A Key to the Icons in *Microbiology: An Evolving Science*

WWW | Weblink icons indicate that there is an author-recommended website related to the topic at hand.

▶⏸ Animation icons in a figure's caption indicate that there is a process animation to further illustrate that particular figure.

Visit Norton StudySpace (wwnorton.com/studyspace) to access these resources and other review material.

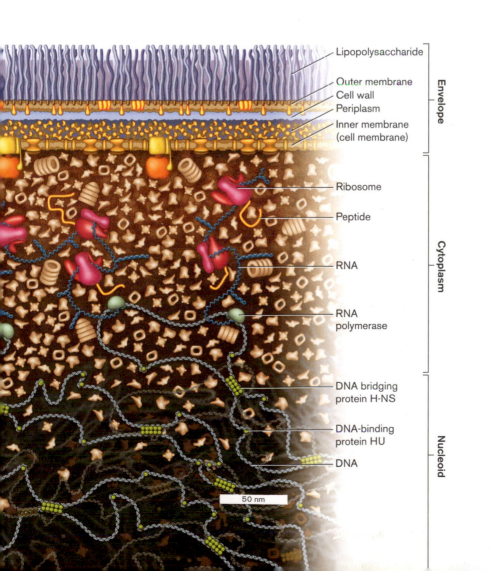

Lipopolysaccharide

Outer membrane
Cell wall
Periplasm
Inner membrane
(cell membrane)

Envelope

Ribosome

Peptide

RNA

RNA
polymerase

Cytoplasm

DNA bridging
protein H-NS

DNA-binding
protein HU

DNA

Nucleoid

50 nm

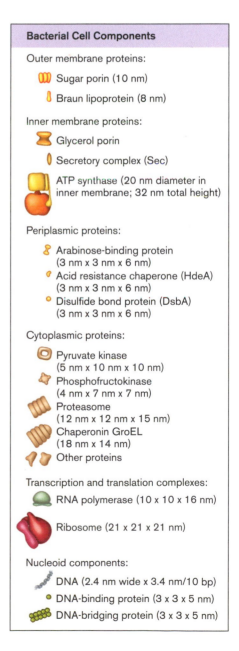

Bacterial Cell Components

Outer membrane proteins:

- Sugar porin (10 nm)
- Braun lipoprotein (8 nm)

Inner membrane proteins:

- Glycerol porin
- Secretory complex (Sec)
- ATP synthase (20 nm diameter in inner membrane; 32 nm total height)

Periplasmic proteins:

- Arabinose-binding protein (3 nm x 3 nm x 6 nm)
- Acid resistance chaperone (HdeA) (3 nm x 3 nm x 6 nm)
- Disulfide bond protein (DsbA) (3 nm x 3 nm x 6 nm)

Cytoplasmic proteins:

- Pyruvate kinase (5 nm x 10 nm x 10 nm)
- Phosphofructokinase (4 nm x 7 nm x 7 nm)
- Proteasome (12 nm x 12 nm x 15 nm)
- Chaperonin GroEL (18 nm x 14 nm)
- Other proteins

Transcription and translation complexes:

- RNA polymerase (10 x 10 x 16 nm)
- Ribosome (21 x 21 x 21 nm)

Nucleoid components:

- DNA (2.4 nm wide x 3.4 nm/10 bp)
- DNA-binding protein (3 x 3 x 5 nm)
- DNA-bridging protein (3 x 3 x 5 nm)

Microbiology

An Evolving Science

Microbiology
An Evolving Science

Joan L. Slonczewski
Kenyon College

John W. Foster
University of South Alabama

Appendices and Glossary by
Kathy M. Gillen
Kenyon College

W·W·NORTON

NEW YORK · LONDON

W. W. Norton & Company has been independent since its founding in 1923, when William Warder Norton and Mary D. Herter Norton first published lectures delivered at the People's Institute, the adult education division of New York City's Cooper Union. The Nortons soon expanded their program beyond the Institute, publishing books by celebrated academics from America and abroad. By mid-century, the two major pillars of Norton's publishing program—trade books and college texts—were firmly established. In the 1950s, the Norton family transferred control of the company to its employees, and today—with a staff of four hundred and a comparable number of trade, college, and professional titles published each year—W. W. Norton & Company stands as the largest and oldest publishing house owned wholly by its employees.

Composition by Precision Graphics
Manufacturing by R. R. Donnelley/Willard
Illustrations by Precision Graphics

Editor: Michael Wright
Developmental editors: Carol Pritchard-Martinez and Philippa Solomon
Senior project editor: Thomas Foley
Copy editor: Janet Greenblatt
Production manager: Christopher Granville
Photography editor: Trish Marx
Marketing manager: Betsy Twitchell
Managing editor, college: Marian Johnson
Science media editor: April Lange
Editorial assistant: Matthew A. Freeman

ISBN: 978-0-393-97857-5

W. W. Norton & Company, Inc., 500 Fifth Avenue, New York, N.Y. 10110
www.wwnorton.com

W. W. Norton & Company Ltd., Castle House, 75/76 Wells Street, London W1T 3QT

1 2 3 4 5 6 7 8 9

Brief Contents

vi

DEDICATION

We dedicate this book to the memory of our doctoral research mentors. Joan's doctoral mentor, Bob Macnab, offered an unfailingly rigorous pursuit of bacterial chemotaxis and physiology, and lasting friendship. John was mentored by Al Moat, a gifted microbial physiologist and humorist who instilled in his neophyte students an appreciation for critical thinking and a love for the science of microbiology.

PART 3
Metabolism and Biochemistry 458

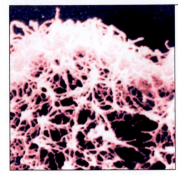

AN INTERVIEW WITH CAROLINE HARWOOD: BACTERIAL METABOLISM DEGRADES POLLUTANTS AND PRODUCES HYDROGEN

PART 4
Microbial Diversity and Ecology   626

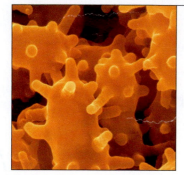

AN INTERVIEW WITH KARL STETTER: ADVENTURES IN MICROBIAL DIVERSITY LEAD TO PRODUCTS IN INDUSTRY

PART 2
Genes and Genomes   218

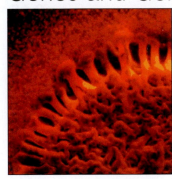

AN INTERVIEW WITH RICHARD LOSICK:
THE THRILL OF DISCOVERY IN MOLECULAR MICROBIOLOGY

Contents

PART 1
The Microbial Cell   2

AN INTERVIEW WITH RITA COLWELL:
THE GLOBAL IMPACT OF MICROBIOLOGY

PART 5
Medicine and Immunology   860

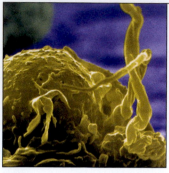

AN INTERVIEW WITH CLIFFORD W. HOUSTON: AN AQUATIC BACTERIUM CAUSES FATAL WOUND INFECTIONS

Preface

Among civilization's greatest achievements are the discovery of microbes and learning how they function. Today, microbiology as a science is evolving rapidly. Emerging species, from *Helicobacter pylori* to ammonia oxidizers, challenge our vision of where microbes can grow, while emerging technologies, from atomic force microscopy to metagenomic sequencing, expand the frontiers of what we can study. As our understanding of microbes and our ability to study them has evolved, what is taught must also evolve. This textbook was designed to present core topics of microbiology in the context of new challenges and opportunities.

Our book gives students and faculty a fresh approach to learning the science of microbiology. A major aim is to balance the coverage of microbial ecology and medical microbiology. We explore the origin of life as a dynamic story of discovery that integrates microfossil data with physiology and molecular biology. This story provides surprising applications in both biotechnology and medicine (Chapter 17, Origins and Evolution). Microbial–host interactions are presented in the context of evolution and ecology, reflecting current discoveries in microbial diversity. For example, *Vibrio cholerae*, the causative agent of cholera, is discussed as part of a complex ecosystem involving invertebrates as well as human hosts (Part 1, Interview with Rita Colwell). Principles of disease are explained in terms of molecular virulence factors that act upon the host cell, including the horizontal transfer of virulence genes that make a pathogen (Chapter 25, Microbial Pathogenesis). Throughout our book, we present the tools of scientific investigation (emphasizing their strengths and limitations) and the excitement of pursuing questions yet to be answered.

We were students when the first exciting reports of gene cloning and the descriptions of molecular machines that compose cells were published. We shared in the excitement surrounding these extraordinary advances, witnessed their impact on the field, and recall how profoundly they inspired us as aspiring scientists. As a result, we believe that conveying the story of scientific advancement and its influence on the way scientists approach research questions, whether classical or modern, is an important motivational and pedagogical tool in presenting fundamental concepts. We present the story of molecular microbiology and microbial ecology in the same spirit as the classical history of Koch and Pasteur, and of Winogradsky and Beijerinck. We drew on all our experience as researchers and educators (and on the input of dozens of colleagues over the past seven years) to create a microbiology text for the twenty-first century.

Major Features

Our book targets the science major in biology, microbiology, or biochemistry. We offer several important improvements over other books written for this audience:

- **Genetics and genomics are presented as the foundation of microbiology.** Molecular genetics and genomics are thoroughly integrated with core topics throughout the book. This approach gives students many advantages, including an understanding of how genomes reveal potential metabolic pathways in diverse organisms, and how genomics and metagenomics reveal the character of microbial communities. Molecular structures and chemical diagrams presented throughout the art program clearly illustrate the connections between molecular genetics, physiology, and pathogenesis.

- **Microbial ecology and medical microbiology receive equal emphasis**, with particular attention paid to the merging of these fields. Throughout the book, phenomena are presented with examples from both ecology and medicine; for example, when discussing horizontal transfer of "genomic islands" we present symbiosis islands associated with nitrogen fixation, as well as pathogenicity islands associated with disease (Chapter 9).

- **Current research examples and tools throughout the text enrich students' understanding of foundational topics.** Every chapter presents numerous current research examples within the up-to-date framework of molecular biology, showing how the latest research extends our knowledge of fundamental topics. For example, in the past two decades, advances in microscopy have reshaped our vision of microbial cells. Chapter 2 is devoted to visualization techniques, from an in-depth treatment of the student's microscope to advanced methods such as atomic force microscopy. Unlike most microbiology textbooks, our text provides size scale information for nearly every micrograph, which is critical when trying to visualize the relationship between different organisms and structures. Examples of current research range from the use of two-hybrid assays to study *Salmonella* virulence proteins to the spectroscopic measurement of carbon flux from microbial communities.

- **Viruses are presented in molecular detail and in ecological perspective.** For example, in marine ecosystems, viruses play key roles in limiting algal populations while selecting for species diversity (Chapter 6). Similarly, a constellation of bacteriophages influences enteric flora. Our coverage of human virology includes the molecular reproductive cycles of herpes, avian influenza, and HIV, including emerging topics such as the role of regulatory proteins in HIV virulence (Chapter 11).

- **Microbial diversity that students can grasp.** We present microbial diversity in a manageable framework that enables students to grasp the essentials of the most commonly presented taxa, devoting one chapter each to bacteria, archaea, and the microbial eukaryotes. At the same time, we emphasize the continual discovery of previously unknown forms such as the nanoarchaea and the marine prochlorophytes. Our book is supported by the on-line *Microbial Biorealm*, an innovative resource on microbial diversity authored by students and their teachers.

- **The physician-scientist's approach to microbial diseases.** Case histories are used to present how a physician-scientist approaches the interplay between the human immune response and microbial diseases. By taking an organ systems approach, we show how a physician actually interacts with the patient, recognizing that patients complain of symptoms, not a species. Ultimately, we let the student in on the clues used to identify infective microbes. The approach stresses the concepts of infectious disease rather than presenting an exhaustive recitation of diseases and microbes.

■ **Scientists pursuing research today are presented alongside the traditional icons.** This approach helps students see that microbiology is an extremely dynamic field of science, full of opportunities for them to do important research as undergraduates or as future graduate students. For example, Chapter 1 not only introduces historical figures such as Koch and Pasteur, but also features genome sequencer Claire Fraser-Liggett, postdoctoral researcher Kazem Kashefi growing a hyperthermophile in an autoclave, and undergraduate students studying acid stress in *E. coli*.

■ **Appendices for students in need of review.** Our book assumes a sophomore-level understanding of introductory biology and chemistry. For those in need of review, two appendices summarize the fundamental structure and function of biological molecules and cells.

Organization

The topics of this book are arranged so that students can progressively develop an understanding of microbiology from key concepts and research tools. The chapters of Part 1 present key foundational topics: history, visualization, the bacterial cell, microbial growth and control, and virology. Chapter 1 discusses the nature of microbes and the history of their discovery, including the key role of microbial genomes. In Chapter 2, basic tools of visualization, from the student's microscope to cryo-EM, provide the foundation for understanding how scientists reveal microbial structure. The basic form and function of bacterial cells emerges in Chapter 3, while Chapters 4 and 5 present core concepts of microbial growth in relation to the environment. Chapter 6 introduces virus structure and culture.

The six chapters in Part 1 present topics treated in more detail in Parts 2 through 5. The topics of nucleoid structure and virus replication introduced in Chapters 3 and 6 lead into Part 2, where Chapters 7 through 12 present modern genetics and genomics. Chapter 11 presents the life cycles of selected viruses in molecular detail. The topics of cell growth and nutrition introduced in Chapter 4 lead into Part 3 (Chapters 13–16), which presents cell metabolism and biochemistry. Diverse forms of metabolism, such as phototrophy and lithotrophy, are explained on a common basis, the fundamental principles of electron transport and energy conservation. These chapters are written in such a way that they can be presented before the genetics material if so desired. Chapter 16 presents food and industrial microbiology, showing how these fields are founded on microbial metabolism. The principles of environmental responses and growth limits introduced in Chapter 5 lead into Part 4 (Chapters 17–22), which explores microbial ecology and diversity. The roles of microbial communities in local ecosystems and global cycling, introduced in Chapter 4, are presented in greater depth in Chapters 21 and 22. And the chapters of Part 5 (Chapters 23–28) present medical and disease microbiology from an investigative perspective, founded on the principles of genetics, metabolism, and microbial ecology.

Special Features

Throughout our book, special features aid student understanding and stimulate inquiry.

ART PROGRAM

The art program offers exceptional depth and clarity, using up-to-date graphical methods to enhance understanding. Key processes are shown in both a simplified version and a more complex version. For example, the Calvin-Benson cycle is introduced with a focus on the incorporation of CO_2 and formation of energy carriers (Fig. 15.5), followed by a more detailed diagram that includes the chemical structures of all intermediates (Fig. 15.7). Overall, our book provides a greater number of figures and photos than our major competitors.

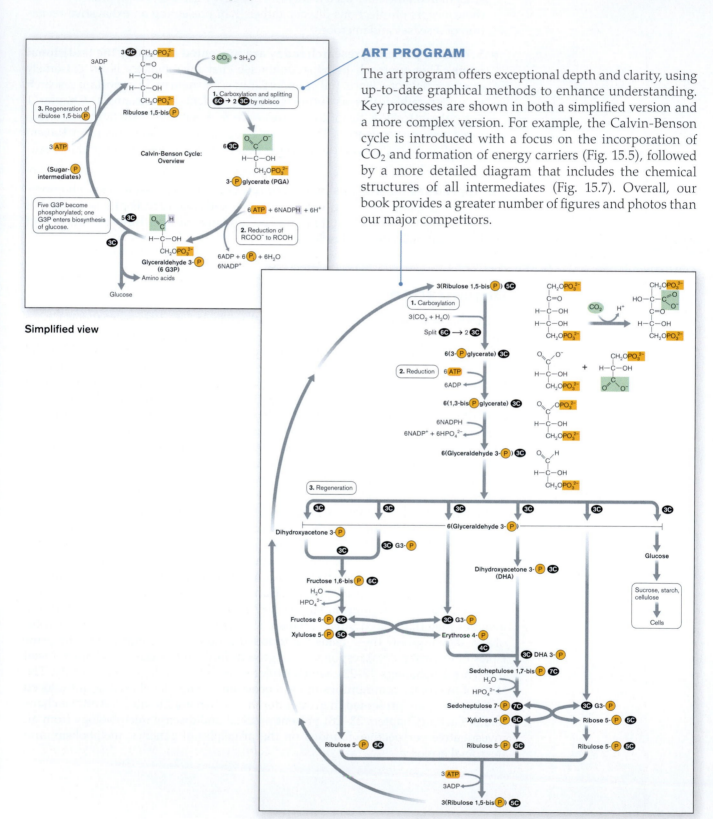

Simplified view

Expanded view

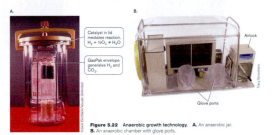

Inset reproduction of page 167:

Part 1 ■ The Microbial Cell   **167**

microbes that can live with or without oxygen. They will grow throughout the tube shown in Figure 5.20. **Facultative anaerobes** (sometimes called aerotolerant) only use fermentation to provide energy but contain superoxide dismutase and catalase (or peroxidase) to protect them from reactive oxygen species. This allows them to grow in oxygen while retaining a fermentation-based metabolism. **Facultative aerobes** (such as *E. coli*) also possess enzymes that destroy toxic oxygen by-products, but have both fermentative *and* aerobic respiratory potential. Whether a member of this group uses aerobic respiration, anaerobic respiration, or fermentation depends on the availability of oxygen and the amount of carbohydrate present. Microorganisms that possess *decreased* levels of superoxide dismutase and/or catalase will be **microaerophilic**, meaning they will grow only at low oxygen concentrations.

The fundamental composition of all cells reflects their evolutionary origin as anaerobes. Lipids, nucleic acids, and amino acids are all highly reduced—which is why our bodies are combustible. We never would have evolved that way if molecular oxygen were present from the beginning. Even today, the majority of all microbes are anaerobic, growing buried in the soil, within our anaerobic digestive tract, or within biofilms on our teeth.

THOUGHT QUESTION 5.6 If anaerobes cannot live in oxygen, how do they incorporate oxygen into their cellular components?

THOUGHT QUESTION 5.7 How can anaerobes grow in the human mouth when there is so much oxygen there?

Culturing Anaerobes in the Laboratory

Many anaerobic bacteria cause horrific human diseases, such as tetanus, botulism, and gangrene. Some of these organisms or their secreted toxins are even potential weapons of terror (for example, *Clostridium botulinum*). Because of their ability to wreak havoc on humans, culturing these microorganisms was an early goal of microbiologists. Despite the difficulties involved, conditions were eventually contrived in which all, or at least most, of the oxygen could be removed from a culture environment.

Three techniques are used today. Special reducing agents (for example, thioglycolate) or enzyme systems (Oxyrase®) that eliminate dissolved oxygen can be added to ordinary liquid media. Anaerobes can then grow beneath the culture surface. A second, very popular, way to culture anaerobes, especially on agar plates, is to use an anaerobe jar (**Fig. 5.22A**). Agar plates streaked with the organism are placed into a sealed jar with a foil packet that releases H_2 and CO_2 gases. A palladium packet hanging from the jar lid catalyzes a reaction between the H_2 and O_2 in the jar to form H_2O and effectively removes O_2 from the chamber. The CO_2 released is required by some reactions to produce key metabolic intermediates. Some microaerophilic microbes, like the pathogens *H. pylori* (the major cause of stomach ulcers) and *Campylobacter jejuni* (a major cause of diarrhea), require low levels of O_2 but elevated amounts of CO_2. These conditions are obtained by using similar gas-generating packets.

For strict anaerobes exquisitely sensitive to oxygen, even more heroic efforts are required to establish an oxygen-free environment. A special anaerobic glove box must be used in which the atmosphere is removed by

Catalyst in lid mediates reaction. $H_2 + \frac{1}{2}O_2 \rightarrow H_2O$

GasPak envelope generates H_2 and CO_2.

Airlock

Glove ports

Figure 5.22 Anaerobic growth technology. **A.** An anaerobic jar. **B.** An anaerobic chamber with glove ports.

Inset reproduction of page 2:

Part 1
The Microbial Cell

AN INTERVIEW WITH
RITA COLWELL: THE GLOBAL IMPACT OF MICROBIOLOGY

Rita Colwell is Distinguished Professor at the University of Maryland and Johns Hopkins University and served as director of the USA National Science Foundation from 1998 to 2004. Colwell's decades of research on *Vibrio cholerae*, the causative agent of cholera, have revealed its natural ecology, its genome sequence, and ways to control it. Colwell originated the concept of viable but nonculturable microorganisms, microbial cells that metabolize but cannot be cultured in the laboratory. She is now chair of the board of Canon US Life Sciences, Inc., and she represents the American Society for Microbiology at the United Nations Educational, Scientific and Cultural Organization (UNESCO).

Rita Colwell, former director of the National Science Foundation.

Why did you decide to make a career in microbiology?
I was first inspired by the report of my college roommate at Purdue University about a wonderful bacteriology professor, Dr. Dorothy Powelson, probably one of only two women at Purdue who were full professors at the time. I enrolled in Powelson's course and was truly inspired by this remarkable woman who was so interested in microbiology and made it fascinating for her students.

How did you choose to study *Vibrio cholerae*? What makes this organism interesting?
I chose to study *Vibrio cholerae* as a result of my having become an "expert" on vibrios through my graduate dissertation on marine microorganisms. Vibrios were the most readily culturable of the marine bacteria and were therefore considered the most dominant. Of course, new information indicates that although vibrios are the dominant bacteria in many estuarine areas, there are other organisms that are very difficult to culture that are important as well.

You led an international collaboration in Bangladesh training women to avoid cholera by filtering water through sari cloth. How did the sari cloth filtration project come about?

When I took my first faculty position at Georgetown University, a friend of mine at NIH, Dr. John Feeley, suggested that I study *Vibrio cholerae*. What makes *V. cholerae* interesting is that it is a human pathogen of extremely great importance, yet resides naturally in estuaries and coastal areas of the world.

What is it like to study this organism?
Vibrio cholerae is naturally occurring (in the environment outside humans) and therefore can never be eradicated; it carries out important functions in the environment, and significant among these is its ability to digest chitin, the structural component of shellfish and many zooplankton. It is at once a "recycling agent" and a public health threat in the form of the massive epidemics of cholera that it causes.

It came about because of collaboration with the International Centre for Diarrhoeal Diseases, Bangladesh, located in Dhaka, Bangladesh, and the Mattlab Field Laboratory, which is located in the village area of Mattlab, Bangladesh. Our work had shown that *Vibrio cholerae* is associated with environmental zooplankton, namely, the copepod. The notion that the copepods are large and could be filtered out and therefore lead to reduced incidence of cholera was a result of my work on the vibrio and the relationships described by my students, notably, Dr. Anwar Huq, who did his thesis on *Vibrio cholerae* attachment to copepods. Anwar Huq is now an associate professor at the University of Maryland.

An important collaborator was Nell Roberts, an outstanding public health microbiologist at Lake Charles, Louisiana, working on public health problems. Nell, Professor Xu (a colleague from Qingdao, China), and I did the critical experiment showing the presence of *Vibrio cholerae* in water from which blue crabs had been harvested—the cause of an outbreak of cholera in Louisiana back in 1982. We were able to use fluorescent antibody to show the presence of the vibrio on copepods in the water.

From there, the idea of sari cloth came about in searching for a simple, inexpensive filter for use by village

2

THOUGHT QUESTIONS

"Thought Questions" throughout the text stimulate students to think critically about their reading. For example, a Thought Question in Chapter 5 (p. 167) asks students to consider how anaerobes incorporate oxygen into their cellular components in spite of their inability to live in oxygen. The question is posed in the context of a discussion of the different levels of oxygen tolerated or required by different types of microbes.

Answers to each Thought Question are provided at the back of the book.

INTERVIEWS WITH PROMINENT SCIENTISTS

Each Part of the book opens with an interview of a prominent microbiologist working today. In each interview, the authors ask the featured scientist questions about everything from how they first became interested in microbiology to how their thought processes and experiments allowed them to make important discoveries. Interviewees include Karl Stetter, the first person to discover living organisms growing at temperatures above 100°C, and Rita Colwell, past director of the National Science Foundation, who used her understanding of the marine ecology of *Vibrio cholerae* to help develop public health measures against cholera in developing countries.

Special Topic 10.2   The Role of Quorum Sensing in Pathogenesis and in Interspecies Communications

Pseudomonas aeruginosa is a human pathogen that commonly infects patients with cystic fibrosis, a genetic disease of the lung. The organism forms a biofilm over affected areas and interferes with lung function. Key to the destruction of host tissues by *P. aeruginosa* are virulence factors such as proteases and other degradative enzymes. But these proteins are not made until cell density is fairly high, a point where the organism might have a chance of overwhelming its host. The organism would not want to make the virulence proteins too early and alert the host to launch an immune response. The induction mechanism involves two interconnected quorum-sensing systems called Las and Rhl, both comprised of regulatory proteins homologous to LuxR and LuxI of *V. fischeri*. Many pathogens besides *Pseudomonas* appear to use chemical signaling to control virulence genes. Genomic analysis has revealed homologs of known quorum-sensing genes in *Salmonella*, *Escherichia*, *Vibrio cholerae*, the plant symbiote *Rhizobium*, and many other microbes.

Some microbial species not only chemically talk among themselves, but appear capable of communicating with other species. *V. harveyi*, for example, uses two different, but converging, quorum-sensing systems to coordinate control of its luciferase. Both sensing pathways are very different from the *V. fischeri* system. One utilizes an acyl homoserine lactone (AHL) as an autoinducer (AI-1) to communicate with other *V. harveyi* cells. The second system involves production of a different autoinducer (AI-2) that contains borate. Because many species appear to produce this second signal molecule, it is thought that mixed populations of microbes use it to "talk" to each other. In the case of *V. harveyi*, specific membrane sensor kinase proteins are used to sense each autoinducer (**Fig. 1**). At low cell densities (no autoinducer), both sensor kinases initiate phosphorylation cascades that converge on a shared response regulator, LuxO, to produce phosphorylated LuxO. Phosphorylated LuxO appears to activate a repressor of the *lux* genes. Thus, at low cell densities, the culture does not display bioluminescence. At high cell density, the autoinducers prevent signal transmission by inhibiting phosphorylation. The cell stops making repressor, which allows another pro-

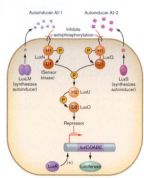

Figure 1   The two quorum-sensing systems of *V. harveyi*.   In the absence of autoinducers (AI-1 and AI-2), both sensor kinases trigger converging phosphorylation cascades that end with the phosphorylation of LuxO. Phosphorylated LuxO (LuxO-P) activates a repressor that inhibits expression of the luciferase genes. As autoinducer concentrations increase, they inhibit autophosphorylation of the sensor kinases and the phosphorylation cascade. As a result, repressor levels decrease, which allows the LuxR protein to activate the *lux* operon.

tein, LuxR (*not* a homolog of the *V. fischeri* LuxR), to activate the *lux* operon. The "lights" are turned on. Bonnie Bassler (**Fig. 2**) and Pete Greenberg (**Fig. 3**) are two of the leading scientists whose studies revealed the complex elegance of quorum sensing in *Vibrio* and *Pseudomonas* species. Other organisms, such as *Salmonella*, have been shown to activate the AI-2 pathway of *V. harveyi*, dramatically supporting the concept of cross-species communication.

A recent report by Ian Joint and his colleagues has shown that bacteria can even communicate across the prokaryotic-eukaryotic boundary. The green seaweed *Enteromorpha* (a eukaryote) produces motile zoospores that explore and attach to *Vibrio anguillarum* bacterial cells in biofilms (**Fig. 4**). They attach and remain there because the bacterial cells produce acetyl homoserine lactone molecules that the zoospores sense. Part of the evidence for this interkingdom communication involved showing that the zoospores would even attach to biofilms of *E. coli* carrying the *Vibrio* genes for the synthesis of acetyl homoserine lactone. The implications of possible inter-

kingdom conversations are staggering. Do our normal flora "speak" to us? Do we "speak" back?

For further discussion of molecular communication between prokaryotes and eukaryotes, see Chapter 21.

Figure 3   Peter Greenberg, one of the pioneers of cell-cell communication research.   Peter Greenberg, first at the University of Iowa and now at the University of Washington, has studied quorum sensing in *Vibrio* species and various other pathogenic bacteria, such as *Pseudomonas*.

Figure 2   **Bonnie Bassler** (left) of Princeton University was instrumental in characterizing interspecies communication between bacteria.

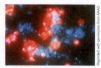

Figure 4   *Enteromorpha* zoospores (red), a type of algae, attach to biofilm-producing bacteria (blue) in response to lactones produced by the bacteria.

SPECIAL TOPICS

Optional "Special Topics" boxes show the process of science and give a human face to the research. Topics are as diverse as scientists discovering "quorum sensing" in pathogenesis (ST 10.2) and undergraduate researchers investigating mycorrhizae in wetland soil (ST 22.1). Whether historical in focus or providing more detail about cutting-edge science, "Special Topics" give students extra background and detail to help them appreciate the dynamic nature of microbiology.

Chapter 3
Cell Structure and Function

Microbial cells face extreme challenges from their environment, enduring rapid changes in temperature and salinity, and pathogens face the chemical defenses of their hosts. To meet these challenges, microbes build complex structures, such as a cell envelope with tensile strength comparable to steel. Within the cytoplasm, molecular devices such as the the ribosome build and expand the cell.

With just a few thousand genes in its genome, how does a bacterial cell grow and reproduce? Bacteria coordinate their DNA replication through the DNA replisome and the cell fission ring. Other devices, such as flagellar propellers, enable microbial cells to compete, to communicate, and even to cooperate in building biofilm communities.

Discoveries of cell form and function have exciting applications for medicine and biotechnology. The structures of ribosomes and cell envelope materials provide targets for new antibiotics. And devices such as the rotary ATP synthase inspire "nanotechnology" the design of molecular machines.

The filamentous cyanobacterium *Anabaena* sp. was engineered to make a cell division protein, FtsZ, fused to green fluorescent protein (GFP). FtsZ-GFP proteins form a ring-like structure around the middle of each cell, where it prepares to divide. Source: Sanner Sakr, et al. 2006. *J. Bacteriol.* 188.

73

5 µm

CHAPTER OPENERS

The title page of each chapter presents an intriguing photo related to a recent research article or current application of the chapter topic. For example, Chapter 3 opens with a fluorescence micrograph of *Anabaena* in which the cell division protein FtsZ fused to "green fluorescent protein" (GFP) fluoresces around the division plane of each cell.

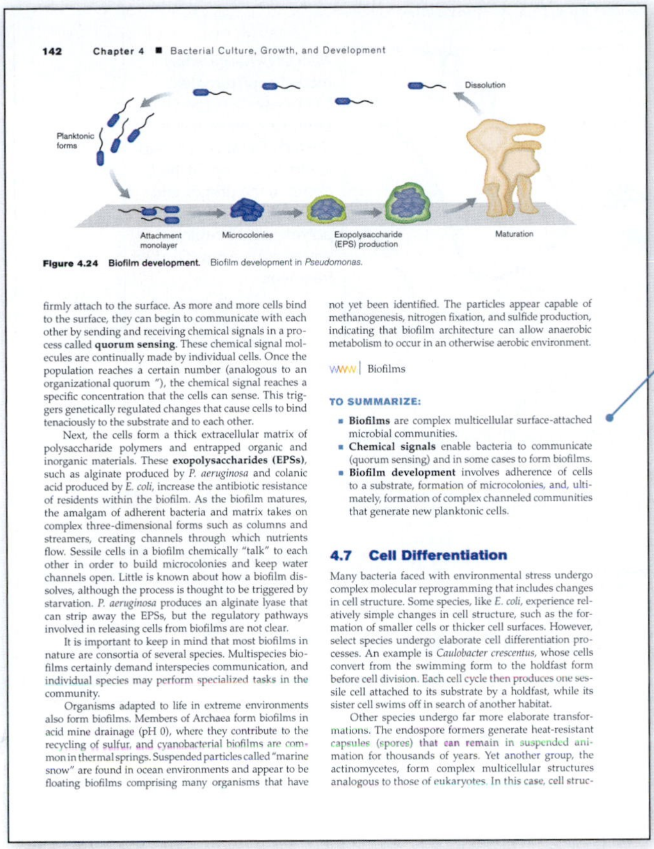

Figure 4.24 Biofilm development. Biofilm development in *Pseudomonas.*

firmly attach to the surface. As more and more cells bind to the surface, they can begin to communicate with each other by sending and receiving chemical signals in a process called **quorum sensing**. These chemical signal molecules are continually made by individual cells. Once the population reaches a certain number (analogous to an organizational quorum "), the chemical signal reaches a specific concentration that the cells can sense. This triggers genetically regulated changes that cause cells to bind tenaciously to the substrate and to each other.

Next, the cells form a thick extracellular matrix of polysaccharide polymers and entrapped organic and inorganic materials. These **exopolysaccharides (EPSs)**, such as alginate produced by *P. aeruginosa* and colanic acid produced by *E. coli*, increase the antibiotic resistance of residents within the biofilm. As the biofilm matures, the amalgam of adherent bacteria and matrix takes on complex three-dimensional forms such as columns and streamers, creating channels through which nutrients flow. Sessile cells in a biofilm chemically "talk" to each other in order to build microcolonies and keep water channels open. Little is known about how a biofilm dissolves, although the process is thought to be triggered by starvation. *P. aeruginosa* produces an alginate lyase that can strip away the EPSs, but the regulatory pathways involved in releasing cells from biofilms are not clear.

It is important to keep in mind that most biofilms in nature are consortia of several species. Multispecies biofilms certainly demand interspecies communication, and individual species may perform specialized tasks in the community.

Organisms adapted to life in extreme environments also form biofilms. Members of Archaea form biofilms in acid mine drainage (pH 0), where they contribute to the recycling of sulfur, and cyanobacterial biofilms are common in thermal springs. Suspended particles called "marine snow" are found in ocean environments and appear to be floating biofilms comprising many organisms that have not yet been identified. The particles appear capable of methanogenesis, nitrogen fixation, and sulfide production, indicating that biofilm architecture can allow anaerobic metabolism to occur in an otherwise aerobic environment.

WWW | Biofilms

TO SUMMARIZE:

- **Biofilms** are complex multicellular surface-attached microbial communities.
- **Chemical signals** enable bacteria to communicate (quorum sensing) and in some cases to form biofilms.
- **Biofilm development** involves adherence of cells to a substrate, formation of microcolonies, and, ultimately, formation of complex channeled communities that generate new planktonic cells.

4.7 Cell Differentiation

Many bacteria faced with environmental stress undergo complex molecular reprogramming that includes changes in cell structure. Some species, like *E. coli*, experience relatively simple changes in cell structure, such as the formation of smaller cells or thicker cell surfaces. However, select species undergo elaborate cell differentiation processes. An example is *Caulobacter crescentus*, whose cells convert from the swimming form to the holdfast form before cell division. Each cell cycle then produces one sessile cell attached to its substrate by a holdfast, while its sister cell swims off in search of another habitat.

Other species undergo far more elaborate transformations. The endospore formers generate heat-resistant capsules (spores) that can remain in suspended animation for thousands of years. Yet another group, the actinomycetes, form complex multicellular structures analogous to those of eukaryotes. In this case, cell struc-

TO SUMMARIZE

This feature ensures that students understand the key concepts of each section before they continue with the reading.

Student Resources

- **StudySpace.** wwnorton.com/studyspace This student website includes multiple-choice quizzes, process animations, vocabulary flashcards, indices of the Weblink reference sites from the text, and prominent links to *Microbial Biorealm*.
- **Process Animations.** Developed specifically for *Microbiology: An Evolving Science*, these animations bring key figures from the text to life, presenting key microbial processes in a dynamic format. The animations can be enlarged to full-screen view, and include VCR-like controls that make it easy to control the pace of animation.
- WWW | **Weblink Icons** throughout the text point students to the student website, which serves as a portal to websites where they can find more information on a host of topics. Each link was reviewed and approved by the authors to ensure that only high-interest, high-quality sites were selected.
- **Microbial Biorealm and Viral Biorealm.** A website maintained at Kenyon College provides information on several hundred genera of microbes and viruses, to which interested students have the opportunity to contribute. Pages are monitored and edited by microbiologists at Kenyon.
- **Ebook.** Same great book at half the price. *Microbiology: An Evolving Science* is also available as an ebook from norton**ebooks**.com. With a Norton ebook, students can electronically highlight text, use sticky notes, and work with fully zoomable images from the book.

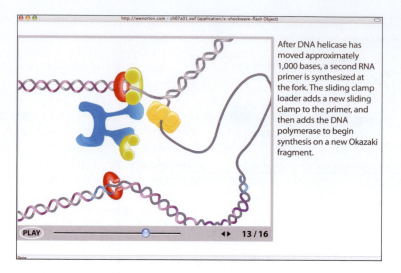

After DNA helicase has moved approximately 1,000 bases, a second RNA primer is synthesized at the fork. The sliding clamp loader adds a new sliding clamp to the primer, and then adds the DNA polymerase to begin synthesis on a new Okazaki fragment.

Instructor Resources

- **Norton Media Library Instructor's CD-ROM:**
  - **Drawn Art and Photographs.** Digital files of all drawn art and most photographs are available to adopters of the text.
  - **Process Animations.** Developed specifically for *Microbiology: An Evolving Science,* these animations bring key figures from the text to life, presenting key microbial processes in a dynamic format. The animations can be enlarged to full-screen view and include VCR-like controls that make it easy for instructors to control the pace of animation during lecture.
  - **Editable PowerPoint Lectures** for each chapter.

- **Norton Resource Library Instructor's Website.** wwnorton.com/instructors Maintained as a service to our adopters, this password-protected instructor website offers book-specific materials for use in class or within WebCT, Blackboard, or course websites. The resources available online are the same as those offered on the Norton Media Library CD-ROM.

- **Instructor's Manual.** The manual includes chapter overviews, answers to end-of-chapter questions, and a test bank of 2,000 questions. Authored by Kathleen Campbell at Emory University.

- **Electronic Test Bank.** The Test Bank includes 2,000 questions in *ExamView Assessment Suite* format.

- **Blackboard Learning System Coursepacks.** These coursepacks include classroom-ready content.

- **Transparencies.** A subset of the figures in the text are available as color acetates.

Acknowledgments

We are very grateful for the help of many people in developing and completing the book. Our first editor at Norton, John Byram, helped us define the aims and scope of the project. Vanessa Drake-Johnson helped us shape the text, supported us in developing a strong art program, and conceived the title. Mike Wright spared no effort to bring the project to completion and to the attention of our colleagues. Our developmental editors, Philippa Solomon and Carol Pritchard-Martinez, contributed greatly to the clarity of presentation. Philippa's strength in chemistry was invaluable in improving our presentation of metabolism. Trish Marx and the photo researchers did a heroic job of tracking down all kinds of images from sources all over the world. Our colleague Kathy Gillen provided exceptional expertise on review topics for the appendices and wrote outstanding review questions for the student website. April Lange's coordination of electronic media development has resulted in a superb suite of resources for students and instructors alike. We thank Kathleen Campbell for authoring an instructor's manual that demonstrates a clear understanding of our goals for the book, and Lisa Rand for editing it. Without Thom Foley's incredible attention to detail, the innumerable moving parts of this book would never have become a finished book. Marian Johnson, Norton's managing editor in the college department, helped coordinate the complex process involved in shaping the manuscript over the years. Chris Granville ably and calmly managed the transformation of manuscript to finished product in record time. Matthew Freeman coordinated the transfer of many drafts among many people. Steve Dunn and Betsy Twitchell have been effective advocates for the book in the marketplace. Finally, we thank Roby Harrington, Drake McFeely, and Julia Reidhead for their support of this book over its many years of development.

For the quality of our illustrations we thank the many artists at Precision Graphics, who developed attractive and accurate representations and showed immense patience in getting the details right. We especially thank Kirsten Dennison for project management; Karen Hawk for the layout of every page in the book; Kim Brucker and Becky Oles for developing the art style and leading the art team; and Simon Shak for his rendering of the molecular models based on PDB files, including some near-impossible structures that we requested.

We thank the numerous colleagues over the years who encouraged us in our project, especially the many attendees at the Microbial Stress Gordon Conferences. We greatly appreciate the insightful reviews and discussions of the manuscript provided by our colleagues, and the many researchers who contributed their micrographs and personal photos. We especially thank the American Society for Microbiology journals for providing many valuable resources. Reviewers Bob Bender, Bob Kadner, and Caroline Harwood offered particularly insightful comments on the metabolism and genetics sections, and James Brown offered invaluable assistance in improving the coverage of microbial evolution. Peter Rich was especially thoughtful in providing materials from the archive of Peter Mitchell. We also thank the following reviewers:

Laurie A. Achenbach, Southern Illinois University, Carbondale
Stephen B. Aley, University of Texas, El Paso
Mary E. Allen, Hartwick College
Shivanthi Anandan, Drexel University
Brandi Baros, Allegheny College
Gail Begley, Northeastern University
Robert A. Bender, University of Michigan
Michael J. Benedik, Texas A&M University
George Bennett, Rice University
Kathleen Bobbitt, Wagner College
James Botsford, New Mexico State University
Nancy Boury, Iowa State University of Science and Technology
Jay Brewster, Pepperdine University
James W. Brown, North Carolina State University
Whitney Brown, Kenyon College undergraduate
Alyssa Bumbaugh, Pennsylvania State University, Altoona
Kathleen Campbell, Emory University
Alana Synhoff Canupp, Paxon School for Advanced Studies, Jacksonville, FL
Jeffrey Cardon, Cornell College

Tyrrell Conway, University of Oklahoma
Vaughn Cooper, University of New Hampshire
Marcia L. Cordts, University of Iowa
James B. Courtright, Marquette University
James F. Curran, Wake Forest University
Paul Dunlap, University of Michigan
David Faguy, University of New Mexico
Bentley A. Fane, University of Arizona
Bruce B. Farnham, Metropolitan State College of Denver
Noah Fierer, University of Colorado, Boulder
Linda E. Fisher, late of the University of Michigan, Dearborn
Robert Gennis, University of Illinois, Urbana-Champaign
Charles Hagedorn, Virginia Polytechnic Institute and State University
Caroline Harwood, University of Washington
Chris Heffelfinger, Yale University graduate student
Joan M. Henson, Montana State University
Michael Ibba, Ohio State University
Nicholas J. Jacobs, Dartmouth College
Douglas I. Johnson, University of Vermont
Robert J. Kadner, late of the University of Virginia
Judith Kandel, California State University, Fullerton
Robert J. Kearns, University of Dayton
Madhukar Khetmalas, University of Central Oklahoma
Dennis J. Kitz, Southern Illinois University, Edwardsville
Janice E. Knepper, Villanova University
Jill Kreiling, Brown University
Donald LeBlanc, Pfizer Global Research and Development (retired)
Robert Lausch, University of South Alabama
Petra Levin, Washington University in St. Louis
Elizabeth A. Machunis-Masuoka, University of Virginia
Stanley Maloy, San Diego State University
John Makemson, Florida International University
Scott B. Mulrooney, Michigan State University
Spencer Nyholm, Harvard University
John E. Oakes, University of South Alabama
Oladele Ogunseitan, University of California, Irvine
Anna R. Oller, University of Central Missouri
Rob U. Onyenwoke, Kenyon College
Michael A. Pfaller, University of Iowa
Joseph Pogliano, University of California, San Diego
Martin Polz, Massachusetts Institute of Technology
Robert K. Poole, University of Sheffield
Edith Porter, California State University, Los Angeles
S. N. Rajagopal, University of Wisconsin, La Crosse
James W. Rohrer, University of South Alabama
Michelle Rondon, University of Wisconsin-Madison
Donna Russo, Drexel University
Pratibha Saxena, University of Texas, Austin
Herb E. Schellhorn, McMaster University
Kurt Schesser, University of Miami
Dennis Schneider, University of Texas, Austin
Margaret Ann Scuderi, Kenyon College
Ann C. Smith Stein, University of Maryland, College Park
John F. Stolz, Duquesne University
Marc E. Tischler, University of Arizona

Monica Tischler, Benedictine University
Beth Traxler, University of Washington
Luc Van Kaer, Vanderbilt University
Lorraine Grace Van Waasbergen, The University of Texas, Arlington
Costantino Vetriani, Rutgers University
Amy Cheng Vollmer, Swarthmore College
Andre Walther, Cedar Crest College
Robert Weldon, University of Nebraska, Lincoln
Christine White-Ziegler, Smith College
Jianping Xu, McMaster University

Finally, we offer special thanks to our families for their support. Joan's husband Michael Barich offered unfailing support, and her son Daniel Barich contributed photo research, as well as filling the indispensable role of technical director for the *Microbial Biorealm* website. John's wife Zarrintaj ("Zari") Aliabadi contributed to the text development, especially the sections on medical microbiology and public health.

To the Reader: Thanks!

We greatly appreciate your selection of this book as your introduction to the science of microbiology. This is a first edition, and as such can certainly benefit from the input of readers. We welcome your comments, especially if you find text or figures that are in error or unclear. Feel free to contact us at the addresses listed below.

Joan L. Slonczewski
slonczewski@kenyon.edu

John W. Foster
jwfoster@jaguar1.usouthal.edu

About the **Authors**

JOAN L. SLONCZEWSKI received her B.A. from Bryn Mawr College and her Ph.D. in Molecular Biophysics and Biochemistry from Yale University, where she studied bacterial motility with Robert M. Macnab. After postdoctoral work at the University of Pennsylvania, she has since taught undergraduate microbiology in the Department of Biology at Kenyon College, where she earned a Silver Medal in the National Professor of the Year program of the Council for the Advancement and Support of Education. She has published numerous research articles with undergraduate coauthors on bacterial pH regulation, and has published five science fiction novels including *A Door into Ocean*, which earned the John W. Campbell Memorial Award. She serves as At-large Member representing Divisions on the Council Policy Committee of the American Society for Microbiology, and is a member of the Editorial Board of the journal *Applied and Environmental Microbiology*.

JOHN W. FOSTER received his B.S. from the Philadelphia College of Pharmacy and Science (now the University of the Sciences in Philadelphia), and his Ph.D. from Hahnemann University (now Drexel University School of Medicine), also in Philadelphia, where he worked with Albert G. Moat. After postdoctoral work at Georgetown University, he joined the Marshall University School of Medicine in West Virginia; he is currently teaching in the Department of Microbiology and Immunology at the University of South Alabama College of Medicine in Mobile, Alabama. Dr. Foster has coauthored three editions of the textbook *Microbial Physiology* and has published over 100 journal articles describing the physiology and genetics of microbial stress responses. He has served as Chair of the Microbial Physiology and Metabolism division of the American Society for Microbiology and is a member of the editorial advisory board of the journal *Molecular Microbiology*.

Microbiology
An Evolving Science

The Microbial Cell

Courtesy of Rita Colwell

Rita Colwell, former director of the National Science Foundation.

AN INTERVIEW WITH

RITA COLWELL: THE GLOBAL IMPACT OF MICROBIOLOGY

Rita Colwell is Distinguished Professor at the University of Maryland and Johns Hopkins University and served as director of the USA National Science Foundation from 1998 to 2004. Colwell's decades of research on *Vibrio cholerae*, the causative agent of cholera, have revealed its natural ecology, its genome sequence, and ways to control it. Colwell originated the concept of viable but nonculturable microorganisms, microbial cells that metabolize but cannot be cultured in the laboratory. She is now chairman of the board of Canon US Life Sciences, Inc., and she represents the American Society for Microbiology at the United Nations Educational, Scientific and Cultural Organization (UNESCO).

Why did you decide to make a career in microbiology?

I was first inspired by the report of my college roommate at Purdue University about a wonderful bacteriology professor, Dr. Dorothy Powelson, probably one of only two women at Purdue who were full professors at the time. I enrolled in Powelson's course and was truly inspired by this remarkable woman who was so interested in microbiology and made it fascinating for her students.

How did you choose to study *Vibrio cholerae*? What makes this organism interesting?

I chose to study *Vibrio cholerae* as a result of my having become an "expert" on vibrios through my graduate dissertation on marine microorganisms. Vibrios were the most readily culturable of the marine bacteria and were therefore considered the most dominant. Of course, new information indicates that although vibrios are the dominant bacteria in many estuarine areas, there are other organisms that are very difficult to culture that are important as well.

When I took my first faculty position at Georgetown University, a friend of mine at NIH, Dr. John Feeley, suggested that I study *Vibrio cholerae*. What makes *V. cholerae* interesting is that it is a human pathogen of extremely great importance, yet resides naturally in estuaries and coastal areas of the world.

What is it like to study this organism?

Vibrio cholerae is naturally occurring (in the environment outside humans) and therefore can never be eradicated; it carries out important functions in the environment, and significant among these is its ability to digest chitin, the structural component of shellfish and many zooplankton. It is at once a "recycling agent" and a public health threat in the form of the massive epidemics of cholera that it causes.

You led an international collaboration in Bangladesh training women to avoid cholera by filtering water through sari cloth. How did the sari cloth filtration project come about?

It came about through collaboration with the International Centre for Diarrhoeal Diseases, Bangladesh, located in Dhaka, Bangladesh, and the Mattlab Field Laboratory, which is located in the village area of Mattlab, Bangladesh. Our work had shown that *Vibrio cholerae* is associated with environmental zooplankton, namely, the copepod. The notion that the copepods are large and could be filtered out and therefore lead to reduced incidence of cholera was a result of my work on the vibrios and the relationships described by my students, notably, Dr. Anwar Huq, who did his thesis on *Vibrio cholerae* attachment to copepods. Anwar Huq is now an associate professor at the University of Maryland.

An important collaborator was Nell Roberts, an outstanding public health microbiologist at Lake Charles, Louisiana, working on public health problems. Nell, Professor Xu (a colleague from Qingdao, China), and I did the critical experiment showing the presence of *Vibrio cholerae* in water from which blue crabs had been harvested—the cause of an outbreak of cholera in Louisiana back in 1982. We were able to use fluorescent antibody to show the presence of the vibrio on copepods in the water.

From there, the idea of sari cloth came about in searching for a very inexpensive filter for use by village

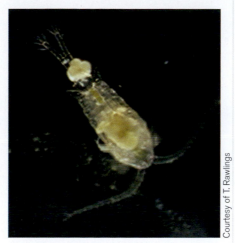

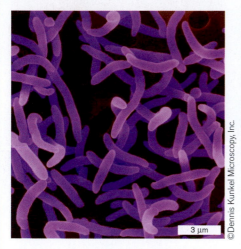

Vibrio cholerae bacteria (left) colonize copepods such as this one (right).

women in Bangladesh. We were able to show that folded sari cloth yielded a 20-micrometer (μm) mesh net. Because plankton are 200 μm or more in size, we could filter them out. The hypothesis that I came up with was that by removing the copepods and associated particulates, we could reduce cholera, which proved to be the case.

What are "viable but nonculturable" organisms?

Viable but nonculturable is a state into which gram-negative microorganisms transform under adverse conditions in the environment. In this state, the bacteria are unable to be cultured, even though they remain viable and potentially pathogenic. Hence, they pose a public health risk, since routine tests done in a bacteri-

A Bangladeshi woman filters water through sari cloth. Colwell's graduate student Anwar Huq compares the filtered and unfiltered water.

ology laboratory would be negative for their presence.

What are the challenges of marine microbiology today? How does marine microbiology impact human health?

The challenges of marine microbiology today are to understand and catalog the extraordinary diversity of marine microorganisms. The world's oceans function in large part as a result of the activities of marine organisms. Marine microbiology impacts human health because of the many pathogens naturally occurring in the environment. But more than that, marine microorganisms may well be the cycling agent that keeps the blue planet inhabitable for humans. Marine microorganisms actively cycle carbon, nitrogen, phosphorus, and other elements in our oceans and even play a role in the weather by producing dimethyl sulfoxide (DMSO), which is involved in cloud formation and moisture condensation.

Why did you move to the National Science Foundation? What difference did you make as a microbiologist heading NSF?

I was asked by the president of the United States to serve as director of the National Science Foundation (NSF). It is a position appointed by

the president and confirmed by the U.S. Senate. As a microbiologist, I was able to bring a molecular understanding of biology to the NSF, while as an interdisciplinary researcher, I was attuned to the needs of all aspects of science, from astronomy to physics. In the biological sciences, my major impact was in launching the Biocomplexity Initiative, which has been enormously productive and continues to yield new information on biological systems, including those of microbiology.

What do you think are the most exciting areas for students entering microbiology today?

Microbial diversity and microbial population studies are two emerging areas of huge interest that will lead to a better understanding of microbial evolution and development.

What advice do you have for today's students?

Develop an expertise as an undergraduate in some area of science, whether it be biology, chemistry, mathematics, physics, or some other area of science or engineering, and be creative and curious about other disciplines. The world of the future will be interdisciplinary and multidisciplinary.

How does your family relate to your work?

I have been happily married ever since I graduated from college! We have two daughters. One is a medical doctor (pediatrician). She recently was named an outstanding physician scholar and voted the best physician in her fellowship class by her colleagues. She also worked in Africa on delivery of health care to women in Tanzania for her PhD. We are equally proud of our other daughter, who earned a PhD in evolutionary biology and now works for the U.S. Geological Survey, cataloging rare plants in Yosemite National Park and Forest.

Chapter 1

Microbial Life: Origin and Discovery

Life on Earth began early in our planet's history with microscopic organisms, or microbes. Microbial life has since shaped our atmosphere, our geology, and the energy cycles of all ecosystems. A human body contains ten times as many microbes as it does human cells, including numerous tiny bacteria on the skin and in the digestive tract. Throughout history, humans have had a hidden partnership with microbes ranging from food production and preservation to mining for precious minerals.

Yet throughout most of our history, humans were unaware that microbes even existed. To study these unseen organisms required a microscope, first developed in the 1600s. In the nineteenth century—the "golden age" of microbiology—microscopes revealed the tiny organisms at work in our bodies and in our ecosystems. The twentieth century saw the rise of microbes as the engines of biotechnology. Microbial discoveries led to recombinant DNA and revealed the secrets of the first sequenced genomes.

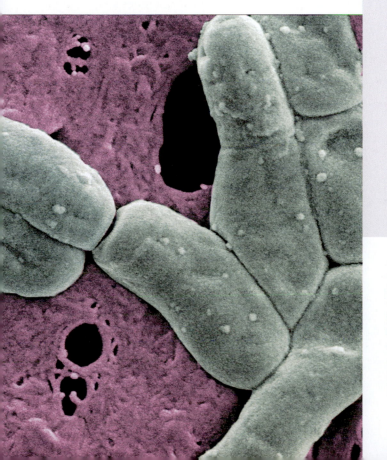

Lactobacillus salivarius bacteria grow normally in human skin, where they produce bacteriocins, compounds that protect us from disease-causing bacteria. Their multi-part genome was sequenced by Marcus Claesson and colleagues. (Claesson, et al. 2006. *Proceedings of the National Academy of Sciences* 103:6718.) Scanning electron micrograph is from Sinead Leahy and D. John, Trinity College, Dublin. Cell length, 1–2 μm.

In 2004, the two Mars Exploration Rovers, *Spirit* and *Opportunity*, landed on the planet Mars (**Fig. 1.1**). The rovers carried scientific instruments to test Martian rocks, to identify minerals and to assess the size and shape of sedimentary particles. The identity of the minerals, as well as their particle structure, could yield clues as to whether the Martian surface had ever been shaped by liquid water. Evidence for water would support the possible existence of living microbes.

Why do we care whether microbes exist on Mars? The discovery of life beyond Earth would fundamentally change how we see our place in the universe. The observation of Martian life could yield clues as to the origin of our own biosphere and expand our knowledge of the capabilities of living cells on our own planet. As of this writing, the existence of microbial life on Mars remains unknown, but here on Earth, many terrestrial microbes remain as mysterious as Mars. Barely 0.1% of the microbes in our biosphere can be cultured in the laboratory; even the digestive tract of a newborn infant contains species of bacteria unknown to science. Our "exploration rovers" for microbiology include, for example, new tools of microscopy and the sequencing of microbial DNA.

On Earth, the microscope reveals microbes throughout our biosphere, from the superheated black smoker vents at the ocean floor to the subzero ice fields of Antarctica. Bacteria such as *Escherichia coli* live in our intestinal tract, while algae and cyanobacteria turn ponds green (**Fig. 1.2**). Protists are the predators of the microscopic world. And viruses such as papillomavirus cause disease, as do many bacteria and protists.

NASA

Figure 1.1   Is there microbial life on Mars?   On February 9, 2004, the Mars Exploration Rover *Spirit* (inset) photographed this windswept surface of the planet Mars. Rock samples were tested for distinctive minerals that are formed by the action of water. The presence of liquid water today would increase the chance that microbial life exists on Mars.

Yet before microscopes were developed in the seventeenth century, we humans were unaware of the unseen living organisms that surround us, that float in the air we breathe and the water we drink, and that inhabit our own bodies. Microbes generate the very air we breathe, including nitrogen gas and much of the oxygen and carbon dioxide. They fix nitrogen into forms used by plants, and they make essential vitamins, such as vitamin B_{12}. Microbes are the primary producers of major food webs, particularly in the oceans; when we eat fish, we indirectly consume tons of algae at the base of the food chain. At the same time, virulent pathogens take our lives. Despite all the advances of modern medicine and public health, microbial disease remains the number one cause of human mortality.

In the twentieth century, the science of microbiology exploded with discoveries, creating entire new fields such as genetic engineering. The promise—and pitfalls—were dramatized by Michael Crichton's best-selling science fiction novel and film, *The Andromeda Strain* (1969; filmed in 1971). In *The Andromeda Strain*, scientists at a top-secret laboratory race to identify a deadly pathogen from outer space—or perhaps from a biowarfare lab (**Fig. 1.3A**). The film prophetically depicts the computerization of medical research, as well as the emergence of pathogens, such as the human immunodeficiency virus (HIV), that can yet defeat the efforts of advanced science.

Today, we discover surprising new kinds of microbes deep underground and in places previously thought uninhabitable, such as the hot springs of Yellowstone National Park (**Fig. 1.3B**). These microbes shape our biosphere and provide new tools that impact human society. For example, the use of heat-resistant bacterial DNA polymerase (a DNA-replicating enzyme) in a technique called the polymerase chain reaction (PCR) allows us to detect minute amounts of DNA in traces of blood or fossil bone. Microbial technologies led us from the discovery of the double helix to the sequence of the human genome, the total genetic information that defines our species.

In this chapter, we introduce the concept of a microbe and the question of how microbial life originated. We then survey the history of human discovery of the role microbes play in disease and in our ecosystems. Finally, we address the exciting century of molecular microbiology, in which microbial genetics and genomics have transformed the face of modern biology and medicine.

1.1   From Germ to Genome: What Is a Microbe?

From early childhood, we hear that we are surrounded by microscopic organisms, or "germs," that we cannot see. What are microbes? Our modern concept of a

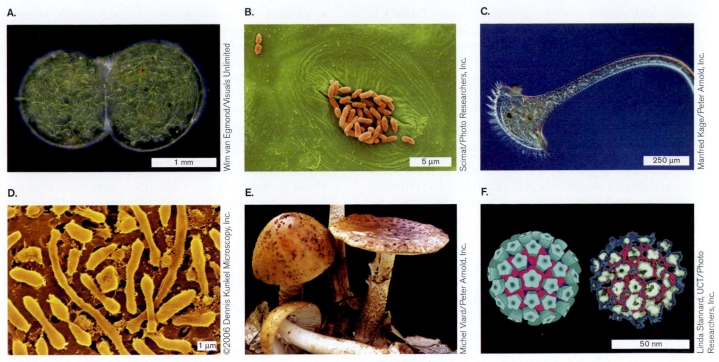

Figure 1.2   Representative microbes.   A. Filamentous cyanobacteria produce oxygen for planet Earth (dark-field light micrograph). **B.** *Escherichia coli* bacteria colonize the stomata of a lettuce leaf cell (scanning electron microscopy). **C.** *Stentor* is a protist, a eukaryotic microbe. Cilia beat food into its mouth. **D.** Halophilic archaea, a form of life distinct from bacteria and eukaryotes, grow at extremely high salt concentration. **E.** Mushrooms are multicellular fungi (eukaryotes). They serve the ecosystem as decomposers. **F.** Papillomavirus causes genital warts, an infectious disease commonly acquired by young adults (model based on electron microscopy).

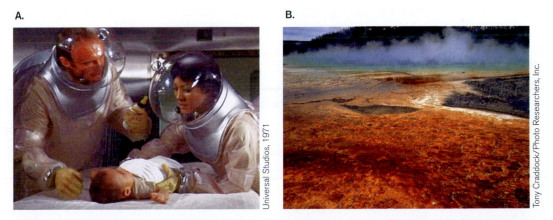

Figure 1.3   Microbial discovery: science fiction and science fact.   A. In *The Andromeda Strain*, medical scientists try to feed a baby who was infected by a deadly pathogen from outer space. While the details of the pathogen are imaginary, the film's approach to identifying the mystery organism captures the spirit of actual investigations of emerging diseases. **B.** Yellowstone National Park hot springs are surrounded by mats of colorful microbes that grow above 80°C in waters containing sulfuric acid. Bacteria discovered at Yellowstone produce enzymes used in polymerase chain reaction (PCR), a technique of DNA amplification.

microbe has deepened through two major research tools: advanced microscopy and the sequencing of genomic DNA. Modern microscopy is covered in Chapter 2, and microbial genetics and genomics are presented in Chapters 6–12.

A Microbe Is a Microscopic Organism

A **microbe** is commonly defined as a living organism that requires a microscope to be seen. Microbial cells range in size from millimeters (mm) down to 0.2 micrometers (μm), and viruses may be tenfold smaller (**Table 1.1**). Some microbes consist of a single cell, the smallest unit of life, a membrane-enclosed compartment of water solution containing molecules that carry out metabolism. Each microbe contains in its genome the capacity to reproduce its own kind.

Our simple definition of a microbe, however, leaves us with contradictions.

▪ **Super-size microbial cells.** Most single-celled organisms require a microscope to render them visible and thus fit the definition of "microbe." Nevertheless, some species of protists and algae, and even some bacterial cells are large enough to see with the naked eye (**Fig. 1.4**). The marine sulfur bacterium *Thiomargarita namibiensis,* called the sulfur pearl of Namibia, grows as large as the head of a fruit fly. Even more surprising, a single-celled plant, the "killer algae" *Caulerpa taxifolia,* spreads through the coastal waters of California. The single cell covers many acres with its leaf-like cell parts.

▪ **Microbial communities.** Many microbes form complex multicellular assemblages, such as mushrooms, kelps, and biofilms. In these structures, cells are differentiated into distinct types that complement each other's function, as in multicellular organisms. And yet, some multicellular worms and arthropods require a microscope to see but are *not* considered microbes.

▪ **Viruses.** A **virus** consists of a noncellular particle containing genetic material that takes over the metabolism of a cell to generate more virus particles. Although viruses are considered microbes, they are not fully functional cells. Some viruses consist of only a few molecular parts, whereas others, such as the Mimivirus infecting amebas (also spelled amoebae) show the size and complexity of a cell.

NOTE: Each section contains questions to think about. These thought questions may have various answers. Possible responses are posted at the back of the book.

THOUGHT QUESTION 1.1 The minimum size of known microbial cells is about 0.2 μm. Could even smaller cells be discovered? What factors may determine the minimum size of a cell?

THOUGHT QUESTION 1.2 If viruses are not functional cells, are they truly "alive"?

In practice, our definition of a microbe derives from tradition as well as genetic considerations. In this book, we consider microbes to include **prokaryotes** (cells lacking a nucleus, including bacteria and archaea) as well as certain classes of **eukaryotes** (cells with a nucleus) that

Table 1.1	Sizes of some microbes.	
Microbe	**Description**	**Approximate size**
Varicella-zoster virus 1	Virus that causes chicken pox and shingles	100 nanometers (nm) = 10^{-7} meter (m)
Prochlorococcus	Photosynthetic marine bacteria	500 nm = 5×10^{-7} m
Rhizobium	Bacteria that fix N_2 in symbiosis with leguminous plants	1 micrometer (μm) = 10^{-6} m
Spirogyra	Filamentous algae found in aquatic habitat	40 μm = 4×10^{-5} m (cell width)
Pelomyxa (an ameba)	Protists found in solid or aquatic habitat	5 millimeters (mm)

A.

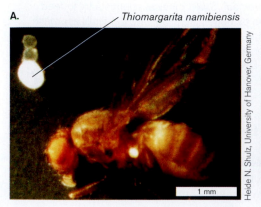

Thiomargarita namibiensis

1 mm

Heide N. Shulz, University of Hanover, Germany

B.

Rachel Woodfield, Merkel & Associates

Figure 1.4 Giant microbial cells. A. The largest known bacterium, *Thiomargarita namibiensis,* a marine sulfur metabolizer, nearly the size of the head of a fruit fly. **B.** "Killer algae," *Caulerpa taxifolia.* All the fronds constitute a single cell, the largest single-celled organism on Earth. Growing off the coast of California. *Source*: A. Reprinted with permission from H. N. Shulz, et al. 1999. *Science* 284(5413):493. © 2005 AAAS.

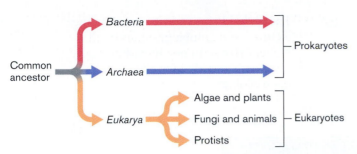

Figure 1.5 Three domains of life. Analysis of DNA sequence reveals the ancient divergence of three domains of living organisms: *Bacteria* and *Archaea* (both prokaryotes) and *Eukarya* (eukaryotes). The color code shown here is used throughout this book to indicate the three domains.

include simple multicellular forms: algae, fungi, and protists (**Fig. 1.5**). The bacteria, archaea, and eukaryotes—known as the three domains—diverged from a common ancestral cell. We also discuss viruses and related infectious particles (Chapters 6 and 11).

> **NOTE:** The formal names of the three domains are ***Bacteria***, ***Archaea***, and ***Eukarya***. Members of these domains are called **bacteria** (singular, **bacterium**), **archaea** (singular, **archaeon**), and **eukaryotes** (singular, **eukaryote**), respectively. The microbiology literature includes alternative spellings for some of these terms, such as "archaean" and "eucaryote."

Microbial Genomes Are Sequenced

Our understanding of microbes has grown tremendously through the study of their genomes. A **genome** is the total genetic information contained in an organism's chromosomal DNA (**Fig. 1.6**). By determining the sequence of genes in a microbe's genome, we learn a lot about how that microbe grows and associates with other species. For

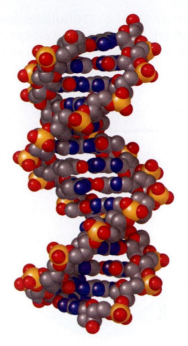

Figure 1.6 DNA. The sequence of base pairs in DNA encodes all the genetic information of an organism.

example, if a microbe's genome includes genes for nitrogenase, a nitrogen-fixing enzyme, that microbe probably can fix nitrogen from the atmosphere into compounds that plants can assimilate into protein. And by comparing DNA sequences, we can measure the degree of relatedness between different species based on the time since they diverged from a common ancestor.

Historically, the first genomes to be sequenced were those of viruses. The first genome whose complete DNA sequence was determined was that of a bacteriophage (a virus that infects bacteria), bacteriophage ϕX174. The DNA sequence of ϕX174 was determined in 1977 by Fred Sanger (**Fig. 1.7A**), who shared the 1980 Nobel Prize in Chemistry with Walter Gilbert and Paul Berg for developing the method of DNA sequence analysis. The

Figure 1.7 Microbial genome sequencers.
A. Fred Sanger, who shared the 1980 Nobel Prize in Chemistry for devising the method of DNA sequence analysis that is the basis of modern genome sequencing. He is reading sequence data from bands of DNA separated by electrophoresis. **B.** Claire Fraser-Liggett, past president of The Institute for Genomic Research (TIGR), which completed the sequences of *H. influenzae* and many other microbial genomes.

genome of bacteriophage φX174 includes over 5,000 base pairs, with just ten genes encoding proteins (**Fig. 1.8A**). Its genome is so compact that several gene sequences actually overlap, sharing the same segment of nucleotides (for example, genes C and K).

WWW | The Nobel Prize website presents the lectures and autobiographies of all Nobel Prize winners, including many who were awarded for advances in microbiology.

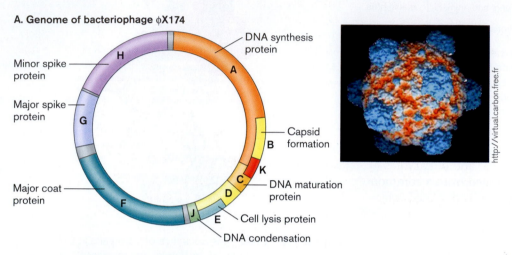

A. Genome of bacteriophage φX174

- Minor spike protein
- Major spike protein
- Major coat protein
- H
- G
- F
- J
- E
- DNA synthesis protein
- A
- B — Capsid formation
- K
- C — DNA maturation protein
- D
- Cell lysis protein
- DNA condensation

http://virtual.carbon.free.fr

B. Genome of *Haemophilus influenzae*

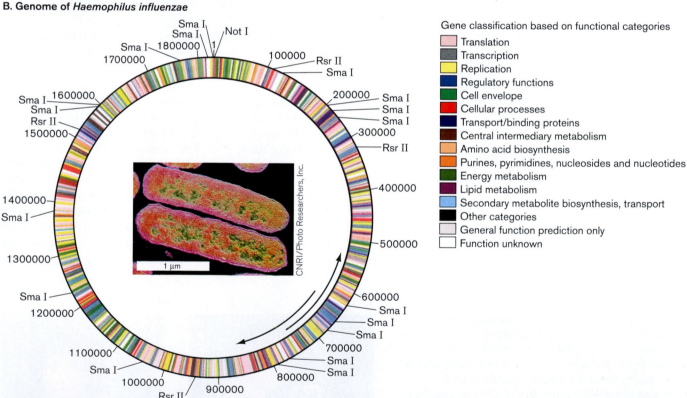

Gene classification based on functional categories

- Translation
- Transcription
- Replication
- Regulatory functions
- Cell envelope
- Cellular processes
- Transport/binding proteins
- Central intermediary metabolism
- Amino acid biosynthesis
- Purines, pyrimidines, nucleosides and nucleotides
- Energy metabolism
- Lipid metabolism
- Secondary metabolite biosynthesis, transport
- Other categories
- General function prediction only
- Function unknown

CNRI/Photo Researchers, Inc.

1 μm

Figure 1.8   The first sequenced genomes.   A. The first organism whose genome sequence was determined was bacteriophage φX174, a virus that grows in *Escherichia coli* (virus diameter, 27 nm). The entire DNA sequence of φX174 contains 5,386 base pairs specifying only ten genes (here labeled A–H and J–K), nine of whose functions have since been determined. Note the highly compact genome, with several overlapping genes. **B.** The genome of *Haemophilus influenzae* Rd, a bacterium that causes ear infections and meningitis, was the first DNA sequence completed for a cellular organism (inset, colorized electron micrograph). The genome of *H. influenzae* contains nearly 2 million base pairs specifying approximately 1,743 genes, which are expressed to make protein and RNA products. The annotated sequence of the genome appears on the website of the National Center for Biotechnology Information. Colored bars indicate gene sequences throughout.

Nearly two decades passed before scientists completed the first genome sequence of a cellular microbe, *Haemophilus influenzae*, a bacterium that causes ear infections and meningitis in children (**Fig. 1.8B**). The strain of *H. influenzae* sequenced has nearly 2 million base pairs, which specify about 1,700 genes. The sequence of *H. influenzae* was determined by a large team of scientists at The Institute for Genomic Research (TIGR) led by Hamilton Smith and Craig Venter, who devised a special computational strategy for assembling large amounts of sequence data. This strategy was later applied to sequencing the human genome. Led by president Claire Fraser-Liggett, (**Fig. 1.7B**), TIGR sequenced the genomes of numerous microbes, such as *Bacillus anthracis*, the bacterium that causes anthrax, and *Colwellia psychrerythraea*, a cold-loving bacterium growing in Antarctic sea ice.

WWW | The National Center for Biotechnology Information (NCBI) provides free access to all published genome sequences.

The growing availability of sequenced genomes at universities and in industry has generated the new field of comparative genomics, which involves the systematic comparison of all genomic sequences of living species, ranging from microbes to *Homo sapiens*. Comparative genomics reveals a set of core genes shared by all organisms, further evidence that all life on Earth shares a common ancestry.

1.2 Microbes Shape Human History

Throughout most of human history, we were unaware of the microbial world. Microorganisms have shaped human culture since our earliest civilizations. Yeasts and bacteria have made foods such as bread and cheese, as well as alcoholic beverages (**Fig. 1.9A**; also discussed in Chapter 16). "Rock-eating" bacteria known as lithotrophs leached copper and other metals from ores exposed by mining, enabling ancient human miners to obtain these metals. The lithotrophic oxidation of minerals for energy generates strong acid, which accelerates breakdown of the ore. Today, about 20% of the world's copper, as well as some uranium and zinc, are produced by bacterial leaching. Unfortunately, microbial acidification also consumes the stone of ancient monuments (**Fig. 1.9B**), a process intensified by airborne acidic pollution. Management of microbial corrosion is an important field of applied microbiology.

As humans became aware of microbes, our relationship with the microbial world changed in important ways (**Table 1.2**, pages 14–15). Early microscopists in the seventeenth and eighteenth centuries formulated key concepts of microbial existence, including their means of reproduction and death. In the nineteenth century, the "golden age" of microbiology, key principles of disease pathology and microbial ecology were established that scientists still use today. This period laid the foundation for modern science, in which genetics and molecular biology provide powerful tools for scientists to manipulate microorganisms for medicine and industry.

Microbial Disease Devastates Human Populations

Throughout history, microbial diseases such as tuberculosis and leprosy have profoundly affected human demographics and cultural practices (**Fig. 1.10**). The bubonic plague, which wiped out a third of Europe's population in the fourteenth century, was caused by *Yersinia pestis*, a bacterium spread by rat fleas. Ironically, the plague-induced population decline enabled the social

Figure 1.9 Production and destruction by microbes.
A. Roquefort cheeses ripening in France. **B.** Statue undergoing decay from the action of lithotrophic microbes. The process is accelerated by acid rain. Cathedral of Cologne, Germany.

A.

Bettmann/Corbis

B.

Johner Images/Getty Images

transformation that led to the Renaissance, a period of unprecedented cultural advancement. In the nineteenth century, the bacterium *Mycobacterium tuberculosis* stalked overcrowded cities, and tuberculosis became so common that the pallid appearance of tubercular patients became a symbol of tragic youth in European literature. Today, societies throughout the world have been profoundly shaped by the epidemic of acquired immunodeficiency syndrome (AIDS), caused by the human immunodeficiency virus (HIV).

Historians traditionally emphasize the role of warfare in shaping human destiny, the brilliance of leaders or the advantage of new technology in determining which civilizations rise or fall. Yet the fate of human societies has often been determined by microbes. For example, much of the native population of North America was exterminated by smallpox unwittingly introduced by European invaders. Throughout history, more soldiers have died of microbial infections than of wounds in battle. The significance of disease in warfare was first recognized by the British nurse and statistician Florence Nightingale (1820–1910) (**Fig. 1.11**).

Better known as the founder of professional nursing, Nightingale also founded the science of medical statistics. She used methods invented by French statisticians to demonstrate the high mortality rate due to disease among British soldiers during the Crimean War. Nightingale's statistics convinced the British government to improve army living conditions and to upgrade the standards of army hospitals. In modern epidemiology, statistical analysis continues to serve as a crucial tool in determining the causes of disease.

WWW | The Centers for Disease Control, in Atlanta, is the global center of modern medical statistics and epidemiology.

Figure 1.10   Microbial disease in history and culture. **A.** Medieval church procession to ward off the Black Death (bubonic plague). **B.** The AIDS Memorial Quilt spread before the Washington Monument. Each panel of the quilt memorializes an individual who died of AIDS.

Microscopes Reveal the Microbial World

The seventeenth century was a time of growing inquiry and excitement about the "natural magic" of science and patterns of our world, such as the laws of gravitation and motion formulated by Isaac Newton (1642–1727). Robert Boyle (1627–1691) performed the first controlled experiments on the chemical conversion of matter. Physicians attempted new treatments for disease involving the application of "stone and minerals" (that is, the application of chemicals), what today we would call chemotherapy. Minds

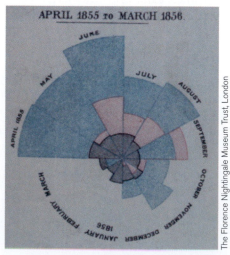

Figure 1.11   Florence Nightingale, founder of medical statistics.   **A.** Florence Nightingale was the first to use medical statistics to demonstrate the significance of mortality due to disease. **B.** Nightingale's polar area chart of mortality data during the Crimean War.

APRIL 1855 TO MARCH 1856.

were open to at least consider the astounding possibility that our surroundings, indeed our very bodies, were inhabited by tiny living beings.

Robert Hooke observes the microscopic world. The first microscopist to publish a systematic study of the world as seen under a microscope was Robert Hooke (1635–1703). As Curator of Experiments to the Royal Society of London, Hooke built the first compound microscope—a magnifying instrument containing two or more lenses that multiply their magnification in series. With his microscope, Hooke observed biological materials such as nematode "vinegar eels," mites, and mold filaments, illustrations of which he published in *Micrographia* (1665), the first publication that illustrated objects observed under a microscope (**Fig. 1.12**).

Hooke was the first to observe distinct units of living material, which he called "cells." Hooke first named the units cells because the shape of hollow cell walls in a slice of cork reminded him of the shape of monks' cells in a monastery. But his crude lenses achieved at best 30-fold power (30×), so he never observed single-celled organisms.

Antoni van Leeuwenhoek observes bacteria with a single lens. Hooke's *Micrographia* inspired other microscopists, including Antoni van Leeuwenhoek (1632–1723), who became the first individual to observe single-celled microbes (**Fig. 1.13A**). As a young man, Leeuwenhoek lived in the Dutch city of Delft, where he worked as a cloth draper, a profession that introduced him to magnifying glasses. (The magnifying glasses were used to inspect the quality of the cloth, enabling the worker to count the number of threads.) Later in life, he took up

Figure 1.12 Hooke's *Micrographia*. An illustration of mold sporangia, drawn by Hooke in 1665, from his observations of objects using a compound microscope.

the hobby of grinding ever stronger lenses to see into the world of the unseen.

Leeuwenhoek ground lenses stronger than Hooke's, which he used to build single-lens magnifiers, complete with sample holder and focus adjustment (**Fig. 1.13B**). First he observed insects, including lice and fleas, then the relatively large single cells of protists and algae, then ultimately bacteria. One day he applied his microscope to observe matter extracted from between his teeth. He wrote, "to my great surprise [I] perceived that the aforesaid matter contained very many small living Animals, which moved themselves very extravagantly."

Over the rest of his life, Leeuwenhoek recorded page after page on the movement of microbes, reporting their size and shape so accurately that in many cases we can determine the species he observed (**Fig. 1.13C**).

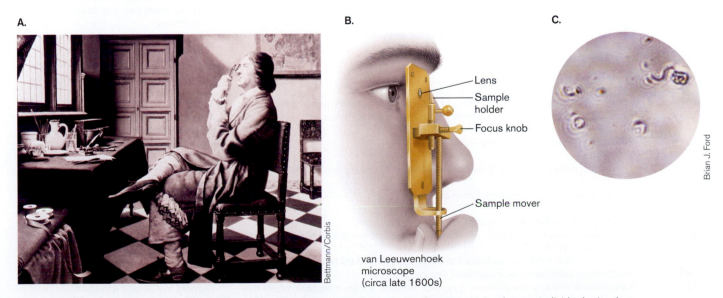

A.

B.

Lens
Sample holder
Focus knob
Sample mover

van Leeuwenhoek microscope (circa late 1600s)

C.

Figure 1.13 Antoni van Leeuwenhoek. A. A portrait of Leeuwenhoek, the first person to observe individual microbes.
B. "Microscope" (magnifying glass) used by Leeuwenhoek. **C.** Spiral bacteria viewed through a replica of Leeuwenhoek's instrument.

Table 1.2 Microbes and human history.

Date	Microbial discovery	Discoverer(s)
Microbes impact human culture without detection		
10,000 BC	Food and drink are produced by microbial fermentation.	Egyptians, Chinese, and others
1,500 BC	Tuberculosis, polio, leprosy, and smallpox are evident in mummies and tomb art.	Egyptians
50 BC	Copper is recovered from mine water acidified by sulfur-oxidizing bacteria.	Roman metal workers under Julius Caesar
1546 AD	Syphilis and other diseases are observed to be contagious.	Girolamo Fracastoro (Padua)
Early microscopy and the origin of microbes		
1676	Microbes are observed under a microscope.	Antoni van Leeuwenhoek (Netherlands)
1688	Spontaneous generation is disproved for maggots.	Francesco Redi (Italian)
1717	Smallpox is prevented by inoculation of pox material, a rudimentary form of immunization.	Turkish women taught Lady Montagu, who brought the practice to England
1765	Microbe growth in organic material is prevented by boiling in a sealed flask.	Lazzaro Spallanzani (Padua)
1798	Cowpox vaccination prevents smallpox.	Edward Jenner (England)
1835	Fungus causes disease in silkworms (first pathogen to be demonstrated in animals).	Agostino Bassi de Lodi (Italy)
1847	Chlorine as antiseptic wash for doctor's hands decreases pathogens.	Ignaz Semmelweis (Hungary)
1881	Bacterial spores survive boiling, but are killed by cyclic boiling and cooling.	John Tyndall (Ireland)
"Golden age" of microbiology: principles and methods established		
1855	Statistical correlation is shown between sanitation and mortality (Crimean War).	Florence Nightingale (England)
1857	Microbial fermentation produces lactic acid or alcohol.	Louis Pasteur (France)
1864	Microbes fail to appear spontaneously, even in the presence of oxygen.	Louis Pasteur (France)
1866	Microbes are defined as a class distinct from animals and plants.	Ernst Haeckel (Germany)
1867	Antisepsis during surgery prevents patient death.	Robert Lister (England)
1877	Bacteria are a causative agent in developing anthrax.	Robert Koch (Germany)
1881	The first artificial vaccine is developed (against anthrax).	Louis Pasteur (France)
1882	First pure culture of colonies on solid medium, *Mycobacterium tuberculosis*.	Robert Koch (Germany)
1884	Koch's postulates are published, based on anthrax and tuberculosis.	Robert Koch (Germany)
1884	Gram stain devised to distinguish bacteria from human cells.	Hans Christian Gram (Netherlands)
1886	Intestinal bacteria include *Escherichia coli*, the future model organism.	Theodor Escherich (Austria)
1889	Bacteria oxidize iron and sulfur (lithotrophy).	Sergei Winogradsky (Russia)
1889	Bacteria isolated from root nodules are proposed to fix nitrogen.	Martinus Beijerinck (Netherlands)
1899	The concept of a virus is proposed to explain tobacco mosaic disease.	Martinus Beijerinck (Netherlands)
Cell biology, biochemistry, and genetics		
1908	Antibiotic chemicals are synthesized and identified (chemotherapy).	Paul Ehrlich (USA)
1911	Cancer in chickens can be caused by a virus.	Peyton Rous (USA)
1917	Bacteriophages are recognized as viruses that infect bacteria.	Frederick Twort (England) and Felix D'Herelle (France)
1924	The ultracentrifuge is invented and used to measure the size of proteins.	Theodor Svedberg (Sweden)
1928	*Streptococcus pneumoniae* bacteria are transformed by a genetic material from dead cells.	Frederick Griffith (England)
1929	Penicillin, the first widely successful antibiotic, is made by a fungus. The molecule is isolated in 1941.	Alexander Fleming (Scotland), Howard Florey (Australia), and Ernst Chain (Germany)
1933–1945	The transmission electron microscope is invented and used to observe cells.	Ernst Ruska and Max Knoll, inventors (Germany); first cells observed by Albert Claude (Belgium), Christian de Duve (Belgium), and George Palade (USA)
1937	The tricarboxylic acid cycle is discovered.	Hans Krebs (England)
1938	The microbial "kingdom" is subdivided into eukaryotes and prokaryotes (Monera).	Herbert Copeland (USA)
1938	*Bacillus thuringiensis* spray is produced as the first bacterial insecticide.	Insecticide manufacturers (France)
1941	One gene encodes one enzyme in *Neurospora*.	George Beadle and Edward Tatum (USA)
1941	Poliovirus is grown in human tissue culture.	John Enders, Thomas Weller, and Frederick Robbins (USA)
1944	The genetic material responsible for transformation of *S. pneumoniae* is DNA.	Oswald Avery, Colin Macleod, and Maclyn McCarty (USA)
1945	Bacteriophage replication mechanism is elucidated.	Salvador Luria (Italy) and Max Delbrück (Germany), working in the USA
1946	Bacteria transfer DNA by conjugation.	Edward Tatum and Joshua Lederberg (USA)
1946–1956	X-ray diffraction crystal structures are obtained for the first complex biological molecules, penicillin and vitamin B_{12}.	Dorothy Hodgkin, J. D. Bernal, and coworkers (England)
1950	Anaerobic culture technique is devised to study anaerobes of the bovine rumen.	Robert Hungate (USA)
1950	Bacteria can carry latent bacteriophages (lysogeny).	André Lwoff (France)
1951	Transposable elements are discovered in maize and later shown in bacteria, where they play key roles in evolution.	Barbara McClintock (USA)
1952	DNA is injected into a cell by a bacteriophage.	Martha Chase and Alfred Hershey (USA)

Table 1.2 Microbes and human history (*continued*)

Date	Microbial discovery	Discoverer(s)
	Molecular biology and recombinant DNA	
1953	The overall structure of DNA is a double helix, based on X-ray diffraction analysis.	Rosalind Franklin and Maurice Wilkins (England)
1953	Double-helical DNA consists of antiparallel chains connected by the hydrogen bonding of AT and GC base pairs.	James Watson (USA) and Francis Crick (England)
1959	Expression of the messenger RNA for the *E. coli lac* operon is regulated by a repressor protein.	Arthur Pardee (England) and François Jacob and Jacques Monod (France)
1960	Radioimmunoassay for detection of biomolecules is developed.	Rosalyn Yalow and Solomon Bernson (USA)
1961	The chemiosmotic hypothesis, which states that biochemical energy is stored in a transmembrane proton gradient, is proposed and tested.	Peter Mitchell and Jennifer Moyle (England)
1966	The genetic code by which DNA information specifies protein sequence is deciphered.	Marshall Nirenberg, H. Gobind Khorana, and others (USA)
1967	Bacteria can grow at temperatures above 80° in hot springs at Yellowstone National Park.	Thomas Brock (USA)
1968	Serial endosymbiosis is proposed to explain the evolution of mitochondria and chloroplasts.	Lynn Margulis (USA)
1969	Retroviruses contain reverse transcriptase, which copies RNA to make DNA.	Howard Temin, David Baltimore, Renato Dulbecco (USA)
1972	Inner and outer membranes of gram-negative bacteria (*Salmonella*) are separated by ultracentrifugation.	Mary Osborn (USA)
1973	A recombinant DNA molecule is created in vitro (in a test tube).	Stanley Cohen, Annie Chang, Robert Helling, and Herbert Boyer (USA)
1974	The bacterial flagellum is driven by a rotary motor.	Howard Berg, Michael Silverman, and Melvin Simon (USA)
1975	mRNA-rRNA base pairing initiates protein synthesis.	Joan Steitz and Karen Jakes (USA) and Lynn Dalgarno and John Shine (Australia)
1975	The dangers of recombinant DNA are assessed at the Asilomar Conference.	Paul Berg, Maxine Singer, and colleagues (USA)
1975	Monoclonal antibodies are produced indefinitely in tissue culture by hybridomas, antibody-producing cells fused to cancer cells.	George Kohler and Cesar Milstein (USA)
1977	A DNA-sequencing method is invented and used to sequence the first genome of a virus.	Fred Sanger, Walter Gilbert, and Allan Maxam (USA)
1977	Archaea are a third domain of life, the others being eukaryotes and bacteria.	Carl Woese (USA)
1978	The first protein catalog is compiled for *E. coli* based on 2D gels.	Fred Neidhart, Peter O'Farrell, and colleagues (USA)
1978	Biofilms are a major form of existence of microbes.	William Costerton and others (Canada)
1979	Smallpox is declared eliminated, the culmination of worldwide efforts of immunology, molecular biology, and public health.	The World Health Organization
	Genomics, structural biology, and molecular ecology	
1981	Invention of the polymerase chain reaction (PCR) makes available large quantities of DNA.	Kary Mullis (USA)
1981–1986	Self-splicing RNA is discovered in the protist *Tetrahymena*, evidence that life could have originated as an "RNA world."	Thomas Cech and Sidney Altman (USA)
1982	Archaea are discovered with optimal growth above 100°C.	Karl Stetter (Germany)
1982	Viable but nonculturable bacteria contribute to ecology and pathology.	Rita Colwell, Norman Pace, and others (USA)
1982	Prions, infectious agents consisting solely of protein, are characterized.	Stanley Prusiner (USA)
1983	Human immunodeficiency virus (HIV) is implicated in the development of AIDS.	Luc Montagnier and colleagues (France)
1983	Genes are introduced into plants by using *Agrobacterium tumefaciens* plasmid vectors.	Eugene Nester, Mary-Dell Chilton, and colleagues at the Monsanto Company (USA)
1984	Acid-resistant *Helicobacter pylori* are discovered in the stomach, where they lead to gastritis.	Barry Marshall and Robin Warren (Australia)
1987–2004	*Geobacter* bacteria that can generate electricity are discovered, and their genomes are sequenced.	Derek Lovley and colleagues (USA)
1988	Earth's smallest and most abundant photosynthesizer is *Prochlorococcus*.	Sallie Chisholm and colleagues (USA)
1993	Giant bacterium (*Epulopiscium*) is identified, large enough to see.	Esther Angert and Norman Pace (USA)
1995	The first genome is sequenced for a cellular organism, *Haemophilus influenzae*.	Craig Venter, Hamilton Smith, Claire Fraser, and others (USA)
2001	The ribosome structure is obtained at near-atomic level by X-ray diffraction.	Marat Yusupov, Harry Noller, and colleagues (USA)
2008	Over 1,000 genome sequences of bacteria and archaea are publicly available.	National Center for Biotechnology Information (USA)

He performed experiments, comparing, for example, the appearance of "small animals" from his teeth before and after drinking hot coffee. The disappearance of microbes from his teeth after drinking a hot beverage suggested that heat killed microbes—a profoundly important principle for the study and control of microbes ever since.

Ironically, Leeuwenhoek is believed to have died of a disease contracted from sheep whose bacteria he observed. Historians have often wondered why it took so many centuries for Leeuwenhoek and his successors to determine the link between microbes and disease. Although observers such as Agostino Bassi de Lodi noted isolated cases of microbes associated with pathology (see Table 1.2), the very ubiquity of microbes—most of them actually harmless—may have obscured their more deadly roles. Also, it was hard to distinguish between microbes and the single-celled components of the human body, such as blood cells and sperm. It was not until the nineteenth century that human tissues could be distinguished from microbial cells by the application of differential chemical stains (discussed in Chapter 2).

> **THOUGHT QUESTION 1.3**   Why do you think it took so long for humans to connect microbes with infectious disease?

Spontaneous Generation: Do Microbes Have Parents?

The observation of microscopic organisms led priests and philosophers to wonder where they came from. In the eighteenth century, scientists and church leaders intensely debated the question of **spontaneous generation**, the theory that living creatures such as maggots could arise spontaneously, without parental organisms. Chemists of the day tended to support spontaneous generation, as it appeared similar to the changes in matter that could occur upon mixture of chemicals. Christian church leaders, however, supported the biblical view that all organisms have "parents" going back to the first week of creation.

The Italian priest Francesco Redi (1626–1697) showed that maggots in decaying meat were the offspring of flies. Meat kept in a sealed container, excluding flies, did not produce maggots. Thus, Redi's experiment argued against spontaneous generation for macroscopic organisms. The meat still putrefied, however, producing microbes that seemed to arise "without parents."

To disprove spontaneous generation of microbes, another Italian priest, Lazzaro Spallanzani (1729–1799),

showed that a sealed flask of meat broth sterilized by boiling failed to grow microbes. Spallanzani also noticed that microbes often appeared in pairs. Were these two parental microbes coupling to produce offspring, or did one microbe become two? By long and tenacious observation, Spallanzani watched a single microbe grow in size until it split in two. Thus, he demonstrated cell fission, the process by which cells arise by the splitting of preexisting cells.

Even Spallanzani's experiments, however, did not put the matter to rest. Proponents of spontaneous generation argued that the microbes in the priest's flask lacked access to oxygen and therefore could not grow. The pursuit of this question was left to future microbiologists, including the famous French microbiologist Louis Pasteur (1822–1895) (**Fig. 1.14A**). In addressing spontaneous generation and related questions, Pasteur and his contemporaries laid the foundations for modern microbiology.

Louis Pasteur reveals the biochemical basis of microbial growth. Pasteur began his scientific career as a chemist and wrote his doctoral thesis on the struc-

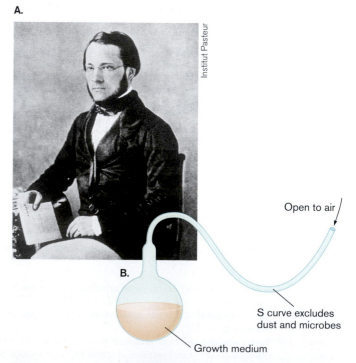

A.

Institut Pasteur

B.

Open to air

S curve excludes dust and microbes

Growth medium

Figure 1.14 Louis Pasteur, founder of medical microbiology. A. Pasteur's contributions to the science of microbiology and immunology earned him lasting fame. **B.** Swan-necked flask. Pasteur showed that in such a flask, after boiling, the contents remain free of microbial growth, despite access to air.

ture of organic crystals. He discovered the fundamental chemical property of chirality, the fact that some organic molecules exist in two forms that differ only by mirror symmetry. In other words, the two structures are mirror images of one another, like the right and left hands. Pasteur found that when microbes were cultured on a nutrient substance containing both mirror forms, only one mirror form was consumed. He concluded that the metabolic preference for one mirror form was a fundamental property of life. Subsequent research has confirmed that most biological molecules, such as DNA and proteins, occur in only one of their mirror forms.

As a chemist, Pasteur was asked to help with a widespread problem encountered by French manufacturers of wine and beer. The production of alcoholic beverages is now known to occur by **fermentation**, a process by which microbes gain energy by converting sugars into alcohol. In the time of Pasteur, however, the conversion of grapes or grain to alcohol was believed to be a spontaneous chemical process. No one could explain why some fermentation mixtures produced vinegar (acetic acid) instead of alcohol. Pasteur discovered that fermentation is actually caused by living yeast, a single-celled fungus. In the absence of oxygen, yeast produces alcohol as a terminal waste product. But when the yeast culture is contaminated with bacteria, the bacteria outgrow the yeast and produce acetic acid instead of alcohol. (Fermentative metabolism is discussed further in Chapter 13.)

Pasteur's work on fermentation led him to test a key claim made by proponents of spontaneous generation. The proponents claimed that Spallanzani's failure to find spontaneous appearance of microbes was due to lack of oxygen. From his studies of yeast fermentation, Pasteur knew that some microbial species do not require oxygen for growth. So he devised an unsealed flask with a long, bent "swan neck" that admitted air but kept the boiled contents free of microbes (**Fig. 1.14B**). The famous swan-necked flasks remained free of microbial growth for many years; but when a flask was tilted to enable contact of broth with microbe-containing dust, growth occurred immediately. Thus, Pasteur disproved that lack of oxygen was the reason for failure of spontaneous generation in Spallanzani's flasks.

But even Pasteur's work did not prove that microbial growth requires preexisting microbes. The Irish scientist John Tyndall (1820–1893) attempted the same experiment as Pasteur, but sometimes found the opposite result. Tyndall found that the broth sometimes gave rise to microbes, no matter how long it was sterilized by boiling. The microbes appear because some kinds of organic matter, particularly hay infusion, are contaminated with a heat-resistant form of bacteria called endospores (or spores). The spore form can only be eliminated by repeated cycles of boiling and resting, in which the spores germinate to the growing, vegetative form that is killed at 100°C.

It was later discovered that endospores could be killed by boiling under pressure, as in a pressure cooker, which generates higher temperatures than can be obtained at atmospheric pressure. The steam pressure device called the **autoclave** became a standard method for the sterilization of materials required for the controlled study of microbes. (Microbial control and antisepsis are discussed further in Chapter 5.)

While spontaneous generation has been discredited as a continual source of microbes, at some point in the past the first living organisms must have originated from nonliving materials. The origin of life is explored in **Special Topic 1.1**.

TO SUMMARIZE:

- **Microbes affected human civilization** for centuries before humans guessed at their existence through their contributions to our environment, food and drink production, and infectious diseases.
- **Robert Hooke and Antoni van Leeuwenhoek** were the first to record observations of microbes through simple microscopes.
- **Spontaneous generation** is the theory that microbes arise spontaneously, without parental organisms. **Lazzaro Spallanzani** showed that microbes arise from preexisting microbes and demonstrated that heat sterilization can prevent microbial growth.
- **Louis Pasteur** discovered the microbial basis of fermentation. He also showed that providing oxygen does not enable spontaneous generation.
- **John Tyndall** showed that repeated cycles of heat were necessary to eliminate spores formed by certain kinds of bacteria.
- **Florence Nightingale** quantified statistically the impact of infectious disease on human populations.

1.3 Medical Microbiology

Over the centuries, thoughtful observers, such as Fracastoro and Agostino Bassi (see Table 1.2), noted a connection between microbes and disease. Ultimately, researchers developed the **germ theory of disease**, the theory that many diseases are caused by microbes.

The first to establish a scientific basis for determining that a specific microbe causes a specific disease was

Special Topic 1.1   How Did Life Originate?

If all life on Earth shares descent from a microbial ancestor, how did the first microbe arise? The earliest fossil evidence of cells in the geological record appears in sedimentary rock that formed as early as 3.8 billion years ago. Although the nature of the earliest reported fossils remains controversial, it is generally accepted that "microfossils" from over 2 billion years ago were formed by living cells. Moreover, the living cells that formed these microfossils looked remarkably similar to bacterial cells today, forming chains of simple rods or spheres (**Fig. 1**).

The exact composition of the first environment for life is controversial. The components of the first living cells may have formed from spontaneous reactions sparked by ultraviolet absorption or electrical discharge. American chemists Stanley Miller (1930–) and Harold C. Urey (1893–1981) argued that the environment of early Earth contained mainly **reduced** compounds—compounds that have a strong tendency to donate electrons, such as ferrous iron, methane, and ammonia. More recent evidence has modified this view, but it is agreed that the strong electron acceptor oxygen gas (O_2) was absent until the first photosynthetic microbes produced it. Today, all our cells are composed of highly reduced molecules

that are readily oxidized by O_2. This seemingly hazardous composition may reflect our cellular origin in the chemically reduced environment of early Earth.

In 1953, Miller attempted to simulate the highly reduced conditions of early Earth to test whether ultraviolet absorption or electrical discharge could cause reactions producing the fundamental components of life (**Fig. 2A**). Miller boiled a solution of water containing hydrogen gas, methane, and ammonia and applied an electrical discharge (comparable to a lightning strike). Astonishingly, the reaction produced a number of amino acids, including glycine, alanine, and aspartic acid. A similar experiment in 1961 by Spanish-American researcher Juan Oró (1923–2004) (**Fig. 2B**) combined hydrogen cyanide and ammonia under electrical discharge to obtain adenine, a fundamental component of DNA and of the energy carrier adenosine triphosphate (ATP).

How could early cells have survived the heat and chemically toxic environment of early Earth? Clues may be found in the survival of archaea that thrive under habitat conditions we consider extreme, such as solutions of boiling sulfuric acid. The specially adapted structures of such microbes may resemble those of the earliest life-forms.

Figure 2   Simulating early Earth's chemistry.
A. Stanley Miller with the apparatus of his early Earth simulation experiment. **B.** Biochemist Juan Oró demonstrated formation of adenine and other biochemicals from reaction conditions found in comets.

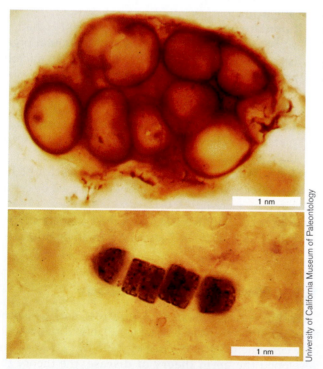

University of California Museum of Paleontology

Figure 1   Evidence of ancient microbial life.
Microfossils of ancient cyanobacteria from the Bitter Springs Formation, Australia, about 850 million years old.

Research since Miller's day has generated as many questions as answers concerning the origin of life. For example:

- Were the reduced forms of early carbon and nitrogen, such as methane (CH_4) and ammonia (NH_4), supplemented by oxidized forms such as CO_2, spewed out by volcanoes? If oxidized carbon and nitrogen were available, different kinds of early-life chemistry may have occurred.
- How did the origin of informational molecules, such as RNA and DNA, coincide with the origin of metabolism? One possibility is that early metabolism was catalyzed by molecules of RNA instead of protein. RNA molecules capable of catalysis, called ribozymes, were discovered in 1982 by Thomas Cech and Sidney Altman, who earned the Nobel Prize in Chemistry in 1989 (**Fig. 3**). The discovery of ribozymes and thousands of small functional RNAs expressed by genomes led to the theory that early organisms were composed primarily of RNA—the so-called "RNA world."
- Geochemical evidence suggests that cells may have originated on Earth as early as 3.8 billion years ago, when Earth was just barely cool enough to allow the existence

of cells. How could cells have formed so quickly? Could the first cells in fact have come from somewhere else?

Most of the molecules that spontaneously formed in Miller's experiments are also found in meteorites and comets. This observation led Oró to propose that the first chemicals of life could have come from outer space, perhaps carried by comets. But could life itself have an extraterrestrial origin? This controversial concept was proposed by British physicists Fred Hoyle (1915–2001) and Chandra Wickramasinghe. Hoyle and Wickramasinghe argued that features of the infrared spectroscopy of interstellar matter might be explained by the existence of microbes in outer space—microbes that could be brought to Earth by comets or meteors. Alternatively, some scientists propose that the first microbes originated on Mars. Because Mars orbits farther out from the sun than Earth, its surface would have cooled before Earth did; therefore, life might have formed on Mars and traveled to Earth on a meteorite. But the "Mars first" explanation says nothing about how life would have arisen on Mars.

Current evidence for the origin and evolution of microbes is discussed in Chapter 17.

A.

Geoffery Wheeler for Howard Hughes Medical Institute

B.

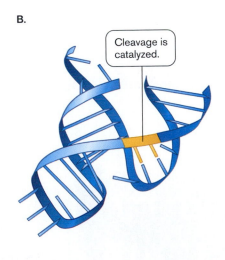

Cleavage is catalyzed.

Figure 3   Tom Cech, discoverer of catalytic RNA.   A. Tom Cech (University of Colorado, Boulder) holding a flask containing *Oxytricha nova*, microbes that make catalytic RNA, the kind of molecule that in early cells may have served both genetic and catalytic functions. **B.** Diagram of a catalytic RNA, where horizontal bars represent bases. The RNA catalyzes cleavage of itself.

the German physician Robert Koch (1843–1910) (**Fig. 1.15**). As a college student, Koch conducted biochemical experiments on his own digestive system. Koch's curiosity about the natural world led him to develop principles and techniques crucial to modern microbial investigation, including the pure-culture technique and the famous Koch's postulates for identifying the causative agent of a disease. He applied his methods to numerous lethal diseases around the world, including anthrax and tuberculo-sis in Europe, malaria in Africa and the East Indies, and bubonic plague in India.

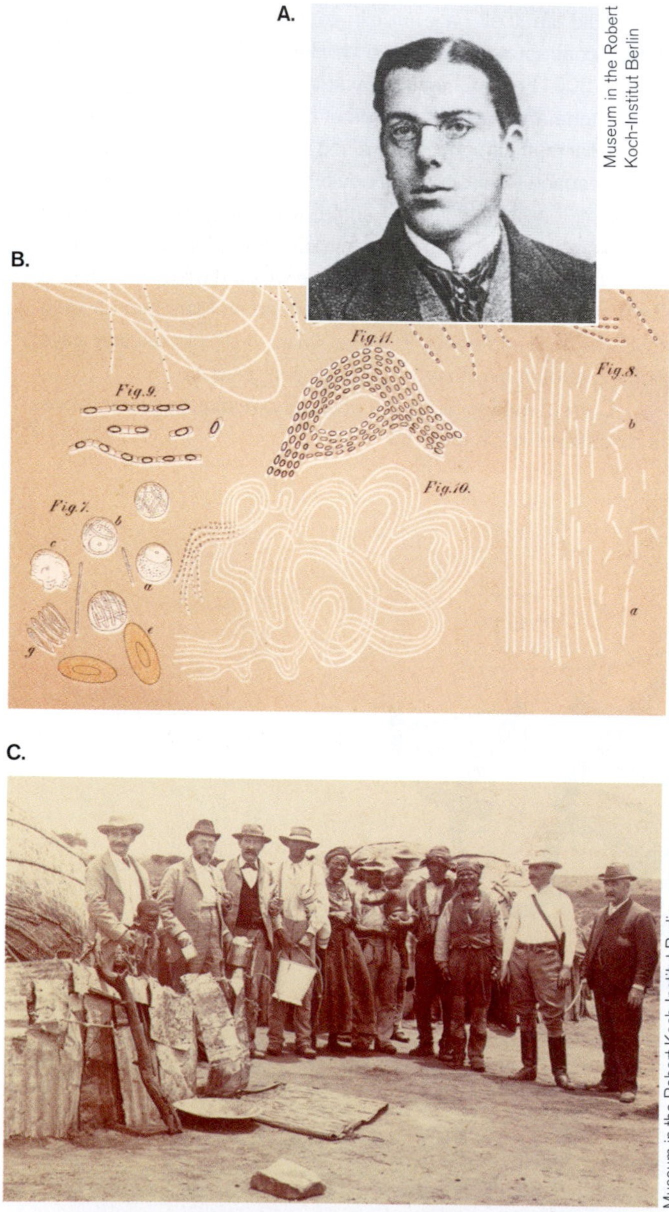

Growth of Microbes in Pure Culture

Unlike Pasteur, who was a university professor, Koch took up a medical practice in a small Polish-German town. To make space in his home for a laboratory to study anthrax and other deadly diseases, his wife curtained off part of his patients' examining room.

Anthrax interested Koch because its epidemics in sheep and cattle caused economic hardship among local farmers. Today, anthrax is no longer a major problem for agriculture, as its transmission is prevented by effective environmental controls and vaccination. It has, however, gained notoriety as a bioterror agent because anthrax bacteria can survive for long periods in the dormant, desiccated form of an endospore. In 2001, anthrax spores sent through the mail contaminated post offices throughout the northeastern United States, as well as an office building of the United States Senate, causing several deaths (**Fig. 1.16**).

To investigate whether anthrax was a transmissible disease, Koch used blood from an anthrax-infected carcass to inoculate a rabbit. When the rabbit died, he used the rabbit's blood to inoculate a second rabbit, which then died in turn. The blood of the unfortunate animal had turned black with long, rod-shaped bacilli. Upon introduction of these bacilli into healthy animals, the ani-

Figure 1.15   Robert Koch, founder of the scientific method of microbiology.   A. Robert Koch as a university student. **B.** Koch's sketch of anthrax bacilli in mouse blood. **C.** Koch (second from left) during his visit to New Guinea to investigate malaria.

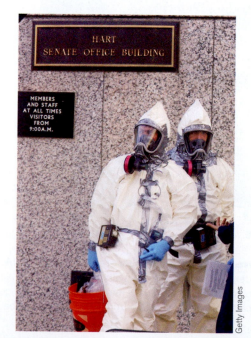

Figure 1.16   Disinfecting the Hart Senate Office Building, Fall 2001.   The building became contaminated with anthrax spores sent through the U.S. mail.

mals became ill with anthrax. Thus, Koch demonstrated an important principle of epidemiology—the **chain of infection**, or transmission of a disease. In retrospect, his choice of anthrax was fortunate, for the microbes generate disease very quickly, multiply in the blood to an extraordinary concentration, and remain infective outside the body for long periods.

Koch and his colleagues then applied their experimental logic and culture methods to a more challenging disease: tuberculosis. In Koch's day, tuberculosis caused one-seventh of reported deaths in Europe; today, tuberculosis bacteria continue to infect millions of people worldwide. Koch's approach to anthrax, however, was less applicable to tuberculosis, a disease that develops slowly after many years of dormancy. Furthermore, the causative bacteria, *Mycobacterium tuberculosis*, are small and difficult to distinguish from human tissue or from different bacteria of similar appearance associated with the human body. How could Koch prove that a particular bacterium caused a particular disease?

What was needed was to isolate a **pure culture** of microorganisms, a culture grown from a single "parental" cell. This had been done by previous researchers using the laborious process of serial dilution of suspended bacteria until a culture tube contained only a single cell. Alternatively, inoculation of a solid surface such as a sliced potato could produce isolated **colonies**, distinct populations of bacteria, each grown from a single cell. For *M. tuberculosis*, Koch inoculated serum, which then formed a solid gel after heating. Later, he refined the solid-substrate technique by adding gelatin to a defined liquid medium, which could then be chilled to form a solid medium in a glass dish. A covered version called the **petri dish** (also called a petri plate) was invented by a colleague, Richard J. Petri (1852–1921). The petri dish consists of a round dish with vertical walls covered by an inverted dish of slightly larger diameter. Today, the petri dish, generally made of disposable plastic, remains an indispensable part of the microbiological laboratory.

Another improvement in solid-substrate culture was the replacement of gelatin with materials that remain solid at higher temperatures, such as the gelling agent **agar** (a polymer of the sugar galactose). The use of agar was recommended by Angelina Hesse (1850–1934), a microscopist and illustrator, to her husband, Walther Hesse (1846–1911), a young medical colleague of Koch's (**Fig. 1.17**). Agar comes from red algae (seaweed) which is used by East Indian birds to build nests; it is the main ingredient in the delicacy "bird's nest soup." Dutch colonists used agar to make jellies and preserves, and a Dutch colonist from Java introduced it to Angelina Hesse. The Hesses used agar to develop the first effective growth medium for tuberculosis bacteria. (Pure culture is discussed further in Chapter 4.)

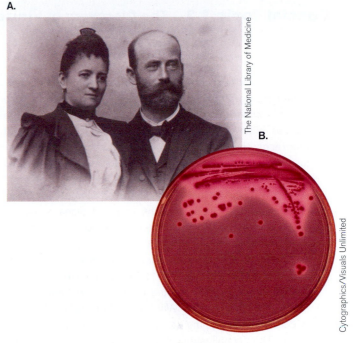

A.

The National Library of Medicine

B.

Cytographics/Visuals Unlimited

Figure 1.17 Angelina and Walther Hesse. A. Portrait of the Hesses, who first used agar to make solid plate media for bacterial growth. **B.** Colonies from a streaked agar plate.

Note that some kinds of microbes cannot be grown in pure culture without other organisms. For example, viruses can be cultured only in the presence of their host cells. The discovery of viruses is explored in **Special Topic 1.2**.

Koch's Postulates Link a Pathogen with a Disease

For his successful determination of the bacterium responsible for tuberculosis, *M. tuberculosis*, Koch was awarded the Nobel Prize in Physiology or Medicine in 1905. Koch formulated his famous set of criteria for establishing a causative link between an infectious agent and a disease (**Fig. 1.18**). These four criteria are known as **Koch's postulates**:

1. The microbe is found in all cases of the disease, but is absent from healthy individuals.
2. The microbe is isolated from the diseased host and grown in pure culture.
3. When the microbe is introduced into a healthy, susceptible host (or animal model), the same disease occurs.
4. The same strain of microbe is obtained from the newly diseased host. When cultured, the strain shows the same characteristics as before.

Special Topic 1.2   The Discovery of Viruses

The discovery of ever smaller and more elusive microorganisms continues to this day. From the nineteenth century on, researchers were puzzled to find contagious diseases whose agent of transmission could pass through a filter of a pore size that blocked known microbial cells—0.1 μm. One of these researchers was the Dutch plant microbiologist Martinus Beijerinck (1851–1931), who studied tobacco mosaic disease, a condition in which the leaves become mottled and the crop yield is decreased or destroyed altogether. Beijerinck concluded that because the agent of disease passed through a filter that retained bacteria, it could not be a bacterial cell.

The filterable agent was ultimately purified by the American scientist Wendell Stanley (1904–1971), who processed 4,000 kilograms (kg) of infected tobacco leaves and crystallized the infective particle. What he had crystallized was the tobacco mosaic virus, the causative agent of tobacco mosaic disease. The crystallization of a virus particle earned Stanley the 1946 Nobel Prize in Chemistry. The fact that an entity capable of biological reproduction could be inert enough to be crystallized amazed scientists and ultimately led to a new, more mechanical view of living organisms.

The individual particle of tobacco mosaic virus consists of a helical tube of protein subunits containing its genetic material coiled within (**Fig. 1**). Stanley thought that the virus was a catalytic protein, but colleagues later determined that it contained RNA as its genetic material. The structure of the coiled RNA was solved through X-ray diffraction crystallography by the British scientist Rosalind Franklin (1920–1958). We now know that all kinds of animals, plants, and microbial cells can be infected by viruses.

Toward the end of the twentieth century, even smaller infective particles were discovered consisting of a single mol-

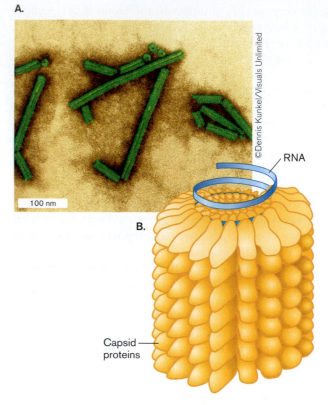

A.

©Dennis Kunkel/Visuals Unlimited

100 nm

RNA

B.

Capsid proteins

Figure 1   Tobacco mosaic virus (TMV). A. Particles of tobacco mosaic virus (colorized transmission EM). **B.** In TMV, a protein capsid surrounds an RNA chromosome.

ecule of RNA (viroids) or of protein (prions). Prions are suspected as a factor in the development of Alzheimer's disease. (The infectious processes of viruses, viroids, and prions are discussed in Chapters 6, 11, and 26.)

Koch's postulates continue to be used to determine whether a given strain of microbe causes a disease. Modern examples include Lyme disease, a tick-borne infection that has become widespread in New England and the Mid-Atlantic states, and hantaviral pneumonia, an emerging disease particularly prevalent among Native Americans in the Southwest. Nevertheless, the postulates remain only a guide; individual diseases and pathogens may confound one or more of the criteria. For example, tuberculosis bacteria are now known to cause symptoms in only 10% of the people infected. If Koch had been able to detect these silent bacilli, they would not have fulfilled his first criterion. In the case of AIDS, the concentration of HIV virus is so low that initially no virus could be detected in patients with fully active symptoms. It took the invention of the polymerase chain reaction (PCR), a method of producing any number of copies of DNA or RNA sequences, to detect the presence of HIV.

Another difficulty with AIDS and many other human diseases is the absence of an animal host that exhibits the same disease. In the case of AIDS, even the chimpanzees, our closest relatives, are not susceptible, although they exhibit a similar disease from a related pathogen, simian immunodeficiency virus (SIV). Experimentation on humans is prohibited, although in rare instances researchers have voluntarily exposed themselves to a proposed pathogen. For example, Australian researcher Barry Marshall ingested *Helicobacter pylori* to convince

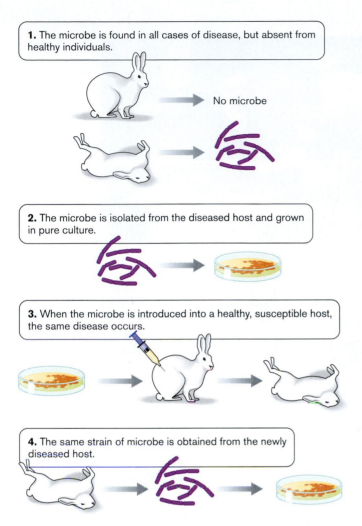

Figure 1.18 Koch's postulates defining the causative agent of a disease.

skeptical colleagues that this organism could colonize the extremely acidic stomach. *H. pylori* turned out to be the causative agent of gastritis and stomach ulcers, conditions that had long been thought to be caused by stress rather than infection. For the discovery of *H. pylori*, Marshall and colleague Robin Warren won the 2005 Nobel Prize in Physiology or Medicine.

THOUGHT QUESTION 1.4 How could you use Koch's postulates to demonstrate the causative agent of influenza? What problems would you need to overcome that were not encountered with anthrax?

Immunization Prevents Disease

Identifying the cause of a disease is, of course, only the first step to developing an effective therapy and prevent-

ing further transmission. Early microbiologists achieved some remarkable insights on how to control pathogens (see Table 1.2).

The first clue as to how to protect an individual from a deadly disease came from the dreaded smallpox. In the eighteenth century, smallpox infected a large fraction of the European population, killing or disfiguring many people. In Turkey, however, the incidence of smallpox was decreased by the practice of deliberately inoculating children with material from smallpox pustules, which contained naturally attenuated virus. Inoculated children usually developed a mild case of the disease and were protected from smallpox thereafter. The practice of smallpox inoculation was introduced from Turkey to Europe in 1717 by Lady Mary Montagu, a smallpox survivor (**Fig. 1.19A**). Stationed in Turkey with her husband, the British ambassador, Lady Montagu learned that many elderly women there had perfected the art of inoculation: "The old woman comes with a nut-shell full of the matter of the best sort of small-pox, and asks what vein you please to have opened." Lady Montagu arranged for the procedure on her own son, then brought the practice back to England where it became widespread.

Preventive inoculation with smallpox was dangerous, however, as some infected individuals still contracted serious disease and were contagious. Thus, doctors continued to seek a better method of prevention. In England, milkmaids claimed that they were protected from smallpox after they contracted cowpox, a related but much milder disease. This claim was confirmed by English physician Edward Jenner (1749–1823), who deliberately infected patients with matter from cowpox lesions (**Fig. 1.19B**). The practice of cowpox inoculation was called **vaccination**, after the Latin word *vacca* for "cow." To this day, cowpox, or vaccinia virus, remains the basis of the modern smallpox vaccine.

Pasteur was aware of vaccination as he studied the course of various diseases in experimental animals. In the spring of 1879, he was studying fowl cholera, a transmissible disease of chickens with a high death rate. He had isolated and cultured the bacteria responsible, but left his work during the summer for a long vacation. No refrigeration was available to preserve cultures, and when he returned to work, the aged bacteria failed to cause disease in his chickens. Pasteur then obtained fresh bacteria from an outbreak of disease elsewhere, as well as some new chickens. But the fresh bacteria also failed to make the original chickens sick (those who had been exposed to the aged bacteria). All of the new chickens, exposed only to the fresh bacteria, contracted the disease. Grasping the clue from his mistake, Pasteur had the insight to recognize that an attenuated (or "weakened") strain of microbe, altered somehow to eliminate its potency to cause disease, could still confer immunity to the virulent disease-causing form.

A.

Hulton-Deutsch Collection/Corbis

B.

Bettmann/Corbis

Figure 1.19 Smallpox vaccination. A. Lady Mary Wortley Montagu, shown in Turkish dress. The artist avoided showing Montagu's facial disfigurement from smallpox. **B.** Dr. Edward Jenner, depicted vaccinating 8-year-old James Phipps with cowpox matter from the hand of milkmaid Sarah Nelmes, who had caught the disease from a cow. **C.** Newspaper cartoon depicting public reaction to cowpox vaccination.

C.

Corbis

Pasteur was the first to recognize the significance of attenuation and extend the principle to other pathogens. We now know that the molecular components of pathogens generate **immunity**, the resistance to a specific disease, by stimulating the **immune system**, an organism's exceedingly complex cellular mechanisms of defense (see Chapter 23). Understanding the immune system awaited the techniques of molecular biology a century later, but nineteenth-century physicians developed several effective examples of **immunization**, the stimulation of an immune response by deliberate inoculation with an attenuated pathogen.

The way to attenuate a strain varies greatly among pathogens. Heat treatment or aging for various periods often turned out to be the most effective approach. In retrospect, the original success of prophylactic smallpox inoculation was probably due to natural attenuation during the time between acquisition of smallpox matter from a diseased individual and inoculation of the healthy patient. A far more elaborate treatment was required for the most famous disease for which Pasteur devised a vaccine: rabies.

The rabid dog loomed large in folklore, and the disease was dreaded for its particularly horrible and inevitable course of death. Pasteur's vaccine for rabies required a highly complex series of heat treatments and repeated inoculations. Its success led to his instant fame (**Fig. 1.20**). Grateful survivors of rabies founded the Pasteur Institute for medical research, one of the world's greatest medical research institutions, whose scientists in the twentieth century discovered the virus HIV, responsible for AIDS.

Antiseptics and Antibiotics Control Pathogens

Before the work of Koch and Pasteur, many patients died of infections transmitted unwittingly by their own doctors. In 1847, Hungarian physician Ignaz Semmelweis (1818–1865) noticed that the death rate of women in childbirth due to puerperal fever was much higher in his own hospital than in a birthing center run by midwives. He guessed that the doctors in his hospital were transmitting pathogens from cadavers that they had dissected. So he ordered the doctors to wash their hands

Figure 1.20 Pasteur cures rabies. Cartoon in French newspaper depicts Pasteur protecting children from rabid dogs.

in chlorine, an **antiseptic** agent (a chemical that kills microbes). The mortality rate fell; but this revelation displeased other doctors, who refused to accept Semmelweis's findings.

In 1865, the British surgeon Joseph Lister (1827–1912) noted that half his amputee patients died of sepsis. Lister knew from Pasteur that microbial contamination might be the cause. So he began experiments to develop the use of antiseptic agents, most successfully carbolic acid, to treat wounds and surgical instruments. After initial resistance, Lister's work, with the support of Pasteur and Koch, drew widespread recognition. In the twentieth century, surgeons developed fully **aseptic** environments for surgery; that is, environments completely free of microbes.

Antibiotics. The problem with most antiseptic chemicals that killed microbes was that if taken internally, they also

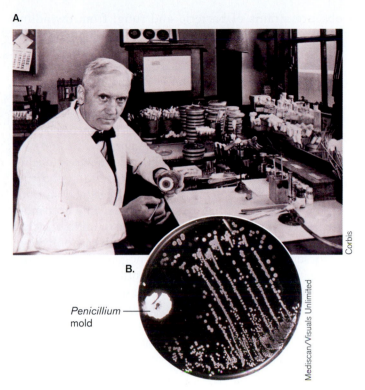

Penicillium mold

Figure 1.21 Alexander Fleming, discoverer of penicillin. A. Alexander Fleming in his laboratory. **B.** Fleming's original plate of bacteria with *Penicillium* mold inhibiting the growth of bacterial colonies.

tended to kill the patients. Researchers sought a "magic bullet," an **antibiotic** molecule that killed microbes alone, leaving their host unharmed.

An important step in the search for antibiotics was the realization that microbes themselves produce antibiotic compounds with highly selective effects. This followed from the famous accidental discovery of penicillin by the English medical researcher Alexander Fleming (1881–1955) (**Fig. 1.21A**). In 1929, Fleming was culturing *Staphylococcus*, which infects wounds. He found that one of his plates of *Staphylococcus* was contaminated with a mold, *Penicillium notatum*, which he noticed was surrounded by a clear region, free of *Staphylococcus* colonies (**Fig. 1.21B**). Following up on this observation, Fleming showed that the mold produced a substance that killed bacteria. We now know this substance as penicillin.

In 1941, British biochemists Howard Florey (1898–1968) and Ernst Chain (1906–1979) purified the antibiotic molecule, which we now know inhibits formation of the bacterial cell wall. Penicillin saved the lives of many Allied troops during World War II, the first war in which an antibiotic became available to soldiers.

The second half of the twentieth century saw the discovery of many new and powerful antibiotics. Most of the new antibiotics, however, were produced by obscure strains of bacteria and fungi from dwindling ecosystems—a circumstance that focused attention on wilderness preservation worldwide. Furthermore, the widespread and often indiscriminate use of antibiotics has selected for pathogens that are antibiotic resistant. As a result, antibiotics have lost their effectiveness against certain strains of major pathogens. For example, multiple-drug-resistant *Mycobacterium tuberculosis* is now a serious threat to public health.

Fortunately, biotechnology provides new approaches to antibiotic development, including genetic engineering of microbial producers and artificial design of antimicrobial chemicals. This industry has become ever more critical because the indiscriminate use of antibiotics has led to a "molecular arms race" in which our only hope is to succeed faster than the pathogens develop resistance. (Microbial biosynthesis of antibiotics is discussed in Chapter 15, and the medical use of antibiotics is discussed in Chapter 27.)

THOUGHT QUESTION 1.5 Why do you think some pathogens generate immunity readily, whereas others evade the immune system?

THOUGHT QUESTION 1.6 How do you think microbes protect themselves from the antibiotics they produce?

TO SUMMARIZE:

- **Robert Koch** devised techniques of pure culture to study a single species of microbe in isolation. A key technique is culture on solid medium using agar, as developed by Angelina and Walther Hesse, in a double-dish container devised by Richard Petri.
- **Koch's postulates** provide a set of criteria to establish a causative link between an infectious agent and a disease.
- **Edward Jenner** established the practice of vaccination, inoculation of cowpox to prevent smallpox. Jenner's discovery was based on earlier observations by Lady Mary Montagu and others that a mild case of smallpox could prevent future cases.
- **Louis Pasteur** developed the first vaccines based on attenuated strains, such as the rabies vaccine.
- **Ignaz Semmelweis** and **Joseph Lister** showed that antiseptics could prevent transmission of pathogens from doctor to patient.
- **Alexander Fleming** discovered that the *Penicillium* mold generated a substance that kills bacteria. **Howard Florey** and **Ernst Chain** purified the substance, penicillin, the first commercial antibiotic to save human lives.

1.4 Microbial Ecology

Koch's growth of microbes in pure culture was a major advance in technology, enabling the systematic study of microbial physiology and biochemistry. In hindsight, this discovery eclipsed the equally important study of microbial ecology. Microbes are responsible for cycling the many minerals essential for all life. Yet barely 0.1% of all microbial species can be cultured in the laboratory—and the remainder make up the majority of Earth's entire biosphere. Only the outer skin of Earth supports complex multicellular organisms. The depths of Earth's crust, to at least 2 miles down, as well as the atmosphere 10 miles out into the stratosphere, remain the domain of microbes. So, to a first approximation, Earth's ecology *is* microbial ecology.

Microbes Support Natural Ecosystems

The first microbiologists to culture microbes in the laboratory selected the kinds of nutrients that feed humans, such as beef broth or potatoes. Some of Koch's contemporaries, however, suspected that other kinds of microbes living in soil or wetlands existed on more exotic fare. Soil samples were known to oxidize hydrogen gas, and this activity was eliminated by treatment with heat or acid, suggesting microbial origin. Ammonia in sewage was oxidized to nitrate, and this process was eliminated by antibacterial treatment. These findings suggested the existence of microbes that "ate" hydrogen gas or ammonia instead of beef or potatoes, but no one could isolate these microbes in culture.

Among the first to study microbes in natural habitats was the Russian scientist Sergei Winogradsky (1856–1953). Winogradsky waded through marshes to discover microbes with metabolism quite alien from human digestion. For example, Winogradsky discovered that species of the bacterium *Beggiatoa* oxidize hydrogen sulfide (H_2S) to sulfuric acid (H_2SO_4). *Beggiatoa* fixes carbon dioxide into biomass without consuming any organic food. Organisms that feed solely on inorganic minerals are known as **chemolithotrophs**, or **lithotrophs**, discussed further in Chapters 4 and 14.

The lithotrophs studied by Winogradsky could not be grown on Koch's plate media containing agar or gelatin. The bacteria that Winogradsky isolated could grow only on inorganic minerals; in fact, some species are actually poisoned by organic food. For example, nitrifiers convert ammonia to nitrate, forming a crucial part of the nitrogen cycle in natural ecosystems. Winogradsky cultured nitrifiers on a totally inorganic solution containing ammonia and silica gel, which supported no other kind of organism. This experiment was an early example of **enrichment culture**, the use of selective growth media that support certain classes of microbial metabolism while excluding others.

Instead of isolating pure colonies, Winogradsky built a model wetland ecosystem containing regions of enrichment for microbes of diverse metabolism. This model is called the **Winogradsky column** (**Fig. 1.22**). The model consists of a glass tube containing mud mixed with shredded newsprint (an organic carbon source) and calcium salts of sulfate and carbonate. After exposure to light for several weeks, several zones of color develop, full of mineral-metabolizing bacteria. At the top, cyanobacteria conduct **photosynthesis**, using light energy to split water and produce molecular oxygen. Below, purple sulfur bacteria use photosynthesis to split hydrogen sulfide, producing sulfur. At the bottom, sulfate reducers produce hydrogen sulfide and precipitate iron.

The gradient from oxygen-rich conditions at the surface to highly reduced conditions below generates a voltage potential, like a battery cell. We now know that the entire Earth's surface acts as a battery—for humans, a potential source of renewable energy.

Geochemical Cycling Depends on Bacteria

Winogradsky and later microbial ecologists showed that bacteria perform unique roles in **geochemical cycling**, the global interconversion of inorganic and organic forms of nitrogen, sulfur, phosphorus, and other minerals. Without these essential conversions (nutrient cycles), no plants or animals could live. Bacteria and archaea *fix nitrogen* (N_2) by reducing it to ammonia (NH_3), the form of nitrogen assimilated by plants. This process is called **nitrogen fixation** (**Fig. 1.23**). Other bacterial species oxidize NH_4^+ in several stages back to nitrogen gas.

Within plant cells, certain bacteria fix nitrogen as **endsymbionts**, organisms living symbiotically inside a larger organism. Endosymbiotic bacteria known as rhizobia induce the roots of legumes to form special nodules to facilitate bacterial nitrogen fixation. Rhizobial endosymbiosis was first observed by the Dutch plant microbiologist Martinus Beijerinck (1851–1931).

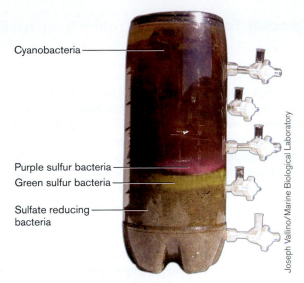

Figure 1.22 Winogradsky column. A wetland model ecosystem designed by Sergei Winogradsky.

Cyanobacteria

Purple sulfur bacteria
Green sulfur bacteria

Sulfate reducing bacteria

Joseph Vallino/Marine Biological Laboratory

THOUGHT QUESTION 1.7 Why don't all living organisms fix their own nitrogen?

Microbial endosymbiosis, in many diverse forms, is widespread in all ecosystems. Many interesting cases involve animal or human hosts (see **Special Topic 1.3**).

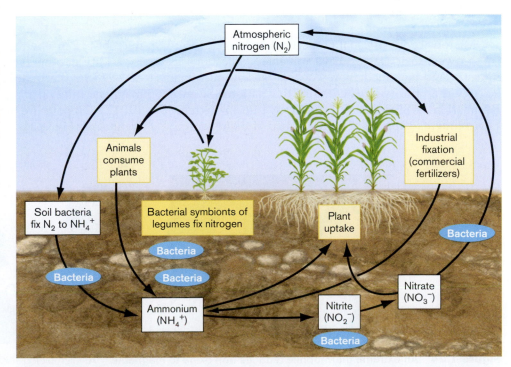

Atmospheric nitrogen (N_2)

Animals consume plants

Industrial fixation (commercial fertilizers)

Soil bacteria fix N_2 to NH_4^+

Bacterial symbionts of legumes fix nitrogen

Plant uptake

Bacteria

Bacteria

Bacteria

Bacteria

Ammonium (NH_4^+)

Nitrite (NO_2^-)

Nitrate (NO_3^-)

Bacteria

Figure 1.23 The global nitrogen cycle. All life depends on these oxidative and reductive conversions of nitrogen—most of which are performed only by microbes.

Special Topic 1.3   Microbial Endosymbionts of Animals

Numerous kinds of endosymbiosis have been found in which microbes make essential nutritional contributions to host animals. Ruminant animals, such as cattle, as well as insects such as termites, require digestive bacteria to break down cellulose and other plant polymers. Even humans obtain about 10% of their nutrition from colonic bacteria. A remarkable form of endosymbiosis is seen in marine animal communities such as those inhabiting thermal vents in the ocean floor, where super-heated water emerges carrying reduced minerals, such as hydrogen sulfide (**Fig. 1**). Near these thermal vents, sulfide-oxidizing bacteria nourish giant worms and clams. The worms have evolved so as to lose their digestive tracts altogether, while their body fluids carry the hydrogen sulfide (highly toxic to most animals) needed for the bacteria to metabolize.

Many invertebrates, such as hydras and corals, harbor endosymbiotic phototrophs that provide products of photosynthesis in return for protection and nutrients. Other kinds of endosymbiosis involve more than nutrition. In the light organs of fish and squid, luminescent bacteria such as *Vibrio fischeri* respond to host signals, causing light emission. Certain sponges maintain endosymbiotic bacteria that produce antibiotics, preventing infection by pathogens. Similarly, some bacteria that normally inhabit the human skin are believed to protect us from infection. Microbial endosymbiosis has even

inspired fictional creations, such as the "breathmicrobes" of Joan Slonczewski's novel *A Door into Ocean*, which acquire oxygen for human hosts swimming underwater. (Microbial endosymbiosis is discussed further in Section 21.2.)

Fritz Heide/WHOI

Figure 1   **Tube worms in a thermal vent ecosystem.** Tube worms do not feed themselves but depend on sulfide-oxidizing symbionts for nutrition.

Microbes with unusual properties, such as the ability to digest toxic wastes, have valuable applications in industry and bioremediation. For this reason, microbial ecology is a priority for funding by the National Science Foundation (NSF). The NSF program Life in Extreme Environments supports research aimed at documenting microbes from environments with extreme heat, salinity, acidity, or other factors. One such microbe is a species of *Geobacter* that reduces rust (iron oxide, Fe_2O_3) to the magnetic mineral magnetite (Fe_3O_4) while growing at 121°C, a temperature high enough to kill all other known organisms (**Fig. 1.24**).

A.

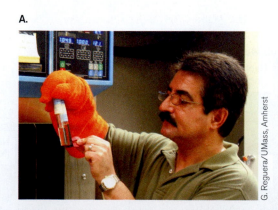

G. Reguera/UMass, Amherst

B.   Bacterium strain 121

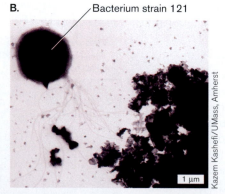

Kazem Kashefi/UMass, Amherst

1 µm

Figure 1.24   An extreme thermophile reduces iron oxide to magnetite.   A. Microbiologist Kazem Kashefi (Michigan State University) pulls a live culture of *Geobacter* out of an autoclave generally used to kill all living organisms at 121°C (250°F). The magnet shows that strain 121 is converting nonmagnetic iron oxide (rust; Fe_2O_3) to the magnetic mineral magnetite, Fe_3O_4. **B.** Strain 121 is a round bacterium with a tuft of rotary flagella (transmission electron micrograph). The black material at right is iron oxide, which feeds the bacteria. *Source:* Part B reprinted with permission from Kashefi, et al. 2003. *Science* 301(5635):934. ©2005 AAAS.

TO SUMMARIZE:

- **Sergei Winogradsky** first developed a system of enrichment culture, the Winogradsky column, to grow microbes from natural environments.
- **Chemolithotrophs (or lithotrophs)** metabolize inorganic minerals, such as ammonia, instead of the organic nutrients used by the microbes isolated by Koch.
- **Geochemical cycling** depends on bacteria and archaea that cycle nitrogen, phosphorus, and other minerals throughout the biosphere.
- **Endosymbionts** are microbes that live within multicellular organisms and provide essential functions for their hosts, such as nitrogen fixation for legume plants.
- **Martinus Beijerinck** first demonstrated that nitrogen-fixing rhizobia grow as endosymbionts within leguminous plants.

1.5 The Microbial Family Tree

The bewildering diversity of microbial life-forms presented nineteenth-century microbiologists with a seemingly impossible task of classification. So little was known about life under the lens that natural scientists despaired of ever learning how to distinguish microbial species. The famous classifier of species, Swedish botanist Carl von Linné (Carolus Linnaeus, 1707–1778), called the microbial world "chaos."

Microbes Are a Challenge to Classify

Two challenges faced would-be taxonomists with respect to microbes. The first was that the resolution of the light microscope allows little more than visualizing the outward shape of microbial cells, and vastly different kinds of microbes look more or less alike (visualization of cells and molecules is discussed in Chapter 2). This challenge was overcome as advances in biochemistry and microscopy made it possible to distinguish microbes based on their metabolism and cell ultrastructure and ultimately on their DNA sequence.

The second challenge was that microbes do not readily fit the classic definition of a species—that is, a group of organisms that interbreed. Unlike multicellular eukaryotes, microbes generally reproduce asexually, with only occasional sexual exchange. When they do exchange genes, they may do so with distantly related species (discussed in Chapter 9). Nevertheless, microbiologists devise working definitions of microbial species that enable us to usefully describe populations while being flexible enough to accommodate continual revision and change. The most useful definitions are based

on genetic similarity. For example, two distinct species generally share no more than 95% similarity of DNA sequence.

NOTE: The actual names of microbial species are often changed to reflect new understanding of genetic relationships. For example, the causative agent of bubonic plague has formerly been called *Bacterium pestis* (1896), *Bacillus pestis* (1900), and *Pasteurella pestis* (1923), but is now called *Yersinia pestis* (1944). The older names, however, still appear in the literature, a point to remember during research. Current and historical nomenclature is compiled at the List of Prokaryotic Names with Standing in Nomenclature.

WWW | List of Prokaryotic Names with Standing in Nomenclature

Microbes Include Eukaryotes and Prokaryotes

In the nineteenth century, taxonomists had no access to DNA information. As they tried to incorporate microbes into the tree of life, they faced a conceptual dilemma in that microbes could not be categorized as either animals or plants, which since ancient times were considered the two "kingdoms" or major categories of life. Taxonomists attempted to apply these categories to microbes—for example, by including algae and fungi with plants. But German naturalist Ernst Haeckel (1834–1919) recognized that microbes differed from both plants and animals in fundamental aspects of their lifestyle, cellular structure, and biochemistry. Haeckel proposed that microscopic organisms constituted a third kind of life—neither animal nor plant—which he called *Monera*.

In the twentieth century, biochemical studies revealed profound distinctions even within the microbial category. In particular, microbes such as protists and algae contain a nucleus bounded by a nuclear membrane, whereas bacteria do not. Herbert Copeland (1902–1968) proposed a system of classification that divided *Monera* into two groups: the eukaryotic protists (protozoa and algae) and the prokaryotic bacteria. Copeland's four-kingdom classification (plants, animals, eukaryotic protists, and prokaryotic bacteria) was later modified by Robert Whittaker (1924–1980) to include fungi as a fifth kingdom of eukaryotic microbes. Whittaker's system thus generated five kingdoms: bacteria, protists, fungi, and the multicellular plants and animals.

Eukaryotes Evolved through Endosymbiosis

The five-kingdom system was modified dramatically by Lynn Margulis, at the University of Massachusetts (**Fig. 1.25**). Margulis tried to explain how it is that eukaryotic cells contain mitochondria and chloroplasts, membranous organelles that possess their own chromosomes. She proposed that eukaryotes evolved by merging with bacteria to form composite cells by intracellular endosymbiosis, in which one cell internalizes another that grows within it. The endosymbiosis may ultimately generate a single organism whose formerly independent members are now incapable of independent existence.

Margulis proposed that early in the history of life, respiring bacteria similar to *E. coli* were engulfed by pre-eukaryotic cells, where they evolved into mitochondria, the eukaryote's respiratory organelle. Similarly, she proposed that a phototroph related to cyanobacteria was taken up by a eukaryote, giving rise to the chloroplasts of phototrophic algae and plants.

The endosymbiosis theory was highly controversial because it implied a **polyphyletic**, or multiple, ancestry of living species, inconsistent with the long-held assumption that species evolve only by divergence from a common ancestor (**monophyletic**). Ultimately, DNA sequence analysis produced compelling evidence of the bacterial origin of mitochondria and chloroplasts. Both these classes of organelles contain circular molecules of DNA, whose sequences show unmistakable homology (similarity) to those of modern bacteria. DNA sequences and other evidence established the common ancestry between mitochondria and respiring bacteria and between chloroplasts and cyanobacteria.

Archaea Differ from Bacteria and Eukaryotes

Gene sequence analysis led to another startling advance in our understanding of the evolution of cells. In 1977, Carl Woese, at the University of Illinois, was studying a group of recently discovered prokaryotes that live in seemingly hostile environments, such as the boiling sulfur springs of Yellowstone, or that exhibit unusual kinds of metabolism, such as production of methane (methanogenesis).

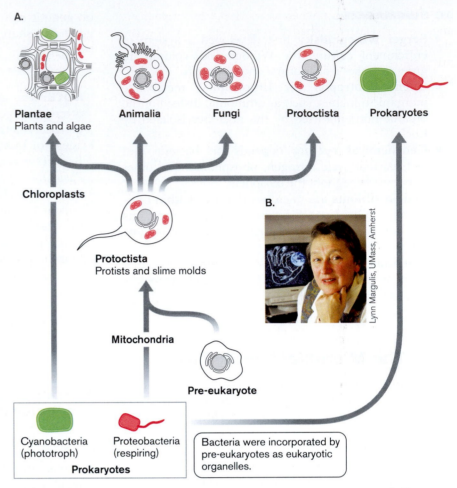

Figure 1.25   Lynn Margulis and the serial endosymbiosis theory.   A. Five-kingdom scheme, modified by the endosymbiosis theory. **B.** Lynn Margulis (University of Massachusetts, Amherst) proposed that organelles evolve through endosymbiosis.

Woese used the sequence of the gene for 16S ribosomal RNA (16S rRNA) as a "molecular clock," a gene whose sequence differences can be used to measure the time since the divergence of two species (discussed in Chapter 17). The divergence of rRNA genes showed that the newly discovered prokaryotes were a distinct form of life, archaea (**Fig. 1.26**). The archaea resemble bacteria in their relatively simple cell structure, in their lack of a nucleus, and in their ability to grow in a wide range of environments. Many archaea, however, grow in environments more extreme than any bacterium, such as 110°C at high pressure. The genetic sequences of archaea differ as much from those of bacteria as from those of eukaryotes; in fact, their gene expression machinery is more similar to that of eukaryotes.

Woese's discovery replaced the classification scheme of five kingdoms with three equally distinct groups, now called the three domains: *Bacteria, Archaea,* and *Eukarya* (**Fig. 1.27**). In the three-domain model, the bacterial ancestor of mitochondria derives from ancient proteobacteria (shaded pink), whereas chloroplasts derive from ancient cyanobacteria (shaded green). The three-domain classification is largely supported by the sequences of microbial genomes, although some horizontal transfer of genes occurs both within and between the domains (discussed in Chapter 17).

> **THOUGHT QUESTION 1.8** What arguments support the classification of *Archaea* as a third domain of life? What arguments support the classification of archaea and bacteria together, as prokaryotes, distinct from eukaryotes?

TO SUMMARIZE:

- **Classifying microbes** was a challenge historically because of the difficulties in observing distinguishing characteristics of different categories.
- **Ernst Haeckel** recognized that microbes constitute a form of life distinct from animals and plants.
- **Herbert Copeland and Robert Whittaker** classified prokaryotes as a form of microbial life distinct from eukaryotic microbes such as protists.

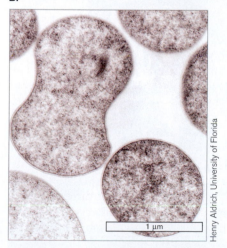

Figure 1.26 Archaea, newly discovered life-forms.
A. Finding archaea in the hot spring Obsidian Pool, at Yellowstone. **B.** *Pyrococcus furiosus*, an organism that lives at temperatures above 100°C (transmission electron micrograph).

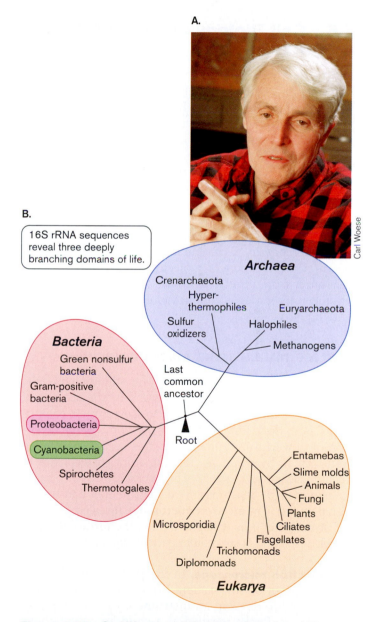

Figure 1.27 Carl Woese and the three domains of life.
A. Carl Woese (University of Illinois at Urbana-Champaign) proposed that archaea constitute a third domain of life. **B.** Three domains, a monophyletic tree, based on 16S rRNA sequences. The length of each branch approximates the time of divergence from the last common ancestor. For an up-to-date tree that synthesizes both approaches, see Chapter 17.

- **Lynn Margulis** proposed that eukaryotic organelles such as mitochondria and chloroplasts evolved by endosymbiosis from prokaryotic cells engulfed by proto-eukaryotes.
- **Carl Woese** discovered a domain of prokaryotes, *Archaea*, whose genetic sequences diverge equally from those of bacteria and those of eukaryotes. Many, though not all, archaea grow in extreme environments.

1.6 Cell Biology and the DNA Revolution

During the twentieth century, amid world wars and societal transformations, the field of microbiology exploded with new knowledge (see Table 1.2). More than 99% of what we know about microbes today was discovered since 1900 by scientists too numerous to cite in this book. Advances in biochemistry and microscopy revealed the fundamental structure and function of cell membranes and proteins. The revelation of the structures of DNA and RNA led to the discovery of the genetic programs of model bacteria such as *E. coli* and the bacteriophage lambda. Beyond microbiology, these advances produced the technology of "recombinant DNA," or genetic engineering, the creation of molecules that combine DNA sequences from unrelated species. These microbial tools offered unprecedented applications to human medicine and industry.

Cell Membranes and Macromolecules

In 1900, the study of cell structure was still limited by the resolution of the light microscope and by the absence of tools that could take apart cells to isolate their components. Both these limitations were overcome by the invention of powerful instruments. Just as society was being transformed by machines ranging from jet airplanes to vacuum cleaners, the study of microbiology was also being transformed by machines. Two instruments had exceptional impact: The **electron microscope** revealed the internal structure of cells (Chapter 2), and the **ultracentrifuge** enabled isolation of subcellular parts (Chapter 3).

The electron microscope. In the 1920s, at the Technical College in Berlin, student Ernst Ruska (1906–1988) was invited to develop an instrument for focusing rays of electrons. Ruska recalled as a child how his father's microscope could magnify fascinating specimens of plants and animals, but that its resolution was limited by the wavelength of light. He was eager to devise lenses that could focus beams of electrons, with wavelengths far smaller than that of light, to reveal living

details never seen before. Ultimately, Ruska built lenses to focus electrons using specially designed electromagnets. Magnetic lenses were used to complete the first electron microscope in 1933 (**Fig. 1.28**). Early transmission electron microscopes achieved about tenfold greater magnification than the light microscope, revealing details such as the ridged shell of a diatom. Further development steadily increased magnification, to as high as a millionfold.

For the first time, cells were seen to be composed of a cytoplasm containing macromolecules and bounded by a phospholipid membrane. By the 1960s, thin sections revealed the entire inner architecture of eukaryotic cells, including intracellular membranes, ribosomes, and organelles such as mitochondria and chloroplasts. Within the smaller cells of bacteria, the DNA-containing nucleoid was revealed, as well as specialized structures, such as photosynthetic organelles called chlorosomes (**Fig. 1.29**).

Subcellular structures, however, raised many questions about cell function that visualization alone could not answer. Biochemists showed how cell function involved numerous chemical transformations mediated by enzymes. A milestone in the study of metabolism was the elucidation by German biochemist Hans Krebs (1900–1981) of the tricarboxylic acid cycle (TCA, or Krebs cycle), by which the products of sugar diges-

Marine Biological Laboratory/WHOI

Figure 1.28 Electron microscopy. An early transmission electron microscope. The tall column contains a series of magnetic lenses.

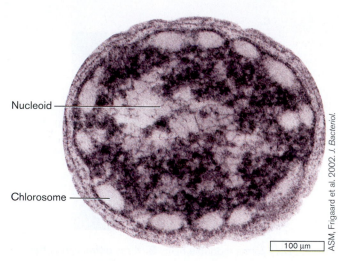

Figure 1.29   Electron microscopy. Transmission electron micrograph of a *Chlorobium* species, a photosynthetic bacterium. The thin section reveals the nucleoid (containing DNA), the light-harvesting chlorosomes, and envelope membranes.

Nucleoid

Chlorosome

100 μm

ASM, Frigaard et al. 2002. J. Bacteriol.

tion are converted to carbon dioxide. The tricarboxylic acid cycle provides energy for many bacteria and for the mitochondria of eukaryotes. But even Krebs understood little of how metabolism is organized within a cell; he and his contemporaries considered the cell a "bag of enzymes." The full significance of cell structure required experiments on isolated parts of cells.

The ultracentrifuge. Whole cells could be separated from the fluid in which they were suspended by centrifugation, the spinning of samples in a rotor so as to subject the cells to centrifugal forces of a few thousand times that of gravity. Some biochemists proposed that even greater centrifugal forces could separate cell components, even macromolecules such as proteins. The Swedish chemist Theodor Svedberg (1884–1971), at the University of Uppsala, set out to build such a machine: the ultracentrifuge.

Svedberg studied the properties of macromolecules such as proteins and polysaccharides. A major challenge was to measure the size of such particles. Svedberg calculated that particle size could be determined based on the rate of the movement of particles in a tube, the subjected to forces of hundreds of thousands times that of gravity. So he designed and built rotors spinning at ever greater rates, rates so high that they required a vacuum to avoid burning up like a space reentry vehicle. Such an advanced centrifuge, used to separate components of cells, is called an ultracentrifuge.

Experiments based on electron microscopy and ultracentrifugation revealed how membranes govern energy transduction within bacteria and within organelles such as mitochondria and chloroplasts. In the 1960s, English biochemists Peter Mitchell (1920–1992) and Jennifer Moyle (1921–) proposed and tested a revolutionary idea called the **chemiosmotic hypothesis**. The chemiosmotic hypothesis states that the redox reactions of the electron transport system store energy in the form of a gradient of protons (hydrogen ions) across a membrane, such as the bacterial cell membrane or the inner membrane of mitochondria. The energy stored in the proton gradient in turn drives the synthesis of ATP.

The chemiosmotic hypothesis earned Mitchell the Nobel Prize in Chemistry in 1978. Nevertheless, the theory took many years to win acceptance by biologists committed to the "bag of enzymes" model of the cell. Ultimately the role of ion gradients across membranes was recognized as fundamental to all living cells. (Ion gradients in membranes are discussed further in Chapters 3 and 4, and their role in energy transduction is presented in Chapter 14.)

Microbial Genetics Leads the DNA Revolution

As the form and function of living cells emerged in the early twentieth century, a largely separate line of research revealed patterns of heredity of cell traits. In eukaryotes, the Mendelian rules of inheritance were rediscovered and connected to the behavior of subcellular structures called chromosomes. Frederick Griffith (1877–1941) showed in 1928 that some unknown substance out of dead bacteria could carry genetic information into living cells, transforming harmless bacteria into a strain capable of killing mice, a process called **transformation**. Some kind of "genetic material" must be inherited to direct the expression of inherited traits, but no one knew what that material was or how its information was expressed.

Then in 1944, Oswald Avery (1877–1955) and colleagues showed that the genetic material responsible for transformation is deoxyribonucleic acid, or DNA. An obscure acidic polymer, DNA was previously thought too uniform in structure to carry information; its precise structure was unknown. As World War II raged among nations, among scientists an epic struggle began: the quest for the structure of DNA.

The discovery of the double helix. The tool of choice for macromolecular structure was X-ray crystallography, a method developed by British physicists in the early 1900s. The field of X-ray analysis included an unusual number of women, including Dorothy Hodgkin (1910–1994), who won a Nobel Prize for the structures of penicillin and vitamin B_{12}. In 1953, crystallographer Rosalind Franklin (1920–1958) joined a laboratory at King's College to study the structure of DNA (**Fig. 1.30A**). As a woman and as a

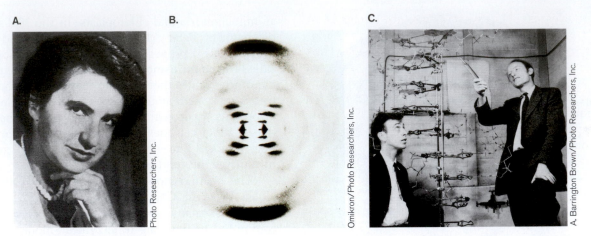

Figure 1.30   The DNA double helix.   A. Rosalind Franklin discovered that DNA forms a double helix. **B.** X-ray diffraction pattern of DNA, obtained by Rosalind Franklin. **C.** James Watson and Francis Crick discovered the complementary pairing between bases of DNA and the antiparallel form of the double helix.

Jew who supported relief work in Palestine, Franklin felt socially isolated at the male-dominated Protestant university. Nevertheless, the exceptional quality of her X-ray micrographs impressed her colleagues. The X-shaped pattern in her micrograph (**Fig. 1.30B**) showed for the first time that the standard B-form DNA was a double helix. Without Franklin's knowledge, her colleague Maurice Wilkins showed her data to a competitor, James Watson (1928–). The pattern led Watson and Francis Crick (1916–2004) to guess that the four bases of the DNA "alphabet" were paired in the interior of Franklin's double helix (**Fig. 1.30C**). They published their model in the journal *Nature*, while denying that they used Franklin's data.

The discovery of the double helix earned Watson, Crick, and Wilkins the 1962 Nobel Prize in Physiology or Medicine. Franklin died of ovarian cancer before the Nobel Prize was awarded. Before her death, however, she turned her efforts to the structure of ribonucleic acid (RNA). She determined the form of the RNA chromosome within tobacco mosaic virus, the first viral RNA to be characterized.

The structure of DNA base pairs led to development of techniques for **DNA sequencing**, the reading of the sequence of DNA base pairs. **Figure 1.31** shows an example of DNA sequence data, where each color represents one of the four bases and each peak represents a DNA fragment terminating in that particular base. The order of fragment lengths yields the sequence of bases in one strand. Reading the sequence enabled microbiologists to determine the beginning and endpoint of genes and ultimately entire genomes.

The DNA revolution began with bacteria. What amazed the world about DNA was that such a simple substance, composed of only four types of subunit, is the genetic material that determines all the different organisms on Earth. The promise of this insight was first fulfilled in bacteria and bacteriophages, whose small genomes and short generation times made key experiments possible. Furthermore, naturally occurring mechanisms of gene exchange in bacteria and viruses provided scientists with the key tools for transferring genes, including those of animals and plants. Consider these examples:

■ **Restriction endonucleases led to recombinant DNA.** Bacteria make **restriction endonucleases**, enzymes that cut DNA at positions determined by specific short base sequences. In nature, restriction

Figure 1.31   A DNA sequence fluorogram.   Sequence obtained by an undergraduate at Kenyon College using an ABI sequencer, from genomic DNA of a microbe isolated from a discarded beverage container. Each colored trace represents the intensity of fluorescence of one of the four bases terminating a chain of DNA.

endonucleases protect bacteria from the foreign DNA of viruses. In the test tube, purified restriction endonucleases were used to "cut and paste" DNA from two unrelated organisms, generating **recombinant DNA**. The construction of artificially recombinant DNA, by "gene cloning," ultimately made it possible to transfer genes between the genomes of virtually all types of organisms, using processes derived from natural phenomena of bacterial transformation.

- **A thermophilic DNA polymerase was used for polymerase chain reaction (PCR) amplification of DNA.** While molecular biologists forged ahead in the laboratory, microbial ecologists discovered new species of bacteria and archaea in increasingly extreme habitats. A hot spring in Yellowstone National Park yielded the bacterium *Thermus aquaticus*, whose DNA polymerase could survive many rounds of cycling to near-boiling temperature. The Taq polymerase formed the basis of a multibillion-dollar industry of PCR amplification of DNA, with applications ranging from genome sequencing to forensic identification.

- **Gene regulation discovered in bacteria provided models for animals and plants.** The information carried by DNA was shown to be expressed by transcription to RNA and then translation of RNA to make proteins. Most of the key discoveries of gene expression were made in bacteria and bacteriophages. The regulation of expression of a bacterial gene was first demonstrated for the *lac* operon, a set of contiguous genes in *E. coli*. Transcription of the *lac* operon was shown to be regulated by a repressor protein binding to DNA. DNA-binding proteins were subsequently demonstrated in all classes of living organisms.

Public response. As early as the 1970s, when the DNA revolution was largely limited to bacteria, its implications drew public concern. The use of recombinant DNA to make hybrid organisms—organisms combining DNA from more than one species—seemed "unnatural," although we now know that interspecies gene transfer occurs ubiquitously in nature. Furthermore, recombinant DNA technology raised the specter of placing deadly genes that produce toxins such as botulin into innocuous human flora such as *E. coli*.

The unknown consequences of recombinant DNA so concerned molecular biologists that they held a conference to assess the dangers and restrict experimentation on recombinant DNA. The conference, led by Paul Berg and Maxine Singer at Asilomar in 1975, was possibly the first time in history that a group of scientists organized and agreed to regulate and restrict their own field.

www | Asilomar conference on the dangers of recombinant DNA

Courtesy of Gabriel Weiss

Figure 1.32 "Protein Jive Sutra." At Stanford University, in 1971, rock dancers represented the process of protein synthesis in the ribosome. The depiction was immortalized in a film produced by chemist Kent Wilson and directed by Gabriel Weiss, with an introduction by Nobel laureate Paul Berg.

On the positive side, the emerging world of molecular biology drew excitement from students at a time of new ideas and social change. In 1971, the newly discovered process of protein translation by the bacterial ribosome inspired a classic cult film, *Protein Synthesis: An Epic on the Cellular Level*, directed by Gabriel Weiss and choreographed by Jackie Bennington, America's 1969 Junior Miss (**Fig. 1.32**). Introduced by Nobel laureate Paul Berg, the film depicts dancers on the Stanford University football field forming the shape of the ribosome while "messenger RNA" and "transfer RNAs" meet to assemble a "protein." The dancers sway to "Protein Jive Sutra," performed by a Haight-Ashbury-inspired rock band. Their excitement echoed that of scientists exploring the extraordinary potential of microbial discovery.

> **THOUGHT QUESTION 1.9** Do you think engineered strains of bacteria should be patentable? What about sequenced genes or genomes?

Microbial Discoveries Transform Medicine and Industry

Twentieth-century microbiology transformed the practice of medicine and generated entire new industries of biotechnology and bioremediation. Following the discovery of penicillin, Americans poured millions of dollars of private and public funds into medical research. The March of Dimes campaign for private donations to prevent polio

Table 1.3   **Fields of research in microbiology.**

Field	Subject of study
Experimental microbiology	Fundamental questions about microbial form and function, genetics, and ecology
Medical microbiology	The mechanism, diagnosis, and treatment of microbial disease
Epidemiology	Distribution and causes of disease in humans, animals, and plants
Immunology	The immune system and other host defenses against infectious disease
Food and industrial microbiology	Microbial and industrial food products, food contamination, and bioremediation
Environmental microbiology	Microbial diversity and microbial processes in natural and artificial environments
Forensic microbiology	Analysis of microbial strains as evidence in criminal investigations
Astrobiology	The origin of life in the universe and the possibility of life outside Earth

led to the successful development of a vaccine that has nearly eliminated the disease. With the end of World War II, research on microbes and other aspects of biology drew increasing financial support from U.S. government agencies, such as the National Institutes of Health and the National Science Foundation, as well as from governments of other countries, particularly the European nations and Japan. Further support has come from private foundations, such as the Pasteur Institute, the Wellcome Trust, and the Howard Hughes Medical Institute.

Research in microbiology finds applications in diverse fields (**Table 1.3**), all of which recruit microbiologists (**Fig. 1.33**). Cloned genes in bacteria produce valuable therapeutic proteins, such as insulin for diabetics. Recombinant viruses make safer vaccines. Bacteria and viral recombinant genomes are used to construct transgenic animals and plants. In environmental science, newly discovered microbes provide new ways to bioremediate wastes and control insect pests. On a global level, the management of our planet's biosphere, with the challenges of pollution and global warming, increasingly depends on our understanding of microbial ecology.

WWW | American Society for Microbiology

Joan Slonczewski, Kenyon College

Figure 1.33   Microbiologists at work. Students at Kenyon College conduct research on bacterial gene expression.

TO SUMMARIZE:

■ **Genetics of bacteria, bacteriophages, and fungi** in the early twentieth century revealed fundamental insights about gene transmission that apply to all organisms.

■ **Structure and function of the genetic material, DNA** was revealed by a series of experiments in the twentieth century.

■ **Molecular microbiology generated key advances**, such as the use of restriction enzymes, the cloning of the first recombinant molecules, and the invention of DNA-sequencing technology.

■ **Genome sequence determination and bioinformatic analysis** became the tools that shape the study of biology in the twenty-first century.

■ **Microbial discoveries transformed medicine and industry.** Biotechnology enables the production of new kinds of pharmaceuticals and industrial products.

Concluding Thoughts

The advances in microbial science raise important questions for society. Who shall bear the growing costs of sophisticated research on microbial pathogens? Should genetically modified bacteria be released into the environment for agriculture to increase crop yield? Should deadly viruses be engineered as vectors for gene therapy?

This book explores the current explosion of knowledge about microbial cells, genetics, and ecology. We introduce the applications of microbial science to human affairs, from medical microbiology to environmental science. Most importantly, we discuss research methods—how scientists make the discoveries that will shape tomorrow's view of microbiology. In the rest of Part 1, Chapter 2 presents the visualization tools that make possible our increasingly detailed view of the structures of cells (Chapter 3) and viruses (Chapter 6). Chapters 4 and 5 introduce microbial nutrition and growth in diverse habitats, including new information derived from the sequencing of microbial genomes. Throughout, we invite readers to share with us the excitement of discovery in microbiology.

CHAPTER REVIEW

Review Questions

1. Explain the apparent contradictions in defining microbiology as the study of microscopic organisms or the study of single-celled organisms.
2. What is the genome of an organism? How do genomes of viruses differ from those of cellular microbes?
3. Under what conditions might microbial life have originated? What evidence supports current views of microbial origin?
4. List the ways in which microbes have affected human life throughout history.
5. Summarize the key experiments and insights that shaped the controversy over spontaneous generation. What key questions were raised, and how were they answered?
6. Explain how microbes are cultured on liquid and solid media. Compare and contrast the culture methods of Koch and Winogradsky. How did their different approaches to microbial culture address different questions in microbiology?
7. Explain how a series of observations of disease transmission led to development of immunization to prevent disease.
8. Summarize key historical developments in our view of microbial taxonomy. What attributes of microbes have made them challenging to classify?
9. Explain how various discoveries in "natural" bacterial genetics were used to develop recombinant DNA technology.

Key Terms

agar (21)
antibiotic (25)
antiseptic (25)
Archaea (9)
archaeon (9)
aseptic (25)
autoclave (17)
Bacteria (9)
bacterium (9)
chain of infection (21)
chemiosmotic hypothesis (33)
chemolithotrophs (26)
colonies (21)
DNA sequencing (34)
electron microscope (32)

endosymbiont (27)
enrichment culture (26)
Eukarya (9)
eukaryote (8)
fermentation (17)
genome (9)
geochemical cycling (27)
germ theory of disease (17)
immune system (24)
immunity (24)
immunization (24)
Koch's postulates (21)
lithotrophs (26)
microbe (7)
monophyletic (30)

nitrogen fixation (27)
petri dish (21)
photosynthesis (27)
polyphyletic (30)
prokaryotes (8)
pure culture (21)
recombinant DNA (35)
reduced (18)
restriction endonuclease (34)
spontaneous generation (16)
transformation (33)
ultracentrifuge (32)
vaccination (23)
virus (8)
Winogradsky column (27)

Recommended Reading

Angert, Esther, Kendall D. Clements, and Norman R. Pace. 1993. The largest bacterium. *Nature* **362**:239–241.

Brock, Thomas D. 1999. *Robert Koch: A Life in Medicine and Bacteriology.* ASM Press, Washington, DC.

Brock, Thomas D. 1999. *Milestones in Microbiology, 1546 to 1940.* ASM Press, Washington, DC.

Doudna, Jennifer A., and Thomas R. Cech. 2004. The chemical repertoire of natural ribozymes. *Nature* **418**:222–228.

Dubos, Rene. 1998. *Pasteur and Modern Science.* Translated by Thomas Brock. ASM Press, Washington, DC.

Fleishmann, Robert D., Mark D. Adams, Owen White, Rebecca A. Clayton, Ewen F. Kirkness, et al. 1995. Whole-genome random sequencing and assembly of *Haemophilus influenzae* Rd. *Science* **269**:496–512.

Hesse, Wolfgang. 1992. Walther and Angelina Hesse—early contributors to bacteriology. *ASM News* **58**:425–428.

Joklik, Wolfgang K., Lars G. Ljungdahl, Alison D. O'Brien, Alexander von Graevenitz, and Charles Yanofsky (eds.). 1999. *Microbiology: A Centenary Perspective.* ASM Press, Washington, DC.

Margulis, Lynn. 1968. Evolutionary criteria in Thallophytes: A radical alternative. *Science* **161**:1020–1022.

Raoult, Didier. 2005. The journey from *Rickettsia* to Mimivirus. *ASM News* **71**:278–284.

Sherman, Irwin W. 2006. *The Power of Plagues.* American Society for Microbiology Press, Washington, DC.

Thomas, Gavin. 2005. Microbes in the air: John Tyndall and the spontaneous generation debate. *Microbiology Today* (Nov. 5): 164–167.

Ward, Naomi, and Claire Fraser. 2005. How genomics has affected the concept of microbiology. *Current Opinion in Microbiology* **8**:564–571.

Westall, Frances. 2005. Life on the early earth: A sedimentary view. *Science* **308**:366–367.

Woese, Carl R., and George E. Fox. 1977. Phylogenetic structure of the prokaryotic domain: The primary kingdoms. *Proceedings of the National Academy of Sciences USA* **74**:5088–5090.

Zuniga, Elina I., Bumsuk Hahm, Kurt H. Edelmann, and Michael B. A. Oldstone. 2005. Immunosuppressive viruses and dendritic cells: A multifront war. *ASM News* **71**:285–290.

Chapter 2

Observing the Microbial Cell

Microscopy reveals the vast realm of bacteria and protists invisible to the unaided eye. The microbial world spans a wide range of size—over several orders of magnitude. For different size ranges, we use different instruments, from the simple bright-field microscope to the electron microscope. The microscope enables us to count the number of microbes in the human bloodstream or in dilute natural environments such as the ocean. It shows how microbes swim and respond to signals such as a new food source.

Advanced forms of microscopy reveal microbes in remarkable detail. Fluorescence microscopy shows how microbial cells develop and reproduce. Electron microscopy explores the cell's interior, showing how all the parts of the cell fit together. The electron microscope is used to build models of viruses, and to probe the subcellular structures of membranes, ribosomes, and flagellar motors. Scanning probe microscopy images live microbes with an unprecedented degree of realism.

Atomic-force microscopic image (AFM) of *Corynebacterium glutamicum* cells, showing the contours of their formidable S-layer. Each cell is 1.0–1.5 μm in length, and about 0.7 μm in diameter. *C. glutamicum* is a major industrial producer of amino acids and vitamins. *Source*: Nicole Hansmeier, et al. 2006. *Microbiology* 152:923–935. Photo from Hansmeier, et al. 2006. *SGM Microbiology* No. 4.

When the microscope of Antoni van Leeuwenhoek (1632–1723) first revealed the tiny life-forms on his teeth, scientists in England refused to believe it until their own instruments showed the same. Van Leeuwenhoek's superior lenses were key to his success. Since the time of van Leeuwenhoek, microscopists have devised ever more powerful instruments to search for microbes in unexpected habitats.

An example of such a habitat is the human stomach, long believed too acidic to harbor life. In the 1980s, however, Australian scientist Barry Marshall reported finding a new species of bacterium in the stomach. The bacteria, *Helicobacter pylori*, proved difficult to isolate and culture, and for over a decade, medical researchers refused to believe Marshall's report. Ultimately, electron microscopy (EM) confirmed the existence of *H. pylori* in the stomach and helped document its role in gastritis and stomach ulcers. The scanning electron micrograph in **Figure 2.1** shows *H. pylori* colonizing the gastric crypt cells. *H. pylori* bacteria are comma-shaped rods with a polar tuft of flagella. The contours of cells and flagella are well resolved by the scanning beam of electrons, an achievement far beyond that possible with a light microscope.

Today, various kinds of microscopes are used in many settings, from hospitals and veterinary clinics to industrial plants and wastewater treatment facilities. This chapter covers two areas:

- **Light microscopy.** The theory and practice of light microscopy are essential for every student and professional observing microorganisms.
- **Advanced tools for research.** Fluorescence microscopy, electron microscopy, and X-ray diffraction crystallography probe ever farther the frontiers of the unseen.

2.1   Observing Microbes

Most microbes are too small to be seen; that is, they are microscopic, requiring the use of a **microscope** to be seen. But why can't we see microbes without magnification? The answer is surprisingly complex. In fact, our definition of "microscopic" is based not on inherent properties of the organism under study, but on the properties of our eyes. What is "microscopic" actually lies in the eye of the beholder.

Resolution of Objects by Our Eyes

The size at which objects become visible depends on the resolution of the observer's eye. **Resolution** is the smallest distance by which two objects can be separated and still be distinguished. In the eyes of humans and other animals, resolution is achieved by focusing an image on a retina packed with light-absorbing photoreceptor cells (**Fig. 2.2**). A group of photoreceptors with their linked neurons forms one unit of detection, comparable to a pixel

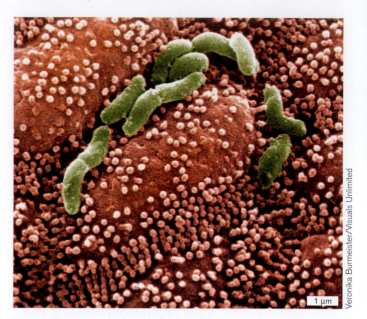

Veronika Burmeister/Visuals Unlimited

Figure 2.1 *Helicobacter pylori* **within the crypt cells of the stomach lining.**   Microscopy demonstrated the presence of *H. pylori*, the causes of stomach ulcers, growing on the lining of the human stomach, a location previously believed too acidic to permit microbial growth. Scanning electron micrograph, colorized to indicate bacteria (green).

on a computer screen. The distance between two retinal "pixels" limits resolution.

The resolution of the human retina (that is, the length of the smallest object most human eyes can see) is about 150 µm, or one-seventh of a millimeter. In the retinas of eagles, photoreceptors are more closely packed, so an eagle can resolve objects eight times as small or eight times as far away as a human can; hence, the phrase "eagle-eyed" means sharp-sighted. On the other hand, insect compound eyes have one-hundredth the resolution of human eyes because their receptor cells are farther apart than ours. If a science-fictional giant ameba had eyes with a retinal resolution of 2 meters (m), it would perceive humans as we do microbes.

NOTE: In this book, we use standard metric units for size:

1 millimeter (mm) = one thousandth of a meter
                     = 10^{-3} m
1 micrometer (µm) = one thousandth of a millimeter
                     = 10^{-6} m
1 nanometer (nm) = one thousandth of a micrometer
                     = 10^{-9} m
1 picometer (pm) = one thousandth of a nanometer
                     = 10^{-12} m

Some authors still use the traditional unit **angstrom (Å)**, which equals a tenth of a nanometer, or 10^{-10} m.

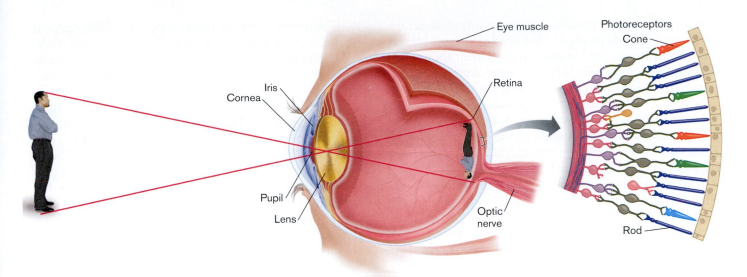

Figure 2.2 Defining the microscopic.   We define what is visible and what is microscopic in terms of the human eye. Within the human eye, the lens focuses an image on the retina. Adjacent light receptors are sufficiently close to resolve the image of a human being, even at a distance where its apparent height shrinks to a millimeter.

Resolution Differs from Detection

Objects of sizes below the resolution limit may yet be *detected*; that is, we can observe their presence as a group. Thus, our eyes can detect a large population of microbes, such as a spot of mold on a piece of bread (about a million cells) or a cloudy tube of bacteria in liquid culture (a million cells per milliliter; **Fig. 2.3A**). **Detection**, the ability to determine the presence of an object, differs from resolution. When the unaided eye detects the presence of mold or bacteria, it cannot resolve individual cells. To resolve bacterial cells, such as those of the wetland phototroph *Rhodospirillum rubrum* (**Fig. 2.3B**), requires magnification. In microscopy, **magnification** means to increase the apparent size of an image so as to resolve smaller separations between objects and thus increase the information obtained by our eyes.

Microbial Size and Shape

Different kinds of microbes differ in size, over a range of several orders of magnitude, or powers of ten (**Fig. 2.4**). Eukaryotic microbes are often sufficiently large (10–100 μm)

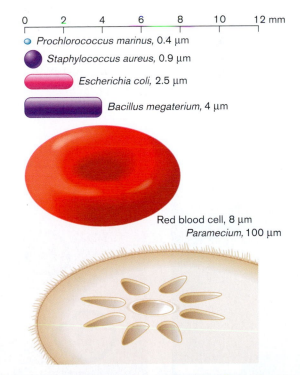

Figure 2.4 Relative sizes of different cells.   Microbial cells come in various sizes, most of which are below the threshold of resolution by the unaided human eye (150 μm). Note that most bacteria are much smaller than the human red blood corpuscle, a small eukaryotic cell.

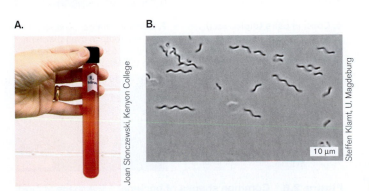

Figure 2.3 Detecting and resolving bacteria.   A. A tube of bacterial culture, *Rhodospirillum rubrum*. Bacterial cells are detected, though not resolved. **B.** Bright-field image. Individual cells of *R. rubrum* are resolved.

that we can resolve their compartmentalized structure under a light microscope. Prokaryotes (bacteria and archaea) are generally smaller (0.4–10 µm); thus, their overall size can be seen, but their internal structures are too small to be resolved. Nevertheless, a few bacterial species, such as *Thiomargarita namibiensis*, are large enough to be seen without a microscope, while many unculturable marine eukaryotes are as small as the smallest bacteria. Thus, the actual size ranges of eukaryotic and prokaryotic microbes overlap substantially.

Many eukaryotes, such as protists and algae, can be resolved under low-power magnification to reveal internal and external structures such as the nucleus, vacuoles, and flagella (**Fig. 2.5**). Protists show complex shapes and appendages. For example, amebas from an aquatic ecosystem show large nuclei and pseudopods to engulf prey (**Fig. 2.5A**). Pseudopods can be seen to move through streaming of their cytoplasm. Another protist readily observed by light microscopy is *Giardia lamblia*, a water contaminant that causes outbreaks of intestinal illness in cities and day-care centers (**Fig. 2.5B**). In the *Giardia* cell, we observe two nuclei and multiple flagella. Eukaryotic flagella propel the cell by a whiplike action. Similarly, we can observe the complex internal patterns of chloroplasts in algae and the exceptional variety of crystalline forms in diatoms. For more on microbial eukaryotes, see Chapter 20.

Prokaryotic cell structures are generally simpler than those of eukaryotes (see Chapter 3). **Figure 2.6** shows representative members of several common cell types, as visualized by light microscopy or by scanning electron microscopy. Note that with light microscopy, the cell shape is barely discernable under the highest power.

With scanning electron microscopy, the shapes appear clearer, but we still see no subcellular structures. Subcellular structures are best visualized with transmission electron microscopy and fluorescence microscopy (discussed below).

Certain shapes of bacteria are common to many taxonomic groups (**Fig. 2.6**). For example, both bacteria and archaea form similarly shaped **rods**, or **bacilli**, and **cocci** (spheres). Thus, rods and spherical shapes have evolved independently within different taxa. On the other hand, a unique bacterial shape is seen in the **spirochetes**, which cause diseases such as syphilis and Lyme disease. Spirochetes possess an elaborate spiral structure with internal axial filaments and flagella as well as an outer sheath (for more on spirochetes, see Section 18.7). Spiral axial filaments are found only within the spirochete group of closely related species.

A. Filamentous rods (bacilli). *Lactobacillus lactis,* gram-positive bacteria (LM).

B. Rods (bacilli). *Lactobacillus acidophilus,* gram-positive bacteria (SEM).

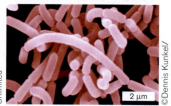

C. Spirochetes. *Borrelia burgdorferi,* causative organism in Lyme disease, among human blood cells (LM).

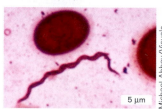

D. Spirochetes. *Leptospira interrogans,* causative organism in leptospirosis in animals and humans (SEM).

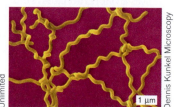

E. Cocci in pairs (diplococci). *Streptococcus pneumoniae,* a cause of pneumonia. Methylene blue stain (LM).

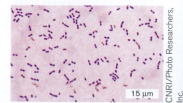

F. Cocci in chains. *Anabaena* spp., filamentous cyanobacteria. Producers for the marine food chain (SEM).

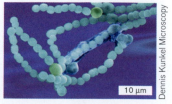

Figure 2.6   Common shapes of bacteria. The shape of most bacterial cells can be discerned with light microscopy (**A, C, E**), but their subcellular structures and surface details cannot be seen. Surface detail is revealed by scanning electron microscopy (SEM) (**B, D, F**). These SEM images are colorized to enhance clarity.

A.

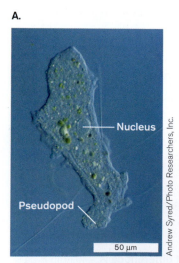

B.

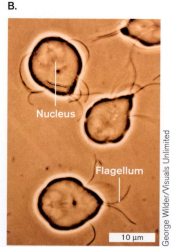

Figure 2.5   Eukaryotic microbial cells. Eukaryotic microbes are large enough that details of internal and external organelles can be seen under a light microscope. **A.** *Amoeba proteus.* **B.** *Giardia lamblia.*

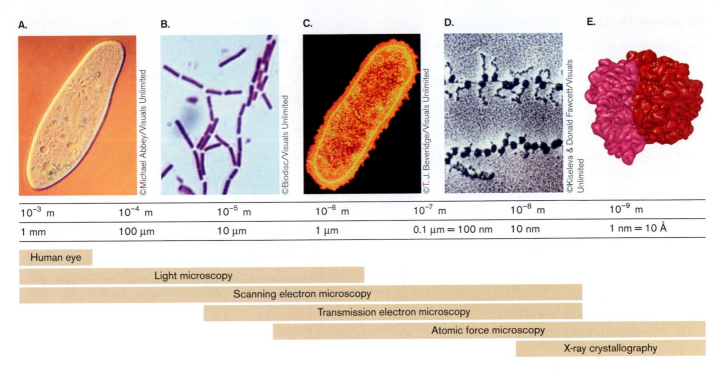

10^{-3} m		10^{-4} m		10^{-5} m		10^{-6} m		10^{-7} m		10^{-8} m		10^{-9} m	
1 mm		100 μm		10 μm		1 μm		0.1 μm = 100 nm	10 nm			1 nm = 10 Å	

Human eye

Light microscopy

Scanning electron microscopy

Transmission electron microscopy

Atomic force microscopy

X-ray crystallography

Figure 2.7   Microscopy and X-ray crystallography, range of resolution.   A. Light microscopy reveals internal structures of a paramecium (a eukaryote). LM, Magnification ×100 at 35 mm size. **B.** *Pseudomonas* sp., rod-shaped bacteria, are barely resolved. LM, Magnification ×1200. **C.** Internal structures of the bacterium *Escherichia coli* are revealed by transmission electron microscopy (TEM). Magnification ×32,000. **D.** Individual ribosomes are attached to the messenger RNA molecules that are translated to make peptides (TEM). Magnification ×150,000. **E.** A ribosome (diameter, 21 nm) cannot be observed directly but can be modeled using X-ray crystallography.

NOTE: The genus name *Bacillus* refers to a specific taxonomic group of bacteria, but the term *bacillus* (plural, bacilli) refers to any rod-shaped bacterium or archaeon.

Microscopy for Different Size Scales

To resolve microbes and microbial structures of different sizes requires different kinds of microscopes. **Figure 2.7** shows the different techniques used to resolve microbes and structures of various sizes. For example, a paramecium can be resolved under a light microscope, but an individual ribosome requires electron microscopy.

■ **Light microscopy** resolves images of individual bacteria based on their absorption of light. The specimen is commonly viewed as a dark object against a light-filled field, or background; this is called **bright-field microscopy** (seen in **Figs. 2.7A, B**). Advanced techniques, based on special properties of light, include dark-field, phase-contrast, and fluorescence microscopy.

■ **Electron microscopy (EM)** uses beams of electrons to resolve details several orders of magnitude smaller than those seen under light microscopy.

In **scanning electron microscopy (SEM)**, the electron beam is scattered from the metal-coated surface of an object, generating an appearance of three-dimensional depth. In **transmission electron microscopy (TEM)** (**Figs. 2.7C, D**), the electron beam travels through the object, where the electrons are absorbed by an electron-dense metal stain.

■ **Atomic force microscopy (AFM)** uses van der Waals forces between a probe and an object to map the three-dimensional topography of a cell.

■ **X-ray crystallography** detects the interference pattern of X-rays entering the crystal lattice of a molecule. From the interference pattern, researchers build a computational model of the structure of the individual molecule, such as a protein or a nucleic acid, or even a molecular complex such as a ribosome (**Fig. 2.7E**).

■ **Nuclear magnetic resonance (NMR)** imaging provides structural information based on the magnetic properties of atomic nuclei.

THOUGHT QUESTION 2.1   You have discovered a new kind of microbe, never observed before. What kind of questions about this microbe might be answered by light microscopy? What questions would be better addressed by electron microscopy?

TO SUMMARIZE:

■ **Detection** is the ability to determine the presence of an object.

■ **Resolution** is the smallest distance by which two objects can be separated and still be distinguished.

■ **Magnification** means an increase in the apparent size of an image so as to resolve smaller separations between objects.

■ **Eukaryotic microbes may be large enough to resolve subcellular structures** under a light microscope, although some eukaryotes are as small as bacteria.

■ **Bacteria and archaea are usually too small for subcellular resolution.** Their shapes include characteristic forms, such as rods and cocci.

■ **Different kinds of microscopy** are required to resolve cells and subcellular structures of different sizes.

WWW | Molecular Expressions: Exploring the World of Optics and Microscopy, Florida State University

2.2 Optics and Properties of Light

Light microscopy directly extends the lens system of our own eyes. Light is part of the spectrum of electromagnetic radiation (**Fig. 2.8**), a form of energy that is propagated as waves associated with electrical and magnetic fields. Regions of the electromagnetic spectrum are defined by wavelength, which for visible light is about 400–750 nm. Radiation of longer wavelengths includes infrared and radio waves, whereas shorter wavelengths include ultraviolet rays and X-rays.

Light Carries Information

All forms of electromagnetic radiation carry information from the objects with which they interact. The information carried by radiation can be used to detect objects; for example, radar (using radio waves) detects a speeding car. All electromagnetic radiation travels through a vacuum at the same speed (*c*), about 3×10^8 meters per second (m/s). The speed of light (*c*) is equal to the wavelength (λ) of the radiation multiplied by its frequency (ν), the number of wave cycles per unit time. Frequency is usually measured in hertz (Hz), reciprocal seconds:

$$c = \lambda\nu$$

Because *c* is constant, the longer the wavelength λ, the lower the frequency ν.

For electromagnetic radiation to resolve an object from neighboring objects or from its surrounding medium, certain conditions must exist:

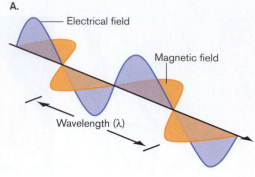

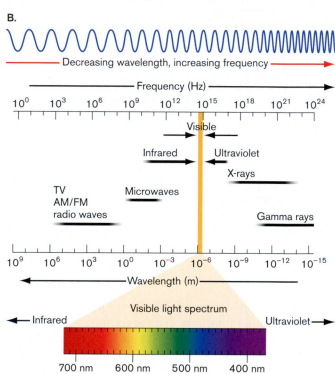

Figure 2.8   Electromagnetic energy.   A. Electromagnetic radiation is composed of electrical and magnetic waves perpendicular to each other. **B.** The electromagnetic spectrum includes the visual range, used in light microscopy.

■ **Contrast between the object and its surroundings. Contrast** is the difference in light and dark. If an object and its surroundings absorb or reflect radiation equally, then the object will be undetectable. Thus, it is hard to observe a cell of transparent cytoplasm floating in water because the aqueous cytoplasm and the extracellular water tend to transmit light similarly, producing little contrast.

■ **Wavelength smaller than the object.** For an object to be resolved, the wavelength of the radiation must be equal to or smaller than the size of the object. If the wavelength of the radiation is larger than the object, then most of the wave's energy will simply pass through it, like an ocean wave passing around a dock post. Thus, radar, with a wavelength of 1–100 centimeters (cm), cannot resolve microbes, though it easily resolves cars and people.

■ **A detector with sufficient resolution for the given wavelength.** The human eye has a retina with photoreceptors that absorb radiation within a narrow range of wavelengths, 400–750 nm (0.40–0.75 μm), which we define as "visible light." But the distance between our retinal photoreceptors is 150 μm, about 500 times the wavelength of light. Thus, our eyes are unable to access all the information contained in the light that enters. In order for us to exploit the full information capacity of light, the light rays from point sources of light must be spread apart sufficiently to be resolved by our retinal photoreceptors. The spreading of the light rays results in magnification of the image.

Light Interacts with an Object

The physical behavior of light in some ways resembles a beam of particles and in other ways a waveform. The particles of light are called photons. Each photon has an associated wavelength that determines how the photon will interact with a given object. The combined properties of particle and wave enable light to interact with an object in several different ways: absorption, reflection, refraction, and scattering.

■ **Absorption** means that the photon's energy is acquired by the absorbing object (**Fig. 2.9A**). The energy is converted to a different form, usually heat (infrared radiation), a form of electromagnetic radiation of longer wavelength than light. When a microbial specimen absorbs light, it can be observed as a dark spot against a bright field, as in bright-field microscopy. Some molecules that absorb light of a specific wavelength reemit energy not as heat, but as light with a longer wavelength; this is called **fluorescence**. Fluorescence is also used in microscopy (discussed later).

■ **Reflection** means that the wave front redirects from the surface of an object at an angle equal to its incident angle (**Fig. 2.9B**). Reflection of light waves is analogous to the reflection of water waves. Reflection from a silvered mirror or a glass surface is used in the optics of microscopy.

■ **Refraction** is the bending of light as it enters a substance that slows its speed (**Fig. 2.9C**). Such a substance is said to be "refractive" and, by definition, has a higher **refractive index** than air. Refraction is the key property that enables a lens to magnify an image.

■ **Scattering** occurs when a portion of the wave front is converted to a spherical wave originating from the object (**Fig. 2.9D**). If a large number of particles

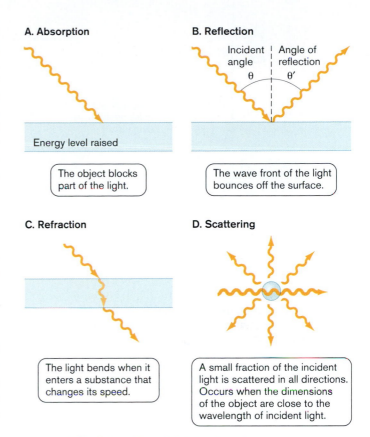

A. Absorption

Energy level raised

The object blocks part of the light.

B. Reflection

Incident angle θ | Angle of reflection θ′

The wave front of the light bounces off the surface.

C. Refraction

The light bends when it enters a substance that changes its speed.

D. Scattering

A small fraction of the incident light is scattered in all directions. Occurs when the dimensions of the object are close to the wavelength of incident light.

Figure 2.9   Interaction of light with matter.

simultaneously scatter light, a haze is observed—for example, the haze of bacteria suspended in a culture tube. Special optical arrangements (such as **dark-field microscopy**, discussed later) can use scattered light to detect (but not resolve) microbial shapes smaller than the wavelength of light.

Refraction Enables Magnification

Magnification requires refraction of light through a medium of high refractive index, such as glass. As a wave front of light enters a refractive material, the region of the wave that first reaches the material is slowed, while the rest of the wave continues at its original speed until it also passes into the refractive material (**Fig. 2.10**). As the entire wave front continues through the refractive material, its path continues at an angle from its original direction. At the opposite face of the refractive material, the wave front resumes its original speed.

> **THOUGHT QUESTION 2.2**   Explain what happens to the refracted light wave as it emerges from a piece of glass of even thickness. How does its new speed and direction compare with its original (incident) speed and direction?

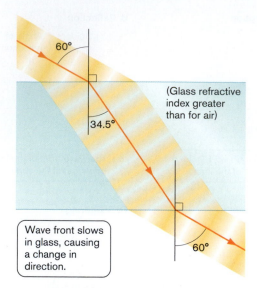

Figure 2.10   Refraction of light waves.   Wave fronts of light shift direction as they enter a substance of higher refractive index, such as glass.

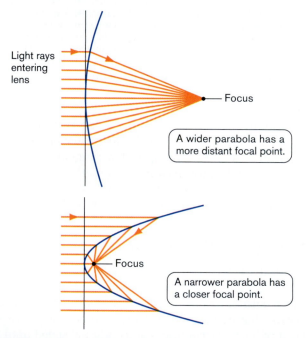

Figure 2.11   The focus of a parabola.   The wider the parabola, the more distant the focal point.

Refraction magnifies an image when light passes through a refractive material shaped so as to spread its rays, widening the wave front. The shape that accomplishes this widening is that of a parabola. Light rays entering a parabolic surface are focused; that is, they are each bent at an angle such that all rays meet at a certain point, called the focus of the parabola (**Fig. 2.11**).

In a lens, a parabolic surface focuses impinging light rays. Parallel light rays entering the lens emerge at angles so as to intersect each other at the **focal point**, equivalent to the focus of the parabola (**Fig. 2.12**). From the focal point, the light rays continue onward in an expanding wave front. This expansion magnifies the image carried by the wave. The distance from the lens to the focal point (called the focal distance) is determined by the degree of parabolic curvature of the lens and by the refractive index of its material.

In **Figure 2.12**, the object under observation is placed just outside the **focal plane**, a plane containing the focal point of the lens. All light rays from the object are bent by the lens, converging at the opposite focal point. At the focal point, the light rays continue through until they converge with nonparallel light rays refracted by the lens. The plane of convergence generates a reversed image of the object. The image is expanded, or magnified, by the spreading out of refracted rays. The distances between parts of the image are enlarged, enabling our eyes to resolve finer details.

www | FSU Tutorial on Magnification

> **THOUGHT QUESTION 2.3**   Parabolic lenses are generally "biconvex," that is, curving outward on both sides. What would happen to parallel light rays that pass through a lens that is concave on both sides? Or a lens that is convex on one side and concave with equal curvature on the other?

Magnification and Resolution

The spreading of light rays does not necessarily increase resolution. For example, an image composed of dots does not gain detail when enlarged on a photocopier, nor does an image composed of pixels gain detail when enlarged on a computer screen. In these cases, resolution fails to increase because the individual details of the image expand in proportion to the expansion of the overall image. Magnification without increase in resolution is called **empty magnification**.

The resolution of detail in microscopy is limited by the degree to which details "expand" with magnification. This expansion of detail is limited by the interference of light rays converging at the focal point. In **interference**, two wave fronts interact by addition (amplitudes in phase) or subtraction (amplitudes out of phase) (**Fig. 2.13A**). The result of interference between two waves is a pattern of alternating zones of constructive and destructive interference (brightness and darkness) (**Fig. 2.13B**).

In theory, a perfect lens that focuses all the light from an object should have no interference as its rays

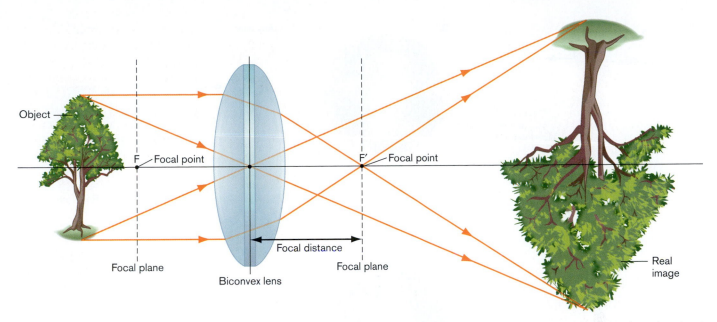

Figure 2.12   Generating an image with a lens.   The object is placed within the focal plane of the lens. All light rays from the object are bent by the lens, converging at the opposite focal point. The light rays continue through the focal point, generating an image of the object that is reversed and magnified.

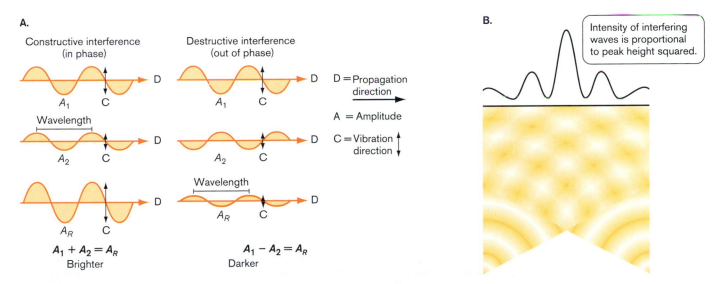

Figure 2.13   Constructive (additive) and destructive (subtractive) interference of light waves.   A. In constructive interference, the peaks of the two wave trains rise together; their amplitudes are additive, creating a wave of greater amplitude. In destructive interference, the peaks of the waves are opposite one another, so their amplitudes cancel, creating a wave of lesser amplitude.
B. Two wave fronts approaching at an angle generate an interference pattern in which intensity alternately increases and decreases.

converge toward the focal point. In fact, however, the outer edges of the lens introduce imperfection. The converging edges of the wave interfere with each other to create alternating regions of light and dark. At the focal point, the interference forms a central sphere of intensity surrounded by a series of concentric spheres of alternating light and dark (**Fig. 2.14A**). The viewing plane intersects the concentric spheres to form an **Airy disk**, a disk containing a bright central peak surrounded by rings

of light and dark. The Airy disk was originally discovered by the astronomer George Airy (1801–1892), who showed that the stars viewed through a telescope could never appear as true point sources, but only as tiny disks of light surrounded by rings. The same principle applies to objects viewed under a microscope.

www | FSU Tutorial on Airy Disk

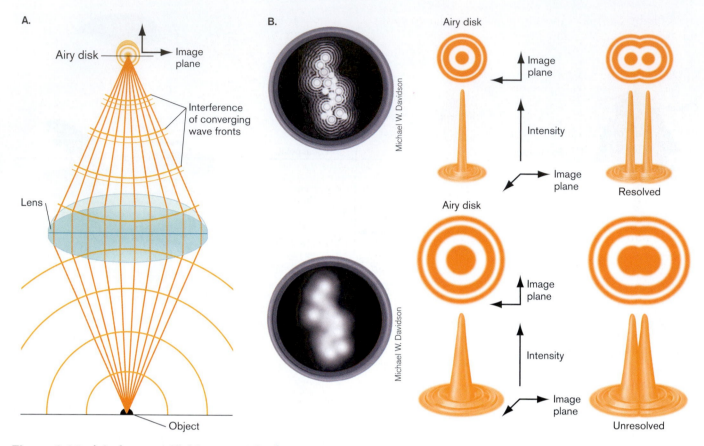

Figure 2.14   Interference of light waves at the focal point generates an Airy disk.   A. The interference of converging spherical wave fronts creates a sphere with a central zone of maximum intensity, surrounded by alternating zones of bright and dark. A plane section through the sphere generates the Airy disk. **B.** The Airy peak and its surrounding rings.

The width of the Airy disk—and hence the limit to resolution of detail—depends on the wavelength of light and the quality of the lens. The shorter the wavelength, the narrower the Airy disk. In addition, the central peak of intensity is sharpest when the specimen is at the focal point, where the specimen is said to be in **focus**.

Suppose an object consists of a collection of point sources of light. Each point source generates an Airy disk of rings surrounding one central peak of intensity. The width of the central peak increases with distance away from the focal point. This width will define the resolution, or separation distance, between any two points of the object (**Fig. 2.14B**). Full resolution is achieved for two points when there is complete separation of the central peaks of the two Airy disks. This resolution distance determines the degree of detail that can be observed along the edge of an object.

TO SUMMARIZE:

■ **Electromagnetic radiation** interacts with an object and acquires information we can use to detect the object.

■ **Contrast** between object and background makes it possible to detect the object and resolve its component parts.

■ **The wavelength of the radiation** must be equal to or smaller than the size of the object if we are to resolve its shape.

■ **Absorption** means that the energy from light (or other electromagnetic radiation) is acquired by the object.

■ **Reflection** means that the wave front bounces off the surface of a particle at an angle equal to its incident angle.

■ **Refraction** is the bending of light as it enters a substance that slows its speed.

■ **Scattering** occurs when a wave front interacts with an object of smaller dimension than the wavelength. Light scattering enables detection of objects whose detail cannot be resolved.

2.3   Bright-Field Microscopy

In bright-field microscopy, an object such as a bacterial cell is perceived as a dark silhouette blocking the passage of light. Details of the dark object are defined by the points of light surrounding its edge.

Increasing Resolution

The optics of a modern bright-field microscope are designed to maximize detail under magnification by a lens. In maximizing the resolution, the following factors need to be considered:

- **Wavelength.** The theoretical maximum magnification can be calculated as the resolution distance of our retina (0.15 mm, or 150,000 nm) divided by the average wavelength of light (550 nm, for green). This gives an approximately 300-fold magnification with increasing resolution. Additional optical factors can bring this magnification closer to 1,000×, although full resolution is rarely reached in practice. Any greater magnification expands only the width of the interference patterns; the image becomes larger but with no greater resolution (empty magnification).

- **Light and contrast.** For any given lens system, there is an optimal amount of light that yields the highest contrast between the dark specimen and the light background. High contrast is needed to achieve the maximum resolution at high magnification.

- **Lens quality.** All lenses possess inherent **aberrations** that detract from perfect parabolic curvature. Instead of trying to grind a "perfect" single lens, modern manufacturers construct microscopes with a series of lenses that multiply each other's magnification and correct for aberrations.

Let us first consider magnification of an image by a single lens. The principles described would hold for any lens. **Figure 2.15** shows an **objective lens**, a lens situated directly above an object or specimen that we wish to observe at high resolution. How can we maximize the resolution of details in the image?

An object at the focal point of a lens sits at the tip of an inverted light cone formed by rays of light from the lens converging at the object. The angle of the light cone is determined by the curvature and refractive index of the lens. The lens fills an aperture, or hole, for the passage of light; and for a given lens, the light cone is defined by an angle θ (theta) projecting from the midline, known as the **angle of aperture**. As θ increases and the horizontal width of the light cone (sin θ) increases, a wider cone of light passes through the specimen. The wider cone of light rays improves the precision of the image by lessening the proportion of edge interference between wave fronts, thereby narrowing the width of the Airy disk interference pattern. Another way to say this is that as sin θ increases, the resolvable distance decreases. Thus, the greater the angle of aperture of the lens, the better the resolution.

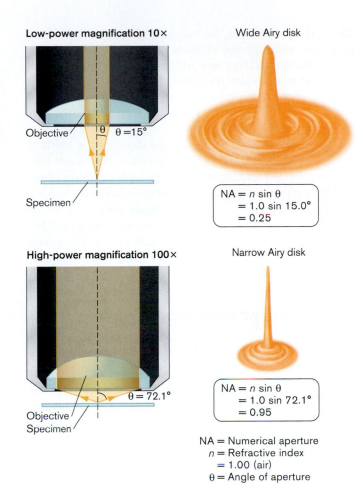

Figure 2.15 Resolution and numerical aperture. The resolution depends on numerical aperture (NA), which equals the refractive index (*n*) of the medium containing the light cone multiplied by the sine of the angle of the light cone (θ). Higher magnification occurs at higher NA.

THOUGHT QUESTION 2.4 In theory, what angle θ would produce the highest resolution? What practical problem would you have in designing a lens to generate this light cone?

Resolution also depends on the refractive index of the medium containing the light cone, which is usually air. The refractive index is the ratio of the speed of light in a vacuum to its speed in another medium. For air, the refractive index (*n*) is taken as 1, whereas lens material has a refractive index greater than 1. The lens bends the light spreading at an angle (θ). The product of the refractive index (*n*) of the medium times sin θ is the **numerical aperture** (NA):

$$NA = n \sin \theta$$

In **Figure 2.15**, we see the calculation of NA for an objective lens of magnification 10×, and for a lens magnification

100×. As NA increases, the magnification power of the lens increases, although the increase is not linear. Note that as the lens strength increases and the light cone widens, the lens must come nearer the object. Defects in lens curvature become more of a problem, and focusing becomes more challenging. As θ becomes very wide, too much of the light from the object is lost owing to refraction at the glass-to-air interface. The greater the refractive index of the medium between the object and the objective lens, the more light can be collected and focused. For the highest-power objective lens, generally 100×, a zone of even refractive index is maintained by insertion of **immersion oil** with a refractive index comparable to that of glass ($n = 1.5$) (**Fig. 2.16**). Immersion oil minimizes loss of light rays at the widest angles and makes it possible to reach 100× magnification with minimal distortion.

The Compound Microscope

The manufacture of higher-power lenses is difficult owing to decreasing tolerance for aberration. For this reason, we rarely observe through a single lens. Instead we use a **compound microscope**, a system of multiple lenses designed to correct or compensate for aberration. A typical arrangement of a compound microscope is shown in **Figure 2.17**. In this arrangement, the light source is placed at the bottom, shining upward through a series of lenses, including the condenser, objective, and ocular lenses.

The **condenser** consists of one or more lenses that focus parallel light rays from the light source onto a small area of the slide. The condensed light rays generate a narrower Airy interference pattern and thus improve the resolution of the objective lens.

Between the light source and the condenser sits a **diaphragm**, a device to vary the diameter of the light column. Lower-power lenses require operation at lower light levels because the excess light makes it impossible to observe the darkening effect of absorbance by the specimen (contrast). Higher-power lenses require more light and thus an open diaphragm.

The nosepiece of a compound microscope typically holds three or four objective lenses of different magnifying power, such as 4×, 10×, 40×, and 100× (requiring immersion oil). These lenses are arranged so as to rotate in turn into the optical column. A high-quality instrument will have the lenses set at different heights from the slide so as to be **parfocal**. In

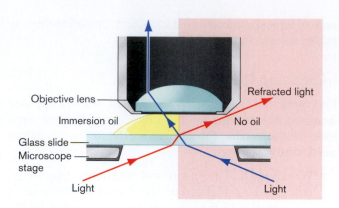

Figure 2.16   Use of immersion oil in microscopy. Immersion oil with a refractive index comparable to that of glass ($n = 1.5$) prevents light rays from bending away from the objective lens. Thus, more light is collected, NA increases, and resolution improves.

a parfocal system, when an object is focused using one lens, it remains in focus, or nearly so, when another lens is rotated to replace the first.

NOTE: Objective lenses can be obtained in several different grades of quality, manufactured with different kinds of correction for aberrations. Lenses should feature at minimum the following corrections: "plan" correction for field curvature, to generate a field that appears flat, and "apochromat" correction for spherical and chromatic aberrations.

The image from the objective lens is amplified by a secondary focusing step through the **ocular lens**. The

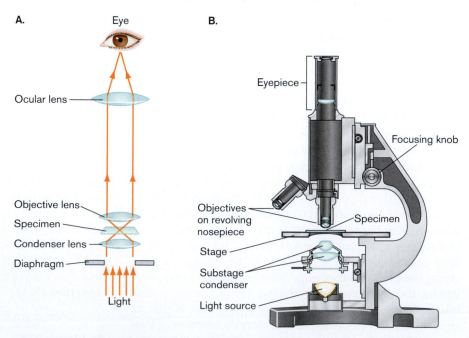

Figure 2.17   The anatomy of a compound microscope.   A. Light path through a compound microscope. **B.** Cutaway view of a compound microscope.

ocular lens sits directly beneath the observer's eye. In the process of magnification, each light ray traces a path toward a position opposite its point of origin, thus creating a mirror-reversed image. This reversal must be kept in mind when exploring a field of cells. Research-grade microscopes include a series of ocular and objective lenses.

The magnification factor of the ocular lens is multiplied by the magnification factor of the objective lens to generate the **total magnification** (power). For example, a 10× ocular times a 100× objective generates 1,000× total magnification.

Observing a specimen under a compound microscope requires several steps:

- **Position the specimen centrally in the optical column.** Only a small area of a slide can be visualized within the field of view of a given lens. The higher the magnification, the smaller the field of view will be seen.
- **Optimize the amount of light.** At lower power, too much light will wash out the light absorption of the specimen without contributing to magnification. At higher power, more light needs to be collected by the condenser. To optimize light, the condenser must be set at the correct vertical position to focus on the specimen, and the diaphragm must be adjusted to transmit the amount of light that produces the best contrast.
- **Focus the objective lens.** The focusing knob permits adjustment of the focal distance between the objective lens and the specimen on the slide so as to bring the specimen into the focal plane. Typically, we focus first using a low-power objective, which generates a greater **depth of field**—that is, a region along the optical column over which the object appears in reasonable focus. After focusing under low power, we can rotate a higher-power lens into view, then perform a smaller adjustment of focus.

Is the Object in Focus?

An object appears in focus (that is, is situated within the focal plane of the lens) when its edge appears sharp and distinct from the background. At higher power, however, recognition of the focal plane is a challenge because of Airy-like interference. The shape of the dark object is actually defined by the points of light surrounding its edge. The partial resolution of these points of light generates extra rings of light surrounding an object whose dimensions are close to the resolution limit.

In **Figure 2.18**, we see microscopic observation of *Rhodospirillum rubrum*, photosynthetic bacteria that contribute to wetland productivity. As cells of *R. rubrum* swim in and out of the focal plane, their appearance changes through optical effects. When a bacterium swims out of the focal plane too close to the lens, resolution declines and the image blurs (**Fig. 2.18A**). When the bacterium swims within the focal plane, its image appears sharp, with a bright line along its edge. A helical cell like *R. rubrum* shows well-focused segments alternating with hazier segments that are too close to the lens (**Fig. 2.18B**). When the cell swims too far past the focal plane, the bright interference lines collapse into the object's silhouette, which now appears bright or "hollow" (**Fig. 2.18C**). In fact, the bacterium is not hollow at all; only its image has changed.

When the cell extends across several focal planes, different portions appear out of focus (either too near or too far from the lens). In addition, when the end of a cell points toward the observer, light travels through the length of the cell before reaching the observer, so the cell absorbs more light and appears dark (see **Fig. 2.18D**). The

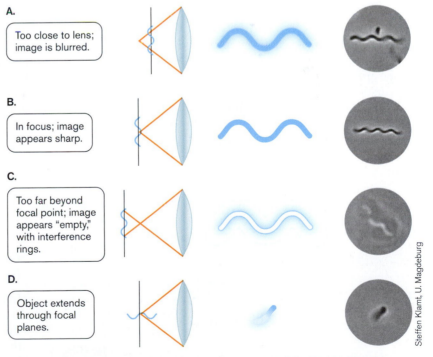

A. Too close to lens; image is blurred.

B. In focus; image appears sharp.

C. Too far beyond focal point; image appears "empty," with interference rings.

D. Object extends through focal planes.

Steffen Klamt, U. Magdeburg

Figure 2.18 *Rhodospirillum rubrum* **observed at different levels of focus. A.** When a bacterium swims too close to the lens, its image blurs. **B.** When the bacterium lies within the focal plane, its image appears sharp. If the width of the cell crosses several focal planes, some parts appear sharp, whereas other parts appear blurred. **C.** When the bacterium lies too far from the lens, its image appears "empty" or "hollow," surrounded by rings of Airy-like interference. **D.** When the spirillum extends through the focal plane, different parts show different focal effects (in focus, too near, or too far).

observation and identification of motile bacteria swimming in and out of the focal plane present a challenge even to experienced microscopists. The higher the magnification, the narrower the depth of the focal plane; thus, when observing swimming organisms, there is a trade-off between magnification and depth of field.

> **THOUGHT QUESTION 2.5**   Under starvation conditions, bacteria such as *Bacillus thuringiensis*, the biological insecticide, repackage their cytoplasm into spores, leaving behind an empty cell wall. Suppose, under a microscope, you observe what appears to be a hollow cell. How can you tell if the cell is indeed hollow or if it is simply out of focus?

Fixation and Staining Improve Resolution and Contrast

The simplest way to observe microbes is to place them in a drop of water on a slide with a coverslip. This is called a **wet mount** preparation. The advantage of the wet mount is that the organism is viewed in as natural a state as possible, without artifacts resulting from chemical treatment; and live behavior such as swimming can be observed. The disadvantage is that most living cells are transparent and therefore show little contrast with the external medium. With limited contrast, the cells can barely be distinguished from background, and both detection and resolution are minimal.

The detection and resolution of cells under a microscope are enhanced by **fixation** and **staining**, procedures that usually kill the cell. Fixation is a process by which cells are made to adhere to a slide in a fixed position. Cells may be fixed with methanol or by heat treatment to denature cellular proteins, whose exposed side chains then adhere to the glass. A stain absorbs much of the incident light, usually over a wavelength range that results in a distinctive color. The use of chemical stains was developed in the nineteenth century, when German chemists used organic synthesis to invent new coloring agents for women's clothing. Clothing was made of natural fibers such as cotton or wool, so a substance that dyed clothing would be likely to react with biological specimens.

How do stains work? Most stain molecules contain conjugated double bonds or aromatic rings that absorb visible light (**Fig. 2.19**) and one or more positive charges that bind to negative charges on the bacterial cell envelope (discussed in Section 3.4). Different stains vary with respect to their strength of binding and the degree of binding to different parts of the cell.

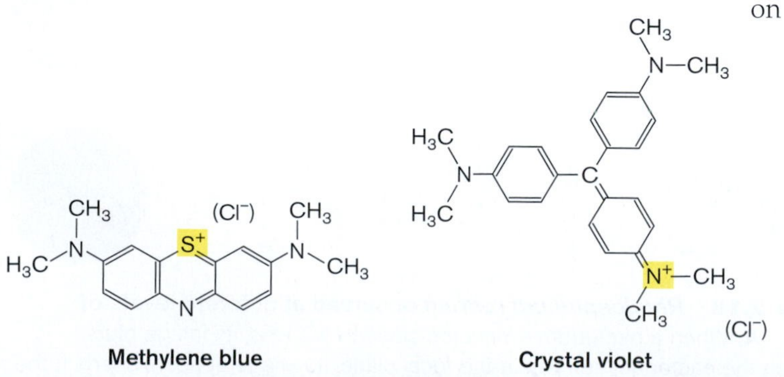

Different Kinds of Stains

A **simple stain** adds dark color specifically to cells, but not to the external medium or surrounding tissue (in the case of pathological samples). The most commonly used simple stain is methylene blue, originally used by Robert Koch to stain bacteria. A typical procedure for fixation and staining is shown in **Figure 2.20**. First, a drop of culture is fixed on a slide by treatment with methanol or by incubation on a slide warmer. Either treatment denatures cell proteins, exposing side chains that bind to the glass. The slide is then flooded with methylene blue solution. The positively charged molecule binds to the negatively charged cell envelope. After excess stain is washed off and the slide has been dried, it is observed under high-power magnification using immersion oil.

A **differential stain** stains one kind of cell but not another. The most famous differential stain is the **Gram stain**, devised in 1884 by the Dutch physician Hans Christian Gram (1853–1938). Gram first used the Gram stain to distinguish pneumococcus (*Streptococcus pneumoniae*) bacteria from human lung tissue. A similar use of the Gram stain is seen in **Figure 2.21A**, where *S. pneumoniae* bacteria appear dark purple against the pink background of human epithelial cells. Not all species of bacteria retain the Gram stain (**Fig. 2.21B**). Different bacterial species are classified gram-positive or gram-negative, depending on whether they retain the purple stain.

Gram Staining Separates Bacteria into Two Classes

In the Gram stain procedure (**Fig. 2.22**), a dye such as crystal violet binds to the bacteria; it also binds to the surface of human cells, but less strongly. After the excess stain is washed off, a **mordant**, or binding agent, is applied. The mordant used is iodine solution, which contains iodide ions (I^-). The iodide complexes with the positively charged crystal violet molecules trapped inside the cells. The crystal violet–iodide complex is now held more strongly within the cell wall. The thicker the cell wall, the more crystal violet–iodide molecules are held.

Figure 2.19   Chemical structure of stains.   Methylene blue and crystal violet are cationic (positively charged) dyes. The positively charged groups react with the bacterial cell envelope, which carries mainly negative charge. Chloride (Cl) is the counter ion.

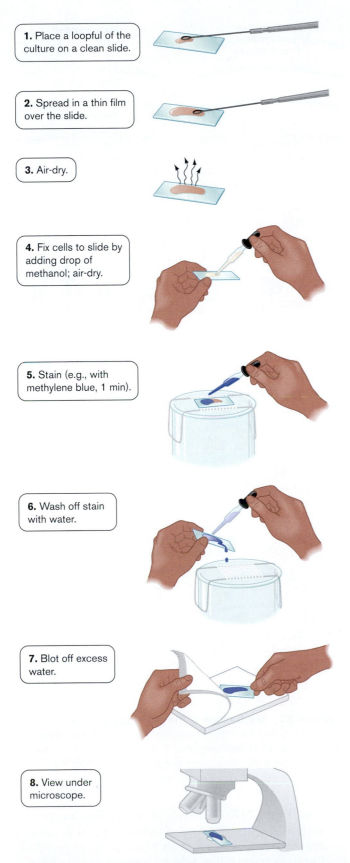

1. Place a loopful of the culture on a clean slide.

2. Spread in a thin film over the slide.

3. Air-dry.

4. Fix cells to slide by adding drop of methanol; air-dry.

5. Stain (e.g., with methylene blue, 1 min).

6. Wash off stain with water.

7. Blot off excess water.

8. View under microscope.

Figure 2.20   Procedure for simple staining with methylene blue.

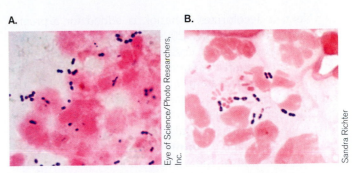

Figure 2.21   Gram staining of bacteria (a type of differential staining).   A. Gram stain of a sputum specimen from a patient with pneumonia, containing gram-positive *Streptococcus pneumoniae* (purple cocci). Cell length, 0.5 to 1.0 μm. **B.** Gram stain of a gingival specimen containing both gram-positive and gram-negative bacteria (purple and pink rods). Cell length, 0.5 to 1.0 μm.

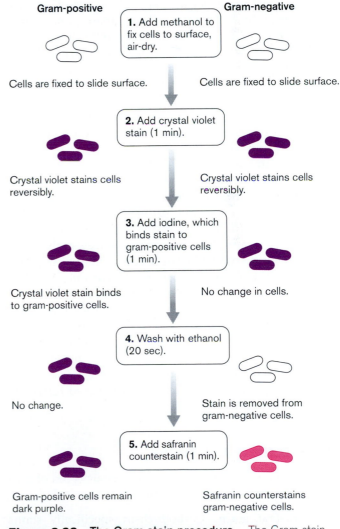

Gram-positive   Gram-negative

1. Add methanol to fix cells to surface, air-dry.

Cells are fixed to slide surface.   Cells are fixed to slide surface.

2. Add crystal violet stain (1 min).

Crystal violet stains cells reversibly.   Crystal violet stains cells reversibly.

3. Add iodine, which binds stain to gram-positive cells (1 min).

Crystal violet stain binds to gram-positive cells.   No change in cells.

4. Wash with ethanol (20 sec).

No change.   Stain is removed from gram-negative cells.

5. Add safranin counterstain (1 min).

Gram-positive cells remain dark purple.   Safranin counterstains gram-negative cells.

Figure 2.22   The Gram stain procedure.   The Gram stain distinguishes between gram-positive cells, with thick cell walls, which retain the crystal violet stain, and gram-negative cells, with thinner cell walls, which lose the crystal violet stain but are counterstained by safranin.

Next, a decolorizer, ethanol, is added for a precise time interval (typically, 20 seconds). The decolorizer removes loosely bound crystal violet–iodide, but gram-positive cells retain the stain tightly. The **gram-positive** cells that retain the stain appear dark purple, while the **gram-negative** cells are colorless. The decolorizer step is critical because if it lasts too long, the gram-positive cells, too, will release their crystal violet stain.

In the final step, a **counterstain**, safranin, is applied. This process allows the visualization of gram-negative material, which is stained pale pink by the safranin.

The Gram stain procedure was originally devised to distinguish bacteria (gram-positive) from human cells (gram-negative). Microscopists soon discovered, however, that many important species, such as the intestinal bacterium *Escherichia coli* and the nitrogen-fixing symbiont *Rhizobium meliloti*, do not retain the Gram stain. It turns out that gram-negative species of bacteria possess a thinner and more porous cell wall than gram-positive species (**Fig. 2.23**). A gram-negative cell wall has only a single layer of peptidoglycan (sugar chains cross-linked by peptides), whereas a gram-positive cell has multiple layers. The multiple layers of peptidoglycan retain enough stain complex so that the cell appears purple.

Among bacteria, the Gram stain distinguishes two groups with distinctive cell wall structures: Proteobacteria, with a thin cell wall plus an outer membrane (gram-negative), and Firmicutes, with a multiple-layered cell wall but no outer membrane (gram-positive). The outer membrane of Proteobacteria (discussed in Chapter 3) often possesses important pathogenic factors, such as the lipopolysaccharide endotoxins that cause toxic shock. During the Gram stain procedure, however, the outer membrane is disrupted by the decolorizer, allowing most of the crystal violet–iodide complex to leak out.

Thus, the Gram stain emerged as a key tool for biochemical identification of species, and it remains essential in the clinical laboratory. Note, however, that still other groups of bacteria and archaea have different kinds of cell walls that may stain either positive or negative and are thus not distinguished by the Gram stain. Prokaryotic diversity and identification are discussed further in Chapters 18 and 19.

Other Differential Stains

Other differential stains applied to various types of prokaryotes are illustrated in **Figure 2.24**. These include:

- **Acid-fast stain.** Carbolfuchsin specifically stains mycolic acids of *Mycobacterium tuberculosis* and *M. leprae*, the causative agents of tuberculosis and leprosy, respectively (**Fig. 2.24A**).
- **Spore stain.** When samples are boiled with malachite green, the stain binds specifically to the endospore

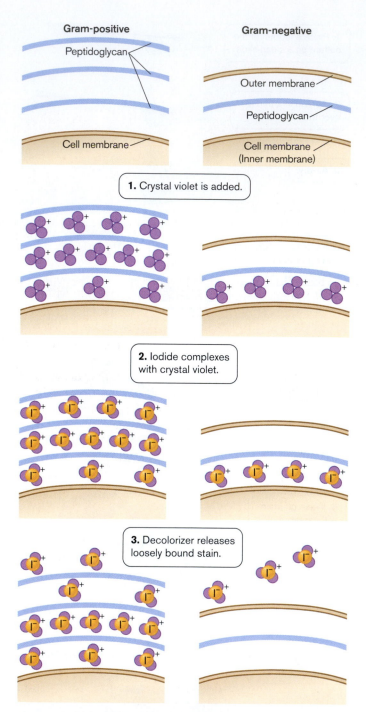

Figure 2.23 Mechanism of the Gram stain. In a gram-positive cell, the crystal violet–iodide complex is retained by multiple layers of peptidoglycan. In a gram-negative cell, the stain leaks out. If the decolorizer is applied for too long, the gram-positive cell will lose its stain as well.

coat (**Fig. 2.24B**). It detects spores of *Bacillus* species such as the insecticide *B. thuringiensis* and *B. anthracis*, the causative agent of anthrax, as well as spores of *Clostridium botulinum*, which produces botulin toxin.

- **Negative stain.** Some bacteria synthesize a capsule of extracellular polysaccharide filaments, which protects

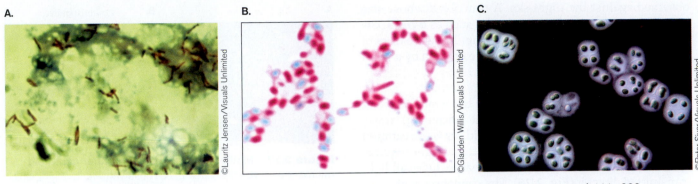

A.

B.

C.

©Lauritz Jensen/Visuals Unlimited

©Gladden Willis/Visuals Unlimited

©Peter Siver/Visuals Unlimited

Figure 2.24   Differential stains.   A. *Mycobacterium tuberculosis*, acid-fast stain (stained cells are red, 1–2 μm long). LM ×600.
B. *Clostridium botulinum*, endospore stain (stained endospores are blue-green, 2 μm long). LM ×400. **C.** *Gloeocapsa*, a colonial blue green alga, whose individual cells are encased in thick sheaths. Staining with India ink makes the sheaths clearly visible. Individual cell size, 3–5 μm.

the cell from predation or from engulfment by white blood cells (discussed in Section 3.4). The capsule is transparent and invisible in suspended cells. However, a suspension of opaque particles such as India ink can be added to darken the surrounding medium. The particles are excluded by the thick polysaccharide capsule, which thus appears clear against the dark background (**Fig. 2.24C**). This is an example of a **negative stain**.

■ **Antibody stains.** Specialized stains utilize monoclonal antibodies to identify precise strains of bacteria or even specific molecular components of cells. The antibody (which binds a specific cell protein) is linked to a reactive enzyme for detection or to a fluorophore (fluorescent molecule) for fluorescence microscopy (see below).

TO SUMMARIZE:

■ **In bright-field microscopy**, resolution depends on:
  ■ The wavelength of light, which limits resolution to about 200 nm.
  ■ The magnifying power of a lens, which depends on its numerical aperture ($n \sin \theta$).
  ■ The position of the focal plane, the location where the specimen is "in focus" (that is, where the sharpest image is obtained).

■ **A compound microscope** achieves magnification and resolution through a series of lenses: the condenser, objective, and ocular lenses.

■ **A wet mount** specimen is the only way to observe living microbes.

■ **Fixation and staining** of a specimen kills it but improves contrast and resolution.

■ **Differential stains** distinguish between different kinds of bacteria with different structural features.

■ **The Gram stain** differentiates between two major bacterial taxa, Proteobacteria (gram-negative) and Firmicutes (gram-positive). Other bacteria and archaea vary in their Gram stain appearance. Eukaryotes stain negative.

2.4   Dark-Field, Phase-Contrast, and Interference Microscopy

Advanced optical techniques enable us to visualize structures that are difficult or impossible to detect under a bright-field microscope, either because their size is below the limit of resolution of light or because their cytoplasm is transparent. These techniques take advantage of special properties of light waves, including scattering and interference patterns.

One application of dark-field illumination is the detection of pathogenic spirochetes, such as *Treponema pallidum*, the causative organism of syphilis. *T. pallidum* cells are so narrow (0.1 μm) that their shape cannot be fully resolved by light microscopy. Nevertheless, the spiral form of *T. pallidum* can be detected by dark-field microscopy (**Fig. 2.25**).

Dark-Field Observation Detects Unresolved Objects

Dark-field optics enables microbes to be visualized as halos of bright light against the darkness, just as stars are

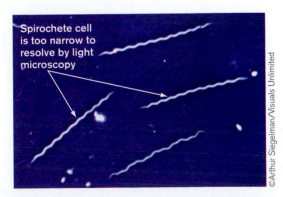

Spirochete cell is too narrow to resolve by light microscopy

©Arthur Siegelman/Visuals Unlimited

Figure 2.25   Dark-field observation of bacteria. *Treponema pallidum* specimen from a patient with syphilis. Note the detection of dust particles. Dark-field LM, ×1,100. Cell length about 10 μm; actual cell width, 0.2 μm.

observed against the night sky. A tiny object whose size is well below the wavelength of light, such as a virus particle, can be detected by light scattering. The wave front of scattered light is spherical, like a wave emitted by a point source (see **Fig. 2.9D**).

Light scattering. The scattered wave has a much smaller amplitude than that of the incident (incoming) wave. Therefore, with ordinary bright-field optics, scattered light is washed out. Detection of scattered light requires a modified condenser arrangement that excludes all light transmitted directly (**Fig. 2.26**). The condenser contains a "spider light stop," an opaque disk held by three "spider legs" across an open ring. The ring permits only a hollow cone of light to focus on the object. The incident hollow cone converges at the object, then generates an inverted hollow cone radiating outward.

The objective lens is positioned in the central region, where it completely misses the directly transmitted light. For this reason, the field appears dark. However, light scattered by the object radiates outward in a spherical wave. A sector of this spherical wave enters the objective lens and is detected as a halo of light.

An intriguing application of dark-field optics is the study of bacterial motility. Motility is important in bacterial diseases such as urethritis, in which the organism needs to swim up through the urethra. The bacterial swimming apparatus consists of helical filaments called

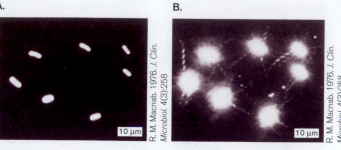

Figure 2.27   Motile bacteria observed under dark-field microscopy.   A. Flagellated *E. coli* observed with low light, which limits scattering. Only cell bodies are detected, no flagella. **B.** The light intensity is increased. Flagella are detected, although their fine structure is not resolved because their width is below the threshold of resolution by light.

flagella, which are rotated by a motor device imbedded in the bacterial cell wall (for flagellar structure, see Section 3.7). The "swimming strokes" of bacteria were first elucidated by Howard Berg (1934–) and Robert Macnab (1940–2003) using dark-field optics to view the helical flagella. The length of flagella is great enough to be resolved with light waves—but their width is not. Thus, dark-field optics is needed to detect and observe the helical flagella (**Fig. 2.27**). Note, however, that the bacterial cell itself appears "overexposed;" its shape is unresolved owing to the high light intensity.

www | Videos of swimming bacteria

THOUGHT QUESTION 2.6   Some early observers claimed that the rotary motions observed in bacterial flagella could not be distinguished from whiplike patterns, comparable to the motion of eukaryotic flagella. Can you imagine an experiment to distinguish the two and prove that the flagella rotate? *Hint*: Bacterial flagella can get "stuck" to the microscope slide or coverslip.

Limitations of dark-field microscopy. A disadvantage of dark-field microscopy is that any tiny particle, including specks of dust, can scatter light and interfere with visualization of the specimen. Unless the medium is extremely clear, it can be difficult to distinguish microbes of interest from particulates. Other methods of contrast enhancement, such as phase contrast and fluorescence, avoid this difficulty.

Phase-Contrast Microscopy

Phase-contrast microscopy exploits differences in refractive index between the cytoplasm and the surrounding medium or between different organelles. This technique is particularly useful for eukaryotic cells, such as algae and protists,

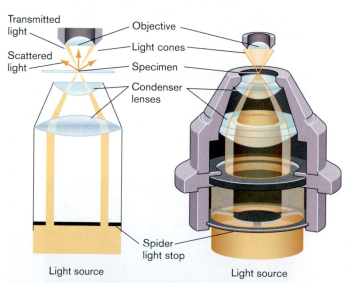

Figure 2.26   A dark-field condenser system, with a spider light stop.   Below the condenser, the spider light stop excludes all but an annular ring of light from the light source. The annular ring converges as a hollow cone of light focused on the specimen. Objects in the specimen scatter light in all directions. The scattered light is collected by the objective lens, but the transmitted light shines outside the range of the objective; thus, in the absence of scattering objects, the field appears dark. Only light scattered by the specimen enters the objective lens.

which contain many intracellular compartments. For example, **Figure 2.28A** shows an image of the protist *Entamoeba histolytica*, a human parasite, obtained by phase-contrast microscopy. In some cells, the nucleus is clearly visible.

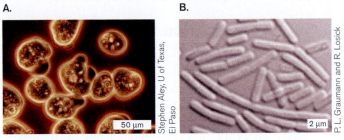

A.

B.

Stephen Aley, U of Texas, El Paso

P. L. Graumann and R. Losick

50 µm

2 µm

Figure 2.28 Phase and interference effects in microscopy.
A. Phase-contrast micrograph of the parasitic protist *Entamoeba histolytica*. **B.** Nomarski interference micrograph of *Bacillus subtilis*. The apparent three-dimensional effect is illusory. Note that one bacterium appears as if crossed through another. *Source*: Part B reprinted by permission from P. L. Graumann and R. Losick. 2001. *J Bacteriol.* 183(13): 4052–4060.

A.

1/2 wavelength

Phase plate

1/4 wavelength

Specimen
Slide

Light source

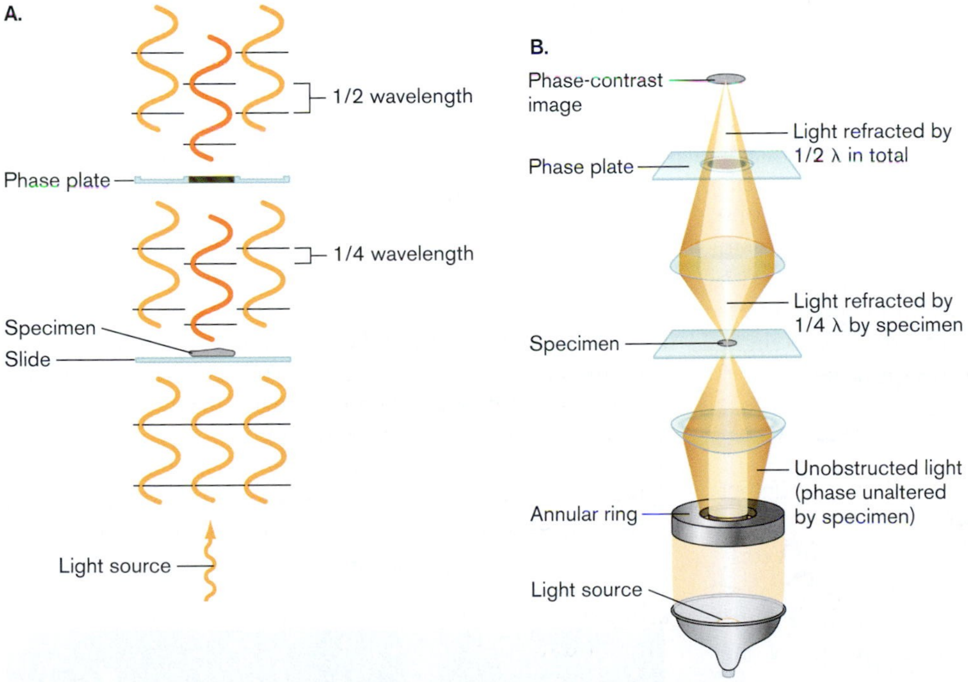

Figure 2.29 Phase-contrast optics. A. The specimen retards light by approximately one-quarter of a wavelength. The phase plate contains a central disk of refractive material that retards light from the specimen by another quarter wavelength, increasing the phase difference to half a wavelength. The light from the specimen and the transmitted light are now fully out of phase, and if they coincide, their waveforms cancel, making the specimen appear dark. **B.** In the phase-contrast microscope, the annular ring forms a hollow cone of light. As the light cone passes through the refractive material of the specimen, it is delayed by about one-quarter of a wavelength, and its path bends inward to the central region. This refracted light is surrounded by the hollow cone of unrefracted light. The light refracted by the specimen enters the lens through the dense central disk of the phase plate, which retards the wave by another quarter wavelength. When the transmitted and refracted light cones rejoin at the focal point, they are out of phase; their amplitudes cancel each other, and that region of the image appears dark against a bright background.

B.

Phase-contrast image

Phase plate

Light refracted by 1/2 λ in total

Specimen

Light refracted by 1/4 λ by specimen

Annular ring

Unobstructed light (phase unaltered by specimen)

Light source

The optical system for phase contrast was invented in the 1930s by the Dutch microscopist Frits Zernike (1888–1966), for which he earned the Nobel Prize in Physics. In this system, slight differences in the refractive index of the various cell components are transformed into differences in the intensity of transmitted light. Zernike's scheme makes use of the fact that living cells have relatively high contrast owing to their high concentration of solutes. Given the size and refractive index of commonly observed cells, light is retarded by approximately one-quarter of a wavelength when it passes through the cell. In other words, after having passed through a cell, light exits the cell about one-quarter of a wavelength behind the phase of light transmitted directly through the medium.

The Zernike optical system is designed to retard the refracted light by an additional one-quarter of a wavelength, so that the light refracted through the cell is slowed by a total of half a wavelength compared with the light transmitted through the medium. When two waves are out of phase by half a wavelength, they produce destructive interference, canceling each other's amplitude (see Fig. 2.13). The result is a region of darkness in the image of the specimen.

As in dark-field microscopy, the light transmitted through the medium in phase-contrast microscopy needs to be separated from the light interacting with the object—in this case, light waves slowed by refraction. This separation is performed by a ring-shaped slit, called an "annular ring," similar in function to the spider light stop. The annular ring stops light from passing directly through the center of the lens system, where the specimen is located, and generates a hollow cone of light, which is focused through the specimen and generates an inverted cone above it (**Fig. 2.29**). Light passing through the specimen, however, is not only retarded; it is also refracted and thus bent into the central region within the inverted cone.

Both the refracted light from the specimen and the outer cone of transmitted light enter the phase plate. The phase plate consists of refractive material that is thinner in the region met by the outer (transmitted) light cone. The refracted light passing through the center of the phase plate is retarded by an additional one-quarter wavelength compared with the transmitted light passing through the thinner region on the outside. The overall difference approximates half a wavelength, so that the two waves are out of phase, thus canceling each other's amplitude. When the light from the inner and outer regions focuses at the ocular lens, the amplitudes of the wave trains cancel and produce a region of darkness. In this system, small differences in refractive index can produce dramatic differences in contrast between the offset phases of light.

WWW | FSU Phase contrast tutorial

Interference Microscopy

Other kinds of optical systems have been devised using light interference to enhance cytoplasmic contrast. **Interference microscopy** enhances contrast by superimposing the image of the specimen on a second beam of light that generates interference fringes. The interference pattern produces an illusion of shadowing across the specimen. For example, the shapes of the *Bacillus subtilis* cells illuminated by interference contrast (**Fig. 2.28B**) are more clearly defined than in conventional bright-field microscopy (for comparison see **Fig. 2.21**).

One optical system for producing interference contrast is based on the Wollaston-Nomarski prism. In this system, light is polarized to obtain waves oriented in one direction. The polarized light is then split by the prism into two separate beams, which are out of phase with each other. The two beams recombine to generate interference patterns whose edges are highly sensitive to slight differences in the refractive index of the specimen.

TO SUMMARIZE:

- **Dark-field microscopy** uses scattered light to detect objects too small to be resolved by light rays.
  Advantage: Extremely small microbes and thin extracellular structures can be detected.
  Limitation: The shape of objects is not resolved. Dust particles easily obscure the image of the specimen.
- **Phase-contrast microscopy** superimposes refracted light and transmitted light shifted out of phase so as to observe differences in refractive index as patterns of light and dark.
  Advantage: Live microbes with transparent cytoplasm, especially eukaryotes, can be observed with high contrast.
  Limitation: Phase contrast is less effective for smaller microbes and for organisms whose cytoplasm has a low refractive index.
- **Interference microscopy** superimposes interference bands on an image, accentuating small differences in refractive index.
  Advantage: The shape of cells can be defined most clearly.
  Limitations: Interference microscopy requires complex optical adjustment and is less effective for organisms with low refractive index.

2.5 Fluorescence Microscopy

In fluorescence microscopy, incident light is absorbed by the specimen and reemitted at a lower energy, thus longer wavelength. Fluorescence microscopy offers a powerful way to detect microbes and subcellular structures while avoiding artifacts caused by dust and other nonspecific materials.

One use of fluorescence microscopy is to assess microbial populations in highly dilute natural environments (**Fig. 2.30**). This technique was used by the Oceanic Microbial Observatory at the Bermuda Biological Station for Research to conduct a long-term study assessing the effects on microbial populations of changing water chemistry due to climate changes such as global warming. Populations of microbes, including viruses, bacteria, and protists, are measured using DNA-specific fluorescence. The advantage of this fluorescence technique is that it detects only live organisms whose DNA is intact, distinguishing them from fine debris in natural environments.

Fluorescence Requires Excitation and Emission at Different Wavelengths

Fluorescence occurs when light of a specific wavelength (the **excitation wavelength**) is absorbed by an atom (or molecule) capable of promoting an electron to an orbital of higher energy (**Fig. 2.31**). Because this higher-energy electron state is unstable, the electron decays to an orbital with a slightly lower energy level through the loss of energy as heat. The electron then falls to its original level by emitting a photon of lower energy and longer wavelength, the **emission wavelength**. The emitted photon has a longer wavelength (less energy) because part of the electron's energy of absorption was lost as heat.

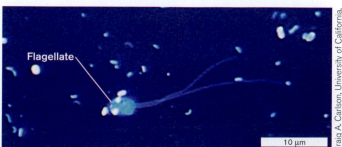

Craig A. Carlson, University of California, Santa Barbara

Figure 2.30 Cell counting using fluorescence microscopy. Bacterioplankton and flagellated protists from the surface waters of the Sargasso Sea near Bermuda are enumerated to investigate their contribution to the oceanic carbon cycle. Live organisms are detected by fluorescence using the DNA-specific fluorophore DAPI.

A.

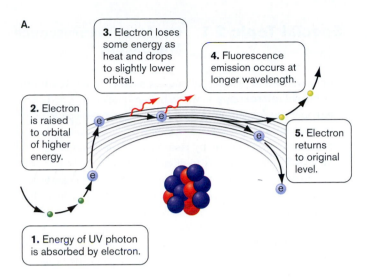

2. Electron is raised to orbital of higher energy.

3. Electron loses some energy as heat and drops to slightly lower orbital.

4. Fluorescence emission occurs at longer wavelength.

5. Electron returns to original level.

1. Energy of UV photon is absorbed by electron.

B.

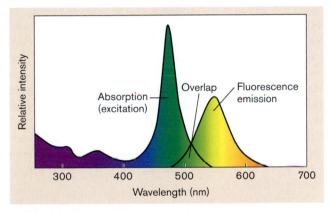

Figure 2.31 Fluorescence. Energy gained from UV absorption is released as heat and as a photon of longer wavelength in the visible region. **A.** Fluorescence on the molecular level. **B.** Comparison of absorption and emission spectra for a fluorophore.

The wavelengths of excitation and emission are determined by the **fluorophore**, the fluorescent molecule used to stain the specimen. For example, the slides for cell counting in the Bermuda study (see **Fig. 2.30**) used the DNA-specific stain 4′,6-diamidino-2-phenylindole (DAPI).

The aromatic groups of DAPI associate exclusively with the base pairs of DNA. Another commonly used DNA-specific fluorophore is acridine orange (**Fig. 2.32**). These fluorophores provide an extremely sensitive means of detecting diverse microorganisms in the environment, including those whose dimensions are too small to be resolved.

The optical system for fluorescence microscopy utilizes filters to limit incident light to the wavelength of excitation and the emitted light to the wavelength of emission. The wavelengths of excitation and emission are determined by the specific fluorophore used.

Fluorophores Can Label Specific Parts of Cells

Fluorescence can be used to label specific parts of cells. The specificity of the fluorophore can be arranged in several ways:

- **Chemical affinity.** Certain fluorophores have chemical affinity for certain classes of biological molecules; for example, the fluorophore acridine specifically binds DNA.
- **Labeled antibodies.** Antibodies that specifically bind a cell component are chemically linked to a fluorophore molecule. The use of antibodies linked to fluorophores is known as immunofluorescence.
- **Gene fusion.** Genetic recombination can be used to create a hybrid gene expressing a protein that generates fluorescence, such as green fluorescent protein (GFP).
- **DNA hybridization.** A short sequence of DNA attached to a fluorophore will hybridize to a specific sequence in the genome, thus labeling one position in the chromosome or nucleoid.

Fluorophore labeling is used to dissect endospore formation in *Bacillus* species (**Fig. 2.33**). The cell envelope and DNA origin of replication are labeled by different fluorophores to track movements of DNA during sporulation. Richard Losick and colleagues applied this technique to show how the DNA origin moves toward one pole of the cell, followed by formation of a septum, a new portion of cell envelope, just behind the developing endospore.

Fluorescein-isothiocyanate (FITC)

Acridine orange (AO)

4′,6-Diamidino-2-phenylindole (DAPI)

Figure 2.32 Fluorescent molecules (fluorophores) commonly used in microscopy. Because of the abundance of conjugated double bonds, these molecules have closely spaced molecular orbitals that give rise to fluorescence.

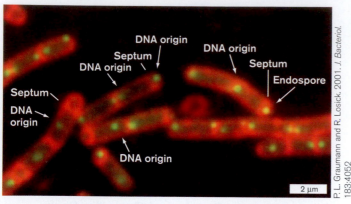

P. L. Graumann and R. Losick. 2001. *J. Bacteriol.* 183:4052

2 µm

Figure 2.33 **Fluorescence micrograph of *Bacillus subtilis* cells during sporulation.** Red fluorescence arises from membrane stained with the dye FM-464. Green fluorescence arises from green fluorescent protein (GFP) bound to the DNA origin of replication. The yellow color occurs where green fluorescence overlaps red.

In the micrograph, the cell envelope is stained red by the fluorescent dye FM-464. The dye FM-464 specifically binds phospholipid membranes. The DNA origin of replication is stained by green fluorescent protein (GFP) fused to a protein that specifically binds the DNA origin of replication. The fluorescence micrographs were taken at two different wavelengths for both excitation and emission to record the two different fluorescent labels. The images were superimposed with two different colors marking envelope and DNA. The result shows how, in sporulation, DNA replication originates not at the midpoint of the cell, as in vegetative growth of most bacteria, but at the poles where the endospore forms.

> **THOUGHT QUESTION 2.7** What experiment could you devise to determine the actual order of events in DNA movement toward the pole during formation of an endospore?
>
> **THOUGHT QUESTION 2.8** Compare and contrast fluorescence microscopy with dark-field microscopy. What similar advantage do they provide, and how do they differ?

Fluorescence microscopy has been used to develop advanced optical systems that reconstruct three-dimensional models of cells. An example is confocal fluorescence microscopy (**Special Topic 2.1**).

TO SUMMARIZE:

■ **Fluorescence microscopy** involves detection of specific cells or cell parts based on fluorescence by a fluorophore.

■ **Cell parts can be labeled** by a fluorophore attached to an antibody stain.

Special Topic 2.1 Confocal Fluorescence Microscopy

An advanced application of fluorescence is **laser scanning confocal microscopy**, in which both excitation light and emitted light are focused together. Confocal microscopy is used to produce images of cells at high resolution, with interference effects decreased by laser optics. The images can be "stacked" computationally to model a cell in three dimensions.

In **Figure 1**, we see a confocal image of human tissue culture cells infected by enteroinvasive *Escherichia coli*. The DNA of bacteria and of the nuclei of HeLa cells (an immortal cancer cell line) are stained blue with DAPI fluorophore. Invading bacteria cause the host to form actin "tails," which help the bacteria move. Tails are stained green by an actin-binding fluorophore, FITC.

In confocal microscopy, a laser beam is focused onto the specimen and scanned across it in two dimensions—that is, in two planes at right angles to each other (**Fig. 2**). The laser beam excites the fluorophore, causing it to emit light at a longer wavelength. The emitted light passes in reverse direction through the objective, where it encounters a dichroic mirror that allows light transmission at the excitation wavelength but reflects light at the wavelength of emission. The reflected emission rays are then focused and pass through a pinhole, which eliminates all unfocused light. A narrow focused beam, representing one "pixel" of image, enters the photomultiplier tube. Scanning across the specimen yields a two-dimensional pattern of pixels that forms an image.

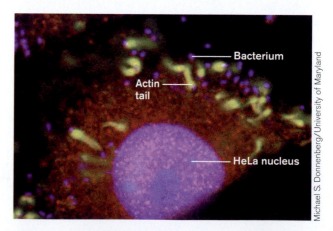

Michael S. Donnenberg/University of Maryland

Figure 1 **Human tissue culture cells infected by enteroinvasive *Escherichia coli*.** DNA of bacteria and of HeLa nuclei are stained blue with DAPI fluorophore. Invading bacteria (blue-stained rods, 1–2 µm long) cause the host to form actin "tails," stained green with an actin-binding FITC fluorophore.

The scanning feature of confocal microscopy is also used to acquire data for high-throughput experiments in which a large number of chemical reactions are arranged in microscopic quantities in an array. An example of an array is a DNA microarray containing probes for all the genes of a genome. Short segments of DNA are arrayed on a microscope slide such that an entire genome of potential protein-encoding genes may be present on a single slide.

One use of arrays is to investigate the mode of action of a new antibiotic molecule by testing which of the microbe's genes are induced or repressed in response to the antibi-otic. The DNA array is hybridized to fluorescence-labeled cDNA copies of messenger RNA from bacteria cultured with or without the antibiotic. The confocal laser then scans the slide to detect binding of the genes to cDNA molecules labeled with a fluorophore. This technique makes it possible to measure transcription of all known genes in the genome and reveal those whose expression varies during exposure to an antibiotic. The antibiotic-induced genes indicate the mode of action of the antibiotic within the cell and aid the design of better antibiotics. DNA microarrays are discussed further in Chapter 12.

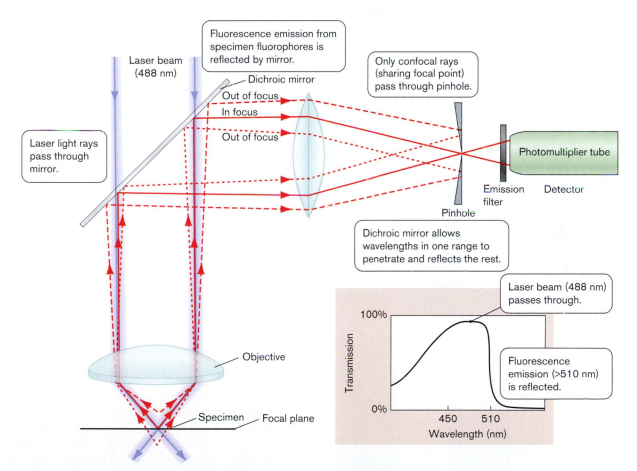

Figure 2 Confocal microscopy.   In confocal optics, the incident laser beam (blue) passes through a dichroic mirror and reaches the specimen, where its absorption leads to fluorescent emission at a longer wavelength. The fluorescent emission travels back to the dichroic mirror, where it is reflected toward the photomultiplier. Only the confocal rays (those emitted from the focal point) pass through the pinhole and reach the photomultiplier. The laser scans across the specimen, generating a pixilated image. The scanned images can be stacked through a series of focal planes to generate a three-dimensional model.

2.6 Electron Microscopy

All cells are built of macromolecular structures. The tool of choice for observing the shape of these macromolecular structures is **electron microscopy (EM)**. In electron microscopy, beams of electrons are focused to generate images of cell membranes, chromosomes, and ribosomes with a resolution a thousand times that of light microscopy. The modern electron microscope was developed in the 1950s and was popularly known as the centerpiece of any biological research program. In Michael Crichton's film, *The Andromeda Strain*, for example, an electron microscope is used to analyze a fictional pathogen from outer space.

The Electron Microscope Focuses Beams of Electrons

How does an electron microscope work? Electrons are ejected from a metal subjected to a voltage potential. The electrons travel in a straight line, like photons. Like photons, electrons interact with matter and carry information about their interaction. And like photons, electrons can exhibit the properties of waves. The wavelength associated with an electron is a 100,000 times smaller than that of a photon; for example, an electron accelerated over a voltage of 100 kilovolts (kV) has a wavelength of 0.0037 nm (0.037 Å) compared with 400–750 nm for visible light. The actual resolution of electrons in microscopy is limited, however, not by the wavelength, but by the aberrations of the lensing

systems used to focus electrons. The magnetic lenses used to focus electrons never achieve the precision required to utilize the full potential resolution of the electron beam.

Electrons are focused by means of a magnetic field directed along the line of travel of the beam (**Fig. 2.34**). As a beam of electrons enters the field, it spirals around the magnetic field lines. The shape of the magnet can be designed to generate field lines that will focus the beam of electrons in a manner analogous to the focusing of photons by a refractive lens. The electron beam, however, forms a spiral because electrons travel around magnetic field lines. Because magnetic lenses generate aberrations, a series of corrective magnetic lenses is generally required to obtain a resolution of about 0.2 nm. This represents a resolution a thousand times greater than the 200-nm resolution of light microscopy.

> **THOUGHT QUESTION 2.9** An electron microscope can be focused at successive powers of magnification, as in a light microscope. At each level, the image rotates at an angle of several degrees. Based on the geometry of the electron beam, as shown in Figure 2.34, why do you think this rotation occurs?

Transmission EM and scanning EM. Two major types of electron microscopy are **transmission electron microscopy (TEM)** and **scanning electron microscopy (SEM)**. In TEM, electrons are transmitted through the specimen as in light microscopy to reveal internal structure. In SEM, the electron beams scan across the surface of the specimen and are reflected to reveal the contours of its three-dimensional surface.

The transmission electron microscope closely parallels the design of a bright-field microscope, including a source of electrons (instead of light), a magnetic condenser lens, a specimen, and an objective lens. A projection lens projects the image onto a fluorescent screen and the final images are obtained by a digital camera (charge-coupled device, CCD) (**Fig. 2.35**). The scanning microscope is arranged somewhat differently than the TEM in that a series of condenser lenses focuses the electron beam onto the surface of the specimen. Reflected electrons are then picked up by a detector (**Fig. 2.36**). In either case, the overall apparatus required for electron microscopy is complex, resembling the bridge of a *Star Trek* spaceship.

Electron Microscopy Requires Specialized Sample Preparation

Electron microscopy poses special problems for biological specimens. With the exception of cryo EM (discussed shortly), the entire optical column must be maintained under vacuum to prevent the electrons from colliding with the gas molecules in air. The requirement for a vacuum precludes the viewing of live specimens, which

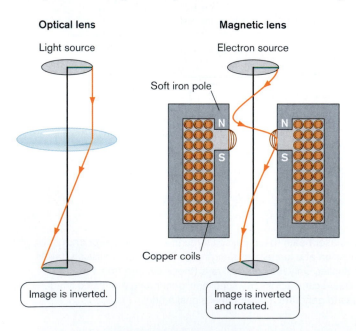

Figure 2.34 A magnetic lens. The beam of electrons spirals around the magnetic field lines. The U-shaped magnet acts as a lens, focusing the spiraling electrons much as a refractive lens focuses light rays.

A.

Light microscopy

Transmission electron microscopy

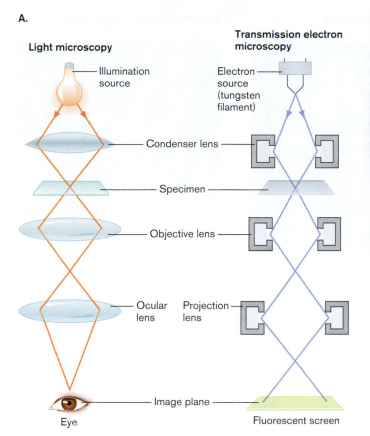

B.

Figure 2.35 Transmission electron microscopy. A. The light source is replaced by an electron source consisting of a high-voltage current applied to a tungsten filament, which gives off electrons when heated. The electron beam is focused by a condenser magnet lens onto the specimen. The specimen image is then magnified by an objective magnetic lens. The projector lens, analogous to the ocular lens of a light microscope, focuses the image on the cathode-ray tube (CRT) screen. *Note*: Because of the greater problem of aberrations in focusing electrons, each magnetic lens shown (condenser, objective, projection) actually represents a series of lenses. **B.** Using an electron microscope.

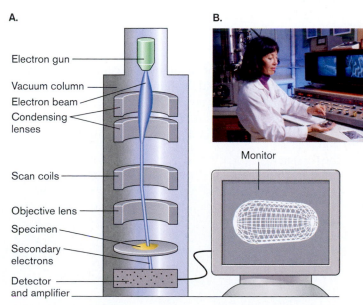

Figure 2.36 Scanning electron microscopy. A. In the scanning electron microscope (SEM), the electron beam is scanned across a specimen coated in gold, which acts as a source of secondary electrons. The incident electron beam ejects secondary electrons toward a detector, generating an image of the surface of the specimen. **B.** Operating an SEM. **C.** Loading a specimen into the vacuum column.

in any case would be quickly destroyed by the electron beam. Moreover, the structure of most specimens lacks sufficient electron density (ability to block electrons) to provide contrast. Thus, the specimen usually requires an electron-dense stain using salts of heavy metals such as gold or uranium.

The specimen can be prepared in one of three ways:

■ **Embedded in a polymer for thin sections.** A special knife called a microtome cuts slices through the specimen, each slice a fraction of a micrometer thick.

■ **Sprayed onto a copper grid.** The electron beam penetrates straight through the entire object, as if it were transparent. This method is effective for virus particles and for isolated macromolecular complexes.

■ **Flash-frozen (for cryo EM).** Samples frozen rapidly in refrigerant provide sufficient contrast for detection by a high-intensity electron beam, a recent innovation.

For sectioned samples or for samples sprayed onto a grid, the specimen is coated with a heavy metal salt such as uranyl acetate. The metal salt is deposited around the biological structures, acting as a negative stain. (For comparison, a negative stain used in light microscopy is the India ink stain, illustrated in Fig. 2.24C.)

In **Figure 2.37**, we see examples of SEM and TEM images from gram-positive rod bacteria. The SEM in **Figure 2.37A** shows an exterior view of the capsules of *M. paratuberculosis*. The image provides far greater magnification of surface detail than does a bright-field view of rod bacteria. In **Figure 2.37B**, flagellar motors have been isolated or "unplugged" from a bacterial cell envelope and spread on a grid for TEM. The micrograph reveals details of the motor, including the axle and individual rings. The TEM of *B. anthracis* in **Figure 2.37C** shows a thin section through a bacillus, including cell wall, membranes, and glycoprotein filaments. The TEM is "transparent," summing electron density throughout the depth of the section.

A. SEM of *Mycobacterium paratuberculosis*

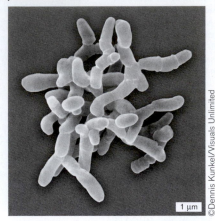

©Dennis Kunkel/Visuals Unlimited

1 μm

B. TEM of flagellar motors from *Salmonella typhimurium*

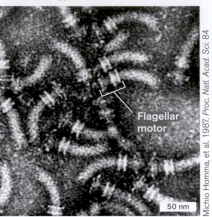

Michio Homma, et al. 1987. *Proc. Natl. Acad. Sci.* 84

Flagellar motor

50 nm

C. TEM of *Bacillus anthracis* showing envelope and cytoplasm

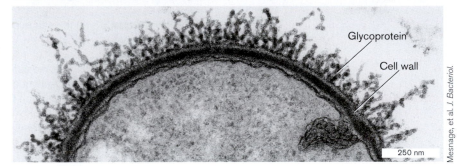

Glycoprotein

Cell wall

Mesnage, et al. *J. Bacteriol.*

250 nm

Figure 2.37 SEM and TEM images of gram-positive rod-shaped bacteria.

THOUGHT QUESTION 2.10 What kind of research questions could you investigate using SEM? What questions would be answered using TEM?

An important limitation of traditional electron microscopy, whether TEM or SEM, is that in most cases it can only be applied to fixed, stained specimens. The fixatives and heavy-atom stains can introduce artifacts into the image, especially at finer details of resolution. In some cases, different preparation procedures have led to substantially different interpretations of subcellular structure.

The development of exceptionally high-strength electron beams now permits low-temperature **cryoelectron microscopy (cryo EM)**. In cryo EM, the specimen is flash-frozen, that is, suspended in water and frozen

rapidly in a refrigerant of high heat capacity (ability to absorb heat). The rapid freezing avoids ice crystallization, leaving the water solvent in a glass-like amorphous phase. The specimen retains water content and thus closely resembles its viable form, although it is still ulti-

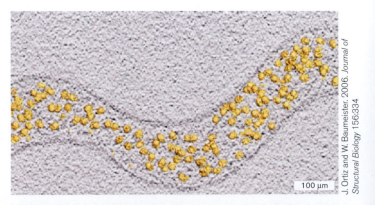

J. Ortiz and W. Baumeister. 2006. *Journal of Structural Biology* 156:334

100 μm

Figure 2.38 Cryoelectron microscopy. Cryoelectron tomography of *Spiroplasma melliferum*. The tomographic slice reveals ribosomes identified using molecular identification software, false-colored yellow.

mately destroyed by electron bombardment. The sample does not require staining because the high-intensity electron beams can detect smaller signals than in earlier instruments. **Figure 2.38** shows a cryo EM of *Spiroplasma melifera*.

Another innovation in cryo EM is **tomography**, the acquisition of projected images from different angles of a transparent specimen. Tomography avoids the need to physically slice the sample for thin-section TEM. The images from EM tomography are combined digitally to visualize the entire object. An example is the tomographic image of the spiral bacterium *Spiroplasma melliferum*, in which individual ribosomes clearly appear (**Fig. 2.38**).

Cryo EM is also used to generate high-resolution models of virus particles and purified structures such as the ribosome (**Special Topic 2.2**). For these small particles, multiple images can be averaged together by computational analysis. The digitally combined images can achieve high resolution, nearly comparable to that of X-ray crystallography.

Microscopy Results Require Careful Interpretation

The images produced by high-level microscopy can be difficult to interpret. For example, an oval that appears hollow might be interpreted as a cell when in fact it represents a deposit of staining material. A microscopic structure that is interpreted incorrectly is termed an **artifact**. Avoiding artifacts is an important concern in microscopy.

Sometimes, published interpretations generate controversy. For example, electron microscopy and immunofluorescence microscopy were used to test for contamination of cultured white blood cells that grew poorly (**Fig. 2.39**). The SEM images were interpreted to show the presence of tiny prokaryotic cells, 200–300 nm in diameter, attached to cultured fibroblast cells (**Fig. 2.39A**). These proposed cells were termed "nanobacteria." Unfortunately, other researchers were unable to confirm these results and suggested that the objects observed might actually be mineral deposits. Another technique, immunofluorescence microscopy, reportedly showed the invasion of cultured cells by nanobacteria (**Fig. 2.39B**). The technique required use of fluorescent-tagged antibodies against nanobacteria, but the actual specificity of the antibodies was unclear. The existence of small infective particles in the blood remains a question of interest to medical researchers.

> **THOUGHT QUESTION 2.11** What kind of experiments could prove or disprove the interpretations of the images of "nanobacteria" in blood plasma?

Emerging Methods of Microscopy

New methods of microscopy are emerging that enable nanoscale observation of cell surfaces, in some cases of living cells suspended in water. A general term for these methods is scanning probe microscopy (SPM). SPM methods differs from light and electron microscopy, in which the sample interacts with a beam of light or electrons. Instead, SPM methods measure a physical interaction, such as the "atomic force" between the sample and a sharp tip. **Atomic force microscopy (AFM)** measures the van der Waals forces between the electron shells of adjacent atoms of the cell surface and the sharp tip. AFM is particularly useful to study the surfaces of live bacteria.

In AFM, an instrument probes the surface of a sample with a sharp tip a couple of micrometers long and often less than 10 nm in diameter (**Fig. 2.40A**). The tip is located at the free end of a lever that is 100–200 µm long. The lever is deflected by van der Waals forces between the tip and the sample surface. Deflection of the lever is measured by a laser beam reflected off a cantilever attached to the tip as the sample scans across. The measured deflections allow a computer to map the topography of cells in liquid medium with a resolution below 1 nm.

In the example in **Figure 2.40B**, obtained by Mark Martin at Occidental College, AFM reveals the outline of

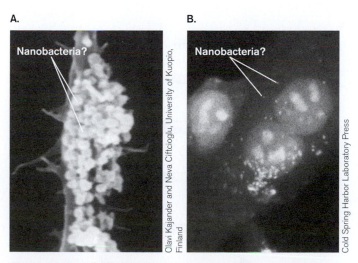

A.

B.

Olavi Kajander and Neva Ciftcioglu, University of Kuopio, Finland

Cold Spring Harbor Laboratory Press

Figure 2.39 Artifacts of microscopy. **A.** Objects attached to the surface of a fibroblast were observed by SEM. The objects (200–300 nm in diameter) were identified as "nanobacteria," exceptionally small bacteria inhabiting human blood plasma, but they are now believed to be mineral deposits. **B.** Cultured human cells observed by immunofluorescence microscopy revealed small particles identified as nanobacteria. This observation, however, has not been reproduced by other researchers.

Special Topic 2.2 **Three-Dimensional Electron Microscopy Solves the Structure of a Major Agricultural Virus**

One of the world's most economically damaging agricultural pathogens is rice dwarf virus. Rice dwarf virus (RDV) is transmitted by leafhopper beetles in China and Southeast Asia, where it systemically infects rice, discoloring the leaves and stunting the plant, devastating the harvest. A retrovirus, RDV has a genome of double-stranded RNA enclosed within an icosahedral capsid of proteins, approximately 70 nm in diameter. Determining the molecular structure of the virus particle may enable us to design antiviral agents.

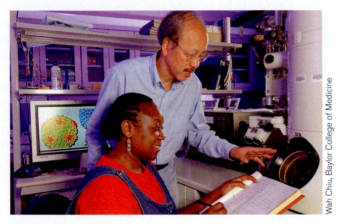

Figure 1 **Three-dimensional cryoelectron microscopy.** Dr. Wah Chiu (standing), at Baylor University, succeeded in imaging the structure of rice dwarf virus using three-dimensional cryoelectron microscopy.

Wah Chiu (**Fig. 1**) and colleagues at the National Center for Macromolecular Imaging, at Baylor University, imaged the structure of RDV at high resolution using three-dimensional cryo EM tomography. Cryo EM is especially important for particles that cannot be crystallized for X-ray diffraction analysis, the most common means of molecular visualization. Because the frozen sample remains hydrated, the biological molecules retain the same conformation as in solution. The sample is imaged without heavy-metal stains, and thus higher resolution is obtained (see **Fig. 2B**). This technique avoids introducing stain artifacts but generates very low contrast. Repeated scans can be summed computationally to obtain an image at higher resolution.

In **Figure 2A**, a cryo EM image reveals the outline of the RDV particles. Each particle was imaged in stereo by rotating the specimen to receive the electron beam at different angles (**Fig. 2B**). In theory, a three-dimensional picture of the virus could be obtained from a stereo pair of images from one particle. In practice, however, one image pair reveals little information because the ratio of signal to noise is so low, so stereo pairs of images were combined from 4,000 individual RDV particles (see **Fig. 2B**). A visual model was generated by computation requiring the use of a supercomputer.

The computational model of RDV reveals in three dimensions the capsid, a protein shell that encloses the virus's RNA

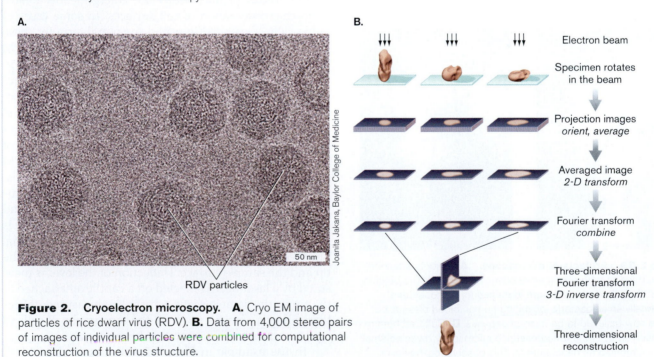

Figure 2. **Cryoelectron microscopy.** **A.** Cryo EM image of particles of rice dwarf virus (RDV). **B.** Data from 4,000 stereo pairs of images of individual particles were combined for computational reconstruction of the virus structure.

chromosome (**Fig. 3A**). The model includes detailed representations of both the outer shell (P8 subunits) and inner shell (P3 subunits). The remarkable resolution of the RDV model provides enough detail to map the alpha helices of a P3 subunit (**Fig. 3B**), although it does not yet reach the atomic resolution of X-ray diffraction. This structure can be used to model molecular interactions for potential antiviral agents that would inactivate the capsid.

A.

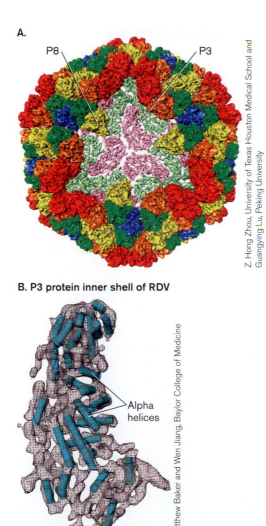

P8   P3

Z. Hong Zhou, University of Texas Houston Medical School and Guangying Lu, Peking University

B. P3 protein inner shell of RDV

Alpha helices

Matthew Baker and Wen Jiang, Baylor College of Medicine

Figure 3   The rice dwarf virus capsid.   A. Model of RDV, at resolution 6.8 Å. The outer shell is composed of 396 P8 subunits (bright colors). A cutaway from the outer shell reveals the inner shell (pale color, within black border) composed of 120 P3 subunits. **B.** The labeled P3 subunit from part A is shown in expanded detail, revealing alpha helices (blue cylinders).

an *E. coli* cell invaded by the parasitic bacterium *Bdellovibrio bacteriovorus*. The cells were observed in water suspension, without stain; and their contours appear without the flattening on a grid required for EM. The infected and uninfected *E. coli* cells differ in texture. This ability to "catch cells in the act" is an advantage of AFM.

A.

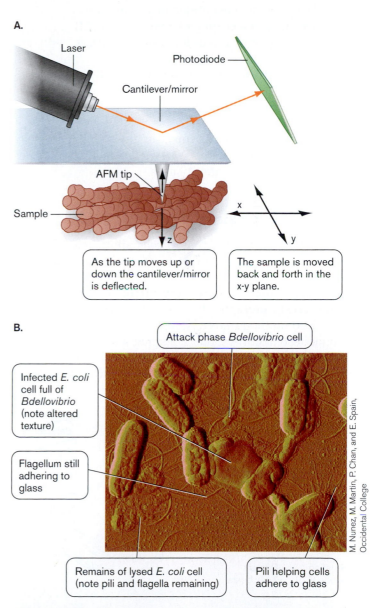

Laser

Photodiode

Cantilever/mirror

AFM tip

Sample

x

z   y

As the tip moves up or down the cantilever/mirror is deflected.

The sample is moved back and forth in the x-y plane.

B.

Attack phase *Bdellovibrio* cell

Infected *E. coli* cell full of *Bdellovibrio* (note altered texture)

Flagellum still adhering to glass

Remains of lysed *E. coli* cell (note pili and flagella remaining)

Pili helping cells adhere to glass

M. Nunez, M. Martin, P. Chan, and E. Spain, Occidental College

Figure 2.40   Atomic force microscopy enables visualization of untreated cells.   A. The atomic force microscope (AFM) has a fine-pointed tip attached to a cantilever that moves over a sample. The tip interacts with the sample surface through atomic force. As the tip is pushed away, or pulled into a depression, the cantilever is deflected. The deflection is measured by a laser light beam focused onto the cantilever and reflected into a photodiode detector. **B.** *Escherichia coli* cells under attack by *Bdellovibrio bacteriovorus*, bacterial predators that invade the periplasmic space of *E. coli*. The untreated cells allow AFM observation of *B. bacteriovorus* cells, *E. coli* cells, pili, and flagella. Each *E. coli* cell is about 2 μm in length.

TO SUMMARIZE:

- **Electron microscopy** is based on the focusing of electron beams on an object stained with a heavy-metal salt that blocks electrons. Much higher resolution is obtained than for light microscopy.
- **Transmission electron microscopy (TEM)** involves electron beam penetration of a thin sample.
- **Scanning electron microscopy (SEM)** involves scanning of a three-dimensional surface with an electron beam.
- **Cryoelectron microscopy (cryo EM)** observes a sample flash-frozen in water solution. Multiple images may be combined digitally to achieve high resolution.
- **Atomic force microscopy**, an emerging alternative to EM, uses van der Waals force measurement to observe cells in water solution.

2.7   Visualizing Molecules

To understand the structure of cells, ultimately we need to isolate the cell's molecules to observe their individual structure and function. The major tool used at present for molecular visualization is **X-ray diffraction analysis**, or **X-ray crystallography**. In some cases, cryo EM modeling based on thousands of samples has also reached near atomic resolution. Another emerging alternative to X-ray diffraction for analysis of small molecules and proteins is **nuclear magnetic resonance (NMR)**. The advantage of NMR is that it presents a dynamic view of molecules in solution, including multiple conformation states.

Unlike microscopy, X-ray diffraction does not present a direct view of a sample, but generates computational models. Dramatic as the models are, they can only represent particular aspects of electron clouds and electron density that are fundamentally "unseeable." That is why molecular structures are represented in different ways that depend on the context—by electron density maps, as models defined by van der Waals radii, or as stick models, for example. Proteins are frequently presented in a cartoon form that shows alpha helix and beta sheet secondary structures.

X-Ray Diffraction Analysis

For substances that can be crystallized, X-ray diffraction makes it possible to fix the position of each individual atom in a molecule, because the wavelengths of X-rays are much shorter than the wavelengths of visible light and are of the same magnitude as the dimensions of atoms. X-ray diffraction, like phase microscopy, involves the principle of wave interference (see **Fig. 2.13**). The interference pattern is generated when a crystal containing many copies of an isolated molecule is bombarded by a beam of X-rays (**Fig. 2.41A**). The wave fronts associated with the X-rays are diffracted as they pass through the crystal, causing interference patterns. In the crystal, the diffraction pattern is generated by a symmetrical array of many sample molecules (**Fig. 2.41B**). The larger the number of copies of the molecule in the array, the narrower the interference pattern and the greater the reso-

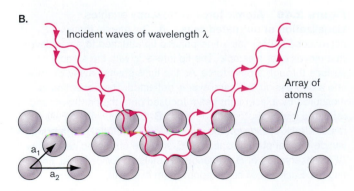

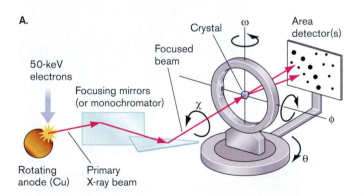

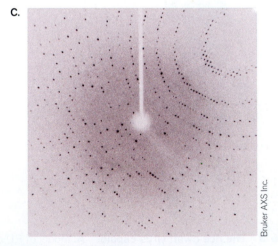

Figure 2.41   Visualizing molecules by X-ray crystallography.   A. Modern apparatus for X-ray crystallography. The X-ray beam is focused onto a crystal, which is rotated over all angles to obtain diffraction patterns. The intensity of the diffracted X-rays is recorded on film or with an electronic detector. **B.** X-rays are diffracted by rows of identical molecules in a crystal. The diffraction pattern is analyzed to generate a model of the individual molecules. **C.** Diffraction pattern from a crystal.

lution of atoms within the molecule. Diffraction patterns obtained from the passage of X-rays through a crystal (**Fig. 2.41C**) can be analyzed by computation to develop a precise structural model for the molecule, detailing the position of every atom in the structure.

The application of X-ray crystallography to complex biological molecules was pioneered by the Irish crystallographer John Bernal (1901–1971) (**Fig. 2.42A**). Bernal was particularly supportive of women students and colleagues, including Rosalind Franklin (1920–1958) who made important discoveries about DNA and RNA, and Nobel laureate Dorothy Crowfoot Hodgkin (1910–1994). Hodgkin won the 1964 Nobel Prize in Chemistry for solving the crystal structures of penicillin and vitamin B_{12} (**Fig. 2.42B**). She later solved one of the first protein structures, that of the hormone insulin.

Today, X-ray data undergo digital analysis to generate sophisticated molecular models, such as the one seen in

Figure 2.43 of anthrax lethal factor, a toxin produced by *B. anthracis* that kills the infected host cells. The model for anthrax lethal factor was encoded in a Protein Data Bank (PDB) text file that specifies coordinates for all atoms of the structure. The Protein Data Bank is a growing world database of solved X-ray structures, freely available on the Internet. Visualization software is used to present the structure as a "ribbon" of amino acid residues, color-coded for secondary structure. In **Figure 2.43**, the pink-colored coils represent alpha helix structures, whereas the blue arrows represent beta sheets (for a review of these secondary structures, see Appendix 1).

WWW | Protein Data Bank: Research Collaboratory for Structural Bioinformatics Protein and Nucleic Acid Databases

WWW | Biomolecules at Kenyon College: molecular tutorials by undergraduate students, on anthrax lethal factor and other proteins; as well as instructions to write your own tutorials.

NOTE: Molecular and cellular biology increasingly rely on visualization in three dimensions. Many of the molecules illustrated in our book are based on structural models deposited in the Protein Data Bank, as indicated by the PDB file code. You may view these structures in 3-D by downloading the PDB file and viewing with a free plug-in such as Jmol.

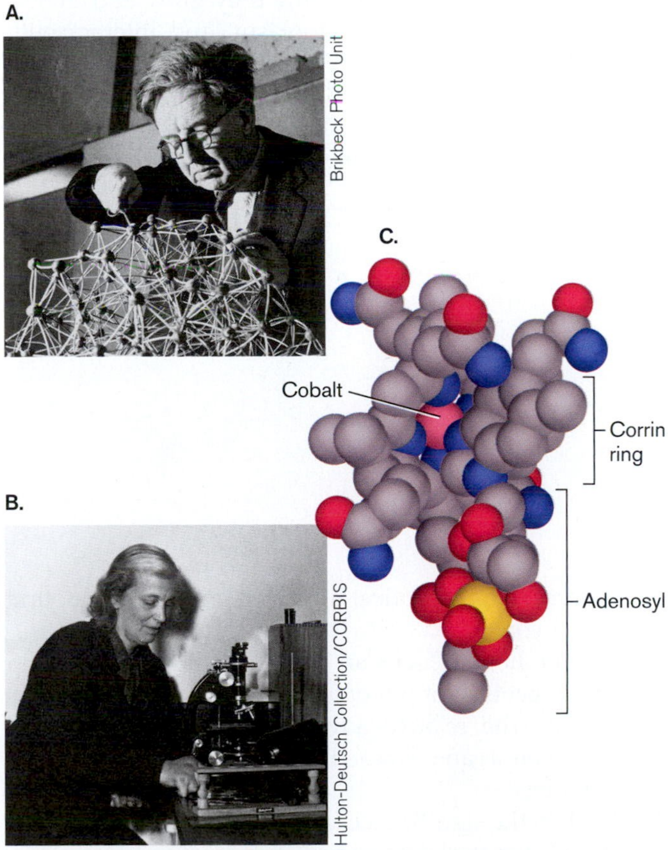

A.

Brikbeck Photo Unit

C.

Cobalt —

— Corrin ring

— Adenosyl

B.

Hulton-Deutsch Collection/CORBIS

Figure 2.42 Pioneering X-ray crystallography. A. John Bernal developed X-ray crystallography to solve the structure of complex biological molecules. **B.** Dorothy Hodgkin (1910–1994) was awarded the 1964 Nobel Prize in Chemistry for her work in X-ray crystallography. **C.** Vitamin B_{12}, whose structure was originally solved by Dorothy Hodgkin. The corrin ring structure (see Chapter 15) is built around an atom of cobalt (pink). Carbon atoms are gray; oxygen, red; nitrogen, blue; phosphorus, yellow. Hydrogen atoms are omitted for clarity.

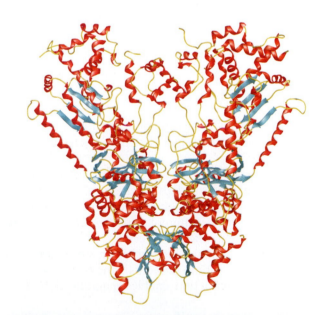

Figure 2.43 X-ray crystallography of a protein complex, anthrax lethal factor.   The toxin consists of a butterfly-shaped dimer of two peptide chains. This cartoon model is based on X-ray crystallographic data, showing alpha helix (pink coils) and beta sheet (orange arrows). (PDB code: 1J7N)

A limitation of X-ray analysis is the unavoidable deterioration of the specimen under bombardment by X-rays. The earliest X-ray diffraction models of molecular complexes such as the ribosome relied heavily on components from thermophiles, microbes that grow at high temperatures. Because these organisms have evolved to grow under higher thermal stress, their macromolecular complexes form more stable crystals than do their homologues in organisms growing at moderate temperatures.

X-ray diffraction analysis of crystals from a wide range of sources was made possible by **cryocrystallography**. In cryocrystallography, as in cryo EM, crystals are frozen rapidly to liquid nitrogen temperature. The frozen crystals have greatly decreased thermal vibrations and diffusion, thus decreasing the radiation damage to the molecules. Models based on cryocrystallography can present multisubunit structures such as the bacterial ribosome complexed with transfer RNAs and messenger RNA. Much of our knowledge of microbial genetics (Chapters 7–12) and metabolism (Chapters 13–16) is based on crystal structures of key macromolecules.

TO SUMMARIZE:

- **X-ray diffraction analysis**, or **X-ray crystallography**, uses X-ray diffraction (interference patterns) from crystallized macromolecules to determine structure at atomic resolution.
- **Cryocrystallography** uses frozen crystals with greatly decreased thermal vibrations and diffusion, enabling the determination of structures of large macromolecular complexes, such as the ribosome.
- **Molecular visualization** by crystallography can only model the "appearance" of a molecule at atomic resolution. Different models emphasize different structural features and levels of resolution.

Concluding Thoughts

The tools of microscopy and molecular visualization described in this chapter have shaped our current understanding of microbial cells—how they grow and divide, organize their DNA and cytoplasm, and interact with other cells. Our current models of cell structure and function are explored in Chapter 3. In Chapter 4, we learn how cells use their structures to obtain energy, reproduce, and develop dormant forms that can remain viable for thousands of years.

CHAPTER REVIEW

Review Questions

1. What principle defines an object as "microscopic"?
2. Explain the difference between detection and resolution.
3. How do eukaryotic and prokaryotic cells differ in appearance under the light microscope?
4. Explain how electromagnetic radiation carries information and why different kinds of radiation can resolve different kinds of objects.
5. Define how light interacts with an object through absorption, reflection, refraction, and scattering.
6. Explain how refraction enables magnification of an image.
7. Explain how magnification increases resolution and why "empty magnification" fails to increase resolution.
8. Explain how angle of aperture and resolution change with increasing lens magnification.
9. Summarize the optical arrangement of a compound microscope.
10. Explain how to focus an object and how to tell when the object is in or out of focus.
11. Explain the relative advantages and limitations of wet mount and stained preparation for observing microbes.
12. Explain the significance (and limitations) of the Gram stain for bacterial taxonomy.
13. Explain the basis of dark-field, phase-contrast, and fluorescence microscopy. Give examples of applications of these advanced techniques.
14. Explain the difference between transmission and scanning electron microscopy and the different applications of each.

Key Terms

aberration (49)
absorption (45)
acid-fast stain (54)
Airy disk (47)
angle of aperture (49)
antibody stain (55)
artifact (65)
atomic force microscopy (AFM) (43, 65)
bacillus (plural, bacilli) (42)
bright-field microscopy (43)
coccus (plural, cocci) (42)
compound microscope (50)
condenser (50)
contrast (44)
counterstain (54)
cryocrystallography (70)
cryoelectron microscopy (cryo EM) (64)
dark-field microscopy (45)
depth of field (51)
detection (41)
diaphragm (50)
differential stain (52)
electron microscopy (43, 62)
emission wavelength (58)

empty magnification (46)
excitation wavelength (58)
fixation (52)
flagella (56)
fluorescence (45)
fluorophore (59)
focal plane (46)
focal point (46)
focus (48)
Gram stain (52)
gram-negative (54)
gram-positive (54)
immersion oil (50)
interference (46)
interference microscopy (58)
laser scanning confocal microscopy (60)
light microscopy (40, 43)
magnification (41)
microscope (40)
mordant (52)
negative stain (54)
nuclear magnetic resonance (NMR) (43, 68)

numerical aperture (NA) (49)
objective lens (49)
ocular lens (50)
parfocal (50)
reflection (45)
refraction (45)
refractive index (45)
resolution (40)
rods (42)
scanning electron microscopy (SEM) (43, 62)
scattering (45)
simple stain (52)
spirochete (42)
spore stain (54)
staining (52)
tomography (65)
total magnification (51)
transmission electron microscopy (TEM) (43, 62)
wet mount (52)
X-ray crystallography (43, 68)
X-ray diffraction analysis (68)

Recommended Reading

Chiu, Wah, et al. 2002. Deriving folds of macromolecular complexes through electron cryomicroscopy and bioinformatics approaches. *Current Opinion in Structural Biology* **12**:263.

Graumann, Peter L., and Richard Losick. 2001. Coupling of asymmetric division to polar placement of replication origin regions in *Bacillus subtilis. Journal of Bacteriology* **183**:4052–4060.

Jiang, Weihang, Juan Chang, Joanita Jakana, et al. 2006. Structure of epsilon15 bacteriophage reveals genome organization and DNA packaging-injection apparatus. *Nature* **439**:612–616.

Lucic, Vladan, Friedrich Förster, and Wolfgang Baumeister. 2005. Structural studies by electron tomography: From cells to molecules. *Annual Review of Biochemistry* **74**:833–865.

Matias, Valério R. F., Ashruf Al-Amoudi, Jacques Dubochet, and Terry J. Beveridge. 2003. Cryo-transmission electron microscopy of frozen-hydrated sections of *Escherichia coli* and *Pseudomonas aeruginosa. Journal of Bacteriology* **185**:6112–6118.

Murphy, Douglas B. 2001. *Fundamentals of Light Microscopy and Electronic Imaging.* Wiley-Liss, Hoboken, N.J.

Popescu, Aurel, and R. J. Doyle. 1996. The Gram stain after more than a century. *Biotechniques in Histochemistry* **71**:145–151.

Chapter 3

Cell Structure and Function

Microbial cells face extreme challenges from their environment, enduring rapid changes in temperature and salinity, and pathogens face the chemical defenses of their hosts. To meet these challenges, microbes build complex structures, such as a cell envelope with tensile strength comparable to steel. Within the cytoplasm, molecular devices such as the the ribosome build and expand the cell.

With just a few thousand genes in its genome, how does a bacterial cell grow and reproduce? Bacteria coordinate their DNA replication through the DNA replisome and the cell fission ring. Other devices, such as flagellar propellers, enable microbial cells to compete, to communicate, and even to cooperate in building biofilm communities.

Discoveries of cell form and function have exciting applications for medicine and biotechnology. The structures of ribosomes and cell envelope materials provide targets for new antibiotics. And devices such as the rotary ATP synthase inspire "nanotechnology," the design of molecular machines.

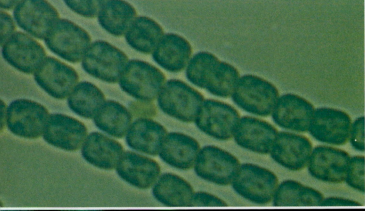

The filamentous cyanobacterium *Anabaena* sp. was engineered to make a cell division protein, FtsZ, fused to green fluorescent protein (GFP). FtsZ-GFP proteins form a ring-like structure around the middle of each cell, where it prepares to divide. *Source:* Samer Sakr, et al. 2006. *J. Bacteriol.* 188.

5 µm

The microbial cell has formidable tasks: to obtain nutrients faster than its competitors, to protect itself from toxins and predators, and to reproduce itself. These tasks are accomplished by the cell's molecular parts.

To study the cell requires microscopy as well as isolating cell parts. Chapter 3 presents an overview of the bacterial cell, then explains how such views derive from cell fractionation and genetic analysis. We explore the structures common to most microbial cells, as well as more specialized devices such as light-harvesting complexes and magnetosomes. We dissect the cell's complex outer layers and see how they interact with the replicat-

ing chromosome to accomplish cell fission. Many of the cell's structures offer targets for antibiotic design as well as opportunities for biotechnology. Overall, this chapter focuses on prokaryotes, with reference to eukaryotes for comparison (eukaryotic microbes are covered in Chapter 20). Our discussion assumes an elementary knowledge of cell biology (reviewed in Appendices 1 and 2).

WWW | **The Microbial Biorealm** online provides a quick, readable reference for microbial species, including those mentioned in this book.

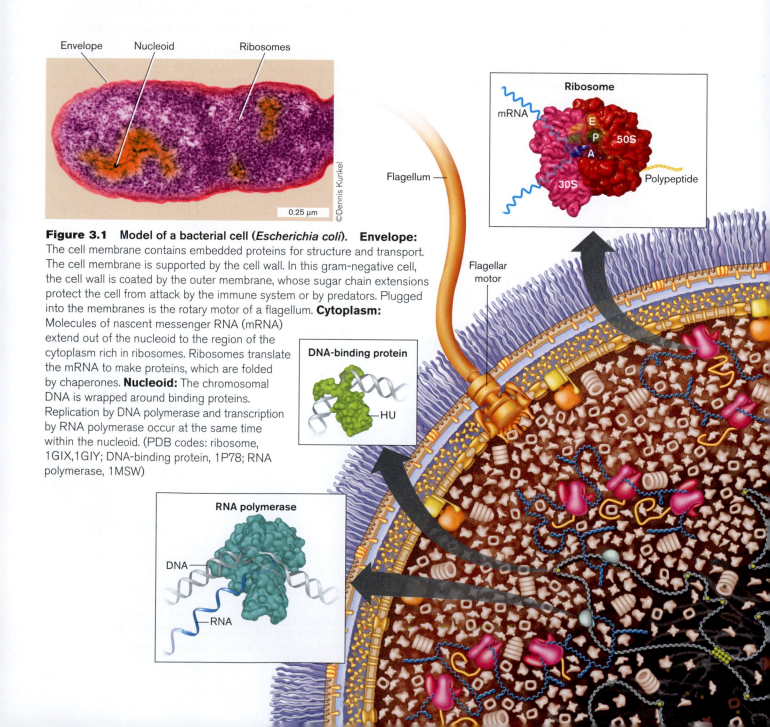

Envelope   Nucleoid   Ribosomes

©Dennis Kunkel

0.25 µm

Figure 3.1   Model of a bacterial cell (*Escherichia coli*).   Envelope: The cell membrane contains embedded proteins for structure and transport. The cell membrane is supported by the cell wall. In this gram-negative cell, the cell wall is coated by the outer membrane, whose sugar chain extensions protect the cell from attack by the immune system or by predators. Plugged into the membranes is the rotary motor of a flagellum. **Cytoplasm:** Molecules of nascent messenger RNA (mRNA) extend out of the nucleoid to the region of the cytoplasm rich in ribosomes. Ribosomes translate the mRNA to make proteins, which are folded by chaperones. **Nucleoid:** The chromosomal DNA is wrapped around binding proteins. Replication by DNA polymerase and transcription by RNA polymerase occur at the same time within the nucleoid. (PDB codes: ribosome, 1GIX,1GIY; DNA-binding protein, 1P78; RNA polymerase, 1MSW)

Flagellum

Ribosome

mRNA

E
P
A
50S

30S

Polypeptide

Flagellar motor

DNA-binding protein

HU

RNA polymerase

DNA

RNA

3.1 The Bacterial Cell: An Overview

Prokaryotic cells show a wide range of form, whose diversity is explored in Chapters 18 and 19. At the same time, most prokaryotes share fundamental traits:

- **Thick, complex outer envelope.** The envelope protects the cell from environmental stress and predators. It also mediates exchange with the environment and communication with other organisms.
- **Compact genome.** Prokaryotic genomes are compact, with relatively little noncoding DNA. Small genomes maximize the production of cells from limited resources.
- **Tightly coordinated cell functions.** The cell's subcellular parts work together in a highly coordinate mechanism. Coordinate action enables high reproduction rate.

In the early twentieth century, the cell was envisioned as a bag of "soup" full of floating ribosomes and enzymes. Modern research shows that in fact the cell's parts fit together in a structure that is ordered, though flexible.

A Model of the Bacterial Cell

Here we present a model of the bacterial cell (**Fig. 3.1**). This model offers an interpretation of how the major components of one bacterial cell fit together. The model represents *Escherichia coli*, but its general features apply to many kinds of bacteria. Remember that we cannot literally "see" the molecules within a cell, but microscopy and subcellular analysis generate a remarkably detailed view.

Within the bacterial cell, the cytoplasm consists of a gel-like network composed of proteins and other macromolecules. The cytoplasm is contained by a **cell membrane** (or **inner membrane**, for a gram-negative cell). The membrane is composed of phospholipids, hydrophobic proteins, and other molecules. The cell membrane prevents

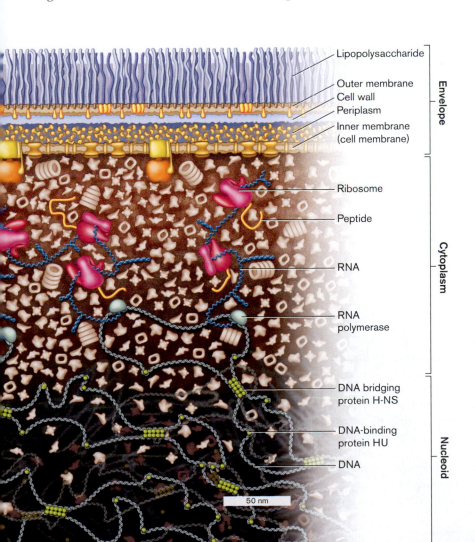

Lipopolysaccharide
Outer membrane
Cell wall
Periplasm
Inner membrane (cell membrane)
Envelope

Ribosome
Peptide
RNA
RNA polymerase
Cytoplasm

DNA bridging protein H-NS
DNA-binding protein HU
DNA
Nucleoid

50 nm

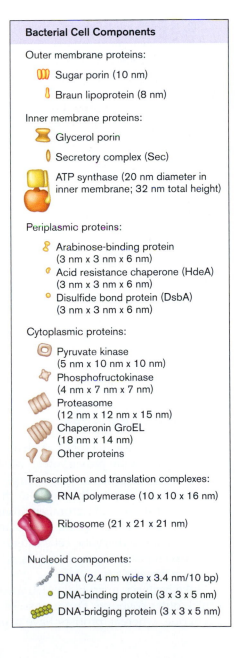

Bacterial Cell Components

Outer membrane proteins:
- Sugar porin (10 nm)
- Braun lipoprotein (8 nm)

Inner membrane proteins:
- Glycerol porin
- Secretory complex (Sec)
- ATP synthase (20 nm diameter in inner membrane; 32 nm total height)

Periplasmic proteins:
- Arabinose-binding protein (3 nm x 3 nm x 6 nm)
- Acid resistance chaperone (HdeA) (3 nm x 3 nm x 6 nm)
- Disulfide bond protein (DsbA) (3 nm x 3 nm x 6 nm)

Cytoplasmic proteins:
- Pyruvate kinase (5 nm x 10 nm x 10 nm)
- Phosphofructokinase (4 nm x 7 nm x 7 nm)
- Proteasome (12 nm x 12 nm x 15 nm)
- Chaperonin GroEL (18 nm x 14 nm)
- Other proteins

Transcription and translation complexes:
- RNA polymerase (10 x 10 x 16 nm)
- Ribosome (21 x 21 x 21 nm)

Nucleoid components:
- DNA (2.4 nm wide x 3.4 nm/10 bp)
- DNA-binding protein (3 x 3 x 5 nm)
- DNA-bridging protein (3 x 3 x 5 nm)

cytoplasmic proteins from leaking out and maintains gradients of ions and nutrients. It is covered by a **cell wall**, a rigid structure composed of polysaccharides linked covalently by peptides (peptidoglycan). The cell wall forms a single cage-like molecule that surrounds the cell.

In gram-negative bacteria such as *E. coli*, the cell wall consists of a single layer of peptidoglycan, a polymer of sugars and peptides. The cell wall extends within the **periplasm**, an aqueous layer containing proteins such as sugar transporters. Outside the cell wall lies an **outer membrane** of phospholipids and **lipopolysaccharides (LPSs)**, a class of lipids attached to long polysaccharides (sugar chains). LPS and other polysaccharides can generate a thick **capsule** surrounding the cell. The capsule polysaccharides form a slippery mucous layer that inhibits phagocytosis by macrophages (see Section 26.3). The cell wall and outer membrane constitute the **envelope**.

The envelope includes cell surface proteins that enable the bacteria to interact with specific host organisms. For example, *E. coli* cell surface proteins enable colonization of the human intestinal epithelium, whereas *Sinorhizobium* cell surface proteins enable colonization of legume plants for nitrogen fixation. Another common external structure is the **flagellum**, a helical protein filament whose rotary motor propels the cell in search of a more favorable environment.

Within the cell, the cell membrane and envelope provide an attachment point for one or more chromosomes. The chromosome is organized within the cytoplasm as a system of looped coils called the **nucleoid**. Unlike the round, compact nucleus of eukaryotic cells, the bacterial nucleoid is not bounded by a membrane, and so the coils of DNA can extend throughout the cytoplasm. Loops of DNA from the nucleoid are transcribed by RNA polymerase to form messenger RNA (mRNA) as well as small functional RNA molecules (sRNA). As the mRNA transcripts grow, they bind ribosomes to start synthesizing polypeptide chains. As the polypeptides grow, protein complexes called chaperones help them fold into their functional conformations.

Microbial Eukaryotes

How do eukaryotic microbes, such as euglenas and amebas, compare with prokaryotes? Most eukaryotic cells that have been studied are much larger than prokaryotes; some amebas are actually visible to the unaided eye. Yet DNA-based surveys of natural environments, such as the ocean and soil, reveal eukaryotes as small as the smallest bacteria. Thus, eukaryotic cells extend over the full size-range encompassed by prokaryotes, in addition to reaching much larger sizes.

Like bacteria, microbial eukaryotes typically possess a thick outer covering. Fungi possess cell walls of polysaccharide, whereas protists have a **pellicle**, consisting of membranous layers reinforced by protein microtubules. Other kinds of eukaryotic cell surface are reinforced by inorganic materials, such as the silicate shells of diatoms.

Eukaryotic microbes contain subcellular organelles composed of membranes that are more extensive and specialized than those of prokaryotes. Examples include the nuclear membrane, golgi, mitochondria, and chloroplasts (reviewed in Appendix 2). These organelles perform some of the physiological functions that are provided by organ systems in multicellular organisms, such as circulation, digestion, and excretion. Mitochondria and chloroplasts are the products of ancestral engulfment of prokaryotic cells, followed by evolution of endosymbiosis, as explained in Chapter 17. Eukaryotic microbial diversity is explored in Chapter 20.

Biochemical Composition of Bacteria

The bacterial cell model indicates shape and size, but tells little about chemical composition. Chemistry explains, for example, why wiping a surface with ethanol eliminates microbial life, whereas water has little effect. Water is a universal constituent of cytoplasm, but is excluded by cell membranes. Ethanol, however, dissolves both polar and nonpolar substances; thus, ethanol disintegrates membranes and destroys the secondary structure of proteins.

All cells share common chemical components:

■ **Water**, the fundamental solvent of life
■ **Essential ions**, such as potassium, magnesium, and chloride ions
■ **Small organic molecules**, such as lipids and sugars that are incorporated into cell structures and that provide nutrition by catabolism
■ **Macromolecules**, such as nucleic acids and proteins, which contain information, catalyze reactions, and mediate transport, among many other functions

Cell composition varies with species, growth phase, and environmental conditions. **Table 3.1** summarizes the chemical components of a cell for the model bacterium *Escherichia coli* during exponential growth.

Small molecules. The *E. coli* cell consists of about 70% water, the essential solvent required to carry out fundamental metabolic reactions and to stabilize proteins. The water solution contains inorganic ions, predominantly potassium, magnesium, and phosphate. Inorganic ions store energy in the form of transmembrane gradients, and they serve essential roles in enzymes. For example, a magnesium ion is required at the active site of RNA polymerase to help catalyze incorporation of ribonucleotides into RNA.

Table 3.1   Molecular composition of a bacterial cell, *Escherichia coli*, during balanced exponential growth.[a]

Component	Percentage of total weight[b]	Approximate number of molecules/cell	Number of different kinds
Water	70%	20,000,000,000	1
Proteins	16%	2,400,000	2,000[c]
RNA: rRNA, tRNA, and other small regulatory RNA (sRNA),	6%	250,000	200
mRNA	0.7%	4,000	2,000[c]
Lipids: phospholipids (membrane)	3%	25,000,000	50
lipopolysaccharide (outer membrane)	1%	1,400,000	1
DNA	1%	2[d]	1
Metabolites and biosynthetic precursors	1.3%	50,000,000	1,000
Peptidoglycan (murein sacculus)	0.8%	1	1
Inorganic ions	0.1%	250,000,000	20
Polyamines (mainly putrescine and spermine)	0.1%	6,700,000	2

[a]Values shown are for a hypothetical "average" cell cultured with aeration in glucose medium with minimal salts at 37°C.

[b]The total weight of the cell (including water) is about 10^{-12} gram (g), or 1 picogram (pg).

[c]The number of kinds of mRNA and of proteins is difficult to estimate because some genes are transcribed at extremely low levels and because RNA and proteins include kinds that are rapidly degraded.

[d]In rapidly growing cells, cell fission typically lags approximately one generation behind DNA replication; hence, two identical DNA copies per cell.

Source: Modified from Neidhardt, F., and H. E. Umbarger. 1996. Chemical composition of *Escherichia coli*, p. 14. In *Escherichia coli* and *Salmonella*: Cellular and Molecular Biology, 2nd ed. ASM Press, Washington, DC.

The cell also contains many kinds of small charged organic molecules, such as phospholipids and enzyme cofactors. A major class of organic cations is the **polyamines**, molecules with multiple amine groups that are positively charged when the pH is near neutral. Polyamines balance the negative charges of the cell's DNA, and they stabilize ribosomes during translation.

Macromolecules. Many cells show similar content of water and small molecules, but their specific character is defined by their macromolecules, especially their proteins. Proteins vary among different species; and a given species makes very different proteins, depending on environmental conditions such as temperature, nutrient levels, and entry into a host organism. Individual proteins, encoded by specific genes, are found in very different amounts, from 10 per cell to 10,000 per cell. The total proteins encoded by a genome, capable of expression in the cell, are known collectively as the **proteome**. Early attempts to define the proteome of a cell were conducted by Fred Neidhardt (**Fig. 3.2**) and colleagues at the University of Michigan, who created the first protein catalog of *E. coli* using **two-dimensional polyacrylamide gel electrophoresis** (**2-D PAGE** or **2-D gels**) (**Fig. 3.3**).

To obtain 2-D gels, cells are lysed by sonication or detergent to release and solubilize as many proteins as possible. The proteins are then subjected to two forms of separation, isoelectric focusing followed by electrophoresis. In **isoelectric focusing**, proteins are placed in a gel that has a pH gradient, and they migrate along the gradi-

Frederick C. Neidhardt, U. of Michigan

Figure 3.2   Proteins of *E. coli*. Fred Neidhardt and colleagues at the University of Michigan, Ann Arbor, used 2-D gel electrophoresis to create the first protein catalog of *E. coli*.

ent until they reach the point where the number of positively charged residues equals the number of negatively charged residues. At this point, their net charge is zero (that is, the pH equals the protein's isoelectric point), and the protein stops moving. In **electrophoresis**, proteins deposited in an electrical field migrate toward a positive electrode. The proteins are coated with a reagent (SDS, or sodium dodecyl sulfate) to confer a uniform negative charge so that their distance of migration depends on molecular weight (roughly a logarithmic function). The proteins are then stained for detection, and the protein spots can be identified, either by N-terminal protein sequencing or by mass spectrometry.

In a 2-D gel of *E. coli* (**Fig. 3.3**), about 500 different proteins can be distinguished. The most highly expressed proteins include ribosomal proteins and translation factors such as elongation factor Tu (EF-Tu). Outer membrane proteins such as OmpX and ProX also show high concentration. Conversely, many proteins of the inner (cell) membrane are too insoluble in water to appear, and some important regulators are present at levels too low to be seen. Nevertheless, 2-D gels are useful for identifying changes in protein expression that occur under different environmental conditions or when bacteria are invading host cells.

Protein synthesis is directed by DNA and RNA. The content of nucleic acids in *E. coli* is nearly 8% by weight, much higher than in multicellular eukaryotes. For microbes, the high nucleic acid content is advantageous, allowing the cell to maximize reproduction of its chromosome while minimizing resources for protein-rich cytoplasm. The high level of nucleic acids is actually toxic to human consumers, who lack the enzymes to digest the uric acid waste product of digested nucleotides. That is why most kinds of bacteria cannot be eaten as a major part of the diet.

Other kinds of macromolecules are found in the cell wall and outer membrane. The bacterial cell wall consists of **peptidoglycan**, an organic polymer that constitutes nearly 1% of the cell mass, approximately the same mass as that of DNA. This shows the importance (for most species) of maintaining turgor pressure in dilute environments, where water would otherwise enter by osmosis, causing osmotic shock.

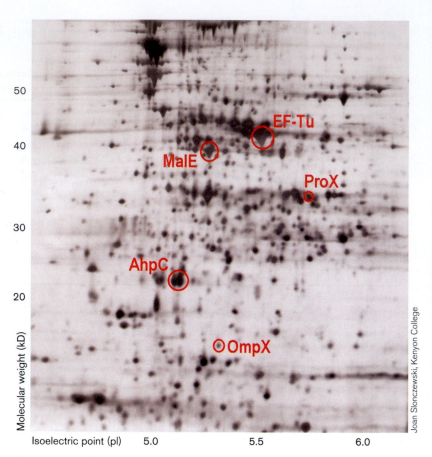

Figure 3.3 **Proteins of *E. coli*.** 2-D gel of proteins of *Escherichia coli* grown aerobically in a casein-yeast extract medium. Proteins were identified by N-terminal sequence and by mass spectroscopy. DnaK is a cytoplasmic chaperone; EF-Tu is an elongation factor for translation; MalE is an outer membrane maltose-binding protein; ProX and OmpX are periplasmic transporters; and AhpC is an antioxidant stress protein.

> **THOUGHT QUESTION 3.1** Which molecules occur in the greatest number in a prokaryotic cell? The smallest number? Why does a cell contain 100 times as many lipid molecules as strands of RNA?

TO SUMMARIZE:

■ **Prokaryotic cells** are protected by a thick cell envelope.
■ **Compact genomes** maximize reproductive potential with minimal resources.
■ **The model bacterial cell** contains a highly ordered cytoplasm in which DNA replication, RNA transcription, and protein synthesis occur coordinately.

■ **The biochemical composition of bacteria** includes relatively high nucleic acid content, as well as proteins, phospholipids, and other organic and inorganic constituents.
■ **Proteins in the cell** vary, depending on the species and environmental conditions.

3.2 How We Study the Parts of Cells

How can we determine how all the macromolecules listed in **Table 3.1** interact and work together? Electron microscopy largely defines how we "see" the cell's interior as a whole. In **Figure 3.1**, we clearly saw major parts of *B. thuringiensis*, such as the endospore and the parasporal crystal. But smaller parts, such as the ribosomes, appear only as small, densely packed particles. Furthermore, even higher-resolution images from electron microscopy cannot tell us the chemical composition of ribosomes or how they function in the living cell.

To study ribosome function requires isolation and analysis of subcellular parts:

- **Subcellular fractionation** enables isolation of cell parts such as ribosomes and membranes so that we can study their form and function.
- **Structural analysis** by X-ray crystallography and related methods reveals the form of cell components.
- **Genetic analysis** dissects the function of cell components based on construction of mutant cells with altered function.

In this section, we present these methods as applied to a key example of cellular function, that of the bacterial ribosome. Each method has its particular strengths and limitations to reveal form and function.

Isolating Parts of Cells

Cellular components, such as ribosomes and flagellar motors, can be isolated readily from cells. The cells must be broken open by techniques that allow subcellular components to remain intact. Here are examples of such techniques:

- **Mild detergent lysis.** Cells can be lysed with a detergent capable of dissolving membranes but not denaturing proteins.
- **Freezing and thawing.** Freezing and then thawing is another way to open cells gently, releasing particles such as ribosomes.
- **Sonication.** Cells can be lysed by intense ultrasonic vibrations above the range of human hearing.
- **Enzymes.** Enzymes such as lysozyme can break the cell wall, allowing the cell to be lysed by mild osmotic shock.

The ultracentrifuge. Different parts of the cell can be isolated by **subcellular fractionation**. A key tool of subcellular fractionation is the **ultracentrifuge**, a device in which solutions containing cell components are rotated in tubes at high speed. The high rotation rate generates centrifugal forces strong enough to separate subcellular particles (**Fig. 3.4A**). The ultracentrifuge was invented

Figure 3.4 Subcellular fractionation by ultracentrifugation. A. An ultracentrifuge is used to fractionate cell components. **B.** *Salmonella* ribosomes (TEM) isolated from the cytoplasm by ultracentrifugation through a linear sucrose density gradient. Polysomes consist of two or more ribosomes attached to mRNA. **C.** High-speed rotation generates high centrifugal forces, measured in units of gravity (*g*). The Svedberg unit (S) offers a measure of particle size based on its rate of travel in a tube subjected to high *g* force. The Svedberg coefficient (number of S units; for example, 30S) is defined in terms of the velocity of the particle in the tube (*v*), the radius of the rotor (*r*), and the rotational velocity (ω). The coefficient of S for a given particle depends on its mass (*m*) and its shape. After centrifugation, fractions are collected from the base of the centrifuge tube. The fractions contain radiolabeled ribosomes. The largest particles (whole ribosomes) sediment near the bottom of the tube, whereas the smaller particles (separate 50S and 30S subunits) appear in upper fractions.

A.

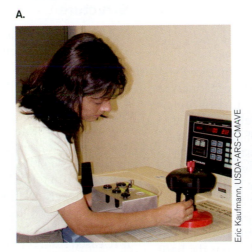

Eric Kaufmann, USDA-ARS-CMAVE

B.

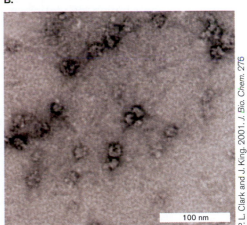

100 nm

P. L. Clark and J. King. 2001. *J. Bio. Chem.* 276

C.

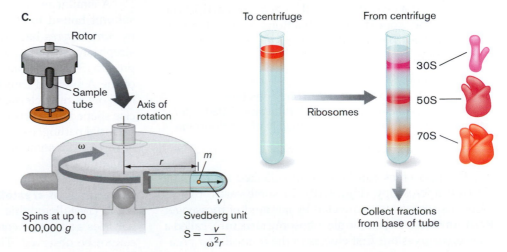

Rotor

Sample tube

Axis of rotation

ω

r

m

v

Spins at up to 100,000 *g*

Svedberg unit

$$S = \frac{v}{\omega^2 r}$$

To centrifuge

From centrifuge

Ribosomes

30S

50S

70S

Collect fractions from base of tube

by the Swedish physicist Theodor Svedberg (1884–1971), who won the 1926 Nobel Prize in Chemistry for the use of ultracentrifugation to separate proteins.

Modern ultracentrifuges have titanium rotors that spin in a vacuum to avoid frictional heating, generating forces up to 100,000 times gravity (100 k g). For fractionation, cells are first lysed by one of the methods previously described to obtain a cell **lysate**, a general term for the contents of a broken cell. The cell lysate is placed in tubes containing a high-density solution, such as sucrose or cesium chloride solution, in which suspended particles sediment slowly. Under a high g force, however, particles sediment at different rates, depending on their size and density.

The **sedimentation rate** is the rate at which particles of a given size and shape travel to the bottom of the tube under centrifugal force. The sedimentation rate is measured by the Svedberg unit, S, which is given by the particle's rate of sedimentation (v), the radius at which the tube rotates (r), and the rotational velocity (ω):

$$S = v/(\omega^2 r)$$

For a given type of particle in suspension, the sedimentation rate also depends on the particle's mass and shape. The contribution of particle mass and shape is defined as its **Svedberg coefficient**; for example, the coefficient of the small subunit of the ribosome is 30, for a sedimentation value of 30S. The value of the Svedberg coefficient increases with the average cross-sectional area of the particle. For bacterial ribosomes, ultracentrifugation yields intact ribosomes (70S) as well as separated ribosomal subunits, the large subunit (50S) and the small subunit (30S). Within cells, ribosomes normally exist as a mixture of joined and separate subunits.

To isolate the ribosomes, a cell lysate is layered onto a tube of sucrose solution, whose density decreases the sedimentation rate and increases the separation of particles of different size (**Fig. 3.4A**). The fractions shown in **Figure 3.4C** were drained sequentially from the base of the tube after centrifugation. The heaviest particle, the 70S ribosome, appears in the fractions nearest the bottom of the tube because it travels fastest under the centrifugal force.

THOUGHT QUESTION 3.2 Why does the Svedberg coefficient of the intact ribosome (70S) differ from the sum of the individual subunits of the ribosome (30S and 50S)?

The ribosomes can be seen in collected fractions by electron microscopy (**Fig. 3.4B**). In some cases, two or more ribosomes are connected by a strand of messenger RNA (mRNA). This multiple-ribosome structure, called a polysome, gives our first clue as to the intracellular orga-

nization of the translation apparatus within a bacterial cell (see Fig. 3.1), where a number of ribosomes translate each mRNA at the same time.

Furthermore, ribosomes isolated by centrifugation can translate messenger RNA in cell-free systems. Experiments in cell-free systems provide the basis of much of our knowledge of protein synthesis (see Chapter 8).

Limitations of subcellular fractionation. Subcellular fractionation yields clues about internal structure but provides little information about processes that require overall integrity of the cell. For example, the role of the transmembrane electrochemical potential, or proton potential, in ATP synthesis was obscured for many years because biochemists were unable to isolate a cytoplasmic complex responsible for it. Transmembrane ion gradients were in fact observed within membrane vesicles— spheres of membrane isolated by cell disintegration and centrifugation. But it was harder to demonstrate that the entire cell membrane of an intact cell supports a proton potential.

Crystallographic Analysis Reveals Structure

Isolated ribosomes can be crystallized for structure analysis by X-ray diffraction crystallography. A crystallographic model of the 30S ribosomal subunit is shown in **Figure 3.5A**. The 30S subunit consists of ribosomal RNA (rRNA) and 27 different proteins, each of which is encoded by a particular gene. The sequences of all the ribosomal genes, with their predicted protein and RNA products, were fitted with X-ray crystallography data to build a three-dimensional model of the 30S subunit. This model includes three bound transfer RNA (tRNA) molecules, each at a key position for translation: the acceptor site (A) for incoming aminoacyl-tRNA; the peptide site (P) attached to the growing peptide chain; and the exit site (E), where tRNA is released from the ribosome.

A similar model was obtained of the 30S ribosomal subunit bound to streptomycin, an antibiotic produced by *Streptomyces* bacteria (**Fig. 3.5B**). In this model, the streptomycin molecule contacts a particular ribosomal protein, S12. The contact with S12 distorts the structure of the A site and thus halts protein synthesis. Only the ribosomes of bacteria, not of eukaryotes, have the precise shape to bind streptomycin; thus, this antibiotic kills bacterial pathogens without harming the human patient. Studying ribosome structure enables us to design new antibiotics.

Limitations of crystallography. As discussed in Chapter 2, crystallographic analysis applies only to isolated particles and under conditions in which its full function cannot be observed. The technique can solve structures

A. The ribosome: x-ray diffraction model

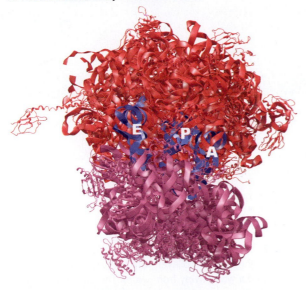

B.

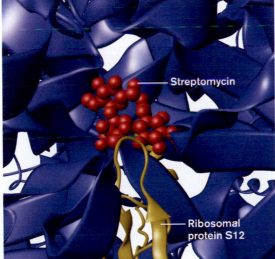

Streptomycin

Ribosomal protein S12

Figure 3.5 **X-ray diffraction models relate structure to function.**
A. The 30S ribosomal subunit, X-ray diffraction model of the 16S RNA (stereo view) showing the acceptor (A) site, the peptide (P) site, and the exit (E) site. To view, place a piece of cardboard vertically between the two images. Position your eyes on opposite sides of the cardboard and force them to focus behind the images. The images will merge and produce a 3-D image. (PDB codes: 1GIX, 1GIY) **B.** When the ribosome is crystallized with streptomycin, the antibiotic molecule binds to protein S12, near the A site. Streptomycin interferes with the tRNA binding to the A site.

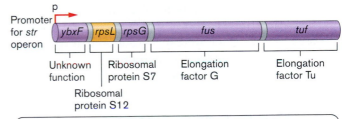

Mutations in the gene *rpsL*, encoding ribosomal protein S12, confer resistance to the antibiotic streptomycin.

Figure 3.6 **Genetic analysis of the streptomycin-resistant ribosome.** The *str* operon has genes encoding components of the ribosome. A mutation in *rpsL*, the gene encoding ribosomal protein S12, alters the protein's interaction with streptomycin. The mutant gene makes the bacteria resistant to the antibiotic.

only for proteins and nucleic acids capable of crystallization. It remains impractical for proteins of flexible, nonrigid structure and for many membrane-soluble proteins.

Genetic Analysis Reveals Function

The crystal structure of a macromolecule only yields clues as to function; it does not show the structure in action. The function of cell components can be dissected by **genetic analysis**. In genetic analysis, mutant strains are selected for loss of a given function, or a strain can be intentionally mutated so as to lose or alter a gene.

An example of genetic analysis is the use of ribosome-encoding genes to show how ribosome func-

tion is blocked by the antibiotic streptomycin. The genes specifying each ribosomal protein and RNA (rRNA) are organized in operons of genes under one promoter, such as the *str* operon (**Fig. 3.6**). The *str* operon was named for streptomycin resistance, a phenotype conferred by mutations in the gene *rpsL*. Mutations conferring streptomycin resistance were known for decades before the structural basis was discovered. These mutations were mapped to the gene *rpsL* encoding the protein S12, whose position in the 30S subunit was shown by X-ray crystallography (see **Fig. 3.5B**). The S12 protein forms part of the ribosome's tRNA acceptor (A) site, which receives the incoming aminoacyl-tRNA. Mutation in the gene encoding S12 generates a protein of altered shape, permitting function in the presence of streptomycin. Thus, genetic analysis and crystallography were combined to demonstrate the precise mode of action of an important antibiotic.

An exciting extension of genetic analysis is the construction of strains with "reporter genes" fused to genes encoding structures of interest. An example of a reporter gene is green fluorescent protein, which fluoresces at the site of the fused protein. Fluorescent reporter genes enable us to observe the function of proteins within a live cell. In Section 3.6, we will discuss the use of fluorescent reporter genes to observe chromosome replication and cell division. The methods of gene manipulation are discussed in Chapters 8 and 12.

> **THOUGHT QUESTION 3.3** What are the advantages and limitations of biochemical and genetic approaches to cell structure and function?

TO SUMMARIZE:

- **Subcellular fractionation** isolates cell parts for structural, biochemical, and genetic analysis.
- **X-ray crystallography** shows the three-dimensional form of cell components at the atomic level.
- **Genetic analysis** shows which genes and proteins are responsible for functions of subcellular complexes such as the ribosome. Mutation of a gene leads to altered function of the cell.

3.3 The Cell Membrane and Transport

The structure that defines the existence of a cell is the cell membrane (**Fig. 3.7**). The cell membrane consists of a phospholipid bilayer containing lipid-soluble proteins. Overall, the membrane serves two kinds of function: It contains the cytoplasm within the external medium, mediating transport between the two; and it carries many proteins with specific functions, such as biosynthetic enzymes and environmental signal receptors.

Membranes Consist of Phospholipids and Proteins

Cell membranes are composed of approximately equal parts phospholipids and proteins. Each phospholipid possesses a charged phosphate-containing "head" that con-

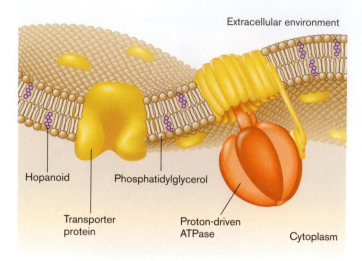

Extracellular environment

Hopanoid

Phosphatidylglycerol

Transporter protein

Proton-driven ATPase

Cytoplasm

Figure 3.7 Bacterial cell membrane. The cell membrane consists of a phospholipid bilayer, with hydrophobic fatty acid chains directed inward, away from water. The bilayer contains stiffening agents such as hopanoids, which serve the same function as cholesterol in eukaryotic membranes. Half the volume of the membrane consists of proteins.

tacts the water interface, as well as a hydrophobic "tail" packed within the bilayer. A phospholipid consists of glycerol with ester links to two fatty acids and a phosphoryl polar head group (for review, see Appendix 2). The polar head group may have a side chain such as ethanolamine in **phosphatidylethanolamine**, the major phospholipid of *E. coli* (**Fig. 3.8**). Lipid biosynthesis is a key process of cells; for example, a bacterial enzyme for fatty acid biosynthesis, enoyl reductase, is the target of triclosan, a common antibacterial additive in detergents and cosmetics.

In the bilayer, all phospholipids face each other tail to tail, keeping their hydrophobic side chains away from the water inside and outside the cell. The two layers of phospholipids in the bilayer are called **leaflets**. One leaflet of phospholipids faces the cell interior, whereas the other faces the exterior. As a whole, the phospholipid bilayer imparts fluidity and gives the membrane a consistent thickness (about 10 nm).

The cell membrane can be thought of as a two-dimensional fluid within which float a vast array of hydrophobic proteins and smaller molecules such as hopanoids (see Fig. 3.7). Membrane proteins serve numerous functions that define the capabilities of the cell, including transport, communication with the environment, and structural support. Examples include:

- **Structural support.** Some membrane proteins anchor together different layers of the cell envelope (see Section 3.4). Other proteins attached to membranes form part of the cytoskeleton. Still others form the base of structures that extend out from the cell, such as pili,

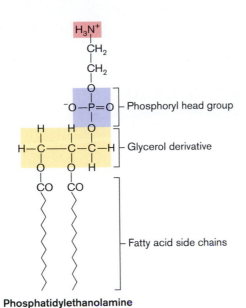

Figure 3.8   Phosphatidylethanolamine. A major bacterial phospholipid consists of glycerol with ester links to two fatty acids and a phosphoethanolamine.

filaments, and flagella; these enable adherence and motility.

■ **Detection of environmental signals.** In *Vibrio cholerae*, the causative agent of cholera, ToxR is a transmembrane protein whose amino-terminal part reaches into the cytoplasm. When ToxR detects acidity and elevated temperature—signs of the host digestive tract—its amino-terminal domain binds to DNA at a sequence that activates expression of cholera toxin and other virulence factors.

■ **Secretion of virulence factors and communication signals.** Membrane proteins form secretion complexes to export toxins and cell signals across the envelope. For example, symbiotic nitrogen-fixing rhizobia require membrane proteins NodI and NodJ to transport nodulation signals out to the host plant roots, where they induce formation of root nodules containing the bacteria.

Proteins embedded within a membrane must have a hydrophobic portion that is soluble within the membrane. Typically, several hydrophobic alpha helices thread back and forth through the membrane. Other regions of peptide extend outside the membrane, containing charged and polar amino acids that interact favorably with water. The combination of hydrophobic and hydrophilic regions effectively locks the protein into the membrane.

Transport across the Cell Membrane

Overall, the cell membrane acts as a barrier to sequester water-soluble proteins of the cytoplasm. The membrane's specific components, particularly proteins, determine which substances cross the membrane between the cytoplasm and the outside. Selective transport is essential for cell survival; it means the ability to acquire scarce nutrients while excluding toxins.

Osmosis. Most cells maintain a concentration of total **solutes** (molecules in solution) that is higher inside the cell than outside. As a result, the internal concentration of water is lower than the concentration outside the cell. Because water can cross the membrane, but its charged solutes cannot, water tends to diffuse across the membrane into the cell, causing expansion of cell volume, in a process called **osmosis.** The resulting pressure on the cell membrane is called **osmotic pressure** (reviewed in Appendix 2). Osmotic pressure will cause a cell to burst, or lyse, in the absence of a countering pressure such as that provided by the cell wall.

Various solutes cross the membrane by different means. Nonspecific transport, or passive diffusion, requires dissolving in the phospholipid bilayer. Specific transport—for example, of nutrients and toxins—requires specific transport proteins. The presence of specific transporters depends on the microbial species and environmental conditions.

Passive diffusion. Small uncharged molecules, such as O_2, CO_2, and water, easily permeate the membrane. Some molecules, such as ethanol, also disrupt the membrane, an action that can make such molecules toxic to cells. By contrast, strongly polar molecules, such as sugars, generally cannot penetrate the hydrophobic interior of the membrane and require transport mediated by specific proteins. Water molecules permeate the membrane, but their rate of passage is increased by protein channels called aquaporins.

Membrane-permeant weak acids and bases. A special case of movement across cell membranes is that of **membrane-permeant weak acids** and **weak bases**, which exist in equilibrium between charged and uncharged forms:

Weak acid: $HA \rightleftharpoons H^+ + A^-$

Weak base: $B + H_2O \rightleftharpoons BH^+ + OH^-$

Weak acids and weak bases cross the membrane in their uncharged form, HA (weak acid) or B (weak base). On the other side, upon reentering aqueous solution, they dissociate (HA to A^- and H^+) or reassociate with H^+ (B to BH^+). In effect, they conduct acid (H^+) or base (OH^-) across the membrane, causing acid or alkali stress. A high-proton concentration outside the cell will increase the amount of uncharged weak acid that can freely enter the cell. Thus, if the H^+ concentration (acidity) outside the cell is higher than inside, it will drive weak acids into the cell.

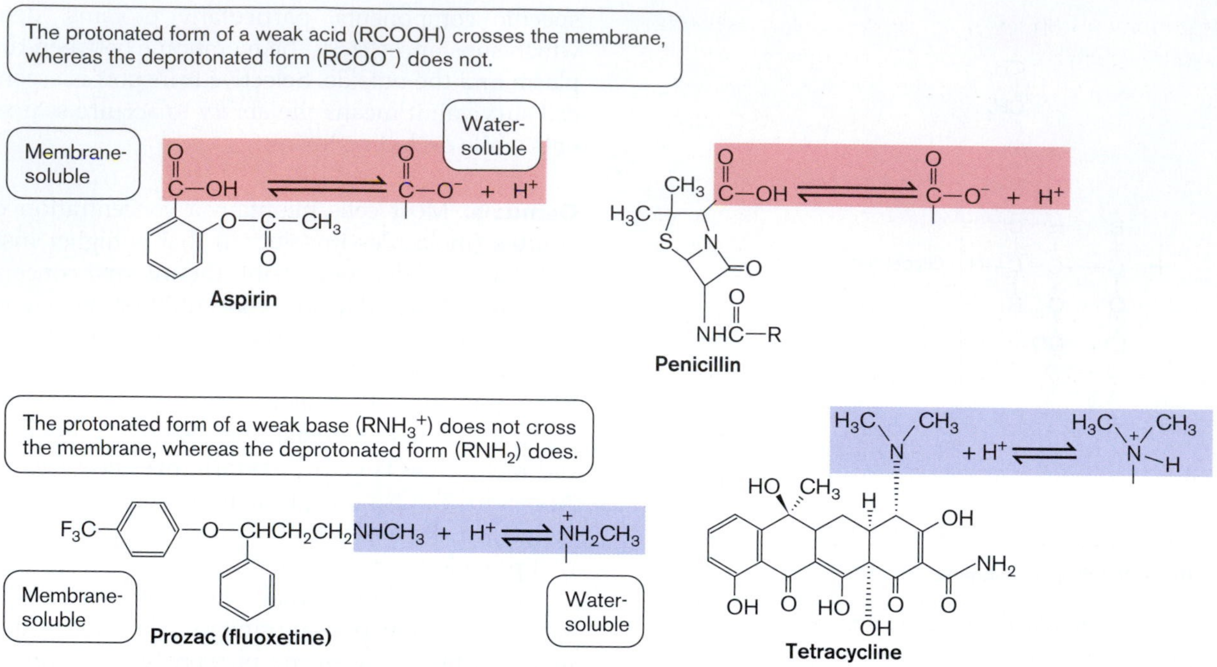

The protonated form of a weak acid (RCOOH) crosses the membrane, whereas the deprotonated form (RCOO⁻) does not.

Aspirin

Penicillin

The protonated form of a weak base (RNH₃⁺) does not cross the membrane, whereas the deprotonated form (RNH₂) does.

Prozac (fluoxetine)

Tetracycline

Figure 3.9   Common drugs are membrane-permeant weak acids and bases. In its charged form, each drug is soluble in the bloodstream. The uncharged form is hydrophobic and penetrates the cell membrane.

Many key substances in cellular metabolism are membrane-permeant weak acids and bases, such as acetic acid. Most pharmaceutical drugs—therapeutic agents delivered to our tissues via the bloodstream—are weak acids or bases whose uncharged forms exist at sufficiently low concentration to cross the membrane without disrupting it. Examples of weak acids that deprotonate (acquiring negative charge) at neutral pH include aspirin (acetylsalicylic acid) and penicillin (**Fig. 3.9**). Examples of weak bases that protonate (acquiring positive charge) at neutral pH include Prozac (fluoxetine) and tetracycline.

> **THOUGHT QUESTION 3.4**   Amino acids have acidic and basic groups that can dissociate. Why are they *not* membrane-permeant weak acids or weak bases? Why do they fail to cross the phospholipid bilayer?

Transport proteins. Proteins enable transport that is highly selective; that is, a given protein will transport only certain nutrients or ions (**Fig. 3.10**). Molecules that carry net charge, such as inorganic ions (Na⁺, Cl⁻), require specific transport by **transport proteins**, or **transporters**. So do organic molecules that carry charge at cytoplasmic pH, such as amino acids.

Transporters are highly specific to different ions and organic substances under different environmental conditions. Much of the character and ecological niche of a microbe—the nutrients it can consume, the drugs it can

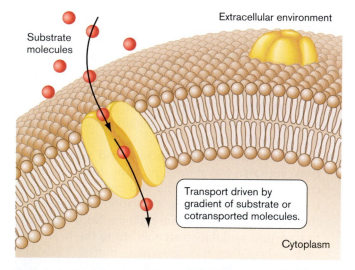

Figure 3.10   Membrane transport proteins. Transport proteins enable the uptake of nutrients into the cell. Uptake can be powered by movement down a gradient of substrate or of a cotransported molecule such as sodium ions. Alternatively, transport may be powered by a coupled chemical reaction that spends energy, such as ATP hydrolysis.

resist—depend on its set of transporters encoded by its genome. Transporters evolve in families common to many species; thus, analysis of known genomes predict the transport capabilities in a newly sequenced genome (discussed in Chapter 8). Organisms that live in complex, changing environments express numerous transporters. For example, the genome of *Streptomyces coelicolor,* a soil-dwelling actinomycete bacterium that produces several antibiotics, shows over 400 different transporters. These include exporters for toxic ions such as arsenate and chromate, uptake of oligosaccharides and peptides, and multiple-drug pumps.

Transport proteins act by various mechanisms, such as a structural channel, or pore, that allows a nutrient molecule to enter the cell or a toxin to be exported. Transport may be passive or active. In **passive transport**, molecules accumulate or dissipate along their concentration gradient. **Active transport**, that is, transport from lower to higher concentration, requires expenditure of energy. The energy for active transport may be obtained by cotransport of another substance down its gradient from higher to lower concentration. For example, many transporters couple uptake of amino acids to uptake of sodium ions. Alternatively, transport may be powered by a coupled chemical reaction that spends energy, such as ATP hydrolysis. Further transport mechanisms are discussed in Chapter 4.

A medically important example of active transport is that of drug efflux proteins powered by the hydrogen ion gradient. Efflux proteins pump antibiotics such as tetracycline out of the bacterial cell, enabling harmful bacteria to grow in the presence of antibiotics. Pathogens and cancer cells evolve multiple drug resistance transporters that enable them to survive chemotherapy.

Transmembrane Ion Gradients Store Energy

Ions such as H^+ or Na^+ usually exist in very different concentrations inside and outside the cell. The ion gradient (ratio of concentrations) across the cell membrane can store energy obtained from metabolizing food or from photosynthesis. For example, membrane proteins such as cytochrome oxidases use energy from respiration to pump hydrogen ions across the cell membrane, generating a hydrogen ion gradient. The hydrogen ion gradient plus the charge difference (voltage potential) across the membrane form an **electrochemical potential**. Because this electrochemical potential includes the hydrogen ion gradient, it is called the proton potential. The proton potential, or proton motive force, drives various cell devices, such as the rotary flagella and efflux pumps that expel antibiotics. Most importantly, the proton potential drives the membrane-bound **ATP synthase (Fig. 3.11)**. The role of the proton potential in metabolism is discussed in detail in Chapters 13 and 14.

The membrane-bound ATP synthase, also called F_1F_o ATP synthase, provides most of the ATP for aerobic respiring cells such as *E. coli;* and essentially the same complex mediates ATP generation in our own mitochondria. The complex of peptides includes a channel (F_o) for

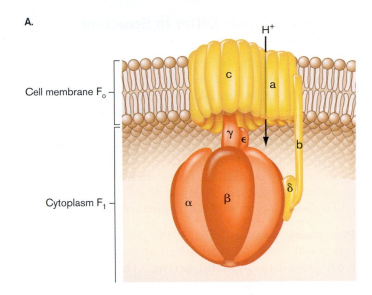

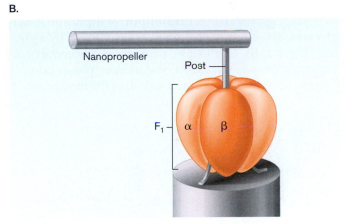

Figure 3.11   Bacterial membrane ATP synthase.   A. The F_1F_o complex of ATP synthase is embedded in the cell membrane. The F_o channel/rotor admits hydrogen ions; the H^+ current flow across the membrane is driven by the concentration and charge differences between inside and outside the cell. The flow of H^+ causes the rotation of F_1. Rotation of F_1 drives conversion of ADP + P_i to ATP. **B.** An artificial "biomolecular motor" was built from an ATPase F_1 unit attached to a nickel post and a nanopropeller.

hydrogen ions, which drive rotation of the ATPase complex (F_1). Rotation of F_1 mediates the formation of ATP.

The idea that a living organism could contain rotating parts was highly controversial when such parts were first discovered in bacterial flagella (discussed in Section 3.7). Before this discovery, scientists had believed that living body parts could not rotate and that only humans had "invented the wheel." The discovery of rotary biomolecules has inspired advances in nanotechnology, the engineering of microscopic devices. For example, a "biomolecular motor" was devised using an ATP synthase F_1 complex to drive a metal submicroscopic propeller (**Fig. 3.11B**). In the future, such biomolecular design may be used to create microscopic robots that enter the bloodstream to perform microsurgery.

Membrane Lipids Differ in Structure and Function

Membranes require a certain uniformity to maintain structural integrity and function. Yet individual membrane lipids differ remarkably in structure. Membrane lipids help determine whether an organism grows in a Yellowstone hot spring or in the human lungs, where it may cause pneumonia. Phospholipids vary with respect to their phosphoryl head groups and with respect to their hydrocarbon side chains. Some membrane lipids lack phosphate altogether, substituting other polar groups; and some replace mobile side chains with fused rings.

Phosphoryl head groups. At the pH of most biological systems, the phosphoryl head group carries net negative charge, called a **phosphatidate** (**Fig. 3.12A**). The negatively charged phosphatidate can contain various organic groups, such as glycerol to form **phosphatidylglycerol**. A more complex phosphatidate is **cardiolipin**, or **diphosphatidylglycerol**, actually a double phospholipid linked by a glycerol (**Fig. 3.12B**). Cardiolipin increases in concentration in bacteria grown to starvation or stationary phase, possibly because its extended structure stabilizes the membrane.

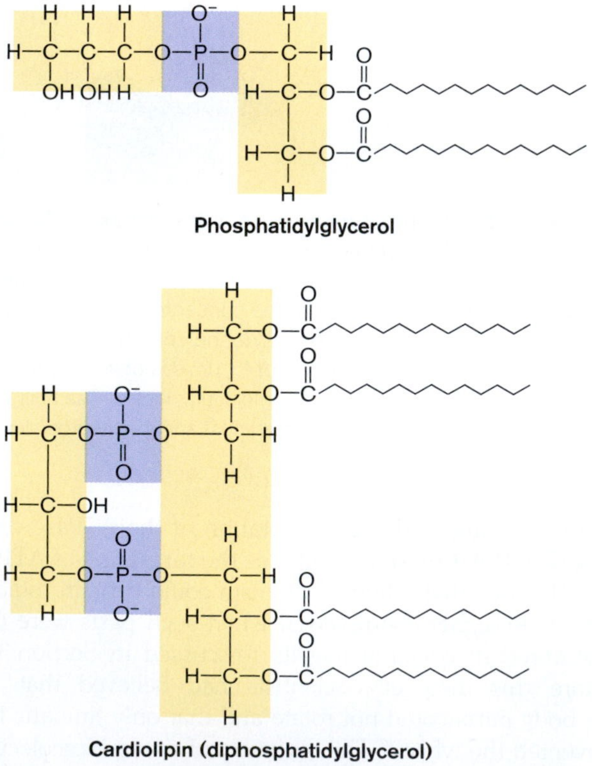

Other phospholipids have a positively charged head group, such as phosphatidylethanolamine. Phospholipids with positive charge or with mixed charges are concentrated in portions of the membrane that interact with DNA, which has negative charge. Also, some transporter proteins require interaction with positively charged phospholipids.

The fatty acid component of phospholipids also varies greatly among bacteria. Certain soil bacteria and pathogens such as *Mycobacterium tuberculosis* have several hundred different kinds of fatty acids. Fatty acid structures that stiffen the membrane are increased under stress conditions, such as starvation or acidity.

Variant fatty acid structures in phospholipids enhance survival under environmental stress. The most common bacterial fatty acids are hydrogenated chains of varying length, typically between 6 and 22 carbons. These chains pack neatly to form a smooth layer. Some fatty acid chains, however, are partly unsaturated (possess one or more carbon-carbon double bonds). If the unsaturated bond is *cis*, meaning that both alkyl chains are on the same side of the bond, the unsaturated chain has a "kink," as in oleic acid (**Fig. 3.13**). The kinked chains do not pack as closely as the straight hydrocarbon chains and thus make the membrane more "fluid." This is why, at room temperature, unsaturated vegetable oils are fluid, whereas highly saturated butterfat is solid. The enhanced fluidity of a kinked phospholipid improves the function of the membrane at low temperature; hence, bacteria can respond to cold and heat by increasing or decreasing their synthesis of unsaturated phospholipids.

Another interesting structural variation is cyclization of part of the chain to create a stiff planar ring with decreased fluidity. The double bond of unsaturated fatty acids can incorporate a carbon from *S*-adenosyl-L-

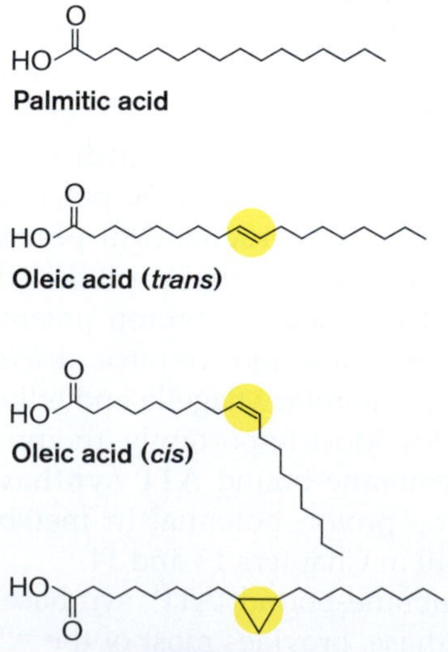

Phosphatidylglycerol

Cardiolipin (diphosphatidylglycerol)

Figure 3.12 Phospholipid head groups. Bacterial membranes contain phospholipids with several kinds of polar head groups. Common examples include phosphatidylglycerol (negatively charged, a phosphatidate), cardiolipin (a double phospholipid joined by a third glycerol), and phosphatidylethanolamine (with a positively charged amine).

Palmitic acid

Oleic acid (*trans*)

Oleic acid (*cis*)

Cyclopropane fatty acid

Figure 3.13 Phospholipid side chains. Bacterial lipid side chains include palmitic acid, oleic acid, and cyclopropane fatty acid.

methionine to form a three-membered ring, becoming a cyclopropane fatty acid (see **Fig. 3.13**). Bacteria convert unsaturated fatty acids to cyclopropane during starvation and acid stress, conditions under which membranes require stiffening. Cyclopropane conversion is an important factor in the pathogenesis of *Mycobacterium tuberculosis* (causative agent of tuberculosis) and in the acid resistance of food-borne toxigenic *E. coli*. Many other structural variants are seen in different species, such as branched chains, hydroxyl and sulfate groups, and polycyclic groups. The less common forms are highly characteristic of particular species and strains. Thus, fatty acid profiles are used to identify certain kinds of pathogens, such as anthrax spores.

Terpene derivatives stiffen membranes. In addition to phospholipids, membranes include planar molecules that fill gaps between hydrocarbon chains (**Fig. 3.14**). These stiff, planar molecules reinforce the membrane, much as steel rods reinforce concrete. In eukaryotic membranes, the reinforcing agents are sterols, such as **cholesterol**. In bacteria, the same function is filled by pentacyclic (five-ring) hydrocarbon derivatives called **hopanoids**, or **hopanes**. Hopanoids appear in geological sediments, where they indicate ancient bacterial decomposition; they provide useful data for petroleum exploration.

Variations in phospholipid side-chain structures reach their extreme in archaea, many of which inhabit the planet's most extreme environments. All archaeal phospholipids replace the ester link between glycerol and fatty acid

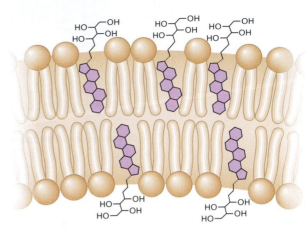

Figure 3.14   Hopanoids add strength to membranes. Hopanoids fit between the fatty acid side chains of membranes and limit their motion, thus stiffening the membrane. A hopanoid has five fused rings and an extended hydroxylated tail.

with an ether link, C—O—C (**Fig. 3.15**). Ethers are much more stable than esters, which hydrolyze easily in water. Another modification is that archaeal hydrocarbon chains are branched **terpenoids**, polymeric structures derived from isoprene, in which every fourth carbon extends a methyl branch. The branches strengthen the membrane by limiting movement of the hydrocarbon chains.

The most extreme hyperthermophiles, which live beneath the ocean at 110°C, have terpenoid chains linked at the tails, creating a tetraether monolayer. In some

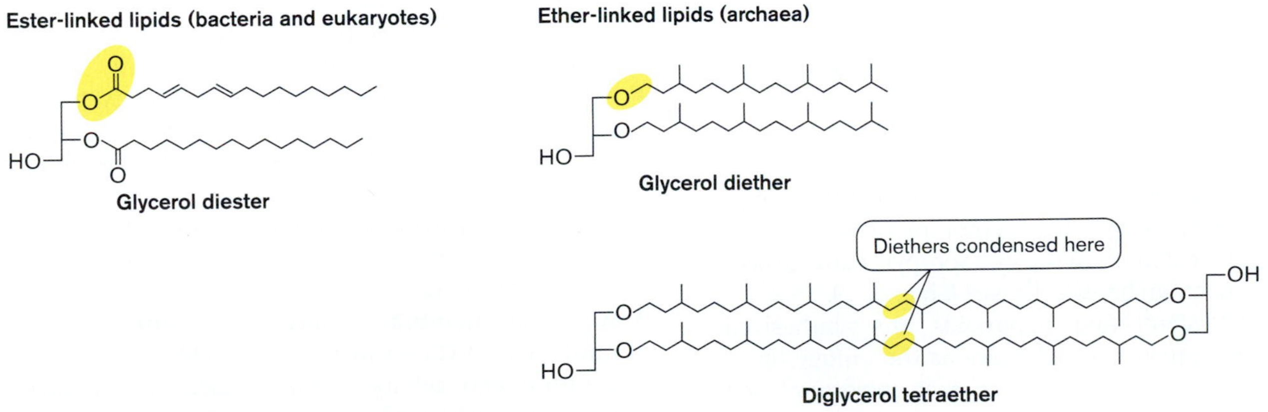

Ester-linked lipids (bacteria and eukaryotes)

Glycerol diester

Ether-linked lipids (archaea)

Glycerol diether

Diethers condensed here

Diglycerol tetraether

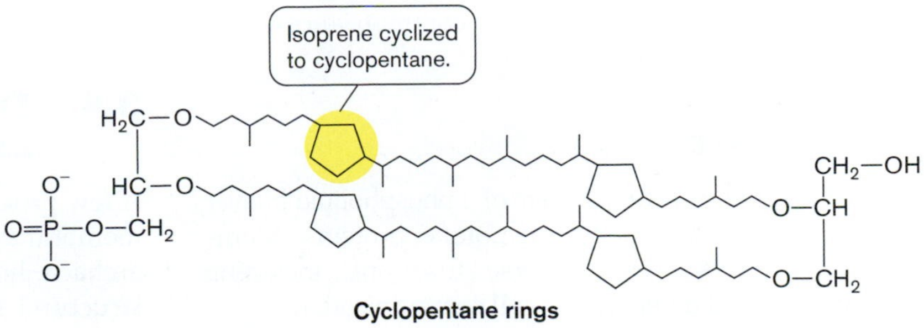

Isoprene cyclized to cyclopentane.

Cyclopentane rings

Figure 3.15   Terpene-derived lipids of archaea. In archaea, the hydrocarbon chains are ether-linked to glycerol, and every fourth carbon has a methyl branch. In some archaea, the tails of the two facing lipids of the bilayer are fused, forming tetraethers; thus, the entire membrane consists of a single layer of molecules (a monolayer).

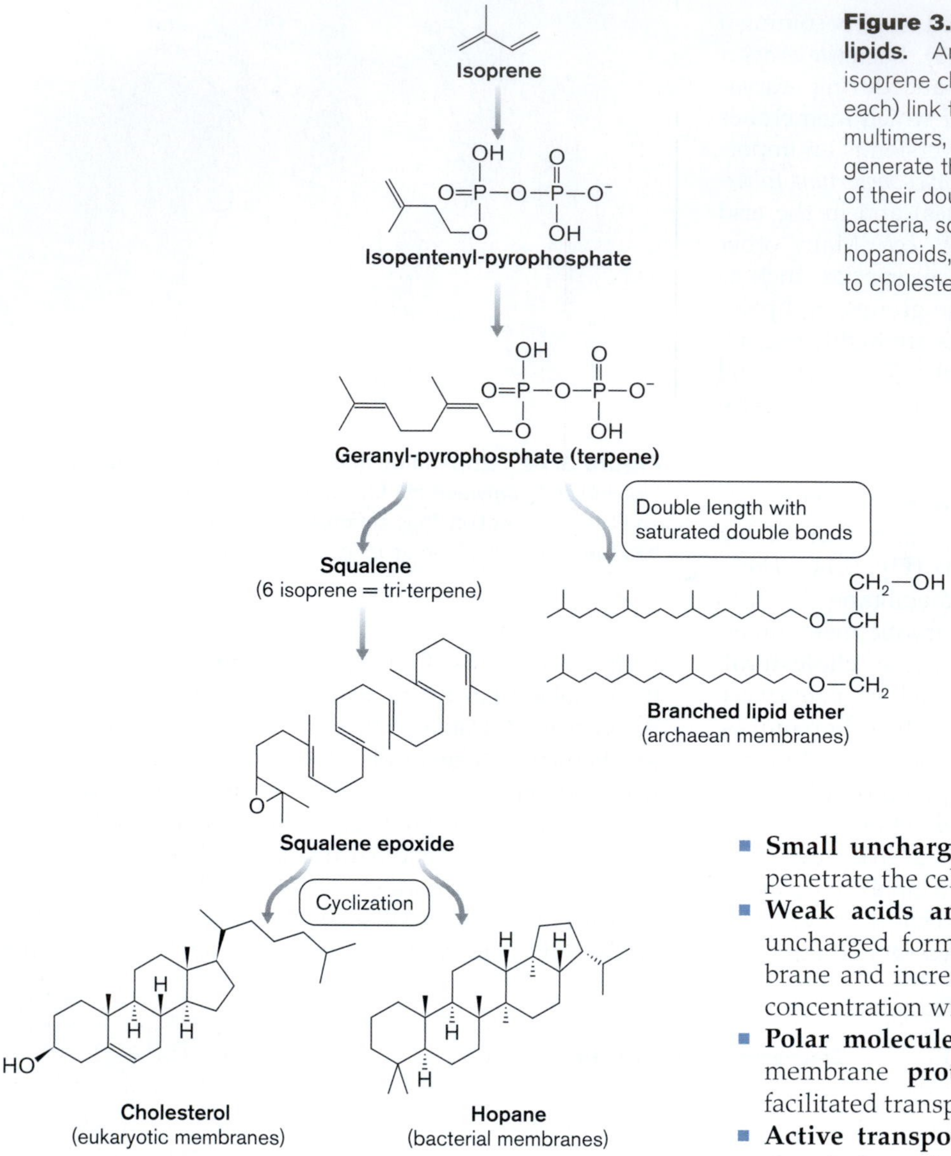

Figure 3.16 Synthesis of terpene-derived lipids. Archaeal lipids are synthesized from isoprene chains. Two isoprene units (five carbons in each) link to form a terpene (ten carbons). Terpenoid multimers, such as the triterpene derivative squalene, generate the lipid chains of archaea, in which most of their double bonds have been hydrogenated. In bacteria, squalene cyclizes to form hopanes and hopanoids, and in eukaryotes, squalene is converted to cholesterol.

species, the terpenoids cyclize to form cyclopentane rings. These planar rings stiffen the membrane under stress to an even greater extent than the cyclopropyl chains of bacteria. For more on archaeal cells, see Chapter 19.

Both cholesterol and hopanoids are synthesized from the same precursor molecules as the unique lipids of archaea (**Fig. 3.16**). It may be that cholesterol and hopanoids persist in bacteria and eukaryotes as derivatives of lipids once possessed by a common ancestor of bacteria, eukaryotes, and archaea.

TO SUMMARIZE:

■ **The cell membrane** consists of a phospholipid bilayer containing hydrophobic membrane proteins. Membrane proteins serve diverse functions, including transport, cell defense, and cell communication.

■ **Small uncharged molecules**, such as oxygen, can penetrate the cell membrane by diffusion.

■ **Weak acids and weak bases** exist partly in an uncharged form that can diffuse across the membrane and increase or decrease, respectively, the H^+ concentration within the cell.

■ **Polar molecules and charged molecules** require membrane **proteins to mediate** transport. Such facilitated transport can be active or passive.

■ **Active transport** requires input of energy from a chemical reaction or from an ion gradient across the membrane. **Ion gradients** generated by membrane pumps store energy for cell functions.

■ **Diverse fatty acids** are found in different microbial species and in microbes grown under different environmental conditions.

■ **Archaeal membranes have ether-linked terpenoids**, which confer increased stability at high temperature and acidity. Some species have cyclized terpenoids, which generate a lipid monolayer.

3.4 The Cell Wall and Outer Layers

A few prokaryotes, such as the mycoplasmas, have a cell membrane with no outer layers. In most bacteria and archaea, however, the cell envelope includes at least one structural supporting layer outside the cell membrane.

A.

J. V. Holtje

5 µm

B.

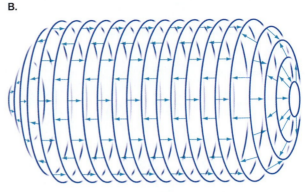

Figure 3.17   The peptidoglycan sacculus.   A. Isolated sacculus from *Escherichia coli* (TEM). **B.** The structure of the sacculus consists of glycan chains (parallel rings) linked by peptides (arrows). The spaces between links are open, porous to large molecules. **C.** Sacculus structure resembles the skeleton of a geodesic dome such as that of Epcot Center at Walt Disney World.

C.

©Kit Kittle/Corbis

The most common structural support is the cell wall. Many species possess additional coverings, such as an outer membrane and capsule.

The Cell Wall Is a Single Molecule

The cell wall confers shape and rigidity to a bacterial cell and helps it withstand the intracellular turgor pressure that can build up as a result of osmotic pressure. The bacterial cell wall, also known as the **sacculus**, consists of a single interlinked molecule that encloses the entire cell. The sacculus has been isolated from *E. coli* and visualized by TEM (**Fig. 3.17A**). In the image shown, the isolated sacculus appears flattened on the sample grid like a deflated balloon. Its geometric structure encloses maximal volume with minimal surface area, analogous to a geodesic dome such as that of Epcot Center. The properties of the cell wall, or sacculus, are largely opposite to those of the membrane: a unimolecular, cage-like structure (**Fig. 3.17B**), highly porous to ions and small organic molecules.

Peptidoglycan structure. Most bacterial cell walls are composed of peptidoglycan, a polymer of peptide-linked chains of amino sugars. Peptidoglycan is synonymous with **murein** ("wall molecule") and consists of parallel polymers of disaccharides cross-linked with peptides of four amino acids, called **glycan** chains (**Fig. 3.18**). Peptidoglycan, or murein, is unique to bacteria, although archaea build analogous structures whose overall physical nature is similar. (Archaeal cell wall structures are presented in Chapter 19.)

Glycan chains are linked by peptide cross-bridges. The long chains of peptidoglycan consist of repeating units of the disaccharide composed of *N*-acetylglucosamine (an amino sugar derivative) and *N*-acetylmuramic acid

(glucosamine plus a lactic acid group; see **Fig. 3.18**). The lactate group of muramic acid forms an amide link with the amino terminus of a short peptide containing four to six amino acid residues. The peptide extension can form **cross-bridges** connecting parallel strands of glycan.

The peptide contains two amino acids in the unusual D mirror form, D-glutamate and D-alanine, a hint that peptidoglycan evolved very early in life's history, before the bias toward the L form had been established. The third amino acid, *m*-diaminopimelic acid, has an extra amine group. It is also used by cells as a biosynthetic precursor of lysine. The fourth and sometimes the fifth amino acids are D-alanine. The second amino group of *m*-diaminopimelic acid is needed to form a cross-link with the COOH of D-alanine from a neighboring peptide. The cross-link forms by removal of a second D-alanine (fifth in the chain). The cross-linked peptides of neighboring glycan strands form the cage of the sacculus.

The details of peptidoglycan structure vary among bacterial species. Some gram-positive species, such as *Staphylococcus aureus* (a cause of toxic shock syndrome) have peptides linked by bridges of pentaglycine instead of the D-alanine link to *m*-diaminopimelic acid. In gram-negative species, the *m*-diaminopimelic acid is linked to the outer membrane, as discussed shortly.

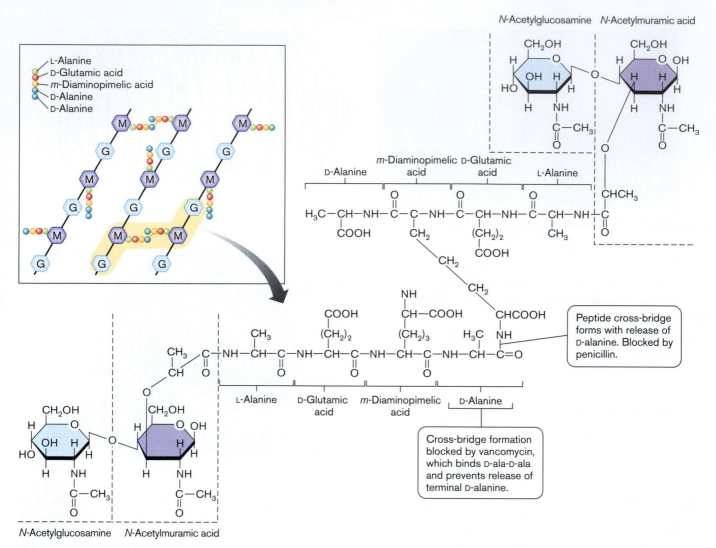

Figure 3.18 Peptidoglycan cross-bridge formation.   A disaccharide unit of glycan has an attached peptide of four to six amino acids. The amino terminus of the peptide forms an amide bond with the lactate group of muramic acid. On the peptide, the extra amino group of *m*-diaminopimelic acid can cross-link to the carboxyl terminus of a neighboring peptide. The addition of D-alanine to the peptide is blocked by vancomycin, and the cross-bridge formation by transpeptidase is blocked by penicillin.

Peptidoglycan synthesis as a target for antibiotics. Synthesis of peptidoglycan requires many genes encoding enzymes to make the special sugars, build the peptides, and seal the cross-bridges. Because peptidoglycan is unique to bacteria, these biosynthetic enzymes make excellent targets for antibiotics (see **Fig. 3.18**). For example, the transpeptidase that cross links the peptides is the target of penicillin. Vancomycin, a major defense against *Clostridium difficile* and drug-resistant staphylococci, prevents cross-bridge formation by binding the terminal D-ala-D-ala dipeptide, thus preventing release of the terminal D-alanine.

Unfortunately, the widespread use of such antibiotics selects for evolution of resistant strains. One of the most common agents of resistance is the enzyme beta-lactamase, which cleaves the lactam ring of penicillin, rendering it ineffective as an inhibitor of transpeptidase.

Strains resistant to vancomycin, on the other hand, contain an altered enzyme that adds lactic acid to the end of the branch peptides in place of the terminal D-alanine. The altered enzyme is no longer blocked by vancomycin. As new forms of drug resistance emerge, researchers continue to seek new antibiotics that target cell wall formation (discussed in Chapter 27).

Gram-Positive and Gram-Negative Bacteria

Most bacteria have additional envelope layers that provide structural support and protection from predators and host defenses (**Fig. 3.19**). Additional molecules are attached to the cell wall and cell membrane, in some cases interpenetrating them. Envelope composition defines two major

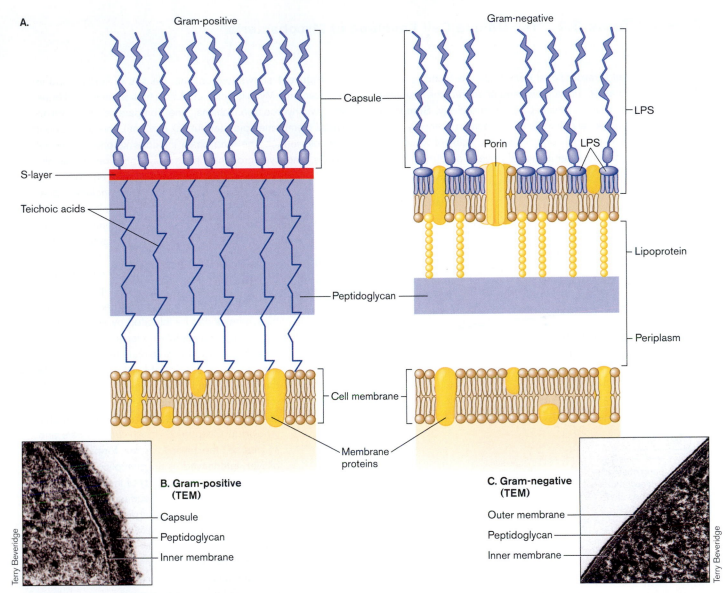

A.

Gram-positive

Gram-negative

S-layer

Teichoic acids

Capsule

LPS

Porin

LPS

Peptidoglycan

Lipoprotein

Periplasm

Cell membrane

Membrane
proteins

Terry Beveridge

**B. Gram-positive
(TEM)**

Capsule

Peptidoglycan

Inner membrane

**C. Gram-negative
(TEM)**

Outer membrane

Peptidoglycan

Inner membrane

Terry Beveridge

Figure 3.19   Cell envelope: gram-positive and gram-negative.   A. The gram-positive cell has a thick cell wall with multiple layers of peptidoglycan, threaded by teichoic acids. The cell wall may be covered by an S-layer. In some gram-positive species, carbohydrate filaments form a capsule. The gram-negative cell has a single layer of peptidoglycan covered by an outer membrane. Some gram-negative species include an S-layer or a capsule (not shown). The cell membrane of gram-negative species is called the inner membrane. **B.** Gram-positive envelope of *Bacillus subtilis* (TEM), showing cell membrane, cell wall, and capsule. **C.** Gram-negative envelope of *Pseudomonas aeruginosa* (TEM), showing inner membrane, thin cell wall in the periplasm, and outer membrane.

categories of bacteria distinguished by the Gram stain (discussed in Chapter 2):

■ **Gram-positive bacteria** have a thick cell wall with multiple layers of peptidoglycan, interpenetrated by teichoic acids. Gram-positive species make up the phylum Firmicutes, such as *Bacillus thuringiensis* and *Streptococcus pyogenes,* the cause of "strep throat."

■ **Gram-negative bacteria** have a thin cell wall (single layer of peptidoglycan) enclosed by an outer membrane. Gram-negative species make up the phylum Proteobacteria, such as *Escherichia coli* and nitrogen-fixing rhizobia.

Outside these two groups, however, many bacterial species do not fit the Gram stain models; consider, for example, the mycobacterial envelope (**Special Topic 3.1**). Archaeal cell envelopes are highly diverse, and they do not fit the Gram models.

THOUGHT QUESTION 3.5   The actual thickness of the cell wall is difficult to determine based solely on electron microscopic observation of the envelope layers. Devise a biochemical experiment that can show the number of layers of peptidoglycan in cells of a given species.

Special Topic 3.1   The Unique Cell Envelope of Mycobacteria

Cell envelopes of exceptional complexity are found in mycobacteria, such as *M. tuberculosis* and *M. leprae*, the causative agents of tuberculosis and leprosy, respectively. Some features of mycobacterial cell walls also appear in actinomycetes, a large and diverse family of soil bacteria, many of which produce antibiotics and other industrially useful products.

Figure 1 shows two unusual classes of lipids found in mycobacteria. Mycolic acids contain a hydroxy acid backbone with two hydrocarbon chains, one comparable in length to typical membrane lipids (about 20 carbons), the other about threefold longer. The long chain includes ketones, methoxyl groups, and cyclopropane rings. Only one type of mycolic acid is shown; hundreds of different forms are known. The phenolic glycolipids include a phenol group, also linked to sugar chains.

The mycobacterial cell wall includes peptidoglycan linked to chains of galactose, called galactans (**Fig. 2**). The galactans are attached to arabinans, polymers of the five-carbon sugar arabinose. The arabinan-galactan polymers are known as arabinogalactans. Arabinogalactan biosynthesis can be inhibited by ethambutol, a major drug against tuberculosis.

The ends of the arabinan chains form ester links to mycolic acids. Mycolic acids provide the basis for acid-fast staining, in which cells retain the dye carbolfuchsin, an important diagnostic test for mycobacteria and actinomycetes (shown in Chapter 2). In *M. tuberculosis* and *M. leprae*, the mycolic acids form a kind of bilayer interleaved with phenolic glycolipids. The extreme hydrophobicity of the phenol derivatives generates a waxy surface that prevents phagocytosis by macrophages.

Overall, the thick, waxy envelope excludes many antibiotics and offers exceptional protection from host defenses, enabling the pathogens of tuberculosis and leprosy to colonize their hosts over long periods. However, the mycobacterial envelope also retards uptake of nutrients. As a result, *M. tuberculosis* and *M. leprae* grow extremely slowly and are a challenge to culture in the laboratory.

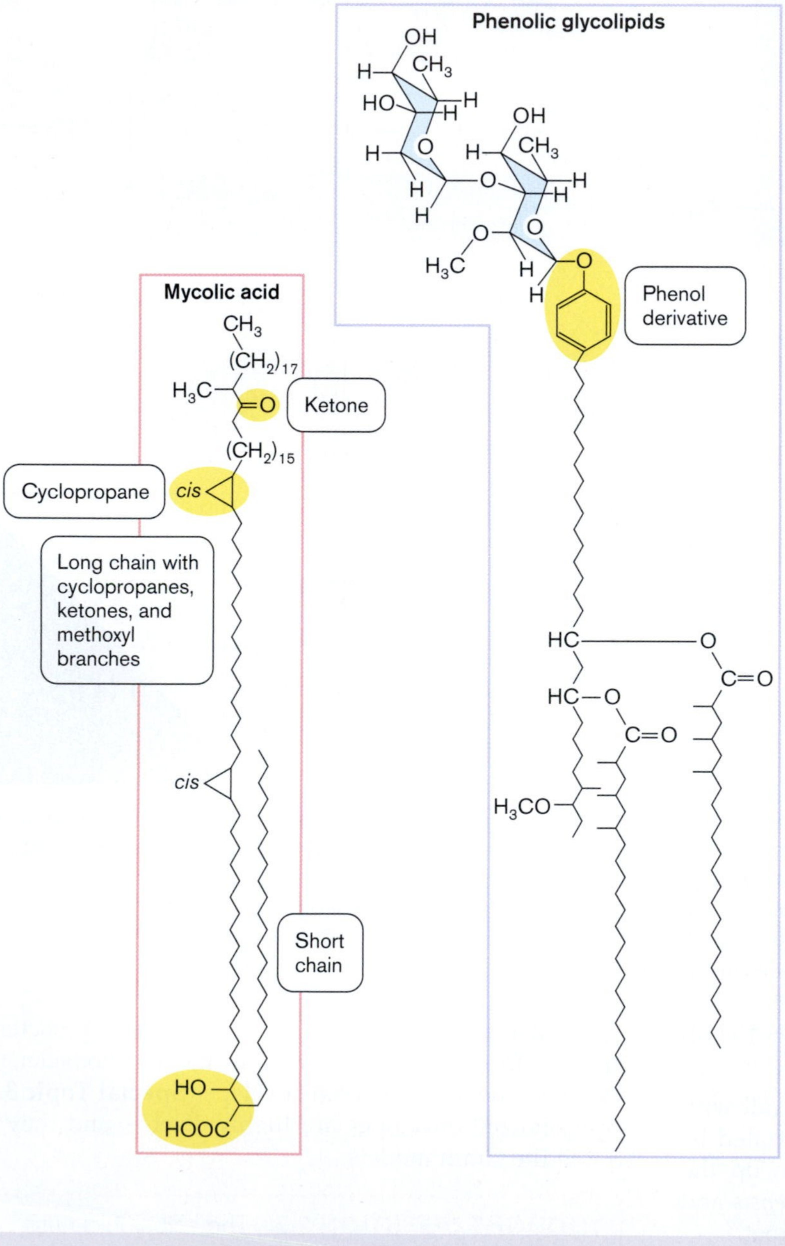

Figure 1   Mycobacterial envelope lipids. *Mycobacterium tuberculosis* possesses unusual envelope components, such as mycolic acids and phenolic glycolipids. These highly hydrophobic molecules form a waxy outer coating that protects cells from host defenses.

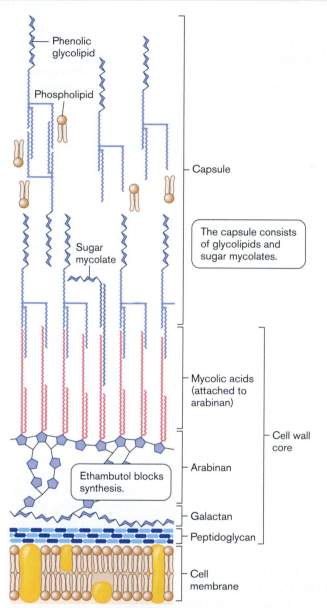

Figure 2 Mycobacterial envelope structure. A complex cell wall includes a peptidoglycan layer linked to a chain of galactose polymer (galactan) and arabinose polymer (arabinan). Arabinan forms ester links to mycolic acids, which form an outer bilayer with phenolic glycolipids. Outside the outer bilayer is a capsule of loosely associated phospholipids and phenolic glycolipids.

The Gram-Positive Cell Envelope

A section of a gram-positive cell envelope is shown in **Figure 3.19**. The gram-positive cell wall consists of multiple layers of peptidoglycan, up to 40 in some species. No space appears between the peptidoglycan layers, although the outer band of envelope appears more dense. The peptidoglycan is reinforced by **teichoic acids** threaded through its multiple layers (**Fig. 3.20**). Teichoic acids are chains of phosphodiester-linked glycerol or ribitol, with sugars or amino acids linked to the middle OH groups. The negatively charged cross-threads of teichoic acids, as well as the overall thickness of the gram-positive cell wall, help retain the Gram stain.

Outside the cell wall, gram-positive cells are often encased in a slippery **capsule** consisting of loosely bound polysaccharides. The capsule can be visualized under the light microscope by using a negative stain consisting of suspended particles of India ink, which the capsule excludes (shown in **Fig. 2.24C**).

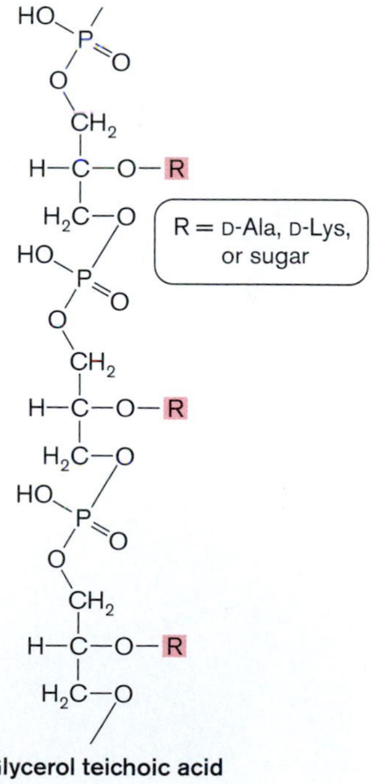

Glycerol teichoic acid

Figure 3.20 Teichoic acids. Teichoic acids in the gram-positive cell wall consist of glycerol or ribitol phosphodiester chains. The middle hydroxyl group of each glycerol or ribitol is typically linked to D-alanine, to D-lysine, or to a sugar such as galactose or N-acetylglucosamine.

The S-layer. An additional protective layer commonly found in gram-positive cells, as well as in archaea, is the surface layer, or **S-layer**. The S-layer is a crystalline layer of thick subunits consisting of protein or glycoprotein (proteins with attached sugars) (**Fig. 3.21**). S-layers are found in most archaea, as well as in many gram-positive and gram-negative bacteria freshly isolated from natural sources.

Each subunit of the S-layer contains a pore large enough to admit a wide range of molecules. The subunits form a smooth layer on the cell wall or outer membrane (see **Fig. 3.21**). The proteins are arranged in a highly ordered array, either hexagonally or tetragonally. They present a formidable physical barrier to predators or parasites. The S-layer is rigid, but it also flexes and allows substances to pass through it in either direction.

If S-layers are common, why were they observed only recently? Bacterial cell structures are generally studied in laboratory cultures. Upon repeated subculturing, however, bacteria lose the genetic ability to synthesize S-layer proteins. Genetic loss of a structure not required in laboratory culture, resulting from accumulation of deleterious mutations, is an example of **reductive evolution**, also known as **degenerative evolution**. Loss of a trait occurs in the absence of selective pressure for genes encoding the trait (discussed in Chapter 17). For example, the mycoplasmas are close relatives of gram-positive bacteria, yet they have permanently lost their cell walls as well as the S-layer. Mycoplasmas have no need for cell walls because they are parasites living in host environments, such as the human lung, where they are protected from osmotic shock.

Figure 3.21 The S-layer. The archaeon *Thermoproteus tenax* has a single tetraether membrane encased by an S-layer (SEM). Note the regular pattern of tiled proteins.

Paul Messner, et al. 1986. *Journal of Bacteriology* 166(3):1046

100 nm

> **THOUGHT QUESTION 3.6** Why would laboratory culture conditions select for evolution of cells lacking an S-layer?
>
> **THOUGHT QUESTION 3.7** Suppose that one cell out of a million has a mutant gene blocking S-layer synthesis, and suppose that the mutant strain can grow twice as fast as the S-layered parent. How many generations would it take for the mutant strain to constitute 90% of the population?

The Gram-Negative Outer Membrane

A gram-negative cell envelope is seen in **Fig. 3.19C**. The thin layer of peptidoglycan is believed to be a single sheet based on calculations of molecular density. The peptidoglycan is covered by an outer membrane. The gram-negative outer membrane confers defensive abilities and toxigenic properties on many pathogens, such as *Salmonella* species and enterohemorrhagic *E. coli* (strains that cause hemorrhaging of the colon). Between the outer and inner (cell) membranes is the periplasm.

Lipoproteins and lipopolysaccharide (LPS). The inward-facing leaflet of the outer membrane has a phospholipid composition similar to that of the cell membrane (in gram-negative species, it is called the **inner membrane** or inner cell membrane). The outer membrane's inward-facing leaflet includes lipoproteins that connect the outer membrane to the peptide bridges of the cell wall. The major lipoprotein is called **murein lipoprotein**, also known as Braun lipoprotein (**Fig. 3.22A**). Murein lipoprotein consists of a protein with an N-terminal cysteine triglyceride inserted in the inward-facing leaflet of the outer membrane. The C-terminal lysine forms a peptide bond with the *m*-diaminopimelic acid of peptidoglycan (murein).

The outward-facing leaflet, however, consists of a special kind of phospholipid called **lipopolysaccharide** (**Fig. 3.22B**). The LPS lipids have shorter fatty acid chains than those of the inner cell membrane, and some are branched. LPS is of crucial medical importance because it acts as an **endotoxin**. An endotoxin is a cell component that is harmless so long as the pathogen remains intact, but when released by a lysed cell, endotoxin overstimulates host defenses, inducing potentially lethal endotoxic shock. Thus, antibiotic treatment of an LPS-containing pathogen can kill the cells but can also lead to death of the patient.

In LPS the fatty acids are esterified to **glucosamine**, an amino sugar found also in peptidoglycan. Two glucosamine dimers each condense with two fatty acid chains, which makes four fatty acid chains in all. Like the glycerol of phospholipid, each glucosamine of LPS connects to a phosphate, whose negative charge interacts with water.

Figure 3.22 Lipoprotein and lipopolysaccharide.
A. Murein lipoprotein consists of a protein with an N-terminal cysteine triglyceride inserted in the inward-facing leaflet of the outer membrane. The protein's C-terminal lysine forms a peptide bond with the *m*-diaminopimelic acid of the peptidoglycan (murein) cell wall. **B.** Lipopolysaccharide (LPS) consists of branched and unbranched short-chain fatty acids linked to a dimer of phosphoglucosamine (also called disaccharide diphosphate). The disaccharide is linked to a core polysaccharide extending out from the cell, which is attached to about 40 repeating units of a distinct polysaccharide known as O antigen.

One of the glucosamines is attached to a **core polysaccharide**, a sugar chain that extends outside the cell. The core polysaccharide consists of about five sugars with side chains such as phosphoethanolamine. It extends to the **O polysaccharides**, a chain of as many as 200 sugars. The O polysaccharide may extend longer than the cell itself. These polysaccharide chains form a layer that helps bacteria resist phagocytosis by white blood cells.

Proteins of the outer membrane. The outer membrane also contains unique proteins not found in the inner membrane. The various layers of the cell envelope need to maintain distinct identities and properties that distinguish them from each other and from the cytoplasm and periplasm. The composition of the inner and outer membranes and aqueous compartments is characterized by ultracentrifugation. Fractions containing various cell components show that each protein is confined to a specific component of the cell (**Table 3.2**). For example, the proton-translocating ATPase is found

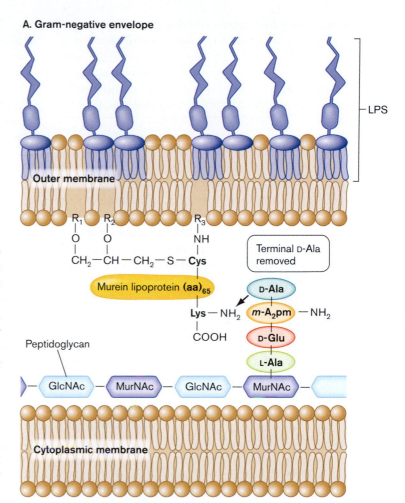

A. Gram-negative envelope

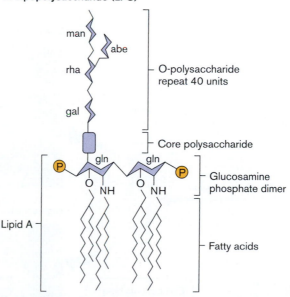

B. Lipopolysaccharide (LPS)

Table 3.2 Subcellular location of proteins in gram-negative bacteria.

Cell fraction	Proteins (examples)
Cytoplasm	Glycolytic enzymes
	Biosynthesis of amino acids
	Cytoplasmic chaperones
Inner membrane (cell membrane)	Proton-translocating ATPase
	Electron transport chain
	Porins for water and glycerol
Periplasm (periplasmic space)	Sugar taxis receptors
	Periplasmic chaperones
Outer membrane	Porins for sugars and peptides
	Sugar-binding proteins

only in the inner membrane fractions, whereas sugar-binding proteins are only in the outer membrane.

THOUGHT QUESTION 3.8   Why would the proteins listed in **Table 3.2** be confined to specific cell fractions? Why could a protein not function throughout the cell?

Outer membrane proteins of pathogens are important targets for vaccine development. For example, the outer membrane proteins were characterized for *Borrelia burgdorferi*, the cause of Lyme disease. In the experiment shown in **Figure 3.23**, fractions of *Borrelia* outer membrane are separated from whole cells by density gradient centrifugation. The sample is located on a tube of sucrose solution whose concentration (and density) increases with depth in the tube. Within the gradient, suspended fragments of membrane migrate to the position where their density equals that of the solution. The outer membrane fractions collect at a density different from that of fractions containing unbroken cells. The gradient contents are then removed sequentially through an opening at the base of the tube. Once isolated, the outer membrane components are separated by gel electrophoresis to reveal distinctive proteins that may serve as vaccine targets.

NOTE: Another type of gradient centrifugation, called equilibrium density gradient centrifugation, uses a cesium chloride solution that forms a gradient when subjected to centrifugal force. Cesium chloride gradients are used most commonly to separate DNA molecules such as plasmids (discussed in Chapter 7).

Outer membranes contain a class of transporters called **porins** that permit entry of nutrients such as sugars and peptides. Outer membrane porins have a distinctive cylinder of beta sheets, also known as a beta barrel. A typical outer membrane porin exists as a trimer of beta barrels, each of which acts as a pore for nutrients. **Figure 3.24** shows a model of the sucrose porin based on a crystal structure.

Cells express different outer membrane porins under different environmental conditions. In a dilute environment, cells express porins of large pore size, maximizing the uptake of nutrients. In a rich environment—for example, within a host—cells downregulate expression of large porins and express porins of smaller pore size, selecting only smaller nutrients and avoiding the uptake of toxins. For example, the porin regulation system of gram-negative bacteria enables them to grow in the intestinal region containing bile salts—a hostile environment for gram-positive bacteria.

Periplasm. The outer membrane is porous to most ions and many small organic molecules, but it prevents passage of proteins and other macromolecules. Thus, the region between the inner and outer membranes of gram-negative cells, including the cell wall, defines a separate membrane-bounded compartment of the cell known as the periplasm (see **Fig. 3.19**). The periplasm contains specific enzymes and nutrient transporters not found within the cytoplasm, such as periplasmic binding protein for sugars, amino acids, or other nutrients. Periplasmic proteins are subjected to fluctuations in pH and salt concentration, because the outer membrane is porous to ions.

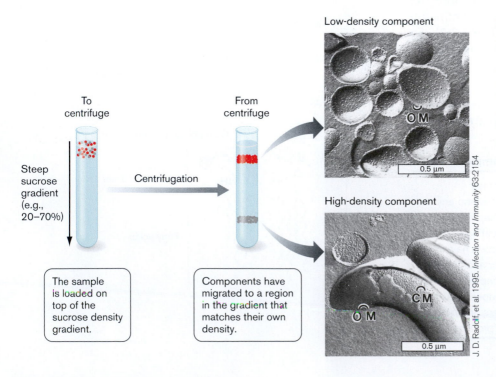

Figure 3.23 Outer membrane analysis by centrifugation. *Borrelia burgdorferi* outer membrane vesicles are separated from the cell by equilibrium density gradient ultracentrifugation. The lighter fractions contain outer membrane vesicles, whereas the denser fractions contain unbroken whole cells (inset images, freeze-fracture TEM).

Low-density component

High-density component

To centrifuge

From centrifuge

Steep sucrose gradient (e.g., 20–70%)

Centrifugation

The sample is loaded on top of the sucrose density gradient.

Components have migrated to a region in the gradient that matches their own density.

0.5 μm

0.5 μm

J. D. Radolf, et al. 1995. *Infection and Immunity* 63:2154

Figure 3.24   Sucrose porin.   Outer membrane porin (OMP) for sucrose transport in *Salmonella typhimurium* (stereo view) based on X-ray crystallography. The porin comprises three beta barrel channels (red, yellow, and blue). Each transports a molecule of sucrose through its channel. To view, place a piece of cardboard vertically between the two images. Position your eyes on opposite sides of the cardboard and force them to focus behind the images. The images will merge and produce a 3-D image. (PDB code: 1A0S)

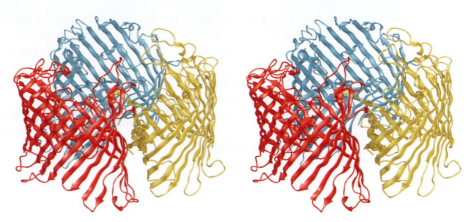

Eukaryotic Microbes: Protection from Osmotic Shock

The problems confronting bacteria, such as adjusting to environmental change, also challenge eukaryotes. A key example is osmotic shock. Eukaryotic microbes possess their own structures to avoid osmotic shock. Algae form cell walls of cellulose fibers, and fungi form walls of chitin. Diatoms form intricate exoskeletons of silicate.

Other eukaryotes, however, such as amebas and ciliated protists, possess a flexible outer coating, the pellicle. The pellicle allows great flexibility of shape; it enables uptake of large particles by endocytosis and of larger objects, even entire cells, by phagocytosis. Thus, eukaryotic cells have the nutritional option of engulfing prey, an option unavailable to prokaryotes and to eukaryotes with cell walls. Eukaryotic microbes that lack a cell wall possess a **contractile vacuole** to pump water out of the cell, avoiding osmotic shock (**Fig. 3.25**). The vacuole takes up water from the cytoplasm through an elaborate network of intracellular channels, and then expels the water through a pore.

> **THOUGHT QUESTION 3.9**   What do you think are the advantages and disadvantages of a contractile vacuole, compared with a cell wall?

TO SUMMARIZE:

- **The cell wall maintains turgor pressure.** The cell wall is porous, but its rigid network of covalent bonds protects the cell from osmotic shock.
- **The gram-positive cell envelope** has multiple layers of peptidoglycan, interpenetrated by teichoic acids.

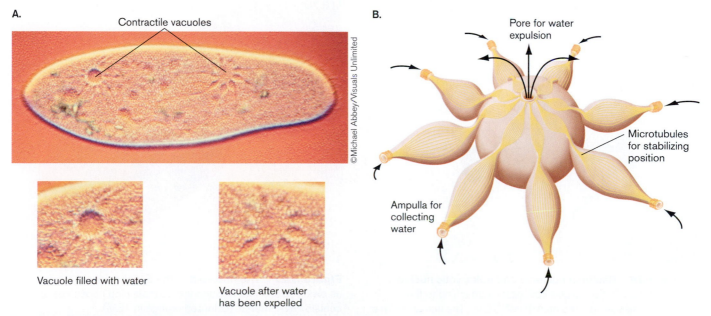

A. Contractile vacuoles

©Michael Abbey/Visuals Unlimited

Vacuole filled with water

Vacuole after water has been expelled

B. Pore for water expulsion

Microtubules for stabilizing position

Ampulla for collecting water

Figure 3.25   Protection from osmotic shock in eukaryotic microbes.   A. A paramecium, showing two contractile vacuoles (phase-contrast micrograph). The inset images show radiating channels that collect water and expel it from the cytoplasm. **B.** Diagram of a contractile vacuole.

- **The S-layer**, composed of proteins, is highly porous, but can prevent phagocytosis and phage infection. In archaea, the S-layer serves the structural function of a cell wall.
- **The capsule**, composed of polysaccharide and glycoprotein filaments, protects cells from phagocytosis. Either gram-positive or gram-negative cells may possess a capsule.
- **The gram-negative outer membrane** regulates nutrient uptake and excludes toxins. The envelope layers include protein pores and transporters of varying selectivity.
- **Eukaryotic microbes** are protected from osmotic shock by polysaccharide cell walls or by a contractile vacuole.

3.5 The Nucleoid and Gene Expression

An important function of the cell envelope is to contain and protect the cell's genome. **Figure 3.26** compares the organization of chromosomal material in enteropathogenic *E. coli* cells with that in a cultured human cell that they have colonized. In this thin-section TEM, each bacterium contains a filamentous nucleoid region that extends through the cytoplasm. In contrast, the nucleus of the eukaryotic cell, only a fraction of which

is visible in the figure, is many times larger than the entire bacterial cell, and the chromosomes it contains are separated from the cytoplasm by the nuclear membrane.

In Prokaryotes, DNA Is Organized in the Nucleoid

All living cells on Earth possess chromosomes consisting of DNA. The genetic functions of microbial DNA are discussed in detail in Chapters 7–12. Here we focus on the physical organization of DNA within the nucleoid of the prokaryotic cell (**Fig. 3.27**).

NOTE: The prokaryotic genome typically consists of a single circular chromosome, but some species have a linear chromosome or multiple chromosomes. In this chapter, we focus on the simple case of a single circular chromosome.

In a bacterial cell, the DNA is organized in loops called domains. All domains connect back to a central point. The central point is the **origin of replication (*ori*)**, which is attached to the cell envelope at a point on the cell's "equator," halfway between the two poles (**Fig. 3.28**). DNA is replicated by DNA polymerase; the process

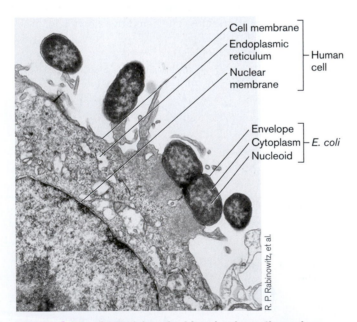

Figure 3.26 Bacterial nucleoid and eukaryotic nucleus. Enteropathogenic *Escherichia coli* bacteria attaching to the surface of a tissue-cultured human cell (TEM). The human cell has a well-defined nucleus delimited by a nuclear membrane, whereas the *E. coli* cells have nuclear material more loosely arranged in the nucleoid region. *E. coli* cells are 1.0–2.0 μm in length.

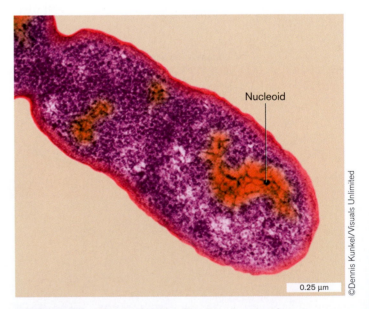

Figure 3.27 The nucleoid. The *E. coli* nucleoid appears as clear regions that exclude the dark-staining ribosomes and contain DNA strands (colorized orange in TEM).

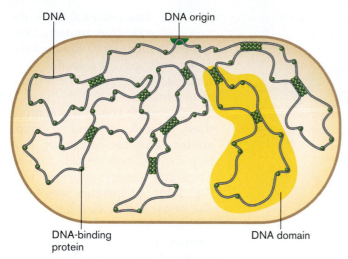

Figure 3.28   Nucleoid organization. The nucleoid forms approximately 50 loops of chromosome called domains, which radiate from the center (shown shaded). Within each domain, the DNA is supercoiled and partly compacted by DNA-binding proteins (shown in green).

is covered in more detail in Chapter 7. At the origin, the DNA double helix begins to unzip, and replication proceeds outward in both directions. Growing cells replicate their DNA continuously, with no resting phase; in fact, the two replicated origins may split again to begin a second round of replication before the first is completed.

> **NOTE:** In biology, the word *domain* is used in several different ways, each referring to a defined portion of a larger entity.
> - **DNA domains** of the nucleoid are distinct loops of DNA that extend from the origin.
> - **Protein domains** are distinct functional or structural regions of a protein.
> - **Taxonomic domains** are genetically distinct classes of organisms, such as bacteria, archaea, and eukaryotes.

How does all the cell's DNA fit neatly into the nucleoid? The DNA is compacted over several levels by supercoiling and by DNA-binding proteins.

Supercoiling. The chromosome includes a number of extra twists, called **supercoils** or superhelical turns, beyond those inherent in the structure of the DNA duplex (double helix). The supercoiling causes portions of DNA to double back and twist upon itself, which results in compaction. Supercoiling is generated by enzymes such as gyrase and maintained by DNA-binding proteins (shown in green in **Fig. 3.28**). The enzyme gyrase is a major target for antibiotics such as quinolones, used, for example, to treat bacterial pneumonia and urinary tract infections.

In bacteria, as in eukaryotic cells, the direction of the extra turns opposes the natural twist of the duplex. Turns that oppose the natural twist are called negative superhelical turns. Negative superhelical turns actually unwind the DNA slightly and help expose the base pairs for transcription. In archaea, however, most supercoiling is in the positive direction, tending to wind the DNA more tightly and therefore increase stability of the duplex. The increased stability of positive supercoils is an advantage for archaea growing in extreme environments, such as temperatures above 100°C.

DNA-binding proteins. Prokaryotic DNA is condensed by winding around various classes of binding proteins (see **Fig. 3.28**). Some binding proteins, such as Hns and Hu, also function as regulators of gene expression. Binding proteins can respond to the state of the cell; for example, under starvation conditions, when most transcription of RNA ceases, the binding protein Dps is used to organize the DNA into a protected crystalline structure. Such "biocrystallization" by Dps and related proteins may be a key to the extraordinary ability of microbes to remain viable for long periods in stationary phase or as endospores.

Transcription and Translation Are Tightly Coupled

The information encoded in DNA is "read" by the processes of transcription and translation to yield gene products. Within the prokaryotic cell, DNA transcription to RNA is coupled tightly to RNA translation to proteins.

DNA transcription to RNA. The initial product of a gene is a single strand of ribonucleic acid (RNA), produced when DNA is transcribed by RNA polymerase (**Fig. 3.29**). In some cases, the newly made RNA has a function of its own, such as one of the RNA components of a ribosome, the cell's protein-making machine. For most genes, however, the RNA is a messenger RNA (mRNA)

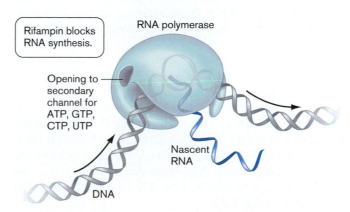

Figure 3.29   DNA transcription to RNA. The DNA information is transcribed into a single strand of RNA.

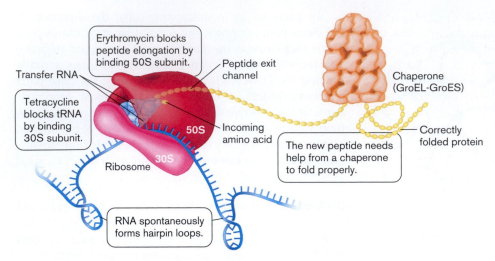

Figure 3.30   Translation and protein folding.   Messenger RNA (mRNA) is translated into a peptide by the ribosome. Nascent polypeptides are folded into their proper conformation, assisted by chaperones.

that immediately binds to a ribosome for translation to generate a polypeptide, a linear sequence of amino acid residues (discussed in Chapter 8).

mRNA translation to protein. A growing bacterial cell typically invests 40% of its energy in translation of mRNA by ribosomes. The ribosome, with its large number of protein and RNA components, is probably the most complex subcellular machine to be discovered (**Fig. 3.30**). Its task of converting the four-letter RNA code into 20 amino acids requires the assistance of 27 different transfer RNAs (tRNAs) and an equal number of aminoacyltransferase enzymes, each encoded by a different gene. Each amino acid is brought by a tRNA to fit into the acceptor site within the ribosome, sandwiched between the 30S and 50S subunits. The amino acid is then transferred onto a peptide chain, where the peptide bond forms. The process of translation is discussed further in Chapter 8.

Because of its complexity, the ribosome is targeted by many kinds of antibiotics. For example, tetracyclines block aminoacyl-tRNA from arriving at the 30S subunit. Erythromycin blocks peptide elongation at the 50S subunit.

Coupling of transcription and translation. In prokaryotes, translation is tightly coupled to transcription; the ribosomes bind to mRNA and begin translation even before the mRNA strand is complete. Thus, a growing bacterial cell is full of mRNA strands dotted with ribosomes. These composite mRNA-ribosome structures are known as **polysomes**. Polysomes can be observed as the mRNA elongates, attached by RNA polymerase to the DNA strand (**Fig. 3.31**).

In rapidly growing bacteria, both transcription and translation occur at top speed while the DNA itself is being

replicated. This remarkable coordination of replication, transcription, and translation is a hallmark of the prokaryotic cell; it explains why some bacteria can double in as little as 10 minutes.

As peptide chains are synthesized, some kinds bind to **chaperones**, enzyme complexes that help fold the new peptide into its functional tertiary structure (see **Fig. 3.30**). Chaperones were first discovered in *E. coli* as "heat-shock proteins," expressed at high temperature to protect proteins from denaturation by heat. Since then, chaperones have been discovered in all organisms, including humans, where chaperone defects are associated with genetic diseases.

In eukaryotes, note that transcription and translation occur only during interphase, when the cell is not undergoing division. Within the nucleus, nascent (elongating) mRNA is translated by ribosomes, which check the transcript for errors and target faulty transcripts for destruction. The majority of eukaryotic translation, however, occurs outside the nucleus, in the cytoplasm.

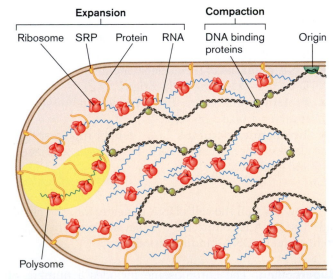

Figure 3.31   Protein synthesis and secretion.   Bacterial transcription of DNA to RNA is coordinated with translation of RNA to make proteins. Even before the RNA chain is complete, ribosomes bind to commence translation. Growing peptides destined for the membrane bind the signal recognition particle (SRP) for insertion into the membrane. The DNA is compacted by binding proteins (shown in green), but it is also pulled outward by the nascent chains of RNA (blue) and membrane-inserted peptides (orange). Each mRNA is translated by multiple ribosomes, called a polysome (highlighted yellow).

Inserting proteins in the membrane. Some of the newly translated proteins are destined for the membrane or for secretion outside the cell. Proteins destined for membrane insertion must be hydrophobic and hence are poorly soluble in the aqueous cytoplasm where translation occurs. How can proteins that are insoluble in water be folded correctly in the cytoplasm? In prokaryotes, membrane and secreted proteins are synthesized in association with the cell membrane (**Fig. 3.31**).

Protein secretion in prokaryotes is managed by several different classes of protein complexes, many of which export toxins and virulence factors, as discussed in Chapter 8. One major secretion system relies on a signal sequence, a short N-terminal sequence rich in hydrophobic residues that heads the coding sequence of a growing polypeptide. As translation begins, the signal sequence binds to a **signal recognition particle (SRP)**, an RNA-protein complex, plus other helper proteins. The SRP complex transfers the polypeptide-ribosome-mRNA complex bearing the signal sequence to the secretory complex within the cell membrane. Translation then continues at the cell membrane, and as it does so, the water-insoluble protein progressively translocates into or across the membrane.

In eukaryotes, membrane proteins are synthesized at the surface of the endoplasmic reticulum (ER). From the ER, eukaryotic proteins are packaged within the Golgi complex for transport to the cell membrane.

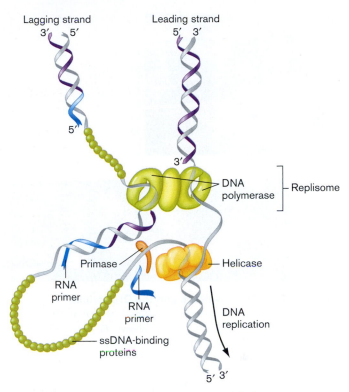

Figure 3.32 DNA replication. DNA replication proceeds by unzipping the helix and synthesizing a complement for each strand. The replisome consists of two DNA polymerase complexes, one of which synthesizes the leading strand, while the other synthesizes the lagging strand. Further details are discussed in Chapter 7.

3.6 Cell Division

Cell division, or cell fission, requires highly coordinated growth and expansion of all the cell's parts. Unlike eukaryotes, prokaryotes synthesize RNA and proteins continually while the cell's DNA undergoes replication. Bacterial DNA replication is coordinated with the expansion of the cell wall and ultimately the separation of the cell into two daughter cells. The DNA replication process is outlined here as it relates to cell division; molecular details are discussed in Chapter 7.

DNA Is Replicated Bidirectionally

In prokaryotes, a circular chromosome begins to replicate at its origin, or *ori* site, a sequence of base pairs on the genetic map. Replication proceeds in both directions (bidirectionally) all around the circle. Some prokaryotes have linear chromosomes, but their replication is poorly understood.

At the origin, the DNA double helix begins to unzip, forming two replication forks. At each replication fork, DNA is synthesized by DNA polymerase (**Fig. 3.32**). The replication fork is propagated by **helicase**, which unwinds the DNA helix ahead of replication, and pri-

mase, which generates RNA primers. The complex of DNA polymerase with its accessory components is called a **replisome**. The replisome actually includes two DNA polymerase enzymes, one to replicate the "leading strand," the other for the "lagging strand." The actual lag time is short compared with the overall time of replication; thus, as the replisome travels along the DNA, it converts one helix into two progeny helices almost simultaneously.

NOTE: Each replisome contains two DNA polymerase complexes (for leading and lagging strands), and each dividing nucleoid requires two replisomes (for bidirectional replication), thus four DNA polymerase complexes overall.

Within the cell, replication proceeds outward in both directions around the genome. Thus, bidirectional replication requires two replisomes, one for each replicating fork. A long-standing question has been: Do the two replisomes move oppositely around the DNA, or do they stay in the middle while the DNA helices slide through them?

A.

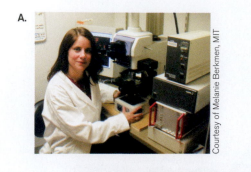

Courtesy of Melanie Berkmen, MIT

B.

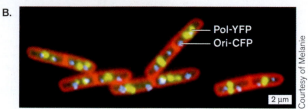

Pol-YFP
Ori-CFP

2 μm

Courtesy of Melanie Berkmen, MIT

Figure 3.33 The replisome and the DNA origin.
A. Melanie Berkmen, working in the laboratory of Alan Grossman, obtains the fluorescence micrographs shown. **B.** Fluorescence microscopy reveals the DNA origin, labeled blue by a protein fused to cyan fluorescent protein, binding at a sequence near the origin (Ori-CFP). Replisomes are labeled yellow by fusion of a DNA polymerase subunit to yellow fluorescent protein (Pol-YFP) in dividing cells of *Bacillus subtilis*. The cell envelope is labeled red with the membrane stain FM4-64.

To answer this question, fluorescence microscopy is used to observe the process of DNA replication within a growing cell of *Bacillus subtilis* (**Figure 3.33**). The DNA origin of replication (*ori*) and the pair of replisomes are labeled by fluorescence. The origin of replication is labeled blue by a hybrid protein fused to a gene encoding cyan fluorescent protein (CFP). The CFP protein binds to a promoter sequence cloned in *B. subtilis* near its origin site. The pair of replisomes are labeled yellow by a hybrid protein expressed from a gene encoding a DNA polymerase subunit fused to a gene encoding yellow fluorescent protein (YFP). The replisomes usually locate together near the center of the cell, but sometimes they separate, visible as two yellow spots.

The fluorescence data are consistent with a model in which two replisomes are located at the midpoint of the growing cell (**Fig. 3.34**⊙⊪). Each of the two replisomes forms a replicating fork that directs two daughter strands of DNA toward opposite poles. The two copies of the DNA origin of replication (green), attached to the cell envelope, move apart as the cell expands. The termination site (red) remains in the middle of the cell.

The two replisomes continue replication at both forks in the middle of the cell. Finally, as the termination site replicates, the two replisomes separate from the DNA. At each new *ori* site, however, two pairs of new replisomes

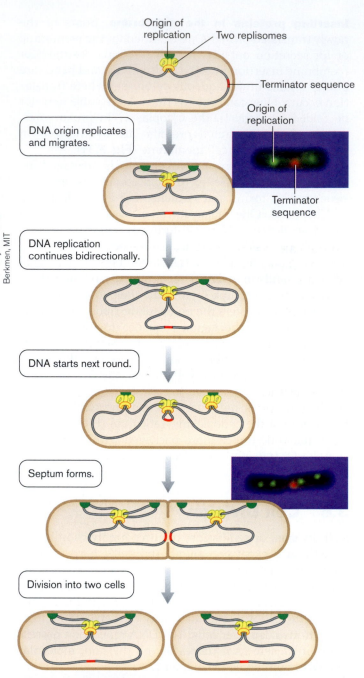

Origin of replication
Two replisomes
Terminator sequence

DNA origin replicates and migrates.

Origin of replication

Terminator sequence

DNA replication continues bidirectionally.

DNA starts next round.

Septum forms.

Division into two cells

Figure 3.34 Replisome movement within a dividing cell.
The DNA origin of replication sites (green) move apart in the expanding cell as the pair of replisomes (yellow) stay near the middle, where they replicate around the entire chromosome, completing the terminator sequence last (red). ⊙⊪ *Source:* Ivy Lau, et al. 2003. *Molecular Microbiology* 49:731.

have formed. The replication of the new *ori* sites begins, sometimes even before termination of the previous round of replication.

Note that the contents of the cytoplasm must expand coordinately with DNA replication for the cell to generate progeny equivalent to the parent. In a rod-shaped

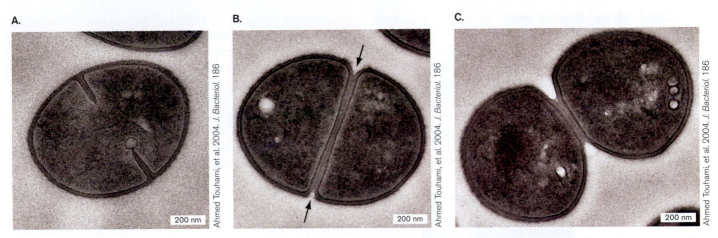

Ahmed Touhami, et al. 2004. *J. Bacteriol.* 186

Ahmed Touhami, et al. 2004. *J. Bacteriol.* 186

Ahmed Touhami, et al. 2004. *J. Bacteriol.* 186

200 nm
200 nm
200 nm

Figure 3.35   Septation in *Staphylococcus aureus*.   A. Furrows appear in the cell envelope, all around the cell equator, as new cell wall grows inward (TEM). **B.** Two new envelope partitions are complete. **C.** The two daughter cells peel apart. The facing halves of each cell contain entirely new cell wall.

cell, the cell envelope and cell wall must elongate as well to maintain progeny of even girth and length.

Septation Completes Cell Division

For the cell to divide, DNA replication must be complete. Replication of the DNA termination site triggers growth of the dividing partition of the envelope, called the **septum**. The septum grows inward from the sides of the cell, at last constricting and sealing off the two daughter cells. This process is called **septation**. Septation and envelope extention require rapid biosynthesis of all envelope components, including membranes and cell wall. The biosynthetic enzymes required are all of great interest as antibiotic targets. Cell wall biosynthesis poses an interesting theoretical problem: How is it possible to expand the covalent network of the sacculus without breaking links to insert new material, thus weakening the wall? The answer remains unclear.

Septation of spherical cells. In spherical cells (cocci), such as *Staphylococcus aureus*, the process of septation generates most of the new cell envelope to enclose the expanding cytoplasm (**Fig. 3.35**). Furrows form in the cell envelope, in a ring all around the cell equator, as new cell wall grows inward. The wall material must compose two separable partitions. When the partitions are complete, the two progeny cells peel apart. The facing halves of each cell consist of entirely new cell wall.

The spatial orientation of septation has a key role in determining the shape and arrangement of cocci (**Fig. 3.36A**). When septation always occurs in parallel planes, such as in *Streptococcus* species, cells form chains. If septation occurs in random orientations, or if cells reassociate loosely after septation, they form compact hexagonal arrays similar to the grape clusters portrayed in classical paintings; hence the Greek-derived term **staphylococci**.

Such clusters are found in colonies of *Staphylococcus aureus*. If subsequent septation occurs at right angles to the previous division, the cells may form tetrads and even cubical octads called sarcinae. Tetrads and sarcinae are formed by *Micrococcus* species (**Fig. 3.36B**).

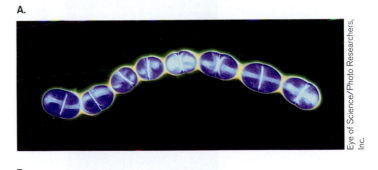

Eye of Science/Photo Researchers, Inc.

Kwangshin Kim/Photo Researchers, Inc.

Figure 3.36   Septation orientation determines the arrangement of progeny cells.   A. A chain of cocci results from septation in the same plane (*Streptococcus*, colorized TEM). **B.** Septation in two planes generates a tetrad (*Micrococcus tetragenus*, negative stain).

Septation of rod-shaped cells. In rod-shaped cells, unlike cocci, cell division requires the envelope to elongate before septation, followed by formation of a new polar envelope for each progeny cell. The process of septation involves an intricate series of molecular signals that is at the frontier of current research.

Certain mutant strains of *E. coli* form long filaments instead of dividing normally. This filamentation results from a failure to form a septum between cells. The mutant genes responsible for this behavior were called *fts* for "filamentation temperature sensitive" because the cells divide normally at the **permissive temperature** but fail to septate at the **restrictive temperature**, forming long filaments.

Some of the *fts* genes encode proteins directly involved in the formation of the septum. The most dramatic example is the protein FtsZ, which assembles to form the "Z ring," a constriction ring around the equator (**Fig. 3.37**). FtsZ is universally found in bacteria and archaea as a key septation protein. FtsZ is also an ancient homolog of tubulin, which is the major component of the mitotic apparatus in eukaryotes. The discovery of FtsZ is interesting because it implies that the processes of mitosis in eukaryotes and cell division in prokaryotes might have evolved from a common process in an ancestral cell. Other bacterial homologs of eukaryotic structural proteins suggest that bacteria possess a kind of cytoskeleton beneath the cell wall (**Special Topic 3.2**).

What signals the cell as to when and where to form the septum? What keeps the cell from septating at positions other than the equator? In fact, defective gene products can lead cells to divide incorrectly with a septum near the pole of the cell. This incorrect septation generates a "minicell," a small cellular compartment with no DNA. Genes whose defects lead to minicells are called *min* genes. In *E. coli*, the protein encoded by *minD* oscillates in rings from pole to pole (**Fig. 3.38**). The Min proteins somehow regulate the formation of the septum, ensuring symmetrical cell division.

Note, however, that normal division of bacteria is not always symmetrical. For example, *Bacillus* species undergo an asymmetrical cell division to form an endospore. In this process, one daughter nucleoid forms an inert endospore capable of remaining dormant but viable for many years (discussed in Chapter 4). Other bacteria expand their cells by polar extension; an example is *Corynebacterium diphtheriae*, the causative agent of diphtheria. Polar extension occurs at variable rates and direction, and thus generates irregularly shaped rods.

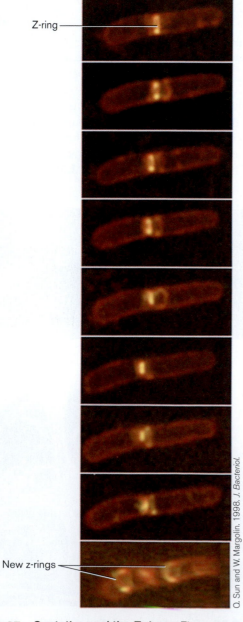

Z-ring

New z-rings

O. Sun and W. Margolin. 1998. *J. Bacteriol.*

Figure 3.37   Septation and the Z ring. Fluorescence microscopy of *E. coli* based on FtsZ-GFP, a genetic fusion of FtsZ with green fluorescent protein (GFP). The Z ring of FtsZ subunits forms around the equator of the constricting cell, as the septum grows inward.

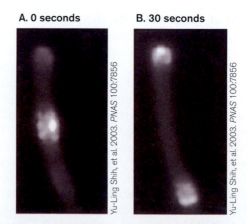

A. 0 seconds   B. 30 seconds

Yu-Ling Shih, et al. 2003. *PNAS* 100:7856

Figure 3.38   MinD protein rings regulate cell division. Fluorescence shows MinD-YFP protein rings in an *E. coli* cell completing division. The rings oscillate from pole to pole (times 0 to 30 s shown).

A major distinction of prokaryotes was thought to be the lack of a cytoskeleton, a structural feature prominent in eukaryotes. Bacterial shape was thought to be maintained by the cell wall. In fact, bacterial shape and cell division require cytoskeletal proteins homologous to eukaryotic cytoskeletal components such as tubulin and actin. The first such protein recognized was the Z ring subunit FtsZ, an analog of tubulin. Further components of the bacterial cytoskeleton were revealed by several means: gene defects that confer the loss of cell shape, fluorescent labeling of the corresponding gene products in wild-type cells, and genomic analysis showing bacterial homologues of eukaryotic cytoskeletal components.

An example of a "shape loss" mutant is shown in **Figure 1**. The bacterium *Shigella* is an enteric pathogen closely related to *E. coli* that normally grows as a rod, or bacillus (see **Fig. 1A**). The opposite poles of the rod bind specific pole proteins, which are marked in the micrographs by fluorescent-labeled antibodies. A mutant strain of *Shigella* was isolated that grows as amorphous spheres (see **Fig. 1B**). Unlike the rod-shaped cells, the spherical mutants show multiple pole-marker proteins at undefined locations. The mutation eliminates expression of *mreB*, encoding the MreB cytoskeletal protein. Sequence analysis shows a strong relationship between MreB and the eukaryotic filament-forming protein actin.

How do MreB proteins maintain the rod shape of the bacterium? Like actin filaments in eukaryotes, MreB proteins form elongated structures beneath the cell membrane. **Figure 2A** shows how MreB proteins polymerize in a helical arrangement around the cell, thus determining the axis of elongation. Because MreB is required for rod elongation, its gene is absent from bacteria that normally grow as cocci, such as *Streptococcus*. In **Figure 2B**, the helical arrangement of MreB is observed by confocal fluorescence microscopy of *Caulobacter crescentus*, a pond-dwelling bacterium. The MreB proteins are labeled by fluorescent antibodies.

Crescent-shaped cells such as *C. crescentus* usually possess a third kind of shape-determining protein, CreS, in addition to FtsZ and MreB (see Fig. 2A). CreS is a homologue of the eukaryotic intermediate filament protein. CreS polymerizes along the inner curve of a crescent cell, as seen in **Figure 2C**. In the micrograph shown, CreS proteins are labeled by recombination of the *creS* gene with a gene encoding the green fluorescent protein (GFP). The result is a gene fusion expressing the hybrid protein CreS-GFP, which fluoresces green. In the confocal fluorescence micrograph, the CreS-GFP appears blue-green, whereas the cell membrane is labeled red by a membrane-binding fluorophore. The membrane-binding fluorophore reveals the overall crescent outline of the cell and the localization of CreS to the cell's inner curve.

Other bacterial proteins contributing to cell shape and division include the Min proteins, required for equal division of cells, and the Par proteins, which segregate replicated plasmids during cell division. In one group of bacteria, the phylum Verrumicrobia, a tubulin homolog has been discovered that forms tubules in vitro. Such tubules may contribute to the unusual extended shapes of Verrumicrobia.

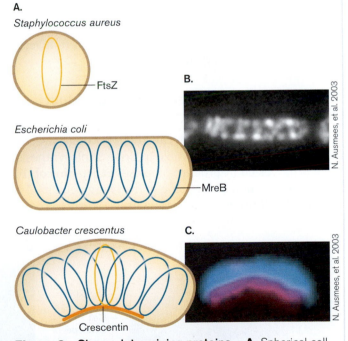

A.
Staphylococcus aureus — FtsZ

Escherichia coli — MreB

Caulobacter crescentus — Crescentin

B.

C.

N. Ausmees, et al. 2003

Figure 2   Shape-determining proteins.   A. Spherical cell size and shape is maintained by FtsZ polymerization to form the Z ring. Elongation of a rod-shaped cell is accomplished by helical polymerization of Mre proteins. Crescent-shaped cells possess a third shape-determining protein, CreS (crescentin), which polymerizes along the inner curve of the crescent. **B.** MreB localizes in a helical coil within *Caulobacter crescentus* cells. Images of MreB immunostaining were obtained by confocal fluorescence microscopy. **C.** Crescentin protein fused to green fluorescent protein GFP (CreS-GFP) localizes to the inner curve of *C. crescentus*. CreS-GFP colocalizes with the membrane-specific stain FM4-64 (red fluorescence).

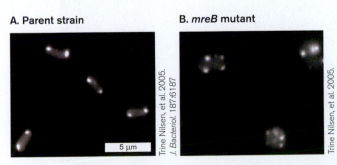

A. Parent strain

B. *mreB* mutant

5 μm

Trine Nilsen, et al. 2005. *J. Bacteriol.* 187:6187

Figure 1   Spherical *mreB* mutants of rod-shaped bacteria.   A. Rod-shaped *Shigella* cells show a pole-associated protein (IcsA-GFP) localized to opposite poles. **B.** Mutant MreB cells form bloated spherical cells with multiple sites localizing IcsA-GFP.

TO SUMMARIZE:

- **The nucleoid** region contains loops of DNA, supercoiled and bound to DNA-binding proteins.
- **Transcription of DNA** occurs in the cytoplasm, often simultaneously with DNA replication.
- **The ribosome translates RNA to make proteins**, which are folded by chaperones and in some cases secreted at the cell membrane.
- **DNA is replicated** bidirectionally by the replisome.
- **Cell expansion and septation** are coordinated with DNA replication.
- **Cell shape** is determined by the FtsZ "Z ring" and other cytoskeletal proteins.

3.7 Specialized Structures

We have introduced the major structures required to contain the cell, organize its contents, maintain its DNA, and express its genes. The cell envelope, the nucleoid, and the gene expression and protein translocation complexes are essential for all living prokaryotes. In addition to these fundamental structures, different species have evolved different kinds of specialized devices adapted to diverse metabolic strategies and environments.

Thylakoids and Carboxysomes Conduct Photosynthesis

Photosynthetic bacteria, also called phototrophs, need to collect as much light as possible to drive photosynthesis. Light is harvested by protein complexes called **phycobilisomes** (see Section 14.6). Some of the energy obtained is stored in the form of storage granules.

To maximize their photosynthetic membranes, phototrophs have evolved specialized systems of extensively folded intracellular membrane called **thylakoids** (**Fig. 3.39A**). Thylakoids consist of layers of folded sheets (lamellae) or tubes of membranes packed with photosynthetic proteins and electron carriers. Cyanobacteria containing thylakoids structurally resemble chloroplasts, which are believed to have evolved from a common ancestor of modern cyanobacteria.

The thylakoids conduct only the "light reaction" of photon absorption and energy storage. The energy obtained is rapidly spent to fix carbon dioxide, which occurs within **carboxysomes**. Carboxysomes are polyhedral, protein-covered bodies packed with the enzyme rubisco for CO_2 fixation.

Aquatic and marine phototrophs, as well as archaea, often possess **gas vesicles** to increase buoyancy and keep themselves high in the water column, near the sunlight (**Fig. 3.39B**). Gas vesicles are specialized vacuoles composed of specific proteins. The vesicles trap and collect gases such as hydrogen or carbon dioxide produced by the cell's metabolism.

Storage Granules

During times of starvation, phototrophs may digest their phycobilisomes for energy and as a source of nitrogen. Alternatively, energy is stored in storage granules composed of glycogen or other polymers, such as poly-hydroxybutyrate (PHB) and poly-3-hydroxyalkanoate (PHA). PHB and PHA polymers are of interest as a biodegradable plastic, which bacteria are engineered to produce industrially. Similar storage granules are also produced by nonphototrophic soil bacteria.

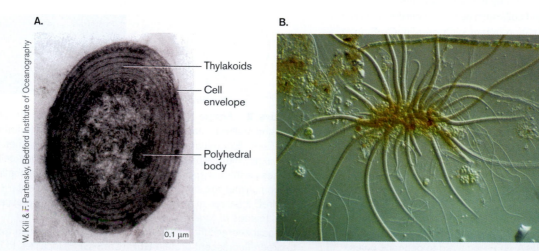

A.

W. Kili & F. Partensky, Bedford Institute of Oceanography

— Thylakoids

— Cell envelope

— Polyhedral body

0.1 μm

B.

©Tom E. Adams/Visuals Unlimited

Figure 3.39 Organelles of phototrophs. **A.** The marine phototroph *Prochlorococcus marinus* (TEM). Beneath the envelope lie the photosynthetic double membranes called thylakoids. Carboxysomes are polyhedral, protein-covered bodies packed with rubisco enzyme for CO_2 fixation. **B.** Filaments (chains of cells) of the cyanobacterium *Planktothrix*. Gas vesicles provide buoyancy, enabling the phototroph to remain at the surface of the water, exposed to light.

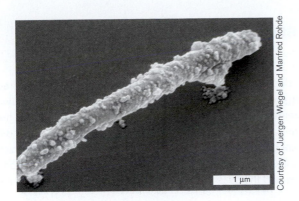

Courtesy of Juergen Wiegel and Manfred Rohde

Figure 3.40 **External sulfur particles.** Sulfur globules dot the surface of *Thermoanaerobacter sulfurigignens*, an anaerobic thermophilic bacterium that gains energy by reducing thiosulfate ($S_2O_3^{2-}$) to elemental sulfur (S^0).

Another type of storage device is sulfur, granules of elemental sulfur produced by purple and green phototrophs through photolysis of hydrogen sulfide (H_2S). Instead of disposing of the sulfur, the bacteria store it in granules, either within the cytoplasm (purple phototrophs) or as "globules" attached outside of the cell. Sulfur-reducing bacteria also make extracellular sulfur globules (**Fig. 3.40**). The sulfur may be usable as an oxidant when reduced substrates become available. Alternatively, the presence of potentially toxic sulfur granules may help cells avoid predation.

Magnetosomes of Magnetite Direct Motility

An unusual structure possessed by magnetotactic species of bacteria (bacteria showing magnetically directed motility) is the magnetosome. **Magnetosomes** are microscopic membrane-bounded crystals of the magnetic mineral magnetite, Fe_3O_4 (**Fig. 3.41**). They are found in anaerobic pond-dwelling organisms such as *Magnetospirillum gryphiswaldense*. The crystals generate a magnetic dipole moment along the length of a bacterium, constraining it to swim along a magnetic field. This magnetic orientation of swimming is called **magnetotaxis**, the ability to sense and respond to magnetism. Magnetotactic bacteria can be collected by placing a magnet within a jar of pond water; bacteria orienting by the field lines collect nearby.

The natural function of magnetosomes appears to be to orient bacterial swimming toward the bottom of the pond. Magnetotactic organisms are anaerobes, which prefer the lower part of the water column, where oxygen concentration is lowest. Because Earth's magnetic field lines point downward in the northern latitudes, bacteria that are magnetotactic swim "downward" toward magnetic north.

> **THOUGHT QUESTION 3.10** How would a magnetotactic species have to behave if it were in the Southern Hemisphere instead of in the Northern Hemisphere?

Magnetotactic bacteria are being studied for their potential applications in wastewater treatment. Through their anaerobic metabolism, some magnetotactic bacteria accumulate high concentrations of toxic metals from the water. They and the toxic metals they scavenge can then be removed by application of a magnetic field to attract and concentrate the bacteria.

Pili and Stalks Enable Attachment

In a favorable habitat, such as a running stream full of fresh nutrients or the epithelial surface of a host, it is advantageous for a cell to adhere to a substrate. **Adherence**, the ability to attach to a substrate, requires specific adherence structures, such as pili (protein filaments). As bacteria grow and proliferate, however, they face the question of whether to stay where they are or leave their present habitat, where nutrients may be depleted and waste products increased. In rapidly changing environments, cell survival requires **motility**, the ability to move and relocate. Motility requires structures such as rotary flagella. Some bacteria have it both ways by producing two kinds of progeny, one adherent, the other motile (see **Special Topic 3.3**).

The most common structures that bacteria use to attach to a substrate are **pili** (singular, **pilus**), or **fimbriae** (singular, **fimbria**), straight filaments of protein monomers called **pilin**. For example, fimbriae attach the

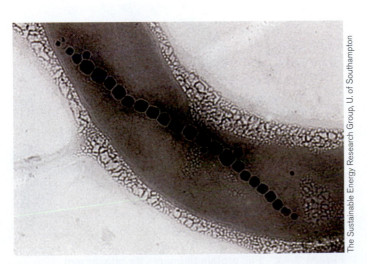

The Sustainable Energy Research Group, U. of Southampton

Figure 3.41 **Magnetotactic bacterium.** A magnetospirillum is a spiral-shaped magnetotactic bacterium. The spiral cell contains a chain of magnetosomes, magnetite crystals (TEM).

Special Topic 3.3   Two Kinds of Progeny: One Stays, One Swims

What if a single species encounters very different environments throughout its life history, some favorable, others unfavorable? Sometimes, adherence might be advantageous; at other times, motility would be desirable. Bacteria such as *E. coli* and *Salmonella* species possess more than one kind of filamentous structure: flagella for motility and pili for attachment once a favorable substrate is reached. An even more dramatic strategy for "having it both ways" is to generate two kinds of daughter cells: one that stays, and one that swims away. This strategy is exemplified by the flagellum-to-stalk transition of the aquatic bacterium *Caulobacter crescentus*.

Caulobacter crescentus exists in two forms, one with a stalk (stalked cell) and one with a single flagellum at one pole (swarmer cell). The swarmer cell swims about freely in an aqueous habitat, such as a pond or a sewage bed. After half an hour of swimming, the swarmer cell undergoes differentiation to shed its flagellum and replace it with a stalk (**Fig. 1**). Once attached, the stalked cell immediately starts to replicate its DNA.

How do the two progeny cells develop different forms? Lucy Shapiro and colleagues at Stanford University showed that the identities of swimmer and stalked cells are determined by the methylation state of their DNA (**Fig. 2**). The original DNA has methyl groups on its cytosine bases, added during the previous round of DNA replication. These methyl groups allow expression of genes needed for a stalked cell. As the DNA now undergoes a new round of replication, each daughter duplex receives a new unmethylated strand, base-paired with the old strand containing methyl groups. The original two parental strands differ in their methylation patterns; thus, the methylation patterns of the two daughter duplexes allow different gene expression in the two daughter cells, one of which is stalked, the other a swarmer. Once DNA synthesis is complete and the cells have differentiated into different forms, the

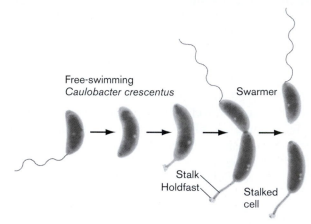

Figure 1   Asymmetrical cell division: A model for development.   A swarmer cell of *C. crescentus* loses its flagellum and grows a stalk. The stalked cell divides to produce a swarmer cell. Photo courtesy of Yves Brun, Indiana University.

---

oral pathogen *Porphyromonas gingivalis* to gum epithelium, where they are associated with periodontal disease (**Fig. 3.42**). A different kind of pili, the **sex pili**, serve to attach a "male" donor cell to a "female" recipient cell for transfer of DNA. This process of DNA transfer is called conjugation. The genetic consequences of conjugation are discussed in Chapter 9.

Another kind of attachment organelle is a membrane-bound extension of the cytoplasm called a **stalk**. The tip of the stalk secretes adhesion factors called **holdfast**, which firmly attach the bacterium in an environment that has proved favorable. The mechanism of stalk and holdfast attachment has been extensively studied in iron-oxidizing bacteria that interfere with mining operations by producing massive biofilms. An example is *Gallionella ferruginea*, an iron-oxidizing species that grows a long stalk (**Fig. 3.43**). The long twisted stalks of adherent *Gallionella* cells become coated by iron hydroxides, and *Gallionella* contributes a major part of the process of biomineralization (biological crystallization of minerals) in iron mines.

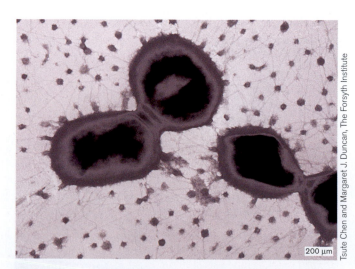

200 μm

Tsute Chen and Margaret J. Duncan, The Forsyth Institute

Figure 3.42   Pili: protein filaments for attachment. *Porphyromonas gingivalis*, a causative agent of gum disease or gingivitis. The *P. gingivalis* cells show fimbriae along with vesicles budding from the cell's outer membrane.

new DNA becomes methylated, restoring the complementary methylation patterns in each cell.

The remaining stalked daughter cell undergoes another round of cell division to produce yet another swarmer cell, as well as a stalked cell. Each swarmer cell eventually converts to a stalked cell and reproduces progeny. By this means, *Caulobacter* can hedge its bets. Half its progeny always remain at a location that, however favorable, may eventually deteriorate, whereas the other half swims away to find a potentially better environment.

Caulobacter represents a simple example of cell differentiation that resembles phenomena seen in more complex developmental systems, such as DNA methylation during cell morphogenesis. Thus, it offers an informative model system for cell differentiation in the development of multicellular organisms.

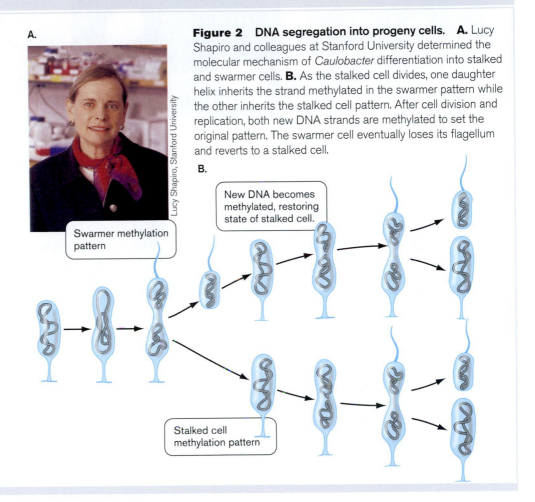

Figure 2 DNA segregation into progeny cells. A. Lucy Shapiro and colleagues at Stanford University determined the molecular mechanism of *Caulobacter* differentiation into stalked and swarmer cells. **B.** As the stalked cell divides, one daughter helix inherits the strand methylated in the swarmer pattern while the other inherits the stalked cell pattern. After cell division and replication, both new DNA strands are methylated to set the original pattern. The swarmer cell eventually loses its flagellum and reverts to a stalked cell.

Lucy Shapiro, Stanford University

Swarmer methylation pattern

New DNA becomes methylated, restoring state of stalked cell.

Stalked cell methylation pattern

Figure 3.43 *Gallionella ferruginea*: **iron-oxidizing stalked bacteria.**

A. The oval cell of *Gallionella ferruginea* generates a long twisted stalk encrusted with iron oxides. **B.** Karen L. Prestegaard, of the University of Maryland, studies iron-oxidizing bacteria attached to iron surfaces, such as this rod in the stream, where the bacteria promote rusting, coating the surface orange.

Eleanora I. Robbins, U.S. Geological Survey

Eleanora I. Robbins, U.S. Geological Survey

Rotary Flagella Enable Motility and Chemotaxis

Bacteria and archaea that are motile generally swim by means of rotary **flagella** (singular, **flagellum**). Flagella are helical propellers that drive the cell forward like the motor of a boat. First shown by Howard Berg at the California Institute of Technology, the bacterial flagellar motor was the first rotary device to be discovered in a living organism.

Different bacterial species have different numbers and arrangements of flagella. **Peritrichous** cells, such as *E. coli* and *Salmonella* species, have flagella randomly distributed around the cell (**Fig. 3.44A**). The

A.

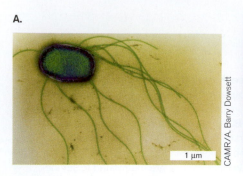

B.

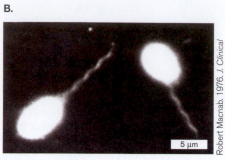

CAMR/A. Barry Dowsett

Robert Macnab. 1976. *J. Clinical Microbiology* 4:258

Figure 3.44   Flagellated *Salmonella* bacteria.   A. *Salmonella enterica* bacterium has multiple flagella (colorized TEM). **B.** The flagella collect in a bundle behind a swimming cell. Under dark-field microscopy, the cell body appears overexposed, about five times as large as the actual cell.

flagella rotate together in a bundle behind the swimming cell (**Fig. 3.44B**) Lophotrichous cells, such as *Rhodospirillum rubrum*, have flagella attached at one or both ends. In monotrichous (polar) species, such as *Pseudomonas aeruginosa*, the cell has a single flagellum at one end.

Each flagellum is a spiral filament of protein monomers called flagellin. The filament is actually rotated by means of a motor driven by the cell's transmembrane proton current, the same proton potential that drives the membrane-bound ATPase. (Alternatively, the motor is driven by a sodium ion potential, particularly in marine bacteria such as *Vibrio cholerae*.) The flagellar motor is embedded in the layers of the cell envelope (**Fig. 3.45**). The motor is observed by electron microscopy, and its parts are defined by genetic analysis of mutants with aberrant motility. The motor actually possesses an axle and rotary parts, all composed of specific proteins. Much of the structure and function of the motor were elucidated by Scottish microbiologist Robert Macnab (1940–2003) at Yale University.

Flagellar motility benefits the cell by causing dispersal of progeny, decreasing competition. In addition, most flagellated cells have an elaborate sensory system that enables them to swim toward favorable environments (attractant signals, such as nutrients) and away from inferior environments (repellent signals, such as waste products). This sensory system is known as **chemotaxis**.

Chemotaxis requires a way for the cell to move toward attractants and away from repellents. This is accomplished by flagellar rotation either clockwise or counterclockwise relative to the cell (**Fig. 3.46A⏺**). When a cell is swimming toward an attractant chemical, the flagella rotate counterclockwise (CCW), enabling the cell to swim smoothly for a long stretch. When the cell veers away from the attractant, receptors send a signal that allows one or more flagella to switch rotation clockwise (CW), against the twist of the helix. This switch in the direction of rotation disrupts the bundle of flagella, causing the cell to tumble briefly, ending up pointed in a random direc-

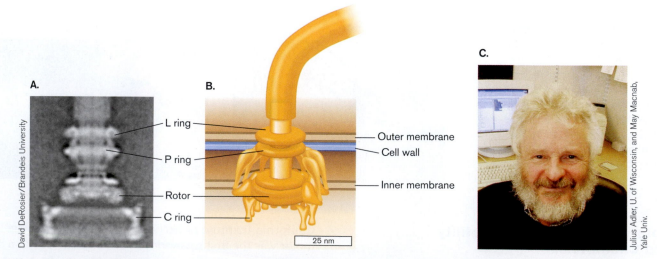

Figure 3.45   The flagellar motor.   A. The basal body, or motor, of the bacterial flagellum (TEM). The image is based on digital reconstruction, in which electron micrographs of purified hook-basal bodies were rotationally averaged. The rings (L, P, and C) correspond to those labeled in the diagram. **B.** Diagram of the flagellar motor, including major protein components. **C.** Robert Macnab (University of Wisconsin) identified many components of the motor and chemotaxis signaling.

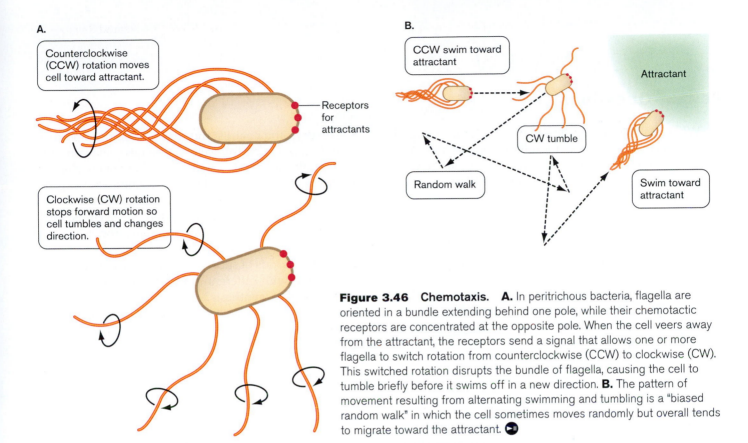

Figure 3.46 Chemotaxis. A. In peritrichous bacteria, flagella are oriented in a bundle extending behind one pole, while their chemotactic receptors are concentrated at the opposite pole. When the cell veers away from the attractant, the receptors send a signal that allows one or more flagella to switch rotation from counterclockwise (CCW) to clockwise (CW). This switched rotation disrupts the bundle of flagella, causing the cell to tumble briefly before it swims off in a new direction. **B.** The pattern of movement resulting from alternating swimming and tumbling is a "biased random walk" in which the cell sometimes moves randomly but overall tends to migrate toward the attractant. ▶❚❚

tion (**Fig. 3.46B**▶❚❚). The cell then swims off in the new direction. The resulting pattern of movement generates a "biased random walk" in which the cell tends to migrate toward the attractant.

> **THOUGHT QUESTION 3.11** Most strains of *E. coli* and *Salmonella* commonly used for genetic research actually lack flagella entirely. Why do you think this is the case? How can a researcher maintain a motile strain?

Note that bacterial flagella differ completely from the whiplike flagella and cilia of eukaryotes. Eukaryotic flagella are much larger structures containing multiple microtubules enclosed by a membrane. They move with a whiplike motion that forms a flat sine wave, propagated by ATP hydrolysis all along the flagellum.

TO SUMMARIZE:

- **Phototrophs** possess thylakoid membrane organelles packed with photosynthetic apparatus and carboxysomes for carbon dioxide fixation. Other subcellular structures may include sulfur granules from H_2S photolysis and gas vesicles for buoyancy in the water column.

- **Storage granules** store polymers for energy.
- **Magnetosomes** orient the swimming of magnetotactic anaerobic bacteria.
- **Adherence structures** enable prokaryotes to remain in an environment with favorable environmental factors. Major adherence structures include pili or fimbriae (protein filaments) and the holdfast (a cell extension).
- **Flagellar motility** occurs by rotary motion of helical flagella.
- **Chemotaxis** involves a biased random walk up a gradient of attractant substance or down a gradient of repellents.

Concluding Thoughts

Research techniques of cell fractionation and biochemistry, together with genetic analysis, continue to reveal the structure and function of cells. The intricate mechanisms derived by microbial evolution challenge the inventers of antibiotics as well as designers of molecular machines. As a journalist observed in *Science*, "When it comes to nanotechnology, physicists, chemists, and materials scientists can't hold a candle to the simplest bacteria."

CHAPTER REVIEW

Review Questions

1. What are the major features of a bacterial cell, and how do they fit together for cell function as a whole?
2. What fundamental traits do most prokaryotes have in common with eukaryotic microbes? What traits are different?
3. Give examples of how our views of ribosome structure and function have emerged from microscopy, cell fractionation, X-ray diffraction crystallography, and genetic analysis. Explain the advantages and limitations of each technique.
4. Outline the structure of the peptidoglycan sacculus, and explain how it expands during growth. Cite two different kinds of experimental data that support our current views of the sacculus.

5. Compare and contrast the structure of gram-positive and gram-negative cell envelopes. Explain the strengths and weaknesses of each kind of envelope.
6. Outline the process of DNA replication, and explain how it is coordinated with cell wall septation.
7. Explain how DNA transcription to RNA is integrated with translation and protein processing and secretion.
8. What are kinds of subcellular structures found in certain cells with different functions, such as magnetotaxis or photosynthesis?
9. Compare and contrast bacterial structures for attachment and motility. Explain the molecular basis of chemotaxis.

Key Terms

active transport (85)
adherence (107)
ATP synthase (85)
capsule (76, 93)
carboxysome (106)
cardiolipin (86)
cell membrane (75)
cell wall (76)
chaperone (100)
chemotaxis (110)
cholesterol (87)
contractile vacuole (97)
core polysaccharide (95)
cross-bridge (89)
diphosphatidylglycerol (86)
electrochemical potential (85)
electrophoresis (77)
endotoxin (94)
envelope (76)
fimbria (plural, fimbriae) (107)
flagellum (plural, flagella) (76, 109)
gas vesicle (106)
genetic analysis (81)
glucosamine (94)
glycan (89)
helicase (101)
holdfast (108)
hopane (hopanoid) (87)
inner membrane (75, 94)
isoelectric focusing (77)

leaflet (82)
lipopolysaccharide (LPS) (76, 94)
lysate (80)
magnetosome (107)
magnetotaxis (107)
membrane-permeant weak acid (83)
membrane-permeant weak base (83)
motility (107)
murein (89)
murein lipoprotein (94)
nucleoid (76)
O polysaccharide (95)
origin of replication (*ori*) (98)
osmosis (83)
osmotic pressure (83)
outer membrane (76)
passive transport (85)
pellicle (76)
peptidoglycan (78)
periplasm (76)
peritrichous (109)
permissive temperature (104)
phosphatidate (86)
phosphatidylethanolamine (82)
phosyphatidylglycerol (86)
phycobilisome (106)
pilin (107)
pilus (plural, pili) (107)
polyamine (77)
polysome (100)

porin (96)
proteome (77)
reductive evolution (degenerative evolution) (94)
replisome (101)
restrictive temperature (104)
sacculus (89)
sedimentation rate (80)
septation (103)
septum (103)
sex pilus (108)
signal recognition particle (SRP) (101)
S-layer (94)
solutes (83)
stalk (108)
staphylococcus (plural, staphylococci) (103)
subcellular fractionation (79)
supercoil (99)
Svedberg coefficient (80)
teichoic acid (93)
terpenoid (87)
thylakoid (106)
transport protein (transporter) (84)
two-dimensional polyacrylamide gel electrophoresis (2-D PAGE, 2-D gels) (77)
ultracentrifuge (79)

Recommended Reading

Ausmees, Nora, Jeffrey R. Kuhn, and Christine Jacobs-Wagner. 2003. The bacterial cytoskeleton: An intermediate filament-like function in cell shape. *Cell* **115:**705–713.

Begic, Sanela, and Elizabeth A. Worobec. 2006. Regulation of *Serratia marcescens ompF* and *ompC* porin genes in response to osmotic stress, salicylate, temperature and pH. *Microbiology* **152:**485–491.

Carter, Andrew P., William M. Clemons, Ditlev E. Brodersen, Robert J. Morgan-Warren, Brian T. Wimberly, and V. Ramakrishnan. 2000. Functional insights from the structure of the 30S ribosomal subunit and its interactions with antibiotics. *Nature* **407:**340–348.

Feucht, Andrea, and Jeff Errington. 2005. *ftsZ* mutations affecting cell division frequency, placement and morphology in *Bacillus subtilis. Microbiology* **151:**2053–2064.

Gitai, Zemer, Natalie Dye, and Lucy Shapiro. 2004. An actin-like gene can determine cell polarity in bacteria. *Proceedings of the National Academy of Sciences USA* **101:**8643–8648.

Komeili, Arash, Zhuo Li, Dianne K. Newman, and Grant J. Jensen. 2006. Magnetosomes are cell membrane invaginations organized by the actin-like protein MamK. *Science* **311:**242–245.

Lemon, Katherine P., and Alan D. Grossman. 1998. Localization of bacterial DNA: Evidence for a factory model of replication. *Science* **282:**1516–1519.

Nilsen, Trine, Arthur W. Yan, Gregory Gale, and Marcia B. Goldberg. 2005. Presence of multiple sites containing polar material in spherical *Escherichia coli* cells that lack MreB. *Journal of Bacteriology* **187:**6187–6196.

Noji, Hiroyuki, Ryohei Yasuda, Masasuke Yoshida, and Kazuhiko Kinoshita, Jr. 1997. F_1F_0 Direct observation of the rotation of F_1-ATPase. *Nature* **386:**299–302.

Parkinson, John S., Peter Ames, and Claudia A. Studdert. 2005. Collaborative signaling by bacterial chemoreceptors. *Current Opinion in Microbiology* **8:**116–121.

Ruiz, Natividad, Daniel Kahne, and Thomas J. Silhavy. 2006. Advances in understanding bacterial outer-membrane biogenesis. *Nature Reviews Microbiology* **4:**57–66.

Schäffer, Christina, and Paul Messner. 2005. The structure of secondary cell wall polymers: How Gram-positive bacteria stick their cell walls together. *Microbiology* **151:**643–651.

Shih, Yu-Ling, Trung Le, and Lawrence Rothfield. 2003. Division site selection in *Escherichia coli* involves dynamic redistribution of Min proteins within coiled structures that extend between the two cell poles. *Proceedings of the National Academy of Sciences USA* **100:**7865–7870.

Sutcliffe, Joyce A. 2005. Improving on nature: Antibiotics that target the ribosome. *Current Opinion in Microbiology* **8**(5):534–542.

Chapter 4

Bacterial Culture, Growth, and Development

The adage "To eat well is to live well" is as true for microbes as it is for humans. Microorganisms constantly struggle to survive in natural habitats because they compete for food. Yet a single microbial pathogen can multiply within the course of a day to cause deadly illness, and a few algae can abruptly bloom and cover the entire surface of a lake. In each of these cases, the population explosion results from the sudden availability of food. Over eons, bacteria have evolved ingenious strategies to find, acquire, and metabolize a wide assortment of potential food sources, ranging from glucose to mothballs. This metabolic diversity arose by necessity, from the need to find new sources of food ignored by competitors.

The remarkable plasticity of microbial genomes enables adaptation to new food sources. During the course of DNA replication, gene systems involved in the use of one food naturally undergo duplications and mutations, some of which sculpt new biochemical pathways capable of metabolizing novel substrates. This genomic flexibility raises hopes that we can engineer microbial biochemistry to remediate pollution and produce tomorrow's wonder drugs.

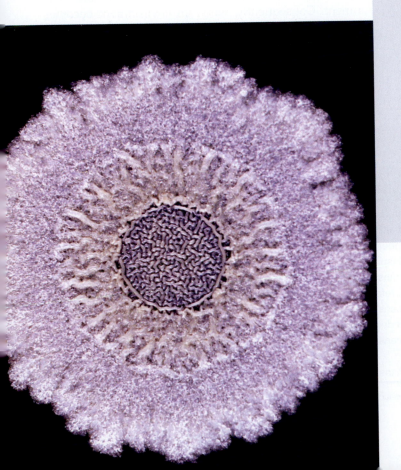

This elaborate bacterial colony is formed by an undomesticated strain of *Bacillus subtilis* isolated from soil. The complex architecture is the result of a tightly regulated developmental process and provides insight into the organization of bacterial communities. *Source:* Cover. 2004. *Journal of Bacteriology* 186(12).

Microorganisms need sources of carbon in order to grow. Understanding how they use food to increase their cell mass and, ultimately, cell number enables us to control their growth and manipulate them to make useful products. Yeast, for example, consume glucose and break it down to ethanol and carbon dioxide gas. These end products are merely waste to the yeast but are extremely important to humans who enjoy beer. Brewers have learned to control the amount of sugar supplied to yeast growing in fermentation vats to produce just the right amount of alcohol and CO_2, which produces beer's bubbling carbonation. Many a home brewer has learned the hard way that providing too much sugar (in malt) will cause yeast to make too much CO_2 gas, turning a homemade fermentor into an unpredictable explosive device.

Learning how bacteria grow has also provided a window through which to view the core processes of life. As a result of studying bacterial growth and nutrition, we now know how DNA replicates, how RNA is made, and how proteins are assembled. We have also learned that the availability of nutrients has influenced the evolution of sophisticated microbial processes designed to avoid or survive starvation. For example, bacteria communicate with each other to build elaborate multicellular reproductive structures, such as fruiting bodies, or complex multispecies biofilms, such as those that erode our teeth and rust the surface of ocean liners.

In this chapter, we consider the microbial biosphere and the basics of bacterial growth. We'll consider a variety of questions: How diverse is this biosphere? Is it true that we have identified only about 0.1% of the bacterial species that inhabit Earth? Where do bacteria get their energy? You might be surprised to learn that some bacteria gain energy from light (photosynthesis), whereas others gain energy from oxidizing sulfur. With so many microbes consuming nutrients, how has nature arranged biosystems to avoid depleting Earth of key compounds?

To understand the biochemistry behind metabolic diversity, we need to grow microbes. How do we ensure their growth and measure it? Some bacteria make elaborate, multicellular structures, whereas others form environmentally resistant fortresses called spores. In this chapter, we discuss how microbes obtain energy and nutrients for growth, and how cell populations develop. A more detailed treatment of mechanisms of energy gain and biosynthesis is presented in Chapters 13–15.

4.1 Microbial Nutrition

Bacterial cells, for all their apparent simplicity, are remarkably complex and efficient replication machines. One cell of *Escherichia coli*, for example, can divide into two cells every 20–30 minutes. At a rate of 30 min per division, one cell could potentially multiply to over 1×10^{14} cells in 24 hours—that is, 100 trillion organisms! Although 100 trillion cells would only weigh about 1 g, after another 24 hours (a total of two days) of replicating every 30 minutes, the mass of cells would explode to 10^{14} g, or 10^7 tons. So why are we not buried under mountains of *E. coli*?

Nutrient Supplies Limit Microbial Growth

One factor limiting growth is that microbes commonly encounter environments where essential nutrients are limited. **Essential nutrients** are those compounds a microbe cannot make itself but must gather from its environment if the cell is to grow and divide. Microbial cells obtain all the essential nutrients for growth from their immediate environment. Consequently, when an environment becomes depleted of one or more essential nutrients, the microorganism will stop growing. How various organisms cope with these periods of starvation until nutrients are restored is a rapidly developing field of microbiology.

All microorganisms require a minimum set of **macronutrients**, nutrients needed in large quantities. Six of these—carbon, nitrogen, phosphorus, hydrogen, oxygen,

---

Table 4.1   Growth factors and natural habitats of organisms associated with disease.

Organism	Diseases	Natural habitats	Growth factors
Shigella	Bloody diarrhea	Humans	Nicotinamide (NAD)[a]
Haemophilus	Meningitis, chancroid	Humans and other animal species, upper respiratory tract	Haemin, NAD
Staphylococcus	Boils, osteomyelitis	Widespread	Complex requirement
Abiotrophia	Osteomyelitis	Humans and other animal species	Vitamin K, cysteine
Legionella	Legionnaires' disease	Soil, refrigeration cooling towers	Cysteine
Bordetella	Whooping cough	Humans and other animal species	Glutamate, proline, cystine
Francisella	Tularemia	Wild deer; rabbits	Complex, cysteine
Mycobacterium	Tuberculosis, leprosy	Humans	Nicotinic acid (NAD),[a] alanine
Streptococcus pyogenes	Pharyngitis, rheumatic fever	Humans	Glutamate, alanine

[a]Both nicotinamide and nicotinic acid are derived from NAD, nicotinamide adenine dinucleotide.

and sulfur—make up the carbohydrates, lipids, nucleic acids, and proteins of the cell. Four other macronutrients are cations whose roles range from serving as enzyme cofactors (Mg^{2+}, Fe^{2+}, and K^+) to acting as regulatory signal molecules (Ca^{2+}). In addition to macronutrients, all cells require very small amounts of certain trace elements, called **micronutrients**. These include cobalt, copper, manganese, molybdenum, nickel, and zinc, which are ubiquitous contaminants on glassware and in water. As a result, these trace elements are not added to laboratory media unless heroic measures have been taken to first remove the elements from the medium. Micronutrients are required by cells because these elements are frequently essential components of enzymes or can themselves be components of cofactors. **Cofactors** are small molecules that fit into specific enzymes and aid in the catalytic process. Cobalt, for example, is part of the cofactor vitamin B_{12}. Some organisms, such as the laboratory "workhorse" bacterium *E. coli*, make all their proteins, nucleic acids, and cell wall and membrane components from this very simple blend of chemical elements and compounds. For many other microbes, this basic set of nutrients is not enough. One example is *Borrelia burgdorferi*, the causative

agent of Lyme disease, which requires an extensive mixture of complex organic supplements to grow.

Microbes Evolved to Grow in Different Environments

Based in part on the ecological niche it inhabits, an organism may have evolved to require additional **growth factors** (**Table 4.1**), specific nutrients that are not required by all cells. For example, why should an organism like *Streptococcus pyogenes* make glutamic acid or alanine if those amino acids are readily available in its normal environment (such as, the human oral cavity)? Because it never needs to make these compounds, *S. pyogenes*, as a matter of efficiency, has lost the genes whose protein products synthesize glutamic acid and alanine. A **defined minimal medium** contains only those compounds needed for an organism to grow (**Table 4.2**). In the case of *S. pyogenes*, this would include glutamic acid and alanine in addition to the macro- and micronutrients mentioned earlier. Other organisms have adapted so well to their natural habitat that we still do not know how to grow them in the laboratory. For example, *Rickettsia prowazekii*, the causative agent of Rocky Mountain

Table 4.2   Composition of commonly used media.

Media	Ingredients per liter		Organisms cultured
Luria Bertani (complex)	Bacto tryptone[a]	10 g	Many gram-negative and gram-positive organisms
	Bacto yeast extract	5 g	
	NaCl	10 g	
	pH 7		
M9 medium (defined)	Glucose	2.0 g	Gram-negative organisms such as *E. coli*
	Na_2HPO_4	6.0 g (42 mM)	
	KH_2PO_4	3.0 g (22 mM)	
	NH_4Cl	1.0 g (19 mM)	
	NaCl	0.5 g (9 mM)	
	$MgCl_2$	2.0 mM	
	$CaCl_2$	0.1 mM	
	pH 7		
Azotobacter medium (defined)	Mannitol	2.0 g	*Azotobacter*
	K_2HPO_4	0.5 g	
	$MgSO_4 \cdot 7H_2O$	0.2 g	
	$FeSO_4 \cdot 7H_2O$	0.1 g	
Sulfur oxidizers (defined)	NH_4Cl	0.52 g	*Thiobacillus thiooxidans*
	KH_2PO_4	0.28 g	
	$MgSO_4 \cdot 7H_2O$	0.25 g	
	$CaCl_2$	0.07 g	
	Elemental sulfur	1.56 g	
	CO_2	5%	
	pH 3		

[a]Bacto tryptone is a pancreatic digest of casein (bovine milk protein).

A.

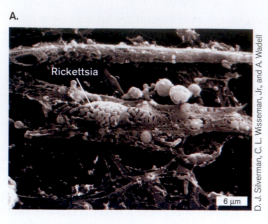

D. J. Silverman, C. L. Wisseman, Jr., and A. Wadell

Rickettsia

6 μm

B.

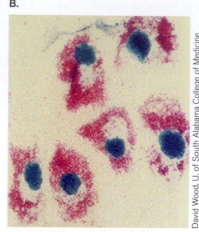

David Wood, U. of South Alabama College of Medicine

Figure 4.1 *Rickettsia prowazekii* **growing within eukaryotic cells.** **A.** *R. prowazekii* growing within the cytoplasm of a chicken embryo fibroblast (SEM). **B.** Giemsa stain of *R. prowazekii* growing within cultured human L cells.

spotted fever, grows only within the cytoplasm of eukaryotic cells (**Fig. 4.1**). This obligate intracellular bacterium lost key pathways needed for independent growth because the host cell supplies them. Despite extensive efforts to grow them outside a host cell (**axenic growth**), cells of *R. prowazekii* have proved uncooperative in this endeavor.

Although it will not grow in a defined medium, we can at least grow *Rickettsia* in the laboratory using eggs or animal cell tissue culture. But it is estimated that 99.9 % of all the microorganisms on Earth cannot be grown in the laboratory at all. How do we even know that these microbes exist if we cannot grow them? Evidence for their existence has been discovered only recently using newly developed tools of molecular biology. All known microorganisms have a set of genes, associated with ribosomes, whose DNA sequences are highly conserved across the phylogenetic tree. A DNA-amplifying procedure called polymerase chain reaction (PCR, described in Section 7.6) can be used to screen for the presence of these genes in soil and water samples. Comparing the DNA sequences of the PCR products with the DNA sequences of similar genes from known organisms reveals that nature harbors many microbes hitherto undiscovered because we cannot grow them in the laboratory. Even though the growth and nutritional requirements of these phantom microbes are unknown, modern genomic techniques can expose their existence. We can gain remarkable insight into the physiology of these "invisible," nonculturable organisms by comparing their gene sequences, mined by PCR, with known gene sequences from culturable organisms that have well-characterized physiologies.

Microbes Build Biomass through Autotrophy or Heterotrophy

Maintaining life on this planet is an amazing process. All of Earth's life-forms are based on carbon, which they acquire by delicately choreographed processes that recycle key nutrients. The carbon cycle, a critical part of this process, involves two counterbalancing metabolic groups of

organisms, heterotrophs and autotrophs. Their pathways of metabolism are discussed in chapters 13 and 14.

Heterotrophs (such as *E. coli*) rely on other organisms to form the organic compounds, such as glucose, that they use as carbon sources. During heterotrophic metabolism, organic carbon sources are disassembled to generate energy and then reassembled to make cell constituents such as proteins and carbohydrates. This process converts a large amount of the organic carbon source to CO_2, which is then released to the atmosphere. Thus, left on their own, heterotrophs would deplete the world of organic carbon sources (converting them to unusable CO_2) and starve to death. For life to continue, CO_2 must be recycled.

Autotrophs assimilate CO_2 as a carbon source, reducing it (adding hydrogen atoms; see Chapter 15) to make complex cell constituents made up of C, H, and O (for example, carbohydrates, which have the general formula CH_2O). These organic compounds can later be used as carbon sources by heterotrophs. As will be discussed below, autotrophs are subclassified, based on how they obtain energy, into **photoautotrophs**, organisms that use light for photosynthesis, and **chemoautotrophs** also known as **chemolithotrophs** or **lithotrophs** (literally, rock-eaters), organisms that gain energy by oxidizing inorganic substances such as iron or ammonia. Both photoautotrophs and chemoautotrophs carry out the autotrophic process of CO_2 fixation, which forms part of the carbon cycle between autotrophs and heterotrophs (**Fig. 4.2**).

> **THOUGHT QUESTION 4.1** In a mixed ecosystem of autotrophs and heterotrophs, what happens when a heterotroph allows the autotroph to grow and begin to make excess organic carbon?

Microbes Obtain Energy through Phototrophy or Chemotrophy

Although the macronutrients mentioned earlier provide the essential building blocks (in other words, carbon) to

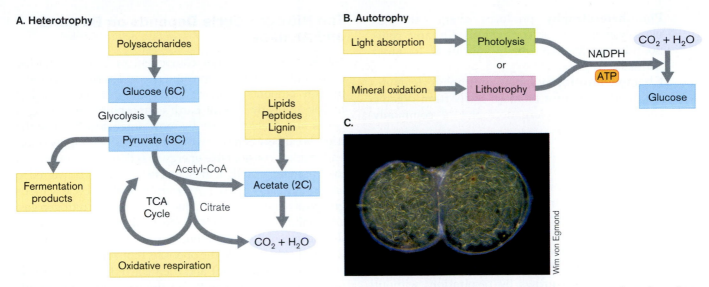

A. Heterotrophy

Polysaccharides

↓

Glucose (6C)

Glycolysis

Pyruvate (3C)

Fermentation products

Acetyl-CoA

TCA Cycle

Citrate

Oxidative respiration

Lipids Peptides Lignin

Acetate (2C)

$CO_2 + H_2O$

B. Autotrophy

Light absorption → Photolysis

or

Mineral oxidation → Lithotrophy

NADPH

ATP

$CO_2 + H_2O$

Glucose

C.

Wim von Egmond

Figure 4.2   The carbon cycle. The carbon cycle requires both autotrophs and heterotrophs. **A.** Heterotrophs gain energy from degrading complex organic compounds, such as polysaccharides, to smaller compounds, such as glucose and pyruvate. The carbon from pyruvate moves through the tricarboxylic acid (TCA) cycle, and CO_2 is released. In the absence of a TCA cycle, the carbon can end up as fermentation products, such as ethanol or acetic acid. **B.** Autotrophs use light energy or energy derived from the oxidation of minerals to capture CO_2 and convert it to complex organic molecules. **C.** *Nostoc* is an autotroph. These cyanobacteria live in a gelatinous sphere. The spheres shown are colonies (approx. 1 mm diameter) that are just beginning to divide. The individual cyanobacteria are the small cells forming the long strands inside the spheres.

make proteins and other cell structures, those synthetic processes require an energy source. Depending on the organism, energy can be obtained from chemical reactions triggered by the absorption of light (**phototrophy**, or photosynthesis) or from oxidation-reduction reactions that transfer electrons from high-energy compounds to make products of lower energy (**chemotrophy**). Chemotrophic organisms include two classes that use different sources of electron donors: chemoautotrophs and chemoheterotrophs. Chemoautotrophs oxidize inorganic chemicals H_2, H_2S, NH_4, NO_2^-, and Fe^{2+} for energy. **Chemoheterotrophs** oxidize organic compounds such as sugars to obtain energy.

> **NOTE:** The following prefixes for "-trophy" terms help distinguish different forms of energy-yielding metabolism.
>
> Carbon source for biomass
>
> Auto- CO_2 is fixed and assembled into organic molecules.
>
> Hetero-   Preformed organic molecules are acquired from outside and assembled.
>
> Energy source
>
> Photo-   Light absorption excites electron to high-energy state.
>
> Chemo-   Chemical electron donors are oxidized.
>
> Electron source
>
> Litho-   Inorganic molecules donate electrons.
>
> Organo-   Organic molecules donate electrons.

In chemotrophy, the amount of energy harvested from oxidizing a compound is directly related to the compound's reduction state. The more reduced the compound is, the more electrons it has to give up and the higher the potential energy yield. A reduced compound, such as glucose, can donate electrons to a less reduced (more oxidized) compound, such as nicotinamide adenine dinucleotide (NAD), releasing energy (in the form of donated electrons) and becoming oxidized in the process. NAD is a cell molecule critical to energy metabolism and is discussed, along with oxidation-reduction reactions, in Chapter 13.

In short, microbes are classified on the basis of their carbon and energy acquisition as follows:

- **Autotrophs.** Autotrophs build biomass by fixing CO_2 into complex organic molecules. Autotrophs gain energy through one of two general metabolic routes that either use or ignore light:

  Photoautotrophy generates energy through light absorption by the photolysis (light-activated breakdown) of H_2O or H_2S. The energy is used to fix CO_2 into biomass.

  Chemoautotrophy (or lithotrophy) produces energy from oxidizing inorganic molecules such as iron, sulfur, or nitrogen. This energy is also used to fix CO_2 into biomass.

- **Heterotrophs.** Heterotrophs break down organic compounds from other organisms to gain energy and to harvest carbon for building their own biomass. Heterotrophic metabolism can be divided into two classes, also based on whether light is involved.

<u>Photoheterotrophy</u> produces energy through photolysis of organic compounds. Organic compounds are broken down and used to build biomass.

<u>Chemoheterotrophy</u> (or organotrophy) yields energy and carbon for biomass solely from organic compounds. Chemoorganotrophy is commonly called heterotrophy.

Note that many species, particularly free-living soil and aquatic bacteria, can utilize more than one of these strategies, depending on environmental conditions. They do this by having multiple gene systems that are expressed under different conditions and whose products carry out different functions. For example, *Rhodospirillum rubrum* grows by photoheterotrophy when light is available and oxygen is absent; but when oxygen is available, the organism grows by respiration, without absorbing light.

The survival and metabolism of any one group of organisms depend on the survival and metabolism of other groups of organisms. Even metazoa (multicellular organisms) rely on microbial metabolism to survive. For example, the cyanobacteria, a type of photosynthetic microorganism that originated 2.5–3.5 billion years ago, produce most of the oxygen we and other metazoa breathe. In fact, cyanobacteria also form the base of the marine food chain. The autotrophic cyanobacteria fix carbon in the ocean and are eaten by heterotrophic protists. The protists then are devoured by fish that produce the CO_2 that is fixed by the cyanobacteria. The cyanobacteria also depend on the heterotrophic bacteria to consume the molecular oxygen that the cyanobacteria produce, since molecular oxygen is toxic to cyanobacteria.

Energy Is Stored for Later Use

Whatever the source, energy once obtained must be converted to a form useful to the cell. This form can be chemical energy, such as that contained in the high-energy phosphate bond in adenosine triphosphate (ATP), or electrochemical energy, where energy is stored in the form of an electrical potential existing between compartments separated by a membrane (see Chapter 14). Energy stored by an electrical potential across the membrane is known as the **membrane potential**. A membrane potential is generated when chemical energy is used to pump protons (and in some cases Na^+) outside of the cell, so that the proton (or Na^+) concentration is greater outside the cell than inside. This ion movement produces an electrical gradient across the cell membrane, making the inside of the cell more negatively charged than the outside. The energy stored in the membrane potential can be used directly to move nutrients into the cell via specific transport proteins (see Section 3.3), to drive motors that rotate flagella, and to drive synthesis of ATP.

The Nitrogen Cycle Depends on Bacteria and Archaea

Nitrogen is an essential component of proteins, nucleic acids, and other cellular constituents, and as such is required in large amounts by living organisms. Nitrogen gas (N_2) makes up nearly 79% of Earth's atmosphere, but nitrogen gas is unavailable for use by most organisms because the triple bond between the two nitrogen atoms is highly stable and requires considerable energy to split them. For nitrogen to be used for growth, it must first be "fixed," or converted to ammonium ions (NH_4^+). As with the carbon cycle, various groups of organisms collaborate to interconvert nitrogen gas, ammonium ions, and nitrate (NO_3^-) ions in what is called the nitrogen cycle. One group "fixes" atmospheric nitrogen while other bacteria do the opposite, transforming ammonia to nitrate (**nitrification**) and then converting nitrate to N_2 (**denitrification**) (**Fig. 4.3**). For the environmental significance of nitrogen metabolism, see Chapter 22.

Nitrogen-fixing bacteria may be free-living in soil or water or they may form symbiotic associations with plants or other organisms. A **symbiont** is an organism that lives in intimate association with a second organism. *Rhizobium*, *Sinorhizobium*, and *Bradyrhizobium* species, for example, are nitrogen-fixing symbionts in leguminous plants such as soybeans, chickpeas, and clover (**Fig. 4.4**). These plant symbionts take atmospheric nitrogen and convert it to the ammonia that the plant needs to form proteins and other essential compounds. This type of beneficial symbiosis is called mutualism. Although symbionts are the most widely known nitrogen-fixing bacteria, the majority of nitrogen in soil and marine environments is fixed by free-living bacteria and archaea.

Once fixed, how does nitrogen get back into the atmosphere? The nitrifying bacteria, such as *Pseudomonas*, *Alcaligenes*, and *Bacillus*, gain energy by oxidizing ammonia to produce nitrate (a form of chemoautotro-

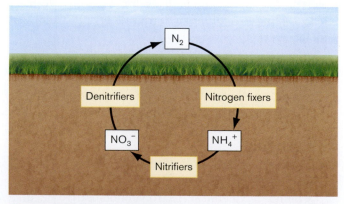

Figure 4.3   The nitrogen cycle. Dinitrogen gas (N_2) is fixed by species of bacteria (nitrogen fixers) that possess the enzyme nitrogenase. Other bacteria oxidize ammonia (NH_4^+) to generate energy. Still others use oxidized forms of nitrogen, such as NO_3^-, as an alternative electron acceptor in place of O_2.

A.

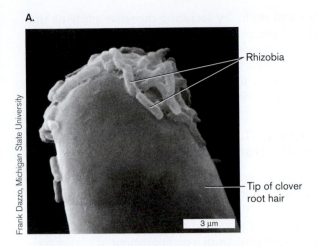

Rhizobia

Tip of clover
root hair

3 μm

Frank Dazzo, Michigan State University

B.

Inga Spence/Photo Researchers, Inc.

Figure 4.4 *Rhizobium* **and a legume. A.** Symbiotic
Rhizobium cells clustered on a clover root tip (SEM). The rhizobia
shown here are clustered on the surface of the root. Soon they
will start to invade the root and begin a symbiotic partnership that
will benefit both organisms. **B.** Root nodules. After the rhizobia
(approx. 0.9 μm × 3 μm) invade the plant root, symbiosis between
plant and microbe produces nodules. Two partly crushed nodules
(arrowheads) are shown with pink-colored contents. This color is
caused by the presence of the pigment leghaemoglobin, which
is found only in the nodules and is not produced by either the
bacterium or the plant when grown alone.

phy). Denitrifying bacteria are heterotrophs that use the
nitrate provided by the nitrifiers to produce a series of
nitrogen compounds, ending with gaseous N_2. The activity of denitrifying bacteria results in a substantial loss of
nitrogen into the atmosphere that roughly balances the
amount of nitrogen fixation that occurs each year. As in
the carbon cycle, this, once again, illustrates how nature
manages to replenish the planet Earth.

Eukaryotic Microbes Include Consumers and Producers

Eukaryotic microbes such as protists and fungi are heterotrophic consumers (discussed in Chapter 20). Protists and

fungi require complex organic molecules for growth; their
lifestyle involves predation, parasitism, or scavenging the
dead. Most protists and fungi need mitochondria, which
are the products of an ancient symbiotic partnering with
internalized prokaryotes (discussed in Section 17.6). Heterotrophic fungi possess the exceptional ability to digest
complex organic compounds such as lignin, a major component of wood and bark. In marine and aquatic systems,
protists form a huge part of the food chain, so that eating
trout or swordfish means, in effect, consuming trillions
of protists.

Just like photosynthetic bacteria, eukaryotic algae
are photoautotrophs that produce biomass through
photosynthesis. In addition to needing chloroplasts for
photosynthesis, however, algae require mitochondria
for energy production. (Recall that chloroplasts, like
mitochondria, evolved from internalized bacteria; see
Section 1.7.) Protist algae such as single-celled *Euglena*
are **mixotrophic**, capable of utilizing photosynthesis or
heterotrophic respiration, depending on environmental
conditions.

TO SUMMARIZE:

- **Microorganisms** require certain essential macro-
  and micronutrients to grow.
- **Microbial genomes** evolve in response to nutrient
  availability.
- **Obligate intracellular bacteria** lose metabolic pathways provided by their hosts and develop requirements for growth factors supplied by their hosts.
- In the **global carbon cycle**, autotrophs use CO_2 as
  a carbon source, either through photosynthesis or
  through lithotrophy.
- **Autotrophs** make complex organic compounds that
  are consumed by heterotrophs.
- **Nitrogen fixers**, **nitrifiers**, and **denitrifiers** contribute to the nitrogen cycle.
- **All the preceding reactions** are carried out by prokaryotes (bacteria or archaea), whereas eukaryotes
  carry out only a limited range of heterotrophic and
  photosynthetic reactions.

4.2 Nutrient Uptake

Whether a microbe is propelled by flagella toward a favorable habitat or, lacking motility, drifts through its environment courtesy of Brownian movement, it must be able
to find nutrients and move them across the membrane
into the cytoplasm. The membrane, however, presents a
daunting obstacle. Membranes are designed to separate
what is outside the cell from what is inside. So, for a cell
to gain sustenance from the environment, the membrane
must be *selectively* permeable to nutrients the cell can use.
A few compounds, such as oxygen and carbon dioxide,

can passively diffuse across the membrane, but most cannot. Selective permeability is achieved in three ways:

- By the use of substrate-specific carrier proteins (**permeases**) in the membrane
- With the aid of dedicated nutrient-binding proteins that patrol the periplasmic space
- Through the action of membrane-spanning protein channels, or pores, that discriminate between substrates

Microbes must also overcome the problem of low nutrient concentrations in the natural environment (for example, lakes or streams). If the intracellular concentration of nutrients were no greater than the extracellular concentration, the cell would remain starved of most nutrients. To solve this dilemma, most organisms have evolved efficient transport systems that concentrate nutrients inside the cell relative to outside. However, moving molecules against a concentration gradient requires some form of energy.

In contrast to environments where nutrients are available but exist at low concentrations (for example, aqueous environments), certain habitats have plenty of nutrients, but those nutrients are locked in a form that cannot be transported into the cell. Starch, a large, complex carbohydrate, is but one example. Many microbes unlock

these nutrient vaults by secreting digestive proteins that break down complex carbohydrates or other molecules into smaller compounds that are easier to transport. The amazing techniques that cells use to extrude these large digestive proteins through the membrane and into their surrounding environment are discussed in Chapter 8.

Facilitated Diffusion: Transport Down a Concentration Gradient

Although most transport systems use cellular energy to bring compounds into the cell, a few do not. Facilitated diffusion is a type of system that simply uses the concentration gradient of a compound to move that compound across the membrane from a compartment of higher concentration to a compartment of lower concentration. The best these systems can do is to equalize the internal and external concentrations of a solute; facilitated transport cannot move a molecule *against* its gradient.

The most important facilitated diffusion transporters are those of the aquaporin family that transport water and small polar molecules such as glycerol. Glycerol transport is performed by an integral membrane protein in *E. coli* called GlpF. The structure of a glycerol channel is shown in **Figure 4.5A**. This protein channel allows passive uptake

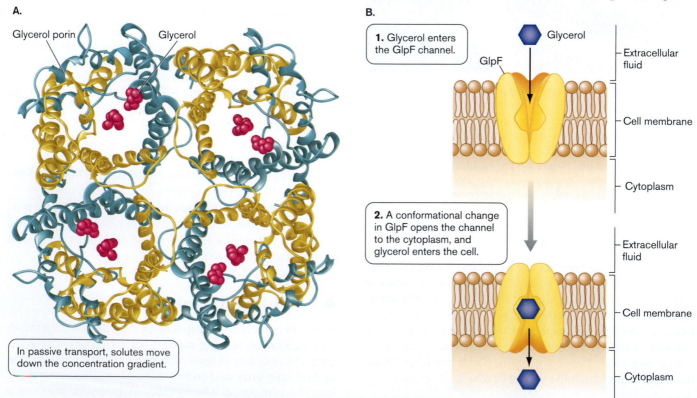

A.

Glycerol porin   Glycerol

In passive transport, solutes move down the concentration gradient.

B.

1. Glycerol enters the GlpF channel.

Glycerol

GlpF

Extracellular fluid

Cell membrane

Cytoplasm

2. A conformational change in GlpF opens the channel to the cytoplasm, and glycerol enters the cell.

Extracellular fluid

Cell membrane

Cytoplasm

Figure 4.5   Facilitated diffusion.   A. The glycerol transporter of *E. coli*, viewed from the external side of the membrane, consists of a tetramer of four dimer channels formed by hydrophobic alpha helices alternating back and forth across the cell membrane. Each channel (blue and yellow) contains a glycerol (magenta). (PDB code: 1FX8) **B.** Facilitated diffusion of glycerol through GlpF. The protein facilitates the movement of the compound from outside the cell, where glycerol is at high concentration, to inside the cell where the concentration of glycerol is low.

of glycerol, a polar molecule useful to cells for energy and for building phospholipids. The glycerol channel complex is viewed from the outer face of the membrane. The complex is a tetramer of four channels. Each channel transports a molecule of glycerol (magenta).

In the membrane, GlpF reversibly assumes two conformations. One form exposes the glycerol-binding site to the external environment, whereas the second form exposes this site to the cytoplasm. When the concentration of glycerol is greater outside than inside the cell, the form with the binding site exposed to the exterior is more likely to find and bind glycerol. After binding glycerol, GlpF changes shape, closing itself to the exterior and opening to the interior (**Fig. 4.5B**). Bound glycerol is then released into the cytoplasm (influx). Once the cytoplasmic concentration of glycerol equals the exterior concentration, bound glycerol can be released to either compartment. If the internal concentration exceeds the external concentration, a situation that can arise if a cell is moved from a high- to a low-glycerol environment, glycerol will move out of the cell (efflux). However, facilitated diffusion normally promotes glycerol influx, because the cell consumes the compound as it enters the cytoplasm, keeping cytoplasmic concentrations of glycerol low.

WWW | GlpF homolog

Active Transport by Membrane Transporters Requires Energy

Most forms of transport expend energy to take up molecules from outside the cell and concentrate them inside. The ability to import nutrients against their natural gradients is critical in aquatic habitats, where nutrient concentrations are low, and in soil habitats, where competition for nutrients is high.

The simplest way to use energy to move molecules across a membrane is to exchange the energy of one chemical gradient for that of another. The most common chemical gradients used are those of ions, particularly the positively charged ions Na^+ and K^+. These ions are kept at different concentrations on either side of the cell membrane. When an ion moves *down* its concentration gradient (from high to low), energy is released. Some transport proteins harness that free energy and use it to drive transport of a second molecule *up*, or against, its concentration gradient. This is called **coupled transport**.

The two types of coupled transport systems are **symport**, where the two molecules travel in the same direction, and **antiport**, in which the actively transported molecule moves in the direction *opposite* to the driving ion. An example of a symporter is the lactose permease

LacY of *E. coli*, one of the first transport proteins whose function was elucidated. This work was carried out by the pioneering membrane biochemist H. Ronald Kaback. LacY moves lactose inward, powered by a proton that is also moving inward (symport). LacY proton-driven transport is said to be **electrogenic** because an unequal distribution of charge results (for example, symport of a neutral lactose molecule with H^+ results in net movement of positive charge).

An example of **electroneutral** coupled transport, in which there is no net transfer of charge, is that of the Na^+/H^+ antiporter. The Na^+/H^+ antiporter couples the export of Na^+ with the import of a proton (antiport). Because molecules of like charge are exchanged, there is no net movement of charge. The mechanism by which Na^+/H^+ antiporters function was elucidated by the Israeli biochemist Etana Padan. Sodium exchange is of universal importance for all organisms and is particularly critical for organisms living in a high-salt habitat (see Section 5.4).

> **THOUGHT QUESTION 4.2** In what situation would antiport and symport be passive rather than active transport?
>
> **THOUGHT QUESTION 4.3** What kind of transporter, other than an antiporter, could produce electroneutral coupled transport?

Symport and antiport transporter proteins function by alternately opening one end or the other of a channel that spans the cell membrane. The channel contains solute-binding sites (**Figs. 4.6A and B**). When the channel is open to the high-concentration side of the membrane, the driving ion (solute) attaches to binding sites. The transport protein then changes shape to open that site to the low-concentration side of the membrane and the ion leaves. When and where the second (cotransported) solute binds depends on whether the transport protein is an antiporter or a symporter.

WWW | Movie on antiport

With all this ion traffic going on across the membrane, a careful accounting must be kept of how many ions are inside relative to outside. The cell must recirculate ions into and out of the cell to maintain certain gradients if the organism is to survive. Because key ATP-producing systems require an electrochemical gradient across the membrane, it is especially important to keep the interior of the cell negatively charged relative to the exterior. However, because the movement of many compounds is coupled to the import of positive ions, the electrochemical gradient will eventually dissipate, or depolarize, unless positive ions are also exported. Depolarization must be avoided

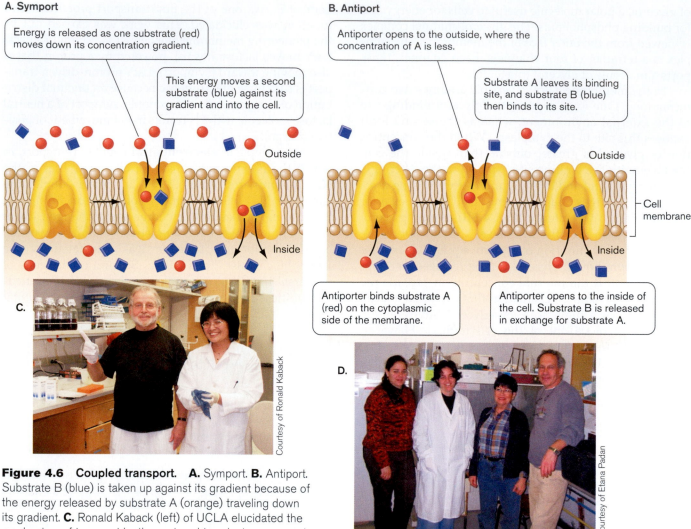

A. Symport

Energy is released as one substrate (red) moves down its concentration gradient.

This energy moves a second substrate (blue) against its gradient and into the cell.

Outside

Inside

B. Antiport

Antiporter opens to the outside, where the concentration of A is less.

Substrate A leaves its binding site, and substrate B (blue) then binds to its site.

Outside

Cell membrane

Inside

Antiporter binds substrate A (red) on the cytoplasmic side of the membrane.

Antiporter opens to the inside of the cell. Substrate B is released in exchange for substrate A.

C.

Courtesy of Ronald Kaback

D.

Courtesy of Etana Padan

Figure 4.6 Coupled transport. A. Symport. **B.** Antiport. Substrate B (blue) is taken up against its gradient because of the energy released by substrate A (orange) traveling down its gradient. **C.** Ronald Kaback (left) of UCLA elucidated the mechanism of transport by the proton-driven lactose symporter LacY. **D.** Etana Padan (second from right) of The Hebrew University of Jerusalem dissected the molecular mechanism of the Na^+/H^+ antiporter NhaA.

because a depolarized cell loses membrane integrity and cannot carry out the simple transport functions needed to sustain life. A healthy cell maintains a proper charge balance using the electron transport chain to move protons out of the cell and by exchanging negatively and positively charged ions as needed.

In addition to the direct linking of ion transport described for symport and antiport systems, ion circuitry is such that movement of one ion can be linked *indirectly* to movement of another molecule. For example, proton concentrations are typically greater outside than inside the cell. The inwardly directed proton gradient is called proton motive force. Proton motive force can impel the exit of Na^+ through the Na^+/H^+ antiporter. The resulting Na^+ gradient can drive the symport of amino acids into the cell. In this case, Na^+ moves back into the cell down its gradient, and the energy released is tied to the import of an amino acid against its gradient.

ABC Transporters Are Powered by ATP

As we pointed out in Section 3.3, a major function of proton transport is to create the proton motive force that powers ATP synthesis (for details on proton motive force, see Chapter 14). The energy stored in ATP, whether generated by the proton current or by cytoplasmic means such as fermentation, can drive membrane transport of nutrients.

The largest family of energy-driven transport systems is the **ATP-binding cassette** superfamily, also known as **ABC transporters**. These transporters are found in bacteria, archaea, and eukaryotes. It is impressive to note that nearly 5% of the *E. coli* genome is dedicated to producing the components of 70 different varieties of uptake and efflux ABC transporters. The uptake ABC transporters are critical for transporting nutrients such as maltose, histidine, arabinose, and galactose. The efflux ABC transporters are generally used as multidrug efflux pumps that allow microbes to survive exposures to hazardous chemicals. *Lactococcus*, for example, can use one pump, LmrP, to export a broad range of antibiotics, including tetracyclines, streptogramins, quinolones, macrolides, and aminoglycosides, conferring resistance to those drugs (see Section 27.6).

An ABC transporter typically consists of two hydrophobic proteins that form a membrane channel and two peripheral cytoplasmic proteins that contain a highly conserved amino acid motif involved with binding ATP. Any conserved amino acid sequence found in a family of proteins is viewed as a "cassette" because the sequence appears to have been inserted into those proteins for specific functions (hence, ATP-binding cassette). The ABC transporter superfamily contains both uptake and efflux transport systems. The uptake systems (but not the efflux systems) possess an additional, extracytoplasmic protein, called a **substrate-binding protein (SBP)**, that initially binds the substrate (also called solute). In gram-negative bacteria, these substrate-binding proteins float in the periplasmic space between the inner and outer membranes. In gram-positive bacteria, which lack an outer membrane, the proteins must be tethered to the cell surface.

ABC transport (**Fig. 4.7**) starts with the SBP snagging the appropriate solute, either as it floats by a gram-positive microbe or as the molecule enters the periplasm of a gram-negative cell. Most substrates nonspecifically enter the periplasm of gram-negative organisms through the outer membrane pores, although some high-molecular-weight substrates, like vitamin B_{12}, require the assistance of a spe-cific outer membrane protein to move the substrate into the periplasm. Because SBPs have a high affinity for their cognate (matched) solutes, their use increases the efficiency of transport when concentrations of solute are low.

Once united with its solute, the SBP binds to the periplasmic face of the channel protein and releases the solute, which now binds to a site on the channel protein. This interaction triggers a structural (or conformational) change in the channel protein that is telegraphed to the nucleotide-binding proteins associated with the cytoplasmic side. On receiving this signal, the nucleotide-binding proteins start hydrolyzing ATP and send a return conformational change through the channel, signaling the channel to open its cytoplasmic side and allow the solute to enter the cell.

The three proteins that constitute ABC transporters appear to have arisen from a common ancestral porter and share a considerable amount of amino acid sequence homology. One unusual feature of ABC transporters is that although the membrane channel components and ATP-hydrolyzing components are generally present as distinctly separate polypeptides in *uptake* transporters, they are fused in *efflux* systems.

Siderophores Are Secreted to Scavenge Iron

Iron, an essential nutrient of most cells, is mostly locked up in nature as $Fe(OH)_3$, which is insoluble and unavailable for transport. Many bacteria and fungi have solved this transport dilemma by synthesizing and secreting specialized molecules called **siderophores** (Greek for "iron bearer") that have a very high affinity for whatever soluble ferric iron is available in the environment. These iron scavenger molecules are produced and sent forth by cells when the intracellular iron concentration is low (**Fig. 4.8**). In most gram-negative organisms, the siderophore binds iron in the environment, and the siderophore-iron complex then attaches to specific receptors in the outer membrane. At this point, either the iron is released directly and is passed to other transport proteins or the complex is transported across the cytoplasmic membrane by a dedicated ABC transporter. The iron is released intracellularly and reduced to Fe^{2+} for biosynthetic use. Other gram-negative microorganisms, such as *Neisseria gonorrhaeae* (the causative agent of gonorrhea), do not use siderophores at all but employ receptors on their surface that bind human iron complexes (for example, transferrin or lactoferrin) and wrest the iron from them.

Group Translocation against a Concentration Gradient

The ABC transporters we have just considered increase concentrations of solute inside the cell relative to the outside concentration. They move nutrients "uphill" against

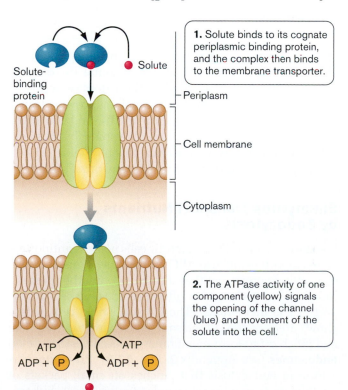

1. Solute binds to its cognate periplasmic binding protein, and the complex then binds to the membrane transporter.

2. The ATPase activity of one component (yellow) signals the opening of the channel (blue) and movement of the solute into the cell.

Figure 4.7   ABC transporters.

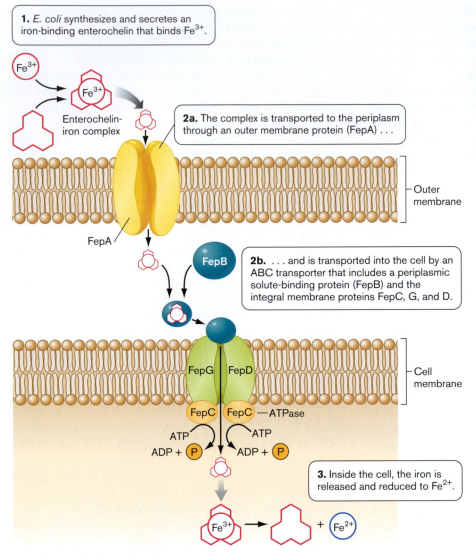

1. *E. coli* synthesizes and secretes an iron-binding enterochelin that binds Fe^{3+}.

Enterochelin-iron complex

2a. The complex is transported to the periplasm through an outer membrane protein (FepA) . . .

Outer membrane

FepA

FepB

2b. . . . and is transported into the cell by an ABC transporter that includes a periplasmic solute-binding protein (FepB) and the integral membrane proteins FepC, G, and D.

FepG   FepD

Cell membrane

FepC   FepC —ATPase

ATP   ATP

ADP + P   ADP + P

3. Inside the cell, the iron is released and reduced to Fe^{2+}.

Fe^{3+} → + Fe^{2+}

Figure 4.8   Siderophores and iron transport.

a concentration gradient. An entirely different system known as **group translocation** cleverly accomplishes the same result but without really moving a substance "uphill." Group translocation alters the substrate during transport by attaching a new group (for example, phosphate) on it. Because the modified nutrient inside the cell is a different chemical entity than the related compound outside, the parent solute entering the cell is always moving *down* its concentration gradient, regardless of how much solute has already been transported. Note that this process uses energy to chemically alter the solute. ABC transporters and group translocation systems both involve active transport, but group translocation systems are *not* ABC transporters.

The **phosphotransferase system (PTS)** is a well-characterized group translocation system present in many bacteria. It uses energy from phosphoenolpyruvate (PEP), an intermediate in glycolysis, to attach a phosphate to specific sugars during their transport into the cell. Glucose, for example, is converted during transport to glucose 6-phosphate. The system has a modular design that accommodates different substrates. There are common elements used by all sugars that are transported by the PTS system and elements that are unique to a given carbohydrate (**Fig. 4.9**). Common elements located in the cytoplasm include Enzyme I (PtsI) and a histidine-rich protein called HPr (PtsH). Enzyme I strips the high-energy phosphate from PEP and passes it to HPr, which in turn distributes the phosphate to various substrate-specific transport proteins called Enzyme II. A typical Enzyme II comprises three domains (A, B, and C) that may be fused together as a single polypeptide or assembled in a variety of combinations. Regardless of the configuration, the phosphorylated HPr (HPr-P) transfers its phosphate to the Enzyme IIA domains/proteins, which relay the phosphate to their cognate Enzyme IIB domains/proteins. Enzyme IIB finally delivers the phosphate to the specific sugar that has been transported into the cell by the Enzyme IIC domain embedded in the cytoplasmic membrane (for example, glucose is transported by Enzyme IIC and converted to glucose 6-phosphate by Enzyme IIB). In Chapter 9, we will see how this physiological system impacts the genetic control of many other systems.

Eukaryotes Transport Nutrients by Endocytosis

Like prokaryotic cells, eukaryotic cells possess antiporters and symporters and use ABC transport systems as multidrug efflux pumps, but they also employ another process, called endocytosis, which often precedes nutrient transport across membranes. The lack of a rigid cell surface allows areas of the cytoplasmic membrane to invaginate and pinch off to form membrane-enclosed vesicles called **endosomes** (see Appendix 2, **Fig. A2.8**). **Pinocytosis** is a form of endocytosis that pinches off small packets of medium to spawn small vacuoles (endosomes). Eukaryotic microbes such as amebas engulf prey bacteria by a

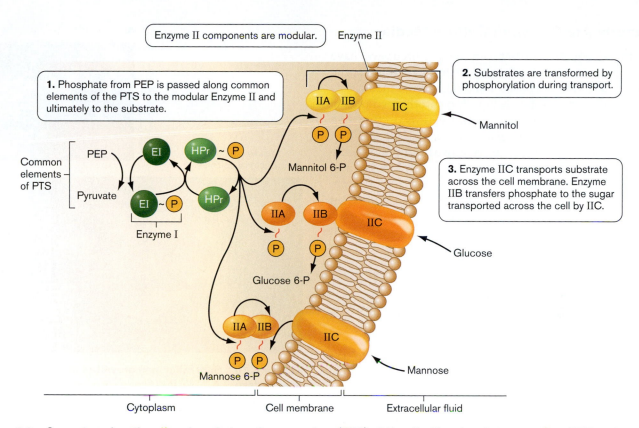

Figure 4.9   Group translocation: the phosphotransferase system (PTS) of *E. coli*.   The phosphate group from PEP is ultimately passed to the substrate during transport. The common elements of the PTS are Enzyme I (PtsI) and HPr (histidine-rich protein; PtsH). Each Enzyme II is specific for a given substrate and consists of modular components. Enzyme II for mannitol is one protein with three domains: A, B, and C. Enzyme II for glucose is really two proteins: One protein contains the A domain, and the B and C domains are joined to form the second protein. Mannose Enzyme II is designed with the opposite arrangement. Its A and B domains are fused into one protein, whereas the membrane protein is simply the C domain. ▶❚

form of endocytosis called **phagocytosis**, producing large vacuoles (**phagosomes**). Lysosomes, cytoplasmic vacuoles that contain various digestive enzymes, then fuse with phagosomes or endosomes, creating phagolysosomes. As a result of the fusion, digestive enzymes gain access to the endocytosed material, degrading it to a form and size that can be transported across the vacuolar membrane and into the cytoplasm.

TO SUMMARIZE:

- **Transport systems** move nutrients across semipermeable membranes.
- **Facilitated diffusion** helps solutes move across a membrane from a region of high concentration to one of lower concentration.
- **Antiporters and symporters** are coupled transport systems in which energy released by moving a driving ion (H⁺ or Na⁺) from a region of high concentration to one of low concentration is harnessed and used to move a solute against its concentration gradient.
- **ABC transporters** use the energy from ATP hydrolysis to move solutes "uphill" against their concentration gradients.

- **Siderophores** are secreted to bind ferric iron and transport it into the cell, where it is reduced to the more useful ferrous form. Siderophore-iron complexes enter cells with the help of ABC transporters.
- **Group translocation systems** chemically modify the solute during transport.
- **Eukaryotes use endocytosis**, a process in which small areas of the cytoplasmic membrane invaginate, trap extracellular fluid, and pinch off into the cytoplasm.

4.3   Culturing Bacteria

Microbes in nature usually exist in complex, multispecies communities, but for detailed studies, they must be grown separately in pure culture. It is nothing short of amazing—and humbling—that after 120 years of trying to grow microbes in the laboratory, we have succeeded in culturing only 0.1% of the microorganisms around us. Since the time of Koch in the late nineteenth century, microbiologists have used the same fundamental techniques to culture bacteria in the laboratory. It is true that improvements have been made, but the vast majority of the microbial world has yet to be tamed.

Bacteria Are Grown in Culture Media

For those organisms that can be cultured, we have access to a variety of culturing techniques that can be used for different purposes. Bacterial culture media may be either liquid or solid. A liquid, or broth, medium, in which organisms can move about freely, is useful when studying the growth characteristics of a single strain of a single species (that is, a **pure culture**). Liquid media are also convenient for examining growth kinetics and microbial biochemistry at different phases of growth. Solid media, usually gelled with agar, are useful when trying to separate mixtures of different organisms as they are found in the natural environment or in clinical specimens.

Pure Cultures Can Be Obtained by Dilution Streaking or by Spread Plate Technique

Solid media are basically liquid media to which a solidifying agent has been added. The most versatile and widely used solidifying agent is agar (for the development of agar medium, see Section 1.5). Derived from seaweed, agar forms an unusual gel that liquefies at 100°C but does not solidify until cooled to about 40°C. Liquefied agar medium poured into shallow, covered petri dishes cools and hardens to provide a large, flat surface on which a mixture of microorganisms can be streaked to separate individual cells. Each cell will divide and grow to form a distinct, visible colony of cells (**Fig. 4.10**). As shown in **Figure 4.11**, a drop of liquid culture is collected using an inoculating loop and streaked across the agar plate surface in a pattern called **dilution streaking**. Organisms fall off the loop as it moves along the agar surface. Toward the end of the streak, few bacteria remain on the loop, so at that point, individual cells will land and stick to different places on the agar surface. If the medium, whether artificial (for example, laboratory medium) or natural (for example, crab carapace) contains the proper nutrients and growth factors, a single cell will multiply into many millions of offspring, forming a **microcolony**. At first visible only under a microscope, the microcolony grows into a visible droplet called a **colony** (**Figs. 4.11B** and **4.12**). A pure culture of the species, or one strain of a species, can be obtained by touching a single colony with a sterile inoculating loop and inserting that loop into fresh liquid medium.

It is important to note that the "one cell equals one colony" paradigm does not hold for all bacteria. Organisms such as *Streptococci* or

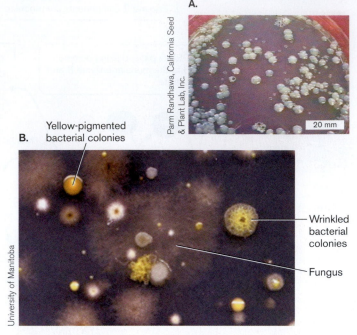

A. Parm Randhawa, California Seed & Plant Lab, Inc.

20 mm

B. University of Manitoba

Yellow-pigmented bacterial colonies

Wrinkled bacterial colonies

Fungus

Figure 4.10   Separation and growth of microbes on an agar surface.   A. Colonies (1 to 5 mm diameter) of *Acidovorax aveane* separated on an agar plate. This organism is a plant pathogen that causes watermelon fruit blotch. **B.** Mixture of yellow-pigmented bacterial colonies, wrinkled bacterial colonies, and fungus separated by dilution on an agar plate.

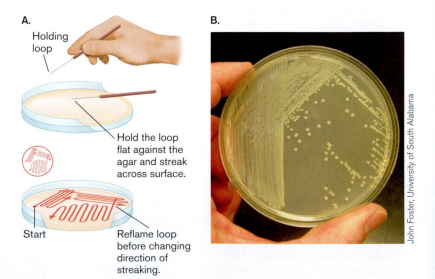

A.

Holding loop

Hold the loop flat against the agar and streak across surface.

Start

Reflame loop before changing direction of streaking.

B. John Foster, University of South Alabama

Figure 4.11   Dilution streaking technique.   A. A liquid culture is sampled with an inoculating loop and streaked across the plate in three to four areas, with the loop flamed between areas. The result is that dragging the loop across the agar diminishes the number of organisms clinging to the loop until only single cells are deposited at a given location. **B.** *Salmonella enterica* culture obtained by dilution streaking.

A.

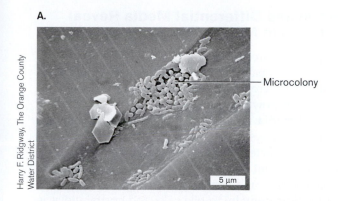

Harry F. Ridgway, The Orange County Water District

Microcolony

5 μm

B.

Douglas Prince, U. of New Hampshire

Figure 4.12   Bacterial colonies.   A. A three-day-old microcolony (biofilm) on the surface of a membrane used to treat municipal wastewater in Fountain Valley, California (SEM). **B.** Bacterial colony attacking a crab carapace.

A.

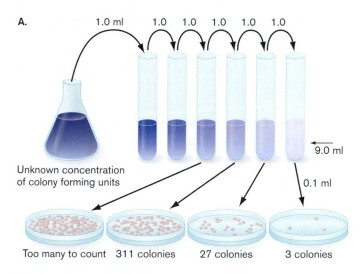

1.0 ml   1.0   1.0   1.0   1.0   1.0

9.0 ml

Unknown concentration of colony forming units

0.1 ml

Too many to count   311 colonies   27 colonies   3 colonies

B.

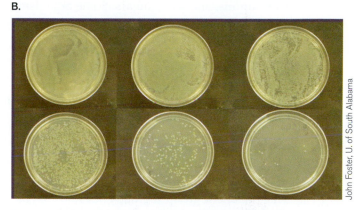

John Foster, U. of South Alabama

**Figure 4.13   Tenfold dilutions, plating, and viable counts.
A.** A culture containing an unknown concentration of cells is serially diluted. One milliliter (ml) of culture is added to 9.0 ml of diluent broth and mixed, and then 1 ml of this $\frac{1}{10}$ dilution is added to another 9.0 ml of diluent (10^{-2} dilution). These steps are repeated for further dilution, each of which lowers the cell number tenfold. After dilution, 1 ml of each dilution is spread onto an agar plate. **B.** Plates prepared as in (A) are incubated at 37°C to yield colonies. By multiplying the number of countable colonies (11 colonies on the 10^{-6} plate) by the reciprocal of the dilution factor you can calculate the number of cells (colony-forming units; CFUs) per milliliter in the original broth tube ($11 \times 10^{6} = 1.1 \times 10^{7}$ CFUs per ml).

Staphylococci usually do not exist as single cells, but occur as chains or clusters of several cells. Thus, a cluster of 10 *Staphyloccocus* cells will only form one colony on an agar medium and is called a colony-forming unit (CFU).

Another way to isolate pure colonies is the **spread plate** technique. Starting from a liquid culture of bacteria, a series of tenfold dilutions is made, and a small amount of each dilution is placed directly on the surface of separate agar plates (**Fig. 4.13**). The sample is spread over the surface of the plate using a heat-sterilized, bent glass rod. The early dilutions, those containing the most bacteria, will produce **confluent** growth that covers the entire agar surface. The later dilutions, containing fewer and fewer organisms, yield individual colonies. As we shall see later, spread plates not only enable us to isolate pure cultures but also can be used to enumerate the number of **viable** bacteria in the original growth tube. A viable organism is one that successfully replicates to form a colony. Thus, each colony on an agar plate represents one viable organism present in the original liquid culture.

There is some disagreement about the use of the term *viable cell*. Is an organism that does not replicate but continues to metabolize viable? Consider, for instance, that human beings who have undergone sterilization are incapable of producing offspring, but they are certainly alive. Nevertheless, based on a tradition dating back to the time of Koch, it remains convention within the scientific community that a "viable" cell reproduces to form a colony on a plate. Organisms that appear to metabolize but for some reason cannot replicate have been referred to as dormant or "**viable but nonculturable**." Increasingly, it seems probable that all viable but nonculturable organisms simply appear that way to us because we have not yet discovered the culture conditions necessary for them to reproduce.

Growth in Complex Media or in Synthetic Media

Bacteria can be grown in nutrient-rich but poorly defined **complex media** or in precisely defined **synthetic media**.

Recipes for complex media usually contain several poorly defined ingredients, such as yeast extract or beef extract, whose exact composition is not known. These additives include a rich variety of amino acids, peptides, nucleosides, vitamins, and some sugars. Some organisms are particularly fastidious, requiring that components of blood be added to a basic complex medium. The complex medium is now called an **enriched medium**.

Complex, or rich, media provide many of the chemical building blocks that a cell would otherwise have to synthesize on its own. For example, instead of making proteins that synthesize tryptophan, all the cell needs is a membrane transport system to harvest prefabricated tryptophan from the medium. Likewise, fastidious organisms that require blood in their media may reclaim the heme released from red blood cells as their own, using it as an enzyme prosthetic group, a group critical to enzyme function (for example, the heme group in cytochromes). All of this saves the scavenging cell a tremendous amount of energy, and as a result, bacteria tend to grow fastest in complex media.

An argument can be made that complex media mimics the rich environment that pathogens encounter in an animal host. However, the metabolism of a microbe growing in a complex medium is hard to characterize. How would you know whether *E. coli* possesses the ability to make tryptophan if the bacterium grows only in complex media? Fortunately, some organisms grow in fully defined synthetic media. In preparing a synthetic medium, one starts with water, then adds various salts, carbon, nitrogen, and energy sources in precise amounts. For self-reliant organisms like *E. coli* or *Bacillus subtilis*, that is all that is needed. Other organisms, such as *Shigella* species or mutant strains of *E. coli* or *B. subtilis*, require additional ingredients to satisfy requirements imposed by the absence of specific metabolic pathways.

THOUGHT QUESTION 4.4   What would be the phenotype (growth characteristic) of a cell that lacks the *trp* genes (genes required for the synthesis of tryptophan)? What would be the phenotype of a cell missing the *lac* genes (genes whose products catabolize the carbohydrate lactose)?

Selective and Differential Media Reveal Differences in Metabolism

Microorganisms are remarkably diverse with respect to their metabolic capabilities and resistance to certain toxic agents. These differences are exploited in **selective media**, which favor the growth of one organism over another, and in **differential media**, which expose biochemical differences between two species that grow equally well. For example, gram-negative bacteria, with their outer membrane, are much more resistant than gram-positive microbes to detergents like bile salts and certain dyes, such as crystal violet. A solid medium containing bile salts and crystal violet is considered selective because it favors the growth of gram-negative organisms over gram-positive ones. On the other hand, *E. coli* and *S. enterica* are both gram-negative, but only *E. coli* can ferment lactose. Colonies of each can be distinguished on solid media containing lactose, the dye neutral red, and a nonfermentable carbon source like peptone. Both organisms grow in the medium, but *E. coli* ferments the lactose and produces acidic end products. The lower pH surrounding an *E. coli* colony causes the dye to enter the cells, turning the colony red. In stark contrast, *S. enterica*, a major cause of diarrhea, will grow on nonfermentable peptides, but, because it cannot ferment lactose, acidic end products are not produced and the colonies remain white, the natural color of the colony. In this example, growth in differential media easily distinguishes colonies of lactose fermenters from nonfermenters.

Several media used in clinical microbiology are both selective and differential. **MacConkey medium**, for example, selects for growth of gram-negative bacte-

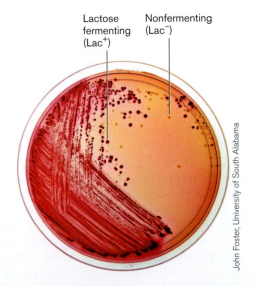

Lactose fermenting (Lac⁺) Nonfermenting (Lac⁻)

John Foster, University of South Alabama

Figure 4.14   MacConkey medium, a culture medium both selective and differential.   Dilution streak of a mixture of Lac⁺ and Lac⁻ bacteria. Only gram-negative bacteria grow on lactose MacConkey (selective). Only a species capable of fermenting lactose produces pink colonies (differential).

ria because it contains bile salts and crystal violet, which prevent the growth of gram-positives. The medium also includes lactose, neutral red, and peptones to differentiate lactose fermenters from nonfermenters. Thus, a culture grown in MacConkey medium will consist of gram-negative organisms that can be identified as lactose fermenters (red colonies) or nonfermenters (uncolored colonies; **Fig. 4.14**). This medium is of particular benefit when diagnosing the etiology (cause) of diarrheal disease because most normal flora are lactose fermenters, whereas two important pathogens, *Salmonella* and *Shigella*, are lactose nonfermenters.

> **THOUGHT QUESTION 4.5**   The addition of sheep blood to agar produces a very rich medium called blood agar. Do you think blood agar can be considered a selective medium? A differential medium? *Hint*: Some bacteria can lyse red blood cells.

TO SUMMARIZE:

- **Microbes in nature** usually exist in complex, multispecies communities, but for detailed studies they must be grown separately in pure culture.
- **Bacteria can be cultured** on solid or liquid media.
- **Minimal defined media** contain only those nutrients essential for growth.
- **Complex, or rich, media** contain many nutrients. Other media exploit specific differences between organisms and can be defined as selective or differential.

4.4   Counting Bacteria

Counting or quantifying organisms invisible to the naked eye is surprisingly difficult because each of the available techniques measures a different physical or biochemical aspect of growth. Thus, a cell density value (given as cells per milliliter) derived from one technique will not necessarily agree with the value obtained by a different method.

Direct Counting Enumerates Both Living and Dead Cells

Microorganisms can be counted directly using a microscope. A dilution of a bacterial culture is placed on a special microscope slide called a hemocytometer (or, more specifically for bacteria, a Petroff-Hausser counting chamber, **Fig. 4.15**). Etched on the surface of the slide is a grid of precise dimensions, and placing a coverslip over the grid creates a space of precise volume. The number of organisms counted within that volume is used to calculate the concentration of cells in the original culture.

However, "seeing" an organism under the microscope does not mean that the organism is alive. Living and dead cells are indistinguishable by this basic approach. Living cells may be distinguished from dead cells by fluorescence microscopy using fluorescent chemical dyes, as discussed in Chapter 2. For example, propidium iodide, a red dye, intercalates between DNA bases but cannot freely penetrate the energized membranes of living cells. Thus, only *dead* cells stain red under a fluorescence scope. Another dye, Syto-9, enters both living and dead cells, staining them both green. By combining Syto-9 with propidium iodide, living and dead cells can be distinguished: Living cells will stain green, whereas dead cells appear orange or yellow because both dyes enter, and Syto-9 (green) plus propidium (red) appears yellow (**Fig. 4.16**).

Direct counting without microscopy can be accomplished by a **Coulter counter**. In the Coulter counter, a microbial culture is forced through a small orifice, through which flows an electrical current. Electrodes placed on both sides measure resistance. Every time

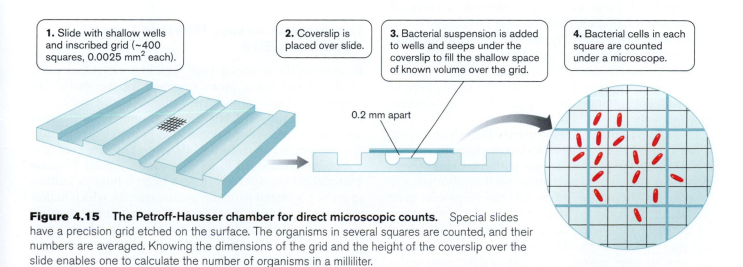

Figure 4.15   The Petroff-Hausser chamber for direct microscopic counts.   Special slides have a precision grid etched on the surface. The organisms in several squares are counted, and their numbers are averaged. Knowing the dimensions of the grid and the height of the coverslip over the slide enables one to calculate the number of organisms in a milliliter.

1. Slide with shallow wells and inscribed grid (~400 squares, 0.0025 mm² each).

2. Coverslip is placed over slide.

3. Bacterial suspension is added to wells and seeps under the coverslip to fill the shallow space of known volume over the grid.

4. Bacterial cells in each square are counted under a microscope.

0.2 mm apart

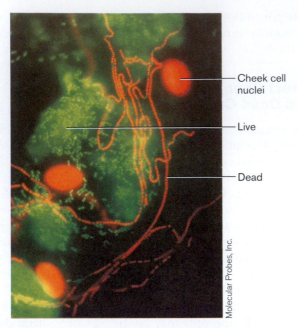

Cheek cell
nuclei

Live

Dead

Molecular Probes, Inc.

Figure 4.16 Live-dead stain. Live and dead bacteria visualized on freshly isolated human cheek epithelial cells using LIVE/DEAD BacLight Bacterial Viability Kit. Dead bacterial cells fluoresce orange or yellow because propidium (red) can enter the cells and intercalate the base pairs of DNA. Live cells fluoresce green because Syto-9 (green) enters the cell. The faint green smears are the outlines of cheek cells.

a cell passes through the orifice, electrical resistance increases, and the cell is counted. The Coulter counter, however, works best with larger eukaryotic cells, such as red blood cells; the instrument is not generally sensitive enough to detect individual bacteria.

An electronic technique more suited to counting and separating bacterial cells requires an instrument called a **fluorescence-activated cell sorter (FACS)**. In the FACS technique, bacterial cells that synthesize a fluorescent protein (see cyan fluorescent protein, Section 3.6) or that have been labeled with a fluorescent antibody or chemical are passed through a small orifice, as in the Coulter counter, and then past a laser (**Fig. 4.17A**). Detectors measure light scatter in the forward direction, a measure of particle size, and to the side, which indicates shape or granularity. In addition, the laser activates the fluorophore in the fluorescent antibody, and a detector measures fluorescence intensity.

The FACS technique can be used to identify and count different populations of cells within a single culture based on cell size and level of fluorescence (that is, one subpopulation may fluoresce more or less than another). For example, by placing the green fluorescent protein gene (*gfp*) under the control of the regulatory DNA sequences of a bacterial gene (creating a gene fusion), researchers can count the cells expressing that gene using FACS analysis; this, in turn, makes it possible to determine under

what conditions that gene is expressed and whether all cells in the population express that gene at the same time and to the same extent (**Fig. 4.17B**).

Viable Counts Estimate the Number of Cells That Can Form Colonies

Viable cells, as noted previously, are those that can replicate and form colonies on a plate. To obtain a viable cell count, dilutions of a liquid culture can be plated directly on an agar surface or added to liquid agar cooled to about 42–45°C. The agar is subsequently poured into an empty Petri plate (**pour plate**), where the agar cools further and solidifies. Because many bacteria resist short exposures to that temperature, individual cells retain the ability to form colonies on, and in, the pour plate. After colonies form, they are counted, and the original cell number is calculated. For example, if 100 colonies are observed in a pour plate made with 100 microliters (μl) of a 10^{-3} dilution of a culture, there are 10^6 organisms per milliliter of the original culture.

Although viable counts are widely used in research, there are problems using this method to measure cell number and determine cultural characteristics. One issue is that colony counting does not reflect cell size or growth stage. Even more problematic is that colony counts usually underestimate the number of living cells in a culture. As noted earlier, metabolically active cells that do not form colonies on agar plates will typically not be counted as alive. Cells damaged for one reason or another, while still alive, may be too compromised to divide. Comparing a viable count with a direct count obtained from a live-dead stain can expose the presence of damaged cells. Organisms such as *Streptococci* pose another problem. These organisms grow in chains. If individual cells in a chain are not separated prior to plating, each colony seen will have formed from a group of cells. This will cause actual cell numbers to be underestimated and is why the results of this counting technique are reported as colony-forming units (CFUs) rather than cells.

Biochemical Assays Measure Overall Population Size

In contrast to methods that visualize individual cells, assays of cell mass, protein content, or metabolic rate measure the overall size of a population of cells. The most straightforward but time-consuming biochemical approach to monitoring population growth is to measure the dry weight of a culture. Cells are collected by centrifugation, washed, dried in an oven, and weighed. Because bacterial cells weigh very little, a large volume of culture must be harvested to obtain measurements, which makes this technique quite insensitive. A more accurate alternative is to measure increases in protein levels, which correlate with increases in cell number. Protein levels are more easily measured and use relatively sensitive assays.

Optical Density: A Rapid Measure of Population Growth

The phenomenon of light scattering was introduced in Chapter 2. We noted that although individual bacteria could not be resolved based on the detection of scattered light by the unaided human eye, the presence of numerous bacteria in a tube of medium could be detected as a cloudy appearance. The decrease in intensity of a light beam due to the scattering of light by a suspension of particles is measured as **optical density**. Note that only particles in suspension, such as cells and macromolecules, can scatter light. True solutions, such as a solution of salt or of glucose, are clear even at high concentration.

The optical density of light scattered by bacteria is a very useful tool for estimating population size. The method is quick and easy, but because light scattering is a complex function of cell number and volume, optical density provides only an approximate result that must be corrected using a standard curve. Typically, a standard curve is obtained that plots viable counts for cultures versus optical densities as measured by a spectrophotometer. Thereafter, the optical density of a growing culture is measured, and the standard curve is used to estimate cell number at any given point during growth.

One inherent problem in estimating cell numbers based on optical density is that a cell's volume can vary, depending on its growth stage, altering its light-scattering properties. Thus, the cell number estimated by the standard curve may deviate from the true number. Another problem is that dead cells also scatter light. Clearly, using optical density to estimate viable count can be misleading, especially when measuring populations in stationary phase.

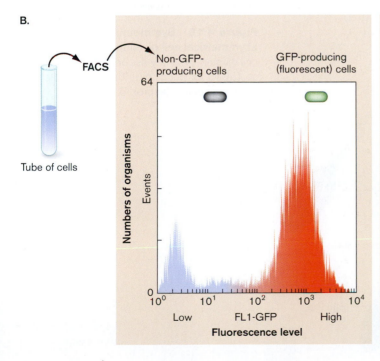

A.

Figure 4.17 Fluorescence-activated cell sorter.
A. Schematic diagram of a FACS apparatus (bidirectional sorting). **B.** Separation of GFP-producing *E. coli* from non-GFP-producing *E. coli*. The low-level fluorescence in the cells on the left is baseline fluorescence (autofluorescence). The scatterplot displays the same FACS data, showing the size distribution of cells (*x*-axis) with respect to level of fluorescence (*y*-axis). The larger cells may be ones that are about to divide.

B.

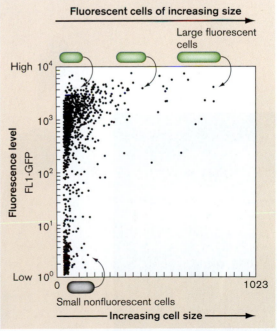

- **Microorganisms in culture may be counted directly** under a microscope, with or without staining, or by using a fluorescence-activated cell sorter.
- **Microorganisms can be counted indirectly**, as in viable counts and measurement of dry weight, protein levels, or optical density.
- **A viable bacterial organism** is defined as being capable of replicating and forming a colony on a solid-medium surface.

4.5   The Growth Cycle

How do microbes grow? What determines their rate of growth? And when does growth start in a nongrowing population? In nature, the answers to these questions are extremely complex. Most microbes in nature do not exist alone, but in complex communities of microbes and multicellular organisms. Studying the growth of multicellular forms requires considering them as a unit. Yet these same multicellular forms can send off **planktonic cells**, free-living organisms that grow and multiply on their own.

Nevertheless, all species at one time or another exhibit both rapid growth and nongrowth, as well as many phases in between. For clarity, we present here the principles of rapid growth, while bearing in mind the actual diversity of growth situations in nature.

The ultimate goal of any species is to make more of its own kind. This fundamental law ensures survival of the species. A typical bacterium (at least the typical bacterium that can be cultured in the laboratory) grows by increasing in length and mass, which facilitates expansion of its nucleoid as its DNA replicates (see Section 3.3). As DNA replication nears completion, the cell, in response to com-plex genetic signals, begins to synthesize an equatorial septum that ultimately separates the two daughter cells. In this overall process called **binary fission**, one parent cell splits into two equal daughter cells (**Fig. 4.18A**). Note, however, that although a majority of culturable bacteria divide symmetrically in two equal halves, some species divide asymmetrically. For example, the bacterium *Caulobacter* forms a stalked cell that remains fixed to a solid surface but reproduces by budding from one end to produce small, unstalked motile cells. The marine organism *Hyphomicrobium* also replicates asymmetrically by budding, releasing a smaller cell from a stalked parent (**Fig. 4.18B**).

Eukaryotic microbes divide by a special form of cell fission involving **mitosis**, the segregation of pairs of chromosomes within the nucleus (see Section 20.2 and Appendix 2). Some eukaryotes also undergo more complex life cycles involving budding and diverse morphological forms. Nevertheless, a large population of eukaryotic microbes will exhibit the same mathematical functions of growth seen in bacteria; indeed, these same growth patterns are seen in large populations of multicellular organisms, including ourselves.

Unlimited Population Growth Is Exponential

The process of reproduction has implications for growth not only of the individual, but also of populations. If we assume that growth occurs without limits, what happens to the population? The unlimited growth of any population obeys a simple law: The **growth rate**, or rate of increase in population numbers or biomass, is *proportional* to the population size at a given time. Such a growth rate is called exponential because it generates an exponential curve, a curve whose slope increases continually.

B.

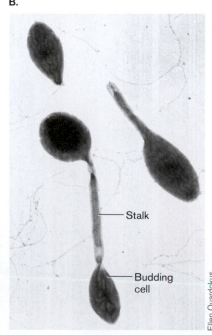

A.

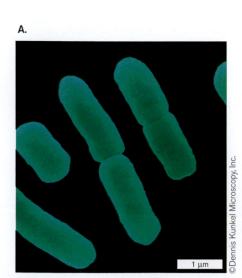

Figure 4.18 Symmetrical and asymmetrical cell division. A. Symmetrical cell division, or binary fission, in *Lactobacillus* sp. (SEM). **B.** Budding of the marine bacterium *Hyphomicrobium* (approx. 4 μm long).

Stalk

Budding cell

1 μm

©Dennis Kunkel Microscopy, Inc.

Ellen Quardokus

How does binary fission of cells generate an exponential curve? If each cell produces two cells per generation, then the population size at any given time is proportional to 2^n, where the exponent n represents the number of generations (replacement of parents by offspring) that have taken place between two time points. Thus, cell number rises exponentially. Many microbes, however, have replication cycles based on numbers other than two. For example, some cyanobacteria form cell aggregates that divide by multiple fission, releasing dozens of daughter cells. The cyanobacterium enlarges without dividing, and then suddenly divides many times without separating. The cell mass breaks open to release hundreds of progeny cells.

It is important to note that simple binary fission is not the only kind of reproduction that would generate an exponential curve. Any generation size will yield exponential growth. For example, the imaginary, softball-sized, fuzzy creatures called "tribbles" created in an episode of the television series *Star Trek* supposedly produced 11 offspring per generation and soon overran the starship. The rate of increase of the tribble population is proportional to 11^n, where n is the number of tribble generations (**Fig. 4.19A**). In comparison with bacterial reproduction, which produces only 2 cells per generation, we might expect the growth *rate* of tribbles to be dramatically faster. In fact, this is the case for the first few generations. But after a few more generations, the bacterial growth curve eventually achieves the same steep slope (that is, rate) as the tribble curve (**Fig. 4.19B**); the two curves are, in fact, graphical representations of the same kind of exponential function.

THOUGHT QUESTION 4.6 A virus such as influenza virus might produce 800 progeny virus particles from one infected host cell. How would you mathematically represent the exponential growth of the virus? What practical factors might limit such growth?

A Constant Generation Time Results in Logarithmic Growth

A variable not accounted for in Figure 4.19B is the length of time from one generation to the next. In an environment with unlimited resources, bacteria divide at a constant interval called the **generation time**. The length of that interval varies with respect to many parameters, including the bacterial species, type of medium, temperature, and pH. The generation time for cells in culture is also known as the **doubling time**, because the population of cells doubles over one generation. For example, one cell of *E. coli* placed into a complex medium will divide every 20 minutes. After 1 hour of growth (three generations), that one cell will have become eight (1 to 2, 2 to 4, 4 to 8). Because cell number (N) *doubles* with each division, the increase in cell number over time is exponential, not linear. A linear increase would occur if cell number rose by a *fixed* amount after every generation (for example, 1 to 2, 2 to 3, 3 to 4).

Starting with any number of organisms (N_0), the number of organisms after n generations will be $N_0 \times 2^n$. For example, a single cell after three generations ($n = 3$) will produce

$$1 \text{ cell} \times 2^3 = 8 \text{ cells}$$

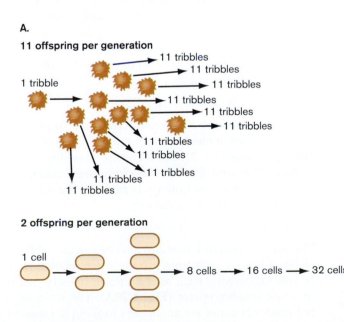

A.

11 offspring per generation

1 tribble

11 tribbles
11 tribbles
11 tribbles
11 tribbles
11 tribbles
11 tribbles
11 tribbles
11 tribbles
11 tribbles
11 tribbles
11 tribbles

2 offspring per generation

1 cell → → → 8 cells → 16 cells → 32 cells

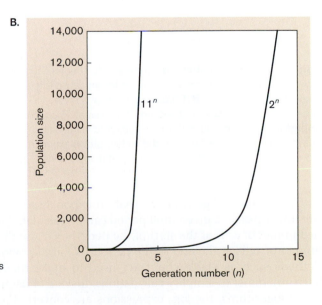

B.

Population size (y-axis): 0, 2,000, 4,000, 6,000, 8,000, 10,000, 12,000, 14,000

Generation number (n) (x-axis): 0, 5, 10, 15

11^n 2^n

Figure 4.19   Reproduction: from one to many.   A. Exponential growth of a hypothetical organism (tribbles from the *Star Trek* series) that produces 11 progeny per generation, compared with an organism that produces 2 offspring per generation. **B.** Growth curves for a population that increases 11-fold per generation and for a population that increases twofold per generation.

A.

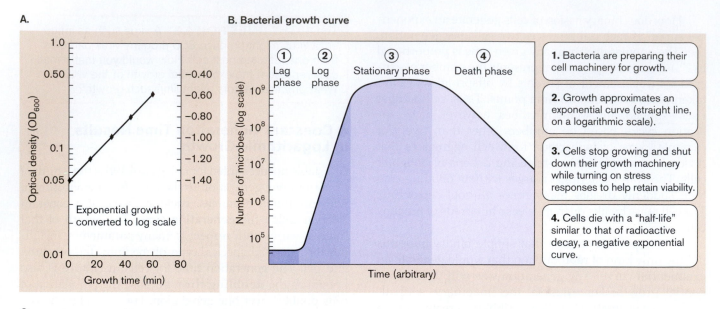

B. Bacterial growth curve

① Lag phase ② Log phase ③ Stationary phase ④ Death phase

1. Bacteria are preparing their cell machinery for growth.

2. Growth approximates an exponential curve (straight line, on a logarithmic scale).

3. Cells stop growing and shut down their growth machinery while turning on stress responses to help retain viability.

4. Cells die with a "half-life" similar to that of radioactive decay, a negative exponential curve.

C.

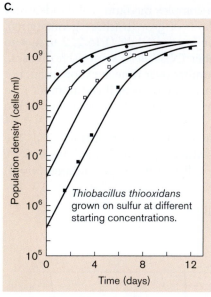

Thiobacillus thiooxidans grown on sulfur at different starting concentrations.

Figure 4.20 Bacterial growth curves. A. Theoretical growth curve of a bacterial suspension measured by optical density at a wavelength of 600 nm. **B.** Phases of bacterial growth in a typical batch culture. **C.** Published growth curves of *Thiobacillus thiooxidans*, an acidophile that oxidizes sulfur to sulfuric acid. Whatever the starting cell density, the culture grows exponentially until it runs out of sulfur; then it enters the stationary phase.
Source: C. Yasuhiro Konishi, et al. 1995. *Applied and Environmental Biology* 61:3617.

The number of generations that an exponential culture undergoes in a given time period can be calculated if the number of cells at the start of the period (N_0) and the number of cells at the end of the period (N_t) are known. Methods such as viable count are used to make those determinations. To simplify calculations, instead of using base 2 logarithms, the $\log_2$ expressions are converted to base 10 through division by a factor of $\log_{10} 2$, which is approximately 0.301. Thus:

$$N_t = N_0 \times 2^n$$

becomes

$$\log_{10} N_t = \log_{10} N_0 + n \log_{10} 2$$
$$n \log_{10} 2 = \log_{10} N_t - \log_{10} N_0$$
$$n = (\log_{10} N_t - \log_{10} N_0) \div \log_{10} 2$$
$$n = \log_{10} (N_t/N_0) \div \log_{10} 2$$
$$n = \log_{10} (N_t/N_0) \div 0.301$$

In practice, exponential growth occurs only for a short period when all nutrients are in full supply and the concentration of waste products has not become a limiting factor.

The rate of exponential growth can be expressed as the **mean growth rate constant** (k), which is the number of generations (n) per unit time (usually generations per hour). Even if we do not know the generation time, we can calculate k if we know the number of organisms at time zero (N_0) and the number of organisms after incubation time t (N_t) as follows:

$$k = n/t = (\log_{10} N_t - \log_{10} N_0) \div 0.301t$$

Thus, a culture with a generation time of 20 minutes (0.33 hour) will have a mean growth rate constant of 3 generations per hour ($k = 3$). For a culture doubling every 120 minutes (2 hours), k is 0.5. Note from these examples that the mean generation time (g) in hours is the reciprocal of the mean growth rate constant:

$$g = 1/k$$

The growth rate constant can also be calculated from the slope of $\log_{10} N$ over time, where N is a relative measure of culture density, such as the optical density measured in a spectrophotometer (**Fig. 4.20A**). The units of N do not matter because we are always looking at ratios of cell numbers relative to an earlier level. For example, we can use the following series of optical density (OD) measurements to measure N:

Time (min)	OD$_{600}$	log$_{10}$ OD$_{600}$
0	0.05	−1.30
15	0.08	−1.10
30	0.13	−0.89
45	0.20	−0.69
60	0.33	−0.48

If we plot log$_{10}$ OD$_{600}$ versus time, we obtain a line with a slope of 0.0136/min. Substituting in equation 4.1, we obtain k, the growth rate constant (doubling time):

$$k = (0.0136/\text{min})(60 \text{ min/h}) \div 0.301$$
$$= 2.7 \text{ doublings per hour}$$

The steeper the slope, the faster the organisms are dividing.

THOUGHT QUESTION 4.7 Suppose 1,000 bacteria are inoculated in a tube of minimal salts medium, where they double once an hour; and 10 bacteria are inoculated into rich medium, where they double in 20 minutes. Which tube will have more bacteria after 2 hours? After 4 hours?

THOUGHT QUESTION 4.8 An exponentially growing culture has an optical density at 600 nm (OD$_{600}$) of 0.2 after 30 minutes and an OD$_{600}$ of 0.8 after 80 minutes. What is the doubling time?

THOUGHT QUESTION 4.9 **Figure 4.20C** shows growth curves for different population densities of *Thiobacillus thiooxidans* when the concentration of sulfur in the medium is constant. Draw the growth curves you would expect to see if the initial population density was constant but the concentration of sulfur varied.

The mathematics of exponential growth is relatively straightforward, but remember that microbes grow differently in pure culture (very rare in nature) than they do in mixed communities, where neighboring cells produce all kinds of substances that may feed or poison other microbes. In mixed communities, the microbes may grow planktonically (floating in liquid), as in the open ocean, or as a biofilm on solid matter suspended in that ocean. In each instance, the mathematics of exponential growth apply at least until the community reaches a density at which different species begin to compete.

THOUGHT QUESTION 4.10 It takes 40 minutes for a typical *E. coli* cell to completely replicate its chromosome and about 20 minutes to prepare for another round of replication. Yet the organism enjoys a 20-minute generation time growing at 37°C in complex medium. How is this possible?

Stages of Growth in Batch Culture

Exponential growth never lasts indefinitely because nutrient consumption and toxic by-products eventually slow the growth rate until it halts altogether. The simplest way to model the effects of changing conditions is to culture bacteria in liquid medium within a closed system, such as a flask or test tube. This is called **batch culture**. In batch culture, no fresh medium is added during incubation; thus, nutrient concentrations decline and waste products accumulate during growth.

These changing conditions profoundly affect bacterial physiology and growth and illustrate the remarkable ability of bacteria to adapt to their environment. As medium conditions deteriorate, alterations occur in membrane composition, cell size, and metabolic pathways, all of which impact generation time. Microbes possess intricate, self-preserving genetic and metabolic mechanisms that slow growth before their cells lose viability. Because many bacteria replicate by binary fission, the plotting of culture growth (as represented by the logarithm of the cell number) versus incubation time allows us to see the effect of changing conditions on generation time and reveals several stages of growth (**Fig. 4.20B**).

Lag phase. Cells transferred from an old culture to fresh growth media need time to detect their environment, express specific genes, and synthesize components needed to institute rapid growth. As a result, bacteria inoculated into fresh media typically experience a lag period, or **lag phase**, where cells do not divide. There are several reasons for this. Cells taken from an aged culture may be damaged and require time for repair. Carbon, nitrogen, or energy sources different from those originally used by the seed culture must be sensed, and the appropriate enzyme systems must be synthesized. The length of lag phase varies, depending on the age of the culture, changes in temperature, and the differences between the new and old media (for example, changes in nutrient levels, pH, and salt concentrations). For example, transferring cells from a complex medium to a fresh complex medium results in a very short lag phase, whereas cells grown in a complex medium and then plunged into a minimal defined medium experience a protracted lag phase, during which time they readjust to synthesize all the amino acids, nucleotides, and other metabolites originally supplied by the complex medium.

Early log, or exponential, phase. Once cells have retooled their physiology to accommodate the new environment, they begin to grow exponentially and enter what is called **exponential**, or **logarithmic (log) phase**. Exponential growth is balanced growth, where all cell components are synthesized at constant rates relative to each other. At this stage, cells are growing and dividing at the maximum rate possible based on the medium and growth conditions provided (such as, temperature, pH, and osmolarity). Cells are largest at this stage of growth. This is the linear part of the growth curve. If cell division were synchronized and all cells divided at the same time, the growth curve during this period would appear as a series of steps with cell numbers doubling instantly after every generation time. But batch cultures are not synchronous. Every cell has an equal generation time, but each cell divides at a slightly different moment, making the cell number rise smoothly.

Cells enjoying balanced, exponential growth are temporarily thrown into metabolic chaos (unbalanced growth) when their medium is abruptly changed. Nutritional downshift (moving cells from a good carbon source such as glucose to a poorer carbon source such as succinate) or nutritional upshift (moving cells to a better carbon source) casts cells into unbalanced growth. Downshift to a carbon source with a lower energy yield not only means that a different set of enzymes must be made and employed to use the carbon source, but also means that the previous high rate of macromolecular synthesis (such as ribosome synthesis) used to support a fast generation time is now too rapid relative to the lower energy yield. Failure to adjust will lead to increased mistakes in RNA, protein, and DNA synthesis, depletion of key energy stores, and ultimately death.

Microbes, however, possess molecular fail-safe systems that immediately dampen rates of macromolecular synthesis until they come into balance with the rest of metabolism. In contrast, nutritional upshift means that the cell will start making more energy (ATP) than it can use at its current rate of macromolecular synthesis. The same fail-safe mechanisms used during downshift will, during upshift, kick macromolecular synthesis into a higher gear, once again establishing balanced growth. Nutritional upshift causes cells to reenter log phase but with a shorter generation time.

Late log phase. As cell density (number of cells per milliliter) rises during log phase, the rate of doubling eventually slows, and a new set of growth-phase–dependent genes are expressed. At this point, some species can also begin to detect the presence of others by sending and receiving chemical signals in a process known as quorum sensing (discussed in Chapter 10).

Stationary phase. Eventually, cell numbers stop rising owing to lack of a key nutrient or buildup of waste products. This occurs for bacteria grown in a complex medium when cell density rises above 10^9 cells per milliliter, but it can occur at lower cell densities if nutrients are limiting. At this point, the growth curve levels off and the culture enters what is called **stationary phase**. In contrast to bacteria, eukaryotic microorganisms, such as protozoa, enter stationary phase at much lower cell numbers, usually around 10^6 organisms per milliliter. The reason for this is simply that eukaryotic cells are bigger than bacteria. Bigger cells use more nutrient and run out of it sooner.

If they did not change their physiology, microbes would be very vulnerable once entering stationary phase. Because cells in stationary phase are not as metabolically nimble as cells in exponential phase, damage from oxygen radicals and toxic by-products of metabolism would read-ily kill them. As an avoidance strategy, some bacteria differentiate into very resistant spores in response to nutrient depletion (see Section 4.7), while other bacteria undergo less dramatic but very effective molecular reprogramming. The microbial model organism *E. coli*, for example, adjusts to stationary phase by decreasing its size, minimizing the volume of its cytoplasm compared with the volume of its nucleoid. Less nutrient is required to sustain the smaller cell. New stress resistance enzymes are also synthesized to handle oxygen radicals, protect DNA and proteins, and increase cell wall strength through increased peptidoglycan cross-linking. As a result, *E. coli* cells in stationary phase become more resistant to heat, osmotic pressure, pH changes, and other stresses that they might encounter while waiting for a new supply of nutrients.

Death phase. Without reprieve in the way of new nutrients, cells in stationary phase will eventually succumb to toxic chemicals present in the environment. Like the growth rate, the **death rate**, the rate at which cells die, is logarithmic. The death rate, however, is a negative exponential function. Recall that the increase in cell number during exponential phase is a positive exponential function of time. In **death phase**, the number of cells that die in a given time period is proportional to the number that existed at the beginning of the time period. Determining microbial death rates is critical to the study of food preservation and to antibiotic development (further discussed in Chapters 5, 16, and 27). Although death curves are basically logarithmic, exact death rates are difficult to define because mutations arise that promote survival, and some cells grow by cannibalizing others. Consequently, the death phase is extremely prolonged. A portion of the cells will often survive for months, years, or even decades.

In **Figure 4.20C**, we see an actual bacterial growth curve from a study of *T. thiooxidans*, an organism that oxidizes sulfur to sulfuric acid and grows below pH 1. Growth curves are shown for several different starting concentrations of bacteria. In each case, the bacteria grow exponentially for several days, until they have exhausted the sulfur in their growth medium. Their growth then slows until they enter stationary phase.

THOUGHT QUESTION 4.11   What can happen to the growth curve when a culture medium contains two carbon sources, one a preferred carbon source of growth-limiting concentration and a second, nonpreferred source? (See Section 10.3.)

THOUGHT QUESTION 4.12   How would you modify the equations describing microbial growth rate to describe the rate of death?

THOUGHT QUESTION 4.13   Why are cells in log phase larger than cells in stationary phase?

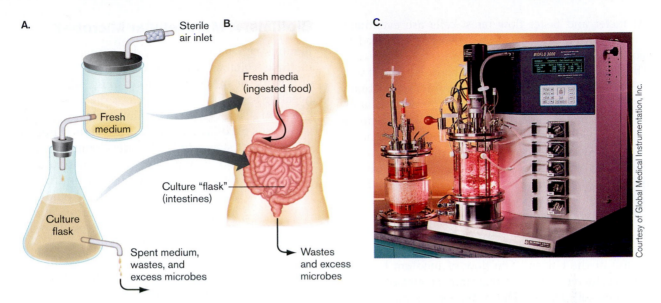

Figure 4.21 Chemostats and continuous culture. A. The basic chemostat ensures logarithmic growth by constantly adding and removing equal amounts of culture media. **B.** Note that the human gastrointestinal tract is engineered much like a chemostat in that new nutrients are always arriving from the throat while equal amounts of bacterial culture exit in fecal waste. **C.** A modern chemostat.

Continuous Culture Maintains Constant Cell Mass during Exponential Growth

In the classic growth curve that develops in closed systems, the exponential phase spans only a few generations. In open systems, however, where fresh medium is continually added to a culture and an equal amount of culture is constantly siphoned off, bacterial populations can be maintained in exponential phase at a constant cell mass for extended periods of time. In this type of growth pattern, known as **continuous culture**, all cells in a population achieve a steady state, which permits detailed analysis of microbial physiology at different growth rates. The **chemostat** is a continuous culture system in which the diluting medium contains a limiting amount of an essential nutrient (**Fig. 4.21**). Increasing the flow rate increases the amount of nutrient available to the microbe. The more nutrient available, the faster a cell's mass will increase. Because cell division is triggered at a defined cell mass, it follows that the growth rate in a chemostat is directly related to the dilution rate, or flow rate (milliliters per hour divided by the vessel volume). The more nutrient a culture receives as a result of increasing flow rate, the faster those cells can replicate (that is, the shorter the generation time).

The complex relationships among dilution rate, cell mass, and generation time in a chemostat are illustrated in **Figure 4.22**. The curves in this figure represent a typical experimental result, which can vary with organism, limiting nutrient, temperature, and so on. Depending on the experimental conditions, the shapes of the curves may change, but the general relationships will remain similar. Note that at moderate dilution rates (defined differently for each species), an increased flow rate of medium

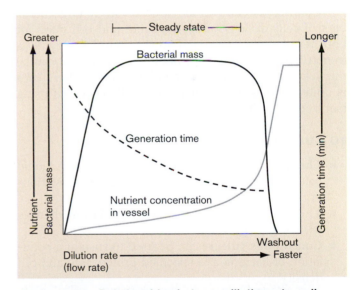

Figure 4.22 Relationships between dilution rate, cell mass, and generation time. As the dilution rate increases in a chemostat (meaning more nutrient is fed to the culture), the generation time decreases (the cells divide more quickly) and the cell mass of the culture increases. This continues until the rate of dilution exceeds the division rate, at which point cells are washed from the vessel faster than they can be replaced by division and the cell mass decreases. The y-axis varies, depending on the curve, as labeled in figure.

through the system increases division rate (thus, generation time decreases as the cells divide faster). A constant cell mass, or density, is maintained over a range of flow rates because the amount of culture (and cells) removed from the vessel exactly compensates for the increased rate of cell division.

At faster and faster flow rates, cells are eventually removed more quickly than they can be replenished by division, so cell density (cell mass) *decreases* in the vessel. This is called washout. Notice, in contrast, that at very *low* dilution (flow) rates, an increase in rate will actually *increase* cell density. This is because at extremely low dilution rates, the nutrient is so limiting that cell mass cannot increase to the point necessary for division. If more nutrient becomes available, the system can maintain a higher cell mass and cell density because the cells are able to divide faster than they are removed. With slow flow rates, the removal of fluid is not a significant factor in determining cell density.

The **turbidostat** is essentially a chemostat in which a photoelectric cell constantly monitors the optical density (turbidity) of the culture. The flow of medium through the vessel is then regulated to maintain a constant turbidity and thus cell density. The turbidostat is most suited to high dilution rates, where the numbers of cells can change quickly and overwhelm a less responsive system; the chemostat, which must be adjusted manually, is better suited at low dilution rates.

Continuous cultures are used to study large numbers of cells at constant growth rate and cell mass for both research and industrial applications. Most bacteria in nature grow at very slow rates, a situation that can be mimicked in a chemostat. The physiology of these cells is quite different from what is typically observed using batch culture.

TO SUMMARIZE:

- **The growth cycle** of organisms grown in liquid batch culture consists of lag phase, log phase, stationary phase, and death phase.
- **The physiology of a bacterial population** changes with growth phase.
- **Continuous culture** can be used to sustain a population of bacteria at a specified growth rate and cell density.

4.6 Biofilms

Bacteria are typically thought of as unicellular; but in nature, many, if not most, bacteria form specialized, surface-attached communities called **biofilms**. Indeed, within aquatic environments, bacteria are found mainly associated with surfaces, a fact that underscores the importance of biofilms in nature. Biofilms also play critical roles in microbial pathogenesis and environmental quality, and cost the nation billions of dollars each year in equipment damage, product contamination, and medical infections. For example, pseudomonad or staphylococcal biofilms can damage ventilators used to assist respiration and can act as direct sources of infection.

Biofilms: A Multicellular Microbe?

Biofilms can be constructed by a single species or by multiple, collaborating species and can form on a range of organic or inorganic surfaces (**Fig. 4.23**). The gram-negative bacterium *Pseudomonas aeruginosa*, for example, can form a single-species biofilm on the lungs of patients with cystic fibrosis or on medical implants (**Special Topic 4.1**). Distinct stages in biofilm development include initiation, maturation, maintenance, and dissolution. Bacterial biofilms form when nutrients are plentiful. The goal is to stay where food is plentiful. Why should a microbe travel off to hunt for food when it is already available? Once nutrients become scarce, however, individuals detach from the community to forage for new sources of nutrient.

Biofilms in nature can take many different forms and serve different functions for different species. In addition, formation of biofilms can be cued by different environmental signals in different species. Such signals include pH, iron concentration, temperature, oxygen availability, and the presence of certain amino acids.

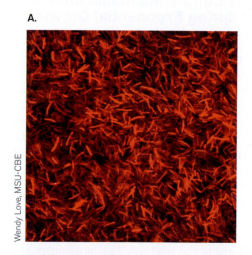

A.

Wendy Love, MSU-CBE

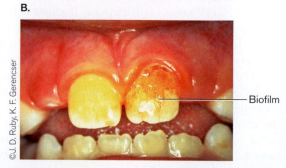

B.

©J. D. Ruby, K. F. Gerencser

— Biofilm

Figure 4.23 Biofilms. A. A thermophilic microbial mat found attached to a rock in Yellowstone National Park (confocal scanning laser microscope image). The autofluorescence of these cyanobacterial cells of the genus *Synechococcus* was generated using a 568-nm krypton laser and a 590LP filter. **B.** Biofilm on a tooth. The biofilm that forms on teeth is called plaque.

Special Topic 4.1 Biofilms, Disease, and Antibiotic Resistance

When causing diseases of plants, animals, or humans, bacteria preferentially exist in surface-attached biofilms. Attachment to host tissues and multicellular growth are important in many situations, from simple wound infections caused by *Staphylococcus aureus* to colonization of the lungs of cystic fibrosis patients by *Pseudomonas aeruginosa*. One characteristic of bacterial biofilms is a marked increase in antibiotic tolerance. The failure to cure infections that stem from bacterial-pathogen biofilm formation is well documented, but the basis of persistence remains unclear. The literature offers different explanations for increased tolerance of antibiotics by microcolonies, including reduced penetration of drug into the microcolony and an altered stress-resistant physiology by the cells in the interior of the colony.

A study of *P. aeruginosa* biofilms clearly demonstrated the importance of multicellular cooperativity of cells within the biofilm. It appears that quorum sensing induces the drug-tolerant state. Cells in the biofilm release a chemical signal molecule called an acyl homoserine lactone (AHL). The population of cells respond to these molecules by increasing resistance to certain antimicrobial agents. As shown in **Figure 1**, cells within a typical biofilm are resistant to the antibiotic tobramycin. The fluorescent stain used in this experiment will stain live cells green and dead cells red. However, simultaneous treatment of *P. aeruginosa* biofilms with a specific quorum-sensing blocker *and* the antibiotic tobramycin led to the effective killing of the cells by the antibiotic. The quorum-sensing blocker, a novel, synthetically produced brominated furanone, did not reduce the viability of the *P. aeruginosa* cells. This result is consistent with the furanone specifically interfering with the AHL quorum-sensing system as seen by a microarray analysis. Hence, the shutdown of the AHL quorum-sensing system in the biofilm cells made them sensitive to the antibiotic.

A. No furanone; 100 μg/ml tobramycin   **B. 10 μg/ml furanone; 100 μg/ml tobramycin**

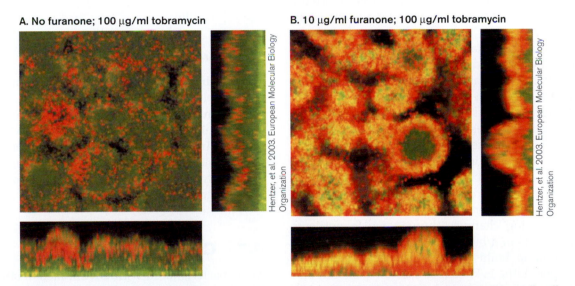

Hentzer, et al. 2003. European Molecular Biology Organization

Figure 1   Resistance of *Pseudomonas aeruginosa* biofilms to the antibiotic tobramycin is mediated by cell-cell signaling.   Tobramycin sensitivity was tested in the presence and absence of the furanone compound C-30, which specifically inhibits cell-cell signaling. *P. aeruginosa* biofilms were grown in the absence (left panel) and presence (right panel) of 10 mM C-30. After three days, the biofilms were exposed to 100 μg/ml tobramycin for 24 hours. Bacterial viability was assayed by staining using the LIVE/DEAD BacLight BacterialViability Kit. Red areas are dead bacteria and green areas are live bacteria. Views are top and cross-sectional. Panels show top and side views of the biofilms.

Nevertheless, a common pattern emerges in the formation of many kinds of biofilms (**Fig. 4.24**).

First, the specific environmental signal induces a genetic program in planktonic cells. The planktonic cells then start to attach to nearby inanimate surfaces by means of flagella, pili, lipopolysaccharides, or other cell surface appendages, and begin to coat that surface with an organic monolayer of polysaccharides or glycoproteins to which more planktonic cells can attach. At this point, cells may move along surfaces using a **twitching motility** that involves the extension and retraction of a specific type of pilus. Ultimately, they stop moving and

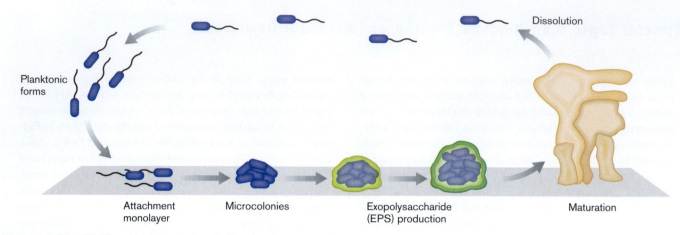

Figure 4.24   Biofilm development.   Biofilm development in *Pseudomonas*.

firmly attach to the surface. As more and more cells bind to the surface, they can begin to communicate with each other by sending and receiving chemical signals in a process called **quorum sensing**. These chemical signal molecules are continually made by individual cells. Once the population reaches a certain number (analogous to an organizational "quorum"), the chemical signal reaches a specific concentration that the cells can sense. This triggers genetically regulated changes that cause cells to bind tenaciously to the substrate and to each other.

Next, the cells form a thick extracellular matrix of polysaccharide polymers and entrapped organic and inorganic materials. These **exopolysaccharides (EPSs)**, such as alginate produced by *P. aeruginosa* and colanic acid produced by *E. coli*, increase the antibiotic resistance of residents within the biofilm. As the biofilm matures, the amalgam of adherent bacteria and matrix takes on complex three-dimensional forms such as columns and streamers, creating channels through which nutrients flow. Sessile cells in a biofilm chemically "talk" to each other in order to build microcolonies and keep water channels open. Little is known about how a biofilm dissolves, although the process is thought to be triggered by starvation. *P. aeruginosa* produces an alginate lyase that can strip away the EPSs, but the regulatory pathways involved in releasing cells from biofilms are not clear.

It is important to keep in mind that most biofilms in nature are consortia of several species. Multispecies biofilms certainly demand interspecies communication, and individual species may perform specialized tasks in the community.

Organisms adapted to life in extreme environments also form biofilms. Members of Archaea form biofilms in acid mine drainage (pH 0), where they contribute to the recycling of sulfur, and cyanobacterial biofilms are common in thermal springs. Suspended particles called "marine snow" are found in ocean environments and appear to be floating biofilms comprising many organisms that have

not yet been identified. The particles appear capable of methanogenesis, nitrogen fixation, and sulfide production, indicating that biofilm architecture can allow anaerobic metabolism to occur in an otherwise aerobic environment.

WWW | Biofilms

TO SUMMARIZE:

- **Biofilms** are complex multicellular surface-attached microbial communities.
- **Chemical signals** enable bacteria to communicate (quorum sensing) and in some cases to form biofilms.
- **Biofilm development** involves adherence of cells to a substrate, formation of microcolonies, and, ultimately, formation of complex channeled communities that generate new planktonic cells.

4.7   Cell Differentiation

Many bacteria faced with environmental stress undergo complex molecular reprogramming that includes changes in cell structure. Some species, like *E. coli*, experience relatively simple changes in cell structure, such as the formation of smaller cells or thicker cell surfaces. However, select species undergo elaborate cell differentiation processes. An example is *Caulobacter crescentus*, whose cells convert from the swimming form to the holdfast form before cell division. Each cell cycle then produces one sessile cell attached to its substrate by a holdfast, while its sister cell swims off in search of another habitat.

Other species undergo far more elaborate transformations. The endospore formers generate heat-resistant capsules (spores) that can remain in suspended animation for thousands of years. Yet another group, the actinomycetes, form complex multicellular structures analogous to those of eukaryotes. In this case, cell structure can change radically; and individual, freewheeling members of a species can relinquish their independence and band together, forming multicellular "organisms." The actinomycete that produces streptomycin is one example. These differentiation programs illustrate the distinction between the presence of genes in a bacterial genome and their activation. Genes for differentiation are expressed only when the cell needs to survive stress. We will discuss gene activation in more detail in Chapter 10.

Eukaryotic microbes also undergo highly complex life cycles. For example, *Dictyostelium discoideum* is a seemingly unremarkable ameba that grows as separate, independent cells. However, when challenged by adverse conditions such as starvation, the microbe secretes chemical signal molecules that choreograph a massive interaction of individuals to form complex multicellular structures. The developmental cycle of this organism may be compared to other eukaryotic microbes that are human parasites (discussed in Chapter 20).

Endospores Are Bacteria in Suspended Animation

Certain gram-positive genera, including important pathogens such as *Clostridium tetani* (tetanus), *Clostridium botulinum* (botulism), and *Bacillus anthracis* (anthrax) have the remarkable ability to develop dormant spores that are heat- and desiccation-resistant. Dessication and heat resistance are properties that make *B. anthracis* spores a potential bioweapon.

Most of our knowledge of bacterial sporulation comes from the gram-positive soil bacterium *B. subtilis*. When growing in rich media, this microbe undergoes normal vegetative growth and can replicate every 30–60 minutes. However, starvation initiates an elaborate 8-hour genetic program that directs an asymmetrical cell division process and ultimately yields a spore (see Section 10.4).

As shown in **Figure 4.25**, sporulation can be divided into discrete stages primarily based on morphological

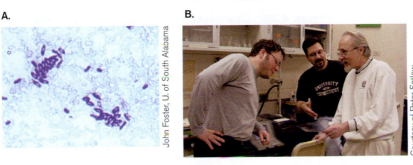

Figure 4.25 Endospore formation.
A. Photomicrograph of *Bacillus subtilis* spores. The cells (approx. 2 μm long) are stained with crystal violet. **B.** Peter Setlow (right) of the University of Connecticut figured out how proteins regulate the process of differentiation of endospores. **C.** Stages of endospore formation.

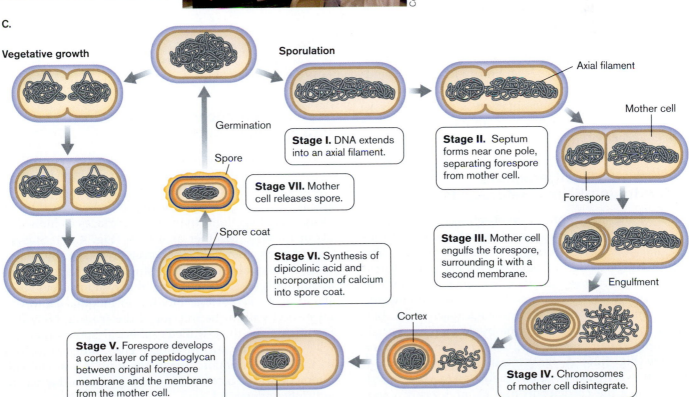

appearance. Stage 0 (not shown) represents the point at which the vegetative cell decides to use one of two potential polar division sites to begin septum formation instead of the central division site used for vegetative growth. Stage I involves replicating and stretching the DNA into a long axial filament that spans the length of the cell. Ultimately, one of the polar division sites wins out, and in stage II, septation occurs, dividing the cell into two unequal compartments, the **forespore**, which will ultimately become the spore, and the larger **mother cell**, from which it is derived. Each compartment contains a chromosome.

In stage III of sporulation, the mother cell membrane engulfs the forespore. Next, the mother cell chromosome is destroyed and a thick peptidoglycan layer (cortex) is placed between the two membranes surrounding the forespore protoplast (stage IV). Layers of coat proteins are then deposited on the outer membrane in stage V. Stage VI completes development of spore resistance to heat and chemical insults. This last process includes the synthesis of dipicolinic acid and the uptake of calcium into the core of the spore. Finally, the mother cell, now called a sporangium, releases the mature spore (stage VII).

Spores are resistant to many environmental stresses that would kill vegetative cells. The nature of this resistance is due, in part, to desiccation of the spore (they have only 10–30% of a vegetative cell's water content). But, as discovered by Peter Setlow and colleagues, spores are also packed with small acid-soluble proteins (SASPs) that bind to and protect DNA. The SASP coat protects the spore's DNA from damage by ultraviolet light and various toxic chemicals.

A fully mature spore can exist in soil for at least 50–100 years and spores have been known to last thousands of years. Once proper nutrient conditions arise, another genetic program, called **germination**, is triggered to wake the dormant cell, dissolve the spore coat, and release a viable vegetative cell.

Cyanobacteria Differentiate into Nitrogen-Fixing Heterocysts

Some of the autotrophic cyanobacteria, such as *Anabaena*, not only make oxygen through photosynthesis but "fix" atmospheric nitrogen to make ammonia. This is surprising because nitrogenase, the enzyme required to fix nitrogen, is very sensitive to oxygen, so one might expect that photosynthesis and nitrogen fixation would be two mutually exclusive physiological activities. *Anabaena* have solved this dilemma by developing specialized cells, called heterocysts, that function in nitrogen fixation (**Fig. 4.26**). A tightly regulated genetic program converts every tenth photosynthetic cell to a heterocyst, which loses the capacity to fix CO_2 and forms a spe-

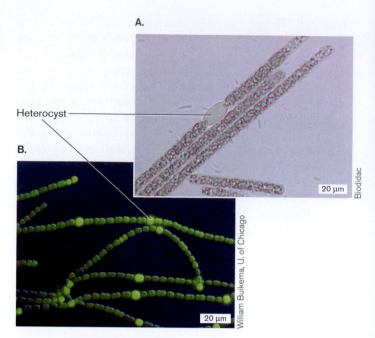

Figure 4.26 Cyanobacteria and heterocyst formation.
A. Light microscope image of cyanobacteria *Phanizomenon*.
B. The cyanobacterial genus *Anabaena*. The expression of genes in heterocysts is different from their expression in other cells. All cells in the figure contain a cyanobacter gene to which the gene for green fluorescent protein (GFP) has been spliced. Only cells that have formed heterocysts are expressing the fused gene, which makes the cell fluoresce bright green.

cialized envelope to limit O_2 access. How this organism produces such a precise spacing of heterocysts is currently the subject of intensive research.

Starvation Induces Differentiation into Fruiting Bodies

Certain species of bacteria, in the microbial equivalent of barn raising, produce architectural marvels called fruiting bodies. The gram-negative species *Myxococcus xanthus* uses a **gliding motility** (involving a type of pilus, not a flagellum) to travel on surfaces as individuals or to move together as a mob (**Fig. 4.27**). Starvation triggers a developmental cycle in which 100,000 or more individuals aggregate, rising into a mound called a fruiting body. At this point, the system resembles a stage in *P. aeruginosa* biofilm formation. However, myxococci within the interior of the fruiting body differentiate into thick-walled, spherical spores that are released into the surroundings. The changes involved in this differentiation process require many cell-cell interactions and a complex genetic program that we do not fully understand.

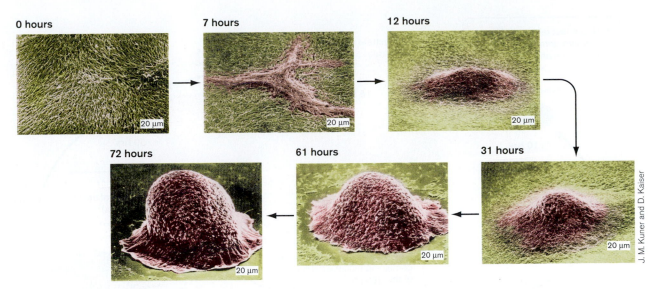

Figure 4.27 *Myxococcus* **swarm erecting a fruiting body.** Approximately 100,000 cells begin to aggregate, and over the course of 72 hours they erect a fruiting body.

Some Bacteria Differentiate to Form Eukaryotic-like Structures

The actinomycetes (see Chapter 19) such as *Streptomyces* are bacteria that form mycelia and sporangia analogous to the filamentous structures of eukaryotic fungi (**Fig. 4.28**). Several developmental programs tied to nutrient availability are at work in this process (**Fig. 4.29**). Under favorable nutrient conditions, a germ tube emerges from a germinating spore, grows from its tip (tip extension), and forms branches that grow along the surface of its food source. This type of growth produces an intertwined net-work of long multinucleate filaments (hyphae) collectively called substrate **mycelia**. After a few days, new genes are activated that cause the hyphae to grow upward, rising above the surface to form aerial mycelia. Compartments at the tips of these aerial hyphae contain 20–30 copies of the genome. Aerial hyphae stop growing as nutrients decline, triggering a developmental program that synthe-sizes antibiotics and produces spores (arthrospores) that are fundamentally different from endospores. This pro-gram lays down multiple septa that subdivide the com-partment into single-genome prespores. The shape of the prespore then changes, its cell wall thickens, and deposits

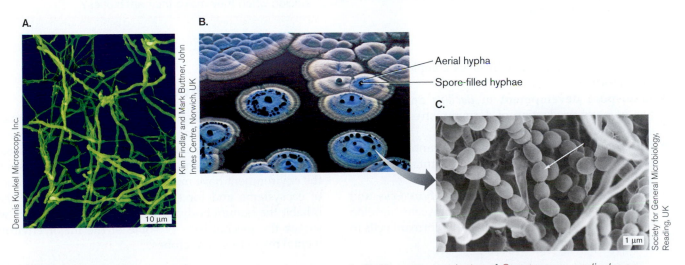

Figure 4.28 **Mycelia.** **A.** *Streptomyces lavendulae* substrate mycelia. **B.** Filamentous colonies of *Streptomyces coelicolor*, an actinomycete known for producing antibiotics such as streptomycin. **C.** *Streptomyces* aerial hyphae. Arrow points to a hyphal spore (approx. 1 μm each).

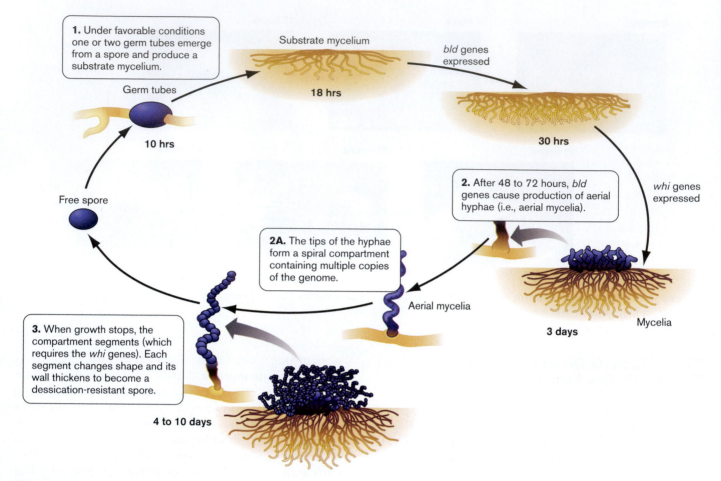

1. Under favorable conditions one or two germ tubes emerge from a spore and produce a substrate mycelium.

Germ tubes

10 hrs

Free spore

Substrate mycelium

18 hrs

bld genes expressed

30 hrs

whi genes expressed

2. After 48 to 72 hours, *bld* genes cause production of aerial hyphae (i.e., aerial mycelia).

2A. The tips of the hyphae form a spiral compartment containing multiple copies of the genome.

Aerial mycelia

Mycelia

3 days

3. When growth stops, the compartment segments (which requires the *whi* genes). Each segment changes shape and its wall thickens to become a dessication-resistant spore.

4 to 10 days

Figure 4.29 Developmental cycle of *Streptomyces coelicolor.*

are made in the spore that increase resistance to desiccation. These organisms are of tremendous interest, both for their ability to make antibiotics and for their fascinating developmental programs.

TO SUMMARIZE:

- **Microbial development** involves complex changes in cell forms.
- **Endospore development** in *Bacillus* and *Clostridium* involves production of dormant, stress-resistant endospores.
- **Heterocyst development** enables cyanobacteria to fix nitrogen anaerobically while maintaining oxygenic photosynthesis.
- **Multicellular fruiting bodies** in *Myxococcus* and *Dictyostelium* and **mycelia** in actinomycetes develop in response to starvation, dispersing dormant cells to new environments.

THOUGHT QUESTION 4.14 How might *Streptomyces* and *Actinomyces* species avoid committing suicide when they make their antibiotics?

Concluding Thoughts

All systems of microbial development share a common theme. They are all triggered in response to a change in their environment, such as depletion of resources, desiccation, or changes in temperature. Microbial responses to the environment have major implications for the function of ecosystems and for the microbial communities that inhabit the human body for good or ill. In Chapter 5, we survey the mechanisms of some of the major environmental responses of microbes.

CHAPTER REVIEW

Review Questions

1. What nutrients do microbes need to grow?
2. Explain the differences between autotrophy, heterotrophy, phototrophy, and chemotrophy.
3. Explain the basics of the carbon and nitrogen cycles.
4. Describe the various mechanisms of transporting nutrients in prokaryotes and eukaryotes. What are facilitated diffusion, coupled transport, ABC transporters, group translocation, and endocytosis?
5. Why is it important to grow bacteria in pure culture?
6. Under what circumstances would you use a selective medium? A differential medium?
7. What are the factors that define the growth phases of bacteria grown in batch culture?
8. Describe the important features of biofilms.
9. Name three kinds of bacteria that undergo differentiation, and give highlights of the differentiation processes.

Key Terms

ABC transporter (124)
antiport (123)
ATP-binding cassette (124)
autotroph (118, 119)
axenic growth (118)
batch culture (137)
binary fission (134)
biofilm (140)
chemoautotroph/chemoautotrophy (118, 119)
chemoheterotroph/chemoheterotrophy (119, 120)
chemolithotroph (118)
chemostat (139)
chemotrophy (119)
cofactor (117)
colony (128)
complex medium (130)
confluent (129)
continuous culture (139)
Coulter counter (131)
coupled transport (123)
death phase (138)
death rate (138)
defined minimal medium (117)
denitrification (120)
differential medium (130)
dilution streaking (128)
doubling time (135)

electrogenic (123)
electroneutral (123)
endosome (126)
enriched medium (130)
essential nutrient (116)
exopolysaccharide (EPS) (142)
exponential phase (137)
fluorescence-activated cell sorter (FACS) (132)
forespore (144)
generation time (135)
germination (144)
gliding motility (144)
group translocation (126)
growth factor (117)
growth rate (134)
heterotroph (118, 119)
lag phase (137)
lithotroph (118)
logarithmic (log) phase (137)
MacConkey medium (130)
macronutrient (116)
mean growth rate constant (136)
membrane potential (120)
microcolony (128)
micronutrient (117)
mitosis (134)
mixotrophic (121)
mother cell (144)

mycelium (145)
nitrification (120)
nitrogen-fixing bacterium (120)
optical density (133)
permease (122)
phagocytosis (127)
phagosome (127)
phosphotransferase system (PTS) (126)
photoautotroph/photoautotrophy (118, 119)
photoheterotrophy (120)
phototrophy (119)
pinocytosis (126)
planktonic cells (134)
pour plate (132)
pure culture (128)
quorum sensing (142)
selective medium (130)
siderophore (125)
spread plate (129)
stationary phase (138)
substrate-binding protein (125)
symbiont (120)
symport (123)
synthetic medium (130)
turbidostat (140)
twitching motility (141)
viable (129)
viable but nonculturable (130)

Recommended Reading

Angert, Esther R. 2005. Alternatives to binary fission in bacteria. *Nature Reviews Microbiology* **3**:214–224 .

Branda, Steven S., Ashlid Vik, Lisa Friedman, and Robert Kolter. 2005. Biofilms: the matrix revisited. *Trends in Microbiology* **13**:20–26.

Davidson, Amy L., and Jue Chen. 2004. ATP-binding cassette transporters in bacteria. *Annual Reviews in Biochemistry* **73**:241–268.

England, Jennifer C., and James W. Gober. 2001. Cell cycle control of cell morphogenesis in *Caulobacter. Current Opinion in Microbiology* **4**:674–680.

Finkel, Steven E., and Robert Kolter. 1999. Evolution of microbial diversity during prolonged starvation. *Proceedings of National Academy of Sciences USA* **96**:4023–4027.

Higgins, Christopher F. 2001. ABC transporters: physiology, structure and mechanism—an overview. *Research in Microbiology* **152**:205–210.

Nystrom, Thomas. 2004. Stationary-phase physiology. *Annual Review of Microbiology* **58**:161–181.

Piggot, Patrick J., and David W. Hilbert. 2004. Sporulation of *Bacillus subtilis. Current Opinion in Microbiology* **7**:579–586.

Skerker, Jeffrey M., and Michael T. Laub, 2004. Cell-cycle progression and the generation of asymmetry in *Caulobacter crescentus. Nature Reviews Microbiology* **2**:325–337.

Chapter 5

Environmental Influences and Control of Microbial Growth

Microbes have both the fastest and the slowest growth rates of any known organism. Some hot-springs bacteria can double in as little as 10 minutes, whereas deep-sea sediment microbes may take as long as 100 years. And while the rapid growth of pathogens underlies the rapid spread of disease, Earth's crust is shaped by microbes that grow very slowly in soil and rock. What determines these differences in growth rate? Nutrition is one factor, but niche-specific physical parameters like temperature, pH, and osmolarity are equally important.

A microbe's physiology is geared to work only within a narrow range of physical parameters. But in nature, the environment can quickly change. Many marine microbes, for instance, can move from deep-sea cold to the searing heat of a thermal vent. How do these organisms survive? Stop-gap measures called stress survival responses help, but a more permanent solution is when a species slowly evolves to thrive, not just survive, in extreme environments. How does the biology of these so-called extremophiles permit growth under conditions that seem uninhabitable? In this chapter, we explore the limits of microbial growth and show how this knowledge has helped us control the microbial world.

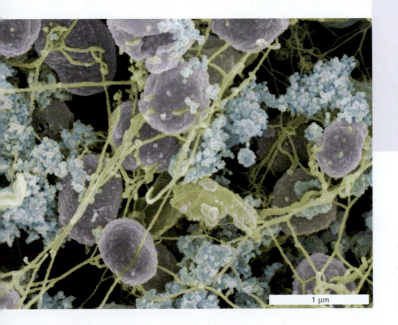

The extremophile *Halorubrum lacusprofundi*, an Antarctic haloarchaeon that forms biofilms at subzero temperatures in 5 M NaCl. Extremophiles are considered model systems for astrobiology.
Source: Shiladitya DasSarma. 2006. *Microbe* 1:120.

In 1998, apple and pear growers in Washington and northern Oregon lost crops worth an estimated $68 million owing to outbreaks of fire blight, a devastating bacterial disease; the causative agent, *Erwinia amylovora*, destroys apple and pear trees, making them appear as if they were torched by fire. As with human disease, we try to control microbial plant infections by controlling the growth of microbes. It should not be surprising, then, that antibiotics are often utilized for this purpose. But while many antibiotics are used to protect humans, only two are currently approved for use on plants: streptomycin and oxytetracycline. The fact that resistance to streptomycin has already been demonstrated in *Erwinia amylovora* underscores the importance of studying all aspects of bacterial growth in the effort to control pathogens of plants, animals, and humans.

We begin this chapter by discussing how physical and chemical changes in the environment modify the growth of different groups of microbes. We will also explore how microorganisms adapt to different environments in ways both transient (involving temporary expression of inactive genes) and permanent (modifications of the gene pool). The permanent genetic changes have led to biological diversity. Finally, we will examine the different ways humans try to limit the growth of microorganisms to protect plants, animals, and ourselves.

As you proceed through this chapter, you will encounter two recurring themes: that different groups of microbes live in vastly different environments and that microbes can respond in diverse ways when confronted by conditions outside their niche, or comfort zone.

5.1 Environmental Limits on Microbial Growth

With our human frame of reference, we tend to think that "normal" growth conditions are those found at sea level with a temperature between 20°C and 40°C, a near-neutral pH, a salt concentration of 0.9%, and ample nutrients. Any ecological niche outside this window is called *extreme* and the organisms inhabiting them **extremophiles** (R. MacElroy first used the term *extremophile* in 1974). Extremophiles are prokaryotes (bacteria and archaea) that are able to grow in extreme environments. For example, one group of organisms can grow at temperatures above boiling, while another group requires a pH 2 acidic environment to grow. Based on our definition of *normal* conditions, conditions on Earth when life began were certainly extreme. Consequently, the earliest microbes likely grew in these extreme environments. Organisms that grow under conditions that seem "normal" to humans likely evolved from an ancient extremophile that gradually adapted as the environment evolved to that of our present-day Earth.

Be aware that multiple extremes in the environment can be encountered simultaneously. For instance, in Yellowstone National Park, one can find an extreme acid pool next to an extreme alkali pool, both at extremely high temperatures. Thus, extremophiles typically evolve to survive multiple extreme environments.

Extremophiles may provide insight into the workings of extraterrestrial microbes we may one day encounter, since outer space certainly qualifies as an extreme environment. Our experiences with extremophiles should alert us to the dangers of underestimating the precautions necessary in handling extraterrestrial samples. For example, we should not assume that irradiation would be sufficient to sterilize samples from future planetary or interstellar missions. Such treatments do not even kill the extremophile *Deinococcus radiodurans* found on Earth.

www | Pyrodictium

How do we even begin to study organisms that grow in boiling water or sulfuric acid solutions or organisms that we cannot even culture in the laboratory? How do we dissect the molecular response of organisms to changes in an environment such as acid rain or to changes in the human body? Genome sequences present new opportunities for investigating these questions.

Bioinformatic analysis allows us to study the biology of organisms that we cannot culture. First, genome sequences from novel organisms are amplified by polymerase chain reaction, or PCR (a technique that multiplies a small sample of DNA; see Section 7.6). PCR amplification can be done directly from the natural environment in which the organisms are found, and the sequences are compared with those of model systems, such as *E. coli*, whose biochemistry is well known. Genomic comparison quickly reveals whether an organism under study may possess specific metabolic pathways and regulatory responses.

Global arrays of DNA and proteins allow us to study the response of an organism to changing environments. Global approaches can reveal in a single experiment the response of all an organism's genes to a single environmental change, such as change in temperature or pH (**Fig. 5.1**). **DNA microarrays** (see Section 10.8) consist of slides with a grid containing DNA probes for every gene in an organism's genome. They are used to assess which genes are expressed to make RNA in a given organism at a given time or under a given condition. **Two-dimensional protein gels** (see Section 3.5) achieve two-dimensional separation of proteins based on differences in each protein's isoelectric point (first dimension) and molecular weight (second dimension). They show which proteins the cell produces when cultured in different environments. Knowing what genes and proteins are expressed in a given situation helps elucidate the molecular strategies that microbes use to grow under different conditions and to defend themselves against environmental stresses.

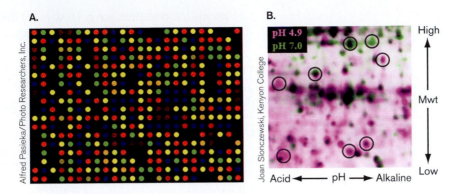

Figure 5.1   Response to environmental stress: global analysis of genes and proteins.   A. This microarray contains DNA probes (small pieces of DNA sequence) representing each gene in the genome. The DNA probes hybridize to preparations of fluorescent-labeled cDNA (complementary DNA) made from whole-cell RNA of cultures grown under different conditions. Each colored disk represents one gene. Red fluorescence indicates RNA expressed under one condition; green indicates expression under the second condition; yellow indicates expression under both conditions. **B.** Two-dimensional gels separate the proteins expressed in the cell. Proteins are separated in one dimension by their charge (isoelectric pH point) and in the second dimension by their molecular weights. Gel images are obtained from cultures grown under two different conditions (in this case, *Escherichia coli* K-12 grown at pH 4.9 versus pH 7.0). Pink spots denote proteins expressed only at pH 4.9; green indicates expression only at pH 7.0. Circled spots were chosen for additional studies.

These techniques of molecular analysis are discussed further in Chapters 8 and 10.

We have already mentioned the fundamental physical conditions (temperature, pH, osmolarity) that define an environment and select for (favor) the growth of specific groups of organisms. And within a given microbial community, each species is further localized to a specific niche defined by a narrower range of environmental factors. The reason for this is that every protein and macromolecular structure within a cell is affected profoundly by changes in environmental conditions and through reaction with the by-products of oxygen consumption. For example, a single enzyme works best under a unique set of temperature, pH, and salt conditions because those conditions allow it to fold into its optimum shape, or conformation. Deviations from these optimal conditions cause the protein to fold a little differently and become less active. While all enzymes within a cell do not boast the same physical optima, these optima must at least be similar and matched to the organism's environment for the organism to function effectively.

As may be apparent from the preceding discussion, there are many different classes of microbes based on their environmental niche. **Table 5.1** summarizes these environmental classes.

Table 5.1   Basic environmental classification of microorganisms.

Environmental parameter	Classification			
Temperature	Hyperthermophile* (growth above 80°C)	Thermophile* (growth between 50°C and 80°C)	Mesophile (growth between 15°C and 45°C)	Psychrophile* (growth below 15°C)
pH	Alkaliphile* (growth above pH 9)	Neutralophile (growth between pH 5–8)	Acidophile* (growth below pH 3)	
Osmolarity	Halophile* (growth in high salt > 2M NaCl)			
Oxygen	Aerobe (growth only in oxygen)	Facultative (growth with or without oxygen)	Microaerophile (growth only in small amounts of oxygen)	Anaerobe (growth only without oxygen)
Pressure	Barophile* (growth at high pressure, greater than 380 atm)		Barotolerant (grown between 10–495 atm)	

*considered extremophiles

5.2 Microbial Responses to Changes in Temperature

Unlike humans (and mammals in general), microbes cannot control their temperature; thus, bacterial cell temperature matches that of the immediate environment. Because temperature affects the average rate of molecular motion, changes in temperature impact every aspect of microbial physiology, including membrane fluidity, nutrient transport, DNA stability, RNA stability, and enzyme structure and function. Every organism has an *optimum* temperature at which it grows most quickly, as well as minimum and maximum temperatures that define the limits of growth. These limits are imposed, in part, by the thousands of proteins in a cell, all of which must function within the same temperature range. The fastest growth rate for a species occurs at temperatures where all of the cell's proteins work most efficiently as a group to produce energy and synthesize cell components. Growth stops when rising temperatures cause critical enzymes or cell structures (such as the cell membrane) to fail. At cold temperatures, growth ceases because enzymatic processes become too sluggish and the cell membrane less fluid. The membrane needs to remain fluid so that it can expand as cells grow larger and so that proteins needed for solute transport can be inserted into the membrane.

Maximum Growth Rate Increases with Temperature

In general, microbes that grow at higher temperatures can achieve higher rates of growth. Remarkably, the relationship between the maximum growth temperature and the growth rate constant *k* (number of generations per hour; see Section 4.5) obeys the Arrhenius equation for simple chemical reactions:

$$\log k = C/T \qquad (5.1)$$

T is the absolute temperature in degrees Kelvin, and *C* is a second constant that combines the gas constant and the average activation energy of cellular reactions. Over a defined temperature range, which differs for each species, growth rate increases (that is, cells divide faster) as temperature increases. If we plot the logarithm of the growth rate versus temperature, we get a straight line whose slope (*C*) can be obtained by substituting in equation 5.1:

$$\log (k_2/k_1) = C/(T_1 - T_2)$$

The general result of the Arrhenius equation is that growth rate roughly doubles for every 10°C rise in temperature (**Fig. 5.2**). This is the same relationship observed for any chemical reaction.

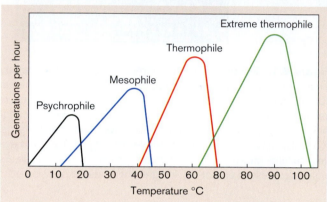

A.

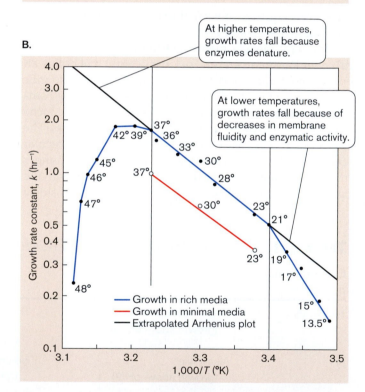

B.

At higher temperatures, growth rates fall because enzymes denature.

At lower temperatures, growth rates fall because of decreases in membrane fluidity and enzymatic activity.

— Growth in rich media
— Growth in minimal media
— Extrapolated Arrhenius plot

Figure 5.2 Relationship between temperature and growth rate. **A.** Relationship between temperature and growth rates of different groups of microbes. Note that the growth rate increases linearly with temperature and obeys the Arrhenius equation. **B.** Growth rate constant (*k*) of the enteric organism *Escherichia coli* is plotted against the inverse of growth temperature on the Kelvin scale (1,000/*T* is used to give a convenient scale on the *x*-axis). This is a more detailed view of a mesophilic growth temperature curve. As temperature rises above or falls below the optimum range, growth rate decreases faster than predicted by the Arrhenius equation. *Source*: B. Sherrie L. Herendeen, et al. 1979. *Journal of Bacteriology* 139:185.

At the upper and lower limits of the growth range, however, the Arrhenius effect breaks down. At high temperatures, critical proteins denature. Lower temperatures decrease membrane fluidity and limit the conformational mobility of enzymes, thereby lowering their activities. As a result, growth fails to occur. The typical growth temperature range spans about 30–40°C, but some organisms have a much narrower range. Within a species, we can identify mutants that are more sensitive to one extreme or the other (heat sensitive or cold sensitive). These mutations often define key molecular components of stress responses, such as the heat-shock proteins (discussed in Chapter 10).

Thermodynamic principles limit a cell's growth to a narrow temperature range. For example, heat increases molecular movement within proteins. Too much or too little movement will interfere with enzymatic reactions. As a result, there is a great biological diversity among microbes as different groups have evolved to grow within very different thermal niches. A species grows within a specific thermal range because its proteins have evolved to tolerate that range. Outside that range, proteins will denature or function too slowly for growth. The upper limit for protists is around 50°C, while some fungi can grow at temperatures as high as 60°C. Prokaryotes, however, have been found to grow at temperatures ranging from below 0°C to above 100°C. Temperatures over 100°C are usually found around thermal vents deep in the ocean, where water temperature can rise to 350°C but the pressure is sufficient to keep water liquid.

> **THOUGHT QUESTION 5.1** Why haven't cells evolved so that all their enzymes have the same temperature optimum? If they did, wouldn't they grow even faster?

Microorganisms Are Classified by Growth Temperature

Based on their ranges of growth temperature, microorganisms can be classified as mesophiles, psychrophiles, or thermophiles (**Fig. 5.2A**).

Mesophiles include the typical "lab rat" microbes, such as *Escherichia coli* and *Bacillus subtilis.* Their growth optima range between 20°C and 40°C, with a minimum of 15°C and a maximum of 45°C. Because they are easy to grow and because human pathogens are mesophiles, much of what we know about protein, membrane, and DNA structure came from studying this group of organisms. Current advances, particularly in obtaining detailed three-dimensional (3-D) views of protein structures, are frequently based on studies of two other classes of organisms whose optimum growth temperature ranges flank

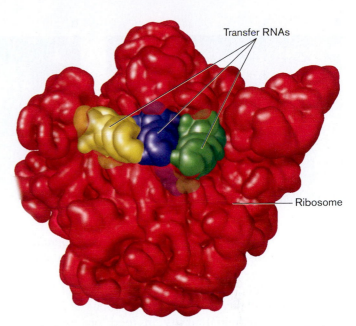

Figure 5.3 Thermophilic proteins. Special properties of thermophilic proteins facilitate 3-D structure analysis, as shown here. The structure of the *Thermus thermophilus* 70S ribosome is shown here at 7.8 Å (or 0.78 nm) resolution. Transfer RNAs (green, blue, and yellow) occupy a cavity between two ribosomal subunits. (PDB codes: 1GIX, 1GIY)

that of the mesophiles, namely **psychrophiles** (on the low temperature side) and **thermophiles** (on the high temperature side; **Fig. 5.3**). Because of their more stable folding structures, proteins from the thermophilic extremophiles are generally easier to crystallize than those of mesophiles or psychrophiles, so it is possible to determine their structures by X-ray crystallography (see Section 2.7).

Psychrophiles are microbes that grow at temperatures as low as 0°C, but their optimum growth temperature is usually around 15°C. Psychrophiles are prominent flora beneath icebergs in the Arctic and Antarctic (**Fig. 5.4**). In addition to true psychrophiles, whose growth optima are below 20°C, there are mesophiles that are cold resistant. For instance, the modern practice of refrigeration has selected for cold-resistant, yet mesophilic, pathogens such as *Listeria monocytogenes* (one cause of food poisoning and septic abortions). Why do these organisms grow so well in the cold? One reason psychrophilic microbes prefer cold is that their proteins are more flexible than those of mesophiles and require less energy (heat) to function. Of course, the downside to the increased flexibility of psychrophilic proteins is that they denature at lower temperatures than their mesophilic counterparts. As a result, psychrophiles grow poorly, if at all, when temperatures rise above 20°C. Another reason psychrophiles favor cold is that their membranes are more fluid at low temperature owing to the high proportion of

A.

B.

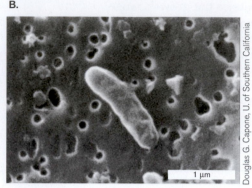

Figure 5.4 **Psychrophilic environment and psychrophiles.** **A.** Iceberg, in which psychrophilic organisms like those shown in (B) can be found. These organisms are unclassified. **B.** Bacteria from South Polar snow (SEM).

unsaturated fatty acids present; at higher temperatures, their membranes are *too* flexible and fail to maintain cell integrity.

Psychrophilic enzymes are of commercial interest because their ability to carry out reactions at low temperature has potential utility in food processing and bioremediation. Many food products require enzymatic processing. Enzymes help brew beer more quickly, break down lactose in milk, and can remove cholesterol from various foods. Production of foods at lower temperatures would be beneficial, since lower processing temperatures would minimize the growth of typical mesophiles that

degrade and spoil food. Another goal is to find organisms that can safely degrade toxic organic contaminants (for example, petroleum) in the cold. Arctic environments are particularly sensitive to pollution because contaminants are slow to degrade in the freezing temperatures. Consequently, the ability to seed Arctic oil spills with psychrophilic organisms armed with petroleum-degrading enzymes could rapidly restore contaminated environments.

Thermophiles (**Fig. 5.5**) are species adapted to growth at high temperature, typically 55°C and higher. **Hyperthermophiles** grow at temperatures as high as

A.

B.

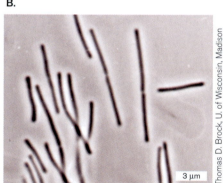

C.

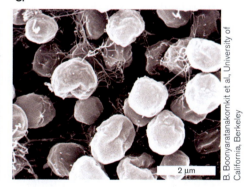

D.

Figure 5.5 **Thermophilic environments and thermophiles.** **A.** Yellowstone National Park hot spring. **B.** *Thermus aquaticus*, a hyperthermophile first isolated at Yellowstone by Thomas Brock. Cell length varies from 3 to 10 μm. **C.** Thermophile *Methanocaldococcus jannaschii* grown at 78°C and 30 psi. **D.** A "smoker," a hydrothermal vent 2 miles deep in the Pacific Ocean. *M. jannaschii* was isolated in 1983 in the area of this vent.

110°C, which occur under extreme pressure (for example, at the ocean floor). These organisms flourish in hot environments such as composts or near thermal vents that penetrate Earth's crust on the ocean floor and on land (for example, hot springs). The thermophile *Thermus aquaticus* was the first source of a DNA polymerase used for PCR amplification of DNA. *T. aquaticus* was discovered in a hot spring at Yellowstone National Park by microbiologist Thomas Brock, a pioneer in the study of thermophilic organisms. Its application to the polymerase chain reaction has revolutionized molecular biology (discussed in Chapter 7).

Extreme thermophiles often have specially adapted membranes and protein sequences. The thermal limits of these structures determine the specific high-temperature ranges in which various species can grow. Because enzymes in thermophiles (thermozymes) do not unfold as easily as mesophilic enzymes, they more easily hold their shape at higher temperatures. Thermophilic enzymes are stable, in part, because they contain relatively low amounts of glycine, a small amino acid that contributes to an enzyme's flexibility. In addition, the amino-termini of proteins in these organisms often are "tied down" by hydrogen bonding to other parts of the protein, making them harder to denature.

Like all microbes, thermophiles have chaperone proteins that help refold other proteins as they undergo thermal denaturation. Thermophile genomes are packed with numerous DNA-binding proteins that stabilize DNA. In addition, these organisms possess special enzymes that function to tightly coil DNA in a way that makes it more thermostable and less likely to denature (think of a coiled phone cord that has twisted and bunched up on itself). Special membranes also help give cells additional stability at high temperatures. Unlike the typical lipid bilayers of mesophiles, the membranes of thermophiles manage to "glue" together parts of the two hydrocarbon layers that point toward each other, making them more stable. They do this by incorporating more saturated linear lipids into their membranes. Saturated lipids form straight hydrocarbon tails that align well with neighboring lipids and form a highly organized structure stable to heat. The membranes of mesophiles are composed mostly of unsaturated lipids that bend against each other and align poorly. This property makes the membranes of mesophiles more fluid at lower temperatures.

High Temperatures Induce the Heat Shock Response

As insurance against extinction, most microorganisms possess elegant genetic programs that remodel their physiology to one that can temporarily survive inhospitable conditions. Rapid temperature changes experienced during growth activate batches of stress response genes, resulting in the **heat shock response** (discussed in Chapter 10). The protein products of these heat-activated genes include chaperones that maintain protein shape and enzymes that change membrane lipid composition. The heat-shock response, first identified in *E. coli* by Yamamori and Yura in 1982, has since been documented in all living organisms examined thus far.

> **THOUGHT QUESTION 5.2** If microbes lack a nervous system, how can they sense a temperature change?

The appearance of different branches of life in the course of evolution reflects, in some ways, the narrowing of tolerance to heat. Different archaeal species, for example, can grow in extremely hot or extremely cold temperatures and some can grow in the middle range. Bacteria, for the most part, tolerate a temperature range that bridges the archaeal extremes. Eukaryotes are less temperature tolerant than bacteria, with individual species capable of growth between 10°C and 65°C. Archaeal species have been found to grow at greater extremes than bacterial species, and bacteria at greater extremes than eukaryotes. As we will see, this evolutionary relationship holds for other environmental conditions.

TO SUMMARIZE:

- **Different species** exhibit different optimal growth values of temperature, pH, and osmolarity.
- **The Arrhenius equation** applies to growth of microorganisms: Growth rate doubles for every 10°C rise in temperature.
- **Membrane fluidity** varies with the composition of lipids in a membrane, which in turn dictates at what temperature an organism can grow.
- **Mesophiles, psychrophiles, and thermophiles** are groups of organisms that grow at moderate, low, and high temperatures, respectively.
- **The heat shock response** produces a series of protective proteins in organisms exposed to temperatures near the upper edge of their growth range.

5.3 Microbial Adaptation to Variations in Pressure

Living creatures at Earth's surface (sea level) are subjected to a pressure of 1 atmosphere (atm), which is equal to 0.101 megapascal (Mpa) or 14 pounds per square inch (psi). At the bottom of the ocean, however, thousands of meters deep, hydrostatic pressure averages a crushing 400 atm and can go as high as 1,000 atm (101 Mpa, or 14,000 psi) in ocean trenches (**Fig. 5.6**). Organisms adapted to

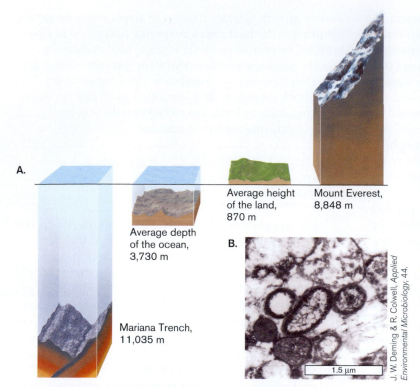

A.

Average height of the land, 870 m

Mount Everest, 8,848 m

Average depth of the ocean, 3,730 m

B.

Mariana Trench, 11,035 m

J. W. Deming & R. Colwell, *Applied Environmental Microbiology*, 44.

1.5 μm

Figure 5.6 Barophilic environments and barophiles.   A. Ocean depths. The deepest part of the ocean is at the bottom of the Mariana Trench, a depression in the floor of the western Pacific Ocean, just east of the Mariana Islands. The Mariana Trench is 1,554 miles long and 44 miles wide. Near its southwestern extremity, 210 miles southwest of Guam, lies the deepest point on Earth. This point, referred to as the "Challenger Deep," plunges to a depth of nearly 7 miles. **B.** Barophile *Shewanella violacea*. Length is approx. 1.5 μm.

designed ribosome structures that can withstand pressures even higher than that.

Specific applications for barotolerant proteins and pressure-regulated genes have not yet been developed. However, many food processes are carried out at high pressure to minimize bacterial contamination (destructive bacteria will not tolerate the pressure), so it is expected that barophiles will ultimately offer useful biotechnology products, such as enzymes that will carry out food processing at high pressure. High-pressure processing has been used on cheeses, yogurt, luncheon meats, and oysters to kill contaminating bacteria without destroying the flavor or texture of the food.

TO SUMMARIZE:

■ **Barophiles** can grow at pressures up to 1,000 atm but fail to grow at low pressures.
■ **Membrane fluidity** can be compromised at high pressures and cold temperatures. Specially designed membranes and protein structures are thought to enable the growth of barophiles.

THOUGHT QUESTION 5.3   What could be a relatively simple way to grow barophiles in the laboratory?

grow at overwhelmingly high pressures are called **barophiles** or **piezophiles**. From the curves in **Figure 5.7**, we can see that barophiles actually *require* elevated pressure to grow, while barotolerant organisms grow well over the range of 1–50 Mpa, but their growth falls off thereafter.

Many barophiles are also psychrophilic because the average temperature at the ocean's floor is 2°C. However, barophilic hyperthermophiles form the basis of thermal vent communities that support symbiotic worms and giant clams (see Chapter 21).

How bacteria can survive pressures of 12,000 psi is still a mystery. It is known that increased hydrostatic pressure and cold temperatures similarly reduce membrane fluidity. Because fluidity of the cell membrane is critical to survival, the phospholipids of deep-sea bacteria commonly have high levels of polyunsaturated fatty acids to increase membrane fluidity. It is thought that in addition to these membrane changes, internal structures must also be pressure adapted. For example, ribosomes in the barosensitive organism *E. coli* dissociate at pressures above 60 Mpa, so barophiles must contain uniquely

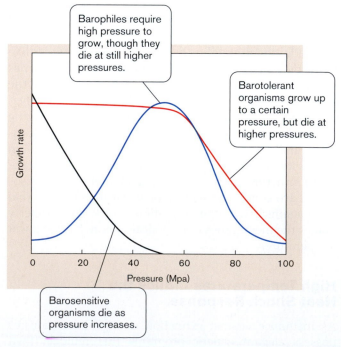

Barophiles require high pressure to grow, though they die at still higher pressures.

Barotolerant organisms grow up to a certain pressure, but die at higher pressures.

Barosensitive organisms die as pressure increases.

Growth rate

Pressure (Mpa)

Figure 5.7   Relationship between growth rate and pressure.

5.4 Microbial Responses to Changes in Water Activity and Salt Concentration

Water is critical to life, but environments differ in terms of how much water is actually available to growing organisms; microbes can only use water that is not bound to ions or other solutes in solution. Water availability is measured as **water activity** (a_w), a quantity approximated by concentration. Because interactions with solutes lower water activity, the more solutes there are in a solution, the less water is available for microbes to use for growth. Water activity is typically measured as the ratio of the solution's vapor pressure relative to that of pure water. A solution is placed in a sealed chamber and the amount of water vapor determined at equilibrium. If the air above the sample is 97% saturated relative to the moisture present over pure water, the relative humidity is 97% and the water activity is 0.97. Most bacteria require water activity to be greater than 0.91 for growth (the water activity of seawater). Fungi can tolerate water activity levels as low as 0.86.

Osmotic Stress

Osmolarity is a measure of the number of solute molecules in a solution and is inversely related to a_w. The more particles there are in a solution, the greater the osmolarity and the lower the water activity (see Appendix 1). Osmolarity is important for the cell because it is related to water activity and also because a semipermeable membrane surrounds microbial cells, so osmolarity inside the cell can be, and often is, different from the osmolarity outside. The principles of physical chemistry dictate that solute concentrations in two chambers separated by a semipermeable membrane will tend to equilibrate. Equilibrating osmolarity across a semipermeable cell membrane, which does not allow the movement of solutes, requires the movement of water. In hypertonic medium, where the external osmolarity is higher than the internal, water will try to *leave* the cell in an attempt to equalize osmolarity across the membrane. In contrast, suspension of a cell in a hypotonic medium (one of lower osmolarity than the cell) will cause an *influx* of water (see Fig. A2.6). This movement of water across cell membranes does not occur primarily by simple diffusion. Special membrane water channels formed by proteins called aquaporins enable water to traverse the membrane much faster than by diffusion and help protect cells against osmotic stress (**Fig. 5.8**). However, too much water moving in or out of a cell is detrimental. Cells may ultimately explode or implode, depending on the direction in which the water moves. Even bacteria with a rigid cell wall suffer. They may not explode like a human cell, but the forces placed on the cell wall are great.

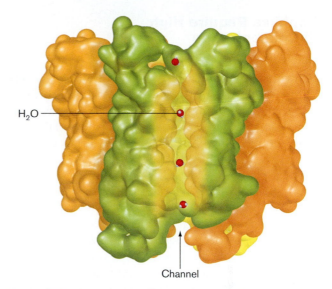

H₂O

Channel

Figure 5.8 Aquaporin. Transverse view of the channel. Complementary halves of the channel are formed by adjacent protein monomers. The curvilinear, size-selective (~4 ± 0.5 nm) core of the channel (~18 nm) is primarily lined by hydrophobic residues. (PDB code: 1J4N)

Cells Minimize Osmotic Stress across Membranes

In addition to moving water, microbes have at least two other mechanisms to minimize osmotic stress across membranes. When stranded in a hypertonic medium (higher osmolarity than the cell), bacteria try to protect their internal water by synthesizing or importing compatible solutes that increase intracellular osmolarity. Compatible solutes are small molecules that do not disrupt normal cell metabolism, even when present at high intracellular concentrations. Increasing intracellular levels of these compounds (for example, proline, glutamic acid, potassium, or betaine) elevates cytoplasmic osmolarity without any detrimental effects, making it unnecessary for water to leave the cell.

Cells also contain pressure-sensitive (mechanosensitive) channels that can be used to leak solutes out of the cell. It is believed that these channels are activated by rising internal pressures in cells immersed in a *hypo*tonic medium. When activated, the channels allow solutes to escape, which lowers internal osmolarity.

Outside their osmotic comfort range—that is, where the aforementioned housekeeping strategies become ineffective at controlling internal osmolarity—microbes launch a global response in which cellular physiology is transformed to tolerate brief encounters with potentially lethal salt concentrations. Some changes are similar to those provoked by heat shock, such as the increased synthesis of chaperones that protect critical cell proteins from denaturation. Other changes include alterations in outer membrane pore composition (for gram-negative organisms).

Halophiles Require High Salt Concentration

Some species of archaea have evolved to require high salt (NaCl) concentration. These are called **halophiles** (**Fig. 5.9**). In striking contrast to most bacteria, which prefer salt concentrations from 0.1 to 1 M (0.2–5% NaCl), the extremely halophilic archaea can grow at an a_w of 0.75 and actually require 2–4 M NaCl (10–20% NaCl) to grow. For comparison, seawater is about 3.5% NaCl. All cells, even halophiles, prefer to keep a relatively low intracellular Na^+ concentration. One reason for this is that some solutes are moved into the cell by symport with Na^+. To achieve a low internal Na^+ concentration, halophilic microbes use special ion pumps to excrete sodium and replace it with other cations, such as potassium, which is a compatible solute. In fact, the proteins and cell components (for example, ribosomes) of halophiles require remarkably high intracellular potassium levels to maintain their structure.

Halophilic algae use a different strategy for dealing with hypertonic media. They redirect photosynthesis to make glycerol (a compatible solute) instead of starch. Increasing glycerol production raises internal osmolarity, which stops water from leaving the cell and prevents dehydration. Although halophilic bacteria appear to have lost ecological flexibility by adapting to such an extreme environment, they have gained access to environments that other organisms are not equipped to use.

> **THOUGHT QUESTION 5.4** How might the concept of water availability be used by the food industry to control spoilage?

TO SUMMARIZE:

- **Water activity** (a_w) is a measure of how much water in a solution is available for a microbe to use.
- **Osmolarity** is a measure of the number of solute molecules in a solution and is inversely related to a_w.
- **Aquaporins** are membrane channel proteins that allow water to move quickly across membranes to equalize internal and external pressures.
- **Compatible solutes** are used to minimize pressure differences across the cell membrane.
- **Mechanosensitive channels** can leak solutes out of the cell when internal pressure rises.
- **Halophilic organisms** grow best at high salt concentration.

5.5 Microbial Responses to Changes in pH

As with salt and temperature, the concentration of hydrogen ions (H^+)—actually, hydronium (H_3O^+)—also has a direct effect on the cell's macromolecular structures. Extreme concentrations of either hydronium or hydroxyl ions (OH^-) in a solution will limit growth. In other words, too much acid or base is harmful to cells. Despite this sensitivity to pH extremes, living cells tolerate a greater range in environmental concentration of H^+ than of virtually any other chemical substance. *E. coli*, for example, tolerates a pH range of 2–9, a 10,000,000-fold difference. For a brief review of pH, refer to Section A1.7.

A.

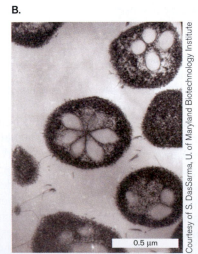

B.

C.

0.5 μm

Wayne P. Armstrong

Courtesy of S. DasSarma, U. of Maryland Biotechnology Institute

Courtesy of S. DasSarma, U. of Maryland Biotechnology Institute

Figure 5.9 **Halophilic salt flats and halophilic bacteria.** **A.** The halophilic salt flats along Highway 50, east of Fallon, Nevada, are colored pinkish red by astronomical numbers of halophilic bacteria. **B.** *Halobacterium* sp. (TEM). Cross section cell width 0.5 to 0.8 μm. **C.** Shiladitya DasSarma and colleagues at the University of Maryland completed the genome sequence of *Halobacterium* species NRC-1. They demonstrated novel features of archaeal genetics, including intriguing similarities with molecular regulatory structures in eukaryotes.

Enzymes Show pH Optima, Minima, and Maxima

The charges on various amino or carboxyl groups within a protein help forge the intramolecular bonds that dictate protein shape and thus protein activity. Because H^+ concentration affects the protonation of these ionizable groups, altering pH can alter the charges on these groups, which in turn changes protein structure and activity. The result is that all enzyme activities exhibit optima, minima, and maxima with regard to pH, much as they do for temperature. As we saw with temperature, groups of microbes have evolved to inhabit diverse niches, for which pH values can range from pH 0 to pH 11.5 (**Fig. 5.10**). However, species differences in optimum growth pH are *not* dictated by the pH limits at which critical cell proteins function.

Generally speaking, the majority of enzymes, regardless of the pH at which their source organism thrives, tend to operate best between pH 5 and 8.5 (which, if you think about it, is still a range in which the hydrogen ion concentration varies more than 1,000-fold), yet many microbes grow in even more acidic or alkaline environments.

Unlike its temperature, the intracellular pH of a microbe, as well as its osmolarity, is not necessarily the same as that of its environment. Biological membranes are relatively impermeable to protons, a fact that allows the cell to maintain an internal pH compatible with protein function when growing in extremely acidic or alkaline environments. When the difference between the intracellular and extracellular pH (ΔpH) is very high, protons can leak through either directly or via proteins that thread the membrane. Excessive influx or efflux of protons can cause problems by altering internal pH.

Membrane-permeant organic acids, also called weak acids (discussed in Chapter 3), can accelerate the leakage of H^+. Unlike H^+, the uncharged form of an organic acid (HA) can freely permeate cell membranes and dissociate intracellularly, releasing a proton that then acidifies internal pH (**Fig. 5.11**). The extent of the pH drop depends on the buffering capacity of the cell's proteins. This shuttling of protons can turn a relatively mild external pH level (say, pH 6) into a deadly acid stress. A naturally occurring example of organic acid stress is the lactic acid produced by lactobacilli during the formation of yogurt. The buildup of lactic acid limits the bacterial growth, leaving yogurt with plenty of food value. The food industry has taken advantage of this phenomenon by preemptively adding citric acid or sorbic acid to certain foods. This allows manufacturers to control microbial growth under pH conditions that do not destroy the

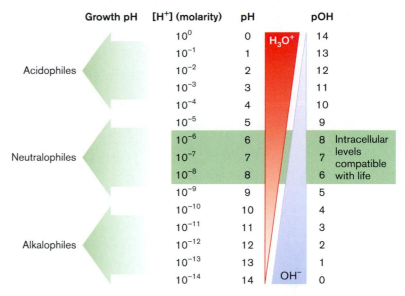

Figure 5.10　Classification of organisms grouped by optimal growth pH.

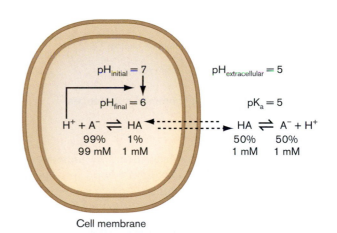

Figure 5.11　**Effect of organic acids on internal pH.** In this contrived example, the organic acid (HA) has an extracellular concentration of 2 mM and has a dissociation constant of pH 5. Because the medium is also pH 5, half of the acid is protonated (undissociated, or un-ionized) and half is dissociated (ionized). The un-ionized form, because it is uncharged, diffuses across the membrane to establish equilibrium between the inside and outside of the cell. However, because the inside of the cell is pH 7 (2 units above the external pH), 99% of the acid will dissociate. This lowers the internal HA concentration, so more HA enters the cell in search of equilibrium. Because neither the ionized form of the acid nor the released proton can diffuse out of the cell, both accumulate. At equilibrium, the concentrations of HA inside and outside the cell are equal, but for every HA that enters the cell, ionization of HA has yielded 99 A^- molecules and an equal number of protons, which lower the internal pH to an extent that depends on the buffering capacity of cellular proteins.

Concentration of
hydrogen ions
compared to
distilled water

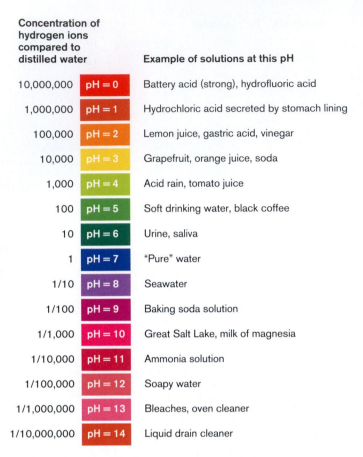

Concentration	pH	Example of solutions at this pH
10,000,000	pH = 0	Battery acid (strong), hydrofluoric acid
1,000,000	pH = 1	Hydrochloric acid secreted by stomach lining
100,000	pH = 2	Lemon juice, gastric acid, vinegar
10,000	pH = 3	Grapefruit, orange juice, soda
1,000	pH = 4	Acid rain, tomato juice
100	pH = 5	Soft drinking water, black coffee
10	pH = 6	Urine, saliva
1	pH = 7	"Pure" water
1/10	pH = 8	Seawater
1/100	pH = 9	Baking soda solution
1/1,000	pH = 10	Great Salt Lake, milk of magnesia
1/10,000	pH = 11	Ammonia solution
1/100,000	pH = 12	Soapy water
1/1,000,000	pH = 13	Bleaches, oven cleaner
1/10,000,000	pH = 14	Liquid drain cleaner

Figure 5.12   pH values of common substances.

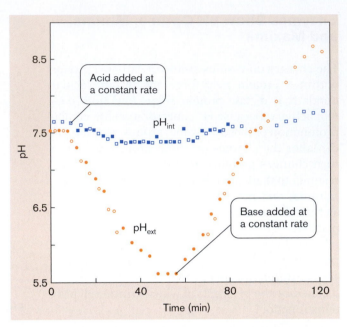

Figure 5.13   Maintaining internal pH (pH homeostasis) over a wide range of external pH.   Internal pH of *Escherichia coli* measured following the addition of acid (HCl, indicated by arrow at $t = 10$ min) to change external pH (circles) and subsequent addition of base (NaOH at $t = 52$ min). In this experiment, internal pH (pH$_{int}$) was determined using nuclear magnetic resonance to measure changes in cellular phosphate (closed circles/squares) and methyl phosphate (open circles/squares). The two phosphate species titrate over different pH ranges. *Source*: Joan L. Slonczewski, et al. 1981. *Proceedings of the National Academy of Sciences USA* 78:6271.

flavor or quality of the food. **Figure 5.12** illustrates the pH values of various everyday items. Food microbiology is further discussed in Chapter 16.

Neutralophiles, Acidophiles, and Alkaliphiles Grow in Different pH Ranges

Cells have evolved to live under different pH conditions not by drastically changing the pH optima of their enzymes but by using novel pH homeostasis strategies that maintain intracellular pH above pH 5 and below pH 8, even when the cell is immersed in pH environments well above or below that range.

There are three classes of organisms marked by the pH of their growth range.

Neutralophiles generally grow between pH 5 and pH 8, and include most human pathogens. Many neutralophiles, including *E. coli* and *Salmonella enterica*, adjust their metabolism to maintain an internal pH slightly above neutrality, which is where their enzymes work best. They maintain this pH even in the presence of moderately acidic or basic external environments (**Fig. 5.13**). Others allow their internal pH to fluctuate with external

pH but usually maintain a pH difference (ΔpH) of about 0.5 pH units across the membrane at the upper and lower limits of growth pH. The ΔpH value is an important component of the transmembrane proton potential, a source of energy for the cell (see Chapter 13).

NOTE: The older term *neutrophile* used for this group of organisms is also similar to the descriptor for a specific type of white blood cell (*neutrophil*). To avoid confusion, the term *neutrophil* should be reserved for the white blood cell and the term *neutralophile* used to designate microbes with growth optima near neutral pH.

Acidophiles are bacteria and archaea that live in acidic environments. They are often chemoautotrophs (lithotrophs) that oxidize reduced metals and generate strong acids such as sulfuric acid. Consequently, they grow between pH 0 and pH 5. The ability to grow at this pH is due partly to altered membrane lipid profiles (high levels of tetraether lipids) that decrease proton permeability as well as to ill-defined proton extrusion mecha-

A.

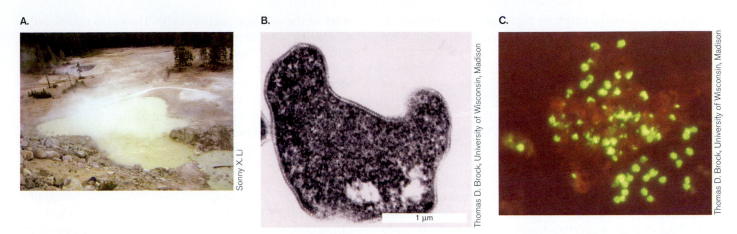

Sonny X. Li

Thomas D. Brock, University of Wisconsin, Madison

Thomas D. Brock, University of Wisconsin, Madison

Figure 5.14   Sulfur Caldron acid spring and *Sulfolobus acidocaldarius*.   A. Sulfur Caldron, in the Mud Volcano area of Yellowstone National Park, is one of the most acidic springs in the park. It is rich in sulfur and in *Sulfolobus*, a bacterium that thrives in hot, acidic waters with temperatures from 60°C to 95°C and a pH of 1–5. **B.** Thin-section electron micrograph of *S. acidocaldarius*. Under the electron microscope, the organisms appear as irregular spheres that are often lobed. **C.** Fluorescent photomicrograph of cells (green) attached to a sulfur crystal. Fimbrial-like appendages (not seen here) have been observed on the cells attached to solid surfaces such as sulfur crystals. Crystal size is approx. 55 mm in real life.

nisms. Often, an organism that is an extremophile with respect to one environmental factor is an extremophile with respect to others. *Sulfolobus acidocaldarius*, for example, is a thermophile and an acidophile (**Fig. 5.14**). It uses sulfur as an energy source and grows in acidic hot springs rich in sulfur.

Alkaliphiles occupy the opposite end of the pH spectrum, growing best at values ranging from pH 9 to pH 11. They are commonly found in saline soda lakes, which have high salt concentrations and pH values (as high as pH 11). Soda lakes, like Lake Magadi in Africa (**Fig. 5.15A**), are steeped in carbonates, which explains their extraordinarily alkaline pH. An alkaliphilic organism first identified in Lake Magadi is *Halobacter salinarum* (also known as a *Natronobacterium gregoryi*), a halophilic archaeon (**Fig. 5.15B**).

The cyanobacterium *Spirulina* is another alkaliphile that grows in soda lakes. Its high concentration of carotene gives the organism a distinctive pink color (note the color of the lake in **Fig. 5.15A**). *Spirulina* is also a major food for the famous pink flamingos indigenous to these African lakes and is, in fact, the reason pink flamingos are pink. After the birds ingest these organisms, digestive

A.

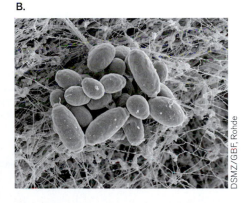

Kazuyoshi Nomachi/Corbis

B.

DSMZ/GBF, Rohde

C.

Wolfgang Kaehler/Corbis

Figure 5.15   A soda lake ecosystem. A. Lake Magadi in Kenya. Its pink color is due to *Spirulina*. **B.** Alkaliphile *Natronobacterium gregoryi*. Cell size approx. 1 × 3 μm. **C.** Pink flamingo colored by ingestion of *Spirulina*.

processes release the carotene pigment to the circulation, which then deposits it in the birds' feathers, turning them pink (**Fig. 5.15C**).

> **NOTE:** Humans also consume *Spirulina* as a health food supplement, but do not turn pink because the cyanobacteria are only a small component of the diet.

The internal enzymes of alkaliphiles, like those of acidophiles, exhibit rather ordinary pH optima (around pH 8). The key to the survival of alkaliphiles is the cell surface barrier that sequesters fragile cytoplasmic enzymes away from harsh extracellular pH. Key structural features of the cell wall, such as the presence of acidic polymers and an excess of hexosamines in the peptidoglycan, appear to be essential. The reason for this is not clear. At the membrane, some alkaliphiles also possess a high level of diether lipids, which are more stable than ester-linked phospholipids, that prevent protons from leaking out of the cell (see Section 3.4).

Because external protons are in such short supply at alkaline pH, most alkaliphiles use a sodium motive force in addition to a proton motive force to do much of the work of the cell (see Section 14.3). They also rely heavily on Na^+/H^+ antiporters (see Section 4.2) to bring protons into the cell, and this keeps the internal pH well below the extremely alkaline external pH. This is partly why many alkaliphiles are resistant to high salt (NaCl) concentrations; sodium ions are expelled while protons are sucked in. Important aspects of sodium circulation in alkaliphiles are depicted in **Figure 5.16**.

In contrast to proteins *within* the cytoplasm, enzymes *secreted* from alkaliphiles are able to work in very alkaline environments. The inclusion of base-resistant enzymes like lipases and cellulases in laundry detergents helps get our "whites whiter and our brights brighter." Other commercially useful alkaliphilic enzymes produce cyclodextrins from starch. The cyclodextrins are complex cyclic carbohydrates whose structure resembles a hollow truncated cone with a hydrophobic (water-hating) core and hydrophilic (water-loving) exterior (**Fig. 5.17**). It is a superb vehicle for drug delivery. For example, cyclodextrins have been used in eyedrops to deliver the antibiotic chloramphenicol. The hydrophobic cavity of cyclodextrin can harbor a poorly soluble drug, while the hydrophilic exterior increases its apparent water solubility. The drug goes into solution more easily, permitting smoother absorption into the body.

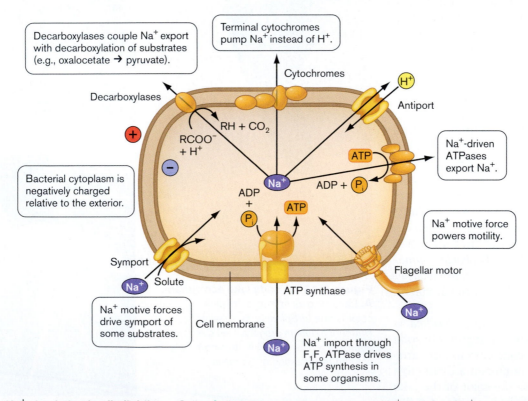

Figure 5.16 Na^+ circulation in alkaliphiles.   Cells of alkaliphiles are designed to use Na^+ in place of H^+ to do the work of the cell. They require an inwardly directed sodium gradient that can be used to rotate flagella, transport nutrient solutes, and generate energy. The Na^+/H^+ antiporter is used to keep internal pH lower than exterior pH. Sodium homeostasis inside the cell is maintained by the net effects of efflux and influx mechanisms.

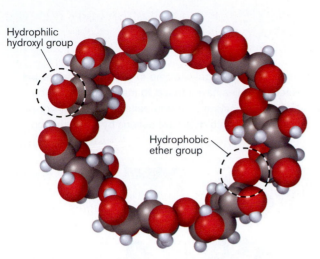

Figure 5.17   A cyclodextrin.   View down the channel in the hydrophobic core lined by C-O-C ether groups, which is surrounded by a hydrophilic exterior covered in -OH hydroxyl groups. The hydrophobic core can harbor drugs that are poorly soluble in water.

THOUGHT QUESTION 5.5   In Chapter 4, we stated that an antiporter couples movement of one ion down its concentration gradient with movement of another molecule uphill, against its gradient. If this is true, how could a Na^+/H^+ antiporter work to bring protons into a haloalkiliphile growing in high salt at pH 10? Since the Na^+ concentration is lower inside the cell than outside and H^+ concentration is higher inside than outside (see Fig. 5.16), both ions are moving against their gradients.

pH Homeostasis and Acid Tolerance Mechanisms Enable Cells to Live in Different pH Environments

When cells are placed in pH conditions below their optimum, protons can enter the cell and lower internal pH to lethal levels. Microbes can prevent the unwanted influx of protons by exchanging extracellular K^+ for intracellular protons when internal pH becomes too low. At the other extreme, under extremely alkaline conditions, the cells can use the Na^+/H^+ antiporters mentioned previously (and in Section 4.2) to recruit protons into the cell in exchange for expelling Na^+ (**Fig. 5.18**). Some organisms can also change the pH of the medium using various amino acid decarboxylases and deaminases. *Helicobacter pylori*, the causative agent of gastric ulcers, employs an exquisitely potent urease to generate massive amounts of ammonia, which neutralizes the acid pH environment. These acid stress and alkaline stress protection systems are usually not made or at least do not become active until the cell encounters an extreme pH.

Many, if not all, microbes also possess an emergency global response system referred to as acid tolerance or acid resistance. In a process analogous to the heat-shock response, bacterial physiology undergoes a major molecular reprogramming in response to hydrogen ion stress. The levels of a large number of proteins increase, while the levels of others decrease. Many of the genes and proteins involved in the acid stress response overlap with other stress response systems, including the heat-shock response. These physiological responses include modifications in membrane lipid composition, enhanced pH homeostasis,

Figure 5.18   Proton circulation and pH homeostasis.   A typical *E. coli* cell uses various proton transport strategies to maintain an internal pH near pH 7.8 in the face of different external pH stresses. Proton pumping through cytochromes also establishes a proton gradient, which drives flagellar rotation and solute transport.

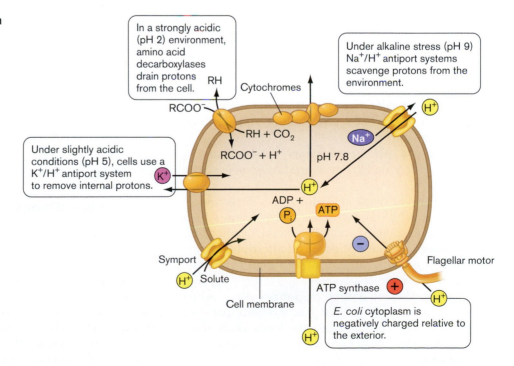

Special Topic 5.1   Signaling Virulence

One of the burning questions in the study of infectious disease is, "How does a microbe know when it has entered a suitable host?" Most, if not all, pathogens express certain genes only *after* infection, so they must sense that something in their environment has changed. A change in environment that causes increased virulence is called a virulence signal. One of these virulence signals is pH. On their way to causing disease, many pathogens travel through acidic host compartments. The most obvious are the stomach, which is extremely acidic (pH 1–3), and infected host cell phagosomes and phagolysosomes, which are mildly acidic (pH 4.5–6). *Salmonella enterica*, a major causative agent of diarrheal disease, provides an intriguing example of virulence that is triggered by low pH. After gaining entrance to the intestine via the mouth and subsequently penetrating intestinal epithelia, this pathogen tricks certain white blood cells, called macrophages, into engulfing the bacterium, placing it within a phagosome (**Fig. 1**). The microbe then waits for the phagosome (initially pH 6) to partially fuse with a lysosome and form a phagolysosome, after which the pH of the compartment drops to about pH 4.5. Numerous bacterial proteins are then expressed, many of which enable *Salmonella* to survive in the macrophage. Drugs that prevent phagolysosome acidification, however, render *Salmonella* extremely susceptible to the various antimicrobial weapons wielded by the macrophage. *Salmonella* "knows" it

is in a macrophage phagolysosome in part because the pH drops. Recent studies have started to map the complex bacterial regulatory circuits involved in sensing acid pH signals and translating them into defensive actions by the pathogen.

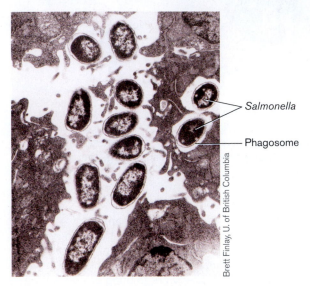

Figure 1   *Salmonella* (1 μm in length) within human epithelial cell phagosomes.

and numerous other changes with unclear purpose. Some pathogens, such as *Salmonella*, sense a change in external pH as part of the signal indicating the bacterium has entered a host cell environment (**Special Topic 5.1**).

TO SUMMARIZE:

- **Hydrogen ion concentration** affects protein structure and function. Thus, enzymes have pH optima, minima, and maxima.
- **Microbes use pH homeostasis mechanisms** to keep their internal pH near neutral when in acidic or alkaline media.
- **The addition of weak acids** to certain foods undermines bacterial pH homeostasis mechanisms, which prevents food spoilage and kills potential pathogens.
- **Neutralophiles, acidophiles, and alkaliphiles** prefer growth under neutral, low, and high pH conditions, respectively.
- **Acid and alkaline stress responses** occur when placing a given species under pH conditions that slow its growth. The cell increases the levels of proteins designed to mediate pH homeostasis and protect cell constituents.

5.6 Microbial Responses to Oxygen and Other Electron Acceptors

Many microorganisms can grow in the presence of molecular oxygen (O_2). Some even use oxygen as a terminal electron acceptor in the **electron transport chain**, a group of membrane proteins (cytochromes) that convert energy trapped in nutrients to a biologically useful form. The use of O_2 as the terminal electron acceptor is called **aerobic respiration** (see Chapter 14).

Oxygen Has Benefits and Risks

Electrons pulled from various energy sources (for example, glucose) possess intrinsic energy, an energy that cytochrome proteins can harness. They do this by extracting the energy in incremental stages and using it to move protons out of the cell. This unequal distribution of H^+ across the membrane produces a transmembrane electrochemical gradient, a sort of biobattery called proton motive force (discussed in Section 14.2). Once the cell has drained as much energy as possible from an electron,

Figure 5.19 Role of oxygen as a terminal electron acceptor in respiration. This pumping of protons out of the cell by cytochromes produces more positive charges outside the cell than inside, creating an electrochemical gradient (also called proton motive force). An electron from a high-energy source is passed from one member of the chain to the next. With each transfer of an electron to the next member of the chain, more energy is extracted and converted to proton motive force. At the end of the chain, the electron must be passed to a final or terminal electron acceptor, clearing the path for the next electron. This net process is called respiration.

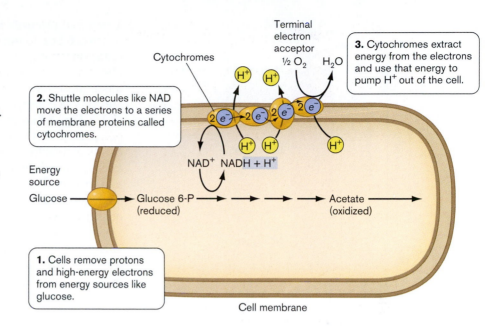

that electron must be passed to a diffusible, final electron acceptor molecule floating in the medium. This clears the way for another electron to be passed down the cytochrome chain (**Fig. 5.19**).

One such terminal electron acceptor is dissolved oxygen. However, oxygen and its breakdown products are dangerously reactive, a serious problem for all cells. As a result, different species have evolved to either tolerate or avoid oxygen all together. **Table 5.2** gives examples of microbes that grow at different levels of oxygen.

Microbes Are Classified by Their Relationship with Oxygen

The relationships between microbes and oxygen are varied. **Figure 5.20** illustrates a test tube with growth medium. The top of the tube, closest to air, is oxygenated; the bottom of the tube has much lower levels of oxygen. Some microbes grow only at the top of the tube, while others prefer to grow at the bottom, the distributions based on each organism's relationship with oxygen. A

Table 5.2 Examples of aerobes and anaerobes.

Aerobic	Anaerobic	Facultative	Microaerophilic
Neisseria spp. Causative organisms of meningitis, gonorrhea	*Azoarcus tolulyticus* Degrades toluene	*Escherichia coli* Normal GI flora; additional pathogenic strains	*Helicobacter pylori* Causative organism of gastric ulcers
Pseudomonas fluorescens Found in soil and water; degrade pollutants such as TNT and aromatic hydrocarbons	*Bacteroides* spp. Normal GI flora	*Saccharomyces cerevisiae* Yeast; used in baking	*Lactobacillus* Ferments milk to form yogurt
Mycobacterium leprae Causative organism of leprosy	*Clostridium* spp. Soil microorganisms; causative agents of tetanus and botulism	*Bacillus anthracis* Causative organism of anthrax	*Campylobacter* spp. Causative organism of gastroenteritis
Azotobacter Soil microorganisms; fix atmospheric nitrogen	*Actinomyces* Soil microrganisms; synthesize antibiotics	*Vibrio cholerae* Causative organism of cholera	*Treponema pallidum* Causative organism of syphilis
Rhizobium spp. Soil microorganisms; plant symbionts	*Desulfovibrio* spp. Reduce sulfate	*Staphylococcus* spp. Found on skin; causative agent of boils	

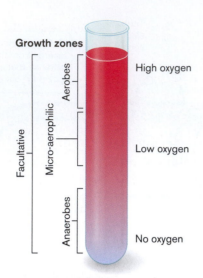

Figure 5.20 **Oxygen-related growth zones in a standing test tube.**

strict aerobe is an organism that not only exists in oxygen but also uses oxygen as a terminal electron acceptor. In fact, a strict aerobe can *only* grow with oxygen present and possesses an aerobic metabolism (aerobic respiration). An aerobe will grow at the top of the tube shown in Figure 5.20. In contrast, a **strict <u>anaerobe</u>** dies in the least bit of oxygen. Strict anaerobes do not use oxygen as an electron acceptor, but this is not why they die in air. They die because this type of microbe is vulnerable to reactive oxygen molecules (also called **reactive oxygen species**; ROS) produced by its own metabolism when it is exposed to oxygen. Anaerobes will grow at the bottom of the tube shown in Figure 5.20.

Do not confuse the ability of an organism to *exist* in the presence of oxygen with its ability to *use* oxygen as

a terminal electron acceptor. Some aerobes can survive in oxygen but do not have the ability to use oxygen. Any organism that possesses NADH dehydrogenase II, aerobe or anaerobe, will, in the presence of oxygen, inadvertently autooxidize the FAD cofactor within the enzyme and produce dangerous amounts of superoxide radicals ($^\bullet O_2^-$; **Fig. 5.21**). Superoxide will degrade to hydrogen peroxide (H_2O_2), another reactive molecule. Iron, present as a cofactor in several enzymes, can then catalyze a reaction with hydrogen peroxide to produce the highly toxic hydroxyl radical ($^\bullet OH$). All of these molecules seriously damage DNA, RNA, proteins, and lipids. Consequently, oxygen is actually an extreme environment in which survival requires special talents. Aerobes, but not anaerobes, destroy reactive oxygen species with the aid of enzymes such as superoxide dismutase (to remove superoxide) and peroxidase and catalase (to remove hydrogen peroxide). Aerobes also have resourceful enzyme systems that detect and repair macromolecules damaged by oxidation.

Anaerobes Ferment and Respire without Oxygen

Anaerobic microbes fall into several categories. Some anaerobes actually do respire using cytochrome systems, but instead of using oxygen, they rely on alternative terminal electron acceptors like nitrate to conduct **anaerobic respiration** and produce energy. Anaerobes of another ilk do not possess cytochromes, cannot respire, and so must rely on carbohydrate **fermentation** for energy (that is, they conduct **fermentative metabolism**). Fermentation involves the production of ATP energy through substrate-level phosphorylation in a process that does not involve cytochromes. **Facultative** organisms are

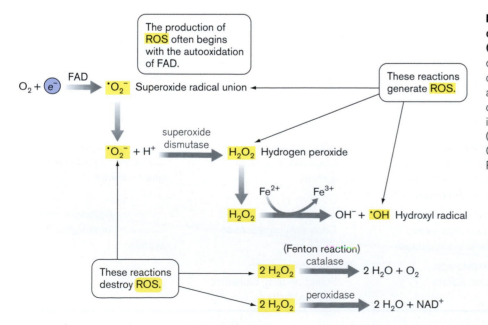

Figure 5.21 **Generation and destruction of reactive oxygen species (ROS).** The autoxidation of flavin adenine dinucleotide (FAD) and the Fenton reaction occur spontaneously to produce superoxide and hydroxyl radicals, respectively. The other reactions require enzymes. FAD is a cofactor for a number of enzymes (for example, NADH dehydrogenase II). Catalase and peroxidase do not produce ROS but detoxify hydrogen peroxide.

microbes that can live with or without oxygen. They will grow throughout the tube shown in Figure 5.20. **Facultative anaerobes** (sometimes called aerotolerant) only use fermentation to provide energy but contain superoxide dismutase and catalase (or peroxidase) to protect them from reactive oxygen species. This allows them to grow in oxygen while retaining a fermentation-based metabolism. **Facultative aerobes** (such as *E. coli*) also possess enzymes that destroy toxic oxygen by-products, but have both fermentative *and* aerobic respiratory potential. Whether a member of this group uses aerobic respiration, anaerobic respiration, or fermentation depends on the availability of oxygen and the amount of carbohydrate present. Microorganisms that possess *decreased* levels of superoxide dismutase and/or catalase will be **microaerophilic**, meaning they will grow only at low oxygen concentrations.

The fundamental composition of all cells reflects their evolutionary origin as anaerobes. Lipids, nucleic acids, and amino acids are all highly reduced—which is why our bodies are combustible. We never would have evolved that way if molecular oxygen were present from the beginning. Even today, the majority of all microbes are anaerobic, growing buried in the soil, within our anaerobic digestive tract, or within biofilms on our teeth.

THOUGHT QUESTION 5.6 If anaerobes cannot live in oxygen, how do they incorporate oxygen into their cellular components?

THOUGHT QUESTION 5.7 How can anaerobes grow in the human mouth when there is so much oxygen there?

Culturing Anaerobes in the Laboratory

Many anaerobic bacteria cause horrific human diseases, such as tetanus, botulism, and gangrene. Some of these organisms or their secreted toxins are even potential weapons of terror (for example, *Clostridium botulinum*). Because of their ability to wreak havoc on humans, culturing these microorganisms was an early goal of microbiologists. Despite the difficulties involved, conditions were eventually contrived in which all, or at least most, of the oxygen could be removed from a culture environment.

Three techniques are used today. Special reducing agents (for example, thioglycolate) or enzyme systems (Oxyrase®) that eliminate dissolved oxygen can be added to ordinary liquid media. Anaerobes can then grow beneath the culture surface. A second, very popular, way to culture anaerobes, especially on agar plates, is to use an anaerobe jar (**Fig. 5.22A**). Agar plates streaked with the organism are placed into a sealed jar with a foil packet that releases H_2 and CO_2 gases. A palladium packet hanging from the jar lid catalyzes a reaction between the H_2 and O_2 in the jar to form H_2O and effectively removes O_2 from the chamber. The CO_2 released is required by some reactions to produce key metabolic intermediates. Some microaerophilic microbes, like the pathogens *H. pylori* (the major cause of stomach ulcers) and *Campylobacter jejuni* (a major cause of diarrhea), require low levels of O_2 but elevated amounts of CO_2. These conditions are obtained by using similar gas-generating packets.

For strict anaerobes exquisitely sensitive to oxygen, even more heroic efforts are required to establish an oxygen-free environment. A special anaerobic glove box must be used in which the atmosphere is removed by

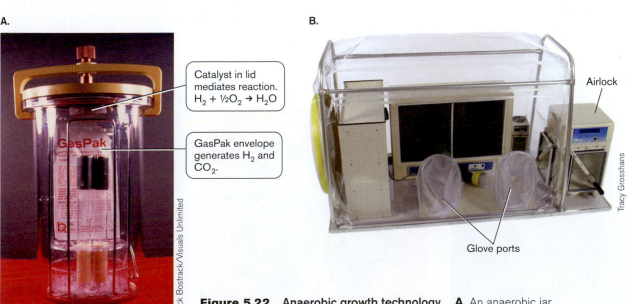

A.

Catalyst in lid mediates reaction. $H_2 + \frac{1}{2}O_2 \rightarrow H_2O$

GasPak envelope generates H_2 and CO_2.

©Jack Bostrack/Visuals Unlimited

B.

Airlock

Glove ports

Tracy Grosshans

Figure 5.22 Anaerobic growth technology. A. An anaerobic jar. **B.** An anaerobic chamber with glove ports.

vacuum and replaced with a precise mixture of N_2 and CO_2 gases (**Fig. 5.22B**).

> **THOUGHT QUESTION 5.8**   What evidence led people to think about looking for anaerobes? *Hint*: Look up Spallanzani, Pasteur, and spontaneous generation on the Web.

TO SUMMARIZE:

- **Oxygen is a benefit** to organisms (aerobes) that can use it as a terminal electron acceptor to extract energy from nutrients.
- **Oxygen is toxic** to all cells (anaerobes) that do not have enzymes capable of efficiently destroying the reactive oxygen species.
- **Anaerobic metabolism** can either be **fermentative** or **respiratory**. Anaerobic respiration requires the organism to possess cytochromes that can use compounds other than oxygen as terminal electron acceptors.
- **Facultative anaerobes** grow either in the presence or in the absence of oxygen, but use fermentation as their primary, if not only, means of gathering energy. These microbes also have enzymes that destroy reactive oxygen species, allowing them to grow in oxygen.
- **Facultative aerobes**, like facultative anaerobes, grow with or without oxygen and have enzymes that destroy reactive oxygen species. In addition, they possess both the ability for fermentative metabolism and the ability to use oxygen as a terminal electron acceptor.

5.7 Microbial Responses to Nutrient Deprivation and Starvation

It is intuitively obvious that limiting the availability of a carbon source or other essential nutrient will limit growth. Not so obvious are the dramatic molecular events that cascade through a cell undergoing nutrient limitation and ultimately starvation. Optimizing growth rate at suboptimal nutrient levels is an important aim of free-living bacteria, given that intestinal, soil, and marine environments rarely offer nutrient excess.

Starvation Slows Cell Metabolism but Activates Key Survival Genes

Numerous gene systems are affected when nutrients decline (see Section 10.3 for details). Growth rate slows, and daughter cells become smaller and begin to experience what is termed a **"starvation" response**, where the microbe senses a dire situation developing but still strives to find new nourishment. The resulting metabolic slowdown generates increased concentrations of critically important small signal molecules such as cyclic AMP and guanosine tetraphosphate (ppGpp) that globally transform gene expression. During this metabolic retooling, transport systems for potential nutrients are produced even if the matching substrates are unavailable. Cells begin to make and store glycogen, presumably as an internal emergency store in case no other nutrient is found. Some organisms growing on nutrient-limited agar plates can even form colonies with intricate geometrical shapes that help the population cope, in some unknown way, with nutrient stress (**Fig. 5.23**). As a cultural environment progressively worsens, the organism must prepare for famine, and many different stress survival genes become active. The products of these genes afford protection against stressors such as reactive oxygen radicals or temperature and pH extremes. No cell can predict the precise stresses it might encounter while incapacitated, so it is advantageous to be prepared for as many as possible. As described in Chapter 4, some species undergo elaborate developmental processes that ultimately produce dormant spores.

A.

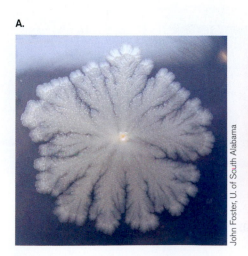

John Foster, U. of South Alabama

B.

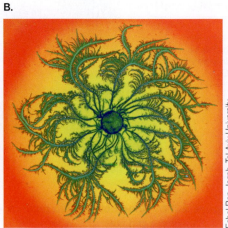

Eshel Ben-Jacob, Tel Aviv University

Figure 5.23   Effects of starvation on colony morphology.   A. Starving *E. coli* colony. **B.** *Pseudomonas dendritiformis* C morphotype, grown on hard agar (1.75%) under starvation conditions. The colony consists of branches with chiral twists (colored green), all with the same handedness. The added coloration indicates time of growth (yellow = oldest; red = most recent).

Some Organisms Only Grow in Nutrient-Poor Media

Most well-studied organisms thrive on organically rich media (more than 2 g of carbon per liter) but struggle to grow at very low nutrient concentration (1–5 mg of carbon per liter). In natural ecosystems, however, the majority of existing microbes appear to be **oligotrophs**, organisms with a high rate of growth at extremely low organic substrate concentrations. Oligotrophs are well suited for growth in the nutrient-poor, or oligotrophic, environments found in certain lakes and streams. Some oligotrophs are actually poisoned by concentrated organics or commit suicide in rich media by overproducing excess toxic hydrogen peroxide. Consequently, they *require* low nutrient levels to survive. In soil, for example, there exist a variety of uncharacterized oligotrophic bacteria that are very sensitive to NaCl and various L-amino acids. How do oligotrophs manage to grow in such impoverished conditions? Some oligotrophic bacteria have thin extensions of their membrane and cell wall called prothecae (stalks) that essentially *expand* the surface area of the cell and increase capacity to transport nutrients. How other oligotrophs grow remains a mystery.

Microbes Encounter Multiple Stresses in Real Life

In keeping with the reductionist approach to science, bacterial stress responses have traditionally been studied in terms of *individual* stresses. *Escherichia coli*, for example, increases synthesis of a specific set of proteins when exposed to high temperature and a different set of proteins when exposed to high salt. There are overlapping members of the two sets; that is, several proteins may be highly expressed under both conditions, but each stress response also includes proteins unique to each stress.

Environmental situations in the world outside of the laboratory, however, can be quite complex, involving multiple, not just single, stresses. An organism could simultaneously undergo carbon starvation in a high salt, low pH environment. A classic study by Kelly Abshire and Frederich Neidhardt examined this situation using the pathogen *Salmonella enterica*, a cause of diarrhea. *S. enterica* invades human macrophage cells and survives in phagocytic vacuoles, where numerous stresses, such as low pH, oxidative stress, and nutrient limitations, are simultaneously imposed on the bacteria. Comparing the proteins synthesized by *Salmonella* growing in this compartment with the proteins synthesized under single stresses in the laboratory revealed an unexpected response pattern. Although many stress-related proteins were induced in the intracellular environment, no one set of stress-induced proteins was induced in its entirety.

Furthermore, several bacterial proteins were induced by growth *only* within the macrophage phagolysosome, suggesting the presence of unknown intracellular stresses. Thus caution is advised when trying to predict cell responses to real-world situations based solely on controlled laboratory studies that alter only single parameters.

Humans Influence Microbial Ecosystems

Natural ecosystems are typically low in nutrients (oligotrophic) but teem with diversity, so that numerous species compete for the same limiting nutrients. Maximum diversity in a given ecosystem is maintained, in part, by the different nutrient-gathering profiles of competing microbes. Take a scenario in which microbe A is better than microbe B at gathering phosphate when phosphate levels are low, but microbe B is superior to A at culling limited quantities of nitrogen (**Fig. 5.24**). Neither organism dominates when both phosphate *and* nitrogen are in short supply. So even though B can harvest more nitrogen than A, it cannot outgrow A, since the phosphate concentration is low. However, if phosphate is suddenly increased so that it is no longer limiting for either organism, the one better adapted to low nitrogen (microbe B) will outgrow and overwhelm the other. The sudden infusion of large quantities of a formerly limiting nutrient, a process called **eutrophication**, can lead to a "bloom" of microbes. Organisms initially held in check by the limiting nutrient now exhibit unrestricted growth, consuming other nutrients to a degree that threatens the existence of competing species.

Humans have caused nutrient pollution in several ways. Runoff from agricultural fields, urban lawns, and golf courses is one source. Untreated or partially treated domestic sewage is another. Sewage was a primary source of phosphorus eutrophication of lakes in the 1960s and 1970s, when detergents contained large amounts of phosphates. The phosphates acted as water softeners to improve cleaning action, but when washed into lakes, they also proved to be powerful stimulants to algal growth (**Fig. 5.25**). The resulting algal "blooms" in many lakes led to oxygen depletion and resultant fish kills. Many native fish species disappeared, to be replaced by species more tolerant of the new conditions.

TO SUMMARIZE:

- **Starvation** is a stress that can elicit a starvation response in many microbes. Enzymes are produced to increase the efficiency of nutrient gathering and to protect cell macromolecules from damage.
- **The starvation response** is usually triggered by the accumulation of small signal molecules such as cyclic AMP or ppGpp.

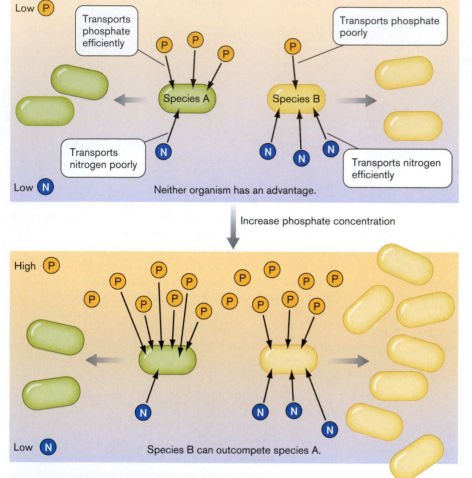

Figure 5.24 Maintaining microbial diversity through nutrient limitation. Species A and species B require both nitrogen and phosphate to grow. *Top*: Neither species dominates because each has a limiting nutrient (N = nitrogen; P = phosphate). *Bottom*: Following phosphate eutrophication, species B will outgrow species A because while both species have enough phosphate, B is better able to assimilate the limited quantity of nitrogen.

Figure 5.25 Eutrophication. Algal bloom in an experimental lake resulting from phosphate eutrophication. A divider curtain separates the lake and prevents mixing of water. The bright green color results from cyanobacteria growing on phosphorus added to the near side of the curtain.

- **Oligotrophs** are organisms that thrive in nutrient-poor conditions.
- **Human activities can cause eutrophication**, which damages delicately balanced ecosystems by introducing nutrients that can allow one member of the ecosystem to flourish at the expense of other species.

5.8 Physical and Chemical Methods of Controlling Microbial Growth

Prevention, control, or elimination of microbes that are potentially harmful to humans is one of the primary goals of our health care system. Within the recent past, infectious disease was an imminent and constant threat to most of the human population. The average family in the United States prior to 1900 had four or five children, in part because parents could expect half of them to succumb to deadly infectious diseases. What today would be a simple infected cut in years past held a serious risk of death, and a trip to the surgeon was tantamount to playing Russian roulette with an unsterilized scalpel. Improvements in sanitation procedures and antiseptics and the advent of antibiotics have, to a large degree, curtailed the incidence and lethal effects of many infectious diseases. Success in this endeavor has played a major role in extending life expectancy and in contributing to the population explosion.

A variety of terms are used to describe antimicrobial control measures. The terms convey subtle, yet vitally important, differences in various control strategies and outcomes.

- **Sterilization** is the process by which *all* living cells, spores, and viruses are destroyed on an object.
- **Disinfection** is the killing, or removal, of *disease-producing* organisms from inanimate surfaces; it does not necessarily result in sterilization. Pathogens are killed, but other microbes may survive.
- **Antisepsis** is similar to disinfection, but it applies to removing pathogens from the surface of living tissues, like the skin. Antiseptic chemicals are usually not as toxic as disinfectants, which frequently damage living tissues.
- **Sanitation**, closely related to disinfection, consists of reducing the microbial population to safe levels and usually involves cleaning an object as well as disinfection.

Antimicrobials can also be classified on the basis of the specific groups of microbes destroyed, leading to the terms *bactericide, algicide, fungicide,* and *virucide.* Furthermore, these agents can be classified either as *-static* (inhibiting growth) or *-cidal* (killing cells). For example, antibacterial agents are called **bacteriostatic** or **bactericidal**. Chemical substances that kill microbes are **germicidal** if they kill pathogens (and many nonpathogens). Germicidal agents do not necessarily kill spores. Although these descriptions place emphasis on killing pathogens, it is also important to note that antimicrobial agents can also kill or prevent the growth of nonpathogens. Many public health standards are based on *total* numbers of microorganisms on an object, regardless of pathogenic potential. For example, to gain public health certification, the restaurants we frequent must demonstrate low numbers of bacteria in their food preparation areas.

Cells Treated with Antimicrobials Die at a Logarithmic Rate

Exposure of microbes to lethal chemicals or conditions does not instantly kill all microorganisms. Microbes die according to a negative exponential curve, where cell numbers are reduced in equal fractions at constant intervals. The efficacy of a given agent or condition is measured as **decimal reduction time (D-value)**, which is the length of time it takes that agent (or condition) to

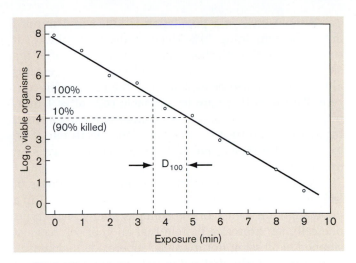

Figure 5.26   The death curve and the determination of D-values. Bacteria were exposed to a temperature of 100°C, and survivors were measured by viable count. The D-value is the time required to kill 90% of cells (that is, for the viable cell count to drop by one $\log_{10}$ unit). In this example, the D-value at 100°C (D_{100}) is approximately 1 minute.

kill 90% of the population (a drop of 1 log unit, or a drop to 10% of the original value). **Figure 5.26** illustrates the exponential death profile of a bacterial culture heated to 100°C. The D-value is a little over 1 minute.

Several factors influence the D-value—that is, the ability of an antimicrobial agent to kill microbes. These include the initial population size (the larger the population, the longer it takes to decrease it to a specific number), population composition (are spores involved?), agent concentration, and duration of exposure. Although the effect of concentration seems intuitively obvious, only over a narrow range is an increase in concentration matched by an increase in death rate. Increases above a certain level might not accelerate killing at all. For example, 70% ethanol is actually better than pure ethanol at killing organisms. This is because some water is needed to help ethanol penetrate cells. The ethanol then dehydrates cell proteins.

So, why is death a logarithmic function? Why don't all cells in a population die instantly when treated with lethal heat or chemicals? The reason is based, in part, on the random probability of the agent causing a lethal "hit" in a given cell. Cells contain thousands of different proteins and thousands of molecules of each. Not all proteins and not all genes in a chromosome are damaged by an agent at the same time. Damage accumulates. Only when enough molecules of an *essential* protein or a gene encoding that protein are damaged will the cell die. Cells that die first are those that accumulate lethal hits early. Members of the population that die later have, by random chance, absorbed more hits on nonessential proteins or genes, sparing the essential ones.

Why, if 90% of a population is killed in 1 minute, aren't the remaining 10% killed in the next minute? It seems logical that all should have perished. Yet after the second minute, 1% of the original population remains alive. This can also be explained by the random hit concept. Although there are fewer viable cells after 1 minute, each has the same random chance of having a lethal hit as when the treatment began. Thus, death rate is an exponential function, much like radioactive decay is an exponential function.

A final consideration involves the overall fitness of individual cells. It is mistaken to assume that all cells in a population are identical. For example, at any given time, one cell may express a protein that another cell has just stopped expressing (for example, superoxide dismutase). In that instant, the first cell might contain a bit more of that protein. If the protein is essential or confers a level of stress protection (such as against superoxide), the cell can absorb more punishment before it is dispatched. The presence of lucky individuals expressing the right repertoire of proteins might also explain why death curves commonly level off after a certain point.

Physical Agents That Kill Microbes

Physical agents are often used to kill microbes or control their growth. Commonly used physical control measures are temperature extremes, pressure (usually combined with temperature), filtration, and irradiation.

High temperature and pressure. Even though microbes were discovered less than 400 years ago, thermal treatment of food products to render them safe has been practiced for over 5,000 years. Moist heat is a much more effective killer than dry heat, thanks to the ability of water to penetrate cells. Many bacteria, for instance, easily withstand 100°C dry heat but not 100°C in boiling water. We humans are not so different, finding it easier to endure a temperature of 32°C (90°F) in dry Arizona than in humid Louisiana.

While boiling water (100°C) can kill most vegetative (actively growing) organisms, spores are built to withstand this abuse, and thermophiles prefer it. Killing spores and thermophiles usually requires combining high pressure and temperature. At high pressure, the boiling point of water rises to a temperature rarely experienced by microbes living at sea level. Even endospores quickly die under these conditions. This combination of pressure and temperature is the principle behind sterilization using the **steam autoclave (Fig. 5.27)**. Standard conditions for steam sterilization are 121°C (250°F) at 15 psi for 20 minutes, a set of conditions that experience has taught us will kill all spores. These are also the conditions produced in pressure cookers used for home canning of vegetables.

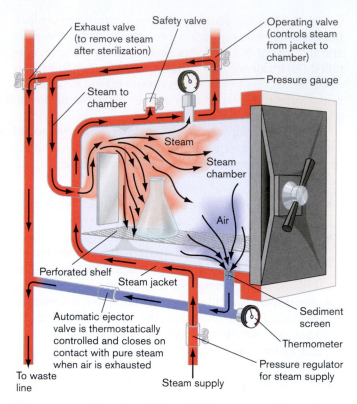

Figure 5.27   Steam autoclave.

Failure to adhere to these heat and pressure parameters can have deadly consequences, even in your own home. For instance, *Clostridium botulinum* is a spore-forming soil microbe that commonly contaminates fruits and vegetables used in home canning. The improper use of a pressure cooker while canning these goods will allow spores of this pathogen to survive. Once the can or jar is cool, the spores will germinate and begin producing their deadly toxin. All of this happens while the canned goods sit on a shelf waiting to be opened and consumed. Once ingested, the toxin makes its way to the nervous system and paralyzes the victim. Several incidents of this disease, called botulism, occur each year in the United States. (For more on food poisoning, see Chapter 16.)

> **THOUGHT QUESTION 5.9**   How would you test the killing efficacy of an autoclave?

The food industry uses several parameters to evaluate the efficiency of heat killing. The D-value has already been described. Two additional measures are **12D**, the amount of time required to kill 10^{12} spores (or reduce a population 12 logs); and the **z-value**, the increase in temperature needed to lower the D-value to $1/10$ its previous value. If, for example, D_{100} (the D-value at 100°C) and D_{110} (the D-value at 110°C) for a given organism are 20 minutes and 2 minutes, respectively, then $12D_{100}$ equals 240 minutes (that is, 20 minutes × 12), and the z-value is 10°C (because a 10°C increase in temperature reduced the D-value to $1/10$, from 20 minutes to 2 minutes). These measurements are extremely important to the canning industry, which must ensure that canned goods do not contain spores of *Clostridium botulinum*, the previously described anaerobic soil microbe that causes the paralyzing food-borne disease botulism.

Because the tastes of certain foods suffer if they are overheated, z-values and 12D-values are used to adjust heating times and temperatures to achieve the same sterilizing result. Take the following example, where D_{121} is 10 minutes and $12D_{121}$ is 120 minutes. Sterilizing food at 121°C for 120 minutes might result in food with a repulsive taste, whereas reducing the temperature and extending the heating time may yield a more palatable product. The D- and z-values are used to adjust conditions for sterilization at a lower temperature. If D_{121} is 15 minutes and the z-value is known to be 10°C, then reducing temperature by 10°C to 110°C will mean D_{111} is 150 minutes (10 times D_{121}). As a result, $12D_{111}$ is 1,800 minutes. Sterilization may take longer, but food quality is likely to remain high, because the sterilizing temperature is lower.

Pasteurization. Originally devised by Louis Pasteur to save products of the French wine industry from devastating bacterial spoilage, **pasteurization** today involves heating a particular food (such as milk) to a specific temperature long enough to kill *Mycobacterium tuberculosis* and *Coxiella burnetii*. These two organisms, the causative agents of tuberculosis and Q fever, respectively, are the most heat-resistant nonspore-forming pathogens known. Many different time and temperature combinations can be used for pasteurization. The LTLT (low-temperature/long-time) process involves bringing milk to a temperature of 63°C (145°F) for 30 minutes. In contrast, the HTST (high-temperature/short-time, also called flash pasteurization) method brings the milk to a temperature of 72°C (161°F) for only 15 seconds. Both processes accomplish the same thing: the destruction of *M. tuberculosis* and *C. burnetii*.

Cold. Low temperatures have two basic purposes in microbiology: to temper growth and to preserve strains. Bacteria not only grow more slowly in cold, but also die more slowly. Refrigeration temperatures (4–8°C, or 39–43°F) are used for food preservation because most pathogens are mesophilic and grow poorly, if at all, at those temperatures. One exception is the gram-positive bacillus *Listeria monocytogenes*, which can grow reasonably well in the cold and causes disease when ingested.

Long-term storage of bacteria usually requires placing solutions in glycerol at very low temperatures (–70°C). Glycerol prevents the production of razor-sharp ice crys-

tals that can pierce cells from without or within. This deep-freezing suspends growth altogether and keeps cells from dying. Another technique called **lyophilization** freeze-dries microbial cultures for long-term storage. In this technique, cultures are quickly frozen at very low temperatures (quick-freezing also limits ice crystal formation) and placed under vacuum, where the resulting sublimation process removes all water from the media and cells, leaving just the cells in the form of a powder. These freeze-dried organisms remain viable for years. Finally, viruses and mammalian cells must be kept at extremely low temperatures (–196°C), submerged in liquid nitrogen. Liquid nitrogen freezes cells so quickly that ice crystals do not have time to form.

Filtration. Filtration through micropore filters with pore sizes of 0.2 μm can remove microbial cells, but not viruses, from solutions. Samples from 1 ml to several liters can be drawn through a membrane filter by vacuum or can be forced through it using a syringe (**Fig. 5.28**). Filter sterilization has the advantage of avoiding heat, which can damage the material sterilized. Of course, since these filters do not trap viruses, the solutions are not really sterile.

Air can also be sterilized by filtration. This forms the basis of several personal protective devices. A surgical mask is a crude example, while **laminar flow biological safety cabinets** are more elaborate (and more effective). These cabinets force air through high-efficiency particulate air (HEPA) filters and remove over 99.9% of airborne particulate material 0.3 μm in size or larger. Biosafety cabinets are critical to protect individuals working with highly pathogenic material (**Fig. 5.29**). Newer technologies have been developed that embed antimicrobial agents or enzymes directly into the fibers of the filter

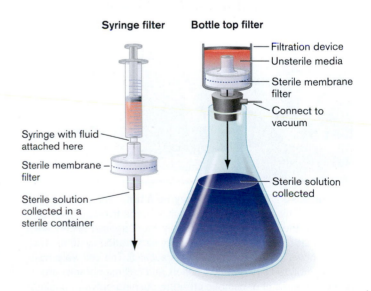

Syringe filter

Bottle top filter

- Filtration device
- Unsterile media
- Sterile membrane filter
- Connect to vacuum
- Sterile solution collected

Syringe with fluid attached here

Sterile membrane filter

Sterile solution collected in a sterile container

Figure 5.28   Membrane filtration apparatus.

A.

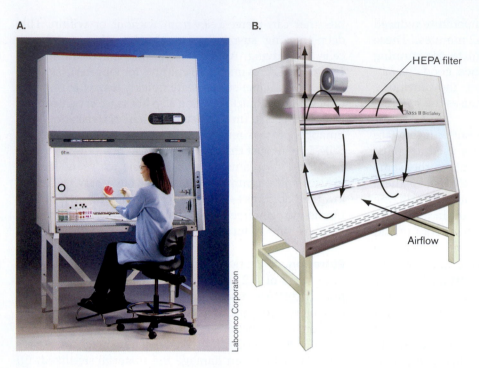

Labconco Corporation

B.

HEPA filter

Class II BioSafety

Airflow

Figure 5.29 **Biological safety cabinet.** Air is drawn into the chamber and continually passed through the HEPA filter to remove hazardous agents. This prevents the escape of aerosolized infectious agents.

(**Fig. 5.30**). Organisms entangled in these fibers are not just trapped; they are attacked by the antimicrobials and lyse.

Irradiation. Public health authorities worldwide are increasingly concerned about food contaminated with pathogenic microorganisms such as *Salmonella* species, *E. coli* O157:H7, *L. monocytogenes*, and *Yersinia enterocolitica*. Irradiation, the bombardment of foods with high-energy electromagnetic radiation, has long been a potent, if politically sensitive, strategy for sterilizing food after harvesting. The food consumed by NASA astronauts, for example, has for some time been sterilized by irradiation as a safeguard against food-borne illness in space.

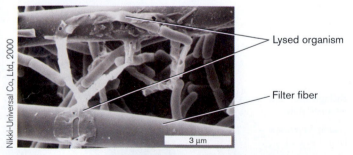

Nikki-Universal Co., Ltd., 2000

Lysed organism

Filter fiber

3 μm

Figure 5.30 **Immobilized enzyme filter.** The primary function of this enzyme filter is to kill airborne microorganisms caught on the surface of the filter to protect against secondary contamination from microorganisms in air filtration systems. The photo shows lysed bacteria (*Bacillus subtilis*). The cell walls have been hydrolyzed by enzymatic action, and cell membranes are broken as a result of osmotic pressure pushing outward against the membrane.

Until recently, public pressure severely limited the use of this technology because of concerns about product safety and exposure of workers. However, the recent use of the U.S. mail service by a bioterrorist to disseminate anthrax spores across the eastern United States has renewed interest in sterilization by irradiation. Foods do not become radioactive when irradiated, but there are concerns that reactive molecules potentially dangerous to humans are produced when high-energy particles are absorbed by the natural chemicals in the food and react to produce toxic substances.

Aside from ultraviolet light, which, owing to its poor penetrating ability, is only useful for surface sterilization, there are three other sources of irradiation: gamma rays, electron beams, and X-rays. Radiation dosage is usually measured in a unit called the gray (Gy), which is the amount of energy transferred to food, microbe, or other substance being irradiated. A single chest X-ray delivers roughly half a milligray (1 mGy = 0.001 Gy). To kill *Salmonella*, freshly slaughtered chicken can be irradiated at up to 4.5 kilograys (kGy), about 7 million times the energy of a single chest X-ray.

When microbes present in food are irradiated, water and other intracellular molecules absorb the energy and create transient reactive chemicals that damage DNA and scramble genetic information. Unless the organism repairs this damage, it will die while trying to replicate. Microbes differ greatly in their sensitivity to irradiation, depending on the size of their DNA, the rate at which they can repair damaged DNA, and other factors. It also matters if the irradiated food is frozen or fresh, as it takes a higher dose of radiation to kill microbes in frozen foods.

The size of the DNA "target" is a major factor in radiation efficacy. Parasites and insect pests, which have large amounts of DNA, are rapidly killed by extremely low doses of radiation, typically with D-values of less than 0.1 kGy (in this instance, the D-value is the *dose* of radiation needed to kill 90% of the organisms). It takes more radiation to kill bacteria (D-values in the range of 0.3–0.7 kGy) because they have less DNA per cell unit (less target per cell). It takes even more radiation to kill a bacterial spore (D-values on the order of 2.8 kGy) because they contain little water, the source of most ionizing damage to DNA. Viral pathogens have the smallest amount of nucleic acid, making them resistant to irradiation doses approved for foods (viruses have D-values of 10 kGy or higher). Prion particles associated with bovine spongiform encephalopathy (BSE, also known as mad cow disease) do not contain nucleic acid and are only inactivated by irradiation at extremely high doses. Thus, irradiation of food is effective in eliminating parasites and bacteria, but is woefully inadequate for eliminating viruses or prions.

Some Bacteria Are Highly Resistant to Physical Control Measures

Deinococcus radiodurans could be nicknamed "Conan the bacterium" and designated as a poster microbe for extremophiles (**Fig. 5.31**). It was discovered in 1956 in a can of meat that spoiled despite having been sterilized by radiation. The microbe has the greatest ability to survive radiation of any known organism. Based on the amount of radiation it can handle, it has been estimated that it could even survive the amount of radiation released in an atomic blast. The bacterium's ability to withstand radiation may have evolved as a side effect of developing resistance to extreme drought, since dehydration and radiation produce similar types of DNA damage.

www | *Deinococcus*

Research by microbiologist John Battista has shown that *D. radiodurans* possesses an unusual capacity for repairing its damaged DNA, although the precise mechanisms remain a mystery. One possible mechanism is that a damaged chromosome in one member of the quartet shown can restore itself using DNA from another member of the quartet. Based on this research, *D. radiodurans* has been genetically engineered to treat radioactive mercury-contaminated waste from nuclear reactors, a process called **bioremediation** (discussed in Chapter 22). The genes for mercury conversion were spliced from a strain of *E. coli* resistant to particularly toxic forms of mercury and inserted into *D. radiodurans*. The genetically altered superbug was able to withstand the ionizing radiation and transform toxic waste into forms that could be removed safely. Fortunately, there is little need to worry about its being a super pathogen, because the organism does not cause disease and is susceptible to antibiotics.

Chemical Agents Supplement Physical Means of Controlling Microbes

Disinfection by physical agents is very effective, but numerous situations arise where their use is impractical (kitchen countertops) or plainly impossible (skin). In these instances, chemical agents are the best approach. There are a number of factors that influence the efficacy of a given chemical agent. These include:

- **The presence of organic matter.** A chemical placed on a dirty surface will bind to the inert organic material present, lowering the agent's effectiveness against microbes. It sometimes is not possible to clean a surface prior to disinfection (as in a blood spill), but the presence of organic material must be factored in when estimating how long to disinfect a surface or object.
- **The kinds of organisms present.** Ideally, the agent should be effective against a broad range of pathogens.
- **Corrosiveness.** The disinfectant should not corrode the surface or, in the case of an antiseptic, damage skin.

Figure 5.31 *Deinococcus radiodurans.* **A.** Based on the amount of radiation this organism can survive, the bacterium could withstand radiation equivalent to that from an atomic blast. The nature of the dark inclusion bodies in three of the four cells in the quartet is not known. **B.** John Battista of Louisiana State University showed that *D. radiodurans* has exceptional capabilities for repairing radiation-damaged DNA.

A.
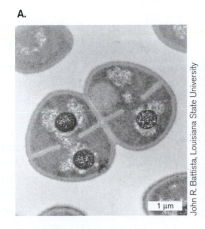
1 μm
John R. Battista, Louisiana State University

B.
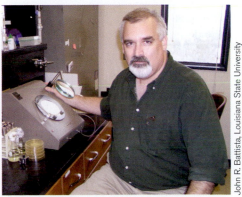
John R. Battista, Louisiana State University

■ **Stability, odor, and surface tension.** The chemical should be stable upon storage, possess a neutral or pleasant odor, and have a low surface tension so it can penetrate cracks and crevices.

The Phenol Coefficient Compares the Effectiveness of Disinfectants

Phenol, first introduced by Joseph Lister in 1867 to reduce the incidence of surgical infections, is no longer used as a disinfectant, but its derivatives, such as cresols and orthophenylphenol, are still in use. The household product Lysol is a mixture of phenolics. Phenolics are useful disinfectants because they denature proteins, are effective in the presence of organic material, and remain active on surfaces long after application.

Today we know phenolics are toxic and should not be used on living tissues. Nevertheless, based on its potency and its history, phenol is the benchmark against which other disinfectants are measured. The **phenol coefficient test** involves inoculating a fixed number of bacteria—for example, *Salmonella typhi* or *Staphylococcus aureus*—into dilutions of the test agent. At timed intervals, samples are withdrawn from each dilution and inoculated into fresh broth (which contains no disinfectant). The phenol coefficient is based on the highest dilution (lowest concentration) of a disinfectant that will kill all the bacteria in a test after 10 minutes of exposure, but leaves survivors after only 5 minutes of exposure. This concentration is known as the maximum effective dilution. Dividing the reciprocal of the maximum effective dilution for the test agent (for example, ethyl alcohol) by the reciprocal of the maximum effective dilution for phenol gives the phenol coefficient (**Table 5.3**). For example, if the maximum effective dilution for agent X is $1/900$ and that of phenol is $1/90$, then the phenol coefficient of X is $900/90 = 10$; the higher the coefficient, the higher the efficacy of the disinfectant.

Table 5.3	Phenol coefficients for various disinfectants.	
Chemical agent	***Staphylococcus aureus***	***Salmonella typhi***
Phenol	1.0	1.0
Chloramine	133.0	100.0
Cresols	2.3	2.3
Ethyl alcohol	6.3	6.3
Formalin	0.3	0.7
Hydrogen peroxide	–	0.001
Lysol	5.0	3.2
Mercury chloride	100.0	143.0
Tincture of iodine	6.3	5.8

Commercial Disinfectants

Ethanol, iodine, chlorine, and surfactants (for example, detergents) are all used to reduce or eliminate microbial content from commercial products (**Fig. 5.32**). The first three are highly reactive compounds that damage proteins, lipids, and DNA. Iodine complexed with an organic carrier forms an iodophor, a compound that is water-soluble, stable, nonstaining, and capable of releasing iodine slowly to avoid skin irritation. Wescodyne and Betadyne (trade names) are iodophors used respectively, for the surgical preparation of skin and for wounds. Chlorine is a disinfectant with universal application. It is recommended for general laboratory and hospital disinfection and kills the HIV virus.

Detergents can also be antimicrobial agents. The hydrophobic and hydrophilic ends of detergent molecules (which makes them amphipathic) will emulsify fat into water. Cationic (positively charged) but not anionic (negatively charged) detergents are useful as disinfectants because the cationic detergents contain positive charges that can gain access to the negatively charged bacterial cell and disrupt membranes. Anionic detergents are not antimicrobial but do help in the mechanical removal of bacteria from surfaces.

Low-molecular-weight aldehydes such as formaldehyde are highly reactive, combining with and inactivating proteins and nucleic acids. This, too, makes them useful disinfectants.

Disposable plasticware like petri dishes, syringes, sutures, and catheters are not amenable to heat sterilization or liquid disinfection. These materials are best sterilized using antimicrobial gases. Ethylene oxide gas (EtO) is a very effective sterilizing agent; it destroys cell proteins, is microbicidal and sporicidal, and rapidly penetrates packing materials, including plastic wraps. Using an instrument resembling an autoclave, EtO at 700 milligrams per liter (mg/L) will sterilize an object after 8 hours at 38°C or 4 hours at 54°C if the relative humidity is kept at 50%. Unfortunately, EtO is explosive. A less hazardous gas sterilant is betapropiolactone. It does not penetrate as well as EtO, but it decomposes after a few hours, which makes it easier to dispose of than EtO.

Antibiotics Selectively Control Bacterial Growth

Antibiotics as made in nature are chemical compounds made by one microbe that selectively kill other microbial species. Naturally occurring antibiotics act like tiny molecular land mines. As a defense against competitors, some organisms secrete antibiotic compounds into their surrounding environment, where they remain until encountered by an intruder. If the interloper is susceptible to the antibiotic, the compound will target specific structures or proteins, disrupting their function. The target cell is either rendered helpless (by bacteriostatic antibiot-

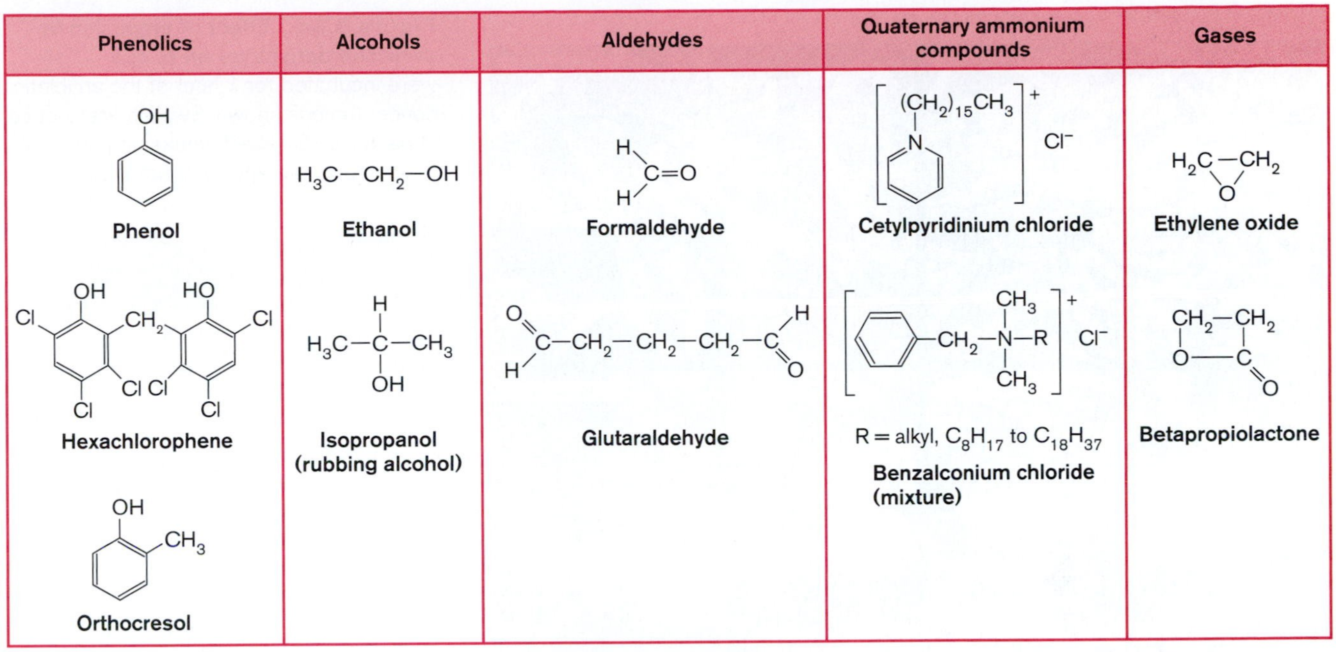

Phenolics
Alcohols
Aldehydes
Quaternary ammonium compounds
Gases
Phenol
Ethanol
Formaldehyde
Cetylpyridinium chloride
Ethylene oxide
Hexachlorophene
Isopropanol (rubbing alcohol)
Glutaraldehyde
R = alkyl, C8H17 to C18H37
Benzalconium chloride (mixture)
Betapropiolactone
Orthocresol

Figure 5.32 Structures of some common disinfectants and antiseptics.

ics) or made nonviable (by bactericidal compounds). (See Chapter 27 for detailed modes of action.) When purified and administered to patients suffering from an infectious disease, these antibiotics can produce seemingly miraculous recoveries.

As we saw in Chapter 1, penicillin, produced by *Penicillium notatum*, was discovered serendipitously in 1929 by Alexander Fleming. **Figure 5.33A** shows a three-dimensional representation of this molecule, which mimics a part of the microbial cell wall. Because of this mimicry, penicillin binds to biosynthetic proteins involved in peptidoglycan synthesis and prevents cell wall formation. The drug is bactericidal because *actively growing* cells lyse without the support of the cell wall (**Fig. 5.34**). Other antibiotics target protein synthesis, DNA replication, cell membranes, and various enzyme reactions. These interactions are described throughout Parts 1–3 of this text.

So how do antibiotic-producing microbes avoid suicide? In some instances, the producing organism lacks the target molecule. *Penicillium* mold, for instance, lacks peptidoglycan and is immune to penicillin by default. Some bacteria produce antimicrobial compounds that target other members of the same species. In this case, the producing organism can modify its own receptors so that they no longer recognize the compound (as with some bacterial colicins). Another strategy is to modify the antibiotic if it reenters the cell. This is the case with streptomycin produced by *Streptomyces griseus.* Streptomycin inhibits bacterial protein synthesis and does not discriminate between the protein synthesis machinery of *Streptomyces*

A.
B.
C.
Antibiotics Save Lives
33
USA
1999
USPS
Christine L. Case/Skyline College

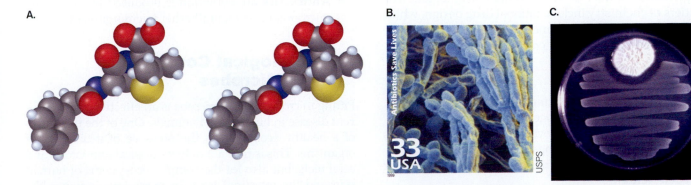

Figure 5.33 Penicillin. A. Three-dimensional and flat views of penicillin G molecule produced by the *Penicillium* mold. To view, place a piece of cardboard vertically between the two images. Position your eyes on opposite sides of the cardboard and force them to focus behind the images. The images will merge and produce a 3-D image. **B.** U.S. postal stamp showing *Penicillium notatum.* **C.** *P. notatum* culture. The mold excretes penicillin, which inhibits the growth of *Staphylococcus aureus.*

A.   B.

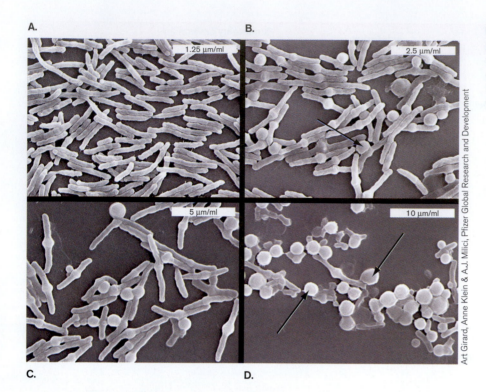

C.   D.

Figure 5.34 **Effect of ampicillin (a penicillin derivative) on *E. coli*.** Cells were incubated for 1 hour at the antibiotic concentrations shown. Swollen areas of cells in panels B–D reflect weakening cell walls. Cells shown are approx. 2 um long.

and that of others. However, while making enzymes that synthesize and secrete streptomycin, *S. griseus* simultaneously makes the enzyme streptomycin-6-kinase, which remains locked in the cell. If any secreted streptomycin reenters the cell, this enzyme renders the drug inactive by attaching a phosphate to it.

Because many microorganisms have become resistant to commonly used antibiotics, pharmaceutical companies are continually using a variety of drug discovery approaches to search for new antibiotics. Traditional procedures include scouring soil and ocean samples collected from all over the world for new antibiotic-producing organisms and chemically redesigning existing antibiotics so that they can bypass microbial resistance strategies. These procedures are now supplemented by "mining the genomes" of microbes for potential drug targets. The frontiers of chemistry include rational drug design, which relies on computer-based methods for predicting the structure and function of potential new antibiotics.

TO SUMMARIZE:

- **Sterilization** kills all living organisms.
- **Disinfection** kills pathogens on inanimate objects.
- **Antisepsis** is the removal of potential pathogens from the surfaces of living tissues.
- **Antimicrobial** compounds can be **bacteriostatic** or **bactericidal.**
- **The D-value** is the time (or dose, in the case of irradiation) it takes an antimicrobial treatment to reduce the numbers of organisms to 10% of the original value.

- **The autoclave** uses high pressure to achieve temperatures that will sterilize objects.
- **The z-value** is a measure of how much more heat is needed to reduce D-value to $1/10$ its original value.
- **Pasteurization** is a heating process designed to kill specific pathogens in milk and other food products.
- **Refrigeration** is used to prevent microbial growth in foods. Extreme cold (freezing) is used to preserve bacteria.
- **Filtration** can remove cells from a solution, but it cannot remove the smallest viruses.
- **Irradiation** can kill pathogens in foods without damaging the food itself.
- **Chemical disinfectants** are compared to one another based on the phenol coefficient.
- **Antibiotics** are compounds produced by one living microorganism that kill other microorganisms.

5.9 Biological Control of Microbes

Pitting microbe against microbe is an effective way to prevent disease in humans and animals. One of the hallmarks of a healthy ecosystem is the presence of a diversity of organisms. This is true not only for tropical rain forests and coral reefs, but also for the complex ecosystems of human skin and the intestinal tract. In these environments, the presence of harmless microbial flora can retard the growth of undesired pathogens. The pathogenic fungus *Phytophthora cinnamomi*, for example, causes root rot in plants but is biologically controlled by fungi belonging to the genus

Myrothecium. Naturally occurring *Staphylococci* on human skin produce short-chain fatty acids that retard the growth of pathogenic strains. Another illustration is the human intestine, which is populated by as many as 500 microbial species. Most of these species are nonpathogenic organisms that exist in symbiosis with their human host. Vigorous competition between members of the normal intestinal flora and the production of permeant weak acids by fermentation help control the growth of numerous pathogens.

Microbial competition has been widely exploited for agricultural purposes and to improve human health in a process known as **probiotics**. In general, a probiotic is a food or supplement that contains live microorganisms and improves intestinal microbial balance. Newborn baby chicks, for instance, are fed a microbial cocktail of normal flora designed to quickly colonize the intestinal tract and prevent colonization by *Salmonella*, a frequent contaminant of factory-farmed chicken. In another example, *Lactobacillus* and *Bifidobacterium* have been used to prevent and treat diarrhea in children.

Russian Nobel laureate Elie Metchnikoff in 1908 first suggested that a high concentration of lactobacilli in intestinal flora was important for health and longevity in humans. Yogurt is an example of a probiotic containing *Lactobacillus acidophilus* and a number of other lactobacilli. It is often recommended as a way to restore a normal balance to gut flora (for example, after it has been disturbed by antibiotic treatment) and appears useful in the treatment of inflammatory bowel disease.

Phage therapy is another biocontrol method first described in 1907 by Felix d'Herelle at France's Pasteur Institute, long before antibiotics were discovered. Bacteriophages are viruses that prey on bacteria (discussed in Section 6.1). Each bacterial species is susceptible to a limited number of specific phages. Because the culmination of a phage infection is often bacterial lysis, it was considered feasible to treat infectious diseases with a phage targeted to the pathogen. At one time, doctors used phages as medical treatment for illnesses ranging from cholera to typhoid fever. In some cases, a liquid containing the phage was poured into an open wound. In other cases, phages were given orally, introduced via aerosol, or injected. Sometimes the treatments worked; sometimes they did not. When antibiotics came into the mainstream, phage therapy largely faded. Now that strains of bacteria resistant to standard antibiotics are on the rise, the idea of phage therapy has enjoyed renewed interest from the worldwide medical community. Several biotechnology companies have even been formed in the United States to develop bacteriophage-based treatments.

TO SUMMARIZE:

- **Biocontrol** is the use of one microbe to control the growth of another.
- **Probiotics** contain certain microbes that, when ingested, aim to restore balance to intestinal flora.
- **Phage therapy** offers a possible alternative to antibiotics in the face of rising antibiotic resistance.

Concluding Thoughts

Microbiology as a science was founded on the need to understand and control microbial growth. The initial impetus was to control the diseases of humans as well as the diseases of plants and animals. But as we will see in later chapters, microbiology has developed into a science that has helped us understand the molecular processes of life. Concepts such as biological diversity, food microbiology, microbial disease, and antibiotics will be revisited in later chapters.

CHAPTER REVIEW

Review Questions

1. Explain the nature of extremophiles and discuss why these organisms are important.
2. What are the parameters that define any growth environment?
3. List and define the classifications used to describe microbes that grow in different physical growth conditions.
4. What do thermophiles have to do with the PCR reaction?
5. Why is water activity important to microbial growth? What changes water activity?
6. How do cells protect themselves from osmotic stress?
7. Why do changes in H^+ concentration affect cell growth?
8. How do acidophiles and alkaliphiles manage to grow at the extremes of pH?
9. Because an organism can live in an oxygenated environment, does that mean that the organism uses oxygen to grow? Because an organism can live in an anaerobic environment, does that mean it cannot use oxygen as an electron acceptor? Why or why not?
10. What happens when a cell exhausts its available nutrients?

11. List and briefly explain the various means by which humans control microbial growth. What is a D-value? What is a phenol coefficient?

12. How do microbes prevent the growth of other microbes?

Key Terms

12D (172)
acidophile (160)
aerobic respiration (164)
alkaliphile (161)
anaerobic respiration (166)
antibiotic (176)
antisepsis (171)
bactericidal (171)
bacteriostatic (171)
barophile (156)
bioremediation (175)
decimal reduction time (D-value) (171)
disinfection (171)
DNA microarray (150)
electron transport chain (164)
eutrophication (169)
extremophile (150)
facultative (166)

facultative aerobe (167)
facultative anaerobe (167)
fermentation (166)
fermentative metabolism (166)
germicidal (171)
halophile (158)
heat shock response (155)
hyperthermophile (154)
laminar flow biological safety cabinet (173)
lyophilization (173)
membrane-permeant organic acid (159)
mesophile (153)
microaerophilic (167)
neutralophile (160)
oligotroph (169)
osmolarity (157)

pasteurization (173)
phage therapy (179)
phenol coefficient test (176)
piezophile (156)
probiotic (179)
psychrophile (153)
reactive oxygen species (166)
sanitation (171)
starvation response (168)
steam autoclave (172)
sterilization (171)
strict aerobe (166)
strict anaerobe (166)
thermophile (153)
two-dimensional protein gels (150)
water activity (157)
z-value (172)

Recommended Reading

Alpuche-Aranda, Celia M., Joel A. Swanson, Wendy P. Loomis, and S. I. Miller. 1992. *Salmonella typhimurium* activates virulence gene transcription within acidified macrophage phagosomes. *Proceedings of the National Academy of Science USA* **89**:10079–10083.

Atomi, Haruyuki. 2005. Recent progress towards the application of hyperthermophiles and their enzymes. *Current Opinion in Chemical Biology* **9**:163–173.

Bang, I. S., Jonathan P. Audia, Y. K. Park, and John W. Foster. 2002. Autoinduction of the *ompR* by acid shock and control of the *Salmonella enterica* acid tolerance response. *Molecular Microbiology* **44**:1235–1250.

Biswas, Biswajit, Sankar Adhya, Paul Washart, et al. 2002. Bacteriophage therapy rescues mice bacteremic from a clinical isolate of vancomycin-resistant *Enterococcus faecium*. *Infection and Immunity* **70**:204–210.

Ferenci, Thomas. 1999. Regulation by nutrient limitation. *Current Opinion in Microbiology* **2**:208–213.

Fredrickson, Jim K., H. M. Kostandarithes, S. W. Li, A. E. Plymale, and Mark J. Daly. 2000. Reduction of Fe(III), Cr(VI), U(VI), and Tc(VII) by *Deinococcus radiodurans* R1. *Applied and Environmental Microbiology* **66**:2006–2011.

Horikoshi, Koki. 1998. Barophiles: Deep-sea microorganisms adapted to an extreme environment. *Current Opinion in Microbiology* **1**:291–295.

Horikoshi, Koki. 1999. Alkaliphiles, some applications of their products for biotechnology. *Microbiology and Molecular Biology Review* **63**:735–750.

MacElroy, Robert D. 1974. Some comments on the evolution of extremophiles. *Biosystems* **6**:74–75.

Makarova, Kira S., L. Aravind, Yuri I. Wolf, Roman L. Tatusov, Kenneth W. Minton, et al. 2001. Genome of the extremely radiation resistant bacterium *Deinococcus radiodurans* viewed from the perspective of comparative genomics. *Microbiology and Molecular Biology Reviews* **65**:44–79.

Maurer, Lisa M., Elizabeth Yohannes, Sandra S. BonDurant, Michael Radmacher, and Joan L. Slonczewski. 2005. pH regulates genes for flagellar motility, catabolism, and oxidative stress in *Escherichia coli* K-12. *Journal of Bacteriology* **187**:304–319.

Peitzman, Steve J. 1969. Felix d'Herelle and bacteriophage therapy. *Transactions & Studies of the College of Physicians of Philadelphia* **37**:115–123.

Rathman, M., Michael D. Sjaastad, and Stan Falkow. 1996. Acidification of phagosomes containing *Salmonella typhimurium* in murine macrophages. *Infection and Immunity* **64**:2765–2773.

Rhen, Mikael, and Charles J. Dorman. 2005. Hierarchical gene regulators adapt *Salmonella enterica* to its host milieus. *International Journal of Medical Microbiology* **294**:487–502.

Saxelin, Maija, Soile Tynkkynen, Tiina Mattila-Sandholm, and Willem M. de Vos. 2005. Probiotic and other functional microbes: From markets to mechanisms. *Current Opinion in Biotechnology* **16**:204–211.

Thomas, D. N., and G. S. Dieckmann. 2002. Antarctic Sea Ice—a habitat for extremophiles. *Science* **295**:641–643.

Chapter 6

Virus Structure and Function

All kinds of cells, including bacteria, eukaryotes, and archaea, can be infected by particles called viruses. Viruses are the smallest known units of reproduction, some with genomes of less than ten genes. Upon entering a cell, a viral genome subverts the cell's machinery to reproduce more virus particles, usually killing the host cell. Alternatively, some viral genomes are copied into the host genome, where they replicate silently with the host. Degenerate viral genomes from ancient viral insertions take up a large part of the human genome.

Virus structure varies in complexity from single RNA molecules to spore-like packages containing multiple layers of protein and membranes. Some viruses resemble degenerate cells. Viral intracellular life cycles enable them to divert complex host cell processes to the formation of new viruses. How can such particles, barely large enough to be detected by an electron microscope, subvert and consume entire cells and multicellular organisms? Alternatively, how can viruses integrate their genomes into the genome of their host, replicating with their host for future generations?

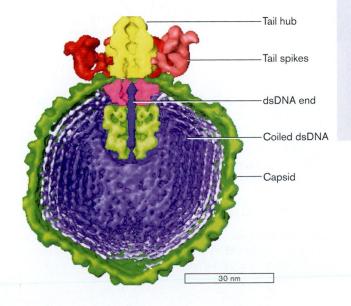

Tail hub

Tail spikes

dsDNA end

Coiled dsDNA

Capsid

30 nm

The bacteriophage epsilon 15 attacks *Salmonella* bacteria, pathogens of the human digestive tract. Such bacteriophages help control our enteric bacterial populations, but they also transfer DNA encoding resistance to antibiotics. The structure of the epsilon 15 bacteriophage was solved by cryo-electron microscopy, revealing details of the tail apparatus that inserts the DNA into the host cell. *Source*: Wen Jiang, et al. 2006. *Nature* 439:612.

In February 2003, travelers from Guangdong, China, began succumbing to a mysterious respiratory syndrome from which one in ten died. The disease spread alarmingly through hospitals in Hong Kong, Vietnam, and Toronto, Canada. Its rapid spread caused thousands of travelers worldwide to cancel plans, while residents of affected cities donned face masks (**Fig. 6.1A**). The disease became known as sudden acute respiratory syndrome (SARS). SARS was caused by a new virus, a member of the coronavirus family, named for their characteristic "corona" of spike proteins (**Fig. 6.1B**). To find clues for the treatment of SARS, the RNA genome of the new virus was sequenced by Canadian researchers with unprecedented speed. Comparison with known viruses showed SARS virus to be related to coronaviruses infecting pigs and birds (**Fig. 6.1C**). So the researchers hypothesized that SARS evolved from an animal virus in the food markets of Guangdong. Viruses evolve new strains even faster than cellular pathogens because of their tiny genomes and little or no proofreading of replication.

Despite their small size, viruses generate frightening epidemics, from smallpox and polio to AIDS and influenza. Other viruses play crucial roles in the environment, particularly in marine ecosystems, where they cycle carbon and curb toxic blooms of algae. Viral predation selects for much of the species diversity of marine protists and invertebrates.

In research, viruses have provided both tools and model systems for our discovery of the fundamental principles of molecular biology. The first genes mapped, the first regulatory switches defined, and the first genomes to be sequenced were all those of viruses. Vectors for gene cloning and gene therapy today continue to be derived from viruses.

In this chapter, we introduce major themes of virus structure and function and the fundamental challenges that all viruses face: genome packaging, cell attachment and entry, and the molecular strategies that enable viruses to divert the metabolism of their host cell. Viruses provide key tools and model systems for molecular biology. Our understanding of viruses, particularly bacterial viruses, called bacteriophages, provides a useful background for the molecular biology we will encounter in Part 2 of this book (**Chapters 7–12**). The molecular biology of viral life cycles is explored further in **Chapter 11**, and viral disease pathology and epidemiology are discussed in **Chapters 25–27**.

6.1   What Is a Virus?

A **virus** is a noncellular particle capable of infecting a host cell, where it reproduces. The virus particle, or **virion**, consists of an infective nucleic acid (DNA or RNA) con-

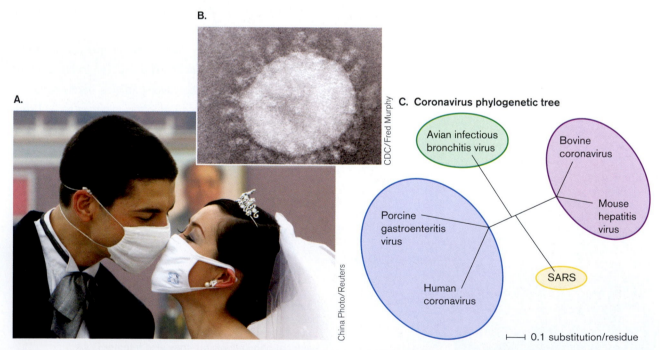

Figure 6.1   Sudden acute respiratory syndrome (SARS): an emerging viral disease.   A. In 2003, as SARS first breaks out, a bride and groom in Hubei, China, wear masks to protect themselves from airborne infection. **B.** A SARS virion (diameter, 78 nm) bristles with spike proteins (TEM). **C.** The family tree of coronavirus shows that the SARS branch is as distant from a human cold-causing coronavirus as from animal coronaviruses.

tained within a protective shell made of protein, called the **capsid**. The capsid usually has a molecular delivery device that enables transfer of the virion's genome into the host cell.

Viruses that infect bacteria are known as **bacteriophages** or **phages**. An example is bacteriophage T2, which infects *Escherichia coli* (**Fig. 6.2A**). The T2 and T4 phages have a capsid with a tail that inserts the viral genome into the host cell, where it directs reproduction of progeny virions. Virions are released when the host cell lyses. As cells lyse, their disappearance can be observed as a **plaque**, a clear spot against a lawn of bacterial cells (**Fig. 6.2B**). Each plaque arises from a single virion or phage particle that lyses a host cell and spreads progeny to infect adjacent cells. Plaques can be counted as representing individual infective virions from a phage suspension.

An example of a virus that infects humans is measles virus. The measles virion has an envelope of membrane that, during infection, fuses with the host cell membrane. After replicating within the infected cell, newly formed measles capsids become enveloped by host cell membrane as they bud out of the host cell (**Fig. 6.2C**). The spreading virus generates a rash of red spots on the skin of infected patients (**Fig. 6.2D**) and is occasionally fatal (one in 500 cases).

Plants are also infected by viruses, such as tobacco mosaic virus (TMV). Within the plant cell, virions accumulate to high numbers (**Fig. 6.2E**) and travel through interconnections to neighboring cells. Infection by tobacco mosaic virus results in mottled leaves and stunted growth (**Fig. 6.2F**). Plant viruses cause major economic losses in agriculture worldwide.

Each species of virus infects a particular group of host species, known as the **host range**. Some viruses can infect only a single species; for example, HIV infects only humans. Close relatives of humans, such as the chimpanzee, are not infected, although they are susceptible to a closely related virus, simian immunodeficiency virus (SIV). On the other hand, the West Nile virus, transmitted by mosquitoes, has a much broader host range, including many species of birds and mammals.

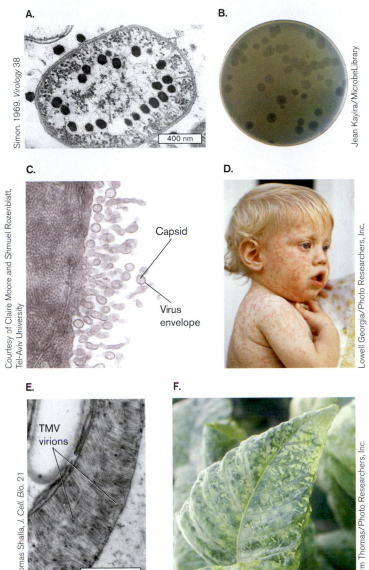

Figure 6.2   Virus infections.   A. Bacteriophage T2 particles form a semicrystalline array within an *E. coli* cell (TEM). **B.** Bacteriophage infection forms plaques of lysed cells on a lawn of bacteria. **C.** Measles virions (diameter, 200–600 nm) bud out of human cells in tissue culture (TEM). **D.** Child infected with measles shows a rash of red spots. **E.** Tobacco leaf section is packed with tobacco mosaic virus particles. **F.** Tobacco leaf infected by tobacco mosaic virus shows mottled appearance.

> **THOUGHT QUESTION 6.1**   Which viruses do you know that have a narrow host range, and which have a broad host range?

www | American Society for Virology

www | International Committee on Taxonomy of Viruses

Viruses Propagate Their Genomic Information

Viral propagation exemplifies the central role of information in biological reproduction. The propagation of viruses is mimicked by the spread of "computer viruses," whose information "infects" computer memory (**Fig. 6.3**).

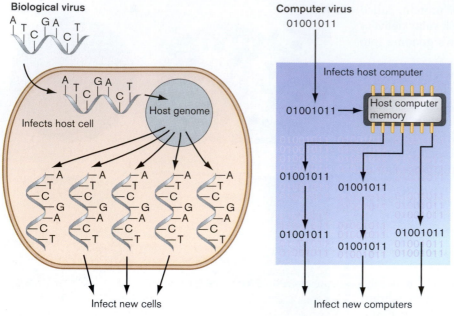

Figure 6.3 **Biological viruses and computer viruses.** From the standpoint of information, the behavior of biological viruses and computer viruses are analogous.

industry; for example, bacteriophages (literally, "bacteria-eaters") infect cultures of *Lactococcus* during the production of yogurt and cheese. Plant pathogens such as cauliflower mosaic virus and rice dwarf virus continue to cause substantial losses in agriculture.

In contrast to our vast arsenal of antibiotics (effective against bacteria), the number of antiviral drugs remains depressingly small. Because the machinery of viral growth is largely that of the host cell, viruses present relatively few targets that can be attacked by antiviral drugs without harming the host. However, an understanding of viral life cycles at the molecular level is now leading to the development of new antiviral drugs, such as AZT and protease inhibitors, which combat HIV.

Despite their lethal potential, viruses have made surprising contributions to medical research. A bacteriophage provides a protein that lyses the cell walls of anthrax bacteria. Other bacteriophages are used as **cloning vectors**, small genomes into which foreign genes can be inserted and cloned for gene technology. Even lethal viruses such as HIV are being developed as vectors for human gene therapy.

Some viruses introduce copies of their own genomes into the host's genome, a process that can mediate evolution of the host genome. Indeed, studies of molecular evolution reveal that viral genomes are the ancestral source of about a tenth of the human genome.

When a biological virus infects a host cell, the information in its genome subverts the host cell machinery to produce multiple copies of the virus; the multiple copies then escape to infect more host cells. Similarly, when a computer virus infects a host computer, its program code subverts the host to produce multiple copies of the virus, which then escape to infect more host computers. Computer viruses generate epidemics analogous to those of biological viruses. The virus's code can even be designed to "mutate" in order to foil the "immune system" of antivirus software.

Viruses Infect All Forms of Life

Viruses are ubiquitous, infecting every taxonomic group of organisms, including bacteria, eukaryotes, and archaea. In marine ecosystems, viruses act as major predators and sequester significant amounts of nutrients. For humans, viruses cause many forms of illness, whose influence on our history and culture would be hard to overstate. More people died of influenza in the global epidemic of 1918 than in the battles of World War I. In the past 30 years, the AIDS pandemic caused by HIV has killed 25 million people worldwide and continues to grow.

Viruses are part of our daily lives. The most frequent infections of college students are due to respiratory pathogens such as rhinovirus (the common cold) and Epstein-Barr virus (infectious mononucleosis), as well as sexually transmitted viruses such as herpes simplex (HSV) and papilloma (genital warts). Viruses also impact human

Viral Genomes

The genome of a virus can be small, encoding fewer than ten genes. In cauliflower mosaic virus, for example, the genome encodes only seven genes (**Fig. 6.4A**), which actually overlap each other in sequence. This overlap in sequence is made possible by the use of different **reading frames**, start positions for translating codons to amino acids. Many viral genomes, such as that of avian leukosis genome, are encoded by RNA. The RNA genome of avian leukosis virus has protein-encoding genes grouped by functional categories of core capsid, replicative enzymes, and envelope proteins.

On the other hand, larger viral genomes, such as that of herpes virus or of bacteriophage T4, have genes dispersed around the chromosome, similar to the genomes of bacteria. The giant Mimivirus, which infects amebas and may cause human pneumonia, is as large as some bacteria

A. Cauliflower mosaic virus genome (DNA)

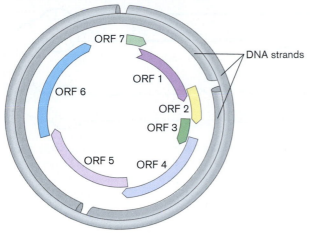

B. Avian leukosis virus genome (RNA)

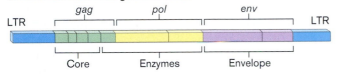

Figure 6.4 Simple viral genomes.   A. Cauliflower mosaic virus has a circular genome of double-stranded DNA, whose strands are interrupted by nicks. The genome encodes seven overlapping genes (open reading frames, ORFs). **B.** Avian leukosis virus: a single-stranded RNA retrovirus resembling eukaryotic mRNA. Three genes (*gag*, *pol*, and *env*) encode polypeptides that are eventually cleaved to form a total of nine functional products. LTR = long terminal repeat.

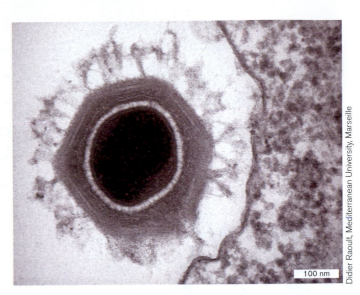

Figure 6.5   Mimivirus infecting an ameba.   About 300 nm across, the Mimivirus is larger than some bacteria.

(**Fig. 6.5**). Furthermore, Mimivirus has a bacterium-sized genome of 1.2 million base pairs. The Mimivirus virion conducts numerous cellular functions, including DNA repair and protein-folding by chaperones. It probably represents a descendant of a cell that parasitized protist cells until most of its cellular traits were lost through degenerative evolution.

Viroids: Infective Genomes with No Capsid

Early in the twentieth century, viruses were believed to be the smallest particles capable of infecting hosts and propagating themselves. Then even smaller virus-like infectious agents were discovered for which the nucleic acid genome is itself the entire infectious particle; there is no protective capsid. Such infectious agents are called **viroids**. Most viroids are RNA molecules that infect plants. An example is the potato spindle tuber viroid. This viroid consists of a circular, single-stranded molecule of RNA that doubles back on itself to form base pairs interrupted by short unpaired loops (**Fig. 6.6A**). The RNA folds up into a globular structure that interacts with host

cell proteins. Most critically, the RNA genome requires a host RNA-dependent RNA polymerase to replicate itself and transcribe its genes. RNA-dependent RNA polymerases occur normally in plant cells, where they contribute to regulation of gene expression.

During viroid infection, the RNA-dependent RNA polymerase replicates progeny copies of the viroid, which encodes no products other than itself. Viroids can cause as much host destruction as "true viruses," and some authors, particularly in plant pathology, classify them as "viruses without capsids."

Some viroids have catalytic ability, comparable to enzymes made of protein. An RNA molecule capable of catalyzing a reaction is called a **ribozyme**. One class of plant-infecting viroids are the **hammerhead ribozymes** (**Fig. 6.6B**). Named for their hammer-shaped tertiary structure, hammerhead ribozymes possess the ability to cleave themselves or other specific RNA molecules. Their ability to cleave very specific RNA sequences has applications in medical research. Hammerhead ribozymes have been engineered to cleave human RNA molecules involved in cancer or in infection by viruses such as HIV. The engineering of ribozymes for medical therapy is an exciting field of biotechnology.

Prions: Infection without Nucleic Acid?

A remarkable class of infectious agents is believed to consist of protein only. These agents, known as **prions**, are believed to be aberrant proteins arising from the host cell. Prions gained notoriety when they were implicated in brain infections such as Creutzfeldt-Jakob disease, popularly known as "mad cow" disease because it may

A. Potato spindle tuber viroid—circular ssRNA

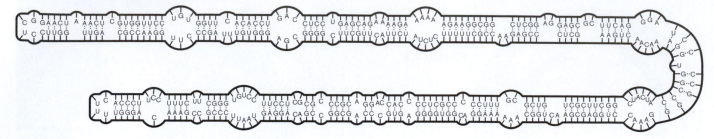

B. Hammerhead ribozyme

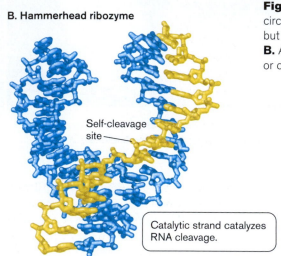

Self-cleavage site

Catalytic strand catalyzes RNA cleavage.

Figure 6.6 **Viroids: infective RNA.** **A.** Potato spindle tuber viroid consists of a circular single-stranded RNA that hybridizes internally. A viroid encodes no genes, but it hijacks the plant cell's RNA-dependent RNA polymerase to replicate itself. **B.** A hammerhead ribozyme catalyzes cleavage of itself (at "self-cleavage site") or of another RNA molecule. (PDB code: 1HMH)

be transmitted through defective proteins in beef from diseased cattle. Other diseases believed to be caused by prion transmission include scrapie, a disease of sheep, and kuru, a degenerative brain disease found in a tribe of people who customarily consumed the brains of deceased relatives.

In prion-associated diseases, the infective agent is unaffected by treatments that destroy RNA or DNA, such as nucleases or UV irradiation. A prion is an aberrant form of a normally occurring cell protein that assumes an abnormal conformation or tertiary structure (**Fig. 6.7A**). The prion form of the protein acts by binding to normally folded proteins of the same class and altering their conformation to that of the prion. The multiplying prion then alters the conformation of other normal subunits, forming harmful aggregates in the cell and ultimately leading to cell death. In the brain, prion-induced cell death leads to tissue deterioration and dementia (**Fig. 6.7B**).

Prion diseases can be initiated by infection with an aberrant protein. More rarely, the cascade of protein misfolding can start with the spontaneous misfolding of an endogenous host protein. The chance of spontaneous unfolding is greatly increased in individuals who inherit certain alleles encoding the protein; thus, spontaneous prion diseases can be inherited genetically. Overall, prion diseases are unique in that they can be transmitted by an infective protein instead of by DNA or RNA; and they propagate conformational change of existing molecules without synthesizing entirely new infective molecules.

A.

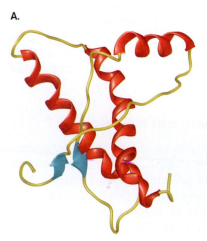

Normal conformation

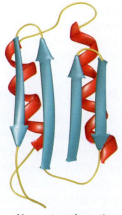

Aberrant conformation

B.

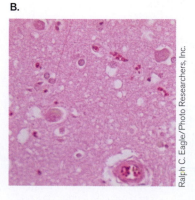

Ralph C. Eagle/Photo Researchers, Inc.

Figure 6.7 **Prion disease.** **A.** The normal conformation of a prion, compared to the abnormal conformation. The abnormal form "recruits" normally folded proteins and changes their conformation into the abnormal form. (PDB code: 1AG2) **B.** Section of a human brain showing "spongiform" holes typical of Creutzfeldt-Jakob disease.

TO SUMMARIZE:

- **Viruses** consist of noncellular particles that infect a host cell and direct its expression apparatus to produce virus particles.
- A **virion** consists of a capsid enclosing a nucleic acid genome. Some viruses are further enclosed by a membrane envelope.
- **All classes of organisms are infected by viruses.** Usually the hosts are limited to a particular host range of closely related strains or species.
- **Viruses contain infective genomes** that take over a cell, reprogramming its cell machinery to make progeny virus particles (virions). Some viral genomes consist of less than 10 genes; others have 100 or 200 genes and may represent degenerate cells.
- **Viroids** that infect plants consist of RNA hairpins with no capsid.
- **Prions** consist of infectious proteins that induce a cell's native proteins to fold incorrectly and impair cell function.

6.2 Virus Structure

A packaged structure of a virus achieves two goals: It keeps the viral genome intact, and it enables infection of the appropriate host cell. First, the stable capsid protects the viral genome from degradation and enables it to be transmitted outside the host. Second, in order for the viral genome to reproduce, the virion must either insert its genome into the host cell or disassemble within the host. In the process, the original particle loses its stable structure and its own identity as such, but it generates numerous progeny virions.

> **THOUGHT QUESTION 6.2** What would happen if a virus particle remained intact within a host cell instead of releasing its genome?

Symmetrical Virus Particles

Different viral species make different forms of capsids. Virus particles may be symmetrical, in which case the capsid is one of two types, icosahedral or filamentous (helical). Each type of capsid exhibits geometrical symmetry. The advantage of symmetry is that it provides a way to form a package out of repeating protein units generated by a small number of genes and encoded by a short chromosomal sequence. The smaller the viral genome, the more genome copies can be synthesized from the host cell's limited supply of nucleotides. Nevertheless, other viruses, such as smallpox virus, are asymmetrical and may have much larger genomes. Large genomes offer a greater range of functions for viral components.

Icosahedral viruses. Many viruses package their genome in an **icosahedral** (20-sided) capsid; examples include poliovirus and the herpes viruses. Icosahedral viral capsids take the form of a polyhedron with 20 identical triangular faces. Bacteriophages often supplement the icosahedral capsid or head coat with an elaborate delivery device. For example, bacteriophage T4 (**Fig. 6.8A**) has a complex structure consisting of an icosahedral headpiece containing the genome, six jointed "legs" that stabilize the structure on the host cell surface, and a neck piece that channels the nucleic acid into the host cell (**Fig. 6.8B**). The structure of phage T4 was first observed by microbiologists during the rise of the NASA space program, and its form was compared to that of the "lunar module" that landed on the moon (**Fig. 6.8C**). Indeed, in the 1960s, the "tailed phages," such as phage T4 and phage lambda, were to molecular biology what the lunar landings were to space exploration.

WWW | **T4 Bacteriophage Cell-Puncturing Device**, a 3-D interactive tutorial

In the capsid, each triangle can be composed of three identical but asymmetrical protein units. An example of an icosahedral capsid is that of the herpes simplex virus (**Fig. 6.9A**). Each triangular face of the capsid is determined by the same genes encoding the same protein subunits. No matter what the pattern of subunits in the triangle, the structure overall exhibits rotational symmetry characteristic of an icosahedron (**Fig. 6.9B**): threefold symmetry around the axis through two opposed triangular faces; fivefold symmetry around an axis through opposite points; and twofold symmetry around an axis through opposite edges. Capsid symmetry is important for structure determination and visualization and for the design of antiviral drugs.

> **THOUGHT QUESTION 6.3** Why do viral capsids take the form of an icosahedron instead of some other polyhedron?

Virus particles can be observed by standard transmission electron microscopy (TEM), but the details of capsid structure as in **Figure 6.9A** require visualization by digital reconstruction of cryo EM (discussed in Section 2.6). Recall from Chapter 2 that in cryo EM, the viral samples for TEM are prepared flash-frozen, preventing formation of ice crystals. Flash freezing enables observation without stain. The electron beams penetrate the object; thus, images of individual capsids actually provide a glimpse of the virus's internal contents. By digitally combining and processing cryo TEM images from a number of capsids, a three-dimensional reconstruction is built for the entire virus particle.

In some icosahedral viruses, the capsid is enclosed in an **envelope** composed of membrane from the host cell in which the virion formed. **Figure 6.10A** shows how herpes virions capture cell membrane to form their envelopes as

A.

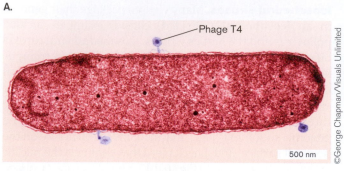

Phage T4

500 nm

©George Chapman/Visuals Unlimited

B.

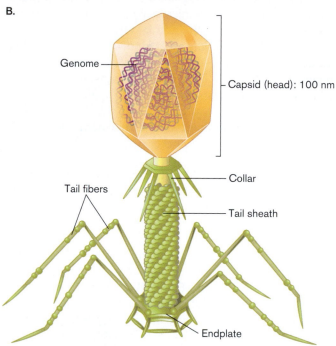

Genome

Capsid (head): 100 nm

Tail fibers

Collar

Tail sheath

Endplate

C.

Bettmann/Corbis

Figure 6.8   Bacteriophage T4 capsid.   A. *E. coli* infected by phage T4 (colorized blue, TEM). **B.** Phage T4 particle with protein capsid containing packaged double-stranded DNA genome. The capsid is attached to a sheath with tail fibers that facilitate attachment to the surface of the host cell. After attachment, the sheath contracts and the core penetrates the cell surface, injecting the phage genome. **C.** The structure of phage T4 resembles an *Apollo* lunar module that landed on the moon.

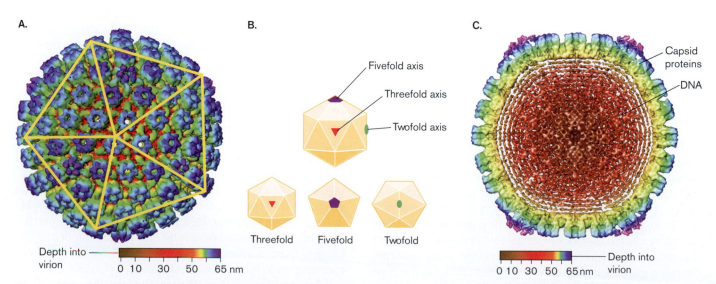

A.

Depth into virion

0 10   30   50 65 nm

B.

Fivefold axis

Threefold axis

Twofold axis

Threefold   Fivefold   Twofold

C.

Capsid proteins

DNA

Depth into virion

0 10   30   50 65 nm

Figure 6.9   Herpes: Icosahedral capsid symmetry.   A. Icosahedral capsid of herpes simplex 1 (HSV-1), envelope removed. Imaging of the capsid structure is based on computational analysis of cryoelectron microscopy (cryo TEM). Images of 146 virus particles were combined digitally to obtain this model of the capsid at 2 nm resolution. **B.** Icosahedral symmetry includes fivefold, threefold, and twofold axes of rotation. **C.** The icosahedral capsid contains spooled DNA. *Source:* Z. H. Zhou, et al. 1999. *J. Virology* 73:3210

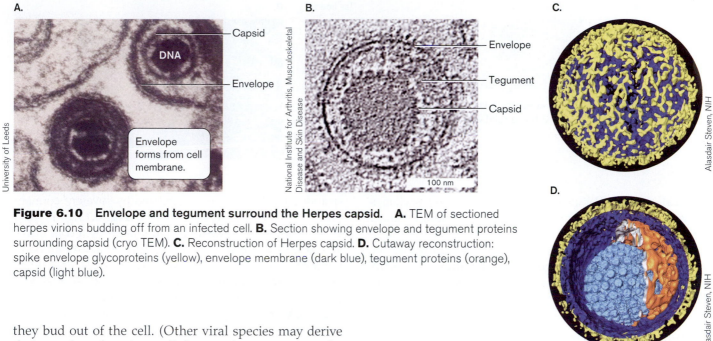

Figure 6.10   Envelope and tegument surround the Herpes capsid.   A. TEM of sectioned herpes virions budding off from an infected cell. **B.** Section showing envelope and tegument proteins surrounding capsid (cryo TEM). **C.** Reconstruction of Herpes capsid. **D.** Cutaway reconstruction: spike envelope glycoproteins (yellow), envelope membrane (dark blue), tegument proteins (orange), capsid (light blue).

they bud out of the cell. (Other viral species may derive their envelope from intracellular membranes, such as the nuclear membrane or endoplasmic reticulum.) The envelope and capsid contents of herpes virus are shown in **Figure 6.10B**. The mature envelope (**Fig. 6.10C**) bristles with glycoprotein **spike proteins** that plug it onto the capsid. The spike proteins enable the virus to attach and infect the next host cell.

Between the envelope and the capsid, additional proteins may be found, called the **tegument** (**Fig. 6.10D**). Tegument proteins, also called **accessory proteins**, serve the virus during the early stages of infection. As the virus enters its host cell, its envelope is dissolved and the tegument proteins are released to start reproduction of new viruses.

NOTE: Distinguish the **viral envelope** (membrane derived from a host cell membrane) from the *bacterial cell envelope* (protective layers outside the bacterial cell membrane). The bacterial envelope is discussed in Chapter 3.

Filamentous viruses. A second major category of virus structure is that of **filamentous viruses**. Filamentous viruses include bacteriophages, such as phage M13 (**Fig. 6.11A**), as well as animal viruses, such as Ebola virus, which causes a swiftly fatal disease of humans and related primates (**Fig. 6.11B**). Filamentous phages have applications in human medicine and industry. Phages similar to M13 infect the gram-positive species *Propionibacterium freudenreichii*, a key fermenting agent for Swiss cheese. Another filamentous phage, CTXφ, integrates its sequence into the genome of *Vibrio cholerae*, where it car-

ries the deadly toxin genes required for cholera. On the other hand, an application in nanoscience is the use of filamentous phages to nucleate the growth of crystalline "nanowires" for electronic devices.

The filamentous bacteriophage M13 (**Fig. 6.11A**) consists of a relatively simple capsid of protein monomers. The monomers are stacked around a coiled genome consisting of a circle of single-stranded DNA. At one end of the filament, short tail fibers mediate specific attachment to the host. Attachment occurs at the F pilus of an *E. coli* bacterium containing an F plasmid that encodes pili.

Filamentous viruses show helical symmetry. The pattern of capsid monomers forms a helical tube around the genome, which usually winds helically within the tube. In a helical capsid, the genome is a single-stranded DNA (as in phage M13) or RNA (as in tobacco mosaic virus). **Figure 6.12A** shows how the RNA strand of tobacco mosaic virus winds in a spiral within a tube of capsid monomers laid down in a spiral array. Such a tube can be imagined as a planar array of subunits that coils around such that each row connects to the row above, generating a spiral (**Fig. 6.12B**). The length of the helical capsid may extend up to 50 times its width, generating a flexible filament.

Unlike the icosahedral capsid, which has a fixed size, the helical capsid can vary in length to accommodate various lengths of nucleic acid. Furthermore, some viruses package several genome segments into separate

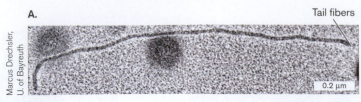

A.

Tail fibers

0.2 μm

Marcus Drechsler, U. of Bayreuth

Figure 6.11   Filamentous viruses.   A. The filamentous bacteriophage M13 has a relatively simple helical capsid that surrounds the genome coiled within (TEM). **B.** Ebola virus filaments.

B.

Frederick A. Murphy, School of Veterinary Medicine, U. of California, Davis

helical capsids. For example, influenza virus packages several different genome segments into separate helical packages of different sizes, contained together within a membrane envelope. Multiple helical packaging enables influenza virus to package different numbers of RNA segments into different virions, thus facilitating rapid evolution of new strains.

Asymmetrical Virus Particles

Some viruses lack a symmetrical capsid. In poxviruses, such as vaccinia and smallpox (**Fig. 6.13A**), the double-

stranded DNA genome is stabilized by covalent connection of its two strands at each end (**Fig. 6.13B**). Instead of a capsid, the DNA is enclosed loosely by a core envelope studded with spike proteins, surrounded by an outer membrane. The core envelope also encloses a large number of accessory proteins needed early in viral infection, such as initiation proteins for transcription of viral genes and RNA-processing enzymes that modify viral mRNA molecules.

Asymmetrical viruses usually contain a large number of accessory proteins. The first viruses to be studied, including TMV and poliovirus, had extremely simple

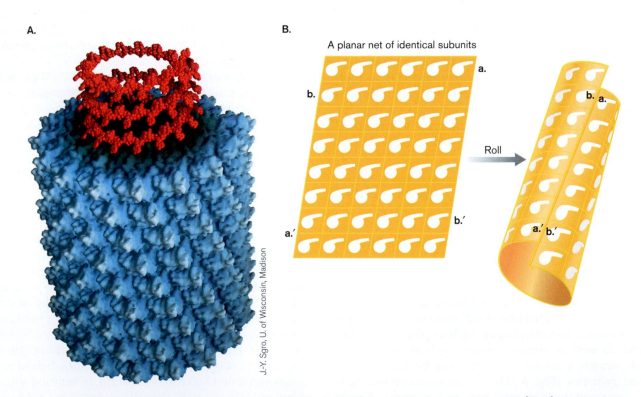

A.

B.

A planar net of identical subunits

a.

b.

a.'

b.'

Roll

b. a.

a.' b.'

J.-Y. Sgro, U. of Wisconsin, Madison

Figure 6.12   Tobacco mosaic virus: helical symmetry.   A. The helical filament of tobacco mosaic virus (TMV) contains a single-stranded RNA genome coiled inside. Image reconstruction is based on X-ray crystallography. **B.** An array corresponding to helical capsid structure can be simulated by rolling up a planar array into a cylinder, then displacing the array along the vertical axis of the cylinder so that the horizontal elements are displaced by one unit.

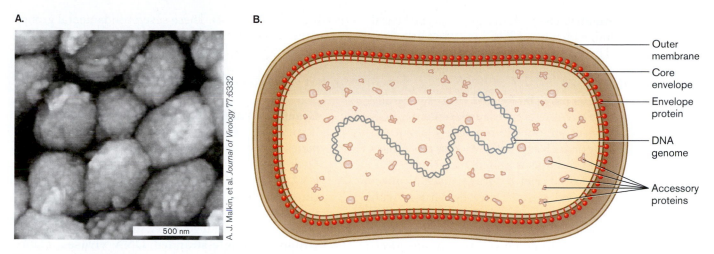

Figure 6.13 Vaccinia pox virus.   A. Vaccinia virions observed in aqueous medium by atomic force microscopy (AFM). **B.** A pox virion includes an outer membrane and a core envelope membrane containing envelope proteins enclosing the double-stranded DNA genome and accessory proteins. The DNA is stabilized by a hairpin loop at each end.

structures that consisted only of nucleic acid and packaging proteins. Based on these models, it was concluded that a virion consists solely of a packaged genome. Further research, however, revealed that other kinds of viruses contain enzymes and regulatory proteins—encoded by the virus, its host, or both. The proteins may be found either inside the capsid or in the tegument between the capsid and the envelope. Examples include the reverse transcriptase and protease enzymes of HIV and the dozen different enzymes contained by vaccinia poxvirus. Large, asymmetrical viruses contain so many enzymes that they appear to have evolved from degenerate cells.

TO SUMMARIZE:

- **The viral capsid** is composed of repeated protein subunits, a structure that maximizes the structural capacity while minimizing the number of genes needed for construction.
- **The capsid packages the viral genome** and delivers it into the host cell.
- **Icosahedral capsids** have regular, icosahedral symmetry.
- **Filamentous (helical) capsids** have uniform width, generating a flexible filamentous virion.
- **Enveloped viruses** consist of a protein capsid and tegument proteins enclosed within membrane derived from the host cell. The envelope includes virus-specific spike proteins.
- **Accessory proteins** are contained within the capsid or as tegument components between the capsid and envelope.

6.3 Viral Genomes and Classification

Living organisms today are classified based on relatedness of their gene sequences. Genetic relatedness can also be used to compare closely related viruses, such as SARS and other coronaviruses. The definition of a virus species, however, is problematic, given the small size and high mutability of viral genomes and the ability of different viruses to recombine their genome segments within an infected host cell. Furthermore, it is not clear whether all viral species are monophyletic, that is, descended from a common ancestor. In fact, it is more likely that different classes of viruses arose from different sources—such as from parasitic cells or from host cell components such as DNA replication enzymes. Viruses are classified based on genome composition, virion structure, and host range.

The International Committee on Taxonomy of Viruses

For purposes of study and communication, a working classification system has been devised by the International Committee on Taxonomy of Viruses (ICTV). The ICTV classification system is based on several criteria:

- **Genome composition.** The nucleic acid of the viral genome can vary remarkably with respect to physical structure: it may consist of DNA or RNA; it may be single- or double-stranded; it may be linear or circular; and it may be whole or **segmented** (that is, divided into separate "chromosomes"). Genomes are classified by the Baltimore method (discussed next).

- **Capsid symmetry.** The protein capsid may be helical or icosahedral, with various levels of symmetry.
- **Envelope.** The presence of a host-derived envelope, and the envelope structure if present, are characteristic of related viruses.
- **Size of the virus particle.** Related viruses generally share the same size range; for example, enteroviruses such as poliovirus are only 30 nm across (about $1/30$ the size of bacteria such as *E. coli*), whereas poxviruses are 200–400 nm, as large as a small bacterium.
- **Host range.** Closely related viruses usually infect the same or related hosts. However, viruses with extremely different hosts can show surprising similarities in genetics and structure. For example, both rabies virus and potato yellow dwarf virus are enveloped, bullet-shaped viruses of the rhabdovirus family.

NOTE: In nomenclature, families of viruses are designated by Latin names with the suffix *-viridae*: for example, *Papovaviridae*. Nevertheless, the common forms of such family names are also used, for example, the papovaviruses. Within a family, a virus species is simply capitalized, as in Papillomavirus.

The Baltimore Virus Classification Based on Genome Structure

Of the classification criteria just described, many virologists consider genome composition the most fundamental; that is, viruses of the same genome class (such as double-stranded DNA) are more likely to share ancestry with each other than with viruses of a different class of genome (such as RNA). In 1971, David Baltimore proposed that the primary distinction among classes of viruses be the genome composition (RNA or DNA) and the route used to express messenger RNA (mRNA). Baltimore, together with Renato Dulbecco and Howard Temin, were awarded the Nobel Prize in Physiology or Medicine in 1975 for discovering how tumor viruses cause cancer.

All cells and viruses need to make messenger RNA (mRNA) to make their fundamental protein components. The production of mRNA from the viral genome is central to a virus's ability to propagate its kind. Cellular genomes always make mRNA by copying double-stranded DNA. For viruses, however, different kinds of genomes require fundamentally different mechanisms to produce mRNA. The different means of mRNA production generate distinct groups of viruses with shared ancestry.

So far, the known mechanisms of replication and mRNA expression define seven fundamental groups of viral species (**Fig. 6.14**). These seven fundamental groups form the basis of the taxonomic survey of viruses in **Table 6.1**.

www | Viral Biorealm

Group I. Double-stranded DNA viruses such as herpes and smallpox make their own DNA polymerase or use that of the host for genome replication. Their genes can be transcribed directly by a standard RNA polymerase, in the same way that a cellular chromosome would be transcribed. The RNA polymerase used can be that of the host cell, or it can be encoded by the viral genome.

Group II. Single-stranded DNA viruses require the host DNA polymerase to generate the complementary DNA strand. The double-stranded DNA can then be transcribed by host RNA polymerase. An example is canine parvovirus.

Group III. Double-stranded RNA viruses, such as the plant pathogenic reoviruses, require a viral **RNA-dependent RNA polymerase** to generate messenger RNA by transcribing directly from the RNA genome. Since the RNA polymerase is required immediately upon infection, such viruses usually package a viral RNA polymerase with their genome before exiting the host cell.

Group IV. (+) sense single-stranded RNA viruses consist of a positive-sense (+) strand, that is, the strand that can serve directly as mRNA to be translated to viral proteins. Replication of the RNA genome, however, requires synthesis of the template (–) strand, that is, the strand complementary to the (+) strand, to form a double-stranded RNA intermediate. Positive-sense (+) RNA viruses are very common. They include the coronaviruses, such as SARS, as well as hepatitis C virus and the flavivirus that causes West Nile encephalitis.

Group V. (–) sense single-stranded RNA viruses such as influenza virus have genomes that consist of template or "negative-sense" RNA. Thus, they need to package a viral RNA-dependent RNA polymerase for transcribing (–) RNA to (+) mRNA. The (–) RNA viral genomes are often segmented, that is, they consist of more than one separate linear chromosome—a key factor in the evolution of killer strains of influenza (see Section 11.4).

Group VI. Retroviruses, or RNA reverse-transcribing viruses such as HIV and feline leukemia virus, have genomes that consist of (+) strand RNA. Instead of RNA polymerase, they package a **reverse transcriptase**, which transcribes the RNA

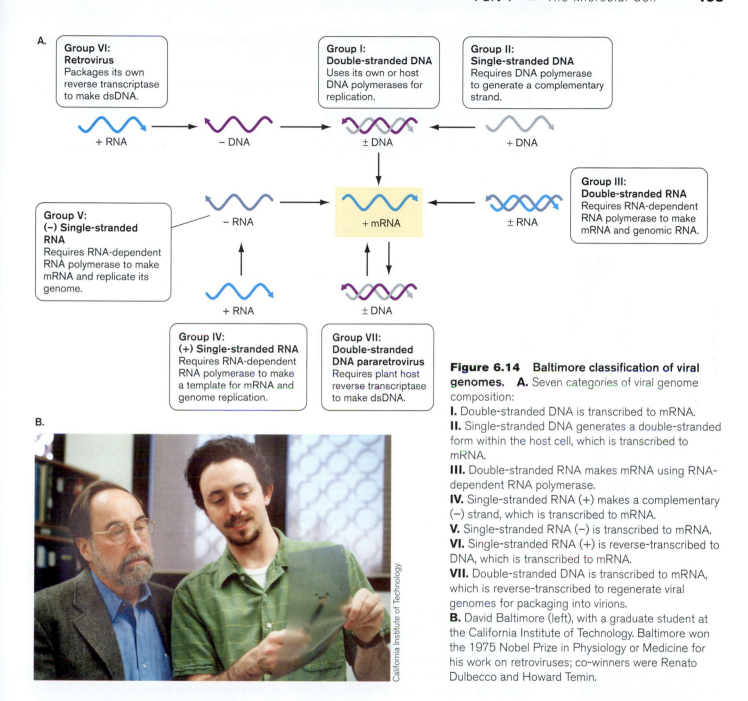

A.

Group VI: Retrovirus
Packages its own reverse transcriptase to make dsDNA.

Group I: Double-stranded DNA
Uses its own or host DNA polymerases for replication.

Group II: Single-stranded DNA
Requires DNA polymerase to generate a complementary strand.

+ RNA   — DNA   ± DNA   + DNA

Group V: (–) Single-stranded RNA
Requires RNA-dependent RNA polymerase to make mRNA and replicate its genome.

Group III: Double-stranded RNA
Requires RNA-dependent RNA polymerase to make mRNA and genomic RNA.

— RNA   + mRNA   ± RNA

+ RNA   ± DNA

Group IV: (+) Single-stranded RNA
Requires RNA-dependent RNA polymerase to make a template for mRNA and genome replication.

Group VII: Double-stranded DNA pararetrovirus
Requires plant host reverse transcriptase to make dsDNA.

B.

California Institute of Technology

Figure 6.14 Baltimore classification of viral genomes. A. Seven categories of viral genome composition:
I. Double-stranded DNA is transcribed to mRNA.
II. Single-stranded DNA generates a double-stranded form within the host cell, which is transcribed to mRNA.
III. Double-stranded RNA makes mRNA using RNA-dependent RNA polymerase.
IV. Single-stranded RNA (+) makes a complementary (–) strand, which is transcribed to mRNA.
V. Single-stranded RNA (–) is transcribed to mRNA.
VI. Single-stranded RNA (+) is reverse-transcribed to DNA, which is transcribed to mRNA.
VII. Double-stranded DNA is transcribed to mRNA, which is reverse-transcribed to regenerate viral genomes for packaging into virions.
B. David Baltimore (left), with a graduate student at the California Institute of Technology. Baltimore won the 1975 Nobel Prize in Physiology or Medicine for his work on retroviruses; co-winners were Renato Dulbecco and Howard Temin.

into a double-stranded DNA (for details, see Section 11.5). The double-stranded DNA is then integrated into the host genome, where it directs the expression of the viral genes.

Group VII. DNA reverse-transcribing viruses, or **pararetroviruses,** have a life cycle that requires reverse transcriptase. Animal viruses such as hepatitis B (a hepadnavirus) first copy their double-stranded DNA genomes into RNA, then reverse-transcribe the

RNA to progeny DNA using a reverse transcriptase packaged in the original virion. In contrast, plant pararetroviruses, such as cauliflower mosaic virus, generate an RNA intermediate that replicates using a reverse transcriptase made by the host cell. Many plant genomes include a gene for reverse transcriptase. Cauliflower mosaic virus is of enormous agricultural significance for its use as a vector to construct pesticide-resistant food crops.

Table 6.1   Groups of viruses—Baltimore classification.

Viruses	Taxonomic group with key traits and examples

Group I. Double-stranded DNA viruses

Replicate using host or viral DNA polymerase.

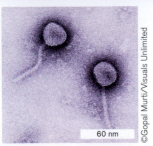

Bacteriophage lambda

©Gopal Murti/Visuals Unlimited

60 nm

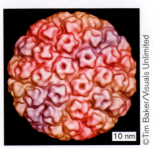

Papillomavirus

©Tim Baker/Visuals Unlimited

10 nm

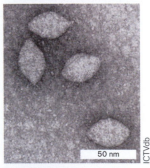

Fusellovirus

ICTVdb

50 nm

Nonenveloped bacteriophages
Structure includes head, neck, and tail.
Myoviridae. Bacteriophage T4 infects *Escherichia coli*.
Siphoviridae. Bacteriophages lambda and Mu infect *E. coli*. Others infect gram-positive hosts such as *Lactobacillus* and *Streptococcus*.
Tectiviridae. Infect enteric bacteria.

Nonenveloped viruses of animals and protists
Adenoviridae. Adenovirus generates tumors in humans.
Papillomaviridae. Papillomavirus causes genital warts.
Phycodnaviridae. Infect chlorella, algal symbionts of paramecia and hydras.
Iridoviridae. Infect insects and amphibians. Regular packaging of capsids in cell confers an iridescent color on the infected cells.
Mimiviridae. The largest known viruses, infect *Acanthameba*.

Enveloped viruses of animals
Herpesviridae. Herpes simplex I and II cause oral and genital herpes, and varicella-zoster virus causes chickenpox.
Poxviridae. Include smallpox and cowpox viruses.
Baculoviridae. Baculoviruses infect insects.

Archaeal viruses
Fuselloviridae. Fusellovirus STIV infects *Sulfolobus* sp., growing at pH 2 at 80°C. Fusellovirus His1 infects *Haloarcules* sp., growing in concentrated salt.
Rudiviridae and *Lipothrixviridae* infect *Sulfolobus* and *Thermoproteus* species.
Gultaviridae. Rod-shaped virus infect *Sulfolobus* sp.
Ampullaviridae. Bottle-shaped viruses infect *Acidianus* sp.
Haloviridae. Haloviruses infect haloarchea such as Haloferax, Halobacterium, and Haloarcula.

Group II. Single-stranded DNA viruses

Genome consists of (+) sense DNA; require a host DNA polymerase to generate the complementary strand; nonenveloped.

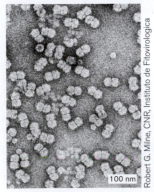

Geminivirus

Robert G. Milne, CNR, Instituto di Fitovirologica Applicata, Torino, Italy

100 nm

Nonenveloped bacteriophages
Inoviridae. Bacteriophage M13 infects *E. coli* and has a slow-release life cycle.
Microviridae. Bacteriophage φX174 infects *E. coli*.

Nonenveloped animal viruses
Parvoviridae. Cause various diseases in cats, pigs, and other animals.
Circoviridae. Infect pigs and birds. These viruses target the lymphoid tissues and cause immunosuppression.

Nonenveloped plant viruses
Geminiviridae. Transmitted by aphids to tomato plants and other important crops. Their virions group in "twins," each member of the pair carrying one DNA circle with part of the genome. Infection requires transmission of both parts.

Table 6.1 Groups of viruses—Baltimore classification (*continued*)

Viruses	Taxonomic group with key traits and examples

Group III. Double-stranded RNA viruses
Require viral RNA-dependent RNA polymerase; usually package the polymerase before exiting host cell.

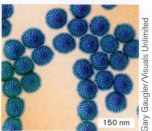

150 nm

Rotavirus

Nonsegmented, enveloped bacteriophages
Cystoviridae. Infect *Pseudomonas* species of bacteria.

Segmented, nonenveloped viruses of animals and plants
Birnaviridae. Infect marine and aquatic fish.
Reoviridae. Orthoreoviruses and rotaviruses infect humans and other vertebrates. Cypovirus infects insects. Fijivirus infects plants. Rice dwarf virus (phytoreovirus) is transmitted by leafhopper beetles and causes major economic damage to rice crops worldwide.
Varicosaviridae. Infect plants.

Group IV. (+) Sense single-stranded RNA viruses
Require viral RNA-dependent RNA polymerase to generate (−) template for progeny (+) genome; usually nonsegmented.

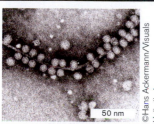

50 nm

Bacteriophage MS2

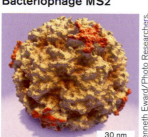

30 nm

Rhinovirus 14

Nonenveloped bacteriophages
Leviviridae. Bacteriophages MS2 and Qβ infect *E. coli.*

Nonenveloped animal and plant viruses
Bromoviridae. Infect many kinds of plants and are often carried by beetles.
Picornaviridae. Poliovirus causes poliomyelitis. Rhinovirus causes the common cold. Apthovirus causes hoof and mouth disease in cattle and other stock.
Tobamoviridae. Tobacco mosaic virus infects plants.
Potyviridae. Viruses such as plum pox virus infect fruits, peanuts, potatoes, and other plants.

Enveloped animal and plant viruses
Coronaviridae. Coronaviruses include SARS and animal viruses.
Flaviviridae. Flaviviruses infecting humans include West Nile virus, yellow fever virus, and hepatitis C virus.
Togaviridae. Include rubella virus and equine encephalitis virus.

Group V. (−) Sense single-stranded RNA viruses
Require viral RNA-dependent RNA transcriptase.

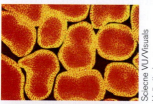

Influenza virions

Ebola virus

Segmented, enveloped viruses
Orthomyxoviridae. Influenza virus causes major epidemics among humans and animals.

Nonsegmented, enveloped viruses
Filoviridae. Ebola virus causes outbreaks among humans and chimpanzees.
Rhabdoviridae. Rabies virus infects mammals.
Paramyxoviridae. Infect humans and cause measles, mumps, and parainfluenza.

Segmented (+/−) strand, enveloped viruses
Arenaviridae. Spread by rodents, arenaviruses cause hemorrhagic fever and lymphocytic choriomeningitis.
Bunyaviridae. Hantaviruses are spread by rodents and infect humans. Tospoviruses are transmitted by thrips, infecting peanuts, onions, garlic, and other plants.

Table 6.1   Groups of viruses—Baltimore classification (*continued*)

Viruses	Taxonomic group with key traits and examples

Group VI. Retroviruses (RNA reverse-transcribing viruses)

Require viral reverse transcriptase to generate DNA copy for integration into host; chromosome package transcriptase before exiting host cell.

100 nm

©Hans Gelderblom/
Visuals Unlimited

HIV I

Retroviridae. Human immunodeficiency virus (HIV) is causing a human pandemic (worldwide epidemic). Feline leukemia virus (FeLV) is endemic among cats.

Group VII. Pararetroviruses (DNA reverse-transcribing viruses)

DNA is transcribed to RNA intermediate; reverse-transcribed to DNA; infect plant cells, which have cytoplasmic reverse transcriptase; or package a viral reverse transcriptase.

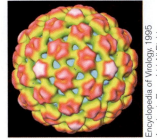

Encyclopedia of Virology. 1995
Academic Press Ltd All Rights
Reserved

Cauliflower mosaic virus

Nonenveloped plant viruses

Caulimoviridae. Transmitted by aphids, cauliflower mosaic virus and related viruses infect cauliflower, broccoli, groundnuts, soybeans, and cassava. The cauliflower mosaic virus promoter sequence is used to construct vectors to put genes into transgenic plants.
Badnaviridae. Badnaviruses infect bananas, cocoa plants, citrus, yams, and sugarcane.

Enveloped animal viruses

Hepadnaviridae. Hepatitis B virus causes widespread disease of the human liver.

Molecular Evolution of Viruses

The phylogeny, or genetic relatedness, of viruses can be determined within families. For example, the herpes family includes double-stranded DNA viruses that cause several human and animal diseases, such as chickenpox, oral and genital herpes infection, and equine herpes respiratory and genital infections. Herpes genomes consist of double-stranded DNA, 120–220 kilobases (kb) encoding about 70–200 genes; an example is that of varicella-zoster virus, the causative agent of chickenpox (**Fig. 6.15A**). The genome includes two "unique" segments of genes, one long and one short (U_L and U_S), joined by two inverted repeats (IRs). Other herpes genomes share similar structure, though they differ in gene order and IR position.

The relatedness of different herpes viruses can be measured by comparing their genome sequences. Comparison is based on **orthologous genes**, or **orthologs**. Orthologs are genes of common ancestry in two genomes that share the same function, a topic discussed in Chapter 8. An example is the ribosomal RNA genes whose sequence is used to measure relatedness of cellular organisms. Viruses have no ribosomal RNA, but closely related viruses share other orthologous genes. In pairs of orthologs, the amount of difference in sequence correlates approximately with the time following divergence from a common ancestor (a topic discussed in Chapter 17). Sequence comparison places herpes into three classes designated alpha, beta, and gamma (**Fig. 6.15B**). The alpha class includes human varicella-zoster virus and the oral and genital herpes viruses (HSV-1 and HSV-2), as well as equine herpes virus. The beta class includes cytomegalovirus, a common cause of congenital infections (present at birth), as well as two less known viruses. The gamma class includes Epstein-Barr virus, the cause of infectious mononucleosis, as well as several viruses of animals.

The more ancient evolution of viruses, however, is difficult to assess based on DNA or RNA sequence. Unlike cells, viruses do not possess genes common to all species, such as the ribosomal RNA genes used to estimate times of divergence of cellular organisms (discussed in Chapter 17). Furthermore, the genomes of viruses are highly mosaic, that is, derived from multiple sources. Mosaic genomes result from recombination between different viruses coinfecting a host.

In some cases, phylogeny is inconsistent with the fundamental chemical composition of the genome. For example, some DNA bacteriophages actually share closer ancestry with RNA bacteriophages than they do with DNA animal viruses. This is because two or more viruses can coinfect a cell and exchange genetic components.

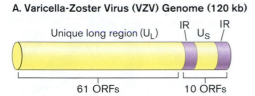

A. Varicella-Zoster Virus (VZV) Genome (120 kb)

Figure 6.15 Phylogeny of herpes viral genomes.
A. Genome structure of human varicella-zoster virus (VZV), the causative agent of chickenpox. **B.** Phylogeny of human and animal herpes viruses, based on whole-genome sequence analysis comparing clusters of orthologous groups of genes. Numbers measure genetic divergence.

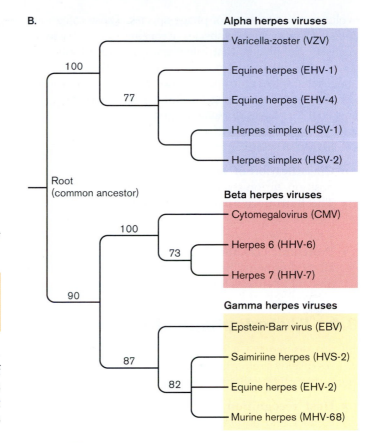

Thus, the genomic content of a virus can be influenced by its host range.

> **THOUGHT QUESTION 6.4** How can viruses with different kinds of genomes (RNA versus DNA) combine and exchange genetic information?

As more viral genomes are sequenced, classification methods have been devised to take advantage of sequence information without requiring gene products common to all species. One promising approach is that of **proteomics**, analysis of the **proteome**, the proteins encoded by genomes (discussed in Chapter 8). Proteins are identified through biochemical analysis of virus particles and through bioinformatic analysis of protein sequences encoded in the genomes. Proteomic analysis is useful for viruses because their small genomes encode a small number of proteins, which can be readily analyzed. Furthermore, because proteomic analysis is based on numerous gene products, it does not require a single gene (such as rRNA) common to all species. Statistical comparison of all proteins generated by a set of viral species reveals underlying degrees of relatedness.

An example of viral classification based on proteomic analysis is that of the "proteomic tree" of bacteriophages proposed by Rohwer and Edwards (**Fig. 6.16**). Unlike earlier trees based on a single common gene sequence, the proteomic tree is based on the statistical comparison of phage protein sequences predicted by genomic DNA of many different species of phages. The proteomic analysis predicts seven major

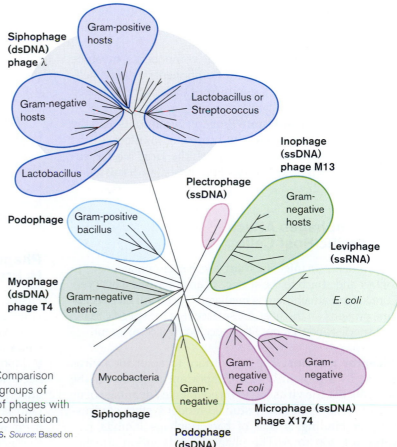

Figure 6.16 The bacteriophage proteomic tree. Comparison of all proteins encoded by each genome predicts distinct groups of bacteriophages. Within each group, there are subgroups of phages with shared hosts, since sharing of hosts facilitates genetic recombination and horizontal transfer of genes between different phages. *Source:* Based on F. Rohwer and R. A. Edwards. 2002. *J. Bacteriol.* 184:4529.

evolutionary categories of phage species. These categories appear to group phage species according to host bacteria. For example, phages that infect gram-negative hosts will show more genetic commonality with each other than with phages that infect gram-positive hosts. Shared hosts have a significant impact on phage evolution because coinfecting a host enables phages to exchange genes.

TO SUMMARIZE:

- **Classification of viruses** is based on genome composition, virion structure, and host range.
- **The Baltimore virus classification** emphasizes the form of the genome (DNA or RNA, single- or double-stranded) and the route to generate messenger RNA.
- **Proteomic classification** includes information from all viral proteins. Statistical analysis reveals common descent of viruses infecting a common host.

6.4 Bacteriophage Life Cycles

All viruses require a host cell for reproduction. While viruses display a remarkable diversity of reproductive strategies, they all face these same needs for host infection:

- **Host recognition and attachment.** Viruses must contact and adhere to a host cell that can support their particular reproductive strategy.
- **Genome entry.** The viral genome must enter the host cell and gain access to the cell's machinery for gene expression.
- **Assembly of virions.** Viral components must be expressed and assembled. Components usually "self-assemble"; that is, the joining of their parts is favored thermodynamically.
- **Exit and transmission.** Progeny virions must exit the host cell, then reach new host cells to infect. In the case of multicellular organisms, the virus must eventually reach other multicellular hosts (as discussed in Chapters 27 and 29).

Bacteriophages Attach to Specific Host Cells

To commence an infectious life cycle, bacteriophages need to contact and attach to the surface of an appropriate host cell. Contact and attachment is mediated by **cell surface receptors**, proteins on the host cell surface that are specific to the host species and that bind to a specific viral component.

Receptor proteins. The cell surface receptor for a virus is actually a protein with an important function for the host cell, but the virus has evolved to take advantage of its existence. An extensively studied model system of virus-receptor binding is that of bacteriophage lambda (see **Table 6.1**, top row). The phage lambda virion attaches

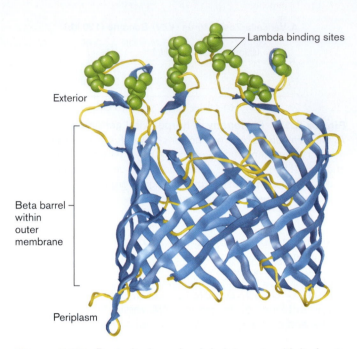

Figure 6.17 **Bacteriophage lambda interacts with its host receptor.** Phage lambda enters *E. coli* by first binding to the maltose porin. The "beta barrel" of maltose porin, shown in blue, is buried in the outer membrane. The phage lambda–binding sites, shown in green, were identified by amino acid substitution mutations that prevent phage binding and confer resistance to lambda on the host cell. (PDB code: 1MAL)

specifically to the maltose porin in the outer membrane of *Escherichia coli* (**Fig. 6.17**). Although the protein is often called "lambda receptor protein," it actually evolved in the host as a way to acquire the sugar maltose to metabolize. Thus, natural selection maintains the maltose porin in *E. coli* despite the danger of phage infection.

The precise domain of the maltose porin that binds to phage lambda was defined experimentally by mutations in *E. coli* that cause amino acid substitution in the protein. Some of the mutant *E. coli* strains were resistant to phage lambda infection. The mutations that conferred host resistance mapped to the domain of maltose porin that binds the phage capsid.

Phage Genomes Direct Host Cells to Produce Progeny Phages

Historically, the life cycles of bacteriophages have provided some of the most fundamental insights in molecular biology. In 1952, Alfred Hershey and Margaret Chase showed that the transmission of DNA by a bacteriophage to a host cell led to production of progeny bacteriophages, thus confirming that DNA is the hereditary material. In 1950, André Lwoff and Antoinette Gutman showed that a phage genome could integrate itself within a bacterial genome—the first recognition that genes could enter and leave a cell's genome. Other

A. Phage T4 DNA insertion

Phage T4 attaches to bacterium.

Sheath contracts and viral DNA enters bacterium.

B. Phage lambda reproductive cycle

Host genome

Phage attaches to host cell and inserts DNA.

Phage particle

Linear dsDNA cyclizes to circular DNA.

Lytic cycle

Viral DNase cleaves host cell DNA. Cell synthesizes capsid proteins.

Cell replicates phage DNA. DNA is packaged into capsids.

Phage lyses cell, and progeny phage are released.

Lysogeny

Phage DNA integrates into host genome to form prophage.

Phage recombines by rejoining the ends of its phosphodiester chain and enters the lytic cycle.

Integrated phage DNA reproduces with host genome.

Stress induces excision of phage DNA.

Integrated phage DNA replicates with host genome.

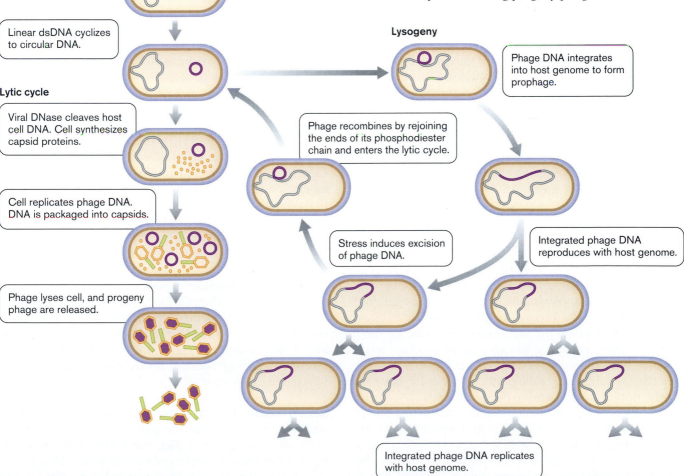

fundamental concepts of the genetic unit and the basis of gene transcription came from experiments on bacteriophages, as discussed in Chapters 7–9.

Most bacteriophages (or phages) insert only their genome into a cell through the cell envelope, thus avoiding the need for the capsid to penetrate the molecular barrier of the cell wall. For example, the phage T4 virion has a neck tube that contracts, bringing the headpiece near the cell surface to insert its DNA (**Fig. 6.18A**⊙). After the genome is inserted, the phage capsid remains outside attached to the cell surface and is termed a "ghost."

The lytic cycle. In a lytic cycle, when a phage particle injects its genome into a cell, it immediately proceeds to reproduce as many progeny phage particles as possible. The process of reproduction involves replicating the phage genome as well as expressing phage mRNA to make enzymes and capsid proteins. Some phages, such as T4, digest the host DNA to increase efficiency of phage production. Finally, the host cell lyses, releasing progeny phages.

Figure 6.18   Bacteriophage reproduction: lysis and lysogeny.   A. Phage T4 attaches to the cell surface by its tail fibers, then contracts to inject its DNA. The empty capsid remains outside as a "ghost." **B.** Lysis occurs when the phage genome reproduces progeny phage particles, as many as possible, then lyses the cell to release them. In phage lambda, lysogeny can occur when the phage genome integrates into that of the host. The phage genome is replicated along with that of the host cell. The phage DNA, however, can direct its own excision by expressing a site-specific DNA recombinase. This excised phage chromosome then initiates a lytic cycle. ⊙

Phage T4 reproduces entirely by the lytic cycle; this is called a **virulent** phage. Other phages, such as phage lambda, have the options of reproducing by lysis or by lysogeny (**Fig. 6.18B**⊜).

Lysis. After phage lambda inserts its DNA, its genes are expressed by the host cell RNA polymerase and ribosomes. "Early genes" are expressed early during the lytic cycle. Other phage-expressed proteins then work together with the cellular enzymes and ribosomes to replicate the phage genome and produce phage capsid proteins. The capsid proteins self-assemble into capsids and package the phage genomes, a process that takes place in defined stages, like a factory assembly line. At last, a "late gene" from the phage genome expresses an enzyme that lyses the host cell wall, releasing the mature virions. Lysis is also referred to as a **burst**, and the number of virus particles released is called the **burst size**.

Lysogeny. A **temperate phage**, such as phage lambda, can infect and lyse cells like a virulent phage, but it also has an alternative pathway: to integrate its genome into that of the host cell (see **Fig. 6.18B**). Phage lambda has a linear genome of double-stranded DNA, which circularizes upon entry into the cell. The circularized genome then recombines into that of the host by **site-specific recombination** of DNA. In site-specific recombination, a recombinase enzyme aligns the phage genome with the host DNA and exchanges the phosphodiester backbone links with those of the host genome. The process of exchange thus integrates the phage genome into that of the host. The integrated phage genome is called a **prophage**.

Integration of the phage genome as a prophage results in **lysogeny**, a condition in which the phage genome is replicated along with that of the host cell as the host reproduces. Implicit in the term *lysogeny*, however, is the ability of such a strain to spontaneously generate a lytic burst of phage. For lysis to occur, the prophage (integrated phage genome) directs its own excision from the host genome by an intramolecular process of site-specific recombination. The two ends of the phage genome exchange their phosphodiester backbone linkages so as to come apart from the host molecule. As the phage DNA exits the host genome, it circularizes and initiates a lytic cycle, destroying the host cell and releasing phage particles.

The "decision" between lysogeny and lysis is determined by proteins that bind DNA and repress transcription of genes for virus replication (see Section 10.7). Exit from lysogeny into lysis can occur at random, or it can be triggered by environmental stress such as UV light, which damages the cell's DNA.

The regulatory switch of lysogeny responds to environmental cues indicating the likelihood that the host cell will survive and continue to propagate the phage genome. If a cell's growth is strong, it is more likely that the phage

DNA will remain inactive, whereas events that threaten host survival will trigger a lytic burst. An analogous phenomenon occurs in animal viral infections such as herpes, in which environmental stress triggers painful outbreaks arising from latent infection, reactivation of a virus that was dormant within cells. The lysogeny of phage lambda provided an early model system for investigating viral persistence in human viral diseases.

Viruses transfer host genes. During the exit from lysogeny or during latent growth of animal viruses, the virus can acquire host genes and pass them on to other host cells. This process of transferring host genes is known as **transduction**. A transducing bacteriophage can pick up a bit of host genome and transfer it to a new host cell. In some forms of transduction, the entire phage genome is replaced by host DNA packaged in the phage capsid, resulting in a virus particle that only transfers host DNA. Host DNA transferred by viruses can become permanently incorporated into the infected host genome. The mechanisms of phage-mediated transduction (generalized and specialized) are discussed in Chapter 9.

The integration and excision of viral genomes make extraordinary contributions to the evolution of host genomes. Lysogenic prophages often contribute key toxins and virulence factors to pathogens: For example, the CTXϕ prophage encodes cholera toxin in *V. cholerae*; the Shiga toxin prophage confers virulence on enteric pathogens *Shigella* and *E. coli* H7:O157; and prophages encode toxins for *Corynebacterium diphtheriae* (diphtheria) and *Clostridium botulinum* (botulism). In natural environments, phage-mediated transduction mediates much of the recombination of bacterial genomes. In the laboratory, the ability of phages to transfer genes provided vectors for recombinant DNA technology, the artificial construction of phages to carry genes from animals and plants for gene cloning.

When viral transfer of genes was first discovered in the 1960s, it was thought that multicellular eukaryotes had a much lower tolerance for genome change. We have since learned, however, that much of the human genome also shows evidence of gene transfer mediated by ancient viruses, including possible retroviruses similar to HIV (discussed in Chapter 11).

Slow release. The **slow-release cycle** differs from lysis and lysogeny in that phage particles reproduce without destroying the host cell (**Fig. 6.19**). Slow release is performed by filamentous phages such as phage M13. In the slow-release life cycle, the single-stranded circular DNA of M13 serves as a template to synthesize a double-stranded intermediate. The double-stranded intermediate slowly generates single-stranded progeny genomes, which are packaged by supercoiling and coating with capsid proteins. The phage particles then extrude through the cell envelope

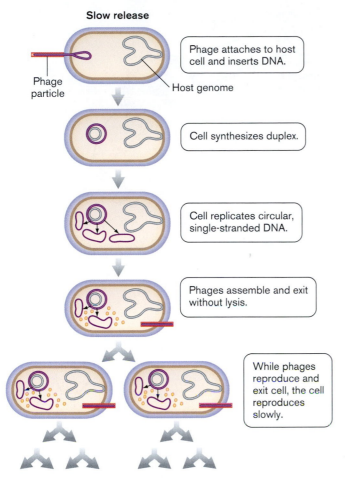

Slow release

Phage attaches to host cell and inserts DNA.

Phage particle

Host genome

Cell synthesizes duplex.

Cell replicates circular, single-stranded DNA.

Phages assemble and exit without lysis.

While phages reproduce and exit cell, the cell reproduces slowly.

Figure 6.19 Bacteriophage life cycle: slow release. In the slow-release life cycle, a filamentous phage produces phage particles without lysing the cell. The host continues to reproduce itself, but more slowly than uninfected cells because much of its resources are being used to make phages.

without lysing the cell. The host cell continues to reproduce, though more slowly than uninfected cells because much of its resources are diverted to virus production.

Some phage genomes consist of RNA. RNA phages include the leviviruses such as phage MS2 and Qβ. The RNA phages replicate their genomes through a double-stranded RNA intermediate, analogous to the double-stranded DNA intermediate of the filamentous phage M13; but RNA phages lyse their hosts. RNA phages have very simple genomes composed of only three or four genes. These genes encode a replicase, a coat protein, and a maturation factor. The RNA genomes of these phages are of historical interest as model systems to study ribosomal translation.

THOUGHT QUESTION 6.5 What are the relative advantages and disadvantages (to a virus) of the slow-release strategy, compared with the strategy of a temperate phage, which alternates between lysis and lysogeny?

TO SUMMARIZE:

- **Host cell surface receptors** mediate attachment of bacteriophages to a cell and confer host specificity.
- **Lytic cycle.** A bacteriophage injects its DNA into a host cell, where it utilizes host gene expression machinery to produce progeny virions.
- **Lysogeny.** Some bacteriophages can copy their genome into that of the host cell, which then replicates the phage genome along with its own. A lysogenic bacterium can initiate a lytic cycle.
- **Slow release.** Some bacteriophages use the host machinery to make progeny that bud from the cell slowly, slowing growth of the host without lysis.

6.5 Animal and Plant Virus Life Cycles

Animal and plant viruses solve problems similar to those faced by bacteriophages: host attachment, genome entry and gene expression, virion assembly, and virion release. The more complex structure of eukaryotic cells, however, leads to greater complexity and diversity of life cycles than is seen in phages. Viral reproduction may involve intracellular compartments such as the nucleus or secretory system and may depend on tissue and organ development in multicellular organisms. Study of animal virus life cycles reveals potential targets for antiviral drugs, such as protease inhibitors for HIV.

NOTE: The virus life cycles in Chapter 6 are simplified. For greater molecular detail, see Chapter 11.

Animal Viruses Bind Receptors and Uncoat Their Genomes

Like bacteriophages, animal viruses evolve to bind specific receptor proteins on their host cell. An example of a human virus-receptor interaction is that of rhinovirus (see Table 6.1, Group IV), which causes the common cold. Rhinovirus attaches to ICAM-1, a human glycoprotein needed for intercellular adhesion (**Fig. 6.20A**). The rhinovirus binds to a domain of ICAM-1 essential for it to bind a lymphocyte protein called integrin.

The host receptors play a key role in determining the host range, the group of host species permitting growth. Within a host, receptor molecules can also determine the **tropism**, or tendency to infect a particular tissue type. Some viruses, such as Ebola virus, exhibit broad tropism, infecting many kinds of host tissues, whereas others, such as papillomavirus, show tropism for only one type, the epithelial tissues. Poliovirus infects only a specific class of human cells that display the immunoglobulin-like receptor protein PVR. Children under age 3 do not yet

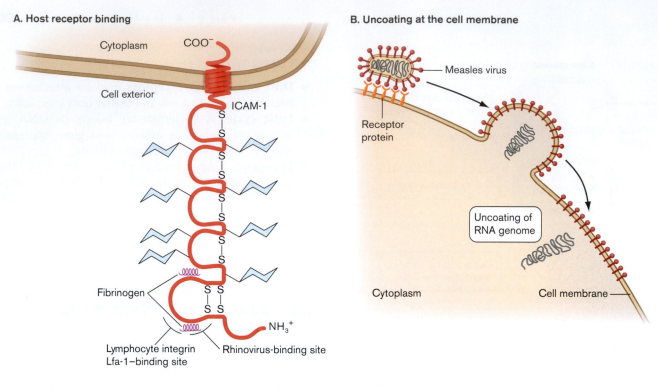

A. Host receptor binding

Cytoplasm

COO⁻

Cell exterior

ICAM-1

Fibrinogen

Lymphocyte integrin Lfa-1–binding site

Rhinovirus-binding site

NH₃⁺

B. Uncoating at the cell membrane

Measles virus

Receptor protein

Uncoating of RNA genome

Cytoplasm

Cell membrane

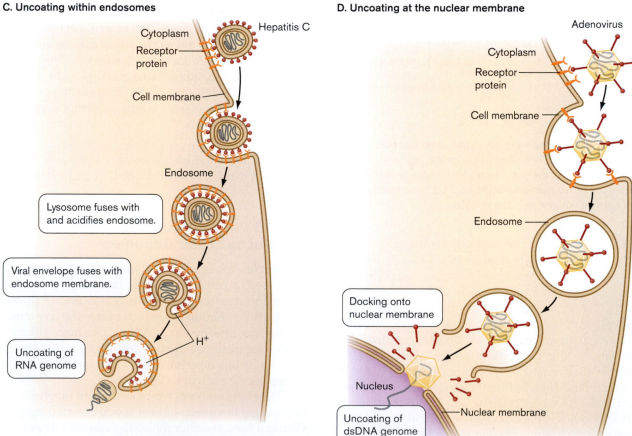

C. Uncoating within endosomes

Hepatitis C

Cytoplasm

Receptor protein

Cell membrane

Endosome

Lysosome fuses with and acidifies endosome.

Viral envelope fuses with endosome membrane.

Uncoating of RNA genome

H⁺

D. Uncoating at the nuclear membrane

Adenovirus

Cytoplasm

Receptor protein

Cell membrane

Endosome

Docking onto nuclear membrane

Nucleus

Nuclear membrane

Uncoating of dsDNA genome

Figure 6.20 Receptor binding and genome uncoating. A. Rhinovirus attaches to the intercellular adhesion molecule (ICAM-1), a glycoprotein required by the host cells to bind a lymphocyte integrin, a cell surface matrix protein required for cell-cell adhesion. After binding a specific receptor on the host cell membrane, an animal virus enters the cell, where its genome is uncoated. **B.** Uncoating at the cell membrane. **C.** Uncoating within endosomes. **D.** Uncoating at the nuclear membrane. *Source:* A. Based on J. Bella, et al. 1998. *PNAS* 95:4140.

express the receptor PVR; thus, they do not experience polio infection of the nervous system. Mice are not normally infected by polio, but when transgenic mice were created to express PVR on their cells, the mice could then be infected.

Different strains of a virus may show radically different host tropisms based on their receptor specificity. An example is that of avian influenza strain H5N1. The H5N1 strain infects birds by binding to a glycoprotein (protein with sugar chains) receptor on cell surfaces of the avian respiratory tract. The H5N1 strain requires a receptor protein with a sialic acid sugar chain terminating in galactose linked at the C-3 position (α2, 3), common in the avian respiratory tract. In humans, however, most nasal upper respiratory cells have receptors with galactose linked at the C-6 position (α2, 6). Cells with the C-3 linkage are more common in the human lower respiratory tract. That is why avian influenza H5N1 infection of humans has been relatively rare. However, only a small mutation in the H5N1 envelope protein could enable it to bind to the (α2, 6) receptor more effectively, allowing rapid transmission between humans.

> **THOUGHT QUESTION 6.6** How could humans evolve to resist rhinovirus infection? Is such evolution likely? Why or why not?

Genome uncoating. Some animal viruses, such as poliovirus, attach to the host cell surface and insert their genome into the host cell, as do bacteriophages. In animal virology, this process is called "extracellular uncoating" of the genome. Most animal viruses, however, enter the host cell as intact virions, then undergo intracellular **uncoating**, a process of virion disassembly in which the genome is released for replication and gene expression.

Uncoating of the genome takes place in several different ways. For example, measles virus, a paramyxovirus, enters the cell by binding host receptor proteins, which causes the viral envelope to fuse with the host cell membrane (**Fig. 6.20B**). The measles RNA genome is then uncoated and released directly into the cytoplasm. Alternatively, a flavivirus such as hepatitis C is taken up by **endocytosis** (**Fig. 6.20C**). In endocytosis, the cell membrane forms a vesicle around the virion and engulfs it, forming an endocytic vesicle. The endocytic vesicle fuses with a lysosome, whose acidic environment activates entry of the capsid into the cytoplasm. The capsid then comes apart, uncoating the viral genome. Still other species, such as adenovirus, enter by endocytosis but require transport to the nucleus (**Fig. 6.20D**). At the nuclear membrane, the virion docks at a nuclear pore and injects its DNA genome into the nucleus. The adenoviral DNA then has access to cellular DNA polymerase and RNA transcriptase.

Animal Virus Life Cycles

The primary factor determining the life cycle of an animal virus is the physical form of the genome. A viral genome made of DNA can utilize some or all of the host replication machinery. If, however, the genome consists of single-stranded RNA, it first needs to synthesize either an RNA-dependent RNA polymerase to generate an RNA template or a reverse transcriptase to generate a DNA template (in the case of retroviruses).

DNA virus life cycle. An example of a double-stranded DNA virus is human papillomavirus (HPV) (see Table 6.1, Group I), the cause of genital warts (**Fig. 6.21A**). HPV is the most common sexually transmitted disease in the United States and one of the most common worldwide. Certain strains infect the skin, whereas others infect the mucous membranes through genital or anal contact (sexual transmission). Like phage lambda, papillomavirus has an active reproduction cycle and a dormant cycle in which the viral genome integrates into that of the host. Genome integration leads to abnormal cell proliferation causing cauliflower-like warts and can progress to cervical or penile cancer.

HPV initially infects basal epithelial cells (**Fig. 6.21B**). In the basal cells, papilloma virions enter the cytoplasm by receptor binding and membrane fusion. The virion then undergoes uncoating, the disintegration of the protein capsid or coat, releasing its circular double-strand genome (**Fig. 6.22**). The uncoated viral genome enters the nucleus, where it gains access to host DNA and RNA polymerases.

The process of viral reproduction is complicated by the developmental progression of basal cells into keratinocytes and ultimately cells to be shed or sloughed off from the surface (see **Fig. 6.21B**). Viral replication is largely inhibited until the basal cells start to differentiate into keratinocytes (mature epithelial cells). Host cell differentiation induces the viral DNA to replicate and undergo transcription by host polymerases. The mRNA transcripts then exit the nuclear pores, as do host mRNAs, for translation in the cytoplasm. The translated capsid proteins, however, return to the nucleus for assembly of the virion. Nuclear virion assembly is typical of DNA viruses (with the exception of poxviruses, which replicate entirely in the cytoplasm).

As the keratinocytes complete differentiation, they start to come apart and are shed from the surface. The DNA virions are released from the cell during this shedding process. Unfortunately, in the basal cells, HPV has an alternative pathway of integrating its genome into that of the host (analogous to phage lysogeny). Genome integration can disrupt the expression of key host genes, causing transformation to cancer.

A.

©Ken Greer/Visuals Unlimited

B.

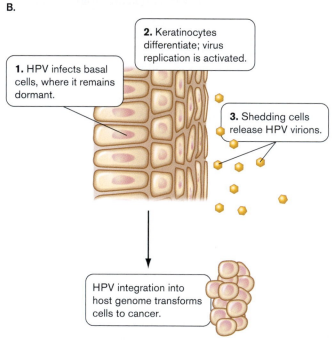

1. HPV infects basal cells, where it remains dormant.

2. Keratinocytes differentiate; virus replication is activated.

3. Shedding cells release HPV virions.

HPV integration into host genome transforms cells to cancer.

Figure 6.21 Human papillomavirus. A. Certain strains of human papillomavirus (HPV) cause warts on the genitals or anus. **B.** HPV infects basal epithelial cells, where the DNA uncoats but remains dormant. As cells differentiate, viral synthesis is activated. Shedding cells then release virions. An alternative fate of HPV is to integrate in the genome of host basal cells, where it leads to carcinogenesis.

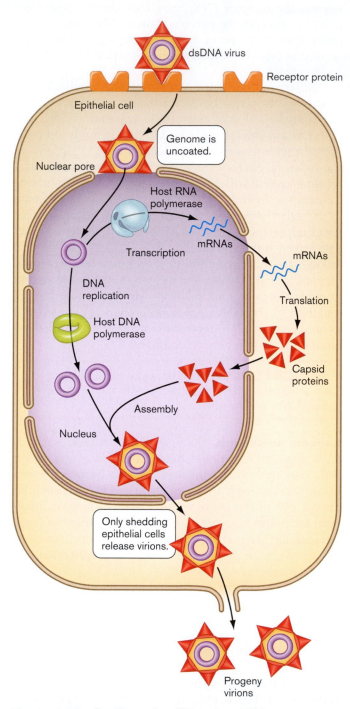

dsDNA virus

Receptor protein

Epithelial cell

Nuclear pore

Genome is uncoated.

Host RNA polymerase

Transcription

mRNAs

mRNAs

Translation

DNA replication

Host DNA polymerase

Capsid proteins

Assembly

Nucleus

Only shedding epithelial cells release virions.

Progeny virions

Figure 6.22 Papillomavirus life cycle. HPV, a double-stranded DNA virus, enters the cytoplasm, where the protein coat disintegrates. The viral DNA enters the nucleus for replication and transcription by host polymerases. Viral mRNA returns to the cytoplasm for translation of capsid proteins, which return to the nucleus for assembly of virions.

RNA virus life cycle. The picornaviruses include polio-virus as well as rhinovirus, which causes the common cold (see Table 6.1, Group IV). Picornavirus genomes contain (+) strand RNA, allowing viral reproduction to occur entirely in the cytoplasm, without any use of DNA.

Picornaviruses bind to a surface receptor, such as ICAM-1 for rhinovirus or the PVR receptor for poliovirus. The (+) strand RNA is uncoated by insertion through the cell membrane into the cytoplasm (**Fig. 6.23**). The role of endocytosis is debated; poliovirus requires no endocytosis, but rhinovirus genome uncoating may require endocytosis and low-pH induction.

After uncoating, a gene in the viral RNA is translated by host ribosomes to make RNA-dependent RNA polymerase. The RNA-dependent RNA polymerase uses the viral RNA template to make (−) strand RNA. The (−) strand RNA then serves as a template for other viral mRNAs as well as progeny genomic RNA. Capsid proteins are translated by host ribosomes, and the capsids self-assemble in the cytoplasm. Virions then assemble at the cell membrane and are released by budding out.

Note that other kinds of RNA virus, such as influenza virus, encapsidate a (−) strand genome. In this case, the (−) strand must serve as template to generate mRNA as well as (−) strand progeny genomes. Influenza virus replication includes other interesting molecular complications, discussed in Chapter 11.

RNA retroviruses. Retroviruses include human immunodeficiency virus (HIV), the causative agent of AIDS, and feline leukemia virus (FeLV), a disease commonly afflicting domestic cats (see Table 6.1, Group VI). A retrovirus uses a reverse transcriptase to make a DNA copy of its RNA genome (**Fig. 6.24**). Instead of being translated from an early gene, the reverse transcriptase is actually carried within the virion, bound to the RNA genome with a primer in place. The virion contains two copies of the HIV genome, each carrying its own reverse transcriptase. After uncoating in the cytoplasm, the viral RNA is copied into double-stranded DNA.

The DNA copy then enters the nucleus, where it integrates by recombining with the host genome. The viral genome then replicates silently in the host cell, generating only a small number of virions without apparent effect on the host. To generate virions, a host RNA polymerase transcribes viral mRNA and viral genomic RNA. The viral mRNA reenters the cytoplasm for translation to produce coat proteins and envelope proteins. The coat proteins are transported by the endoplasmic reticulum (ER) to the cell membrane, where virions self-assemble and bud out.

At a certain point, the host cell can suddenly begin to generate large numbers of virions. The cause of accelerated reproduction is poorly understood, although it

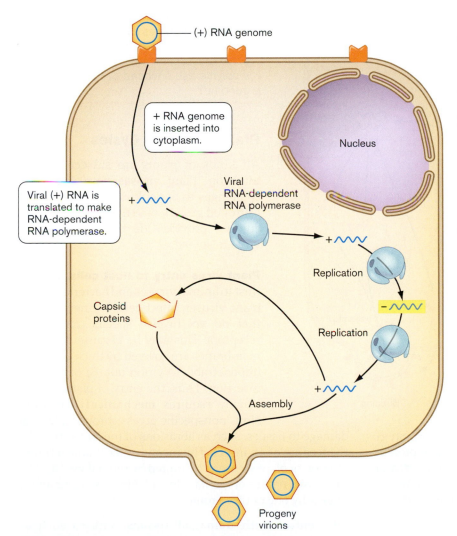

Figure 6.23 Picornavirus life cycle. A picornavirus inserts its (+) strand RNA into the cell. Reproduction occurs entirely in the cytoplasm. A key step is the early translation of a viral gene to make RNA-dependent RNA polymerase. The RNA-dependent RNA polymerase uses the picornavirus RNA template to make (−) strand RNA, which then serves as a template for other viral mRNAs as well as progeny genomic RNA.

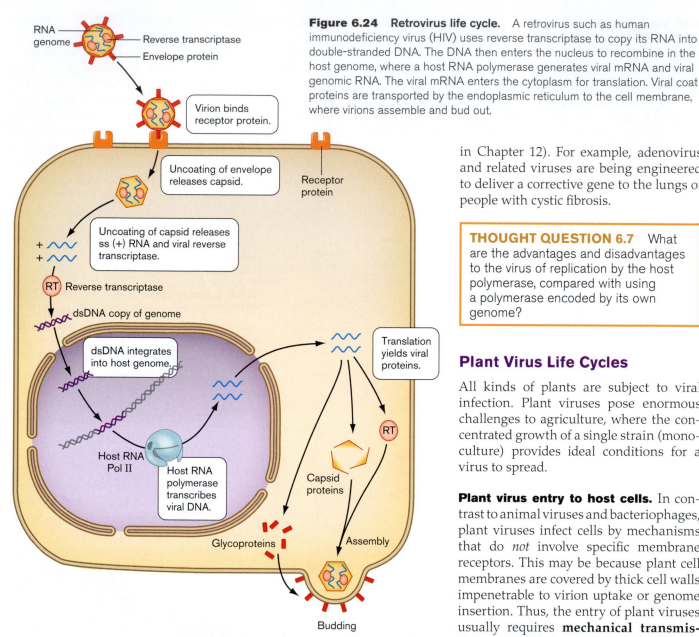

Figure 6.24 Retrovirus life cycle. A retrovirus such as human immunodeficiency virus (HIV) uses reverse transcriptase to copy its RNA into double-stranded DNA. The DNA then enters the nucleus to recombine in the host genome, where a host RNA polymerase generates viral mRNA and viral genomic RNA. The viral mRNA enters the cytoplasm for translation. Viral coat proteins are transported by the endoplasmic reticulum to the cell membrane, where virions assemble and bud out.

in Chapter 12). For example, adenovirus and related viruses are being engineered to deliver a corrective gene to the lungs of people with cystic fibrosis.

THOUGHT QUESTION 6.7 What are the advantages and disadvantages to the virus of replication by the host polymerase, compared with using a polymerase encoded by its own genome?

Plant Virus Life Cycles

All kinds of plants are subject to viral infection. Plant viruses pose enormous challenges to agriculture, where the concentrated growth of a single strain (monoculture) provides ideal conditions for a virus to spread.

Plant virus entry to host cells. In contrast to animal viruses and bacteriophages, plant viruses infect cells by mechanisms that do *not* involve specific membrane receptors. This may be because plant cell membranes are covered by thick cell walls impenetrable to virion uptake or genome insertion. Thus, the entry of plant viruses usually requires **mechanical transmission**, nonspecific access through physical damage to tissues, such as abrasions of the leaf surface by the feeding of an insect. The means of mechanical transmission of plant viruses are limited by the cell wall and by the sessile nature of plants. Most plant viruses gain entry to cells by one of three routes:

can involve stress conditions such as poor health or pregnancy. The mechanism of control involves several protein regulators encoded by the virus. The full process and regulation of HIV reproduction is exceedingly complex; further molecular details are discussed in Chapter 11.

Oncogenic viruses. DNA viruses such as HPV and adenovirus can insert their genomes into the chromosome of a host cell in a manner analogous to that of a lysogenic bacteriophage. In some cases, the transfer of host genes to abnormal chromosome locations can generate **oncogenes**, genes whose expression at inappropriate times causes uncontrolled proliferation of the cell and ultimately cancer. Viruses that carry oncogenes are known as **oncogenic viruses**. This capacity for gene transfer may be manipulated artificially for gene therapy (discussed

- **Contact with damaged tissues.** Viruses such as tobacco mosaic virus appear to require nonspecific entry into broken cells.
- **Transmission by an animal vector.** Insects and nematodes transmit many kinds of plant viruses. For example, the geminiviruses are inoculated into cells by plant-eating insects such as aphids, beetles, or grasshoppers.
- **Transmission through seed.** Some plant viruses enter the seed and infect the next generation.

Figure 6.25   Plum pox is caused by potyvirus.   A. Potyvirus, a filamentous (+) strand RNA virus, approximately 800 nm in length (TEM). **B.** Potyvirus is transmitted by aphids, which suck the plant sap and release potyvirus into the damaged tissues. **C.** Streaking of flowers caused by potyvirus infection. **D.** Ring-shaped "pox" appear on the infected fruit and the stone inside.

An economically important plant virus is the potyvirus plum pox virus (see Table 6.1, Group IV), a major pathogen of plums, peaches, and other stone fruits. Plum pox virus, a (+) strand RNA virus, is transmitted by aphids (**Fig. 6.25**). Following infection, the spread of the virus generates streaked leaves and flowers as well as ring-shaped "pox" marks on the surfaces of the fruit and of the stone within.

Plant virus transmission through plasmodesmata. Within a plant, the thick cell walls prevent a lytic burst or budding out of virions. Instead, plant virions spread to uninfected cells by traveling through **plasmodesmata** (singular, **plasmodesma**). Plasmodesmata are membrane channels that connect adjacent plant cells (**Fig. 6.26**). The outer channel connects the cell membranes of the two cells, whereas the inner channel connects the endoplasmic reticulum.

Passage through the plasmodesmata requires action by **movement proteins** whose expression is directed by the viral genome. In some cases, the movement proteins transmit the entire plant virion; in other cases, only the nucleic acid itself is small enough to pass through. The independent movement and transmission of plant viral genomes suggest infective strategies in common with those of viroids, which lack capsids altogether.

DNA pararetroviruses. Pararetroviruses possess a DNA genome that requires transcription to RNA in the cytoplasm, followed by reverse transcription to form DNA genomes for progeny virions (**Fig. 6.27**). While some pararetroviruses infect humans, such as hepadnavirus, the best-known pararetrovirus is cauliflower mosaic virus, or caulimovirus (see Table 6.1, Group VII). Cauli-

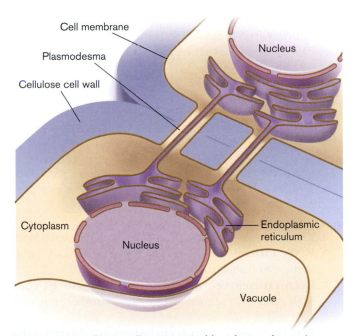

Figure 6.26   Plant cells connected by plasmodesmata. Plasmodesmata offer a route for plant viruses to reach uninfected cells.

movirus is an important tool for biotechnology because it has a highly efficient promoter for gene transcription, allowing high-level expression of cloned genes. Vectors derived from caulimoviruses are often used to construct transgenic plants.

Caulimovirus is transmitted by secretions from an insect whose bite damages plant tissues, providing access to the cytoplasm. The caulimovirus genome moves from the cytoplasm to the cell nucleus through a nuclear pore. Within the nucleus, two promoters on its DNA genome direct transcription to RNA. The two RNA transcripts exit the nucleus for translation by host ribosomes to make viral proteins. A host reverse transcriptase, present in plant cells, generates DNA viral genomes. After virions are assembled in the cytoplasm, movement proteins help transfer them through plasmodesmata into an adjacent cell.

A caulimovirus promoter sequence is commonly used in gene transfer vectors for plant biotechnology. The advantage of viral promoters is their highly efficient transcription of a transferred gene, such as one conferring pesticide resistance. In the field, 10% of cruciferous vegetables are typically infected with caulimovirus. Some critics of gene technology are concerned that the prevalence of the caulimovirus promoter in transgenic crops may lead to the evolution of new pararetroviruses.

TO SUMMARIZE:

- **Host cell surface receptors** mediate animal virus attachment to a cell and confer host specificity and tropism.
- **Animal DNA viruses** either inject their genome or enter the host cell by endocytosis. The viral genome requires uncoating for gene expression.
- **RNA viruses** use an RNA-dependent RNA polymerase to transcribe their messenger RNA.
- **Retroviruses** use a reverse transcriptase to copy their genomic sequence into the chromosome of a host cell.
- **Plant viruses** enter host cells by transmission through a wounded cell surface or an animal vector. Plant viruses travel to adjacent cells through plasmodesmata.
- **Pararetroviruses** contain DNA genomes but generate an RNA intermediate that requires reverse transcription to DNA for progeny virions.

6.6 Culturing Viruses

To study any living microorganism, we must culture it in the laboratory. Culturing viruses provides large numbers of virions and their components for analysis. A complica-

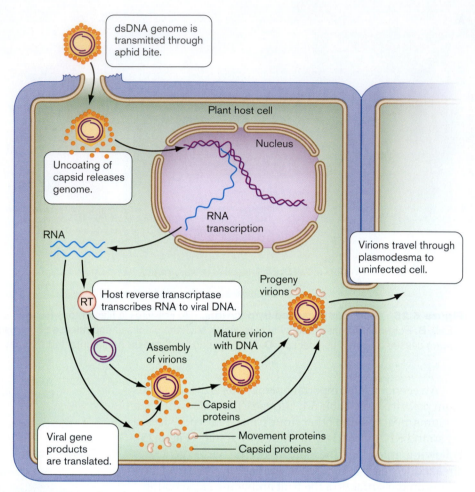

dsDNA genome is transmitted through aphid bite.

Plant host cell

Nucleus

Uncoating of capsid releases genome.

RNA transcription

RNA

Virions travel through plasmodesma to uninfected cell.

RT Host reverse transcriptase transcribes RNA to viral DNA.

Progeny virions

Mature virion with DNA

Assembly of virions

Capsid proteins

Movement proteins

Capsid proteins

Viral gene products are translated.

Figure 6.27 Caulimovirus life cycle.   The cauliflower mosaic virus, a caulimovirus, uses host RNA polymerase to copy its DNA into RNA and uses host reverse transcriptase to make DNA copies. The DNA genomes are packaged into progeny virions that use virus-encoded movement proteins to travel through plasmodesmata to an adjacent uninfected cell.

tion of virus culture is the need to grow the virus within a host cell. Therefore, any virus culture system must be a double culture of host cells plus viruses. Culturing viruses of multicellular animals and plants creates additional complications, as viruses show tropism for particular tissues or organs.

Batch Culture of Bacteriophages in Large Populations

Batch culture, or culture in liquid medium, enables growth of a large population of viruses for study. Bacteriophages can be inoculated into a growing culture of bacteria, usually in a culture tube or a flask. The culture fluid is then sampled over time and assayed for phage particles. The growth pattern usually takes the form of a step curve (**Fig. 6.28**).

To observe one cycle of phage reproduction, phages are added to host cells at a multiplicity of infection (MOI, ratio of phage to cells) such that every host cell is infected. The

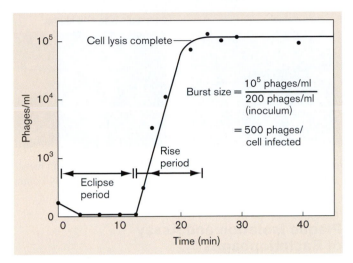

Figure 6.28 **One-step growth curve of a bacteriophage.** After initial infection of a liquid culture of host cells, the titer of virus drops near zero as all virions attach to the host. During the eclipse period, progeny phages are being assembled within the cell. As cells lyse (the rise period), virions are released until they reach the final plateau. The infectious cycle is typically complete within less than an hour.

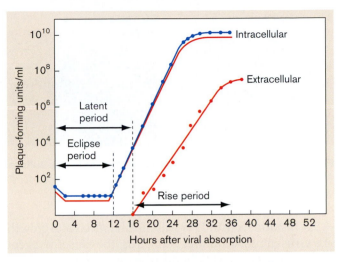

Figure 6.29 **One-step growth curve for a virus.** The titer of extracellular virus drops near zero during the latent period, as all virions adsorb to the host. Then progeny virions begin to emerge by budding out from the infected cell. The growth curve of a rapidly replicating animal virus typically takes hours to level off; the "burst" event is not defined as clearly as for phages.

phage particles immediately adsorb to surface receptors of host cells and inject their DNA. As a result, virions are virtually undetectable in the growth medium for a short period after infection; this is called the **eclipse period**.

For some species, it is possible to distinguish between the eclipse period and a **latent period**, which includes the eclipse period plus the time during which progeny viruses have been formed but remain trapped within the cell. The latent period is particularly significant in animal viruses, where large numbers of virions usually generate progeny through budding out of the host cell (**Fig. 6.29**).

> **NOTE:** The *latent period* of a lytic virus is the period between initial phage-host contact and the first appearance of progeny phage. This must be distinguished from the *latent infection* of a virus that maintains its genome within a host cell without reproducing virions.

As cells begin to lyse and liberate progeny viruses, the culture enters the **rise period**, in which virus particles are appearing in the growth medium. The rise period ends when all the progeny viruses have been liberated from their host cells. If the number of viruses that go on to reinoculate further host cells is small, then the virus concentration at the end point divided by the original concentration of inoculated phage approximates the **burst size**; that is, the number of viruses produced per infected host cell. The burst size may be estimated by dividing the concentration of progeny virions by the concentration of

inoculated virions, assuming that all the original virions infect a cell.

The burst size, together with the cell density prior to lysis, determines the concentration of the resultant suspension of virus particles, called a **lysate**. In the case of bacteriophages, a lysate of phage particles can be extremely stable, remaining infective at room temperature for many years. Eukaryotic viruses tend to be less stable and need to be maintained in culture or deep freeze.

> **THOUGHT QUESTION 6.8** Why does bacteriophage reproduction give a step curve, whereas cellular reproduction generates an exponential growth curve? Could you design an experiment in which viruses generate an exponential curve? Under what conditions does the growth of cellular microbes give rise to a step curve?

Animal Viruses Are Grown in Tissue Culture

In the case of animal and plant viruses, the multicellular nature of the host is an important factor in the pathology and transmission of the pathogen (discussed in Chapters 26 and 27). Animal viruses can be cultured within whole animals by serial inoculation, where virus is transferred from an infected animal to an uninfected one. Culture within animals ensures that the virus strain maintains its original **virulence** (ability to cause disease). But the process is expensive and laborious, involving large-scale use of animals.

0 h uninfected	5.5 h cells round up	8 h cells detach	24 h cells lyse and clump
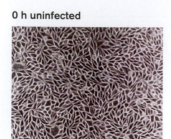	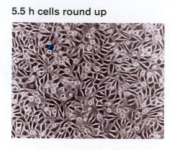	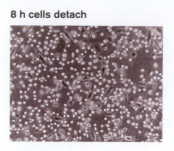	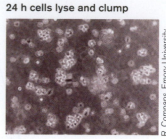

R. Compans, Emory University, School of Medicine

Figure 6.30   **Poliovirus replication in human tissue culture.**   Before infection (0 hours), the cultured cells grow in a smooth layer. At 5.5 hours after infection, cells are starting to retract and round up. At 8 hours, infected cells detach from the culture dish. By 24 hours, cells have lysed or in some cases clumped with other cells.

A historic event in 1949 was the first successful growth of a virus in tissue culture. Poliovirus, the causative agent of the devastating childhood disease poliomyelitis, was grown in human cell tissue culture (**Fig. 6.30**) by John F. Enders, Thomas J. Weller, and Frederick Robbins at the Children's Hospital in Boston. As heralded that year in *Scientific American*, "It means the end of the 'monkey era' in poliomyelitis research. . . . Tissue-culture methods have provided virologists with a simple in vitro method for testing a multitude of chemical and antibiotic agents." Since then, tissue culture has remained the most effective way to study the molecular biology of animal and plant viruses and to develop vaccines and antiviral agents.

Some viruses can be grown in a tissue culture of cells growing confluently on a surface. The fluid bathing the tissue layer is sampled for virus concentration. As in the case of bacteriophage batch culture, we can define an eclipse period, a latent period before appearance of the first progeny virions in the culture fluid, and a rise period. The intracellular completion of virion assembly within the cell is important for transmission of viral infection because some viruses have the ability to spread their virions from cell to cell without ever existing as such outside. In this way, they can avoid detection by the immune system to which they might be vulnerable.

In tissue culture, the time course of animal virus replication is usually much longer (hours or days) than that of bacteriophages (typically less than an hour under optimal conditions). The burst size, however, of animal viruses is typically several orders of magnitude larger than that of phages. The reason for the larger burst size is probably that the volume of the eukaryotic host cell is usually much larger than that of a bacterial host, thus providing a larger supply of materials to build virions.

Not all animal viruses exhibit growth that can be represented by a step curve. Some species, called **slow viruses**, bud off virions relatively slowly, without immediately lysing the host cell. A well-known family of slow viruses is the **lentiviruses** (literally, "slow viruses"), a group of retroviruses that includes HIV. HIV and related retroviruses are known for their long incubation periods, in some cases many years, during which time extremely low concentrations of virions are produced by infected cells.

Plaque Isolation and Assay of Bacteriophages

For the investigation of cellular microbes, an important tool is the culturing of individual colonies on a solid substrate that prevents dispersal throughout the medium, as described in Chapters 1 and 4. Plate culture of colonies enables us to isolate a population of microbes descended from a common progenitor. But viruses cannot be isolated as "colonies." The reason is that although viruses can be obtained at incredibly high concentrations, they disperse in suspension. Even on a solid medium, viruses never form a solid visible mass comparable to the mass of cells that constitutes a cellular colony.

In viral plate culture, viruses from a single progenitor lyse their surrounding host cells, creating a clear area called a plaque. The **plaque assay** for lytic bacteriophages was invented early in the twentieth century by the French microbiologist Felix d'Herelle, who was the first to recognize that bacterial cells can be infected by bacteriophages.

To perform a plaque assay of bacteriophages, a diluted suspension of bacteriophages is mixed with bacterial cells in soft agar, and the mixture is then poured over a nutrient agar plate (**Fig. 6.31**). If no bacteriophages are present, the bacteria grow homogeneously as an opaque sheet over the surface (confluent growth). The presence of a phage is detected as a plaque, a clear circle seemingly cut out of the bacterial growth. A plaque is formed when a single virus infects a cell, replicates, and spreads progeny phages to adjacent cells, killing them as well (**Fig. 6.32**). In contrast to the clear plaques produced by lytic phages (as shown in the figure), temperate (lysogenic) phages make cloudy plaques containing lysogenized viable cells. The prophage in the host genome protects the cells from subsequent infection and lysis by another phage.

Plaques offer a convenient way to isolate a recombinant DNA molecule contained in a bacteriophage vector. In **Figure 6.32B**, the blue plaques result from the phage vector, a derivative of phage M13 that carries a gene encoding the enzyme β-galactosidase. This enzyme converts a colorless compound into a blue dye. When the indicator gene is interrupted by an inserted recombinant gene, the phage produces white plaques, which indicate the successful production of recombinant DNA phages.

Plaques can be counted and used to calculate the concentration of phage particles, or **plaque-forming units (PFUs)**, in a given suspension of liquid culture. The liquid culture can be analyzed by serial dilution in the same way one would analyze a suspension of bacteria.

Plaque Isolation and Assay of Animal Viruses

For animal viruses, the plaque assay has to be modified because it requires infection of cells in tissue culture. Tissue culture usually involves growth of cells in a monolayer on the surface of a dish containing fluid medium, which would quickly disperse any viruses released by lysed cells. To solve this problem, in 1952 Renato Dulbecco, at the California Institute of Technology, modified the tissue culture procedure for plaque assays (**Fig. 6.33A**). In Dulbecco's method, the tissue culture with liquid medium is first inoculated with virus. After sufficient time to allow for viral attachment to cells, the fluid is removed and replaced by a gel medium. The gel retards dispersal of viruses from infected cells, and as the host cells die, plaques can be observed. **Figure 6.33B** shows a plate culture of adenovirus.

Animal viruses that do not kill their host cells require a different kind of assay based on identification of a **focus** (plural, **foci**), a group of cells infected by the virus. In a **fluorescent-focus assay**, the infected cells under the gel overlay are incubated for a sufficient period to allow production of progeny virions. The plasma membranes of the cells are then made permeable by treatment with an organic solvent, and an antiviral antibody is added. Unattached antibodies are then washed away, and a second antibody is added that recognizes the first antibody molecule. The second antibody is conjugated to a fluorophore whose fluorescence reveals

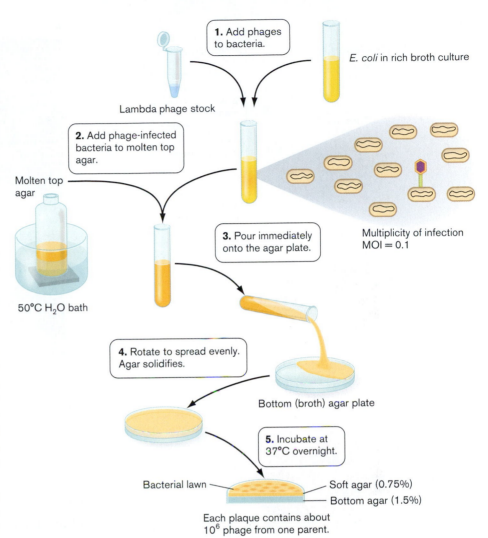

1. Add phages to bacteria.

E. coli in rich broth culture

Lambda phage stock

2. Add phage-infected bacteria to molten top agar.

Molten top agar

50°C H₂O bath

3. Pour immediately onto the agar plate.

Multiplicity of infection MOI = 0.1

4. Rotate to spread evenly. Agar solidifies.

Bottom (broth) agar plate

5. Incubate at 37°C overnight.

Bacterial lawn — Soft agar (0.75%) — Bottom agar (1.5%)

Each plaque contains about 10⁶ phage from one parent.

Figure 6.31   Plating a phage suspension to count isolated plaques.   A suspension of bacteria in rich broth culture is inoculated with a low proportion of phage particles (multiplicity of infection is approximately 0.1). This means that only a few of the bacteria become infected immediately, while the rest continue to grow. Each plaque arises from a single infected bacterium that bursts, its phage particles diffusing to infect neighboring cells.

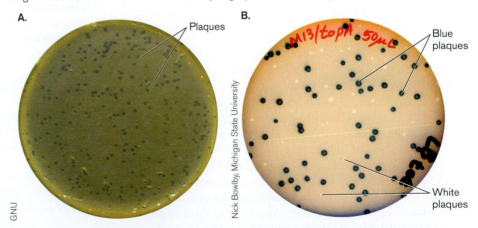

A.   Plaques

GNU

B.   M13/topA 50μC   Blue plaques

Nick Bowlby, Michigan State University

White plaques

Figure 6.32   Phage plaques on a lawn of bacteria.   A. Phage lambda plaques on a lawn of *Escherichia coli* K-12. **B.** Plaques of recombinant phage M13 on *E. coli*. The original phage expresses β-galactosidase, an enzyme that makes a blue product (blue plaques). White plaques are produced by phage particles whose genome is recombinant (contains a cloned gene interrupting the gene for β-galactosidase).

A.

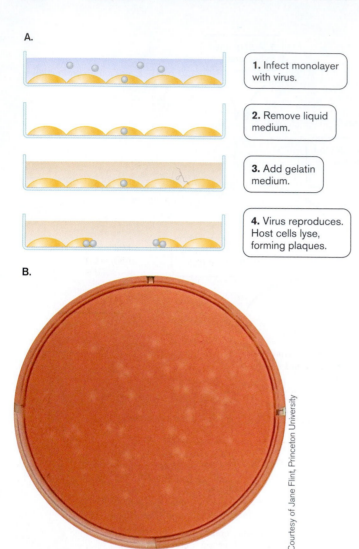

1. Infect monolayer with virus.

2. Remove liquid medium.

3. Add gelatin medium.

4. Virus reproduces. Host cells lyse, forming plaques.

B.

Courtesy of Jane Flint, Princeton University

Figure 6.33   Plate culture of animal viruses.   A. Modified plaque assay for animal viruses. The gelled medium retards dispersal of progeny virions from infected cells, restricting new infections to neighbor cells. The result is a visible clearing of cells (a plaque) in the monolayer. **B.** Plaque assay in which a serial dilution of adenovirus suspension was plated on a monolayer of cells in tissue culture.

A.

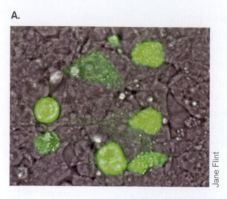

Jane Flint

B.

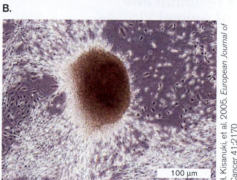

100 μm

H. Kisanuki, et al. 2005. *European Journal of Cancer* 41:2170

Figure 6.34   Focus assays of animal viruses.   A. GFP fluorescent-focus assay. **B.** Transformed-focus (cancer-forming) assay of an oncogenic virus. Cells infected by an oncogenic virus have produced three foci of transformed cells. Transformed cells grow in an uncontrolled manner, which can produce a tumor.

TO SUMMARIZE:

- **Culturing viruses** requires growth in host cells. Bacteriophages may be cultured either in batch culture or as isolated plaques on a bacterial lawn.
- **Batch culture** of viruses generates a step curve.
- **Animal viruses** are cultured within animals or as plaques in a tissue culture.
- **Fluorescent-focus assays** reveal foci of virus-infected cells.
- **Oncogenic viruses** are cultured as foci of cancer-transformed host cells.

6.7   Viral Ecology

Viruses exist naturally within host organisms in complex ecosystems. Viral persistence in natural ecosystems is significant for human health and for agricultural plants and animals. Persistence in wild populations provides a reservoir of infection that hinders eradication of a pathogen, especially when it is associated with an insect vector (carrier organism) such as ticks or mosquitoes.

Environmental Change Leads to Emergence of Viral Pathogens

Changing distribution patterns of insect vectors and animal hosts can generate new epidemics of a pathogen in

foci of cells, each of which has arisen from a single virion in the original inoculum.

Another kind of fluorescent-focus assay uses green fluorescent protein (GFP). The viruses are genetically engineered to contain a GFP gene fusion, a labeling technique discussed in Chapter 3. The GFP fluorescence reveals foci of virus-infected cells (**Fig. 6.34A**).

A focus assay can also be used to isolate oncogenic viruses. Oncogenic, or cancer-causing, viruses actually **transform** their host cells into cancer cells. Cancer cells lose contact inhibition; they grow up in piles instead of remaining in the normal monolayer. These piles of transformed cells, or transformed foci, can easily be visualized and counted. This is known as the **transformed-focus assay** (**Fig. 6.34B**).

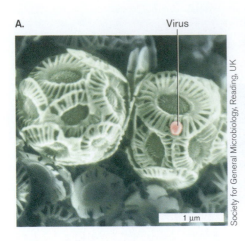

A.

Virus

1 μm

Society for General Microbiology, Reading, UK

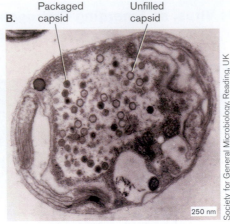

B.

Packaged capsid

Unfilled capsid

250 nm

Society for General Microbiology, Reading, UK

Figure 6.35   Viruses infect algae.
A. A virus attaches to the surface of a marine phytoplankton, *Emiliana huxleyi* (SEM). **B.** Progeny virions assembling within the phytoplankton *Pavlova vivescens*.

regions where the virus could not spread before. Such changes in distribution can be brought about by many factors, including global climate change (**Special Topic 6.1**).

Some emerging viruses arise as variants of endemic milder pathogens. Viruses long associated with a host, such as the common cold viruses (rhinoviruses), tend to have evolved a moderate disease state that provides ample opportunities for host transmission. A virus that "jumps" from an animal host, however, may cause a more acute syndrome with higher mortality. The best-known case is the exceptionally virulent emerging strains of influenza, which generally result from intracellular recombination of human strains with strains from pigs or ducks (discussed in Chapter 11). Wild animals consumed for food introduce entirely new species of viruses to human populations. For example, HIV is believed to have emerged from apes whose meat was consumed by humans.

Viruses Have Important Roles in Ecosystems

Viruses fill important niches in all ecosystems. Two important roles of viruses are:

- **Limiting host population density.** An increase in host population density increases the rate of transmission of viral pathogens. As the host population declines, natural selection results in individuals that are less susceptible. Thus, viruses can limit host density without extinction of the host. An example is the myxovirus introduced into Australia to curb the population of rabbits (which had been previously introduced by British colonists). The myxovirus ultimately maintained the rabbits at a reduced density, less destructive to the ecosystem.

- **Selecting for host diversity.** Each viral species has a limited host range and requires a critical population density to sustain the chain of infection. The virus limits its host species to a population density far lower than that sustainable by the available resources, thus preventing the dominance of any one species. In this way, the presence of viruses fosters the evolution of many distinct host species. The role of viruses

in diversity is exemplified in marine phytoplankton, which are highly susceptible to viruses and display an exceptional range of species diversity.

In the oceans, viruses are the most numerous and genetically diverse forms of life. The number of bacteriophages and algal viruses can reach 10^7 (10 million) per milliliter. **Figure 6.35** shows examples of marine viruses infecting phytoplankton such as the algae *Emiliana huxleyi* and *Pavlova vivescens*. When marine algae overgrow, generating an algal bloom, viruses play a decisive role in controlling the bloom, such as the overgrowth of *Emiliana huxleyi* (**Figure 6.36**). Consumer organisms apparently cannot grow fast enough to control such blooms, but

Plymouth

Society for General Microbiology, Reading, UK

Figure 6.36   Viral control of marine algal blooms.
Phytoplankton bloom of *E. huxleyi* in the English Channel, satellite image, July 30, 1999. Algal populations are generally controlled by viruses. *Source:* From the Plymouth Marine Laboratory Remote Sensing Group, based on data from the NERC Dundee Satellite Receiving Station.

Special Topic 6.1   West Nile Virus, an Emerging Pathogen

In the spring of 1999, New Yorkers noticed an unusual number of dead crows and other local birds. Then people began falling ill with an unfamiliar form of encephalitis (brain inflammation) caused by an avian virus never before seen in North America. The virus, a (+) single-stranded RNA virus, comes from the flavivirus family, which includes yellow fever and dengue (breakbone fever). These pathogens are all endemic to warmer environments, such as the Mediterranean and North Africa—hence the name, West Nile virus. Like yellow fever, a disease made infamous by its devastation of the southern United States in the nineteenth century, West Nile virus is transmitted between humans and animals by mosquitoes (**Fig. 1**).

Why the sudden appearance of West Nile virus in the United States? The answer is not known for certain, but some climatologists argue that West Nile and other emerging pathogens arise from unusual weather patterns related to global climate change (**Fig. 2**). West Nile virus is generally found in species of mosquitoes that fare poorly in the cold of winter. In 1998–1999, however, the unusually warm winter enabled a few mosquitoes carrying the virus to survive until the spring. The spring that followed was drier than usual, forcing birds to spend more time at dwindling water pools, where mosquitoes were concentrated. The July heat wave then increased the rate of viral replication in the mosquitoes, which reached unusual population densities later that summer. As the cycle continued, mosquitoes reinfected birds, which then infected more mosquitoes. Eventually, the cycle included mosquitoes infecting humans.

As of this writing, West Nile virus has spread to most of the United States and appears to be endemic to North America. Meanwhile, research continues to determine whether clearer links can be established between climate changes and emerging pathogens.

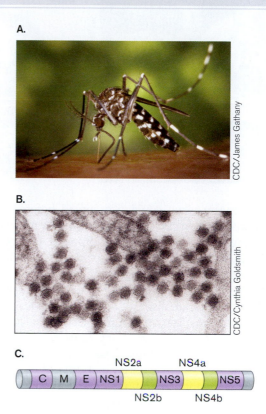

Figure 1   West Nile virus.   A. The mosquito carries West Nile virus. **B.** West Nile virus particles (50 nm in diameter) forming within an infected cell (TEM). **C.** Like hepatitis C, West Nile virus is a flavivirus, with a (+) strand RNA of 10,000 nucleotides. The genome encodes ten proteins, three of which are structural proteins (C, M, and E) and seven of which are nonstructural proteins (NS1, NS2a, NS2b, NS3, NS4a, NS4b, and NS5).

---

viruses spread rapidly through the population. By lysing the algae as they grow, viruses return algal carbon and minerals to the surface water before the algae starve to death and their bodies sink to less productive depths.

On a global scale, marine viruses play a significant role in the carbon balance. Models of ocean carbon flux typically emphasize the consumption of phytoplankton by grazers and the consumption of grazers by carnivores

(**Fig. 6.37**). At each level, some of the carbon fixed as biomass sinks to the ocean floor, removed from the carbon cycle. But at each level, viral infection and lysis converts the bodies of phytoplankton, grazers, and carnivores into detritus consisting of small organic particles. The organic material is consumed by bacteria, whose rapid respiration releases much CO_2. Thus, carbon is diverted from the ocean sink into CO_2 returning to the atmo-

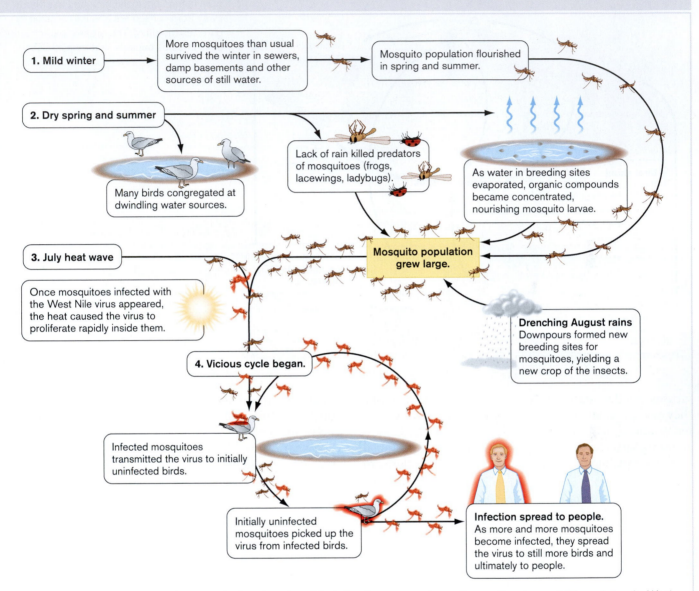

Figure 2   An emerging pathogen.   In the changing global climate, an emerging pathogen threatens wildlife and people. West Nile virus, originally found in Africa and the Mediterranean, is now spreading through the Western Hemisphere, probably owing to an increase in warm weather. The spread of the virus involves transfer between mosquitoes and avian, animal, or human hosts.
Source: Adapted from P. Epstein. 2000. *Scientific American,* August: 36–43.

sphere—a factor that must be considered in models of global warming.

TO SUMMARIZE:

■ **Environmental change** results in new emerging viral pathogens.
■ **Host population density** is limited by virus infection, and host genetic diversity is increased.

■ **Marine viruses** are the major consumers of phytoplankton. Viral activity substatially impacts the global carbon balance.

Concluding Thoughts

In this chapter, we have covered general approaches to studying the structure and function of all kinds of

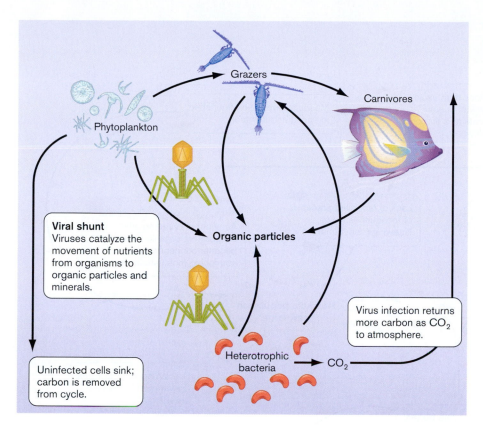

Figure 6.37   Marine viruses recycle carbon.   Viruses divert the flow of carbon from marine phytoplankton and bacteria away from consumers and into organic particles (detritus). The organic particles are consumed by bacteria that respire CO_2.

viruses. As devastating as viruses can be to their hosts, they also provide engines of genomic change (see Chapter 9) and ecological balance (see Chapter 21). Understanding virus function prepares us to discuss microbial genetics in Chapters 7–10. Chapter 11 then explores in depth the molecular basis of viral genomes and reproduction. Molecular virology is a field of growing importance for medical and agricultural research, for genetic engineering and gene therapy, and for understanding natural ecosystems.

CHAPTER REVIEW

Review Questions

1. Compare and contrast the form of icosahedral and filamentous (helical) viruses, citing specific examples.
2. How do viral genomes gain entry into cells in bacteria, plants, and animals?
3. Explain the structure and function of the seven Baltimore groups of viral genomes.
4. How do viral genomes interact with host genomes, and what are the consequences for host evolution?
5. Compare and contrast the lytic, lysogenic, and slow-release life cycles of bacteriophages. What are the strengths and limitations of each?
6. Compare and contrast the life cycles of RNA viruses and DNA viruses in animal hosts. What are the strengths and limitations of each?
7. Explain the plate titer procedure for enumerating viable bacteriophages. How must this procedure be modified to titer animal viruses? Oncogenic viruses?
8. Explain the generation of the step curve of virus proliferation. Why is virus proliferation generally observed as a single step, or generation, in contrast to the life cycles of cellular microbes, outlined in Chapter 4?
9. Explain the key contributions of viruses to natural ecosystems. What may happen in an ecosystem where viruses are absent or fail to cause significant infection?

Key Terms

accessory protein (189)
bacteriophage (phage) (183)
burst (200)
burst size (200, 209)
capsid (183)
cell surface receptor (198)
cloning vector (184)
DNA reverse-transcribing virus (193)
double-stranded DNA virus (192)
double-stranded RNA virus (192)
eclipse period (209)
endocytosis (203)
envelope (187)
filamentous virus (189)
fluorescent-focus assay (211)
focus (211)
hammerhead ribozyme (185)
host range (183)
icosahedral (187)
latent period (209)
lentivirus (210)
lysate (209)
lysogeny (200)

mechanical transmission (206)
(−) sense single-stranded RNA virus (192)
movement protein (207)
oncogene (206)
oncogenic virus (206)
ortholog (196)
orthologous gene (196)
pararetrovirus (193)
plaque (183)
plaque assay (210)
plaque-forming unit (PFU) (211)
plasmodesma (207)
(+) sense single-stranded RNA virus (192)
prion (185)
prophage (200)
proteome (197)
proteomics (197)
reading frame (184)
retrovirus (192)
reverse transcriptase (192)
ribozyme (185)

rise period (209)
RNA reverse-transcribing virus (192)
RNA-dependent RNA polymerase (192)
segmented genome (191)
single-stranded DNA virus (192)
site-specific recombination (200)
slow-release cycle (200)
slow virus (210)
spike protein (189)
tegument (189)
temperate phage (200)
transduction (200)
transform (212)
transformed-focus assay (212)
tropism (201)
uncoating (203)
viral envelope (189)
virion (182)
viroid (185)
virulence (209)
virulent (200)
virus (182)

Recommended Reading

Baltimore, David. 1971. Expression of animal virus genomes. *Bacteriology Reviews* **35**:235–241.

Edwards, Robert A., and Forest Rohwer. 2005. Viral metagenomics. *Nature Reviews Microbiology* **3**:504–510.

Flint, S. Jane, Lynn W. Enquist, Vincent R. Racaniello, and A. M. Skalka. 2003. *Principles of Virology: Molecular Biology, Pathogenesis, and Control of Animal Viruses.* 2nd ed. ASM Press, Washington, DC.

Gromeier, Matthias. 2002. Viruses for treating cancer. *ASM News* **68**:438.

Grünewald, Kay, Prashant Desai, Dennis C. Winkler, J. Bernard Heymann, David M. Belnap, et al. 2003. Three-dimensional structure of Herpes Simplex Virus from cryo-electron tomography. *Science* **302**:1396–1398.

Hull, Roger. 2001. *Matthews' Plant Virology.* Academic Press, New York.

Moineau, Sylvain, et al. 2002. Phages of lactic acid bacteria: From genomics to industrial applications. *ASM News* **68**:388–391.

Ptashne, Mark. 2004. *Genetic Switch: Phage Lambda Revisited.* Cold Spring Harbor Laboratory Press, Cold Spring Harbor, NY.

Raoult, Didier, Stéphane Audic, Catherine Robert, Chantel Abergel, and Patricia Renesto. 2004. The 1.2-megabase genome sequence of Mimivirus, *Science* **306**:1344–1350.

Suttle, Curtis A. 2005. Viruses in the sea. *Nature* **437**:356–361.

Suttle, Curtis A. 2007. Marine viruses—major players in the global ecosystem. *Nature Reviews Microbiology* **5**:801–812.

Wilson, Willie. 2002. Giant algal viruses: Lubricating the great engines of planetary control. *Microbiology Today* **29**(November):180.

Part 2

Genes and Genomes

Courtesy of Richard Losick

AN INTERVIEW WITH

RICHARD LOSICK: THE THRILL OF DISCOVERY IN MOLECULAR MICROBIOLOGY

Richard Losick has taught at Harvard for more than three decades, and his creative teaching was honored when he was named a Howard Hughes Medical Institute (HHMI) Professor. A member of the National Academy of Sciences, Losick has authored more than 200 research articles on bacterial genetics and cell biology. He discovered how *Bacillus* sporulation is controlled by alternative sigma factors that regulate transcription. To dissect sporulation and the cell division cycle, he uses fluorescent proteins to visualize individual subcellular parts of bacteria.

Richard Losick, Harvard College Professor. Losick has made major discoveries on the molecular and cellular biology of *Bacillus subtilis*.

Why did you decide to make a career in microbiology? Did you begin research as an undergraduate?

When I was in grade school, I used to read science books and do experiments on my own. I thought that school was boring, and I didn't realize that science was a subject that I would learn about later in school!

When I was an undergraduate at Princeton, I joined the laboratory of a newly arrived professor, Charles Gilvarg, who studied amino acid biosynthesis in bacteria. I fell in love with bacteria from that experience, and I have remained fascinated by bacteria ever since.

Why do you study *Bacillus subtilis*? How is your work relevant to the insecticide *B. thuringiensis* and the human pathogen *B. anthracis*?

Bacillus subtilis is extraordinarily rich in its biology. Its ability to metamorphose into a spore is endlessly fascinating. *B. subtilis* has the capacity to differentiate into specialized cells for DNA uptake (competence), to swim, to swarm on surfaces, to cannibalize sibling cells, and to form architecturally elaborate, multicellular communities. At the same time, its property of genetic competence (gene transfer between cells) makes it exceptionally facile for molecular genetic manipulation.

B. thuringiensis and *B. anthracis* are less favorable than *B. subtilis* for genetic manipulation, but the main features of spore formation are very similar in all three organisms. The crystal toxin of *B. thuringiensis* that has useful insecticidal properties is produced under sporulation control. The remarkable environmental resistance properties of *B. subtilis* spores are shared with its evil cousin *B. anthracis* and help us understand the challenge posed by the bioterrorism agent.

Your career has spanned a generation of revolutionary developments in genetics and molecular biology, such as the discovery of the sigma factors that regulate transcription of many genes. Would you describe how one of these developments came about?

After winning my PhD from MIT, I came to Harvard to work on a project involving the effect of a virus on bacterial cell membranes. The sigma subunit of *E. coli* RNA polymerase had just been discovered (by Dick Burgess and Andrew Travers at Harvard and by John Dunn and Ekkehard Bautz at Rutgers), and people were excited about the possible existence of alternative sigma factors (proteins that regulate large groups of genes). Meanwhile, A. L. Sonenshein was studying how certain bacterial viruses (bacteriophages) get trapped during spore formation by *Bacillus subtilis*. The phage is virulent but is shut off transcriptionally upon entering a sporulating cell. The phages are unable to grow in cells that have started to sporulate and instead become trapped in what will become the spore. So the poor spore, after waiting for long periods of time to germinate, gets killed upon germination by the enemy within.

Putting two and two together, we reasoned that maybe changes in the RNA polymerase during entry into sporulation could help explain the shutdown in viral gene expression. We got so excited about this idea that we began to work on RNA polymerase full time. This eventually led to the discovery that sporulation is governed by a cascade of bacterial sigma factors.

Meanwhile, in collaboration with Jan Pero, we found that phage SP01 encodes and produces two alternative sigma factors upon infecting growing cells. These regulatory proteins, the first examples of alternative sigmas, drive the phage program of gene transcription. Since then, alternative sigmas have emerged as a widespread

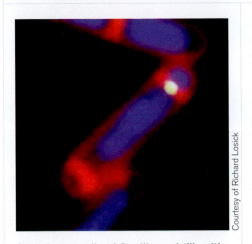

Sporulating cells of *Bacillus subtilis* with molecular parts labeled by fluorophores. Red fluorescence (FM4-64) labels the cell membrane; blue (DAPI) the DNA; and green (green fluorescent protein) a DNA pump that translocates a chromosome across the division septum from the mother cell (the big cell) into the forespore (the small cell on the upper right). (Cell length, 2–3 μm.)

Courtesy of Richard Losick

theme in the control of gene sets in many different kinds of bacteria.

What is the significance of sporulation as a developmental process?

Sporulation is an extremely attractive system for understanding cellular differentiation because it is both complicated enough to mimic features of development in higher organisms and accessible enough to be subject to genetic, biochemical, and cytological manipulation. Sporulating cells undergo a true process of cellular differentiation, in which the two cellular compartments of the sporangium communicate with each other in a back-and-forth manner by intricate signaling systems.

How was the technology developed to visualize chromosome dynamics in *Bacillus*?

In the old days, bacteria were viewed as an amorphous vessel with enzymes floating around inside. Now we understand that bacteria are highly organized, that proteins often have subcellular addresses in bacteria; thus, cytology has become an important aspect of bacterial research.

One of the keys to this revolution was the introduction of green fluorescent protein (GFP) as a tool for visualizing the subcellular localization of proteins in cells. Andrew Murray and Andrew Belmont figured out that GFP could also be harnessed to visualize the location of sites on DNA by attaching the fluorescent protein to a protein that binds to specific sequences in DNA. Murray and Belmont used this trick to visualize centromeres in yeast. Andrew Wright and I extended this methodology to visualizing sites on chromosomes in bacteria. Traditional thinking was that origins tended to remain in cell middles, but we found that replication origins moved toward the cell poles during chromosome segregation, and with GFP we could visualize this in living cells.

How do students get involved with your research? Do you work with undergraduates as well as graduate students?

Yes, I always have undergraduates as well as graduate students in my lab, and both have contributed to some of our most significant findings. One special undergraduate was Aurelio Teleman, who was both a brilliant student and a gifted investigator. Working with a graduate student, Teleman helped create and exploit the system for visualizing the movement of a specific site in the bacterial chromosome that had been tagged with tandem lactose operon operators. He is now at the European Molecular Biology Laboratory.

As an educator, what innovations have you developed?

I enjoy the challenge of explaining complicated concepts and of making large-class instruction lively and interactive. I run a program for placing freshmen from disadvantaged backgrounds in laboratories of faculty for long-term research mentoring. Such students are often at greatest risk for dropping out of science, but a long-term relationship with a laboratory that begins early during the college years can keep these students (indeed all students) excited about science throughout college and beyond.

I believe that a positive experience in experimental work can represent one of the most memorable aspects of undergraduate education. And its benefit is not limited to future scientists. A science major with an associated experience in discovery research is excellent preparation for a wide variety of careers, including medicine, business, law, public policy, journalism, and education.

What advice do you have for today's students?

Join a lab! In science you can learn something that nobody else on Earth ever knew before. There's a special thrill about learning how living things function—the inner workings of living things. If you learn some new aspect of it, even if it's a tiny part, you've advanced knowledge for the whole of humankind.

Losick teaching undergraduates at Harvard. Harvard Gazette staff photo by Kris Snibbe.

Photo by Kris Snibbe

Chapter 7

Genomes and Chromosomes

A genome is all the genetic information that defines a species. We have known that chromosomes consist of DNA for over a hundred years, yet we have only recently been able to sequence complete genomes. Now we are in the postgenomic era in which biological research is driven, in large part, by knowledge of complete genome sequences.

Microbial genomes consist of one or more chromosomes made of DNA. Many bacteria have a single, circular chromosome, but some species have multiple chromosomes or even a mixture of linear and circular chromosomes. For example, the genome of *Sinorhizobium melliloti*, a major contributor to nitrogen fixation, consists of three circular chromosomes.

This chapter addresses basic as well as some advanced issues of DNA and genomes. What is DNA? How is it packaged in the cell? Because DNA is essential to life, nature must have devised a remarkable machine to replicate it. But how does this machine work? And do different species use different machines? Microbiology has played a leading role in answering these difficult questions—opening the door to post-genomic studies of all living things.

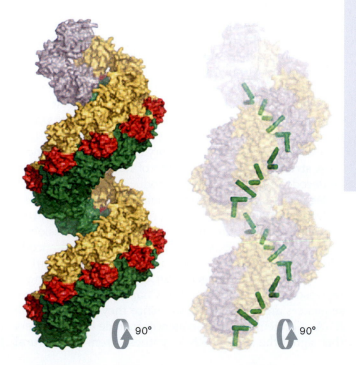

All cells accurately time the initiation of chromosome replication to cell division. In bacteria, DnaA protein accomplishes this by accumulating during growth and binding to the origin of replication, forming a right-handed helical filament. The figure shows four DnaA tetramers bound to origin DNA. On the left, the initial monomer of is colored grey to distinguish a single subunit of the filament. Gold, green and red mark different domains of subsequent monomers. The image on the right uses different colors to show individual monomers. The torsional strain caused by the filament may destabilize the adjacent DNA to allow assembly of the replication machine.
Source: Melissa L. Mott and James M. Berger. 2007. Nature Reviews Microbiology 5:343–354.

221

What makes each species unique is its genome, the sum total of its genes. The human genome is also supplemented by the genomes of all the microbes in the digestive tract. Collectively, intestinal microbes contain 100 times more genes than the human genome and provide key metabolic capabilities that humans lack, such as digesting complex plant materials. In the mid-twentieth century, one intestinal bacterium, *Escherichia coli*, became the focus of efforts to understand genes and genetics. The choice of *E. coli* was fortunate, as its genetics proved to be especially pliable.

In one remarkable experiment, a cell of *E. coli* was lysed, releasing its chromosome for electron microscopy. What spewed out of this single cell was a strand of DNA 1,500 times longer than *E. coli* itself. Scientists wondered how this enormous molecule could fit into a single cell, and how an enzyme could duplicate that much DNA (over 4.6 million base pairs) without the whole DNA molecule getting tangled up (**Fig. 7.1**). And how did it manage to do this in under 20 minutes, the doubling time of the organism? Half a century later, those questions continue to intrigue us.

Chapter 7 explores the nature of DNA, the structure of prokaryotic genomes, and the mechanisms used for their replication. We also examine how microbial enzymes that manipulate DNA were used to develop the fundamental techniques of present-day biotechnology, such as DNA restriction analysis, DNA sequencing, and the polymerase chain reaction (PCR). The expression of genes to make RNA and protein is discussed in Chapter 8. The remaining chapters of Part 2 present gene recombination and transmission (Chapter 9), the regulation of gene expression (Chapter 10), the specialized genetic mechanisms of viruses (Chapter 11), and the use of microbial genetic tools for research and biotechnology (Chapter 12). From the advances of the last 50 years, a new molecular and genomic perspective of biology has emerged.

7.1 DNA: The Genetic Material

The ability of plants and animals to transfer genetic traits was known well before we knew that DNA was the carrier of genetic information. Early researchers understood the Mendelian model of gene inheritance and how it dictated the passage of traits from parent to offspring in eukaryotes. This type of gene transfer is known as **vertical transmission**, as from parent to child. Bacteria, however, appeared to operate differently, because in addition to vertical transmission, they were able to conduct **horizontal transmission**—the transfer of small pieces of genetic information from one cell into another. The study of horizontal gene transfer in bacteria ultimately led to the discovery of horizontal transfer throughout animals and plants, albeit on a slower time scale because of their longer generation times. Mechanisms of horizontal transmission are discussed in Chapter 9.

Our path to understanding DNA began in 1928 with the discovery by Fred Griffith (1881–1941) that some unknown substance from dead cells of a virulent strain of *Streptococcus pneumoniae* could transfer virulence to living but otherwise harmless strains when both strains were coinjected into mice. This form of horizontal gene transfer, called transformation, led to the discovery in 1944 that genetic information is embedded in the base sequence of deoxyribonucleic acid (DNA) (see Section 1.6). A more contemporary example of transformation is the toxic shock gene that can be moved horizontally among strains of *Staphylococcus aureus*. As a result of these and many other experiments, we now appreciate that chromosomes are made of contiguous packets of information, called genes.

Genes are units of information composed of a sequence of DNA nucleotides of four different types, adenine (A), guanine (G), thymine (T), and cytosine (C). (For review at the level of introductory biology, see Appendix 1.) A **structural gene** is a string of nucleotides that can be decoded by an enzyme to produce a functional RNA molecule. A structural gene usually produces an RNA molecule that in turn encodes a protein. A **DNA control sequence**, on the other hand, regulates the expression of a structural gene. DNA control sequences do not encode RNA but regulate the expression (RNA production) of an adjacent structural gene, for example, by binding a repressor protein (discussed in Chapter 8). DNA control sequences are not really genes, since they do not encode RNA or protein. Control sequences include the promoters that signal the start of RNA synthesis from a structural

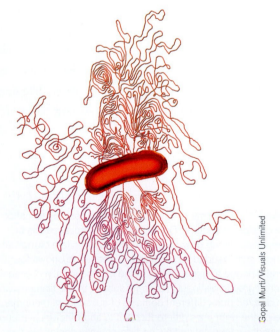

Figure 7.1 Osmotically disrupted bacterial cell with its DNA released.

Gopal Murti/Visuals Unlimited

gene and the binding sites for regulatory proteins that can activate or inactivate that promoter.

As noted previously, the entire genetic complement of DNA in a cell that defines it as a species is called its genome. Our primary goal in this chapter is to convey what is known about how genomes are maintained and replicated. Information about how computers are used to mine the informational content of DNA sequences and genes is described more fully in Chapters 8 and 10.

Bacterial Genomes Can Be Circular or Linear

In the early twentieth century, the chromosomes of bacteria, unlike those of eukaryotes, could not be observed by light microscopy. The reason for this is that bacteria, unlike eukaryotes, do not undergo mitosis, a process in which chromosomes are condensed and thickened about 1,000-fold. Important clues to bacterial chromosome structure were, however, gleaned from painstaking genetic studies. In the 1950s, it was discovered that **conjugation**, a horizontal gene transfer mechanism requiring cell-to-cell contact, could transfer large segments of the bacterial chromosome—not all at once, as in the established Mendelian model of plants and animals, but sequentially over a period of time (it takes 100 minutes to move the entire *E. coli* chromosome from one cell to another).

Thus, even though the bacterial chromosome could not be seen, conjugation allowed genes to be mapped relative to each other based on time of transfer. For example, a donor strain that can synthesize the amino acids alanine and proline can directly transfer the responsible genes to a recipient cell defective in those genes. The transfer process is nonspecific; any gene can be transferred in this way. Completion of transfer also requires **recombination**, in which the donor DNA fragment replaces the recipient DNA fragment. Successful transfer of the genes for amino acid synthesis enables the formerly defective recipient to form colonies on minimal media lacking either amino acid. But because transfer occurs over time from a fixed starting point (that is, not all genes are transferred at the same time), it might take one gene 10 minutes of cell contact to be transferred while the second gene, farther away from the starting point of DNA transfer, takes an additional 20 minutes.

Knowing that eukaryotic chromosomes are linear, scientists initially expected that bacterial chromosomes would be linear, too. However, the early genetic maps drawn from conjugation experiments just would not fit together in a manner consistent with a linear model—for the simple reason that the bacterial chromosome in *E. coli*, the organism under study at the time, is circular. We now know that many microbes have circular chromosomes, although some, like the Lyme disease agent *Borrelia burgdorferi*, have linear chromosomes and others,

like *Agrobacterium tumefaciens*, have a mixture of circular and linear chromosomes (**Table 7.1**).

It is instructive to compare genome sizes across the phylogenetic tree, from viruses to humans. The size range is enormous. In general, the simpler the organism, the smaller its genome. At some point, as DNA content is trimmed, the organism loses independence and *must* parasitize another organism. This begs the question, what constitutes a "minimal" genome? How "lean" can a chromosome become and still support independent growth?

TO SUMMARIZE:

- **A genome** is all the genetic information that defines a species.
- **Prokaryotic genomes** are made up of chromosomes and plasmids made of DNA.
- **Prokaryotic chromosomes** can be circular or linear, as can plasmids.
- **Functional units** of DNA sequences include structural genes and regulatory sequences.

7.2 Genome Organization

New techniques for constructing physical maps of genomes and determining the sequences of whole genomes have revealed tremendous diversity in the size and organization of prokaryotic genomes.

Genomes Vary in Size

Bacterial chromosomes range from 580 kilobase pairs (kbp) to 9,400 kbp, and archaeal chromosomes range from 935 kbp to 6,500 kbp. For comparison, eukaryotic chromosomes range from 2,900 kbp (microsporidia) to over 4,000,000 kbp (humans).

> **NOTE:** The designation kbp refers to the length of a double-stranded DNA molecule. A 1,000-base-pair double-stranded genome is 1 kbp in size. In contrast, a 1,000-*base* single-stranded genome (as in some viruses) is referred to as 1 kb.

The smallest cellular genomes identified thus far are those of *Mycoplasma*. These pathogens rely on their host environment for many products but can still grow outside a host cell. The complete genome of *Mycoplasma genitalium* consists of only 580 kbp and encodes 480 proteins. *M. genitalium* lacks the genes required for many biosynthetic functions. These organisms cannot synthesize amino acids or construct cell walls, and they do not have a functional tricarboxylic acid (TCA) cycle. In contrast, free-living bacteria that can grow in soil have larger genomes and dedicate many genes

Table 7.1   Genomes of representative bacteria and archaea.

Species (strain)	Genomic chromosome(s)* (kilobase pairs, kbp)	Plasmid(s)* (kbp)	Total (kbp)
Bacteria			
Mycobacterium tuberculosis Tuberculosis	4,400		4,400
Mycoplasma genitalium Normal flora, human skin	580		580
Burkholderia cepacia	3,870 + 3,217 + 876	93	8,056
Escherichia coli K-12 (W3110) Model strain for *E. coli* proteomics	4,600		4,600
Anabaena species (PCC 7120) Cyanobacteria: Major photosynthetic producer of carbon source for aquatic ecosystems	6,370	110 + 190 + 410	7,080
Borrelia burgdorferi Lyme disease	911	21 plasmids with sizes between 9–58	>1,250
Agrobacterium tumefaciens Tumors in plants; genetic engineering vector	2,840 + 2,070	214 + 542	5,666
Archaea			
Methanocaldococcus jannaschii Methanogen from thermal vent	1,660	16 + 58	1,734
Haloarcula marismortui Halophile from volcanic vent	3,130 + 288	33 + 33 + 39 + 50 + 155 + 132 + 410	4,270

1,000 kbp   500 kbp

*Colored schematics indicate relative sizes of genomic elements and whether these are circular or linear. Size bars are provided under each column.

to the synthesis or acquisition of amino acids or TCA cycle intermediates. Even different strains of one species, such as *Salmonella enterica*, may vary considerably in gene distribution (**Fig. 7.2**).

Another feature that distinguishes prokaryotic and eukaryotic genomes is the amount of so-called noncoding DNA. Many, but not all, eukaryotes contain huge amounts of noncoding DNA scattered between genes. In some species (such as humans), over 90% of the total DNA is noncoding. These noncoding regions include enhancer sequences needed to drive transcription of eukaryotic promoters and DNA expanses that separate enhancers. Enhancer sequences can function at large distances from the gene they regulate. A **promoter** is the DNA sequence immediately in front of a gene that is needed to activate the gene's expression. Most of the noncoding spacers appear to be remnants of genes lost over the course of evolution and pieces of defunct viral genomes. Noncoding regions may, however, provide raw material for future evolution.

Figure 7.2 **The genome of *Salmonella enterica* serovar Typhimurium LT2.** The circular chromosome. Base pairs are indicated around the perimeter. The two outer multicolored circles indicate genes encoded on the separate strands, which means they are transcribed in opposite directions. Color codings represent homologies between eight compared species. Blue indicates genes present only in *S. typhimurium* LT2. Orange indicates that the gene has a close homolog in all eight genomes compared. Green indicates genes with a close homolog in at least one other *Salmonella* (*S. typhi*, *S. paratyphi A*, *S. paratyphi B*, *S. arizonae*, or *S. bongori*) but not in *E. coli* K12, *E. coli* O157:H7, and *Klebsiella pneumoniae*. Gray indicates other combinations. The black inner circle is the GC content of the LT2 DNA (peaks pointing outward indicate GC-rich areas).

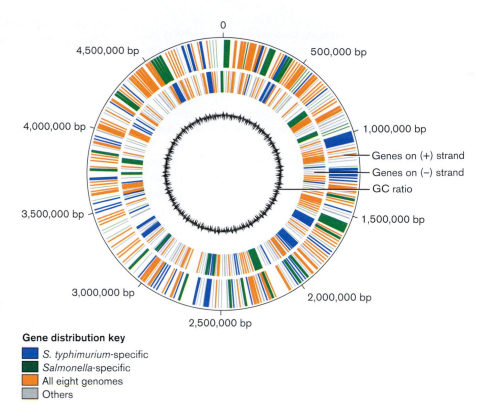

Gene distribution key

- ■ *S. typhimurium*-specific
- ■ *Salmonella*-specific
- ■ All eight genomes
- ■ Others

In contrast to many eukaryotes, prokaryotes tend to have very little noncoding DNA (typically less than 15% of the genome). The completed genome sequence of the bacterium *Mycobacterium leprae*, however, has a substantial amount (about 50%) of DNA with no known or predicted function, and this noncoding DNA may represent degenerate or inactivated genes. Still, this is less than the percentage of noncoding DNA seen in most eukaryotes.

Genes Are Organized into Functional Units

In the simplest case, a gene can stand alone, operating independently of other genes. The RNA produced from a stand-alone gene is said to be monocistronic, coding for one protein. Alternatively, a gene may exist in tandem with other genes in a unit called an **operon** (**Fig. 7.3A**). All genes in an operon are situated head to tail on the chromosome and are controlled by a single regulatory sequence located in front of the first gene. The single RNA molecule produced from the operon contains all the information from all the genes in that operon.

On a functional level, a collection of genes and operons at multiple positions on the chromosome can hold membership in a **regulon** when they have a unified biochemical purpose (such as amino acid biosynthesis; **Fig. 7.3B**). The various mechanisms regulating expression of these functional genetic units are discussed in Chapter 10.

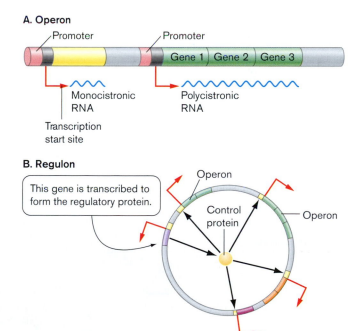

Figure 7.3 **Gene organization.** **A.** Diagram of a segment of DNA containing a single gene producing a monocistronic message (codes for one protein) and an operon of three genes producing a polycistronic message (from which three different proteins are made, one corresponding to each gene). **B.** Diagram of a circular bacterial genome, containing genes and operons coordinately controlled by a regulatory protein.

NOTE: In bacteria, the names of genes are given as a three-letter abbreviation of the encoded enzyme's name (for example, the gene *dam* encodes <u>d</u>eoxy-<u>a</u>denosine <u>m</u>ethylase) or the function of related genes (for example, genes designated *proA*, *proB*, and *proC* encode enzymes involved in <u>pro</u>line biosynthesis). Bacterial gene names are written in italics with lowercase letters. For example, the genes involved in catabolizing lactose are the *lac* genes. If several genes are involved in the pathway, a fourth letter, capitalized, is used. Thus, the three genes *lacZ*, *lacY*, and *lacA* are all associated with lactose catabolism. When speaking of a gene product, a nonitalic (roman) font is used, and the first letter is capitalized. Thus, *lacZ* is the gene, and LacZ is the protein product of that gene.

www | Microbial Genome Database

DNA Function Depends on Chemical Structure

DNA is composed of four different nucleotides linked by a phosphodiester backbone (**Fig. 7.4A**). Each nucleotide consists of a nitrogenous base attached through a ring of nitrogen to carbon 1 of 2-deoxyribose in the phosphodiester backbone. The 2-deoxy position that distinguishes DNA from RNA is circled in the figure. A **phosphodiester bond** (marked) joins adjacent deoxyribose molecules in DNA to form the phosphodiester backbone. Phosphodiester bonds link the 3' carbon of one ribose to the 5' carbon of the next ribose. The two backbones are **antiparallel** so that at either end of the DNA molecule, one strand ends with a 3' hydroxyl group and the opposite strand ends with a 5' phosphate. This antiparallel arrangement is necessary so that complementary bases protruding from the two strands can pair properly via hydrogen bonding. Base pairing is not possible if one tries to model DNA strands in a parallel arrangement.

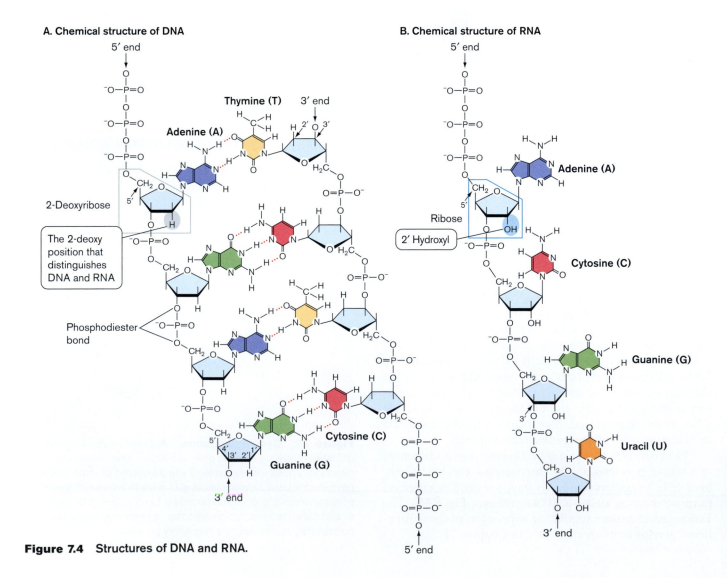

Figure 7.4   Structures of DNA and RNA.

The nitrogenous bases in DNA are planar hetero-aromatic structures stacked perpendicular to the phosphodiester backbone and parallel to each other (see **Fig. 7.4A**). **Purines** (bicyclic bases, adenine or guanine) pair with **pyrimidines** (monocyclic bases, thymine or cytosine). Under physiological conditions of salt (about 0.85%) and pH (pH 7.8), the hydrogen bonding of the bases permits adenine to pair only with thymine (via two hydrogen bonds) and likewise guanine with cytosine (via three hydrogen bonds). These complementary base interactions enable the two phosphodiester backbones to wrap around each other to form the classic double helix, or duplex.

The thousands of H-bonds that form between purines and pyrimidines along the interior of a DNA duplex (**Fig. 7.5**) make the bonding of the two complementary strands of DNA highly specific, so that a duplex is formed only between complementary strands.

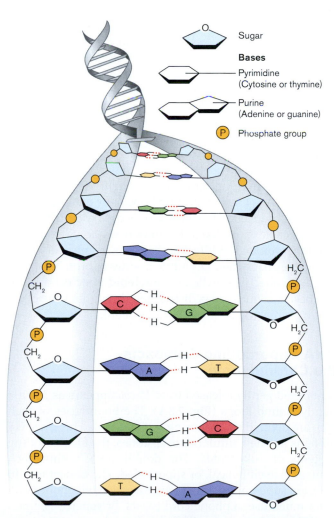

Figure 7.5   Progressive magnification of the DNA helix. The top half of the molecule illustrates the typical double-helix structure that becomes magnified in the lower half to show individual bases and hydrogen bonding.

Although H-bonds govern the specificity of strand pairing, the thermal stability of the helix is predominantly due to the stacking of the hydrophobic base pairs. Stacking creates interactions between these base pairs so that water and ions can interact with the negatively charged, hydrophilic phosphate backbone but not with the interior of the double helix.

> **THOUGHT QUESTION 7.1**   What do you think happens to two single-stranded DNA molecules isolated from *different* genes when they are mixed together at very high concentrations of salt?

At high temperatures (50–90°C), the hydrogen bonds in DNA break, and the duplex falls apart, or **denatures**, into two single strands. The temperature required to denature a DNA molecule depends on the GC/AT ratio of a sequence. More energy is required to break the three hydrogen bonds of a GC base pair than the two bonds of an AT base pair. Thus, DNA with a high GC content requires a higher denaturing temperature than a similar-sized DNA with a lower GC content. The black center ring in **Figure 7.2** illustrates how the GC content of bases can change around a single chromosome.

When DNA has been heated to the point of strand separation, lowering the temperature permits the two single strands to find each other and reanneal into a stable double helix. The kinetics of DNA denaturation is much faster than that of renaturation, since the latter is a random, hit-or-miss process of complementary sequences finding each other. This melting/reannealing property of DNA is exploited in a number of molecular techniques (see the discussion of the polymerase chain reaction in Special Topic 7.3). Note, however, that bacteria and archaea growing at extreme pH or temperature protect their DNA from denaturation through the use of remarkable DNA-binding proteins. The role of DNA-binding proteins in microbial survival is the subject of considerable research.

> **THOUGHT QUESTION 7.2**   How do you think the kinetics of denaturation and renaturation are dependent on DNA concentration?

Ribonucleic Acid (RNA) Differs Slightly from DNA

As we learned in Chapter 3, the growing cell continually makes temporary copies of its genes in the form of RNA molecules that direct the synthesis of proteins. DNA and RNA are chemically similar, except that in RNA the sugar ribose replaces deoxyribose and the pyrimidine base uracil replaces thymine (**Fig. 7.4B**). Functionally,

A. Space-filled   **B. Surface**

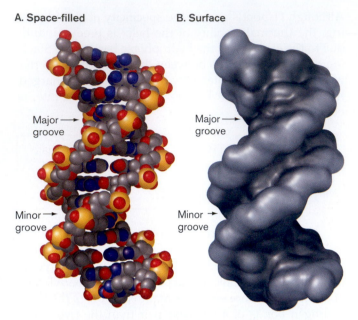

Figure 7.6   Models of DNA.   A. Space-filling model of DNA. **B.** DNA surface modeled using nuclear magnetic resonance. (PDB code: 1K8J)

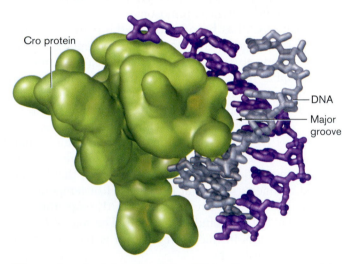

Figure 7.7   Protein recognizing DNA.   Cro repressor protein binds DNA within the major groove. (PDB code: 6CRO)

these two differences prevent enzymes meant to work on DNA, such as DNA polymerases, from acting on RNA. However, uracil can still base-pair with adenine, which means that hybrid RNA-DNA double-stranded molecules (**hybridization**) can form when base sequences are complementary. In fact, hybridization is a necessary step in the decoding of genes to make proteins. Also, note that DNA in a cell is usually double-stranded, whereas RNA is usually single-stranded.

As seen in the space-filling model of DNA in **Figure 7.6A** and the contour map in **Figure 7.6B**, the B form of the DNA double helix has grooves: the wider major

groove and the narrow minor groove. The two grooves are generated by the angles at which the paired bases meet each other. These grooves provide DNA-binding proteins access to base sequences buried in the center of the molecule, so that proteins can interact with the bases without the strands being separated (**Fig. 7.7**).

Bacterial Chromosomes Are Compacted into a Nucleoid

The chromosome of *E. coli* has over 4.6 million bases in one strand, or over 9 million counting both strands. This is a huge molecule. In fact, this one molecule contributes greatly to the overall negative charge of the cytoplasm. This is because at the normal pH of the cell (pH 7.8), all the phosphates in the backbone (9 million) are unprotonated and negatively charged. You might wonder how the cell handles this macromolecule.

In **Figure 7.1**, we see DNA spewing out of a damaged bacterial cell. Laid out, the chromosome is 1,500 times longer than the cell. It is obvious from this photomicrograph that an intact, healthy cell must compact a huge bundle of DNA into a very small volume. DNA is the second largest molecule in the cell (only peptidoglycan is larger) and comprises a large portion of a bacterial cell's dry mass, about 3–4%. While packaging 3% of a cell's dry weight may not seem like a challenge, realize that DNA is further confined only to ribosome-free areas of the cell, so the chromosome-packing density reaches about 15 mg/ml. In a test tube, DNA at 15 mg/ml is almost a gel, so how can anything move inside a cell? And how does all this DNA keep from getting hopelessly entangled?

As introduced in Chapter 3, cells pack their DNA into a manageable form that still allows ready access to DNA-binding proteins. While bacteria lack a nuclear membrane, they pack their DNA into a series of protein-bound domains collectively called the **nucleoid** (see Section 3.5). The bacterial nucleoid is distributed throughout the cytoplasm, unlike the compact nucleus of eukaryotes.

DNA Supercoiling Compacts the Chromosome

A nucleoid gently released from *E. coli* appears as 30–100 tightly wound loops (**Fig. 7.8A**). The boundaries of each loop are defined by anchoring proteins called histone-like proteins for their similarity to histones, the DNA-binding proteins of eukaryotes. The double helix within each domain is itself helical, or supercoiled. The easiest way to envision supercoiling is to picture a coiled telephone cord. After much use, a phone cord twists, or supercoils, upon itself. Note that supercoiled phone cords are quite compact, taking up less space than a relaxed cord. Circular DNA works the same way, a property used by the cell to pack its chromosome.

Figure 7.8 **Bacterial nucleoid.**
Nucleoid showing domain loops after gentle release from cells. The single-strand nick unwinds (relaxes) only one loop.

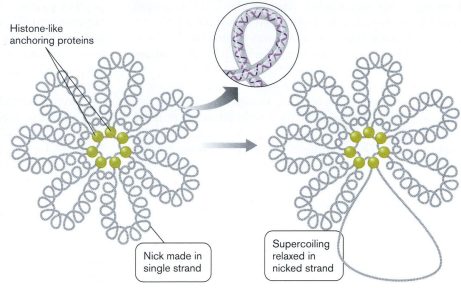

Histone-like anchoring proteins

Nick made in single strand

Supercoiling relaxed in nicked strand

Remarkably, the supercoiling of one domain of DNA is maintained independently of other loops. The independence of supercoiled domains was demonstrated by introducing one single-strand nick in the phospho-diester backbone of one domain (see **Fig. 7.8**). You can do this by adding very small amounts of a nuclease (an enzyme that cleaves a nucleic acid). The ends of the broken strand, driven by the energy inherent in the supercoil, rotate about the unbroken complementary strand of the duplex and relax the supercoil. However, supercoils were removed from only the one domain. How is this possible if the chromosome is one circular molecule? The unaffected chromosomal domains remained supercoiled because they were constrained at their bases by anchoring proteins such as Hu and HNS (histone-like proteins) that prevent rotation.

How does DNA achieve the supercoiled state? The bacterial cell produces enzymes that can twist DNA into supercoils and relieve supercoils. A single twist introduced into a 300-bp circular DNA molecule creates a single supercoil as shown in **Figure 7.9**. An enzyme makes a double-stranded break at one point in the circle, passes another part of the DNA through the break, and reseals it. This produces the same result as if one end of the broken circle were twisted one full turn. Twisting in the opposite direction of the helical turn tightens the helix by adding more turns (overwinding). Think of the phone cord again. Look down the length of the cord from one end. If the cord turns left to right (that is, clockwise) down its length, twist it from one end right to left (counterclockwise). This *increases* the number of twists.

The resulting torsional stress of overwinding is relieved when the DNA (or phone cord) subsequently twists upon itself, introducing positive supercoils. In contrast, negative supercoils are formed if one end of a DNA molecule is turned in the *same* direction as the helix (thereby underwinding the DNA). In terms of the phone cord, if the cord naturally turns clockwise down its length, turn one end clockwise several more times. Torsional stress results in this case, as well, because the maneuver tries to *decrease* the number of turns in the helix. To maintain the same number of turns, the molecule must supercoil in the opposite direction and form negative supercoils that reduce the torsional strain. The nucleoids of bacteria and most archaea, as well as the nuclear DNA of eukaryotes, are

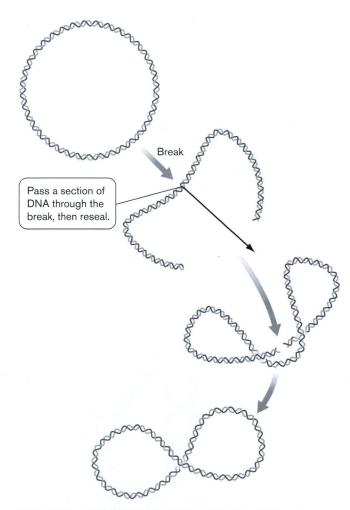

Break

Pass a section of DNA through the break, then reseal.

Figure 7.9 **Supercoiling of 300-bp circular DNA.** A supercoil can be introduced into a double-stranded circular DNA molecule by (1) cleaving both strands at one site in the molecule, (2) passing an intact part of the molecule between ends of the cut site, and (3) reconnecting the free ends.

kept *negatively* supercoiled. Because the DNA is underwound, the two strands of negatively supercoiled DNA are easier to separate than positively supercoiled DNA. This is important for transcription enzymes like RNA polymerase that must separate strands of DNA to make RNA.

Note, however, that some archaeal species living in acid at high temperature have nucleoids that are *positively* supercoiled to keep DNA double-stranded in these inhospitable environments (discussed below). Positively supercoiled DNA is harder to denature because it takes excess energy to separate overwound DNA.

Topoisomerases Regulate the Supercoiling of DNA

Supercoiling changes the topology of DNA. Topology is a description of how spatial features of an object are connected to each other. Thus, enzymes that change DNA supercoiling are called **topoisomerases**. To maintain proper DNA supercoiling levels, a cell must delicately balance the activities of two types of topoisomerases. Type I topoisomerases cleave only one strand of a double helix, while type II enzymes cleave both strands. Type I enzymes are generally used to relieve or unwind supercoils, while type II enzymes use energy to introduce them. **Figures 7.10** and **7.11** illustrate the mechanisms used by type I and type II topoisomerases, respectively. Type I enzymes are usually single proteins, while type II enzymes have multiple subunits. An example of a type II topoisomerase

is DNA gyrase, whose function is to introduce negative supercoils in DNA (see **Fig. 7.11**). The active gyrase complex is a tetramer composed of two GyrA and two GyrB proteins. **Figure 7.12** shows a three-dimensional representation of DNA gyrase in the midst of generating a supercoil.

Enzymes that make or manage bacterial DNA, RNA, and proteins are common targets for antibiotics. For instance, the **quinolone antibiotics** specifically target bacterial type II topoisomerases. A modern quinolone, ciprofloxacin, was the treatment of choice for anthrax pneumonia during the 2001 anthrax attacks. The patriarchs of this drug family, nalidixic and oxolinic acids, were used to map *gyrA* and *gyrB*, the first drug resistance genes identified in *E. coli*. The modern successors of these drugs, the fluoroquinolones, are among the most widely used antimicrobials in the world. These drugs do not block topoisomerase action but stabilize the complex in which DNA gyrase is covalently attached to DNA (see **Fig. 7.11**). This forms a physical barrier in front of the DNA replication complex, and the bacterial cell dies.

Extreme thermophiles (hyperthermophilic archaea) possess an unusual gyrase called reverse DNA gyrase. In contrast to the DNA gyrase from mesophiles, reverse gyrase introduces positive supercoils into the chromosome. It is proposed that tightening the coil helps protect the chromosome against thermal denaturation. Because the DNA has *extra* turns, it takes more energy (heat) to separate the strands.

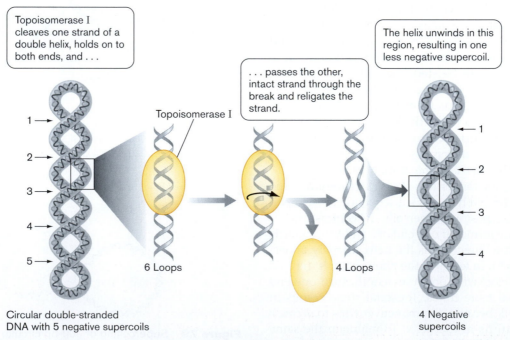

Figure 7.10   Mechanism of action for type I topoisomerases (Topo I of *E. coli*). Topoisomerase I relaxes a supercoiled DNA molecule. From left to right: Circular, supercoiled, double-stranded DNA is nicked by topoisomerase I, unwound, and released with one less superturn.

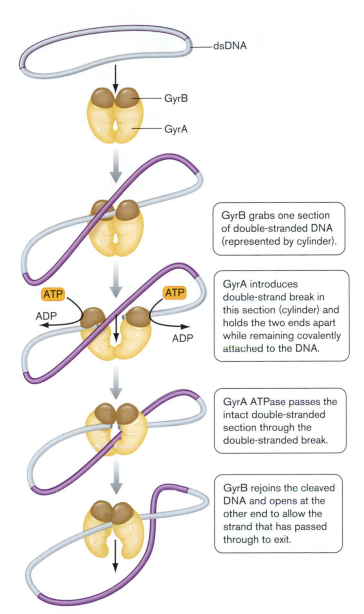

Figure 7.11   Mechanism of action for type II topoisomerases (DNA gyrase of *E. coli*).   The gyrase enzyme grabs DNA and, in an ATP-dependent process, introduces a double-strand break, passes another part of the double helix through the break, and then reseals the break. The result is the introduction of a negative supercoil.

The following boxes appear alongside the figure:

- GyrB grabs one section of double-stranded DNA (represented by cylinder).
- GyrA introduces double-strand break in this section (cylinder) and holds the two ends apart while remaining covalently attached to the DNA.
- GyrA ATPase passes the intact double-stranded section through the double-stranded break.
- GyrB rejoins the cleaved DNA and opens at the other end to allow the strand that has passed through to exit.

THOUGHT QUESTION 7.3   DNA gyrase is essential to cell viability. Why, then, are nalidixic acid–resistant cells that contain mutations in *gyrA* still viable?

THOUGHT QUESTION 7.4   Bacterial cells contain many enzymes that can degrade linear DNA. How, then, do linear chromosomes in organisms like *Borrelia burgdorferi* (the causative agent in Lyme disease) avoid degradation?

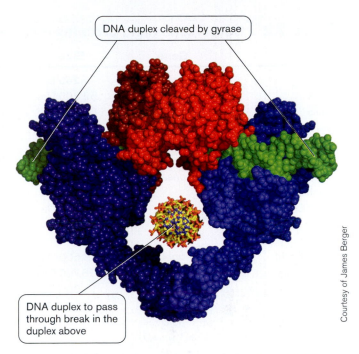

Courtesy of James Berger

Figure 7.12   Three-dimensional representation of DNA gyrase.   The 3-D model of gyrase was determined using data from X-ray crystallographic studies (blue and red regions). The gyrase complex is modeled in the process of gripping a broken DNA duplex (shown in green) and transporting a second duplex (the multicolored rosette).

Labels in figure: DNA duplex cleaved by gyrase; DNA duplex to pass through break in the duplex above.

TO SUMMARIZE:

- **The smallest genome known** for a free-living microbe encodes a possible 480 proteins.
- **Noncoding DNA** can constitute a large amount of a eukaryotic genome, while prokaryotes have very little noncoding DNA.
- **DNA is composed of two antiparallel chains** of purine and pyrimidine nucleotides in which phosphate links the 5′ carbon of one nucleotide with the 3′ carbon of the next in the chain. This forms a double helix containing a deep major groove and a more shallow minor groove.
- **Hydrogen bonding and interactions between the stacked bases** hold together complementary strands of DNA.
- **Supercoiling** by topoisomerases compacts DNA into an organized nucleoid.
- Bacteria, eukaryotes and most archaea possess **negatively supercoiled DNA**. Archaea living in extreme environments have **positively supercoiled genomes**.
- **Type I topoisomerases** cleave one strand of a DNA molecule and *relieve* supercoiling; **type II enzymes** cleave both strands of DNA and use ATP to *introduce* supercoils.

Table 7.2 **Some genes and proteins involved in DNA replication of *E. coli*.**

Protein	Gene	Size	Function
	oriC	245 bp	Origin of replication
DnaA	*dnaA*	52,300 Da	Initiation, binds *oriC*, DnaB loading
DnaB	*dnaB*	52,200 Da	Helicase (hexamer), prepriming priming, DNA-dependent rNTPase
DnaC	*dnaC*	27,800 Da	Loading factor for DnaB
Pol III (alpha)	*dnaE*	129,700 Da	DNA Pol III holoenzyme, elongation
Primase (DnaG)	*dnaG*	65,400 Da	Priming, RNA primer synthesis, rifampin resistant RNA polymerase
Gamma and tau subunits	*dnaX*	47,500 Da; 71,000 Da	Synthesis, part of the gamma complex, promotes dimerization of Pol III
Beta (EFI)	*dnaN*	40,400 Da	Beta clamp, processivity
Single-stranded binding protein	*ssb*	19,000 Da	Helix destabilizing
Delta (EFIII)	*holA*	38,500 Da	Part of gamma complex, clamp loading
Delta prime	*holB*	36,700 Da	Part of gamma complex, clamp loading
Psi	*holD*	15,000 Da	Part of gamma complex, clamp loading
Chi	*holC*	16,500 Da	Part of gamma complex, clamp loading
Theta	*holE*	8,700 Da	Pol III dimerization
Epsilon subunit	*dnaQ (mutD)*	27,000 Da	Proofreading, 3'-to-5' exonuclease
DNA gyrase (subunit beta)	*gyrB*	89,800 Da	Relaxation of supercoils
DNA Pol II	*polB*	120,000 Da	DNA repair
RNA Pol, beta subunit	*rpoB*	150,000 Da	Initiation of replication, also bulk RNA synthesis
Pol I	*polA*	103,000 Da	Gap filling
DNA ligase	*ligA*	73,400 Da	Joins phosphodiester ends

7.3 DNA Replication

Microbial DNA needs to replicate itself as accurately and as quickly as possible so that the organism can grow and compete with other species. Replication efficiency is one reason why bacterial pathogens such as *Salmonella* can cause disease so quickly after ingestion. In this respect, bacteria differ from multicellular organisms, which need to regulate cell division carefully within their tissues; unregulated growth within tissues leads to cancer. The process of bacterial replication involves an amazing number of proteins and genes coming together in a complex machine (**Table 7.2**). Its operation is all the more remarkable considering that some bacteria, such as thermophilic *Bacillus* species that live in hot springs, can double in less than 15 minutes.

The molecular details of bacterial DNA replication are important because they provide targets for new antibiotics, as well as tools for biotechnology such as the polymerase chain reaction (PCR). In addition, the proteins of DNA replication have homologs in the human genome, in which defects are the basis of inherited human diseases such as xeroderma pigmentosum, and can cause a predisposition to certain cancers.

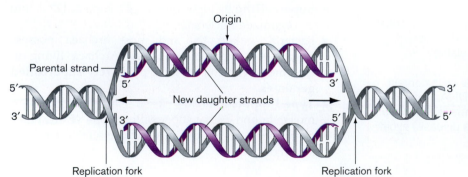

Figure 7.13 Semiconservative replication. A replication bubble with two replication forks. Parental strands are gray, and newly synthesized daughter strands are purple. Replication is called semiconservative because one parental strand is conserved and inherited by each daughter cell genome. It is called bidirectional because it begins at a fixed origin and progresses in opposite directions.

Overview of Bacterial DNA Replication

To replicate a molecule containing millions of base pairs poses formidable challenges. How does replication begin and end? How can two identical copies be generated? How is accuracy checked and maintained?

Semiconservative replication. Replication of cellular DNA in most cases is **semiconservative**, meaning that each daughter cell receives one parental strand and one newly synthesized strand (**Fig. 7.13**). At the **replication fork**, the advancing DNA synthesis machine separates the parental strands while extending the new, growing strands. The semiconservative mechanism provides a means for each daughter duplex to be checked for accuracy, based on its parent strand.

Enzymes that synthesize DNA or RNA can only connect nucleotides together in a 5′-to-3′ direction. That is, every newly made strand begins with a 5′ triphosphate and ends with a 3′ hydroxyl group (see **Fig. 7.4A**). A polymerase (a chain-lengthening enzyme complex) fastens the 5′ alpha-phosphate of an incoming nucleoside triphosphate to the 3′ hydroxyl end of the growing chain, thus forming a phosphodiester bond. (The alpha-phosphate is the phosphate closest to the sugar base.) This 5′-to-3′ enzymatic constraint produces an interesting mechanistic puzzle. If polymerases can only synthesize DNA in a 5′-to-3′ direction and the two phosphodiester backbones of the double helix are antiparallel, how are both strands of a moving replication fork synthesized simultaneously? One strand presents no problem because it is synthesized in a 5′-to-3′ direction toward the fork, but synthesizing the other new strand in a 5′-to-3′ direction would seem to dictate that it move *away* from the fork (**Fig. 7.14**). How is this possible? The answer was revealed in Chapter 3 but is discussed in greater detail here.

> **THOUGHT QUESTION 7.5** Suppose you have the following capabilities: You can label DNA in a bacterium by growing cells in medium containing either nitrogen ^{14}N or the heavier isotope ^{15}N; you can isolate pure DNA from the organism; and you can subject DNA to centrifugation in a cesium chloride solution, a solution that forms a density gradient when subjected to centrifugal force, thereby separating the light (^{14}N) and heavy (^{15}N) forms of DNA to different locations in the test tube. Given these capabilities, how might you prove that DNA replication is semiconservative?

The process of DNA replication is divided into three phases: (1) initiation, which is the melting (unwinding) of the helix and the loading of the DNA polymerase enzyme complex; (2) elongation, which is the sequential addition of deoxyribonucleotides from deoxynucleotide triphosphates, followed by proofreading; and (3) termination,

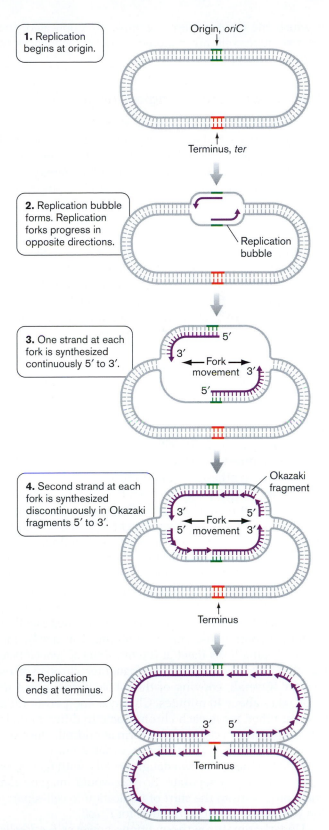

Figure 7.14   Comparing direction of fork movement with direction of DNA synthesis.

in which the DNA duplex is completely duplicated, the negative supercoils are restored, and key sequences of new DNA are methylated!

Replication from a Single Origin

Replication begins at a single defined DNA sequence called the **origin** (*oriC*). Following initiation, a circular microbial chromosome replicates *bidirectionally* (see **Fig. 7.14**, step 1) until it terminates at defined **termination** (*ter*) **sites** located on the opposite side of the molecule. Once the process has begun, the cell is committed to completing a full round of DNA synthesis. As a result, the decision of when to start copying the genome is critical. If it starts too soon, the cell accumulates unneeded chromosome copies; if it starts too late, the dividing cell's septum "guillotines" the chromosome, killing both daughter cells. Consequently, elaborate fail-safe mechanisms link the initiation of DNA replication with cell mass, generation time, and cellular health, making the timing of initiation remarkably precise.

Fundamentals of DNA replication. The basic process of chromosome replication is outlined in **Figure 7.14**. After initiation of replication, a replication bubble forms at the origin. The bubble contains two replication forks that move in opposite directions around the chromosome (**Fig. 7.14**, step 2). DNA polymerases synthesize DNA in a 5'-to-3' direction. Thus, at each fork, one new DNA strand can be synthesized continuously until the terminus region (**Fig. 7.14**, step 3). However, because the two DNA strands are antiparallel and the DNA polymerases only synthesize 5'-to-3', the other daughter strand has to be synthesized discontinuously, in stages—seemingly backward relative to the moving fork (**Fig. 7.14**, step 4). The fragments of DNA formed on this discontinuously synthesized strand are called **Okazaki fragments**, after the scientist who discovered them. As we will discuss later, the Okazaki fragments are progressively stitched together to make a continuous, unbroken strand. Ultimately, the two replicating forks meet at the terminus sequence (see **Fig. 7.14**, step 5) and the two daughter chromosomes separate. Overall, copying of the whole chromosome in *E. coli* takes about 40 minutes. Chromosome-partitioning processes then move each chromosome to different ends of the cell so that a cell wall can form at midcell. Once the cell wall is complete (this time is variable but generally about 20 minutes), the two daughter cells, with their new chromosomes, can separate. So you would imagine the whole process from the start of replication to cell separation would take about 60 minutes for *E. coli*.

Under optimal growth conditions, however, *E. coli* cells divide in 20 minutes. How is this possible if the process of replication takes at least 40 minutes? The answer is that a partially replicated chromosome can start new rounds of replication at the two daughter origins even before the

first round is complete. This overlapping of generations enables the cell to accommodate the 40-minute DNA replication time within a 20-minute generation time. During these rapid cell divisions, the DNA-partitioning mechanisms ensure that the two actively replicating chromosomes move to different ends of the cell, which keeps them from being severed when the cell septum forms.

Now let's examine each step in molecular detail to answer some important questions about mechanism.

Initiation of Replication

What determines when replication begins? Initiation is controlled by DNA methylation, and by the binding of a specific initiator protein to the origin sequence. Further molecular events load the elaborate DNA polymerase complex and generate the first RNA primer for the new DNA strand.

DNA Methylation Controls Timing

The chromosomal origin of *E. coli* is a sequence of 245 base pairs designated *oriC*. It is subject to a critical molecular control mechanism that dictates the precise timing of replication initiation. Initiation of replication at *oriC* is activated by one protein, DnaA, and inhibited by another, SeqA. Immediately after a cell has divided, the level of active DnaA (DnaA bound to ATP) is low, and the inhibitor SeqA will bind to *oriC* and prevent ill-timed initiations (before the cell has grown enough to divide again).

How does SeqA know to bind just after the origin has replicated? The key is DNA methylation. *E. coli* uses the enzyme deoxyadenosine methylase (Dam) to attach a methyl group to the adenine residue at position N-6. The methylated adenines appear in all GATC sequences. GATC sequences (the recognition sites of Dam methylase) are scattered throughout the chromosome and occur on both strands. However, just after the origin has replicated, there is a short lag before the newly synthesized strand is methylated. As a result, the origin is temporarily hemimethylated, a situation in which only one of the two complementary strands is methylated. Because SeqA has a high affinity for hemimethylated origins, this inhibitor will bind most tightly immediately after the origin has been replicated and prevent another initiation event. Eventually, the Dam methylase will methylate the new strand and decrease SeqA binding.

The replication initiator protein, DnaA. The onset of the initiation phase is determined by the concentration of the replication initiator protein, DnaA. DnaA protein recognizes specific 9-bp repeats at *oriC*. As the cell grows, the level of active DnaA initiator protein rises until it is sufficient to bind to a series of 9-bp repeats at *oriC* (**Fig. 7.15●**, step 1). DnaA actually binds as a complex with ATP (DnaA-ATP). This binding initiates the assembly of

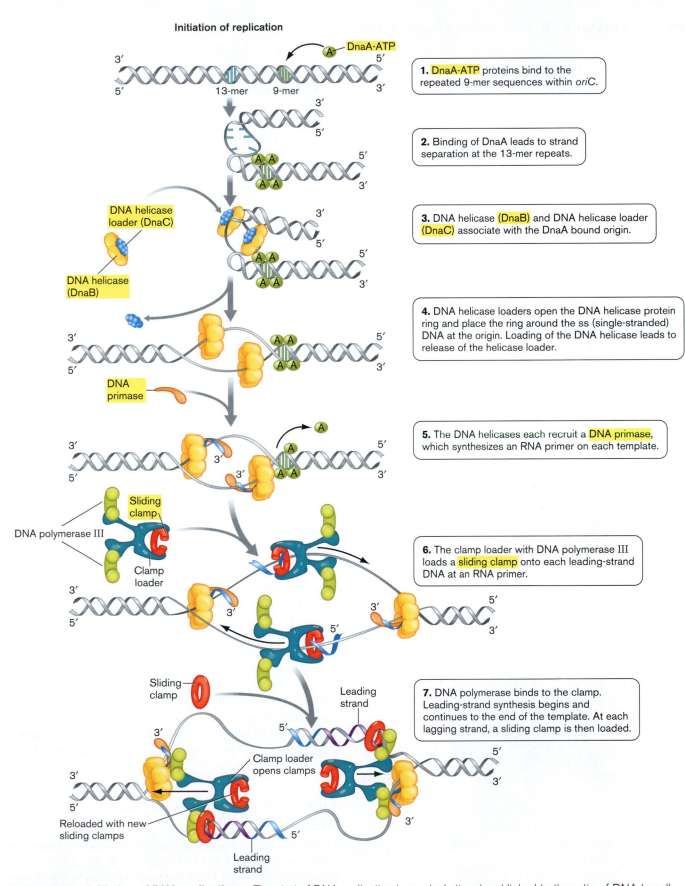

Initiation of replication

1. DnaA-ATP proteins bind to the repeated 9-mer sequences within *oriC*.

2. Binding of DnaA leads to strand separation at the 13-mer repeats.

3. DNA helicase (DnaB) and DNA helicase loader (DnaC) associate with the DnaA bound origin.

4. DNA helicase loaders open the DNA helicase protein ring and place the ring around the ss (single-stranded) DNA at the origin. Loading of the DNA helicase leads to release of the helicase loader.

5. The DNA helicases each recruit a DNA primase, which synthesizes an RNA primer on each template.

6. The clamp loader with DNA polymerase III loads a sliding clamp onto each leading-strand DNA at an RNA primer.

7. DNA polymerase binds to the clamp. Leading-strand synthesis begins and continues to the end of the template. At each lagging strand, a sliding clamp is then loaded.

Figure 7.15 Initiation of DNA replication. The start of DNA replication is precisely timed and linked to the ratio of DNA to cell mass. In *E. coli*, the initiator protein DnaA accumulates during growth and then triggers the initiation of replication. It begins with DnaA-ATP complexes binding to 9-mer (9-bp) repeats upstream of the origin. This binding (along with other proteins not shown), first causes the DNA to loop in preparation for melting open by the helicase (DnaB). ▶❚❚

a membrane-bound replication hyperstructure, a complex assembly of numerous proteins forming a functional unit, at midcell (at the cell equator).

The origin sequence (Ori), after moving through the replication complex, cannot trigger another round of replication because of inhibition by SeqA and decreasing levels of unbound DnaA-ATP. Another round of replication can only begin after (1) the origin becomes fully methylated, (2) SeqA dissociates, and (3) the DnaA-ATP concentration rises.

Initiation Requires RNA Polymerases

An unexpected feature of DNA replication is that its initiation actually requires RNA polymerases. One RNA polymerase helps separate the strands of the DNA helix at the origin, and a second RNA polymerase produces the *primer*, or starter, fragments needed to synthesize new DNA. The first of these polymerases (the housekeeping RNA polymerase used to make most of the RNA in the cell; discussed in Chapter 8) transcribes DNA at *oriC* to produce RNA that helps separate the two DNA strands (**Fig. 7.15**, step 2, this RNA polymerase is not shown). This separation allows a special DNA helicase (protein DnaB) in association with a DNA helicase loader (DnaC) to bind the two replication forks formed during bidirectional replication (**Fig. 7.15**, step 3, and see **Fig. 7.16**).

As the lead protein of the replication machine, the helicase (DnaB) uses energy from ATP hydrolysis to unwind the DNA helix as the DNA moves into the DNA polymerase replicating complex. The ringlike DnaB is

assembled around one DNA strand at each replication fork. After loading DnaB at the origin, DnaC is released (**Fig. 7.15**, step 4). Coincident with the unwinding of DNA, small single-strand DNA binding proteins (ssb, seen in **Fig. 7.17**) coat the exposed single-stranded DNA, protecting it from nuclease activities patrolling the cell.

DNA-dependent DNA polymerases possess the unique ability to "read" the nucleotide sequence of a DNA template and synthesize a complementary DNA strand. The discovery of this activity earned Arthur Kornberg (1918–) the Nobel Prize in 1959. As remarkable as these enzymes are, *no* DNA polymerase can start synthesizing DNA unless there is a preexisting DNA or RNA fragment to extend—a primer fragment. The primer fragment possesses a 3′ OH end that can receive incoming deoxynucleotides. Consequently, once DnaB (helicase) is bound to DNA, the next step in initiation is to make RNA primers at each fork (**Fig. 7.15**, step 5). In contrast to DNA polymerases, RNA polymerases *can* synthesize RNA without a primer. A specific RNA polymerase called DNA **primase** (DnaG) synthesizes short (10–12 nucleotides) RNA primers at the origin, which launches DNA replication. One primase is loaded at each of the two replication forks. Note that primase is different from the RNA polymerase involved in initially separating the DNA strands at the origin.

A Sliding Clamp Tethers DNA Polymerase to DNA

At this point, the DNA is almost ready for DNA polymerase. But first, a **sliding clamp** protein (the beta subunit) must be loaded to keep the DNA polymerase affixed to the DNA (**Fig. 7.15**, step 6). Without this clamp, DNA polymerase would frequently "fall off" the DNA molecule (see **Special Topic 7.1**). A multisubunit complex (called the clamp-loading complex) places the beta clamp, along with an attached pair of DNA polymerase molecules, onto DNA. DNA polymerase (specifically DNA Pol III; discussed later) can then bind to the 3′ OH terminus of the primer RNA molecule and begin to synthesize new DNA (**Fig. 7.15**, step 7). **Figure 7.16** presents the molecular structure of the beta clamp loaded onto DNA.

Elongation: DNA Polymerases Synthesize DNA

Escherichia coli contains five different DNA polymerase proteins designated Pol I through Pol V. All polymerases catalyze synthesis of DNA in the 5′-to-3′ direction. However, only the replication polymerases Pol III and Pol I participate directly in chromosome replication. The other polymerases conduct operations to rescue stalled replication forks and repair DNA damage.

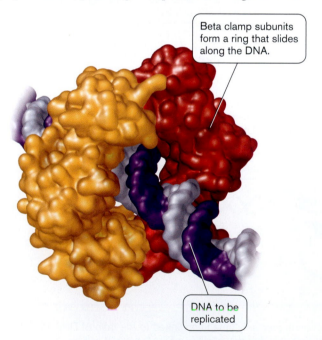

Beta clamp subunits form a ring that slides along the DNA.

DNA to be replicated

Figure 7.16 DNA replication: The sliding clamp of *E. coli*. The clamp dimer encircles the gray and purple DNA strand. (PDB codes: 1OK7, 1K8J)

Michael O'Donnell and colleagues in 1991 performed an elegant experiment that supports the sliding clamp model. O'Donnell hypothesized that if the beta subunit were really a sliding clamp, then once it was loaded onto a small *circular* DNA molecule, the protein would never come off. It would continually slide around the circle. However, the protein complex would easily slide off a linear DNA molecule.

The experiment required detecting differences in molecular weight (or size) between DNA and DNA-protein complexes. Free protein is smaller than the DNA-protein complex formed when the same protein binds to DNA. To measure these size differences, various mixtures of DNA and ^{3}H-labeled clamp proteins were passed through a gel filtration column, a column filled with tiny, hollow gel beads, each of which contains channels that serve as molecular sieves (**Fig. 1A**). The smaller the molecule, the more easily it can enter the bead channel. Thus, small molecules become temporarily trapped in each bead, taking a long time to move from the top of the gel column to the bottom. Larger molecules, however, are too large to enter the beads. They flow quickly around the beads and exit the column first. Fractions of defined volume are collected over

time. Proteins collected in early fractions are larger than those collected in later fractions. Consequently, a DNA–clamp protein complex, being large, will move more quickly through the gel and will appear in the early fractions. Unbound protein will travel more slowly and appear in later fractions.

The beta clamp complex was loaded onto a circular DNA molecule and onto the same molecule cut with the restriction enzyme *Sma*I. The separate mixtures were passed through identical gel filtration columns. Fractions were collected and counted for radioactivity (see **Fig. 1B**). Note that the protein loaded onto the circular molecule eluted in a much earlier fraction (fraction 12) than did protein loaded onto the linear molecule (fraction 25). The earlier elution indicated that the protein did not slide off the circular molecule and was maintained as a large complex. The late-elution profile of the protein loaded onto the linear molecule indicated that the protein was free to slide off the DNA and failed to maintain a large complex. The bottom panel, however, shows that a protein (for example, eukaryotic EBNA1) bound to the ends of the linear molecule will block exit of the beta clamp. This is evident in the fact that both the circular and linear DNA molecules eluted in the same fractions.

A.

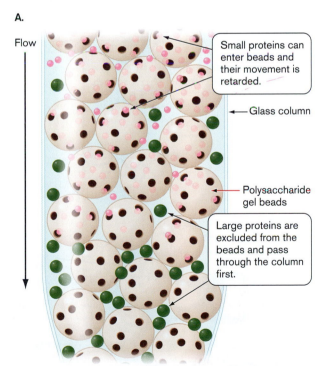

B.

Figure 1   Gel filtration evidence supports the sliding clamp model.   A. Principle of gel filtration. Porous beads separate small molecules from large molecules because small molecules are temporarily trapped inside the beads. **B.** *Top:* ^{3}H-labeled beta clamp dimers were added to nicked, circular DNA (red line) or linear DNA (black line). After passing through separate gel filtration columns, radioactivity was measured in the material that eluted from the column. The results show that the clamp cannot leave the circular model, so the large protein–DNA complex elutes from the column in an early fraction (fraction 12). The second red peak eluting at fraction 25 is protein that did not load onto DNA. Protein loaded on linear DNA, however, slides off of the molecule and does not form a large complex. *Bottom:* This is the same experiment, but the DNA contained two EBNA1 binding sites. Clamp protein was loaded onto circular DNA (red line) or the same DNA cut between the sites with *Eco*RV (linear). EBNA1 protein blocks the ends and prevents the beta clamp from sliding off the linear molecule. Note that the beta clamp mixed with linear DNA (black line) in the bottom panel eluted earlier from the column than in the top panel. Again, the peaks at fraction 28 reflect protein that did not load onto DNA. *Source:* B. P. Todd Stukenberg, et al. 1991. *Journal of Biological Chemistry* 266:11328.

Elongation of DNA synthesis

1. The leading-strand DNA Pol III enzyme replicates the leading strand. SSBs cover and protect the unreplicated single strand. The DNA helicase remains on the lagging strand, unwinding the dsDNA moving into the replisome complex.

2. Lagging-strand DNA polymerase synthesizes the lagging strand. The lagging strand loops out after passing through the polymerase.

3. After DNA helicase has moved approximately 1,000 bases, a second RNA primer is synthesized on each lagging strand.

4. When the lagging-strand polymerase bumps into the 5′ end of a previously synthesized fragment, the DNA polymerase is released and the clamp is disengaged.

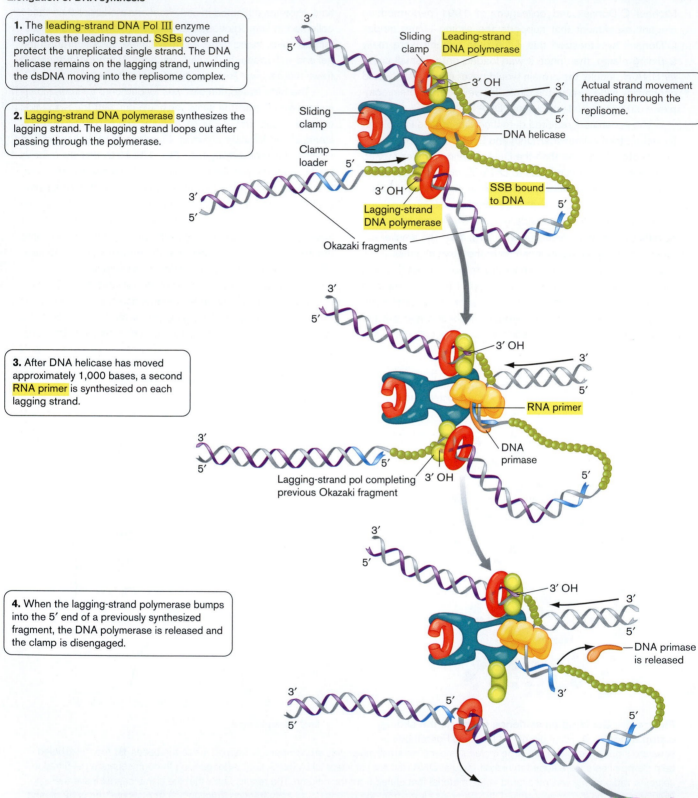

Sliding clamp
Leading-strand DNA polymerase
3′ OH
Sliding clamp
Clamp loader
DNA helicase
SSB bound to DNA
Lagging-strand DNA polymerase
Okazaki fragments

Actual strand movement threading through the replisome.

3′ OH
RNA primer
DNA primase
Lagging-strand pol completing previous Okazaki fragment
3′ OH

3′ OH
DNA primase is released

Figure 7.17 **A. The DNA polymerase dimer acting at a replication fork.** Both the leading and lagging strands are synthesized simultaneously in the 5′-to-3′ direction. For clarity, the beta clamp on the lagging strand is shown on the opposite side of Pol III as compared to its position on the leading strand. If the clamp was placed on the left side of polymerase, as it is shown in many models, the lagging strand would have to completely loop around the polymerase and enter the polymerase from the left.

DNA polymerase III. The main replication polymerase, Pol III, is a complex, multicomponent enzyme. Because of its complexity, it is often referred to as a "molecular machine." The DNA synthesis activity of Pol III is held in the alpha subunit of the complex, while other subunits are used for improving fidelity (accuracy of replication) and processivity (a measure of how long the polymerase remains attached to, and replicates, a template). The Pol III epsilon subunit (DnaQ), for example, contains a **proofreading** activity that prevents mistakes and improves fidelity.

Proofreading activities within DNA polymerases scan for mispaired bases that have been inappropriately added to a growing chain. Mispaired bases are detected by their increased mobility. A mispaired base mistakenly linked by a phosphodiester bond to a growing DNA chain is more mobile than the correct base because it does not hydrogen-bond to the template base. This motion halts DNA elongation by Pol III because the base is not properly positioned at the enzyme's active site. This stalling of Pol III activity triggers an intrinsic 3'-to-5' **exonuclease** activity in the epsilon subunit. Exonucleases degrade DNA starting from either the 5' end or the 3' end. The exonuclease activity of Pol III cleaves the phosphodiester bond, releasing the improperly paired base from the growing chain. Once the wayward mispaired base has been excised, Pol III activity can proceed.

Both DNA Strands Are Elongated Simultaneously

After initiation, each replication fork contains one elongating 5'-to-3' strand called the leading strand (look back at **Fig. 7.15**, step 7). But how is the opposite strand at each fork replicated? There are no known DNA polymerases capable of synthesizing DNA in the 3'-to-5' direction, which would seem to be needed if both strands are to be synthesized simultaneously. DNA synthesis of one strand continuing all the way back to the origin is not a solution because it would leave the unreplicated strand at each fork exposed to possible degradation for too long and would double the time needed to complete DNA replication. The cell has solved this dilemma by coordinating the activity of *two* DNA Pol III enzymes in one complex, one for each strand. The two associated Pol III complexes together are called the replisome. As the dsDNA unwinds at the fork, the problem strand loops out and primase (DnaG) synthesizes a primer. The second Pol III enzyme binds to the primed section of the loop and synthesizes DNA in the 5'-to-3' direction (imagine the lower template strand in **Fig. 7.17A** threading from left to right, through the polymerase ring). All the while the second polymerase is moving along in tandem with the first polymerase (on the leading strand) relative to the fork (**Fig. 7.17A**, step 1). Realize that in actuality, the replisome remains at a fixed, midcell location in the cell, probably attached to the membrane, and the template DNA threads through it (discussed later).

Note that simultaneous extension of the two strands requires that synthesis of the looped strand *lag* behind the leading strand and that new RNA primers be synthesized periodically by primase (DnaG) every thousand bases or so. Thus, the lagging strand is synthesized discontinuously, in pieces called Okazaki fragments, while the leading strand can be synthesized continuously. As the leading strand moves forward,

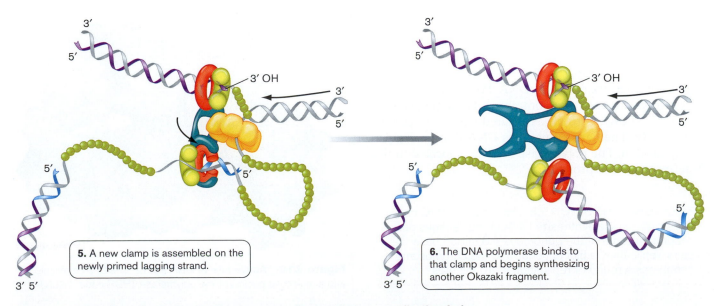

5. A new clamp is assembled on the newly primed lagging strand.

6. The DNA polymerase binds to that clamp and begins synthesizing another Okazaki fragment.

Figure 7.17 (*continued*)   B. The DNA polymerase dimer acting at a replication fork.

advancing the fork, there remains a long stretch of lagging strand complementary to the already replicated leading strand. This lagging strand remains single-stranded but protected by single-stranded DNA binding proteins (SSBs) (**Fig. 7.17A**, step 2). After about 1,000 bases, DNA primase reenters and synthesizes a new RNA primer in anticipation of lagging strand DNA synthesis (**Fig. 7.17A**, step 3). At some point, the lagging-strand polymerase bumps into the 5′ end of the previously synthesized fragment. This interaction causes DNA polymerase to disengage from that strand (**Fig. 7.17A**, step 4), and the clamp loader loads a new clamp near the new RNA primer (**Fig. 7.17B**, step 5). The DNA polymerase binds to that clamp and begins synthesizing another Okazaki fragment (**Fig. 7.17B**, step 6). This process repeats every 1,000 bases or so around the chromosome.

DNA polymerase I. Discontinuous DNA synthesis results in a daughter strand containing long stretches of DNA punctuated by tiny patches of RNA primers. This RNA must be replaced with DNA to maintain chromosome integrity. To remove the RNA, cells typically use an RNase enzyme specific for RNA-DNA hybrid molecules (called RNase H). A DNA Pol I enzyme enters after the RNase and synthesizes a DNA patch using the 3′ OH end of the preexisting DNA fragment as a priming site (**Fig. 7.18**). When DNA Pol I reaches the next fragment, the enzyme removes the 5′ nucleotide and resynthesizes it. This process of replicating DNA increases accuracy and decreases mutations.

Once DNA Pol I stops synthesizing, it cannot join the 3′ OH of the last added nucleotide with the 5′ phosphate of the abutting fragment. The resulting nick in the phosphodiester backbone is repaired by **DNA ligase**, which in *E. coli* uses energy gained by cleaving nicotinamide adenine dinucleotide (NAD) to create the phosphodiester bond (see **Fig. 7.18**). DNA ligase from eukaryotes and some other microbes uses ATP rather than NAD in this capacity.

A cleavage fragment of DNA Pol I, which can be made in the laboratory, containing DNA polymerase activity is shown with an associated DNA molecule in **Figure 7.19**. The figure illustrates DNA filling a crevice in the protein. The polymerase activity is located in the N terminus of the protein.

> **THOUGHT QUESTION 7.6** How fast does *E. coli* DNA polymerase synthesize DNA (in nucleotides per second), given that the genome is 4.6 million base pairs, replication is bidirectional, and the chromosome completes a round of replication in 40 minutes?

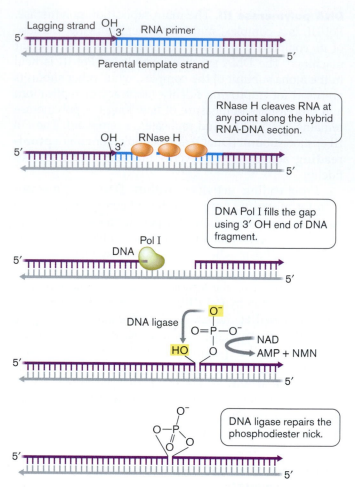

Figure 7.18 RNase H removing the RNA primer. RNase H cleaves the RNA primer (blue). DNA polymerase I uses the preexisting 3′ OH end of the DNA fragment to fill the gap. Finally, DNA ligase repairs the phosphodiester nick using energy derived from the cleavage of NAD.

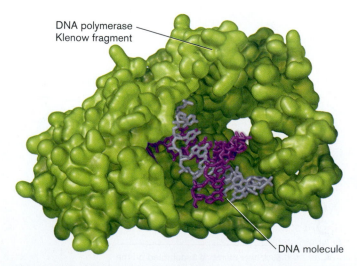

Figure 7.19 Active site of DNA Pol I. The DNA molecule fits into a molecular pocket of the polymerase. (PDB code: 1KLN)

DNA Replication Generates Supercoils

As template DNA is threaded through the replisome, the helicase continually pulls apart the two strands of the DNA helix. This causes the DNA ahead of the fork to twist, introducing positive supercoils. (Try this yourself. Twist two pieces of string together, staple one end of the pair to a piece of cardboard, and then pull the two strands apart from the free end. Notice the supercoiling that takes place beyond, or downstream of, the moving fork.) The increasing torsional stress in the chromosome could stop replication by making strand separation more and more difficult. What prevents the buildup of torsional stress is the DNA gyrase that is located ahead of the fork, removing the positive supercoils as they form.

Another topological question, noted earlier, is how DNA polymerase maneuvers through the cell. Although many descriptions of replication give the impression that the polymerase complex travels around the chromosome like a train along a track, DNA is actually fed through a stationary pair of replisomes (each a double Pol III complex) located at the cell membrane. This was shown by Katherine Lemon and Alan Grossman using *Bacillus subtilis*. The replisomes in this organism were each tagged with green fluorescent protein, and the location of the complex was monitored in replicating cells using fluorescence microscopy. If the replisomes moved like a train on a track, the polymerase-GFP protein would be found at different positions in each cell. Instead, however, in every replicating cell, replisomes were observed as distinct fluorescent foci located at or near midcell (**Fig. 7.20**). Cellular DNA stained with a blue fluorescent dye (DAPI) clearly occupied most of the cytoplasmic space.

Terminating Replication

Bidirectional replication of a circular bacterial chromosome results in the two membrane-associated replication machines attempting to replicate through the same DNA sequences 180° from the origin—that is, halfway around the chromosome. What tells the polymerases to stop? There are at least two and as many as six terminator sequences (*ter*) on the chromosome that polymerases enter but rarely, if ever, leave (**Fig. 7.21A**). One set of terminators deals with the clockwise-replicating polymerase, while the other set halts DNA polymerases replicating counterclockwise relative to the origin. A protein called Tus (<u>t</u>erminus <u>u</u>tilization <u>s</u>ubstance) binds to these sequences and acts as a counter-helicase when it comes in contact with an advancing helicase (DnaB). The bound Tus protein effectively halts polymerase movement. Six terminator sites ensure that the polymerase complex does not escape and continue replicating DNA.

Once DNA polymerases have completed duplicating the chromosome and have been removed at the *ter* sites, the cell is still faced with what could be called a "knotty" problem. Because of the topology of the chromosome, the two daughter molecules will appear as a **catenane** of linked rings after replication (**Fig. 7.21B**). The cell resolves this structure using enzymes called

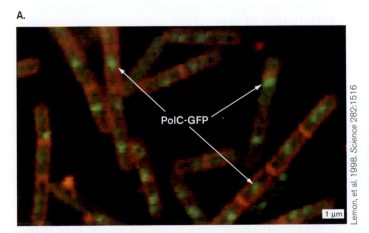
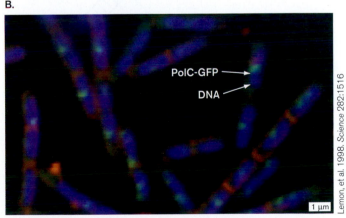

A. PolC-GFP

Lemon, et al. 1998. *Science* 282:1516

1 µm

B. PolC-GFP

DNA

Lemon, et al. 1998. *Science* 282:1516

1 µm

Figure 7.20 Location of replicative DNA polymerase in living cells.   *Bacillus subtilis* cells containing PolC-GFP (DNA polymerase III tagged with green fluorescent protein) were grown at 30°C in defined minimal medium. Images were captured with a cooled charge-coupled device (CCD) camera. **A.** PolC-GFP is seen localized as discrete green foci. Membranes were stained orange with vital membrane stain FM4-64. **B.** Using the same cells, DNA was stained blue with DAPI (blue) and the image overlayed with the one in part A. This image shows that DNA is distributed throughout the cell while PolC-GFP is localized to midcell.

A.

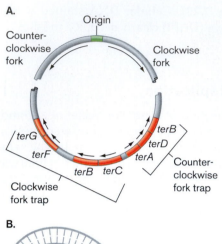

B.

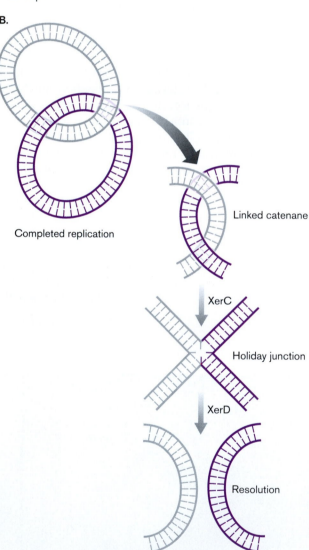

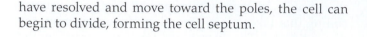

Figure 7.21   Terminating replication of the chromosome.   A. Terminator regions for DNA replication on the *E. coli* chromosome. Replication forks moving clockwise are trapped by *terG*, *terF*, *terC*, and *terB*. Counterclockwise moving forks are trapped by *terA*, *terD*, and *terB*. **B.** Resolution of DNA replication catenanes. On the left is a linked chromosome catenane. The highlighted area is enlarged to the right. XerC and XerD catalyze a breaking and rejoining that passes the chromosomes through each other, resolving the link.

have resolved and move toward the poles, the cell can begin to divide, forming the cell septum.

THOUGHT QUESTION 7.7   Individual cells in a population of *E. coli* typically initiate replication at different times (asynchronous replication). However, depriving the population of a required amino acid can synchronize reproduction of the population. What happens is that ongoing rounds of DNA synthesis finish, but new rounds do not begin. Replication stops until the amino acid is once again added to the medium, an action that triggers simultaneous initiation in all cells; that is, reproduction of the population becomes synchronized. Why?

THOUGHT QUESTION 7.8   The antibiotic rifampin inhibits transcription by RNA polymerase, but not by primase (DnaG). What happens to DNA synthesis if rifampin is added to a synchronous culture?

www | DNA replication in archaea

TO SUMMARIZE:

- **Replication is semiconservative**, with newly synthesized strands lengthening in a 5′-to-3′ direction. It involves initiation, elongation, and termination.
- **Initiation of replication** occurs from a fixed DNA origin attached to the cell membrane. Initiation depends on the mass and size of the growing cell. It is controlled by the accumulation of initiator and repressor proteins and by methylation at the origin.
- **Elongation** requires that primase (DnaG) lay down an RNA primer, that DNA polymerase III act as a dimer at each replication fork, and that a sliding clamp keep DNA Pol III attached to the template DNA molecule.
- **The 3′-to-5′ proofreading** activity of Pol III corrects accidental errors during polymerization.
- **DNA ligase** joins Okazaki fragments.
- **Two replisomes, each containing a pair of DNA Pol III complexes**, are fixed at the membrane, and DNA feeds through them.
- **Termination** involves stopping replication forks halfway around the chromosome at *ter* sites that inhibit helicase (DnaB) activity.

XerC and XerD that recognize a specific site (called *dif*) on both DNA molecules and catalyze a series of cutting and rejoining steps that essentially pass one molecule through the other. Once the two daughter chromosomes

■ **Ringed catenanes** formed at the completion of replication are separated by the proteins XerC and XerD.

7.4 Plasmids and Bacteriophages

Two kinds of extragenomic DNA molecules can interact with bacterial genomes: **plasmids** and the genomes of bacteriophages, viruses that infect bacterial cells. Plasmid-encoded functions can contribute to the physiology of the cell, and in some cases the plasmid or phage DNA itself will integrate into the bacterial genome (see Section 9.2). Eukaryotic genomes and the genomes of eukaryote-specific plasmids and viruses are similarly subject to sharing of proteins and chromosomal interactions, such as insertions. Viruses of eukaryotes and their applications in genetic engineering are discussed in greater detail in Chapters 11 and 12.

Plasmids Replicate Autonomously within Cells

Plasmids are much smaller than chromosomes (**Fig. 7.22**) and are found in archaea and bacteria as well as in eukaryotic microbes. Plasmids are usually circular, and circular plasmids, like circular chromosomes, are typically negatively supercoiled. Replication of many of these extrachromosomal elements is not tied to chromosome replication. Each plasmid contains its own origin sequence for DNA replication, but only a few of the genes needed for replication. Thus, even when the timing of plasmid replication is not linked to that of the chromosome, many of the proteins used for plasmid replication are actually host enzymes. Which host proteins are used depends on the plasmid.

Plasmids can replicate in two different ways. Bidirectional replication, as described in **Figure 7.23A**, occurs in two directions simultaneously, while rolling-circle replication (**Fig. 7.23B**) is unidirectional. In rolling-circle replication, a replication initiator (RepA, encoded by a plasmid gene) binds to the origin of replication, nicks one strand, and holds on to one end (5′ PO₄) of it while the other end (3′ OH) serves as a primer for host DNA polymerase to replicate the intact, complementary strand. The RepA protein recruits a helicase that unwinds DNA, which becomes coated by single-strand binding proteins. As replication proceeds, the nicked strand progressively peels off until replication is complete. Then the two ends of the nicked single strand are rejoined by the Rep protein and released. A complementary strand is replicated by host enzymes to produce a double-stranded molecule. Although most known plasmids use one or the other strategy, a few can use either bidirectional or unidirection replication, depending on the cell circumstance.

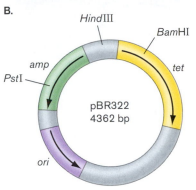

Figure 7.22   Plasmid map.   A. Note the huge difference in size between a circular plasmid DNA molecule and chromosomal DNA after both are gently released from a cell (approx. 1 μm). The arrow points to the circular plasmid. **B.** Map of plasmid pBR322. This plasmid contains an origin of replication (*ori*) and genes encoding resistance to ampicillin (*amp*) and tetracycline (*tet*). The locations of three unique restriction sites are shown (*Hind*III, *Bam*HI, and *Pst*I).

If plasmids replicate autonomously, can cells easily lose their plasmids? Plasmids come equipped with self-preservation genes that help maintain the plasmid in the host (**Special Topic 7.2**).

Plasmids can ensure their inheritance by carrying genes whose functions benefit the host bacterium under certain conditions. For instance, bacterial plasmids can carry genes responsible for antibiotic resistance (discussed in Chapter 28). As long as the antibiotic is present in the environment, any cell that loses the plasmid will be killed or stop growing. Antibiotic resistance plasmids benefit bacteria, but they are a major problem for modern hospitals, where plasmids carrying multiple drug resistance genes are transmitted from harmless bacteria into pathogens. On the other hand, plasmids (such as pBR322) containing drug resistance genes are the workhorses of genetic technology and have benefited society tremendously.

Other kinds of host survival genes carried by plasmids include genes providing resistance to toxic metals,

A.

B.

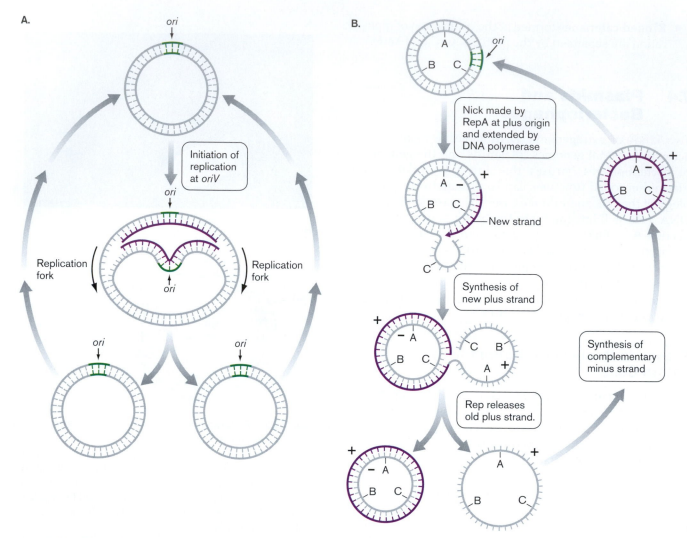

Figure 7.23 Plasmid replication: bidirectional versus rolling circle. A. DNA replication begins at a fixed origin (*ori*) and proceeds bidirectionally. **B.** Replication begins at a fixed origin but only moves in one direction. As the polymerase moves, it displaces one strand as it replicates the other. At the end of a round, the polymerase releases the displaced strand as a single-stranded circle that is then replicated.

genes encoding toxins that aid pathogenesis, and genes encoding proteins that enable symbiosis. These will be discussed in later chapters. Most of the genes involved in the nitrogen-fixing symbiosis of *Rhizobium*, for example, are plasmid-borne. Far from being freeloaders, plasmids often contribute significantly to the physiology of an organism.

Plasmids Are Transmitted between Cells

Some plasmid molecules are self-transferable via conjugation, a process that involves cell-to-cell contact to move the plasmid from a donor cell to a recipient (as discussed in Section 9.2). Other plasmids are incapable of conjugation (nontransmissible). A third group can be transferred only if a self-transferable plasmid resides in the same cell.

In this case, the conjugation mechanism produced by one plasmid will act on the other plasmid. Plasmids released from dead cells can also be taken up intact by some bacteria in a process called transformation. Finally, plasmid transmission in nature can occur by the accidental packaging of plasmids into bacteriophage head coats—in other words, by bacteriophage transduction (discussed in Section 9.2).

Replication Mechanisms of Bacteriophages

Bacteriophages show diverse mechanisms of DNA replication, which may differ considerably from those of cells. For instance, many phage genomes replicate unidirectionally rather than bidirectionally, and they may copy

Special Topic 7.2   Plasmid Partitioning and Addiction

By virtue of sheer numbers in a cell, *high-copy-number* plasmids ensure their inheritance into daughter cells through simple diffusion (**Fig. 1**). *Low-copy-number* plasmids, however, require dedicated partitioning mechanisms to guarantee that each daughter cell receives a copy prior to cell division.

Some enterprising plasmids even have addiction systems designed to kill cells that lose the plasmid. A typical addiction system includes a long-lived toxin and an antidote protein. The antidote is very labile and in need of constant synthesis if it is to neutralize the more stable toxin. Consequently, failure to inherit the plasmid means that the antidote can no longer be made. The antidote that remains will decay, leaving active toxin to kill the errant, plasmid-less cell. Why is this necessary? What happens if a cell division produces a cell with no plasmid? The cell without the plasmid may be at a competitive advantage for growth because maintaining a plasmid drains energy. Thus, the cells without plasmids could overgrow the plasmid-containing cells, eventually eliminating them from the population. Pres-

ence of an addiction molecule ensures that the plasmid is maintained in a population even if that plasmid does not provide any other selective advantage to the cell.

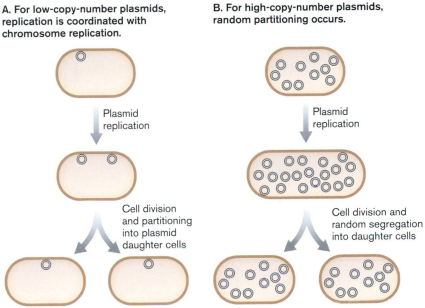

A. For low-copy-number plasmids, replication is coordinated with chromosome replication.

Plasmid replication

Cell division and partitioning into plasmid daughter cells

B. For high-copy-number plasmids, random partitioning occurs.

Plasmid replication

Cell division and random segregation into daughter cells

Figure 1   Partitioning of low-copy-number and high-copy-number plasmids.

only one strand. Often a phage genome encodes specialized polymerase enzymes to conduct its own replication, coordinated with host components (see Section 7.3). Representative phage replication pathways are shown in **Figure 7.24**.

For example, the linear DNA duplex of phage T4 replicates bidirectionally from multiple origins using T4 gene products such as gene product 43, the T4 DNA polymerase. An interesting feature of T4 is that each capsid packs a linear DNA molecule containing redundant sequences at each end. Consequently, the genome is said to be terminally redundant. As indicated in Figure 7.24A, genes at one end of the packaged DNA are identical to genes at the other end. After the T4 DNA has replicated, it must be packaged into new capsids. The packaging apparatus, however, requires a long concatemer of DNA in which multiple genomes are joined end to end. The concatemer is made when the terminally redundant ends of individual copies of T4 DNA recombine to make a single DNA molecule (**Fig. 7.24A**). From the linear T4 DNA concatemer the individual genomes are packaged into head coats, then cleaved in such a way that 3% of each

encapsidated DNA is redundant. Because each cleavage occurs farther along the genome, every T4 DNA molecule has a different terminal repeat. The genome is linear, but when the genes in a population of T4 phages are examined as a whole, the terminal repeats make the genome seem circular.

For other mechanisms of bacteriophage and viral genome replication, see Chapters 6 and 11.

TO SUMMARIZE:

- **Plasmids are autonomously replicating** circular or linear DNA molecules that are part of a cell's genome and can be transferred to other cells.
- **Plasmids replicate** by rolling-circle and/or bidirectional mechanisms.
- **Plasmids can be transferred** between cells.
- **Phage can replicate** by forming linear concatemers that are cut up to fill capsid heads. Some phage incorporate single-stranded DNA or RNA into their capsids.

A.

Phage: T4
Form of genome: Linear DNA,
double strand, permuted map

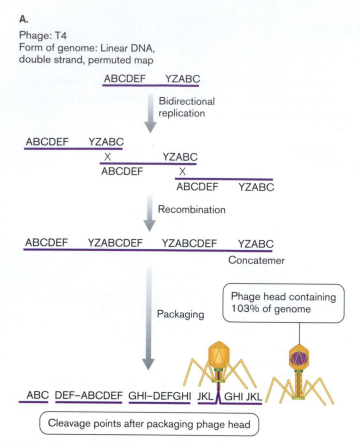

B.

Phage: Lambda
Form of genome: Linear DNA,
double strand, linear map

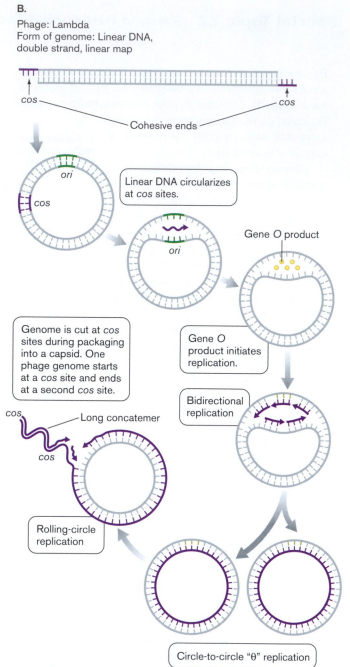

Figure 7.24   Genome replication in bacteriophages.
A. Phage T4 replicates by the rolling-circle method.
Initially, the linear genome circularizes, then replicates as
a linear concatemer. From the concatemer, the individual
genomes are packaged into head coats, then cleaved.
Because the phage head carries 103% of the genome,
each encapsidated DNA contains an end-duplicated 3%
of its genome. **B.** Phage lambda replicates its double-
stranded DNA in two stages. In the first stage, each
strand serves as a template for a complementary strand,
replicated in opposite directions (bidirectional rope). In
the second stage, rolling-circle replication generates a
concatemer, which is precisely cleaved at *cos* sites.

7.5   Eukaryotic Chromosomes: Comparison with Prokaryotes

The chromosomes of eukaryotic microbes such as protists and algae have much in common with those of
prokaryotes. Both consist of double-stranded DNA, and
both usually replicate by bidirectional replication. Yet
important differences are revealed as well—differences
in genome structure. Eukaryotic chromosomes are linear,
contained within a nucleus, and their replication involves
the process of mitosis.

Eukaryotic Genomes Are Large and Linear

Overall, the genomes of eukaryotic nuclei are larger than
those of bacteria, sometimes by several orders of magnitude (see Chapter 18). In addition, most eukaryotic
chromosomes are linear, whereas most (though not all)
bacterial chromosomes are circular. Eukaryotic chromosomes require segregation by mitosis, in order to ensure
that each daughter cell receives the correct combination
of daughter chromosomes.

Because their genomes are linear, eukaryotes require a special process to duplicate the chromosome ends, providing a primer for the lagging strand. The primer is provided by a special enzyme called telomerase. At each end of the chromosome (called a telomere) there often is not enough room on the lagging strand to add an appropriate RNA primer. So after each round of replication, the chromosome would be a little shorter and information would eventually be lost. Telomerase is actually a reverse transcriptase, an enzyme that reads RNA as a template to synthesize a complementary DNA molecule. Telomerase uses an intrinsic RNA (an RNA that is part of the enzyme) as a template to add DNA repeat sequences to the end of a lagging strand forming the telomere. The telomere allows for an RNA primer to be synthesized so that the end of the chromosome can be replicated. In this way, genetic information is not lost during division.

Telomerase may have evolved from the ancient progenitor cells that contained RNA rather than DNA genomes (see Section 17.2). For cells with RNA genomes to evolve modern chromosomes made of DNA, a reverse transcriptase must have been necessary. The reverse transcriptase may also be the evolutionary source of retroviruses (see Section 11.5).

Eukaryotic cells pack their DNA within the confines of a nucleus, using a series of proteins called **histones**. Histones are rich in arginine and lysine, and therefore, are positively charged, basic proteins that easily bind to the negatively charged DNA. The DNA becomes wrapped around the histones to form units called nucleosomes. For protection and condensation, all eukaryotic chromosomes are packaged by histones, proteins that play a regulatory role through methylation and acetylation. Bacteria, too, have DNA-packaging proteins, but they are more diverse and less essential for function (see Sections 3.5 and 7.2).

The detailed structure of the genomes of eukaryotes and prokaryotes reveal surprising differences (**Fig. 7.25**). The genome of the bacterium *E. coli* is packed with genes (green) encoding proteins or RNA molecules. These genes are separated by very little unused or noncoding DNA (purple) and with only an occasional mobile element, such as an insertion sequence, that can move from one DNA molecule to another. Most prokaryotic genes are organized in coordinately regulated operons such as *thrABC*, which encodes three enzymes needed to synthesize the amino acid threonine (see Section 10.1). In contrast, 95% of the human genome consists of noncoding sequences composed largely of regulatory sequences and the fossil genomes of ancient viruses. The coding genes are separated by large stretches of noncoding sequences and usually are not situated together in operons. Moreover, coding genes in the human genome are interrupted by **introns** (DNA within a gene that is not part of the coding sequence for a protein; shown as yellow in the figure) and ancient gene duplications that have decayed into nonfunctional, vestigial **pseudogenes** (orange). Bacteria also have pseudogenes, but they account for much less of the genome than in eukaryotes. The more rapid replication of bacteria causes pseudogenes to be lost more quickly.

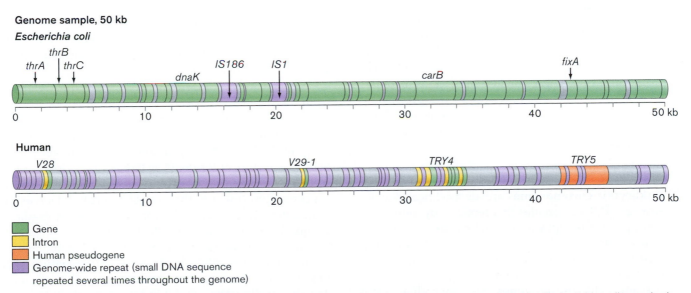

Figure 7.25 Genome structure in a prokaryote and in a eukaryote. The genome of the prokaryote *Escherichia coli* is packed with coding genes (green), with very little unused sequence between them (purple), only an occasional mobile element, such as an insertion sequence (IS). Most genes are organized in operons coordinately regulated by a single regulatory molecule such as *thrABC*. By contrast, the human genome consists of 95% noncoding sequences consisting largely of regulatory sequences and the decayed genomes of ancient viruses. The coding genes are separated by large stretches and usually are not situated together in operons. Moreover, coding genes are interrupted by introns (yellow). Ancient gene duplications have decayed into pseudogenes (orange).

The amount of noncoding DNA present in a eukaryote varies with the complexity of the eukaryotic species (for example, yeasts contain a much lower percentage of noncoding DNA than humans). It is believed that eukaryotes have retained noncoding DNA in their genomes because they rely heavily on it somehow for survival.

By themselves, promoters of eukaryotic genes have a very low activity and require enhancer DNA sequences to drive transcriptional activity. Enhancer sequences act at large distances from promoters, and scientists suspect that once enhancers became important, it was necessary to place enough DNA between them to reduce activation of other, unrelated but adjacent promotors.

Archaeal Genomes Combine Features of Bacteria and Eukaryotes

Like bacteria, archaea have polygenic operons, and their reproduction is predominantly asexual. Archaea are true prokaryotes in that their cells lack a nuclear membrane. On the other hand, in most species of archaea, the structures of the DNA-packing proteins, RNA polymerase, and ribosomal components more closely resemble those of eukaryotes. Even the DNA polymerase and the origin recognition sequence show greater similarity to those of eukaryotes. Finally, it should be noted that archaeal genomes encode certain unique components such as the metabolic pathway of methanogenesis. (For more on archaea, see Chapter 19.)

What experimental data allow us to make such comparisons? Overwhelmingly, we rely on new data from the growing number of microbial genomes sequenced. Comparison of genomes has revolutionized our understanding of evolutionary relationships among microbes.

We will now examine the tools of DNA sequence analysis that have made these studies possible. The most important of these tools—restriction mapping, DNA sequencing, and amplification by the polymerase chain reaction—actually harness the molecular apparatus used by bacteria to replicate or protect their own chromosomes.

TO SUMMARIZE:

■ **Eukaryotic chromosomes** are always linear double-stranded DNA molecules that replicate by mitosis.

■ A **reverse transcriptase** called telomerase is needed to finish replicating the ends of a eukaryotic chromosome.

■ **Histones** (eukaryotic DNA-packing proteins) play a critical role in forming chromosomes.

■ **Introns and pseudogenes** are noncoding DNA sequences that make up a large portion of eukaryotic chromosomes.

■ **Archaeal chromosomes** resemble those of bacteria in size and shape, but archaeal DNA polymerases are more closely related to eukaryotic enzymes.

7.6 DNA Sequence Analysis

We have just described the core concepts of DNA structure, packaging, and replication. This knowledge is crucial to understanding genomics and the fundamentals of genomic analysis. It is now appropriate to discuss the basic techniques used to manipulate DNA. These include isolating genomic DNA from cells, snipping out DNA fragments with surgical precision, splicing them into plasmid vehicles, and reading their nucleotide sequences. These are the techniques that drove the genomic revolution.

DNA Isolation and Purification

The chemical uniformity of DNA means that simple and reliable purification methods can be used to isolate it. A variety of techniques are used to extract DNA from bacterial cells. The cells may be lysed using lysozyme to degrade the peptidoglycan of the cell wall, followed by treatment with detergents to dissolve the cell membranes. The next step is to remove most proteins by precipitating them in a high salt solution. These proteins are removed by centrifugation, and the cleared lysate containing DNA is passed through a column containing a silica resin that specifically binds DNA. The remaining proteins are washed out of the column because they do not stick, and the DNA is eluted with water. The DNA is then concentrated via alcohol precipitation (ethanol or isopropanol). DNA may be precipitated from aqueous solution by adding ethanol and salt because the charge density of the phosphates makes the molecule particularly insoluble in nonaqueous solvents. The precipitated DNA is then redissolved in water or a weak buffer. At this point, the extracted DNA can be examined with a variety of analytical tools.

Plasmid or chromosomal DNA molecules released from lysed cells can also be purified using a technique known as equilibrium density gradient centrifugation. (The principles of centrifugation are discussed in Chapter 3.) The density gradient is typically generated using a cesium chloride (CsCl) solution. Under high centrifugal force, the heavy Cs^+ atoms will increase in concentration towards the end of the tube, decreasing their concentration near the top. This forms a gradient at equilibrium ranging from the least dense at the top of the tube to the densest at the bottom. DNA molecules dissolved in the solution prior to centrifugation will migrate along the gradient according to each molecule's density. Adding the DNA-binding dye ethidium bromide to the mix will magnify these differences. Because the dye binds in between bases (intercalates), it binds more to linear DNA (such as fragmented chromosomal DNA) than it does to supercoiled molecular plasmids. The different molecules will stop moving down the gradient at different points where their apparent densities match the density of the Cs⁺ gradient. No matter how long you spin, the bands stop at one density point, and they are extremely sharp. In this way, CsCl density gradient centrifugation differs from the sucrose gradient centrifugation, described in Chapter 3, where the gradient is prepared in a test tube *before* centrifugation. Besides isolating DNA molecules, CsCl density gradient centrifugation can be used to purify DNA/protein complexes and virus particles.

www | DNA purification protocols

Restriction Enzyme Digestion

Although the method has largely been supplanted in recent years by PCR techniques (**Special Topic 7.3**), cloning genes for study typically required cutting the chromosome into distinct pieces, or fragments, that could be selectively "plucked" from the sea of other genes composing the chromosome. Fortunately, bacteria themselves provided a way to produce these small fragments (discussed in Section 9.2). The bacterial proteins involved are called **restriction enzymes**, which naturally function to clip the "foreign DNA" of invading plasmids and phages at specific points. Different species of bacteria use restriction enzymes that recognize different DNA sequences.

The most useful restriction enzymes for molecular biology recognize specific base sequences (usually four to six bases in length) in a DNA molecule and cleave both phosphodiester backbones at sites either near or within the center of the site. The cut may produce blunt ends or staggered ends; in staggered ends, the top strand is cut at one end of the site and the bottom strand cut at the other (**Fig. 7.26A**). Staggered ends are also called *cohesive ends* because the protruding strand of one of those ends can base-pair with complementary protruding strands from any DNA fragment cut with the same restriction enzyme, regardless of the source organism. This ability of cohesive ends from different organisms to base-pair forms the basis of recombinant DNA technology.

Notice in **Figure 7.26A** that each recognition sequence, also called a restriction site, is a **palindrome** in which the top and bottom strands read the same in the 5′-to-3′ direction. This is typical of the sequences recognized by one class of restriction enzymes (there are three general classes). But how do bacteria making these scissor-like DNA enzymes keep from destroying their own DNA? They protect themselves by producing a companion methylating enzyme that recognizes the same DNA sequence recognized by the restriction enzyme and methylates a base in both strands of the

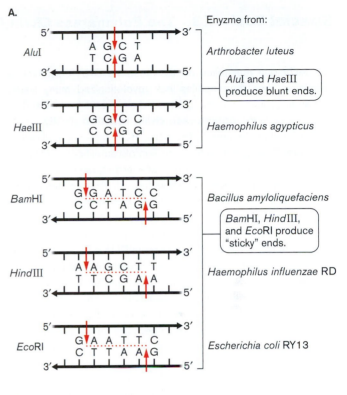

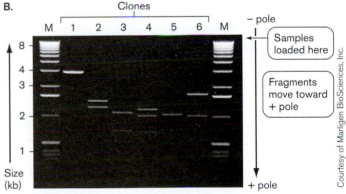

Figure 7.26 Bacterial DNA restriction enzymes. A. Target sequences of sample enzymes. The names of these enzymes reflect the genus and species of the source organism. For example, *Eco*RI comes from *Escherichia coli*. **B.** Agarose gel size analysis of *Eco*RI-restricted DNA fragments. M refers to marker fragments of known size. Smaller fragments move toward the bottom.

recognition sequence. The restriction enzyme cannot cut a site if either one or both of the strands are methylated. Thus, newly replicated double-stranded molecules are protected, since one strand, the template, remains methylated at all times. Phage DNA that originates from a cell with one type of restriction modification system will *not* be protected once it infects a cell with a different restriction modification system because the phage will not be methylated in the right places. As such, it faces a greater chance of destruction than of undergoing protective methylation.

Special Topic 7.3   The Polymerase Chain Reaction

A technique that relies on thermostable DNA polymerases isolated from thermophiles has revolutionized many fields, including biological research, medicine, and forensics. The technique is the **polymerase chain reaction (PCR)**, briefly described in Chapter 1. The basic PCR process, outlined in **Figure 1**, can produce over a millionfold amplification of target DNA. Using PCR, one can produce large quantities of a specific DNA sequence within a few hours. In this technique, specific oligonucleotide primers (usually between 20 and 30 bp) are annealed to known DNA sequences flanking the target gene. A thermostable DNA polymerase uses these primers to replicate the target DNA. Heat-stable DNA polymerases must be used in this process because the reaction mixture must be subjected to repeated cycles of heating to 95°C (to separate DNA strands so that they are available for primer annealing), cooling to 55°C (to allow primer annealing), and heating to 72°C (the optimum reaction temperature for the polymerase). The polymerase Taq, from the thermophile *Thermus aquaticus*, is often used for this purpose (polymerases from other thermophiles are also used). Because PCR reactions require 25–30 heating and cooling cycles, a machine called a thermocycler is used to reproducibly and rapidly deliver these cycles.

This basic PCR technique has been modified to serve many different purposes. Primers can be engineered to con-

tain specific restriction sites that simplify subsequent cloning. **Cloning** is the process by which DNA from one source is spliced into DNA from another source. **Restriction sites** are small (4–8 bp) sequences that are recognized and cleaved by endonucleases called restriction enzymes. If the primers used for PCR are highly specific for a gene that is present in only one microorganism, PCR can be used to detect the presence of that organism in a complex environment, such as the presence of the pathogen *E. coli* O157:H7 in hamburger.

PCR has profoundly impacted human society. By making it possible to amplify the tiniest amounts of DNA contaminating a crime scene, PCR has changed our judicial system by providing conclusive evidence in court cases where no evidence would have existed previously. And there may come a day when individual human genomes are sequenced as a standard medical test—with profound ethical and societal implications. The hopes and fears raised by the invention of PCR-based genomic sequencing inspired the science fiction film *GATTACA*, directed by Andrew Niccol, depicting an imaginary future in which everyone's destiny is determined by his or her DNA sequence. This knowledge could impact which jobs are available to someone, whether insurance coverage can be withheld, and even whether two individuals can marry.

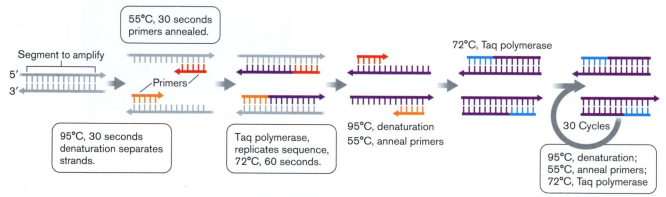

Figure 1   The polymerase chain reaction.   1. DNA template is heated to 95°C (30 seconds) to separate strands. 2. The reaction is cooled to 55°C for 30 seconds, which allows amplifying DNA primers to anneal. 3. Within seconds, the temperature is raised to 72°C, optimal for Taq polymerase activity. Polymerization is allowed to proceed for about 60 seconds, and then the strands are separated again at 95°C to prepare for another cycle. Each polymerization step increases (amplifies) the target sequence until, ultimately, only the fragment bounded by the primers is amplified. Thus, from a single DNA molecule, potentially 10^{30} copies of a fragment can be made.

Restriction enzymes from hundreds of bacteria are now commercially available for analyzing DNA. A few examples of such enzymes are shown in **Figure 7.26A**. Agarose gels can be used to analyze the DNA fragments obtained by treatment with these enzymes. In **Figure 7.26B**, each lane represents a different, homogeneous

population of DNA molecules cut with the same restriction enzyme. Because DNA is negatively charged, all DNA molecules travel to the positive pole during electrophoresis. Pore sizes in the agarose are such that they allow small molecules to speed through the gel, while larger molecules take longer to move. Thus, DNA fragments in this gel separate on the basis of size—the smaller the fragment is, the farther it travels down the lane. The inserted DNA fragments were of different sizes and so travel different distances in the gel. The sum of the sizes of fragments in each lane yields the size of the original, intact, uncut molecule.

Cloning: The Birth of Recombinant DNA Technology

The ability to clone genes rests on several seemingly unrelated findings. The discovery of restriction enzymes in the late 1960s was initially considered interesting but esoteric, and its significance was grossly underestimated at the time. Only later would the full impact of these enzymes on biological research be realized. The beginning of the recombinant DNA revolution can be traced to 1972. It was known at that time that plasmids were small circular DNA molecules capable of independent replication in bacterial cells. Some of the known plasmids contained a single site for certain restriction enzymes. As already noted, many restriction enzymes generate cohesive ends that can base-pair with the ends of any DNA cut with the same enzyme.

The significance of these facts went unrecognized until Stanley Cohen, Herb Boyer, and Stanley Falkow were relaxing with colleagues in a Waikiki deli after a scientific meeting in 1972 (**Figs. 7.27A** and **B**). In something of a "eureka" moment, the scientists suddenly realized that it might be possible to use a restriction enzyme (such as *Eco*RI) to cut a piece of DNA from one organism's chromosome and graft it in vitro into a plasmid cut in a single place with the same enzyme (**Fig. 7.27C**). They also knew that DNA ligase (see earlier discussion) would seal the fragment to the plasmid and form a new artificial

A.

B.

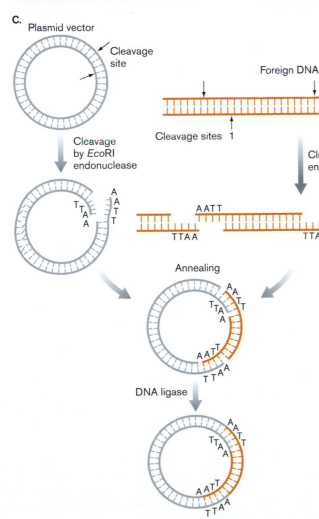

Figure 7.27   Formation of recombinant DNA olecules.
A. Stanley Cohen of Stanford University and **B.** Herbert Boyer of the University of California, San Francisco, two of the founders of recombinant DNA technology. **C.** A plasmid vector is cut with a restriction enzyme (*Eco*RI) that produces cohesive ends. The same enzyme is used to digest foreign DNA, producing a series of fragments all with identical cohesive ends. The cut vector and foreign DNA fragments are mixed. Cohesive ends of the foreign DNA and vector anneal to form a chimeric molecule containing two nicks. DNA ligase is used to seal the nicks and connect the phosphodiester backbones.

DNA molecule. This gene, as part of the recombinant molecule, could then be introduced into *E. coli* by transformation (discussed in Section 9.2) and be produced in large numbers through plasmid replication. Three months later, the strategy worked. This seminal work was published by Boyer and Cohen along with students Annie Chang and Robert Helling. Over the next decade, the technique became known as *gene cloning*, an immensely powerful technology that thrust biology into the genomic era.

www | Cloning strategies

One problem with earlier cloning techniques is that *E. coli* plasmid origins will not function (that is, will not initiate replication) in most other bacterial species, but especially eukaryotes or archaea. This means that cloned genes from these groups of organisms could be studied only in *E. coli*. The problem with studying these foreign genes in *E. coli* is that the processing of RNA or proteins naturally carried out in the eukaryote, for instance, will not be possible in *E. coli* because *E. coli* does not possess those systems. Nor can function of these eukaryotic proteins be tested in the prokaryotic cell. The solution to these problems is to use a **shuttle vector**. A shuttle vector is a plasmid that contains a replication origin compatible with *E. coli* and a second origin that will allow the plasmid to replicate in an unrelated bacterial species, or in a eukaryote or archaeon. In this way, the cloned gene can be genetically manipulated in *E. coli* using well-characterized molecular protocols (that is, creating specific mutations) and then be reintroduced into the source organism to study function.

THOUGHT QUESTION 7.9 Knowing that *E. coli* possesses restriction enzymes that cleave DNA from other species, how is it possible to clone a gene from one organism to another without the cloned gene sequence being degraded?

THOUGHT QUESTION 7.10 PCR is a powerful technique, but one can easily contaminate a sample and amplify the wrong DNA—perhaps sending an innocent person to jail. What might you do to minimize this possibility?

DNA Sequencing Based on Sanger Dideoxynucleotides

Even with the development of gene cloning, our advance into the genomic age would not have been possible without a rapid way to sequence large pieces of DNA. The most commonly used DNA-sequencing method relies on the Sanger dideoxy strategy. Ingenious in its simplicity, the dideoxy method relies on the fact that the 3′ hydroxyl

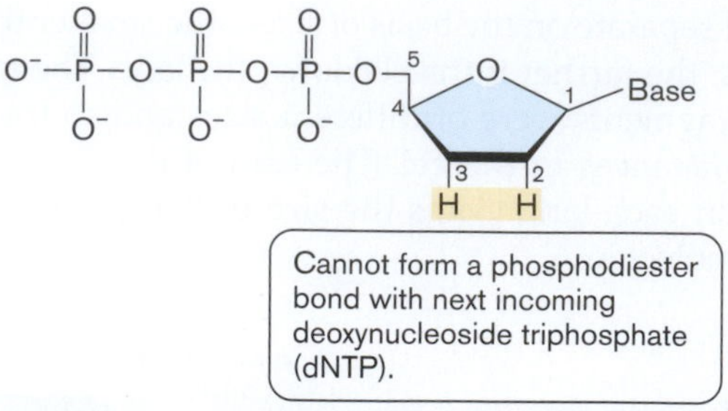

Figure 7.28   Dideoxy nucleotide. DNA nucleotides lack the 2′ OH group on ribose but retain the 3′ OH needed for DNA synthesis. Dideoxy nucleotides lack both the 2′ OH and the 3′ OH. Thus, the 5′ phosphate of a dideoxy nucleoside triphosphate can still form a phosphodiester bond to a growing chain but lacks the 3′ OH acceptor for a new incoming nucleotide.

group on 2′-deoxyribonucleotides (shown in **Fig. 7.3A**) is absolutely required for a DNA chain to grow. Thus, incorporation of a 2′,3′-dideoxynucleotide (**Fig. 7.28**) into a growing chain prevents further elongation because without the 3′ OH, extension of the phosphodiester backbone is impossible.

The dideoxy method begins with a short (20–30-bp) oligonucleotide DNA primer molecule designed to anneal to the plasmid vector at a site immediately adjacent to the cloned insert to be sequenced. The mixture is first heated to 95°C to separate the DNA strands and then cooled, which allows the primer to bind. DNA polymerase can then build on this primer and synthesize DNA using the cloned insert as a template. A small amount of dideoxyadenosine triphosphate (dideoxy ATP) is included in the DNA synthesis reaction, which already contains all four normal deoxyribonucleotides. Because normal 2′-deoxyadenosine is present at high concentration, chain elongation usually proceeds normally. However, the sequencing reaction contains just enough dideoxy ATP to make DNA polymerase occasionally substitute the dead-end base for the natural one, at which point the chain stops growing. The result is a population of DNA strands of varying size, each one truncated at a different adenine position.

Now imagine using four different dideoxynucleotides corresponding to A, T, C, and G, with each dideoxy base tagged with a different-colored fluorescent dye (**Fig. 7.29A**). The result will be a series of different-sized strands tagged at their 3′ end with different colors, depending on the base incorporated. Analyzing the results of this reaction by electrophoresis in a single lane of a polyacrylamide gel will separate the various fragments based on their size. A laser and detector positioned at the bottom of the gel reads the individual fragments as they pass (**Fig. 7.29B**). Because we know the color of each tagged base, we can use a computer to print a series of colored peaks whose order corresponds to the template DNA sequence (**Figs. 7.29C** and **7.30**). Although the fig-

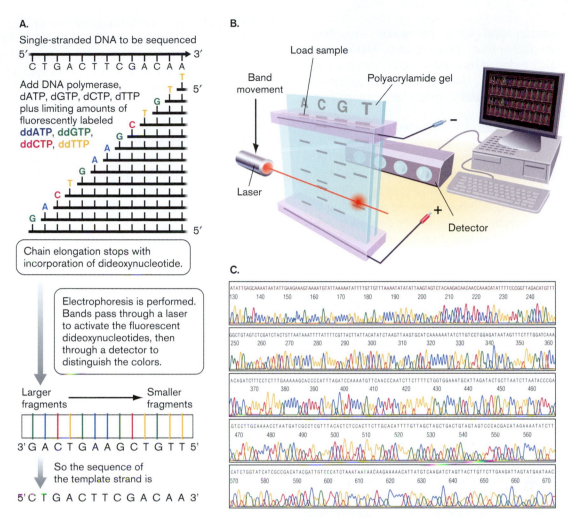

A.

Single-stranded DNA to be sequenced

5' ──────────── 3'
C T G A C T T C G A C A A

Add DNA polymerase,
dATP, dGTP, dCTP, dTTTP
plus limiting amounts of
fluorescently labeled
ddATP, **ddGTP**,
ddCTP, **ddTTP**

Chain elongation stops with
incorporation of dideoxynucleotide.

Electrophoresis is performed.
Bands pass through a laser
to activate the fluorescent
dideoxynucleotides, then
through a detector to
distinguish the colors.

Larger → Smaller
fragments   fragments

3' G A C T G A A G C T G T T 5'

So the sequence of
the template strand is

5' C T G A C T T C G A C A A 3'

B.

Load sample

Band movement

Polyacrylamide gel

A C G T

Laser

Detector

C.

Figure 7.29   DNA sequencing using fluorescently tagged dideoxynucleotides to randomly stop chain elongation.   The steps consist of synthesizing tagged strands using dideoxynucleotides and then separating the strands. **A, B.** The reaction products are separated based on size using a polyacrylamide gel apparatus that includes a laser and detector to specifically identify the different tagged fragments. **C.** The bases are then read and printed out as differently colored peaks. ▶❙❙

ure depicts a standard polyacrylamide slab gel, the rapid automated DNA sequencers used today for large-scale projects utilize small capillary tube gels.

Sequencing an Entire Genome

Sequencing a genome begins by indiscriminately cloning all the fragments of chromosomal DNA that have been randomly generated, either by restriction digestion (for small fragments) or by sonic breakage (for large pieces). The fragments can either be cloned into conventional plasmid vectors, which may only hold 40–50 kbp of insert DNA, or be cloned into super vectors known as bacterial artificial chromosomes (BACs) that can harbor inserts of 150–1,000 kbp. This random "shotgun" cloning approach must be done in such a way as to ensure overlapping cloned fragments. This is important when finally reassembling the DNA sequences read from various fragments into one contiguous sequence representing the entire chromosome.

Keith Weller, USDA

Figure 7.30   Rapid DNA sequencing using an automated DNA sequencer.   Cletus Kurtzman inspects a yeast DNA sequence from a previous run while Larry Tjarks loads samples of new sequencing reactions at the National Center for Agricultural Utilization Research. The lanes on the screen represent the sequences of different DNA molecules.

NOTE: Sonication and some restriction enzymes generate blunt, rather than cohesive ends. These molecules can still be cloned into linearized plasmids that also contain blunt ends. However, the efficiency of cloning blunt end molecules is much lower than with cohesive ends.

Next, each clone is sequenced using primers that originate from a sequence in the vector. The vector provides a clear starting point and a known sequence to design a primer. Typically, 500–1,000 bases can be read using the Sanger sequencing technique previously described. New primers are then designed based on the previously read sequence, and the sequencing team continues to "walk" along the fragment until its sequencing is complete. An alternative to "walking" along the chromosome is to break large (greater than 5 kb) cloned fragments into many overlapping, smaller fragments, which are then subcloned into another plasmid vector. The entire sequence of the smaller subclones can usually be determined using the vector primers. It is important to remember that both complementary strands must be sequenced in order to verify a fragment's sequence and that the region should be sequenced at least two or three times.

Once the fragment sequences are verified, overlapping sequences from different fragments are identified using supercomputers and assembled in silico (in a computer) into **contigs**, which are contiguous sequences spanning several fragments. Ideally, each chromosome of a genome will assemble into a single contig. However, if some gaps remain between contigs, focused sequencing can span them.

Due to constantly improving technologies, the length of time it takes to sequence an entire genome has shortened remarkably over the past decade. It originally took many years to sequence the 4.6-million-bp genome of *E. coli*. Today it would take less than a month. The top DNA sequencing laboratories can sequence nearly 10 million bases per month.

How the information from sequences is mined will be discussed more in Chapter 8. Suffice it to say, many important discoveries have resulted from this technology. A particularly intriguing recent discovery was finding the entire genome of the bacterium *Wolbachia* (which infects 20% of the world's insect population) is embedded within the genome of *Drosophila*, the common fruit fly. The evolutionary implications of this are under intense investigation.

TO SUMMARIZE:

- **DNA restriction enzymes** used for DNA analysis cleave DNA at specific recognition sequences, which are usually 4–6 bp in length and produce either a blunt or a staggered cut.
- **Agarose electrophoresis** will separate DNA molecules based on their size.
- **Restriction enzyme digested DNA** molecules were first cloned into plasmids in the early 1970s.
- **The most commonly used DNA sequencing** technique (Sanger technique) uses nucleotides that lack both the 2′ OH and 3′ OH groups to terminate chain elongation.
- **Entire genomes** are sequenced by shotgun cloning of overlapping DNA fragments, each of which is sequenced. Overlapping regions are matched and joined by computer analysis until the entire genome is reconstructed.

Concluding Thoughts

We used to think of bacteria as simple, single-celled organisms. While they are single cells, they are far from simple. Complexity is evident in the elegant mechanisms they use to organize and replicate their genomes. The next three chapters will expand on this idea and discuss the way bacteria make proteins, regulate genes, exchange DNA, and, in the process, evolve. Knowing how a bacterial cell replicates and genetically controls the repertoire of available proteins and enzymes provides a unique perspective from which to study the physiology of microbial growth, explored in Part 3 of this textbook. Furthermore, understanding the bacterial genome allows us to explore the physiology of pathogenic as well as nonculturable organisms and gain greater insight into the evolutionary diversity of species that we will discuss in Part 4.

CHAPTER REVIEW

Review Questions

1. What is the difference between vertical and horizontal gene transmission?
2. Explain the structural types of bacterial genomes. What is a structural gene?
3. Describe the four functional levels of gene organization.
4. What are the differences between DNA and RNA?

5. Explain DNA supercoiling. Why is it important to microbial genomes?
6. Discuss the mechanisms of topoisomerases. What drug targets a topoisomerase?
7. What are the basic mechanisms involved in DNA replication?
8. How does the bacterial cell regulate the initiation of chromosome replication?
9. What is the gamma complex? Primase? DnaB? DnaC? DNA proofreading?
10. How is the problem of replicating both strands at a replication fork solved?
11. What is a catenane? What does it have to do with DNA replication?
12. Explain the polymerase chain reaction.
13. What is rolling-circle replication?
14. What is DNA restriction and modification? Why is it important to bacteria? Why is it important to forensic scientists?
15. Explain the basic technique of DNA sequencing.

Key Terms

antiparallel (226)
catenane (241)
cloning (250)
conjugation (223)
contig (254)
denature (227)
DNA control sequence (222)
DNA ligase (240)
enhancer (224)
exonuclease (239)
histone (247)
horizontal transmission (222)
hybridization (228)
intron (247)

Okazaki fragments (234)
operon (225)
origin (*oriC*) (234)
palindrome (249)
phosphodiester bond (226)
plasmid (243)
polymerase chain reaction (PCR) (250)
primase (236)
promoter (224)
proofreading (239)
pseudogene (247)
purine (227)
pyrimidine (227)
quinolone antibiotic (230)

recombination (223)
regulon (225)
replication fork (233)
restriction enzyme (249)
restriction site (250)
semiconservative (233)
shuttle vector (252)
sliding clamp (236)
structural gene (222)
termination (*ter*) site (234)
topoisomerase (230)
vertical transmission (222)

Recommended Reading

Actis, Luis, Marcelo E. Tolmasky, and Jorge H. Crosa. 1998. Bacterial plasmids: Replication of extrachromosomal genetic elements encoding resistance to antimicrobial compounds. *Frontiers in Bioscience* 3:D43–62.

Amick, Jean D., and Yves V. Brun. 2001. Anatomy of a bacterial cell cycle. *Genome Biology* 2:1020.1–1020.4.

Boeneman, Kelly, and Elliott Crooke. 2005. Chromosomal replication and the cell membrane. *Current Opinion in Microbiology* 8:143–148.

Cohen, Stanley N., Annie C. Chang, Herbert W. Boyer, and Robert B. Helling. 1973. Construction of biologically functional bacterial plasmids in vitro. *Proceedings of the National Academy of Science* 70:3240–3244.

Dame, Remus T. 2005. The role of nucleoid-associated proteins in the organization and compaction of bacterial chromatin. *Molecular Microbiology* 56:858–870.

Falkow, Stanley. 2001. I'll have chopped liver please, or how I learned to love the clone. *ASM News* 67:555.

Forterre, Patrick, Simon Gribaldo, Daniele Gadelle, and Marie-Claude Serre. 2007. Origin and evolution of DNA topoisomerases. *Biochimie* 89:427–446.

Giovannoni, Stephen J., H. James Tripp, Scott Givan, Mircea Podar, Kevin L. Vergin, et al. 2005. Genome streamlining in a cosmopolitan oceanic bacterium. *Science* 309:1242–1245.

Khodursky, Arkady B., and Nicholas Cozzarelli. 1998. The mechanism of inhibition of topoisomerase IV by quinolone antibacterials. *Journal of Biological Chemistry* 273:27668–27677.

Lemon, Katherine P., and Alan D. Grossman. 1998. Localization of bacterial DNA polymerase: Evidence for a factory model of replication. *Science* 282:1516–1519.

Mott, Melissa L., and James M. Berger. 2007. DNA replication initiation: Mechanisms and regulation in bacteria. *Nature Reviews Microbiology* 5:343–354.

Myllykallio, Hannu, Philippe Lopez, Purificación López-Garcia, Roland Heilig, William Saurin, et al. 2000. Bacterial mode of replication with eukaryotic-like machinery in a hyperthermophilic archaeon. *Science* 288:2212–2215.

Rothfield, L.awrence, Sheryl Justice, and Jorge Garcia-Lara. 1999. Bacterial cell division. *Annual Review of Genetics* **33:**423–428.

Stukenberg, P. Todd, Patricia S. Studwell-Vaughn, and Mike O'Donnell. 1991. Mechanism of the sliding β clamp of DNA polymerase III holoenzyme. *Journal of Biological Chemistry* **266:**11328–11334.

Thanbichler, Martin, Patrick H. Viollier, and Lucy Shapiro. 2005. The structure and function of the bacterial chromosome. *Current Opinion in Genetic Development* **115:**153–162.

Vicente, Miguel, Ana Isabel Rico, Rocio Martinez-Arteaga, and Jesus Mingorance. 2006. Septum enlightenment: Assembly of bacterial division proteins. *Journal of Bacteriology* **188:**19–27.

Chapter 8

Transcription, Translation, and Bioinformatics

The cell accesses the vast store of data in its genome through transcription, which is the reading of a DNA template to make an RNA copy, and translation, the decoding of RNA to assemble protein. The molecular machines that carry out these processes have been studied relentlessly, yet they still hold secrets.

After translation, each polypeptide must be properly folded and placed at the correct cellular or extracellular location. How does the cell do this? And what does the cell do with proteins it no longer needs? As we will discuss, elegant chaperone pathways help fold proteins; complex transport systems move proteins out of the cytoplasm; and regulated proteolytic systems properly dispose of unneeded or damaged proteins. What emerges is a picture of remarkable biomolecular integration, homeostatically controlled to maintain balanced growth and ensure survival.

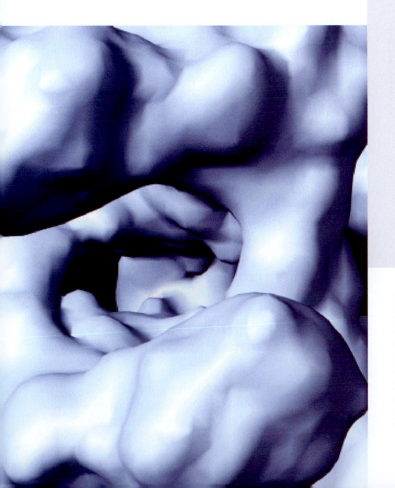

The *Escherichia coli* DegP protein is a periplasmic protease that degrades misfolded proteins including "sentry" regulators that control certain stress responses. It has been hypothesized that unfolded polypeptides are threaded in an extended conformation into the cage to access the proteolytic sites. The image is a close-up view of the inner cavity of DegP. *Source*: Ahmad Jomaa, et al. 2007. *Journal of Bacteriology* 189:706; © 2007, American Society for Microbiology. All Rights Reserved.

By 1960, it was clear that the genetic code was embedded in DNA, but the code itself remained a mystery, as was the mechanism by which the code produced protein. RNA was a suspected, but unproven, intermediate. Marshall Nirenberg and Heinrich Matthaei, a postdoctoral student of Nirenberg's at the National Institutes of Health, set out in 1959–1960 to design a cell-free system (a cell lysate of *E. coli*) to test the hypothesis (**Fig. 8.1A**). Key to their investigation was the use of synthetic RNA molecules containing simple, known, repeated sequences such as poly-A (consisting only of adenylic acid), poly-U (poly-uridylic acid), poly-AAU, and poly-AAAAU. These RNA molecules were tested in a cell-free protein-synthesizing system to see if the repetitive sequence might direct incorporation of a specific amino acid into a protein. Each synthetic polynucleotide was tested in the presence of a radiolabeled amino acid. If the radioactive amino acid was incorporated into a polypeptide, the polypeptide would also be radioactive. On the morning of May 27, 1960, the results of experiment 27Q indicated that the poly-U RNA specified the assembly of radioactive poly-phenylalanine. It was the first break in the genetic code.

B.

A.

Courtesy of National Library of Medicine/
National Institute of Health

Courtesy of Maxine Singer

Figure 8.1 Many scientists contributed to understanding the genetic code. A. Heinrich Matthaei (left) and Marshall Nirenberg (right) were first to crack the genetic code. An early hand-drawn model of what would eventually be known as translation is behind them. **B.** Maxine Singer, a key contributor to the genetic code experiments, also helped develop guidelines for recombinant DNA research.

Nirenberg reported the result at a conference in Moscow at the height of the Cold War. News that Nirenberg's poly-U experiment had determined the first "word" of the genetic code was an international media event. In January 1962, the *Chicago Sun-Times* announced: "No stronger proof of the universality of all life has been developed since Charles Darwin's *The Origin of Species* . . . In the far future, the . . . hereditary lineup will be so well known that science may deal with the aberrations of DNA . . . that produce cancer, aging, and other weaknesses of the flesh." This statement was truly prophetic, for understanding the genetic code was the first step in discovering the basis of many genetic diseases. For his work, Nirenberg, along with Har Gobind Khohara (who developed methods for making synthetic nucleic acids) and Robert Holley (who solved the structure of yeast transfer RNA), won the 1968 Nobel Prize in Physiology or Medicine. Maxine Singer, who also played an important role in deciphering the genetic code by synthesizing the RNA molecules used in the Nirenberg experiments, became a leader in developing guidelines for recombinant DNA research (**Fig. 8.1B**).

Since Nirenberg's breakthrough, we have learned a tremendous amount about genes and proteins and, in this electronic age, we can use the information in ways never before possible. A new discipline called bioinformatics uses powerful computer technologies to store and analyze gene and protein sequences. Bioinformatic programs can compare a new gene sequence with hundreds of thousands of known genes. These comparisons help predict for any given microbe what genes are present, what proteins are made, and even what food it consumes. Patterns in DNA sequences across species also allow us to pose new questions about microbial life, disease, and evolution.

In this chapter, we use *E. coli* as a paradigm to explore the way microbes interpret the nucleotide sequence of DNA and convert it to proteins. From there we look at what the cell does with those proteins once they are made. Specific proteins are targeted for movement into the periplasm, while other proteins must be inserted into membranes. We also show how damaged proteins are selectively degraded. We conclude with bioinformatics, now an essential tool of the new microbiology. Chapters 9 and 10 then discuss how microbes respond to their environment by controlling the expression of their genes.

8.1 RNA Polymerases and Sigma Factors

To survive and reproduce, every cell needs to form proteins using the information encoded within DNA. Chromosomal DNA is large and cumbersome, so the first step in the process is to make multiple copies of the information in snippets of RNA that can move around the cell and, like disposable photocopies, be destroyed once the encoded protein is no longer needed. This copying of the DNA to RNA is called **transcription**. Much of what we have learned about transcription has come from studying *E. coli*. However, bioinformatic analyses indicate that the mechanisms and components of transcription in other bacterial species are similar to those characterized in *E. coli*.

RNA Polymerase Transcribes DNA to RNA

An enzyme complex called RNA polymerase, also known as DNA-dependent RNA polymerase, carries out the process of transcription, making RNA copies (called **transcripts**) of a DNA template. The DNA **template strand** specifies the base sequence of the new complementary strand of RNA. Transcription involves (1) initiation, which is the binding of RNA polymerase to the beginning of the gene, followed by melting of the helix and synthesis of the first nucleotide of the RNA; (2) elongation, the sequential addition of ribonucleotides from nucleoside triphosphates; and (3) termination, in which

sequences at the end of the gene trigger release of the polymerase and the completed RNA molecule.

RNA polymerase in bacteria consists of a "core polymerase," which is a protein complex containing the *essential* components required for elongation of the RNA chain, plus a **sigma factor**, a protein needed only for initiation of RNA synthesis, not for its elongation. Core polymerase plus sigma factor together are called "holoenzyme."

Core RNA polymerase is a complex of four different subunits: two alpha (α) subunits, one beta (β) subunit, and one beta-prime (β') subunit for every core enzyme (**Fig. 8.2**). The beta-prime subunit contains the Mg^{2+}-containing catalytic site for RNA synthesis, as well as sites for the ribonucleotide and DNA substrates and the RNA products. The three-dimensional structure of RNA polymerase resembles a hand, whose "fingers" consist of the beta and beta-prime subunits (**Fig. 8.3**). DNA fits into a cleft formed by the beta and beta-prime subunits (**Fig. 8.2**). The alpha subunit assembles the other two subunits into a functional complex and communicates through physical "touch" with

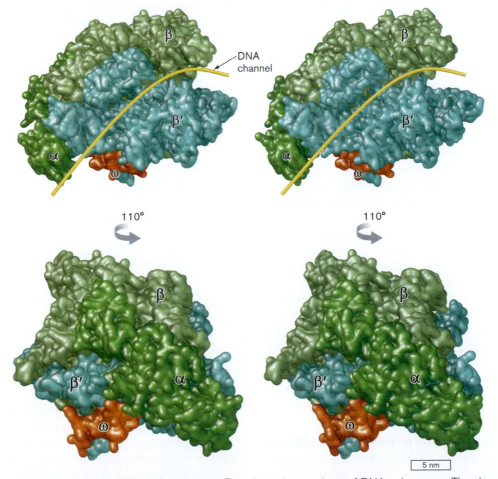

Figure 8.2　Subunit structure of core RNA polymerase.　Two stereo-image views of RNA polymerase. The channel for the DNA template is shown by the yellow line. The molecule is rotated as shown. Subunits are color-coded. Omega is a subunit of unclear function sometimes associated with RNA polymerase. To view the stereo-images, place a piece of cardboard vertically between the two images. Position your eyes on opposite sides of the cardboard and force them to focus behind the images. The images will merge and produce a 3-D image. (PDB code: 1HQM) *Source*: Robert D. Finn, Elena V. Orlova, Brent Gowen, Martin Buck, and Marin van Heel. 2000. *Escherichia coli* RNA polymerase core and holoenzyme structures. *EMBO J.* 19(24):6833–6844.

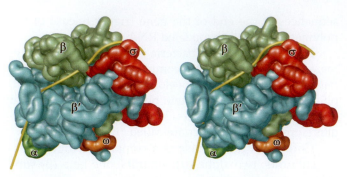

Figure 8.3 Three-dimensional structure of RNA polymerase. RNA polymerase holoenzyme. Note the open channel able to accept DNA. To view the stereo-image, follow the instructions from Figure 8.2. (PDB code: 1L9U)

various regulatory proteins that can bind DNA. The resulting protein-protein communications instruct RNA polymerase in what to do after binding DNA.

Promoter binding. Where does a gene begin and end? The helical structure of a DNA sequence appears largely homogeneous compared to proteins, which bend and fold in complex ways. In fact, without sigma factor, the core RNA polymerase binds and releases DNA at random. Yet there must be a mechanism that tells RNA polymerase where a gene starts, because random transcription is extremely wasteful.

The proteins responsible for guiding RNA polymerase to genes are the sigma (σ) factors (see **Fig. 8.3B**). A sigma factor binds RNA polymerase through the alpha subunit and then helps the core enzyme detect a specific DNA sequence, called a **promoter**, which signals the beginning of a gene. We will discuss how sigma recognizes a promoter later.

A single bacterial species can make several different sigma factors, with each sigma factor helping core RNA polymerase find the start of a different subset of genes. However, a single core polymerase complex can bind only one sigma factor at a time. By acting as a sort of "seeing-eye" protein, sigma factors help RNA polymerase recognize different classes of promoter sequences. The specific sigma factor used to initiate transcription of a given gene will vary, depending on the gene and on the environmental signals needed to initiate transcription of that gene. However, regardless of which sigma factor is used, core RNA polymerase is required for all gene transcription.

Sigma factors regulate major physiological responses. Individual sigma factors coordinately control genes involved in nitrogen metabolism, flagellar synthesis, heat stress, starvation, sporulation, and many other physiological responses. So far, over 100 sigma factor genes have been sequenced from numerous species. While they all have different amino acid sequences, they also show sequence similarities that make them recognizable as sigma factors. The DNA sequence recognized by a given sigma factor can be determined by comparing known promoter sequences of different genes whose expression requires the same sigma. The DNA sequence similarities seen when comparing these different promoters defines a **consensus sequence** likely recognized by the sigma factor (**Fig. 8.4A**). A consensus sequence consists of the most likely base (or bases) at each position of the predicted promoter. Note that while the promoter is a dsDNA sequence, convention is to present the promoter as the ssDNA sequence of the sense strand, which matches the RNA product sequence.

The best consensus sequences are based on a large set of promoters that use the same sigma factor, or other regulator. Some positions in a consensus sequence are highly conserved—that is, the same base is found in that position in every promoter. Other, less conserved positions can be occupied by different bases. Few promoters will actually have the most common base at every position. Even highly efficient promoters usually differ from the consensus at one or two positions.

Microbes are guarded by an array of protein sensors that continually sample the internal and external environments. These sensors "look" for chemical deficiencies or dangers and, when triggered, direct the cell to increase synthesis or decrease destruction of the appropriate sigma factor. Accumulation of a specialty sigma factor dislodges other sigma factors from core polymerase and redirects RNA polymerase to the promoters of genes whose products can best solve the problem sensed. Regulation of multiple genes by a single sigma factor enables the cell to coordinately control those genes simply by regulating when a given sigma factor accumulates (Section 10.4). This control is critical to survival in constantly changing environments.

> **THOUGHT QUESTION 8.1** If each sigma factor recognizes a different promoter, how does the cell manage to transcribe genes that respond to multiple stresses, each involving a different sigma factor?

Every cell has a "housekeeping" sigma factor that keeps essential genes and pathways operating. In the case of *E. coli* and other gram-negative rod-shaped bacteria, that factor is sigma-70, so named because it is a 70-kilodalton (kDa) protein (its gene designation is *rpoD*). Genes recognized by sigma-70 all contain similar promoter consensus sequences consisting of two parts. The DNA base corresponding to the start of the RNA transcript is called nucleotide +1 (+1 nt). Relative to this landmark, the consensus promoter sequences are characteristically centered at −10 and −35 nt before the start of transcription (see **Fig. 8.4**). Other sigma factors typically recognize different

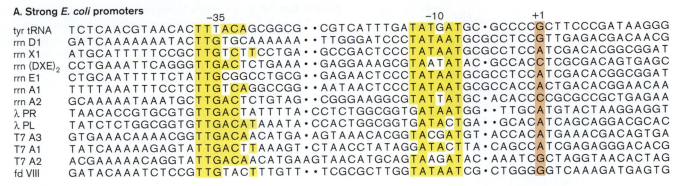

A. Strong E. coli promoters

```
                    −35                              −10              +1
tyr tRNA   TCTCAACGTAACACTTTACAGCGGCG··CGTCATTTGATATGATGC·GCCCCGCTTCCCGATAAGGG
rrn D1     GATCAAAAAAATACTTGTGCAAAAAA··TTGGGATCCCTATAATGCGCCTCCGTTGAGACGCAACG
rrn X1     ATGCATTTTTCCGCTTGTCTTCCTGA··GCCGACTCCCTATAATGCGCCTCCATCGACACGGCGGAT
rrn (DXE)2 CCTGAAATTCAGGGTTGACTCTGAAA··GAGGAAAGCGTAATATAC·GCCACCTCGCGACAGTGAGC
rrn E1     CTGCAATTTTTCTATTGCGGCCTGCG··GAGAACTCCCTATAATGCGCCTCCATCGACACGGCGGAT
rrn A1     TTTTAAATTTCCTCTTGTCAGGCCGG··AATAACTCCCTATAATGCGCCACCACTGACACGGAACAA
rrn A2     GCAAAAATAAATGCTTGACTCTGTAG··CGGGAAGGCGTATTATGC·ACACCCCGCGCCGCTGAGAA
λ PR       TAACACCGTGCGTGTTGACTATTTTA·CCTCTGGCGGTGATAATGG··TTGCATGTACTAAGGAGGT
λ PL       TATCTCTGGCGGTGTTGACATAAATA·CCACTGGCGGTGATACTGA··GCACATCAGCAGGACGCAC
T7 A3      GTGAAACAAAACGGTTGACAACATGA·AGTAAACACGGTACGATGT·ACCACATGAAACGACAGTGA
T7 A1      TATCAAAAAGAGTATTGACTTAAAGT·CTAACCTATAGGATACTTA·CAGCCATCGAGAGGGACACG
T7 A2      ACGAAAAACAGGTATTGACAACATGAAGTAACATGCAGTAAGATAC·AAATCGCTAGGTAACACTAG
fd VIII    GATACAAATCTCCGTTGTACTTTGTT··TCGCGCTTGGTATAATCG·CTGGGGGTCAAAGATGAGTG
```

B. Consensus sequences of σ⁷⁰ promoters

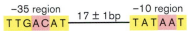

```
−35 region              −10 region
TTGACAT   17 ± 1bp   TATAAT
```

C. Sequence of lac promoter

```
−35 region              −10 region
TTTACAC                 TATGTT
   │ │                     │ │
   A A                     A A
  └───┘                   └───┘
  Down                     Up
```

Figure 8.4   **−10 and −35 sequences of E. coli promoters. A.** Alignment of sigma-70 (σ⁷⁰)-dependent promoters from different genes is used to generate a consensus sequence in panel B (yellow indicates conserved nucleotides, brown denotes transcript start site). **B.** Consensus sequences of σ⁷⁰-dependent promoters (red-shaded letters indicate nucleotide positions where different promoters show a high degree of variability). **C.** Mutations in the *lac* promoter that affect promoter strength (*lac* genes encode proteins used to metabolize the carbohydrate lactose). Some mutations can cause decreased transcription (called down mutations), while others cause increased transcription (up mutations). **D.** Some E. coli promoter sequences recognized by different sigma factors.

D.

Sigma factor	Promoter recognized	Promoter consensus sequence	
		−35 region	−10 region
RpoD σ⁷⁰	Most genes	TTGACAT	TATAAT
RpoH σ³²	Heat-shock-induced genes	TCTCNCCCTTGAA	CCCCATNTA
RpoF σ²⁸	Genes for motility and chemotaxis	CTAAA	CCGATAT
RpoS σ³⁸	Stationary phase and stress response genes	TTGACA	TCTATACTT
		−24 region	−12 region
RpoN σ⁵⁴	Genes for nitrogen metabolism and other functions	CTGGNA	TTGCA

consensus sequences at one or both of these positions (or at nearby locations in some cases); or they shape the overall polymerase complex to bind the promoter.

How do sigma factors, or any other DNA-binding protein for that matter, recognize specific DNA sequences when the DNA is a double helix? The phosphodiester backbone is quite uniform, and the bases appear inaccessible owing to base pairing. Recognition of DNA sequences is possible, however, because the alpha-helical portions of DNA-binding proteins recognize base side groups (noncovalent binding) that protrude from the base into the major and minor grooves of DNA. **Figure 8.5A** illustrates how portions of sigma-70 from *E. coli* wrap around DNA allowing certain parts of the protein to fit into DNA grooves.

A key to understanding how sigma factors recognize different promoters is their structure. Based on amino acid sequence comparisons, sigma factors generally contain four highly conserved regions (**Figs. 8.5A** and **B**). Part of region 2 of the sigma-70 family recognizes −10 sequences, while region 4 recognizes the −35 sites. A part of region 1 helps separate the DNA strands in preparation for RNA synthesis. The structure of the holoenzyme RNA polymerase complex positioned at a promoter is shown in **Figure 8.5C**.

THOUGHT QUESTION 8.2   With respect to two different sigma factors with different promoter recognition sequences, predict what would happen to the overall gene expression profile in the cell if one sigma factor were artificially overexpressed? Could there be a detrimental effect on growth?

THOUGHT QUESTION 8.3   Why might some genes contain multiple promoters, each one specific for a different sigma factor?

A.

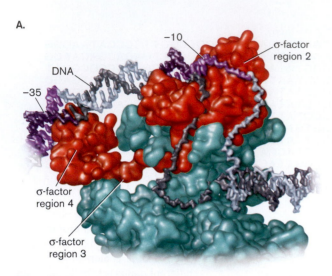

Figure 8.5 **A.** *E. coli* sigma factor binds to two sequences in a DNA promoter. Region 2 of sigma binds to the −10 region of promoters, while region 4 binds to the −35 promoter region. Cyan marks part of core RNA-P. Purple on DNA denotes the −10 and −35 promoter regions. **B.** Sigma-70 amino acid homology and domain interactions with different regions of the promoter sequence. Conserved amino acid regions within examples of the sigma-70 family. **C.** Taq holoenzyme/promoter DNA complex. Double-stranded DNA is shown as atoms. The possible locations of the two alpha subunit C-terminal domains (drawn as gray spheres, labeled I and II) on the DNA UP elements is illustrated. The UP element is a DNA sequence that increases transcription. (PDB codes: 1L9Z, 1K8J) *Source*: A. Seth A. Darst. 2001. Bacterial RNA polymerase. *Current Opinion in Structural Biology* 11:155–162.

B.

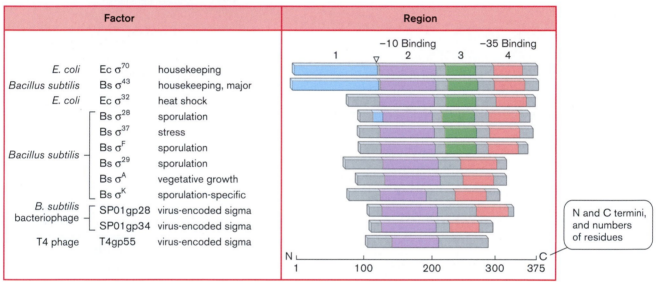

C.

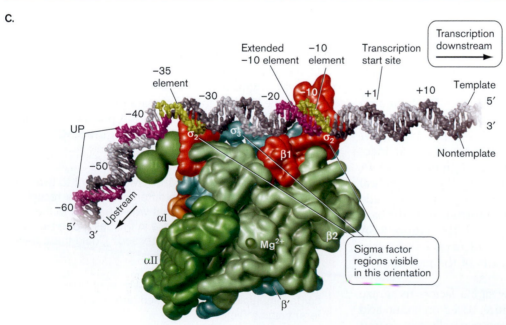

TO SUMMARIZE:

- **RNA polymerase holoenzyme** consisting of core RNA polymerase and a sigma factor executes transcription of a DNA template strand.
- **Sigma factors** help core RNA polymerase locate consensus promoter sequences near the beginning of a gene. The sequences identified and bound by *E. coli* sigma-70 are located –10 and –35 bp upstream of the transcription start site.
- **Dynamic changes in gene expression** result when the relative levels of different sigma factors change.

8.2 Transcription Initiation, Elongation, and Termination

Sigma binding is the first step of transcription initiation. Once the promoter has been activated by binding with its sigma factor, RNA polymerase completes the process of

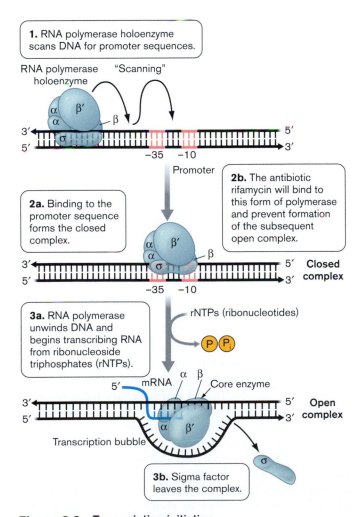

Figure 8.6 Transcription initiation.

initiation, in which the closed RNA polymerase complex becomes an open complex. In elongation, the RNA chain is extended, followed by termination, in which the RNA polymerase comes apart from the DNA.

Initiation of Transcription

RNA polymerase constantly scans DNA for promoter sequences (**Fig. 8.6**, step 1). It binds DNA loosely and comes off repeatedly. Once bound to the promoter, RNA polymerase holoenzyme forms a loosely bound, closed complex with DNA, which remains annealed and double-stranded (that is, unmelted) (step 2). To successfully transcribe a gene, this closed complex must become an open complex through the unwinding of one helical turn, which causes DNA to become unpaired in this area (step 3). After promoter unwinding, RNA polymerase in the open complex becomes tightly bound to DNA. Region 1 of the sigma factor is important for DNA unwinding, as is a "rudder" complex in β' that separates the two stands (shown in **Fig. 8.7A**).

The open complex form of RNA polymerase begins transcription. The first ribonucleoside triphosphate (rNTP) of the new RNA chain is usually a purine. It base-pairs to the DNA template designated position +1, marking the start of the gene. As the enzyme complex moves along the template, subsequent NTPs diffuse through a channel in the polymerase and into position at the DNA template (see **Fig. 8.7B**). After the first base is in place, each subsequent ribonucleoside triphosphate transfers a nucleoside monophosphate to the growing chain while releasing a pyrophosphate:

$$\text{RNA—3'—OH} + \text{}^-\text{O—P—O—P—O—P—O—nucleoside—3'—OH}$$

$$\downarrow$$

$$\text{RNA—3'—O—P—O—nucleoside—3'—OH} + \text{}^-\text{O—P—O—P—O}^-$$

The energy released by cleaving the rNTP triphosphate groups (phosphoric anhydride bond) is used to create the phosphodiester link to the growing polynucleotide chain.

As seen in **Figure 8.7**, the β' subunit forms part of the pocket, or cleft, in which the DNA template is read. Another part of the β' cleft is involved in hydrolyzing ribonucleoside triphosphates. Deep at the base of the cleft is the active site of polymerization, defined by three evolutionarily conserved aspartate residues that chelate (that is, bond with) two magnesium ions (Mg^{2+}). As we saw for DNA polymerases, these metal ions play a key role in catalyzing the polymerization reaction.

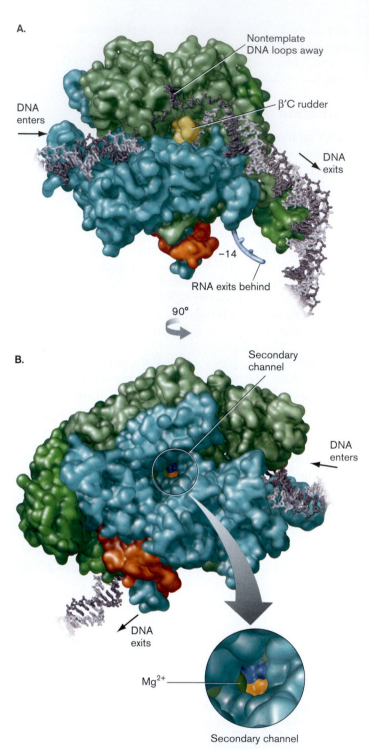

A.

Nontemplate
DNA loops away

DNA
enters

β'C rudder

DNA
exits

−14

RNA exits behind

90°

B.

Secondary
channel

DNA
enters

DNA
exits

Mg²⁺

Secondary channel

Figure 8.7   Three-dimensional view of transcription by *Thermus aquaticus* **(Taq) RNA polymerase showing DNA, RNA, and the position of the rudder.**   The process is shown from two different angles. Transcription (polymerase movement on the DNA) is going from right to left in panel A and the opposite in panel B. Color-coded molecular surfaces are as follows: beta subunit, green; beta G flap in red, beta-prime C rudder in gold; alpha and omega, not seen in this view; the DNA template (template strand, purple; nontemplate strand, gray); and the RNA transcript (blue). The directions of the entering downstream DNA and exiting upstream DNA are indicated (black arrows). **A.** A view perpendicular to the main active site channel, which runs roughly horizontal. Parts of the protein structure are colored. **B.** In this view, the structure is rotated 90° relative to view (A). With the entering DNA removed for clarity, Mg²⁺ is seen in the catalytic site; the secondary channel where nucleotides (orange) enter and a part of the RNA strand (blue) are also seen. (PDB codes: 1L9Z, 1K8J) *Source: A. Seth A. Darst. 2001. Bacterial RNA polymerase. Current Opinion in Structural Biology 11:155–162.*

the template, synthesizing RNA at approximately 45 bases per second. The unwinding of DNA ahead of the moving complex forms a 17-bp transcription bubble. Because of this unwinding, positive supercoils are formed ahead of the advancing bubble. These positive supercoils are removed by enzymes that return the DNA to its normal negative supercoiled state (see Section 7.2).

Termination of transcription. We have learned how RNA polymerase recognizes the beginning of a gene and transcribes RNA, but how does it know when to stop? Again, the secret is in the sequence. All bacterial genes use one of two known transcription termination signals. One termination mechanism, called **Rho-dependent**, relies on a protein called Rho and an ill-defined sequence at the 3′ end of the gene that appears to be a strong pause site. Pause sites are areas of DNA that slow or stall RNA polymerase movement. Some pause sites occur within the gene's **open reading frame (ORF)**, the part of a gene that actually encodes a protein, to allow time for the ribosome to catch up to a transcribing RNA polymerase (discussed in Section 8.3). The transcription termination pause site, however, is located after the ORF, beyond the translation stop codon, because if transcription were to cease before the ribosome reaches the translation stop codon, an incomplete protein would be made. Rho factor binds to an exposed region of RNA after the ORF at GC-rich sequences that lack obvious secondary structure. This is the transcription terminator pause site. Rho monomers assemble as a hexamer around the RNA (**Fig. 8.8A**). Then, like a raft pulled to shore by winding a rope around a piling, Rho pulls itself to the paused RNA

Elongation of RNA transcripts. The sigma factor remains associated with the transcribing complex until about nine bases have been joined; then it dissociates (see **Fig. 8.6**). The newly liberated sigma factor can then reassociate with an unbound core RNA polymerase to direct another round of promoter binding. Meanwhile, the original RNA polymerase continues to move along

A. Rho-dependent termination

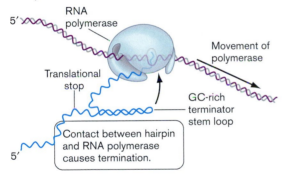

Rho binds to GC-rich regions.

Translational stop

5′

RNA polymerase

Pause site

Translational stop

5′

5′

RNA wraps around Rho hexamer, pulling Rho toward RNA polymerase.

5′

Contact between Rho and RNA polymerase causes termination.

B. Rho-independent termination

RNA polymerase

5′

Movement of polymerase

Translational stop

GC-rich terminator stem loop

Contact between hairpin and RNA polymerase causes termination.

5′

Figure 8.8 **Transcription termination.**

poly-A base pairs at the 3′ terminus contain only two hydrogen bonds per pair and are usually weak (that is, easier to melt). The melting of the hybrid molecule releases the transcript and halts transcription. The pause in polymerase movement, and thus transcription, is important to prevent the formation of tighter base pairing downstream of the UA region.

Antibiotic Resistance Helped Reveal Components of RNA Synthesis Machines

How did scientists determine the details of RNA and protein synthesis? The processes were observed in cell-free systems. For example, transcription and translation were observed in the test tube using purified RNA polymerase and ribosomes. In addition, the various protein subunits of RNA polymerase were separated from each other and mixed together again to make an active complex.

One particularly clever strategy was based on the genetics of antibiotic resistance. We now know that many antibiotics target specific components of the bacterial transcription and translation machinery. However, organisms, such as *E. coli* mutate at low frequency to become resistant to a given antibiotic. The resistance mutation often affects the part of the RNA or protein synthesis machine that binds to the antibiotic. The effect is a decreased affinity of that altered protein for the antibiotic.

The subunits that conferred resistance to particular antibiotics were determined in cell-free systems. For example, rifamycin is an antibiotic that inhibits transcription (**Special Topic 8.1**). To determine which RNA polymerase subunit was targeted by the drug, RNA polymerases from rifamycin-sensitive and -resistant strains were purified, the component parts of the polymerase were separated, and then a chimeric RNA polymerase was reassembled using all but one of the protein subunits from the sensitive strain. The missing subunit was supplied by the resistant strain and the polymerase preparation tested in vitro for an ability to make RNA in the presence of rifamycin. This will happen only when the subunit conveying resistance has been added—in this case, the beta subunit. Subsequent in vitro assays and X-ray crystallography used this information to learn more about the enzymatic mechanism of this subunit.

Different Classes of RNA Have Different Functions

Now that we have shown how RNA is synthesized, it is important to note that not all RNA molecules are translated to protein. Six known classes of RNA are made. Each

polymerase by wrapping downstream RNA around itself via an intrinsic ATPase activity. Once Rho touches the polymerase, an RNA-DNA helicase activity built into Rho appears to unwind the RNA-DNA heteroduplex, which releases the completed RNA molecule and frees the RNA polymerase.

The second type of termination event, called **Rho-independent**, occurs in the absence of Rho or any other protein cofactors. Rho-independent termination requires a GC-rich region of RNA roughly 20 bp upstream from the 3′ terminus as well as four to eight consecutive uridine residues at the terminus. The GC-rich sequence contains complementary bases and forms an RNA stem, or **stem loop**, a structure that "grabs" RNA polymerase, causing it to pause (**Fig. 8.8B**). While the polymerase is paused, the DNA-RNA duplex is weakened because the poly-U,

Table 8.1 Classes of RNA in *E. coli*.[a]

RNA class	Function	Number of types	Average size	Approx. half-life	Unusual bases
mRNA (messenger RNA)	Encodes protein	Thousands	1,500 nt	3–5 minutes	No
rRNA (ribosomal RNA)	Synthesizes protein	Three	5S, 120 nt; 16S, 1,542 nt; 23S, 2,905 nt	Hours	Yes
tRNA (transfer RNA)	Shuttles amino acids	27 typeset (86 genes)	80 nt	Hours	Yes
sRNA (small RNA)	Controls translation, controls transcription	20–30	<100 nt	Variable	No
tmRNA (properties of transfer and messenger RNA)	Frees ribosomes stuck on damaged mRNA	Roughly one per species	300–400 nt	3–5 minutes	No
Catalytic RNA	Carries out enzymatic reactions (e.g., RNase P)	??	Varies	3–5 minutes	No

[a]Six major classes of RNA exist in all bacteria. They differ based on their function in the cell, the number of types, average sizes, half-lives, and whether they contain modified bases.

is designed for a different purpose (**Table 8.1**). RNA molecules that encode proteins are called **messenger RNA (mRNA)**. Molecules of mRNA average 1,000–1,500 bases in length but can be much longer or shorter, depending on the size of the protein they encode. Another class of RNA forms the scaffolding on which ribosomes are built and is called **ribosomal RNA (rRNA)**. As will be discussed later, rRNA also forms the catalytic center of the ribosome. **Transfer RNA (tRNA)** molecules ferry amino acids to the ribosome. A unique property of rRNA and tRNA is the presence of unusual modified bases not found in other types of RNA.

The fourth important class of RNA is called "**small RNA" (sRNA)**. Molecules of sRNA do not encode proteins but are used to *regulate* the stability or translation of specific mRNAs into proteins. As will be discussed later in more detail, all mRNA molecules contain untranslated leader sequences that precede the actual coding region. Complementary sequences within some of these RNA leader regions snap back on themselves to form double-stranded stems and stem loop structures that can obstruct ribosome access and limit translation. The regulatory sRNAs are thought to disrupt or stabilize intrastrand stem structures by base-pairing to key regions within the mRNA molecule.

The final two classes of RNA are **tmRNA**, which has properties of both tRNA and mRNA, and **catalytic RNA**. Catalytic RNA is usually found associated with proteins in which the enzymatic (catalytic) activity actually resides in the RNA portion of the complex rather than in the protein.

RNA Stability

Once released from RNA polymerase, most prokaryotic mRNA transcripts are doomed to a short existence owing to their degradation by various intracellular RNases. This is especially true of transcripts that encode proteins (mRNA). The rationale for this is that the cell must be prepared to react rapidly to a changing environment, quickly halting synthesis of superfluous or even detrimental proteins. An effective way to do this is to rapidly destroy the mRNAs that encode those proteins. Rapid mRNA degradation makes it necessary to continually transcribe those genes as long as the need for their products exists.

RNA stability is measured in terms of half-life, which is the length of time the cell needs to degrade half the molecules of a given mRNA species. The average half-life for mRNA is 1–3 minutes but can be as little as 15 seconds. Other RNAs can be fairly stable, with half-lives of 10–20 minutes or longer. However, different kinds of RNAs differ drastically with respect to stability. The mRNAs are usually unstable compared with rRNAs and tRNAs, which contain modified bases less susceptible to RNase digestion. These more stable RNA molecules have half-lives of hours.

TO SUMMARIZE:

- **A closed complex** occurs on initial binding of RNA polymerase to promoter DNA.
- **Opening (melting) of the DNA** strands produces a "bubble" of DNA around the polymerase and forms the open complex.

- **Rho-dependent and Rho-independent** mechanisms mediate transcription termination.
- **Half-lives of RNA** species vary within the cell.
- **Messenger RNA** molecules encode proteins.
- **Ribosomal RNA** are stable molecules found in ribosomes and contain unusual modified bases.
- **Transfer RNA** are stable molecules that bind amino acids and contain modified bases.
- **Small RNA** molecules can regulate gene expression, have catalytic activity (catalytic RNA), or function as a combination of tRNA and mRNA (tmRNA).

8.3 Translation of RNA to Protein

Once a gene has been copied into mRNA, the next stage is **translation**, the decoding of the RNA message to synthesize protein. An mRNA molecule can be thought of as a sentence in which triplets of nucleotides, called **codons**, represent individual words, or amino acids (**Fig. 8.9**). Ribosomes are the machines that read the language of mRNA and convert, or translate, it to protein. Numerous steps are involved in translation, the first of which is the search by ribosomes for the beginning of an RNA coding region. Because the code consists of triplet codons, a ribosome must start translating at precisely the right base (in the right frame) or the product will be gibberish. Before we discuss how the ribosome finds the right reading frame, let's review the code itself and the major players in the translation process.

The Code and tRNA Molecules

When learning a new language, we need a dictionary that converts words from one language into the other, in this case from RNA to protein. Through painstaking efforts of Marshall Nirenberg and colleagues, the molecular code was broken, and it was found that each codon (a triplet of nucleotides) represents an individual amino acid. Remarkably, and with few exceptions, the code operates universally across species. (Some exceptions: UGA is normally read as a stop codon but encodes tryptophan in vertebrate mitochondria, *Saccharomyces*, mycoplasmas; and *Candida albicans* read CUG for serine instead of leucine.) On examining the code presented in **Figure 8.9**, we can see that most amino acids have multiple codon synonyms. Glycine, for example, is translated from four different codons, leucine from six, whereas methionine and tryptophan each have only one

					2nd Base		
RNA Codon	Amino Acid	RNA Codon	Amino Acid	RNA Codon	Amino Acid	RNA Codon	Amino Acid
U U U	Phe	U C U	Ser	U A U	Tyr	U G U	Cys
U U C	Phe	U C C	Ser	U A C	Tyr	U G C	Cys
U U A	Leu	U C A	Ser	U A A	Stop	U G A	Stop
U U G	Leu	U C G	Ser	U A G	Stop	U G G	Trp
C U U	Leu	C C U	Pro	C A U	His	C G U	Arg
C U C	Leu	C C C	Pro	C A C	His	C G C	Arg
C U A	Leu	C C A	Pro	C A A	Gln	C G A	Arg
C U G	Leu	C C G	Pro	C A G	Gln	C G G	Arg
A U U	Ile	A C U	Thr	A A U	Asn	A G U	Ser
A U C	Ile	A C C	Thr	A A C	Asn	A G C	Ser
A U A	Ile	A C A	Thr	A A A	Lys	A G A	Arg
A U G	Met	A C G	Thr	A A G	Lys	A G G	Arg
G U U	Val	G C U	Ala	G A U	Asp	G G U	Gly
G U C	Val	G C C	Ala	G A C	Asp	G G C	Gly
G U A	Val	G C A	Ala	G A A	Glu	G G A	Gly
G U G	Val	G C G	Ala	G A G	Glu	G G G	Gly

1st Base (left margin)   3rd Base (right margin)

Figure 8.9 The genetic code. Codons within a single box encode the same amino acid. Color-highlighted amino acids are encoded by codons in two boxes. Stop codons are highlighted pink. Often, single letter abbreviations for amino acids are used to convey protein sequences, Ala (A); Arg (R); Asn (N); Asp (D); Cys (C); Gln (Q); Glu (E); Gly (G); His (H); Ile (I); Leu (L); Lys (K); Met (M); Phe (F); Pro (P); Ser (S); Thr (T); Trp (W); Tyr (Y); Val (V).

Special Topic 8.1   Antibiotics That Affect Transcription

To be useful in medicine, all antibiotics need to meet two fundamental criteria: They must kill or retard the growth of a pathogen, and they must not harm the host. Antibiotics used against bacteria, for instance, must *selectively* attack features of bacterial targets not shared with eukaryotes. Many antibiotics possess highly specific modes of action, binding to and altering the activity of one specific "machine" of one type of cell. Thus, the quest to discover how antibiotics worked ultimately revealed the molecular mechanics of life.

One example of this is the antibiotic rifamycin B (**Fig. 1A**), produced by the actinomycete *Streptomyces mediterranei*

(**Fig. 1B**). Rifamycin B selectively targets bacterial RNA polymerase. Rifamycin binds to the bacterial RNA polymerase beta subunit near the Mg^{2+} active site (**Fig. 1C**) and sterically blocks phosphodiester bond formation during initiation but does not prevent RNA polymerase from binding to promoters and does not inhibit complexes already transcribing DNA (that is, elongation). The semisynthetic derivative of this drug, rifampin, is used for treating tuberculosis and leprosy and is given to contacts (family members or roommates) of individuals with bacterial meningitis caused by *Neisseria meningitidis*.

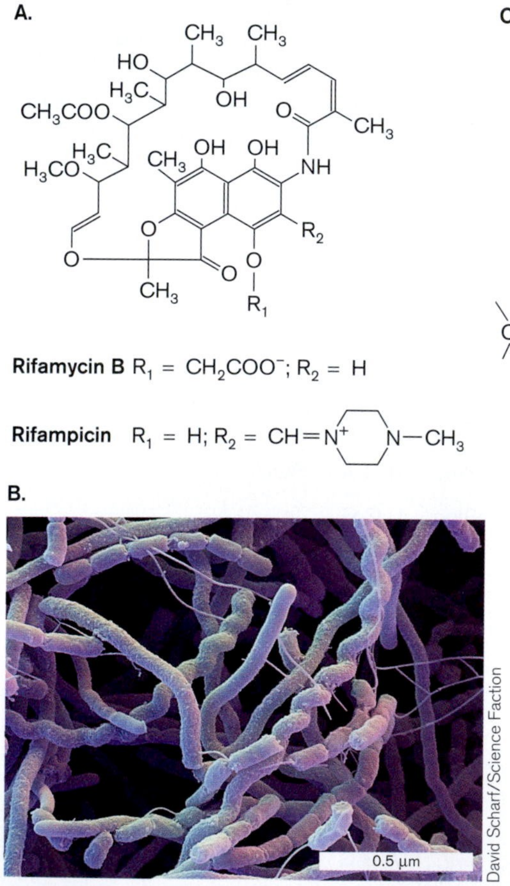

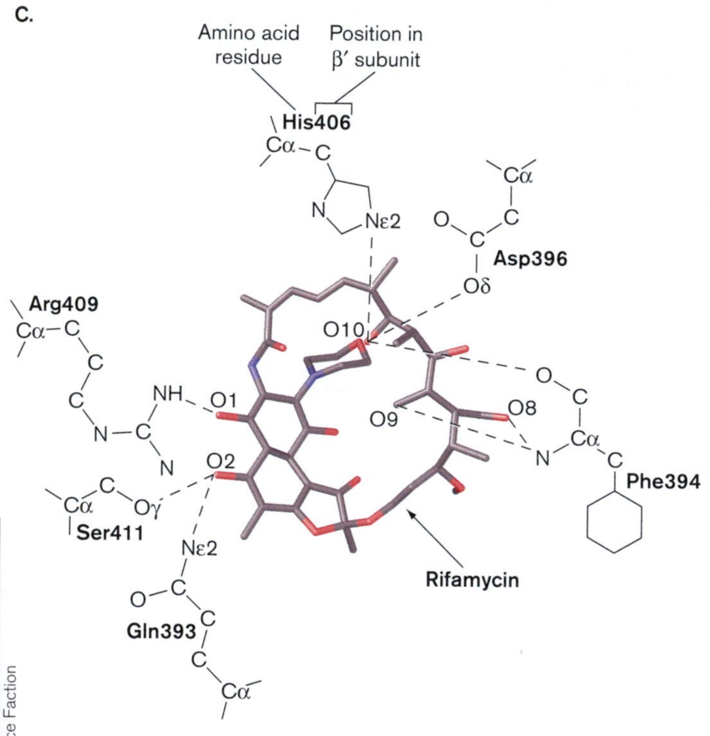

Figure 1   Structure and mode of action of rifamycin. A. Structure of rifamycin. R groups indicated are added to alter the structure and pharmacology of the basic structure. **B.** Electron micrograph of *Streptomyces*. **C.** Contact points between rifamycin and residues in the beta-prime subunit of RNA polymerase.

David Scharf/Science Faction

codon. Because more than one codon can encode the same amino acid, the code is said to be degenerate or redundant. Notice that in almost every case, synonymous codons differ only in the *last* base. How the cell handles this degeneracy (redundancy) in the code will be explained later.

Only 61 out of a possible 64 codons specify amino acids. Four of these act as codons that can mark the begin-

ning of the protein "sentence." In order of preference, they are AUG (90%), GUG (8.9%), UUG (1%), and CUG (0.1%). Due to their inefficiency in starting translation, the use of a rare start codon, such as CUG, limits the translation of any ORF in which they are found. Although these **start codons** are used to begin all proteins, they are not restricted to that role; they also encode amino acids found in the middle of

Actinomycin D (**Fig. 2**) is another antibiotic produced by a *Streptomyces* species. Its phenoxizone ring is a planar structure that "squeezes," or intercalates, in between GC base pairs within DNA. The side chains then extend in opposite directions along the minor groove (see **Fig. 2B**). Because it mimics a DNA base, actinomycin D blocks the elongation phase of transcription; but because it binds any DNA, it is not selective for prokaryotes. It can be used to treat human cancers because cancer cells replicate rapidly, but has severe side effects due to its inhibition of DNA synthesis in normal cells. (Further discussion of antibiotics can be found in Chapter 27.)

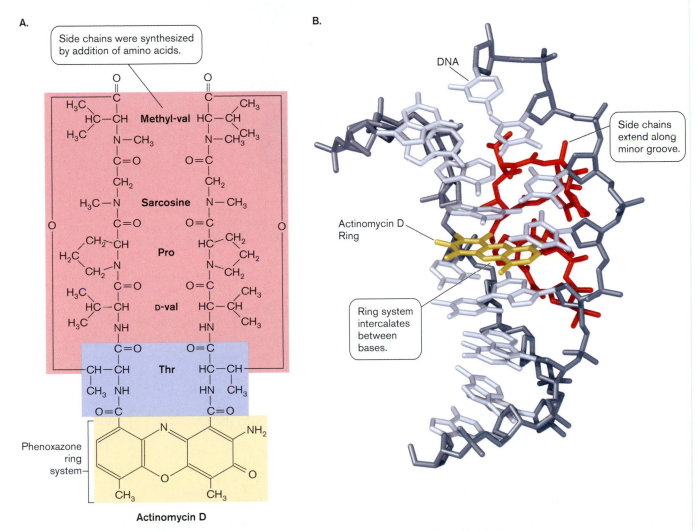

A.

Side chains were synthesized by addition of amino acids.

Methyl-val

Sarcosine

Pro

D-val

Thr

Phenoxazone ring system

Actinomycin D

B.

DNA

Side chains extend along minor groove.

Actinomycin D Ring

Ring system intercalates between bases.

Figure 2 Structure (A) and mode of action (B) of actinomycin D. (PDB code for B: 1dsc)

coding sequences. Consequently, there must be another mRNA feature that signifies whether an AUG codon, for example, marks the start of a protein or resides internally, where it codes for an amino acid (discussed below).

The other three triplets (UAA, UAG, and UGA), however, are equally important, for they tell the ribosome when to stop reading a gene sentence. Called **stop codons**, they trigger a series of events to be described later that dismantle ribosomes from mRNA and release a completed protein.

THOUGHT QUESTION 8.4 How might the redundancy of the genetic code be used to establish evolutionary relationships?

A.

5′ GCGGAUUUAGCUCAGDDGGGAGAGCMCCAGACUGAAYAUCUGGAGMUCCUGUGTPCGAUCCACAAUUCGCACCA 3′

B.

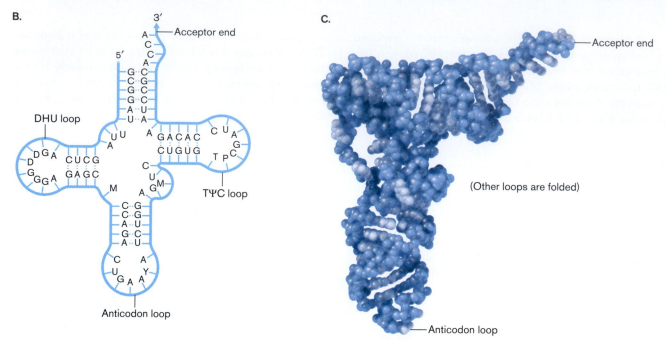

C.

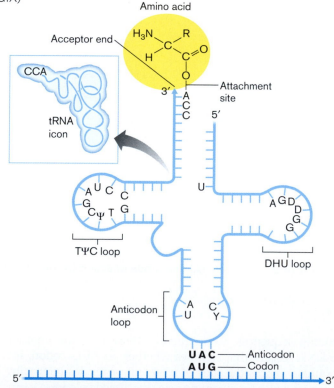

Figure 8.10 Transfer RNA. A. Primary sequence. The letters D, M, Y, T, and P stand for modified bases found in tRNA. **B.** Cloverleaf structure. DHU (or D) is dihydroxyuracil, which only occurs in this loop; TΨC are thymine, pseudouracil, and cytosine bases that occur as a triplet in this loop **C.** Three-dimensional structures. The DHU and TΨC loops are named for the modified nucleotides that are characteristically found there. The anticodon loop binds to the codon, while the acceptor end binds to the amino acid. (PDB code: 1GIX)

The decoder (or adaptor) molecules that convert the language of RNA (codons) to the language of proteins (amino acids) are the tRNAs. Their job is to travel through the cell and return to the ribosome with amino acids in tow. Approximately 80 bases in length, a tRNA laid flat appears as a cloverleaf with three loops (**Figs. 8.10A** and **B**). In nature, however, the molecule folds into a characteristic "boomerang-like" three-dimensional structure (**Fig. 8.10C**). The very bottom or middle loop of all tRNA molecules (as depicted in **Figs. 8.10B** and **C**) harbors an **anticodon** triplet that base-pairs with codons in mRNA (**Fig. 8.11**). As a result, this loop is called the anticodon loop. Notice that codon and anticodon pairings are aligned in an antiparallel manner. Most tRNA molecules begin with a 5′ G, and all end with a 3′ CCA, to which amino acids attach (see **Figs. 8.10** and **8.11**). Because of this, the 3′ end of the molecule is called the acceptor end.

Transfer RNA molecules contain a large number of unusual, modified bases, which accounts for the strange letter codes seen in **Figure 8.10A** (for example, D, M, Y, and T). Some of the odd bases used are diagrammed in **Figure 8.12**. For example, wybutosine (with the symbol Y) has three rings instead of the two found in a normal purine. During transcription of the tRNA genes, normal, unmodi-

Figure 8.11 Codon-anticodon pairing. The tRNA anticodon consists of three nucleotides at the base of the anticodon loop. The anticodon hydrogen-bonds with the mRNA codon in an antiparallel fashion.

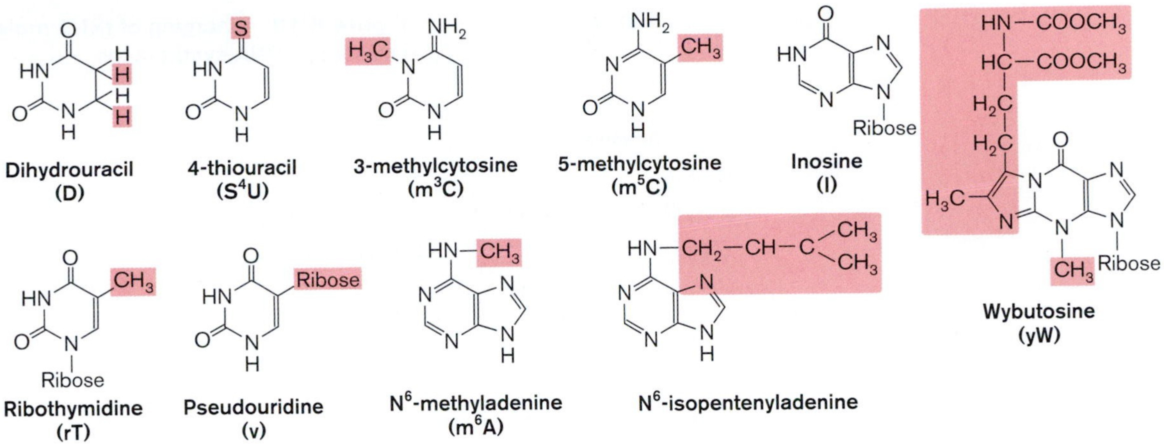

Figure 8.12 **Modified bases present in tRNA and rRNA molecules.** Modifications are shown in red.

fied bases are incorporated, but some of these are modified later by specific enzymes. The remarkable stability of tRNA molecules is explained, in part, by these unusual bases because they are poor substrates for RNases.

When looking at the cloverleaf structure of tRNA, two of the loops are named after modifications that are inveriently present in those loops. One is called the TPC (or TΨC) loop because this loop in every tRNA has the nucleotide triplet thymidine, pseudouridine (Ψ), and cytidine. The other loop always contains dihydroxyuridine and is called the DHU loop. These loops are recognized by the enzymes that match tRNA to the proper amino acid.

As noted previously, the code is redundant in that many amino acids have codon synonyms. Redundancy occurs primarily in the third position of the codon (for example, UU$\underline{U}$ and UU$\underline{C}$ both encode phenylalanine). This is the result of a "wobble" in the first position of the anticodon, which corresponds to the third position of the codon (remember, as with DNA, base-pairing between RNA strands is antiparallel). The wobble is due in part to the curvature of the anticodon loop and to the use of an unusual base (inosine) at this position in some tRNA molecules. The wobble structure allows one anticodon to pair with several codons differing only in the third position.

Aminoacyl-tRNA Transferases Attach Amino Acids to tRNA

When a tRNA GCG anticodon, for instance, pairs with a CGC codon, the ribosome has no way of checking that the tRNA is attached (that is, charged) to the "correct" amino acid (arginine in this case). Consequently, each tRNA must be charged with the proper amino acid *before* it encounters the ribosome. How are amino acids correctly matched to tRNA molecules and affixed to their 3' ends? The charging of tRNAs is carried out by a set of enzymes called **aminoacyl-tRNA transferases**. There are generally 20 of these "match and attach" proteins in

the cell, one for each amino acid. Some bacteria, however, have only 18, choosing instead to modify already charged glutamyl-tRNA and aspartyl-tRNA by adding an amine to make glutamine and asparagine derivatives. Each aminoacyl-tRNA transferase enzyme has a specific binding site for its cognate (matched) amino acid. Each enzyme also has a site that recognizes the target tRNA and an active site that joins the carboxyl group of the amino acid to the 3' OH (class II synthetases) or 2' OH (class 1 synthetases) of the tRNA by forming an ester (**Fig. 8.13**).

The specificity of the aminoacyl-tRNA transferase is based on recognition of the anticodon loop as well as other features of the tRNA structure such as the TΨC and DHU loops. It is very important not only that each enzyme recognize its own tRNA but that it does not bind to any other tRNA. Thus, each tRNA has its own set of interaction sites that only match the proper aminoacyl-tRNA transferase.

The Ribosome, a Translation Machine

The decoding process of the ribosome may be compared to a language translation machine that converts one language to another. The ribosome can be viewed as translating the language of the mRNA code into sensible protein sequences that conduct the activities of the cell. Ribosomes are composed of two complex subunits, each of which includes rRNA and protein components. In prokaryotes, the subunits are named 30S and 50S for their "size" in Svedberg units. A Svedberg unit reflects the rate at which a molecule sediments under the centrifugal force of a centrifuge (discussed in Chapter 3). Within the living cell, the two subunits exist separately but come together on mRNA to form the translating 70S ribosome (**Fig. 8.14**). Note that Svedberg units are not directly additive, since they represent a rate of sedimentation, not a weight.

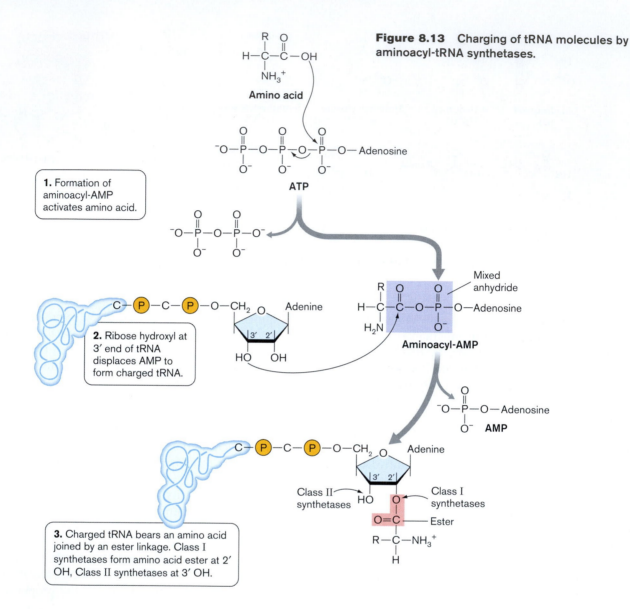

Figure 8.13 Charging of tRNA molecules by aminoacyl-tRNA synthetases.

Amino acid

ATP

1. Formation of aminoacyl-AMP activates amino acid.

Mixed anhydride

Aminoacyl-AMP

2. Ribose hydroxyl at 3′ end of tRNA displaces AMP to form charged tRNA.

AMP

Adenine

Class II synthetases

Class I synthetases

Ester

3. Charged tRNA bears an amino acid joined by an ester linkage. Class I synthetases form amino acid ester at 2′ OH, Class II synthetases at 3′ OH.

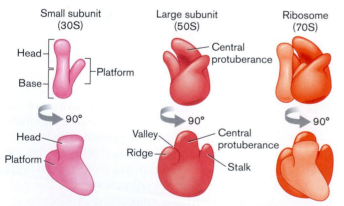

Figure 8.14 **Schematic ribosome structure.** Each panel shows two rotated views of the 30S subunit, 50S subunit, and 70S ribosome complex, respectively. Note that the platform of the 30S subunit fits into the valley of the 50S subunit when forming the 70S ribosome.

Small subunit (30S)

Head — Platform

Base

90°

Head

Platform

Large subunit (50S)

Central protuberance

90°

Valley — Central protuberance

Ridge — Stalk

Ribosome (70S)

90°

The smaller, 30S subunit (900,000 Da) contains 21 ribosomal proteins (named S1–S21; S = small) assembled around one 16S rRNA molecule. The 50S subunit (1.6 million Da) consists of 31 proteins (designated L1–L34; three numbers were not used; L = large) formed around two rRNA molecules (5S and 23S). **Figure 8.15** presents a three-dimensional spatial arrangement of rRNA and protein in the 50S subunit. Note that the majority of the ribosome is RNA.

The typical *E. coli* cell has approximately 18,000 ribosomes. The ribosome is a molecular machine, whose production, together with that of other translation components, can take up 40% of the energy of a growing bacterium. How does such a complex molecular machine get built? Assembly of the ribosome starts with the transcription of ribosomal RNA genes (DNA), sometimes referred to as rDNA. Although there are only three forms of rRNA

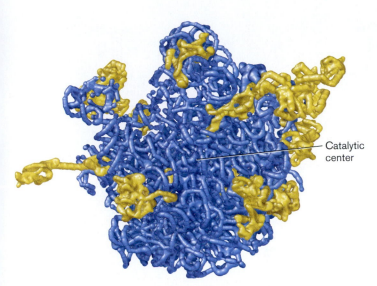

Figure 8.15   RNA: protein interfaces in the large (50S) ribosome subunit.   rRNA is in blue, proteins are gold. (PDB code: 1GIY)

in a prokaryotic cell (16S, 23S, and 5S), many microbes possess multiple copies of the genes encoding them. Arranged together as packaged sets on the chromosome, the 16S, 23S, and 5S rRNAs are initially transcribed as a single RNA molecule from a common DNA promoter, as illustrated in **Figure 8.16**.

In the initial RNA transcript, there are also leaders, spacers, and trailer sequences that often contain tRNA sequences. This polycistronic message is the starting

point from which the ribosomes are built. A **cistron** is a name used to define a functional unit of RNA; a **polycistronic** RNA is a single RNA molecule that contains information from several contiguous genes. In a carefully choreographed series of steps, the rRNA precursor molecule is surgically processed by various RNases that trim away excess RNA. (The RNA stem structures represented in the figure are specifically required substrates for some RNases.) Although shown as forming two-dimensional loops in **Figure 8.17**, the rRNA molecules actually form complex, three-dimensional folded structures. As with tRNA molecules, many bases in rRNA are enzymatically modified during processing to enhance the molecule's rigidity and stability.

Simultaneous with rRNA transcription and processing, the ribosomal proteins begin to assemble, finding their places nestled in the secondary rRNA structures (see **Fig. 8.17**). Ribosomal proteins bind sequentially in waves, the first wave providing new binding sites for a second wave of different proteins. Thus, the ribosome is built by precise, timed molecular interactions that occur between RNA sequences, between RNA and protein, and between protein and protein.

THOUGHT QUESTION 8.5   Synthesis of ribosomal RNAs and ribosomal proteins causes a major energy drain on the cell. How might the cell regulate synthesis of these molecules when growth rate slows as a result of amino acid limitation?

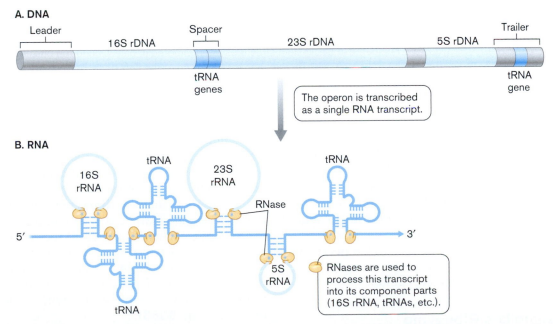

Figure 8.16   Processing of ribosomal RNA and tRNA.   A. The genes encoding rRNAs in *E. coli* exist in operons that include several tRNA genes. *E. coli* contains seven such operons in its chromosome, all of which make the same rRNA molecules but differ in which tRNA molecules are present. **B.** Processing of precursor rRNA molecules. Specific RNases cleave the transcript at precise places to release 16S, 23S, and 5S rRNA and the interspersed tRNA molecules.

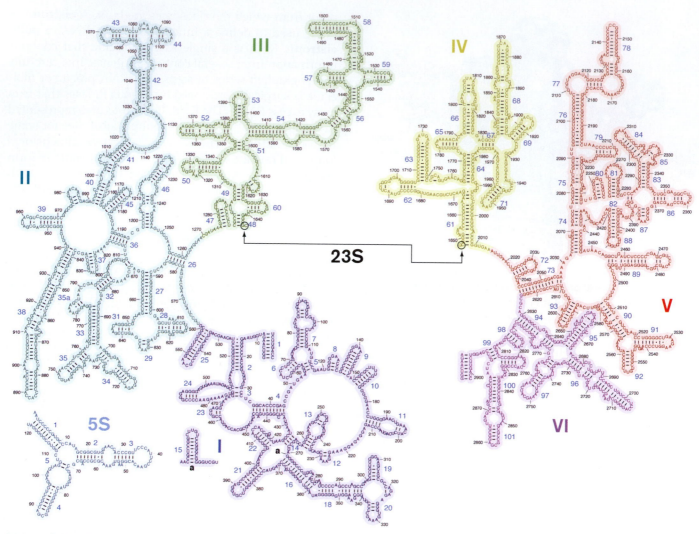

Figure 8.17   Secondary structure of the 5S and 23S rRNA.   Note the many double-stranded hairpin structures that fold into domains (roman numerals). Nucleotides are numbered in black; stem structures are numbered in blue. During the building of the ribosome, proteins bind to specific sites on the RNA molecule. The proteins bind to some of the secondary and tertiary rRNA structures and help form the key tertiary 23S structure. Domains where various ribosomal proteins bind are indicated by different colors. The 23S RNA is split where indicated to provide space for all secondary structures.

The 70S ribosome harbors three binding sites for tRNA (**Fig. 8.18A**). During translation, each tRNA molecule moves through the sites in an assembly-line fashion, first entering at one site, then moving progressively through the others before being jettisoned from the ribosome. The first position is the aminoacyl-tRNA **acceptor site (A site)**, which binds an incoming aminoacyl-tRNA. The growing peptide is attached to tRNA berthed in the **peptidyl-tRNA site (P site)**. Lastly, a tRNA recently stripped of the polypeptide is located in the **exit site (E site)**.

The Ribosome Is a Ribozyme

The ribosome's specialty is to make the peptide bonds that stitch amino acids together. This is carried out by a remarkable enzymatic activity called **peptidyl-transferase**, present in the 50S subunit. Contrary to expectation, the peptidyltransferase activity resides not in ribosomal proteins, but in the rRNA. Peptidyl-transferase is actually a **ribozyme** (an RNA molecule that carries out catalytic activity), and it is part of 23S rRNA highly conserved across all species. Proteins surrounding this active center offer structural assistance to ensure that the RNA is folded properly. The catalytic mechanism of peptide bond formation appears to involve a key adenine residue within the 23S rRNA molecule (**Fig. 8.18B**).

The sequences of ribosomal RNAs from all microbes are very similar, in large part because rRNA plays an important catalytic role in the activities of ribo-

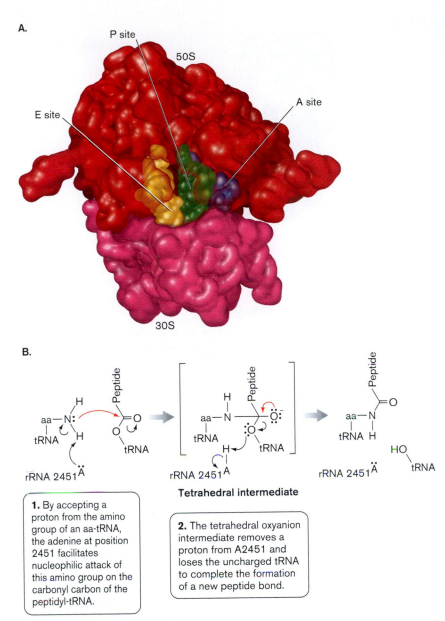

A.

P site
50S
E site
A site
30S

B.

1. By accepting a proton from the amino group of an aa-tRNA, the adenine at position 2451 facilitates nucleophilic attack of this amino group on the carbonyl carbon of the peptidyl-tRNA.

Tetrahedral intermediate

2. The tetrahedral oxyanion intermediate removes a proton from A2451 and loses the uncharged tRNA to complete the formation of a new peptide bond.

Figure 8.18   Binding of tRNA. A. X-ray crystallographic model of *Thermus aquaticus* ribosome with associated tRNAs. 50S is red, 30S is violet, and tRNAs in the A, P, and E sites are purple, green, and gold, respectively. (PDB codes: 1GIX, 1GIY) **B.** Proposed general acid-base mechanism for peptidyltransferase. Adenine at position 2451 on 23S rRNA receives a proton from the amino group in an amino acyl (aa)-tRNA. This allows the amino nitrogen to launch a nucleophilic attack on the carbonyl carbon of the peptidyl-tRNA, forming a tetrahedral intermediate. The oxyanion then accepts a proton from A2451, forming a new peptide bond and an uncharged tRNA.

somes. But there are differences in rRNA sequences that increase in relation to the evolutionary distance between species. As a result, rRNA serves as a molecular clock that measures the approximate time since two species diverged.

How Do Ribosomes Find the Right Reading Frame?

Every mRNA has three potential reading frames, depending on which base the ribosome happens to start with. Having a one in three chance of randomly picking the right frame on the template, a ribosome that chooses the wrong frame would produce proteins with totally different amino acid sequences. In addition, a shifted reading frame often generates an inadvertent stop codon that prematurely terminates the peptide; thus, the translated protein would not even have the right length. So how does the ribosome find the right frame?

The key is that each mRNA contains additional sequences upstream of the segment that encodes protein. Upstream, untranslated leader RNA plays a critical role in directing the prokaryotic ribosome to the right place. The upstream leader RNA contains a purine-rich sequence with the consensus 5'-AGGAGGU-3' located four to eight bases upstream of the start codon in *E. coli*. This upstream sequence is called the **ribosome-binding site** or **Shine-Dalgarno sequence** after the scientists who discovered it in 1974, Lynn Dalgarno and her student John Shine (Australian National University). The Shine-Dalgarno site is complementary to a sequence at the 3' end of 16S rRNA (5'-ACCUCCU-3'), found in the 30S ribosomal subunit. Binding of mRNA to this site positions the start codon precisely in the ribosome P site, ready to pair with an N-formylmethionyl-tRNA.

NOTE: Eukaryotic ribosomes have a different mechanism for finding the start codon. They generally start translating at the first AUG after the 5' cap. The 5' cap is a 7-methylguanosine added post-transcriptionally to the 5' end of eukaryotic mRNA molecules. The normal role of the cap is thought to be as a signal to transport RNA out of the nucleus and/or to stabilize mRNA.

Although Dalgarno and Shine predicted the importance of the ribosome-binding site in 1974, experimental evidence that this mRNA region truly bound 16S rRNA was lacking. The evidence was supplied in 1975 by Joan

Steitz in an elegant yet simple experiment (**Fig. 8.19A**). A radioactive, 30-nt, ^{32}P-labeled RNA fragment was taken from the 5′ end of an mRNA molecule and added to ribosomes. The 5′ end of the mRNA is where the initiation sequences should be. Steitz then treated the ribosome initiation complex with colicin E3, a toxin made by some bacteria as a defense against others. Colicin E3 cleaves the large 16S rRNA molecule once, about 50 nt from the 3′ end, to yield a small, manageable piece that would easily enter a gel. The ribosomal proteins were then peeled away using the detergent SDS (sodium dodecyl sulfate), and the mixture was subjected to electrophoresis through a polyacrylamide gel to separate different RNA bands by size. The gel was dried and placed onto X-ray film to visualize the RNA bands. SDS does not disrupt RNA-RNA interactions; consequently, if the 3′ end of 16S rRNA oriented the mRNA message for translation by binding to the 5′ end of the mRNA, a 50-nt + 30-nt (rRNA-^{32}P labeled mRNA fragment) hybrid molecule would exist and, based on its size, travel to a specific position in the gel. When the film was developed, the radioactive band was exactly where Steitz predicted a 50 nt–30 nt hybrid should be (**Fig. 8.19B**).

The Three Stages of Protein Synthesis

Like transcription, peptide synthesis is a three-step process. Once the ribosome has been properly positioned on the message, peptide synthesis can begin. Peptide synthesis has three phases: initiation, which brings the two ribosomal subunits together, placing the first amino acid in position; elongation, which sequentially adds amino acids as directed by the mRNA transcript; and termination. Note that there are several steps in the process that can be inhibited by certain antibiotics (**Special Topic 8.2** on page 281).

Initiation of translation. In bacteria, initiation of protein synthesis requires three small proteins called initiation factors (IF1, IF2, and IF3). In archaea (for example, *Methanococcus jannaschii*), initiation is much more complex, involving six different IF proteins. For simplicity, we will limit discussion to the process in bacteria, such as *E. coli*.

IF3 first brings mRNA and the 30S ribosome subunit together, allowing the ribosome-binding site to find its complementary site on 16S rRNA (**Fig. 8.20** ●). Next, IF1 binds and blocks the A site. IF2 bound to GTP then escorts the initiator N-formylmethionyl-tRNA (f-met-tRNA) to the start codon located at what will be the P site. N-formylmethionyl-tRNA binds to all start codons and is the only aminoacyl-tRNA to bind directly to the P site. Once the initiator tRNA is in place, IF3 is released. The 50S subunit then docks to the 30S subunit, GTP is

A.

Courtesy of Joan Seitz

B.

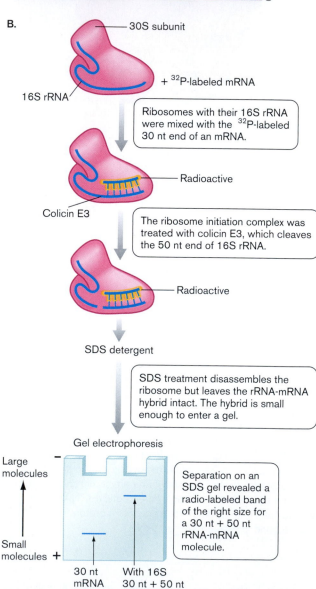

Figure 8.19 Evidence for the ribosome-binding site on mRNA. A. Joan Steitz (second from left) at Yale University with some of her students. **B.** Outline of the key experiment.

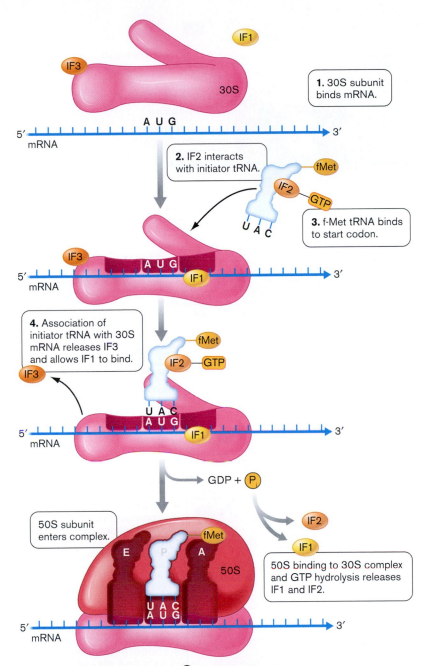

Figure 8.20 **Translation initiation.** See text for details. ⏸

hydrolyzed, and IF1 and IF2 are released. The ribosome is now "locked and loaded."

Elongation of the peptide. Elongation involves repetition of three basic steps: an aminoacyl-tRNA binds to the A, or acceptor, site; peptide bond formation occurs between the new amino acid and the growing peptide prepositioned in the P site; and the message must move by one codon (**Fig. 8.21**⏸). First, an elongation factor (EF-Tu) associates with GTP to form a complex (EF-Tu-GTP) that binds to free, charged aminoacyl-

tRNAs (except for the initiator tRNA). Sequences within the 50S rRNA recognize features of the EF-Tu-GTP-aminoacyl-tRNA complex, guiding it into the A site (**Fig. 8.21**, step 1). Correct selection of tRNA is guided in large part by codon-anticodon pairing, but conformational changes sensed by various ribosomal proteins customize the fit. If the fit is not perfect, the tRNA is rejected.

Once an aminoacyl-tRNA is in the A site (step 2), a peptide bond is formed between the amino acid moored there and the terminal amino acid linked to tRNA in the

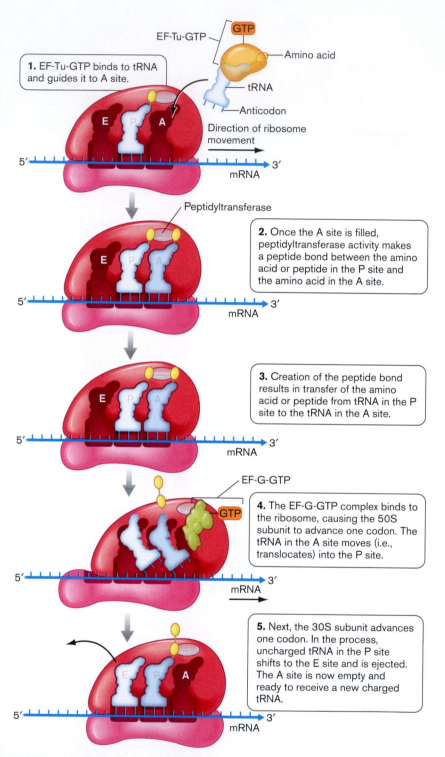

Figure 8.21 Elongation of peptide. 🔊

P site (step 3). Simultaneously, a GTP is hydrolyzed and the resulting EF-Tu-GDP is expelled. Peptide bond formation effectively transfers the peptide from the tRNA in the P site to tRNA in the A site. Finally, for protein synthesis to continue, the ribosome must advance by one codon, which moves the peptidyl-tRNA from the A site into the P site, leaving the A site vacant. The process, called **translocation**, involves another elongation factor, EF-G, associated with GTP. EF-G-GTP binds to the ribosome, GTP is hydrolyzed, and the 50S subunit

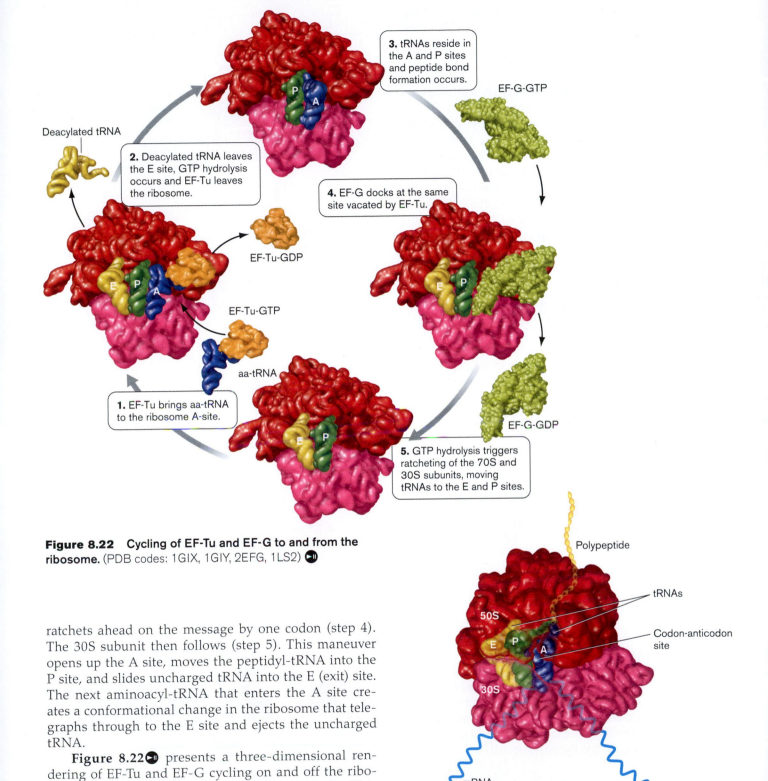

3. tRNAs reside in the A and P sites and peptide bond formation occurs.

EF-G-GTP

2. Deacylated tRNA leaves the E site, GTP hydrolysis occurs and EF-Tu leaves the ribosome.

Deacylated tRNA

EF-Tu-GDP

EF-Tu-GTP

aa-tRNA

1. EF-Tu brings aa-tRNA to the ribosome A-site.

4. EF-G docks at the same site vacated by EF-Tu.

EF-G-GDP

5. GTP hydrolysis triggers ratcheting of the 70S and 30S subunits, moving tRNAs to the E and P sites.

Figure 8.22 Cycling of EF-Tu and EF-G to and from the ribosome. (PDB codes: 1GIX, 1GIY, 2EFG, 1LS2) ◐❚❚

ratchets ahead on the message by one codon (step 4). The 30S subunit then follows (step 5). This maneuver opens up the A site, moves the peptidyl-tRNA into the P site, and slides uncharged tRNA into the E (exit) site. The next aminoacyl-tRNA that enters the A site creates a conformational change in the ribosome that telegraphs through to the E site and ejects the uncharged tRNA.

Figure 8.22◐❚❚ presents a three-dimensional rendering of EF-Tu and EF-G cycling on and off the ribosome. Note that both factors bind to the same site. This reflects a structural similarity (or molecular mimicry) between EF-G-GTP and the EF-Tu-GTP-aminoacyl-tRNA complex. Because both factors bind to the same site, they cannot do so simultaneously, so they must cycle on and off the ribosome sequentially. **Figure 8.23**◐❚❚ illustrates the path of mRNA into the ribosome

Polypeptide

tRNAs

Codon-anticodon site

50S

E P A

30S

mRNA

5′

3′

Figure 8.23 Orientation of tRNA molecules within the ribosome and tracks of mRNA and nascent polypeptide. Note that the mRNA travels along the 30S subunit, and the growing peptide exits from a channel formed in the 50S subunit. (PDB codes: 1GIX, 1GIY) ◐❚❚

and the nascent (that is, incomplete) peptide emerging from it. While this process seems incredibly complex, the ribosomes of *E. coli* manage to link together 16 amino acids per second.

Termination of translation. Eventually, the ribosome arrives at the end of the coding region, but not the end of the RNA. As noted previously, the end of the coding region is marked by one of three stop codons. Formation of the last peptide bond and the subsequent translocation of mRNA leads to ejection of tRNA in the E site (**Fig. 8.24**, step 1) and brings the stop codon into the A site (**Fig. 8.24**, step 2). No tRNA binds, but one of two **release factors** (RF1 or RF2) will enter and activate the peptidyltransferase (step 2). This cuts the bond tethering the completed peptide to tRNA in the P site (step 3).

With the protein released, the ribosome must disassemble. RF3 causes RF1 or RF2 to depart the ribosome (step 4). Then ribosome recycling factor (RRF), along with EF-G, binds at the A site, and an accompanying GTP hydrolysis undocks the two ribosomal subunits (step 5). IF3 then reenters the 30S subunit to replace the remaining uncharged tRNA and mRNA (step 6). The liberated ribosomal subunits are now free to diffuse through the cell, ready to bind yet another mRNA and begin the translation sequence anew.

> **THOUGHT QUESTION 8.6**   While working as a member of a pharmaceutical company's drug discovery team, you have found that a soil microbe snatched from the jungles of South America produces an antibiotic that will kill even the most deadly, drug-resistant form of *Enterococcus faecalis*, which is an important cause of heart valve vegetations in bacterial endocarditis. Your experiments indicate that the compound stops protein synthesis. How could you more precisely determine the antibiotic's mode of action?

Many genes in bacterial chromosomes are arranged and transcribed in tandem and produce a polycistronic mRNA (see Section 7.2). It should be remembered that the beginning of each gene in a polycistronic mRNA carries its own ribosome-binding site. This means that different ribosomes can bind simultaneously to the start of each cistron within a polycistronic mRNA.

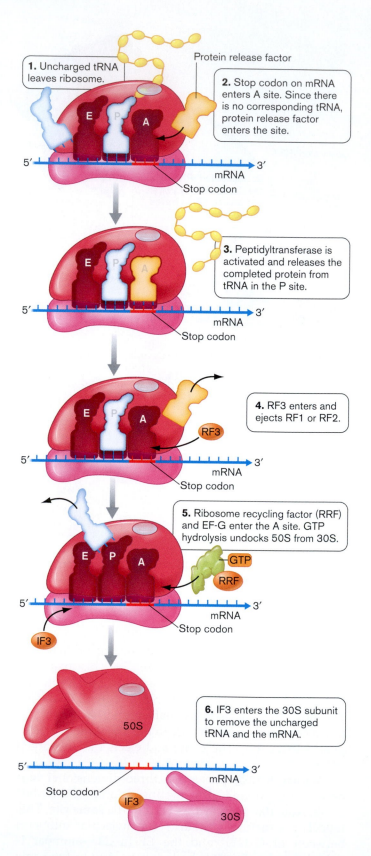

1. Uncharged tRNA leaves ribosome.

Protein release factor

2. Stop codon on mRNA enters A site. Since there is no corresponding tRNA, protein release factor enters the site.

mRNA
Stop codon

3. Peptidyltransferase is activated and releases the completed protein from tRNA in the P site.

mRNA
Stop codon

4. RF3 enters and ejects RF1 or RF2.

RF3

mRNA
Stop codon

5. Ribosome recycling factor (RRF) and EF-G enter the A site. GTP hydrolysis undocks 50S from 30S.

GTP
RRF

IF3

mRNA
Stop codon

6. IF3 enters the 30S subunit to remove the uncharged tRNA and the mRNA.

50S

mRNA
Stop codon

IF3

30S

Figure 8.24   Translation termination.   The completed protein is released and the ribosome subunits are recycled.

Special Topic 8.2   Antibiotics That Affect Translation

Streptomycin (**Fig. 1A**), a well-known member of the aminoglycoside family of antibiotics, is produced by a species of *Streptomyces*. The drug targets bacterial small ribosomal subunits by binding to a region of 16S rRNA that forms part of the decoding A site and to protein S12, a protein critical for maintaining the specificity of codon-anticodon binding. At high concentrations, streptomycin binds to the 30S-mRNA-tRNA initiation complex and prevents binding of the 50S subunit. It stops the initiation of translation. At low concentrations, translation can go on, but the A site becomes "sloppy," permitting illicit codon-anticodon matchups that result in a mistranslated protein sequence. Bacteria usually become resistant to streptomycin by spontaneously mutating the S12 gene (*rpsL*), although mutations in 16S rRNA can also prevent binding. The altered S12 protein maintains function but will not bind streptomycin. Some bacteria gain resistance by acquiring an aminoglycoside phosphotransferase that modifies streptomycin so that it cannot bind to its target. Additional mechanisms are discussed in Chapter 27. Other therapeutically important aminoglycosides include gentamicin, kanamycin, and amikacin.

Tetracycline (**Fig. 1B**) also targets the 30S ribosomal subunit, but instead of preventing formation of the 70S complex (as with streptomycin), it outright prevents binding of aminoacyl-tRNA to the A site and stops protein synthesis. It binds to 16S rRNA, right above the A site for incoming tRNA. Resistance to tetracycline can be conferred by an efflux transport system that effectively removes the antibiotic from the bacterial cell. Genes encoding this resistance are usually carried on mobile genetic elements like plasmids that can be passed from cell to cell (see Section 9.2). Other resistance mechanisms are described in Chapter 27.

Another species of *Streptomyces* produces chloramphenicol (**Fig. 1C**), which attacks the 50S subunit. It binds to 23S rRNA at the peptidyltransferase active site and inhibits peptide bond formation. Resistance to this drug involves the ability to synthesize an enzyme, chloramphenicol acetyltransferase, that destroys chloramphenicol activity by modifying the chemical.

Erythromycin, made by *Streptomyces erythrens*, is one of a large group of related antibiotics called macrolides whose hallmark is a large lactone ring of 12–22 carbon molecules (**Fig. 1D**). They all affix themselves to the 50S subunit by binding to protein L15 and 23S rRNA in the peptidyltransferase cavity. This triggers an abortive translocation step that evicts peptidyl-tRNA from the P site while preventing peptide bond formation. In some organisms, mutational alterations in L15 decrease macrolide binding and confer resistance to these drugs. Other microbes reduce erythromycin binding by methylating the relevant area of 23S rRNA using an enzyme produced from another mobile genetic element.

Other translation targeting antibiotics interfere with mRNA binding to the ribosome (kasugamycin), prevent translocation by targeting EF-G (fusidic acid), or use **molecular mimicry** to trick peptidyltransferase into action without having a bonafide tRNA in the A site (puromycin). We chronicle the discovery and use of antibiotics more completely in Chapter 27.

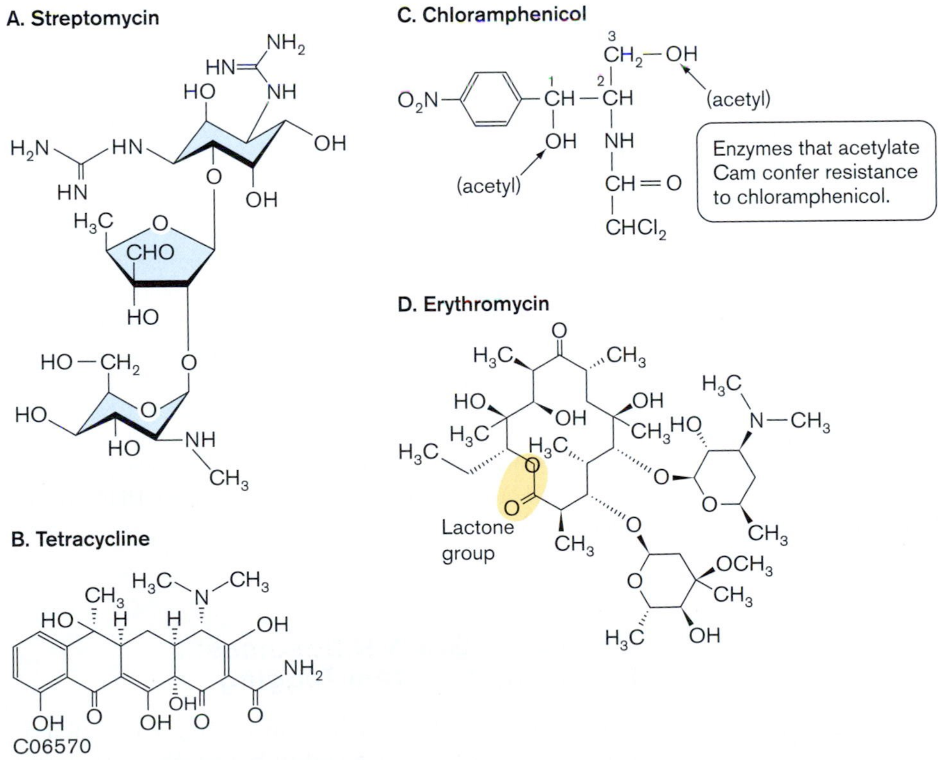

Figure 1   Antibiotics that inhibit protein synthesis in bacteria.

Prokaryotic Transcription and Translation Are Coupled

Ribosomes bind to the 5′ end of mRNA and begin translating protein even before RNA polymerase has finished transcribing the mRNA (**Fig. 8.25**). This is called coupled transcription and translation. Transcriptional-translational coupling in prokaryotes provides an opportunity to use translation as a means of regulating transcription, a process that is explained further in Chapter 10.

Eukaryotic microbes, on the other hand, use separate cellular compartments to carry out most of their transcription and translation. They transcribe genes in the nucleus and, after splicing and other forms of processing, transport mRNA to the cytoplasm, where the majority of translation occurs. However, a small amount of translation is carried out in the eukaryotic nucleus, where it is also coupled to transcription.

In prokaryotes, coupled transcription and translation presents a potential problem. Ribosomes generally travel along mRNA more slowly than mRNA is generated by RNA polymerase, so RNA polymerase can potentially scoot ahead of the ribosome and leave large tracts of RNA unprotected and susceptible to nucleases. Unintentional cleavage of the RNA between the ribosome and polymerase would separate the two macromolecular machines and destroy the message. The cell handles this problem by modulating transcriptional speed. The rate of RNA synthesis averages about 45 nt per second, which roughly equals the average rate of translation (16 amino acids per second). However, these speeds are not constant. For example, RNA polymerase pauses momentarily at sites rich in GC content (GC base pairs, with three hydrogen bonds, are harder to melt than AT pairs with only two hydrogen bonds). Once sigma factor exits the transcription complex, proteins called NusA and NusG enter the complex and can actually lengthen the pause in RNA polymerase activity to allow time for the trailing protein-synthesizing ribosome to catch up to the polymerase. Pausing allows the ribosome to follow RNA polymerase closely and protect the RNA.

> **THOUGHT QUESTION 8.7** How might one gene code for two proteins with different amino acid sequences?
>
> **THOUGHT QUESTION 8.8** Why involve RNA in protein synthesis? Why not translate directly from DNA?
>
> **THOUGHT QUESTION 8.9** Codon 45 of a 90-codon gene was changed into a translation stop codon. This produced a shortened, truncated protein. What kind of mutant cell could produce a full-length protein from the gene *without* removing the stop codon?

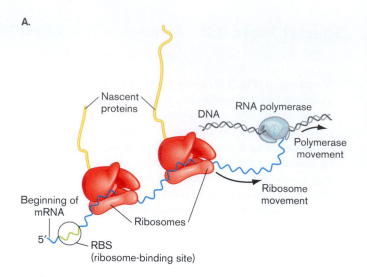

A.

B.

Figure 8.25   Coupled transcription and translation in prokaryotes.   A. Ribosomes attach at mRNA ribosome-binding sites and start synthesizing protein before transcription of the gene is complete. **B.** Electron micrograph of a polysome (ribosome is 21 nm across). Several ribosomes may translate a single mRNA molecule at the same time. This structure is called a polysome. The beginning (5′ end) of the mRNA is to the right (at the arrow), and the 3′ end is to the left. Note that the synthesized protein molecule grows longer and longer the closer the ribosome gets to the 3′ end of the mRNA. The protein molecule is most clearly seen at the end of the mRNA (to the upper right).

Unsticking Stuck Ribosomes: tmRNA and Protein Tagging

Sometimes an RNase will shear off the 3′ end of a message, removing the stop codon before the translation is complete. What happens after a ribosome has finished translating this damaged mRNA molecule? Without a stop codon, there is nothing to trigger ribosome release when the ribosome reaches the end of the message. So the ribosome is stuck at the end of the mRNA with a peptidyl-tRNA in the ribosome P site.

The answer to this problem is a translation rescue molecule, the previously mentioned tmRNA, which has the properties of both tRNA and mRNA. One end of the tmRNA molecule has an attached amino acid and a folded structure that resembles a tRNA. This aminoacyl end acts like tRNA, entering the unoccupied ribosome A site where a peptide bond then forms between the stalled polypeptide (in the P site) and the emergency amino acid on the tmRNA. Another section of the tmRNA is then read as mRNA, which adds a tag of 12 or so amino acids to the now dislodged peptide (SsrA in **Fig. 8.26A**). These amino acids, called a proteolysis tag, predestine the protein for destruction. A stop codon present in tmRNA triggers peptide release and ribosome disassembly. A helper protein (SspB) recognizes the proteolysis tag and brings the useless aberrant polypeptide to the ClpXP protease (discussed shortly) for degradation. **Figure 8.26B** shows this process for tmRNA from *E. coli*.

TO SUMMARIZE:

- **Triplet nucleotide codons** in mRNA encode specific amino acids. **tRNA molecules** interpret the genetic code and bring specific amino acids to the A (acceptor) site in the ribosome.
- **Specific codons** mark the beginning and end of a gene, but the Shine-Dalgarno sequence in mRNA located before the start codon helps the ribosome find the correct reading frame in the mRNA.
- **Initiation of protein synthesis** in bacteria requires three initiation factors that bring the ribosomal subunits together on an mRNA molecule.
- **Peptidyltransferase** activity of the ribosome is carried out by ribosomal RNA, not protein. **Elongation of translation** occurs when the ribosome ratchets one codon length along the mRNA.
- **Translation terminates** upon reaching a stop codon. The ribosome pauses because it cannot find an appropriate tRNA, and a release factor enters the A site and triggers peptidyltransferase activity. This frees the completed protein from tRNA in the A site.
- **Ribosome release factor** and EF-G bind to the A site to dissociate the two ribosomal subunits from the mRNA.
- **Transcription and translation** in prokaryotes are coupled.
- **RNA polymerase pauses** during transcription to allow the slower translating ribosomes to stay close. This minimizes exposure of mRNA to degradative cellular enzymes.
- **tmRNA** rescues ribosomes stuck on damaged mRNA that lacks a stop codon.

8.4 Protein Modification and Folding

For many proteins, translation is not the last step in producing a functional molecule. Cell function often requires that a protein be modified *after* translation to achieve an appropriate three-dimensional structure or to regulate the activity of the protein. There are many ways of modifying the primary, secondary, or tertiary structure of proteins after the primary protein sequence has been assembled by the ribosomes.

Protein Structure May Be Modified after Translation

Completed proteins released from the ribosome contain N-formylmethionine at the N-terminus (as previously described). With some proteins, the N-formyl group is surgically removed from the N terminus with methionine deformylase, leaving methionine. Alternatively, methionyl aminopeptidase will remove the whole amino acid. It is of interest to note that N-formylmethionine (f-Met) peptides are produced only by bacteria and mitochondria, not by archaea or by cytoplasmic ribosomes of eukaryotes. Therefore, during the immune response, white blood cells detect N-formyl-Met (f-Met) peptides as a sign of invading bacteria or of necrotic (dying) host cells releasing mitochondria. The f-Met peptides are detected at incredibly low concentrations, around 10^{-12} M.

Some proteins undergo other types of processing in which acetyl groups or AMP can be attached to change their functions, and proteolytic cleavages may either activate or inactivate a protein. Examples of modified bacterial enzymes include glutamine synthetase (adenylylation), isocitrate dehydrogenase (phosphorylation), and a variety of ribosomal proteins (acetylation). These groups directly regulate enzyme activity (glutamine synthetase and isocitrate dehydrogenase) or alter tertiary structure (ribosomal proteins).

Proteins Must Be Correctly Folded

Christian Anfinsen (1916–1995) won the Nobel Prize in 1972 for demonstrating that for some proteins, folding is governed solely by the protein itself. In other words, the optimal three-dimensional structure of a protein is determined solely by the linear sequence of amino acid residues. But three decades later, other scientists discovered that folding of many proteins requires assistance from other proteins. These helper proteins are called **chaperones** (or chaperonins). Chaperones associate with target proteins during some phase of the folding process, then dissociate, usually after folding of the target protein is completed. Although there is some specificity, a

A.

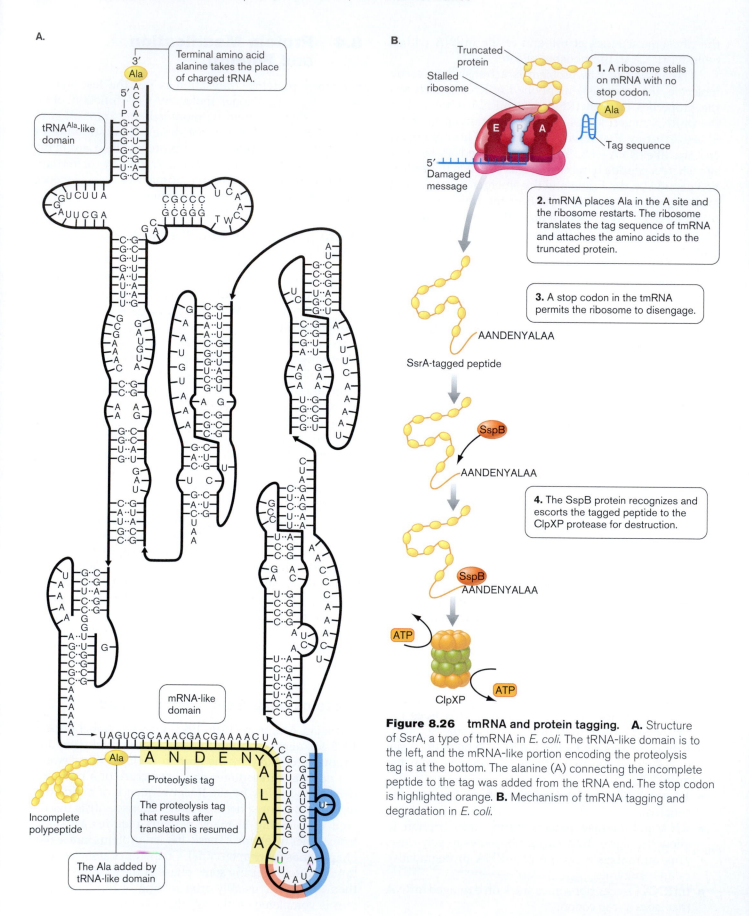

Terminal amino acid alanine takes the place of charged tRNA.

tRNA^Ala-like domain

mRNA-like domain

Incomplete polypeptide

Proteolysis tag

The proteolysis tag that results after translation is resumed

The Ala added by tRNA-like domain

B.

Truncated protein

Stalled ribosome

5′ Damaged message

1. A ribosome stalls on mRNA with no stop codon.

Tag sequence

2. tmRNA places Ala in the A site and the ribosome restarts. The ribosome translates the tag sequence of tmRNA and attaches the amino acids to the truncated protein.

3. A stop codon in the tmRNA permits the ribosome to disengage.

AANDENYALAA

SsrA-tagged peptide

4. The SspB protein recognizes and escorts the tagged peptide to the ClpXP protease for destruction.

AANDENYALAA

ClpXP

Figure 8.26   tmRNA and protein tagging.   A. Structure of SsrA, a type of tmRNA in *E. coli*. The tRNA-like domain is to the left, and the mRNA-like portion encoding the proteolysis tag is at the bottom. The alanine (A) connecting the incomplete peptide to the tag was added from the tRNA end. The stop codon is highlighted orange. **B.** Mechanism of tmRNA tagging and degradation in *E. coli*.

given chaperone can help fold many different types of proteins.

The major chaperone family in most species includes GroEL (60 kDa), GroES (10 kDa), DnaK (70 kDa), and trigger factor (48 kDa). Because their levels in *E. coli* increase in response to high temperature stress, these chaperones were originally named **heat-shock proteins (HSPs)** and are, in fact, more resistant to heat denaturation than the average protein. Representatives of these chaperones are found in all species. As a result, DnaK examples throughout nature are called HSP70s, and GroEL homologs are called HSP60s. Chaperones increase in response to heat stress because they are needed to help refold heat-damaged proteins.

The GroEL and GroES chaperones form a stacked ring with a hollow center (**Figs. 8.27** and **8.28**). The chaperoned protein fits inside. The small capping protein GroES controls entrance to the chamber (**Fig. 8.28A**). Cycles of ATP binding and hydrolysis cause conformational changes within the chamber that can reconfigure target proteins. DnaK (HSP70) chaperones have a very different structure (**Fig. 8.28B**). They do not form rings like the GroEL and GroES chaperones, but can clamp down on a peptide to assist folding. Proteins emerging from a prokaryotic ribosome enter a folding pathway that involves a hierarchy of these chaperones.

TO SUMMARIZE:

■ **Protein modifications** are made after translation.
■ **The N-terminal amino acid** (N-formylmethionine) can be removed by methionine deformylase.
■ **An inactive precursor protein** can be cleaved into a smaller active protein, or other groups can be added to the protein (for example, phosphate or AMP).
■ **Chaperone proteins** help translated proteins fold properly.

8.5 Secretion: Protein Traffic Control

Microorganisms, especially gram-negative bacteria, face an intriguing challenge in delivering proteins to different target locations in the cell. Recall that gram-negative microbes are surrounded by two layers of membrane (the inner membrane, or cell membrane, and an outer membrane) between which lies a periplasmic

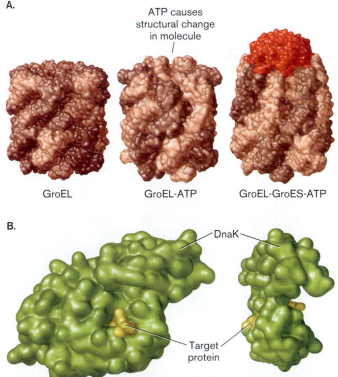

A.

ATP causes structural change in molecule

GroEL   GroEL-ATP   GroEL-GroES-ATP

B.

DnaK

Target protein

Figure 8.28 GroEL-GroES and DnaK (HSP70) structures. A. Three-dimensional reconstructions of GroEL, GroEL-ATP, and GroEL-GroES-ATP from cryo-EM. The GroES ring is seen as a disk above GroEL. (PDB codes: 1SS8, 2C7E, 1PCQ) **B.** DnaK clamping down on a peptide (yellow). (PDB code: 1DKX)

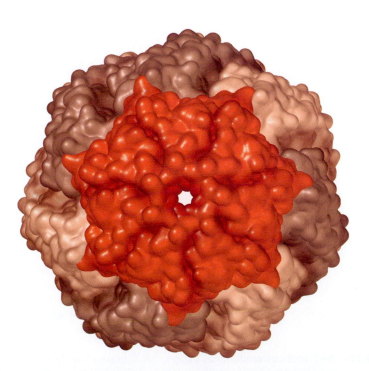

Figure 8.27 Top view of the GroEL-GroES complex of *E. coli.* The heat-shock chaperone-10 (equivalent to GroES) is shown in red, which is tacked on a GroEL ring. The chaperoned protein fits inside the open hole. (PDB code: 1PCQ)

space (see Section 3.4). Many proteins are specifically destined for one or another of these cell compartments. Other proteins are secreted completely out of the cell into the surrounding environment (for example, hemolysins that lyse red blood cells). But how do these diverse proteins know where to go? Protein traffic out of the cell is directed by an elaborate set of **protein secretion systems**. Each system selectively delivers a set of proteins originally made in the cytoplasm to various extracytoplasmic locations.

The term *secretion* is used to describe movement of a protein out of the cytoplasm. There are protein secretion systems that move proteins to the periplasm, others that move proteins to the outer membrane, and still others that deliver proteins across both of the membranes and into the surrounding environment. An added complication of protein export is that periplasmic proteins are usually delivered unfolded into the periplasm and require another set of chaperones to fold properly in this cell compartment.

Protein Export Out of the Cytoplasm

Proteins destined for the bacterial cell membrane or envelope regions (periplasm, outer membrane, or extracellular spaces) require special export systems. These systems manage to move hydrophilic proteins through one or more hydrophobic membrane barriers. Proteins meant for the inner membrane (for example, cytochromes) are tagged, as part of the open reading frame, with very hydrophobic N-terminal **signal sequences** of 15–30 amino acids. Signal sequences tether nascent proteins to the membrane and confer conformations that allow the proteins to melt into the fabric of the membrane. Inner membrane proteins also contain hydrophobic transmembrane spanning regions (20–25 amino acids) that aid in this insertion process. These hydrophobic regions are important because they are compatible with the hydrophobicity of the membrane itself. A nutrient transport protein often has 12 such membrane-spanning regions, which weave back and forth across the membrane.

One special export system begins with a complex called the **signal recognition particle (SRP)**, which targets proteins for inner membrane insertion. A second export mechanism uses a protein called trigger factor for proteins destined for the periplasm. (Trigger factor was mentioned

earlier as a chaperone.) These two protein traffic pathways converge on a general secretion complex composed of three proteins, collectively called the **SecYEG translocon**, embedded in the cell (inner) membrane.

Protein Export to the Cell Membrane

The pathway leading proteins to the inner (cell) membrane begins with SRP. In *E. coli*, SRP consists of a 54-kDa protein Ffh complexed with a small RNA molecule (*ffs*). SRP binds to the signal sequences of integral cell membrane proteins as they are being translated (**Fig. 8.29**) and halts further translation in the cytoplasm. The nascent protein with its paralyzed nontranslating ribosome is delivered to the membrane-bound protein FtsY, where translation resumes. The partially translated protein is now subject to one of two fates: It may be cotranslationally inserted directly into the cell membrane, meaning that the protein is inserted even as it is still being translated. Alternatively, the protein may be completely synthesized, after which it is delivered to the

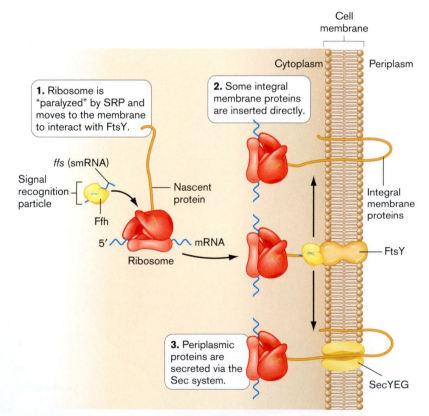

Figure 8.29 SRP and cotranslational export. A ribosome "paralyzed" by SRP does not resume translating protein until encountering FtsY in the membrane. Translation can then recommence. Some proteins designated for integral membrane location are inserted directly (top). Other integral membrane proteins, and proteins destined for the periplasm, are inserted or secreted via the Sec system (bottom).

SecYEG translocon for insertion. The route to membrane insertion that occurs depends on the protein.

Protein Export to the Periplasm: The Sec-Dependent General Secretion Pathway

The periplasm contains important proteins that bind nutrients for transport into the cell and other proteins that carry out enzymatic reactions. For example, one form of superoxide dismutase (SOD), an enzyme that degrades superoxide, is a periplasmic protein in *Salmonella enterica* and other gram-negative bacteria. Many periplasmic proteins, such as SOD and maltose-binding protein (which imports the sugar maltose), are delivered to the periplasm by a common pathway called the general secretion pathway.

There are several steps in the general secretion pathway. First, the peptide is completely translated in the cytoplasm (**Fig. 8.30**, step 1). The completed presecretion protein is then captured by a piloting protein called SecB (step 2 and **Fig. 8.31**) and delivered in an unfolded state to SecA, a protein peripherally associated with the membrane-spanning SecYEG translocon (step 3). SecB discourages folding in the cytoplasm. This assists in secretion because sliding an unfolded protein through a membrane is far easier than trying to deliver a folded one.

The SecA ATPase appears to act like a plunger (step 4). It inserts deep into the SecYEG channel, shoving about 20 amino acids of the target export protein into the channel. ATP hydrolysis causes SecA to release the protein and withdraw (step 5). At this point, SecA can bind fresh ATP, rebind the target protein, and reinsert, pushing another 20 amino acids through. Proteins needed in the periplasm have cleavable signal sequences at their amino-terminal ends. Immediately following translocation into the periplasm, periplasmic signal peptidases (LepB is one of several examples in *E. coli*) snip off the amino-terminal signal sequence of the protein. This allows release of the mature protein into the periplasm (step 6). Signal peptidases will not, however, cleave signals from proteins destined to stay embedded within the membrane (integral membrane proteins).

Periplasmic proteins delivered by the Sec system arrive unfolded and inactive. Because the folding chaperones mentioned earlier are cytoplasmic, periplasmic proteins need another set of dedicated periplasmic chap-

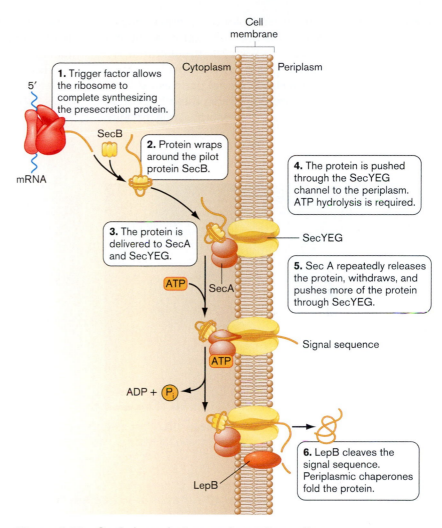

1. Trigger factor allows the ribosome to complete synthesizing the presecretion protein.

2. Protein wraps around the pilot protein SecB.

3. The protein is delivered to SecA and SecYEG.

4. The protein is pushed through the SecYEG channel to the periplasm. ATP hydrolysis is required.

5. Sec A repeatedly releases the protein, withdraws, and pushes more of the protein through SecYEG.

6. LepB cleaves the signal sequence. Periplasmic chaperones fold the protein.

Cell membrane

Cytoplasm — Periplasm

SecB · mRNA · 5′ · ATP · SecA · ADP + P$_i$ · SecYEG · Signal sequence · LepB

Figure 8.30 SecA-dependent general secretion pathway.

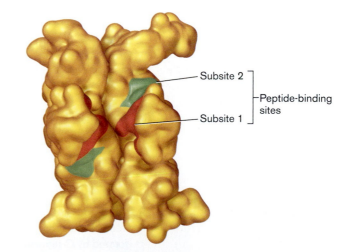

Subsite 2
Subsite 1
Peptide-binding sites

Figure 8.31 Proposed peptide-binding channel of SecB. The exposed surfaces encompassing the two proposed peptide-binding subsites are highlighted. The presecretion protein is wrapped around these sites in SecB. (PDB code: 1FX3)

erones to guide their tertiary folding. Another problem with periplasmic proteins is that the oxidizing environment of aerobic cells can oxidize cysteines within a protein and produce inappropriate cysteine disulfide bonds that destroy enzyme function. Special periplasmic disulfide reductases are required to reverse these S—S bonds and form two SH groups. Many periplasmic proteins,

however, need certain disulfide bonds to be active. The periplasm also contains a disulfide bond catalyst (DsbA) to make those bonds.

Eukaryotic microbes such as the yeast *Saccharomyces cerevisiae* also possess secretion systems that move proteins to the membrane and beyond. However, the eukaryotic Sec systems are more complex and are evolutionarily

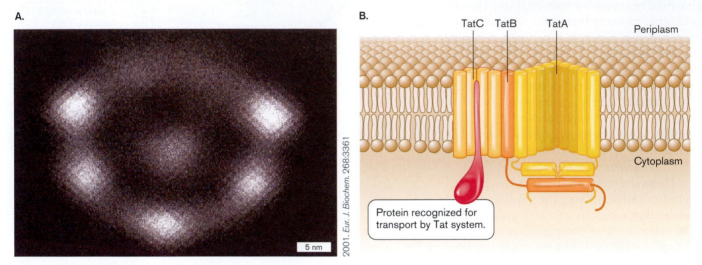

Figure 8.32   The twin arginine translocase.   A. A negative stain electron microscopic image of the purified TatAB complex. **B.** A commonly accepted model for the Tat protein translocase, which includes proteins TatA, TatB, and TatC.

Table 8.2   Comparison of mechanisms that secrete proteins across the outer membrane.[a]

Type	Mechanism	Structure	Location of protein substrate	Location of secretory signals	Number of components
I	Coupled to TolC	ABC type	Cytoplasm	N terminus, not cleaved	3
II	Extending and contracting	Pilus-like structure	Periplasm	Cleaved N terminus for Sec-dependent transport	12–16
III	Injecting proteins directly into host eukaryotic cell cytoplasm	Syringe, related to flagella biogenesis	Cytoplasm	None	20
IV	Conjugation-like	Multicomponent	Some are cytoplasmic and are moved directly to outside of cell; others use Sec system to get to periplasm first	None	8 or 9
V	Autotransport	Self-transporting channel in outer membrane formed by C-terminal domain	Periplasm	N terminus, cleaved	1

[a]The five classes are grouped on the basis of their structure. Types II, III, and IV are evolutionarily related to mechanisms that assemble pili (type II) or flagella (type III) or that carry out conjugation (type IV). Some systems pick up protein substrates from the cytoplasm and transport them across both membranes. Substrates for other systems are collected in the periplasm. Substrates also differ as to the presence of N-terminal signal sequences.

distinct from bacterial Sec systems. Archaea secretion systems are actually more similar to those of eukaryotes.

Export of Prefolded Proteins to the Periplasm

In a dramatic departure from Sec-dependent transport systems, some proteins like TorA, a component of an anaerobic respiratory chain, can be transported fully folded across the membrane to their periplasmic destination. These proteins contain the amino acid motif RRXFXK within their N-terminal signal sequence (R is shorthand for arginine, F is phenylalanine, and X stands for any amino acid). This sequence, called the twin arginine motif because it begins with two arginines, targets the protein to the membrane-bound **twin arginine translocase (TAT)**, a transport protein that ships fully folded proteins across the cell membrane to the periplasm (**Fig. 8.32**). Whereas Sec-dependent transport is ATP driven, the TAT system is powered by proton motive force (see Chapters 4 and 12).

Journeys through the Outer Membrane

For many reasons, bacteria need to export proteins completely out of the cell and into their surrounding environment. Some exported proteins digest extracellular peptides for carbon and nitrogen sources, whereas others act as free-floating toxins that bind and kill host cells. Still others are injected directly into eukaryotic cells by pathogenic or symbiotic microbes to commandeer host metabolic processes. Five elegant secretion systems, identified as type I–type V, have evolved to ship these proteins out of the cell (**Table 8.2**). A few start with the Sec system just to get the protein into the periplasm, where dedicated outer membrane systems take over and complete export. Other systems provide nonstop service, delivering the protein directly from the cytoplasm to the extracellular space. The diversity of system design is impressive. It is the result, in some instances, of selective evolutionary pressures appropriating established cell processes (for example, pilus assembly). This occurs through the accidental duplication of one set of genes followed by random mutations that innovatively redesign the duplicated set for a new purpose. We know this because the footprints of genetic divergence have been left behind in the DNA sequence. Type I secretion will be described here. Other systems will be covered during the discussion of pathogenesis in Chapter 25.

Type I Protein Secretion

Chapter 4 describes the family of ATP-binding cassette (ABC) influx transporters whose signature is an amino acid motif that binds ATP. Similar ABC transporters function in the opposite direction to export various toxins, proteases, and lipases, as well as antimicrobial drugs (multidrug efflux transporters). These ABC transporters are the simplest of the protein secretion systems and make up what is called type I protein secretion (one example is the Hly system that secretes hemolysin; **Fig. 8.33**). Type I systems all have three protein components, one of which contains an ATP-binding cassette. One component is an outer membrane channel, the second is an ABC protein at the inner membrane, and the third is a periplasmic protein lashed to the inner membrane.

Proteins secreted through type I systems never contact the periplasm because they pass through a continu-

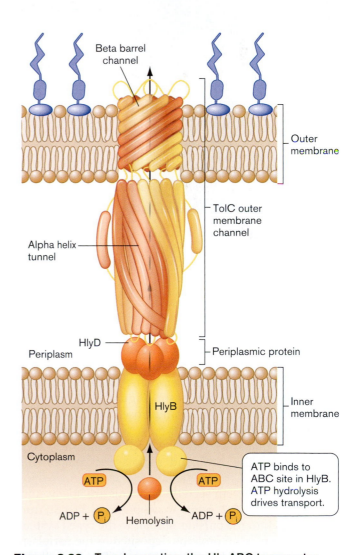

Figure 8.33 Type I secretion, the Hly ABC transporter. Hemolysin (HlyA) is transported directly from the cytoplasm into the extracellular medium through a multicomponent ABC transport system. The HlyB and D proteins are dedicated to HlyA transport. TolC is shared with other transport systems. Not drawn to scale. *Source*: Modified from Moat, Foster, and Spector. 2002. *Microbial Physiology* 4th ed. Wiley-Liss.

A.

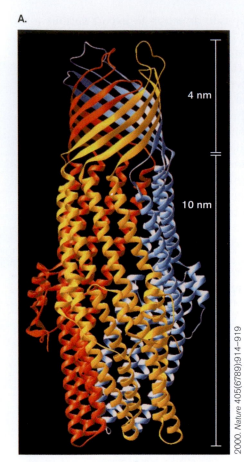

4 nm

10 nm

2000. Nature 405(6789):914–919

B.

C.

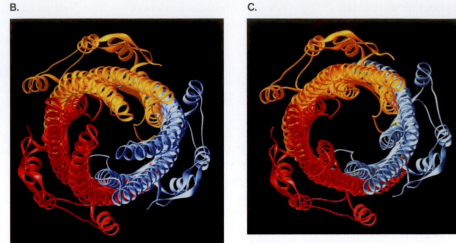

Figure 8.34   TolC protein structure.   A. The beta barrel channel spans the outer membrane (not shown), and the alpha-helical tunnel extends into the periplasm. The inner membrane would be located at the bottom of this figure. Three monomers make up the channel. **B.** A view of the crystal structure from the periplasmic end of the molecule, looking down the threefold symmetry axis. The alpha-helical tunnel is closed. **C.** A hypothetical model of how the tunnel might be opened, with the same orientation as in (B).

ous channel that extends from the cytoplasm to the outer membrane. The inner membrane and periplasmic subunits are generally substrate specific, but numerous ABC export systems share the channel protein TolC. TolC is an intriguing protein composed of a beta-barrel channel embedded in the outer membrane and an alpha-helix tunnel spanning the periplasm. **Figure 8.34** illustrates how the tunnel might open and close. The type I transport shown in **Figure 8.33** exports a hemolysin (HlyA) from *E. coli* that lyses red blood cell membranes. HylB and HylD are the ABC and periplasmic components, respectively.

Four other protein secretion systems are briefly summarized in **Table 8.2**. Some move proteins directly from the cytoplasm to the outside, similar to the type I system, while others pick up proteins deposited in the periplasm by the Sec system. It is important to note that they all play important roles in microbial pathogenesis and are more fully discussed in Chapter 25.

www | Chaperone-assisted protein folding

TO SUMMARIZE:

- **Special protein export mechanisms** are used to move proteins to the inner membrane, the periplasm, the outer membrane, and the extracellular surroundings.
- **N-terminal amino acid signal sequences** help target membrane proteins to the membrane.
- **The general secretory system** involving the SecYEG translocon can move unfolded proteins to the inner membrane or periplasm.
- **Signal recognition protein (SRP)** pauses translation of a subset of proteins that will be placed into the membrane.
- **SecB** binds to certain proteins that will eventually end up in the periplasm. SecB protein pilots these unfolded proteins to the SecYEG translocon.
- **The twin arginine translocase** can move a subset of folded proteins across the inner membrane and into the periplasm.
- **Type I secretion systems** are ATP-binding cassette (ABC) mechanisms that move certain secreted proteins directly from the cytoplasm to the extracellular environment.

8.6   Protein Degradation: Cleaning House

What happens when a cell no longer needs a specific protein—or when a cell synthesizes a protein with incorrect amino acids? These useless or disabled proteins must be degraded to maintain cellular health. In addition, because cellular needs are constantly changing, proteins no longer needed can produce deleterious effects on the cell. This is particularly true of regulatory proteins, whose concentrations must change with time or in response to alterations in the cellular condition.

Many normal proteins contain degradation signals called **degrons** that dictate the stability of a protein. The **N-terminal rule** describes one type of degron. The rule states that the N-terminal amino acid of a protein directly correlates with its stability. For example, proteins beginning with arginine, lysine, or phenylalanine experience a short half-life (2 minutes or less), while proteins with aspartic acid, glutamic acid, or cysteine in the lead position enjoy a much longer half-life (more than 10 hours). Why this correlation exists is not clear.

Abnormally folded proteins are recognized by proteases in part because regions that are normally buried within the protein's three-dimensional structure become exposed. The protein is progressively degraded into smaller and smaller pieces by a series of these proteases. Initial cuts usually involve ATP-dependent endoproteases like Lon protein or ClpP followed by digestion with tripeptidases and dipeptidases. Endopeptidases cleave proteins somewhere within the sequence but not from the ends of the sequence. Many peptidases use ATP hydrolysis to help unfold the target protein prior to digestion. Unfolding is necessary for the protein to slide into a barrel-shaped protease such as ClpAP, ClpXP, or the 28S proteasome of eukaryotic microbes (**Fig. 8.35**). Proteasomes are protein-degrading organelles of eukaryotes. The protein-degrading enzymes are classified as serine, cysteine, or threonine proteases, based on the key residue in their active sites.

There is a striking resemblance between prokaryotic and eukaryotic ATP-dependent proteases (**Fig. 8.36**). The core particle of the eukaryotic proteasome contains 2 copies of 14 different proteins assembled in four stacked rings, each containing 7 proteins (heteroheptameric). The core particle is capped at the top and bottom by the regulatory particles, also made from 14 proteins that are different from those in the core. Six of those proteins are ATPases, while others recognize a ubiquitin peptide tag (see later) placed on doomed proteins. The purpose of this cap is to unfold the target protein and inject it into the 20S proteasome barrel. Archaea also contain a proteasome, but it comprises 14 copies of 2 different proteins assembled in four stacked rings, each containing 7 proteins (homoheptameric). Unlike eukaryotic proteasomes, there is no regulatory particle, and substrate proteins are not ubiquitinated (see **Special Topic 8.3**).

Bacteria contain **Clp proteases**, which contain a proteolytic core made of two homoheptameric rings of the protein ClpP. These proteases also have interchangeable homohexameric ATPase caps made of ClpX, ClpA, ClpB, or ClpC, each of which recognizes different substrates. The accessory proteins recognize and present different substrate proteins to the ClpP protease and thereby regulate which proteins are degraded. **Figure 8.36** shows a side-by-side comparison of the bacterial Clp proteases and the eukaryotic 26S proteasome.

How Do Cells Cope with Stress-Damaged Proteins?

Microbes are constantly exposed to environmental insults such as high temperature or pH extremes that damage proteins and cause them to misfold. As an energy cost-saving device and to prevent interruption of protein func-

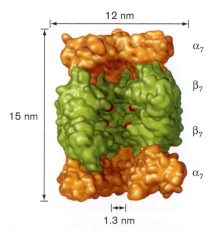

Figure 8.35 Protein degradation machines. Predicted structure of the 20S proteasome from the methanoarchaeon *Methanosarcina thermophila*. The active sites involved in peptide bond cleavage are indicated in red. (PDB code: 1GOU)

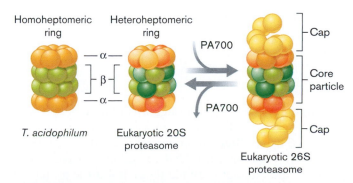

Figure 8.36 Proteolysis degrading machines from prokaryotes and eukaryotes. Structural comparison of the 20S proteasome from an archeaon (*Thermoplasma acidophilum*), a eukaryote, and the 26S eukaryotic proteasome.

Special Topic 8.3 Ubiquitination: A Ticket to the Proteasome

In eukaryotes, protein degradation occurs through organelles called proteasomes. However, proteins are not degraded randomly; they are targeted to the proteasome by the carrier protein **ubiquitin (Ub)**, a small, 76-amino-acid protein found only

in eukaryotic organisms. Among eukaryotes, ubiquitin is highly conserved and found ubiquitously throughout the cell (hence its name). Following ATP hydrolysis, Ub is attached (conjugated) to proteins through a covalent bond between glycine at the C-terminal end of Ub and the side chains of lysine on target proteins. Single Ub molecules can be conjugated to proteins, but more commonly, Ub chains are attached. Ubiquitination of proteasomal substrate proteins is performed by a complex system of three enzymes, ubiquitin-activating (E1) enzyme, ubiquitin-conjugating (E2) enzyme, and substrate recognition protein (E3 enzyme) (**Fig. 1**).

Ubiquitination controls protein turnover by closely regulating the degradation of specific proteins. But what determines if a protein gets marked by Ub? Several factors are involved. First, ubiquitination can be triggered when a substrate recognition protein encounters an N-degron (already discussed) or a PEST sequence, a short stretch of about eight amino acids enriched with proline (P), glutamic acid (E), serine (S), and threonine (T). Other signals buried in the hydrophobic core can do the same thing if they are exposed to the surface when the protein is abnormal or partially unfolded.

The 26S proteasome is the structure that actually carries out degradation. Without Ub, proteins interact with the proteasome, but quickly dissociate. The role of Ub in degradation is to stabilize the interaction between proteasomes and their target substrate proteins. This increases the likelihood that the proteasome will degrade the protein.

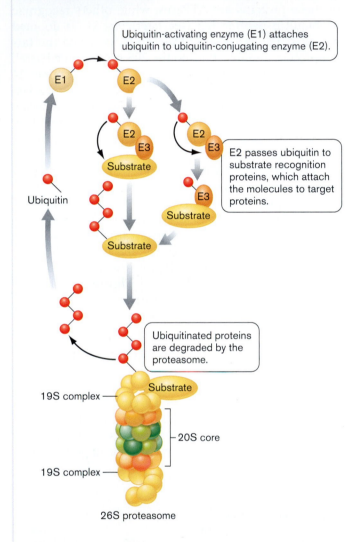

Ubiquitin-activating enzyme (E1) attaches ubiquitin to ubiquitin-conjugating enzyme (E2).

E2 passes ubiquitin to substrate recognition proteins, which attach the molecules to target proteins.

Ubiquitin

Ubiquitinated proteins are degraded by the proteasome.

19S complex — Substrate
20S core
19S complex
26S proteasome

Figure 1 Protein degradation and ubiquitin. Most proteins that are degraded by 26S proteasomes are tagged for destruction by the addition of polyubiquitin chains. Ubiquitin is a small 76-amino-acid protein that is highly conserved in eukaryotic cells. Ubiquitination of proteasomal substrate proteins is performed by a complex system consisting of ubiquitin activating (E1) enzymes, ubiquitin-conjugating (E2) enzymes and substrate recognition proteins (E3 enzymes).

tion, injured proteins go through a kind of triage process that evaluates whether they are salvageable or must be destroyed before they can endanger the cell. Chaperones constantly hunt for misfolded (or otherwise damaged) proteins and attempt to refold them. But if the protein is released from a chaperone and remains misfolded, it can, by chance, either reengage the chaperone or bind a protease that destroys it (**Fig. 8.37**). This fold-or-destroy triage system is essential if a microbe is to survive environmental stress.

TO SUMMARIZE:

- **All proteins** are eventually degraded.
- **The N-terminal rule** is one type of degradation signal (degron) that marks the half-life of a protein (how long it takes 50% of the protein to degrade).
- **ATP-dependent proteases** such as Lon or ClpP usually initiate degradation of a large protein.
- **Ubiquitination** marks eukaryotic proteins for degradation by the proteasome (see **Special Topic 8.3**).

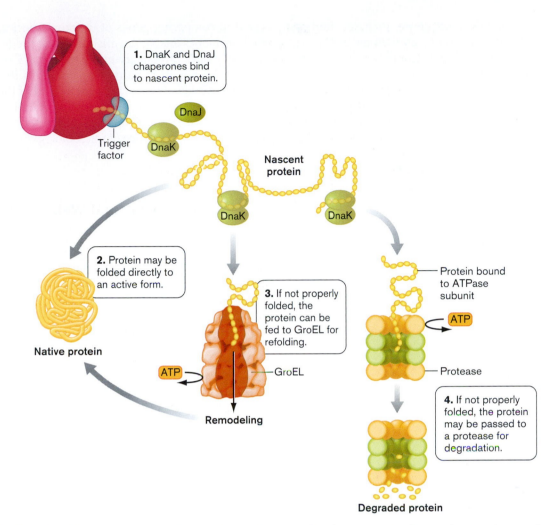

Figure 8.37   Protein folding versus degradation triage pathways.   The figure depicts what can happen to a newly synthesized protein. However, a protein that unfolds as a result of environmental stress (for instance, heat) will undergo the same triage process.

■ **Damaged proteins** randomly enter chaperone-based refolding pathways or degradation pathways until the protein is repaired or destroyed.

8.7 Bioinformatics: Mining the Genomes

We have just described how the information within a genome is deciphered by the cell to produce proteins. But the experiments described cannot be performed in the vast majority of microbes, which are unculturable. In Chapter 7, we discussed how we can quickly and efficiently sequence entire genomes (see Section 7.6). This knowledge, combined with our understanding of protein structure (amino acid sequences) and the relationships between protein structure and function, has brought us to the postgenomic era. We can now call on the vast store of information gath-

ered over the last century to make predictions about the genetics and physiology of microbes even when we cannot grow them in the laboratory. The following sections reveal how these predictions are made.

Annotating the Genome Sequence

Chapter 7 explained how the DNA sequence of a genome is obtained. Knowing the sequence is just a first step; meaningful information comes from **annotation** of the DNA sequence—that is, understanding what the sequence means. Annotation is analogous to identifying separate sentences and words in an unknown language. For most languages, you would look for punctuation marks like periods or exclamation points to identify sentences and then scan for blank spaces between letters to signify the beginning and end of a word. You may not initially know what the sentence says, but defining what makes a sentence is the first step.

Annotation of a DNA sequence includes finding the start and stop sites of genes, as well as predicting the function of the gene product. Computers use established rules to mark potential genes—open reading frames (ORFs). Then similarities are sought between the deduced amino acid sequences of those ORFs and the sequences of proteins with known functions. Similarities are used to infer the function of the unknown ORF. Genes encoding transfer RNA and ribosomal RNAs do not encode proteins but can be identified because of the conservation of these sequences across vast phylogenetic distances. The results of annotation can be represented in many different forms. **Figure 8.38** shows a commonly used method of presenting the results, a circular display map—in this case of the *Mycoplasma mycoides* genome—with genes color-coded by predicted function.

Since 1998, the complete genomes of over 225 microbial species have been published, and many others have been partially sequenced. This wealth of information has spawned a new discipline called **bioinformatics** dedicated to comparing genes of different species. Data from bioinformatics enable scientists to make predictions about an organism's physiology and evolutionary development, even if the organism cannot be grown in the laboratory. The magnitude of this achievement, and of the job ahead, can be appreciated if we recall that fewer than 1% of microorganisms can be cultured. Bioinformatics has forever changed how the science of microbiology is conducted.

www | Sequenced Genomes

Computer Analysis and Web Science

Lengthy DNA sequences are analyzed by computer programs such as ORFfinder. The program uses the universal genetic code to deduce all possible protein sequences that could be formed in all reading frames on RNA molecules transcribed from either direction on the chromosome (**Fig. 8.39A**). An open reading frame (ORF), the equivalent of a sentence in our analogy, is defined

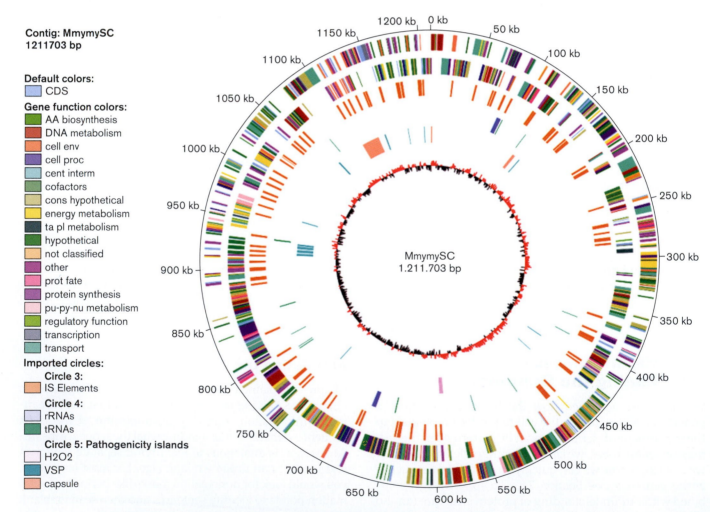

Figure 8.38   Circular display map of the *Mycoplasma mycoides* genome (1,211,703 bp).   Color code indicates gene clustering by function. The two outer rings represent genes transcribed from different DNA strands. The innermost circle uses peaks to show GC content levels above or below 50%.

as a DNA sequence that can potentially encode a string of amino acids of minimum length—say, for example, 50 residues. The 50 residues of the ORF could encode a 5,500-Da (5.5-kDa) protein, since the average weight of an amino acid is 110 Da. Each ORF begins with a translation start codon (usually ATG or, more rarely, GTG or TTG). A translation start codon marks where ribosomes start to read a messenger RNA molecule. An ORF ends with a termination codon (in the DNA, these are TAA, TAG, or TGA), which, as UAA, UAG, and UGA in RNA, signals the ribosome to stop translation. Additionally, the computer can identify an ORF by looking for potential ribosome-binding sites upstream of the start codon, but these ribosome-binding sites may differ between species, so finding one is not essential to declaring the presence of an ORF.

While the preceding methods work for prokaryotes, identifying ORFs in eukaryotes is more difficult. In eukaryotes, most genes contain **introns** (long noncoding sequences that occur in the middle of genes), as do the initial mRNA products produced from those genes (**Fig. 8.39B**). The vast majority of known prokaryotic genes do not contain introns. Introns serve several regulatory func-

tions in eukaryotes that determine whether a protein product is made. However, the eukaryotic cell must use special splicing mechanisms to remove the intron sequences and rejoin the protein-encoding sequences, called **exons**, of the mRNA prior to translation. (In eukaryotic organisms, the mRNA transcribed directly from DNA is thus called the **preliminary mRNA transcript** or **pre-mRNA**; the "final" mRNA transcript is produced when the introns have been spliced out.) Because there are few identifying sequence characteristics that mark an intron, extremely sophisticated computer analysis is needed to determine where to remove introns in order to derive the ORF coding sequence of a eukaryotic gene.

Once an ORF is identified, computers use mathematical algorithms to determine whether the protein predicted by the ORF resembles any other protein deposited in the worldwide databases or, even better, resembles proteins of known function. By "resemble" we mean that the query protein possesses amino acid sequences that are identical or functionally similar to those found in other proteins. In a functionally similar sequence, certain amino acids are replaced by similar amino acids—for example, isoleucine for leucine. We know from consider-

Figure 8.39 Predicting open reading frames (ORFs) in a DNA sequence.
A. Each predicted ORF in this 1,600-bp bacterial sequence begins with an AUG or GUG (potential translation start codons in mRNA) and ends with a translation terminator codon. The computer can translate in all six reading frames (3 on top strand, 3 on bottom strand). Determining which is the real ORF requires additional information, such as potential ribosome-binding sites and transcriptional start site. **B.** In eukaryotes, finding ORFs is complicated by the presence of noncoding DNA sequences called introns. The actual ORF (exons) is disrupted by introns, which must be spliced out of the primary RNA message before the final mRNA leaves the nucleus.

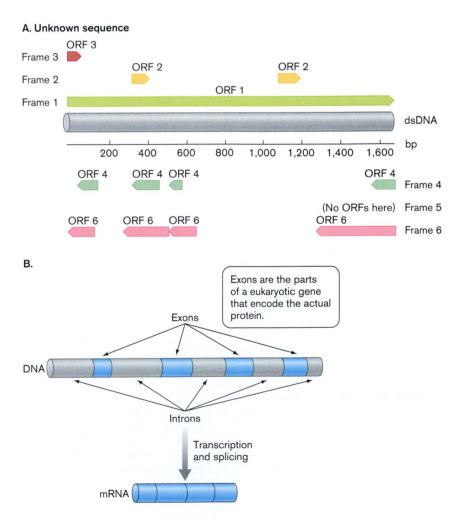

able experimental precedent that proteins with the same function, regardless of species, usually have critical amino acid sequences in common because they evolved from the same ancestral sequence. Sequences common to two different protein or DNA molecules are called homologous sequences.

Figure 8.40 shows a sequence alignment between eight examples of dihydrolipoamide dehydrogenases from different species and illustrates how this alignment was used to partially characterize the function of a virulence gene in an important pathogen. Ordinarily, dihydrolipoamide dehydrogenase is integrally important to the tricarboxylic acid cycle, which extracts energy and carbon from carbohydrate food sources. Note the highlighted areas in the figure; these represent regions of identity or similarity at similar positions. Identity means the same amino acid is present in each protein, whereas similarity means that chemically related amino acids are substituted (for example, valine is substituted by the similar leucine, or glutamate is substituted by aspartate). Also notice that the top sequence in the figure is an unknown ORF from the *Streptococcus pneumoniae* genome, an organism that causes pneumonia. The computer program flagged the sequence similarities between this unknown ORF and the known proteins shown in the figure. Based on that homology, the ORF was deduced to encode a dihydrolipoamide dehydrogenase, a prediction confirmed by mutating the gene and by looking directly for the biochemical activity. However, *S. pneumoniae* are strictly fermentative organisms that do not have a complete TCA cycle. Nevertheless, deleting the gene decreased virulence of this pathogen in a mouse model, suggesting an important function for the enzyme unrelated to the TCA cycle. In this and many other instances, bioinformatics provides important insights into microbial physiology.

Alignments can be done either with base sequences in DNA or with protein sequences deduced from the DNA sequence. Because of the degeneracy of the code, however, it is usually easier to pick up evolutionary relationships between genes from distantly related species by comparing their protein products rather than the DNA sequence. Remember, one amino acid can be encoded by more than one DNA codon. So a DNA strand (nontemplate) encoding the peptide Ala-Leu-Ser could be 5′ GCU CCU UCC or 5′ GCA CCG UCA. It is easier to see the homology using the deduced peptide. On the other hand, not all genes encode proteins.

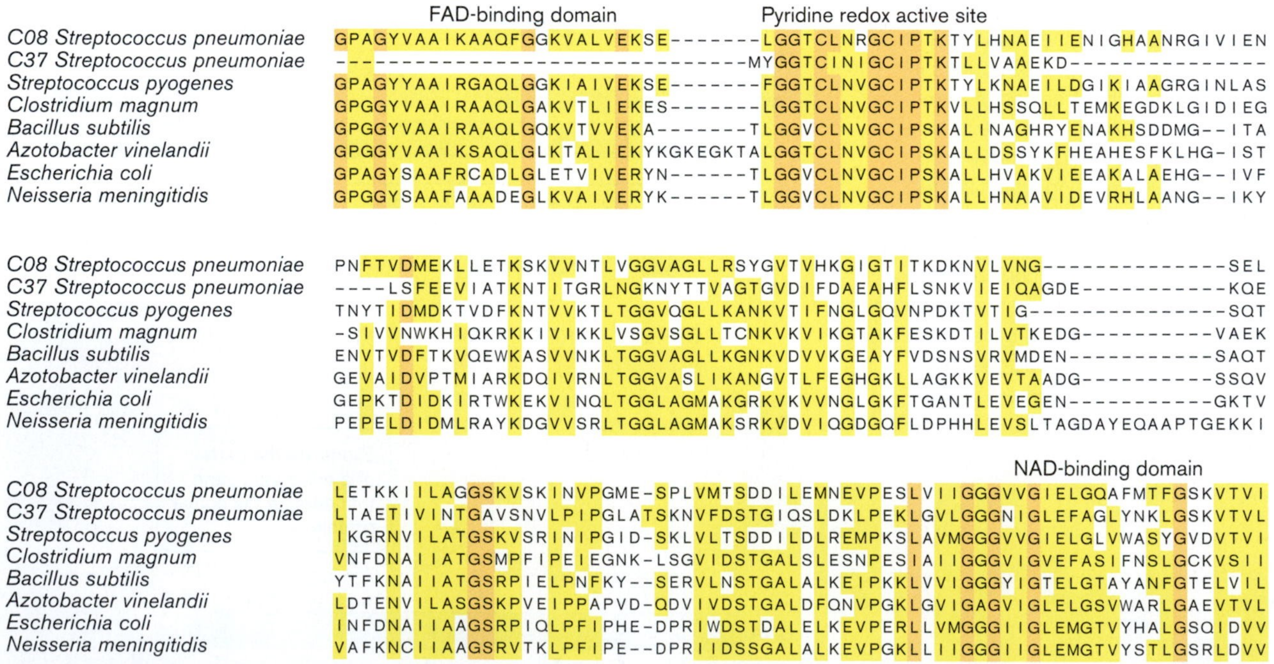

Figure 8.40 Multiple sequence alignment. The program CLUSTALW (available online) was used to align the amino acid sequence of dihydrolipoamide dehydrogenases from *Streptococcus pneumoniae* C08 ORF with selected orthologs from other known species (only a portion of the alignment is shown). These proteins are called orthologs because they have similar sequences and carry out the same biochemical reaction. Top sequence in each tier is a different portion of the C08 sequence. Amino acid sequences of over 51% similarity are shaded. Highly conserved residues where at least five residues are the same or similar amino acids are shaded yellow. Very highly conserved residues where at least seven out of eight are identical are shaded brown. Deletions (sequences missing in one or more orthologs) are indicated as dots.

Special Topic 8.4   What Is the Minimal Genome?

Defining life is more than a philosophical exercise. Attempts to address the topic have touched almost every scientific discipline. From a genetic viewpoint, we could try to define what makes life by determining the minimum number of genes needed to sustain the life of a cell. Mycoplasmas, bacteria that lack cell walls, have long been viewed as minimal organisms. (Viral DNA is not considered to represent a minimal genome because viruses are not cells and are incapable of independent growth and reproduction.) The *Mycoplasma* species with the smallest genome is *Mycoplasma genitalium* (**Fig. 1**), whose chromosome encodes a mere 480 proteins, compared with over 4,000 for *E. coli*. The small size of the *M. genitalium* genome is thought to reflect its parasitic lifestyle. For example, the many genes required to synthesize amino acids, make cell wall components, and run a citric acid cycle have been lost from this organism, presumably because the host cell supplies these products.

Based on this finding should we conclude that 480 genes is the minimum needed to sustain cellular life? Any effort to define such a minimum depends on what we define as "living." How can we know which, if any, of the 480 genes "enhance" life rather than being absolutely essential? Removing even nonessential genes can reduce the fitness of an organism and result in its eventual extinction. For example, microbes with small genomes often have a slower growth

rate than their relatives with larger genomes and exhibit a greater dependence on outside assistance. At some point during genome reduction, the ability to maintain active cultures will become impossible. That point may differ, however, depending on the environment. Environmental factors will dictate whether a given gene is essential for survival. For example, a fish requires gills to live; humans do not. Thus, the evolutionary niche occupied by a microbe will influence which genes are retained in its genome. An organism that lives in acidic environments will require genes whose products will protect the cell against low pH. Those genes are dispensable to an organism that grows only at neutral pH.

In experiments to determine the minimal genome required to support growth of *M. genitalium*, scientists have used transposon mutagenesis, a technique in which pieces of DNA (transposons) are inserted into the the cell's DNA to disrupt gene function (described in Section 9.6). This technique makes it possible to randomly knock out genes and see which ones could be neutralized without losing cell viability. Based on a statistical treatment of their results, these investigators concluded that 129 genes appeared dispensable (they could be knocked out and the cell would still grow). This does not mean that the minimal genome is 351 genes, since it is impossible to know if any of the paralogous genes present among the 129 dispensables might carry out identical (redundant) functions such that losing one or the other would not affect viability. However, losing both simultaneously would be lethal.

It is sobering to realize that one-third of the 351 essential genes in *M. genitalium* are of undefined function. It brings into question a basic assumption held by many that the fundamental mechanisms of life have, for the most part, been recognized. On the other hand, it may be that many of those 100 or so genes with unannotated function encode activities that are well characterized in other organisms but are simply not recognized by sequence as being orthologs in *M. genitalium*. Some may be specialized for colonization or parasitism. Whether these genes encode unknown fundamental life processes or not, there is much to be done before claiming a clear understanding of the simplest cell. The clearest solution to determining the minimum number of genes required to support life would to be to build a genome from scratch in the test tube, place it into a cell whose genome had been excised, and then see if that cell can survive.

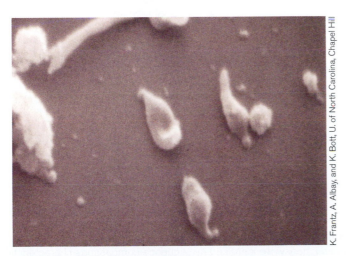

K. Frantz, A. Albay, and K. Bott, U. of North Carolina, Chapel Hill

Figure 1   *Mycoplasma genitalium.*   Scanning electron micrograph.

Prime examples are the genes encoding stable ribosomal RNA molecules. The DNA sequences for these ribosomal RNAs are highly conserved across species, and their similarities reveal homology even between the three ancestral kingdoms of bacteria, archaea, and eukarya.

> **NOTE:** Do not confuse template and nontemplate strands of DNA for a given gene. The template strand, read 3′-to-5′ by RNA polymerase, is transcribed to make the 5′-to-3′ RNA transcript. The nontemplate DNA strand, also called the sense strand, has the *same* sequence as the RNA transcript, with the substitution of T's for U's.

Homologs, Orthologs, and Paralogs

Homologies found between genes or proteins suggest an evolutionary relatedness. Genes or proteins that are **homologous** probably evolved from a common ancestral gene. Homologous genes or proteins can be classified as either orthologous or paralogous. **Orthologous** genes have functions essentially the same but occur in two or more different species. For example, the gene for glutamine synthase (*gltB*) in *E. coli* is orthologous to *gltB* in *Vibrio cholerae* and to *gltB* in the gram-positive bacterium *Bacillus subtilis*. **Paralogous** genes arise by duplication within the same species (or progenitor) but evolve to carry out different functions (**Fig. 8.41**). For example, genes encoding type III protein secretion systems, which export virulence proteins, have evolved from paralogous genes encoding flagellar biogenesis. The two sets of genes share a common ancestor but have since evolved completely different functions (for example, secretion and motility). Paralogous genes are maintained in a microbial genome because their distinct functions contribute to the organism's adaptive

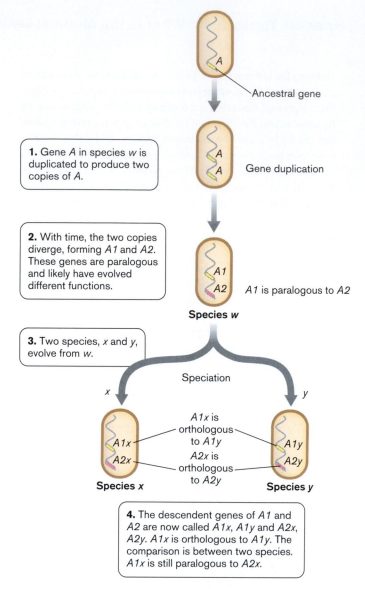

1. Gene *A* in species *w* is duplicated to produce two copies of *A*.

2. With time, the two copies diverge, forming *A1* and *A2*. These genes are paralogous and likely have evolved different functions.

3. Two species, *x* and *y*, evolve from *w*.

4. The descendent genes of *A1* and *A2* are now called *A1x*, *A1y* and *A2x*, *A2y*. *A1x* is orthologous to *A1y*. The comparison is between two species. *A1x* is still paralogous to *A2x*.

Figure 8.41   Paralogous versus orthologous genes. An ancestral gene can undergo a duplication to evolve an orthologous or paralogous gene.

Figure 8.42   Genomic predictions of solute transport and metabolic pathways of *Helicobacter pylori* strain 26695.   The large rectangle represents a cell. The transporters arrayed around the periphery were identified by sequence comparisons characteristic to transporters in gram-negative bacteria. The ABC transporters (multicomponent) are predicted to transport oligopeptides, dipeptides, proline, glutamine, molybdenum, and iron III. There are also predicted P-type ATPases that extrude toxic metals from the cell and a glutathione-regulated potassium-efflux protein (encoded by *kefB*). An integrated view of the main components of *H. pylori* central metabolism is presented within the rectangle. This scheme is based on using glucose as the sole carbohydrate source. Urease, a multisubunit enzyme crucial for survival of *H. pylori* at acid pH, is indicated as a complex (purple circle). A question mark is attached to pathways that could not be completely elucidated. A red arrow represents pathways or steps for which no enzymes were predicted from the sequence. Pathways for macromolecular biosynthesis (RNA, DNA, and fatty acids) were found but are not shown. Other abbreviations for *H. pylori* genes: *ackA*, acetate kinase; *acnB*, aconitase B; *aspC*, aspartate aminotransferase; *dld*, D-lactate dehydrogenase; *gdhA*, glutamate dehydrogenase; *glnA*, glutamine synthetase; *gltA*, citrate synthase; *hydABC*, hydrogenase complex; *icd*, isocitrate dehydrogenase; *pfl*, pyruvate formate lyase; *por*, pyruvate ferredoxin oxidoreductase; *ppc*, phosphoenolpyruvate carboxylase; *pps*, phosphoenolpyruvate synthase; *pta*, phosphate acetyltransferase; *gldD*, glycerol-3-phosphate dehydrogenase; NDH-1, NADH–ubiquinone oxidoreductase complex.

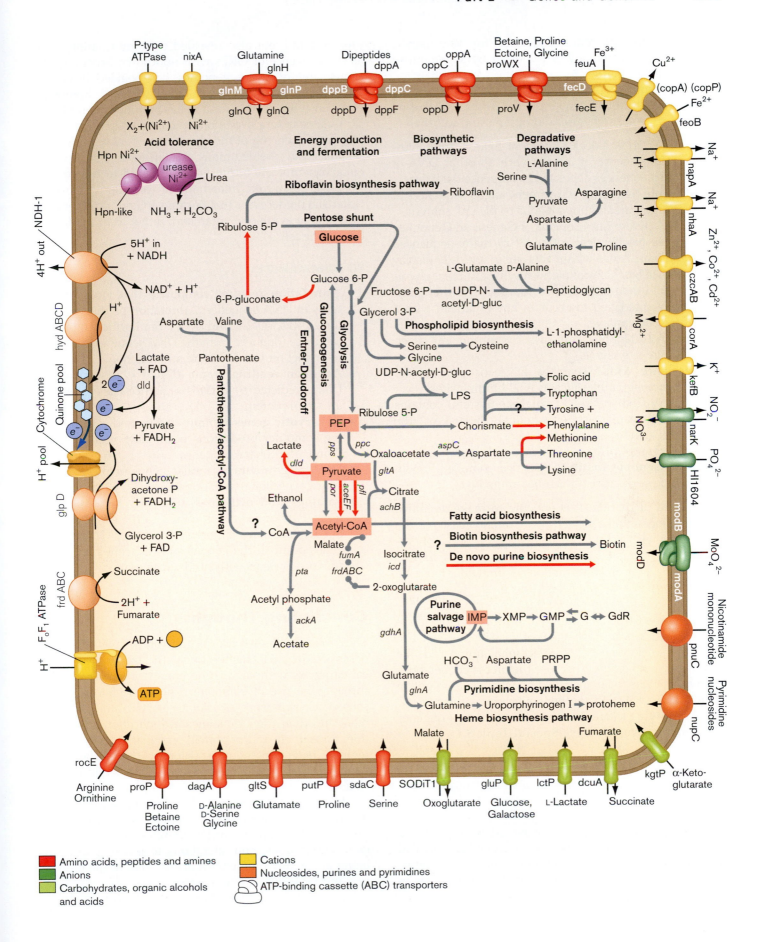

potential. Duplicate genes of identical function usually result in loss of one copy or the other by degenerative evolution (see Chapter 9).

Many computer programs and resources used to analyze DNA and protein sequences are freely available on the Web. Here are some useful programs and websites:

- **BLAST** (**B**asic **L**ocal **A**lignment **S**earch **T**ool) compares a sequence of interest with all other DNA or protein sequences deposited in sequence databases.
- **Multiple Sequence Alignment** aligns sequences of genes identified as homologous by BLAST analysis.
- **KEGG** (**K**yoto **E**ncyclopedia of **G**enes and **G**enomes) outlines biochemical pathways in many sequenced organisms. Areas within this site graphically display reference biochemical pathways (for example, glycolysis) and then indicate which proteins in that pathway are predicted to occur in any organism with a sequenced genome.
- **PUMA2** identifies pathways containing any given substrate or enzyme.
- **Sequence Motif** searches DNA or proteins for sequence signatures such as ATP-binding sites.
- **ExPASy** (**Ex**pert **P**rotein **A**nalysis **Sy**stem) contains many molecular tools, including Swiss-Prot, the definitive index of known proteins.
- **Colibri** and **Colibase** focus on *E. coli* genomics.
- **Joint Genome Institute** has also compiled known genome sequences of microbes and eukaryotes.

A word of caution. It is enticing to make definitive proclamations about gene or protein function based on the computer analysis of a DNA sequence. As good as a prediction may seem, it is only a prediction, a well-educated guess. Biochemical confirmation of function must be made, if possible. Nevertheless, the predictions we can make are powerful. In one recent example, a metabolic model was constructed by genome annotation for *Helicobacter pylori*, the causative agent of gastritis (ulcers). The *H. pylori* model was used to predict what amino acids the organism would require (because the predicted metabolic pathway appears to be missing) and what genes might be essential (**Fig. 8.42**). In another application, some investigators are trying to predict the smallest number of genes required for a living cell (see **Special Topic 8.4**).

> **THOUGHT QUESTION 8.10** An ORF 1,200 bp in length could encode a protein of what size and molecular weight?

The information age has spawned what could be called Web biology, an *in silico* science that has provided new insight into evolution, physiology, and pathogenesis. Bioinformatics has not only confirmed what we know in greater detail, but also revealed new information, such as the existence of ORFs with no known function. The sequencing of entire genomes has enabled a greater understanding of sequence similarity or homology between genes or proteins and has provided a foundation for understanding evolutionary relationships. For example, the sequencing of the alpha proteobacterium *Rickettsia prowazekii* revealed many similarities to genes within eukaryotic mitochondrial DNA. This led to the conclusion that eukaryotic mitochondria were derived from a rickettsial predecessor that became an endosymbiont (an intracellular symbiotic organism), rather than food, for a eukaryotic cell.

TO SUMMARIZE:

- **Annotation** requires computers that look for patterns in DNA sequence. Annotation predicts regulators, ORFs, rDNA, and tRNA. Similarities in protein sequence (deduced from the DNA sequence) are used to predict protein structure and function.
- **An open reading frame (ORF)** is a sequence of DNA predicted by various sequence cues to encode an actual protein.
- **Eukaryotic genes** contain introns and exons, making computer predictions of an ORF more difficult than with prokaryotic genes.
- **DNA alignments** of similar genes or proteins can reveal evolutionary relationships.
- **Paralogs and orthologs** arise from gene duplications. Paralogous genes coexist in the same genome but usually have different functions. Orthologous genes occur in the genomes of different species but produce proteins with similar functions.

Concluding Thoughts

The efficient cell must use the macromolecular processes described in this chapter to synthesize its biochemical pathways without wasting energy. Microbes do this through the use of elaborate control mechanisms that sense the organism's physiological state and environment and then trigger changes in replication, transcription, translation, and/or protein processing. How bacteria regulate gene expression in response to environmental stimuli, including threats to survival, will be discussed in Chapter 10.

In Chapter 9, we explore how natural selection randomly redesigns genomes to adapt to ecological niches. Microbes use a variety of DNA exchange mechanisms, gene duplications, and alterations to evolve into forms better adapted to their environments and, in the process, may produce entirely new species.

CHAPTER REVIEW

Review Questions

1. What are some characteristics of an ORF?
2. What is a DNA sequence alignment, and what can it tell you?
3. Describe the differences between a pair of orthologous genes and a pair of paralogous genes.
4. How can bioinformatics predict a metabolic pathway for an organism that cannot be grown in the laboratory?
5. What defines a promoter?
6. What are sigma factors, and what role do they play in gene expression?
7. Describe the three stages of transcription.
8. Explain degeneracy of the genetic code. What is the wobble in codon-anticodon recognition?
9. Describe the stages of protein synthesis. Why is the ribosome called a ribozyme?
10. Discuss some antibiotics that affect transcription or translation.
11. What is meant by coupled transcription and translation? Does this occur in eukaryotic cells?
12. How do bacterial cells release ribosomes that are stuck onto damaged mRNA molecules lacking termination codons?
13. What can happen to misfolded proteins?
14. Why are only certain proteins secreted from the bacterial cell? What are some secretion mechanisms?
15. Compare protein degradation in eukaryotes and prokaryotes.
16. What is annotation? How does it apply to bioinformatics?

Key Terms

acceptor site (A site) (274)
aminoacyl-tRNA transferase (271)
annotation (293)
anticodon (270)
bioinformatics (294)
catalytic RNA (266)
chaperone (283)
cistron (273)
Clp protease (291)
codon (267)
consensus sequence (260)
degron (291)
exit site (E site) (274)
exon (295)
heat-shock protein (285)
homologous (297)
intron (295)
messenger RNA (mRNA) (266)

molecular mimicry (281)
N-terminal rule (291)
open reading frame (ORF) (264)
orthologous (297)
paralogous (297)
peptidyltransferase (274)
peptidyl-tRNA site (P site) (274)
polycistronic (273)
preliminary mRNA transcript
   (pre-mRNA) (295)
promoter (260)
protein secretion system (286)
release factor (280)
Rho-dependent (264)
Rho-independent (265)
ribosomal RNA (rRNA) (266)
ribosome-binding site (275)
ribozyme (274)

SecYEG translocon (286)
Shine-Dalgarno sequence (275)
sigma factor (259)
signal recognition particle (SRP) (286)
signal sequence (286)
"small RNA" (sRNA) (266)
start codon (268)
stem loop (265)
stop codon (269)
template strand (259)
tmRNA (266)
transcript (259)
transcription (259)
transfer RNA (tRNA) (266)
translation (267)
translocation (278)
twin arginine translocase (TAT) (289)
ubiquitin (Ub) (292)

Recommended Reading

Buckanan, Susan K. 2001. Type I secretion and multidrug efflux; transport through the TolC channel-tunnel. *Trends in Biochemical Science* **26:**3–6.

Buttner, Daniela, and Ulla Bonas. 2002. Port of entry—the type III secretion translocon. *Trends in Microbiology* **10:**186–192.

Christie, Peter J. 2001. Type IV secretion: intercellular transfer of macromolecules by systems ancestrally related to conjugation machines. *Molecular Microbiology* **40:**294–305.

Darst, Seth. 2001. Bacterial RNA polymerase. *Current Opinion in Structural Biology* **11:**155–162.

Gowrishankar, Jayaraman, and Rajendran Harinarayanan. 2004. Why is transcription coupled to translation in bacteria? *Molecular Microbiology* **54:**598–603.

Hutchison, Clyde, Scott Peterson, Steven R. Gill, Robin T. Cline, Owen White, et al. 1999. Global transposon mutagenesis and a minimal *Mycoplasma* genome. *Science* **286:**2165–2169. [Online.] http://www.tigr.org/about/minimalgenes.pdf.

Kaczanowska, Magdalena, and Monica Rydén-Aulin. 2007. Ribosome biogenesis and the translation process in *Escherichia coli. Microbiology and Molecular Biology Reviews.* **71:**477–494.

Korzheva, Natalia, Arkady Mustaev, Maxim Kozlov, Arun Malhotch, Vadim Nikiforov, et al. 2000. A structural model of transcription elongation. *Science* **289:**619.

Murakami, Katsuhiko S., Shoko Masuda, Elizabeth A. Campbell, Oriana Muzzin, and Seth Darst. 2002. Structural basis of transcription initiation: RNA polymerase holoenzyme-DNA complex. *Science* **296:**1285–1290.

Peterson, Scott N., and Claire M. Fraser. 2001. The complexity of simplicity. *Genome Biology* **2:**2002.2–2002.8.

Pugsley, Antony P., Olivera Francetic, Arnold J. Driessen, and Victor de Lorenzo. 2004. Getting out: protein traffic in bacteria. *Molecular Microbiology* **52:**3–11.

Santangelo, Thomas J., and John N. Reeve. 2006. Archaeal RNA polymerase is sensitive to intrinsic termination directed by transcribed and remote sequences. *Journal of Molecular Biology* **355:**196–210.

Steitz, Joan A., and Karen Jakes. 1975. How ribosomes select initiator regions in mRNA: base pair formation between the 3′ terminus of 16S rRNA and the mRNA during initiation of protein synthesis in *Escherichia coli. Proceedings of the National Academy of Science* **72:**4734–4738.

Ward, Naomi, and Claire M. Fraser. 2005. How genomics has affected the concept of Microbiology. *Current Opinion in Microbiology* **8:**564–571.

Wilson, Daniel N., and Knud H. Nierhous. 2003. The ribosome through the looking glass. *Angewandte Chemie International Edition* **42:**3464–3486.

Chapter 9

Gene Transfer, Mutations, and Genome Evolution

DNA was once thought to be a monolithic "master molecule" that governs all traits of an organism while remaining unchanged itself. We now know that DNA sequences change over generations through various mutations, DNA rearrangements, and gene transfers between species. Bacterial and archaeal genomes experience a rapid flux, sometimes shuttling large clusters of genes between members of different taxonomic kingdoms. All this inter- and intraspecies DNA traffic has led ecologists to expand the definition of the "microbial genome" to include DNA in the cell plus all the DNA out in the environment to which the microbe has potential access.

A number of important questions come to mind when we contemplate the consequences of DNA plasticity. For example, because the processes of adding, subtracting, and mutating genes are all processes that come at some cost, can the cell protect itself from undergoing too many drastic changes? Are some mutations good and others bad? Why would nature allow this much genetic uncertainty? This chapter explores these long-standing evolutionary questions and shows how microbial genomes continually change.

No UV

B. With UV

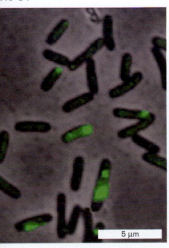
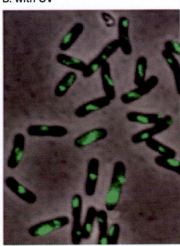

5 μm

Visualization of *E. coli* responding to ultraviolet light. A gene fusion between the recombinant gene *recA* and green fluorescent protein (GFP) was introduced into *E. coli*. When DNA is damaged by UV light, the cell increases the production of the RecA-GFP fusion protein as part of the SOS response. Panel A shows unexposed cells; Panel B shows the increase in RecA-GFP binding to DNA 20 minutes after UV exposure. *Source*: Renzette, et al. 2005. *Molecular Microbiology* 57:1074.

We think of evolution as taking place over thousands if not millions of years. However, John Sullivan and Clive Ronson, in New Zealand, witnessed evolution taking place within a mere decade. These scientists were studying the microbe *Mesorhizobium loti*, a bacterial species that establishes symbiotic relationships with plants by forming nitrogen-fixing nodules on the plant's roots. The genes encoding symbiosis are located as a group on the mesorhizobial chromosome. In addition to these symbiotic rhizobia, there are many nonsymbiotic species that are incapable of forging this relationship. In a remarkable experiment, Sullivan and Ronson inoculated a single strain of *M. loti* into an area of land devoid of natural nodulating rhizobia. Seven years later, they discovered that the area contained many genetically diverse symbiotic mesorhizobia now able to nodulate the flowering plant *Lotus corniculatus*. These microbes did not exist before the experiment. The scientists found that the 500-kbp genome segment encoding symbiosis had somehow made its way from *M. loti* into these other bacteria—essentially creating new species. How did this happen so quickly?

In this chapter, we address this question beginning with a discussion of the mechanisms that mediate transient as well as heritable DNA movements between species. Next we turn to a description of the competing processes of mutagenesis and mutation repair in the context of self-preservation and evolutionary change. The chapter closes with an examination of how all these processes collaborate over millennia to remodel genomes and build biological diversity.

9.1 The Mosaic Nature of Genomes

Genomic analysis has revealed that over the millennia, microbes have undergone extensive gene loss and gain. Archaea, for example, arose from a common eukaryotic/archaeal phylogenetic branch and as a result possess many traits in common with eukaryotes, such as the structure and function of their DNA and RNA polymerases. However, many archaeal genes whose products are involved with intermediary metabolism look purely bacterial. In fact, 37% of the proteins found in the archaeon *Methanococcus jannaschii* are found in all three domains—Archaea, Eukarya, and Bacteria. Another 26% are otherwise found only in bacteria, while a mere 5% are confined to archaea and eukaryotes. This clearly suggests that archaea enjoy a mixed heritage.

Another surprise arising from bioinformatic studies is the mosaic nature of the *E. coli* genome and, indeed, of all microbial genomes. Though we have intensively studied *E. coli* for over 100 years, we now find that the organism's DNA is rife with pathogenicity islands, fitness islands, inversions, deletions (when compared to similar species), paralogous genes, and orthologous genes. How did all this genomic blending happen? The answer appears to involve heavy gene traffic between species (horizontal gene transfer), recombination events occurring within a species, and a variety of mutagenic and DNA repair strategies. All of these processes accelerated natural selection, where trial and error helped mold genomes.

Why was this gene movement unexpected? Sometimes a powerful idea in biology actually delays discovery of the next step. For example, James Watson's vision of DNA as the "master molecule" and the concept of the single-rooted evolutionary tree led to so many important concepts in molecular biology that observations of more fluid gene exchange were ignored or resisted at first. However, the history of microbiology provided many clues as to its existence.

9.2 Gene Transfer: Transformation, Conjugation, and Transduction

In 1928, a perceptive English medical officer, Fred Griffith (1879–1941), found that he could kill mice by injecting them with *dead* cells of a virulent pneumococcus (*Streptococcus pneumoniae*), a cause of pneumonia, together with *live* cells of a nonvirulent mutant. Even more extraordinary was the fact that the bacteria recovered from the dead mice were of the live, *virulent* type. Were the dead bacteria brought back to life? Unfortunately, Griffith was killed by a German bomb during an air raid on London in 1941 and never learned the answer.

In a landmark series of experiments published in 1944, Oswald Avery (1877–1955), Colin MacLeod (1909–1972), and Maclyn MacCarty (1911–2005) proved that Griffith's experiment was not a case of reviving dead cells, but involved the transfer of DNA released from the virulent, dead strain *into* the harmless living strain of *S. pneumoniae*, an event that transformed the live strain into a killer. The process of importing free DNA into bacterial cells is now known as **transformation**.

Transformation provided the first clue that gene exchange can occur in microorganisms. But why bacteria carry out this and other forms of gene transfer was not fully appreciated until much later. Recent genomic studies indicate that the fundamental purpose of bacterial gene transfer is to acquire genes that might be useful as the environment changes. Before we can discuss how bacterial genomes evolve, we need to understand the various gene exchange mechanisms available to them. Following the explanations of gene exchange, we will discuss the mechanisms that incorporate newly acquired DNA into genomes (for example, recombination).

Gene Exchange by Transformation

Many bacteria have the ability to import DNA fragments and plasmids released from nearby dead cells via the process of transformation. Natural transformation is a property inherent to many species, conferred by specific protein complexes. Examples include gram-positive bacteria like *Streptococcus* and *Bacillus*, as well as gram-negative species such as *Haemophilus* and *Neisseria* species. Transformation is a natural part of their growth cycles.

Other bacteria, however, such as *E. coli* and *Salmonella*, do not possess the equipment needed to import DNA naturally. They require artificial manipulations to drive DNA into the cell. These laboratory techniques include perturbing the membrane by chemical ($CaCl_2$) and electrical (**electroporation**) methods. $CaCl_2$ alters the membrane, which makes these cells electrocompetent and allows DNA to pass. Electroporation, on the other hand, uses a brief electrical pulse to "shoot" DNA across the membrane.

Why do species such as *Neisseria* species undergo natural transformation? First, species that indiscriminately import DNA may use the transformed DNA as food. Second, the more finicky species that transform only compatible DNA sequences may use DNA released from dead compatriots to repair their own damaged genomes. Finally, transformation may play an evolutionary role, enabling species to adjust to new environments by acquiring new genes from other species in a process called **horizontal gene transfer**. Horizontal gene transfer is distinguished from vertical transfer, the generational passing of genes from parent to offspring, as in cell division. If horizontal acquisition of a gene system improves the competitiveness of the cell, the new genes will be retained and the descendants will have evolved into a new kind of organism. Pathogens such as *Neisseria gonorrhoeae*, the cause of gonorrhea, may have used transformation to acquire new genes whose protein products now help the organism evade the host immune system.

Gram-positive organisms transform DNA using a translocasome complex. Natural transformation in gram-positive organisms typically involves the growth-phase-dependent assembly of a **translocasome** complex across the cell membrane (**Fig. 9.1**). The translocasome is composed of a binding protein that captures extracellular DNA floating in the environment, plus proteins that form a transmembrane pore. The complex also includes a nuclease that degrades one strand of a double-stranded DNA molecule while pulling the other strand intact

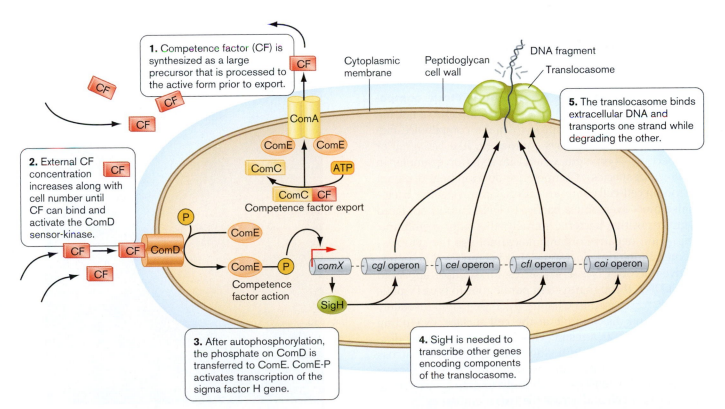

Figure 9.1 Transformation in *Streptococcus*.   The process of transformation in this organism begins with the synthesis of a signal molecule (competence factor) and concludes with the import of a single-stranded DNA strand through a translocasome complex.

through the pore and into the cell. Once inside, the strand can be incorporated into the chromosome by recombination, a process we will discuss in Section 9.3.

Once the translocasome is assembled, the cell is **competent**, meaning that it can import free DNA fragments and incorporate them into its genome. What triggers growth-stage-dependent competence? For some gram-positive bacteria, competence for transformation is generated by a chemical conversation that takes place between members of the culture. Every individual in a growing population produces and secretes a small, 15–20 amino acid peptide generically called **competence factor** that accumulates in the medium until it induces a genetic program that makes the population competent (**Fig. 9.1**, step 1). The sequence of the competence factor peptide is unique to each species, as are the specifics of the induction process. For *Streptococcus pneumoniae*, as the population increases—that is, as the cell density increases—so does the level of competence factor in the medium (step 2).

Above a certain concentration threshold, competence factor is able to bind to a sensory protein built into the cell membrane (ComD for *S. pneumoniae*). This binding begins what is called a phosphorylation cascade. In the phosphorylation cascade, the sensory protein uses ATP to autophosphorylate and then passes the phosphate to a cytoplasmic regulatory protein, ComE, that stimulates expression of a novel sigma factor, SigH (step 3). The resulting sigma factor is specifically used to transcribe genes encoding the translocasome (step 4). The protein products of these genes are assembled at the membrane, and the cell becomes competent (step 5).

Gram-negative species transform without competence factors. Gram-negative species capable of natural transformation do not appear to make competence factors. Either they are always competent, like *Neisseria*, or they become competent when starved (for example, in *Haemophilus* species). Competence development does not depend on cell-cell communication. Gram-negative bacteria also must overcome a barrier to DNA import not faced by gram-positive bacteria, namely, the outer membrane. Thus, gram-negative organisms cannot use the cytoplasmic membrane translocasome that is employed by gram-positive microbes. *Neisseria* species, for example, appear to import DNA through a system paralogous to type IV pilus assembly (discussed in Chapter 8). Type IV pili reversibly assemble and disassemble at the cell surface, expanding and contracting as a result. This happens constantly during the growth of a culture and can help cells move across solid surfaces. During transformation in species of *Neisseria*, disassembly of a pilus is thought to drag transforming DNA into the cell and across the two membranes.

In another departure from the gram-positive example, transformation in species of *Haemophilus* and *Neisseria* is species and sequence specific, thereby limiting gene exchange between different genera. Specificity is due to particular sequences in DNA that are recognized by part of the uptake apparatus. However, not all gram-negative competence systems display such specificity; *Acinetobacter calcoaceticus,* a lung pathogen, is able to take up DNA from any source at very high frequency. This is possible because the DNA-binding part of its transformation system does not need to recognize a specific nucleotide sequence.

Gene Transfer by Conjugation

Conjugation, or "bacterial sex," requires cell-cell contact typically initiated by a special pilus protruding from a donor cell (**Fig. 9.2**). Conjugation occurs in many species of bacteria and archaea, even in hyperthermophiles such as *Sulfolobus* species. In nature, mixed-species biofilms are thought to be a way in which cross-species DNA transfer may even occur, albeit at a low frequency. Some species, especially gram-positives, use nonpilus attachment proteins. In the case of *E. coli*, the tip of the specialized sex pilus attaches to a receptor on the recipient cell and then contracts, drawing the two cells closer. The two cell envelopes fuse and generate a conjugation complex, similar to the transformation complex, through which single-stranded DNA passes from donor to recipient. The DNA passes across the membranes through a conjugation complex.

Bacterial conjugation requires the presence of special *transferable* plasmids that usually contain all the genes

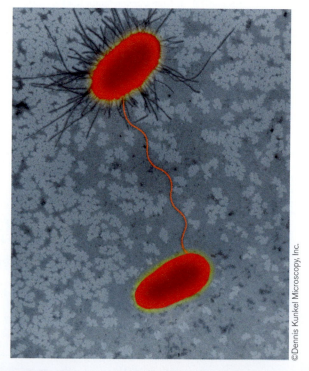

©Dennis Kunkel Microscopy, Inc.

Figure 9.2   Sex pilus connecting two *E. coli* cells.   The top cell with pili is the donor. The smaller, bottom cell is the recipient.

needed for pilus formation and DNA export. A well-studied, transferable plasmid in *E. coli* is called **fertility factor (F factor)** (Fig. 9.3). Not only does this plasmid transfer itself to a recipient cell, but if it integrates into the chromosome of its host, it can also transfer host genes. F factor contains two replication origins, *oriV* and *oriT*, located at different positions on the plasmid. The origin *oriV* is used to replicate and maintain the plasmid in non-conjugating cells, whereas *oriT* is used only to replicate DNA during DNA transfer.

Conjugation begins with cell-to-cell contact between a donor cell, called the **F⁺ cell**, which carries the plasmid, and a recipient **F⁻ cell** (step 1). Formation of the fused membrane conjugation conduit (step 2) triggers synthesis of an F factor helicase/endonuclease (encoded on the plasmid) that nicks the phosphodiester backbone at *oriT* (step 3). DNA polymerase III (Pol III) is then recruited to *oriT*, where replication begins (step 4). In contrast to bidirectional replication used to duplicate the chromosome, replication for conjugal transfer

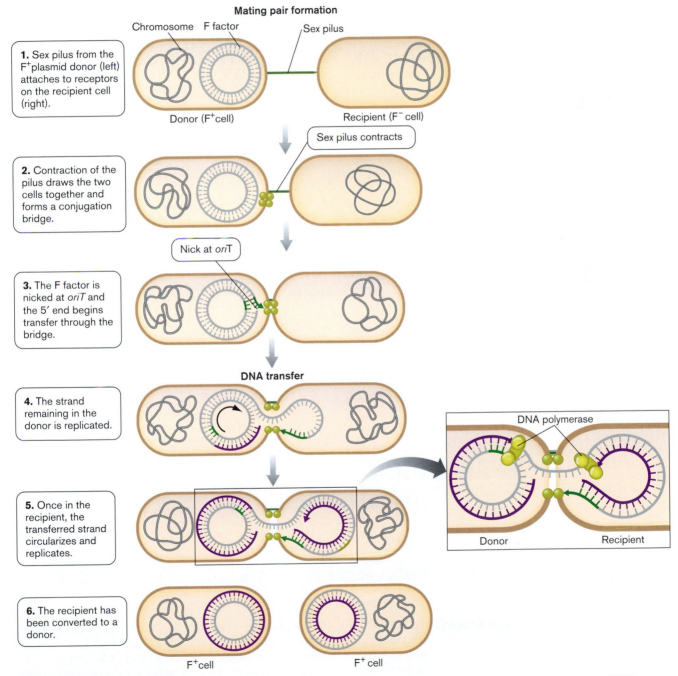

Figure 9.3   The conjugation process.   Some plasmids, such as F factor in *E. coli*, can mediate cell-to-cell transfer of DNA.

occurs unidirectionally by rolling-circle replication (step 5) (see Section 7.4).

The 5′ end of the nicked strand of F factor moves through a pore in the conjugation complex by a process driven by cellular DNA polymerase III, which synthesizes a replacement strand in the donor cell using the intact strand as a template (see blowup in step 5). The intact strand remains in the donor as the transferred strand enters the recipient. The strand entering the recipient is also converted to a double-stranded molecule by DNA synthesis (see Section 7.3). The last portion of F factor moved into the recipient is *oriT*. Once transferred, the F factor circularizes to re-form a plasmid, thereby converting the F⁻ recipient cell to a new F⁺ donor cell (step 6). The conjugation complex spontaneously comes apart, and the membranes seal.

Is it possible for two donors to exchange DNA with each other? The answer is generally no. Donors have

mechanisms in place to prevent this. For example, they make a cell surface protein encoded by the F factor that inhibits formation of a conjugation complex with another donor possessing the same protein. This mechanism prevents the pointless transfer of a plasmid to a cell that already has that plasmid.

The F-factor plasmid can integrate into the chromosome. Once inside the recipient cell, F factor can remain an independent plasmid or be integrated into the chromosome of the recipient cell (**Fig. 9.4**). Integration of the plasmid into the genome involves recombination between two circular DNA duplexes, a process explained in Section 9.3. Because the plasmid can exist in extrachromosomal and integrated forms, it is sometimes called an **episome** (from the Greek *epi*, "over," and *some*, "chromosome").

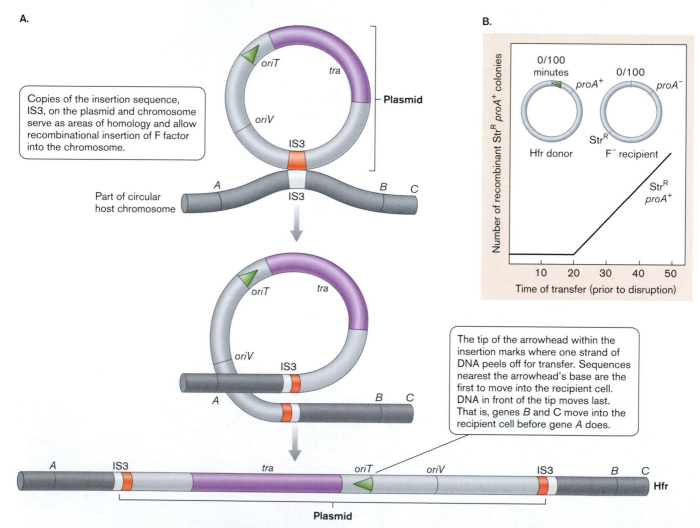

Figure 9.4 Hfr formation and gene mapping. A. The making of an Hfr by F-factor integration. Regions of sequence homology between an F factor and the host chromosome are recognized by host cell recombination proteins that then integrate F factor into the host chromosome. The transfer genes (*tra*), and replication origins are shown (vegetative = *oriV*; transfer = *oriT*). **B.** Conjugational mapping of genes. The figure shows Hfr donor and recipient chromosomes. Recipient requires proline added to media in order to form colonies. The graph illustrates that it takes 20 minutes of conjugation to start seeing ProA⁺ recombinants.

When the chromosome has integrated an F factor, the cell is designated an **Hfr**, or high-frequency recombination, strain. The term *high frequency* refers to the fact that there are more cells capable of transferring *chromosomal* DNA in an Hfr population than in an F⁺ culture where very few Hfr cells are present. It does not refer to the speed with which individual DNA molecules are transferred. Because the F factor is inserted into the chromosome, an Hfr cell is capable of transferring all or part of the chromosome into a recipient cell. Essentially, the entire chromosome becomes the F factor. However, the integrated F factor is actually the last bit of DNA transferred during Hfr conjugation. Because it takes nearly 100 minutes to transfer the entire *E. coli* chromosome (as opposed to only 5 minutes for the free F plasmid), the integrated F factor is rarely transferred before the conjugation bridge breaks, and the recipient almost never becomes an F⁺ or Hfr cell.

Mapping gene position by conjugation. Earlier scientists exploited Hfr integration to begin mapping the locations of genes on the *E. coli* chromosome. Because it takes nearly 100 minutes for an Hfr to transfer the entire *E. coli* chromosome, it was possible to "map" genes based on how long it took to transfer different genes to a recipient. Note that an F factor can integrate at different locations on the chromosome and in different orientations. The resulting Hfr strains will transfer different chromosomal genes early during conjugation. Timing of Hfr gene transfer was used to construct the original gene maps of *E. coli*, which were expressed in minutes. The gene map was divided into 100 equal segments of about 44 kbp each.

The mapping process required mating an *E. coli* Hfr strain with an F⁻ recipient that contained a mutation in the gene to be mapped. It is important in any genetic cross that the mutation create an observable phenotype. Say, for example, we wanted to map the gene *proA*, which is needed to synthesize the amino acid proline. Cells defective in *proA* will not grow on minimal medium unless proline is added. However, if the mutated *proA* gene is repaired by conjugation and subsequent recombination between the donor and recipient DNA molecules, the recombinant cell will grow and form a colony on agar medium *lacking* proline. It is also important that the F⁻ recipient be resistant to an antibiotic such as streptomycin.

An Hfr culture that is streptomycin sensitive is mixed with the streptomycin-resistant *proA* mutant culture (**Fig. 9.4B**). Then, at timed intervals, an aliquot (that is, a measured sample) of the mating mixture is removed and the conjugation bridges broken by blending in a blender. This is called interrupted mating. The longer the mating goes before breaking the bridges, the more of the chromosome is transferred from the Hfr into the recipient. At each interval, the disrupted mixture is plated onto a selective agar lacking proline but containing streptomycin. The *only* cell that can grow on this medium is a recombinant in which the *proA*⁺ gene transferred from the Hfr strain and recombined into the chromosome of the streptomycin-resistant F⁻ cell. Until the gene transfers, no colonies will form. Because the position of the integrated F factor on the chromosome is fixed, the farther the wild-type gene is from the insertion site, the longer it will take for the gene to move into the mutant. In the example, the *proA*⁺ gene is located 20 map minutes from the location of the F-insertion.

An integrated F factor can dis-integrate from the chromosome. Another type of gene shuffling can occur when an integrated F factor subsequently excises from the chromosome. Once an F factor has been integrated into a host chromosome, it can also be excised via host recombination mechanisms. Usually, it is excised completely and restored to its original form. Occasionally, however, it is excised along with some DNA from the host chromosome, creating a product that will be an F-factor plasmid that also contains some of the chromosomal DNA. The derivative F plasmid that contains host DNA is called an **F-prime (F′) plasmid** or F′ factor (**Fig. 9.5**).

In contrast to chromosomal genes whose transfer is directed by Hfr, genes hitchhiking on an F′ plasmid do not have to recombine into the recipient chromosome to be maintained. The extra genes can be expressed as part of the F′ plasmid. This establishes what is called a **merodiploid** situation in which a conjugal recipient of the F′ factor contains two copies of those few genes—one set on the chromosome, the other on the F′ factor. This is useful to the cell because the second copy of the gene could serve as the raw material needed to evolve a new gene. It is also useful to the geneticist because it allows one to ask whether a mutant allele of a gene is dominant over a normal, or wild-type, allele. Merodiploids can also help geneticists understand gene organization.

There are many different types of plasmids in the microbial world. Most are not transferable by themselves. However, one group of plasmids that cannot transfer themselves can be *mobilized* if a transferable plasmid is also present in the same cell. Mobilizable plasmids usually contain an *oriT*-like DNA replication origin recognized by the conjugation apparatus of the transferable plasmid. As a result, when the transferable plasmid begins conjugating, so does the mobilizable plasmid. This is one way in which antibiotic resistance genes on plasmids called R factors can be spread throughout a microbial population.

THOUGHT QUESTION 9.1 Transfer of an F factor from an F⁺ cell to an F⁻ cell converts the recipient to F⁺. Why does transfer of an Hfr not do the same?

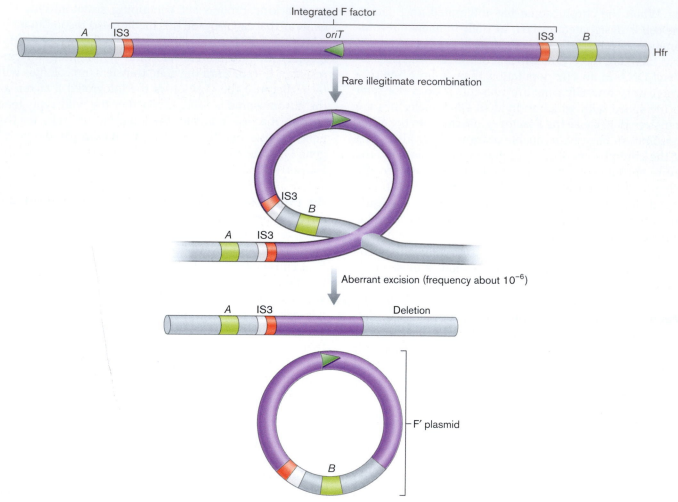

Figure 9.5 **Formation of an F′ factor.** A rare, illegitimate recombination event in an Hfr cell occurs between sequences within the integrated F factor and sequences in the host chromosome proper. The result is excision (deletion) of the gene from the host chromosome and production of a plasmid containing F factor and host genes. This F′-factor plasmid is capable of transferring to a new cell via conjugation.

Bacterial pheromones can promote conjugation. Conjugation can, in certain species, be promoted by chemical communication between cells. One example is plasmid pAD1 in *Enterococcus faecalis*. This plasmid produces a protein called aggregation substance, which catalyzes cell-cell contact and formation of the conjugation complex. Ordinarily, this gene and others on the plasmid are inactive. Potential recipients, however, try to entice donors to mate with them by producing small behavior-altering molecules called pheromones. The chemical signal molecules released by bacteria are small peptides that enter the donor cell through oligopeptide permeases present in the membrane. Once inside the donor cell, the pheromone stimulates transcription of the *pAD1* transfer genes. Pheromone-responsive conjugation thus far is unique to enterococci. Most known conjugation systems do not use pheromones to stimulate transfer.

Microbial transfer of genes into eukaryotes. In eukaryotes, sexual exchange of genes usually occurs only within a single species. Microbes, on the other hand, are more promiscuous. Mating among different microbial genera is common and perhaps even desirable from an evolutionary viewpoint. For example, *Salmonella* can conduct interspecies mating with *E. coli*. Sharing genes between species allows each one to sample genes from the other and keep genes that increase fitness.

What may be surprising is that some bacteria can actually transfer genes across biological domains. One striking example is *Agrobacterium tumefaciens*, which causes crown gall disease in plants. This gram-negative plant pathogen, common to the rhizosphere (the area around root surfaces), contains a tumor-inducing plasmid (Ti) that can be transferred via conjugation to plant cells. These bacteria detect and swim toward phenolic wound compounds released by damaged plant cells.

A.

Robert Calentine/Visuals Unlimited

B.

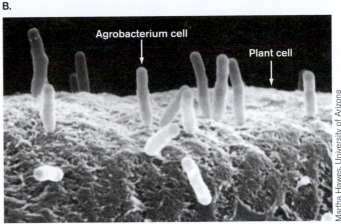

Agrobacterium cell

Plant cell

Martha Hawes, University of Arizona

Figure 9.6 An example of gene transfer between bacteria and plants. A. Crown gall disease tumor caused by the bacterium *Agrobacterium tumefaciens*. **B.** Electron micrograph of *A. tumefaciens* attached to plant cells.

The phenolic compounds signal where the microbe can gain access to plant cells. Subsequent transfer, integration, and expression of the Ti plasmid in the plant cell genome triggers the release of plant hormones that stimulate tumorous growth of the plant (**Fig. 9.6**). Plant cells within the tumor release amino acid derivatives called "opines" that the microbe can then use as a source of carbon and nitrogen.

The unique mode of action of *A. tumefaciens* has made this bacterium an indispensable tool for plant breeding. Any desired genes, such as insecticidal toxin genes (like those produced by *Bacillus thuringiensis*) or herbicide resistance genes, can be engineered into the bacterial plasmid DNA and thereby inserted into the plant genome (discussed in Chapter 16). The use of *Agrobacterium* not only shortens the conventional plant-breeding process, but also allows entirely new (nonplant) genes to be engineered into crops.

NOTE: Conjugation in bacteria and archaea is mechanistically different from conjugation among eukaryotic microbes. Some eukaryotic microbes exchange nuclei through a structure very different from that of bacteria, though also called a conjugation bridge (discussed in Chapter 20).

Gene Exchange by Bacteriophage Transduction

Bacteriophages harbor their own genomes, separate from that of their host cells. However, bacteriophages also can *accidentally* move bacterial genes between cells as an offshoot of the phage life cycle (see Sections 6.4 and 7.4). The process in which bacteriophages carry payloads of host DNA from one cell to another is known as **transduction**. There are two basic types of transduction: generalized and specialized. **Generalized transduction** can take any gene from a donor cell and transfer it to a recipient cell. In contrast, **specialized transduction** (also known as restricted transduction) can only transfer a few closely linked genes between cells.

Bacteriophages capable of generalized transduction have trouble distinguishing their own DNA from that of the host when attempting to package DNA into their capsids, so pieces of bacterial host DNA accidentally become packaged in the phage capsid. In the case of *Salmonella* P22 phage, the packaging system recognizes a certain DNA sequence on P22 DNA called a *pac* site. During rolling-circle replication, P22 DNA forms long concatemers containing many P22 genomes arranged in tandem. The *pac* site defines the ends of the phage genome, marking where the packaging system cuts the P22 DNA and starts packaging DNA into the next empty phage head. However, certain DNA sequences on the *Salmonella* chromosome also "look" like *pac* sites. As a result, the packaging system sometimes mistakenly packages host DNA instead of phage DNA.

The result of this mistake is that 1% of the phage particles in any population of P22 do not contain phage

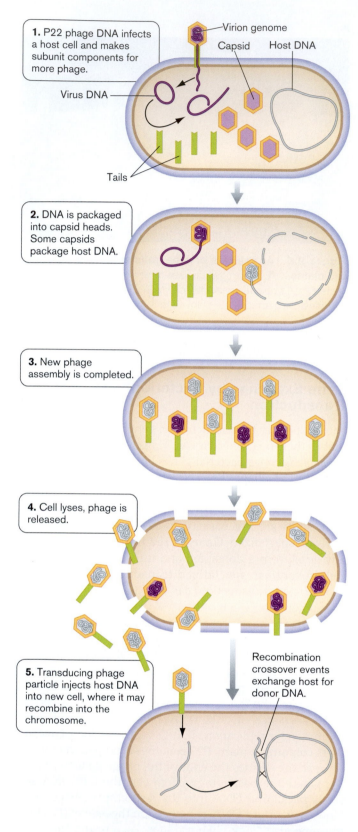

1. P22 phage DNA infects a host cell and makes subunit components for more phage.

Virion genome

Capsid   Host DNA

Virus DNA

Tails

2. DNA is packaged into capsid heads. Some capsids package host DNA.

3. New phage assembly is completed.

4. Cell lyses, phage is released.

5. Transducing phage particle injects host DNA into new cell, where it may recombine into the chromosome.

Recombination crossover events exchange host for donor DNA.

Figure 9.7 Generalized transduction. Generalized transduction involves the movement of any segment of donor chromosome to a recipient cell by phage vectors. The number of genes transferred in any one phage capsid is limited, however, to what can fit in the phage head.

DNA but carry host DNA plucked from around the chromosome (**Fig. 9.7**). The phages that carry host DNA are called **transducing particles**. Any one transducing particle will only contain one segment of host DNA, but different particles in the phage population will contain different segments of host DNA. When a transducing particle injects its DNA into a cell, no new phages are made, but the highjacked host DNA can recombine, or exchange, with sequences in the host chromosome of the newly infected cell, changing the genetic makeup of the recipient.

> **THOUGHT QUESTION 9.2** Would it be easier to demonstrate generalized transduction using a temperate phage (which generates a lysogen) or a lytic (virulent) phage?

Specialized or restricted transduction is a phage-mediated gene transfer mechanism that resembles the formation of the F' factors previously described. In contrast to generalized transduction, specialized transduction can only move a limited number of host genes. *E. coli* phage lambda is the classic example of specialized transduction. To understand this gene exchange process, however, you must first understand the basic biology of this phage (discussed in Chapter 6). When lambda phage DNA first enters the cell, it is linear. It circularizes at cohesive ends called *cos* sites. Then a small DNA sequence in lambda called *attP* can specifically recombine with a small host DNA sequence (called *attB*) located between the *gal* (galactose catabolism) and *bio* (biotin synthesis) genes on the *E. coli* chromosome (**Fig. 9.8**, step 1). This process, carried out by the phage integrase protein, produces a chromosome with an integrated phage genome, referred to as a prophage, flanked by chimeric *att* sites called *attL* and *attR*. They are called chimeric because each one is made half from the bacterial and half from the phage *att* sites. Prophage DNA remains latent until something happens to the cell to activate it.

The process of specialized transduction begins with this prophage DNA. The majority of the time, when the prophage is reactivated, recombination mechanisms involving phage proteins can excise the lambda DNA precisely, so that the chromosome and viral DNAs are restored to their native states. The viral DNA will then replicate and make more phage particles containing normal phage DNA. However, on rare occasions, improper excision can take place between host DNA sequences that lay adjacent to the phage insertion site (*attB* in **Fig. 9.8**, step 2) and similar DNA sequences within the prophage. Improper, or aberrant, excision between virus and host sequences yields a virus that will lack a few viral genes (tail genes in step 3) but will include host genes lying adjacent to the phage attachment site (the galactose utilization gene *gal* in **Fig.**

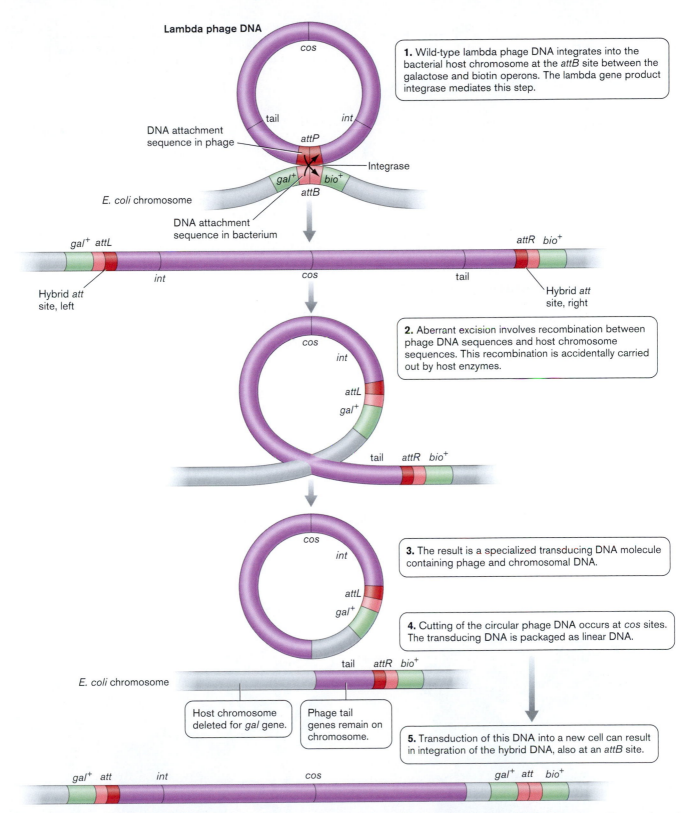

Figure 9.8   Specialized transduction is restricted to moving host genes flanking the phage attachment site.   The number of genes transferred is limited, based on the size of the phage head. The resulting transductant chromosome becomes partially diploid for the transferred gene, in this case *gal*.

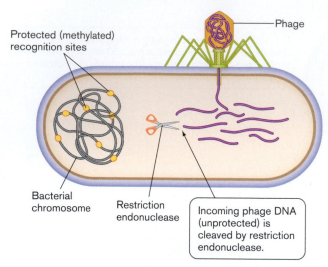

Figure 9.9   **Restriction of invading phage DNA.**   Phage DNA is injected into a host where restriction enzymes can digest it. Host DNA is protected because specific methylations of its own DNA prevent the enzymes from cutting it.

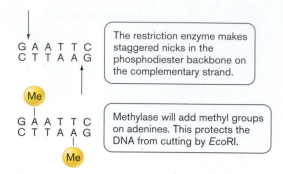

Figure 9.10   *Eco*RI **restriction site.**   *Eco*RI is an example of a class II restriction-modification system. Shown here are cleavage (top) and methyl modification (bottom). The DNA sequence shown is specifically recognized by the endonuclease and methylase enzymes.

9.8, step 3). Aberrant excision is accidentally carried out by host recombination enzymes. With some phages, the specialized transducing particles can replicate unaided. However, in **Figure 9.8** the result is a defective, specialized transducing phage DNA (lambda d*gal*, or λd*gal*). Specialized transducing phage particles that are defective cannot replicate by themselves but require the presence of a helper phage to supply missing gene products.

Once formed, specialized transducing phages can deliver the hybrid DNA molecule to a new recipient cell.

This is the transduction process. Once in that cell, the phage DNA can integrate into the host *attB* site, carrying the donor host gene(s) with it (steps 4 and 5). This will establish another merodiploid situation where the new recipient contains two copies of a host gene, one originally present on its chromosome and one brought in by the transducing DNA.

Genes other than *gal* or *bio* can also be moved by specialized transduction if lambda is used to infect mutant host strains lacking the normal *att* site. Because the primary integration site is gone, lambda DNA can integrate at low frequency to other, secondary attachment sites somewhere else in the chromosome. Improper excision at one of those locations will package adjacent host DNA.

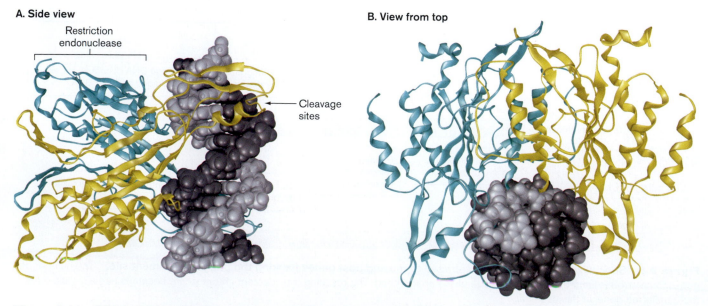

Figure 9.11   **DNA restriction enzyme attached to a DNA restriction sequence.**   **A.** In the side view, the part of the protein shown in gold is about to cut the dark gray DNA backbone at the top. The light blue part of the protein will cut the light gray DNA at the bottom. **B.** Viewed from the top, the enzyme is seen to hold the DNA backbone. (PDB code: 1QRI)

DNA Restriction and Modification Systems Prevent Promiscuous DNA Transfer

There are dangers to the cell associated with the indiscriminate transfer of DNA between bacteria. The most obvious risk involves bacteriophages whose goal is to replicate at the expense of target cells. A bacterium that can digest invading phage DNA while protecting its own chromosome would have a far better chance of surviving in nature than cells unable to make this distinction. However, beyond the threat of overt destruction by phages, the unrestricted incorporation of foreign DNA can be an energetic drain on the cell. Genes encoding competing or useless products or products that lack regulatory restraints would squander resources and lower the overall fitness of the cell. As a result, bacteria have developed a kind of "safe-sex" approach to gene exchange. It is an imperfect approach, however, that still leaves room for beneficial genetic exchanges.

This protection system, called restriction and modification, involves the enzymatic cleavage (restriction) of alien DNA and the protective methylation (modification) of self DNA (**Fig. 9.9**). Most bacteria produce DNA **restriction endonucleases** (also known as restriction enzymes), enzymes that recognize specific short DNA sequences (known as recognition sites) and cleave DNA at or near those sequences (**Fig. 9.10**). A structural model of the restriction enzyme *Eco*RI cleaving its target DNA sequence is shown in **Figure 9.11**.

Because the producing organism's own chromosomes contain these same restriction sequences, how do bacteria avoid committing suicide? They protect themselves by producing matching modification enzymes that use S-adenosyl methionine to attach methyl groups to those same sequences. A three-dimensional model of a methylation enzyme interacting with a DNA recognition site is shown in **Fig. 9.12**. This modification makes the sequence invisible to the cognate (matched) restriction enzyme. Only one strand of the sequence needs to

be methylated to protect the duplex from cleavage; thus, even newly replicated, and consequently hemimethylated (only one strand methylated), DNA sequences are invisible to the restriction enzyme.

There are three main types of restriction endonucleases based on how many subunits there are in the enzyme, what cofactors are required, and where the DNA cleavage sites are located relative to the recognition site (**Table 9.1**). Types I and III restriction enzymes have their restriction and modification activities in one multifunctional protein and cleave DNA some distance away from the recognition site. Type II restriction enzymes, those used most often for cloning, only possess endonuclease activity. A separate modification protein carries out methylation of the restriction site. Type II restriction enzymes generally recognize palandromic

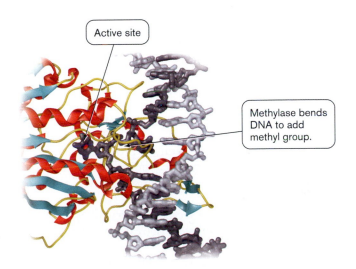

Figure 9.12   DNA methylase used for DNA modification. The DNA methylase adds a methyl group to a restriction site, changing the site's conformation so that it is no longer recognized by the restriction enzyme. The methylase bends the DNA helix to bring the target base into the active site (left side). (PDB code: 1DCT)

Table 9.1	**Main types of restriction-modification systems.**		
Characteristics	**Type I**	**Type II**	**Type III**
Restriction and modification activities	Present in one multifunctional enzyme	Separate methylase and restriction enzymes	Present in one multifunctional enzyme
Subunits	Three different subunits	One or two identical subunits per activity	Two different subunits
Recognition site	5–7 bp asymmetrical	4–6 bp, palindromic	5–7 bp asymmetrical
Cleavage and modification sites	Random, 100 bp or more from recognition site	At or near recognition site	24–26 bp from recognition site
Examples	*Eco*K in *E. coli*; *Sty*LTIII in *Salmonella enterica*	*Eco*RI in *E. coli*; *Hind*III in *Haemophilus influenzae*	*Eco*571 in *E. coli*; *Bce*SI in *Bacillus cereus*

DNA sequences and cleave at those sites. **Figure 9.11** illustrates how the class II restriction enzyme *Eco*RI holds the DNA as it creates a staggered cut at the cleavage (restriction) site.

THOUGHT QUESTION 9.3   How do you think phage DNA containing restriction sites evade the restriction-modification screening system of its host?

THOUGHT QUESTION 9.4   Type I restriction-modification system genes are often designated *hsdR*, *hsdM*, and *hsdS*, which code, respectively, for the restriction (HsdR), modification (HsdM), and sequence recognition (HsdS) proteins. Can a plasmid grown in a wild-type strain be used to transform a strain defective in *hsdR*? An *hsdS* mutant? Can a plasmid grown in an *hsdM* mutant strain be transformed into an *hsdR* defective strain?

TO SUMMARIZE:

- **Transformation** is the uptake by living cells of free-floating DNA from dead, lysed cells.
- **Competence** for transformation in some organisms involves a genetically programmed physiological change.
- **Conjugation** is a DNA transfer process mediated by a transferable plasmid that requires cell-cell contact and formation of a protein complex between mating cells.
- **Pheromone peptides** secreted by gram-positive cocci such as *Enterococcus faecalis* activate transfer genes in nearby donor cells.
- **Some bacteria can transfer DNA across phylogenetic domains.** For example, *Agrobacterium tumefaciens* conjugates with plant cells.
- **Transduction** is the process whereby bacteriophages transfer fragments of bacterial DNA from one bacterium to another. Generalized transduction is when a phage preparation can move any gene in a bacterial genome to another bacterium. Specialized transduction is when a phage can only move a limited number of bacterial genes.
- **Restriction endonucleases** protect bacteria from invasion by foreign DNA. Restriction-modification enzymes methylate restriction target sites in the host DNA to prevent self-digestion.

9.3 Recombination

Once a new piece of DNA has entered a cell through one of the gene exchange systems, what happens to it? The answer to that question depends on the nature of the acquired DNA. If the DNA is a plasmid, capable of autonomous replication, it can coexist in the cell separate from the host chromosome. If the DNA is not capable of autonomous replication, it is usually degraded by nucleases. Alternatively, it may be incorporated into the chromosome through recombination.

Two different DNA molecules in a cell can recombine by one of several mechanisms. **Generalized recombination** requires that the two recombining molecules have a considerable stretch of homologous DNA sequence. In contrast to generalized recombination, **site-specific recombination** occurs via a mechanism that requires very little sequence homology between the recombining DNA molecules, but does require a short (10–20 bp) sequence recognized by the recombination enzyme.

Generalized Recombination Requires RecA

Enzymes participating in generalized recombination are able to find and align homologous stretches of DNA in two different DNA molecules and catalyze an exchange of strands. The result can be a **cointegrate** molecule where a single recombination event (that is, a single crossover) joins the two participating DNA molecules. For example, when two circular DNA molecules are joined by a single recombination event, the two molecules are added together and form one large circular molecule (for example, **Fig. 9.4**; also see preceding discussion of F factor or lambda phage). A second crossover separated by some distance from the first, however, will lead to an equal exchange of DNA in which neither molecule increases in length.

Why is recombination advantageous? There are three probable functions for generalized recombination in the microbial cell:

- Recombination probably first evolved as an internal method of DNA repair, useful to fix mutations or restart stalled replication forks. This role does not involve foreign DNA.
- Cells with damaged chromosomes use DNA donated by others of the same species to repair their damaged genes.
- Recombination is also part of a "self-improvement" program that samples genes from other organisms for an ability to enhance the competitive fitness of the cell.

The central, but by no means only, player in generalized recombination is a protein called RecA. RecA molecules are also called synaptases because they are able to scan DNA molecules for homology and align the homologous regions, forming a triplex DNA molecule, or synapse (**Fig. 9.13**). Homologs of RecA are found in many other species.

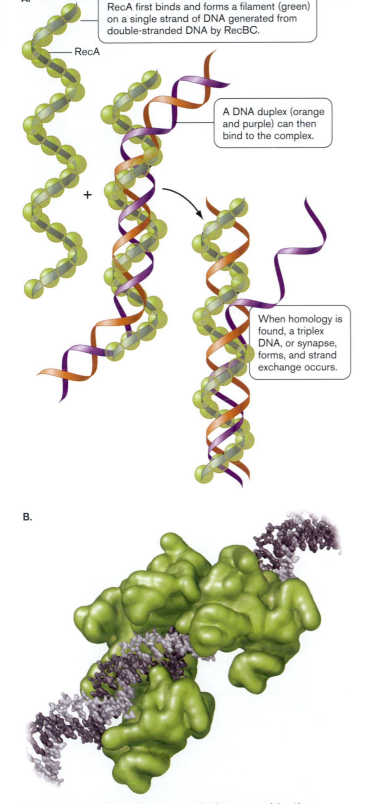

A.

RecA first binds and forms a filament (green) on a single strand of DNA generated from double-stranded DNA by RecBC.

RecA

A DNA duplex (orange and purple) can then bind to the complex.

+

When homology is found, a triplex DNA, or synapse, forms, and strand exchange occurs.

B.

Figure 9.13   RecA protein catalyzing recombination.
A. The standard mechanism of recombination. **B.** A molecular model of RecA protein filament catalyzing recombination. The DNA is light gray and dark gray. (PDB codes: 1N03, 1K8J)

The clearest model of RecA-mediated recombination comes from *E. coli* (**Figs. 9.13** and **9.14 ◑**). Before RecA can find homology between two DNA molecules, the donor double-stranded DNA molecule must be converted to a single strand. *E. coli* has two enzyme systems that can prepare the single-stranded template: the RecBCD and RecJ systems. The RecBCD complex, for example, enters at the end of a DNA fragment and begins to unwind it (**Fig. 9.14**, steps 1 and 2). The complex changes activity when it encounters 8-bp sequences called **Chi** that are scattered throughout most DNA molecules. RecBCD nicks DNA at a Chi site and continues unwinding the strand. RecA then loads onto the single strand as a filament (step 2).

RecA filaments coat a stretch of single-strand DNA, then RecA scans double-stranded DNA, looking for stretches of homology. When it finds homology (minimum required is about 50 bp), RecA catalyzes strand invasion and **D-loop formation** (**Fig. 9.14**, step 3). In strand invasion, the donor single-strand DNA invades the homologous region in the double-strand DNA recipient molecule and displaces its like strand. A drawing of the resulting triplex molecule resembles the letter "D" (a "D-loop," shown in **Fig. 9.14**, step 3). At this point, a single-strand crossover has been made. The single-strand crossover creates a molecule with four duplex ends.

Other recombination proteins, RuvA and B, assemble at the crossover point and pull donor and recipient DNA strands in opposite directions, extending the invasion, in a process called strand assimilation (**Fig. 9.14**, step 4, and **Fig. 9.15**) or branch migration. This process extends the base pairing between homologous donor and recipient strands.

Ultimately, the end of the displaced strand at the D loop is cleaved, possibly by RecBCD (**Fig. 9.14**, step 5), and ligated to the donor strand (step 6). When spread out on an electron microscope grid, such a structure resembles a cross (step 7 and also seen in **Fig. 9.15**) that is known as a **Holliday junction**, after the scientist (Robin Holliday) who first proposed it. The final step is resolution of the crossover Holliday junction. RuvC protein cleaves across the junction in one of two directions. To visualize this, we mentally rotate the right half of the Holliday structure as shown in **Figure 9.14**, step 7. RuvC can then cleave the junction in the horizontal or vertical direction across the two strands and then religate them as shown in step 8 to give the alternative products shown. The product on the right has exchanged only a small segment of DNA, while the other product represents a true crossover.

The results of the crossover are shown in step 9. If donor and recipient DNA both started as circular molecules in step 1, the product would form a cointegrate circle (shown to the left). If, however, we are viewing the end result of transduction, where the recipient is a

Figure 9.14   Mechanism of generalized recombination. ▶❙❙

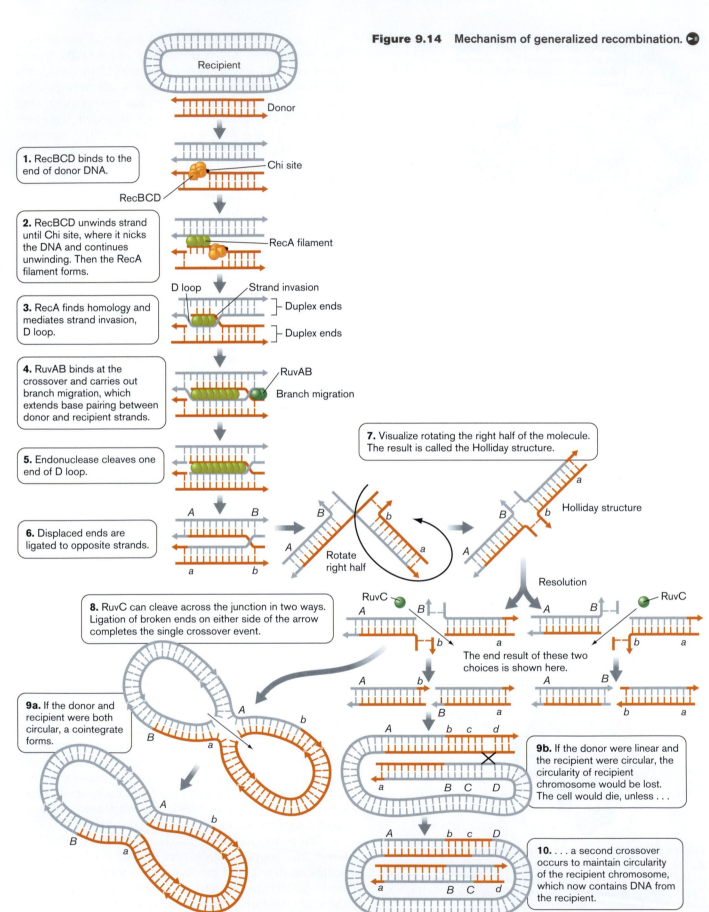

1. RecBCD binds to the end of donor DNA.

2. RecBCD unwinds strand until Chi site, where it nicks the DNA and continues unwinding. Then the RecA filament forms.

3. RecA finds homology and mediates strand invasion, D loop.

4. RuvAB binds at the crossover and carries out branch migration, which extends base pairing between donor and recipient strands.

5. Endonuclease cleaves one end of D loop.

6. Displaced ends are ligated to opposite strands.

7. Visualize rotating the right half of the molecule. The result is called the Holliday structure.

8. RuvC can cleave across the junction in two ways. Ligation of broken ends on either side of the arrow completes the single crossover event.

9a. If the donor and recipient were both circular, a cointegrate forms.

9b. If the donor were linear and the recipient were circular, the circularity of recipient chromosome would be lost. The cell would die, unless . . .

10. . . . a second crossover occurs to maintain circularity of the recipient chromosome, which now contains DNA from the recipient.

Recipient

Donor

Chi site

RecBCD

RecA filament

D loop   Strand invasion

Duplex ends

Duplex ends

RuvAB

Branch migration

Holliday structure

Rotate right half

Resolution

RuvC

RuvC

The end result of these two choices is shown here.

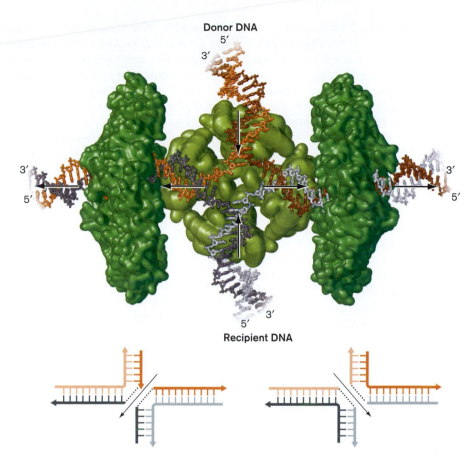

Figure 9.15 RuvAB proteins catalyzing branch migration. At the Holliday junction, these proteins pull matched parental strands in opposite directions. Vertical DNA molecules represent the two different parental helices (donor, recipient). The top vertical donor helix strands are marked as light orange/dark orange, and the bottom recipient helix is marked light gray/gray. Arrows mark movement of the helices. This movement elongates the region of base pairing between donor and recipient strands (horizontal). Bottom illustrations reveal alternative ways that RuvC can resolve this junction. (PDB codes: 1IN4, 1BDX, 3CRX)

circular chromosome and the donor is a double-stranded linear fragment, the recombinant molecule would appear exactly as the right-hand structure shown in **Figure 9.14**, step 9. As you can see, the bacterial chromosome would lose its integrity, no longer forming a circle. To restore circular integrity to the chromosome, a second crossover event would have to occur downstream of the first (bottom of **Fig. 9.14**, step 10). Note that any genes residing between the first and second crossovers are simultaneously exchanged.

One consequence of this type of double recombination is that multiple genes clustered next to each other on a chromosome can be recombined as a group into a recipient's genome. In the case of transduction, a piece of DNA from the donor bacterium replaces a segment of DNA in the recipient host. One gene in the segment is said to be **cotransduced** with another. The closer the two genes are physically on the chromosome, the higher the probability they will be cotransduced. This is because the closer two genes are to each other, the less likely a crossover will occur in between them.

> **THOUGHT QUESTION 9.5** In a transductional cross between an $A^+B^+C^+$ genotype donor and an $A^-B^-C^-$ genotype recipient, 100 A^+ recombinants were selected. Of those 100, 15% were also B^+, while 75% were C^+. Is gene B or C closer to gene A?

The benefit of recombination to bacteria can be demonstrated using a culture of *E. coli* that contains a defective *lacZ* gene. The *lacZ*⁺ gene product (β-galactosidase) is normally used to catabolize the carbohydrate lactose. If lactose is the only carbohydrate available, the *lacZ* mutant cell culture fails to grow. However, if a transducing phage lysate grown on a *lacZ*⁺ strain is added to the *lacZ* mutant cells, phage particles containing the functional *lacZ*⁺ can inject the gene into the mutant (transduction), where recombination systems can exchange it for the defective gene. (Note that the defect may only be a single base change, so there is plenty of homology left for RecA to recognize.) The new recombinant cell has now been converted to *lacZ*⁺ and can grow on lactose.

When trying to understand recombination (and gene exchange overall), it is important to remember that the processes are generally random. The lucky bacterium that gets the right piece of DNA and recombines the right segment of that DNA will benefit. This is why a genetic experiment usually requires screening hundreds of millions of cells to find a few recombinants that grow into visible colonies.

Deinococcus Uses RecA to Repair Fragmented Chromosomes

As first noted in Chapter 5, *Deinococcus radiodurans* is a tough microbe. It can endure 3,000 Gy of gamma irradiation and survive. That is 3,000 times the lethal dose for a human. Although *Deinococcus* chromosomal DNA is fragmented by irradiation, the organism can reassemble the pieces of DNA in the correct order to re-form the chromosome, a process that is even more amazing considering that the organism possesses two *different* chromosomes.

The *Deinococcus* protein that mediates repair is a homolog of RecA, but the sequence of strand exchange is the exact inverse of the established RecA pathway (see previous discussion). Most RecA homologs first form filaments on single-stranded DNA, then take up homologous double-stranded DNA. The *Deinococcus* RecA forms a filament on double-stranded DNA and then incorporates a homologous single-stranded molecule. *Deinococcus* RecA does not need ssDNA to bind. Instead, *Deinococcus* RecA is thought to bind the double-stranded fragments of irradiated DNA and splice homologous ends together to reconstruct the chromosome. This assumes that at least two chromosome copies are present per cell so that overlapping fragments are generated during irradiation. Unlike most bacteria, *Deinococcus* keeps several copies of each chromosome, making it more likely that it will find an intact copy of any given stretch of its sequence. It is a remarkable evolutionary adaptation that has allowed this microbe to inhabit and survive extremely stressful environments, such as the extreme dryness of the desert and the upper atmosphere exposed to cosmic radiation.

Site-Specific Recombination Is RecA Independent

In contrast to generalized recombination mechanisms that require RecA protein and can recombine any region of the chromosome, site-specific recombination does not utilize RecA and only moves a limited number of genes. This form of recombination involves very short regions of homology between donor and target DNA molecules. Dedicated enzyme systems specifically recognize those sequences and catalyze a crossover between them to produce a cointegrate molecule. The integration of phage lambda, described in Section 9.2, is one example of site-specific recombination. A short, 15-bp sequence called the *att* site is present on both the lambda phage DNA and on the *E. coli* chromosome. A phage-encoded protein called **integrase** engages the two *att* sites and, through a strand breakage and rejoining mechanism, combines the two molecules into one (see **Fig. 9.8**).

Other examples of site-specific recombination are flagellar **phase variation** in the pathogen *Salmonella enterica* and phase variation in the expression of type I pili in *E. coli*. In phase variation, the sequence of a segment of DNA is inverted (flipped), resulting in on-or-off regulation of adjacent gene expression. Phase variation is frequently employed by pathogens to evade the host immune system by changing expression of cell surface proteins. For *S. enterica,* as the immune system starts to make antibodies against one form of flagellar protein, the organism switches to make a different form not recognized by the antibody. The mechanism involves a "flippable" DNA cassette that sits adjacent to a flagellar gene on the chromosome. In one orientation, the flagellar gene is expressed; in the opposite orientation, it is not. For more on flagellar phase variation, see Section 10.6.

TO SUMMARIZE:

- **Recombination** is the process by which DNA sequences can be exchanged between DNA molecules.
- **General recombination** involves large regions of sequence homology between recombining DNA molecules.
- **RecA synaptase** mediates generalized recombination.
- **The extremely radiation-resistant *Deinococcus radiodurans*** is thought to use RecA protein to patch together homologous ends of fragmented DNA in a way that reconstructs the chromosome after extreme radiation damage.
- **Site-specific recombination** requires little homology between donor and recipient DNA molecules. Site-specific recombination enables phage DNA to integrate into bacterial chromosomes and is a process that can turn on or turn off certain genes, as in flagellar phase variation in *Salmonella*.

9.4 Mutations

Any permanent, heritable alteration in a DNA sequence, whether harmful, beneficial, or neutral, is called a **mutation**. There are two basic requirements to produce a heritable mutation: there must be a change in the base sequence and the cell must fail to repair the change before the next round of replication. Repair mechanisms will be discussed in Section 9.5. Although it is tempting to think all mutations destroy a protein's activity, this is not the case. A given mutation may or may not affect the informational content or phenotype of the organism.

Mutations come in several different physical and structural forms:

- A **point mutation** is a change in a single nucleotide (**Fig. 9.16 A** and **B**). Among point mutations, changing a purine to a different purine or a pyrimidine to a different pyrimidine is called a **transition**, while swapping a purine for a pyrimidine (and vice versa) is a **transversion**.
- **Insertions** and **deletions** involve, respectively, the addition or subtraction of one or more nucleotides (**Figs. 9.16C** and **D**). This makes the sequence either longer or shorter than it was originally.
- An **inversion** occurs when a fragment of DNA is flipped in orientation relative to DNA on either side (**Fig. 9.16E**).

Mutations can also be categorized into several informational classes. Mutations that do not change the amino acid sequence of a translated open reading frame (ORF) are called **silent mutations**. For example, a point mutation changing TTT to TTC in the sense DNA strand (corresponding to a UUU-to-UUC codon change in mRNA) still codes for a phenylalanine. (Remember, RNA polymerase reads from the DNA template strand to make RNA, whereas the complementary, or sense, DNA strand has the same sequence as the mRNA.) Thus, even though the DNA sequence has changed, the protein sequence remains the same. This is also called a synonymous substitution. However, if the UUU codon were changed to a UUA (a U-to-A transversion), the protein would have a leucine where a phenylalanine had been (see **Fig. 9.16A**). This type of mutation is a **missense mutation** because the amino acid sequence of the protein has changed.

The amino acid substitution resulting from a missense mutation may or may not alter protein function. The outcome depends on the structural importance of the original amino acid and how close in structure the replacement amino acid is to the original. Missense mutations result in either conservative amino acid replacements, where the new amino acid is structurally similar to the original (for example, leucine is substituted for isoleucine), or nonconservative replacements, in which a very different amino acid is substituted (for example, tyrosine for alanine). A missense change may decrease or eliminate the activity of the protein (a **loss-of-function mutation**), or it may make the protein more active. It could even gain a new activity such as an expanded substrate specificity or a completely different substrate specificity (these are called **gain-of-function mutations**).

A **nonsense mutation** is a point mutation that changes an amino acid codon into a translation termination codon—for example, UCA (serine) to UAA. The result will be a truncated protein most likely lacking any function. A mutation that eliminates function is known as

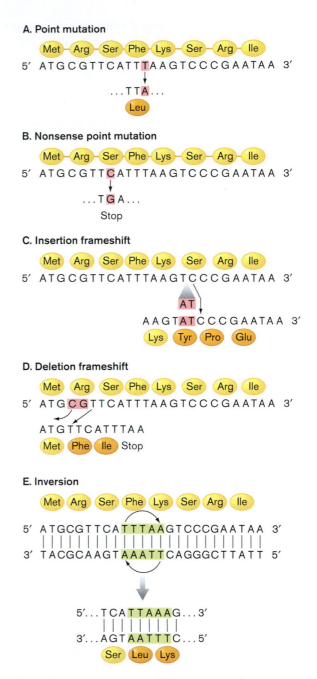

Figure 9.16 Changes in a DNA sequence that result in different classes of mutations. **A.** Missense mutation. The T to A transversion is a point mutation that converts the phenylalanine codon TTT to the leucine codon TTA. **B.** Nonsense point mutation. The C to G transversion converts the TCA codon, encoding serine, to the TGA stop codon. **C.** Insertion frameshift. The addition of AT into the middle of the TCC serine codon creates a shift in the reading frame that changes all the downstream amino acids. **D.** Deletion frameshift. Removing two nucleotides from the arginine codon also causes a frameshift. **E.** Inversion. Rotating a DNA sequence 180° relative to adjacent sequences.

a **knockout mutation** (see **Fig. 9.16B**). Knockout mutations can include multiple-base insertions and deletions as well as nonsense mutations.

Insertions and deletions can alter the reading frame of the DNA sequence (see **Figs. 9.16C** and **D**). Remarkable as it is, the ribosome simply reads RNA sequences one codon at a time, stringing amino acids together in the process. It translates each codon "word" but cannot understand the overall protein "sentence." It does not recognize when bases have been added or removed through mutation; instead, it keeps on reading the sequence in triplets. If the number of bases inserted or deleted is not a multiple of three, the ribosome will read the wrong triplets. The result is a **frameshift mutation**, which produces a garbled protein "sentence." This often results in the ribosome encountering a premature stop codon, originally in a different reading frame. If the insertion or deletion involves multiples of three bases, the reading frame is not changed but one or more amino acids are added or removed.

An inversion mutation involves the flipping of a DNA sequence (**Fig. 9.16E**). Imagine the highlighted sequence rotating 180° while the adjacent sequences in black remain stationary. The rotation would allow retention of 5'-to-3' polarity in the new molecule, but if the inversion occurred within a gene, it would likely change the codons in the area and alter the resulting protein. What is shown in the figure is a small inversion; however, inversions often involve large tracts of DNA encompassing several genes. If an entire gene with its promoter inverts, the gene will likely remain functional, and its encoded protein may very well still be made. Inversions occur within a genome as a result of recombination events between similar DNA sequences or as a consequence of mobile genetic elements jumping between different areas of a genome.

Mutations can affect both the genotype and the phenotype of an organism. The genotype of an organism reflects its genetic makeup. Regardless of whether a mutation causes a change in a biochemical trait, every mutation causes a change in the genotype, that is, the makeup of the genome. In contrast to genotype, phenotype only comprises observable characteristics, such as biochemical, morphological, or growth traits. For instance, consider the arginine biosynthesis gene *argA*. The product of this gene allows an organism to make enough arginine to support growth in media devoid of arginine. Any mutation in *argA* is a genotypic change, but if the result is a synonomous amino acid substitution, there is no phenotypic change—the organism can still make its own arginine. However, if the mutation destroys the activity of the *argA* product, the microbe can no longer make arginine, and the amino acid must be supplied in the medium. This is a phenotypic change.

It is also important to realize that small mutations, such as a single base substitution, can have large effects on phenotype. Conversely, large mutations—for instance, the insertion of a 5-kb transposon between two genes—may have little or no effect. To illustrate, consider that a single point mutation in the *hpr* gene of the phosphotransferase sugar transport system (discussed in Section 4.2) will render a bacterium incapable of growing on many sugars. However, inserting 5 kb of DNA just past the *hpr* stop codon will have no effect on cell growth. For mutations, it's often not the size that counts, it's the location.

Mutations Arise by Diverse Mechanisms

Although DNA is quite stable, especially compared to mRNA which is rapidly degraded, it is susceptible to damage inflicted by a variety of physical and chemical

Table 9.2   **Mutagenic agents and their effects.**

Mutagenic agent	Effects
Chemical agent	
Base analog	Substitutes "look-alike" molecule for the normal nitrogenous base during DNA replication: point mutation
Examples: caffeine, 5-bromouracil	
Alkylating agent	Adds an alkyl group, such as methyl group ($-CH_3$), to nitrogenous base, resulting in incorrect pairing: point mutation
Example: nitrosoguanidine	
Deaminating agent	Removes an amino group ($-NH_2$) from a nitrogenous base: point mutation
Examples: nitrous acid, nitrates, nitrites	
Acridine derivative	Inserts (intercalates) into DNA ladder between backbones to form a new rung, distorting the helix: can cause frameshift mutations
Examples: acridine dyes, quinacrine	
Electromagnetic radiation	
Ultraviolet rays	Link adjacent pyrimidines to each other, as in thymine dimer formation, and thereby impairs replication; lethal if not repaired
X-rays and gamma rays	Ionize and break molecules in cells to form free radicals, which in turn break DNA; lethal if not repaired

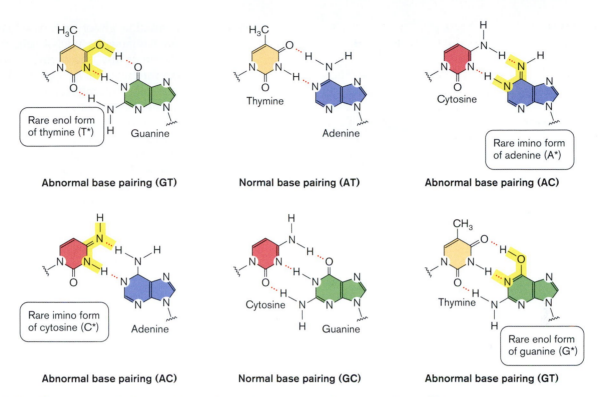

Figure 9.17   Rare tautomeric forms of bases have altered base-pairing properties.   Cytosine, which normally base pairs with guanine, will, in its imino form, pair with adenine. The imino form of adenine will also base-pair with cytosine, rather than its normal complementary base thymine. Likewise, the enol form of thymine base-pairs with guanine, while the enol form of guanine binds thymine. These tautomeric transitions can lead to permanent mutations.

agents. For example, irradiation by X-rays can cause a massive number of DNA strand breaks. When this happens, the integrity of the chromosome is lost and the cell dies. Other agents can directly modify the bases in DNA while leaving its overall structure intact. In this case, the modified bases have altered hydrogen bond base-pairing properties that result in the incorporation of an inappropriate base during replication. When that happens, a mutation occurs.

Mutations can be caused by **mutagens**, chemical agents that can damage DNA (**Table 9.2**). Even in the absence of a mutagen, mutations arise spontaneously. Because DNA proofreading and repair pathways are so efficient, spontaneous mutations are rare, having a frequency of occurrence ranging from 10^{-6} to 10^{-8} per generation.

Spontaneous mutations in a genome arise for many reasons—for example, due to tautomeric shifts in the chemical structure of the bases as shown in **Figure 9.17**. Tautomeric shifts involve a change in the bonding properties of amino ($-NH_2$) and keto ($=O$) groups. Normally, the amino and keto forms predominate (over 85%), but when an amino group, for example, shifts to an imino ($=NH$) group, base pairing changes. A cystosine, which normally base-pairs with guanine, will, in its rare imino form, base-pair with adenine. Tautomeric shifts that occur during DNA replication will increase the number

of mutational events. Even though the replication apparatus is very accurate with various proofreading and repair functions, such as the 3′-to-5′ exonuclease activity of DNA polymerase III (see DnaQ in Section 7.3), mistakes do occur at a very low rate.

Besides misincorporation mistakes, naturally occurring intracellular chemical reactions with water can damage DNA. These endogenous reactions are an important source of spontaneous mutations. For example, cytosine deaminates spontaneously to yield uracil (**Fig. 9.18**). This results in a GC to AT transition. In addition, purines are particularly susceptible to spontaneous loss from DNA by breakage of the glycosidic bond connecting the base to the sugar backbone (**Fig. 9.19**). The result of this loss is the formation of an **apurinic site** (one missing a purine

Figure 9.18   Spontaneous deamination of cytosine. Oxidative deamination changes cytosine to uracil, which will base-pair with adenine. This creates a transition mutation.

Depurination of DNA

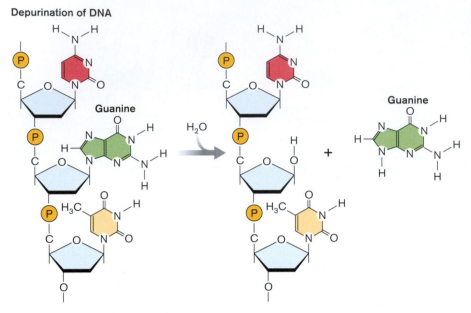

Figure 9.19 **Spontaneous formation of an apurinic site.**

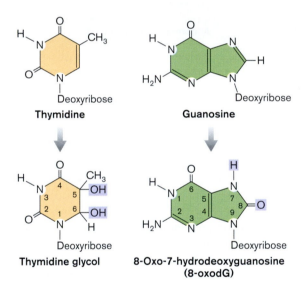

Figure 9.20 **Examples of damage caused by reactive oxygen species.** Growth in the presence of oxygen leads to the production of reactive oxygen species that can modify nucleotide residues. The modifications can interfere with polymerase function and stop replication or interfere with the transcription of affected genes. The blue highlighted groups are modifications to thymidine and guanosine residues.

base) in the DNA. This would obviously hinder transcription and replication.

DNA damage can occur as a result of metabolic activities of the cell that produce intermediates such as hydrogen peroxide (H_2O_2), superoxide radicals ($^\bullet O_2^-$), and hydroxyl radicals ($^\bullet OH$), collectively called reactive oxygen species. Even though bacteria have biochemical mechanisms to detoxify reactive oxygen species, sometimes the systems are overwhelmed. Oxidative damage causes the production of

thymidine glycol or 8-oxo-7-dehydro-deoxyguanosine in DNA (**Fig. 9.20**).

Naturally occurring intracellular methylation agents (for example, S-adenosylmethionine) can spontaneously methylate DNA to produce a variety of altered bases. The spontaneous methylation of the N-7 position of guanine, for example, weakens the glycosidic bond and results in spontaneous loss of the base (apurinic site formed) or opening of the imidazole ring forming a methyl formamide pyrimidine. In addition to mispairing, some of these spontaneous events can lead to major chromosomal rearrangements, such as duplications, inversions, and deletions. Mutagens tend to increase the mutation rate by increasing the number of mistakes in a DNA molecule as well as by inducing repair pathways that themselves introduce mutations.

Ultraviolet (UV) light produces a striking structural alteration in DNA molecules because pyrimidines are highly susceptible to UV radiation. The energy absorbed by a pyrimidine hit with UV light boosts the energy of its electrons to the point where the molecule is unstable. If two pyrimidines are neighbors on a single DNA strand, their energized electrons can react to form a four-membered cyclobutane ring. The result is a pyrimidine dimer that will block replication and transcription (**Fig. 9.21**).

Clearly, there are many ways DNA can be damaged. However, the cell has many opportunities to repair that damage before it becomes fixed as a mutation. But the repair mechanisms are not perfect. These missed opportunities contribute heavily to the formation of heritable mutations and, thus, evolution. Although most mutations decrease the fitness of a species, some can improve fitness by, among other things, enabling an organism to use, or compete for, available food sources with greater efficiency.

Mutation Rates and Frequencies Are Determined Experimentally

The **mutation rate** for a given gene is defined as the number of mutations formed per cell doubling. It can be estimated according to the formula $a = m$ per number of cell generations, where a is the mutation rate and m is the number of mutations that occur as the number of cells increases from the initial number N_0 to N (see Chapter 4). The number of cell generations is equal to ($\log_{10} N_t - \log_{10} N_0$) ÷ 0.301 (see Section 4.5).

A less complicated calculation is **mutation frequency**. Mutation frequency tells you how many mutant cells are present in a population. It is calculated simply as the ratio of mutants per total cells in the population.

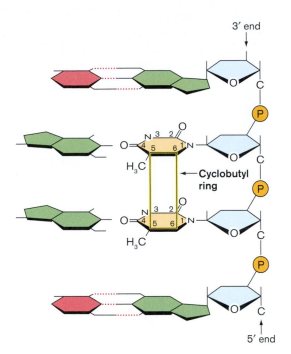

Figure 9.21 **Production of a of pyrimidine dimer.** The energy from ultraviolet light irradiation can be absorbed by pyrimidine molecules. The excited electrons of carbons 5 and 6 on adjacent pyrimidines can then be shared to create a four-member cyclobutane ring between adjacent pyrimidines. The pyrimidine dimer acts as a block to replication and transcription.

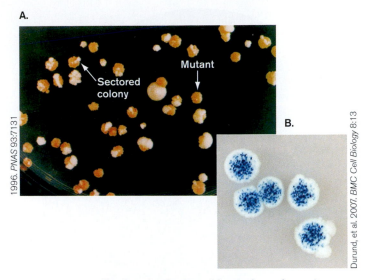

1996. PNAS 93:7131

Durund, et al. 2007. BMC Cell Biology 8:13

Figure 9.22 **Sectored colonies of bacteria and yeast.** **A.** Sectored colonies of yeast. On this medium, normal yeast cells form white colonies. The occasional mutant cell that develops as the colony grows will replicate as part of the colony but forms a reddish-orange sector. **B.** *E. coli* mutator strain. Cells are lactose-negative mutants that do not ferment lactose and do not turn the indicator blue. However, because the strain has a mutation in one of the mutator genes, there is a high rate of reversion to the lactose fermentor strain, which shows up as blue papillae on the white colony.

The colonies in **Figure 9.22** graphically illustrate the result of a high mutation rate. Single yeast cells placed on an agar surface multiplied to form the colonies shown (**Fig. 9.22A**). As the yeast cell replicates to form a colony, a spontaneous mutant may develop at some point. The mutant yeast is unable to metabolize a compound which then accumulates and turns the cell red. As the mutant cell increases in number alongside normal cells, a sectored colony forms.

In the case of the bacteria, the medium contains an indicator that turns blue if the organism can ferment lactose (**Fig. 9.22B**). This is a useful tool to follow the development of mutations in the gene needed to ferment lactose. Because the original cell in this case was Lac⁻ (due to a mutation in that gene), most of the colony is white. However, the strain is a mutator strain defective in a DNA repair gene. This caused a high rate of reversion to Lac⁺. Each cell that reverted to become a lactose fermenter produced offspring that can turn the indicator blue. These mutants appear as blue papillae on the surface of the white colony. Each papilla represents a separate mutational event.

THOUGHT QUESTION 9.6 Calculate the mutation rate for a gene in which the number of mutations increases from 0 to 50 as cell number increases from 10^1 to 10^8.

Mutagens Can Be Identified Using Bacterial "Guinea Pigs"

In humans, many nonmutagenic chemicals can be transformed into potent mutagens by the liver. The liver is the chief organ for detoxifying the body, a task accomplished by chemically modifying foreign chemicals by the action of liver enzymes. In a world where new chemical entities are constantly being introduced, it is important to determine which ones are potential mutagens. Conventional methods require testing these compounds on large populations of animals. Bruce Ames and his colleagues invented a simple alternative that uses bacteria as a rapid initial screen, for which Ames received the National Medal of Science. The method, called the Ames test, relies on a mutant of *Salmonella enterica* that is defective in the *hisG* gene, whose product is involved in histidine biosynthesis (**Fig. 9.23**). The *hisG* mutant cannot grow on a minimal defined medium lacking histidine. However, if a reversion mutation occurs in the *hisG* gene, restoring the gene to its original functional state, the new mutant cell will form a colony, even in the absence of histidine. This method is called a reversion test.

Ames used this *his* reversion test to screen compounds for potential mutagenicity. If a chemical is able to mutate one gene (*hisG*), it has the potential to affect any gene. As shown in **Figure 9.23,** a mutagen-containing

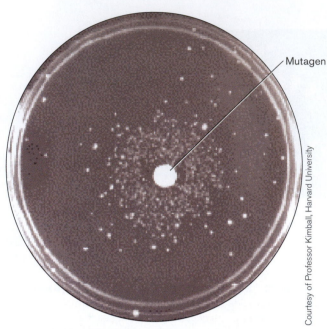

Courtesy of Professor Kimball, Harvard University

Figure 9.23 Basic Ames test for mutagenesis. A mutation in the *hisG* gene of *Salmonella enterica* creates a histidine auxotroph, which is a strain that requires histidine to grow. When plated onto medium lacking histidine, only a few spontaneous mutations (reversions) occur that reverse the mutation so the cell can make histidine again. These cells form colonies on medium lacking histidine. A paper disk containing a mutagen placed in the center of the plate will progressively, over time, increase the number of prototrophic revertants seen orbiting the disk.

disk is placed in the middle of an agar plate spread with the original *his* mutant. As the mutagen diffuses into the medium and causes reversion mutations, colonies start to appear, forming a ring around the disk. More revertants are produced the longer the plate is incubated.

How can this technique be used to screen for hidden mutagens that require processing by mammalian enzymes? The basic technique was modified by treating a potential mutagen with a rat liver extract before applying the mixture to bacteria for the *his* reversion test (**Fig. 9.24**). In the absence of liver enzymes, these compounds are not mutagenic. If the enzymes of the liver, the main detoxifying organ of the body, convert the compound to a mutagen (it does this accidentally—not on purpose), His⁺ revertant colonies will be seen. This method has helped accelerate the process of drug discovery by providing an inexpensive preliminary screen for weeding out mutagenic chemicals before the more expensive animal testing is undertaken.

The most recent assays for mutagenicity involve transgenic mice. Mice are engineered to have the *lacZ* gene, for instance, inserted into their chromosomes. The *lacZ* gene encodes an enzyme that is easily assayed, so it is useful as a reporter for mutagenic activity. This reporter gene

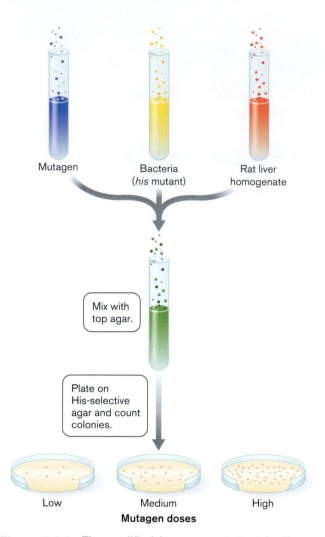

Figure 9.24 The modified Ames assay to test for the mutagenic properties of chemical processed through the liver. The potential mutagen, *his*-mutant bacteria, and liver homogenate are combined and mixed with agar. The combination is poured into a Petri plate. If enzymes from the liver extract modify the test compound and convert it to a mutagen, increasing numbers of His⁺ revertants will be observed with increasing doses of mutagen.

will be distributed throughout the mouse in all organs. A potential mutagen is administered to the mouse, and after some time the mouse is sacrificed, organs are removed, and the DNA is extracted and examined for mutations arising in the *lacZ* gene. This approach allows scientists to track where the mutagen is distributed in the mouse and to determine whether certain organs convert harmless precursor chemicals to dangerous mutagenic compounds.

TO SUMMARIZE:

- **A mutation** is any heritable change in DNA sequence, regardless of whether there is a change in gene function.

- **Genotype** reflects the genetic makeup of an organism, whereas **phenotype** reflects its physical traits.
- **Classes of mutations** include silent mutations, missense mutations, nonsense mutations, point mutations, insertions, deletions, and frameshift mutations.
- **Spontaneous mutations** reflect tautomeric shifts in DNA nucleotides during replication, accidental incorporation of noncomplementary nucleotides during replication, or "natural" levels of chemical or physical (irradiation) mutagens in the environment.
- **Chemical mutagens** can alter purine and pyrimidine structure and change base-pairing properties.
- **Bacterial cultures** can be used to assess the mutagenicity of a chemical.

9.5 DNA Repair

Microorganisms are equipped with a variety of molecular tools that repair DNA damage before the damage becomes a heritable mutation (see **Table 9.3**). The type of repair mechanism used (and when it is used) depends on two things: the type of mutation needing repair and the extent of damage involved. Some mechanisms excise whole fragments of DNA that contain damaged bases; others precisely excise the damaged bases or directly reverse the damage. After extensive damage, special "emergency" DNA polymerases are expressed that sacrifice replication accuracy to rescue the damaged genome. Whether damage is introduced by mutagens or by inaccurate DNA synthesis, microbial survival depends on the ability to repair DNA.

First we cover the **error-proof repair** pathways that prevent mutations. These include methyl mismatch repair, photoreactivation, nucleotide excision repair, base excision repair, and recombinational repair. We will then discuss **error-prone repair** pathways. These pathways risk introducing mutations and operate only when damage is so severe that the cell has no other choice but to die.

Error-Proof Repair Pathways

Mismatch repair. What happens if DNA polymerase simply makes a mistake and incorporates a normal but incorrect base? Although proofreading functions in DNA polymerases are designed to prevent the misincorporation of bases during replication, misincorporation does occur. Curiously, far fewer mutations arise than one would predict based on the inherent error rate of DNA polymerase III (approximately 1 mistake per 10^6 bases synthesized, after proofreading). However, the mutation rate in a live cell is actually only 10^{-9} to 10^{-10} per base pair replicated. How can a cell repair a mutation after it has been introduced during replication? One method is **methyl mismatch repair**, which is based on recognition of the methylation pattern in DNA bases. As discussed in Section 7.3, many bacteria tag their parental DNA by methylating it at specific sites. In E. coli, for example, deoxyadenosine methylase (Dam) methylates the palindromic sequence GATC to produce GA^{ME}TC. The Dam methylase does this soon, but not immediately, after replication of a DNA sequence.

When DNA polymerase misincorporates a base during replication, a mismatch forms between the incorrect base in the newly synthesized, but unmethylated strand and the correct base residing in the parental, methylated strand. Methyl-directed mismatch repair enzymes (MutS, MutL, and MutH) bind to the mismatch. MutS first binds to the mismatch and recruits MutL and MutA. MutL recognizes the methylated strand (GATC-Me), brings it in a loop to meet MutS and MutH, and then cleaves the unmethylated strand containing the mutation, near the GATC (**Fig. 9.25**). A DNA helicase called UvrD then

Table 9.3   Basic types of DNA repair.

System	Genes	Mutations recognized	Repair mechanism	Result
Photoreactivation	*phrB*	Pyrimidine dimers	Cyclobutane ring cleaved	Accurate
Nucleotide excision	*uvrABCD*	Helical destabilization, e.g., *pyrimidine* dimers	Patch of nucleotides excised	Accurate
Base excision	*fpg, ung, tag, mutY, nfo*	Various modified bases	Glycosylases remove base from phosphodiester backbone; apurinic (AP) sites formed	Accurate
Methyl mismatch	*mutHSL, dam*	Transitions, transversions	Nick on nonmethylated strand, excision of nucleotides	Accurate
Recombination	*recA*	Single-strand gaps	Recombination	Accurate
Translesion bypass synthesis	*umuDC*	Gaps	Part of SOS sytem	Error-prone (generates mutations)

unwinds the cleaved strand, exposing it to a variety of exonucleases. The result is a gap that is filled in by DNA polymerase I (Pol I) and sealed by DNA ligase.

The methyl-directed mismatch repair proteins (and genes) are called Mut (and *mut*) because a high mutation

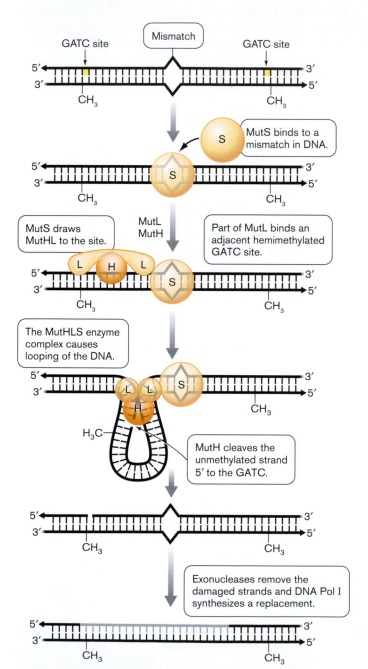

Figure 9.25   Methyl mismatch repair. The bacterial cell, in this case *E. coli*, can use specific methylations on DNA to recognize parental DNA strands for preferential DNA repair. Newly replicated strands are not immediately methylated. So when a mismatch is found, the mismatch repair system views the newly synthesized strand as suspect and replaces the section of unmethylated DNA encompassing the mismatch. The steps following cleavage are similar to what is shown in Fig. 9.20 (steps 7 and 8).

rate results in strains that are defective in one of these proteins. A bacterial strain with a high mutation rate is called a **mutator strain**.

NOTE: Not all DNA methylations serve as signals for methyl mismatch repair. DNA methylation by restriction-modification systems, for example, prevent cleavage by DNA restriction enzymes but are not used to designate parental DNA after replication.

Photoreactivation and nucleotide excision. DNA repair mechanisms are not influenced by DNA methylation. Many microbes in the environment commonly encounter ultraviolet light. The pyrimidine dimers that form as a result of ultraviolet irradiation can be repaired by a light-activated mechanism called **photoreactivation**. In photoreactivation, the enzyme photolyase binds to the dimer and cleaves the cyclobutane ring linking the two adjacent, damaged nucleotides. The damage is repaired without excising any bases.

While photolyase is specific for pyrimidine dimer repair and requires light, a system called **nucleotide excision repair (NER)** operates in the dark (or light) and is used to excise other kinds of damaged DNA as well as pyrimidine dimers. In this system, a three-subunit exonuclease (UvrABC) excises a patch of 12–13 nucleotides that includes the dimer (**Fig. 9.26**). The basic mechanism involves UvrA, associated with UvrB, recognizing the damaged base (steps 1–3). UvrA is ejected (step 4), and UvrB binds UvrC (step 5), which actually cleaves the damaged strand at two sites flanking the damaged base (step 6). The small fragment is removed and the gap repaired by DNA polymerase I (steps 7 and 8). Nucleotide excision repair is also used to excise other kinds of damaged DNA besides pyrimidine dimers.

Nucleotide excision repair can be enhanced by ongoing RNA transcription. Bacteria as well as eukaryotic cells preferentially repair genes that are being transcribed. The preferential repair of transcriptionally active genes makes sense because a cell unable to transcribe such a gene due to recent DNA damage would be at a survival disadvantage. An RNA polymerase that encounters a UV dimer or other unrecognizable base on a template DNA strand will stall during transcription. The stalled RNA polymerase is then recognized by a protein that mediates a process called transcription-coupled repair. The transcription-coupled repair protein then snares a nearby UvrAB to begin nucleotide excision repair.

Base excision repair snips damaged bases from DNA. Although nucleotide excision repair can recognize and repair several types of damage, it does not work in all instances. Another error-proof process known as **base**

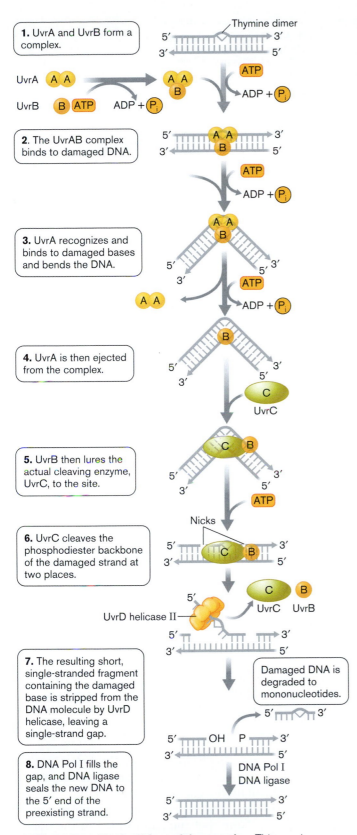

Figure 9.26 Nucleotide excision repair. This repair mechanism cuts out a single-stranded segment of DNA containing the mutation and uses DNA replication to repair the gap.

The following labels appear in the figure:

1. UvrA and UvrB form a complex.

UvrA [A A]
UvrB [B ATP] ADP + Pi

Thymine dimer
5' 3'
3' 5'
ATP
ADP + Pi

2. The UvrAB complex binds to damaged DNA.

ATP
ADP + Pi

3. UvrA recognizes and binds to damaged bases and bends the DNA.

ATP
ADP + Pi

4. UvrA is then ejected from the complex.

UvrC

5. UvrB then lures the actual cleaving enzyme, UvrC, to the site.

ATP

Nicks

6. UvrC cleaves the phosphodiester backbone of the damaged strand at two places.

UvrD helicase II
UvrC UvrB

7. The resulting short, single-stranded fragment containing the damaged base is stripped from the DNA molecule by UvrD helicase, leaving a single-strand gap.

Damaged DNA is degraded to mononucleotides.

OH P

DNA Pol I
DNA ligase

8. DNA Pol I fills the gap, and DNA ligase seals the new DNA to the 5' end of the preexisting strand.

excision repair (**BER**) employs a battery of glycosylase enzymes that can recognize and clip certain damaged bases from the phosphodiester backbone. Uracil DNA glycosylase, hypoxanthine DNA glycosylase, and 3-methyladenine glycosylase recognize, respectively, uracil, hypoxanthine, and 3-methyladenine when these bases are present in DNA. Uracil can be found in DNA either as a spontaneous deamination product of cytosine (see **Fig. 9.18**) or as a result of inappropriate incorporation during replication. Spontaneous deamination of adenine residues produces hypoxanthine. Uracil and hypoxanthine have base-pairing properties very different from those of the original bases, so their formation can lead to mutations. Mutagens such as methyl methane sulfonate can alkylate adenine to produce the adduct 3-methyladenine. A 3-methyladenine residue will totally block DNA replication, making this a lethal form of damage. Consequently, repairing these mutations is vital for survival.

The glycosylases previously noted cleave the bond connecting the damaged base to deoxyribose in the phosphodiester backbone (**Fig. 9.27**, step 1). The result is an intact phosphodiester backbone missing a base (an abasic site). The site is called an **AP site** because it is missing a purine (it is apurinic) or a pyrimidine (it is apyrimidinic). The next step in base excision repair involves **AP endonucleases** that specifically cleave the phosphodiester backbone at AP sites (step 2). The 5'-to-3' exonuclease activity of the gap-filling DNA polymerase I will degrade the cleaved strand downstream of the AP site and at the same time synthesize in its stead a replacement strand containing the proper base (step 3). DNA ligase seals the remaining nick and the repair process is complete (step 4).

> **THOUGHT QUESTION 9.7** It has been reported that hypermutable bacterial strains are overrepresented in clinical isolates. Out of 500 isolates of *Haemophilus influenzae*, for example, 2–3% were mutator strains having mutation rates 100–1,000 times higher than lab reference strains. Why might mutator strains be beneficial to pathogens?

All of the repair processes described thus far (mismatch repair, photoreactivation, nucleotide excision repair, and base excision repair) are error-proof pathways that rely on the presence of a good template strand opposite the damaged strand. Replication of the template strand is used to replace the damaged bases accurately. But what happens when both strands are damaged or when there is only one strand?

Recombinational repair. Figure 9.28 illustrates one repair mechanism that occurs when DNA replication takes place before a UV dimer can be excised by nucleotide excision repair. Because DNA polymerase III (the main

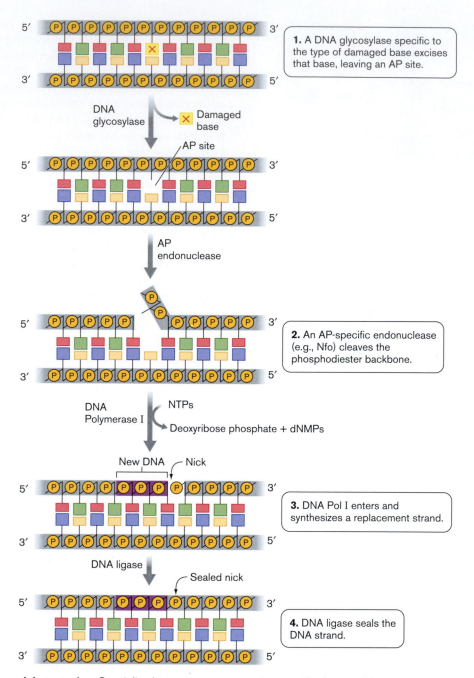

1. A DNA glycosylase specific to the type of damaged base excises that base, leaving an AP site.

DNA glycosylase → Damaged base

AP site

AP endonuclease

2. An AP-specific endonuclease (e.g., Nfo) cleaves the phosphodiester backbone.

DNA Polymerase I — NTPs

Deoxyribose phosphate + dNMPs

New DNA — Nick

3. DNA Pol I enters and synthesizes a replacement strand.

DNA ligase — Sealed nick

4. DNA ligase seals the DNA strand.

Figure 9.27   Base excision repair.   Specialized enzymes can recognize specific damaged bases and remove them from DNA without breaking the phosphodiester backbone. The result is an abasic site (AP; apurinic or apyrimidinic) that can be recognized and cleaved by a specific AP endonuclease. This allows DNA polymerase I to synthesize a replacement strand containing the proper base.

polymerase for chromosome replication) cannot decode a dimer, it skips over the damaged area and restarts with a new RNA primer, leaving a gap opposite the dimer. At this point, the Uvr complex cannot excise the UV dimer because chromosome integrity would be lost. However, the RecA protein, involved with recombination, can bind to the gap and initiate a genetic exchange in which a piece of the undamaged, properly replicated strand is spliced into the gap. This is called **recombinational repair** and

mostly occurs near replication forks. Once the gap is filled, the Uvr complex can remove the dimer and DNA polymerase I will fill the gap in the other strand.

The recombinational repair system is an error-proof repair pathway. Note that the gap formed in the donor strand is easily replaced by the gap-filling DNA polymerase I. Recombinational repair is not limited to pyrimidine dimers. It will work on any damage that causes gaps during replication.

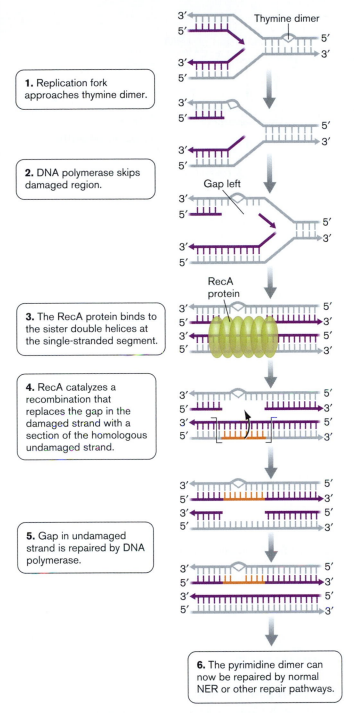

1. Replication fork approaches thymine dimer.

2. DNA polymerase skips damaged region.

3. The RecA protein binds to the sister double helices at the single-stranded segment.

4. RecA catalyzes a recombination that replaces the gap in the damaged strand with a section of the homologous undamaged strand.

5. Gap in undamaged strand is repaired by DNA polymerase.

6. The pyrimidine dimer can now be repaired by normal NER or other repair pathways.

Figure 9.28 Recombinational repair. Recombination can be used to repair DNA damage of replicating DNA when one daughter strand is undamaged. A single-stranded segment of the undamaged daughter strand can be used to replace a gap in the damaged daughter strand.

Error-Prone DNA Repair

SOS "Save Our Ship" repair. When DNA damage is extensive, none of the repair strategies that excise pieces of DNA or that rely on one of the two strands remaining undamaged can work without destroying the chromosome. The cell must take more drastic measures known as the **SOS response**, a coordinated cellular response to DNA damage that, in order to save the cell, can introduce mutations into severely damaged DNA. So in this case, repair refers to saving the integrity of the circular chromosome even if incorrect bases are introduced. In the SOS system, RecA protein appears to sense when DNA damage is extensive based on the level of single-stranded DNA available. For example, excessive ultraviolet irradiation will produce numerous ssDNA gaps because DNA polymerase cannot replicate through pyrimidine dimers.

Formation of RecA filaments (see **Fig. 9.13**) on ssDNA activates a second function of RecA called coprotease activity that stimulates autodigestion of the LexA repressor, a protein that normally binds to DNA at the promoters of DNA repair genes and prevents their transcription (**Fig. 9.29**). Cleavage of LexA unleashes expression of these genes to make DNA repair enzymes. Among these enzymes are two "sloppy" DNA polymerases that lack proofreading activity: UmuDC, also called DNA polymerase V (Pol V), and DinB, also called DNA polymerase IV (Pol IV) (**Fig. 9.29**). The UmuDC enzyme is perfect for replicating through damaged bases (a process called translesion replication) because it sacrifices accuracy for continuity. When the enzyme encounters an undecipherable damaged base, it will insert whatever nucleotide is available. This, of course, will lead to numerous mutations, but the benefit is that the cell has a chance to live if it can tolerate the mutations. This high mutation frequency is a calculated risk. However, the cell really has no other option, since housekeeping DNA polymerases like Pol III cannot move through damaged areas of chromosomal DNA. The choice really is "mutate or die." (Note that the term *housekeeping* is applied to proteins or enzymes that keep the cell running at all times.)

When the cell is severely compromised by mutation, it temporarily stops dividing. Like a pit stop during an auto race, this pause in cell division allows time for repair enzymes to fix the damage. The product of another SOS-regulated gene called *sulA* causes this pause by binding to the FtsZ cell division protein, keeping it from initiating cell division (see Section 3.6). Once the damage has been repaired, RecA coprotease is inactivated and LexA repressor accumulates, turning off all the SOS genes, including *sulA*. The lingering SulA protein is then degraded by a protease called Lon. The protein is called Lon because when it is missing, SulA inappropriately accumulates, inhibits cell division, and produces *lon*g filamentous cells.

Human homologs of bacterial repair genes. About 30% of *E. coli* genes have human homologs; the functions of the human genes may be similar to those of

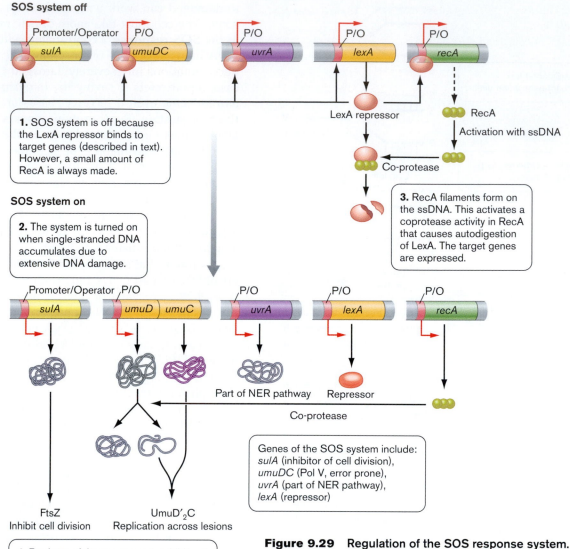

SOS system off

1. SOS system is off because the LexA repressor binds to target genes (described in text). However, a small amount of RecA is always made.

LexA repressor

RecA
Activation with ssDNA

Co-protease

SOS system on

2. The system is turned on when single-stranded DNA accumulates due to extensive DNA damage.

3. RecA filaments form on the ssDNA. This activates a coprotease activity in RecA that causes autodigestion of LexA. The target genes are expressed.

Part of NER pathway   Repressor

Co-protease

FtsZ
Inhibit cell division

UmuD′$_2$C
Replication across lesions

Genes of the SOS system include:
sulA (inhibitor of cell division),
umuDC (Pol V, error prone),
uvrA (part of NER pathway),
lexA (repressor)

4. Products of the target genes inhibit cell division, mediate replication across lesions (Pol V), and participate in nucleotide excision repair.

Figure 9.29 Regulation of the SOS response system. The emergency DNA repair system known as the SOS response is induced when there is extensive DNA damage. The system is not a single repair mechanism but a set of different mechanisms that collaborate to rescue the cell.

E. coli or they may have newly acquired functions. These genes often turn out to play key roles in human genetic diseases. An example is *mutS*, of the methyl mismatch DNA repair system. In humans, the ancestor of *mutS* evolved to have multiple repair functions as well as participate in postmeiotic segregation. Defects in this repair mechanism have been linked to increased risk of certain colon cancers. Numerous other human genetic diseases are caused by mutations in homologs of bacterial repair genes. For example, defective excision repair genes in humans cause xeroderma pigmentosa, a disease that causes blindness and skin cancers. A deficiency in transcription-coupled repair causes Cockayne syndrome in humans, a devastating neurodegenerative disease.

TO SUMMARIZE:

- **DNA repair pathways** in microorganisms include error-proof and error-prone mechanisms.
- **Methyl mismatch repair** uses methylation of the parental DNA strand to discriminate it from newly replicated DNA. The premise is that the parental strand will contain the proper DNA sequence.
- **Photoreactivation** cleaves the cyclobutane rings of pyrimidine dimers.
- **Nucleotide excision repair** (NER) clips out a patch of single-stranded DNA containing certain types of damaged bases.
- **Base excision repair** (BER) excises structurally altered bases without cleaving the phosphodiester

backbone. The resulting AP (apurinic/apyrimidinic) site is targeted by AP nucleases.

- **Recombinational repair** takes place at a replication fork. A "good" strand of DNA is used to replace a homologous damaged strand.
- **Extensive DNA damage leads to induction of the SOS response** with increased levels of the error-proof repair systems, as well as error-prone translesion bypass DNA polymerases that introduce mutations.

9.6 Mobile Genetic Elements

In 1948, Barbara McClintock (1902–1992) noticed that certain genetic traits of corn defied the laws of Mendelian inheritance. The genes responsible for these traits, sometimes called "jumping genes," seemed to hop from one chromosome to another. Although McClintock's theories were provocative at the time, we now know that these types of genes, referred to as **transposable elements**, exist in virtually all life-forms and can move both within and between chromosomes.

Since McClintock's work, transposable elements have been most widely studied in bacteria. Nancy Kleckner, Russel Chan, Bik-Kwoon Tye, and David Botstein described the presence of transposable elements in bacteria in 1975 (**Fig. 9.30**). In this classic study, a transposable element carrying a tetracycline resistance gene (Tc) was found inserted into the P22 phage genome. P22 is a lysogenic phage that infects *Salmonella*. Recall that a lysogenic phage can integrate into the host chromosome and remain dormant for long periods of time.

The experiment used a defective P22 phage (carrying the Tc resistance gene) that could only replicate in certain strains of *Salmonella*. When this phage was used to infect nonpermissive strains of *Salmonella* (strains that would not allow the phage to grow or lysogenize), the *Salmonella* still became tetracycline resistant at a high frequency. Many of those tetracycline-resistant colonies also developed an auxotrophic requirement for an amino acid, nucleic acid base, or vitamin. The simple explanation was that the tetracycline resistance (Tc^R) element translocated ("jumped") out of the P22 DNA and into one of a number of locations on the *Salmonella* chromosome, sometimes inserting within a structural gene, thereby creating a mutation and phenotypic change. Today we know that these mobile DNA elements have contributed to remodeling of genomes throughout the evolutionary development of all species.

> **NOTE:** The symbol :: is a convention representing insertion. The element appearing after the two colons has been inserted into the element noted in front of the two colons. For example, *proA*::Tn*10* means Tn*10* (a transposable element) is inserted into the *proA* gene involved in proline synthesis.

Transposable elements are not autonomous; unlike plasmids, they are incapable of existing outside of a larger DNA molecule. They exist only as hitchhikers integrated into some other DNA molecule (**Fig. 9.31 ◉**). The DNA sequence of a transposable element includes a gene encoding a **transposase**—an enzyme that catalyzes the transfer or copying of the element from one DNA molecule into another. A simple element (700–1500 bp) called an **insertion sequence (IS)** element consists of a transposase gene flanked by short inverted repeat sequences that are targets of the transposase (see **Fig. 9.31**).

> **NOTE:** An inverted repeat is a DNA sequence identical to another downstream sequence. The repeats have reversed sequences and are separated from each other by intervening sequences:
>
> 5'-AATCGAT . . . . ATCGATT-3'
>
> When no nucleotides intervene between the inverted sequences, it is called a palindrome.

Figure 9.30   A. Nancy Kleckner (Harvard University) and B. David Botstein (Princeton University).   Two of the discoverers of bacterial transposition.

A.

©Marcus Halevi

B.

Courtesy Vera Bremmer and David Botstein

Transposition is the process of moving a transposable element *within* or *between* DNA molecules. During transposition, a short DNA sequence on the target DNA molecule is duplicated so that one copy of the sequence will flank each end of the element (**Fig. 9.31**). Note that the transposase randomly selects one of many possible target sequences where it will move the insertion element.

IS elements transfer by one of two mechanisms: replicative or nonreplicative transposition (**Fig. 9.32**⏵).

In nonreplicative transposition, the insertion sequence excises itself out of one host DNA while integrating into the next DNA. In replicative tranposition, the sequence copies itself into the new host DNA while the original copy remains within the original host.

Nonreplicative Transposition

The nonreplicative model of transposition is shown in **Figure 9.33**⏵. The transposase protein binds to the inverted repeat ends of the transposable element and to the target DNA, forming a **transpososome** complex (**Fig. 9.33**, step 1). The transposase cuts the phosphodiester backbone (step 2), severing one strand at one end of the insertion sequence and the other strand at the other end. The 3′ OH ends of the IS element then attack the unnicked strands in a transesterification reaction that produces hairpin structures (step 3). In this instance, joining the ends of the double-stranded molecule produces a single strand with a hairpin.

After the host carrier DNA is ejected from the transpososome, the hairpin ends are renicked and the 3′ OH ends attack the target DNA molecule in a staggered manner (step 4). The element has successfully "jumped" from one molecule to another without replicating. Note that the target DNA segment replicates as a result of the insertion process, producing a copy of the target sequence at each end of the IS element (step 5).

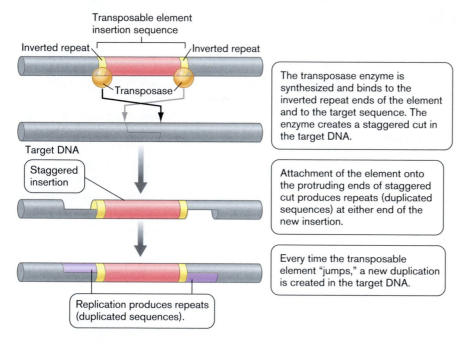

Figure 9.31   Basic transposition and origin of target site duplication. Enzymes that catalyze transposition generate duplications in the target site by ligating the ends of the insertion element to the long ends of a staggered cut at the target DNA site. ⏵

Replicative Transposition

Figure 9.34 shows how a transposable element in a plasmid can bring about cointegration of the plasmid into a target DNA molecule during replicative transposition. As a result of replication of the IS element, the cointegrate molecule will contain two copies of the element flanking the integrated plasmid. Thus, transposition can provide an organism with a whole set of genes that might extend its metabolic capacity, provide drug resistance, or serve as the starting point for gene divergence through mutation. Do not be misled; transposition is not a process designed to help the host; rather, it evolved to ensure survival of the transposon. The benefit to the host is accidental.

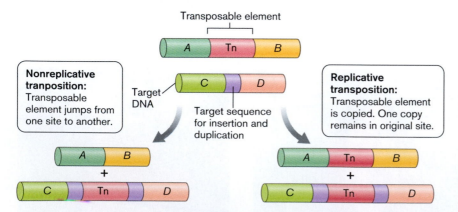

Figure 9.32   Products of nonreplicative and replicative transposition. Nonreplicative transposition moves an insertion element from one DNA site to another without leaving a copy of the element at the original site. Replicative transposition leaves the element at the original site and moves a replicated copy to the new site. ⏵

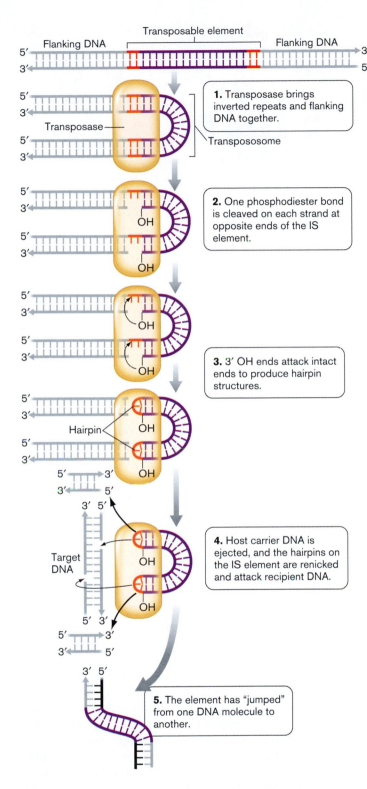

Figure 9.33   Nonreplicative transposition.   The transpososome complex includes the transposase binding to the ends of the transposable element and the target DNA. ▶️

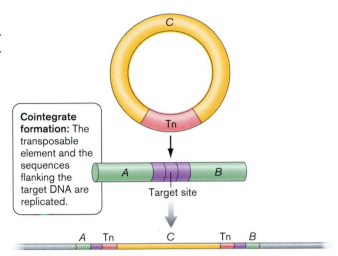

Cointegrate formation: The transposable element and the sequences flanking the target DNA are replicated.

Figure 9.34   Transposition-mediated cointegrate formation between two DNA elements.   During replicative transposition, the insertion element is copied and simultaneously ligated to the target site. This creates a transient cointegrate molecule where donor DNA and target DNA become one molecule.

Transposons are one type of transposable element. They are more complex than simple insertion elements, since they carry other genes in addition to those required for transposition. They come in different configurations. A **composite transposon** typically consists of two insertion sequences that flank an antibiotic resistance gene, or one or more catabolic genes (for example, benzene catabolism). Often the interior inverted repeats of the IS element have degenerated, so the transposase primarily acts on the two outermost inverted repeats, causing the whole transposon to move as one unit. Tn*10* is one such example (**Fig. 9.35**). Some transposons also include regulatory genes, such as a repressor that inhibits transcription of the transposase.

Complex transposons have, as the name suggests, a more complex organization. Tn3, for instance, not only possesses transposase and antibiotic resistance genes, but includes a gene whose product, called resolvase, specifically resolves (unlinks) the cointegrate produced by replicative transposition (**Fig. 9.34**). The transposase acts on the inverted repeat ends of the element, causing simultaneous insertion and replication. Resolvase is a site-specific recombinase that mediates recombination at the internal Tn3 *res* sites. The result of this recombination is restoration of the two original circular replicons, each one now containing a copy of the transposon.

All of the transposition events mentioned so far involve transfer between DNA molecules within the same cell. Some transposons, called **conjugative transposons**, are able to transfer from one cell to another by conjugation. One example is the transposon Tn*916* of

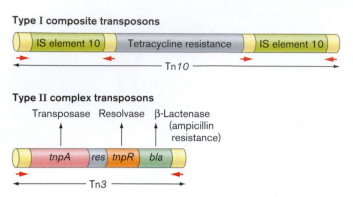

Figure 9.35   Examples of composite and complex transposons. In Tn*10*, the yellow fragments represent inverted repeats and the arrows depict orientation. TnpR is a repressor of the Tn*3 tnpA* transposase gene and is the enzyme that recognizes the *res* site required to resolve cointegrates formed during transposition.

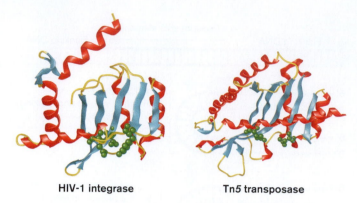

Figure 9.36   Comparison of the catalytic regions for HIV-1 integrase and Tn5 transposase. Note the similarity in structures. The alpha helices are shown in red, the beta sheets are blue, and the catalytic residues are green. Both enzymes contain a five-strand mixed beta sheet and the DDE motif (aspartate/aspartate/glutamate) of catalytic residues. (PDB codes: 1BIH, 1MUH)

Enterococci, which encodes tetracycline resistance. These conjugative transposons resemble plasmids in that they form a circular intermediate just prior to conjugation, but they do not replicate autonomously. They must be part of a larger self-replicating DNA entity (for example, plasmid, chromosome, or phage). The transfer mechanism of conjugative transposons also resembles that of temperate phage in that these elements excise from donor DNA and integrate into the recipient chromosomes. Tn*916* and some other conjugative transposons are designed to transfer when a population is facing danger. For example, in the absence of tetracycline, only a few cells in a population may have Tn*916*. However, the presence of just a small amount of tetracycline triggers conjugative transfer of the element, spreading it throughout the population.

The study of transposons may provide insight into the workings of the human immunodeficiency virus (HIV) that causes AIDS. The structure of transposase from transposon Tn*5* bears a striking similarity to the HIV integrase needed to embed the reverse-transcribed HIV viral DNA into the human genome (**Fig. 9.36**; see Section 11.5). Another type of mobile element called an integron has evolved in bacteria to capture DNA cassettes that encode antibiotic resistance genes (**Special Topic 9.1**).

TO SUMMARIZE:

■ **Transposons and insertion sequences** ("jumping genes") move from one DNA molecule to another, usually without replicating separately (that is, they are not plasmids).

■ **Transposase** is an enzyme that forms a transpososome complex with the transposon and target DNAs.

■ **Transposons move** by nonreplicative or replicative mechanisms.

■ **Transposons can carry a variety of genes**, including antibiotic resistance genes.

9.7   Genome Evolution

"Nothing makes sense in biology except in the light of evolution," said biologist Theodor Dobzhansky (1900–1975). This statement, made in 1972, rings especially true now that we have access to rapid sequencing of whole genomes. A sense of evolution, of what went before, pervades efforts to interpret the results of the many genome projects. Microbial genomes are dynamic entities that continually evolve. Genes can be removed (deletion), added (insertion), rearranged (recombination), or divided to the point where the genome has only a slight resemblance to what it once was. Evolution of a genome is a random process driven by natural selection. The process of microbial evolution is discussed in Chapter 17. Here we note the molecular mechanisms that drive genome change.

The result of genome evolution can be seen in the genomes of *Escherichia coli*. Several strains of *E. coli* have been sequenced. The first was the K12 strain (called MG1655). It contains 4,639,221 base pairs (compared with the 3 billion base pairs of the human genome). Of

> **THOUGHT QUESTION 9.8** Diagram how a composite transposon like Tn*10* can generate inversions or deletions in target DNA during transposition. *Hint:* This occurs when the transposon tries to jump from one site to another in the same chromosome using the ends of the inverted repeats closest to the tetracycline resistance gene (see **Fig. 9.35**). If you draw it right, you will see that in the end, the tetracycline gene is lost.

this genome, 87.8% encode proteins and 0.8% code for stable RNAs such as tRNA and rRNA used in translation. DNA with no known function encompasses another 0.7%. Approximately 11% of the chromosome is involved with various forms of regulation. Even though *E. coli* is the best-studied organism on the planet, 28% of the 4,288 open reading frames still remain a mystery to us, having no known or annotated function.

Examining the sequences more closely reveals how the current organism we call *E. coli* came about through the evolution of an ancestral genome. Two basic processes are thought to contribute to genome remodeling: horizontal gene transfers and duplications followed by functional divergence through mutation.

Horizontal Gene Transfer Occurs via Transformation, Conjugation, and Transduction

Microbial genomes evolve by randomly assembling an eclectic array of genes from many sources. For example, *E. coli* strain O157:H7, the culprit in several fatal outbreaks of food-borne disease in the United States and Europe, contains 1,387 genes lacking in the strain K12. These additional genes represent about 25% of the O157:H7 genome and encode virulence factors, metabolic pathways, and prophages (phage genes integrated into host chromosomes), all of which were acquired from other species. Until 1990, it was thought that bacterial genomes were rather static; one strain of *E. coli* was thought to be very similar to any other strain. But because of genomic comparisons like this, we now know that enterobacteria are subject to a great deal more interspecies recombination than previously thought.

Evidence for horizontal transfer. More evidence of gene shuffling comes from comparing the proportion of GC base pairs along the chromosome (also known as GC content). On average, the *E. coli* genome consists of 50.8% GC pairs. But 15% of the K12 genome and 26% of the O157:H7 genome show GC proportions that differ significantly from the rest of the genome and show a different codon usage, suggesting that these genes came from other bacterial lines or species and were acquired by *E. coli* more recently. These horizontal gene transfers, occurring via conjugation, transduction, or transformation, are estimated to happen at the rate of about 16 kbp, or 0.4% of the *E. coli* genome, every million years. A sudden change in the GC content within a chromosome is the equivalent of an evolutionary "footprint" marking a horizontal gene transfer.

Horizontal gene transfer appears to be a primary force in microbial evolution. Some estimates place the level of recombination between species to be about 100 times greater than the rate of mutation. In *E. coli*, it has been estimated that nearly 20% of its genome may have originated in other microbes. Gene acquisitions performed by *E. coli* have played an important role in the bacterium's ecological plasticity and contributed to its success as a pathogen.

Genome rearrangement is not restricted to *E. coli*, of course. All life forms enjoy this form of genetic "gambling." It is thought that every species of microbe has a *core* gene pool that includes the minimal set of genes needed for independent growth and replication. Every genome is also thought to possess a *flexible* gene pool that is not common to all strains within the species (**Table 9.4**). Therefore, differences in genome size between strains are usually thought to reflect variations in the flexible gene pool brought about by loss (deletion) or gain (insertion) of chromosomal DNA.

Genomic islands are large, transferred genetic elements containing a large number of genes that serve a common function. For example, a **pathogenicity island** is a genomic island that encodes virulence factors. They are found in bacteria pathogenic for plants and animals, but not in related nonpathogenic strains. *Salmonella enterica* var

---

Table 9.4 **Composition of bacterial genomes.**

Core gene pool		Flexible gene pool	
DNA elements	**Encoded features**	**DNA elements**	**Encoded features**
Chromosomes and some plasmids	Ribosomes	Genomic islands	Pathogenicity
	Cell envelope	Genomic islets	Antibiotic resistance
	Key metabolic pathways	Bacteriophages	Secretion
	DNA replication	Plasmids	Symbiosis
	Nucleotide turnover	Integrons	Degradation
		Transposons	Secondary metabolism; restriction-modification; transposases/integrases

Source: From Dobrindt, U., U. Hentschel, J. B. Kaper, and J. Hacker. 2002. Genome plasticity in pathogenic and nonpathogenic enterobacteria. *Current Topics in Microbiology and Immunology* 264:157–175.

Special Topic 9.1 Integrons and Gene Capture

Antibiotic-resistant pathogens are a growing problem throughout the world. One factor contributing to this situation is the movement of drug resistance transposons; another is the existence of **integrons** scattered among various microbial genomes. Integrons are large and complex mobile elements that can accept, or capture, a shopping cart full of different antibiotic resistance gene cassettes to use later when needed. Integrons come in various forms, but in general they

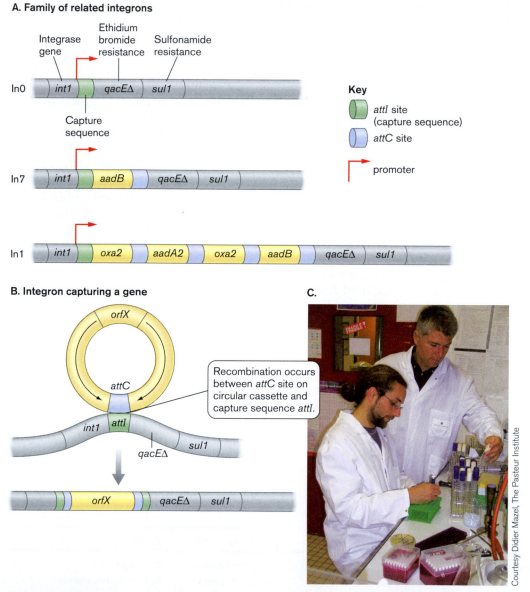

Figure 1 Examples of integrons. A. Family of related integrons. **B.** An integron capturing a gene. **C.** Didier Mazel (right) from the Pasteur Institute studies the molecular mechanism of integron cassette transfer. Here he is seen helping PhD student Guillaume Cambray.

consist of structural genes linked to a 59-bp element (an *attC* site) and assemble as cassettes at an attachment, or capture, site (*attI*). The homologous *attI* and *attC* DNA attachment sites are used to capture the antibiotic resistance cassettes. A strong promoter next to *attI* will drive expression of an adjacent, integrated antibiotic resistance gene. Note that strong promoters initiate transcription more frequently than do weak promoters. Strong promoters have −10 and/or −35 sequences that are closer to the ideal consensus sequence for RNA polymerase binding.

There are at least three classes of integrons based on the type of integrase gene they possess. The most widely distributed integrons are of the first class. All class 1 integrons have, at their 5′ ends, an integrase gene, a capture sequence called *attI* that captures the antibiotic cassettes, and the promoter P next to *attI* (**Fig. 1A**). At the 3′ ends many contain an ethidium bromide resistance locus (*qacEDΔ*) and a sulfonamide resistance gene (*sulI*). The antibiotic resistance genes captured by integrons are located on circular gene cassettes. Although these cassettes can exist free of a chromosome, unlike plasmids they cannot replicate, nor do the antibiotic genes come equipped with a promoter. The cassettes do contain *attC* sequences that allow them to be captured through recombination at *attI* in the chromosomal integron (**Fig. 1B**). Once captured, the cassette closest to the promoter will be expressed to the highest extent. However, the cassettes can be "shuffled" in their position by excision and reintegration. Through this shuffling process, an integron containing several antibiotic resistance genes can, though trial and error, move the resistance gene that best combats the current antimicrobial challenge to a position where it will be more highly expressed (**Fig. 1C**). *Vibrio cholerae*, for example, possesses a super integron carrying over 150 cassettes, each with a variant of an *attC* site.

Typhimurium, for example, contains three pathogenicity islands, each with a role specific to a different stage of disease. Pathogenicity islands are often flanked by boundary regions such as direct repeats or insertion elements, usually found near tRNA genes (**Fig. 9.37**). The association of pathogenicity islands with tRNA genes may reflect the fact that some tRNA genes serve as attachment sites for the integration of some phage DNAs or transposons. Also, some tRNA genes are dispensable since there are often multiple copies of a given tRNA gene within a genome. The pathogenicity island as a whole is often genetically unstable, subject to further rearrangements or deletions. In addition to virulence genes, pathogenicity islands

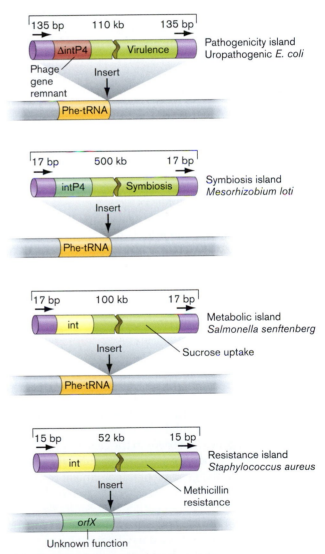

Figure 9.37 Genomic islands from different microorganisms. Pathogenicity islands are often found inserted adjacent to tRNA genes and have direct repeats at their ends (arrows). Functions for each island are shown above the insert.

contain mobility genes encoding transposases, integrases, or other enzymes associated with recombination.

Another kind of genomic island helps a microbe enter into symbiotic relationships. This is called a mobile symbiosis island. For example, *Mesorhizobium loti* is a symbiotic bacterium that lives in nodules on the roots of legumes. This root nodule bacterium derives nourishment from the plant in exchange for fixing nitrogen from the atmosphere that the plant can use. Several species of rhizobia carry the genes for nodulation and nitrogen fixation on large plasmids. Others, like *Bradyrhizobium* species, have symbiosis genes as part of their genomes but clustered in a way that suggests they were once derived from horizontal transfer. These gene clusters appear to have lost whatever mobility they once had. The symbiosis genes of *M. loti,* however, are contained on a mobile symbiosis island. These were the genes horizontally transferred in the Sullivan and Ronson field experiment described in the introduction to this chapter.

Pathogenicity islands and symbiosis islands alike encode remarkable protein secretion systems (called type III and type IV secretion systems) capable of injecting effector proteins directly from the bacterium straight into a eukaryotic cell. Once inside the target cell, these effector proteins alter the function of eukaryotic cell proteins, making the plant, in the case of symbiosis, more accepting of the symbiont. Type III and IV secretion systems are discussed further in Chapter 25.

Other genomic islands are called fitness islands. Fitness islands encode functions that are important to survival in the environment. *E. coli*, for example, possesses an acid fitness island that enables the organism to survive at pH 2, a condition found in the human stomach (see Section 12.1).

Genome Evolution Includes Genome Reduction

It is important to note that evolution toward pathogenicity involves gene loss as well as gene acquisition. Large-scale loss of genes through evolution is known as genome reduction. For example, pathogenic shigellae (the cause of bacillary dysentery) exhibit chromosomal "black holes," regions lacking genes that occur in the closely related *E. coli*. Absence of these genes is required for a fully virulent phenotype. Similarly, the genome of the organism responsible for whooping cough, *Bordetella pertussis*, is 100 kb smaller than the genome of the less pathogenic *Bordetella bronchiseptica*, another example of evolution by genome reduction. Genome reduction is discussed in Chapter 17.

Sometimes, evidence of genome reduction in the making can be observed. Many genomes contain pseudogenes, genes that by homology appear to encode an enzyme but are nonfunctional because a portion is missing through deletion. These appear to be the remnants of genes that were useful to an ancestral species but made superfluous when one of its descendants adapted to a new environmen-

tal niche. When evolutionary pressure to retain a functional form of the gene no longer exists, the imperative to repair mutations within that gene also evaporates. The pseudogene, therefore, is in the process of being eliminated.

Duplications and Divergence Generate New Genes

Gene duplication is the most important mechanism for generating new genes and biochemical processes. Duplication frees a gene from its previous functional constraints and allows divergent evolution through mutation. Duplications can arise in several ways, including transposition of a transposable element and recombination after replication between multiple direct or inverted repeat sequences located in the chromosome. **Figure 9.38** shows a duplication formed by recombination between direct repeats of chromosomal sequences. Superfamilies of proteins arising from divergent evolution share structural and functional features but may catalyze different reactions. One example is the ABC superfamily of ABC transporters. Homology between members of the ABC family may be as little as 10%, suggesting that once an enzyme adopts a new function, sequence diverges rapidly.

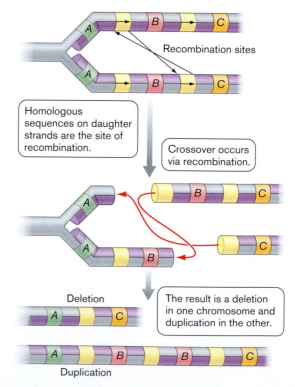

Recombination sites

Homologous sequences on daughter strands are the site of recombination.

Crossover occurs via recombination.

Deletion

The result is a deletion in one chromosome and duplication in the other.

Duplication

Figure 9.38 Duplication during replication. Replicating DNA with three genes *A*, *B*, and *C*. Yellow segments denote areas with sequence homology. Arrows represent orientation. Crossover lines indicate a recombination event between homologous areas in different places on daughter strands. Red arrows indicate where the strands reconnect after recombination.

TO SUMMARIZE:

- **Horizontal gene transfers** between species are thought to have occurred by conjugation, transformation, and transduction.
- A DNA sequence with a **GC content** different from that of flanking chromosomal DNA is one sign of horizontal gene transfer.
- **Pathogenicity islands** are the result of horizontal gene transfers that improve the pathogenicity of a recipient. Fitness islands are similar to pathogenicity islands but improve environmental survival characteristics of the recipient.
- **Superfamilies of functional proteins** result from gene duplications and mutations that cause divergent evolution of function.

THOUGHT QUESTION 9.9   The heat-shock gene *dnaK* in *E. coli* encodes a chaperone (Hsp70), homologs of which are found in all three domains of life. However, there is no other phylogenetic evidence that a *dnaK*-like gene was present in the most recent common ancestor of all organisms. For instance, some species of archaea do not encode a *dnaK* homolog, while other archaeal species do. In fact, molecular phylogenies indicate that the archaeal homologs are quite closely related to those of bacteria. It seems the most recent common ancestor of all *dnaK* (or *hsp70*) genes existed more recently than the common ancestor of all organisms. That being the case, how do you suppose *hsp70* genes arose in archaea?

Concluding Thoughts

Fortuitous movement of genes between species through conjugation (cell-cell contact), transformation (DNA uptake), transduction (phage vector), and transposition (jumping genes) has contributed in a fundamental way to evolution, causing evolutionary leaps and enabling ancestral bacteria to develop, among other things, symbiotic and pathogenic mechanisms that opened up new ecological niches. It has become apparent that DNA repair mechanisms, while designed to prevent the establishment of deleterious mutations, nevertheless allow some mutations to occur. An imperfect DNA repair mechanism offers the cell the opportunity to throw the genetic dice. The vast majority of mutations will be detrimental, but a key few may benefit the organism and enable the variant strain to outcompete its neighbors in its ecosystem. A serious question is raised by our new awareness of genetic mobility: How do we now define a species? For instance, there are numerous strains of *Helicobacter pylori* where large differences in gene sequence and organization are evident. Should each be called a different species? An arbitrary setting of ribosomal RNA gene divergence is used to assign species, but some wonder whether this is still a valid landmark. For further discussion of microbial evolution, see Chapter 17.

CHAPTER REVIEW

Review Questions

1. What are the basic ways microorganisms exchange DNA?
2. Discuss horizontal versus vertical gene transfer.
3. Describe competence and how it comes about in a population.
4. What is an F factor, and how does it (and other factors like it) contribute to gene exchange?
5. What is a merodiploid? Explain how a cell can become a merodiploid.
6. Compare the differences between the activities of competence factor made by *Streptococcus* and pheromones made by *Enterococcus*.
7. What does microbial gene exchange have to do with the plant disease called crown gall disease?
8. Compare specialized versus generalized transduction.
9. Discuss how bacteria protect themselves from invading bacteriophages.

10. What is the value of recombination to a species?
11. List the major proteins that contribute to recombination, and specify their roles in the process.
12. Define Holliday structure, strand assimilation, cointegrate, and cotransduction.
13. List and explain the different types of mutations.
14. How are genotype and phenotype different?
15. Describe six different DNA repair mechanisms. Which ones contribute to mutations?
16. Explain the basic process of transposition. Why are insertion sequences always flanked by direct repeats of host DNA? How are transposons different from plasmids?
17. What is a pathogenicity island? A fitness island? What are their characteristics?

Key Terms

AP endonuclease (329)
AP site (329)
apurinic site (323)
base excision repair (BER) (328–329)
Chi (317)
cointegrate (316)
competence factor (306)
competent (306)
complex transposon (335)
composite transposon (335)
conjugative transposon (335)
cotransduced (319)
deletion (321)
D-loop formation (317)
electroporation (305)
episome (308)
error-prone repair (327)
error-proof repair (327)
F⁻ cell (307)
F⁺ cell (307)
fertility factor (F factor) (307)
F-prime (F′) factor (309)
F-prime (F′) plasmid (309)

frameshift mutation (322)
gain-of-function mutation (321)
generalized recombination (316)
generalized transduction (311)
genomic island (337)
Hfr (309)
Holliday junction (317)
horizontal gene transfer (305)
insertion mutation (321)
insertion sequence (IS) (333)
integrase (320)
integron (338)
inversion (321)
knockout mutation (322)
loss-of-function mutation (321)
merodiploid (309)
methyl mismatch repair (327)
missense mutation (321)
mutagen (323)
mutation (320)
mutation frequency (324)
mutation rate (324)
mutator strain (328)

nonsense mutation (321)
nucleotide excision repair (NER) (328)
pathogenicity island (337)
phase variation (320)
photoreactivation (328)
point mutation (321)
recombinational repair (330)
restriction endonuclease (315)
silent mutation (321)
site-specific recombination (316)
SOS response (331)
specialized transduction (311)
transducing particles (312)
transduction (311)
transformation (304)
transition mutation (321)
translocasome (305)
transposable element (333)
transposase (333)
transposition (334)
transposon (335)
transpososome (334)
transversion (321)

Recommended Reading

Bridges, Bryn A. 2005. Error-prone DNA repair and transle-sion synthesis: Focus on the replication fork. *DNA Repair* **4**:618–619.

Chen, Inês, Peter J. Christie, and David Dubnau. 2005. The ins and outs of DNA transfer in bacteria. *Science* **310**:1456–1460.

Davidson, Tonje, and Tone Tønjum. 2006. Meningococcal genome dynamics. *Nature Reviews Microbiology* **4**:11–22.

Dobrindt, Ulrich, Ute Hentschel, James B. Kaper, and Jörg Hacker. 2002. Genome plasticity in pathogenic and non-pathogenic enterobacteria. *Current Topics in Microbiology and Immunology* **264**:157–175.

Doyle, Marie, Maria Fookes, Al Ivens, Michael W. Mangan, John Wain, and Charles J. Dorman. 2007. An H-NS-like stealth protein aids horizontal DNA transmission in bacteria. *Science* **5**:654–666.

Frost, Laura S., Raphael Leplae, Anne O. Summers, and Ariane Toussaint. 2005. Mobile genetic elements: The agents of open source evolution. *Nature Reviews Microbiology* **3**:722–732.

Iyer, Ravi R., Anna Pluciennik, Vickers Burdett, and Paul L. Modrich. 2006. DNA mismatch repair: Functions and mechanisms. *Chemical Reviews* **106**:302–323.

Kim, Jong-Il, and Michael M. Cox. 2002. The RecA proteins of *Deinococcus radiodurans* and *Escherichia coli* promote DNA

exchange via inverse pathways. *Proceedings of the National Academy of Science* **12**:7917–7921.

Kleckner, Nancy, Rodney K. Chan, Bik K. Tye, and David Botstein. 1975. Mutagenesis by insertion of a drug resistance element carrying an inverted repetition. *Journal of Molecular Biology* **97**:561–575.

Koo, Jovanka T., Juno Choe, and Steve L. Moseley. 2004. HrpA, a DEAH-box RNA helicase, is involved in mRNA processing of a fimbrial operon in *Escherichia coli*. *Molecular Microbiology* **52**:1813–1826.

Mazel, Didier. 2007. Integrons agents of bacterial evolution. *Nature Reviews Microbiology* **4**:608–620.

Murray, Noreen E. 2000. Type I restriction systems: Sophisticated molecular machines (a legacy of Bertani and Weigle). *Microbiology and Molecular Biology Reviews* **64**:412–434.

Schilling, Cristophe H., Markus W. Covert, Iman Famili, George M. Church, Jeremy S. Edwards, and Bernhard O. Palsson. 2002. Genome scale metabolic model of *Helicobacter pylori* 26695. *Journal of Bacteriology* **184**:4582–4593.

Sullivan, John T., and Clive W. Ronson. 1998. Evolution of rhizobia by acquisition of a 500-kb symbiosis island that integrates into a phe-tRNA gene. *Proceedings of the National Academy of Science* **95**:5145–5149.

Thomas, Christopher, and Kaare M. Neilsen. 2005. Mechanisms of, and barriers to, horizontal gene transfer between bacteria. *Nature Reviews Microbiology* **3:**711–721.

Tippin, Brigette, Phuong Pham, and Myron F. Goodman. 2004. Error-prone replication for better or worse. *Trends in Microbiology* **12:**288–295.

Truglio, James J., Deborah L. Croteau, Bennett Van Houten, and Caroline Kisker. 2006. Prokaryotic nucleotide excision repair: The UrrABC system. *Chemical Reviews* **106:**233–252.

Yang, Wei. 2006. Poor base stacking at DNA lesions may initiate recognition by many repair proteins. *DNA Repair* **5:**654–666.

Chapter 10

Molecular Regulation

A bacterial genome encodes thousands of different proteins needed to handle many different environmental contingencies. Some proteins, such as RNA polymerase, are always required for growth; these are called "housekeeping proteins." Most proteins, however, are needed only under a limited set of conditions. Proteins that degrade the sugar lactose, for example, are only useful when lactose is present. Likewise, *Vibrio cholerae* needs to make cholera toxin only when it's growing inside the human body, not when it's living in the ocean. To compete successfully with others, the microbe will not waste energy making unneeded proteins. Cells achieve molecular efficiency using elegant control systems that selectively increase or decrease gene transcription, mRNA translation, or mRNA degradation, as well as by degrading or sequestering regulatory proteins. But how does a cell "know" when to alter expression?

This chapter describes the variety of mechanisms used by bacteria to sense changes in internal and external environments, and adjust gene expression accordingly to maintain biochemical homeostasis and achieve optimum growth.

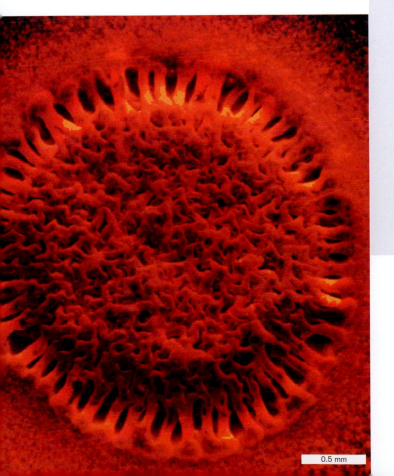

Pseudomonas aeruginosa cells form a wrinkled aggregate (3 mm in diameter) protruding above the surface of swimming motility agar. Autoaggregation is caused by activation of the regulatory protein WspR, which controls expression of cell surface factors, including aggregative fimbriae and exopolysaccharides. Source: David A. D'Argenio, et al. 2002. *Journal of Bacteriology* 184(23):6481.

0.5 mm

During the day, the bobtailed squid, *Euprymna scolopes*, remains buried in the sand of shallow reef flats around Hawaii. After sunset, the animal emerges from its hiding place and begins its search for food. As it swims in the moonlit night, its light organ projects light downward in an apparent attempt to camouflage the squid from predatory fish swimming below. Looking up, the fish would not see the squid, but would instead see what appeared to be stars. What does this have to do with microbiology? Inside the light organ are luminescent bacteria called *Vibrio fischeri*. The bacteria, not the squid, produce the light. However, these microbes do not *constantly* glow. They only light up when their cell density rises above a certain level, at which time the genes needed to make light are "turned on." This intricate level of genetic control by such a small, seemingly simple bacterium is truly amazing.

Microbes use numerous mechanisms to sense their internal and external environments and then translate that information into action. The cell's surface, for example, contains an array of sensing proteins that monitor osmolarity, pH, temperature, and the chemical content of the surroundings or that detect the presence of a host or competitor. When triggered, a sensor phosphorylates itself using ATP and transfers the phosphate to another protein called the response regulator. Response regulators are typically DNA-binding proteins that, when phosphorylated, activate or inactivate the expression of functionally related sets of target genes.

Even more elaborate are the sigma factor cascades and protein localization mechanisms that control developmental processes such as sporulation or asymmetrical cell division (as in the stalked bacterium *Caulobacter*, discussed in Section 4.7). Other systems, such as quorum sensing, in which bacteria secrete and sense chemical signal molecules, make it possible for members of microbial communities to communicate and cooperate. All these mechanisms orchestrate the synthesis of waves of proteins whose concentrations change with changing environments. Even virulence genes have tapped into these sensing systems, using them to determine whether the microbe has entered a susceptible host. This chapter explores the fundamental principles of gene regulation in microbes and discusses how individual systems are woven into a global regulatory network that interconnects many processes throughout the cell.

Why are there so many different mechanisms for controlling microbial gene expression? Part of the answer is evolutionary: If it works, it stays. Another reason is timing: Different mechanisms act over different time scales. Some changes in an environment may require a very rapid increase in the level of a specific protein; other changes may be more forgiving of a delay. Still other situations require sequential expression. Sporulation, for instance, is a differentiation pathway in which it is critical that certain proteins are not made until other prerequisite proteins have been produced. A feature common to all control mechanisms, however, is an ability to sense that something inside or around the cell has changed.

10.1 Regulating Gene Expression

A cell needs to monitor two different compartments in order to know when to alter gene expression and adjust its physiology: the cytoplasm within the cell and the environment outside. Intracellularly, the concentrations of vitamins, amino acids, and nucleotides must be sufficient and balanced to supply the biosynthetic and energetic needs of the cell. This usually requires controlling de novo (new) synthesis of these compounds. The cell must also know what carbon and energy sources are present within the cell in order to assemble the proper catabolic pathway.

The microbe also needs to know whether conditions outside the cell are hazardous, and it needs to distinguish, based on the combination of chemicals present, whether it is floating in a pond of water, in a gastrointestinal tract, or within a mammalian host cell. Once the environment is sensed, the cell can change the repertoire of genes it expresses to meet its needs or to protect itself.

Cells use different mechanisms to sense and respond to conditions within and outside the cell membrane. Sensing conditions within the cell is relatively straightforward. **Regulatory proteins**, also called **regulators**, bind specific small-molecular-weight compounds called ligands to determine their concentrations (**Fig. 10.1A**). Different regulators bind different ligands. For example, one regulator will bind a carbohydrate ligand and alert the cell that a new potential carbon source is available, while a different regulator will sense if there is enough of the amino acid tryptophan present in the cytoplasm to carry out protein synthesis. The ligand, once bound, then alters the ability of the regulatory protein to latch onto specific DNA regulatory sequences located near the promoters of target genes. These DNA regulatory sequences are called **operators** (or operator sequences) if binding down-regulates expression of the target genes; and **activators** (or activator sequences) if binding increases expression. In this way, regulatory proteins can then activate or dampen transcription, depending on the gene. Note that the term "operator" is used generically by some investigators to indicate *any* DNA sequence that binds an activator or a repressor protein. To avoid confusion, we will avoid that usage.

General Concepts of Transcriptional Control by Regulatory Proteins

Genes encoding regulatory proteins are usually, but not always, transcribed separately from the target gene (**Fig. 10.1A**). As noted earlier, regulatory proteins come in two forms: repressors and activators. Repressor proteins bind to operator sequences and prevent the transcription of target genes, an event known as **repression**. However, repression happens in one of two ways, depending on the

**Figure 10.1   General aspects of transcriptional regulation
by repressor and activator proteins.   A.** Schematic drawing of
a regulatory system. The regulatory gene may be located near or
quite far away from the target gene on the chromosome. When it is
located nearby, it is often, but not always, transcribed separately from
the target gene. The product of the regulatory gene (the regulatory
protein) binds to DNA sequences near the promoter of the target
gene and controls whether or not transcription occurs. **B.** Repressor
proteins bind to operator sequences and prevent transcription.
One type of repressor (scenario 1) releases from the operator
after binding a chemical ligand (called an inducer) that is specific
for that repressor. The target gene is thereby induced. The second
type of repressor (scenario 2) must bind a chemical ligand, called a
corepressor, *before* it binds to operator DNA. **C.** Activator proteins
generally bind to specific chemical ligands in the cytoplasm before
the protein can bind to activator DNA sequences near target genes.

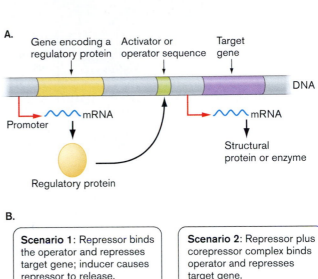

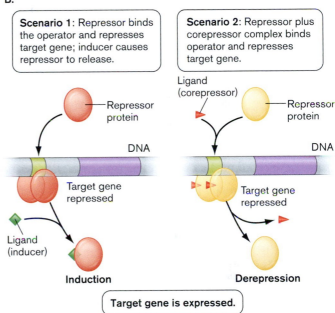

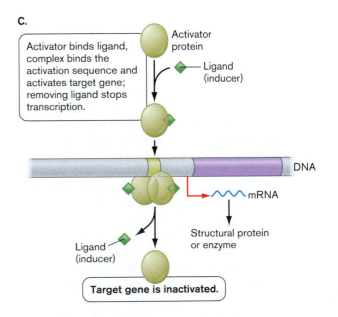

repressor. Some repressors bind operator DNA by themselves and prevent transcription (**Fig. 10.1B**, scenario 1).
Expression of the target gene requires that a specific
ligand, called an **inducer**, bind to the repressor, causing it
to release from the operator. Because a small inducer molecule is required, the increased expression of the target
gene is called **induction**. The lactose operon, discussed
later, is one example of an inducible system.

Other repressor proteins do not bind well to operator DNA unless they first bind a small ligand called a
corepressor (**Fig. 10.1B**, scenario 2). As the ligand disappears from the cell, it is no longer available to bind to
the repressor. When this happens, the repressor releases
from the operator and the target gene is expressed. This
process is called **derepression** rather than induction. The
tryptophan operon, also discussed later, is an example of
a repression/derepression system.

Activator proteins, on the other hand, generally bind
poorly to activator DNA sequences unless an inducer is present (**Fig. 10.1C**). Most activator proteins, once bound to DNA,
directly contact an RNA polymerase molecule stuck at the
nearby promoter, spurring it to initiate transcription. When
the intracellular concentration of inducer falls, the activator
protein (without inducer) either leaves the DNA or moves to
a nearby site from which it can no longer contact RNA polymerase. Thus, transcription of that target gene stops.

Sensing the Extracellular Environment

Sensing what goes on outside the cell is more challenging because intracellular regulatory proteins cannot
reach through the membrane and touch what is outside.
A common mechanism used by gram-positive and gram-
negative organisms to transmit information from outside
to inside the cell relies on a series of two-protein phophor-
ylation relay systems. Each protein pair will activate different sets of genes. The first protein in each relay, the **sensor
kinase** protein, spans the membrane (**Fig. 10.2**). One end
of this sensor protein contacts the outside environment

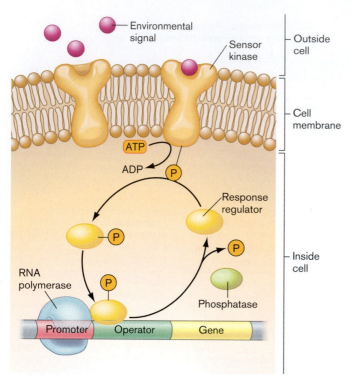

Figure 10.2 Two-component signal transduction systems sense the external environment. A transmembrane sensor kinase protein senses an environmental condition outside the cell (as in gram-positive bacteria) or in the periplasm (as in gram-negative bacteria). Binding of the environmental signal triggers autophosphorylation of the sensor kinase, as shown in the example, and in other instances prevents it. The phosphorylated sensor kinase will then transfer the phosphate to a cognate response regulator protein in the cytoplasm. The response regulator then binds to target operator or activator DNA sequences and inhibits or stimulates gene expression. A phosphatase will downregulate the system by removing the phosphate.

(or periplasm), and the other end protrudes into the cytoplasm. Activating the external sensory domain of the molecule triggers a conformational change in the cytoplasmic part, called the kinase domain. Each sensor protein recognizes a different molecule or condition (for example, PhoQ in *Salmonella* senses magnesium). The altered kinase domain activates a self-phosphorylation reaction that uses ATP to place a phosphate on a specific conserved histidine residue located in a different part of the protein. Then, like two relay runners exchanging a baton, the phosphorylated sensor kinase protein passes the phosphate to a cognate (matched) cytoplasmic protein called a **response regulator**. This transfer, called transphosphorylation, occurs at a specific glutamate residue within the response regulator. The phosphorylated response regulator commonly binds to regulatory DNA sequences in front of one or more specific genes and activates or represses expression. Regulatory relays of this type are called **two-component signal transduction systems**.

Microbes Control Gene Expression at Several Levels

Hundreds of intracellular and extracellular sensing mechanisms monitor the overall health of the cell and its environment. But how do these systems control gene expression?

It is important to note that expression of genes and operons and their products (mRNA, which produces the protein product, or small functional RNAs) can be controlled at various levels. Different levels of control offer different advantages to the cell. The major levels of control can be categorized as DNA sequence controls, transcriptional controls, translational controls, and post-translational modifications. In general, DNA sequence-level control is the most drastic and the least reversible, whereas control at the protein (translational/posttranslational) level offers the most rapid and reversible control. Examples of these control levels are summarized here and then discussed in detail later in the chapter.

- **Alterations of DNA sequence.** Some microbes can program the mutation of DNA so as to activate or disable a particular gene. An example of such a mechanism is phase variation, in which DNA rearrangement involving the reversible flipping of a DNA segment enables a pathogen to turn on or turn off expression of cell surface proteins to evade the host immune system.
- **Control of transcription and mRNA stability.** Many types of gene regulation occur at the level of transcription. The most common mechanisms of transcriptional control in prokaryotes involve operons. Recall that an operon is a string of two or more genes in a chromosome that are expressed from a common promoter in front of the first gene in the operon. Sets of genes in operons are coordinately regulated by protein repressors, activators, and sigma factors, as well as by small RNAs (sRNA). Coordinate regulation means that the expression of all members of the gene set increase or decrease simultaneously. The regulatory proteins recognize and bind to sequences of DNA usually located within 100–200 bp upstream of the regulated genes. Binding of these regulatory molecules determines how much mRNA is made. **Repressor** proteins, when bound to DNA, *prevent* transcription, while activators stimulate it. Levels of specific mRNA molecules are also regulated by RNase activity, which degrades some mRNA molecules as fast as they are transcribed. In some, but not all, instances, sRNA molecules bind RNA transcripts and help or hinder degradation.
- **Translational control.** Translation by ribosomes can be regulated by *translation initiation sequences* in the mRNA, which recognize specific translational repressor proteins. Translational control mechanisms are often coupled to transcriptional mechanisms, offering "fine-tuning" of operon control. Attenuation, for

instance, is a mechanism that uses *translation* to sense the level of an amino acid in the cell and then increase or decrease transcription of the genes encoding the enzymes that synthesize the amino acid.

■ **Posttranslational control.** Once proteins are made, their activity can be controlled by modifying their structure—for example, by protein cleavage, phosphorylation, methylation, or acetylation.

In recent years, studies of bacterial biology have revealed **integrated control circuits** in which all the previously described levels of control work together to coordinately regulate systems throughout the cell. We will discuss examples of integrated control circuits, such as the phage lambda lysis/lysogeny decision and the metabolic/genetic system of nitrogen regulation (see Section 10.7). But first we will explore model examples of control. Most of these examples are at the transcriptional level, which is the level best understood.

TO SUMMARIZE:

■ **Regulatory proteins** help a cell sense changes in its internal environment and alter gene expression to match.

■ **Repressor and activator proteins** bind to operator and activator DNA sequences, respectively, in front of target genes. Repressors prevent transcription, whereas activator proteins stimulate transcription.

■ **Two-component signal transduction** systems also help the cell sense and respond to its environment, both inside and outside.

■ **Gene expression** can be controlled at several levels: DNA sequence, transcription, translation, or by modifying the protein posttranslationally.

■ **Integrated control circuits** connect many individual regulatory systems to coordinate regulation throughout the cell.

10.2 Paradigm of the Lactose Operon

In 1961, French scientists Jacques Monod (1910–1976) and François Jacob (1920–) proposed the revolutionary idea that genes could be regulated (**Figs. 10.3A** and **B**). They, among others, noticed that the enzyme used by *Escherichia coli* to consume the carbohydrate lactose was produced *only* when lactose was added to growth media. The term *induction* was coined to describe this phenomenon. The enzymes required to metabolize glucose, however, were different. They were always (that is, *constitutively*) present. The groundbreaking work of Jacob and Monod, and of André Lwoff (**Fig. 10.3C**) for his study of phage lysogeny, won them the Nobel Prize and launched the field of gene regulation, a scientific realm where discoveries continue to surprise us. Still, it took many years after Monod and Jacob's discovery to learn exactly how the lactose-degrading enzyme β-galactosidase is induced. Many of the concepts presented here hold for numerous bacterial gene systems.

Lactose Induces Expression of the *lacZYA* Operon

Figure 10.4A shows the genes in *E. coli* that encode the simple regulatory circuit for lactose catabolism. Catabolism is a general term to describe the degradation of an organic food source. The genes *lacZ*, *lacY*, and *lacA* form an **operon**, which is a group of genes cotranscribed from

A.

B.

C.

Figure 10.3   Discoverers of gene regulation.   Jacques Monod (**A**), François Jacob (**B**), and André Lwoff (**C**) all worked at the Pasteur Institute and won the Nobel Prize in 1965 in Physiology or Medicine for their groundbreaking work on induction and gene regulation.

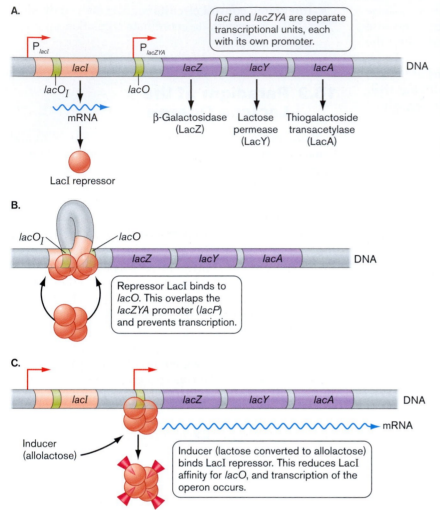

A.

lacI and *lacZYA* are separate transcriptional units, each with its own promoter.

β-Galactosidase (LacZ)

Lactose permease (LacY)

Thiogalactoside transacetylase (LacA)

LacI repressor

B.

Repressor LacI binds to *lacO*. This overlaps the *lacZYA* promoter (*lacP*) and prevents transcription.

C.

Inducer (allolactose)

Inducer (lactose converted to allolactose) binds LacI repressor. This reduces LacI affinity for *lacO*, and transcription of the operon occurs.

D.

Binding site for CRP

Promoter (binding site for RNA polymerase)

Operator (binding site for *lac* repressor)

Start

−35   −10   +1

AATTAATGTGAGTTAGCTCACTCATTAGGCACCCCAGGCTTTACACTTTATGCTTCCGGCTCGTATGTTGTGTGGAATTGTGAGCGGATAACAATTTCACACAGGAAACAGCTATG
TTAATTACACTCAATCGAGTGAGTAATCCGTGGGGTCCGAAATGTGAAATACGAAGGCCGTGCATACAACACACCTTAACACTCGCCTATTGTTAAAGTGTGTCCTTTGTCGATAC

Start of *Z* gene

mRNA

Z gene encodes β-Galactosidase

Figure 10.4 Transcriptional induction of the lactose operon.   A. Organization of the operon. Bent arrows mark promoters. Green boxes indicate LacI protein-binding sites on DNA. **B.** The LacI tetrameric repressor binds to specific DNA sites (the operator; *lacO*) **C.** The inducer allolactose (an altered form of lactose made by low levels of β-galactosidase) removes the repressor LacI and allows expression of *lacZYA*. **D.** DNA sequence of *lac* control region. Boxes enclose sequences important to binding of CRP (defined later), RNA polymerase, and LacI. ▶❚❚

a common promoter. The *lacY* gene encodes lactose permease (LacY), an integral membrane protein that imports (transports) lactose from the extracellular environment. The product of *lacZ*, β-galactosidase (LacZ), intracellularly cleaves the disaccharide lactose into its component parts, glucose and galactose (**Fig. 10.5**). Both of these sugars are subsequently degraded by the enzymes of glycolysis (Section 13.4). To break down lactose in this way requires synthesis of both β-galactosidase and lactose permease; without these two gene products, the catabolic energy of lactose is unavailable to the cell.

The role of *lacA*, encoding thiogalactoside transacetylase (LacA), is unclear, but it is not needed to ferment lactose. One function may be to detoxify a harmful by-product of lactose metabolism. As we can see, even

well-studied systems such as the *lac* operon still pose unanswered questions.

NOTE: Lactose operon, *lac* operon, and *lacZYA* operon all refer to the same system.

THOUGHT QUESTION 10.1   If the gene *lacZ* has a nonsense mutation in its open reading frame, will *lacY* be translated?

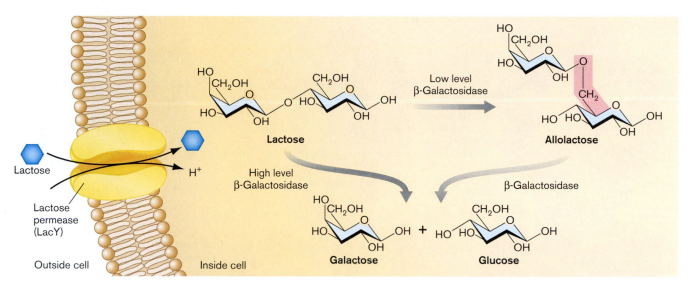

Figure 10.5 Lactose transport and catabolism. A dedicated lactose permease uses proton motive force to move lactose (and a proton) into the cell. Once there, the enzyme β-galactosidase can cleave the disaccharide into its component parts, galactose and glucose, or alter the linkage between the monosaccharides to produce allolactose, an important chemical needed to induce the genes that encode this pathway.

In the absence of lactose, the *lac* operon is transcribed at extremely low levels, generating less than ten molecules of β-galactosidase per cell. The reason is that transcription of *lacZYA* is repressed by the protein product of the regulator gene, *lacI*. The gene *lacI* is situated immediately upstream of *lacZYA* but is transcribed from a different promoter. A tetramer of LacI repressor protein forms in the cell and binds to two regions of DNA called operators. One operator sequence is called *lacO*, which partially overlaps the *lacZYA* promoter (P$_{lacZYA}$). The second operator site occurs within *lacI* and is called *lacO*$_I$ (**Fig. 10.4A**). The *lac* operators control whether the three structural genes are transcribed. The LacI tetramer simultaneously binds the *lacO* and *lacO*$_I$ operators (a dimer at each site). This causes the intervening DNA to loop out (**Figs. 10.4B** and **10.6A**) and prevents RNA polymerase from continuing transcription into the structural *lacZYA* genes. (A third binding site for LacI exists but is not important to our discussion here.)

Once lactose is added to the medium, the *lac* operon is expressed at 100-fold higher levels. How does lactose get into the cell to induce the operon if the operon encoding its transport protein is not expressed? It turns out that even when the *lacZYA* operon is uninduced, it is constitutively expressed at a low level. This means that lactose can be transported into the cell in low amounts because of constant low-level expression of the lactose transporter LacY. The tiny amount of β-galactosidase (LacZ) that is expressed in uninduced cells is also important. At this low level, the enzyme does not completely *cleave* the glycosidic bond of lactose but rearranges it to form **allolactose**, the form of lactose that activates the operon (the structure of which can be seen in **Fig. 10.5**). Allolactose binds the repressor and "unlocks" the protein, so it is released

from the operator (**Fig. 10.4C**). Once this happens, sigma D (the housekeeping sigma factor of *E. coli*, not shown in the figure) can find the *lac* promoter sequences and enable RNA polymerase to initiate transcription of the *lacZYA* structural genes. The roles of cAMP and CRP will be described later.

Glucose Represses the lac Operon

What happens if, in addition to lactose, the medium also contains an alternative carbon source, such as glucose? Enzymes for glucose catabolism (glycolysis) are always produced at high levels because glucose is the favored catabolite (carbon source), providing the quickest source of energy. Many carbohydrates such as lactose must first be converted to glucose to be catabolized. So, should *E. coli* forestall induction of the *lac* operon while glucose is present, in the interest of greater efficiency? In fact, this is precisely what happens. The repression of an operon enabling catabolism of one nutrient by the presence of a more favorable catabolite (commonly glucose) is known as **catabolite repression**.

When glucose and lactose are both present in the medium, cells grow by breaking down glucose until the glucose is depleted; then growth stops. At this point, the cells sense a need for carbon and induce the *lacZYA* operon (because lactose is present). This allows them to begin consuming lactose and reinitiate growth. The biphasic curve of a culture growing on two carbon sources is often called **diauxic growth** (**Fig. 10.7A**). But, if lactose was present from the start, why was *lacZYA* turned off? The answer is that glucose actively *prevents* the induction of *lacZYA*. **Figure 10.7B** illustrates that even when *lacZYA* is already induced, adding glucose stops (or represses) induction.

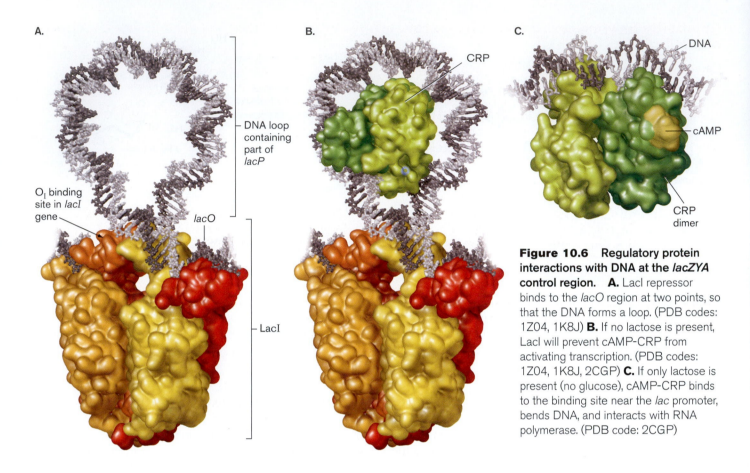

A.

B.

C.

O$_I$ binding site in *lacI* gene

lacO

LacI

DNA loop containing part of *lacP*

CRP

DNA

cAMP

CRP dimer

Figure 10.6 Regulatory protein interactions with DNA at the *lacZYA* control region. **A.** LacI repressor binds to the *lacO* region at two points, so that the DNA forms a loop. (PDB codes: 1Z04, 1K8J) **B.** If no lactose is present, LacI will prevent cAMP-CRP from activating transcription. (PDB codes: 1Z04, 1K8J, 2CGP) **C.** If only lactose is present (no glucose), cAMP-CRP binds to the binding site near the *lac* promoter, bends DNA, and interacts with RNA polymerase. (PDB code: 2CGP)

A.

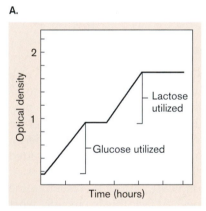

B.

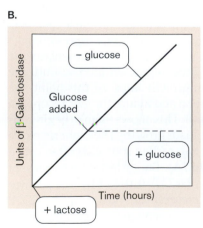

Figure 10.7 Catabolite repression of the *lacZYA* operon. **A.** The diauxic growth curve of *E. coli* growing on a mixture of glucose and lactose. **B.** Glucose repression of LacZ (β-galactosidase) production. Lactose was added at the beginning of the experiment into parallel cultures, and β-galactosidase activity was measured. At the point indicated, glucose was added to one culture. The synthesis of β-galactosidase continues to increase in the culture without glucose but stops in the culture with glucose. **C.** Although the inducer allolactose removes the repressor LacI and allows expression of *lacZYA*, maximum expression requires the presence of cAMP and cAMP receptor protein (CRP, also known as CAP for *c*atabolite *a*ctivator *p*rotein), which bind to a separate site at the *lacZYA* promoter. Bound CRP can interact with RNA polymerase and increase the rate of transcription *initiation*.

C.

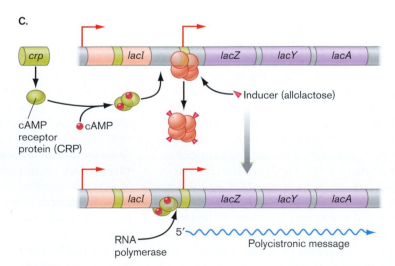

Catabolite Repression Changes the Level of Cyclic AMP

Catabolite repression of the lactose operon by glucose occurs because growth of *E. coli* on glucose prevents the synthesis of the nucleotide cyclic AMP (cAMP). This molecule, which activates numerous operons, including *lacZYA*, is a derivative of AMP in which the 5′ phosphate is linked to the 3′ OH group of the ribose, making a cyclic bond. Cyclic AMP controls the expression of many genes by combining with a dimeric regulatory protein called cAMP receptor protein (CRP). The cAMP-CRP complex binds to specific DNA sequences located near many bacterial genes and modifies their transcription, usually acting as an activator. This is the case for the *lacZYA* operon (**Figs. 10.7C** and **10.4C**).

> **NOTE:** The original name for CRP was catabolite activator for protein (CAP), given to reflect its role in catabolite repression. Though some investigators still use the CAP designation, the term CRP is generally preferred today. Thus, the gene encoding the protein is called *crp*.

Glucose inhibits cAMP production by a clever mechanism that ties the synthesis of cAMP, catalyzed by adenylate cyclase (**Fig. 10.8**), to the transport of glucose. Many sugars, including glucose, are transported by the multicomponent phosphotransferase system (PTS; see Section 4.2). This system transfers a phosphate group from phosphoenolpyruvate (PEP), along a series of protein intermediates, ending with different membrane transport proteins that import specific carbohydrates. The carbohydrate-specific membrane proteins pass the phosphate to the sugar during transport. The glucose-specific PTS relay includes a cytoplasmic protein called IIA^{Glc} and the integral membrane protein IIBC^{Glc}. Protein IIA^{Glc} is the key to glucose repression.

> **NOTE:** IIA^{Glc} and IIBC^{Glc} are also referred to as Enz IIA and Enz IIBC in the literature.

Glucose is changed to glucose 6-phosphate during its transport through the PTS system. Thus, as the cell imports glucose, phosphate is continually removed from IIA^{Glc}, making unphosphorylated IIA^{Glc} the predominant form of the protein. This is important because unphosphorylated IIA^{Glc} interacts with and inhibits adenylate cyclase activity (**Fig. 10.8A**). The net effect of inhibiting adenylate cyclase is that cAMP levels fall. This happens because cAMP synthesis stops and because cAMP is continually degraded by cAMP phosphodiesterase. This is the basis of glucose repression.

If, however, glucose is not in the medium and therefore is unavailable for transport, the phosphorylated form of IIA^{Glc} accumulates because there is no sugar to receive the phosphate (**Fig. 10.8B**). Because the phosphorylated protein IIA^{Glc}-P does *not* interfere with adenylate cyclase, cAMP levels rise and complex with CRP, enabling CRP-dependent, catabolite-repressible genes such as the *lac* operon to be expressed. Thus, catabolite repression by glucose is linked to glucose transport in *E. coli*.

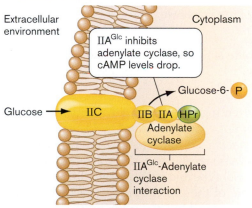

A. Glucose present

IIA^{Glc} inhibits adenylate cyclase, so cAMP levels drop.

B. Glucose absent

Figure 10.8 Control of adenylate cyclase via the phosphotransferase system. A. When glucose is present, the phosphorylated forms of IIA^{Glc} are low because glucose siphons off the phosphate. IIA^{Glc} then interacts with and inhibits adenylate cyclase activity. **B.** In the absence of glucose, the phosphorylated forms of glucose-specific IIA^{Glc} and IIBC^{Glc} accumulate because they cannot pass the phosphate to substrate (there is no glucose). Adenylate cyclase functions in this situation to produce cAMP. The inset on the right shows the conversion of ATP to cyclic AMP by adenylate cyclase. The figure illustrates control over adenylate cyclase activity.

In contrast to gram-negative bacteria, cAMP is not generally found in gram-positive bacteria. Catabolite repression in these organisms is mediated by other mechanisms.

Activation of Transcription by cAMP-CRP

How does the cAMP-CRP complex ultimately activate the expression of *lacZYA*? RNA polymerase cannot easily form an open complex and is essentially "stuck" at *lacP* and other CRP-dependent promoters, even in the absence of a repressor. This problem is overcome when the cAMP-CRP complex binds to a 22-bp DNA sequence located –60 bp from the start of transcription (just upstream of the *lacZYA* operator), which causes the DNA to curve (see **Figs. 10.4D** and **10.6C**). CRP can then directly interact with RNA polymerase bound at the *lac* promoter.

Interaction with RNA polymerase at the *lac* promoter and at the promoters of several other genes

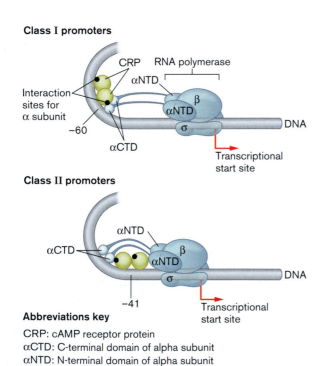

Class I promoters

Interaction sites for α subunit

CRP   RNA polymerase
αNTD
β
αNTD
–60
σ
αCTD
DNA
Transcriptional start site

Class II promoters

αNTD
αCTD
β
αNTD
σ
–41
DNA
Transcriptional start site

Abbreviations key

CRP: cAMP receptor protein
αCTD: C-terminal domain of alpha subunit
αNTD: N-terminal domain of alpha subunit

Figure 10.9   CRP interactions with RNA polymerase. There are three classes of CRP-dependent promoters, which differ as to where CRP binds relative to the promoter region. Class I promoters, like the one at the *lacZYA* operon, possess a CRP-binding site positioned around –60 bp from the transcriptional start. CRP on these promoters interact with the alpha subunit C-terminal domain (αCTD) of RNA polymerase. The CRP-binding site on class II promoters (–41 bp) overlaps the RNA polymerase (RNAP)-binding site, allowing CRP to interact with the C-terminal and N-terminal domains of the alpha subunit. CRP binds the third class of promoter (not shown) at –90 bp. *Source:* Adapted from Moat, Foster, and Spector. 2002. *Microbial Physiology,* 4th ed.

typically occurs at the C-terminal domain of the RNA polymerase alpha subunit (**Fig. 10.9**, class I promoters). With some other promoters affected by cAMP, CRP can also engage the N-terminal domain of the alpha subunit (Fig. 10.9, class II promoters). These protein-protein interactions stimulate RNA polymerase bound at the promoter to initiate transcription. Notice in Figure 10.6B that when neither glucose nor lactose is present, cAMP-CRP cannot by itself activate *lacZYA* transcription because it binds within the DNA loop formed by the LacI tetramer, so there is no room for RNA polymerase to bind to DNA.

There are three classes of promoters requiring CRP activation, with each class differing in where CRP binds. CRP binds class I, II, and III promoters at –60 bp, –41 bp, and –90 bp, respectively, from the start of transcription. The differences in binding location reflect the amount of "encouragement" a particular RNA polymerase–promoter complex must receive from CRP to change from a closed promoter complex to the open promoter complex necessary for transcription. The C terminus of the RNA polymerase alpha subunit can bind DNA sequences near certain promoters and inhibit RNA polymerase movement. When cAMP-CRP touches regions in the C terminus of the alpha subunit, this inhibition is relieved. This contact is enough to stimulate class I promoters (such as *lacP*). In class II promoters, however, where CRP binding overlaps the –35 promoter region, CRP also touches the N terminus of the alpha subunit, which, along with contact at the C terminus, is needed to stimulate transcription. CRP essentially strengthens a weak –35 site. Transcription of class III promoters requires activator proteins in addition to CRP (see the discussion of the arabinose operon in Section 10.3).

WWW | Catabolite activator protein, Biomolecules at Kenyon

NOTE: In presenting a genotype, gene names written *with* a superscript + are considered wild-type. Gene names written *without* a superscript + are considered mutant genes. Also recall that writing a gene name in roman (not italic) type and with a capital first letter is really the name of the protein product of that gene. Thus, *lacZ* is the gene, but LacZ is the protein.

THOUGHT QUESTION 10.2   Null mutations completely eliminate the function of a mutated gene. Predict the effect of the following null mutations on the induction of β-galactosidase by lactose and predict whether the *lacZ* gene is expressed at high or low levels: *lacI, lacO, lacP, crp, cya* (the gene encoding adenylate cyclase). Also, what effect will those mutations have on catabolite repression?

THOUGHT QUESTION 10.3 Predict what will happen to the expression of *lacZ* under the following partial diploid conditions (see Section 9.1). The genotypes of these strains are presented as follows: chromosomal genes/plasmid gene. (a) *lacI lacO⁺P⁺Z⁺Y⁺A⁺*/plasmid *lacI⁺*; (b) *lacO lacI⁺P⁺Z⁺Y⁺A⁺*/plasmid *lacO⁺*; (c) *crp lacI⁺O⁺P⁺Z⁺Y⁺A⁺*/plasmid *crp⁺*.

Glucose Also Causes Inducer Exclusion

Failure of lactose to induce *lacZYA* during growth on glucose (**Fig. 10.7B**) is partly due to the effect of the PTS protein IIA^Glc on cAMP levels, but it is not the whole story. Growth on glucose also keeps lactose out of the cell. This phenomenon is known as **inducer exclusion**. As already noted, glucose transport results in the accumulation of the unphosphorylated PTS protein IIA^Glc in the membrane. The unphosphorylated form of the protein binds to and inhibits the activity of certain non-PTS carbohydrate permeases, like LacY, the transporter for lactose. Shutting transport down keeps the inducer out, which, in the case of lactose, prevents *lacZYA* expression. The precise nature of the communication between these two proteins is not well understood.

THOUGHT QUESTION 10.4 Researchers often use isopropyl-β-D-thiogalactopyranoside (IPTG) rather than lactose to induce the *lacZYA* operon. IPTG structure resembles lactose, which is why it can interact with the LacI repressor, but it is not degraded by β-galactosidase. Why do you think the use of IPTG is preferred in these studies?

TO SUMMARIZE:

- **The lactose utilization *lacZYA* operon** of *E. coli* was the first gene regulatory system described.
- **LacI binds as a tetramer** to the operator region and represses the *lac* operon by preventing open complex formation by RNA polymerase.
- **β-galactosidase (LacZ)**, when at low concentration, cleaves and rearranges lactose to make the inducer allolactose.
- **Allolactose complexes to LacI**, reducing repressor affinity for the operator and allowing induction of the operon.
- **The cAMP-CRP complex** binds to the *lac* operator region and stimulates RNA polymerase through CRP interaction with the C-terminal domain of the RNA polymerase alpha subunit. cAMP-CRP regulates many types of operons.
- **Catabolite repression** is a process whereby a preferred carbon source, such as glucose, will prevent induction of the *lac* operon by lactose. Catabolite repression works by lowering cAMP levels.

- **Glucose transport through the phosphotransferase system** inhibits adenylate cyclase, the enzyme that makes cAMP, and inhibits function of the LacY permease.

10.3 Other Systems of Operon Control

The lactose operon is not the only model of regulation. Many other mechanisms that regulate genes and operons are utilized by bacteria. In this section, we discuss proteins that have dual regulatory functions—that perform as both activators (positive control) and repressors (negative control). We will also see how the coupling of transcription and translation is used to regulate gene expression by attenuation and how an idling ribosome sends a chemical signal that affects the expression of many genes and operons.

Repression of Anabolic (Biosynthetic) Pathways

Many different gene systems use repressor proteins to inhibit transcription. Repressing biosynthetic pathways is fundamentally different from repressing catabolic (degradative) systems. Repressor proteins that control catabolic pathways, such as lactose degradation, typically bind the initial substrate or a closely related product (for example, allolactose in the case of the *lac* operon). Binding the substrate decreases repressor protein affinity for operator DNA. Thus, increased concentration of the substrate or inducer actually removes the repressor from the operator and *derepresses* expression of the operon. This derepression makes sense because the cell wants to make the enzymes that use the substrate as a carbon and energy source.

In contrast, genes encoding biosynthetic enzymes are regulated by repressors (called inactive aporepressors) that must bind the end product of the pathway (for example, tryptophan for the *trp* operon) to become active repressors. The end product that binds the aporepressor is called a corepressor. Binding of the corepressor (end product) to the repressor increases the repressor's affinity for the operator sequence upstream of the target gene or operon. As the concentration of end product, such as an amino acid or nucleotide, increases in the cell beyond that needed to support growth, the cell will shut the biosynthetic pathway down and not waste energy making a superfluous pathway or compound.

The AraC Regulator Activates and Represses Transcription

The *lac* operon regulatory paradigm is a wonderful way to regulate gene systems involved in catabolism. What could

Special Topic 10.1   How Do We Study Protein-DNA Binding?

A long series of painstaking experiments led to the discovery of how the *lac* operon is regulated, but current techniques have made it much easier to study how DNA and proteins interact and how these interactions control gene expression. Three experimental techniques are widely used to show whether a protein binds DNA and to determine where it binds.

Sequence Analysis

DNA sites that bind regulatory proteins typically exhibit a sequence symmetry that allows unambiguous recognition. Dimers of a regulatory protein bind to the DNA, with each member of the duo binding half of the symmetrical DNA sequence. Symmetry usually involves an inverted repeat (**Fig. 1**). Once a potential regulatory protein and a presumed target gene have been identified, the DNA sequence extending 200–300 bp upstream of the target gene's transcriptional start site is examined using computer programs designed to reveal possible binding sites. Finding an inverted repeat within 100–200 bases of a promoter suggests the presence of a regulatory binding site. Finding very similar sequences in front of two or more different promoters controlled by the same regulatory protein is stronger evidence that, in fact, the sequences represent regulatory binding sites. Proof ultimately requires showing that the protein will bind a piece of DNA containing this sequence. Protein-DNA binding can be demonstrated in several ways, as described below.

Electrophoretic Mobility Shift Assay

DNA fragments containing the suspected binding site are generated by PCR and radiolabeled. The labeled target fragment is then mixed in vitro with purified regulatory protein and subjected to electrophoresis through a nondenaturing polyacrylamide gel. The fragment alone will travel quickly through the gel matrix, but if a DNA-protein complex forms in the lane loaded with protein *and* the DNA fragment, it will travel more slowly due to its larger size. Thus, a shift in position of a DNA band indicates DNA-protein binding. This technique is called an **electrophoretic mobility shift assay (EMSA)** or gel shift (**Fig. 2**).

DNA Footprinting

Once a DNA fragment that binds the regulator has been identified, it is useful to determine the specific DNA sequences

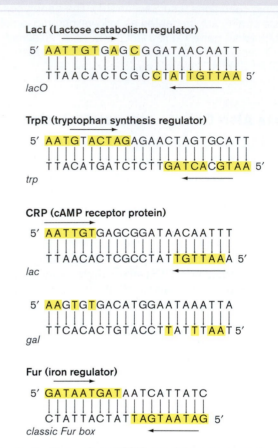

Figure 1   Examples of DNA regulatory sequences. The sequences shown are located upstream of the genes noted in italics. Inverted repeats are shown in yellow. Note that in some inverted repeats, the occasional base is not repeated (only those bases that repeat are highlighted in the diagram). Arrows indicate direction of symmetry.

that contact the protein. One way to define where a regulatory protein contacts DNA is to allow the protein to bind the DNA and then to treat the complex with a DNase (**Fig. 3**). The area of DNA in close contact with the regulatory protein will be protected from the cleaving enzyme.

The basic procedure is to radiolabel the 5′ end of *one* strand before adding DNase. Just enough DNase is added so that each individual DNA molecule is cut only once. The result-

be better than having a substrate induce expression of the operon needed to catabolize it? How about a regulator that can repress *or* activate gene expression, depending on whether substrate is available? This would provide

very tight control over the synthesis of catabolic enzymes and enable an accelerated induction. One such regulator, called AraC, regulates the genes encoding arabinose catabolism. When arabinose is absent, AraC represses

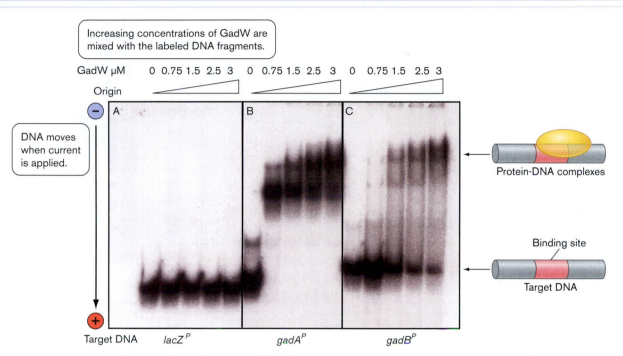

Figure 2   Electrophoretic mobility shift assay (EMSA).   To test DNA-protein interactions, target DNA fragments (roughly 50–1,000 bp in length) are labeled at the 5′ end with a radioactive phosphate group (^{32}P) from ATP. The fragment may then be visualized using X-ray film. Varying concentrations of the protein are mixed with the labeled DNA fragments and loaded at the top of a polyacrylamide gel. Because DNA is negatively charged, it will move toward the positive pole (bottom of gel) once a current is applied. Free (unbound) DNA will travel most quickly down the gel, while DNA bound to protein is larger and travels more slowly. This figure illustrates the addition of increasing concentrations of a protein regulator called GadW to three DNA fragments: the promoter regions of *lacZ* (panel **A**) and the two genes encoding glutamate decarboxylase, *gadA* (panel **B**) and *gadB* (panel **C**). Glutamate decarboxylase is an enzyme that contributes to *E. coli*'s resistance to extreme acid (pH 2) environments. After electrophoresis, the gel is dried by vacuum and exposed to X-ray film. No shift in fragment mobility is seen in panel A, meaning GadW does not bind to the promoter for *lacZ*. The progressively slower band movements seen in panels B and C with increasing concentrations of GadW indicate that this regulator will bind the *gad* promoters. *Source:* A–C. ©American Society for Microbiology, *Journal of Bacteriology.* 2002. 184:7001–7012.

ing fragments of varying size are then separated by gel electrophoresis and visualized by autoradiography. A DNA sequence ladder of the same strand is run parallel to the DNase-treated lanes. (See DNA sequencing, Fig. 7.29.) A DNase-treated band and a DNA sequence band that are the same size will end with the same base. So a band in the DNase lane whose sequence matches in size a G (guanosine)-ending band in the DNA sequence lane will itself end in G. Ideally, unprotected

DNA cut with DNase will yield as many fragments as it has bases. DNA protected by a binding protein, however, will leave a "footprint" where the binding protein comes in close proximity to DNA. Because DNase cannot cut DNA protected by the protein, bands generated by cuts in the region of the DNA that binds to the protein will disappear. Thus, by observing the DNA sequencing lanes, the sequence of DNA covered or protected by the protein can be determined.

(*continued on next page*)

expression of the genes that break down arabinose; when arabinose is present, AraC activates these same genes. The products of these genes ultimately convert the five carbon sugar L-arabinose to D-xylulose 5-phosphate, an interme-

diate in the pentose phosphate shunt, a pathway that provides reducing energy for biosynthesis (see Section 15.2). It turns out that many regulatory systems, including some virulence regulators, use AraC-type regulation.

Special Topic 10.1 How Do We Study Protein-DNA Binding? (*continued*)

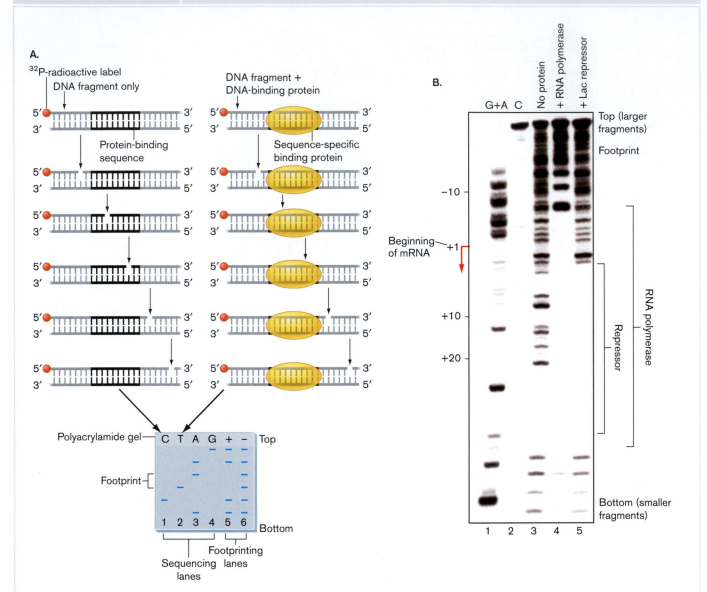

Figure 3 DNA footprinting. A. DNA fragments are labeled with ^{32}P at just one 5′ end to tag one strand. The DNA–binding protein (for example, a repressor or activator protein) is added to the appropriate tube (right side). Mixtures with binding protein (protected) and without binding protein (unprotected) are treated with a DNase. The area of DNA bound by the regulatory protein will be protected from DNase digestion. Just enough DNase is added to nick each DNA molecule only once (indicated by vertical arrows). Each molecule will be a different size, depending on where the enzyme cut. The mixtures are loaded onto a polyacrylamide gel (bottom) to separate the fragments by size. The original uncut fragment is sequenced and loaded into adjacent lanes (lanes 1–4). The size of each footprint fragment (lanes 5 and 6) can be compared with the sequencing ladder to determine at which base the DNase has cut. **B.** An actual footprint showing where RNA polymerase and LacI bind to the *lacP* partial sequence (lanes 1 and 2). In the figure, lane 1 shows fragments that end in G or A and lane 2 shows fragments that end in C. The unprotected digest of the DNA fragment is in lane 3, the DNA fragment plus RNA polymerase is in lane 4, and the DNA fragment plus LacI repressor protein is in lane 5. Note that the area protected by the repressor overlaps the area protected by RNA polymerase. From this, one can conclude that repressor bound to the DNA will prevent RNA polymerase access to this promoter. *Source*: B. Albert Schmitz and David J. Galas, Department de Biologie Moleculaire, University de Geneve.

The *ara* operon. Part of the operon encoding the arabinose degradation enzymes is shown in **Figure 10.10A**. Note that the *ara* operon is transcribed divergently from a central promoter region. The structural genes (*araBAD*) are transcribed in one direction and the regulator gene, *araC*, is transcribed in the opposite direction. The product of *araC* is a 33-kDa protein (AraC) with two domains: the C-terminal end and the N-terminal end (**Fig. 10.10B**). The C-terminal end contains a DNA-binding domain that contains two helix-turn-helix motifs. Helix-turn-helix motifs are made up of two alpha helices separated by a short amino acid chain that allows a bend. Usually, the helix closer to the C

terminus fits into the major groove of the target DNA. The C-terminal DNA-binding domain is attached by a flexible linker peptide to an N-terminal dimerization domain. AraC forms a dimer in vivo that can assume one of two conformations depending on whether arabinose is available. When arabinose is absent, the dimer exists in a rigid, elongated form that represses expression of the machinery for arabinose degradation (*araBAD*). When arabinose is present, it binds to the dimer, which then assumes a more compact form that activates expression of *araBAD* (**Fig. 10.10C**).

The flexible N-terminal arm of an AraC monomer is the key to determining what type of dimer forms. When

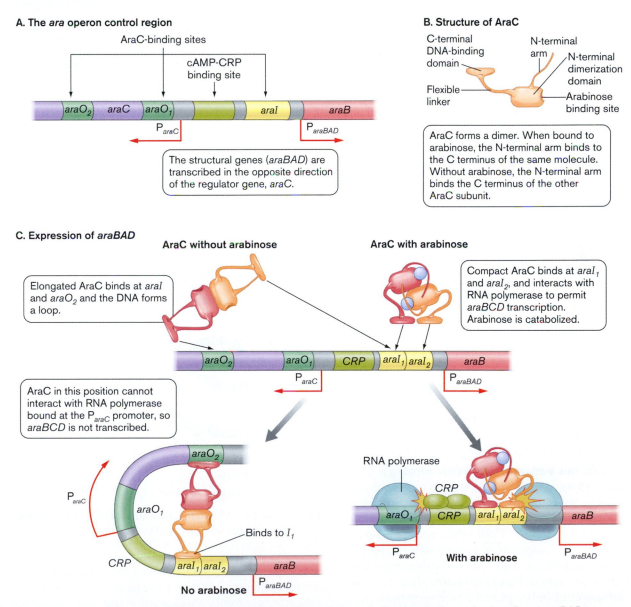

Figure 10.10   Regulation of *araBAD* operon by the AraC regulator.   A. Divergently transcribed *araC* and *araBAD* operons showing binding sites for the AraC and CRP proteins. The region shown is about 400 bp. **B.** View of AraC showing dimerization and DNA-binding domains. **C.** Alternative conformations of AraC dimers. The different conformations change the location of where the dimer can bind DNA. (Note that O_1 and O_2 are operators, and I_1 and I_2 are other DNA-binding sites that, when occupied by AraC, induce expression.)

arabinose is not present, the N-terminal arm of an AraC monomer binds to the C domain of the same molecule, forming a rigid, extended structure. However, when arabinose docks to its binding site in the AraC protein, the N arm can bend and bind to the C domain of its partner protein, producing a more compact structure. **Figure 10.10C** shows the alternative dimerization configurations of the AraC N-terminal domain in the absence and presence of L-arabinose.

So how does AraC act as both repressor and activator? First, unlike the *lac* repressor, AraC does not function by denying RNA polymerase access to the promoter. Instead, AraC represses *araBAD* expression by producing a looped DNA structure that limits activation by cAMP-CRP, which, as discussed in Section 10.2, is a global regulator of many bacterial genes. Then, when placed in the right position on the DNA, AraC, as well as CRP, can activate *araBAD* transcription by physically contacting an RNA polymerase complex that is already bound to the promoter.

There are four DNA sites to which AraC can bind, as illustrated in **Figure 10.10C**. With arabinose absent, the elongated, repressor form of the AraC dimer (AraC$_2$) can only bind the widely separated sites I$_1$ and O$_2$, causing the intervening DNA to form a loop. This positions AraC too far from the *araBAD* promoter to ever interact with RNA polymerase. The loop also prevents CRP from interacting with RNA polymerase. However, the compact, inducer form of AraC$_2$, formed by the binding of arabinose, can bind to the DNA sites I$_1$ and I$_2$ (I represents inducer site), which are closer to the promoter. This places AraC closer to an RNA polymerase idling on the *araBAD* promoter and allows the two complexes to touch. This contact releases RNA polymerase from the promoter and allows it to move into and transcribe the *araBAD* structural genes. This position also allows bound CRP to activate expression. **Figure 10.11** illustrates the structural effect of arabinose on the N-terminal dimerization domain of AraC. The binding of arabinose to AraC engineers dimerization through interactions between coiled-coil alpha-helical regions of the protein.

Advantages of the activator-repressor strategy. This system offers an advantage over a simple repressor circuit. In the *lac* operon, LacI repressor must fully dissociate from operator DNA during induction and ends up dispersing throughout the cell, often nonspecifically binding to other DNA sequences. Reestablishing repression requires random diffusion of LacI protein back to the *lacO* DNA sequence. This takes some time because random diffusion of bulky proteins in the cytoplasm is slow. In contrast, AraC protein rarely leaves the vicinity of the *ara* operon. The response of AraC to varying levels of arabinose allows AraC to simply shuttle back

and forth between repressor and activator sites, shortening the delay between induction and repression. The inducer arabinose is relatively small; thus, it can diffuse quickly throughout the cell to find AraC positioned at the promoter.

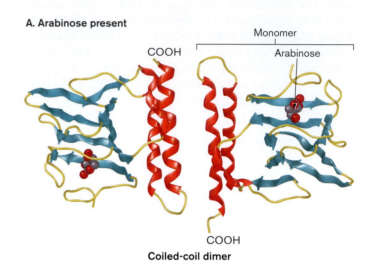

A. Arabinose present

Monomer

Arabinose

COOH

COOH

Coiled-coil dimer

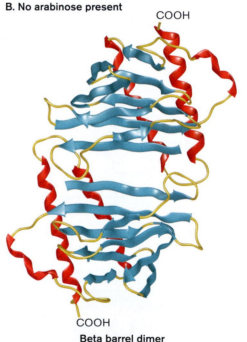

B. No arabinose present

COOH

COOH

Beta barrel dimer

Figure 10.11 Molecular structure of the AraC binding domain. Figures show alternative binding conformations of the dimerization domain only. **A.** Dimer formed by the interaction of coiled-coil regions in the presence of arabinose. (PDB code: 2ARC) **B.** Alternative dimer formed by the interaction between the beta barrel regions in the absence of arabinose. Note the helix-turn-helix motif (helixes shown in red). This is a common structure present in many DNA-binding proteins. The helix-turn-helix is able to fit into the major groove of a DNA molecule. (PDB code: 2ARC)

AraC-like Regulators: A Diverse Group of Intracellular Sensors

Computer analysis of numerous microbial genomes reveals a family of over 800 regulators that share a 100-amino-acid region of homology with AraC and the closely related XylS activator (xylose catabolism). This signature sequence, found at the C-terminal end of AraC, forms an independently folding domain that contains two helix-turn-helix DNA-binding motifs. The AraC/XylS family members regulate a variety of different cell functions, including carbon metabolism and virulence, as well as many responses to environmental conditions.

Although family resemblance is strong at the DNA-binding domains of these AraC-like proteins, homology usually evaporates at the other end of the protein, known as the dimerization domain. The nonfamily domain is where these proteins appear to bind or respond to some ligand (for example, arabinose). However, for many, if not most, of the AraC/XylS family members, the identity of the ligand remains a mystery.

In Attenuation, Translation Regulates Transcription

As noted previously, many amino acid biosynthetic pathways are controlled by transcriptional repression in which a repressor protein binds to a DNA operator sequence to prevent transcription. For instance, when internal tryptophan levels exceed cellular needs, the excess tryptophan (acting as a corepressor) will bind to an inactive aporepressor protein, TrpR, converting it to an active holorepressor (**Fig. 10.12**). TrpR holorepressor then binds to an operator DNA sequence positioned upstream of the tryptophan operon (*trp* operon), which encodes the enzymes required for tryptophan biosynthesis. Holorepressor bound to the *trp* operator represses expression by blocking

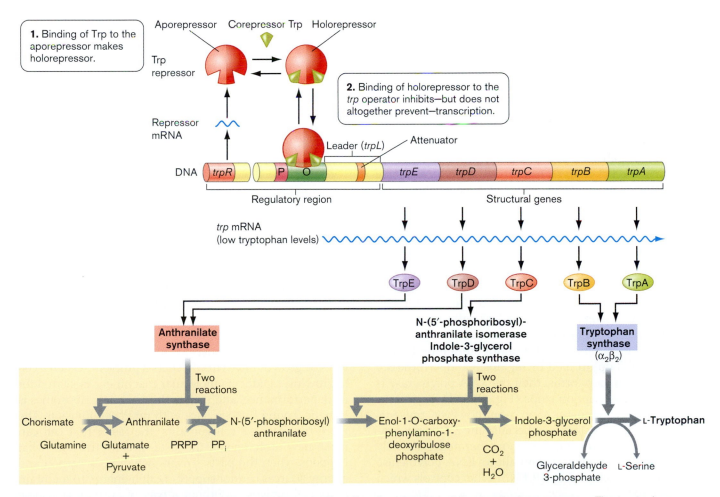

Figure 10.12 The tryptophan biosynthetic pathway in *E. coli* and repression of the tryptophan operon. The tryptophan biosynthetic enzymes and their encoding genes. TrpR aporepressor (inactive repressor) binds excess tryptophan when intracellular concentration exceeds need. The holorepressor (active repressor) then binds to the *trp* operator and prevents transcription. Note the long polycistronic message in blue. Repression lowers expression about 100-fold.

RNA polymerase. But repression is not the whole story in regulating the *trp* operon. Many amino acid biosynthetic operons, including the *trp* operon, have adopted a second strategy for downregulating tryptophan synthesis, which can be used alone or in conjunction with repression. This second mechanism is called **transcriptional attenuation**. Attenuation uses the ribosome as a sensor of amino acid levels. It is a transcriptional control mechanism in which the ability to *translate* part of an mRNA determines whether the RNA polymerase that began transcribing the operon is allowed to continue downstream transcription. Because attenuation halts transcription in progress, it affords an even quicker response to changing amino acid levels than simple repression.

Transcriptional attenuation was discovered by Charles Yanofsky and his colleagues at Stanford University. While examining the beginning of the *trp* operon in *E. coli*, they discovered an odd DNA region located between the *trp* operator and the first structural gene, *trpE*. This region, called the **leader sequence**, does not directly control the biosynthesis of tryptophan, but instead determines whether or not RNA polymerase, already authorized to begin transcription, is granted permission to proceed into the *trp* structural genes. The leader sequence encodes a peptide, but the peptide has no enzymatic function. The importance of the leader sequence lies in a pair of tryptophan codons (UGGUGG) imbedded within it. The act of translating those codons couples the intracellular level of tryptophan (measured as charged tryptophanyl-tRNA) to transcription. If the level of tryptophan is sufficient to maintain a level of charged tryptophanyl-tRNA adequate to support growth at a rapid growth rate, the mRNA of the leader region jettisons RNA polymerase before it reaches *trpE*, the first structural gene involved in the biosynthesis of tryptophan.

The attenuation mechanism hinges on four complementary nucleotide stretches within the leader mRNA. These regions, numbered 1 through 4, can base pair to form competing stem loop structures (**Fig. 10.13A** ⬤). Two of the stem loop structures are critical to the mechanism. These are the **anti-attenuator stem loop** formed by regions 2 and 3 and the **attenuator stem loop** (or terminator stem loop) formed by regions 3 and 4. If the 3:4 attenuator stem loop forms, then the RNA polymerase is ejected and transcription stops (**Fig. 10.13B**). Formation of the 2:3 anti-attenuator stem loop, however, prevents formation of the 3:4 stem loop because the 2:3 stem is longer than the 3:4 stem and, thus, more thermodynamically stable. The anti-attenuator stem allows RNA polymerase to transcribe into *trpE* (**Fig. 10.13C**). What controls which stem loop forms?

High tryptophan levels. When the cell is replete with charged tryptophanyl-tRNA and needs no more, the ribosome quickly translates through the key Trp codons

and runs into a translation stop codon between regions 1 and 2. The ribosome in this position envelops region 2 and prevents formation of the 2:3 stem. As a result, once RNA polymerase transcribes through region 4, the 3:4 attenuator stem snaps together. This structure interacts with the RNA polymerase ahead of it and halts transcription. As you would expect, the ribosome dissociates after reaching the region 2 stop codon, but because the 3:4 stem loop is already in place, a 2:3 stem loop does not form. Ribosome release, therefore, leads to formation of a 1:2 stem structure, precluding all possibility of regions 2 and 3 annealing.

Low tryptophan levels. However, if the level of charged tryptophanyl-tRNA is low, the ribosome following behind RNA polymerase stalls over the tryptophan codons. Because these codons occur right before region 1, the ribosome does not cover and protect region 2. As soon as RNA polymerase transcribes region 3, the 2:3 anti-attenuator stem loop forms and precludes formation of the 3:4 attenuator stem loop. The result is that RNA polymerase can continue into the structural genes, and ultimately more tryptophan is made. Transcriptional attenuation is a common regulatory strategy used to control many operons that code for amino acid biosynthesis.

> **NOTE:** Even though translation is part of the attenuation control mechanism, attenuation is not considered translational control. This is because RNA polymerase, rather than the ribosome, is the target of the control. Translational control of gene expression will be discussed later.

When Energy Sources Dwindle, Ribosome Synthesis Slows

During transitions from nutrient-rich to nutrient-poor conditions, microbes must contend with dramatic fluctuations in growth rate. This presents a problem. When a cell is growing rapidly, its molecular machinery is geared for peak performance, and the synthesis of new ribosomes is frenetic, trying to keep pace with rapid cell division. The more ribosomes a cell contains, the faster that cell can make new proteins and the faster it can grow. But what happens when the party's over—when poor carbon and energy sources cannot supply enough energy to maintain rapid cell division? Without a way of curbing ribosome construction, cells would soon fill with idle ribosomes. Under these conditions, bacteria undergo a process called the **stringent response**. The stringent response causes a decrease in the number of rRNA transcripts made for

A. Stem loop structures in attenuator region

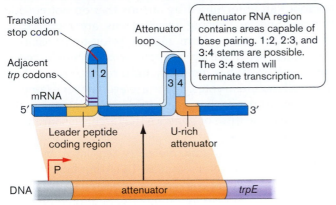

Translation stop codon

Adjacent *trp* codons

mRNA

5′

Attenuator loop

Attenuator RNA region contains areas capable of base pairing. 1:2, 2:3, and 3:4 stems are possible. The 3:4 stem will terminate transcription.

3′

Leader peptide coding region

U-rich attenuator

P

DNA | attenuator | *trpE*

Figure 10.13 The transcriptional attenuation mechanism at the *trp* operon. A. Relationship between the mRNA attenuator region and encoding DNA. **B.** Attenuation when *E. coli* is growing in high tryptophan concentrations. **C.** Transcriptional read through when *E. coli* is growing in low tryptophan concentrations. **D.** Charles Yanofsky (Stanford University) was instrumental in discovering attenuation and several other gene regulatory mechanisms while studying tryptophan metabolism. Here he is seen receiving the National Medal of Science from President George W. Bush in 2003.

B. High tryptophan levels

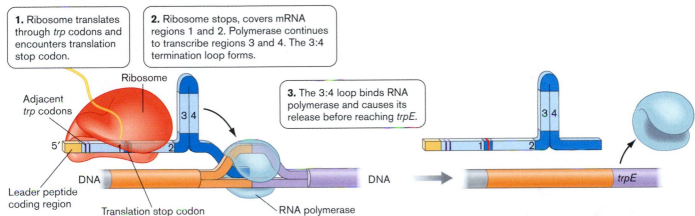

1. Ribosome translates through *trp* codons and encounters translation stop codon.

2. Ribosome stops, covers mRNA regions 1 and 2. Polymerase continues to transcribe regions 3 and 4. The 3:4 termination loop forms.

3. The 3:4 loop binds RNA polymerase and causes its release before reaching *trpE*.

Ribosome

Adjacent *trp* codons

5′

DNA

Leader peptide coding region

Translation stop codon

RNA polymerase

DNA

trpE

C. Low tryptophan levels

1. Ribosome translates leader.

2. Scarce tRNA^trp makes ribosome stall at *trp* codons. Polymerase continues through attenuator.

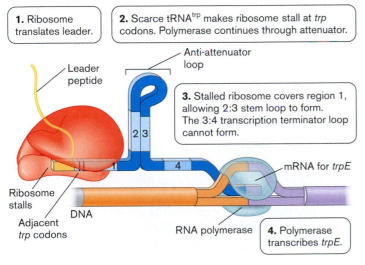

Anti-attenuator loop

Leader peptide

3. Stalled ribosome covers region 1, allowing 2:3 stem loop to form. The 3:4 transcription terminator loop cannot form.

mRNA for *trpE*

Ribosome stalls

DNA

Adjacent *trp* codons

RNA polymerase

4. Polymerase transcribes *trpE*.

D.

AP photo/J. Scott Applewhite

ribosome assembly and alters the expression of numerous other genes.

In the stringent-response strategy, idling ribosomes produce a signal molecule called guanosine tetraphosphate (ppGpp) that interacts with RNA polymerase and lowers its ability to transcribe genes encoding ribosomal

RNA (**Fig. 10.14**). How is ppGpp made? When there is an uncharged tRNA bound at the ribosome A site, which can happen during amino acid starvation, a ribosome-associated protein called RelA transfers phosphate from ATP to GTP to form ppGpp. This signal nucleotide interacts with the beta subunit of RNA polymerase and

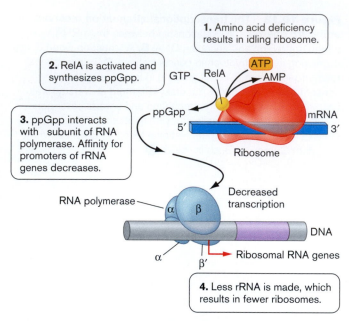

1. Amino acid deficiency results in idling ribosome.

2. RelA is activated and synthesizes ppGpp.

3. ppGpp interacts with subunit of RNA polymerase. Affinity for promoters of rRNA genes decreases.

4. Less rRNA is made, which results in fewer ribosomes.

GTP RelA ATP AMP ppGpp 5′ mRNA 3′ Ribosome

RNA polymerase α β Decreased transcription DNA α β′ Ribosomal RNA genes

Figure 10.14   Ribosome-dependent synthesis of guanosine tetraphosphate and the stringent response.

diminishes its recognition of promoters for operons producing rRNA and tRNA. The result is the downregulation of rRNA and tRNA synthesis. The less rRNA scaffolding there is available for building ribosomes, the fewer ribosomes will be produced.

This raises another question. Even though the synthesis of ribosomal RNA has been curtailed, won't the cell continue to waste resources on the synthesis of ribosomal proteins? It turns out that some ribosomal proteins can bind to the mRNA that encodes them and inhibit translation. So when there is less rRNA in the cell, free ribosomal proteins accumulate in the cytoplasm, unassociated with ribosomes. These excess ribosomal proteins begin to bind to their own mRNA molecules and inhibit the translation of their own coding region, as well as the coding regions of other ribosomal proteins residing on the same polycistronic mRNA. This is an example of **translational control**, where regulation affects translation of an mRNA by ribosomes rather than transcription by RNA polymerase.

In Eukaryotes, Most Genes Are Controlled Individually

There are a number of differences between gene regulation in prokaryotes and eukaryotes. In eukaryotes, most genes are present as isolated sequences encoding a single polypeptide, rather than as multigene operons. Eukaryotic microbes, like multicellular animals and plants, possess three different RNA polymerases: RNA polymerases I and III (pol I and pol III) to transcribe small functional RNAs and RNA polymerase II (pol II) for mRNA. Prokaryotes generally use a single RNA

polymerase. Protein-coding sequences of eukaryotic microbes are interrupted by introns that do not encode protein. Special splicing mechanisms are required to remove the introns from mRNA prior to translating the exons that make up a single protein. Single-celled eukaryotic microbes generally have fewer introns than multicellular eukaryotes. There are even fewer examples of introns in the prokaryotic world. Some introns occur in archaea, a feature of genetics that archaea share with eukaryotes.

Transcription in eukaryotes is regulated by general transcription factors such as TATA-binding protein, which are considered comparable to bacterial sigma factors because they help locate eukaryotic promoters; and specialized transcription factors, which are comparable to the activator and repressor proteins of bacteria. Both general and specialized transcription factors of eukaryotes bind regulator DNA sequences, including **enhancers** and **silencers**, that activate or repress RNA transcription, respectively. Enhancers are regions of DNA that may be located thousands of base pairs away from the gene they control. Binding of transcriptional factors to enhancer sequences increases the rate of transcription of the gene. Enhancers can be located upstream, downstream, or even within the gene they control. Silencers are control regions of DNA that, like enhancers, may be located thousands of base pairs away from the gene they control. When transcription factors bind to silencers, expression of the gene they control is repressed.

Translational control in eukaryotes is poorly understood. One form of control involving translation is that of nonsense-mediated mRNA decay, which terminates transcription in the nucleus. As the mRNA is being transcribed, nuclear ribosomes begin to translate the message. If a ribosome runs into a stop codon (resulting from faulty RNA transcription), transcription terminates and the faulty RNA is degraded.

TO SUMMARIZE:

■ **DNA-binding proteins** often recognize symmetrical DNA sequences.

■ **Many anabolic pathway genes** (for example, for amino acid biosynthesis) are repressed by the end product of the pathway (for example, the amino acid), which binds to a corepressor that inhibits transcription.

■ **AraC-like proteins** are a large family of regulators, present in many bacterial species, that can activate or repress operons by assuming different conformations.

■ **Attenuation** is a transcriptional regulatory mechanism in which translation of a leader peptide affects transcription of a downstream structural gene.

■ **Idling ribosomes** synthesize the signal molecule ppGpp, which interacts with the beta subunit of RNA polymerase and decreases the affinity of RNA polymerase for ribosomal RNA and a variety of other genes needed for rapid growth. The result, called the **stringent response**, is lower transcription of these genes, reduced levels of rRNA, and a decrease in the number of ribosomes.

10.4 Sigma Factor Regulation

DNA repressor and activator proteins are great for controlling the expression of individual genes and operons and are even used to coordinate expression of regulons, genes scattered around the chromosome that have related functions (such as tyrosine biosynthesis). In many cases, such as in the stringent response, bacteria need to coordinately activate large sets of genes and operons of seemingly disparate function that nevertheless collaborate to aid survival in particularly hostile environmental situations. One way to indirectly regulate expression of large sets of genes is to first regulate the synthesis or activity of the sigma factor that directs the expression of all those genes.

Many bacteria employ alternative sigma factors to direct transcription of distinct sets of genes (see Section 8.2). For example, many gram-negative bacteria, such as *E. coli* or *S. enterica*, use the sigma factor called sigma S (also called sigma-38, RpoS, σ^S or σ^{38}) to initiate the transcription of a variety of genes associated with survival in stationary phase. The level of sigma S is low in exponentially growing cells but rises dramatically as cells enter stationary phase. There are myriad ways to control whether a sigma factor gene is expressed and, if expressed, whether its product accumulates or functions.

Sigma Factor Activity Can Be Regulated

The cell can regulate an enzymatic reaction by changing the amount of the enzyme in a cell or by changing the *activity* of an enzyme already present. The same is true for sigma factors. Some microbes use **anti-sigma factor** proteins to inhibit sigma factor activity. Anti-sigma factor proteins target specific sigma factors, blocking their access to core RNA polymerase (that is, blocking their activity). This strategy prevents expression of that sigma factor's target genes.

Salmonella uses an anti-sigma factor (FlgM) to time events involved in constructing flagella. The transcription factor sigma F (also called sigma-28, RpoF, σ^F, or σ^{28}) is required to synthesize proteins used in the last stages of flagellar biosynthesis. The anti-sigma factor FlgM, however, keeps sigma F function at bay until membrane assembly of the flagellar basal bodies is complete. Once completed, the basal body selectively secretes the anti-sigma factor from the cell. This frees intracellular sigma F to direct transcription of the final set of flagellar assembly genes.

But what happens when the blocked sigma factor is suddenly needed? How do cells counter anti-sigma factors? Some systems counter anti-sigma factors with **anti-anti-sigma factors** that bind the blocking protein, like a decoy, and free the sigma factor to join core RNA polymerase. Other systems link the liberation of a bound sigma factor with a cell cycle event or with the assembly of a structure. *Bacillus* species, for instance, produce spores. During the process of making a spore, the cell couples formation of a forespore crosswall (see Section 5.8) to the activation of an anti-anti-sigma factor, which triggers the release of a whole cascade of new sigma factors needed only to complete spore formation (discussed later in this section).

Sigma Factor Translation Is Regulated by mRNA Secondary Structure and by Proteolysis

The synthesis or accumulation of sigma factors can also be regulated. The regulation of the heat-shock sigma factor sigma H (also called sigma-32, RpoH, σ^H, or σ^{32}) in *E. coli* illustrates how the synthesis of a sigma factor can be controlled at the level of translation (**Fig. 10.15**). Excessive heat, above 42°C for *E. coli*, will cause proteins to denature and membrane structure to deteriorate. All cells subjected to heat above their comfort zones (optimal growth range) will express a set of proteins called heat-shock proteins. These proteins include chaperones that refold damaged proteins and a variety of other proteins that affect DNA and membrane integrity. The transcription of many *E. coli* heat-shock genes requires the sigma factor sigma H. So one of the first consequences of growth at elevated temperature is an increase in the amount of sigma H present in the cell.

The level of sigma H is regulated by two temperature-dependent mechanisms; one of these controls the rate of sigma H synthesis, while the other determines its rate of proteolysis. The gene encoding sigma H is *rpoH*. At 30°C, *rpoH* mRNA adopts a secondary structure at the 5′ end that buries a ribosome-binding site, so *rpoH* mRNA is poorly translated. A sudden rise in temperature melts this secondary structure and exposes the ribosome-binding site, allowing translation to occur more readily. Thus, heat shock increases sigma H synthesis, which in turn increases transcription of the heat-shock genes whose products include chaperones and proteases.

Sigma Factor Levels Are Regulated by Proteolytic Degradation

Proteolysis is another mechanism used to control the level of certain sigma factors. This works by limiting sigma factor accumulation. For example, at 30°C, the *rpoH* message is poorly translated owing to secondary structure, as previously described, but some sigma H protein is made. To prevent inappropriate expression of the sigma H–dependent heat-shock genes at 32°C, the DnaK, DnaJ, GrpE chaperone system interacts with sigma H and shuttles it to various

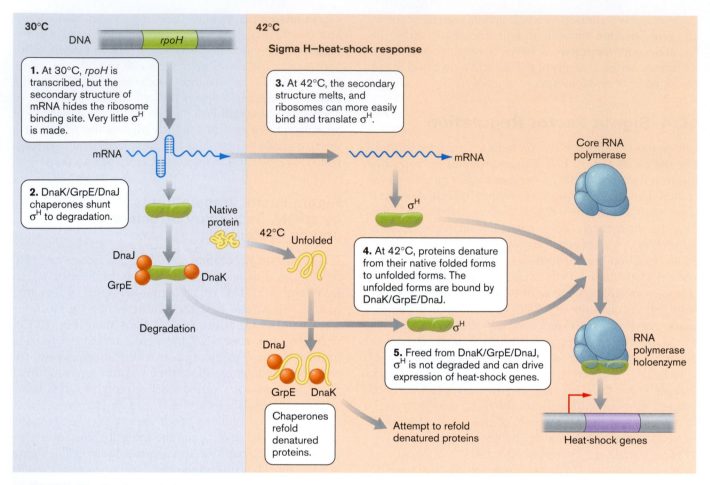

Figure 10.15 The heat-shock response of *E. coli*. Two mechanisms control sigma H levels. The small amount of sigma H that can be made at 30°C is met by the DnaK/GrpE/DnaJ chaperones and shuttled toward degradation. At 42°C, however, misfolded cytoplasmic proteins siphon off the chaperone trio and release sigma H to direct transcription of the heat-shock genes.

proteases for digestion (see **Fig. 10.15**). At 42°C, however, proteolysis of sigma H decreases, and sigma H is allowed to accumulate. Decreased degradation occurs because at the higher temperature, the chaperones are siphoned away from sigma H by the many other heat-denatured proteins formed. The new goal of the chaperones is to refold and rescue those damaged proteins. Chaperone redeployment frees sigma H to transcribe the heat-shock genes, which include the chaperone genes *dnaK*, *dnaJ*, and *grpE*. Thus, as the temperature rises, the amount of sigma H is increased by two temperature-dependent mechanisms: One increases translation by exposing the ribosome-binding site, while the second redeploys chaperones that direct its proteolysis.

In Sporulation, Different Sigma Factors Are Activated in the Mother Cell and Forespore

Bacillus and *Clostridium* species produce spores when nutrients become scarce. This requires an asymmetrical cell division in which one of the compartments becomes the spore (see Section 4.7). The regulation of this system is quite

complex, involving programs of transcription in the mother cell that are separate from those in the forespore compartment. How is this possible? The answer is that at the time of septum formation, when the forespore is first produced, only 30% of the chromosome (the part near the replication origin) is actually inside the forespore. The remainder of the chromosome is slowly pumped in over a 15-minute period. The sporulating cell takes advantage of this delay to selectively express different genes in different compartments.

For example, distinctly different transcriptional programs are directed by sigma F in the forespore and sigma E (σ^E) in the mother cell. Soon after formation of the septum that divides the forespore from the mother cell, sigma F becomes associated with RNA polymerase to direct the transcription of select genes in the forespore while sigma E directs transcription of specific genes in the mother cell. Later, sigma G and sigma K associate with RNA polymerase to direct transcription of the next developmental sequence in the forespore and mother cell compartments, respectively.

How does the cell selectively direct transcription in the forespore? It starts with an intricate timing mechanism that involves anti-sigma and anti-anti-sigma factors (**Fig. 10.16**). Like all sporulation sigma factors, sigma F is inactive when first synthesized. It is inactive because it binds to an anti-sigma F protein dubbed SpoIIAB that is cosynthesized with sigma F in the predivisional cell (**Fig. 10.16A**, step 1). After division, both proteins become equally partitioned between the two compartments. However, in addition to an anti-sigma, there is an anti-anti-sigma F protein, called SpoIIAA (AA for anti-anti), which is also equally distributed (steps 2 and 3). Anti-anti factor frees sigma F from anti-sigma F, which is degraded by another protein complex, the ClpPC protease. This is not a problem in the mother cell because that compartment just makes more AB (anti-sigma), so sigma F is never active.

But all of the previous events occur in both compartments (forespore and mother cell). So why is sigma F activation limited to the forespore? Recall that at the time of septum formation, one chromosome is slowly pumped into the forespore. The gene encoding the AB anti-sigma factor is located far from the origin and, as such, is not in the part of the genome trapped in the forespore immediately after septal formation (**Fig. 10.16B**, step 2). Thus, no new AB (anti-sigma) is made in the forespore during this time, and what was there is inactivated by anti-anti factor and degraded by ClpPC (**Fig. 10.16B**, steps 3 and 4). This depletion of anti-sigma in the forespore efficiently frees sigma F to act only in the forespore compartment. Liberated sigma F transcribes the genes needed for subsequent sporulation steps; these genes reside in the part of the genome trapped early on in the forespore.

Elegant proof for this model was obtained in the laboratory of Richard Losick, a leading scientist in the field of microbial development. In Losick's experiment, the gene for SpoIIAB (anti-sigma) was moved from its normal location to a position near the origin, where it stopped sporulation. Because its new position was near the origin,

A. Vegetative cell—σ^F is inactive

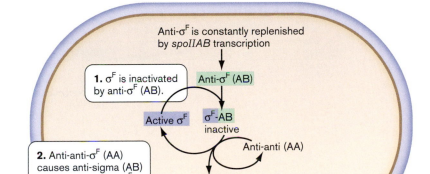

Figure 10.16   Regulating sporulation by genetic asymmetry. A. Summary of mechanisms that control sigma F activity in growing cells of *Bacillus subtilis*. The interactions between sigma F, anti-sigma F (AB), anti-anti-sigma (AA), and SpoIIE phosphatase collectively keep sigma F inactive. **B.** Activation of sigma F in the forespore occurs as the forespore chromosome is pumped into the forespore. Activation of sigma F leads to synthesis of sigma G in the forespore (not shown), which directs the next stage of spore development.

B. Forespore—σ^F is active

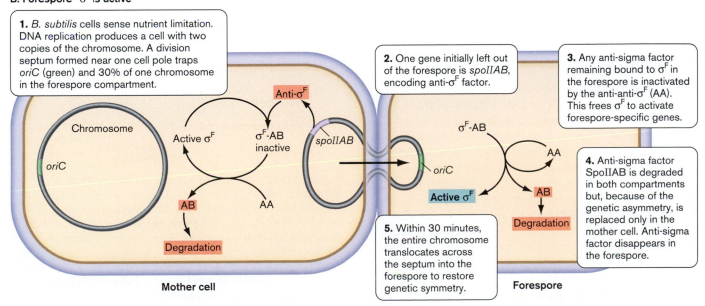

the anti-sigma F gene entered the forespore early and continued to replenish the anti-sigma F that was lost by degradation.

The simple, unequal distribution of the chromosome during septal formation is heavily exploited during sporulation to trigger differential gene expression in two different cell compartments.

> **THOUGHT QUESTION 10.5** Predict the phenotype of a *spoIIAA* mutant that completely lacks this anti-anti sigma factor (see **Fig. 10.16**).

TO SUMMARIZE:

- **Changing the synthesis or activity of alternative sigma factors** is used to coordinately regulate sets of related genes.
- **Sigma factors can be controlled** by altered transcription, translation, proteolysis, and anti-sigma factors.
- **Secondary structures** at the 5′ end of mRNA can obscure access to ribosome-binding sites. Conditions that melt the secondary structures will increase translation of the sigma factor.
- **Chaperones** can direct certain sigma factors toward degradation by proteases. Conditions that draw those chaperones away from the sigma factor will allow the sigma factor to accumulate.
- **Sporulation** relies on the hierarchical activation of a series of different sigma factors.
- **Temporary asymmetrical distribution** of the chromosome between the forespore and the mother cell results in differential activation of compartment-specific alternative sigma factors.
- **Sigma F is a forespore-specific sigma factor** preferentially activated by temporary asymmetrical distribution of the chromosome during the early stages of sporulation.

10.5 Small Regulatory RNAs

A considerable fraction of a bacterial chromosome does not encode mRNA, rRNA, or tRNA. Many of these intergenic regions encode small untranslated RNA (sRNA) molecules that carry out a variety of biological functions. Examples of sRNA can be found in many different organisms, ranging from bacteria to mammalian cells. Many act as regulators of gene expression at a posttranscriptional level, either by interacting with proteins or by acting as **antisense RNAs** that bind to complementary sequences of target transcripts and stimulate or prevent translation. Regulatory RNAs are involved in the control of a variety of processes, such as plasmid replication, transposition in both prokaryotes and eukaryotes, phage development, viral replication, bacterial virulence, envi-

ronmental stress responses, and developmental control in lower eukaryotes.

An example of regulation by an sRNA occurs within the Fur (ferric uptake regulator) **regulon** of *E. coli* (**Fig. 10.17**). (A regulon is made up of genes that, despite being scattered around the chromosome, are controlled by the same regulatory protein). Iron is an important component of living systems. However, too much iron can be detrimental to the cell by increasing oxidative stress. Fur is a repressor protein that regulates genes whose products scavenge iron from the environment or store iron in the cell (see Section 4.2). When intracellular iron levels are high, Fur represses expression of scavenging genes but induces production of iron storage proteins (**Fig. 10.17**). Fur represses gene expression by directly binding to spe-

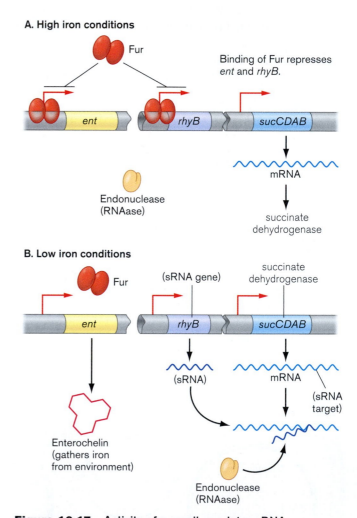

Figure 10.17 Activity of a small regulatory RNA molecule. A. When iron levels are high, Fur repressor protein binds to the *ent* and *rhyB* Fur box DNA sequence (a short specific DNA sequence in front of the genes regulated by Fur) and represses their expressions. Enterochelin is no longer made, but the *sucCDBA* message encoding succinate dehydrogenase can be translated. **B.** Under low iron conditions, the small RNA *rhyB* is expressed. RhyB sRNA binds to the *sucCDAB* message and renders it susceptible to an RNase.

Figure 10.18 **A.** Susan Gottesman (center) and **B.** Gisela Storz, both from the National Institutes of Health, played instrumental roles in establishing the importance of sRNA molecules in bacteria.

cific DNA sequences in front of target genes. But how does Fur activate other genes? One mechanism involves controlling production of a small RNA called RhyB (90 nt) (see **Fig. 10.17**). When iron levels are low, RhyB sRNA downregulates production of several iron storage and iron-using proteins (such as, succinate dehydrogenase). In this way, the cell prevents the shuttling of limited iron resources in wasteful directions. The cell needs to use the available iron, not store it. When iron is plentiful, however, Fur will directly repress *rhyB* and indirectly promote expression of the storage genes. Succinate dehydrogenase, an iron-containing enzyme, can also be made, which enables the use of succinate as a carbon and energy source.

What is the advantage to the cell of using sRNA to control expression? Using sRNA does not require protein synthesis and could be one of the most economical and efficient ways to globally repress genes. Global regulation refers to the coordinated control of many genes and regulons. Because sRNA molecules typically act on preexisting messages, they should also work more quickly than mechanisms that regulate transcription, which require degradation of preexisting mRNA before synthesis of the protein stops.

sRNA Genes Are Hard to Identify

How common are regulatory sRNA molecules? How can you look at a genome and predict what sequences encode sRNA genes? Identifying sRNA genes requires a different approach than identifying ORFs, which is relatively easy to do. ORFs have a translational start codon heading up a string of amino acid codons; sRNAs do not. The translational start codon of an ORF is preceded by a potential ribosome-binding site, which is itself preceded by a potential –10/–35 (or similar) promoter site sequences. Genes encoding sRNA will not have the ribosome-binding site but should have a recognizable promoter (see Sections 8.1 and 8.2). The promoter is not, however, enough to mark a potential sRNA gene. This is because all genomes contain

sequences that resemble promoters yet fail to bind RNA polymerase.

The laboratories of Susan Gottesman and Gisela Storz at NIH (**Fig. 10.18**) discovered that genes encoding known sRNA molecules always occur in intergenic regions between genes encoding ORFs and exhibit high degrees of homology between related species (greater than 80%). The researchers used this knowledge to compare the intergenic regions of *E. coli* with those of *Salmonella* and *Klebsiella*. After eliminating tRNAs, rRNAs, and other repetitive elements, they ended up with 295 possible sRNA candidates. A microarray experiment (see Section 10.9) using oligonucleotide probes to intergenic regions was then used to determine if any of the predicted sRNA molecules were actually made. Using this strategy, 23 new sRNA species were identified (**Fig. 10.19**). It appears that sRNA genes are common and integral players in gene control.

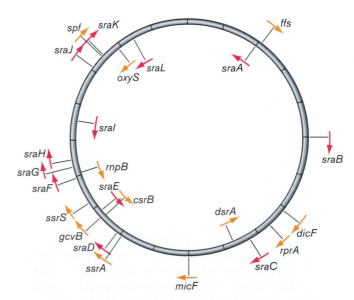

Figure 10.19 **Predicted sRNA genes in *E. coli*.** Location of various sRNA genes identified in the *E. coli* genome. Orange arrows depict sRNAs of proven function; red arrows indicate predicted sRNAs of unknown function.

Compared with prokaryotes, eukaryotic microbes possess an even broader array of small functional RNA molecules, including those of the spliceosome and of various cytoplasmic protein-RNA complexes. The genomes of eukaryotes such as yeast reveal thousands of potential sRNA sequences. The significance of sRNA discovery was highlighted by the decision of the journal *Science* to designate sRNA as the "Molecule of the Year" in 2002.

TO SUMMARIZE:

■ **Small regulatory RNAs** found within bacterial intergenic regions regulate the transcription or stability of specific mRNA molecules.

■ **The antisense nature** of sRNA allows these molecules to bind target mRNA. Binding can either stabilize the target mRNA or make it susceptible to degradation.

■ **Identifying sRNAs** is more complex than identifying ORFs.

■ **Eukaryotic microbes** use small interfering RNA to silence genes.

10.6 DNA Rearrangements: Phase Variation by Shifty Pathogens

All of the operon mechanisms previously described involve binding reactions between DNA, RNA, and proteins. More drastic means of control involve altering the DNA sequence itself.

Like a chameleon changing its color, some microbes use gene regulation to periodically change their appearance in a process called **phase variation**. Phase variation involves changing the amino acid composition of a particular protein on the bacterial surface. Bacteria that infect mammals, for example, must contend with immune systems whose mission is to destroy any and all invaders (discussed in Chapter 24). A central feature of immunity is the production of antibodies that recognize and bind foreign structures like microbial proteins and lipopolysaccharide (LPS). An infection will trigger the production of antibodies that are specific to the invader's component parts, such as pili, flagella, and LPS. Antibodies that bind to these microbial surface structures are especially useful for clearing an infection. Some pathogens, however, confound the immune system by changing the composition of these structures while the infection is in progress. This "shape-shifting" by the microbe, called **immune avoidance**, renders useless those antibodies specific for the old structure, and the embattled immune system must start all over making new antibodies. As a result, the course of infection is prolonged.

Gene Inversion Acts as an On-Off Switch

Flagellar phase variation in the gram-negative bacterium *Salmonella enterica* is a classic example of the regulatory genre known as **gene inversion**. Gene inversion is a recombinational event that flips the orientation of a gene or DNA segment in the chromosome, thereby turning that gene, or an adjacent gene, on or off. A major cause of diarrhea, *S. enterica* periodically changes the type of protein, called flagellin, used to make its flagella. Each organism has two genes, widely separated on the chromosome, that encode different forms of flagellin. The mechanism of the switch is a site-specific DNA recombination that turns off one gene while turning on the other. The target of the switch is a 993-bp DNA fragment (or cassette) called the H region that contains an outwardly directed promoter and a gene called *hin*, whose product, Hin recombinase (also called Hin invertase), mediates the recombination (**Fig. 10.20**). In antigenic parlance, the term *H antigen* refers to flagella, so the acronym Hin stands for *H in*version. The DNA cassette is flanked by short (26-bp) inverted repeats called *hixL* (left) and *hixR* (right).

NOTE: An **inverted repeat** is a sequence found in identical (but inverted) forms at two sites on the same double helix (for example, 5'-ATCGATC-GnnnnnnnnnnnnnnnCGATCGAT-3'). A **direct repeat** is a sequence found in identical form at two sites on the same double helix (for example, 5'-ATCGATCGnnnnnnnnnnnnnnnnATCGATCG-3'. A **tandem repeat** is a direct repeat without any intervening DNA sequence (for example, ATCGATCGATCGATCGATCGATCG).

WWW | Hin recombinase molecular tutorial, Biomolecules at Kenyon

The Hin recombinase collaborates with other less specific DNA remodeling proteins, such as Fis, to engineer an association between the 26-bp left (*hixL*) and right (*hixR*) ends of the invertible DNA element. The two ends, each bound to a Hin monomer, are brought together by Hin-Hin protein interactions. DNA within the cassette then forms a loop. Hin cuts within the center of each *hix* site, producing staggered ends. An exchange of Hin subunits is thought to lead to strand inversion, so that the orientation of the DNA cassette is reversed relative to the flanking DNA on either side.

In one orientation, the outwardly directed promoter of the H region directs expression of H2 flagellin, encoded by *fljB*, and a repressor, FljA, that prevents transcription of the other flagellin gene, *fliC* (see **Fig. 10.20A**). After the switch, however, the promoter points in the wrong direction, so there is no produc-

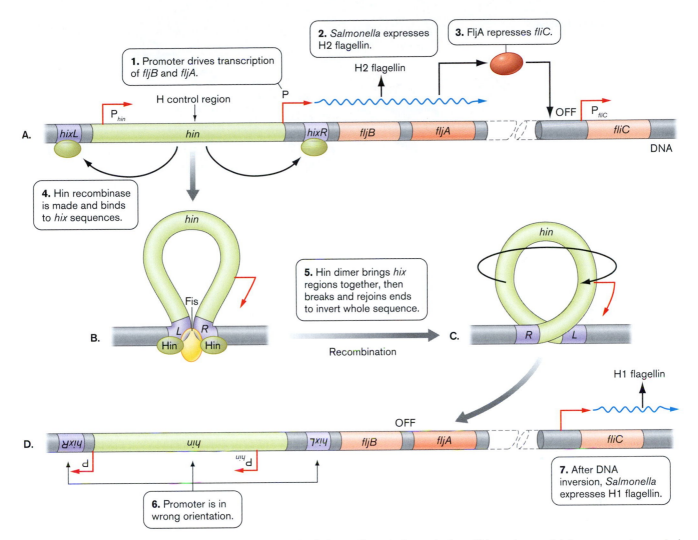

Figure 10.20   Phase variation of flagellar proteins in *Salmonella enterica*.   An invertible region containing a promoter controls the expression of two unlinked flagellar protein genes. In one orientation (**A**), the promoter drives synthesis of H2 flagellin (*fljB*) and a repressor (FljA) of the H1 flagellin gene (*fliC*). Action by the Hin recombinase causes the segment to invert (**B** and **C**), thereby removing the promoter. Because the repressor FljA is not formed, the gene for H1 can be expressed (**D**).

tion of H2 flagellin or FljA, the repressor of *fliC* (see **Fig. 10.20D**). Lacking this repressor, the *fliC* flagellin gene is expressed. Thus, H1 flagellin (present in phase 1 cells) is synthesized instead of H2 flagellin, which is present in phase 2 cells. The amino acid sequences, and thus the antigenicities, of the two flagellar proteins are different, so this mechanism allows *Salmonella* to change its appearance to a host immune system. In each generation, the rate of the reversible switch varies from about one cell in 10^3 to one in 10^5. Note, however, that this switch would not accomplish much if the infecting population of bacteria started out mixed—that is, producing both types of flagella. The initial infection must be of one phenotype.

THOUGHT QUESTION 10.6   What is the phenotype of a *fljA* mutant? A *fliC* mutant?

Slipped-Strand Mispairing Is Another Mechanism of Phase Variation

A different type of phase variation occurs within the *opa* genes of *Neisseria gonorrhoeae*, the causative agent of gonorrhea. The *opa* multigene family encodes 11 related outer membrane proteins that help the microbe adhere to the host cell epithelium. Opa refers to the *opaque* appearance of the bacterial colonies that result from the presence of these proteins on the cell surface. Variable numbers of the 5-bp repeat CTCTT are interspersed within the signal peptide regions of each of the genes for the outer membrane proteins. As discussed in Section 8.4, signal peptide sequences mark some membrane proteins for membrane insertion. Mispairing occurs during DNA replication when one of the strands, either the growing strand or the template strand, slips and pairs with the wrong repeat on the other strand (**Fig. 10.21**). If the newly synthesized strand slips back rela-

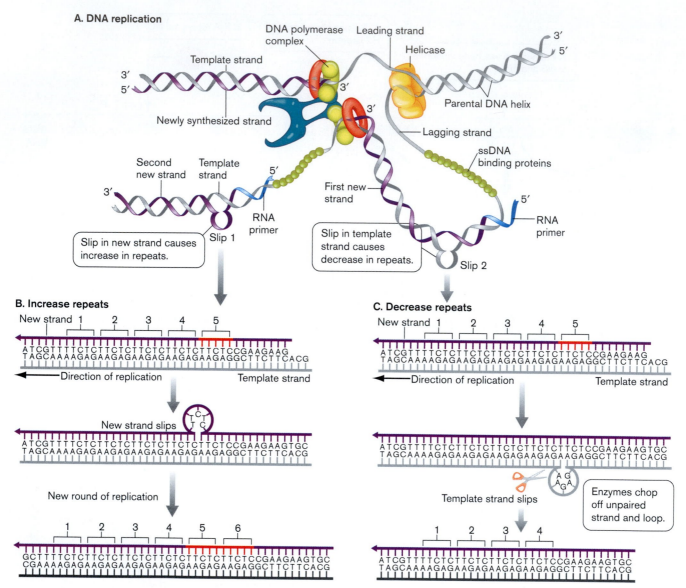

Figure 10.21   Slipped-strand mispairing.   A. A DNA polymerase dimer moving along a replication fork. Slipping events (1 and 2) are portrayed on the lagging strand. **B.** The series of events that occur if the newly synthesized strand slips back relative to the template strand. When this new dsDNA, containing the looped-out slipped strand, is subsequently replicated, one of the daughter dsDNA molecules will have an extra repeat because of the slippage. **C.** What happens if, prior to replication, the original template strand slips back relative to the newly synthesized complementary strand. If repair enzymes get to the loop prior to replication, the loop is excised and a pentanucleotide repeat is removed. Subsequent daughter strands will be shortened.

tive to the template strand, one extra copy of the repeated unit is inserted in the growing strand (panel A). If instead the template strand slips back relative to the growing strand, one copy is deleted in the growing strand (panel C).

The effect of slipped-strand mispairing is to change the number of bases in a run of repeats. Because the length of the repeated sequence is not an even multiple of three, its insertion or deletion within the coding region of a gene generates a frameshift. The frameshift causes the gene to encode a run of amino acids completely different from that encoded by the original sequence (see Section 9.4). For *opa*, the number of CTCTT repeats in the amino-terminal portion of the gene determines whether the full-length protein is made (in which case the *opa* gene is considered on) or not (in which case the gene is off). One repeat more or less means a shift in the reading frame of the *opa* message so that a translation stop codon materializes, terminating production of that protein. The frequency of such slippage is very high; each time the bacteria divide, approximately one out of every 100 or 1,000 daughter cells will carry a mutation that changes the number of CTCTT repeats.

Each *opa* gene switches on or off independently. As a result, none, one, or several Opa polypeptides can be expressed simultaneously. Because no regulatory proteins are made, this is a wonderful low-maintenance way to evade the immune system.

Eukaryotic microbes, especially pathogenic sporozoans, possess elaborate mechanisms of phase variation. The trypanosome that causes "sleeping sickness" undergoes extensive genetic shuffling and mutation of its coat proteins over successive generations, essentially overwhelming the host immune system by presenting every possible form of antigen.

SUMMARY:

- **Phase variation** occurs when a bacterial protein structure reversibly changes from one form to another as a result of a genetic mechanism.
- **An invertible promoter switch** regulates two genes of *Salmonella enterica* encoding two alternative structural types of flagellin.
- **Slippage while replicating** through repetitive DNA sequences, a process called **slip-strand mispairing**, can reversibly activate or inactivate a series of outer membrane proteins in *Neisseria gonorrhoeae*.

10.7 Integrated Control Circuits

The regulatory circuits outlined to this point have mostly involved single-switch mechanisms. But microbes build on these mechanisms by combining multiple switches with redundant feedback controls that can form a kind of integrated gene circuit. Integrated circuits can send a virus down alternate lifestyle paths (lysis or lysogeny), couple the genetic and biochemical control of a metabolic pathway (ammonia assimilation), or time the events of a developmental cycle such as sporulation. The deeper we delve into the regulatory workings that govern the expression of a cell's genome, the more clearly we see the hierarchy of control. Groups of simple control circuits are collectively regulated by other, more complex circuits, which, in turn, are governed by an even higher regulatory authority. One illustration of this was already presented, the role of cAMP and CRP in regulating lactose and arabinose catabolism. The *lacZYA* and *araBAD* operons are each controlled separately by dedicated regulatory switches—LacI and AraC, respectively. At a higher level, however, *both* operons are coregulated by cAMP and CRP in response to energy status (whether or not glucose is present and utilized). The systems we are about to discuss represent increasingly complex examples of nature's inventiveness and need for control.

The Phage Lambda Lysis/Lysogeny "Decision" Is to Kill or Not to Kill

Some bacterial viruses, known as temperate or lysogenic bacteriophages, have a "decision" to make after they infect a cell (discussed in Chapter 6). They can rapidly replicate intracellularly, fill the cell with progeny, and then cause the cell to lyse by destroying the cell membrane. This process is called the **lytic** life cycle of the phage.

Alternatively, lysogenic bacteriophages can remain in a dormant state after infecting a bacterial cell. This generally happens when times are good and the host grows quickly. The infecting viral DNA keeps a low profile and does not kill the host. Depending on the phage, it quietly hides by integrating into the host chromosome or masquerades as a harmless plasmid. The dormant form of the virus is called a **prophage**, and the host bacterial cell is called a **lysogen**. This lifestyle "choice" of the phage is termed **lysogeny** (Section 6.4). Lysogeny is good for the phage because if it quickly killed all the hosts in the area, the phage population could perish if no new host was found. Lysogeny can also be good for the bacterium if the prophage contains genes that contribute to bacterial virulence. However, prophages are not committed to a lysogenic existence forever. If the lysogenic cell gets into trouble (for example, by starvation or DNA damage), prophage DNA will begin replicating to produce viral progeny that kill the host and "abandon ship." The prophage switches to a lytic life cycle in the hope of finding a new, healthy host. Thus, lysogenic bacteriophages could be viewed as sleeping assassins. **Figure 10.22** illustrates the type of plaques formed by lysogenic and lytic variants of a lambdoid phage.

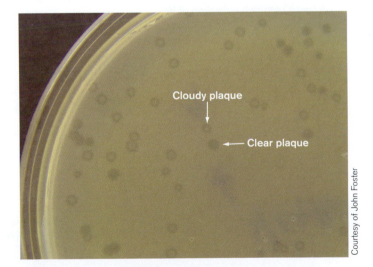

Courtesy of John Foster

Figure 10.22 Lysogenic plaques versus lytic plaques. A lambda-like phage (phage P22) was diluted and plated onto a lawn of *Salmonella enterica*. Two types of plaques are evident. Those with cloudy centers were made by P22 capable of lysogenizing the cells (cloudy plaques). The cloudy centers are due to the growth of lysogenic cells containing quiescent prophage DNA. A ring of lysis surrounds the cloudy center because low-level release of virus particles from lysogenic cells results in infection and lysis of surrounding cells. Plaques with clear centers are made by phages defective in the P22 homolog of the CI repressor (clear plaques). These phages are programmed for lysis.

The phage regulatory circuit that best illustrates this lysis/lysogeny decision is that of phage lambda, an *E. coli* virus. The decision awaits the outcome of a competition between two phage repressors, CI and Cro. If CI repressor wins, lambda DNA becomes a quiescent prophage that integrates into the host chromosome and expresses only one gene, *cI*. If Cro prevails, progeny phage are made and the cell is doomed.

It is important to understand the rationale for the lambda decision-making process before we discuss the process itself. When several lambda phages infect a single cell (a high multiplicity of infection), CI repressor tends to win and the cell becomes a lysogen. Lysogeny makes sense because the multiple infection of each cell shows that there are already more phages than there are bac-

teria. If progeny phage were released, there would be no other bacteria to infect. In contrast, if there are more bacteria than phage (low multiplicity infection), it makes sense for lambda to make progeny phage to infect the available hosts. So in this event, the genetic mechanisms are designed for Cro to win, initiating a lytic cycle.

The region of lambda DNA involved in the lysis/lysogeny control circuit is shown in **Figure 10.23A**. When lambda phage DNA is first injected into the cell, transcription initiates at two promoters, the leftward promoter P_L and rightward promoter P_R. Transcription from the P_L and P_R early promoters yields proteins needed for the transcription of later stage genes required for replication and capsid synthesis, but also encode the repressor Cro, as well as the proteins CII and CIII. CII protein can

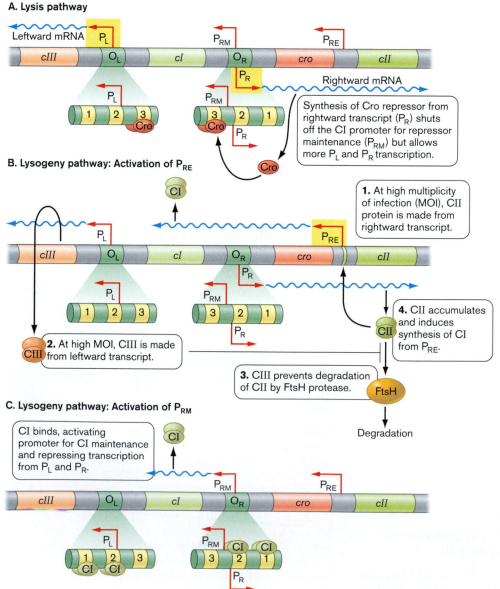

A. Lysis pathway

Leftward mRNA

Synthesis of Cro repressor from rightward transcript (P_R) shuts off the CI promoter for repressor maintenance (P_{RM}) but allows more P_L and P_R transcription.

B. Lysogeny pathway: Activation of P_{RE}

1. At high multiplicity of infection (MOI), CII protein is made from rightward transcript.

2. At high MOI, CIII is made from leftward transcript.

3. CIII prevents degradation of CII by FtsH protease.

4. CII accumulates and induces synthesis of CI from P_{RE}.

Degradation

C. Lysogeny pathway: Activation of P_{RM}

CI binds, activating promoter for CI maintenance and repressing transcription from P_L and P_R.

Figure 10.23 The phage lambda lysis/lysogeny decision. **A.** Lysis pathway. The right and left operators, O_L and O_R, possess three binding sites each for the CI and Cro repressors. Synthesis of Cro repressor shuts off the CI promoter for repressor maintenance. Phage chooses lysis. **B.** At high MOI, CII protein is made and triggers the CI promoter for repressor establishment. CI is made. **C.** Once made, CI binds to sites 1 and 2 in the operator regions, turning on the promoter for CI maintenance while repressing the leftward and rightward promoters needed to make more phage particles. The phage chooses lysogeny.

activate transcription of the *cI* gene from the promoter for CI repressor establishment, called P_{RE}. The host protease FtsH, however, degrades CII (**Fig. 10.23B**). Thus, at low multiplicity of infection, where at most only one phage genome is injected into any one cell, CII never achieves a concentration high enough to trigger production of the repressor CI, and the phage will head toward lysis rather than lysogeny. A few other events that we will not discuss must occur before the phage is committed to replicate.

While the CI repressor is struggling to be made, the repressor Cro can bind to regions within the left and right operator regions (O_L and O_R) (**Fig. 10.23A**). These operators control transcription from the left and right promoters. Note in the figure that each operator is made up of three binding sites, each of which can bind Cro or CI. Cro binds with highest affinity to region 3, which overlaps another promoter for *cI* called the promoter for repressor maintenance (P_{RM}). Cro, even at low levels, will repress this promoter. At this point, CI, the repressor that can prevent transcription of later stage promoters, is not made and the phage has committed to lysis.

High multiplicity of infection, where multiple phage genomes infect a single cell, results in a larger amount of CII being made in part because of multiple gene copies available for transcription (**Fig. 10.23B**). In general, the more copies of a gene there are, the more of that gene's protein product can be made. The concentration of CII rises even more because its degradation by the host protease FtsH is inhibited by CIII, which is also synthesized in larger amounts at high multiplicity of infection. CIII protein binds to and inhibits FtsH activity.

Because CII is not degraded by FtsH, it will achieve a concentration that can initiate *cI* transcription from P_{RE}. The CI repressor resulting from this transcription then binds preferentially to sites 1 and 2 within the operators O_R and O_L. This shuts down transcription from the two primary phage promoters P_L and P_R (**Fig. 10.23C**). CI binding to O_{R2} also activates *cI* transcription from P_{RM}. This promoter is required to maintain long-term expression of CI because when CI binds to P_R sites 1 and 2, transcription from P_{RE} stops because CII does not accumulate. Now all transcription, except for that of the CI repressor, is terminated. The cell has committed to lysogeny.

Many breakthroughs in molecular biology have resulted from the study of phage lambda. One of the most significant was the revelation of how repressors bind to DNA and control genes. As an undergraduate student in 1960, Mark Ptashne became entranced by the bold hypothesis made by Jacob and Monod regarding regulation of cell behavior. Some vague molecule called a "repressor" controlled the expression of genes. Was it protein? Did it bind DNA? These were critical questions no one was addressing. Told that his quest was too challenging for a graduate student, Ptashne had to wait until his postdoctoral years to search for his holy grail. Using CI repressor as the model, he discovered that CI is a protein that dimerizes and, in that form, binds to a specific DNA operator site (**Fig. 10.24**). His group was the first to demonstrate what others had predicted. Ptashne and his colleagues at Harvard University also discovered within the CI repressor

Figure 10.24 Structure of the Lambda repressor. A. Mark Ptashne (currently at Memorial Sloan Kettering Cancer Center). **B.** Dimer of lambda CI repressor binding to DNA. Note the helix-turn-helix motif located in two successive major grooves. Helices are gold and purple; the turn is green. (PDB code: 1LMB)

Courtesy of Mark Ptashne

the classic helix-turn-helix DNA-binding motif that is common to many DNA-binding proteins.

THOUGHT QUESTION 10.7 The plaques of normal phage lambda appear as clear rings with cloudy centers of lysogenized cells. What do you think the plaque will look like in a *cI* mutant? A *cII* mutant? A *cro* mutant?

THOUGHT QUESTION 10.8 Predict why UV irradiation can activate lambda prophage. *Hint:* What other host system is activated by UV?

Coupling Metabolic and Genetic Control: Nitrogen Regulation

Another mechanism by which levels of key compounds in the cell are regulated is to link biochemical control of a metabolic pathway (that is, whether an enzyme is active or inactive) to the genetic regulation of that pathway (which determines whether the enzyme is made or not). The assimilation of nitrogen is a classic example of coupled regulation (see Section 15.5).

All cells require nitrogen to grow. Thus, when nitrogen becomes limiting, bacteria activate pathways to gather nitrogen from the environment. The indicator of nitrogen abundance turns out to be the cellular levels of two amino acids, glutamate and glutamine. Actually, it is the intracellular ratio between alpha-ketoglutarate (the precursor of glutamic acid) and glutamine levels, but to make the discussion easier we will use glutamate/glutamine levels. Put simply, when glutamine is in excess, the cell has plenty of nitrogen; when glutamate (or alpha-ketoglutarate) is in abundance, the organism is nitrogen starved.

During times of nitrogen abundance, cells accumulate a store of nitrogen via glutamine synthetase (GlnA). Glutamine synthetase uses the energy released from ATP hydrolysis to assimilate NH_4^+ into glutamic acid and produce glutamine (**Fig. 10.25A**, step 1). However, if the cell makes too much active glutamine synthetase, then all of the cell's glutamate will be converted to glutamine, and not enough glutamate will remain for protein synthesis. To prevent this, glutamine, when present at high levels relative to glutamic acid, signals the cell to stop making GlnA and inactivate whatever GlnA is already present. Thus, the glutamate to glutamine ratio is conserved. *E. coli* uses a single coupling protein called GlnB to regulate both the transcription of *glnA* and the biochemical activity of its product, glutamine synthetase. GlnB allows cells to balance the activation/inactivation of glutamine synthetase activity with its rate of synthesis.

We will start by explaining transcriptional control of the system and then show how GlnB links transcriptional and biochemical controls. Genetic control of *glnA*

transcription is managed by the products of the *nitrogen regulator* genes *ntrB* and *ntrC* and a dedicated 54-kDa sigma factor, sigma N (also called sigma-54, RpoN, σ^N, or σ^{54}), encoded by the gene *rpoN*. NtrB and NtrC form a two-component signal transduction system. NtrB is a sensor kinase that phosphorylates the response regulator NtrC when glutamine levels are low (**Fig. 10.25A**, step 2). The phosphorylated form of NtrC (NtrC-P) induces expression of *glnA* and increases the level of glutamine synthetase in the cell (**Fig. 10.25A**, step 3). The NtrC-P regulator binds not to an operator region but to an enhancer sequence. Unlike operators, which must lie close to the promoter they control, enhancers can influence target gene expression from great distances (1 or 2 kb). Bound to its enhancer sequence, NtrC-P activates transcription of *glnA* through direct interactions with sigma N RNA polymerase bound at the *glnA* promoter resulting in a DNA loop (**Fig. 10.26**). As a result, glutamine synthetase is made, nitrogen (in the form of ammonia) is assimilated, and glutamine is produced.

NOTE: Unlike the DNA loop formed by AraC, enhancer sequences can be moved experimentally to great distances from the promoter and still affect expression of the promoter.

As already noted, a dire consequence of unrestricted glutamine production is the depletion of intracellular glutamate levels relative to glutamine levels. Ideally, when glutamine levels are in excess, the cell would stop making GlnA and would inactivate whatever GlnA is already made. That way the cell can preserve the glutamate/glutamine balance. Conversely, when glutamine levels are limiting, the cell would resume making GlnA and make sure it is active. The GlnB regulator coordinates these genetic and biochemical controls by modulating the phosphorylation of NtrC (thereby influencing transcription of *glnA*) and by altering GlnA activity.

The direction of control centers on whether or not a UMP group is attached to GlnB (uridylylation). When glutamine levels are low, GlnD protein uridylylates GlnB (**Fig. 10.25B**, step 4). GlnB-UMP allows *glnA* transcription and activates pre-existing GlnA enzyme (**Fig. 10.25B**, step 5). Conversely, when internal stores of glutamine are high relative to glutamate, GlnD removes UMP from GlnB (**Fig. 10.25B**, step 6). GlnB then halts *glnA* transcription and inactivates pre-existing GlnA enzyme.

To carry out these functions, GlnB interacts with two proteins, the NtrB sensor kinase and another protein called GlnE (**Fig. 10.25B**). GlnB (the form when glutamine levels are high) causes NtrB to dephosphorylate NtrC, which effectively halts *glnA* transcription. GlnB also compels GlnE to covalently add an AMP to GlnA (a process called adenylylation). This modification inac-

Figure 10.25 Regulation of nitrogen metabolism. A. Genetic control. NtrB-NtrC form a cytoplasmic two-component signal transduction system that controls *glnA* transcription. **B.** Biochemical control. GlnB coordinates the function of NtrB (transcriptional control) and GlnE (metabolic control).

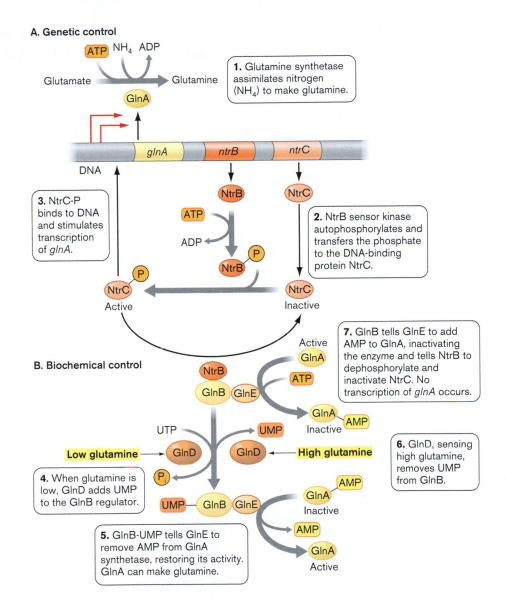

tivates whatever GlnA remains in the cell. The result is that ammonia is no longer assimilated and the glutamate/glutamine ratio is protected.

When glutamine levels are low, relative to glutamate, GlnB-UMP allows the phosphorylation of NtrC by NtrB. So NtrC-P is made and stimulates the synthesis of glutamine synthetase. In addition, GlnB-UMP causes GlnE protein to remove AMP from preexisting glutamine synthetase, thus activating it. The result is that transcription of *glnA* is enhanced, glutamine synthetase is activated, and ammonia is assimilated to make glutamine once again.

THOUGHT QUESTION 10.9 Predict the phenotype of a *glnB* mutant. Will it be a glutamine auxotroph or not? What about a *glnD* or a *glnE* mutant?

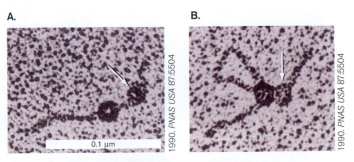

1990. *PNAS USA 87*:5504

1990. *PNAS USA 87*:5504

Figure 10.26 Visualization of NtrC function. NtrC binds to an enhancer sequence a great distance from the promoter of *glnA*. **A.** Electron micrograph of the NtrC protein (gray circle marked by arrow) and sigma-54 RNA polymerase (black circle) bound to their respective sites on the DNA but not in contact. **B.** Looping of the DNA brings the proteins in contact. In this configuration, NtrC will activate transcription.

TO SUMMARIZE:

■ **Bacterial genes are regulated by a hierarchy of regulators** that form integrated gene circuits.

■ **Gene switches** can regulate "decision-making" processes in microbes.

■ **Two repressors (CI and Cro)** made by phage lambda of *E. coli* control whether the phage travels through the lytic (Cro dominates) or lysogenic (CI dominates) pathway of replication. Both repressors bind similar phage DNA sequences but do so with different affinities. Cro prevents CI synthesis but allows expression of other genes needed for lysis. CI prevents expression of all lambda genes except its own. Multiplicity of infection determines which repressor dominates and whether the phage DNA will replicate to produce progeny or will become a prophage.

■ **The study of lambda CI** repressor led to the discovery of the helix-turn-helix motif common in many DNA-binding proteins. The helices bind to appropriate sequences in the grooves of target DNA.

■ **The NtrB-NtrC two-component signal transduction system** regulates nitrogen assimilation by altering transcription of the gene encoding glutamine synthetase (*glnA*).

■ **The protein GlnB** links the biochemical control and genetic regulation of nitrogen assimilation.

10.8 Quorum Sensing: Chemical Conversations

A discovery that fundamentally changed the way we think about microbes was made while studying *Vibrio fischeri*, a peculiar marine microorganism that colonizes the light organ of the Hawaiian squid (*Euprymna scolopes*) (**Fig. 10.27**). As discussed at the beginning of this chapter, *V. fischeri* is bioluminescent, but only glows at high cell densities—a situation achieved naturally in the squid's colonized light organ or artificially in the test tube (**Figs. 10.27C and D**). What accounts for the dependence of gene expression on cell density? How do cells *know* they are crowded?

The phenomenon was originally dubbed "**quorum sensing**" because it seemed akin to parliamentary rules of order that require a minimum number of members (a quorum) to be present at a meeting in order to conduct business. In the microbial world, however, gene regulation is only loosely associated with actual cell numbers. Induction of a quorum-sensing gene system really requires the accumulation of a secreted small molecule called an **autoinducer** (usually a homoserine lactone, although grampositive organisms are known to use short peptides). The more cells there are in a given space, the faster the critical level of autoinducer is reached.

At a certain extracellular concentration, the secreted autoinducer reenters cells and binds to a regulatory molecule. In the case of *V. fischeri*, the regulatory molecule is LuxR (**Fig. 10.28**). The LuxR-autoinducer complex activates transcription of the luciferase target genes responsible for bioluminescence. Luciferase is the enzyme responsible for light production in *V. fischeri*.

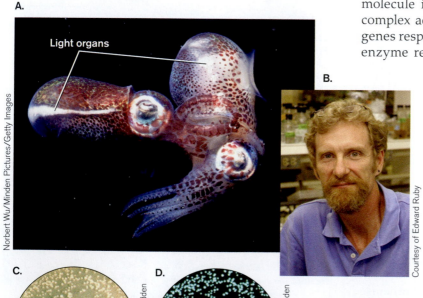

A.

Light organs

Norbert Wu/Minden Pictures/Getty Images

B.

Courtesy of Edward Ruby

C.

Courtesy of Professor Madden

D.

Courtesy of Professor Madden

Figure 10.27   Visual demonstration of quorum sensing.   A. The organism *Vibrio fischeri* colonizes the light organ of the Hawaiian squid (*Euprymna scolopes*). Shown is a pair mating at night. **B.** Edward Ruby, at the University of Hawaii, has studied various aspects of the symbiotic relationship between *V. fischeri* and its squid host. **C.** When colonies of the luminescent bacterium *V. fischeri* are observed in a well-lit place, the light emitted by bacteria is not visible. **D.** If the same colonies are viewed in darkness, the intensity of luminescence is remarkable.

One can bypass the apparent cell density requirement by simply adding purified autoinducer to a low-density cell culture. Many other microbes use these chemical languages to coordinate behavior of the population (**Table 10.1**). Quorum sensing is used by pathogens to time the production of virulence factors and can even be used to communicate between different species (**Special Topic 10.2**).

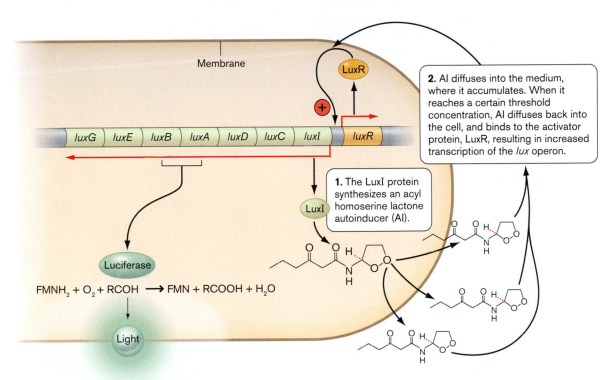

2. AI diffuses into the medium, where it accumulates. When it reaches a certain threshold concentration, AI diffuses back into the cell, and binds to the activator protein, LuxR, resulting in increased transcription of the *lux* operon.

1. The LuxI protein synthesizes an acyl homoserine lactone autoinducer (AI).

$$FMNH_2 + O_2 + RCOH \longrightarrow FMN + RCOOH + H_2O$$

Figure 10.28 Microbial communication through quorum sensing. The *lux* system of *Vibrio fischeri* mediates that organism's bioluminescence. Synthesis and accumulation of an autoinducer (AI) triggers expression of the *lux* operon. The greater the cell number and the smaller the container, the faster AI will accumulate. The resulting luciferase enzymes catalyze bioluminescence. The luciferase reaction, catalysed by LuxA and LuxB, uses oxygen and reduced flavin mononucleotide to oxidize a long-chain aldehyde (RCHO) and in the process produces blue-green light. Other *lux* gene products are involved in synthesis of the aldehyde.

Table 10.1 Examples of microbial quorum-sensing systems.

System	Organism	Autoinducer family	Function
LuxR/LuxI	*Vibrio fischeri*	Homoserine lactone	Bioluminescence
LusQ/LuxS	*Vibrio harveyi*	Furanosyl borate diester	Bioluminescence
LuxN/LuxLM	*Vibrio harveyi*	Homoserine lactone	Bioluminescence
LasR/LasI	*Pseudomonas aeruginosa*	Homoserine lactone	Exoenzyme production
Rhl	*Pseudomonas aeruginosa*	Homoserine lactose	Exoenzyme production
Agr	*Staphylococcus*	Peptide	Extracellular toxins
StrR	*Streptomyces griseus*	γ-butyrolactone	Aerial hyphae; antibiotic production
TraR/TraI	*Agrobacterium tumefaciens*	Homoserine lactone	Conjugation
YpeR/YpeI	*Yersinia pestis*	Unknown	Unknown
SdiA	*E. coli*	Unknown	Cell division

Structures of different auto inducer families

Homoserine lactone family   γ-Butyrolactone   Furanosyl borate diester

Special Topic 10.2   The Role of Quorum Sensing in Pathogenesis and in Interspecies Communications

Pseudomonas aeruginosa is a human pathogen that commonly infects patients with cystic fibrosis, a genetic disease of the lung. The organism forms a biofilm over affected areas and interferes with lung function. Key to the destruction of host tissues by *P. aeruginosa* are virulence factors such as proteases and other degradative enzymes. But these proteins are not made until cell density is fairly high, a point where the organism might have a chance of overwhelming its host. The organism would not want to make the virulence proteins too early and alert the host to launch an immune response. The induction mechanism involves two interconnected quorum-sensing systems called Las and Rhl, both comprised of regulatory proteins homologous to LuxR and LuxI of *V. fischeri*. Many pathogens besides *Pseudomonas* appear to use chemical signaling to control virulence genes. Genomic analysis has revealed homologs of known quorum-sensing genes in *Salmonella*, *Escherichia*, *Vibrio cholerae*, the plant symbiote *Rhizobium*, and many other microbes.

Some microbial species not only chemically talk among themselves, but appear capable of communicating with other species. *V. harveyi*, for example, uses two different, but converging, quorum-sensing systems to coordinate control of its luciferase. Both sensing pathways are very different from the *V. fischeri* system. One utilizes an acyl homoserine lactone (AHL) as an autoinducer (AI-1) to communicate with other *V. harveyi* cells. The second system involves production of a different autoinducer (AI-2) that contains borate. Because many species appear to produce this second signal molecule, it is thought that mixed populations of microbes use it to "talk" to each other. In the case of *V. harveyi*, specific membrane sensor kinase proteins are used to sense each autoinducer (**Fig. 1**). At low cell densities (no autoinducer), both sensor kinases initiate phosphorylation cascades that converge on a shared response regulator, LuxO, to produce phosphorylated LuxO. Phosphorylated LuxO appears to activate a repressor of the *lux* genes. Thus, at low cell densities, the culture does not display bioluminescence. At high cell density, the autoinducers prevent signal transmission by inhibiting phosphorylation. The cell stops making repressor, which allows another pro-

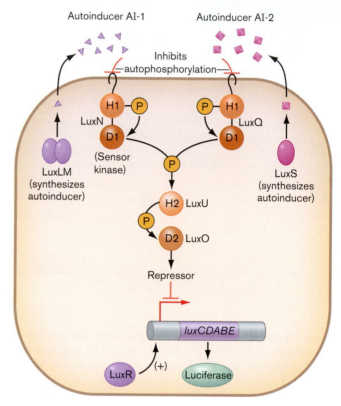

Figure 1   The two quorum-sensing systems of *V. harveyi*.   In the absence of autoinducers (AI-1 and AI-2), both sensor kinases trigger converging phosphorylation cascades that end with the phosphorylation of LuxO. Phosphorylated LuxO (LuxO-P) activates a repressor that inhibits expression of the luciferase genes. As autoinducer concentrations increase, they inhibit autophosphorylation of the sensor kinases and the phosphorylation cascade. As a result, repressor levels decrease, which allows the LuxR protein to activate the *lux* operon.

This knowledge begs the question, Why would a squid want to harbor this bioluminescent microbe? Buried in the sand by day, this animal emerges at night from its safe hiding place to hunt for food. In moonlight, the squid would appear as a dark silhouette from below, marking it as easy prey for predators. It is thought that the squid camouflages itself by projecting light downward from its light organ, giving the appearance of celestial stars as perceived by predators below.

THOUGHT QUESTION 10.10   Genes encoding luciferase can be used as "reporters." What do you think would happen if the promoter for an SOS response gene were fused to the luciferase open reading frame?

THOUGHT QUESTION 10.11   What do you think would happen if a culture were coinoculated with a *V. fischeri luxI* mutant and a *luxA* mutant, neither of which produces light?

tein, LuxR (*not* a homolog of the *V. fischeri* LuxR), to activate the *lux* operon. The "lights" are turned on. Bonnie Bassler (**Fig. 2**) and Pete Greenberg (**Fig. 3**) are two of the leading scientists whose studies revealed the complex elegance of quorum sensing in *Vibrio* and *Pseudomonas* species. Other organisms, such as *Salmonella*, have been shown to activate the AI-2 pathway of *V. harveyi*, dramatically supporting the concept of cross-species communication.

A recent report by Ian Joint and his colleagues has shown that bacteria can even communicate across the prokaryotic-eukaryotic boundary. The green seaweed *Enteromorpha* (a eukaryote) produces motile zoospores that explore and attach to *Vibrio anguillarum* bacterial cells in biofilms (**Fig. 4**). They attach and remain there because the bacterial cells produce acetyl homoserine lactone molecules that the zoospores sense. Part of the evidence for this interkingdom communication involved showing that the zoospores would even attach to biofilms of *E. coli* carrying the *Vibrio* genes for the synthesis of acetyl homoserine lactone. The implications of possible interkingdom conversations are staggering. Do our normal flora "speak" to us? Do we "speak" back?

For further discussion of molecular communication between prokaryotes and eukaryotes, see Chapter 21.

Figure 3 **Peter Greenberg, one of the pioneers of cell-cell communication research.** Peter Greenberg, first at the University of Iowa and now at the University of Washington, has studied quorum sensing in *Vibrio* species and various other pathogenic bacteria, such as *Pseudomonas*.

Courtesy of University of Iowa Medical Photography Unit

Figure 2 **Bonnie Bassler** (left) of Princeton University was instrumental in characterizing interspecies communication between bacteria.

©Denise Applewhite

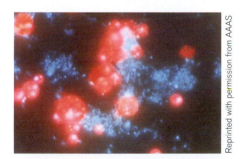

Reprinted with permission from AAAS

Figure 4 *Enteromorpha* **zoospores (red),** a type of algae, attach to biofilm-producing bacteria (blue) in response to lactones produced by the bacteria.

10.9 Genomics and Proteomics: Tools of the Future

Most of the studies exposing the regulatory systems discussed in this chapter were accomplished by examining only a small number of genes at any one time. Now, however, new technologies enable scientists to examine thousands of genes or proteins in a single experiment.

Gene Arrays Provide Global Snapshots of Gene Expression

A pebble tossed into a pond makes a wave that spreads and ultimately disturbs the entire pond. Similarly, altering one biochemical pathway in a cell produces a wave effect that impacts, to a greater or lesser degree, all other pathways. We are just now beginning to develop the tools needed to see these molecular "waves." One pow erful technique, called DNA microarrays, can simultaneously examine the expression of every gene in a cell. For an organism like *Salmonella typhi*, the causative agent of typhoid fever, this requires seeing over 4,600 genes. What is amazing is that all this can be done on a slide the size of a postage stamp.

The DNA microarray technique uses a tool called a **DNA microchip**, where DNA fragments from every ORF in a genome are affixed to separate locations on a solid support surface, producing a grid, or array (**Fig. 10.29**). The DNA fragments are generated by PCR or by in vitro chemical synthesis. The key to the technique is

Figure 10.29   DNA microarray technology.   This procedure is used to determine transcript levels produced from every gene in a cell (transcriptome). It assesses the relative level of each transcript produced by cells grown in different conditions or between mutant and wild-type strains of bacteria. **A.** In the basic procedure, RNA is extracted from cultures grown under different conditions (pH 7 versus pH 4.5 in this example). The RNA molecules from the two cultures are converted to DNA with reverse transcriptase (RT), then amplified and quantitatively tagged with different-colored fluorescent tags using PCR techniques. **B.** DNA fragments of each ORF in the bacterial genome are spotted by robotics to a glass slide (DNA microchip). The tagged (fluorescent) RT/PCR mixtures are then applied to the DNA microchip and allowed to anneal. Each tagged RT/PCR fragment will only anneal to the spot corresponding to its gene. **C.** Laser scanning of the annealed chip will reveal red spots if that gene was expressed mostly under pH 4.5 conditions, green spots if it was expressed mostly at pH 7, and yellow or orange if it was equally expressed under both conditions (an equal mix of red tag and green tag).

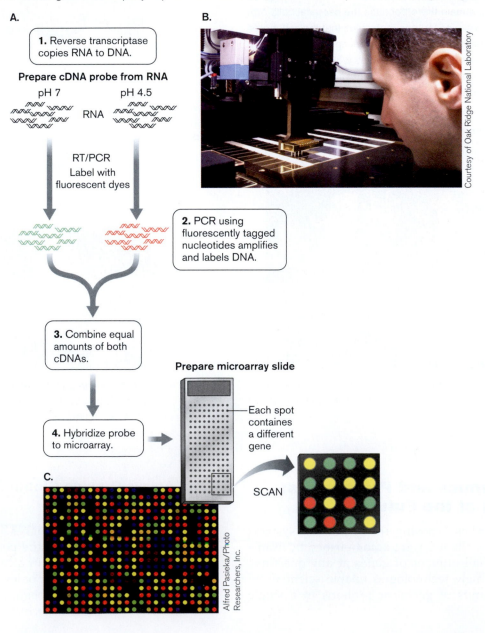

A.

1. Reverse transcriptase copies RNA to DNA.

Prepare cDNA probe from RNA

pH 7   pH 4.5

RNA

RT/PCR
Label with
fluorescent dyes

2. PCR using fluorescently tagged nucleotides amplifies and labels DNA.

3. Combine equal amounts of both cDNAs.

4. Hybridize probe to microarray.

Prepare microarray slide

Each spot containes a different gene

B.

Courtesy of Oak Ridge National Laboratory

C.

Alfred Pasieka/Photo Researchers, Inc.

SCAN

that only one strand of a double helix is actually fixed to the slide. Heating the slide breaks the hydrogen bonds holding down the complementary strands. The released strands are washed off, leaving the tethered strands free to anneal with any other complementary DNA that approaches. Think of these strands as a thousand tiny pieces of flypaper, with each piece engineered to catch a different type of fly.

The DNA microchip can be used to analyze RNA extracted from microbes grown under two different conditions. The ultimate goal is to compare the relative expressions of each gene under the two sets of environmental conditions; for example, at a pH of 7 versus a pH of 4.5. The latter is a pH that bacteria like *Salmonella* might encounter in eukaryotic host cell vesicles. After extraction, the two RNA samples are treated separately with the enzyme reverse transcriptase, which makes a DNA called **complementary DNA**, or **cDNA**, from RNA templates. Fluorescently tagged nucleotides are used in the reaction so that the product cDNAs also carry a fluorescent tag. How much of one cDNA is made depends on how much mRNA was initially present. Different colored tags (usually red and green) are used for each sample. In the example shown in **Figure 10.29**, the cDNA from the pH 7 and pH 4.5 cultures were labeled red and green, respectively.

The microchip is then flooded with a mixture of the two fluorescently tagged batches of cDNA, allowing the individual cDNA molecules to find and bind their mates on the organized grid. This is the hybridization step. After hybridization, a laser scans the slide and a fluorescence detector reads emissions. Composite images of the two scans, one for red and one for green, are made and analyzed by computer. If a given gene is expressed equally under both test conditions, the corresponding spot on the chip will contain equal amounts of red and green, a result that produces a yellow color. If a gene is expressed more during growth at pH 7, the spot will glow green. If expressed more at pH 4.5, it will fluoresce red. The result is a comprehensive view of what has been called the cell's **transcriptome**—all the cell's expressed mRNAs (**Fig. 10.29C**).

Two-Dimensional Gels Provide a Global View of Protein Expression

If all the expressed mRNAs in a microbe make up its transcriptome, then all of the expressed proteins form its **proteome**. Unlike the DNA genome, which is fixed, the transcriptome and proteome continually change in response to a changing environment. Gene array technology can capture what happens transcriptionally, but gene arrays do not necessarily reveal which proteins are expressed because stress often causes the translation of a message to change without changing its transcription. Thus, a significant increase in the level of a protein can occur without any change in the amount of mRNA. In addition, posttranslational modifications of proteins (for example, acetylation, phosphorylation) can take place in response to different environments. The powerful technique of two-dimensional gel electrophoresis (introduced in Section 3.5), is used to capture and view fluctuations in the proteome.

The first step in the process involves isoelectric focusing (IEF), where proteins in a cell extract are sorted based on their individual charges. Every protein is adorned with ionizable side groups (amino and carboxyl) that can impart charge. The degree of protonation (ionization) of each group is affected by pH and by the dissociation constant (pK) of the group. Generally, an ionizable group will be protonated at pH values below its pK. At pH values below their intrinsic pKs, amino groups become positively charged ($R-NH_3^+$), whereas carboxyl groups are neutral (R-COOH). When the pH rises above a group's pK, the amino groups will be neutral (NH_2) and carboxyl groups become negatively charged (COO^-). Proteins have hundreds of amino and carboxyl groups in different ratios, so for each protein, there will be a specific pH, called the **isoelectric point**, where the plus and minus charges cancel out—that is, where the net charge on the protein is zero.

Isoelectric focusing takes advantage of this phenomenon. Cell proteins are applied to a gel containing an immobilized pH gradient, called an IPG strip, and are subjected to an electrical field (**Fig. 10.30**). Negatively charged proteins move toward the positive pole until they reach a pH in the gel where the charges cancel out. Without a charge, the protein does not move under the influence of the electrical field. Positively charged proteins behave similarly, but move toward the negative pole.

Proteins with very different molecular weights can share the same isoelectric point. As a result, one band on an IEF gel could contain ten proteins. The goal of the second dimension, therefore, is to further separate proteins based on their molecular weights. To accomplish this, the IEF gel is transferred to the top of a sodium dodecyl sulfate (SDS) polyacrylamide gel (**Fig. 10.30D**). SDS is a detergent that forms positive and negative ions in solution. The negatively charged ion (dodecyl sulfate) has a hydrophobic end that coats proteins, giving all of them a negative charge. Polyacrylamide is a porous gel whose pore size varies in relationship to the acrylamide concentration and the amount of cross-linking used to make it a gel. When placed in an electrical field, the proteins (which are all negatively charged by SDS) are pulled toward the positive pole, located at the bottom of the gel. Small proteins easily slip through the tiny polyacrylamide pores and race to the positive electrode. The larger proteins find it more difficult to squeeze through the pores and are slow to move. The result is a gradient of proteins ranging

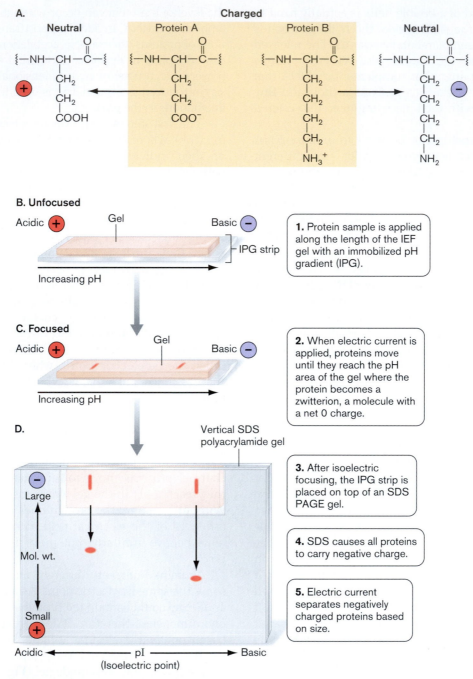

Figure 10.30 Two-dimensional analysis of cell proteins. **A.** General movement of negatively and positively charged proteins in an electrical current applied to an IPG strip. Once a protein reaches a position corresponding to its isoelectric point, movement stops and the protein focuses in the area. **B.** Sample applied to an IPG strip. **C.** The IPG strip after focusing. **D.** The focused IPG strip is layered onto a standard SDS polyacrylamide gel and separated by size. The result is separation of the proteins in a two-dimensional pattern. *Source*: Modified from Moat, Foster, and Spector. 2002. *Microbial Physiology*, 4th ed.

from the largest at the top of the gel to the smallest at the bottom.

When combined with isoelectric focusing, the two procedures display all the cell's proteins in a two-dimensional array, as shown in **Figure 10.31**. If the proteins are radioactively labeled before analysis, they can be visualized by autoradiography, or the proteins can be stained with fluorescent dyes after separation and read by a laser scanner. Subsequent computer analysis of the proteome patterns obtained from cells grown under two different conditions will reveal those proteins whose levels increase or decrease in response to the changing environ-

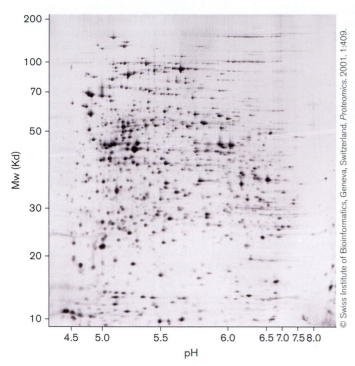

© Swiss Institute of Bioinformatics, Geneva, Switzerland. *Proteomics.* 2001. 1:409.

Figure 10.31   Proteome of *Escherichia coli*. Each spot represents a different protein.

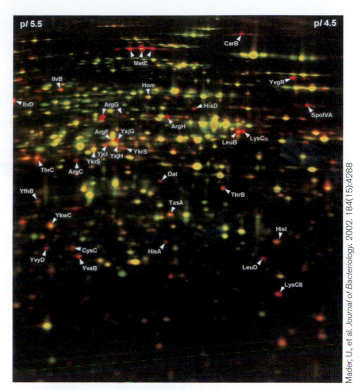

Mader, U., et al. *Journal of Bacteriology.* 2002. 184(15):4288

Figure 10.32   Proteomic profile of *B. subtilis* cells grown in minimal media with and without mixed amino acid supplementation (casamino acids). The IEF gradient used in the first dimension was pH 4.5–5.5; this represents only a part of the entire proteome. The figure is the result of dual-channel image analysis of silver-stained gels. A computer assigns the color red to proteins expressed in minimal media and the color green to proteins expressed in amino-acid-supplemented media. If the proteins are expressed under both conditions, the red and green colors combine to form yellow or orange.

ment. **Figure 10.32** shows a dual-channel image analysis that compares *Bacillus subtilis* proteomes from cultures grown in a minimal glucose medium to that of cultures grown in the same medium but supplemented with a rich assortment of amino acids. The different colors indicate whether the level of a protein is higher, lower, or the same in the two cultures.

Identifying Proteins Plucked from a Two-Dimensional Gel

The true power of proteomics becomes evident when it is linked to genomics. Knowing the complete sequence of an organism's chromosome makes it easier to identify proteins following growth under any experimental condition. For example, levels of certain proteins increase when a bioremediating microbe is grown on benzene instead of glucose. How does one determine which proteins are induced? Those proteins, and the genes encoding them, can be identified from a two-dimensional gel if the organism's genome is sequenced. This procedure is shown schematically in **Figure 10.33**. The protein spot of interest is excised from the gel and digested into peptide fragments with a protease. The peptide fragment mixture is then analyzed by mass spectrometry, which determines the precise molecular weight of each fragment. Combining the molecular weight of all the fragments gives the molecular weight of the whole protein. Computer programs match the experimentally determined molecular

weight of the protein (based on the sum of its peptide masses) with the calculated molecular weights of all ORFs predicted from the genome sequence. A match indicates identity.

> **THOUGHT QUESTION 10.12**   Why might the mass of a protein excised from a gel fail to match any of the ORF molecular weights predicted by the genome sequence?

This technology has been used for many microbial applications. One fascinating example centers on the transcriptional events that occur during the developmental cycle of *Caulobacter crescentus*. This organism progresses from a nonreplicating swimming form (swarmer cell) to a sessile stalked form that can replicate to make more swarming offspring. DNA microarray and proteomic analysis have revealed several potential regulators of this complex developmental process.

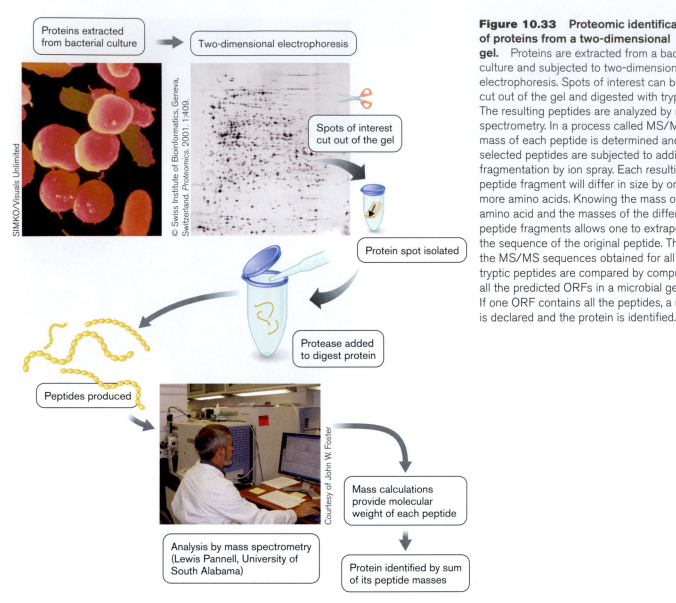

Figure 10.33 Proteomic identification of proteins from a two-dimensional gel. Proteins are extracted from a bacterial culture and subjected to two-dimensional electrophoresis. Spots of interest can be cut out of the gel and digested with trypsin. The resulting peptides are analyzed by mass spectrometry. In a process called MS/MS, the mass of each peptide is determined and then selected peptides are subjected to additional fragmentation by ion spray. Each resulting peptide fragment will differ in size by one or more amino acids. Knowing the mass of each amino acid and the masses of the different peptide fragments allows one to extrapolate the sequence of the original peptide. Then the MS/MS sequences obtained for all of the tryptic peptides are compared by computer to all the predicted ORFs in a microbial genome. If one ORF contains all the peptides, a match is declared and the protein is identified.

Image labels:
- Proteins extracted from bacterial culture
- Two-dimensional electrophoresis
- SIMKO/Visuals Unlimited
- © Swiss Institute of Bioinformatics, Geneva, Switzerland. *Proteomics.* 2001. 1:409.
- Spots of interest cut out of the gel
- Protein spot isolated
- Protease added to digest protein
- Peptides produced
- Courtesy of John W. Foster
- Analysis by mass spectrometry (Lewis Pannell, University of South Alabama)
- Mass calculations provide molecular weight of each peptide
- Protein identified by sum of its peptide masses

TO SUMMARIZE:

- **The transcriptome and proteome** constitute all of a cell's mRNA molecules and proteins, respectively. Transcriptomes and proteomes change as environmental conditions change.
- **A DNA microarray chip** is used to monitor the levels of individual mRNA molecules made during growth.
- **Two-dimensional gels** separate proteins by isoelectric point and molecular weight. They offer a snapshot of the proteome at any given point of growth or under any given growth condition.
- **Mass spectrometry** can determine the exact molecular weight of a protein (or the sequence of a component peptide) taken from an otherwise anonymous spot on a two-dimensional gel. The exact molecular weight or peptide sequence is compared against the database of ORFs deduced from the genomic sequence of the organism. A successful match, therefore, determines the protein's identity.

Concluding Thoughts

The systems discussed in this chapter are but a small sampling of the known gene regulatory systems. Many more await discovery. How the cell coordinates all these systems is still a matter of conjecture. Scientists, like genetic cartographers, are now trying to map all the regulatory circuits of *E. coli*, as well as to identify novel regulatory strategies used by other organisms. The structural biology of regulatory protein-protein interactions and protein-DNA binding are certainly the research waves of the future. For though we know the "what and where" of many molecular interactions, we remain, for the most part, ignorant of *how* they occur structurally.

CHAPTER REVIEW

Review Questions

1. List regulatory mechanisms discussed in this chapter that work at the DNA level, transcription level, translational level, and posttranslational level.
2. How does lactose induce the *lacZYA* operon?
3. If *lacY* is not induced unless lactose is present, how does external lactose induce the system?
4. How does tryptophan repress the tryptophan operon?
5. Describe a two-component signal transduction system.
6. Discuss how glucose impacts the utilization of lactose as a carbon source.
7. Name a regulatory protein that can activate and repress an operon's expression. How does it do that?
8. What is DNA footprinting?
9. What is the regulatory mechanism that uses translation to control transcription? How does it work?
10. When growth of *E. coli* slows, how does the cell slow its production of ribosomes?
11. Describe four ways that sigma factor production/activity can be regulated.
12. What are small regulatory RNA molecules?
13. How do bacterial cells couple genetic control and biochemical control in terms of nitrogen assimilation?
14. What is quorum sensing?
15. Describe DNA microarrays and two-dimensional gels. Can you think of a situation where a protein could be shown to increase in response to a stress but that same stress will not increase synthesis of the mRNA encoding the protein?

Key Terms

activator (346)
allolactose (351)
anti-anti-sigma factor (365)
anti-attenuator stem loop (362)
antisense RNA (368)
anti-sigma factor (365)
attenuator stem loop (362)
autoinducer (378)
catabolite repression (351)
complementary DNA (cDNA) (383)
corepressor (347)
derepression (347)
diauxic growth (351)
direct repeat (370)
DNA microchip (382)
electrophoretic mobility shift assay (EMSA) (356)

enhancer (364)
gene inversion (370)
immune avoidance (370)
inducer (347)
inducer exclusion (355)
induction (347)
integrated control circuit (349)
inverted repeat (370)
isoelectric focusing (383)
isoelectric point (383)
leader sequence (362)
lysogen (373)
lysogeny (373)
lytic (373)
operator (346)
operon (349)
phase variation (370)

prophage (373)
proteome (383)
quorum sensing (378)
regulatory protein (regulator) (346)
regulon (368)
repression (346)
repressor (348)
response regulator (348)
sensor kinase (347)
silencer (364)
stringent response (362, 365)
tandem repeat (370)
transcriptional attenuation (362)
transcriptome (383)
translational control (364)
two-component signal transduction system (348)

Recommended Reading

Camilli, Andrew, and Bonnie L. Bassler. 2006. Bacterial small-molecule signalling pathways. *Science* **311**:1113–1116.

Eagan, Susan M. 2002. Growing repertoire of AraC/XylS activators. *Journal of Bacteriology* **184**:5529–5532.

Fang, Ferric C. 2005. Sigma cascades in prokaryotic regulatory networks. *Proceedings of the National Academy of Science USA* **102**:4933–4934.

Fujita, Masaya, and Richard Losik. 2002. An investigation into the compartmentalization of the sporulation transcription factor σE in *Bacillus subtilis*. *Molecular Microbiology* **43**:27–38.

Henke, Jennifer M., and Bonnie L. Bassler. 2004. Bacterial social engagements. *Trends in Cell Biology* **14**:648–656.

Joint, Ian, Karen Tait, Maureen E. Callow, James A. Callow, Debra Milton, et al. 2002. Cell-to-cell-communication across the prokaryotic-eukaryotic boundary. *Science* **298**:1207.

Lewis, Mitchell. 2005. The *lac* repressor. *Critical Reviews in Biology* **328**:521–548.

Magnusson, Lisa U., Anne Farewell, and Thomas Nyström. 2005. ppGpp: A global regulator in *Escherichia coli*. *Trends in Microbiology* **13**:236–242.

Majdalani, Nadim, Carin K. Vanderpool, and Susan Gottesman. 2005. Bacterial small RNA regulators. *Critical Reviews in Biochem Mol Biol* **40**:93–113.

Merino, Enrique, and Charles Yanofsky. 2005. Transcription attenuation: A highly conserved regulatory strategy used by bacteria. *Trends in Genetics* **21**:260–264.

Pardee, Arthur B., Francois Jacob, and Jacques Monod. 1959. The genetic control and cytoplasmic expression of inducibility in the synthesis of β-galactosidase by *E. coli*. *Journal of Molecular Biology* **1**:165–178.

Piggot, Patrick J., and David W. Hilbert. 2004. Sporulation of *Bacillus subtilis*. *Current Opinions in Microbiology* **7**:579–586.

Wassarman, Karen M., Francis Repoila, Carsten Rosenow, Gisela Storz, and Susan Gottesman. 2001. Identification of novel small RNAs using comparative genomics and microarrays. *Genes and Development* **15**:1637–1651.

Chapter 11

Viral Molecular Biology

Virus structure ranges from a single RNA molecule to a spore-like package containing multiple layers of protein capsid and membranes. When a virus infects a cell, its reproduction may require a genome of less than 10 genes or more than 200. Viral reproduction varies from the relatively simple lytic program of bacteriophage T4 to the elaborate multi-organelle cycle of herpes viruses. Viruses such as herpes sequester their genome within that of the host. Latent viral genomes have contributed surprisingly to human evolution.

The molecular mechanisms of viruses raise intriguing questions: How do viruses trick host cell defenses into permitting their entry and replication? How did viruses originate: as free-living cells that took up predatory lifestyles or as host cell components that went wrong? Viral molecular biology offers clues to these questions. The molecular mechanisms of viral infection reveal targets for antiviral drugs.

Avian influenza H5N1 hemagglutinin, a viral envelope protein, with map of mutated amino acid residues. The circled area marks the site that binds the cell receptor. Mutations in this site may enable the influenza virus to infect humans efficiently and be transmitted rapidly. (PDB code: 2FK0)
Source: James Stevens, et al. 2006. *Science* 312:409.

In the 1990s, as the AIDS epidemic spread around the globe, health care workers noticed certain individuals who remained free of HIV over many years of exposure. These people proved to have a defective gene that encodes a macrophage cell surface protein, CCR5 (**Fig. 11.1**). Individuals who carry a single copy of the gene defect, known as deletion 32 (CCR5-Δ32), possess only one functional copy of CCR5, encoded by the homologous chromosome. The CCR5-Δ32 heterozygotes show delayed onset of AIDS, and individuals with two copies of the CCR5 defect resist infection by the most common strains of HIV.

The basis of the CCR5-Δ32 resistance to HIV is that the virus envelope proteins need to recognize the specific protein sequence of CCR5 in order to attach and enter the host cell. The single CCR5 defect decreases the number of CCR5 receptor proteins and thus the efficiency of infection. The defect on both homologous chromosomes eliminates the protein and prevents virus attachment altogether. This protein-protein interaction offers hope for developing methods to prevent or cure AIDS. It

may be possible to transfer the CCR5-Δ32 trait into the genome of a patient's macrophages, which will then resist HIV infection.

The HIV receptor defect is just one example of the promise of viral molecular biology for medical research and biotechnology. In Chapter 10 we discussed how molecular mechanisms such as repressors and DNA inversions regulate the function of microbial cells. Chapter 11 presents the molecular mechanisms of viral infection of cells. We assume a fundamental understanding of the nature of viruses, as presented in Chapter 6. Here we explore in greater depth the molecular basis of infection. First we consider two kinds of bacteriophages: a tailed phage and a filamentous phage. We then examine four viruses infecting humans, which include RNA and DNA genomes as well as a retrovirus. We see how all viral infection mechanisms share common themes of host recognition and co-opting cellular metabolism. At the same time, differences in genome structure dictate fundamentally different molecular requirements for infection.

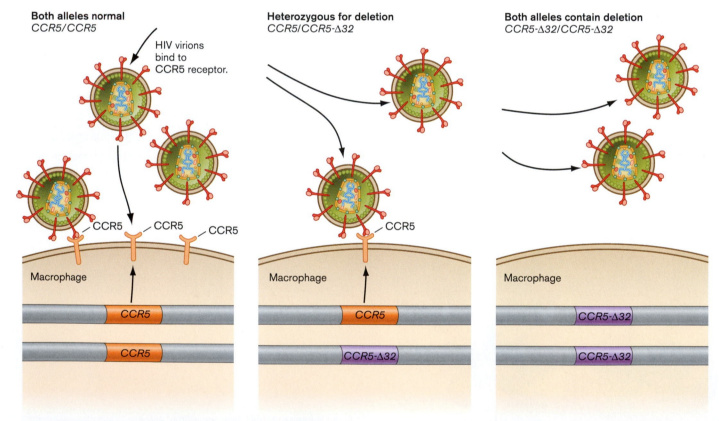

Figure 11.1   Resistance to HIV: a molecular mechanism.   HIV normally binds to the CCR5 surface protein of immune cells. Individuals heterozygous for a CCR5 deletion show delayed onset of AIDS. Individuals homozygous for the CCR5 defect show full protection from the major strains of HIV, apparently because the virus fails to bind to its target host cell.

This chapter emphasizes the fundamental biology of viral infection and propagation. The consequences of viral disease for the host organism are presented in Chapter 26.

www | The Viral Biorealm

11.1 Phage T4: The Classic Molecular Model

Bacteriophages are of interest for several reasons. Historically, phages provided the first living systems simple enough to dissect at the molecular level. Phages yielded fundamental discoveries in gene regulation, including the classic case of lambda lysogeny (discussed in Section 10.7). The first model for the genetic analysis of development was a phage system, the assembly of the "tailed phage" T4 (*Myoviridae*). A different kind of phage, the filamentous phage M13 (*Inoviridae*), provided key cloning vectors for DNA sequencing in the 1970s. Today, phages provide new tools for

antibacterial agents and new ways to make vaccines. They are also used to assemble wires in microscopic devices.

Here we present the replicative cycles of phages T4 and M13. These two phages offer interesting contrasts in their very different strategies of host infection.

Phage T4 Structure

Phage T4 may be the most complicated noncellular reproductive unit yet characterized. The intricate virion requires 20 gene products for assembly (**Fig. 11.2**). The T4 genome of 166 kilobase pairs (kbp) specifies 170 genes, many of which encode metabolic enzymes that duplicate host cell functions (**Fig. 11.3**). This viral genome almost looks like a cell genome in miniature. Thus, T4 has served as a model to probe the fundamental molecular processes of transcription and gene recombination as well as the global function of cells. The mutation strategies devised to dissect T4 function provided the basis for dissecting molecular development in animals.

www | 3-D tutorial on the T4 tail injector

The phage T4 virion contains numerous specialized protein subunits. The overall structure includes a polyhedral head, or capsid, containing its chromosome, a linear double strand of DNA (see **Fig. 11.2**) attached to a delivery device called the "tail;" hence the term **tailed phage**. The head is connected by a narrow neck tube to an internal tube that extends down within the sheath. The sheath plus the internal tube make up the tail. From the tail extend two sets of fibers, six connected at the neck and six at the baseplate. The elegant mechanical structure and function of this "molecular motor" was an early inspiration for nanotechnology.

Adsorption to Host and Delivery of DNA

Viral infection requires a molecular "fit" between a viral component and one or more molecules of the host cell surface. Phage T4 infects strains of *Escherichia coli* whose outer membrane possesses lipopolysaccharide (LPS) and the porin OmpC. The phage adsorbs (attaches) to the cell surface by contact between its tail fibers and the outer membrane (**Fig. 11.4**). Once several of the tail fibers are anchored to the outer membrane by LPS, the phage baseplate makes contact. The contact

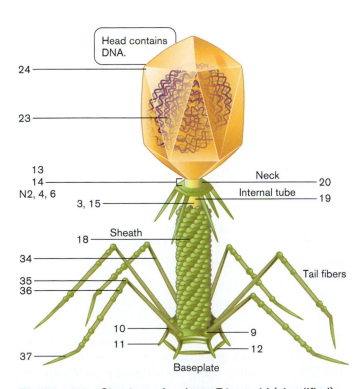

Figure 11.2　Structure of a phage T4 capsid (simplified). Proteins are assigned to specific genes (numbered) that are mapped on the chromosome in Figure 11.3.

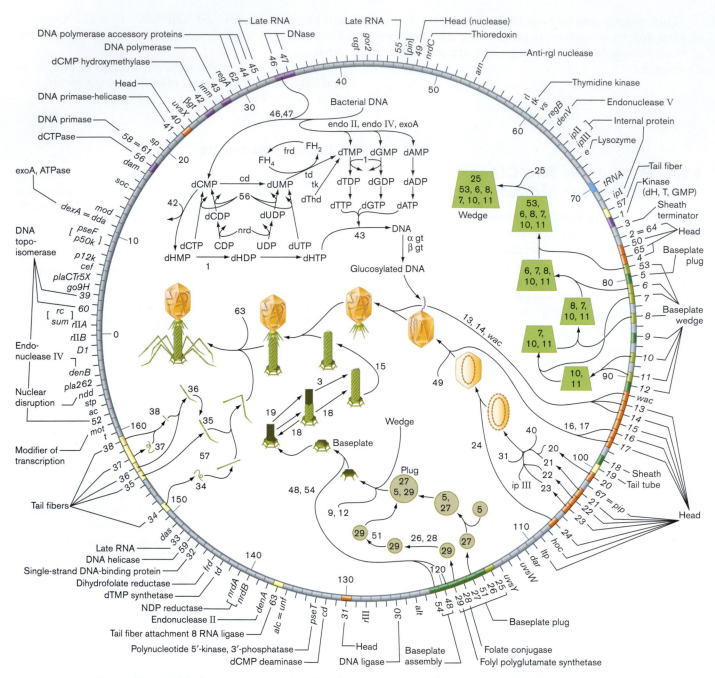

Figure 11.3 **Phage T4 genome showing functions of mapped genes.** The gene numbers correspond to those assigned to steps of assembly in Figure 11.2. *Source*: Based on Matthews, et al. *Bacteriophage T4*, ASM Press, 1983.

between baseplate and cell outer membrane generates a conformational change in the tail. The outer tube of the tail, the sheath, is a helical tube of subunits of protein P18. The sheath contracts by shortening and widening, enabling the internal tube to push through, like a molecular syringe.

The internal tube of the tail is capped by the needle-like "injector," a unique trimer composed of proteins P27 and P5. The injector structure was first determined by

Michael Rossman and colleagues at Purdue (**Fig. 11.5A**). Rossman's lab developed techniques of X-ray analysis and cryo EM to visualize many challenging viral structures, including whole virions such as rhinovirus (cold virus), dengue fever flavivirus, and canine parvovirus. These viral structures may now be used to test new antiviral drugs.

The P5 proteins form three intertwined strands in a "beta helix." As the sheath surrounding the internal

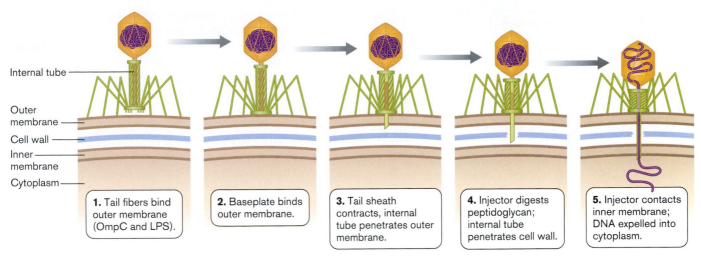

Internal tube

Outer membrane

Cell wall

Inner membrane

Cytoplasm

1. Tail fibers bind outer membrane (OmpC and LPS).

2. Baseplate binds outer membrane.

3. Tail sheath contracts, internal tube penetrates outer membrane.

4. Injector digests peptidoglycan; internal tube penetrates cell wall.

5. Injector contacts inner membrane; DNA expelled into cytoplasm.

Figure 11.4   Phage T4 adsorption and DNA injection.   First the tail fibers contact the outer membrane; then the baseplate completes the connection. The sheath contracts, and the internal tube (composed of protein 19) penetrates the outer membrane, followed by digestion of the peptidoglycan. The tube extends to the inner membrane, where DNA is released into the cytoplasm.

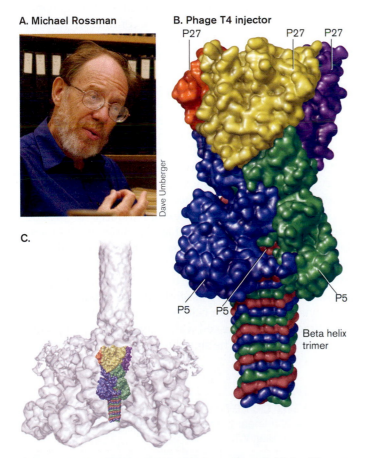

A. Michael Rossman

Dave Umberger

B. Phage T4 injector

P27   P27   P27

P5   P5

P5

Beta helix trimer

C.

Figure 11.5   Injector device of phage T4.   A. Michael Rossman and colleagues at Purdue University deciphered the structure of the phage T4 injector. They have also solved the structures of numerous whole viruses, such as the first rhinovirus (cold virus), dengue fever flavivirus, and canine parvovirus. **B.** Molecular model of the injector for the phage DNA. (PDB code: 1K28) **C.** Position of the injector within the baseplate, using low-resolution X-ray data. *Source*: Based on Rossman, et al. 2004. *Current Opinion in Structural Biology* 4:171–180.

tube contracts, the injector pokes through the host outer membrane and penetrates the cell wall (**Figs. 11.5B and C**). The peptidoglycan surrounding the injector is then digested by protein P5, whose globular alpha-helical domain has lysozyme activity. The hole in the cell wall enables the entire tail to penetrate the cell wall and inner membrane. When the end of the tail tube penetrates the inner membrane, the phage DNA is released through it into the cytoplasm.

Virulent Replication Destroys the Host Cell

Phage T4 is fully virulent; that is, its only reproductive option is to assemble progeny virions while destroying a host cell (**Fig. 11.6A**). By contrast, other tailed phages, such as phage lambda can undergo lysogeny, the integration of their genome into that of their host cell. The integrated phage genome then replicates passively with the host. Lysogeny is described in Chapter 6, and the genetic regulation of phage lambda is presented in Chapters 7 and 10.

When phage T4 DNA enters the host cytoplasm, a set of phage genes are activated for transcription. The phage genes are transcribed by the host cell RNA polymerase and translated by the host ribosomes. These host components, however, are supplemented by phage-encoded components such as tRNAs.

Phage genes expressed early in the infectious cycle are called **early genes**. The early-gene products include proteins needed to cleave host DNA, thus halting host macromolecular synthesis. In addition, several phage DNA synthesis enzymes replace key enzymes of the host.

A.

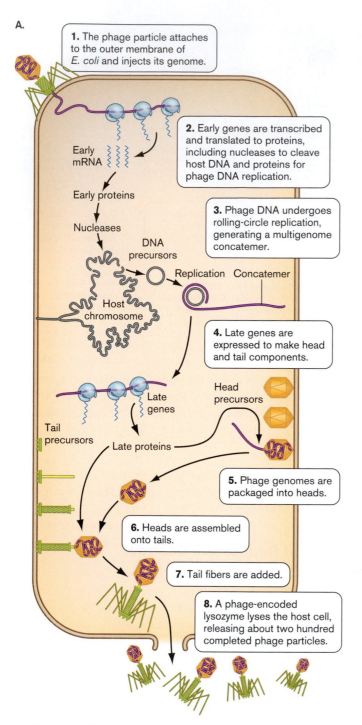

1. The phage particle attaches to the outer membrane of *E. coli* and injects its genome.

Early mRNA

2. Early genes are transcribed and translated to proteins, including nucleases to cleave host DNA and proteins for phage DNA replication.

Early proteins

Nucleases

DNA precursors

3. Phage DNA undergoes rolling-circle replication, generating a multigenome concatemer.

Replication   Concatemer

Host chromosome

4. Late genes are expressed to make head and tail components.

Late genes

Head precursors

Tail precursors

Late proteins

5. Phage genomes are packaged into heads.

6. Heads are assembled onto tails.

7. Tail fibers are added.

8. A phage-encoded lysozyme lyses the host cell, releasing about two hundred completed phage particles.

B. 5-Hydroxymethylcytosine

Figure 11.6   Replicative cycle of phage T4.   A. The phage particle attaches to the surface of *E. coli* and injects its genome. The genome is reproduced and packaged into progeny virions that are released upon lysis of the host cell. **B.** 5-Hydroxymethylcytosine replaces cytosine during DNA replication by T4-encoded DNA polymerase.

One T4 enzyme increases the rate of phage DNA replication tenfold. Another enzyme replaces cytosine with the modified base 5-hydroxymethylcytosine (**Fig. 11.6B**), which substitutes for cytosine throughout the phage chromosome. This modification of phage DNA prevents cleavage by endonucleases.

> **THOUGHT QUESTION 11.1**   Why would phage T4 production require a tenfold higher rate of DNA replication than does the host cell? Why would the phage substitute all of its cytosine with an unusual base that requires greater energy to synthesize?

Phage T4 DNA is synthesized by rolling-circle replication (described in Chapter 7). First, the linear DNA duplex of phage T4 circularizes within its host cell by recombination between the terminally redundant ends. The circularized duplex then replicates by the rolling-circle method, generating a linear concatemer in which multiple genomes are joined end to end (**Fig. 11.7**). The concatemer serves as template to synthesize the complementary strand. Out of the linear T4 DNA concatemer, the individual genomes are packaged into head coats. Each head coat actually contains sufficient volume to pack 3% more DNA than contained in a phage genome; thus, when the DNA duplexes extending from each head coat are cleaved, each phage DNA contains 3% terminal duplication (terminal redundancy). Inclusion of the terminal duplication causes each cleavage to occur another 3% farther along the genome. Thus, every T4 DNA molecule ends up with a different terminal repeat at some point throughout the genome.

Ultimately, the **late genes** are induced to produce the capsid and tail proteins that assemble on the membrane to make mature phage. During **assembly**, the phage genomes are stuffed into heads, the filled heads are attached to tails, and finally the various tail, collar, and capsid substructures are assembled into about 200 complete phage particles per cell. At last a late gene encodes a lysozyme to digest the cell wall, releasing the phages into their surroundings. The signal for lysis and phage release remains unknown.

> **THOUGHT QUESTION 11.2**   The phage T4 chromosome is a linear piece of DNA, yet the genomic map of T4 is circular. Why?

The temporal separation of early genes and late genes is common to the replicative cycles of all kinds of phages as well as viruses of eukaryotes. Viruses differ, however, in the size of their genomes; some, such as phage M13 (discussed shortly), contain only a small number of genes, whereas others, such as phage T4, encode products that duplicate host functions of DNA synthesis, ribosomal proteins, tRNA, as well as biosynthetic enzymes such as dihydrofolate reductase. Some phage-encoded enzymes, such as the T4 DNA polymerase, modify a cellular function to favor phage reproduction.

One hypothesis proposed for the evolution of T4 is that the phage arose by degeneration of a cellular parasite that ultimately lost its cytoplasm but retained metabolic components. An alternative hypothesis is that the phage T4 genome acquired many host genes through recombination with scraps of the host genome. The acquired genes then evolved new functions that increased the efficiency of phage assembly; for example, dihydrofolate reductase is used to assemble folate into the baseplate of the T4 injector.

Phage Particles Self-Assemble

The assembly of phage T4 particles within the host cytoplasm offers exceptional opportunities to visualize a molecular pathway. Each phage particle is assembled by convergence of three pathways involving the head coat, the tail, and the tail fibers (**Fig. 11.8**). All of these stages can be isolated and observed within infected cells by electron microscopy (**Fig. 11.9**). But with all these many steps, how did we determine the order of assembly and identify the genes responsible?

The stages of phage assembly were discovered using strains that carry mutations in various genes encoding proteins essential for development (see **Fig. 11.8**). In some cases, defective assembly leads to a bizarrely altered shape of the particle, such as a phage with a "giant head" (**Fig. 11.9A**). In other cases, a defective gene simply prevents progression in the pathway, halting assembly at an unfinished stage. For example, **Figure 11.9B** shows the phage particles found in a cell infected by a phage defective for gene 23. The unfinished particles consist of tail structures only. This occurs because gene 23 is required for assembly of the head, which must be completed before attachment to the tail. In **Figure 11.9C**, which shows phages defective for gene 19, we see only baseplates because the product of gene 19 is needed to build the tail upon the baseplate. In this case, assembled heads are found near the membrane of the cell (not shown).

> **THOUGHT QUESTION 11.3** Which numbered genes from Figure 11.8 might be responsible for each of the defective phage populations presented in Figure 11.9D?

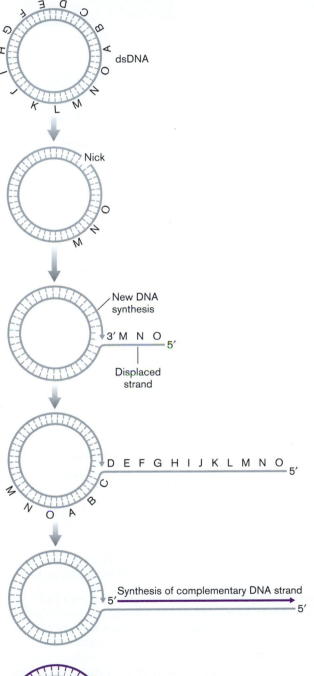

Figure 11.7 Rolling-circle replication of phage T4. Initially, the linear genome circularizes, then replicates as a concatemer. Out of the concatemer, the individual genomes are packaged into head coats, then cleaved. Each encapsidated DNA contains an end-duplicated 3% of its genome.

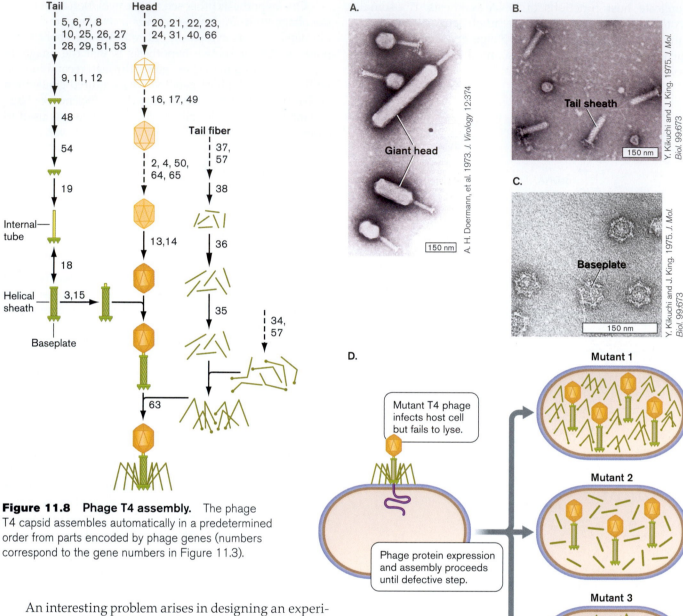

Figure 11.8 **Phage T4 assembly.** The phage T4 capsid assembles automatically in a predetermined order from parts encoded by phage genes (numbers correspond to the gene numbers in Figure 11.3).

Figure 11.9 **Developmental mutants of Phage T4.** Analyzing such mutants made it possible to decipher the genetic pathway governing assembly of the structure. **A.** Mutations in developmental genes lead to variant capsid morphology such as "giant head." **B.** Gene 23 defect results in tails without heads. **C.** Gene 19 defect results in baseplates without tails. **D.** Design of experiment using temperature-dependent mutations to identify steps of assembly.

An interesting problem arises in designing an experimental strategy: How do we grow populations of phages whose particles fail to complete assembly? The answer lies in using **conditional lethal mutations**, that is, mutations that cause a lethal defect under one growth condition but permit growth under a second condition. Two classes of conditional lethal mutant strains of phage T4 were obtained through mutagenesis by Robert Edgar and Richard Epstein in the early 1960s. These two classes are:

■ **Temperature-sensitive mutants.** Mutants that are temperature sensitive can grow at one temperature but fail to grow at another. For example, a point mutation may result in a protein product that is stable at 25°C but denatures at 42°C, so that the phage can reproduce at 25°C but not at 42°C.

■ **Nonsense mutations countered by suppressor tRNA.** A mutant that contains a nonsense codon (stop codon) in the middle of a gene prematurely terminates translation of the protein. Such a strain, however, may be grown in a strain of *E. coli* that makes a certain type of mutant tRNA. The mutant tRNA contains an anticodon that matches the stop codon sequence, which allows the tRNA to place an amino acid into the growing protein. This kind of tRNA is called a **nonsense suppressor** because the tRNA anticodon mutation suppresses the effect of the nonsense mutation in the phage.

Through many years of analysis of phage mutants, Edgar and Epstein and their colleagues showed that conditional lethal mutations could be used to dissect a developmental pathway as complex as that of phage T4. Their work laid the foundation for the molecular analysis of development in animals and plants and the discovery of the molecular basis of inherited diseases. With whole-genome sequences in hand, we use conditional lethal mutations and suppressor mutations to decipher how all the gene products work together. For example, genetic analysis of development in the fruit fly *Drosophila* uses temperature-sensitive mutations to dissect mechanisms of gene regulation.

TO SUMMARIZE:

- **The phage T4 virion** consists of a head containing its DNA genome and accessory proteins; a tail composed of an internal tube with a sheath; and tail fibers.
- **Adsorption of T4** occurs by binding of tail fibers to bacterial outer membrane receptors, followed by binding of the tail baseplate.
- **Injection of T4 DNA** through the internal tube leads to lytic infection. Early genes encode a DNA polymerase and a DNase to cleave host DNA.
- **Rolling-circle replication** generates progeny genomes linked in a concatemer. The concatemer is cut with an offset, so that each linear genome has a slight overlap, cut at a different position in the sequence.
- **Self-assembly of virions**, including packaging of DNA into the head, occurs in arrays beneath the inner membrane.
- **Cell lysis** is caused by a lysozyme expressed by late genes in the T4 genome.
- **The phage T4 assembly process was dissected using temperature-sensitive mutants and nonsense mutants with suppressor tRNA.** This strategy of genetic dissection was used as a model for experiments on animal development.

11.2 The Filamentous Phage M13

Filamentous phages consist of an extended, flexible tube of variable length. An exciting use of the filamentous phage M13 is to grow microscopic metal "nanowires" that conduct electricity. The wires are made using a mutant phage whose coat proteins bind metals such as gold and cobalt. Around the phage filaments form microscopic tubes of metal, which can be used in the manufacture of tiny batteries.

The filamentous phages, like the tailed phages, must construct progeny using the materials and metabolism of a host cell. Yet the means by which they achieve this goal differ greatly from those of T4. Where T4 featured remarkable complexity in its capsid, the structure of M13 is a simple helical filament (**Fig. 11.10**). Instead of the 170 genes of the T4 genome, M13 manages just as successfully with 11, including some that overlap (**Fig. 11.11**). Instead of lysing its host, M13 allows the host to continue growing, although more slowly than uninfected cells.

Filamentous Phage Structure

The M13 capsid consists of a flexible protein cylinder that is about 150 times as long as it is wide (**Fig. 11.10A**). The cylinder is assembled around a supercoiled single-stranded circle of DNA. The advantage of the filamentous structure is that it requires a relatively small number of different protein subunits encoded in the genome. Similar flexible filamentous structures have evolved independently in animal and plant viruses; compare, for example,

A. M13 bacteriophage

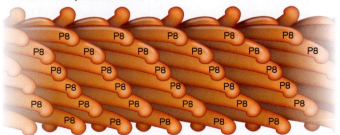

Protein 3

Carolos F. Barbas, et al. 1991. *PNAS* 79:78

B. Filament of protein 8 subunits

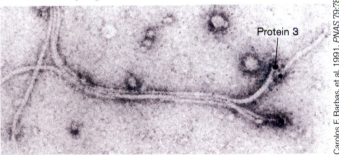

Figure 11.10   A filamentous phage: Page M13.   A. M13 filamentous phage particles (8 nm wide, 900 nm long) with tail protein P3 gene fused to a gene encoding anti-tetanus-toxoid antibody. The protein P3-antibody was stained with tetanus toxoid-colloidal gold, observed as a black sphere at the tip of the phage filament (TEM). **B.** Model of filamentous phage structure consisting of a helical tube of protein 8 subunits, based on X-ray crystallography.

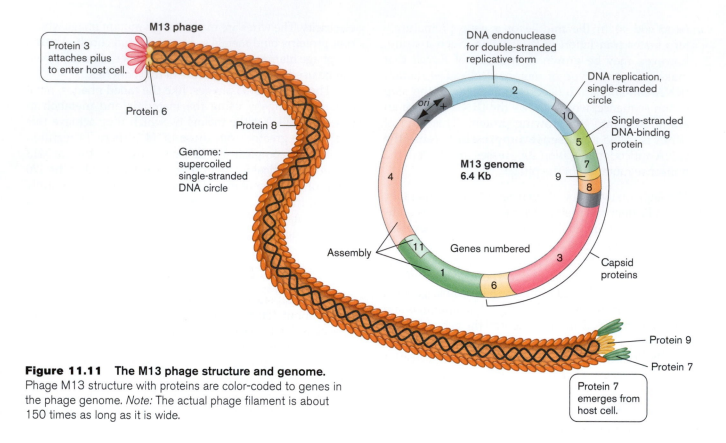

Figure 11.11 **The M13 phage structure and genome.** Phage M13 structure with proteins are color-coded to genes in the phage genome. *Note:* The actual phage filament is about 150 times as long as it is wide.

Ebola filovirus and tobacco mosaic virus (see Section 6.2). Filamentous animal viruses, however, show a virulent infectious cycle completely different from the slow release of filamentous phages.

Each subunit of the M13 filament consists of a short alpha-helical peptide encoded by gene 8. The protein P8 subunits have positively charged lysine residues at the internal end to bind the negatively charged DNA. The subunits extend outward at a shallow angle, looking like bunches of bananas stacked as a long tube (**Fig. 11.10B**). The tube is tipped at one end by five copies of protein P3, a much larger and more flexible protein that binds the cell surface receptor during phage adsorption (see **Fig. 11.11**). At the opposite end of the tube are proteins P7 and P9, which enable a progeny phage particle to emerge from the cell.

The simple, flexible nature of the filamentous phage has proved useful for development of the technology of **phage display**. Phage display is used in drug discovery to identify molecular interactions between protein domains. In phage display, a gene encoding a key portion of a protein domain (usually a short peptide) is cloned at the terminus of the sequence for one of the phage coat proteins, such as protein P8 or protein P3 (see **Fig. 11.11**). The recombinant phage now displays the desired protein domain on its coat, where it can be recognized by antibodies or can selectively bind to a target molecule. This technique can be used to screen a library of recombinant phages for binding partners to an antibody or to a target regulatory protein, such

as a cell cycle regulator involved in tumorigenesis. Phage display is discussed further in Chapter 12.

Adsorption to Host and Delivery of DNA

Phage M13 infects only F$^+$ strains of *E. coli*, which possess F pili. The phage adsorbs by binding of the tip protein P3 to an F pilus (**Fig. 11.12**). The pilus contracts, bringing the phage into contact with a **coreceptor**, or secondary receptor protein, TolA. The TolA complex is an important structural anchor of the *E. coli* cell envelope. Coreceptors for host attachment are a common feature of viruses, including animal viruses such as influenza virus and HIV-1 (discussed later).

As the M13 filament contacts the cell, it transfers its DNA across the cell envelope by an unknown mechanism. Unlike phage T4, the empty M13 filament does not remain intact outside the cell. Instead, the filament subunits disassemble and insert into the inner membrane, where they eventually join the pool of newly synthesized subunits to form progeny phages. Because most M13 assembly steps involve the membrane, we know less about the details of M13 assembly than we do about assembly of T4, whose components are all cytoplasmic and therefore easier to study.

M13 Reproduction Does Not Kill the Host

As the infecting capsid dissolves in the inner membrane, the phage DNA (single-stranded circle) enters the cytoplasm. Phage DNA is replicated by host enzymes while

Figure 11.12   M13 adsorption to the pilus and TolA.   Attachment to pilus of F⁺ cell.

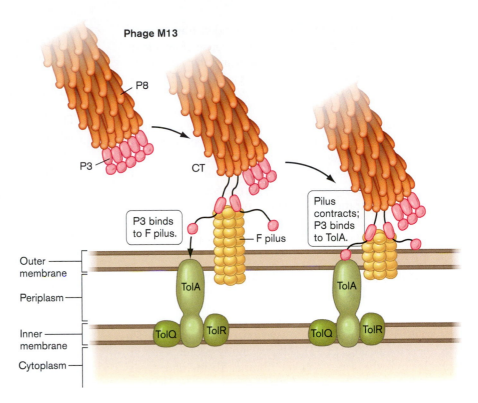

host cell growth continues. Replication of the single-stranded genome starts with generation of a double-stranded replicative form (**Fig. 11.13A**). The replicative form then undergoes rolling-circle replication of the (+) strand, like phage T4. The growing concatemer (multiple repeats of the minus strand) is protected by subunits of protein P5, a single-strand binding protein. As each genome length is completed, it is nicked and circularized without overhang. The circular DNA is packaged with capsid proteins at the cell membrane.

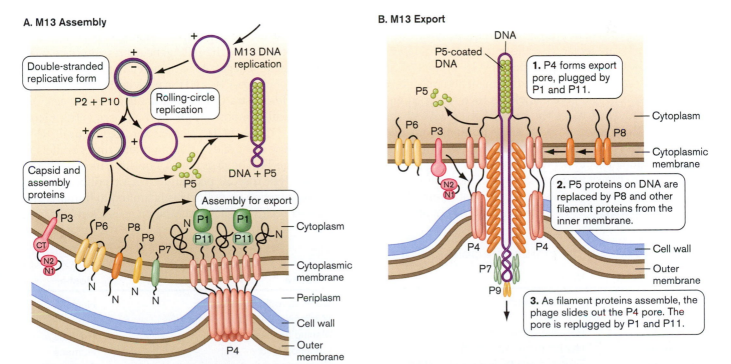

Figure 11.13   Assembly and export of M13 progeny phages.   A. Replication of phage M13. Most assembly is conducted by phage proteins within the cell membrane. **B.** Assembly for export through the pore made of P4 monomers. P4 makes a channel used to export filamentous phages without lysing the bacterial host. The single-stranded DNA of the filamentous virus is coated by P8 subunits as it passes through the membrane, then passed through the P4 channel to emerge as mature, infectious phages.

The single-stranded to double-stranded transition of phage M13 was important historically because of its role in DNA sequencing. When DNA-sequencing methods were first invented by Sanger, they required a template that was single-stranded. But recombinant DNA techniques required double-stranded DNA for restriction enzyme cleavage and ligation. The most effective way to obtain large quantities of a single-strand template was to splice a gene sequence within a double-stranded M13 replicative form. The replicative form was then used to transform cells that subsequently produced M13 phages carrying single-stranded DNA. Today, a single-stranded template is no longer required for PCR-based sequencing, but M13-based vectors remain useful for special applications, such as phage display.

Filamentous Phages Self-Assemble at the Inner Membrane

The M13 genome expresses packaging proteins, including protein P8 as well as the specialized tip proteins P3, P7, and P9 (see **Fig. 11.11**). All packaging proteins insert initially into the inner cell membrane. To mediate their assembly, a pore complex forms, composed of protein P4. P4 monomers extend through all layers of the envelope and assemble in a ring to build the pore. The positive charges on the N-terminal end of P4 help attract the negatively charged DNA into the pore. The pore guides the assembly of P8 monomers (also positively charged) and other packaging proteins around the DNA (see **Fig. 11.13**). To avoid lysing the host cell, the pore is plugged by proteins P1 and P11 until phage assembly begins.

As the P5-coated chromosome enters the pore, the protein P5 subunits are replaced by the growing tube of P8 proteins. The final step of phage export involves capping with protein P3, which is needed for attachment and infection of the next host cell. The pore complex is then recapped with P1 and P11 to avoid lysis. The production of M13 phages slows growth of the host because of resource consumption, but it does not destroy the host cell; in fact, two of the phage's 11 gene products (about one-sixth of its genome) are devoted to preventing host cell lysis. The evolved strategy of M13 is to maintain its current host at minimal cost, rather than to maximize its immediate number of progeny.

> **THOUGHT QUESTION 11.4**   Compare and contrast the reproductive strategies of phages T4 and M13.

TO SUMMARIZE:

- **The filamentous phage M13** has a filamentous capsid of variable length, able to package genomes of variable size; this is an advantage for cloning vectors.
- **Phage M13 adsorbs to F pili.** The pilus contracts, bringing the phage into contact with the host outer membrane. The filament penetrates to the inner membrane, where the filament subunits fuse and dissolve into the membrane.
- **Host enzymes replicate phage DNA and express phage genes.**
- **Phage M13 replicates without destroying the host cell.**
- **The DNA genome is a single-stranded DNA circle.** It must be replicated to a double-stranded circle for synthesis of progeny single-stranded genomes.
- **M13 virions are assembled at the inner membrane** for export through a phage-encoded pore complex.

11.3 A (+) Strand RNA Virus: Polio

Compared to bacteriophages, viruses that infect animal cells usually have the simpler task of crossing the eukaryotic cell membrane. Within the eukaryotic cell, however, the interior structure is generally more complex than that of prokaryotes. To use the cell's gene expression system, an animal virus must negotiate various cellular compartments, such as endocytic vesicles and the nucleus. Different viruses use very different strategies to manipulate the host cell.

A medically important group of viruses is the **picornaviruses** (**Table 11.1**), a class of icosahedral RNA viruses whose life cycle has some interesting parallels to that of phage T4. The picornavirus **poliovirus** is the cause of the paralytic disease **poliomyelitis**, or polio. The host range for infection by poliovirus is limited to humans and our closest relatives, such as chimpanzees. Closely related

Table 11.1   **Picornavirus species (examples).**

Genus	Virus	Disease	Host
Enterovirus	Coxsackievirus	Common cold, myocarditis	Humans
	Echovirus	Meningitis, paralysis, encephalitis	Humans
	Poliovirus (PV)	Meningitis, paralysis	Humans
Rhinovirus	Human rhinovirus (HRV)	Common cold	Humans
Aphthovirus	Foot-and-mouth disease virus (FMDV)	Foot-and-mouth disease	Cattle, swine
Hepatovirus	Hepatitis A virus (HAV)	Hepatitis	Humans
Cardiovirus	Encephalomyocardiovirus	Encephalitis, myocarditis	Mice
Erbovirus	Equine rhinotracheitis B virus (ERBV)	Rhinotracheitis (respiratory infection)	Horses

picornaviruses such as the **rhinoviruses** and coxsackie-virus cause the common cold. A major veterinary concern is aphthovirus, the cause of foot-and-mouth disease in cattle and swine. In 2001, an outbreak of foot-and-mouth disease in the United Kingdom caused major economic and societal disruption.

Poliovirus is famous historically for the outbreaks of poliomyelitis that swept the United States during the first half of the twentieth century (**Fig. 11.14**). The virus enters orally from fecal sources such as contaminated water. In most cases, poliovirus infection causes a mild gastrointestinal disease, but the virus can invade motor neurons of the spinal cord and lower brain, resulting in paralysis. Tens of thousands of Americans were infected, including many young children. Many had few or no symptoms but could transmit the disease to others. About 1% became paralyzed, of whom about 15% died. Decades later, paralyzed individuals may suffer post-polio syndrome, a condition in which muscles that were previously affected by the polio infection experience atrophy and pain. Today, polio infection has nearly been wiped out worldwide, although outbreaks still occur due to low levels of vaccination in isolated communities such as the Amish and in countries of Southeast Asia and Africa.

President Franklin Roosevelt, himself a victim of the disease, established a national foundation called the March of Dimes to develop a vaccine. Support from the March of Dimes led Jonas Salk and colleagues at the University of Pittsburgh to develop the first vaccine against polio in 1952. The spectacular success of the March of Dimes set the pattern for future public support for research on more challenging diseases, such as cancer and AIDS.

Despite the success of the polio vaccine, questions remain. The vaccine is not perfect; could better vaccines avoid the one-in-a-million deaths actually caused by the vaccine? What is the cause of the post-polio syndrome that affects some former polio victims 30–40 years after their illness? Are there positive applications of poliovirus, perhaps as a gene therapy vector or even as a treatment for brain tumors? To address these questions an understanding of the virus and its host cell at the molecular level is required.

Poliovirus Structure and (+) Strand RNA Genome

The structure of poliovirus is so compact and symmetrical that it was crystallized early in the twentieth century. Its three-dimensional structure was modeled by X-ray crystallography and by digital reconstruction based on cryo EM. Like many viruses, the capsid is fundamentally icosahedral (20-sided) (discussed in Chapter 6). Its form can be visualized as 60 triangular faces, each composed of proteins VP1, VP2, and VP3 (**Fig. 11.15A**). The three proteins arise from one long "polyprotein" or polypeptide translated as a unit, then cleaved into three pieces. The polyhedral form enables a small number of genes to encode a relatively large capsid. For this reason, capsids of icosahedral symmetry are common among animal and plant viruses, including, for example, the rhinoviruses (cold viruses), aphthovirus (foot-and-mouth disease), herpes viruses, and rice yellow mottle virus.

Each group of three triangular faces forms a pentamer. Beneath the pentamers of VP1, VP2, and VP3 lie subunits of a fourth protein, VP4. VP4 subunits coat the interior, strengthen the structure, and help package the RNA genome.

The RNA genome of poliovirus consists of a coding (+) strand. Its 3′ OH end has a poly-A tail, like eukaryotic mRNA, but its 5′ OH end is capped with a special viral protein, VPg.

Virus Attachment and Genome Entry

The poliovirus attaches to the host at a species-specific receptor protein location. The virus can only infect host cells that contain a **poliovirus receptor (PVR)**, a cell membrane protein found only in humans and close relatives such as chimpanzees. The structure of PVR protein suggests host

Figure 11.14   Polio in the United States.   A. Board of Health quarantine notice, around 1900, warning people to stay away from a polio-infected household. **B.** The most famous polio victim, President Franklin Delano Roosevelt, with his granddaughter Ruthie Bie and his dog Fala. Roosevelt's legs were paralyzed permanently, and he used a wheelchair throughout his three-plus terms as president. (He died a year after his fourth reelection.)

A. Quarantine notice

March of Dimes archive

B. President Franklin D. Roosevelt

Franklin D. Roosevelt Library

A.

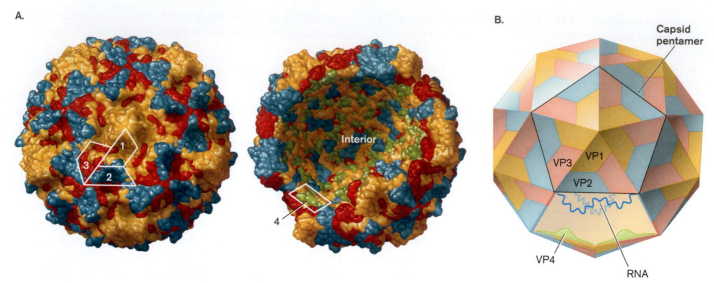

B.

Figure 11.15   Poliovirus structure.   A. Digital reconstruction of the poliovirus capsid based on TEM. *Left:* The icosahedral structure is composed of external capsid proteins VP1 (yellow), VP2, (blue), and VP3 (red). The three proteins form a structural unit, five of which form a pentamer (B). *Right:* Beneath the pentamer, VP4 subunits (green) coat the interior. The interior normally contains the RNA chromosome (not shown). (PDB code: 1ASJ) **B.** Schematic of capsid proteins forming a pentamer.

function related to inflammation or cell adhesion, but the details are unknown. A transgenic mouse expressing a recombinant human gene for PVR can be infected by polio-virus, generating all the major symptoms of poliomyelitis. The polio mouse was the first example of a transgenic animal created as a model system for a human disease.

Poliovirus receptors diffuse freely within the host cell membrane. PVR can bind the polio virion at any vertex of the capsid pentamer (a pentamer is shown in **Fig. 11.15B**). PVR actually binds to adjacent VP2 and VP3 proteins surrounding the VP1 subunits, as shown in **Figure 11.16**. When poliovirus attaches to a cell, multiple receptor molecules diffuse near it and eventually bind to several faces of the capsid. As the poliovirus receptors bind, the membrane forms a cup around the capsid. Unlike other viruses, such as influenza virus, poliovirus requires only the earliest stage of endocytosis; the virion delivers its RNA to the cytoplasm immediately after internalization beneath the cell membrane. Thus, poliovirus infection is insensitive to drugs such as bafilomycin A1, which prevents infection by viruses that enter cells through endosomes.

As the poliovirus interacts with the poliovirus receptors, its pentamer structures shift so as to extrude the N-terminal ends of the five VP1 peptides (**Fig. 11.16**). The five VP1 molecules of the pentamer poke through the cell membrane together, generating a pore through which the chromosome passes into the cytoplasm. The transfer of the poliovirus chromosome parallels that of phage T4 in that the nucleic acid is injected across the cell membrane through a viral device while the capsid proteins are left outside.

The viral chromosome, a (+) strand RNA with a protein VPg attached to the 5′ end, is said to be uncoated.

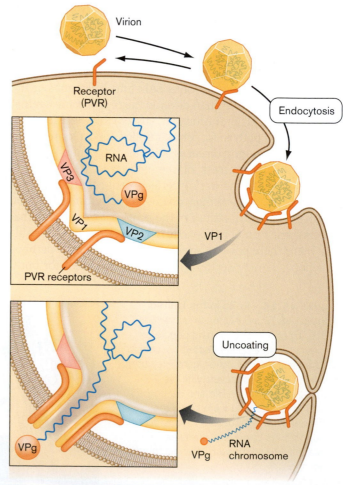

Figure 11.16   Poliovirus attachment to the host cell.   The poliovirus binds PVR at the VP2 and VP3 subunits. A conformational change in the VP1 subunits allows insertion of genome.

The uncoated genome is now ready to direct formation of progeny virions.

Gene Expression and Replicative Cycle

The uncoated viral RNA has two distinct roles in producing progeny virions: (1) mRNA translation by host ribosomes and (2) template replication of progeny RNA genomes. These two processes occur in separate compartments within the host cells.

Translation to make peptides. In the cytoplasm, the RNA is translated to make three large precursor peptides, P1, P2, and P3 (**Fig. 11.17A**). All three peptides are

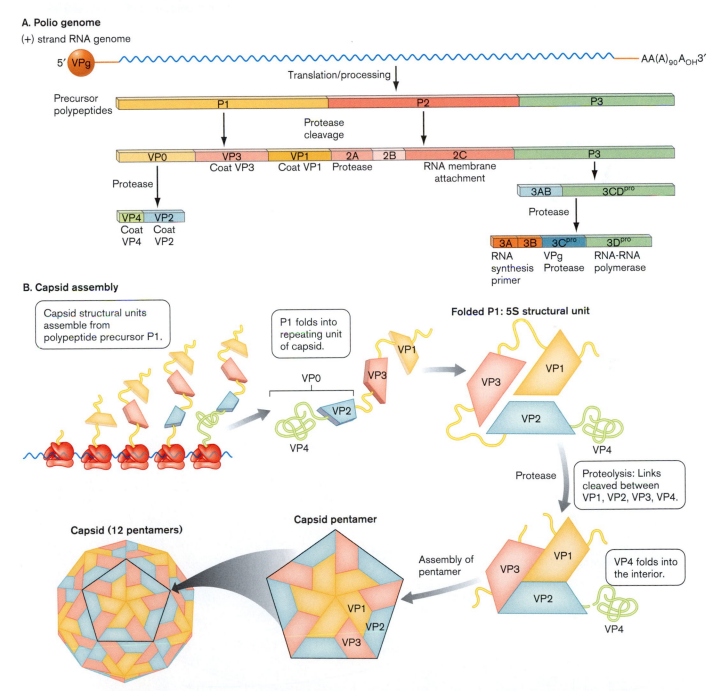

Figure 11.17   Polio genome and capsid assembly.   A. The (+) strand RNA genome encodes three precursor polypeptides, P1, P2, and P3. These precursors are cleaved by proteases to generate the virus capsid proteins (VP1, VP2, and VP3) as well as the RNA-dependent RNA polymerase, proteases, and other assembly proteins. **B.** The capsid proteins are assembled together into the 5S structural unit before cleavage by protease. Five 5S units are packed into a pentamer, 12 of which assemble to build the icosahedral capsid.

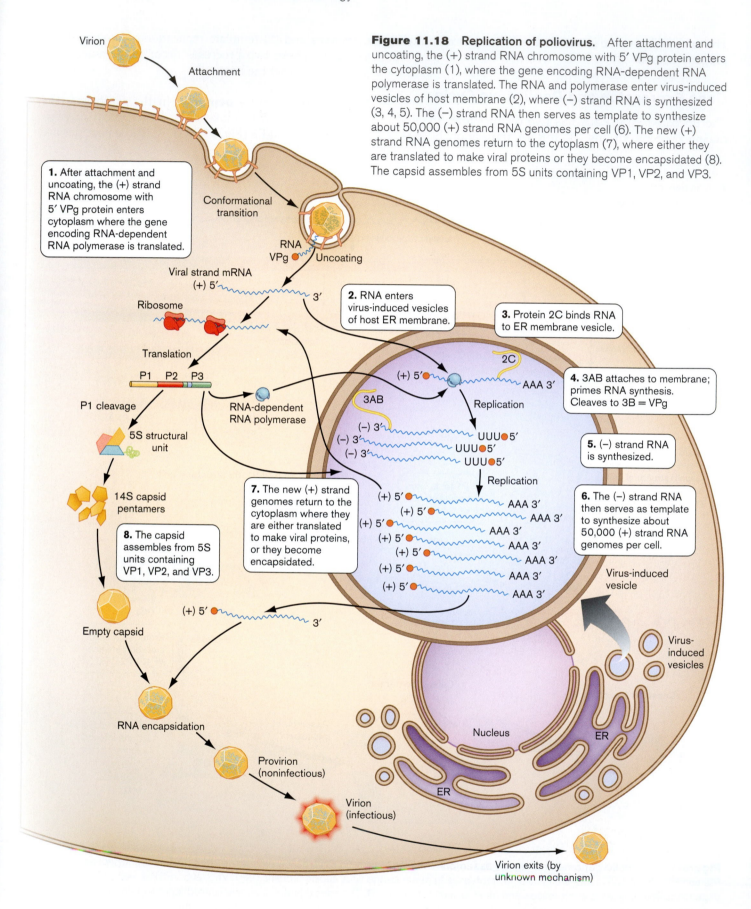

Figure 11.18 Replication of poliovirus. After attachment and uncoating, the (+) strand RNA chromosome with 5′ VPg protein enters the cytoplasm (1), where the gene encoding RNA-dependent RNA polymerase is translated. The RNA and polymerase enter virus-induced vesicles of host membrane (2), where (−) strand RNA is synthesized (3, 4, 5). The (−) strand RNA then serves as template to synthesize about 50,000 (+) strand RNA genomes per cell (6). The new (+) strand RNA genomes return to the cytoplasm (7), where either they are translated to make viral proteins or they become encapsidated (8). The capsid assembles from 5S units containing VP1, VP2, and VP3.

1. After attachment and uncoating, the (+) strand RNA chromosome with 5′ VPg protein enters cytoplasm where the gene encoding RNA-dependent RNA polymerase is translated.

2. RNA enters virus-induced vesicles of host ER membrane.

3. Protein 2C binds RNA to ER membrane vesicle.

4. 3AB attaches to membrane; primes RNA synthesis. Cleaves to 3B = VPg

5. (−) strand RNA is synthesized.

6. The (−) strand RNA then serves as template to synthesize about 50,000 (+) strand RNA genomes per cell.

7. The new (+) strand genomes return to the cytoplasm where they are either translated to make viral proteins, or they become encapsidated.

8. The capsid assembles from 5S units containing VP1, VP2, and VP3.

Virion

Attachment

Conformational transition

RNA VPg Uncoating

Viral strand mRNA (+) 5′ — 3′

Ribosome

Translation

P1 P2 P3

P1 cleavage

5S structural unit

14S capsid pentamers

Empty capsid

RNA-dependent RNA polymerase

RNA encapsidation

Provirion (noninfectious)

Virion (infectious)

(+) 5′ — 3′

2C

(+) 5′ AAA 3′

Replication

3AB

(−) 3′

(−) 3′ UUU 5′
(−) 3′ UUU 5′
UUU 5′

Replication

(+) 5′ AAA 3′
(+) 5′ AAA 3′
(+) 5′ AAA 3′
(+) 5′ AAA 3′
(+) 5′ AAA 3′
(+) 5′ AAA 3′
(+) 5′ AAA 3′

Virus-induced vesicle

Virus-induced vesicles

Nucleus

ER

ER

Virion exits (by unknown mechanism)

eventually cleaved by one of two viral proteases (products 2A and 3C) to generate a total of 11 different protein products. The P1 precursor forms the repeating unit of the isosahedral capsid; each unit is ultimately cleaved into four peptides. The P2 precursor is cleaved to form a two-subunit protease and an RNA-membrane interaction protein. The P3 subunit is cleaved to form RNA-dependent RNA polymerase, the Vpg primer for RNA synthesis, and another protease.

The capsid proteins ultimately arise from cleavage of P1. Cleavage, however, does not occur immediately (**Fig. 11.17B**). Instead, the peptide P1 (including sequences V4–V2–V3–V1) first folds into the tertiary structure of the capsid's major repeating unit. The repeating unit is called the 5S structural unit, referring to its size in Svedberg units of sedimentation rate. After folding together, each 5S unit then undergoes proteolysis at the junctions between the four subunits (V1 through V4). The 5S units then join together in groups of five to form pentamers, 12 of which ultimately assemble around a viral RNA chromosome.

RNA genome replication. In the overall cycle of replication (**Fig. 11.18**), the RNA chromosome serves as mRNA for peptide translation (step 1) and also as a template for RNA synthesis by RNA-dependent RNA polymerase (translated from the genomic RNA as part of precursor peptide P3). A theoretical problem is: How do the two molecular machines (ribosomes and RNA polymerase) avoid colliding with each other? The problem is solved by containing the chromosome and polymerase within special vesicles formed out of the ER, induced by the poliovirus (step 2). The virus-induced vesicles contain no ribosomes, so only RNA synthesis occurs.

Within the virus-induced vesicles, the (+) strand RNA is bound to the vesicle membrane by protein 2C (step 3) and protein 3AB (step 4). Protein 3AB primes RNA synthesis, then it is cleaved to form VPg. The (+) strand RNA first serves as template to synthesize (–) strands of RNA (step 5). The (–) strands then serve as templates to make new (+) strands (step 6). For each process of RNA synthesis, the template RNA is bound at the membrane by protein 2C. The daughter strand synthesis is primed by a tyrosine-OH of protein 3AB, which binds to the 3′ end of the template. Protein primers are unique to viruses; in uninfected host cells, all DNA synthesis is primed by the 3′ OH end of RNA. Protein primers, however, are common in many kinds of viruses.

After synthesis of (+) strand RNA is complete, the membrane-bound 3AB protein is cleaved, leaving the 3B portion (now designated VPg) attached to the 5′ OH end of the RNA. Polioviral RNA synthesis is extremely efficient, generating about 50,000 copies per host cell. The (+) strand RNAs exit the ER (step 7), where some of them serve as mRNA for further translation to peptides, while others are packaged into capsids (step 8) to complete the poliovirus particles.

Viral Exit and Dissemination

How mature polio virions exit the cell is poorly understood, but their release is rapid and destroys the cell. Rapid, destructive release of virions is typical of species with nonenveloped capsids. The cytopathic effects of release are compounded by the host's immune response, which includes inflammation that further damages cells.

Animal viruses, unlike bacteriophages, require a means of transmission not only from cell to cell but also from host to host. To complete its infectious cycle, the animal virus may need to infect several different kinds of cells in different tissues. Poliovirus initially infects epithelial cells of the gut lining (**Fig. 11.19**). Some of the progeny virions remain in the gut, sustaining infectious cycles for weeks or months. In most cases, the infection remains confined to the gut; but in some cases, virions move to the gut-associated lymphoid tissue, leading to

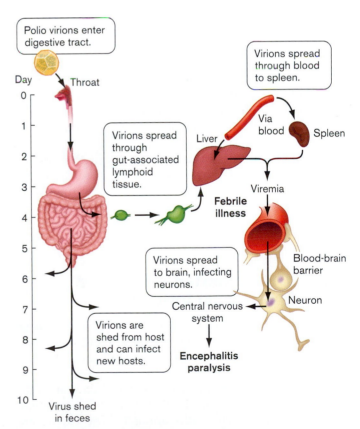

Figure 11.19 Poliovirus dissemination. Polio virions initially infect the gut epithelium, followed by dissemination through the lymphatic system and blood. Some virions cross the blood-brain barrier to infect neurons, causing paralysis.

dissemination throughout the lymphatic system and associated organs. Large numbers of virus are released into the blood, a condition known as **viremia**. Viremia facilitates dissemination across the blood-brain barrier, leading to infection of neurons. The destroyed neurons cannot be regrown; thus, nerve loss results in paralysis.

Throughout the course of infection, polio virions continue to be released from the host via the digestive tract. The process of viral release is known as **virus shedding**. Poliovirus can be shed from the gastrointestinal tract for several weeks after symptoms have ended, thus enhancing transmission to the next host.

TO SUMMARIZE:

- **Poliovirus**, a picornavirus, is icosahedral. Its genome consists of single-stranded (+) strand RNA.
- **Polio virions attach to poliovirus (PVR) receptors.** Conformational change in the capsid allows DNA insertion for uncoating in the cytoplasm.
- **Poliovirus RNA is replicated in the ER by RNA-dependent RNA polymerase.** Genome replication is primed by virus-encoded protein primer 3AB, which is cleaved to 3B (VPg).
- **(+) strand RNA serves as a template for (–) strand RNA.** The (–) strand RNA then serves as template for (+) strand RNA, which is either translated to proteins or encapsidated into progeny virions.
- **RNA for polio gene expression is exported** to the cytoplasm for translation by host ribosomes. Virions are packaged in the cytoplasm and exported by an unknown mechanism.
- **Poliovirus enters the body through the gastrointestinal tract.** In severe cases, the virus disseminates through the lymph and blood, crossing the blood-brain barrier to infect neurons and cause paralysis.

11.4 A Segmented (–) Strand RNA Virus: Influenza

Some viruses package a (–) strand RNA genome, whose replication involves a more complex replicative cycle than that of (+) strand RNA viruses. Influenza A virus, an orthomyxovirus, is one of the most common life-threatening viruses in the United States. Each year, influenza A infects approximately 10% of the U.S. population, causing about 20,000 deaths annually. The elderly are most susceptible, but in an epidemic year, mortality rises among young people.

Influenza shows a cyclic appearance of extremely virulent strains that cause pandemic mortality, such as the famous pandemic of 1918, which infected 20% of the

world's population and killed more people than World War I. The 1918 strain arose as a mutant form of an influenza strain infecting birds. In 2006, a similar avian influenza strain emerged as a potential source of the world's next influenza pandemic. As of this writing, however, the avian strain has not yet mutated to a form readily transmitted between humans.

Highly virulent strains of influenza usually result from reassortment between distantly related strains. The reassortment process is enhanced by a particular feature of their molecular biology, namely, the **segmented genome**. A segmented genome of a virus consists of multiple separate nucleic acids, like the multiple chromosomes of a eukaryotic cell. The influenza genome consists of eight separate linear (–) strands of RNA, which can combine from different strains to generate a novel hybrid strain.

www | CDC page on influenza virus

Influenza Virion Structure and Genome

The influenza virion, unlike that of polio, has no geometrical capsid (**Fig. 11.20A**). Its RNA chromosome segments are loosely contained by a shell of **matrix proteins** (M1). Matrix proteins enclose the core or capsid, supplementing its connection to the membrane envelope (**Fig. 11.21A**). The envelope derives from the phospholipid membrane of the host cell, which incorporates viral proteins such as hemagglutinin (HA) and neuraminidase (NA). These viral envelope proteins peg the membrane to the enclosed capsid, maintaining its structure. Thus, mutations in the gene encoding neuraminidase produce aberrant virions with highly irregular shapes (**Fig. 11.20B**). Neuraminidase can be blocked by the antiviral agent Tamiflu (oseltamivir), the main drug available to treat influenza.

A.

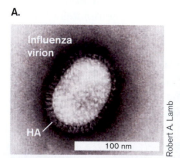

B.

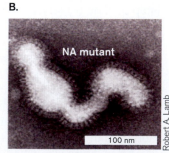

Figure 11.20 Influenza A virions. A. Influenza A virions (TEM) consist of an amorphous membrane envelope surrounding packaged segments of the RNA chromosome. The brush-like border coating the envelope contains hemagglutinin (HA) and neuraminidase (NA). **B.** A distorted virion results from a mutation that truncates the internal domain of NA. *Source:* Hong Jin, et al. 1997. *EMBO Journal* 16:1236.

A.

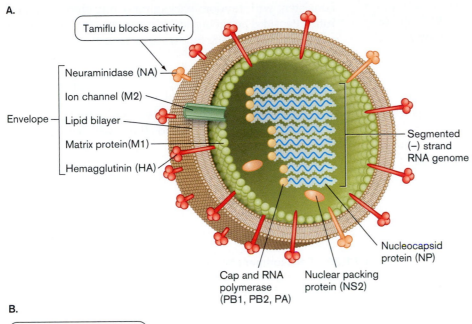

Tamiflu blocks activity.

Envelope
- Neuraminidase (NA)
- Ion channel (M2)
- Lipid bilayer
- Matrix protein (M1)
- Hemagglutinin (HA)

Segmented (−) strand RNA genome

Nucleocapsid protein (NP)

Cap and RNA polymerase (PB1, PB2, PA)

Nuclear packing protein (NS2)

Figure 11.21 Influenza A: structure and genome. A. Diagram of virion structure showing envelope (tan), envelope proteins, matrix protein (green), RNA segments with attached polymerase, and the nuclear packing protein NS2. **B.** Genome of influenza A includes eight RNA segments, each encoding one protein. Each (−) strand segment must be transcribed to a (+) strand mRNA with a host-derived 5′ cap and viral RNA polymerase, and 3′ poly-A tail. Two of the segments (7 and 8) undergo additional processing to express two additional proteins (M2 and NS2).

B.

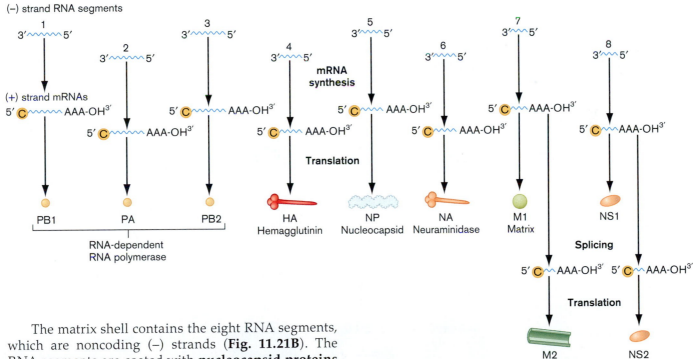

Influenza A (orthomyxovirus) genome consists of 8 segments (−) strand RNA.

(−) strand RNA segments

(+) strand mRNAs

mRNA synthesis

Translation

PB1   PA   PB2

RNA-dependent RNA polymerase

HA Hemagglutinin   NP Nucleocapsid   NA Neuraminidase   M1 Matrix   NS1

Splicing

Translation

M2 Ion channel   NS2

The matrix shell contains the eight RNA segments, which are noncoding (−) strands (**Fig. 11.21B**). The RNA segments are coated with **nucleocapsid proteins (NP)**. The term "nucleocapsid" refers generally to proteins coating a viral genome and packaged within or as part of the viral capsid. Each NP-coated RNA segment also possesses a bound RNA-dependent RNA polymerase complex. Thus, unlike poliovirus and most bacteriophages, the influenza virion contains more than its genome; it also contains its own RNA polymerase enzyme, bound to each RNA segment and poised for synthesis. During infection each (−) strand RNA segment must be transcribed to a (+) strand mRNA for eukaryotic translation.

THOUGHT QUESTION 11.5 Why does influenza virus have to provide its polymerase ready-made, whereas poliovirus does not?

NOTE: In figures of this chapter, AAA-OH$^{3'}$ represents a poly-A tail consisting of a variable number of adenine nucleotides with a 3' hydroxyl end.

The segmented genome of influenza poses a major problem for virus assembly: Given that viral infection requires all eight segments, how does the assembly mechanism package exactly eight segments, one of each? In fact, the segments are packaged at random, resulting in a vast majority of defective particles. The fraction of influenza particles observed to be capable of infection is consistent with random packaging of any eight segments into a capsid. The theoretical fraction of infectious virions would be given by the following equation:

$$\frac{8!}{8^8} = \frac{8 \times 7 \times 6 \times 5 \times 4 \times 3 \times 2 \times 1}{16,777,216} = \frac{1}{416}$$

One in 400 may sound tremendously inefficient, but if 10,000 virions are produced, there will be more than enough infective particles to propagate the virus. Furthermore, animal viruses require a relatively small number of progeny to ensure infecting the neighboring cells of a multicellular tissue, as compared to phages or viruses of single-celled organisms. Thus, influenza virus may avoid

A.

James Gathany/CDC

Figure 11.22   Reassortment between human and avian strains generates exceptionally virulent strains.   A. The reconstructed strain of the 1918 Pandemic Influenza Virus is studied by Dr. Terrence Tumpey, microbiologist at the Centers for Disease Control and Prevention, Atlanta. The 1918 strain may hold clues to combatting the "avian influenza" strain today. **B.** Major epidemics of exceptionally virulent strains arise from reassortment with chromosomal segments from strains that evolved within ducks or swine, agricultural animals that live in proximity to humans. The segments are labeled as in Figure 11.21. The 1918 pandemic may actually have derived all eight segments from an avian strain. The 1957 strain incorporated avian segments 2, 4, and 5; subsequently, the 1968 strain replaced segments 2 and 4 with those of yet another avian strain. The numbers following H and N refer to different alleles of the genes encoding hemagglutinin and neuraminidase, respectively.

B.

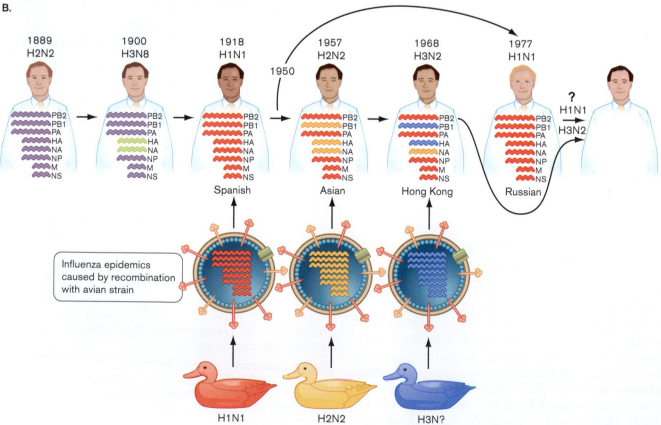

the energetic expense of an accurate packaging mechanism for its segmented genome.

THOUGHT QUESTION 11.6 Explain why the expression $(8!/8^8)$ approximates the proportion of infective particles of influenza virus. What assumption might be changed to make the proportion greater or less?

The key advantage of a segmented genome is that it facilitates recombination between two strains coinfecting the same cell. Coinfection can generate an instant new strain that evades a host immune system. Even strains that infect animals such as ducks or swine may reassort their segments with those of a human virus (**Fig. 11.22**). Influx of genes from a distantly related strain can sharply increase virulence and mortality. Major epidemics of exceptionally virulent influenza arise from reassortment with chromosomal segments from strains that evolved within ducks or swine, agricultural animals that live in close proximity to humans.

Figure 11.22B shows how various combinations of human and avian genome segments led to new allelic combinations of the envelope proteins hemagglutinin (H numbers) and neuraminidase (N numbers). These variant envelope proteins can foil the immune system; for example, a human immune system is unlikely to have developed antibodies against an avian envelope protein. Based on sequence analysis, the 1918 pandemic virus may actually have derived all eight segments from an avian strain. Subsequently, the 1957 virulent strain incorporated avian segments 2, 4, and 5, whereas the 1968 virulent strain replaced segments 2 and 4 with those of yet another avian strain.

Overall, the opportunities for reassortment among influenza strains make it difficult to generate vaccines and predict which strains to vaccinate against in a given flu season. Nevertheless, alert clinicians and effective public health reporting enable physicians to predict lesser epidemics and protect many people by vaccination, as discussed in Chapter 28. The rarer pandemics involving exceptionally virulent avian-human recombinant strains remain a major challenge to public health.

Attachment and Entry to the Host Cell

An influenza virion attaches to a cell when its hemagglutinin envelope protein binds to a host cell receptor protein that contains polysaccharide terminating with sialic acid. The precise structure of the host receptor binding the hemagglutinin may determine whether a strain such as avian influenza H5N1 will spread directly between humans. For example, the sialic acid connection in the receptor polysaccharide can involve different OH groups of the sugar galactose: a linkage to the OH-3 (α2,3) or to

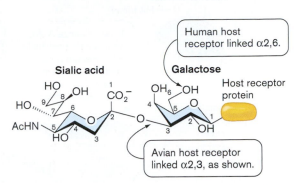

Figure 11.23 Sialic acid in influenza receptor. The avian host receptor polysaccharide contains sialic acid with an α2,3 bond to galactose, whereas the human receptor in the upper respiratory tract has an α2,6 bond.

the OH-6 (α2,6) (**Fig. 11.23**). The influenza strain H5N1 recognizes mainly the α2,3-linked protein, found in birds. In humans, the upper respiratory tract contains mainly α2,6-linked receptors; α2,3-linked receptors are found only deeper within the lungs. Thus, avian influenza strain H5N1 is rarely transmitted between humans. However, more rapid transmission could arise from a mutation in the avian hemagglutinin that binds the receptor, leading to an influenza pandemic.

The hemagglutinin complex consists of a trimer of subunits, each of which has an N-terminal **fusion peptide**. A fusion peptide is a portion of an envelope protein that changes conformation so as to facilitate envelope fusion with the host cell membrane.

Before membrane contact, the fusion peptides are buried within the core of the hemagglutinin (HA) trimer (**Fig. 11.24**). The HA C-terminal domains each bind a sialic acid receptor in the host membrane. When the virion is taken up by endocytosis, the endocytic vesicle fuses with a lysosome and its interior acidifies. The lowered pH induces a conformational change shifting the C-terminal ends back and the N-terminal fusion peptides outward to face the vesicle membrane. The peptides extend into the membrane, where they mediate fusion between viral and host membranes. The fusion process expels the contents of the virion into the host cytoplasm, where the RNA segments become uncoated. Unlike poliovirus, which injects only its chromosome with an attached peptide, influenza virus releases several enzymes and structural proteins along with its genetic material.

Replicative Cycle of Influenza A

The replication of influenza virus is more complex than that of poliovirus because the viral components travel in and out of the nucleus. In addition, envelope proteins require transport through the ER and Golgi to the cell membrane (**Fig. 11.25**). By contrast, for poliovirus no envelope is assembled.

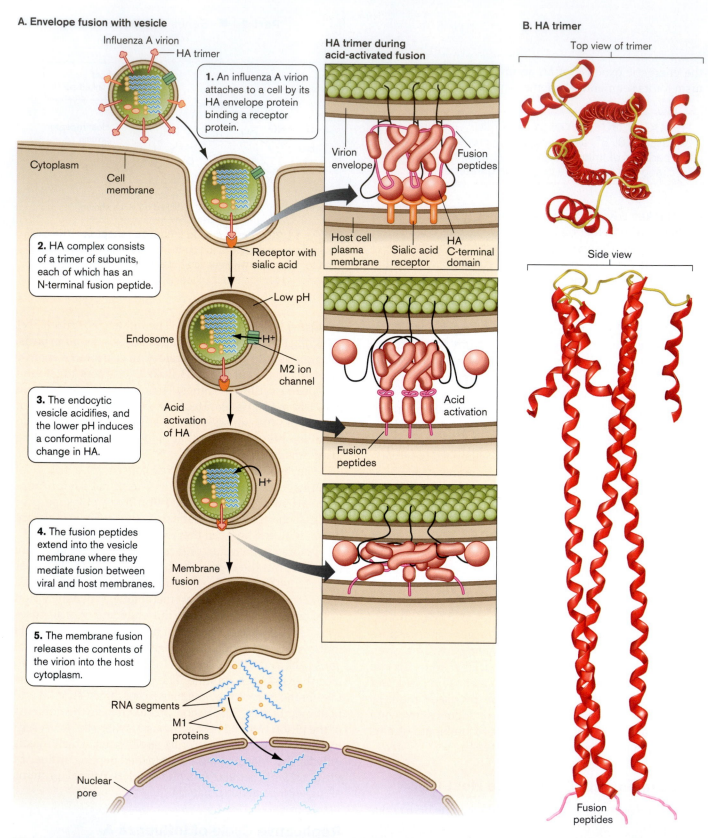

A. Envelope fusion with vesicle

Influenza A virion
HA trimer

1. An influenza A virion attaches to a cell by its HA envelope protein binding a receptor protein.

Cytoplasm
Cell membrane

2. HA complex consists of a trimer of subunits, each of which has an N-terminal fusion peptide.

Receptor with sialic acid

Low pH

Endosome

H+

M2 ion channel

3. The endocytic vesicle acidifies, and the lower pH induces a conformational change in HA.

Acid activation of HA

H+

4. The fusion peptides extend into the vesicle membrane where they mediate fusion between viral and host membranes.

Membrane fusion

5. The membrane fusion releases the contents of the virion into the host cytoplasm.

RNA segments

M1 proteins

Nuclear pore

HA trimer during acid-activated fusion

Virion envelope
Fusion peptides

Host cell plasma membrane
Sialic acid receptor
HA C-terminal domain

Acid activation

Fusion peptides

B. HA trimer

Top view of trimer

Side view

Fusion peptides

Figure 11.24 Entry and acid activation of influenza A virion. A. The HA envelope protein of an influenza A virion attaches to a host cell (1) binding a sialic acid receptor protein. HA complex consists of a trimer of subunits, each of which has an N-terminal fusion peptide. When the virion is taken up by endocytosis, the interior of the endocytic vesicle acidifies, and the lower pH induces a conformational change in HA (2, 3). The receptor-binding domains fold back, and the fusion peptides extend into the vesicle membrane, where they mediate fusion between viral and host membranes. The membrane fusion releases the contents of the virion into the host cytoplasm (4). **B.** Model of the core portion of HA trimer (top view and side view). (PDB code: 1HTM) ⏯

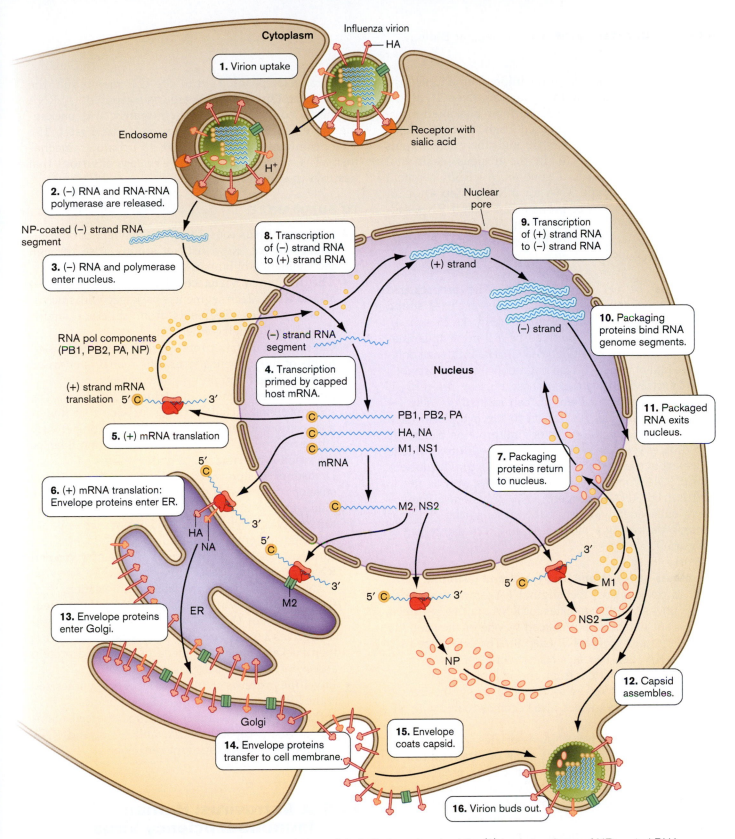

Figure 11.25 Replication of influenza virus. HA-mediated attachment and uptake (1) leads to release of NP-coated RNA segments in the cytoplasm (2). The (−) strands enter the nucleus (3), where the viral polymerase makes (+) strand RNA (4). RNA synthesis is primed by "capped" fragments of host mRNA, obtained by "cap snatching." The (+) strand RNA segments return to the cytoplasm (5) for translation to viral proteins (6, 7). Additional (+) strand RNA is made to serve as a template for (−) strands, primed by NP protein (8, 9). Proteins MI, NS1, and NS2, as well as newly synthesized RNA-dependent RNA polymerase components (PA, PB1, and PB2), return to the nucleus to bind the progeny (−) strand segments (10). Protein-packaged RNA segments then return to the cytoplasm (11) where the capsid assembles (12). Envelope proteins (M2, NA, and HA) are synthesized at the ER (6), where they are glycosylated by host enzymes and transferred to the Golgi (13) for export to the cell membrane (14). At the cell membrane, the packaged (−) RNA segments are enveloped by host membrane containing the envelope proteins (15). ▶️

After the influenza virion undergoes endocytosis (step 1) and release into the cytoplasm (step 2), all the viral (−) RNA segments with their prepackaged primers and polymerases are released. Each NP-coated RNA segment individually passes through the nuclear pores into the nucleus (step 3). In the nucleus, each viral polymerase attached to a genomic (−) RNA segment synthesizes (+) strand RNA for mRNA (step 4) or for templates to synthesize progeny (−) strand RNA segments (steps 8, 9). The processes of mRNA synthesis and template synthesis for progeny genomic RNA are primed differently.

Synthesis of (+) strand mRNA. The prepackaged RNA-dependent RNA polymerase uses each (−) strand RNA segment to synthesize mRNA (**Fig. 11.25**, step 4). The mRNA synthesis is primed by a 5-methylguanine-"capped" RNA fragment (labeled C). The influenza polymerase had obtained the cap fragments from the previous host by cleaving them from host nuclear pre-mRNA, a process quaintly known as "cap snatching." The (+) strand mRNA molecules return to the cytoplasm (step 5) for translation to all types of viral proteins. Segments encoding envelope proteins attach to the ER for ultimate transport to the host cell membrane (step 6). The nucleocapsid RNA-packaging protein (NP), as well as newly synthesized RNA-dependent RNA polymerase components, subsequently return to the nucleus (step 7). Other genome-packaging proteins (M1 and N2) also return to the nucleus.

Synthesis of (+) strand and (−) strand genomic RNA. Back in the nucleus, the original (−) strand RNA segments now serve as templates for RNA synthesis primed *not* by cap snatching, but instead by one of the subunits of nucleocapsid protein NP (step 8). The NP-primed (+) strand RNA then becomes coated with the newly made NP subunits imported from the cytoplasm. The NP-coated (+) strand serves as template to synthesize (−) strand RNA (step 9), which also becomes coated with NP (step 10).

The NP-coated (−) RNA associates with a newly made polymerase for a future cycle of viral replication. The RNA is then complexed with matrix protein (M1) and nuclear packaging protein (N2), proteins that were imported from the cytoplasm earlier. At last, the fully packaged (−) RNA segments exit the nucleus to the cytoplasm (step 11), where they approach the cell membrane for packaging into progeny virions (step 12).

Envelope synthesis and assembly. The envelope proteins synthesized at the ER include hemagglutinin (HA) and neuraminidase (NA). Within the ER lumen, these proteins are glycosylated by host enzymes, then transferred to the Golgi (step 13) for export to the cell

membrane (step 14). Within the cell membrane, the envelope proteins assemble around a group of (−) RNA segments complexed with their matrix and packaging proteins, completing the virion particle (step 15). The virion buds out (step 16) by an unknown mechanism. The net result is a massive release of virions that destroys the host cell.

TO SUMMARIZE:

- **Influenza virus** causes periodic pandemics of respiratory disease. New virulent strains arise through recombination of human and avian strains.
- **The influenza virus consists of segmented (−) strand RNA.** Each segment is packaged with nucleocapsid proteins. Segments from different strains recombine through coinfection.
- **Capsid and matrix proteins** enclose the RNA segments of the influenza virus. The matrix is enclosed by an envelope containing spike proteins.
- **Spike proteins mediate virion attachment.** The spike proteins include a fusion peptide that undergoes conformational change to cause fusion between viral envelope and host cell membrane. For influenza, the virion is internalized by endocytosis.
- **Lysosome fusion with endosomes** triggers viral envelope fusion with the endosome membrane. The viral genome and proteins are then released into the cytoplasm. Viral (−) strand RNA segments are uncoated and enter the nucleus.
- **Influenza mRNA synthesis is primed by capped RNA fragments**, cleaved from host mRNA. The viral mRNAs return to the cytoplasm for translation.
- **Genomic RNA synthesis is primed by the nucleocapsid protein (NP).** First, (+) strand RNA is synthesized as a template for (−) RNA strands, which are then packaged in newly made nucleocapsid protein and exported to the cytoplasm.
- **Envelope proteins** of influenza virus are synthesized at the ER for transport to the cell membrane.
- **Influenza viral assembly** occurs at the cell membrane, where capsid, matrix, and (−) strand RNA components are packaged into envelope.

11.5 A Retrovirus: Human Immunodeficiency Virus

So far we have discussed (+) strand RNA viruses, which need to generate a (−) strand intermediate as a template both for (+) strand mRNA and for progeny genomes; and (−) strand viruses, which can generate mRNA directly but need to make a (+) strand template to generate progeny virions. Another major class of RNA viruses is **retroviruses**. Retroviruses are so called because they reverse the normal order of synthesis to copy their RNA into a double-stranded DNA, which is then integrated into the host genome. As introduced in Chapter 6, the RNA-to-DNA synthesis requires a key enzyme, reverse transcriptase (RT), the target of the major anti-AIDS drug azidothymidine (AZT). Retroviruses form a large family of viruses that infect animals; for examples, see **Table 11.2**.

www | Reverse Transcriptase Tutorial, Biomolecules at Kenyon

The most famous of the retroviruses is **human immunodeficiency virus (HIV)**, the causative agent of **acquired immunodeficiency syndrome (AIDS)**. In 2005, according to the United Nations, nearly 40 million people globally were estimated to be living with HIV. That year, the AIDS epidemic claimed more than 3 million lives, and close to 5 million people became newly infected with HIV.

www | UNAIDS

Besides HIV, several other retroviruses have significant historical importance and human impact. **Rous sarcoma virus (RSV)** was the first virus demonstrated to cause cancer. Indeed, the DNA insertion mechanism of retroviruses is ideally suited to alter host gene regulation, either by decoupling a host gene from its regulatory sites or by insertion of an oncogene (an altered host gene that causes cancer). Another retrovirus, **feline leukemia virus (FeLV)** is a major veterinary pathogen. FeLV remains America's number one killer of outdoor cats, despite the availability of an effective vaccine. Related to FeLV is primate T-lymphotrophic virus (PTLV-1), formerly called human T-cell leukemia virus (HTLV). PTLV-1 was the first virus shown to involve

reverse transcription, discovered by American virologist Robert Gallo in 1980. Gallo's studies laid the foundation for the discovery by French virologist Luc Montagnier of HIV and its role in AIDS.

www | Feline Leukemia Virus, Cornell Feline Health Center

www | HIV InSite, University of California at San Francisco

Human Immunodeficiency Virus Causes AIDS

HIV-1 is a **lentivirus,** a genus of retroviruses that evolved from viruses infecting African monkeys. Two strains are recognized: HIV-1, the cause of most infections at present, and HIV-2, a strain that appears to have evolved from simian immunodeficiency virus (SIV). The virus is transmitted through blood and through genital or oral-genital contact. The HIV life cycle is that of a "slow virus" that can hide in the host cell for many years, with only gradual production of virions. Eventually, however, the virus destroys the body's T leukocytes, leaving the host defenseless against many organisms that normally would be harmless.

The first cases of AIDS in the United States were reported in 1981. Twenty-six years later, HIV infects over a million Americans, and more than half a million have died of AIDS. Worldwide, HIV infects one in every 100 adults, equally among women and men. In the developed countries, treatment effectively prolongs the life of people with AIDS; but treatment is too expensive for the majority of those infected worldwide. Countries in southern Africa have already experienced population decline. In Swaziland, for example, 40% of adults test positive for

Table 11.2 Retroviruses of animals (examples).

Genus	Virus	Disease	Host
Alpharetrovirus	Avian leukosis virus (ALV)	Leukemia	Birds
Alpharetrovirus	Rous sarcoma virus (RSV)	Leukemia	Birds
Betaretrovirus	Mouse mammary tumor virus (MMTV)	Mammary tumor	Mice
Gammaretrovirus	Feline leukemia virus (FeLV)	Lymphoma, immune deficiency	Cats
	Moloney murine leukemia virus (MMLV)	Leukemia	Mice
Deltaretrovirus	Bovine leukemia virus (BLV)	Leukemia	Cattle
	Primate T-lymphotrophic virus (PTLV-1) [formerly Human T-cell leukemia (HTLV)]	Leukemia	Humans
Epsilonretrovirus	Walleye dermal sarcoma virus (WDSV)	Sarcoma	Fish
Lentivirus	Human immunodeficiency virus (HIV-1, HIV-2)	AIDS	Humans
	Simian immunodeficiency virus (SIV)	Simian AIDS	Monkeys
	Equine infectious anemia virus (EIAV)	Anemia	Horses
	Maedi-Visna virus (MV)	Neurological disease	Sheep

A. *And the Band Played On*

Photos12.com/Collection Cinéma

B. Luc Montagnier and Robert Gallo

John Mottern/AFP/Getty Images

Figure 11.26 HIV discovery, in fact and film. A. The film *And the Band Played On* depicts the first decade of the AIDS epidemic. In this scene, French scientists observe HIV grown in T-lymphocyte tissue culture. **B.** Luc Montagnier and Robert Gallo signing agreement to collaborate on development of an AIDS vaccine, 2002.

HIV. Despite all we know, AIDS continues to grow in North America and Europe, and particularly high rates of increase are seen in Eastern Europe and central Asia. As of this writing, there is no cure and no vaccine.

A historical view of AIDS in the United States emerges in the book *And the Band Played On* by Randy Shilts, adapted as an award-winning film in 1993 (**Fig. 11.26A**). The book and film show how American society failed for many years to grasp the significance of AIDS because the syndrome first appeared in societal groups considered marginal (homosexual men and certain ethnic immigrants), although it spread to all social classes. Also, the virus proved extremely difficult to detect and grow in culture. The discovery of HIV sparked controversy in the scientific community, as Gallo failed to acknowledge his use of a virus-producing cell line from Montagnier, the first to isolate HIV-1. Since that time, the two scientists and many others have collaborated to

develop a test for HIV-1 infection and to search for a vaccine (**Fig. 11.26B**).

The lack of a cure for AIDS should not surprise us, since only a small number of effective antiviral drugs have been found. But why can we find no vaccine when safe and effective vaccines were obtained long ago for major killers such as smallpox and polio and even for the feline retrovirus FeLV? The answers to this question are complex. They include:

- **High mutation rate.** The mutation rate of all retroviruses is exceptionally high because of the high error rate of reverse transcriptase. In addition, the lentiviruses (including HIV) appear to have evolved selectively high mutation rates for envelope proteins that are exposed to the immune system.
- **Complex regulation of replication.** The lentiviruses express a greater number of regulator proteins

A. HIV virions

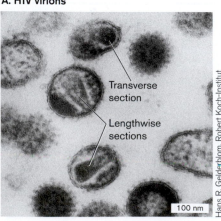

Transverse section

Lengthwise sections

100 nm

Hans R. Gelderblom, Robert Koch-Institut

B. Virion budding out

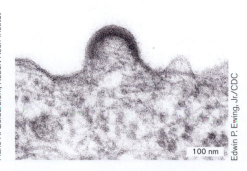

100 nm

Edwin P. Ewing, Jr./CDC

Figure 11.27 HIV-1 virus particles. A. HIV virions from infected human lymphoma tissue culture (thin section, TEM). Individual virions were sliced through at different angles: Transverse section through the cone-shaped capsid appears round, whereas longitudinal (lengthwise) sections appear rod-shaped or triangular. **B.** Budding virions show thickened membrane as spike proteins concentrate.

than other retroviruses. These complex regulatory options may enable HIV to hide itself more effectively within host cells than other retroviruses such as FeLV.

While a cure remains elusive, studying the molecular biology of HIV has led to successful treatments that greatly extend the life span and improve the quality of life for infected individuals.

HIV Structure and Genome

The structure of HIV as visualized by TEM consists of an electron-dense **core particle** (or capsid) surrounded by a phospholipid envelope (**Fig. 11.27A**). In some sections, the core appears round, whereas other sections cut longitudinally reveal an elongated shape, like a cone or cylinder. The thickening of the membrane around the core indicates the presence of **spike proteins**, which peg the membrane to the matrix, as in influenza virus. The envelope forms around the core by budding out of the host cell membrane (**Fig. 11.27B**), a process that does not rapidly lyse the cell but has other devastating consequences for cell function. The conical core is composed of capsid subunits (CA) whose arrangement is partly icosahedral (**Fig. 11.28A**). The capsid subunits enclose two identical copies of the RNA genome plus reverse transcriptase and other enzymes.

Figure 11.28 HIV-1 structure and genome. **A.** Internal structure of virion, color-coded to match the genome. In the genome sequence, the staggered levels indicate three different reading frames. Each virion contains two copies of the RNA genome plus multiple copies of reverse transcriptase (RT) and protease (PR) enclosed within a conical capsid (CA subunits plus subunits of a host protein, cyclophilin A). The capsid is surrounded by a matrix (MA subunits), which reinforces the host-derived phospholipid membrane, pegged by spike proteins (SU, TM). The genome also encodes six accessory proteins that are expressed within the infected host cell and regulate the replicative cycle. **B.** Flossie Wong-Staal, pioneering AIDS researcher at University of California, San Diego, was the first to clone the HIV genome. Wong-Staal now pursues gene therapy approaches to AIDS prevention and develops lentiviral gene vectors.

Each virion contains two copies of the RNA genome, coated with nucleocapsid proteins (NC) similar in function to those of influenza virus (**Fig. 11.28A**). For priming, each RNA is complexed with a tRNA derived from the previously infected host cell. The primed and packaged RNA is contained within the capsid or core composed of capsid subunits (CA). The capsid also contains about 50 copies of reverse transcriptase (RT) and protease (PR), as well as a DNA integration factor (IN). Unique to strain HIV-1, subunits of a host chaperone named cyclophilin A are incorporated into the structure, about one for every ten capsid subunits. An HIV-1 mutant that fails to incorporate cyclophilin A can attach to a host cell and insert its capsid, but the capsid fails to come apart, and infection is halted.

The capsid is surrounded by a matrix (MA subunits), which reinforces the host-derived phospholipid membrane. The membrane is pegged by spike proteins composed of the envelope subunits TM and SU. As in influenza virus, the spike proteins play crucial roles in host attachment and entry.

The genome of HIV was first cloned for molecular study by Flossie Wong-Staal, now at the University of California at San Diego (**Fig. 11.28B**). Born in China in 1947, Wong-Staal immigrated to the United States, then worked with Gallo on the early discoveries of HIV. Wong-Staal now pursues gene therapy approaches to combat HIV infection and is developing retroviral vectors for human gene therapy.

The HIV genome encodes three main open reading frames: *gag*, *pol*, and *env*. The *gag* sequence encodes capsid and nucleocapsid proteins; *pol* encodes reverse transcriptase (RT), integrase, and protease; and *env* encodes envelope proteins. During infection, each reading frame is transcribed and translated as a "polyprotein;" then, at subsequent stages, the polyprotein is cleaved by proteases to form the mature products. The three polyprotein reading frames are found in all retroviruses, but each retroviral species processes their products differently.

In HIV-1, the *gag* and *pol* sequences overlap, and *env* overlaps with genes encoding **accessory proteins**, a term for proteins that modify and regulate retroviral infection. The accessory proteins are expressed within the infected host cell and regulate the replicative cycle (**Table 11.3**). For example, Tat protein activates transcription of the viral genome. The HIV-1 genome encodes at least six accessory proteins—a greater number than any other retrovirus. They are major targets for research and drug discovery aimed at preventing HIV proliferation.

Table 11.3 Accessory proteins of HIV-1.

Protein	Function	Effect of mutation
Vif	Virion component: • Associates with cytoplasmic filaments • Required for infectivity of progeny virions	Virions produced are noninfective
Vpr	Virion component: • Transcription factor, activates HIV transcription during G_2 phase of cell cycle; arrests T-cell growth • Imports DNA across nuclear membrane; avoids need to infect rapidly dividing cells in which mitosis dissolves the nucleus	Lower production of virions
Nef	Virion component: • Internalizes and degrades CD4 receptors • Controls signal transduction to downregulate immune response by T cells • Enhances infectivity of virus particle • Accelerates progression from infection to disease syndrome (AIDS)	Slower progression to AIDS
Vpu	Membrane protein: • Degrades CD4, releasing bound spike proteins • Promotes virion assembly and export	Host cell dies early; lower production of virions
Rev	Nuclear phosphoprotein, combines with host cell proteins: • Stabilizes certain mRNAs in nucleus • Export of mRNA out of nucleus into the cytoplasm • By promoting mRNA export, induces shift from latent phase to virion-producing phase	Failure of infection
Tat	Transcription factor: • Binds TAR site on nascent RNA to activate transcription • Associates with histone acetylases and kinases to activate transcription of integrated viral DNA	Failure of chromosome replication

HIV Attachment and Entry to the Host Cell

Like other viruses, HIV needs to recognize specific receptor molecules on the surface of its target cells. The primary receptor for HIV is the CD4 surface protein on CD4 T lymphocytes (T cells). The normal function of CD4 surface proteins is to connect the T cell with an antigen-presenting cell, which activates the T cell to turn on production of antibodies (discussed in Section 24.5). Note, however, that CD4 proteins appear on many other cell types, such as microglia (macrophage-like cells in the central nervous system) and Langerhans cells (immune cells of the epidermis). Their presence may make other cells susceptible to infection by HIV.

Spike proteins mediate membrane fusion. The binding of HIV to CD4 receptors involves the envelope spike protein SU (**Fig. 11.29A**). Spike proteins are the main external proteins accessible to the host immune system. The binding of SU to CD4 requires several crucial domains of the SU peptide sequence (constant domains C1–C5; **Fig. 11.29B**). The constant domains show the same sequence in most isolates of HIV. In contrast, the variable domains of SU (V1–V5) show highly variable sequences. Apparently, the viral genome tolerates a wider range of variation in parts of the protein not directly involved in receptor binding. Variations in the nonbinding domains helps the virus evade the host immune system.

HIV attachment to the cell membrane requires a fusion peptide rearrangement similar to that of influenza virus. When SU binds to CD4, the spike transmembrane component (TM) unfolds and extends its fusion peptide into the host cell membrane (see **Fig. 11.29A**). In addition, SU binds to secondary receptors in the membrane called **chemokine receptors (CCR)**. After attachment of the spike protein SU to receptors and the insertion of the TM fusion peptide, the HIV-1 envelope fuses with the plasma membrane.

Chemokine receptors such as CCR5 participate in cell signaling for host defenses (discussed in Chapter 24). The requirement for CCR attachment varies among different strains of HIV. Some chemokine receptors are found on neurons, and their involvement in HIV infection may be responsible for the neurological disorders seen in AIDS. As described at the beginning of the chapter, people with a defective gene for CCR5 are resistant to infection by HIV-1. Researchers used this knowledge to develop a transgenic therapy for HIV-1-positive volunteers. From the HIV-infected individuals, CD4 T cells were removed and placed into culture. The genome of the cultured T cells was engineered to contain a defective CCR5 gene; thus, the T cells no longer expressed CCR5 on their

A. HIV spike proteins bind host receptors

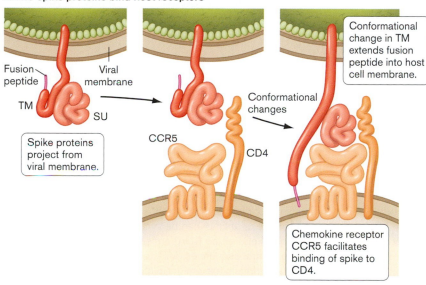

Fusion peptide

Viral membrane

TM

SU

Spike proteins project from viral membrane.

CCR5

CD4

Conformational changes

Conformational change in TM extends fusion peptide into host cell membrane.

Chemokine receptor CCR5 facilitates binding of spike to CD4.

B. SU constant regions bind CD4

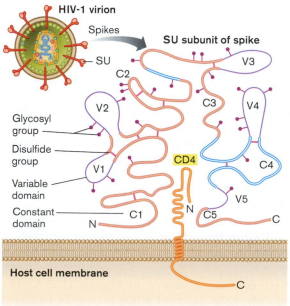

HIV-1 virion

Spikes

SU

SU subunit of spike

C2

V3

V2

C3

V4

Glycosyl group

Disulfide group

Variable domain

Constant domain

V1

CD4

C4

V5

C1

N

C5

N

C

Host cell membrane

C

Figure 11.29 HIV attachment to host cell. A. The SU subunit of the spike complex attaches to the receptor, CD4 cell surface protein. **B.** The region of SU that binds the CD4 receptor contains constant regions (C1–C5) and variable regions (V1–V5) that differ in amino acid sequence in different HIV-1 isolates. The constant regions directly interact with the CD4 receptor. The variability of the V regions enables the virus to evade the host immune system.

surface. The CCR5-negative T cells were then returned to the volunteers, where they produced a resurgence of HIV-1-resistant T cells in the blood. The transgenic therapy may prove useful in maintaining the immune systems of HIV-positive patients. It may not, however, prove useful for all strains of HIV, and it does not cure the infection. Nevertheless, it represents hope for future molecular research to overcome AIDS.

After the HIV envelope fuses with the cell membrane, the HIV core directly enters the cytoplasm. The HIV entry mechanism differs from that of influenza virus, in which endocytosis and lysosome fusion are required to open the capsid and release the genome into the cell. The HIV core (composed of CA and the host-derived cyclophilin A) dissolves, releasing the twin RNA genomes along with associated viral enzymes into the cytoplasm.

> **THOUGHT QUESTION 11.7** How do attachment and entry of HIV resemble attachment and entry of influenza virus? How do attachment and entry differ between these two viruses?

Retroviral RNA Genome

The twin RNA genomes each possess a 5' "cap" and a 3' poly-A "tail" that probably enable them to mimic host nuclear mRNA. Each RNA is hybridized to a tRNA primer for DNA synthesis. The primer is a lysine-specific tRNA from the previous infected cell (**Fig. 11.30**). The primer might be expected to hybridize at the 3' end of the template, where its 3' OH "points" toward the opposite end, positioned to synthesize all the way down. Surprisingly, however, the 3' OH end of the tRNA actually binds near the 5' end, where initially it can only generate a brief sequence. These early sequences bind key regulatory factors for transcription and for DNA insertion into the host genome.

> **THOUGHT QUESTION 11.8** Compare and contrast the priming of chromosome replication in HIV with the priming mechanisms of poliovirus and influenza virus.

Reverse Transcriptase Copies RNA to Double-Stranded DNA

The defining component of a retrovirus is **reverse transcriptase (RT)**, the enzyme that synthesizes the double-stranded DNA copy of the genome (**Fig. 11.31A**). Reverse transcriptase is the target of the first clinically useful drug to treat HIV infection, the nucleotide analog **azidothymidine (AZT)**. The reverse transcriptase complex actually possesses three different activities:

■ **DNA synthesis from the RNA template.** Synthesis of DNA is first primed by the host tRNA, which

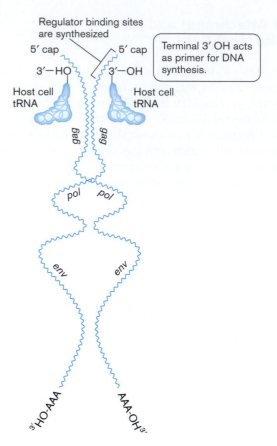

Figure 11.30 Structure of the RNA dimer. The RNA dimer packed in the virion includes two host tRNA molecules, each hybridized so as to present a 3' OH for priming synthesis of DNA. The primer hybridizes so near the 5' end of the chromosome that only a short piece of DNA can be synthesized (see Figure 11.28). The capped 5' end of the RNA forms several hairpin loops containing control sequences: TAR, to bind Tat protein, which induces transcription; a signal for 3' polyadenylation; PBS, the primer-binding site for host tRNA; DIS, site of contact for RNA dimer; a splice site to generate shortened transcripts for *env* and for accessory protein genes; a signal required for core packaging; and the AUG start codon for the *gag* transcript.

was hybridized to the chromosome within the virion.

■ **RNA degradation.** After DNA synthesis, the template RNA is gradually removed through an RNase activity of the RT complex. Removal of RNA enables replacement of the entire original RNA template by DNA.

■ **DNA-dependent DNA synthesis.** To make the DNA complementary strand replacing the RNA, the RT needs to use the newly made DNA as its template. Thus, RT has the rare ability to use either DNA or RNA as template.

The RT complex is shown in **Figure 11.31A**. The RNA template with its short RNA primer is threaded

A. Reverse transcriptase

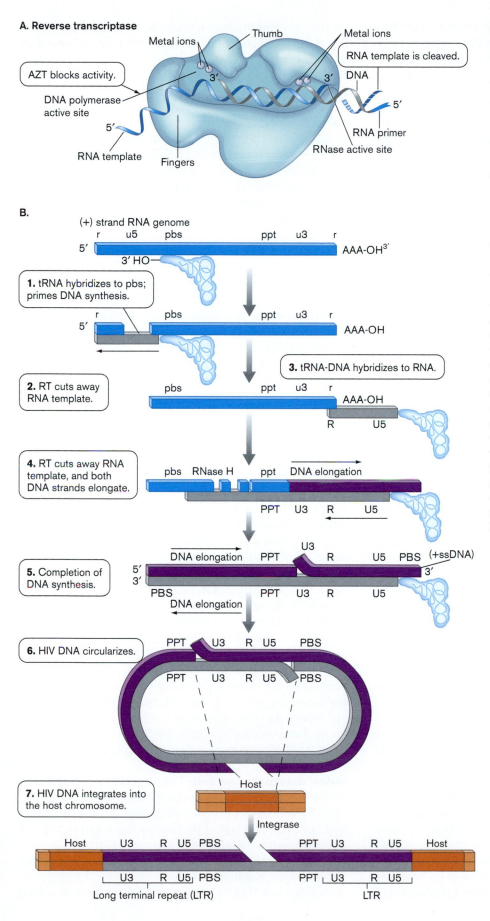

B.

1. tRNA hybridizes to pbs; primes DNA synthesis.

2. RT cuts away RNA template.

3. tRNA-DNA hybridizes to RNA.

4. RT cuts away RNA template, and both DNA strands elongate.

5. Completion of DNA synthesis.

6. HIV DNA circularizes.

7. HIV DNA integrates into the host chromosome.

Figure 11.31 Reverse transcription of the HIV genome and integration into host DNA. A. The RNA template with its short RNA primer is threaded through the RT complex between the "thumb" and "fingers," a configuration typical of other RNA polymerases. The RT complex adds successive deoxynucleotides from dNTPs, starting at the 3′ OH end of the RNA primer (as in regular DNA synthesis). As DNA elongates, the RNA template is cleaved from behind. **B.** The tRNA primer (1) initiates a short sequence of DNA complementary to the 5′ end of the HIV chromosome (2). The corresponding template is then degraded. The original RNA template has repeated ends (r), so the exposed DNA end (extended tRNA primer of the second copy) then hybridizes to the 3′ end (3). DNA elongation now extends along the rest of the chromosome (4), up to the polymerase-binding site for mRNA transcription (PBS). DNA completion is followed by degradation of the remaining RNA template. Complementary DNA is then synthesized up to the PPT-U3 junction (5). The completed double-stranded DNA copy of the HIV genome, plus a short repeat of the 5′ end (U3 R U5), circularizes (6) and integrates into the host chromosome (7), mediated by integrase (IN). The repeat of the 5′ end generates a long terminal repeat (LTR) flanked by short repeats of host sequence.

through the RT complex between the "thumb" and "fingers," a configuration typical of other RNA polymerases (discussed in Chapter 8). The RT complex adds successive deoxynucleotides from dNTPs starting at the 3' OH end of the RNA primer. As DNA elongates, however, the RNA template is cleaved from behind by the RT complex. Thus, the new DNA actually replaces preexisting RNA sequence. This "destructive replication" is unique to retroviruses. The details are important, as they suggest possible targets for new antiviral drugs.

www | Reverse Transcriptase tutorial, Biomolecules at Kenyon

Initiation of reverse transcription. Figure 11.31B outlines the entire process of reverse transcription in HIV-1, from primed RNA template to integration of duplex DNA. First, the host-derived tRNA primer initiates synthesis of a DNA strand complementary to the RNA chromosome. DNA synthesis is directed toward the 5' end of the HIV chromosome, generating a short segment (step 1). The original RNA template for this short segment is then degraded, leaving only the DNA extension off the tRNA primer (step 2).

The new DNA primes the second template. The original RNA template had repeated ends (labeled r, lowercase for RNA), and the exposed DNA copy of the 5' end has a complementary sequence (R, uppercase for DNA). The R DNA from the second tRNA extension hybridizes to the 3' end of the original RNA (step 3). From the hybridized DNA, elongation now extends along the rest of the chromosome (step 4), up to the polymerase-binding site for mRNA transcription (pbs). DNA completion is followed by degradation of the remaining RNA template, except for occasional short fragments to serve as primers, such as the polypurine tract (ppt). A complementary DNA strand is then synthesized through the PPT primer, leaving a nick ahead (step 5).

DNA integration into the host genome. The final step of genome processing requires integration into host DNA. The double-stranded DNA copy of the HIV genome (plus a short repeat of the 5' end) circularizes (step 6). The circular molecule then integrates into the host chromosome by site-specific recombination (step 7), mediated by the HIV integrase protein (IN), generating an integrated viral genome, or **provirus**. The provirus contains a repeat of the 5' end at its 3' end; this pair of repeats is called a **long terminal repeat (LTR)**. Proviral sequences can now be expressed, directing production of progeny virions.

An alternative to virion production is that the integrated HIV genome can lie dormant, like the lambda prophage in *E. coli* (discussed in Chapter 6). The viral genome is replicated passively within the genome of its host cell, hiding for many years with only infrequent production of virions. The few virions shed by the patient, however, can infect an unsuspecting individual who shares sexual or blood contact.

Replicative Cycle of HIV

The steps of HIV replication are outlined in **Figure 11.32**. The main points of viral entry and replication are typical of retroviruses. HIV, however, has an exceptionally large number of accessory proteins that govern the level of virus production and the duration of the quiescent phase, when the integrated chromosome replicates with the host cell.

Synthesis of HIV mRNA and progeny genomic RNA. After the HIV virion attaches to the host receptors, its envelope fuses with the host membrane (step 1). Unlike influenza virus, the HIV core dissolves in the cytoplasm directly, without endocytosis (step 2). The RNA chromosomes are then reverse-transcribed to make double-stranded DNA (step 3). The double-stranded DNA enters the nucleus through a nuclear pore (step 4), a key step facilitated by Vpr accessory protein. Vpr enables infection of nondividing cells, which only lentiviruses can do; other retroviruses, such as those causing lymphoma, must infect dividing cells, in which the nuclear membrane dissolves during mitosis.

Upon entering the nucleus, the DNA copy of the HIV genome integrates its sequence as a provirus at a random position in a host chromosome (step 5). Within the nucleus, full-length RNA transcripts are made by host RNA polymerase II, including a 5' cap and a 3' poly-A tail (step 6). Some of the RNAs exit the nucleus (step 7) to serve as mRNA for translation of polyproteins. Polyproteins are translated in alternative versions, such as Gag-Pol. Other full-length RNA transcripts exit the nucleus to form RNA dimers for progeny virions (step 8). Still other RNA transcripts within the nucleus are cut and spliced to complete the *env* gene sequence for translation of envelope proteins (step 9).

The Env (envelope) proteins are made within the endoplasmic reticulum (ER) (step 10). They pass through the Golgi for glycosylation and packaging (step 11) and are exported to the cell membrane (step 12). At the membrane, Env proteins plug into the core particle as it forms from the RNA dimers plus Gag-Pol peptides (step 13).

Virion assembly and exit. The assembly and packaging of core particles into virions (step 14) is mediated by Vif, one of the unique HIV accessory proteins. As the virion buds off, the protease (PR) cleaves the Gag-Pol peptide to complete the maturation of the core structure containing Gag subunits as well as RNA polymerase (RT) (step 15). The Gag subunits now form the conical core structure. Proteases that cleave Gag-Pol offer important drug

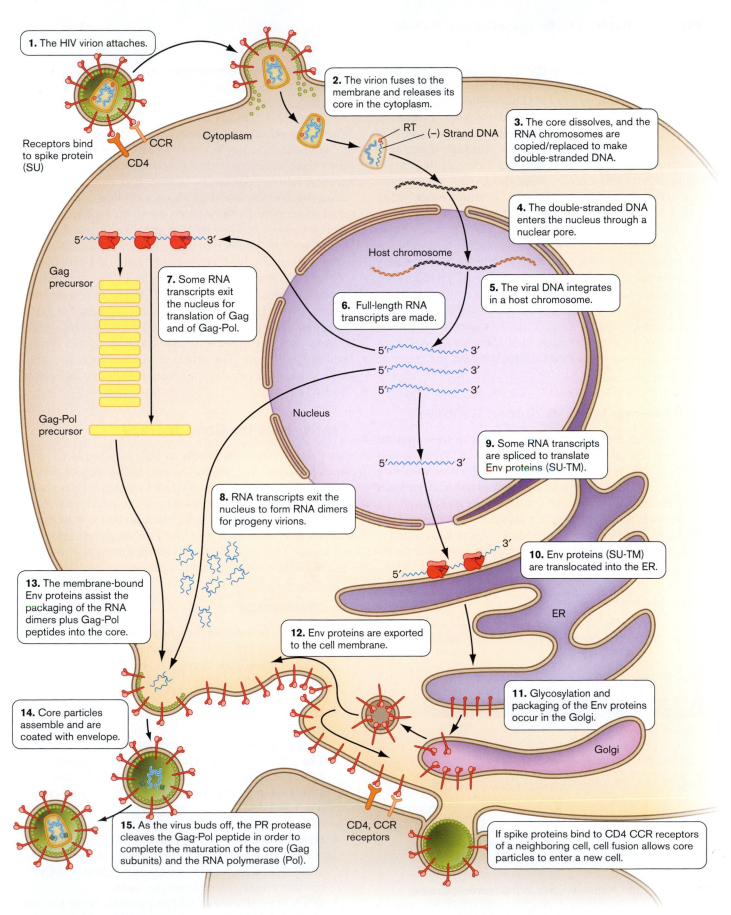

Figure 11.32 HIV replicative cycle. The HIV virion attaches its receptor and fuses with the host cell membrane, releasing its contents in the cytoplasm to undergo a replicative cycle. ▶‖

The following labels appear within the figure:

1. The HIV virion attaches.

2. The virion fuses to the membrane and releases its core in the cytoplasm.

3. The core dissolves, and the RNA chromosomes are copied/replaced to make double-stranded DNA.

4. The double-stranded DNA enters the nucleus through a nuclear pore.

5. The viral DNA integrates in a host chromosome.

6. Full-length RNA transcripts are made.

7. Some RNA transcripts exit the nucleus for translation of Gag and of Gag-Pol.

8. RNA transcripts exit the nucleus to form RNA dimers for progeny virions.

9. Some RNA transcripts are spliced to translate Env proteins (SU-TM).

10. Env proteins (SU-TM) are translocated into the ER.

11. Glycosylation and packaging of the Env proteins occur in the Golgi.

12. Env proteins are exported to the cell membrane.

13. The membrane-bound Env proteins assist the packaging of the RNA dimers plus Gag-Pol peptides into the core.

14. Core particles assemble and are coated with envelope.

15. As the virus buds off, the PR protease cleaves the Gag-Pol peptide in order to complete the maturation of the core (Gag subunits) and the RNA polymerase (Pol).

If spike proteins bind to CD4 CCR receptors of a neighboring cell, cell fusion allows core particles to enter a new cell.

Receptors bind to spike protein (SU)

CD4   CCR   Cytoplasm   RT   (–) Strand DNA

Gag precursor   Gag-Pol precursor

Host chromosome   Nucleus   ER   Golgi

CD4, CCR receptors

targets, which have led to the development of anti-HIV drugs known as **protease inhibitors**.

An alternative to viral budding is cell fusion, mediated by binding of Env in the membrane to CD4 receptors on a neighboring cell. The two cells then fuse, and HIV core particles can enter the new cell through their fused cytoplasm. The fusion of many cells can create a giant multinucleate cell called a syncytium. Cell fusion with formation of syncytia enables HIV to infect neighboring cells without exposure to the immune system.

The intricate scheme in **Figure 11.32** actually omits many functions of HIV accessory proteins that enhance the virulence of HIV infection (**Table 11.3**). Accessory proteins are the subject of intensive research. Mutation of genes for accessory proteins often decreases virulence; thus, these proteins are important potential targets for chemotherapy. For example, the Nef accessory protein accelerates progression from HIV infection to AIDS, so a drug that inactivates Nef might prevent the onset of AIDS.

Retroelements in the Human Genome

Suppose that an integrated HIV genome were to mutate and lose the ability to produce progeny. What would happen to its genome? The integrated genome would be "trapped" within a cell; and over many generations in its host, it would inevitably accumulate more mutations.

In fact, the human genome shows evidence for numerous remains of decaying retroviral genomes, collectively known as retroelements (**Fig. 11.33**). Some retroelements

are **endogenous retroviruses**, sequences that contain all the genomic elements of a retrovirus, including *gag*, *env*, and *pol* genes, yet never generate virions. Presumably, they lost this ability by mutation. Other elements known as **retrotransposons** retain only partial retroviral elements but may maintain a reverse transcriptase to copy themselves into other genomic locations. An example of a retrotransposon is the well-known Alu sequence, a short sequence found in about a million copies in the human genome. In some cases, a retrotransposon such as Alu can interrupt a key human gene, leading to a genetic defect such as a defective lipoprotein receptor associated with abnormally high cholesterol level and heart failure. Still other retroelements, known as LINEs (long interspersed nuclear repeats) and SINEs (short interspersed nuclear repeats), show more vestigial remnants of retroviral genomes. Amazingly, about half the sequence of the human genome appears to have originated from viruses and retroelements. The origin of viruses, including their interaction with cellular evolution, is explored in **Special Topic 11.1**.

Lentiviruses for Gene Therapy

How could a virus as deadly as HIV be used to improve human health? In fact, the exceptional ability of lentiviruses to deliver their own DNA to human cells makes them promising candidates for gene therapy. Lentiviruses are particularly promising for their ability to infect nonmitotic cells. Research is focused on engineering lentiviral vectors that can transfer a recombinant gene to a host cell but fail to generate progeny virions.

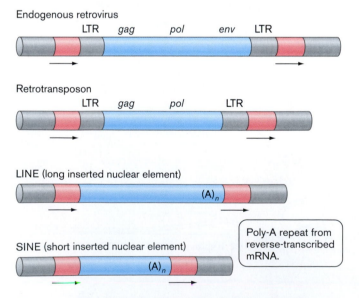

Figure 11.33   Retroelements in the human genome. Endogenous retroviruses and other retroelements in the human genome may arise from progressive degeneration of ancestral retroviruses, or they may be progenitors of new retroviruses.

TO SUMMARIZE:

- **Human immunodeficiency virus (HIV)** causes an ongoing epidemic of acquired immunodeficiency syndrome (AIDS). There is no cure, although molecular biology has led to drugs that extend life expectancy.
- **HIV is a retrovirus** whose RNA genome is reverse-transcribed into double-stranded DNA, which integrates into the DNA of the host cell.
- **The HIV core particle** contains two copies of its RNA genome, each bound to a primer (host tRNA) and reverse transcriptase (RT). The core, or capsid, is surrounded by an envelope containing spike proteins.
- **HIV binds the CD4 receptor** of T lymphocytes with coreceptor CCR5. Following virion-receptor binding and envelope-membrane fusion, HIV virions are released into the cytoplasm.
- **DNA is synthesized in the nucleus from the retroviral RNA**, primed by the tRNA, and synthesized by reverse transcriptase. **RNA degradation**

allows formation of a double-stranded DNA. The retroviral **DNA integrates** into the host genome.

- **Retroviral mRNAs are exported to the cytoplasm** for translation. Envelope proteins are translated at the ER and exported to the cell membrane.

- **Retroviral assembly occurs at the cell membrane**, where virions are released slowly, without lysis. Alternatively, the accumulation of retroviral proteins in the cell membrane leads to cell fusion, forming syncytia.

- **Accessory proteins regulate virion formation** and the latent phase, in which double-stranded DNA persists without reproduction of progeny virions.

- **Ancient retroviral sequences** persist within animal genomes, including the human genome.

11.6 A DNA Virus: Herpes Simplex

Many important viruses of humans and other animals contain genomes of double-stranded DNA (**Table 11.4**). DNA viruses include the causative agents of well-known diseases such as smallpox, chickenpox, and infectious mononucleosis (mono). Most DNA viruses are considerably larger than RNA viruses and encode a wider range of viral enzymes; for example, the vaccinia genome encodes nearly 200 different proteins. The complexity of viruses

such as vaccinia and herpes approaches that of small prokaryotic cells—a fact with interesting implications for viral evolution (see **Special Topic 11.1**).

Replication of viral DNA occurs by mechanisms similar to replication of prokaryotic and phage genomes, either bidirectionally from an origin of replication (as in bacteria) or by the rolling-circle method (as in phages such as T4). As for cellular genomes, DNA replication requires more than a polymerase; enzymes such as helicase, primase, and single-stranded binding proteins are also needed. Viral species differ as to the source of their replication components. SV40 and Epstein-Barr virus rely entirely on cellular components, whereas poxviruses rely entirely on viral components for DNA replication. Other species, such as adenovirus and papillomavirus, combine viral and cellular components.

Herpes Simplex Virus Infects the Oral or Genital Mucosa

An important example of a DNA virus is **herpes simplex virus (HSV)**. Strains HSV-1 and HSV-2 cause one of the most common infections in the United States. Approximately 60% of Americans acquire herpes simplex, usually HSV-1, in epithelial lesions commonly known as cold sores. About 30–60% acquire genital herpes, usually HSV-2, through sexual contact (oral or vaginal). Genital herpes

Table 11.4 **DNA viruses of animals (examples).**

Virus	DNA replication	Disease	Host
Adenovirus (many strains)	Viral DNA polymerase, single-strand binding protein, and protein primer	Enteritis or respiratory diseases	Humans, other mammals, birds
Papovavirus—simian virus 40 (SV40)	Cellular DNA polymerases	Asymptomatic	Monkeys
Herpes viruses			
Herpes simplex virus 1 and 2	All viral components (DNA polymerase, primase, etc.)	Epithelial and genital lesions, latency in neurons	Humans
Varicella-zoster virus	Viral components	Chickenpox, shingles	Humans
Epstein-Barr virus	All cellular components (DNA polymerase, etc.)	Infectious mononucleosis, Hodgkin's lymphoma	Humans
Other strains		Epithelial lesions, cancer	Monkeys, cattle, horses
Papillomavirus			
Human papillomaviruses (many strains)	Viral DNA helicase; cellular polymerase	Genital warts, cervical and penile cancer, cutaneous warts	Humans
Other papillomaviruses		Warts, cancer	Rabbits, cattle, sheep
Poxviruses	All viral components		
Variola virus major		Smallpox	Humans
Vaccinia virus		Cowpox	Cattle, humans
Other poxviruses		Monkeypox	Monkeys, camels, birds

Special Topic 11.1 How Did Viruses Originate?

The origin of viruses remains a mystery. Did viruses evolve within cells? Or could they possibly have predated cells? Unlike cellular organisms, viruses are too small to have left fossil traces that we can detect. Viral genomes show evidence of so much horizontal transfer that lineages are impossible to trace back over time. Unlike cells, which all possess ribosomal RNA genes with a common ancestor, all virus species share no one gene in common. No viral sequence data point to a first common ancestor, nor to a common means of origin.

Three theories are proposed to explain the origin of viruses (**Fig. 1**).

Reductive Evolution

Reductive evolution leads to gradual loss of unselected traits. For example, pathogens such as *Helicobacter* and *Mycoplasma* have lost many genes encoding metabolic functions and stress responses that are unnecessary when the host organism provides these functions. Suppose such a cell were to become an intracellular parasite, increasingly dependent on host functions. Ultimately, it might lose all but a few metabolic genes, losing functions for genome replication with the rest of its functions provided by the host.

Some viruses show signs of degeneration from ancestral cells. Phage T4, the poxviruses, and herpes viruses possess large genomes of 80–200 genes. Their virions include enzymes and other proteins that provide some of the virus's own metabolic functions. The size of a poxvirus approaches half a micrometer—well within the size range of prokaryotic cells (**Fig. 2**). A vaccinia particle actually looks like a miniature cell, with its multilayered envelope containing a DNA chromosome surrounded by enzymes.

Recently, an even larger "cellular virus" was discovered, called the Mimivirus because it mimics a cell (discussed in Chapter 6). Originally discovered within amebas, the Mimivirus was at first mistaken for a bacterium. In fact, Mimivirus consists of a giant icosahedral particle with a diameter of

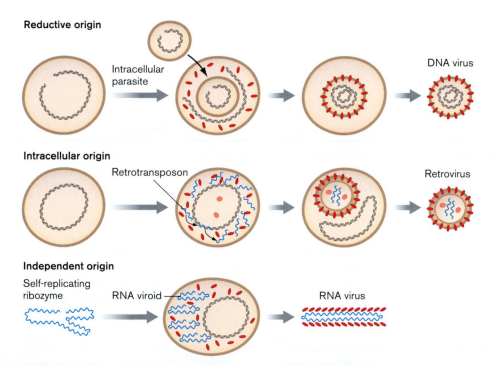

Figure 1 Theories of viral origin. Three theories have been proposed to account for the origin of viruses. According to the theory of reductive origin, viruses evolved by degenerative evolution from intracellular parasitic cells. According to the theory of intracellular origin, viruses evolved from functional parts of cells that acquired an ability to reproduce themselves uncontrolled by the cell. The theory of independent origin holds that viroid nucleic acids could have evolved outside of cells during the RNA world and acquired the ability to infect cells.

400 nm. Its genome is a circle of double-stranded DNA, about 1.2 megabases (Mb). The genome was sequenced by Jean-Michel Claverie and colleagues at the Institut de Biologie Structurale et Microbiologie, Marseille. It includes 1,262 open reading frames—a greater number than the genomes of several prokaryotes, such as mycoplasmas. It is the first viral genome known to include protein translation components such as aminoacyl-tRNA synthetases and elongation and termination factors. Mimivirus may represent a "missing link" between bacteria and viruses.

Intracellular Origin

Other viruses, particularly retroviruses, support the idea of intracellular origin. Retroviruses require reverse transcriptase to convert their RNA information to DNA. When retroviruses were first discovered, reverse transcriptase was considered an anomaly because no such enzyme had been reported in host cells. We have since discovered specialized reverse transcriptases in both plant and animal cells. In retrospect, we now believe that a reverse transcriptase must have existed very early in the evolution of all cells because something had to copy the information from the ancient "RNA world" cell into the DNA of modern genomes.

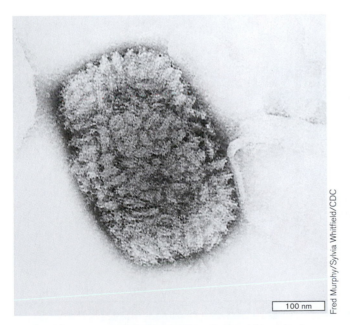

Fred Murphy/Sylvia Whitfield/CDC

100 nm

Figure 2   Large cell-like viruses.   Support for reductive origin of some viruses can be seen in the size and complexity of viruses such as vaccinia, the infective agent of cowpox.

In animal cells, a native type of reverse transcriptase called telomerase serves to replicate the ends of linear chromosomes. The sequence of telomerase shows homology with that of known reverse transcriptases of retroviruses. Mammalian genomes show repeated elements of DNA that have arisen from mistaken reverse copying of host RNA molecules back into the DNA genome. One can imagine that mutation of a cellular reverse transcriptase might lead to intracellular evolution of a retrovirus.

Independent Origin

Some of the simpler viral elements might have evolved by themselves in the RNA world, first as free-living entities, later as stowaways within host cells. Perhaps the best clue for such an origin is that of the viroid ribozymes that infect plant cells. Viroids are self-contained RNA molecules that possess catalytic function. Present-day viroids rely entirely on host cell metabolism; but one could imagine that viroids might have evolved from an ancestral RNA molecule that managed its own self-replication.

An intriguing possibility is that some viruses not only originated as independent cells, but were the first to evolve DNA, and they contributed DNA to all living cells. The viral origin of DNA is proposed by evolutionary biologists, including Patrick Forterre, at the University of Paris-Sud in Orsay, France. The earliest cells are believed to have had genomes of RNA, the simplest nucleic acid. These RNA-world cells may have been parasitized by viruses that evolved DNA chromosomes to avoid cleavage by host cellular enzymes. If a viral DNA chromosome became latent in a host cell (as do herpes viruses, for example), it might eventually acquire genes from the host genome through recombination events. Any host gene transferred to the DNA chromosome would be favored in evolution because DNA is more stable than RNA. Eventually, according to this hypothesis, the entire host genome would end up transferred to the DNA chromosome, and all cells would have genomes of DNA (as indeed cells do today). The hypothesis of viral origin of DNA remains unproven but inspires exciting research to test its implications.

The origin of viruses has important implications for the principles of evolution, including origins of the earliest cells. In addition, the medical importance of studying viral origins is that it yields insights into virus function, suggesting pharmacological applications. For example, ribozymes are being studied for possible use in therapeutic regulation of cancer gene expression.

causes recurrent eruptions of infection in the reproductive tract (**Fig. 11.34**). Many of those infected are unaware of symptoms, but they can still transmit the disease to others.

Herpes simplex virus typically infects cells of the oral or genital mucosa, causing ulcerated sores. The primary infection is epithelial, followed by latent infection within

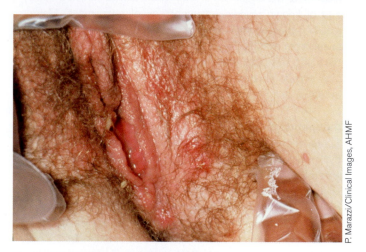

Figure 11.34 Genital herpes infection. Lesions of the vulva caused by HSV-2 infection.

neurons of the ganglia. A common site of infection is the trigeminal ganglion, which processes nerve impulses between the face and eyes and the brain stem.

The latent infection of the ganglia later leads to new outbreaks of virus, often triggered by stress such as menstruation, sunlight exposure, or depression of the immune system. Progeny virions travel back down the dendrites to the epithelia, causing lytic infection. In the trigeminal ganglion, herpes reactivation can lead to eye disease or lethal brain infection. In most cases, herpes symptoms can be controlled by antiviral agents such as acyclovir (discussed in Chapter 26). There is no cure or prevention of future outbreaks. In pregnant women, HSV can be transmitted to the fetus, with serious complications for the child.

Herpes simplex virus is closely related to varicella-zoster virus (VZV), the cause of chickenpox, also an epithelial infection. Varicella, too, can hide in ganglial neurons, emerging decades later to cause skin lesions called shingles.

Herpes Virus Structure

The herpes virion comprises a double-stranded DNA chromosome packed within an icosahedral capsid (**Fig. 11.35A**). The capsid is surrounded by **tegument**, a collec-

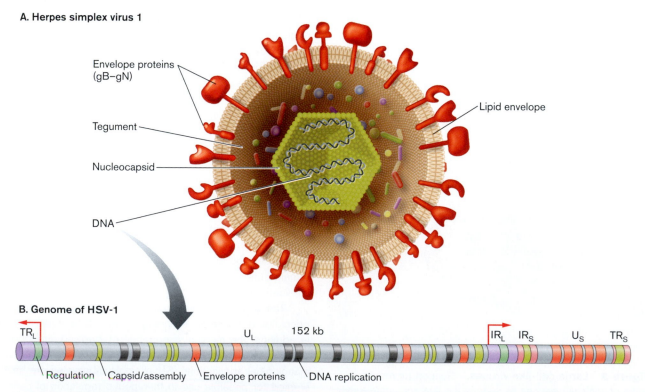

A. Herpes simplex virus 1

Envelope proteins (gB–gN)

Tegument

Nucleocapsid

DNA

Lipid envelope

B. Genome of HSV-1

TR_L U_L 152 kb IR_L IR_S U_S TR_S

Regulation   Capsid/assembly   Envelope proteins   DNA replication

Figure 11.35 Herpes simplex virus 1: virion and genome. A. HSV-1 virion consists of a double-stranded DNA chromosome packaged within an icosahedral capsid. The capsid is surrounded by tegument, a collection of 15 different kinds of virus-encoded proteins. The tegument is contained within a host-derived membrane envelope including several kinds of envelope proteins. **B.** The genome of HSV 1 spans 152,000 base pairs, encoding more than 70 gene products. The HSV sequence consists of two segments, long (U_L) and short (U_S). Each segment contains a unique region (U_L or U_S) flanked by two inverted repeat regions, terminal (TR_L or TR_S) and internal (IR_L or IR_S) where the two segments meet.

tion of about 15 different kinds of virus-encoded proteins. The tegument is contained within a host-derived membrane envelope with several kinds of spike proteins. The HSV-1 genome spans 152 kbp, encoding more than 70 gene products (**Fig. 11.35B**). The sequence includes two unique segments, long (U_L) and short (U_S), each flanked by a terminal repeat (TR_L or TR_S) and an internal repeat (IR_L or IR_S). Within the host, the genome circularizes, so the genetic linkage map appears circular.

Attachment and Entry of Herpes Simplex

Unlike poliovirus and HIV, the herpes virion can bind to several alternative receptor molecules in the host cell membrane, such as a homolog of tumor necrosis factor receptor called HveA or intercellular adhesion molecules called nectins (**Fig. 11.36** ▶⏸, step 1). As in the case of HIV, the entire herpes capsid enters the cytoplasm (step 2); but unlike HIV, whose capsid dissolves, the herpes capsid travels down a scaffold of microtubules (step 3) to the nuclear membrane. During this stage, the virion host shut-off factor (Vhs) degrades host mRNA, while the virion protein VP16 protects viral mRNA. At a nuclear pore, the herpes chromosome is inserted into the nucleus (step 4). The DNA then circularizes (step 5) to form a plasmid-like intermediate.

The herpes genome now takes one of two alternative directions: expression of mRNA for proteins of the infectious cycle or expression of mRNA encoding LAT proteins to maintain latency (step 6). If the latent course is taken, the DNA circle can persist within the cell for decades before switching to lytic infection. Latent infection most commonly occurs in nerve cells, such as those of the trigeminal ganglion.

Replication of Herpes Simplex Virus

In the nucleus, herpes DNA is transcribed to mRNA by host RNA polymerase II (**Fig. 11.36**). If mRNA for lytic infection is produced, it leaves the nucleus to be translated by ribosomes. Many different mRNAs are produced and exported, including those required for early and intermediate stages of infection (step 7). The translated proteins return to the nucleus for packaging within capsids.

To generate progeny genomes, the circular DNA is replicated by viral enzymes including DNA polymerase, single-strand binding protein, and a proofreading endonuclease (step 8). Additional enzymes are provided by the host cell. Replication occurs by the rolling-circle method, generating a concatemer similar to that of phage T4. Unlike T4, however, the herpes DNA is eventually cut into segments defined by the terminal repeat sequences.

The newly synthesized DNA expresses late-stage mRNA (step 9), which exits the nucleus for translation. Translated envelope proteins are inserted into the ER membrane, through which they migrate to the nuclear membrane (step 10). Other late proteins reenter the nucleus for assembly into capsids containing DNA genomes (step 11). The envelope may form from the nuclear membrane (step 12), although some evidence supports formation in the ER or Golgi. The virions are then transported through the ER (step 13) to the Golgi, and ultimately to the cell membrane (step 14). Finally, the cell membrane releases mature virions through exocytosis. Rapid release destroys the cell, causing the characteristic sores of herpes infection.

> **THOUGHT QUESTION 11.9**   Compare and contrast the fate of the HSV chromosome with that of the chromosome of HIV.

TO SUMMARIZE:

- **Herpes simplex** causes recurring eruptions of sores in the oral or genital mucosa. Initial transmission is by oral or genital contact, followed by eruptions from reactivated virus latent in ganglial neurons.
- **The herpes virion** contains a double-stranded DNA genome, packed in an icosahedral capsid. The capsid is surrounded by numerous matrix proteins and by an envelope.
- **HSV attachment** may involve several alternative receptors. **A microtubular scaffold** transports the herpes virions to the nucleus, where the DNA genome is inserted. The DNA circularizes for transcription.
- **DNA genomes of HSV** are synthesized by the rolling-circle method, using viral DNA polymerase supplemented by viral and host-generated components.
- **Infectious mRNA expression** leads to production of capsid, matrix, and envelope proteins for assembly of HSV. Alternatively, **LAT protein expression** leads to latent infection, usually in nerve cells, where the DNA persists silently for months or years.
- **HSV assembly** occurs at the nuclear membrane or other membranes. The virions are released from the cell by exocytosis. Rapid release leads to mucosal pathology.

Viral Infection Mechanisms: Commonalities and Diversity

In considering the mechanisms of viral infection, a picture emerges of interesting commonalities as well as intriguing diversity:

- **Viral infection requires host membrane receptors.** Different viruses evolve to take advantage of different cell membrane proteins in order to recognize and gain entry into host cells. Mutation of host receptors confers resistance to the virus.

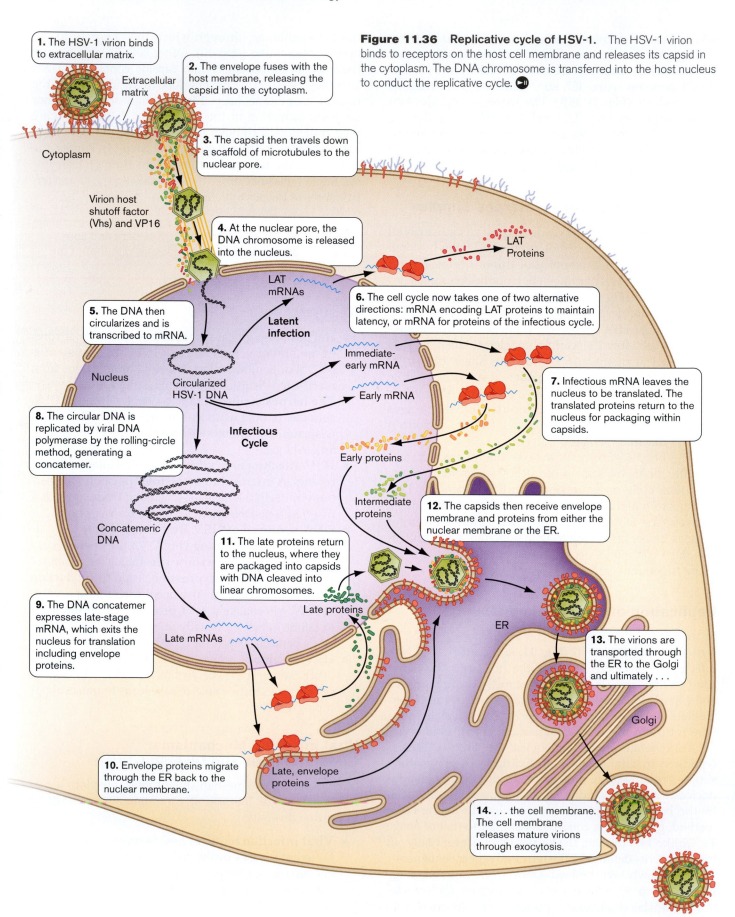

1. The HSV-1 virion binds to extracellular matrix.

2. The envelope fuses with the host membrane, releasing the capsid into the cytoplasm.

Figure 11.36 **Replicative cycle of HSV-1.** The HSV-1 virion binds to receptors on the host cell membrane and releases its capsid in the cytoplasm. The DNA chromosome is transferred into the host nucleus to conduct the replicative cycle. ▶❚❚

Extracellular matrix

Cytoplasm

3. The capsid then travels down a scaffold of microtubules to the nuclear pore.

Virion host shutoff factor (Vhs) and VP16

4. At the nuclear pore, the DNA chromosome is released into the nucleus.

LAT Proteins

LAT mRNAs

Latent infection

6. The cell cycle now takes one of two alternative directions: mRNA encoding LAT proteins to maintain latency, or mRNA for proteins of the infectious cycle.

5. The DNA then circularizes and is transcribed to mRNA.

Nucleus

Circularized HSV-1 DNA

Immediate-early mRNA

Early mRNA

7. Infectious mRNA leaves the nucleus to be translated. The translated proteins return to the nucleus for packaging within capsids.

8. The circular DNA is replicated by viral DNA polymerase by the rolling-circle method, generating a concatemer.

Infectious Cycle

Early proteins

Intermediate proteins

12. The capsids then receive envelope membrane and proteins from either the nuclear membrane or the ER.

Concatemeric DNA

11. The late proteins return to the nucleus, where they are packaged into capsids with DNA cleaved into linear chromosomes.

Late proteins

9. The DNA concatemer expresses late-stage mRNA, which exits the nucleus for translation including envelope proteins.

Late mRNAs

ER

13. The virions are transported through the ER to the Golgi and ultimately . . .

Golgi

10. Envelope proteins migrate through the ER back to the nuclear membrane.

Late, envelope proteins

14. . . . the cell membrane. The cell membrane releases mature virions through exocytosis.

- **Viruses recruit host cells to replicate their genomes.** Viral genomes are remarkably diverse, including RNA or DNA, single- or double-stranded, linear or circular. Replication within the host cell occurs by a variety of methods, with varying dependence on viral and host enzymes.

- **Virus particles (and their genomes) range in size from a few components to assemblages approaching the complexity of cells.** The advantage of simplicity is the minimal requirement for resources. The advantage of complexity is the fine-tuning of infection mechanisms for evading host defenses.

Concluding Thoughts

The study of viral molecular biology raises intriguing questions about the nature of viral replication and about viral origins (see **Special Topic 11.1**). Research in virology offers hope for new drugs and cures for humanity's worst plagues, as well as devastating diseases of agricultural plants and animals. At the same time, it is sobering to note that despite the enormous volumes we now know about viruses such as HIV and influenza, the AIDS pandemic continues to grow, and we face the likely emergence of a new deadly flu strain. Molecular research can succeed only in partnership with epidemiology and public health (discussed in Chapter 28).

CHAPTER REVIEW

Review Questions

1. How does phage T4 attach to its correct host cell and insert its genome for replication?
2. How does phage M13 maximize its reproductive options while avoiding loss of its host cell?
3. How does poliovirus compartmentalize the expression and replication of its genome? How are the products coordinated to complete viral assembly?
4. How do influenza virions gain access to the host cytoplasm? Explain the role of fusion peptides.
5. How does influenza virus manage the replication and packaging of its segmented genome? What is the consequence of genome segmentation for virus evolution?
6. How does HIV provide ready-made components for replication of its genome? How does reverse transcriptase convert the single-stranded RNA genome to double-stranded DNA?
7. What is the role of protease in HIV replication? What is the significance of protease for AIDS therapy?
8. How did ancient retroviruses participate in evolution of the human genome?
9. How does herpes virus compartmentalize the expression and replication of its DNA genome?
10. Compare and contrast the needs of DNA genome replication with those of RNA genome replication.

Key Terms

accessory protein (416)
acquired immunodeficiency syndrome (AIDS) (413)
assembly (394)
azidothymidine (AZT) (418)
chemokine receptor (CCR) (417)
conditional lethal mutation (396)
core particle (415)
coreceptor (398)
dissemination (406)
early gene (393)
endogenous retrovirus (422)
feline leukemia virus (FeLV) (413)
fusion peptide (409)

herpes simplex virus (HSV) (423)
human immunodeficiency virus (HIV) (413)
late gene (394)
lentivirus (413)
long terminal repeat (LTR) (420)
matrix protein (406)
nonsense suppressor (397)
nucleocapsid protein (NP) (407)
phage display (398)
picornavirus (400)
poliomyelitis (polio) (400)
poliovirus (400)
poliovirus receptor (Pvr) (401)

protease inhibitor (422)
provirus (420)
retrotransposon (422)
retrovirus (411)
reverse transcriptase (RT) (418)
rhinovirus (401)
Rous sarcoma virus (RSV) (413)
segmented genome (406)
spike protein (415)
tailed phage (391)
tegument (426)
viremia (406)
virus shedding (406)

Recommended Reading

Crotty, Shane, Maria-Carla Saleh, Leonid Gitlin, Oren Beske, and Raul Andino. 2004. The poliovirus replication machinery can escape inhibition by an antiviral drug that targets a host cell protein. *Journal of Virology* **78:**3378–3386.

Dufresne, Andrew T., and Matthias Gromeier. 2006. Understanding polio: New insights from a cold virus. *Microbe Magazine* **1:**13–18.

Flint, S. Jane, Lynn W. Enquist, Robert M. Krug, Vincent R. Racaniello, and Anna Marie Skalka. 2004. *Principles of Virology: Molecular Biology, Pathogenesis and Control of Animal Viruses,* 2nd ed. ASM Press, Washington, DC.

Frampton, A. R., Jr., W. F. Goins, K. Nakano, E. A. Burton, and J. C. Glorioso. 2005. HSV trafficking and development of gene therapy vectors with applications in the nervous system. *Gene Therapy* **12:**891–901.

Girones, Rosina. 2006. Tracking viruses that contaminate environments. *Microbe Magazine* **1:**19–25.

Levine, Bruce L., Wendy B. Bernstein, Naomi E. Aronson, Katia Schlienger, et al. 2002. Adoptive transfer of costimulated CD4+ T cells induces expansion of peripheral T cells and decreased CCR5 expression in HIV infection. *Nature Medicine* **8:**47–53.

Nam, Ki Tae, Dong-Wan Kim, Pil J. Yoo, Chung-Yi Chiang, Nonglak Meethong, et al. 2006. Virus-enabled synthesis and assembly of nanowires for lithium ion battery electrodes. *Science* **312:**885–888.

Nicola, Anthony V., Anna M. McEvoy, and Stephen E. Straus. 2003. Roles for endocytosis and low pH in Herpes Simplex Virus entry into HeLa and Chinese hamster ovary cells. *Journal of Virology* **77:**5324–5332.

Raoult, Didier, Stéphane Audic, Catherine Robert, Chantal Abergel, Patricia Renesto, et al. 2004. The 1.2-Megabase genome sequence of Mimivirus. *Science* **306:**1344–1350.

Rossman, Michael G., Vadim V. Mesyanzhinov, Fumio Arisaka, and Petr G. Leiman. 2004. The bacteriophage T4 DNA injection machine. *Current Opinion in Structural Biology* **14:**171–180.

Schuch, Raymond, Daniel Nelson, and Vincent A. Fischetti. 2002. A bacteriolytic agent that detects and kills *Bacillus anthracis. Nature* **418:**884–889.

Slusarczyk, Janusz (editor in chief). 2002–present. *HIV & AIDS Review: International Journal of HIV-Related Problems.*

Stevens, James, Ola Blixt, Terrence M. Tumpey, Jeffrey K. Taubenberger, James C. Paulson, and Ian A. Wilson. 2006. Structure and receptor specificity of the hemagglutinin from an H5N1 influenza virus. *Science* **312:**404–410.

van Riel, Debby, Vincent J. Munster, Emmie de Wit, Guus F. Rimmelzwaan, Ron A. M. Fouchier, et al. 2006. H5N1 virus attachment to lower respiratory tract. *Science* **312:**399.

Chapter 12

Molecular Techniques and Biotechnology

Technologies that are developed to answer basic questions in biology often yield useful consumer products. This is especially evident in today's world of genetic engineering, where the term *biotechnology* has become ubiquitous. The science of biotechnology uses living organisms or their products to improve human health. Although we think of it as a new field, the earliest biotechnologists actually lived about 10,000 years ago. They learned to use yeast to make bread and alcohol. They also unwittingly used naturally existing bacteria to make cheeses and yogurts. The field changed little over the millennia, until about 60 years ago, when the Scotsman Alexander Fleming discovered antibiotics. That watershed event, along with unraveling the structure of DNA and deciphering the genetic code, ushered in the modern high-tech era of biotechnology. We have since learned a great deal about the molecular biology of the microbes our ancestors used. Now, through gene-splicing techniques, we can force bacterial cells to produce human hormones; make fruits and vegetables produce vaccines; and create biochemical pathways to synthesize therapeutic molecules that never before existed.

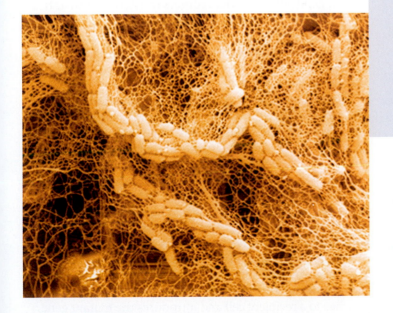

Scanning electron micrograph of an *Alcanivorax borkumensis* SK2 mutant overproducing polyhydroxyalkanoate during growth on n-octadecane (0.8 to 1.2 micrometers in length). Disrupting the acyl-coenzyme A thioesterase gene results in the overproduction and extracellular deposition of this biotechnologically important biopolymer used in sutures, cardiovascular patches, skin substitutes, and dozens of other applications.
Source: Heinrich Lünsdorf. 2006. *Journal of Bacteriology* 188(24):3452 (HZI-Helmholtz Centre for Infection Research).

Today's farmers must spray huge amounts of anti-fungal agents on their crops to combat potato blight disease—the cause of the famous Irish potato famine that wiped out 30% of the Irish population in the nineteenth century. Spraying plants as many as 25 times a season with chemicals, the typical U.S. potato farmer spends about $250 per acre to fight the disease—a huge incentive to find less expensive means of prevention. But the problem may be solved by biotechnology. Scientists discovered a wild Mexican potato that was completely resistant to *Phytophthora infestans*, the cause of potato blight disease. Although the Mexican potato had no commercial value (it is not edible), the gene conveying resistance was genetically engineered into commercial potato breeds, making them utterly resistant to the fungus. This discovery may eventually improve the health of potato crops (research is ongoing) and will certainly contribute to our understanding of disease resistance in plants. This is but one example of how biotechnology can potentially improve human existence.

Of course, with technological power comes a sobering responsibility, for if nature has taught us anything, it has taught us that it cannot be tampered with without cost. For instance, what would happen if an *Escherichia coli* that produced human growth hormone managed to colonize the human gut? Or, worse yet, one that was engineered to make botulism toxin. While these possibilities may seem unlikely, future advances in biotechnology must be guided by serious considerations of bioethics and an eye for unintended consequences.

We begin our discussion of biotechnology by explaining the techniques of modern molecular biology that are used to probe the inner workings of microbes. A research case history will illustrate how many of these techniques are applied to a single problem. Other applications will be discussed at the end of the chapter.

12.1 Basic Tools of Biotech: A Research Case Study

Biotechnology encompasses numerous molecular techniques developed over the last few decades. We need to understand these techniques to fully appreciate the potential of biotechnology. Some of these techniques were described in earlier chapters: DNA sequencing was introduced in Section 7.6; gene array technology and the two-dimensional PAGE analysis of proteomes were covered in Section 10.9. In this chapter, we present several other essential molecular tools in the context of a basic research question concerning a remarkable microbial stress response that allows *E. coli* to survive in the highly acidic conditions of the human stomach.

Few microbes that grow at neutral pH can survive at pH 2, a degree of acidity equivalent to 10 mM HCl. In the stomach, such a highly acidic environment kills most ingested pathogens, but strains of the intestinal bacterium *E. coli*, such as enterohemorrhagic *E. coli* O157:H7, can survive at pH 2 for hours. This means that a person needs to ingest only a few cells of *E. coli* to become colonized or infected. Thus, the organism is said to have a low infectious dose. What makes *E. coli* so special? The molecular tools that we will describe here helped reveal the novel mechanisms of acid resistance used by this microbe and the regulatory networks that control them. Our story of these acid resistance mechanisms begins with the discovery of acid-sensitive *E. coli* mutants that failed to survive at pH 2.

Investigating acid resistance in *E. coli* called for various techniques to determine which genes were defective in the acid-sensitive mutants and computer-assisted analysis of those genes to gain insight into their possible function (see Section 12.2). In Section 12.3, we will see how regulation of each gene was investigated by looking at their RNA transcripts (Northern blots) and protein products (Western blots). We will describe methods used to purify possible regulatory proteins and strategies used to prove that proteins regulating acid resistance bind to target DNA sequences. We will also discuss techniques that allow us to map protein-protein interactions in the cell, design novel vaccines to immunize against disease, and even carry out evolution in the test tube.

12.2 Genetic Analyses

Transposons Inactivate and Mark Target Genes

An early step commonly taken to gain insight into a biological process involves isolating mutants defective in that process. Such was the case with *E. coli* acid resistance. The initial goal was to inactivate relevant genes and tag them with an antibiotic resistance marker. In other words, a known gene whose product confers resistance to an antibiotic (that is, an antibiotic resistance gene) is inserted into the middle of a target gene sequence. This process inactivates the target gene and marks the mutation with antibiotic resistance. As a result, the mutant cell becomes antibiotic resistant.

Once constructed, the insertion mutation can be moved from one strain of *E. coli* to another by transduction. Transduction is the process whereby bacteriophage package bacterial DNA segments into capsids, rather than phage DNA, and then transfers the host DNA to another cell. Transducing phage is grown on the mutant bacterial strain (the donor) that contains the tagged host gene. Some of the progeny phage in the resulting lysate will contain the tagged gene. The phage lysate is then mixed with antibiotic-sensitive cells (the recipient) and the mixture plated onto agar medium containing the antibiotic. The only way a colony can develop is if a phage particle containing the marked (antibiotic resistance) gene attaches to a recipient cell and introduces the mutant gene into the cytoplasm. There the donor DNA can recombine into the recipient's genome in exchange for the wild-type gene. The recipient's genome remains identical in all other aspects except it now contains the mutant gene and is resistant to the antibiotic.

To generate mutations in genes needed to confer acid resistance, researchers infected a population of *E. coli* with bacteriophages containing a tetracycline resistance transposon. The bacteriophage was a mutant that fails to replicate in most strains of *E. coli*, so for an infected cell to become tetracycline resistant, the transposon had to "hop," or transpose, out of the phage DNA and into the host chromosome (see Section 9.6). Because target sequences are randomly selected during transposition, each tetracycline-resistant cell contained a transposon inserted into a different gene. As each cell grew into a colony, the colony was picked with a sterile toothpick and transferred into a well of a microtiter plate containing growth medium (**Fig. 12.1**).

After about 10,000 colonies were transferred to microtiter dishes and grown, a multipronged replicator device (a square with 48 metal prongs arranged in 6 rows of 8) was used to transfer small amounts of each culture to new microtiter plate wells that contained a pH 2 medium. After several hours of incubation, the multipronged replicator was again used to sterilely transfer a sample of the pH 2 culture from each well to an agar plate. Most of the insertion mutants survived at pH 2 and were able to form colonies on agar. These insertions occurred in genes unimportant for acid resistance. A few mutants, however, were killed. The transposon in these acid-sensitive mutants had inserted into a gene required for acid resistance, rendering it inactive. The acid-sensitive mutants were retrieved from the original growth medium microtiter plates and subjected to further analysis.

THOUGHT QUESTION 12.1   In this postgenomic era, how might you generally design a strategy to construct a strain of *E. coli* that cannot synthesize histidine?

DNA Sequencing of Insertion Sites and Computer Analysis

Once mutants in a project are identified, it is important to know in what genes the insertions occurred. Subsequent bioinformatic analysis of the gene may then link function to phenotype. In our example, an oligonucleotide primer (a short, single-stranded DNA molecule) that anneals to one end of the transposon was used to sequence across the insertion joint and into the adjacent *E. coli* DNA. The DNA sequences obtained were subjected to computer-based homology searches with the sequenced genome of *E. coli*. The matches obtained indicated that insertions occurred in four genes. Two of these genes encode iso-

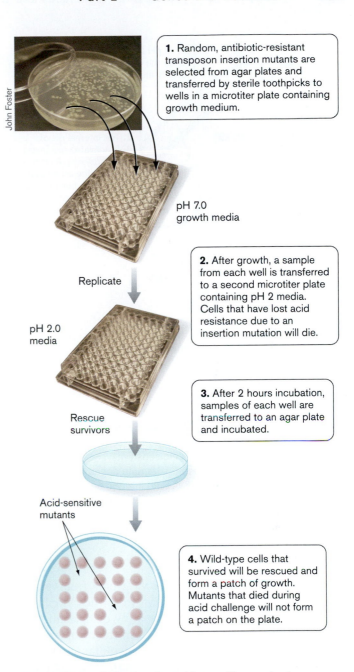

1. Random, antibiotic-resistant transposon insertion mutants are selected from agar plates and transferred by sterile toothpicks to wells in a microtiter plate containing growth medium.

pH 7.0 growth media

Replicate

pH 2.0 media

2. After growth, a sample from each well is transferred to a second microtiter plate containing pH 2 media. Cells that have lost acid resistance due to an insertion mutation will die.

3. After 2 hours incubation, samples of each well are transferred to an agar plate and incubated.

Rescue survivors

Acid-sensitive mutants

4. Wild-type cells that survived will be rescued and form a patch of growth. Mutants that died during acid challenge will not form a patch on the plate.

Figure 12.1   Selection for acid-sensitive mutants. Mutagenesis with transposons led to the discovery of several *E. coli* genes involved in acid resistance. The technique is applicable to many other biochemical systems.

forms of glutamate decarboxylase (the isoforms differ by only a few amino acids), which converts glutamic acid to gamma-aminobutyric acid (γ-aminobutyric acid), known as GABA; a third gene produces a glutamate/GABA antiporter; and a fourth gene encodes a putative DNA-binding protein with no known function, although it was annotated as a possible regulator.

The nature of the gene's functions suggests how the system might protect cells submerged in acid (**Fig. 12.2**). When *E. coli* is placed into a pH 2 environment, the cytoplasmic pH falls to dangerously low levels. Glutamate

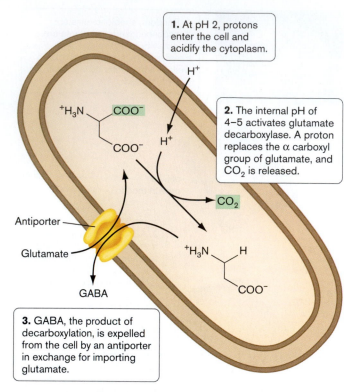

1. At pH 2, protons enter the cell and acidify the cytoplasm.

2. The internal pH of 4–5 activates glutamate decarboxylase. A proton replaces the α carboxyl group of glutamate, and CO_2 is released.

Antiporter

Glutamate

GABA

3. GABA, the product of decarboxylation, is expelled from the cell by an antiporter in exchange for importing glutamate.

Figure 12.2 Proposed model of acid resistance in *E. coli.* DNA sequence data identified the location of the transposon insertions leading to acid sensitivity. The genes identified included two isoforms of glutamate decarboxylase and a putative antiporter that brings glutamate into the cell in exchange for the decarboxylation product gamma-aminobutyric acid (GABA), which is transported out of the cell. The combined action of decarboxylase and antiporter would remove H^+ from the cell.

decarboxylase removes a carboxyl group from glutamate and replaces it with a proton from the cytoplasm, thereby decreasing acidity and elevating cytoplasmic pH. But to continue draining protons from the cytoplasm, *E. coli* must bring in more glutamate. The glutamate/GABA antiporter does this by bringing in glutamate in exchange for expelling the decarboxylation end product, GABA.

Exploring Gene Regulation: Reporter Fusions

Another important question in our example was whether any of the acid resistance genes were regulated in response to changes in the environment. Are they always expressed, and is the cell ever-ready to encounter acid stress? Or does the cell have sensing mechanisms that anticipate when it might encounter life-threatening low pH and only then express acid resistance genes? Several experimental strategies can be used to determine if a given gene is regulated in response to changes in the environment. Recall that the cell can control production of a protein at several levels.

The cell can regulate transcription, translation, and the stability of the mRNA or protein. How do microbiologists differentiate between these possibilities or determine if there is any regulation in the first place?

One technique used to examine gene regulation involves linking (or fusing) the promoter of the gene of interest to what is called a **reporter gene**, such as *lacZ* (beta-galactosidase) or *gfp* (green fluorescent protein). Reporter genes encode proteins whose activities are easily assayed. The result is called a **reporter fusion**. Reporter fusions facilitate monitoring the expression of genes whose products are difficult to measure directly. In the case of *lacZ*, beta-galactosidase (β-galactosidase) converts the substrate *o*-nitrophenyl galactoside (ONPG) to a yellow product that is easily measured with a spectrophotometer.

Two general types of reporter fusions are used to investigate gene regulation in bacteria: **operon fusions** and **gene fusions**. Operon fusions will only reveal transcriptional control of the target gene, whereas gene fusions expose both transcriptional and translational controls. Both types of fusions involve linking a reporter gene lacking its own promoter to the 3′ end of a target gene that still possesses its promoter (**Fig. 12.3**). The result is an artificial operon. Because there is only one promoter, only one mRNA transcript is made encompassing the target and reporter gene transcripts. The 5′ end of the newly engineered polycistronic mRNA will contain the target gene transcript followed at the 3′ end by the reporter gene transcript. When the reporter gene is inserted into the middle of the target gene, as shown in Figure 12.3, the downstream part of the target gene is not transcribed because of a transcriptional terminator located at the end of the reporter.

In both types of fusions, any factor that controls expression of the target gene promoter will also control the production of the reporter. The difference between operon fusions (also known as transcriptional fusions) and gene fusions (also known as protein or translational fusions) is whether the reporter gene carries its own ribosome-binding site. In an operon fusion, the reporter gene carries its own ribosome-binding site, but in a gene fusion, it does not. Because the reporter gene in an operon fusion maintains its own ribosome-binding site, translation of the reporter message occurs independently of the fused, target gene's message. As a result, operon fusions will only reflect transcriptional control at the promoter, not control at the level of translation.

In a gene fusion the reporter is missing *both* its promoter and ribosome-binding site. So for the reporter protein to be synthesized, the reporter gene must be fused to the target gene in the proper codon reading frame (**Fig. 12.3C**). Now, not only will the mRNA messages from the target and reporter genes be fused, but the peptides of the truncated target gene and the reporter gene will also be linked. In this case, anything controlling the transcription *or* translation of the target gene will also control reporter protein levels.

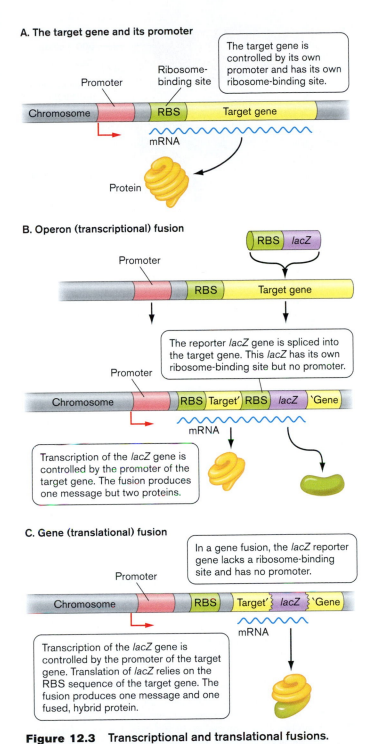

A. The target gene and its promoter

Promoter

Ribosome-binding site

The target gene is controlled by its own promoter and has its own ribosome-binding site.

Chromosome RBS Target gene

mRNA

Protein

B. Operon (transcriptional) fusion

RBS lacZ

Promoter

RBS Target gene

The reporter lacZ gene is spliced into the target gene. This lacZ has its own ribosome-binding site but no promoter.

Promoter

Chromosome RBS Target' RBS lacZ 'Gene

mRNA

Transcription of the lacZ gene is controlled by the promoter of the target gene. The fusion produces one message but two proteins.

C. Gene (translational) fusion

In a gene fusion, the lacZ reporter gene lacks a ribosome-binding site and has no promoter.

Promoter

Chromosome RBS Target' lacZ 'Gene

mRNA

Transcription of the lacZ gene is controlled by the promoter of the target gene. Translation of lacZ relies on the RBS sequence of the target gene. The fusion produces one message and one fused, hybrid protein.

Figure 12.3 Transcriptional and translational fusions. Aspects of a gene's regulation can be monitored by fusing a second gene, one that produces an easily assayed protein, to the gene under study. **A.** The gene targeted for construction of a reporter fusion. **B.** Construction of an operon fusion. **C.** Construction of a gene fusion. The lacZ open reading frame must be fused in-frame to the amino-terminal end of the target gene.

Each type of fusion can be constructed in vitro using plasmids and then transferred into recipient cells in a manner that will promote exchange of the reporter fusion for the resident gene. Once constructed, these fusions allow the researcher to monitor how different growth conditions influence the transcription or translation of the gene. They can also be used to test whether mutations in known or suspected regulators affect expression of the gene. In addition, these fusions provide convenient phenotypes (for example, lactose fermentation or fluorescence) that can be exploited in mutant hunts to screen for new regulators.

In our acid resistance example, a strain engineered to have a *gadA-lacZ* fusion would highly express that gene when growing on pH 5.5 agar media. If that medium contained X-Gal, a chemical indicator of LacZ activity, the colonies would appear dark blue. β-galactosidase cleaves a bond in the colorless X-Gal molecule and releases a blue chemical that stays in the colony. One can then search for regulatory genes required to activate *gadA* by subjecting the *gadA-lacZ* strain to random transposon mutagenesis and plating the insertion mutants onto pH 5.5 X-Gal medium. The medium includes an antibiotic that only allows growth of insertion mutants whose transposons contain an antibiotic resistance gene. An insertion into a gene encoding a positive regulator of *gadA-lacZ* will then produce a white (LacZ⁻) colony, distinguishable from the blue LacZ⁺ colonies.

> **THOUGHT QUESTION 12.2** You have monitored expression of a gene fusion where *lacZ* is fused to the glutamate decarboxylase gene, *gadA*. You find that there is a 50-fold increase in expression of *gadA* (as measured for β-galactosidase) when the cells carrying this fusion are grown in media at pH 5.5 as compared to cells grown at pH 8. What additional experiment must you do to tell if the control is transcriptional or translational?

TO SUMMARIZE:

- **Transposons** can be used to mutate genes and mark the defective gene with antibiotic resistance.
- **The location of a transposon insertion** can be determined by sequencing from the end of the transposon across the insertion junction. The junction sequence is examined by online comparative BLAST analysis to identify the gene in the genome.
- **Annotations of the gene** may provide clues as to the function of the protein.
- **Regulation of a gene** can be determined by fusing that gene to an easily assayed reporter gene.
- **Operon fusions** reflect transcriptional control of the subject gene.
- **Gene fusions** (also called translational fusions) reflect transcriptional and translational control.

12.3 Molecular Analyses

There are many situations and numerous organisms in which creating a fusion is not possible. One must then use more direct measures to monitor changes in mRNA or protein levels. This section will describe the Northern, Southern, and Western blot procedures that accomplish these goals and will describe methods to examine protein-DNA interactions as well as rapidly purify proteins.

Northern Blots Help Visualize Specific mRNA Messages

Edwin M. Southern, in the 1970s, invented a method, now called the **Southern blot**, for identifying a particular sequence of DNA in a complex mixture of DNA fragments. Scientists later used a similar technique to examine RNA fragments and named their method the **Northern blot** in counterpoint to the Southern blot.

Northern blots are used to analyze the presence, size, and processing of a specific RNA molecule in a cell extract. In the Northern blot technique, RNA is extracted from the cell. Special precautions must be taken to avoid contaminating these preparations with RNases that are ubiquitous to the laboratory environment. The RNA is then fractionated by electrophoresis on an agarose-formaldehyde gel. Formaldehyde helps keep the RNA denatured and unkinked by preventing base pairing, so the molecules can be separated by size. The separated fragments are transferred by simple capillary action onto a nylon or nitrocellulose membrane, creating the blot.

The diagram in **Figure 12.4A** illustrates how buffer is drawn up through the agarose gel and then through the facing membrane. RNA (or DNA) travels with the buffer out of the gel and is deposited on the membrane in exactly the same pattern as it was displayed in the gel. Once transferred, the RNA is fixed to the membrane by UV cross-linking.

Once the RNA is transferred, the membrane is hybridized (or "probed") with a small DNA fragment specific for a given mRNA. This is done under conditions that allow annealing between the DNA probe and matching RNA fragment in a process called **hybridization**. The probe can be labeled with radioactivity, a fluorescent dye, or biotin. The biotin is detected later by a chemiluminescent enzyme assay. RNA fragments to which a radiolabeled probe has bound are visualized by exposing the membrane to X-ray film, followed by photographic development of the film, otherwise known as **autoradiography**. Alternatively, the membrane is subjected to phosphoimaging. (A phosphoimager is a machine that records the energy emanated as light or radioactivity from gels or membranes.)

The molecular weights of bands visualized by autoradiography or phosphoimaging can be determined by comparing their locations on the gel with the locations of

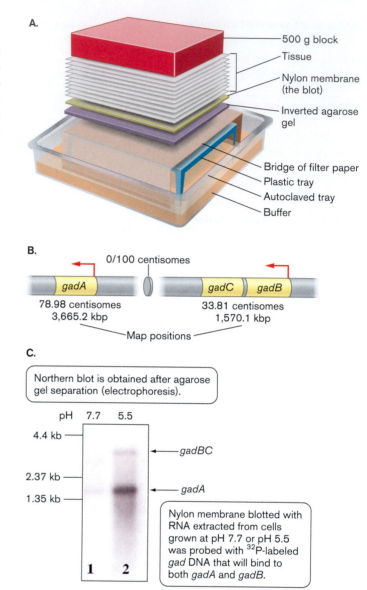

A.
- 500 g block
- Tissue
- Nylon membrane (the blot)
- Inverted agarose gel
- Bridge of filter paper
- Plastic tray
- Autoclaved tray
- Buffer

B.
0/100 centisomes

gadA

gadC   gadB

78.98 centisomes
3,665.2 kbp

33.81 centisomes
1,570.1 kbp

Map positions

C.

Northern blot is obtained after agarose gel separation (electrophoresis).

pH   7.7   5.5

4.4 kb —

2.37 kb —

1.35 kb —

← gadBC

← gadA

Nylon membrane blotted with RNA extracted from cells grown at pH 7.7 or pH 5.5 was probed with ^{32}P-labeled gad DNA that will bind to both gadA and gadB.

1   **2**

Figure 12.4   Northern blot to view mRNA levels.   Northern blots can be used to monitor the quantity and breakdown of RNA. **A.** Apparatus to perform Northern blot. Capillary action draws buffer through the gel and carries the RNA upward to the membrane, which binds the RNA. **B.** Organization of genes responsible for acid resistance in *E. coli.* Genes *gadA* and *gadB* encode isoforms of glutamate decarboxylase. Gene *gadC* encodes the putative antiporter. Map positions are given in centisomes and kpb. **C.** Northern blot illustrates that *gadB* and *gadC* are transcribed as an operon, while *gadA*, located elsewhere in the genome, is transcribed separately. The data also illustrate that *gadA* and *gadBC* mRNA accumulate after growth at pH 5.5, probably due to increased transcription. *Source:* From Zhuo Ma, et al. 2003. *Molecular Microbiology* 49:1309–1320.

standard RNA molecular weight markers run in a parallel lane. This technique can be used to determine the presence of any given mRNA. By estimating the size of the transcript, the method can also be used to evaluate whether two adjacent genes are transcribed as an operon.

In our example, how are the *E. coli* acid resistance genes regulated? **Figure 12.4B** shows that the *gadA* gene is located at a position on the genome different from those of the *gadB* and *gadC* genes, which together form an operon. The Northern blot in **Figure 12.4C** reveals that when cells are grown under acidic conditions (low pH), the *gadA* gene and the operon *gadBC* are both induced. The RNA gel was probed with a fragment of DNA that can bind mRNA from either the *gadA* or *gadB* gene. Because *gadB* and *gadC* form an operon, a longer mRNA molecule appears. The RNA bands are more intense in the lane from acid-grown cultures (pH 5.5) than in the lane from cultures grown at high pH (pH 7.7). The results suggest that the transcription of these genes is induced by acid. A current alternative procedure for quantitating levels (but not sizes) of specific RNA molecules is real time quantitative PCR, which is discussed in Section 12.4.

Southern Blots Help Visualize Specific DNA Sequences

The Southern blot technique for DNA uses a probing strategy similar to that of Northern blots to detect the presence of specific bands of DNA. The basic difference between the Southern and Northern blot techniques is the starting material. Southern blots begin with restriction endonuclease digestion of genomic DNA to produce discrete DNA fragments that are then separated by electrophoresis in an agarose or acrylamide gel. The DNA fragments are denatured to single strands then blotted onto a membrane. A DNA probe is generated from one bacterial species by PCR, although other techniques can be used, and labelled with a radioactive or nonradioactive tag. This probe is hybridized to the DNA blot and the hybridized bands visualized by autoradiography or phosphoimaging. Standardized double-stranded DNA markers are used to help determine sample fragment sizes. Southern blotting is used, for example, to detect the presence of specific genes among different species or strains of a single species.

Western Blots Help Visualize Proteins

An alternative approach to examining gene regulation is to detect the protein products themselves; the technique used is called the **Western blot**. Antibodies that specifically bind a given protein can be used to detect those proteins within a complex mixture of different proteins. To carry this out, the protein extract is subjected to SDS polyacrylamide gel electrophoresis (discussed in Section 10.9). The protein bands are electrophoretically transferred from the gel to a membrane made of polyvinylidene fluoride (PVDF), which tightly binds proteins (**Fig. 12.5A**). The membrane is then probed with a primary antibody directed against the specific protein. These antibodies are made by injecting the protein into animals and then extracting the serum after a few months.

Depending on the source of the primary antibody (mouse, rabbit, or other), the membrane is then probed with

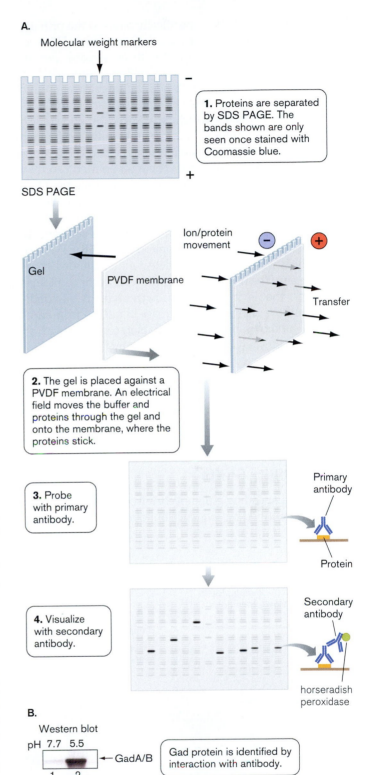

A.

Molecular weight markers

1. Proteins are separated by SDS PAGE. The bands shown are only seen once stained with Coomassie blue.

SDS PAGE

Gel

PVDF membrane

Ion/protein movement

Transfer

2. The gel is placed against a PVDF membrane. An electrical field moves the buffer and proteins through the gel and onto the membrane, where the proteins stick.

3. Probe with primary antibody.

Primary antibody

Protein

4. Visualize with secondary antibody.

Secondary antibody

horseradish peroxidase

B.

Western blot

pH 7.7 5.5

← GadA/B

1   2

Gad protein is identified by interaction with antibody.

Figure 12.5 Western blot analysis. Specific proteins within a mixture of many proteins can be identified by Western blot. **A.** The Western blot procedure. **B.** Western blot analysis of glutamate decarboxylase content of *E. coli* grown at pH 7.7 versus pH 5.5. The production of the protein is induced by growth under acidic conditions. *Source: From Zhuo Ma, et al. 2003. Molecular Microbiology 49:1309.*

a secondary antibody that specifically binds to the primary antibody (for example, anti-mouse IgG antibody). This secondary antibody may be tagged with horseradish peroxidase, for instance. The result is an antibody "sandwich" in which the primary antibody binds the target protein and the secondary antibody binds the primary antibody. Adding hydrogen peroxide, which is oxidized by horseradish peroxidase and luminol (a chemical that emits light when oxidized) sets off a luminescent reaction wherever an antibody sandwich has assembled. The light emitted can then be detected by autoradiography or phosphoimaging. Only protein bands to which the primary antibody has bound will be detected. The technique is used to determine if, and at what levels, specific proteins are present in a cell extract.

In our acid resistance example (**Fig. 12.5B**), Western blot analysis was performed to detect glutamate decarboxylase protein. The blot shows that the level of this protein is increased—that is, induced by growth under acidic conditions. Thus, the acid-induced increase in mRNA from transcription (as shown by the Northern blot) correlates with an increase in translated protein (as shown by Western blot).

> **THOUGHT QUESTION 12.3**   What would you conclude if the Northern blot of the *gad* genes showed no difference in mRNA levels in cells grown at pH 7.7 and pH 5.5 but the Western blot showed more protein at pH 5.5 than at pH 7.7?

DNA Mobility Shifts

Once a gene encoding a putative transcriptional regulatory protein is identified by sequence annotation, proof of func-

tion requires a demonstration that the protein will bind to a DNA sequence present in the target promoter area. One approach is to add purified protein to the putative target DNA fragment and see if the protein causes a shift in the electrophoretic mobility of that fragment in a gel. An example of this method, as applied to our acid resistance problem, is illustrated in **Figure 12.6A**. Linear DNA molecules travel in agarose or polyacrylamide gels at an inverse rate relative to their size; larger molecules travel more slowly than smaller ones. If a protein binds to a DNA molecule, the complex is larger than the DNA alone and will travel even more slowly in the gel. This method of analyzing protein-DNA binding is called the **electrophoretic mobility shift assay (EMSA)** or gel shift (first introduced in Section 10.2).

In our research example, one of the genes that affected acid resistance (*yhiE*) was annotated in the database as a potential transcriptional regulator. The encoded protein was purified and tested by EMSA for its ability to bind to a DNA sequence conserved within the promoter regions of the two potential targets, *gadA* and *gadBC*. As shown in **Figure 12.6B**, the purified YhiE protein, renamed GadE because of its role in regulating glutamic acid decarboxylase expression, did bind the *gadA* sequence and retarded its mobility. Thus, GadE appears to be a true regulator of the acid resistance genes. It appears to work in collaboration with the GadW protein shown in Special Topic 10.1, Figure 2.

Purifying Proteins in One Step

How do researchers purify proteins today? In the past, protein purification was a long, complex process involving

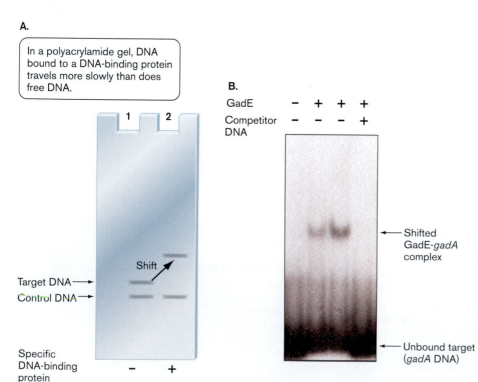

A.

In a polyacrylamide gel, DNA bound to a DNA-binding protein travels more slowly than does free DNA.

Target DNA →
Control DNA →
Shift

Specific DNA-binding protein   −   +

B.

GadE   −   +   +   +
Competitor   −   −   −   +
DNA

→ Shifted GadE-*gadA* complex

→ Unbound target (*gadA* DNA)

Figure 12.6   Electrophoretic mobility shift assay (EMSA).   Protein-DNA interactions can be monitored by EMSA. **A.** Diagrammatic representation of EMSA showing slowed mobility of a hypothetical DNA-protein complex in a polyacrylamide gel. **B.** Gel shift experiment showing that the GadE regulatory protein binds to a radiolabeled promoter fragment of *gadA*. Lane 1, radiolabeled target DNA only; Lanes 2 and 3 contain target DNA and GadE protein. Lane 4 is the same as Lane 3 but includes excess unlabeled promoter fragment. The unlabeled fragments effectively outcompete labeled fragments for the protein, so no shift in the radiolabeled fragment is seen. This provides a check on the specificity of the binding. *Source:* From Zhuo Ma, et al. 2003. *Molecular Microbiology* 49:1309.

numerous columns and time-consuming assays. Now, a desired protein with a known sequence can be purified in only a week or so. The trick is to take the gene encoding the protein and fuse it to a DNA sequence encoding a peptide tag that has a strong binding affinity to some small ligand molecule. The target ligand is attached to beads, and the cell extract containing the tagged protein is passed over the beads. Only the tagged protein, also called a fusion protein, will bind to the beads. The other proteins will pass through the column. This technique allows researchers to essentially "fish" the tagged protein from a complex mixture of proteins present in a cell extract.

There are several commonly used peptide tags. One tag is a series of six histidines, called a His_6 tag, which tightly binds to nickel. A commercially available plasmid used to make His-tagged proteins is shown in **Figure 12.7A**.

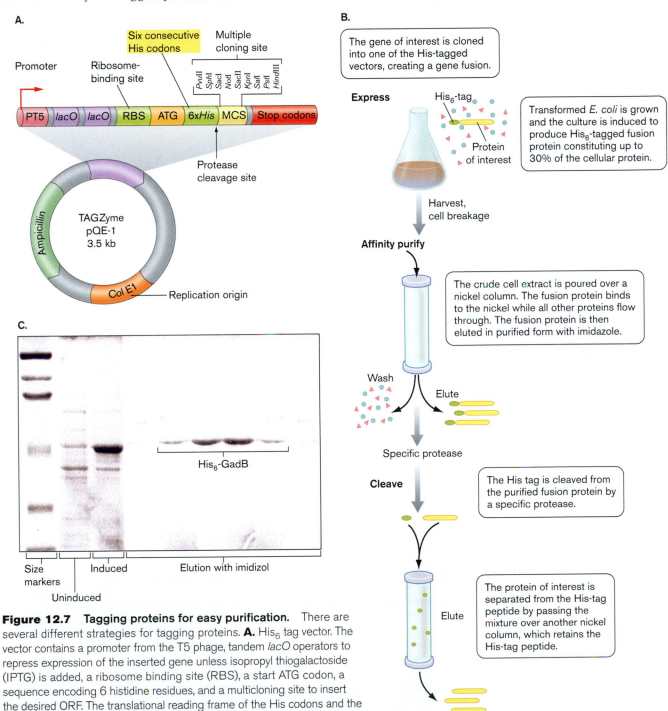

Figure 12.7   Tagging proteins for easy purification.   There are several different strategies for tagging proteins. **A.** His_6 tag vector. The vector contains a promoter from the T5 phage, tandem *lacO* operators to repress expression of the inserted gene unless isopropyl thiogalactoside (IPTG) is added, a ribosome binding site (RBS), a start ATG codon, a sequence encoding 6 histidine residues, and a multicloning site to insert the desired ORF. The translational reading frame of the His codons and the codons of the ORF must be the same to create a protein fusion. **B.** Purification of the fusion protein on a nickel column. **C.** SDS polyacrylamide gel electrophoresis of fractions from the purification of His_6-GadB. Lane 1: protein molecular weight markers. Lane 2: uninduced cells. Lane 3: induced cells. Lanes 4–9: His_6-GadB eluted from nickel column with imidazole.

Source: C. From Zhuo Ma, et al. 2003. *Molecular Microbiology* 49:1309.

These plasmids typically include an inducible promoter to control when the fusion protein is produced. To purify the protein produced, nickel-coated beads are loaded into a column, and then a crude extract containing the tagged protein is added to the top (**Fig. 12.7B**). Proteins that do not bind to the nickel (that is, those without the His$_6$ tag) are eluted with buffer. The tagged protein molecules that did bind are then eluted from the column with an elution buffer containing imidazole. Imidazole has a stronger affinity for the nickel and will displace the His$_6$-tagged protein from the beads. Analysis of the progressive purification of GadB by the His$_6$ tag fusion method is shown in **Figure 12.7C**.

The other commonly used tag is maltose-binding protein (MBP), which binds to amylose. The MBP-GadE protein in our research example (Fig. 12.6) was made by genetically fusing maltose-binding protein to the amino terminus of GadE.

Primer Extension Analysis Identifies Transcriptional Start Sites

For many genes, transcription can occur from multiple promoters to produce transcripts of different sizes. The transcripts all *end* at the same terminator but *begin* at different defined nucleotides located upstream of the struc-

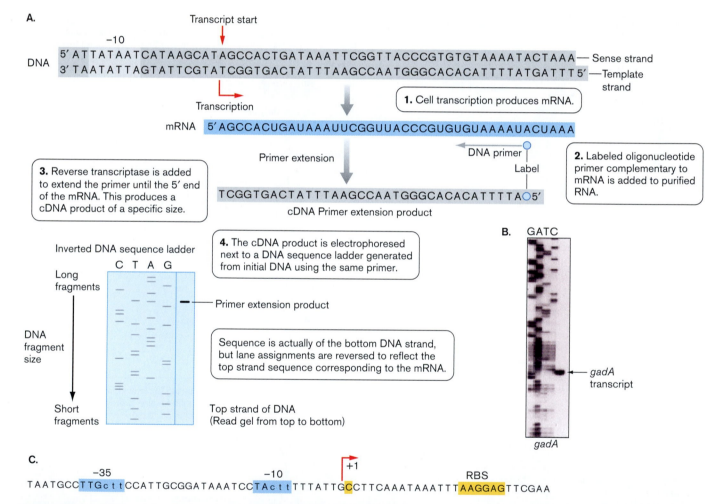

Figure 12.8 Primer extension analysis to determine transcriptional start sites. **A.** Top sequence represents the DNA sequence of an imaginary promoter. The mRNA product is shown below. After separating the RNA from DNA, a DNA oligonucleotide primer that binds to the mRNA at a defined place is added, followed by a reverse transcriptase reaction. Reverse transcriptase produces a cDNA primer extension product that ends at the beginning (5′ end) of the mRNA. The product is run in a polyacrylamide gel next to a sequencing ladder of the original DNA made with the same primer used in the reverse transcriptase reaction. The primer extension product will migrate at the same position as the base that represents the start site of transcription. **B.** Primer extension (lane on far right) showing the transcriptional start site of the *gadA* gene (arrow) involved in *E. coli* acid resistance. **C.** The sequence of the promoter is shown with −10, −35, and transcriptional start site marked. Lowercase letters indicate bases that differ from consensus −10 and −35 sequences. Note that the first base of the mRNA transcript is marked C (cytosine). This corresponds to a G (guanosine) in the gel sequence ladder (panel B). This is because the DNA sequence obtained using the primer is actually that of the strand complementary to the message. *Source: De Biase, et al. 1999. Molecular Microbiology 32:1198.*

tural gene. These different promoters are useful because they can respond to different cell signals (for example, high temperature vs. acidic pH vs. nitrogen limitation).

To begin the search for upstream promoter sequences, we need to define where each transcript begins. If there are different transcriptional start sites, there are multiple promoters. One method to determine transcript length is called **primer extension** (**Fig. 12.8**). Knowing the sequence of the gene, the researcher designs a single DNA primer that will anneal to the mRNA of interest near to the suspected start site. The primer is used in a reverse transcription reaction (reverse transcriptase makes DNA from RNA) that extends the primer to the 5′ end (the start) of the message. This produces a complementary DNA (cDNA) of a precise length. The cDNA is radiolabeled by using radioactive nucleotides in the reaction mix.

The key to using the primer extension technique is that the same primer is also used in a DNA-sequencing reaction with the template DNA (see **Fig. 12.8A**). The cDNA primer extension fragment is then run on a poly-acrylamide gel alongside a DNA-sequencing reaction of the gene fragment. The size of the primer extension cDNA fragment will be identical to the size of one of the "rungs" on the DNA sequence ladder. That "rung" represents the base in the sequence that correlates to the transcriptional start site. The cDNA from the transcripts of genes that have multiple promoters will produce primer extension fragments of varying sizes that comigrate with different-sized sequencing fragments.

Primer extension analysis of one of the *E. coli* acid resistance genes, *gadA*, is illustrated in **Figure 12.8B**. The primer extension analysis indicates that the transcript begins opposite a G in the sequence of the complementary DNA strand. This corresponds to a C in the sense strand of DNA shown in **Figure 12.8C**. The sense strand has the same sequence as the mRNA. Thus, the promoter sequences for *gadA* reside at –10 and –35 bp from the transcript start site. Once the promoter is identified, we can begin to search for nearby DNA sequences that control expression of the promoter.

DNA Protection Analysis Identifies DNA-Protein Contacts

An electrophoretic mobility shift assay will identify fragments of DNA to which a transcriptional regulatory protein will bind, but it does not actually reveal which nucleotides in the sequence are bound. A first step in determining potential DNA-protein contacts on DNA is to see what bases the binding protein can protect from nuclease digestion. The protected bases represent what is called the "footprint" of the binding protein.

Because a regulatory protein covers DNA at its binding site sequence, an enzyme that might attack the sequence cannot get close enough to do damage. Deoxy-ribonuclease I (DNase I), one of many different types of DNases, is one such enzyme. It cleaves phosphodiester bonds between bases. **Figure 12.9** presents the fundamental method that uses DNase I to determine the protective "footprint" a regulatory protein leaves on a DNA sequence. In this method, DNase I is added to a DNA fragment at a concentration that can nick each molecule only once during the time of incubation. When the DNA is later denatured by heat, arrays of single-stranded fragments are produced, each fragment being of a different size, depending on where the nick occurred. Because only one end of one strand of the DNA is labeled, only those fragments containing that label will be seen following polyacrylamide gel electrophoresis and autoradiography.

The bases protected by a hypothetical protein are shown in **Figure 12.9A** and **B**, while **Figure 12.9C** presents the actual footprint of GadX, another regulator of *E. coli* acid resistance. The regulatory protein GadX covers a region around the promoter of *gadA* and *gadB* (only *gadA* is shown). The region is immediately upstream of the promoter. It is thought that both regulatory proteins GadE (previously described) and GadX communicate with RNA polymerase at the promoter, allowing it to transcribe the acid resistance genes.

DNA protection assays do not definitively show contact points between protein and DNA. Areas of the protein only have to be close enough to the DNA to hinder nuclease access. Protein-DNA interaction sites can be better defined by other, more difficult methods that cause a cross-link between amino acids of the protein and the specific bases they contact in the DNA. Nevertheless, the DNA protection assay remains a very useful tool.

> **THOUGHT QUESTION 12.4** Another regulator of glutamate decarboxylase (Gad) production does not affect the production of *gad* mRNA but is required to accumulate Gad protein. What two regulatory mechanisms might account for this phenotype?

Two-Hybrid Analysis Reveals Protein-Protein Interactions

Many cellular proteins interact with and influence the function of one or more other proteins in the cell. For example, anti-sigma factors help control the timing of sporulation in *Bacillus* by interacting with sigma proteins. Protein-protein interactions, therefore, constitute an essential aspect of the normal workings of a living cell. Unraveling the various interactions between individual proteins is an invaluable way of understanding protein function.

It is important to probe protein-protein interactions in vivo, where the proteins actually work. An ingenious tool to measure in vivo protein-protein interaction is the two-hybrid system. Two-hybrid analysis can be used to mine for unknown "prey" proteins that interact with a known "bait" protein.

A. DNase footprint assay, part 1

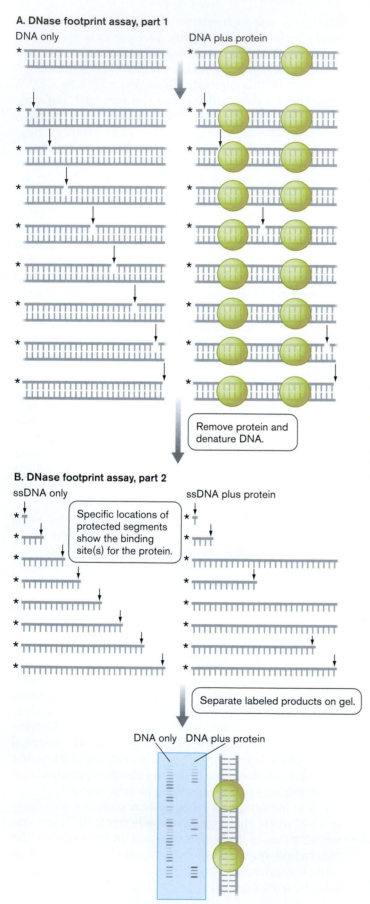

Remove protein and denature DNA.

B. DNase footprint assay, part 2

ssDNA only

Specific locations of protected segments show the binding site(s) for the protein.

ssDNA plus protein

Separate labeled products on gel.

DNA only   DNA plus protein

While there are many types of two-hybrid techniques, a classic example is the **yeast two-hybrid system** (**Fig. 12.10**). Yeast two-hybrid analysis involves fusing genes encoding two potentially interacting proteins to separated components of the yeast GAL4 transcriptional factor (**Fig. 12.10A**). The GAL4 transcription factor is composed of two domains: an activation domain and a sequence-

Figure 12.9 **DNA protection assay using DNase I.** Protein-DNA interactions can be revealed by an ability of the protein to protect DNA from nuclease digestion. **A.** DNase I will randomly nick unprotected double-stranded DNA. Conditions are designed so that there will be only one nick per molecule of DNA. To facilitate tracking, only one end of one strand is labeled (asterisk). *Left:* Random cleavage of unprotected DNA. *Right:* DNA-binding proteins block DNase I activity. **B.** The protein is removed, and the strands are separated by heat denaturation and run on polyacrylamide gels. By comparing the two lanes, one can see the "footprint" where the DNA-binding protein protected the fragment from nuclease digestion. **C.** Gel showing DNase I footprint where a regulator of *E. coli* acid resistance, GadX, binds to the *gadA* promoter region. The actual protein used was an MBP (maltose-binding protein)-GadX fusion protein. Numbers to the left indicate the number of bases relative to the transcriptional start (+1). The first two lanes represent the fragment's DNA sequence (G and A lanes only). Numbers above the gel indicate increasing amounts (picomoles) of GadX regulatory protein added. Orange boxes marked with roman numerals on the right indicate areas protected by GadX. The red box marks the sequence bound by the other regulator, GadE. The bent arrow denotes the transcriptional start site. *Source: Angela Tramonti, et al. 2002. Journal of Bacteriology 184:2603.*

C.

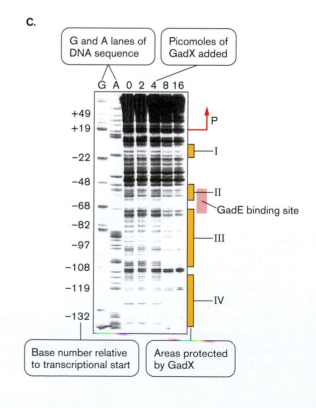

G and A lanes of DNA sequence

Picomoles of GadX added

Base number relative to transcriptional start

Areas protected by GadX

specific DNA-binding domain. The DNA-binding domain binds to the yeast GAL1 gene promoter region, positioning the GAL4 activation domain to bind RNA polymerase and stimulate GAL1 transcription. If the domains are separated into two different peptides, GAL1 is not activated. However, when separated, both domains can still work together if each domain is fused to one member of a pair of other proteins that interact.

This strategy is exploited in the yeast two-hybrid system. The target gene is a chromosomal *GAL1-lacZ* fusion gene, although other reporter genes can be used. The *lacZ* gene encodes beta-galactosidase, whose enzymatic activity is easily measured. When the two resulting hybrid proteins interact via the two target protein domains, the GAL4 DNA-binding and activation domains are brought together. Thus assembled, the two-hybrid GAL4 complex will bind and activate transcription of *GAL1-lacZ*. Expression of *GAL1-lacZ* is then visualized on agar plates containing a substrate of beta-galactosidase, X-Gal. X-Gal is a colorless substrate that, when cleaved, produces a blue product.

The power of this technique is that it can also be used to find unknown proteins that interact with a known protein. A known protein is fused to one of the GAL4 domains and used as "bait" to find the "prey" proteins expressed by randomly cloned "prey" genes fused to the other GAL4 domain. Plasmids containing the bait and prey fusions are transformed into yeast cells and the transformants plated onto a medium containing an indicator of beta-galactosidase activity. After plating, colonies that express beta-galactosidase are the result of protein-protein interactions occurring between the bait and prey fusion proteins. Sequencing the DNA insertion can then identify the gene encoding the "prey" protein. An example of two-hybrid analysis is shown in **Figure 12.10B**. The two-hybrid approach could be used in our acid resistance research example to learn if any of the proteins involved interact with any other proteins in the cell.

> **THOUGHT QUESTION 12.5** How could you use two-hybrid analysis to determine if *three* proteins can interact? You suspect that proteins A and C can simultaneously bind to protein B and that this interaction then allows A to interact with C in the complex.

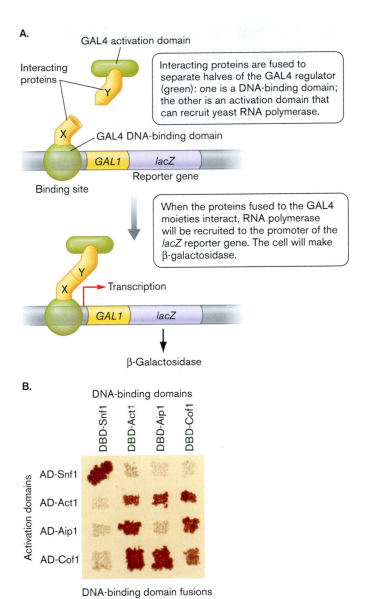

Figure 12.10 Detecting protein-protein interactions: the yeast two-hybrid system. A. Interacting proteins are fused to separate halves of the GAL4 regulator. **B.** Examples of strains containing interacting proteins. Each patch of growth represents a different pair of proteins fused to the GAL4 domains. The interaction activates synthesis of beta-galactosidase and turns colonies red on this indicator medium. *Source*: B. David Amberg. 1999. *Journal of Cell Biology* 145(6):1251.

TO SUMMARIZE:

- **Northern blot technology** uses labeled DNA from a specific gene to probe the levels and sizes of RNA made from that gene.
- **Southern blot technology** uses labeled DNA from a specific gene to probe other genomes for the presence of that gene.
- **Western blot technology** uses antibodies to specific proteins to detect the quantity and size of those proteins in cell extracts.

- **Electrophoretic mobility shift assays** (EMSA) examine protein interactions with DNA.
- **Fusion proteins** are made to easily purify proteins for biotechnology purposes.
- **Primer extension analysis** uses labeled DNA probes and reverse transcriptase to identify the 5′ end of a transcript and thus the transcriptional start site of a gene.
- **DNase protection assays** reveal the bases in a DNA sequence protected by a DNA-binding protein.
- **Two-hybrid analysis** uses hybrid proteins to detect protein-protein interactions.

12.4 "Global" Questions of Cell Physiology

In addition to answering specific questions of cell physiology, biotechnology can be used to look into broad—even global—questions of whole-cell physiology and evolution. When a cell encounters a change in its environment, it is rare that only one gene or protein becomes affected. Rather, the impact on the cell radiates along and between pathways affecting many different genes and proteins. Tracking these global regulatory influences can show how a species adjusts its entire physiology to better tolerate, or even flourish in, a new environment.

This section describes a number of these incredible tools. For example, how can we map protein-protein interactions throughout an entire cell? At the other extreme, how can we follow the intracellular movement of a single protein as it does its job? We will also describe ingenious PCR strategies that allow the unequivocal identification of individual species in a complex microbial community.

Mapping Protein-Protein Interactions

One example of global analysis of cell physiology involves a previously described tool that examines protein-protein interactions—the yeast two-hybrid system. Micheline Fromont-Racine and colleagues at the Pasteur Institute in Paris developed a high-throughput, genome-wide version of the yeast two-hybrid system to create protein interaction maps for whole cells (**Fig. 12.11**). The automated version of the procedure is rapidly becoming the method of choice for mapping whole proteomes. With a yeast cell mating procedure that increases screening efficiency, Fromont-Racine and colleagues used their complex yeast genomic library of 5×10^6 clones to test 7×10^8 interactions against 15 proteins.

The method identified and classified 170 potential interactors, including approximately 70 proteins of previously unknown function. More than 25% of the interactors are probably biologically relevant, although one cannot be sure without additional in vivo testing. The other 75% may be spurious interactions due to the unnatural overexpression of the two proteins in the test cells. Overexpression of two proteins could magnify low-affinity interactions that would not take place when those proteins are present at lower, more physiological concentrations.

The achievements of this group have opened the way to the systematic analysis of the protein interaction networks of the 6,000 open reading frames of the yeast proteome. Another European team (Hybrigenics, in Paris) has adapted the Fromont-Racine procedure to analyze a bacterial genome and has linked half of the 1,600 proteins of the ulcer-provoking bacterium *Helicobacter pylori* into partial protein interaction maps.

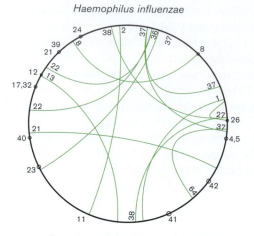

Haemophilus influenzae

Figure 12.11 Protein-protein interaction maps.
Representation of a protein interaction map for the most likely interactions in *H. influenzae*. The large black circle represents the genome. 0° corresponds to the first base pair and 360° to the last base pair of the genome. Predicted protein interactions are indicated by linking the circular map positions of the genes (each gene named by an arbitrary number) involved. In cases of neighboring genes (<5°), a small circle indicates the predicted interaction between two genes at that region; otherwise, an arc links the two genes in question.

Tracking Cells and Cell Constituents with Light

The use of the green fluorescent protein GFP to track the location of DNA polymerase in living cells of *B. subtilis* is described in Section 7.3. The original technology was limited, however, by the single color of fluorescence, green, emitted by GFP. More recently, mutations in the *gfp* gene have been introduced to make derivative proteins that fluoresce at different wavelengths (that is, emit different colors). For example, YFP and CFP fluoresce yellow and blue (cyan), respectively. **Figure 12.12** illustrates how different-colored GFP proteins were used to show that type IV pili responsible for twitching motility are also required for biofilm development. Type IV pili contribute to biofilm formation because the pili can attach to many solid surfaces.

GFP technology can also be used to target how microbial proteins move through eukaryotic cells. For instance, the potato virus X (or potexvirus) is a positive-strand RNA virus (a positive-strand RNA is equivalent to mRNA). One viral protein called TBGp3 is required for the virus to move between plant cells. Investigators fused GFP to the TBGp3 protein and asked where the protein ended up in the plant leaves. They delivered the genes to plant cells by bombarding leaves with the fusion plasmids. As shown in **Figure 12.13**, GFP alone was diffusely located throughout the leaf. The GFP-TBGp3 fusion protein, however, inserted only into the reticulated network of veins. Altering specific residues within the protein via mutation prevented this localization and allowed the fusion protein to remain diffused throughout the plant

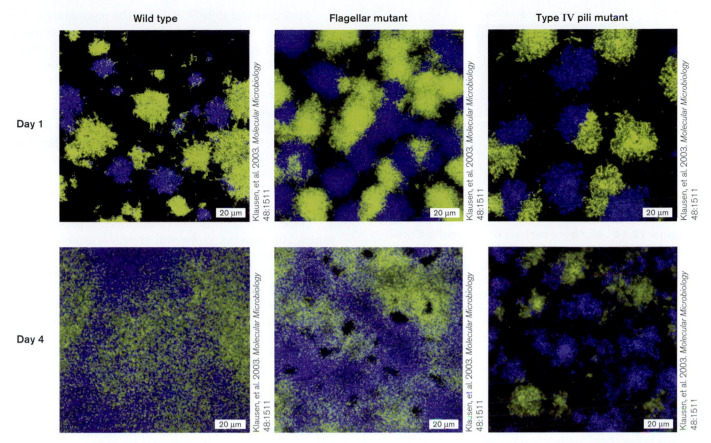

Figure 12.12   The use of fluorescently tagged cells to demonstrate that twitching motility is important for biofilm formation. Different strains of *Pseudomonas aeruginosa* were tagged with two different versions of green fluorescent protein, one that fluoresces yellow-green (YFP) and one that emits blue, or cyan (CFP). The two cell types were mixed, and their ability to form biofilms was determined by observing the spreading out and merging of microcolonies. Top panel shows microcolonies after one day of growth, while bottom panels are after four days of growth. Wild type and mutants lacking flagella merged well by day four, but mutants lacking type IV pili, which are responsible for twitching motility, do not merge.

cells. These results suggest that this protein helps move the virus through the plant's reticulated network.

Multiplex and Real-Time PCR

Food safety is a major concern for human health. Screening a food product for multiple pathogens one pathogen at a time can be laborious and costly. A technique that can screen for several organisms simultaneously would save time and money and thereby help keep food costs down. One such technique is PCR. Different species of microbes are distinguished by specific genes that define them. Pairs of oligonucleotide probes (that is, DNA primers) that amplify species-specific genes can be combined in a single PCR reaction called **multiplex PCR** (**Fig. 12.14**). For multiplex PCR to be successful, the oligonucleotide pairs cannot bind

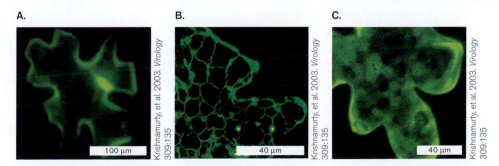

Figure 12.13   Using GFP protein to locate the potato virus X protein TBGp3 in the reticulated plant network.   Tagging a protein of interest with GFP allows one to determine the cellular location of that protein. **A.** Plasmid containing GFP only. Fluorescence is distributed throughout cell. **B.** Plasmid containing a GFP-TBGp3 fusion. Fluorescence is restricted to reticulated network. **C.** Plasmid containing mutant GFP-TBGp3 fusion gene. Mutant protein is not properly targeted to the reticulated network.

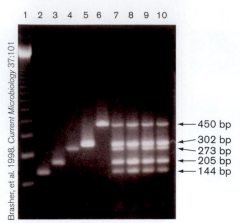

Figure 12.14 Multiplex analysis of pathogenic *E. coli*, *Salmonella*, and *Vibrio* species. PCR products are shown in this agarose gel. Lane 1 shows the 123-bp DNA ladder used to estimate sizes of the amplified fragments. Each of the other lanes represents a multiplex PCR reaction that can detect segments of five different genes representing different bacterial pathogens. Lanes 2 through 6 used the multiplex oligonucleotide primers to detect single strains. Lane 2 detects the *uidA* (beta-glucuronidase) gene from *E. coli* (144 bp amplicon); lane 3, the *cth* (cytolysin hemolysin) gene from *V. vulnificus* (205 bp); lane 4, the *invA* invasion gene from *S. typhimurium* (273 bp); lane 5, the *ctx* (cholera toxin) gene from *V. cholerae* (302 bp); and lane 6 the *tl* hemolysin gene from *V. parahaemolyticus* (450 bp). Lanes 7 through 10 are replicates in which the multiplex reaction was tested on mixture of all five species.

to each other or inadvertently amplify some other chromosomal gene. In addition, the **amplicon** (amplified product) sizes for each pair must be clearly different so that they can be separated on a gel. If these conditions are met, finding a given fragment will unambiguously demonstrate that the corresponding organism was present in the material tested.

Another revolutionary modification of the PCR technique is called **real-time PCR**. Real-time PCR is a fluorescence assay with the ability to monitor the progress of PCR as it occurs (that is, in real time). Data are collected throughout the entire PCR process rather than just at the end of the reaction. A major advantage of real-time PCR is that it can be used to quantitate the level of DNA or RNA in a sample.

Real-time PCR changes how we approach PCR-based quantitation of DNA and RNA. In real-time PCR, DNA is quantified based on how long it takes to *first* detect an amplified product while the polymerase chain reaction is still running rather than how much product is formed after a fixed number of cycles, the endpoint technique typically used in the past. The higher the starting copy number of the nucleic acid target, the sooner a significant increase in fluorescence is observed; thus, the time at which fluorescence increases will indicate the amount of nucleic acid in the original sample. In contrast, the endpoint assay only measures the amount of accumulated PCR product at the end of the whole PCR cycle.

How does real-time PCR work? The procedure relies on an oligonucleotide probe that contains a reporter fluorescent dye on the 5' end and a quencher dye on the 3' end. As long as the probe remains intact, the quencher dye reduces the fluorescence emitted by the reporter dye by a process called **fluorescence resonance energy transfer (FRET)**. In other words, the quencher absorbs the energy emitted by the fluorescent dye. The reporter probe's function is not to act as a primer for DNA synthesis, but to act as an indicator (or reporter) of whether Taq polymerase has synthesized DNA from an upstream primer (**Fig. 12.15**). Thus, the reporter probe is designed to anneal in a target DNA sequence downstream of where the DNA synthesis primer anneals.

If the DNA sequence to be amplified is present in the sample, the reporter probe will anneal downstream from where the primer probe anneals. Taq polymerase will begin to synthesize DNA from the upstream primer. However, Taq polymerase also has 5'-to-3' exonuclease activity. So, as Taq approaches the reporter probe, it will cleave the probe while the polymerase continues to synthesize DNA. Cleavage of the reporter probe separates the reporter dye from the quencher dye, increasing the fluorescence emitted from the reporter dye. The reporter probe is destroyed and thus removed from the template during this process. Meanwhile, primer extension by Taq polymerase continues to the end of the template. The template is amplified. After each annealing cycle, more reporter probe binds to the newly made templates and becomes cleaved with Taq polymerase during each round of polymerization. As a result, fluorescence continues to increase as the amount of template increases.

> **THOUGHT QUESTION 12.6** How would you have to modify real-time PCR to quantitate the level of a specific mRNA?

TO SUMMARIZE:

- **Automated yeast two-hybrid systems** expose interactions between many proteins in the cell, which can be used to build protein interaction maps.
- **Green fluorescent protein (GFP)** fused to a cell protein allows tracking of that protein's location or movement in a cell.
- **Multiplex PCR** is used to detect multiple organisms in complex environmental or food samples.
- **Real-time PCR** is used to quantitate specific DNA or RNA molecules present in cell extracts.

12.5 Biotechniques of Artificial Evolution

Previous chapters have discussed the principles that drive evolution of genomes in nature. But with all we know about manipulating DNA, can we speed the pro-

Figure 12.15   Real-time PCR.
A. The Taq DNA polymerase extends an upstream primer and reaches the downstream reporter probe where a 5′-to-3′ exonuclease activity degrades the probe, releasing the fluorescent dye from the vicinity of the quencher so it can now fluoresce. **B.** Amplification plot of *Rhodococcus* with primer BPH4. Different numbers of DNA copies were used as follows: 10^7, 10^6, 10^5, 10^4, 10^3 copies per reaction mixture, and no-template control. The more DNA copies present in the initial reaction, the sooner (represented in numbers of cycles) fluorescence becomes detectable above background.

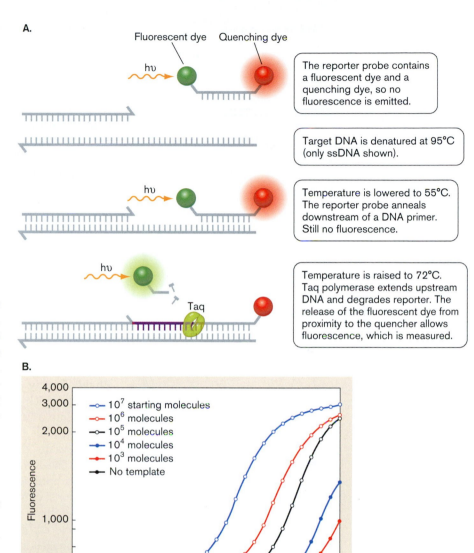

cess artificially in the laboratory to create a better enzyme? This section will describe how DNA sequences can be "shuffled" in vitro to evolve new genes not present in nature and how permutations of proteins can be expressed on phage surfaces so that the best proteins for a given purpose may be selected.

DNA Shuffling Creates In Vitro Evolution

Natural evolution takes centuries if not millennia to produce a new gene. We can now artificially evolve new genes in hours by a process called **DNA shuffling**. The basis of DNA shuffling is to mix and match fragments of different examples (or orthologs) of a given gene to make a new functional gene.

The DNA of each gene is digested with the enzyme DNase I into random fragments 10–50 bp in length. These fragments are then mixed and heated to separate the strands, then made to reanneal by lowering the temperature. This is done in the presence of a DNA polymerase. The separated strands of DNA pair up according to complementary base sequences in homologous DNA. The 3′ ends of overlapped fragments will "prime" the synthesis of DNA, for which the unpaired, opposite strand is used as a template, until the complete double-stranded DNA is restored. The fragments produced from the orthologs will reanneal randomly, but their order is ultimately based on overlapping homology. The DNA polymerase enzymes essentially fill in the gaps in the two strands. A simplified diagram of what is achieved in a typical experiment is shown in **Figure 12.16**.

DNA shuffling, therefore, is a method for quickly evolving new genes, operons, and even whole viruses for the acquisition of any desired properties. This has been used, for example, to generate new retroviruses that have altered tissue specificities. The process has also been used to create new cytokines. Cytokines are small immunomodulary peptides, produced by mammalian cells, that activate or inhibit functions of cells of the immune system (discussed in Chapters 23 and 24).

DNA shuffling of a family of more than 20 interferon-alpha genes (IFN-alpha, or IFN-α, is one type of cytokine) was used to derive variants that have increased antiviral and antiproliferative (inhibiting cell division) activities in mouse cells. Normally, the effects of a particular IFN are species specific, so there is minimal cross-species reactivity. For example, human IFN works on human cells, and mouse

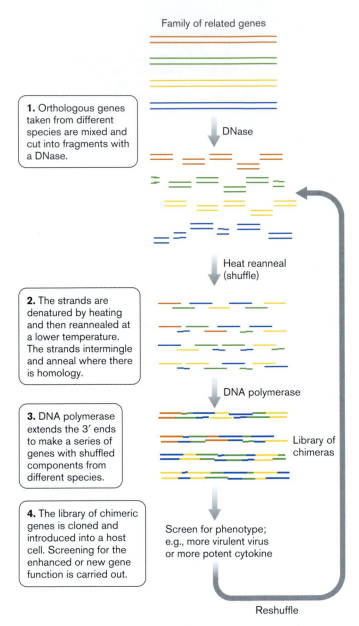

Family of related genes

1. Orthologous genes taken from different species are mixed and cut into fragments with a DNase.

DNase

Heat reanneal (shuffle)

2. The strands are denatured by heating and then reannealed at a lower temperature. The strands intermingle and anneal where there is homology.

DNA polymerase

3. DNA polymerase extends the 3′ ends to make a series of genes with shuffled components from different species.

Library of chimeras

4. The library of chimeric genes is cloned and introduced into a host cell. Screening for the enhanced or new gene function is carried out.

Screen for phenotype; e.g., more virulent virus or more potent cytokine

Reshuffle

Figure 12.16   DNA shuffling: In vitro evolution. Starting with four genes from different sources, DNA shuffling can create a new gene with enhanced functions.

so much more potent. Is it only due to enhanced receptor interactions or are there other reasons? DNA microarray analysis of cells that have been treated with these highly evolved cytokines may even reveal previously unrecognized target genes activated by cytokines.

In an experiment that shuffled retroviral genomes, six mouse leukemia viruses (MLVs) were recombined in a single round, giving a library of 5 million replicating recombinant viruses. Among them were completely new viruses capable of infecting Chinese hamster ovary cells, a feat none of the original MLVs could achieve. There is no doubt, however, that the shuffling procedure itself is potentially dangerous. Some of the recombinant viruses could well be lethal. Consequently, the potential misuse of this technology should not be dismissed. Nevertheless, many products developed from DNA-shuffling techniques will be beneficial for human life. These may include new cancer treatments, new vaccines, and new enzymes used in food making.

Directed Evolution through Phage Display Technology

Phage display, briefly described in Section 11.2, is a technique in which DNA sequences that encode peptides are cloned into a phage capsid gene. These peptides are then synthesized as part of the capsid protein and, upon phage assembly, are "displayed" on the surface of the phage (**Fig. 12.17**). The DNA sequences cloned can code for random peptides, antibodies cloned from natural sources, or libraries (collections) of mutant enzymes that one wishes to study.

The phage vectors used most often are the filamentous bacteriophages fd or M13 (discussed in Section 11.2). These phages infect *E. coli* strains containing the F plasmid (the phages attach to the sex pili) and replicate without killing the host cell. The phage particle consists of a single-stranded DNA molecule encapsulated in a long cylindrical capsid made up of five proteins. The major protein—the product of gene 8 (gp8)—is a small 5.2-kDa protein that forms the cylinder of the capsid. The other coat proteins (gp3, gp6, gp7, and gp9) cap the ends of the cylinder. The product of gene 3—present in three to five copies—is responsible for phage infectivity. Most phage display libraries have been created by cloning into gene 3.

Peptides displayed on filamentous phages have been used extensively to select peptide ligands or antibodies with high affinities for specific receptors or antigenic structures known as haptens (see Section 24.2). Recombinant phages that display high-affinity binding peptides are isolated from libraries in a process called biopanning (**Fig. 12.17B**). In biopanning, the target ligand—for example, a human virus attachment protein—is attached to a plastic plate or other support, and phages displaying random peptides are added. The phages that do not stick to the human virus protein are washed away. Those that do bind are subsequently released by adding free ligand (for example, human

IFN works on mouse cells. In the shuffling experiment, the most active human IFN clone that was generated also worked on mouse cells 285,000-fold better than the original IFN. The resulting IFN is a chimeric protein derived from shuffling DNA fragments encoding parts of two or more natural IFNs. Sequence analysis of the human-mouse IFN chimeras showed that the sequence of the carboxyl-terminal residues at positions 121–125 is responsible for the unusually high activity in mouse cells. On the basis of functional experiments and modeling, the residues in this region are proposed to interact with the mouse IFN-alpha receptor.

The implications of this approach are profound. It will be of great interest to determine why the shuffled IFNs are

A.

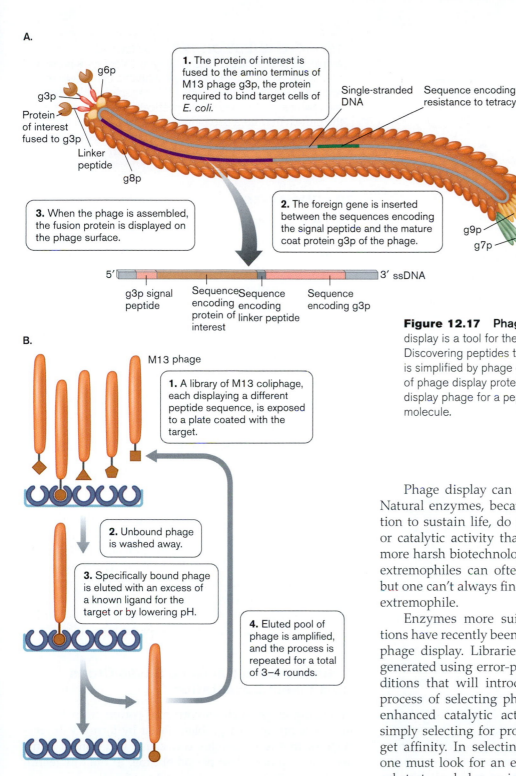

1. The protein of interest is fused to the amino terminus of M13 phage g3p, the protein required to bind target cells of *E. coli.*

Single-stranded DNA

Sequence encoding resistance to tetracycline

g6p
g3p
Protein of interest fused to g3p
Linker peptide
g8p

g9p
g7p

3. When the phage is assembled, the fusion protein is displayed on the phage surface.

2. The foreign gene is inserted between the sequences encoding the signal peptide and the mature coat protein g3p of the phage.

5′ 3′ ssDNA

g3p signal peptide | Sequence encoding protein of interest | Sequence encoding linker peptide | Sequence encoding g3p

B.

M13 phage

1. A library of M13 coliphage, each displaying a different peptide sequence, is exposed to a plate coated with the target.

2. Unbound phage is washed away.

3. Specifically bound phage is eluted with an excess of a known ligand for the target or by lowering pH.

4. Eluted pool of phage is amplified, and the process is repeated for a total of 3–4 rounds.

Figure 12.17 Phage display technology. Phage display is a tool for the directed evolution of peptides. Discovering peptides that bind to specific target proteins is simplified by phage display techniques. **A.** Construction of phage display protein. **B.** Biopanning a library of M13 display phage for a peptide that specifically binds to a target molecule.

Phage display can also help retool entire enzymes. Natural enzymes, because they were selected by evolution to sustain life, do not necessarily have the stability or catalytic activity that would make them suitable for more harsh biotechnological applications. Enzymes from extremophiles can often be used in these applications, but one can't always find an appropriate enzyme from an extremophile.

Enzymes more suited to biotechnological applications have recently been engineered using whole-enzyme phage display. Libraries of phage enzyme mutants are generated using error-prone PCR (PCR done under conditions that will introduce occasional mutations). The process of selecting phages that display enzymes with enhanced catalytic activity is more demanding than simply selecting for protein variants with increased target affinity. In selecting for increased enzyme activity, one must look for an enzyme with a higher affinity for substrate and also an increase in the ability to catalyze a chemical reaction.

An experiment to prove that this in vitro evolution technique would work involved making a more effective *Bacillus cereus* metallo-beta-lactamase (**Fig. 12.18**). Beta-lactamase (β-lactamase) is an enzyme that cleaves the antibiotic penicillin. The selection process combined affinity chromatography with what is called catalytic elution. The protocol had three basic steps: (1) Phage-displayed

virus protein). This process is analogous to the way miners would pan rivers for gold. Several rounds of biopanning are required to obtain high-affinity clones. Following each round, the selected phages are enriched by repropagation in *E. coli.* Phage display was successfully used to identify high-affinity peptides that bind to attachment proteins of some eukaryotic viruses. These virus-binding peptides effectively block viral infection.

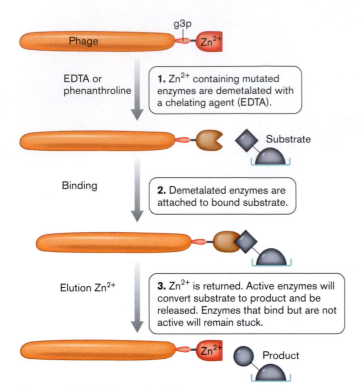

Figure 12.18   Example of phage enzyme display.
Phage display was used to obtain a more active Zn^{2+}-binding protein. Randomly altered versions of the metalloenzyme beta-lactamase are displayed on phage. The Zn^{2+} from the prosthetic group is removed and the phage library added to plates containing bound substrate (penicillin). Only phage displaying high-affinity versions of the enzyme will attach. Returning Zn^{2+} will restore activity to versions of the enzymes that retained their catalytic activity. Cleavage of the substrate releases phages that display an active enzyme with high affinity. Those few phages can be repropagated to high numbers and repanned for further purification. *Source*: A. Fernandez-Gacio, et al. 2003. *Trends in Biotechnology* 21:408–414.

mutant enzymes were inactivated by extracting the catalytic metal ion; (2) the demetalated phage enzymes were adsorbed onto a support coated with penicillin, which bound the enzyme at the active site (only a phage with high affinity enzymes will attach); and (3) phages that displayed active enzymes on their capsids were selectively eluted by adding back the metal ion. The selective elution method works because adding back the metal cofactor reactivates the enzymes, which then convert the immobilized penicillin to a product. Because the enzyme has a lower affinity for the *product*, the phage enzyme is released and can be collected. The process eliminates mutant proteins that don't bind at all or that bind but have lost catalytic activity. This protocol succeeded in extracting several beta-lactamase enzymes with higher enzyme activity from a library of 5×10^6 clones generated by error-prone PCR. The purpose of this exercise was not to make a more active beta-lactamase, but to offer proof that in vitro evolution of an enzyme using phage display was possible.

THOUGHT QUESTION 12.7   You have just employed phage display technology to select for a protein that tightly binds and blocks the eukaryotic cell receptor targeted by anthrax toxin. When this receptor is blocked, the toxin cannot get into the target cell. You suspect that the protein may be a useful treatment for anthrax. How would you recover the gene following phage display and then express and purify the protein product?

TO SUMMARIZE:

- **DNA shuffling**—shuffling of DNA fragments made from orthologous genes (genes encoding similar proteins from different species)—can be used in vitro to create new genes whose protein products have new properties.
- **Phage display** is another form of in vitro evolution, where DNA sequences encoding random peptides are cloned into phage capsid protein genes. Phage particles displaying peptides with a desired binding property are pulled from the general phage population by biopanning.

12.6 Applied Microbial Biotechnology

It is remarkable how many useful products can be imagined and are being developed using the tools of biotechnology. You will be surprised to learn that bananas, as one example, are being engineered as edible vaccine delivery systems. In another development, we are finding that vaccines can be delivered not only by injecting a protein but by simply injecting DNA, which expresses the antigenic protein. We even have technology whereby a genetically engineered virus can repair a human genetic defect. These are all remarkable additions to our scientific toolbox and are already helping improve human life.

A Simple Bacterial Gene Saves Crops from Insect Devastation

Protecting crops such as corn and cotton from hungry insects is an age-old problem faced by farmers. Insects such as moths or beetles can cause widespread damage to a crop in a short period of time. In the nineteenth and twentieth centuries, chemical pesticides were widely used to kill these pests. While generally effective, chemical pesticides can contaminate soil and groundwater and can impact human health. But in 1911, a microorganism was discovered that led to a new way of protecting crops. The organism was *Bacillus thuringiensis*, a close relative of *Bacillus anthracis*.

Bacillus thuringiensis sporulates on the surfaces of plants. However, in addition to the spore, the sporulat-

A.

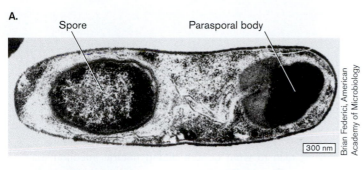

Spore   Parasporal body

300 nm

Brian Federici, American Academy of Microbiology

B.

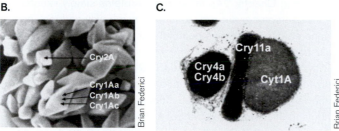

Cry2A
Cry1Aa
Cry1Ab
Cry1Ac

Brian Federici

C.

Cry11a
Cry4a
Cry4b
Cyt1A

Brian Federici

Figure 12.19   The insecticidal bacterium *Bacillus thuringiensis*. A. A cell of *B. thuringiensis* in the process of sporulation, during which insecticidal proteins are produced and crystallize. **B.** Parasporal crystals from *B. thuringiensis* subsp. Kurstaki. Four major crystal proteins are formed (Cry1aa, Cry1ab, Cry1ac, and Cry2aa). They are used to control caterpillar pests. **C.** The parasporal crystal of *B. thuringiensis* subsp. israelensis contains four insecticidal proteins and is used to control mosquito and blackfly larvae.

ing cell makes a separate parasporal body that cradles the spore (**Fig. 12.19**). The parasporal body contains crystallized proteins that are toxic to insects feeding on the plant. A single subspecies of *B. thuringiensis* can produce multiple insecticidal proteins. Different subspecies (there are 34 of them) can produce different combinations of these proteins. After the insect or insect larva ingests the insecticidal protein crystals, the alkaline environment in the insect midgut dissolves the crystals, and insect proteases inadvertently activate the proteins. Activated insecticidal proteins insert into the membrane of the midgut cells and form pores that lead to loss of membrane potential, cell lysis, and death of the insect through starvation. Farmers have taken to spreading *B. thuringiensis* directly on their crops, where even dead cells are effective.

B. thuringiensis is generally considered a highly beneficial pesticide with few downsides. Unlike most insecticides, *B. thuringiensis* insecticides do not have a broad spectrum of activity, so they do not kill animals or beneficial insects—including the natural enemies of harmful insects (predators and parasites) as well as beneficial pollinators, such as honeybees. Therefore, *B. thuringiensis* integrates well with other natural crop controls. Perhaps the major advantage is that *B. thuringiensis* is essentially nontoxic to people, pets, and wildlife, so it is safe to use

on food crops or in other sensitive sites where chemical pesticides can cause adverse effects.

While spreading *B. thuringiensis* on crops is helpful, an even more effective means of delivering the toxin has been developed. The genes encoding insecticidal proteins have been spliced right into the genomes of certain plants, such as cotton and corn. The gene is placed under the control of a plant promoter, so that the transgenic plant makes its own insecticide. **Table 12.1** presents the timeline of events leading up to this technological advance. As of 2003, 72% of the cotton, and 40% of the corn planted in the United States contain *B. thuringiensis* genes.

Although there have been no reports of adverse effects on humans, these proteins could cause allergic reactions in some people if they were ingested (much like peanuts, citrus, or shellfish). As a result of this concern, the use of transgenic corn expressing insecticidal proteins has been limited to cattle feed. In the bovine digestive tract, the protein is thoroughly digested to individual amino acids; thus, the protein does not show up in meat products produced from cattle and cannot cause an allergic reaction.

THOUGHT QUESTION 12.8   Would insect resistance to an insecticidal protein be a concern when developing a transgenic plant? Why? How would you design a transgenic plant to limit the possibility of insects developing resistance?

Table 12.1 | Timeline in the development of *B. thuringiensis* as an insecticide.

Over a century after its discovery, *B. thuringiensis* is a widely used natural insecticide.

1901	Ishiwata Shigetane discovers a disease of silkworms caused by a previously undescribed bacterium.
1911	E. Berliner names the organism causing the silkworm disease *Bacillus thuringiensis (Bt)*.
1938	The first commercial products that contain *Bt* are marketed in France.
1954	T. Angus shows that crystalline proteins are responsible for the insecticidal action of *Bt*.
1983	Scientists demonstrate that plants can be genetically engineered.
1987	Plants engineered to contain the *Bt* gene are demonstrated to be pest resistant.
1996	*Bt* cotton reaches the U.S. market.
2001	69% of the cotton and 26% of the corn planted in the United States contain *Bt* genes.
2003	72% of cotton and 40% of corn contain *Bt* genes.

Special Topic 12.1   DNA Vaccines

Since the time of Edward Jenner, vaccines used to prevent microbial diseases have consisted of incapacitated microbes or inactivated microbial proteins (for example, inactivated toxins). These cellular or protein vaccines stimulate the immune system to make antibodies and immune cells that destroy the invader (see Chapters 23 and 24). Now, a new type of vaccine is being tested. It involves injecting DNA encoding bacterial proteins into humans.

To make the vaccine, DNA encoding a protein is isolated from an infectious organism such as a virus or bacterium and is altered to place eukaryotic promoters in front of the gene or genes. The modified DNA is then integrated into a plasmid vector. Quite amazingly, if this plasmid DNA is injected into a human cell, the injected DNA's genes will be expressed inside that cell and produce the microbial protein. At present, the muscle cell is the most common target for injection. The presence of these foreign proteins inside human cells causes the immune system to launch a powerful protective immune response against the microbial protein. This immune response will also protect against the live infectious organism.

But why make a DNA vaccine when other, traditional vaccines already work? One problem is that live, attenuated bacteria, even with minimal disease-producing potential, can still cause illness and in some cases serious disease. Selecting and expressing only one or two proteins from an infecting organism (provided those proteins are not toxins) enables us to avoid the risk of disease altogether. So why not just administer the protein directly? While this does work for many vaccines, the best way to generate a potent immune response is if the protein is actively synthesized within a host cell. The

reason for this is that a small amount of any foreign protein synthesized in a host cell will be degraded and the pieces presented on the surface of that cell. Specialized immune cells surveying the body see these proteins, recognize that they are foreign, and initiate an immune reaction. This process is akin to what occurs during natural infection (discussed in Chapter 24). In contrast, proteins simply injected into a person are not processed the same way and may not generate a response as complete or long-lasting. **Figure 1** illustrates how a DNA vaccine will protect against infection with herpes simplex virus 2 (HSV-2). **Figure 1A** shows a newborn with an HSV-2 infection. The graph in **Figure 1B** shows the results of a mouse model examining HSV-2 infection and the protective effect of a DNA vaccine.

How is the DNA vaccine delivered? A modified air pistol called a gene gun can blast DNA-coated beads into muscle. A more intriguing delivery system places the recombinant DNA vaccine vector into attenuated strains of intracellular bacteria like *Salmonella* and lets the microorganism deliver the plasmid into cells (**Fig. 1C**). For *Salmonella*, this system will specifically deliver vaccine to the intestinal mucosa, where the salmonellae invade. Production of vaccine proteins in these host cells yields a powerful immune response.

We should stress that DNA vaccines are still experimental, although their future looks promising. For example, one report has used the herpes simplex virus type 2 glycoprotein D gene to immunize mice against genital herpes. Future application of this type of vaccine to numerous human diseases would eliminate considerable pain and suffering, both physical and emotional.

Vaccine Delivery Can Be Accomplished with Fruits and Vegetables

Starting with the war against smallpox in the late 1700s, vaccines have been used to inoculate humans against many terrible diseases (see Chapters 25 and 27). However, the cost of producing them is enormous, and the price, of course, is passed on to those who are vaccinated or to governments that conduct vaccination programs. This becomes problematic for the world's impoverished

nations, and as a result, 20% of the world's infants do not receive proper vaccination, resulting in over 2 million preventable deaths annually.

An amazing solution to this problem is the invention of edible vaccines. In these vaccines, the gene encoding the vaccine protein is inserted into the genome of a fruit or vegetable native to the country where vaccination is desired. As in the case of transgenic plants encoding *Bt* toxins, the vaccine gene must be placed under the control of a well-expressed plant promoter. Their presence in var-

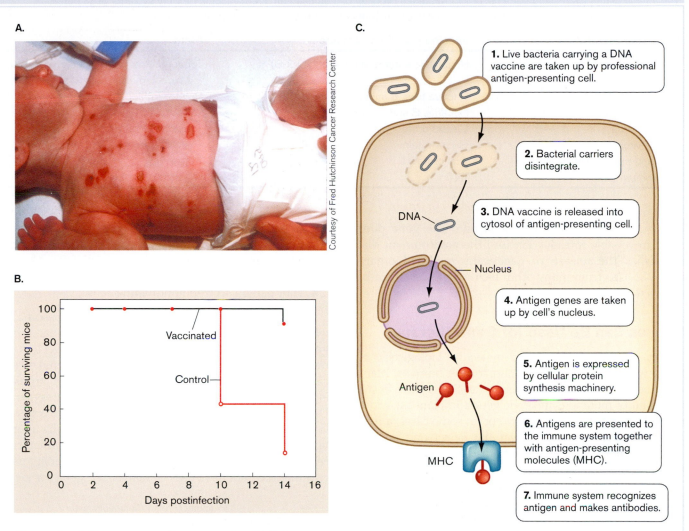

A.

Courtesy of Fred Hutchinson Cancer Research Center

B.

C.

1. Live bacteria carrying a DNA vaccine are taken up by professional antigen-presenting cell.

2. Bacterial carriers disintegrate.

DNA

3. DNA vaccine is released into cytosol of antigen-presenting cell.

Nucleus

4. Antigen genes are taken up by cell's nucleus.

5. Antigen is expressed by cellular protein synthesis machinery.

Antigen

6. Antigens are presented to the immune system together with antigen-presenting molecules (MHC).

MHC

7. Immune system recognizes antigen and makes antibodies.

Figure 1   DNA vaccines.   Directly injecting DNA that encodes a virulence protein into an animal can protect against infection. **A.** View of a newborn infected with HSV-2. **B.** Survival data for mice vaccinated with DNA encoding HSV-2 glycoprotein D show that vaccinated mice are protected from a lethal dose of HSV-2. Control mice ($n = 30$) were vaccinated with plasmid vector (no HSV gene). Test mice ($n = 40$) were vaccinated with plasmid containing the gene for HSV-2 glycoprotein D. Three weeks later, the mice were infected with a lethal dose of HSV-2. Survival was followed over 14 days. **C.** Example of bacterial vaccine delivery.

ious poor countries makes bananas, rice, corn, wheat, and soy the best choices for edible vaccine development. The agricultural vaccine machines can be grown and simply distributed to the population, without the need to handle needles or subject young patients to the pain of injection. Examples of edible vaccines under study include potatoes with hepatitis B surface antigen, *E. coli* labile toxin, and even human papillomavirus.

Actually, transgenic plants expressing bacterial or viral virulence proteins can be used in two ways: They can be direct vaccine delivery systems, as through edible fruits and vegetables (such as potatoes, tomatoes, and corn); or, as with transgenic tobacco crops, they can be a cheap way to grow large amounts of the vaccine protein that can be extracted later and used in more conventional ways. Other plants used to make vaccines are cited in **Table 12.2**.

Although this technology seems an ideal solution to vaccination, concerns persist. A key issue is that any release of genetically engineered microbes into the envi-

Table 12.2 **Potential vaccine antigens expressed in plants.**

Transgenic plants expressing bacterial or viral virulence proteins can be used as direct vaccine delivery systems (e.g., potatoes, tomatoes, corn) or as an inexpensive way to grow large amounts of the vaccine protein (tobacco crops).

Source of the protein	Vaccine protein or peptide	Plant
Enterotoxigenic *E. coli*	Heat-labile enterotoxin B subunit (LT-B)	Tobacco, Potato, Tobacco chloroplast, Maize kernels
Vibrio cholerae	Cholera toxin B subunit (CT-B)	Potato, Tobacco chloroplast
Hepatitis B virus	Hepatitis B surface antigen	Tobacco, Potato
Norwalk virus	Capsid protein (NVCP)	Tobacco, Potato
Rabies virus	Glycoprotein	Tomato
Foot-and-mouth disease virus	Viral protein I	Alfalfa
Clostridium tetani	TetC	Tobacco chloroplast

ronment introduces the transgene into the gene pool. This irreversible introduction into the environment distinguishes the release of genetically modified organisms (GMOs) from many other technologies that can be reversed.

Viruses Can Be Used for Gene Therapy

A long-standing challenge for molecular biologists is to find a way to correct genetic mutations in humans. Many diseases, such as cystic fibrosis or Tay-Sachs disease, are the result of mutations in single genes. Implanting a copy of the correct gene in an embryo or, in some cases, a fully developed human would repair the defect and prevent development of disease. Cystic fibrosis, for example, is the result of a defect in a chloride channel gene (Cystic Fibrosis Transmembrane Regulator, CFTR) in the lungs. Introducing a normal copy of the gene into lung cells would prevent the disease. This type of treatment is called gene therapy.

But how do you get good genes into cells? One way is by cloning a copy of the gene into a viral vector. When that virus infects the cell, the good gene goes with it. Of course, not just any virus can serve as a vector. The virus DNA chosen must be one that enters the cell nucleus. For this reason, retroviruses are often used, since they efficiently enter the nucleus. The ideal vector would also integrate into the DNA of the nucleus, making the replacement permanent. However, the insertion process, which appears random, could mutate other genes. Another obvious requirement is that the virus be stripped of its disease potential.

Adenovirus and adeno-associated virus are vectors that have been employed in several gene therapy trials (**Fig. 12.20**). Although they don't have all the ideal properties just mentioned, they do have several others that make them attractive:

- Adenoviruses are double-stranded genomes, so cloning is easy.

- Adeno-associated viruses having a linear, single strand genome, can infect many different cell types and enter the nucleus.
- They don't cause severe disease.
- They can insert into nondividing cells.

These viruses have drawbacks, however, some of which may be corrected by modifying the adenovirus genome. One problem is that many adenovirus vectors are mutated in genes that, after infection, might make cells cancerous or lead to cell death. Another major problem is that an adenovirus does not permanently integrate its DNA into cells. As a result, the therapeutic gene could be lost when it disintegrates because the genetically altered virus will not replicate. To circumvent these difficulties, Mark Kay and his colleagues at Stanford University used two modified adenoviruses to deliver the gene for factor

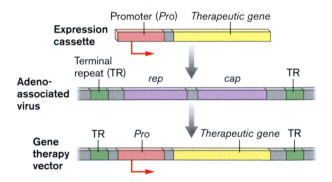

Figure 12.20 **Basic construction of a gene therapy vector.** The adeno-associated virus is a small linear single-stranded virus consisting of two genomic regions: *rep*, encoding proteins involved in replication and integration; and *cap*, encoding viral structural proteins. These regions are flanked on either side by a terminal repeat (TR). Reconstruction of the virus requires replacement of the *rep-cap* sequence with an expression cassette carrying a promoter (Pro) and the therapeutic gene. Because the genes required for replication were removed, replication of the vector virus requires cotransfection of the host cell with a helper adenovirus. The helper virus supplies the missing functions.

IX into mouse liver cells. People with a defective factor IX gene suffer from type B hemophilia. The gene delivery system resulted in permanent integration of the gene into the mouse cell genome while avoiding the integration of potentially cancerous adenovirus genes.

The strategy they used is illustrated in **Figure 12.21**. One adenovirus was genetically engineered to contain the normal factor IX gene. The integrated cassette was flanked first by inverted repeat DNA sequences that are targets for a specific transposase and second by sequences (called FRT) recognized by a recombinase called Flp (vector 1). They essentially constructed a minitransposon cassette that contained the therapeutic gene (factor IX). The second adenovirus was a helper virus that contained the transposase gene and the yeast gene for Flp recombinase (vector 2, not shown in figure). Neither adenovirus vector could, by itself, integrate into the genome. The goal was to have both viruses infect the same cell. The Flp recombinase made from vector 2 would excise the factor IX cassette from Vector 1 and circularize it to make a DNA molecule devoid of adenovirus genes (steps 1 and 2). The transposase made from vector 2, which acts only on circular molecules, would then catalyze transposition of the factor IX cassette from the circular molecule into the mouse genome (step 3). (See Chapter 9 for mechanisms of transposition and recombination.) When the scientists injected the two adenovirus vectors into mouse livers, they wound up with liver cells containing the factor IX gene integrated into the mammalian genome, exactly as planned.

Many other ideas to deliver genes have been developed and are being tested. For example, DNA sequences that target DNA to specific cellular locations in the cell (for example, nucleus and mitochondria) are attached to the gene payload. DNA is administered either by gene gun or by electroporation (see Section 9.2) to specific organs. The presence of the targeting DNA sequence ensures that the payload gene goes to the correct location.

Another amazing development is the use of modified human immunodeficiency viruses (HIV), a lentivirus, as gene delivery systems. The vector does not carry genes it needs to replicate, but will reverse transcribe its RNA engineered to encode therapeutic products, and integrate the cDNA into the genome of the host cell. The power of these vectors is that they can integrate into the genome of dividing or non-dividing cells. They are being developed as gene therapies for metabolic diseases, viral infections (HIV itself), hemophilia, cancer, and a host of other conditions.

TO SUMMARIZE:

- **Genes encoding insecticidal proteins** from *Bacillus thuringienesis* have been engineered into plant genomes to protect crops from insect devastation.
- **Proteins from pathogenic bacteria and viruses** cloned into fruits and vegetables may provide a cost-effective means to vaccinate large populations.
- **Genetically engineered viruses** are being tested as gene therapy vehicles to deliver DNA to correct inherited diseases in humans.

Figure 12.21 Use of an adenovirus gene delivery system. Overview of a strategy to transpose the factor IX gene into the genome of mouse liver cells. The adenovirus used in this case is a double-stranded DNA virus. *Source:* From Yant, et al. 2002. *Nature Biotechnology* 20:999–1004.

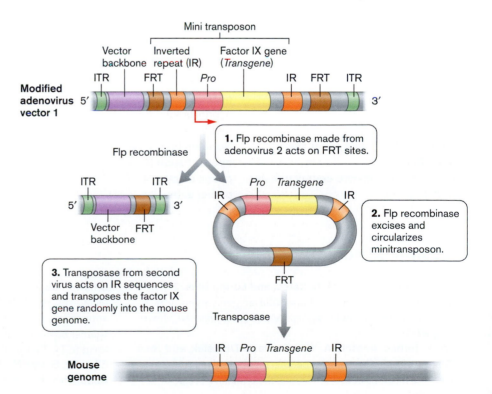

Concluding Thoughts

This chapter has discussed only some of the available molecular techniques and developing technologies used to probe biological processes and manipulate them to our benefit. Keep them in mind as we progress deeper into the physiology of microbial growth and the mechanisms of microbial pathogenesis. Many of the approaches described here were used to generate the knowledge presented in later chapters.

CHAPTER REVIEW

Review Questions

1. What are important considerations when designing a mutant selection strategy?
2. Once the DNA sequence of a gene is known, how does one gain clues as to possible function of the gene product?
3. What are the key features of reporter genes used to measure transcriptional control? Translational control?
4. What is the difference between a Northern blot and a Southern blot?
5. Explain the principle of a Western blot.
6. What techniques are used to determine if a protein binds to a DNA sequence?
7. Explain the important points for constructing a plasmid that makes His$_6$-tagged protein.
8. Explain how one can determine the start site of a transcript.
9. How can protein-protein interactions be determined in vitro?
10. Describe two-hybrid analysis. What are bait and prey proteins?
11. How can the cellular location of a protein be tracked in vivo?
12. Explain how PCR can be used in a single reaction to identify which of two species of pathogenic bacteria are present in a food sample.
13. Discuss how gene shuffling or phage display might be used to create a more toxic toxin. What are the ramifications of these technologies?
14. Discuss some ways biotechnology has helped the agricultural industry. How has the field impacted vaccine delivery or our approaches to the treatment of metabolic diseases in humans?

Key Terms

amplicon (446)
autoradiography (436)
DNA shuffling (447)
electrophoretic mobility shift assay (EMSA) (438)
fluorescence resonance energy transfer (FRET) (446)
gene fusion (434)
hybridization (436)
multiplex PCR (445)
Northern blot (436)
operon fusion (434)
phage display (448)
primer extension (441)
real-time PCR (446)
reporter fusion (434)
reporter gene (434)
Southern blot (436)
Western blot (437)
yeast two-hybrid system (442)

Recommended Reading

Baldwin, Brett R., Cindy H. Nakatsu, and Loring Nies. 2003. Detection and enumeration of aromatic oxygenase genes by multiplex and real-time PCR. *Applied and Environmental Microbiology* **69**:3350–3358.

Branyik, Tomás, António A. Vicente, Pavel Dostálek, and José A. Teixeira. 2005. Continuous beer fermentation using immobilized yeast cell bioreactor systems. *Biotechnology Progress* **21**:653–663.

Chikwamba, Rachel, Joan Cunnick, Diane Hathaway, Jennifer McMurray, Hugh Mason, and Kan Wang. 2002. A functional antigen in a practical crop: LT-B producing maize protects mice against *Escherichia coli* heat labile enterotoxin (LT) and cholera toxin (CT). *Transgenic Research* **11**:479–493.

Elbashir, Sadya M., Jens Harborth, Winfried Lendeckel, Abdullah Yalcin, Klaus Weber, and Thomas Tuschl. 2001.

Duplexes of 21-nucleotide RNAs mediate RNA interference in cultured mammalian cells. *Nature* **411**:494–498.

Fernandez-Gacio, Ana, Marilyne Uguen, and Jacques Fastrez. 2003. Phage display as a tool for the directed evolution of enzymes. *Trends in Biotechnology* 21:408–414.

Ferrer, Manuel, Olga Golyshina, Anocibar Beloqui, Peter N. Golyshin. 2007. Mining enzymes from extreme environments. *Current Opinion in Microbiology.* **10**:207–214.

Gebhard, John R., Jr., J. Zhu, X. Cao, J. Minnick, and B. A. Araneo. 2000. DNA immunization utilizing a herpes simplex virus type 2 myogenic DNA vaccine protects mice from mortality and prevents genital herpes. *Vaccine* 18:1837–1846.

Graumann, Klaus, and Andreas Premstaller. 2006. Manufacturing of recombinant therapeutic proteins in microbial systems. *Biotechnology Journal* **1**:164–186.

Hartley, Oliver, and Robin E. Offord. 2005. Engineering chemokines to develop optimized HIV inhibitors. *Current Protein and Peptide Science* 3:207–219.

Klausen, Mikkel, Arne Heydorn, Paula Ragas, Lotte Lambertsen, Anders Aaes-Jorgensen, et al. 2003. Biofilm formation by *Pseudomonas aeruginosa* wild type, flagella and type IV pili mutants. *Molecular Microbiology* 48:1511–1524.

Krishnamurthy, Konduru, Marty Heppler, Ruchira Mitra, Elison Blancaflor, Mark Payton, et al. 2003. The Potato virus X TGBp3 protein associates with the ER network for virus cell-to-cell movement. *Virology* **309**:135–151.

Ma, Zhuo, Shimei Gong, Hope Richard, Donald L. Tucker, Tyrrell Conway, and John W. Foster. 2003. GadE (YhiE) activates glutamate-decarboxylase-dependent acid resistance in *Escherichia coli* K12. *Molecular Microbiology* 49:1309–1320.

Olivares, Eric C., Roger P. Hollis, Thomas W. Chalberg, Leonard Meuse, Mark A. Kay, and Michele P. Calos. 2002. Site-specific genomic integration produces therapeutic Factor IX levels in mice. *Nature Biotechnology* **20**:1124–1128 .

Petri, Ralf, and Claudia Schmidt-Dannert. 2004. Dealing with complexity: evolutionary engineering and genome shuffling. *Current Opinions in Biotechnology* **15**:298–304.

Rabaey, Korneel, and Willy Verstraete. 2005. Microbial fuel cells: novel biotechnology for energy generation. *Trends in Biotechnology* **23**:291–298.

Subbara, Kanta, and Jacqueline M. Katz. 2004. Influenza vaccines generated by reverse genetics. *Current Topics in Microbiology and Immunology* **283**:313–342.

Tramonti, Angela, Paolo Visca, Michele De Canio, Maurizio Falconi, and Daniella De Biase. 2002. Functional characterization and regulation of *gadX*, a gene encoding an AraC/XylS-like transcriptional activator of the *Escherichia coli* glutamic acid decarboxylase system. *Journal of Bacteriology* **184**:2603–2613.

Part 3

Metabolism and Biochemistry

Courtesy of Caroline Harwood

Caroline Harwood, professor of microbiology at the University of Washington.

AN INTERVIEW WITH

CAROLINE HARWOOD: BACTERIAL METABOLISM DEGRADES POLLUTANTS AND PRODUCES HYDROGEN

Caroline Harwood taught microbiology from 1988 to 2004 as a professor at the University of Iowa and since 2005 at the University of Washington. Harwood showed how soil bacteria catabolize some of nature's toughest compounds, such as lignin components, as well as related compounds that pollute our environment. She led the project to sequence the genome of *Rhodopseudomonas palustris*, a gram-negative purple bacterium capable of photosynthesis, heterotrophy, and hydrogen production. The diverse metabolism of *R. palustris* may be manipulated for bioremediation and production of hydrogen fuel. Harwood serves as editor for the journal *Applied and Environmental Microbiology*.

Why did you decide to make a career in microbiology?

I was interested in science since I was very young, but there were very few examples of women scientists at that time. I was fortunate to have been given excellent educational opportunities by my family. After college I had a chance to participate in an intensive summer course called "Microbial Diversity," given at the Marine Biological Laboratory at Woods Hole, Massachusetts. This was my first real exposure to bacteria and to microbiology, and I became captivated by the metabolic diversity of bacteria. Years later I took a turn as the director of this same course at Woods Hole.

Why did you study aromatic metabolism? What is significant about this process in microbes?

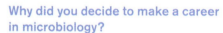

Aromatic compounds first caught my attention when I took organic chemistry. It was then that I became interested in the unique chemical properties of the benzene ring. I was

also aware that a huge proportion of plant biomass—specifically the lignin component of wood—is a polymer of aromatic compounds and that these compounds must be degraded by living organisms in order for Earth's carbon cycle to operate. And of course, many environmental pollutants have benzene rings as part of their structure.

In graduate school, I found that biochemistry and microbiology textbooks included very little mention of aromatic compound degradation. Some might have concluded from this that the topic of aromatic compound degradation was not important, but I realized that bacteria and fungi can't read and so they didn't know this. In fact, aromatic compounds serve as the principal source of carbon and energy for many kinds of soil microbes. I could see an opportunity for discovery, so I went to do postdoctoral research with L. Nicholas Ornston at Yale, a world expert on bacterial degradation of aromatic compounds.

Why do you study the soil bacterium *Rhodopseudomonas palustris*?

At the time I was a postdoctoral fellow, it was generally believed that only aerobic microbes could degrade aromatic compounds because one of the principal degradative enzymes inserts oxygen into the aromatic ring. But most soil microbes grow anaerobically, without oxygen. One day at lunch I wondered if aromatic compounds might also be degraded under anaerobic conditions. My postdoctoral advisor said yes, they could, but that not much was known. After lunch I went to the library and found that a few papers had been written on anaerobic aromatic degradation by a photosynthetic bacterium named *Rhodopseudomonas palustris*. At about that time, I moved to Cornell University and found a biochemistry professor, Jane Gibson, who showed me how to grow *R. palustris*. We collaborated on studies to work out the metabolic pathway for anaerobic benzoate degradation.

What did you find was unusual about *R. palustris* metabolism?

At first, we worked with *R. palustris* because it served as an excellent model organism to study anaerobic degradation of aromatic compounds. But when we saw its genome sequence, we realized that *R. palustris* was even more metabolically versatile than we had thought. It has genes for four dif-

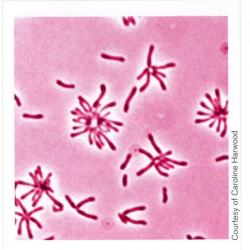

***Rhodopseudomonas palustris* bacteria.**
The bacteria grow in rosette-shaped clusters
(cell length, 2–4 μm; LM).

ferent aerobic benzene ring cleavage
pathways in addition to the anaerobic
pathway that we had been studying.
It has five different sets of "light-harvesting" genes, enabling photosynthesis at low levels of light. Finally, from
the genome sequence, we realized
that *R. palustris* encodes three different nitrogenase enzymes for converting nitrogen gas to ammonia and
hydrogen. *Rhodopseudomonas palustris*
is a remarkable microbe that can grow
using all the major modes of energy
metabolism that exist on Earth.

**Why does *R. palustris* produce
hydrogen gas? What are the potential applications of biogenic hydrogen for energy?**
R. palustris produces hydrogen as an
unintended product of nitrogen fixation.
For humans, hydrogen is a clean-burn-
ing form of energy that can be easily
converted to electricity in hydrogen fuel
cells. It can be used as automobile fuel
and as fuel for any electricity-driven
process. At present, the major commercial routes for hydrogen production
depend on fossil fuels and petroleum
products. Biogenic hydrogen, derived
from microbes, does not depend on
these dwindling resources, but instead
depends on light from the sun and biomass, a renewable resource.

**Would you describe your experience
working with graduate students?**
I would be nothing without the
graduate students that I've had the
privilege of working with. The best
graduate students make one see
things in new and different ways
and will drag you into new areas of
inquiry. Through the process of mentoring graduate students and helping
them to become more critical think-
ers and careful experimenters, I have
become a better scientist myself.
Working with graduate students is
one of the greatest joys of my job.

**As an educator, how has your
research impacted your teaching?**
I try to communicate the thrill of the
hunt in research and to make students realize that research is a creative process and a social process that
depends on teamwork among laboratory workers. The arena that I favor
for teaching undergraduate students
is the teaching laboratory. It has a
more relaxed pace, and teaching laboratories give me a chance to get to
know students on an individual basis,
to see their personalities and how
they interact with each other.

**What have you learned through edit-
ing a research journal?**
As a journal editor I have learned that
people react to constructive criticisms
from peer reviewers in surprisingly
varied ways. Most people are gracious
and respond by trying to improve
their papers and their work; these
investigators have been good role
models for me. But others have taught
me what not to do. I also have a better
appreciation for what it takes to make
an original and significant contribution to the scientific literature.

**What do you think are the most
exciting areas for students entering
microbiology today?**
Advances in genome sequencing and
in imaging make possible sophisticated studies of interactions between
bacteria and viruses and their human
or animal hosts. In the near future, it
will be possible to study the activities
of single bacterial cells and of complex bacterial communities.

**How does your family relate to your
career in microbiology?**
I grew up as one of six children and
part of a large extended family, and
I learned the value of family support
networks. My experiences and atti-
tudes about large families apply to my
interactions with the members of my
laboratory. The line between family
and work is sometimes a bit blurred in
my case because my husband is also a
microbiology professor. It has been a
help to have a spouse who completely
understands my job and my passion.

**What advice do you have for today's
students?**
Pursue what interests you. Follow
your passions. This is the most certain
pathway to success and happiness.

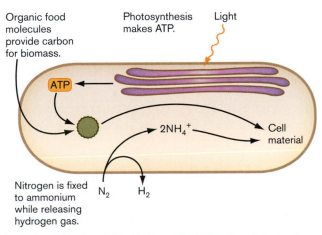

Organic food molecules provide carbon for biomass.

Photosynthesis makes ATP. Light

ATP

Nitrogen is fixed to ammonium while releasing hydrogen gas. N_2 H_2

$2NH_4^+$ → Cell material

***R. palustris* conducts photoheterotrophy and fixes nitrogen, while producing hydrogen
gas.** The bacteria make ATP through photosynthesis, but use organic molecules from the soil
to build biomass. They fix nitrogen gas into ammonium ion, generating hydrogen
gas as a by-product. *Source:* Modified from Larimer, et al. 2005. *Nature Biotechnology* 22:57–61.

Chapter 13

Energetics and Catabolism

To grow, all cells need energy. Energy comes from chemical reactions that transfer electrons within a molecule or between two molecules. For microbes, the variety of such reactions is limitless; virtually any kind of molecule in our biosphere can yield energy for some kind of microbe, from hydrogen gas to chlorinated pollutants.

To capture and use energy, microbes must regulate their energy-yielding reactions and couple them to biosynthesis, using enzymes. Enzymes direct the transfer of energy onto carriers such as ATP. ATP acts as an energy currency that is generated by energy-yielding reactions and spent by reactions that build the cell.

A major class of energy-yielding reactions used by microbes involves the breakdown of complex molecules into smaller ones, a process called catabolism. Microbes catabolize chemicals within our own digestive tracts and all around us, in the soil and water. Collectively, microbes show astonishing potential to catabolize nearly any organic substance, including petroleum, hard woods, and synthetic polymers. The products of their catabolism range from ethanol to the generation of electricity in fuel cells.

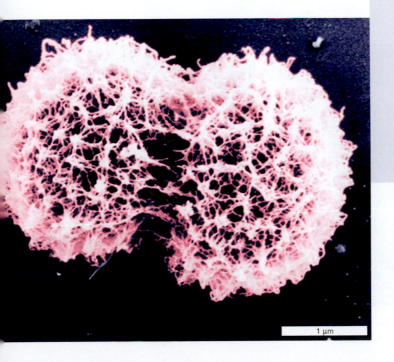

1 μm

A new species of bacteria was discovered to catabolize naphthalene, a complex aromatic molecule that is a common environmental pollutant from creosote-treated wood. The bacterium, called strain CJ2, is a beta-proteobacterium closely related to *Polaromonas vacuolata*. Its rough surface texture consists of extracellular polysaccharides. *Source*: Che O. Jeon, et al. 2003. *Proceedings of the National Academy of Sciences USA* 100:13591.

All growing microbes gain energy from their surroundings. One place to see microbial energetics at work is in composting, a process by which we dispose of food and yard waste. Composting is done by individuals or at community composting centers (**Fig. 13.1**). The U.S. Army uses microbial composting to remediate sediments containing explosives such as trinitrotoluene (TNT). In composting, nitrogen-rich manure or food waste is mixed with cellulose-rich yard waste or newspaper. The mixed compost materials are aerated to provide oxygen for the oxidative breakdown of food by bacteria and fungi. The breakdown of complex molecules, such as cellulose, to smaller molecules, such as carbon dioxide, is called **catabolism**. Catabolism provides energy for **anabolism**, the reactions that build microbial cells. But some of the energy is always released as heat. In a compost pile, heat is produced faster than it dissipates, and the temperature typically rises to 60°C. As temperature rises, thermophilic species of *Bacillus* and *Thermus* take over. The thermophiles metabolize even faster, and the heat they release maintains the compost at a high temperature that favors their growth. The heat of compost can actually be used by other organisms, such as certain nesting birds that use compost to warm their eggs.

In Part III of this book we explore microbial metabolism and biochemistry. **Chapter 13** explains how microbes gain energy from chemical reactions, based on the transfer of electrons between molecules. This chapter emphasizes catabolism, the breakdown of complex organic molecules with transfer of their electrons to an electron acceptor. Inorganic reactions such as metal oxidation can also yield energy for microbes, as shown in Chapter 14. Chapter 14 explores electron transport in greater depth, through pathways of organic respiration (oxidation of organic nutrients), lithotrophy (oxidation of *in*organic nutrients), and photosynthesis. The energy from all these pathways is used to build cells by pathways of anabolism (biosynthesis), the focus of Chapter 15. Finally, Chapter 16 explores commercial applications of microbial metabolism to produce food and beverages as well as industrial products and pharmaceuticals.

The major kinds of energy-yielding reactions discussed in Chapters 13 and 14 are summarized in **Table 13.1**. Chapter 13 emphasizes reactions of organic compounds donating electrons to yield energy, known as **organotrophy**. For each class of metabolism, Table 13.1 shows only a few examples. Microbes use many energy-yielding reactions unavailable to animals or plants. For example, *Pseudomonas* and related species of soil bacteria can gain energy by catabolizing camphor, benzene, or chlorinated aromatic pollutants. On the other hand, the archaeon *Pyrodictium occultum* gains energy by oxidizing hydrogen gas with sulfur, an inorganic reaction.

> **NOTE: Organotrophs** use preformed organic compounds to yield energy. Most organotrophs are also **heterotrophs**, organisms that use preformed organic compounds for biosynthesis. Introductory texts often equate heterotrophy with organotrophy, although in rare cases an organism shows one but not the other.

A.

B.

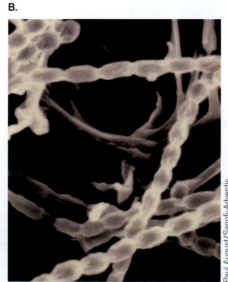

C.

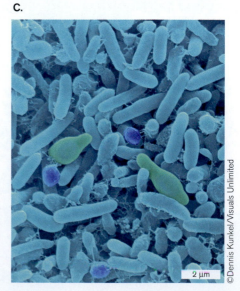

Figure 13.1   Microbial composting.   A. Disposing of yard waste in the community compost site. **B.** Actinomycete bacteria digest wood and newspaper (filament width, 0.5–1.0 µm; SEM). **C.** Thermophiles such as *Alicyclobacillus* take over compost digestion at 50–60°C (colorized SEM).

Table 13.1 Energy acquisition in bacteria and archaea.

Compounds metabolized	Class of metabolism	Chemistry of energy acquisition	Use of oxygen	Systems for energy acquisition
Organotrophy Organic compounds donate electrons	**Fermentation** Catabolism	$C_6H_{12}O_6 \rightarrow 2C_3H_6O_3$ (or other small molecules)	Anaerobic	Glycolysis and other catabolism
	Organic respiration Catabolism with inorganic or small organic electron acceptor	$C_6H_{12}O_6 + 6H_2O + 6O_2 \rightarrow$ $6CO_2 + 12H_2O$ Electron acceptor is O_2	Aerobic	Glycolysis and other catabolism TCA cycle Electron transport system
		$C_6H_{12}O_6 + 6H_2O + 12NO_3^- \rightarrow$ $6CO_2 + 12H_2O + 12\ NO_2^-$ Electron acceptor is NO_3^-, SO_4^{2-}, or oxidized organic molecules	Anaerobic	
Lithotrophy Inorganic compounds donate electrons	**Lithotrophy or chemoautotrophy**	Electron donor for respiration is H_2, Fe^{2+}, H_2S, NH_4^+ Electron acceptor is O_2 or NO_3^-	Aerobic or Anaerobic	Electron transport system
	Methanogenesis	Electron donor is H_2, CH_3OH, CH_3NH_2 Electron acceptor is CO_2 $CO_2 + 4H_2 \rightarrow CH_4 + 2H_2O$	Anaerobic	Methanogenesis
	Photoautotrophy	Photolysis of H_2O $6CO_2 + 12H_2O \rightarrow$ $C_6H_{12}O_6 + 6H_2O + 6O_2$	Aerobic (oxygenic)	Photosystems I *and* II
		Photolysis of H_2S, HS^-, or Fe^{2+} $6CO_2 + 12H_2S \rightarrow$ $C_6H_{12}O_6 + 6H_2O + 12S$	Anaerobic (anoxygenic)	Photosystem I *or* II
Light absorption supplements use of organic compounds	**Photoheterotrophy** Catabolism with light absorption	Photolysis of H_2S, HS^-, or small organic molecules H^+ pump or Na^+ pump supplements catabolism		Photosystem I *or* II Bacteriorhodopsin or proteorhodopsin

Figure 13.2 Living organisms acquire energy to build cells. Early twentieth-century biologists wondered how the "spontaneous" growth of cells could obey the laws of conservation of energy and increase of entropy.

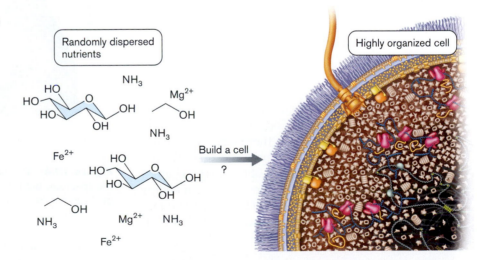

13.1 Energy and Entropy: Building a Cell

Every form of life, from a composting microbe to a human body, uses energy. **Energy** is the ability to do work, such as flagellar propulsion or the growth of a cell. Energy is used to organize proteins and to main-tain ion gradients across the cell membrane. Yet the laws of thermodynamics tell us that systems tend to become less ordered and that **entropy**, the disorder, or random-ness, of the universe, always increases. So how do cells assemble simple, disordered molecules into complex, ordered forms (**Fig. 13.2**)? How does life build order out of disorder?

Microbes Use Energy to Build Order

To explain how microbes build order requires a closer look at the thermodynamic relationship between energy and entropy. Gaining energy enables organisms to grow, thus increasing their structural order. As order increases, the cell is said to decrease in entropy, or disorder. But the decrease in entropy is local to the cell, and temporary. Ultimately, the cell's energy must be spent as heat, which radiates away, causing entropy to increase. In other words, the local, temporary gain of energy enables a cell to grow. Continued growth requires continual gain of energy and continual radiation of heat.

What is true of the cell holds as well for the entire biosphere. In Earth's biosphere, the total metabolism of all life must ultimately dissipate most energy as heat. Biological heat production is not always obvious because soil and water provide a tremendous heat sink. But overall, Earth's biosphere behaves as a giant thermal reactor (**Fig. 13.3**). As solar radiation reaches Earth, a small fraction is captured by photosynthetic microbes and plants. The fraction captured is largely in the range of visible light (**Fig. 13.4**), the range of wavelengths in which the photon energies are appropriate for the controlled formation and dissociation of molecular bonds. Microbial and plant photosynthesis generates biomass, which is consumed by heterotrophs and decomposers. The consumers store a small fraction of their energy in biomass. At each successive level, the majority of energy is lost, radiated from Earth as heat. Despite growth of living organisms on Earth, the universe as a whole becomes more disordered.

> **THOUGHT QUESTION 13.1** Why can't photosynthesis be driven by solar microwaves or by X-rays?

Gibbs Free Energy Change Includes Heat and Entropy

To provide energy to a cell, a biochemical reaction must go forward from reactants to products. The direction of a reaction can be predicted by a thermodynamic quantity known as the **Gibbs free energy change**, ΔG, also known as free energy change or Gibbs energy change. The ΔG value of a reaction determines how much energy is potentially available to do work, such as to drive rotary flagella, to build a cell wall, or to store accurate information in DNA.

ΔG includes enthalpy and entropy. The free energy change ΔG has two components:

- ΔH = change in **enthalpy**, the heat energy absorbed or released as reactants become products at constant pressure. When reactants absorb heat from their surroundings as they convert to products, ΔH is positive.

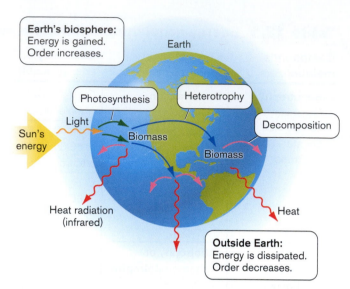

Figure 13.3 Solar energy. Solar radiation reaches Earth, where a small fraction is captured by photosynthetic microbes and plants. The microbial and plant biomass enters heterotrophs and decomposers, which convert a small fraction to biomass at each successive level. At each level, the majority of energy is lost, radiated from Earth as heat.

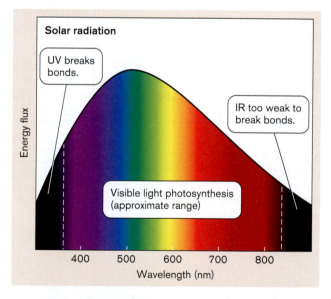

Figure 13.4 The solar spectrum. The sun radiates across the spectrum, but the intensity of solar radiation reaching Earth peaks in the range of visible light.

When, instead, heat energy is released, then ΔH is negative. Release of heat (negative value of ΔH) can yield energy for the cell to use.

- ΔS = change in **entropy**, or disorder. Entropy is based on the number of states of a system, such as the number of possible conformations of a molecule. If a cellular reaction splits one molecule into two, all else being equal, entropy increases; the system is more disordered, and ΔS is positive. If, however, two molecules become one, then order is increased overall, and ΔS is negative.

The relationship between ΔH and ΔS is given by the Gibbs-Helmholtz equation:

$$\Delta G = \Delta H - T\Delta S$$

The value of free energy, ΔG, determines whether a process may go forward spontaneously. Negative values of ΔG mean that heat is released and/or order is decreased. If ΔG is negative, the process may go forward, whereas positive values mean that the reaction will go in reverse.

The overall sign of ΔG depends on its two components: ΔH, the absorption or release of heat energy, and $-T\Delta S$, the product of entropy change (ΔS) and temperature (T). In living organisms, a sufficiently negative ΔH (energy lost as heat) often overrides $-T\Delta S$, the term for increase in order (negative value of ΔS, positive value of $-T\Delta S$). Thus, a living organism, whose development entails increasing order and decreasing ΔS, can grow so long as the sum of its metabolism has a sufficiently negative value of ΔH. The negative ΔH includes energy gained by the cell as well as energy lost to the surroundings as heat. The heat loss associated with ΔH is obvious in a compost pile, and it occurs as well for all living organisms and communities.

Negative ΔG Drives a Reaction Forward

An example of a thermodynamically favored reaction is the oxidation of hydrogen gas to form water. Hydrogen gas is oxidized for energy by many kinds of bacteria in soil and water. For example, hydrogenotrophic bacteria of the genus *Ralstonia* have been isolated from ultrapure water used for nuclear fuel storage, where radioactive ionization generates H_2. The chemical reaction of hydrogen oxidation is:

$$2H_2 + O_2 \rightarrow 2H_2O$$

In this reaction, two molecules of hydrogen gas donate four electrons to oxygen, forming water. For this reaction, at standard temperature (298 K) and pressure (sea level), $\Delta H = -484$ kJ/mol (kilojoules per mole). The ΔH is strongly negative (much heat is released) because the bonds of the product H_2O are much more stable than those of the substrates H_2 and O_2. However, entropy decreases because the three molecules are replaced by two, a more ordered state. Thus, ΔS is negative ($\Delta S = -0.0886$ kJ/mol · K); and in the Gibbs equation, the negative sign on the entropy term $-T\Delta S$ makes its contribution to ΔG positive, unfavorable for reaction. So which term wins, ΔH or $-T\Delta S$?

$$\Delta G = \Delta H - T\Delta S$$
$$= -484 \text{ kJ/mol} - (298 \text{ K})(-0.0886 \text{ kJ/mol} \cdot \text{K})$$
$$= -484 \text{ kJ/mol} + 27 \text{ kJ/mol}$$
$$= -457 \text{ kJ/mol}$$

Overall, ΔG is negative, and so bacteria with the appropriate enzyme pathways can use the reaction of hydrogen gas with oxygen to provide energy.

> **NOTE: Joules (J)** are the standard SI unit to denote energy. 1 kilojoule (kJ) = 1,000 joules. Another unit commonly used is kilocalories (kcal). The conversion factor is: 1 kJ = 0.239 kcal.

TO SUMMARIZE:

- **Energy** enables cells to build ordered structures out of simple molecules from their environment.
- **The free energy change (ΔG)** includes enthalpy (ΔH), the heat energy absorbed or released, and $-T\Delta S$, the product of temperature and entropy change (ΔS).
- **Negative values of ΔG** show that a reaction may drive the cell's metabolism. The sign of ΔG depends on the relative magnitude of ΔH and $-T\Delta S$.

13.2 Energy and Entropy in Biochemical Reactions

The energy required for all growth processes is supplied by biochemical reactions. But which direction will a biochemical reaction go? The values of ΔH and $-T\Delta S$ determine which reactions will proceed and whether they can provide energy under actual conditions of the cell. The overall free energy change ΔG determines which "foods" a microbe can eat or, more precisely, which reactions between available molecules can be harnessed for microbial growth.

Enthalpy and Entropy in Microbial Metabolism

The relative contributions of ΔH and $-T\Delta S$ indicate the roles of chemistry and temperature in microbial metabolism.

- **ΔH-driven reactions release heat.** The ΔH term dominates bioenergetic reactions involving strong oxidants such as O_2 (reactions such as glucose respiration and photosynthesis). ΔH-driven reactions release a lot of heat, causing, for example, the rise in temperature of an aerated compost pile.
- **ΔS-driven reactions depend on temperature.** The entropy-driven component of Gibbs energy, $-T\Delta S$, grows larger as temperature increases. This term dominates reactions that generate a larger number of products than reactants but have a relatively small change in oxidation state. An example is glucose fermentation to ethanol and CO_2 during production of alcoholic beverages. Fermentation reactions produce less heat than oxidative catabolism, and their energy yield increases with environmental temperature.

An extreme case of entropy-driven metabolism is that of acetate conversion to methane and CO_2, conducted by soil archaea such as *Methanosarcina barkeri*. This methanogen actually cools its environment, absorbing heat to form products of higher chemical energy. Its growth is driven by the increase in entropy.

Measuring ΔH and ΔS. The actual ΔH and $-T\Delta S$ of biochemical reactions can be determined by calorimetry, measuring the heat released or absorbed during a reaction. One type of calorimetry is isothermal titration calorimetry, in which heat release is measured as the amount of energy needed to cool the system while keeping the temperature constant. Observing the reaction at different fixed temperatures reveals the temperature dependence of the reaction and hence the relative contribution of the entropic term $-T\Delta S$. A common application of isothermal titration calorimetry is to study the ΔH and $-T\Delta S$ of a reaction in which a drug binds to its target receptor protein, such as an inhibitor binding to the protease of HIV (human immunodeficiency virus). The ΔH and $-T\Delta S$ of the inhibitor-binding reaction indicate its binding strength, an important factor in predicting its efficacy in treating patients.

Factors That Determine the Direction of a Reaction

The ΔG for a given reaction is determined by several factors:

- **Molecular stability.** When reactants combine to form products with more stable bonds, the reaction has a negative ΔH. For example, the reaction of a sugar with oxygen has a negative ΔH because the relatively unstable oxygen molecules are reduced to H_2O.
- **Entropy increase.** Reactions in which a complex molecule is broken down to a greater number of smaller molecules increase entropy (positive ΔS, negative $-T\Delta S$). Entropy also increases with formation of gaseous products, such as CO_2, or of carboxylic acids, such as lactic acid, in which H^+ dissociates from a carboxylate ion ($RCOO^-$).
- **Concentrations and environmental factors.** The direction of reaction depends not only on the intrinsic properties of the reactants, but on environmental factors such as temperature, pressure, and reactant concentration. A large excess of reactants over products makes ΔG more negative (favoring the forward reaction). An excess of product makes ΔG more positive (favoring the reverse reaction).

Standard reaction conditions. The thermodynamic values reported in data tables hold only for isolated reactions under specified conditions that differ greatly from those of living cells. These conditions include temperature, ionic strength, and gas pressure (in the case of gaseous components, such as CO_2), as well as the concentrations of reactants and products. To enable comparison, thermodynamic values are commonly presented under standard conditions for temperature, pressure, and concentration. The standard conditions define a **standard Gibbs free energy change, $\Delta G°$.** The standard conditions for $\Delta G°$ are as follows:

- The temperature is 298 K (25°C).
- The pressure is 1 atm (standard atmospheric pressure).
- All concentrations of substrates and products are 1 molar.

These standard conditions also apply to changes in enthalpy ($\Delta H°$) and entropy ($\Delta S°$), which contribute to $\Delta G°$. In biochemistry, an additional standard condition is that of pH 7, as living cells commonly maintain their cytoplasm within a unit of neutral pH. The thermodynamic terms in biochemistry are thus designated $\Delta G°'$, $\Delta H°'$, and $-T\Delta S°'$.

NOTE: Remember to distinguish these forms of the Gibbs free energy term:

ΔG = Change in Gibbs free energy for a reaction under defined conditions.

$\Delta G°$ = ΔG at standard conditions of temperature (298 K) and pressure (sea level), with all reactants and products at a concentration of 1 M.

$\Delta G°'$ = ΔG at standard temperature, pressure, and concentrations for $\Delta G°$, plus the additional condition of pH 7 (H^+ concentration of 10^{-7} M) commonly specified by biochemists.

Concentrations affect ΔG. In living cells, the concentrations of reactants and products usually differ from 1 molar; thus, the actual ΔG differs from $\Delta G°$ or $\Delta G°'$. Consider a reaction in which reactants A and B are converted to products C and D, as well as conversion in reverse

$$A + B \rightleftharpoons C + D$$

Higher concentration of reactants (A or B) drives the reaction forward, whereas higher concentration of products drives it in reverse. So, ΔG includes the ratio of products to reactants:

$$\Delta G = \Delta G° + RT \ln \frac{[C]\,[D]}{[A]\,[B]}$$

$$= \Delta G° + 2.303\, RT \log \frac{[C]\,[D]}{[A]\,[B]}$$

Table 13.2 Effect of the concentration ratio on ΔG.

Ratio of products to reactants: $\dfrac{[C][D]}{[A][B]}$	Change in ΔG (kJ/mol) at standard temperature (298 K) and atmospheric pressure	Result of change from standard concentrations
10^{-4}	−23	Products increase
10^{-2}	−11	Products increase
1	0	ΔG = ΔG°
10^{2}	+11	Reactants increase
10^{4}	+23	Reactants increase

where R is the gas constant (R = 8.315 J/mol) and T is the absolute temperature in Kelvins (298 K for 25°C). The factor 2.303 converts the logarithm of the ratio of products to reactants from base e to base 10.

The effect of the concentration ratio on ΔG is shown in **Table 13.2**. In reactions at medium temperature (25–40°C), a hundredfold increase in the ratio of products to reactants adds about 11 kJ/mol to ΔG. ΔG is then less negative and the reaction less favorable. On the other hand, a hundredfold decrease in the concentration ratio makes ΔG more negative by –11 kJ/mol.

In some environments, a highly negative concentration term can override a positive ΔG°, resulting in a reaction with negative ΔG that microbes can use for energy. For example, in an iron mine the high concentration of reduced iron favors iron-oxidizing microbes. Alternatively, a high temperature may increase the magnitude of the term 2.303 RT log [products]/[reactants] when it is negative, until it overrides a positive ΔG°. This occurs in thermophiles such as *Sulfolobus*, which metabolize sulfur at 90°C by reactions unfavorable at lower temperatures.

Another way the direction of reaction may change is for a second metabolic process to remove one of the reaction's products (C or D) as fast as it is produced. The total ΔG for both reactions could then be negative, even if the first reaction had a positive ΔG. In **glycolysis**, the pathway of glucose breakdown to pyruvate, many of the individual reactions have ΔG°′ values smaller than 5 kJ/mol. Their actual direction of reaction in the cell depends on concentrations of products and reactants. Glycolytic reactions with near-zero ΔG°′ are reversible and can in fact participate in the biosynthetic pathway of gluconeogenesis (glucose biosynthesis).

Energy and Entropy in Concentration Gradients

A growing cell needs to obtain essential molecules from outside, such as sugars and amino acids, and inorganic ions. The difference in concentration of a nutrient inside a cell versus outside generates a concentration gradient that stores energy.

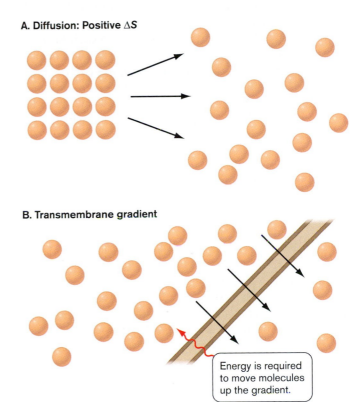

A. Diffusion: Positive ΔS

B. Transmembrane gradient

Energy is required to move molecules up the gradient.

Figure 13.5 **Diffusion and transport.** **A.** Water-soluble molecules diffuse to uniform concentration throughout the solution: ΔS is positive, −TΔS negative; the entropy term favors the process. **B.** If a membrane separating two compartments is permeable, molecules move from a compartment with high concentration to one with low concentration. Energy is required to move molecules up their concentration gradient.

In water solution, a dissolved substance diffuses by random movements until its distribution has the same concentration throughout (**Fig. 13.5A**). The random distribution of molecules at uniform concentration represents the state of greatest entropy. Diffusion in the environment ultimately brings nutrients into contact with microbial cells, even cells that lack chemotactic motility to hunt for food. For example, sugars in the food we eat diffuse through our saliva to reach the bacteria growing in biofilms on our teeth.

The cell membrane contains transporters for useful molecules, allowing the nutrient molecules to cross. Entropy favors their movement from higher to lower concentration (**Fig. 13.5B**). In most environments, however, the nutrients are at lower concentrations than inside the cell. To obtain these molecules from outside, the bacterial cell must transport them against their gradient—that is, from lower to higher concentration—increasing the concentration difference and thus decreasing entropy. Uptake against a gradient requires an energy source to power transport proteins embedded in the membrane. The transporters must spend energy to compensate for the decrease in entropy when a cell transports the nutrient against its concentration gradient.

To form a chemical gradient requires expenditure of energy, so establishing a gradient stores energy. The energy stored in a gradient across a cell membrane can be interconverted with the energy of biochemical reactions such as sugar catabolism. For example, respiration generates a proton gradient (chemical gradient of H^+), which then drives the ATP synthase to synthesize ATP. The interconversion of gradient energy with chemical reaction energy was first known as the "chemiosmotic hypothesis," earning the 1978 Nobel Prize in Chemistry for Peter Mitchell (discussed in Chapter 14).

Microbial Thermodynamics Includes Other Factors

Within living cells, numerous other factors influence the Gibbs energy values of microbial growth. The Gibbs equation as presented assumes constant concentrations of reactants and products. In living cells, however, the reactant concentrations are continually changing. To account for changing concentrations requires a level of analysis more complex than we can present here.

An important source of concentration change is the coupling of energy-generating reactions to the energy-spending reactions of biosynthesis. The energy-spending reactions subtract from the observed free energy release of a growing microbe. A model proposed by Urs von Stockar and colleagues predicts that microbes optimize their coupling of catabolism with biosynthesis to maintain a net Gibbs energy of growth within a range of 300–450 kJ/mol per carbon of substrate catabolized. Different forms of catabolism, however, offer very different yields of biomass (cell mass synthesized per unit substrate). For example, the oxidative respiration of glucose yields five times the biomass produced by glucose fermentation, although respiring and fermenting cells show similar net Gibbs energy of growth.

A surprising discovery has been that many bacteria and archaea in natural environments grow extremely slowly, at values of ΔG approaching zero (that is, thermodynamic equilibrium). When the actual ΔG (under actual reaction conditions) equals zero, a reaction proceeds equally forward and reverse, and there is no net change in energy. At equilibrium, the ratio of product and reactant concentrations exactly cancels $\Delta G°$ or $\Delta G°'$:

$$\Delta G = 0 = \Delta G° + 2.303 \; RT \log \frac{[C] \; [D]}{[A] \; [B]}$$

$$\Delta G° = -2.303 \; RT \log \frac{[C] \; [D]}{[A] \; [B]}$$

Living cells can never grow exactly at equilibrium ($\Delta G = 0$), but some gain energy from reaction pathways whose Gibbs free energy just exceeds equilibrium. Michael McInerney, Derek Lovley, and colleagues have identified soil bacteria and archaea that gain energy from metabolic pathways with ΔG values as small as –20 kJ/mol, less than one ATP's worth of energy per cycle. An example of an energy-yielding pathway with near-zero ΔG is CO_2 reduction to methane (a form of methanogenesis). How such pathways are coupled to biosynthesis is poorly understood. The discovery of low-ΔG energetics has opened new possibilities for industrial microbiology previously thought impossible, such as the anaerobic digestion of complex organic pollutants in contaminated soil.

THOUGHT QUESTION 13.2 Methanogens are a small proportion of the microbial community in soil, and the ΔG of methanogenesis is small. Yet methanogens produce large volumes of methane, large enough to contribute significantly as a greenhouse gas to global warming. Why would methanogens produce a relatively large quantity of waste product?

THOUGHT QUESTION 13.3 The soil bacterium *Geobacter* can metabolize acetic acid by the following reaction:

$$CH_3COOH + 2H_2O \rightarrow 2CO_2 + 4H_2$$

Yet the $\Delta G°'$ is +95 kJ/mol. What could happen in the soil enabling *Geobacter* to grow?

TO SUMMARIZE:

- **The direction of reaction** is determined by the molecular stability of reactants and products and by entropy change associated with number of products, gaseous products, or products with multiple forms.
- **Concentrations of reactants and products affect ΔG.** The lower the concentration ratio of products to reactants, the more negative the value of ΔG.
- **A concentration gradient stores energy.** Solutes run down their concentration gradient unless energy is applied to reverse the flow.

A. ADP phosphorylation to ATP

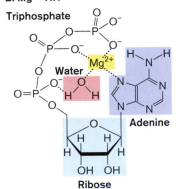

Figure 13.6   ADP plus inorganic phosphate makes ATP. The reaction requires energy input (positive ΔG) because the negatively charged oxygens of the phosphates are forced into proximity. **A.** The chemical reaction phosphorylating ADP to ATP. **B.** Model of Mg^{2+}-ATP. The multiple negative charges of ATP are stabilized by complexing with magnesium ion plus a water molecule.

- **Cellular thermodynamics involves** changing reactant and product concentrations and coupling of energy-yielding reactions with energy-spending reactions.
- **Reaction pathways with ΔG near zero** can drive microbial growth.

13.3 Energy Carriers and Electron Transfer

Our ΔG equations show only the total energy of a reaction such as glucose oxidation. If all the energy were released at once, however, it would dissipate as heat without building biomass. In living cells, glucose oxidation never occurs in one step. Instead, the energy yield is divided among a large number of stepwise reactions with smaller energy changes. In this way, the cell can be thought of as "making change" by converting a large energy source to numerous smaller sources that can be "spent" conveniently for cell function and biosynthesis. The "spending" of energy is controlled by enzymes that couple all energy-providing reactions to specific energy-spending reactions.

Energy Carriers Gain and Release Energy

Many of the cell's energy transfer reactions involve **energy carriers**. Examples of energy carriers are ATP (adenosine triphosphate) and NADH (nicotinamide adenine dinucleotide). Energy carriers are molecules that gain and release small amounts of energy in reversible reactions. Energy carriers are used to transfer energy in a wide range of biochemical reactions.

Some energy carriers, such as NADH, transfer energy associated with electrons received from a food molecule. A molecule that transfers, or "donates," electrons to another molecule is called an **electron donor** or a reducing agent; a molecule that receives, or "accepts," electrons is called an **electron acceptor**. For example, during glucose catabolism, a molecule of glyceraldehyde 3-phosphate transfers a pair of electrons ($2e^-$) with a hydrogen ion to NAD^+, forming NADH. NAD^+ is an electron acceptor that receives the electrons; it then becomes the electron donor, NADH. Electron donors such as NADH mediate electron transfer from reduced food molecules to a terminal electron acceptor such as oxygen. Energy carriers that transfer electrons are needed for all energy-yielding pathways and for biosynthesis of cell components such as amino acids and lipids.

ATP Carries Energy through Phosphorylation

Adenosine triphosphate ATP (**Fig. 13.6A**) is composed of a base (adenine), a sugar (ribose), and three phosphates. Note that adenine-ribose-phosphate is equivalent to a nucleotide "link" of RNA. The base adenine is a fundamental molecule of life, one that forms spontaneously from methane and ammonia in experiments simulating the origin of life on early Earth. Like the sugar ribose, ATP is an ancient component of cells, found in all living organisms.

Note that under physiological conditions, ATP always complexes with Mg^{2+} (**Fig. 13.6B**). The magnesium cation partly neutralizes the negative charges of the ATP phosphates, stabilizing the structure in solution. Most enzyme-binding sites for ATP actually bind Mg^{2+}-ATP. This is one reason magnesium is an essential nutrient for all living cells.

ADP phosphorylation to ATP. During cell metabolism, ATP is generated by **phosphorylation**, the condensation of inorganic phosphate with **adenosine diphosphate (ADP)**:

The phosphorylation of ADP to form ATP requires energy input (positive ΔG).

Why does ATP formation require energy? The inorganic phosphate has four oxygen atoms that share a negative charge. When phosphate reacts with another phosphate to form a bond, the charged oxygens of adjacent phosphates are forced into proximity despite charge repulsion. The two phosphoanhydride bonds in ATP show charge repulsion, associated with a large negative ΔG of hydrolysis. Hydrolysis of each bond yields energy; thus, they are referred to as high-energy bonds. The formation and hydrolysis of ATP are shown as

$$\text{A-P}\sim\text{P} + \text{H}^+ + \text{P}_i \rightleftharpoons \text{A-P}\sim\text{P}\sim\text{P} + \text{H}_2\text{O}$$

where ~ designates each "high-energy" bond and P_i designates inorganic phosphate; or in more concise shorthand,

$$\text{ADP} + \text{P}_i \rightleftharpoons \text{ATP} + \text{H}_2\text{O}$$

ATP transfers energy. ATP can transfer energy to cell processes in three different ways: hydrolysis releasing phosphate; hydrolysis releasing pyrophosphate (diphosphate); and phosphorylation of an organic molecule. Each process serves different functions in the cell.

■ **Hydrolysis releasing phosphate.** The **hydrolysis** of ATP at the terminal phosphate consumes H_2O to produce ADP and P_i. Hydrolysis of the phosphodiester bond releases energy. The energy released by ATP hydrolysis can be transferred to a coupled reaction of biosynthesis, such as building an amino acid. The two reactions are coupled by an enzyme with binding sites specific for ATP and the substrate.

■ **Hydrolysis releasing pyrophosphate.** ATP can hydrolyze at the middle phosphate, releasing pyrophosphate (PP_i). The pyrophosphate usually hydrolyzes shortly afterward to make 2 P_i; thus, approximately twice as much energy is spent as in release of 1 P_i from ATP. The advantage of pyrophosphate release is that it drives a reaction strongly forward. Pyrophosphate release occurs in reactions that must avoid reversal—for example, the incorporation of nucleotides into growing chains of RNA.

■ **Phosphorylation of an organic molecule.** ATP can transfer its phosphate to the hydroxyl group of a molecule such as glucose to activate the substrate for a subsequent rapid reaction. No inorganic phosphate appears, and no water molecule is consumed:

ATP + glucose → ADP + glucose 6-phosphate

Some enzymes catalyze ATP transfer of phosphate to activate sugar molecules for catabolism. Other enzymes couple the phosphorylation of a sugar to its transport across the cell membrane; consequently, these enzymes are known as the **phosphotransferase (PTS)** family. The PTS enzymes play a critical role in determining which nutrients from the environment a microbe can acquire and catabolize (discussed in Chapter 4).

Note that besides ATP, other nucleotides carry energy. Guanidine triphosphate (GTP) provides energy for ribosome elongation of proteins. And the phosphodiester bonds of all four nucleotide triphosphates, as well as their corresponding deoxyribonucleotide triphosphates, carry energy for their own incorporation into RNA and DNA, respectively.

ATP produced by glucose catabolism. A large number of ATPs can be formed by coupling ATP synthesis to the step-by-step breakdown and oxidation of a food molecule such as glucose. In theory, complete oxidation of glucose through respiration can produce as many as 38 ATP molecules. The overall $\Delta G^{\circ\prime}$ of the coupled reactions is

$$\text{C}_6\text{H}_{12}\text{O}_6 + 6\text{H}_2\text{O} + 6\text{O}_2 \rightarrow$$
$$12\text{H}_2\text{O} + 6\text{CO}_2 \qquad \Delta G^{\circ\prime} = -2{,}878 \text{ kJ/mol}$$
$$\underline{38\,[\text{ADP} + \text{P}_i \rightarrow \text{ATP} + \text{H}_2\text{O}] \qquad \Delta G^{\circ\prime} = 38 \times (31 \text{ kJ/mol})}$$
$$\text{Net } \Delta G^{\circ\prime} = -1{,}700 \text{ kJ/mol}$$

The difference in $\Delta G^{\circ\prime}$ for the coupled reactions is the energy lost as heat and entropy, in this case $-1{,}700$ kJ/mol ($-2{,}978$ kJ/mol $+ 1{,}178$ kJ/mol). Thus, the maximal efficiency of energy capture by ATP is about 40%, a level that may be approached by highly efficient systems such as mitochondria. When ΔG values are corrected for cellular concentrations of reactants and products, the actual efficiency appears to be greater than 50%. For comparison, the efficiency of a typical machine such as an internal combustion engine is about 20%.

Under conditions such as low oxygen concentration, much smaller amounts of ATP are made per molecule of glucose. On the other hand, the vast majority of microbial catabolism in soil and water involves nonsugar substrates consumed by reactions at near-zero values of ΔG, which require multiple cycles to make a single molecule of ATP. These near-equilibrium reactions remain poorly understood, but their efficiency may approach 100%.

THOUGHT QUESTION 13.4   After ATP has been dephosphorylated to ADP, how else might ADP provide energy for the cell?

THOUGHT QUESTION 13.5   Linking an amino acid to its cognate tRNA is driven by ATP hydrolysis to AMP plus pyrophosphate. Why release PP_i instead of P_i?

THOUGHT QUESTION 13.6   In the microbial community of the bovine rumen, the actual ΔG value has been calculated for glucose fermentation to acetate:

$$C_6H_{12}O_6 + 2H_2O \rightarrow 2C_2H_3O_2^- + 2H^+ + 4H_2 + 2CO_2$$
$$\Delta G = -318 \text{ kJ/mol}$$

If the actual ΔG for ATP formation is 44 kJ/mol and each glucose fermentation yields four molecules of ATP, what is the thermodynamic efficiency of energy gain? Where does the lost energy go?

NADH Carries Energy and Electrons

A major energy carrier that donates and accepts electrons is **nicotinamide adenine dinucleotide** (**NADH**, reduced; **NAD⁺**, oxidized). NADH carries two or three times as much energy as ATP, depending on cell conditions. It is used to carry electrons from breakdown products of glucose. The oxidized form NAD^+ receives two electrons ($2e^-$) plus a proton (H^+) from a food molecule; a second H^+ from the food molecule enters water solution. Overall, reduction of NAD^+ consumes two hydrogen atoms to make NADH:

$$NAD^+ + 2H^+ + 2e^- \rightarrow NADH + H^+ \qquad \Delta G^{\circ\prime} = 62 \text{ kJ/mol}$$

For this reaction, $\Delta G^{\circ\prime}$ is positive; therefore, it requires input of energy from the food molecule. The reduced energy carrier NADH can then reverse the above reaction by donating two electrons ($2e^-$) to another molecule, regenerating NAD^+.

NOTE: In this book, our discussion of metabolism makes the following assumptions:

- A proton is equivalent to a hydrogen ion (H^+). In water solution, H^+ combines with H_2O to form hydronium (H_3O^+), but for clarity we use H^+.
- An atom of hydrogen removed from a C–H bond consists of a proton (H^+) plus an electron (e^-). In a reaction, the proton and electron may be transferred to one molecule or to different molecules.

NADH structure and function. NAD^+ consists of an ADP molecule attached to a nicotinamide group instead of a third phosphate. The nicotinamide group, like a ribonucleotide, contains a nitrogenous base attached to a sugar phosphate. In NAD^+, the dehydrogenated (oxidized) nicotinamide has a ring structure (shaded pink in **Fig. 13.7A**) that forms a stable cation.

NAD^+ is a relatively stable structure because the ring electrons are **aromatic**; that is, the bonding electrons delocalize equally around the ring, as in benzene. Aromatic rings that contain noncarbon atoms are said to be **heteroaromatic**. Many biologically active molecules are heteroaromatic, including adenine and other nucleotide bases. A heteroaromatic ring is stable, but its disruption requires less energy than the disruption of benzene. Thus, it is possible to disrupt the ring by adding two electrons with a proton, eliminating one double bond. The donation of electrons eliminates the ring's aromaticity, generating NADH, which carries energy in an amount useful for cell reactions.

The electrons transferred to NADH eventually must be put somewhere else, onto the next electron acceptor. NADH cannot build up in the cell; it must always be recycled to NAD^+, to continue oxidizing food molecules. One way the energy stored by NADH can be spent is to transfer $2H^+ + 2e^-$ onto a product of catabolism. For example, in ethanolic fermentation to make wine or beer, NADH reduces pyruvate to ethanol. In this case, however, the energy is lost to the cell. Alternatively, NADH can transfer its electrons to one of a series of electron carrier molecules known as the **electron transport system (ETS)**.

The Electron Transport System (ETS)

The electron transport system (also known as the electron transport chain) stores energy from electron transfer as ion gradients across the membrane of the cell or an organelle. The ETS enables production of ATP. The ETS includes a series of proteins and small organic molecules that can be reduced and cyclically reoxidized. At the end of the series of oxidation-reduction transfers, the pair of

A. NAD⁺ (NADP⁺)

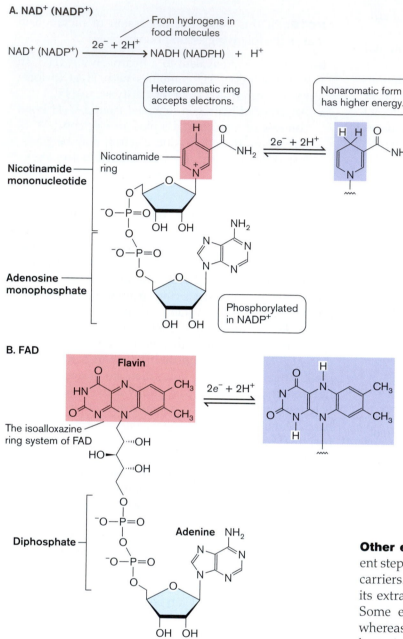

From hydrogens in food molecules

$$NAD^+ (NADP^+) \xrightarrow{2e^- + 2H^+} NADH (NADPH) + H^+$$

Heteroaromatic ring accepts electrons.

Nonaromatic form has higher energy.

$$2e^- + 2H^+$$

Nicotinamide mononucleotide

Nicotinamide ring

Adenosine monophosphate

Phosphorylated in NADP⁺

B. FAD

Flavin

$$2e^- + 2H^+$$

The isoalloxazine ring system of FAD

Diphosphate

Adenine

Ribose

Figure 13.7   Reduction of NAD⁺ and FAD.
A. In the reduction of NAD⁺, the nicotinamide ring (shaded pink) loses a double bond as two electrons are gained from an electron donor. Two hydrogen atoms are consumed; one bonds to NADH, while the other ionizes. **B.** In FAD, the flavin isoalloxazine ring system gains two electrons associated with two hydrogens.

The total energy spent by NADH oxidation through the ETS is –220 kJ/mol (–62 kJ/mol –158 kJ/mol). This energy is converted to transmembrane proton potential, which then drives nutrient transport, motility, and synthesis of ATP. The ETS and the proton potential are discussed in detail in Chapter 14.

Many microbes can reduce terminal electron acceptors other than O_2, such as nitrate (NO_3^-) or pyruvate. Humans, unlike most microbes, have no alternative electron acceptors; thus, we need to breathe oxygen. When no terminal electron acceptor is available to accept electrons, NADH builds up to unfavorably high levels, leaving no more NAD⁺ available to be reduced. When muscle tissues run out of oxygen, they transfer electrons from NADH back to the products of glucose catabolism, producing lactic acid. The buildup of lactic acid drives down pH, causing the muscle to cramp.

Other energy carriers that transfer electrons. Different steps of metabolism utilize different but related energy carriers. For example, NADPH differs from NADH only in its extra phosphate attached to the 2′ carbon of adenine. Some enzymes can utilize both NADPH and NADH, whereas other enzymes use only one or the other. In many bacteria, NADPH is used for biosynthesis, whereas NADH feeds the ETS to yield energy. Other species, particularly phototrophs, use NADH and NADPH interchangeably.

Another related coenzyme is **flavin adenine dinucleotide (FADH₂)**, in which flavin substitutes for nicotinamide. The flavin nucleotide includes a ring structure whose aromaticity is eliminated by receiving two electrons (**Fig. 13.7B**). The redox function of the flavin isoalloxazine ring system is similar to that of NADH:

$$FAD^+ + 2H^+ + 2e^- \rightarrow FADH_2$$

Like NADH, FADH₂ donates $2e^-$ to an electron acceptor. FADH₂ is a weaker electron donor than NADH, but when combined with a strong electron acceptor such as O_2, electron transfer occurs and significant energy is released ($\Delta G^{\circ\prime} = -158$ kJ/mol).

electrons are transferred to a **terminal electron acceptor** whose product leaves the cell. For example, molecular oxygen (O_2) as a terminal electron acceptor is reduced to H_2O.

The reaction of O_2 reduction to H_2O is coupled to oxidation of NADH or another reduced energy carrier:

$$NADH + H^+ \rightarrow$$
$$NAD^+ + 2H^+ + 2e^- \qquad \Delta G^{\circ\prime} = -62 \text{ kJ/mol}$$

$$\tfrac{1}{2}O_2 + 2H^+ + 2e^- \rightarrow H_2O \qquad \Delta G^{\circ\prime} = -158 \text{ kJ/mol}$$

$$NADH + H^+ + \tfrac{1}{2}O_2 \rightarrow$$
$$NAD^+ + H_2O \qquad \Delta G^{\circ\prime} = -220 \text{ kJ/mol}$$

For several reasons, different kinds of reactions use different energy carriers:

- **Different amounts of energy.** Biochemical reactions yield different amounts of energy—that is, different values of ΔG. Suppose that a reaction can provide more than enough energy to generate ATP from ADP (31 kJ), but not quite enough to generate NADH from NAD⁺ (62 kJ). An example is the conversion of succinate to fumarate in the TCA cycle (see Section 13.6). This reaction provides the energy to reduce FAD to FADH₂, whose oxidation by O₂ can yield two molecules of ATP. Thus, the use of FADH₂ enables the cell to make more efficient use of its food than if generation of ATP or NADH were the only choices.

- **Different redox levels.** Food molecules may have more or fewer electrons (level of reduction/oxidation) than those associated with the cell structure. For example, lipids are more highly reduced than glucose. Thus, lipid catabolism requires a greater proportion of electron-accepting energy carriers (such as NAD⁺ or NADP⁺) than glucose catabolism and makes relatively few ATP molecules directly. A combination of energy carriers that do or do not change redox state enables cells to balance their electrons while acquiring energy.

- **Regulation and specificity.** Specific energy carriers can direct metabolites into different pathways serving different functions. For example, in many bacteria, NADH is directed into the ETS, whereas NADPH, the 2′-phosphorylated form of NADH, is directed into biosynthesis of cell components such as amino acids and lipids.

The concentrations and reduction level of energy carriers provide much information on the state of a cell, such as the effects of environmental stress on cell metabolism. Energy carriers can be observed in living cells by the technique of nuclear magnetic resonance (NMR) spectroscopy (see **Special Topic 13.1**). The use of NMR to observe living cells led to the development of magnetic resonance imaging (MRI) to observe the entire human body.

Enzymes Catalyze Metabolic Reactions

In living cells, each reaction must occur only as needed, in the right amount at the right time. The rate of reaction is determined by the **activation energy** (E_a), the input energy needed to generate the high-energy transition state on the way to products (**Fig. 13.8**). Most biochemical reactions have an activation energy that exceeds the kinetic energy of the reactant molecules colliding. Thus, no matter how negative the ΔG, the reaction will proceed only when the activation energy is lowered by interaction with a catalyst, an agent that participates in a reaction without being consumed.

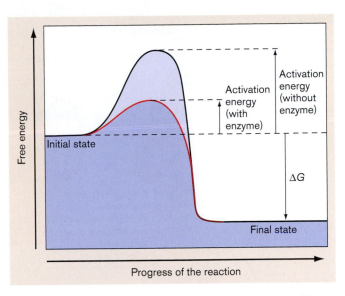

Figure 13.8 Enzyme lowers the activation energy of the transition state. In the presence of an enzyme, the activation energy of the reactants is decreased, allowing rapid conversion of reactants to products.

Biological reactions are catalyzed by **enzymes**, structures composed of protein (or, in some cases, RNA) that bind substrates of a specific reaction. The enzyme lowers the activation energy by bringing the substrates in proximity to one another and by correctly orienting them to react. In some cases, enzymes provide a reactive amino acid residue to participate in a transition state between reactants and products. Microbial enzymes have growing importance in industry; they are used for food production, fabric treatment, and drug therapies (discussed in Chapter 16).

Enzymes couple specific energy-yielding reactions (such as those of glucose breakdown) with the cell's reactions requiring energy, such as making ATP. An example of coupled reactions is shown in **Figure 13.9**. The substrate **phosphoenolpyruvate (PEP)**, a phosphorylated breakdown product of glucose, is converted to pyruvic acid while the phosphate is added to ADP to generate ATP:

Phosphoenolpyruvate

Pyruvate

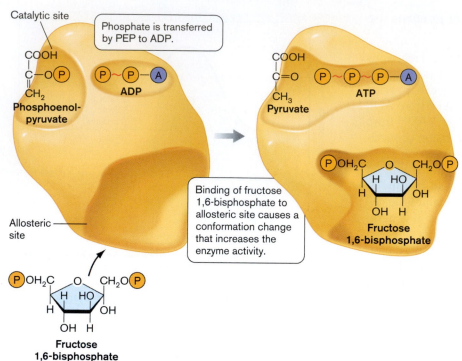

Catalytic site

Phosphate is transferred by PEP to ADP.

COOH
|
C—O—(P)
|
CH₂

**Phosphoenol-
pyruvate**

(P)~(P)—(A)
ADP

Allosteric
site

Binding of fructose 1,6-bisphosphate to allosteric site causes a conformation change that increases the enzyme activity.

(P)OH₂C   O   CH₂O(P)
H   HO
H   OH
OH H

**Fructose
1,6-bisphosphate**

COOH
|
C=O
|
CH₃

Pyruvate

(P)~(P)~(P)—(A)
ATP

(P)OH₂C   O   CH₂O(P)
H   HO
H   OH
OH H

**Fructose
1,6-bisphosphate**

Figure 13.9 The enzyme pyruvate kinase. Pyruvate kinase catalyzes the transfer of a phosphoryl group from PEP to ADP, generating pyruvate and ATP. The enzyme possesses separate binding sites for substrates and allosteric effectors.

The $\Delta G°'$ of phosphate cleavage from PEP is −62 kJ/mol, whereas the $\Delta G°'$ of ATP formation is only 31 kJ/mol. Thus, the net $\Delta G°'$ is negative (−31 kJ/mol), and the reaction goes forward to pyruvate. But a high activation energy makes the reaction extremely slow; it essentially never occurs on the timescale of life. The reaction occurs only when the two reacting substrates (PEP and ADP) are coupled by the enzyme **pyruvate kinase** (**Fig. 13.10**). Pyruvate kinase has specific binding sites for each of its substrates, PEP and ADP. The ADP and PEP are brought together by the enzyme and positioned so as to lower the activation energy of phosphate transfer.

In addition to the catalytic site, the enzyme has an **allosteric site** (a regulatory site distinct from the substrate-binding site) for its activator fructose 1,6-bisphosphate. The difference between a substrate-binding site and an allosteric site can be seen in the molecular model, based on X-ray crystallography (**Fig. 13.10**). The allosteric site is found at a distance from the substrate-binding site, but its interaction with the regulator, fructose 1,6-bisphosphate, alters the conformation of the entire enzyme, increasing the rate of reaction. As we shall see, fructose 1,6-bisphosphate is a central intermediate of glycolysis; thus, it makes sense that as this molecule builds up, it activates pyruvate kinase to remove products farther down the chain of catabolic reactions, maintaining the steady flow of metabolites.

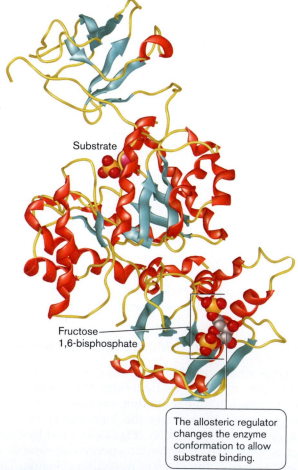

Substrate

Fructose
1,6-bisphosphate

The allosteric regulator changes the enzyme conformation to allow substrate binding.

Figure 13.10 Structure of pyruvate kinase. The molecular model of pyruvate kinase, based on a crystal structure of the enzyme bound to a substrate analog and an allosteric effector, fructose 1,6-bisphosphate. (PDB code: 1A3W)

Special Topic 13.1   Observing Energy Carriers in Living Cells

Energy carriers lead a dynamic existence, often with rapid turnover. How are energy carriers measured in living cells? One way is to use nuclear magnetic resonance (NMR) spectroscopy (**Fig. 1**). The use of NMR for real-time observation of live organisms was pioneered in the 1970s by Robert Shulman and colleagues at Yale University. Further advances led to magnetic resonance imaging (MRI), a technique in which the human body is scanned for NMR signals to build a three-dimensional image. MRI is now the method of choice for noninvasive imaging of patients with tumors, stroke, or other internal defects.

NMR spectroscopy gives detailed information about molecules containing nuclei of odd atomic number or mass number, which possess nuclear magnetic moment. Nuclei particularly useful in biochemistry include ^1H, ^{13}C, and ^{31}P. NMR detects nuclear spin transitions associated with the nuclear magnetic moment of nuclei suspended within the intense magnetic field of a superconducting magnet. The spin transition occurs as the nuclear magnetic moment of each atom interacts with electromagnetic radiation, producing a signal.

Within live cells, NMR can be used to observe concentrations of molecules with extremely high turnover, such as ATP and NADH (**Fig. 1**). These energy carriers are detected based on NMR signals from their phosphate groups. The naturally occurring phosphorus isotope ^{31}P has a nuclear spin, which generates the nuclear magnetic moment required for an NMR signal. The ^{31}P NMR spectrum of live *E. coli* cells reveals several peaks corresponding to cytoplasmic levels of nucleotide triphosphates (primarily ATP) and NADH. Each phosphorus signal position is "shifted" by shielding from electrons in molecular orbitals. The shift differs for each phosphorus in the molecule.

In **Figure 1**, the levels of phosphorylated energy carriers were used to assess the overall energy levels of *E. coli* cells exposed to tellurite, a toxic chemical. The ^{31}P NMR spectrum reveals three shift peaks corresponding to the three phosphates of ATP and the two of ADP. The middle phosphate of ATP (ATP β) generates a well-isolated peak indicating ATP concentration. As the overall energy level of the cell rises upon addition of glucose, the ATP β peak increases. Thus, the concentration of phosphorylated energy carriers serves as a measure of energy flux within the cell. Similar measurements for human tissues are used to study the biochemistry of athletic performance.

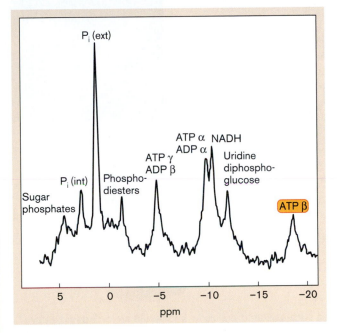

Figure 1   Phosphorylated energy carriers observed by ^{31}**P NMR.** The ^{31}P NMR spectrum reveals peaks corresponding to cytoplasmic levels of nucleotide triphosphates (primarily ATP) and NADH. The alpha-phosphate links directly to the nucleotide; the beta-phosphate lies between two phosphates; and the gamma-phosphate is terminal. The beta-phosphate of ATP (ATP β) generates a well-isolated peak indicating ATP concentration. *Source: Elke M. Lohmeier-Vogel, et al. 2004. Applied and Environmental Microbiology 70:7342.*

THOUGHT QUESTION 13.7   What would happen to the cell if pyruvate kinase catalyzed PEP conversion to pyruvate but failed to couple this reaction to ATP production?

Note that pyruvate kinase is actually named for its reverse reaction, the use of ATP to phosphorylate pyruvate to PEP. This may be because the activity of the purified enzyme was originally observed in the reverse direction. Enzymes can catalyze both forward and reverse reactions; the predominating direction depends on the concentrations of substrate and products, which determine ΔG, and on allosteric regulators.

TO SUMMARIZE:

- **Catabolic pathways organize the breakdown of large molecules** in a series of sequential steps coupled to reactions that store energy in small carriers such as ATP and NADH.
- **ATP and other nucleotide triphosphates store energy** in the form of phosphodiester bonds.
- **NADH, NADPH, and FADH each store energy** associated with an electron pair that carries reducing power.
- **Enzymes catalyze reactions by lowering the ΔG** required to reach the transition state. They couple energy transfer reactions to specific reactions of biosynthesis and cell function.

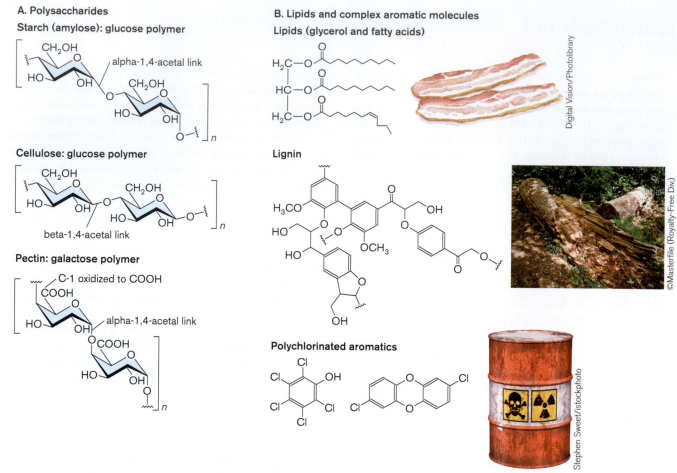

Figure 13.11 **Carbon sources for catabolism.** **A.** Polysaccharides such as starch, cellulose, and pectin are broken down by microbes to glucose, which undergoes glycolysis. **B.** Lipids and complex aromatic structures are broken down to acetate, which enters the TCA cycle. Halogenated aromatic pollutants are broken down by white rot fungi, a promising agent of bioremediation.

13.4 Catabolism: The Microbial Buffet

In the early twentieth century, it was thought that microbes could catabolize a limited subset of naturally occurring organic molecules, such as sugars. Molecules not known to be catabolized were termed "xenobiotics," especially if they were "synthetic" products of human industry. We have since found that virtually any organic molecule can be catabolized by a microbe that has evolved the appropriate enzymes. Catabolism also plays key roles in microbial disease; for example, the causative agent of acne, *Propionibacterium acnes*, degrades skin cell components such as sialic acids, matrix molecules, lipids, and pore-forming factors.

Note that besides releasing energy, the breakdown of organic food molecules provides substrates for biosynthesis. Biosynthesis is covered in Chapter 15.

Classes of Catabolism

Microbial pathways of catabolism are summarized in **Table 13.1**. Most catabolic pathways fall into one of three classes:

■ **Fermentation.** First identified by Louis Pasteur as "la vie sans air," or "life without air," fermentation is the partial breakdown of organic food without net transfer of electrons to an inorganic terminal electron acceptor; thus, it can take place in the absence of oxygen. Fermentation has a favorable ΔG owing to breakdown of a large molecule to several smaller products, which are usually more stable as well. Types of fermentation pathways include ethanolic fermentation, producing alcoholic beverages, and lactic acid fermentation by lactic acid bacteria, producing cheese and yogurt.

■ **Respiration.** Respiration combines catabolic breakdown of organic molecules with electron transfer to a terminal electron acceptor such as oxygen. Respiration yields far more energy from catabolism than does fermentation. For humans, respiration is synonymous with breathing; but in the absence of O_2, many microbes use alternative electron acceptors, such as nitrate or sulfate. Thus, microbes—unlike humans and other animals—have the capacity for anaerobic respiration.

■ **Photoheterotrophy.** Some microbes conduct catabolism with a "boost" from light absorption. In photolytic catabolism, or photoheterotrophy, light absorption by chlorophyll drives the photolysis of an organic molecule such as succinate. Photoheterotrophy is common in marine and freshwater bacteria (discussed in Chapter 14).

Diverse Substrates for Catabolism

Microbes catabolize many different kinds of substrates, or catabolites (**Fig. 13.11**). While in principle virtually any organic constituent may be catabolized, certain kinds of substrates are widely used. Many of these substrates are of economic and environmental significance and generate products important to our nutrition and technology.

Carbohydrate catabolism. Historically, the catabolism of carbohydrates (sugars and sugar polymers) has been important because of its relevance to human digestion and to the microbial production of food and drink (presented in Chapter 16). Glucose as such is rarely available to microbes, except to those pathogens growing within a host. But the pathways of glucose catabolism provide entry points for a diverse range of food molecules found in the environment, such as sugar chains, known as **polysaccharides**. For example, the cellulose of plant cell walls, the starch of potatoes, the pectin of fruit—all contain sugar chains that are broken down by microbial enzymes, first to short chains (oligosaccharides), then to two-sugar units (disaccharides), and then to monosaccharides such as glucose or fructose.

Even tiny differences between sugars can mean the difference between growth or starvation for a microbe. **Figure 13.12B** shows ^{13}C NMR spectra of two monosaccharides, glucose and galactose, that differ in the position of the hydroxyl group on a single carbon, C-4. This small difference is important because different microbial

Figure 13.12 Different sugars distinguished by NMR. A. Mo Hunsen (left), chemist at Kenyon College, demonstrates the use of the Bruker 300-MHz NMR. A sample is inserted into a superconducting magnet. NMR spectroscopy gives detailed information about molecules containing nuclei of odd atomic number or mass number, which possess nuclear magnetic moment. The nuclear magnetic moment of each atom interacts with electromagnetic radiation, producing a signal. **B.** ^{13}C NMR spectra distinguish two highly similar sugars, glucose and galactose, which are carbon sources for microbial catabolism. Within each sugar, the signal of each ^{13}C nucleus has a "shift" in position based on its shielding by electrons in the molecule. Because glucose and galactose each exist as an equilibrium mixture of two different isomers, each carbon is associated with two peaks with different shift positions.

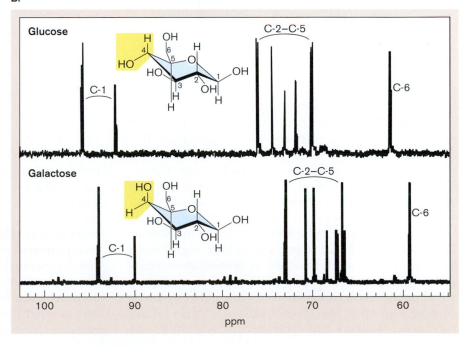

A.

Courtesy of Mo Hunsen

B.

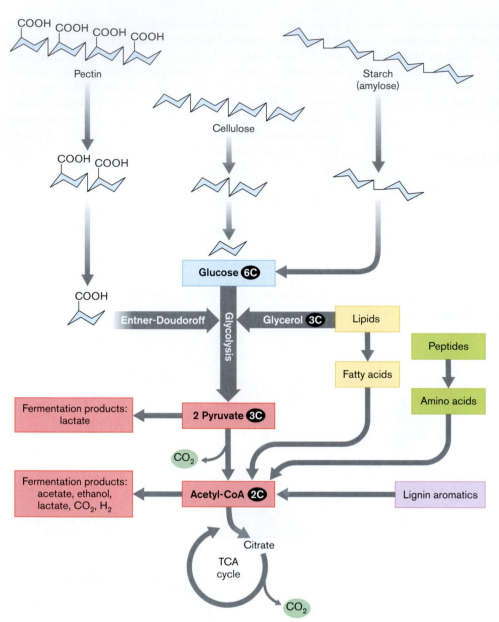

Figure 13.13 Many carbon sources enter central pathways of catabolism. Carbohydrates are broken down by specific enzymes to disaccharides, then to monosaccharides such as glucose. Glucose and sugar acids are converted to pyruvate, which yields acetyl groups. Acetyl groups or acetate are also the breakdown products of fatty acids, amino acids, and complex aromatic plant materials such as lignin.

enzymes (encoded by different genes) have evolved specifically to catabolize different sugars.

Polysaccharides are hydrolyzed to products that enter central catabolic pathways such as glycolysis (**Fig. 13.13**). In some sugars, the aldehyde is replaced by a hydroxyl (sorbitol, mannitol) or a carboxylate (gluconate, glucuronate). These sugar derivatives may be taken up by specific transporters and converted to glucose by specific enzymes. Alternatively, the sugar acids (carboxylate sugars) can be broken down by the Entner-Doudoroff pathway, an important variant of glycolysis. Pyruvate is commonly reduced to fermentation products such as lactate or ethanol. In the presence of a terminal electron acceptor such as oxygen, pyruvate yields an acetyl group to the tricarboxylic acid cycle (TCA), also known as the Krebs cycle or citric acid cycle. The

TCA cycle transfers electrons to the electron transport system, and ultimately the terminal electron acceptor. Glycolysis and the TCA cycle are presented in Sections 13.5 and 13.6.

Different species differ profoundly in their abilities to digest particular polysaccharides. The starches (glucans) and pectins (pectic acids) are the most widely digested. Cellulose forms particularly stable fibers that are digestible by some bacteria and fungi, but not by animals. For this reason, some animals, such as cattle and termites, require bacteria such as *Ruminococcus* and *Succinomonas* living within their digestive tracts to ferment cellulose from grasses or wood. Even humans derive about a tenth of our caloric intake from catabolism of plant fibers by bacteria such as *Bacteroides thetaiomicron*. Microbiologists consider the human gut flora a functional part of the human body, and their genomes are functionally part of the human "metagenome." Medical conditions such as obesity may depend in part on the catabolic activities of gut flora.

In a complex environment such as the intestinal lumen, with numerous potential substrates for catabolism, organisms select preferred substrates based on availability and energy efficiency. Selection occurs by regulating gene expression, as discussed in Chapter 10. For example, in *E. coli*, the sugar lactose induces transcription of genes that encode beta-galactosidase (*lacZ*) and lactose permease (*lacY*). In the presence of glucose, however, a preferred carbon source, *lac* transcription is halted. Halting *lac* transcription enables preferential catabolism of glucose. The process of prioritized consumption of substrates is known as **catabolite repression**.

Catabolism of lipids and amino acids. Many bacteria respire lipids from sources such as milk, animal fats, and nuts; this causes the rancid odor of partly oxidized lipids in meat or butter. Lipids are catabolized by hydrolysis to glycerol and fatty acids (see **Fig. 13.13**). Glycerol, a three-carbon triol, can enter catabolism either as a three-carbon intermediate of glycolysis or through breakdown to acetate, entering the TCA cycle. Fatty acids undergo oxidative breakdown to acetyl groups.

Amino acids are the building blocks of proteins, but when present in excess of the cell's needs, they are catabolized to provide energy. Specific enzymes catalyze the early steps in the degradation of each amino acid until products are formed that can enter common pathways of carbohydrate catabolism or the TCA cycle. The initial step of amino acid degradation is one of two kinds: decarboxylation (removal of CO_2) to produce an amine (**Fig. 13.14**); or deamination (removal of NH_3) to produce a carboxylic acid. Carboxylic acids are degraded through the TCA cycle. Amine products, such as cadaverine and putrescine, are often excreted, causing the noxious odor of decomposing flesh. In general, microbes favor decarboxylation during growth at low pH, because the amine products buffer the cell against acidity. Growth at higher pH favors deamination, because the acidic products buffer the cell against alkalinization.

The initial steps of amino acid catabolism are regulated genetically by multiple factors, including substrate availability and environmental conditions such as pH. In *E. coli*, low pH induces expression of several amino acid decarboxylases that release amines raising pH. An example is the *cadBA* operon encoding lysine permease and lysine decarboxylase (see **Fig. 13.14**). Expression of *cadBA* is induced by lysine binding to the membrane-bound regulator

LysP; by acid, which changes the conformation of regulator CadC; and by low oxygen, which increases transcription by unknown means.

Other reactions of amino acids generate small amounts of products with intense odors and flavors. Products of amino acid catabolism confer some of the distinctive flavors of fermented foods and beverages. For example, catabolites of aspartate and methionine generate the flavors of cheese (**Special Topic 13.2**). Microbial food production is discussed in Chapter 16.

Aromatic catabolism. Aromatic compounds are more difficult to digest than sugars because of the exceptional stability of aromatic ring structures. Yet many bacteria metabolize benzene derivatives, and even polycyclic aromatic molecules, either partly or all the way to CO_2. A particularly important aromatic substance found in nature is **lignin** (see **Fig. 13.11B**), which forms the key structural support of trees and woody stems. A discouragingly complex molecule, lignin is made from sugars oxidized to benzene rings, with ether connections that are difficult for enzymes to break down. Certain fungi and soil bacteria have evolved to digest lignin, ultimately directing its carbon into the TCA cycle.

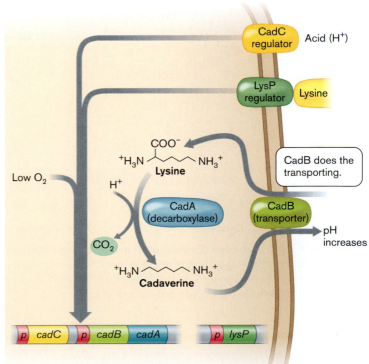

Figure 13.14   Lysine catabolism responds to the environment. Transcription of the operon encoding lysine transporter (CadB) and lysine decarboxylase (CadA) is induced by acid, lysine, and low oxygen concentration.

Special Topic 13.2   Swiss Cheese: A Product of Bacterial Catabolism

Swiss cheese is famous for its large, round holes, known as "eyes," as well as for its distinctive flavor (**Fig. 1A**). Both the eyes and the flavor derive from common pathways of bacterial fermentation that occur during cheese production. The original Swiss cheese, or Emmentaler, was made in the Emmen valley from cows pastured in the mountains. Today, related "Swiss-type" cheeses such as Jarlsburg are produced in large quantities throughout Europe and America.

Emmentaler cheese production involves two fermentation stages. The first stage, at high temperature, is dominated

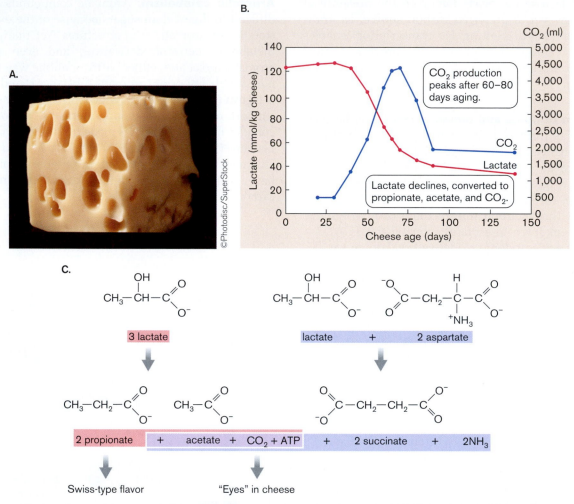

Figure 1   Swiss cheese (Emmentaler) involves fermentation of lactate to propionate.   A. Swiss cheese (Emmentaler). **B.** As a Swiss cheese ages, lactate declines and CO_2 rises, gradually forming the "eyes" (holes). **C.** Lactate is fermented by *Propionibacterium freudenreichii* to propionate, acetate, and CO_2. Concurrent fermentation of lactate and aspartate generates additional CO_2, increasing the size and number of eyes.

Today, the environment contains increasing amounts of aromatics, including growing amounts of halogenated aromatics, such as polychlorinated biphenyls. These compounds, produced for herbicides and other industrial uses, are the source of highly toxic dioxins. These substances concentrate in the food chain, reaching levels potentially dangerous to human consumers. Halogenated aromatics turn out to be catabolized by a number of soil microbes, which offer promising candidates for bioremediation of polluted environments (described in Section 13.7).

by thermophilic lactic acid bacteria such as *Lactobacillus helveticus* and *Streptococcus salivarius*. Lactic acid bacteria ferment the milk sugar lactose, a disaccharide of glucose and galactose, generating mainly lactic acid. The high temperature ensures rapid buildup of acid and retards growth of harmful species from the unpasteurized milk.

The second fermentation stage is unique to Swiss-type cheeses. The fermented milk curd is cooled and inoculated with *Propionibacterium freudenreichii*, a species that converts lactate to propionate, acetate, and CO_2 (**Figs. 1B** and **C**). This propionic fermentation generates another ATP, in addition to the ATPs of glycolysis. Propionic fermentation proceeds at cool temperatures, very slowly, over many months. During this period as the cheese is "aged," lactate declines while CO_2 increases. The gradual CO_2 production forms the eyes, or holes, for which Swiss cheese is known. The propionate contributes to the distinctive flavor.

The optimal production of eyes and flavor depends on several side pathways of fermentation of amino acids. Fermentation of aspartate plus lactate generates additional CO_2, thus increasing the size of the eyes, and increases the overall efficiency of lactate fermentation. Thus, addition of aspartate as well as use of a strongly aspartate-fermenting bacterial strain can enhance the quality of the cheese.

The full flavor of a cheese results from a large number of minor fermentation pathways that produce trace amounts of aldehydes, esters, alcohols, and ketones. One substance that contributes to the Swiss-type flavor is methional, a derivative of methionine (**Fig. 2**). During the initial fermentation by lactic acid bacteria, small amounts of methionine are deaminated and decarboxylated to methional, a strong odorant. An alternative side reaction generates methanethiol, an odorant more characteristic of cheddar cheese. The basis for favoring one pathway or another remains unknown, but these side reactions are responsible for generating many distinctive varieties of cheeses. Microbial food production is discussed further in Chapter 16.

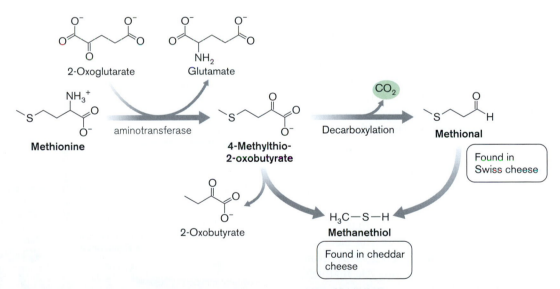

Figure 2   Methionine is fermented to flavor molecules.   Methionine undergoes deamination and oxidation, followed by decarboxylation to methional, an aldehyde odorant characteristic of Swiss cheese. Alternative reactions generate methanethiol, more typical of cheddar.

A given species of bacteria or fungi may catabolize thousands of different carbon sources, each requiring specific transporters and enzymes for initial breakdown. In most cases, these products ultimately funnel into one of a few common pathways of metabolism. The remainder of this chapter presents key pathways in detail: glycolysis and other pathways of glucose catabolism; the TCA cycle; and the catechol pathway of benzoate catabolism. These pathways play key roles in medical microbiology and in industrial fields such as bioremediation.

TO SUMMARIZE:

- **Fermentation, respiration, and photoheterotrophy are forms of catabolism.** In fermentation, the catabolite is broken down to smaller molecules without an inorganic electron acceptor. Respiration requires an inorganic terminal electron acceptor such as O_2 or nitrate. In photoheterotrophy, catabolism is supplemented by light absorption.
- **Carbohydrates or polysaccharides are broken down to disaccharides, then to monosaccharides.** Sugars and sugar derivatives, such as amines and acids, are catabolized to pyruvate.
- **Pyruvate and other intermediary products of sugar catabolism are fermented, or they are further catabolized to CO_2 and H_2O through the TCA cycle** (in the presence of a terminal electron acceptor).
- **Lipids and amino acids are catabolized to glycerol and acetate** as well as other metabolic intermediates.
- **Aromatic compounds such as lignin and benzoate derivatives** are catabolized to acetate through different pathways, such as the catechol pathway.

13.5 Glucose Breakdown and Fermentation

Glucose catabolism is important not only as a widespread source of energy, but also as a source of key substrates for biosynthesis such as five-carbon sugars to build nucleic acids (discussed in Chapter 15). Glucose and related sugars are catabolized through a series of phosphorylated sugar derivatives. A common theme in sugar catabolism is the split of a six-carbon substrate into two three-carbon products. The three-carbon products may yield two molecules of pyruvate:

$$C_6H_{12}O_6 \rightarrow 2C_3H_4O_3 + 4H \text{ (on NADH)}$$

Under anaerobiosis—the prevailing condition of many microbial habitats—the pyruvate obtained from sugar breakdown must be converted to forms that receive electrons from NADH, in order to restore the electron-accepting form NAD^+. Different microbes reduce pyruvate to different end products of fermentation. Alternatively, through anaerobic respiration, NADH may reduce an electron acceptor such as oxygen or nitrate, allowing pyruvate to feed into the TCA cycle (discussed in Section 13.6).

NOTE: The carboxylic acid intermediates of metabolism exist in equilibrium with their dissociated, or ionized, form, identified by the suffix *-ate*. For example, lactic acid dissociates to lactate; acetic acid dissociates to acetate. We use the "-ate" terms for acids whose ionized form predominates under typical cell conditions (around pH 7).

To catabolize glucose, bacteria and archaea use three main routes to pyruvate (**Fig. 13.15**):

- **Glycolysis** or the **Embden-Meyerhof-Parnas (EMP) pathway**, in which glucose 6-phosphate isomerizes to fructose 6-phosphate, ultimately yielding two molecules of pyruvate. EMP is used by many bacteria, eukaryotes, and archaea. From each glucose, the pathway generates net 2ATP and 2NADH.
- **Entner-Doudoroff (ED) pathway**, in which glucose 6-phosphate is oxidized to 6-phosphogluconate, a phosphorylated sugar acid. Alternatively, sugar acids may be converted directly to 6-phosphogluconate. Sugar acids often derive from intestinal mucus, and the ED pathway is essential for enteric bacteria to colonize the intestinal epithelium. The ED pathway generates only one ATP, one NADH, and one NADPH.
- **Pentose phosphate shunt (PPS)**, in which glucose 6-phosphate is oxidized to 6-phosphogluconate, then decarboxylated to a five-carbon sugar (pentose), ribulose 5-phosphate. The PPS produces sugars of three to seven carbons, which serve as precursors for biosynthesis or convert to pyruvate as needed. The PPS generates one ATP plus 2NADPH, the reducing cofactor most commonly associated with biosynthesis.

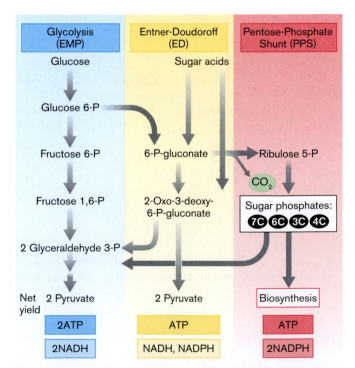

Figure 13.15   From glucose to pyruvate: three pathways. The Embden-Meyerhof-Parnas (EMP) pathway of glycolysis, the Entner-Doudoroff (ED) pathway, and the pentose-phosphate shunt (PPS) each catabolize carbohydrates by related but different routes.

The Embden-Meyerhof-Parnas Pathway Generates ATP and NADH

The Embden-Meyerhof-Parnas (EMP) pathway, or glycolysis, is the form of glucose catabolism most commonly studied in introductory biology; it is central for animals and plants as well as many bacteria. In the EMP pathway, one molecule of D-glucose undergoes stepwise breakdown to two molecules of pyruvic acid (or its anion, pyruvate) (**Fig. 13.16**). The breakdown of glucose takes place in two stages. In the first stage, the glucose molecule is primed for breakdown by two steps of sugar phosphorylation by ATP. Each ATP phosphotransfer step requires input of Gibbs energy. The phosphoryl groups tag two sides of the glucose for splitting into two three-carbon molecules of glyceraldehyde 3-phosphate (G3P). In the second stage, each glyceraldehyde 3-phosphate is oxidized by NAD^+ through steps leading to pyruvate. Each conversion of glyceraldehyde 3-phosphate to pyruvate forms two molecules of ATP, one from dephosphorylation of the substrate and one from the addition of inorganic phosphate. The net gain of energy carriers is two molecules of NADH plus two molecules of ATP.

Most of the conversion steps are associated with a small change in energy, so small that the sign of ΔG depends on the concentrations of substrates or products; thus, some individual steps are reversible. In the cytoplasm, however, as intermediate products form, they are quickly consumed by the next step, and so the pathway flows in one direction. The direction of flow is determined by the key irreversible reactions that consume ATP. These steps "prime" the pathway by spending energy.

Phosphorylation and splitting of glucose. In the first stage of the EMP pathway, the six-carbon sugar is activated by two phosphorylation steps (**Fig. 13.17**). The first phosphoryl group (phosphate) is added at carbon 6 of glucose. (In some species of bacteria, the first phosphoryl group is added by phosphoenolpyruvate instead of ATP, but the net effect is the same.) The next enzyme-catalyzed step, rearrangement of glucose 6-phosphate to fructose 6-phosphate, involves no significant change in energy, but prepares the sugar to receive the second phosphoryl group, yielding fructose 1,6-bisphosphate.

The sugar then splits into two three-carbon sugars (trioses), each tagged with one of the two phosphates. The splitting of this molecule has a favorable entropy change (ΔS), but its chemical change (ΔH) is unfavorable, largely canceling out the energy yield. The two triose phosphates, glyceraldehyde 3-phosphate and dihydroxyacetone phosphate, have nearly the same energy value, so an enzyme interconverts them reversibly. Interconversion is necessary because only glyceraldehyde 3-phosphate proceeds further in the pathway.

> **THOUGHT QUESTION 13.8** Some bacteria make an enzyme, dihydroxyacetone kinase, that phosphorylates dihydroxyacetone to dihydroxyacetone phosphate. Why would this enzyme be useful?

> **NOTE:** In Chapters 13–16, every substrate conversion shown requires catalysis by an enzyme. For the EMP pathway, the enzymes are shown, but for other pathways, the enzyme names are omitted.

Figure 13.16 Energy changes during the Embden-Meyerhof-Parnas pathway of glycolysis. Glucose is activated through two substrate phosphorylations by ATP. The breakdown of glucose to two molecules of pyruvate is coupled to net production of two ATP and two NADH.

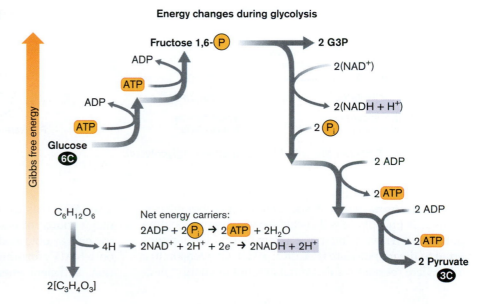

Energy changes during glycolysis

Net energy carriers:
$$2ADP + 2P_i \rightarrow 2ATP + 2H_2O$$
$$2NAD^+ + 2H^+ + 2e^- \rightarrow 2NADH + 2H^+$$

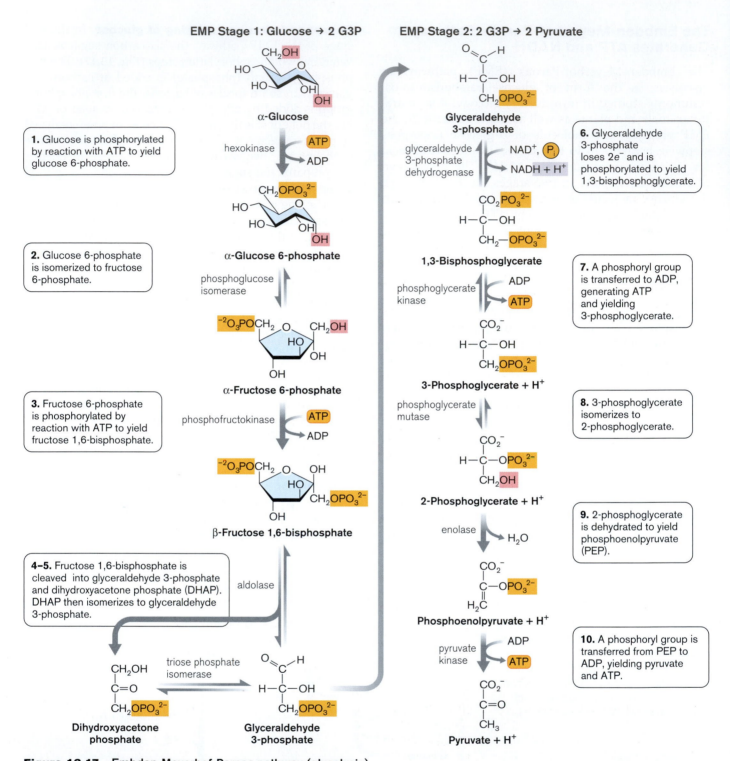

Figure 13.17   Embden-Meyerhof-Parnas pathway (glycolysis).

ATP generation. In glycolysis, each three-carbon glyceraldehyde 3-phosphate is directed into an energy-yielding pathway to pyruvate. First, the aldehyde (R-CHO) is converted to the carboxylate (R-COO⁻ + H⁺). Conversion to a carboxylate releases a substantial amount of energy (neg-

ative ΔG). This oxidation of the aldehyde represents the major source of energy in glycolysis—the step at which the energy obtained is used to transfer a pair of electrons onto NAD⁺, forming NADH with an ionized H⁺. In addition, sufficient energy is released to add a phosphoryl group (from inorganic phosphate, P_i) to the carboxylate, generating 1,3-bisphosphoglycerate.

In subsequent steps, the added phosphoryl group will be transferred to ADP, yielding the one net ATP generated per pyruvate (two per glucose). The transfer of a phosphoryl group from an organic substrate to make ATP is called **substrate-level phosphorylation**. With subtraction of the initial two ATP molecules invested, the net energy carriers gained from each glucose are as follows:

$$2NAD^+ \rightarrow 2NADH + 2H^+$$

$$2ADP + P_i \rightarrow 2ATP + 2H_2O$$

> **THOUGHT QUESTION 13.9** In the EMP pathway, how are the two water molecules generated?

Regulation of glycolysis. Enzymes of catabolism are regulated at the level of transcription of the enzyme. In addition, the activities of certain enzymes in long pathways require allosteric regulation. Allosteric regulation by enzyme substrates and products ensures that excess intermediates do not build up and avoids releasing more energy than the cell can use at a given time. Glycolysis is regulated so that its reactions go forward only when the cell needs energy, not when the cell is trying to synthesize glucose. The regulation occurs at steps where the products are consumed so rapidly that their forward reaction is effectively irreversible. Irreversible steps are shown as unidirectional arrows in **Figure 13.17**.

Enzymes that catalyze irreversible steps are regulated so as to maintain consistent levels of intermediates in the pathway. The most important irreversible reaction in glycolysis is the phosphorylation of fructose 6-phosphate to fructose 1,6-bisphosphate, mediated by the enzyme phosphofructokinase. This enzyme is activated allosterically by ADP and inhibited by ATP or by the alternative phosphoryl donor, phosphoenolpyruvate.

What happens when the cell needs to reverse glycolysis in order to make glucose? Most of the intermediate reactions are reversible, and so the same enzymes can be used for biosynthesis. Pathways that participate in both catabolism (breakdown) and anabolism (biosynthesis) are called **amphibolic**.

An amphibolic pathway such as glycolysis includes enzymes such as phosphoglucose isomerase that can run in either direction, but also requires key enzymes that operate only in the catabolic direction or in the anabolic direction. For example, in glycolysis the ATP phosphorylation of fructose 6-phosphate is catalyzed by the enzyme phosphofructokinase, whereas in biosynthesis this step is reversed by a different enzyme, fructose bisphosphatase. Instead of regenerating ATP, fructose bisphosphatase removes the second phosphoryl group as inorganic phosphate—a step releasing energy and thus driving the whole pathway in reverse (toward biosynthesis of sugar). The two enzymes are regulated differently; the catabolic enzyme phosphofructokinase is activated by ADP, a signal of energy need, whereas the biosynthetic enzyme is inhibited by such signals.

The Entner-Doudoroff Pathway Catabolizes Sugar Acids

The Entner-Doudoroff (ED) pathway offers a slightly different route to catabolize sugars as well as sugar acids (sugars with acidic side chains). The ED pathway was originally studied for its role in production of the Mexican beverage *pulque*, or "cactus beer," by *Zymomonas* fermentation of the blue agave plant. More recently, Tyrrell Conway and colleagues have found genes encoding the Entner-Doudoroff enzymes in the genomes of numerous bacteria and archaea. In the human colon, the ED pathway enables *E. coli* and other enteric bacteria to feed on mucus secreted by the intestinal epithelium (**Fig. 13.18**). Some gut flora, such as *Bacteroides thetaiotaomicron*, actually induce colonic production of the mucus that they consume. These bacteria that "farm" intestinal mucus may enhance human health by preventing colonization by pathogens.

The ED pathway probably evolved earlier than the EMP pathway, as it involves fewer substrate phosphorylation steps and produces less ATP, and it is found in a wider range of prokaryotes. As in the EMP pathway, glucose is phosphorylated to glucose 6-phosphate (**Fig. 13.19**). The next step, however, involves oxidation by NAD^+ at carbon 1, with loss of two hydrogens to form 6-phosphogluconate. Gluconate is a sugar acid found in intestinal mucus; it can be phosphorylated to enter the ED pathway.

The hydrogens and electrons from glucose 6-phosphate are transferred to $NADP^+$, instead of the NAD^+ as in the EMP pathway. This step differs from the EMP pathway in two respects: The carrier used is $NADP^+$ instead of NAD^+, and the electron transfer occurs early,

A. **B.**

10 μm

Paul Cohen, U. Rhode Island and Tyrrell Conway, U. Oklahoma

Courtesy of Tyrrell Conway

Figure 13.18 Intestinal bacteria use the Entner-Doudoroff pathway. A. Intestinal *E. coli* (orange) primarily feed on gluconate from mucus secretions (fluorescence micrograph). **B.** Tyrrell Conway, at the University of Oklahoma, used genomics and genetic analysis to dissect the role of the Entner-Doudoroff pathway in the enteric bacterial catabolism of sugar acids from intestinal mucus.

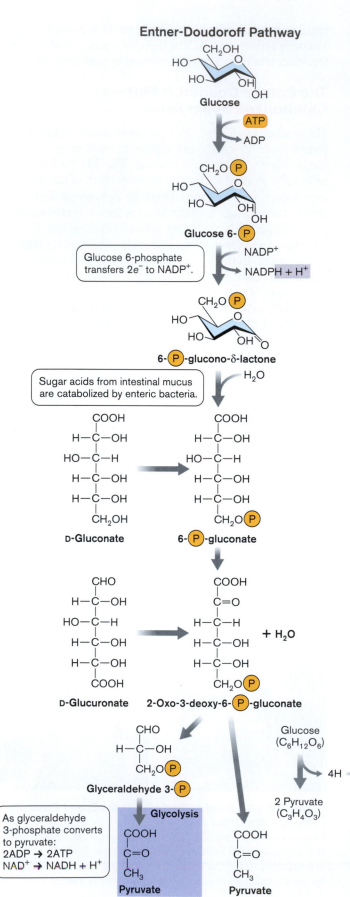

Entner-Doudoroff Pathway

Glucose 6-phosphate transfers 2e⁻ to NADP⁺.

Sugar acids from intestinal mucus are catabolized by enteric bacteria.

As glyceraldehyde 3-phosphate converts to pyruvate:
2ADP → 2ATP
NAD⁺ → NADH + H⁺

without a second ATP-consuming phosphorylation step. When the six-carbon substrate is eventually split into two three-carbon products, one of the three-carbon products is glyceraldehyde 3-phosphate, which enters the second stage of glycolysis. NADH is made, and two ATP are made by substrate-level phosphorylation. The remaining three-carbon product, however, is pyruvate. This one-step production of pyruvate short-circuits the catabolic pathway, missing the formation of an ATP. The unused potential energy is released as waste heat.

The net ATP gain from the Entner-Doudoroff pathway is only one ATP per molecule of glucose, half that of the EMP pathway (see page 483 and **Fig. 13.15**). The electrons transferred, however, are equivalent: Instead of two molecules of NADH, the Entner-Doudoroff pathway generates one NADH and one NADPH.

THOUGHT QUESTION 13.10   Explain why the ED pathway generates only one ATP, whereas the EMP pathway generates two.

The Pentose Phosphate Shunt Yields NADPH and Substrates for Biosynthesis

A third pathway of glucose catabolism is the pentose phosphate shunt (PPS), which forms the key intermediate **ribulose 5-phosphate**, a five-carbon sugar. The pentose phosphate shunt generates one ATP with no NADH, but two NADPH for biosynthesis (**Fig. 13.20**). In addition, PPS generates a more complex series of intermediates than the EMP or ED pathway, which can be redirected as substrates for biosynthesis of diverse cell components such as amino acids and vitamins.

The pentose phosphate shunt starts like the Entner-Doudoroff pathway: Glucose 6-phosphate gives up two electrons to form NADPH and is oxidized to 6-phosphogluconate. Thus, only one net ATP is gained from breakdown to pyruvate. The next step involves a second oxidation to NADPH, with loss of a carbon as CO_2. The loss of CO_2 generates the five-carbon sugar ribulose 5-phosphate (hence the pathway name "pentose phosphate"). In succeeding steps, pairs of sugars, such as

Net energy carriers:
$$ADP + P_i \rightarrow ATP + H_2O$$
$$NADP^+ + 2H^+ + 2e^- \rightarrow NADPH + H^+$$
$$NAD^+ + 2H^+ + 2e^- \rightarrow NADH + H^+$$

Figure 13.19   Entner-Doudoroff pathway. Glucose 6-phosphate is oxidized to 6-phosphogluconate, with one pair of electrons transferred to NADPH. The 6-phosphogluconate is dehydrated and cleaved to form one pyruvate plus one glyceraldehyde 3-phosphate that enters the EMP pathway to pyruvate. Phosphoryl groups are shown as (P).

sedohepulose 7-phosphate and glyceraldehyde 3-phosphate, exchange short carbon chains, giving rise to sugar phosphates of various lengths—for example, ribose 5-phosphate and erythrose 4-phosphate, which are precursors of purines and aromatic amino acids, respectively. Alternatively, if these routes to biosynthesis are not taken, the intermediates convert to fructose 6-phosphate and reenter the EMP pathway, where ATP and NADH are produced.

Fermentation Completes Catabolism and Yields Useful Products

None of the pathways from glucose to pyruvate constitutes a completed pathway of catabolism, because NADH and NADPH remain to be recycled. In the absence of oxygen or other electron acceptors, heterotrophic cells must transfer the hydrogens from NADH + H$^+$ back onto the

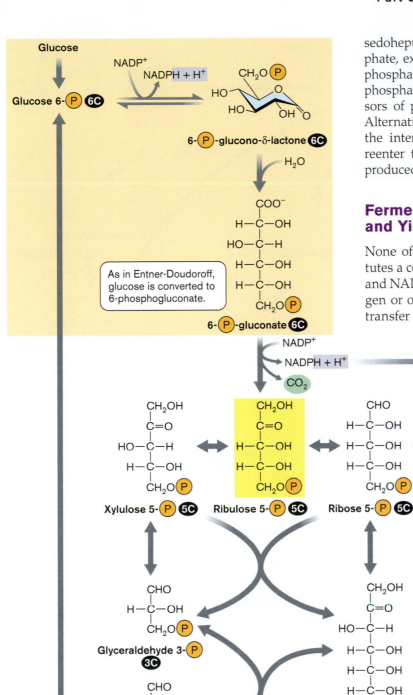

Figure 13.20 Pentose phosphate shunt pathway of glucose catabolism. Like the Entner-Doudoroff pathway, the pentose phosphate shunt forms 6-phosphogluconate. One CO_2 is released, and one NADPH is produced for biosynthesis. The pathway can generate ribose 5-phosphate for purine synthesis or erythrose 4-phosphate to synthesize aromatic amino acids.

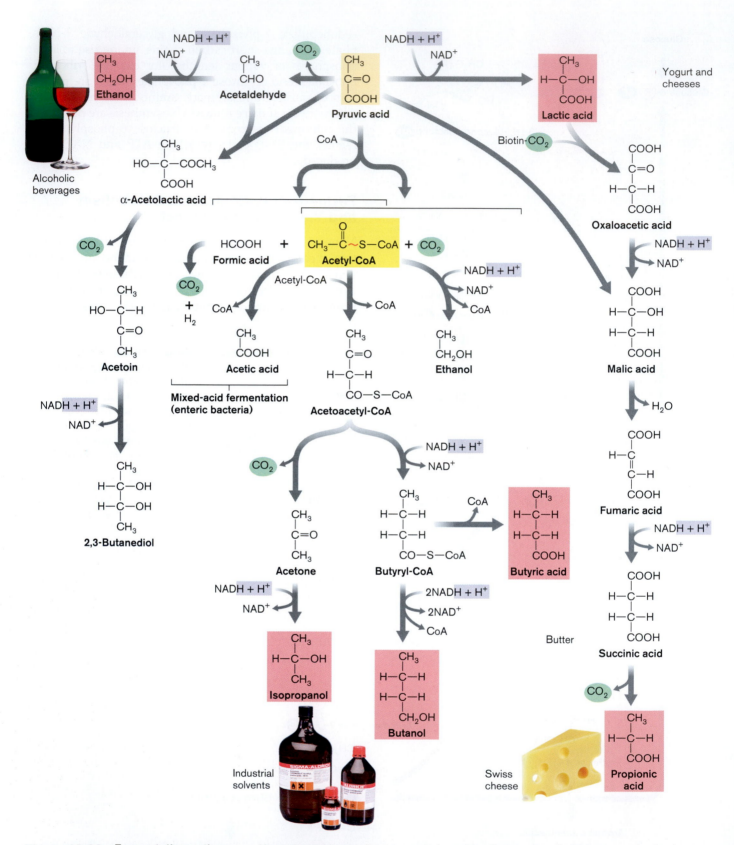

Figure 13.21   Fermentation pathways.   Alternative pathways from pyruvate to end products, many of which we use for food or industry. Different species conduct different portions of the pathways shown. *Sources:* Wine: ©Food Collection/SuperStock; Yogurt: John E. Kelly/ Jupiterimages; Butter: D. Hurst/Alamy; Solvents: Sigma-Aldrich Corporation; Cheese: ©Lee Hacker/Alamy.

products of pyruvate, forming partly oxidized fermentation products with the same redox level (balance of O and H) as the original glucose. For example, glucose may be fermented to two molecules of lactic acid (**lactate fermentation**) or two molecules of ethanol plus two CO_2 (**ethanolic fermentation**) or one lactic acid, one ethanol, and one CO_2 (**heterolactic fermentation**). These fermentation products are then excreted from the cell.

Most fermentation pathways do not generate ATP beyond that produced by substrate-level phosphorylation, the direct transfer of a phosphate group from an organic phosphate to ADP. An example is the final step of glycolysis, catalyzed by pyruvate kinase (**Fig. 13.17**). Much of the energy available from glucose remains unspent or is lost as heat. Nevertheless, fermentation is essential for microbes in environments such as anaerobic soil or animal digestive tracts. Even aerated cultures of bacteria start fermenting once their demand for oxygen exceeds the rate of oxygen dissolving in water. Microbes compensate for the low efficiency of fermentation by consuming large quantities of substrate.

Fermentation pathways. The large quantities of substrate consumed in fermentation generate large amounts of waste products to be excreted. Numerous different pathways have evolved to dispose of the waste in different forms (**Fig. 13.21**). *E. coli* ferments mainly by the mixed-acid pathway, yielding acetate, formate, lactate, and succinate, as well as ethanol, H_2, and CO_2. The mix of products varies with pH; low pH favors ethanol and lactate (which minimize acidification) in preference to formate and acetate. *Clostridium* species produce alcohols (butanol, isopropanol), whereas *Porphyromonas gingivalis*, a cause of periodontal disease, produces short-chain acids (propionate, butyrate).

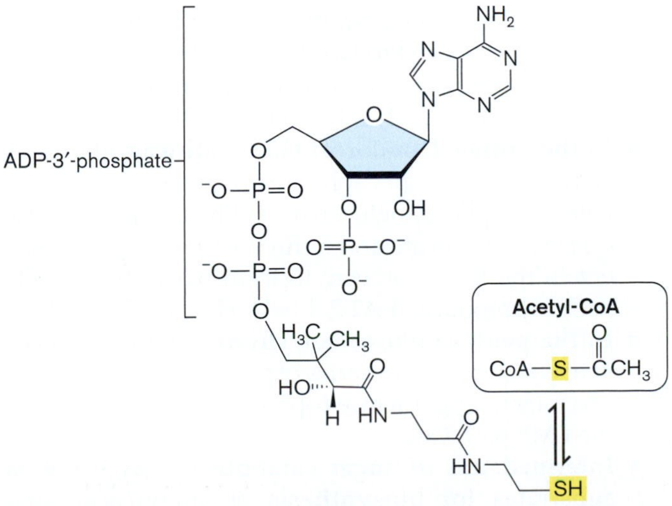

Figure 13.22   Structure of coenzyme A.   The thiol (SH) forms an ester link with the COOH of acetic acid, generating acetyl-CoA.

Many fermentation products share key intermediates, such as **acetyl-CoA**. Acetyl-CoA is a versatile two-carbon intermediate, the "Lego block" of metabolism. It consists of an acetyl group esterified to **coenzyme A** (**CoA**; Fig. 13.22), a famous coenzyme whose discovery won Fritz Lipmann the 1953 Nobel Prize in Medicine with Hans Krebs. CoA has a thiol (–SH) that exchanges its hydrogen for an acyl group, thus activating the molecule for transfer in various metabolic pathways.

> **THOUGHT QUESTION 13.11**   How does the structure of coenzyme A resemble that of NADH? How does it differ?

In the mixed-acid fermentation pathway, typical of *E. coli* and *Salmonella*, the enzyme pyruvate formate lyase splits pyruvate to form acetyl-CoA plus formate:

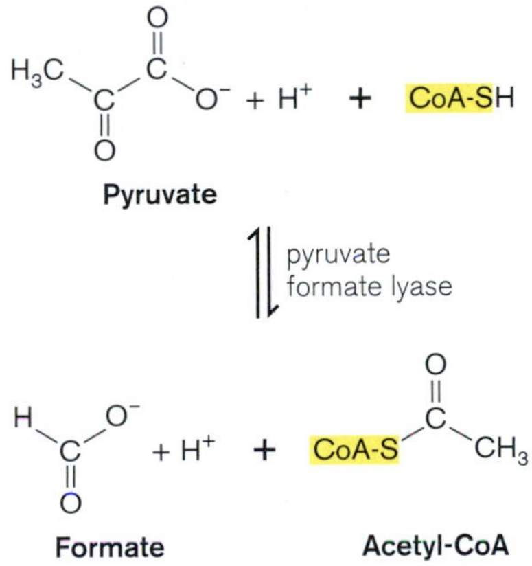

The SH group of CoA is indicated to emphasize its role in accepting the acetyl group to form acetyl-CoA. The hydrogen ($H^+ + e^-$) of the SH group is transferred onto the carboxyl carbon of pyruvate, generating formate.

Acetyl-CoA can be converted to different fermentation products by several different pathways. The simplest is water exchange with CoA, yielding acetate:

$$CH_3CO-S-CoA + H_2O \rightleftharpoons$$
$$CH_3COO^- + H^+ + HS-CoA$$

Incorporation of water restores the hydrogen to the thiol of CoA and the OH to acetate. The acetate thus formed may then be excreted by the cell.

Note that the excreted acids and alcohols are readily recovered by cells when an electron acceptor becomes available for oxidation or when these fermentation products are needed as building blocks for biosynthesis. Alternatively, the excreted "wastes" may be utilized by other species capable of fermenting them further.

Food and industrial applications. The "waste products" of fermentation retain much of their organic structure and food value. Thus, fermentation products have proved extremely useful in human culture and technology. For thousands of years, ethanolic fermentation by yeast has been used to produce wine and beer, while lactate fermentation has been used to produce yogurt and cheese. The minor product butyric acid (butyrate) lends taste to butter. Similarly, propionic acid (propionate) provides the distinctive flavor of Swiss cheese, while CO_2 gas generates the holes, or "eyes" (see **Special Topic 13.2**).

In the chemical industry, microbial fermentation produces industrial solvents such as butanol and acetone. Acetone production had historic impact during the World War I, when Britain needed a source of acetone to manufacture gunpowder. Acetone and butanol were produced as fermentation products by *Clostridium acetobutylicum*, a bacterium identified by the biochemist Chaim Weitzmann. Weitzmann was a Russian-born Jew who sought a Jewish homeland in Palestine. At the end of the war, Weitzmann's process for acetone production helped earn him the British government's support for the national homeland of Israel, where the biochemist later became the country's first president.

Today, microbial fermentation produces many "commodity chemicals" such as ethanol, butanol, and glycerol. Compared to industrial alternatives, such as production from petroleum, microbial culture is environmentally desirable and energy efficient. Furthermore, microbes are "smart" producers of small but complex pharmaceuticals such as vitamins, amino acids, and antibiotics, molecules that would require many steps of organic synthesis. Applications of microbial biosynthesis are discussed further in Chapters 15 and 16.

Diagnostic applications. Another important application of fermentation lies in diagnostic microbiology. To quickly identify the microbe causing a disease and prescribe an effective antibiotic, hospitals use rapid and inexpensive biochemical tests. A general test for fermentation pathways is the **phenol red broth test** (**Fig. 13.23A**). Phenol red is a pH indicator, which is orange-red at neutral pH. It turns yellow in medium acidified by fermentation acids (below pH 6.8) and red at higher pH (above pH 7.4). A culture of *Escherichia coli*, which ferments quickly, turns phenol red to a medium yellow after 24 hours. *Alcaligenes faecalis*, meanwhile, ferments poorly on sugar but converts peptides in the broth to alkaline amines; this culture turns deep red.

More specific tests depend on the microbe's ability to ferment specific sugars. Different species possess different enzymes able to convert different sugars and sugar derivatives to glucose, which then enters the common fermentation pathway. A well-known example is the use of sorbitol MacConkey agar to test for *E. coli* O157:H7, a lethal pathogen contaminating ground beef and cider. The pathogen lacks the ability to ferment sorbitol, so failure to ferment sorbitol indicates a high probability that the strain is *E. coli* O157:H7. On sorbitol MacConkey agar, bacteria that fer-

A. Phenol red test	B. Sorbitol MacConkey agar

Gas

Figure 13.23   Clinical tests based on fermentation.
A. Phenol red broth test. Organisms are cultured in a tube of broth containing phenol red. The dye turns yellow when protonated at low pH. Phenol red is orange-red at neutral pH, yellow at lower pH (acid), and red at higher pH (base). A small inverted tube (Durham tube) collects gaseous fermentation products (CO_2 and H_2). Left to right: *Escherichia coli* gives acidic fermentation products (yellow) and gas in Durham tube; *Alcaligenes faecalis* does not ferment, tube turns red without gas; uninoculated control, red. **B.** Sorbitol fermentation test for pathogen *E. coli* O157:H7. White colonies (strain O157:H7) fail to ferment sorbitol, unlike red colonies (nonpathogenic *E. coli*).

ment sorbitol during growth produce acids. The acidity causes a dye to turn red (the opposite of the phenol red test). Failure to ferment sorbitol (observed as pale colonies) indicates high probability of *E. coli* O157:H7 (**Fig. 13.23B**).

TO SUMMARIZE:

■ **Glucose catabolism** occurs by three related pathways of stepwise degradation of glucose to pyruvate. In the absence of oxygen, completion of catabolism requires generation of fermentation products.

■ **In the Embden-Meyerhof-Parnas (EMP) pathway**, glucose is activated by two substrate phosphorylations, then cleaved to two three-carbon sugars. Both sugars eventually are converted to pyruvate. The pathway produces 2 ATPs and 2 NADHs.

■ **In the Entner-Doudoroff (ED) pathway**, glucose is activated by one phosphorylation, then dehydrogenated to 6-phosphogluconate. 6-Phosphogluconate is cleaved to pyruvate and a three-carbon sugar, which enters the EMP pathway to form pyruvate. The ED pathway produces 1 ATP, 1 NADH, and 1 NADPH.

■ **In the pentose phosphate shunt**, 6-phosphogluconate is converted to a series of sugars of carbon length 3 to 7, including fructose 6-phosphate, which reenters the EMP pathway.

■ **Intermediates of sugar catabolism may serve as substrates for biosynthesis** of amino acids and other cell components.

■ **Glucose catabolism is reversible**, enabling cells to build glucose from small molecules. Glucose biosynthesis uses some enzymes of glycolysis in reverse, but also requires enzymes that bypass irreversible steps in the catabolic pathway.

■ **Fermentation is the completion of catabolism** *without* the electron transport system and a terminal electron acceptor. The electrons from NADH are restored to pyruvate or its products in reactions that generate fermentation products, including alcohols and carboxylates as well as H_2 and CO_2. Fermentation has applications in food, industrial, and diagnostic microbiology.

■ **Acetyl-CoA is a key intermediate** in several fermentation pathways.

13.6 The Tricarboxylic Acid (TCA) Cycle

In the presence of O_2 or another terminal electron acceptor, the products of sugar breakdown can be catabolized to CO_2 and H_2O through the **tricarboxylic acid (TCA) cycle**. The TCA cycle is also known as the Krebs cycle, named for Hans Krebs (1900–1981) who shared the 1953 Nobel Prize in Medicine with Fritz Lipmann. Krebs and his colleagues at Sheffield University, England, studied catabolism by observing the oxidizing activities of crude enzyme preparations from sources such as pigeon breast muscle, beef liver, and cucumber seeds. In all of these animal and plant tissues, the TCA cycle is conducted by mitochondria, using virtually the same process as their bacterial ancestors.

Glucose catabolism connects with the TCA cycle through pyruvate breakdown to acetyl-CoA and CO_2.

Recall from Section 13.4 that acetyl-CoA is also generated from many other catabolic pathways, including those for breakdown of fatty acids, amino acids, and even benzoate derivatives. Regardless of its source, the acetyl-CoA enters the TCA cycle by condensing with the four-carbon intermediate oxaloacetate to form citrate (**Fig. 13.24**). Citrate undergoes two steps of oxidative decarboxylation, in which CO_2 is released and two hydrogens with electrons are transferred to produce NADH + H^+ or $FADH_2$. The TCA cycle, in whole or in part, is found in all microbial species except for degenerately evolved pathogens dependent on host metabolism.

We present first the connecting step between pyruvate and acetyl-CoA, followed by the details of the TCA cycle. Different species of bacteria and archaea use at least ten known variations on the TCA cycle, conducted by diverse species under various environmental conditions. You will be relieved to hear that we present only one, the Krebs TCA pathway best known in gram-negative and gram-positive bacteria. Variant pathways are detailed in online resources such as the KEGG Pathway Database.

www | KEGG

Pyruvate Dehydrogenase Connects Sugar Catabolism to the TCA Cycle

Pyruvate is converted to acetyl-CoA through removal of CO_2 and transfer of $2e^-$ onto NAD^+. The removal of CO_2 and transfer of two electrons is known as oxidative decarboxylation. The oxidative decarboxylation of pyruvate,

Figure 13.24 Acetyl-CoA feeds into the TCA cycle. Pyruvate undergoes oxidative decarboxylation and incorporates CoA to form acetyl-CoA. Depending on the state of the cell, acetyl-CoA is converted to acetate for excretion or it enters the TCA cycle.

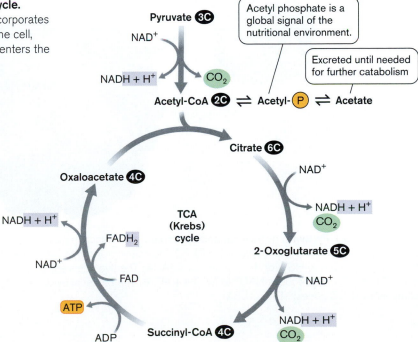

A. Pyruvate is converted to acetyl-CoA

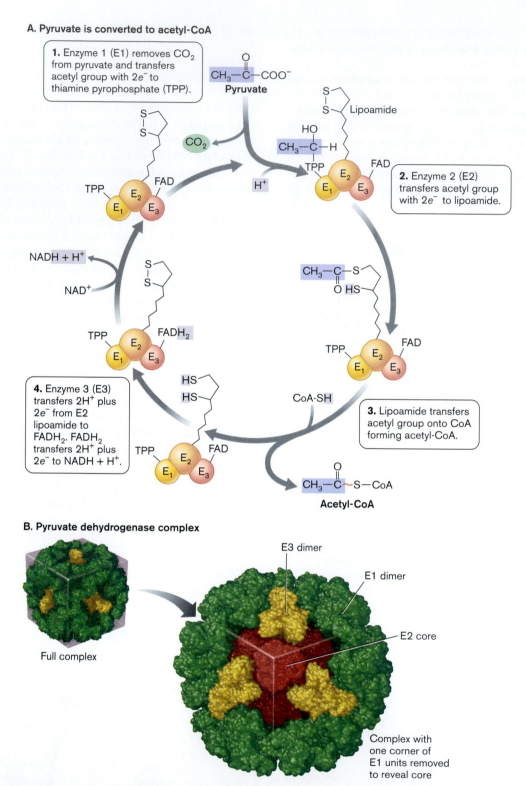

1. Enzyme 1 (E1) removes CO_2 from pyruvate and transfers acetyl group with $2e^-$ to thiamine pyrophosphate (TPP).

2. Enzyme 2 (E2) transfers acetyl group with $2e^-$ to lipoamide.

3. Lipoamide transfers acetyl group onto CoA forming acetyl-CoA.

4. Enzyme 3 (E3) transfers $2H^+$ plus $2e^-$ from E2 lipoamide to $FADH_2$. $FADH_2$ transfers $2H^+$ plus $2e^-$ to NADH + H^+.

B. Pyruvate dehydrogenase complex

Full complex

E3 dimer

E1 dimer

E2 core

Complex with one corner of E1 units removed to reveal core

Figure 13.25   Pyruvate dehydrogenase complex.   A. Reaction mechanism: (1) Enzyme 1 acquires the acetyl group of pyruvate, condensed to thiamine pyrophosphate (TPP). (2) The acetyl group with $2e^-$ is transferred to the lipoamide of enzyme 2. (3) The acetyl group is transferred to HS-CoA to form acetyl-CoA. (4) The two thiols are oxidized to disulfide, and their $2e^- + 2H^+$ reduce FAD to make $FADH_2$. $FADH_2$ reduces NAD^+ to NADH + H^+. **B.** The bacterial pyruvate dehydrogenase complex contains multiple copies of each component. (PDB codes: 1EAA, 1L8A)

coupled to CoA incorporation, is performed by an unusually large multisubunit enzyme called the **pyruvate dehydrogenase complex**, or **PDC** (**Fig. 13.25**). PDC is a key component of metabolism in bacteria and mitochondria, the first molecular player to direct sugar catabolism into respiration. In human mitochondria, defects in PDC are associated with myocardial malfunction and heart failure, maple syrup urine disease, and neurodegeneration.

Reaction mechanism of PDC. The overall reaction catalyzed by PDC is:

$$CH_3COCOO^- + H^+ + HS\text{-}CoA \rightarrow$$
$$CH_3CO\text{-}S\text{-}CoA + CO_2 + 2H^+ + 2e^-$$

$$NAD^+ + 2H^+ + 2e^- \rightarrow NADH + H^+$$

The removal of stable CO_2 yields energy for the electron transfer to NADH. The protons dissociated from pyruvate and thiol (SH) of CoA ($2H^+$ in total) are balanced by the net gain of protons by $NADH + H^+$.

The coupling of all the molecular transfers actually requires three distinct enzyme components within the complex, designated E1, E2, and E3 (see **Fig. 13.25**). Each of these enzymes possesses a distinct coenzyme to mediate electron transfer.

The bacterial pyruvate decarboxylase complex contains multiple copies of all three enzymes (**Fig. 13.25B**). Central to the structure is a core containing 24 E2 proteins surrounded by 12 E1 dimers and 6 E3 dimers. Mitochondrial complexes are even larger and more intricate. The reason for this degree of complexity is unknown, but it is clear that organisms invest considerable energy and material in the construction and function of PDC.

Regulation of PDC. The activity of PDC is increased by high concentrations of its substrates, CoA and NAD^+, and inhibited by its products acetyl-CoA and NADH. The product, acetyl-CoA, may enter one of several pathways. In *E. coli*, when glucose is plentiful, acetyl-CoA is mostly converted to acetate via the intermediate acetyl phosphate (see **Fig. 13.24**). Acetyl phosphate is a global signal molecule that indicates to the cell the quantity and quality of carbon source available. As glucose decreases, the cell starts to take back acetate, converting it back to acetyl-CoA for entry into the TCA cycle.

At the level of gene expression, PDC responds to environmental conditions. As would be expected, PDC gene expression is repressed by carbon starvation and at low oxygen. More surprising, expression is strongly induced both at high pH and at low pH, compared to growth at pH 7. The basis for extreme-pH induction is unknown.

> **THOUGHT QUESTION 13.12** Compare the reactions catalyzed by pyruvate dehydrogenase and pyruvate formate lyase (see Section 13.4). What conditions favor each reaction, and why?

Acetyl-CoA Enters the TCA Cycle

Acetyl-CoA enters the TCA cycle by condensing the acetyl group with oxaloacetate, a four-carbon dicarboxylate (double acid). The condensation yields citrate, a six-carbon tricarboxylate (**Fig. 13.26**). An advantage of intermediates with two or more acidic groups is that the concentration of the fully protonated form is extremely low; thus, the molecule is unlikely to be lost from the cell by diffusion across the membrane, as are monocarboxylic acids, such as acetate. Through the rest of the cycle, citrate loses two carbons as CO_2 by a series of reactions that transfer increments of energy to 3NADH, $FADH_2$, and ATP. Each reaction step couples energy-yielding to energy-storing events.

Step 1. As the acetyl group condenses with oxaloacetate, the removal of CoA consumes a molecule of H_2O to restore HS-CoA. The removal of HS-CoA yields energy to incorporate acetate into oxaloacetate, forming citrate.

Step 2. Citrate undergoes two rearrangements with little energy change to form isocitrate. Isocitrate then undergoes oxidative decarboxylation. As we saw for pyruvate, removal of CO_2 yields energy to transfer $2H^+ + 2e^-$ to form $NADH + H^+$, producing 2-oxoglutarate (alpha-ketoglutarate).

Step 3. 2-oxoglutarate undergoes oxidative decarboxylation to release CO_2 and make another $NADH + H^+$. In this case, CoA is incorporated, making succinyl-CoA.

Step 4. Succinyl-CoA releases CoA, providing energy to phosphorylate ADP to ATP. To form fumarate, $2H^+ + 2e^-$ are transferred to FAD to form $FADH_2$, a reaction involving negligible free energy change.

Step 5. Fumarate incorporates water across its double bond, forming the hydroxy acid malate. The increasing stability from fumarate to malate and from malate to oxaloacetate releases enough energy to form the final $NADH + H^+$. Oxaloacetate is the intermediate of lowest energy in the cycle, now ready to accept the next acetyl-CoA.

> **NOTE:** Many textbooks state that the enzyme catalyzing step 4, succinyl-CoA synthetase, phosphorylates $GDP \rightarrow GTP$. According to the primary literature, ADP phosphorylation predominates in *E. coli* (Margaret Birney et al., 1996), whereas in *Pseudomonas* species, various nucleotide diphosphates are phosphorylated (Vinayak Kapatral et al., 2000). Human mitochondria have two forms of the enzyme, which form ATP and GTP, respectively (David Lambeth et al., 2004).

Observing the TCA cycle intermediates. How were all the TCA intermediates identified? The main experimental approach available to Krebs and his contemporaries was

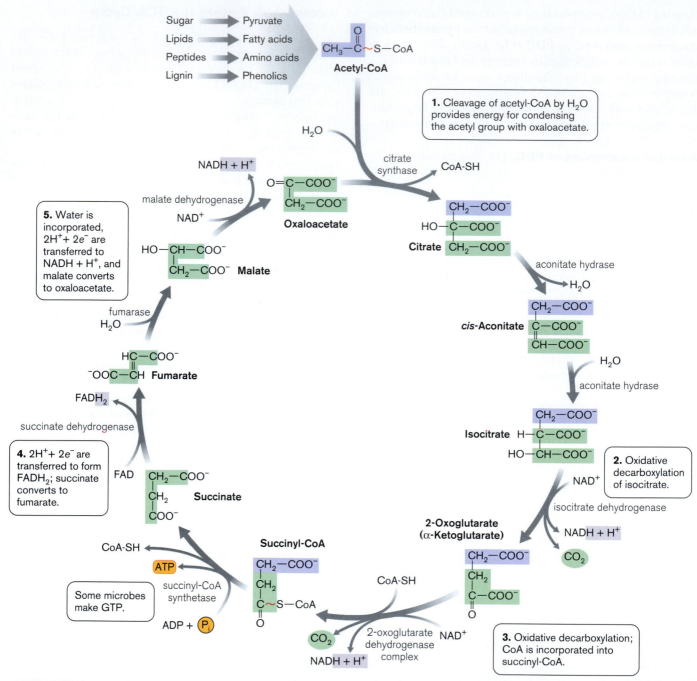

Figure 13.26 The tricarboxylic acid (TCA) cycle. Acetyl-CoA derived from pyruvate and other catabolic pathways enters the TCA cycle. Blue highlights show the fate of labeled acetate incorporated into the TCA cycle. (1) The acetyl group condenses with four-carbon oxaloacetate to produce citrate, a tricarboxylic acid. (2) Citrate rearranges to isocitrate, which is decarboxylated and transfers $2H^+ + 2e^-$ to form NADH + H^+. (3) 2-Oxoglutarate is decarboxylated and transfers $2H^+ + 2e^-$ to form NADH + H^+ while incorporating CoA to form succinyl-CoA. (4) Succinate is symmetrical; thus, the "old" and "new" carbons are now indistinguishable. Succinate transfers $2H^+ + 2e^-$ to form $FADH_2$, yielding fumarate. (5) Water is incorporated, and $2H^+ + 2e^-$ is transferred to form NADH + H^+, yielding oxaloacetate, ready to take up another acetyl group.

to guess at dozens of short-chain acids known to exist in cells, then add each individually to an enzyme preparation and test for TCA cycle activity. In aerobic organisms, the TCA cycle is tightly tied to respiration, so an increase in uptake of oxygen signaled a TCA intermediate. A major experimental advance was the use of tracer isotopes such as ^{14}C, discovered in the 1930s.

Radioisotope tracers answered an important question about the TCA cycle. As the two acetyl carbons cycle through, which two carbons of each intermediate are removed as CO_2? Are the acetyl carbons lost first, or those of the original oxaloacetate? This question was answered using substrates radiolabeled with ^{14}C (**Fig. 13.26**, highlighted in blue). In the experiment, bacteria are fed ^{14}C-radiolabeled acetate, which enters the TCA cycle. The radiolabeled carbons are captured by the TCA intermediates and retained into the next cycle, whereas two COOH carbons from the original oxaloacetate are lost. Thus, the four-carbon intermediate does not recycle intact, but breaks down and re-forms each time it passes through the cycle.

After loss of the second CO_2, the third CH_2 of the carbon skeleton is oxidized to COOH, generating succinate. Succinate is a symmetrical molecule; thus, the former identities of the acetyl and oxaloacetate moieties are now erased, and the carbons are equally likely to disappear from either half of the molecule. Further conversion steps regenerate oxaloacetate—half its carbon from the acetyl group and half from the original oxaloacetate.

> **THOUGHT QUESTION 13.13** Suppose a cell is pulse-labeled with ^{14}C-acetate (fed the ^{14}C label briefly, then "chased" with unlabeled acetate). Can you predict what will happen to the level of radioactivity observed in isolated TCA intermediates? Plot a curve showing your predicted level of radioactivity as a function of number of rounds of the cycle.

The TCA cycle and oxidative phosphorylation. In all, each acetate generates three NADH molecules, one FADH$_2$, and one ATP; and all the carbons from pyruvate (ultimately from glucose) have been released as waste CO_2. From the standpoint of the carbon skeleton, the glucose breakdown is now complete. But do we have a completed metabolic pathway? No, because all of the NADH and FADH$_2$ need to be recycled by donating their electrons onto a terminal electron acceptor.

The process of electron transfer from NADH and FADH$_2$ is mediated by a series of cell membrane proteins called the electron transport system (ETS) or electron transport chain (introduced in Section 13.3). The membrane proteins use the energy of electron transfer to pump

protons, generating a gradient of hydrogen ions across the membrane (**Fig. 13.27**), a process discussed in detail in Chapter 14. Assuming 3 ATPs generated per NADH and 2 ATPs per FADH$_2$, the hydrogen ion gradient then drives the membrane ATP synthase to synthesize as many as 34 ATPs. Another 4 ATPs came from glucose breakdown and the TCA cycle (38 total per glucose). Under actual conditions, however, fewer ATPs are made; about 20 ATP per glucose are made by a well-aerated culture of *E. coli*.

The overall process of electron transport and ATP generation is termed **oxidative phosphorylation**. The overall process of oxidative catabolism from substrate breakdown to oxidative phosphorylation is called respiration. The overall equation for the respiration of glucose is

$$C_6H_{12}O_6 + 6H_2O + 6O_2 \rightarrow 12H_2O + 6CO_2$$

Glucose respiration can generate a relatively large number of ATPs per glucose, far more than fermentation. In bacteria, however, the actual number of ATPs generated varies widely with availability of carbon source and oxygen. For example, as oxygen decreases in the environment, the ability to oxidize NADH decreases, so the cell may make only one or two ATPs per NADH (discussed in Chapter 14).

Regulation of the TCA cycle. The enzymes of the TCA cycle are regulated extensively by substrate induction and product inhibition, and their expression is induced at high oxygen and carbon source availability. At very low levels of glucose, alternative pathways are favored, particularly the **glyoxylate bypass**. The glyoxylate bypass consists of two enzymes that divert isocitrate to glyoxylate and incorporate a second acetyl-CoA to form malate:

The glyoxylate bypass cuts out all the generation of ATP, FADH$_2$, and NADH, with the exception of one NADH from malate to oxaloacetate. Thus, limited energy is released; but, like the low gear of a bicycle, the cycle keeps turning under tough conditions. The glyoxylate bypass can also be used to regenerate TCA cycle intermediates as needed, a process known as anaplerosis (discussed in Chapter 15).

TCA cycle intermediates as substrates for biosynthesis. Analysis of pathway evolution indicates that the TCA cycle originally evolved to provide substrates for biosynthesis. For example, the TCA cycle intermediate

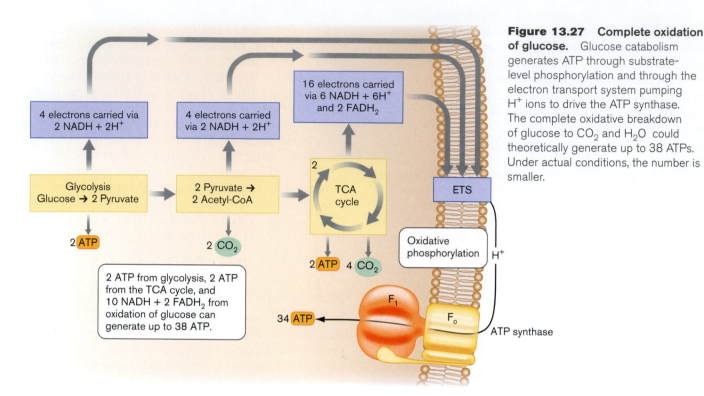

Figure 13.27 Complete oxidation of glucose. Glucose catabolism generates ATP through substrate-level phosphorylation and through the electron transport system pumping H^+ ions to drive the ATP synthase. The complete oxidative breakdown of glucose to CO_2 and H_2O could theoretically generate up to 38 ATPs. Under actual conditions, the number is smaller.

2-oxoglutarate (alpha-ketoglutarate) is aminated to form glutamate, which leads to glutamine. Oxaloacetate is aminated to form aspartate, entering pathways to purines and pyrimidines as well. The TCA cycle, like glycolysis, is an amphibolic pathway that provides substrates for biosynthesis. Many bacteria use the TCA cycle and glycolytic enzymes to build their sugars and amino acids, as discussed in Chapter 15. Others, such as *Treponema pallidum*, the cause of syphilis, have lost the TCA cycle by reductive evolution. They must obtain amino acids from their host organism (see **Special Topic 13.3**).

TO SUMMARIZE:

■ **The pyruvate dehydrogenase complex (PDC)** removes CO_2 from pyruvate, generating acetyl-CoA. Two electrons are transferred to NAD^+, forming NADH/H^+. PDC activity is a key control point of metabolism, induced when carbon sources are plentiful and repressed under carbon starvation and low oxygen.

■ **The tricarboxylic acid cycle (TCA cycle)** converts the acetyl group to $2CO_2$ and $2H_2O$ in the presence of a terminal electron acceptor such as O_2 to receive the electrons associated with the hydrogen atoms.

■ **Acetyl-CoA enters the TCA cycle by condensing with oxaloacetate to form citrate.** A series of enzymes sequentially remove carbon dioxide and water molecules and generate 2NADH, $FADH_2$, and ATP. Each reaction step couples energy-yielding to energy-storing events.

■ **NADH and $FADH_2$ transfer electrons to the electron transport system** and ultimately the terminal electron acceptor such as O_2. Electron transport generates a proton gradient that powers the membrane-bound ATPase to make ATP.

■ **Respiration** consists of substrate catabolism plus oxidative phosphorylation, the process of electron transport and ATP generation.

13.7 Aromatic Catabolism

In forest environments, bacteria and fungi catabolize aromatic molecules such as benzene derivatives. Aromatic catabolism is crucial to recycle lignin, the main component of woody biomass. Today, aromatic catabolism has growing applications for bioremediation of soil pollutants such as nitrate explosives, aniline dyes, and industrial solvents such as toluene. Aromatic molecules are notoriously difficult to catabolize because of the stability associated with the benzene ring. The benzene ring breaks down slowly in natural environments and was once thought to require aerobic digestion. We now know, however, that soil bacteria catabolize a wide range of aromatic molecules, anaerobically as well as aerobically. These microbial processes have major industrial and environmental importance. Here we explore bacterial catabolism of benzene derivatives such as benzoate (benzene carboxylate). Benzoate has a central role in aromatic catabolism, comparable to that of glucose in sugar catabolism.

Benzoate Derivatives Are Degraded to Catechol

Bacteria such as *Pseudomonas* and *Rhodococcus* species degrade benzoate and related molecules aerobically or anaerobically. The catabolic enzymes are often encoded on a large plasmid, of 100–450 kb. The aerobic pathways (**Fig. 13.28**) share key features:

- Removal of substituents such as chlorides or nitrates
- Oxidation to form the dihydroxy derivative catechol
- Ring cleavage and catabolism to acetyl-CoA, which may enter the TCA cycle

Aerobic benzoate catabolism. Aerobic degradation typically proceeds through the key intermediate **catechol**, a benzene ring bearing two adjacent hydroxyl groups. Benzoate and various related compounds, such as toluene (methyl benzene) and nitrobenzene, are each converted to catechol by a specific **dioxygenase**, an enzyme that coordinately oxygenates two adjacent ring carbons.

The intermediate catechol then undergoes another key oxidation by a **catechol dioxygenase**, which adds two more oxygens while cleaving the ring. In different bacterial species, the enzyme may oxidize either the 1, 2 positions, or the 2, 3 positions as shown. The product is either a monocarboxylate or a dicarboxylate. In either case, the carbon skeleton is further oxidized, and its constituents are directed into the TCA cycle for full degradation to CO_2.

Genetic analysis of toluene catabolism. Many pathways of aromatic catabolism are still being discovered. To study unknown metabolism, we must identify the genes that encode the enzymes responsible. These genes lie waiting in the sequenced microbial genomes, where a large proportion of genes remains unidentified. How are their functions discovered?

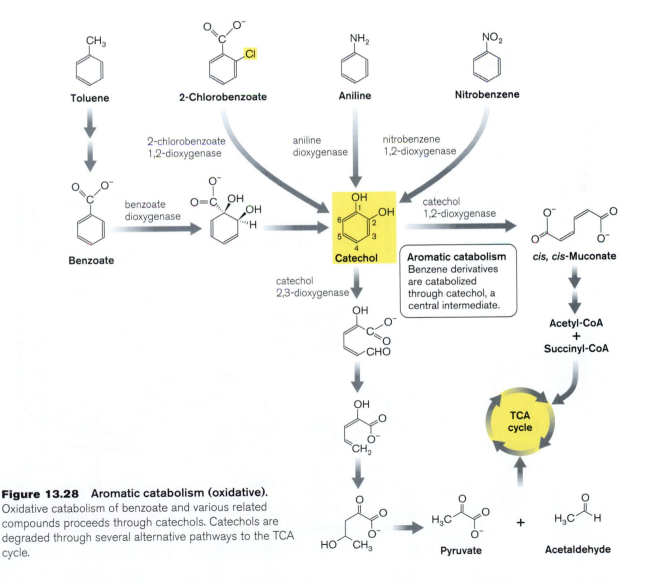

Figure 13.28 Aromatic catabolism (oxidative). Oxidative catabolism of benzoate and various related compounds proceeds through catechols. Catechols are degraded through several alternative pathways to the TCA cycle.

Special Topic 13.3 **Genomic Analysis of Metabolism**

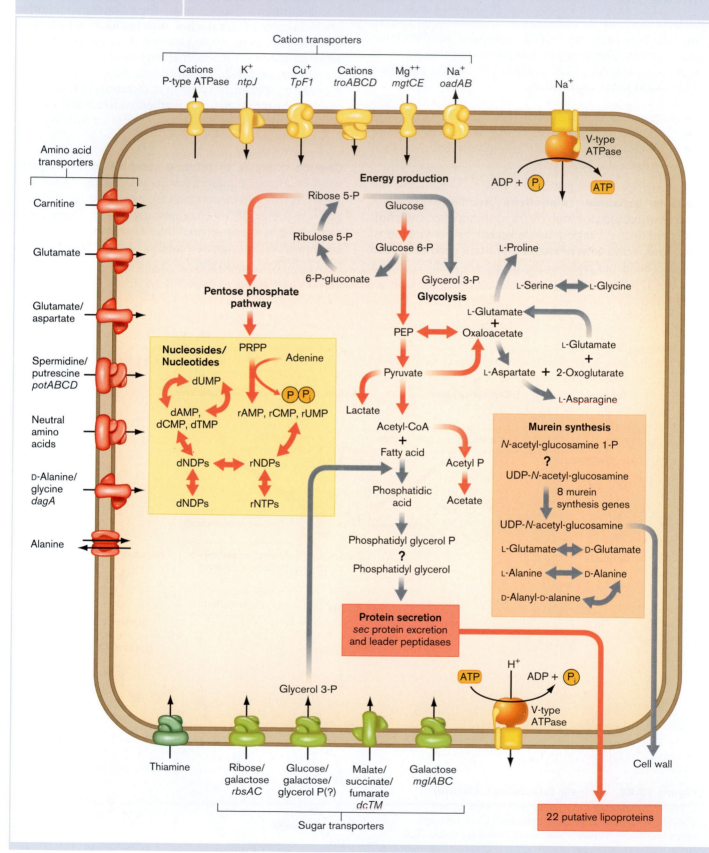

The dissection of the TCA cycle required decades of painstaking experiments in biochemistry and genetic analysis on model organisms such as *E. coli*. Now that so much metabolism is understood in model organisms, much of the metabolic capacity of other organisms can be deduced by comparison of their genomes. Genome analysis can be especially useful in characterizing pathogens that are difficult to culture and study, such as *Treponema pallidum*, the causative agent of syphilis.

The sequence of a microbial genome typically reveals a high percentage of genes whose function can be identified based on homology with known genes in other species (discussed in Chapter 8). For example, the components of the pyruvate dehydrogenase complex show highly similar sequences in all organisms. Thus, it is possible to survey an annotated microbial genome and predict much of the microbe's metabolic capacities.

The genome of *T. pallidum* was sequenced by Claire Fraser-Liggett and colleagues at The Institute for Genomic Research (TIGR) (**Fig. 1**). The genome encodes key enzymes of glycolysis (red arrows, central) as well as those of fermentation to lactate or acetate. However, enzymes of the TCA cycle are completely absent, as are the proteins of electron transport. Thus, *T. pallidum* can only generate ATP by substrate-level phosphorylation. The genome does encode a small number of sugar transporters (yellow symbols) for entry to glycolysis, as well as amino acid transporters (red symbols). It does not encode enzymes of amino acid catabolism or of their biosynthesis. Thus, the organism can ferment a limited number of sugars but can neither catabolize nor form its own amino acids. It must depend on its host to supply many fundamental nutrients. The general loss of major metabolic systems from *T. pallidum* is typical of ancient pathogens that have evolved over a long period in obligate association with their host.

The genes encoding metabolism can be identified through use of clone libraries, as discussed in Chapter 12. For example, clone complementation was used by Hyung-Yeel Kahng and colleagues to reveal new genes encoding degradation of toluene and its derivatives associated with petroleum pollution (**Fig. 13.29**).

Kahng and coworkers obtained a strain of the soil bacterium *Burkholderia*, which degrades toluene, benzene, and chlorobenzene. A genomic library was generated to contain plasmids carrying cloned segments from throughout the *Burkholderia* genome. The library of clones was used to transform a strain of *Pseudomonas aeruginosa* that cannot degrade the toluene derivatives. Each transformed strain contained a plasmid with a different segment of the *Burkholderia* genome. One clone, pHYK2000, conferred the ability to catabolize toluene, benzoate, and chlorobenzoate. The plasmid in this clone was mapped by restriction enzymes to delimit the range of DNA sequence needed for catabolism. Ultimately, the plasmid was shown to contain two previously unknown operons for aromatic catabolism, *tbc1* and *tbc2*. Some of the genes in the operons showed homology to known genes in other organisms that encode oxygenases for aromatic catabolism. Other *Burkholderia* genes showed little relatedness to known genes. The functions of their gene products were tested subsequently by biochemical assays. These genes point to a potential new microbial route for biodegradation of petroleum waste.

Anaerobic benzoate catabolism. Anaerobic catabolism is especially desirable for cleanup of contamination underground, where oxygen levels are extremely low. Yet the pathways of anaerobic degradation of aromatic molecules remain poorly understood. An anaerobic pathway of

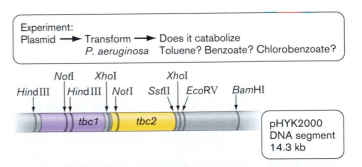

Figure 1   Genome of *Treponema pallidum* reveals metabolic pathways.   The *Treponema* genome encodes key enzymes of glycolysis and fermentation (red arrows, central). However, enzymes of the TCA cycle and electron transport are absent. Thus, *Treponema* is capable of little or no oxidative respiration. The genome does show sugar transporters (yellow symbols) and amino acid transporters (red symbols) but no enzymes for amino acid catabolic or biosynthetic pathways.

Figure 13.29   Complementation analysis reveals a toluene catabolism gene cluster.   A cloned plasmid pHYK2000 containing a DNA segment from a genomic library of *Burkholderia* species was transformed into a *Pseudomonas aeruginosa* strain, where it conferred the ability to catabolize toluene, benzoate, and chlorobenzoate. The plasmid was mapped by restriction enzymes and shown to contain two previously unknown operons for aromatic catabolism, *tbc1* and *tbc2*.
Source: Modified from Hyung-Yeel Kahng, et al. 2001. *Appl. Env. Microbiol.* 67:4805.

benzoate catabolism was observed in the soil bacterium *Azoarcus evansii* (**Fig. 13.30**). In *Azoarcus*, under anaerobic conditions, benzoate undergoes reductive degradation instead of oxidation. First, the benzoate is activated by CoA, then it is reduced by NADPH. The NADPH reduction eliminates aromaticity before ring cleavage.

The reduction of benzoyl-CoA by NADPH was demonstrated by Christa Ebenau-Jehle and colleagues at the University of Freiburg by separating radiolabeled intermediates. Ring carbons of benzoyl-CoA were labeled with ^{14}C, and reaction products from *Azoarcus* species were analyzed using thin-layer chromatography (TLC), a method that separates molecules based on different solubilities in a solvent mixture (**Fig. 13.30B**).

The chromatogram shows radiolabeled spots of [ring-^{14}C] benzoyl-CoA and its products, which are carried different distances by the chromatographic solvent. In the presence of NADPH, the labeled benzoyl-CoA steadily disappears, while products of catabolism emerge. If the nonreductive $NADP^+$ substitutes for NADPH, no loss of

benzoyl-CoA occurs. Thus, benzoyl-CoA degradation was shown to require NADPH. This anaerobic route of aromatic degradation differs from that of the oxidative process, in which benzoate is oxidized and decarboxylated to catechol.

Polynuclear Aromatic Hydrocarbons (PAHs) Are Degraded Slowly

A particularly challenging class of catabolic substrates is the polynuclear aromatic hydrocarbons (PAHs). The multiple fused rings of PAH compounds may take years to biodegrade in natural environments. PAHs are found in coal tar products such as creosote, a preservative widely used to protect wood from insects and fungal decomposition. Creosote leaks from wood structures such as railway ties (**Figs. 13.31A** and **B**) and contaminates wetland environments. Many PAHs are teratogenic and carcinogenic, and their increase in our environment is of growing concern.

The pathways and even the microbial species responsible for PAH degradation in nature remain poorly known. In one study, the biodegradation of a PAH compound in creosote was observed by Sergey Selifonov and colleagues at the University of Minnesota. Selifonov used ^{13}C NMR to track the catabolism by *Pseudomonas* strain A2279 of ^{13}C-labeled acenaphthene, a tricyclic naphthalene derivative (**Fig. 13.31C**). The compound was labeled at a single aliphatic position so that its resonance would appear as a single peak separate from the range of aromatic carbon, where natural-abundance ^{13}C peaks appear. New peaks appeared, which were identified as various oxidized products of acenaphthene. The enzymes responsible for these products remain to be discovered. Furthermore, the completion of degradation remains unclear. It is likely that such complex molecules require a consortium of different species to complete their degradation to acetate.

A.

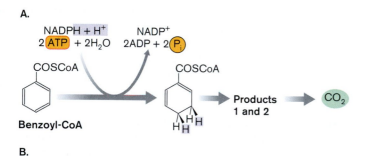

B.

Figure 13.30 **Anaerobic catabolism of benzoate: analysis by thin-layer chromatography.** **A.** Under anaerobic conditions, benzoate is activated by coenzyme A, then reduced (hydrogenated) by NADPH. **B.** The reduction of benzoyl-CoA in *Azoarcus evansii* was observed by thin-layer chromatography (TLC). In the presence of the reducing agent NADPH, the spot due to benzoyl-CoA disappears, while those due to two products of catabolism increase. In the presence of $NADP^+$, with no reducing power, no reaction occurs. *Source:* Modified from Christa Ebenau-Jehle, et al. 2003. *J. Bacteriol.* 185:6119.

> **THOUGHT QUESTION 13.14** How might the enzymes that catalyze acenaphthene catabolism be discovered?

TO SUMMARIZE:

- **Catabolism of aromatic molecules by bacteria and fungi recycles lignin** and other important substances within ecosystems. Toxic pollutants are also degraded.
- **Benzoate undergoes aerobic catabolism to catechol.** The catechol ring is cleaved, generating acetyl-CoA, which enters the TCA cycle.
- **Anaerobic catabolism of benzoate involves activation by HS-CoA and reduction of $NADP^+$ to NADPH.**

Figure 13.31 PAH biodegradation.
A. Railroad ties are a source of the pollutant creosote, which contains a mixture of polycyclic aromatic compounds. **B.** Typical structures of polyaromatic aromatic hydrocarbons (PAH). **C.** A ^{13}C-labeled polycyclic PAH, acenaphthene, is included in a creosote sample undergoing catabolism by *Pseudomonas* sp. strain A2279. After seven days, the oxidized digestion products are revealed by characteristic resonance peaks in the ^{13}C NMR spectrum. *Source*: C. Modified from Sergey A. Selifonov, et al. 1998. *Appl. Envir. Microbiol.* 64:1447.

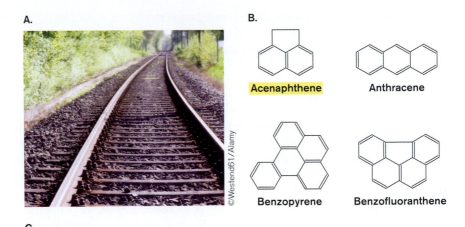

A.

©Westend61/Alamy

B.

Acenaphthene   Anthracene

Benzopyrene   Benzofluoranthene

C.

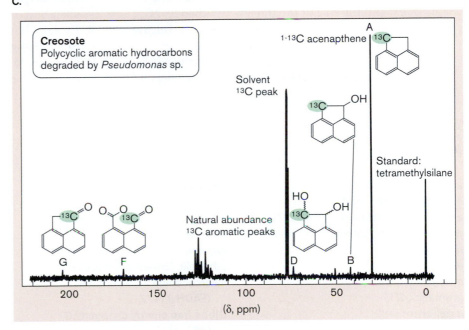

■ **Polynuclear aromatic hydrocarbons (PAHs) are degraded slowly by microbes.** PAH catabolism has exciting potential applications for bioremediation of toxic pollutants.

■ **Aromatic catabolism by environmental microbes is an important field of research.** Research techniques include genetic and genomic analysis, chromatography of radiolabeled substrates, and NMR spectroscopy.

Concluding Thoughts

In this chapter, we have seen how energy-yielding reactions enable living cells to generate order from disorder. In fact, the result of cell metabolism increases entropy throughout the universe, but enables local growth of complexity. Living organisms acquire energy by trans-

ferring energy from reactions with negative ΔG to cell-building reactions with positive ΔG. The fundamental energy-yielding pathways of Earth's biosphere are those of photosynthesis; and the biomass generated through photosynthesis provides materials for catabolism. Much of human civilization has been built on harnessing microbial catabolism for waste treatment, food production, and biotechnology.

All forms of metabolism involve chemical exchange of electrons through oxidation and reduction. In Chapter 14, we focus on the transport of electrons between donor and acceptor molecules, through membrane-embedded complexes that generate ion gradients across the membrane. This transport of electrons is fundamental to respiration, the oxidative completion of catabolism to CO_2 and water. It also underlies the autotrophic means of building biomass: the gain of energy from light (phototrophy) and from mineral oxidation (lithotrophy.)

CHAPTER REVIEW

Review Questions

1. Why must the biosphere continually take up energy from outside? Why can't all the energy be recycled among organisms, like the fundamental elements of matter?

2. Explain how a biochemical reaction can be driven by change in enthalpy, ΔH. Explain how a different reaction can be driven by change in entropy, ΔS. In each case, explain the role of the free energy change, ΔG.

3. Why do some biochemical reactions only release energy above a threshold temperature?

4. How do organisms determine which of their catabolic pathways to use? How does catabolism depend on environmental factors?

5. Beer is produced by yeast fermentation of grain to ethanol. Why must the process of beer production be anaerobic? Why are such large quantities of ethanol produced, with relatively small production of yeast biomass?

6. Explain the three different routes to catabolize glucose to pyruvate. Why is it necessary to start by spending one or two molecules of ATP?

7. Explain how the TCA cycle incorporates an acetyl group. How are the two CO_2 molecules removed?

8. Compare and contrast aerobic and anaerobic processes of benzoate catabolism.

Key Terms

acetyl-CoA (489)
activation energy (473)
adenosine diphosphate (ADP) (470)
adenosine triphosphate (ATP) (469)
allosteric site (474)
amphibolic (485)
anabolism (462)
aromatic (471)
catabolism (462)
catabolite repression (479)
catechol dioxygenase (497)
coenzyme A (CoA) (489)
dioxygenase (497)
electron acceptor (469)
electron donor (469)
electron transport system (ETS) (471)
Embden-Myerhof-Parnas (EMP)
    pathway (482)
energy (463)

energy carrier (469)
enthalpy (464)
Entner-Doudoroff (ED) pathway (482)
entropy (463, 464)
enzyme (473)
ethanolic fermentation (489)
flavin adenine dinucleotide
    (FADH$_2$) (472)
Gibbs free energy change (ΔG)
    (464, 466)
glycolysis (467)
glyoxalate bypass (495)
heteroaromatic (471)
heterolactic fermentation (489)
heterotrophy, heterotroph (462)
hydrolysis (470)
joules (J) (465)
lactate fermentation (489)
lignin (479)

nicotinamide adenine dinucleotide
    (NADH, NAD$^+$) (471)
organotrophy, organotroph (462)
oxidative phosphorylation (495)
pentose phosphate shunt (PPS) (482)
phenol red broth test (490)
phosphoenolpyruvate (PEP) (473)
phosphorylation (470)
phosphotransferase (PTS) (470)
polysaccharide (477)
pyruvate dehydrogenase complex
    (PDC) (493)
pyruvate kinase (474)
ribulose 5-phosphate (486)
substrate-level phosphorylation (485)
terminal electron acceptor (472)
tricarboxylic acid (TCA) cycle (491)

Recommended Reading

Bäckhed, Fredrik, Ruth E. Ley, Justin L. Sonnenburg, Daniel A. Peterson, and Jeffrey I. Gordon. 2005. Host-bacterial mutualism in the human intestine. *Science* 307:915–920.

Birney, Margaret, Hong-Duck Um, and Claudette Klein. 1996. Novel mechanisms of *Escherichia coli* succinyl-coenzyme A synthetase regulation. *Journal of Bacteriology* 178:2883–2889.

Brüggemann, Holger, Anke Henne, Frank Hoster, Heiko Liesegang, Arnim Wiezer, et al. 2004. The complete genome sequence of *Propionibacterium acnes*, a commensal of human skin. *Science* 305:671–673.

Bunge, Michael, Lorenz Adrian, Angelika Kraus, Matthias Opel, Wilhelm G. Lorenz, et al. 2003. Reductive dehalogenation of chlorinated dioxins by an anaerobic bacterium. *Nature* **421:**357–360.

Ebenau-Jehle, Christa, Matthias Boll, and Georg Fuchs. 2003. 2-Oxoglutarate:NADP$^+$ oxidoreductase in *Azoarcus evansii*: Properties and function in electron transfer reactions in aromatic ring reduction. *Journal of Bacteriology* **185:**6119–6129.

Fischer, Eliane, and Uwe Sauer. 2003. A novel metabolic cycle catalyzes glucose oxidation and anaplerosis in hungry *Escherichia coli*. *Journal of Biological Chemistry* **278:**46446–46451.

Fraser, Claire M., Steven J. Norris, George M. Weinstock, Owen White, Granger G. Sutton, et al. 1998. Complete genome sequence of *Treponema pallidum*, the syphilis spirochete. *Science* **281:**375–388.

Head, Ian M., D. Martin Jones, and Wilfred F. M. Röling. 2006. Marine microorganisms make a meal of oil. *Nature Reviews Microbiology* **4:**173–182.

Jackson, Bradley E., and Michael J. McInerney. 2002. Anaerobic microbial metabolism can proceed close to thermodynamic limits. *Nature* **415:**454–456.

Kahng, Hyung-Yeel, Juliana C. Malinverni, Michelle M. Majko, and Jerome J. Kukor. 2001. Genetic and functional analysis of the *tbc* operons for catabolism of alkyl- and chloroaromatic compounds in *Burkholderia* sp. strain JS150. *Applied and Environmental Microbiology* **67:**4805–4816.

Kapatral, Vinayak, Xiaowen Bina, and A. M. Chakrabarty. 2000. Succinyl coenzyme A synthetase of *Pseudomonas aeruginosa* with a broad specificity for nucleoside triphosphate (NTP) synthesis modulates specificity for NTP synthesis by the 12-kilodalton form of nucleoside diphosphate kinase. *Journal of Bacteriology* **182:**1333–1339.

Lambeth, David O., Kristin N. Tews, Steven Adkins, Dean Frohlich, and Barry I. Milavetz. 2004. Expression of two succinyl-CoA synthetases with different nucleotide speci-

ficies in mammalian tissues. *Journal of Biological Chemistry* **279:**36621–36624.

Lohmeier-Vogel, Elke M., Shiela Ung, and Raymond J. Turner. 2004. In vivo ^{31}P nuclear magnetic resonance investigation of tellurite toxicity in *Escherichia coli*. *Applied and Environmental Microbiology* **70:**7342–7347.

Maurer, Lisa M., Elizabeth Yohannes, Sandra S. BonDurant, Michael Radmacher, and Joan L. Slonczewski. 2005. pH regulates genes for flagellar motility, catabolism, and oxidative stress in *Escherichia coli* K-12. *Journal of Bacteriology* **187:**304–319.

Peekhaus, Norbert, and Tyrrell Conway. 1998. What's for dinner? Entner-Doudoroff metabolism in *Escherichia coli*. *Journal of Bacteriology* **180:**3495–3502.

Sonnenburg, Justin L., Jian Xu, Douglas D. Leip, Chien-Huan Chen, Benjamin P. Westover, et al. 2005. Glycan foraging in vivo by an intestine-adapted bacterial symbiont. *Science* **307:**1955–1959.

von Stockar, Urs, Thomas Maskow, Jingsong Liu, Ian W. Marison, and Rodrigo Patiño. 2006. Thermodynamics of microbial growth and metabolism: an analysis of the current situation. *Journal of Biotechnology* **121:**517–533

Wolfe, Alan J. 2005. The acetate switch. *Microbiology and Molecular Biology Reviews* **69:**12–50.

Xu, Jian, Magnus K. Bjursell, Jason Himrod, Su Deng, Lynn K. Carmichael, et al. 2003. A genomic view of the human-*Bacteroides thetaiotaomicron* symbiosis. *Science* **299:**2074–2075.

Young, David Young, Donna Parke, and L. Nicholas Ornston. 2005. Opportunities for genetic investigation afforded by *Acinetobacter Baylyi*, a nutritionally versatile bacterial species that is highly competent for natural transformation. *Annual Review of Microbiology* **59:**519–551.

Chapter 14

Respiration, Lithotrophy, and Photolysis

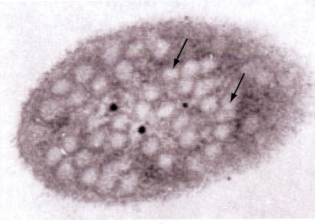

$+O_2$ $-O_2$

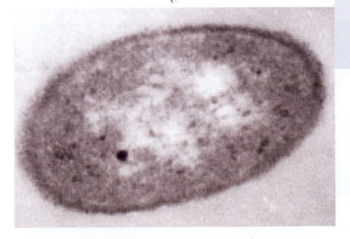

Microbes transfer energy by moving electrons—the equivalent of an electric current. Electrons move from reduced food molecules onto energy carriers, from energy carriers onto membrane proteins called cytochromes, and from cytochromes onto oxygen or oxidized minerals. Some bacteria can actually donate electrons to an electrode and power a fuel cell. In the soil, bacteria donate electrons to metals such as iron, gold, and uranium. Bacteria in Earth's crust have deposited most of the iron mined today, and possibly the gold as well.

A special kind of electron current flows between proteins and cofactors of an electron transport system in a cell membrane. The electron transport system generates a "proton motive force" that drives protons across the membrane. The proton motive force stores energy to make ATP. Similar electron transport systems power our own mitochondria and the photosynthetic membranes of plant chloroplasts.

Rhodobacter sphaeroides grows by oxidative respiration (lower panel), but under oxygen limitation the bacterium forms tubes of photosynthetic membranes to harvest light (upper panel, shown by arrows; cell width, approx. 0.5 μm). The genes regulating photosynthetic membrane formation were revealed by microarray and genomic analysis.
Source: Mao, et al. 2005. *Microbiology* 151:3197. Image by Chris Mackenzie; copyright 2005 Nature Publishing Group.

Electric current in living organisms has fascinated scientists ever since the eighteenth century when Luigi Galvani (1737–1798) showed that a voltage potential caused a dead frog limb to flex. The idea of biological electricity inspired Mary Shelley's famous science fiction novel *Frankenstein: or, the Modern Prometheus*, in which a physician is imagined to "create life" by jolting body parts with an electric shock.

Today, electric currents and voltage potentials are understood to be fundamental aspects of all cells. Nevertheless, it came as a surprise to discover that bacteria could directly power an electrical fuel cell. Bacteria commonly found in soil can oxidize organic nutrients using iron ions (ferric ion, Fe^{3+}) instead of oxygen. The organic molecules transfer electrons to the iron. But Fe^{3+} is usually present as insoluble grains of iron oxide, outside the cell; so how do cells reach the iron? Derek Lovley and colleagues at the University of Massachussetts, Amherst, wondered how bacteria gained access to iron oxide—and whether they might also transfer electrons to an iron electrode. They found that *Geobacter metallireducens* transfers electrons to a metal electrode, generating enough power to run an electric clock. A related species of *Geobacter* may transfer electrons through its pili (**Fig. 14.1**). The pili act as nanowires conducting electricity to iron oxide, thus generating a charge difference across the bacterial membranes.

The source of electrons for *Geobacter* metabolism can include hard-to-digest organic molecules, even aromatic compounds such as benzoates that often contaminate soil. Thus, bacteria may combine electricity generation with organic waste remediation. In fact, the bacterial community of our entire biosphere acts as a battery in which underground bacteria continually transfer electrons from Earth's reduced interior to the oxidizing atmosphere. Useful electricity can be obtained simply by placing electrodes at appropriate depths in the ocean floor. While the amount of power obtained at present is small, in the future subterranean microbes may contribute significantly to our energy supply.

As we saw in Chapter 13, biochemical reactions can provide energy in different ways (see **Table 13.1**). Catabolism provides energy through breakdown of complex molecules (increasing entropy) and through electron transfer converting molecules to more stable forms (releasing heat energy). Electron transfer also yields energy from inorganic minerals (lithotrophy) and from light capture (phototrophy). Microbial electron transfer reactions have profound consequences for ecosystems; they acidify or alkalinize soil and water, and they can contribute or remove nitrogen and minerals.

Chapter 14 develops a unified view of electron transfer reactions, or oxidation-reduction reactions, in respiration, lithotrophy, and phototrophy. In all kinds of metabolism, some of the energy from electron transfer is stored in the form of an electrochemical potential (voltage) across a membrane. The voltage potential includes a concentration gradient of ions (H^+ or Na^+) plus the charge difference across the membrane. Voltage potentials drive ATP synthesis and nutrient uptake, mediate pathogenesis of infective microbes, and shape the chemistry of our environment.

14.1 Electron Transport Systems

As we learned in Chapter 13, energy to support life is obtained through reactions that convert substrates to products of lower energy—that is, reactions in which ΔG is negative. Most such reactions involve transfer of electrons from a reduced **electron donor** to an oxidized **electron acceptor**. The simplest path of electron flow occurs in fermentation, where electrons from the fermented substrate, such as glucose, are transferred onto NAD^+ to make NADH, then returned to the glucose breakdown products. More complex kinds of metabolism, such as aerobic respiration, transfer electrons through a series of membrane-soluble carriers called an **electron transport system (ETS)**, also known as an **electron transport chain**. A membrane-embedded ETS, unlike cytoplas-

A.

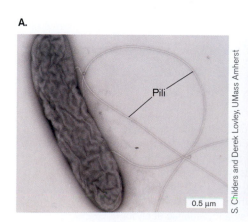

Pili

0.5 μm

S. Childers and Derek Lovley, UMass Amherst

B.

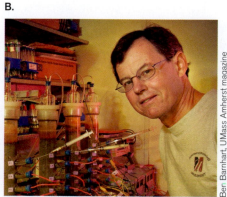

Ben Barnhart, UMass Amherst magazine

Figure 14.1   Electricity from iron-reducing bacteria.   A. *Geobacter metallireducens*, a bacterium that transfer electrons to iron through pili (TEM). **B.** Derek Lovley, University of Massachusetts at Amherst, culturing *G. metallireducens*. Lovley discovered that *G. metallireducens* could donate electrons to a metal electrode, generating electricity to power electromechanical devices.

mic redox reactions, can convert its energy into an ion potential or electrochemical potential between two compartments separated by the membrane. The ion potential is most commonly a proton (H^+) potential, also known as proton motive force (PMF). The proton motive force drives essential cell processes such as synthesis of ATP (discussed in Section 14.2).

Electron Transport Systems Use Different Electron Donors and Acceptors

Different types of metabolism use electron transport systems with different components embedded in the bacterial membrane. Each ETS must accept electrons from an initial electron donor consisting of an organic "food" molecule, such as glucose, or a reduced mineral, such as Fe^{2+}. The ETS proteins and cofactors act sequentially as electron donors and acceptors, whose oxidation-reduction reactions are coupled to proton transport across the membrane. Each ETS finally transfers its electrons to a terminal electron acceptor molecule, whose product may exit the cell.

Metabolism using an ETS is classified based on the nature of the initial electron donors and terminal electron acceptors. In this chapter, we cover three major classes of prokaryotic energy acquisition involving an ETS (their relationships are summarized in Table 13.1):

- **Respiration** is the oxidation of electron donors such as sugars, lipids, or amino acids to CO_2 and H_2O. (Oxidation of minerals is a form of respiration called lithotrophy, discussed below.) Organic molecules donate electrons associated with hydrogen atoms, a process known as dehydrogenation. Additional hydrogen atoms are donated by water molecules. The oxidant (terminal electron acceptor) may be O_2, for aerobic respiration, or an anaerobic alternative, for anaerobic respiration. The anaerobic electron acceptor may be inorganic (such as NO_3^-) or organic (such as fumarate, $^-O_2C-CH=CH-CO_2^-$).
- **Lithotrophy** (or **chemolithotrophy**, or chemoautotrophy) is the oxidation of inorganic electron donors, such as Fe^{2+} or H_2. The electron acceptor may be O_2 or an anaerobic alternative, such as NO_3^-. A few inorganic electron donors, such as H_2, can be used with organic electron acceptors. Most lithotrophs are autotrophs, fixing CO_2 for biosynthesis (discussed in Chapter 15).
- **Phototrophy** involves light capture by chlorophyll, usually coupled to splitting of H_2S or H_2O (photolysis) or organic molecules (photoheterotrophy). Alternatively, light absorption drives a proton pump without photolysis. Photoautotrophs couple photolysis to CO_2 fixation for biomass (discussed in Chapter 15). Photoheterotrophs use photolysis to supplement catabolism.

NOTE: Distinguish these prefix terms:

Organo-	Organic electron donors for energy
Litho-	Inorganic electron donors for energy
Chemo-	Chemical reactions yield energy
Photo-	Light absorption drives or supplements chemical reactions for energy
Hetero-	Organic molecules for carbon source
Auto-	Inorganic carbon source, usually CO_2

An Electron Transport System Stores Energy

In Chapter 13, we showed how the free energy change ΔG determines whether energy can be obtained from a given reaction of substrates to products. Similar considerations of ΔG govern the reactions of electron flow through an ETS and the storage of energy in transmembrane ion gradients.

The reduction potential. To obtain energy, biochemical reactions require a negative ΔG overall. In oxidation-reduction reactions (redox reactions), the ΔG values are proportional to the **reduction potential** (E) between the oxidized form of a molecule (electron acceptor) and its reduced form (electron donor). The reduction potential represents the tendency of a molecule to accept electrons, measured in volts (V) or millivolts (mV). The oxidized and reduced states together form a **redox couple**. An example of a redox couple is the electron acceptor O_2 and the electron donor H_2O. Because O_2 is a strong electron acceptor, the redox couple O_2/H_2O has a high positive value of E.

Standard values of E at equilibrium ($E°$) are known as the **standard reduction potential**. The reduction potential $E°'$ (assuming a concentration of 1 M for all components at pH 7) for H_2 formation from $2H^+ + 2e^-$ is -0.42 volts (V). This large negative reduction potential means that it takes a lot of energy to add the two electrons to $2H^+$. Thus, the reverse reaction donating $2e^-$ from H_2 to an appropriate electron acceptor yields a lot of energy. Hydrogen-oxidizing bacteria grow in our intestines, using H_2 made by fermenting bacteria.

Reduction potentials represent a form of the standard free energy change, $\Delta G°'$. The value of $\Delta G°'$ (in kJ/mol) between reactants and products is given by

$$\Delta G°' = -nFE°'$$

where n is the number of electrons transferred, F is Faraday's constant ($F = 96.4$ kJ/V · mol) and $E°'$ is the reduction potential at 1 M concentration, 25°C, and pH 7. Note

Table 14.1 "Electron tower" of standard reduction potentials.[a]

Electron acceptor	→	electron donor	$E^{o\prime}$ (mV)[b]
$CO_2 + 4H^+ + 4e^-$	→	$[CH_2O]$ glucose + H_2O	−430
$2H^+ + 2e^-$	→	H_2	−420
$NAD^+ + 2H^+ + 2e^-$	→	$NADH + H^+$	−320
$S^0 + H^+ + e^-$	→	HS^-	−278
$CO_2 + 2H^+ + 3H_2 + 2e^-$	→	$CH_4 + 2H_2O$	−240
$SO_4^{2-} + 10H^+ + 8e^-$	→	$H_2S + 4H_2O$	−220
$FAD + 2H^+ + 2e^-$	→	$FADH_2$	−220[c]
$FMN + 2H^+ + 2e^-$	→	$FMNH_2$	−190
Menaquinone + $2H^+ + 2e^-$	→	Menaquinol	−74
Fumarate + $2H^+ + 2e^-$	→	Succinate	33
Ubiquinone + $2H^+ + 2e^-$	→	Ubiquinol	+110
$NO_3^- + 2H^+ + 2e^-$	→	$NO_2^- + H_2O$	+420
$NO_2^- + 8H^+ + 6e^-$	→	$NH_4^+ + 2H_2O$	+440
$MnO_2 + 4H^+ + 2e^-$	→	$Mn^{2+} + 2H_2O$	+460
$NO_3^- + 6H^+ + 5e^-$	→	$\frac{1}{2}N_2 + 3H_2O$	+740
$Fe^{3+} + e^-$	→	Fe^{2+} (at pH 2)	+770
$\frac{1}{2}O_2 + 2H^+ + 2e^-$	→	H_2O	+820

[a]Values taken from Rudolf K. Thauer, Kurt Jungermann, and Karl Decker. 1977. Energy conservation in chemotrophic anaerobic bacteria. *Bacteriological Reviews* 41:100–180.

[b]Voltage potentials for redox couples when all species are 1 M at pH 7.

[c]This value is for free FAD; FAD bound to a specific flavoprotein has a different $E^{o\prime}$ that depends on its protein environment.

that the reaction is favored by positive values of E, which yield negative values of ΔG.

> **NOTE:** A higher reducing potential (more positive $E^{o\prime}$) means a stronger electron acceptor (stronger oxidant). A lower reducing potential (more negative $E^{o\prime}$) means a stronger electron donor (stronger reductant).

The standard reduction potential $E^{o\prime}$ gives the potential difference between the oxidized and reduced states at equilibrium, when both oxidized and reduced species are at equal concentrations, namely, 1 M. Assuming equal concentrations at 1 M, we may compare $E^{o\prime}$ of various molecules available in the environment for microbial respiration. Such a comparison generates an "electron tower" representing the reduction potential $E^{o\prime}$ of molecules that may act as electron acceptors or donors (**Table 14.1**). The oxidized state of each redox couple in the table is shown in *red*, the reduced state in blue.

In the electron tower, the more negative values represent couples (half reactions) with a stronger electron donor (H_2, NADH), whereas the more positive values represent stronger electron acceptors (O_2, NO_3^-). For example, in the redox couple $NAD^+/NADH + H^+$, the oxidized form is NAD^+ and the reduced form is NADH. The

reduction potential is strongly negative (−320 mV), so NADH is a strong electron donor; that is, the reverse reaction, $NADH + H^+ \rightarrow NAD^+$ goes forward releasing a lot of energy (+320 mV).

A complete redox reaction combines two redox couples, one accepting electrons (arrow forward), the other donating electrons (arrow reversed). Redox couples with more negative values of $E^{o\prime}$ (higher in the tower) can provide electron donors (blue column) for electron acceptors with more positive values of $E^{o\prime}$ (lower in the tower, red column). The $E^{o\prime}$ for the overall reaction is then given by the subtraction of the donor $E^{o\prime}$ from the acceptor $E^{o\prime}$. A positive value of $E^{o\prime}$ means that the reaction may proceed to provide energy. For example, the oxidation of NADH by O_2 results from combining two couples:

Acceptor:
$$\frac{1}{2}O_2 + 2e^- + 2H^+ \rightarrow H_2O \qquad E^{o\prime} = 820 \text{ mV}$$

Donor:
$$NADH + H^+ \rightarrow NAD^+ + 2e^- + 2H^+ \qquad E^{o\prime} = -320 \text{ mV}$$

$$NADH + H^+ + \frac{1}{2}O_2 \rightarrow H_2O + NAD^+ \qquad E^{o\prime} = 1{,}140 \text{ mV}$$

The aerobic oxidation of NADH pairs a strong electron donor (NADH) with a strong electron acceptor (O_2). Thus, NADH oxidation via the ETS provides the cell with a huge amount of potential energy (comparable to a 1-V battery cell) to make ATP and generate ion gradients. NADH is oxidized by many electron transport systems of bacteria and archaea, as well as mitochondria.

Bacteria and archaea have evolved many alternative donor-acceptor systems; the one they use depends on what their environment provides. For example, in the absence of oxygen, *Escherichia coli* can oxidize NADH with fumarate, which is reduced to succinate. From **Table 14.1**, we find that

$$NADH + H^+ + \text{fumarate} \rightarrow NAD^+ + \text{succinate}$$

$$E^{o\prime} = 320 \text{ mV} + 33 \text{ mV} = 353 \text{ mV}$$

When oxygen reappears in the environment, the succinate can donate electrons back to fumarate and water:

$$\frac{1}{2}O_2 + 2H^+ + \text{succinate} \rightarrow H_2O + \text{fumarate}$$

$$E^{o\prime} = 820 \text{ mV} - 33 \text{ mV} = 787 \text{ mV}$$

> **THOUGHT QUESTION 14.1** *Pseudomonas fluorescens*, a soil bacterium, oxidizes NADH with nitrate (NO_3^-) at neutral pH. What is the value of $E^{o\prime}$?

Concentrations of electron donors and acceptors. The standard reduction potential $E^{o\prime}$ assumes 1 M concentrations of reactants and products at pH 7. Actual values of E within cells depend on actual reactant concentrations. In Chapter 13, we noted that ΔG includes a term incorporating the ratio of product concentrations (C and D) to reactant concentrations (A and B):

$$\Delta G = \Delta G^{\circ\prime} + 2.303\, RT \log \frac{[C][D]}{[A][B]}$$

where R is the gas constant (8.315 J/mol), and T is the temperature in Kelvins. The reduction potential E of a redox reaction depends on the ratio of reduced products to oxidized reactants:

$$E(mV) = \frac{-\Delta G}{nF}$$

$$= E^{\circ\prime} - \frac{2.303\, RT}{nF} \times \log \frac{[C][D](\text{reduced})}{[A][B](\text{oxidized})}$$

$$= E^{\circ\prime} - 60 \log \frac{[C][D](\text{reduced})}{[A][B](\text{oxidized})}$$

where n is the number of electrons transferred and F is the Faraday constant. As the ratio of reduced products to oxidized reactants increases, E decreases; in other words, there is less potential to do work. At moderate temperatures (25–40°C), the constants in the concentration term (2.303 RT/F) combine to yield approximately 60 millivolts (mV) per electron for a tenfold ratio of products to reactants. Thus, a tenfold ratio of products to reactants subtracts about 60 mV from ΔE (decreases energy yield), whereas a tenfold excess of reactants adds 60 mV to E (increases energy yield).

A high concentration of an electron donor can drive metabolism using reactions with small values of $E^{\circ\prime}$. For example, *Gallionella* species use the reduction potential between Fe^{3+}/Fe^{2+} and O_2/H_2O to oxidize Fe^{2+}. *Gallionella* bacteria grow in drainage from iron mines, where high concentrations of reduced iron (Fe^{2+}) support growth of thick bacterial biofilms.

> **THOUGHT QUESTION 14.2** Could a bacterium obtain energy from succinate as an electron donor with nitrate (NO_3^-) as an electron acceptor?

Electron Transport Systems Function within a Membrane

An electron transport system transfers electrons onto membrane-embedded carriers in a series of increasing reduction potential. Some of these steps yield energy to pump ions across the membrane. To store this energy, the ETS must maintain an ion gradient across a membrane that fully separates two aqueous compartments. For example, a bacterial cell membrane separates the cytoplasm and external medium. Gram-negative bacteria such as *E. coli* have their ETS in the inner cell membrane, which separates the cytoplasm from the periplasm. The gram-negative outer membrane surrounding the periplasm is permeable to protons and other small molecules; thus, the outer membrane does not store energy.

Figure 14.2A shows the inner membrane of *Sulfurospirillum barnesii*, a respiring bacterium isolated from farm drainage in Nevada. *S. barnesii* can respire using O_2 as a terminal electron acceptor—or it can use one of several alternative electron acceptors, such as selenium oxyanions (SeO_3^{2-} or SeO_4^{2-}). The oxyanions are reduced to pure selenium (Se), which collects in solid granules within the cell. Whether with O_2 or with an alternative electron acceptor, electrons are transferred from a reduced nutrient molecule (electron donor) through ETS components that pump protons across the membrane.

A similar structure is found in the ETS of our own mitochondria (**Fig. 14.2B**). Mitochondrial respiration uses only O_2 as a terminal electron acceptor. The ETS proteins are embedded in folds of the mitochondrial inner membrane called cristae. The inner membrane separates the inner mitochondrial space from the intermembrane space (between the inner and outer membranes). The mitochondrial inner membrane, including its electron

Figure 14.2 Respiratory membranes.
A. Electron transport occurs in the inner membrane of *Sulfurospirillum barnesii* (TEM). Reduction of selenium oxyanions (SeO_3^{2-} or SeO_4^{2-}) generates granules of selenium (Se^0). **B.** Mitochondria in a human pancreatic acinar cell (TEM). The inner membrane containing the electron transport system is organized in pockets called cristae, about 40 nm thick.

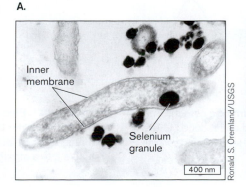

Inner membrane

Selenium granule

400 nm

Ronald S. Oremland/USGS

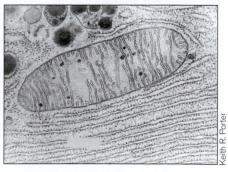

Keith R. Porter

transport proteins, evolved from the cell membrane of an endosymbiotic bacterial ancestor. Among modern bacteria, the closest relatives of mitochondria are the rickettsias, obligate intracellular parasites that cause serious diseases, such as Rocky Mountain spotted fever. Like mitochondria, the rickettsias show evidence of extensive evolutionary degeneration of genes no longer required for intracellular survival. Mitochondrial evolution is discussed further in Chapter 17.

The electron carrier molecules include proteins and small organic molecules called cofactors, which are loosely or tightly bound to the proteins. The protein components of an ETS were first discovered in the 1930s at Cambridge University by Russian scientist David Keilin (1887–1963). Keilin was an entomologist who studied insect mitochondria. The mitochondrial inner membranes contained proteins called **cytochromes**, which were named for their deep colors, typically red to brown.

In prokaryotes, cytochromes are often found in the cell membrane. **Figure 14.3** shows the absorbance spectrum of a cytochrome from the cell membrane of *Haloferax volcanii*, a halophilic archaeon isolated from the Dead Sea. In the reduced cytochrome, light absorption peaks in the blue range (440 nm) and in the red (607 nm). Upon oxidation, however, the cytochrome loses its absorption peak at 607 nm, and the 440-nm peak shifts to a shorter wavelength; its reflected color shifts red. Different species of bacteria make many different cytochromes, with different absorption peaks, but all cytochromes show spectral change with change in reduction state.

A membrane ETS typically includes several different cytochromes with different reduction potentials. Keilin proposed that the cytochromes pass electrons sequentially from each protein complex to the next stronger electron acceptor, with each step providing a small amount of energy to the organism (**Fig. 14.4**). The electron transport proteins are now generally known as **oxidoreductases** because they oxidize one substrate (removing electrons) and reduce another (donating electrons). Thus, they couple different half-reactions in the electron tower (**Table 14.1**). Oxidoreductases consist of multiple-protein complexes that include cytochromes as well as noncytochrome proteins. The structure and function of ETS oxidoreductase complexes are discussed in Section 14.3.

TO SUMMARIZE:

- **An electron transport chain (ETS)** consists of a series of electron carriers that sequentially transfer electrons to the carrier of next higher reduction potential (that is, next stronger electron acceptor). Electron flow through the ETS begins with an initial electron donor from outside the cell and ultimately transfers all electrons to a terminal electron acceptor that leaves the cell.

- **The reduction potential E** for a complete redox reaction must be positive to yield energy for metabolism. The standard reduction potential $E^{\circ\prime}$ indicates the voltage potential at which both members of a redox couple are present in equal concentration.

- **Concentrations of electron donors and acceptors** in the environment influence the actual reduction potential E experienced by the cell.

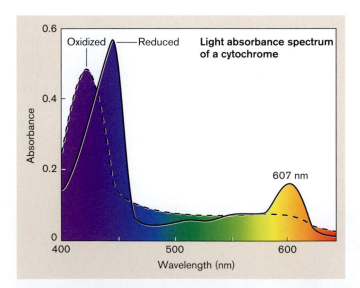

Figure 14.3 Light absorbance spectrum of a cytochrome. Absorption peaks shift between the oxidized and reduced forms of a cytochrome from the electron transport system of *Haloferax volcanii*, a halophilic archaeon.

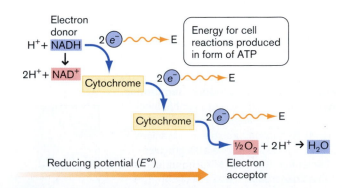

Figure 14.4 Electron transport system. Keilin's model for electron transfer through cytochromes. Each cytochrome in turn receives the electrons from a stronger electron donor and transfers it to a stronger electron acceptor.

- **The ETS is embedded in a membrane that separates two compartments.** Two aqueous compartments must be separate to maintain an ion gradient generated by the ETS.
- **The ETS is composed of protein complexes and cofactors.** Protein complexes called oxidoreductases include cytochromes and noncytochrome proteins. Cytochromes are colored proteins whose absorbance spectrum shifts when there is a change in redox state.

14.2 The Proton Motive Force

The sequential transfer of electrons from one ETS protein to the next yields energy to pump ions (in most cases H^+) across the membrane. Proton pumping generates a **proton motive force** (or **proton potential**) composed of the H^+ concentration difference plus the charge difference across the membrane. The proton motive force (PMF) can be used to drive many different cell processes, including ATP synthesis, flagellar rotation, and nutrient transport.

The Electron Transport System Pumps Protons

The ion pumped by most ETS complexes is H^+. In this book, we use *hydrogen ions* and H^+ interchangeably with *protons*. In fact, we do not know the precise form of the ion that crosses the membrane. In water solution, H^+ never occurs as such because it combines with a water molecule to form hydronium ion, H_3O^+. Current data, however, are consistent with the passage of isolated hydrogen nuclei (protons) across the proton pumps of the electron transport system.

The proton pump generates an H^+ concentration difference across the membrane. Also, the separation of charge generates a charge difference across the membrane. The H^+ concentration gradient (ΔpH) plus the charge difference ($\Delta\psi$) make a proton potential (Δp), or proton motive force. The proton potential (in volts) equals the electrochemical proton gradient (in joules) multiplied by Faraday's constant.

The discovery of the proton potential radically changed the field of biochemistry. Early in the twentieth century, Keilin and other scientists knew that the energy acquired by electron transport proteins was used to make ATP, but they did not know how. Most were convinced that electron transport was somehow directly coupled to ATP synthesis. The actual means of coupling was discovered by a student of Keilin's, Peter Mitchell (1920–1992). In 1961, Mitchell proposed an astonishing explanation

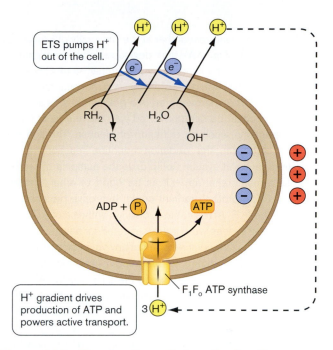

Figure 14.5 The chemiosmotic hypothesis. The chemiosmotic hypothesis states that the electron transport system pumps protons out of the cell. The resulting electrochemical gradient of protons (proton motive force) drives conversion of ADP to ATP through ATP synthase, as well as transport of various substrates such as sugars.

for the coupling of electron transport to ATP synthesis (**Fig. 14.5**). The **chemiosmotic hypothesis** states that the energy from electron transfer between membrane proteins is used to pump protons across the membrane, accumulating a higher H^+ concentration in the compartment outside. The pump generates the proton motive force Δp, which stores energy that can be used to make ATP (discussed in Section 14.3).

At first, many biologists found it difficult to understand how a proton potential could be coupled to ATP synthesis at an enzyme complex separate from the ETS, at a distant location on the membrane. The explanation is that the H^+ concentration gradient and charge difference exist everywhere on the membrane separating the two compartments. Because the membrane is impermeable to hydrogen ions, the H^+ current can only flow back through a proton-driven complex such as the membrane ATP synthase (discussed in section 14.3). In effect, the proton potential is a "proton battery," analogous to the electron potential of an electrical battery. The concept of cellular electricity proved highly controversial among scientists, who devised many experiments to test it before it was accepted (see **Special Topic 14.1**).

In 1961, as a young researcher at the University of Edinburgh, Peter Mitchell (**Fig. 1A**) first developed the chemiosmotic theory. He developed the idea of vectorial metabolism—that metabolic energy could be stored in a directional form, in an ion gradient. The energy stored in a gradient of ions across a membrane could be interconverted with the energy of chemical interactions, such as oxidation of glucose. Through vectorial metabolism, the oxidation-reduction reactions of the electron transport system could drive the pumping of protons across a membrane, as well as transport of other ions and nutrients. The proton gradient could then drive protons back through the F_1F_o ATP synthase, driving synthesis of ATP.

Mitchell's ideas were largely rejected for several reasons. Established colleagues had spent many years discovering the molecular "intermediates" of metabolism, and they expected to find one or two more between the ETS and the ATP synthase. Furthermore, most biochemists lacked sufficient training in physics to appreciate Mitchell's ideas, and Mitchell's communication skills did not cross the gap. Eventually, private family wealth enabled Mitchell to leave Edinburgh and form the Glynn Research Foundation. The foundation supported a bioenergetics laboratory at Glynn House, a country estate in Cornwall that Mitchell restored and equipped (**Fig. 1C**). At Glynn, Mitchell developed a research program to test various principles and predictions of the chemiosmotic theory.

During his graduate studies at Cambridge, Mitchell had collaborated with an outstanding experimentalist, Jennifer Moyle (**Fig. 1B**). Moyle was a student of the Cambridge biochemist Marjorie Stephenson, the first woman to be selected as a Fellow of the Royal Society. Stephenson suggested that Moyle work with Mitchell, and it was Moyle who performed most of the early experiments testing the chemiosmotic theory. Moyle continued working with Mitchell at Edinburgh, and she moved with him to Glynn, where they worked together for the next 30 years.

In the 1960s, Mitchell and Moyle devised several experiments to test the chemiosmotic theory. A key requirement was to show that the ETS generates a proton potential. Moyle showed that respiration of mitochondria was associated with proton efflux: Mitochondria isolated from rat livers were exposed to oxygen, causing efflux of hydrogen ions. The H^+ efflux occurred as electrons were transferred across the ETS and hydrogen ions were expelled by the proton pumps. In Moyle's experiments, the number of protons extruded per electron transferred down the ETS was consistent with the chemiosmotic model. Similar results were obtained using vesicles made first from mitochondrial membranes and later from the membranes of chloroplasts and bacteria.

A greater challenge was to demonstrate that a proton motive force could in fact drive the ATP synthase and other ion transporters without any undiscovered intermediate. Many researchers attempted to show this, often with conflicting results. The first successful experiment was reported in 1966 by André Jagendorf, a colleague from Johns Hopkins University who had visited Mitchell at Glynn. Jagendorf and his postdoctoral fellow Ernest Uribe tested the effect of a proton gradient imposed on spinach chloroplasts (**Fig. 2A**). Chloroplasts, unlike mitochondria and bacteria, contain ATP synthase directed outward (that is, driven by proton flow from inside to outside). The chloroplasts were partly opened by osmotic shock, then suspended in medium containing concentrated hydrogen ions (pH 4). With the osmotic balance restored, the chloroplast membranes closed again, with increased [H^+] trapped inside. When the pH outside was raised to pH 8.3 (that is, lower [H^+]), protons from the acidic interior of the

A.

Courtesy of Peter Rich, U. of London

B.

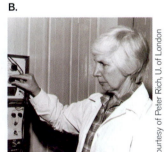

Courtesy of Peter Rich, U. of London

C.

Courtesy of Peter Rich, U. of London

Figure 1   Peter Mitchell and Jennifer Moyle proposed and tested the chemiosmotic hypothesis.   A. Peter Mitchell conducts a test of the chemiosmotic hypothesis. **B.** Jennifer Moyle uses an oxygen electrode for an experiment testing the prediction that a transmembrane voltage potential stores energy as a proton gradient. **C.** Glynn House, the Regency mansion in Cornwall, England, that Mitchell restored for his laboratory, is shown here in 1990 during a conference in Mitchell's honor, attended by five Nobel Prize winners.

vesicles flowed out through the ATP synthase, and ADP and inorganic phosphate were converted to ATP. Jagendorf interpreted this result as consistent with Mitchell's chemiosmotic hypothesis for the mechanism of phosphorylation.

A. A pH difference, ΔpH, drives ATP synthesis.

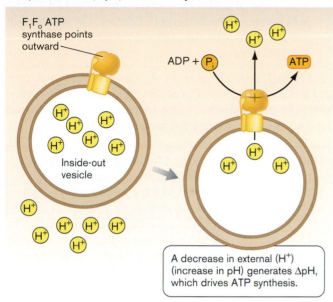

A decrease in external (H^+) (increase in pH) generates ΔpH, which drives ATP synthesis.

B. A charge difference, Δψ, drives ATP synthesis.

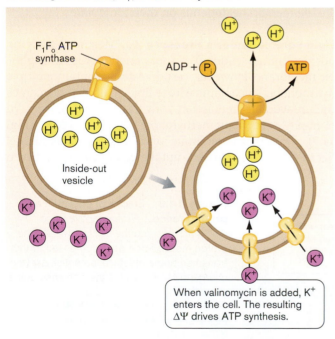

When valinomycin is added, K^+ enters the cell. The resulting ΔΨ drives ATP synthesis.

Figure 2   Either ΔpH or Δψ drives ATP synthesis.   A. A pH difference imposed across the chloroplast inner membrane drives a proton current through an outwardly directed ATP synthase. **B.** A charge difference generated by K^+ influx drives a proton current through ATP synthase.

Other researchers showed that ATP synthesis could be driven by a charge difference across a vesicle membrane (**Fig. 2B**). An artificial electrical potential (Δψ) was applied by loading vesicles with potassium ions (K^+). The potassium ions were conducted across the membrane by an ionophore, a small organic molecule that binds to an ion and solubilizes it in the membrane. The K^+ ionophore was the antibiotic valinomycin, a cyclic peptide of 12 amino acids produced by the bacterium *Streptomyces fulvissimus*. Valinomycin specifically binds K^+, while its hydrophobic side chains solubilize it in the membrane; thus the molecule conducts K^+ across the membrane down its concentration gradient (see **Fig. 2B**). In the experiment shown, the vesicles are "inside-out" bacterial membrane vesicles, in which the ATP synthase points outward instead of inward. The K^+ influx adds positive charge, which drives H^+ out through ATP synthase, catalyzing formation of ATP. Thus, Δψ drives formation of ATP.

The preceding experiments, however, could not definitively rule out the existence of an undetected component of the membrane that somehow energized ATP synthase. A more compelling experiment was performed in 1975 by Efrem Racker and colleagues at Cornell University. Racker developed a process to make artificial vesicles called "liposomes," composed of purified phospholipids in the absence of any membrane proteins. He then purified a proton pump called bacteriorhodopsin from a halophilic archaeon, a *Halobacterium* species. Bacteriorhodopsin acts as a single-protein proton pump driven by light absorption. The bacteriorhodopsin was combined with phospholipids and with ATP synthase purified from mitochondria to obtain "reconstituted liposomes" with functional bacteriorhodopsin and mitochondrial ATP synthase. When the liposomes were exposed to light, the bacteriorhodopsin pumped protons and the ATP synthase made ATP. Thus, a proton pump and a proton-driven ATP synthase from two completely different organisms could work together in a membrane, connected solely by the proton motive force.

In 1978, Mitchell received the Nobel Prize in Chemistry, and the scientific community fully embraced the role of ion currents in metabolism. Today the chemiosmotic principle is firmly established as the basis of energy transduction in respiration, lithotrophy, and photosynthesis. Nevertheless, interesting questions remain, some of which were addressed by Peter Rich (now at University College, London), who succeeded Mitchell as director of research at Glynn in 1987. Rich introduced modern forms of spectrophotometry to investigate the mechanisms of proton transport by specific ETS components, such as cytochrome oxidase. Other researchers are now investigating how extreme acidophiles and alkaliphiles manage to grow at pH values as low as pH 1 or as high as pH 14 while maintaining internal pH within two units of neutrality. Medical researchers, meanwhile, investigate how pathogenic bacteria and tumor cells use proton-driven efflux pumps to expel antibiotics from the cell.

The Proton Potential Includes the Electrical Potential and the H⁺ Concentration Gradient

When protons are pumped across the membrane, energy is stored in two different but energetically equivalent forms: the separation of charge, or electrical potential (because H⁺ carries a positive charge), and the gradient of H⁺ concentration, or pH difference, as shown in **Figure 14.6**. Either form (or both) can drive Δp-dependent cell processes.

Thus, the Δp contains two different components:

- **The electrical potential** ($\Delta\psi$) arises from the separation of charge between the cytoplasm (more negative) and the solution outside the cell membrane (more positive). For many bacteria, this "battery" potential is about −50 to −150 mV.
- **The pH difference** (ΔpH) is the log ratio of external to internal chemical concentrations of H⁺. For example, if the bacterial internal pH is 7.5 and the external pH is 6.5, the ΔpH is 1.0, and the ratio of $[H^+]_{out}$ to $[H^+]_{in}$ is 10. A ΔpH of 1.0 corresponds to a proton chemical concentration potential of approximately −60 mV.

The relationship between electrical and chemical components of the proton potential Δp is given by:

$$\Delta p = \Delta\psi - (2.3RT/F)\,\Delta\text{pH}$$

or approximately

$$\Delta p = \Delta\psi - 60\Delta\text{pH}$$

For cells grown at neutral pH, all three terms (Δp, $\Delta\psi$, and -60ΔpH) usually have a negative value, meaning that their force drives protons inward from outside.

In living cells, the relative contributions of $\Delta\psi$ and -60ΔpH vary depending on other sources of charge difference and protons, as shown in **Figure 14.6**. Both the $\Delta\psi$ and ΔpH components of Δp are influenced by other factors besides ETS proton transport. For example:

- **The charge difference $\Delta\psi$** includes charges on other ions, such as K⁺ and Na⁺. These ions are pumped or exchanged by ion-specific membrane transport proteins.
- **The pH difference ΔpH** is affected by metabolic generation of acids, such as fermentation acids, or by pH changes outside the cell. Permeant acids can run down the gradient through the membrane, collapsing the ΔpH to zero.

Overall, the cell uses its various membrane pumps and metabolic pathways to adjust and maintain its proton motive force at a size sufficient to drive ATP synthesis but not so great as to disrupt the membrane.

> **THOUGHT QUESTION 14.3**   What do you think happens to $\Delta\psi$ as the cell's external pH increases or decreases? What happens to Δp in bacteria growing at a pH two or three units above their internal pH?

How are the Δp and its two components actually measured? In larger cells, such as those of eukaryotes, a microelectrode can be inserted into the cell to measure voltage potentials directly. Most bacterial cells are too small to insert electrodes, but $\Delta\psi$ and ΔpH can be measured by the uptake of molecules whose ability to permeate the membrane depends on their ionic state.

Lipid-soluble cations measure $\Delta\psi$. Most ions carrying positive charge (cations) cannot penetrate the hydrophobic membrane. However, certain cations contain multiple hydrophobic side chains whose lipid solubility overcomes the unfavorable interaction with the positive charge. An example is triphenylmethylphosphonium (TPMP) (**Fig. 14.7**). TPMP can penetrate and cross the membrane. Because no H⁺ is carried, uptake of the cation is driven

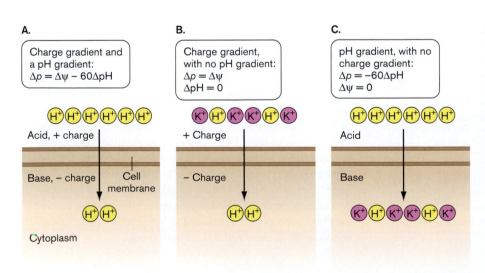

A.

Charge gradient and a pH gradient:
$\Delta p = \Delta\psi - 60\Delta$pH

Acid, + charge

Base, − charge   Cell membrane

Cytoplasm

B.

Charge gradient, with no pH gradient:
$\Delta p = \Delta\psi$
ΔpH = 0

+ Charge

− Charge

C.

pH gradient, with no charge gradient:
$\Delta p = -60\Delta$pH
$\Delta\psi = 0$

Acid

Base

Figure 14.6   Electrical potential and pH difference. Proton motive force drives protons into the cell (arrow). **A.** The proton motive force Δp is composed of the transmembrane electrical potential, $\Delta\psi$ (the charge difference), plus the transmembrane pH difference, ΔpH (the chemical concentration gradient of H⁺). **B.** If the pH inside and outside the cell are equal, then $\Delta p = \Delta\psi$. **C.** If the electrical charge inside and out are equal, then $\Delta p = -60\Delta$pH.

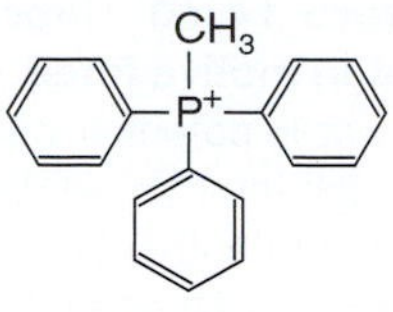

CH_3

Triphenylmethylphosphonium ion (TPMP)

Figure 14.7   A lipid-soluble cation.
Triphenylmethylphosphonium ion (TPMP) is soluble in the
membrane because its positive charge on the phosphorus is
surrounded by hydrophobic groups.

solely by the charge difference (usually negative inside
the cell) and thus affects only the electrical potential. A
radiolabeled TPMP will accumulate in the cell in proportion to the charge difference (**Fig. 14.8**). Thus, the cellular
uptake of a lipid-soluble cation such as TPMP can be used
to measure $\Delta\psi$.

Membrane-permeant weak acids measure ΔpH. Weak
acids such as benzoic acid cross the membrane primarily
in the uncharged form (HA) (**Fig. 14.9A**), as discussed in
Chapters 3 and 4. Many fermentation acids, such as acetate
and lactate, are membrane-permeant weak acids. At low
concentration, the weak acid acts as a proton carrier, soluble
in the membrane. Inside the cell, at higher pH, the HA dissociates to H^+ plus A^-. The acid accumulates inside until the
pH inside falls to the level of the pH outside. Thus, uptake
of a radiolabeled weak acid can be used to measure ΔpH.

Some weak acids cross the membrane in both charged
and uncharged forms. The weak acid 2,4-dinitrophenol
(DNP) dissociates to an anion whose charge is relatively
evenly distributed around the molecule; thus, it remains
overall hydrophobic enough to penetrate the membrane
(**Fig. 14.9B**). Because both protonated (uncharged) and

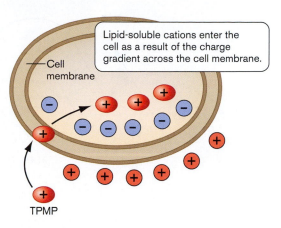

Lipid-soluble cations enter the
cell as a result of the charge
gradient across the cell membrane.

Cell
membrane

TPMP

Figure 14.8   Measuring the electrical potential, $\Delta\psi$.
Uptake of lipid-soluble cations is proportional to the
transmembrane charge difference, or electrical potential, $\Delta\psi$.

deprotonated (negatively charged) forms cross the membrane, DNP can cyclically bring protons into the cell and
collapse both $\Delta\psi$ and ΔpH (**Fig. 14.9C**). Such molecules
are called **uncouplers** because they uncouple electron
transport from ATP synthesis through the membrane ATP
synthase. Uncouplers are highly toxic to bacteria, and in
the early twentieth century, they were considered for use
as antibiotics. Unfortunately, they are also highly toxic to
human cells. All cells need to maintain ion gradients and
voltage potentials.

The Proton Potential Drives
Many Cell Functions

Besides ATP synthesis, Δp drives many cell processes
directly, such as the rotation of flagellar motors (**Fig.
14.10**). Proton transport is also coupled to transport of ions

**Figure 14.9   Membrane-
permeant weak acids
and uncouplers.   A.** The
transmembrane pH difference,
ΔpH, drives uptake of a
membrane-permeant weak acid
such as benzoic acid, which
dissolves in the membrane only
in its protonated form. **B.** If an
acid such as 2,4-dinitrophenol
crosses the membrane in both
protonated and unprotonated
forms, it cycles H^+ both ways
until the entire proton motive
force is dissipated. **C.** A weak
acid and an uncoupler cross the
membrane.

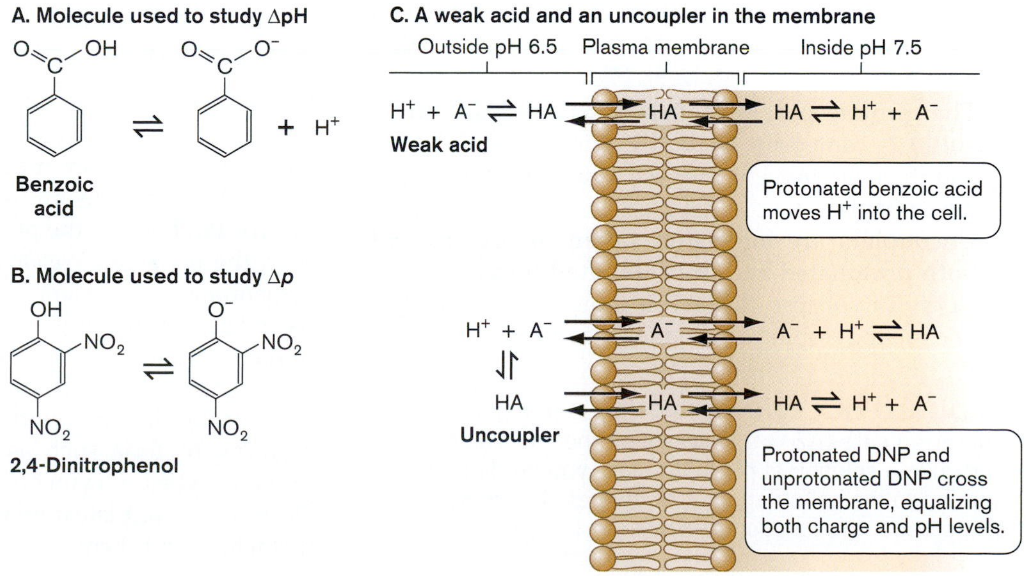

A. Molecule used to study ΔpH

Benzoic
acid

B. Molecule used to study Δp

2,4-Dinitrophenol

C. A weak acid and an uncoupler in the membrane

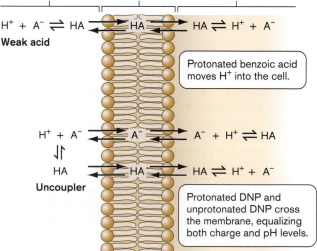

Outside pH 6.5   Plasma membrane   Inside pH 7.5

$H^+ + A^- \rightleftharpoons HA$ HA $HA \rightleftharpoons H^+ + A^-$
Weak acid

Protonated benzoic acid
moves H^+ into the cell.

$H^+ + A^-$ A^- $A^- + H^+ \rightleftharpoons HA$

HA   HA $HA \rightleftharpoons H^+ + A^-$
Uncoupler

Protonated DNP and
unprotonated DNP cross
the membrane, equalizing
both charge and pH levels.

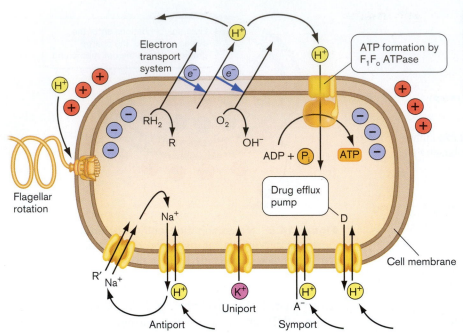

Figure 14.10   Processes driven by the proton motive force. Processes powered by proton potential include ATP synthesis through the F_1F_o ATPase, flagellar rotation, uptake of nutrients, and efflux of toxic drugs. R refers to an organic nutrient.

such as K^+ or Na^+ through parallel transport (symport) or oppositely directed transport (antiport), as discussed in Chapters 3 and 4. Ion flux then drives uptake of nutrients such as amino acids or efflux of molecules such as antibacterial drugs. Drug efflux pumps are a major problem in hospitals, where the pump proteins are encoded by genes carried on plasmids that spread among virulent strains.

TO SUMMARIZE:

■ **The ETS complexes generate a proton motive force** (Δp). The force is usually directed inward, driving protons into the cell.

■ **The proton potential drives ATP synthase** and other functions, such as ion transport and flagellar rotation.

■ **The proton potential (Δp, measured in millivolts)** is composed of the electrical potential ($\Delta\psi$) and the hydrogen ion chemical gradient (ΔpH): $\Delta p = \Delta\psi - 60\Delta pH$.

■ **Uncouplers are molecules taken up by cells in both protonated and unprotonated forms.** Uncouplers can collapse the entire proton potential, thus uncoupling respiration from ATP synthesis.

THOUGHT QUESTION 14.4   Suppose that de-energized cells ($\Delta p = 0$) with an internal pH 7.6 are placed in a solution at pH 6. What do you predict will happen to the cell's flagella? What does this demonstrate about the function of Δp?

14.3 The Respiratory ETS and ATP Synthase

In the respiratory electron transport system (ETS), a series of carrier molecules harvest the reducing potential of electrons in small steps. The respiratory ETS most commonly presented is that of the mitochondrial inner membrane, in which NADH and $FADH_2$ transfer electrons ultimately to O_2, producing H_2O. Microbes, however, have evolved a diverse array of electron carriers matched to diverse electron donors and acceptors.

Cofactors Provide Small Energy Transitions

ETS proteins such as cytochromes must associate electron transfer with small energy transitions. The small energy transitions are mediated by cofactors, small molecules that associate strongly or loosely with the protein. **Figure 14.11** shows the protein structure of a cytochrome from the pathogen *Pseudomonas aeruginosa*. The protein, cytochrome *c*, transfers electrons for nitrate respiration. Its reddish color derives from the buried cofactor called **heme**, a ring of conjugated double bonds surrounding an iron atom. The heme plays a key role in acquiring and transferring electrons, with an Fe^{2+}/Fe^{3+} transition. Metal-reducing bacteria such as *Geobacter* make hundreds of different types of cytochromes in their periplasm, where electrons accumulate until the bacterium finds an oxidized metal to accept them.

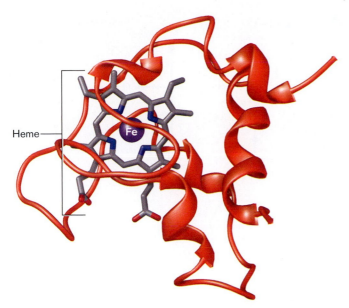

Figure 14.11 Cytochrome c. Portion of the protein structure of cytochrome c containing one heme cofactor, from the pathogen *Pseudomonas aeruginosa*. (PDB code: 2PAC)

Cofactors such as heme provide the capacity for small, reversible redox changes (**Fig. 14.12**). Examples include flavin mononucleotide (FMN) and ubiquinone. The structure of each cofactor must allow transition of an electron between orbitals with closely spaced energy levels to avoid "spending it all in one place." If all energy

were spent in one transition, most of it would be lost as heat instead of conversion to several small processes, such as pumping H^+ across the membrane.

Small energy transitions are typically accomplished by these kinds of molecular structures:

■ Metal ions such as iron or copper, coordinated (and hence held in place) with amino acid residues. Iron is often coordinated by sulfur atoms of cysteine residues in the protein; examples shown in **Figure 14.12B** are [2Fe-2S] and [4Fe-4S]. Transition metals make useful electron carriers because their outer electron shell has several closely spaced energy levels, facilitating small energy transitions.

■ Conjugated double bonds and heteroaromatic rings, such as the nicotinamide ring of $NAD^+/NADH$, also provide narrowly spaced energy transitions. Membrane-soluble carriers such as **quinones** (reduced to **quinols**) allow even smaller energy transitions than does $NAD^+/NADH$.

The major protein complexes of electron transport each contain one or more reaction centers containing either metal ions or conjugated double bonds or both. In FMN, the conjugated double bonds allow a small energy transition (**Fig. 14.12A**). In iron sulfur clusters (Fe_2S_2 and Fe_4S_4), the metal atoms provide the site for a small energy transition, its size dependent on the cluster's connections within the associated protein (**Fig. 14.12B**). The heme group found in cytochromes and other oxidoreductases contain

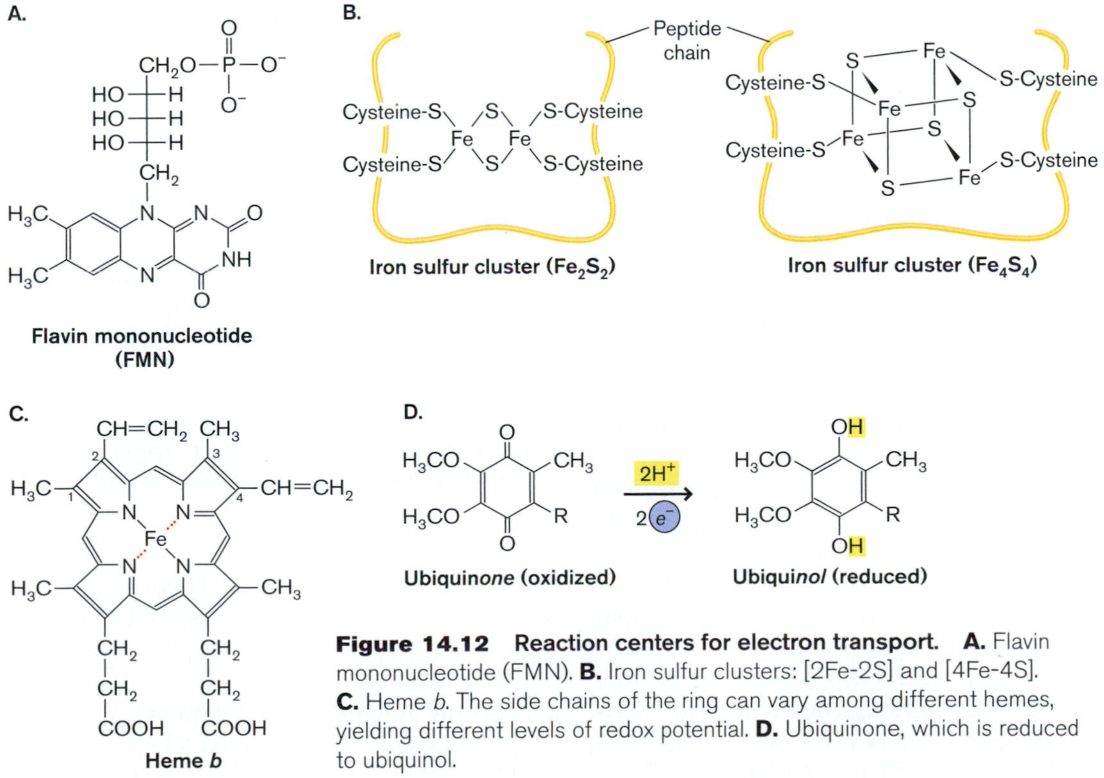

A.

Flavin mononucleotide
(FMN)

B.

Iron sulfur cluster (Fe_2S_2)

Iron sulfur cluster (Fe_4S_4)

C.

Heme *b*

D.

Ubiquin*one* (oxidized)   Ubiquin*ol* (reduced)

Figure 14.12 Reaction centers for electron transport. A. Flavin mononucleotide (FMN). **B.** Iron sulfur clusters: [2Fe-2S] and [4Fe-4S]. **C.** Heme *b*. The side chains of the ring can vary among different hemes, yielding different levels of redox potential. **D.** Ubiquinone, which is reduced to ubiquinol.

extensive conjugated double bonds, coordinated around the metal Fe^{3+} (**Fig. 14.12C**). Reduction by a transferred electron converts the iron to Fe^{2+}. The branching of side chains on the ring varies among different hemes, altering the magnitude of $E°'$ for the redox couple Fe^{3+}/Fe^{2+}.

Some electron carriers are small molecules that associate loosely with a protein complex, then come apart to diffuse freely throughout the membrane. Mobile electron carriers include quinones such as ubiquinone (**Fig. 14.12D**), which are reduced by hydrogenation to quinols:

$$Q + 2H^+ + 2e^- \rightarrow QH_2$$

Quinols carry electrons and protons laterally within the membrane between the proton-pumping protein complexes of the ETS. After transferring their electrons to the next protein complex, they revert to quinones, capable of accepting electrons again.

> **NOTE:** *Dehydrogenases, reductases,* and *oxidases* are all **oxidoreductases**, which oxidize one substrate (remove electrons) and reduce another (donate electrons). Oxidoreductases that accept electrons from NADH or $FADH_2$ are called dehydrogenases because their reaction releases hydrogen ions.

Oxidoreductase Protein Complexes

Here we present the molecular structures of a typical bacterial ETS conducting respiration. A respiratory electron transport system includes at least three functional components: a substrate oxidoreductase (or dehydrogenase), a mobile electron carrier, and a terminal oxidase. In some cases, alternative forms for each component allow access to alternative electron-donating substrates and terminal electron acceptors.

Substrate oxidoreductase. An ETS begins with an oxidoreductase that receives a pair of electrons from an organic substrate such as NADH, or an inorganic substrate such as H_2 or Fe^{2+}. Note that NADH arises through removal of $2H^+$ from an organic product of catabolism (designated RH_2) (**Fig. 14.13A**). The $2H^+$ are ultimately balanced by $2H^+$ from the cytoplasm combining with O_2 (or another terminal electron acceptor) at the end of the ETS.

NADH donates electrons to **NADH dehydrogenase (NADH:quinone oxidoreductase, NDH-1)** (**Fig. 14.13A**). In bacteria, the NDH-1 complex includes 14 different subunits; in mitochondria, there are 46, one of the most elaborate membrane complexes known. A bacterial complex includes the cofactor FMN as well as several iron sulfur clusters (typically $7Fe_4S_4$ and $2Fe_2S_2$). The cofactors "hand off" electrons to each other through adjacent connections; see, for example, the placement of FMN and the first Fe_4S_4 within the peptide coils of NDH-1 (**Fig. 14.13B**).

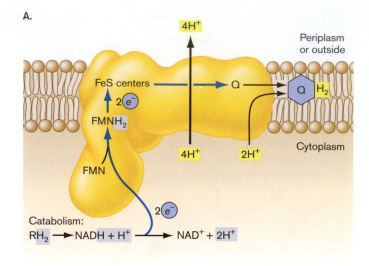

A.

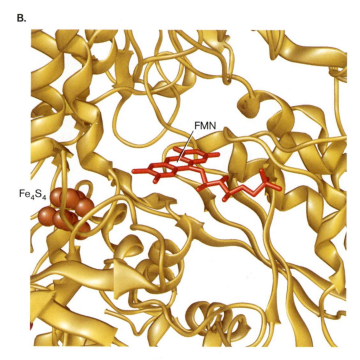

B.

Figure 14.13   NADH:quinone oxidoreductase complex (NDH-1).   A. NDH-1 transfers two electrons from NADH onto FMN ($FMNH_2$), through several FeS centers, ultimately onto a quinone, Q (forming quinol, QH_2). The energy from oxidizing NADH is coupled to pumping an additional $4H^+$ across the cell membrane. **B.** Within the NDH-1 complex, FMN lies adjacent to the first iron sulfur center, Fe_4S_4. (PDB code: 1NOX)
Source: B. L. A. Sazanov and P. Hinchliffe. 2006. Science 311:1430–1436.

Each electron from NADH travels individually through FMN and the iron sulfur series. At the end of the chain, the electrons are transferred to a quinone, which picks up $2H^+$ from solution and is thus reduced to quinol. Quinones are generally designated Q, and quinols QH_2.

Elsewhere within the NDH-1 complex, the oxidation of NADH and reduction of Q to QH_2 is coupled to pumping of up to 4H⁺ across the membrane, using the energy yielded by oxidation of NADH. Note that the 4H⁺ pumped across the membrane are distinct from the 2H⁺ transferred onto the quinone (Q → QH₂). In some halophilic bacteria, 4Na⁺ are pumped instead of 4H⁺. The mechanism of the coupling reaction remains unknown, although it is of interest because of its contribution to the proton potential. In human mitochondria, the NADH dehydrogenase (aka complex I) is critical for health; genetic defects in complex 1 are associated with diseases such as Parkinson's disease and some forms of diabetes.

> **NOTE:** In our figures, protons that are formed and consumed without affecting the transmembrane potential are shaded gray (see **Fig. 14.13A**). Protons that cross the membrane by the end of the ETS are highlighted yellow.

Not all substrate oxidoreductases pump protons. For example, *E. coli* has an alternative NADH dehydrogenase (NDH-2) that transfers two electrons to Q without pumping additional protons across the membrane. (The unused energy is lost as heat.) NDH-2 functions during rapid growth, when the cell must limit its proton potential to avoid membrane breakdown. Other complexes, such as succinate dehydrogenase, transfer electrons from substrates that lack sufficient energy to pump extra protons.

Quinone pool. A quinone can receive $2e^-$ from the substrate oxidoreductase. When it picks up 2H⁺ from solution, the quinone is converted to a quinol (**Fig. 14.12D**). The quinols diffuse within the membrane and carry reduction energy to other ETS components. After transferring $2e^-$ to the next protein complex, 2H⁺ are released in solution, usually across the membrane (**Fig. 14.14**). Thus quinol transfer contributes 2H⁺ to the transmembrane proton potential. The reoxidized carriers then recycle back as quinones.

Each quinone can bind to a substrate dehydrogenase, pick up a pair of electrons and hydrogen ions, then diffuse away and carry the electrons to a reductase. The quinones and quinols, referred to as the **quinone pool**, diffuse freely within the phospholipids of the membrane. Thus, the quinones/quinols are able to transfer electrons between many different redox enzymes.

Note that quinones with different side chains have different names, such as ubiquinone and menaquinone. For clarity, we refer to all as quinones (Q) and their reduced forms, quinols (QH₂).

Terminal oxidase. A terminal oxidase complex receives electrons from a quinol (QH₂) and transfers them to a terminal electron acceptor, such as O_2 or NO_3^- (**Fig. 14.14**). The complex usually includes a cytochrome that accepts electrons from quinols. Cytochromes of comparable function are designated by letters, such as cytochrome *b* (*E. coli* and mitochondria) and cytochrome *c* (mitochondria). The

Figure 14.14 Cytochrome *bo* quinol oxidase complex. Each quinol (QH₂) transfers two electrons to heme *b*. The two H⁺ from each quinol are expelled to the periplasm (or outside the bacterial cell). The $2e^-$ from the quinols are transferred to heme o_3, a step coupled to pumping of 2H⁺ across the membrane. At heme o_3, $2e^-$ combines with 2H⁺ plus an oxygen atom from O_2 to form water. *Source:* Based on the structure determined by Jeff Abramson, et al. 2000. *Nature Structural Biology* 7:910–917.

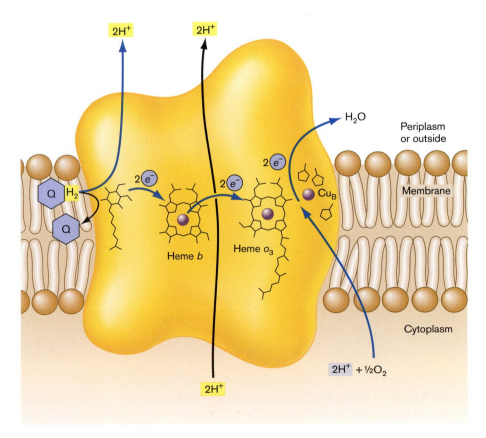

cytochrome is bound to an oxidase complex containing a series of electron-transferring carriers: two iron-centered hemes and three copper atoms. This unique center couples electron transfer and proton pumping.

The *E. coli* cytochrome *bo* quinol oxidase consists of cytochrome *b* plus oxidase complex *o*. The cytochrome *b* subunit receives two electrons from a quinol ($QH_2 \rightarrow Q$) and releases the $2H^+$ out to the periplasm. Each electron from quinol travels through the two hemes of the oxidase complex. Thus, because $2H^+$ were consumed from the cytoplasm at the NADH oxidoreductase (**Fig. 14.13**) and $2H^+$ were released outside by a quinol, there is net increase of $2H^+$ outside the cell. In addition, the transfer of $2e^-$ between the two oxidase hemes is coupled to pumping $2H^+$ from the cytoplasm across to the periplasm (**Fig. 14.14**).

The second heme of the oxidase (heme o_3) receives an atom of oxygen from O_2. Each oxygen atom in turn receives two electrons and combines with two protons ($2H^+$) from the cytoplasm to form H_2O. The $2H^+$ consumed balances the $2H^+$ released by catabolism to make $NADH + H^+$.

Note that the oxidase is conventionally shown as obtaining two electrons from the cytoplasm and donating them to $\frac{1}{2}O_2$, although, of course, $\frac{1}{2}O_2$ does not exist as such. A full reaction cycle of cytochrome *bo* quinol oxidase actually puts four electrons from two quinols (originally 2NADH) onto O_2, taking up $4H^+$ from the cytoplasm to make two molecules of H_2O.

Besides cytochrome *bo* quinol oxidase, bacteria express different terminal oxidases that differ with respect to the ratio of cytoplasmic protons pumped to electrons transferred. For example, the alternative cytochrome *bd* oxidase reduces O_2 to water but pumps no extra protons. Cytochrome *bd* is favored when external pH is already low and cell growth would be impaired by further acidification.

Electron Transport Pathways for Respiration

As we have seen, a respiratory ETS includes at least three phases of electron transfer: (1) Substrate electrons are donated to an oxidoreductase; (2) the electrons are transferred to a quinone, which is reduced to a quinol; and (3) the quinol electrons are transferred to a terminal oxidoreductase while its $2H^+$ are released outside. Both enzymes and quinones show considerable complexity and diversity among different species in different environments. **Figure 14.15** summarizes one example of a complete ETS for oxidation of NADH by $\frac{1}{2}O_2$ in the inner membrane of *E. coli*.

The entering carrier NADH carries two electrons with protons obtained from catabolized food molecules. The two electrons ($2e^-$) from NDH-1 and two protons ($2H^+$) from solution are transferred onto Q (quinone), converting it to QH_2 (quinol). The electrons transferred

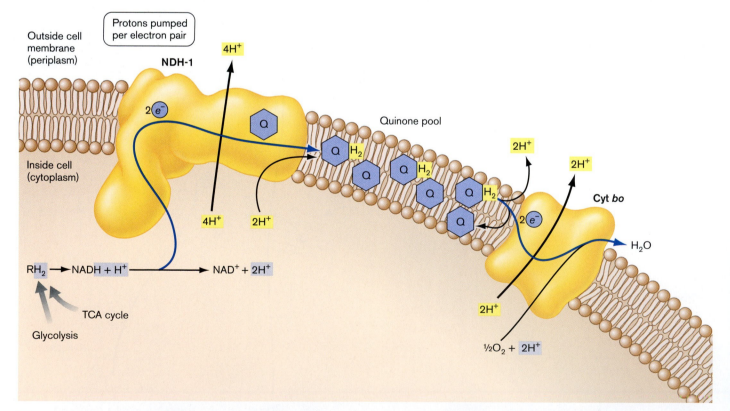

Figure 14.15   A bacterial ETS for aerobic NADH oxidation.   In *E. coli*, electrons from NDH-1 are transferred to quinones, generating quinols. The quinols transfer electrons onto Cyt *bo*. For each NADH oxidized, up to $8H^+$ may be pumped across the membrane. ▶❚❚

from NADH include sufficient energy to pump up to $4H^+$ across the membrane. The exact number depends on cell conditions, such as the concentrations of NADH and the terminal electron acceptor.

The hydrogenated QH_2 diffuses within the membrane until it reaches a terminal oxidase complex, such as the cytochrome *bo* quinol oxidase. The $2H^+$ from QH_2 are released outside the cell, while the $2e^-$ enter the oxidase, reducing the two hemes. The two electrons join $2H^+$ from the cytoplasm, combining with an oxygen atom to make H_2O. The reaction is coupled to pumping of $2H^+$ across the membrane plus a net increase of $2H^+$ outside through redox reactions.

Overall, the oxidation of NADH exports about $8H^+$ per $2e^-$ transferred through the ETS to make H_2O. The export of protons generates a proton potential Δp.

NOTE: Protons (hydrogen ions, H^+) can have three different fates in the ETS:

1. Protons are pumped across the membrane by an oxidoreductase complex; contributes Δp.
2. Protons are consumed from the cytoplasm by quinone/quinol, then released across the membrane; contributes Δp.
3. Protons are consumed by combining with the terminal electron acceptor (O_2). Balances protons released by catabolism; does not affect Δp.

THOUGHT QUESTION 14.5 What is the advantage of the oxidoreductase transferring electrons to a pool of mobile quinones, which then reduce the terminal reductase (cytochrome complex)? Why does each oxidoreductase not interact directly with a cytochrome complex?

THOUGHT QUESTION 14.6 Why are most electron transport proteins fixed within the cell membrane? What would happen if they "got loose" in aqueous solution?

Environmental modulation of the ETS. The ETS just described represents optimal conditions, when food and oxygen are unlimited. What happens in tough conditions, when food (that is, electron donors) or oxygen is scarce? Under varying environmental conditions, bacteria adjust the efficiency of their ETS by expressing alternative oxidoreductases. This adjustment depends on available concentrations of electron donors (such as NADH) and electron acceptors (such as O_2). Available concentrations of substrates determine the actual reduction potential E for NADH oxidation and hence the proton motive force Δp.

The overall equation for NADH oxidation by O_2 is

$$2NADH + 2H^+ + O_2 \rightarrow 2NAD^+ + 2H_2O$$

The reduction potential E depends on the concentrations of reactants and products.

$$E = E^{\circ\prime} - \frac{2.303\,RT}{nF} \times \log \frac{[NAD^+]^2}{[NADH]^2[H^+]^2[O_2]}$$

At low concentration of NADH or O_2 or at high pH (low H^+ concentration), the concentration term is negative, so the amount of energy released by the oxidation of NADH by O_2 is decreased. This means that fewer protons can be pumped per electron. Thus, it is advantageous for *E. coli* to have a range of options for its electron transport, pumping two to eight protons per NADH as needed.

The efficiency of the bacterial ETS is adjusted by making different NADH dehydrogenases and terminal oxidases under different environmental conditions (**Fig. 14.16**). In *E. coli*, the ETS components are regulated as follows:

- **High oxygen level.** When O_2 concentration is high, some electrons are donated to NDH-2 instead of NDH-1, allowing NAD^+ to be recycled without pumping extra protons. This is necessary to avoid generating too much negative potential and too high a pH in the cytoplasm.
- **Low oxygen level.** Cytochrome *bo* quinol oxidase requires a high concentration of O_2 to pump protons. At low $[O_2]$, electrons are transferred instead from cytochrome *bd* quinol oxidase, which has a higher

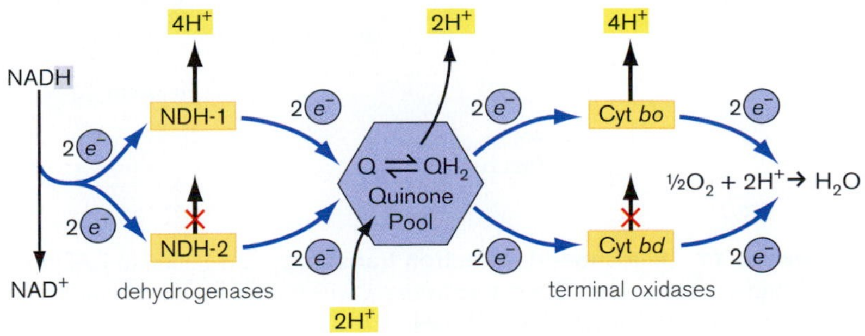

Figure 14.16 Alternative substrate dehydrogenases and terminal oxidases.
NADH:quinone oxidoreductase complex (NDH-1) pumps $2H^+$ whereas NDH-2 pumps none. Cyt *bo* pumps an extra proton per electron, whereas Cyt *bd* only exports protons from QH_2 and consumes $2H^+$ to make H_2O.

affinity for O_2. However, the *bd* oxidase cannot pump protons. So at low oxygen level, expression of NDH-2 is repressed, allowing maximal proton pumping by NDH-1.

■ **High external pH.** When external pH exceeds the pH of the cytoplasm, the cell needs to minimize proton export to maintain a constant internal pH. So the *bd* oxidase is expressed in preference to the proton-pumping *bo* oxidase.

> **THOUGHT QUESTION 14.7**   Why would bacteria evolve dehydrogenases that fail to pump protons even when sufficient donor potential exists?
>
> **THOUGHT QUESTION 14.8**   Experiments suggest that NADH dehydrogenase 2 is more often coupled with cytochrome *bo*, and dehydrogenase 1 more often with cytochrome *bd*. Why might this be the case?

Bacteria also have alternative oxidoreductases to serve different electron donors and acceptors. Some enzymes take electrons from donors lacking the potential of NADH; for example, succinate dehydrogenase catalyzes the one step of the TCA cycle that yields $FADH_2$, a step that provides not quite enough energy to produce NADH (discussed in Chapter 13). Succinate dehydrogenase is

the only TCA enzyme embedded in the membrane as an ETS component. Other complexes donate electrons to anaerobic electron acceptors such as nitrate, or they accept electrons from inorganic electron donors (discussed in Sections 14.4 and 14.5).

Mitochondrial respiration. In contrast to bacteria, mitochondria have only a single ETS, optimized for a relatively uniform intracellular environment (**Fig. 14.17**). Protected by the eukaryotic cytoplasm, mitochondria do not need to maintain alternative versions of their respiratory ETS. Instead, a set of just four electron-carrying complexes has evolved so as to maximize the energy obtained from NADH, minimizing energy lost as heat. This highly efficient mitochondrial respiration is nearly universal throughout the cells of animals, plants, and most eukaryotic microbes.

The mitochondrial ETS differs from that of *E. coli* in these respects:

■ **An intermediate cytochrome oxidase complex.** Besides NADH dehydrogenase (complex I) and cytochrome *c* oxidase (complex IV), both of which have homologs in *E. coli*, mitochondria show an intermediate step of electron transfer to ubiquinol-cytochrome *c* oxidoreductase (complex III, shown in **Fig. 14.17**). The intermediate electron transfer step pumps an

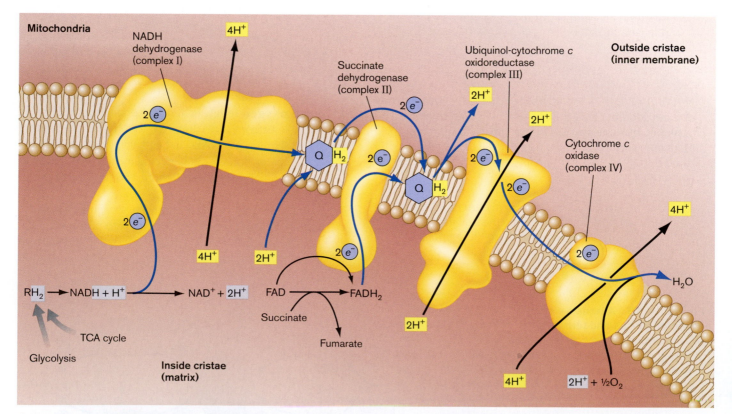

Figure 14.17   Mitochondrial electron transport.   In addition to NADH dehydrogenase and a terminal cytochrome oxidase, mitochondria possess cytochrome *bc* oxidase, which provides an intermediate electron transfer step. As a result, mitochondrial membranes export 10–12 H^+ per NADH.

additional $2H^+$. Another $2e^-$ may come from succinate dehydrogenase (complex II), which forms $FADH_2$ through the TCA cycle.

- **The mitochondrial ETS pumps more protons per NADH.** As many as 10–12 protons may be pumped per NADH, in contrast to 2–8 protons in *E. coli*.
- **Homologous complexes have numerous extra subunits.** The mitochondrial homologs of bacterial enzymes generally have evolved additional nonhomologous subunits specific to eukaryotes. For example, the homologous subunits of cytochrome *c* complex are enveloped by a series of eukaryotic-specific proteins.

The Proton Potential Drives ATP Synthesis

The proton potential drives synthesis of ATP by the membrane ATP synthase, also known as the F_1F_o ATP synthase. The proton-driven synthesis of ATP completes the cycle of oxidative phosphorylation, in which hydrogen atoms from organic foods are transferred to $NADH + H^+$. The same type of ATPase can use the proton potential generated by lithotrophy (see Section 14.5) or by phototrophy (see Section 14.6).

The F_1F_o ATP Synthase

The F_1F_o ATP synthase is a protein complex highly conserved in the bacterial cell membrane, the mitochondrial inner membrane, and the chloroplast thylakoid membrane. An elegant molecular machine, the ATP synthase is composed of two rotating complexes, F_o and F_1 (**Fig. 14.18**). The F_o complex translocates protons across the membrane. Twelve *c* subunits form a cylinder embedded in the membrane, stabilized by subunits *a* and *b*. The F_1 complex consists of six alternating subunits of types alpha and beta surrounding a gamma subunit that acts as a drive shaft. Each "third" of the F_1 complex (an alpha plus a beta subunit) interconverts $ADP + P_i$ with $ATP + H_2O$. The gamma subunit connects the tripartite "knob" of F_1 to the membrane-bound F_o. Proton transport through F_o drives ATP synthesis by F_1.

One proton at a time enters the subunit *a* channel and moves into a subunit *c* of F_o (**Fig. 14.19** ▶❚). The proton potential directed inward ensures that protons more often enter from the outside than from the cytoplasm. Each entering proton causes a subunit *c* to rotate around the axle, with release of a bound proton to the cytoplasm. The flux of three protons through F_o is coupled to forming one molecule of ATP by one alpha-beta-gamma unit of F_1. During each cycle generating ATP, the 12 c subunits of the F_o rotor rotate in the membrane one-third of one turn relative to the F_1 in the cytoplasm.

Note that the generation of ATP is completely reversible, so that ATP hydrolysis by F_1 can pump protons back through F_o across the membrane. In the absence of a pro-

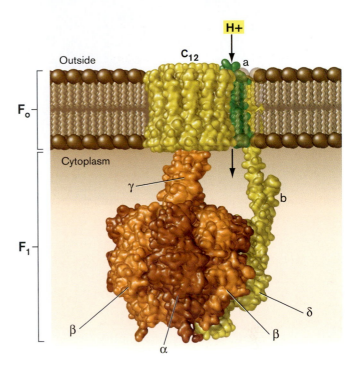

Figure 14.18 Bacterial membrane-bound ATP synthase (F_1F_o ATPase). The F_1F_o complex plugged into the *E. coli* plasma membrane. The three pairs of alpha and beta subunits rotate around the gamma axle, catalyzing formation and hydrolysis of ATP. The ring of *c* subunits rotates while translocating three protons. (PDB codes: 1B9U, 1C17, 1E79, 2CLY)

ton potential, a high ATP concentration can drive the F_o in reverse, actually pumping protons to generate Δp. This reversal of ATP synthase is used, for example, by *Enterococcus faecalis*, enteric gram-positive cocci that have no electron transport system but generate ATP entirely by fermentation and substrate phosphorylation. *E. faecalis* can operate the membrane-bound F_1F_o ATP synthase in reverse, consuming ATP and thus generating a proton potential for nutrient uptake and ion transport.

How do alkaliphilic bacteria make ATP and maintain a proton potential when growing in a high pH environment, such as the Dead Sea? Their ATP synthase looks surprisingly similar to that of bacteria living at neutral pH, but its peptide sequences differ at key amino acid positions (see **Special Topic 14.2**).

THOUGHT QUESTION 14.9 How might the bacterial cell compensate for environmental conditions in which Δp falls below 100 mV, a suboptimal level for ATP synthesis?

THOUGHT QUESTION 14.10 Would *E. coli* be able to grow in the presence of an uncoupler that eliminates the proton potential supporting ATP synthesis?

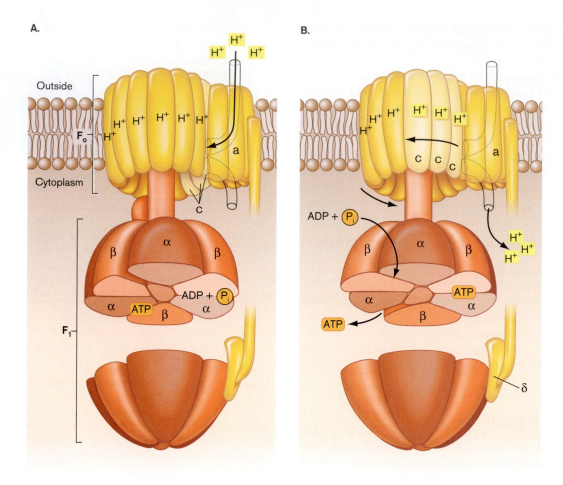

Figure 14.19 **H⁺ flux drives ATP synthesis.** **A.** Three protons enter *c* subunits of the F_o complex. **B.** The three *c* subunits rotate ⅓ turn relative to the F_1. Flux of three protons through F_o is coupled to F_1 converting ADP + P_i to ATP. ◐ *Source: Animation by Don Nicholson.*

Na⁺ Pumps: An Alternative to H⁺ Pumps

While a proton potential provides primary energy storage for most species, some bacteria generate an additional potential of sodium ions. A sodium motive force (ΔNa^+) is analogous to the proton motive force in that it includes the electrical potential $\Delta\psi$ plus the sodium ion concentration gradient (log ratio of the Na^+ concentration difference across the membrane). In extreme halophilic archaea, which grow in concentrated NaCl, the sodium potential entirely substitutes for the proton potential to drive ATP synthesis. These "haloarchaea" make use of the high external Na^+ concentration to store energy in the form of a sodium potential.

In some bacteria, an ETS oxidoreductase pumps Na^+ instead of H^+. For example, the proton-pumping NADH dehydrogenase can be supplemented by an NADH dehydrogenase that pumps Na^+. This primary sodium pump is found in many pathogens, including *Vibrio cholerae*, the cause of cholera, and *Yersinia pestis*, the cause of bubonic plague. These pathogens use the sodium-rich circulatory fluids of their hosts to store energy in a sodium potential.

TO SUMMARIZE:

- **Electron carriers** containing metal ions and/or conjugated double-bonded ring structures are used for electron transfer. Cytochrome *c* quinol oxidase has two hemes and three copper ions.
- **A substrate dehydrogenase** receives a pair of electrons from a particular reduced substrate such as NADH. NADH dehydrogenase (NADH:quinone oxidoreductase) typically has an FMN carrier and nine iron sulfur clusters.
- **Quinones receive electrons** from the substrate dehydrogenase and become hydrogenated to quinols. Typically, a quinol receives $2H^+$ from the cytoplasm, then releases $2H^+$ across the membrane upon transfer of $2e^-$ to an electron acceptor complex.
- **Protons are pumped** by substrate dehydrogenases (oxidoreductases) and terminal oxidases. The bacterial cytochrome *bo* oxidase pumps $4H^+$ across the membrane, coupled to transfer of $4e^-$ onto O_2 to generate $2H_2O$.
- **Protons are consumed** by combining with the terminal electron acceptor, such as combining with oxygen to make H_2O.

- **The number of protons pumped by a bacterial ETS** is determined by environmental conditions, such as the concentrations of substrate and terminal electron acceptor.
- **The proton potential drives ATP synthesis** through the membrane-bound F_1F_o ATP synthase. Three protons drive each F_1F_o cycle, synthesizing one molecule of ATP.
- **An Na^+ potential (ΔNa^+) can drive ATP synthesis**, replacing or supplementing the proton potential for some bacteria and archaea.

14.4 Anaerobic Respiration

Nearly all multicellular animals and plants require electron transport to oxygen. Bacteria and archaea, however, sample an extraordinary menu of terminal electron acceptors, including metals, oxidized ions of nitrogen and sulfur, and chlorinated organic molecules. This **anaerobic respiration** generally occurs in environments where oxygen is scarce, such as in wetland soil and water and within the human digestive tract.

Alternative Electron Acceptors and Electron Donors

Respiratory bacteria usually possess several different terminal oxidoreductases to reduce alternative electron acceptors, as shown for *E. coli* in **Figure 14.20**. These enzymes, although comparable to cytochrome quinol oxidases, are conventionally termed "reductases" to emphasize reduction of the alternative electron acceptor. Some of the electron acceptors are inorganic, such as nitrate (NO_3^-) reduced to nitrite (NO_2^-), or NO_2^- reduced to NO. Others are organic products of catabolism; for example,

the TCA cycle intermediate fumarate can be reduced to succinate. Organic electron acceptors play important roles in food decomposition. For example, the substance trimethylamine oxide, used by fish as an osmoprotectant against sea salt, is reduced by bacteria to trimethylamine—the main cause of the "fishy" smell.

At the top of the ETS, alternative dehydrogenases receive electrons from different organic electron donors as well as from molecular hydrogen (H_2). The various dehydrogenases all "connect" to various terminal reductases through the pool of quinones. Note that the various electron donors differ greatly in their reduction potential and hence in their capacity to generate proton potential (see **Table 14.1**). In a given environment, the strongest electron donor and the strongest electron acceptor available may be used. The best donor and best acceptor usually induce expression of genes encoding their respective redox enzymes. For example, in the presence of nitrate, genes encoding nitrate reductase are expressed. At the same time, nitrate represses the expression of reductases for poorer electron acceptors, such as fumarate. (The mechanisms of gene induction and repression are discussed in Chapter 10.)

Nitrogen and Sulfur: Oxidized Forms Accept Electrons

Oxidized forms of nitrogen and sulfur, usually anions such as nitrate or sulfate, can accept electrons from many species of soil and water bacteria. The nitrogen series offers an abundant source of strong electron acceptors. Reduction of oxidized states of nitrogen for energy yield is called **dissimilatory denitrification**. Dissimilatory nitrate reduction contributes to respiration, whereas assimilatory reduction of nitrate leads to ammonia for fixation into biomass (discussed in Chapters 15 and 22).

Figure 14.20 Alternative electron donors and electron acceptors. *Escherichia coli* can oxidize various foods (electron donors) while reducing various electron acceptors. Each electron donor may utilize alternative dehydrogenases, depending on the environmental conditions. Each dehydrogenase sends protons through the quinol pool to the terminal reductases.

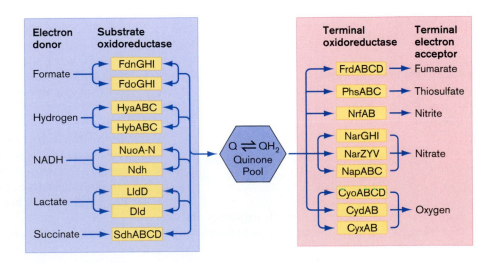

Special Topic 14.2   ATP Synthesis at High pH

Alkaliphilic bacteria isolated from soda lakes, such as Lake Natron in Tanzania, grow at pH values as high as pH 14 while maintaining internal pH at 9–10. Thus, their ΔpH is inverted, with 10,000-fold higher H^+ concentration inside the cell. How do these alkaliphiles maintain a proton motive force driving protons inward? Maintaining internal pH several units lower than outside the cell seems incompatible with generation of a proton potential that runs ATP synthase. So how do these organisms make ATP?

One model to explain ATP synthase coupling at high pH was proposed by Terry Krulwich, at Mount Sinai School of Medicine (**Fig. 1**). Krulwich proposed that protons pumped by the electron transport system are sequestered just outside the membrane and directed into the ATP synthase (**Fig. 1B**). This model (as yet unproven) would require a cell compartment, or trap, for protons outside the cell membrane or some kind of direct connection between the ETS proton pumps and the ATP synthase.

A.

B.

Courtesy of Terry A. Krulwich

Figure 1   Alkaliphiles maintain an inverted pH gradient. At external pH several units higher than internal pH, how do alkaliphilic bacteria make ATP? **A.** Terry Krulwich, researcher on alkaliphilic bacteria. **B.** Proposed model for how protons pumped by the electron transport system might be sequestered just outside the membrane and directed into the ATPase.

Oxidized forms of nitrogen. In the nitrogen redox series, a given oxidation state can serve as an acceptor in one redox couple but as a donor in the next. The redox couples are summarized here:

$$\underset{\text{Nitrate}}{NO_3^-} \xrightarrow{2e^-} \underset{\text{Nitrite}}{NO_2^-} \xrightarrow{e^-} \underset{\substack{\text{Nitric}\\\text{oxide}}}{NO} \xrightarrow{e^-} \underset{\substack{\text{Nitrous}\\\text{oxide}}}{\tfrac{1}{2}N_2O} \xrightarrow{e^-} \underset{\substack{\text{Nitrogen}\\\text{gas}}}{\tfrac{1}{2}N_2}$$

In general, a single bacterial species expresses the terminal reductase for only one or two transformations in the series, such as reduction of nitrate and nitrite, while growing in communities that include species conducting the next steps. Each nitrogen state requires a specialized reductase, such as nitrate reductase or nitrite reductase, to receive electrons from the ETS. The full series of nitrate reduction to N_2 plays a crucial role in producing the nitrogen gas of Earth's atmosphere (discussed in Chapter 22).

An alternative option for many soil bacteria, such as *Bacillus* species, is to reduce nitrite to ammonium ion:

$$NO_2^- + 8H^+ + 6e^- \rightarrow NH_4^+ + 2H_2O$$

The structure of the hypothetical proton coupling connection is unclear, but one prediction from the model is that the membrane-bound F_o component of ATP synthase, which transduces proton flow, would have a form specialized for alkaliphilic conditions. Krulwich and colleagues compared the peptide sequences of F_o from an alkaliphilic species, *Bacillus pseudofirmus* OF4, with that of a neutralophilic species of the same genus, *B. megaterium*. Six amino acid differences were observed in the sequences of the membrane-bound *a* and *c* subunits, which mediate the proton flux.

Mutant strains of the alkaliphile *B. pseudofirmus* were obtained, each of which had an amino acid substitution at one of the sites 1–6 OF4 (**Fig. 2**). Four of the respective mutants failed to synthesize ATP when grown at high pH (mutation sites 1, 2, 4, and 6). This observation shows that the alkaliphile has evolved a form of ATP synthase specifically adapted to function at high external pH. It is consistent with a model that the alkaliphile's cell membrane contains a special proton compartment or coupling device to drive ATP synthesis, although other explanations are possible.

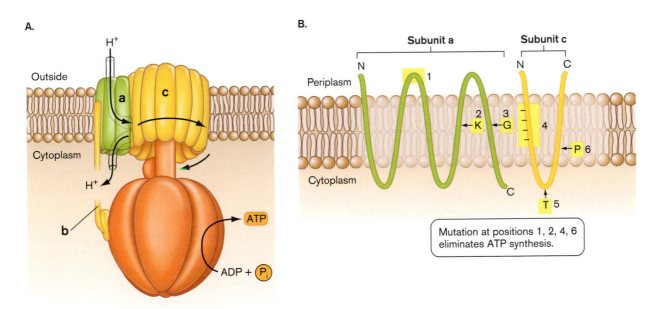

Figure 2   Mutations that inactivate ATP synthase in an alkaliphile.   A. ATP synthase structure in the alkaliphile *B. pseudofirmus* OF4, showing membrane location of subunits *a* and *c*. **B.** Location of mutant residues in subunits *a* and *c* that inactivate the ATP synthase at high pH.

Dissimilatory nitrate and nitrate reduction to ammonia can increase the soil pH, precipitating metals such as iron.

The presence of a particular reductase can be used as a diagnostic indicator for clinical isolates of bacteria. For example, the chemical test for nitrate reduction is a key step in the diagnosis of *Neisseria gonorrheae*, the causative agent of gonorrhea. *N. gonorrheae* happens to lack the terminal reductase for nitrate; hence, it tests negative, whereas several closely related species test positive.

Respiration using oxidized forms of nitrogen is widespread among bacteria and has been found in a growing number of archaea. Most eukaryotes must breathe oxygen, but some eukaryotic microbes respire on nitrate and nitrite. In anaerobic soil, many yeasts and filamentous fungi can reduce nitrate to nitrite, and nitrite to nitrous oxide. In the digestive tract of termites and of ruminant vertebrates, many fungi and protists grow anaerobically.

Oxidized forms of sulfur. A similar series of oxidized sulfur molecules serve as electron acceptors for bacteria that synthesize appropriate reductases:

$$\underset{\text{Sulfate}}{SO_4^{2-}} \xrightarrow{2e^-} \underset{\text{Sulfite}}{SO_3^{2-}} \xrightarrow{2e^-} \underset{\substack{\text{Thiosulfate}}}{\tfrac{1}{2}S_2O_3^{2-}} \xrightarrow{2e^-} \underset{\substack{\text{Elemental}\\\text{Sulfur}}}{S^0} \xrightarrow{2e^-} \underset{\substack{\text{Hydrogen}\\\text{Sulfide}}}{H_2S}$$

Their redox potentials are generally lower than those for oxidized nitrogen forms. Nevertheless, sulfate and sulfite receive electrons from many kinds of electron donors, including acetate, hydrocarbons, and H_2. Sulfate-reducing bacteria and archaea are widespread in marine habitats, from the arctic waters to submarine thermal vents. The ubiquity of sulfate reduction may be related to the high sulfate content of seawater, in which SO_4^{2-} is the most common anion after chloride.

Dissimilatory Metal Reduction

An important class of anaerobic respiration involves the reduction of metal cations, or **dissimilatory metal reduction**. The term *dissimilatory* indicates that the metal reduced as a terminal electron acceptor is excluded from the cell. This is in contrast to minerals reduced for the purpose of incorporation into cell components (assimilatory metal reduction). The metals most commonly reduced through anaerobic respiration are iron ($Fe^{3+} \rightarrow Fe^{2+}$) and manganese ($Mn^{4+} \rightarrow Mn^{2+}$), but virtually any metal with multiple redox states can be reduced or oxidized by some bacteria.

Metal electron acceptors require a specific oxidoreductase (or reductase). Metal ions in solution interact directly with reductases; but oxidized metals such as Fe^{3+} are often barely soluble in water. How does a bacterial reductase interact with an insoluble source of metal outside the cell? One way that soil bacteria reduce metals is to "shuttle" electrons using quinone-like degradation products of lignin, a complex aromatic substance that forms the bulk of wood and woody stems. Lignin degradation products are also called "humics" from their presence in humus, the organic components of soil. Humics accumulate in anaerobic environments, where decomposition is slow. Some are quinones (or their reduced counterparts, quinols). The extracellular quinols can shuttle electrons from the cell's ETS to extracellular metals.

Anoxic or anaerobic environments, such as the bottom of a lake or wetland sediment, offer a series of different electron acceptors (**Fig. 14.21**). The stronger electron acceptors will be consumed, in turn, by species that have the terminal oxidases to use them. For example, as oxygen grows scarce, nitrate is reduced to nitrogen gas ($NO_3^- \rightarrow N_2$) by nitrifying bacteria. As nitrate is used up, other species reduce manganese ($Mn^{4+} \rightarrow$

Mn^{2+}), iron ($Fe^{3+} \rightarrow Fe^{2+}$), and sulfate ($SO_4^{2-} \rightarrow H_2S$). At the bottom of the lake or sediment, carbon dioxide is reduced to methane by methanogenesis. Methanogenesis is conducted by archaea (see Section 14.5). Microbial reduction of minerals plays a critical role in every ecosystem and participates in the geochemical cycling of elements throughout Earth's biosphere (discussed in Chapter 22).

The reduction of Fe^{3+} is of great interest from the standpoint of early evolution of the biosphere. The biosphere of early Earth lacked molecular oxygen but had plenty of iron, the most abundant element in our planet. Most of the iron would have existed in the reduced, soluble form, Fe^{2+}. Without the ozone layer, however, ultraviolet rays may have reacted with Fe^{2+}, removing an electron to produce Fe^{3+}. Fe^{3+} could have been used by early hydrogenotrophs and even phototrophs as a terminal electron acceptor. Cycles of microbial iron reduction and oxidation are believed to have generated the major iron ores we mine for iron metal worldwide (discussed in Chapter 17).

A particularly interesting family of metal-reducing microbes is Geobacteraceae, whose members reduce metals such as iron and manganese as well as uranium ($U^{6+} \rightarrow U^{4+}$). Reduced uranium is insoluble in water, precipitating as uranite (UO_2). *Geobacter metallireducens* can use uranium reduction to respire on simple organic substrates such as acetate, providing a way to remediate uranium-contaminated water. The U.S. Department of Energy has used uranium-reducing bacteria for remediation at Rifle, Colorado, where uranium contamination has threatened the Colorado River. In this process, acetate is pumped into the water table, where *G. metallireducens* oxi-

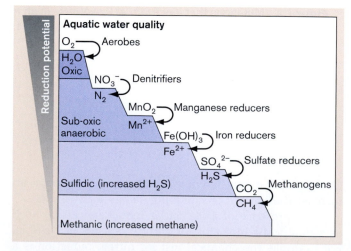

Figure 14.21   Anaerobic respiration in a lake water column. As each successive terminal electron acceptor is used up, its reduced form appears. The next best electron acceptor is then used, generally by a different species of microbe.

dizes the substrate to CO_2 by reducing U^{6+} to U^{4+}. The reduced uranium then precipitates out of the water into the soil, where it can be collected and removed, while the cleansed water flows through.

> **THOUGHT QUESTION 14.11** The proposed scheme for uranium removal requires injection of acetate under highly anoxic conditions, with less than 1 part per million (ppm) dissolved oxygen. Why must the acetate be anoxic?

Another intriguing type of metal reduction associated with anaerobic respiration is that of gold ($Au^{+3} \rightarrow Au^0$). The gold ion is soluble in water, but gold metal precipitates on the surface of gold-reducing bacteria. It is speculated that bacterial gold respiration may have formed much of the gold metal deposits found on Earth.

TO SUMMARIZE:

- **Anaerobic terminal electron acceptors**, such as nitrogen and sulfur oxyanions, oxidized metal cations, and oxidized organic substrates, accept electrons from a specific reductase complex of an ETS.
- **Nitrate** is successively reduced by bacteria to nitrite, nitric oxide, nitrous oxide, and ultimately nitrogen gas. Alternatively, nitrate and nitrite may be reduced to ammonium ion, a product that alkalinizes the environment.
- **Sulfate** is successively reduced by bacteria to sulfite, thiosulfate, elemental sulfur, and hydrogen sulfide. Sulfate reducers are especially prevalent in seawater.
- **Oxidized metal ions** such as Fe^{3+} and Mn^{4+} are reduced by bacteria in soil and aquatic habitats, a form of anaerobic respiration also known as dissimilatory metal reduction.
- **The geochemistry of natural environments** is largely shaped by anaerobic bacteria and archaea.

14.5 Lithotrophy and Methanogenesis

Many reduced minerals can serve as electron donors for an ETS, in the energy-yielding form of metabolism known as lithotrophy (or chemolithotrophy). Lithotrophy is distinguished from organotrophy, metabolism that derives energy from breakdown of organic compounds (compounds with at least one carbon-carbon bond.) Only prokaryotes can grow by metabolizing inorganic compounds without photosynthesis. Thus, bacteria and archaea fill many key niches in ecosystems that eukaryotes cannot.

As summarized in the previous chapter in Table 13.1, lithotrophs may share common components of ETS with organotrophs. The electron-accepting oxidoreductase, however, needs to be specialized for the electron donor, such as H_2, NH_4^+, or Fe^{2+}. Most inorganic substrates other than H_2 are relatively poor electron donors compared to organic donors such as glucose. Therefore, the terminal electron acceptor is usually a strong oxidant, such as O_2, NO_3^-, or Fe^{3+}. Obligate lithotrophs (organisms that conduct only lithotrophy) consume no organic carbon source; they build biomass by fixing CO_2. Other bacteria, however, possess alternative metabolic pathways for lithotrophy and heterotrophy.

Nitrogen Oxidation

A form of lithotrophy of great importance to the environment is oxidation of nitrogen compounds:

$$NH_4^+ \xrightarrow{\frac{1}{2}O_2} NH_2OH \xrightarrow{O_2} HNO_2 \xrightarrow{\frac{1}{2}O_2} HNO_3$$

Ammonium Hydroxylamine Nitrous acid (nitrite) Nitric acid (nitrate)

Reduced forms of nitrogen, such as ammonium ion derived from fertilizers, support growth of **nitrifiers**, bacteria that generate nitrites or nitrates (forming nitrous acid or nitric acid, respectively). Acid production can degrade environmental quality. In soil treated with artificial fertilizers, nitrifiers decrease the reduced nitrogen content and produce toxic concentrations of nitrites, which leach into groundwater. Still, nitrifiers can be useful in sewage treatment, where they eliminate ammonia that would harm aquatic life (discussed in Chapter 22).

Ammonia and nitrite are relatively poor electron donors compared to organic molecules. Thus, their oxidation through the ETS pumps fewer protons, and the bacteria must cycle relatively large quantities of substrates to grow. Nevertheless, where reduced carbon supplies are exhausted, ammonium is oxidized rapidly.

What happens to ammonium from detritus that accumulates in anaerobic regions, such as the bottom of lakes? Surprisingly, ammonium can yield energy through oxidation by nitrite produced from nitrate respiration:

$$NH_4^+ + NO_2^- \rightarrow N_2 + 2H_2O \qquad E^{\circ\prime} = 430 \text{ mV}$$

Under conditions of high ammonium and extremely low oxygen, this metabolism supports ample growth of bacteria. Known as the **anammox reaction**, anaerobic ammonium oxidation plays a major role in wastewater treatment, and it cycles tremendous amounts of nitrogen back to the atmosphere. Anammox is conducted by planctomycetes, irregularly shaped bacteria with unusual nucleus-like membranes enclosing their nucleoid (discussed in Chapter 18).

Sulfur and Metal Oxidation

Major sources of lithotrophic electron donors are minerals containing reduced sulfur, such as hydrogen sulfide and sulfides of iron and copper. As we saw for nitrogen compounds, each sulfur compound that undergoes partial oxidation may serve as an electron donor and be further oxidized.

$$H_2S \xrightarrow{\frac{1}{2}O_2} S^0 \xrightarrow{\frac{1}{2}O_2} \frac{1}{2}S_2O_3^{2-} \xrightarrow{O_2 + H_2O} H_2SO_4$$

Hydrogen sulfide   Elemental sulfur   Thiosulfate   Sulfuric acid

Sulfur oxidation produces the strong acid sulfuric acid, which dissociates to produce an extremely high H^+ concentration. In the early twentieth century, no one would have believed that living organisms could live in concentrated sulfuric acid, much less produce it. The *Star Trek* science fiction episode "The Devil in the Dark" portrayed an imaginary alien creature, the Horta, that produced sulfuric acid in order to eat its way through solid rock (**Fig. 14.22A**). In actuality, of course, no creatures beyond the size of a microbe are yet known to grow in sulfuric acid. But archaea such as *Sulfolobus* species oxidize hydrogen sulfide to sulfuric acid and grow at pH 2, often in hot springs at near-boiling temperatures (**Figs. 14.22 B** and **C**). *Sulfolobus* makes irregular Horta-shaped cells without even a cell wall to maintain shape; how it protects its cytoplasm from disintegration is not yet understood.

The result of microbial sulfur oxidation can be severe environmental acidification, causing deterioration of concrete structures and stone monuments. The problem is compounded by sulfur oxidation in the presence of iron, such as in iron mine drainage. For example, the sulfur-oxidizing archaeon *Ferroplasma acidarmanus* was discovered in the huge abandoned mine Iron Mountain, in northern California, by Katrina Edwards of the Woods Hole Oceanographic Institution and colleagues from the University of Wisconsin–Madison. *Ferroplasma* oxidizes ferrous sulfide (FeS_2) with ferric iron (Fe^{3+}) and water:

$$FeS_2 + 14Fe^{3+} + 8H_2O \rightarrow 15Fe^{2+} + 2SO_4^{2-} + 16H^+$$

This reaction generates large quantities of sulfuric acid. The acidity of the mine water is near pH 0, one of the most acidic environments found on Earth. As *Ferroplasma* grows, it forms biofilms of thick streamers in the mine drainage, which poisons aquatic streams.

Anaerobic reactions between sulfur and iron cause hidden hazards for human technology, such as the corrosion of steel in underwater bridge supports. Anaerobic corrosion was long considered a mystery, since iron was known to rust through spontaneous oxidation by O_2. In anaerobic conditions, however, corrosion can be caused by two alternative pathways involving sulfur-reducing bacteria (**Fig. 14.23**). In one pathway, the bacteria reduce elemental sulfur (S^0) with H_2 to hydrogen sulfide (H_2S).

A. *Star Trek*: An imaginary creature produces H_2SO_4

Paramount/CBS Photo Archive

B. A hot spring supports sulfur-oxidizing archaea

© Getty Images

C. *Sulfolobus acidocaldarius*

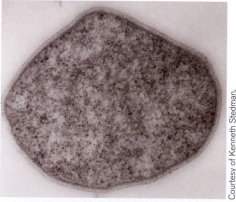

Courtesy of Kenneth Stedman, Portland State University

Figure 14.22   Organisms produce sulfuric acid: science and science fiction.   A. In the *Star Trek* episode "The Devil in the Dark," starship officers encounter an imaginary creature called the Horta that tunnels through rock by producing sulfuric acid. **B.** Volcanic rocks and hot springs support growth of sulfur-oxidizing thermophilic archaea such as *Sulfolobus* species whose growth at pH 2 colors the rocks. **C.** *Sulfolobus acidocaldarius* grows as irregular spheres about 1 μm in diameter (TEM).

Figure 14.23 Anaerobic iron corrosion. A. Anaerobic corrosion of iron is accelerated by sulfur-reducing bacteria. Alternatively, in other bacteria, iron reduces sulfate. **B.** Sulfate-reducing bacteria corrode iron. *Source: A. Derek Lovley, Environmental Microbe-Metal Interactions, p. 163.*

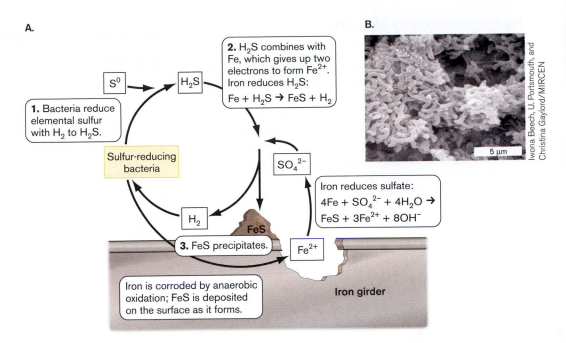

A.

2. H_2S combines with Fe, which gives up two electrons to form Fe^{2+}. Iron reduces H_2S:
$Fe + H_2S \rightarrow FeS + H_2$

1. Bacteria reduce elemental sulfur with H_2 to H_2S.

S^0 → H_2S

Sulfur-reducing bacteria

SO_4^{2-}

Iron reduces sulfate:
$4Fe + SO_4^{2-} + 4H_2O \rightarrow FeS + 3Fe^{2+} + 8OH^-$

H_2

FeS

3. FeS precipitates.

Fe^{2+}

Iron is corroded by anaerobic oxidation; FeS is deposited on the surface as it forms.

Iron girder

B.

5 μm

Iwona Beech, U. Portsmouth, and Christina Gaylord/MIRCEN

H_2S then combines with iron metal (Fe^0), which gives up two electrons to form Fe^{2+}, precipitating as iron sulfide (FeS). The displaced $2H^+$ combine with the $2e^-$ from iron, regenerating H_2—now available to reduce sulfur once again. In an alternative mechanism, bacteria use Fe^0 to reduce sulfate directly to FeS. These damaging processes may resemble the iron-based metabolism of Earth's most ancient life-forms.

Nevertheless, like the imaginary Horta that ended up helping miners with their excavations, acid-producing microbes are now used to supplement commercial mining. Lithotrophs such as *Thiobacillus ferrooxidans* oxidize sulfides of iron and copper found in minerals such as chalcopyrite ($CuFeS_2$), chalcocite (Cu_2S), and covellite (CuS). The oxidation of Cu^+ to Cu^{2+}, as well as the acidification resulting from production of sulfate, dissolves the metal from the rock. Oxidation of Cu^+ can occur either aerobically with O_2 or anaerobically with NO_3^- present in the soil. Other metals oxidized by *T. ferrooxidans* include selenium, antimony, molybdenum, and uranium. The process

of metal dissolution from ores is called **leaching**. Leaching of minerals has been a part of mining since ancient times, long before the existence of microbes was known. Today, over 10% of the copper supply in the United States is provided by microbial leaching from ores too low in copper to smelt directly (**Fig. 14.24**). A similar process is being developed to mine gold, and special strains of *T. ferrooxidans* are being engineered to optimize gold recovery.

In the absence of sulfur, reduced metal ions such as Fe^{2+} and Mn^{2+} provide energy through oxidation by O_2 or NO_3^-. These lithotrophic processes occur commonly in soil where weathering exposes reduced minerals. They generate metal ions with higher oxidation states (such as Fe^{3+} or Mn^{4+}), which other bacteria use for anaerobic respiration. Environments such as ponds and wetlands that experience frequent shifts between oxygen availability and oxygen depletion are likely to host a variety of metal-oxidizing lithotrophs as well as metal-reducing anaerobic heterotrophs. Lithotrophy in ecology is discussed further in Chapters 21 and 22.

Figure 14.24 Copper mining. A. Bingham Canyon copper mine near Salt Lake City, where copper is leached from low-grade ores using *Thiobacillus ferrooxidans*. **B.** *Thiobacillus ferrooxidans*, a gram-negative rod that oxidizes copper and iron sulfides (TEM).

A. A copper mine

Robert Harding/Alamy

B. *Thiobacillus ferrooxidans*

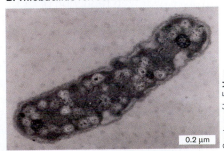

0.2 μm

Courtesy of L. E. Murr

Hydrogenotrophy Uses H$_2$ as an Electron Donor

The use of molecular hydrogen (H$_2$) as an electron donor for electron acceptors is known as **hydrogenotrophy** (**Table 14.2**). An example is oxidation of H$_2$ by sulfur to form H$_2$S, performed by *Pyrodictium brockii*, a species of archaea that grows at thermal vents above 100°C. Hydrogen gas is a stronger electron donor than most organic foods, so it can be oxidized with the full range of electron acceptors. It is available in anaerobic communities as a fermentation product.

Hydrogenotrophy is hard to categorize. Because H$_2$ is inorganic, oxidation by O$_2$ is considered lithotrophy; yet many species that oxidize H$_2$ with molecular oxygen are heterotrophs that also respire on organic foods. When hydrogen reduces an organic electron acceptor, such as fumarate, the process is considered a form of fermentation or (more correctly) anaerobic respiration. When hydrogen reduces a mineral such as sulfur or sulfate, the process is anaerobic lithotrophy.

A form of hydrogenotrophy with enormous potential for bioremediation is **dehalorespiration**, the reduction of halogenated organic molecules by H$_2$ (**Fig. 14.25A**). Chlorinated molecules such as chlorobenzenes, perchloroethene, and polyvinylchloride (known as PVC) are highly toxic environmental pollutants. In dehalorespiration, the chlorine is removed as chloride anion and replaced by hydrogen, a reaction requiring input of two electrons from H$_2$. Many soil bacteria conduct dehalorespiration; a particularly unusual cell wall-less bacterium was discovered to dechlorinate chlorobenzene, a highly stable aromatic molecule (**Fig. 14.25B**).

Methanogenesis and Methane Oxidation

Hydrogen is such a strong electron donor that it can even reduce the highly stable carbon dioxide to methane. Reduction of CO$_2$ and other single-carbon compounds to methane is called **methanogenesis**. Methanogenesis supports a major group of archaea known as **methanogens**, many of which grow solely by autotrophy, generating methane. The availability of methane then provides a niche for methane-oxidizing bacteria.

Methanogenesis. The simplest form of methanogenesis involves hydrogen reduction of CO$_2$:

$$CO_2 + 4H_2 \rightarrow CH_4 + 2H_2O$$

Because both sides of the equation contain a weak electron acceptor (CO$_2$, H$_2$O) and a strong electron donor (H$_2$, CH$_4$), it is surprising that such a reaction can yield energy for growth, but in the presence of sufficient carbon dioxide and hydrogen, enormous communities of methanogens are supported. Such conditions prevail wherever bacteria grow by fermentation, and their gaseous products

Table 14.2 Hydrogenotrophy: examples.

Specific reaction	General description
$O_2 + 2H_2 \rightarrow 2H_2O$	Aerobic oxidation of H$_2$
Fumarate + H$_2$ → succinate	Organic + H$_2$ → organic
$2CO_2 + 4H_2 \rightarrow CH_3COOH + 2H_2O$	Mineral + H$_2$ → organic
$2H^+ + SO_4^{2-} + 4H_2 \rightarrow H_2S + 4H_2O$	Mineral + H$_2$ → mineral
$CO_2 + 4H_2 \rightarrow CH_4 + 2H_2O$	Methanogenesis— only in archaea

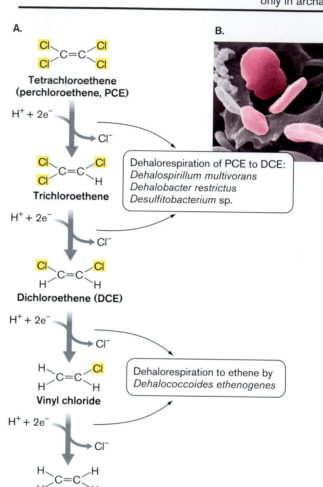

Figure 14.25   Dehalorespiration.   A. Reduction of a chlorinated substrate by H$_2$ yields energy for dehalorespiring bacteria while dechlorinating a toxic pollutant (tetrachloroethene, also known as perchloroethene) to a nontoxic form (ethene). **B.** *Dehalococcoides* strain, CBDB1, a cell wall-less bacterium that reductively dechlorinates chlorobenzene; cell diameter, about 1 µm (SEM).

are trapped, such as in landfills, where methane can be harvested as natural gas, as well as in the digestive systems of cattle and humans. Variants of methanogenesis also include pathways by which H$_2$ reduces various one-carbon and two-carbon molecules, such as methanol, methylamine, and acetic acid. Methanogens include a variety of archaeal species nearly as diverse as an entire domain.

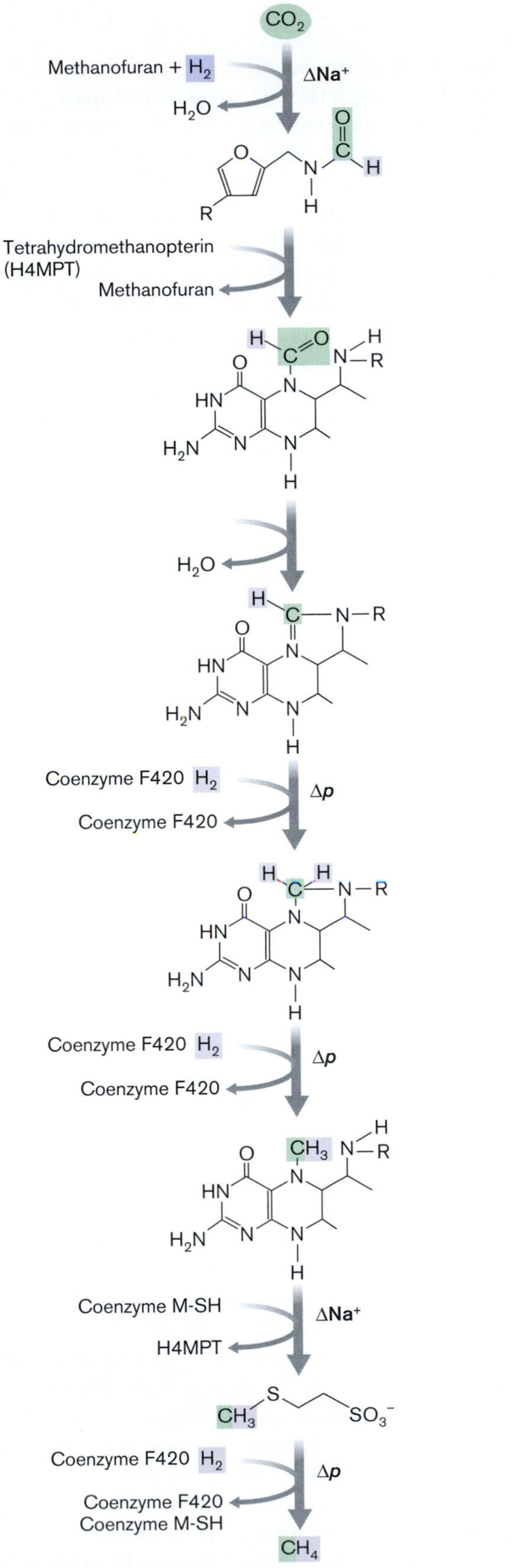

Figure 14.26 Methanogenesis. A methanogen reduces carbon dioxide with hydrogen, generating methane. The incorporation of hydrogen is driven both by a proton potential and by a sodium potential (discussed in Chapter 19).

A simplified pathway of methanogenesis from CO_2 is shown in **Figure 14.26**. The CO_2 undergoes stepwise hydrogenation, and each oxygen is reduced to water. The increasingly reduced carbon is transferred through a series of unique cofactors (methanofuran, tetrahydromethoperin, coenzyme M-SH). Three of the hydrogenation steps involve a membrane ETS complex including the carrier coenzyme F_{420}, whose reaction generates a proton potential. Note, however, that the initial hydrogenation step and the final step of methane production generate a transmembrane sodium potential (ΔNa^+). The use of a sodium potential is consistent with genetic evidence that methanogens evolved from ancient halophiles, organisms living in a high-salt environment (discussed in Chapter 19).

THOUGHT QUESTION 14.12 Hydrogen gas is so light that it rapidly escapes from Earth. Where does all the hydrogen come from to be used for hydrogenotrophy and methanogenesis?

Methane oxidation. Methanogenesis provides a niche for methanotrophs, bacteria and archaea that oxidize methane with a terminal electron acceptor. In deep-ocean sediment, the activity of methanogens is so high that enormous quantities of methane become trapped on the seafloor in the form of water-based crystals known as methane hydrates. If all the methane were to be released at once, it would greatly accelerate global warming. How is this methane recycled? Until recently, it was thought that most methane oxidation, or **methanotrophy**, required O_2; indeed, many bacteria are aerobic methanotrophs. The deep-ocean sediments reveal other species of bacteria and archaea that oxidize methane using nitrate or sulfate. This anaerobic methane oxidation may be critical for the global carbon cycle, as it suggests a mechanism for removal of submarine methane.

NOTE: Distinguish among methanogens, which generate methane; methanotrophs, which oxidize methane; and methylotrophs, which oxidize single-carbon compounds other than methane, such as methanol or methylamine.

TO SUMMARIZE:

- **Lithotrophy (chemolithotrophy)** is the acquisition of energy by oxidation of inorganic electron donors.
- **Sulfur oxidation** includes oxidation of H_2S to sulfur or to sulfuric acid by sulfur-oxidizing bacteria. Sulfuric acid production leads to extreme acidification, damaging stone structures and poisoning mine drainage.
- **Iron oxidation** often accompanies sulfur oxidation.

- **Nitrogen oxidation** includes successive oxidation of ammonia to hydroxylamine, nitrous acid, and nitric acid. Anammox, the anaerobic oxidation of ammonium by nitrate, generates nitrogen gas from anoxic environments.

- **Hydrogenotrophy** uses hydrogen gas as an electron donor. Hydrogen (H_2) has sufficient reducing potential to donate electrons to nearly all biological electron acceptors, including chlorinated organic molecules (through dehalorespiration).

- **Methanogenesis** occurs by oxidation of H_2 by carbon dioxide to form methane. Methanogenesis is performed only by the methanogen group of archaea.

- **Methane oxidizers** use O_2, nitrate, or sulfate to oxidize the methane produced by methanotrophs.

14.6 Phototrophy

On Earth today, the ultimate source of electrons driving metabolism is phototrophy, the harnessing of photoexcited electrons to power cell growth. Every year, photosynthesis converts more than 10% of atmospheric carbon dioxide to biomass, most of which then feeds microbial and animal heterotrophs. Most of Earth's photosynthetic production, especially in the oceans, comes from microbes (discussed in Chapter 21).

In phototrophy, the energy of a photoexcited electron is used to pump protons. Different kinds of phototrophy include the bacteriorhodopsin proton pump; the single-cycle chlorophyll-based photosystems I and II; and the double-cycle Z pathway of oxygenic photosynthesis in cyanobacteria and chloroplasts.

Bacteriorhodopsin: A Simple Light Pump

In most ecosystems, the dominant source of carbon and energy is photosynthesis based on chlorophyll. At the same time, many halophilic archaea and marine bacteria supplement their metabolism with a simpler, more ancient form of phototrophy based on a single-protein light-driven proton pump, **bacteriorhodopsin**. We present bacteriorhodopsin first as a simple case to grasp, followed by the more complex ETS-based photolysis in bacteria and chloroplasts.

Bacteriorhodopsin. Bacteriorhodopsin is a small membrane protein commonly found in halophilic archaea (or haloarchaea) such as *Halobacterium salinarum*, a single-celled archaeon that grows in evaporating salt flats containing concentrated NaCl. For many years, bacteriorhodopsin-like proton pumps were thought to be limited to extreme halophilic archaea. As bacterial genomes were sequenced, however, homologs of the protein appeared in several species of gram-negative bacteria or proteobacteria; the homologs were termed **proteorhodopsin**. In 2005, Oded Béjà and colleagues from Israel, Austria, Korea, and the United States surveyed the genomes of unculturable bacteria from the upper waters of the Mediterranean and Red Seas. They found that 13% of the marine bacteria contain proteorhodopsins, accounting for a substantial—and previously unrecognized—fraction of marine phototrophy.

Bacteriorhodopsin absorbs light with a broad peak in the green range; thus, the organisms containing large amounts of bacteriorhodopsin reflect blue and red, appearing purple. The protein consists of seven hydrophobic alpha helices surrounding a molecule of **retinal**, the same cofactor bound to light-absorbing opsins in the vertebrate retina (**Fig. 14.27A**). In bacteriorhodopsin, the retinal is attached to the nitrogen end of a lysine residue (**Fig. 14.27B**).

Retinal has a series of conjugated double bonds that absorb visible light. Upon absorbing a photon, an electron in one of the double bonds is excited to a higher energy level. As the electron falls back to the ground state, the bond shifts position from *trans* (substituents pointing opposite) to *cis* (substituents pointing in the same direction). This change in shape of the retinal alters the conformation of the entire protein, causing it to pick up a proton from the cytoplasm. Eventually, the retinal switches back to its original *trans* configuration. The reversion to *trans* is coupled to the release of the proton from the opposite end of the protein facing outside of the cell. Thus, photoexcitation of bacteriorhodopsin is coupled to the pumping of one H^+ across the membrane.

The proton gradient generated by bacteriorhodopsin drives ATP synthesis by a typical F_1F_o ATP synthase. Light capture by bacteriorhodopsin supplements, but does not replace, catabolism for energy and heterotrophy for carbon source. The combination of light absorption and heterotrophy is called **photoheterotrophy**.

Purple membrane captures light rays. One problem every phototroph needs to solve is how to "capture" light rays. For chemotrophs, food molecules diffuse in solution and can be picked up by receptors for transport into a cell. In phototrophy, however, a photon impinges on one point of the cell, where it either is absorbed or passes through. Thus, the only way to absorb a high percentage of photons is to spread light-absorbing pigments over a wide surface area. To maximize light absorption, *Halobacterium salinarum* archaea pack their entire cell membranes with bacteriorhodopsin. The protein forms trimers that pack in hexagonal arrays, forming the so-called "purple membrane" (**Fig. 14.28**).

While the bacteriorhodopsin cycle is much simpler than the chlorophyll-based photosynthesis (discussed below), it nevertheless illustrates several principles that apply to more complex forms of phototrophy:

- A photoreceptor absorbs light, causing excitation of an electron to a higher energy level, followed by return to the ground state.
- To maximize light collection, large numbers of photoreceptors are packed throughout a membrane.
- The photocycle (absorption and relaxation of the light-absorbing molecule) is coupled to energy storage in the form of a proton gradient.

NOTE: Distinguish these terms: **Phototrophy** covers all forms of energy-yielding metabolism that involve absorption of light energy. **Photolysis** means light absorption coupled to splitting an electron from a molecule. **Photosynthesis** means photolysis with CO_2 fixation and biosynthesis. Thus the bacteriorhodopsin light cycle is a form of phototrophy, but it is *not* photolysis or photosynthesis.

Photolysis: Light Absorption Separates an Electron from Chlorophyll

The energy-yielding phase of photosynthesis is **photolysis**, in which the photoexcitation of a chlorophyll leads to electron transfer through an ETS. The elucidation of photolysis in bacteria and chloroplasts—the fundamental source of energy for Earth's biosphere—was one of the most exciting projects of the twentieth century. Among hundreds of important contributors, we note two major figures: Roger Stanier (1916–1982) and Germaine Cohen-Bazire (1920–2001) (**Figs. 14.29A and B**). Stanier, a Canadian microbial physiologist at the University of California, Berkeley, clarified the nature of cyanobacteria as phototrophic prokaryotes distinct from eukaryotic algae, and he helped distinguish the photolytic cycles of cyanobacteria from those of sulfur-metabolizing purple

A.

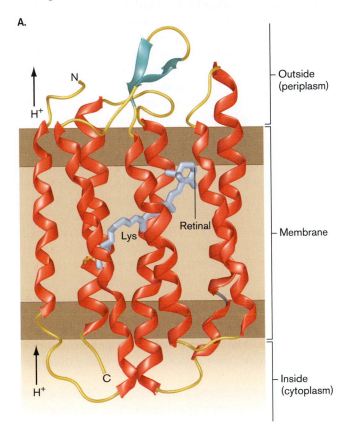

B.

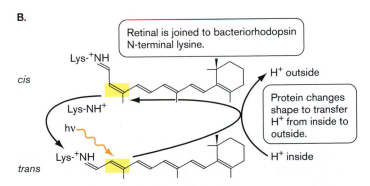

Figure 14.27 The light-driven cycle of bacteriorhodopsin.
A. Bacteriorhodopsin contains seven alpha helices that span the membrane in alternating directions and surround a molecule of retinal, which is linked to a lysine residue. (PDB code: 1FBB)
B. A photon (hν) is absorbed by retinal, which shifts configuration from *trans* to *cis*. The cycle of excitation and relaxation back to the *trans* form is coupled to pumping of $1H^+$ from the cytoplasm across the membrane.

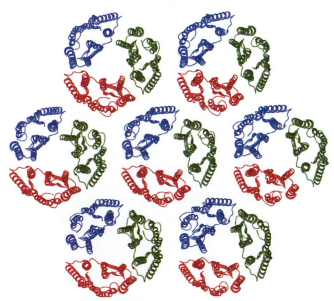

Figure 14.28 Bacteriorhodopsin purple membrane.
Trimers of bacteriorhodopsin (monomers shown red, blue, and green) are packed in hexagonal arrays, forming the purple membrane. (PDB code: 1MOL) *Source:* Purple Membrane: Theoretical Biophysics Group, VMD Image Gallery, NIH Resource for Macromolecular Modeling and Bioinformatics.

A. Roger Stanier

Stanier Institute

B. Germaine Cohen-Bazire

Stanier Institute

C. *Synechococcus elongatus*

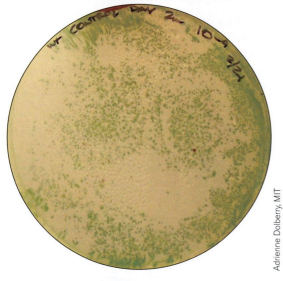

Adrienne Dolberry, MIT

Figure 14.29 Roger Stanier and Germaine Cohen-Bazire studied cyanobacterial photosynthesis. A. Roger Stanier pioneered the physiology of cyanobacteria. **B.** Germain Cohen-Bazire performed the first studies of genetic regulation of bacterial photosynthesis. **C.** Green colonies of the cyanobacterium *Synechococcus elongatus*.

bacteria. He married Cohen-Bazire, a French bacterial geneticist who had studied *lac* operon regulation with Nobel laureate Jacques Monod at the Institut Pasteur in Paris. Cohen-Bazire applied her genetics skills to phototrophs, and she conducted the first genetic analysis of photosynthesis in purple bacteria and cyanobacteria.

Cyanobacteria, the only oxygen-producing bacteria, appear green, like algae or plants (**Fig. 14.29C**). Cyanobacteria include a wide range of species, such as the ocean's major producers, the submicroscopic *Prochlorococcus marinus*, barely visible under a light microscope. Other cyanobacteria have cells as large as algae and form complex developmental structures with important symbiotic associations (see Chapters 19 and 21). Cyanobacteria are among the most successful and diverse groups of life on Earth. Cyanobacteria, and the chloroplasts of algae and plants produce all the oxygen available for aerobic life.

Overview of photolysis. The energy for photosynthesis derives from the photoexcitation of a light-absorbing pigment. Photoexcitation leads to photolysis, the light-driven separation of an electron from a molecule coupled to an ETS. The components of the ETS are often homologous to those of respiratory electron transport, and they share common electron carriers such as iron sulfur clusters.

In plant chloroplasts, photolysis is also known as the "light reaction," coupled to the "dark reaction" of carbon dioxide fixation. In green plants and cyanobacteria, photolysis is tightly coupled to the "dark reaction" in which CO_2 is fixed into biomass. We discuss CO_2 fixation with other biosynthetic pathways in Chapter 15. Note, however, that many of the sulfur- or organic-based bacterial phototrophs, such as *Rhodospirillum rubrum*, combine photolysis with heterotrophy instead of CO_2 fixation.

In ETS-based photosynthesis, photoexcitation leads to actual charge separation—that is, separation of an electron from a donor molecule such as H_2O or H_2S. Each electron is then transferred to an ETS, whose components show common ancestry with respiratory ETS proteins. The photolytic ETS generates a proton potential and the reduced cofactor NADPH. The proton potential drives ATP synthesis through an F_1F_o ATPase, similar or identical to the one for respiration.

Chlorophylls absorb light. The main light-absorbing pigments are **chlorophylls**. Each type of chlorophyll contains a characteristic **chromophore**, a light-absorbing electron carrier. The chlorophyll chromophore consists of a heteroaromatic ring complexed to a magnesium ion (Mg^{2+}) (**Fig. 14.30**). As we saw for ETS electron carriers, metal ions and aromatic bonds offer electrons with relatively narrow energy transitions. The chromophore absorbs a photon through a reversible energy transition, such that the chlorophyll can alternate between excited and relaxed states.

Chlorophyll molecules differ slightly in their substituent groups around the ring; for example, chlorophyll *a* of chloroplasts has a methyl group in ring II, whereas chlorophyll *b* has an aldehyde. These slight differences alter their absorption spectra (**Fig. 14.31A**). Chlorophyll *a* absorption peaks in the red, whereas chlorophyll *b* absorbs mainly blue. Both chlorophylls *a* and *b* are made by chloroplasts and by cyanobacteria, their nearest bacterial relatives. Because they absorb red and blue, they reflect the middle range of the spectrum and so appear green.

By contrast, the chlorophylls of anaerobic phototrophs, or "purple bacteria," such as *Rhodobacter* and *Rhodospirillum*, absorb most strongly in the far-red (infrared) and in some cases ultraviolet range (**Fig. 14.31B**). These chlorophylls are specifically named **bacteriochlorophylls**. In purple bacteria, bacteriochlorophylls are supplemented by accessory pigments called **carotenoids**, which absorb

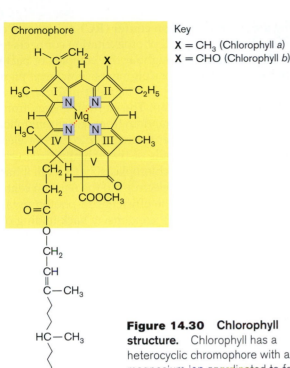

Figure 14.30 **Chlorophyll structure.** Chlorophyll has a heterocyclic chromophore with a magnesium ion coordinated to four nitrogens. Different chlorophyll types differ mainly in small substituents of the chromophore; for example, an aldehyde replaces the ring II methyl group in chlorophyll a to make chlorophyll b.

light of green wavelengths and transfer the energy to bacteriochlorophyll. The combination of green-absorbing carotenoids and infrared-absorbing bacteriochlorophylls makes cultures appear deep purple or brown.

The infrared radiation absorbed by bacteriochlorophylls is too weak to permit splitting H_2O to produce oxygen. Thus, purple bacteria are limited to photolysis of H_2S and small organic molecules; most are photoheterotrophs. On the other hand, infrared rays are available in water below the oxygenic phototrophs absorbing red and blue. Thus, anaerobic phototrophs grow at depths where their more high-powered oxygenic relatives do not.

NOTE: Distinguish among these classes of photopigments: **bacteriorhodopsin**, a retinal-containing proton pump; **chlorophyll**, a charge-separating photopigment (usually referring to chloroplasts and cyanobacteria); **bacteriochlorophyll**, a charge-separating chlorophyll of anaerobic purple and green bacteria (also known generically as chlorophyll); and **carotenoid**, an accessory pigment that absorbs the midrange of light wavelengths but does not directly conduct photolysis.

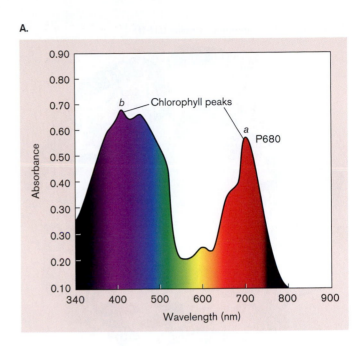

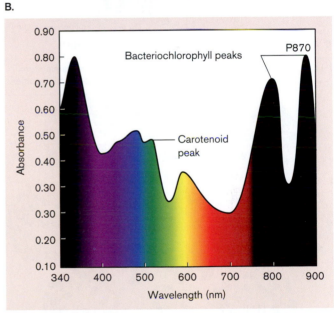

Figure 14.31 **Absorption spectra of photosynthetic pigments.** **A.** Absorption by chloroplasts including chlorophyll a (P680), chlorophyll b, and carotenoid accessory pigments. The middle range (green) is reflected. **B.** Absorption by purple photosynthetic bacteria, including bacteriochlorophyll (P870) and carotenoids. The main absorption is infrared (range 750–900 nm); therefore, the bacteria appear purple or brown.

Antenna complex and reaction center. To maximize light collection, many molecules of chlorophyll are grouped in an **antenna complex**. These antenna complexes are arranged like a satellite dish within the plane of the membrane in an elaborate cluster around accessory proteins (**Figure 14.32A**). The clusters then associate in

A. Light-harvesting antenna complex (LH-II)

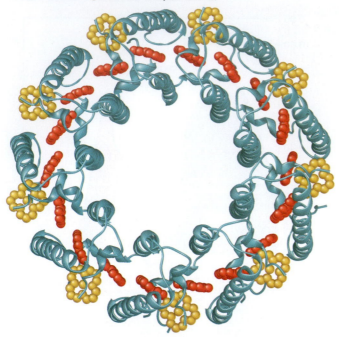

B. Antenna complexes surround the reaction center (RC)

LH-II harvests electrons.

LH-I directs electrons into the reaction center.

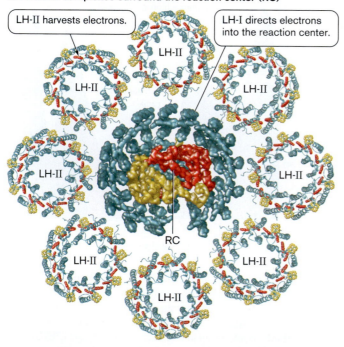

Figure 14.32   Antenna complexes surround the reaction center.   **A.** The antenna complex of *Rhodopseudomonas viridis* contains 9 chlorophylls with chromophores facing parallel to the membrane (gold) and 18 with chromophores facing out from the ring (red). **B.** Multiple rings of chlorophyll-protein antenna complexes surround the reaction center, like a funnel collecting photons. LH-II is the accessory light-harvesting antenna complex; LH-I is the central complex directing electrons into the reaction center (RC). (PDB codes: 1PYH, 2FKW) *Source:* Quantum Biology of the PSU, NIH Resource for Macromolecular Modeling and Bioinformatics—BChl antenna.

a ring around the **reaction center (RC)**, the protein complex in which chlorophyll photoexcitation connects to the ETS (**Fig. 14.32B**). Throughout the complex of bacteriochlorophylls and accessory pigments, whichever pigment molecule happens to be in "the right place at the right time" captures the photon. The electron energy from the photon then transfers at random from one chromophore to the next, until it arrives at the reaction center for electron transfer to the ETS. Details of the antenna structure differ in cyanobacteria and chloroplasts, but overall it has the same effect of funneling electrons to the reaction center.

In purple bacteria, the efficiency of photon uptake is increased even more by the extensive back folding of the photosynthetic membranes in oval pockets, stacked like pita breads (**Fig. 14.33A**). These oval pockets, also typical of chloroplasts, are called **thylakoids**. The extensive packing of thylakoids gives an incident photon hundreds of chances to meet a chlorophyll at just the right angle for absorption.

The thylakoids are connected by tubular extensions, so that overall there exists one interior space, the **lumen**, separated topologically from the regular cytoplasm, or **stroma**. Protons are pumped from the stroma across the thylakoid membrane into the lumen. The F_1F_o complex is embedded in the thylakoid, where it makes ATP using the proton current running through it into the cell. The F_1 knob of ATP synthase appears to face "outward" in photosynthetic organelles (as opposed to "inward" in respiratory chains). In each case, however, the proton current and ATP motor face in the same direction with respect to the cytoplasm (or stroma). The proton potential is more negative in the cytoplasm (or stroma), thus drawing protons through the ATPase to generate ATP.

The photolytic electron transport system. In photolysis, the absorption of light by chlorophyll or bacteriochlorophyll drives the separation of an electron from a molecule. The photolyzed molecule may be the chlorophyll itself, as in *Rhodospirillum* or *Rhodobacter*, or it may be a hydrogen donor such as H_2S or H_2O, depending on the photosystem of a given bacterial species. The excited electron enters a membrane-embedded chain of oxidoreductases and quinones/quinols, as we saw for the ETS of respiration and lithotrophy.

Diverse kinds of photosynthesis in different environmental niches include oxygenic, sulfur-based, iron-dependent, and even heterotrophic photolysis. Nevertheless, all forms of photolysis share a common design (**Fig. 14.34**):

1. **Antenna system.** A complex of chlorophylls and accessory pigments in the photosynthetic membrane collects photons. Energy from each photoexcited electron is transferred to the reaction center.

2. Reaction center complex. A quantum of light energy absorbed by chlorophyll raises an electron to a higher-energy orbital. In the reaction center, the energy from chlorophyll photoexcitation is used to separate an elec- tron from a small molecule such as H_2S (photosystem I) or from bacteriochlorophyll (photosystem II).

3. Electron transport system. Each photoexcited electron enters an ETS. In PS I, electrons separated from H_2O or H_2S are transferred to NAD^+ to form NADPH. In PS II, the electron separated from bacteriochlorophyll is replaced by an electron returned from the ETS. In the oxygenic Z pathway (H_2O photolysis), electrons flow from PS II into PS I, ultimately releasing O_2 from H_2O.

4. Energy carriers. In PS I, electrons are used to make NADPH. In PS II, electron transfer provides energy to pump protons and drive synthesis of ATP. The Z pathway makes both NADPH and ATP, which are used to fix CO_2.

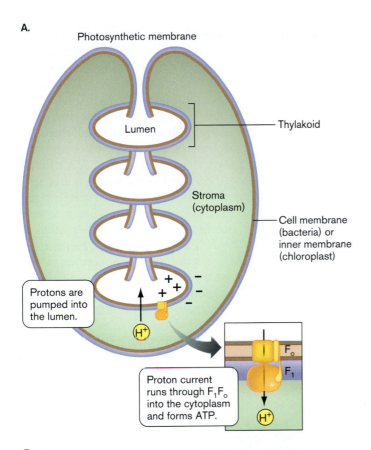

A.

Photosynthetic membrane

Lumen — Thylakoid

Stroma (cytoplasm)

Cell membrane (bacteria) or inner membrane (chloroplast)

Protons are pumped into the lumen.

H^+

Proton current runs through F_1F_o into the cytoplasm and forms ATP.

F_o
F_1

H^+

B.

R. Howard Berg/Visuals Unlimited

Figure 14.33 Photosynthetic membranes. A. The photosynthetic membranes of bacteria and chloroplasts appear as hollow disks with tubular interconnections. The disks are called thylakoids. Topologically, the membrane separates the cytoplasm (stroma) from the interior space (lumen). **B.** TEM section of a chloroplast, showing the stacked thylakoids.

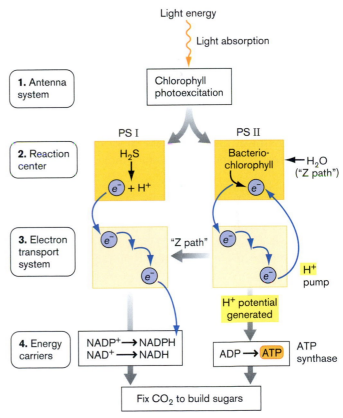

Light energy

Light absorption

1. Antenna system

Chlorophyll photoexcitation

PS I   PS II

2. Reaction center

H_2S Bacterio-chlorophyll   ← H_2O ("Z path")

e^- + H^+ e^-

3. Electron transport system

e^- "Z path" e^-

e^- e^-

H^+ pump

H^+ potential generated

4. Energy carriers

$NADP^+ \rightarrow NADPH$
$NAD^+ \rightarrow NADH$

$ADP \rightarrow ATP$ ATP synthase

Fix CO_2 to build sugars

Figure 14.34 Energy transformation in photosynthesis. Light energy is absorbed by the antenna system of chlorophylls and proteins, which channel the energy to a single reaction center. The excitation of the reaction center drives splitting of a hydrogen atom from a donor such as H_2S or succinate (anaerobic photosynthesis) or H_2O (oxygenic photosynthesis). Each electron from the hydrogen atom is transferred to an electron transport system, where it may drive pumping of protons or formation of a reduced cofactor (NADH or NADPH). The proton gradient drives synthesis of ATP.

Electrons Are Transferred through Photosystems I and II

The steps of photolysis and electron transport occur in three different kinds of systems, in different classes of bacteria:

- **Anaerobic photosystem I** separates electrons associated with hydrogens from H_2S or an organic electron donor such as succinate or from reduced iron (Fe^{2+}). Found, for example, in chlorobia, "green sulfur" bacteria.
- **Anaerobic photosystem II** separates an electron from bacteriochlorophyll. Found, for example, in alpha-proteobacteria, "purple nonsulfur" bacteria.
- **Oxygenic Z pathway** includes homologs of photosystems I and II. Two pairs of electrons are separated from two water molecules to generate O_2. Found in cyanobacteria and in the chloroplasts of green plants.

The components of photosystems I and II (PS I and PS II) show homology with each other, implying common ancestry. Each system runs anaerobically, producing sulfur or oxidized organic by-products, but not O_2. Each photosystem shows more recent homology with the respective PS I and PS II components of the oxygenic Z pathway (so called because the electron path through a diagram of the two photosystems traces a Z). The Z pathway ultimately generates O_2—the only biochemical reaction known to produce molecular oxygen.

NOTE: In photolysis, each quantum of light excites a single electron. Some ETS components, such as the quinones/quinols ($Q \rightleftharpoons QH_2$) actually process two of these electrons in completing their redox cycle. The single-electron intermediate states of quinones are not shown.

Photosystem I in chlorobia. In bacteria such as *Chlorobium* species, light is absorbed primarily by bacteriochlorophyll P840, named for its peak absorption at 840 nm. The wavelength range for PS I typically extends into the near infrared (750–850 nm). Some chlorobia actually conduct photosynthesis using thermal radiation from deep-sea thermal vents.

In the PS I reaction center, the light energy from the antenna complex is used to separate an electron from H_2S or an organic electron donor (**Fig. 14.35A**). The high-potential electron is transferred by a quinone, phylloquinone (PQ) to **ferredoxin**, an FeS protein. Ferredoxin transfers the electron to the enzyme ferredoxin NAD-reductase; a pair of these electrons then reduce NAD^+ or the energetically equivalent $NADP^+$. The reduced carrier (NADH or NADPH) provides reductive energy for CO_2 fixation and biosynthesis (to be discussed in Chapter 15). Bacteria using PS I also generate a net proton gradient by consuming H^+ inside and generating H^+ outside the cell, thus providing proton motive force to drive ATP synthesis.

Photosystem II in alpha-proteobacteria. Phototrophic alpha-proteobacteria such as *Rhodospirillum rubrum* are typically found in wetlands and streams, where they capture light not used by other phototrophs. The wavelength absorbed by bacteriochlorophyll P870 lies so far into the infrared that the photon energy is insufficient to split hydrogenated substrates. Instead, the light energy at the reaction center separates an electron from the bacteriochlorophyll itself (**Fig. 14.36**). The high-energy electrons are then transferred by quinones to a terminal cytochrome oxidoreductase, as in the respiratory ETS.

The reduction potential, however, is too small to reduce $NADP^+$ to NADPH. Instead, as in respiration, the quinols pass their electrons to cytochromes, while transferring $2H^+$ across the membrane. No extra protons are pumped, but the proton potential drives synthesis of ATP. When the electrons reach cytochrome c, they flow back to bacteriochlorophyll, where they can be reexcited by photon energy from the antenna complex. Because the electron path traces back to its source, the ETS of photosystem II leading to ATP synthesis is called **cyclic photophosphorylation**.

Photosystem II, unlike photosystem I, provides no direct way to make NADH or NADPH for reductive biosynthesis such as CO_2 fixation. To make NADPH, purple bacteria obtain hydrogen from molecules such as H_2, H_2S, or organic acids in a pathway outside photolysis. The mechanism is poorly understood, but it involves reverse electron flow through the ETS, driven by the proton potential.

THOUGHT QUESTION 14.13 Suppose you discover bacteria that require a high concentration of Fe^{2+} for photosynthesis. Can you hypothesize what the role of Fe^{2+} may be? How would you test your hypothesis?

Oxygenic photolysis. The "Z" pathway of photolysis found in cyanobacteria and chloroplasts combines key features of both PS I and II (**Fig. 14.37**). Both reaction centers, however, contain chlorophylls that absorb at shorter wavelengths (higher energy) than those of the respective purple or green homologs: P680 instead of P870 (PS II), and P700 instead of P840 (PS I). Thus, the cyanobacterial reaction centers can split water and yield

A. Photosystem I

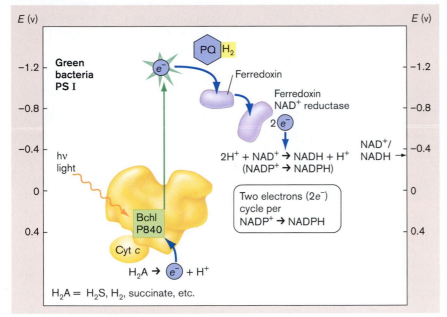

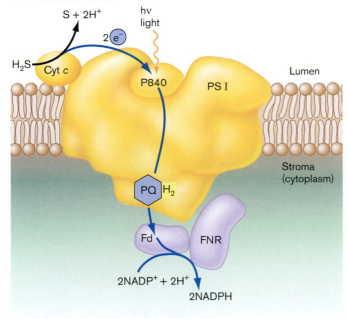

B. PS I reaction center

Figure 14.35 Photosystem I separates electrons from sulfides and organic molecules. A. In green sulfur bacteria, red light photoexcites bacteriochlorophyll P840 (Bchl P840; photosystem I). Photoexitation of P840 enables splitting (photolysis) of two hydrogens from H_2S or an organic electron donor. The $2e^-$ are transferred by a quinone (phylloquinone, PQ) to ferredoxin (Fd). Ferredoxin is oxidized by ferredoxin NADH oxidoreductase (FNR), transferring the $2e^-$ to NAD^+ (or the energetically equivalent $NADP^+$) to form NADH (or NADPH). **B.** The PS I reaction center within a photosynthetic membrane.

$4e^-$ transferred to carriers, and formation of O_2. The net reaction forming oxygen is

$$2H_2O \rightarrow 4H^+ + 4e^- + O_2$$

Through the ETS, the electrons are transferred to quinones. As in respiration, each $2e^-$ reduction of quinone to quinol requires pickup of $2H^+$ from the stroma (equivalent to cytoplasm). Thus the four electrons transferred generate a net change in the proton gradient of $4H^+$. Furthermore, the energy of electron transfer to cytochrome bf enables pumping of an additional $2H^+$ across the membrane. Thus, in all, for each conversion of $2H_2O$ to O_2, the net protons transferred across the thylakoid membrane include $4H^+$ (water photolysis) plus $4 \times 2H^+$ (quinones to quinols) through cytochrome bf, to yield a total of $12H^+$ for the proton gradient. The proton gradient drives the synthesis of approximately 3ATP per O_2 formed.

The electrons from PS II do not cycle back to the PS II reaction center, as they do in purple bacteria. Instead they are transferred to PS I by a protein called plastocyanin. The energy of the electron transferred by plastocyanin is augmented through absorption of a second photon by the chlorophyll of PS I. Subsequent electron flow through ferredoxin can now generate NADH or NADPH. If the reduction potential is too large, however, particularly under intense light, some of the electron flow instead cycles back to cytochrome bf.

greater energy overall. Oxygenic phototrophs dominate the shallow water depths, whereas anaerobes grow at lower depths using light at wavelengths unused by cyanobacteria and algae near the surface.

The P680 reaction center of photosystem II possesses sufficient potential to split H_2O (see **Fig. 14.37**). The entire cycle of water splitting involves $4H^+$ removed from $2H_2O$,

A. Photosystem II

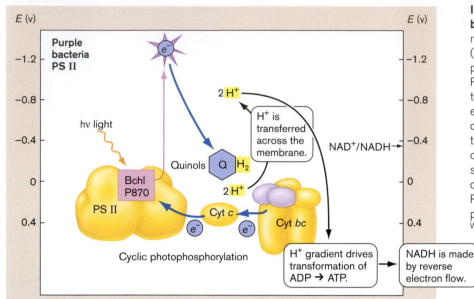

Figure 14.36 **Photosystem II separates an electron from bacteriochlorophyll.** **A.** In the PS II reaction center, bacteriochlorophyll P870 (Bchl P870; photosystem II) absorbs primarily infrared light. Bacteriochlorophyll P870 donates an energized electron to a quinone (Q). Two of these donated electrons complete the conversion of quinone to quinol (QH$_2$). Electrons flow through cytochrome bc, coupled to pumping of protons. The proton potential drives synthesis of ATP. The cytochrome bc complex transfers the electrons back to P870. **B.** The PS II reaction center within a photosynthetic membrane. *Source*: A. Based on Whitmarsh and Govindjee.

B. PS II reaction center and F$_1$F$_o$ ATP synthase

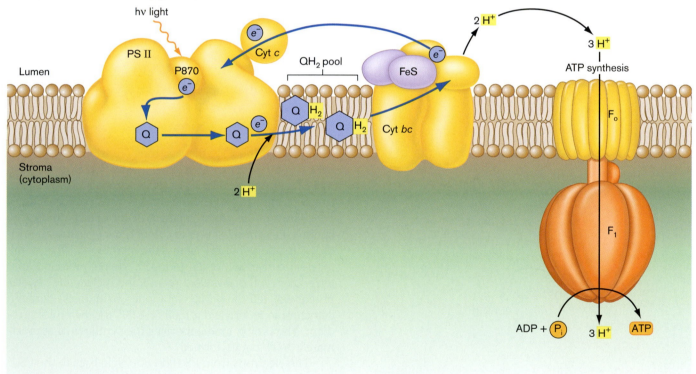

The overall equation for energy yield of oxygenic photolysis can be represented as

$$2H_2O + 2NADP^+ + 3[ADP + P_i] \rightarrow$$
$$O_2 + 2[NADPH + H^+] + 3ATP + 3H_2O$$

To generate one molecule of glucose by CO$_2$ fixation, we need six rounds of the photolysis equation—one per CO$_2$ molecule "fixed" into sugar (C$_6$H$_{12}$O$_6$)

(discussed in Chapter 15). The CO$_2$ fixation equation works out to

$$12H_2O + 12NADP^+ + 18[ADP + P_i] \rightarrow$$
$$6O_2 + 12[NADPH + H^+] + 18ATP + 18H_2O$$

$$6CO_2 + 12[NADPH + H^+] + 18ATP + 18H_2O \rightarrow$$
$$\underline{C_6H_{12}O_6 + 6H_2O + 12\ NADP^+ + 18\ [ADP + P_i]}$$
$$12H_2O + 6CO_2 \rightarrow C_6H_{12}O_6 + 6H_2O + 6O_2$$

A. Z pathway of oxygenic photosynthesis

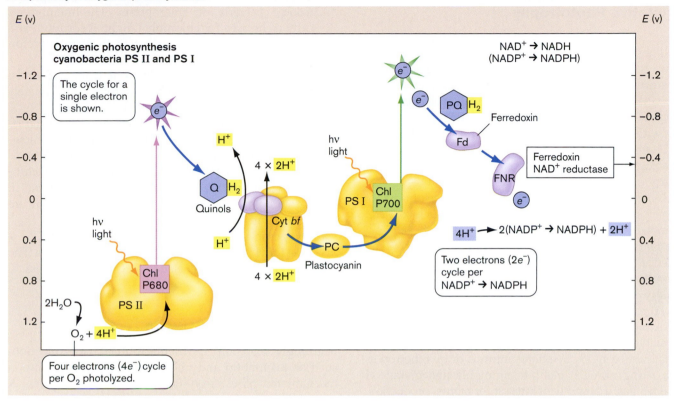

B. Z pathway reaction complexes and ATP synthase

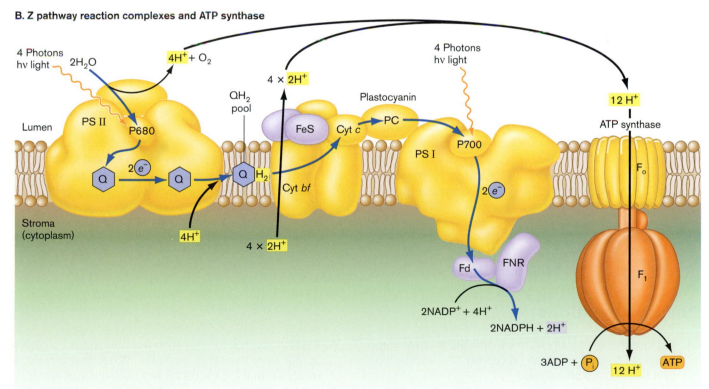

Figure 14.37 Oxygenic photosynthesis in cyanobacteria and chloroplasts. A. H_2O is photolyzed via the Z pathway of PS II and PS I. The $2e^-$ from water is transferred to QH_2, which pumps $2H^+$ across the membrane to cytochrome *bf*, which pumps $4H^+$; and via plastocyanin to a PS I containing chlorophyll P700. A second photon excites P700, enabling transfer of $2e^-$ to ferredoxin, thence to $NADP^+ \rightarrow NADPH$. **B.** The Z pathway within a photosynthetic membrane. ▶️ *Source:* A. Based on Whitmarsh and Govindjee. B. Based on crystallographic data from thermophilic cyanobacteria: From Kurisu, et al. 2003. *Science* 302:1009.

Combined Metabolism: Facultative Phototrophy

Biology courses often give the impression that heterotrophy, lithotrophy, and photosynthesis are discrete metabolic pathways utilized by completely different organisms. In fact, however, a microbial species rarely depends on a single option. Many bacteria found in soil or water use alternative metabolic options such as respiration and photosynthesis, often with shared components. The combined pathways of a facultative phototroph possess useful properties for biotechnology and environmental remediation.

An example of a facultative phototroph is *Rhodopseudomonas palustris*, a purple alpha-proteobacterium that grows in soil and water samples throughout our biosphere. The metabolic diversity of *R. palustris* was revealed by Caroline Harwood and colleagues by analysis of its genome sequence, using methods described in Chapter 8. The bacterium that "eats everything," *R. palustris* catabolizes sugars, fatty acids, and lignin components, including benzene rings, through aerobic or anaerobic respiration. It degrades toxic compounds such as 3-chlorobenzoate, which makes it valuable as an agent of bioremediation. Alternatively, *R. palustris* fixes CO_2 by photosynthesis or by lithotrophy, while fixing nitrogen gas into ammonia and giving off hydrogen gas, a property of interest for biofuel production. Besides all this metabolic versatility, *R. palustris* possesses at least seven different multidrug resistance pumps that may unfortunately be acquired by human pathogens (a less happy accomplishment, from the human point of view).

TO SUMMARIZE:

- **Bacteriorhodopsin** is a light-driven proton pump that supplements heterotrophy in haloarchaea. The homolog, proteorhodopsin, is found in marine proteobacteria.
- **The antenna complex** of chlorophylls and other photopigments captures light for transfer to the reaction center in chlorophyll-based photosynthesis.
- **Thylakoids** are folded membranes within phototrophic bacteria or chloroplasts. The membranes extend the area for chlorophyll light absorption, and they separate two compartments to form a proton gradient.
- **Photosystem I** separates electrons from H_2S or an organic electron donor. The electrons are ultimately transferred to NADH or NADPH.
- **Photosystem II** separates an electron from bacteriochlorophyll. An electron ultimately returns to bacteriochlorophyll through cyclic photophosphorylation.
- **The oxygenic Z pathway in cyanobacteria and chloroplasts** includes homologs of photosystems I and II. Eight photons are absorbed and two electron pairs are removed from $2H_2O$, ultimately producing O_2.
- **Oxygenic photosynthesis generates 3ATP + 2NADPH** per $2H_2O$ photolyzed and O_2 produced. The ATP and NADPH are used to fix CO_2 into biomass.

Concluding Thoughts

Seemingly disparate biochemical means of nutrition, such as respiration and photosynthesis, share common mechanisms of electron flow and proton transfer. A recurring theme is that all forms of metabolism involve charge separation and electron transfer. As molecules are rearranged by transfer of electrons from one substrate to another, energy is obtained and harnessed to the formation of ion gradients and energy carriers. The energy carriers always need to be balanced between redox-neutral carriers, such as ATP, and reducing carriers, such as NADH or NADPH, for biosynthesis (as discussed in Chapter 15).

Whatever the means of obtaining energy, ultimately the energy must be applied to biosynthesis. Chapter 15 presents how microbes spend their energy to construct the most fundamental "nuts and bolts" of their cells.

CHAPTER REVIEW

Review Questions

1. Explain the source of electrons and the sink for electrons (terminal electron acceptor) in respiration, lithotrophy, and photolysis.
2. How do bacteria combine redox couples for a metabolic reaction that yields energy? Cite examples, calculating the reduction potential.
3. How do environmental conditions affect the reduction potential of a metabolic reaction?
4. Explain the role of cytochromes and redox cofactors in electron transport systems. What features of a molecule make it useful for redox biochemistry?
5. Explain how a proton potential is composed of a chemical concentration difference plus a charge difference. Explain how each component of Δp can drive a cellular reaction.

6. Explain the role of the substrate dehydrogenase (oxidoreductase), the quinones, and the cytochrome oxidase (oxidoreductase) in the respiratory ETS.
7. Compare the ETS function in lithotrophy with that in respiration.
8. Summarize the inorganic redox couples that can be used in anaerobic respiration and those that can be used in lithotrophy. What constraints determine whether a given molecule can serve as electron acceptor or as electron donor?

9. How do diverse forms of anaerobic respiration and lithotrophy contribute to ecosystems?
10. Explain the differences and common features of bacteriorhodopsin phototrophy and chlorophyll phototrophy.
11. Explain the differences and common features of photosystems I and II. Explain how the two photosystems combine in the Z pathway. Why can the Z pathway generate oxygen, whereas PS I and PS II cannot?

Key Terms

anaerobic respiration (525)
anammox reaction (529)
antenna complex (537)
bacteriochlorophyll (536, 537)
bacteriorhodopsin (534, 537)
carotenoid (536, 537)
chemiosmotic hypothesis (511)
chlorophyll (536, 537)
chromophore (536)
cyclic photophosphorylation (540)
cytochrome (510)
dehalorespiration (532)
dissimilatory denitrification (525)
dissimilatory metal reduction (528)
electron acceptor (506)
electron donor (506)
electron transport chain (506)
electron transport system (ETS) (506)

ferredoxin (540)
heme (516)
hydrogenotrophy (532)
leaching (531)
lithotrophy (chemolithotrophy) (507)
lumen (538)
methanogen (532)
methanogenesis (532)
methanotrophy (533)
NADH dehydrogenase (518)
NADH:quinone oxidoreductase (NDH-1) (518)
nitrifier (529)
oxidoreductase (510, 518)
oxygenic Z pathway (540)
photoheterotrophy (534)
photolysis (535)
photosynthesis (535)

photosystem I (540)
photosystem II (540)
phototrophy (507, 535)
proteorhodopsin (534)
proton motive force (511)
proton potential (511)
quinol (517)
quinone (517)
quinone pool (519)
reaction center (RC) (538)
redox couple (507)
reduction potential (E) (507)
respiration (507)
retinal (534)
standard reduction potential ($E°$) (507)
stroma (538)
thylakoid (538)
uncoupler (515)

Recommended Reading

Adrian, Lorenz, Ulrich Szewzyk, Jörg Wecke, and Helmut Görisch. 2000. Bacterial dehalorespiration with chlorinated benzenes. *Nature* **408**:580–583.

Beatty, J. Thomas, Jörg Overmann, Michael T. Lince, Ann K. Manske, Andrew S. Lang, et al. 2005. An obligately photosynthetic bacterial anaerobe from a deep-sea hydrothermal vent. *Proceedings of the National Academy of Science USA* **102**:9306–9310.

Belevich, Ilya, Michael I. Verkhovsky, and Mårten Wikström. 2006. Proton-coupled electron transfer drives the proton pump of cytochrome *c* oxidase. *Nature* **440**:829–832.

Dinh, Hang T., Jan Kuever, Marc Mumann, Achim W. Hassel, Martin Stratmann, and Friedrich. 2004. Iron corrosion by novel anaerobic microorganisms. *Nature* **427**:829–832.

Ferreira, Kristina N., Tina M. Iverson, Karim Maghlaoui, James Barber, and So Iwata. 2004. Architecture of the photosynthetic oxygen-evolving center. *Science* **303**:1831–1838.

Gennis, Robert B., and Valley Stewart. 1996. Respiration. *In* Ecosal, online; also in F. C. Neidhart (ed.), *Escherichia coli and Salmonella typhimurium.* 2nd ed. American Society for Microbiology Press, Washington, D.C.

Karl, David M. 2002. Hidden in a sea of microbes. *Nature* **415**:590–591.

Kashefi, Kazem, Jason M. Tor, Kelly P. Nevin, and Derek R. Lovley. Reductive precipitation of gold by dissimilatory Fe(III)-reducing bacteria and archaea. 2001. *Applied and Environmental Microbiology* **67**:3275–3279.

Lovley, Derek R. 2002. Dissimiliatory metal reduction: From early life to bioremediation. *ASM News* **68**:231–237.

Newman, Dianne, and Jillian Banfield. 2002. Geomicrobiology: How molecular-scale interactions underpin biogeochemical systems. *Science* **296**:1071.

Reguera, Gemma, Kevin D. McCarthy, Teena Mehta, Julie S. Nicoll, Mark T. Tuominen, and Derek R. Lovley. 2005. Extracellular electron transfer via microbial nanowires. *Nature* **435:**1098–1101.

Rich, Peter R. 2003. The molecular machinery of Keilin's repiratory chain. *Biochemical Society Transactions* **31:**1095–1105.

Roszak, Aleksander W., Tina D. Howard, June Southall, Alastair T. Gardiner, Christopher J. Law, et al. 2003. Crystal structure of the RC-LH1 core complex from *Rhodopseudomonas palustris*. *Science* **302:**1969–1972.

Sazanov, Leonid A., and Philip Hinchliffe. 2006. Structure of the hydrophilic domain of respiratory complex I from *Thermus thermophilus*. *Science* **311:**1430–1436.

Strous, Marc, Eric Pelletier, Sophie Mangenot, Thomas Rattei, Angelika Lehner, et al. 2006. Deciphering the evolution and metabolism of an anammox bacterium from a community genome. *Nature* **440:**790–794.

Wang, Zhen X., David B. Hicks, Arthur A. Guffanti, Katisha Baldwin, and Terry A. Krulwich. 2004. Replacement of amino acid sequence features of a and c subunits of ATP synthases of alkaliphilic *Bacillus* with the *Bacillus* consensus sequence results in defective oxidative phosphorylation and non-fermentative growth at pH 10.5. *Journal of Biological Chemistry* **279:**26546–26554.

Chapter 15

Biosynthesis

How do microbes build their cells? Autotrophic bacteria and archaea build themselves entirely from carbon dioxide and nitrogen gas plus a few salts. From these simple molecules, microbial enzymes construct amino acids, the cell wall and envelope, and all the machinery of the cell. Some microbes produce complex secondary products with valuable properties, such as antibiotics. In the laboratory, microbial biosynthesis can be engineered to make cloned proteins, pesticides, and industrial reagents.

Once the key elements of carbon and nitrogen are incorporated into small molecules, more complex structures are built using many enzymes in intricate pathways. How do cells organize their biosynthesis to build precisely the forms they need? How do they avoid wasting energy on excess production? These questions are answered by the use of radioisotope tracers, genetic analysis of mutants, and the decoding of genomes. When a new species is discovered, its genome can reveal its capacities for biosynthesis, including hints of new pharmaceuticals.

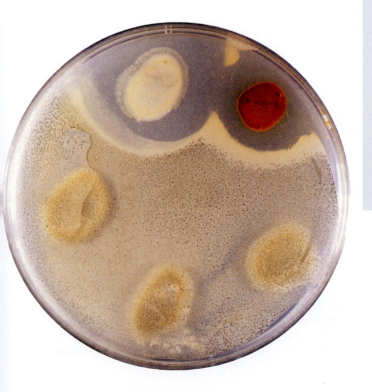

Streptomyces bacteria synthesize hundreds of kinds of antibiotics starting from simple organic compounds. A newly discovered antibiotic produced by *Streptomyces platensis* inhibits fatty acid biosynthesis. The antibiotic, platensimycin, kills methicillin-resistant strains of *Staphylococcus aureus* (MRSA), a cause of deadly skin infections. Platensimycin was discovered by Jun Wang, et al. 2006. *Nature* 441:358. *Source:* © Brad Mogen/Visuals Unlimited.

Microbes synthesize complex products such as vitamin B_{12} (**Fig. 15.1**), an essential cofactor for human enzymes. Vitamin B_{12} deficiency causes anemia, the failure to make sufficient red blood cells. In developed countries, virtually all our dietary B_{12} is produced by bacteria in the digestive tract of animals whose meat and dairy products we consume. Our own intestinal bacteria produce only small amounts of B_{12}, which are insufficient for our needs. Thus, strict vegetarians require vitamin B_{12} as a supplement, produced industrially by bacteria such as the Swiss cheese fermenter *Propionibacterium freudenreichii*. In vegan diets of traditional cultures, fermenting bacteria in unrefrigerated foods provide this essential vitamin.

The biosynthesis of vitamin B_{12} requires about 25 genes, each specifying a different enzyme to catalyze an essential step—a large amount of DNA to maintain error-free in the genome. Smaller carbon skeletons, such as succinyl coenzyme A (succinyl-CoA) and glycine, must be synthesized and joined to form a tetrapyrrole ring. Expression of all those enzymes, as well as the steps they catalyze, entails a substantial cost in energy. The energy is obtained by microbes as described in Chapters 13 and 14.

Chapter 15 shows how microbes assimilate the key elements carbon and nitrogen to build complex biomolecules, and how they regulate biosynthesis to make only the products they need. Microbial products find key applications as antibiotics, vitamins, and industrial chemicals.

15.1 Overview of Biosynthesis

Biosynthesis is the building of complex biomolecules, also known as **anabolism**, the reverse of catabolism. Products of biosynthesis include the common building blocks of cells, such as amino acids and sugars, as well as secondary products to be exported from the cell, such as antibiotics and signaling molecules. Biosynthesis requires a source of essential elements, such as carbon, oxygen, hydrogen, and nitrogen. The central element carbon is obtained either through CO_2 fixation (autotrophy) or through catabolism (heterotrophy). In either case, the carbon and associated atoms (oxygen and hydrogen, for example) must be formed into various small molecules that serve as substrates or building blocks for structural and functional parts of the cell.

Substrates for Biosynthesis Are Metabolic Intermediates

Many substrates for biosynthesis arise from the catabolic pathways discussed in Chapter 13 (**Fig. 15.2**). For example, the succinyl-CoA molecules used to build the tetrapyrrole ring of vitamin B_{12} come directly from the tricarboxylic acid (TCA) cycle, while the glycines derive from 3-phosphoglycerate, an intermediate of glycolysis. Glycerol 3-phosphate provides the glyceride backbone of lipids. Pyruvate forms the backbone of several amino acids with aliphatic side chains, whereas erythrose 4-phosphate contributes to the ring structures of aromatic amino acids. Other amino acids derive their carbon skeleton from TCA cycle intermediates oxaloacetate and 2-oxoglutarate, which incorporate nitrogen in the form of ammonium ion.

Biosynthesis Spends Energy

Biosynthesis entails substantial costs to the organism. The organism's genome must maintain the DNA that encodes all the enzymes that catalyze all the steps of the pathway; a mutation in any one enzyme may negate the entire process.

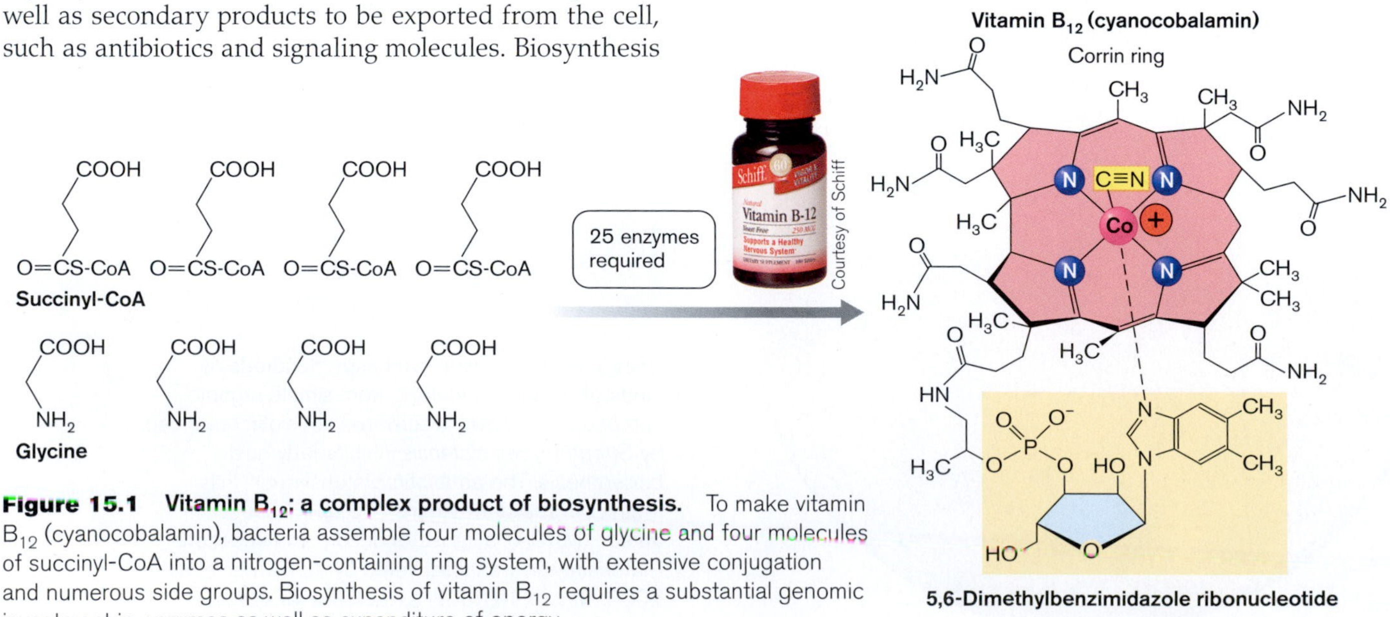

Figure 15.1 **Vitamin B_{12}: a complex product of biosynthesis.** To make vitamin B_{12} (cyanocobalamin), bacteria assemble four molecules of glycine and four molecules of succinyl-CoA into a nitrogen-containing ring system, with extensive conjugation and numerous side groups. Biosynthesis of vitamin B_{12} requires a substantial genomic investment in enzymes as well as expenditure of energy.

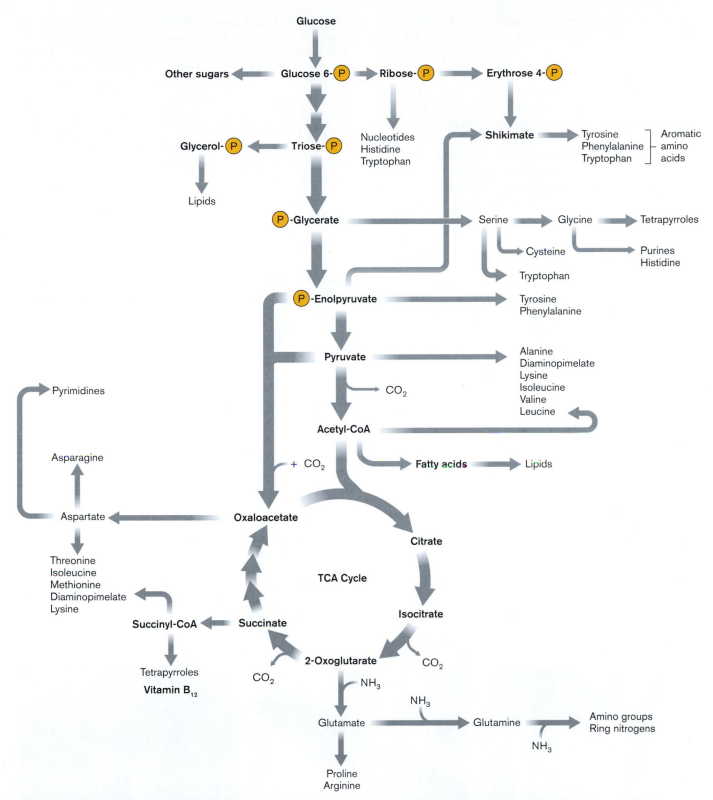

Figure 15.2 Substrates for biosynthesis. Substrates for biosynthesis of lipids and amino acids derive from the amphibolic pathways of glycolysis and the TCA cycle. Acetyl–CoA links these two pathways and serves as a substrate for many products of biosynthesis.

Enzyme activity requires substantial expenditure of energy in the form of ATP, NADPH, and other energy stores. The genomic and energetic costs lead different microbes to evolve different strategies to control these costs:

■ **Regulation.** Biosynthesis is regulated at several levels by the enzyme products. In this way, microbes avoid making more than they need under given environmental conditions. Regulation occurs at the levels of transcription and translation of enzymes, as well as through feedback inhibition of enzyme activity.

■ **Genome degeneration.** Each species makes an evolutionary "choice" whether to maintain the expense of a particular assimilatory or biosynthetic pathway, such as N_2 fixation, or to lose the pathway and become dependent on other species in the environment. Parasitic microbes that grow only inside a host cell show extensive loss of biosynthetic genes.

■ **Secondary products.** Free-living bacteria often produce secondary products that are not essential nutrients but enhance nutrient uptake or inhibit competitors. Many secondary products are antibiotics, such as streptomycin, produced by the actinomycete *Streptomyces coelicolor.*

First, we present the pathways by which autotrophs assimilate carbon and nitrogen into simple building blocks, such as acetyl-CoA. Then we will present pathways of construction of the key parts of the cell, such as fatty acids and amino acids. Finally, we will consider exciting recent discoveries in the biosynthesis of secondary products such as antibiotics.

15.2 CO₂ Fixation: The Calvin Cycle

The fundamental significance of **carbon dioxide fixation** by green plants and algae was recognized in the early twentieth century. "Fixation" refers to the covalent incorporation of a small molecule into larger biochemical material. The bulk of biomass on Earth consists of carbon fixed by chloroplasts and bacteria through the **reductive pentose phosphate cycle**, which recycles a pentose phosphate intermediate. The cycle is also known as the **Calvin cycle**, for which Melvin Calvin (1911–1997) was awarded the 1961 Nobel Prize in Chemistry. The mechanism of the pentose phosphate cycle was solved by Calvin with colleagues Andrew Benson and James Bassham, at the University of California, Berkeley. The importance of the Calvin cycle to the global ecosystem can scarcely be overestimated; it plays a major role in removing atmospheric CO_2. The Calvin cycle's responsiveness to CO_2, temperature, and other factors must be considered in all models of global warming.

NOTE: The Calvin cycle is also known as the **Calvin-Benson cycle**, the **Calvin-Benson-Bassham cycle**, or the **CBB cycle**. This book uses the terms *Calvin cycle* and *CBB cycle.*

The Calvin cycle is performed by several categories of organisms (**Fig. 15.3**):

■ **Oxygenic phototrophic bacteria**, mainly cyanobacteria, fix CO_2 by the Calvin cycle coupled to oxygenic photosynthesis. Cyanobacteria are believed to generate the majority of oxygen gas in Earth's atmosphere.

■ **Chloroplasts** of algae and multicellular plants use the Calvin cycle as the "dark reaction" or "light-independent reaction" of oxygenic photosynthesis.

■ **Facultatively anaerobic purple bacteria**, including sulfur oxidizers and photoheterotrophs such as *Rhodospirillum* and *Rhodobacter*, use the Calvin cycle to fix CO_2. These bacteria also obtain carbon through catabolism.

■ **Lithotrophic bacteria** fix CO_2 through the Calvin cycle, using NADPH and ATP provided by the O_2 oxidation of minerals.

A.

Michael Clayton, U. Wisconsin, Madison

B.

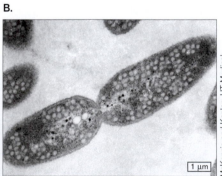

McKenzie and Kaplan, UT-Medical School, Houston

1 μm

C.

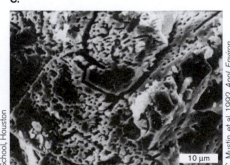

C. Mustin, et al. 1992. *Appl. Environ. Microbiol.* 38:1175

10 μm

Figure 15.3 **Aerobic and facultatively anaerobic phototrophs use the Calvin cycle.** **A.** *Oscillatoria*, a filamentous cyanobacterium, fixes CO_2 through oxygenic photosynthesis. Filaments are about 8 μm wide (light micrograph). **B.** *Rhodobacter sphaeroides*, purple photoheterotrophs with spherical photosynthetic membranes (TEM). **C.** The iron and sulfur bacterium *Thiobacillus ferrooxidans* oxidizes the sulfur in pyrite (FeS_2), using the Calvin cycle to fix CO_2 (TEM).

To date, the Calvin cycle has not been found among archaea and is largely absent in obligate anaerobic bacteria (**Table 15.1**). Archaea and anaerobes fix carbon by several different pathways, as described in Section 15.2.

^{14}C Labeling and Chromatography Revealed Calvin Cycle Intermediates

Early in the twentieth century, biochemists tried to figure out the mechanism of CO_2 fixation, believing that agricultural photosynthesis could be made more efficient. With the tools then available, however, researchers had no hope of sorting out the intermediate products through which CO_2 was fixed. Elucidating the pathways of CO_2 fixation required development of two fundamental tools: tracer radioisotopes, which are specific compounds labeled with radioactivity; and paper chromatography, a means of separating labeled compounds based on differential migration in a solvent. Both isotope labeling and chromatographic separation remain key tools of biochemical analysis today, enhanced by modern developments.

Carbon isotope labeling. As a means of tracking the conversion of intermediates within a biochemical pathway, specific atoms within molecules can be "labeled" for detection. Atoms are labeled by substituting an uncommon isotope, a version of an element containing a number of neutrons different from that of the most common isotope found in nature. Chemically, the isotopes of an element behave nearly the same in biochemical reactions, although enzymes may show a slight preference for one isotope.

The number of neutrons added to the number of protons (atomic number) yields the mass number of the isotope. For example, the predominant isotope of carbon, ^{12}C (carbon = 12), contains six protons and six neutrons, adding up to a mass number of twelve. The next most abundant isotope, ^{13}C, contains seven neutrons. An isotopic label needs to be detected based on its physical properties, such as its mass (by mass spectroscopy) or its nuclear magnetic resonance (by NMR). The most sensitive property for detection and the most widely available to researchers is based on radioactive decay. Radioactive decay requires an unstable isotope (radioisotope), such as ^{14}C.

In the 1930s, when Calvin started his research, few useful radioisotopes were known for biological molecules. The discovery of ^{14}C by Martin Kamen (1913–2002) in 1940 revolutionized biochemistry, enabling discovery of all kinds of cellular metabolism (**Special Topic 15.1**).

Paper chromatography separates intermediates of CO_2 fixation. When ^{14}C-labeled substrates became available in the 1940s, a method was needed to isolate and characterize the unknown intermediates. The technique of **paper chromatography** was developed by Calvin for his studies of carbon fixation. In paper chromatography, a mixture of chemicals is spotted on a special paper, one end of which is immersed in a carrier solvent (**Fig. 15.4A**). The solvent is then drawn up through the paper by capillary action. Because the various chemical components of the solution differ in their solubility, they travel in the solvent at different rates. After, the paper is dried and turned at a right angle. A second solvent is then added and drawn through in the crosswise direction. In the second solvent, the chemicals travel at different rates than in the first solvent, thus separating the chemicals in two dimensions. The chance of two different kinds of molecule migrating the same distance in both solvents is very low.

Figure 15.4B shows one of Calvin's original chromatographic separations. The alga *Chlorella* was exposed to light so it could conduct photosynthesis. $^{14}CO_2$ was added. Then samples of the alga were killed at specific times by plunging them into boiling alcohol. Five sec-

Table 15.1   Carbon dioxide fixation pathways.

Pathway	Bacteria	Archaea	Eukaryotes
	Organisms in which pathways occur		
	Aerobes and facultative anaerobes (Section 15.2)		
Calvin cycle	Cyanobacteria; purple phototrophs; lithotrophs	Rubisco homologs appear, but their function is unclear	Chloroplasts
	Anaerobes and archaea (Section 15.3)		
Reductive (reverse) TCA cycle	Green sulfur phototrophs (*Chlorobium*); thermophilic epsilon-*Proteobacteria*	Hyperthermophilic sulfur oxidizers (*Thermoproteus* and *Pyrobaculum*)	Anaplerotic reactions fix CO_2 to regenerate TCA intermediates
Reductive acetyl-CoA pathway	Anaerobes: acetogenic bacteria and sulfate reducers	Methanogens; other anaerobes	None known
3-Hydroxypropionate cycle	Green phototrophs (*Chloroflexus*)	Aerobic sulfur oxidizers (*Sulfolobus*)	None known

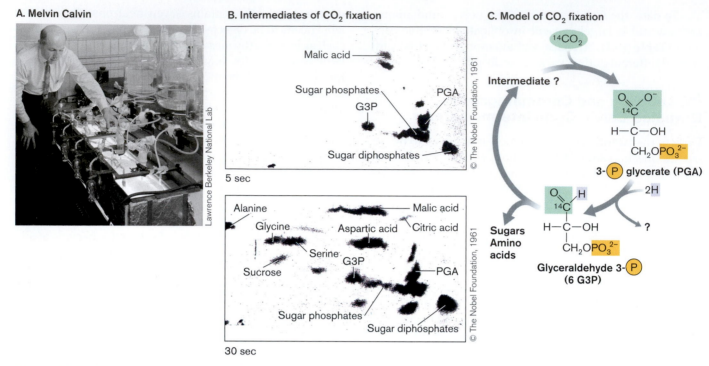

A. Melvin Calvin

B. Intermediates of CO₂ fixation

5 sec

30 sec

C. Model of CO₂ fixation

Figure 15.4 **Discovery of the Calvin cycle.** **A.** Melvin Calvin won the Nobel Prize for elucidating CO_2 fixation through the Calvin cycle. **B.** Paper chromatography reveals early intermediate products. Five seconds after addition of $^{14}CO_2$ to photosynthesizing *Chlorella*, the main spots showing radioactivity contain 3-phosphoglycerate (PGA) and glyceraldehyde 3-phosphate (G3P). Thirty seconds after $^{14}CO_2$ addition, radiolabel has entered subsequent products such as amino acids. **C.** Model of CO_2 fixation suggested by the chromatogram of ^{14}C-labeled intermediates.

onds after the addition of $^{14}CO_2$, the main spots showing radioactivity correspond to 3-phosphoglycerate (PGA) and glyceraldehyde 3-phosphate (G3P), early intermediates of CO_2 fixation. Thirty seconds after $^{14}CO_2$ was added, radiolabel is detected in subsequent products such as amino acids.

The chromatography data illustrate the extraordinary speed of CO_2 fixation, while revealing a glimpse of its earliest intermediates. At even earlier times (2 seconds) only PGA shows radiolabel. **Figure 15.4C** shows the portion of the cycle deduced from the data: CO_2 is incorporated to form PGA, then is hydrogenated to G3P by an unknown reducing agent (later shown to be NADPH). In this particular experiment, the key intermediate that first assimilates CO_2 remains undetected. The key intermediate was later identified as ribulose 1,5-bisphosphate, also known for its role in the pentose phosphate pathway of glucose catabolism (discussed in Section 13.5).

> **THOUGHT QUESTION 15.1** Propose a simple experiment to reveal the key intermediate to receive CO_2.

Since Calvin's time, chromatographic separation has been developed into ever more sophisticated forms. High-performance liquid chromatography (HPLC) provides extremely high-resolution separation of chemical products through columns packed with beads of various physical properties.

Overview of the Calvin Cycle

Decades of biochemical and genetic experiments established the details of CO_2 fixation in the Calvin cycle. An overview is shown in **Figure 15.5**; greater detail appears in **Figure 15.7**.

In each "turn" of the cycle, one molecule of CO_2 is condensed (combined, forming a new C–C bond) with the five-carbon sugar ribulose 1,5-bisphosphate. The resulting six-carbon intermediate splits into two molecules of 3-phosphoglycerate (PGA). The fixed CO_2 ultimately ends up as a carbon of glyceraldehyde 3-phosphate (G3P), reduced by $2H + 2e^-$ from NADPH + H^+. Recall from Chapter 13 that NADPH is a phosphorylated derivative of NADH commonly associated with biosynthesis.

How does the fixed CO_2 become one "corner" of a glucose molecule? For every three turns of the cycle, fixing three molecules of CO_2, the cycle feeds one molecule of G3P ($C_3H_5O_3$–PO_3^{2-}) into biosynthesis:

$$3CO_2 + 6NADPH + 6H^+ + 9ATP + 9H_2O \longrightarrow$$
$$C_3H_5O_3\text{–}P + 6NADP^+ + 9ADP + 8P_i$$

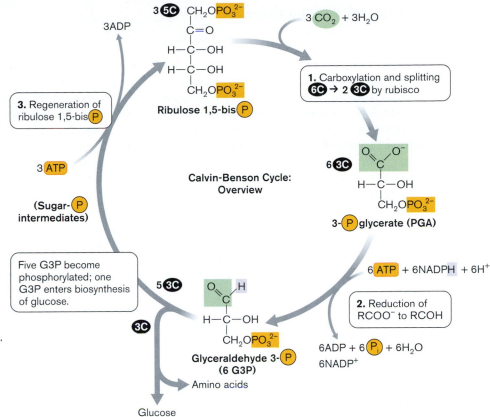

Figure 15.5 **The Calvin cycle: overview.** The Calvin cycle condenses CO_2 and H_2O with the intermediate ribulose 1,5-bisphosphate. Overall, three molecules each of CO_2 and of H_2O are fixed and split into six molecules of 3-phosphoglycerate (PGA), which are reduced by 6NADPH (with 6ATP) to six glyceraldehyde 3-phosphate (G3P). An additional 3ATP are consumed during sugar exchange reactions to regenerate three molecules of the recycled 5C intermediate ribulose 1,6-bisphosphate.

The H_2O and the phosphoryl group of G3P are ultimately recycled during biosynthetic assimilation of G3P. Two molecules of G3P may condense (that is, form a new C–C bond) in a pathway to form glucose. The overall condensation of $6CO_2 \longrightarrow 2$ G3P $\longrightarrow$ glucose yields:

$$6CO_2 + 12NADPH + 12H^+ + 18ATP + 18H_2O \longrightarrow$$
$$C_6H_{12}O_6 + 12NADP^+ + 18ADP + 18P_i$$

Alternatively, instead of forming glucose, G3P can enter biosynthesis of amino acids, vitamins, and other essential components of cells. Glyceraldehyde 3-phosphate is the fundamental unit of carbon assimilation into biomass.

The Calvin Cycle Requires Reduction and Regeneration of Ribulose 1,5-Bisphosphate

Each "turn" of the Calvin cycle has three main phases: carboxylation and splitting into two three-carbon (3C) intermediates; reduction of two molecules of PGA to two molecules of G3P; and regeneration of ribulose 1,5-bisphosphate (see **Fig. 15.5**):

1. **Carboxylation and splitting: 6C $\longrightarrow$ 2[3C].** Ribulose 1,5-bisphosphate condenses with CO_2 and H_2O, mediated by ribulose 1,5-bisphosphate carbon dioxide reductase/oxidase, generally referred to by the

acronym **rubisco**. Rubisco generates a six-carbon intermediate, which immediately hydrolyzes (splits into two parts by incorporating H_2O). The split produces two molecules of PGA, one of which contains the CO_2 fixed by this cycle.

2. **Reduction of PGA to G3P.** The carboxyl group of each PGA molecule is phosphorylated by ATP. The phosphorylated carboxyl group is then hydrolyzed and reduced by NADPH, forming G3P.

3. **Regeneration of ribulose 1,5-bisphosphate.** Of every six G3P, resulting from three cycles of fixing CO_2, five G3P enter a complex series of reactions (including hydrolysis of 3ATP) to regenerate three molecules of ribulose 1,5-bisphosphate. The net conversion of five G3P molecules to three molecules of ribulose 1,5-biphosphate yields two H_2O, restoring two of the 3H_2O fixed with 3CO_2. The remaining sixth G3P exits the cycle, available to be used in the biosynthesis of sugars and amino acids. Thus, three fixed carbons lead to one three-carbon product.

Overall, each CO_2 fixed sends one carbon into biosynthesis and regenerates one ribulose 1,5-bisphosphate.

The full Calvin cycle occurs only in bacteria and in plastid organelles (chloroplasts) that evolved from bacteria. It has not been found in archaea or in the cytoplasm of eukaryotes. Thus, the Calvin cycle appears

Special Topic 15.1 The Discovery of ^{14}C

In 1937, a young physicist from the University of Chicago, Martin Kamen (1913–2002), arrived at the University of California at Berkeley to study nuclear reactions. Berkeley possessed one of the world's largest cyclotrons, an instrument in which charged particles are accelerated to high speeds by passage through a voltage potential (**Fig. 1**). One use of a cyclotron was to generate special isotopes of many elements by particle collision. A charged particle collides with a target atom, whose nucleus then undergoes a reaction to form a new isotope. For example, a deuteron (proton plus neutron, mass 2) can collide with ^{10}B (boron 10: five protons, five neutrons) to produce ^{11}C (six protons, five neutrons) plus a neutron. The radioactive isotope ^{11}C decays to ^{10}B with a half-life of 21 minutes.

Kamen and a colleague, chemist Samuel Ruben, were assigned the task of providing radioisotopes requested by physicians for use in cancer therapy, an arrangement that generated funding for the cyclotron. As a result, Kamen and Ruben heard from biochemists about the need for a tracer isotope for carbon. At that time, the only radioactive isotope of carbon that Kamen and Ruben knew how to produce was the short-lived ^{11}C. But it occurred to them that with a ready supply of ^{11}C at their cyclotron, they themselves could attempt tracer experiments in physiology. They focused on a key question: the identity of the first compound to assimilate CO_2 during photosynthesis.

Kamen and Ruben designed experiments in which ^{11}C-radiolabeled CO_2 was assimilated by photosynthesis in *Chlorella*, a species of alga that is a food source for aquarium fish (**Fig. 2**). But the short half-life of ^{11}C made it difficult to complete an experiment while enough radioactivity remained to be detected. As the day's supply of ^{11}C was prepared by the cyclotron, the researchers in the biochemistry laboratory felt "like sprinters at the starting gate," according to Kamen: "Anyone looking in . . . when an experiment was in progress would have had the impression of three madmen hopping about in an insane asylum, what with the frenzied activity punctuated by loud classical music from the radio monitor, and Sam's yells to get on with it and hand him samples while he sat at the counter table, feverishly taking background and sample counts."

With their limited technology, Kamen and colleagues failed to identify intermediates, but they made another key discovery: Fixation of CO_2 occurs in organisms that lack photosynthesis. Colleagues had shown that an anaerobic bacterium fermented glycerol to succinate, a process that appeared to involve the addition of a fourth carbon from CO_2. So Kamen and Ruben used ^{11}C to test for assimilation by various bac-

A.

B.

Lawrence Berkeley National Lab

Lawrence Berkeley National Lab

Figure 1 Radioactive isotopes of carbon. A. The 60-inch cyclotron at the Radiation Laboratory at Berkeley in 1939. Neutrons generated by the cyclotron bombarded stable atoms to generate radioactive isotopes. **B.** Martin Kamen at the cyclotron in 1941.

to have evolved after the divergence of the three domains of life. The relative uniformity of the pathway across species, compared with the diversity of anaerobic pathways of CO_2 fixation, supports the view of late emergence. Nevertheless, some archaeal genomes do show a homolog of rubisco. The function of archaeal rubisco is not yet understood.

Ribulose 1,5-Bisphosphate Fixes CO_2 and Is Regenerated

Rubisco is found in large quantities in CO_2-fixing cells; it is one of the most prevalent proteins on Earth. Its characteristics determine the rate and efficiency of CO_2 fixation. The structure of rubisco is highly conserved across

teria, by yeast, and even by rat liver and muscle. All of these organisms assimilated radioactive CO_2. It was ultimately established that all organisms—including humans—fix some CO_2 as part of their central metabolism, usually by reversal of steps in the TCA cycle.

Kamen and his colleagues sought a longer-lived radio-isotope of carbon. Theoretical predictions suggested that a much longer half-life would be exhibited by the isotope ^{14}C. But ^{14}C had yet to be generated and isolated from the cyclotron. Kamen recalled observing unusual collision trails in a sample of nitrogen gas (^{14}N) exposed to neutrons from the cyclotron. Each collision of ^{14}N with a neutron formed a long trail characteristic of a positron (an emitted proton), attached to a short trail expected for a more massive product. The more massive product, Kamen guessed, was ^{14}C.

The isotope ^{14}C proved to have a half-life of 5,700 years. This stability made ^{14}C exactly the kind of tracer needed for experiments in biology. In 1945, Kamen reported the first biological experiment utilizing ^{14}C to investigate the assimilation of CO_2 by the bacterium *Clostridium thermoaceticum*. The bacteria were shown to assimilate two molecules of CO_2 to form acetic acid.

Kamen's own use of ^{14}C for tracer studies was cut short in 1944 by his dismissal from Berkeley stemming from charges that he was a security risk. During the war years, the cyclotron was devoted to weapons development, and all researchers were under tight scrutiny. Kamen faced a decade of accusations, including an appearance before the House Un-American Activities Committee, before he cleared himself of all charges. By that time, however, the fundamental questions of CO_2 fixation in oxygenic photosynthesis had been answered by Melvin Calvin and colleagues, using ^{14}C back at Berkeley. Kamen eventually discovered other kinds of CO_2 fixation in anaerobic bacteria, and he received the Enrico Fermi Award in 1996 for his original discovery of ^{14}C.

Today, short-lived radioisotopes such as ^{11}C find new applications in medical research. For human subjects, short-lived isotopes are desirable because their rapid decay limits exposure to harmful radiation. ^{11}C is particularly useful in positron-emission tomography (PET), a technique in which a molecule labeled with ^{11}C is followed in the body based on detection of emitted positrons. PET can be used to follow the fate of neurotransmitters and psychotropic substances in the brain.

Figure 2 The first carbon radioisotope experiments in biology.
A. Sam Ruben, chemist (seated), with plant physiologist Zev Hassid, demonstrating equipment for their ^{11}C studies of photosynthesis in *Chlorella*.
B. The alga *Chlorella*, in which the first ^{11}C-label studies were performed.

A.

Hansel Mieth/Time Life/Getty Images

B.

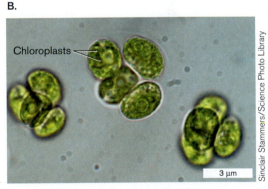

Chloroplasts

3 μm

Sinclair Stammers/Science Photo Library

bacterial and chloroplast domains. It consists of two types of subunits, designated small and large (**Fig. 15.6A**). The large (L) subunit contains the active (catalytic) site. Different species contain different multiples of the small and large subunits: eight (in lithotrophs, green phototrophs, and some algae), six (in plant chloroplasts and some algae), or two (in purple phototrophs). Nevertheless,

the fundamental mechanism of CO_2 fixation in all these organisms appears to be similar.

Mechanism of rubisco. Initially, rubisco adds CO_2 to ribulose 1,5-bisphosphate to give an unstable six-carbon intermediate that remains bound to the enzyme (**Fig. 15.6B and C**). The intermediate hydrolyzes into two molecules

A.

B.

C.

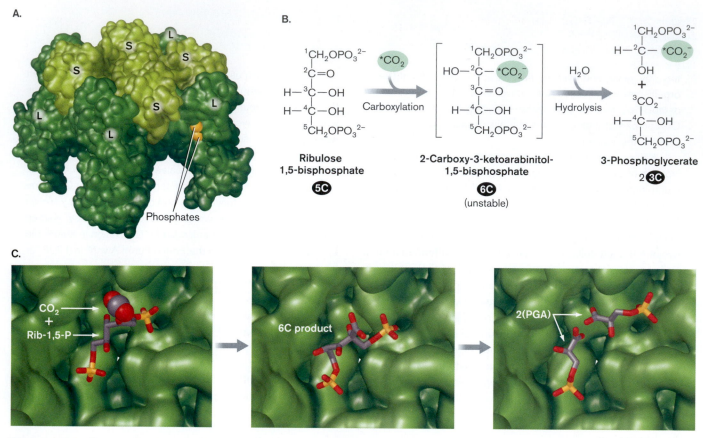

Figure 15.6   The mechanism of rubisco.   A. Rubisco from *Alcaligenes eutrophus* consists of eight large subunits (L) crowned by eight small subunits (S). Only the upper four L and S subunits are shown. On each L subunit, a catalytic site contains two phosphates (orange) that compete with the substrates for binding. (PDB code: 1BXN) **B.** Rubisco adds CO_2 to ribulose-1,5-bisphosphate to give an unstable six-carbon intermediate bound to the enzyme. The bound intermediate hydrolyzes to form two molecules of 3-phosphoglycerate (PGA). Both PGA molecules enter the cellular pool of PGA and behave equivalently in the rest of the cycle. **C.** The mechanism of CO_2 fixation at the catalytic site. CO_2 adds to the ketone, generating the 6C intermediate, which splits into two molecules of PGA. The two phosphates appear to act as anchors for the molecule(s) throughout the CO_2 assimilation and cleavage. (PDB code: 1RUS)

of PGA, each held at the active site by its phosphate. Only one molecule of PGA contains a carbon from the newly fixed CO_2; nevertheless, as both molecules dissociate from the enzyme, they enter the cellular pool of PGA and behave equivalently in the rest of the cycle.

Rubisco has a high affinity for CO_2, and the typical concentration of rubisco active sites (one per large subunit, six per complex) in plant chloroplasts is 4 mM, about 500 times greater than the concentration of CO_2. Thus, considerable energy is invested in CO_2 absorption. Yet the efficiency of carbon fixation by rubisco is lowered by the existence of a competing reaction with O_2 that leads to 2-phosphoglycolate instead of 3-phosphoglycerate. Researchers are trying to engineer a more efficient oxygen-resistant rubisco enzyme to improve efficiency of crop yields.

THOUGHT QUESTION 15.2   Speculate why rubisco catalyzes a competing reaction with oxygen. Why might researchers be unsuccessful in attempting to engineer rubisco without this reaction?

Regeneration of ribulose 1,5-bisphosphate. Overall, each glyceraldehyde 3-phosphate (G3P) arises from three rounds of CO_2 fixation by ribulose 1,5-bisphosphate. Thus, to maintain the cycle, for each G3P provided to biosynthesis, five other molecules of G3P must be recycled into three ribulose 1,5-bisphosphate. Details of regenerating the intermediate are shown in the lower half of **Figure 15.7**.

Figure 15.7 The Calvin cycle in detail.
This diagram shows the fixation of CO_2, reduction of 3-phosphoglycerate to glyceraldehyde 3-phosphate (G3P), and conversion of five G3P into three ribulose 1,5-phosphate.

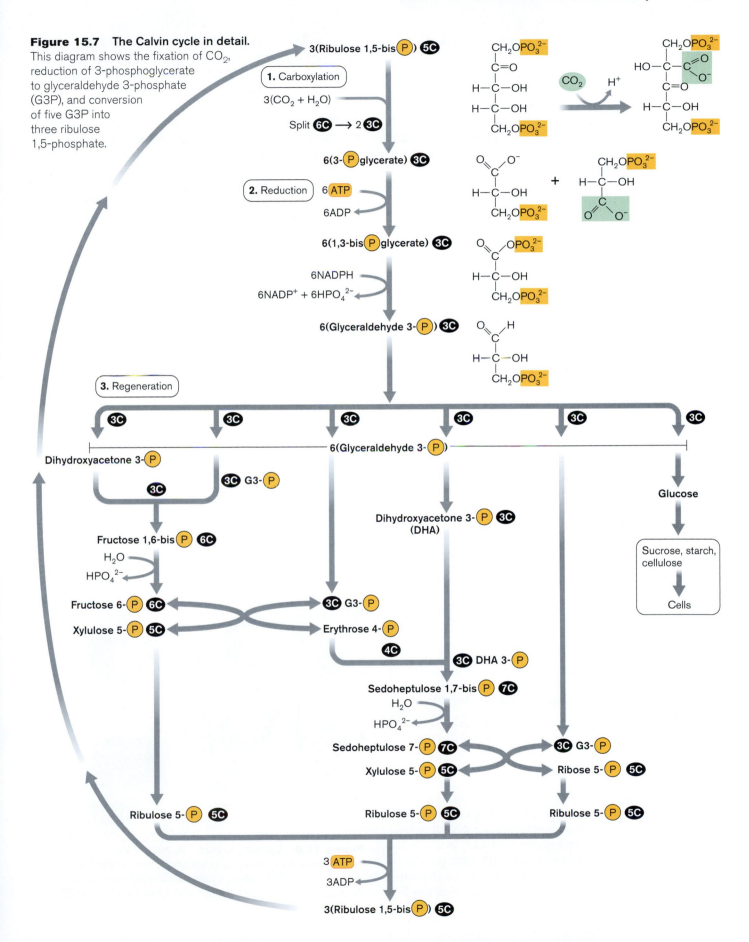

The regeneration pathway is summarized here:

- Two molecules of G3P condense to form the six-carbon sugar fructose 6-phosphate.
- Fructose 6-phosphate condenses with a third G3P. The nine-carbon molecule splits to form a five-carbon sugar (xylulose 5-phosphate) and a four-carbon sugar (erythrose 4-phosphate). Xylulose 5-phosphate rearranges to ribulose 5-phosphate.
- Erythrose 4-phosphate condenses with a fourth G3P (rearranged to dihydroxyacetone 3-phosphate) to make a seven-carbon sugar.
- The seven-carbon sugar condenses with the fifth G3P and splits into two five-carbon sugars (ribulose 5-phosphate via xylulose 5-phosphate and ribose 5-phosphate).
- Each of the three five-carbon sugars receives a second phosphoryl group from ATP, generating ribulose 1,5-bisphosphate. Each ribulose 1,5-bisphosphate is now ready for rubisco to fix another CO_2.

Why would a cycle have evolved requiring so many enzymatic steps to so many different intermediates? First, a cycle with many steps breaks down the energy flow into numerous reversible conversions with near-zero values of ΔG. The nearer to equilibrium, the more energy is conserved in the conversion. Second, the multiple different intermediates provide substrates for biosynthesis. For example, some molecules of erythrose 4-phosphate and ribose 5-phosphate are withdrawn from the cycle to build aromatic amino acids and nucleotides.

THOUGHT QUESTION 15.3 Why does ribulose 1,5-bisphosphate have to contain two phosphoryl groups, whereas the other intermediates of the Calvin cycle contain only one?

THOUGHT QUESTION 15.4 Which catabolic pathway (see Chapter 13) includes some of the same sugar-phosphate intermediates as those of the Calvin cycle? What might this suggest about the evolution of the two pathways?

Regulation of the Calvin Cycle

The Calvin cycle is tightly organized and regulated within cells. Expression of the CO_2-fixing enzymes varies with CO_2 concentration, light levels (for phototrophs), and temperature. The concentration of CO_2 is a special problem because CO_2 diffuses readily through phospholipid membranes. Thus, cells cannot concentrate this substrate across the cell membrane to reach the level needed

to drive rubisco. The gas concentration problem is solved by some species through special carbonate-concentrating structures. Other bacteria use alternative CO_2-fixing systems adapted to different CO_2 concentrations.

Carboxysomes. Many organisms that fix CO_2 contain the rubisco complex within subcellular structures called **carboxysomes** (**Fig. 15.8**). A carboxysome consists of a polyhedral shell of protein subunits surrounding packed molecules of rubisco. The protein shell sequesters carbon dioxide, permitting less leakage than a membrane-bounded organelle. Mutant strains of bacteria lacking carboxysomes can fix CO_2 only at high concentration (5%), much higher than the atmospheric CO_2 concentration (0.037%). Carboxysomes are found within CO_2-fixing lithotrophs, as well as within cyanobacteria and chloroplasts.

Regulation of gene expression. When CO_2 levels decline, genes are induced to express transmembrane transporters of CO_2 and bicarbonate (HCO_3^-) (**Fig. 15.9**). The inducible genes encode transport systems known collectively as the **CO_2-concentrating mechanism (CCM)**. The CCM enables rapid concentration of HCO_3^- into the carboxysome, where it is dehydrated to CO_2 by the enzyme carbonic anhydrase. The CO_2 is then fixed by rubisco to PGA, which exits the carboxysome for further conversion in the cytoplasm.

The CO_2-concentrating mechanism is regulated genetically. The CCM genes encode a low-affinity CO_2 transporter (NdhF4), a high-affinity CO_2 transporter (NdhF3), and two high-affinity HCO_3^- transporters

A.

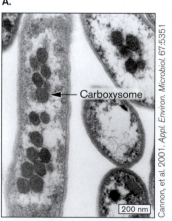

Carboxysome

200 nm

Cannon, et al. 2001. *Appl. Environ. Microbiol.* 67:5351

B.

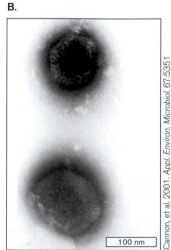

100 nm

Cannon, et al. 2001. *Appl. Environ. Microbiol.* 67:5351

Figure 15.8   Carboxysomes. A. *Halothiobacillus neapolitanus*, a sulfur-oxidizing lithotroph; thin section (TEM) showing polyhedral carboxysomes. **B.** Carboxysome isolated from *Synecchococcus* species packed with rubisco complexes (TEM).

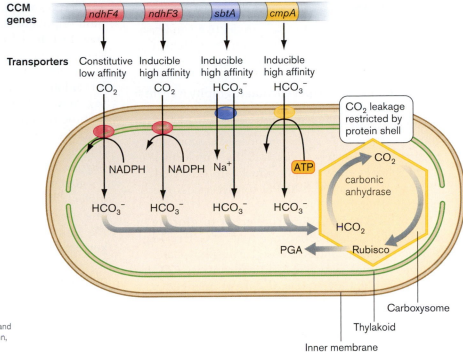

Figure 15.9   The CO$_2$-concentrating mechanism (CCM) in cyanobacteria. CO$_2$ and HCO$_3^-$ are brought into the cell by transport complexes in the inner membrane or thylakoid membranes of *Synechocystis* species. The HCO$_3^-$ is concentrated in the carboxysome and converted to CO$_2$ by carbonic anhydrase. *Source:* Model from Dean Price and colleagues, Australian National University at Canberra; McGinn, et al. 2003. *Plant Physiology* 132:218.

(CmpABCD and SbtA). The low-affinity CO$_2$ transporter is effective only when CO$_2$ concentration is high. By contrast, the high-affinity transporters are induced by CO$_2$ starvation.

Induction of transporters by low CO$_2$ levels is shown by reverse transcriptase PCR (RT-PCR), a technique for detection of minute quantities of mRNA using the enzyme reverse transcriptase followed by polymerase chain reaction, a technique discussed in Chapter 12, **Fig. 12.15.** In RT-PCR, the RNA transcripts are reverse-transcribed to DNA; the DNA is then amplified by PCR to generate a detectable product. RNA transcripts are obtained for genes expressed before and after CO$_2$ limitation. The *sbtA* and *cmpA* transcripts, encoding HCO$_3^-$ transporters, show particularly strong induction by CO$_2$ (see **Fig. 15.10**). By contrast, the constitutive low-affinity CO$_2$ transporter NdhF4 shows no induction.

Alternative CO$_2$ fixation systems. Instead of carboxysomes, some phototrophs possess entire alternative systems of CO$_2$ fixation adapted to different levels of CO$_2$. For example, the genome of the purple phototroph *Rhodobacter sphaeroides* includes two unlinked operons, each encoding a different set of CO$_2$ fixation genes, called form I and form II. Form I and form II each encode all the key enzymes such as rubisco. When CO$_2$ is limiting, form I enzymes work best; when CO$_2$ levels are saturating, form II enzymes work best.

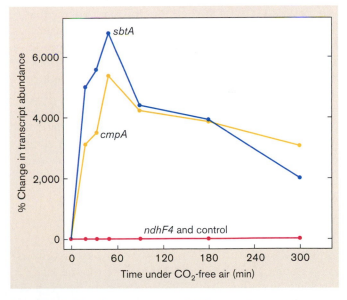

Figure 15.10   Time course of CCM gene transcription. When CO$_2$ is limiting, *Synechocystis* genes encoding the carbon uptake complexes show induced transcription. Induction is measured by reverse transcriptase PCR (RT-PCR). The *sbtA* and *cmpA* genes show especially high induction, whereas the constitutive system (*ndhF4*) shows no induction.

TO SUMMARIZE:

- **The Calvin cycle** fixes CO_2 by reductive condensation with ribulose 1,5-bisphosphate. The Calvin cycle is used by cyanobacteria and chloroplasts and by some lithotrophs and photoheterotrophs.
- **^{14}C radiolabel and paper chromatography** were first used to identify intermediates of the Calvin cycle.
- **Rubisco** catalyzes the condensation of CO_2 with ribulose 1,5-bisphosphate. The six-carbon intermediate immediately splits into two molecules of 3-phosphoglycerate (PGA), which are activated by ATP and reduced by NADPH to glyceraldehyde 3-phosphate (G3P).
- **One of every six G3P is converted to glucose** or amino acids. The other five molecules of G3P undergo reactions to regenerate ribulose 1-5-bisphosphate.
- **Carboxysomes sequester and concentrate CO_2** for fixation by the Calvin cycle. CO_2 levels regulate gene expression for carboxysome transporters and the Calvin cycle.

15.3 CO_2 Fixation in Anaerobes and Archaea

We have seen how aerobic organisms and facultative anaerobes fix CO_2 by the Calvin cycle. Some anaerobes, however, fix CO_2 by alternative means (see **Table 15.1**). The most ancient pathways, from the standpoint of evolution and phylogenetic distribution, are the reductive, or reverse, TCA cycle of anaerobic phototrophs and the reductive acetyl-CoA pathway of methanogens. Anaerobic CO_2 fixation plays an essential role in soil, aquatic, and wetland ecosystems, as well as in the digestive systems of animals such as cattle.

The Reductive, or Reverse, TCA Cycle

Portions of the TCA cycle (discussed in Section 13.6) are found in all major groups of organisms. Most of the individual reaction steps of the TCA cycle are reversible, enabling the assimilation of small amounts of CO_2. Furthermore, all organisms, including humans, fix small amounts of CO_2 that regenerate TCA cycle intermediates. These regeneration steps are called **anaplerotic reactions**. Common anaplerotic reactions are the formation of oxaloacetate from phosphoenolpyruvate (PEP), catalyzed by PEP carboxylase, and the formation of malate from pyruvate, catalyzed by malate dehydrogenase.

In some anaerobic bacteria and archaea, the entire TCA cycle functions in "reverse," reducing CO_2 to generate acetyl-CoA and build sugars (**Fig. 15.11A**). The **reductive** or **reverse TCA cycle** is used by bacteria such

as *Chlorobium tepidum*, a thermophilic green sulfur bacterium originally isolated from a New Zealand hot spring (**Fig. 15.11B**). *Chlorobium* conducts anoxygenic photosynthesis by photolyzing H_2S to produce elemental sulfur, which collects in extracellular granules. Sulfur granule formation is of interest for the development of a process to remove H_2S from sulfide-generating industries such as refining of coal and oil. The reductive TCA cycle is also used by epsilon-proteobacteria of hydrothermal vent communities and by sulfur-reducing archaea, such as *Thermoproteus* and *Pyrobaculum*.

Reversal of the TCA cycle uses four or five ATP molecules to fix four molecules of CO_2 and generates one molecule of oxaloacetate. The enzymes used are the same as for the "forward" cycle, except for three key enzymes that spend ATP to drive the reaction in reverse: ATP citrate lyase, 2-oxoglutarate:ferredoxin oxidoreductase, and fumarate reductase. For example, the enzyme ATP citrate lyase catalyzes the cleavage of citrate into acetyl-CoA and oxaloacetate. CO_2 assimilation requires reduction by NADPH. The reductive TCA cycle is believed to be the most ancient means of CO_2 fixation and amino acid biosynthesis, the original cycle of biomass generation in the ancestors of all three living domains.

From CO_2 to acetyl-CoA, the reverse TCA cycle essentially reverses the complementary reaction of catabolism:

$$2CO_2 + 2ATP + 8[H] + H\text{-}SCoA \longrightarrow$$
$$CH_3CO\text{-}SCoA + 3H_2O + 2ADP + 2P_i$$

The coenzyme A is recycled as the acetyl group enters biosynthesis. For example, acetyl-CoA can assimilate another CO_2 with reduction to pyruvate. Pyruvate then builds up to glucose and related sugars by **gluconeogenesis** or reverse glycolysis, with expenditure of ATP and NADPH. Alternatively, acetyl-CoA can enter biosynthetic pathways to produce fatty acids or amino acids.

Unlike the Calvin cycle, the reverse TCA cycle provides several different molecular intermediates that can assimilate CO_2. Here are the key steps:

- Succinyl-CoA assimilates CO_2 to form 2-oxoglutarate (alpha-ketoglutarate).
- 2-oxoglutarate assimilates CO_2 to form isocitrate.
- Acetyl-CoA (produced by the reverse TCA cycle) assimilates CO_2 to form pyruvate.

Each CO_2 assimilation requires one or more reduction steps. 2-oxoglutarate is reduced by the protein ferredoxin (FDH_2). Ferredoxin also mediates acetyl-CoA reduction to pyruvate. Other reduction steps may be accomplished by NADPH or NADH.

A.

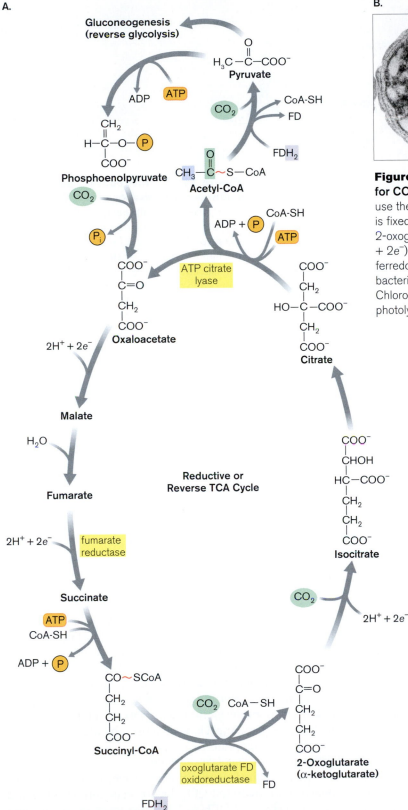

Gluconeogenesis
(reverse glycolysis)

$$H_3C-\overset{\overset{\displaystyle O}{\|}}{C}-COO^-$$
Pyruvate

ADP   ATP

CO_2

CoA-SH
FD

FDH_2

$$H-\overset{\overset{\displaystyle CH_2}{\|}}{\underset{\underset{\displaystyle COO^-}{|}}{C}}-O-P$$
Phosphoenolpyruvate

$$CH_3-\overset{\overset{\displaystyle O}{\|}}{C}\sim S-CoA$$
Acetyl-CoA

CO_2

P_i

ADP + P   CoA-SH

ATP

$$\overset{\displaystyle COO^-}{\underset{\displaystyle COO^-}{\overset{\displaystyle |}{\underset{\displaystyle |}{\overset{\displaystyle C=O}{\underset{\displaystyle CH_2}{|}}}}}}$$
Oxaloacetate

==ATP citrate lyase==

$$\overset{\displaystyle COO^-}{\underset{\displaystyle COO^-}{\overset{\displaystyle |}{\underset{\displaystyle |}{\overset{\displaystyle CH_2}{\underset{\displaystyle CH_2}{HO-C-COO^-}}}}}}$$
Citrate

$2H^+ + 2e^-$

Malate

H_2O

Fumarate

**Reductive or
Reverse TCA Cycle**

$$\overset{\displaystyle COO^-}{\underset{\displaystyle COO^-}{\overset{\displaystyle |}{\underset{\displaystyle |}{\overset{\displaystyle CHOH}{\underset{\displaystyle CH_2}{\overset{\displaystyle HC-COO^-}{|}}}}}}}$$
Isocitrate

$2H^+ + 2e^-$ ==fumarate reductase==

Succinate

ATP
CoA-SH

ADP + P

$$\overset{\displaystyle CO\sim SCoA}{\underset{\displaystyle COO^-}{\overset{\displaystyle |}{\underset{\displaystyle |}{\overset{\displaystyle CH_2}{\underset{\displaystyle CH_2}{|}}}}}}$$
Succinyl-CoA

CO_2 CoA—SH

$$\overset{\displaystyle COO^-}{\underset{\displaystyle COO^-}{\overset{\displaystyle |}{\underset{\displaystyle |}{\overset{\displaystyle C=O}{\underset{\displaystyle CH_2}{\overset{\displaystyle CH_2}{|}}}}}}}$$
**2-Oxoglutarate
(α-ketoglutarate)**

CO_2

$2H^+ + 2e^-$

==oxoglutarate FD
oxidoreductase==

FD

FDH_2

B.

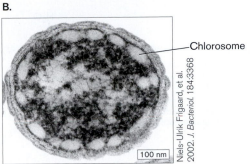

Chlorosome

100 nm

Niels-Ulrik Frigaard, et al.
2002. *J. Bacteriol.* 184:3368

Figure 15.11   The reductive, or reverse, TCA cycle for CO_2 fixation.   A. Anaerobic phototrophs and archaea use the reverse TCA cycle to fix carbon into biomass. CO_2 is fixed by several intermediates including succinyl-CoA, 2-oxoglutarate, and acetyl-CoA. Reduction (addition of $2H^+ + 2e^-$) is performed by NADPH or NADH and by reduced ferredoxin (FDH_2). **B.** *Chlorobium tepidum*, a green sulfur bacterium that fixes CO_2 by the reductive TCA cycle (TEM). Chlorosome membranes contain the reaction centers of photolysis.

Acetyl-CoA Pathway

Yet another route for CO_2 assimilation into acetyl-CoA and pyruvate is the **reductive acetyl-CoA pathway** (**Fig. 15.12A**). In the acetyl-CoA pathway, two CO_2 molecules are condensed through converging pathways to form the acetyl group of acetyl-CoA. The acetyl-CoA pathway is used by anaerobic soil bacteria, such as *Clostridium thermo-aceticum*, and by most autotrophic sulfate reducers, such as *Desulfobacterium autotrophicum*. It also provides the main source of biomass for methanogens. Most of the CO_2 absorbed by methanogens is used to yield energy by reducing CO_2 to methane (discussed in Chapter 14). However, 5% of the CO_2 absorbed by a methanogen enters the acetyl-CoA pathway for biosynthesis, generating the entire biomass of the organism.

The substrates and ultimate products of the acetyl-CoA pathway are the same as those of the reverse TCA cycle, except that the reducing agent is H_2 instead of NADPH. The acetyl-CoA pathway is slightly more efficient than the TCA cycle, requiring one less ATP:

$$2CO_2 + ATP + 4H_2 + \text{H-SCoA} \longrightarrow$$
$$CH_3\text{CO-SCoA} + 3H_2O + ADP + P_i$$

Nevertheless, the form of the pathway and its intermediate compounds are completely different from those of the TCA cycle. The

A.

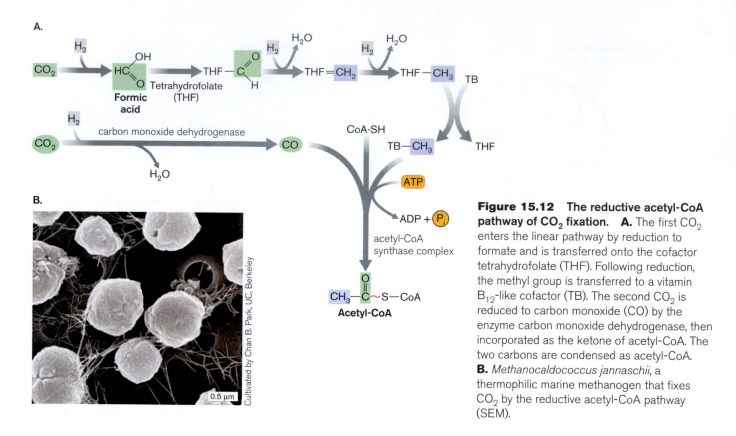

Cultivated by Chan B. Park, UC, Berkeley

Figure 15.12   The reductive acetyl-CoA pathway of CO_2 fixation.   A. The first CO_2 enters the linear pathway by reduction to formate and is transferred onto the cofactor tetrahydrofolate (THF). Following reduction, the methyl group is transferred to a vitamin B_{12}-like cofactor (TB). The second CO_2 is reduced to carbon monoxide (CO) by the enzyme carbon monoxide dehydrogenase, then incorporated as the ketone of acetyl-CoA. The two carbons are condensed as acetyl-CoA. **B.** *Methanocaldococcus jannaschii*, a thermophilic marine methanogen that fixes CO_2 by the reductive acetyl-CoA pathway (SEM).

reductive acetyl-CoA pathway is linear, with no recycled intermediates.

Reduction of the first CO_2. The first CO_2 enters the linear pathway by reduction to formate. The formate is transferred onto tetrahydrofolate (THF), a reduced form of **folate**. Folate (folic acid) is a heteroaromatic cofactor, comparable to NADH. Folate is required by many living organisms and is an essential vitamin for humans. It is particularly important for fetal brain development.

The carbon carried by reduced folate (THF) is reduced in successive steps to a methyl group. In methanogens, a THF methyl group is converted to methane for net energy acquisition. An occasional methyl group, however, is fixed into biomass. For incorporation into biomass, the methyl must be transferred from THF onto a different cofactor, a specialized tetrapyrrole (TB) related to vitamin B_{12}.

Reductive CO_2 incorporation into acetyl-CoA. The second CO_2 is reduced to carbon monoxide (CO) by the enzyme carbon monoxide dehydrogenase. This same enzyme is used in reverse reaction by some lithotrophs to gain energy from carbon monoxide as an electron donor. For fixation, the CO is condensed with the methyl group carried by TB to form acetyl-CoA. The acetyl-CoA then enters pathways of biosynthesis, as it does when formed by the reverse TCA cycle.

The fixation of CO_2 and the release of methane by methanogens is a major concern for cattle farming.

Within the rumen, bacteria convert feedstock into CO_2 and H_2; and methanogens divert a significant portion of these gases into methane. The methane then escapes through burping and flatulence. If methanogen growth could be prevented, the CO_2 and H_2 could be fixed by other microbes into fermentation acids assimilated by the rumen. Beef production would be more efficient, with less release of the greenhouse gas methane.

The 3-Hydroxypropionate Cycle

New pathways of carbon fixation continue to be discovered. In the **3-hydroxypropionate cycle**, CO_2 is fixed by acetyl-CoA into 3-hydroxypropionate (**Fig. 15.13A**). The cycle was discovered in the green photoheterotroph *Chloroflexus aurantiacus* (**Fig. 15.13B**). *Chloroflexus* bacteria are filamentous thermophiles that absorb light at longer wavelengths than cyanobacteria and often grow below cyanobacteria in thick mats of biofilm. The 3-hydroxypropionate cycle is also used by archaea, such as the thermoacidophile *Acidianus brierleyi* and the aerobic sulfur oxidizer *Sulfolobus metallicus*.

In the 3-hydroxypropionate cycle, acetyl-CoA condenses with hydrated CO_2 (bicarbonate ion, HCO_3^-) and is reduced by 2 NADPH to 3-hydroxypropionate. Through several intermediates, including succinyl-CoA, condensation with a second molecule of HCO_3^- yields malyl-CoA. These intermediates were detected by [14C] acetate incorporation into the metabolism of *Chloroflexus* (**Fig. 15.13C**).

A.

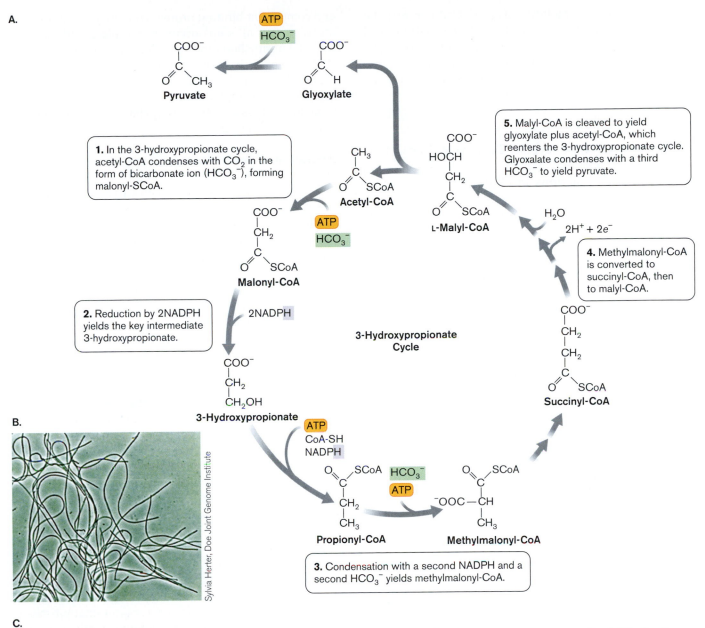

1. In the 3-hydroxypropionate cycle, acetyl-CoA condenses with CO₂ in the form of bicarbonate ion (HCO₃⁻), forming malonyl-SCoA.

2. Reduction by 2NADPH yields the key intermediate 3-hydroxypropionate.

3. Condensation with a second NADPH and a second HCO₃⁻ yields methylmalonyl-CoA.

4. Methylmalonyl-CoA is converted to succinyl-CoA, then to malyl-CoA.

5. Malyl-CoA is cleaved to yield glyoxylate plus acetyl-CoA, which reenters the 3-hydroxypropionate cycle. Glyoxalate condenses with a third HCO₃⁻ to yield pyruvate.

3-Hydroxypropionate Cycle

Pyruvate · Glyoxylate · Acetyl-CoA · Malonyl-CoA · 3-Hydroxypropionate · Propionyl-CoA · Methylmalonyl-CoA · Succinyl-CoA · L-Malyl-CoA

B.

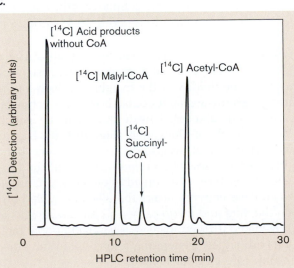

Sylvia Herter, Doe Joint Genome Institute

C.

[¹⁴C] Detection (arbitrary units)

[¹⁴C] Acid products without CoA

[¹⁴C] Malyl-CoA

[¹⁴C] Acetyl-CoA

[¹⁴C] Succinyl-CoA

HPLC retention time (min)

Figure 15.13 The 3-hydroxypropionate cycle of CO₂ fixation.
A. In the 3-hydroxypropionate cycle, acetyl-CoA condenses with CO₂ in the form of bicarbonate ion (HCO₃⁻), forming malonyl-SCoA (1). Reduction by 2NADPH yields the key intermediate 3-hydroxypropionate (2). Condensation with a second NADPH (3) and a second HCO₃⁻ yields methylmalonyl-CoA. Step 4 converts methylmalonyl-CoA to malyl-CoA. In step 5, malyl-CoA is cleaved to yield glyoxyalate plus acetyl-CoA, which reenters the 3-hydroxypropionate cycle. Glyoxalate condenses with a third HCO₃⁻ to yield pyruvate. **B.** *Chloroflexus*, a green phototroph that performs the 3-hydroxypropionate cycle. **C.** Products of [¹⁴C] acetate incorporate into the 3-hydroxypropionate cycle. After 1 minute, samples were separated by HPLC (high-performance liquid chromatography). The nonpolar coenzyme A esters of each acid separate in the HPLC solvent. *Source:* A. Modified from Silke Friedmann, et al. 2006. *J. Bacteriol.* 188:2646. C. Herter, et al. 2002. *J. Biol. Chem.* 277:20277.

Malyl-CoA releases a molecule of acetyl-CoA to renew the cycle, plus glyoxalate. Glyoxalate condenses with a third HCO_3^- to yield pyruvate, which enters standard routes for biosynthesis of sugars and amino acids. Thus in all, three molecules of CO_2 are fixed into one molecule of pyruvate, which serves as a substrate for biosynthesis.

TO SUMMARIZE:

- **The reductive, or reverse, TCA cycle** fixes CO_2 in some anaerobes and archaea.
- **Anaplerotic reactions** in all organisms regenerate TCA cycle intermediates by fixing CO_2.
- **The acetyl-CoA pathway** in anaerobic bacteria and methanogens fixes CO_2 by condensation to form acetyl-CoA.
- **The 3-hydroxypropionate cycle** in chlorobia and in some hyperthermophilic archaea fixes CO_2 in a cycle generating the intermediate 3-hydroxypropionate.

15.4 Biosynthesis of Fatty Acids and Polyesters

The biosynthesis of cell parts begins with small, versatile substrates such as acetyl-CoA, a product of catabolic pathways such as the TCA cycle and benzoate catabolism. A key pathway of biosynthesis from acetyl-CoA generates fatty acids. Fatty acids condense with glycerol to form the phospholipids of cell membranes; they also form the lipid components of envelope proteins and are converted to enzyme cofactors. Fatty acid biosynthesis provides targets for antimicrobial drugs such as triclosan, an inhibitor of enoyl-ACP reductase. Related condensation pathways yield industrial materials, such as polyesters.

Fatty Acids Are Built from Repeating Units

Fatty acids exemplify the construction of cell components from repeating units (**Fig. 15.14**). The construction of molecules based on repeating units requires a cyclic pathway that feeds its products back repeatedly as substrates for further synthesis. The advantage of a cyclic process is that large polymers can be made using a limited number of enzymes; for example, the mycolic acids of *Mycobacterium tuberculosis* are extended to lengths of over 100 carbons.

While fatty acid structures used by cells include many different forms (discussed in Chapter 3), their synthesis always begins with the successive addition of two-carbon acetyl groups. Thus, most fatty acids contain an even number of carbon atoms.

The fatty acid synthase complex. The cyclic process of fatty acid synthesis is managed by the **fatty acid synthase complex**. The complex contains all the enzymes

and component binding proteins bound together in proximity, so that all steps occur in one place without "losing" the unfinished molecule. The main bacterial version of this complex is designated FASII. All components are essential for viability and offer drug targets. Analogous multienzyme complexes are used by actinomycete bacteria to synthesize other long-chain products such as polyketide antibiotics.

Activation of acetyl-CoA. Most acetyl groups within the cell are "tagged" with coenzyme A (CoA), a cofactor that directs acetate into catabolic pathways such as the TCA cycle. To redirect acetyl groups into fatty acid biosynthesis (**Fig. 15.14**), acetyl CoA molecules are first tagged at the "back end" by condensation with CO_2 (step 1), catalyzed by acetyl-CoA carboxylase. The addition of CO_2 in this step does not "count" as carbon fixation, because the CO_2 added does not permanently add to the carbon skeleton. Instead, like phosphate or CoA, the CO_2 exists to be displaced when its function is no longer required. The CO_2-tagged acetyl-ACP is now called malonyl-CoA. The coenzyme A is replaced by **acyl carrier protein (ACP)**, making malonyl-ACP (**Fig. 15.14A**, step 2).

Malonyl-ACP condenses with the growing chain. In step 3, malonyl-ACP hooks onto the head of an acetyl-ACP or a longer-growing chain. The process of hooking onto the new unit involves displacement of CO_2 from the back end of malonyl-ACP, which replaces ACP from the front end of the growing chain. The growing chain now contains a ketone from the former acetyl group. The ketone must be reduced to CH_2 by 2 NADPH, gaining $4H^+ + 4e^-$ (steps 4–6). The chain is now fully hydrogenated (saturated).

Once hydrogenated, the chain is ready to take on the next malonyl-ACP, which, as before, loses CO_2 and replaces the ACP from the front of the growing chain (return to step 3). Successive addition can continue many times to build a saturated fatty acid of indefinite length. The regulation of fatty acid chain length is poorly understood.

Unsaturation. Certain fatty acids require unsaturated "kinks" in the chain to serve a structural purpose, such as to increase the fluidity of the membrane. Such a kink is created by an unsaturated double bond between adjacent carbons (an alkene). Unsaturation can be generated during a cycle of elongation (**Fig. 15.14B**). During the first four cycles of acetyl addition, an alkene bond forms between the *second* and *third* carbons of the fatty acid. After the fifth two-carbon addition, however, a special dehydratase enzyme forms the alkene double bond between the *third* and *fourth* carbons. This alkene fails to be hydrogenated further. Instead, several additional malonyl units are added, generating a long-chain fatty acid with a "kink" of a *cis* double bond.

A.

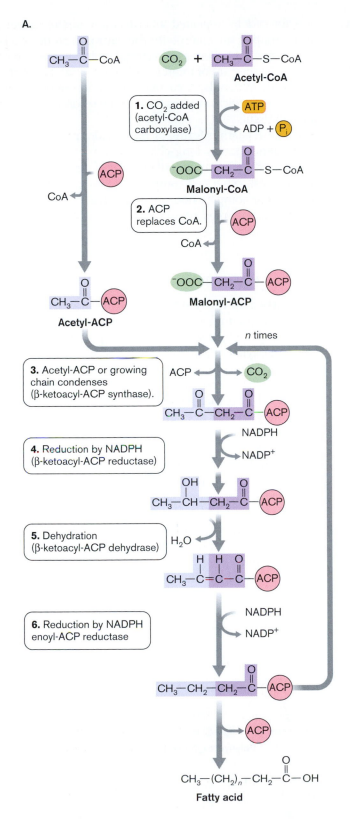

B.

Figure 15.14 **Fatty acid biosynthesis. A.** Stepwise elongation of a saturated fatty acid. Units of acetyl-CoA are carboxylated to malonyl-CoA, then successively condense to form the long chain of a fatty acid. Each two-carbon unit requires reduction by 2 NADPH. **B.** An alkene "kink" can be left unsaturated during chain extension. A special dehydratase enzyme forms the double bond between the third and fourth carbons, rather than the third and second. In this position, the double bond escapes reduction as the chain lengthens through subsequent cycles of malonyl addition.

Regulation of fatty acid synthesis. The synthesis of fatty acids consumes enormous quantities of reducing energy; thus, it must be regulated closely to avoid waste. From a structural standpoint, production of fatty acids incorporated into membranes must be balanced with growth of the cytoplasm. Furthermore, the many different variants of fatty acids are regulated in response to particular environmental needs. In *E. coli,* key points of regulation include the following:

■ **Acetyl-CoA carboxylase represses its own transcription.** Transcription of an operon encoding two subunits of acetyl-CoA carboxylase (AccB, AccC) is repressed by one of its subunits, protein AccB. As AccB increases in concentration, it binds the promoter of the *accBC* operon, repressing further transcription. Thus, initiation of fatty acid biosynthesis (**Fig. 15.14**, step 1) is always limited by the amount of AccB-AccC enzyme present.

■ **Starvation blocks fatty acid biosynthesis.** Starvation for carbon sources blocks fatty acid biosynthesis through the "stringent response." Blockage is mediated by the polyphosphorylated nucleotide ppGpp, the global regulator of the stringent response (discussed in Section 10.3).

■ **Temperature regulates fatty acid composition.** Bacterial fatty acid composition is regulated by environmental factors such as temperature. In *E. coli*, low temperature favors unsaturated fatty acids because they are less rigid and maintain membrane flexibility. Low temperature induces expression of the gene *fabA* encoding the dehydratase enzyme that desaturates the fatty acid bond. As the dehydratase activity is increased, more unsaturated fatty acids are made.

THOUGHT QUESTION 15.5   For a given species, uniform thickness of a cell membrane requires uniform chain length of its fatty acids. How do you think chain length may be regulated?

The biosynthesis of fatty acids provides a model for study of the related biosynthesis of a major family of pharmaceuticals, the polyketides. A well-known polyketide is erythromycin, a broad-spectrum antibiotic frequently prescribed for pneumonia and other serious bacterial infections. Polyketides are commonly produced by actinomycetes, particularly species of *Streptomyces* (**Special Topic 15.2**). Antibiotics are produced as **secondary metabolites**, or **secondary products**—molecules not essential for survival, but which can enhance competition with other strains or species in a crowded natural environment. These secondary products accumulate to especially high levels during stationary phase, when growth of the cell is greatly slowed.

Bacterial Polyesters

Cyclic elongation routes comparable to fatty acid biosynthesis are used to build polymers for energy storage. For example, many bacteria synthesize polyesters (**Fig. 15.15**). The term *polyester* indicates the multiple ester groups that are formed by repeated esterification of the carboxylic acid group of the chain with the hydroxyl group of a new alkanoate unit. Because bacterial polyesters are biodegradable, they are of interest for commercial development as plastic materials. A current application of polyalkanoates is for surgical sutures that can be resorbed within the healed tissue.

The repeating forms of fatty acids, polyketides, and polyesters enables long-chain molecules to be synthesized with a relatively small number of different enzymes. By contrast, other cell components, such as amino acids, have more complex nonrepeating structures whose biosynthetic enzymes require long operons (presented in Section 15.6).

TO SUMMARIZE:

■ **Fatty acid biosynthesis** involves successive condensation of malonyl-ACP groups formed from acetyl groups tagged with acyl carrier protein (ACP) and a carboxylate. Each successive malonyl group is transferred onto the growing acyl chain, with release of CO_2. The condensation reaction is catalyzed by acetyl-CoA carboxylase.

■ **The growing acyl chain is dehydrogenated.** Each added unit is hydrogenated by two molecules of NADPH unless an unsaturated kink is required.

■ **Some fatty acids are partly unsaturated.** An unsaturated kink may be generated by a special dehydratase, which forms the alkene double bond between the third and fourth carbons.

■ **Fatty acid biosynthesis is regulated** by the levels of acetyl-CoA carboxylase and by the stringent response to carbon starvation. Bond saturation is regulated by temperature and other environmental factors.

■ **Polyesters** for energy storage are synthesized by cyclic elongation of polyalkanoates.

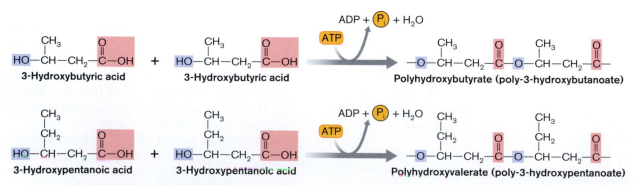

Figure 15.15   Microbial polyester: polyhydroxyalkanoates.   Some bacterial species produce natural polyesters for energy storage. The term *polyester* refers to the multiple ester groups produced by successive esterification of the carboxylic acid group of the chain with the hydroxyl group of a new hydroxy acid unit.

A majority of known antibiotics and antitumor drugs are derived from secondary products of *Streptomyces* bacteria. Many are polyketides, such as erythromycin, originally isolated from *S. erythreus;* ivermectin, a broad-spectrum antiparasitic drug used for heartworm and river blindness, isolated from *S. avermitilis;* and the experimental antitumor agent C-1027, isolated from *S. globisporus* (**Fig. 1**).

Like fatty acids, polyketides are built by the successive condensation of malonyl-ACP units; but each malonyl group carries a unique extension, or R group. In erythromycin, the R groups are all methyl groups. In other polyketides, the R groups range from hydroxyl groups and amides to aromatic rings, some of which undergo secondary reactions and interconnections.

How is the order of different R groups determined? The polyketide is actually constructed on a "modular enzyme" (**Fig. 2**). A modular enzyme contains a series of active sites, or modules, each of which catalyzes the addition of one of a series of units. Each module contains all the enzyme activities needed to add the unit and recognizes a unit with one specific R group. The modular polyketide synthase may consist of a single protein with a series of domains; or it may consist of a complex of several protein subunits. Either the multidomain enzyme or the complex acts as an assembly line to generate the specific polyketide.

Figures 2A and **B** show the chain-extension synthesis of a polyketide. The initial two domains of enzyme 1 are acyltransferase (AT) and an acyl carrier protein (ACP) domain, analogous to the ACP of fatty acid biosynthesis but specialized to accept one component of the polyketide. The acyltransferase transfers the acyl group (R_1-acetyl) from R_1-acetyl-CoA onto the ACP, with release of CoA (**Fig. 2A**). The R_1-acetyl is then transferred to a ketosynthase domain (KS).

Meanwhile, a malonyl group with its own R group (R_2-malonyl) has been transferred by a second AT onto the second ACP domain (**Fig. 2B**). The R_1-acetyl group then leaves KS and condenses with R_2-malonyl-ACP, releasing CO_2.

The R_1R_2 acyl chain subsequently undergoes a series of extensions by R-malonyl groups (**Fig. 2C**). For erythromycin, all R groups are methyl; other polyketides have R groups that are more complex. The initial AT and ACP domains

constitute a "loading module" for the first acyl group, whereas subsequent sets of domains constitute "extender modules" that each add a different R-malonyl group. Extender modules can include secondary activities such as dehydratase (DH, removes OH) and ketoreductase (KR, hydrogenates a ketone to OH). In principle, the modular approach can construct a limitless range of products with diverse antibiotic properties.

Upon completion, elongation is terminated by the thioesterase (TE), which hydrolyzes the thioester bond to the final

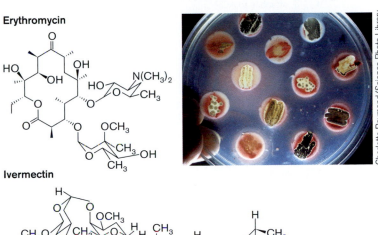

Erythromycin

Ivermectin

C-1027

Figure 1   **Polyketide drugs from *Streptomyces* species.**
Polyketide drugs include the antibiotic erythromycin, the antiparasitic drug ivermectin, and the antitumor agent C-1027. Inset: Colonies of *Streptomyces* species produce numerous secondary products.

Charlotte Raymond/Science Photo Library

(*continued on next page*)

Special Topic 15.2 Polyketide Drugs Are Synthesized by Multienzyme Factories (*continued*)

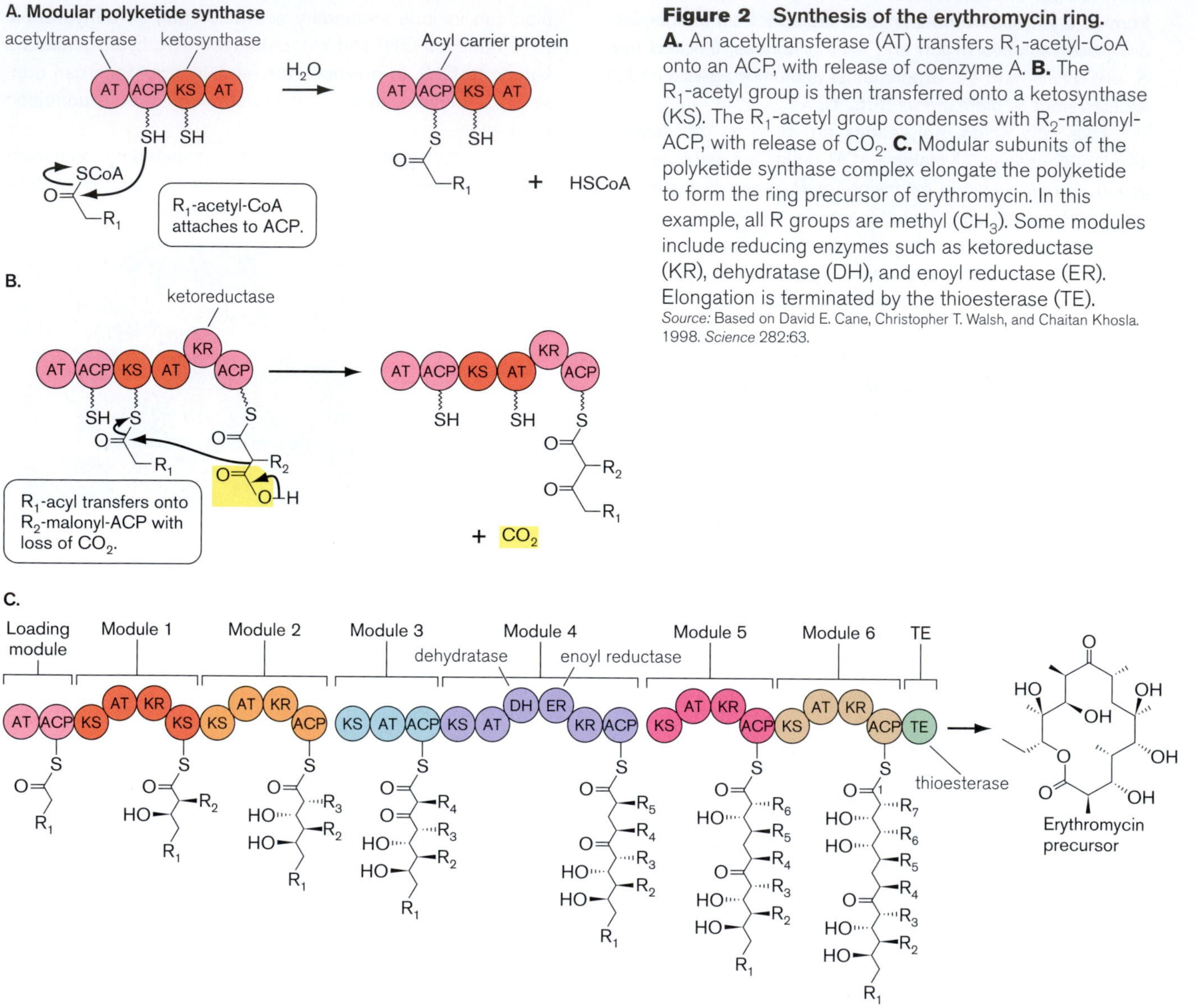

A. Modular polyketide synthase

B.

C.

Figure 2 Synthesis of the erythromycin ring.
A. An acetyltransferase (AT) transfers R_1-acetyl-CoA onto an ACP, with release of coenzyme A. **B.** The R_1-acetyl group is then transferred onto a ketosynthase (KS). The R_1-acetyl group condenses with R_2-malonyl-ACP, with release of CO_2. **C.** Modular subunits of the polyketide synthase complex elongate the polyketide to form the ring precursor of erythromycin. In this example, all R groups are methyl (CH_3). Some modules include reducing enzymes such as ketoreductase (KR), dehydratase (DH), and enoyl reductase (ER). Elongation is terminated by the thioesterase (TE).
Source: Based on David E. Cane, Christopher T. Walsh, and Chaitan Khosla. 1998. *Science* 282:63.

ACP. Additional enzymes (not shown) are required to link extra components, such as sugars, to the polyketide core, completing the product.

Bacteria in nature produce numerous kinds of antibiotics we have yet to discover. One way to discover and produce new polyketide antibiotics is by genome scanning of new bacteria isolates (**Fig. 3**). The scanning of a large number of unsequenced genomes enables cloning of biosynthesis operons for hundreds of new secondary products. The pharmaceutical activities of these products can then be tested.

Genome scanning was used by Ben Shen at the University of Wisconsin–Madison, to discover the potent antitumor agent C-1027. C-1027 is one of an unusual family of polyketides called enediynes, notable for the cyclized core containing two carbon-carbon triple bonds (see **Fig. 1**). The process of genome scanning is based on two assumptions. First, genes for biosynthesis of secondary metabolites tend to be variants of genes for known products. Second, the genes for synthesis of a particular product usually map together as a result of being transferred as a cluster from another organism.

Figure 3   Genome scanning to identify a biosynthesis cluster for a new antibiotic.   Biosynthesis genes occur in clusters in a bacterial genome. To identify an unknown cluster whose genes are related to known clusters: 1. Amplify gene sequence tags (GST) from an unsequenced actinomycete genome. Compare GST sequences from the new genome with a database of known antibiotic biosynthesis gene clusters. Identify promising GSTs. 2. Hybridize GST sequences to cloned plasmids containing sequences adjacent to the GST. Subclone and obtain the adjacent sequences, which may contain a new biosynthesis cluster. 3. Assemble the sequence of the new gene cluster for biosynthesis. *Source:* Based on Zazopoulos, et al. 2003. *Nature Biotechnology* 21:187.

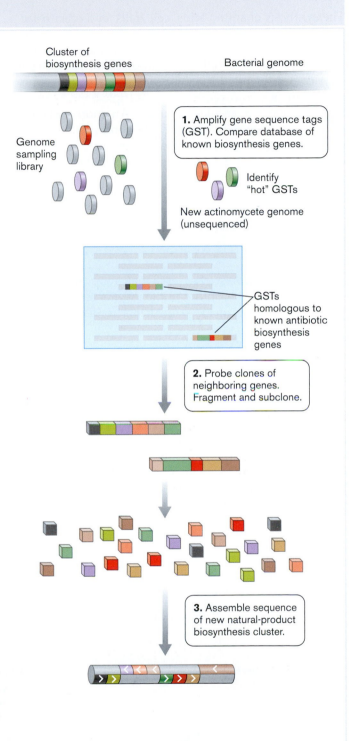

Cluster of biosynthesis genes

Bacterial genome

Genome sampling library

1. Amplify gene sequence tags (GST). Compare database of known biosynthesis genes.

Identify "hot" GSTs

New actinomycete genome (unsequenced)

GSTs homologous to known antibiotic biosynthesis genes

2. Probe clones of neighboring genes. Fragment and subclone.

3. Assemble sequence of new natural-product biosynthesis cluster.

Thus, genome scanning is designed to isolate an unknown cluster whose genes are related to known clusters in other organisms. The result is valuable because often a variant of a known antibiotic proves to be active against pathogens resistant to the original substance, or it may have fewer side effects in the host.

Step 1 of genome scanning is to amplify gene sequence tags (GSTs) from an isolated bacterium. A gene sequence tag is a short sequence of DNA bounded by primer ends homologous to known biosynthesis genes (see **Fig. 3**). The primer ends are then used to amplify and sequence the GST. The GSTs are amplified by PCR from plasmid clones in a genomic library of a suspected drug-producing organism. Next (step 2), the GSTs are compared with a database of known biosynthesis genes. The "hot" GSTs (those confirmed as biosynthetic homologs) are then used (step 3) to probe clones containing genes adjacent to the GST sequences. The adjacent genes may contain completely novel biosynthesis components. To identify these components, DNA is sequenced upstream and downstream of the GST. Finally, the sequences are assembled to reveal the biosynthesis gene cluster. The sequence is annotated and deposited in the database for future genome scanning.

To identify new polyketide antibiotics, including C-1027, Shen scanned genomic libraries of 50 different actinomycete strains for biosynthetic clusters encoding natural polyketides not known previously. Eight of the strains revealed novel antibiotic compounds. These new products show varying degrees of toxicity and antibacterial action; they are presently under testing for medical use.

15.5 Nitrogen Fixation

The synthesis of amino acids, cell walls, and other cofactors requires an additional element we have not yet considered, namely, nitrogen. In principle, nitrogen should be more accessible to cells than carbon, since nitrogen gas (N_2) comprises more than three-quarters of our atmosphere. In fact, however, N_2, with its triple bond, is one of the most stable molecules in nature. An enormous input of energy is required to reduce nitrogen to ammonia for assimilation into carbon skeletons. Early in evolution, all cells may have fixed their own N_2, but today only certain species of bacteria and archaea retain the ability. All other organisms depend on reduced or oxidized forms of nitrogen, which ultimately derive from N_2. Thus, all living organisms, directly or indirectly, depend on N_2-fixing prokaryotes.

> **NOTE:** Prokaryotes play essential roles in global cycling of nitrogen, sulfur, and phosphorus. The geochemical cycling of these elements is discussed in Chapter 22.

Nitrogen Is Assimilated in Different Forms

To assimilate into biomass, any form of nitrogen must be fully reduced to ammonia (NH_3), which at pH 7 is mostly protonated to ammonium ion, NH_4^+. Unlike carbon, nitrogen rarely appears in oxidized form in complex biomolecules. Inorganic forms, such as nitric oxide (NO),

are used as defense mechanisms against invading pathogens (see Chapter 24) or as signaling molecules. In macromolecules, however, virtually all the nitrogen is reduced; organic compounds containing oxidized nitrogen are generally toxic.

While N_2 is the ultimate source and sink of biospheric nitrogen, several oxidized or reduced forms occur in the environment (**Fig. 15.16**). Most free-living bacteria can acquire nitrate (NO_3^-) or nitrite (NO_2^-) for reduction to ammonium ion, although they repress these energy-expensive pathways when ammonium ion is present. Even nitrogen-fixing legume symbionts, such as *Rhizobium*, can utilize nitrate.

In natural environments, most potential sources of nitrogen are subject to competition from dissimilatory metabolism in which the molecule is oxidized or reduced for energy, as discussed in Chapter 14. For example, anaerobic respirers convert nitrate and nitrite to N_2 (denitrification), whereas lithotrophs oxidize NH_4^+ to nitrite and nitrate (nitrification). An important consequence of nitrification is that most of the commercial ammonia fertilizer spread on agricultural fields is soon oxidized by lithotrophs to nitrates and nitrites. High concentrations of nitrites in water are harmful because they combine with hemoglobin, generating a form that cannot take up oxygen.

Nitrogen Fixation: Early Discoveries

The first nitrogen fixers discovered by Martinus Beijerinck and colleagues in the late 1800s were soil and wetland bacteria such as *Beggiatoa* and *Azotobacter*. Species of *Rhizobium* were discovered to conduct nitrogen as endosymbionts of

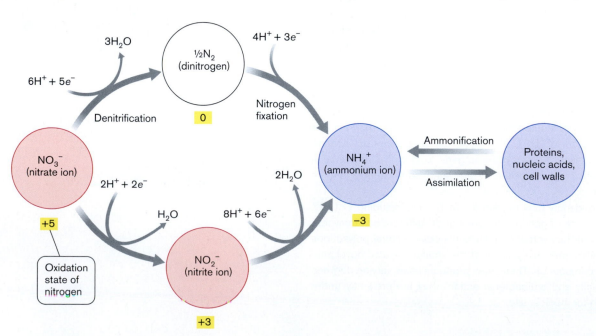

Figure 15.16 Nitrogen assimilation. Different oxidation states of nitrogen require different amounts of reducing energy for assimilation into biomass.

leguminous plants (see Chapter 23). For several decades, it was believed that N_2 was fixed only by a few special bacteria in the soil or in symbiotic plant bacteria. But in the 1940s, Martin Kamen and colleagues noticed that phototrophs such as *Rhodospirillum rubrum* produced hydrogen gas—a known by-product of nitrogen fixation. Nitrogen fixation was hard to demonstrate reliably in the laboratory because at that time researchers did not know that in *R. rubrum* nitrogen fixation was repressed by the presence of alternative nitrogen sources, such as ammonia. When traces of ammonia and other nitrogen sources were eliminated, Kamen's student Herta Bregoff found that photosynthetic bacteria actually fix nitrogen. The fixation of nitrogen was confirmed by experiments showing uptake of the radio-isotope ^{15}N into biomass. In fact, N_2 is fixed by most phototrophic bacteria (green and purple bacteria, as well as cyanobacteria) and by many archaea. Marine cyanobacteria fix a large proportion of both the nitrogen and carbon dioxide assimilated by our biosphere.

By the end of the twentieth century, about half the nitrogen in the biosphere originated from human industry. Worldwide, factories fix enormous quantities of N_2 into ammonia for agricultural fertilizers. Industrial nitrogen fixation uses the **Haber process**, in which nitrogen gas is hydrogenated by methane (natural gas) under extreme heat and pressure. The anthropogenic (human-caused) influx of nitrogen into the biosphere is an astonishing example of the influence of human society on our planet.

The Mechanism of Nitrogen Fixation

In living cells, nitrogen fixation is an enormously energy-intensive process. The mechanism is largely conserved across species:

$$N_2 + 8H^+ + 8e^- + 16ATP \longrightarrow$$
$$2NH_3 + H_2 + 16ADP + 16P_i$$

The total energy investment includes approximately three ATP-equivalents per electron (assuming reduction by NADPH), plus sixteen ATP molecules as shown. That makes $24 + 16 = 40ATP$ in all, more than that provided by complete oxidation of glucose. The production of H_2 is surprising, as it consumes extra ATP. Hydrogen loss results from initiating the cycle of nitrogen reduction (discussed shortly). Some bacteria have secondary reactions to reclaim the lost hydrogen with part of its lost energy.

NOTE: While nitrogen fixation is commonly represented as producing ammonia (NH_3), under most conditions of living cells the predominant form is protonated, ammonium ion (NH_4^+).

Nitrogenase reaction mechanism. The overall conversion of nitrogen gas to two molecules of ammonia is cat-alyzed in four cycles by **nitrogenase**, an enzyme highly conserved in nearly all nitrogen-fixing species. Like rubisco, nitrogenase probably evolved once in a common ancestor of all nitrogen-fixing organisms.

The mechanism of nitrogenase is of intense interest to agricultural scientists because of the potential benefits of improving efficiency of plant growth and of extending nitrogen-fixing symbionts to nonleguminous plants such as corn. In 1960, scientists at the DuPont Laboratory first isolated the nitrogenase complex from a bacterium, *Clostridium pasteurianum*. They measured its activity by incorporation of radiolabeled nitrogen, $^{15}N_2$, into $^{15}NH_3$. Since then, the detailed structure of nitrogenase has been solved (**Fig. 15.17**).

The active nitrogenase complex includes two kinds of subunits encoded by different genes: Fe protein (shaded green in **Fig. 15.17**), containing a 4Fe-4S center, and FeMo protein (shaded cyan), containing iron and molybdenum. The Fe protein contains a typical 4Fe-4S structure to facilitate electron transfer. As we learned in Chapter 14, metal atoms help to transfer electrons because their orbitals are closely spaced and transitions between these orbitals involve relatively small amounts of energy. The electrons are funneled through a second FeS cluster (the P cluster) to the FeMo (iron-molybdenum) cluster. The FeMo cluster is an unusual structure characteristic of nitrogenase, in which trios of sulfur atoms alternate with trios of iron atoms. One end of the cluster is capped with another iron, and the other end is capped with an atom of molybdenum (Mo). A consequence of the nitrogenase structure is that most nitrogen-fixing organisms (and their plant hosts, for leguminous symbionts) require the element molybdenum for growth. Some bacteria make an alternative nitrogenase that substitutes vanadium for molybdenum.

Four cycles of reduction. Nitrogen fixation requires four reduction cycles through nitrogenase (**Fig. 15.18**).

To initiate the first cycle of N_2 reduction, Fe protein acquires $2e^-$ from an electron transport protein such as a ferredoxin or a flavodoxin (step 1). Energy is provided by spending four molecules of ATP. The reduced Fe protein then transfers each electron to the FeMo center. The FeMo protein binds $2H^+$, which is reduced to H_2 (step 2). Only then does N_2 bind to the active site, by displacing the H_2 (step 3).

In the next reduction cycle, two electrons are transferred to Fe protein, where they reduce the iron. The reduced Fe protein transfers each electron to the FeMo center, near the binding site for N_2. The electron transfer requires expenditure of 2ATP per electron, or 4ATP per $2e^-$. Two hydrogen ions join the N_2 and receive the two electrons, forming $HN=NH$.

A third pair of electrons enters the Fe protein, which then joins another $2H^+$, reducing N_2 to H_2N-NH_2. The cycle again requires hydrolysis of 4ATP. A final cycle of

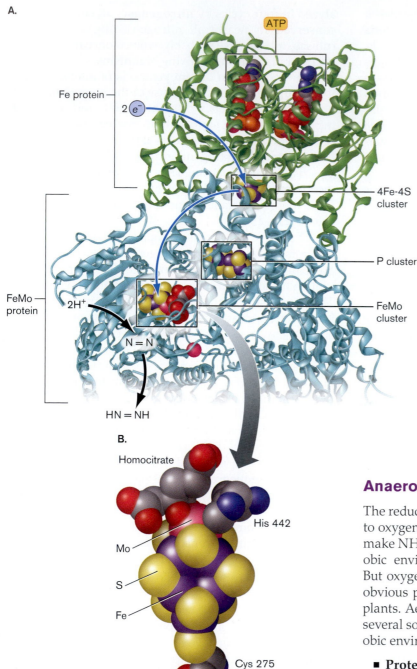

A.

Fe protein

ATP

2 e^-

4Fe-4S cluster

FeMo protein

P cluster

2H$^+$

FeMo cluster

N = N

HN = NH

B.

Homocitrate

His 442

Mo

S

Fe

Cys 275

Figure 15.17   Structure of the nitrogenase complex.
A. In each active site of the Fe-protein-FeMo-protein, the Fe protein binds ATP and receives electrons from the electron donors. The electrons are subsequently channeled down through the 4Fe-4S cluster and the P cluster (FeS cluster) to the FeMo cluster, where they reduce the N$_2$. (PDB code: 1N2C) **B.** The metal cluster (Fe$_7$S$_9$Mo) is held in place by coordination with a molecule of homocitrate plus two amino acid residues of nitrogenase, His 442 and Cys 275.

electron transfer and H$^+$ uptake reduces H$_2$N–NH$_2$ to 2NH$_3$. At typical cytoplasmic pH values (near pH 7), NH$_3$ is protonated to NH$_4^+$.

The loss of H$_2$ during nitrogen fixation is puzzling because it represents lost energy in a highly energy-intensive process. The actual fate of the H$_2$ varies among species. *Klebsiella pneumoniae*, a gram-negative organism in which nitrogen regulation has been much studied, gives off the H$_2$ without further reaction. On the other hand, *Azotobacter* uses an irreversible hydrogenase enzyme to convert H$_2$ back to 2H$^+$ and recover the transferred electrons. In leguminous rhizobial symbionts such as *Sinorhizobium* species, H$_2$ recovery varies surprisingly among strains and can affect the efficiency of plant growth.

The process of nitrogen fixation is expensive, both in terms of protein synthesis and in terms of reducing energy and ATP. Even bacteria capable of nitrogen fixation repress the process in the presence of other nitrogen sources, such as nitrate (NO$_3^-$), nitrite (NO$_2^-$), ammonia (NH$_3$), or nitrogenous organic molecules acquired from dead cells.

Anaerobiosis and N$_2$ Fixation

The reductive action of nitrogenase is extremely sensitive to oxygen because of the large reducing power needed to make NH$_4^+$. Thus, cells can fix nitrogen only in an anaerobic environment. This is no problem for anaerobes. But oxygenic phototrophs such as cyanobacteria face an obvious problem, as do bacterial symbionts of oxygenic plants. Aerobic and oxygenic organisms have developed several solutions to the problem of fixing nitrogen in aerobic environments.

- **Protective proteins.** Aerobic *Azotobacter* species synthesize protective proteins that stabilize nitrogenase and prevent attack by oxygen. For rhizobia, on the other hand, their plant hosts produce leghemoglobin (named for "legume" plants), a form of hemoglobin that sequesters oxygen away from the bacteria.
- **Temporal separation of photosynthesis and N$_2$ fixation.** Some species of cyanobacteria fix nitrogen only at night, when photosynthesis does not occur and thus no oxygen is made.
- **Specialized cells for N$_2$ fixation.** Filamentous cyanobacteria form specialized nitrogen-fixing cells called **heterocysts** (**Fig. 15.19**). The heterocysts lose their photosynthetic capacity entirely and specialize in

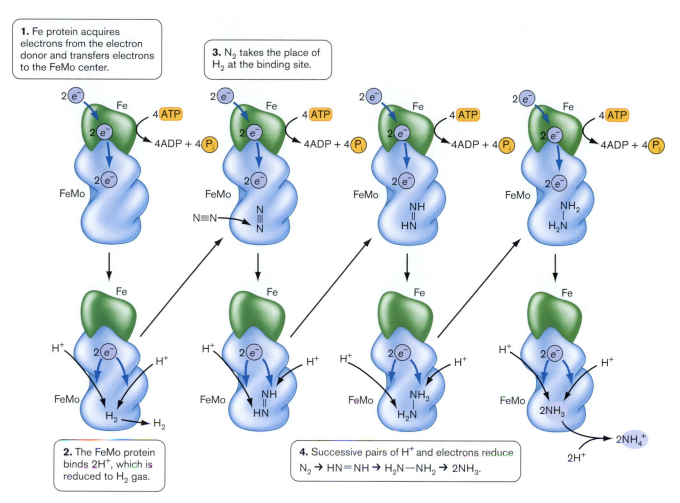

1. Fe protein acquires electrons from the electron donor and transfers electrons to the FeMo center.

3. N$_2$ takes the place of H$_2$ at the binding site.

2. The FeMo protein binds 2H$^+$, which is reduced to H$_2$ gas.

4. Successive pairs of H$^+$ and electrons reduce N$_2$ → HN=NH → H$_2$N—NH$_2$ → 2NH$_3$.

Figure 15.18   Nitrogen fixation by nitrogenase.   The enzyme nitrogenase successively reduces nitrogen by electron transfer, ATP hydrolysis, and H$^+$ incorporation. The nitrogenase complex includes two classes of subunits encoded by different genes: Fe protein (shaded green) and MoFe protein (shaded blue).

Figure 15.19   Cyanobacteria form heterocysts to fix nitrogen under anaerobic conditions.   A. *Anabaena spiroides*, a filamentous cyanobacterium, segregates N$_2$ fixation in heterocysts, cells that do not photosynthesize and thus maintain anoxic conditions. **B.** Autofluorescence reveals chlorophyll, which is absent in the heterocyst.

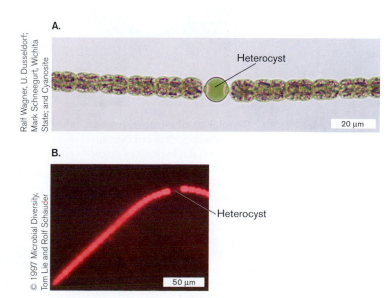

A.

Heterocyst

20 μm

B.

Heterocyst

50 μm

nitrogen fixation; their loss of photosynthesis is revealed by the absence of chlorophyll autofluorescence (**Fig. 15.19B**). Heterocyst development is directed by a complex genetic program, induced by nitrogen starvation. In natural environments, the heterocysts leak organic acids that attract heterotrophic bacteria, which use up all the oxygen around the heterocyst, generating ideal anaerobic conditions for nitrogen fixation.

Molecular Regulation of N_2 Fixation by NtrC and Sigma-54

Nitrogen fixation costs substantial energy, and therefore the process is highly regulated. Oxygen and NH_4^+ availability regulate expression of the *nif* genes (*nifHDKTY*) encoding nitrogenase and other nitrogen-fixation proteins. Regulation of *nif* is mediated by several molecular regulators, including a nitrogen starvation sigma factor (sigma N) and the NtrB-NtrC two-component signal transduction system (discussed in Chapter 10). The NtrB-NtrC system has been dissected in *Klebisiella pneumoniae* by Sydney Kustu and colleagues at the University of California, Berkeley (**Fig. 15.20**).

The NtrC protein regulates nitrogenase expression in response to NH_4^+ concentration (**Fig. 15.20B**). When

cellular levels of NH_4^+ are high, no activators bind the *nifHDKTY* promoter, and the nitrogen fixation genes are not expressed. When NH_4^+ is low, however, NtrC is phosphorylated by NtrB. The NtrC-P phosphoprotein now binds its DNA site to activate expression of *nifLA*, making NifL and NifA proteins. Expression of *nifLA* is coactivated by the sigma 54. Sigma 54 and the NifA protein together activate expression of *nifHDKTY*. Thus, low NH_4^+ concentration turns on nitrogen fixation.

Nitrogen fixation cannot occur with oxygen, so high oxygen levels block nitrogenase expression. When oxygen is high, NifA binds NifL and is prevented from binding the nitrogenase promoter. Low oxygen allows NifA to bind the promoter and express nitrogenase to fix nitrogen. Overall, NtrC governs response to NH_4^+, whereas NifA governs response to oxygen. Response to multiple environmental signals is typical of energy-expensive biosynthetic pathways.

TO SUMMARIZE:

■ **Oxidized or reduced forms of nitrogen**, such as nitrate, nitrite, and ammonium ions, can be assimilated by bacteria and plants. Assimilation competes with dissimilatory reactions that obtain energy.

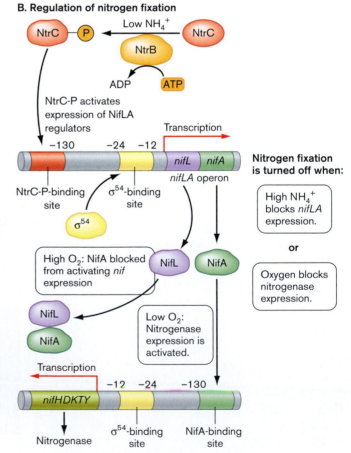

Figure 15.20   Regulation of nitrogen fixation. A. Sydney Kustu has made important discoveries about NtrC and other forms of nitrogen regulation in the gram-negative soil bacterium *Klebsiella pneumoniae*. **B.** When cellular levels of NH_4^+ are low, NtrC is phosphorylated and activates expression of *nifLA*. Expression is coactivated by the nitrogen starvation sigma factor, sigma 54. The NifA protein (and sigma 54) activate expression of *nif* genes encoding nitrogenase. When oxygen levels are high, however, nitrogen fixation cannot occur, so NifA is blocked by binding NifL. (Numbers refer to positions upstream of the translation site.)

■ **Nitrogen gas (N₂) is fixed into ammonium ion** only by some species of bacteria and archaea, never by eukaryotes.

■ **Nitrogenase enzyme** includes a protein containing an iron-sulfur core (Fe protein) and a protein containing a complex of molybdenum, iron, and sulfur (FeMo protein). Electrons acquired by Fe protein (with energy from ATP) are transferred to FeMo protein to reduce nitrogen.

■ **Four cycles of reduction by NADPH** or an equivalent reductant are required to reduce one molecule of N_2 to two molecules of NH_3. At neutral pH, NH_3 is protonated to NH_4^+.

■ **Oxygen inhibits nitrogen fixation.** Bacteria and plants have various means of separating nitrogen fixation from aerobic respiration, such as heterocyst development or temporal separation.

■ **Regulation of nitrogen fixation** by nitrogen and oxygen levels occurs via NtrC and sigma 54 regulation of transcription.

15.6 Biosynthesis of Amino Acids and Nitrogenous Bases

Microbes need amino acids to make proteins and cell walls, as well as nitrogenous bases to form nucleic acids. When possible, they obtain these molecules from their environment through membrane-bound transporters. But competition for such valuable nutrients is high, especially for free-living microbes in soil or water. Most free-living microbes and plants possess the ability to make all the standard amino acids and bases of the genetic code, as well as nonstandard variants used for cell walls and transfer RNAs. Microbial biosynthesis of amino acids "essential" for animals is used for industrial production of food supplements.

Diversity and Complexity in Amino Acid Synthesis

Like fatty acid biosynthesis, synthesis of amino acids and nitrogenous bases requires the input of large amounts of reducing energy. These compounds pose additional challenges because of their unique and diversified forms, which cannot be made by the cyclic processes that generate molecules from repeating units. Synthesis of complex, asymmetrical molecules such as amino acids requires many different conversions, each mediated by a different enzyme. Nevertheless, some economy is gained by arrangement of branched pathways in which early intermediates are utilized to form several products (**Fig. 15.21**). For example, oxaloacetate is converted to aspartate, which can be converted to five other amino acids.

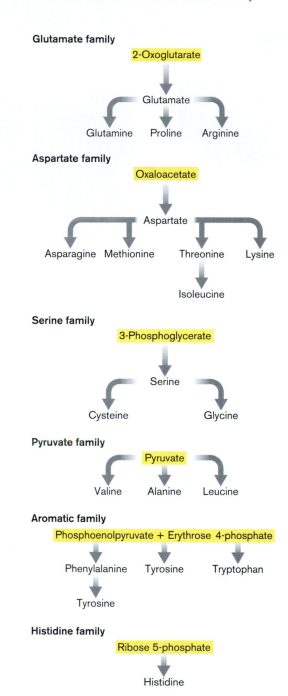

Figure 15.21   Major pathways of amino acid biosynthesis. Some amino acids arise from key metabolic intermediates (highlighted), whereas others must be synthesized out of other amino acids. Note alternative pathways (see Figure 15.2).

Amino Acids Arise from Metabolic Intermediates

The carbon skeletons of amino acids arise from diverse intermediates of metabolism (see **Fig. 15.21**). Note that certain amino acids arise directly from key metabolic intermediates (for example, glutamate from 2-oxoglutarate), whereas others must be synthesized from preformed

Rodney Start/Museum Victoria, Melbourne Museum

Figure 15.22   The Murchison meteorite. Fragments of a meteorite that fell in Murchison, Australia, in 1969 were shown to contain the five fundamental amino acids of biosynthetic pathways (glutamate, aspartate, valine, alanine, and glycine).

As in fatty acid biosynthesis, precursor molecules are channeled into amino acid biosynthesis by specialized cofactors and reducing energy carriers such as NADPH. Unlike fatty acids, however, amino acids must assimilate another key ingredient: nitrogen.

Amino Acids Form through Assimilation of NH_4^+

The NH_4^+ produced by N_2 fixation or by nitrate reduction is the key source of nitrogen for biosynthesis. But NH_4^+ always exists in equilibrium with NH_3, which is toxic to cells. Moreover, NH_3 travels freely through membranes, making it difficult to store NH_4^+ within a cell. Deprotonation to NH_3 increases as pH rises; by pH 9.2, deprotonation reaches 50%. Even at neutral pH, a very small equilibrium concentration of NH_3 can drain NH_4^+ out of the cell. Thus, cells avoid storing high levels of NH_4^+; instead, it is incorporated immediately into organic products.

amino acids (for example, glutamine, proline, and arginine from glutamate). Some amino acids can arise from more than one source; for example, leucine and isoleucine can be made from succinate as well as from pyruvate.

It has been hypothesized that the amino acids that arise in just one or two steps from central intermediates are more ancient in cell evolution than those requiring more complex pathways. Five of these "ancient" amino acids—glutamate, aspartate, valine, alanine, and glycine—are the same as those detected in meteorites, whose composition resembles that of prebiotic Earth (**Fig. 15.22**). These same five amino acids also appear in early-Earth simulation experiments in which methane, ammonia, and water are heated under reducing conditions and subjected to electrical discharge. Thus, we speculate that the first amino acids that early cells evolved to make were the same as those that arose spontaneously in the prebiotic chemistry of our planet.

2-oxoglutarate and glutamate condense with NH_4^+. In most bacteria, the main route for NH_4^+ assimilation is by condensing NH_4^+ with 2-oxoglutarate to form glutamate, or with glutamate to form glutamine (**Fig. 15.23**). Three key enzymes interconvert these substrates:

■ **Glutamate dehydrogenase (GDH)**, actually named for its reverse activity, condenses NH_4^+ with 2-oxoglutarate to form glutamate. The condensation requires reduction by NADPH.

■ **Glutamine synthetase (GS)** condenses a second NH_4^+ with glutamate to form glutamine, a process driven by spending one ATP.

■ **Glutamate synthase (GOGAT, glutamine: 2-oxoglutarate amidotransferase)** converts 2-oxoglutarate plus glutamine into two molecules of glutamate. Different variants of this enzyme use different reducing agents: NADPH, NADH, or ferredoxin. The NADPH variant is shown.

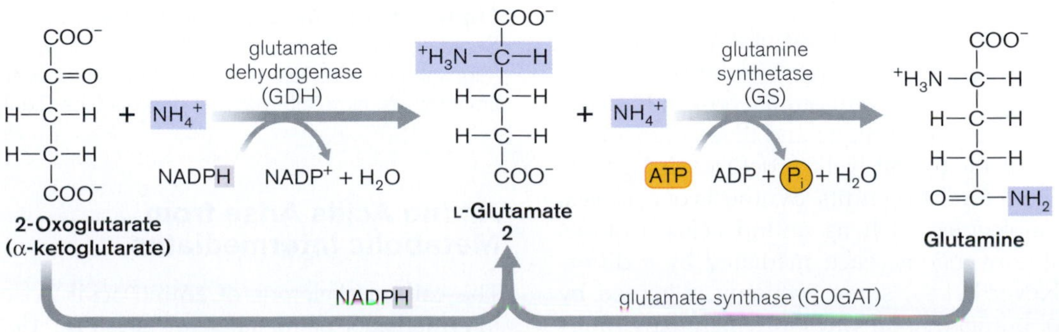

Figure 15.23   Assimilation of NH_4^+ into glutamate and glutamine. The key TCA cycle intermediate 2-oxoglutarate (alpha-ketoglutarate) incorporates one molecule of NH_4^+ at the ketone to form glutamate. Glutamine can combine with oxoglutarate to produce two molecules of glutamate. These reactions provide sources of nitrogen to feed other pathways of amino acid synthesis.

All three substrates thus exchange amino groups readily. High nitrogen levels induce GDH to take up NH_4^+, but repress GS and GOGAT. GS has a higher affinity than GDH for NH_4^+. Low nitrogen levels induce GS (to make glutamine) and GOGAT (to convert some glutamine to glutamate as needed).

Both glutamate and glutamine contribute an amine, as well as their carbon skeletons, to the synthesis of other amino acids in the biosynthetic "tree." The transfer of ammonia between two metabolites such as glutamate and glutamine is called **transamination**. Transamination also occurs between many other pairs of amino acids and metabolic intermediates. For example, glutamate transfers NH_3 to oxaloacetate, making aspartate and 2-oxoglutarate; the reaction can also be reversed. In another example, valine transfers an amine group to pyruvate, generating alanine and 2-ketoisovalerate.

> **THOUGHT QUESTION 15.6** Suggest two reasons why transamination is advantageous to cells.

The cellular levels of glutamate and glutamine act as indicators of nitrogen availability. When mineral forms such as NH_4^+ and nitrate are absent, some microbes seek organic sources of nitrogen in their environment. During nitrogen starvation, NtrC is phosphorylated and induces expression of glutamine synthetase as well as nitrogenase. NtrC phosphate also induces a high-affinity ammonia transporter and transporters for organic sources of nitrogen, such as amino acids, oligopeptides, and cell wall fragments containing amino sugars. All these molecules can be "scavenged" to obtain nitrogen for biosynthesis.

Building Complex Amino Acids Requires Numerous Enzymes

Seven "fundamental" amino acids show relatively simple biosynthetic pathways. Others require longer pathways involving numerous enzymes. An amino acid produced by a complex pathway is arginine (**Fig. 15.24**). Bacterial arginine biosynthesis generally involves about a dozen different enzymes distributed among four to eight operons. Operon expression is regulated by an arginine repressor that binds the promoter of each operon to prevent transcription in the presence of sufficient arginine for cell needs. In some species, the arginine repressor also activates enzymes of arginine catabolism, making the excess amino acid available as a carbon source.

> **THOUGHT QUESTION 15.7** Which energy carriers (and how many) are needed to make arginine from 2-oxoglutarate? (See Fig. 15.24.)

Arginine biosynthesis. Arginine synthesis begins with the condensation of glutamate with acetyl-CoA (**Fig. 15.24**, step 1). Three more amine groups are transferred subsequently by glutamine and aspartate. A second glutamate transfers an amino group to the arginine precursor (step 2) while its own carbon skeleton cycles back as 2-oxoglutarate (alpha-ketoglutarate). This transfer of NH_4^+ between two organic intermediates is an example of transamination. The original acetyl group from acetyl-CoA is then hydrolyzed, producing ornithine. Ornithine is a central intermediate, used by cells to synthesize proline and various polyamines, as well as arginine.

Ornithine receives a third amino group (step 3) from glutamine via conversion of CO_2 to carbamoyl phosphate [$H_2N\text{-}CO(PO_4)^-$]:

$$CO_2 + H_2O \longrightarrow H^+ + HCO_3^-$$

$$HCO_3^- + ATP + glutamine \longrightarrow$$
$$H_2N\text{-}CO(PO_4)^{2-} + ADP + glutamate$$

Carbamoyl phosphate provides a way to assimilate one nitrogen plus one carbon into a structure.

The fourth nitrogen is acquired from aspartate in a two-step process requiring ATP release of pyrophosphate (PPi) (step 4). The release of fumarate, a TCA cycle intermediate, yields arginine.

Aromatic amino acids. The complexity of aromatic amino acids requires particularly energy-expensive biosynthesis. The number of different enzymes required, however, is minimized by the presence of a common core pathway branching to several different amino acids (**Fig. 15.25**).

The three aromatic amino acids, phenylalanine, tyrosine, and tryptophan, each require a common precursor, chorismate. The pathway to chorismate starts with simple glycolytic intermediates (phosphoenolpyruvate and erythrose 4-phosphate), but its synthesis requires 10 enzymes encoded in a single operon, *aroABCDEFGHKL*. From chorismate to phenylalanine or tyrosine requires an additional three enzymatic steps. From chorismate to tryptophan, the most complex amino acid specified by the genetic code, requires another 5 enzymes, expressed by *trpABCDE*. Thus, a total of at least 15 enzymes encoded by different genes are required to make tryptophan out of simple substrates.

Not surprisingly, the presence of tryptophan in the cell effectively represses expression of its own biosynthetic enzymes. Repression of the *trp* operon includes two mechanisms: the Trp repressor, effective over moderate to high levels of tryptophan; and an RNA loop mechanism called attenuation, sensitive to lower levels of tryptophan (discussed in Chapter 10). Attenuation involves the destabilization of the transcription complex by formation of a specific stem loop on the nascent *trp* mRNA. The mechanism of attenuation in *E. coli* was discovered in 1981 by Charles Yanofsky, winner of the National Medal of Science. Several other amino acids commonly show

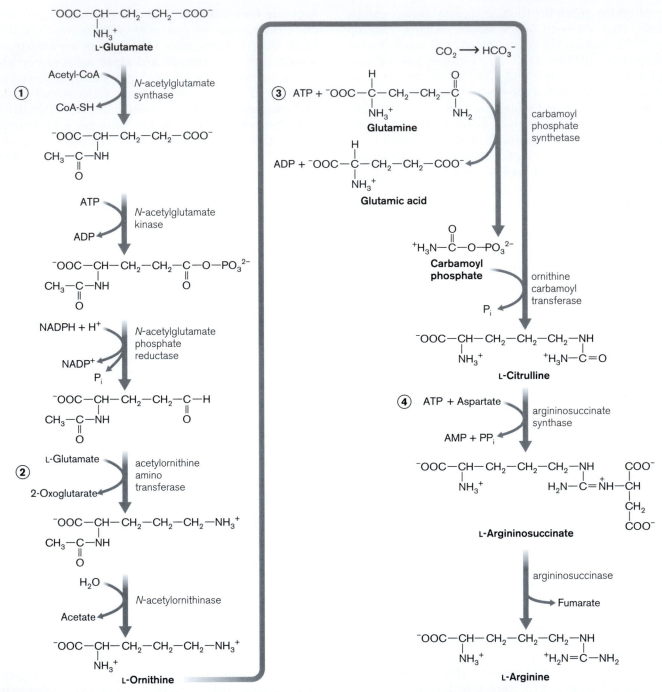

Figure 15.24 Arginine biosynthesis in *E. coli*. Arginine biosynthesis involves condensation of two glutamate with acetyl-CoA, as well as nitrogen donation by glutamate and glutamine.

attenuation in bacteria, particularly histidine, threonine, phenylalanine, valine, and leucine.

Purine and Pyrimidine Synthesis Requires Multiple Enzyme Pathways

Nitrogenous bases include purines and pyrimidines, the essential coding components of DNA and RNA. The accuracy of the entire genetic code requires accurate syn-

thesis of these bases. Besides their role in nucleic acids, purine and pyrimidine nucleotides such as ATP serve as energy carriers.

Bases are built onto ribose 5-phosphate. The purine and pyrimidine bases are not built as isolated units; rather they are constructed on a ribose 5-phosphate substrate, forming a nucleotide (sugar-base-phosphate). Note that ribonucleotides are synthesized first and that these are

Figure 15.25 Biosynthesis of aromatic amino acids. Aromatic amino acids are assembled out of various carbon skeletons. In *E. coli*, the pathways to chorismate and tryptophan are encoded by operons of contiguous genes, *aro* and *trp*.

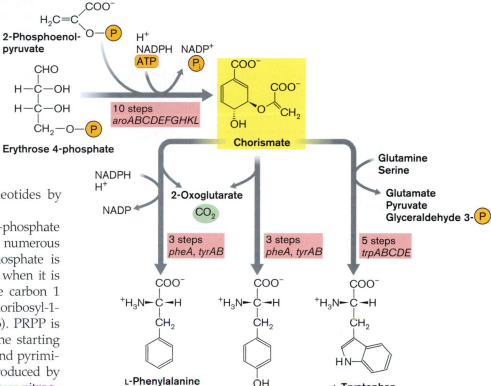

then converted to deoxyribonucleotides by enzymatic removal of the 2′ OH.

As we have seen, ribose 5-phosphate and related sugars participate in numerous metabolic pathways. Ribose 5-phosphate is directed into nucleotide synthesis when it is tagged with pyrophosphate at the carbon 1 (C-1) position, forming 5-phosphoribosyl-1-pyrophosphate (PRPP) (**Fig. 15.26**). PRPP is a major metabolic intermediate, the starting point for the synthesis of purine and pyrimidine nucleotides. PRPP is also produced by "scavenger" pathways, in which excess nitrogenous bases are broken down and recycled. Its synthesis is regulated by feedback inhibition to avoid overproduction of purines. In humans, overproduction of PRPP leads to gout, a condition in which the purine breakdown product uric acid precipitates in the joints, causing painful swelling of wrists and feet.

Purine synthesis. The hydrolysis of PRPP releases pyrophosphate, thus spending two high-energy phosphate bonds. This irreversible reaction drives forward the subsequent reactions to make purine. To construct the purine, the pyrophosphate at carbon 1 (C-1) is replaced by an amine from glutamine. The purine is built up by addition of a series of single-carbon groups (formyl, methyl, and CO_2) alternating with nitrogens from glutamine and carbamoyl phosphate. The number of single-carbon assimilations, including assimilation of CO_2, is striking. It may reflect the ancient origin of the purine synthetic pathway, which evolved in CO_2-fixing autotrophs.

The first purine constructed is inosine monophosphate, the "wobble" purine found at the third position of transfer RNAs. Inosine monophosphate is subsequently converted to adenine monophosphate or guanine monophosphate.

Pyrimidine synthesis. The pyrimidine is built by a slightly different route. Instead of building up the base on the ribose, the six-membered pyrimidine ring forms separately from aspartate plus carbamoyl phosphate. The pyrimidine ring then displaces the pyrophosphate of

PRPP, attaching by a nitrogen to the ribosyl carbon 1. The first pyrimidine built is uracil, which can be converted to cytosine or thymine.

> **THOUGHT QUESTION 15.8** Why are purines and pyrimidines not synthesized separately from the sugar phosphate?
>
> **THOUGHT QUESTION 15.9** Why are the ribosyl nucleotides synthesized first, then converted to deoxyribonucleotides as necessary? What does this suggest about the evolution of nucleic acids?

Nonribosomal Peptide Antibiotics

In addition to the standard amino acids used by ribosomes, there exist hundreds of different amino acids used by cells for functions such as the cross-bridges of peptidoglycan (discussed in Chapter 3). Some of these are made by one-step modification of standard amino acids, such as epimerization (altering the chirality from L to D) or halogenation with chlorine or fluorine. Others have complex carbon skeletons, including unsaturated bonds and epoxides (three-membered rings):

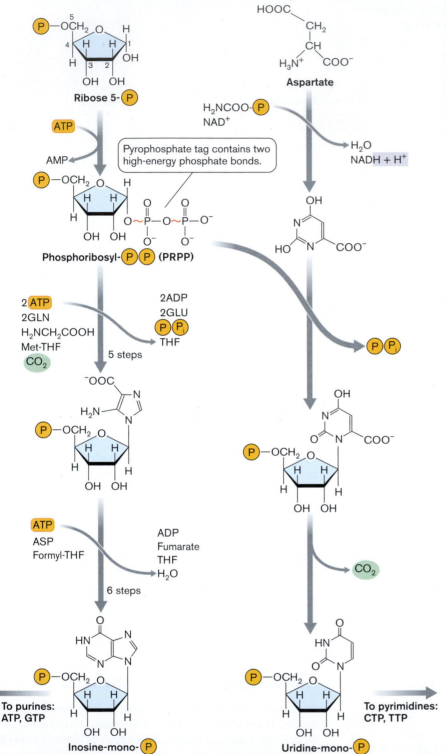

Figure 15.26 Biosynthesis of purines and pyrimidines. Both purines and pyrimidines are built onto ribose 5-phosphate. Ribose 5-phosphate is directed into purine and pyrimidine biosynthesis by the activating step of ATP ⟶ AMP, converting the sugar to phosphoribosyl-PP (PRPP). The purine ring is built up out of successive additions of amines, one-carbon units, plus a formyl group from formyl tetrahydrofolate (formyl-THF). The purine ring is modified to yield the corresponding adenine and guanine compounds. Likewise, the pyrimidine ring is modified to cytidine triphosphate or thymidine triphosphate.

analogous to those that build polyketides. (See **Special Topic 15.3**, on the biosynthesis of vancomycin.)

TO SUMMARIZE:

- **Amino acid biosynthesis** requires a large number of different enzymes to catalyze many unique conversions. Structurally related amino acids branch from a common early pathway.
- **Metabolic intermediates** from glycolysis and the TCA cycle initiate amino acid biosynthetic pathways.
- **Ammonium ion is assimilated by TCA intermediates**, such as oxaloacetate into glutamate. Glutamate assimilates ammonium ion to form glutamine. Transamination is the donation of NH_4^+ from one amino acid to another, such as glutamine transferring ammonia to oxaloacetate to make aspartate.
- **Arginine biosynthesis** requires multiple steps of NH_3 transfer and carbon skeleton condensation.
- **Aromatic amino acids** are built from a common pathway that branches out. Their biosynthesis is regulated tightly at both transcriptional and translational levels.
- **Purines are built as nucleotides** attached to a ribose phosphate. Several single-carbon groups are assimilated, including CO_2, a phenomenon suggesting an ancient pathway. Pyrimidines are made from aspartate, then added onto PRPP.
- **Nonribosomal peptide antibiotics** are built by modular enzymes analogous to those that build polyketides.

Both standard and nonstandard amino acids are used by actinomycetes to build secondary metabolites with antimicrobial activity. These secondary metabolites are called **nonribosomal peptide antibiotics** because their peptide backbone is constructed by an enzyme without ribosomes. Nonribosomal peptide antibiotics are synthesized by modular "assembly-line" enzymes,

Heme *b*

Chlorophyll

R = —CH₃
(Chlorophyll *a*)

R = —CHO
(Chlorophyll *b*)

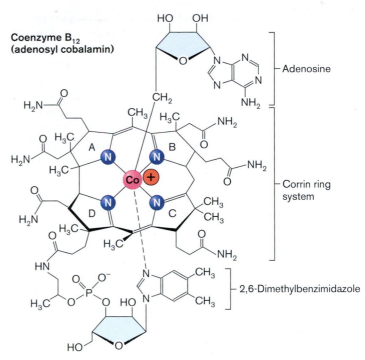

Coenzyme B₁₂
(adenosyl cobalamin)

Adenosine

Corrin ring system

2,6-Dimethylbenzimidazole

Figure 15.27　Tetrapyrroles. Tetrapyrrole derivatives include hemes, chlorophylls, and coenzyme B₁₂. Different variations of the ring enable coordination to different metal atoms.

tions comes from its multiple conjugated double bonds, with many narrowly spaced energy transitions. The size of these transitions and the wavelength of light absorption are altered incrementally by reduction or oxidation at specific positions or by addition of particular substituents around the ring. For example, the modified ring structure of chlorophyll *a* enables the molecule to absorb red and blue light for photosynthesis. Small differences in ring structure enable different tetrapyrroles to coordinate to different metal atoms, such as iron, magnesium, or cobalt.

Of all the essential primary products that cells need to make, the tetrapyrroles are among the most complicated and require the greatest number of distinct enzymatic steps. The tetrapyrrole system consists of four five-membered rings, each including an atom of nitrogen. Its double bonds are highly conjugated around the system. The nitrogens are directed inward, where their electron pairs are well positioned to complex a metal ion.

Biosynthesis of the Tetrapyrrole Ring System

Tetrapyrroles arise from condensation of four **pyrrole** rings. The five-membered pyrrole ring has three substituent groups, one of which (—CH₂NH₂) serves as a linker in

15.7 Biosynthesis of Tetrapyrroles

The **tetrapyrrole** family includes some of the best-known and essential cofactors of energy transduction and biosynthesis, such as chlorophylls, hemes of cytochromes and hemoglobin, and coenzyme B₁₂ (**Fig. 15.27**). The versatility of the tetrapyrrole for redox reac-

condensation of the tetrapyrrole. The individual pyrrole rings are generated by one of two different pathways (**Fig. 15.28**), found in different organisms: the glutamate pathway, in most bacteria as well as in plants; and the glycine-succinate pathway, in purple bacteria, animals, and fungi. The glutamate pathway is unusual in that it starts from a glutamine residue attached to a transfer RNA, generally used in protein synthesis. In either case, condensation and rearrangement generates the linear molecule 5-aminolevulinic acid (ALA). Two molecules of ALA cyclize to form the ring of porphobilinogen.

Once the pyrroles are formed, an enzyme condenses four of them in a chain, removing each amine as NH_4^+. The four linked pyrroles are conventionally labeled A through D. When the chain cyclizes, an unusual isomerization occurs in which the D pyrrole is inverted. The mechanism of the D-ring isomerization is unknown, but it imparts a critical asymmetry to the molecule as a whole. This first cyclized tetrapyrrole is called uroporphyrinogen III.

Uroporphyrinogen III forms the foundation of most biologically active tetrapyrroles. By inspection of functional tetrapyrroles, we can see the subtle modifications that distinguish them (see **Fig. 15.27**). Chlorophyll and coenzyme B_{12} (the active form of vitamin B_{12}, in which adenosine replaces cyanide) each have an extra-long side chain at the D ring, leading to completely different functional groups. The ring systems of both heme and chlorophyll are more unsaturated than the original ring system found in uroporphyrinogen III; the extra conjugated double bonds facilitate energy transitions in these electron carriers. The cobalamin core of coenzyme B_{12} has a more hydrogenated D ring and one less carbon (a carbon is lost in the connection between rings A and D).

The synthesis of tetrapyrroles is highly regulated by nutritional needs. For example, purple photoheterotrophs such as *Rhodospirillum rubrum*, switch off synthesis of bacteriochlorophylls and other photopigments in the presence of oxygen, preferring aerobic respiration to phototrophy. Thus, when exposed to oxygen, the bacteria change color from purple-red to white.

Figure 15.28 Biosynthesis of tetrapyrroles.   The fundamental tetrapyrrole ring arises from simple TCA intermediates and amino acids. Two amino acids condense to form 5-aminolevulinic acid (ALA), which cyclizes to form porphobilinogen, a pyrrole ring with three substituents—two ending in COOH, one in NH_2. Four of the pyrroles link together, then cyclize to form the corrin ring system. The final step of cyclization includes reversal of the D-ring linkage to yield uroporphyrinogen III, which enters various pathways to form different products.

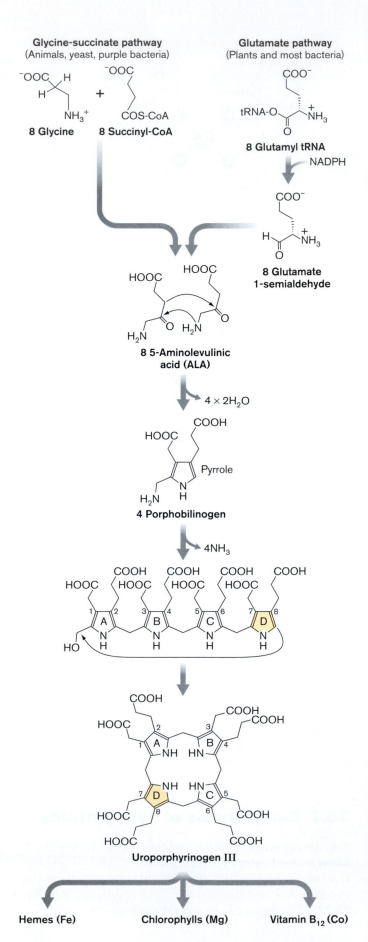

Riboswitch Regulation

Biosynthesis of tetrapyrroles is enormously energy intensive and subject to sensitive means of regulation. An interesting form of regulation is the **riboswitch** mechanism of regulating the biosynthesis of coenzyme B_{12}. A riboswitch is an RNA sequence capable of forming an elaborate stem loop structure at the sequence upstream of a biosynthetic operon (**Fig. 15.29A**). The stem loop structure is stabilized by the operon's product, in this case coenzyme B_{12}. Riboswitch regulation occurs for several ancient cofactors, including riboflavin, thiamin, and S-adenosylmethionine. It is unusual in requiring no protein repressor. The riboswitch suggests a way of regulating gene expression in the earliest cells of the proposed "RNA world," cells that lacked DNA as well as protein. (The RNA world theory of the origin of life is discussed in Chapter 17.)

Riboswitches can act either to block translation or to prematurely terminate transcription. **Figure 15.29B** shows one example: translational repression of the cobalamin operon by coenzyme B_{12}. When coenzyme B_{12} levels are high, the riboswitch stem loop structure allows formation of a neighboring stem loop, called the sequestrator. The sequestrator contains the translational start site, consisting of the Shine-Dalgarno sequence plus AUG (discussed in Chapter 8). In effect, the start site is "sequestered" by hybridization within the stem loop. When B_{12} is scarce, however, it dissociates from the riboswitch. The riboswitch stem loops come apart, exposing new sequences for potential hybridization. The exposed sequences form a new stem loop, longer and more stable than that of the sequestrator. The Shine-Dalgarno and AUG sequences are now exposed to bind the ribosome and initiate translation of biosynthetic enzymes for cobalamin.

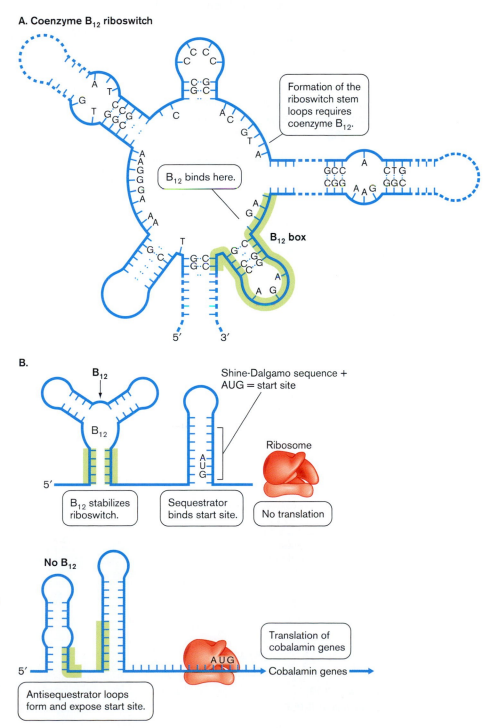

Figure 15.29 The B_{12} riboswitch. A. A riboswitch upstream of the cobalamin operon. Formation of the riboswitch stem loops requires coenzyme B_{12}. **B.** The B_{12} riboswitch allows formation of the sequestrator stem loop, which sequesters the translation start site, preventing translation of cobalamin biosynthesis genes. In the absence of coenzyme B_{12}, the antisequestrator stem loops form instead of the sequestrator, permitting the ribosome to bind and translate genes.

Special Topic 15.3 Modular Biosynthesis of Vancomycin

A major nonribosomal peptide antibiotic is vancomycin, the drug of last resort for life-threatening *Clostridium difficile* and for "flesh-eating" methicillin-resistant staphylococcal infections (MRSA) (**Fig. 1**). Vancomycin has a peptide backbone (shaded) equivalent to that of ribosomal peptide, but its aminoacyl residues are nonstandard and show atypical secondary connections.

Vancomycin is produced by *Streptomyces orientalis*, an actinomycete isolated from the soil of India and Indonesia. Like polyketides, polypeptide antibiotics are synthesized by a modular enzyme complex. The vancomycin peptide backbone is built by a nonribosomal peptide synthetase (NRPS) containing seven repeating modules. Each module includes the following key domains (**Fig. 2**):

- **Domain A: Adenylation and transfer.** The A domain catalyzes adenylation (addition of adenosine monophosphate, AMP) to the carboxylate of the amino acid. The adenylation reaction is driven by ATP releasing pyrophospate (PP$_i$). Next, release of AMP drives transfer of the aminoacyl group onto a sulfur atom of polypeptide carrier protein (S-PCP). Each S-PCP recognizes only one specific aminoacyl group.

Figure 1   Vancomycin: A nonribosomal peptide antibiotic. The peptide backbone is shaded. *Source:* Christopher Walsh. 2003. *Antibiotics: Actions, Origins, Resistance.* ASM Press.

A.

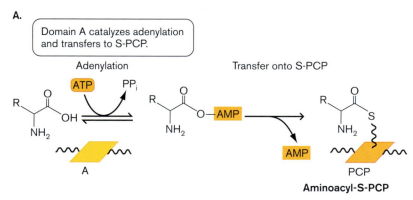

Figure 2   Elongation of a nonribosomal peptide.   A. Enzyme domain A adenylates (activates) an amino acid, then transfers it onto S-PCP (peptidyl carrier protein). **B.** Domain C catalyzes transfer of the aminoacyl group to form a peptide bond. In subsequent rounds, the nascent peptide is transferred onto the next amino acid. *Source:* Christopher Walsh. 2003. *Antibiotics: Actions, Origins, Resistance.* ASM Press.

B.

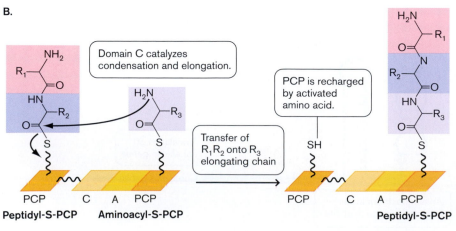

- **Domain C: Condensation and elongation.** The C domain catalyzes transfer of the peptidyl-S-PCP from one module onto the amine of the next aminoacyl-S-PCP. In each subsequent round, the nascent peptide is transferred onto the amino group of the next aminoacyl PCP. As each peptide travels down all seven modules, it accretes seven aminoacyl groups in all.
- **Domain E: Epimerization.** Some (not all) modules include an E domain to epimerize the aminoacyl group (change its configuration from L to D).

The vancomycin NRPS comprises three multidomain proteins (CepA, B, C), each providing one to three modules for chain extension (**Fig. 3**). The seven aminoacyl groups include leucine, asparagine, three tyrosines, and two hydroxyphenylglycines; hydroxyphenylglycine is a nonstandard derivative of glycine. Once the seven-member peptide is formed, additional enzymes (not shown) catalyze chlorination, cross-linking of hydroxyls, and sugar transfer to complete the structure shown in **Figure 1**. In all, vancomycin biosynthesis requires about 30 different genes, as predicted by bioinformatic analysis of the vancomycin gene cluster.

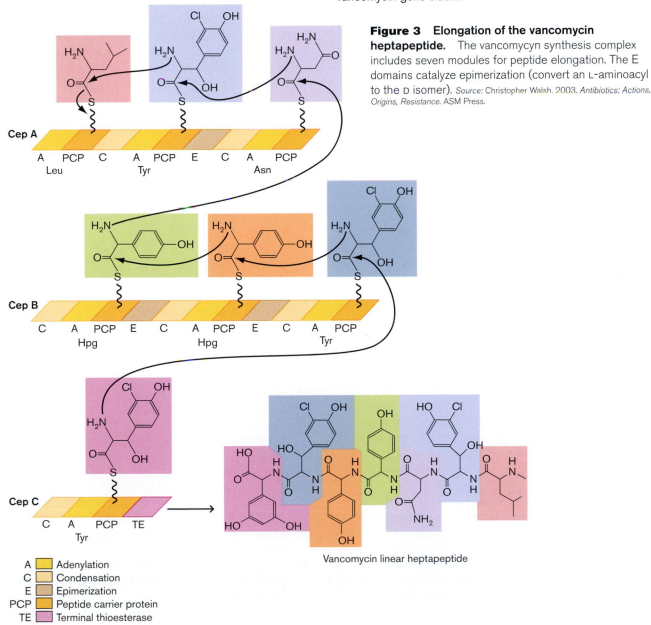

Figure 3 Elongation of the vancomycin heptapeptide. The vancomycyn synthesis complex includes seven modules for peptide elongation. The E domains catalyze epimerization (convert an L-aminoacyl to the D isomer). *Source:* Christopher Walsh. 2003. *Antibiotics: Actions, Origins, Resistance.* ASM Press.

TO SUMMARIZE:

■ **Tetrapyrroles** are conjugated ring systems made up of pyrroles, each five-membered ring including one nitrogen positioned to coordinate a central metal ion. Tetrapyrrole derivatives include chlorophylls, hemes, hemoglobin, and cyanocobalamin.

■ **The glutamate pathway or the glycine-succinate pathway** condenses eight molecules of 5-aminolevulinic acid into four linked pyrroles. The linked pyrroles are cyclized to uroporphyrinogen III, the foundation of most biologically active tetrapyrrole derivatives.

■ **Riboswitch regulation** limits the synthesis of the tetrapyrrole derivative cobalamin (vitamin B_{12} precursor). A riboswitch is a regulatory RNA sequence (upstream of the operon) that forms a stem loop structure stabilized by the operon's product. A riboswitch limits transcription or translation of genes encoding biosynthetic enzymes.

Concluding Thoughts

In living cells, no enzymatic pathway occurs in isolation. All pathways of energy acquisition and biosynthesis occur together, sharing common substrates and products. For example, acetyl-CoA is produced by glycolysis or by fatty acid degradation; it is then utilized by pathways that synthesize fatty acids, amino acids, bases, and tetrapyrroles. Tetrapyrrole biosynthesis requires products of the TCA cycle. At the same time, the TCA cycle is needed to provide energy through catabolism. The microbial cell regulates these competing needs on an extremely rapid timescale; within seconds of the disappearance or appearance of a nutrient, such as an amino acid, its biosynthesis is turned on or shut off.

A given microbe needs to build as much biomass as it can by the most efficient means available. For a pathogen growing within a host, this means utilizing preformed host compounds for rapid growth and replication. On the other hand, free-living microbes face intense competition for organic nutrients; they need to fix essential elements into readily accessible forms. Many soil and marine microbes build the most complicated biological molecules out of single-carbon and single-nitrogen sources, although they scavenge preformed organic substrates when available.

Microbial metabolism has many applications in food preparation and preservation, and in industrial production of antibiotics and other pharmaceuticals. The principles of food and industrial microbiology are discussed in Chapter 16. Later, in Chapters 21 and 22, we see how microbial metabolism within populations and communities contributes to global cycles of Earth's biosphere.

CHAPTER REVIEW

Review Questions

1. What are the sources of substrates for biosynthesis? From what kind of pathways do they arise?
2. How do microbial species economize by synthesizing only the products they need? Cite long-term as well as short-term mechanisms.
3. Compare and contrast the different cycles of carbon dioxide fixation. What classes of organisms conduct each type?
4. How was carbon isotope labeling used to identify intermediates of the Calvin cycle?
5. How is ribulose 1,5-bisphosphate consumed and re-formed through the Calvin cycle? What key products emerge to form sugars and amino acids?
6. How do oxygenic phototrophs maintain CO_2 at sufficient levels to conduct the Calvin cycle?
7. Explain the cyclic process of chain extension in fatty acid biosynthesis. Explain the generation of occasional unsaturated "kinks" in the chain.
8. Explain the different kinds of regulation of fatty acid biosynthesis.
9. What are the different environmental sources of nitrogen? Explain how and why microbes use these different sources.
10. Explain the process of nitrogenase converting N_2 to $2NH_4^+$. Why is H_2 produced?
11. Explain the different ways that microbes maintain anaerobic conditions for nitrogenase.
12. Explain the molecular basis of regulation of nitrogen fixation and nitrogen scavenging.
13. Compare and contrast the general scheme of biosynthesis of amino acids with that of fatty acids.
14. Outline the interconversions of 2-oxoglutarate, glutamate, and glutamine that provide nitrogen for amino acid biosynthesis.
15. Compare and contrast the processes of purine and pyrimidine biosynthesis. What is the role of the sugar ribose in each case?
16. How does the biosynthesis of tetrapyrroles generate diverse products from common small organic substrates?

Key Terms

acyl carrier protein (ACP) (564)
anabolism (548)
anaplerotic reaction (560)
biosynthesis (548)
Calvin cycle (Calvin-Benson cycle, Calvin-Benson-Bassham cycle, CBB cycle) (550)
carbon dioxide fixation (550)
carboxysome (558)
chloroplast (550)
CO_2-concentrating mechanism (CCM) (558)

fatty acid synthase complex (564)
folate (562)
gluconeogenesis (560)
glutamate dehydrogenase (GDH) (576)
glutamate synthase (GOGAT) (576)
glutamine synthetase (GS) (576)
Haber process (571)
heterocyst (572)
3-hydroxypropionate cycle (562)
nitrogenase (571)
nonribosomal peptide antibiotic (580)
paper chromatography (551)

pyrrole (581)
reductive acetyl-CoA pathway (561)
reductive pentose phosphate cycle (550)
reductive (reverse) TCA cycle (560)
riboswitch (583)
rubisco (553)
secondary metabolite (566)
secondary product (550, 566)
tetrapyrrole (581)
transamination (577)

Recommended Reading

Ahlert, Joachim, Erica Shepard, Natalia Lomovskaya, Emmanuel Zazopoulos, Alfredo Staffa, et al. 2002. The calicheamicin gene cluster and its iterative type i enediyne PKS. *Science* **297**:1173–1176.

Andersson, Sive G. E., and C. G. Kurland. 1990. Codon preferences in free-living microorganisms. *Microbiological Reviews* **54**:198–210.

Cane, David E., Christopher T. Walsh, and Chaitan Khosla. 1998. Harnessing the biosynthetic code: Combinations, permutations, and mutations. *Science* **282**:63–68.

Eisen, Jonathan A., et al. 2002. The complete genome sequence of *Chlorobium tepidum* TLS, a photosynthetic, anaerobic, green-sulfur bacterium. *Proceedings of the National Academy of Sciences USA* **99**:9509–9514.

Gollnick, Paul, Paul Babitzke, Alfred Antson, and Charles Yanofsky. 2005. Complexity in regulation of tryptophan biosynthesis in *Bacillus subtilis*. *Annual Review of Genetics* **39**:47–68.

Hansen, S., V. B. Vollan, E. Hough, and K. Andersen. 1999. The crystal structure of rubisco from *Alcaligenes eutrophus* reveals a novel central eight-stranded beta-barrel formed by beta-strands from four subunits. *Journal of Molecular Biology* **288**:609–621.

Heath, Richard J., J. Ronald Rubin, Debra R. Holland, Erli Zhang, Mark E. Snow, and Charles O. Rock. 1999. Mechanism of triclosan inhibition of bacterial fatty acid synthesis. *Journal of Biological Chemistry* **274**:11110–11114.

Herter, Sylvia, Jan Farfsing, Nasser Gad'On, Christoph Rieder, Wolfgang Eisenreich, et al. 2001. Autotrophic CO_2 fixation by *Chloroflexus aurantiacus:* Study of glyoxylate formation and assimilation via the 3-hydroxypropionate cycle. *Journal of Bacteriology* **183**:4305–4316.

Hügler, Michael, Carl O. Wirsen, Georg Fuchs, Craig D. Taylor, and Stefan M. Sievert. 2005. Evidence for autotrophic CO_2 fixation via the reductive tricarboxylic acid cycle by members of the ε subdivision of *Proteobacteria. Journal of Bacteriology* **2187**:3020–3027.

Kamen, Martin. 1985. *Radiant Science, Dark Politics. A Memoir of the Nuclear Age.* University of California Press, Berkeley.

Liu, Wen, Steven D. Christenson, Scott Standage, and Ben Shen. 2002. Biosynthesis of the enediyne antitumor antibiotic C-1027. *Science* **297**:1170–1173.

Valbuzzi, Angela, and Charles Yanofsky. 2001. *B. subtilis* regulatory protein TRAP. *Science* **293**:2057–2059.

Zazopoulos, Emmanuel, Kexue Huang, Alfredo Staffa, Wen Liu, Brian O. Bachmann, et al. 2003. A genomic-guided approach for discovering and expressing cryptic metabolic pathways. *Nature Biotechnology* **21**:187–190.

Zhang, Yong-Mei, Stephen W. White, and Charles O. Rock. 2006. Inhibiting bacterial fatty acid synthesis. *Journal of Biological Chemistry* **281**:17541–17544.

Zimmer, Daniel P., Eric Soupene, Haidy L. Lee, Volker F. Wendisch, Arkady B. Khodursky, et al. 2000. Nitrogen regulatory protein C-controlled genes of *Escherichia coli:* scavenging as a defense against nitrogen limitation. *Proceedings of the National Academy of Sciences USA* **97**:14674–14679.

Chapter 16

Food and Industrial Microbiology

Microbes have nourished humans for centuries, generating cheese, bread, wine and beer, tempeh, and soy sauce. And yet, from the moment of harvest, we humans compete with microbes for our food. Microbes from the food's surface or from the air colonize food, and their uncontrolled growth can render the food rancid or putrid. Historically, the need for food storage and preservation has led to practices such as drying, smoking, and adding spices, all of which retard microbial growth.

The principles of food microbiology have been extended to a much broader field of industrial microbiology. Industrial microbiology includes the development of microbial products such as antibiotics and enzymes, as well as transgenic microbes that produce human proteins such as insulin and growth hormones. Many products derive from extremophiles, organisms adapted to extreme environments, whose enzymes can withstand industrial conditions. Industrial microbiology faces special challenges with regard to microbial growth conditions, scaled-up fermentation, genetic optimization of production, and safety testing.

Chocolate production requires fermentation of cocoa beans by yeasts, lactic acid bacteria, and acetic acid bacteria, a process analyzed by Rosane Schwan at the Federal University of Lavras, Brazil. The beans are fermented by microbes naturally present in pulp of the cocoa fruit. The fermented beans are dried and processed to make cocoa and cocoa butter. (Schwan, R. F., and A. E. Wheals. 2004. *Critical Reviews in Food Science and Nutrition* 44:205)
Source: Alain Caste/Stockford.

The fermentation industry reaches back to 5000 BCE, the date of pottery jars excavated from a Neolithic mud-brick kitchen in the Zagros Mountains of modern Iran (**Fig. 16.1A**). The jars contain residue showing chemical traces of grapes fermented to wine. Besides winemaking, Neolithic people leavened bread, another staple food requiring microbial growth. For thousands of years, microbial fermentation has been a daily part of human life and commerce.

Today, microbial growth drives companies such as Genencor International, now earning $400 million in annual revenues from engineering industrial enzymes. Genencor delivers 250 products ranging from contact lens cleansing agents to new fashion finishes for denim at manufacturing centers in the United States, Europe, Argentina, and China. Most of Genencor's product enzymes are made by bacteria grown in giant fermentation vessels (**Fig. 16.1B**). Founded as a joint venture by Genentech and Corning, Genencor's success heralded a growing investment by biotech start-ups as well as traditional chemical companies in the fermentation industry—the use of microbes to produce industrial products.

The fermentation industry is based on long-standing microbial associations with human food. Until the relatively recent invention of steam-pressure sterilization, all foods contained live microbes. Microbial metabolism could spoil food; or it could improve the food by adding flavor, preserving valuable nutrients, and preventing growth of pathogens. As we learned in Chapter 1, one of the first great microbiologists, Louis Pasteur, began his career as a chemist investigating fermentation in winemaking. In Chapter 16, we see how the many kinds of microbial biochemistry presented in Chapters 13–15 give rise to the characteristics of food so familiar to us, from the taste of cheese and chocolate to the rising of bread dough and the physiological effects of alcoholic beverages. Industrial research continues to improve food through the fields of food biochemistry, discovering the molecular basis for the flavor and texture of microbial foods; food preservation, eliminating undesirable microbial decay by preservative methods; and food engineering, improving the quality, shelf life, and taste of microbial foods.

Beyond food, the commercial applications of microbes extend to industrial microbiology, the use of microbes to generate useful products of all kinds. Industrial microbiology generates enzymes, chemical feedstocks, fuels, and pharmaceuticals. New products are developed through genomic engineering, and the technologies presented in Chapter 12. Increasingly, research microbiologists reach beyond the laboratory bench to invent new uses for microbial products—and start their own companies to develop and market them.

16.1 Microbes as Food

Certain kinds of microbial bodies have long been eaten as food, especially the fruiting bodies of fungi and the fronds of marine algae. Single-celled algae and cyanobacteria are also used as food supplements. These microbes can provide important sources of protein, vitamins, and minerals.

Edible Fungi: From Survival Food to Gourmet Treat

In the children's classic *Homer Price*, by Robert McCloskey, settlers on the Ohio frontier save themselves from starvation when they discover "forty-two pounds of edible fungus, in the wilderness a-growin'." Fungal **fruiting bodies**, multicellular reproductive structures that generate spores, are commonly known as mushrooms (see Chapter 18). Mushrooms offer a flavorful source of protein and minerals, albeit at the risk of consuming the deadly toxins produced by a few species. The protein content of edible mushrooms can be as high as 25% dry weight, comparable to that of whole milk (percent dry weight), and includes all essential dietary amino acids.

Mushrooms contributed to the survival of pre-industrial humans, while killing those unlucky enough to consume varieties that were poisonous. Less than 1% of mushrooms are poisonous, but those few are deadly, such as the amanita, or "destroying angel," which produces toxins including the RNA polymerase inhibitor

A.

Courtesy The University of Pennsylvania Museum

B.

Genencor International

Figure 16.1 The fermentation industry: then and now. A. A Neolithic jar that contained wine, from 5000 BCE. Excavated at Hajji Firuz Tepe, Iran. **B.** An industrial fermentation apparatus for growing microbes to produce enzyme products, at Genencor International.

alpha-amanitin. People who ingest the amanita typically die of liver failure.

On the other hand, certain mushrooms actually invite predation in order to disperse their spores. For example, the underground truffles prized in European cooking produce odorant molecules that attract animals to dig them up. Truffle odorants include sex pheromones such as androstenol, found in human perspiration.

Many varieties of edible mushrooms are farmed and marketed for human fare. **Figure 16.2** shows the culturing of *Agaricus bisporus*, the mushroom variety most commonly sold in the United States as button mushrooms and portobellos. Button mushrooms are harvested at an early stage, whereas portobellos are harvested later, when the gills are fully exposed and some of the moisture has evaporated. The decrease in moisture in portobellos concentrates the flavor and creates a dense, meaty texture; the mushrooms are often served in gourmet sandwiches as a vegetarian alternative to hamburger. *Agaricus* culture was first developed around 1700 in France, where the mushrooms were grown in underground caves. In modern mushroom farms, *Agaricus* mushrooms are cultured on composted horse manure or chicken manure in chambers controlled for temperature and humidity.

Other mushroom varieties from China and Japan are grown on logs or wooden blocks. Wood-grown mushrooms include the strong-flavored *Lentinula edodes* (black forest mushroom, or shiitake), *Pleurotus* (oyster mushrooms), and *Flammulina velutipes* (enoki mushrooms) with long thin white stalks and delicate flavor.

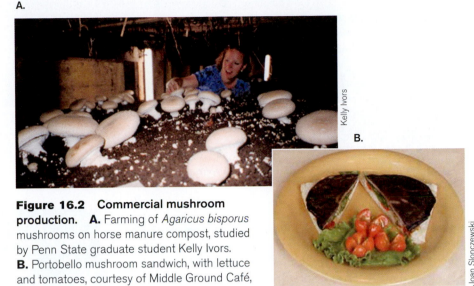

Figure 16.2   Commercial mushroom production.   A. Farming of *Agaricus bisporus* mushrooms on horse manure compost, studied by Penn State graduate student Kelly Ivors.
B. Portobello mushroom sandwich, with lettuce and tomatoes, courtesy of Middle Ground Café, Gambier, Ohio.

Kelly Ivors

Joan Slonczewski

are harvested for processing into sheets. The sheets are toasted, turning dark green.

Another group of seaweeds, the "brown algae," or kelp, are grown for food or to extract food additives. Kelp of the genus *Laminari* are cultured as a source of the flavor additive monosodium glutamate (MSG). A different genus, the giant kelp *Macrocystis*, is grown off the coast of California for production of the polymer alginate, a common thickener for ice cream, salad dressings, and paints. Another additive produced from brown algae is carrageenan, an emulsifier (combiner of oil and water solutions) used in commercial pies and toothpastes.

Edible Bacteria and Yeasts: Single-Celled Protein and Vitamins

Since prehistoric times, consumption of single-celled fungi or yeasts has provided a supplemental source of

Edible Algae: Wrapping Sushi

Several kinds of seaweed (marine algae) are cultivated, most notably in Japan. The red alga *Porphyra*, a eukaryotic "true alga" (discussed in Chapter 20), forms large multicellular fronds cultured for **nori** (**Fig. 16.3**). Nori is best known for its use in wrapping rice, fish, and vegetables to form sushi.

For nori production, the red algae are grown as "seeds," or starter cultures, in enclosed tanks. The starter cultures are then distributed on nets in a protected coastal area, usually an estuary. The cultures grow until they hang heavy from the nets, when they

Figure 16.3   Nori production for sushi.   A. Nori grows from nets in seawater. **B.** Toasted sheets of nori wrap rice, vegetables, and fish to make sushi.

Courtesy of Tao Nature International

Klaus Arras/Stockford

A.

B.

C.

Figure 16.4   Single-celled protein from _Spirulina_.   A. Farming _Spirulina_. **B.** _Spirulina_ processed into protein-rich food supplements. **C.** _Spirulina_ cells (LM ×250).

protein and vitamins. Yeasts such as _Saccharomyces_ species were grown to high concentration in fermented milks and grain beverages. Traditional beers contained only a low percentage of alcohol with a thick suspension of nutrient-rich yeasts. Especially important was the content of vitamin B_{12}, an essential substance for the human diet that cannot be obtained from plant sources.

Few bacteria are edible as isolated organisms, mainly because their small cells contain a relatively high proportion of DNA and RNA. The high nucleic acid content is a problem because nucleic acids contain purines, which the human digestive system converts to uric acid. Because we humans lack the enzyme urate oxidase, the uric acid cannot be metabolized. Consumption of more than 2 g per day of nucleic acids causes uric acid to precipitate, causing painful conditions such as gout and kidney stones. For this reason, most bacteria cannot be consumed in quantity, except as a minor component of food mass, as in fermented foods.

An exception is the cyanobacterium _Spirulina_ (**Fig. 16.4**). _Spirulina_ consists of spiral-shaped cells that grow photosynthetically in fresh water. _Spirulina_ is sold as a food additive, rich in protein, vitamin B_{12}, and minerals. It also contains antioxidant substances that may prevent cancer. _Spirulina_ is grown with illumination in special ponds lined for food production. The final product is collected and vacuum-dried to form a dark green powder of flour-like consistency.

In the food industry, _Spirulina_ is classified as a form of **single-celled protein**, a term for edible microbes of high food value. Other kinds of single-celled protein include eukaryotes such as yeast and algae. The twentieth century saw the development of single-celled protein as a food source for impoverished populations. The yeast _Saccharomyces cerevesiae_ was grown for protein by Germany during World War I, using inexpensive molasses for culture; and during World War II, _Candida albicans_ was grown on paper mill wastes. Single-celled protein was later promoted by Western countries as a food for rapidly expanding populations of developing countries. This idea inspired the science fiction film _Soylent Green_ (1973), in which people of a future overpopulated Earth are forced

Figure 16.5   _Soylent Green_ (1973).   Science fiction film about a future Earth whose overgrown population is forced to eat a form of food called "Soylent," supposedly based on soybeans and single-celled protein.

to eat "Soylent," a food supposedly based on soybeans and single-celled protein though its true source was recycled humans (**Fig. 16.5**).

TO SUMMARIZE:

- **Fungal fruiting bodies** such as mushrooms and truffles are consumed as a protein-rich food.
- **Edible algae** include nori (toasted red algae), used to wrap sushi, and kelp (brown algae), which provides food additives.
- _Spirulina_ is edible cyanobacteria, a source of single-celled protein. Most bacteria, however, are inedible because of their overly high concentration of nucleic acids.
- **Yeasts** have been grown as an economical protein and vitamin supplement.

16.2 Fermented Foods: An Overview

Virtually all human cultures have developed varieties of **fermented foods**, food products that are modified biochemically by microbial growth. The purposes of food fermentation include:

- **To preserve food.** Certain microbes, particularly the lactobacilli, metabolize only a narrow range of nutrients before their waste products build up and inhibit further growth. Typically, the waste fermentation products that limit growth are carboxylic acids, ammonia (alkaline), or alcohol. Buildup of these substances renders the product stable for much longer than the original food substrate.
- **To improve digestibility.** Microbial action breaks down fibrous macromolecules and tenderizes the product, making it easier for humans to digest. Meat and vegetable products are tenderized by fermentation.
- **To add nutrients and flavors.** Microbial metabolism generates vitamins, particularly vitamin B_{12}, as well as flavor molecules such as esters and sulfur compounds.

Different societies have devised thousands of different kinds of fermented foods. Examples are given in **Table 16.1**. Fermented foods that are produced commercially include dairy products such as cheese and yogurt, soy products such as miso (from Japan) and tempeh (from Indonesia), vegetable products such as sauerkraut and kimchi, and various forms of cured meats and sausages. Alcoholic beverages are made from grapes and other fruits (wine), grains (beer and liquor), and cacti (tequila). Other kinds of foods require microbial treatment for special purposes, such as leavening by yeast (for bread), or cocoa bean fermentation (for chocolate). Besides commercial production, numerous fermented products are homemade by traditional methods thousands of years old. Such products are known as **traditional fermented foods**. Occasionally, a traditional fermented food enters commercial production and becomes widespread. For example, soy sauce, a traditional Japanese product, was marketed by the Kikkoman company and achieved global distribution in the twentieth century.

The nature of fermented foods depends on the quality of the fermented substrate as well as on the microbial species and the type of biochemistry performed. Traditional fermented foods usually depend on **indigenous flora**, that is, microbes found naturally in association with the food substrate; or on starter cultures derived from a previous fermentation, as in yogurt or sourdough fermentation. Commercial fermented foods use highly engineered microbial strains to inoculate their cultures, although in some cases indigenous flora still participate. For example, wines and cheeses aged in the same caves for centuries often include fermenting organisms that persist in the air and the containers used.

Major classes of fermentation reactions are summarized in **Figure 16.6**. The most common conversions involve anaerobic fermentation of glucose, as discussed in Chapter 13. Glucose is fermented to lactic acid (**lactic acid fermentation**) in cheeses and sausages, primarily by lactic acid bacteria such as *Lactobacillus*. A second-stage fermentation of lactic acid to propionic acid (**propionic acid fermentation**) by *Propionibacterium* generates the special flavor of Swiss and related cheeses. Some kinds of vegetable fermentation, as in sauerkraut, involve production of lactic acid, ethanol, and carbon dioxide (**heterolactic fermentation**) by *Leuconostoc*. Fermentation to ethanol plus carbon dioxide without lactic acid (**ethanolic fermentation**) is conducted by yeast during bread leavening and production of alcoholic beverages.

In some food products, particularly those fermented by *Bacillus* species, proteolysis and amino acid catabolism generate ammonia in amounts that raise pH (**alkaline fermentation**). For example, alkaline fermentation forms the soybean product natto. Other products require the growth of mold, such as the mold-spiked Roquefort cheese and the soy product tempeh. Mold growth requires some oxygen for aerobic respiration. Respiration must be limited, however, to avoid excessive decomposition of food substrate and loss of food value.

Note that the conversions cited here include only the major reactions in achieving the food product. In addition, thousands of minor or secondary reactions occur, some of

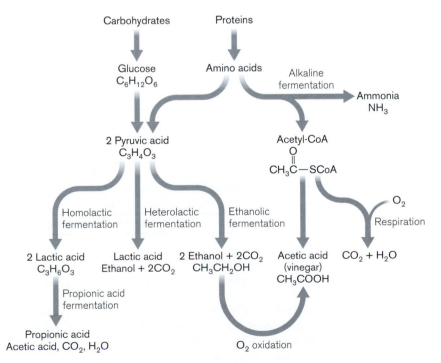

Figure 16.6 Major chemical conversions in fermented foods.

Table 16.1 Fermented foods and beverages.

Product	Description	Microbial genera	Geographic origin
Acidic fermentation of dairy, meat, and fish			
Buttermilk	Bovine milk, lactic fermented	Lactococcus	Europe, Asia
Yogurt	Bovine milk, lactic fermented and coagulated	*Lactobacillus, Streptococcus*	Europe, Asia
Kefir	Bovine or sheep milk, mixed fermentation, acidic with some alcohol	*Lactobacillus, Streptococcus,* yeasts, others	Eastern Europe, Russia
Sour cream	Bovine cream, lactic fermented	Lactococcus	Europe, Asia
Cheese (many kinds)	Milk (bovine, sheep, or goat) lactic fermented, coagulated, and pressed; in some cases cooked; mold ripened (spiked or coated)	Acid fermentation: *Lactobacillus, Streptococcus, Propionibacterium* Mold ripening: *Penicillium*	Europe, Asia
Sausage	Ground beef and/or pork encased with starter culture, lactic fermented, dried, or smoked	*Lactobacillus, Pediococcus, Staphylococcus,* others	Europe, Asia
Fermented fish	Many kinds of fish, mixed fermentation, acid and amines produced	Unknown	Africa, Asia
Acidic fermentation of vegetables			
Tempeh	Soybean cakes, fungal fermentation	*Rhizopus oligosporus*	Indonesia
Miso	Soy and rice paste, fungal fermentation	*Aspergillus*	Japan
Soy sauce	Extract of soy and wheat, fungal fermentation, brined, bacterial fermentation	*Aspergillus,* followed by halotolerant bacteria	China, Japan
Kimchi	Cabbage, peppers, and other vegetables, with fish paste, brined; container is buried	*Leuconostoc,* other bacteria	Korea
Sauerkraut	Cabbage fermented, making lactic and acetic acids and CO_2	*Leuconostoc, Pediococcus, Lactobacillus*	Europe
Pickled foods	Cucumbers, carrots, fish; brined, then fermented	*Leuconostoc, Pediococcus, Lactobacillus*	Europe, Asia
Kenkey	Maize, fermented, wrapped in banana leaves and cooked	Unknown	Western Africa
Chocolate	Cocoa beans soaked and fermented before processing to chocolate	*Lactobacillus, Bacillus, Saccharomyces*	South America
Alkaline fermentation			
Pidan	Duck eggs, coated in lime (CaO), aged, producing ammonia and sulfur odorants	*Bacillus*	Japan, China
Natto	Whole soybeans, fermented	*Bacillus natto*	Japan, China
Dawadawa	Locust beans, fermented	*Bacillus*	Africa
Ogiri	Melon seed paste, fermented	*Bacillus*	Africa
Leavened bread dough			
Yeast breads	Ground grain, dough leavened by yeast	*Saccharomyces*	Europe, Asia, America
Sourdough	Ground grain, dough leavened by starter culture from previous dough	*Saccharomyces, Torulopsis, Candida*	
Injera	Ground *teff* grain, dough leavened and fermented three days by organisms from the grain	*Candida*	Ethiopia
Alcoholic fermentation			
Wine	Grape juice, yeast fermented, followed by malolactic fermentation	*Saccharomyces, Oenococcus*	Europe, Asia, America
Beer	Barley and hops, yeast fermented	*Saccharomyces*	Europe, Asia, America
Sake	Rice extract, yeast fermented	*Saccharomyces*	Japan
Tequila	Blue agave, yeast fermented and distilled	*Saccharomyces*	Mexico
Whiskey	Barley or other grains or potatoes, fermented and distilled	*Saccharomyces*	United Kingdom, America, Japan

which produce tiny amounts of potent odorants and flavors. While these flavor molecules have less nutritional consequence than the main fermentation products, they are responsible for the complex, "sophisticated" taste for which fine cheeses, wines, and soy products are known.

> **THOUGHT QUESTION 16.1**   Why do the lipid components of food experience relatively little breakdown during fermentation?
>
> **THOUGHT QUESTION 16.2**   Why does oxygen allow excessive breakdown of food, compared with anaerobic processes?

TO SUMMARIZE:

- **Fermentation of food** enhances preservation, digestibility, nutrient content, and flavor.
- **Acidic fermentations** lead to organic acid fermentation products, such as lactate and propionate.
- **Alkaline fermentations** produce ammonia and break down proteins to peptides.
- **Ethanolic fermentation** produces ethanol and carbon dioxide.
- **Lipids are relatively stable** under anaerobic conditions of fermentation.

16.3 Acidic and Alkaline Fermented Foods

Many food fermentations produce acids or bases. An acid or base serves as an effective preservative because the pH change is unlikely to be reversed, and because animal or plant bodies grown at near-neutral pH are unlikely to support growth of acidophiles or alkaliphiles, which grow at extreme pH conditions.

Acidic Fermentation of Dairy Products

The conversion of milk to solid or semisolid fermented products dates far back in human civilization. The practice of milk fermentation probably arose among herders who collected the milk of their pack animals but had no way to prevent the rapid growth of bacteria. The milk had to be stored in a portable container such as the stomach of a slaughtered animal. After hours of travel, the combined action of lactic acid-producing bacteria and stomach enzymes caused the coagulation of milk proteins into **curd**. The curd naturally separated from the liquid portion, called **whey**. Both "curds and whey" can be eaten, as in the nursery rhyme "Little Miss Muffet." The curds, however, are particularly valuable for their concentrated protein content.

Curd formation. A **cheese** is any milk product from a mammal (usually cow, sheep, or goat) in which the milk protein coagulates to form a semisolid curd. Curd formation results from two kinds of processes: acidification, usually as a result of the microbial production of lactic acid, and treatment by proteolytic enzymes such as rennet. The curd may then be separated and processed to varying degrees, depending on the type of cheese.

How does milk coagulate? The major organic components of cow's milk are milk fat (about 4% unless skimmed), protein (3.3%), and the sugar lactose (4.7%). Milk starts out at about pH 6.6, very slightly acidic. At this pH, the milk proteins are completely soluble in water; otherwise, they would clog the animal's udder as the milk comes out. Fermentation generally begins with bacteria such as *Lactobacillus* and *Streptococcus* (**Fig. 16.7**). As bacteria ferment lactose to lactic acid, the pH starts to decline. The dissociation constant of lactic acid ($pK_a = 3.9$) allows greater deprotonation than other fermentation products such as acetate ($pK_a = 4.8$). Thus, lactic acid rapidly acidifies the milk product to levels that halt further growth of bacteria. Halting bacterial growth minimizes the oxidation of amino acids, thus maintaining food quality.

Milk contains micelles (suspended droplets) of hydrophobic proteins called **caseins**. As the pH of milk declines below pH 5, the acidic amino acid residues of caseins become protonated, eventually destablilizing the tertiary structure. As the casein molecules unfold (or "denature"), they expose hydrophobic residues that regain stability by interacting with other hydrophobic molecules. The intermolecular interaction of caseins generates a gel-like network throughout the milk, trapping other substances such as droplets of milk fat. This protein network is responsible for the semisolid texture of **yogurt**, a simple product of milk acidified by lactic acid bacteria.

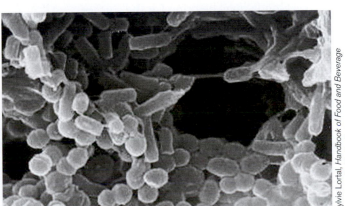

Sylvie Lortal, Handbook of Food and Beverage Fermentation Technology

Figure 16.7   Bacterial community within Emmentaler cheese (SEM).   Bacterial species include *Lactobacillus helveticus* (rods, 2.0–4.0 μm in length) and *Streptococcus thermophilus* (cocci).

In most kinds of cheese formation, a further step of casein coagulation is accomplished by proteases such as **rennet**. Rennet derives from the fourth stomach of a calf, although modern versions are made by genetically engineered bacteria. Calf rennet includes two proteolytic enzymes, chymosin and pepsin. Chymosin specifically cleaves casein into two parts, one of which is charged and water-soluble, the other hydrophobic. The hydrophobic portion forms a firmer curd than intact casein and is responsible for the harder texture of solid cheeses. The water-soluble portion, about one-third of the total casein, enters the whey and is lost from the curd. Processing of some cheese varieties includes heating to high temperature, which denatures even the whey protein so it is retained in the curd.

Varieties of cheese. An extraordinary number of cheese varieties have been devised (**Fig. 16.8**). These fall into several categories, based on particular steps in their production.

- **Soft, unripened cheeses**, such as cottage cheese and ricotta, are coagulated by bacterial action, without rennet. The curd is cooked slightly, and the whey is partly drained, but their water content is 55% or greater. These cheeses spoil easily; there are no steps of aging, or **ripening**.
- **Semihard, ripened cheeses**, such as Muenster and Roquefort, include rennet for firmer coagulation, and the curd is cooked down to a water content of 45–55%. The cheese is aged for several months.
- **Hard cheeses**, such as Swiss cheese and cheddar, are concentrated to even lower water content. Extra-hard varieties, such as Parmesan and Romano, have a water content as low as 20%. These cheeses are aged for many months, even several years.
- **Brined cheeses**, such as feta, are permeated with brine (concentrated salt), which limits further bacterial growth and develops flavor. Harder cheeses, such as Gouda, may be brined at the surface.
- **Mold-ripened cheeses** are inoculated with mold spores that germinate and grow during the ripen-

ing, or aging, process to contribute texture and flavor. The mold may be inoculated on the surface, to form a crust (as in Brie and Camembert), or it may be spiked deep into the cheese (as in blue cheese or Roquefort).

Cheese production. Commercial production of cheese involves a standard series of steps (**Fig. 16.9**). At each of these steps, choices of treatment lead to very different varieties. Key steps are illustrated for the example of Gouda cheese in **Fig. 16.10**.

In the first step, the milk is filtered to remove particulate objects, such as straw, and microfiltered or centrifuged to remove potentially pathogenic bacteria and spores. Most modern production includes flash pasteurization (brief heating to 72°C), although some traditional cheeses continue to be made from unpasteurized milk. Unpasteurized milk in cheese has been linked to illness, particularly from *Listeria*, bacteria that grow at typical refrigeration temperatures.

The fermenting microbes are added as a **starter culture**. The starter was traditionally derived from a sample of previous fermentation, in which case the flora are undefined. Commercial cheese production now uses defined species. In all but the soft cheeses, bacterial coagulation and curd formation are supplemented by rennet or by genetically engineered proteases.

The solid curd is then cut, or **cheddared** (hence the name "cheddar" cheese). The finer the pieces, the more whey can be pressed out and the harder the cheese produced. Curd is then heat treated, with or without the whey; if whey is included, more protein is retained. Brining at this stage leads to a salty cheese, such as feta.

The pressed curd is then shaped into a mold, which determines the ultimate shape of the cheese. Before ripening (or aging), the cheese may be floated in brine to generate a rind; or it may be coated or spiked with a *Penicillium* mold. The ripening period then allows development of flavor. Texture also changes; for example, where fermentation has produced CO_2, the trapped gas forms "eyes," or holes.

A.   **B.**   **C.**   **D.**

Figure 16.8 Cheese varieties.   A. Cottage cheese, an unripened perishable cheese. **B.** Feta cheese, a soft cheese from goat's milk, preserved in brine. **C.** Emmentaler Swiss cheese, with eyes produced by carbon dioxide fermentation. **D.** Roquefort, a medium-hard cheese ripened by spiking with *Penicillium roqueforti*.

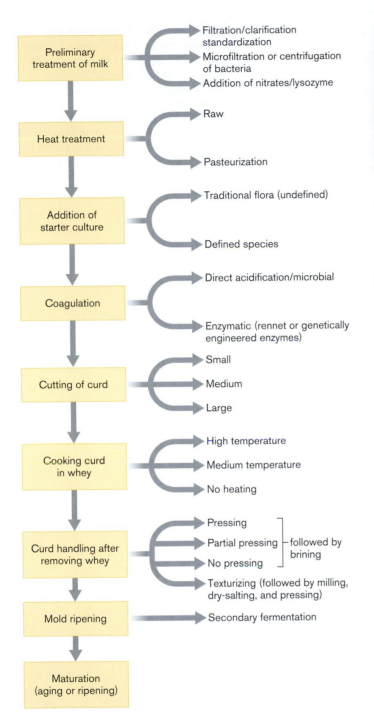

Figure 16.9 **Flow chart for cheese production.** The alternative procedures shown on the right produce different varieties.

A. Milk fermentation

B. Cheddaring the curd

C. Shaping the curd

D. Brining

Figure 16.10 **Cheese production.** Gouda cheese is produced at the Henri Willig factory, the Netherlands. **A.** Milk is poured into a fermentation tub with a bacterial starter culture and rennet. **B.** The milk curd is cut, or "cheddared." **C.** The curds are shaped into round molds, then pressed to remove whey. **D.** The solidified curds are floated in brine to form the characteristic rind of Gouda cheese. The cheese then dries and ripens on the shelf.

Flavor generation in cheese. In all fermented foods, microbial metabolism generates by-products that confer a characteristic aroma and flavor. In some cases, particular species confer distinctive flavors; for example, *Propionibacterium* ferments lactate or pyruvate to propionate, a distinctive flavor component of Swiss cheese. All bacteria generate a surprising range of side reactions, forming trace products that confer distinctive flavors. For example, while *Lactobacillus* converts most of the lactose to lactic acid, a small fraction of the pyruvate is converted to acetoin, acetaldehyde, or acetic acid, which contribute flavor. Most amino acids are retained intact; but traces are converted to flavorful alcohols, esters, and sulfur compounds (**Fig. 16.11**). For example, methanethiol (CH_3SH) contributes to the desirable flavor of cheddar cheese. Lipids are not significantly metabolized by *Lactobacilli*, but in mold-ripened cheeses such as Camembert, *Penicillium* oxidizes a small amount of the lipids to flavorful methylketones, alcohols, and lactones.

THOUGHT QUESTION 16.3 In an outbreak of listeriosis from unpasteurized cheese, only the refrigerated cheeses were found to cause disease. Why would this be the case?

THOUGHT QUESTION 16.4 Cow's milk contains 4% lipid (butterfat). What happens to the lipid during cheese production?

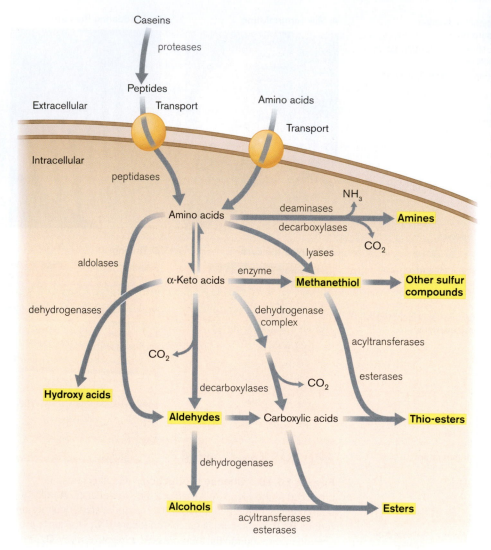

Figure 16.11 Flavor generation from amino acid catabolism. Casein catabolism generates flavor molecules (highlighted). Extracellular enzymes break down casein into peptides and amino acids, which are taken into the bacterial cell by membrane transporters. The amino acids are fermented to volatile alcohols and esters. In some cases, they combine with sulfur to form methanethiol and other sulfur-containing odorants characteristic of cheese.

Acid Fermentation of Vegetables

Many kinds of vegetable products are based on microbial fermentation. Commercial products marketed globally include pickles, soy sauce, and sauerkraut. Other products provide staple foods for particular nations or regions, such as Indonesian tempeh and Korean kimchi.

Soy fermentation. Soybeans offer one of the best sources of vegetable protein and are indispensable for the diet of millions of people, particularly in Southeast Asia. In North America, soy products are important for vegetarian consumption and as a milk substitute, as well as for animal feed. But soybeans also contain substances that decrease their nutritive value. Phytate, or inositol hexaphosphate, chelates minerals, such as iron, inhibiting their absorption by the intestine. Lectins are plant glycoproteins that bind to cell surface glycoproteins within the human body. At high concentration, lectins may upset digestion and induce autoimmune diseases. Protease inhibitors interfere with

chymotrypsin and trypsin, thus decreasing the amount of protein that can be obtained from the soy food.

All of these drawbacks of soybeans are diminished by microbial fermentation, while the protein content remains comparable to that of the unfermented bean (40%). A variety of fermented soy foods have been developed. Most soy fermentation involves mold growth, supplemented by bacteria that contribute vitamins, including vitamin B_{12}.

A major fermented soy product is **tempeh**, a staple food of Indonesia, the world's fifth most populous country, as well as of other countries of Southeast Asia (**Fig. 16.12**). Tempeh consists of soybeans fermented by *Rhizopus oligosporus*, a common bread mold. Besides decreasing the negative factors of soy, the mold growth breaks down proteins into more digestible peptides and amino acids. During World War II, tempeh was fed to American prisoners of war held by the Japanese. The tempeh was later credited with saving the lives of prisoners whose dysentery and malnutrition impaired their ability to absorb intact proteins.

Figure 16.12   Tempeh, a mold-fermented soy product.   A. Fried tempeh. **B.** *Rhizopus oligosporus* mold, used to make tempeh.

A.

Unicorn Productions

B.

Gregory G. Dimijan/Photo Researchers, Inc.

Tempeh is commonly produced in home-based factories in Indonesia. The soybeans are soaked in water overnight, allowing initial fermentation by naturally present lactic acid bacteria; in some cases, a crude "starter" may be introduced from the water of soybeans soaked previously. This pre-fermentation allows bacterial generation of vitamins and produces mild acid that promotes growth of mold. The soaked beans are then dehulled, cooked, and cooled to room temperature for inoculation with *R. oligosporus* spores or with a previous tempeh culture. The inoculated beans are wrapped in banana leaves or in perforated plastic bags, then allowed to incubate for two days. The mold grows as a white mycelium that permeates the beans, joining them into a solid cake. The final product has a mushroom-like taste and is often served fried or grilled like a hamburger.

Other soy products undergo acidic fermentation by the mold *Aspergillus oryzae*. The Japanese condiment **miso** is made from ground soy and rice, salted and fermented for two months by *A. oryzae*. **Soy sauce** is made from jiang, a Chinese condiment similar to miso in which the rice starter culture is replaced by wheat. The fermentation generates glutamic acid, a flavor-enhancing compound known popularly in the form of its salt—monosodium glutamate, or MSG.

Fermentation of cabbage and other vegetables. Various leaf vegetables are fermented by traditional societies, originally as a means of storage over the winter months. In Europe and North America, the best-known fermented products include sauerkraut and pickles. Sauerkraut production involves heterolactic fermentation by *Leuconostoc mesenteroides*. In heterolactic fermentation, each fermented sugar molecule yields lactic acid as well as ethanol and carbon dioxide. The culture is used to inoculate shredded cabbage, which is layered in alternation with salt. The salt helps limit the species and extent of microbial growth. A similar brine-enhanced fermentation process is used to pickle cucumbers, olives, and other vegetables.

An important food based on brine-fermented cabbage is Korean **kimchi** (**Fig. 16.13**). Kimchi is prepared from Chinese cabbage, salted and layered with radishes, peppers, onions, and other vegetables. The vegetables are covered with a paste of fish, rice, and chili peppers. Pickled seafoods such as shrimp or oysters may be included. The entire mixture is stored in a pot, traditionally buried underground for several months. The main fermentation organism is *Leuconostoc mesenteroides*, although *Streptococcus* and *Lactobacillus* species participate.

One of the most complex of all food fermentations is that of cocoa beans. Cocoa and coffee beans both require fermentation within the juice of the fruit before the beans are dried and processed. In the case of cocoa, at least three different kinds of fermentation must be performed by three different kinds of microorganisms (see **Special Topic 16.1**).

Alkaline Fermentation: Natto and Pidan

In Western countries, food-associated fermentation is almost synonymous with acidification. In Southeast Asia and in Africa, however, many food products involve

A.

B.

Alamy

Courtesy of David Jemison/*Life in Asia*

Figure 16.13   Kimchi.   A. To make kimchi, cabbage leaves are layered alternating with chili paste containing salted fish and vegetables. **B.** The salted layers are packed to be buried and aged for two months.

Special Topic 16.1   Chocolate: The Mystery Fermentation

Chocolate, the product of the cocoa bean *Theobroma cacao*, or "food of the gods," requires one of the most complex fermentations of any food. For all the commercialization of chocolate production, totaling 2.5 billion kilos per annum worldwide, no "starter culture" has yet been standardized to ferment the cocoa bean. Instead, the beans harvested in Africa or South America are heaped in mounds upon plantain leaves for fermentation by indigenous microorganisms, essentially the same way cocoa has been processed for thousands

of years (**Fig. 1**). The fermentation must occur immediately where the beans are harvested; they cannot be exported and fermented later.

The microbial fermentation actually occurs outside the cocoa bean, within the pulp that clings to the beans after they are removed from the cocoa fruit. The pulp contains 1% pectin, a complex branched polysaccharide, plus a rich supply of amino acids and minerals. These nutrients support growth of many kinds of microbes. Brazilian microbiologist Rosane Schwan, of

A.

Superstock

B.

ICCO photo

Figure 1   Cocoa beans.
A. Cocoa fruit, showing beans encased in mucilage. **B.** Heap of beans covered by plantain leaves. Fruit pulp ferments, liquefies, and drains away, while the beans acidify and turn brown.

A.

Courtesy of Rosane Schwan

B.

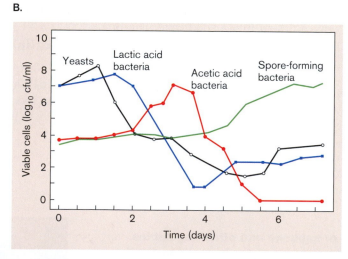

Figure 2   Microbial succession during cocoa pulp fermentation.   A. Rosane Schwan harvests cocoa beans for her research on cocoa fermentation. **B.** Yeasts generate alcohol, which lactic acid bacteria convert to lactate, and acetic acid bacteria oxidize to CO_2. *Source*: B. From R. Schwan.

increased pH. Such fermentations typically release small amounts of ammonia, which raises the pH to about 8, retarding growth of all but alkali-tolerant bacteria. The fermenting bacteria are usually *Bacillus*, aerobic species tolerant of moderate alkali and capable of extensive proteolysis and amino acid decomposition. The fermentation needs to be controlled to limit the loss of protein content, but the end result is a highly stable food product.

Natto. The Japanese soybean product **natto** is prepared by a process similar to that for tempeh. Soybeans are washed, preformented, and cooked briefly before incubation with the starter organism *Bacillus natto*. The fermenting beans are incubated in a shallow, ventilated container at a slightly raised temperature (40°C).

B. natto secretes numerous extracellular enzymes, including proteases, amylases, and phytases. These enzymes

the Federal University of Lavras, Brazil, analyzed the microbial community (**Fig. 2**). Schwan defined three stages of succession: yeasts, lactic acid bacteria, and acetic acid bacteria.

Yeast fermentation (anaerobic). The pulp initially is full of citric acid (pH 3.6). The acidity favors growth of yeasts, including *Candida, Kloekera,* and *Saccharomyces.* The yeasts consume citric acid, increasing pH. They also degrade pectin into glucose and fructose, allowing the pulp to liquefy.

As the liquefied pulp drains, its remaining sugars are fermented to ethanol, CO_2, and acetate. These products eventually rise to levels that inhibit yeast growth. The ethanol and acetate penetrate the bean embryo, preventing germination and disrupting cell membranes. The disrupted cells release enzymes that generate key flavor molecules of chocolate by hydrolysis of proteins to produce hydrophobic peptides. Other products include psychoactive molecules, such as theobromine, a stimulant similar to caffeine, and 2-phenylethylamine, a neurotransmitter associated with the pleasure response (**Fig. 3**).

Lactic acid bacteria (anaerobic). The consumption of citric acid by yeast increases the pH to pH 4.2, encouraging growth of lactic acid bacteria. *Lactobacillus* species convert citric acid to acetate, CO_2, and lactic acid. As fermentable substrates disappear, however, lactic acid bacteria are inhibited.

Acetic acid bacteria (aerobic). After one or two days, the beans are overturned and mixed periodically to permit access to oxygen. Aerobic bacteria such as *Acetobacter* now convert the ethanol and acids to CO_2. The consumption of acids neutralizes undesirable acidity. The bean is also penetrated by oxygen, which oxidizes key cocoa components, such as polyphenols. Polyphenol oxidation generates the brown color of cocoa and contributes flavor.

Oxidative respiration within the mound of beans generates heat faster than it can dissipate, causing temperature to rise as high as 50°C, which halts fermentation. After fermentation and pulp drainage, the beans are dried and roasted. The roasting process completes the transformation of cocoa substances that contribute flavor. Without the preceding fermentation process, no flavor would develop. Cocoa liquor and cocoa butter are extracted from the beans, then recombined with sugar and other components to make "cocoa mass" (**Fig. 4**). The cocoa mass is stirred for several days to achieve a smooth texture; then it is molded into the decorative forms known as chocolate.

The quality of chocolate rests largely on the original process of pulp fermentation, whose details remain poorly understood. Schwan is working to develop a defined starter culture, a community of microbes to produce a predictable high quality of flavor. At present, however, one of the world's most refined and highly prized commercial food products still depends on the indigenous microbial fermentation of cocoa pulp.

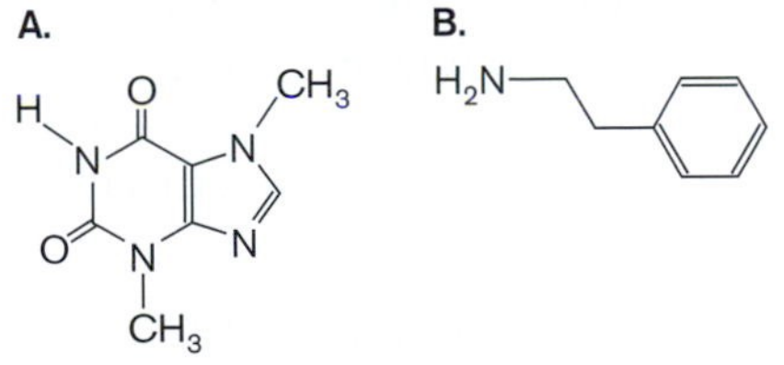

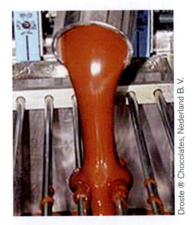

Droste ® Chocolates, Nederland B. V.

Figure 4. Chocolate manufacture. Cocoa mass contains cocoa butter and liquor extracted from the cocoa beans, mixed with sugar and other ingredients.

A.   **B.**

Figure 3   Psychoactive products of cocoa fermentation. A. Theobromine, a stimulant. **B.** 2-Phenylethylamine, an antidepressant.

decrease the undesirable components of soy, such as phytates and lectins, while liberating more easily digestible peptides and amino acids. Some of the amino acids are deaminated, generating ammonia; a well-ventilated natto chamber allows most of this gas to escape. In addition, *B. natto* synthesizes extracellular polymers such as polyglutamate (a peptide chain consisting exclusively of glutamic acid residues). Polyglutamate generates long elastic strings that bind the beans together. The stretching of these strings from chopsticks is considered a sign of a good natto (**Fig. 16.14A**).

Alkaline fermented vegetables. In Africa, numerous vegetable products are based on alkaline fermentation predominantly by *Bacillus* species. An example is dawadawa, a paste of fermented locust beans common in western Africa. The locust beans are washed and supplemented with potash, or potassium hydroxide, originally obtained from wood ashes. This addition of alkali retards growth of bacteria other than *Bacillus* species, which predominate at higher pH. The beans are fermented by indigenous bacteria (bacteria already present in the beans). The fermented beans are sun-dried, releasing most of the ammonia, and pounded into cakes for storage. Similar alkali-fermented vegetables include ogiri, from melon seeds, and ugba, from oil beans.

Pidan. An ancient means of preserving eggs has produced the famous Chinese delicacy pidan, or "thousand-year egg," now a favorite at dim sum restaurants (**Fig. 16.14B**). To make pidan, duck eggs are covered with a mixture of brewed tea, lime (CaO), and sodium carbonate (Na_2CO_3). The lime and sodium carbonate react to form NaOH, which penetrates the eggshells, raising pH and coagulating the egg white proteins. The eggs are buried in mud for several months, during which time the combined action of alkali and *Bacillus* fermentation generates dark colors and interesting flavors. Hydrogen sulfide and acetaldehyde contribute to flavor generation.

A.

Courtesy of Matt Wegener

B.

Eleanor Nakama-Mitsunaga, Honolulu Star-Bulletin

Figure 16.14 Alkaline-fermented foods.   A. Natto consists of soybeans fermented by *Bacillus natto*. The fermentation generates long strings of polyglutamate. **B.** Pidan, or "thousand-year egg," consists of duck eggs coagulated by sodium hydroxide and fermented by *Bacillus* species. One egg is cut open, revealing the transformed yolk, which develops a greenish color.

- **Vegetables** are fermented and brined to make sauerkraut and pickles. Cabbage and supplementary foods are fermented and brined to make kimchi.
- **Alkaline fermented vegetables** include the soy product natto, the egg product pidan, and the locust bean product dawadawa. The main fermenting organisms are *Bacillus* species.

> **THOUGHT QUESTION 16.5**   In traditional fermented foods, without pure starter cultures, what determines the kind of fermentation that occurs?

TO SUMMARIZE:

- **Milk curd** forms by lactic acid fermentation and rennet proteolysis, rendering casein insoluble. The cleaved peptides coagulate to form a semisolid curd. The main fermentative organisms are lactic acid bacteria.
- **Cheese varieties** include unripened cheese; semihard and hard cheeses that are cooked down and ripened; brined cheeses; and mold-ripened cheeses.
- **Cheese flavors** are generated by minor side products of fermentation, such as alcohols, esters, and sulfur compounds.
- **Soy fermentation** to tempeh and other products improves digestibility and decreases undesirable soy components such as phytates and lectins. The fermentative agent of tempeh is the bread mold *Rhizopus oligosporus*.

16.4 Ethanolic Fermentation: Bread and Wine

Some of our most nutritionally significant and culturally important foods, including most bread and alcoholic beverages, derive from ethanolic fermentation by yeast fungi. Ethanolic fermentation converts pyruvate to ethanol and carbon dioxide:

$$C_3H_4O_3 \longrightarrow CH_3CH_2OH + CO_2$$

The most prominent yeast used is *Saccharomyces cerevisiae*, known as baker's yeast or brewer's yeast (**Fig. 16.15**). A hardy organism, *S. cerevisiae* easily survives on a grocery shelf for home use and is genetically tractable for fundamental research. The yeast has been studied since the time of Pasteur, who used it to prove the biological basis of fermentation. *S. cerevisiae* today is a major model system of cell biology, yielding the molecular secrets of human cancer and other diseases.

Bread making depends on carbon dioxide to generate air spaces that **leaven** the dough, making its substance easier to chew and digest. The small amount of ethanol produced is eliminated during baking. For alcoholic bever-

Figure 16.15 Baker's yeast, the "champion" fermenter. A. *Saccharomyces cerevisiae* cells budding; some show bud scars (SEM). **B.** *S. cerevisiae* is used to study the function of human proteins such as α-synuclein (green fluorescence), which plays a role in Parkinson's disease. Yeast cells engineered to express one copy of the gene (left panel) show the protein normally within their cell membrane. Two gene copies (right panel) cause the protein to clump and kill the cells.

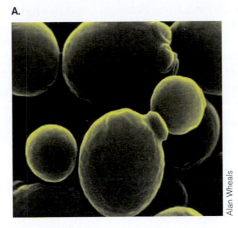
A.

Alan Wheals

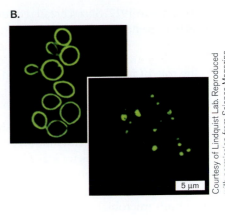
B.

5 µm

Courtesy of Lindquist Lab. Reproduced with permission from Science Magazine.

ages, however, ethanol is the key product, accompanied by carbon dioxide bubbles for "fizz," known as carbonation.

Bread Dough Is Leavened by Microbial CO_2

Bread is made in many different forms (**Fig. 16.16**) and from diverse kinds of flour, or ground grain. The earliest breads probably arose from grain mush naturally contaminated by yeast. Later, bread makers learned to include yeast left over from wine or beer production as a starter culture.

Yeast bread production. The preparation of all forms of yeast bread requires the same fundamental steps. A starter culture of yeast is included in the dough. The yeast can be commercial baker's yeast, or it can be **sourdough** starter, an undefined yeast population derived from a previous batch of dough. The dough is kneaded to develop a fine network of air pockets and allowed to rise, expanding with production of the carbon dioxide gas (**Fig. 16.17**). The finest-textured breads are made from wheat flour, which contains gluten, a substance that forms a fine molecular network supporting the rising dough.

Modern bread sold commercially is produced on an industrial scale. The dough is made in huge vats, then cut into regular chunks that are placed into the bread pans. The loaves are baked under hot-air convection, a system that greatly decreases the baking time. In the United States, most commercial bread is sliced mechanically, an invention that

A.

Susan C. Bourgoin/Food Pix/Jupiter Images

B.

Susan C. Bourgoin/Food Pix/Jupiter Images

Guy Cali and Associates/Stockford

Figure 16.16 Yeast bread. Many varieties of bread are made.

Figure 16.17 Making bread. A. Kneading the dough. **B.** As yeast fermentation generates carbon dioxide gas, the dough rises.

dates back to 1912. Sliced bread requires preservative chemicals to prevent growth of mold on the exposed interior.

> **THOUGHT QUESTION 16.6** Compare and contrast the role of fermenting organisms in the production of cheese and bread.

Injera: extended fermentation. Most kinds of bread involve only a short fermentation period, just long enough to produce enough gas for leavening. A prolonged fermentation, with more extensive microbial activity, occurs in the dough for an Ethiopian bread called **injera** (**Fig. 16.18**). The high microbial content provides a substantial source of vitamins not found in quick-rising breads.

Injera is made from teff, a grain with small, round kernels that grows in Ethiopia at high altitude. Teff contains no gluten, so it cannot rise as much as wheat flour, but it

A.

Jim Sugar/Corbis

B.

Alamy

Figure 16.18 Injera. A. After three days of fermentation, the injera dough is baked in a ceramic pan upon a "Mirte" charcoal stove. **B.** Injera forms an edible tablecloth for a variety of Ethiopian foods.

makes a kind of flatbread. The dough is spread into a wide pancake, then allowed to ferment for three days, based on organisms present in the grain and air. The fermentation includes a succession of species, usually dominated by the yeast *Candida*. The extended fermentation generates a complex range of by-products that confer exceptional flavors in the baked product. Injera forms the basis of an entire Ethiopian meal, served with other food items placed upon it as an edible tablecloth. Diners wrap samples of each food in a fold of injera and consume them together.

Alcoholic Beverages: Beer and Wine

Ethanolic fermentation of grain or fruit was important to early civilizations because it provided a drink free of waterborne pathogens. Traditional forms of beer also provided essential vitamins in the unfiltered yeasts.

Ethanol is unique among fermentation products in that it provides a significant source of caloric intake, but it is also a toxin that impairs mental function. A modest level of ethanol enters the human circulation naturally from intestinal flora, equivalent to a fraction of a drink per day. The human liver produces the enzyme **alcohol dehydrogenase**, which detoxifies ethanol. This enzyme in a healthy liver can metabolize small amounts of alcohol without harm. However, excess alcohol consumption can overload the liver's capacity for detoxification and permanently damage the liver and brain.

Beer: alcoholic fermentation of grain. Beer production is one of the most ancient fermentation practices and is depicted in the statuary of ancient Egyptian tombs dated to 5,000 years ago (**Fig. 16.19A**). The earliest Sumerian beers were made from bread soaked in water and fermented. Today, most beer is produced commercially by fermenting barley using giant vats (**Fig. 16.19B**). Production of high-quality beer involves complex processing with many steps, including the germination of barley grains, the mashing in water and cooking, and the introduction of hops for flavor (**Special Topic 16.2**).

Before fermentation, the barley starch is broken down to maltose by enzymes from the grain. Most of the maltose is fermented by yeast to ethanol and carbon dioxide. However, minor side products contribute flavors—or unpleasant off-flavors if present in too great an amount (**Fig. 16.20**). Some of the off-flavors result from the presence of oxygen, which is needed because yeast is not a true anaerobe; it requires aerobic metabolism to synthesize some of its cell components. The presence of oxygen diverts some pyruvate into acetaldehyde, most of which converts to ethanol. Some of the acetaldehyde, however, remains unconverted, and some is converted to diacetyl. Both acetaldehyde and diacetyl cause off-flavors.

Most pyruvate undergoes ethanolic fermentation, but a small fraction is drawn off to make amino acids via TCA cycle intermediates, as discussed in Chapter 15. The 2-oxo

Figure 16.19 Beer production: ancient and modern. A. Making beer in ancient Egypt, circa 3000 BCE. The mash is stirred in earthen jars. **B.** Fermenters in a modern brewery.

acids of the TCA cycle are analogous to pyruvate, with the methyl group replaced by extended carbon chains (R group). A tiny amount of the 2-oxo acids is converted to long-chain alcohols, which add desirable flavor to beer.

THOUGHT QUESTION 16.7 Compare and contrast the role of low-concentration by-products in the production of cheese and beer.

Wine: alcoholic fermentation of fruit. The fermentation of fruit gives rise to wine, another class of alcoholic products of enormous historical and cultural significance. Grapes produce the best-known wines, but wines and distilled liquors are also made from apples, plums, and other fruits. The key difference between fermentation of fruits and fermentation of grains is the exceptionally high monosaccharide content in fruits. Grape juice, for example, can contain concentrations of glucose and

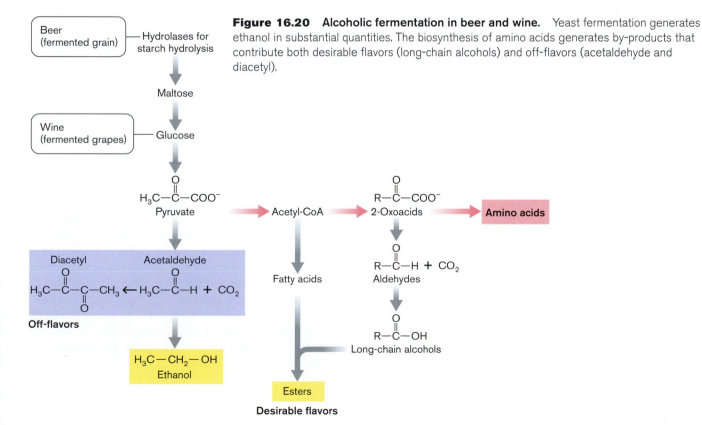

Figure 16.20 Alcoholic fermentation in beer and wine. Yeast fermentation generates ethanol in substantial quantities. The biosynthesis of amino acids generates by-products that contribute both desirable flavors (long-chain alcohols) and off-flavors (acetaldehyde and diacetyl).

Special Topic 16.2   Beer Is Made from Barley and Hops

Beer production encompasses five main stages (**Fig. 1**). These include malting, the germination of barley grains; mashing, a stepwise heating process to promote starch hydrolysis; wort boiling with hops; fermentation; and postfermentation treatments.

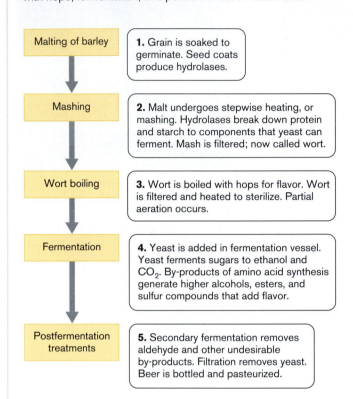

Malting of barley	**1.** Grain is soaked to germinate. Seed coats produce hydrolases.
Mashing	**2.** Malt undergoes stepwise heating, or mashing. Hydrolases break down protein and starch to components that yeast can ferment. Mash is filtered; now called wort.
Wort boiling	**3.** Wort is boiled with hops for flavor. Wort is filtered and heated to sterilize. Partial aeration occurs.
Fermentation	**4.** Yeast is added in fermentation vessel. Yeast ferments sugars to ethanol and CO_2. By-products of amino acid synthesis generate higher alcohols, esters, and sulfur compounds that add flavor.
Postfermentation treatments	**5.** Secondary fermentation removes aldehyde and other undesirable by-products. Filtration removes yeast. Beer is bottled and pasteurized.

Figure 1   Process of beer production.   Barley is malted, milled, and mashed to break down long-chain sugars and proteins into short chains and monomers that yeast can digest, forming wort. The wort is fermented by yeast to make beer.

Malting of barley. The barley grains are malted, or soaked in water, to encourage germination (**Fig. 2**). During germination, the endosperm (stored food) of the grain secretes gibberellin, a hormone that stimulates the growth of rootlets and the emerging stem. Gibberellin induces the aleurone (lining of the endosperm) to produce hydrolases, enzymes that break down starch to maltose and proteins to amino acids for use by the growing plant. The hydrolases actually become activated during the second stage, called mashing, when the grains are crushed and stirred in huge vats of water.

Mashing. While the grain is mashed, the temperature is raised in steps, each of which optimizes the activity of a different hydrolase (**Fig. 3**). At 52°C, the protein hydrolases are activated. Then at 68°C, the starch hydrolases convert long-chain sugars to the disaccharide maltose, which can support yeast fermentation. The final temperature (77°C) inactivates all enzymes; then the mash is cooled, pressed, and filtered. The liquid filtrate of the mash is called wort.

Wort boiling. The wort is supplemented with hops, a traditional herb used for centuries in Europe to contribute a distinctive flavor of beer. After boiling with hops, the wort is again filtered.

Fermentation. The wort is inoculated with a special strain of *Saccharomyces cerevisiae* known as brewer's yeast. The yeast conducts ethanolic fermentation on the maltose from hydrolyzed starch. At the same time, minor by-products, such as long-chain alcohols, impart good flavors. The time of fermentation is a key factor in the quality of the beer.

---

fructose as high as 15%. With simple sugars available, yeast can set to fermentation immediately, without need for preliminary breakdown of long-chain carbohydrates, as in the malting and mashing of beer.

Most modern wine production uses strains of the grape *Vitis vinifera*. The grapes are crushed to release juices, usually in the presence of antioxidants such as sulfur dioxide (**Fig. 16.21**). For white wine, the skins are removed before juice is fermented. For red wine, the skins are included in early fermentation to extract the red and purple anthocyanin pigments as well as phenolic flavor compounds. The first few days of fermentation are dominated by indigenous species of yeast naturally present on the grapes, such as *Kloeckera* and *Hanseniaspora* species. Commercial producers usually inoculate with standard *S. cerevisiae*, whose population dominates the late stage

of fermentation (6–20 days). The yeast growth ends once the ethanol level reaches about 15%; to achieve higher alcohol content, distillation is required.

After fermentation, the wine is drained, or "racked," from the sediment of grape and yeast material, the lees. The liquid may be further clarified by centrifugation. Then it is stored for 2–3 weeks in tanks or barrels. During storage, a second stage of fermentation may be performed, called **malolactic fermentation**. In malolactic fermentation, the wine is seeded with *Oenococcus oeni* bacteria, which ferment L-malate (deprotonated L-malic acid), a side product from glucose fermentation. The L-malate is decarboxylated to L- or D-lactate:

$$^-OOC-CHOH-CH_2-COO^- \longrightarrow$$
$$CH_3-CHOH-COO^- + CO_2$$

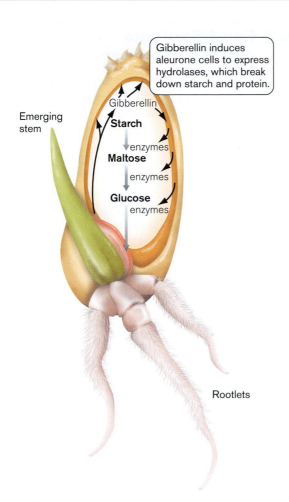

Figure 2 **Malting the grain.** Malting involves germination of the grain to induce expression of hydrolase enzymes.

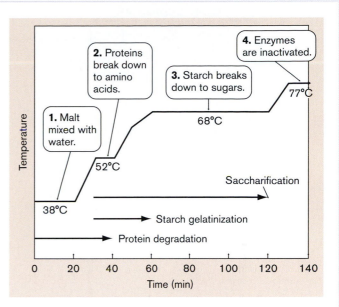

Figure 3 **Mashing the wort.** During mashing, stepwise heating activates the hydrolases to break down carbohydrates and proteins.

Postfermentation treatment. This includes filtering of the wort to remove the bulk of the yeast. The product, now recognizably beer, still contains undesirable levels of acetaldehyde and diacetyl generated by partial oxidation. These oxidized by-products can be reduced by the few remaining yeast cells during a period of secondary fermentation. During secondary fermentation, oxygen is completely excluded and the temperature is decreased to 15°C or lower. The best German beers are aged at 2°C for several months.

By removal of a carboxylate group as CO_2, malolactic fermentation decreases acidity of the wine, improving flavor. It also removes a potential carbon source whose later fermentation by undesirable organisms could spoil the wine.

As in the case of beer production, yeast fermentation of wine produces numerous minor products contributing flavor, such as long-chain alcohols and esters. At the same time, overgrowth of yeast or the growth of undesired species can produce excess amounts of these compounds, such as sulfides and phenolics, giving rise to off-flavors. Some undesired species require oxygen exposure, whereas others can grow during storage and bottling. The balance of microbial populations is challenging to control and has a major role in determining the quality of a given wine vintage.

TO SUMMARIZE:

- **Bread is leavened** by yeasts conducting limited ethanolic fermentation, producing enough carbon dioxide gas to expand the dough.
- **Injera** bread dough undergoes more extensive fermentation by indigenous organisms, and as a result generates multiple flavors.
- **Beer** derives from alcoholic fermentation of grain. Barley grains are germinated, allowing enzymes to break down the starch to maltose for yeast fermentation.
- **Secondary products of grain fermentation**, such as long-chain alcohols and esters, generate the special flavors of beer.
- **Wine** derives from alcoholic fermentation of fruit, most commonly grapes. The grape sugar (glucose) is

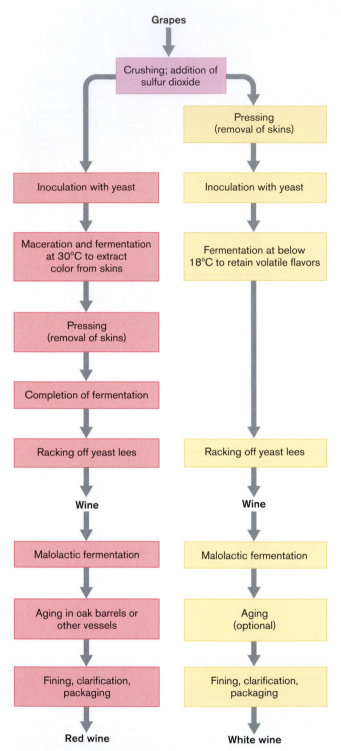

Grapes

↓

Crushing; addition of sulfur dioxide

↓ (left branch)   ↓ (right branch)

Right branch (white wine):

Pressing (removal of skins)

↓

Inoculation with yeast

↓

Fermentation at below 18°C to retain volatile flavors

↓

Racking off yeast lees

↓

Wine

↓

Malolactic fermentation

↓

Aging (optional)

↓

Fining, clarification, packaging

↓

White wine

Left branch (red wine):

Inoculation with yeast

↓

Maceration and fermentation at 30°C to extract color from skins

↓

Pressing (removal of skins)

↓

Completion of fermentation

↓

Racking off yeast lees

↓

Wine

↓

Malolactic fermentation

↓

Aging in oak barrels or other vessels

↓

Fining, clarification, packaging

↓

Red wine

Figure 16.21   Production of red and white wines. For red wine, the grapes are fermented with the skins at a temperature that increases extraction of color and tannins. For white wine, the skins are removed before addition of yeast starter, and the temperature is kept low to retain volatile flavors. Both kinds of wine usually undergo malolactic fermentation by *Oenococcus oeni*, special lactic acid bacteria that consume malic acid. *Source:* G. H. Fleet. Wine. Chapter 37 in *Food Microbiology*, ASM Press.

fermented by yeast to alcohol. A secondary product, malate, undergoes malolactic fermentation by *Oenococcus oeni* bacteria.

16.5 Food Spoilage and Preservation

We humans have always competed with microbes for our food. When early humans killed an animal, microbes commenced immediately to consume its flesh. Because meat perished so fast, it made economic sense to share the kill immediately and consume all as soon as possible. Vegetables might last longer, but eventually succumbed to mold and rot. Later societies developed preservation methods, such as drying, smoking, and canning, that enabled humans to survive winters and dry seasons on stored food.

Modern food preservation depends on **antimicrobial agents**, chemical substances that either kill microbes or slow their growth, as well as physical preservative measures, such as treatment with heat and pressure. These general principles of microbial control are described in Chapter 5. Here we focus on microbial contamination and food preservation from the perspective of the food industry.

Food Spoilage and Food Contamination

After food is harvested, several kinds of chemical changes occur. Some begin instantly, whereas others take several days to develop. Some changes may be considered desirable, such as meat tenderizing, whereas others, such as putrefaction, render food unfit for consumption. The major classes of food change include:

■ **Enzymatic processes.** Following the death of an animal, its flesh undergoes proteolysis by its own enzymes. Limited proteolysis tenderizes meat. Plants after harvest undergo other changes; for example, in harvested corn, the sugar rapidly converts to starch.

■ **Chemical reactions with the environment.** The most common abiotic chemical reactions involve oxidation by air—for example, lipid autoxidation, which generates rancid odors.

■ **Microbiological processes.** Microbes from the surface of the food begin to consume it, some immediately, others later in succession, generating a wide range of chemical products. In meat, internal organs of the digestive tract are an important source of microbial decay.

Microbial activity can aid food production, but it can also have various undesirable effects. Two different classes of microbial effects are distinguished: food spoilage and food contamination with pathogens.

Food spoilage refers to microbial changes that render a product obviously unfit or unpalatable for consumption.

For example, rancid milk or putrefied meats are unpalatable and contain metabolic products that may be deleterious to human health, such as oxidized fatty acids or organic amines. Even in these cases, however, the definition of spoilage partly depends on cultural practice. What is sour milk to one person may be buttermilk to another; what one society considers spoiled meat, another considers merely aged.

Different pathways of microbial metabolism lead to different kinds of spoilage. Sour flavors result from acidic fermentation products, as in sour milk. Alkaline products generate bitter flavor. Oxidation, particularly of fats, causes **rancidity**, whereas general decomposition of proteins and amino acids leads to **putrefaction**. The particularly noxious odors of putrefaction derive from amino acid breakdown products that often have apt names, such as the amines cadaverine and putrescine and the aromatic product skatole.

Food contamination, or **food poisoning**, refers to the presence of microbial pathogens that cause human disease; for example, rotaviruses that cause gastrointestinal illness. Pathogens usually go unnoticed as food is consumed because their numbers are very low, and they may not even grow in the food. Even the freshest-appearing food may cause serious illness if it has been contaminated with a small number of pathogens.

How Food Spoils

Different foods spoil in different ways, depending on their nutrient content, the microbial species, and environmental factors such as temperature. **Table 16.2** summarizes common forms of spoilage.

Dairy products. Milk and other dairy products contain carbon sources, such as lactose, protein, and fat. In fresh milk, the nutrient most available for microbial catabolism is lactose, which commonly supports anaerobic fermentation to sour milk. Fermentation by the right mix of microbes, however, leads to yogurt and cheese production, as previously described.

Under certain conditions, bitter off-flavors may be produced by bacterial degradation of proteins. The release of amines causes a rise in pH. Protein degradation is most commonly caused by psychrophiles, species that grow well at cold temperatures, such as those of refrigeration.

Cheeses are less susceptible than milk to general spoilage because of their solid structure and lowered water activity. However, cheeses can grow mold on their surface. Historically, the surface growth of *Penicillium* strains led to invention of new kinds of cheeses. Other kinds of mold, however, such as *Aspergillus*, produce toxins and undesirable flavors.

Meat and poultry. Meat in the slaughterhouse is easily contaminated with bacteria from hide, hooves, and intestinal contents. Muscle tissue offers high water content, which supports microbial growth as well as rich nutrients, including glycogen, peptides, and amino acids. The breakdown of peptides and amino acids produces the undesirable odorants that define spoilage (for example, cadaverine and putrescine).

Meat also contains fat, or adipose tissue, but the lipids are largely unavailable to microbial action because they consist of insoluble fat (triacylglycerides). Instead, meat lipids commonly spoil abiotically by autoxidation (reaction with oxygen) of unsaturated fatty acids, independent of microbial activity. Thus, when meat is exposed to air during storage, it turns rancid, particularly meats such as pork, which contains highly unsaturated lipids. Autoxidation can be prevented by anaerobic storage, such as vacuum packing, which also prevents growth of aerobic microorganisms. The absence of the suppressed organisms, however, favors growth of lactic acid bacteria and anaerobes such as *Brochothrix thermosphacta*. These organisms generate short-chain fatty acids, which taste sour.

In industrialized societies, the most significant determinant of microbial populations in meat spoilage is the practice of refrigeration. Refrigeration prolongs the shelf life of meat because contaminating microbes are predominantly mesophilic (grow at moderate temperatures, as discussed in Chapter 5). But ultimately, the few psychrotrophs initially present do grow; typically, these are *Pseudomonas* species. The pseudomonads are also favored by the low pH of meat (pH 5.5–7.0), which results from the accumulation of lactic acid in the muscle.

Seafood. Fish and other seafood contain substantial amounts of protein and lipids, as well as amines such as trimethylamine oxide. Fish spoils more rapidly than meat and poultry for several reasons. First, fish do not thermoregulate, and they inhabit relatively low-temperature environments. Because fish grow in low-temperature environments, their surface microorganisms tend to be psychrotrophic and thus grow well under refrigeration. In addition, marine fish contain high levels of the osmoprotectant trimethylamine oxide (TMAO), which bacteria reduce to trimethylamine, a volatile amine responsible for the "fishy" smell of seafood. Finally, the rapid microbial breakdown of proteins and amino acids leads to foul-smelling amines and sulfur compounds, such as hydrogen sulfide and dimethyl sulfide.

> **THOUGHT QUESTION 16.8** Why would bacteria convert trimethylamine oxide (TMAO) to trimethylamine? Would this kind of spoilage be prevented by exclusion of oxygen?

Table 16.2 Food spoilage (examples).

Food product	Signs of spoilage	Microbial cause
Milk and dairy products		
Milk	Sour flavor	Lactic acid bacteria produce lactic and acetic acids.
Milk	Coagulation	Lactic acid bacteria produce proteases that destabilize casein and lower pH, causing coagulation.
Milk	Bitter flavor	Psychrophilic bacteria degrade proteins and amino acids.
Cheese	Open texture, fissures	Lactic acid bacteria produce carbon dioxide.
Cheese	Discoloration and colonies	Molds such as *Penicillium* and *Aspergillus* grow on the cheese.
Meat and poultry		
Meat and poultry	Rancid flavor	Psychrotrophic bacteria produce free fatty acids, which become oxidized.
Meat and poultry	Putrefaction	*Pseudomonas* and other aerobes degrade amino acids, producing amines and sulfides.
Meat and poultry	Discolored patches	Molds such as *Mucor* and *Penicillium* grow on the surface.
Eggs	Pink or greenish egg white	*Pseudomonas* and related bacteria grow on albumin, producing water-soluble pigments.
Eggs	Sulfurous odor	Bacterial growth on albumin releases hydrogen sulfide.
Seafood		
Fish	Fishy smell	Anaerobic psychrophiles such as *Photobacterium* convert trimethylamine oxide (TMAO) to trimethylamine.
Fish	Odor of putrefaction	*Pseudomonas* and other gram-negative species degrade amino acids, producing amines and sulfides.
Shellfish	Odor of putrefaction	*Vibrio* and other marine bacteria decompose the protein.
Fruits, vegetables, and grains		
Plants before harvest	Rotting or wilting	Plant pathogens, most commonly fungi such as *Alternaria, Aspergillus,* and *Penicillium*
Stored plant foods	Rotting or wilting	Molds or bacteria produce degradative enzymes, such as pectinases and cellulases.
Apples, pears, cherries	Geosmin off-flavor	*Penicillium* mold
Peeled oranges	Discoloration and off-flavor	*Enterobacter* and *Pseudomonas* spp.
Pasteurized fruit juices	Medicine-like phenolic off-flavor	Acid- and heat-tolerant spore former, *Alicyclobacillus* sp., produces 2-methoxyphenol (guaiacol).
Bread	Ropiness	*Bacillus* spp.
Bread	Red discoloration	*Serratia marcescens*

Plant foods. Fruits, vegetables, and grains spoil differently than animal foods because of their high carbohydrate content and their relatively low water content. The low water content of plant foods usually translates into considerably longer shelf life than for animal-based foods. Carbohydrates favor microbial fermentation to acids or alcohols that limit further decomposition, and this microbial action can be managed to produce fermented foods, as described in Section 16.2.

A distinctive role is assigned to plant pathogens, which rarely infect humans but may destroy the plant before harvest. Most plant pathogens are fungi, although some are bacteria, such as *Erwinia* species. Historically, plant pathogens have caused major agricultural catastrophes, such as the Irish potato famine, caused by a fungus-like pathogen. Plant pathogens continue to devastate local economies and cause shortages worldwide; a contemporary example is the witches' broom fungus *Crinipellis perniciosa*, which causes a serious fungal disease of cocoa trees that has drastically cut Latin American cocoa production.

After harvest, various molds and bacteria can soften and wilt plant foods by producing enzymes that degrade the pectins and celluloses that give plants their structure. In general, the more processed the food, the greater the opportunities for spoilage. For example, citrus fruits generally last for several weeks, but peeled oranges are susceptible to spoilage by gram-negative bacteria.

Baked bread usually resists spoilage except for surface molds. However, in rare cases, improperly baked bread can show contamination. The appearance of red bread, caused by the red bacterium *Serratia marcescens*, is believed to have been responsible for the "blood" observed in communion bread during a Catholic mass in the Italian town of Bolsena in 1263, an event that became known as the Miracle of Bolsena.

Human Pathogens Spread through Food

The U.S. Centers for Disease Control and Prevention estimate that there are 76 million cases of gastrointestinal illness a year, usually spread through water or food. Thus, one in four Americans experiences gastrointestinal illness in a given year. Intestinal pathogens spread readily because many pathogens can be transmitted through food without any outward sign that the food is spoiled.

Food-borne pathogens typically arise from a range of sources; see, for example, the diagram of transmission of *Listeria monocytogenes*, a psychrotrophic pathogen that invades the cells of the intestinal epithelium, causing listeriosis (**Fig. 16.22**). (Psychrotrophic organisms grow optimally at moderate temperatures but also grow slowly at lower temperatures, typically 0–30°C.) *Listeria* can be transferred from soil and feed to cattle, whose manure then cycles it back to soil. From cattle, the pathogen contaminates milk and meat, where it can eventually infect human consumers. Because *Listeria* is a psychrotroph, it outcompetes other food-borne bacteria under refrigeration.

The U.S. Public Health Service judges the importance of food-borne illnesses based on their incidence and/or severity (**Table 16.3**). For example, the Norwalk-type viruses, or noroviruses, infect 180,000 Americans per year; their spread is very difficult to control, especially in close quarters, such as a cruise ship, where an outbreak can easily infect a large proportion of passengers. Fortunately, the course of illness, however unpleasant, is short, with few complications.

On the other hand, the spore-forming pathogen *Clostridium botulinum* causes only about 100 cases of botulism per year. The incidence is relatively low, but if untreated, the fatality rate is 50%. In this case, the low incidence of botulism actually enhances its danger because the condition is likely to go undiagnosed.

What distinguishes a pathogen from a spoilage organism? Pathogens possess highly specific mechanisms for host colonization, as discussed in Chapter 25. **Figure 16.23** shows intestinal crypt cells covered with *Escherichia coli* O157:H7 bacteria, an emergent pathogen first recognized in fast-food hamburgers.

Bacterial factors that contribute to disease are often encoded together in the genome in a region known as a **pathogenicity island**. A pathogenicity island consists of a set of genes and operons that function coordinately (as discussed in Chapter 9). The colocalization of the genes enables transfer of virulence capability to other species as pathogens evolve. **Figure 16.24** shows an example of a pathogenicity island in *Salmonella*. Four of its operons contribute to the Type III secretion complex, which

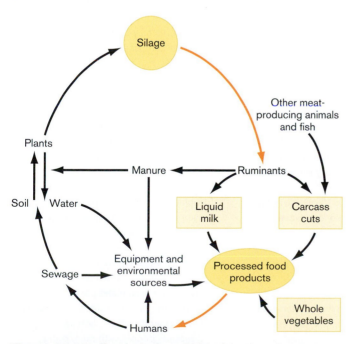

Figure 16.22 Transmission of *Listeria monocytogenes*. Transmission occurs through various routes, including passage through food products. Colored arrows indicate transmission of disease.

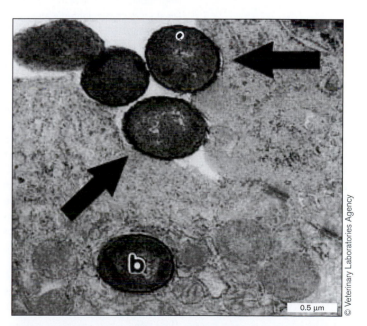

Figure 16.23 Intestinal crypt cells with adherent bacteria, strain *Escherichia coli* O157:H7. A gnotobiotic (germ-free) piglet was infected with the bacteria (arrows). A thin section was stained with toluidine blue.

Table 16.3 Food-borne pathogens in the United States[a].

Pathogen	Incidence and transmission	Course of illness
Norovirus (Norwalk and Norwalk-like viruses)	Most common cause of diarrhea; also called "stomach flu" (no connection with influenza). 180,000 cases per year are estimated. Transmitted mainly by virus-contaminated food and water. Infection rates are highest under conditions of crowding in close quarters, such as inside a ship or a nursing home.	Disease lasts for 1 or 2 days. Includes vomiting, diarrhea, and abdominal pain; headache and low-grade fever may occur.
Salmonella sp.	Most common food-borne cause of death; more than 1 million cases per year; estimated 600 deaths per year. Transmission nearly always through food—raw, undercooked, or recontaminated after cooking, especially eggs, poultry, and meat; also contaminates dairy products, seafood, fruits, and vegetables.	Causes gastrointestinal disease that includes diarrhea, fever, and abdominal cramps lasting 4–7 days. Fatal cases are most common in immunocompromised patients.
Campylobacter	More than 1 million cases of campylobacteriosis per year; estimated 100 deaths per year. Grows in poultry without causing symptoms. Transmission is mainly through raw and undercooked poultry; contaminates half of poultry sold. Occurs less often in dairy products or in foods contaminated after cooking.	In humans, usually causes severe bloody diarrhea, fever, and abdominal cramps lasting 7 days. Fatal cases are most common in immunocompromised patients.
Escherichia coli O157:H7	An emerging pathogen, first recognized in a hamburger outbreak in 1982; now known to infect 73,000 yearly, including 60 deaths per year. Grows in cattle without causing symptoms. Transmitted most often through ground beef; also through unpasteurized cider and from salad items.	In humans, usually causes severe bloody diarrhea and abdominal cramps lasting 5–10 days. About 5% of patients, especially children and elderly, develop hemolytic uremic syndrome, in which the red blood cells are destroyed and the kidneys fail.
Clostridium botulinum	Causes about 100 cases per year of botulism, with a 50% fatality rate if untreated. C. botulinum grows in improperly home-canned foods, more rarely in commercially canned low-acid foods and improperly stored leftovers such as baked potatoes. Spores occur in honey, endangering infants under 2 years of age.	Botulinum toxin from growing bacteria causes progressive paralysis, with blurred vision, drooping eyelids, slurred speech, difficulty swallowing, and muscle weakness. Infant botulism causes lethargy and impaired muscle tone, leading to paralysis.
Listeria monocytogenes	Listeria bacteria grow in animals without symptoms. Animal feces may contaminate water, which is then used to wash vegetables. Transmission occurs mainly through vegetables washed in contaminated water and through soft cheeses. Listeria is psychrotrophic, growing at refrigeration temperatures.	Listeriosis involves fever, muscle aches, and sometimes gastrointestinal symptoms. In pregnant women, symptoms may be mild but lead to serious complications for the unborn child.
Shigella	Shigella infects about 18,000 people a year in the United States; in developing countries, Shigella infections are endemic in most communities. Transmission occurs through fecal-oral contact or from foods washed in contaminated water.	Shigellosis involves gastrointestinal symptoms such as diarrhea, fever, and stomach cramps, usually lasting 7–10 days. Complications are rare.
Staphylococcus aureus	S. aureus is best known as a skin infection transmitted through open wounds. However, S. aureus can also be transmitted through high-protein foods such as ham, dairy products, and cream pastries.	S. aureus causes toxic shock syndrome. Can also cause food poisoning via pre-formed toxins.
Toxoplasma gondii	T. gondii is a parasite believed to infect 60,000 people annually, most with no symptoms. In a few cases, serious disease results. T. gondii is transmitted through contact with feces of infected animals, particularly cats, or through contaminated foods such as pork.	Toxoplasmosis causes mild flu-like symptoms; but in pregnant women, its transmission to the unborn child can lead to severe neurological defects, including death. Neurological complications also occur in immunocompromised patients.
Vibrio vulnificus	V. vulnificus is a free-living marine organism that contaminates seafood or open wounds. About 40 cases per year are reported.	V. vulnificus can infect the bloodstream, causing septic shock. Mainly threatens people with preexisting conditions such as liver disease.

[a]Ten major food-borne pathogens highlighted by the U.S. Public Health Service (USHPS).

Pathogenicity island SPl2 in the *Salmonella* genome

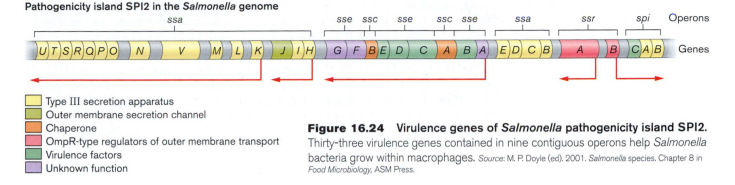

Type III secretion apparatus
Outer membrane secretion channel
Chaperone
OmpR-type regulators of outer membrane transport
Virulence factors
Unknown function

Figure 16.24   Virulence genes of *Salmonella* pathogenicity island SPl2. Thirty-three virulence genes contained in nine contiguous operons help *Salmonella* bacteria grow within macrophages. *Source*: M. P. Doyle (ed). 2001. *Salmonella* species. Chapter 8 in *Food Microbiology*, ASM Press.

secretes toxins and host colonization factors (discussed in Chapter 25). Other genes encode outer membrane proteins that counteract host defenses, as well as regulators of virulence gene expression.

The most dangerous consequence of infection by food-borne pathogens is the production of a potentially fatal toxin. For example, *E. coli* O157:H7 infection of the intestines can be overcome, but the bacteria produce Shiga toxin, which can destroy the kidneys. In cases of adult botulism, *Clostridium botulinum* does not usually grow within the patient; the botulinum toxin comes from bacteria that grew previously in improperly sterilized food. The botulinum toxin has a highly specific effect, inhibiting synaptic vesicle fusion in the terminal of peripheral motor neurons (**Fig. 16.25**). Synaptic inhibition prevents activation of muscle cells, causing paralysis. Microbial toxins are discussed further in Chapter 25.

Food Preservation: Physical and Chemical Methods

Cultural practices and cuisines have long evolved so as to limit food spoilage. Such practices include cooking (heat treatment), addition of spices (chemical preservation), and fermentation (partial microbial digestion). In modern commercial food production, spoilage and contamination are prevented by numerous methods based on fundamental principles of physics and biochemistry that limit microbial growth (discussed in Chapter 5).

Physical means of preservation include:

■ **Dehydration and freeze-drying.** Removal of water prevents microbial growth. Water is removed either by application of heat or by freezing under vacuum (known as **freeze-drying**, or **lyophilization**). Drying is especially effective for vegetables and pasta. The disadvantage of drying is that some nutrients are broken down.

■ **Refrigeration and freezing.** Refrigeration temperature (typically –2°C to 16°C) slows microbial growth, as shown in an experiment comparing bacterial growth in ground beef at different temperatures (**Fig. 16.26**). Nevertheless, refrigeration also selects for psychrotrophs. Freezing halts the growth of most microbes, but preexisting contaminant strains often survive to grow again when the food is thawed. This is why deep-frozen turkeys, for example, can still

A.

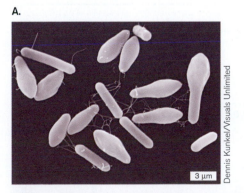

Figure 16.25   *Clostridium botulinum* produces botulinum toxin.   A. Club-shaped morphology of *C. botulinum* cells containing endospores. **B.** Botulinum toxin inhibits synaptic vesicle fusion in the terminal of a peripheral motor neuron, preventing activation of the muscle cell.

B.

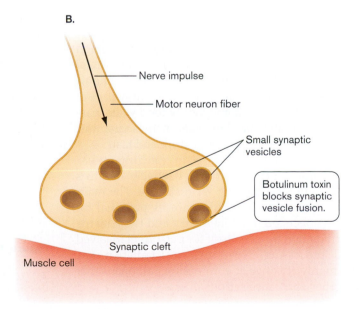

Nerve impulse

Motor neuron fiber

Small synaptic vesicles

Botulinum toxin blocks synaptic vesicle fusion.

Synaptic cleft

Muscle cell

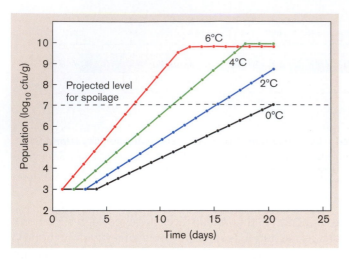

Figure 16.26 Bacterial growth in ground beef. The growth rate of total aerobic bacteria in ground beef declines at lower storage temperatures. *Source*: M. P. Doyle (ed). 2001. Meat, Poultry, and Seafood. Chapter 5 in *Food Microbiology*, ASM Press.

cause *Salmonella* poisoning, especially if the interior is not fully thawed before roasting.

■ **Controlled or modified atmosphere.** Food can be packed under vacuum, or stored under atmospheres with decreased oxygen or increased CO_2. Controlled atmospheres limit abiotic oxidation as well as microbial growth. For example, CO_2 storage is particularly effective for extending the shelf life of apples.

■ **Pasteurization.** Invented by Louis Pasteur, pasteurization is a short-term heat treatment designed to decrease microbial contamination with minimal effect on food value and texture. For example, milk is commonly pasteurized at 63°C for 30 minutes followed by quick cooling to 4°C. Pasteurization is most effective for extending the shelf life of liquid foods with consistent, well-understood microbial flora, such as milk and fruit juices.

■ **Canning.** In canning, the most widespread and effective means of long-term food storage, food is cooked under pressure to attain a temperature high enough to destroy endospores (typically 121°C). Commercial canning effectively eliminates microbial contaminants, except in very rare cases. The main drawback of canning is that it incurs some loss of food value, particularly that of labile biochemicals such as vitamins, as well as loss of desirable food texture and taste.

■ **Ionizing radiation.** Exposure to ionizing radiation, known as food irradiation, effectively sterilizes many kinds of food for long-term storage. The main concerns about food irradiation are its potential for unknown effects on food chemistry and the hazards of the irradiation process itself for personnel involved in food processing. Nevertheless, irradiation has proved highly effective at eliminating pathogens that would otherwise cause serious illness.

Often, two or more means of preservation are used in combination, such as acid treatment and refrigeration. For example, **Figure 16.27** shows results of a typical experiment measuring the effect of pH on microbial survival in refrigerated food—in this case, *E. coli* O157:H7 in Greek eggplant salad. Note the critical threshold pH required to decrease bacterial counts. At pH 4, about the pH of lemon juice, the bacteria show a steep exponential death curve, whereas at pH 4.5, the bacteria remain viable for many days.

Many kinds of chemicals are used to preserve foods. Major classes of chemical preservatives include:

■ **Acids.** While microbial fermentation can preserve foods by acidification, an alternative approach is to add acids directly. Organic acids commonly used to preserve food include benzoic acid, sorbic acid, and propionic acid. The acids are generally added as salts: sodium benzoate, potassium sorbate, sodium propionate. These acids act by crossing the cell membrane in the protonated form, then releasing their protons at the higher intracellular pH. For this reason, they work best in foods that already have moderate acidity (pH 5–6), such as dried fruits and processed cheeses.

■ **Esters.** The esters of organic acids often show antimicrobial activity, whose basis is poorly understood. Examples include fatty acid esters and parabens (benzoic acid esters). They are used to preserve processed cheeses and vegetables.

■ **Other organic compounds.** Numerous organic compounds, both traditional and synthetic, have antimicrobial properties. For example, cinnamon and cloves contain the benzene derivative eugenol, a potent antimicrobial agent.

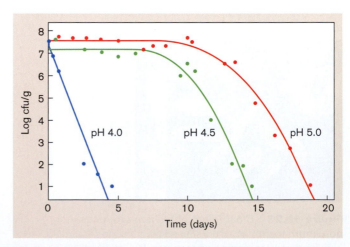

Figure 16.27 Bacteria die sooner at lower pH. Survival curves of *E. coli* O157:H7 in eggplant salad stored at 5°C at pH 4.0 (•), pH 4.5 (•), and pH 5.0 (•). *Source*: Panagiotis N. Skandamis and George-John E. Nychas. 2000. *Applied and Environmental Microbiology* 66:1646.

■ **Inorganic compounds.** Inorganic food preservatives include salts, such as phosphates, nitrites, and sulfites. Nitrites and sulfites inhibit aerobic respiration of bacteria, and their effectiveness is enhanced at low pH. These substances, however, may have harmful effects on humans; nitrites can be converted to toxic nitrosamines, and sulfites cause allergic reactions in some people.

THOUGHT QUESTION 16.9 Is it possible for physical or chemical preservation methods to completely eliminate microbes from food? Explain.

TO SUMMARIZE:

■ **Food spoilage** refers to chemical changes that render food unfit for consumption. Food spoils through degradation by enzymes within the food, through spontaneous chemical reactions, and through microbial metabolism.

■ **Food contamination, or food poisoning**, refers to the presence of microbial pathogens that cause human disease, or toxins produced by microbial growth. Food harvesting, processing, and shared consumption are all activities that spread pathogens.

■ **Dairy products** can be soured by excessive fermentation or made bitter by bacterial proteolysis.

■ **Meat and poultry** are putrefied by decarboxylating bacteria, which produce amines with noxious odors.

■ **Fish and other seafood** spoil rapidly because their unsaturated fatty acids rapidly oxidize; they harbor psychrotrophic bacteria that grow under refrigeration; and their TMAO is reduced by bacteria to the fishy-smelling trimethylamine.

■ **Vegetables** spoil by excess growth of bacteria and molds. Plant pathogens destroy food crops before harvest.

■ **Food preservation** includes physical treatments, such as freezing and canning, as well as the addition of chemical preservatives, such as benzoates and nitrites.

16.6 Industrial Microbiology

The production and preservation of food is only one field of **industrial microbiology**, the commercial exploitation of microbes. Industrial microbiology is commonly understood to include a broad range of commercial products relevant to microbes, including vaccines and clinical devices; industrial solvents and pharmaceuticals; bioinformatic analysis of genomes; and genetically modified plants and animals using microbial vectors. Another growing field of industrial microbiology is bioremediation and environmental engineering, covered in Chapters 21 and 22.

Industrial Microbiology Aims for Commercial Success

The practical application of a microbial product or device may arise out of an industrial laboratory, or it may be conceived by a research scientist with the aim of meeting a compelling need in society. For example, **Special Topic 16.3** profiles a biotechnology company formed to develop innovative products for the prevention, testing, and treatment of tuberculosis. In all cases, however, the key goal is success in the marketplace—that is, to generate a product that achieves adoption by customers in preference to alternative technologies. The product's sales must cover the costs of raw materials and production and (in a for-profit company) generate a profit for the shareholders. Success requires:

■ **Identifying a useful product.** Possible products include small molecules, such as antibiotics; human proteins from cloned genes; or proteins from microbes with useful properties, such as thermostability.

■ **Isolating a microbe to produce the product.** A novel product, such as an antibiotic, is generally discovered in a naturally occurring microbe. The genes encoding product biosynthesis are then cloned into an industrial vector.

■ **Scaling up production in quantity.** The engineered microbial strain containing the vector with its cloned genes must be grown on an industrial scale, and the product must be isolated and purified.

■ **Developing a business plan.** The scientist-entrepreneur must obtain partners skilled at industrial management, finance, and marketing. Patents must be filed to protect the intellectual property rights.

■ **Safety and efficacy testing.** Human consumption or consumer use require many levels of clinical testing for approval by government agencies.

■ **Effective marketing.** The benefits of the new product must be communicated effectively to convince customers of its superiority to current products or processes.

Failure at any of the preceding tasks spells doom for the product. Thus, a prudent business plan includes multiple alternative products in development. Although the failure rate of new products is high, all the products we use had to overcome these risks.

Molecular Products from Human or Microbial Sources

Increasingly, microbes are grown to produce a commercially valuable chemical substance, such as a vitamin, an

Special Topic 16.3 Start-Up Companies Take On Tuberculosis

In the 1990s, newspaper headlines announced the resurgence of tuberculosis (TB) in American cities. Carol Nacy was then a chief scientific officer at a small biotechnology firm, following a research career at the Walter Reed Army Institute of Research and a term as president of the American Society for Microbiology (**Fig. 1**). In 1996, Nacy was invited by the National Institutes of Health to assist in a review of tuberculosis research grant support to U.S. universities. She was surprised to discover the lack of attention to TB, a disease that is the world's number one killer of women aged 15–44 and the leading killer of men after traffic accidents. The United States spends $1 billion yearly to treat 16,000 active cases; yet the standard antibiotics for tuberculosis were developed before 1970, and the main diagnostic test available (the tuberculin skin test) dates to 1880, the time of Robert Koch. The best available vaccine, the bacille Calmette-Guérin (BCG) attenuated-live vaccine, is only 50% effective.

Despite the clear needs, tuberculosis has been of little interest to industrial research or development. So Nacy decided to found two new companies to address TB: a nonprofit medical research company to develop vaccines and

Figure 1 Carol Nacy developed two companies to fight tuberculosis.

© Joanne Lawton/Washington Business Journal

conduct clinical trials and a for-profit company to develop innovative drugs and treatment devices.

A nonprofit company for vaccine development. The nonprofit company Aeras Global TB Vaccine Foundation, now directed by Jerald Sadoff, received $100 million in grants from the Bill and Melinda Gates Foundation—more than doubling the amount spent worldwide on TB vaccines in 1999. Additional funding came from the U.S. Centers for Disease Control and Prevention and from the government of Denmark. The Aeras foundation established a clinical trial site in Cape Town, South Africa. South Africa has the highest rate of pediatric TB in the world: over 500 cases per 100,000 persons. Aeras trained South African investigators and support staff in the art of clinical trials, while helping community health care workers to serve the local community. They helped test and counsel the community for HIV, a major risk factor for tuberculosis, and they vaccinated thousands of babies with BCG vaccine.

By 2005, Aeras had 12 different experimental TB vaccines in clinical and preclinical trials. The experimental vaccines are based on leading-edge principles of molecular biology, such as a recombinant vaccinia virus expressing antigenic proteins from *Mycobacterium tuberculosis*; recombinant fusion proteins from *M. tuberculosis*; and a mutant strain of the bacterium deleted for genes encoding two transcriptional sigma factors. An advanced "DNA vaccine" was developed consisting of a recombinant DNA vector designed to express antigenic proteins within host cells. The fundamental principles behind these experimental vaccines are introduced in Chapters 11 and 12.

A for-profit company develops drugs and devices. Nacy's for-profit company, Sequella, Inc., targeted innovative ideas with high risk but high potential to improve performance, rapidity, and safety of diagnosing and treating TB infections. Nacy and her cofounders scanned the academic community for novel ideas that had succeeded against TB in "proof of principle" animal models. The most promising ideas were developed for improved antibiotics, rapid and less invasive tests for TB exposure, and devices to measure the extent of pulmonary infection. Because any one idea had a high risk of failure, multiple prospects were pursued in each category.

industrial solvent, or an enzyme (**Table 16.4**). Each product requires a gene or operon of genes encoding either the product itself or the enzymes for the product's biosynthesis. We may distinguish between two fundamentally different sources of products: cloned genes from human, animal, or plant sources; and native microbial products, often from newly discovered species in extreme environments. Cloned human genes typically encode a protein

of valuable function in the human body. For example, the Claragen company produces the recombinant human protein CC10, a lung development protein often deficient in the lungs of premature infants. The recombinant protein, produced and purified from a recombinant bacterium, can be provided to the infant to reduce lung inflammation.

Cloning of gene products in recombinant organisms is discussed in Chapter 12. Here we discuss obtaining

Several Sequella products have since reached advanced clinical trials. Their lead product is the Transdermal Patch Test, a diagnostic transdermal patch to distinguish between active TB and prior TB infection or BCG vaccination (**Fig. 2**). Prior to the transdermal patch, it has been difficult to determine whether an individual who tests positive for TB antibodies actually has active infection or has simply been exposed to the bacillus in the past. The transdermal product allows antigens to penetrate the skin without needle injection, and it produces results more reliable than the standard tuberculin skin test.

For treatment of TB, several promising new antibiotics are in the pipeline. The most promising antibiotic, SQ109, was discovered by high through-put screening of a chemical library, in collaboration with Clifford Barry at the National Institutes of Health. The chemical library consisted of over 60,000 analogs of a known TB antibiotic, ethambutol (**Fig. 3**). Ethambutol is part of the current standard course of treatment for TB, whose six-month time course has a poor compliance rate. It is hoped that improved drugs will shorten the time course and improve compliance, thereby decreasing the appearance of drug-resistant strains.

The analog molecules were selected based on a common diamine core, with different combinations of side chains. The 60,000 compounds in the library were subjected to combinatorial screening, a mathematically intensive analysis based on numerous tests. Of the compounds tested, 2,796 showed activity against *M. tuberculosis* in the test tube. The 69 best compounds were tested for cytotoxicity in tissue culture, activity in TB-infected macrophages, and activity in infected ani-

mals. The compound with the greatest efficacy and fewest side effects was SQ109, a molecule with an unusual cage-like side group of three fused rings. SQ109 is now in preclinical trials for toxicology and dose range. With these products, Sequella has positioned itself to develop more effective vaccines, diagnostics, and treatments to alleviate the global burden of tuberculosis—and to make a handsome profit for investors.

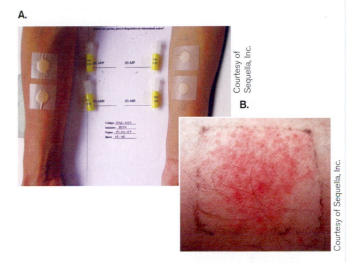

Figure 2   The transdermal patch reveals active TB infection.   A. The transdermal patch administers an antigen test reagent without requiring needle injection. **B.** A positive test result for active TB.

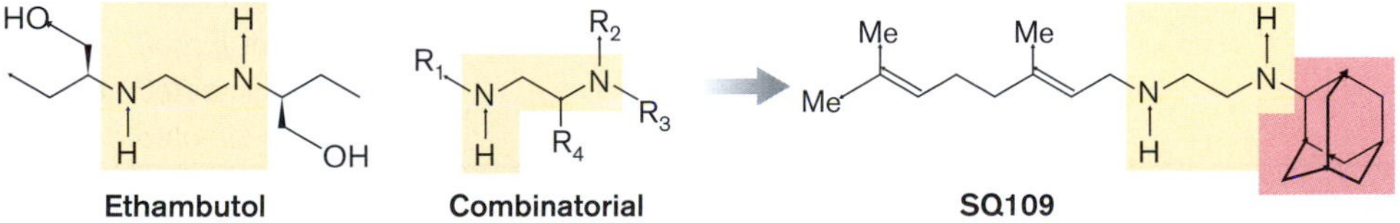

Ethambutol

Combinatorial library of 63,238 diamines

SQ109

Figure 3   A new antibiotic for TB is obtained by screening ethambutol analogs.   A combinatorial library of compounds containing the ethambutol diamine core (yellow) was screened for antibacterial effect against *M. tuberculosis*. The most promising agent screened was SQ109, with an unusual carbon-cage side group (pink).

products from native microbial sources. For example, the Dutch company Novozymes markets over 600 microbial enzymes, including Stainzyme, a laundry detergent acting at lower temperatures, an oxidase to control bleaching of paper pulp, and proteases for leather treatment. Novozymes also produces numerous enzymes as biocatalysts for "green chemistry." Green chemistry refers to environmentally friendly procedures for reactions in

organic chemistry, typically using water solution in place of petroleum-derived organic solvents.

Bioprospecting.   To identify new microbial products, chemical companies screen thousands of microbial strains from diverse ecosystems. The search for organisms with potential commercial applications is called **bioprospecting**. Bioprospecting can be done anywhere,

Table 16.4 **Commercial products from microbes.**

Class of product	Product	Producer microbe[a]
Agricultural products	Gibberellin (plant hormone)	*Fusarium moniliforme* (fungus)
	Fungicides	*Coniothyrium minitans* (fungus)
	Insecticides (live pathogens)	*Bacillus thuringiensis*
Enzymes	Alpha-amylase	*Bacillus subtilis*
	Amyloglucosidase	*Aspergillus niger*
	Lactase (beta-galactosidase)	*Kluyveromyces lactis* (fungus)
	Lipases	*Candida cylindraceae* (yeast)
	Alkaline protease	*Aspergillus oryzae* (fungus)
Food supplements	L-Lysine	*Brevibacterium lactofermentum*
	L-Tryptophan	*Klebiella aerogenes*
	Monosodium glutamate (MSG)	*Corynebacterium ammoniagenes*
	Vitamin B$_{12}$	*Pseudomonas denitrificans*
	Vitamin C (ascorbic acid)	*Acetobacter suboxidans*
Fuels and solvents	Acetone	*Clostridium* spp.
	Butanol	*Clostridium acetobutylicum*
	Glycerol	*Zygosaccharomyces rouxii*
	Methane (natural gas)	Methanogens (archaea)
Organic acids	Acetic acid	*Acetobacter xylinum*
	Citric acid	*Aspergillus niger* (fungus)
	Lactic acid	*Lactobacillus delbruckii*
Pharmaceuticals: antibiotics	Streptomycin	*Streptomyces griseus*
	Penicillin	*Penicillium chrysogenum* (fungus)
	Erythromycin	*Saccharopolyspora erythraea*
	Tetracycline	*Streptomyces aureofaciens*
Pharmaceuticals: other drugs	Lysergic acid (LSD, hallucinogen)	*Claviceps paspali* (fungus)
	Cyclosporin (immunosuppressant)	*Trichoderma polysporum*
	Steroids	*Arthrobacter* spp.
Pharmaceuticals: cloned human proteins	Insulin	Recombinant *Escherichia coli*
	Interferon	Recombinant *E. coli* or *Saccharomyces cerevisiae*
	Human growth hormone	Recombinant *E. coli*
	CC10 lung development protein	Recombinant *E. coli*
	Antibodies	Baculovirus-infected caterpillars
	Cancer regulators	Baculovirus-infected caterpillars

[a]Bacteria, unless stated otherwise

Source: M. J. Waites. 2001. *Industrial Microbiology,* Blackwell Science, Table 4.1, pp. 76–77.

from one's backyard to Yellowstone National Park. Unique, endangered ecosystems are the most promising sources of previously unknown microbial strains with valuable properties. Extreme environments, such as the hot springs of Yellowstone, are particularly promising because their products may tolerate conditions of temperature or pH required for industrial use. Psychrophiles from extremely cold environments, such as Antarctica, are a useful source for enzymes that do not require heating for activity, such as the Stainzyme laundry detergent.

An important aspect of bioprospecting is "mining the genome." Once a promising source strain is obtained, its genome is sequenced to investigate the genetic sequences that encode the useful product and regulate its expression. The operons encoding the product (or enzymes for its production) can then be cloned for optimal production (see Chapter 12). The cloned genes are transferred into an **industrial strain**, a strain whose growth characteristics are well studied and optimized for industrial production. An industrial strain must possess the following attributes:

■ **Genetic stability and manipulation.** The industrial strain must reproduce reliably, without major DNA rearrangements. It must also have an efficient gene transfer system by which vectors can introduce genes of interest into its genome.

- **Inexpensive growth requirements.** Industrial strains grow on low-cost carbon sources with minimal special needs, such as vitamins, and at easily maintained conditions of temperature and gases.
- **Safety.** Industrial strains are nonpathogenic and do not produce toxic byproducts.
- **High level of product expression.** The strain or recombinant vector must possess an efficient gene expression system to generate the desired product as a high proportion of its cell mass.
- **Ready harvesting of product.** Either the product must be secreted by the cell or, if the product is intracellular, the cells must be easily breakable to liberate the product.

Species for industrial strains include the bacterium *Bacillus subtilis*, the yeast *Candida utilis*, and the filamentous fungus *Aspergillus niger*. Each of these species is safe, grows to high density on inexpensive carbon sources, such as molasses, and expresses desired products at high concentration.

> **THOUGHT QUESTION 16.10** Why would different industrial strains or species be used to express different kinds of cloned products?

Fermentation Systems

Commercial success requires optimizing every detail of the fermentation system. "Fermentation" in industrial terms refers not just to anaerobic metabolism, but to all means of growth of microbes on an industrial scale. In an **industrial fermenter**, the growth vessel and all its environmental supports, such as temperature control and oxygenation, must be scaled up to thousands of liters (**Fig. 16.28**). This increase in scale generates many problems of quality control, such as maintaining even temperature, pH, and oxygenation throughout the vessel and minimizing foaming of the culture liquid. A small change in any of the growth factors can impact production costs and profit margin. Another major concern is to avoid contamination by other organisms.

The fermenter is the core of the first half of industrial production, known as **upstream processing** (**Fig. 16.29**). Upstream processing refers to the culturing of the industrial microbe to produce large quantities of product. All aspects of the process must be controlled so as to maximize the final concentration of product, which in most cases peaks at a specific time in the microbial growth cycle. Following microbial growth, the culture must be harvested and the product purified. These processes constitute **downstream processing**. The first step of downstream processing is to separate the microbial cells from

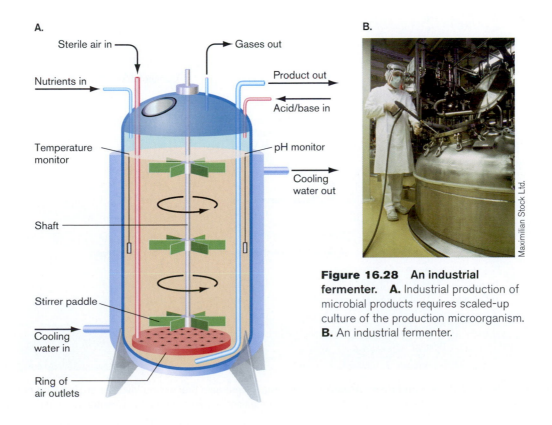

A.

Sterile air in — Gases out
Nutrients in — Product out
Acid/base in
Temperature monitor — pH monitor
Cooling water out
Shaft
Stirrer paddle
Cooling water in
Ring of air outlets

B.

Maximilian Stock Ltd.

Figure 16.28   An industrial fermenter.   A. Industrial production of microbial products requires scaled-up culture of the production microorganism. **B.** An industrial fermenter.

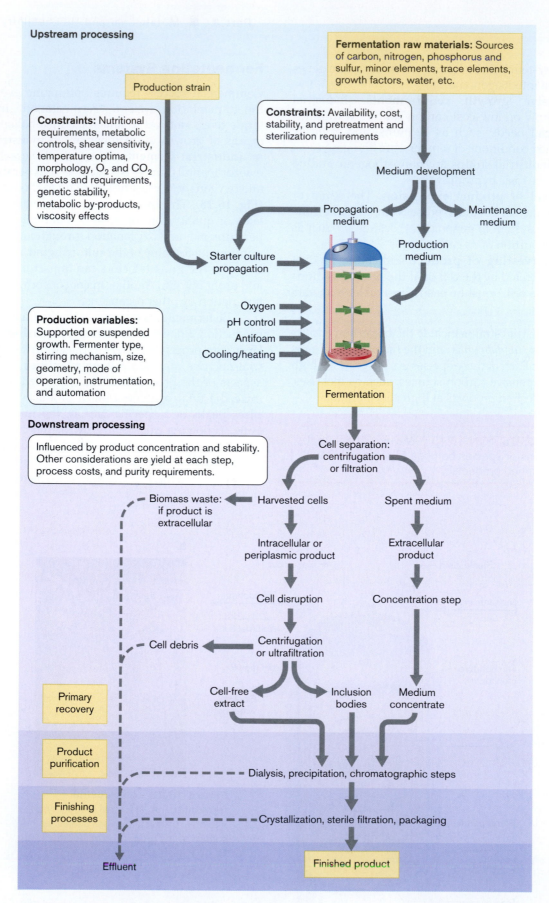

Upstream processing

Fermentation raw materials: Sources of carbon, nitrogen, phosphorus and sulfur, minor elements, trace elements, growth factors, water, etc.

Production strain

Constraints: Nutritional requirements, metabolic controls, shear sensitivity, temperature optima, morphology, O_2 and CO_2 effects and requirements, genetic stability, metabolic by-products, viscosity effects

Constraints: Availability, cost, stability, and pretreatment and sterilization requirements

Medium development

Propagation medium

Maintenance medium

Production medium

Starter culture propagation

Production variables: Supported or suspended growth. Fermenter type, stirring mechanism, size, geometry, mode of operation, instrumentation, and automation

Oxygen
pH control
Antifoam
Cooling/heating

Fermentation

Downstream processing

Influenced by product concentration and stability. Other considerations are yield at each step, process costs, and purity requirements.

Cell separation: centrifugation or filtration

Biomass waste: if product is extracellular

Harvested cells

Spent medium

Intracellular or periplasmic product

Extracellular product

Cell disruption

Concentration step

Cell debris

Centrifugation or ultrafiltration

Primary recovery

Cell-free extract

Inclusion bodies

Medium concentrate

Product purification

Dialysis, precipitation, chromatographic steps

Finishing processes

Crystallization, sterile filtration, packaging

Effluent

Finished product

Figure 16.29   Details of upstream and downstream processing.   Upstream processing involves engineering of the microbial strain and large-scale growth to generate the product. Downstream processing involves product concentration and purification. *Source*: M. J. Waites. 2001. *Industrial Microbiology*, Blackwell Science.

the culture fluid, by centrifugation or by filtration. Next, the **primary recovery** of product follows one of two different pathways, depending on whether the product is maintained within the cells or secreted into the culture fluid. Many kinds of subsequent purification and finishing steps are necessary before the product has acceptable quality for its desired use. Again, failure of any detail can render the entire process unusable.

Products designed for human internal use face formidable hurdles in clinical testing and finally approval by the appropriate regulatory agency, such as the U.S. Food and Drug Administration (FDA). After millions of dollars are invested in process development, the product may still fail this final test and never come to market. Not surprisingly, a pharmaceutical company must research thousands of potential products before achieving one that makes a profit. The consumer cost inevitably includes the development costs not only of the one successful product, such as recombinant insulin, but also of all the products that failed.

Production in Animal or Plant Systems

Some kinds of products require subtle kinds of processing that occur correctly only within animal or plant cells. For example, protein products may require posttranslational modifications such as glycosylation (attachment of polysaccharide chains). In such cases, the expressed genes must be transferred from a bacterial or viral vector into an animal or plant tissue culture or to a transgenic animal or plant. The microbially transformed animal or plant may itself be the industrial product. We present two examples of microbial production involving multicellular hosts: that of the plant vector host *Agrobacterium tumefaciens* and that of the baculovirus insect vector.

***Agrobacterium tumefaciens* engineers plants.** Bacteria and viruses provide the vectors for transferring genes into multicellular organisms, as discussed in Chapter 12. A bacterium of major industrial importance is *Agrobacterium tumefaciens*, a tumor-inducing plant pathogen that conducts natural genetic engineering on dicot plants (**Fig. 16.30**). Long before scientists invented "recombinant DNA," *A. tumefaciens* had evolved a gene transfer system by which it induces infected plant cells to generate food molecules to feed the pathogen. This highly efficient gene transfer system is readily modified to insert genes of interest, such as herbicide resistance, into plant genomes.

Tumorigenic strains of *A. tumefaciens* possess a special plasmid for engineering plant cells, called the **Ti plasmid** (tumor-inducing plasmid). The Ti plasmid is transferred

into a plant cell through a process mediated by bacterial proteins, similar to conjugation (discussed in Chapter 7). The Ti plasmid includes the *vir* operon, which encodes virulence genes, as well as T-DNA (transferable DNA), a sequence of genes that will be transferred to the host plant and recombined into its genome. T-DNA encodes tumor induction genes as well as enzymes for biosynthesis of a carbon and nitrogen source called an opine. Opines are specialized amino acids made by a one-step synthesis from arginine, typically by amination of a central metabolite such as pyruvate or 2-oxoglutarate. A given strain of *A. tumefaciens* typically provides one type of opine synthesis enzyme and has the ability to metabolize the corresponding opine.

The *vir* gene products from the Ti plasmid detect the presence of a plant host and stimulate plasmid transfer (see **Fig. 16.30**). In the cell envelope, VirA protein detects a chemical signal from a wounded plant, which is capable of being infected. VirA then activates VirG to induce expression of other *vir* genes, encoding endonucleases VirD1 and VirD2. The VirD endonucleases cleave the left and right ends of the T-DNA and direct its transfer into the plant cell. Within the plant cell nucleus, the T-DNA becomes integrated into the plant genome, where it induces opine production. The opine-producing cells proliferate, forming a tumor.

For industrial use, a recombinant strain of *A. tumefaciens* has the Ti plasmid divided into two separate plasmids. One contains the *vir* operon conducting DNA transfer, whereas the other plasmid contains T-DNA with its left and right ends intact but most of its genes substituted by a desired recombinant gene. Upon infection, the virulence system induces transfer of the T-DNA to the plant cell without any tumor-inducing genes, allowing genomic integration of the recombinant gene without tumor induction or opine production.

> **THOUGHT QUESTION 16.11** Why would an herbicide resistance gene be desirable in an agricultural plant? What long-term problems might be caused by microbial transfer of herbicide resistance genes into plant genomes?

Baculovirus in insects. Insect viruses such as baculovirus provide a surprisingly practical means of generating large amounts of product. The larvae, or caterpillars, grow rapidly on inexpensive substrates, accumulating large quantities of biomass. The most effective host is generally a strain of an aggressive pest such as *Trichoplusia ni*, the cabbage looper, an insect pest that feeds on the underside of cabbage leaves. Virus-infected caterpillars are used to

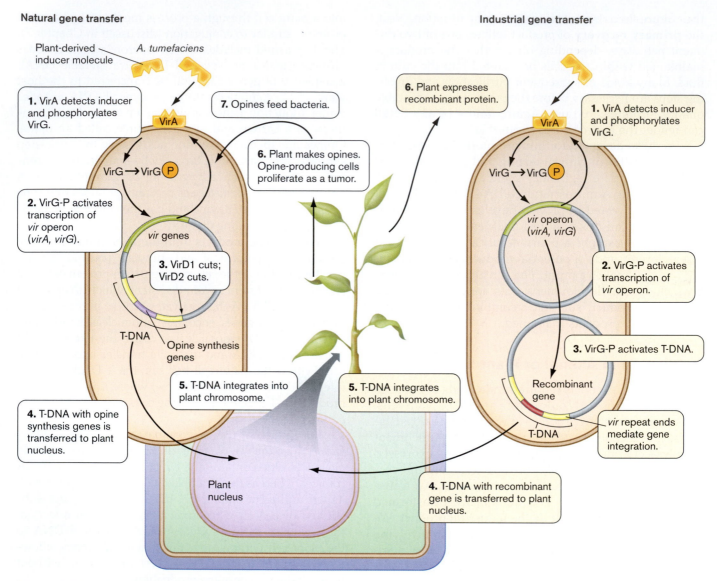

Natural gene transfer

Plant-derived inducer molecule

A. tumefaciens

1. VirA detects inducer and phosphorylates VirG.

VirA

VirG → VirG ⓟ

2. VirG-P activates transcription of *vir* operon (*virA*, *virG*).

vir genes

3. VirD1 cuts; VirD2 cuts.

T-DNA

Opine synthesis genes

7. Opines feed bacteria.

6. Plant makes opines. Opine-producing cells proliferate as a tumor.

5. T-DNA integrates into plant chromosome.

4. T-DNA with opine synthesis genes is transferred to plant nucleus.

Plant nucleus

6. Plant expresses recombinant protein.

5. T-DNA integrates into plant chromosome.

Industrial gene transfer

1. VirA detects inducer and phosphorylates VirG.

VirA

VirG → VirG ⓟ

vir operon (*virA*, *virG*)

2. VirG-P activates transcription of *vir* operon.

3. VirG-P activates T-DNA.

Recombinant gene

vir repeat ends mediate gene integration.

T-DNA

4. T-DNA with recombinant gene is transferred to plant nucleus.

Figure 16.30 *Agrobacterium tumefaciens*: **a natural gene transfer vector for plants.** Left: *A. tumefaciens* transfers T-DNA containing opine synthesis genes into a plant, which then produces opines to feed the bacteria. Right: *A. tumefaciens* can be engineered to transfer T-DNA containing a recombinant gene of interest into the plant genome. A recombinant strain of *A. tumefaciens* has the Ti plasmid divided into two separate plasmids, one containing the *vir* operon conducting DNA transfer, the other containing T-DNA with most of its genes substituted by a desired recombinant gene. ⏵❚❚

produce human proteins such as antibodies, growth factors, and antimicrobial peptides.

A baculovirus expression system (**Fig. 16.31**) was developed by insect virus researchers Max Summers and Gale Smith at Texas A&M University. This expression system involves two kinds of microbial vector: a bacterial plasmid carrying an engineered product gene and an insect virus. Baculoviruses are circular double-stranded DNA viruses that can infect a wide range of insects and insect cell lines in tissue culture. In the baculovirus expression system, the gene of interest, encoding, for example, a vaccine antigen, is first cloned into a standard

bacterial plasmid, called the transfer vector. The transfer vector contains an antibiotic resistance gene, a cloning site for gene insertion, and a promoter from the polyhedrin gene, a baculovirus gene promoter that induces high-level expression in infected insect cells. The transfer vector plus linearized baculovirus DNA are cotransfected (introduced) into insect cells in tissue culture. For the virus to reproduce, it must recombine with the transfer vector, restoring the circularity of its genome. All the progeny virus particles now contain the recombinant gene, and when replicated, they will express it to make high levels of recombinant protein.

Baculovirus expression system for eukaryotic gene products

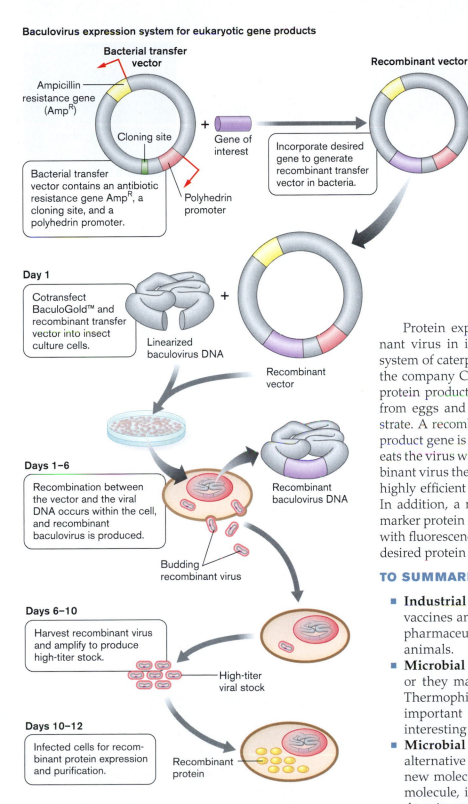

Figure 16.31 Baculovirus expression system for production in insects. A gene cloned in a bacterial transfer vector is recombined with an insect virus to express a protein product in an insect tissue culture. *Source*: BaculoGold™.

Bacterial transfer vector

Ampicillin resistance gene (Amp^R)

Cloning site

Bacterial transfer vector contains an antibiotic resistance gene Amp^R, a cloning site, and a polyhedrin promoter.

Gene of interest

Polyhedrin promoter

Incorporate desired gene to generate recombinant transfer vector in bacteria.

Recombinant vector

Day 1

Cotransfect BaculoGold™ and recombinant transfer vector into insect culture cells.

Linearized baculovirus DNA

Recombinant vector

Days 1–6

Recombination between the vector and the viral DNA occurs within the cell, and recombinant baculovirus is produced.

Recombinant baculovirus DNA

Budding recombinant virus

Days 6–10

Harvest recombinant virus and amplify to produce high-titer stock.

High-titer viral stock

Days 10–12

Infected cells for recombinant protein expression and purification.

Recombinant protein

Protein expression requires growth of the recombinant virus in insect caterpillars. A remarkably efficient system of caterpillar protein production was developed by the company Chesapeake PERL, Inc. (**Fig. 16.32**). In the protein production system, *Trichoplusia* caterpillars hatch from eggs and grow within chambers full of food substrate. A recombinant baculovirus containing the desired product gene is sprayed into each chamber. The caterpillar eats the virus with its food, becoming infected. The recombinant virus then directs synthesis of its product, using the highly efficient promoter from the viral polyhedrin gene. In addition, a recombinant gene expresses a fluorescent marker protein such as DsRed. When the caterpillars glow with fluorescence, this signals that they are producing the desired protein and are ready for harvesting.

TO SUMMARIZE:

- **Industrial microbiology** includes the production of vaccines and clinical devices, industrial solvents and pharmaceuticals, and genetically modified plants and animals.
- **Microbial product molecules** may be indigenous or they may be cloned from nonmicrobial sources. Thermophiles and psychrophiles are particularly important sources of new strains with potentially interesting new properties.
- **Microbial products must be competitive** with alternative technologies. Developing a competitive new molecular product requires identifying a useful molecule, isolating and engineering a strain to produce it, scaling up for production in quantity, developing a business plan, and testing for safety.
- **Bioprospecting** is the mass screening of new microbial strains for potentially valuable protein and small-molecule products.

A.

Food containing baculovirus

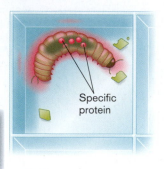

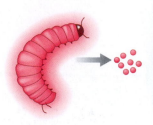

B. Glowing larvae

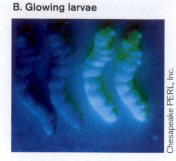

Specific protein

Week 1: Infection
The caterpillar hatches from the egg. It sits amid food in a container made of 56 individual cells, each of which contains one caterpillar. At week's end, a recombinant insect virus encoding a specific protein is sprayed into each cell.

Week 2: Production
As the caterpillar grows, it eats the virus-infested food and becomes infected. The recombinant virus directs it to produce the specific protein as well as a protein that causes it to glow.

End of week 2: Harvest
The caterpillar emits an intense glow signaling it's ready for harvesting. The caterpillars are ground up, and the specific proteins are separated and purified from caterpillar cadavers.

Figure 16.32   Insect caterpillars are engineered to produce a valuable protein.   A. Caterpillars of *Trichoplusia ni* are infected with recombinant baculovirus, which produces a desired protein product as well as a fluorescent marker protein such as DsRed, indicating successful infection. **B.** The infected caterpillars ready for harvest fluoresce; compare with uninfected controls (dark). *Source:* Photo courtesy of Chesapeake PERL, Inc., using a DsRed recombinant virus provided by Kevin O'Connell and Patricia Anderson of the U.S. Army Edgewood Chemical Biological Center.

- **Industrial strains**, commonly *E. coli* or *B. subtilis*, are used to incorporate the newly discovered genes into a host microbe.
- **Upstream processing** refers to the culturing of the industrial microbe to produce large quantities of product. **Downstream processing** involves product recovery and purification.
- **Posttranscriptional processing of human or plant genes** may be facilitated by use of a transgenic source or a virus-infected animal host.

Concluding Thoughts

Overall, we see how microbes from our environment contribute products for human use, from foods and vitamin supplements to industrial enzymes and highly specific protein therapeutics. These microbial applications incorporate much of the natural metabolism and biochemistry introduced in Chapters 13–15. Human inventiveness and industrial and financial management must meet many challenges to develop new microbial products.

CHAPTER REVIEW

Review Questions

1. What kinds of microbes are consumed as food? Why are most bacteria inedible for humans?
2. What are the advantages of fermentation for a food product?
3. What are the differences between traditional fermented foods and commercial fermented foods?
4. How do acidic fermentations contribute to the formation of different kinds of cheeses? What is the role of different kinds of metabolism performed by different microbial species?
5. Compare and contrast acidic and alkaline fermentation processes. What different kinds of foods are produced?
6. Compare and contrast the role of ethanolic fermentation in bread making and winemaking.
7. How do relatively minor fermentation reactions contribute to the flavor of food?

8. Explain the differences between food spoilage and food poisoning.
9. What are the most important food-borne pathogens, based on infection rates? Based on mortality rates?
10. What are the major means of preserving food? Compare and contrast their strengths and limitations.
11. What tasks must be accomplished to develop a microbial product for commercial marketing?
12. What are the sources of potential new microbial products? What is the difference between a source strain and an industrial strain?
13. Explain upstream processing and downstream processing.
14. Explain why genes encoding industrially useful products may be transferred into plant or animal systems for production. What are the roles of microbes in these systems?

Key Terms

alkaline fermentation (593)
alcohol dehydrogenase (604)
antimicrobial agent (608)
bioprospecting (617)
caseins (595)
cheddared (595)
cheese (595)
curd (595)
downstream processing (619)
ethanolic fermentation (593)
fermented food (592)
food poisoning (609)
food spoilage (608)
freeze-drying (613)
fruiting body (590)
heterolactic fermentation (593)

indigenous flora (593)
industrial fermenter (619)
industrial microbiology (615)
industrial strain (618)
injera (604)
kimchi (599)
lactic acid fermentation (593)
leaven (602)
lyophilization (613)
malolactic fermentation (606)
miso (599)
natto (600)
nori (591)
pathogenicity island (611)
primary recovery (621)
propionic acid fermentation (593)

putrefaction (609)
rancidity (609)
rennet (596)
ripening (596)
single-cell protein (592)
sourdough (603)
soy sauce (599)
starter culture (596)
tempeh (598)
Ti plasmid (621)
traditional fermented food (593)
upstream processing (619)
whey (595)
yogurt (595)

Recommended Reading

Chang, Shu-Ting, and Philip G. Miles. 2004. *Mushrooms. Cultivation, Nutritional Value, Medicinal Effect, and Environmental Impact.* 2nd ed. CRC Press, New York.

Doyle, Michael P., Larry R. Beuchat, and Thomas J. Montville. 2001. *Food Microbiology. Fundamentals and Frontiers.* ASM Press, Washington, DC.

Giraffa, Giorgio. 2004. Studying the dynamics of microbial populations during food fermentation. *FEMS Microbiological Reviews* **28**:251–260.

Hui, Y. H., Lisbeth Meunier-Goddick, Åse S. Hansen, Jytte Josephsen, Wai-Kit Nip, et al. (eds). 2004. *Handbook of Food and Beverage Fermentation Technology.* Marcel Dekker, New York.

Marilley, L., and M. G. Casey. 2004. Flavours of cheese products: metabolic pathways, analytical tools and identification of producing strains. *International Journal of Food Microbiology* **90**:139–159.

Mills, David A., Helen Rawsthorne, C. Parker, D. Tamir, and K. Makarova. 2005. Genomic analysis of *Oenococcus oeni* PSU-1 and its relevance to winemaking. *FEMS Microbiological Reviews* **29**:465–475.

Schwan, Rosane F. 1998. Cocoa fermentations conducted with a defined microbial cocktail inoculum. *Applied Environmental Microbiology* **64**:1477–1483.

Schwan, Rosane F., and Alan E. Wheals. 2004. The microbiology of cocoa fermentation and its role in chocolate quality. *Critical Reviews in Food Science and Nutrition* **44**:205–221.

Steinkraus, Keith H., ed.-in-chief. 1983. *Handbook of Indigenous Fermented Foods.* Marcel Dekker, Inc., New York.

Part 4

Microbial Diversity and Ecology

Karl Stetter collecting samples from volcanic hot springs in Siberia.

AN INTERVIEW WITH

KARL STETTER: ADVENTURES IN MICROBIAL DIVERSITY LEAD TO PRODUCTS IN INDUSTRY

Karl Stetter, professor at the University of Regensburg in Germany, was the first person to discover living organisms growing at temperatures above 100°C. Hyperthermophiles grow in hot springs and at thermal vents in the ocean floor. Such organisms possess exceptionally stable enzymes with enormous potential uses in industry. In 2004, Stetter cofounded the Diversa Company to isolate and develop such enzymes. For example, Diversa developed an enzyme for bleaching paper from a microbe isolated at a geyser in Russia. Such industrial enzymes often replace polluting processes with processes that are more environmentally friendly.

Why did you decide to make a career in microbiology?
I had been fascinated about microbes since I was a little child, when my parents had given me a simple microscope.

What enabled you to discover the hyperthermophile *Pyrodictium occultum*?
Just believe the laws of physics and chemistry and never believe prejudices (such as, "If it would exist, other people would have discovered it already"). Finding life above 100°C was a big surprise. I had to be perfectly sure about my results and confirmed them several times. It finally paid off, and this topic became my first *Nature* paper.

Would you describe the discovery that led to your 1982 paper ("Ultrathin mycelia-forming organisms from submarine volcanic areas having an optimum growth temperature of 105°C," *Nature* 300[1982]:258–260)?

This paper raised a lot of public interest; I even had to give my first (completely unexpected!) press conference. Here is how my discovery came about: During my family holidays in 1981, in order to check my hypothesis of life possibly existing above 100°C (under overpressure to keep the water liquid), I did some scuba sampling at the shallow hot vents off the coast of Vulcano, Italy. This had been my first attempt in underwater sampling. To do this, I developed some special sampling gear, which I used and improved on all my later diving expeditions (electronic waterproof thermometers with sturdy, very long probes, syringes with enlarged inlets, and so on). During our holidays, my wife and my little daughter Sabine had fun supporting me in our little rubber boat, processing the still-boiling hot samples that I brought up to the surface. The samples were kept anaerobic by injecting a reducing agent, a mixture of sodium dithionite and sodium sulfide.

Back in the Regensburg lab, together with my technician, we began extensive cultivation attempts, which lasted for several months before the exciting bugs finally grew up. For the first time, from several samples at 100°C incubation temperature, disk-shaped cells grew up within two days. As we found out later, their fastest growth was in superheated water at 105°C, and their upper limit for growth temperature was even at 110°C!

What is it like for students working with extremophiles in the laboratory? How do you modify laboratory practice for the culture and observation of these organisms?
It's great fun for students to culture hyperthermophiles, as it is for me. It's great to culture organisms that had been thought to be impossible. Of course, some safety briefing is essential to avoid lab accidents. Every culture bottle under pressure in my lab is housed in an explosion-safe cage. It's an incredible thrill being the first to see a novel bug growing!

What have we learned from the genome sequences of hyperthermophiles? What do they tell us about microbial evolution?
This question would require a whole book of answers. Here is my answer in a nutshell: Hyper-

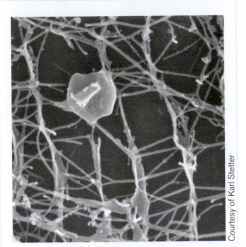

Pyrodictium occultum, the first hyperthermophile discovered. The disk-shaped archaeal cells form complex networks of interconnecting filaments. (Cell diameter, 2–4 μm.)

thermophiles harbor rather small genomes. Interestingly, their DNA double helix is not stabilized by an elevated GC/AT content of its DNA. Instead, all hyperthermophiles possess reverse gyrase to stabilize DNA. The enzymes and other proteins are structurally related to their homologs in mesophilic (moderate-growth-temperature) organisms.

A lot of horizontal gene transfer goes on in hyperthermophiles. The information-processing machinery in hyperthermophiles (small-subunit rRNA, and so on) appears to be still rather primitive. In the small-subunit rRNA–based tree of life, hyperthermophiles include all the deep-branching lineages of organisms that diverged earliest.

You also discovered the unusual hyperthermophilic symbiont *Nanoarchaeum*, which lives at marine vents at 70–98°C associated with another archaeon, *Ignicoccus*. *Nanoarchaeum* cells are only 400 nm in diameter and lack cell walls, their cell membranes covered only by a protein S-layer. They require cell-cell contact with an actively growing *Ignicoccus* cell in order to grow. What do we know

about how these two organisms affect each other's growth?
Nanoarchaeum definitively needs *Ignicoccus* to grow. But *Ignicoccus* grows without *Nanoarchaeum*. This is all just under laboratory conditions. So far, we are pretty much away from a deeper understanding of this exciting symbiosis, which most likely has existed since the dawn of life.

Why is the Diversa Company so interested in your extremophiles?
From its onset, Diversa has been interested in extremophiles, with the view of developing sturdy enzymes as industrial catalysts. Diversa has already developed several promising products, such as stock food additives, enzymes for bleaching paper pulp, and oil recovery enzymes.

What do you think are the most exciting areas for students entering microbiology today? What advice do you have for today's students?
I think our hyperthermophile research at present is only the tip of a damned hot iceberg! But truly,

there are so many exciting questions open in all kinds of fields in microbiology. Students should follow their passion and try to do an excellent job in whatever field they choose. And if they don't mind burning their fingers once in a while . . . well, then hyperthermophiles may be the right choice for them!

After sharing your early discoveries, how does your family relate to your work today?
My wife Heidi is a microbiologist and a high school teacher of chemistry and biology. Our oldest daughter Sabine is now an attorney in criminal law. Our son Florian is a very successful actor, on the stage and in movies.

What are your pursuits outside science?
In addition to my number one hobby—cultivating hot bugs—I grow tropical orchids in my greenhouse (about 500 species) and drive a hot car (Ferrari) at 190 miles per hour over the German Autobahn.

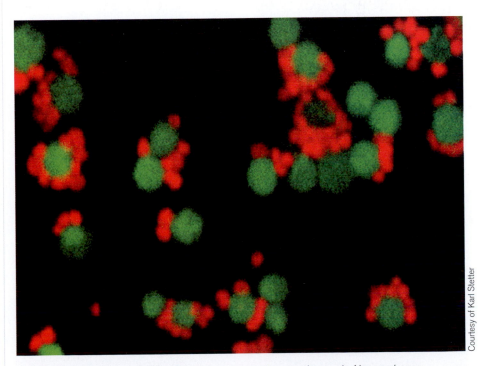

Symbiotic hyperthermophiles. This fluorescence micrograph reveals *Nanoarchaeum equitans* (red), the smallest archaeon known, as an obligate symbiont of another archaeon, *Ignicoccus* (green). These archaea were found at submarine thermal vents north of Iceland.

Chapter 17

Origins and Evolution

Microbial life may have appeared as early as 3.8 billion years ago, soon after our planet Earth formed out of dust from the young sun. Since then, microbes have diverged into new forms adapted to diverse ways of life, from psychrophiles beneath the ice of Antarctica to anaerobes in the human colon. Descendants of early microbes include all living plants and animals, including ourselves. How did microbes originate on Earth? How did their metabolism shape the chemistry of Earth's crust and atmosphere?

DNA sequence-based taxonomy has revealed vast numbers of previously unknown microbial species. How do we classify them? Some evidence, particularly the sequences of rRNA genes, support a traditional branched tree of life. Other gene sequences, however, support a more complex picture, including gene transfer between distant branches of the tree. Many microbes have evolved mutualistic partnerships with other living things, such as bacteria that fix nitrogen within plant cells. Still other microbes evolve as pathogens whose adaptations enable them to live at the expense of their host and cause disease.

A diver is sectioning a stromatolite, a layered community of cyanobacteria in a tidal pool in the Bahamas. Fossil stromatolites have been dated at 3.4 billion years ago, among the earliest forms of life on Earth. *Source: Abigail C. Allwood, et al. 2006. Nature 441:714.*

For centuries, observers of the natural world have wondered where life came from. As early as 1802, the naturalist Erasmus Darwin, grandfather of Charles Darwin, wrote:

Organic life beneath the shoreless waves
Was born and nurs'd in ocean's pearly caves;
First forms minute, unseen by spheric glass,
Move on the mud, or pierce the watery mass;
These, as successive generations bloom,
New powers acquire and larger limbs assume . . .

Thus, nineteenth-century biologists developed the idea that all living organisms had evolved from microbes, perhaps even from cells too small to be seen with the "spheric glass" of a microscope. Even without modern tools of isotope analysis and genetics, thoughtful observers recognized the overwhelming commonalities among all living cells, such as the membrane-bound compartment of cytoplasm and common metabolic pathways such as sugar metabolism. Today, lines of evidence from geology, biochemistry, and genetics overwhelmingly support the microbial origin of life.

What did early life look like? The earliest forms of life for which we have clear fossil evidence are bacterial communities called **stromatolites** (**Fig. 17.1A**). A stromatolite is a bulbous mass of sedimentary layers of limestone (calcium carbonate, $CaCO_3$) accreted by microbes over years, even centuries. Within the outer layers, microbes grow as a "microbial mat," a kind of biofilm. The outermost layers of the mat contain oxygenic phototrophs, such as diatoms and filamentous cyanobacteria, that exude bubbles of oxygen. A few millimeters below the surface, red light supports bacteria photolyzing H_2S

to sulfate, which is then reduced by still lower layers of sulfate-reducing bacteria. Stromatolites today survive mainly in isolated pools whose high salt concentration excludes predators, as in Hamlyn Bay, Australia. But 3 billion years ago, in the absence of predators, stromatolites covered shallow seas all over Earth. Fossil stromatolites appear in ancient rock formations such as the 3.4-billion-year-old Strelley Pool Chert, a sedimentary formation in Pilbara Craton, Australia (**Fig. 17.1B**). Their layers preserve the wavy form of the microbial mats, remarkably similar to living stromatolites. Stromatolites are studied by NASA for clues as to how life may appear on other planets.

Questions about life's origin have long sparked controversy. In eighteenth-century Europe, the idea of "spontaneous generation" of microbes was considered so dangerous that priests and monks performed scientific experiments to disprove it (described in Chapter 1). The priests ultimately won their scientific point: All microbes today have "parents"—that is, preexisting microbes. But these experiments did not address the origin of the first living cells or how early life gave rise to modern species. This concept was so controversial that in the early twentieth century, laws forbade teaching it in American public schools. In 1929, such a law in Tennessee was put to the test by high school teacher John Scopes in what became the famous Scopes trial (**Fig. 17.2**). According to a 14-year-old student who testified at the trial, Scopes taught that "there was a little germ of one cell organism formed, and this organism kept evolving . . . and from this was man." A large body of evidence now supports the view that microbes living more than 3 billion years ago gave rise to all animals and plants as well as to all modern microbes.

A.

B.

© Roger Garwood & Trish Ainslie/Corbis

Francois Gohier/Photo Researchers, Inc.

1 cm

Figure 17.1   Stromatolites: an ancient life-form.   A. Cyanobacteria form colonial stromatolites, the present-day structures believed to resemble most closely the earliest forms of life on Earth, Hamlyn Bay, Western Australia. **B.** Section through a 3.4-billion-year-old stromatolite from the Strelley Pool Chert. Pilbara Craton, Australia.

Figure 17.2 Objection to microbial ancestry. In 1925, outside the Scopes trial in Dayton, Tennessee, demonstrators opposed the teaching that all life—including humans—evolved from a microbe.

Hulton-Deutsch Collection/Corbis

Chapter 17 explores the evidence from geochemistry and molecular biology for the nature of the earliest cells, as well as the challenges in interpreting data from so long ago. We explore how molecular techniques reveal deep similarities among all life-forms: the core macromolecular apparatus of DNA, RNA, and proteins, expressed by genomes that share a common set of ancestral genes. We focus on three major concerns of microbial evolution:

■ The origin of life on Earth and the nature of the earliest cells.
■ The divergence of microbes from common ancestors.
■ Gene transfer and symbiosis as agents of evolution.

17.1 Origins of Life

Before the first cells could evolve, several fundamental conditions were required. These conditions included:

■ **Essential elements.** Because all life on Earth is composed of molecules, the origin of life required the fundamental elements that compose organic molecules.
■ **Continual source of energy.** The generation of life requires continual input of energy, which ultimately is dissipated as heat. The main source of energy for life is the sun.
■ **Temperature range permitting liquid water.** Above 150°C, life's macromolecules fall apart; below the freezing point of water, metabolic reactions cease. Maintaining the relatively narrow temperature range conducive to life depends on the nature of our sun,

our planet's distance from the sun, and the heat-trapping capacity of our atmosphere.

Elements of Life

For life to arise and grow, elements such as carbon and oxygen needed to be available on Earth. The planet Earth coalesced during formation of the solar system 4.5 Gyr (gigayears, or billions of years) ago. Central to the solar system is our sun, a "yellow" star of medium size and surface temperature (5,770 K). The sun's surface temperature generates electromagnetic radiation across the spectrum, peaking in the range of visible light. As we learned in Chapter 13, the photon energies of visible light are sufficient to drive photosynthesis but not so energetic that they destroy biomolecules. Thus, the stellar class of our sun makes organic life possible.

NOTE: A billion means the number 10^9 in American English; but in British English, a billion is 10^{12}. For clarity, we designate 10^9 years as a gigayear, or **Gyr**. A million years (10^6) is a "megayear," or **Myr**.

The sun's surface temperature and luminosity are generated by nuclear fusion reactions in which hydrogen nuclei fuse to form helium. Yet 2% of the solar mass consists of heavier elements, such as carbon, nitrogen, and oxygen, as well as traces of iron and other metals—elements that compose Earth, including its living organisms. Where did these heavier elements come from? To answer this question, we must look to other stars in the universe at different stages of their development (**Fig. 17.3**).

Elements of life formed within stars. Throughout the universe, young stars such as our sun fuse hydrogen to form helium. As stars age, they use up all their hydrogen. With hydrogen gone, the aging star contracts and its temperature rises, enabling helium nuclei to fuse forming carbon (see **Fig. 17.3**). Carbon drives a cyclic nuclear reaction, the CNO cycle, to form isotopes of nitrogen and oxygen. Further nuclear reactions generate heavier elements through iron (Fe). Thus, the major elements of biomolecules were formed within stars that aged before the birth of our solar system.

The later nuclear reactions of aging stars generate heavier nuclei up to iron. The aging star expands, forming a red giant (see **Fig. 17.3**). When a star of sufficient mass expands (a supergiant), it explodes as a **supernova** (**Fig. 17.4**). The explosion of a supernova generates in a brief time all the heaviest elements and ejects the entire contents of the star at near-light speed. Billions of years before our sun was born, the first stars aged and died, spreading

Life cycle of a massive star

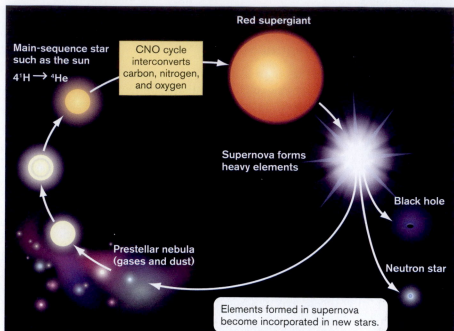

Figure 17.3 **Stellar origin of atomic nuclei that form living organisms.** In young stars, hydrogen nuclei fuse to form helium. In older stars, fusion of helium forms carbon, nitrogen, oxygen, and all the heavier elements up to iron. Massive stars explode as supernovae, spreading all the elements of the periodic table across space. These elements are picked up by newly forming stars, such as our own sun.

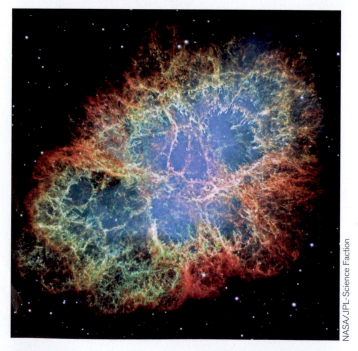

Figure 17.4 **Crab Nebula, the remnant of a supernova.** Observed by radioastronomy.

all the elements of the periodic table across the universe. Some of these elements coalesced with our sun and formed the planets of our solar system. In effect, all life on Earth is made of stardust, the remains of stars long gone.

THOUGHT QUESTION 17.1   What would have happened to life on Earth if the sun were a different stellar class, substantially hotter or colder than it is now?

Elemental composition of Earth. When our solar system formed, individual planets coalesced out of matter attracted by the force of gravity. Because of Earth's small size, most of the hydrogen gas escaped Earth's gravity very early. The most abundant dense component of Earth was iron (**Fig. 17.5**). Much of Earth's iron sank to the center to form the core. The core is surrounded by a mantle, composed primarily of iron combined with less dense crystalline minerals such as silicates of iron and magnesium, $(Fe, Mg)_2SiO_4$. The mantle is coated by Earth's thin outer crust. The crust is composed primarily of silicon dioxide, SiO_2, also known as quartz or chert. Crustal rock includes smaller amounts of numerous minerals, including the carbonates and nitrates that provided the essential elements for life. Overall, the crust shows a redox gradient, reducing in the interior and oxidizing at the surface.

The crust, also called the lithosphere, provides a habitat for microbes down to surprising depths, such as within gold mines excavated to 3 kilometer (km). Some **endolithic** microbes (microbes living within rock) metabolize by oxidizing electron donors generated through decay of radioactive metals. The discovery of endolithic organisms deep in the crust was of great interest to NASA scientists

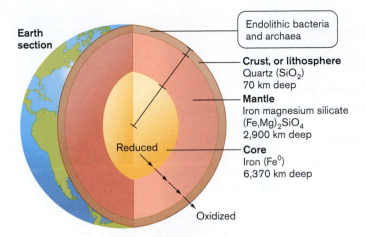

Figure 17.5 Geological composition of Earth. The cross section of Earth shows the core, the mantle, and the thin outer crust. The core and mantle are rich in iron; oxygen content increases toward the crust. The crust is composed primarily of silicates such as quartz (SiO_2). Crustal rock supports endolithic prokaryotes.

seeking life on Mars, whose crust might provide a similar habitat to Earth's.

The outer surface of the crust supports the remainder of the **biosphere**, the sum total of all life on Earth. The biosphere generates oxidants (electron acceptors), most notably O_2. Oxygen-breathing organisms can live only on the outer surface, where O_2 is produced by photosynthesis.

Earth's atmosphere. From the crust and the mantle of early Earth, volcanic activity released gases such as carbon dioxide and nitrogen, which formed Earth's first atmosphere, while volcanic water vapor formed the ocean. The composition of this first atmosphere, before life evolved, looked much like that of Mars: thin, about 1% as dense as that of Earth today, consisting primarily of CO_2. But unlike Mars, Earth developed living organisms that filled the atmosphere with gaseous N_2 and O_2 and that continue to produce these gases today. Organisms also produce CO_2, as well as fixing it into biomass. Some CO_2 and N_2 arise from geological sources such as volcanoes, but their contribution is small compared to that of biological cycles (discussed in Chapter 22). The overall composition of Earth's atmosphere is determined by living organisms, primarily microbes.

Temperature. Another important aspect of Earth's habitat, determined by the atmospheric density and composition, is temperature. Atmospheric gases absorb light and convert the energy to heat, raising the temperature of the surface and atmosphere. This rise in temperature is known as the **greenhouse effect**. Because car-

bon dioxide is an especially potent greenhouse gas, the CO_2-rich atmosphere of early Earth could have heated the planet to temperatures approaching that of Venus, eliminating the possibility of life. Instead, microbial consumption of CO_2 and generation of nitrogen and oxygen gases limited Earth's surface temperature to an average of 13°C. The cooling effect may have led to an ice age, possibly forestalled by rising methane from methanogens. One way or another, the history of Earth's atmosphere is intimately related to the history of microbial evolution.

Geological Evidence for Early Life

The first period of Earth's existence, ranging 4.5–3.8 Gyr ago, is designated the **Hadean eon** (**Fig. 17.6**), named for Hades, the ancient Greek world of the dead. During the Hadean eon, repeated bombardment by meteorites vaporized the oceans, which then cooled and recondensed. Meteor bombardment may have killed off incipient life more than once before living microbes finally became established. Still, scientists speculate whether some forms of life might have survived Hadean conditions, perhaps growing 3 km below the Earth's surface. Like the dead spirits imagined by the Greeks to have populated Hades, Earth's earliest cells might have reached deep enough within the crust that they were protected from the heat and vaporization at the surface. As of this writing, however, no geological evidence supports the existence of life earlier than 3.8 Gyr ago.

The Archaean era. The earliest geological evidence for life dates to 3.8–2.5 Gyr ago, in the **Archaean eon** (**Fig. 17.6**). In the Archaean, meteor bombardment was less frequent, and the Earth's crust had become solid. The Archaean marked the first period with stable oceans containing the key ingredient of life: liquid water. Water is a key medium for life because it remains liquid over a wide range of temperatures and because it dissolves a wide range of inorganic and organic chemicals. Rock strata dating to the Archaean era reveal the first evidence of living organisms and their metabolic processes.

NOTE: The term *Archaean* refers to the earliest eon in which life existed, whereas *archaeal* is the adjective referring to the taxonomic domain Archaea. The domain Archaea (originally, Archaebacteria) was named by Carl Woese based on his theory that the species of this domain most closely resembled the earliest life-forms of the Archaean eon. In fact, early life may have encompassed diverse traits later associated with archaea, bacteria, and eukaryotes.

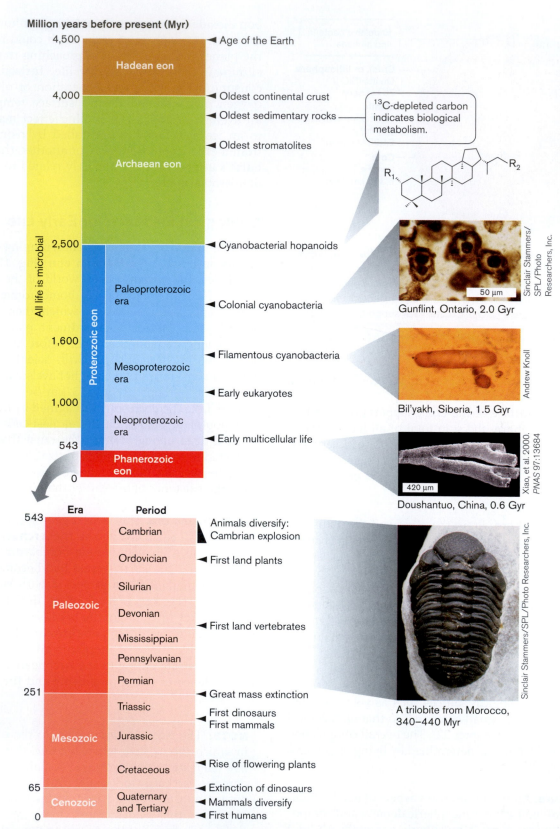

Million years before present (Myr)

4,500 — Age of the Earth
Hadean eon

4,000 — Oldest continental crust
— Oldest sedimentary rocks

Archaean eon
— Oldest stromatolites

¹³C-depleted carbon indicates biological metabolism.

2,500 — Cyanobacterial hopanoids

Paleoproterozoic era
— Colonial cyanobacteria

All life is microbial

Gunflint, Ontario, 2.0 Gyr
50 µm

1,600
Proterozoic eon
— Filamentous cyanobacteria

Mesoproterozoic era
— Early eukaryotes

Bil'yakh, Siberia, 1.5 Gyr

1,000
Neoproterozoic era
— Early multicellular life

543
Phanerozoic eon
0

Doushantuo, China, 0.6 Gyr
420 µm

Era	Period	
543	Cambrian	Animals diversify: Cambrian explosion
Paleozoic	Ordovician	First land plants
	Silurian	
	Devonian	First land vertebrates
	Mississippian	
	Pennsylvanian	
	Permian	
251	Triassic	Great mass extinction
Mesozoic	Jurassic	First dinosaurs / First mammals
	Cretaceous	Rise of flowering plants
65 Cenozoic	Quaternary and Tertiary	Extinction of dinosaurs / Mammals diversify / First humans
0		

A trilobite from Morocco, 340–440 Myr

Figure 17.6 Geological evidence for early life. The geological record shows evidence of microbial life early in Earth's history, 3 Gyr before the first multicellular forms.

How and when did living cells arise out of inert materials? Without a time machine to take us back 4 Gyr, we must rely on evidence from Earth's geology. Interpreting geology is a challenge because most forms of evidence for early life are indirect and subject to multiple interpretations. The farther back in time, the more change has occurred to the rock strata and the greater the difficulties. One way to meet this challenge, however, is to compare the results from different kinds of evidence (**Table 17.1**). If two or more kinds of evidence (such as microfossils and isotope ratios) point to life in the same location, the conclusion is strengthened.

Stromatolites. Fossil stromatolites are layers of carbonate or silicate rock that resemble modern living stromatolites. Presumably, the fossils formed as layers of phototrophic microbial communities grew and died, their

form filled in by calcium carbonate or silica. Fossil stromatolites accepted by geologists date as early as 3.4 Gyr ago (see **Fig. 17.1**). These rock formations appear remarkably similar to the layered forms of stromatolites today. The ancient rock, however, is too deformed to reveal the detailed structure of cells, and the biological origin of such fossils is questioned by some researchers.

Microfossils. The most convincing evidence is the visual appearance of **microfossils**, microscopic fossils in which calcium carbonate deposits have filled in the form of ancient microbial cells (**Fig. 17.7**). Microfossils are dated based on the age of the rock formation in which they are found, which in turn is based on evidence such as radioisotope decay. Convincing microfossils need to show regular three-dimensional patterns of cells that cannot be ascribed to abiotic (nonbiological) causes.

Table 17.1 Geological evidence of early life.

Type of evidence	Advantages	Limitations
Stromatolites. Layers of phototrophic microbial communities grew and died, their form filled in by calcium carbonate or silica.	Fossil stromatolites can be observed in the oldest rock of the Archaean eon. Their distinctive shapes resemble those of modern living stromatolites.	Some layered formations attributed to stromatolites have been shown to be generated by abiotic (nonbiological) processes.
Microfossils. Early microbial cells decayed, and their form was filled in by calcium carbonate or silica. The size and shape of microfossils resemble that of modern cells.	Microfossils are visible and measurable under a microscope, offering direct evidence of cellular form.	Microscopic rock formations require subjective interpretation. Some formations may result from abiotic processes.
Isotope ratios. Microbes fix $^{12}CO_2$ more readily than $^{13}CO_2$. Thus, limestone depleted of ^{13}C must have come from living cells. Similarly, sulfate reduced by sulfate-respiring bacteria shows depletion of ^{34}S compared with ^{32}S.	Isotope ratios offer objective measurement of a highly reproducible physical quantity. They provide the best evidence for dating the earliest life. Isotope ratios generated by key biochemical reactions can calibrate the timeline of phylogenetic trees.	We cannot prove absolutely that no abiotic process could generate a given isotope ratio. Isotope ratios tell us nothing about the form of early life or how it evolved.
Biosignatures. Certain organic molecules found in sedimentary rock are known to be formed only by microbes. These molecules are used as biosignatures.	Biosignatures such as hopanoids are complex molecules and highly specific to different life-forms.	In the oldest of rocks, organic molecules are eliminated by metamorphic processes.
Oxidation state. The oxidation state of metals such as iron and uranium indicates the level of O_2 available when the rock formed. Banded iron formations (BIF) suggest oxidation by microbial phototrophs that intermittently produced oxygen.	Oxidized metals offer evidence of microbial processes even in highly deformed rock.	It is hard to rule out abiotic causes of oxidation. If biological processes were responsible, the kind of metabolism is not revealed.

Microfossils

A. Filamentous prokaryotes

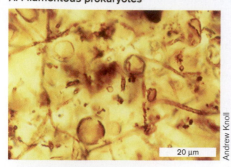

20 μm

Andrew Knoll

C. Colonial cyanobacteria

20 μm

© Hans Hofmann, McGill University, Montreal

E. Algae (eukaryote)

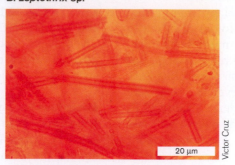

25 μm

Nicholas Butterfield, UK

Modern species

B. *Leptothrix* sp.

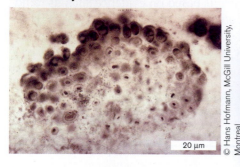

20 μm

Victor Cruz

D. *Entophysalis* sp.

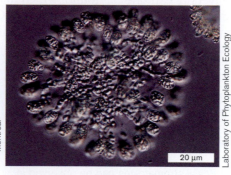

20 μm

Laboratory of Phytoplankton Ecology

F. *Bangia* sp., red algae

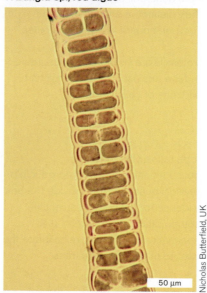

50 μm

Nicholas Butterfield, UK

Figure 17.7   Microfossils compared with modern bacteria.   A. Filamentous prokaryotes, 2.0 Gyr, from Gunflint formation, Ontario, Canada. **B.** Modern *Leptothrix* sp. filamentous bacteria. **C.** Colonial cyanobacteria, about 2.0 Gyr, from Belcher Islands, Canada. **D.** Modern *Entophysalis* cyanobacteria. **E.** Filamentous algae, 1.2 Gyr, from arctic Canada. **F.** Modern red algae, a eukaryote, *Bangia* species.

The earliest convincing microfossils are dated at 2.0 Gyr. Microfossils dated to 2.0 Gyr include filamentous prokaryotes in the Gunflint formation, Ontario, Canada (**Fig. 17.7A**). The Gunflint outcrops consist of chert, a kind of silicate formed by precipitation from an ancient sea. The sea was rich in carbonates and reduced iron, a good combination for redox metabolism. Some of the fossils resemble the form of filamentous iron-metabolizing bacteria today, such as *Leptothrix* species (**Fig. 17.7B**). Other microfossils dated at 2.0 Gyr resemble colonial cyanobacteria (compare **Fig. 17.7C** with **17.7D**). More recent strata, dated at 1.2 Gyr, contain larger fossil cells comparable to those of modern eukaryotes such as algae (**Figs. 17.7E** and **F**).

If life existed in more ancient times, such as the Archaean eon, where are the microfossils? Archaean rock is metamorphic, greatly modified by temperature and pressure. The macroscopic contours of stromatolites can be identified, but microfossil interpretation is highly

Figure 17.8   Microfossils or artifacts?   Structures originally identified as cyanobacterial microfossils from western Australia, dated to 3.85 Gyr. Further testing indicates that the structures are nonbiological artifacts.

Brasier, et al. 2002. *Nature* 416: 76

controversial. For example, microfossils of cyanobacteria dated to 3.85 Gyr by William Schopf in the early 1990s were accepted and described in many textbooks (**Fig. 17.8**). The 3.85-Gyr fossils have since been reinterpreted by Martin Brazier and colleagues as nonbiogenic artifacts (caused by abiotic processes). The form of the proposed Archaean microfossils is less regular and convincing than the form of later specimens, particularly when observed at different angles not shown in the original publication.

Another kind of evidence is that of a **biosignature** or **biological signature**, a chemical indicator of life. Biosignatures have been found that are even earlier than the oldest fossils. Their significance is limited, however, as it is hard to rule out nonbiogenic explanations, so researchers seek additional, corroborating evidence based on independent principles.

Isotope ratios. An **isotope ratio** may serve as a biosignature if the ratio between certain isotopes of a given element is altered by biological activity. Enzymatic reactions, unlike abiotic processes, are so selective for their substrates that their rates may differ for molecules containing different isotopes. For example, the carbon-fixing enzyme rubisco, found in chloroplasts, preferentially fixes CO_2 containing ^{12}C rather than ^{13}C. The carbon dioxide fixed into microbial cells eventually is converted to calcium carbonate in sedimentary rock. Thus, the calcium carbonate deposited by CO_2-fixing autotrophs (such as cyanobacteria) shows lower ^{13}C content than calcium carbonate deposited by abiotic processes (**Fig. 17.9B**). The difference, $\delta^{13}C$, is defined by the fractional difference (in parts per thousand) between the $^{13}C/^{12}C$ ratios in a sample versus a standard inorganic rock:

$$\delta^{13}C = \frac{^{13}C/^{12}C \text{ (experimental)} - {}^{13}C/^{12}C \text{ (standard)}}{^{13}C/^{12}C \text{ (standard)}} \times 1,000$$

Typical $\delta^{13}C$ values are shown in **Figure 17.9B**. Organisms on land and sea show $\delta^{13}C$ values of −10 to −30 parts per thousand, a significant ^{13}C depletion through carbon fixation. A comparable $\delta^{13}C$ is observed in fossil fuels, which formed from plant and animal bodies decomposed by bacteria.

A. Minik Rosing

BBC

Figure 17.9   Biosignatures of early life.   A. Minik Rosing (left), in Greenland, shows the Isua rocks whose carbon isotope ratios indicate photosynthesis at 3.8 Gyr. **B.** ^{13}C isotope depletion (negative $\delta^{13}C$) occurs in biomass as a result of the Calvin-Benson cycle. Negative $\delta^{13}C$ is observed at 3.7 Gyr in sedimentary graphite, which may derive from sedimented phototrophs. Little or no isotope depletion is seen in carbonate rock, which has no biological origin. **C.** 2-methylhopane, a biological signature of cyanobacteria, is found in rock strata dated at 2.5 Gyr. The 2-methyl group is highlighted.

B. ^{13}C **isotope depletion**

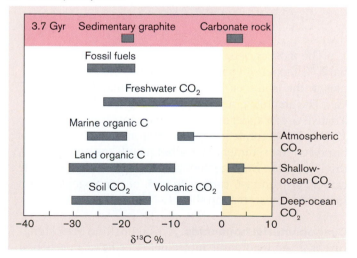

C. 2-Methylhopane

A.

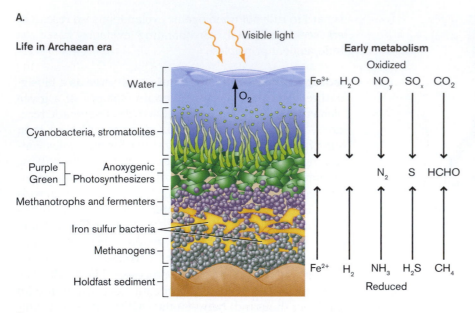

B.

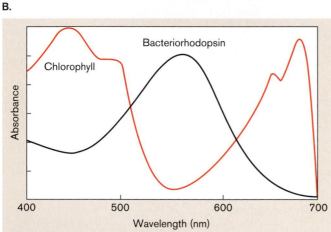

Figure 17.10   Early metabolism.
A. Early metabolism could have been based on various reactions between oxidized minerals that diffuse down from the air and water and reduced minerals in the sediment, upwelling from hydrothermal vents. **B.** Early photosynthesis may have resembled that of haloarchaea, whose bacteriorhodopsin absorbs light in the central range of sunlight (yellow-green, 500–600 μm), in contrast to the chlorophyll of cyanobacteria and plants, which absorbs blue and red. *Source:* B. Shil DasSarma, U. Maryland.

The most ancient mineral samples showing a substantial $\delta^{13}C$ (about −18 parts per thousand) are graphite granules in the Isua rock bed of West Greenland, dated at 3.7 Gyr. The graphite granules were analyzed by Minik Rosing, a native Greenlander at the Danish Lithosphere Center (**Fig. 17.9A**). The graphite grains derive from microbial remains buried within sediment that subsequently metamorphosed, driving out the water content but leaving behind the telltale carbon. By contrast, carbonate rock of nonbiogenic origin from the same formation shows a $\delta^{13}C$ near zero.

Cyanobacterial hopanoids. A different kind of biosignature is given by organic molecules specific to a particular life form. Certain organic molecules may last within rock for hundreds of millions of years. A particularly durable class of molecules are membrane lipids. Recall from Chapter 3 that bacterial cell membranes contain steroid-like molecules called hopanoids (**Fig. 17.9C**). A hopanoid consists of four or five fused rings of hydrocarbon with variable side groups, depending on the bacterial species.

The 2-methylhopanoids, containing a methyl side chain, are highly specific to cyanobacteria and are found in no other class of organism today. They are considered a biosignature of cyanobacteria.

The hopanoid derivative 2-methylhopane is found in sedimentary rock of the Hamersley Basin of Western Australia, dated to 2.5 Gyr. This highly specific biosignature is strong evidence that cyanobacteria existed by the end of the Archaean eon. Furthermore, because cyanobacterial photosynthesis is a relatively advanced form of metabolism, the finding suggests that more primitive microbes must have evolved earlier.

Metabolism of the First Cells

How did the earliest life-forms metabolize without oxygen gas to respire and without the complex machinery of photosynthesis? The nature of the first metabolism is unknown, but geochemistry and modern metabolism suggest several possibilities (**Fig. 17.10A**).

- **Oxidation-reduction reactions.** The early oceans contained oxidized forms of nitrogen, sulfur, and iron that could interact with reduced minerals from the crust. For example, nitrate (NO_3^-) or sulfate (SO_4^{2-}) could be reduced by hydrogen gas to yield energy (hydrogenotrophy, discussed in Chapter 14). The oxidized molecules were generated by reactions driven by ultraviolet radiation, which penetrated the atmosphere in the absence of the ozone layer. Sulfur isotope ratios ($^{34}S/^{32}S$) suggest the growth of sulfate-reducing bacteria as early as 3.47 Gyr ago.

- **Light-driven ion pumps.** A simple light-driven pump, such as the bacteriorhodopsin of haloarchaea (halophilic archaea), could have conducted the first kind of phototrophy. The absorption spectrum of bacteriorhodopsin matches the peak of solar radiation reaching the upper layers of ocean (**Fig. 17.10B**), in contrast to the chlorophylls of cyanobacteria and plants, which absorb the outer ranges of blue and red. Perhaps cyanobacteria evolved in the presence of haloarchaea, filling the unexploited photochemical niches.

■ **Methanogenesis.** Climate models of early Earth suggest a methane atmosphere, produced by methanogenic archaea. Methanogenesis involves reaction of H_2 and CO_2 producing CH_4 and H_2O. Methanogens show highly divergent genomes, a finding that suggests early evolution of their common ancestor. Their biochemistry and evolution (discussed in Chapter 19) are consistent with proposed models of ancient life.

Oxygen from Cyanobacteria Appeared Gradually

An extraordinary event in the planet's history was the evolution of the first oxygenic phototrophs, cyanobacteria that split water to form O_2. The entry of O_2 into Earth's biosphere is often portrayed as a sudden event that would have been disastrous to microbial populations that lacked defenses against its toxicity. In fact, geological evidence shows that oxygen arose gradually in the oceans, starting about 2 Gyr ago, and may have arisen and disappeared numerous times before reaching a high steady-state level in our atmosphere.

Banded iron formations. Evidence for oxygen in the biosphere comes from the oxidation state of minerals, particularly those containing iron. The bulk of crustal iron is in the reduced form (Fe^{2+}), which is soluble in water and reached high concentrations in the anoxic early oceans. Sedimentary rock, however, contains many fine layers of oxidized iron (Fe^{3+}), which is insoluble and forms a precipitate, such as iron oxide (Fe_2O_3). The layers of iron oxide suggest periods of alternating oxygen-rich and anoxic conditions. These layered iron minerals are called **banded iron formations (BIFs)** (**Fig. 17.11A**). A common form of banded iron consists of gray layers of silica oxide (SiO_2) alternating with layers colored red by iron

oxides and iron oxyhydroxides [$FeO_x(OH)_y$]. Banded iron formations are widespread around the world and provide our major sources of iron ore (**Fig. 17.11B**).

Banded iron formations are often found in rock strata containing signs of past life such as ^{13}C depletion and other biomarkers. For example, the Isua formations (Greenland) and Hamersley formations (western Australia), which both show ^{13}C depletion, also contain extensive banded iron. Calculations show that the layers of oxidized iron could result from biological metabolism involving iron oxidation. One possibility is that the iron was oxidized by chemolithotrophs, using molecular oxygen produced by cyanobacteria. The Archaean and early Proterozoic eons experienced fluctuating levels of molecular oxygen in the atmosphere. These fluctuations could have led to oscillating levels of iron oxide thus producing bands in the sediment, as microbes used up all the oxygen.

Alternatively, Dianne Newman, at the California Institute of Technology, proposes that the iron oxides arose directly from anaerobic photosynthesis, in which the reduced iron served as the electron donor (**Fig. 17.12**). Photosynthetic oxidation of Fe^{2+} to Fe^{3+}, or "photoferrotrophy," could have occurred in cycles until the marine iron was all oxidized, generating sedimentary layers of iron oxides and iron oxyhydroxides. This argument was supported by Newman's discovery of iron phototrophy by modern purple bacteria such as *Rhodopseudomonas palustris*.

By 2.3 Gyr ago, the prevalence of oxidized iron and other minerals indicates the steady rise of oxygen from photosynthesis in Earth's atmosphere. All the dissolved Fe^{2+} from the ocean floor was oxidized, leaving the oceans in the iron-poor state that persists today. Oxygen as the most efficient electron acceptor enabled the evolution of aerobic respiratory bacteria. Aerobic bacteria gave rise to mitochondria, which enabled the evolution of eukaryotes and ultimately multicellular organisms (**Fig. 17.13**).

Remarkably, modern cells are still composed primarily of reduced molecules, highly reactive with oxygen, a relic of the time when our ancestral cells evolved in the absence of oxygen. The conditions under which such cells may have evolved can be simulated in the laboratory— conditions under which some of life's most common

Figure 17.11 Banded iron formations. A. Banded iron formation in ancient sedimentary rock. Its main component is chert, a form of quartz (silica oxide, SiO_2) with layers colored red by iron oxide (Fe_2O_3). **B.** The BHP Iron Ore Mine at Newman, western Australia.

A.

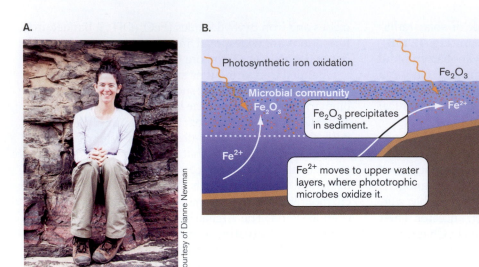

Courtesy of Dianne Newman

Photosynthetic iron oxidation

Fe_2O_3

Microbial community
Fe_2O_3 → Fe^{2+}

Fe_2O_3 precipitates in sediment.

Fe^{2+}

Fe^{2+} moves to upper water layers, where phototrophic microbes oxidize it.

Figure 17.12 **Iron phototrophy.**
A. Dianne Newman at the California Institute of Technology proposes that early iron phototrophs caused the iron oxide deposition generating banded iron formations.
B. Photosynthetic oxidation of Fe^{2+} to Fe^{3+} may have generated sedimentary layers of Fe_2O_3 and $FeO_x(OH)_4$.

Origin and evolution of life on Earth

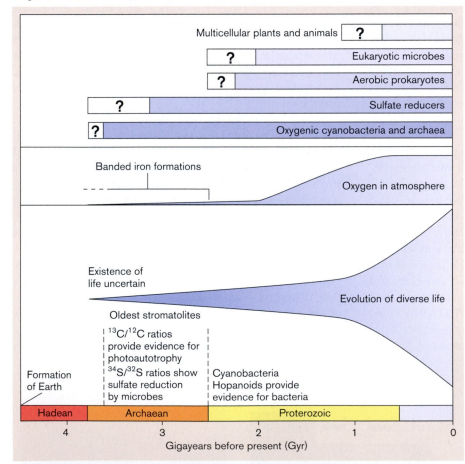

Figure 17.13 **Proposed timeline for the origin and evolution of life.** The planet Earth formed during the Hadean eon (about 4.5 Gyr ago). The environment was largely reducing until cyanobacteria pumped O_2 into the atmosphere. When the O_2 level reached sufficient levels (about 0.6 Gyr ago), multicellular animals and plants evolved. Question marks designate periods when evidence for given life-forms is uncertain.

molecules, such as adenine and simple amino acids, form spontaneously. These early-Earth simulation experiments can never prove the actual conditions under which life began, but they can suggest testable models with intriguing implications.

TO SUMMARIZE:

- **Elements of life** were formed through nuclear reactions within stars that exploded into supernovae before the birth of our own sun.
- **Reduced molecules** compose Earth's interior. Oxidized minerals are found only near the surface. Early Earth had no molecular oxygen (O_2).
- **Archaean rocks show evidence for life** based on fossil stromatolites, isotope ratios, and chemical biosignatures. Fossil stromatolites appear in chert formations formed 3.4 Gyr ago. Isotope ratios for carbon indicate photosynthesis at 3.7 Gyr ago and sulfate reduction at 3.47 Gyr. Cyanobacterial hopanoids appear at 2.5 Gyr ago.
- **Microfossils of filamentous and colonial prokaryotes** date to 2.0 Gyr ago. At 1.2 Gyr, larger fossil cells resemble those of modern eukaryotes.
- **Early metabolism** involved anaerobic oxidation-reduction reactions. Likely forms of early metabolism include sulfate respiration, light-driven ion pumps, iron phototrophy, and methanogenesis.

■ **Banded iron formations** reflect the cyclic increase and decrease of oxygen produced by cyanobacteria and consumed through reaction with reduced iron. After all the ocean's iron was oxidized, oxygen increased gradually in the atmosphere.

17.2 Models for Early Life

Various models have been proposed to explain how the first life-forms originated from nonliving materials and how they replicated and evolved. Models for early life attempt to address the following questions: In what kind of environment did the first cells form? What kind of metabolism did the first cells use to generate energy? What was their hereditary material?

Models for Origin of the First Cells

Models for early life include:

■ **The prebiotic soup.** Organic building blocks of life could arise **abiotically** (in the absence of life) out of simple reduced chemicals such as ammonia and methane. This "prebiotic soup" could have generated complex macromolecules that eventually acquired the apparatus needed for self-replication and membrane compartmentalization.

■ **Metabolist models.** The central components of intermediary metabolism, including the TCA cycle to generate amino acids, arose from self-sustaining chemical reactions based on inorganic chemicals. These abiotic reactions then acquired self-replicating macromolecules and membranes.

■ **The RNA world.** Proposed by Francis Crick in the 1960s, the **RNA world** is a model of early life in which RNA performed all the informational and catalytic roles of today's DNA and proteins. The concept of an RNA world derives support from the emerging sequences of genomes, which reveal thousands of catalytic and structural RNAs.

The prebiotic soup. In the mid-twentieth century, biochemists Aleksandr Oparin, at Moscow University, and Stanley Miller and Harold Urey, at the University of Chicago, showed that organic building blocks of life could arise abiotically out of a mixture of water and reduced chemicals, including CH_4, NH_3, and H_2 (**Fig. 17.14**). The mixture was subjected to an electrical discharge, similar to the lightning discharges that arise from volcanic eruptions, which would have been common in the late Hadean or early Archaean eons. The chemical reaction produced fundamental amino acids such as glycine and alanine. Similar experiments by Juan Oró, at the University of Houston, showed the formation of adenine by condensation of ammonia and methane. The same amino acids and nucleic bases are found in meteorites, which are believed to retain the chemistry of the early solar system "frozen" in time.

The original model conditions for the prebiotic soup assumed that molecules in the early Archaean ocean were largely reduced, with little or no oxygen present in the atmosphere. More recent geochemical evidence suggests that the early ocean actually included oxidized forms of nitrogen, sulfur, and iron that arose through reactions driven by ultraviolet radiation, which penetrated the atmosphere in the absence of the ozone layer. These oxidized minerals could have reacted with the reduced crustal minerals, releasing energy to drive production of more complex biomolecules.

Forming the first cell, or "proto-cell," must have required enclosing the first biochemical reactants within a membrane-like compartment. Such compartments can arise spontaneously from fatty acid derivatives such as fatty acid glycerol esters. The fatty acid derivatives are "amphipathic," that is, they posses both hydrophobic portions that associate together as well as hydrophilic portions that associate with water. In water, the fatty acid derivatives collect in "micelles," small round aggregates in which the hydrophobic portions associate in the interior and the hydrophilic portions associate with water. Under certain conditions, micelles can aggregate to form hollow spheres of membrane. The membrane spheres can

Figure 17.14 The prebiotic soup model for the origin of life. A. In the prebiotic soup, inorganic molecules could have reacted to form complex macromolecules that eventually acquired the apparatus for self-replication and membrane compartmentalization. **B.** Lightning accompanies eruption of the Galunggung volcano in West Java, Indonesia, 1982. The first biomolecules may have formed as a result of lightning triggered by volcanic eruption.

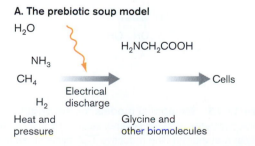

A. The prebiotic soup model

H_2O

NH_3

CH_4

H_2

Heat and pressure

Electrical discharge

H_2NCH_2COOH

Glycine and other biomolecules

Cells

B.

R. Hadian, U.S. Geological Survey

be made to enclose molecules such as RNA, suggesting a primitive cell-like form. These spontaneous processes of membrane formation suggest models for how the first living cells may have arisen.

Metabolist model. Other models attempt to explain the origin of biosynthesis based on CO_2 fixation, the fundamental metabolism of all life today. Proponents of a **metabolist model**, including Harold Morowitz and Günter Wächterhäuser, at George Mason University, propose that CO_2-based metabolism originated through self-sustaining reactions (**Fig. 17.15A**). Simulation experiments suggest that such premetabolic reactions could have been catalyzed by metal sulfides prevalent in the early ocean. For example, abiotic polymerization of CO_2 into TCA cycle intermediates may be catalyzed by FeS.

How did simple inorganic reactions lead to biochemical cycles involving more complex biopolymers such as nucleic acids and proteins? A major challenge is to explain the genetic code by which nucleic acid codons are assigned to unique amino acids. The genetic code may have arisen from an earlier pretranslational mechanism of amino acid biosynthesis (**Fig. 17.15B**). In this mechanism, proposed by Morowitz and colleagues, amino acids were originally synthesized from TCA cycle acids complexed to a dinucleotide. The dinucleotide later evolved into the first two nucleotides of the codon specifying the amino acid. The proposed dinucleotide association explains certain features of the genetic code, such as the fact that most amino acids specified by a codon starting with the same

nucleotide are synthesized from the same TCA cycle acid (discussed in Chapter 15).

The RNA world. Neither the prebiotic soup model nor the metabolist models account for the evolution of macromolecules that encode complex information, such as nucleic acids and proteins. A candidate for life's first "informational molecule" is RNA. RNA is a relatively simple biomolecule, with only 4 different "letters," compared to the 20 standard amino acids of proteins. Its purine bases, especially adenine, have been shown to arise spontaneously from ammonia and carbon dioxide under conditions believed to resemble those of the Archaean eon. Its ribose sugar is a fundamental building block of living cells, with key roles in numerous biochemical pathways, including the Calvin-Benson cycle and the synthesis of deoxyribose for DNA. Compared to DNA, RNA is easier to form and degrade, and its pyrimidine base uracil is formed early by biochemical pathways; only later is it transformed to the thymine used by DNA.

Most importantly, RNA molecules have been shown to possess catalytic properties analogous to those of proteins. Catalytic RNA molecules are called **ribozymes**. The first ribozyme, discovered by Nobel laureate Tom Cech in the protist *Tetrahymena*, can splice introns in mRNA. Other ribozymes actually catalyze synthesis of complementary strands of RNA, suggesting a model for early replication of RNA chromosomes. The most elaborate example of catalytic RNA is found in the ribosome. In the ribosome, X-ray crystallography reveals that the

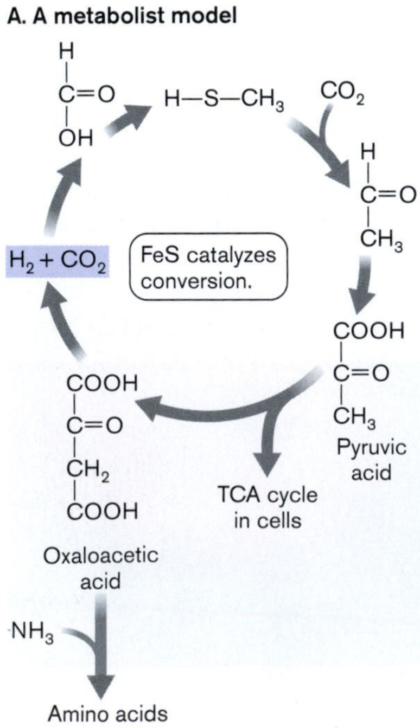

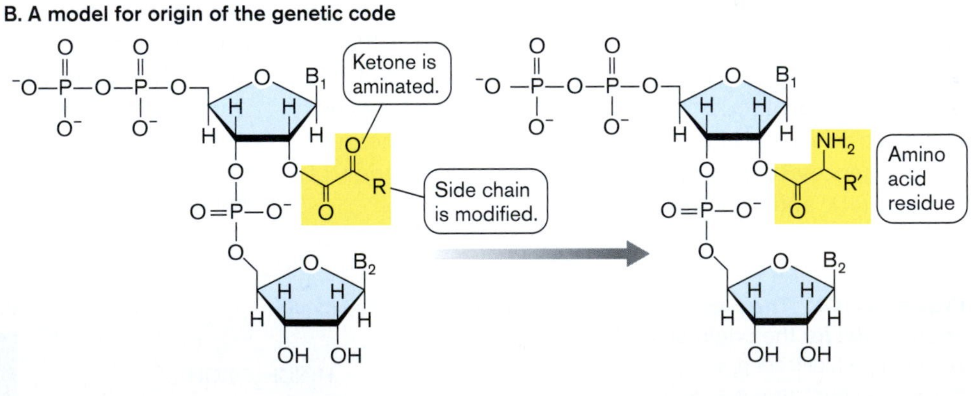

A. A metabolist model

B. A model for origin of the genetic code

Figure 17.15 **Metabolist models for the origin of life.** **A.** Self-sustaining abiotic chemical reactions, such as polymerization of CO_2 and H_2, could have formed the basis of cellular metabolism. The reductive TCA cycle arose out of intermediates with small thermodynamic transitions to CO_2. **B.** The genetic code may have originated from synthesis of an amino acid complexed to a dinucleotide. This dinucleotide could later have evolved into the first two nucleotides of the codon specifying the amino acid.

key steps of protein synthesis, such as peptide bond formation, are actually catalyzed by the RNA components, not proteins (discussed in Chapter 8). The ribosomal proteins possess relatively little catalytic function; their main role seems to be protection and structural support of the RNA.

Thus, in the earliest cells, RNA might have fulfilled all the functions today filled by DNA and proteins, including information storage, replication, and catalysis (**Fig. 17.16**). This model is known as the "RNA world." The prominent function of RNA in the ribosome, one of life's most ancient and conserved molecular machines, leads to a model for the transition from an RNA world to the modern cell, with unexpected relevance for modern medicine (**Special Topic 17.1**).

The RNA world model explains the central role of RNA throughout the history of living cells. Yet the role of RNA offers little clue as to the origins of cell compartmentalization and metabolism. No one model of life's origin yet addresses all the requirements for a living cell—metabolism, membrane compartmentalization, and hereditary material. Each model does, however, offer important insights into the evolutionary history and mechanisms of life today.

> **THOUGHT QUESTION 17.2** Outline the strengths and limitations of each model of the origin of living cells. Which aspects of living cells does each model explain?

Unresolved Questions about Early Life

Overall, geology and biochemistry provide compelling evidence that organisms resembling today's cyanobacteria lived on Earth at least 2.5 Gyr ago, possibly 3.7 Gyr ago, and that bacteria with anaerobic metabolism evolved as early or earlier. Many intriguing questions remain. We outline here three unresolved questions regarding the temperature of early Earth, the role of methane in the early atmosphere, and the actual source of Earth's first cells.

Thermophile or psychrophile? The apparent existence of life so soon after Earth cooled suggests a thermophilic origin. Thermophily is supported by the fact that in the domains Bacteria and Archaea, the deepest-branching species (that is, species that diverged the earliest from others in the domain) are thermophiles. Such organisms could have thrived at hydrothermal vents, which offer a continual supply of H_2S and carbonates.

On the other hand, after meteoric bombardment abated, early Earth should have become glacially cold. In the Archaean, solar radiation was 20–30% less intense than it is today, and the thin CO_2 atmosphere was insufficient to increase the temperature by a greenhouse effect. A colder habitat would support psychrophiles. Psychrophiles might have had an advantage in an "RNA world," given the thermal instability of RNA compared to DNA and proteins.

A world of methane? Among Earth's earliest life-forms were methanogens (methane producers). Methanogens are one of the most widely divergent groups of organisms and persist today in environments ranging from anaerobic sediment to the human intestine. Methanogenesis requires only carbon dioxide and hydrogen gas, which would have been plentiful in early anoxic sediment. The production of methane, an extremely potent greenhouse gas, could have greatly increased Earth's temperature during the Archaean. Thus, methanogenesis could explain how Earth escaped the permanent freeze of Mars. Overheating would have been halted when the oxygen gas produced by cyanobacteria enabled growth of methanotrophs, bacteria that oxidize methane. The decline of methane and the rise of CO_2 then brought about relative thermal stability.

The debate over temperature and climate of early Earth has interesting implications for Earth today when we again face the prospect of massive global climate change. Human agriculture favors explosive growth of methanogens, which threaten to accelerate global warming faster than the biosphere can moderate it. Understanding the climate of early Earth may help us better understand and manage our own climate, as discussed in Chapter 22.

Origin on Earth or elsewhere? Twenty years ago, the accepted picture of life's origin was still one of anaerobic early metabolism that gave rise much later to oxygenic and aerobic metabolism. But emerging evidence from fossils and geochemistry has inexorably pushed back the earliest known dates for several kinds of metabolism closer to 3.7 Gyr. This implies that as soon as Earth cooled to a temperature suitable for life, all the fundamental components of cells evolved almost immediately. How could

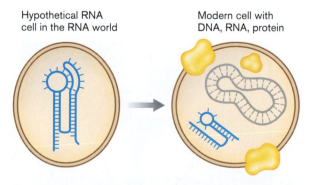

Hypothetical RNA cell in the RNA world

Modern cell with DNA, RNA, protein

Figure 17.16   The RNA world model for the origin of life. In the RNA world, RNA molecules performed all the structural and catalytic functions performed today by proteins and DNA.

Special Topic 17.1   The RNA World: Clues for Modern Medicine

An intriguing question is how the cells of the hypothetical RNA world evolved to use proteins. Tom Cech, Sidney Altman, and colleagues propose that the earliest RNA components of cells acquired peptide components to enhance stability, decreasing the tendency of RNA to hydrolyze (**Fig. 1**). As the peptide components increased through natural selection, the RNA component may have shrunk by reductive evolution (the loss of a trait in the absence of selection pressure) until all that remained was one or two nucleotides. Dinucleotide cofactors such as NADH persist in enzymes today, perhaps representing vestigial remnants of the RNA world.

Furthermore, the persistence of ribonucleotides in molecular mechanisms of gene regulation argues for an early central role of RNA in cell function. Bacterial operons for ribosomal components and for purine and pyrimidine biosynthesis are regulated not by DNA-binding proteins but by RNA molecules known as riboswitches, that interact with the DNA promoters (discussed in Chapter 15). Finally, many viruses have RNA genomes and ribozymes that interact with host cells in unique ways, as discussed in Chapters 6 and 11.

If cells acquired proteins through evolution, how did they acquire DNA? A controversial proposal by Patrick Forterre, at the University of Paris-Sud in Orsay, France, is that DNA was acquired through viruses. As discussed in Chapters 6 and 11, many kinds of viruses are capable of inserting their genomes into a host cell for replication along with host chromosomes. Viruses could have evolved DNA as a modified form of chromosome that avoided cleavage by protective enzymes of the RNA-genomic host. A viral DNA molecule within a host cell could eventually acquire genes from the host RNA chromosome by recombination or reverse transcription. Host genes converted to DNA would tend to outlast their RNA counterparts because of DNA's greater stability. Eventually, according to Forterre's model, the entire host cell genome would be converted to DNA.

The study of life's origin some billions of years ago may seem distant from modern human concerns. Yet investigation of early-life phenomena such as ribozymes has led to important developments in medical research. Jennifer Doudna and colleagues discovered the function of an unusual RNA component of hepatitis C virus (**Fig. 2**). The hepatitis viral mRNA possesses a specialized stem loop structure called an internal ribosome entry site (IRES). The IRES binds to a host ribosome and shuts down host protein synthesis while facilitating translation of the viral mRNA. The IRES site then catalyzes cleavage of the RNA to complete the formation of viral genomes by rolling circle replication.

A.

David Lilley, University of Dundee, UK

Figure 1   Transition from the RNA world to modern cells.   A. Nobel laureate Tom Cech (left) proposed a model for the RNA world transition to protein-based cells. He is shown here discussing a student's poster at an international workshop on ribozymes in Dundee, Scotland; 2001. **B.** Earliest cells may have been composed of RNA enzymes (ribozymes). As the RNA cells evolved, the ribozymes acquired protein subunits that eventually assumed most of the catalytic functions. Remnants of the original RNA may persist as nucleotide cofactors such as NADH. **C.** Nobel laureate Sidney Altman (right) studies a genomic sequence with a student at Yale University. Source: B. *Yale Bulletin* 2002–2003 Vol. 2, Number 1.

B.

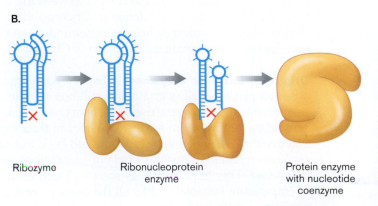

Ribozyme   Ribonucleoprotein   Protein enzyme
                  enzyme   with nucleotide
                                             coenzyme

C.

© 2001 Yale University

Doudna published the first structure of an IRES bound to a ribosome initiation site (**Fig. 2B**). The IRES binds specifically to the ribosome at the mRNA recognition groove, so as to facilitate transcription of viral templates while excluding those of the host. Its tertiary structure positions a cytosine just right so as to catalyze hydrolysis of the RNA backbone (**Fig. 2C**).

The study of viral ribosomes has led to design of artificial ribozymes for chemotherapy. One goal is to design a ribozyme to cleave a gene transcript essential in cancer cells. The potential applications of artificial ribozymes have led to the start-up of several ribozyme companies; for example, Ribozyme Pharmaceuticals, Inc. (RPI) conducts clinical trials on ribozymes to treat cancer.

A.

Courtesy of Ronald Sutherland, UC, Berkeley

B.

Figure 2 **A viral ribozyme.** **A.** Jennifer Doudna elucidates the structure and function of ribozymal components of viruses such as hepatitis C virus. **B.** The internal ribosome entry site (IRES) of the hepatitis C viral mRNA, showing the catalytic site. (PDB code: 1VC5) **C.** The IRES is folded such that specific cytosine is positioned to catalyze hydrolysis of the RNA backbone. *Source*: B. J. Doudna and T. Cech. 2002. *Nature* 418:222.

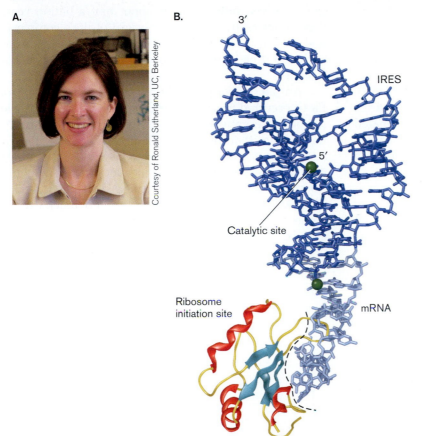

life, with all its diverse kinds of metabolism, have arisen so quickly?

The mystery deepens as we combine geochemistry with **phylogeny**, the measurement of genetic relatedness (discussed in the next section). Australian geochemist Roger Buick and colleagues at the University of Washington discovered early evidence of sulfate respiration based on isotope ratios of sulfur (**Fig. 17.17A**). Sulfate reducers use sulfate (SO_4^{2-}) as an electron acceptor to catabolize organic compounds in the absence of O_2 (as discussed in Chapter 14). The sulfate reductase enzyme preferentially acts on ^{32}S, generating sulfide depleted for ^{34}S. Buick discovered ^{34}S-depleted sulfides trapped in rock crystals of formations in western Australia dated to 3.47 Gyr.

The early date was surprising because sulfate respiration is performed by a class of bacteria that evolved relatively late, based on DNA sequence analysis (**Fig. 17.17B**). Sequence similarity data show that sulfate respirers diverged from other bacteria well after bacteria diverged from archaea and eukaryotes. So if the sulfate respirers had already diverged by 3.47 Gyr ago, when did bacteria diverge from archaea?

DNA sequence analysis shows that the common ancestor of bacteria, eukaryotes, and archaea already possessed most of the complexity of the organisms that diverged from it. Furthermore, data from Buick's laboratory in 2006 indicates biosignatures of eukaryotes as early as 2.4 Gyr ago, far earlier than was thought possible.

If the rate of evolution were constant, the three major domains would have diverged before the formation of Earth. In other words, life would have evolved elsewhere, and then come to Earth.

The idea that life-forms originated elsewhere and "seeded" life on Earth is called **panspermia**. Theories of panspermia remain highly speculative. One hypothesis is that microbial life originated on Mars and was then carried to Earth on meteorites. As the solar system formed, Mars would have cooled sooner than Earth, and as a smaller planet, its weaker gravity would have generated less bombardment by meteorites. Martian rocks ejected into space by meteor impact have reached the Earth, and calculations based on simulated space habitats show that microbes could survive such a journey. A Martian origin, however, only gains us about half a billion years; it does not really explain the origin of life's complexity and diversity. Did life-forms come from still farther away, perhaps borne on interstellar dust from some other solar system?

An alternative possibility is that early evolution on Earth occurred much faster than later evolution. Simpler reproductive entities such as viruses mutate and evolve much faster than modern cells. If early cells mutated as fast as viruses, they would have evolved faster than indicated by molecular clock molecules such as small-subunit ribosomal RNA (a method discussed in Section 17.3). Thus, for example, the common ancestor of Buick's sulfate reducers and cyanobacteria may have arisen on Earth but evolved much faster than later bacteria. The answers are unknown.

> **THOUGHT QUESTION 17.3**   Suppose living organisms were to be found on Mars. How might such a find shed light on the origin and evolution of life on Earth?

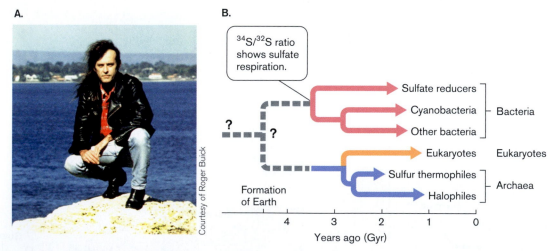

Figure 17.17 Chemical biosignatures calibrate microbial divergence.   A. Roger Buick, Australian geochemist, explores isotope ratios and fossil evidence for early microbial life. Besides science, he likes "listening to punk bands, reading interminable Russian novels and arguing politics ferociously." **B.** The divergence of major life-forms is based on comparing sequences of small-subunit rRNA. Discovery of ^{34}S-depleted rock formations 3.5 Gyr old suggests that sulfate respirers evolved by that time. Other biosignatures calibrate the divergence of eukaryotes at 2.70 Gyr. *Source*: B. Based on Shen, et al. 2001. *Nature* 410:77–81.

TO SUMMARIZE:

- **Prebiotic soup models** propose that the fundamental biochemicals of life arose spontaneously through condensation of reduced inorganic molecules.

- **Metabolist models** propose that components of intermediary metabolism arose from self-sustaining chemical reactions that connected nucleotides with amino acids, forming the basis of the genetic code.

- **The RNA world model** proposes that in the first cells, RNA performed all the informational and catalytic roles of today's DNA and proteins.

- **Thermophile or psychrophile?** Classic models of early life assume thermophily, but Earth may actually have been cold when the first cells originated.

- **A world of methane?** If the first cells were methanogens, methane production could have led to the first greenhouse effect, warming the Earth and enabling evolution of other kinds of life.

- **Origin on Earth or elsewhere?** Isotope ratios suggest the presence of complex metabolism by 3.7 Gyr ago, shortly after Earth cooled (3.9 Gyr ago). Simpler cells existing before 3.9 Gyr ago may have evolved much faster than life today, or they may have come to Earth from another location, where life evolved earlier. These hypotheses are highly speculative.

17.3 Microbial Taxonomy

Once living cells arose, they evolved and diverged, generating all the many diverse forms of modern life, an ongoing process that continues as populations become extinct while new groups emerge. The description of distinct life-forms and their organization into different categories (**taxa**, singular **taxon**) is called **taxonomy**. Taxonomy includes **classification**, the recognition of different classes of life; **nomenclature**, the naming of different classes; and **identification**, the recognition of the class of a given microbe isolated in pure culture. In this section, we consider how different kinds of microbes are classified and identified, based on DNA sequence as well as phenotypic traits. DNA sequences often bring surprises, upsetting traditional assumptions about life's family tree.

Classification and Nomenclature

Classification generates a hierarchy of taxa (groups of related organisms) based on successively narrow criteria. The fundamental basis of modern taxonomy is DNA sequence similarity; but the use of DNA arises within a long historical tradition of phenotypic description that shapes the views and practice of microbial taxonomy. Historically, taxa have been defined and named based on a combination of genetic and phenotypic traits. The nomenclature of microbes remains surprisingly fluid; as new traits are identified and the genetic sequence of a microbe is established, species are all too frequently renamed. Current information on microbial names is available through online databases such as List of Prokaryotic Names with Standing in Nomenclature.

WWW | List of Prokaryotic Names with Standing in Nomenclature

Traditional classification designates levels of taxonomic hierarchy, or **rank**, such as phylum, class, order, and family. Some levels of rank are designated by certain suffixes: for example, -*ales* (order) and -*aceae* (family). The ultimate designation of a type of organism is that of **species**, which includes the capitalized **genus name** (group of closely related species) followed by the uncapitalized **species name**, for example, *Streptomyces coelicolor*.

The conventions of nomenclature and levels of rank vary widely across different taxa, as seen in **Table 17.2**. Species with a long history of study tend to show many levels of rank in the literature, whereas species discovered recently show relatively few, regardless of genetic diversity. *Streptomyces coelicolor* is a well-studied member of the actinomycetes, filamentous Gram-positive bacteria that produce many complex natural products, including antibiotics. The actinomycetes are subdivided into classes, orders, and families, as well as sublevels of the traditional ranks. Compare the actinomycetes with a species of a more recently described class, the hyperthermophilic methanogen *Methanocaldococcus jannaschii* (formerly *Methanococcus jannaschii*). *M. jannaschii* is defined only by the minimal levels of rank, without subclasses or suborders.

Still more recent, a group of mostly unculturable marine species, the SAR11 cluster, was found to comprise approximately 25% of all marine microbial cells. SAR11 was originally defined based on its small-subunit rRNA sequence of uncultured environmental samples. Surprisingly, the rRNA sequence places these marine oligotrophs in the order Rickettsiales, most of whose known members are obligate intracellular parasites, such as *Rickettsia*, the cause of Rocky Mountain spotted fever. The SAR11 cluster has not been classified below the rank of order, except for one species cultured in 2002, *Pelagibacter ubique*.

NOTE: Taxonomic categories generally have two forms, formal and informal. The formal term is capitalized, with a latinized suffix: Actinomycetales, *Pseudomonas*, *Micrococcus*. The informal term is lowercase, in some cases with an anglicized ending; and informal references to genera are not italicized: actinomycete, pseudomonad, micrococci.

Table 17.2   Taxonomic hierarchy of classification.

Taxon rank	A long-studied taxon	A less-studied taxon	An uncultivated environmental sample
Domain	Bacteria	Archaea	Bacteria
Division	Actinobacteria	Euryarchaeota	Proteobacteria
(phylum)	Filamentous gram-positive	Methanogens and halophiles	Purple bacteria and relatives; gram-negative
Class	Actinobacteria	Methanococci	Alpha Proteobacteria
	High GC gram-positive	Methanogens	Gram-negative bacteria
Subclass	Actinobacteridae		
Order	Actinomycetales	Methanococcales	Rickettsiales
	Filamentous; acid-fast stain	Methanogenic cocci	Includes intracellular bacteria
Suborder	Streptomycineae		
Family	Streptomycetaceae	Methanocaldococcaceae	SAR11 cluster
	Filamentous; hyphae produce spores	Thermophilic methanogens	Nonculturable planktonic marine bacteria
Genus	*Streptomyces*	*Methanocaldococcus*	*Pelagibacter*
Species	*S. coelicolor*	*M. jannaschii*	*P. ubique*
(date first described)	(1908)	(1984)	(2002)

Nongenetic Categories for Medicine and Ecology

Genetic relatedness is the standard for classifying and naming organisms in all fields of biology. At the same time, several nongenetic systems of categorization serve a practical purpose in certain fields. These systems include:

- **Phenotypic categories for identification.** Categories such as pigmentation or cell shape (rod or coccus) may have minor genetic significance but are useful for practical identification of organisms isolated from field or clinical sources.
- **Ecological categories.** In ecology, the niche filled by an organism may be more important than its phylogeny. For example, cyanobacteria and sequoia trees both fill the role of photosynthetic producer of biomass. Both are primary producers, a trophic category of organisms that feed other organisms in the food web. Ecological categories are discussed further in Chapter 21.
- **Disease categories.** In medical microbiology, microorganisms are categorized according to the type of disease they cause or the host organ system they inhabit. For example, *Mycoplasma pneumoniae* and influenza virus are both pulmonary pathogens, whereas *Escherichia coli* and *Bacteroides thetaiotaomicron* are both normal gut flora. Disease categories are discussed further in Chapters 25 and 26.

Defining a Species

What is a species? Among eukaryotes, a species is defined by the principle that members of different species do not normally interbreed with each other. Thus, the failure to interbreed is the traditional property defining species. Microbes, however, and especially prokaryotes, reproduce asexually, so interbreeding is not a basis for classification. At the same time, microbes show much evidence of **horizontal transfer** of genetic material (also known as **lateral transfer**), that is, gene flow between distantly related species. Horizontal transfer is particularly problematic in natural communities where distantly related organisms interact. In the laboratory, on the other hand, one population may give rise to many different isolated strains that acquire different mutations during laboratory cultivation, such as resistance to an antibiotic. Even populations frozen away in laboratory stocks may show selective survival of genetic variants.

Microbial species are defined based on DNA relatedness. Members of a common species are generally defined as a value of 70% or greater genomic sequence similarity based on DNA-DNA hybridization, and 97% ribosomal small subunit sequence similarity. However, definitions of a species are debated. For example, the genus *Salmonella* historically was subdivided into hundreds of species and subspecies. In 1987, a study based on DNA analysis proposed renaming all *Salmonella* bacteria as members of a single species, *Salmonella enterica*, subdivided by subspecies and serovars (strains expressing envelope proteins with different antigenicity). Clinicians objected, however, insisting on retaining species designation for certain strains that cause distinctive diseases, such as *Salmonella typhi*, which causes typhoid fever.

A commonly accepted set of names for species and higher taxa is essential for research and communication.

The official rules for naming species and taxa are determined by the International Committee on Systematics of Prokaryotes (ICSP). ICSP establishes minimal criteria for designating species, genera, classes, and other taxa of bacteria and archaea. To establish a new species, a new form of microbe must be isolated and grown in pure culture, known as an **isolate.** The isolate's unique genetic and phenotypic traits are published, with a proposed species name designated "*Candidatus*" for **candidate species.** An example of a candidate species is *Candidatus "Nitrosopumilus maritimus,"* a marine archaeon that oxidizes ammonia, first isolated in 2005. The candidate species becomes accepted as an official species upon publication in the *International Journal of Systematic and Evolutionary Microbiology,* the official journal of record for novel prokaryotic taxa.

All accepted taxonomic categories and species descriptions are compiled in *Bergey's Manual of Determinative Bacteriology,* a multivolume reference work of prokaryotic taxonomy. Up-to-date taxonomy of prokaryotes as well as eukaryotes is maintained in the interactive online database National Center for Biotechnology Information (NCBI) under the Taxonomy Browser, supported by the U.S. government.

We will revisit the question of defining bacterial species at the end of Section 17.5, after considering the molecular basis of phylogeny and gene sequence relatedness.

Identification

Once a species has been described and classified, we need a way to identify future members of the species isolated from natural environments. Identification poses special difficulties for microbes, which by definition are invisible to the unaided eye. Even under the electron microscope, thousands of divergent species may possess similar shape and form.

Because the definition of species is based on its genetic sequence, the most consistent and straightforward way to identify an isolate is to sequence part or all of its genome or to attempt hybridization of its DNA with a labeled sequence probe (see Chapter 7). In clinical practice, such DNA-based methods are used increasingly. Nevertheless, even in the clinical lab—as well as in field and environmental microbiology—it is convenient to narrow down the possibilities using various easily determined traits, such as cell shape, staining properties, and metabolic reactions. Thus, practical identification is based on a combination of phylogeny (relatedness based on DNA sequence divergence) and phenetic or phenotypic traits.

A common strategy of practical identification is the **dichotomous key**, in which a series of yes/no decisions successively narrows down the possible categories of species. **Figure 17.18** shows an example of a dichotomous key used to identify filamentous bacteria found during wastewater treatment. Filamentous bacteria are important in wastewater because their entangled filaments interfere with the settling of bacteria out of the treated water. Note that in the key shown, most of the traits are phenotypic, such as cell size and motility. Most of the branch decisions have two choices, although one juncture (cell motility and shape) has four. The key identifies some organisms down to the species level (*Sphaerotilus natans*), others to the genus level (*Flexibacter*), and others only to numbered samples whose characterization remains incomplete (0914, 021N).

> **THOUGHT QUESTION 17.4** Based on **Figure 17.18**, what would be the identification of a straight nonsheathed gram-negative bacterium that has sulfur granules, is motile, and is 1.0 μm wide? What would happen if you happened to assign the bacterium a width of 0.9 μm?

A disadvantage of the dichotomous key is that it requires a series of steps, each of which takes time. In the clinical setting, time is critical in identifying a potential pathogen and prescribing appropriate treatment. An alternative means of identification in the clinical setting is the probabilistic indicator. A **probabilistic indicator** is a battery of biochemical tests performed simultaneously on an isolated strain (**Fig. 17.19**). The indicator requires a predefined database of known bacteria from a well-studied habitat, such as Gram-negative bacteria from the human intestinal tract. To build such a database, numerous strains of each species must be isolated in pure culture and tested for all the traits in the database. The fraction of isolates that test positive for each trait, for each species, is noted in the database as the probability of obtaining a positive result. The probability of a negative result would be 1 minus the probability of a positive result. For example, if 97 out of 100 strains of *Salmonella enterica* test positive for ornithine decarboxylase, the probability of a positive result is 0.97, and the probability of a negative result is 0.03.

The database can be used to identify a newly isolated specimen from a patient. When the test results are obtained, the probabilities for all results, negative or positive, for a given species are then multiplied to generate a probability score for the species. For each species in the database, a score is computed, and the scores are sorted. The top-scoring species is then taken as the most probable identification.

A simplified example of indicator results appears in **Figure 17.19B.** Most of the tests generate pH-dependent color changes from products of bacterial catabolism, such as fermentation of sugars to acid or decarboxylation of

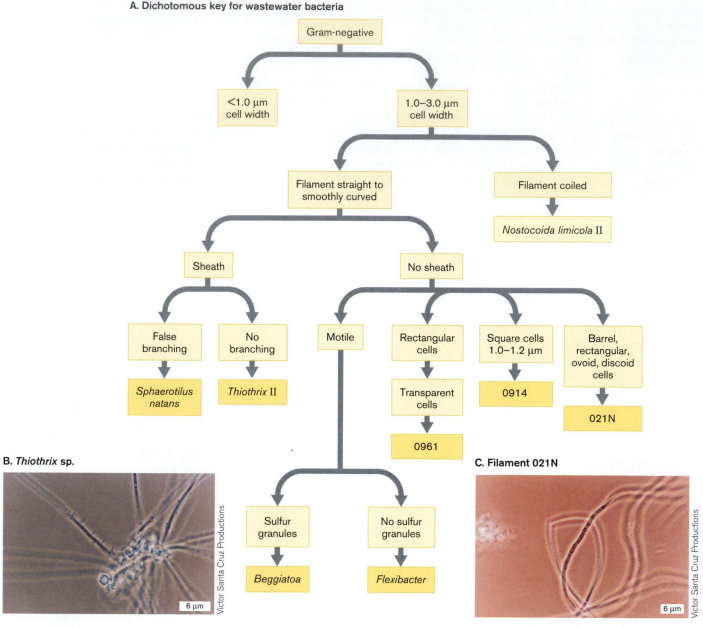

Figure 17.18   Dichotomous key for wastewater bacteria.   A. A section of a dichotomous key for identifying gram-negative bacteria from wastewater. The designations 0961, 0914, and 021N refer to unidentified species. **B.** *Thiothrix* species, a filamentous wastewater bacterium. **C.** Filament 021N, an unclassified organism that forms starburst filaments. *Source*: Santa Cruz Database Central, Wastewater organisms.

amino acids to generate alkaline amines. For each species in the database, the probabilities of all the test results are multiplied; for example, for *Salmonella enterica*:

$$\text{Score} = (0.99)(0.98)(0.97)(0.95)(1 - 0.01)$$
$$(1 - 0.01)(0.95)(1 - 0.96)$$

$$= 3.3 \times 10^{-2}$$

The advantage of the probabilistic indicator is its inherent redundancy; if an isolate gives one atypical result for its species, it may still show the highest probability score in the database.

THOUGHT QUESTION 17.5   In **Figure 17.19B**, were any test results atypical for *Salmonella enterica*? What would happen if another atypical result had been obtained? What would happen if the actual isolate had not been one of those in the database?

Note that both the dichotomous key and the probabilistic indicator assume a known set of organisms. To explore and characterize a completely unknown organism requires phylogenetic analysis—when possible, the sequencing of the organism's entire genome.

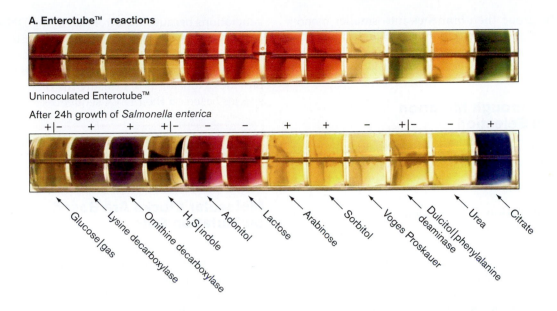

A. Enterotube™ reactions

Uninoculated Enterotube™

After 24h growth of *Salmonella enterica*

| +|– | + | + | +|– | – | – | + | + | – | +|– | – | + |

Glucose|gas — Lysine decarboxylase — Ornithine decarboxylase — H₂S|indole — Adonitol — Lactose — Arabinose — Sorbitol — Voges Proskauer — Dulcitol|phenylalanine deaminase — Urea — Citrate

B.

Species	D-Glucose acid	Lysine decarbox.	Ornithine decarbox.	H₂S	Adonitol ferment	Lactose ferment	D-Sorbitol ferment	Dulcitol ferment	Score
Test Results:	+	+	+	+	–	–	+	–	
Salmonella enterica	0.99	0.98	0.97	0.95	0.01	0.01	0.95	0.96	3.3 E-2
Proteus mirabilis	0.99	0.01	0.99	0.98	0.01	0.02	0.01	0.01	9.2 E-5
Yersinia enterocolitica	0.99	0.01	0.95	0.01	0.01	0.05	0.99	0.01	8.7 E-5
Enterobacter aerogenes	0.99	0.98	0.98	0.01	0.98	0.95	0.99	0.05	8.9 E-6
Klebsiella pneumoniae	0.99	0.98	0.01	0.01	0.90	0.98	0.99	0.30	1.3 E-7

Figure 17.19 The Enterotube™ II, a probabilistic indicator to identify enteric gram-negative pathogens. **A.** The upper tube shows appearance of the uninoculated control. The lower tube shows test results for *Salmonella enterica*. **B.** A simplified indicator table in which the probabilities are multiplied, yielding a probability score for each organism in the database. *Source*: Joan Slonczewski and BD Diagnostics.

TO SUMMARIZE:

- **Taxonomy** is the description and organization of life-forms into classes (taxa). Taxonomy includes classification, nomenclature, and identification.
- **Classification** is traditionally based on a hierarchy of ranks. Groups of organisms long studied tend to have many ranks, whereas recent isolates have few.
- **DNA sequence relatedness** defines microbial taxa. Below genus level, however, the definition of bacterial species can be problematic.
- **Practical identification** is based on phenotypic and genetic traits. Methods of identification include the dichotomous key and the probabilistic test battery. Both methods assume a predefined set of organisms.

- **Nongenetic categories** based on ecological niche or host organ infected are useful in ecology or medicine, respectively.

17.4 Microbial Divergence and Phylogeny

The unifying assumption of modern biology is genetic relatedness, or molecular phylogeny. Phylogeny generates a series of branching groups of related organisms called **clades**. Each clade is a **monophyletic group**—that is, a group of species that share a common ancestor not shared by any species outside the clade. Each monophyletic group then branches into smaller monophyletic groups and ultimately species. The full description of branching divergence of a species is called its phylogeny.

Divergence through Mutation and Natural Selection

Populations of organisms diverge from each other through several fundamental mechanisms of evolution. These include:

- **Random mutation.** DNA sequence changes through rare mistakes (in prokaryotes, typically one out of a million base pairs) as the chromosome replicates. Replication errors result in mutation. Most mutations are neutral; that is, they have no effect on gene function.
- **Natural selection.** In a given environment, natural selection favors organisms that produce greater numbers of offspring in that environment. Genes encoding traits under selection pressure may show mutation frequencies much higher or lower than the random mutation rate. Genes under selection pressure do not serve as molecular clocks.
- **Reductive evolution (degenerative evolution).** In the absence of selection for a trait, the genes encoding the trait accumulate mutations without affecting the organism's reproductive success. Because mutations that decrease function are more common than mutations that improve function, accumulating mutations without selection pressure leads to decline and ultimately loss of the trait. The loss or mutation of DNA encoding unselected traits is called **reductive evolution** (or degenerative evolution).

Random mutations with neutral effects that are not subject to selection tend to accumulate at a steady rate over generations because the error rate of the DNA replication machinery stays about the same. The resulting "genetic drift" causes sequences in separate populations to diverge over time. The constancy of mutation rate (within limits) provides a tool for us to measure the time of divergence of species based on their DNA sequences.

> **THOUGHT QUESTION 17.6** What kind of DNA sequence changes have no effect on gene function?

Molecular Clocks Are Based on Mutation Rate

An important conceptual advance of the twentieth century was that information contained in a macromolecule such as protein or DNA could measure the history of a species. The temporal information contained in a macromolecular sequence is called a **molecular clock**. Molecular clocks have revolutionized our understanding of the emergence of all living organisms, including human beings. The first molecular clocks, based on protein sequencing and DNA hybridization, were developed in the 1960s. Subsequently, the sequences of ribosomal RNA (rRNA) were used by Carl Woese to reveal the divergence of three domains of life. The rRNA sequence is particularly useful because ribosome structure and function are highly similar across all organisms. Genomic sequences now offer many other genes to measure divergence at different levels of classification.

A molecular clock is based on the acquisition of new random mutations in each round of DNA replication. In **Figure 17.20**, each offspring in generation 2 acquires two new mutations; their sequences now differ by 25%. In the next generation, each individual propagates the earlier mutant sequences while acquiring two more random mutations. Strain 3A now differs by 50% from strains 3B

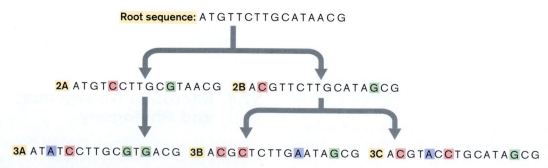

Figure 17.20 The molecular clock. As genetic molecules reproduce, the number of mutations accumulated at random is proportional to the number of generations and thus the time since divergence. Sequence label numbers indicate generations; letters identify individual organisms.

and 3C, which differ by 25% from each other. (Actual mutation frequencies, of course, are much lower, about one base per million per generation.) The key assumption is that the chromosomes of offspring acquire a consistent number of random mutations from their parents, and therefore the number of sequence differences between two species should be proportional to the time of divergence between them.

In practice, the molecular clock works best for a particular gene sequence with the following features:

- The gene has the same function across all species compared. That is, all versions are orthologous; they have not evolved to serve different functions. Functional difference may lead to different rates of change.
- The generation time is the same for all species compared. Shorter generation times (more frequent reproductive cycles) lead to overestimates of the overall time of divergence because of the increased opportunity for DNA mutation.
- The average mutation rate remains constant among species and across generations. If different species mutate at different rates, species with more rapid rates of mutation will appear to have diverged over a longer time than is actually the case.

Genes that show the most consistent measures of evolutionary time encode components of the transcription and translation apparatus, such as ribosomal RNA and proteins, tRNA, and RNA polymerase. The most widely used molecular clock is the gene encoding the **small-subunit rRNA** (**Fig. 17.21**). The small-subunit rRNA is also known by its sedimentation coefficient: 16S rRNA (bacteria) or 18S rRNA (eukaryotes). (Sedimentation coefficients are discussed in Chapter 3.)

The use of a molecular clock requires the alignment of homologous sequences in divergent species or strains (**Fig. 17.22**). Alignment refers to the correlation of portions of two gene sequences that diverged from a common ancestral sequence. In practice, correct alignment requires difficult decisions based on rule-of-thumb assumptions about mutation rates. These assumptions include:

- **Minimum number of changes.** The best alignment between two sequences is that which assumes the smallest number of mutational changes.
- **Loop positions change more frequently than stems.** Bases found in single-stranded loops of RNA, and the lengths of these loops, are more likely to change than bases found in double-stranded stems. This is because a base change in one strand of a loop necessitates a second mutation in the opposite strand to maintain base pairing in the stem. Intramo-

lecular base pairing is required for RNA structure in solution.

- **Functional sequences change more slowly.** Functional sequences change more slowly than portions of the molecule without essential catalytic function. Within small-subunit rRNA, sequences essential to the function of translation remain remarkably consistent across all three domains of life. These regions can be used to devise primer pairs for PCR amplification of rDNA from unknown organisms.
- **Key positions in the sequence are conserved.** Key positions in the sequence are conserved within a particular class of organisms, but differ between classes.

THOUGHT QUESTION 17.7   Solve the three alignment problems shown in **Figure 17.22**. How could errors or ambiguities in alignment lead to mistakes in interpreting the molecular clock?

Phylogenetic Trees

Once homologous sequences are aligned, the frequency of differences between them can be used to generate a **phylogenetic tree** that estimates the relative amounts of evolutionary divergence between them. All phylogenetic trees depend on complex mathematical analysis to measure degrees of divergence and determine the most probable tree that connects present sequences with their common ancestor. Two common approaches are:

- **Maximum parsimony.** Evolutionary distances are computed for all pairs of taxa based on numbers of nucleotide or amino acid substitution between them. A proposed common ancestor, or "ancestral state," is reconstructed. All possible trees comparing relative time of divergence from the common ancestor are computed. The "best fit" tree is defined as that which requires the smallest number of changes to fit the data (that is, the most "parsimonious"). A limitation of parsimony reconstruction is that more than one tree may produce results consistent with the data.
- **Maximum likelihood.** For each possible tree, one calculates the likelihood (probability) that such a tree would have produced the observed divergence data. The probability of given substitution events is based on complex statistical calculations. Maximum likelihood methods require large amounts of computation but obtain the most information from the data, usually generating one tree or a small set of probable trees.

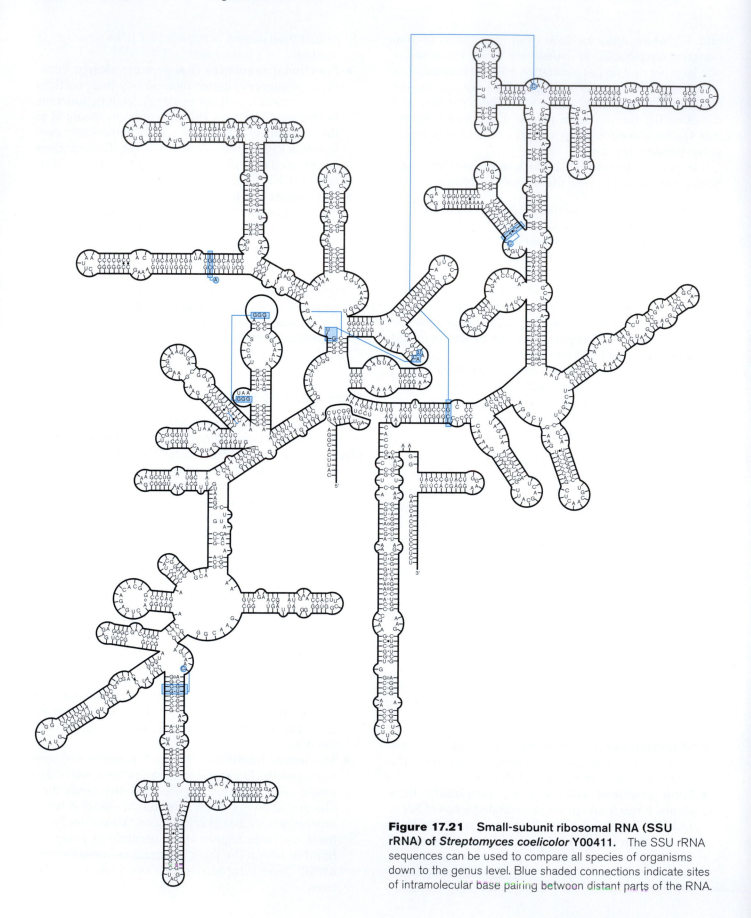

Figure 17.21   Small-subunit ribosomal RNA (SSU rRNA) of *Streptomyces coelicolor* Y00411.   The SSU rRNA sequences can be used to compare all species of organisms down to the genus level. Blue shaded connections indicate sites of intramolecular base pairing between distant parts of the RNA.

Figure 17.22 RNA sequence comparison by phylogeny exercises.
A. Homologous RNA sequences from different species can be aligned at regions of similarity. The best alignment is that which minimizes mismatches. **B.** Knowledge of the secondary structure can help predict alignments. Inset: James W. Brown, author of these phylogeny exercises, samples diverse thermophiles from Obsidian Pool, Yellowstone National Park. **C.** A new archaeal isolate can be assigned to a division based on key positions in the sequence. (Dots between bases indicate partial hydrogen bonding.)

RNA phylogeny exercises

A. Add the new sequence from organism F to this alignment:

Seq A   CCCCAGCUUCGGCUGGGGGAGG
Seq B   CCUUAGCGAAAGCU–AAGGAGG
Seq C   CCUCAGCGUGAGCU–GAGGAGG
Seq D   CCCAAGCUUU–GCUAUGGGAGG
Seq E   CC––AGCUUUGGCU–––GGAGG
Seq F   CCAAGCGAGAGCUUGGAGG

Courtesy of James W. Brown

B. Make an alignment of these RNAs:

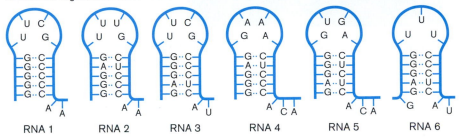

RNA 1   RNA 2   RNA 3   RNA 4   RNA 5   RNA 6

C. Based on its rRNA sequence, to which archaeal group does isolate X most likely belong?

Reference R2 RNA   Isolate X R2 RNA

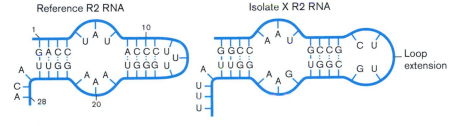

Position number	Crenarchaea	Euryarchaea	Korarchaea
4	C or U	C	U
5	U	U	A
8	A	G	C
11	C	G	G
15	A or G	C	U
18	U	U	C
19	A	G	C or U
20	A	A	U
21	A	A	C or U
22	G	G	A

A portion of a computed phylogenetic tree is shown in **Figure 17.23**. The sequence data were obtained from PCR-amplified sequences of small-subunit rRNA from isolates in Obsidian Pool, a thermal spring at Yellowstone. The sequences were analyzed by Susan Barns and colleagues at Indiana University in the laboratory of Norman Pace, a pioneering investigator of extreme habitats. Note that some of the Obsidian Pool samples are known species, such as the thermophile *Pyrodictium occultum,* whereas others are uncharacterized organisms given alphanumeric designations, their existence known only by the rRNA sequences reported here. In fact, any microbial habitat surveyed by PCR, including a "soap scum" biofilm on a shower curtain, will yield previously unknown species (**Special Topic 17.2**).

The Obsidian Pool rRNA sequences were aligned based on the principles shown in **Figure 17.22**, and divergence distances were analyzed by maximum likelihood. In each tree, the total distance (in terms of base substitution frequency) between any two sequences is approximated by the distance from each sequence and its branch point. The tree at left (**Fig. 17.23B**) is an **unrooted tree**; it shows only the relative distances between different sequences, without indicating which of these diverged earliest from the common ancestor. To root the tree (at right), the researchers identified a sequence well outside the selected cluster of isolates (portion of tree not shown). The root is the position from which the most different sequence first diverged from the rest.

The Obsidian Pool example reveals an important limitation of phylogenetic trees: Not all the branches add up to the same total length. In other words, some lineages accumulate mutations faster than others. The difference in rate arises from differences in mutation rate and from differences in generation time between organisms whose sequences are compared. Without calibration data outside the tree, approximations need to be made to estimate the actual divergence times. Thus, our molecular clocks are inevitably distorted, especially for distantly diverged organisms with disparate mutation rates. The distortion is reminiscent of the famous painting by Salvador Dali, *Persistence of Memory* (**Fig. 17.24**), in which warped watch dials represent the biological nature of memory.

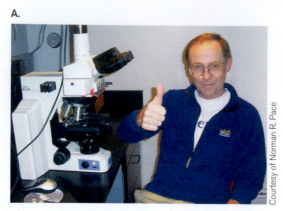

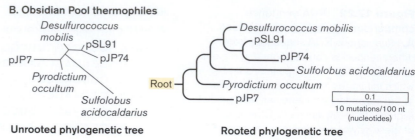

B. Obsidian Pool thermophiles

Unrooted phylogenetic tree

Rooted phylogenetic tree

Figure 17.23   Phylogenetic trees: rooted and unrooted.   A. Norman R. Pace was the first to characterize the sequence phylogeny of thermophiles from high-temperature environments. **B.** Comparison of rRNA sequences was used to generate a phylogenetic tree for thermophiles isolated from Obsidian Pool in Yellowstone National Park. A phylogenetic tree shows the degrees of relatedness among taxa based on the number of molecular substitutions on each branch (see scale bar). An unrooted tree does not indicate the position of a common ancestor for all taxa. A rooted tree indicates the position of a common ancestor. *Source*: B. Susan M. Barns, et al. 1996. *PNAS* 93:9188.

A second problem is the need to calibrate the tree, that is, to relate mutation frequency to units of time. To indicate actual time since divergence, an external measure of time is required. A tree can be calibrated if some kind of fossil evidence or geological record exists to confirm at least one branch point of the tree. An example of a calibrated tree appears in **Figure 17.17**, in which a chemical biosignature (the sulfur isotope ratio) was used to set a minimum time since the divergence of the sulfate-reducing clade from other lineages of bacteria. Another method of calibration is to correlate the divergence of microbial species growing only within particular host species with the divergences of their hosts, based on the host fossil record.

> **THOUGHT QUESTION 17.8**   What are the major sources of error in constructing phylogenetic trees?

Divergence of Three Domains of Life

Carl Woese first used small-subunit rRNA phylogeny to reveal the existence of a third domain of life, Archaea, roughly as distant from prokaryotes as from eukaryotes (**Fig. 17.25**). How was an entire domain of life missed in the past? Many archaea grow in habitats previously thought inhospitable for life; for example, *Thermoplasma* species grow at 60°C at pH 2 (**Fig. 17.26**). Others, such as methanogens and halophiles, were long known to microbial ecologists, but without tools for genetic analysis they were simply classified among bacteria.

Rooting the tree of life. Where is the root of the tree, the position of the last universal common ancestor of all life-forms? Which of the three domains (bacteria, archaea, eukaryotes) first diverged from the other two?

Figure 17.24   A distorted measure of time.   *Persistence of Memory*, by Salvador Dali, suggests that all organisms measure time through clocks that are distorted. In fact, the molecular clocks of DNA are distorted by differences in mutation rate and generation time.

The question has profound importance for biology, as the research community defines "model systems" for study based on their commonalities with organisms of importance to humans. For example, investigators of intron splicing in archaea argue that archaea represent a model system for related processes in complex eukaryotes such as humans.

The root, however, can only be found by measuring divergence relative to an outside group of organisms. Because no "outgroup" exists for the tree of all life, how is the tree rooted? One approach is to compare a pair of homologous genes within one organism, homologs that diverged from a common ancestral gene and acquired distinct functions (paralogs). The pair of paralogs chosen for analysis must have diverged within the common

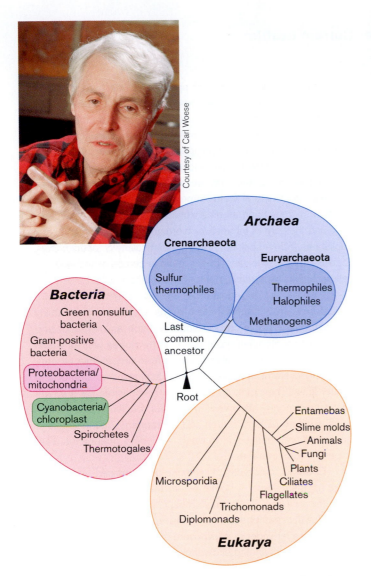

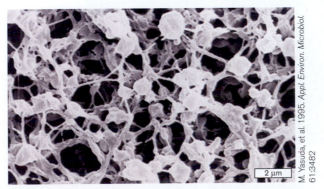

ancestral cell *before* the divergence of the three domains of life. The early divergence is determined by comparing the pair's sequence similarity in many different organisms and constructing a tree of gene phylogeny. A limitation of this approach is that genes with divergent function have been subjected to natural selection in addition to random variation.

As of this writing, there is no general agreement as to which of the three domains diverged first from the other two. While the position of the root remains controversial, the three major divisions of life based on rRNA phylogeny have been confirmed by sequence data from the hundred-plus genomes published to date. Furthermore, structural and physiological comparison of the three domains largely confirms the rRNA tree (**Table 17.3**). Note first that all living cells on Earth share profound similarities. All cells consist of membrane-bound compartments that shelter the same fundamental apparatus of cell production, the DNA-RNA-protein machine. The fundamental components of this machine appear to have evolved before the three domains diverged from their last universal common ancestor. From a molecular standpoint, all cells on Earth appear more similar than they are different.

Nonetheless, important differences emerge between each pair of domains; indeed, each domain shows distinctive traits absent or scarce in the other two. We summarize here the major features common to most members of each domain. Specific classes and species within each domain are explored in Chapters 18–20, and on the Microbial Biorealm website.

Archaea, bacteria, and eukaryotes. Of the three domains, the eukaryotes stand out as having a nucleus and other complex membranous organelles (**Table 17.3**). Eukaryotic organelles include mitochondria and chloroplasts, which evolved from internalized bacteria. Bacteria and archaea are prokaryotes; they possess no nucleus and have relatively simple intracellular membranes. Their size is limited by diffusion across the cell membrane, with occasional exceptions, such as the "giant bacterium" *Epulopiscium fishelsoni*. The larger size and complexity of eukaryotic cells generally requires the most high-powered sources of energy, such as aerobic respiration and oxygenic photosynthesis, although some protists conduct fermentation. By contrast, the prokaryotes (bacteria and archaea) employ a wider range of metabolic alternatives, including lithotrophy and anaerobic respiration. Finally, eukaryotic plants and animals have attained a degree of multicellular complexity unknown in the prokaryotic domains.

On the other hand, eukaryotes share key traits with archaea that distinguish both from bacteria. The core information machinery of eukaryotes more closely resembles that of archaea. The two domains share closely

Figure 17.25 Three domains of life.   Carl Woese (inset) used SSU rRNA sequencing to reveal three equally distinct domains of life: Bacteria, Eukarya (eukaryotes), and Archaea. The tree is "unrooted," since it does not show which of the three domains diverged first from the common ancestor.

Figure 17.26 *Thermoplasma*.   This archaeon lives at 60°C at pH 2—with no cell wall, only a cell membrane.

Special Topic 17.2   Phylogeny of a Shower Curtain Biofilm

The deep sea and the tropical forest receive much attention as habitats to discover new exotic life-forms. But novel microorganisms can be discovered closer to home—in the soil, on the roots of grass, within our own digestive tract. Even within our own homes lurk microbial communities that include potentially dangerous pathogens. Such organisms were discovered by Norman Pace and colleagues at San Diego State University and at the University of Colorado, taking a break from their submarine studies to explore a domestic aquatic habitat: the "soap scum" biofilm on a shower curtain (**Fig. 1**).

Domestic water sources are well known as a source of pathogens such as *Legionella pneumophila*, the cause of Legionellosis, a deadly form of pneumonia. Pathogens persist in biofilms growing within plumbing and air-conditioning lines. Lesser known reservoirs for microbes are the biofilms that collect on household surfaces such as shower curtains, which are often neglected during cleaning. Viable cells of the shower curtain biofilm are revealed by fluorescence microscopy using the DAPI stain for DNA (**Fig. 1B**).

Pace's students Scott Kelley and Ulrike Thiesen obtained biofilm from a shower curtain and used it to extract DNA. The biofilm DNA was amplified for sequences transcribed to rRNA using the PCR technique. The PCR amplification was conducted with a pair of sequence primers based on portions of the small-subunit rRNA sequence that are extremely conserved across all known species of bacteria. The region between the primers, however, is known to vary among species; thus, the DNA amplified should show distinguishing features that identify the source organism.

The shower curtain biofilm revealed a wide range of species, including 117 unique sequences for small-subunit rRNA.

The most abundant groups of species include two genera of Alpha-Proteobacteria, *Sphingomonas* and *Methylobacterium*. *Sphingomonas* is named for its outer membrane content of sphingolipids, lipids containing an amide link to fatty acid. The phylogeny of the shower curtain sphingomonads is shown in **Figure 2**. *Sphingomonas* species are known to grow in a wide range of soil and water habitats and are occasionally isolated as yellow colonies from natural water supplies. They catabolize complex carbon sources such as dibenzofuran and hexachlorocyclohexane and thus are of interest for bioremediation of environmental pollutants. Some species, however, particularly *S. paucimobilis*, cause opportunistic infections in immune-compromised individuals, including bacteremia, peritonitis, and abscesses.

Species of the next-most abundant genus, *Methylobacterium*, are versatile heterotrophs known for their ability to metabolize single-carbon sources such as methanol and methylamine. Species grow in soil and water as well as within plant tissues; they are also isolated from automobile air-conditioning systems, printing paper machines, and dental unit water lines. Their pink color may be responsible for the pink color commonly observed in shower biofilms. Some species, such as *M. extorquens* and *M. zatmanii,* cause illness in immune-compromised individuals, including pneumonia, skin ulcers, and bacteremia.

Growing numbers of immune-compromised individuals care for themselves at home, where they need to control their own microbiological exposure. Thus, it is of concern to discover opportunistic pathogens in a home setting. The authors conclude with a reminder that "exposure can be minimized by regular cleaning or by changing shower curtains."

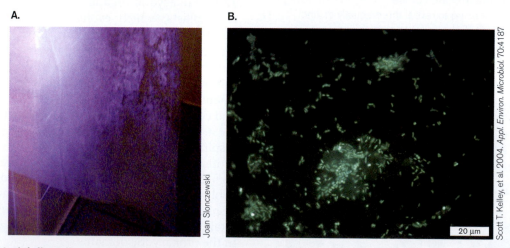

Figure 1   Bacterial diversity on a shower curtain.   A. Shower curtain "soap scum" consists of a biofilm. **B.** Biofilm from a shower curtain is visualized by epifluorescence microscopy using DAPI stain.

Joan Slonczewski

Scott T. Kelley, et al. 2004. *Appl. Environ. Microbiol. 70*:4187

20 μm

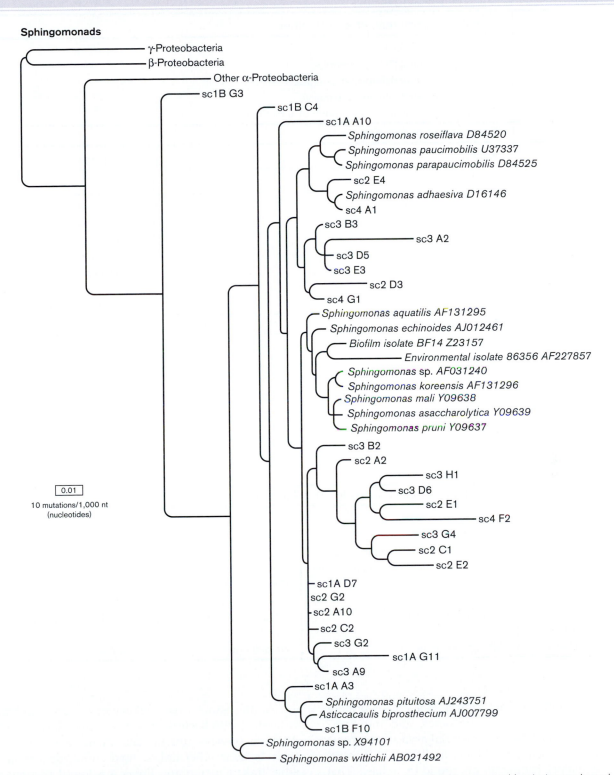

Sphingomonads

Figure 2 *Sphingomonas* **phylogeny from a shower curtain biofilm.** Phylogeny of sphingomonad bacteria was based on small-subunit RNA gene sequence comparison.

Table 17.3 **Three domains of life.**

Characteristic	Traits of living organisms		
	All cells on Earth resemble each other		
Chromosomal material	Double-stranded DNA		
RNA transcription	Common ancestral RNA polymerase		
Translation	Common ancestral rRNAs and elongation factors		
Protein	Common ancestral functional domains		
Cell structure	Aqueous cell compartment bounded by a membrane		
	Comparison of domains		
	Bacteria	**Archaea**	**Eukaryotes**
	Archaea resemble bacteria		
Cell volume	1–100 μm³ (usually)		1–10⁶ μm³
DNA chromosome	Circular (usually)		Linear
DNA organization	Nucleoid		Nucleus with membrane
Gene organization	Multigene operons		Single genes
Metabolism	Denitrification, N₂ fixation, lithotrophy, respiration, and fermentation		Respiration and fermentation
Multicellularity	Simple		Simple or complex
		Archaea resemble eukaryotes	
Intron splicing	Introns are rare	Introns are common	
RNA polymerase	Bacterial	Eukaryotic form	
Transcription factors	Bacterial	Eukaryotic form	
Ribosome sensitivity to chloramphenicol, kanamycin, and streptomycin	Sensitive	Resistant	
Translation initiator	Formylmethionine	Methionine (except mitochondria use formylmethionine)	
Cell wall	Peptidoglycan	Pseudopeptidoglycan or other polymer; or protein S-layer	
	Bacteria resemble eukaryotes and differ from archaea		
Methanogenesis	No	Yes	No
Thermophilic growth	Up to 90°C	Up to 120°C	Up to 70°C
Photosynthesis	Many species; bacteriochlorophyll Proteorhodopsin derived from archea	Haloarchaea only; bacteriorhodopsin	Many species; chlorophyll (bacterial origin)
Chlorophyll light absorption	Red and blue	Green (central range of solar spectrum)	Red and blue (chloroplasts of bacterial origin)
Membrane lipids (major)	Ester-linked fatty acids	Ether-linked isoprenoid	Ester-linked fatty acids
Pathogens infecting animals or plants	Many pathogens	No pathogens	Many pathogens

related components of the central DNA-RNA-protein machine: RNA polymerase, ribosomes, and transcription factors. Even such hallmarks of Eukarya as intragenic introns and the splicing machinery and the "RNA interference" regulatory complexes are found in archaea. This explains why rRNA trees place archaea closer to eukaryotes than to bacteria.

At the same time, eukaryotes share fundamental structures with bacteria that differ from those in archaea. Archaea possess unique cell membrane components, such as their ether-linked lipids. Outside the archaea, ether-linked membrane lipids are found in only a few deep-branching bacterial species that share habitat (and exchange genes) with hyperthermophilic archaea. Only the domain Archaea includes species capable of growth in the most "extreme" environments of temperature (above 110°C or below –20°C) and pH (below pH 1), although other archaeal species grow well at mesophilic temperatures in soil or water. Perhaps the most striking distinction of archaea is the complete absence of known pathogens of animals or plants. Even the many methanogens that live within animal digestive tracts have never been shown to cause disease.

Overall, the three-domain phylogeny divides life usefully into three distinctive groups. Yet the tree also shows signs of gene flow unaccounted for by monophyletic descent. The eukaryotes contain mitochondria derived from entire assimilated prokaryotes, whose genomes persist within the organelle. And pathogenic bacteria such as *Agrobacterium* species transfer DNA into the genomes of plants. Moreover, sequenced genomes reveal evidence of gene transfer between bacteria and archaea sharing high-temperature habitats. What if the tree of life is not strictly monophyletic?

TO SUMMARIZE:

- **Phylogeny is the divergence of related organisms.** Organisms diverge through random mutation, natural selection, and reductive evolution.
- **Molecular clocks are based on mutation rate.** Given a constant mutation rate and generation time, the degree of difference between two DNA sequences correlates with the time since the two sequences diverged from a common ancestor.
- **Different sequences diverge at different rates.** Under selection pressure, the actual divergence rates of sequences depend on structure and function of the RNA or protein products.
- **Phylogenetic trees are calculated based on sequence analysis.** The more different the two sequences are, the longer the branch representing time since divergence from the common ancestor. Rooting a tree requires comparison with an outgroup.
- **The tree of life diverges to three domains: Bacteria, Archaea, and Eukarya.** Eukaryotes are distinguished by the nucleus, lacking in prokaryotes (archaea and bacteria). Archaea possess ether-linked isoprenoid lipids rare or absent in bacteria and eukaryotes, and they are never pathogens. The machinery of archaeal gene expression resembles that of eukaryotes.

17.5 Horizontal Gene Transfer

In retrospect, the very first demonstration of hereditary material, the transformation of avirulent *Pneumococcus* to virulence in 1928, involved horizontal transfer. Horizontal transfer is the acquisition of a piece of DNA from another cell, as distinguished from **vertical transfer**, the transmission of an entire genome from parent to offspring. Among bacteria, DNA is transferred horizontally by plasmids, transposable elements, and bacteriophages, as well as through the process of transformation, as discussed in Chapter 9. For example, drug resistance genes are transferred from harmless flora to pathogens of distantly related species. Another example is the transfer of genes encoding light-driven proton pumps (bacteriorhodopsin) from halophilic archaea into marine bacteria. Such transfer events are relatively rare, occurring perhaps once in a million generations. But over time, the number of such "rare" events can accumulate.

Genomic Evidence for Gene Transfer

A strong sign of horizontal transfer is a DNA sequence whose **GC/AT ratio** (proportion of GC and AT base pairs) differs from the rest of the genome. A surprising proportion of genomic DNA can show "spikes" of GC contact indicating origin elsewhere, even from species of a different kingdom. For example, *Archaeoglobus fulgidus*, a sulfur-oxidizing archaeon that grows in petroleum deposits, possesses numerous genes for lipid catabolism closely related to counterparts in bacteria. The presence of such genes in an archaeon fits no phylogenetic tree unless horizontal transfer is assumed. In other archaea, particularly hyperthermophiles *Pyrococcus* and *Aeropyrum*, 10–20% of the genes appear to be of bacterial origin. On the other hand, the thermophilic bacterium *Thermotoga maritima* shows a number of genes transferred from archaea.

Between more closely related taxa, transfer occurs even faster. For example, the *Escherichia coli* genome has acquired about 18% of its genes from other species after its relatively recent divergence from the close relative *Salmonella enterica*. Some medically important genera, such as *Neisseria* (causing gonorrhea and meningitis), are particularly "recombinogenic." Rapid gene exchange enables pathogens to avoid the host immune system by expressing novel proteins not recognized by host antibodies (discussed in Chapter 23).

Eukaryotic genomes include many genes acquired from bacteria, most of them from the endosymbiotic ancestors of their mitochondria and chloroplasts. And the content of their noncoding intergenic sequences and introns derives largely from ancient viruses and retroposons (discussed in Chapter 11). In 1999, Ford Doolittle, at Dalhousie University, redrew the standard tree of life with a bewildering array of cross-cutting lineages (**Fig. 17.27**). Perhaps the "last common ancestor" was actually a "last common community" of diverse life-forms that contributed different parts of our genetic legacy.

A. Monophyletic tree: Limited lateral transfer

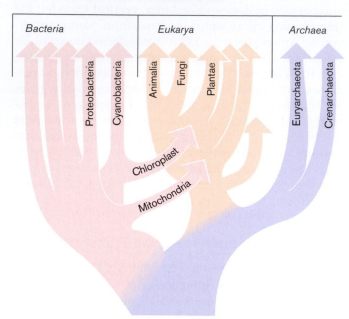

Last common ancestor

B. Lateral transfer obscures phylogeny

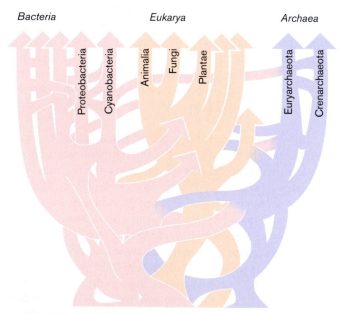

Last common community

Figure 17.27 Vertical and horizontal gene transfer.
A. The traditional view of phylogeny holds that the vast majority of gene transfer is vertical and that most lineages are monophyletic. Only occasional transfers occur, such as that of mitochondria and chloroplasts from bacteria into eukaryotes.
B. An alternative view argues that lateral (or horizontal) gene transfer is so provalent among microbes that it obscures monophyletic distinctions between taxa. *Source*: W. Ford Doolittle. 1999. *Science* 284:2126.

> **THOUGHT QUESTION 17.9** What are the limits of evidence for horizontal gene transfer in ancestral genomes? What alternative interpretation might be offered?

Reconciling Vertical and Horizontal Transfer

A promising model to sort out vertical and horizontal gene transfer was proposed by James Lake and colleagues at the University of California Los Angeles and developed further by Doolittle. Lake argues that the genome contains two kinds of genes that evolve differently. **Informational genes** specify products essential for transcription and translation of information-rich macromolecules. Such products include RNA polymerase as well as ribosomal RNAs and elongation factors. Informational genes need to interact directly in complex ways with large numbers of cellular components; thus, their capacity for horizontal transfer is limited. On the other hand, **operational genes** are those whose products govern metabolism, stress response, and pathogenicity. Operational genes function with relative independence from other cell components and consequently undergo horizontal transfer more easily than informational genes. Operational gene products also tend to meet conditional needs, dependent on the availability of nutrients and other environmental conditions, whereas informational gene products have functions essential to cell growth under every condition.

The concept of differentially transferable genes does much to clarify genome-based phylogeny. New phylogenetic trees are drawn based on sets of informational genes. These multiple-gene trees largely confirm the results of the small-subunit rRNA tree of life. At the same time, we increasingly find that significant traits of species arise from horizontal transfer. In particular, pathogens show extensive horizontal transfer of virulence genes and genes encoding resistance to host defenses and antibiotics. Groups of such genes are often transferred together on plasmids or on **genomic islands**, regions of DNA of foreign origin that confer special properties such as virulence or nitrogen fixation (discussed in Chapter 9). Examples of a virulence plasmid and a genomic island are found in the deadly pathogen *E. coli* O157:H7 (**Special Topic 17.3**).

A balanced view of vertical and horizontal transfer is shown in **Figure 17.28**. The row of arrows designates vertical transfer of the bulk of the genome as two species diverge. Black lines represent core informational genes, such as ribosomal RNA, whereas gray indicates genes with more flexible function. At various levels, colored lines indicate where genes enter each lineage. Gray lines

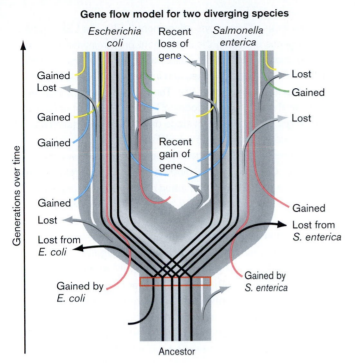

Gene flow model for two diverging species

Figure 17.28 Genes enter and leave genomes. Both rRNA trees and whole-genome trees consistently reflect monophyletic descent (black lineages) down to the genus level. Below the genus level, monophyletic descent may be obscured by high rates of horizontal transfer (colored lines entering, gray lines departing). White spaces indicate genes lost by reductive evolution.

peel out, indicating loss of a gene. Overall, the vertical lineage persists for most of the core informational genes (black lines), while horizontally acquired genes enter and leave.

Species: A Molecular Definition

Our changing view of molecular phylogeny requires us to revisit the concept of species. As genome sequence data became available, scientists hoped that quantitative measures of divergence could provide a consistent basis for defining species of asexually reproducing microbes. But in some organisms, such as *Helicobacter pylori*, the genomes of different strains that cause the same disease (gastritis) differ by as much as 7%. On the other hand, strains of *Bacillus* with nearly identical genomes cause completely different diseases, such as anthrax (*B. anthracis*) and caterpillar infection (the biological pesticide *B. thuringiensis*).

A working definition of microbial species includes two perspectives: phylogeny (based on DNA relatedness) and ecology (based on shared traits and ecological niche).

Phylogenetic relatedness. A species is a group of individuals that share relatedness of a key set of "housekeeping genes," typically informational genes such as ribosomal and transcriptional components. Ideally, these genes should all be orthologs (genes with a common origin and function), not paralogs (which diverged from a common ancestor but now differ in function). Within a genus, species that cannot be distinguished based on small-subunit rRNA alone may be defined by analysis of multiple genes. For example, analysis of multiple gene loci effectively distinguishes *Neisseria meningitidis* (the cause of meningitis) from *N. lactamica*, a harmless resident of the nasopharynx. The multigene approach has proved successful even in the case of highly "recombinogenic" organisms known to acquire and rearrange genes readily.

Ecological niche. Besides a high degree of genomic relatedness, a species should include individuals that share common traits and an ecological niche. Shared traits should include cell shape and nutritional requirements, and there should be a common habitat and life history (for example, causing the same disease). By this criterion, highly divergent strains of *H. pylori* causing gastritis make up one species, whereas *B. anthracis* (anthrax) and *B. thuringiensis* (caterpillar infection) are different species despite their high degree of genetic similarity. In 2005, Dirk Gevers and colleagues from several countries proposed a three-part basis for classifying microbial species: rRNA sequence analysis identifies the family or genus; multiple-sequence analysis or whole-genome analysis distinguishes putative species; and the species distinction is confirmed based on ecological niche or disease involvement.

TO SUMMARIZE:

■ **Horizontal transfer of genes occurs between different species.** Horizontal transfer is most frequent between closely related species or between distantly related species that share a common habitat.

■ **Genomes include informational genes and operational genes.** Informational genes involve central processes of gene expression; they tend to be transferred vertically. Operational genes involve metabolic processes that function independently of other components. They are more likely to be transferred horizontally.

■ **Horizontal transfer is important for adaptation to new environments and for pathogenesis.** Gene transfer among pathogenic and nonpathogenic strains leads to emergence of new pathogens.

■ **The definition of microbial species requires complex criteria.** A microbial species may be defined based on rRNA sequence, multilocus sequence analysis, and ecological niche or pathogenesis.

Microbial evolution is essential to pathogenesis (discussed in Chapter 25). Evolution provides clues to the source of emerging pathogens that cause new diseases, such as *E. coli* O157:H7, the "hamburger *E. coli*." The original appearance of *E. coli* O157:H7 in foods came as a surprise, especially given how fast the strain caused death in healthy children (**Fig. 1A**). The strain commonly contaminates grain-fed beef and the surfaces of vegetables grown near pastured cattle or manure.

Despite its genetic similarities to harmless enteric *E. coli,* the O157:H7 strain clearly possesses several new genes that contribute to virulence. Some of the virulence genes specific to strain O156:H7 (for example, genes *bfp* and *per*) are carried in a distinctive plasmid. Other virulence genes (such as the LEE genes) are found in the core genome. Both kinds of virulence genes contribute to the invasion cycle (**Fig. 1B**). First, the bacteria attach to the intestinal mucosa via special pili encoded by the "bundle-forming pilus" gene *bfp*. The attached bacteria then express proteins causing irritation and effacement (removal) of projecting villi. This "attaching and effacing" behavior is mediated by a genetic region called the locus of enterocyte effacement (LEE). The LEE region includes genes encoding the attachment protein intimin (*eae*) in the outer membrane and the intimin attachment protein Tir (*tir*), which the *E. coli* actually makes and inserts in the host cell membrane (discussed in Chapter 25). Then each bacterium induces the intestinal cell to form a "pedestal" stiffened by host actin filaments (**Figs. 1B** and **C**). Each stage of colonization requires specific virulence factors, such as the LEE proteins.

But where did the LEE genes come from, and how did they make the *E. coli* pathogen so different from its close relatives in the normal human flora? Completion of the genome sequences of O157:H7 and of *E. coli* K-12 revealed an astonishing 25% more genes in the pathogen. These novel genes are not scattered at random around the chromosome; instead, they appear in genomic islands, large chunks transferred together (red segments in **Fig. 2**). Other genomic islands (yellow) are present at the same location in both strains, but differ in sequence.

The chunks of contiguous transferred genes show GC content and tRNA codon usage substantially different from that of the genomic sequences shared with *E. coli* K-12. Thus, they appear to represent horizontal transfer from distantly related species. This horizontal transfer brings an entire system of gene products conferring a mechanism of pathogenesis, such as the LEE system for intestinal villus effacement. Such a transferred set of virulence genes is known as a "pathogenicity island" (discussed in Chapter 9). Pathogenicity islands undergo horizontal transfer among different organisms, generating new pathogenic strains.

A.

Reuters/Corbis

B.

Figure 1   **An emerging pathogen, *E. coli* O157:H7.**   **A.** A child infected by *E. coli* O157:H7 in one of North America's biggest epidemics is evacuated to a hospital. **B.** The cycle of colonization. *E. coli* proteins Bfp and DsbA make pili to attach microvilli and initiate colonization. The LEE proteins cause effacement of microvilli and polymerize actin. Eae (intimin) mediates attachment and pedestal formation. **C.** The *E. coli* cells induce the intestinal mucosa to form "pedestals" on which the bacteria attach, drawing nourishment from the host cell. *Source*: B. Valerie Souza, et al. 2002. *American Scientist* 90:332.

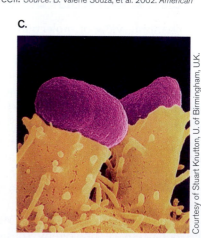

Courtesy of Stuart Knutton, U. of Birmingham, U.K.

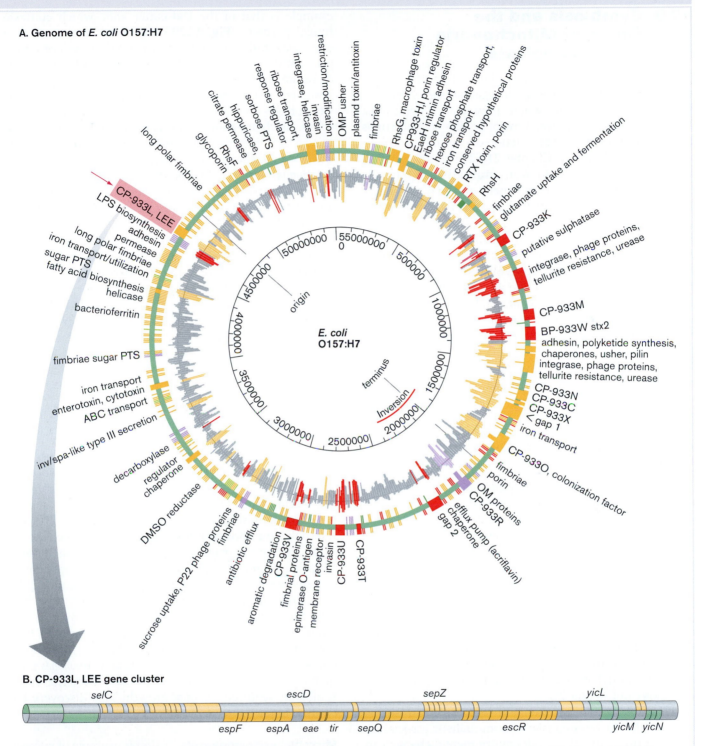

A. Genome of *E. coli* O157:H7

B. CP-933L, LEE gene cluster

Figure 2   Horizontal transfer of virulence genes.   A. Genome comparison between *E. coli* strains of K-12 and O157:H7. Outer circle shows DNA sequence shared between the two strains (green); horizontally acquired islands of sequence specific to O157:H7 (red); islands at the same location in O157:H7 and K-12 (yellow). Second circle shows the GC content calculated for each gene longer than 100 codons, plotted around the mean value for the whole genome, and color-coded like outer circle. Innermost circle gives the scale in base pairs. **B.** Close-up view of the LEE gene cluster. *Source:* Nicole T. Perna, et al. 2001. *Nature* 409:529–533.

17.6 Symbiosis and the Origin of Mitochondria and Chloroplasts

So far, we have largely considered evolution of a species in isolation. In fact, however, all organisms evolve in the presence of others, with whom they share interactions, both positive and negative. A major engine of evolution is **symbiosis**, the intimate association of two unrelated species (discussed fully in Chapter 21). The word *symbiosis* is popularly understood to mean **mutualism**, a relationship in which both partners benefit and may absolutely require each other. Biologists, however, recognize **parasitism**, in which one partner is harmed, as a relationship equally as intimate as mutualism; both are forms of symbiosis. Intimate relationships between species, either negative or positive, lead to **coevolution**, the evolution of two species in response to one another, showing parallel phylogeny. Coevolution commonly involves reductive evolution, where one partner species loses traits provided by the other partner. For example, the intracellular bacterium *Wolbachia*, found within nematodes that cause river blindness, has lost the ability to produce many nutrients, while retaining the ability to make purines and pyrimidines required by their host cell.

Microbial Interactions Lead to Coevolution

Interactions of microbial populations with different organisms include a range of positive through negative relationships (discussed in Chapter 21). A famous example of mutualism is that of bacterial nitrogen fixation, in which rhizobia form intracellular "bacteroids" within legume tissues. Both rhizobia and their plant hosts are highly evolved to respond to each other chemically and develop the nitrogen-fixing system. The plants provide protection and nutrients from photosynthesis while receiving fixed nitrogen from the bacteria. Another case of mutualism is that of lichens, multicellular systems composed of fungi and algae. These two cases and other forms of mutualism are discussed from the perspective of ecology in Chapter 21.

The distinction between mutualist and parasite is often subtle. Lichens consist of a mutualistic association between fungi and algae; but environmental change can convert the fungus to a parasite. On the other hand, parasitic microbes may coevolve with a host to the point that each depends on the other for optimal health. For example, the incidence of certain allergies and immune disorders, such as multiple sclerosis, appears to correlate with lack of exposure to pathogens and parasites.

In natural habitats, microbial symbiosis may involve multiple partners, both positive and negative. A classic example is that of the leaf-cutter ants, which cultivate fungal partners (**Fig. 17.29A**). Leaf-cutter ants arrange cut leaves in underground burrows for growth of a fungus that the ant is adapted to consume. Each ant species associates with one species of fungus. A third "partner," however, is a parasitic fungus of the genus *Escovopsis*, which feeds on the cultivated fungus and harms the ant colony. Each ant-fungus pair is parasitized by an *Escovopsis* species unique to that pair. To counteract the parasite, however, the ant supports growth of a fourth partner, *Streptomyces* bacteria that produce potent antifungal antibiotics. The bacteria grow on a specific structure on the ant cuticle, evolved for the specific function of hosting the bacteria (**Fig. 17.29B**).

Three of the four partners of the leaf-cutter system show extensive evidence of coevolution (**Fig. 17.29C**). The molecular phylogeny of both cultivated and parasitic fungal species tracks the divergence of the ant species. For example, the fungus strains from *Atta* and *Acromyrmex* host species are highly diverged from the fungi of *Trachymyrmex* hosts. The bacterial phylogeny is as yet unknown, but it is likely that the *Streptomyces* species as well has coevolved with its three partners.

Endosymbiosis Leads to Reductive Evolution and Obligate Association

The most intimate kind of symbiosis is **endosymbiosis**, in which one partner population grows within the body of another organism. Endosymbiosis includes communities of microbes within the digestive tract of animals, such as the human intestinal flora (discussed in Chapter 21). The internalized **endosymbiont** can also be intracellular, as in the case of rhizobial bacteroids within legume tissues. Rhizobia retain the genetic capacity for independence, growing readily in soil. Other intracellular endosymbionts, however, become wholly dependent on their host cells. Such endosymbionts undergo drastic reductive evolution, evolving ever deeper interdependence with their host cells. A surprising number of human invertebrate parasites, such as filarial nematodes and *Anopheles* mosquitoes, have been discovered to carry bacterial endosymbionts required for host growth. This discovery has exciting implications for treatment.

Microbial endosymbionts. A simple example of intracellular endosymbiosis is that of the alga *Chlorella* growing within *Paramecium bursaria* (**Fig. 17.30**). The algae conduct photosynthesis and provide nutrients to the paramecium, which in turn shelters the algae from predators and viruses. The relationship is highly specific in that only certain species of algae and paramecia participate, and it is highly controlled in that the algal growth is limited to a population that avoids harming the host. This relation-

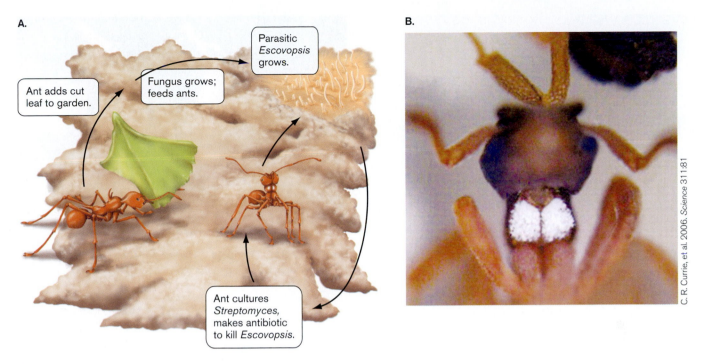

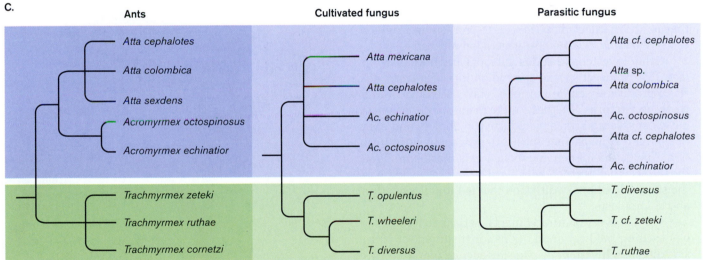

Figure 17.29 Ant fungal gardens: a four-way symbiosis. A. Attine ants cut leaves to cultivate mutualistic fungi, which feed the ants. When the parasitic fungus *Escovopsis* sp. grows, ants grow *Streptomyces* sp. on their bodies to produce antibiotics. **B.** Colonies of *Streptomyces* sp. on an ant cuticle. **C.** Coevolution of ants, cultivated fungus, and parasitic fungus. In most cases, the divergence of the ant species is paralleled by the divergence of their symbionts. Strains are designated based on the host ant gardens where they were isolated. *Source: A, C. Cameron R. Currie, et al. 2003. Science 229:386.*

ship may give clues to how the bacterial ancestor of chloroplasts began its intracellular existence.

The algal symbiosis, however, is reversible in that *Chlorella* retains its ability to multiply outside the paramecium. Moreover, under conditions of starvation in the absence of light, the paramecium may start to digest its endosymbionts as prey. Thus, the nature of the symbiosis (mutualistic or predatory) depends on the environment.

Bacterial endosymbionts of invertebrates. Many invertebrate animals, often themselves parasites of animals or plants, possess obligate bacterial endosymbionts. In some cases the endosymbiont is a parasite, such as the bacterium *Wolbachia pipientis* that grows within cells of the fruit fly *Drosophila*. *Wolbachia* bacteria are related to the rickettsias, obligate intracellular pathogens carried by arthropods and transmitted to humans. The *Wolbachia* strains that infect *Drosophila* cells are transmitted only through

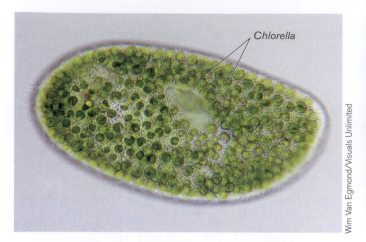

Figure 17.30 Endosymbiosis. *Paramecium bursaria*, a ciliate protist with endosymbiotic Chlorella algae. Cell length of *Paramecium*: 100 to 150 μm (LM).

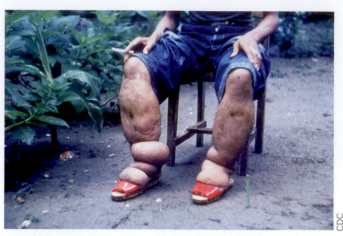

Figure 17.31 Filariasis. Patient suffering from filariasis, a form also known as elephantiasis.

egg cells; they cannot exist outside the insect. The bacteria have evolved ways to manipulate *Drosophila* reproduction so as to enhance their own transmission; for example, they feminize infected males so as to produce eggs, which carry *Wolbachia* into the next generation of flies.

Other invertebrate endosymbionts are mutualists. In fact, 15% of insect species depend on intracellular bacteria to produce essential nutrients, such as certain amino acids or vitamins. In these mutualisms, both partners have lost essential traits by reductive evolution, and each now requires the partner species to provide the lost function. The degree of genome reduction is most extreme in the intracellular partner, which can be seen as heading in the same evolutionary direction that led to mitochondria and chloroplasts.

The discovery of intracellular bacteria in invertebrates has exciting medical implications for treatment of parasitic diseases. Invertebrate parasites such as filarial worms invade human lymph nodes, causing forms of disease (filariasis) that are notoriously difficult to treat. A form of filariasis is elephantiasis, in which a limb expands with huge numbers of worms (**Fig. 17.31**). Filariasis afflicts more than 120 million people worldwide, largely in the Indian subcontinent and in Africa.

The filarial worms are nematodes that harbor *Wolbachia* endosymbionts (**Figs. 17.32A** and **B**). These *Wolbachia* strains differ from those that parasitize insects. *Wolbachia* may have entered the nematode originally as a pathogen or parasite, then persisted based on its metabolic contributions to the host. The nematode endosymbiont strains are mutualists; their presence is required for the nematode's embryonic development. The bacteria are found within tissue layers beneath the nematodes' skin and within the uterine tubes of females, where they enter the developing offspring (**Fig. 17.32B**). When human patients infected by the nematodes are treated with anti-

A.

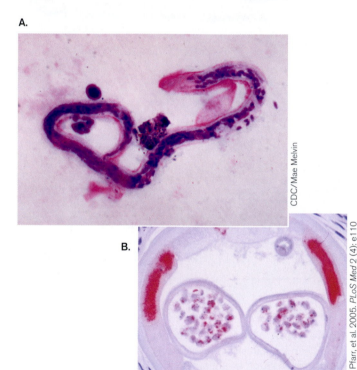

B.

Figure 17.32 The filarial endosymbiont *Wolbachia*.
A. Nematode worm *Brugya malayi*, a cause of filariasis.
B. Nematode cross section showing *Wolbachia* bacteria (stained pink) within the dermis and the uterine tubes.

biotics such as tetracycline, the bacteria disappear from worm tissues. The worm burden gradually decreases, and no offspring are produced. Antibacterial antibiotics eliminate the worms sooner and more completely than treatment with anti-nematode agents.

The genome of a *Wolbachia* strain from the filarial nematode *Brugia malayi* reveals extensive reductive evolution (**Fig. 17.33**). With barely a million base pairs, the

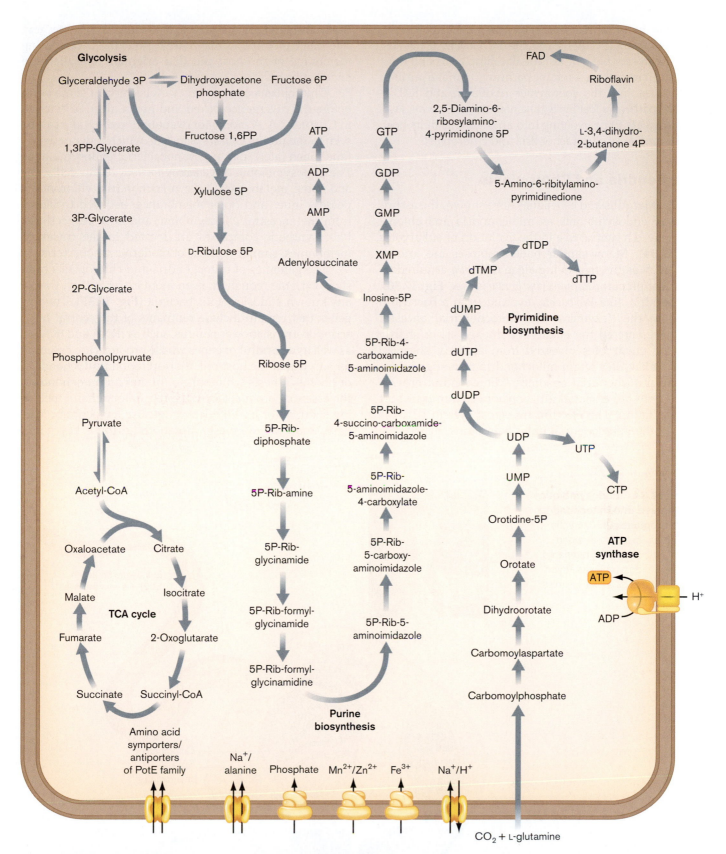

Figure 17.33 Metabolic pathways retained in the *Wolbachia* genome. Reductive evolution has eliminated much of the *Wolbachia* genome, but it retains metabolic pathways to make purines, pyrimidines, and vitamins such as riboflavin for its nematode host.

Wolbachia genome has lost many metabolic pathways, barely retaining glycolysis and the TCA cycle. It has lost the pathways for biosynthesis of all amino acids and most vitamins. It nonetheless retains pathways to make purines, pyrimidines, and the coenzymes riboflavin and FAD—essential pathways lost by its host nematode. Overall, *Wolbachia* appears to be evolving into an organelle of its host, like the ancestors of mitochondria and chloroplasts.

Mitochondria and Choroplasts

As Lynn Margulis and colleagues have shown, the assimilation of endosymbionts as mitochondria and chloroplasts played a central role in the evolution of eukaryotes (**Fig. 17.34**). Many similar endosymbioses are known today, such as the free-living algae *Chlorella* acquired by the protist predator *Paramecium bursaria* (see **Fig. 17.30**). Mitochondria, like *Wolbachia*, evolved from a bacterium related to the rickettsias. The mitochondrial ancestor must have entered the eukaryotic lineage as, or shortly after, the eukaryotes diverged from archaea, since all known eukaryotes retain mitochondria or vestigial remnants of mitochondrial genomes. Mitochondria provide the cell with the essential functions of electron transport and respiration. The electron transport system (ETS) is found in the mitochondrial inner membrane, believed to

derive from the cell membrane of the ancestral bacterium. The outer membrane may derive from the invaginating membrane of the host cell that originally engulfed the endosymbiont.

Chloroplasts arose from cyanobacteria at some point before the divergence of red and green algae (discussed in Chapter 20). A model for cyanobacterial uptake can be seen in the protist *Glaucocystophyta*, which independently (and much later) took up cyanobacterial endosymbionts. The endosymbionts of *Glaucocystophyta* retain cell walls and some metabolism. Like mitochondria, chloroplasts possess inner and outer membranes, believed to derive from the ancestral endosymbiont and host, respectively. Photosynthetic complexes are located in the thylakoid membranes, similar to those of modern cyanobacteria.

The genomes of mitochondria and chloroplasts both show extreme reduction, even more extreme than that of any known endosymbiotic bacteria (**Fig. 17.35**). The few genes that remain include remnants of the central transcription-translation apparatus, such as rRNA and tRNAs, as well as a handful of genes whose products are essential for survival of the host cell: for respiration (mitochondria) or photosynthesis (chloroplast). In human mitochondria, the essential products form the respiratory chain, including subunits of NADH dehydrogenase, cytochrome oxidase, and ATP synthase. Mutations in these key genes

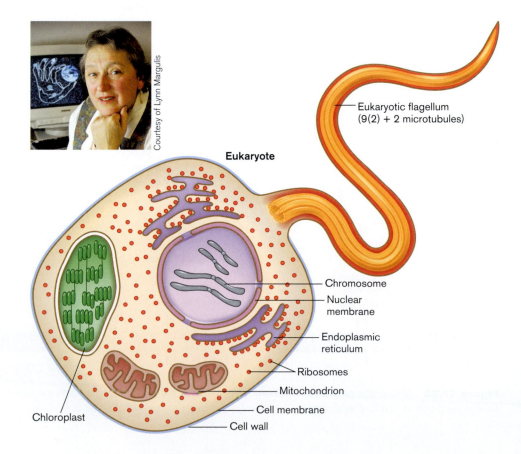

Figure 17.34 Endosymbiotic cells evolved into mitochondria and chloroplasts. Eukaryotic cells contain mitochondria and chloroplasts, organellar remnants of ancient endosymbioses. (Lynn Margulis, inset.) *Source:* Margulis. 1993.

Courtesy of Lynn Margulis

Eukaryote

Eukaryotic flagellum (9(2) + 2 microtubules)

Chromosome

Nuclear membrane

Endoplasmic reticulum

Ribosomes

Mitochondrion

Cell membrane

Cell wall

Chloroplast

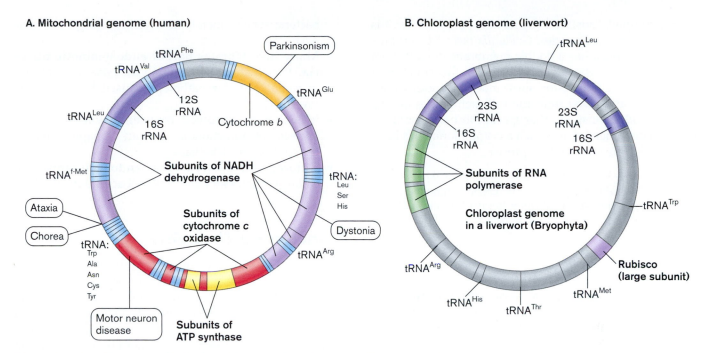

A. Mitochondrial genome (human)

B. Chloroplast genome (liverwort)

Figure 17.35 Genomes of mitochondria and chloroplasts. A. The mitochondrial genome retains large and small subunit rRNAs, five tRNA genes, plus subunits of the respiratory electron transport chain. Bubbles indicate human diseases associated with mitochondrial defects. **B.** The chloroplast genome retains large and small subunit rRNAs (23S and 16S), several tRNA and RNA polymerase genes, plus rubisco and components of photosystems I and II (dispersed around the circle, not labeled in figure).

lead to serious diseases; for example, damage to mitochondrial genes of respiration is associated with motor neuron disease, parkinsonism, and forms of ataxia.

But thousands of genes encoding ETS subunits, as well as other essential parts of mitochondria, have migrated from the mitochondrion to the nucleus. Some of these genes show tissue-specific expression, resulting in different mitochondrial types associated with different tissues. Thus, some mitochondrial defects are actually inherited through the nuclear genome. The mitochondria have evolved as integral parts of the host cell.

In chloroplasts, the organellar genome encodes essential products for photosynthesis, including photosystems I and II and the ATP synthase. The chloroplast genome shown in **Fig. 17.35B** retains the large subunit of rubisco, whereas the gene encoding the small subunit has migrated to the nucleus.

Remarkably, the process of **symbiogenesis**, the generation of new symbiotic associations, continues in a number of protist species. Protist-algae called cryptomonads result from symbiogenesis in which an alga (containing a chloroplast) was engulfed by an ancestral protist. The cryptomonad cell contains the degenerate remains of the algal endosymbiont (**Fig. 17.36**). The algal mitochondrion was lost through reductive evolution, and the nucleus shrank to a **nucleomorph**, the vestigial remains of a nucleus containing a small amount of the chromosomal DNA of the original algal nucleus. But the algal chloroplast

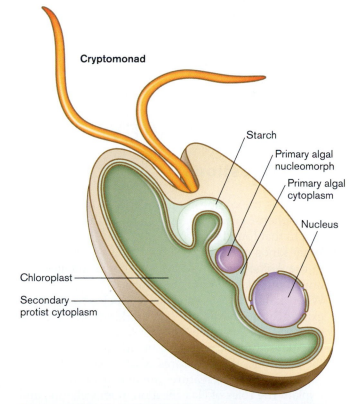

Figure 17.36 A cryptomonad results from secondary endosymbiosis. The cryptomonad *Guillardia theta* contains the primary algal chromosome surrounded by the remains of the primary algal cytoplasm and nucleus, which is shrunk to a nucleomorph.
Source: Modified from Paul R. Gilson. 2001. *Genome Biology* 2:1022.1.

was maintained, "enslaved" by its new host. The result is a new cryptomonad species, *Guillardia theta*. The cryptomonad chloroplast has a double membrane derived from the original cyanobacterial ancestor of the chloroplast and from the algal ancestor, surrounded by another double membrane derived from the cell membranes of alga and a secondary host. In other species, tertiary symbiosis has been documented, in which a cryptomonad has been swallowed in turn by another protist.

> **THOUGHT QUESTION 17.10** Besides mitochondria and chloroplasts, what other kinds of entities within cells might have evolved from endosymbionts?

TO SUMMARIZE:

- **Symbiosis is the intimate association of two unrelated species.** A symbiosis in which both partners benefit is called mutualism. If one partner benefits while harming the other, this is called **parasitism**.
- **Symbiotic partners undergo coevolution**, the evolution of two species in response to one another. Coevolution involves reductive (degenerative) evolution, in which each partner species loses some functions that the other partner provides.
- **An endosymbiont lives inside a much larger host species.** Many microbial cells harbor endosymbiotic

bacteria whose metabolism yields energy for their hosts.
- **Many invertebrates harbor endosymbiotic bacteria.** The bacteria are required for host survival and in some cases for pathology caused by a parasitic invertebrate.
- **Mitochondria evolved from endosymbionts.** The ancestor of mitochondria was an alphaproteobacterium related to rickettsias.
- **Chloroplasts evolved from endosymbionts.** The chloroplast ancestor was a cyanobacterium.

Concluding Thoughts

Over 3.85 Gyr ago, microbial communities had evolved, their strata building rock layers that persist today. From these or other microbes, all subsequent life evolved. It is hard to say which is more astonishing: the overall commonalities of all living cells, including membrane-bounded support systems for genomes of 3.85 billion years of shared ancestry, or the subsequent evolution of organisms with vastly different adaptations to exploit every possible niche of our planet. The next three chapters explore these diverse adaptations: Chapter 18, bacterial diversity; Chapter 19, archaeal diversity, and Chapter 20, diversity among microbial eukaryotes including fungi, algae, and protists.

CHAPTER REVIEW

Review Questions

1. What was the composition of Earth's early crust and atmosphere? What processes changed their composition to that found today?
2. What kinds of evidence support the presence of life in the Archaean eon? What are the advantages and limitations of each kind of evidence?
3. What kinds of metabolism are believed to have existed in Archaean life? What kinds of evidence support their existence?
4. Compare and contrast three models for the origin of the first cells. What features of life does each model explain, and which features are unexplained?
5. Explain the roles of classification, nomenclature, and identification for microbial taxonomy.
6. Why is the definition of species in prokaryotes more problematic than for eukaryotes? What is generally considered the present basis for defining prokaryotic species?

7. Discuss the roles of mutation, natural selection, and reductive evolution in the divergence of microbial species. Cite specific examples.
8. Explain the basis of a "molecular clock" for measuring microbial evolution. What fundamental properties must be met by a gene to function as a microbial clock? What are the limitations of a molecular clock?
9. Explain the basis of a phylogenetic tree. Why is the fundamental tree at the divergence of bacteria, archaea, and eukaryotes unrooted?
10. How does horizontal gene transfer determine genomic content? What kinds of genes are likely to undergo horizontal transfer?
11. Explain how endosymbiosis can lead to obligate association. Explain how reductive evolution and gene transfer lead to the evolution of organelles that are inseparable from host cells.

Key Terms

abiotic (641)
Archaean eon (633)
banded iron formation (BIF) (639)
biosignature (biological signature) (637)
biosphere (633)
candidate species (649)
clade (651)
classification (647)
coevolution (666)
dichotomous key (649)
endolithic (632)
endosymbiont (666)
endosymbiosis (666)
GC/AT ratio (661)
genomic island (662)
genus name (647)
greenhouse effect (633)
Gyr (631)
Hadean eon (633)

horizontal transfer (lateral transfer) (648)
identification (647)
informational gene (662)
isolate (649)
isotope ratio (637)
lateral transfer (648)
metabolist model (642)
microfossil (635)
molecular clock (652)
monophyletic group (651)
mutualism (666)
Myr (631)
nomenclature (647)
nucleomorph (671)
operational gene (662)
panspermia (646)
parasitism (666)
phylogenetic tree (653)

phylogeny (649)
probabilistic indicator (649)
rank (647)
reductive evolution (652)
ribozyme (642)
RNA world (641)
small-subunit rRNA (653)
species (647)
species name (647)
stromatolite (630)
supernova (631)
symbiogenesis (671)
symbiosis (666)
taxon (647)
taxonomy (647)
unrooted tree (655)
vertical transfer (661)

Recommended Reading

Bada, Jeffrey. 2004. How life began on Earth: A status report. *Earth and Planetary Science Letters* **226**:1–15.

Barns, Susan M., Charles F. Delwiche, Jeffery D. Palmer, and Norman R. Pace. 1996. Perspectives on archaeal diversity, thermophily and monophyly from environmental rRNA sequences. *Proceedings of the National Academy of Sciences USA* **93**:9188–9193.

Brasier, Martin D., Owen R. Green, Andrew P. Jephcoat, Annette K. Kleppe, Martin J. Van Kranendonk, et al. 2002. Questioning the evidence for Earth's oldest fossils. *Nature* **416**:76–81.

Carroll, Sean B. 2001. Chance and necessity; the evolution of morphological complexity and diversity. *Nature* **409**:1102–1109.

Currie, Cameron R., et al. 2003. Ancient tripartite coevolution in the attine ant-microbe symbiosis. *Science* **299**:386.

Daubin, Vincent, Manolo Gouy, and Guy Perrière. 2002. A phylogenomic approach to bacterial phylogeny: Evidence of a core of genes sharing a common history. *Genome Research* **12**:1080–1090.

Doolittle, W. Ford. 1999. Phylogenetic classification and the universal tree. *Science* **284**:2124.

Gevers, Dirk, Frederick M. Cohan, Jeffrey G. Lawrence, Brian G. Spratt, Tom Coenye, et al. 2005. Re-evaluating prokaryotic species. *Nature Reviews Microbiology* **3**:733–739.

Gilson, Paul R. 2001. Nucleomorph genomes: Much ado about practically nothing. *Genome Biology* **2**:1022.1–1022.5.

Hanage, William P., et al. 2005. Fuzzy species among recombinogenic bacteria. *BMC Biology* **3**:6.

Hanczyc, Martin M., Shelly M. Fujikawa, and Jack W. Szostak. 2003. Experimental models of primative cellular compartments: Encapsulation, growth, and division. *Science* **302**:618–622.

Jiao, Yongqin, Andreas Kappler, Laura R. Croal, and Dianne K. Newman. 2005. Isolation and characterization of a genetically-tractable photoautotrophic Fe(II)-oxidizing bacterium, *Rhodopseudomonas palustris* strain TIE-1. *Applied and Environmental Microbiology*, **71**(8):4487–4496.

Kasting, James F., and Janet L. Siefert. 2002. Life and the evolution of earth's atmosphere. *Science* **296**:1066–1067.

Kelley, Scott T., Ulrike Theisen, Largus T. Angenent, Allison St. Amand, and Norman R. Pace. 2004. Molecular analysis of shower curtain biofilm microbes. *Applied and Environmental Microbiology* **70**:4187–4192.

Moser, Duane P., Thomas M. Gihring, Fred J. Brockman, James K. Fredrickson, David L. Balkwill, et al. 2005. *Desulfotomaculum* and *Methanobacterium* spp. dominate a 4- to 5-kilometer-deep fault. *Applied and Environmental Microbiology* **71**:8773–8783.

Ochman, Howard. 2005. Genomes on the shrink. *Proceedings of the National Academy of Sciences USA* **102**:11959–11960.

Salzberg, Steven L., Julie C. Dunning Hotopp, Arthur L. Delcher, Mihai Pop, Douglas R. Smith, et al. 2005. Serendipitous discovery of *Wolbachia* genomes in multiple *Drosophila* species. *Genome Biology* **6**:R23.

Shen, Yanan, Roger Buick, and Donald E. Canfield. 2001. Isotopic evidence for microbial sulphate reduction in the early Archaean era. *Nature* **410**:77–81.

Souza, Valeria, Amanda Castillo, and Luis Equiarte. 2002. Evolutionary ecology of *E. coli*. *American Scientist* **90**:332.

Wernegreen, Jennifer J. 2005. Endosymbiosis: Lessons in conflict resolution. *PLoS Biology* **2**:307–311.

Yasuda, Moriyoshi, Hiroshi Oyaizu, Akihiko Yamagishi, and Tairo Oshima. 1995. Morphological variation of new *Thermoplasma acidophilum* isolates from Japanese hot springs. *Applied and Environmental Microbiology* **61**:3482–3485.

Chapter 18

Bacterial Diversity

Bacteria vary tremendously in their cell structure and metabolism. They include heterotrophs, phototrophs, and lithotrophs, with some species that could be classified as all three. There are obligate aerobes, anaerobes, and microaerophiles. Their cell shapes include rods, cocci, spirals, and budding forms. Ecologically, bacteria include mutualists, pathogens, and organisms that cannot be cultured in our laboratories. New species, even entire new phyla of bacteria, are continually discovered in soil and water, in our homes, and even within our own bodies.

How do we begin to make sense of the thousands of different kinds of bacteria? This chapter introduces the major categories of bacteria about which we know most. These include the gram-positive rods and cocci, the gram-negative proteobacteria, anaerobes, phototrophs, and spirochetes. There are cell wall-less chlamydiae and planctomycetes and deep-branching thermophiles. Even now, all the species we know represent but a tiny fraction of the diverse bacterial species growing in nature.

1 μm

Verrucomicrobium spinosum, isolated by Heinz Schlesner at Christian Albrechts University. Verrucomicrobia, or "bumpy bacteria," metabolize sugars in soil and water and in the human digestive tract. These bacteria were barely known until recently, but DNA analysis shows they may account for 5% of all bacteria. The sequencing of their genomes is supported by the U.S. Department of Energy Joint Genome Institute, in collaboration with German and Australian scientists (2006). Photo courtesy of Dennis Kunkel/Phototake.

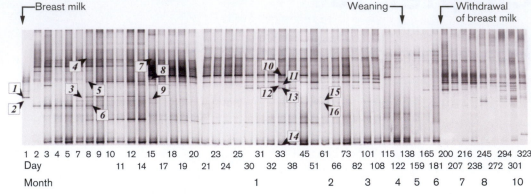

B. Electrophoresis of 16S rRNA genes

Breast milk

Weaning

Withdrawal of breast milk

Day	1 2 3 4 5 7 8 9 10 12 15 18 20 23 25 31 33 45 61 73 101 115 138 165 200 216 245 294 323
	11 14 17 19 21 24 30 32 38 51 66 82 108 122 159 181 207 238 272 301
Month	1 2 3 4 5 6 7 8 10

A. Intestinal bacteria from a human infant: Nearest relatives

Species	%
1. *Escherichia coli*	97
2. *Veillonella dispar*	97
3. *Streptococcus thermophilus*	99
4. *Enterococcus raffinosus*	96
5. *Ruminococcus gnavus*	96
6. *Enterococcus avium*	99
7. *Streptococcus salivarius*	96
8. *Clostridium paraputrificum*	91
9. *Veillonella atypical*	93
10. *Enterobacter aerogenes*	98
11. *Clostridium neonatale*	97
12. *Clostridium neonatale*	93
13. *Clostridium neonatale*	98
14. *Bifidobacterium breve*	98
15. *Veillonella dispar*	96
16. *Streptococcus salivarius*	98

Figure 18.1 **Bacterial succession in the digestive tract of a newborn baby.** **A.** A nursing infant shows a unique combination of bacterial species. Comparative sequence analysis reveals the closest relative of each species isolated. **B.** From a newborn baby over time, fecal samples were tested. PCR-amplified sequences of 16S (SSU) rRNA genes were separated by denaturing gradient gel electrophoresis (DGGE). The patterns of bands change as species disappear and new ones appear, with the greatest changes occurring at weaning from breast milk to solid food. Band numbers refer to isolates for which 16S rRNA was sequenced (list at left). *Source*: B. Christine Favier, et al. *Applied and Environmental Microbiology* 68:219.

Bacteria have evolved a bewildering array of life-forms that colonize every habitat on Earth, from the dunes of the Sahara desert to the ice-buried Lake Vostok of Antarctica. Every day we discover new species never before recorded, not just from exotic environments, but from habitats closer to home. Consider, for example, the digestive tract of a newborn infant (**Fig. 18.1**). Dutch researcher Antoon Akkermans and colleagues identified bacterial species from the feces of an infant during the first year of life. The bacterial species were identified using polymerase chain reaction (PCR) to amplify small-subunit rRNA gene sequences directly, without culturing, as discussed in Chapter 17. The rRNA sequences from the infant's fecal sample were separated by denaturing gradient gel electrophoresis (DGGE), a technique in which the cloned DNA products denature and form secondary structures that influence their banding patterns in the gel. As each infant grows, the DGGE bands change as bacterial species disappear and new ones appear. The species distribution changes the most during months 4–6 as the infant is weaned from breast milk onto foods containing very different carbon sources. Some of the DNA samples show close sequence homology with well-characterized anaerobic genera such as *Bifidobacterium*, a gram-posi-

tive rod that ferments without gas production. The lack of gas production by *Bifidobacterium* helps prevent colic (gas pains). Other gram-positive rods were identified as clostridia, gram-positive spore formers that include harmless species as well as pathogens causing tetanus and botulism. But more than half of the fecal isolates showed rRNA with less than 97% similarity to any known sequence, a criterion for distinct species. Thus, even a human infant constitutes a microbial frontier full of unknown kinds of bacteria.

Characterizing bacteria is a formidable task, as even in familiar habitats the vast majority of species remain unknown and new phyla are discovered daily. At the same time, new fields of genomics, microscopy, and culture techniques clarify our view of major groups of bacteria with physiological, ecological, and medical significance. Chapter 18 presents these taxa in a format that emphasizes phylogeny (evolutionary relatedness) as well as key traits such as the gram-positive cell wall of Firmicutes and the oxygenic photosynthesis of cyanobacteria. For each taxon, a few key species are introduced to represent the spectrum of diversity that spans even a group of close relatives. For more information on particular species, consult the following public online resources:

- **The National Center for Biotechnology Information (NCBI) Taxonomy Database**, funded by the U.S. government, lists the full taxonomy of biological species reported in the literature and an ever-growing database of sequenced genomes.

www | National Center for Biotechnology Information (NCBI)

- **The Microbial Biorealm** at Kenyon College is a student-edited online encyclopedia of microbial genera. Entries briefly summarize the taxonomy, genome structure, physiology, and societal relevance of each species.

www | Microbial Biorealm

NOTE: Bacteria are also known as eubacteria or "true bacteria," to distinguish them from archaea, many of which were originally described as bacteria. Current literature equates bacteria with eubacteria. Archaea are discussed in Chapter 19.

18.1 Bacterial Diversity at a Glance

To survey bacterial diversity in one chapter is like touring all the countries of a continent in a single day. Like countries, bacterial taxa have complex traits and histories and often contested borders. At the same time, bacteria as a whole share major commonalities that offer an important perspective from which to appreciate their differences.

Common Traits of Bacteria

Chapter 17 summarized the differences and similarities between the three major domains—Bacteria, Archaea, and Eukarya (see **Table 17.3**). A profound commonality of bacteria is their central apparatus of gene expression, particularly their RNA polymerase and their ribosomal RNAs and translation factors. The bacterial gene expression complexes differ more from those of Archaea or Eukarya than those of Archaea or Eukarya differ from each other. This subtle point of molecular biology has a profound consequence for human medicine and agriculture: It underlies the selective activity of a wide array of antibiotics, such as streptomycin, that attack only bacterial pathogens, without harming the animal or plant host.

Another fundamental trait of bacteria with useful consequences in controlling pathogens is that most bacterial cells possess a cell wall of peptidoglycan comprising a single covalent molecule (**Fig. 18.2**; discussed in Chapter 3). Peptidoglycan is composed of disaccharide-peptide chains that can cross-link in three dimensions; key enzymes that build the peptide links are blocked by antibiotics such as penicillin and vancomycin. Variant forms of peptidoglycan distinguish different species. For instance, the gram-positive pathogen *Staphylococcus aureus* has cell wall peptides cross-linked by pentaglycine (a chain of five glycine residues). Some species such as mycoplasmas lack peptidoglycan altogether, but they arose by reductive evolution from bacteria that possess it.

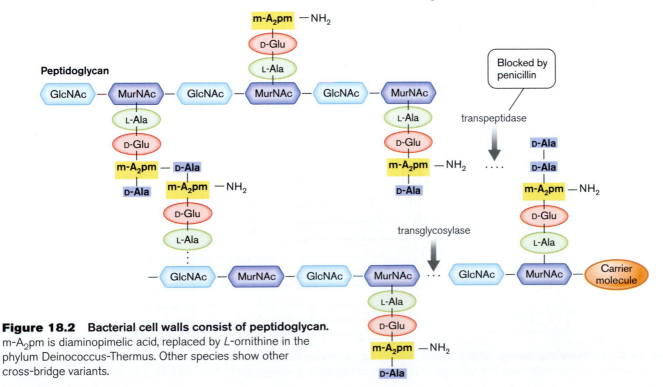

Figure 18.2   Bacterial cell walls consist of peptidoglycan. m-A$_2$pm is diaminopimelic acid, replaced by *L*-ornithine in the phylum Deinococcus-Thermus. Other species show other cross-bridge variants.

By contrast, most archaea lack cell walls, although methanogens have independently evolved an analogous sugar-peptide structure called pseudopeptidoglycan (discussed in Chapter 19). Eukaryotes such as fungi and plants have cell walls of polysaccharides such as cellulose.

The Bacterial Tree

The phylogeny of known bacteria is presented in **Figure 18.3**. The names of well-studied phyla are lettered blue. A **phylum** is defined as a group of organisms sharing a common ancestor that diverged early from other bacteria based on rRNA sequence (discussed in Chapter 17). Phyla and other major divisions are also defined on the basis of historical convention and consensus of the research community.

A bacterial phylum generally comprises species that share key traits as well as ancestry. While sharing key traits, the member species often show remarkable diversity in other ways. Some phyla, such as Cyanobacteria, share a unique form of metabolism (oxygenic photosynthesis), yet have evolved many diverse cell shapes and grow in diverse habitats. Other phyla, such as Spirochetes, share a unique cell structure while diverging in habitat and metabolism.

Table 18.1 introduces the bacterial species that are most frequently encountered in the literature. Bear in mind, however, that more than 50 other bacterial phyla are known to date and that deep-branching clades continue to be discovered.

Deep-branching thermophiles. The most deeply branching bacterial phyla (that is, those that appear to have diverged earliest) include hyperthermophiles that share physiology and habitat with archaea. For example, *Aquifex* grows at 95°C and possesses traits typical of archaea, such as ether-linked membrane lipids. Some "archaeal" traits of bacteria arose from a common ancestor of bacteria and archaea, whereas others arose through lateral transfer of genes from archaea to bacteria sharing the high-temperature habitat. In other aspects of structure and physiology, the deep-branching phyla diverge from each other. Order Aquificales consists of gram-negative rods that oxidize hydrogen, whereas order Thermotogales is made up of sheathed gram-negative rods that are obligate anaerobic heterotrophs. Deinococcus-Thermus includes moderately thermophilic gram-negative heterotrophs, often filamentous. *Thermus aquaticus* was the first source of heat-resistant DNA polymerase for PCR.

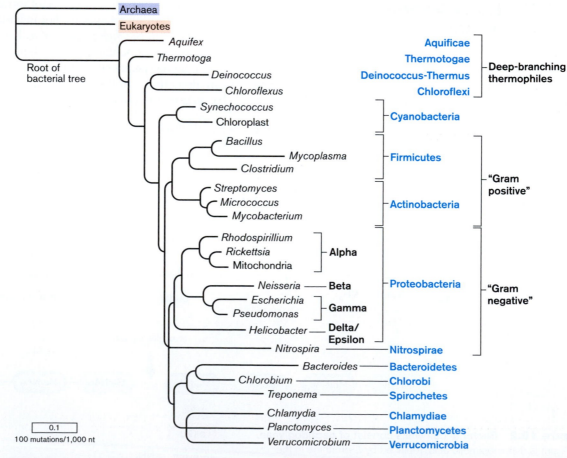

Figure 18.3 **Bacterial phylogeny.** A phylogenetic tree of representative Bacteria based on 16S rRNA sequence comparison. The tree is rooted with respect to Archaea and Eukarya. Bold labels correspond to major groups in Table 18.1. Phylum names are lettered blue.

Table 18.1   Representative groups of bacteria.

Note: Each trait applies to *most* members of the taxon described, but exceptions have evolved.

Blue-lettered terms indicate phyla.

• **Bulleted terms are representative orders** within the phylum.

Deep-branching Thermophiles

Thermophilic bacteria that diverged early from the archaea and eukaryotes. Many genes transferred laterally from archaea.

Aquifex

K. O. Stetter & R. Rachel, U. of Regensburg

Aquificae. Hyperthermophiles (70–95°C). Oxidize H₂, H₂S, or thiosulfate.
• **Aquificales.** Outer membrane; cells may be single with flagella or grow in filaments. *Aquifex, Thermocrinis.*

Thermotogae. Thermophiles or hyperthermophiles (55–100°C).
• **Thermotogales.** Outer membrane with large periplasm; anaerobic heterotrophs. *Petrotoga, Thermotoga.*

Deinococcus

Dennis Kunkel/Visuals Unlimited

Deinococcus-Thermus. Peptidoglycan contains ornithine. Diverse growth temperatures.
• **Deinococcales.** Thick envelope; stains gram-positive. Not thermophilic, but extremely resistant to ionizing radiation and to desiccation. *Deinococcus.*
• **Thermales.** Species are filamentous clustered cells or single-celled. Grow at 70–75°C. *Thermus aquaticus* is the source of Taq polymerase for PCR.

Chloroflexi. Filamentous green nonsulfur bacteria.
• **Chloroflexales.** Filamentous phototrophs. Most contain photosystem II in chlorosomes. *Chloroflexus aurantiacus* is thermophilic; forms mats in hot springs.

Chloroflexus aurantiacus

DOE-Herter/Visuals Unlimited

Cyanobacteria

Oxygenic photoautotrophs with thylakoid membranes. Share ancestry with chloroplasts.

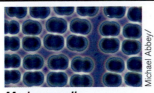

Merismopedia

Michael Abbey/Visuals Unlimited

Cyanobacteria. Oxygenic photoautotrophs with thylakoid membranes. Fix CO₂ using rubisco. Often mutualists. Share ancestry with chloroplasts.
• **Chroococcales.** Square colonies based on two division planes.
  Chroococcus. Single, double, or quartet of cells. Grow in pond sediment.
  Merismopedia. Platelike colonies of elliptical cells.
  Synechococcus. Single or double cells. Marine producer.

Anabaena

©Carolina Biological Supply/Visuals Unlimited

• **Nostocales.** Filamentous chains with N₂-fixing heterocysts. Often grow symbiotically with other microbes, corals, or plants.
  Anabaena. Aquatic. Grow in association with red water fern (*Azolla*).
  Nostoc. Aquatic. Grow independently; or mutualistically with fungi, as lichens; or as endosymbionts of *Gunnera* plant cells.
• **Oscillatoriales.** Filamentous chains with motile hormogonia (short chains).
  Oscillatoria. Aquatic or marine. Grow independently; or as sponge endosymbionts.
  Spirulina. Aquatic. Farmed as food supplement.
  Trichodesmium. Major marine producer. Forms "blooms" of overgrowth.
• **Pleurocapsales.** Globular colonies; reproduce through baeocytes. *Pleurocapsa, Myxosarcina.*
• **Prochlorales.** Tiny single cells, elliptical or spherical.
  Prochlorococcus. Most abundant marine producer. One of the smallest cells of a free-living microbe, with a highly reduced genome.
  Prochloron. Tropical marine producer; endosymbiont of sea squirt.

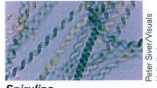

Spirulina

Peter Siver/Visuals Unlimited

(continued)

Table 18.1	**Representative groups of bacteria (continued)**

Firmicutes and Actinobacteria (gram-positive)

Gram-positive. Peptidoglycan multiple layers, cross-linked by teichoic acids. Aerobes and facultative anaerobes.

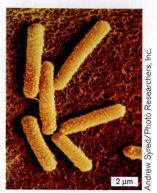

Bacillus subtilis

Mycoplasma genitalium

Firmicutes. Low GC, gram-positive rods and cocci.
- **Bacillales.** Aerobic or facultative anaerobes.
  - *Bacillus.* Endospore-forming rods. Obligate aerobes. Grow in soil. *B. subtilis* is gram-positive model system; *B. anthracis* causes anthrax. Include extremophiles: *B. alkalophilus, B. thermophilus, B. halodurans.*
  - *Listeria.* Non-spore-forming rods. *L. monocytogenes* is an enteric intracellular pathogen.
  - *Staphylococcus.* Non-spore-forming cocci; hexagonal clusters. *S. aureus* causes toxic shock and lethal flesh infections.
- **Clostridiales.** Anaerobic rods.
  - *Carboxydothermus.* Oxidizes carbon monoxide (CO) to CO_2. Forms endospores.
  - *Clostridium.* Forms endospores. *C. botulinum* and *C. difficile* are pathogens. *C. acetobytylicum* generates butanol.
  - *Dehalobacter.* Non-spore-forming rods. Dechlorinate chloroethenes.
  - *Epulopiscium.* Exceptionally large cells. Reproduce by "live birth."
  - *Heliobacterium, Heliophilum.* Endospore-forming photoheterotrophs; use a distinctive bacteriochlorophyll *g.*
  - *Metabacterium.* Exceptionally large cells; form multiple endospores. *M. polyspora.*
  - *Ruminococcus.* Digestive flora of ruminant animals.
- **Lactobacillales.** Facultative anaerobes. Ferment producing lactic acid.
  - *Enterococcus.* Enteric cocci.
  - *Lactobacillus.* Non-spore-forming rods. *L. acidophilus* is used for dairy culture.
  - *Lactococcus.* Non-spore-forming cocci. Used for dairy culture.
  - *Streptococcus.* Non-spore-forming cocci; chains. Human throat flora. Group A streptococci cause "strep throat" and scarlet fever.
- **Mollicutes** (class). Lacks cell wall and S-layer; requires an animal host. *Mycoplasma genitalium* has one of smallest known genomes. *M. pneumoniae* causes pneumonia.

Streptomyces

Micrococcus luteus

Actinobacteria. High GC, gram-positive bacteria with moderate salt tolerance.
- **Actinomycetales.**
  - **Actinomycetes.** Filamentous, producing aerial hyphae and spores.
    - *Streptomyces* species produce many antibiotics. *S. coelicolor, S. griseus.*
    - *Frankia* **species** fix nitrogen for plants.
    - *Actinomyces.* *A. israelii* causes actinomycosis.
  - **Corynebacteraceae.** Irregular rods. *Corynebacterium diphtheriae* causes diphtheria.
  - **Mycobacteriaceae.** Exceptionally thick cell envelope holds acid-fast stain.
    - *Mycobacterium.* *M. tuberculosis* causes tuberculosis; *M. leprae* causes leprosy.
  - **Micrococcaceae.** Nonfilamentous soil bacteria; mostly obligate aerobes.
    - *Arthrobacter.* Rods. Respire on oxygen or on chlorinated aromatics.
    - *Micrococcus.* Aerobic cocci such as *M. luteus.* Grow in soil.
- **Bifidobacteriales.** Rods; fermenting without gas production. Commonly found among the enteric flora of breast-fed infants.
- **Propionibacteria.** Facultative anaerobic rods. Propionic acid fermentation. *Propionibacterium acnes* causes acne; *P. freudenreichii* makes Swiss cheese.

Proteobacteria and Nitrospirae (gram-negative)

Gram-negative. Outer membrane contains LPS. Diverse metabolism. Lineage includes mitochondria.

Proteobacteria. Gram-negative. Diverse cell forms and metabolism.
Alpha Proteobacteria (class). Heterotrophic rods or spirilla with diverse metabolism.
- **Caulobacterales.** Aquatic oligotrophs. Crescent-shaped cells with stalk; develop flagellar form.
  - *Caulobacter.* Other oligotrophs with variable shape (lineage uncertain): *Hyphomicrobium, Seliberia.*

Table 18.1 Representative groups of bacteria (*continued*)

- **Nitrobacter.** Nitrite-oxidizing lithotrophs. *Nitrobacter winogradskyi.*
- **Rhizobiales.** Plant mutualists and pathogens, methyl oxidizers, and animal pathogens.
    - *Agrobacterium. A. tumefaciens* causes plant tumors; transgenic plant vector.
    - *Rhizobium* group. Related to *Agrobacterium. Rhizobium, Bradyrhizobium, Sinorhizobium* fix nitrogen intracellularly in legumes.
    - *Brucella.* Animal pathogens. *B. abortus.*
    - *Methylobacterium.* Grow on single-carbon compounds. Opportunistic pathogens found in shower-curtain soap scum.
    - *Rhodopseudomonas, Rhodomicrobium.* Soil photoheterotrophs.
- **Rhodobacterales, Rhodospirales.** Photoheterotrophs using sulfides and organic substrates. Unicellular with flagella. *Rhodobacter sphaeroides, Rhodospirillum rubrum, Rhodopseudomonas palustris.* Some non-phototrophs: *Acetobacter* makes vinegar.
- **Rickettsia.** Includes intracellular parasites; lineage includes mitochondria.
    - *Rickettsia.* Intracellular parasites. *R. rickettsii* causes Rocky Mountain spotted fever; *R. prowazekii* causes typhus.
    - **SAR11 cluster.** Marine photoheterotrophs use proteorhodopsin. *Pelagibacter.*
- **Sphingomonadales.** Catabolizes complex organics; useful for bioremediation. Some species are opportunistic pathogens. *Sphingomonas.* Also aerobic photoheterotrophs: *Citromicrobium, Erythromicrobium.*

Beta Proteobacteria (class). Phototrophs, heterotrophs, and lithotrophs.
- **Burkholderiales.** Soil bacteria and opportunistic pathogens. *Burkholderia pseudomallei* causes melioidiosis in humans and farm animals.
- **Hydrogenophilales.** Lithotrophs.
    - *Thiobacillus. T. ferrooxidans* oxidizes iron and sulfur.
- **Neisseriales.** Mucous membrane normal flora and pathogens. Aerobic or microaerophilic diplococci. *Neisseria gonorrhoeae* causes gonorrhea; *N. meningitidis* causes meningitis.
- **Nitrosomonadales.** Lithotrophs. *Nitrosomonas europea* oxidizes ammonia.
- **Rhodocyclales.** Soil heterotrophs and photoheterotrophs. *Azoarcus evansii* catabolizes complex aromatic molecules. *Rhodocyclus* species are purple photoheterotrophs.

Gamma Proteobacteria (class). Facultative anaerobes and lithotrophs.
- **Aeromonadales.** Aquatic heterotrophs. *Aeromonas hydrophila* infects fish; causes opportunistic infections in humans.
- **Chromatiales.** Lithotrophs (*Nitrococcus* oxidizes ammonia). Sulfur and iron phototrophs (*Chromatium*). Nitrite phototrophs (*Thiocapsa*).
- **Enterobacterales.** Enterobacteraceae. Facultative anaerobes; colonize the human colon.
    - *Escherichia coli.* Strains include normal flora and enteric pathogens.
    - *Salmonella.* Pathogens. *S. enterica* causes gastrointestinal infection. *S. typhi* causes typhoid fever.
    - *Yersinia.* Pathogens. *Y. pestis* causes bubonic plague.
- **Legionellales.** Waterborne intracellular pathogens. *Legionella pneumophila* causes legionellosis pneumonia.
- **Oceanospirilles.** Marine heterotrophs, including degraders of petroleum. *Alcanivorax.*
- **Pseudomonadales.** Rods, usually aerobic or respire on nitrate; catabolize aromatics. *Pseudomonas aeruginosa* opportunistically infects lungs of cystic fibrosis patients. *P. fluorescens* suppresses plant diseases.
- **Thiotrichales.** Lithotrophs and heterotrophs. *Beggiatoa alba* and *Thiomargarita namibiensis* oxidize sulfur. *Cycloclasticus* catabolizes polycyclic aromatic hydrocarbons in petroleum.
- **Vibrionales.** Marine heterotrophs. *Vibrio cholerae* causes cholera.
- **SAR86 cluster.** Marine photoheterotrophs use proteorhodopsin.

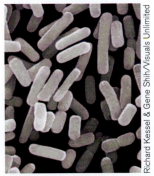
Escherichia coli

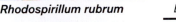
0.5 μm

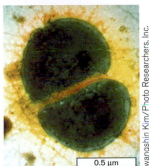
Neisseria gonorrhoeae

Rhodospirillum rubrum

David M. Phillips/Visuals Unlimited

Kwangshin Kim/Photo Researchers, Inc.

Richard Kessel & Gene Shih/Visuals Unlimited

(continued)

Table 18.1   Representative groups of bacteria (*continued*)

Myxococcus

Michiel Vos, U. of Oxford

Delta Proteobacteria (class). Diverse form and metabolism.
- **Bdellovibrionales.** Bacterial intracellular predators. *Bdellovibrio bacteriovorus* parasitizes and consumes *E. coli*.
- **Desulfobacterales.** Reduces sulfate. *Desulfobacter* spp.
- **Desulfuromonadales.** Lithotrophs; reduce or oxidize sulfur or metals.
    Geobacter metallireducens reduces iron oxide.
    Desulfuromonas reduces elemental sulfur.
- **Myxococcales.** Gliding bacteria that form fruiting bodies. *Myxococcus* spp.

Epsilon Proteobacteria (class).
- **Campylobacterales.** Spirillar pathogens.
    Campylobacter. *C. jejuni* causes food poisoning.
    Helicobacter. *H. pylori* causes gastritis.
    Thiovulum. Oxidizes sulfides, producing sulfur granules.

Thiovulum

Carl O. Wirsen and Holger W. Jannasch. 1978. *Journal of Bacteriology* 136:765

Nitrospirae. Nitrite oxidizers. Obligate aerobes.
- **Nitrospirales.** Tight spirillum. In soil and water, oxidizes NO_2^- to NO_3^-. *Nitrospira*; *Leptospirillum*. Includes acidophilic Fe oxidizers.

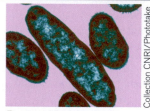

Nitrospira

S. Ehrich, et al. 1995. *Arch. Microbiol.* 164:16

1 μm

Bacteroidetes/Chlorobi Group

Gram-negative. Obligate anaerobes and/or green sulfur phototrophs.

Bacteroides

Collection CNRI/Phototake

Bacteroidetes. Obligate anaerobes. Heterotrophs that feed on diverse carbon sources.
- **Bacteroidales.** Obligate anaerobes, soil or human flora.
    Bacteroides. Enteric. *B. thetaiotaomicron* digests complex plant carbohydrates in gut.
        Bacteroides species escaping the colon cause abscesses.
    Porphyromonas. Includes gingival pathogens such as *P. gingivalis*.
- **Flavobacteriales.** Heterotrophs in soil and water; opportunistic pathogens. *Flavobacterium psychrophilum* infects freshwater trout fish.

Chlorobi. Green sulfur-oxidizing phototrophs.
- **Chlorobiales.** Conducts anaerobic photolysis of H_2S, absorbing primarily red and infrared. *Chlorobium tepidum*.

Spirochetes (Spirochaeta)

Narrow coiled cell encased by sheath, which encloses polar flagella.

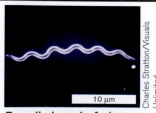

Borrelia burgdorferi

10 μm

Charles Stratton/Visuals Unlimited

Spirochetes. Narrow coiled cell with axial filaments, encased by sheath. Polar flagella beneath sheath double back around cell. Often symbiotic or pathogenic.
- **Spirochaetales.** Aquatic, free-living or pathogenic.
    Borrelia. *B. burgdorferi* causes Lyme disease, transmitted by ticks.
    Hollandina. Termite gut endosymbionts.
    Leptospira. L-shaped animal pathogen; causes leptospirosis.
    Treponema. *T. pallidum* causes syphilis.
    Spirochaeta. Aquatic, free-living.

Table 18.1 Representative groups of bacteria (*continued*)

Chlamydiae, Planctomycetes, and Verrucomicrobia

Irregular cells lacking peptidoglycan, with subcellular structures analogous to eukaryotes.

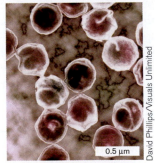

0.5 µm

David Phillips/Visuals Unlimited

Chlamydia trachomatis

Chlamydiales. Intracellular cell wall-less pathogens of animals or protists.

- **Chlamydiales.** Reticulate bodies form multiple spore-like transfer particles (elementary bodies) that infect the next host.

  Chlamydia. *C. trachomatis* causes sexually transmitted disease and trachoma (eye infection).

  Chlamydophila. *C. pneumoniae* causes pneumonia; *C. abortus* causes spontaneous abortion in animals; *C. psittaci* causes psittacosis in birds and humans.

  Parachlamdia. Chlamydia-like species that infect free-living amebas.

Planctomycetes. Nucleoid has double membrane analogous to eukaryotic nuclear membrane and other intracytoplasmic subcellular membrane compartments.

- **Planctomycetales.** Aquatic or marine. Flexible cell shape.

  Brocadia. *B. anammoxidans* anaerobically oxidizes ammonium, using a membrane-bound organelle, the anammoxosome.

  Gemmata. Nuclear double membrane, surrounded by additional intracellular membrane.

  Pirellula. Only one intracellular membrane. Has fimbriae and single flagellum. Marine habitat.

  Planctomyces. One intracellular membrane. Halophile.

Verrucomicrobia. Stalk-like appendages contain actin filaments in addition to peptidoglycan. Aquatic oligotrophs.

- **Verrumicrobiales.**

  Prosthecobacter. Each cell has a polar cytoplasmic extension called a prostheca.

  Verrucomicrobium. Stellate cells have multiple prosthecae.

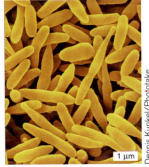

1 µm

Dennis Kunkel/Phototake

Prosthecobacter fusiformis

By contrast, members of the closely related order Deinococcales grow in the middle range of temperature and stain gram-positive. *Deinoccus radiodurans* shows extreme resistance to ionizing radiation and desiccation (discussed in Chapters 5 and 9). The phylum Chloroflexi consists of thermophilic green phototrophs that often share habitat with thermophilic cyanobacteria.

Cyanobacteria. The **Cyanobacteria** are a deeply branching phylum of profound importance for all ecosystems. Cyanobacteria conduct photosynthesis by splitting water or H_2S and fixing CO_2 (discussed in Chapter 14). Only cyanobacteria and chloroplasts (in eukaryotes) can produce oxygen gas. They have a unique two-photosystem apparatus for oxygenic photosynthesis arranged in lamellar arrays of membranes called thylakoids. Unique to cyanobacteria and chloroplasts are the chlorophylls, most notably chlorophylls *a* and *b*; these are distinct from the "bacteriochlorophylls" used by nonoxygenic bacterial phototrophs. Cyanobacteria share a common metabolism, yet the cell shape and organization of cyanobacterial species show an immense range of different forms, including chains of cells, square arrays, and globular colonies. Many cyanobacteria grow in seawater or fresh water, whereas others grow in mutualism with a eukaryotic host.

Gram-positive bacteria. Two major phyla, **Firmicutes** and **Actinobacteria**, are called "the gram-positive bacteria" because most members stain gram-positive. Some Firmicutes fail to stain gram-positive because they lack cell walls (the mycoplasmas), whereas some Actinomycetes possess a thick waxy coat that excludes the gram stain. Most Firmicutes and Actinobacteria have an exceptionally thick cell wall, with several layers of peptidoglycan generally interpenetrated by supporting molecules such as teichoic acids or mycolic acids. The thick, reinforced cell wall is what retains the Gram stain (discussed in Chapter 2). In addition, most gram-positive species possess a well-developed S-layer of protein with glycan strands. By contrast, in other bacterial phyla the S-layer is either absent or present in diminished form.

Firmicutes and Actinobacteria are distinguished genetically based on their **GC content**, that is, the proportion of their genomes consisting of guanine-cytosine base pairs (as opposed to adenine-thymine). The Firmicutes (low GC) generally grow as well-defined rods or cocci, isolated or in simple filaments consisting of cells that divide but remain attached end to end. Many Firmicutes form **endospores**, inert heat-resistant spores that can remain viable for thousands of years. Endospores are the most durable type of spore formed by bacteria. The

Actinobacteria have a relatively high GC content. Actinobacteria (high GC) include the **actinomycetes** (order Actinomycetales), which undergo complex life cycles forming filamentous hyphae and arthrospores. Other groups closely related to actinomycetes grow as isolated rods and cocci, often with variable shape, such as the corynebacteria. Actinomycete relatives include the well-known causative agents of tuberculosis (*Mycobacterium tuberculosis*) and leprosy (*M. leprae*).

Gram-negative bacteria. **Proteobacteria** and **Nitrospirae** are called "the gram-negative bacteria" because their single layer of peptidoglycan fails to retain the Gram stain. (Note, however, that members of other phyla also stain gram-negative.) Species of Proteobacteria and of Nitrospirae have a complex outer membrane consisting of lipopolysaccharides (LPS) and porins. Some porins in the outer membrane are homologs of firmicute S-layer proteins, which suggests that the gram-negative outer membrane may have evolved from an S-layer. All Proteobacteria have in common the LPS outer membrane, but they show an immense range of shape (rods, cocci, spirals, and filaments) and metabolism (heterotrophy, lithotrophy, and anaerobic phototrophy). Most species are aerobic, facultative, or microaerophilic.

Proteobacteria include five major classes, labeled Alpha, Beta, Gamma, Delta, and Epsilon. (Some sources merge the Greek letter into the name, as in Alphaproteobacteria, Betaproteobacteria, and so on.) Proteobacteria include famous model organisms and pathogens, such as *Escherichia coli*, *Salmonella typhimurium*, and *Yersinia pestis*. Many are human or animal symbionts, either as mutualists or as pathogens—including *Rickettsia*, the genus most closely related to mitochondria. Members of the gram-negative phylum Nitrospirae largely resemble proteobacteria in form, though they diverged earlier. Species of Nitrospirae (such as *Nitrospira* spp.) oxidize nitrite to nitrate, a lithotrophic conversion essential for ecosystems.

Bacteroidetes and Chlorobi. The phyla **Bacteroidetes** and **Chlorobi** stain gram-negative, but nearly all their members are obligate anaerobes. Within Bacteroidetes, *Bacteroides* species ferment complex carbohydrates, serving as the major mutualists of the human gut. Chlorobi species are also obligate anaerobes, but they are "green sulfur" phototrophs that photolyze sulfides or H_2.

Spirochetes. The **Spirochetes** (Spirochaetes) exhibit a unique and complex cellular form of a flexible, extended spiral, resembling a telephone cord. The cytoplasm and cell membrane are contained within an outer membrane called the sheath. Between the sheath and the cell membrane extend flagella, doubled back from each pole. The rotation of these flagella is coordinated so as to twist and

flex the helical body, generating motility and chemotaxis. Spirochetes include many free-living forms in aquatic systems, as well as digestive endosymbionts and pathogens.

Irregular shaped bacteria. Three phyla of bacteria display remarkable adaptations suggestive of eukaryotes: **Chlamydiae**, **Verrucomicrobia**, and **Planctomycetes**. Most members of these groups have lost their peptidoglycan cell walls, but they show complex structural adaptations and developmental forms. The Chlamydiae (best-known genus *Chlamydia*) consists of intracellular parasites that lose most of their cell envelope during intracellular growth. The replicating parasites generate multiple spore-like "elementary bodies" that escape to infect the next host. Verrucomicrobia, by contrast, are free-living aquatic bacteria with wart-like protruding structures containing actin. Planctomycetes, also free-living, have stalked cells that reproduce by budding. Each planctomycete cell contains an extra double membrane surrounding its nucleoid, analogous to a eukaryotic nuclear membrane, though it evolved independently.

THOUGHT QUESTION 18.1   Which taxonomic groups in **Table 18.1** stain gram-positive and which gram-negative? Which group contains both gram-positive and gram-negative species? For which groups is the Gram stain undefined, and why?

THOUGHT QUESTION 18.2   Which groups of species share common structure and physiology within the group? Which groups show extreme structural and physiological diversity?

NOTE: Formal names for taxa such as phyla, orders, and genera are capitalized (phylum Cyanobacteria, genus *Streptococcus*. Informal forms are lowercase roman (cyanobacteria, streptococci), as are adjectival forms (cyanobacterial, streptococcal).

New Isolates: Unclassified and Uncultured Bacteria

The seven groups outlined in **Table 18.1** represent those bacteria of greatest historical interest and those best characterized by microbiologists. As new organisms emerge, they may be known initially only by their habitat and their small-subunit rRNA sequence. Such isolates require description of essential cell structure and metabolism before designation as new species. "Incompletely described" samples are designated as follows:

- An **unclassified organism** or **uncultured organism** is assigned to a taxonomic rank based on rRNA sequence but not yet grown in pure culture; for example, an "uncultured actinomycete."

- An **environmental sample** is designated by its habitat and assigned a rank based on rRNA. Examples of environmentally defined actinomycetes in the NCBI database include "oil-degrading bacterium AOB1" and "glacier bacterium FJS11."

- A species with some physiological characterization beyond DNA sequence may be published with a provisional status of **candidate species**, designated by the prefatory term *Candidatus*. For example, *Candidatus Nostocoida limicola* was published as "a filamentous bacterium from activated sludge."

Figure 18.4 Deep-branching thermophiles.
A. Mass of pink filamentous *Thermocrinis ruber* in a hot spring at near-boiling temperature.
B. *T. ruber* filamentous cells (SEM).

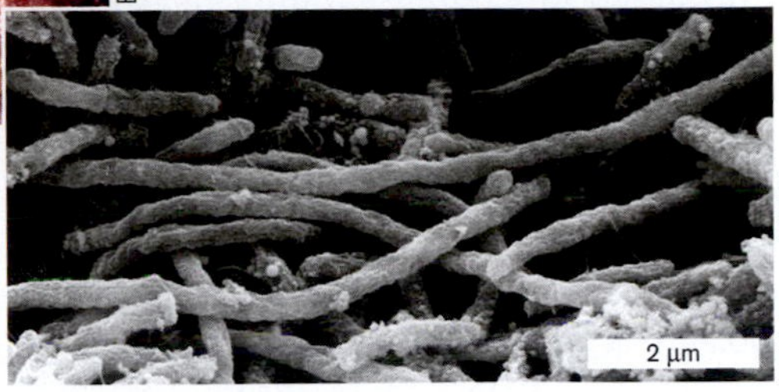

Following our brief "one-day tour" of the bacterial domain, we now explore some of the diverse species within the major groups. Each section presents species from one of the seven major groups outlined in **Table 18.1**.

NOTE: The primary organizing principle used in Chapters 18–20 is that of phylogeny based on DNA relatedness. Characteristic traits described for each branch apply to the majority of its known species, though exceptions are discovered daily, such as members of Spirochetes that lack spiral form.

18.2 Deep-Branching Thermophiles

Several of the most deeply branching phyla show traits similar to those of archaeal extremophiles and share high-temperature habitats such as hot springs. The precise nature of their relationship with archaea is controversial, however, because these organisms also show the fastest doubling rates of all cells (in some cases less than 10 minutes) and high rates of mutation, which may have "accelerated" their molecular clock. Thus, the deep-branching thermophiles may appear to have diverged from other bacteria earlier than they in fact did.

Aquificales and Thermotogales: The Most Extreme Bacterial Hyperthermophiles

The order Aquificales includes bacteria growing at high temperatures, up to 95°C. Most members of this group are hydrogenotrophs, oxidizing hydrogen gas with molecular oxygen to make water. *Aquifex pyrophilus*, a flagellated rod (see **Table 18.1**), was first discovered

by extremophile microbiologist Karl Stetter at Regensburg University in a submarine hydrothermal vent north of Iceland. Stetter named it *Aquifex*, Latin for "water maker," in reference to its metabolism of oxidizing hydrogen gas with O_2 to form water. *Aquifex* species are obligate autotrophs, fixing CO_2 into biomass using the reverse TCA cycle.

Other genera of Aquificales form filamentous mats in hydrothermal springs. *Thermocrinis ruber* was discovered by Thomas Brock at the University of Wisconsin-Madison as a mat of pink filamentous streamers ("pink filaments") growing at 82–88°C in the outflow channel of Octopus Spring, in Yellowstone National Park (**Fig. 18.4**). As the filaments grow, they break off flagellated single cells that grow new filaments. *T. ruber* gains energy by oxidizing hydrogen, thiosulfate, and elemental sulfur as well as small organic molecules such as formate.

Another deep-branching order of thermophiles (growing at 50–80°C) is Thermotogales. *Thermotoga maritima* was isolated from a geothermal vent in Volcano, Italy. Their cells have a loosely bound sheath, or "toga," for which the genus is named. They use anaerobic respiration, reducing elemental sulfur, thiosulfate, and sulfite to hydrogen sulfide.

The genomes of *T. maritima* and of *Aquifex aeolicus* reveal a surprisingly high content of genes transferred horizontally from archaea, well after the bacteria diverged from their common ancestor. Nearly a quarter of the *T. maritima* genome appears to be of archaeal origin, including fundamental enzymes such as glutamate synthase and inositol 1-phosphate synthase.

THOUGHT QUESTION 18.3 What taxonomic questions are raised by the apparent high rate of gene transfer between archaea and thermophilic bacteria?

Thermus and *Deinococcus*

The phylum Deinococcus-Thermus shares a unique structural trait, the substitution of *L*-ornithine for diaminopimelic acid in the peptidoglycan cross-bridge. *Thermus* species (growing at 70–75°C) are heterotrophs commonly isolated from hot tap water. *Deinococcus* species, however, are not thermophilic. *D. radiodurans* bacteria resist extremely high doses of ionizing radiation (discussed in Chapters 5 and 9). An otherwise ordinary heterotroph, *D. radiodurans* was originally isolated from cans of meat supposedly sterilized with a dose of several megarads. The natural habitat of *Deinococcus* species is not known, but the genus also resists extreme desiccation; it may have evolved as a soil organism capable of surviving drought. *Deinococcus* species, unlike *Thermus* species, have a thick cell wall that stains gram-positive, as well as an S-layer.

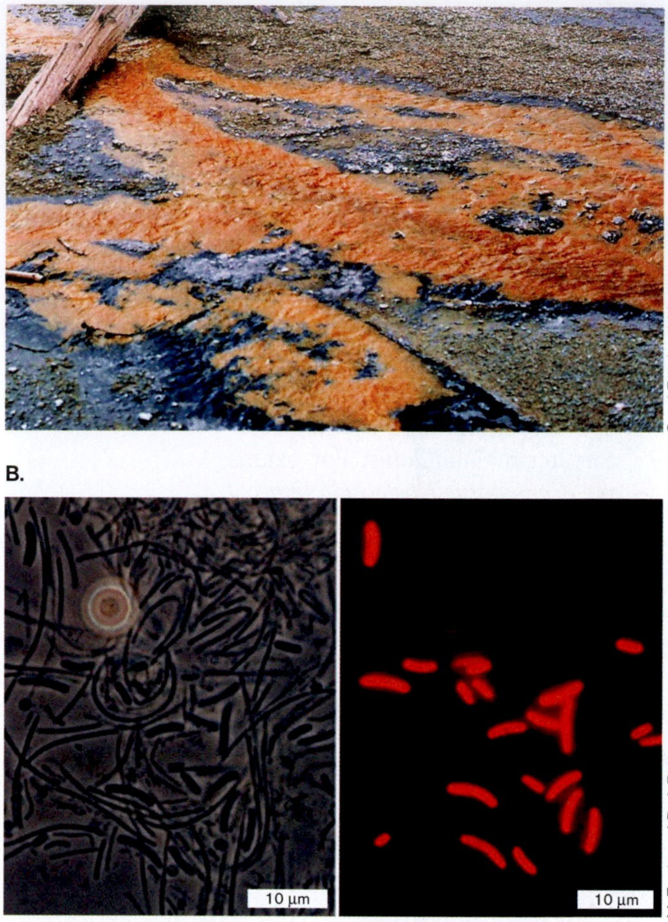

Chloroflexi: Thermophilic Green Photoheterotrophic Bacteria

Chloroflexi bacteria are filamentous photoheterotrophs, absorbing light to split organic molecules. Photosystem II (PS II) generates ATP through cyclic electron flow (discussed in Chapter 14). Chloroflexi species conduct nonphotosynthetic heterotrophy in the presence of oxygen. Most are moderate thermophiles, often found in hot springs in association with thermophilic cyanobacteria. *Chloroflexus aurantiacus* and thermophilic cyanobacteria form massive microbial mats in the hot springs of Yellowstone (**Fig. 18.5A**).

Chloroflexi are informally called "green nonsulfur bacteria" to distinguish them from Chlorobi, a different phylum of green phototrophs that photolyze sulfur compounds. Some chloroflexi appear green because their bacteriochlorophylls (Bchl) absorb primarily blue and red, allowing transmission of green (**Table 18.2**, top row). Other species, however, appear red or yellow owing to accessory pigments. A more inclusive term for this group is filamentous anoxygenic phototrophs, distinguishing them from other photoheterotrophs that mainly grow as single cells.

Chloroflexus species contain their photosynthetic apparatus within membranous organelles called **chlorosomes**. The chlorosome contains Bchl *c* and Bchl *d*, complexed with Bchl *a* at the plasma membrane. Other species of Chloroflexi, however, such as *Roseiflexus* and *Helothrix*, lack chlorosomes. Chloroflexi are the main filamentous phototrophs outside the Cyanobacteria.

Table 18.2 summarizes key features of phototrophy in clades throughout the bacterial domain; we shall return to this table as a guide throughout the chapter. All light-harvesting complexes descend from one of two ancestral

Figure 18.5   *Chloroflexus*: A thermophilic phototroph. A. Hot spring bacterial mat contains *Chloroflexus* and thermophilic cyanobacteria, Yellowstone National Park. **B.** *C. aurantiacus* filaments often combine with *Synechococcus* (cyanobacteria) to form a microbial mat. Red autofluorescence reveals the cyanobacteria.

sources: the chlorophyll/bacteriochlorophyll with electron transport (PS I, PS II) or the proteorhodopsin proton pump. It is not clear whether one or both ancestral photosystems were present in the ancestor of all bacteria or whether extensive horizontal transfer distributed the photosystems among divergent clades.

NOTE: Distinguish phylum Chloroflexi (filamentous photoheterotrophs) from phylum Chlorobi (sulfur photoautotrophs of the Bacteroidetes/Chlorobi group). Both are phototrophs containing green photopigment complexes arranged in chlorosomes, but they are deeply divergent genetically.

Table 18.2 **Phototrophic bacteria.**

Taxon	Energy generation	Cell structure and photopigments	Absorption spectrum (whole cell)	Reaction center
Chloroflexi "Green nonsulfur" *Chloroflexus*	$+O_2$ Heterotrophy $-O_2$ Photoheterotrophy	Chlorosome, Bchl a; Bchl *c/d*		PS II
Cyanobacteria "Blue-greens" *Anabaena*	$+O_2$ Oxygenic phototrophy $-O_2$ Photolithoautotrophy on reduced sulfur	Thylakoid; Chl *a/b*		PS I and PS II
Firmicutes "Sun bacteria" *Heliobacterium*	$+O_2$ $-O_2$ Photoheterotrophy	Forms endospore; Bchl g		PS I
Proteobacteria "Purple nonsulfur" (Alpha and Beta classes) Bchl a *Rhodospirillum*	$+O_2$ Heterotrophy $-O_2$ Photoheterotrophy	Bchl a		PS II
Bchl b *Blastochloris*	$+O_2$ Heterotrophy $-O_2$ Photoheterotrophy	Bchl b		PS II
"Purple sulfur" (Gamma class) *Chromatium*	$+O_2$ $-O_2$ Photolithoautotrophy on reduced sulfur	Bchl a/b		PS II
"Proteorhodopsin" (Alpha and Gamma classes) *Pelagibacter* (SAR11) SAR86	$+O_2$ Heterotrophy $-O_2$ Photoheterotrophy	Proteorhodopsin		PR
Chlorobi "Green sulfur" *Chlorobium*	$+O_2$ $-O_2$ Photolithoautotrophy on reduced sulfur	Chlorosome, Bchl a; Bchl *c/d/e*		PS I

Sources: J. Overman and F. Garcia-Pichel. 2001. The phototrophic way of life. *Prokaryotes*, Springer; Oded Beha, et al. 2001. *Nature* 411:786.

TO SUMMARIZE:

- **Deep-branching thermophiles** such as *Aquifex* and *Thermatoga* species share traits and habitats with thermophilic archaea. They show extensive transfer of archaeal genes.
- **Deinococcus-Thermus** bacteria have ornithine in their peptidoglycan cross-bridges. *Deinococcus* bacteria are highly resistant to ionizing radiation, and they can be isolated from irradiated foods. *Thermus* bacteria are moderate thermophiles isolated from hot-water taps.
- **Chloroflexi** bacteria are deep-branching thermophilic photoheterotrophs, occurring as filamentous mats in association with thermophilic cyanobacteria in hot springs.

18.3 Cyanobacteria: Oxygenic Phototrophs

Cyanobacteria (or Cyanophyta) are the only bacteria that produce oxygen gas. The phylum is named for the blue phycocyanin accessory pigments possessed by some genera, giving them a bluish tint. Cyanobacteria commonly appear green due to the predominant blue and red absorption by chlorophylls *a* and *b* (see **Table 18.2**). Essentially the same chlorophylls are found in plant chloroplasts. Some cyanobacteria, however, appear red due to the accessory pigment phycoerythrin. Cyanobacteria are the only prokaryotes that use both photosystems I and II, photolyzing water to produce oxygen (see **Table 18.2**; details of oxygenic photosynthesis are given in Chapter 14). Under anoxic conditions, most cyanobacteria can also photolyze hydrogen, reduced sulfur compounds, and organic compounds.

The success of oxygenic phototrophy is shown by its fundamental similarity across all members of Cyanobacteria, which account for about a quarter of all of Earth's biomass production and vary widely in cell form and

behavior as well as genetic diversity. Even two filamentous genera such as *Synechococcus* and *Anabaena* share only 37% of their genes. Various cyanobacteria are found in all habitats, from tropical soils to Antarctica. Cyanobacteria include species edible for humans, in products such as *Spirulina* salad or *Nostoc* soup (discussed in Chapter 16).

Cyanobacterial Cell Structure

The photosynthetic apparatus of cyanobacteria is organized within thylakoids, pockets of membrane resembling flattened spheres packed with photosynthetic apparatus. The thylakoids may be distributed through the cell, as in "traditional" genera such as *Nostoc* (**Fig. 18.6A**), or they may encircle the cell in concentric layers, as in the more recently discovered marine species *Prochlorococcus marinus* (**Fig. 18.6B**). *Prochlorococcus* is one of the smallest and most abundant oxygen producers in the biosphere. In both large cells and small, the thylakoids are completely separate from the plasma membrane, unlike the attached chlorosomes of Chloroflexi and the plasma membrane extensions of "purple" proteobacteria (see **Table 18.2**). Cyanobacterial thylakoids resemble the thylakoids of eukaryotic chloroplasts; they are the most complex and specialized form of photosynthetic apparatus.

Cyanobacteria have several other subcellular structures. **Carboxysomes** (also known as polyhedral bodies) conduct CO_2 fixation. Cyanobacteria store energy-rich compounds in lipid bodies. To maintain height in the water column and consequently access to sunlight, cyanobacteria have **gas vesicles** whose buoyancy enables cells to float. Their external structures include a thick peptidoglycan cell wall, similar to that of gram-positive cells, plus several external layers that vary with different species. Many species move by "gliding," a form of motility poorly understood.

Besides fixing CO_2, most cyanobacteria fix N_2. Because nitrogen fixation requires the absence of oxygen, cyanobacteria have to solve the problem of maintaining anaerobic biochemistry while producing huge quanitities

A.

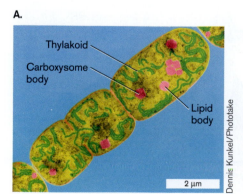

Thylakoid

Carboxysome body

Lipid body

2 μm

Dennis Kunkel/Phototake

B.

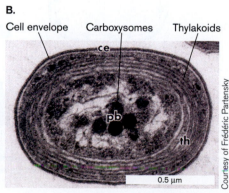

Cell envelope   Carboxysomes   Thylakoids

ce

pb

th

0.5 μm

Courtesy of Frédéric Partensky

Figure 18.6 Cyanobacterial cell structure. A. Intracellular organelles of *Nostoc*, a typical filamentous cyanobacterium (colorized TEM). **B.** Intracellular organelles of *Prochlorococcus*, a prochlorophyte cyanobacterium, the smallest known phototroph. *Prochlorococcus* accounts for 40–50% of marine phototrophic biomass.

of highly toxic O_2. Different species solve this problem in different ways:

- Formation of specialized nitrogen-fixing cells called **heterocysts**.
- Temporal separation, alternating between photosynthesis during daylight and nitrogen fixation at night.
- Accumulation of large aggregates of cells in which the interior becomes sufficiently anaerobic for nitrogenase to function, while the exterior continues oxygenic photosynthesis.
- Symbiotic association with microbes or plants that consume oxygen or otherwise maintain anoxic conditions.

Filamentous and Colonial Cyanobacteria

Cyanobacteria have evolved several major categories of form. Single-celled forms include *Synechococcus* and *Prochlorococcus*, the most abundant phototrophs in the oceans. Other genera, such as *Oscillatoria*, generate long filaments of hundreds or even thousands of cells, which can associate in thick biofilms (**Fig. 18.7**). *Oscillatoria* cells are stacked like plates, wider than they are long. To disseminate their cells beyond the biofilm, the filaments produce **hormogonia**, short motile chains of three to five cells. Other filamentous genera, such as *Nostoc*, arrange their filaments in balls of mucilage, possibly to protect from grazing (**Fig. 18.8**). Most filamentous species develop heterocysts to fix nitrogen (**Fig. 18.8B**). The function of heterocysts is discussed in Chapter 11.

Under environmental stress, such as light limitation or phosphate starvation, filamentous cyanobacteria such as *Anabaena* form specialized spore cells called **akinetes**. An akinete forms as a long, oval cell adjacent to a heterocyst, where it stores nitrogen and develops a thickened envelope. Like other types of spores, akinetes resist desiccation and remain viable for long periods. **Table 18.3** compares the properties of akinetes with those of other spore types (which are discussed under Firmicutes and other taxa). Akinetes lie dormant but viable until improved conditions permit germination and growth of new vegetative filaments. In lake water, akinete germination may cause toxic blooms of *Anabaena*.

> **THOUGHT QUESTION 18.4** What are the relative advantages and disadvantages of propagation by hormogonia, as compared with akinetes?

Nonfilamentous species divide to form small groups or larger colonies (**Fig. 18.9**). *Gloeocapsa* and *Chroococcus* form doublets or quartets, encased in a thick protective mucous slime (**Fig. 18.9A**). Still others, such as *Merismopedia*, continue cell division in two planes, extending to form long square sheets of attached cells (**Fig. 18.9B**).

A. *Oscillatoria*

B. Dome cell

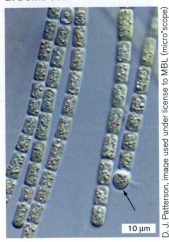

Figure 18.7 Filamentous cyanobacteria. A. *Oscillatoria* forms large filaments whose cells are wider than their length. **B.** Cyanobacteria filament terminates in a dome cell.

A.

B.

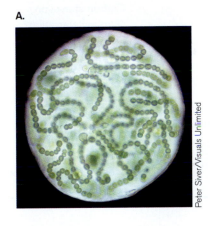

Heterocysts

Figure 18.8 Clumped filamentous cyanobacteria. A. *Nostoc* forms filaments that clump together in large balls of mucilage. **B.** *Nostoc* makes heterocysts (round cells approximately 5 µm in diameter).

Colonial genera such as *Myxosarcina* and *Pleurocapsa* reproduce by multiple fission, producing large cell aggregates (**Figs. 18.9C** and **D**). As the aggregate matures, some cells continue to divide and release unicellular structures called **baeocytes**. Each baeocyte reproduces and develops into a new cell aggregate. The aggregate group maintains anoxic conditions at the center for nitrogen fixation. Both filamentous and colonial cyanobacteria fill key roles in diverse habitats (**Special Topic 18.1**).

> **THOUGHT QUESTION 18.5** What are the relative advantages and disadvantages of the different strategies for maintaining separation of nitrogen fixation and photosynthesis?

Table 18.3   Spore types in bacteria.

Spore type	Bacteria that produce the spore	Initiation of spore formation	Formation of the spore	Properties of the spore
Akinete	Filamentous cyanobacteria	Light limitation Cold temperature Phosphate starvation	Adjacent to a heterocyst, the akinete develops as a large oval cell with a thickened cell wall and multilayered envelope.	Desiccation resistant Cold resistant Viable for decades
Arthrospore	Actinomycetes	Carbon starvation Phosphate starvation	At the tip of an aerial mycelium, cells undergo vegetative division. Cells develop thick coat and pinch off as arthrospores.	Desiccation resistant Heat resistant
Elementary body	Chlamydiae	Completion of intracellular life cycle	Intracellular chlamydia reticular bodies replicate, then develop into elementary bodies whose envelope contains outer membrane proteins cross-linked by disulfide bonds. Elementary bodies survive outside the host cell.	Survives outside host Desiccation resistant
Endospore	Firmicutes	Carbon starvation Nitrogen starvation Phosphate starvation Low pH Inducers: purine nucleosides, peptide antibiotics	Individual cell develops motherspore and forespore. Motherspore contributes DNA and nutrients to the forespore, which develops into an endospore with a spore coat reinforced by keratin and calcium dipicolinate.	Highly desiccation resistant Highly heat resistant Viable for centuries
Myxospore	Myxobacteria	Nutrient starvation Heat shock Inducers: glycerol, dimethyl sulfoxide, or phenethyl alcohol	Myxobacteria aggregate to form a fruiting body in which vegetative cell division forms a mass of myxospores.	Desiccation resistant UV resistant

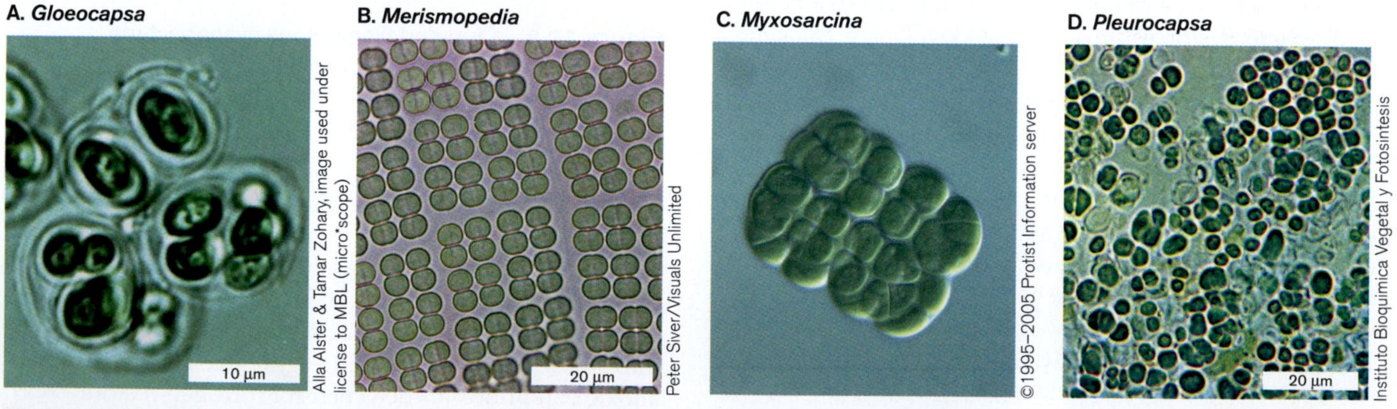

A. Gloeocapsa   **B. Merismopedia**   **C. Myxosarcina**   **D. Pleurocapsa**

10 μm   20 μm   20 μm

Alla Alster & Tamar Zohary, image used under license to MBL (micro*scope)

Peter Siver/Visuals Unlimited

© 1995–2005 Protist Information server

Instituto Bioquimica Vegetal y Fotosintesis

Figure 18.9   Single-celled and colonial cyanobacteria.   A. *Gloeocapsa* is surrounded by mucus. Cells grow as single cells, doublets, or quartets. **B.** *Merismopedia* forms extended quartets, octets, and so on. **C.** *Myxosarcina* cells divide in three planes and form amorphous masses of cells. **D.** *Pleurocapsa* forms enormous aggregates that release baeocytes.

Special Topic 18.1   Cyanobacterial Communities: From Ocean to Animal

Cyanobacterial communities play key roles in many ecosystems. Their role as primary producers in marine habitats is highlighted when an environmental change triggers a bloom in population. For example, the colonial cyanobacteria *Trichodesmium* forms giant blooms visible from outer space, covering many square kilometers of ocean surface (**Fig. 1**). Such a bloom can be triggered by an influx of iron carried by wind from a dust storm blowing off the Sahara desert. The concentrated *Trichodesmium* converts nitrogen to forms that promote growth of eukaryotic algae, which prove toxic to consumers such as fish and manatees.

Other cyanobacteria grow in biofilms, containing either a single species or multiple species. In salt marshes and sand flats, cyanobacteria participate in multilayered microbial mats, such as the section shown in **Figure 2**, cut from the sand flats of Great Sippewissett Salt Marsh, on Cape Cod, Massachusetts. The high concentration of sulfides in the sediment support growth of high populations of sulfur phototrophs. Typically, cyanobacteria and eukaryotic algae such as diatoms form the upper green layer. Below, the purple layer consists of "purple sulfur" proteobacteria, whose photopigments (primarily Bchl *a*) absorb at longer wavelengths (discussed in Section 18.5). The pale-colored layer below the purple layer consists of proteobacteria with Bchl *b*, which absorbs farther into the infrared.

Cyanobacteria participate in many kinds of mutualism with animals, plants, fungi, and protists. Sponges growing on coral reefs may have endosymbiotic cyanobacteria that provide the sponge with nutrients from photosynthesis. The products of photosynthesis supplement the nutrients obtained by the sponge from its filter feeding, and these extra nutrients greatly augment the sponge growth rate within the competi-

tive coral reef environment. For example, the sponge *Lamellodysidea* (**Fig. 3**) may gain up to 80% of its nutritional needs from photosynthesis by *Oscillatoria* and *Synechocystis* bacteria growing in the channels between sponge epithelial tissues. As sponge species have evolved, their microbial endosymbionts have evolved in parallel and remain specific to each species of sponge. Sponge symbionts may be associated with the production of pharmaceutically active compounds.

Cyanobacteria and diatoms
Purple sulfur proteobacteria
Long-wavelength purple sulfur bacteria

2.54 cm

Jörg Overmann and Ferrau Garcia-Pichel

Figure 2   Cyanobacteria in microbial mats.   Cutaway through a multilayered microbial mat from the sand flats of Great Sippewissett Salt Marsh (Cape Cod, Massachusetts). Cyanobacteria and diatoms form the upper green layer, above layers of purple sulfur proteobacteria.

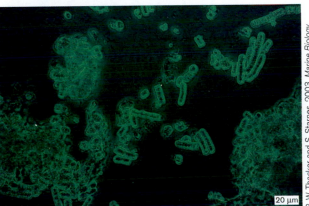

R. W. Thacker and S. Starnes. 2003. *Marine Biology* 142:643

Figure 3   Coral reef sponges contain endosymbiotic cyanobacteria.   *Oscillatoria spongeliae* cyanobacteria, in a "squash preparation" from the marine sponge *Lamellodysidea herbacea* (phase contrast with green filter).

Horn Point Laboratory, Thomas C. Malone

Figure 1   Marine cyanobacterial bloom.   A ship plows through an ocean bloom of *Trichodesmium*.

TO SUMMARIZE:

- **Cyanobacteria** are the only oxygenic prokaryotes. They conduct photosynthesis in thylakoids, fix CO_2 in carboxysomes, and maintain buoyancy using gas vesicles. They exhibit gliding motility.
- **Single-celled cyanobacteria** such as *Prochlorococcus* are among the smallest and most abundant phototrophic producers in the oceans.
- **Filamentous cyanobacteria** such as *Nostoc* and *Oscillatoria* are common in aquatic systems. Filaments form heterocysts to fix nitrogen, and reproduce by hormogonia or by akinetes.
- **Colonial cyanobacteria** such as *Myxosarcina* produce large cell aggregates with an anaerobic core for nitrogen fixation. The colonies reproduce through baeocytes.
- **Symbiotic associations** of cyanobacteria are formed with animals, fungi, and plants.

18.4 Gram-Positive Firmicutes and Actinobacteria

Gram-positive bacteria were formerly all classified as Firmicutes, the "tough skin" bacteria, and older sources designate Actinobacteria within this phylum. Today, as shown in **Table 18.1**, the phylum Firmicutes is defined to contain "low GC" species, whereas Actinobacteria are the "high GC" species. They form two distinct phylogenetic branches. Species in both groups have thick cell walls that

retain the Gram stain and partly exclude antibiotics (**Fig. 18.10**; discussed in Chapter 3).

Many firmicutes, such as *Bacillus* and *Clostridium* species, survive unfavorable environmental conditions by forming durable endospores. Non-spore-formers such as *Lactobacillus* and *Streptococcus* may have evolved from a common Firmicutes ancestor that formed endospores. In many cases, the machinery to form endospores is revealed in the sequenced genomes of organisms believed to be non-spore-formers such as *Carboxydothermus* sp., soil bacteria that oxidize CO (carbon monoxide) to CO_2. On the other hand, actinobacteria of the order Actinomycetales (actinomycetes) such as *Streptomyces* do not form endospores, but they develop filaments dispersing arthrospores (see **Table 18.3**).

Firmicutes Include Endospore-Forming Rods

Endospore-forming bacteria are common in soil and air, as their spore forms resist desiccation and can remain viable in a dormant state for thousands of years. The best-known orders are Bacillales, mainly aerobic respirers, and Clostridiales, obligate anaerobes. Both groups include species of environmental and economic importance, as well as causative agents of well-known diseases.

Bacillales, genus *Bacillus*. *Bacillus* was one of the first bacterial genera to be classified in the nineteenth cen-

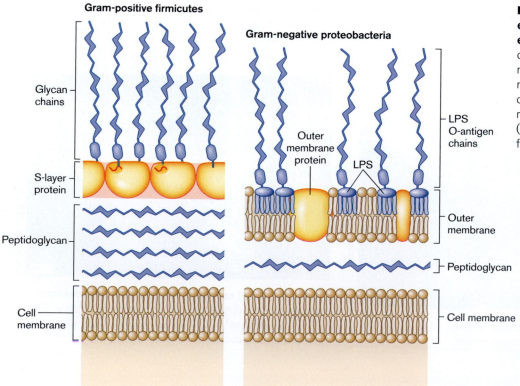

Figure 18.10 Gram-positive envelope and gram-negative envelope. The gram-positive cell has an S-layer and a multiplex cell wall. The gram-negative cell has a single layer of peptidoglycan and an outer membrane containing proteins (OMPs) that may have evolved from ancestral S-layer proteins.

tury. Colonies are readily isolated on a nutrient agar plate exposed to air. *Bacillus* species can be isolated from soil or food by suspending a sample in water and heating at 80°C for half an hour. Vegetative cells (that is, cells undergoing binary fission) and non-spore-formers are killed at that temperature. The remaining endospores will germinate and grow on a beef broth agar plate at 25–30°C. *Bacillus* species isolated in this way include over a thousand characterized strains, all but a few of them harmless to humans.

The large, rod-shaped **vegetative cells** (growing and replicating form) of *Bacillus* species are easily stained and visualized. As the vegetative cells run short of nutrients on an agar plate, they start to produce inert endospores. The resulting pattern of vegetative cells, hollowed out after loss of spores, is highly distinctive in stained specimens (**Fig. 18.11A**). The life cycle of endospore production involves a coordinated developmental plan, in which cell fission directed by FtsZ occurs at the pole instead of at the cell equator (**Figs. 18.11B** and **C**). The polar compartment develops as the **forespore**, directed by unique regulatory proteins such as SpoIIE (**Fig. 18.11B**). The larger compartment, called the **motherspore**, provides DNA and nutrients to the growing forespore and disintegrates after release of the mature endospore. When the released endospore encounters favorable conditions of moisture and nutrients, it germinates and restarts vegetative growth. Further details of sporulation are discussed in Chapter 4.

A species of particular scientific importance is *Bacillus subtilis*, the best-studied gram-positive organism, a "model system" for Firmicutes. The *B. subtilis* genome sequence reveals a large number of transporters for carbon sources and many secretory complexes for industrially important enzymes and drug resistance proteins. It also reveals several integrated prophages (integrated phage genomes). Phage genomes contribute to bacterial evolution by transferring genes between different strains and species, as discussed in Chapter 17. Like *E. coli*, *B. subtilis* is studied as a model system with respect to stress response. Enormous changes in protein expression accompany starvation and general stress conditions, which ultimately cause cells to sporulate.

Figure 18.11 *Bacillus* species: Gram-positive endospore formers. **A.** Gram-stained *Bacillus* species, sporulating culture. Dark blue stain is lost by cells that have expelled their spores. **B.** Correlated fluorescence imaging of membrane migration, protein translocation, and chromosome localization during *Bacillus subtilis* sporulation. Membranes were stained with red fluorescent FM 4-64. Chromosomes were localized with the blue fluorescent nuclear counterstain DAPI. The small green fluorescent patches indicate the localization of a green fluorescent protein gene fusion to SpoIIIE, a protein essential for both initial membrane fusion and forespore engulfment. Progression of the engulfment is shown from left to right. **C.** Cycles of binary fission and sporulation in *Bacillus* species.

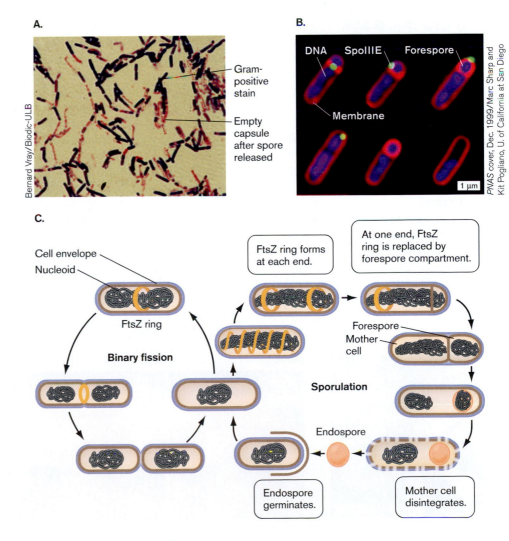

A.

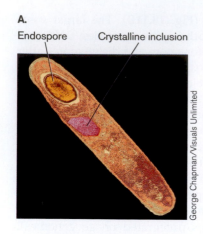

Endospore   Crystalline inclusion

George Chapman/Visuals Unlimited

B.

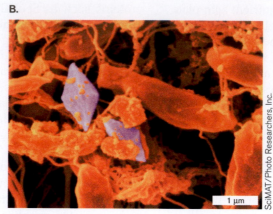

1 μm

SciMAT/Photo Researchers, Inc.

Figure 18.12 *Bacillus thuringiensis:* **the biological insecticide. A.** Sporulating cell of *B. thuringiensis*, showing crystalline inclusion of insecticidal toxin (colorized TEM). **B.** Toxin crystals (colorized blue) embedded in *B. thuringiensis* colony.

A *Bacillus* species of great economic importance is *B. thuringiensis*, the most successful biological control agent yet produced (**Fig. 18.12**). *B. thuringiensis* was originally discovered in 1901 by Japanese bacteriologist Shigetane Ishiwatari as a cause of disease in silkworms. The organism proved easy to culture on agar-based medium, and its spores are now applied as an insecticide against the gypsy moth caterpillar. During sporulation, *B. thuringiensis* generates a diamond-shaped crystal adjacent to its endospore (**Fig. 18.12A**). The crystal contains an insecticidal protein, known as delta-endotoxin. The crystals are ultimately released from the cells (**Fig. 18.12B**). The toxin is activated only at high pH in the digestive tract of insect larvae; thus, it is completely safe for animals with acidic digestive tracts.

Nevertheless, controversy arose when the gene encoding the toxin was introduced transgenically into plants, with the aim of creating insect-resistant crops. The concern is that the *Bt* gene might escape into wild plants and endanger beneficial or harmless species of moths and butterflies.

Many *Bacillus* species are extremophiles, growing at high pH (*B. alkalophilus*), at high temperature (*B. thermophilus*), or in high salt (*B. halodurans*). Some species combine alkaliphily with thermophily, as in the case of the "alkalithermophile" *B. alkalophilus*. Genomic research has focused on the surprisingly subtle differences that distinguish an extremophile from a closely related mesophile. For example, thermostability of proteins can be determined by comparing the protein sequences of a thermophile with those of a related mesophile. The thermophilic sequence shows specific patterns of amino acid residues that confer thermostability. Such amino acid substitutions provide useful information for industrial engineering of enzymes.

Clostridiales, genus *Clostridium*. The anaerobic family of spore formers includes the genus *Clostridium*, species of which cause botulism (*C. botulinum*) and tetanus (*C. tetani*) (**Fig. 18.13A**). The botulism toxin, botulinum, or "botox," is famous for its therapeutic use to relax muscle spasms and to smooth wrinkles in skin (**Fig. 18.13B**; discussed in Chapter 26). Other species, such as *C. acetobutylicum*, have economic importance as producers of industrial solvents such as butanol and acetone (for industrial microbiology, see Chapter 16). The butanol pathway is an example of the wide range of diverse fermentative strategies found among clostridia. Phylogenetically, the genus *Clostridium* is less tightly defined than *Bacillus*, whose members all show highly similar DNA sequence, whereas *Clostridium* species diverge throughout the phylogeny of Firmicutes.

Clostridium cells sporulate in a form distinct from that of *Bacillus* (see **Fig. 18.13A**). The growing endospore swells the end of the cell, forming a "drumstick" appear-

A. *Clostridium* sp.

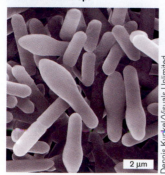

2 μm

Dennis Kunkel/Visuals Unlimited

B. Botox treatment

Before   After

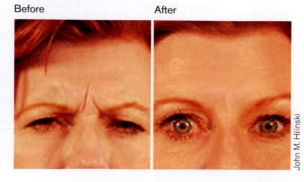

John M. Hiinski

Figure 18.13 *Clostridium* **species: spore-forming anaerobes. A.** As *Clostridium* cells sporulate, the endospore swells, forming a characteristic "drumstick" appearance. **B.** The deadly botulinum toxin (botox) from *C. botulinum* is used to relax muscle spasms. This woman was unable to open her eyelids fully (left frame), a condition that was cured by injection of botox (right frame).

ance. Mature clostridial spores are generally less heat resistant than the spores of *Bacillus* and thus more difficult to isolate in pure culture from natural environments. Nevertheless, *Clostridium* spores are found in soil and water, ready to germinate and grow when the environment becomes anoxic. Many harmless species are found in the human colon, particularly in infants (see **Fig. 18.1**). Pathogenic *C. botulinum* can grow within the colon of very young infants, and infant botulism has been implicated in some cases of sudden infant death syndrome. Another species, *C. difficile*, is emerging in hospitals as a life-threatening intestinal pathogen, resistant to most antibiotics. *C. difficile* emerges in patients treated with antibiotics that eliminate normal enteric bacteria, allowing growth of the pathogen.

Related to *Clostridium* are the Heliobacteraceae or "solar bacteria," the only known phototrophs in Firmicutes (see **Table 18.2**, third row). Their name derives from their yellowish color caused by the shift of their red peak absorbance into the infrared, which allows transmission of red light plus green light (perceived as yellow). The heliobacteria are photoheterotrophs, with a PS I reaction center providing a modest boost of energy to supplement heterotrophic metabolism of simple organic compounds. Most heliobacterial cells are elongated rods with peritrichous flagella, and they form endospores—the only known endospore-forming phototrophs.

> **NOTE:** Distinguish *Heliobacterium* from *Helicobacter*, a helical-shaped species of the gammaproteobacteria.

Variant sporulation and "live birth." The order Clostridiales includes species of exceptionally large bacteria that grow only within the digestive tract of specific animal hosts. These species have evolved intriguing variations of the endospore former life cycle, as revealed by Esther Angert and colleagues at Cornell University (**Fig. 18.14A**). *Metabacterium polyspora* (size 15–20 μm) grows throughout the digestive tract of guinea pigs (**Fig. 18.14B**). Endospores ingested from feces germinate in the upper intestine but rarely undergo binary fission. Instead, the growing cell forms forespores at both poles (**Fig. 18.14C**). The forespores actually multiply within the motherspore to form several endospores, which are released in the colon before defecation.

An even larger enteric endosymbiont is *Epulopiscium fishelsoni*, found in the digestive tract of sturgeonfish (**Fig. 18.15**). These bacteria are large enough to be seen by eye, about the size of the period at the end of this sentence. As in *M. polyspora*, Angert showed that *E. fishelsoni* reproduction is synchronized with the digestive cycle of its host, but it has gone even further in transformation of the sporulation cycle. Binary fission is eliminated; the cell must fission at both poles. Each polar fission generates

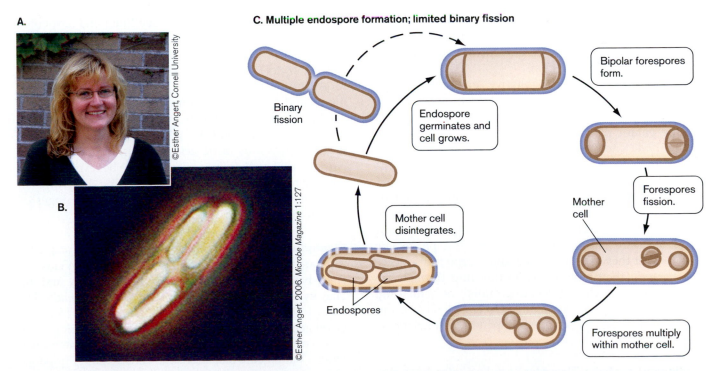

Figure 18.14 Multiple endospore formation. **A.** Esther Angert, at Cornell University, characterized unusual forms of sporulation and reproduction in exceptionally large firmicute bacteria. **B.** *Metabacterium polyspora* forms multiple endospores (cell length 15–20 μm; phase contrast LM). **C.** A forespore forms at each pole. Forespores fission and multiply within the motherspore, then are released. Germinated cells undergo limited or no binary fission.

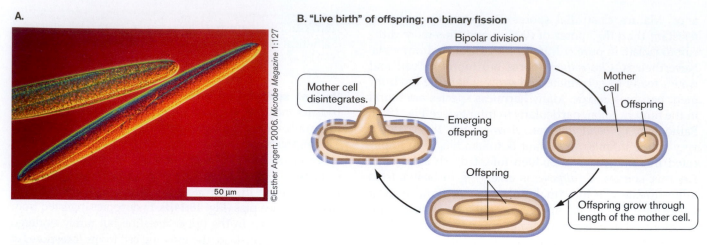

Figure 18.15 "Live birth" in *Epulopiscium*. A. *Epulopiscium fishelsoni* forms offspring cells that grow internally. **B.** An offspring cell forms by fission at each pole. The cells grow internally until released. No binary fission occurs outside the mother cell.

an intracellular daughter cell that grows to nearly the full length of the mother cell. The two intracellular offspring ultimately emerge in "live birth" from the mother cell, which then disintegrates.

Non-spore-forming Firmicutes

Both Bacillales and Clostridiales, as well as other gram-positive orders, include many non-spore-forming rods and cocci. In these taxa, the endospore-forming system was probably lost by reductive evolution. Non-spore-forming firmicutes include human pathogens such as *Listeria* and *Streptococcus* species, as well as lactic acid bacteria that are important for food production.

Listeria species are facultative anaerobic bacilli, named for the British surgeon Joseph Lister (1827–1912), who first promoted antisepsis during surgery. They include enteric pathogens such as *L. monocytogenes* that contaminate cheese and sauerkraut. Unlike other food-associated organisms, *Listeria* grows at temperatures as low as 4°C. Under preindustrial conditions of food preparation, *L. monocytogenes* was generally outcompeted by other flora, but the era of refrigeration has brought about the emergence of *Listeria* as a serious pathogen. *L. monocytogenes* cells are taken up by macrophages into phagocytic vesicles; but they avoid digestion and escape the vesicles. The bacteria then multiply while traveling through the host cytoplasm, while generating "tails" of actin (**Fig. 18.16**). The actin tails eventually project *Listeria* cells out of the original cell and enable it to penetrate a neighboring host cell.

Lactic acid bacteria. The order Lactobacillales, known as lactic acid bacteria, are aerotolerant (capable of growth in the presence of oxygen), but most are obligate fermenters; that is, they generate ATP by substrate-level phosphory-

lation but cannot use oxygen to respire. They ferment primarily by converting sugars to lactic acid (a fermentation pathway discussed in Chapter 13). As the acid builds up, the pH decreases until it halts bacterial growth; thus, the carbon source retains much of its food value for human consumption. This is the basis of yogurt and cheese production. *Lactococcus* and *Lactobacillus* species are extremely important for the dairy industry (discussed in Chapter 16). Other common genera of lactic acid bacteria include *Leuconostoc*, often responsible for spoilage of meat, and *Pediococcus*, found in sauerkraut and fermented bean products as well as meat products such as sausage.

The shape of lactate-producing bacteria varies among species, from long, thin rods to curved rods and cocci. Most lactic acid bacteria have fastidious growth requirements and need many amino acids and vitamins. They can be isolated from pasture grasses incubated anaerobically in moderate acid (pH 5). The human intestinal flora include species of lactic acid bacteria. Certain species, particularly *Lactobacillus acidophilus*, are believed to play a positive role in human health by inhibiting the growth of pathogens. For this reason *L. acidophilus* may be ingested as a probiotic therapy.

***Staphylococcus* and *Streptococcus*.** The staphylococci are facultative anaerobic cocci that grow in clusters, often hexagonal in arrangement (**Figs. 18.17A** and **B**). They include common skin flora such as *S. epidermidis*. The staphylococci are generally salt tolerant, and their fermentation generates short-chain fatty acids that inhibit growth of skin pathogens. Certain species, however, are themselves serious pathogens. *Staphylococcus aureus* causes impetigo and toxic shock syndrome, as well as pneumonia, mastitis, osteomyelitis, and other diseases (discussed in Chapter 26). It is a major cause of nosocomial (hospital-acquired) infections, especially contamination of surgical wounds.

Figure 18.16 *Listeria monocytogenes*: **Intracellular pathogen travels on tails of actin.** **A.** Fluorescent antibodies mark the tails of polymerized actin (green) behind the *Listeria* cell bodies (red) traveling within an infected macrophage (fluorescence micrograph). **B.** Invading bacteria encapsulate themselves in actin. Actin tails propel them through the host cytoplasm and out through the cell membrane to invade a neighboring cell.
Source: B. U. South Carolina, Microbiology and Immunology On Line.

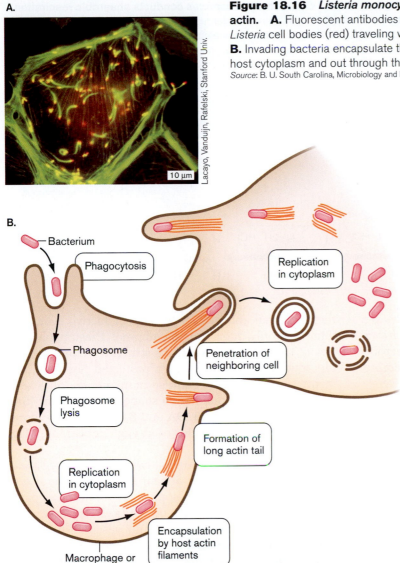

The most dangerous strains, now resistant to most known antibiotics, are termed MRSA (methicillin-resistant *S. aureus*).

Streptococcus species generally form chains instead of clusters, as their cells divide in a single plane (**Figs. 18.17C** and **D**). They are aerotolerant (grow in the presence of oxygen) but metabolize by fermentation. Many live on oral or dental surfaces, where they are responsible for caries (tooth decay). Their fermentation of sugars produces such high concentrations of lactic acid that the pH at the tooth surface can fall to pH 4. *Streptococcus* species cause many serious diseases, including pneumonia (*S. pneumoniae*), strep throat, erysipelas, and scarlet fever (*S. pyogenes* or group A strep).

The streptococci are less salt tolerant than *Staphylococcus* species, which tolerate as much as 5–15% NaCl. Another genus whose size and fermentative metabolism resembles that of streptococci is *Enterococcus*. *Enterococcus faecalis* is a common member of the intestinal flora, and related strains are enteric pathogens.

Anaerobic dechlorinators. Some firmicutes from the soil show promising abilities to degrade chlorinated pollutants, such as drycleaning solvents that are biodegraded very

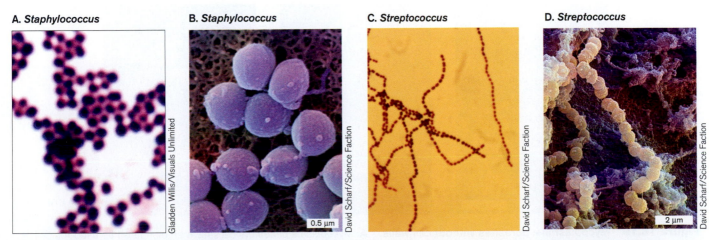

Figure 18.17 **Staphylococci and streptococci.** **A.** *Staphylococcus* species (cell size 0.5–1.0 μm; Gram stain). **B.** *Staphylococcus* species (colorized SEM). **C.** *Streptococcus* species (cell size 0.5–1.0 μm; Gram stain). **D.** *Streptococcus* species (colorized SEM).

A.

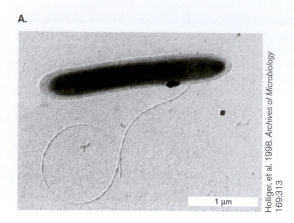

Holliger, et al. 1998. *Archives of Microbiology* 169:313

1 μm

Figure 18.18 *Dehalobacter restrictus* **conducts anaerobic respiration by dechlorination.** **A.** *Dehalobacter restrictus*, a flagellated rod related to clostridia. **B.** *D. restrictus* donates electrons to remove chlorine from tetrachloroethene, a major industrial pollutant. The ultimate product, ethene, is much less harmful in the environment.

B. Dehalogenation of tetrachloroethene by *Dehalobacter restrictus*

slowly in the environment. The chlorinated molecules are reduced as alternative electron acceptors. *Dehalobacter restrictus*, a flagellated rod, was isolated as an anaerobe capable of respiring by donating electrons to chlorine atoms in tetrachloroethene (**Fig. 18.18**). Thus, the discovery of *D. restrictus* has promising potential for bioremediation of chlorinated pollutants.

While genetic analysis places *D. restrictus* among the clostridia, the bacterium actually stains gram-negative, perhaps because its peptidoglycan layer is relatively thin. Nevertheless, the species shows no gram-negative outer membrane. It does possess a thick S-layer of hexagonally tiled proteins, typical of gram-positive bacteria.

Mollicutes lack a cell wall. Organisms of the class Mollicutes (named in Latin, "soft skin") have completely lost

their cell wall and S-layer through reductive evolution, retaining only their cell membrane. Presumably, the loss of these energy-expensive structures enhanced the reproductive rate of cells in a protected host environment. The Mollicutes comprise many genera of flexible wall-less cells that maintain a shape through some kind of cytoskeleton (**Figs. 18.19A and B**). On agar they form colonies of a characteristic "fried-egg" appearance (**Fig. 18.19C**).

The best-known genus of Mollicutes is *Mycoplasma*, although its species now appear to fall in several distantly related branches. Mycoplasmas are found as parasites of every known class of multicellular organism, including vertebrates, insects, and vascular plants; in humans, they cause pneumonia and meningitis. Some *Mycoplasma* species remain adherent to the host cell,

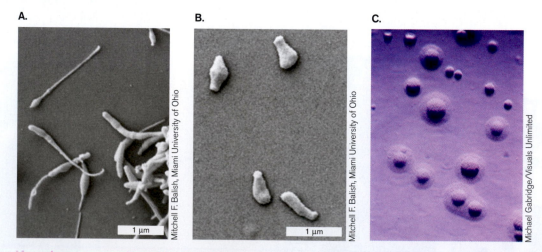

Figure 18.19 **Mycoplasmas: parasites without cell walls.** **A.** *Mycoplasma penetrans* cells have an elongated tip used for attachment to the host (SEM). **B.** *Mycoplasma mobile* (SEM). **C.** Mycoplasmas cultured on agar show a typical "fried-egg" shape of colony.

whereas others penetrate and grow intracellularly. Most mycoplasma cells have a rounded cell shape with one or two extended tips. In *Mycoplasma penetrans*, an opportunistic pathogen infecting AIDS patients, the cell's attachment tip is coated with adhesion molecules that enable attachment to a host cell surface (**Fig. 18.19A**). The attachment tip penetrates deep into epithelial tissues. By contrast, the fish pathogen *M. mobile* has no attachment tip, but glides along a surface at a rate of seven cell lengths per second (**Fig. 18.19B**). The basis of mycoplasma motility is not understood, although it may involve gliding or cytoskeletal contraction. In some species, motility involves a specialized cell tip called the "terminal organelle," but species that lack this organelle are equally motile. Species of *Spiroplasma* have a spiral shape and undergo corkscrew motion; the basis for this motion is also unknown.

Mycoplasma genomes are among the smallest known cellular organisms, and they lack biosynthetic pathways for amino acids and phospholipids, which must be acquired from the host. Mycoplasmas have unique nutritional requirements, such as cholesterol, a membrane component typical of eukaryotes but rare for prokaryotes. For these reasons, mycoplasmas are difficult to grow in pure culture, although they readily infect tissue cultures. In fact, mycoplasma contamination of tissue culture is so prevalent that it has compromised major studies of cancer and AIDS.

Actinomycetes Form Multicellular Filaments

The phylum Actinobacteria, the "high GC gram-positives," comprises several orders, including the orders Actinomycetales and Bifidobacteria. Members of the order Actinomycetales are "actinomycetes." Actinomycetes can be identified under the microscope by the **acid-fast stain,** a procedure in which cells are penetrated with a dye that is retained under treatment with acid alcohol (discussed in Chapter 2). The acid-fast property is associated with unusual cell wall lipids, such as mycolic acids. Actinomycetes include filamentous spore formers, such as *Streptomyces* and *Frankia*, as well as short-chain organisms, such as *Mycobacterium*, and irregularly shaped cells, such as *Corynebacterium*.

***Streptomyces*: filamentous spore formers.** *Streptomyces* bacteria form multicellular filaments that generate dispersable spores. The streptomycete life cycle is discussed in Chapter 4. Streptomycetes play a major role in the ecosystems of soil (discussed in Chapter 21). Decaying *Streptomyces* cells produce the compound **geosmin,** which causes the characteristic odor of soil and can affect the taste of drinking water. In culture, the best-known species, *S. coelicolor* (Latin, "sky color"), forms strikingly blue colonies (**Fig. 18.20A**). Other species produce filaments that are red, orange, green, or gray, depending on

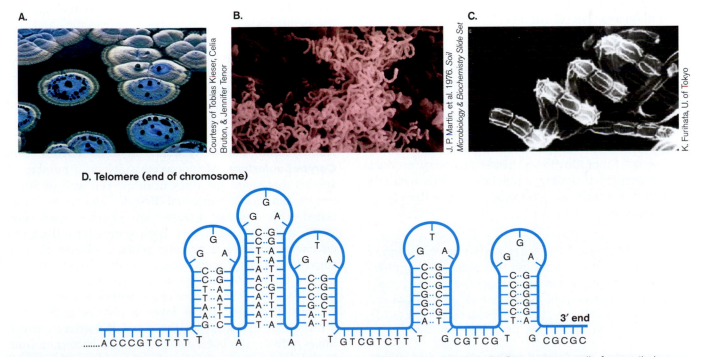

Figure 18.20 *Streptomyces* **bacteria. A.** Colonies of *S. coelicolor* show sky-blue mycelia. **B.** *Streptomyces* cells form coiled filaments (filament width, about 0.5 μm; SEM). **C.** Close-up of a coiled filament, showing individual cells (SEM). **D.** Hairpin-looped telomere end of the linear chromosome of *S. griseus*. Source: D. Prokaryotes, Springer.

their distinctive products of metabolism, many of which are antibiotics.

THOUGHT QUESTION 18.6 Why would *Streptomyces* produce antibiotics targeting other bacteria?

Streptomyces species are obligate aerobes, requiring access to air to complete their life cycle. When a *Streptomyces* spore germinates, it forms extended branched filaments into the substrate called **vegetative mycelia**, branched filaments of actively dividing cells. Some of the filaments then grow upward into the air, where they develop into **aerial mycelia**. The aerial mycelia in some species grow in tightly coiled spirals (**Figs. 18.20B and C**). As the mycelial colony runs out of nutrients, older cells of the filament age and lyse, releasing nutrients absorbed by the younger cells. The nutrients also attract other scavengers, which may be killed by antibiotics produced by the aging streptomycete cells. The dead scavengers, too, release nutrients that feed the growing tip of the mycelium.

As mycelia mature, they fragment into smaller cells called **arthrospores**. Arthrospores are vegetative cells, not dormant like endospores (**Table 18.3**). The arthrospores separate and are dispersed by the wind, enabling them to colonize a new location. Streptomycete mycelia can be obtained from natural habitats by burying a glass slide in soil, then waiting several days for spores to germinate, covering the slide with mycelia. They are challenging to isolate in pure culture, however, because their coiled filaments trap cells of other bacteria.

The *S. coelicolor* genome contains over 8 million base pairs, one of the largest prokaryotic genomes. Streptomycete chromosomes are linear with special "telomeres," single-stranded end sequences that double back to form hairpin loops (**Fig. 18.20D**). Much of the lengthy genome of a streptomycete encodes catabolism of a rich array of diverse organic components of decaying plant and animal matter, including even lignin. Other genes encode extensive operons for production of diverse secondary products (see Chapter 15), including antibiotics. More than half the antibiotics currently used in medicine derive from *Streptomyces* species.

NOTE: Distinguish the order Actinomycetales from the genus *Actinomyces* within this order. Distinguish *Streptomyces*, filamentous rod-shaped actinomycetes, from nonactinomycete *Streptococcus*, gram-positive cocci that form short unbranched chains.

Actinomycetes associated with animals and plants.

Most actinomycetes, such as *Actinomyces* and *Streptomyces*, have no direct association with animals, either posi-

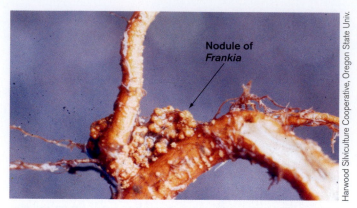

Figure 18.21 *Frankia* species associate endophytically with plants. Root of alder tree (*Alnus glutinosa*) bears the orange-yellow-colored nodules (arrowhead) containing *Frankia*.

tive or negative, but there are important exceptions. For example, *Actinomyces* species cause **actinomycosis**, a condition of skin abscesses in humans and cattle. A species of *Streptomyces* maintains a mutualism with leaf-cutter ants. The ants culture the bacteria on special organs to produce antibiotics against parasites of their fungal gardens (discussed in Chapter 17).

Actinomycetes have many mutually beneficial associations with plants. For example, *Frankia* associates with the alder tree, which develops orange-yellow colored nodules on its roots to fix nitrogen (**Fig. 18.21**). *Frankia* species live as **endophytes**, endosymbionts of vascular plants. This nitrogen fixation mutualism is analogous to the better-known symbiosis between legumes and rhizobia (see under Alpha Proteobacteria). Other *Frankia* species benefit wheat by conferring resistance to major fungal pathogens, such as "take-all" disease, as well as to certain parasites and insects. A few actinomycetes cause plant diseases. For example, potato scab disease is caused by *Streptomyces scabies*.

Nonmycelial actinomycetes: *Mycobacterium* and *Corynebacterium*. Actinomycetes include a number of species that share the thick acid-fast cell wall of *Streptomyces* but lack the mycelial lifestyle. Two genera associated with dreaded diseases are *Mycobacterium* (**Fig. 18.22**) and *Corynebacterium*. Both genera have thick cell envelopes containing **mycolic acids**, a diverse class of sugar-linked fatty acids (discussed in Chapter 3). The mycolic acids of *M. tuberculosis* are extremely diverse and include some of the longest-chain acids known, up to 90 carbons. The thick, dense layer of mycolic acids forms a waxy coat that impedes the entry of nutrients, which limits growth rate; but it also protects the bacterium from host defenses and antibiotics. For this reason, the cure of tuberculosis requires an exceptionally long course of antibiotic therapy.

A.

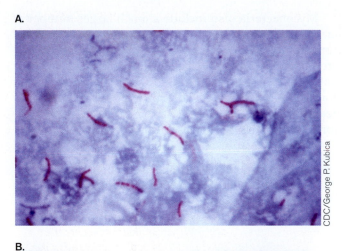

CDC/George P. Kubica

B.

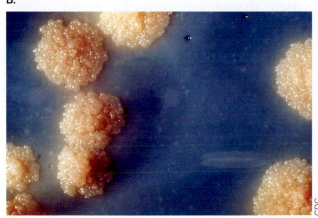

CDC

C.

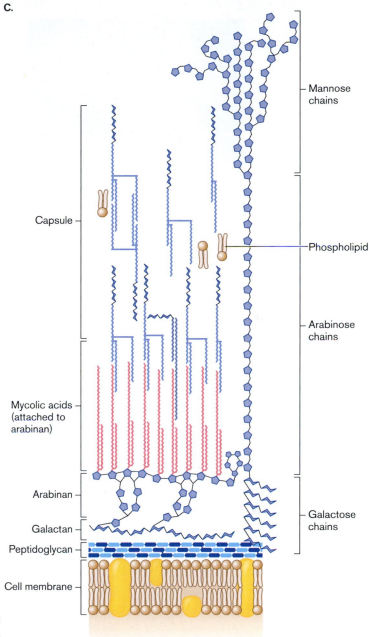

Mannose chains

Capsule

Phospholipid

Arabinose chains

Mycolic acids (attached to arabinan)

Arabinan

Galactose chains

Galactan

Peptidoglycan

Cell membrane

Figure 18.22 *Mycobacterium tuberculosis* **causes tuberculosis. A.** Acid-fast stain of tissue sample containing *M. tuberculosis* (chains of pink rods, 2–4 μm in length). **B.** Crinkled appearance of *M. tuberculosis* colonies. **C.** The cell wall of *M. tuberculosis* is impregnated with mycolic acids linked to chains of arabinose, galactose, and mannose.

Mycobacterium includes the species *M. tuberculosis* and *M. leprae*, as well as lesser-known pathogens such as *M. ulcerans*. Cells of *M. tuberculosis* can be detected by the acid-fast stain as tiny rods associated with sloughed cells in sputum (**Fig. 18.22A**). The organism forms crinkled colonies after two weeks of growth on agar-based media (**Fig. 18.22B**).

The closely related species *M. leprae* causes the disfiguring disease leprosy (**Fig. 18.23A**). The species has one of the longest known doubling times of any pathogen (about 14 days), and it can take a year to grow enough cells in the laboratory for observation. Lower temperature is preferred; for this reason, leprosy attacks the extremities. Culture on artificial media is impossible; the bacteria can only be grown within low-temperature animals, such as armadillos, or within genetically immune-deficient mice.

The genomes have been sequenced both for *M. tuberculosis* and *M. leprae*. *M. tuberculosis* has surprisingly few recognizable pathogenicity genes but a large number of environmental stress components, including 16 environmental sigma factors as well as 250 genes for its complex lipid metabolism. Over half the *M. leprae* genome consists of pseudogenes, homologs of *M. tuberculosis* genes undergoing reductive evolution (**Fig. 18.23B**). Thus, *M. leprae* appears to be an evolving pathogen "caught in the act" of losing many genes no longer needed in its sheltered host environment. How it lost the need for so many genes preserved in *M. tuberculosis* remains a mystery.

A.

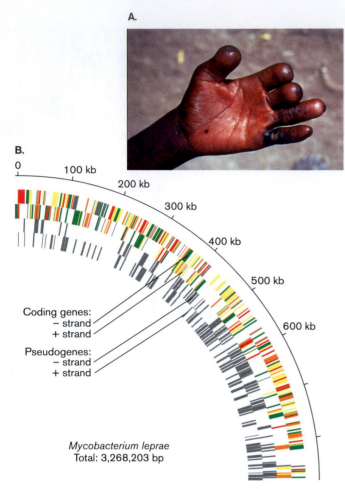

B.

Coding genes:
— strand
+ strand
Pseudogenes:
— strand
+ strand

Mycobacterium leprae
Total: 3,268,203 bp

C. James Webb/Phototake

Figure 18.23 *Mycobacterium leprae* **causes leprosy.**
A. Hand disfigured by leprosy. **B.** The genome of *M. leprae*
shows a high content of decaying pseudogenes (gray bars),
most of which correspond to functional genes in the genome of
M. tuberculosis.

Mycobacteria also include a much larger number of
harmless commensals, such as *M. smegmatis*, isolated
from human skin. Species of mycobacteria can be isolated from soil and water, as well as from various animal sources. Their culture is difficult because of their
slow growth rates, but isolation can be enhanced by
treatment with a base (NaOH or KOH) at concentrations that kill most other bacteria.

NOTE: Distinguish *Mycobacterium*, rods whose cell
walls contain mycolic acids, from *Mycoplasma*, cell
wall-less firmicutes related to *Bacillus* and *Clostridium*.

Irregularly shaped actinomycetes. Several nonmycelial actinomycetes show unusual cell shapes. Members of the genus *Corynebacterium* include soil bacteria
as well as pathogens such as *C. diphtheriae*, the cause
of the lung disease diphtheria. *Corynebacterium* species grow as irregularly shaped rods, which may divide
by a "half-snapping" mechanism in which one side of
the cell remains attached like a hinge (**Fig. 18.24A**).
Related soil bacteria include the genera *Nocardia* and
Rhodococcus.

Soil bacteria of the genus *Arthrobacter* exhibit an
unusual cell cycle in which coccoid stationary-phase cells
sprout into rods, which eventually run out of nutrients
and revert to the coccoid form. The growing rods form
irregular branched filaments (**Fig. 18.24B**). An *Arthrobacter* species was discovered conducting anaerobic respiration by reduction of hexavalent chromium (Cr^{6+}), a
toxic metal pollutant, to a less toxic oxidation state. Now
Arthrobacter shows potential as an agent of bioremediation of hexavalent chromium.

A. *Corynebacterium diphtheriae*

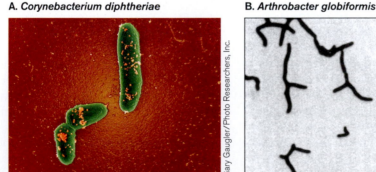

Gary Gaugler/Photo Researchers, Inc.

B. *Arthrobacter globiformis*

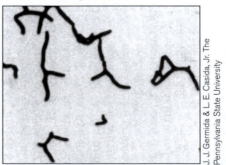

J. J. Germida & L. E. Casida, Jr. The
Pennsylvania State University

Figure 18.24   Irregularly shaped actinomycetes: *Corynebacterium* and *Arthrobacter*.   A. *Corynebacterium diphtheriae* (cells
1–2 µm in length) divide by snapping off one side while remaining attached at the other; the result is a typical V-shape or "Chinese
letter" arrangement (colorized SEM). **B.** *Arthrobacter globiformis* cultures form coccoid cells in stationary phase. With added nutrients, the
coccoid cells grow out as rods.

Figure 18.25 *Micrococcus* species grow in tetrads or sarcinae. **A.** *M. luteus* growing in tetrads (negative or indirect stain with nigrosin). **B.** *M. luteus* bacteria grow in clumps when cultured on a solid substrate (SEM).

A.

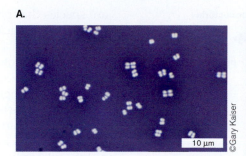

B.

Micrococcaceae. Relatives of *Arthrobacter* are nonfilamentous cocci such as *Micrococcus*. *Micrococcus luteus* is one of the most ubiquitous of soil bacteria, appearing readily as yellow colonies on agar plates exposed to air. Historically the genus *Micrococcus* in the family Micrococcaceae was classified with *Staphylococcus*, but its DNA sequence data now place *Micrococcus* and most of the Micrococcaceae within the actinomycetes.

Micrococci are aerobic heterotrophs, commonly isolated from air and dust, although their habitat of choice is the human skin. Micrococci grow in square or cuboid formations, dividing in two or three planes (**Fig. 18.25**); cuboid clusters are known as **sarcinae**. *M. luteus* is harmless to humans and grows well at room temperature, so it makes an excellent laboratory organism for observation by students.

TO SUMMARIZE:

- **Firmicutes**, the low GC gram-positive bacteria, include endospore-forming genera such as *Bacillus* and *Clostridium*. The cycle of endospore formation was probably present in the common ancestor of this phylum.
- **Nonsporulating firmicutes** include pathogenic rods such as *Listeria* as well as food-producing bacteria such as *Lactobacillus* and *Lactococcus*.
- *Staphylococcus* and *Streptococcus* are gram-positive cocci that include normal human flora as well as serious pathogens causing toxic shock syndrome, pneumonia, and scarlet fever.
- **Mycoplasmas** belong phylogenetically to Firmicutes but lack the cell wall and S-layer. They have flexible cytoskeletons and show ameboid motility. Species cause diseases such as meningitis and pneumonia.
- **Actinomycetes** (order Actinomycetales) include mycelial spore-forming soil bacteria, such as *Streptomyces*. Other actinomycetes have irregularly shaped cells, such as *Clostridium* species that cause tetanus and botulism. Actinomycetes stain acid-fast.
- **Mycobacteria** are actinomycete rods whose cell envelope contains a diverse assemblage of complex mycolic acids. Mycobacterial species cause tuberculosis and leprosy. They stain acid-fast.

18.5 Gram-Negative Proteobacteria and Nitrospirae

The Proteobacteria, with all their diversity of form and metabolism, share a common structure—their triple-layered, gram-negative cell envelope, which consists of outer membrane, peptidoglycan cell wall permeated by the periplasm, and inner membrane (plasma membrane) (see **Fig. 18.10**). The outer membrane is packed with receptor proteins and porins, comprising two-thirds the mass of the membrane. The outer membrane lipids contain long sugar polymer extensions (LPS). In pathogens, LPS repels phagocytosis and has toxic effects when released by dying cells.

The Protean Metabolism of Proteobacteria

Proteobacteria, the "protean" or "many-formed" bacteria, comprise at least five classes, labeled Alpha through Epsilon (see **Table 18.1**). Even within each closely related group, we see nearly as wide a range of cell shape and metabolic strategies as we see in Proteobacteria as a whole.

A closer look at proteobacterial metabolism shows that processes that seem very different on the surface actually connect through linked modules (**Fig. 18.26**; metabolism is discussed in detail in Chapter 14). Many proteobacteria are "photoheterolithotrophs" capable of nearly all the fundamental classes of metabolism, depending on environmental conditions, such as availability of light, oxygen, and nutrients. An example is *Rhodopseudomonas palustris*, the species of Alpha Proteobacteria characterized and sequenced by Caroline Harwood at the University of Washington. In such a "protean" organism, the core of all proteobacterial energy acquisition is a respiratory chain of electron donors and acceptors. Electrons may enter the chain from a photoexcited chlorophyll, from organic electron donors such as sugars or benzoates, or from a mineral electron donor such as reduced sulfur or iron (lithotrophy). In facultative anaerobes, the electron acceptor may be oxygen, or it may be a mineral such as nitrate or sulfate (anaerobic respiration).

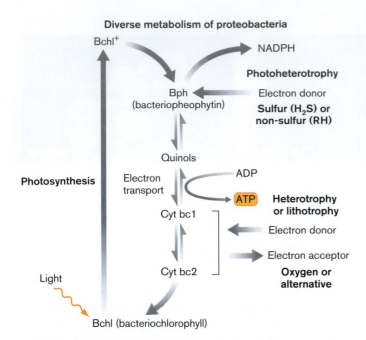

Diverse metabolism of proteobacteria

Figure 18.26   The protean metabolism of Proteobacteria. In many gram-negative species, metabolic diversity arises through minor "add-ons" of biochemical modules such as light absorption by bacteriochlorophyll; use of sulfide or organic electron donors; and use of oxygen or alternative (anaerobic) electron acceptors.

Photoheterotrophy. Light-supplemented heterotrophy occurs in Alpha, Beta, and Gamma Proteobacteria (see **Table 18.2**). This ubiquity suggests that the common ancestor of the phylum was a photoheterotroph. At the same time, evidence supports horizontal transfer of photosystems among proteobacterial branches. The evidence is particularly strong in the case of **proteorhodopsin**, a homolog of the retinal protein light-driven proton pump first characterized in halophilic archaea (discussed in Chapter 19). In any case, **Table 18.2** shows the remarkable diversity of light absorption niches that have evolved. The proteorhodopsin light pump in *Pelagibacter* spp. absorbs green and yellow (wavelength 500–600 nm), the mid-range of solar radiation reaching Earth's surface. The various bacteriochlorophylls of Proteobacteria generally peak in two ranges, in the blue and in the red or infrared. Bacteriochlorophyll *b*, in *Blastochloris* spp., peaks well beyond 1,000 nm. Species with different photopigments often grow together in stratified layers of sediment or wetland, the infrared absorbers below, where they capture the longer wavelengths "left over" from the shorter-wavelength absorbers above.

In the literature, proteobacterial phototrophs are historically called "purple bacteria." The actual colors of all the phototrophs range from purple-red through yellow and brown. "Purple sulfur bacteria" are species that photolyze reduced forms of sulfur, such as H_2S, HS^-, S^0, or S_2O_3 (thiosulfate), whereas "purple nonsulfur bacteria"

are species that photolyze small organic molecules such as short-chain acids. However, it turns out that many "purple bacteria" photolyze both sulfur and organic molecules. Most proteobacterial phototrophs conduct photosynthesis only in the absence of oxygen, reverting to aerobic heterotrophy when oxygen is present. Some, however, are obligate anaerobes. Still others, found in marine environments, surprisingly photolyze sulfide or organic compounds only in the presence of oxygen. Making sense of this extraordinary diversity is the job of marine and aquatic microbiologists, particularly those calculating global cycles of carbon and oxygen (discussed in Chapters 21 and 22).

> **THOUGHT QUESTION 18.7** Why might genes for the proteorhodopsin retinal photopump be more likely to transfer horizontally than the bacteriochlorophyll-based photosystems PS I and PS II?

Lithotrophy. The use of inorganic electron donors recurs throughout the phylogeny of bacteria, with notable diversity among Proteobacteria (**Table 18.4**). Many lithotrophic reactions generate acid, which conveniently breaks down rock containing additional reduced minerals. Bacterial and archaeal lithotrophy has tremendous significance for global cycling of elements in the biosphere (discussed in Chapter 22). Some lithotrophs are autotrophs, fixing carbon dioxide into biomolecules by the Calvin-Benson cycle (in carboxysomes) or the reverse TCA cycle. Others have alternative pathways of heterotrophy when organic foods are available. The ability to oxidize or reduce minerals evolved multiple times in different clades as an offshoot of the general electron transport chain.

We now survey species of the major classes of Proteobacteria and the gram-negative phylum Nitrospira.

Alpha Proteobacteria: Photoheterotrophs, Methylotrophs, and Intracellular Endosymbionts

The class Alpha Proteobacteria includes photoheterotrophs and heterotrophs, as well as bacteria metabolizing single-carbon compounds. A remarkable array of intracellular mutualists and pathogens arise in this group.

Photoheterotrophs. The Alpha class photoheterotrophs are generally unicellular. Their cell shapes range from flagellated spirilla to rounded rods (*Rhodobacter sphaeroides*), wide spirals (*Rhodospirillum rubrum*), stalked cells (*Rhodomicrobium*) (**Figs. 18.27A** and **B**). The TEM of *R. sphaeroides* shows how the cell is packed with photomembranes (membranes containing the photosynthetic complex) arranged in folds or tubes invaginated from the cytoplasmic membrane. These intracellular photomem-

Table 18.4 Lithotrophic reactions in bacteria (examples).

Class	Example species	Reaction	$\Delta G^{\circ\prime}/2e^-$ (kJ)	Class of reaction
Alpha Proteobacteria	Nitrobacter winogradskii	$NO_2^- + \frac{1}{2}O_2 \longrightarrow NO_3^-$	−54	Nitrite oxidation (nitrification to nitrate)
	Paracoccus pantrophus	$H_2S + 2O_2 \longrightarrow 2H^+ + SO_4^{2-}$		Sulfide oxidation to sulfate
Beta Proteobacteria	Nitrosomonas europea	$NH_3 + O_2 \longrightarrow HNO_2^- + 3H^+ + 2e^-$		Ammonia oxidation (nitrification to nitrite)
	Ralstonia eutropha	$2H_2 + \frac{1}{2}O_2 \longrightarrow 2H_2O$	−237	Hydrogen oxidation
	Gallionella ferruginea Thiobacillus ferrooxidans	$4FeS_2 + 15O_2 + 14H_2O \longrightarrow$ $4Fe(OH)_3 + 16H^+ + 8SO_4^{2-}$	−164	Iron sulfur oxidation
	Thiobacillus denitrificans	$3S^0 + 4NO_3^- \longrightarrow 3SO_4^{2-} + 2N_2$		Anaerobic sulfur oxidation
	Thiobacillus thiooxidans	$2S^0 + 3O_2 + 2H_2O \longrightarrow$ $2SO_4^{2-} + 4H^+$	−196	Sulfur oxidation
Gamma Proteobacteria	Methylococcus capsulatus	$CH_4 + 2O_2 \longrightarrow HCO_3^- + H^+ + H_2O$	−203	Methanotrophy
	Beggiatoa alba	$2H_2S + O_2 \longrightarrow 2S^0 + 2H_2O$	−210	Sulfide oxidation
	Chromatium	$4Fe^{2+} + CO_2 + 11H_2O + h\nu$ $\longrightarrow 4Fe(OH)_3 + [CH_2O] + 8H^+$		Iron phototrophy (photoferrotrophy)
Delta Proteobacteria	Desulfovibrio	$4S^0 + 4H_2O \longrightarrow$ $SO_4^{2-} + 3HS^- + 5H^+$	−11.3 (at pH 8)	Sulfur oxidation
Nitrospirae	Nitrospira	$NO_2^- + \frac{1}{2}O_2 \longrightarrow NO_3^-$	−54	Nitrite oxidation (nitrification to nitrate)
Planctomycetes	Brocadia anammoxidans	$NH_4^+ + NO_2^- \longrightarrow N_2 + 2H_2O$	−238	Anaerobic ammonium oxidation (anammox)

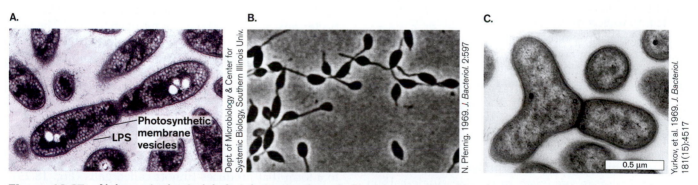

Figure 18.27 **Alpha proteobacterial photoheterotrophs.** **A.** *Rhodobacter sphaeroides*, showing intracellular photosynthetic membranes and outer membrane LPS filaments (TEM). **B.** *Rhodomicrobium acidophila* with stalked cells (length 2–7 µm; phase contrast). **C.** *Citromicrobium* species, an aerobic photoheterotroph, forms highly pleomorphic shapes, including the Y shape (TEM).

branes expand the surface area for photon capture; but during growth with oxygen, the photomembranes disappear. The outer surface of the cell shows LPS filaments extending from the outer membrane.

The metabolism of *R. rubrum* has been studied extensively. Anaerobically, *R. rubrum* bacteria grow dark red because they synthesize membrane containing Bchl *a*, which absorbs green and infrared and reflects primarily red. Under aerobic conditions, however, *R. rubrum* fails to synthesize Bchl *a* and the cells grow white as

their metabolism switches to straight heterotrophy. A surprising capacity is their ability to oxidize carbon monoxide (CO), a common by-product of photosynthesis. CO oxidation is now thought to occur in a number of proteobacteria.

Alpha photoheterotrophs requiring oxygen for phototrophy (but *not* photolyzing water for O_2) were found unexpectedly in genomic surveys of marine bacteria. Originally thought to be straight heterotrophs, these organisms revealed genes encoding bacteriochlorophyll.

Their aerobic heterotrophy was shown to be driven by light absorption. The source of the oxygen requirement is unclear, as their photosystems do not involve oxygen. Examples of O_2-requiring phototrophs have since been found in most of the proteobacterial clades, and species have been isolated from all major habitats including freshwater, marine, and soil ecosystems. Some show unusual cell morphology, such as the Y shape of *Citromicrobium* (**Fig. 18.27C**).

The aerobic photoheterotroph *Erithromicrobium ramosus* can reduce toxic metal compounds such as tellurite (TeO_3^{2-}). The cells generate tellurium crystals that take up 30% of their cell weight, providing a convenient way to remove tellurite from liquid waste. Other toxic metals reduced by aerobic photoheterotrophs include selenium and arsenic. Thus, these obscure bacteria now have a promising future in remediation of metal-contaminated industrial wastes.

Aquatic and soil oligotrophs. Alpha Proteobacteria also include many nonphototrophic heterotrophs of soil and water. Many aquatic and soil heterotrophs are oligotrophs adapted to extremely low nutrient concentrations; an example is *Caulobacter crescentus*, the stalk-to-flagellum organism discussed in Chapters 3 and 4. Oligotrophic bacteria often have unusual extended shapes enhancing nutrient uptake, such as the starlike cell aggregates of *Seliberia stellata*. In *Seliberia* species, the individual tightly coiled rods generate oval or spherical reproductive cells by a budding process. The budding reproductive cells germinate into rods, which then form new aggregates.

Some heterotrophs common in soil are pathogens. *Brucella* sp. is an intracellular pathogen of animals that can also infect humans. It is classed as a potential bioterrorism agent. The soil bacterium pathogen *Granulobacter bethesdensis* is associated with chronic granulomatous disease, an inherited disorder of the phagocyte oxidase system that leaves patients susceptible to infection.

Methylotrophy and methanotrophy. Methylotrophy is the ability of an organism to oxidize single-carbon compounds such as methanol, methylamine, or methane. The Alpha Proteobacteria include several genera of methylotrophs, which are found in all environments, including soil, freshwater, and the ocean. Most methylotrophs can grow on both single-carbon and organic compounds. One such versatile genus, *Methylobacterium*, is equally at home in soil and water, on plant surfaces, and as a contaminant of facial creams and purified water for silicon chip manufacture. Other species are restricted to single-carbon compounds, incapable of metabolizing organic compounds with carbon-carbon bonds. Organisms that grow solely on methane (CH_4) are called **methanotrophs**.

Methane-oxidizing bacteria of both Alpha and Gamma classes contribute to aquatic ecosystems, serving as major food sources for zooplankton. They eliminate much of the methane produced by methanogens before it reaches the atmosphere, where it has a potent greenhouse effect (see Chapter 22).

Endosymbionts: mutualists and parasites. The Alpha Proteobacteria include many highly evolved intracellular symbionts, some mutualists, others parasites. As isolated bacteria, they are generally rod-shaped with aerobic metabolism, but their shape is transformed within the host cell. Intracellularly, they need to solve special problems, such as the exclusion of oxygen from nitrogen fixation, a process requiring anaerobiosis—or, in the case of pathogens, the need to resist host defenses.

Nitrogen-fixing endosymbionts of plants include genera such as *Rhizobium*, *Bradyrhizobium*, and *Sinorhizobium*. The nomenclature has undergone many changes, but the species are still generally referred to as rhizobia. Rhizobia can live freely in the soil, but they prefer to colonize plants, usually legumes such as peas or alfalfa, where they form distinctive nodule structures. Each species of bacteria colonizes a particular host range for infection. Complex chemosensory processes attract the bacteria to the surface of the host root, where their presence induces root hairs to form curls around the bacteria (**Fig. 18.28**). The curling of the root hairs enables the bacteria to form infection threads, chains of bacteria that invade the root cells. Within the host cells, the bacteria lose their cell wall and become **bacteroids**, specialized cells for nitrogen fixation. The host plant cells provide nutrients as well as protective components such as **leghemoglobin**, an oxygen-binding protein that maintains anaerobiosis within infected cells. Leghemoglobin turns the nodule interior pink (**Fig. 18.28A**).

Agrobacterium. The genus *Agrobacterium* includes plant pathogens closely related to the rhizobia. *Agrobacterium tumefaciens* is known for its ability to convert host cells to a form that produces tumors called galls (**Fig. 18.29**). The ability of *A. tumefaciens* to insert its DNA into plant genomes has made it a major tool for plant biotechnology (see Chapter 16).

Rickettsias: intracellular pathogens and mitochondria. Short coccoid rods, rickettsias lack flagella and can only grow within a host cell. The best-known rickettsia is *Rickettsia rickettsii*, the cause of Rocky Mountain spotted fever, a disease spread by ticks in the northwestern United States. *Rickettsia* species parasitize human endothelial cells (**Fig. 18.30**). The bacteria induce phagocytosis, then dissolve the phagocytic vesicle and escape into the cytoplasm. Some rickettsias propel themselves

A.

B.

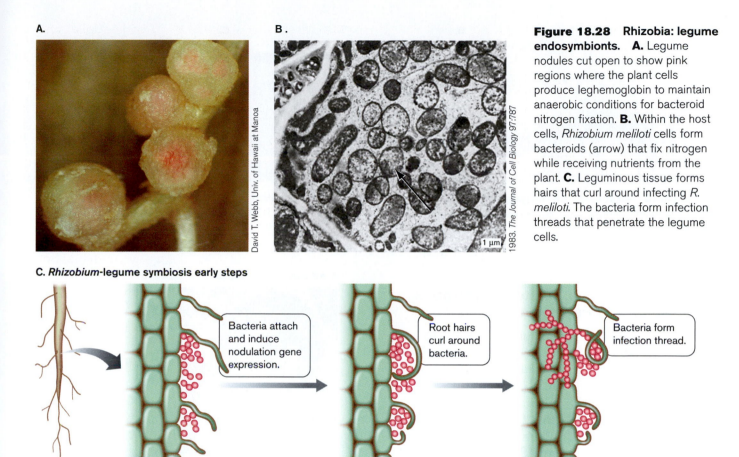

David T. Webb, Univ. of Hawaii at Manoa

1983. *The Journal of Cell Biology* 97:787

1 μm

Figure 18.28   Rhizobia: legume endosymbionts.   A. Legume nodules cut open to show pink regions where the plant cells produce leghemoglobin to maintain anaerobic conditions for bacteroid nitrogen fixation. **B.** Within the host cells, *Rhizobium meliloti* cells form bacteroids (arrow) that fix nitrogen while receiving nutrients from the plant. **C.** Leguminous tissue forms hairs that curl around infecting *R. meliloti*. The bacteria form infection threads that penetrate the legume cells.

C. *Rhizobium*-legume symbiosis early steps

Bacteria attach and induce nodulation gene expression.

Root hairs curl around bacteria.

Bacteria form infection thread.

through the host cell by polymerizing cytoplasmic actin behind them, a process similar to that of *Listeria* (described under Firmicutes). The actin tails eventually project outward as filopodia, extensions of host cytoplasm and membrane that protect the bacteria from host

defenses while enabling them to invade adjacent cells. The rickettsias include other intracellular pathogens, such as *Coxiella burnetii*, the causative agent of human Q fever.

Genetic analysis shows that the rickettsia clade includes mitochondria. All eukaryotic mitochondria appear to be descendants of an ancient rickettsial parasite whose respiratory apparatus ultimately became essential to power eukaryotic cells.

Genomic analysis of natural ecosystems continues to reveal surprises. We can now sequence a "community genome," or **metagenome**, of numerous organisms from a given microbial community. In 2004, Craig Venter and colleagues from several institutions sequenced a billion base pairs of the metagenome from the Sargasso Sea near Bermuda. Numerous uncultured species were revealed, including free-living rickettsias. Some of these rickettsias, such as *Pelagibacter ubique* (isolate SAR11), contain proteorhodopsin, a bacterial version of the retinal-based light-driven proton pump found in halophilic archaea (discussed in Chapter 14). It was unexpected to find a nonparasitic marine rickettsia, let alone one with an archaeal form of photosynthesis.

Nigel Cattlin/Photo Researchers, Inc.

Figure 18.29   Plant gall induced by *Agrobacterium tumefaciens*. Crown gall on Chrysanthemum plant.

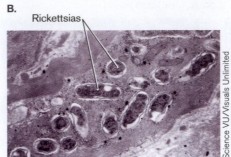
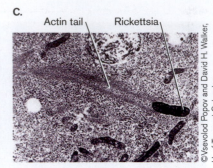

A.

B. Rickettsias

C. Actin tail   Rickettsia

James L. Castner/Visuals Unlimited

Science VU/Visuals Unlimited

©Vsevolod Popov and David H. Walker, U. of Texas at Galveston

Figure 18.30   Rickettsias are obligate intracellular parasites.   A. The Rocky Mountain tick carries *Rickettsia rickettsii*, the cause of Rocky Mountain spotted fever. **B.** *Rickettsia* species parasitize human endothelial cells (TEM). **C.** The rickettsias propel themselves through the host cell by polymerizing cytoplasmic actin behind them (TEM).

Beta Proteobacteria: Photoheterotrophs, Lithotrophs, and Pathogens

The Beta Proteobacteria include photoheterotrophs such as *Rhodocyclus*, as well as a diverse range of lithotrophs (**Table 18.4**). Several important pathogens infect humans and other animals.

Lithotrophs: nitrifiers and sulfur oxidizers. An important group of nitrogen lithotrophs is the **nitrifiers**. Nitrifiers oxidize ammonia (NH_3) to nitrite (NO_2^-) (**Fig. 18.31A**) or nitrite to nitrate (NO_3^-). Typically, different species conduct the two reactions separately while coexisting in the soil and water. Nitrifiers are of enormous economic and practical importance for wastewater treatment, as they decrease the reduced nitrogen content of sewage. Special systems have been developed to retain nitrifier bacteria behind filters as one stage of water treatment (**Fig. 18.31B**). In the system shown, nitrifying bacteria are encapsulated in pellets to retain them within the bioreactor while the treated water flows through a filter.

The most commonly isolated ammonia oxidizers are *Nitrosomonas* species of the Beta class. *Nitrosomonas* cells conduct electron transport through extensive internal membranes, either stacked or invaginated. Related genera such as *Nitrosolobus* and *Nitrosovibrio* have also been described, some with peritrichous or polar flagella. One ammonia-oxidizing genus of the Gamma group has been found, *Nitrosococcus*. Major nitrite oxidizers also include the nonproteobacterial phylum Nitrospirae. Although genetically outside the Proteobacteria, the unusual spiral-shaped cells of Nitrospirae have a gram-negative cell envelope.

Another important group of Beta Proteobacteria is the sulfur and iron oxidizers, such as the genus *Thiobacillus* (**Fig. 18.32A**). Most are short rods or vibrios (comma-shaped). *Thiobacillus* and other sulfur-oxidizing genera can undergo a number of different reactions oxidizing

A. *Nitrosomonas europea*

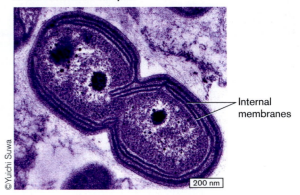

©Yuichi Suwa

Internal membranes

200 nm

B. Encapsulated nitrifiers

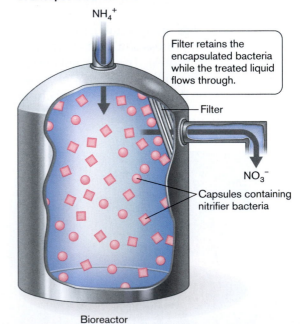

NH_4^+

Filter retains the encapsulated bacteria while the treated liquid flows through.

Filter

NO_3^-

Capsules containing nitrifier bacteria

Bioreactor

Figure 18.31   Nitrifiers.   A. *Nitrosomonas europea*, of the Beta Proteobacteria, oxidizes ammonia to nitrite (cell length 2–4 μm; TEM). Internal membranes contain the electron transport complexes. **B.** Wastewater treatment uses nitrifier bacteria to remove ammonia.

A.

5 μm

P. Romano, et al. 2001. *J. Chem. Tech. Biotech.* 76:723

B.

Photo courtesy of Kennecott Utah Copper, Salt Lake City, UT

Figure 18.32 *Thiobacillus*: iron oxidizers. **A.** *Thiobacillus ferrooxidans* leaches copper and iron from molybdenite ore (SEM). **B.** In a copper mine, *Thiobacillus* species can oxidize copper ores, leaching the copper into solution for retrieval.

H_2S to S^0 and S^0 to SO_4^{2-} (see **Table 18.4**). Sulfate production creates an environment acidic enough to erode stone monuments and the interior surface of concrete sewer pipes. Sulfur oxidation is often coupled to oxidation of iron, $Fe^{2+} \longrightarrow Fe^{3+}$. The bacterium *T. ferrooxidans* is known for its role in acidification of mine water, where it contributes to leaching of iron, copper, and other minerals (**Fig. 18.32B**).

Pathogens. The Beta class includes aerobic heterotrophic cocci such as *Neisseria.* The cocci of *Neisseria* species form distinctive pairs known as **diplococci** (shown in **Table 18.1**). Most *Neisseria* species are harmless commensals of the nasal or oral mucosa, but *N. gonorrhoeae* causes the sexually transmitted disease gonorrhea. *N. gonorrhoeae* is actually a microaerophile, requiring a narrow range of oxygen concentration; it has fastidious growth requirements, necessitating cultivation on special blood-based medium. A related organism, *N. meningitidis*, may be carried asymptomatically by a

quarter of the human population, but occasionally it causes meningitis, which can be fatal. Other neisserias, such as *N. sicca*, are easily isolated from skin and often presented as an unknown for identification by undergraduates.

Other members of the Beta class are important animal and plant pathogens, such as *Burkholderia. B. cepacia* was originally isolated from onions as a cause of bulb rot; it is now known to be a major opportunistic invader of the lungs of cystic fibrosis patients.

Gamma Proteobacteria: Photolithotrophs, Enteric Flora, and Pathogens

The Gamma Proteobacteria are well known for the Enterobacteriaceae family of facultative anaerobes found in the human colon, including important pathogens. In the environment, marine species catabolize complex organic pollutants, such as petroleum hydrocarbons. Gamma Proteobacteria also include unusual phototrophs, some of which oxidize iron and nitrite.

Chromatium: sulfur and iron phototrophs. The Gamma group phototrophs, such as *Chromatium* species (**Fig. 18.33**), mainly utilize sulfide and produce sulfur, which is deposited as intracellular granules visible within the cytoplasm. Their phototrophy is entirely anaerobic. Some of the Gamma group are true autotrophs and do not use organic substrates. Some of these actually conduct phototrophy using iron (Fe^{2+}) to donate electrons. Iron phototrophy, or "photoferrotrophy," is considered an intriguing possibility for metabolism of early life (discussed in Chapter 17). *Thiocapsa* uses NO_2^-, the first nitrogen-based phototroph discovered.

Sulfur lithotrophs. The sulfur-oxidizing genus *Beggiatoa* (Gamma class) was one of the first kinds of lithotroph described by pioneer microbial ecologist Sergei Winogradsky (discussed in Chapter 1). *Beggiatoa* species oxidize H_2S to elemental sulfur, which collects as sulfur granules within the periplasm. *Beggiatoa* also stores carbon in cytoplasmic granules of polyhydroxybutyrate, a common strategy among proteobacteria. The cells of *Beggiatoa* grow as extended filaments with visible sulfur granules, forming biofilms on sulfide-rich sediment.

Enteric fermenters and respirers. The family Enterobacteriaceae, facultative anaerobes of the Gamma subdivision, include some of the most intensively studied species of all bacteria. Species are readily isolated from the contents of the human digestive tract and easily grown on laboratory media based on human food. The best-known species of Enterobacteriaceae—indeed, the best studied of all bacterial species—is the model organism *Escherichia*

A.

B.

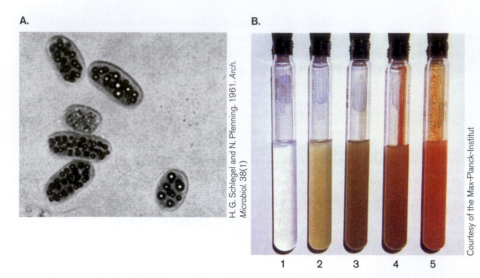

H. G. Schlegel and N. Pfenning. 1961. *Arch. Microbiol.* 38(1)

1 2 3 4 5

Courtesy of the Max-Planck-Institut

Figure 18.33 *Chromatium*: **Sulfur and iron phototrophs.** **A.** *Chromatium* forms single-flagellated rods full of sulfur granules. **B.** Photoferrotrophy: Under illumination, color develops over time (1–5) as a *Chromatium* isolate oxidizes Fe^{2+} to Fe^{3+}.

coli. Strains of *E. coli* grow normally in the human intestine, feeding on our mucous secretions and producing vitamins, such as vitamin K. But other strains, such as *E. coli* O157:H7, cause serious illness. A large proportion of the world's children die of *E. coli*-related intestinal illness before the age of 5.

The Enterobacteriaceae are gram-negative rods, although as nutrients diminish, their size dwindles almost to a coccoid form. They grow singly, in chains, or in biofilms. Many species are motile, with numerous flagella. Most strains grow well with or without oxygen, by either respiration (aerobic or anaerobic) or fermentation. They ferment rapidly on carbohydrates, generating fermentation acids, ethanol, and gases (CO_2 plus H_2) in varying proportions, depending on the species. Their presence in the intestine supports the growth of organisms utilizing these gases, including methanogens (discussed in Chapter 19.) Many strains form biofilms. Biofilm formation explains the persistence of drug-resistant infections, such as those associated with urinary catheters in long-term hospital patients.

THOUGHT QUESTION 18.8 Why do you think it took many years of study to realize that *Escherichia coli* and other Proteobacteria can grow as a biofilm?

Because of their ease of cultivation and their long study in clinical laboratories, many genera of Enterobacteraceae are familiar to students in introductory microbiology laboratory courses. A common laboratory exercise is to distinguish *Enterobacter* and *Klebsiella* species from *E. coli* by their fermentation to the pH-neutral product butanediol, which tests positive in the Voges-Proskauer test. (Fermentation is discussed in Chapter 13.) *Enterobacter* species occur more frequently in freshwater streams than in the human body, although a few species colonize the intestine and cause illness.

Proteus mirabilis and *P. vulgaris* cause bladder and kidney infections, particularly as a complication of surgical catheterization. *Proteus* species are heavily flagellated and display a remarkable **swarming** behavior (**Fig. 18.34**). Upon an

A.

B.

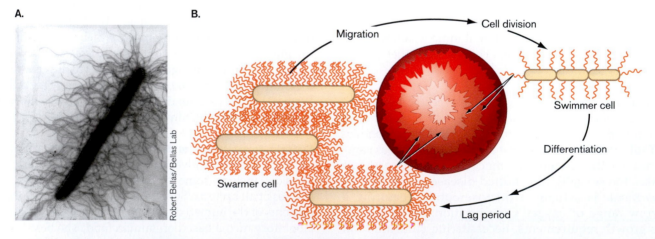

Robert Bellas/Bellas Lab

Migration

Cell division

Swimmer cell

Differentiation

Lag period

Swarmer cell

Figure 18.34 **The enteric rod *Proteus mirabilis*: Isolated swimmer or cooperative swarmer.** **A.** A thickly flagellated swarmer cell (TEM). Cell length can reach 20 μm. **B.** Waves of migration of *P. mirabilis* through blood agar.

A.

B.

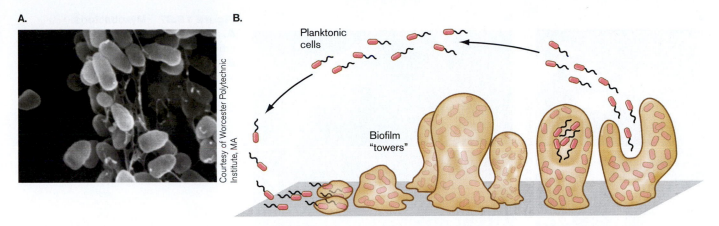

Planktonic cells

Biofilm "towers"

Courtesy of Worcester Polytechnic Institute, MA

Figure 18.35 *Pseudomonas* **species form biofilms.** **A.** *Pseudomonas fluorescens* biofilm on a plant surface. **B.** Biofilm development in *Pseudomonas* species.

environmental signal, the flagellated rods grow into long-chain swarmer cells. The swarmers gather together forming "rafts" that swim together, growing into a complex biofilm.

Besides symbionts of animals, Enterobacteraceae include plant pathogens, such as *Erwinia* species. *Erwinia* species cause wilts, galls, and necrosis of a wide variety of plants, including bananas, tomatoes, and orchids (see Chapter 23).

Aerobic rods. Closely related to Enterobacteraceae are several genera of rod-shaped bacteria that are obligate respirers. Many are obligate aerobic respirers (requiring O_2 for growth), although some can use alternative electron acceptors such as nitrate. These genera metabolize an extraordinary range of natural compounds, including aromatic derivatives of lignin; thus, they have important roles in natural recycling and soil turnover.

The Pseudomonadaceae are a large and amorphous group. As the "pseudo" prefix suggests, their taxonomic unit is poorly defined and includes species whose DNA sequence has necessitated their reassignment to other groups. The Gamma class pseudomonads, such as *Pseudomonas aeruginosa* and *P. fluorescens*, respire on oxygen or nitrate and are vigorous swimmers with single or multiple polar flagella. *P. aeruginosa* can swim throughout a standard agar plate (much to the chagrin of students attempting to isolate colonies). Nevertheless, under appropriate environmental conditions, pseudomonad cells give up their motility and develop biofilms (**Fig. 18.35**; discussed in Chapter 4). Biofilms of *P. aeruginosa* cause lethal infections of the pulmonary lining of cystic fibrosis patients.

Some pseudomonad pathogens have the unusual ability to infect both plants and animals. For example, *P. aeruginosa* commonly infects plants as well as humans. *P. fluorescens* infects seedlings and causes rotting of citrus fruit, while it also appears as an opportunistic pathogen of immunocompromised cancer patients. Other pseudomonad species, however, are harmless residents of soil or sewage.

Legionella pneumophila is a well-publicized pathogen related to the pseudomonads (class Legionellales). Incapable of growth on sugars, *L. pneumophila* requires oxygen to respire on amino acids. The organism exhibits an unusual dual lifestyle, alternating between intracellular growth within human macrophages and intracellular growth within freshwater amebas (**Fig. 18.36**). Growth within amebas facilitates transmission through aerosols into the

Figure 18.36 *Legionella pneumophila* **colonizes an ameba.** **A.** *Legionella pneumophila* cell caught by an ameba's pseudopod. **B.** *L. pneumophila* cells have colonized the ameba.

A.

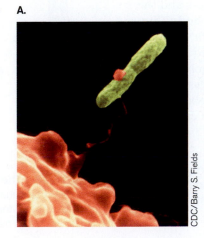

CDC/Barry S. Fields

B.

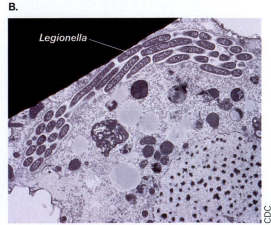

Legionella

CDC

A.

Wolgemuth, et al. 2002. *Current Microbiology* 12:369

10 μm

B.
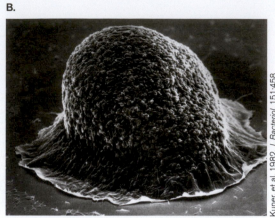

Kuner, et al. 1982. *J. Bacteriol.* 151:458

Figure 18.37 Myxobacteria.
A. *Myxococcus xanthus* cells glide while producing trails of slime (dark-field LM).
B. Upon starvation, cells come together to generate a fruiting body (100–200 μm across) packed with spherical myxospores (SEM).

human lung. *L. pneumophila* is an emerging pathogen that took advantage of our changing lifestyle (the prevalence of large-scale air-conditioning units with unfiltered water).

> **NOTE:** Distinguish myxobacteria from *Mycobacterium*, the gram-positive genus that includes the causative agents of tuberculosis and leprosy.

Delta Proteobacteria: Lithotrophs and Multicellular Communities

The Delta Proteobacteria include important sulfur and iron reducers such as *Geobacter metallireduscens*. Other species have complex life cycles that include multicellular developmental forms. The best-studied example is the myxobacteria. Myxobacteria have exceptionally large genomes, such as that of *Sorangium cellulosum* (12.6 Mb).

Myxobacteria. The myxobacteria, such as *Myxococcus xanthus*, are free-living soil bacteria that can grow as isolated cells or come together to form a differentiated structure for the purpose of reproduction (**Fig. 18.37**). When nutrients are plentiful, the myxobacteria grow and divide as individual cells. As nutrients run out, the cells begin to attract each other, moving into parallel formations. Myxobacteria have no flagella, but they move along a surface, a form of motility called gliding.

The formations develop into a bulging mass and the aggregating cells coalesce to form a **fruiting body** (**Fig. 18.38**). Fruiting body formation also occurs in slime molds, a class of eukaryotic microbe (discussed in Chapter 20). The myxobacteria, however, have evolved their process independently. Within the *Myxococcus* fruiting body develop durable spherical cells called **myxospores**. Ultimately, the myxospores are released to be carried on the wind to a more favorable location.

> **THOUGHT QUESTION 18.9** Compare and contrast the formation of cyanobacterial akinetes; firmicute endospores; actinomycete arthrospores; and myxococcal myxospores.

Bdellovibrios parasitize bacteria. Bacteria, like eukaryotic cells, can be parasitized or preyed on by smaller bacteria. The Delta class includes *Bdellovibrio* species, which attack proteobacterial host cells. The structure of the "attack cell" is a small comma-shaped rod with a single flagellum. The attack cell attaches to the envelope of its

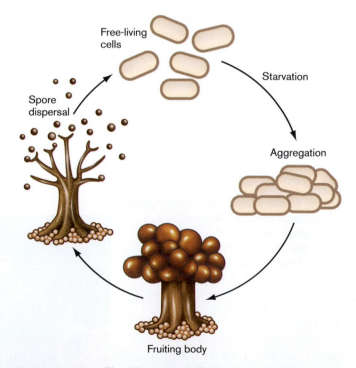

Free-living cells

Starvation

Spore dispersal

Aggregation

Fruiting body

Figure 18.38 The life cycle of *Myxococcus xanthus*.
Starving myxobacteria glide toward each other to aggregate. The aggregation generates a fruiting body with bulges packed with small, spherical myxospores. *Source:* Dale Kaiser.

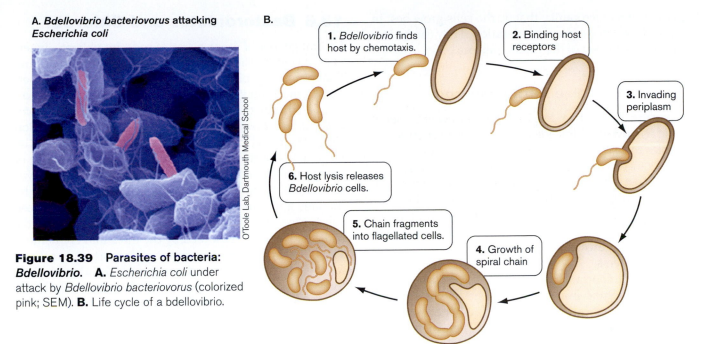

A. *Bdellovibrio bacteriovorus* attacking
Escherichia coli

B.

1. *Bdellovibrio* finds host by chemotaxis.

2. Binding host receptors

3. Invading periplasm

6. Host lysis releases *Bdellovibrio* cells.

5. Chain fragments into flagellated cells.

4. Growth of spiral chain

O'Toole Lab, Dartmouth Medical School

Figure 18.39   Parasites of bacteria:
Bdellovibrio. **A.** *Escherichia coli* under
attack by *Bdellovibrio bacteriovorus* (colorized
pink; SEM). **B.** Life cycle of a bdellovibrio.

host, then penetrates into the periplasm, where it uses host resources to grow (**Fig. 18.39**). The growing cell produces enzymes that cross the inner membrane to degrade host macromolecules and make their components available to the bdellovibrio. The entire host cell loses its shape and becomes a protective incubator for the predator. This stage is called the bdelloplast.

Within the periplasm, the invading bdellovibrio elongates as a spiral filament while replicating several copies of its DNA. When most nutrients have been exhausted, the filament septates into multiple short cells. The cells develop flagella, and the bdelloplast bursts, releasing the newly formed attack cells.

Bdellovibrios can be isolated from sewage, soil, or marine water—any environmental source of gram-negative prey bacteria. Different species infect *E. coli* and *Pseudomonas*, as well as *Agrobacterium* and *Rhizobium* in their free-living state. They can be cultured as plaques on a top-agar plate containing host bacteria, the same procedure used to isolate bacteriophages (discussed in Chapter 6).

Epsilon Proteobacteria: Microaerophilic Helical Pathogens

Epsilon Proteobacteria include the genera *Campylobacter* and *Helicobacter*. *Helicobacter pylori* is well known today as the causative agent of gastritis and stomach ulcers (**Fig. 18.40**). Yet until recently, most microbiologists believed that bacteria could not live in the acidic stomach. The discovery of *H. pylori* as the cause of gastritis earned the Nobel Prize in Medicine, in 2005, for Barry Marshall and J. Robin Warren, of the University of Western Australia.

H. pylori and related species grow primarily on the stomach epithelium, at about pH 6, which is less acidic than the gastric contents (pH 2–4). The bacteria bury themselves in the epithelial layer and neutralize their acidic surroundings by secreting urease enzyme, which converts urea to ammonia and carbon dioxide. *Helicobacter* species form wide spiral cells (spirillum) with an unusual grouping of flagella at one end. The metabolism of

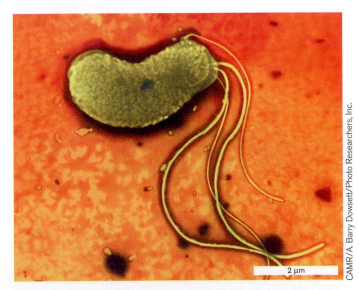

CAMR/A. Barry Dowsett/Photo Researchers, Inc.

2 μm

Figure 18.40   *Helicobacter*, a neutralophile growing within the acidic stomach. *H. pylori*, a short spirillum with unusual knobbed flagella projecting from one end.

H. pylori is microaerophilic; that is, the cells grow best in a narrow range of oxygen concentration. They can be isolated through biopsy of the gastric mucosa.

Nitrospirae Oxidize Nitrite and Iron

The phylum Nitrospirae consists of gram-negative spiral bacteria that oxidize nitrite ion to nitrate. Their cell structure resembles that of the Proteobacteria, although their phylogenetic branch is deep enough for assignment to a separate phylum. Most species, such as *Nitrospira* sp. (see **Table 18.1**), are true lithotrophs or autotrophs, fixing carbon in the form of carbon dioxide or carbonate using carboxysomes. *Nitrospira* species are generally found in freshwater or salt water. Their removal of excess nitrite makes a key contribution to aquatic ecosystems.

Another important genus is *Leptospirillum*, which includes acidophilic iron oxidizers. They are also strict autotrophs, fixing carbon using Fe^{2+} as their electron donor and O_2 as the electron acceptor. Their metabolism generates acid, contributing to acid mine drainage in iron mines in Iron Mountain, California, where they grow in massive pink biofilms.

TO SUMMARIZE:

- **Proteobacteria** stain gram-negative, with a thin cell wall and an LPS outer membrane. They show wide diversity of form and metabolism, including phototrophy, lithotrophy, and heterotrophy on diverse organic substrates.
- **Alpha Proteobacteria** include photoheterotrophs (such as *Rhodospirillum*) and heterotrophs as well as methylotrophs. They include intracellular mutualists, such as rhizobia, and pathogens, such as the rickettsias. Rickettsias share ancestry with mitochondria.
- **Beta Proteobacteria** include photoheterotrophs (*Rhodocyclus*) as well as nitrifiers (*Nitrosomonas*) and sulfur iron oxidizers (*Thiobacillus*). Pathogenic diplococci include *Neisseria gonorrhea*, the cause of gonorrhea.
- **Gamma Proteobacteria** include sulfur and iron photolithotrophs (*Chromatium*) as well as members of Enterobacteraceae found in the human colon. Intracellular pathogens include *Salmonella* and *Legionella*. The pseudomonads, aerobic rods, can respire on a wide range of complex organic substrates.
- **Delta Proteobacteria** include sulfur and iron reducers (*Geobacter*), fruiting body bacteria (*Myxobacteria*), and bacterial predators (*Bdellovibrio*).
- **Epsilon Proteobacteria** are spirillar pathogens such as *Helicobacter pylori*, the cause of gastritis.
- **Nitrospirae** are gram-negative spiral bacteria that oxidize nitrite to nitrate (*Nitrospira*).

18.6 Bacteroidetes and Chlorobi

Several genera of gram-negative rods are obligate anaerobes, including *Bacteroides, Cytophaga, Flavobacterium,* and *Chlorobium*. Despite their gram-negative cell envelopes, these genera are genetically distant from Proteobacteria.

Bacteroides species, such as *B. fragilis* and *B. thetaiotaomicron*, are the major inhabitants of the human colon (**Fig. 18.41**). Their envelope polysaccharides help the bacteria evade the immune system. Their main source of energy is fermentation of a wide range of sugar derivatives from plant material, compounds indigestible by humans and potentially toxic. *Bacteroides* converts these substances into simple sugars and fermentation acids, some of which are absorbed by the intestinal epithelium. Thus, *Bacteroides* species serve two important functions for their host: They breakdown potential toxins in plant food, and their fermentation products make up as much as 15% of the caloric value we obtain from food. A third benefit of *Bacteroides* is their ability to remove side chains from bile acids, enabling return of bile acids to the hepatic circulation. In effect, our *Bacteroides* population constitutes a functional organ of the human body.

Bacteroides species cause trouble, however, when they reach parts of the body not designed to host them. During abdominal surgery, bacteria can escape the colon and invade the surrounding tissues. The displaced bacteria can form an abscess, a localized mass of bacteria and pus contained in a cavity of dead tissue. The interior of the abscess is anaerobic and often impenetrable to antibiotics.

Chlorobium species, known informally as green sulfur bacteria, are genetically related to *Bacteroides*, but their metabolism is quite different. They are strict photolithotrophs, using PS I to split electrons from H_2, H_2S, or other reduced sulfur compounds (**Fig. 18.42A**; see

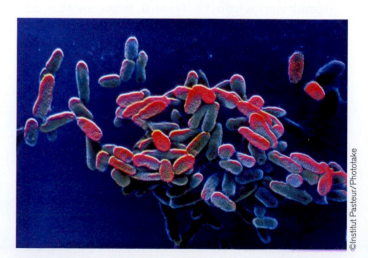

©Institut Pasteur/Phototake

Figure 18.41 *Bacteroides fragilis* **cells colonize the human colon.** *B. fragilis* causes intestinal infections (cell length 1–2 μm; colorized SEM).

A.

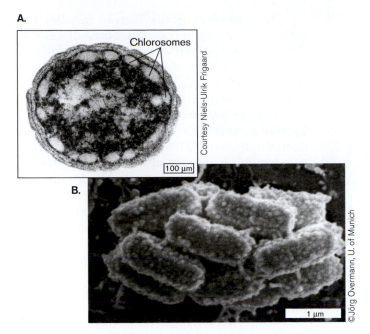

Chlorosomes

Courtesy Niels-Ulrik Frigaard

100 μm

B.

©Jörg Overmann, U. of Munich

1 μm

Figure 18.42 *Chlorobium*: gram-negative "green bacteria." **A.** *Chlorobium tepidum* cell containing chlorosomes. **B.** *Chlorobium* sp. cells covered with sulfur globules; cells form a cluster surrounding a non-phototrophic symbiotic bacterium.

Table 18.2). During the oxidation of sulfide, *Chlorobium* species deposit elemental sulfur extracellularly, forming attached sulfur globules (**Fig. 18.42B**). In contrast, the sulfur granules of Gamma Proteobacteria such as *Chromatium* are deposited internally.

Chlorobium species require extensive membrane systems full of photopigments for light absorption. The chlorophyll reaction centers of *Chlorobium* are contained within chlorosomes associated with the cytoplasmic membrane (**Fig. 18.42**), similar to chlorosomes of the deep-branching phylum Chloroflexi. The photopigment of *Chlorobium* is predominantly Bchl *c*, which absorbs in the blue (460 nm) and near-infrared (750 nm), thus reflecting the middle range, brownish green.

TO SUMMARIZE:

- ■ *Bacteroides* bacteria are anaerobes that ferment complex plant materials in the human colon. They may enter body tissues through wounds and cause abscesses.
- ■ *Chlorobium* bacteria are green sulfur phototrophs, obligate anaerobes incapable of heterotrophy.

18.7 Spirochetes: Sheathed Spiral Cells with Internalized Flagella

A unique phylum of heterotrophic bacteria is Spirochetes (or Spirochaetes). Different species of spirochetes conduct a broad range of heterotrophy, from aerobic to anaerobic; but all spirochetes have in common a distinctive cell structure consisting of a long, tight spiral, flexible like a telephone cord (**Fig. 18.43**). In many species, the spiral is so thin that its width cannot be resolved by bright-field microscopy, and the organisms can pass through a filter of pore size 2 μm.

Cell Structure of Spirochetes

The spirochete cell is surrounded by a thick outer sheath of lipopolysaccharides and proteins. The spirochete sheath is similar to a proteobacterial outer membrane, except that the periplasmic space completely separates the sheath from the plasma membrane. At each end of the cell, one or more polar flagella extend and double back around the cell body within the periplasmic space (**Fig. 18.44**). The periplasmic flagella rotate on proton-driven motors, as do regular flagella; but because they twine back around the cell body, their rotation forces the entire cell to twist around, corkscrewing through the medium.

This corkscrew motion of the cell body turns out to have a physical advantage in highly viscous environments, such as human mucous secretions or agar culture

A.

B.

C.

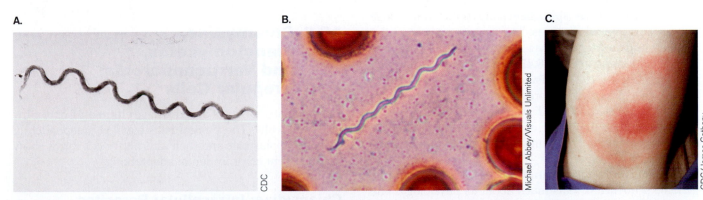

CDC

Michael Abbey/Visuals Unlimited

CDC/James Gathany

Figure 18.43 Spirochetes. A. *Treponema pallidum*: the causative agent of syphilis (cell length 5–10 μm, width 0.2 μm; TEM). **B.** *Borrelia recurrentis* cell in a blood film (stained micrograph). The organism causes relapsing fever. **C.** Lesion caused by tick-borne infection by *Borrelia burgdorferi*, a cause of Lyme disease.

A.

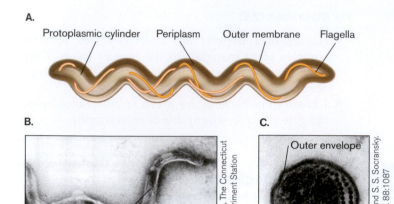

Protoplasmic cylinder   Periplasm   Outer membrane   Flagella

B.

Flagella

0.1 μm

John F. Anderson, The Connecticut
Agricultural Experiment Station

C.

Outer envelope

Axial filament

M. A. Listgarten and S. S. Socransky.
1964. *J. Bacteriol.* 88:1087

Figure 18.44   Spirochete structure.   A. Spirochete cell structure, showing the arrangement of periplasmic flagella. **B.** Spirochete with two periplasmic flagella, isolated from mouse blood (TEM). **C.** Cross section through a spirochete, showing outer envelope and axial filaments (TEM). The unidentified spirochete was obtained from a human lesion of acute necrotizing ulcerative gingivitis.

medium. Few nonspirochetes can swim through agar at standard culture concentrations (1.5% agar); thus, growth within agar provides a way to isolate anaerobic spirochetes from environmental sources. When an environmental sample is inoculated into agar, other bacteria grow and concentrate at the injection point, whereas spirochetes migrate outward in a "veil" through the agar.

Spirochete Diversity

Spirochetes grow in a wide range of habitats, from ponds and streams to the digestive tract of animals. Aquatic systems carry free-living sugar fermenters of the genus *Spirochaeta*. A particularly interesting spirochete community is found in the termite gut, where the organisms form elaborate symbiotic associations with protists and assist digestion of cellulose.

The best-known spirochetes are those that cause human and animal diseases. The causative agent of the sexually transmitted disease syphilis is the spirochete *Treponema pallidum* (**Fig. 18.43A**). The cell of *T. pallidum* is too narrow (0.2 μm) to visualize by bright-field microscopy; either dark-field, fluorescence, or electron microscopy must be used. While *T. pallidum* is a highly virulent pathogen, other members of the genus *Treponema* are normal residents of the human and animal oral, intestinal, and genital regions.

T. pallidum is not culturable in the laboratory, but its genome sequence reveals much about its physiology. The genome of 1.1 million base pairs is highly degenerate, lacking nearly all components of biosynthesis and of the Krebs cycle; its only system for ATP production is glycolysis.

The spirochete genus *Borrelia* includes two pathogens causing serious tick-borne diseases in the United States. *Borrelia recurrentis* causes relapsing fever, and *B. burgdorferi* causes Lyme disease (**Figs. 18.43B** and **C**). *Borrelia* species have been cultured, although they grow very slowly.

A unique feature of *Borrelia* species is their multipartite genomes. Each species possesses a linear main chromosome of less than a million base pairs, plus a number of linear and circular plasmids. *B. burgdorferi*, for example, has a linear chromosome of 910,725 base pairs, with at least 17 linear and circular plasmids that total an additional 533,000 base pairs. The reason for this unusual fragmentation of *Borrelia* genomes is unknown.

Spirochetes include important pathogens of animals, such as *Leptospira*, the cause of leptospirosis, a form of nephritis (kidney inflammation) with complications in the liver and other organs. *Leptospira* cells are known for the peculiar L shape of the cell, as each end of the spirochete turns out at an angle. The genus *Treponema* also includes the causative agent of digital dermatitis in cattle and sheep. As always, however, the pathogenic species are far outnumbered by harmless flora.

TO SUMMARIZE:

- **The spirochete cell** is a tight coil, surrounded by a sheath and periplasmic space containing periplasmic flagella.
- **Motility** occurs by a flexing motion caused by rotation of the periplasmic flagella, propagated the length of the coil.
- **Spirochetes grow in diverse habitats.** Some are free-living fermenters in water or soil. Others are pathogens, such as *T. pallidum*, the cause of syphilis. Still others are endosymbionts of an animal digestive tract, such as the termite gut.

18.8 Chlamydiae, Planctomycetes, and Verrucomicrobia: Irregular Cells

Several related phyla of bacteria have eliminated or diminished cell walls. These organisms evolved independently of the mycoplasmas, and their alternative cell forms show very different environmental adaptations.

Chlamydiae: Intracellular Parasites

In the phylum Chlamydiae, the genera *Chlamydia* and *Chlamydophila* evolved a complex developmental life cycle of parasitizing host cells. *Chlamydia trachomatis* is the causative agent of a major sexually transmitted disease in the United States. It is also a cause of trachoma, an eye disease dating back to records in ancient Egypt. The related species *Chlamydophila pneumoniae* causes pneumonia and has been implicated in cardiovascular disease.

Chlamydiae alternate between two developmental stages with different functions: elementary bodies and reticulate bodies (**Fig. 18.45A**). The form of chlamydia transmitted outside host cells is called an **elementary body**. Elementary bodies resemble endospores in that they are metabolically inert, with a compacted chromosome. While lacking a cell wall, they possess an outer membrane whose proteins are cross-linked by disulfide bonds, making a tough coat that provides osmotic stability. The elementary body adheres to a host cell surface and is endocytosed (**Fig. 18.45B**).

To reproduce, the elementary body must transform itself into a **reticulate body**, named for the netlike appearance of its uncondensed DNA. The reticulate body has active metabolism and divides rapidly, but outside the cell it is incapable of infection and vulnerable to osmotic shock. To complete the infectious cycle, therefore, the reticulate bodies must develop into new elementary bodies before exiting the host. When the host cell lyses, the elementary bodies are released to infect new cells. Chlamydiae infect a wide range of host cell types, from respiratory epithelium to macrophages.

Planctomycetes: A Nucleus-like Compartment

Another cell wall-less group, the planctomycetes, evolved largely as free-living organisms. Planctomycetes are oligotrophs, heterotrophs requiring nutrients at extremely low concentration. They grow in aquatic, marine, and saline environments. Their mechanism of osmoregulation remains poorly understood.

The planctomycete cells possess multiple internal membrane compartments of unknown function (**Fig. 18.46**). An internal membrane just inside the cell membrane divides the cytoplasm into concentric portions. A double membrane surrounds the entire nucleoid, analogous to the double membrane surrounding the eukaryotic nucleus. The purpose of this segregation is unclear, however, as ribosomal translation appears to occur equally both inside and outside the planctomycete "nucleus." It remains

a remarkable example of independent analogous evolution between bacteria and the eukaryotes.

Like eukaryotic protists (discussed in Chapter 20), the planctomycetes have flexible cell bodies that can assume diverse forms. For example, some *Planctomyces* species have rotary flagella, whereas *P. bekefii* cells have stalks that attach to each other to generate a starlike aggregate (**Fig. 18.46C**). Most planctomycetes reproduce by budding, a strategy typical of eukaryotic yeasts.

A.

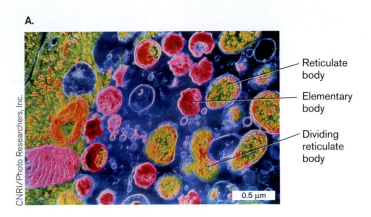

B. *Chlamydia* **developmental cycle**

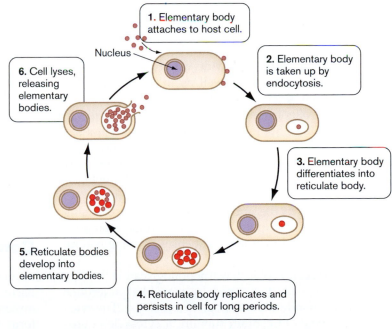

Figure 18.45 *Chlamydia* **life cycle. A.** *Chlamydia trachomatis* multiplying within a human cell (colorized TEM). Infected cell (equivalent to step 5 in part B) contains reticulate bodies (yellow) growing and dividing, as well as newly formed elementary bodies (red). **B.** *Chlamydia* species persist outside the host as a spore-like "elementary body" that is metabolically inactive and contains compacted DNA. Upon endocytosis, the elementary body avoids lysosomal fusion and develops into a "reticulate body" in which the DNA, now uncondensed, has a reticular (netlike) appearance. The reticulate body replicates within the host cytoplasm, then develops into new elementary bodies that are released when the cell lyses.

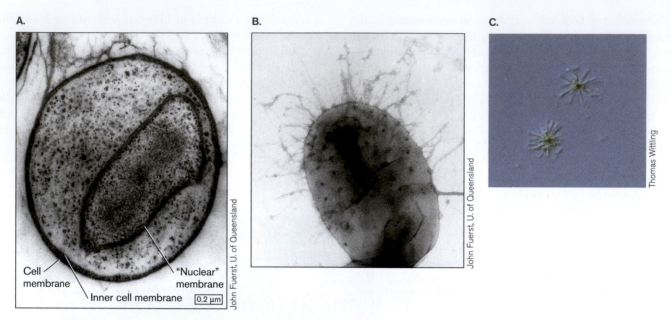

Figure 18.46 *Planctomyces*: bacteria with a "nuclear membrane." **A.** Section through *Gemmata obscuriglobus* showing membrane compartmentalization (TEM). The DNA is contained within a double membrane analogous to a eukaryotic nuclear membrane. **B.** A swarmer cell of *Planctomyces* species with multiple flagella. **C.** *Planctomyces bekefii* cells form stalks that attach to each other at the center to yield a starlike form.

Verrucomicrobia: Wrinkled Microbes

Verrucomicrobia, "wrinkled microbes," are irregularly shaped bacteria found in a wide variety of aquatic and terrestrial environments as well as in the mammalian gastrointestinal tract (see **Table 18.1**). Most verrucomicrobia are oligotrophs, growing heterotrophically in low-salt habitats. Some are ectosymbionts of protists, attached to the cell surface of the eukaryote, where they eject harpoon-like objects. They are rarely cultured, and until recently they failed to show up in PCR amplification of natural isolates because their rDNA sequences are poorly amplified by the standard primer sequences. Using appropriate primer pairs, Verrucomicrobia comprise 5% of all surveyed microbial sequences in some natural environments.

The verrucomicrobia have peptidoglycan cell walls, but their shape is dominated by wart-like projections containing a poorly characterized cytoskeleton. The cytoskeleton appears to contain tubulin, a cytoskeletal protein previously believed to exist only in eukaryotes. In 2002, genes encoding tubulin were found in the partly sequenced genome of *Prosthecobacter dejongeii*, a free-living member of Verrucomicrobia. The genes appear so similar to those of eukaryotes that they must have undergone horizontal transfer from a eukaryotic genome. This horizontal transfer of a eukaryotic trait may be contrasted with the independent development of a nucleus-like structure in planctomycetes.

TO SUMMARIZE:

- **Chlamydiae** are obligate intracellular parasites that undergo a complex developmental progression, culminating in a spore form called an elementary body that can be transmitted outside the host cell. Chlamydiae lack cell walls.
- **Planctomycetes** lack cell walls, and they have evolved a membrane enclosing the nucleoid, analogous to the eukaryotic nuclear membrane.
- **Verrucomicrobia** have cell projections containing tubulin. Their tubulin genes are believed to have arisen through horizontal transfer from a eukaryote.

Concluding Thoughts

Several themes emerge in the diversity of the domain Bacteria. Some phyla, such as Proteobacteria and Actinomycetes, have evolved highly diverse metabolism and cell form. Others show remarkable uniformity in cell structure (Spirochetes) or in metabolism (Cyanobacteria). At the same time, our attempts to generalize about any given clade inevitably run up against the unexpected appearance of exceptions, such as the heliobacteria, photosynthetic endospore formers that unaccountably branch from clostridia. Given the range of bacterial phenotypes, it is hard to imagine that yet more diverse forms of microbial life exist, but they do, as we shall find among the archaea (Chapter 19) and the microbial eukaryotes (Chapter 20).

CHAPTER REVIEW

Review Questions

1. Which deep-branching phyla include hyperthermophiles? Why is the actual branch position of these groups controversial?
2. Compare and contrast Chloroflexi and Cyanobacteria with respect to habitat, cell structure, and means of photosynthesis.
3. Compare and contrast the colonial and filamentous cyanobacteria with respect to life cycle and means of nitrogen fixation.
4. Name and describe three genera of Firmicutes that form endospores. Discuss the difference between single and multiple spore formers. Explain the existence of non-spore-forming species of Firmicutes.
5. Compare and contrast firmicute endospores and actinomycete arthrospores with respect to their means of production, their resistance properties, and their dispersal mechanism.
6. Describe the diverse kinds of metabolism available to different species of Proteobacteria. Explain how it is possible for some species to perform many different kinds of energy-gaining metabolism.
7. What do species of Bacteroidetes and Chlorobi have in common, and how do they differ?
8. Explain the structure and mechanism of motility typical in species of Spirochetes.
9. Discuss the properties of as many intracellular parasites as you recall from various phyla. What do they have in common, and how do they differ?
10. Describe the unique cell structure of *Planctomyces*. Explain how the traits of planctomycete cells appear analogous to aspects of eukaryotic cells.
11. How do microbiologists seek to find previously unknown kinds of bacteria? What techniques reveal unknown microorganisms?

Key Terms

acid-fast stain (699)
Actinobacteria (683)
actinomycetes (684)
actinomycosis (700)
aerial mycelium (700)
akinete (689)
arthrospore (700)
bacteroid (706)
Bacteroidetes (684)
baeocytes (689)
candidate species (685)
carboxysome (688)
Chlamydia (684)
Chlorobi (684)
chlorosome (686)
Cyanobacteria (683)
diplococcus (709)

elementary body (717)
endophyte (700)
endospore (683)
environmental sample (685)
Firmicutes (683)
forespore (693)
fruiting body (712)
gas vesicle (688)
GC content (683)
geosmin (699)
heterocyst (689)
hormogonium (689)
leghemoglobin (706)
metagenome (707)
methanotroph (706)
methylotrophy (706)
motherspore (693)

mycolic acid (700)
myxospore (712)
nitrifier (708)
Nitrospirae (684)
phylum (678)
Planctomycetes (684)
Proteobacteria (684)
proteorhodopsin (704)
reticulate body (717)
sarcina (703)
Spirochetes (684)
swarming (710)
vegetative cell (693)
vegetative mycelium (700)
Verrucomicrobia (684)

Recommended Reading

Anderson, Gregory G., et al. 2003. Intracellular bacterial biofilm-like pods in urinary tract infections. *Science* **301**:105–107.

Andersson, Siv G. E., Alireza Zomorodipour, Jan O. Andersson, Thomas Sicheritz-Pontén, U. Cecilia M. Alsmark, et al. 1998. The genome sequence of *Rickettsia prowazekii* and the origin of mitochondria. *Nature* **396**:133–140.

Angert, Esther. 2006. Beyond binary fission: Some bacteria reproduce by alternative means. *Microbe Magazine* **1**:127–131.

Beatty, J. Thomas, Jörg Overmann, Michael T. Lince, Ann K. Manske, Andrew S. Lang, et al. 2005. An obligately photosynthetic bacterial anaerobe from a deep-sea hydrothermal vent. *Proceedings of the National Academy of Sciences* **102**:9306–9310.

Béjà, Oded, L. Aravind, Eugene V. Koonin, Marcelino T. Suzuki, Andrew Hadd, et al. 2000. Bacterial rhodopsin: Evidence for a new type of phototrophy in the sea. *Science* **289**:1902–1906.

Bern, Marshall, and David Goldberg. 2005. Automatic selection of representative proteins for bacterial phylogeny. *BMC Evolutionary Biology* **5**:34.

Chouari, Rakia, Denis Le Paslier, Patrick Daegelen, Philippe Ginestet, Jean Weissenbach, and Abdelghani Sghir. 2003. Molecular evidence for novel planctomycete diversity in a municipal wastewater treatment plant. *Applied and Environmental Microbiology* **69**:7354–7363.

Clark, Robert W., Nicholas D. Lanz, Andrea J. Lee, Robert L. Kerby, Gary P. Roberts, and Judith N. Burstyn. 2006. Unexpected NO-dependent DNA binding by the CooA homolog from *Carboxydothermus hydrogenoformans*. *Proceedings of the National Academy of Sciences* **103**:891–896.

Daims, Holger, Jeppe L. Nielsen, Per H. Nielsen, Karl-Heinz Schleifer, and Michael Wagner. 2001. In situ characterization of *Nitrospira*-like nitrite-oxidizing bacteria active in wastewater treatment plants. *Applied and Environmental Microbiology* **67**:5273–5284.

D'Hondt, Steven, Bo Barker Jørgensen, D. Jay Miller, Anja Batzke, Ruth Blake, et al. 2004. Distributions of microbial activities in deep subseafloor sediments. *Science* **306**:2216–2221.

Dover, Lynn G., A. M. Cerdeno-Tarraga, M. J. Pallen, J. Parkhill, and Gurdyal S. Besra. 2004. Comparative cell wall core biosynthesis in the mycolated pathogens, *Mycobacterium tuberculosis* and *Corynebacterium diphtheriae*. *FEMS Microbiological Reviews.* **28**:225–250.

Ehrenreich, Armin, and Friedrich Widdel. 1994. Anaerobic oxidation of ferrous iron by purple bacteria, a new type of phototrophic metabolism. *Applied and Environmental Microbiology* **60**:4517–4526.

Favier, Christine F., Elaine E. Vaughan, Willem M. De Vos, and Antoon D. L. Akkermans. 2002. Molecular monitoring of succession of bacterial communities in human neonates. *Applied and Environmental Microbiology* **68**:219–226.

Frigaard, Niels-Ulrik, Asuncion Martinez, Tracy J. Mincer, and Edward F. DeLong. 2006. Proteorhodopsin lateral gene transfer between marine planktonic *Bacteria* and *Archaea*. *Nature* **439**:847–850.

Giovannoni, Stephen J., H. James Tripp, Scott Givan, Mircea Podar, Kevin L. Vergin, et al. 2005. Genome streamlining in a cosmopolitan oceanic bacterium. *Science* **309**:1242–1245.

Jenkins, Cheryl, Ram Samudrala, Iain Anderson, Brian P. Hedlund, Giulio Petroni, et al. 2002. Genes for the cytoskeletal protein tubulin in the bacterial genus *Prosthecobacter*. *Proceedings of the National Academy of Sciences* **99**:17049–17054.

Krinos, Corinna M., Michael J. Coyne, Katja G. Weinacht, Arthur O. Tzianabos, Dennis L. Kasper, and Laurie E. Comstock. 2001. Extensive surface diversity of a commensal microorganism by multiple DNA inversions. *Nature* **414**:555–558.

Schlieper, Daniel, María A. Oliva, José M. Andreu, and Jan Löwe. 2005. Structure of bacterial tubulin BtubA/B: Evidence for horizontal gene transfer. *Proceedings of the National Academy of Sciences* **102**:9170–9175.

Sunnenshine, Rebecca H., and L. Clifford McDonald. 2006. *Clostridium difficile*-associated disease: New challenges from an established pathogen. *Cleveland Clinic Journal of Medicine* **73**:187–197.

Venter, J. Craig, Karin Remington, John F. Heidelberg, Aaron L. Halpern, Doug Rusch, et al. 2004. Environmental genome shotgun sequencing of the Sargasso Sea. *Science* **304**:66–74.

Wang, Jenny, Cheryl Jenkins, Richard I. Webb, and John A. Fuerst. 2002. Isolation of Gemmata-like and isosphaera-like planctomycete bacteria from soil and freshwater. *Applied and Environmental Microbiology* **68**:417–422.

Wu, Martin, Qinghu Ren, A. Scott Durkin, Sean C. Daugherty, Lauren M. Brinkac, et al. 2005. Life in hot carbon monoxide: The complete genome sequence of *Carboxydothermus hydrogenoformans* Z-2901. *PLoS Genetics* **1**:e65.

Zahran, Hamdi H. 1999. *Rhizobium*-legume symbiosis and nitrogen fixation under severe conditions and in an arid climate. *Microbiology and Molecular Biology Reviews 1999* **63**:968–989.

Chapter 19

Archaeal Diversity

Archaea are as different from the "true" bacteria as they are from eukaryotes. These highly diverse organisms include hyperthermophiles, inhabiting the hottest environments on Earth, as well as Arctic and Antarctic psychrophiles growing beneath sea ice and in subglacial lakes. Many of the hyperthermophiles experience extreme pressure or acidity as well as extreme temperature. Most grow without molecular oxygen, as did early life on Earth. Their proposed resemblance to the earliest life-forms earned this group the name Archaea.

Still other kinds of archaea grow in soil or water, even within our own digestive tract. The haloarchaea grow in concentrated brine, coloring salt lakes red. Some have forms of metabolism unique to archaea, such as methanogenesis, the reduction of CO_2 or other small organic molecules to methane. Methanogens grow in the digestive tracts of animals and humans. Yet surprisingly, the archaeal domain seems to lack pathogens. No archaeon has yet been shown to cause disease in animals or plants.

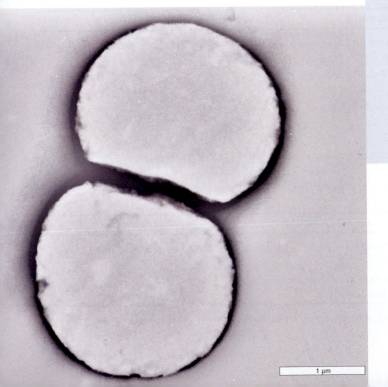

1 μm

A hyperacidophile, the archaeon *Picrophilus torridus* grows at pH 0. It was first isolated from a solfatara (volcanic steam vent) in Hokkaido, northern Japan. *P. torridus* can grow in IM sulfuric acid at temperatures up to 65°C. Its genome was sequenced by Wolfgang Liebl and colleagues at the University of Goettingen. *Source*: O. Fütterer, et al. 2004. Genome sequence of *Picrophilus torridus* and its implications for life around pH 0. *Proceedings of the National Academy of Sciences USA* 101:9091–9096. Photo courtesy of Wolfgang Liebl.

Archaea are prokaryotic microbes best known for their dominance of superheated habitats such as the hot springs of Yellowstone National Park and thermal vents at the ocean floor. Yet genetic surveys based on rRNA genes reveal archaeal species in the coldest habitats as well as the hottest. Ace Lake, Antarctica, is a well-studied cold habitat with temperatures in the range of 14–24°C and bottom anoxic layers never warmer than 2°C (**Fig. 19.1**). The lake supports a wide range of psychrophilic microbes, including methanogens and other archaea. Some of these psychrophilic archaea show surprising relatedness to the sulfur-metabolizing thermophiles found at thermal vents.

Overall, archaeal species grow in habitats at a wider range of temperature and other environmental factors than either bacteria or eukaryotes. Archaea grow in alkaline soda lakes, as well as highly acidic mine runoff streams; within oxygenated hot springs, as well as within the anaerobic human digestive tract. Extremophiles are of interest

Ricardo Cavicchioli. 2006. *Nature Reviews Microbiology* 4:331

Figure 19.1 **Ace Lake, Antarctica.** Psychrophilic archaea and other microbes are studied here.

Table 19.1 Archaeal traits absent from bacteria and eukaryotes.

Traits of archaea	Alternative traits of bacteria and/or eukaryotes	Archaea showing the trait
Cell envelope		
Membrane lipids: isoprenoid L-glycerol ethers or diethers	Membrane lipids: D-glycerol hydrocarbon diesters	All archaea
Membrane lipid chains stiffen by covalent cross-links or by forming pentacyclic rings.	Chains stiffen by unsaturation.	Most archaea
S-Layer of glycoprotein	Murein sacculus	Hyperthermophilic Crenarchaeota
S-Layer of protein, methanochondroitin, or sulfated polysaccharide	Murein sacculus	Methanogens and haloarchaea
Pseudomurein sacculus contains talosaminuronic acid; peptide bridges contain only L-amino acids.	Murein	Methanobacteriales, Methanopyrales
Metabolism		
Nonphosphorylated intermediates of sugar catabolism and synthesis (Embden-Meyerhof glycolysis in some cases)	Embden-Meyerhof pathway of glycolysis, with phosphorylation mediated by NAD or NADP (Entner-Doudoroff in some cases)	All archaea
Methanogenesis from H_2 and CO_2; or from CO, methanol, methyl sulfides, formate, or acetate	Anaerobic metabolism such as fermentation. No methane production.	Methanogens
Archaeal coenzymes: coenzymes F_{420}, F_{430}, and coenzyme M	Flavin mononucleotide, coenzyme A, others	Most methanogens
Retinal-associated light-driven membrane pumps for H^+ or Na^+	Chlorophyll-based photosynthesis	Haloarchaea
Nucleic acid structure and function		
Positive superturns generated by reverse gyrase protect DNA from extreme acid.	Negative superturns generated by gyrase	Hyperthermophilic archaea
Unique base structures in tRNA, such as the guanine analog archaeosine	tRNA bases, such as queosine, found only in bacteria and eukaryotes	Most archaea

Sources: O. Kandler and H. König. 1998. *Cell. Mol. Life Sci.* 54:305–308; C. Bullock. 2000. *Biochem. and Mol. Biol. Ed.* 28:186–191.

to biotechnology companies because their exceptionally thermostable enzymes can catalyze reactions under industrially useful conditions, such as high salt or acidity or temperatures above the boiling point of water.

While dominating certain extreme habitats, archaea are also abundant in "moderate" habitats, such as the open ocean, the soil, and the surface of plant roots. Furthermore, species of archaea display unique traits, such as energy-yielding methanogenesis and cyclic diether membranes, found only in the archaeal domain, not in bacteria or eukaryotes. Perhaps the most compelling unique feature of archaea, from the human point of view, is the complete absence of known archaeal pathogens of animals or plants. A few reports suggest the presence of methanogens associated with periodontal disease, but even in these cases, no causal relationship has been demonstrated. This lack of pathogenicity is particularly striking given the intimate association of many archaea with animal digestive systems, with marine sponges, and with the roots of plants.

In this chapter, we explore the diversity of archaeal form and function. We emphasize the unique structures and metabolism of archaeal cells. While surveying their major taxonomic categories, we introduce research techniques used to study organisms in extreme environments and those with unique forms of metabolism such as methanogenesis.

As we saw for bacteria in Chapter 18, new species of archaea continue to be discovered daily. For more detailed coverage of particular species, we recommend the following online resources: Microbial Biorealm, and the National Center for Biological Information (NCBI) Taxonomy Database.

19.1 Archaeal Traits and Diversity

The Archaea show key features unique to this domain, as well as intriguing traits shared by eukaryotes, such as transcription factors required by eukaryotic polymerase II. Archaea show distinctive forms, from intricate cell networks to square-shaped cells. Here we summarize key traits of archaea, then outline the major phylogenetic divisions.

Archaeal Cell Structure and Metabolism

Distinctive features of archaea, sometimes called "archaeal signatures," include components of the cell membrane and envelope and certain metabolic pathways to gain energy (**Table 19.1**).

Isoprenoid membranes and cell wall. The most distinctive structure of archaea is their membrane

(**Fig. 19.2**). The membrane lipids of archaea differ profoundly from those of bacteria and eukaryotes, with the exception of a few thermophilic bacteria that show evidence of horizontal transfer of genes conferring synthesis of archaeal-type lipids. Archaeal lipid structure includes:

- **L-Glycerol.** Archaeal membrane lipids incorporate L-glycerol, rather than the mirror-symmetrical form D-glycerol, which is used by bacteria and eukaryotes. The two chiral forms show similar thermal stability, but their biochemistry requires different enzymes,

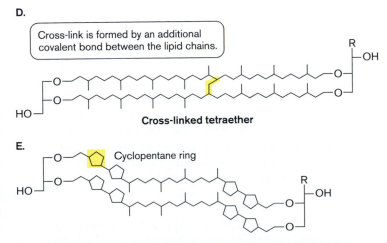

Figure 19.2   Branched-chain ether lipids are characteristic of archaeal membranes.   Archaeal membranes show diverse forms of cyclization that offer different levels of protection from extremely high temperature or acidity. *Source*: R. M. Daniel and D. A. Cowan. 2000. *Cellular and Molecular Life Sciences* 57:250–264.

and thus they represent a deep divergence in ancestry. In some archaea, the glycerol is extended by six carbons, forming nonitol (nine OH groups).

- **Ether linkage.** The glycerol units are linked to side chains by ether links (R–O–R) instead of the ester links (R–COO–R) found in bacteria and eukaryotes (**Fig. 19.2A**). Ether links are much more stable than esters; in other words, their breakage requires greater input of energy.

- **Isoprenoid chains.** The side chains of archaeal lipids are branched at every fourth carbon. The methyl branches arise by condensation (C–C bond formation) of units of isoprene (see **Fig. 19.2A**). Condensed isoprene chains are called **isoprenoid** or diphytanyl chains; thus, the overall lipid is diphytanylglycerol diether. Isoprenoid branched chains increase membrane stability.

- **Tetraether-linked lipids.** In some hyperthermophiles, the ends of side chains are linked covalently, either to each other (**Fig. 19.2B**) or to a lipid on the opposite side of the membrane (**Fig. 19.2C**). Two lipid chains crosslinked across the membrane form a **tetraether**, so called because the complex contains four ether links in all. In some cases, an additional covalent bond links the two linked pairs of side chains across the middle (**Fig. 19.2D**).

- **Cyclopentane rings.** Unlike bacteria and eukaryotes, archaea lack unsaturated side chains that decrease mobility. Instead, in some species, the lipid's methyl branches cyclize, forming cyclopentane rings (**Fig. 19.2E**).

Archaea show distinctive envelope structures, such as the cell wall and S-layer. Some methanogens possess a cell wall comparable to the bacterial sacculus, a unimolecular cage of sugar chains linked by peptide bridges (discussed in Chapter 3). But the sacculus of methanogens is composed not of murein, but of a related macromolecule called **pseudomurein** or **pseudopeptidoglycan**. Pseudopeptidoglycan contains chains of alternating sugar derivatives including N-acetylglucosamine, as in bacterial peptidoglycan. The bacterial N-acetylmuramic acid, however, is replaced by a related sugar, N-acetyltalosaminuronic acid; and the sugar linkage is $\beta(1,4)$ instead of $\beta(1,3)$. As a result, these archaea are resistant to lysozyme, which degrades bacterial cell walls at the sugar linkage. The peptide cross-bridges of pseudopeptidoglycan differ as well, causing resistance to penicillin.

Other species of archaea, especially members of Crenarchaeota, possess no cell wall at all, only an S-layer composed of proteins. The shape of these cells is more flexible than that of most bacterial cells, which maintain turgor pressure against a sacculus. How archaeal cells maintain turgor pressure without a cell wall is unknown.

Unique metabolic pathways. Archaea show many distinctive metabolic pathways. Glucose is catabolized by several variants of the Entner-Doudoroff (ED) and Embden-Meyerhof-Parnas (EMP) pathways that rarely or never occur in bacteria (**Fig. 19.3**). For example, the sulfur thermophiles *Sulfolobus* and *Thermoplasma* convert glucose to gluconate without phosphorylation, ultimately generating pyruvate with no net production of ATP. On the other hand, halophilic archaea such as *Halobacterium* phosphorylate the dehydrated product of gluconate (2-oxo-3-deoxygluconate), thus channeling into the "standard" ED pathway. The ED pathway generates one molecule of pyruvate and one molecule of 3-phosphoglycerate, which produces one net ATP through the second stage of the EMP pathway. A unique variant of the EMP pathway is seen in the vent thermophile *Pyrococcus furiosus*, which oxidizes glyceraldehyde 3-phosphate using ferredoxin instead of NAD and avoids phosphorylation. The reduced ferredoxin is then used to reduce $2H^+$ to H_2 in an energy-yielding reaction.

The energy-yielding process of **methanogenesis** occurs only in archaea. Methane production by methanogenic archaea (methanogens) makes a growing contribution to global warming, as discussed in Chapter 22. Methanogenesis includes a unique pathway of carbon fixation, called the **carbon monoxide reductase pathway** because the key enzyme can fix CO as well as CO_2. A CO group is fixed into acetyl-CoA by condensation with a methyl group generated by methanogenesis. Acetyl-CoA then enters the TCA cycle. Methanogenesis also uses unique cofactors that largely replace NAD, FAD, and FMN (discussed in Section 19.4).

The only form of phototrophy in archaea is that of retinal-based ion pumps. Membrane protein complexes known as **bacteriorhodopsin** and halorhodopsin contain retinal and pumps for H^+ or Na^+ (discussed in Chapter 14). Bacteriorhodopsins are found in haloarchaea (halobacteria) such as *Halobacterium halobium*; they were named before *Halobacterium* was known to be an archaeon (discussed in Section 19.5). By contrast, the chlorophyll-based phototrophy found in bacteria and plants is completely unknown in archaea. Light-driven ion pumps were believed to be unique to archaea, until bacteriorhodopsin homologs were discovered in marine bacteria. These bacterial H^+ pumps, called proteorhodopsins, are believed to have evolved by horizontal gene transfer from archaea.

Nucleic acid structure. A distinctive feature of archaeal chromosome function is the "reverse gyrase" enzyme found in hyperthermophiles. All bacteria and eukaryotes have gyrase to maintain their DNA in an "underwound" negatively supercoiled state, but hyperthermophiles with reverse gyrase maintain positive supercoiling. The positive superturns "overwind" the DNA, enhancing stability and preventing the helix from melting into separate strands at high temperature.

Figure 19.3 Glucose catabolism in archaea. *Sulfolobus* and *Thermoplasma* species catabolize glucose to pyruvate by a modified ED pathway without phosphorylating glucose and produce no net ATP. *Halobacterium* species phosphorylate 2-keto-3-deoxygluconate and produce 1 net ATP by the EMP stage 2 pathway. *Pyrococcus furiosus* oxidizes glyceraldehyde 3-phosphate using ferredoxin instead of NAD$^+$ and avoids phosphorylation. *Source*: Hatim Ahmed, et al. 2005. *Biochemistry Journal* 390:529–540.

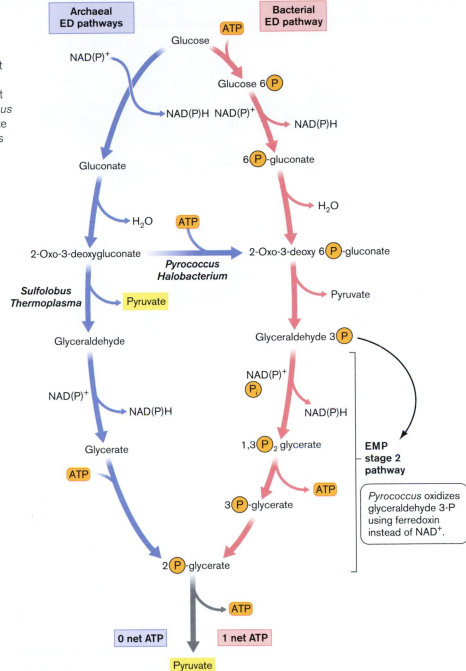

Archaea also have distinctive modified bases in their tRNA molecules. In particular, the guanosine analog archaeosine (7-formamidino-7-deazaguanosine) is used by nearly all archaea but no bacteria or eukaryotes. Other unusual tRNA bases, such as queuosine, are found only in bacteria and eukaryotes, not in archaea.

Gene Regulation Resembles That of the Eukaryotic Nucleus

The genomes of archaea generally resemble those of bacteria in size and gene density, and genes of related func-tion are often arranged in operons. Certain tRNA gene sequences, however, are interrupted by nontranscribed sequences called introns, similar to the tRNA introns found in eukaryotes. Furthermore, the archaeal appara-tus for DNA and RNA polymerases, transcription factors, and protein synthesis show remarkable similarity to those of eukaryotes. **Figure 19.4A** compares the components of an archaeal RNA polymerase (from *Sulfolobus*) with those of eukaryotic RNA polymerase II. The archaeal polymerase possesses two transcription factors (regula-tory protein components) that are found in eukaryotes: TATA-binding protein (TBP) and transcription factor B

A.

Archaeal RNA polymerase

Eukaryotic RNA polymerase II

B.

John Reeve

C.

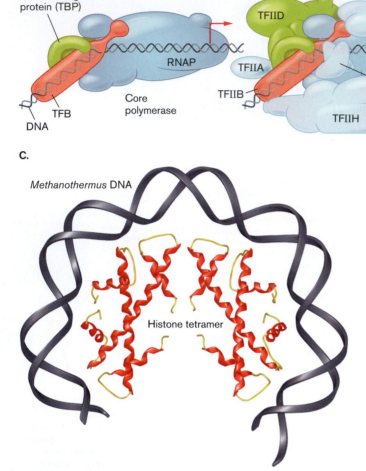

Methanothermus DNA

Histone tetramer

Figure 19.4   Central genetic molecules of Archaea resemble those of Eukarya.   A. The core RNA polymerase subunits (RNAP) of the archaeon *Sulfolobus* show homology to those of the eukaryotic RNA polymerase II (RNAPII). The TATA-binding protein (TBP) and transcription factor B (TFB) also show homology to eukaryotic counterparts. **B.** The laboratory research group of John Reeve (left), at Ohio State University, working on the molecular biology of methanogens. **C.** Archaeal histones are found in Euryarchaeota such as the methanogens. *Methanothermus fervidus* DNA binds to histone tetramers that share descent with the histones of eukaryotes. The histone tetramer binds to specific AT-rich sequences. The peptide sequences of the archaeal histones show significant homology to the sequence of a eukaryotic histone. (PDB code: 1A7W)
Source: Kathryn A. Bailey, et al. 2002. *PNAS* 277:9293.

(TFIIB). By contrast, bacteria have no homologs of these factors. A consequence of the eukaryotic-like transcription and translation in archaea is that archaea are resistant to antibacterial antibiotics that target transcription and translation.

Another eukaryotic structure for which archaeal homologs were discovered is that of **histones**, the fundamental packaging proteins of DNA. The histone complex found in eukaryotic chromosomes contains a histone $(H3 + H4)_2$ tetramer flanked by two histone (H2A + H2B) dimers. The archaeal homologs form a $(H3 + H4)_2$ tetramer, with no H2A + H2B homologs. John Reeve and colleagues at Ohio State University (**Fig. 19.4B**) showed that isolated DNA of specific sequences can be bound and curved around a histone tetramer from the methanogen *Methanothermus fervidus* (**Fig. 19.4C**). The DNA has to bind AT-rich sequences that specifically fit the histone complex. The key sequences of the histones (the ones that bind to DNA) show homology to eukaryotic histones. Histones have since been found in many species of archaea.

Phylogeny of Archaea

The domain Archaea includes several deeply branching phyla, which are also called divisions (**Fig. 19.5**). The two best-characterized phyla are **Crenarchaeota** (known for sulfur thermophiles and marine mesophiles) and **Euryarchaeota** (known for methanogens, halophiles, and extreme acidophiles). Major groups of species are outlined in **Table 19.2**.

Crenarchaeota. The Crenarchaeota include well-studied thermophiles that metabolize sulfur, either by anaerobic reduction (such as by H_2 to form H_2S) or by aerobic oxidation (by O_2 to form sulfuric acid). Anaerobic sulfur metabolizers include moderate thermophiles (growth range about 60–80°C) as well as hyperthermophiles (90–120°C). Many of the hyperthermophiles are also barophiles, growing under high pressure at hydrothermal vents on the ocean floor; an example is *Pyrodictium abyssi*. In addition, however, ribosomal RNA gene probes reveal numerous unculturable Crenarchaeota species that are mesophiles (grow at medium temperature) or psychrophiles (grow at low temperature). Thus, the species of Crenarchaeota encompass the widest range of growth temperature of any division of life.

Figure 19.5   Archaeal divergence.
Major clades diverged within the phyla (divisions) Crenarchaeota and Euryarchaeota. Other phyla are less well characterized. Divergence was measured based on small-subunit rRNA (16S rRNA) sequences.
Sources: Susan Barns, et al.1996. *PNAS* 93:9188; O. Nercessian, et al. 2003. *Environ. Microbiol.* 5:492; E. F. DeLong, et al. 1994. *Nature* 371:695.

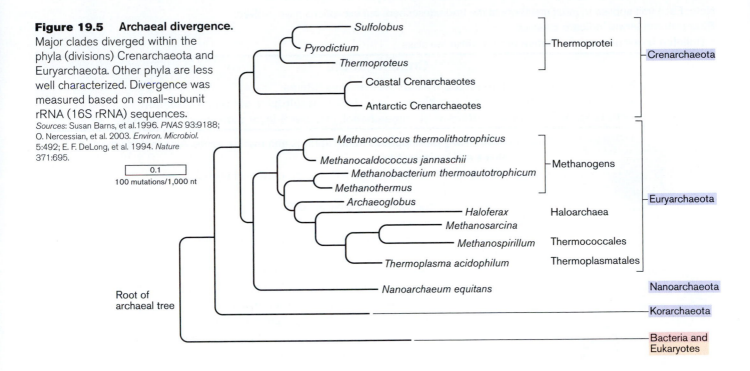

Euryarchaeota. The phylum Euryarchaeota has a broader range of metabolism than Crenarchaeota. The most highly divergent group of Euryarchaeota, and the group most widespread in natural ecosystems, is the methanogens. Methanogens serve a key energetic role in ecosystems by offering an anaerobic mechanism to remove excess H_2 and other small-molecule reductants. Despite their common energetic pathway, methanogens show a wide range of cell form and environmental adaptations. One methanogenic clade branches to the Haloarchaea, extreme halophiles that are the only form of life to grow in concentrated brine (NaCl). Most haloarchaea, such as *Halobacterium* species NRC-1, are photoheterotrophs that can supplement their metabolism with light-driven ion pumps. Still other genera, such as *Archaeoglobus*, show mixed physiology, including components of methanogenesis linked to other pathways, such as glycolysis. The euryarchaeotes include mesophiles as well as thermophiles, such as the order Thermococcales, but few psychrophiles have been found so far. With respect to pH, the euryarchaeotes show the widest range of any clade, from *Ferroplasma* (the order Thermococcales) growing at pH 0 to *Natronococcus* growing at pH 10.

Emerging phyla. Besides the two major phyla Crenarchaeota and Euryarchaeota, we continue to discover new isolates that branch near the root of the tree. Most such isolates are unculturable, known only by their rRNA sequence. One deeply branching group obtained from a Yellowstone hot spring was designated **Korarchaeota**. Another group, **Nanoarchaeota**, includes the tiny hyperthermophile *Nanoarchaeum*, which grows symbiotically

on the larger cells of a crenarchaeote, *Ignicoccus*. A genome for *Nanoarchaeum* has been sequenced, and it has proved to be one of the smallest of archaeal genomes.

> **THOUGHT QUESTION 19.1**   If two deeply diverging clades each show a wide range of growth temperature, what does this suggest about the evolution of thermophily or psychrophily?

TO SUMMARIZE:

- **Archaeal membranes are composed of L-glycerol diether or tetraether lipids**, with isoprenoid side chains that may include cross-links or pentacyclic rings.
- **Many archaea have no cell wall, only an S-layer.** Some methanogens have cell walls of pseudomurein.
- **Glucose is catabolized by variants of the ED pathway.** Other metabolic pathways found in archaea include methanogenesis and retinal-associated light-driven ion pumps such as bacteriorhodopsin.
- **Central genetic functions of archaea resemble those of eukaryotes**, as seen in the structure of DNA and RNA polymerases and of histone-like DNA-binding proteins.
- **Two major phyla or divisions of archaea are Crenarchaeota and Euryarchaeota.** Crenarchaeota includes sulfur hyperthermophiles as well as mesophiles, while Euryarchaeota includes methanogens, halophiles, and acidophiles. Each group includes thermophiles as well as mesophiles.

Table 19.2 Representative groups of archaea.

Note: Each trait applies to *most* members of the taxon described, but exceptions have evolved.
Blue-lettered terms indicate classes.
• **Bulleted terms are representative orders** within the class.

Crenarchaeota

Temperature: Hyperthermophiles and psychrophiles, as well as mesophiles
Metabolism: Sulfur and hydrogen oxidation, aerobic and anaerobic heterotrophy, ammonia oxidation
Envelope: Membrane lipids include tetraethers with crenarchaeol. Flexible S-layer surrounds membrane.

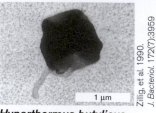

1 µm

Zillig, et al. 1990. *J. Bacteriol.* 172(7):3959

Hyperthermus butylicus

1 µm

Oliver Meckes/Nicole Ottawa

Sulfolobus sp.

Thermoprotei. Many thermophiles in hot springs and marine vents. Also include marine mesophiles and psychrophiles.
• **Caldisphaerales.** Thermoacidophilic heterotrophs grow in hot springs.
  Caldisphaera spp.
• **Cenarchaeales.** Sponge symbionts; grow at 10°C.
  Cenarchaeum symbiosum.
• **Desulfurococcales.** Anaerobic sulfur reduction with organic electron donors. Irregularly shaped cells with glycoprotein S-layer; no cell wall.
  Aeropyrum pernix, Desulfurococcus mobilis, Hyperthermus butylicus.
  Pyrodictium abyssi and *P. occultum* grow at marine thermal vents, up to 110°C. Form three-dimensional network of cells and extracellular cannulae.
  Pyrolobus fumarii grows at marine thermal vents, up to 113°C.
  Ignicoccus islandicus. Unique periplasmic space contains membrane vesicles.
  Thermosphaera. Heterotrophic; inhibited by sulfur.
• **Nitrosopumilales.** Ammonia oxidizers, marine.
  Nitrosopumilus maritimus.
• **Sulfolobales.** Aerobic acidophiles; moderate thermophiles. Oxidize H_2S to H_2SO_4.
  Sulfolobus, Sulfurisphaera, Acidianus.
• **Thermoproteales**
  Pyrobaculum, Thermoproteus, Vulcanisaeta.
Marine pelagic plankton. Mesophiles of open ocean; includes heterotrophs and autotrophs; uncultured and largely uncharacterized.
Psychrophilic marine Crenarchaeota. Abundant in Antarctic ocean and in Arctic sea ice.
Marine benthic anaerobes. Anaerobic heterotrophs in seafloor sediment; psychrophiles. Some reduce sulfate; others conduct anaerobic methanotrophy (oxidation of methane) in association with methanogens.
Soil and plant root–associated Crenarchaeota. Many uncharacterized species.

Euryarchaeota

Temperature: Mesophiles and thermophiles; fewer psychrophiles
Metabolism: Methanogenesis, halophilic photoheterotrophs, and sulfur and hydrogen oxidizers; acidophiles and alkaliphiles
Envelope: Methanogens and halophiles have rigid cell walls of glycans, glycoproteins, or pseudopeptidoglycan.

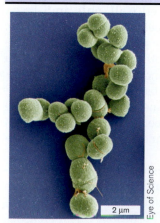

2 µm

Eye of Science

Halobacteriales

Archaeoglobi. Hyperthermophiles; metabolize sulfur and use reverse methanogenesis pathway.
• **Archaeoglobales.** Sulfate oxidation of H_2 or organic hydrogen donors. *Archaeoglobus fulgidus.*

Halobacteria (Haloarchaea). Halophiles; grow in brine (concentrated NaCl). Conduct photoheterotrophy, with light-driven H^+ pump and Cl^- pump.
• **Halobacteriales**
  Haloarcula, Halobacterium, Halococcus, Haloferax. Mesophiles at neutral pH. Grow in salterns and salt lakes. *Haloarcula marismortui; Halobacterium salinarum.*
  Haloquadra (Haloquadratum). Square-shaped mesophilic halophiles.
  Halorubrum lacusprofundi. Psychrophiles grow in subglacial Antarctic salt lakes.
  Natronococcus, Natronomonas. Alkaliphiles, grow above pH 9 in soda lakes.

Table 19.2 **Representative groups of archaea (*continued*)**

Methanogens (four classes). Generate methane from CO_2 and H_2, formate, acetate, other small molecules. Strict anaerobes. Must associate with bacteria producing their substrates.

Cell walls of pseudopeptidoglycan or polysaccharides.

Wide variety of shapes: rods, filaments, cocci, and coccoid clusters.

Wide range of anaerobic habitats: wetlands, landfills, and intestinal tracts of animals. Psychrophiles in deep-ocean floor sediment generate methane hydrates.

- **Methanobacteriales.** Lacks cytochromes; reduces CO_2, formate, or methanol with H_2.
  Methanobacterium spp. have peptidoglycan cell walls.
  Methanobrevibacter smithii and *Methanosphaera stadtmanae* inhabit human digestive tract. Increase caloric output of human food by reducing methanol, a breakdown product of pectin.
  Methanothermus fervidus. Marine vent thermophiles.
- **Methanomicrobiales, Methanococcales, and Methanopyrales.** Lack cytochromes; reduces CO_2, formate, or methanol with H_2.
  Methanoculleus nigri.
  Methanocaldococcus jannaschii. Vent thermophile.
  Methanogenium frigidum. Psychrophile.
  Methanopyrus kandleri. Marine vent thermophile.
- **Methanosarcinales.** Possesses cytochromes; reduces methylamines and acetate (as well as CO_2 and formate) with H_2.
  Methanosarcina acetivorans. Globular colonies; cell walls of sulfated polysaccharides.
  Methanosaeta concilii. Filamentous colonies.
  Methanococcoides alaskense. Arctic marine habitat.
  Methanohalophilus. Halophile. Inhabits salt lakes, often at high pH.

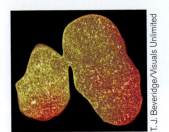

Methanoculleus nigri

T. J. Beveridge/Visuals Unlimited

Methanosarcina mazei

©Ralph Robinson/Visuals Unlimited

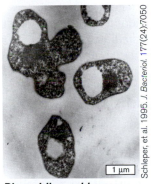

Picrophilus oshimae

Schleper, et al. 1995. *J. Bacteriol.* 177(24):7050

1 μm

Thermococci. Hyperthermophiles (grow above 100°C) and barophiles (up to 200 atmospheres pressure). Anaerobes; reduce sulfur.

- **Thermococcales.**
  Pyrococcus abyssi, P. furiosus. Use S^0 to oxidize H_2 or organic hydrocarbons. *Thermococcus* species are closely related.

Thermoplasmata. Extreme acidophiles; oxidize sulfur from pyrite (FeS_2), generating sulfuric acid. Mesophiles or moderate thermophiles.

- **Thermoplasmatales.**
  Ferroplasma acidiphilum and *F. acidarmanus* grow at 37–50°C. Oxidize sulfur from FeS_2, generating ambient pH as low as pH 0. No cell wall.
  Picrophilus torridus grows above 60°C. Oxidizes sulfur, generating acid. Possesses cell wall.
  Thermoplasma acidophilum grows at 59°C and pH 2.

Korarchaeota

Deep-branching hyperthermophiles

Korarchaeote OPF1-KOR; SRI-306. rRNA sequences branch near the root of the archaeal tree. Original isolate was from Yellowstone hot springs.

Nanoarchaeota

Hyperthermophilic symbionts; cell size less than 400 μm

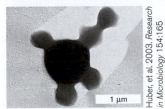

Nanoarchaeum equitans (attached to surface of *Ignicoccus* sp.)

Huber, et al. 2003. *Research in Microbiology* 154:165

1 μm

Nanoarchaeum equitans. Grows attached to cells of *Ignicoccus* sp. rRNA sequence deeply branches from common ancestor of archaea. Small genome contains only 490 kb. Grows in hydrothermal vents, up to 98°C. Uncultured; metabolism unknown.

19.2 Crenarchaeota: Hyperthermophiles

The name Crenarchaeota means "scalloped archaea," derived from the amorphous scalloped cell shapes of the first hyperthermophiles discovered. Crenarchaeotes synthesize a distinctive tetraether lipid, called crenarchaeol, containing occasional six-membered cyclic rings (**Fig. 19.6**). Crenarchaeol is found in varying amounts in all known crenarchaeotes; thus, the molecule serves as a physiological biosignature for this division.

The best-studied crenarchaeotes are members of the class Thermoprotei, which includes sulfur oxidizers, aerobic sulfur reducers, and heterotrophs (**Table 19.2**). The first members described were thermophiles; mesophiles and psychrophiles are now known as well.

Figure 19.7   Thermophiles colonize the edge of a hot spring.   Morning Glory hot springs, Yellowstone National Park. Yellow-orange regions indicate growth of thermophiles.

Habitats for Thermophiles

Thermophiles and hyperthermophiles commonly grow in hot springs and geysers, such as those of Yellowstone National Park (**Fig. 19.7**) or the Solfatara volcanic area near Naples, Italy. A hot spring occurs where water seeps underground above a magma chamber, which heats the water to near boiling. The heated water expands and is forced upward through fissures, coming out in a heated spring. In a geyser, the water is heated under pressure. As the water escapes upward, it turns into steam, which expands and jets upward, falling into a heated pool. These heated pools and their surrounding edges generate extreme ranges of temperature, mineral content, and acidity. They support a diverse range of microbial life, including thermophilic cyanobacteria and firmicutes as well as archaea.

Several features of hot springs and geysers are important for thermophiles.

- **Reduced minerals.** The heated water dissolves high concentrations of sulfides and other reduced minerals. When the water emerges and cools, the minerals precipitate. These reduced minerals serve as rich energy sources for autotrophs.
- **Low oxygen content.** At higher temperatures, the oxygen concentration of water is decreased. Therefore,

hyperthermophiles tend to be anaerobic, although there are important exceptions, such as *Sulfolobus*.

- **Steep temperature gradients.** The temperature of the water falls dramatically within a short distance from the source, creating a steep gradient. Different species of thermophiles are adapted to different temperatures and grow in separate patches at the different temperatures, causing a variegated pattern.
- **Acidity.** In some cases, hot-spring environments show extreme acidity. The acidity results from oxidation of sulfur or iron in reactions that generate strong inorganic acids such as sulfuric acid (H_2SO_4).

A major subcategory of volcanic hot-spring habitats is that of submarine hydrothermal vents on the ocean floor. So-called "vent" thermophiles must evolve adaptations to high pressure under several kilometers of ocean. Pressure increases by approximately 100 atm per kilometer of ocean depth. Organisms that only grow at high pressure are called **barophiles** (discussed in Chapter 5).

Desulfurococcales: Reducing Sulfur from Hot Springs

Species of the order Desulfurococcales show distinctive cell structures and forms of metabolism (**Fig. 19.8A; Table 19.3**). All possess elaborate S-layers but lack a cell wall; their diphytanylglycerol membranes contain a combination of diethers and tetraethers. Many take advantage of the high temperatures that increase the thermodynamic favorability of sulfur redox reactions. An example is *Desulfurococcus mobilis*, a flagellated coccoid cell isolated from hot springs; it grows optimally at 85°C (see **Fig. 19.8A**). *D. mobilis* respires anaerobically by reducing elemental sulfur (S^0) to sulfide (HS^-). Sulfur

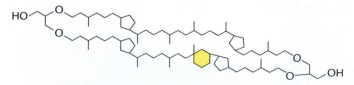

Figure 19.6   Crenarchaeol: a biosignature for crenarchaeotes.   Crenarchaeol is a glycerol diphytanyl diether containing a six-membered cyclic ring.

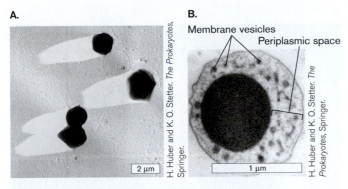

A.

B.
Membrane vesicles
Periplasmic space

H. Huber and K. O. Stetter. *The Prokaryotes,* Springer.

2 μm

1 μm

H. Huber and K. O. Stetter. *The Prokaryotes,* Springer.

Figure 19.8   Hyperthermophilic crenarchaeotes.
A. *Desulfurococcus mobilis* (shadow EM). **B.** *Ignicoccus islandicus* (TEM). Its unique periplasmic space contains membrane vesicles.

reduction is coupled to oxidizing small organic molecules such as sugars.

Another flagellated coccus, *Ignicoccus islandicus*, has an unusual cell architecture (**Fig. 19.8B**). *Ignicoccus* possesses an outer membrane surrounding its cytoplasmic membrane, with a large periplasmic space between them. The periplasmic space contains membrane-bound vesicles of unknown function. No other species is known to have a periplasmic space containing vesicles. The evolution of this structure, similar to the extra membranes of the bacterial genus *Planctomyces* (see Chapter 18), suggests a model for an intermediate stage of evolution of the eukaryotic nucleus.

Ignicoccus islandicus, unlike *Desulfurococcus*, is a marine organism, growing at temperatures as high as 98°C. It is a lithotroph, oxidizing hydrogen with sulfur:

$$H_2 + S^0 \longrightarrow H_2S$$

Most of the cultured species of Desulfurococcales are obligate anaerobes. An exception is *Aeropyrum pernix*, one of the first archaea whose genome was sequenced. *A. pernix* is an aerobic heterotroph, respiring with O_2 on complex compounds during growth at 70–100°C.

Barophilic hyperthermophiles. The most extreme hyperthermophiles are barophiles adapted to grow near hydrothermal vents at the ocean floor. The high pressure beneath several kilometers of ocean allows water to remain liquid at temperatures above 100°C; the highest known temperature for growth of an organism is 121°C. Isolation and study of such organisms require specialized research apparatus (**Special Topic 19.1**).

A common feature of thermal vents is the **black smoker** (**Fig. 19.9**). A black smoker is a chimneylike structure resulting from the upwelling of seawater superheated by an undersea magma chamber. As in a geyser aboveground, the heated water is forced upward through a small opening. Because the thermal vent is under steam pressure, the water can reach temperatures of over 400°C, enabling it to dissolve high concentrations of minerals such as iron II sulfide (FeS). When the rising water escapes, however, it immediately cools, depositing iron sulfide around the edge of the vent chimney and precipitating iron sulfide particles that cloud the water; hence the term *black smoker*. While no organism can grow at 400°C, various species of archaea are adapted to grow in the range of 100–120°C, where the vent stream meets the seawater and minerals precipitate (**Fig. 19.9B**).

Vent-adapted crenarchaeotes include *Pyrodictium abyssi* (**Fig. 19.10**), *P. occultum*, and *P. brockii*; the latter is named for Thomas Brock of the University of Wisconsin, Madison, a pioneering researcher of hyperthermophiles.

Table 19.3   Hyperthermophilic Crenarchaeota.

Species	Growth temperature (°C)	Growth pH	Cell shape	Metabolism
Aeropyrum pernix	70–100°	pH 5–9	Cocci	O_2 respiration
Desulfurococcus mobilis	78–87°	pH 6	Flagellated cocci	Anaerobic S^0 respiration or fermentation
Ignicoccus islandicus	70–98°	pH 5–7	Flagellated cocci with periplasmic space	Anaerobic lithotrophy, S^0 oxidation of H_2
Thermosphaera aggregans	65–90°	pH 5–7	Flagellated cocci in aggregates	Anaerobic fermentation
Pyrodictium brockii	85–110°	pH 5–7	Disks linked by cannulae	Anaerobic oxidation of H_2 by S^0 or $S_2O_3^{2-}$ or fermentation
Pyrodictium abyssi	80–110°	pH 5–7	Disks linked by cannulae	Anaerobic fermentation
Sulfolobus solfatericus	50–87°	pH 2–4	Irregular cocci	O_2 respiration on S^0, producing H_2SO_4

A.

Michael Perfit, U. of Florida, and NOAA VENTS program

Figure 19.9 **Extreme temperature and pressure: black smoker vents.** **A.** Black smoker vents with steam escaping from "chimneys" of sulfide minerals at the Juan de Fuca Ridge on the ocean floor. **B.** Different parts of the smoker vent system support different classes of archaea.

B.

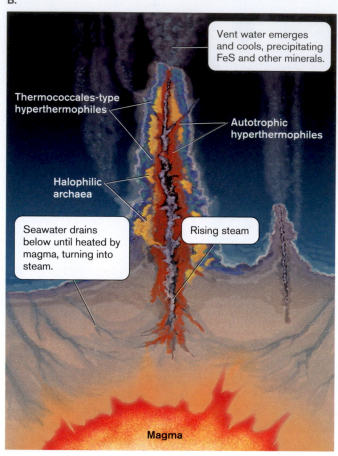

Vent water emerges and cools, precipitating FeS and other minerals.

Thermococcales-type hyperthermophiles

Autotrophic hyperthermophiles

Halophilic archaea

Seawater drains below until heated by magma, turning into steam.

Rising steam

Magma

For energy, *Pyrodictium* species reduce sulfur to H₂S, either with molecular hydrogen or with organic compounds. A membrane-bound sulfur-reducing complex and a proton-translocating ATPase have been isolated from *P. abyssi*. The complexes are extremely heat stable, exhibiting a temperature optimum of 100°C.

Pyrodictium species grow as flat, disk-shaped cells that can be as thin as 0.1 μm. The cells contain a periplasm and outer membrane with an S-layer that is coated with zinc sulfide, presumably precipitated from the vent minerals. The cell disks are interconnected by cytoplasmic extensions called **cannulae**. The cannulae can extend to more than 0.1 mm, forming complex networks of connections (**Fig. 19.10A**); in liquid culture, the networks grow into white balls up to 10 mm in diameter. A thin-section TEM of a *Pyrodictium* cell appears to show that the cannulae bridge the cytoplasm between cells (**Fig. 19.10B**).

What happens when a *Pyrodictium* cell divides? Cells of *P. abyssi* generate new cannulae as they undergo fission (**Fig. 19.11**). Some of the new cannulae form as loops connecting the two daughter cells while simultaneously

A.   **B.**

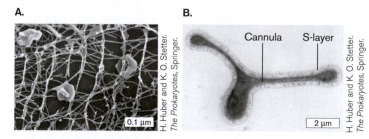

Cannula   S-layer

H. Huber and K. O. Stetter. *The Prokaryotes*, Springer.

Figure 19.10 *Pyrodictium abyssi* **growing as networks of cells linked by cannulae.** **A.** *P. abyssi* (SEM). **B.** Thin section through three linked cells of *P. abyssi* (TEM).

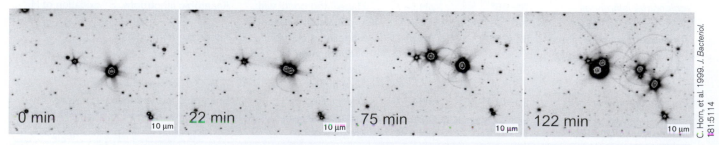

0 min   22 min   75 min   122 min

C. Horn, et al. 1999. *J. Bacteriol.* 181:5114

Figure 19.11 *Pyrodictium abyssi* **undergoing cell division.** Cells of *P. abyssi* generate new interconnecting cannulae as they divide.

Special Topic 19.1　Research on Deep-Sea Hyperthermophiles

To study hyperthermophiles from black smoker vents requires specialized equipment and methods. The isolation of such organisms is a challenge because their habitats endanger our own survival. Undersea vent systems must be

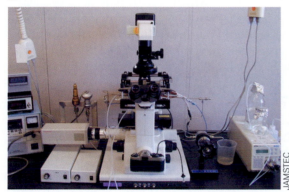

Figure 1　Robotic sampling from a black smoker vent. Engineer Gene Massion, from the Monterey Bay Aquarium Research Institute, deploys the submersible Environmental Sample Processor with robotic collection arm at an ocean site off the coast of Maine. *Source*: Courtesy of Monterey Bay Aquarium Research Institute.

approached by a special submersible device with a robotic arm. An example is the Environmental Sample Processor from the Monterey Bay Aquarium Research Institute (**Fig. 1**). The robotic system samples temperature and other properties of fluid emerging from a black smoker hydrothermal vent. It can then sample organisms for study. An advanced version of this device can actually process the organism's DNA. Thus, the DNA can be obtained from vent-adapted microbes that could not survive transfer to a laboratory at sea level. The robotic sample processor is supported by NASA as a model for a future space probe to explore one of Jupiter's satellites, Europa, considered a possible source of extraterrestrial life.

　Organisms that do survive transport to sea level must nonetheless be maintained at high pressure and temperature to ensure viability. In the laboratory, all devices for microscopy and cultivation (if possible) must also be maintained under pressure and at high temperature. **Figure 2A** shows a microscope system with a high-temperature and high-pressure cell installed, from the Marine-Earth Data and Information Department of the Japan Agency for Marine-Earth Science and Technology. Organisms are cultured in a pressurized cell such as the Deep Aquarium (**Fig. 2B**). The culture must be provided with reduced minerals and gases needed for growth of vent microbes.

A.

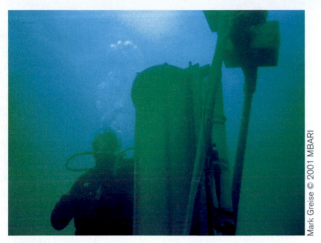

B.

Figure 2　Observation and culture of hyperthermophiles.　A. Microscope system with high-temperature and high-pressure cell installed. **B.** The "Deep Aquarium," a pressurized device for cultivation of vent organisms. *Source*: Images courtesy of Japan Agency for Marine-Earth Science and Technology.

pushing the two cells apart. In this fashion, the cell division process expands the cell network.

NOTE: Two genera of vent thermophiles have similar names but only distant genetic relatedness: *Pyrodictium abyssi*, a crenarchaeote, and *Pyrococcus abyssi*, a euryarchaeote.

The intricate cell network of *P. abyssi* is an example of a single-species biofilm. Other forms of single-cell and multicellular biofilms are found at hydrothermal vents. *Thermosphaera aggregans* forms colonies so tightly bound that they cannot be dissociated by protease treatment or sonication (**Fig. 19.12A**). Biofilms of other unidentified hyperthermophiles line the chimneys of black smoker vents (**Fig. 19.12B**).

A.

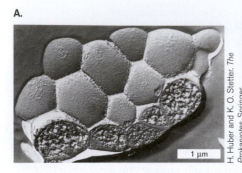

H. Huber and K. O. Stetter. *The Prokaryotes,* Springer

B.

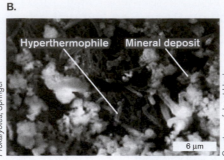

Hyperthermophile   Mineral deposit

Courtesy of Joost Hoek

Figure 19.12 Hyperthermal biofilms. A. Freeze-etched sample of *Thermosphaera aggregans* reveals a dense aggregate of cells (SEM). **B.** Biofilm of rod-shaped hyperthermophiles mingled with mineral sulfides at a black smoker vent, Central Indian Ridge.

> **THOUGHT QUESTION 19.2** What might be the advantages of flagellar motility for a hyperthermophile living in a thermal spring or in a black smoker vent? What would be the advantages of growth in a biofilm?

Sulfolobales: High Temperature and Extreme Acid

The crenarchaeote order Sulfolobales includes species that respire by oxidizing sulfur (instead of reducing it, like *Desulfurococcus*). These organisms, such as *Sulfolobus* species, grow at 80–90°C within hot springs and solfateras (volcanic vents that emit only gases). Ken Stedman and colleagues at Portland State University study *Sulfolobus solfataricus*, a species that grows at 80°C and pH 3 (**Fig. 19.13**). *S. solfataricus* oxidizes H$_2$S to sulfuric acid:

$$2S^0 + 3O_2 + 2H_2O \longrightarrow 2H_2SO_4 \longrightarrow 4H^+ + 2SO_4^{2-}$$

As a result of sulfuric acid production, the pH of the organism's surroundings falls to pH 2–3, effectively excluding all but acidophiles. While other archaea grow at higher temperatures (up to 120°C) or lower pH (below pH 0), *Sul-*

Image courtesy of Ken Stedman

Figure 19.13 Isolating *Sulfolobus* from Yellowstone. A researcher holds a collecting tube at length to obtain samples from a steaming spring, Rabbit Creek, at Yellowstone National Park.

folobus is of interest as a "double extremophile," requiring both high temperature and extreme acidity simultaneously.

Sulfolobus cells have a membrane composed mainly of tetraethers with cyclopentane rings (see **Fig. 19.2E**). Tetraether membranes are commonly seen in acidophilic thermophiles, probably because they are exceptionally impermeable to protons. Remarkably, *Sulfolobus* has no cell wall, only an S-layer of glycoprotein (**Fig. 19.14A**). Like all archaea, these organisms are nonpathogenic to animals and plants, yet they secret toxins deadly to competitor strains of *Sulfolobus*.

Sulfolobus species are not obligate autotrophs; they can also grow heterotrophically on sugars or amino acids. In fact, many species are easily cultured in tryptone broth at 80°C, pH 3. Their internal pH is typically pH 6.5; thus, they maintain more than three units of pH difference across their membrane. The full metabolic potential of this organism is revealed by annotation of its genome, which reveals homologs of sugar and amino acid transporters as well as the non-ATP-forming Entner-Doudoroff pathway of glucose catabolism. Genes are present for enzymes to oxidize HS$^-$ and S$_2$O$_3^{2-}$ as well as S^0. The main redox carrier for respiration appears to be ferredoxin (instead of NADH, which is relatively unstable at high temperature).

> **THOUGHT QUESTION 19.3** What problem with cell biochemistry is faced by acidophiles that conduct heterotrophic metabolism?

Because *Sulfolobus* species are readily cultured, their metabolism and ecology have been studied extensively, revealing unexpected features. Some species show multiple origins of replication of their DNA; for example, *S. solfataricus* and *S. acidocalderius* each use three active replication origins. The presence of multiple origins in some archaea may be related to the closer affinity of archaeal DNA management to that of eukaryotes.

Another interesting discovery was that of archaeal viruses. *Sulfolobus* species are attacked by fuselloviruses, the best-studied class of archaeal viruses to date (**Fig. 19.14B**). Fuselloviruses generally resemble bacteriophages

Figure 19.14 *Sulfolobus.*
A. *Sulfolobus* cell with thick S-layer.
B. Fusellovirus, a double-cone-shaped virus that infects *S. solfataricus.*

A.

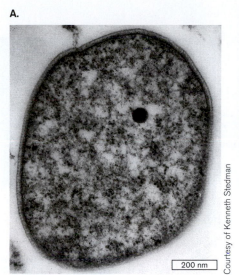

Courtesy of Kenneth Stedman

200 nm

B.

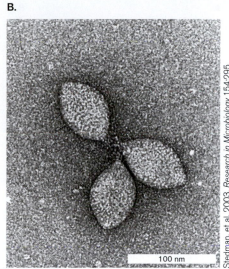

Stedman, et al. 2003. *Research in Microbiology* 154:295

100 nm

in size and function, but their capsid is spindle-shaped, forming a cone at each end. Spindle-shaped capsids are common in archaeal viruses, but never seen in bacteriophages. Since the discovery of fuselloviruses, numerous other archaeal viruses have been discovered in high-temperature environments, including icosahedral, tailed, and filamentous forms. All have in common genomes consisting of double-stranded DNA. This suggests that only double-stranded DNA is stable enough for virus particles to persist at high temperature. Viruses of mesophilic archaea are only beginning to be explored.

> **THOUGHT QUESTION 19.4** What conclusions might be drawn if viruses of mesophilic archaea are found to have RNA genomes? What if they all have DNA genomes only?

Other Thermophilic Crenarchaeotes

The class Thermoprotei includes two other orders of thermophiles (as well as mesophiles and psychrophiles; see next section). Caldisphaerales is an order of thermoacidophiles first isolated from a Philippine hot spring at Mount Maquiling. Members of Caldisphaerales, such as *Caldisphaera,* typically grow at pH 3 up to 80°C. Unlike *Sulfolobus* species, *Caldisphaera* species are anaerobes or microaerophiles, tolerating only low concentrations of oxygen. They grow by fermentation or anaerobic respiration. Another order, **Thermoproteales**, includes hyperthermoacidophiles isolated from marine vents. *Thermoproteus* species grow up to 97°C at pH values lower than pH 3. Their rod-shaped cells are less than 0.3 μm in length, one of the smallest cell types known. Their metabolism is autotrophic, gaining energy by reducing sulfur with H_2 to H_2S. A related species, *Pyrobaculum islandicum,* grows in temperatures up to 103°C, although at less acidic pH (pH 5–7). New thermophilic crenarchaeotes continue to be discovered.

TO SUMMARIZE:

- **Crenarchaeol** is a tetraether containing a six-membered ring, found in varying amounts in all known species of the division Crenarchaeota.
- **Habitats for hyperthermophiles** include hot springs and submarine hydrothermal vents. Vent organisms are barophiles as well as thermophiles.
- **Desulfurococcales includes diverse thermophiles.** Most are anaerobes that use sulfur to oxidize hydrogen or organic molecules.
- *Pyrodictium* **species show unusual networked cells.** Disk-shaped cells are interconnected by cytoplasmic bridges called cannulae.
- **Sulfolobus species are aerobic thermoacidophiles.** *Sulfolobus* species oxidize sulfur to sulfuric acid or catabolize organic compounds.
- **Caldisphaerales and Thermoproteales include anaerobic hyperthermophilic acidophiles.**

19.3 Crenarchaeota: Mesophiles and Psychrophiles

For two decades, crenarchaeotes were thought to be largely restricted to high temperatures. In 1992, however, fluorescent DNA probes based on rRNA sequences revealed mesophilic archaea growing throughout the ocean. Edward DeLong and colleagues first found archaea in the Pacific Ocean off the coast of North America; a survey in 2001 of marine archaea at the Hawaii Time-Series Station found high numbers of crenarchaeotes (**Fig. 19.15**). The abundance of crenarchaeotes varied according to season

A.

Courtesy of Ed DeLong

B.

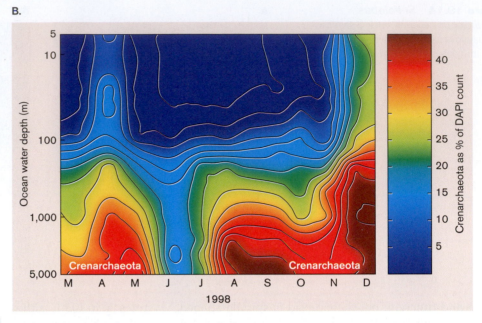

Figure 19.15 Crenarchaeota in the Pacific Ocean. A. Ed DeLong, now at Massachussetts Institute of Technology, discovers marine crenarchaeotes. **B.** The proportion of Crenarchaeota (color profile) is shown as a function of depth and season. Crenarchaeotes were identified by RNA hybridization to an rRNA sequence DNA probe containing DAPI, a fluorophore. *Source*: B. Markus Karner, et al. *Nature* 409:507.

and increased with depth, typically comprising 40% of the total microbial population at depths of 1,000 m, where temperatures are cold. The high proportion of ribosomal RNA in these cells indicates that they are metabolically active, even at the relatively cold temperatures of the ocean. These uncultivated mesophiles and psychrophiles are probably the predominant species of Crenarchaeota on Earth. In addition, crenarchaeotes are now known to be present in many dryland environments as well, including soil and the surfaces of plant roots.

Psychrophilic Crenarchaeotes

Some of the psychrophilic crenarchaeotes found by DeLong and colleagues exhibit a remarkable symbiosis with marine animals such as the sponge (**Fig. 19.16**). The crenarchaeote *Cenarchaeum symbiosum* inhabits the sponge *Axinella mexicana* (**Fig. 19.16B**). Like other marine

crenarchaeotes, *C. symbiosum* has yet to be grown in pure culture in the laboratory. The contributions of the microbe to its sponge host are unknown, but the sponge and its symbiotic crenarchaeotes can be cocultured in an aquarium for many years.

Even colder habitats were explored by Alison Murray and colleagues at the Desert Research Institute, who found psychrophilic crenarchaeotes growing in sea ice off Antarctica (**Figs. 19.17** and **19.18**). Archaea and bacteria were collected from ice and seawater at a temperature of –1.8°C; the water remains liquid below zero because of the high salt concentration. Cells were concentrated by filtration, and concentrated samples were analyzed to quantify the number of cells and classify them. The microbes collected were identified based on fluorescence microscopy using probes of PCR-amplified rDNA, revealing both bacteria and archaea. Typically, 20% of the isolates are crenarchaeotes; occasional euryarchaeotes are found as well.

A.

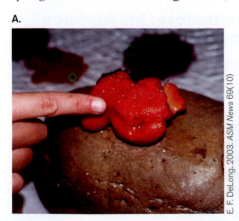

E. F. DeLong. 2003. *ASM News* 69(10)

B.

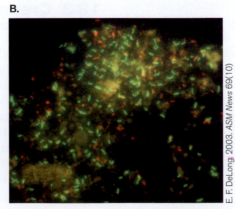

E. F. DeLong. 2003. *ASM News* 69(10)

Figure 19.16 Symbiosis between a mesophilic Crenarchaeote and a sponge. A. *Cenarchaeum symbiosum* inhabits the sponge *Axinella mexicana*. **B.** Differential fluorescent staining of *C. symbiosum* present in sponge tissue reveals that bacterial DNA (stained with DAPI, green) is segregated from host cell ribosomes (stained with an rRNA probe, red).

Figure 19.17   Collecting psychrophiles in Antarctica.   A. Alison Murray, from the Desert Research Institute, studies psychrophilic microbes in Antarctica. **B.** Microbes are collected off Bonaparte Point, Antarctica, from seawater samples down to 45 meters. **C.** Antarctic seawater samples are filtered to concentrate psychrophilic archaea and bacteria.

Figure 19.18   Identifying Antarctic psychrophiles.   A. Antarctic psychrophiles are observed by epifluorescence microscopy, with DAPI stain for DNA. The small spherical cells are approximately 0.5 μm in diameter. **B.** DNA of a psychrophilic archaeon is isolated and genes are identified. Microarray hybridization identifies which genes are turned on (expressed) or turned off (not expressed).

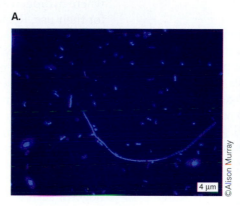

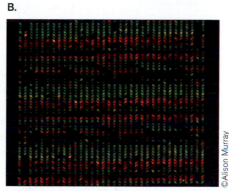

Most psychrophilic crenarchaeotes remain uncultivated, but the new technologies of genomics and micromanipulation now make it possible to study uncultivated organisms. The DNA of a single cell taken directly from the field can be amplified by PCR to generate a DNA library for genomic sequencing. Microarrays show which genes are expressed in these organisms and provide insights as to the molecular basis of life at cold temperatures.

Yet another important psychrophilic environment is the marine benthos, or seafloor sediment. At temperatures of about 2°C, seafloor strata show crenarchaeotes that are anaerobic heterotrophs, sulfate reducers, and anaerobic methanotrophs (methane oxidizers). Methanotrophs are particularly crucial in recycling the methane produced by seafloor methanogens (euryarchaeotes, discussed in Section 19.4).

Ammonia-Oxidizing Crenarchaeotes Affect Global Nitrogen

From a global standpoint, the most significant crenarchaeotes discovered recently may be the marine ammonia oxidizers of the order Nitrosopumilales. The first known ammonia-oxidizing crenarchaeote, *Nitrosopumilus maritimus*, was isolated from a marine aquarium in 2005 by David Stahl and colleagues at the University of Washington, Seattle. The organism gains energy by aerobically oxidizing ammonia to nitrite:

$$2NH_3 + 3O_2 \longrightarrow 2NO_2^- + 2H_2O + 2H^+$$

In 2006, Ann Pearson and colleagues from Harvard University and Scripps Institution of Oceanography showed that crenarchaeote ammonia oxidizers are widespread throughout the oceans. They are also common in soil. Ammonia-oxidizing crenarchaeotes fix carbon dioxide as well as ammonia, thus potentially making a substantial contribution to global cycles of carbon and nitrogen.

TO SUMMARIZE:

- **Oceans, soil, plant roots, and animals** provide habitats for mesophilic and psychrophilic crenarchaeotes.
- **Psychrophiles in marine sediment** play a key role in recycling methane hydrates produced by methanogens.
- **Ammonia-oxidizing crenarchaeotes** contribute to global nitrogen cycling.

19.4 Euryarchaeota: Methanogens

The second major branch of known archaea is Euryarchaeota, the "broad-ranging archaea" (**Table 19.2**). The euryarchaeotes are dominated by the **methanogens**, a vast array of species that derive energy through reactions producing methane. Biogenic methane has actually been observed since 1776, when the Italian priest Carlo Campi and physicist Alessandro Volta (1745–1827) investigated bubbles of "combustible air" from a wetland lake. Volta wrote:

> Being in a little boat on Lake Maggiore, and passing close to an area covered with reeds, I started to poke and stir the bottom with my cane. So much air emerged that I decided to collect a quantity in a large glass container. . . . This air burns with a beautiful blue flame.

Volta noted that methane arose from wetlands containing water-saturated decaying plant material. In 1882, the German medical researcher Hermann Von Tappeiner (1847–1927) combined plant materials with ruminant stomach contents (a source of methanogens) and showed that both components were essential to produce methane. During the late nineteenth century, in England, methane was collected from manure and sewage and used as a fuel for streetlamps.

Methanogens: Different Paths to Methane

Different species generate methane from different substrates, such as CO_2 and H_2 or small organic compounds, generally fermentation products of bacteria. Each pathway of methanogenesis generates a small free energy change ($\Delta G°'$) that is just enough to drive processes of carbon fixation and biosynthesis. All types of methanogenesis are poisoned by molecular oxygen and therefore require extreme anaerobiosis. Major substrates and reactions include:

Carbon dioxide: $CO_2 + 4H_2 \longrightarrow CH_4 + 2H_2O$

Formic acid: $4CHOOH \longrightarrow CH_4 + 3CO_2 + 2H_2O$

Acetic acid: $CH_3COOH \longrightarrow CH_4 + CO_2$

Methanol: $4CH_3OH \longrightarrow 3CH_4 + CO_2 + 2H_2O$

Methylamine:
$4CH_3NH_2 + 2H_2O \longrightarrow 3CH_4 + CO_2 + 4NH_3$

Dimethyl sulfide:
$2(CH_3)_2S + 2H_2O \longrightarrow 3CH_4 + CO_2 + 2H_2S$

In CO_2 reduction, the most common form of methanogenesis, carbon dioxide plus molecular hydrogen combine to form water and methane. CO_2-reducing methanogens are autotrophs, growing solely on CO_2 and H_2 with a source of nitrogen and other minerals. Methanogenesis from other carbon sources, such as formate, acetate, or methanol, is heterotrophic, and generates CO_2 as a product in addition to methane. The nitrogen-containing substrate methylamine generates ammonia in addition to methane and CO_2, whereas the sulfur-containing substrate dimethyl sulfide generates hydrogen sulfide.

Note that only a narrow range of substrates supports methanogenesis. For unknown reasons, most methanogens lack the vast array of alternative pathways found in soil bacteria such as *Rhodopseudomonas* or *Streptomyces*. Thus, methanogens generally require close association with bacterial partners to provide their substrates. They usually grow in habitats with minimal resource flux, where hydrogen and carbon dioxide gases can be trapped for their use. Removal of these gases then enhances bacterial metabolism. This kind of metabolic partnership is called "syntrophy."

Many kinds of methanogens can be cultured in the laboratory. Their culture poses special challenges because the reactions of methanogenesis are halted by oxygen. To exclude oxygen, all growth and manipulation of cultures must be performed in an anaerobic chamber. Furthermore, methanogens that use CO_2 and H_2 as substrates must receive a steady supply of these gases. Even more challenging is the culture of hyperthermophilic methanogens, such as *Methanopyrus*, a vent hyperthermophile that grows at 98°C. For these organisms, both high temperature and high pressure must be maintained.

Thermophiles, mesophiles, and psychrophiles. Methanogens include four classes, of which five major orders are shown (**Table 19.4**). The amount of divergence within an order can be nearly as great as that between orders. Particularly striking is the wide range of growth temperatures among closely related species. Thermophiles and even hyperthermophiles branch from closely related mesophiles; for example, the order Methanobacteriales includes *Methanobrevibacter ruminantium* (37–39°C), *Methanobacterium thermoautotrophicum* (50–75°C), and *Methanothermus fervidus* (60–97°C).

Methanogens Show Diverse Cell Forms

Despite their metabolic similarity, methanogens display an astonishing diversity of form, perhaps as diverse as the entire domain of bacteria (**Fig. 19.19**). For example, *Methanocaldococcus janaschii* cells grow as cocci with numerous flagella attached to one side, while *Methanosarcina barkeri* form peach-shaped cocci lacking flagella. *Methanothermus fervidus* cells are short, fat rods without flagella, and

Table 19.4 Methanogens.

Representative species	Growth temperature (°C)	Growth pH	Cell shape	Substrates for methanogenesis
Methanobacteriales				
Methanobrevibacter ruminantium	37–39°	pH 6–9	Chains of short rods	H_2 and CO_2
Methanobacterium thermoautotrophicum	50–75°	pH 7–8	Filaments of long rods	H_2 and CO_2, formate
Methanothermus fervidus	60–97°	pH 6–7	Rods	H_2 and CO_2
Methanosarcinales				
Methanosarcina barkeri	20–50°	pH 5–7	Aggregates of cocci	H_2 and CO_2, methanol, methylamine, acetate
Methanosaeta concilii	10–45°	pH 6–8	Filaments of rods	Acetate
Methanohalophilus zhilinae	45°	pH 8–10	Cocci	Methanol, methylamine, dimethyl sulfide
Methanococcales				
Methanococcus vaniellii	20–40°	pH 7–9	Cocci with flagella	H_2 and CO_2
Methanocaldococcus jannaschii	48–94°	pH 6–7	Cocci with flagella	H_2 and CO_2
Methanomicrobiales				
Methanoculleus olentangii	30–50°	pH 6–8	Cocci	Acetate, complex nutrients
Methanospirillum hungatei	20–45°	pH 6–7	Spirillum	Acetate
Methanopyrales				
Methanopyrus kandleri	84–110°	pH 6–8	Long rods (2–14 μm)	H_2 and CO_2

Source: *The Prokaryotes*, Springer.

A. *Methanosarcina barkeri*

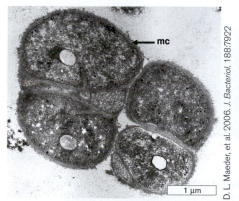

B. *Methanothermus fervidus*

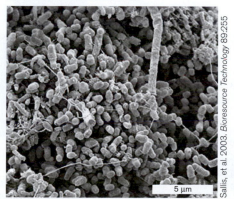

C. *Methanobacterium thermoautotrophicum*

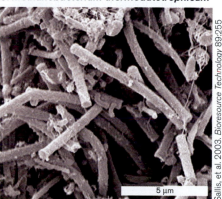

Figure 19.19 Methanogens show a wide range of shapes. A. *Methanosarcina barkeri*, a lobed coccus form lacking flagella (TEM). **B.** *Methanothermus fervidus*, a short bacillus (SEM). **C.** *Methanobacterium thermoautotrophicum*, an elongate bacillus (SEM).

Methanobacterium thermoautotrophicum grows as elongate rods reminiscent of the bacterial genus *Bacillus*. Still others, such as *Methanospirillum hungatei*, form wide spirals. The morphological diversity of methanogens may be explained in part by their rigid cell walls that can maintain a distinctive shape.

The composition of methanogen cell walls is much more diverse than those of bacteria. *Methanobacterium* species have a cell wall composed of pseudopeptidoglycan, a structure in which chains of alternating amino sugars are linked by peptide cross-bridges analogous to those of peptidoglycan. On the other hand, *Methanosarcina* species have a cell wall composed of sulfated polysaccharides. The genera *Methanomicrobium* and *Methanococcus* have protein cell walls.

Filamentous methanogens form chains of large cells similar to those of filamentous cyanobacteria. Filamentous methanogens, such as *Methanosaeta*, perform a key function in the treatment of sewage waste (**Fig. 19.20**). In waste treatment, the raw sewage first undergoes aerobic respiration by bacteria, converting the organic materials to small molecules such as CO_2 and acetate, followed by anaerobic digestion of the remainder. The bacteria performing anaerobic decomposition become trapped in filaments of *Methanosaeta*, which convert bacterial fermentation products such as acetate into methane and CO_2. The methanogenic filaments serve a key function by trapping bacteria into granules that settle out from the liquid.

Anaerobic Habitats for Methanogens

Methanogens grow in soil, in animal digestive tracts, and in marine floor sediment. A major methanogenic environment is the anaerobic soil of wetlands (**Fig. 19.21A**). The wetlands that generate the most methane are typically disturbed or artificial wetlands, particularly rice paddies, which contain high levels of added fertilizer that bacteria convert to the substrates used by methanogens. From

the standpoint of the organism, methane is an incidental by-product, but this product has great significance for our biosphere as a greenhouse gas (discussed in Chapter 22).

Methanogenesis in soil and landfills. Chinese farmers have found that one way to decrease methane production is to drain the soils used for rice production. Changsheng Li and colleagues at the University of New Hampshire showed in experimental plots that drained soil produces less methane. The drained soil receives more oxygen, which blocks methanogenesis. Draining soil also stimulates rice root development and accelerates decomposition of organic matter in the soil to release nitrogen. Sufficiently dry rice paddy soil can actually become a sink for atmospheric methane, thanks to methane-oxidizing bacteria.

Another major source of methanogens is landfills (**Figs. 19.21B** and **C**). Landfills such as those outside New York City are among the largest structures on Earth made by humans. They are rich in organic wastes, which bacteria ferment to CO_2, H_2, and short-chain organic molecules that methanogens convert to methane. As a result, large amounts of gas can build up and spontaneously combust, causing explosions. To avoid explosions, the methane needs to be piped out. In some cases, as in the Los Reales landfill in Tucson, Arizona, it can be collected and used to generate electricity.

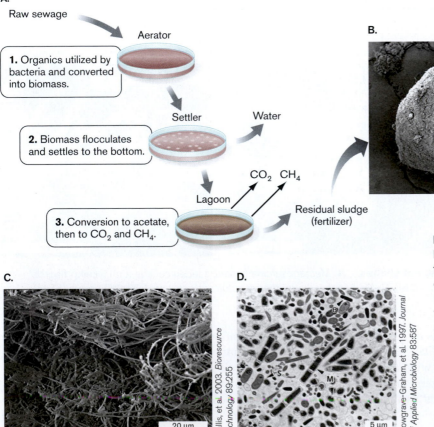

A.

Raw sewage

Aerator

1. Organics utilized by bacteria and converted into biomass.

Settler   Water

2. Biomass flocculates and settles to the bottom.

CO_2 CH_4

Lagoon

3. Conversion to acetate, then to CO_2 and CH_4.

Residual sludge (fertilizer)

B.

Sallis, et al. 2003. *Bioresource Technology* 89:255

200 μm

C.

20 μm

Sallis, et al. 2003. *Bioresource Technology* 89:255

D.

5 μm

Howgrave-Graham, et al. 1997. *Journal of Applied Microbiology* 83:587

Figure 19.20   Filamentous methanogens bind bacterial communities in waste treatment. A. Raw sewage is aerated and decomposed by bacteria, followed by anaerobic incubation. Under anaerobiosis, the bacterial waste products are converted to methane and CO_2. The bacteria are packed together by filamentous methanogens to form sludge. **B.** Sludge forms granules packed with bacteria and methanogens. **C.** Granule is packed with microbial filaments, probably *Methanosaeta* sp. **D.** Slice through a sludge granule (TEM) reveals mixed archaea and bacteria, probably Methanothrix (M), Methanobrevibacter (B), and Syntrophobacter (S).

Figure 19.21   Terrestrial habitats for methanogens.   A. Chinese farmers plant rice in waterlogged soil, an ideal anaerobic habitat for methanogens. **B.** The Puente Hills Landfill Gas Recovery Facility in California. Methane gas from the landfill is burned to generate electricity. **C.** Methane gas from a landfill near Kearny, New Jersey, is pumped through a series of pipes and recovery wells.

Digestive methanogenic symbionts. Numerous methanogens also grow within the digestive fermentation chambers of animals such as termites and cattle (**Fig. 19.22A**; discussed in Chapter 21). Termites have to aerate their mounds continually in order to remove methane; when a rainfall temporarily clogs the mound, it can be ignited by lightning and explode. Cattle support methanogenesis within their rumen and reticulum; a common veterinary trick is to insert a cannula and light a flame on the escaping gas. Bovine methanogenesis diverts carbon from meat production, and it makes a significant contribution to global methane.

Research on rumen microbiology focuses on attempts to suppress growth of methanogens. For example, Steven Ragsdale and his graduate student Bree DeMontigny at the University of Nebraska, Lincoln, test potential chemical inhibitors of methanogenesis on samples of rumen fluid (**Fig. 19.22B**). The rumen fluid is maintained in vials that serve as artificial rumens. The artificial rumens allow researchers to assess how well the proposed methane-blocking compounds might perform in an actual bovine rumen.

Human digestion. Recent research reveals key contributions of methanogens to human digestion. Most humans test positive for methanogens, primarily the species *Methanobrevibacter smithii* and *Methanosphaera stadtmanae*. These archaea constitute about 10% of gut anaerobes. *M. smithii* increases the efficiency of digestion by consuming excess reduced products (formate and H_2) from bacterial mixed-acid fermentation (discussed in Chapter 13). High levels of H_2 inhibit bacterial NADH dehydrogenases and thus decrease the proton potential for ATP production. *M. stadtmanae* consumes methanol, a potentially toxic byproduct of bacterial degradation of pectin, a major fruit polysaccharide. Thus, methanogens may enhance the growth of human enteric bacteria.

If methanogens influence the fermentation efficiency of gut flora, do they affect the amount of caloric content

Figure 19.22   Methane production in the bovine rumen.
A. Methanogens grow with the community of fermentative bacteria in the rumen and reticulum of ruminating animals. **B.** Bree DeMontigny, a graduate student in animal science working with Steven Ragsdale, pipets rumen fluid samples into vials that serve as "artificial rumens." The vials will be incubated and then assayed for the amount of methane formed.

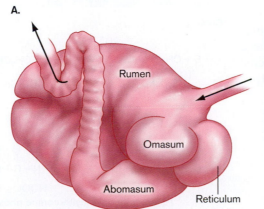

Rumen

Omasum

Abomasum

Reticulum

we obtain from food? In 2006, Buck Samuel and Jeffrey Gordon at Washington University, St. Louis, conducted an experiment to test the effect of methanogenesis on energy balance in gnotobiotic (germ-free) mice. The gnotobiotic mice were colonized with *Bacteroides thetaiotaomicron*, an anaerobic bacterium that ferments complex polysaccharides from the plants we consume. The mice were colonized in the presence or absence of *M. smithii*, which consumes H_2 through methanogenesis. Mice colonized with both *B. thetaiotaomicron* and *M. smithii* were found to make more fat after consuming the same quantity of food as those colonized solely with *B. thetaiotaomicron*. This result is in contrast with the case of the bovine rumen, in which methanogenesis draws off so much carbon as to decrease fermentation efficiency. The results suggest interesting hypotheses for future study of weight control in humans.

Figure 19.23   Methane hydrate. A crystalline form of methane and water generated by benthic methanogens; a potential source of fuel.

Methane hydrates on the seafloor. Psychrophilic marine methanogens grow at or beneath the seafloor. These seabed methanogens generate large volumes of methane that seep up slowly from the sediment. Under the great pressure of the deep ocean, the methane becomes trapped as **methane gas hydrates**, which are crystalline materials in which methane molecules are surrounded by a cage of water molecules (**Fig. 19.23**). The methane hydrates accumulate in vast quantities. If a large part of this methane were to be released, it could greatly accelerate global warming.

Methane hydrates are also of interest as a potential source of fuel to be mined for the generation of electricity.

Biochemistry of Methanogenesis

With all the roles that methanogens play in our environment and in human digestion, much effort goes into better understanding of how methanogenesis works. Knowledge of methanogenic biochemistry may help us develop controls for these processes, such as methanogen-specific antibiotics to minimize methane output from cattle.

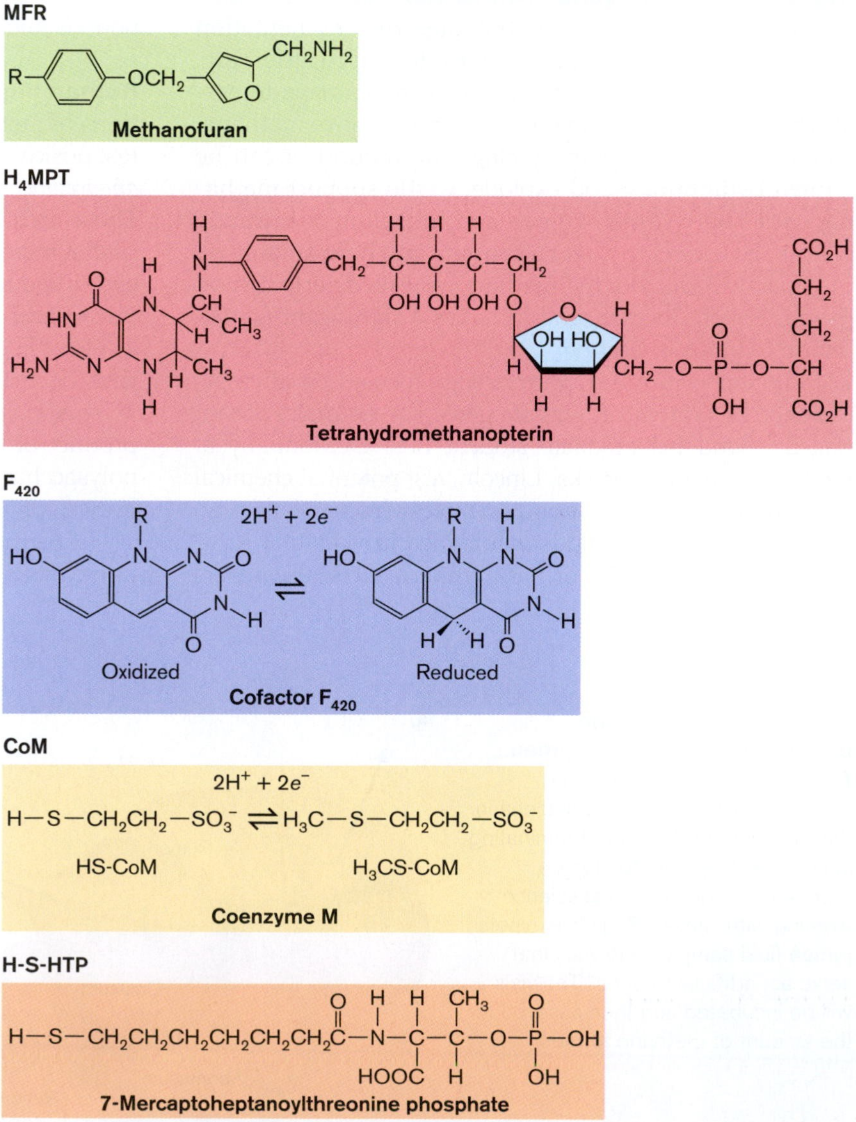

Figure 19.24   Cofactors for methanogenesis. Cofactors specific for methanogenesis transfer the hydrogens and the increasingly reduced carbon to each enzyme in the pathway.

MFR

Methanofuran

H₄MPT

Tetrahydromethanopterin

F₄₂₀

$2H^+ + 2e^-$

Oxidized   Reduced

Cofactor F₄₂₀

CoM

$2H^+ + 2e^-$

$H-S-CH_2CH_2-SO_3^- \rightleftharpoons H_3C-S-CH_2CH_2-SO_3^-$

HS-CoM   H₃CS-CoM

Coenzyme M

H-S-HTP

$H-S-CH_2CH_2CH_2CH_2CH_2C-N-C-C-O-P-OH$

7-Mercaptoheptanoylthreonine phosphate

Cofactors for methanogenesis. The process of methanogenesis uses a series of specific cofactors to carry each carbon from CO_2 (or other substrates) as it becomes progressively reduced by hydrogen. The hydrogen atoms also require redox carriers. Most of the cofactors are unique to methanogens, although the general types of structure resemble redox cofactors we have seen in Chapter 14 (**Fig. 19.24**). For example, cofactor F_{420} is a heteroaromatic molecule (containing nitrogens in the aromatic rings) that undergoes a redox transition by acquiring or releasing two hydrogens, a transition similar to that of the nicotinamide ring of NADH. HS-coenzyme M plays a role similar to that of HS-CoA.

Methanogenesis from CO_2. The formation of methane from CO_2 and H_2 (**Fig. 19.25**) is technically a form of anaerobic respiration in which H_2 is the electron donor and CO_2 is the terminal electron acceptor (discussed in Chapter 14). The process fixes CO_2 onto the cofactor methanofuran (MFR), then passes the carbon stepwise

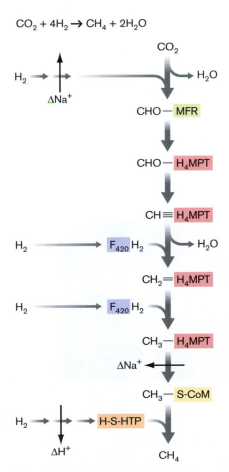

$$CO_2 + 4H_2 \rightarrow CH_4 + 2H_2O$$

Figure 19.25 Methanogenesis from CO_2 and H_2. Carbon dioxide is hydrogenated and deoxygenated in a stepwise fashion. The initial incorporation of H_2 requires a coupled sodium potential (ΔNa^+). A later step generates proton potential (ΔH^+). All steps of the reaction require specific enzymes (not shown).

from one cofactor to the next, each time losing an oxygen to form water or gaining a hydrogen carried by another cofactor.

The first step in the conversion of CO_2 and H_2 to methane is fixing the carbon onto methanofuran (see **Fig. 19.25**). This requires placement of protons onto the oxygen to form water. The flow of protons may be coupled to the transmembrane gradient of sodium ion, which is required for this step. Most methanogens require Na^+ for growth, unlike bacteria, many of which can grow without sodium. A higher external Na^+ concentration provides a sodium potential (ΔNa^+) that drives the two steps of methanogenesis. The sodium requirement of methanogens is something they share with another division of Euryarchaeota, the halophiles (discussed in Section 19.5).

Another important feature of methanogenesis is that the enzymes that catalyze each step require several transition metals. For example, the hydrogenase that reduces F_{420} requires both nickel and iron. The enzyme that catalyzes the reaction

$$CO_2 + \text{methanofuran} + 3H \longrightarrow \\ \text{CHO-methanofuran} + H_2O$$

requires either molybdenum or tungsten, depending on the species; some species have two alternative enzymes, depending on which metal is available. Another metal, cobalt, is required for a B_{12}–related cofactor that participates in methane production from methanol and methylamines.

> **THOUGHT QUESTION 19.5** What do the multiple metal requirements suggest about the evolution of methanogens?

Methanogenesis from acetate. Methane production from acetate is particularly important for wastewater treatment, where most of the substrate consists of short-chain bacterial fermentation products. The acetate methanogenesis pathway is not fully understood, but the initial incorporation of H_2 probably requires a coupled gradient of sodium ion, as it does for CO_2 methanogenesis. The two carbons from acetate enter different pathways, which eventually converge:

$$CH_3\text{-S-CoM} + \text{H-S-HTP} \longrightarrow \\ CH_4 + \text{CoM-S-S-HTP}$$

TO SUMMARIZE:

- **Methanogens gain energy through redox reactions that generate methane** from CO_2 and H_2, formate, and acetate. They require association with bacterial species that generate the needed substrates.

- **Methanogens have rigid cell walls of diverse composition in different species**, including pseudopeptidoglycan, protein, and sulfated polysaccharides.
- **Species of methanogens show a wide range of different shapes**, including rods (single or filamentous), cocci (single or clumped), and spirals.
- **Methanogens inhabit anaerobic environments** such as wetland soil, marine benthic sediment, and animal digestive organs.
- **Biochemical pathways of methanogenesis** involve transfer of the increasingly reduced carbon to cofactors that are unique to methanogens.

19.5 Euryarchaeota: Halophiles

In the saturated salt water of Utah's Great Salt Lake or Israel's Dead Sea, swimmers float with their heads well above the dense brine; but few living things can grow (**Fig. 19.26**). The main inhabitants of high salt environments are extremely halophilic archaea, or **haloarchaea**. Most haloarchaea are members of **Halobacteriales**, an order that was named before the archaea were classified as distinct from bacteria. The order Halobacteriales includes species of haloarchaea that diverge among the methanogens (see **Fig. 19.5**). The halophiles, however, do not conduct methanogenesis, but grow as photoheterotrophs, using light energy to drive a retinal-based ion pump to establish a proton potential. Their photopigments color salterns, brine pools evaporated to mine salt (**Fig. 19.27**).

The halophilic archaea require at least 1.5 M NaCl or equivalent ionic strength, and most grow optimally at near saturation (about 4.3 M, seven times the concentration of seawater). Most are colored red by bacterioruberin,

which protects cells from damage by light (**Fig. 19.28**). Thus, their growth colors their habitat red.

Haloarchaeal Form and Physiology

Halophilic microbes need some way to maintain turgor pressure—that is, to avoid cell shrinkage as cytoplasmic water runs down the osmotic gradient toward external high salt. Most bacterial halophiles compensate for high external salt by uptake or synthesis of other kinds of osmolites, such as small organic molecules. Haloarchaea,

Courtesy of Don Green Photography and Morton Salt

Figure 19.27   A saltern for salt production.   Aerial view of a solar saltern facility in Grantsville, Utah.

Priya DasSarma

A.

B.

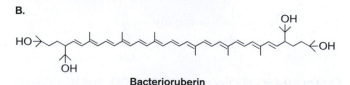
Bacterioruberin

Figure 19.28   Red pigmentation.   A. Brine inclusions in a salt crystal colored by haloarchaea. **B.** The red pigment bacterioruberin protects cells from damage by light.

Alison Wright

Figure 19.26   A high-salt environment.   A swimmer floats with her head up in the highly concentrated brine of the Dead Sea, in Israel.

however, adapt to high external NaCl by maintaining a high intracellular concentration of potassium chloride (about 4 M KCl). Potassium ion concentrations are moderately high in most microbial cells (commonly 200 mM), but the exceptionally high KCl concentration within haloarchaea requires major physiological adaptations:

- **High GC content of DNA.** High salt concentration decreases the fidelity in base pairing of DNA. Because the triple hydrogen bonds of GC pairs hold more strongly, the exceptionally high GC content of haloarchaea (above 60% for most species) may protect their DNA from denaturation in high salt.
- **Acidic proteins.** Most haloarchaeal proteins are highly acidic, with an exceptional density of negative charges at their surface. The high negative charge maintains a layer of water in the form of hydrated potassium ions (K^+). This unusual hydration layer keeps the acidic proteins soluble in the high salt in the cytoplasm.

The high salt content of haloarchaea provides a convenient way to lyse cell contents for analysis: Simply transfer cells into low-salt buffer, and they fall apart owing to osmotic shock. This technique provides a quick way for beginning students to isolate DNA. This and other techniques have made the organism *Halobacterium* a model system for molecular biology education (**Special Topic 19.2**).

The properties of diverse halophiles are presented in **Table 19.5**. The haloarchaea are generally uniform with respect to temperature range (mesophilic). With respect to pH, two categories are found: halophiles that grow at neutral pH and those that are alkaliphiles, growing in soda lakes above pH 9.

With respect to cell shape, haloarchaea display considerable diversity. Some species form symmetrical rods, as in *Halobacterium* NRC-1, the model system studied by Shiladitya DasSarma and colleagues at the University of Maryland. Other species form pleomorphic cells, flattened like pancakes; and still others form regular cocci (*Haloferax*, **Fig. 19.29A**). A few species grow as flattened

Table 19.5 Haloarchaea.

Representative species	NaCl range (M)	Growth temperature (°C)	Growth pH	Cell shape	Location of isolate
Halobacterium salinarum	3.0–5.2	35–50°	pH 7	Rods	Salted cow hide
Halorubrum lacusprofundi	1.5–5.2	1–44°	pH 7	Long rods (12 μm)	Deep Lake, Antarctica
Haloarcula valismortis	3.5–4.3	40°	pH 7.5	Pleomorphic rods	Salt pools, Death Valley, CA
Haloarcula quadrata	2.7–4.3	53°	pH 7	Square flat; pleomorphic	Sabkha, Sinai, Egypt
Halococcus morrhuae	2.5–5.2	30–45°	pH 7	Cocci	Dead Sea, Israel
Haloferax volcanii	1.5–5.2	40°	pH 7	Pleomorphic flat	Dead Sea, Israel
Natronobacterium gregoryi	2.5–5.2	37°	pH 9.5	Rods	Lake Magadi, Kenya
Natronococcus occultus	1.4–5.2	30–45°	pH 9.5	Cocci	Lake Magadi, Kenya
Natronomonas pharaonis	2.0–5.2	45°	pH 9–10	Rods	Wadi Natrun, Egypt

Sources: Aharon Oren, The Order Halobacteriales, *The Prokaryotes,* Springer; *H. lacusprofundi* growth temperature, from Cavicchioli. 2006. *Nature Reviews Microbiology* 4:331.

A. Haloferax dombrowskii

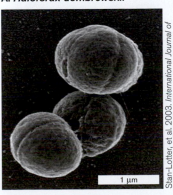

Stan-Lotter, et al. 2003. *International Journal of Astrobiology* 1(4):271, ©Cambridge University Press

B. Square haloarchaea

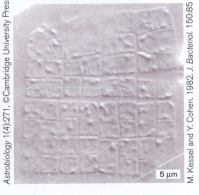

M. Kessel and Y. Cohen. 1982. *J. Bacteriol.* 150:851

C. Haloquadra walsbyi

W. Stoeckenius. 1981. *J. Bacteriol.* 148:352

Figure 19.29 Diverse haloarchaea. A. *Haloferax dombrowskii* (SEM). **B.** Square haloarchaea during their sixth round of cell division, which alternates in vertical and horizontal directions. (Photomicrograph, Nomarski optics.) **C.** Cross-sectioned cells of square haloarchaea show their thinness (TEM).

A halophilic archaeon, *Halobacterium* NRC-1, has found surprising applications as a teaching tool for precollege education. Teaching microbiology and biotechnological concepts has become crucially relevant at all grade levels. Microbes relate to many precollege content areas, including hygiene, food production and preservation, environmental science, genetics, and biotechnology. But many of the bacteria traditionally used by educators are reported to cause occasional illness in immunocompromised individuals, such as children with diabetes. Archaea such as *Halobacterium* have never caused a documented illness in humans.

Halobacterium has further advantages for education. It grows readily at room temperature on inexpensive broth medium supplemented with concentrated NaCl, which excludes most contaminants. It produces attractive colonies with mutants that show interesting color phenotypes (**Fig. 1**). And its cells lyse readily in water for instant extraction of DNA as well as other cellular components, such as gas vesicles and purple membrane.

For use in high schools, teachers require training and a kit of materials and instructions. Educator Priya DasSarma has developed such a series of kits and leads workshops to train teachers (**Fig. 2**). DasSarma and her husband, Shiladitya DasSarma, at the University of Maryland, have studied *Halobacterium* for two decades, developing the tools of genetics and molecular biology to investigate halobacterial physiology, including sequencing its genome. Now Priya DasSarma has developed a set of hands-on laboratory exercises that enable students to frame experimental questions and test them. Available from Carolina Biological Supply Company, the kit's exercises teach a wide range of concepts in microbiology, such as exponential growth and competition, colony formation, mutation, motility, DNA function, antibiotic sensitivity and resistance, bioinformatics, and genetic transformation and complementation.

In DasSarma's workshop, students and teachers learn to use the microbe for inquiry-based teaching by asking fundamental questions about what the students can see. Why do cells turn from a beautiful pink to a deep red when put under pressure, such as by centrifugation? The pink color is caused by light scattering from intracellular gas vesicles and light absorption by red carotenoid pigments; gas vesicles collapse under pressure, revealing the bright red pigment. The purpose of the gas vesicles is to keep the cells afloat in their natural habitat, such as the Great Salt Lake, where cells use sunlight for energy. On agar plates, why are some *Halobacterium* colonies pink and others red, while others are sectored like a pie? Pink colonies contain gas vesicles that scatter light, whereas a red colony derives from a cell containing an insertion element in a gas vesicle gene. Red sectors arise from spontaneous gas vesicle mutations in a fraction of cells within a colony; the mutants' descendants grow outward in a red wedge.

For advanced students, the *Halobacterium* cultures can be grown and lysed in distilled water and the DNA spooled out. The DNA is then run on electrophoretic gels to reveal plasmids. In other experiments, plasmid DNA can be used to transform a genetic strain to a new genotype. The availability of a completely sequenced genome permits a wide range of teaching applications.

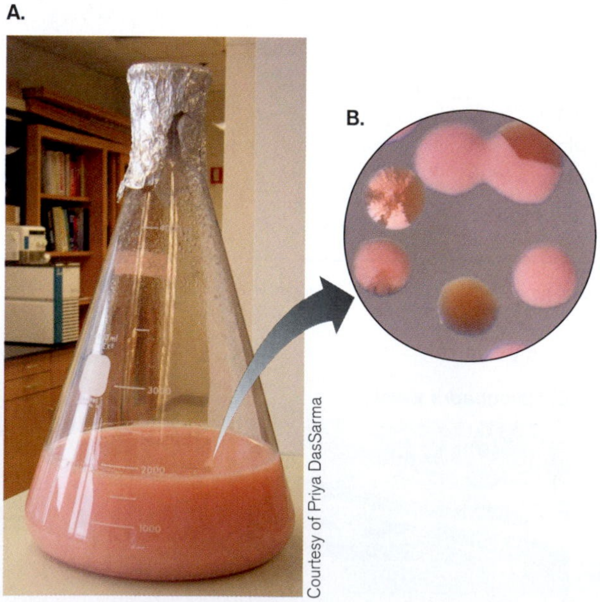

A.

B.

Courtesy of Priya DasSarma

Figure 1   *Halobacterium* in the high school classroom.
A. Flask of pink *Halobacterium* NRC-1 liquid culture in the DasSarma laboratory. **B.** Pink, opaque (wild type); red, translucent (lacking gas vesicles); and sectored colonies of *Halobacterium*.

Courtesy of Priya DasSarma

Figure 2   High school teachers use *Halobacterium*.
Educator Priya DasSarma (right) leads a workshop including teachers Janet Kovach (left) and Christina K. Schwalm. The workshop trains teachers to use *Halobacterium* in the classroom.

squares, the only microbial cells known to be square (*Haloquadra walsbyi*) (**Figs. 19.29B, C**). The basis for maintaining the various shapes is unknown. In oligotrophic (low-nutrient) environments, it is likely that the greater surface-to-volume ratio gives flattened cells a competitive edge in obtaining nutrients.

The cell envelopes of most haloarchaea contain rigid cell walls of glycoprotein, as do some of the methanogens. Most species possess flagella for phototaxis and **gas vesicles** for maintaining their buoyancy in the upper layer of the water column. Unlike phospholipid vesicles, the gas vesicles are made entirely of protein, and they are filled with air.

Different Hypersaline Habitats Support Different Haloarchaea

Not all hypersaline (high-salt) habitats are alike. They differ with respect to pH, temperature, and the presence of other minerals, such as magnesium ion. Different kinds of hypersaline habitat support different species of haloarchaea (**Table 19.5**). Major types of habitat include:

- **Thalassic lakes.** Thalassic lakes (Greek *thalassa*, meaning "ocean"), such as the Great Salt Lake, in Utah, contain saturated salts with essentially the same ionic proportions as the ocean: Na^+ and Cl^- (NaCl) predominate, followed by Mg^{2+}, K^+ and SO_4^{2-}. Thalassic lakes support genera such as *Halobacterium*.
- **Athalassic lakes.** Athalassic lakes, such as the Dead Sea in Israel, contain higher proportions of magnesium ions. These habitats favor genera such as *Haloarcula*, which require 100 mM Mg^{2+} for growth and grow best at magnesium concentrations above 1 M.
- **Solar salterns.** These are artificial pools of brine, or saturated NaCl, that evaporate in the sun, precipitating halite (salt crystals) for commercial production. Commercial evaporation pools for salt actually benefit from the red microbes, whose light absorption accelerates heating and evaporation.
- **Brine pools beneath the ocean.** Undersea brine pools collect near geothermal vents, as in the Gulf of Mexico, Mediterranean Sea, and Red Sea. These hypersaline regions contain salts brought up by vent water. They support hyperthermophilic halophiles, most of which have not yet been cultured.
- **Alkaline soda lakes** such as Lake Magadi in Kenya, where carbonate salts drive the pH above pH 9. These lakes support alkaliphilic haloarchaea such as *Natronobacterium*.
- **Antarctic brine lakes.** Reaching temperatures as low as –18°C, Antarctic brine lakes support growth of cold-adapted halophiles such as *Halorubrum lacusprofundi*.

- **Underground salt deposits** contain micropockets of salt-saturated water where halophiles may survive for thousands of years.
- **Salted foods** such as meat and fish and salt-cured hides can be spoiled by species such as *Halobacterium*.

Retinal-Based Photoheterotrophy

Most haloarchaea are photoheterotrophs, in which light-directed energy acquisition supplements respiration on complex carbon sources. Respiration occurs with oxygen or anaerobically with nitrate. A typical habitat for haloarchaea starts out as a pool with moderate salt content, containing a range of bacterial and archaeal species with varied tolerance to salt. As the pool evaporates, the salt concentrates, and the various bacteria die and lyse, releasing cell components that the haloarchaea can metabolize. The halophiles supplement their utilization of organic substrate energy by using light-driven ion pumps.

Light is captured by retinal-containing transmembrane proteins called **bacteriorhodopsin** (**BR**), and the chloride pump **halorhodopsin** (**HR**) (**Fig. 19.30**). (The role of light-driven ion pumps in phototrophy is discussed in Chapter 14.) The proton pump bacteriorhodopsin was named to indicate that it was a prokaryotic version of the better-known eukaryotic retinal rhodopsins; the name was chosen before the domain Archaea was recognized. The chloride pump halorhodopsin was discovered and named in reference to the chloride ion (a halide.) The two proteins are homologs with similar structure and mechanism. In the cell membrane, they form complexes that aggregate in patches called purple membrane.

Each bacteriorhodopsin proton pump contains seven alpha helices that traverse the membrane, surrounding a buried molecule of retinal, the same light-absorbing molecule found in photoreceptors of the human retina. Light absorption triggers a conformational change that enables a proton from the cytoplasmic face to be picked up by an aspartate residue. Subsequent proton transfer through the molecule leads to release of the proton outside the cell. Halorhodopsin has a similar structure, in which chloride (instead of H^+) is pumped *into* the cell instead of outward.

The net result of proton transfer by bacteriorhodopsin is generation of a proton motive force that can run a proton-driven ATPase, storing energy as ATP (**Fig. 19.30B●**). The halorhodopsin pumps chloride ions in the reverse direction, into the cell. Because chloride is negatively charged, this contributes to the proton motive force. Light-driven chloride pumps are unique to haloarchaea. Unimolecular proton pumps driven by light were thought to be restricted to haloarchaea until homologs

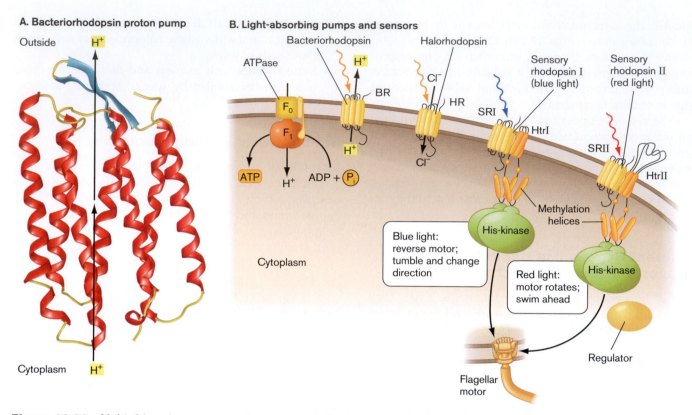

Figure 19.30   Light-driven ion pumps and sensors.   A. Bacteriorhodopsin absorbs light and pumps H⁺ out of the cell, whereas light-activated halorhodopsin pumps chloride into the cell. (PDB code: 1QK0) **B.** Rhodopsin family molecules in the membrane of *Halobacterium*: bacteriorhodopsin (BR), light-driven proton extrusion; halorhodopsin (HR), light-driven chloride intake; sensory rhodopsins I and II (SRI and SRII), with their signal transduction proteins HtrI and HtrII. The sensory transduction proteins phosphorylate and dephosphorylate a protein regulating the direction of rotation of the flagellar motor. ▶❙❙

were found to be widespread among marine species of gammaproteobacteria. These proteorhodopsins appear to have been obtained through horizontal transmission from archaea (discussed in Chapter 14).

Haloarchaea also possess homologs of bacteriorhodopsin that serve as sensory devices, the sensory rhodopsins I and II. These proteins, too, each have seven alpha helices containing retinal. When the sensory rhodopsins absorb light, they signal the cell to swim using its flagella (discussed in Chapter 3). The activated sensory rhodopsins direct the cell to swim toward red light, the optimum range for photosynthesis, and away from blue to ultraviolet light, which causes photooxidative damage to DNA. The rhodopsins signal through the chemotaxis machinery to the flagellar motor, using histidine kinase enzymes that phosphorylate regulatory proteins (see **Fig. 19.30B**). This response to light, or phototaxis, is an important model for archaeal molecular regulation.

The photosynthesis and phototaxis systems can be seen in the context of the overall metabolic map of *Halobacterium* NRC-1 (**Fig. 19.31**). The map is based on its genome sequence, published in 2000 by Wailap Ng and colleagues at the University of Maryland (the first halo-archaeal genome to be completed). In the metabolic map, the proton-translocating ATPase is seen to be driven by the proton potential generated through bacteriorhodopsin. Alternatively, a proton potential can be generated by the respiratory chain, through consumption of organic foods. Glucose catabolism occurs by the semiphosphorylated Entner-Doudoroff pathway, in which the intermediate 2-keto-3-deoxygluconate is formed and then phosphorylated, completing the pathway with one net molecule of ATP produced.

The *Halobacterium* genome reveals homologs for a number of genes known to encode transporters of organic food molecules such as sugars and amino acids. In addition, the genome shows a number of cation and anion transporters. Some transporters move in needed molecules such as phosphate, whereas others pump out harmful substances such as arsenate. Efflux of inorganic toxins is particularly critical for growth in evaporating brine, where the concentrations of trace metals are increasing.

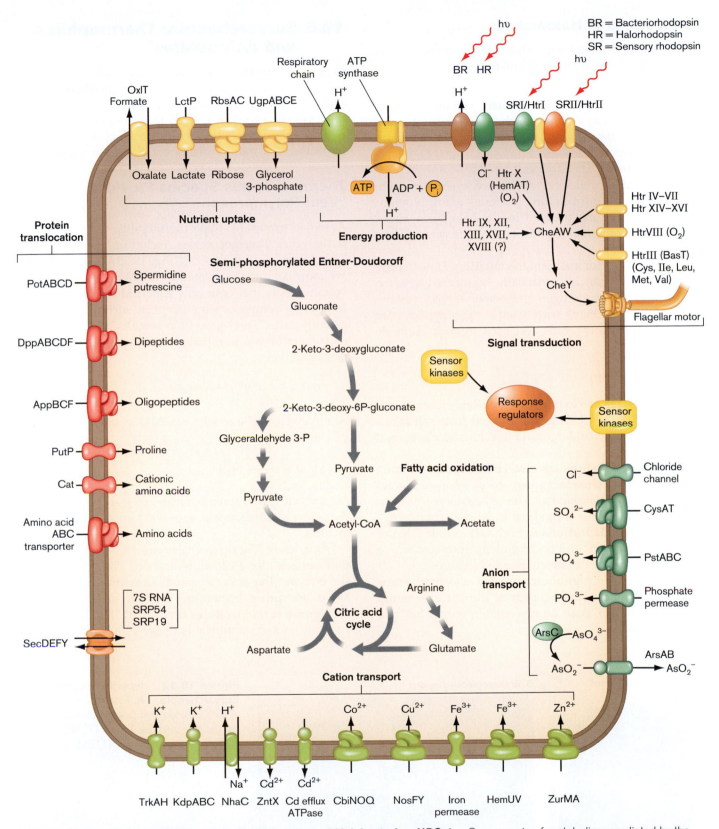

Figure 19.31 Metabolic pathways from the genome of *Halobacterium* NRC-1. Components of metabolism predicted by the genome, including light-driven energy circuit and sensory apparatus. *Source*: W. V. Ng, et al. 2000. *PNAS* 97:12176.

Applications of Haloarchaea

Haloarchaeal proteins and other cell components have found applications in research and industry. Some examples include:

- **Bacteriorhodopsin photoswitch.** The use of bacteriorhodopsin as a photoswitch is a model system for nanoscale experimental electronic circuits and memory devices.
- **Antibiotics.** Haloarchaea produce antibacterial agents called halocins, which accelerate the death and lysis of surrounding bacteria, whose components can be metabolized by the halophiles. Halocins have potential applications in antibacterial drug development.
- **Gas vesicles for antigen presentation.** Haloarchaea gas vesicles can be genetically engineered to serve as vehicles for antigen presentation. Gas vesicles containing proteins with fused antigen peptides are purified from the cells and used to elicit antibody response.

TO SUMMARIZE:

- **Haloarchaea are extreme halophiles**, growing in at least 1.5 M NaCl. They are isolated from salt lakes, solar salterns, underground salt micropockets, and salted foods.
- **Halobacteria show diverse cell shapes**, including slender rods (*Halobacterium*), cocci (*Haloferax*), and flat squares (*Haloquadra*). The cell envelopes of most haloarchaea contain rigid cell walls of glycoprotein.
- **Molecular adaptations to high salt** include DNA of high GC content and acidic proteins (proteins with a high number of negatively charged residues).
- **Retinal-based photoheterotrophy** involves the proton pump bacteriorhodopsin or the chloride pump halorhodopsin. Both pumps contain retinal for light absorption.

19.6 Euryarchaeota: Thermophiles and Acidophiles

Besides the methanogens and halophiles, the phylum Euryarchaeota includes a number of hyperthermophiles that superficially resemble the crenarchaeotes, although their genetic divergence from these organisms is deep. Some of the euryarchaeotes are acidophiles that grow under the most extreme acidic conditions on Earth.

Thermococcales Species Provide "Vent Polymerase"

A major group of hyperthermophilic euryarchaeotes is the order Thermococcales, including genera such as *Thermococcus* and *Pyrococcus* (**Fig. 19.32A**). These genera are most commonly isolated from submarine solfataras, volcanic vents that emit only gases. Most are anaerobes fermenting complex carbon sources, such as peptides or carbohydrates. Their growth is accelerated by using elemental sulfur as a terminal electron acceptor for anaerobic respiration:

$$2H^+ + 2e^- + S^0 \longrightarrow H_2S$$

Alternatively, these species can oxidize molecular hydrogen with sulfur:

$$H_2 + S^0 \longrightarrow H_2S$$

Most species grow at temperatures well above 90°C; *Pyrococcus woesei*, for example, grows at temperatures as high as 105°C.

Species of Thermococcales are the source of "vent polymerases" for PCR amplification. They now replace the enzyme Taq polymerase obtained from the Yellowstone hot-spring bacterium *Thermus aquaticus*, which only grows at temperatures up to 90°C. The Taq enzyme has only limited stability at or above 95°C, but DNA poly-

A. *Pyrococcus horikoshii*

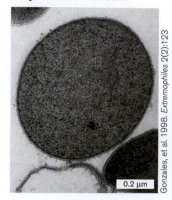

Gonzales, et al. 1998. *Extremophiles* 2(2):123

0.2 µm

B. *Archaeoglobus fulgidus*

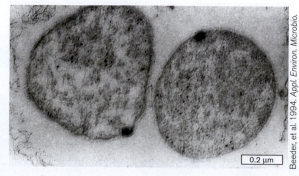

Beeder, et al. 1994. *Appl. Environ. Microbio.* April: 1227

0.2 µm

Figure 19.32 Hyperthermophilic Euryarchaeota. A. *Pyrococcus horikoshii* (TEM), isolated from a hydrothermal vent at the Okinawa Trough (TEM). **B.** *Archaeoglobus fulgidus* (TEM).

merases with greater stability at higher temperatures and increased accuracy in replication have been produced from vent-dwelling archaea such as *Pyrococcus furiosus* and *Thermococcus litoralis*. These "vent polymerases" now allow higher-temperature denaturation and synthesis of GC-rich sequences that were difficult to amplify with the original enzyme.

The first member of Thermococcales whose genome was sequenced was *Pyrococcus abyssi*. (Remember to distinguish *Pyrococcus abyssi* from the crenarchaeote *Pyrodictium abyssi*.) The sequence was completed in 1999 by Daniel Prieur, Patrick Forterre, and colleagues at Genoscope, the French National Sequencing Center. The genome of *Pyrococcus abyssi* reveals an especially high number of eukaryotic homologs for DNA replication, transcription, and protein translation. Examples include the Eukarya-like primase, helicase, and endonuclease for generation of Okazaki fragments in the lagging strand of DNA synthesis. At the same time, the genome also shows bacterial homologs for cell division and DNA repair. Thus, *P. abyssi* offers a striking view of mixed heritage from the common ancestor of the three domains.

Pyrococcus species possess enzymes conducting most of the classic conversions of the EMP pathway of glycolysis; but several steps use enzymes unrelated to those in bacteria. Examples include two ADP-dependent enzymes, glucokinase and phosphofructokinase, as well as the phosphate-independent enzyme glyceraldehyde 3-phosphate ferredoxin oxidoreductase. The glyceraldehyde 3-phosphate conversion is unique in that it sidesteps the use of inorganic phosphate, required at this step by bacterial glycolysis (see **Fig. 19.3**).

Pyrococcus and other members of Thermococcales have enzymes requiring tungsten, a metal requirement rare outside the archaea. (Tungsten is present in elevated concentrations at hydrothermal vents.) Another unusual feature of *Pyrococcus* is that both proton motive force generation and the ATPase appear to involve H^+/Na^+ antiport. This dependence on sodium is a trait shared with the methanogens and the halophiles.

> **THOUGHT QUESTION 19.6** Compare and contrast the metabolic options available for *Pyrococcus* and for the crenarchaeote *Sulfolobus*.

Archaeoglobus Runs Methanogenesis in Reverse

A hyperthermophile with unusual metabolic characteristics is *Archaeoglobus fulgidus*, of the order Archaeoglobales (**Fig. 19.32B**). The genome of *A. fulgidis* reveals an exceptionally complex array of metabolic pathways,

including an acetyl-CoA degradation pathway involving "reverse methanogenesis" (**Fig. 19.33**). All the enzymes and cofactors of a methanogen are present, but instead of generating methane from CO_2, they run in reverse, converting the methyl group of acetate to CO_2. The energy required for reversal is gained through fermentation of carbohydrates and through anaerobic respiration with sulfate, which is successively reduced to sulfite and sulfide. This system provides more energy than methanogenesis when organic substrates are available.

Thermoplasmatales

The order **Thermoplasmatales** includes thermophiles and acidophiles with no cell wall and no S-layer—only a plasma membrane. How they maintain their cells against extreme conditions is poorly understood. An example is the moderate thermophile isolated from self-heating coal refuse piles, *Thermoplasma acidophilum*, growing at 59°C and pH 2. The cells have flagella and are motile, but it is unclear how the flagella maintain a torque against the membrane without a rigid envelope for support. The metabolism of *T. acidophilum*, like that of *Pyrococcus*, is based on S^0 respiration of organic molecules.

The genome sequence of *T. acidophilum* contains just 1.5 million bp, one of the smallest known for a free-living organism. It shows substantial evidence of lateral transfer from the crenarchaeote *Sulfolobus*, a hyperthermophile that shares the same range of habitats. Lateral transfer between distant relatives turns out to be common among the hyperthermophiles, causing some controversies in their classification.

Another acidophilic genus of the order Thermoplasmatales is *Ferroplasma*. *Ferroplasma* species are found in mines containing iron pyrite ore (FeS_2). Using dissolved Fe^{3+} as an oxidizing agent in the presence of H_2O, they oxidize the sulfur to sulfuric acid. The chemical equation is

$$FeS_2 + 14Fe^{3+} + 8H_2O \longrightarrow 15Fe^{2+} + 2SO_4^{2-} + 16H^+$$

The reaction generates pH values below pH 0 (1 M H^+). This degree of acidity is enough to dissolve a metal shovel within a day.

The amorphous cells of *Ferroplasma acidarmanus* grow in biofilms that form long streamers into streams draining from the mine (**Fig. 19.34**). The oxidative disintegration of iron-bearing ores can be useful for leaching of minerals, but it also causes acid mine drainage into aquatic systems (discussed in Chapter 22).

> **THOUGHT QUESTION 19.7** Compare and contrast sulfur metabolism in *Pyrococcus* and *Ferroplasma*.

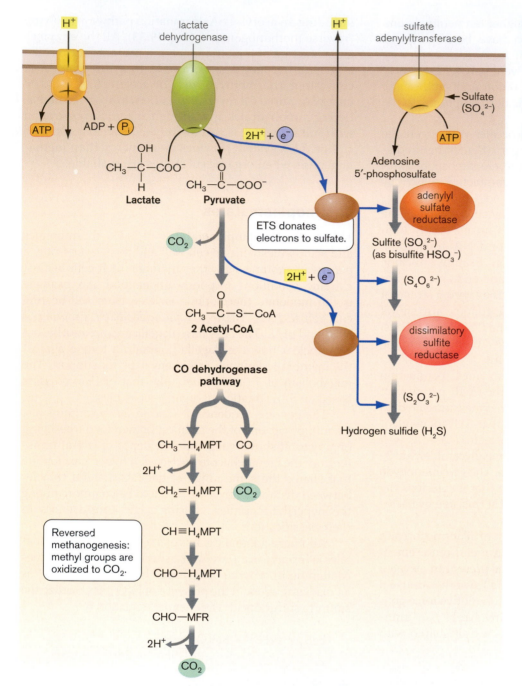

Figure 19.33 Metabolism of *Archaeoglobus fulgidus*. *A. fulgidus* gains energy by sulfate respiration on small organic molecules such as lactate. This process drives a unique acetyl-CoA degradation pathway involving reversal of methanogenesis. *Source:* H.-P. Klenk, et al. 1997. *Nature* 390:364.

Labels in figure:

H⁺

lactate dehydrogenase

H⁺

sulfate adenylyltransferase

ATP   ADP + Pᵢ

2H⁺ + e⁻

Sulfate (SO₄²⁻)

ATP

$$CH_3-\overset{OH}{\underset{H}{C}}-COO^-$$

Lactate

$$CH_3-\overset{O}{C}-COO^-$$

Pyruvate

Adenosine 5′-phosphosulfate

adenylyl sulfate reductase

ETS donates electrons to sulfate.

CO₂

Sulfite (SO₃²⁻) (as bisulfite HSO₃⁻)

2H⁺ + e⁻

(S₄O₆²⁻)

$$CH_3-\overset{O}{C}-S-CoA$$

2 Acetyl-CoA

dissimilatory sulfite reductase

CO dehydrogenase pathway

(S₂O₃²⁻)

Hydrogen sulfide (H₂S)

CH₃—H₄MPT   CO

2H⁺

CH₂=H₄MPT   CO₂

CH≡H₄MPT

Reversed methanogenesis: methyl groups are oxidized to CO₂.

CHO—H₄MPT

CHO—MFR

2H⁺

CO₂

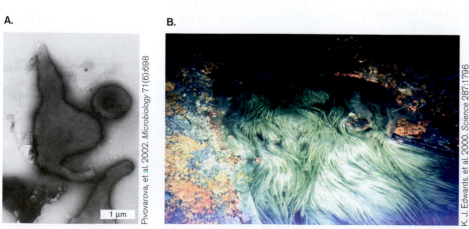

A. (TEM image)

Pivovarova, et al. 2002. *Microbiology* 71(6):698

1 μm

B.

K. J. Edwards, et al. 2000. *Science* 287:1796

Figure 19.34 The extreme acidophile: *Ferroplasma*.
A. *Ferroplasma acidiphilum* grows at pH 0 (TEM). **B.** Streamers of *F. acidarmanus* anchored to deposits of pyrite within the Iron Mountain mine. The stream is about a meter across, its water about pH 0. This level of acidity will dissolve a metal shovel in a day.

TO SUMMARIZE:

- **Thermococcales includes hyperthermophiles and acidophiles.** Most species use sulfur to oxidize complex organic substrates.
- *Pyrococcus* and *Thermococcus* **are the source of vent polymerases** for polymerase chain reaction (PCR).
- **Tungsten is commonly used by enzymes** of species of Thermococcales.
- *Archaeoglobus* **species (order Archaeoglobales) run methanogenesis in reverse.** The methyl group of acetate is oxidized to CO_2.
- **Thermoplasmatales includes extreme acidophiles.** *Ferroplasma* oxidizes iron pyrite ore (FeS_2) in a process that generates concentrated sulfuric acid, causing acid mine drainage.

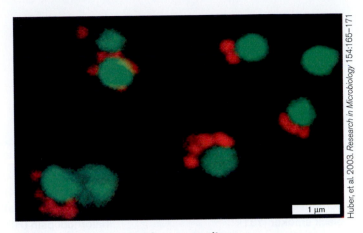

Huber, et al. 2003. *Research in Microbiology* 154:165–171

Figure 19.35 *Nanoarchaeum equitans*, a hyperthermophilic symbiont of *Ignicoccus*. Fluorescent probes to the DNA of *Ignicoccus* (green) and *N. equitans* (red) distinguish the two species.

19.7 Nanoarchaeota and Other Emerging Divisions

New archaeal species continue to be discovered through PCR-amplified rDNA probes. Some of their sequences diverge more deeply than the divergence between Crenarchaeota and Euryarchaeota. Most such strains are uncultivated, and little is known of them other than their rRNA sequence. Nevertheless, new techniques of genome sequencing enable us in some cases to read most of a genome from a single cell. The genomes of such mystery strains may provide clues as to their needs for cultivation.

Deep-Branching Hyperthermophiles

A deep-branching division is Korarchaeota, which includes hyperthermophiles isolated by Susan Barns and Norman Pace from Obsidian Pool at Yellowstone. Unfortunately, the isolates remain unculturable, and nothing more is known of them beyond their rDNA sequence.

Another deep-branching clade is Nanoarchaeota, represented so far by the species *Nanoarchaeum equitans* (**Fig. 19.35**). *N. equitans* shows about equal sequence similarity to both Crenarchaeota and Euryarchaeota and thus appears to represent a distinct division. The organism consists of exceptionally small cells that are obligate symbionts of the crenarchaeote *Ignicoccus*. The nano-

archaeal genome is exceptionally small (under 500 kb), typical of a dependent organism that has lost numerous genes through degenerative evolution and dependence on a host. It has not been established whether *N. equitans* is parasitic on *Ignicoccus* or whether it makes a metabolic contribution to its host.

Concluding Thoughts

Overall, species of archaea inhabit a wider range of environments than either bacteria or eukaryotes, from extreme heat to extreme cold, from high pH to extreme acid. Their metabolism includes unique capacities, such as methanogenesis. Nevertheless, it is important to remember that we probably know less about the actual scope of Archaea than we do about the other two domains because so many members remain uncultured. We recognize two major divisions, the crenarchaeotes, including sulfur-metabolizing hyperthermophiles as well as marine mesophiles and psychrophiles; and the euryarchaeotes, including methanogens, halophiles, and sulfur-metabolizing extremophiles. New deep-branching isolates continue to be discovered, potentially representing entire new divisions. Much of what we call Archaea remains to be explored.

CHAPTER REVIEW

Review Questions

1. What distinctive structures are seen in the archaeal cell membrane and envelope?
2. Which aspects of archaeal genetics resemble that of bacteria, and which aspects have more in common with eukaryotes?
3. Compare and contrast the two major divisions of archaea, Crenarchaeota and Euryarchaeota.
4. Outline the genetic phylogeny and key traits of these groups of archaea: haloarchaea, methanogens, Thermococcales, Thermoplasmatales.

5. What are some specific physiological adaptations found in hyperthermophiles? Psychrophiles? Extreme acidophiles?

6. Outline three different specific types of mutualism involving an archaeal symbiont.

7. Explain how different archaea contribute to cycling of nitrogen and sulfur in ecosystems.

8. Explain what is known and what is unknown about these groups of archaea: marine pelagic crenarchaeotes; marine benthic anaerobes; soil and plant root–associated crenarchaeotes. What kinds of experiments may reveal further traits of these organisms?

Key Terms

bacteriorhodopsin (BR) (724, 747)
barophile (730)
black smoker (731)
cannula (732)
carbon monoxide reductase pathway (724)
Crenarchaeota (726)
Euryarchaeota (726)
gas vesicle (747)

haloarchaeon (744)
Halobacteria or Haloarchaea (744)
halorhodopsin (HR) (747)
histone (726)
isoprenoid (724)
Korarchaeota (727)
methane gas hydrate (742)
methanogen (738)
methanogenesis (724)

Nanoarchaeota (727)
pseudomurein (724)
pseudopeptidoglycan (724)
tetraether (724)
Thermococcales (750)
Thermoproteales (735)
Thermoplasmatales (751)

Recommended Reading

Barns, Susan M., Charles F. Delwiche, Jeffrey D. Palmer, and Norman R. Pace. 1996. Perspectives on archaeal diversity, thermophily and monophyly from environmental rRNA sequences. *Proceedings of the National Academy of Sciences USA* **93**:9188–9193.

Biddle, Jennifer F., Julius S. Lipp, Mark A. Lever, Karen G. Lloyd, Ketil B. Sørensen, et al. 2006. Heterotrophic Archaea dominate sedimentary subsurface ecosystems off Peru. *Proceedings of the National Academy of Sciences USA* **103**:3846–3851.

Brochier, Céline, Patrick Forterre, and Simonetta Gribaldo. 2005. An emerging phylogenetic core of Archaea: Phylogenies of transcription and translation machineries converge following addition of new genome sequences. *BMC Evolutionary Biology* **5**:36.

Cavicchioli, Ricardo. 2006. Cold-adapted archaea. *Nature Reviews Microbiology* **4**:331–343.

Daniel, R. M., and D. A. Cowan. 2000. Biomolecular stability and life at high temperatures. *Cellular and Molecular Life Sciences* **57**:250–264.

DasSarma, Shiladitya. 2007. Extreme microbes. *American Scientist* **95**:224–231.

Edwards, Katrina J., Philip L. Bond, Thomas M. Gihring, and Jillian F. Banfield. 2000. An archaeal extreme acidophile involved in acid mine drainage. *Science* **287**:1796–1799.

Ferry, James G., ed. 1993. *Methanogenesis.* Chapman & Hall, Philadelphia.

Golyshina, Olga V., and Kenneth N. Timmis. 2005. *Ferroplasma* and relatives, recently discovered cell wall–lacking archaea making a living in extremely acid heavy metal-rich environments. *Environmental Microbiology* **7**:1277–1288.

Ingalls, Anitra E., Sunita R. Shah, Roberta L. Hansman, Lihini I. Aluwihare, Guaciara M. Santos, et al. 2006. Quantifying archaeal community autotrophy in the mesopelagic ocean using natural radiocarbon. *Proceedings of the National Academy of Sciences USA* **103**:6442–6447.

Karner, Markus B., Edward F. DeLong, and David M. Karl. 2001. Archaeal dominance in the mesopelagic zone of the Pacific Ocean. *Nature* **409**:507–510.

Karr, Elizabeth A., Joshua M. Ng, Sara M. Belchik, W. Matthew Sattley, Michael T. Madigan, and Laurie A. Achenbach. 2006. Biodiversity of methanogenic and other archaea in the permanently frozen Lake Fryxell, Antarctica. *Applied and Environmental Microbiology* **72**:1663–1666.

Lepp, Paul W., Mary M. Brinig, Cleber C. Ouverney, Katherine Palm, Gary C. Armitage, and David A. Relman. 2006. Methanogenic Archaea and human periodontal disease. *Proceedings of the National Academy of Sciences USA* **101**:6176–6181.

Ng, Wailap V., Sean P. Kennedy, Gregory G. Mahairas, Brian Berquist, Min Pan, et al. 2000. Genome sequence of *Halobacterium* sp. NRC-1. *Proceedings of the National Academy of Sciences USA* **97**:12176–12181.

Reysenbach, Anna-Louise, Yitai Liu, Amy B. Banta, Terry J. Beveridge, Julie D. Kirshtein, et al. 2006. A ubiquitous thermoacidophilic archaeon from deep-sea hydrothermal vents. *Nature* **442**:444–447.

Ruepp, Andreas, Werner Graml, Martha-Leticia Santos-Martinez, Kristin K. Koretke, Craig Volker, et al. 2000. The genome sequence of the thermoacidophilic scavenger *Thermoplasma acidophilum. Nature* **407**:508.

Samuel, Buck S., and Jeffrey I. Gordon. 2006. A humanized gnotobiotic mouse model of host–archaeal–bacterial mutualism. *Proceedings of the National Academy of Sciences USA* **103**:10011–10016.

Sapra, Rajat, Karine Bagramyan, and Michael W. W. Adams. 2003. A simple energy-conserving system: Proton reduction coupled to proton translocation. *Proceedings of the National Academy of Sciences USA* **100**:7545–7550.

Chapter 20

Eukaryotic Diversity

The domain Eukarya encompasses a breathtaking range of size and form, from giant whales and sequoias to microbial fungi, algae, and protists. Some eukaryotic microbes are single cells as small as the smallest bacteria.

Fungi include multicellular forms such as mushrooms, as well as unicellular yeasts and filamentous *Penicillium* and *Neurospora*. Fungal filaments interconnect the roots of forest plants, forming a vast underground network of nutrients. Algae conduct photosynthesis using chloroplasts. Algae include broad sheets of kelp as well as unicellular phytoplankton. Many algae turn out to be secondary endosymbionts—protists whose ancestors engulfed preexisting algae whole, only to assimilate them and utilize their prey's chloroplasts.

Protists include amebas, zigzag-swimming euglenas with their chloroplasts, paramecia with hundreds of cilia, and stalked vorticellae whose rings of cilia draw prey toward the mouth. Most protists are free-living, but some are deadly parasites, such as the trypanosome of sleeping sickness, and the plasmodium of malaria.

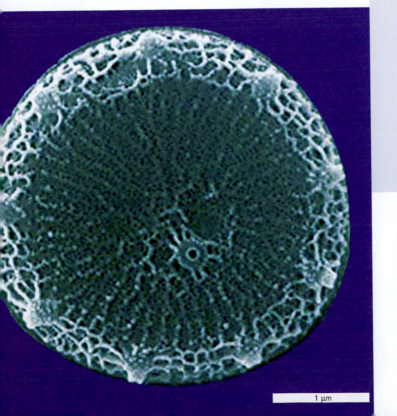

Diatoms are algae with silicate shells of intricate symmetry. They account for 40% of marine carbon productivity. *Thalassiosira pseudonana* was the first diatom whose genome was sequenced, by Virginia Armbrust and colleagues through an international collaboration involving twenty-four institutions (2004). The structure of diatoms such as *T. pseudonana* is now used for model systems of silica-based nanotechnology. *Source*: SEM image courtesy of the U.S. Department of Energy Joint Genome Institute.

1 μm

When we think of eukaryotes, we think first of plants and animals consisting of complex multicellular bodies. Macroscopic multicellular eukaryotes provide most of the food we consume, from breakfast cereal to beef. But eukaryotes also include microbes, such as algae and protists. While we rarely eat these directly, marine and aquatic algae and protists form the core of a vast food web supporting fish and other metazoan seafood, a major food source of protein. A drop of pond water reveals these microbial eukaryotes in their extraordinary diversity of form (**Fig. 20.1**): filamentous algae with spiral choroplasts, silica-shelled diatoms and radiolarians with intricate symmetry, predators with engulfing mouths, and parasites with syringe-like injectors. Of the three domains of life, Eukarya shows the greatest range of size and shape.

In Chapter 20, we explore the diversity of eukaryotic microbes, including essential partners in ecosystems as well as parasites that cause devastating pathology. Recall that the structure of eukaryotic cells is defined by the presence of the nucleus and other membrane-bound organelles, which enable eukaryotic cells to grow a thousandfold larger than those of prokaryotes (for review, see Appendix 2). Despite their extraordinary range of form and cell structure, the metabolism of eukaryotes is less diverse than that of either bacteria or archaea. Most eukaryotes are either oxygenic phototrophs or heterotrophs growing on complex carbon sources. All descended from an ancestral cell that incorporated a bacterial endosymbiont giving rise to mitochondria, the source of aerobic respiration.

This chapter presents the phylogeny of major groups of eukaryotes, including the branching of animals and plants from the microbial family tree. We explore the form and function of key microbes, including fungi, amebas and slime molds, algae, and various classes of parasites. As in Chapters 18 (Bacteria) and 19 (Archaea), we can show only representative examples of major clades of Eukarya. For further exploration, we recommend the NCBI Taxonomy Database and the Microbial Biorealm.

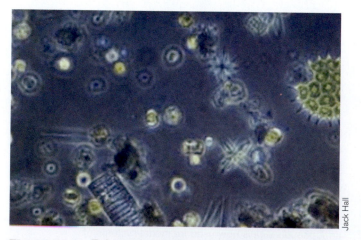

Figure 20.1 Eukaryotic microbes in a drop of pond water. Eukaryotic microbes include phototrophs (producers) and heterotrophs (consumers) that fill key niches in the food web.

20.1 Phylogeny of Eukaryotes

Our view of the domain Eukarya, including its relation to Bacteria and Archaea, has undergone several transitions over the past century, as discussed in Chapter 1. Classification of eukaryotes has always been a challenge for several reasons. Complex eukaryotic cells frequently lose structures through reductive (degenerative) evolution. Thus, for example, a clade originally characterized by possession of flagella often includes members that lack flagella. In addition, superficially similar forms of organisms have evolved independently in distantly related taxa; this is called convergent evolution. For example, the "water molds" that grow on aquarium fish superficially resemble fungi, but they actually evolved in a clade that includes brown algae and diatoms.

Modern classification is based on DNA sequence. The size and complexity of eukaryotic genomes has hindered the completion of their genomes. Nevertheless, genomic data brings us near a consensus view of eukaryotic descent, including the branching of our own human species.

Historical Overview of Eukaryotes

For most of human history, life was understood in terms of macroscopic multicellular eukaryotes: animals (creatures that move to obtain food) and plants (rooted organisms that grow in the sun). Fungi, which lack photosynthesis, were nonetheless considered a form of plant because they grow on the soil or other substrate. Thus, **mycology**, the study of fungi, was often included with botany, the study of plants. But the basis of fungal growth was poorly understood, and their mystery was often associated with magic, as seen in the Victorian painting in **Figure 20.2A**, depicting fairies upon a "fairy ring" of mushrooms.

In the eighteenth and nineteenth centuries, microscopists came to recognize microscopic forms of fungi such as hyphae (filaments of cells) and unicellular yeasts. Furthermore, unicellular protozoa (originally considered "microscopic animals") and algae (considered microscopic plants) became a hugely popular subject for observation (**Fig. 20.2B**). The cellular dimensions of eukaryotes are typically ten- to a hundred-fold larger than those of bacteria, and their form and motility offer intriguing subjects for observation. They are nonetheless microscopic in that visualization of their details requires light microscopy. And so, unicellular and microscopic forms of fungi, protozoa, and algae came to be included in the subject of microbiology.

By the mid-twentieth century, a growing number of biochemical and molecular discoveries led us to redefine these organisms. For example, the motile organisms defined as protozoa often contain chloroplasts and fit the classification of algae. On the other hand, slime molds, originally classified with fungi, show form and motility more typical of protozoa. In 1866 Ernst Haeckel at the University of Jena classified protozoa and motile forms of algae as **protists**. Lynn Margulis, at the University of Massachusetts, Amherst, in 1978, defined "protoctista" as

A.

B.

The MBLWHOI Library

Figure 20.2   Historical views of fungi and protozoa (protists).   A. "A fairy ring," a Victorian painting of fairies upon mushrooms, by Walter Jenks Morgan. Courtesy of the Leicester Gallery. **B.** A nineteenth-century depiction of ciliated protists, by Rudolph Leuckart.

the protists plus slime molds and oomycetes, organisms previously classified with fungi. Today, the term *protist* remains the most useful, although it now includes several distantly related clades (**Fig. 20.3**).

The shifting definitions of microbial eukaryotes are sorted out by molecular phylogeny. Eukaryotic genomes are harder to sequence because they are typically severalfold larger than those of prokaryotes, and 50–90% of their DNA consists of noncoding sequences. Thus, eukaryotic

genomes take longer to sequence and are more challenging to annotate than those of prokaryotes.

Furthermore, the evolution of eukaryotes includes multiple events of **endosymbiosis**, in which an engulfed cell evolved into an essential organelle. An endosymbiotic incorporation of a proteobacterium by the ancestor of all eukaryotes gave rise to mitochondria. A later incorporation of a cyanobacterium by the ancestor of plants and algae gave rise to chloroplasts. Much later, several lineages

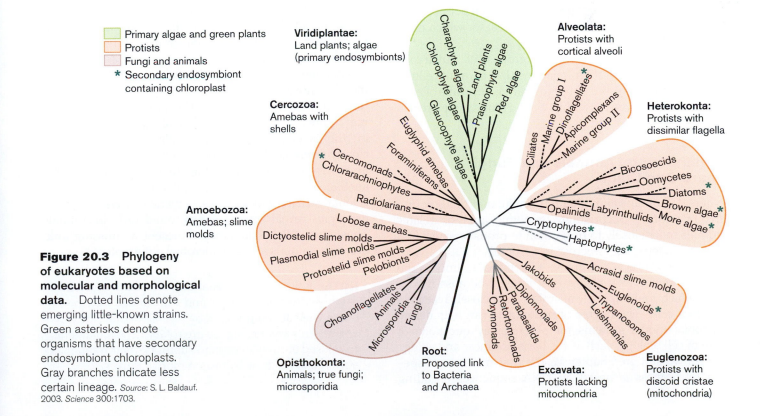

Figure 20.3   Phylogeny of eukaryotes based on molecular and morphological data. Dotted lines denote emerging little-known strains. Green asterisks denote organisms that have secondary endosymbiont chloroplasts. Gray branches indicate less certain lineage. *Source:* S. L. Baldauf. 2003. *Science* 300:1703.

of protists took up chloroplast-bearing algae, which now show varying stages of evolution as organelles. The lineage of algae derived from a single endosymbiotic event are **primary endosymbionts** (the Viridiplantae in **Fig. 18.3**), while the protists that later incorporated algae are **secondary endosymbionts** (green asterisk in **Fig. 18.3**).

While uncertainties remain, a consensus view of eukaryotic phylogeny has emerged based on a combination of DNA sequence comparison, protein trees, and the appearance of specific gene fusions and deletions. In the current consensus model:

- **Fungi** refer to the "true fungi," or Eumycota. Fungi, multicellular animals, and a few protist-like organisms such as microsporidia form a clade called Opisthokonta (**opisthokonts**).
- **Plants** include land plants and primary endosymbiont algae, known as Viridiplantae.
- **Protists** include all other eukaryotic microbes, such as amebas and ciliates, as well as secondary endosymbiont algae and certain organisms formerly classed with fungi (slime molds and oomycetes). Protists include about six to eight clades as distinct from each other as they are from Opisthokonta or Viridiplantae.

The consensus phylogeny based on DNA sequence analysis is shown in **Figure 20.3**. The distinct clades of protists include amebas and slime molds (Amoebozoa), shelled amebas (Cercozoa), ciliates and dinoflagellates (Alveolata), oomycetes, brown algae, and diatoms (Heterokonta), and a growing number of other groups not as well characterized (such as Excavata and Euglenozoa). The "rank" of these major groups is less well defined than for major groups of bacteria and archaea, for which more sequence data are available.

Table 20.1 outlines major groups of eukaryotic microbes, including their relationships to multicellular animals and plants. We now summarize key traits of each major group. Diversity within each group is explored in the sections that follow.

Opisthokonts: Animals and Fungi Share One Clade

Where do humans and other multicellular animals fit into this taxonomy? Surprisingly, genomic analysis relates animals more closely to fungi than to protists, which had been considered the "microscopic animals."

The position of animals ("metazoa") among the eukaryotic microbial clades is of interest because it suggests which contemporary microbes most closely resemble our own cells. The degree of relatedness can help define microbial model systems for probing key questions of human cell biology, such as the basis of cancer and aging.

As we saw in Chapter 17, the first measure of species relatedness is based on DNA sequences encoding ribosomal RNA, particularly small-subunit rRNA. For eukaryotes, however, rRNA sequences yield ambiguous results, in part because of differing rates of change among different clades. So researchers have searched the genomes for stronger clues to relative divergence—that is, which clade diverged earliest from the others. A particularly strong clue would be the appearance of a gene insertion or deletion unique to a particular clade. Such a sequence indicates that all species containing the gene insertion or deletion must have diverged after their common ancestor branched from the outgroup.

Animals and fungi share several key gene insertions and deletions, as discovered by Sandra Baldauf and colleagues at the University of York. Baldauf focused on a short sequence insertion in a key gene encoding a protein translation factor, elongation factor 1α (EF-1α) (**Fig. 20.4**). The inserted DNA, encoding 12 amino acids, appears in all sequenced genomes of animals and fungi and is absent from all plants and most protists. It is present in a few organisms, such as microsporidians, that were previously considered protists. This gene sequence, as well as several others, support the grouping of animals, fungi, and microsporidians together in one clade, the opisthokonts (see **Fig. 20.3**).

Animal and fungal cells also share a structural feature distinguishing them from protists: the presence of an unpaired flagellum. Both animals and fungi include species showing a uniflagellar stage in their life cycle (in contrast to protists, most of which have flagella in pairs). In the case of humans, the uniflagellar stage is the spermatozoan. Similarly, some species of fungi generate uniflagellar reproductive cells called **zoospores**. As a whole, the clade including single-flagellar members is termed "opisthokont" based on the Greek words meaning "backward spear," referring to the appearance of the backward-pointing flagellum.

Among opisthokonts, the microbes that diverged most recently from animals (600 million years ago) appear to be the choanoflagellates. The prefix *choano-*, meaning "funnel," refers to the collar of filaments surrounding the flagellum. The collared cells of choanoflagellates closely resemble the choanocyte cells of colonial sponges, an ancient form of animal (**Fig. 20.5**). Furthermore, choanoflagellates form colonies that crudely resemble the colonial structure of sponges. Genetic studies of choanoflagellates reveal several genes found only in animals. Thus, choanoflagellates may represent a "missing link" between animals and the microbial eukaryotes.

> **NOTE:** Eukaryotic flagella are whiplike organelles composed of microtubules and surrounded by a membrane, their action powered by ATP along the entire filament. Prokaryotic flagella are rotary helical filaments composed entirely of protein subunits; their rotation is powered at the base by proton motive force.

EF-1α peptide sequence alignment

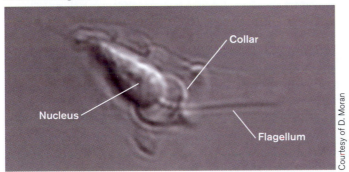

		1 9 7	Insertion within EF gene	2 6 5
Animals	Homo	P I SGWNGDNMLEPSAN–MPWFKG	WKVTRKDG–––––NASG	TTLLEALDCILPPTRPTDKPLRLPL
	Drosophila	P I SGWHGDNMLEPSTN–MPWFKG	WEVGRKEG–––––NADG	KTLVDALDAILPPARPTDKALRLPL
	Acmaea	P I SGYNGDNMLEKSPN–MPWYKG	WKVEQKDDKGNASTVTG	DTLTQALDSIQPPKRPTDKALRLPL
	Caenorhabditis	P I SGFNGDNMLEVSSN–MPWFKG	WAVERKEG–––––NASG	KTLLEALDSIIPPQRPTDRPLRLPL
	Hydra	PVSGWHGDNMIEPSPN–MSWYKG	WEVEYKDTG––––KHTG	KTLLEALDNIPLPARPSSKPLRLPL
	Anemonia	P I SGWHGDNMLEKSDK–MPWWNG	FELFNKSQG––––SKTG	TTLFDGLDDINVPSRPTDKALRLPL
Fungi	Sordaria	P I SGFNGDNMLEASTN–CPWYKG	WEKETKAG–––––KSTG	KTLLEAIDAIEQPKRPTDKPLRLPL
	Triochoderma	P I SGFNGDNMLTPSTN–CPWYKG	WEKETKAG–––––KFTG	KTLLEAIDSIEPPKRPTDKPLRLPL
	Histoplasma	P I SGFEGDNMLEPSPN–CTWYKG	WNKETASG–––––KSSG	KTLLDAIDAIEPPTRPTDKPLRLPL
	Schizosacch.	PVSGFQGDNMIEPTTN–MPWYQG	WQKETKAG––––VVKG	KTLLEAIDSIEPPARPTDKPLRLPL
	Candida	P I SGWNGDNMIEPSTN–CPWYKG	WEKETKSG––––KVTG	KTLLEAIDAIEPPTRPTDKPLRLPL
Microsporidians	Glugea	P I SGYLGINIVEKGDK–FEWFKG	WKPV–SGA–––––GDSI	FTLEGALNSQIPPPRPIDKPLRMPI
Plants	Triticum	P I SGFEGDNMIERSTN–LDWYKG	–––––––––––––––––	PTLLEALDQINEPKRPSDKPLRLPL
	Arabidopsis	P I SGFEGDNMIERSTN–LDWYKG	–––––––––––––––––	PTLLEALDQINEPKRPSDKPLRLPL
Protists	Dictyostelium	P I SGWNGDNMLERSDK–MEWYKG	–––––––––––––––––	RTLLEALDAIVEPKRPHDKPLRIPL
	Plasmodium	P I SGFEGDNLIEKSDK–TPWYKG	–––––––––––––––––	RTLIEALDTNQPPKRPYDKPLRIPL
	Entamoeba	P I SGFQGDNMIEPSTN–MPWYKG	–––––––––––––––––	PTLIGALDSVTPPERPVDKPLRLPL
	Trypanosoma	P I SGWQGDNMIEKSEK–MPWYKG	–––––––––––––––––	PTLLEALDMLEPPVRPSDKPLRLPL
	Euglena	P I SGWGDNMIEASEN–MGWYKG	–––––––––––––––––	LTLIGALDNLEPPKRPSDKPLRLPL
	Giardia	PTSGWTGDNIMEKSDK–MPWYKG	–––––––––––––––––	PCLIDAIDGLKAPKRPTDKPLRLPI

(Opisthokonts brackets Animals, Fungi, and Microsporidians.)

Figure 20.4 Alignment of peptide sequences for EF-1α reveals insertion in opisthokont species. An insertion of 11–17 amino acid residues appears in the EF-1α sequence of animals, fungi, and microsporidians (known collectively as opisthokonts) but not in plants or most protists. This finding suggests that opisthokonts have a common ancestor that diverged from the eukaryote lineage before divergence of the plants and nonopisthokont protists. *Source:* S. L. Baldauf. 1999. *American Naturalist* 154:S178.

A. Choanoflagellate

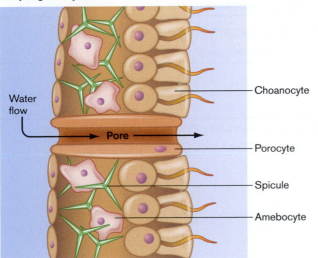

Courtesy of D. Moran

Collar
Nucleus
Flagellum

B. Sponge body wall

Water flow
Pore
Choanocyte
Porocyte
Spicule
Amebocyte

C. Sponge

Andrew J. Martinez/Photo Researchers, Inc.

Figure 20.5 Choanoflagellates resemble sponge choanocytes. **A.** A choanoflagellate, *Acanthocorbis unguiculata* (light micrograph; cell length, approximately 6 μm). **B.** Sponge choanocytes resemble choanoflagellates. Within the sponge, choanocytes assist the circulation of water and uptake of nutrients. **C.** Branching tube sponge (*Pseudoceratina crassa*).

Table 20.1 Representative groups of eukaryotes.

Note: Each trait applies to *most* members of the taxon described, but exceptions have evolved.
Blue-lettered terms indicate phyla or comparable rank.
• Bulleted terms are representative groups within the phylum.

Opisthokonta (fungi and metazoan animals)

Single basal flagellum on reproductive cells (usually).
Includes complex macroscopic multicellular organisms, such as humans.

Human spermatozoan

Eye of Science

Animals. Metazoa. Multicellular organisms with motile cells and body parts.
• **Colonial animals:** Sponges, jellyfish
• **Invertebrates:** Hydra, mollusks, arthropods, worms.
• **Vertebrates:** Fish, amphibians, reptiles, birds, mammals, including *Homo sapiens*

Choanoflagellata. Single flagellum with collar of microvilli.
Resemble sponge choanocytes. Possible link to common ancestor of metazoan (multicellular) animals.

Choanoflagellate

©1994–2000 by C. J. O'Kelly and T. Littlejohn/U. of Montreal

Fungi (Eumycota). Cells form hyphae with cell walls of chitin.
• **Chytridiomycota:** The deepest-branching fungal clade. Zoospores (motile gametes) with a single flagellum resemble the gametes of animals. Saprophytes or anaerobic rumen fungi. *Allomyces.*
  Batrachochytrium dendrobatidis, frog pathogen
  Neocallomastix, bovine rumen digestive endosymbiont
• **Ascomycota:** Fruiting bodies form asci containing haploid ascospores.
  Filamentous species: *Neurospora, Aspergillus, Penicillium*
  Yeasts: Unicellular species have lost mycelial stages.
  Saccharomyces cerevisiae ferments beer and bread dough;
  Candida albicans and *Pneumocystis carinii* infect immunocompromised patients.
  Stachybotrys, known as "black mold," contaminates homes.
  Morels and truffles: Large fruiting bodies are highly valued foods.
• **Basidiomycota:** Basidiospores form primary mycelium; fuses to form secondary mycelium. Generates large fruiting body called a mushroom. Examples: *Lycoperdon* is edible; *Amanita* is extremely poisonous.
  Cryptococcus neoformans is a yeast-form basidiomycete, an opportunistic pathogen.
• **Zygomycota:** Nonmotile gametes grow toward each other and fuse to form the zygote (zygospore). Saprophytes or insect parasites. Some form mycorrhizae, a mutualistic association with plant roots.
• **Lichens:** Mutualistic association between an ascomycete and green algae (*Trebouxia*) or cyanobacteria (*Nostoc*).

Zoospores *Allomyces*

J. C. Clark/California State Polytechnic Univ, Pomona

Aspergillus **mold** *versicolor*

©Dennis Kunkel/Visuals Unlimited

Microsporidia. Protist parasites that inject spores through tube into host cells, causing microsporidiosis.
• *Encephalitozoon* species: Commonly infect AIDS patients.
Other protists: Corallochytrium, Ichthyosporea, and Nucleariidae

Viridiplantae (primary endosymbiotic algae and plants)

Include algae and multicellular plants. Chloroplasts arose from a primary endosymbiont.

Charophyte

John Clegg/Science Photo Library

Plants. Adapted to growth on land.
Nonvascular plants: Mosses, ferns
Vascular plants: Gymnosperms, angiosperms

Charophyta (stoneworts). Multicellular algae with rhizoids that adhere to sediment. Form green mats with crust of calcium carbonate; hence called "stonewort."

Table 20.1 **Representative groups of eukaryotes (continued)**

Cymopolia barbata

S. Berger/Heidelberger Institut

Chlorophyta (green algae). Chlorophyll a confers green color. Inhabit upper waters.

Unicellular with paired flagella:

Chlamydomonas is a unicellular green alga; a model system for research on algae.

Volvox forms colonies of flagellated cells.

Multicellular:

Ulva grows in large sheets.

Spirogyra forms chains of cells.

Cymopolia forms calcified stalks with filaments.

Palmaria palmata

Edward Kinsman/Photo Researchers, Inc.

Rhodophyta (red algae). Phycoerythrin obscures chlorophyll, colors the algae red. Absorption of blue-green light enables colonization of deeper waters.

Porphyra form sheets edible by humans.

Sebdenia and *Plocamium* form branched fronds.

Mesophyllum forms coralline algae, hardened by calcium carbonate crust; resembles coral.

Amoebozoa or Lobosea (protists: amebas and slime molds)

Lobe-shaped (lobose) pseudopods driven by sol-gel transition of actin filaments.

Chaos carolinense
(size: 1–2 mm)

©Michael Abbey/Visuals Unlimited

Amebas. Unicellular. No microtubules to define shape. Life cycle is primarily asexual. **Predators in soil or water.**

Ameba proteus, Pelomixa, and *Chaos chaos* are giant free-living amebas in soil and water; they consume small invertebrates.

Entamoeba histolytica is an intestinal parasite.

Acanthamoeba is a soil predator, an opportunistic parasite causing meningitis.

Mycetozoa. Slime molds. Upon starvation, amebas aggregate to form a fruiting body, which undergoes meiosis and produces spores.

- **Cellular slime mold.** Multicellular fruiting body alternates with single-celled amebas.

  Dictyostelium discoideum. Provides an important model system for multicellular development.

- **Plasmodial slime mold.** Multinucleate plasmodium alternates with single-celled amebas.

  Physarum polycephalum. In aqueous environment, amebas generate flagella.

Cercozoa (protists: amebas with filament-shaped pseudopods)

Filament-shaped (filose) pseudopods. Some species have a test (shell) of silica or other inorganic materials.

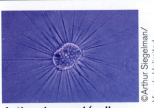

Actinophrys sol (cell size 30–90 μm)

©Arthur Siegelman/ Visuals Unlimited

Foraminifera form spiral tests. Fossil "forams" are an indicator of petroleum deposits.

Radiolaria form thin pseudopods (filopodia) reinforced by microtubules, radiating starlike from the center. Some possess inorganic skeletons or spicules.

Euglyphida are amebas with a test.

Chlorarachniophyceae have algal endosymbionts.

Alveolata (protists with cortical alveoli)

Cortex contains flattened vesicles called alveoli, reinforced below by lateral microtubules.

Vorticella (length 1 mm)

©Wim van Egmond/ Visuals Unlimited

Ciliophora. Common aquatic predators. Reproduce by conjugation, in which micronuclei are exchanged, then regenerate macronucleus for gene expression. Macronuclear DNA may be cut into thousands of segments.

Paramecium, Spirostomum, and *Oxytricha* are covered with cilia.

Didinium has two equatorial rings of flagella.

Vorticella and *Stentor* are stalked, with a mouth ringed by cilia.

Suctorians such as *Acineta* have knobbed tentacles.

(continued)

Table 20.1 **Representative groups of eukaryotes** (*continued*)

Ceratium tripos and
Ceratium furca

100 μm

©Dennis Kunkel/Visuals Unlimited

Dinoflagellata. Secondary or tertiary endosymbiont algae, from ancestral engulfment of red algae and diatoms. Cortical alveoli contain stiff plates. Pair of flagella, one of which wraps around the cell.

> **Free-living marine and aquatic dinoflagellates** such as *Peridinium* spp. supplement their photosynthesis with predation. Some produce bioluminesence. Blooms of marine dinoflagellates generate the "red tide."
>
> **Zooxanthellae** such as *Symbiodinium* spp. are endosymbionts of corals, providing essential nutrition through photosynthesis. "Coral bleaching" is a serious condition in which the zooxanthellae are expelled under environmental stress. Zooxanthellae also inhabit clams, flatworms, mollusks, and jellyfish.

Apicomplexa (formerly sporozoa). Parasites with complex life cycles. Lack flagella or cilia; possess apical complex for invasion of host cells. Vestigial chloroplasts.

> *Plasmodium falciparum* causes malaria.
> *Toxoplasma gondii* causes feline-transmitted toxoplasmosis.
> *Cryptosporidium parvum* is a water-borne opportunistic parasite.

Heterokonta (protists with pair of dissimilar flagella)

Paired flagella of dissimilar form, one much shorter than the other.

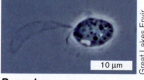

Diatom

©Wim van Egmond/
Visuals Unlimited

Paraphysomonas sp.

10 μm

Great Lakes Envir.
Lab/NOAA

Secondary endosymbiont algae. Free-living in marine and aquatic systems.
> **Mixotrophic; combine photrophy with heterotrophy.**
> - **Bacillariophyta:** Diatoms. Possess intricate silicate shells with radial symmetry (centric) or bilateral symmetry (pennate).
> - **Phaeophyceae:** "Brown algae" such as kelps and Sargassum weed. Stems and fronds extend for many meters.
> - **Chrysophyceae:** "Golden algae," pale-colored flagellates such as *Ochromonas* spp. and *Paraphysomonas* spp.
> - **Xanthophyceae:** "Yellow-green algae" and other less-studied forms of phytoplankton.

Oomycetes. Water molds. Superficially resemble fungi; often infect fish or plants. *Phytophthora infestans* destroyed potato crops and caused the Great Irish Famine.

Euglenozoa or Discicristata (protists with disk-shaped cristae)

Usually possess a deep feeding groove. Disk-shaped cristae of mitochondria.

Euglena gracilis

©Michael Abbey/
Visuals Unlimited

Euglenida. Free-living flagellates. Some contain a secondary endosymbiotic chloroplast. *Euglena gracilis* is a common aquatic flagellate.

Trypanosoma. Parasites. Cause sleeping sickness and Chagas' disease.

Leishmania. Parasites. Cause leishmaniasis.

Excavata (protists lacking mitochondria)

Parasitic flagellates that have lost mitochondria and Golgi through degenerative evolution. Usually possess a deep feeding groove.

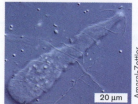

Pyrsonympha sp.

20 μm

L. Amaral-Zettler,
L. Olendzenski, and
D. J. Patterson.
MBL
(micro*scope)

Diplomonadida and Retortamonadida. Human intestinal parasites. Highly degenerated cells, lacking mitochondria. *Giardia lamblia.* Cause of diarrhea from contaminated drinking water.

Oxymonadida and Parabasalidea. Symbionts of termite gut. *Pyrsonympha.*

Fungi (Eumycota) consist of cells with chitinous cell walls that grow in chains called **hyphae** (singular, hypha). Fungi range from single-celled organisms, such as yeast, to complex multicellular forms, such as mushrooms. The deepest-branching clade of fungi, Chytridiomycota, produces the uniflagellar zoospores that define opisthokonts. Other fungi, however, generate nonmotile reproductive cells, a result of reductive evolution.

Several taxa that historically were grouped with fungi (for example, slime molds) are now classified genetically with protist clades. Slime molds generate populations of cells that migrate into a unified structure called a **fruiting body** for production of reproductive cells. Slime molds are now grouped with amebas (Amebozoa, discussed in Section 20.4). Water molds, plant and animal pathogens of the family Oomycetes, are now recognized as protists (see Section 20.2).

Algae Evolved by Engulfing Phototrophs

Algae are commonly defined as single-celled plants and simple multicellular plants lacking true stems, roots, and leaves. Algal cells contain **chloroplasts** or plastids, membrane-bounded organelles of photosynthesis derived from a cyanobacterium (**Fig. 20.6**). In Earth's biosphere, algae plus bacterial phototrophs form the foundation of all marine and aquatic ecosystems, producing the majority of oxygen and biomass available for Earth's consumers.

The "green plants" (Viridiplantae) include primary endosymbiont algae descended from a common ancestor containing a chloroplast. The chloroplasts of primary endosymbionts are enclosed by two membranes (**Fig. 20.6A**), the inner membrane (from the ancestral phototroph's cell membrane) and the outer membrane (from the host cell membrane as it enclosed its prey). Both **green algae** (**chlorophytes**) and **red algae** (**rhodophytes**) are primary endosymbionts. Their plastids have diverged from their common ancestor to utilize pigments absorbing different ranges of the light spectrum. Genetically, the green algae are the most closely allied with plants.

Several major taxa traditionally considered to be algae turn out to be secondary endosymbionts derived from protist hosts. The symbiotic history is most evident in the **cryptophyte algae** (Fig. 20.6B), which still retain a vestigial nucleus, or **nucleomorph**, derived from the engulfed cell. Their plastid is surrounded by two extra membranes, one from the primary endosymbiont (engulfed alga) cell membrane and one from the secondary host. Other secondary endosymbiont algae include the **chrysophytes**, such as kelps and diatoms. The **dinoflagellates**, photoheterotrophs of the protist clade Alveolata, are secondary or tertiary endosymbionts, decended from a flagellate that consumed one or more types of algae. Dinoflagellates also engage in "kleptoplasty," or "plastid stealing," in which the plastid of a digested prey is retained long enough to derive some photosynthetic energy but ultimately consumed. The variety of endosymbiosis among protists provides clues as to how the original chloroplast evolved within the ancestral algae.

Protists Form Many Divergent Clades

The protists are now known to include several distantly related categories of eukaryotes (see **Table 20.1**). All protists are heterotrophs, commonly predators or parasites, although many also conduct photosynthesis as secondary endosymbionts (algae). Protists are important producers and consumers in marine, aquatic, and soil food webs. In ecology, phototrophic protists are termed phytoplankton and heterotrophs are termed zooplankton, although many in fact are "mixotrophs" that act as both producers and consumers.

Amebas are unicellular organisms of highly variable shape that form **pseudopods**, locomotory extensions of cytoplasm bounded by the cell membrane. Their size can reach several millimeters, visible to the unaided eye. There are two major groups of amebas: Amoebozoa and Cercozoa (see **Table 20.1**). The Amoebozoa or Lobosea, the most familiar kind of amebas, have lobed pseudopods, pseudopods that extend lobes of cytoplasm through cytoplasmic streaming. Most lobed amebas are free-living in aquatic habitats, but some cause human diseases such as meningitis. Lobed amebas also include slime molds, in which individual amebas converge to form a fruiting body. The second group of amebas, the Cercozoa, have

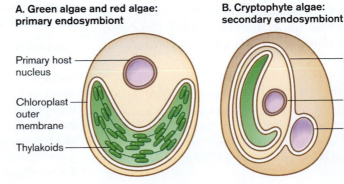

Figure 20.6 Chloroplast evolution: primary and secondary endosymbiosis. A. Green algae (Chlorophyta) and red algae (Rhodophyta) contain chloroplasts (green) that evolved from engulfed cyanobacteria. Primary host cytoplasm is colored yellow; primary host nuclear membrane is colored brown. **B.** Cryptophyte algae contain chloroplast (green), vestigial primary host cytoplasm (yellow), and vestigial nucleus or nucleomorph (brown) from the engulfed primary endosymbiont.

A. Green algae and red algae: primary endosymbiont

Primary host nucleus

Chloroplast outer membrane

Thylakoids

B. Cryptophyte algae: secondary endosymbiont

Membrane derived from engulfed algal cell membrane

Nucleomorph

Cryptophyte nucleus

thin, filamentous pseudopods, often radially arranged like a star, as in heliozoan amebas. Some cercozoans, such as the foraminiferans, form inorganic shells called tests. Fossil foraminiferan tests are common in rock formations derived from ancient seas. Foraminiferan shells formed the white cliffs of Dover in Britain and the stone used to build the Egyptian pyramids.

The **Alveolates** include ciliated protists (ciliates), dino-flagellates, and apicomplexans. Alveolates are known for their complex outer covering, or cortex. The cortex contains networks of vesicles called **cortical alveoli** (**Fig. 20.7A**). Alveoli store calcium ion and in some species grow protective plates. Organisms equipped with paired cilia or flagella are known as **flagellates** and **ciliates**, respectively. Flagella and cilia are essentially equivalent organelles composed of microtubules and enveloped by the cell membrane (**Fig. 20.7B**), although cilia are shorter and more numerous than flagella, and usually cover a broad surface. Alveolata also includes a major group of parasites that no longer possess flagella, the **apicomplexans**. A well-known apicomplexan parasite is *Plasmodium falciparum*, which causes malaria.

Heterokonts (the Heterokonta) are named for their pairs of differently shaped flagella (see **Table 20.1**). Flagellated members of this clade possess two flagella of unequal length and different structure. They include many voracious zooplankton. Nonflagellated heterokonts include the oomycetes, or water molds (formerly classed with fungi), as well as diatoms and kelps (secondary endosymbiotic algae with chloroplasts). **Diatoms** are single cells with unique bipartite shells that fit together like a Petri dish. The shells of diatoms form an infinite variety of different patterns for different species. **Kelps**, also known as brown algae, extend multicellular sheets floating at the water's surface. The kelp known as sargassum weed is famous for its growth in the Sargasso Sea.

Some protists, such as the Euglenozoa or Discicristata (discicristates) and the Excavata (excavates), show extensive evolutionary reduction. The Euglenozoa have mitochondria with distinctive disk-shaped cristae (membrane pockets). They include euglenas, free-living flagellates with a relatively simple cortex. Other Euglenozoa are parasites such as the trypanosomes that cause sleeping sickness and Chagas' disease. The Excavata lack mitochondria altogether, although traces of mitochondrial genes persist in their nuclei. Excavata include the diplomonads such as *Giardia lamblia*, a common contaminant of water supplies.

NOTE: "Algae" may refer to primary algae, or "true algae," as well as secondary endosymbiont algae that derive from protist clades. Fungi (the "true fungi," or Eumycota) are opisthokonts. Several fungus-like organisms have been reclassified within protist clades.

A. Cortex of a paramecium

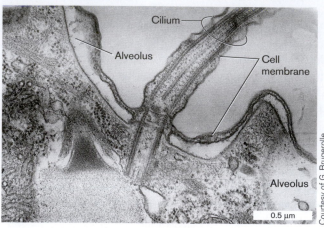

Courtesy of G. Brugerolle

B. Flagellum or cilium

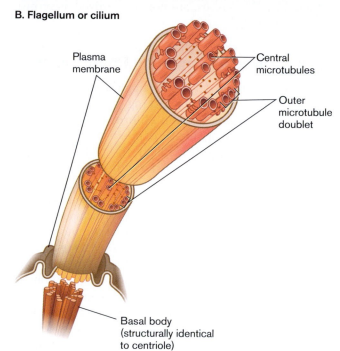

Figure 20.7   Alveolates have alveoli in their cortex.
A. Cortex of *Paramecium tetraurelia* with cilium and alveoli.
B. Flagella and cilia are composed of doublet microtubules and enveloped in cell membrane.

Newly Discovered Eukaryotes as Small as Bacteria

Genes from natural communities continually reveal new species of microbial eukaryotes in previously unknown divisions. Many of the new isolates are single cells as small as bacteria, designated nanoeukaryotes (3–10 μm) and picoeukaryotes (0.5–3 μm). Their genomes in some cases appear comparably downsized; for example, the picoeukaryote *Ostreococcus tauri*, a chlorophyte alga, has a genome of only 8 Mb, barely twice that of the bacterium *Escherichia coli*. Genetic analysis shows that these miniaturized eukaryotes branch deeply from all groups in the phylogenetic tree, potentially doubling the known number of major eukaryotic taxa.

Furthermore, eukaryotes are emerging in "extreme" environments previously believed to be restricted to bacteria and archaea, ranging from Antarctic sea ice to the hyperacidic Tinto River in Spain. The next decade may substantially reshape our overall understanding of the domain Eukarya.

TO SUMMARIZE:

- **Opisthokonta** includes true fungi (Eumycota) and multicellular animals (Metazoa), as well as certain kinds of protists.
- **Viridiplantae** includes green plants and primary endosymbiont algae.
- **Several groups of protists** are Amoebozoa, the unshelled amebas; Cercozoa, shelled amebas; Alveolata, ciliates and flagellates with complex cortical structure; Heterokonta, the kelps, diatoms, and flagellates with nonequivalent paired flagella; and Discicristata and Excavata, primarily parasites showing evolutionary reduction.
- **Many protists are phototrophs as well as heterotrophs**, based on secondary or tertiary endosymbiosis derived from engulfed algae.

20.2 Fungi

Fungi provide essential support for all communities of multicellular organisms. Fungi recycle the biomass of wood and leaves, including substances such as lignin, which other organisms may be unable to digest. Underground fungal filaments called mycorrhizae extend the root systems of most plants, forming a nutritional "internet" that interconnects the plant community (discussed in Chapter 21). Within animal digestive systems, fungi ferment plant materials. On the other hand, pathogenic fungi infect plants and animals, and they contribute to the death of immunocompromised human patients. Still other fungi produce antibiotics such as penicillin, as well as food products such as wine and cheeses (discussed in Chapter 16).

Subtle details of the fungal cell are exploited as targets for antifungal agents. For example, fungal membranes contain ergosterol, an analog of cholesterol not found in animals or plants. Inhibitors of ergosterol biosynthesis, such as the triazoles, are used to treat fungal infections.

The following traits are common to most fungal cells:

- **Absorptive nutrition.** Most fungi cannot ingest particulate food, as do protists, because their cell walls cannot part and re-form, like the flexible pellicle of amebas and ciliates. Instead they secrete digestive enzymes, then absorb the broken-down molecules from their environment.
- **Hyphae.** Most fungi grow by extending multinucleate cell filaments called hyphae (**Fig. 20.8A**). As a hypha extends, its nuclei divide mitotically without cell division, generating a multinucleate cell. Hyphae grow by cytoplasmic extension and branching. A branched mass of extending hyphae is called a **mycelium**.
- **Cell walls containing chitin.** Chitin is an acetylated amino-polysaccharide of immense tensile strength, stronger than steel (**Fig. 20.8B**). Its strength derives from multiple hydrogen bonds between fibers. Chitinous cell walls enable fungi to penetrate plant or animal cells, including tough materials such as wood. Inhibitors of chitin synthesis, such as the polyoxins and nikkomycins, are used as antibiotics against fungal infections.

Fungal Hyphae Absorb Nutrients

At the growing tip of a hypha, the cell membrane expands by incorporating vesicles generated by the endoplasmic reticulum (**Fig. 20.9A**). The vesicle fusion provides phospholipids and proteins to extend the membrane surface area, allowing the cytoplasm to expand rapidly. Expansion of cell membrane and cytoplasm is driven by uptake of hydrogen ions in exchange for potassium ions (**Fig. 20.9B**). Molecules that deregulate K^+/H^+ exchange, such as nystatin, serve as antifungal agents.

Figure 20.8 Fungi grow hyphae with cell walls of chitin.   A. Fungal hyphae extend and form branches, generating a mycelium. **B.** Chitin consists of beta-linked polymers of *N*-acetylglucosamine.

A. Hyphae

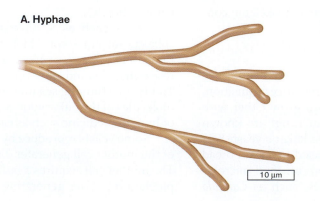

10 μm

B. Chitin

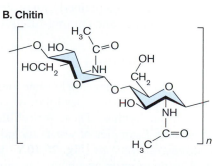

A.

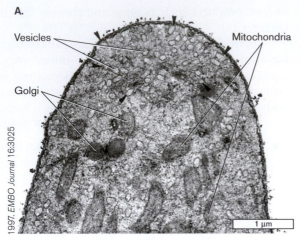

1997. *EMBO Journal* 16:3025

Figure 20.9 **Cellular basis of hyphal extension.**
A. Section through growing tip of a hypha (TEM). Vesicles collect at the tip, where they fuse into the cell membrane, enabling extension. **B.** Hyphal extension is regulated by uptake of protons and extrusion of potassium ions at the tip of the apical growth zone. The absorption zone takes in nutrients, while the storage zone synthesizes and stores cell constituents.

B.

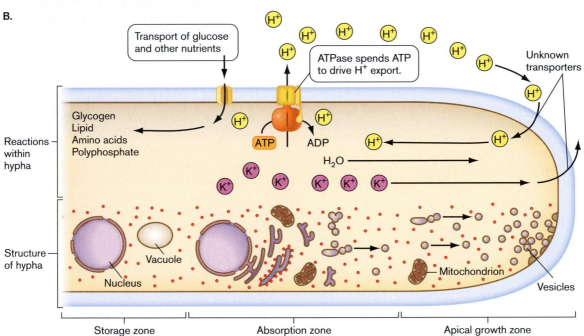

Just behind the hypha's growing tip lies its absorption zone. The absorption zone takes in nutrients from the surrounding medium, such as the cytoplasm of an invaded animal cell. Behind the absorption zone, the older part of the hypha collects and stores nutrients. The tubular shape of the hypha is maintained by a protein cytoskeleton consisting of alpha- and gamma-tubulin.

Unicellular Fungi

Despite the advantages of hyphae for resource acquisition, various species are unicellular, or can grow either with or without hyphae. These unicellular fungi are known as **yeasts**. Examples of yeasts include baker's yeast, *Saccharomyces cerevisiae* (**Fig. 20.10A**), used to leaven bread and to brew wine and beer (for more on food microbiology, see Chapter 16). Other yeasts, such as *Candida*

albicans (**Fig. 20.10B**), are important members of human vaginal flora. Some are important opportunistic pathogens, occurring frequently in AIDS patients; for example, *Pneumocystis carinii* is a yeast-form ascomycete, whereas *Cryptococcus parvum* is a yeast-form basidiomycete (discussed shortly).

Yeasts such as *S. cerevisiae* provide major research subjects for eukaryotic biology. For example, the Nobel Prize in Physiology or Medicine in 2001 was won by yeast researchers Leland Hartwell and Paul Nurse (shared with Tim Hunt, studying sea urchins). Their work revealed key molecules of the eukaryotic cell cycle, relevant to concerns of human medicine such as cancer.

Some yeasts reproduce by **budding**, in which mitosis of the mother cell generates daughter cells of smaller size. The mother cell acquires a bud scar where the smaller one pinched off. After generating a limited number of buds,

A. *Saccharomyces cerevisiae*

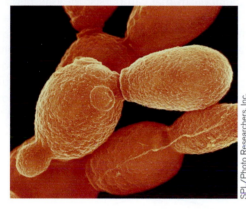

Bud

Bud scar

©Dennis Kunkel/Visuals Unlimited

B. *Candida albicans*

SPL/Photo Researchers, Inc.

C. Yeast alternation of generations

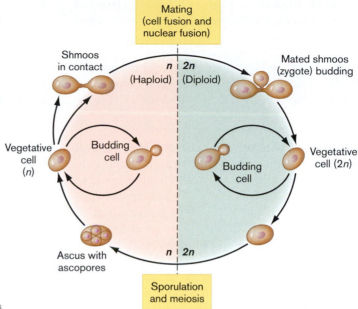

Figure 20.10   Yeast are nonmycelial fungi.   A. Baker's yeast, *Saccharomyces cerevisiae*, reproduces by budding. Note bud scar next to budding daughter cell. SEM. **B.** *Candida albicans*, a common inhabitant of the human vaginal tract, alternates between unicellular and filamentous forms (SEM). **C.** In the life cycle of baker's yeast, haploid cells reproduce many generations by budding.

the mother cell senesces and dies. Thus, yeasts provide a unicellular model system for the process of aging.

> **THOUGHT QUESTION 20.1**   Why would yeasts remain unicellular? What are the relative advantages and limitations of hyphae?

Some yeasts are asexual, whereas others can undergo sexual **alternation of generations** (**Fig. 20.10C**). This life cycle alternates between generation of a haploid population, with a single copy of each chromosome (n), and a diploid population, with a diploid chromosome number ($2n$). In the haploid generation, haploid spores divide and proliferate by mitosis, forming a haploid mycelium. Under environmental signals such as starvation, mating factors induce the haploid cells to differentiate into gamete forms called "shmoos." Gametes of two different mating types fuse, and their nuclei combine to form a zygote. In the diploid generation, the zygote divides mitotically, generating a population of diploids that appear superficially similar to haploid cells. Under stress, particularly desiccation, the diploids undergo meiosis. Meiosis generates an **ascus** that contains four haploid spores.

Many fungi and protists undergo modified versions of alternation of generations, utilizing a wide variety of haploid and diploid structures to accomplish essentially the same genetic tasks. In many fungi, the haploid form predominates; for example, ascomycetes such as *Aspergillus* and *Neurospora* form mainly haploid mycelia. Some fungi lack sexual reproduction entirely. An example is the famous *Penicillium* mold, an asexual ascomycete, from which the antibiotic penicillin is derived. Species that lack sexual life cycles were traditionally classified as "imperfect" fungi. We now know that such species in fact evolved independently from several different fungal clades. Today, asexual species are called **mitosporic fungi** because they generate spores only by mitosis.

> **THOUGHT QUESTION 20.2**   Why would some fungi have lost their sexual life cycle? What are the advantages and limitations of sexual reproduction?

Fungi: Mycelia, Mushrooms, and Mycorrhizae

Different species of fungi show vastly different forms, from the familiar mushrooms, fruiting bodies that can weigh several pounds, to the mycelia of pathogens and the symbiotic partners of algae in lichens. Major phyla of fungi include Chytridiomycota, Zygomycota, Ascomycota, and Basidiomycota. A phylum formerly classified with fungi, Oomycota (Oomycetes), has been reclassified as heterokont protists.

> **NOTE:** The major groups of fungi are also named with the alternative suffix -etes: Chytridiomycetes, Zygomycetes, Ascomycetes, Basidiomycetes.

Chytridiomycota: motile zoospores. The deepest branching clade of fungi is Chytridiomycota (the chytrids), which share with animals and choanoflagellates the motile, flagellated reproductive form known as a zoospore. The zoospore form has been lost by other fungi.

Chytrid species include bovine rumen inhabitants whose hyphae penetrate tough plant material, facilitating digestion. An example is *Neocallomastix* species (**Fig. 20.11A**). Unlike most fungi, *Neocallomastix* is an obligate anaerobe whose mitochondria have evolved into **hydrogenosomes**, organelles that ferment carbohydrates in a pathway generating H_2. Hydrogenosomes are a unique adaptation of certain anaerobic fungi and protists.

Other chytrids are aerobic animal pathogens. **Figure 20.11B** shows the skin of a frog infected by the chytridiomycete *Batrachochytrium dendrobatidis*. *B. dendrobatidis* has caused a widespread die-off of frogs in Central and South America, in an epidemic associated with global warming. The mycelium of *B. dendrobatidis* grows within the frog skin, producing capsules full of diploid zoospores called zoosporangia. Each zoosporangium protrudes through the skin surface, ready to expel zoospores in search of a new host.

The life cycle of a chytrid includes both haploid (gametophyte) and diploid (sporophyte) mycelia (**Fig. 20.11C**). Haploid mycelia produce motile gametes that detect each other by sex-specific attractants. The gametes fuse to produce a motile zygote. The zygote forms a cyst, a cell with arrested metabolism that can persist for long periods. In a favorable environment, the cyst germinates to form a diploid mycelium, or sporophyte. The sporophyte generates zoosporangia full of zoospores. There are two alternative forms of zoosporangia: those that produce diploid zoospores, which form cysts and regenerate the diploid mycelium, and those that undergo meiosis to produce haploid zoospores. The haploid zoospores generate a haploid mycelium (gametophyte) capable of producing haploid gametes.

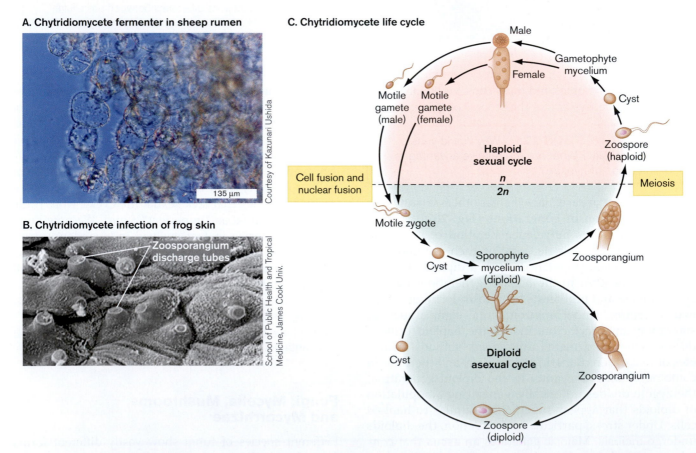

A. Chytridiomycete fermenter in sheep rumen

135 μm

Courtesy of Kazunari Ushida

B. Chytridiomycete infection of frog skin

Zoosporangium discharge tubes

School of Public Health and Tropical Medicine, James Cook Univ.

C. Chytridiomycete life cycle

Male
Gametophyte mycelium
Female
Motile gamete (male)
Motile gamete (female)
Cyst
Haploid sexual cycle
Zoospore (haploid)
Cell fusion and nuclear fusion
n
$2n$
Meiosis
Motile zygote
Cyst
Sporophyte mycelium (diploid)
Zoosporangium
Cyst
Diploid asexual cycle
Zoosporangium
Zoospore (diploid)

Figure 20.11 Chytridiomycete form and life cycle. A. Chytrids such as *Neocallomastix* species ferment complex plant material within the bovine rumen, providing nutrition for the host. **B.** A pathogenic chytrid, *Batrachochytrium dendrobatidis*, infects the skin of a frog. Frog skin cells are penetrated by discharge tubes of diploid zoosporangia about to release zoospores. **C.** Life cycle of a chytrid.

Zygomycota: nonmotile sporangia. The **zygomycetes** and other non-chytridiomycete fungi generate nonmotile spores. Nonmotile spores require transport by air or water or ballistic expulsion (expulsion under pressure) from a spore-bearing organ, called the **sporangium**. A common zygomycete is the bread mold *Rhizopus* (**Fig. 20.12A**). Most zygomycetes, such as *Mucor* species, are soil molds that decompose plant material or other fungi or the droppings of animals (**Fig. 20.12B**). These modest molds fill important niches in all terrestrial ecosystems. Others, particularly *Glomales* species, form mutualistic associations with plant roots known as **vesicular-arbuscular mycorrhizae**. Mycorrhizae expand the roots' absorptive capacity (see Chapter 21). Most trees and crop plants require mycorrhizae for optimal growth.

The life cycle of a zygomycete parallels that of a chytrid in its alternative options of haploid (*n*) and diploid (*2n*) forms (**Fig. 20.12C**). The mechanics differ, however, owing to the lack of motile gametes. The haploid spore (sporangiospore) is disseminated through air currents. A sporangiospore does not directly undergo sexual reproduction; it grows into a haploid mycelium. The haploid mycelium then forms special hyphae whose tips differentiate into gamete cells. The gametes cannot separate from the filament; instead, two gamete-bearing hyphae must grow toward each other in order to fuse and form a **zygospore**. The zygospore undergoes meiosis and generates the sporangium, a haploid structure that releases **sporangiospores**.

> **THOUGHT QUESTION 20.3** What are the advantages and limitations of motile gametes, as compared to nonmotile spores?

Ascomycota: mycelia with paired nuclei. The **ascomycete** fungi are famous in the history of science as well as in the culinary arts (**Fig. 20.13**). The bread mold *Neurospora* was used by George Beadle and Edward Tatum in the 1940s to formulate the one gene–one protein theory. In *Neurospora*, meiosis produces pods (asci) of **ascospores** aligned in rows that reflect the ordered tetrads of meiotic division. The tetrad patterns were used by geneticists to demonstrate the segregation and independent assortment of chromosomes. In other species, by contrast, the asci are packed in large mushroom-like fruiting bodies known as morels (*Morchella hortensis*) and truffles (*Tuber aestivum*). The ascospores of such fruiting bodies are spread by animals attracted by their delicious flavor. Human collectors traditionally use muzzled pigs to detect and unearth the famous underground truffles.

A. Bread mold, *Rhizopus*

Gregory G. Dimijian

B. *Mucor* **diploid hyphae form zygospores**

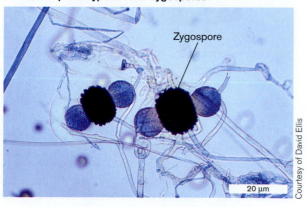

Zygospore

20 μm

Courtesy of David Ellis

C. Zygomycete life cycle

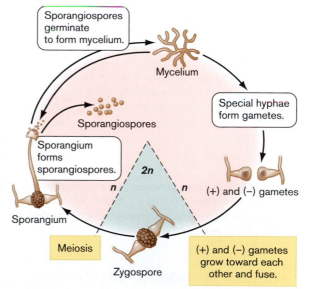

Sporangiospores germinate to form mycelium.

Mycelium

Special hyphae form gametes.

Sporangiospores

Sporangium forms sporangiospores.

2n

n   *n*

(+) and (−) gametes

Sporangium

Meiosis

Zygospore

(+) and (−) gametes grow toward each other and fuse.

Figure 20.12 Zygomycete fungi form nonmotile sporangia. A. *Rhizopus* (bread mold) haploid sporangia contain sporangiospores. **B.** Diploid hyphae of *Mucor* species terminate in zygospores. **C.** The life cycle of zygomycetes involves primarily haploid mycelia. Special hyphae form gametes at their tips. Gametes of different mating types fuse together to form the zygospore. The zygospore undergoes meiosis, regenerating haploid cells that form sporangia. The sporangia release sporangiospores, which germinate to form new mycelia.

A. Morel (an ascomycete fruiting body)

B. Asci containing ascospores

C. Ascomycete life cycle

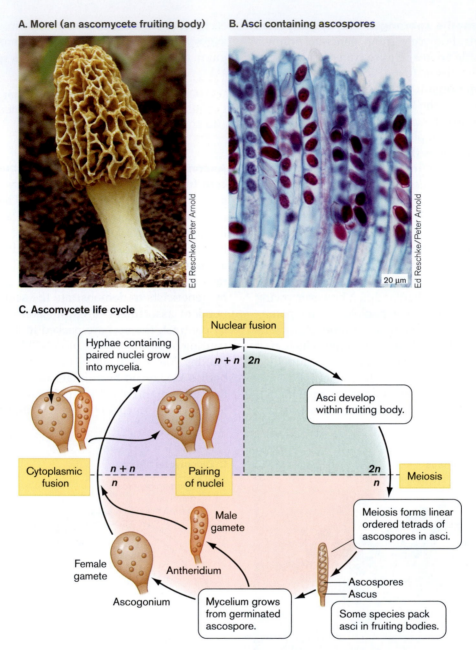

Figure 20.13 Ascomycetes produce large fruiting bodies. A. The culinary delicacies known as morels are fruiting bodies of the species *Morchella hortensis*. The dark pits of the morel are lined with asci. **B.** Ascomycete asci containing ascospores (stained red). **C.** The life cycle of an ascomycete alternates between the diploid and haploid forms. The diploid mycelium produces asci, within which the haploid ascospores are formed.

The ascomycete life cycle (**Fig. 20.13C**) includes a phase in which each cell possesses a pair of separate nuclei, one from each parent (chromosome number is designated $n + n$). The "dikaryotic" (paired-nuclei) phase is generated by haploid mycelia in which male and female reproductive structures fuse, followed by migration of all the male nuclei into the female structure. The paired nuclei then undergo several rounds of mitotic division while migrating into the growing mycelium. In the mycelial tips, the paired nuclei finally fuse (becoming $2n$) and the mycelial tips develop into asci. Each ascus then undergoes meiosis in which the haploid products segregate in the same order that the meiotic chromosomes separated.

Some ascomycetes, such as *Aspergillus* and *Penicillium* species, form small asexual fruiting bodies called conidiophores for airborne spore dispersal (**Fig. 20.14**). While *Penicillium* is known for antibiotic production, *Aspergillus* is a growing medical problem as an opportunistic pathogen of immunocompromised patients. Conidiospore-forming ascomycetes are the major form

Figure 20.14 *Aspergillus.* **A.** *Aspergillus* forms a microscopic asexual fruiting structure called a conidiophore, containing spores in its spherical tip. **B.** Colony of *Aspergillus nidulans* on an agar plate.

A.

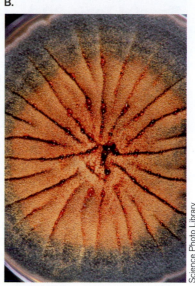

B.

David M. Phillips/Visuals Unlimited

Science Photo Library

of mold associated with dampness in human dwellings; they caused massive damage to homes flooded in the wake of Hurricane Katrina (**Special Topic 20.1**).

Basidiomycetes: cells with paired nuclei form mushrooms. The **basidiomycetes** are known for the exceptional size and diversity of their aboveground fruiting bodies, known as "true" mushrooms (**Fig. 20.15**). Mushrooms produce some of the world's deadliest poisons, such as alpha-amanitin, which inhibits RNA polymerase II. Alpha-amanitin is produced by the amanita, or "destroying angel," a taste of which is usually fatal. Other mushroom species include some of the world's most prized culinary delights, such as the portobello. Many grow in soil, while others, such as *Piptoporus*, grow on tree bark. Some mushrooms have evolved elaborate insect-attracting structures and odors, such as the "starfish stinkhorn," with its ring of bright red horns.

Figure 20.15 Mushrooms and other basidiomycetes form large, complex fruiting bodies. **A.** The amanita makes one of the most dangerous toxins known, alpha-amanitin, an inhibitor of RNA polymerase II. **B.** *Aleuria* mushrooms releasing spores from their gills, to be carried on the wind. **C.** The "starfish stinkhorn" forms a smelly mass of spores that attracts insects for dispersal.

A. *Amanita*

©Forest Buchanan/Visuals Unlimited

B. *Aleuria*

©Bruce Fuhrer

C. Starfish stinkhorn

©Bruce Fuhrer

Special Topic 20.1　Mold after Hurricane Katrina

In the wake of Hurricane Katrina in 2005, much of the city of New Orleans was flooded. Entire neighborhoods were submerged for days or weeks. The flooding of homes full of drywall and organic materials made ideal conditions for the growth of "mold," which generally consists of airborne ascomycete fungi. Mold grew not only on materials submerged, but also on the surfaces above, exposed to water-saturated air. **Figure 1** shows how mold grew up to three feet above the floodline (the highest level submerged) within the kitchen of a flooded home. Unfortunately, most homeowner insurance policies only covered damage "up to the floodline."

Mold growing on damp surfaces releases spores into the air. Inhaled mold spores cause respiratory problems, even at relatively low levels (such as 2,000 spores per cubic meter). In New Orleans, after the waters receded, the mold counts in air within homes reached as high as 650,000 spores per cubic meter. Such spore levels cause asthma attacks and hypersensitivity pneumonitis, a condition in which fever rises and the lungs fail.

All remediation workers were advised to wear an N95 mask respirator. The Natural Resources Defense Council (NRDC) found that even after flooded homes underwent full removal of contaminated items, including furniture, carpets, and drywall down to the studs, airborne spore counts still reached 70,000.

The most common types of mold found by the NRDC researchers were ascomycetes such as *Cladosporium* (**Fig. 2**). *Cladosporium* species form black or dark green colonies of conidia that readily break off and become airborne. Other ascomycetes found at high levels included *Aspergillus* and *Penicillium* species. These organisms occur at low levels even in clean air. Their spores cause allergy attacks and infect immunocompromised individuals. Some ascomycete molds produce mycotoxins, fungal toxins that cause long-term health effects. In addition, some flooded homes showed significant levels of *Stachybotrys* species, known as "black mold," a possible cause of neurological problems and immune suppression. Once such molds establish growth, their eradication is a major challenge.

Figure 1　Mold grows above the floodline.　An undergraduate student volunteer indicates mold colonies growing above the floodline on a kitchen wall in a New Orleans home, six months after flooding.

Figure 2　*Cladosporium* mold.　Conidiophores (branched hyphae) producing conidia (asexual spores) of *Cladosporium cladosporioides* (spore size 5–20 µm; light micrograph).

The mushroom itself is only the fruiting body of the basidiomycete, whose life cycle involves transitions between n, $n + n$, and $2n$ (**Fig. 20.16A**) similar to those of ascomycetes (see **Fig. 20.13C**). In the basidiomycete, however, the fruiting body consists largely of cells with paired nuclei ($n + n$). A few of the paired nuclei fuse to form diploid cells ($2n$) called basidia, which line the gills of the mushroom. The basidia undergo meiosis to form haploid **basidiospores** (*n*). Some types of basidia can release basidiospores under pressure, whereas other basidiospores are transmitted by wind (see **Fig. 20.15B**).

The basidiospores germinate to form underground mycelium. This haploid "primary mycelium" generates gametes that ultimately fuse to form *n* + *n* "secondary mycelium." The primary and secondary mycelia may radiate underground, unseen, until their tips generate mushrooms aboveground, at points approximately equidistant from the origin. The result is a mysterious "fairy ring" of mushrooms (see **Fig. 20.2A**). In fact, these invisible underground hyphae of basidiomycetes and zygomycetes contribute mycorrhizae to extend the root systems of trees (**Fig. 20.16B**).

Oomycota superficially resemble fungi. The **oomycetes**, or water molds, were originally classified as fungi based on the superficial similarity of their infective filaments to fungal hyphae, and the fungus-like appearance of their infections of plants and animals. The term *oomycete* means "egg fungi," referring to the large round egg-like form of the female gamete (**Fig. 20.17A**). The oomycete genus *Phytophthora* ("plant destroyer") includes serious plant pathogens. Destruction of potato crops by *Phytophthora infestans* was largely responsible for the Great Irish Famine in the 1840s, in which a million people died and many others migrated to the United States. Today, another species of *Phytophthora* causes "sudden oak death," the loss of tens of thousands of oaks and other trees on the west coast of the United States (**Fig. 20.17B**).

Despite superficial similarities between oomycetes and fungi, their cell forms and DNA sequences show fundamental differences. Oomycete cell walls contain no chitin, being composed primarily of glucans and cellulose. DNA analysis shows that oomycetes are heterokonts, a class of protists with pairs of dissimilar flagella (Heterokonta, major category in Table 20.1). The heterokont cell form is seen in the primary and secondary zoospores (**Fig. 20.17C**). Oomycetes and fungi offer an interesting case of convergent evolution.

A. Basidiomycete (mushroom) life cycle

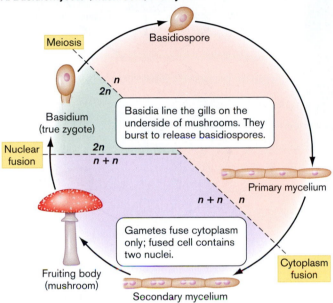

B.

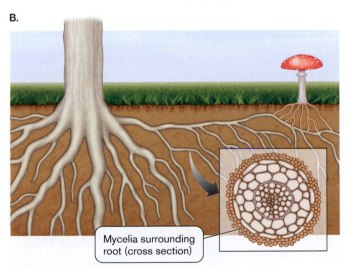

Mycelia surrounding root (cross section)

Figure 20.16   Mushroom life cycle.   A. Haploid basidiospores generate primary mycelium underground, where they form gametes. Gametes of opposite mating types fuse their cytoplasm only, forming secondary mycelium. The parental nuclei remain separate throughout many generations of mitosis during development of the fruiting body or mushroom. As the mushroom matures, the basidia undergo nuclear fusion and meiosis, forming progeny basidiospores. **B.** The underground secondary mycelia of some mushrooms form mycorrhizae with tree roots. Mycorrhizae enhance and extend the absorptive power of the roots.

> **THOUGHT QUESTION 20.4**   Compare the life cycle of an oomycete (**Fig. 20.17C**) with that of a chytridiomycete (**Fig. 20.11C**). How are they similar, and how do they differ?

TO SUMMARIZE:

- **Fungi form hyphae with cell walls of chitin.** Hyphae absorb nutrients from decaying organisms or from infected hosts. Some fungi remain unicellular; these are called yeasts, or mitosporic fungi.
- **Chytridiomycete fungi have motile zoospores.** Motile reproductive forms are a trait shared with animals. Flagellar motility has been lost by other fungi through reductive evolution.
- **Zygomycete fungi form haploid mycelia.** Hyphal tips differentiate into gametes and grow toward each other to undergo sexual reproduction. Some zygomycetes form mycorrhizae that connect the roots of plants.

Figure 20.17 Oomycetes are fungus-like pathogens. **A.** *Phytophthora infestans*, cause of potato late blight. **B.** Oak tree in California dying of *Phytophthora ramosum* infection, known as "sudden oak death." **C.** Life cycle of an oomycete. Primary and secondary zoospores show the pair of differently shaped flagella typical of heterokont protists.

A.

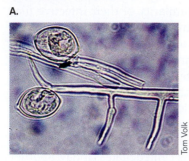

Tom Volk

B.

©Joseph O'Brien, USDA Forest Service

C.

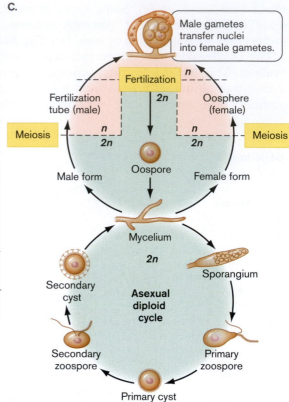

- **Ascomycete fungal mycelia form paired nuclei.** Within the hyphal cells, the paired nuclei fuse, followed by meiosis and development of ascospores. Some ascomycetes form fruiting bodies called conidia.
- **Basidiomycete fungi form mushrooms.** Cells with paired nuclei (secondary mycelium) form large fruiting bodies called mushrooms. The paired nuclei fuse to form the diploid basidium, which generates haploid basidiospores. The basidiospores develop underground hyphae or mycorrhizae that interconnect plant roots.
- **Oomycetes are heterokont protists.** Oomycete cells lack chitin, and their reproductive forms include motile cells with pairs of dissimilar flagella. They form infective filaments that superficially resemble fungi.

20.3 Algae

Algae serve as primary producers in all ecosystems, most crucially for aquatic and marine habitats. In aquatic and marine ecology, the algae, together with photosynthetic bacteria, are known as **phytoplankton** (see Chapter 21). All algae possess chloroplasts. The primary endosymbiotic algae, or "true algae," are products of a single ancestral endosymbiosis that also gave rise to land plants. The biochemistry and cell structures of true algae and land plants are similar; thus, the study of photosynthesis in the alga *Chlorella* by Martin Kamin, Melvin Calvin, and others has advanced our understanding of plant physiology (discussed in Chapter 15).

Other photosynthetic eukaryotes, or secondary endosymbiotic algae, arose from protists that engulfed a primary or secondary endosymbiont. Secondary endosymbionts show "mixotrophic" nutrition, involving both phototrophy and heterotrophy. For example, dinoflagellate "algae" are voracious predators on smaller protists. The heterokont algae, diatoms and kelps, are covered in this section. Dinoflagellates are covered under alveolates (Section 20.5).

Green Algae, or Chlorophyta

The primary endosymbiotic algae include two major clades: Chlorophyta, or green algae, and Rhodophyta, or red algae, although not all members of each group appear green or red, respectively. In most cases, the chlorophytes appear green because the color of their chlorophyll *a* is not obscured by secondary pigments.

Unicellular green algae. An important model system for genetics and phototaxis is *Chlamydomonas reinhardtii*, a unicellular alga common in freshwater systems as well as Antarctic pools. The genetics of cell cycle regulation in *C. reinhardtii* provides clues to the role of homologous regulators in human tumor cells.

C. reinhardtii possesses a symmetrical pair of flagella, a common pattern for green and red algae and their gametes (**Fig. 20.18A**). The alga swims by beating its flagella

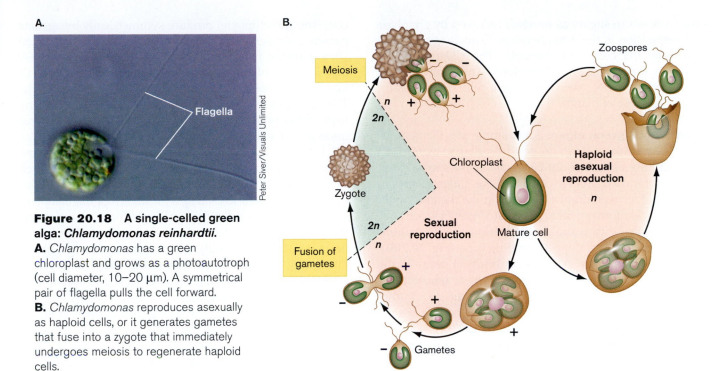

Figure 20.18   A single-celled green alga: *Chlamydomonas reinhardtii*.
A. *Chlamydomonas* has a green chloroplast and grows as a photoautotroph (cell diameter, 10–20 μm). A symmetrical pair of flagella pulls the cell forward.
B. *Chlamydomonas* reproduces asexually as haploid cells, or it generates gametes that fuse into a zygote that immediately undergoes meiosis to regenerate haploid cells.

back toward the cell, like a breast stroke. *Chlamydomonas* cells are mostly haploid, reproducing by asexual cell division (**Fig. 20.18B**). Sexual reproduction occurs by fusion of opposite mating types to form a zygote, which loses flagella and forms a spiny protective coat. The zygote undergoes meiosis to regenerate haploid cells.

The cell ultrastructure of *Chlamydomonas* is typical of algal cells (**Fig. 20.19**). The nucleus is cupped by a single chloroplast, which is surrounded by a double membrane. The double membrane indicates a primary endosymbiont; the inner membrane derives from the ancestral bacterium, and the outer membrane derives from the engulfing host. Within the chloroplast is a pyrenoid, an organelle that concentrates bicarbonate (HCO_3^-) and converts it to CO_2 for fixation. The pyrenoid is surrounded by one or more starch bodies for energy storage. The starch is

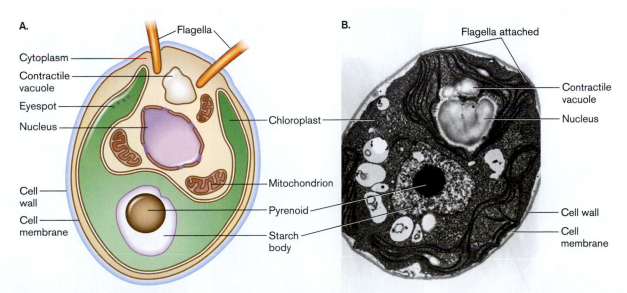

Figure 20.19   Cell structure of *Chlamydomonas reinhardtii*.   A. The cell membrane is surrounded by a cell wall of cellulose and glycoproteins. The single chloroplast fills much of the cell and wraps around the nucleus. Within the chloroplast lies a pyrenoid, a structure for concentrating bicarbonate ion for conversion to CO_2. The pyrenoid is surrounded by starch bodies that store high-energy compounds. A contractile vacuole maintains constant osmotic pressure. **B.** Electron micrograph of *C. reinhardtii* shows the nucleus, contractile vacuole, chloroplast, pyrenoid, and other organelles. *Source:* Ohad, et al. 1967. Rockefeller U. Press.

broken down to sugars as needed, followed by glycolysis and respiration in the mitochondria. Osmolarity is maintained by the contractile vacuole. The *Chlamydomonas* cell is encased in a cell wall predominantly composed of glycoprotein. Other green algae have cellulose cell walls, similar to those of plants.

Filamentous green algae. An example of a filamentous alga is *Spirogyra*, a common pond dweller known for its spiral chloroplasts (**Fig. 20.20A**). *Spirogyra* species form long, unbranched chains of cells. Like *Chlamydomonas*, the haploid form predominates; but unlike the unicellular alga, *Spirogyra* forms neither flagellated gametes nor zoospores. Instead, its sexual reproduction requires alignment of two stalks of opposite mating type (**Fig. 20.20B**). Cells **conjugate** (form cytoplasmic bridges) between the two stalks. The "male" gametes are those whose cytoplasm inserts through the conjugation bridge to join that of the "female" gamete. As the gametes fuse, a row of empty cell walls is left behind (**Fig. 20.20C**). The zygote eventually hatches and germinates a new chain of cells.

Sheet-forming green algae. Some forms of marine algae grow in undulating sheets. An example familiar to beach bathers is the "sea lettuce" *Ulva* (**Fig. 20.21A**). Sheets of *Ulva* can extend over many square meters, although they are only two cells thick (see **Fig. 20.21A** inset). The *Ulva* life cycle consists of classic alternation of generations between haploid and diploid forms (**Fig. 20.21B**). Haploid sheets of

cells (the gametophyte) produce symmetrically biflagellate gametes, similar to unicellular *Chlamydomonas*. But when *Ulva* gametes fuse, the zygote grows into an immense diploid sheet of cells. This sporophyte (diploid multicellular body) appears similar in form to the gametophyte. The sporophyte eventually undergoes meiosis, releasing haploid zoospores with two pairs of flagella. The zoospores undergo mitosis and regenerate the gametophyte.

Like the fungi, marine and freshwater algae show many complex and beautiful forms. Colonial algae such as *Volvox* (**Fig. 20.22A**) generate geodesic spheres of biflagellate cells similar to *Chlamydomonas*. Their flagella point outward from the colony, propelling it forward and drawing nutrients across its surface. Each cell of *Volvox* connects by cytoplasmic bridges to five or six of its neighbors. The colony reproduces by generating daughter colonies within the sphere, which grow until the outer sphere falls apart, liberating the daughters.

Other green algae, such as *Cymopolia*, form long calcified stalks punctuated by tufts of green filaments, where most of the photosynthesis occurs (**Fig. 20.22B**). Yet another group, the **siphonous algae**, are known for forming long, siphon-like tubes and fronds without cell partitions—in effect, giant organisms that are unicellular (**Fig. 20.22C**). The siphonous species *Caulerpa taxifolia* has become a notorious pest, destroying coastal communities in the Mediterranean Sea and off the coast of California. Originally bred for aquariums, *C. taxifolia* is exceptionally cold resistant and produces toxins that prevent grazing.

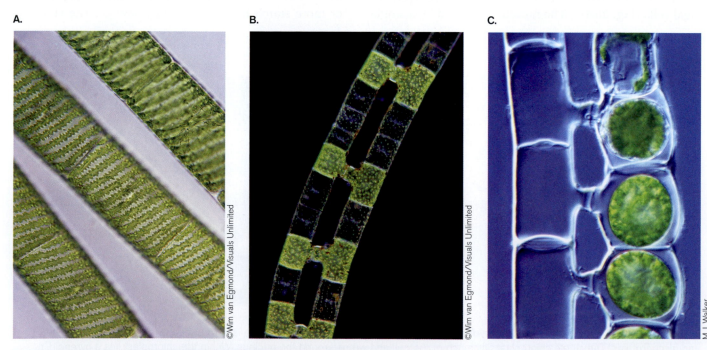

A. **B.** **C.**

Figure 20.20 Stalks with spiral chloroplasts: *Spirogyra*. A. *Spirogyra* species grow in long multicellular filaments, typically 25 µm in width and several centimeters long. Each cell contains one or more chloroplasts that spiral around the cytoplasm. **B.** Sexual reproduction involves conjugation between cells of two mating types. The cytoplasm from each male cell exits from its cell wall and enters the female cell. **C.** Gamete fusion is complete, generating zygotic spores.

A.

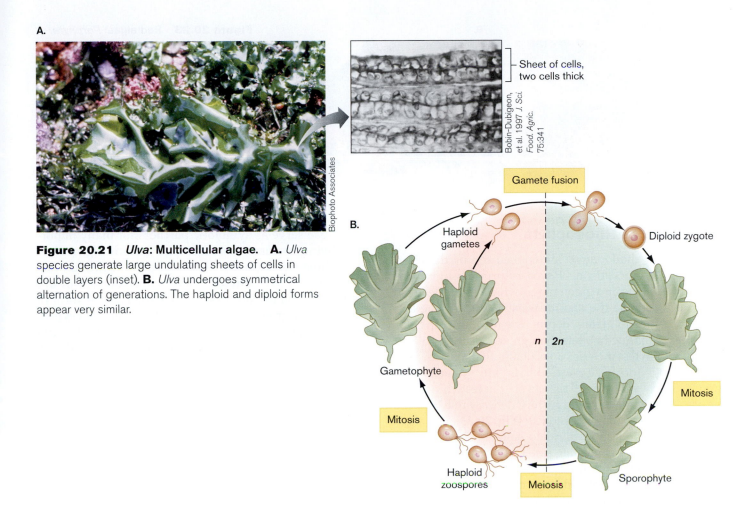

Sheet of cells, two cells thick

Bobin-Dubigeon, et al. 1997 *J. Sci. Food. Agric.* 75:341

Figure 20.21 *Ulva*: **Multicellular algae. A.** *Ulva* species generate large undulating sheets of cells in double layers (inset). **B.** *Ulva* undergoes symmetrical alternation of generations. The haploid and diploid forms appear very similar.

B.

Gamete fusion

Haploid gametes

Diploid zygote

Gametophyte

n | *2n*

Mitosis

Mitosis

Haploid zoospores

Meiosis

Sporophyte

A. *Volvox*

B. *Cymopolia*

C. *Caulerpa*

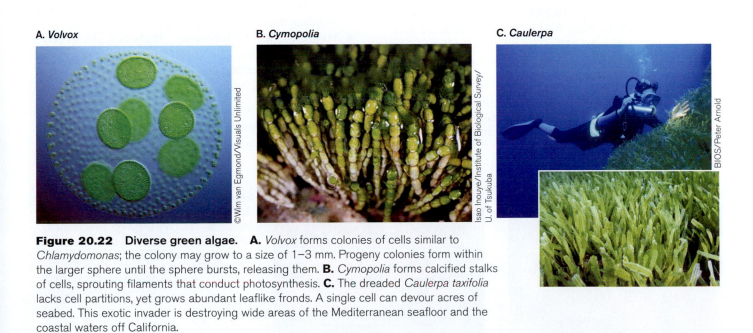

©Wim van Egmond/Visuals Unlimited

Isao Inouye/Institute of Biological Survey/ U. of Tsukuba

BIOS/Peter Arnold

Figure 20.22   Diverse green algae. A. *Volvox* forms colonies of cells similar to *Chlamydomonas*; the colony may grow to a size of 1–3 mm. Progeny colonies form within the larger sphere until the sphere bursts, releasing them. **B.** *Cymopolia* forms calcified stalks of cells, sprouting filaments that conduct photosynthesis. **C.** The dreaded *Caulerpa taxifolia* lacks cell partitions, yet grows abundant leaflike fronds. A single cell can devour acres of seabed. This exotic invader is destroying wide areas of the Mediterranean seafloor and the coastal waters off California.

A.

©Univ. of California Museum of Paleontology

B.

©FoodCollection/SuperStock

Figure 20.23   Red algae: *Porphyra.*
A. *Porphyra* forms large multicellular sheets. **B.** Known as *nori*, the harvested and dried red sheets are used to wrap sushi, a Japanese delicacy.

Red Algae, or Rhodophyta

Red algae, or rhodophytes are colored red by the photopigment phycoerythrin. Phycoerythrin absorbs efficiently in the blue and green range, which green algae fail to absorb. Because the shorter wavelengths of blue and green penetrate to greater depths, the red algae can colonize deeper marine habitats than green algae.

Rhodophytes include unicellular, filamentous, and multicellular forms. Several kinds are human food sources. *Porphyra* forms large sheets harvested in Japan for use as nori, for wrapping sushi, a delicacy that includes rice, vegetables, and uncooked fish (**Fig. 20.23**). Red algae contain valuable polymers called sulfated polygalactans (sugar polymers with sulfate side chains). Sulfated polygalactans include agar, used to solidify microbial growth media; agarose, a processed sugar derivative used to form electrophoretic gels; and carrageenan, an additive used in processed foods.

Rhodophytes such as *Plocamium* form delicate branched filaments (**Fig. 20.24A**). Still others, the **coralline algae**, calcify their fronds into hardened shapes similar to corals (**Fig. 20.24B**). The calcified bodies of corallines grow at greater depths than corals, and they often form the foundation for coral reefs.

> **THOUGHT QUESTION 20.5**   How are coralline algae able to grow at greater depths than coral?

Algae serve as symbiotic partners for many important living systems. For example, certain algae grow in intimate association with fungi to form unified structures called **lichens**, important colonizers of dry and cold habitats (discussed in Chapter 21). A form of algal-fungal ground cover similar to lichens is **cryptogamic crust**, common on desert soil. Still other algae grow within the cells of paramecia, hydras, and corals, providing photosynthetic nutrition in exchange for protection.

Secondary Endosymbiotic Algae: Diatoms and Kelps

Several phototrophs long classified as algae have since been reclassified as protists that once engulfed a primary endosymbiotic alga. Their cell structure shows two traits that distinguish them from primary endosymbionts:

- **More than two membranes surround the chloroplast.** The extra membranes derive from the cell membrane of the engulfed alga.
- **Mixotrophic metabolism includes heterotrophy.** By contrast, the primary endosymbionts are near-obligate autotrophs, catabolizing only the simplest substrates, such as acetate.

Several major groups of secondary-symbiont algae are heterokonts (stramenopiles). These include the diatoms, phylum Bacillaraceae; the brown algae, or Phaeophy-

A. *Plocamium*

©Seatrends

B. *Mesophyllum*

©Keoki Stender

Figure 20.24   Diverse red algae.
A. *Plocamium* forms delicate feather-like fronds. **B.** The coralline alga *Mesophyllum* forms calcified fronds that provide a foundation for coral reefs.

ceae, such as kelps; the golden algae, or Chrysophyceae, mainly pale-colored flagellates; the yellow-green algae, or Xanthophyceae; and other less studied forms of phytoplankton. Flagellated species of heterokont algae always show a pair of differently shaped flagella. Typically, one flagellum is brush-like with side branches, while the other is shorter, often wrapped around the cell, its function uncertain.

Diatoms, or Bacillariophyta. Diatoms are unicellular heterokonts found ubiquitously in aquatic and marine waters. The cell grows a unique kind of bipartite shell called a **frustule**. The frustule is composed of silica (cross-linked silicon dioxide, SiO_2). Frustules of decomposed diatoms eventually sediment on the ocean floor, where they contribute to sedimentary rocks or wash up to shore as beach sand. Their diverse species show high sensitivity to environmental parameters such as pH, and the frequency of their shells in sediment can be used to track a lake's environmental history.

Frustules of different species form an extraordinary range of shapes with intricate pore formations (**Fig. 20.25A**). The shapes fall into two classes, either centric, with radial symmetry, or pennate, with bilateral symmetry. All, however, pose a unique challenge

to cell division: As the diatom grows and fissions, each daughter cell receives one parental half of the frustule while forming a new half fitting *within* the parental half, like the bottom dish of a Petri plate (**Fig. 20.25B**). Thus, each generation results in an inexorable decline in size of the organism. As its cell size reaches a critical point, the diatom must undergo meiosis to generate gametes. When the gametes fuse, they form a zygote that expands to the original size of the diatom before forming the first frustule.

Brown algae, or Phaeophyceae. Brown algae, such as kelps, possess vacuoles of leucocin, an oily lipid for energy storage whose color gives the organism a brown or yellow tint. Kelps are familiar to ocean bathers as the long, dark brown blades that root near the beach until the surf rips the blades off and tosses them ashore (**Fig. 20.26A**). Kelps support important communities of multicellular organisms, known as kelp forests. The most famous kelp forests are those of the unrooted **sargassum weeds**, which float upon the Sargasso Sea. Sargassum consists of stalks with photosynthetic blades and round gas bladders to keep the organism afloat (**Fig. 20.26B**). Sargassum supports a complex food chain of invertebrate and vertebrate animals, including worms, crabs, and fish.

A.

©Dennis Kunkel/Phototake

Figure 20.25 Diatoms. A. Diatoms of various species (colorized SEM). **B.** Life cycle of a centric diatom. Asexual cell division requires repeated formation of an in-fitting frustule half, resulting in progressive decrease in size. At a certain size limit, the diatom must undergo meiosis to form flagellated gametes. As gametes fuse, they regenerate to give a diatom of the original size.

B.

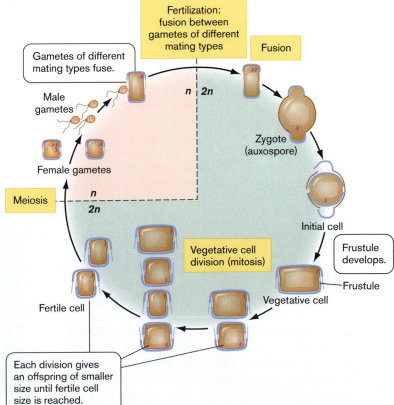

A. Kelp

©Mark Spencer/AUSCAPE/Minden Pictures

B. Sargassum weed

Visuals Unlimited

Figure 20.26   Kelp forests.   A. Kelps form broad leaflike sheets attached to the seabed. **B.** *Sargassum natans* forms the basis of the Sargasso Sea. The brown alga forms stalks with leaflike blades and round gas bladders to keep the alga afloat.

TO SUMMARIZE:

- **Chlorophyta (green algae)** absorb red and blue light and grow near the top of the water column. Green algae include unicellular, filamentous, and sheet forms.
- **Rhodophyta (red algae)** have the accessory photopigment phycoerythrin, which absorbs green and longer-wavelength blue light, enabling growth at greater depths. Red algae include species of diverse forms, many of which are edible for humans.
- **Secondary endosymbiotic algae** are derived from protists that had engulfed primary symbiotic algae. They are mixotrophs, combining phototrophy and heterotrophy.
- **Diatoms are heterokonts with silicate shells called frustules.** Diatoms replicate by an unusual division cycle generating successively smaller frustules.
- **Kelps are heterokonts that grow in long sheetlike fronds.** Kelps play an important role in the ecology of the open ocean as well as the ecology of marine beaches.

20.4 Amebas and Slime Molds

The ameba is familiar to most of us as an apparently amorphous form of microscopic life, capable of engulfing and consuming prey in a dramatic fashion (**Fig. 20.27**). While the ameba's shape is exceptionally variable, the pseudopods, or "false feet," that it extends, far from being amorphous, are complex structures that undertake highly controlled and specific movements. Most amebas are free-living predators in soil or water. They extend in size up to 5 mm, large enough to consume bacteria, algae, ciliates,

smaller amebas, and even invertebrates such as rotifers. A few are dangerous parasites of humans or animals. Furthermore, free-living amebas can harbor bacterial pathogens such as *Legionella pneumophila*, which contaminates water supplies and air ducts. The bacteria cause legionellosis, an often fatal form of pneumonia. The host ameba enables the pathogen's persistence and transmission to human hosts.

> **THOUGHT QUESTION 20.6**   What might happen when an ameba phagocytoses algae?

> **NOTE:** The taxonomy of amebas and slime molds remains problematic, with diverse views as to the number of clades, their relatedness, and their degree of divergence.

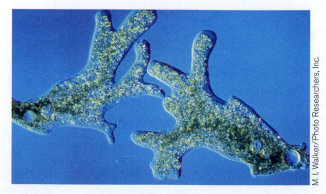

M. I. Walker/Photo Researchers, Inc.

Figure 20.27   Amebas with lobe-shaped pseudopods. *Amoeba proteus* moves by extending its pseudopods.

Diverse Kinds of Amebas

Ameba-like cell forms occur in many different species. Some species persist as an ameba throughout all or most of the life cycle. Others can convert to flagellates, particularly when the habitat fills with water, a low-viscosity condition favoring flagellar motility. Still other protist species, such as dinoflagellates, never become fully amebic, but they can extend a pseudopod to engulf prey.

Different kinds of amebas have different kinds of pseudopods. The "classic" ameba (phylum Amoebozoa or Lobosea) has lobe-shaped pseudopods (**Fig. 20.27**). Lobe-shaped pseudopods have the most variable shape. A different form is the sheetlike pseudopod, or **lamellar pseudopod**. Lamellar pseudopods are extended by dinoflagellates. Similar lamellar pseudopods are generated by human white blood cells such as leukocytes. Finally, needlelike pseudopods, or filopodia, are thin extensions reinforced by microtubules. Filopodia are made by the shelled amebas (phylum Cercozoa), such as Foraminifera (spiral shells) and Radiolaria (radial form).

Amebas Move by Cytoplasmic Streaming into Pseudopods

The extension of lobe-shaped and lamellar pseudopods has been studied closely for its relevance to human white blood cells (for more on white blood cells as host defenses, see Chapter 23). The mechanism of pseudopod motility remains poorly understood, but it is known to involve a sol-gel transition between cortical cytoplasm (just beneath the cell surface) and the cytoplasm of the deeper interior (**Fig. 20.28**). The tip of a pseudopod contains a gel of polymerized actin beneath its cell membrane. From the center of the ameba, liquid cytoplasm (sol) containing actin subunits streams forward along microtubular "tracks," powered by ATP hydrolysis. The actin subunits stream into the pseudopod, where they polymerize, forming a gel. The gel region grows, pushing the membrane forward and extending the pseudopod. As the gel is pushed backward, it resolubilizes to continue the cycle.

Ameba Reproduction and Genetics

Amebas can be uninuclear or multinuclear. They are usually haploid and reproduce asexually by nuclear mitosis, without dissolution of the nuclear membrane, followed by fission of the cytoplasm. Some species do have reproductive alternatives, such as cyst formation, gamete fusion and meiosis, and even growth of flagella in a favorable habitat.

> **THOUGHT QUESTION 20.7** What kind of habitats would favor a flagellated ameba?

Ameba genetics is poorly understood, but the sequence of at least one genome is near completion, that of the intestinal parasite *Entamoeba histolytica*. The strain being sequenced contains 20 Mb of DNA in 14 chromosomes; some of these are linear, whereas others are circular. Closely related strains show considerable variation in organization, suggesting that ameba genomes (like their cytoplasm) undergo extensive rearrangement.

Slime Molds Generate Multicellular or Multinucleate Fruiting Bodies

Some amebas conduct a life cycle in which thousands of individuals aggregate into a complex differentiated fruiting body (**Fig. 20.29A**). Such an organism is called a **slime mold**. Slime molds, as the name implies, were originally classed with fungi because their fruiting bodies superficially resemble fungal reproductive forms. There are two kinds of slime molds: **cellular slime molds**, in which the individual cells remain cellular, and **plasmodial slime molds**, in which the mass of aggregating cells becomes a multinucleate single cell.

Figure 20.28 Pseudopod motility.
A pseudopod extends by flow of liquid cytoplasm (sol state) followed by actin polymerization (gel state). As actin polymerizes, the cell rotates down toward the substrate like a tank tread.

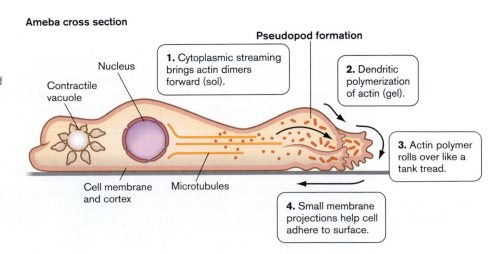

Ameba cross section

Pseudopod formation

1. Cytoplasmic streaming brings actin dimers forward (sol).

2. Dendritic polymerization of actin (gel).

3. Actin polymer rolls over like a tank tread.

4. Small membrane projections help cell adhere to surface.

Nucleus

Contractile vacuole

Cell membrane and cortex

Microtubules

A.

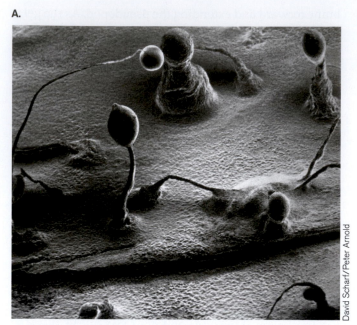

David Scharf/Peter Arnold

Figure 20.29   A cellular slime mold, *Dictyostelium discoideum*.   A. Fruiting bodies of *D. discoideum* (composite light micrograph). **B.** Life cycle of *D. discoideum*.

B.

Fruiting body

Cyst

24 hr

Meosis

Ameba

n $2n$ n

Macrocyst

Gamete fusion

Starvation

0 hr

4 hr

cAMP waves

Aggregation

8 hr

18 hr

16 hr

14 hr

12 hr

Slug

Tipped aggregate

as well. Successive waves of cyclic AMP continue to attract thousands of amebas to the center, where they pile on top of each other to form a slug as long as 1 mm. The slug then migrates, attracted by light and warmth, to find an appropriate place to form a fruiting body and disperse its spores.

The slug at last differentiates into a fruiting body, a spherical sporangium supported upon a stalk of largely empty cells that emerges from a basal disk. The sporangium then releases spores (also called cysts), which are dispersed on air currents and can remain viable for several years. When a spore detects chemical signals from bacteria, it germinates an ameba to feed on them.

Note that the entire reproductive cycle just described is asexual; the amebas and their differentiated structures remain haploid throughout. *D. discoideum* amebas do have a sexual alternative in which cells of opposite mating type can fuse to form a diploid zygote, then undergo meiosis, restoring haploid amebas.

By contrast, the amebas of a plasmodial slime mold, such as *Physarum polycephalum*, aggregate only following gamete fusion to form diploid cells. The diploids then aggregate and become transformed into **plasmodium**, a giant multinucleate cell that can spread over an area of many square centimeters. Out of the plasmodium arise fruiting bodies whose sporangia undergo meiosis, producing haploid spores. A large plasmodium can occasionally be seen as a yellow mass of slime spreading over decaying wood.

NOTE: Distinguish the term plasmodium (a large multinucleate cell) from the genus *Plasmodium* (an apicomplexan parasite, such as *Plasmodium falciparum*, which causes malaria.)

A well-studied example of a cellular slime mold is *Dictyostelium discoideum*, historically an important model system for multicellular development (**Fig. 20.29**). *D. discoideum* amebas are relatively small, about 10 μm, but large enough to consume bacteria. They can be cocultured on a plate with *Escherichia coli*. As the haploid amebas consume bacteria, they divide asexually until their food runs out. At this point, a few amebas begin to emit the aggregation signal molecule cyclic AMP. An ameba emiting cyclic AMP attracts other amebas nearby, which move toward the center and begin emitting cyclic AMP

A. Radiolarian tests

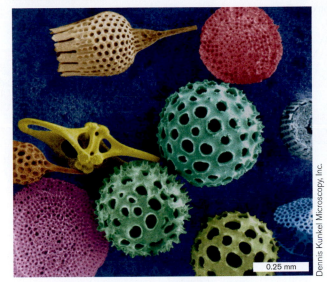

0.25 mm

Dennis Kunkel Microscopy, Inc.

B. A foraminiferan

C. Hemleben/©Cushman Foundation for Foraminiferal Research, Inc.

Figure 20.30   Filamentous and shelled amebas.
A. Shells (tests) of radiolarians (colorized SEM). **B.** A live foraminiferan, *Globigerinella aequilateralis* (darkfield).

Filamentous and Shelled Amebas

A distinct clade of amebas (phylum Cercozoa) form needlelike pseudopods. Most of these amebas are encased by mineral shells called **tests** (**Fig. 20.30A**). A major class of shelled amebas is the **radiolarians**, whose shells are made of silica perforated with numerous holes through which pseudopods appear to radiate in all directions. Many different kinds of radiolarians exist today, and many can be recognized in fossil rock. A second class of shelled amebas is the **foraminiferans** (**Fig. 20.30B**). The foraminiferans, or forams, generate shells of calcium carbonate as chambers laid down in helical succession. Their pseudopods all extend from one opening in the most recent chamber.

Radiolarians and forams grow in marine and aquatic habitats. Their shells make up a substantial part of reef formations, sedimentary rock, and beach sand. Forams, in particular, play a key role in geological surveys as indicators of petroleum deposits.

TO SUMMARIZE:

- **Amebas** have an amorphous form that moves using pseudopods. In different species, pseudopods are lobe-shaped, lamellar, or filamentous (filopodia).
- **Cytoplasmic streaming** through cycles of actin polymerization and depolymerization drives the extension and retraction of pseudopods.
- **Slime molds** show an asexual reproductive cycle in which individual amebas attract each other and aggregate to form a slug. The slug develops into a fruiting body that forms spores.

- **Radiolarians** have silicate shells penetrated by filamentous pseudopods.
- **Foraminiferans** have calcium carbonate shells with helical arrangement of chambers, the most recent of which opens to extend filamentous pseudopods.

20.5 Alveolates: Ciliates, Dinoflagellates, and Apicomplexans

The Alveolata include voracious predators such as the ciliated protists (**Fig. 20.31A**). Alveolates are named for the flattened vacuoles called alveoli within their outer cortex (**Fig. 20.31B**; see also **Fig. 20.7A**). Some alveoli contain plates of stiff material, such as protein, polysaccharide, or minerals. Besides alveoli, most alveolate protists possess other kinds of cortical organelles such as extrusomes for delivery of enzymes or toxins, bands of microtubules for reinforcement, and whiplike cilia or flagella. The alveolate cell form is highly structured, in contrast to the amorphous shape of amebas. Major groups of alveolates include ciliates, dinoflagellates, and apicomplexans.

Ciliates Possess Numerous Cilia for Motility and Prey Capture

A diverse group of alveolates known as Ciliophora, or ciliates, possess large numbers of cilia, short projections containing 9+2 microtubules, whose whiplike action is driven by ATP (for review, see Appendix 2). The cilia

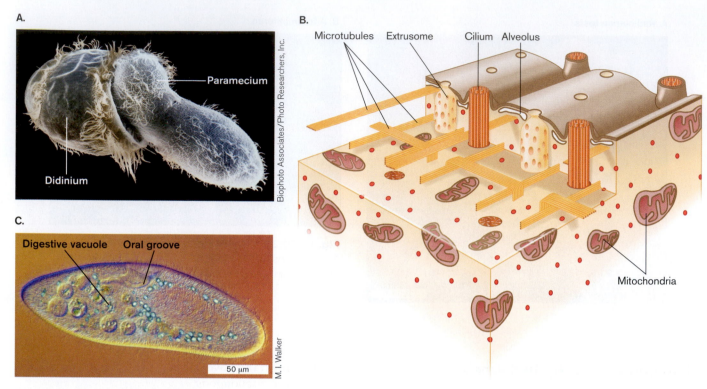

Figure 20.31   Ciliated protists.   A. *Didinium* consuming *Paramecium* (SEM). **B.** Cortical structure of a ciliate (colorized phase contrast). Beneath the outer membrane lie flattened sacs of fluid called alveoli. The cilia, composed of 9+2 microtubules, are rooted in a complex network of lateral microtubules. Extrusomes secrete various enzymes or toxins. **C.** Intracellular organelles of a paramecium include the contractile vacuole, digestive vacuoles, and the oral groove for ingestion.

beat in coordinated waves that maximize the efficiency of motility. Cilia serve two functions:

- **Cell propulsion.** Coordinated waves of beating cilia, usually covering the cell surface, propel the cell forward.
- **Acquiring food.** By generating water currents into the mouth of the cell, cilia, usually arranged in a ring around the mouth, bring food into the cell.

Ciliate cell structure. *Paramecium* is one of the most studied ciliates. Paramecia feed on bacteria and in turn are consumed by larger ciliates such as *Didinium* (**Fig. 20.31A**). They can also take up smaller particles through endocytosis by specialized pores in their cortex, called parasomal sacs.

The cell structure of *Paramecium* includes an **oral groove** for uptake of food driven by the beating cilia (**Fig. 20.31C**). Once ingested through the oral groove, a food particle travels within the digestive vacuole in a circuit around the cell. The digestive vacuole ultimately empties into the cytoproct, a specialized vacuole for the discharge of waste outside the cell. Paramecia maintain osmotic balance by means of a **contractile vacuole**, a vacuole that withdraws water from the cytoplasm to shrink it or con-

tracts to expand it. Contractile vacuoles are widespread among protists and algae, but their mode of action has been most studied in paramecia.

Genetics and reproduction. Most ciliates have a complex genetic system involving one or more **micronuclei** and **macronuclei**. The micronucleus contains a diploid set of chromosomes that undergoes meiosis for sexual exchange. For gene expression, however, the micronucleus generates hundreds of copies of its DNA within a macronucleus. This phenomenon has surprising applications for the study of human aging (**Special Topic 20.2**).

The macronucleus is formed by genetic rearrangement and excision of some micronuclear DNA. Only macronuclear genes are transcribed to RNA and translated to protein. When a ciliate reproduces asexually, the micronucleus undergoes mitosis, whereas the macronucleus divides by a different mechanism that is poorly understood. Cell division occurs across the long axis, necessitating generation of a new oral groove for the posterior daughter cell and a new cytoproct for the anterior daughter cell—again, a process poorly understood.

Most ciliates are diploid and never produce haploid gamete cells. Instead, they reproduce sexually

Special Topic 20.2   A Ciliate Model for Human Aging

An unusual model for a molecular process of human aging is seen in the ciliate *Oxytricha fallax*. Like all ciliates, *Oxytricha* cells exchange their micronuclei during sexual reproduction by conjugation (**Fig. 1A**). After transfer, when the new micronuclei form, the chromosomes of the micronucleus are subsequently replicated into a thousand copies, forming a macronucleus to serve gene expression.

As the macronucleus forms, the chromosomes fragment into many pieces. During this process, virtually all noncoding DNA, about 95% of the genome, is eliminated. In *Oxytricha*, the fragmentation is particularly extreme; the genome is cut into tens of thousands of pieces until each cut-down chromosome contains only one to three genes. And all the tens of thousands of cut-down chromosomes must possess new **telomeres**, special end sequences of DNA that protect the chromosome termini. A telomere is a special form of DNA in which the 3′ OH end of the double helix extends beyond the 5′ OH end and doubles back to form base pairs that enable "filling in" of the last portion of the helix.

In multicellular animals, telomeres usually shorten throughout the life of an individual. Telomere shortening is associated with chromosome damage as part of the process of aging. Thus, telomeres are of vital importance to research on human aging.

With its abundant supply of telomeres, the obscure protist *Oxytricha* serves as a key model system for telomere research. The three-dimensional structure of *Oxytricha* telomeres has been determined by George Thomas and colleagues at the University of Missouri, using X-ray crystallography, and by Juli Feigon and colleagues at the University of California, Los Angeles, using nuclear magnetic resonance (NMR) spectroscopy (techniques discussed in Chapter 2). The structure consists of a quadruple helix in which the end of each DNA strand doubles back, and four guanines form base pairs with each other (**Figs. 1B** and **C**). The resulting knot-like structure protects the chromosomal end from degradation by exonucleases. Similar structures have since been observed for human telomeres. The structure of telomeres yields clues to the function of telomerases in maintaining the length of our chromosomes.

A.

©2000–2003, José de Ondarza

C.

B.

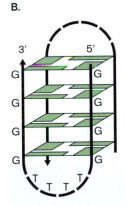

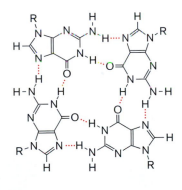

Figure 1 *Oxytricha*, **a model for the study of telomeres.** **A.** Conjugating pair of *Oxytricha fallax*. **B.** *Left*: George Thomas's proposed secondary structure of the *Oxytricha* telomere. *Right*: Cross section through the quadruple-stranded structure shows base pairing of four guanines. **C.** Space-filling model of the *Oxytricha* telomere quadruplex. (PDB code: 156D) *Source*: B. Kraft, et al. 2002. *Nucleic Acids Research* 30:3981.

Figure 20.32 Conjugation. A. Two paramecia conjugating (light micrograph). **B.** In conjugation, two paramecia of opposite mating type form a cytoplasmic bridge. The 2*n* micronucleus of each cell undergoes meiosis. Each macronucleus, as well as three out of four meiotic products, disintegrates. The haploid micronuclei undergo mitosis forming two daughter micronuclei. Daughter nuclei from each cell are exchanged across the cytoplasmic bridge, then fuse with their respective counterparts, restoring 2*n* micronuclei. The cells come apart, and each micronucleus generates a new macronucleus.

A. Conjugating paramecia

©Michael Abbey/Visuals Unlimited

B. Conjugation between ciliates

Micronucleus (2*n*)
Macronucleus

Conjugation

Meiosis

Macronuclei disintegrate.

Three of four meiotic products disintegrate, leaving one micronucleus (*n*).

Haploid micronuclei replicate and divide (mitosis).

Haploid micronuclei exchange.

Haploid micronuclei fuse, becoming diploid (2*n*).

Macronucleus re-forms.

by exchanging haploid micronuclei. Micronuclei are exchanged through **conjugation**, a process in which two cells of opposite mating type join by a cytoplasmic bridge and exchange nuclear products of meiosis (**Fig. 20.32**). During fusion, the micronucleus of each cell undergoes meiosis to form four haploid nuclei. Three out of four of the haploid nuclei disintegrate, as does the entire macronucleus. The haploid micronuclei then undergo mitosis, and one of each daughter nuclei is exchanged across the cytoplasmic bridge. Each transferred nucleus then fuses with its haploid counterpart, restoring diploidy. The two cells come apart, and each recombined micronucleus generates a new macronucleus.

A. *Stentor* (stalked ciliate)

Courtesy of the Olympus Microscopy Resource Center/Hoffman Gallery

B. *Acineta* (suctorian)

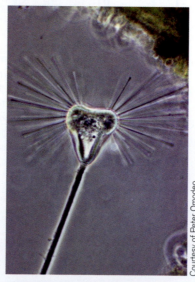

Courtesy of Peter Omodeo

Figure 20.33 Stalked ciliates. A. *Stentor*, a ciliate with a flexible stalk, 1.5–2.0 mm in length (phase contrast). The oral ring of cilia creates currents drawing food into the mouth. **B.** The suctorian *Acineta* replaces cilia with knobbed tentacles (light micrograph).

Stalked ciliates. Some ciliates adhere to a substrate and use their cilia primarily to obtain prey (**Fig. 20.33A**). **Stalked ciliates** such as *Stentor* and *Vorticella* possess a ring of cilia surrounding a large mouth. The ciliary beat is specialized to draw large currents of water and whatever prey it carries. Stalked ciliates are commonly found in pond sediment and in wastewater during biological treatment by microbial digestion, where they are attached to flocs of filamentous bacteria.

Another group of stalked ciliates, the **suctorians** (**Fig. 20.33B**), possess cilia for only a short period after a daughter cell is released by the stalked cell. The daughter cell swims by ciliary motion until it finds a good habitat to settle, whereupon its cilia are replaced by knobbed tentacles similar to the filopodia of shelled amebas. Suctorians prey on swimming ciliates such as paramecia.

Dinoflagellates Are Phototrophs and Predators

The dinoflagellates (class Dinophyceae) are a major group of marine phytoplankton, essential to marine food chains. Like ciliates, they are highly motile, but instead of numerous short cilia, dinoflagellates possess just two long flagella, one of which wraps along a crevice encircling the cell (**Fig. 20.34A**). Dinoflagellates are secondary or tertiary algal endosymbionts. They have a chloroplast derived from a red alga, which in some species was later replaced by a heterokont alga, itself a secondary endosymbiont (**Fig. 20.34C**). Some dinoflagellates possess carotenoid pigments that confer a red color. Blooms of red dinoflagellates cause the

A. *Gymnodium* (dinoflagellate)

David M. Phillips/Visuals Unlimited

B. Red tide

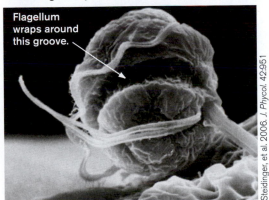

Bill Bachman/Photo Researchers, Inc.

C. Dinoflagellate cell structure

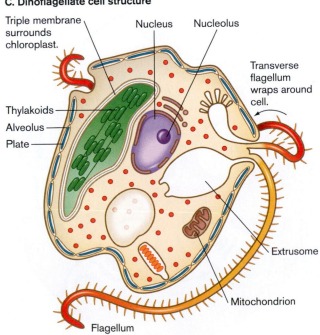

Triple membrane surrounds chloroplast.
Nucleus
Nucleolus
Transverse flagellum wraps around cell.
Thylakoids
Alveolus
Plate
Extrusome
Mitochondrion
Flagellum

D. Dinoflagellate predation

Flagellum wraps around this groove.

Steidinger, et al. 2006. *J. Phycol.* 42:951

Figure 20.34 Dinoflagellates. A. *Gymnodium* sp., a dinoflagellate, with flagellum wrapped around the cell (colorized SEM). **B.** "Red tide," caused by bloom of dinoflagellates. **C.** Diagram of a dinoflagellate. Protective plates of protein or calcified polysaccharides are formed within cortical alveoli. The chloroplast is surrounded by a triple membrane. **D.** Dinoflagellate phagocytosing a smaller dinoflagellate (SEM).

famous red tide, which may have inspired the biblical story of the plague in which water turns to blood (**Fig. 20.34B**).

The armor-plated appearance of a dinoflagellate results from its stiff alveolar plates, composed of proteins or calcified polysaccharides (**Fig. 20.34C**). The complex outer cortex includes various extrusomes (organelles that extrude a defensive substance) and endocytic pores, as well as a species-specific pattern of alveolar plates. Dinoflagellates supplement their photosynthesis by predation, extending a lamellar pseudopod to engulf prey (**Fig. 20.34D**). Some dinoflagellates have evolved to lose their chloroplasts altogether, becoming obligate predators or parasites.

Some phototrophic dinoflagellates inhabit other organisms as endosymbionts. Their hosts include shelled amebas, sponges, sea anemones, and most importantly, reef-building corals. Coral endosymbionts, known as **zooxanthellae**, are vital to reef growth. The zooxanthellae are temperature sensitive, and their health is endangered by global warming. Rising temperatures in the ocean lead to coral bleaching (the expulsion of zooxanthellae), after which the coral dies.

Apicomplexans Are Specialized Parasites

Apicomplexans form a major group of parasites of humans and other animals. They are characterized by the presence of the **apical complex**, a highly specialized structure that

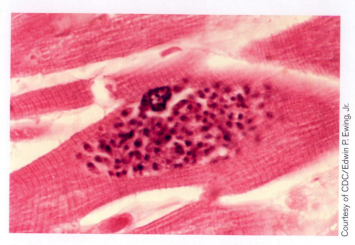

Courtesy of CDC/Edwin P. Ewing, Jr.

Figure 20.35   *Toxoplasma gondii*: An apicomplexan parasite.   Numerous *T. gondii* parasites infecting myocardium (a cell of heart muscle; stained tissue section, light micrograph).

facilitates entry of the parasite into a host cell. An example of an apicomplexan is *Toxoplasma gondii*, a parasite commonly carried by cats and transmissible to humans, where it can harm a developing fetus (**Fig. 20.35**). Like the ciliates and dinoflagellates, apicomplexans possess an elaborate cortex composed of alveoli, pores, and microtubules. But as parasites, apicomplexans have undergone extensive reduc-

A. *Plasmodium falciparum* infects red blood cells

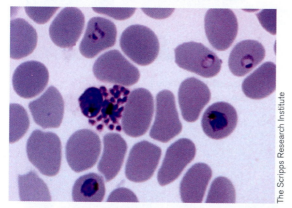

The Scripps Research Institute

Figure 20.36   *Plasmodium falciparum*, a cause of malaria.   A. Red blood cells infected with *Plasmodium falciparum*, which is stained purple with a dye that interacts with DNA (light micrograph). One red blood cell can be seen bursting, unleashing parasites on surrounding cells. **B.** Merozoite form of *P. falciparum* invades a red blood cell. The apical complex facilitates invasion, then dissolves as the merozoite transforms into an intracellular form.

B. Malarial merozoite invading a red blood cell

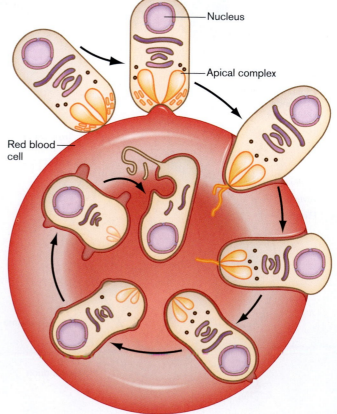

Nucleus

Apical complex

Red blood cell

tive evolution, losing their flagella or cilia. They possess a unique organelle called the **apicoplast**, derived by extensive genetic reduction from an endosymbiotic chloroplast. No capacity for photosynthesis remains, but the apicoplast provides one essential function in fatty acid metabolism.

The best known apicomplexan is *Plasmodium falciparum*, the main causative agent of **malaria**, the most important parasitic disease of humans worldwide (**Fig. 20.36A**). *P. falciparum* is carried by mosquitoes, which transmit the parasite to humans when the insect's proboscis penetrates the skin. The disease is endemic in areas inhabited by 40% of the world's population; it infects hundreds of millions of people and kills more than a million African children each year.

The transmitted parasites invade the liver, then develop into the **merozoite** form that invades red blood cells (**Fig. 20.36B**). The merozoite first contacts a red blood cell through interaction between its surface proteins on the apical complex and specific host receptors. The apical complex consists of a pair of secretory organelles capped by the "conoid," a ring of microtubules that injects enzymes that aid entry of the parasite. The cone-like tip of the apical complex penetrates the host cell,

enabling secretion of lipids and proteins that facilitate invasion. Eventually, the entire merozoite enters the host cell, leaving no traces of the parasite on the host cell surface. Thus, the internalized parasite becomes invisible to the immune system until its progeny burst out.

P. falciparum acquires resistance rapidly and no longer responds to drugs such as quinine that nearly eliminated the disease half a century ago. The life cycle and molecular properties of *P. falciparum* have been studied extensively for clues as to the development of new antibiotics and vaccines. The elaborate life cycle of *P. falciparum* and other parasites involves several common features:

- **Schizogony**, mitotic reproduction of a diploid or haploid form to achieve a large population within a host tissue.
- **Gamogony**, the differentiation of haploid cells into male and female gametes capable of fertilization.
- **Sporogony**, mitotic reproduction and development of the diploid zygote into a spore-like form transmissible to the next host.

In the case of malaria (**Fig. 20.37**), whip-shaped sporozoites injected by the mosquito invade the liver,

Figure 20.37   Malaria: cycle of *P. falciparum* transmission between mosquito and human.

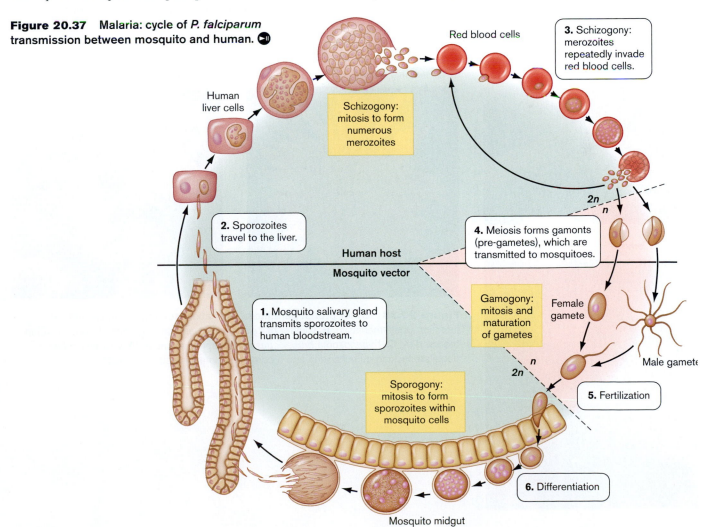

Red blood cells

3. Schizogony: merozoites repeatedly invade red blood cells.

Human liver cells

Schizogony: mitosis to form numerous merozoites

2n

n

2. Sporozoites travel to the liver.

4. Meiosis forms gamonts (pre-gametes), which are transmitted to mosquitoes.

Human host

Mosquito vector

Gamogony: mitosis and maturation of gametes

Female gamete

1. Mosquito salivary gland transmits sporozoites to human bloodstream.

n

Male gamete

2n

Sporogony: mitosis to form sporozoites within mosquito cells

5. Fertilization

6. Differentiation

Mosquito midgut

where they undergo schizogony in the form of merozoites. The merozoites invade the red blood cells, where they feed on hemoglobin. The red cells burst, liberating progeny merozoites that invade another round of red cells. The bursting of red blood cells also releases cell fragments that trigger the cyclic fevers characteristic of malaria.

Some of the merozoites in the bloodstream undergo meiosis, generating pre-gamete cells, or gamonts. The gamonts multiply and mature (a process called gamogony), then they are acquired by bloodsucking mosquitoes. In the mosquito's midgut, the gamonts develop into female eggs and flagella-like male cells. The male cells fertilize the egg cells, and the resulting zygotes undergo mitosis (sporogony) and differentiate into sporozoites, which enter the salivary gland for transmission to the next human host.

The nuclear genome of *P. falciparum* consists of 24 Mb contained in 14 chromosomes. Sequence annotation and expression studies predict 5,300 protein-encoding ORFs, comparable to the number in a yeast genome. The parasite has lost many genes encoding enzymes and transporters while expanding its repertoire of proteins involved in antigenic diversity. In addition, the parasite contains two smaller nonnuclear genomes: that of its mitochondria and that of the chloroplast-derived apicoplast.

Unique metabolic and transport components emerge from its genome that offer promising targets for drug design. For example, the fatty acid biosynthesis occurring within the apicoplast is targeted by triclosan and other antibiotics. Other promising targets for antimalarial drugs are the unique proteases required to digest hemoglobin within the *P. falciparum* food vacuole.

TO SUMMARIZE:

- **Ciliates are covered with numerous cilia.** Cilia provide motility and help capture prey. Ciliates undergo complex reproductive cycles involving exchange of micronuclei through conjugation.
- **Dinoflagellates are phototrophic predators.** Dinoflagellates are tertiary endosymbiotic algae. Their alveoli contain calcified plates; they have paired flagella, one of which is used for propulsion. Predation occurs by extension of a lamellar pseudopod.
- **Apicomplexans are parasites that penetrate host cells.** The apicoplast is a specialized organ for cell penetration. Apicomplexans such as *Plasmodium falciparum* conduct complex life cycles involving several different cell forms within mammalian and arthropod hosts.

A. *Leishmania* **infection**

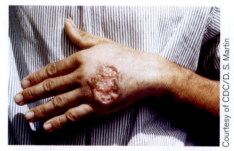

Courtesy of CDC/D. S. Martin

D. *Giardia lamblia*

©Dennis Kunkel Microscopy, Inc.

B. *Leishmania major*

©Dennis Kunkel

C. *Trypanosoma brucei*

Eye of Science/Photo Researchers, Inc.

Figure 20.38   Other protist parasites.   A. Patient suffering from *Leishmania* infection (Leishmaniasis). **B.** Cluster of *Leishmania major* undergoing schizogony within the sand fly. **C.** *Trypanosoma brucei*, cause of sleeping sickness. **D.** *Giardia lamblia*, a flagellate lacking mitochondria and a common human intestinal parasite.

20.6 Trypanosomes, Microsporidia, and Excavates

Several other major groups of protists are known mainly for organisms that are human parasites. These parasites cause some of the most gruesome and debilitating conditions known to humanity; for example, *Leishmania major*, the cause of leishmaniasis, causes swelling and decay of the extremities and eventually death (**Figs. 20.38A** and **B**). The leishmanias are related to **trypanosomes**, ruffle-shaped cells with a cortical skeleton of microtubules culminating in a long flagellum. *Trypanosoma brucei*, carried by the tsetse fly, causes African sleeping sickness (**Fig. 20.38C**), whereas *T. cruzi*, carried by the bedbug, causes Chagas' disease, a debilitating infection of the internal organs. Chagas' disease is prevalent in South and Central America, and global warming is expected to expand its range north. Like apicomplexans, trypanosomes show genetic evidence of descent from chloroplast-bearing flagellates, in this case Euglenida (a clade distinct from Alveolata). Euglenida includes *Euglena gracilis*, a chloroplast-containing flagellate that is frequently presented to students for microscopic observation.

The rise of AIDS and the growth of elderly and immunocompromised populations have led to the emergence of new opportunistic pathogens. An example is found in Microsporidia, a clade that falls within the Opisthokonta, including animals and fungi (see **Fig. 20.3**). The microsporidian *Encephalitozoon intestinalis* is now a common intestinal pathogen among susceptible populations.

Analogous to apicomplexans, the microsporidians possess a specialized invasion complex that enables entry to host cells, typically macrophages.

Excavata, a highly degenerate group of parasites, includes the intestinal parasite *Giardia lamblia*, an organism familiar in North America, Europe, and Russia (**Fig. 20.38D**). Giardia is a frequent nemesis of day-care centers as well as a contaminant of aquatic streams frequented by bears and other wildlife. *Giardia* occasionally contaminates community water supplies and has become endemic in some Russian cities. *Giardia* and other excavates are noted for their anaerobic metabolism and their complete absence of mitochondria, reflecting their absolute dependence on the anaerobic intestinal environment of their hosts.

Concluding Thoughts

The microbial eukaryotes, including fungi, algae, and many kinds of protists, serve many diverse roles in our ecosystems. A survey of protists may leave an impression that most protist clades exist primarily to parasitize humans. However, recent genetic surveys based on DNA sequence analysis of environmental communities suggest the existence of twice as many protist clades in nature as we have studied to date. Most of these unknown clades have no direct connection with humans, yet they may fill crucial niches in the ecosystems on which human existence depends. The remaining chapters in Part 4 emphasize the interconnections among many kinds of microbes that form the foundations of Earth's biosphere.

CHAPTER REVIEW

Review Questions

1. Discuss the evidence for the branching of fungi and animals within one clade, the Opisthokonta, which is distinct from algae and protists.

2. How do primary symbiont algae differ from secondary and tertiary symbiont algae? Compare with respect to cell structure and nutritional options.

3. Compare and contrast the molecular basis of motility in amebas and ciliates. Cite particular species.

4. Compare and contrast the traits of these heterokont protists: kelps, diatoms, and dinoflagellates. Compare with respect to cell structure, colony organization, and nutritional options.

5. Summarize the key traits of fungi. What do fungi have in common with protists, and how do they differ?

6. Outline the life cycles of the major phyla of fungi: Chytridiomycetes, Zygomycetes, Ascomycetes, and Basidiomycetes. Explain their ecological significance.

7. Compare and contrast the traits of green and red algae (Chlorophyta and Rhodophyta, respectively).

8. Outline the life cycle of the slime mold *Dictiostelium discoidium*. Compare and contrast its features with that of fungi that produce fruiting bodies, such as Basidiomycetes.

9. Outline the complex parasitic life cycles of an apicomplexan parasite and a trypanosome. Cite evidence of reductive evolution, as well as of evolution of elaborate specialized structures to facilitate the parasite's life cycle.

Key Terms

alga (763)
alternation of generations (767)
alveolate (764)
ameba (763)
apical complex (788)
apicomplexan (764)
apicoplast (789)
ascomycete (769)
ascospore (769)
ascus (767)
basidiomycete (771)
basidiospore (773)
budding (766)
cellular slime mold (781)
chloroplast (763)
chrysophyte (763)
ciliate (764)
conjugate (776)
conjugation (786)
contractile vacuole (784)
coralline alga (778)
cortical alveolus (764)
cryptogamic crust (778)
cryptophyte alga (763)
diatom (764)
dinoflagellate (763)

endosymbiosis (757)
flagellate (764)
foraminiferan (783)
fruiting body (763)
frustule (779)
fungus, pl. fungi (761)
gamogony (789)
green alga (chlorophyte) (763)
heterokont (764)
hydrogenosome (768)
hypha (763, 765)
kelp (764)
lamellar pseudopod (781)
lichen (778)
macronucleus (784)
malaria (789)
merozoite (789)
micronucleus (784)
mitosporic fungus (767)
mycelium (765)
mycology (756)
nucleomorph (763)
oomycete (774)
opisthokont (761)
oral groove (784)
phytoplankton (774)

plasmodial slime mold (781)
plasmodium (782)
primary endosymbiont (761)
protist (756, 761)
pseudopod (763)
radiolarian (783)
red alga (rhodophyte) (763)
sargassum weed (779)
schizogony (789)
secondary endosymbiont (761)
siphonous alga (776)
slime mold (781)
sporangiospore (769)
sporangium (769)
sporogony (789)
stalked ciliate (787)
suctorian (787)
telomere (785)
test (783)
trypanosome (791)
vesicular-arbuscular mycorrhizae (769)
yeast (766)
zoospore (761)
zooxanthella (788)
zygomycete (769)
zygospore (769)

Recommended Reading

Amaral-Zettler, Linda A., Felipe Gomez, Erik R. Zettler, Brendan G. Keenan, Ricardo Amils, and Mitchell L. Sogin. 2002. Eukaryotic biodiversity and physiology at acidic extremes: Spain's Tinto River. Eukaryotic diversity in Spain's River of Fire. *Nature* **417:**137.

Armbrust, E. Virginia, John A. Berges, Chris Bowler, Beverley R. Green, Diego Martinez, Nicholas H. Putnam, et al. 2004. The genome of the diatom *Thalassiosira pseudonana*: ecology, evolution, and metabolism science. *Science* **306:**79–86.

Baldauf, Sandra L. 2003. The deep roots of eukaryotes. *Science* **300:**1703–1706.

Baum, Jake, Anthony T. Papenfuss, Buzz Baum, Terence P. Speed, and Alan F. Cowman. 2006. Regulation of apicomplexan actin-based motility. *Nature Reviews Microbiology* **4:**621–628.

Carlile, Michael J., Sarah C. Watkinson, and Graham W. Gooday. 2001. *The Fungi.* 2nd ed. San Diego: Academic Press.

Dayel, Mark J., and R. Dyche Mullins. 2004. Activation of Arp2/3 complex: Addition of the first subunit of the new filament by a WASP protein triggers rapid ATP hydrolysis on Arp2. *PLoS Biology* **2:**E91.

Gardner, Malcolm J., Shamira J. Shallom, Jane M. Carlton, Steven L. Salzberg, Vishvanath Nene, et al. 2002. The genome sequence of the human malaria parasite *Plasmodium falciparum*. *Nature* **419:**498–511.

Graham, Linda E., and Lee W. Wilcox. 2000. *Algae*. Englewood Cliffs, NJ: Prentice-Hall.

Haldar, Kasturi, Sophien Kamoun, N. Luisa Hiller, Souvik Bhattacharje, and Christiaan van Ooij. 2006. Common infection strategies of pathogenic eukaryotes. *Nature Reviews Microbiology* **4:**922–931.

Mochizuki, Kazufumi, Noah A. Fine, Toshitaka Fujisawa, and Martin A. Gorovsky. 2002. Analysis of a piwi-related gene implicates small RNAs in genome rearrangement in *Tetrahymena*. *Cell* **110:**689–699.

Pounds, J. Alan, Martin R. Bustamante, Luis A. Coloma, Jamie A. Consuegra, Michael P. L. Fogden, et al. 2007. Widespread amphibian extinctions from epidemic disease driven by global warming. *Nature* **439:**161–167.

Schultze, Peter, Flint W. Smith, and Juli Feigon. 1994. Refined solution structure of the dimeric quadruplex formed from the *Oxytricha* telomeric oligonucleotide d(GGGGTTTTGGGG). *Structure* **2:**221–233.

Chapter 21

Microbial Ecology

Of all organisms, microbes grow in the widest range of habitats, from Antarctic lakes to the acid drainage of iron mines, as well as the air, water, and soil that surround us. Microbial communities form the foundation of Earth's biosphere, shaping the environments inhabited by plants and animals. In the ocean, vast quantities of microbes produce the biomass that ultimately feeds fish and humans. In forests and fields, microbes are actually the main consumers; they decompose the majority of plant material, generating fertile soil. Through their biochemistry, diverse microbes largely determine the quality of soil, air, and water for human life.

In all ecosystems, microbes associate with animals and plants. Microbes form elaborate symbiotic systems, such as mycorrhizae, underground networks of fungal hyphae that support and connect the roots of plants throughout forest and field. Rhizobia fix nitrogen for legumes, whereas algae photosynthesize for coral reefs. Anaerobes digest complex plant fibers for termites, cattle, even humans. This chapter explores how microbes interact with each other and with their many diverse habitats on Earth.

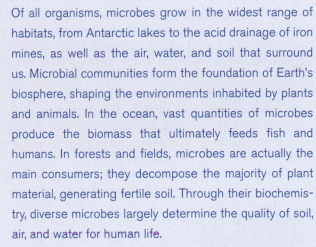

Erik Zinser, a postdoctoral fellow with Sallie Chisolm at the Massachusetts Institute of Technology, charted the distribution of the tiny cyanobacterium *Prochlorococcus* across the Atlantic Ocean. *Prochlorococcus* is one of Earth's most abundant producers of biomass and of oxygen in our atmosphere. Zinser is now at the University of Tennessee, Knoxville. Source: Johnson, et al. 2006. *Science* 311:1737.

The enteric disease cholera causes hundreds of thousands of deaths annually in India, Bangladesh, and other countries. The causative bacterium, *Vibrio cholerae*, inhabits both ocean and freshwater, contaminating the drinking water of millions of people too impoverished to boil it. Microbial ecologist Anwar Huq, from Bangladesh, worked with Rita Colwell, director of the U.S. National Science Foundation, to devise a simple way to decrease the incidence of cholera, based on the ecology of *V. cholerae*. Huq found that *V. cholerae* cells rarely exist in free suspension; rather, they colonize the surfaces of tiny invertebrates called copepods (**Fig. 21.1**). The copepods actually depend on these bacteria to eat through the chitin of their egg cases, releasing their young.

The copepods are much larger than individual bacteria—large enough to be excluded by a simple cloth filter. Most filter materials would be too expensive for villagers, but Huq found an alternative: the sari cloth worn by Bangladeshi women. A study showed that villagers could halve the incidence of cholera by filtering all their water through several layers of folded cloth from old saris. Although the sari cloth could not remove isolated bacterial cells, it removed the copepods that *V. cholerae* inhabited. This case shows how public health can benefit from understanding microbial ecology.

This chapter shows how microbes interact with each other in communities of organisms. Microbial communities critically impact other organisms in all habitats, from oceans to forests (**Fig. 21.2**). Microbes recycle organic material into aquatic and terrestrial ecosystems, providing resources for plants and animals. Our agricultural productivity depends in part on positive and negative effects of microbes in the soil and air. And deep below Earth's surface, microbial communities shape crustal rock.

Figure 21.1   Copepods carrying *Vibrio cholerae*.   Copepod with case of eggs, to be "hatched" by *V. cholerae* bacteria.

Following this chapter, Chapter 22 shows how microbial geochemistry shapes the global biosphere as a whole and how microbes in our soil, air, and water determine Earth's environmental quality. As a scientist at the U.S. Department of Energy's Microbial Genome Program observed, "Microbes rule. They operate this planet, we don't."

21.1 Microbes in Ecosystems

"No man is an island," nor is any microbe. All organisms exist within **ecosystems** consisting of populations of species plus their habitat or environment. A **population** is a group of individuals of one species living in a common location. Within an ecosystem, each population of organisms fills a specific **niche**. The niche is a set of traits that defines the conditions for an organism to live and the

A.   **B.**   **C.**

Figure 21.2   Microbes shape all environments.   A. Microbes recycle organic material into aquatic and terrestrial habitats, providing resources for plants and animals. **B.** Agricultural productivity depends on positive and negative effects of microbes in the soil and air. **C.** Deep below the surface, microbial activity shapes crustal rock.

organism's relations with other members of the ecosystem. For example, the niche of *Synechococcus*, a cyanobacterium, is that of a free-living filamentous organism that fixes CO_2 into biomass while producing molecular oxygen utilized by swarms of heterotrophic bacilli. The habitat of *Synechococcus* is the upper water layer of the ocean; its biomass provides food for protist predators, which in turn feed invertebrates and fish.

Microbes Fill Unique Niches in Ecosystems

The essential role of microorganisms in all ecosystems was formulated by the Dutch microbiologist Cornelis B. Van Niel (1897–1985). Van Niel was the first to show that bacteria in the soil and water can conduct photosynthesis without producing oxygen, using electron donors such as H_2S. This surprising discovery revealed one of many kinds of metabolism unique to microbes but unknown in plants or animals. Other unique forms of microbial metabolism (discussed in Chapters 13–15) include nitrogen fixation and the degradation of lignin by bacteria and fungi. Moreover, microbial metabolism provides ecosystems with their sole source of key elements, such as sulfur, phosphorus, and iron.

Unique microbial metabolism provides unique roles for microbes in ecosystems. Van Niel expressed this principle in two key postulates of microbial ecology:

■ **Every component or product of living cells can be used as a source of carbon or energy by a microorganism found somewhere in the biosphere.** Any cell component, such as sugars and amino acids, or cell product, such as fermentation acids and H_2S, can participate in some kind of energy-yielding reaction. If an energy-yielding reaction exists, some microbe must have evolved to use it.

■ **Microbes are found in every environment on Earth.** Every possible habitat for life supports microbes. In fact, the largest part of the biosphere (below Earth's surface) is colonized solely by microbes.

The Van Niel postulates imply an extraordinary variety of microbial species using different energy-yielding functions. But what determines the contribution of each microbe? What a microbe gives to and takes from its ecosystem depends on two things:

■ **The microbial genome.** Each genome provides a set of metabolic enzymes available to interconvert molecules. For example, the genomes of pseudomonads encode enzymes for catabolizing complex aromatic molecules, whereas genomes of cyanobacteria encode the apparatus of oxygenic photosynthesis.

Microbial genomes share content through mobile genes (acquired by transfer mechanisms, discussed in Chapter 9).

■ **Environmental factors.** The environment's physical and chemical factors, such as temperature, nutrient supplies, oxygen availability, and pH, determine the range of metabolic options. For example, if H_2 is available in the presence of O_2, *Rhodopseudomonas palustris* will oxidize hydrogen to water; on the other hand, if O_2 is absent but light is available, *R. palustris* will use anaerobic photosynthesis, splitting H_2S to make sulfur. Each organism's environment includes other organisms in the community. For example, the H_2 oxidized by *R. palustris* is provided by fermenting bacteria.

All organisms depend, directly or indirectly, on the presence of other organisms. Some populations cooperate by reciprocally supplying each other's needs. Cooperation by organisms may be incidental, as in the case of hydrogen-oxidizing bacteria using H_2 from fermenters; or it may involve mutualism, a highly developed partnership in which two species have coevolved to support each other (discussed in Section 21.2).

Carbon Assimilation and Dissimilation: The Food Web

The interactions between microbes and their ecosystems include two common roles of metabolic input and output, often referred to as assimilation and dissimilation, respectively. Most of these metabolic processes were discussed in Chapters 13, 14, and 15, but the ecological viewpoint offers a different perspective.

Assimilation refers to processes by which organisms acquire an element, such as carbon from CO_2, to build into body parts. When the environment lacks organic compounds containing an element such as nitrogen or phosphorus, microbes may assimilate the element from mineral sources. Major forms of assimilation include carbon dioxide fixation and nitrogen fixation. Organisms that produce biomass from inorganic carbon (usually CO_2 or bicarbonate ions) are called **primary producers**. Producers are a key determinant of productivity for other members of the ecosystem.

Dissimilation is the process of breaking down organic nutrients to inorganic minerals such as CO_2 and NO_2^-, usually through oxidation. Microbial dissimilation releases minerals for uptake by plants, and it provides the basis of wastewater treatment (discussed in Chapter 22). On the other hand, microbial dissimilation of an energy-rich nitrogenous fertilizer, such as N_2, causes nitrogen to be lost from the soil.

This chapter covers microbial assimilation and dissimilation of carbon and nitrogen in association with the plants and animals of an ecosystem. The cycles of other key elements, and their effects on the global biosphere, are explored in Chapter 22.

> **THOUGHT QUESTION 21.1** From Chapters 13–15, give examples of microbial metabolism that fit patterns of assimilation and dissimilation.

The major interactions among organisms in the biosphere are dominated by the production and transformation of **biomass**, the bodies of living organisms. To obtain energy and materials for biomass, all organisms participate in the **food web** (**Fig. 21.3**). A food web describes the ways in which various organisms consume each other as well as each other's products. Levels of consumption are called **trophic levels**. Each trophic level represents the consumption of biomass of organisms from the previous level, lower in the food web. At each trophic level, the fraction of biomass retained by the consumer is small; most is dissipated as CO_2 to provide energy.

Every food web depends on primary producers for two things:

■ **Absorbing energy from outside the ecosystem.** In most cases, the ultimate source of energy is the sun, and the producers are phototrophs.

■ **Assimilating minerals into biomass.** The biomass of producers is then passed on to subsequent trophic levels.

The majority of carbon in Earth's biosphere is assimilated by oxygen-producing phototrophs such as cyanobacteria, algae, and plants. Certain important ecosystems are founded on carbon fixation by lithotrophs—for example, the hydrothermal vent communities, in which bacteria oxidize hydrogen sulfide to fix CO_2, capturing both gases as they well up from Earth's crust. The vent communities, however, require oxygen generated by phototrophs from above.

In addition to producers, all ecosystems include **consumers** that acquire nutrients from producers and ultimately dissimilate biomass by respiration, returning carbon back to the atmosphere (**Fig. 21.3A**). Consumers constitute several trophic levels based on their distance from the primary producers. The first level of consumers, generally called **grazers**, directly feed on producers. Grazers usually convert 90% of the carbon back to CO_2 through respiratory metabolism, yielding energy. The next level of consumers, often called **predators**, feed on the grazers, again converting 90% back to atmospheric CO_2. In microbial ecosystems, the trophic relationships are often highly complex, as a given species may act both as producer and consumer.

At each trophic level, some of the organisms die, and their bodies are consumed by **decomposers**, returning carbon and minerals back to the environment for use by producers. Decomposers are invariably microbes (fungi or bacteria). Decomposers have particularly versatile digestive enzymes capable of breaking down complex molecules such as lignin. Without decomposers, carbon and minerals needed by phototrophs would be locked away by ever-increasing mounds of dead biomass. Instead, all biomass is recycled, and all the energy acquired by ecosystems is eventually converted to heat.

Note that in the function of ecosystems, phylogenetic distinctions between bacteria, archaea, and eukaryotes are of less significance than the biological and biochemical consequences of an organism's presence in the community; that is, what does the organism produce or consume? Thus, in this chapter, we place greater emphasis on trophic roles than on phylogenetic distinctions.

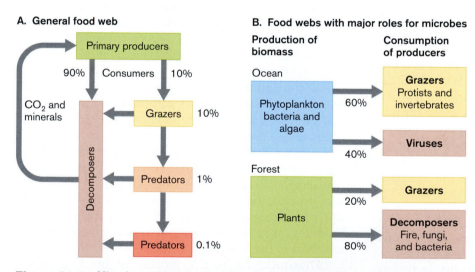

Figure 21.3   Microbes within food webs.   A. The general view of a food web includes primary producers that fix CO_2 into biomass, grazers (primary consumers) that feed on primary producers, and predators that consume the grazers and smaller predators. Decomposers consume dead bodies of organisms from every level, recirculating their minerals to the producers. Percentages indicate fraction of original CO_2 converted to biomass at each trophic level. **B.** In different habitats, the relative significance of the various trophic component varies. In marine ecosystems, the primary producers are bacteria, archaea, and algae. About 10–40% of consumption is by viruses, which mineralize both producer and consumer microbes. In a forest, the major producers are trees, much of whose biomass is poorly digested by grazers; 80% of consumption occurs by fungal and bacterial decomposition. In some forests, the major decomposer is fire.

Table 21.1 **Aerobic and anaerobic metabolism.**

Element	Oxidized by O$_2$ (lithotrophy)	Reduced by CHO* or H$_2$ (anaerobic respiration)
Nitrogen	$NH_3 + O_2 \longrightarrow NO_3^-$	$NO_3^- + CHO \longrightarrow N_2$
Manganese	$Mn^{2+} + O_2 \longrightarrow Mn^{4+}$	$Mn^{4+} + CHO \longrightarrow Mn^{2+}$
Iron	$Fe^{2+} + O_2 \longrightarrow Fe^{3+}$	$Fe^{3+} + CHO \longrightarrow Fe^{2+}$
Sulfur	$H_2S + O_2 \longrightarrow SO_4^{2-}$	$SO_4^{2-} + CHO \longrightarrow H_2S$
Carbon	$CH_4 + O_2 \longrightarrow CO_2$	$CO_2 + H_2 \longrightarrow CH_4$

*CHO = organic material.

The relative significance of microbial and multicellular producers and consumers varies considerably among different habitats. A major difference appears between marine and terrestrial ecosystems (**Fig. 21.3B**). In the oceans, most CO$_2$ fixation and biomass production are performed by the smallest inhabitants, phototrophic bacteria. The major consumers are protists and viruses; in terms of bulk biomass, multicellular organisms barely contribute. In terrestrial ecosystems, by contrast, the major primary producers and fixers of CO$_2$ are multicellular plants. Plants generate **detritus**, discarded biomass such as leaves and stems, that requires decomposition by fungi and bacteria. These differences between the food webs of ocean and dry land explain why most of the food we harvest from the ocean consists of predators at the higher trophic levels (fish), whereas most food harvested on land consists of producers and first-level consumers (plants and herbivores). Fish depend on a vast invisible food web of microbes, and thus their numbers remain limited despite the seemingly huge volume of ocean. Thus, marine fish populations are especially vulnerable to overharvesting and to any environmental change that disrupts the food chain.

Oxygen and Other Electron Acceptors

The availability of oxygen and other electron acceptors is the most important factor that determines the kind of assimilation and dissimilation occurring in a given environment. Examples of aerobic and anaerobic metabolism are presented in **Table 21.1**. In aerobic environments, microbes use molecular oxygen as an electron acceptor to respire on organic compounds (abbreviated CHO) produced by other organisms (discussed in Chapters 13 and 14). Aerobic respiration on organic compounds is highly dissimilatory in that it tends to break compounds down to CO$_2$. Microbes also use oxygen to respire on reduced minerals such as NH$_3$, H$_2$S, and Fe^{2+} (lithotrophy). Lithotrophy is coupled to CO$_2$ fixation and is therefore assimilatory metabolism.

In anaerobic environments, such as deep soil, microbes use minerals such as Fe^{3+} and NO$_2^-$ to oxidize organic compounds supplied by other organisms (anaerobic respiration). Other anaerobes use these electron acceptors to oxidize reduced minerals. Anaerobic environments allow much slower rates of assimilation and dissimilation than occurs in the presence of oxygen. The total volume of anaerobic microbial communities, however, far exceeds that of the oxygenated biosphere.

Temperature, Salinity, and pH

Other abiotic factors can profoundly affect the environment for microbes and other members of the food chain, either directly or by impacting other factors. The effects are most significant for **extremophiles**, species that grow in environments considered extreme by human standards (**Table 21.2**; discussed in Chapter 5). Temperature limits

Table 21.2 **Extremophiles.**

Class of extremophile	Typical environmental conditions for growth
Acidophile	Acidic environments at or below pH 3
Alkaliphile	Alkaline environments at a range of pH 9–14
Barophile	High pressure, usually at ocean floor, above 380 atm
Endoliths	Within rock crystals down to a depth of 3 km
Halophile	High salt, typically above 2 M NaCl
Hyperthermophile	Extreme high temperature, above 80°C
Oligotroph	Low carbon concentration, below 1 ppm
Psychrophile	Low temperature, below 15°C
Thermophile	Moderately high temperature, 50–80°C
Xerophile	Dessication, water activity below 0.8

the rate of metabolism: Higher temperatures, particularly those found in hot springs, enable the fastest growth rates measured (doubling times as short as 10 minutes for some hyperthermophiles). On the other hand, temperature limits the oxygen concentration in water, so hyperthermophiles often use sulfur instead.

High salt concentration limits the growth of microbes adapted to freshwater conditions. On the other hand, many microbial species have adapted to high salinity (they are called halophiles). The haloarchaea, for instance, bloom in population as a body of water shrinks and becomes hypersaline.

Acidity is important geologically because a high concentration of hydrogen ions accelerates the release of reduced minerals from exposed rock. Extreme acidity is often produced by lithotrophic acidophiles (discussed in Chapter 14). The increasing acidity enhances further release of minerals to be oxidized, while excluding acid-sensitive competitors. Other kinds of habitat, such as soda lakes, show extreme alkalinity resulting from high sodium carbonate concentration.

TO SUMMARIZE:

- **Microbial populations fill unique niches in ecosystems.** Every environment on Earth (to depths of 3 km below the surface) contains microbes. Every chemical reaction that may yield free energy can be utilized by some kind of microbe.
- **Microbes fix or assimilate essential elements into biomass**, which recycles within ecosystems. Important elements, such as nitrogen, are fixed solely by prokaryotes.
- **Primary producers fix single-carbon units, usually CO_2.** Microbial primary producers include algae, cyanobacteria, and lithotrophs.
- **In a food web, consumers catabolize the bodies of producers**, generating CO_2 and releasing heat energy. Dissimilation is the process of breaking down nutrients to inorganic minerals such as CO_2 and NO_2^-, usually through oxidation.
- **Microbial activity depends on levels of oxygen, carbon, nitrogen, and other essential elements**, as well as environmental factors such as temperature, salinity, and pH. The largest volume of our biosphere contains anaerobic bacteria and archaea. Beneath Earth's surface, anaerobic forms of metabolism predominate.

21.2 Microbial Symbiosis

Crucial relationships with microbes may occur at a distance; for example, the oxygen gas generated by marine cyanobacteria is breathed by organisms around the globe. Other kinds of interactions, however, require intimate association between microbes or between a microbial

population and animals or plants. Intimate association between organisms of different species is called **symbiosis**. Symbiotic associations include a full range of positive through negative relationships (**Table 21.3**).

Mutualism Involves Partner Species That Require Each Other

The most highly developed forms of symbiosis involve partner species that require each other for survival, a phenomenon termed **mutualism**. A highly evolved form of mutualism is exemplified by **lichens** (**Fig. 21.4**). Lichens consist of an intimate symbiosis between a fungus and an alga or cyanobacterium—sometimes both. The highly evolved association usually requires compatible partner species. The alga or bacterium provides photosynthetic nutrition, while the fungus provides minerals and protection. Lichens grow very slowly, but they tolerate extreme dessication.

A.

B.

Figure 21.4 Lichens on tombstone. A. Tombstone at Kenyon College cemetery encrusted with lichens (pale blue) and mosses (dark green), a nonvascular plant. **B.** Close-up of lichens on tombstone.

Table 21.3 Types of interactions involving microbial species.

Type of interaction	Effects of interaction	Example
Mutualism	Two organisms grow in an intimate species-specific relationship in which both partner species benefit and may fail to grow independently.	Lichens consist of fungi and algae (in some cases cyanobacteria) growing together in a complex layered structure. Both species only grow in the presence of the other.
Synergism	Both species benefit through growth in proximity, but the partners are easily separated and either partner can grow independently of the other.	Human colonic bacteria produce H_2 and CO_2 as fermentation products, which methanogenic archaea metabolize to methane. The methanogens gain energy, and the bacteria benefit energetically through the steady-state removal of the end products of their metabolism.
Commensalism	One species benefits, while the partner species neither benefits nor is harmed.	In wetlands, *Beggiatoa* bacteria oxidize H_2S for energy. Removal of H_2S from the environment enables growth of other microbes for whom H_2S is toxic. The other microbes are not known to benefit *Beggiatoa*.
Amensalism	One species benefits by harming another. The relationship is nonspecific.	On the human skin, *Staphylococcus epidermidis* and related species produce short-chain fatty acids whose uptake inhibits growth of yeasts and pathogens. The staphylococci benefit by outcompeting inhibited microbes for scarce water and nutrients.
Parasitism	One species (the parasite) benefits at the expense of the other (the host). The relationship is usually obligatory for the parasite.	*Legionella pneumophila*, the cause of legionellosis pneumonia in humans, parasitizes amebas in natural aquatic habitats. Within the human body, *L. pneumophila* parasitizes macrophages.

Lichens show a surprising variety of form. Different species may form a flat crust, branched filaments (**Fig. 21.5A**), or leaflike lobes. A cross section of the leaflike lichen *Lobaria pulminaria* reveals a layer of algae (green) covered by fungal mycelium (white), which probably protects the algae from excessive photodamage (**Fig. 21.5B**). In addition to the algae, this lichen includes patches of cyanobacteria—a three-way mutualism. For dispersal, the lichen forms asexual clumps of algae wrapped in fungal mycelium. The clumps flake off and are carried by wind to new locations.

Other forms of mutualism involve a plant or animal host. A mutualism particularly crucial to ecosystems is that of bacterial nitrogen fixation. Many cyanobacteria fix nitrogen in association with specific plants, similar to the legume symbiosis in which rhizobial bacteria such as *Sinorhizobium* species form intracellular "bacteroids" within plant cells (discussed in Section 21.5). Both the

Figure 21.5 Lichen form and structure.   A. *Cladina evansii*, known as "deer moss," a branch-like lichen, growing on oak leaves and sandy soil in Florida. **B.** Section through lichen (*Lobaria pulminaria*) shows fungal, algal, and bacterial symbionts.

A. "Deer moss" lichen

Stephen & Sylvia Sharnoff, *Lichens of N. America*

B. Lichen cross section

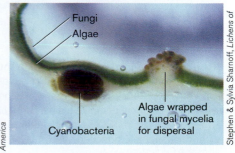

Fungi
Algae
Cyanobacteria
Algae wrapped in fungal mycelia for dispersal

Stephen & Sylvia Sharnoff, *Lichens of N. America*

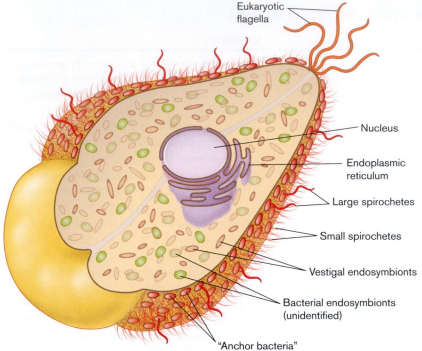

B.

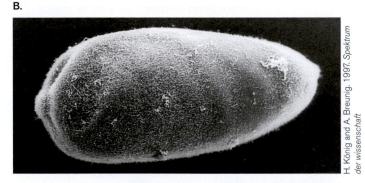

H. König and A. Breunig. 1997. *Spektrum der wissenschaft*

C.

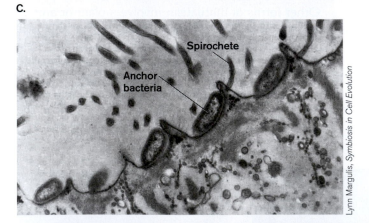

Lynn Margulis, *Symbiosis in Cell Evolution*

Figure 21.6 *Mixotricha paradoxa*: a multiple symbiont.
A. The protist *Mixotricha paradoxa* possesses attached spirochetes (large and small species), "anchor bacteria," and two kind of bacterial endosymbionts. **B.** *M. paradoxa* (length, 300–500 µm; SEM). **C.** TEM section through pellicle including anchor bacteria and attached spirochetes. *Source*: Lynn Margulis, *Symbiosis in Cell Evolution*, 1993.

cyanobacteria and their plant hosts are highly evolved to respond to each other chemically and develop the nitrogen-fixing system. The plants provide protection and nutrients from photosynthesis, while receiving fixed nitrogen from the bacteria.

Many animal hosts require endosymbiotic bacteria to digest plant material such as cellulose. Digestive symbiosis is well known in termites, whose bacteria digest wood, and in ruminants such as cattle. Less known is that many primates, such as gorillas, depend on bacterial fermentation. Even humans derive about 15% of their caloric intake from bacterial fermentation.

A particularly complex digestive endosymbiont is *Mixotricha paradoxa* (**Fig. 21.6**), a large ciliate (as long as half a millimeter) that grows in the termite gut. The *Mixotricha* cell contains several kinds of bacterial endosymbionts. Intracellular bacteria digest cellulose from wood consumed by the termite. There are also organelles that appear to be vestigial remnants of another endosymbiont, diminished by reductive evolution. On the protist's surface, four kinds of bacteria are attached. Two of the attached species are spirochetes, one significantly larger than the other. The spirochetes extend from the protist's cell membrane; they are flagellated, and their flagellar motility propels the protist cell. Two other species of "anchor bacteria" are gram-negative rods attached to knobs of the protist surface; their function is unknown. All members of the partnership appear to have evolved an obligate relationship.

Symbiosis May Involve Cooperation or Parasitism

Less intimate cooperation is called **synergism**, in which both species benefit but can grow independently and show less specific cellular communication (**Table 21.3**). For example, human colonic bacteria produce fermentation products that colonic methanogens metabolize to methane. The methanogens gain energy, and the fermenting bacteria benefit energetically through the steady-state removal of their end products. This kind of metabolic cooperation is also known as **syntrophy**, or "feeding together." In other cases, one species derives benefit from another without return; for example, some wetland bacteria derive benefit from *Beggiatoa* because *Beggiatoa* bacteria oxidize H_2S, which inhibits growth of other species. An interaction that benefits one partner only is called **commensalism**. Commensalism is difficult to define in practice since "commensal" microbes often provide a hidden benefit to their host. For example, gut bacteria such as Bacteroides species were considered

Labels on figure A (clockwise): Eukaryotic flagella; Nucleus; Endoplasmic reticulum; Large spirochetes; Small spirochetes; Vestigal endosymbionts; Bacterial endosymbionts (unidentified); "Anchor bacteria"

Labels on figure C: Spirochete; Anchor bacteria

commensals until it was discovered that their metabolism aids human digestion.

An interaction that harms one partner nonspecifically, without an intimate symbiosis, is called **amensalism**. An example of amensalism is actinomycete production of antimicrobial peptides that kill surrounding bacteria. The dead bacterial components are then catabolized by the actinomycete. Finally, **parasitism** is an intimate relationship in which one member (the parasite) benefits while harming a specific host. Many microbes have evolved specialized relationships as parasites, including intracellular parasitic bacteria such as the rickettsias, which cause diseases such as Rocky Mountain spotted fever.

The distinction between mutualist and parasite is often subtle. Lichens consist of a mutualistic association between fungus and algae; but environmental change can convert the fungus to a parasite. On the other hand, parasitic microbes may coevolve with a host to the point that each depends on the other for optimal health. For example, the incidence of certain allergies and immune disorders appears to correlate with lack of exposure to pathogens and parasites.

TO SUMMARIZE:

- **Symbiosis** is an intimate association between organisms of different species.
- **Mutualism** is a form of symbiosis in which each partner species benefits from the other. The relationship is usually obligatory for growth of one or both partners. Examples of mutualism include lichens (algae or cyanobacteria and fungi) and termite gut communities of cellulose-digesting bacteria and protists.
- **Parasitism** is a form of symbiosis in which one species grows at the expense of another, usually a much larger host organism.

21.3 Marine and Aquatic Microbiology

The oceans cover more than two-thirds of the Earth's surface, reaching depths of several kilometers and forming an immense habitat (**Fig. 21.7**). Marine water is distinguished by its salt concentration, averaging 3.5%. The major ions are Na^+ and Cl^-, with significant levels of sulfate, iodide, and bromide. Marine salt concentration is high enough to prevent growth of many aquatic and terrestrial bacteria such as *E. coli*, although salt-tolerant organisms such as *Vibrio cholerae* grow well over a broad range of salt concentration.

Marine Habitats

Marine waters vary considerably with respect to temperature, pressure, light penetration, and concentration

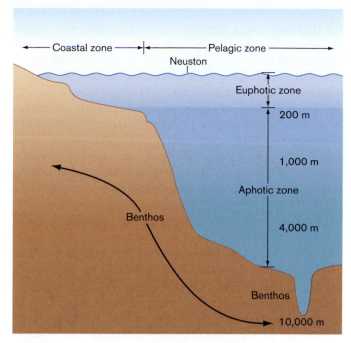

Figure 21.7   Regions of marine habitat. The marine habitat subdivides into several categories. The coastal shelf region is defined as the water extending from the shoreline out to a depth of 200 m. The open ocean includes several depth regions: the neuston, the microscopic interface between water and air; the euphotic zone of light penetration, where phototrophs can grow, down to 100–200 m; the aphotic zone, which supports only heterotrophs and lithotrophs; and the benthos, at the ocean floor.

of organic matter. In the open ocean, the water column (known as the **pelagic habitat**) is subdivided into distinct regions (see **Fig. 21.7**):

- **Neuston (about 10 μm).** The **neuston** is the air-water interface. Although extremely thin, the neuston layer contains the highest concentration of microbes. Many algae and protists have evolved so as to "hang" from the layer of surface tension.
- **Euphotic zone (100–200 m).** The **euphotic zone**, or **photic zone**, is the upper part of the water column that receives light for phototrophs. In the open ocean, the euphotic region extends down a couple of hundred meters, whereas at the **coastal shelf** (<200 m to floor), a higher concentration of silt and organisms reduces the photic zone to as little as 1 m.
- **Aphotic zone.** Below the reach of light, in the **aphotic zone**, only heterotrophs and lithotrophs can grow.
- **Benthos.** The **benthos** includes the region where the water column meets the ocean floor, as well as sediment below the surface. Organisms that live in the benthos, such as those of thermal vent communities, are called **benthic organisms**.

Another important determinant of marine habitat is the **thermocline**, a depth at which temperature decreases steeply and water density increases. A thermocline typically exists in an unmixed region. At the thermocline, a population of heterotrophs will peak, feeding on organic matter that settles from above.

Coastal regions show the highest concentration of nutrients and living organisms and the least light penetration. The open ocean is largely oligotrophic (extremely low concentration of nutrients and organisms). The concentration of heterotrophic microorganisms determines the **biochemical oxygen demand (BOD)**, the amount of oxygen removed from the water by aerobic respiration. The open ocean has such a low concentration of organisms that the BOD is extremely low; therefore, the dissolved oxygen content is high. This explains why enough oxygen reaches the ocean floor to serve lithotrophs growing at hydrogen sulfide thermal vents.

To nineteenth-century microbiologists, the oceans appeared virtually free of bacteria. Trained in the tradition of Robert Koch (discussed in Chapter 1), microbiologists attempted to isolate marine bacteria by plate culture, considered the definitive way to study a microbial species. But the colonies that grew on traditional plate media were extremely few. Then in 1959, the German microbial ecologist Holger Jannasch (1927–1998; **Fig. 21.8A**) showed that many more bacteria could be seen by light microscopy than could be grown on plates. Later, with colleagues at the Woods Hole Oceanographic Institution, Jannasch went on to study life at the ocean floor, using the famous submersible vessel *Alvin* (**Fig. 21.8B**). They documented the bacteria and archaea of biofilms at black smoker thermal vents such as those of Guaymas Basin, in the Gulf of California (**Fig. 21.8C**).

The crew of the *Alvin* discovered entire ecosystems based on symbiotic sulfide-oxidizing bacteria adapted to the enormous pressure under many kilometers of seawater. They made important discoveries about the effect of deep-sea pressure on microbial decomposition. An accident that left the *Alvin* stranded on the ocean floor led to observation of an abandoned sandwich that lasted several months with minimal decomposition. This and similar observations showed that our familiar microbial decomposers grow extremely poorly under deep-sea conditions—a discovery with implications for the fate of human garbage dumped in the open sea.

Because marine bacteria proved difficult to grow in culture, Jannasch's initial observations sparked decades of controversy. How could the organisms be alive if they were not cultured? Other researchers, including Rita Colwell, showed that even nonculturable bacteria could show metabolic activities. These elusive bacteria were described as **viable but nonculturable (VBNC)**. Viable bacteria can be nonculturable in the laboratory for many reasons, such as lack of unknown complex growth requirements, the presence of inhibitory molecules, or unfavorable environmental factors such as temperature and pressure. We now know that the vast majority of bacteria alive in the biosphere are nonculturable under current laboratory conditions.

A. Holger Jannasch

B. Submersible *Alvin*

C. Black smoker chimney

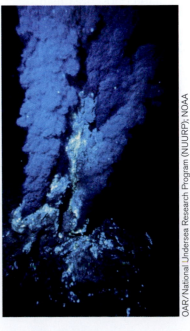

Figure 21.8 Holger Jannasch discovered unculturable marine bacteria. A. Holger Jannasch pioneered the study of marine microbes in the open sea and on the sea floor. **B.** The submersible *Alvin* is being lowered from the ship *Lulu*, a marine research vessel of the Woods Hole Oceanographic Institution. **C.** A black smoker chimney at the Guaymas Basin hydrothermal vent site in the Gulf of California.

Nonculturable bacteria are observed and studied by PCR amplification of their DNA and by epifluorescence microscopy.

Measuring Planktonic Communities

In marine science, the term **plankton** refers to organisms that float passively in water (**Fig. 21.9A**). The term is used loosely, as it includes motile microorganisms. Microbial plankton are known by different prefixes based on their approximate size ranges (**Table 21.4**): **microplankton**, 20–200 μm, consisting of relatively large ciliated protists and algae; **nanoplankton**, 2–20 μm, consisting of smaller algae and flagellated protists as well as filamentous cyanobacteria; and **picoplankton**, 0.3–2 μm, consisting of primarily bacterial phototrophs, heterotrophs, and lithotrophs. The picoplankton include the smallest living cells known. In addition to these different classes of floating cells, marine ecologists refer to marine viruses as **femtoplankton**, the smallest detectable particles capable of reproduction.

Counting and Identifying Marine Plankton

From an ecological perspective, how do we define a planktonic population? This is challenging because not all marine microbes float independently. Many colonize suspended inorganic particles known as **marine snow** (**Fig. 21.9B**). Until recently, the ecological significance of these miniature biofilm-like particles of marine snow was discounted because they appeared so dilute. In fact, perhaps half of all marine bacteria are associated with particulate substrates. As observed by Farooq Azam at the Scripps Institution of Oceanography, "Seawater is an organic matter continuum, a gel of tangled polymers with embedded strings, sheets, and bundles of fibrils and particles, including living organisms, as 'hotspots'." Marine snow adds a new dimension to our view of the oligotrophic marine habitat, in that the ocean as a whole has a low average nutrient concentration, but marine waters include a suspension of concentrated tangles of nutrient-rich substrates.

Population size can be defined in different ways: the number of individual reproductive units, total organic biomass available to consumers, or the rate of productivity or assimilation of key nutrients such as carbon dioxide. All ways of measuring these quantities present challenges; there is no one right method, but different approaches serve to answer different questions.

Table 21.4 Microbial plankton: size categories.

Size category	Microbial group	Size range (μm)
Microplankton	Eukaryotes	20–200
	Microalgae—phytoplankton	
	Heterotrophs, ciliates—zooplankton	
	Mixotrophs and symbiotic protists	
	Microscopic multicellular plants and invertebrates	
Nanoplankton	Prokaryotes: Cyanobacteria (filamentous)	2–200
	Eukaryotes	2–20
	Nanoalgae	
	Heterotrophs and mixotrophs—flagellates	
Picoplankton	Prokaryotes	
	Bacteria	0.5–1.0
	Phototrophs (phytoplankton)	
	Prochlorophytes	0.5–2.0
	Cyanobacteria (coccoid)	1.0–2.0
	Lithotrophs (oxidize NH_4^+, H_2S, etc.)	0.3–1.0
	Heterotrophs (zooplankton)	0.3–1.0
	Archaea	
	Eukaryotes	
	Picoalgae—phytoflagellates	1.0–2.0
	Heterotrophic protists—zooflagellates	1.0–2.0
Femtoplankton	Viruses	0.01–0.02

Source: Kirchman, D. 2000. *Microbial Ecology of the Ocean.*

A. Phytoplankton

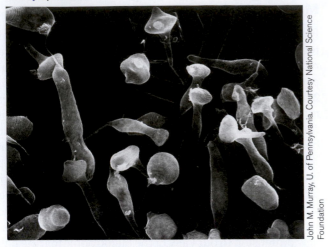

John M. Murray, U. of Pennsylvania. Courtesy National Science Foundation

B. Marine snow

Susumo Honjo/Woods Hole Oceanographic Institute

Figure 21.9   Pelagic phytoplankton.   A. Euglenophytes, a form of phytoplankton (phototrophs). **B.** Marine snow, particles of organic and inorganic matter colonized by bacteria (dark-field light micrograph).

Enumeration by epifluorescence microscopy. Marine microbes of all sizes, even viruses, can be counted under the microscope using a DNA-intercalating fluorescent dye (**Fig. 21.10**). Recall that fluorescence enables detection even of particles whose size is below the resolution limit defined by the wavelength of light (discussed in Chapter 2). Epifluorescence microscopy with DNA-binding fluorophores is used to detect planktonic microbes (**Fig. 21.11A**). The fluorescent dye DAPI reveals extremely small bacteria and algae (large spots) and viruses (small spots). For differential detection of specific classes or species of microbe, the fluorescent dye can contain a DNA probe, a short DNA sequence that can hybridize only to DNA of a particular species.

Biomass. The amount of biomass in a population can be determined by standard chemical assays of protein and other forms of organic matter. Net biomass of a population, however, does not indicate productivity within an ecosystem because it misses the amount of carbon cycled through respiration. Marine microbes have an extremely

rapid rate of turnover, but what appears to be a small population may nonetheless conduct tremendous rates of carbon fixation into biomass, which is rapidly consumed by the next trophic level.

Incorporation of radiolabeled substrates. The measured amount of biomass does not indicate the rate of biomass production. The rate of production can be estimated based on the cells' incorporation of a radiolabeled substrate. Uptake of $^{14}CO_2$ indicates the rate of carbon fixation—a highly significant property for study of global warming. Uptake of ^{14}C-thymidine measures the rate of DNA synthesis, an indicator of the rate of cell division. These procedures can be conducted in seawater under native conditions, without requiring laboratory cultivation.

A limitation of these methods is that addition of the radiolabeled substrate may raise a nutrient concentration to artificially high levels that distort the naturally occurring rates of activity. In the case of labeled thymidine, another limitation is that not all growing cells incorporate exogenous thymidine into their DNA.

Mapping marine phytoplankton. Marine phytoplankton produce a substantial part of the world's oxygen and consume much of atmospheric CO_2. How do we measure marine phytoplankton and calculate their metabolic contributions? Teams of microbiologists, including undergraduate students, conduct such studies at the Bermuda Biological Station for Research, based in the Sargasso Sea southeast of Bermuda. At the Bermuda station, a long-term study of marine ecology known as the Bermuda Atlantic Time-Series (BATS) measures seasonal and long-term changes in microbial production, populations, and diversity (**Fig. 21.10A**). The BATS study, led by Stephen Giovannoni and Craig Carlson, tracks how marine ecosystems respond to environmental disturbances such as the wind patterns of El Niño and global warming.

The BATS researchers use epifluorescence microscopy to detect and count bacteria, most of which cannot be cultured in the laboratory. The bacteria are stained with DAPI, a DNA-binding fluorescent stain that indicates live organisms with intact DNA (**Fig. 21.11A**). The number of DAPI-stained organisms is counted in samples of seawater from a series of depths (0–300 m). Depth profiles over time further reveal the seasonal dynamics of the microbial community (**Fig. 21.11B**). At Bermuda, the highest populations of bacteria (coded red) are generally found at depths of 40–100 m, peaking during the summer. Overall, the BATS results can help improve our models of productivity of the world's oceans.

In another experiment at the Bermuda station, a group of undergraduates mapped the bacterial community over the depth series and compared it with a profile of ^{14}C-thymidine uptake as an indicator of DNA synthesis (**Fig. 21.12**). The profiles showed that while the bacterial concentration

A.

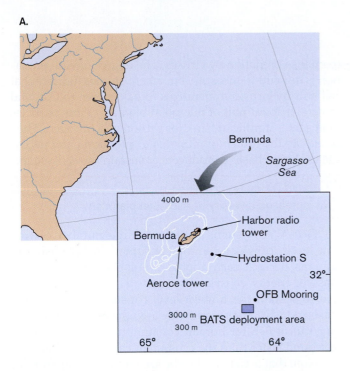

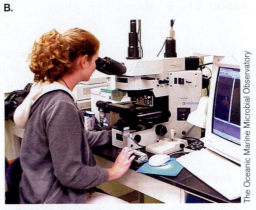

Figure 21.10 Mapping Bermuda phytoplankton.
A. The Bermuda Atlantic Time-Series (BATS) is an ongoing study of the marine microbial community of the Sargasso Sea. **B.** An undergraduate student from the Marine Microbial Ecology Summer Course at the Bermuda Biological Station uses epifluorescence microscopy to measure concentrations of picoplankton and viruses.

Figure 21.11 Epifluorescence observation of picoplankton.
A. Picoplankton, mainly bacteria, observed as part of the BATS (epifluorescence micrograph). The bacteria (large spots) and viruses (small spots) are stained with DAPI, a fluorescent dye that intercalates DNA. **B.** Concentration of bacteria at one BATS site as a function of depth and time over the course of a year, observed by epifluorescence using DAPI DNA stain.

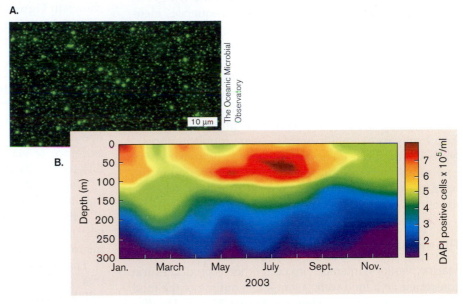

Figure 21.12 Depth profiles of marine bacteria. **A.** The bacterial concentration is plotted as a function of water depth. **B.** The rate of DNA synthesis, a measure of biomass production, is determined by uptake of radiolabeled thymidine. Thymidine incorporation is plotted as a function of water depth.

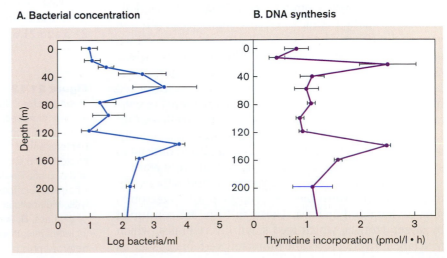

peaks at about 60 m, the biochemical activity (DNA synthesis) peaks at a shallower depth (about 20 m). This occurs because the photosynthesis is most active near the surface, with greatest light exposure. The phototrophic cells, however, are lysed and degraded rapidly by viruses. Thus, the phototrophs near the surface cycle tremendous amounts of carbon relative to their net biomass.

Metagenomic analysis of microbial DNA. The behavior of marine plankton, such as its interactions with marine animals and its production and consumption of carbon, depends upon its species. The vast majority of microbial species can only be identified based on their DNA. The DNA of marine microbial populations is now characterized by metagenomics, the sequencing of "community DNA." For metagenomic analysis, the DNA is obtained from organisms taken directly from their habitat without culturing.

In 2007 a team led by Craig Venter of the J. Craig Venter Institute sequenced over six billion base pairs of DNA from ocean water samples taken from around the globe. The research team estimated that their sequences identified 25,000 different microbial species per liter of seawater. Their survey revealed many new gene sequences, doubling the number of known protein types for all living organisms. Many protein classes believed to be found in only one domain (bacteria, archaea, or eukaryotes) were discovered in members of another domain. For example, proteorhodopsins, light-driven ion pumps originally found only in halophilic archaea, were found in numerous bacterial species. The bacterial proteorhodopsins have evolved diverse spectral sensitivities; those found in the open ocean are optimized to absorb blue light, whereas those in coastal regions absorb green light.

Planktonic Food Webs

Marine plankton interact with each other as producers and consumers in a food web. A simplified outline of the food web in the open ocean is shown in **Figure 21.13A**. The diagram of a food web is often called a "spaghetti diagram" because so many trophic interactions cross each other in various directions. The phototrophic producers are known as **phytoplankton**. Examples of phytoplankton include tiny *Prochlorococcus*, predominating in the upper photic layer, and *Synechococcus*, in a slightly deeper layer reached by longer wavelengths of light. In addition, there are myriad forms of algae, such as diatoms and dinoflagellates. Our knowledge of planktonic communities is evolving rapidly, as many marine phytoplankton are so small and difficult to study that major new species are discovered every year.

Bacterial phototrophs (picoplankton and nanoplankton) are mainly consumed by protists and by heterotrophic bacteria and archaea. These, in turn, are consumed by larger protists, which feed small invertebrates, which feed larger invertebrates and ultimately vertebrates such as fish. All levels of microbial plankton, however, are subject to intense predation by viruses. The degree of viral predation is difficult to measure, but recent studies indicate that cell lysis by viruses mineralizes (breaks down to CO_2 and minerals) about half of microbial biomass. Virus particles represent a significant sink for carbon and nitrogen. They accelerate the return of minerals to producers and necessitate a larger base of producers to sustain the ecosystem.

> **THOUGHT QUESTION 21.3**   How do viruses select for increased diversity of microbial plankton?

Ocean ecosystems contain a large number of trophic levels, of which the top consumers include fish and humans. Because each trophic level spends 90% of its biomass for energy, ecosystems require a huge lower foundation to sustain the highest-level consumers. This is one reason why fisheries worldwide are now in danger of running out of fish for human consumption: Fish are being harvested faster than the ecosystem can replace them.

Note that actual food webs are far more complex than shown in **Figure 21.13A**. One source of complexity is that many "algal protists" such as chrysophytes (golden algae) and dinoflagellates are actually **mixotrophs**, organisms conducting both photosynthetic and heterotrophic metabolism. Mixotrophs include secondary endosymbiont algae, such as Sargassum weed, as well as protists containing cyanobacterial or algal endosymbionts. A scheme for a food web that includes mixotrophs is shown in **Figure 21.13B**. The scheme is further complicated by the existence of elaborate mutualistic associations between different kinds of protists, algae, and bacteria.

Finally, it must be remembered that the microbial flora includes numerous unique species associated within or adherent to macroscopic organisms, including invertebrates, fish, and marine mammals. An example of a microbe found solely within a specialized animal host is *Epulopiscium fishelsoni*, a highly unusual bacterium that "gives birth" to progeny formed within the parent cell. *E. fishelsoni* grows only within the digestive tract of the Red Sea surgeonfish. Microbial communities within plants and animals are discussed in detail in Sections 21.5 and 21.6.

Figure 21.13   The pelagic marine food web.   A. In the marine water column, the vast majority of carbon transfer occurs among microbes. Arrows lead from organisms consumed and point to their consumers. The primary producers are phototrophic bacteria and algae, with a smaller contribution from lithotrophic bacteria and archaea. Grazers include heterotrophic bacteria and protists. Predators include protist flagellates and ciliates. About 40% of bacterial and protist biomass is degraded by viral lysis. Multicellular organisms cycle a relatively small fraction of biomass. **B.** A more complex view of the marine food web includes mixotrophs, organisms that combine phototrophic and heterotrophic nutrition, and symbiotic associations between protists and phototrophic bacteria or algae.

A. Marine food web

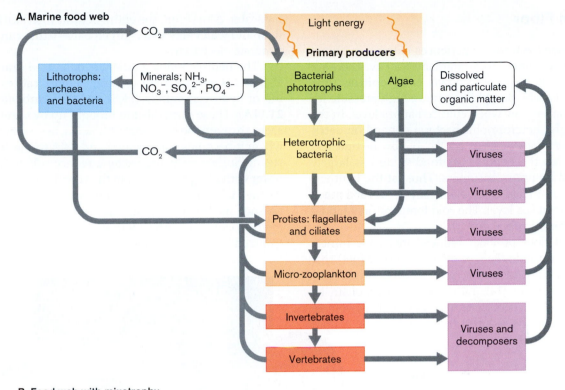

B. Food web with mixotrophy

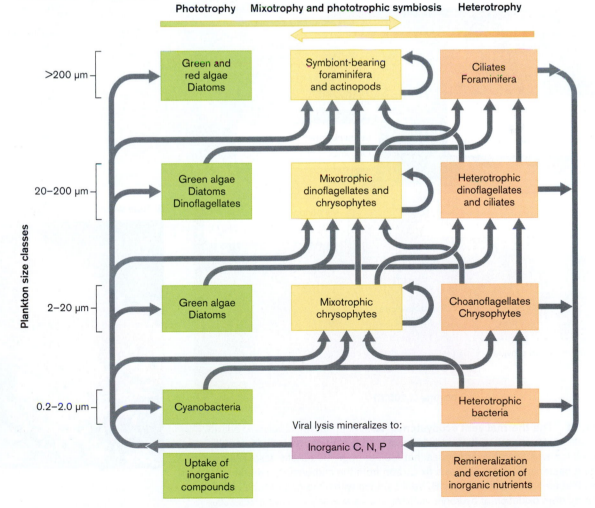

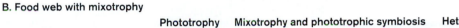

The Ocean Floor

The ocean floor (benthos) experiences extreme pressure from the water column. Most organisms that live there are pressure-dependent species, known as **barophiles** (discussed in Chapter 5). Barophiles require high pressure for growth, failing to grow when cultured at sea level. In the absence of light, heterotrophic bacteria depend on detritus from above as their carbon source; but by the time any material reaches the benthos, its food value has largely been depleted by prior consumers. Thus, at the benthos, with less food available, microbial decay occurs at a much lower rate than at sea level. The cold temperatures (about 10°C) also limit microbial growth.

The ocean floor provides reduced inorganic minerals such as iron sulfide (FeS) and manganese (Mn^{2+}) that can combine with dissolved oxygen to drive lithotrophy (discussed in Chapter 14). Lithotrophic metabolism by microbes in the sediment generates a permanent voltage potential between the reduced sediment and the oxidiz-

ing water. As a result, the entire benthic interface between floor and water acts as a charged battery that can actually generate electricity.

The benthic redox gradient is enhanced dramatically at thermal vents, where volcanic activity causes upwelling of hydrogen sulfide (H_2S), H_2, and carbonates (**Fig. 21.14A**). These minerals are brought up by seawater that seeps through the sediment until it reaches a magma pool, where it becomes superheated and rises to the surface as steam. Sulfate-reducing bacteria reduce sulfate from seawater with H_2 upwelling from the vent fluids, to form H_2S. As the H_2S rises, it is oxidized by sulfur-oxidizing bacteria such as *Thiomicrospira*. The vent waters also support methanogens, and the methane they produce becomes oxidized by methanotrophs.

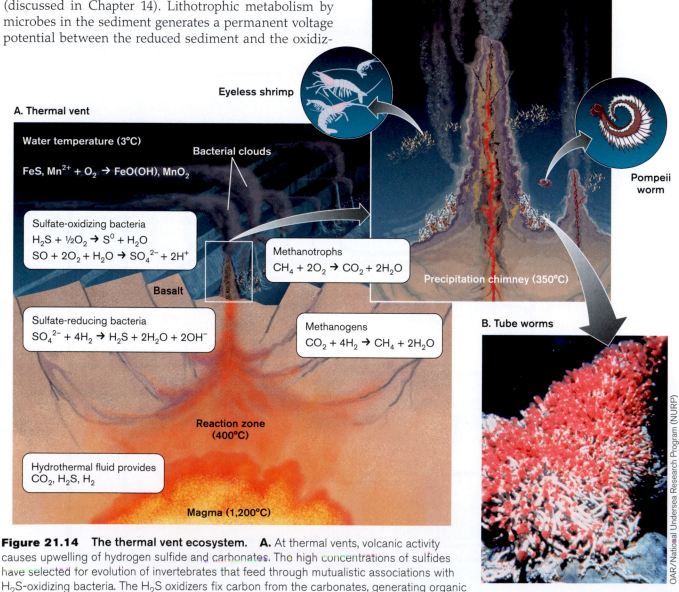

A. Thermal vent

Water temperature (3°C)

$FeS, Mn^{2+} + O_2 \rightarrow FeO(OH), MnO_2$

Bacterial clouds

Eyeless shrimp

Sulfate-oxidizing bacteria
$H_2S + \frac{1}{2}O_2 \rightarrow S^0 + H_2O$
$SO + 2O_2 + H_2O \rightarrow SO_4^{2-} + 2H^+$

Basalt

Methanotrophs
$CH_4 + 2O_2 \rightarrow CO_2 + 2H_2O$

Sulfate-reducing bacteria
$SO_4^{2-} + 4H_2 \rightarrow H_2S + 2H_2O + 2OH^-$

Methanogens
$CO_2 + 4H_2 \rightarrow CH_4 + 2H_2O$

Reaction zone (400°C)

Hydrothermal fluid provides
CO_2, H_2S, H_2

Magma (1,200°C)

Pompeii worm

Precipitation chimney (350°C)

B. Tube worms

OAR/National Undersea Research Program (NURP)

Figure 21.14   The thermal vent ecosystem.   A. At thermal vents, volcanic activity causes upwelling of hydrogen sulfide and carbonates. The high concentrations of sulfides have selected for evolution of invertebrates that feed through mutualistic associations with H_2S-oxidizing bacteria. The H_2S oxidizers fix carbon from the carbonates, generating organic metabolites that feed their animal hosts. **B.** Vent bacteria within the gills of tube worms provide energy from oxidation of hydrogen sulfide. *Source:* A. Maier, et al. *Environmental Microbiology.*

The high concentrations of sulfides at thermal vents have selected for a remarkable evolution of invertebrate species that feed through mutualistic associations with H₂S-oxidizing bacteria. The H₂S oxidizers fix carbon from the carbonates, generating organic metabolites that feed their animal hosts—species of worms, anemones, and giant clams, all closely related to surface-dwelling species that would be poisoned by H₂S. The tube worm *Riftia* is colored bright red by a pigment carrying H₂S and O₂ in its circulatory fluid (**Fig. 21.14B**). The worm has evolved such a complete dependence on its symbionts that it has lost its own digestive tract.

Freshwater Microbial Communities

Many features of marine habitats also apply to freshwater aquatic systems, such as lakes and rivers. Freshwater habi-

tats, of course, contain much lower salt, usually less than 0.1%. The study of aquatic systems is called limnology. Like the marine water column, the profile of a lake water column is defined by depth, aeration, and temperature (**Fig. 21.15**).

Large, undisturbed lakes are usually **oligotrophic**, with dilute concentrations of nutrients and microbes (**Fig. 21.15A**). The warm upper water layer above the thermocline is called the epilimnion. The epilimnion water is warm, well mixed, and oxygenated relative to the lower layers of water that are isolated from the atmosphere. It supports oxygenic phototrophs such as algae and cyanobacteria. Below the epilimnion lies a thermocline, a steep transition zone to colder, denser water below.

The epilimnion reaches only about 10 m in depth, much shallower than the marine euphotic zone. In a lake, the depth of light penetration varies greatly, depending on the concentration of microbes and particulate matter.

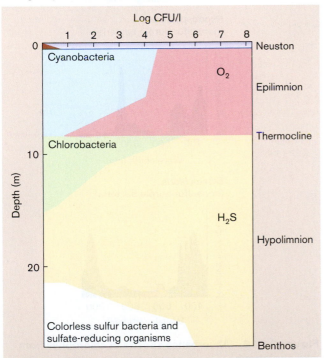

A. Oligotrophic lake (dilute nutrients)

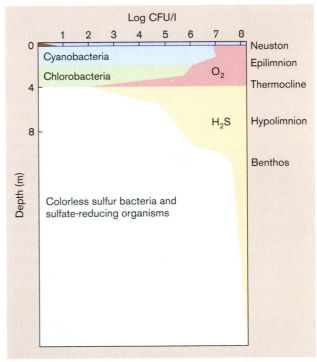

B. Eutrophic lake (rich nutrients)

C. Algal bloom

Figure 21.15 Aquatic microbial metabolism in lakes. A. Depth profile of an oligotrophic lake. The oxygenated zone (epilimnion) reaches to the thermocline, below which grow only anaerobes. **B.** Depth profile of a eutrophic lake. The oxygenated epilimnion is much shallower, and microbial concentrations are tenfold higher than in an oligotrophic lake. **C.** Algal bloom due to excess nitrogen from septic systems releasing into the pond.

At the edge of the lake, where the upper layer becomes shallow enough for rooted plants, is the **littoral zone**. In the deeper water, below the epilimnion, lies the hypolimnion, a region that becomes anoxic in lakes that are eutrophic (rich in organic nutrients).

When deeper water is anoxic, it supports only anaerobic microbes. These include a series of anaerobic phototrophs that do not produce O_2. Enough light may penetrate to support anaerobic H_2S-splitting phototrophs such as *Chlorobium* and *Rhodopseudomonas*. Although H_2S photolysis provides less energy than oxygenic H_2O photolysis, these bacteria have evolved in a productive niche by using chlorophyll pigments whose spectrum extends into the infrared (**Fig. 21.16**). Light in the red and infrared portions of the spectrum extends the deepest into the lake column. Light with wavelengths in this range cannot be used by oxygenic phototrophs because the photon energy is insufficient to split water. The less efficient H_2S photolyzers, however, can harness the energy of red and infrared radiation (discussed in Chapter 14). The green and purple phototrophs overlap metabolically with anaerobic heterotrophs and lithotrophs that reduce oxidized minerals. Some bacteria, such as *Rhodospirillum rubrum*, can grow anaerobically with or without light; others grow only by anaerobic catabolism, unassisted by light.

At the bottom of the water column, the water meets the sediment (benthos). In the benthic sediment, gradients develop in which successive electron acceptors become reduced by anaerobic respirers and lithotrophs (**Fig. 21.17**). Electron acceptors that provide the most energy are consumed first; as each in turn is depleted, the electron acceptor with the most energy is consumed next. First, molecular oxygen is used to oxidize organic material and reduced minerals such as NH_4^+. As molecular oxygen falls off, bacteria use nitrate (NO_3^-) from oxidized ammonium ion as an electron acceptor to respire on remaining organic material. As the nitrate is used up, still other bacteria use manganese (Mn^{4+}) as an electron acceptor, followed by iron (Fe^{3+}) and sulfate (SO_4^{2-}). Reduction of sulfate leads to H_2S, which eventually returns to the upper layers supporting anaerobic photolysis. Reduction of CO_2 by H_2 produces methane (CH_4). Methane collects below and sometimes ignites when it escapes to the surface. Methane is of global concern as a potent "greenhouse gas" (discussed in Chapter 22).

NOTE: Certain minerals and organic molecules occur in equilibrium between ionized and un-ionized states over the range of pH typical of most common habitats (pH 4–9). Examples include ammonia (NH_3), which protonates to ammonium ion (NH_4^+), and organic acids such as acetic acid (CH_3COOH), which deprotonates to acetate (CH_3COO^-). In this chapter, we refer to the form most prevalent at pH 7 unless stated otherwise.

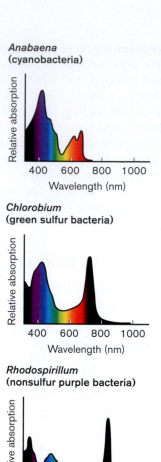

Anabaena
(cyanobacteria)

Chlorobium
(green sulfur bacteria)

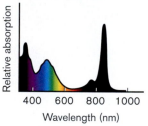

Rhodospirillum
(nonsulfur purple bacteria)

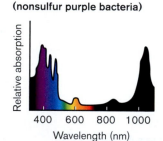

Blastochloris
(nonsulfur purple bacteria)

Figure 21.16 Absorbance spectra of chlorophyll from lake phototrophs. In the upper waters, algae and cyanobacteria absorb primarily blue and red. Below, where red has been absorbed by microorganisms above, anaerobic phototrophs such as *Blastochloris* absorb infrared (wavelengths beyond 750 nm).

Throughout the water column, fungi and protists serve as grazers and predators on algae and bacteria, respectively. They also interact with invertebrates and fish as parasites. Other important consumers are the viruses, which lyse about half of microbial populations in lakes, as they do in the ocean. Viruses serve an important function by limiting the number of microbes and keeping the water clear enough for light to penetrate.

Note that all fish and invertebrates need access to the epilimnion for oxygen. Only anaerobic microbes can

Benthic microbial redox transformations

Carbon

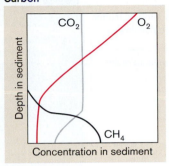

Nitrogen

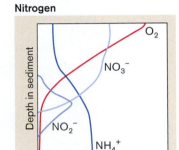

Sulfur

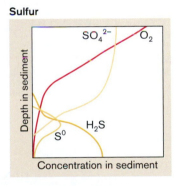

Figure 21.17   Redox gradients in the benthic sediment.
At the top of the sediment interface with the water column, minerals are first oxidized lithotrophically by O_2; then, as O_2 declines, the oxidized minerals are used as alternative electron acceptors for anaerobic respiration. The nitrate, nitrite, and sulfur concentrations fall as these species are used as alternative electron acceptors for anaerobic respiration.

grow entirely in the anoxic hypolimnion. Thus, problems arise when effluents entering a lake carry high concentrations of nutrients causing overgrowth of phototrophs. Initially, the phototrophs increase the oxygen content of the epilimnion. But, the cell mass of the phototrophs sinks to the bottom, where overgrowth of heterotrophs depletes oxygen, causing **eutrophication** (**Fig. 21.15B**). So the anaerobic zone near the bottom of the lake grows larger, meaning that the boundary between the aerobic and anaerobic zones rises. In a eutrophied lake, fish die off owing to lack of oxygen, which heterotrophic microbes have consumed. Such a lake is said to have a high level of biochemical oxygen demand (BOD).

Common causes of eutrophication include:

- **Phosphates.** Because phosphorus is commonly a **limiting nutrient** (nutrient in shortest supply) for algae, addition of phosphates from detergents and fertilizers can lead to an **algal bloom** in which algae overgrow the water surface. The algae die, and their consumption by heterotrophic bacteria depletes oxygen, suffocating fish.
- **Nitrogen** from sewage effluents can lead to algal blooms by relieving nitrogen limitation.
- **Organic pollutants** from sewage effluents overfeed heterotrophic bacteria, depleting the epilimnion of oxygen.

A **eutrophic lake** is one in which the lower layers have become depleted of oxygen as a result of overgrowth of microbial producers. Thermal stratification may break up, and the lake may mix one to several times per year, thus reoxygenating the entire lake. A permanently eutrophic lake typically supports ten times the microbial concentrations of an oligotrophic lake (see **Figs. 21.15A** and **B**), but shows greatly decreased animal life.

TO SUMMARIZE:

- **The euphotic zone of the ocean** is the upper part of the water column that receives light for phototrophs. Below, in the aphotic zone, only heterotrophs and lithotrophs can grow. The benthos includes the region where the water column meets the ocean floor, as well as sediment below the surface.
- **Plankton are small floating organisms**, including swimming microbes. Phytoplankton are phototrophs such as cyanobacteria and algae. Microbial consumers include protists and viruses. Many marine protists are mixotrophs (producers and consumers in one).
- **Picoplankton** are measured by epifluorescence microscopy. Their biochemical rate of production is measured by uptake of radiolabeled nutrients.
- **New microbial species** are discovered by metagenomic analysis.
- **Benthic microbes are barophiles.** The seafloor supports psychrophiles, whereas hydrothermal vents support thermophiles. Vent microbes include sulfur-oxidizing bacteria, sulfur-reducing bacteria, and methanogens. Bacteria that oxidize H_2S feed symbiotic animals such as tube worms.
- **Aquatic freshwater lakes** exhibit stratified water levels. As depth increases, minerals become increasingly reduced. Anaerobic forms of metabolism predominate, with the more favorable alternative electron acceptors used in turn.
- **Lakes may be eutrophic or oligotrophic.** Eutrophic lakes may show such high biochemical oxygen demand (BOD) that they cannot support vertebrate life.

21.4 Soil and Subsurface Microbiology

In contrast to the ocean, where the base of producers is almost entirely microbial, the major producers of terrestrial ecosystems are macroscopic plants. Plants vary greatly in size and form, from mosses and bryophytes to prairie grasses and forest trees, but all terrestrial plants are rooted in **soil**. Soil is a complex mixture of decaying organic and mineral matter that feeds vast communities of microbes—arguably the most complex microbial ecosystem on our planet. And soil-based agriculture is the major source of food for our planet's human inhabitants. The qualities of a given soil—oxygenated or water saturated, acidic or alkaline, salty or fresh, nutrient-rich or -poor—define what food can be grown and whether the human community will eat or starve.

Soil Microbiology

The general structure of soil (**Fig. 21.18**) includes a series of layers called **horizons** that arise as a result of soil-forming factors such as rainfall, temperature variation, wind, and biological activity. Note that the soils of different habitats, such as prairie, forest, and desert, vary greatly as to the depth and quality of each layer.

The surface layer of soil we see is the organic horizon. The organic horizon consists of dark, organic detritus, such as shreds of leaves fallen from plants. The detritus of the organic horizon is in the earliest stages of decomposition by microbes, primarily fungi and bacteria such as actinomycetes. Early-stage decomposition is defined loosely as a state in which the origin of the detritus may be still recognizable.

Beneath the organic horizon lies the lighter colored aerated horizon, in which organic particles in more advanced stages of decomposition combine with minerals from rock at lower levels. In the aerated horizon, the source of the organic particles is no longer recognizable, and decomposers have broken down some of the more difficult-to-digest plant structural components such as lignin. (Lignin is a complex aromatic molecule whose digestion is discussed later in this section.) Such material is often sold by garden stores as peat or topsoil.

In well-drained soil, both the organic and aerated horizons are full of oxygen as well as nutrients liberated by the decomposers and used by plants. In these layers, soil consists of a complex assemblage of organic and inorganic particles (**Fig. 21.19A**). In between soil particles are air spaces that provide access to oxygen. Each particle of soil supports miniature colonies, biofilms, and filaments of bacteria and fungi that interact with each other and with the roots of plants. An interaction of fundamental significance is that of fungi with plants in the symbiotic system known as mycorrhizae (discussed shortly).

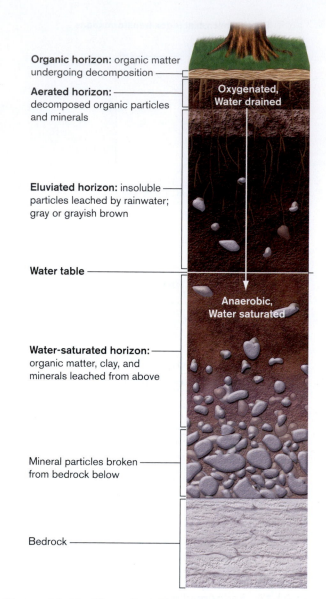

Figure 21.18 The soil profile. Soil forms layers in which decomposing organic material predominates at the top, minerals toward the bottom, at bedrock. The top layers are aerated, providing heterotrophs with access to O_2, whereas the bottom layers are water-saturated and anaerobic.

Below the aerated horizon, the eluviated horizon experiences periods of water saturation from rain. Rainwater leaches (dissolves and removes) some of the organic and mineral nutrients from the upper layers. Below the eluviated horizon lie increasing proportions of minerals and rock fragments broken off from bedrock below. These lower layers experience increasing water saturation, forming the **water table**. Prolonged water saturation generates anoxic conditions. This anoxic, water-saturated region contains mainly lithotrophs and anaerobic heterotrophs.

The soil layers finally end at bedrock, a source of mineral nutrients such as carbonates and iron. Interestingly, bedrock is permeated with microbes. Core samples show

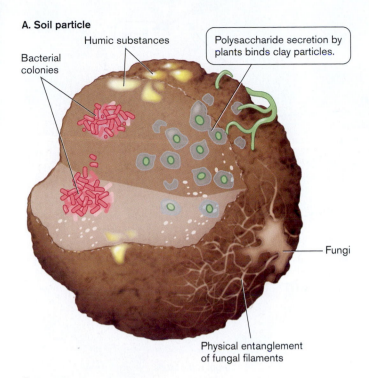

A. Soil particle

Humic substances

Bacterial colonies

Polysaccharide secretion by plants binds clay particles.

Fungi

Physical entanglement of fungal filaments

B.

Li-Hung Lin

Figure 21.19   Microbes in soil and rock.   A. Soil particles support growth of complex assemblages of microbes. A soil particle contains bacterial colonies, biofilm associations, and microbes associated with fungi and plant roots. **B.** Endolithic bacteria grow 3.5 km underground in this South African gold mine.

that crustal rock as deep as 3 km down contains **endoliths**, bacteria growing within the very crystals of solid rock (**Fig. 21.19B**). What energy source feeds microbes trapped within rock? For some endoliths, a surprising answer may be the radioactive decay of uranium. Uranium-238 decay generates hydrogen radicals that combine to form hydrogen gas. The hydrogen gas combines with CO_2 from carbonate rock to feed methanogens and other endolithic lithotrophs. Still other endoliths are consumers that feed on the lithotrophs or the organic compounds they produce.

The Soil Food Web

The top horizons of soil feature a food web of extraordinary complexity (**Fig. 21.20**). The major producers are green plants, whose leaves generate detritus and whose root systems serve predators, scavengers, and mutualists. Some carbon is also fixed by lithotrophs oxidizing reduced nitrogen (NH_3), hydrogen sulfide (H_2S), and iron (Fe^{2+}). In well-aerated soil, however, the proportion of carbon fixed by lithotrophs is small.

Plant matter is decomposed by many species of fungi, such as *Mycena* species, and by bacteria such as the actinomycetes. The actinomycetes include *Streptomyces*, a genus famous for the production of antibiotics and for generating chemicals whose odors give soil its characteristic smell (**Fig. 21.21A**). Besides fallen leaves, another source of organic matter from plants is the **rhizosphere**, the region of soil surrounding plant roots. The rhizosphere contains proteins and sugars released by roots, as well as sloughed-off plant cells. These materials feed large numbers of bacteria, which then cycle minerals back to the plant. Bacteria in the rhizosphere may also discourage growth of plant pathogens. Another habitat colonized by bacteria is the surface of fungi, whose filaments are considerably larger than bacterial cells (**Fig. 21.21B**).

At the next trophic level, bacteria feeding on leaf detritus and root exudates are then preyed on by protists and nematodes (**Figs. 21.21C** and **D**). Diverse predators such as nematodes, fungi, and protists exhibit different preferences for bacteria, fungi, or protists as prey. *Vampirella* protists drill holes in fungal hyphae to suck out their nutrients. Parasitic fungi prey on plants or invertebrates; some actually capture and strangle nematodes. Besides predation, there exist many complex forms of mutualism, such as lichens (discussed earlier). Another mutualistic interaction of fungi is that of **mycorrhizae**, fungal infections of plants in which the fungal hyphae extend the root surface area and increase the root's ability to absorb nutrients.

Microbes also cooperate to form complex communities such as microbial mats and biofilms. The microbes ultimately feed invertebrates, which then feed larger invertebrates and vertebrate predators. Some predators, such as earthworms and burrowing animals, enhance the soil quality by turning over the matter, aerating the soil particles and helping to mix the organic matter from above with the mineral particles from below.

Decomposition of Lignin to Humus

A critical role of fungal decomposers (also known as **saprophytes**) is the breakdown of extremely complex structural components of vascular plants such as grass and trees. Trees, in particular, accumulate vast stores of biomass in

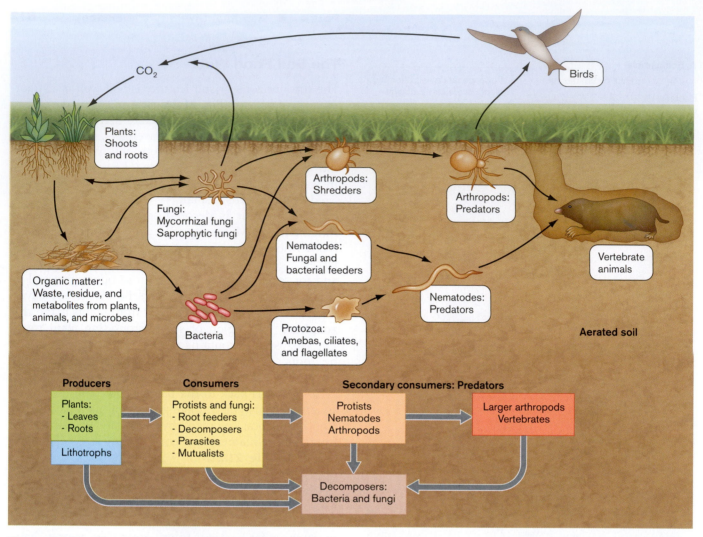

Figure 21.20   The soil food web. Plants are the major producers, although some production also occurs from lithotrophs such as ammonia oxidizers. Detritus from plants is decomposed by fungi and bacteria, which feed protists and small invertebrates such as nematodes. Protists and small invertebrates are consumed by larger invertebrates and vertebrate animals.

A. *Streptomyces griseus*

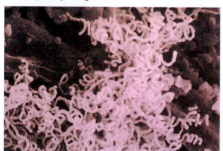

D. Mites feed on decaying leaves

B. Bacteria grow on fungi

C. Nematode feeds on bacteria

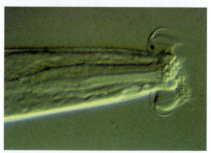

Figure 21.21   Soil microbes.   A. Hyphae of actinomycete bacteria, *Streptomyces griseus* (filament width 1 µm; SEM). *Streptomyces* bacteria give the soil its characteristic odor. **B.** Bacteria grow on the surfaces of fungi (hyphal width 10–20 µm). **C.** A bacteria-feeding nematode has special mouthparts to consume bacteria. Other species of nematodes graze on plant roots or fungi. **D.** Mites feed on dead leaves decayed by fungi and bacteria. *Source:* A–D. Reprinted from *The Soil Biology Primer* with permission. ©2000 Soil and Water Conservation Society.

forms that are difficult to digest, such as lignin. Fungal and bacterial decomposers possess enzyme systems to degrade lignin and other complex components of plants. Examples of decomposers include white rot fungi (**Fig. 21.22A**) and actinomycete soil bacteria. The prevalence of lignin is one reason that decomposition by fungi plays a much larger role in terrestrial ecosystems than in marine ecosystems.

Lignin is a highly complex and diverse covalent polymer composed of interlinked phenolic groups (benzene rings with OH or related oxygen-bearing side groups; **Fig. 21.22B**). These phenolic polymers can be broken down

A. White rot fungi

B. Lignin

Figure 21.22 Microbial formation of humus. A. *Mycena haematopus*, a white rot fungus, growing on a willow log. The fungus degrades lignin. **B.** Lignin is a complex organic polymer that is a component of wood and bark, one of the most abundant polymers on Earth.

relatively rapidly to smaller units, some containing only a single phenol or benzoic acid. This first phase of degradation (about 50%) occurs within a year of deposition in the soil. The remaining phenolics, however, may degrade less than 5% per year; and some samples dated by ^{14}C isotope ratios have been shown to last 2,000 years. These phenolic molecules are called **humic material** or **humus**. Because of its slow degradation, humic material provides a steady slow-release supply of nutrients for plant growth. But forests whose rate of microbial decomposition is particularly low—for example, the New Jersey Pine Barrens—ultimately depend on fire to clear the mounting layers of humus.

Lignin is also a major constituent of newsprint and other kinds of paper disposed of by composting or in landfills. Because of the low degradation rate, particularly in anoxic soil, newsprint in landfills can maintain its structure for decades. Archaeology of landfills outside New York City has revealed readable headlines from papers dating to the 1940s.

Microbes Associated with Roots

The presence of plant roots provides yet another level of complexity to soil communities (**Fig. 21.23**). Plant roots influence the surrounding soil by taking up nutrients and by secreting organic substances and molecules that modulate their surroundings. The environment surrounding a plant root can be further subdivided into two categories: the **rhizoplane**, the root surface; and the **rhizosphere**, the region of soil outside the root surface but still influenced by plant exudates (materials secreted by the plant). Particular bacterial species are adapted to these environments. For example, in anaerobic wetland soil, the rhizoplane and rhizosphere of plant roots provide oxygen for methanotrophs that oxidize methane produced by methanogens.

The rhizoplane and rhizosphere also provide the environment for symbiotic fungi that generate mycorrhizae. At least 80% of plants in nature, including 90% of forest trees, require mycorrhizae for growth.

Mycorrhizae: The Fungal Internet

The function and significance of mycorrhizae for plant growth is just beginning to be understood. Mycorrhizae (from *myco*, fungal, and *rhiza* root) consist of fungal mycelia that associate intimately with the roots of plants, extending access to minerals while obtaining in return the energy-rich products of plant photosynthesis. Mycorrhizae were first discovered in the 1880s by German truffle hunters who sought to cultivate the prized delicacy, the fruiting body of an ascomycete (for review of fungi, see Chapter 20). The propagation of truffles was investigated by mycologist A. B. Frank at the Landwirtschaftlichen Hochschule, Berlin. To Frank's surprise, he found that the truffles extended their mycelia far beyond the site of the

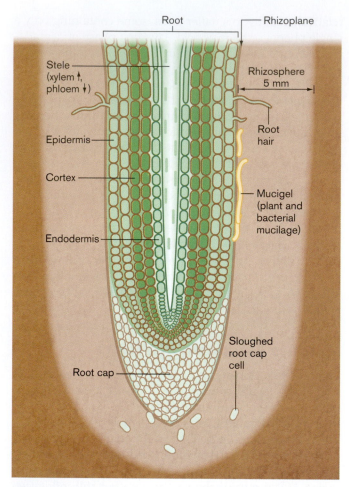

Figure 21.23   Plant roots create particular habitats for microbes.   The rhizoplane is the region of soil directly contacting the plant root surface. The rhizosphere is the soil outside the rhizoplane that receives substances from the root, such as mucilage, sloughed cells, and exudates.

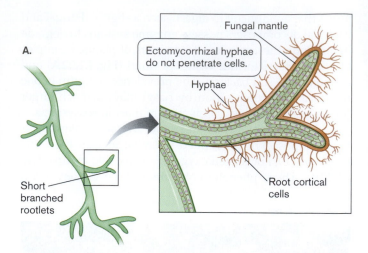

Figure 21.24   Ectomycorrhizae: Fungi colonize the surface of the rootlet.   A. Plant root cross section. Ectomycorrhizae extend hyphae from the root surface. **B.** Ectomycorrhizal hyphae from a rootlet (light micrograph).

fruiting body, and that the mycelia formed an impenetrable tangle with plant roots. Frank called the tangled mycelia "fungus-root" or mycorrhizae. A century later, we are beginning to appreciate that these mysterious fungal-root tangles offer a vast interconnected network for exchange of nutrients among fungi and many different plants, like an internet connecting countless sites.

Two different kinds of mycorrhizae are observed: ectomycorrhizae and endomycorrhizae. **Ectomycorrhizae** colonize the rhizoplane, the surface of plant rootlets, the most distal part of plant roots (**Fig. 21.24**). The fungal mycelia never penetrate the root; instead, they form a thick mantle surrounding the root, then extend long mycelia away from the root to absorb nutrients. Numerous kinds of fungi form ectomycorrhizae, including ascomycetes, such as truffles, and basidiomycetes, known by their mushrooms, such as stinkhorns. **Endomycorrhizae** form a more intimate association, in which the fungal filaments penetrate plant cells.

A typical "large plant, small plant" experiment demonstrates the impact of mycorrhizae (**Fig. 21.25**). In this experiment, pine seedlings of *Pinus densiflora* were grown with or without inoculation with the ectomycorrhizal fungus *Pisolithus tinctorius*. The seedlings inoculated with the fungus clearly grew faster and larger than those without. Furthermore, analysis of root/shoot ratios shows that plants grown with mycorrhizae invest less of their body mass in roots and more in the aboveground stems and leaves—an important consideration for agriculture, in which the aboveground plant is usually the part harvested.

Unlike the ectomycorrhizae, the endomycorrhizae generate an intimate association in which fungal hyphae penetrate plant cells deep within the cortex (**Fig. 21.26**). The penetrating hyphae form knobbed branches that resemble

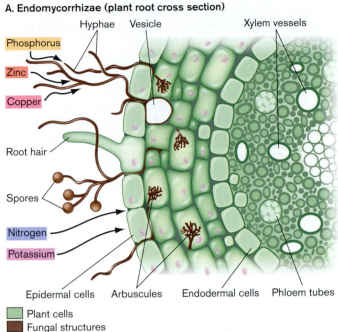

A. Endomycorrhizae (plant root cross section)

Hyphae   Vesicle   Xylem vessels

Phosphorus

Zinc

Copper

Root hair

Spores

Nitrogen

Potassium

Epidermal cells   Arbuscules   Endodermal cells   Phloem tubes

☐ Plant cells
☐ Fungal structures

B. Arbuscule within a root cell

Mark Brundett, CSIRO Forestry and Forest Products

Figure 21.26   Endomycorrhizae: Fungi invade root cells, forming arbuscules.   A. Plant root cross section. Endomycorrhizal hyphae penetrate cells deep within the root cortex. **B.** The penetrating hypha forms an arbuscule within a root cell.

Michael Amaranthus

Figure 21.25   Pine seedlings grown with or without mycorrhizal fungi.

microscopic "trees," or **arbuscules** within the root cells. Some of the hyphae form specialized vesicles within the plant that store nutrients. Another name for this kind of mycorrhizae is **vesicular-arbuscular mycorrhizae (VAM)**.

Endomycorrhizae are more specialized than ecto-mycorrhizae. They comprise a relatively small number of fungal species, such as members of the zygomycete family Glomales, and they show obligate dependence on their host plants. Their presence in nature and their importance in the ecosystem, however, are actually greater. Endo-mycorrhizal species exist entirely underground (they do not form mushrooms), and they completely lack sexual cycles. They may acquire 25% of the photosynthetic product of their hosts in exchange for tremendous expansion of access to soil resources.

Mycorrhizae greatly enhance the plant's uptake of water as well as minerals such as nitrogen and phosphorus. In addition, the hyphae sequester toxins, and they actually distribute organic substances from one plant to another. Mycorrhizae may join many different plants, even different species, in a vast nutrient-sharing network.

The reliance of plants on mycorrhizae may explain why some plants, such as orchids, grow better in nature than in sterilized potting soil, which lacks mycorrhizal fungi.

THOUGHT QUESTION 21.4   Design an experiment to test the hypothesis that the presence of mycorrhizae enhances plant growth in nature.

Wetland Soils

So far, we have considered the interface between ground and water (the benthic sediment beneath oceans and lakes) and the interface between ground and air (aerated soil). An interesting case that combines the two is that of wetlands (**Fig. 21.27A**). A **wetland** is defined as a region of land that undergoes seasonal fluctuations in water level, so that sometimes the land is dry and oxygenated, at other times water saturated and anaerobic. Wetlands provide many crucial functions; for example, the Everglades filter much of the water supply for Florida communities.

Wetland soil that undergoes such periods of anoxic water saturation is known as **hydric soil**. Hydric soil is characterized by "mottles," patterns of color and paleness (**Fig. 21.27B**). The reddish brown portions (for example, surrounding a plant root) result from oxidized iron (Fe^{3+}). The gray-colored "gley" portions indicate loss of iron in its water-soluble reduced form (Fe^{2+}) generated by anaerobic respiration.

Soil becomes anoxic when the rate of oxygen diffusion is too low to support aerobic metabolism. Anaerobic metabolism allows much lower rates of production than metabolism in the presence of oxygen, as anaerobes use oxidants of lower redox potential and limited quantity, such as sulfate and nitrate. Many kinds of anaerobic bacteria inhabit wetlands. For example, denitrifiers (bacteria using nitrate to oxidize organic food) remove nitrate from water before it enters the water table—one of the ways that wetlands protect our water supply.

The alternative oxidants (electron acceptors) are always in limited supply, so further catabolism occurs by fermentation. Fermentation allows only incomplete breakdown of food molecules, generating a rich and diverse supply of nutrients for a variety of consumers, including aerobic organisms when the water recedes. Thus, despite its lower overall productivity, anaerobic soil contributes to the nutritional diversity of wetland ecosystems. The relatively slow rate of decomposition can lead to accumulation of high levels of organic carbon, particularly rich for plant growth.

The anoxic conditions of wetlands also favor methanogenesis. Methanogenesis is performed solely by archaea (discussed in Chapters 13 and 17). Methanogenesis occurs when fermenting bacteria generate H_2, CO_2, and other one- or two-carbon substrates that methanogens convert to methane. Methane is a more potent greenhouse gas than CO_2, and although the current methane concentration in our atmosphere is low, it is rising exponentially.

In water-filled anaerobic soils, particularly those of rice fields, significant quantities of methane escape to the air through air-conducting channels in the roots of the rice plant. The roots of rice plants, like those of other wetland vascular plants, contain channels for oxygen to reach the roots. These same channels, however, allow methane to escape from anaerobic soil. Fortunately, the rhizosphere of the roots supports methanotrophs that oxidize methane, so research is being done to maximize methanotroph activity. There is some evidence that natural wetlands actually take in more carbon than they put out, whereas disturbed wetlands (wetlands altered by human activity) generate net efflux of CO_2 and CH_4. The role of wetlands in global cycling is discussed further in Chapter 22.

Dry Land: Lichens and Cryptogamic Crust

The opposite of wetlands is the dry, nutrient-poor habitats that support extremely hardy microbial communities such as lichens. Lichens are essential for boreal (northern) forests, where they cover the majority of the ground and serve as food for grazing animals (**Fig. 21.28A**). In winter, lichens are a major food source for caribou, which dig beneath the snow to obtain them (**Fig. 21.28B**). Native populations of North America often use lichens for medicines and to dye cloth (**Fig. 21.28C**). Many lichens are found to contain antibacterial agents.

More southern climates, such as the dry hot deserts of the western United States, host a slow-forming type of soil called **cryptogamic crust** (**Fig. 21.29**).

Courtesy of Chien-Lu Ping, Univ. of Alaska, Anchorage

Figure 21.27 Wetland soil. A. A true wetland experiences periods of water saturation alternating with dry soil. Soil that is water saturated becomes anaerobic because O_2 diffuses slowly through water. **B.** Alternating periods of water saturation and dryness generate an appearance of mottled color in soil. The red-orange colored portions around root holes result from oxidized iron (Fe^{3+}), whereas the gray portions ("gley") indicate loss of iron in its water-soluble reduced form (Fe^{2+}) generated by anaerobic respiration.

A. Lichen ground cover

Stephen & Sylvia Sharnoff

B. Caribou feed on lichens

Fran Mauer, U.S. Fish & Wildlife Services

C. Blankets dyed with lichens

Stephen & Sylvia Sharnoff

Figure 21.28 Boreal forest ecosystems and native cultures depend on lichens. **A.** Lichens cover most of the ground in boreal forests. **B.** Caribou dig up lichens beneath the snow. **C.** Chilkat Tlingit dancing blankets are traditionally dyed with lichens.

A.

Christopher Prince

B.

Biological Soils

Figure 21.29 Cryptogamic crust. **A.** Cryptogamic crust interspersed by vegetation, Utah's Capitol Reef National Park. **B.** Close-up view of cryptogamic crust just after rainfall.

(*Crypto* means "hidden," referring to the hidden nature of this ground cover.) Cryptogamic crust includes a complex, spongy association of lichens, mosses, and nonlichenous algae and fungi. The crust takes many years to grow a centimeter thick and is easily destroyed by bikes or hiking boots—a problem for desert parks such Utah's Capitol Reef National Park. The crust plays a vital role in desert ecosystems by holding moisture, protecting plant seeds for germination, and preventing erosion.

TO SUMMARIZE:

- **The uppermost horizons of soil** consist of detritus, largely aerated. Below the aerated layers, the eluviated horizon experiences water saturation. Lower layers are anoxic.
- **The soil food web** includes a complex range of microbial producers, consumers, predators, decomposers, and mutualists. Microbial mats and biofilms are complex multispecies communities of bacteria, fungi, and other microbes.
- **Fungi decompose lignin**, a complex aromatic tree component that is challenging to digest. Lignin decomposition forms humus.

- **Fungi form symbiotic associations** with plant roots called mycorrhizae. Mycorrhizae transport soil nutrients among many different kinds of plants.
- **Wetland soils** alternate aerated (dry) with anoxic (water-saturated) conditions. Anoxic wetland soil favors methanogenesis. Wetland soils are among the most productive ecosystems.
- **Dry soils** depend on lichens.

21.5 Microbial Communities within Plants

Specialized habitats for microbes exist within the bodies of plants and animals. Multicellular plants and animals coevolve with the microbes in their environment, and in many cases their microbes can exist nowhere else. The microbial partners may grow as mutualists, commensals, or parasites. A form of bacteria-plant mutualism critical for agriculture is that of nitrogen fixation by rhizobia, a group of alpha proteobacteria related to *E. coli* and other gram-negative species. Major rhizobial genera include *Rhizobium*, *Bradyrhizobium*, and *Sinorhizobium*.

Rhizobia Fix Nitrogen for Legumes

Rhizobia, associated with legumes such as peas and beans, fix more nitrogen than the plants absorb from soil, actually increasing the soil's nitrogen content. For this reason, farmers often alternate crops such as corn with soybeans to restore nitrogen to the soil. The rhizobial bacteria develop specialized forms within plant cells, called **bacteroids**. Bacteroids lack cell walls and are unable to reproduce; their function is specialized for nitrogen fixation. The rhizobial infection of root hairs induces formation of nodules within which the nitrogen-fixing bacterioids are sequestered (**Fig. 21.30A**).

The initiation, development, and maintenance of the rhizobia-legume symbiosis poses intriguing questions of genetic regulation. How does the association begin? How do host plant and bacterium recognize each other as suitable partners? The legume exudes signal molecules called **flavonoids** into its rhizosphere. Flavonoids resemble steroid hormones such as estrogen and have similar effects on animals; they are also called phytoestrogens. The flavonoids are detected by rhizobial bacteria, which respond by chemotaxis, swimming toward the root surface.

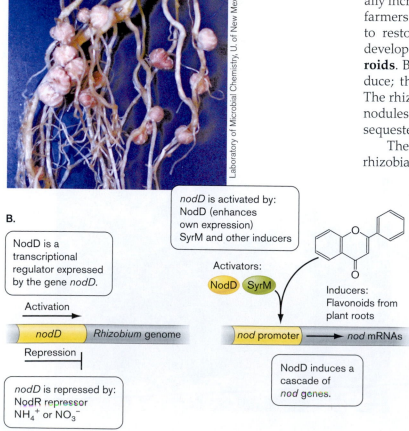

A.

Laboratory of Microbial Chemistry, U. of New Mexico

B.

NodD is a transcriptional regulator expressed by the gene *nodD*.

Activation →

nodD   *Rhizobium* genome

Repression ⊣

nodD is repressed by:
NodR repressor
NH_4^+ or NO_3^-

nodD is activated by:
NodD (enhances own expression)
SyrM and other inducers

Activators:
NodD   SyrM

Inducers:
Flavonoids from plant roots

nod promoter → *nod* mRNAs

NodD induces a cascade of *nod* genes.

Figure 21.30   Rhizobia induce legume roots to form nitrogen-fixing nodules.   A. *Rhizobium* nodules on pea plant roots. **B.** Rhizobial infection and nodulation of a root hair is mediated by flavonoid induction of the bacterial *nod* genes.

A.

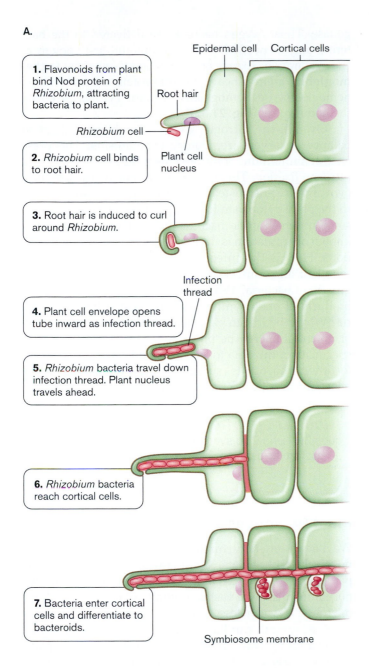

1. Flavonoids from plant bind Nod protein of *Rhizobium*, attracting bacteria to plant.

Epidermal cell   Cortical cells

Root hair

Rhizobium cell

Plant cell nucleus

2. *Rhizobium* cell binds to root hair.

3. Root hair is induced to curl around *Rhizobium*.

Infection thread

4. Plant cell envelope opens tube inward as infection thread.

5. *Rhizobium* bacteria travel down infection thread. Plant nucleus travels ahead.

6. *Rhizobium* bacteria reach cortical cells.

7. Bacteria enter cortical cells and differentiate to bacteroids.

Symbiosome membrane

B.   **C.**

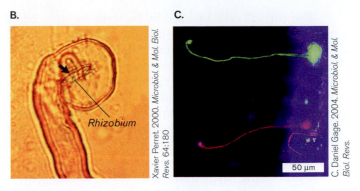

Rhizobium

Xavier Perret. 2000. Microbiol. & Mol. Biol. Revs. 64:180

50 µm

C. Daniel Gage. 2004. Microbiol. & Mol. Biol. Revs.

Figure 21.31   The infection thread.   A. Rhizobia are attracted to the legume by chemotaxis toward exuded flavonoids. A bacterium induces an epidermal root hair to curl around it and take it up into a tube of plant cell wall material. The bacteria entering through the plant tube form an infection thread. The thread eventually penetrates cortical cells, where the bacteria lose their cell walls and become nitrogen-fixing bacteroids. **B.** Root hair curling around *Rhizobium*, and the formation of an infection thread (light microscopy). **C.** Infection threads invading the cortex (fluorescence microscopy). Bacteria are labeled with transgenes expressing either DsRed (pink) or green fluorescent protein (green).

> **THOUGHT QUESTION 21.5**   High levels of nitrate or ammonium ion corepress the *nod* genes through NodR. Why?

Within the bacterial cell, the signal molecules induce transcription of genes involved in nodule formation, called *nod* genes (**Fig. 21.30B**). The overall regulatory protein NodD then amplifies its own subsequent expression, subject to limitation by the repressor NodR. This highly sensitive feedback loop leads to coordinated expression of proteins that induce root infection, plant cell proliferation, and nodule development. One of the functions of the *nod* genes is to synthesize Nod factors, molecules composed of chitin with lipid attachments. Nod factors communicate with the host plant and help establish species specificity between bacterium and host.

The entry of the bacteria into the host involves a fascinating interplay between bacterial and plant cells, whose mechanism remains largely unknown (**Fig. 21.31A**). First, a bacterium is attracted by flavonoids to the surface of a root hair extended by a root epidermal cell. The presence of the bacterium somehow induces the root hair to curl around it and ultimately surround the bacterium with plant cell envelope. The bacterium then induces growth of a tube poking into the plant cell. The tube growth is directed by the plant nucleus, which migrates toward the plant cortex. As the tube grows, bacteria proliferate, forming a column of cells that projects down the tube. This column of cells is known as the **infection thread**. The infection thread can be visualized by light microscopy (**Fig. 21.31B**).

As the infection thread develops, signals from the bacteria induce the cortical cells (below the epidermis) to prepare to receive the bacteria. The bacteria induce further tube formation into the cortical cells and continue penetration as they grow. The penetration of the infection thread into the cortex appears strikingly in fluorescence micrographs (**Fig. 21.31C**) in which the bacteria are engineered to express a fluorescent protein.

The cortical cells invaded by bacteria are induced to proliferate in an organized manner, forming nodules. Within the nodules, most of the infecting bacteria differentiate into wall-less bacteroids that will fix nitrogen (**Fig. 21.32A**). A few bacteria fail to differentiate; their fate is unclear. The bacteroids remain sequestered within a sac of plant-derived membrane known as the **symbiosome**. The symbiosome membrane contains special transporters that mediate the exchange of nutrients between the bacteroid and its host cell, sustaining bacteroid metabolism while preventing harm to the host.

From the plant cytoplasm, the bacteroid receives catabolites such as malate, which enter the TCA cycle and donate electrons for respiration. The oxygen for respiration comes from the plant's photosynthesis (**Fig. 21.33**). About a fifth of the plant's photosynthesis is required to fix nitrogen. As discussed in Chapter 15, nitrogen is fixed by the nitrogenase enzyme into ammonium ions:

$$N_2 + 8H + 2H^+ \longrightarrow 2NH_4^+ + H_2$$

The reaction requires expenditure of 8 NADPH or NADH plus 16 ATP, which are generated by respiration. But respiration requires oxygen, which poisons nitrogenase. Thus, oxygen needs to be delivered to the bacteroid only as needed, and in an amount just enough to run respiration. Excess oxygen is sequestered by **leghemoglobin**, an iron-bearing plant protein related to blood hemoglobin. Leghemoglobin in plant cytoplasm colors the nodule pink (**Fig. 21.32B**).

Overall, the nitrogen fixation symbiosis is kept in balance by several regulatory mechanisms. The presence of ammonium or nitrate ions inhibits symbiosis and nitrogen fixation. The bacteroids cannot synthesize their own amino acids; instead, they must provide ammonium to the plant cytoplasm for assimilation into amino acids, some of which cycle back to the bacteroid. But what is known of regulation is dwarfed by the unanswered questions. How is the infection thread formed? How does the plant allow infection while preventing uncontrolled growth of bacteria? What determines how much of plant products of photosynthesis are harvested by the bacteria? Why is hydrogen gas released, and how can this loss of potential energy be prevented? What limits the host specificity of rhizobia to legumes, and can it be extended to other crop plants, such as corn? Research on these questions is critical for agriculture.

> **THOUGHT QUESTION 21.6** One unanswered question is, how do symbiotic rhizobia reproduce? Why do bacteroids develop if they cannot proliferate?

Plant Pathogens

We have seen how many bacteria and fungi interact positively with plants, but many other species act as pathogens. In any environment, pathogens are always outnumbered by the vast community of neutral or helpful microbes. Nevertheless, when a pathogen does colonize a plant, its growth can have effects ranging from minimal to devastating (**Fig. 21.34**). A relatively harmless plant virus was associated with a famous historical phenomenon, the tulip craze in the Netherlands in the sixteenth century. The virus caused streaking of tulip petals, a pattern much admired by tulip fanciers. Other viruses, however, can cause devastating blights and epidemics. (For more on viruses, see Chapters 6 and 11.)

A pathogenic relative of rhizobia is the bacterium *Agrobacterium tumefasciens*, whose DNA transforms plant cells to form crown gall tumors (**Fig. 21.34B**). *A. tumefasciens* has an unusually broad host range, and its natural genetic transformation system has been applied widely for commercial plant engineering (discussed in Chapter 16). The tumors remain largely confined and have relatively little effect on plant growth. Other bacterial pathogens, particularly species of *Erwinia* and *Xanthomonas*, severely damage plants.

A. Bacteroids with plant cells

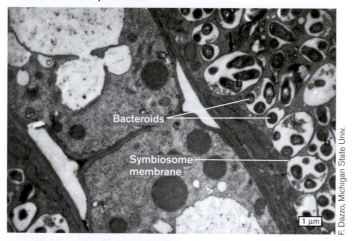

Bacteroids

Symbiosome membrane

1 μm

F. Dazzo, Michigan State Univ.

B. Nodules contain leghemoglobin

Stephanie Batchelet

Figure 21.32 Bacteroids form within nodules.
A. *Rhizobium* cells within plant cytoplasm lose their cell walls and become bacteroids, contained within the plant-generated symbiosome membrane. Bacteroid-containing cells form nodules. **B.** Within a sectioned nodule, the pink color indicates leghemoglobin.

Figure 21.33 Energy and oxygen regulation during nitrogen fixation. The bacteroid receives photosynthetic products from the plant, such as malate, and oxygen to generate ATP for nitrogen fixation. The amount of oxygen is regulated closely by leghemoglobin. The bacteroid provides nitrogen (fixed as ammonium ion) to the plant cell.

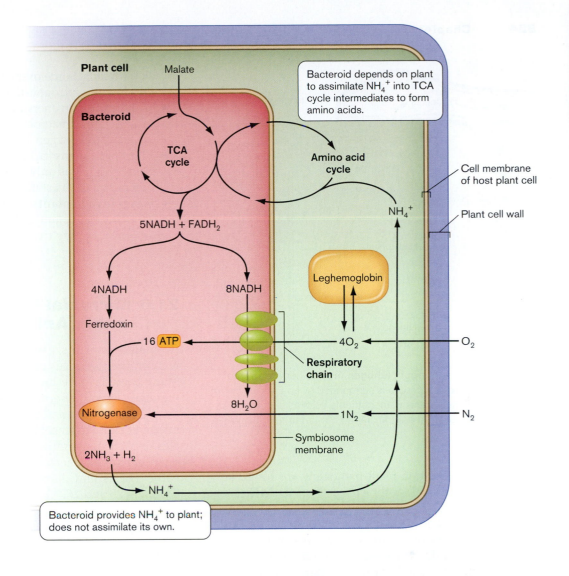

Plant cell

Malate

Bacteroid

Bacteroid depends on plant to assimilate NH_4^+ into TCA cycle intermediates to form amino acids.

TCA cycle

Amino acid cycle

NH_4^+

Cell membrane of host plant cell

Plant cell wall

$5NADH + FADH_2$

$4NADH$

$8NADH$

Leghemoglobin

Ferredoxin

16 ATP

$4O_2$

O_2

Respiratory chain

$8H_2O$

Nitrogenase

$1N_2$

N_2

$2NH_3 + H_2$

Symbiosome membrane

NH_4^+

Bacteroid provides NH_4^+ to plant; does not assimilate its own.

A. Virus infection of tulips

Courtesy of Kathy Nichols

B. Crown gall on rose stem

U. of Nebraska, Lincoln/Dept. of Plant Pathology

C. Anthracnose fungus on white oak leaf

Paula Flynn, Iowa State Extension Service

D. Dutch elm disease

R. Scott Cameron, International Paper, U.S.

Figure 21.34 Plant diseases range from innocuous to devastating. A. Striped tulips result from a virus. **B.** Crown gall tumor on rose stem caused by *Agrobacterium tumefaciens*. **C.** White oak leaf spotted by anthracnose fungus. **D.** Elm tree wilted by Dutch elm disease, caused by the fungus *Ophiostoma novo-ulmi*.

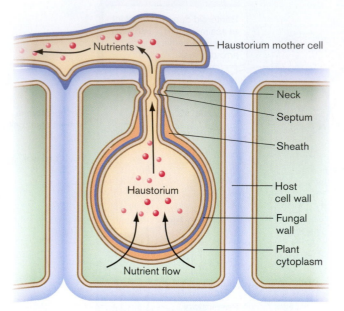

Figure 21.35 A fungal pathogen inserts haustoria into plant cells.   The haustorium is surrounded by an invagination of the plant's cell membrane. Plant nutrients such as sucrose flow into the haustorium and are transferred out to the fungal mycelia.

The most common plant pathogens are fungi. Fungus diseases such as anthracnose (**Fig. 21.34C**) cause substantial losses in agriculture, affecting cucumbers, tomatoes, and other vegetables. Dutch elm disease, which has wiped out nearly all the native elms of the United States (**Fig. 21.34D**), is caused by the fungus *Ophiostoma novo-ulmi*. The fungus is carried by bark beetles, which bore into the xylem, damaging the plant's transport vessels and allowing access to fungal spores.

Some fungal pathogens generate specialized structures to acquire nutrients from plants. As a hypha grows across the plant epidermis, its tip can penetrate the plant cell wall, followed by ingrowth of a bulbous extension called a **haustorium** (**Fig. 21.35**). The haustorium never penetrates the plant cell membrane, thus avoiding leakage and loss of plant cytoplasm. Instead, it causes the membrane to invaginate, while expanding into the volume of the plant cell. The haustorium takes up nutrients such as sucrose, generated by adjacent chloroplasts. Depending on the species of fungus, haustorial parasitism can lead to mild growth retardation or it can rapidly kill the plant.

> **THOUGHT QUESTION 21.7**   Compare and contrast the processes of plant infection by rhizobia and by fungal haustoria.

TO SUMMARIZE:

■ **Rhizobia induce legume roots** to nodulate for nitrogen fixation. The bacteria enter the root as inva-

sion threads. Some of the bacteria enter root cells and develop into nitrogen-fixing bacteroids.

■ **Bacteroids** gain energy from plant cell respiration but must remain anaerobic. The anaerobic state is protected by the oxygen-binding protein leghemoglobin.

■ **Plant pathogens** include bacteria, fungi, and viruses. Some pathogens only mildly affect the plant, whereas others cause devastation.

■ **Fungi invade plants using haustoria.** The haustoria grow into the plant cell by penetrating the cell wall and invaginating the cell membrane to avoid leakage of cytoplasm.

21.6 Microbial Communities within Animals

Animals, like plants, associate with communities of bacteria on their surfaces or within certain internal organs. Many bacteria have beneficial effects, such as enhancing digestion or generating protective substances in the skin. A relatively small proportion of animal-associated microbes cause disease. We discuss here just a few examples of particular significance for ecology and veterinary science. Human microflora are discussed in Chapter 23. Pathogenesis and disease in humans are discussed in Chapters 25 and 26.

Zooxanthellae in Corals and Sea Anemones

Many invertebrate animals acquire endosymbiotic algae, most commonly dinoflagellates. These endosymbionts are called **zooxanthellae**. The algae receive protection from predators, while the animal receives photosynthetic products. Examples include hydras, anemones, clams, and corals (**Fig. 21.36A**). Coral zooxanthellae are extremely important to the biosphere because healthy coral is responsible for reef formation and much of the biological productivity of coastal shelf ecosystems. The slight rise in temperature that has occurred from global warming has already caused severe problems with **coral bleaching**, in which the algal symbionts die or are expelled. The coral turns white and soon dies unless its symbionts return.

The most common algal partners are members of the dinoflagellate genus *Symbiodinium* (**Fig. 21.36B**). A given coral or anemone may harbor several different species of *Symbiodinium*, which show different preferences for light or shade and different tolerances for temperature change. Studies of coral bleaching due to temperature increase suggest that corals containing diverse species of symbionts are more likely to survive, because one of their species may happen to be resistant to a rise in temperature.

Figure 21.36   Endosymbiotic algae.
A. Brain coral (*Diploria* sp.) containing zooxanthellae. **B.** *Symbiodinium* sp. are zooxanthellae (symbiotic dinoflagellate algae) of coral.

A.

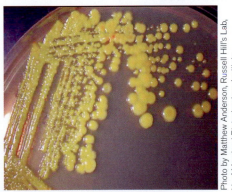

Becky A. Dayhuff, National Oceanic Atmospheric Administration/Dept. of Commerce

B.

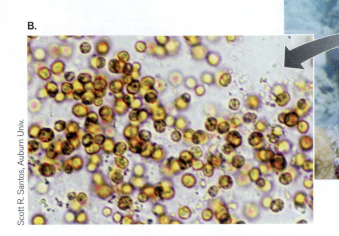

Scott R. Santos, Auburn Univ.

Sponge Communities

Animals often harbor extremely diverse communities of microbes whose specific contributions are poorly understood—for example, bacteria associated with marine sponges (**Fig. 21.37**). In 2001, Russell Hill and colleagues at the University of Maryland, Baltimore, characterized the bacterial species associated with the sponge

Rhopaloeides odorabile, which grows off the Australian Great Barrier Reef. Using 16S rDNA probes, they identified novel species of delta-proteobacteria, planctomycetes, flavibacteria, green sulfur and nonsulfur bacteria, as well as deep-branching isolates of no known major clade. Most interesting, they discovered a novel actinomycete that produces antimicrobial substances. Hill hypothesizes that the sponge harbors these bacteria to defend it from pathogens, which rapidly consume sponges that lack their own microbial community. Sponge bacteria, therefore, may offer a wealth of previously unknown antibiotics.

A. Sponge containing actinomycetes

Nicole Webster, Australian Institute of Marine Science

B. Actinomycete isolated from sponge

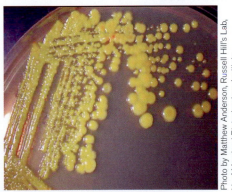

Photo by Matthew Anderson, Russell Hill's Lab, U. of Maryland Biotechnology Institute

Figure 21.37   Sponge bacterial communities.   A. The sponge *Rhopaloeides odorabile* contains actinomycetes that produce novel antimicrobial compounds. **B.** A sponge-associated actinomycete growing on an agar plate.

A.

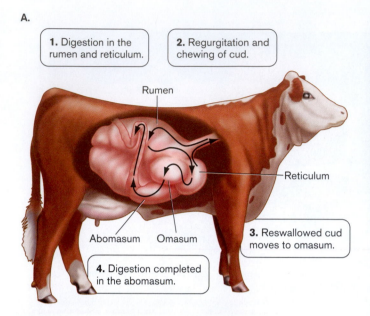

1. Digestion in the rumen and reticulum.

2. Regurgitation and chewing of cud.

Rumen

Reticulum

Abomasum   Omasum

3. Reswallowed cud moves to omasum.

4. Digestion completed in the abomasum.

B.

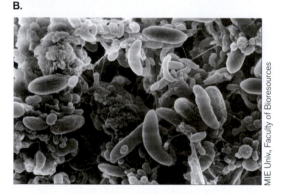

MIE Univ, Faculty of Bioresources

Digestive Communities

Some of the most complex and nutritionally important communities of microbes are those found within animal digestive systems. Digestive chambers, such as the stomach and intestines, often support thousands of species of bacteria, protists, and archaea unique to the animal. Among insects, termites harbor thousands of unique cellulolytic species, including spirochetes and multipartner symbiotic organisms such as *Mixotricha*. The human colon contains numerous fermenters and methanogens, some of which feed on intestinal mucus, whereas others digest complex plant fibers, providing up to 15% of our caloric intake. On the other hand, the extremely acidic human stomach supports a very different community, including unique acid-tolerant spirilla such as *Helicobacter pylori*, a cause of gastric ulcers.

The most extensively studied digestive communities are those of ruminant animals, such as cattle, sheep, llamas, and caribou. Throughout most of human civilization, ruminants have provided protein-rich food, textile fibers, and mechanical work. A historical reference is the biblical

C.

Photo courtesy of James B. Russell, Cornell Veterinary School

Figure 21.38   The bovine rumen.   A. The rumen is one of four chambers of the bovine stomach. **B.** Bacteria growing within the rumen. **C.** Todd Callaway, Francisco Diez Gonzales, James B. Russell, and Minas Kizoulis study microbial diversity and metabolism of the bovine rumen.

injunction to consume an animal that "is cleft-footed and chews the cud"—that is, "ruminates," or redigests its food in the fermentation chamber known as the **rumen** (**Figs. 21.38A** and **B**).

The microbial community of the rumen enables herbivores to acquire nutrition from complex plant fibers that the animal could not otherwise digest. From a genomic standpoint, such an arrangement makes evolutionary sense. If the animal had to digest all the diverse polysaccharide chains encountered in nature, its own genome would have to encode a wide array of different enzyme systems. Instead, the ruminant relies on diverse microbial species to conduct various kinds of digestion. Organisms that partly digest a substrate (for instance, converting sugars to lactate) provide a substrate that others can further metabolize. This process is known as syntrophy. Microbial fermentation under the anaerobic conditions of the rumen generates short-chain fatty acids that are absorbed by the ruminant intestinal lining and digested to completion through the animal's respiration.

THOUGHT QUESTION 21.8   Why does ruminant fermentation leave food value for the animal host? How is the animal able to obtain nourishment from waste products that the microbes could not use?

The most intensively studied ruminants are cattle. The bovine stomach has four chambers. The rumen initially digests the feed, then passes it to the reticulum. The reticulum breaks the feed into smaller pieces and traps indigestible objects, such as stones or nails. After initial digestion, feed is regurgitated for rechewing, then returned to the rumen, by far the largest of the chambers.

In the rumen, feed is reduced to small particles and fermented slowly by thousands of species of microbes. The partially digested feed passes to the omasum, which absorbs water and short-chain acids produced by fermentation. The abomasum then decreases pH and secretes enzymes to digest proteins before sending its contents to the colon for further nutrient absorption and waste excretion.

While rumen digestion has been studied since the 1830s, major advances in understanding rumen fermentation were first achieved in 1966 by Robert Hungate (1906–2004) at the University of California, Davis, who pioneered techniques of anaerobic microbiology. An example of Hungate's methods still in use today is that of obtaining anaerobic cultures from fistulated cattle—that is, cattle in which an artificial connection is made between the rumen and the animal's exterior (**Special Topic 21.1**). The full diversity of ruminal flora was revealed by James Russell and colleagues at Cornell Veterinary School (**Fig. 21.38C**), whose rRNA studies showed hundreds of isolates from all three microbial domains—archaea, prokaryotes, and eukaryotes—including several major groups within each domain. Different species fill different niches in ruminal metabolism (**Fig. 21.39**).

In nature, the most important role for ruminants is that of breaking down cellulose and complex plant fibers such as lignins. Cattle grown on relatively poor forage (that is, forage high in complex plant content) show a high proportion of ruminal fungi, the chytridiomycetes (discussed in Chapter 20). Chytridiomycete mycelia appear on ruminal food particles, and their motile zoospores—formerly mistaken for protists—swim through rumen fluid. On the other hand, cattle fed a high cellulose diet, such as hay, grow faster and show cellulolytic bacteria such as *Ruminococcus albus* and *Fibrobacter flavefaciens*.

Metabolism of cellulolytic bacteria requires the presence of small amounts of branched-chain fatty acids. The branched-chain acids turn out to be produced by amino acid fermenters such as *Megasphaera elsdenii* and *Peptostreptococcus anaerobius*. However, too much degradation of amino acids can lead to overproduction of ammonia, poisoning the animal; thus, the protein content of cattle feed must be limited.

Another problem with ruminal fermentation is the frequent production of H_2 and CO_2. Hydrogen production is hard to avoid because the quantity of electron donors (reduced food molecules) greatly exceeds that of the available electron acceptors. The hydrogen and CO_2

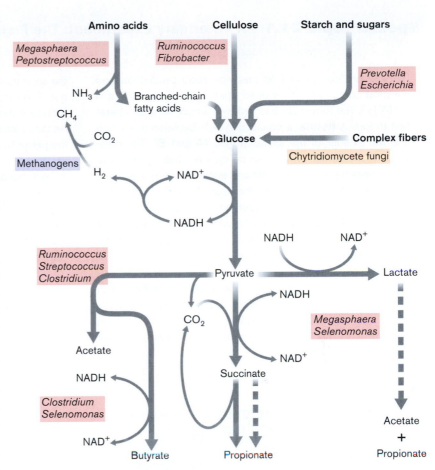

Figure 21.39   Ruminal metabolism.   *Source*: Based on Russell and Rychlik. 2001. *Science* 292:1119.

from fermentation support methanogens, wasting valuable carbon from feed and contributing a substantial part of global methane emissions. So much methane is formed by the rumen that a cannula inserted into the rumen liberates enough of the gas to light a flame.

> **THOUGHT QUESTION 21.9**   How do you think cattle feed might be altered or supplemented to decrease methane production?

Another concern with modern cattle rearing is the shift in feed from hay to grain, whose higher content of starch leads to more rapid digestion and faster growth of the animal. Unfortunately, rapid starch digestion favors fermenters such as *Prevotella* species and *E. coli*, which generate higher acid levels and gases, leading to "starch bloat." Furthermore, rumen acidity selects for acid-resistant pathogens such as the *E. coli* strain O157:H7. Pathogenic *E. coli* contaminates meat and produce in North America and is now a leading cause of serious illness in young people.

Special Topic 21.1 A Veterinary Experiment: The Fistulated Cow

Ruminal ecology is vital to maximize dairy production and minimize contamination by deadly pathogens such as *E. coli* O157:H7. The rumen is observed by a simple surgical operation to form a **fistula**, a connecting hole between the rumen and the exterior of the animal (**Figs. 1A** and **B**). The presence of the fistula, which can be capped and reopened at will, causes the animal no harm and ensures it several years of comfortable existence beyond the usual life of dairy cattle. The fistula can be used to insert experimental feed materials, such as grain or nitrogen supplements, and it can be used to test parameters of rumen fluid, such as pH and microbial content. Study of fistulated cattle is a common exercise for students of veterinary medicine.

An experiment using the fistula is shown in **Figure 2**. In the experiment shown, James Russell and colleagues tested the hypothesis that increasing proportions of grain in cattle feed result in increased acid production and favor the appearance of acid-resistant bacterial species such as *E. coli*. The top part of the figure shows the pH of ruminal fluid (obtained through the fistula) as well as fluid from the colon. As feed includes higher proportions of grain (horizontal axis), the pH of the rumen declines, as does that of the colon. As pH declines, the number of *E. coli* bacteria increases greatly, in part because they are more acid resistant than other species.

These results suggest that the high-grain content in modern cattle feed has contributed to the emergence of pathogenic strains of *E. coli*. One response among consumers has been a renewed interest in hay-fed or free-range cattle. On the other hand, it is likely that fiber-rich diets produce even more methane than grain-rich diets supplemented with antimethanogen agents. Further research on rumen flora will be needed to optimize cattle production while minimizing the growth of pathogens and methanogens.

A.

U. of California, Davis, Veterinary School

B.

U. of California, Davis, Veterinary School

Figure 1 The fistulated cow. A. A cow carries a surgical fistula, an artificial opening into the rumen. **B.** The fistula is uncapped, exposing the contents of the rumen.

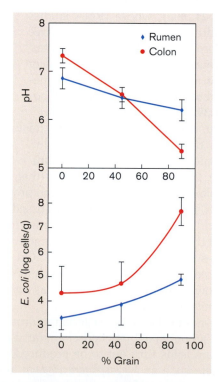

Figure 2 Rumen pH and bacterial contents. An experiment by James B. Russell and colleagues, in which the fistula was used to test pH and *E. coli* bacterial content of the rumen. As the percentage of grain in feed rises, pH declines and *E. coli* bacteria increase. *Source: Francisco Diez-Gonzalez, et al. 1998. Science 281:1666.*

TO SUMMARIZE:

- **Animals harbor microbial communities.** Microbes populate the skin, where they provide protection from pathogens. Other species grow in the animal's digestive system, where they enhance nutrition.
- **Corals and sea anemones harbor zooxanthellae.** Zooxanthellae are algae that provide products of photosynthesis in exchange for a protected habitat.
- **The rumen of ruminant animals is a complex microbial digestive chamber.** Rumen microbes, including bacteria, protists, and fungi, digest complex plant materials. The microbial digestion generates short-chain fatty acids that are absorbed by the intestinal lining.

Concluding Thoughts

In this chapter, we have seen how microbes colonize a vast array of habitats ranging from the deserts and oceans to the roots of plants and the digestive tracts of animals. Microbes are found in the upper layers of our planet's atmosphere, down to the deepest rock strata we can reach. Wherever found, microbes both respond to and modify the environment that surrounds them. While this chapter has focused on food webs in local habitats, Chapter 22 takes a global perspective of microbial ecology and its roles in the cycling of essential elements, such as nitrogen and iron, in Earth's biosphere.

CHAPTER REVIEW

Review Questions

1. What unique functions do microbes perform in ecosystems?
2. Explain the difference between carbon assimilation and dissimilation.
3. What kinds of microbial metabolism are favored in aerated environments? In anoxic environments?
4. What are examples of microbial producers in ecosystems? Include phototrophs as well as lithotrophs. Can a microbe be both a producer and a consumer? Explain.
5. Explain the microbial relationships in various forms of symbiosis, including mutualism, commensalism, and parasitism. Outline an example of each, detailing the contributions of each partner.
6. Compare and contrast the marine food web with the soil food web. What kinds of organisms are the producers and consumers? How many trophic levels are typically found?

7. Explain the role of biofilms in marine and soil habitats. Are these habitats typically uniform or patchy?
8. Compare and contrast the roles of microbes in photic and aphotic marine communities.
9. Compare and contrast the microbial activities in aerated and waterlogged soil.
10. Explain the different methods of quantifying microbial communities, including the advantages and limitations of each.
11. Explain how anaerobic microbial metabolism can enrich soil for plant cultivation.
12. What are mycorrhizae, and how are they important for plant growth?
13. Explain how bacterial mutualists fix nitrogen for plants.

Key Terms

algal bloom (811)
amensalism (801)
aphotic zone (801)
arbuscule (817)
assimilation (795)
bacteroid (820)
barophile (808)
benthic organism (801)

benthos (801)
biochemical oxygen demand (BOD) (802)
biomass (796)
coastal shelf (801)
commensalism (800)
consumer (796)
coral bleaching (824)

cryptogamic crust (818)
decomposer (796)
detritus (797)
dissimilation (795)
ecosystem (794)
ectomycorrhizae (816)
endolith (813)
endomycorrhizae (816)

euphotic (photic) zone (801)
eutrophic lake (811)
eutrophication (811)
extremophile (797)
femtoplankton (803)
fistula (828)
flavonoid (820)
food web (796)
grazer (796)
haustorium (824)
horizon (812)
humic material (815)
humus (815)
hydric soil (818)
infection thread (821)
leghemoglobin (822)
lichen (798)
lignin (815)

limiting nutrient (811)
littoral zone (810)
marine snow (803)
microplankton (803)
mixotroph (806)
mutualism (798)
mycorrhiza (813)
nanoplankton (803)
neuston (801)
niche (794)
oligotrophic lake (809)
parasitism (801)
pelagic habitat (801)
phytoplankton (806)
picoplankton (803)
plankton (803)
population (794)
predator (796)

primary producer (795)
rhizoplane (815)
rhizosphere (813, 815)
rumen (826)
saphrophyte (813)
soil (812)
symbiosis (798)
symbiosome (822)
synergism (800)
syntrophy (800)
thermocline (802)
trophic level (796)
vesicular-arbuscular mycorrhiza (VAM) (817)
viable but nonculturable (VBNC) (802)
water table (812)
wetland (818)
zooxanthellae (824)

Recommended Reading

Brugerolle, Guy. 2004. Devescovinid features, a remarkable surface cytoskeleton, and epibiotic bacteria revisited in *Mixotricha paradoxa*, a parabasalid flagellate. *Protoplasma* 224:49–59.

Colwell, Rita R., Anwar Huq, M. Sirajul Islam, K. M. A. Aziz, M. Yunus, et al. 2003. Reduction of cholera in Bangladeshi villages by simple filtration. *Proceedings of the National Academy of Science USA* 100:1051–1055.

Diez-Gonzalez, Francisco, Todd R. Callaway, Menas G. Kizoulis, and James B. Russell. 1998. Grain feeding and the dissemination of acid-resistant *Escherichia coli* from cattle. *Science* 281:1666–1668.

Fierer, Noah, and Robert B. Jackson. 2006. The diversity and biogeography of soil bacterial communities. *Proceedings of the National Academy of Science USA* 103:626–631.

Gage, Daniel J. 2004. Infection and invasion of roots by symbiotic, nitrogen-fixing rhizobia during nodulation of temperate legumes. *Microbiology and Molecular Biology Reviews* 68:280–300.

Johnson, Zackary I., Erik R. Zinser, Allison Coe, Nathan P. McNulty, E. Malcolm S. Woodward, and Sallie W. Chisholm. 2006. Niche partitioning among *Prochlorococcus* ecotypes along ocean-scale environmental gradients. *Science* 311:1737–1740.

Karl, David M. 2007. Microbial oceanography: Paradigms, processes, and promise. *Nature Reviews Microbiology* 5:759–769.

Kirchman, David. 2000. *Microbial Ecology of the Oceans.* Wiley-Liss, New York.

Konopka, Allan. 2006. Microbial ecology: Searching for principles. *Microbe* 1:175–179.

Newman, Dianne K., and Jillian F. Banfield. 2002. Geomicrobiology: How molecular-scale interactions underpin biogeochemical systems. *Science* 296:1071–1077.

Ogunseitan, Oladele. 2005. *Microbial Diversity.* Blackwell Publishing, Boston, MA.

Rusch, Douglas B., Aaron L. Halpern, Granger Sutton, Karla B. Heidelberg, Shannon Williamson, et al. 2007. The Sorcerer II global ocean sampling expedition: Northwest Atlantic through eastern tropical Pacific. *Public Library of Science Biology* **5(3)**:e77.

Russell, James B., and Jennifer L. Rychlik. 2001. Factors that alter rumen microbial ecology. *Science* 292:1119–1122.

Walker, Jeffrey J., John R. Spear, and Norman R. Pace. 2005. Geobiology of a microbial endolithic community in the Yellowstone geothermal environment. *Nature* 434:1011–1014.

Webster, Nicole S., Kate J. Wilson, Linda L. Blackall, and Russell T. Hill. 2001. Phylogenetic diversity of bacteria associated with the marine sponge *Rhopaloeides odorabile. Applied and Environmental Microbiology* 67:434–444.

Young, I. M., and J. W. Crawford. 2004. Interactions and self-organization in the soil-microbe complex. *Science* 304:1634–1637.

Chapter 22

Microbes and the Global Environment

Microbes throughout the biosphere recycle carbon, nitrogen, sulfur, and other elements essential for all life. Through their biochemical transformations, both helpful and harmful, diverse microbial activities largely determine the quality of soil, air, and water for human life. Today, all of these geochemical cycles are altered profoundly by human activity. The burning of fossil fuels, which were generated millions of years ago by subterranean microbes, releases quantities of carbon dioxide too great to be absorbed by marine bacteria and algae. Growing rice production increases the release of methane by methanogens that thrive in submerged rice paddies. Both carbon dioxide and methane are "greenhouse gases" that lead to global warming. Will we humans learn to manage the perturbations of our own biosphere, from pollution to global warming?

From local wastewater treatment to the control of greenhouse gases, microbes are our hidden partners on Earth. And Earth's microbial cycles lead us to wonder whether biospheres exist on other worlds. Our neighbor planet Mars shows tantalizing hints of water and the possibility of hidden microbial life.

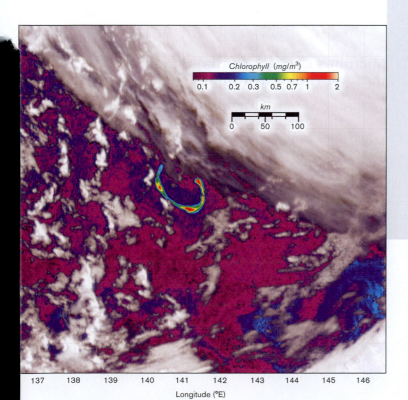

Satellite image of chlorophyll concentration in the waters of the Southern Ocean, due to phytoplankton bloom following an iron enrichment experiment. Green through orange color indicates overgrowth of phytoplankton where iron was added. Iron limits phytoplankton productivity in much of the world's ocean waters. Thus, iron controls algal blooms, which affect the carbon cycle and may influence global warming. Photo provided by Norman Kuring, NASA. *Source*: Philip W. Boyd, et al. 2007. *Science* 315:612.

Two billion years ago, ancient cyanobacteria acquired the ability to photolyze water and produce molecular oxygen. Oxygen is a powerful oxidant, lethal to most life in that anoxic era, when all living organisms were anaerobic microbes. Massive extinctions must have occurred until a lucky few species evolved mechanisms of antioxidant protection. Since then microbes have shaped our biosphere by generating oxygen, by fixing gaseous nitrogen and returning it to the atmosphere, and by fixing and producing carbon dioxide. These and countless other microbial processes not only have made the Earth's atmosphere what it is but continue to play a crucial role in its homeostasis.

During the past century, key aspects of global biospheric chemistry have been perturbed by human technology. Already 15% of the nitrogen in the biosphere comes from anthropogenic (human-generated) sources. And increases in atmospheric CO_2 and CH_4, important greenhouse gases, correlate with the rise of the industrial age (**Fig. 22.1**). These gases are produced by the biosphere by bacteria, fungi, and protists, as well as by human technology. Human burning of petroleum releases CO_2 from the product of ancient bacterial cells. Much of global carbon dioxide is fixed by phototrophic prokaryotes and algae as well as land plants. But the added CO_2 production by human industry has outpaced the rate of CO_2 fixation.

Can we adapt to our changing biosphere, just as microbes respond and adapt to environmental change? Concern over the effect of increasing greenhouse gases on global climate change led the United Nations, meeting in Kyoto in 1997, to adopt the Kyoto Protocol for reduction of industrial emissions of CO_2, CH_4, and other gases that contribute to global warming. Over 160 nations adopted the protocol, although the United States declined, contending that it posed a disproportionate economic burden on developed nations. Will the human-induced global warming cause mass extinctions of a majority of Earth's species—as did the rise of ancient cyanobacteria? Or can we use our knowledge of microbial ecology to channel microbial activities into recovering the balance—for example, by increasing microbial CO_2 fixation?

Throughout most of this book, we present microbial biochemistry in the context of growth of individual organisms. In this chapter, we show how the collective metabolic activities of microbial populations contribute to global cycles of elements throughout Earth's biosphere. We consider the ways that we humans enlist microbes to manage our environment. Finally, our awareness of global microbiology has renewed our quest for life beyond Earth.

22.1 Biogeochemical Cycles

Microbes possess exceptional abilities to interconvert molecules. Nearly any kind of molecule can be metabolized by some species somewhere if the reaction provides energy or a useful nutrient. But even microbes face the limits of the actual elements themselves: No microbe, nor any living creature, can convert one element to another. Living organisms conduct chemical reactions, not nuclear reactions—although microbes that live deep underground may obtain chemical energy from hydrogen ions generated by uranium decay.

Microbes Cycle Essential Elements through Their Environment

Because organisms on Earth cannot create their own elements, they need to get them from their environment. While small amounts of matter enter the biosphere from outer space, and gases continually escape the planet, the rate of these changes is tiny compared to the rates of living processes. So organisms acquire their elements either from nonliving components of their environment, such as by fixing atmospheric CO_2, or from other organisms, by grazing, predation, or decomposition. Furthermore, all organisms recycle their components back to the biosphere. The partners in this recycling include **abiotic** entities such as air, water, and minerals, as well as **biotic** entities such as predators and decomposers. Collectively, the metabolic interactions of microbial communites with the biotic and abiotic components of their ecosystems are known as **biogeochemistry** or **geomicrobiology**.

Which elements need to be recycled and made available for life? In most organisms, six elements predomi-

Atmospheric carbon dioxide (CO_2) concentrations (1750 to present)

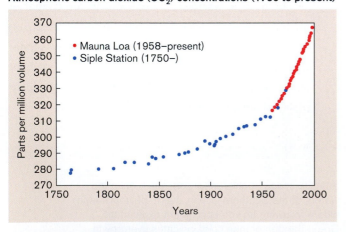

Figure 22.1   Global carbon dioxide levels.   Atmospheric CO_2 levels since 1750 reveal the increase of this greenhouse gas accompanying the rise in industrial consumption of fossil fuels. Microbial CO_2 fixation helps limit CO_2 increase in the atmosphere. *Source:* World Resources Institute.

Table 22.1 Typical elemental composition of gram-negative bacteria.

Element	Dry weight, %
Carbon	50
Oxygen	20
Nitrogen	14
Hydrogen	8
Phosphorus	3
Sulfur	1
Potassium	1
Sodium	1
Calcium	0.5
Magnesium	0.5
Chlorine	0.5
Iron	0.2
All others	0.3

nate: carbon, oxygen, nitrogen, hydrogen, phosphorus, and sulfur (discussed in Chapter 4). **Table 22.1** summarizes the elemental composition typical of gram-negative bacteria. In other microbes, elemental proportions vary with species and growth conditions. Some species include inorganic components such as the silicate shells of diatoms or the sulfur granules of phototrophic bacteria.

All of these elements flow in **biogeochemical cycles** of nutrients throughout the biotic and abiotic components of the biosphere. The environmental levels of key elements can limit biological productivity. For example, the concentration of iron limits the marine populations of phytoplankton, which are dependent on iron supplied by wind currents. Other elements, such as zinc and copper, are **micronutrients**, nutrients required in still smaller amounts. Micronutrients, too, must be cycled, although their flux is challenging to measure.

Sources and Sinks of Essential Elements

Biogeochemical cycles include both biological components (such as phototrophs that consume CO_2 and heterotrophs that produce CO_2) and geological components (such as volcanoes that produce CO_2 and oceans that absorb CO_2). The major parts of the biosphere containing significant amounts of an element needed for life are called reservoirs of that element (**Table 22.2**). Each reservoir acts both as a source of that element for living organisms and as a sink to which the element returns. For example, the ocean is an important reservoir for carbon (CO_2 in equilibrium with HCO_3^-, bicarbonate ion). The ocean's carbon cycles rapidly; thus, the ocean serves as both a source and a sink for carbon.

As elements cycle from sources to sinks, microbial metabolism generates a series of redox changes. The major oxidation states of carbon, nitrogen, and sulfur

Table 22.2 Global reservoirs of carbon, nitrogen, and sulfur.[a]

Reservoir	Carbon	Rate of cycling	Nitrogen	Rate of cycling	Sulfur	Rate of cycling
Atmosphere	700 (CO_2)	Fast	3,900,000 (N_2)	Slow	0.0014 (SO_2, H_2S)	Fast
Ocean						
Biomass	4	Slow	0.5	Fast	0.15	Fast
Organic molecules	2,100	Slow	300	Fast	–	–
Inorganic molecules	38,000 (HCO_3^-, CO_3^{2-})	Fast	20,000 (N_2) 690 (NO_3^-, NO_2^-, NH_4^+)	Slow Fast	1,200,000 (SO_4^{2-})	Slow
Land						
Biomass	500	Fast	25	Fast	8.5	Fast
Organic matter (soil)	1,200	Fast	110	Slow	16	Fast
Crust (below land and ocean)	120,000,000	Slow	770,000	Slow	18,000,000	Slow
Fossil fuel (oil, coal, natural gas)	13,000	Fast				

[a]Units are 10^9 metric tons, or 10^{12} kg.

Source: R. Maier, et al. 2000. *Environmental Microbiology*. Academic Press: San Diego.

Table 22.3 Oxidation states of cycled compounds.

Oxidation state	Carbon		Nitrogen		Sulfur	
−4	CH_4	Methane				
−3			NH_3, NH_4^+	Ammonia, ammonium ion		
−2	CH_3OH, $(CH_2)_n$	Methanol, hydrocarbon	H_2N-NH_2	Hydrazine	H_2S, S^{2-}	Sulfides
−1			NH_2OH	Hydroxylamine		
0	$(CH_2O)_n$	Carbohydrate	N_2	Nitrogen	S^0	Elemental sulfur
+1			N_2O	Nitrous oxide		
+2	HCOOH	Formic acid	NO	Nitric oxide	$S_2O_3^{2-}$	Thiosulfate
+3			HNO_2, NO_2^-	Nitrous acid, nitrite ion		
+4	CO_2, HCO_3^-	Carbon dioxide, bicarbonate ion	NO_2	Nitrogen dioxide	SO_3^{2-}	Sulfite
+5			HNO_3, NO_3^-	Nitric acid, nitrate ion		
+6					H_2SO_4, SO_4^{2-}	Sulfuric acid, sulfate ion

are summarized in **Table 22.3**. Biospheric carbon can be found as CH_4 generated by methanogens (completely reduced, −4), as CO_2 produced by respiration and fermentation (completely oxidized, +4), or as one of various intermediate states of oxidation. Nitrogen is excreted by many organisms in its most reduced form, ammonia (NH_3), which is protonated to ammonium ion (NH_4^+). Ammonium is oxidized by lithotrophic bacteria through several stages to nitrite (NO_2^-) and nitrate (NO_3^-), which serve as terminal electron acceptors for anaerobic respiration. Sulfides (H_2S, HS^-) serve as electron donors for respiration and sulfur phototrophy, whereas oxidized forms, such as sulfite (SO_3^{2-}) and sulfate (SO_4^{2-}), serve in anaerobic respiration.

THOUGHT QUESTION 22.1 Why is oxidation state critical for the acquisition, usability, and potential toxicity of cycled compounds? Cite examples based on your study of microbial metabolism.

How do we study microbial cycling on a global scale? How do we figure out whether ecosystems are net sources or sinks of CO_2? Does microbial activity enhance or limit availability of nitrogen? Rates of flux of elements in the biosphere are very difficult to measure, yet the questions have enormous political and economic implications.

To measure environmental carbon, nitrogen, and other elements, various methods are used. These methods fall under the following categories:

■ **Chemical and spectroscopic analysis.** Bulk quantities of CO_2, nitrates, and other chemicals can be determined by sophisticated chemical instrumentation. Atmospheric CO_2 is measured by infrared absorption spectroscopy, applied to samples from towers such as those of the NASA FLUXNET study (**Fig. 22.2**). Gas chromatography is used to separate and quantitate various gases, including oxygen, nitrogen, sulfur dioxide, and carbon monoxide. Mass spectroscopy detects extremely small quantities of different molecules, even distinguishing different isotopes of elements.

■ **Radioisotope incorporation.** The influx and efflux of CO_2 can be measured by the uptake of ^{14}C-labeled substrates in a small, controlled model ecosystem called a **mesocosm**. Alternatively, CO_2 flux can be measured with radioisotope tracers in the field using a field chamber.

■ **Stable isotope ratios.** Some enzyme reactions show a preference for one isotope over another, such as ^{14}N versus ^{15}N. For example, denitrifiers (bacteria that metabolize nitrate) strongly prefer the ^{14}N isotope, leaving behind nitrate enriched in ^{15}N. The $^{14}N/^{15}N$ ratio is measured using mass spectroscopy. Measuring nitrogen isotope ratios can indicate whether denitrifiers could have conducted metabolism in the sample.

least accessible to the biosphere as a whole. Crustal rock yields carbon only to organisms at the surface and to subsurface microbes that grow extremely slowly. The carbon reservoir that cycles most rapidly is that of the atmosphere, a source of CO_2 for photosynthesis and lithotrophy. The atmosphere also acts as a sink for CO_2 produced by heterotrophy and by geological outgassing from volcanoes.

The atmospheric reservoir is much smaller than other sources, such as the oceans, crustal rock, and fossil fuels. For this reason, the industrial burning of fossil fuels has perturbed the balance between atmospheric CO_2 and other reservoirs, such as the ocean. The ocean actually absorbs a good part of the extra CO_2, converting it to carbonates. In addition, marine phototrophs trap a substantial amount of carbon in biomass, which is then consumed by protists and sinks to lower layers. But despite these buffering effects of the ocean, atmospheric CO_2 continues to increase at a rate of about 1% per year. The CO_2 traps solar radiation as heat, a process known as the **greenhouse effect**. CO_2 is one of several greenhouse gases contributing to global warming, the overall rise in temperature of our biosphere over the past hundred years.

Note, however, that the fact that CO_2 is a greenhouse gas does not make it inherently "bad" for the environment. In fact, if heterotrophic production of CO_2 were to cease altogether, phototrophs would run out of CO_2 in roughly 300 years despite the vast quantities of carbon present in the ocean and crust. Thus, both CO_2 fixers and heterotrophs need each other.

Figure 22.2 Measuring flux of elements in the biosphere. An American FLUXNET tower at Austin Cary Management Forest is used for atmospheric CO_2 sampling as part of a global effort to monitor carbon flux.

TO SUMMARIZE:

- **Microbes cycle essential elements in the biosphere.** Many key cycling reactions are performed only by bacteria and archaea.
- **Elements cycle between organisms and abiotic sources and sinks.** The most accessible source of carbon and nitrogen is the atmosphere. The Earth's crust stores large amounts of key elements, but their availability to organisms is limited.
- **Environmental flux of elements is measured through biochemistry.** Methods include infrared and mass spectroscopy, gas chromatography, radioisotope incorporation, and the measurement of isotope ratios.

22.2 The Carbon Cycle

The foundation of all food webs involves influx and efflux of carbon. The major reservoirs of carbon are shown in Table 22.2. In theory, carbonate rock forms the largest reservoir of carbon. But Earth's crust is the source

Carbon Cycles Depend on Oxygen

The global cycle of carbon in the biosphere is closely linked to the cycles of oxygen and hydrogen, elements to which most carbon is bonded (**Fig. 22.3A**). Overall, carbon cycles between carbon dioxide (CO_2) and various reduced forms of carbon, including biomass (living material). Note that the results of carbon cycling differ greatly, depending on the presence of molecular oxygen.

Aerobic carbon cycling. Aerated ecosystems, such as the photic zone of oceans and the oxygenated surface of terrestrial habitats, are dominated by the photosynthetic fixation of CO_2 into biomass, designated by the shorthand [CH_2O]. Phototrophs include bacteria and protists, as well as plants. Aerobic CO_2 fixation is accompanied by production of O_2. The O_2 is then used by heterotrophs (such as bacteria, protists, and animals) to convert [CH_2O] back to CO_2. In the presence of light, a net excess of O_2 is produced; in the ocean's photic zone, the rate of photosynthesis exceeds the rate of respiration (**Fig. 22.3B**).

A.

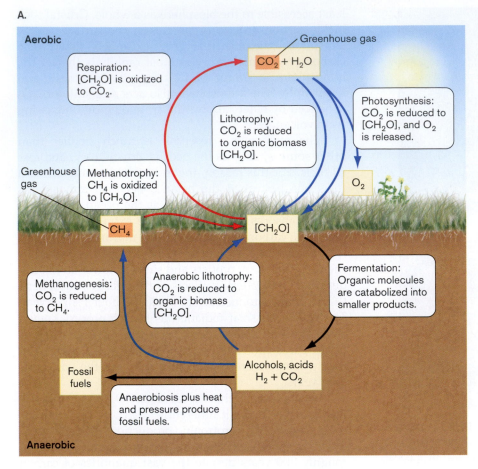

Figure 22.3   The carbon cycle: aerobic and anaerobic.   A. Aerobic and anaerobic conversions of carbon. Blue = reduction of carbon; red = oxidation of carbon; black = fermentation; $[CH_2O]$ = organic biomass. In an aerobic environment, photosynthesis generates molecular oxygen (O_2), which enables the most efficient metabolism by heterotrophs and lithotrophs. In an anaerobic environment, photosynthesis generates only oxidized minerals, which support limited anaerobic respiration. Fermentation generates organic carbon products as well as CO_2 and H_2. In the absence of oxygen, methanogens convert carbon dioxide (CO_2) and molecular hydrogen (H_2) to methane (CH_4), one of the most potent greenhouse gases. **B.** The aerated upper water layer of a pond or ocean. The layer of water in which enough light penetrates to power more photosynthesis than heterotrophy is called the photic zone.

B.

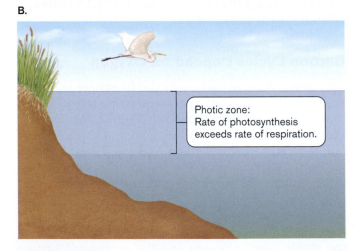

Anaerobic carbon cycling. Anoxic environments support lower rates of biomass production than do oxygen-rich environments, as they depend on oxidants of lower redox potential and limited quantity such as Fe^{3+}. Anaerobic conversion of CO_2 to biomass is mainly done by bacteria and archaea. Vast permanently anaerobic habitats extend several kilometers below Earth's surface, encompassing greater volume than the rest of the biosphere put together. In these habitats, endolithic bacteria inhabit the interstices of rock crystals (discussed in Chapter 21).

In soil and water, anaerobic metabolism includes fermentation of organic carbon sources, as well as respiration and lithotrophy with alternative electron acceptors such as nitrate, ferric iron (Fe^{3+}), and sulfate. Anaerobic decomposition by microbes is one stage in the formation of fossil fuels such as oil and natural gas (primarily methane). In soil, anoxic conditions (extremely low in O_2) favor incomplete breakdown of organic material. This characteristic of anaerobic soil actually enriches ecosystems, particularly those of wetlands, which undergo periodic cycles of aeration and hydration. The partly decomposed matter becomes available for further decomposition with oxygen. Unfortunately, anoxic environments near the surface also favor production of methane from the H_2, CO_2, and other fermentation products of anaerobes.

Biomass is also produced through lithotrophy—the oxidation of hydrogen, hydrogen sulfide, ferrous iron (Fe^{2+}) and other reduced minerals, and even carbon monoxide. Lithotrophy is especially prominent in soil and in weathered areas of crustal rock. Lithotrophy is performed solely by bacteria and archaea, essential microbial partners in these ecosystems.

The Global Carbon Balance

The global balance of biological CO_2 fixation and release largely determines the atmospheric level of CO_2. Since the beginning of the industrial age, however, the release of CO_2 has accelerated significantly. The major part of this increase comes from the combustion of **fossil fuels**, which adds about 6×10^{15} g of carbon annually to the atmosphere. Fossil fuels are the product of microbial anaerobic digestion of plant and animal remains, reduced to hydrocarbons by the pressure and heat of Earth's crust. When burned as fuel, carbon that had accumulated over millions of years is rapidly returned to the atmosphere as CO_2. Some of the CO_2 flux is compensated by increased CO_2 fixation and ocean absorption, but there remains a net flux of 4×10^{15} g of carbon annually.

Another factor in the increase of atmospheric CO_2 is deforestation, which adds about 2×10^{15} g carbon annually to the atmosphere, or approximately a third as much as burning fossil fuels. As forests are cut, CO_2 is released through wood burning and microbial decay. The role of microbes in CO_2 release, as compared to wood burning, is challenging to measure. One study focused on carbon flux in the forests of Amazonia, the Amazon river watershed. The Amazon watershed may account for as much as 10% of global carbon flux from deforestation. In this study, researchers mapped various processes of carbon uptake and efflux throughout the region (**Fig. 22.4**). Processes that generate CO_2 include burning of trees (red line) and microbial decay of cut trees (blue line). Removal of CO_2 from the atmosphere occurs by regrowth of trees (green line). Since 1978, the major contribution to total net carbon efflux has been microbial decomposition of felled trees. Thus, more CO_2 has been released by tree cutting and microbial decay than by burning.

The contribution of ocean ecosystems to carbon flux is even more difficult to measure. New molecular genetic surveys of the upper ocean are revealing previously unrecognized communities of microbial phototrophs, such as submicroscopic cyanobacteria. In the deep ocean, similar studies reveal major new groups of lithotrophic producers, such as anaerobic ammonia oxidizers ("anammox" bacteria). A key question is the fate of the benthic methane hydrates accumulated through methanogenesis. At present, bacteria are believed to oxidize most of the methane released; but a sudden release from methane hydrates could alter the carbon balance and accelerate global warming.

Studies of carbon and oxygen flux based on isotope ratios suggest a much greater level of biological production than earlier estimates. It now appears that the ocean's microbial communities may be responsible for more than half of the global biological uptake of carbon from the atmosphere. This kind of data is crucial for predicting the global effects of changes in atmospheric CO_2.

From a political standpoint, ecosystems such as forests that act as **carbon sinks** by fixing carbon into stable

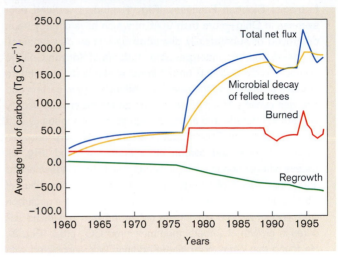

Carbon flux out of Amazonian forest

Figure 22.4   Carbon flux from Amazonian forest during 1962–1998. Carbon flux (CO_2 efflux to the atmosphere and uptake by forests) was measured from the Brazilian Amazon forest watershed. Total net carbon flux (blue line) includes contributions from microbial decay (yellow line), burning of trees (red line), and regrowth of trees (green line). Tg C yr^{-1} = teragrams of carbon per year (1 Tg = 10^{12} g). *Source:* Based on R. A. Houghton, et al. 2000. *Nature* 403:301.

biomass are considered desirable because they lessen the rate of CO_2 input into the atmosphere. Ecosystems that act as sources of carbon dioxide may be viewed with disfavor because they contribute to global warming. But what if CO_2-generating ecosystems provide other environmental benefits? For example, wetlands are among Earth's most productive ecosystems, supporting vast amounts of plant and animal life, but they may also release significant amounts of CO_2 and methane (see **Special Topic 22.1**).

TO SUMMARIZE:

- **The most accessible reservoir for carbon is the atmosphere (CO_2).** Atmospheric carbon is severely perturbed by the burning of fossil fuels.
- **Cyanobacteria and other phytoplankton cycle much of the CO_2 in the biosphere.** Oxygen produced by phototrophs is used by heterotrophs and lithotrophs.
- **Carbon cycling is linked to the cycling of hydrogen and oxygen.**
- **Anaerobic environments cycle carbon through bacteria and archaea.** Bacteria conduct fermentation and anaerobic respiration. Methanogens produce methane, much of which is oxidized by methanotrophs.
- **Microbial decomposition returns CO_2 to the atmosphere.** Microbial decomposition is a major contributor to the accelerated CO_2 flux associated with deforestation.

Ecologist Siobhan Fennessy (**Fig. 1**) studies wetland ecosystems in Ohio—more than 90% of which have already been destroyed or substantially disturbed by human development. Fennessy and her undergraduate students at Kenyon College compare disturbed and undisturbed wetlands to investigate species diversity and the role that wetlands play in carbon and nitrogen cycling. Much of the carbon and nitrogen is cycled by anaerobic bacteria and archaea such as methanogens.

Student John Tisdale addressed the relationship between nitrogen loading and carbon efflux from a wetland (**Fig. 2**). Excess nitrogen loading into wetlands results from industrial and agricultural runoff of nitrogenous fertilizers. Does this influx of nitrogen to a nitrogen-limited ecosystem affect the rate of carbon flux? Tisdale applied ammonium nitrate fertilizer to wetland soil cores in growth chambers, then measured CO_2 efflux (**Fig. 3**). He found that ammonium nitrate addition was associated with increased CO_2 efflux, presumably caused by the increased respiration by heterotrophs. Moreover, addition of CO_2 together with nitrate had a synergistic effect, augmenting the fertilizer-induced

efflux of CO_2. Elevated atmospheric CO_2 is known to increase carbon assimilation by plants, causing an increase in fine root growth and mycorrhizal association. This suggests the possibility of a positive feedback in the carbon cycle: higher CO_2 availability increases plant growth, which leads to higher rates of CO_2 output from the soil, further increasing atmospheric concentrations. Soil carbon flux is one of the largest fluxes in the global carbon cycle, so even small changes in soil flux can lead to substantial carbon inputs to the atmosphere. Thus, the rise in greenhouse gases may lead to significant changes in carbon dynamics.

Today's land developers are required to replace lost wetlands with "created wetlands." But are natural wetlands replaceable? Student Dan Gustafson addressed the survival of mycorrhizae (symbiotic fungal extensions of plant roots) in the anaerobic soil of well-established wetlands. Mycorrhizae are important for productivity of natural ecosystems as well as human agriculture. Gustafson proposed that mycorrhizae require oxygen provided by the oxygenated rhizosphere surrounding the roots of wetland plants. This hypothesis was tested by cutting off the oxygen moving down through the root channels. Oxygen levels were measured by determining soil redox potentials, which decrease when oxygen is depleted. In low oxygen, mycorrhizae failed to grow. Thus, the experiment showed that the mycorrhizae indeed require oxygen provided by the plant roots, even in anaerobic soil. Subsequent field studies showed that "created" wetlands fail to support as much mycorrhizae as natural wetlands do. This delicate mutualism between fungus and plant requires a long-established wetland habitat, a system that could take many decades to form in a "created wetland."

Figure 1 Siobhan Fennessy studies the effect of human disturbance on Ohio wetlands.

Figure 2 Fennessy and Kenyon College undergraduates John Tisdale and Dan Gustafson collect soil cores to study carbon and nitrogen flux.

Figure 3 Field monitoring stations in an Ohio wetland. The stations are used to study carbon flux and water levels. The white PVC pipes are wells with attachments for CO_2 flux measurements using a portable infrared gas analyzer.

22.3 The Hydrologic Cycle and Wastewater Treatment

The fate and distribution of complex carbon compounds is largely a function of the **hydrologic cycle**, the cyclic exchange of water between atmospheric water vapor and Earth's ecosystems (**Fig. 22.5A**). A vast reservoir of water is supplied by the ocean. In the hydrologic cycle, water precipitates as rain, which is drawn by gravity into groundwater, rivers, lakes, and, ultimately, the ocean. All along this route, of course, evaporation returns water to the air. Human communities interact with the hydrologic cycle by drawing water for drinking and other purposes, and by returning wastewater. Wastewater requires cleansing of organic contaminants, including key forms of treatment by microbes.

Biochemical Oxygen Demand

A major problem in aquatic ecosystems is organic contamination. Water passing through the ground and aquatic ecosystems carries organic carbon material from humus, sewage, and fertilizer runoff. A sudden influx of rich carbon substrates accelerates respiration by aquatic microbes. Microbial respiration then competes with that of fish, invertebrates, and amphibians for the limited supply of oxygen dissolved in water, raising the **biochemical oxygen demand (BOD)**, also called the biological oxygen demand (BOD). The higher the concentration of organic substances, the higher the BOD arising from microbial oxygen consumption.

High BOD can cause a massive die-off of fish and other aquatic animals. Thus, a routine part of monitoring the health of lakes and streams is the measurement of BOD. BOD is measured by observing the rate of oxygen uptake in a water sample by a defined set of heterotrophic bacteria (**Fig. 22.5B**). Oxygen uptake is observed in a BOD meter (**Fig. 22.5C**), which detects dissolved oxygen in water. The rate of decrease of dissolved oxygen measured by the BOD meter is approximately proportional to the amount of dissolved organic matter available for respiration.

A. Hydrologic cycle

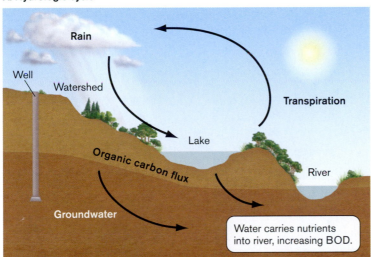

B. BOD measurement

Joel Gordon, Pittsfield wastewater treatment plant

Figure 22.5 The hydrologic cycle interacts with the carbon cycle. **A.** The hydrologic cycle carries bacteria and organic carbon into groundwater and aquatic systems. **B.** Bottled water samples are measured for dissolved oxygen over time; the rate of decrease of dissolved oxygen indicates biological oxygen demand (BOD). The rate of decrease of dissolved oxygen in water samples is approximately proportional to the concentration of organic matter available for respiration. **C.** A microprocessor-controlled Biox-1010 BOD analyzer measures rate of respiration. Water samples are mixed with a concentrated microbial biomass, and a dissolved-oxygen sensor measures small rates of oxygen decrease over time.

C. BOD analyzer

Endress & Hauser, Inc.

Until recently, BOD was considered a local issue, affecting the health of lakes and rivers in a community. Today, however, large regions of ocean have become **dead zones**, or **zones of hypoxia**, devoid of most fishes and invertebrates. A well-known dead zone is a region in the Gulf of Mexico off the coast of Louisiana where the Mississippi River releases about 40% of the U.S. drainage to the sea (**Fig. 22.6**). Over its long, meandering course, which includes inputs from the Ohio and Missouri rivers, the Mississippi builds up high levels of organic pollutants, as well as nitrates from agricultural fertilizer. When these nitrogen-rich substances flow rapidly out to the Gulf in the spring, they lift the nitrogen limitation on algal growth and feed massive algal blooms. The algae then crash, and their sedimenting bodies are consumed by heterotrophic bacteria. The heterotrophs use up the available oxygen, causing **hypoxia**. Hypoxia kills off the fish, shellfish, and crustaceans over a region equivalent in size to the state of New Jersey.

Dead zones now occur along the coasts of industrial and developing countries throughout the world, including, for example, India and Australia. Dead zones contribute to the crash of fisheries worldwide, removing a critical food resource for human populations. Plans to clean up dead zones face daunting costs. For example, in 2000, the Mississippi River/Gulf of Mexico Watershed Nutrient Task Force agreed on a plan to restore 30% of the Gulf dead zone by the year 2015, at a cost of $1 billion per year.

The only way to avoid such dead zones is if all the communities throughout the river drainage areas treat their wastes to eliminate nitrogenous wastes before disposal. There are two common approaches to community wastewater treatment, both of which involve microbial partners: wastewater treatment plants and wetland filtration. Both approaches depend on microbes to remove organic carbon and nitrogen from water before it returns to aquatic systems and ultimately the ocean.

Wastewater Treatment by Microbial Communities

In industrialized nations, all municipal communities use some form of **wastewater treatment** (**Fig. 22.7**). The purpose of wastewater treatment is to decrease the BOD and the level of human pathogens before water is returned to the local rivers.

A. Mississippi River Basin with Gulf of Mexico hypoxia

B. Dead zone

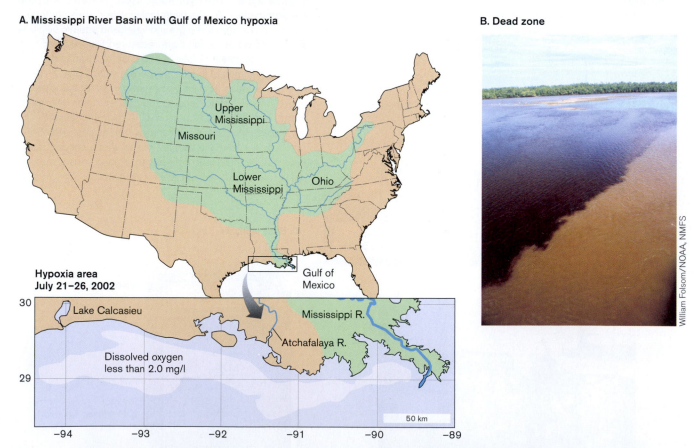

Figure 22.6 The dead zone in the Gulf of Mexico. A. Map of the Mississippi River drainage, which empties into the Gulf of Mexico. Every summer, drainage high in organic carbon and nitrogen causes algal blooms, leading to hypoxia and death of fish. **B.** The edge of the dead zone (right) off the Gulf Coast lies west of the Mississippi outlet. *Source:* A. National Oceanographic and Atmospheric Administration.

A.

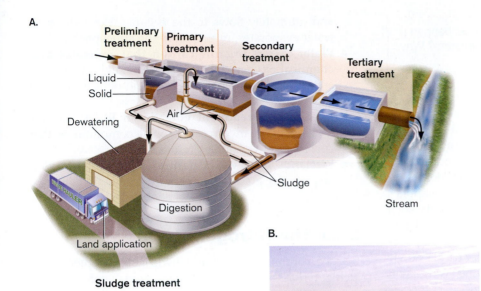

Preliminary treatment
Primary treatment
Secondary treatment
Tertiary treatment

Liquid
Solid
Dewatering
Air
Sludge
Stream
Digestion
Land application

Sludge treatment

Figure 22.7 Wastewater treatment plant. A. In a typical municipal treatment plant, wastewater undergoes primary treatment (filtering and settling), secondary treatment (microbial decomposition), and tertiary treatment (chlorination or other chemical treatments). **B.** Aeration basin for secondary treatment.

B.

City of Huntsville, AL, Water Pollution Control

The wastewater treatment plant is the final destination for all household and industrial liquid wastes passing through the municipal sewage system. A typical plant includes the following stages of treatment:

■ **Preliminary treatment** consists of screens that remove solid debris, such as sticks, dead animals, and feminine hygiene items.
■ **Primary treatment** includes fine screens and sedimentation tanks that remove insoluble particles. The particles eventually are recombined with the solid products of wastewater treatment (known as **sludge**). The sludge ultimately is used for fertilizer or landfill.

■ **Secondary treatment** consists primarily of microbial ecosystems that decompose the soluble organic content of wastewater. The microbes generate particulate flocs of biofilm. The flocs are sedimented as sludge, also known as activated sludge owing to their microbial activity.
■ **Tertiary (advanced) treatment** includes chemical applications such as chlorination to eliminate pathogens.

The microbial ecosystems of secondary treatment require continual aeration to maximize breakdown of molecules to carbon dioxide and nitrates. The microbes typically include bacilli such as *Zooglia*, *Flavobacterium*, and *Pseudomonas*, as well as filamentous species such as *Nocardia* (**Fig. 22.8A**). Optimal treatment depends on the ratio of filamentous to single-celled bacteria: enough filaments to hold together the flocs for sedimentation, but not so many as to trap air and cause flocs to float and foam, preventing sedimentation.

Besides bacteria, the ecosystem of activated sludge includes filamentous methanogens (discussed in Chapter 19), which metabolize short-chain molecules such as acetate within the anaerobic interior of flocs. Thus, wastewater treatment generates methane, often in quantities sufficient for recovery as fuel.

In addition, the bacteria are preyed upon by protists such as stalked ciliates (**Fig. 22.8B**), swimming ciliates, and amebas, as well as invertebrates such as rotifers and nematodes. The predators serve a valuable function of limiting the planktonic population of bacteria, enabling the bulk of the biomass to be removed by sedimentation.

Figure 22.8 Microbes in wastewater treatment. A. *Nocardia* sp. filamentous bacteria from flocs formed during secondary treatment (light micrograph). **B.** Flocs with a stalked ciliate, which preys on bacteria (light micrograph).

A. *Nocardia* sp.

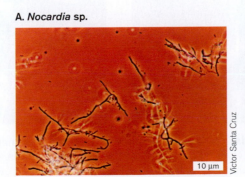

10 μm

Victor Santa Cruz

B. Flocs

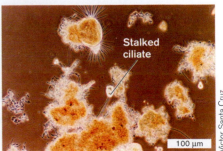

Stalked ciliate

100 μm

Victor Santa Cruz

THOUGHT QUESTION 22.2 What would happen if wastewater treatment lacked microbial predators? Why would the result be harmful?

Wastewater treatment plants are remarkably effective at converting human wastes to ecologically safe water and ultimately human drinking water. The plants are, however, impractical for purifying the runoff from large agricultural operations. For large-scale alternatives to treatment plants, communities and agricultural operations are looking to **wetland restoration**. Much of our current water supplies are already filtered and purified by natural wetlands, such as the Florida Everglades (**Fig. 22.9A**). Wetlands act to remove nitrogen through the action of denitrifying bacteria. In wetlands, rainwater and river water trickle slowly through vast stretches of soil, where microbial conversion acts as the foundation of a macroscopic ecosystem including trees and vertebrates. Thus, much of the carbon and nitrogen is fixed into valuable biomass. In the Everglade wetlands, the remaining water filters slowly through limestone into underground aquifers, which ultimately provide water through wells to human communities.

Some agricultural operations are building artificial wetlands to replace treatment plants. **Figure 22.9B** shows a hog farm where a series of terraced wetlands was built to drain liquified manure. The wetlands were found to produce fewer odors and to remove organics more efficiently and at a lower cost than traditional filtration plants.

TO SUMMARIZE:

- **The hydrological cycle is the cyclic exchange of water between the atmosphere and the biosphere.** Water precipitates as rain, which enters the ground and ultimately flows to the oceans. Along the way, water evaporates, returning to the atmosphere.
- **Water carries organic carbon that generates biochemical oxygen demand (BOD).** High BOD accelerates heterotrophic respiration and depletes oxygen needed by fish.
- **Wastewater treatment cuts down BOD.** Secondary treatment involves microbial communities that decompose the soluble organic content.
- **Wetlands filter water naturally.** Wetland filtration helps purify groundwater entering aquifers.

22.4 The Nitrogen Cycle

Besides carbon, oxygen, and hydrogen, another major element that cycles largely by microbial conversion is nitrogen (**Fig. 22.10**). The nitrogen cycle is notable in two respects:

- **Many oxidation states.** A larger number of oxidation states exist for biological nitrogen than for any other major biological element (see **Table 22.3**). Conversion between these oxidation states requires metabolic processes such as N_2 fixation, lithotrophy, and anaerobic respiration.
- **Dependence on prokaryotes.** Many steps of the nitrogen cycle require prokaryotes—that is, bacteria and archaea. Without bacteria and archaea, the nitrogen cycle would not exist.

THOUGHT QUESTION 22.3 How many kinds of biomolecules can you recall that contain nitrogen? What are the usual oxidation states for nitrogen?

A. Everglades

Everglades National Park

B. Artificial wetlands

Tim McCabe USDA/NRCS

Figure 22.9 Water filtration by wetlands. A. The marshes of the Everglades act as natural filters for the aquifers of southern Florida. **B.** A filtering system installed by Steve Kerns on his hog farm in Taylor County, Iowa. A series of hillside terraces form wetlands containing bacteria that purify hog manure and wastewater.

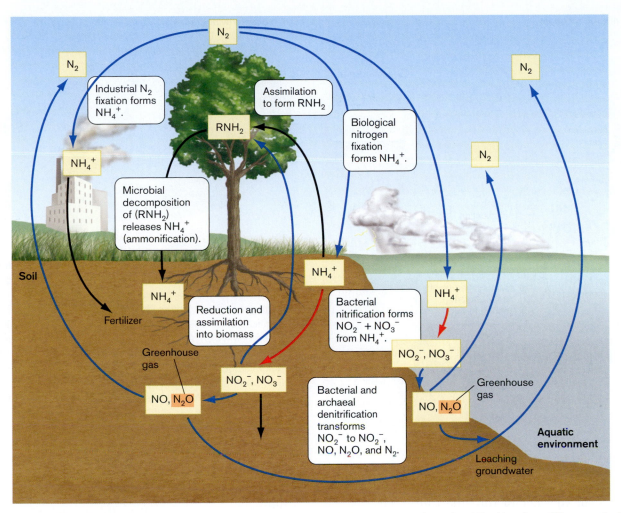

Figure 22.10   The global nitrogen cycle.   Prokaryotic conversions of nitrogen occur throughout the biosphere. Blue = reduction; red = oxidation; black = redox-neutral.

Sources of Nitrogen

Where is nitrogen found on Earth? A significant amount of nitrogen is found in Earth's crust, in the form of ammonium salts in rock (see **Table 22.2**). This form is largely inaccessible to microbes. The major accessible source of nitrogen is the atmosphere, of which dinitrogen gas (N_2) constitutes 79%. However, N_2 is a highly stable molecule that requires enormous input of reducing energy before assimilation is possible (see Chapter 14). Thus, for many natural ecosystems, and most forms of agriculture, nitrogen is the limiting nutrient for primary productivity.

Until recently in Earth's history, nitrogen was fixed entirely by nitrogen-fixing bacteria and archaea. In the twentieth century, however, the Haber process (see Chapter 14) was invented for artificial nitrogen fixation to generate fertilizers for agriculture. The process was devised by German chemist Fritz Haber, who won the 1918 Nobel Prize in Chemistry. Today, the Haber process for producing fertilizers accounts for more than half of nitrogen fixation in the entire biosphere, while other human activities contribute substantially to oxidized forms of nitrogen such as nitrous oxide (NO_2), a potent greenhouse gas. The nitrogen cycle is the most perturbed of the major biogeochemical cycles.

The "Nitrogen Triangle"

The numerous oxidation states available for microbial metabolism of nitrogen generate a complex cycle of conversions. One way to organize the complexity is to envision a "nitrogen triangle" whose corners include three key forms: atmospheric N_2; the reduced forms NH_3 and NH_4^+; and the oxidized forms NO_2^- and NO_3^- (**Fig. 22.11A**). At the base of the triangle, both reduced and oxidized forms of nitrogen are assimilated into biomass.

A. The nitrogen triangle

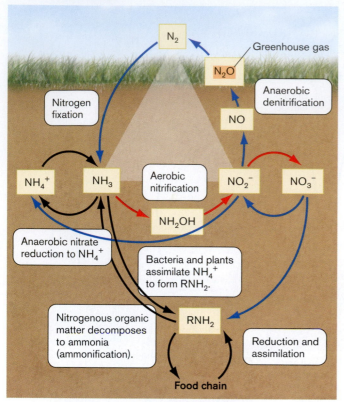

B. *Nitrobacter winogradskyi*

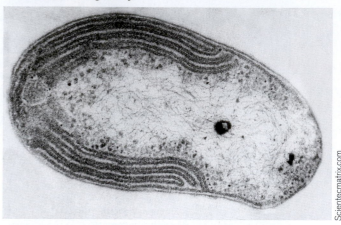

Figure 22.11 The nitrogen cycle: fixation, nitrification, denitrification. A. The "nitrogen triangle" involves nitrogen fixation and assimilation, reductive dissimilation of nitrate (denitrification; blue), and oxidation (nitrification; red). Denitrification includes production of the potent greenhouse gas nitrous oxide (N_2O). Assimilation into biomass is often reductive; virtually all nitrogen in biomolecules is highly reduced. Oxidation of ammonia generates nitrites and nitrates, whose runoff can pollute water supplies. **B.** *Nitrobacter winogradskyi* oxidizes nitrite to nitrate (TEM). Folded layers of membrane contain the electron transport complexes.

We shall consider in turn the three "legs" of the nitrogen triangle:

- **N_2 fixation** to ammonium (NH_4^+), a form assimilated into biomass by microbes and plants.
- **Nitrification** or oxidation of ammonia to nitrite (NO_2^-) and nitrate (NO_3^-), which are also assimilated by microbes and plants.
- **Denitrification** of NO_2^- and NO_3^- back to N_2 (or, in carbon-rich habitats, reduction to NH_3).

Nitrogen fixation. The main avenue for entry of nitrogen into the biosphere is **nitrogen fixation**—specifically, fixation of dinitrogen, or nitrogen gas (N_2) into NH_3, protonated to NH_4^+. Ammonium ion is rapidly assimilated by bacteria and plants, typically through combination with Krebs cycle intermediates to form key amino acids such as glutamate (discussed in Chapter 14). The ability to assimilate NH_4^+ into nitrogenous organic molecules is found in virtually all primary producers.

Fixation of nitrogen requires enormous energy because the triple bond of N_2 is exceptionally stable. Breaking the triple bond to generate ammonia requires a series of reduction steps involving high input of energy:

Dinitrogen

$$N{\equiv}N \xrightarrow{} HN{=}NH \xrightarrow{} H_2N{-}NH_2$$
$$2H^+ + 2e^- 2H^+ + 2e^-$$

$$\xrightarrow{} \text{Ammonia} \quad \text{Ammonium}$$
$$2H^+ + 2e^- \quad 2NH_3 \rightleftharpoons 2NH_4^+$$
$$2H^+$$

All the intermediate reactions of nitrogen fixation are tightly coupled, so the intermediate compounds rarely become available for other uses. The final product, ammonia, exists in ionic equilibrium with ammonium ion:

$$NH_3 + H_2O \rightleftharpoons NH_4^+ + OH^-$$

Fixation of nitrogen is catalyzed by the enzyme nitrogenase (discussed in Chapter 15). The highly reductive reactions of nitrogenase require exclusion of oxygen; yet they also require tremendous input of energy, usually provided by aerobic metabolism. Therefore, nitrogen-fixing bacteria generally separate anaerobic nitrogen fixation from aerobic metabolism by one of several mechanisms, such as the heterocysts of cyanobacteria (discussed in Chapter 14).

Given the energy expense and the need to exclude oxygen, many species of bacteria as well as all eukaryotes have lost the nitrogen fixation pathway by degenerative evolution. But all ecosystems, both marine and terrestrial, include some species of bacteria and archaea that fix N_2 into ammonia. Nitrogen-fixing bacteria in the soil include obligate anaerobes, such as *Clostridium* species, and facultative gram-negative enteric species of *Klebsiella*, *Salmonella*, and *Escherichia* (though not the commonly studied laboratory strains of *E. coli*), as well as obligate respirers such as *Pseudomonas*. In ocean and in freshwater systems, cyanobacteria are the major nitrogen fixers. After fixation, all these organisms assimilate the reduced nitrogen into essential components of their cells. Within an ecosystem, nitrogen fixers ultimately make the reduced nitrogen available for assimilation by nonfixing microbes and plants, either directly through symbiotic association (such as that of rhizobia and legumes) or indirectly through predation (marine cyanobacteria) and decomposition (soil bacteria).

If nitrogen fixation is ubiquitous in soil and water, then why is nitrogen limiting? Nitrogen fixation is extremely energy intensive; thus, the rate of fixation usually fails to meet the potential demand of other members of the ecosystem. The one exception is legume symbiosis with rhizobia, which provide ample nitrogen for their hosts.

In agriculture, symbiotic nitrogen fixation by rhizobial bacteria increases the yield of crops such as soybeans (**Fig. 22.12A**). Soybeans can derive as much as 50% of their nitrogen from microbial symbionts. To enhance colonization by nitrogen fixers, farmers apply molecules called isoflavonoids, which mimic the natural plant-derived attractants for rhizobia.

Nitrification. Free ammonia in soil or water is quickly oxidized for energy by **nitrifiers**, bacterial species that possess enzymes for oxidation of ammonia to nitrite (NO_2^-), or of nitrite to nitrate (NO_3^-). This process is called **nitrification**. Nitrification of ammonia is a form of lithotrophy, an energy generation pathway involving oxidation of minerals. The nitrification pathway generates the red leg of the triangle in Figure 22.11A. The pathway of nitrification includes:

$$\underset{+\ \frac{1}{2}O_2}{\underset{NH_3}{Ammonia}} \longrightarrow \underset{+\ O_2}{NH_2OH} \longrightarrow \underset{+\ \frac{1}{2}O_2}{\underset{NO_2^-}{Nitrite}} \longrightarrow \underset{NO_3^-}{Nitrate}$$

The pathway includes two separate energy generation mechanisms, utilized by different species: (1) ammonia through NH_2OH to nitrite, and (2) nitrite to nitrate. Typically, soil contains both kinds of microbes living together, a collaboration ensuring complete conversion to nitrate. Nitrifying genera include *Nitrosomonas*, which oxidizes ammonia to nitrite, and *Nitrobacter* (**Fig. 22.11B**) and *Nitrospira*, which oxidize nitrite to nitrate.

THOUGHT QUESTION 22.4 In the laboratory, which kind of bacteria would likely grow on artificial medium including NH_2OH as the energy source: *Nitrosomonas* or *Nitrobacter*?

A. Soybeans grow with symbiotic rhizobia

©Thomas Hovland/Alamy

B. Nitrate contamination of groundwater

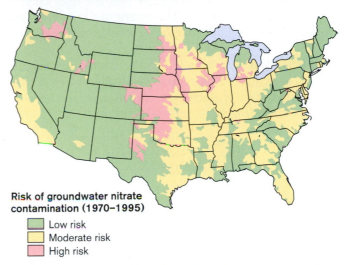

Risk of groundwater nitrate contamination (1970–1995)

- ▢ Low risk
- ▢ Moderate risk
- ▢ High risk

Figure 22.12 **Agricultural benefits and consequences of microbial nitrogen metabolism.** **A.** Soybean field in Maine. Rhizobial nitrogen fixation enhances growth of soybeans and other major crops. **B.** Nitrate in drinking water is especially prevalent in agricultural regions of the United States. Ammonification of fertilizer, followed by nitrification, generates nitrate and nitrite. Nitrate and nitrite runoff from oxidized nitrogenous fertilizers pollutes streams and groundwater.

When nitrate is produced in the soil, it is assimilated by plants and bacteria nearly as quickly as ammonium ion, although extra energy is needed to reduce nitrate to NH_4^+ for incorporation into biomass. Nitrate assimilation to biomass is called **assimilatory nitrate reduction**. In agriculture, intensive fertilization generates a large excess of ammonia resulting from **ammonification**, the conversion of organic nitrogen to ammonia. Ammonia is oxidized rapidly by lithotrophic bacteria (nitrifiers). The excess ammonia leads to a buildup of nitrites and nitrates via nitrification faster than they can be assimilated. The nitrites and nitrates are highly soluble in water, and they readily diffuse into aquatic systems. Nitrate influx into lakes leads to eutrophication as the nitrogen limit is lifted and algal blooms cause a rise in BOD. Nitrate influx also causes formation of toxic nitrosamines.

Consumption of nitrate in drinking water can lead to methemoglobinemia, a blood disorder in which hemoglobin is inactivated. Methemoglobinemia occurs in infants whose stomachs are not yet acidic enough to inhibit growth of bacteria that convert nitrate to nitrite. Nitrite oxidizes the iron in hemoglobin, eliminating its capacity to carry oxygen. The failure to carry oxygen leads to a bluish appearance, one cause of "blue baby syndrome." Nitrite-induced blue baby syndrome is a problem in intensively cultivated agricultural regions, such as Kansas and Nebraska (**Fig. 22.12B**).

Recently, another role was discovered for nitrite: as an electron donor for photosynthesis. Certain gamma-protobacteria, related to *Thiocapsa*, can use light energy to oxidize nitrite to nitrate. The extent of nitrogen-based photosynthesis in global nitrogen cycles is unknown.

Denitrification. The blue leg of the triangle in **Figure 22.11A** is generated by reduction of nitrate and nitrite through a series of decreased oxidation states back to atmospheric nitrogen:

$$\underset{\substack{2NO_3^- \\ 4H^+ + e^-}}{Nitrate} \longrightarrow \underset{\substack{2NO_2^- \\ 2H^+ + 2e^-}}{Nitrite} \longrightarrow \underset{\substack{2NO \\ 2H^+ + 2e^-}}{\substack{Nitric \\ oxide}} \longrightarrow \underset{\substack{N_2O \\ 2H^+ + 2e^-}}{\substack{Nitrous \\ oxide}} \longrightarrow \underset{N_2}{Dinitrogen}$$

This process of denitrification, also known as **dissimilatory nitrate reduction**, is performed primarily by anaerobic respirers. (In contrast, assimilatory nitrate reduction incorporates the nitrogen into biomass.) Anaerobic respirers conduct electron transport using an oxidized

form of nitrogen as an alternative electron acceptor in the absence of O_2 (discussed in Chapter 14). Many bacterial species in soil and water can respire using one or more of the oxidized nitrogen intermediates. Nitrate respiration is, however, repressed in the presence of oxygen, a more favorable electron acceptor; therefore, it is limited to anoxic habitats.

In undisturbed environments, the products of nitrate respiration rarely build up to significant levels. Heavy fertilization, however, causes buildup of excess nitrate and so increases the environmental rate of denitrification. During this process, some of the nitrogen escapes as nitrous oxide gas (N_2O). A highly potent greenhouse gas, N_2O generates 200 times the warming effect of CO_2; thus, relatively small amounts of N_2O can make a disproportionate contribution to global warming. Furthermore, N_2O in the upper atmosphere reacts catalytically with ozone, depleting the ozone layer. Thus, the atmospheric effects of bacterial denitrification are a serious concern in agricultural and waste treatment processes.

Nitrous oxide also builds up in marine dead zones. A recent study by Syed Wajih Naqvi and colleagues at the National Institute of Oceanography (Goa, India) investigated denitrification in the zone of hypoxia off the Indian coast (**Fig. 22.13A**). Under anoxic conditions, nitrate is used for anaerobic respiration. The study revealed unexpectedly high levels of N_2O production at a series of stations within the zone. In an experiment, introduction of nitrate into water samples from the dead zone led quickly to production of nitrite and N_2O (**Fig. 22.13B**), a process almost certainly caused by denitrifying bacteria. Thus, denitrifying bacteria in polluted ocean waters may contribute significantly to global warming.

A.

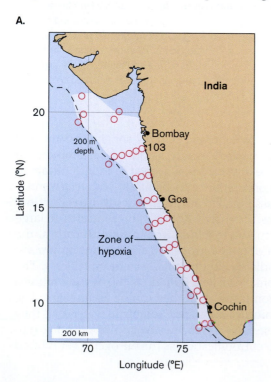

B.

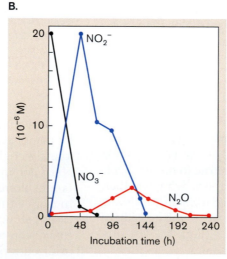

Figure 22.13 N_2O production from a coastal dead zone.
A. Zone of hypoxia off the coast of India. Circles represent stations for sample collection. **B.** Levels of NO_3^-, NO_2^-, and N_2O following addition of NO_3^- to a water sample from the zone of hypoxia. The sequential rise of NO_2^- and N_2O indicates metabolism of denitrifying bacteria. *Source*: Based on S. W. A. Naqvi, et al. 2000. *Nature* 408:346.

Dissimilatory nitrate reduction to ammonia. Most environmental nitrate and nitrite are reduced via N_2O as described previously. Certain conditions, however, favor the alternate route of dissimilatory nitrate reduction to ammonia (DNRA) (see **Fig. 22.11A**). Nitrate reduction to ammonia is a form of anaerobic lithotrophy or hydrogenotrophy, in which nitrate serves as an electron acceptor and hydrogen gas (H_2) is the electron donor:

$$NO_3^- + 4H_2 + H^+ \longrightarrow NH_3 + 3H_2O$$

Bacteria reduce nitrate to ammonia mainly in anaerobic environments rich in organic carbon and H_2 generated by fermentation, but low in reduced nitrogen, such as sewage sludge and stagnant water. Another carbon-rich habitat favoring this pathway is the rumen, the digestive tract of cattle, goats, and other ruminant animals (see Chapter 21). Ruminants consume grasses whose cellulose requires digestion by microbial endosymbionts. They also depend on their microbial endosymbionts to assimilate nitrate into ammonia and synthesize amino acids.

> **THOUGHT QUESTION 22.5** The nitrogen cycle has to be linked with the carbon cycle, since both contribute to biomass. How might the carbon cycle of an ecosystem be affected by increased input of nitrogen?

TO SUMMARIZE:

- **Nitrogen in ecosystems is found in a wide range of oxidation states.** Interconversion of most of these states requires prokaryotes.
- **The main source and sink of nitrogen is the atmosphere.** Nitrogen gas is fixed into ammonium ion by some bacteria and archaea. Denitrifying bacteria reduce NO_3^- successively back to N_2 and return it to the atmosphere.
- **Nitrogen fixation is conducted by symbiotic bacteria in association with specific plants.** Legume-associated nitrogen fixation is critical for agriculture.
- **Most NH_4^+ in the soil and water is oxidized to nitrate.** Nitrifying bacteria gain energy through lithotrophy.
- **Nitrate can be reduced to ammonia under anoxic conditions.** Especially in the deep ocean, NO_3^- may be reduced by hydrogen gas to ammonia.

22.5 Sulfur, Phosphorus, and Metals

Besides carbon and nitrogen, many other elements participate in biochemical cycles that have important consequences for the biosphere as well as for human environments. Sulfur undergoes a "triangle" of redox conversions analogous to that of nitrogen. Phosphorus, unlike other biological elements, is generally assimilated in the oxidized state, and phosphate is often a limiting nutrient for plants. Iron cycles in complex interactions with sulfur and phosphate. Beyond these macronutrients, many toxic metals in the environment, such as mercury and arsenic, are either bioactivated or detoxified by microbes.

The Sulfur Cycle

Sulfur is a major component of biomass, including proteins and cofactors. Like carbon and nitrogen, sulfur can be assimilated in either mineral or organic form. At the same time, assimilation competes with dissimilation. Reduced and oxidized forms of sulfur (comparable to those of nitrogen; see **Table 22.3**) offer electron donors and acceptors for dissimilatory reactions that generate energy. Reduced forms of sulfur, such as H_2S, can also participate in photolysis in reactions analogous to those of H_2O. Oxidized forms, such as sulfate, serve as anaerobic electron acceptors. As with nitrogen, most of these redox reactions are performed solely by bacteria and archaea. The biochemistry of sulfur cycling has important environmental consequences, such as the corrosion of concrete and iron.

Where is sulfur found? A remarkably large source of sulfur is marine sulfate (see **Table 22.2**). In the ocean, sulfate is the second most common anion after chloride. Marine sulfate turns over slowly and constitutes an essentially limitless supply. Sulfur is similarly plentiful in freshwater sediments. Thus, sulfur is rarely a limiting nutrient.

While the amount of sulfur in ocean water is relatively large compared to that of nitrogen or carbon, the amount in the atmosphere, mainly sulfur dioxide, is small. Nevertheless, these small amounts of sulfur compounds generate toxic effects as well as acidic pollution. Thus, both biochemical and industrial sources of atmospheric sulfur are of concern.

Competing assimilatory and dissimilatory sulfur reactions in the biosphere form a "sulfur triangle" between H_2S, S^0, and SO_4^{2-} (**Fig. 22.14A**). The oxidation and reduction pathways are analogous to those of nitrogen. Sulfur pathways include further options of anaerobic phototrophy. In an aquatic system, H_2S arises from spring waters welling up from the sediment and from decomposition of detritus. In decomposition, anaerobic respirers such as *Desulfovibrio* species convert sulfate to sulfur, then to H_2S. As H_2S rises to the oxygenated surface water, it is readily oxidized by sulfur bacteria such as *Thiobacillus* species (**Figs. 22.14B** and **C**). Microbial sulfur oxidation is helpful for environments because it removes H_2S, which is highly toxic to most nonsulfur bacteria and plants. Because of the toxicity of H_2S, autotrophs more readily assimilate SO_4^{2-} into biomass, despite the extra energy needed to reduce SO_4^{2-}.

A. The sulfur triangle

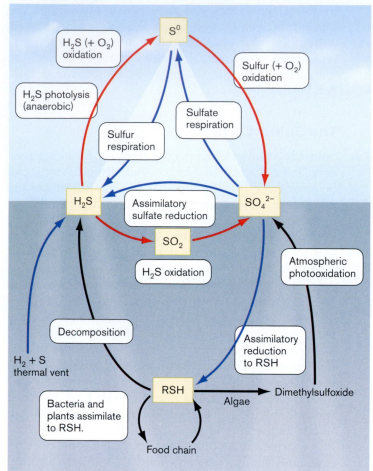

B. White sulfur bacteria

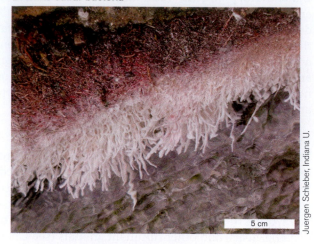

Juergen Schieber, Indiana U.

C. Bacteria containing sulfur granules

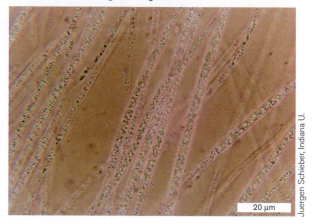

Juergen Schieber, Indiana U.

Figure 22.14   The sulfur cycle.   A. The "sulfur triangle." In the presence of oxygen, H_2S is oxidized to sulfur dioxide, then sulfate. Anaerobically, H_2S may be photolyzed to sulfur. Sulfate serves as an electron acceptor for anaerobic respiration, or it may be assimilated into biomass, organic molecules with reduced sulfur groups (RSH). Elemental sulfur may be oxidized to sulfate by lithotrophs. **B.** In springs containing H_2S, white sulfur bacteria (H_2S oxidizers) form microbial mats. **C.** Close-up of white sulfur bacteria. Within their filaments are intracellular sulfur granules formed by oxidation of H_2S to S^0.

An alternate route for H_2S oxidation, if light is available, is photolysis by anaerobic phototrophs such as *Rhodopseudomonas* species. Some phototrophic bacteria further oxidize the S^0 generated to sulfate. In some sulfur-rich lakes, such as the Russian Lake Sernoye, underground springs pump so much H_2S that most of the sulfur is photolytically converted to S^0, which forms up to 5% of the sediment. This elemental sulfur can be mined commercially.

In thermal vent ecosystems, several sulfur-based reactions drive metabolism. For example, thermal vent archaea such as *Pyrodictium* species use sulfur to oxidize hydrogen gas to H_2S:

$$H_2 + S^0 \longrightarrow H_2S$$

This form of sulfur-based hydrogenotrophy is enhanced by the vent conditions of extreme pressure and temperature

(100°C), conditions under which elemental sulfur exists in a molten state more accessible to microbes than solid sulfur.

Decomposition of biomass generates various organic sulfur compounds, many of which are volatile. Odors of certain microbial sulfur products contribute to the smell of rotting eggs, but others enhance the taste of cheeses (discussed in Chapter 16). In marine environments, algae excrete substantial quantities of dimethyl sulfide, which escapes to the atmosphere, providing part of the characteristic "smell of the sea." Dimethyl sulfide is believed to be the major biogenic source of atmospheric sulfur. Dimethyl sulfide and other organic sulfur compounds in the atmosphere are subject to photochemical oxidation to SO_2 and SO_4^{2-}, a nonbiogenic process.

Sulfur is metabolized underground by bacteria beneath deposits of fossil fuels such as oil and coal (**Fig. 22.15A**). Oil and coal contain sulfur in the form of

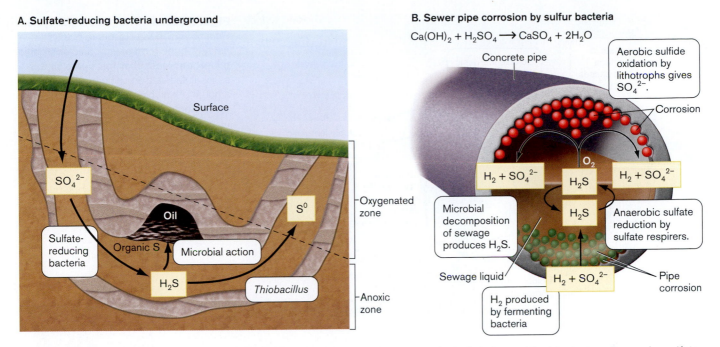

Figure 22.15   Consequences of the sulfur cycle.   A. Oil- or coal-bearing geological strata provide rich electron donors for sulfate-reducing bacteria (sulfate respirers). Eventually, various forms of sulfur contaminate the oil or coal, which, when burned as fuel, generate first SO_2 and ultimately sulfuric acid (acid rain). **B.** In a sewer pipe, sulfate-reducing bacteria (sulfate respirers) generate H_2S. Sulfur-oxidizing bacteria then oxidize H_2S to sulfate in the form of sulfuric acid. The sulfuric acid reacts with calcium hydroxide in the concrete, thus corroding the interior surface of the pipe.

SO_4^{2-} and provide a carbon source for sulfate respirers (bacteria that respire using sulfate as a terminal electron acceptor). The products of sulfate respiration include S^0 and organic sulfur. When the fuel is burned, these forms of sulfur cause severe pollution. Before burning, however, the S^0 and organic sulfur can be oxidized by sulfur-oxidizing bacteria. Fuel processors are now turning to sulfur-oxidizing microbes for experimental use in "desulfuration," the removal of sulfur from coal.

A habitat that exhibits the entire range of sulfur oxidation states is that of a sewage pipe. In a concrete sewer pipe, the alternation between anaerobic and oxygenated sulfur biochemistry generates severe corrosion (**Fig. 22.15B**). The microbial decomposition of sewage yields large quantities of toxic H_2S, which then volatilize to high levels that endanger sewer workers. The H_2S is then oxidized to sulfuric acid by *Thiobacillus ferrooxidans*, a bacterium that colonizes the surface of the concrete. The sulfuric acid (H_2SO_4) decreases the pH at the concrete surface to pH 2. In the concrete surface, sulfuric acid converts calcium hydroxide to soluble calcium sulfate. Over several years, this corrosion can cut the thickness of a sewer pipe by half.

Sulfur metabolism shows important connections with metabolism of metals. Many sulfur-oxidizing bacteria, such as *Thiobacillus ferrooxidans*, oxidize iron as well (discussed shortly).

> **THOUGHT QUESTION 22.6**   Compare and contrast the cycling of nitrogen and sulfur. How are the cycles similar? How are they different?

The Phosphate Cycle

Phosphorus is a macronutrient required for assimilation into organic biomass. Phosphate is a fundamental component of nucleic acids, phospholipids, and phosphorylated proteins. Unlike sulfur and nitrogen, which are found in several different oxidation states, phosphorus is cycled almost entirely in the fully oxidized state of phosphate (**Fig. 22.16**). The absence of fully reduced phosphorus in ecosystems may be due to the fact that reduced phosphorus (phosphine, PH_3) undergoes spontaneous combustion in the presence of oxygen. Nevertheless, some anaerobic decomposers may use phosphate as a terminal electron acceptor, reducing it to phosphine. In marshes and graveyards, where extensive decomposition occurs, phosphine has been observed emanating from the ground, where it ignites with a green glow. Microbial respiration might be responsible for such "ghostly" apparitions.

Although phosphate is abundant in Earth's crust, its availability in ecosystems is limited by its tendency to precipitate with calcium, magnesium, and iron ions. Thus, dissolved phosphate in water and soil is often a limiting

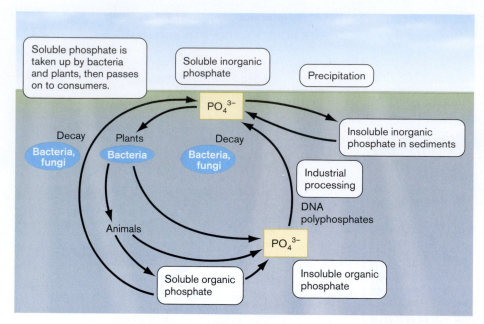

Figure 22.16   The phosphate cycle. Phosphorus in the biosphere occurs entirely in the form of inorganic or organic phosphate. Most phosphate precipitates as insoluble salts in sediment. The small amount of soluble phosphate is taken up by plants and bacteria, then passed on to consumers. Decomposers return phosphate to the environment.

nutrient for productivity. In natural ecosystems, whatever phosphate is available is taken up rapidly by bacteria and phytoplankton, then consumed by grazers and predators and dispersed by decomposers at each level. The importance of phosphate limitation in the oceans was underscored by the Sargasso Sea collection of genomes sequenced by Craig Venter and colleagues (discussed in Chapter 21). In the genomes they assembled, they found a large number of genes encoding diverse systems of phosphate acquisition.

In agriculture, phosphate is often added as a fertilizer. Phosphate fertilizer is obtained by treating calcium phosphate rock with sulfuric acid, producing calcium sulfate (gypsum) and phosphoric acid. Excess phosphate from fertilizer or industry may drain into streams and lakes where phosphorus is the limiting element. The sudden influx of phosphate causes an algal bloom. The overgrowth of algae leads to overgrowth of heterotrophs, depletion of oxygen, and destruction of the food chain.

THOUGHT QUESTION 22.7   Compare and contrast the cycling of nitrogen and phosphorus. How are the cycles similar? How are they different?

The Iron Cycle

Iron is one of a large number of micronutrients, elements that form a negligible part of biomass but are essential for growth of organisms. Nearly all organisms require iron, which acts as a cofactor for enzymes and is an essential component of oxygen carrier molecules such as hemoglobin. Other common micronutrients required for enzymes and cofactors are zinc, copper, and selenium.

While iron is a major component of Earth's crust, its availability to organisms is limited by its extremely low solubility in the oxidized form. In microbial biochemistry, the oxidized form, ferric ion (Fe^{3+}), interconverts with the reduced form, ferrous ion (Fe^{2+}). In the presence of oxygen, iron metal (Fe^0) "rusts" and is therefore available to organisms mainly as Fe^{3+}. Ferric ion is especially insoluble at high pH, precipitating as ferric hydroxide [$Fe(OH)_3$] or ferric phosphate ($FePO_4$) (**Fig. 22.17A**).

In iron mines, where pyrite (FeS_2) is exposed to air, spontaneous oxidation produces sulfuric acid:

$$2FeS_2 + 7O_2 + 2H_2O + 4H^+ \longrightarrow 2Fe^{2+} + 4H_2SO_4$$

This reaction drives the growth of lithotrophs such as *Thiobacillus ferrooxidans*, which greatly increases the rate of sulfuric acid production, leading to acid mine drainage. Where a source of concentrated reduced iron from acid mine drainage enters an aquatic system, it is oxidized by lithotrophs to ferric hydroxide, forming an orange precipitate, seen in home plumbing (**Fig. 22.17B**). Iron mine drainage causes severe pollution of streams (**Fig. 22.18**).

In anoxic habitats, such as benthic sediment, Fe^{3+} is largely reduced to Fe^{2+} by anaerobic respiration. Reduction leads to loss of the reddish color, generating gray-colored sediment, as in wetland soil (discussed in Chapter 21). The reduction of Fe^{3+} to Fe^{2+} is one of the enriching contributions of anaerobic sediment in aquatic wetlands and coastal estuaries, because reduced iron is more available to plants and bacteria. Oxidized iron, in aerobic soils, can only be taken up by bacteria. Bacteria synthesize special iron uptake systems, including molecules called **siderophores** that bind ferric ion outside the cell (discussed in Chapter 4). Bacterial iron then becomes available to consumers in the ecosystem.

A. The iron cycle

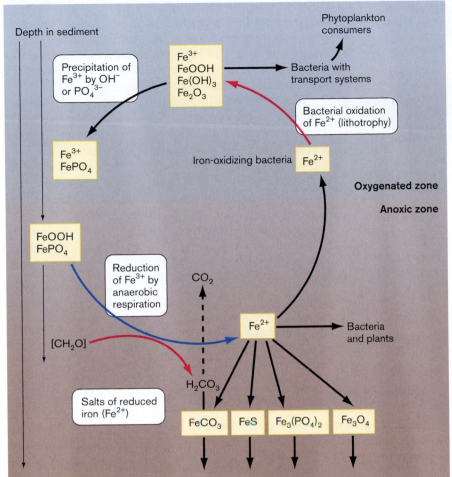

B. Iron precipitation due to lithotrophy

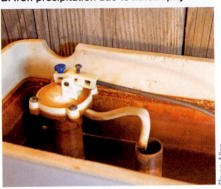

Bryan Allen

Figure 22.17   The iron cycle.   A. Ferric ion (oxidized iron, Fe^{3+}) precipitates with hydroxide or phosphate. Only bacteria can assimilate Fe^{3+}, which is then passed on to consumers. In anaerobic sediment, bacterial respiration converts Fe^{3+} to Fe^{2+}, a more soluble form of iron available to plants. **B.** Home plumbing shows signs of lithotrophic oxidation of Fe^{2+} to Fe^{3+}, which forms an orange precipitate.

Thomas R. Fletcher/Alamy

Figure 22.18   Mine drainage enters a stream.   Lithotrophic oxidation of reduced iron yields the orange material polluting this stream in Preston County, West Virginia.

Iron cycling often connects with sulfur in ways that can prove unfortunate for human engineering. The most serious problem is anaerobic corrosion of iron. Iron corrosion occurs spontaneously in the presence of oxygen. In anoxic environments, however, little spontaneous corrosion occurs. Instead, iron is corroded by sulfate-reducing bacteria (**Fig. 22.19**). Bacteria such as *Desulfovibrio* and *Desulfobacterium* can grow on the iron surface, forming an anaerobic biofilm. Within the biofilm, iron is oxidized to Fe^{2+}, and sulfate dissolved in the water is reduced to ferrous sulfide:

$$4Fe + SO_4^{2-} + 4H_2O \longrightarrow FeS + 3Fe^{2+} + 8OH^-$$

The FeS flakes away, exposing more iron to react with water, resulting in cyclic corrosion.

In oceans, iron is limiting because the benthos is so distant that its iron is largely inaccessible. The main

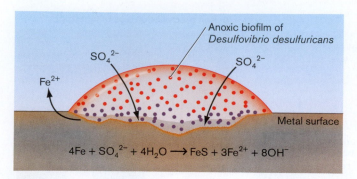

Figure 22.19   Anaerobic corrosion.   Iron-sulfur bacteria convert Fe⁰ to FeS. FeS flakes off, exposing iron to further oxidation and rapid corrosion. *Source*: H. T. Dinh, et al. 2004. *Nature* 427:829–832.

source of marine iron is eolian (wind-borne) dust from dry land. Most of the wind-borne iron is particulate, oxidized, and unavailable to eukaryotic phytoplankton (algae). Thus, bacteria that acquire and reduce ferric ion provide the main entry of iron into the ecosystem. This may explain why so many marine algae are mixotrophs: Although their photosynthesis can fix plenty of carbon for biomass, they must consume bacteria as a source of iron.

Iron limitation in the ocean affects other cycles, particularly those of nitrogen and phosphate. Evidence suggests that in the North Pacific, lack of iron limits the growth of nitrogen fixers, thus setting the limit for nitrogen assimilation into biomass. In the North Atlantic, on the other hand, iron is more available, so nitrogen fixation is not limited; instead, phosphate becomes limiting.

The role of iron limitation has led to speculation that we could remove excess CO_2 from the atmosphere by fertilizing the ocean with iron, accelerating growth of marine phytoplankton. An experiment to demonstrate the feasibility of iron fertilization was conducted by members of the Planktos Foundation, a private organization supported by Canadian rock star Neil Young. In this experiment, a large quantity of iron-rich ore was dumped in marine waters off the Hawaiian Islands. Sure enough, a substantial algal bloom followed. Calculations showed that the algal bloom was large enough to fix significant CO_2.

THOUGHT QUESTION 22.8   Do you think that iron fertilization would succeed in maintaining lowered atmospheric CO_2? What consequences might you predict for widespread iron fertilization?

Table 22.4   Microbial metabolism of metals.

Metal	Major conversions	Microbial genera (examples)	Effects on environment
Manganese, Mn	$Mn^{2+} \longrightarrow Mn^{4+}$ $Mn^{4+} \longrightarrow Mn^{2+}$	*Hyphomicrobium, Arthrobacter, Bacillus, Geobacter, Pseudomonas*	Mn is a trace element required for enzymes Anaerobic respiration in sediment
Mercury, Hg	$Hg^{2+} \longrightarrow Hg^0$ $Hg^{2+} \longrightarrow (CH_3)Hg^+$ $(CH_3)_2Hg^+ \longrightarrow (CH_3)_2Hg$	*Thiobacillus* *Desulfovibrio*	Hg^0 volatilizes; little harm $(CH_3)Hg^+$ is a severe neurotoxin; ingested through contaminated fish $(CH_3)_2Hg$ volatilizes; poisonous but does not accumulate
Arsenic, As	$AsO_2^- \longrightarrow AsO_4^{3-}$ $AsO_4^{3-} \longrightarrow AsO_2^-$ $AsO_4^{3-} \longrightarrow (CH_3)_3As$	*Alcaligenes, Pseudomonas* *Bacillus, Chrysiogenes, Pyrobaculum* *Candida* (a fungus), *Scopulariopsis* (a fungus)	AsO_2^- (arsenite) and AsO_4^{3-} (arsenate) both poison cell metabolism Arsenate as terminal e^- acceptor Methylarsines poison; inhalation of moldy wallpaper with arsenic pigment
Chromium, Cr	$CrO_4^{2-} \longrightarrow Cr^{3+}$	*Aeromonas, Arthrobacter, Desulfovibrio*	CrO_4^{2-} [Cr(VI)] is mutagenic and carcinogenic; as e^- acceptor, reduced to Cr^{3+} [Cr(III)], less toxic
Vanadium, V	$VO_3^- \longrightarrow VO(OH)$	*Veillonella, Desulfovibrio, Clostridium*	V is a trace element for some nitrogenases and invertebrate blood pigments VO_3^- (vanadate) is oxidized as an e^- donor
Selenium, Se	$Se^0 \longrightarrow SeO_3^{2-}$	*Bacillus, Micrococcus*	Se is a human trace element, but toxic at higher doses
Uranium, U	$UO_2^{2+} \longrightarrow UO_2$	*Veillonella, Shewanella, Geobacter*	UO_2^{2+} [U(VI)] is highly soluble; microbes reduce as e^- acceptor to uranium dioxide, UO_2 [U(IV)], insoluble; used for cleanup of radioactive uranium contamination

The iron fertilization experiment was highly controversial, as ecologists pointed out numerous unforeseen consequences of such an attempt. While iron fertilization is unlikely to prove feasible, other means of positive engagement with the biosphere need to be considered to counteract the increasingly negative impacts of human civilization on biogeochemical cycles. Our view of Earth's biosphere increasingly is moving toward **environmental management**—the concept that wilderness as such no longer exists, only a biosphere to be managed for better or worse. Microbes will have many roles as partners in environmental management.

Other Metals in the Environment

Another concern for environmental management is that of toxic metals such as mercury and arsenic. Besides iron, numerous trace metals interact with bacterial species as either electron donors or acceptors (**Table 22.4**). Bacterial conversion of metals can either produce toxic species or remove toxic metals from ecosystems. For example, chromium-6 [Cr(VI)] is an extremely toxic pollutant that can be reduced by soil bacteria to the much less toxic Cr(III).

Mercury contamination. A well-known case of metal toxicity mediated by bacteria is that of mercury (**Fig. 22.20**). Mercury is found in small quantities in nature, but human production of pesticides puts out 10,000 metric tons annually, in addition to

3,000 metric tons released by the burning of fossil fuels. Elemental mercury (Hg^0) has limited toxicity; but in the environment, mercury is methylated by bacteria, forming methylmercury, $(CH_3)Hg^+$, which is highly toxic. When bacteria containing methylmercury are consumed by grazers and predators, the mercury concentration increases as a proportion of biomass until it may reach levels toxic to animals such as fish and shellfish. When these animals are consumed by humans, we may experience extreme toxicity, particularly neurological effects.

Note that methylation of mercury is just one of the reactions bacteria may conduct. Anaerobic respirers may reduce Hg^{2+}, while others may demethylate methylmercury to the less toxic Hg^{2+}. Addition of a second methyl group produces the extremely toxic and volatile compound dimethylmercury [$(CH_3)_2Hg$] that dissipates initially, but can be taken up by organisms.

A. Mercury conversions in the environment

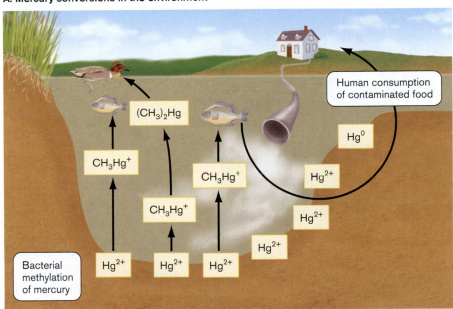

B.

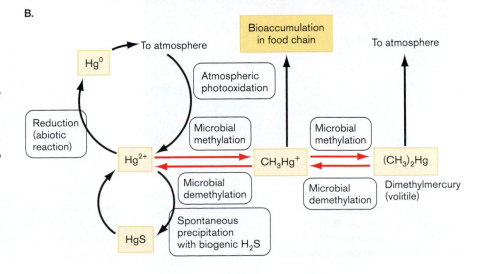

Figure 22.20   Mercury conversions include microbial metabolism.
A. Mercury methylated by bacteria is accumulated by consumers across the food web, including, ultimately, humans who eat contaminated fish or shellfish.
B. Various reactions of mercury are mediated by bacteria. Bacteria may modify the metal through lithotrophy or use it to oxidize H_2S. Methylation of ionic mercury (Hg^{2+}) is a common response, probably to detoxify it for the bacterium. Unfortunately, methylmercury (CH_3Hg^+) is the most hazardous form for consumers because it sticks within sediment and dissolves in membrane lipids. Addition of a second methyl group forms dimethylmercury [$(CH_3)_2Hg$], a volatile, highly toxic compound.

Uranium contamination. Increasingly, bacteria are being harnessed for metal removal and detoxification technologies. An example is the use of microbes to reduce radioactive uranium to an insoluble form, which is easier to remove from contaminated water or soil, such as that of abandoned uranium mines (**Fig. 22.21**). A strain of *Desulfovibrio desulfuricans* reduces U(VI) to U(IV), an insoluble form. *Desulfovibrio* is better known for anaerobic H_2S generation associated with corrosion of sewer pipes and oil wells (discussed earlier). For metal removal, however, the organism can use U(VI) oxide for anaerobic respiration, with an electron donor such as acetate or sugar. By pumping acetate or crude sugars such as molasses into contaminated soil, U(VI) is reduced to insoluble U(IV), which is immobilized in the soil and prevented from migrating further into the water table.

A. *Desulfovibrio desulfuricans*

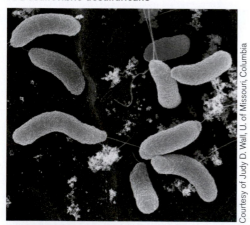

Courtesy of Judy D. Wall, U. of Missouri, Columbia

B. Uranium mine

Kevin Moloney/The New York Times/Redux

Figure 22.21 **Bacteria remove uranium contamination.** **A.** *Desulfovibrio desulfuricans* reduces U(VI) with an organic electron donor (cell length, 1–2 μm; SEM). Because the resulting U(IV) is insoluble, bacterial reduction can remove uranium from contaminated wastewater. **B.** A uranium mine in Moab, Utah, is 200 times more radioactive than the public exposure limit. Such sites may someday be decontaminated by bacteria.

TO SUMMARIZE:

- **A major source of sulfur is marine sulfate.**
- **Oxidized and reduced forms of sulfur are cycled in ecosystems.** Sulfate and sulfite serve as electron acceptors for respiration. Hydrogen sulfide serves as an electron donor. Sulfur cycling participates in acid mine drainage and pipe erosion.
- **Phosphorus cycles primarily in the fully oxidized form (phosphate).** Phosphate limits growth of algae in some aquatic and marine systems.
- **Iron cycles in oxidized and reduced forms.** Oxidized iron (Fe^{3+}) serves as a terminal electron acceptor in anaerobic soil and water. Reduced iron (Fe^{2+}) from rock is oxidized through weathering or mining. Bacterial lithotrophy accelerates iron oxidation, leading to acidification.
- **Metal toxins such as mercury and uranium can be metabolized by bacteria.** Bacterial metabolism may increase toxicity (as in the case of mercury), or it may convert the metal to an oxidation state with changed solubility, enhancing decontamination.

22.6 Astrobiology

As we come to appreciate the ubiquitous contributions of microbes to shaping our planet, in all its diverse habitats, increasingly we wonder whether microbes exist on worlds beyond Earth. Is Earth unique in supporting life; or have living cells evolved as well on Mars or Venus or on Jupiter's planet-sized moons? **Astrobiology** is the study of life in the universe, including its origin and possible existence outside Earth. The discovery of life beyond Earth would arguably be the most significant advance in science since a human set foot on the moon.

If the same physical and chemical laws govern the universe everywhere, then it is hard to suppose that only one of the billions of stars would have a planet supporting life. On the other hand, we have no idea of the number of planets and/or planetary histories that are or were capable of developing and sustaining a biosphere. As Isaac Asimov said, "There are two possibilities. Maybe we're alone. Maybe we're not. Both are equally frightening."

If life exists elsewhere, is it built on the same fundamental elements as ours? Many lines of evidence suggest that the biochemistry of life elsewhere would resemble that of Earth. Terrestrial life is founded on macroelements in the first two rows of the periodic table, including carbon, nitrogen, oxygen, phosphorus, and sulfur (see **Table 22.1**). The valence numbers of these elements from the middle of the periodic table enable them to form complex molecular structures with strong covalent bonds. The same fundamental molecules that appear in early-Earth simulation experiments, such as adenine and glycine, also appear in meteorites. Thus, we suspect that the fundamental building blocks for biochemistry are universal. On the other hand, if life could indeed be founded on some other basis, how would we recognize it?

If life is found on other planets, it will almost certainly include microbes. In fact, a case can be made that the majority of biospheres in our galaxy would consist entirely of microbial life, since microbes inhabit a wider range of conditions than multicellular plants and animals. Even on Earth itself, the largest bulk of our biosphere—including deep sediments and rock strata—consists of microbial ecosystems.

Did Mars Once Support a Biosphere?

For several reasons, the most studied candidate for extraterrestrial life has been the planet Mars.

- **Geology.** Of all the solar planets, Mars seems the most similar to Earth in its topography; indeed, some areas of Mars remarkably resemble desertscapes on Earth. Martian rock contains the fundamental elements needed for life.
- **Day and year length.** Mars has a day length similar to that of Earth, and a year only twice as long as ours.
- **Temperature.** The average temperature on Mars is 220 K (–53°C), too cold for most biochemistry on Earth, but its temperature rises above freezing at the equator. By contrast, the torrid heat of Venus (460°C) would exclude stable macromolecules.
- **Atmosphere.** The overall atmospheric pressure on Mars, 6 mbar, is barely a hundredth that of Earth (1,013 mbar) and lacks molecular oxygen. Thus, aerobes could not grow. But the Martian atmosphere does include carbon dioxide, actually at 20 times the CO_2 content on Earth, so there would be plenty for photoautotrophic production of biomass.
- **Water.** No liquid water exists at the Martian surface, but evidence supports the existence of water in the past. Liquid water may yet exist deep underground, supporting life-forms similar to the endoliths of Earth's crustal rock.

The existence of liquid water is a key question because on Earth, wherever liquid water exists, even brine (concentrated salt) at –20°C, there we find microbial life. Evidence for liquid water, past or present, was a priority for NASA's highly successful Mars Rover mission of 2004. The rovers obtained several types of mineral deposits that on Earth are formed only through crystallization from water, such as jarosite, a form of iron sulfate hydroxide. Another mineral found was smooth spherical deposits of hematite (Fe_2O_3, iron oxide) about 2 mm in diameter, nicknamed "blueberries." Again, these minerals are believed to arise only out of water solution.

Other evidence for water on Mars comes from geological formations surveyed by the Mars Reconnaissance

Figure 22.22   Evidence of past water on Mars. Sedimentary deposits in one of Mars's deepest canyons, Candor Chasma, mapped by the Mars Reconnaissance Orbiter in 2007. The layered patterns are best explained by a model involving fluid flow, such as the flow of water. *Source:* From C. H. Okubo and Alfred S. McEwen. 2007. *Science* 315:983.

Orbiter in 2007. The Orbiter mapped sedimentary deposits in one of Mars's deepest canyons, Candor Chasma (**Fig. 22.22**). The layered patterns are best explained by a model involving fluid flow, such as the flow of water.

If life once existed on Mars, what became of it? Two main possibilities are considered:

- **Life developed and existed until the planet froze.** Under this scenario, life originated much as it did on Earth, during a time of heavy bombardment from space and outgassing of nitrogen and carbon dioxide. Unfortunately, however, life failed to generate sufficient atmosphere for a greenhouse effect to sustain temperate conditions. No oxygenic organisms produced molecular oxygen and an ozone layer.
- **Life developed and still exists underground.** In the absence of an ozone layer, the Martian surface is sterilized by cosmic radiation. Nevertheless, microbes may yet exist deep underground, similar to the endolithic prokaryotes on Earth. Endoliths metabolize within interstitial water and are protected from cosmic radiation.

Do Biosignatures Indicate Life?

If microbes do exist on Mars, they should be detectable. But the detection of unknown life, even on Earth, remains a challenge. The advent of PCR sequence detection of species based on ribosomal genes has revealed thousands of unknown species, most of which cannot be grown or recognized by other methods. Other species may well be missed if their rRNA sequences fail to amplify with our probes. On Mars, assuming life evolved independently of Earth life-forms, we would have no way to define sequence probes for detection.

Instead, researchers try to define **biosignatures**, chemical and physical signs that could only have been created by life. Most of the proposed biosignatures are based on types of evidence for life on Earth, either as fossils of ancient life (discussed in Chapter 17) or signs of current life in extreme habitats. Proposed biosignatures include the following:

- **Microfossils.** The mineralization of microbial cells leads to formation of structures that can be visualized under a microcope. The cell fossils must be sufficiently distinct to establish that no abiotic process could have formed them.
- **Isotope ratios.** Certain biochemical reactions preferentially use one isotope of an atom over another; for example, rubisco, the key enzyme of carbon dioxide fixation, uses ^{12}C in preference to ^{13}C. Thus, photosynthetic use of ^{12}C can decrease the ratio of $^{12}C/^{13}C$ in subsequent carbonate deposits. Isotope ratios of nitrogen, oxygen, and sulfur are also used as biosignatures.
- **Mineral deposits.** Certain mineral formations are only observed to be caused by microbial activity. For example, insoluble manganese oxides such as Mn_2O_3 are almost always the result of microbial oxidation of reduced manganese. Reduced manganese ions are very stable, and their abiotic oxidation rate is extremely slow.
- **Metabolic activity.** Samples of soil can be incubated with radioactive tracer substances such as $^{14}CO_2$ and tested for metabolic conversion or incorporation into biomass. It must be established that the conversion could not have occurred abiotically and that no organisms from Earth were present.

Various kinds of evidence for life on Mars have been reported, such as possible microfossils within a Martian meteorite that landed in Antarctica. Metabolic activity was tested in samples obtained by the NASA Viking Lander in 1975. As of this writing, however, no evidence has proved conclusive for active or past living microorganisms on Mars.

Could Mars Be Terraformed to Support Earth Life?

If no life exists—or if it exists only in the form of microbes deep underground—should we consider human intervention, or **terraforming**, to make Mars habitable for life from Earth? Scenarios for terraforming, long explored in science fiction, are now receiving serious thought among space scientists. Terraforming Mars would require increasing the temperature and air pressure. The temperature of the atmosphere might be increased by release of greenhouse gases. In addition, sufficient carbon dioxide and nitrogen gases must be made available from the Mars surface rock. In principle, microbes from Earth could be seeded to grow and generate an atmosphere containing nitrogen, oxygen, and CO_2.

In favor of terraforming, it is argued that Mars offers enormous natural resources of potential benefit for humanity, especially as terrestrial resources are used up. Human settlements on Mars would be a major step forward for space exploration. On the other hand, it is argued that the planet Mars is a natural monument, a place with its own right to exist as such. It should be allowed to remain in its natural state for future generations to appreciate. As a practical matter, terraforming remains unfeasible for the near future. For example, the amount of chlorofluorocarbons calculated to raise Martian temperature is 100 times greater than our global capacity to produce such substances.

Does Europa Have an Ocean?

Farther out in the solar system, surprising candidates for microbial life are the moons of Jupiter. In 2000, the Galileo space probe passed several of Jupiter's moons, including Ganymede, Callisto, and Europa (**Fig. 22.23**). While their distance from the sun results in extreme cold, these bodies receive extra heat from friction generated by tidal forces from the giant planet Jupiter. In the case of Europa, the tidal forces have been calculated to provide enough heat to liquefy water without boiling it off. Furthermore, measurement of Europa's magnetic field suggests that its composition includes a dense iron core surrounded by 15% water. Most of the water must be locked in ice, but tidal heating could melt enough water for an underlying ocean of brine. On Earth, similar brine lakes beneath the ice of Antarctica harbor halophilic archaea that grow at −20°C.

If such oceans do exist, how could they support life without photosynthesis? Photosynthesis is impossible at Jupiter's distance from the sun. However, an alternative source of chemical energy might be the influx of charged particles accelerated in Jupiter's magnetic field. Charged particles entering Europa's ice can react with water to form hydrogen peroxide (H_2O_2). The H_2O_2 then breaks down, releasing molecular oxygen. Oxygen reaching the brine layer below could combine with electron donors from crustal vents and power metabolism. Alternatively, molecular hydrogen (H_2) could be generated from water ionized by decay of radioisotopes. A similar source of H_2

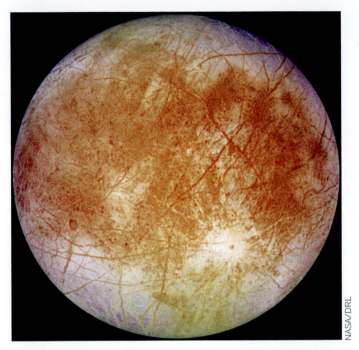

Figure 22.23   Jupiter's moon Europa.   Does life exist beneath Europa's ice, in a salty sea?

Figure 22.24   A stellar nursery.   Nebula RCW-49, in the Milky Way, contains more than 300 newborn stars, from which NASA's Spitzer Space Telescope detected spectroscopic signals of common organic constituents of life (infrared photograph).

for life has been proposed to occur on Earth in rock strata several kilometers below the surface, where it may support lithotrophs. The hydrogen could then combine with oxygen gas or other oxidants to power life. These schemes are highly speculative—but tantalizing enough to encourage further NASA missions to take a closer look at Europa and its sister moons.

Does Life Exist on Planets of Distant Stars?

The past decade of astronomy has seen an extraordinary growth in our knowledge of solar systems beyond our own. Now that we know so many other stars possess planets, we can only wonder whether they, too, possess biospheres of life-forms we would recognize.

Figure 22.24 shows an infrared photograph through the Spitzer Telescope of nebula RCW-49, a gaseous cloud full of newborn stars. Recall from Chapter 17 that as stars form, they take up dust of supernovae that includes all the elements needed to form biomolecules. Spectroscopic observation of RCW-49 reveals stars surrounded by disks coalescing into planets. The planetary disks contain icy particles full of organic molecules such as methanol, glycine, and ethylene glycol, a reduced form of sugar. Could there be biospheres in the making?

TO SUMMARIZE:

- **Astrobiology is the study of life in the universe including possible habitats outside Earth.**

- **The search for extraterrestrial life is based on methods similar to those used to seek early life on Earth.** Methods include chemical and physical biosignatures, isotope ratios, microfossils, and metabolic activity.
- **Mars is the planet whose geology most closely resembles that of Earth.** Geological features strongly support the past existence of flowing water, a prerequisite for microbial life.
- **Jupiter's moon Europa is proposed as another possible site for life.** Europa is bathed in a sea of brine similar to terrestrial habitats for halophiles.

Concluding Thoughts

Our observations of distant molecules, as well as our analysis of meteorites (discussed in Chapter 17), suggest that the biomolecules of Earth fit into a universal pattern of interstellar chemistry. Whether life exists elsewhere or we are alone, we must remember that our Earth is the only place we know of at this time that can support humans and the forms of life we require for our own survival. The function of our entire biosphere depends on our microbial partners to cycle key elements and acquire energy to drive the food web. For the first time in history, human technology now rivals the ability of microbes to alter fundamental cycles of biogeochemistry. But to manage and moderate our alterations—for our own survival and that of the biosphere—our fate still depends on the microbes.

CHAPTER REVIEW

Review Questions

1. Explain the major sources and sinks of carbon, nitrogen, and sulfur. Which sources recycle rapidly, and why?
2. Explain how the carbon cycle differs in oxygenated and anoxic environments.
3. Explain two different chemical methods of measuring the environmental levels of carbon and nitrogen.
4. Outline the hydrological cycle. Explain the role of biochemical oxygen demand (BOD) in water quality and how it may be perturbed by human pollution.
5. Outline the function of a wastewater treatment plant. Include the phases of primary, secondary, and tertiary treatment. Explain the roles of microbes in these phases of water treatment.
6. Outline the main transformations of the nitrogen cycle. Which reactions are carried out only by microbes?
7. Outline the main transformations of the sulfur cycle. Which features are comparable to the nitrogen cycle, and which are unique to the sulfur cycle?
8. For the iron cycle, explain aerobic and anaerobic processes of microbial transformation. Explain how microbial iron transformation may be linked to sulfur transformation.
9. Explain how bacteria may convert metals to toxic forms. Alternatively, explain how bacteria may detoxify metal-polluted sediment.
10. Offer several arguments for and against the existence of life on planets beyond Earth.

Key Terms

abiotic (832)
ammonification (845)
assimilatory nitrate reduction (845)
astrobiology (854)
biochemical oxygen demand (BOD) (839)
biogeochemical cycle (833)
biogeochemistry (832)
biosignature (856)
biotic (832)
carbon sink (837)
dead zone (840)

denitrification (844)
dissimilatory nitrate reduction (846)
environmental management (853)
fossil fuel (837)
geomicrobiology (832)
greenhouse effect (835)
hydrological cycle (839)
hypoxia (840)
mesocosm (834)
micronutrient (833)
nitrification (844, 845)
nitrifier (845)

nitrogen fixation (844)
preliminary treatment (841)
primary treatment (841)
secondary treatment (841)
siderophore (850)
sludge (841)
terraforming (856)
tertiary (advanced) treatment (841)
wastewater treatment (840)
wetland restoration (842)
zone of hypoxia (840)

Recommended Reading

Azam, Farooq. 1998. Microbial control of oceanic carbon flux: The plot thickens. *Science* **280:**694–696.

Boyd, P. W., Tim Jickells, Clarice S. Law, Stacy Blain, E. A. Boyle, et al. 2007. Mesoscale iron enrichment experiments 1993–2005: Synthesis and future directions. *Science* **315:**612–616.

Chyba, Christopher F., and Kevin P. Hand. 2001. Planetary science. Life without photosynthesis. *Science* **292:**2026–2027.

Dinh, Hang T., Jan Kuever, Marc Mußmann, Achim W. Hassel, Martin Stratmann, et al. 2004. Iron corrosion by novel anaerobic microorganisms. *Nature* **427:**829–832.

Houghton, Richard A., D. L. Skole, Carlos A. Nobre, J. L. Hackler, K. T. Lawrence, et al. 2000. Annual fluxes of carbon from deforestation and regrowth in the Brazilian Amazon. *Nature* **403:**301–304.

Jakosky, Bruce. 2002. *The Search for Life on Other Planets.* Cambridge University Press.

Kerr, Richard A. 1997. Life goes to extremes in the deep Earth—and elsewhere? *Science* **276:**703–704.

Maranger, Roxanne, D. F. Bird, and Neil M. Price. 1998. Iron acquisition by photosynthetic marine phytoplankton from ingested bacteria. *Nature* **396:**248–251.

Mori, Tadahiro, Tsuguhiro Nonaka, Kazue Tazaki, Makoto Koga, Yasuo Hikosaka, and S. Nota. 1992. Interactions of nutrients, moisture and pH on microbial corrosion of concrete sewer pipes. *Water Research* **26:**29–37.

Naqvi, Syed Wajih A., D. A. Jayakumar, P. V. Narvekar, H. Naik, V. V. S. S. Sarma, et al. 2000. Increased marine production of N_2O due to intensifying anoxia on the Indian continental shelf. *Nature* **408:**346–349.

Okubo, Chris H., and Alfred S. McEwen. 2007. Fracture-controlled paleo-fluid flow in Candor Chasma, Mars. *Science* **315:**983–985.

Wu, Jingfeng, William Sunda, Edward A. Boyle, and David M. Karl. 2000. Phosphate depletion in the western North Atlantic Ocean. *Science* **289:**759–762.

Part 5

Medicine and Immunology

Courtesy of Clifford W. Houston

AN INTERVIEW WITH

CLIFFORD W. HOUSTON: AN AQUATIC BACTERIUM CAUSES FATAL WOUND INFECTIONS

Clifford W. Houston has taught medical microbiology for over two decades at the University of Texas Medical Branch (UTMB). Houston developed assays to detect toxins from group A streptococci and from *Salmonella* species. He showed how *Aeromonas* bacteria acquired from waterways can fatally infect wounds in humans. A leader in microbial education, he developed innovative outreach programs for UTMB, chaired the Education Board of the American Society for Microbiology (ASM), and was elected president of ASM in 2006. As deputy associate administrator for education at the National Aeronautics and Space Administration (NASA) in 2003, he addressed the medical challenges faced by astronauts and funded programs to inspire the next generation of explorers.

Clifford Houston, professor of microbiology and immunology and associate vice president for educational outreach at the University of Texas Medical Branch at Galveston.

When did you first become interested in science?

I am a native of Oklahoma City, Oklahoma, and the first member of my family to attend college. My parents always encouraged me to obtain as much education as possible. My interest in science began when I did my first science fair project in elementary school. My teacher was very encouraging and gave me confidence that I could do the project. When I was in junior high school, I transformed the basement of my parents' home into a research laboratory.

How did you decide to make a career in microbiology?

My first undergraduate course in microbiology had a significant impact on my life. I was inspired so much by the subject that I immediately changed my undergraduate major from biology to microbiology. It was my good fortune to have the opportunity to conduct my predoctoral

studies in the laboratory of Joseph J. Ferretti, whose expertise is streptococcal genetics. In Ferretti's laboratory, I worked on the kinetics and regulation of type A streptococcal exotoxin during the growth of *Streptococcus pyogenes*. This work resulted in one of my first publications in *Infection and Immunity*. Ferretti's mentorship sustained my career interest in determining the role that bacterial toxins play in the pathogenesis of disease. I acquired my expertise in working with gram-negative bacteria from Johnny Peterson as a postdoctoral fellow in his laboratory. As a faculty member, I was honored to have Ashok Chopra as a postdoctoral fellow, who conducted much of our work on *Aeromonas*.

You developed sensitive assays for detection of toxins from group A streptococci and from *Salmonella*. What are the special challenges of assay development for pathogens?

During the mid to late 1970s, I modified the enzyme-linked immunosorbent assay (ELISA) to detect bacterial toxins that play a role in the pathogenesis of disease. The major challenges in developing this type of assay are sensitivity and specificity. I succeeded in developing an extremely sensitive assay (capable of detecting nanogram quantities), as well as an assay specific enough to avoid a high frequency of false positive reactions. The ability to detect these toxins can be used to link a pathogen to causation of human disease.

How did you choose to study *Aeromonas hydrophila*?

We were surveying the cultural filtrates of a variety of bacterial isolates in the laboratory employing an ELISA toxin assay using cholera antitoxin. To our surprise, the *Aeromonas* cultural filtrates tested positive in the ELISA, which meant that some antigen in the cultural filtrate cross-reacts with the cholera toxin.

Aeromonas is associated with wound infections and causes gastroenteritis. We found that a cytotoxic enterotoxin related to cholera toxin is associated with tissue damage in cell cultures and rabbit skin testing, as well as fluid accumulation in rabbit intestinal loops. We also showed that *Aeromonas* with mutations in the cytotoxic enterotoxin gene has diminished virulence.

Aeromonas sp. bacteria adhering to human epithelial cells. (Cell length, 2–4 μm.)

Is *Aeromonas* an opportunistic pathogen? Why are opportunistic pathogens a growing topic of concern for research and for public health?

Some species of *Aeromonas* may be pathogens, while others may be opportunists. Opportunistic microorganisms are capable of causing disease only when the conditions are favorable to the bacteria, such as when the host's immune system has been weakened as a result of stress, radiation treatment, chemotherapy, or viral diseases. Opportunistic pathogens are a growing concern because they may live in the host for years, waiting for the opportunity to cause disease.

How does a patient contract illness from *Aeromonas*?

Aeromonas hydrophila originally was thought to infect only cold-blooded animals, such as frogs, snakes, and fish. However, there are many isolated cases of *Aeromonas* infections reported in the literature resulting from individuals whose skin has been cut, abraded, and/or punctured coming in contact with water contaminated with *Aeromonas* species. One case involved a motorcycle accident on the side of a highway in which the rider received a compound fracture in one of his legs and decided to put his injured leg into nearby pond water, which unfortunately contained *Aeromonas* species. From the pond, the microorganism entered his bloodstream, and the motorcyclist died in a hospital 48 hours later.

It is possible to acquire an *Aeromonas* infection from infected fish. For example, people have been infected by contaminated fish hooks or by handling infected fish when their hands that have cuts or abrasions. In addition, humans can get gastroenteritis by ingestion of *Aeromonas*-contaminated food.

What are your goals as an educator in microbiology?

My major goal is to expose young precollege and college students to the field of microbiology and to direct more students toward careers in the biomedical sciences. My Educational Outreach Office offers a summer research program for high school students. The students work in the laboratory of research faculty at the University of Texas Medical Branch. The students work 40 hours per week for seven weeks during the summer, get paid a stipend, and present a poster on their work at the end of the program. More than 95% of these students go on to college.

How did microbiology fit into your work at NASA?

I served on a team of NASA experts whose task was to plan the details on returning to the moon and on to Mars. Microbiology will play an important role in our plans because NASA must be careful not to contaminate the moon or Mars with microorganisms from Earth, and vice versa. We are concerned with the impact of long-term space travel on the astronauts' immune system and their susceptibility to infection. Also, there is concern about the impact of long-term exposure to UV light in causing mutations in harmless bacteria that make them virulent or antibiotic resistant.

Do you think the astronauts will find extraterrestrial microorganisms?

Recent data indicate that there may be water on Mars, which suggests the possibility that life may exist on the planet. Water will be examined for microorganisms as a first sign of life on the planet. Plans for isolation and containment have been discussed to prevent transporting Earth microorganisms to another planet because mutations might occur during the journey and the microorganisms could become more virulent. We also need to take precautions to prevent transporting extraterrestrial microorganisms back to Earth.

What do you think are the most exciting areas for students entering microbiology today?

The most exciting areas are astrobiology and biophysics. Both fields have relevance to microbiology. Another exciting area is biodefense. This field will play a significant role in the future protection of our country against biological warfare. Finally, another important field to remember is bioethics, which will always be a compelling career choice.

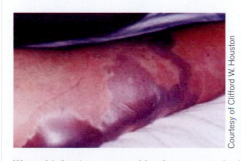
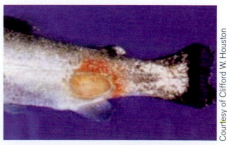

Wound infections caused by *Aeromonas hydrophila*, in a human patient (left) and in a fish (right).

Chapter 23

Human Microflora and Nonspecific Host Defenses

The human body teems with microbial hitchhikers. Most are harmless as long as they stay where they belong, such as on the skin or in the intestines, and some are even beneficial to their unsuspecting hosts. Consequently, organisms that are part of the normal human microflora typically exist in a symbiotic relationship with us. Of course, the human body is also under constant attack from microbial invaders that inhabit the surrounding environment. Fortunately, we have developed a series of barriers and elaborate fail-safe mechanisms that keep normal flora and invading pathogens at bay. Obstacles such as skin and stomach acid will repel most microorganisms; these are considered nonspecific defenses. But for those microbes able to breach these barriers, there await adaptive and nonadaptive immune defenses.

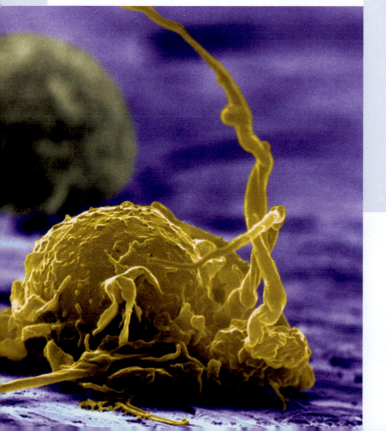

A human macrophage (10–15 μm diameter, colorized SEM). Found mainly in the liver, spleen, and lymph nodes, macrophages are a part of the immune system. Their primary function is to ingest (phagocytosis) and destroy anything they do not recognize as healthy tissue. Magnification ×7000.
Source: David M. Phillips/Photo Researchers, Inc.

Imagine two vacationing tourists, unknown to each other, each visiting the Alabama gulf coast at the same time. Both decide to take a swim in the Gulf of Mexico. One, a 12-year-old girl, has a cut on her leg caused by the sharp edge of a scooter. The other vacationer, a man 66 years of age, has recently cut his arm on a nail protruding from a wooden handrail. Within moments of entering the Gulf of Mexico, several small gram-negative microbes invade both of their bodies through those cuts. Four days later, the young girl is heading back to Birmingham in the back of the family car, oblivious to the battle recently waged in her bloodstream. At the same time, the man lies dead of an aggressive blood infection.

Both swimmers were attacked by the same pathogen, *Vibrio vulnificus*. Why did one live and the other die? As is the case for most healthy people, a variety of nonspecific, innate immune factors present in the girl's body killed the invading pathogen before it could multiply. The man, on the other hand, was an alcoholic with liver disease. He was deficient in several of those defense mechanisms—mechanisms that made the difference between life and death. What are these powerful innate factors? How can they be nonspecific, able to attack many different microbes without ever previously encountering any of them? And why do they kill the invaders and not our own cells?

Before exploring these questions of immunity and self-defense, we must first understand that each of us is a self-contained ecosystem, home to numerous microbes inhabiting all sorts of body niches. This ecological relationship helped shape our immune system. The day-to-day presence of these microbial guests also sharpens our immunity, but it is a benefit that comes with considerable risk.

23.1 Human Microflora: Location and Shifting Composition

From the moment of our birth to the time of our death, we are constantly exposed to microbes. As a result, humans—and all other higher life-forms—are heavily populated with bacteria. Colonization typically occurs where our body interfaces with the external environment (for example, mouth, skin, and parts of the genitourinary tract). The more sequestered body sites are sterile. For example, most internal organs, the blood, and the cerebrospinal fluid should not harbor any bacteria. Microbes discovered at these sites usually have deleterious effects on the host. Their presence means an infection is under way. Bacteria normally found at various nonsterile body sites (such as the intestine) are called **commensal organisms** (commensal, from the Latin "to share a table"). By strict definition, a commensal is an organism that derives benefit from the host but does not harm the host. However, many organisms originally defined as commensal were later shown to

benefit the host in relationships that are called mutualistic (see Section 17.6). The organisms, however, are still called commensal, especially in human hosts. A remarkable variety of bacterial species populate these sites, coexisting with each other in delicate, balanced ecosystems. In this section, we describe some of these organisms and why they populate different sites. **Figure 23.1** illustrates various human body sites colonized by bacteria and provides examples of the bacteria found there.

First, we will consider microbe-human interactions that take place in the host's outermost host layers—the skin and the eye—followed by the environments of the nose and throat. We will then take up the relatively commensal-free respiratory tract and the commensal-rich digestive tract. Finally, normal flora of the genitourinary tract, with its widely diverse environments, is described. **Table 23.1** lists the four prominent ecosystems of humans along with the number of microorganisms that typically inhabit them (called the **bioburden**). It includes the ratio of aerobes to anaerobes and the initial origins of resident species. Note that the composition of microflora will vary throughout life as a result of continual reseeding from the environment. Viruses are not included in this list because they are not considered commensal organisms.

Skin

The human adult, on average, is covered with 2 m^2 (over 9 ft^2) of skin (**epidermis**) populated by 10^{12} microorganisms. As with all colonized body sites, there are resident (normal) and transient microflora. But even a resident microbe exhibits diversity in that different strains colonize at different times. Thus, our microflora does not comprise a static population but represents a vibrantly changing milieu of strains and species.

Several features of epidermis make it difficult to colonize. Large expanses are subject to drying, although some harbor enough moisture to support microbial growth; these areas include scalp, ear, armpit, and genital and anal regions. The skin also has an acidic pH (pH 4–6) owing to the secretion of organic acids by oil and sweat glands. As noted in Section 5.5, organic acids inhibit microbial growth by lowering bacterial cytoplasmic pH. Epidermal secretions are also high in salt and low in water activity (see Section 5.4), and they contain enzymes such as lysozyme that degrade bacterial peptidoglycan.

Despite these hurdles, many species of bacteria manage to colonize the epidermal habitat. Most of these organisms are gram-positive because they tend to be more resistant to salt and dryness. *Staphylococcus epidermidis*, various *Bacillus* species, and yeast such as *Candida* are common examples of normal skin flora. Though normal, they are not always innocuous. One member of the skin's resident microflora, the gram-positive, anaerobic rod *Propionibacterium*

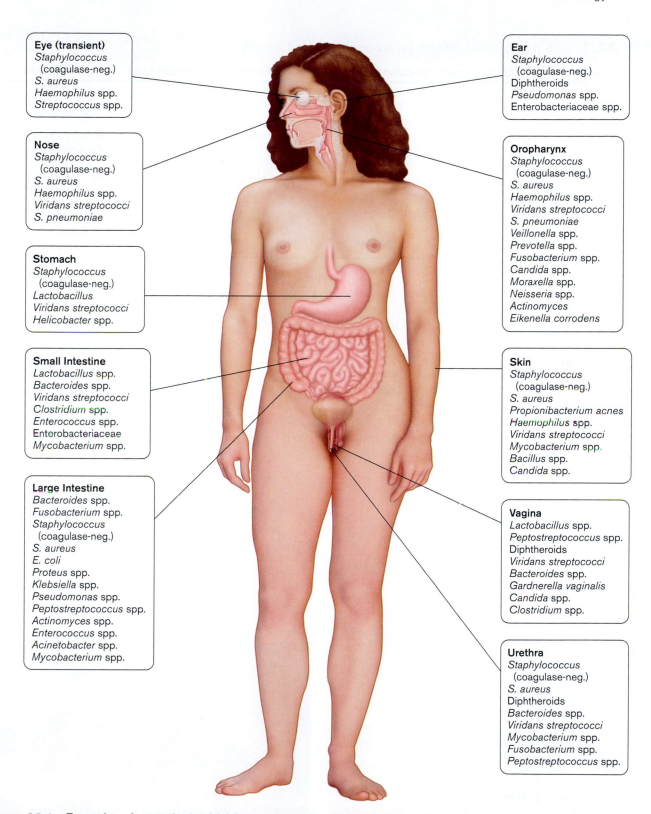

Eye (transient)
Staphylococcus
  (coagulase-neg.)
S. aureus
Haemophilus spp.
Streptococcus spp.

Nose
Staphylococcus
  (coagulase-neg.)
S. aureus
Haemophilus spp.
Viridans streptococci
S. pneumoniae

Stomach
Staphylococcus
  (coagulase-neg.)
Lactobacillus
Viridans streptococci
Helicobacter spp.

Small Intestine
Lactobacillus spp.
Bacteroides spp.
Viridans streptococci
Clostridium spp.
Enterococcus spp.
Enterobacteriaceae
Mycobacterium spp.

Large Intestine
Bacteroides spp.
Fusobacterium spp.
Staphylococcus
  (coagulase-neg.)
S. aureus
E. coli
Proteus spp.
Klebsiella spp.
Pseudomonas spp.
Peptostreptococcus spp.
Actinomyces spp.
Enterococcus spp.
Acinetobacter spp.
Mycobacterium spp.

Ear
Staphylococcus
  (coagulase-neg.)
Diphtheroids
Pseudomonas spp.
Enterobacteriaceae spp.

Oropharynx
Staphylococcus
  (coagulase-neg.)
S. aureus
Haemophilus spp.
Viridans streptococci
S. pneumoniae
Veillonella spp.
Prevotella spp.
Fusobacterium spp.
Candida spp.
Moraxella spp.
Neisseria spp.
Actinomyces
Eikenella corrodens

Skin
Staphylococcus
  (coagulase-neg.)
S. aureus
Propionibacterium acnes
Haemophilus spp.
Viridans streptococci
Mycobacterium spp.
Bacillus spp.
Candida spp.

Vagina
Lactobacillus spp.
Peptostreptococcus spp.
Diphtheroids
Viridans streptococci
Bacteroides spp.
Gardnerella vaginalis
Candida spp.
Clostridium spp.

Urethra
Staphylococcus
  (coagulase-neg.)
S. aureus
Diphtheroids
Bacteroides spp.
Viridans streptococci
Mycobacterium spp.
Fusobacterium spp.
Peptostreptococcus spp.

Figure 23.1   Examples of normal microbial flora present at various colonizing sites.   *Staphylococcus* species are differentiated by their differing abilities to produce coagulase, an enzyme that coagulates serum. *S. aureus* is generally coagulase-positive (there are some coagulase-negative mutants), whereas other species of staphylococci normally found on skin are coagulase-negative. Some organisms, *S. pneumoniae*, for example, are pathogens but can be found as normal flora in some individuals.

Table 23.1 The prominent bacterial ecosystems of humans.

Microbial reservoirs	Total "bioburden" or colony-forming units	Aerobe-to-anaerobe ratio	How acquired
Skin	10^{4-6}/sweat gland	1:10	Birth canal Oral environments
Mouth	10^{6-8}	1:10	Birth canal Caregiver
Genitourinary tract	10^{8-9}/vagina/ureter	1:100	Surrounding external environment
Intestine	10^{11}/cm^3	1:1,000	Baby formula Mother Caregiver

acnes, causes acne, a very visible plague of adolescence. Curiously, this is the same genus (but different species) that makes Swiss cheese (see Section 16.2). Increased hormonal activity during the teen years stimulates oil production by the sebaceous glands (**Fig. 23.2**). *P. acnes* readily degrades the triglycerides in this oil, turning them into free fatty acids that then promote inflammation of the gland. One consequence of the inflammatory response is the formation of a blackhead, a plug of fluid and keratin that forms in the gland duct. The result is the typical skin eruptions of acne. Because of its microbial basis, treatments for acne include tetracycline to kill the microbe.

Eye

Because the eye is exposed to the outside environment, one might expect it to be heavily colonized. Actually, this is not the case; colonization is inhibited by the presence of antimicrobial factors such as lysozyme in the tears that continually rinse the eye surface (**conjunctiva**). Despite this, a few commensal bacteria can be found on the conjunctiva. Skin flora such as *S. epidermidis* and diphtheroids (gram-positive rods that look like clubs) as well as some gram-negative rods such as *Escherichia coli*, *Klebsiella*, and *Proteus* manage to at least temporarily make the eye their home without causing damage. Occasionally, bacteria such as *Streptococcus pneumoniae* and *Haemophilus influenzae*, as well as some viruses, can cause an ocular disease known as pinkeye, in which the eye becomes reddened with a watery discharge.

Oral and Nasal Cavities

Within hours of birth, a human infant's mouth becomes colonized with nonpathogenic *Neisseria* species (gram-negative cocci), *Streptococcus*, *Actinomyces*, *Lactobacillus* (all gram-positive), and some yeasts. These organisms come from the environment surrounding the newborn, such as

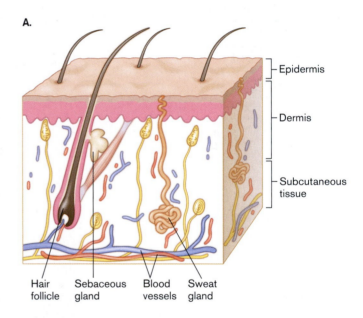

A.

Epidermis

Dermis

Subcutaneous tissue

Hair follicle   Sebaceous gland   Blood vessels   Sweat gland

B.

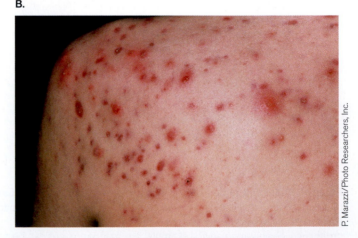

P. Marazzi/Photo Researchers, Inc.

Figure 23.2 **Microbiology of skin and development of acne.** **A.** Location of sebaceous glands in skin. **B.** Acne.

the mother's skin and garments. As teeth emerge in the newborn, the anaerobic nature of the space between teeth and gums supports the growth of anaerobes such as *Prevotella* and *Fusobacterium* (**Fig. 23.3**). Whatever the organism, colonizers of the oral cavity must be able to adhere to surfaces, like teeth and gums, to avoid mechanical removal and flushing into the acidic stomach. The teeth and gingival crevices are colonized by over 500 species of bacteria.

Organisms like *Streptococcus mutans*, which attaches to tooth enamel, and *Streptococcus salivarius*, which binds gingival surfaces, form a glycocalyx that enables them to firmly adhere to oral surfaces and to each other. They are but two of the microbes that lead to dental plaque formation. The acidic fermentation products of these organisms cause demineralization of teeth and dental caries (that is, tooth decay).

Important microbial habitats of the throat include the **nasopharynx**, which is the area leading from the nose to the oral cavity, and the **oropharynx**, which lies between the soft palate and the upper edge of the epiglottis (see **Fig. 23.3A**). Organisms like *Staphylococcus aureus* and *S. epidermidis* populate these sites. The nasopharynx and oropharynx can also harbor relatively harmless streptococci such as *S. salivarius* and *S. oralis*, as well as *S. mutans*, a major cause of tooth decay. Other oropharyngeal organisms include a large number of diphtheroids and the small gram-negative rod *Moraxella catarrhalis*. Within the tonsillar crypts (small pits along the tonsil surface) lie anaerobic species such as *Prevotella*, *Porphyromonas*, and *Fusobacterium*.

Even though the oral microflora is normally harmless, it does hold potential for disease. Dental procedures, for instance, will often cause these organisms to enter the bloodstream, producing what is called **bacteremia** (infection of the bloodstream). Normal immune mechanisms typically clear these transient bacteremias quite easily; but in patients who have a heart mitral valve prolapse (heart murmur), microbes can become trapped in the defective valve and form bacterial vegetations. Vegetations are formed by a large number of bacterial cells encased within glycocalyx, a polysaccharide or peptide polymer secreted by the organism, and fibrin, produced by clotting blood. Because the onset of disease is often insidious (slow), it is called subacute bacterial endocarditis (**Fig. 23.4**). Once ensconced within a vegetation, the microbes are extremely difficult to kill with antibiotics.

THOUGHT QUESTION 23.1 How can an anaerobic microorganism grow on skin or in the mouth, both of which are exposed to air?

THOUGHT QUESTION 23.2 What can a person with mitral valve prolapse do to prevent formation of subacute bacterial endocarditis when visiting the dentist?

A. Oral and nasal cavities

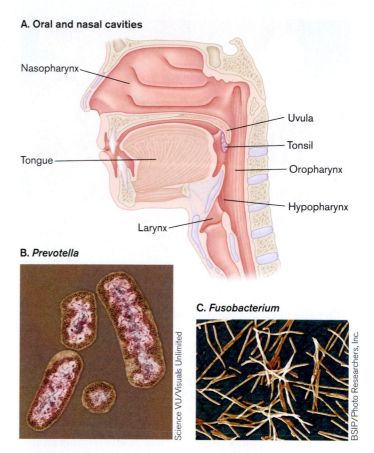

B. *Prevotella*

C. *Fusobacterium*

Science VU/Visuals Unlimited

BSIP/Photo Researchers, Inc.

D. Periodontal disease

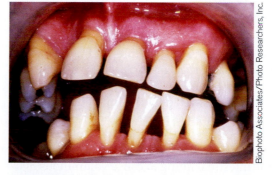

Biophoto Associates/Photo Researchers, Inc.

Figure 23.3 Examples of anaerobic oral flora and periodontal disease. A. Structures of the oral and nasal cavities. **B.** *Prevotella* (colorized TEM, each cell approx. 2 μm long). **C.** *Fusobacterium* (SEM, each cell approx. 1 to 5 μm long). **D.** Symptoms of periodontal disease include red swollen gums, bleeding gums, gum shrinkage, and teeth drifting apart.

The Respiratory Tract

The lungs and trachea do not harbor normal microflora. Many organisms entering the nasopharynx are trapped in the nose by cilia that beat toward the pharynx. Their ultimate destination is the acidic stomach and death. Microorganisms that make it into the trachea are trapped by mucus produced by ciliated epithelial cells that line the airways leading to the lung, and the cilia themselves

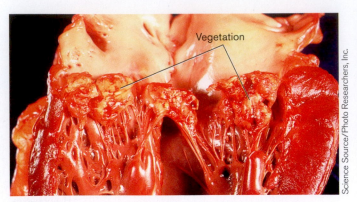

Figure 23.4 **Gross pathology of subacute bacterial endocarditis involving the mitral valve.** The left ventricle of the heart has been opened to show mitral valve fibrin vegetations due to infection. These growths are not present in a normal heart.

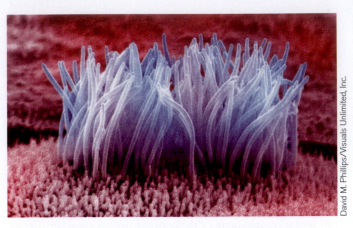

Figure 23.5 **Mucociliary elevator.** Movement of these hairlike cilia ushers particles up and out of the trachea and lungs (colorized SEM, diameter between 0.5 to 1 μm).

usher the microbes up and away from the lungs. The ciliated mucous lining of the trachea, bronchi, and bronchioles makes up the **mucociliary elevator** (**Fig. 23.5**). The mucociliary elevator constantly sweeps foreign particles up and out of the lung. This is extremely important for preventing respiratory infections. When the mucociliary elevator fails or is overwhelmed by inhaling too many infectious microbes, infections such as the common cold (for example, rhinovirus) or pneumonia (for example, *Streptococcus pneumoniae*) can result.

Stomach

We have known for a hundred years that the stomach contents are acidic and that gastric acidity can kill bacteria. Just how important that acidity is for protection against microbes is illustrated by the infection caused by *Vibrio cholerae,* the causative agent of cholera. Cholera is a severe diarrheal disease endemic to many of the poorer countries of the world. The toxin produced by the organism acts on the intestinal lining to cause voluminous diarrhea—as much as 10 liters a day. Death can occur rapidly as a result of dehydration and shock. Although cholera actually affects the intestines, not the stomach, the bacteria must survive passage through the stomach to reach the intestines.

In the middle of the last century, the U.S. government was concerned that troops dispatched to endemic countries could develop cholera and become incapacitated in large numbers, so the U.S. Army decided to test potential cholera vaccines for their efficacy. The vaccines were living cultures of genetically weakened cholera bacteria that could not cause serious disease. Living strains were preferred because they were expected to grow in the gastrointestinal tract and stimulate natural mucosal immunity (discussed later and in Section 24.3). But after adminis-

tration of huge numbers of virulent organisms to healthy volunteers (as a control group), very few contracted cholera. This was confusing because hundreds of thousands of malnourished people in India easily contract the disease.

Why did volunteers given live cholera generally fail to develop the disease? Variations in resistance to cholera were discovered to be related to differences in stomach acidity. The organism *V. cholerae* is extremely sensitive to low pH; even a pH of 4 will readily kill it. Because the pH in the stomachs of healthy volunteers was well below 4, it easily killed *V. cholerae*. Malnourished people, however, suffer from achlorhydria (loss of stomach acid). This permissive environment gives ingested microbes time to enter the less acidic intestine, where they can thrive and cause disease. Cholera epidemics in which thousands of people die, even today, typically occur among poor, malnourished populations. Because the U.S. populace is better nourished, cholera is not a major problem in the United States.

Although the contents of the stomach are very acidic, the mucous lining of the stomach is much less so. It is there that some bacteria can take refuge. In fact, the stomach harbors a diverse microflora (as detected using cultural and molecular techniques such as polymerase chain reaction (PCR) identification of 16S rDNA). A classic stomach pathogen is *Helicobacter pylori*. Although this organism has a remarkable ability to resist acid pH (it survives at pH 1 using the enzyme urease to generate ammonia), it will not grow at that pH. It can, however, grow and divide in the mucous lining of the stomach, where the pH is closer to 5 or 6 (**Fig. 23.6**). The U.S. Centers for Disease Control and Prevention (CDC) estimates that *Helicobacter* colonizes the stomachs of half the world's population. Most of the time, *Helicobacter* does not cause any apparent problem, but on occasion, the organism can produce gastric ulcers and even cancer.

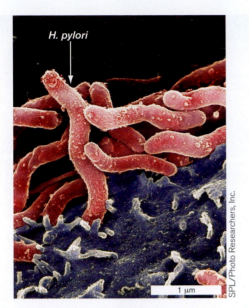

Figure 23.6 *Helicobacter pylori* **defies stomach acidity.** *Helicobacter pylori* (colorized SEM) growing in the mucus on stomach epithelium. *Helicobacter pylori* bacteria attach to gastric epithelial cells and induce specific changes in cellular function, such as an increase in the expression of laminin receptor 1, a protein associated with malignancy.

The Human Intestine

The gastrointestinal tract beyond the stomach consists of an extremely long tube made of several sections, each of which provides a uniquely different environment that supports the growth of different bacterial species (see **Fig. 23.1**). Pancreatic secretions (at pH 10) enter the intestine at a point just past the stomach and raise the intestinal pH to about pH 8, which immediately relieves the acid stress placed on microbes. The relatively high pH and bile content of the duodenum and jejunum allow colonization by only a few resident, mostly gram-positive, bacteria (enterococci, lactobacilli, and diphtheroids). These particular gram-positive organisms possess a bile salt hydrolase that helps them grow in the presence of intestinal bile salts. The distal parts of the human intestine (ileum and colon) have a slightly acidic pH (pH 5–7) and a lower concentration of bile salts, conditions that support a more vibrant ecosystem. The intestine actually contains roughly 10^9–10^{11} bacteria per gram of feces (**Fig. 23.7**). This generally anaerobic environment is populated by both anaerobic and facultative microbes in a ratio of 1,000 anaerobes to one facultative organism. Why is the intestine anaerobic? The small amount of oxygen that diffuses from the intestinal wall into the lumen is immediately consumed by the facultative microbes, which render the environment anaerobic.

One such anaerobe, *Bacteroides thetaiotaomicron*, provides us with a great benefit. Much of this organism's genome is dedicated to taking many of the complex car-

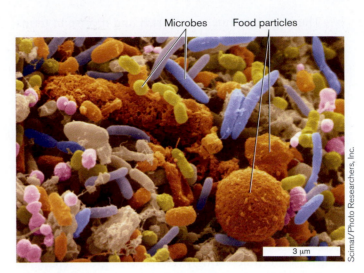

Figure 23.7 Human feces showing food particles and resident bacteria (colorized SEM).

bohydrates we eat and breaking them down into products that can be absorbed by the body. We humans actually absorb 15–20% of our daily caloric intake in this way.

Only recently have we really begun to ascertain what microbes inhabit our intestines (mainly because we cannot culture them). Through the use of culture-independent techniques (like ribotyping), we now have a sense of the identity, diversity, and variation among individuals of the human microflora. For example, of 395 unique "phylotypes" recently identified in humans, 60% were unknown at the species level, 80% have never been cultured, and huge differences in the population of microbes were observed between individuals.

All resident organisms normally present in the intestine are innocuous and considered **normal flora**. But if they accidentally escape into nearby tissues, they can cause serious disease. Intestinal bacteria, encompassing 400–500 different species, have special talents that allow them to bind and colonize the intestinal cell wall. One reason why such a large and eclectic mix of species is supported is that different bacteria attach to different host cell receptors. Another reason is that many different food sources are available to support diverse groups of microbes.

Organisms that inhabit the intestine include the family Enterobacteriaceae, of which *E. coli* is a member, and many anaerobes such as *Bacteroides* (gram-negative rods), *Peptostreptococcus* (gram-positive cocci), and *Clostridium* (gram-positive, spore-forming rods). In fact, the vast majority of our intestinal flora consists of the phyla Bacteroidetes and Firmicutes, of which *Bacteroides* species and *Clostridium* species, respectively, are members. Besides bacteria, other common and innocuous inhabitants are the yeast *Candida albicans* and protozoa such as *Trichomonas hominis* and *Entamoeba hartmanni*.

The large intestine is both rich and diverse in terms of nutrient availability. The majority of intestinal bacteria require a fermentable carbohydrate for growth, so it has generally been assumed that carbohydrate metabolism is necessary for colonization by most species. In the large intestine, carbohydrates are derived from partially digested food and from host secretions. Naturally occurring sugar acids, such as galacturonate from pectin and gluconate and ketogluconate from muscle tissues, are present in the foods we eat. The mucous layer covering epithelial tissues is also recognized as an important source of carbohydrates in the intestine. Mucus is a complex gel of glycoproteins and glycolipids. The sugar substituents of mucus include N-acetylglucosamine, N-acetylgalactosamine, galactose, fucose, sialic acids, and lesser amounts of glucuronate and galacturonate. Interestingly, *E. coli* will grow rapidly in mucus by catabolizing gluconate, a component of secreted mucus, using the Entner-Doudoroff pathway (see Section 13.5). The organism grows very poorly in feces itself. Thus, *E. coli* doesn't grow on what we *eat* but on what we *secrete*.

Despite this buffet-like assortment of food, many species still end up vying for the same nutrient, so competition for food is fierce. Perturbing the balance of power with antibiotics, diet, or stress can lead to disease.

Although we have considerable knowledge of the gastrointestinal microflora, there remains much to learn. For example, many of the species that inhabit the intestine have never been cultured in the laboratory. Molecular techniques such as PCR are now being used to identify the nonculturable organisms in this environment.

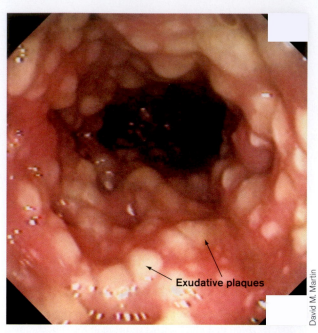

Figure 23.8 Pseudomembranous enterocolitis caused by *Clostridium difficile*. The organism is often a part of the normal intestinal microflora, but antibiotic treatment, especially following surgery, can kill off competing microbes, leaving *C. difficile* to grow unabated. A toxin produced by the organism growing at the epithelial surface damages and kills host cells, causing exudative plaques to form on the intestinal wall. The small plaques eventually coalesce to form a large pseudomembrane that can slough off into the intestinal contents.

Exudative plaques

David M. Martin

> **THOUGHT QUESTION 23.3** Why do many gram-positive microbes that grow on the skin, such as *S. epidermidis*, grow poorly or not at all in the gut?
>
> **THOUGHT QUESTION 23.4** How might normal flora escape the intestine and cause disease at other body sites?

Probiotics. As already noted, the microbial composition of the intestine is complex but balanced. It is complex because there are many species residing there. It is balanced because these species manage to coexist without eliminating each other. The rival organisms help prevent infection by competing with pathogenic species for nutrients and can actually contribute to nutrition (for example, *E. coli* makes vitamin B$_{12}$, whereas other organisms help digest complex substrates). Emotional stress, a change in diet, or antibiotic therapy can throw off that balance and lead to poor digestion or disease, such as pseudomembranous enterocolitis (**Fig. 23.8**). Some people favor restoring the natural microbial balance by orally ingesting living microbes such as the lactobacilli present in yogurt. Taking these supplements, called **probiotics** (from the Greek meaning "for life"), is thought to restore balance to the microbial community and return the host to good health. The most commonly used probiotic genera are *Lactobacillus* and *Bifidobacterium* (**Fig. 23.9**). The potential mechanisms by which they improve intestinal health include competitive bacterial interactions (normal flora prevent growth of pathogenic bacteria), production of antimicrobial compounds, and immune modulation (an ability to change the activity of cells of the immune system). The emerging use of probiotics in several gastrointestinal disorders (including inflammatory bowel disease, IBS) has led to increased interest in their use in patients. Early results are promising, but much of the research remains controversial.

> **THOUGHT QUESTION 23.5** Why can the colon be considered a fermenter?

Genitourinary Tract

A significant portion of the genitourinary tract is usually free from microbes. These areas include the kidneys, which remove waste products from the blood; ureters, which remove the urine from the kidneys; and the urinary bladder, which holds urine until it is excreted. The distal

Figure 23.9 Commonly used probiotic microorganisms. Colorized SEMs of **A.** *Bifidobacterium* (note the "Y" shape of some cells) and **B.** *Lactobacillus acidophilus.*

A.

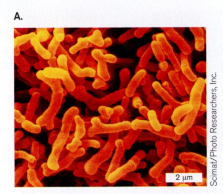

B.

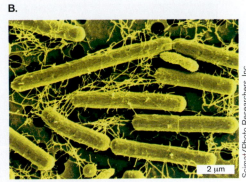

urethra, however, because of its proximity to the outside world, normally contains *S. epidermidis, Enterococcus* species, and some members of Enterobacteriaceae. Some of these organisms, if they can make their way into the bladder (for example, via catheterization), will cause bladder disease (known as **urinary tract infection**, or **UTI**).

The large surface area and associated secretions of the female genital tract make it a rich environment for microbes. Composition of the vaginal microflora changes with the menstrual cycle, owing to changing nutrients and pH. The mildly acidic nature of vaginal secretions (approximately pH 4.5) discourages the growth of many microbes. As a result, the acid-tolerant *Lactobacillus acidophilus* is among the most populous vaginal species. As with all microbial ecosystems associated with the human body, the balance between competing species is crucial to preventing infection. Antibiotic therapy to treat an infection can also cause the loss of normal flora. In the vagina, this allows infection by *C. albicans* (yeast infection), which is not susceptible to antibiotics designed to kill bacteria.

TO SUMMARIZE:

- The **normal microflora** present on body surfaces constantly changes.
- **Normal flora can cause disease** if microbes breach the surfaces they colonize and gain access to the circulation or deeper tissues.
- **Colonization of skin** by microbes is difficult owing to surface dryness, an acidic pH, high salinity, and the presence of degradative enzymes.
- **Skin flora** consists primarily of gram-positive microbes, including *Propionibacterium acnes*, which can cause acne.
- **Microbe-free areas of the body** are the lungs, cerebrospinal fluid, and bladder. The eyes are generally microbe-free, although some microbes can be present for short periods.
- **Oral and nasal surfaces** are colonized by aerobic and anaerobic microbes.
- **The intestine** is populated by 10^9–10^{11} microbes per gram of feces in a ratio of approximately 1,000 anaerobes to 1 aerobe.

23.2 Risks and Benefits of Harboring Microbial Populations

It is enticing to portray colonizing bacteria as hostile armies encamped at our body's gates (also called portals of entry), just waiting for the opportunity to invade. But bacteria do not possess hostile intent, only the need to find food. The microbes of our normal microflora, unlike true pathogens, have not evolved mechanisms to specifically breach human defenses. For the most part, the needs of commensal organisms are met at the colonizing site. However, accidental penetration beyond these sites by normal flora can cause disease. A cancerous lesion in the colon, for example, can provide a passageway for microbes to enter deeper tissues, where they can begin to grow unabated. *Bacteroides fragilis*, a harmless anaerobe in the gut, can invade tissues through surgical wounds, causing abscesses that persist (and even lead to gangrene) after abdominal surgery (**Fig. 23.10**).

By and large, the barriers of defense work well to prevent incursions by normal flora or, in the event of an incursion, to kill the invader. These defenses sometimes break down in persons with an immune system compromised by medical treatments (such as anticancer drugs) or by disease (such as a deficiency in complement factors, discussed later). Such a person is described as a **compromised host** and can be repeatedly infected by normal flora. Organisms causing disease in this situation are called **opportunistic pathogens**.

Commensal species are often beneficial (mutualistic). They can interfere with the colonization of pathogens by competing for attachment receptors on host cells, competing for food sources, and synthesizing antimicrobial compounds such as fermentation end products. For example, the acidic fermentation products produced by lactobacilli in the female genitourinary tract help maintain a low pH. The low pH dissuades colonization by various pathogenic microbes.

Commensal organisms also enhance function of the immune system. Bacterial proteins such as catalase can

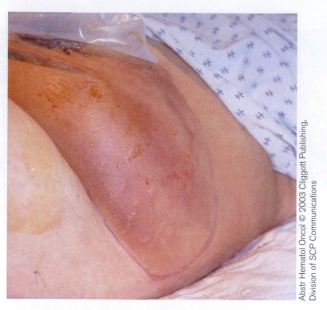

Abstr Hematol Oncol © 2003 Cliggott Publishing, Division of SCP Communications

Figure 23.10 Anaerobic gas gangrene of the abdominal wall caused by *Bacteroides*. This infection was an unfortunate result of bowel surgery. Organisms escaped from the intestine and initiated infection in the abdominal wall.

act as immunomodulins. **Immunomodulins** made by normal flora growing on mucosal surfaces modify the secretion of host proteins, such as cytokines and tumor necrosis factor, which influence the immune response. **Cytokines** are small secreted host proteins that bind to various cells of the immune system and regulate the extent and duration of responses by those cells. (Cytokines are discussed in Sections 23.5 and 24.6.)

Some toxic bacterial products are also "double-edged swords." **Enterotoxins**, for example, are proteins produced by some gram-negative pathogens that damage the small intestine of the host and cause diarrhea. But they may also protect against colorectal cancer by activating membrane calcium channels in intestinal epithelial cells. Increasing membrane conductance for calcium turns on an antiproliferative pathway that provides resistance to colon cancer.

A recently discovered benefit of commensal organisms has been their development as vaccine delivery systems. Select commensal species can be genetically engineered to produce proteins from pathogenic species on their cell surfaces. Colonization of a host by the modified commensal strain will elicit an immune response to the cloned protein and provide immunity to the pathogen. Commensal microbes such as *S. salivarius* (part of the oral and nasopharyngeal microflora), *Lactococcus* species (residents of the intestinal tract), and *Lactobacillus* species (present in the urogenital and rectal tracts) are being exploited in this way. For example, if one wished to provide protection against *Streptococcus pyogenes*, the causative agent of strep throat, an oral commensal organism could be engineered to present the conserved portion of the M protein (the streptococcal attachment pilus) to the oral mucosa. The resulting local immune response could protect the individual from strep throat—at least a strep throat caused by that M-protein-bearing strain of *S. pyogenes* (there are at least 80 different M-protein-type strains). This approach can be adapted, in theory, to virtually any pathogen that enters through or colonizes a given mucosal surface on any human or animal. It is important, however, to choose proteins from the pathogen that will not also confer pathogenicity on the commensal strain in which it is expressed.

The benefit that commensal organisms offer to their hosts is especially apparent in gnotobiotic animals. A **gnotobiotic animal** is an animal that is germ-free or one in which *all* the microbial species present are *known*. Developing a gnotobiotic colony of animals involves delivering offspring by cesarean section under aseptic conditions in an isolator. The newborn is moved to a separate isolator where all entering air, water, and food are sterilized. Once gnotobiotic animals are established, the colony is maintained by normal mating between the members. Germ-free animals, however, often have poorly developed immune systems, lower cardiac output, and thin intestinal walls, and they are more susceptible to infection by pathogens. The lesson from these animals is that the presence of normal flora continually challenges the immune system and keeps it active.

TO SUMMARIZE:

- **Opportunistic pathogens** infect only compromised hosts.
- **Benefits of commensal microbes** include interfering with pathogen colonization, the production of immunomodulatory proteins, and potential use as vaccine delivery vehicles.
- **Gnotobiotic animals** are germ-free or colonized with a known set of microbes.

23.3 Overview of the Immune System

Because we are surrounded by and host numerous bacteria, how is it we are not constantly infected? Here we begin our discussion of the many layers of protection designed to prevent infection and disease. As you will see, there are numerous physical and chemical barriers that provide an effective first line of defense against infection by invading microbes. As such, they play a critical role in managing our resident ecosystems. But these barriers are not unbreachable. Organisms can still slip through. Consequently, humans, as well as other mammals, have a more aggressive defense called the immune system. The **immune system** is an integrated system of organs, tissues, cells, and cell products that differentiates self from nonself and neutralizes potentially pathogenic organisms or substances. This complex collection of cells and soluble proteins is capable of responding to nearly any foreign molecular structure.

Immunity Is of Two Types: Innate and Adaptive

There are two broad types of immunity: **nonadaptive immunity**, often called **innate immunity**, and **adaptive immunity**. Innate and adaptive immunity have several key differences. Innate immunity includes physical barriers such as skin, chemical barriers such as stomach acid, and relatively nonspecific cellular responses to infection that engage if the physical and chemical barriers are breached. Innate immunity is essentially "hardwired" into the body. It is present at birth, so that mechanisms of innate immunity exist before the body ever encounters a microbe. The protection they afford is nonspecific, capable of blocking or attacking many different types of foreign substances and organisms.

In contrast to innate immunity, adaptive immunity is designed to react to very specific structures called **antigens**. An antigen is any chemical, compound, or structure foreign to the body that elicits an immune response. Adaptive immune responses to a specific antigen do not occur until the body "sees" that antigen. Once activated, adaptive immune mechanisms can recognize at least 10^{10} different antigenic structures and specifically launch a directed attack against each one. Once such an attack has been activated, the organism keeps a "memory" of the exposure in the form of specific memory cells, and an encounter with the microbe years later will reactivate the memory cells specific to the antigen.

The two types of immunity are illustrated by the response of the immune system to infection by the microorganism *Neisseria gonorrhoeae*, which causes the sexually transmitted disease gonorrhea. A component of the innate immune response is **complement**, comprised of several soluble protein factors constantly present in the blood. Within moments of an initial infection, complement attacks the bacterial membrane. Complement proteins form holes in bacterial membranes, thereby killing the microbe (see Section 23.8). On the other hand, the adaptive immune response will generate specific antibodies made to the cells of *N. gonorrhoeae* that escaped the innate mechanisms. These antibodies, however, are not made until well after the organism infects a person. Together, innate and adaptive immunity can help ward off disease caused by this organism. Note that contact with an infectious agent does not guarantee that a person will actually contract the disease. If the number of infecting organisms is small and the immune system (innate and adaptive) is effective, the individual may not develop disease (see chapter introduc-

tion), although a person known to have been exposed to certain microorganisms will be treated with antibiotics as a preventive measure. Chapter 26 will more fully discuss the difference between being infected and having a disease.

Any microbe that launches a successful attack on a human or other animal and causes disease must first breach the host's physical and chemical barriers to gain entrance to the body. It must then survive the innate defense mechanisms and begin to multiply. Finally, the microbe must surmount the last line of defense, adaptive immunity, which begins to respond as the microbe struggles to overcome innate immune defenses. The rest of this chapter will discuss the various innate defense mechanisms. Adaptive immunity is discussed in Chapter 24.

Although innate and adaptive immunity are often treated as separate entities, certain kinds of cells and organs play a role in both types of immunity. We will thus introduce here the various cells and organs of the immune system as a whole before focusing on innate immunity. Our coverage of innate immunity will include physical and chemical barriers to infection, inflammation and nonspecific killing through phagocytosis, interferon, natural killer cells, and complement.

Cells of the Innate and Adaptive Immune Systems

Blood is composed of red blood cells, white blood cells, and platelets (**Fig. 23.11**). The many types of white blood cells are formed by differentiation of stem cells produced in bone marrow (**Fig. 23.12**). Among these white blood cells are various components of innate immunity that

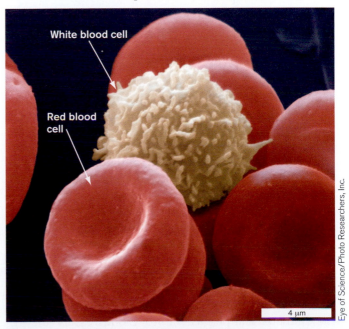

White blood cell

Red blood cell

4 μm

Eye of Science/Photo Researchers, Inc.

Figure 23.11   Red and white blood cells. This scanning electron micrograph illustrates the relative sizes and three-dimensional morphologies of these cell types.

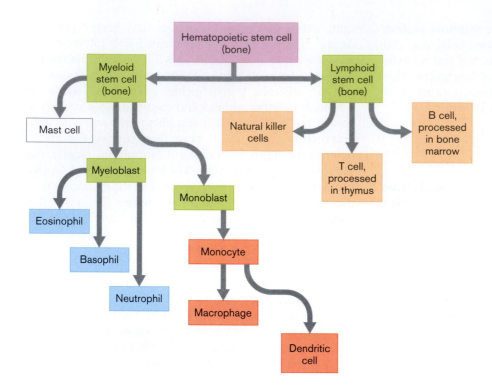

Figure 23.12   Development of white blood cell components of the immune system.   Pleuripotent stem cells in bone marrow divide to form two lineages. One lineage consists of the myeloid stem cells, which develop into polymorphonuclear leukocytes (PMNs) and monocytes. These cells primarily function as part of innate immunity. The other branch consists of the lymphoid stem cells, which ultimately form natural killer cells, B cells, and T cells. Final maturation into B cells and T cells (the principle cells involved in adaptive immunity) occurs in the bone marrow and thymus, respectively. Colors indicate a group of differentiated cells that arise from the same progenitor.

differentiated from myeloid stem cells. These differentiated cells include:

- Polymorphonuclear leukocytes (PMNs)
- Monocytes
- Macrophages
- Dendritic cells
- Mast cells

PMNs, also called granulocytes, have multilobed nuclei, differentiate from an intermediate cell called the myeloblast, and contain enzyme-rich lysosome organelles (**Fig. 23.13B**). PMNs are of several types named for their different staining characteristics. Each cell type has a different function. **Neutrophils**, making up the vast majority of white cells in the blood, can engulf microbes by phagocytosis and kill them by fusing enzyme-gorged lysosomes with the phagosomes containing the organism (**Fig. 23.14**). **Phagosomes** are the vacuoles formed around a microorganism as it is engulfed by a host cell. Enzymes spilling from the lysosome into the phagosome will destroy various components of the microbe and, ultimately, the microbe itself. **Basophils**, which stain with basic dyes, and **eosinophils**, which stain with the acidic dye eosin, do not phagocytize microbes but release products, such as major basic protein, that are toxic to the microbe (**Fig. 23.13D**). These blood cells also release chemical mediators (called vasoactive agents) that affect the diameter and permeability of blood vessels (the significance of which is discussed in Section 23.5). **Mast cells** are similar to basophils in structure but differentiate in a lineage separate from PMNs. Unlike PMNs, mast

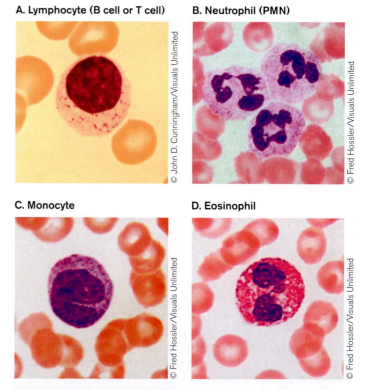

A. Lymphocyte (B cell or T cell)   **B. Neutrophil (PMN)**

C. Monocyte   **D. Eosinophil**

Figure 23.13   Types of white blood cells.   A. Lymphocyte (B cell or T cell). **B.** Neutrophil (PMN). **C.** Monocyte. **D.** Eosinophil.

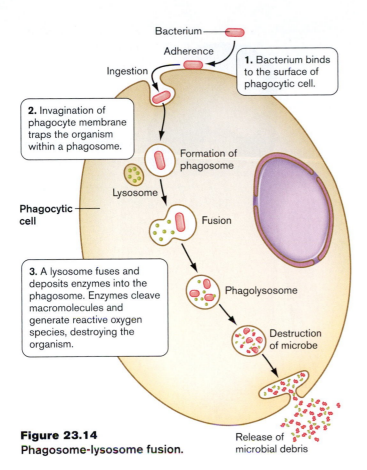

Bacterium

Adherence

1. Bacterium binds to the surface of phagocytic cell.

Ingestion

2. Invagination of phagocyte membrane traps the organism within a phagosome.

Formation of phagosome

Lysosome

Phagocytic cell

Fusion

3. A lysosome fuses and deposits enzymes into the phagosome. Enzymes cleave macromolecules and generate reactive oxygen species, destroying the organism.

Phagolysosome

Destruction of microbe

Figure 23.14
Phagosome-lysosome fusion.

Release of microbial debris

A. Macrophage

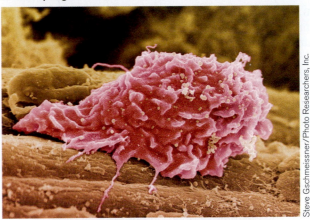

B. Dendritic cell

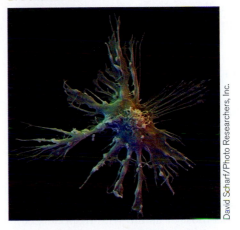

C. Macrophage engulfing bacteria

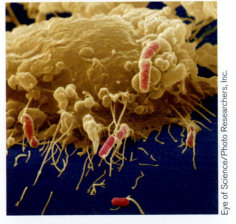

Figure 23.15 **The innate immune system depends on white blood cells called macrophages and dendritic cells.** **A.** A macrophage (20 μm long) at the site of a skin wound (colorized SEM). **B.** Dendritic cell (colorized SEM). **C.** Membrane protrusions from a macrophage detecting and engulfing bacteria (pink; *E. coli*, 1.5 μm long). (Colorized SEM.)

cells are residents of connective tissues and mucosa and do not circulate in the bloodstream. Basophils and mast cells also contain high-affinity receptors for a class of antibody called immunoglobin E (IgE) associated with allergic responses (detailed in Section 24.8).

Monocytes (**Fig. 23.13C**) are white blood cells with a single nucleus (not multilobed like a PMN); they engulf (phagocytize) foreign material. Monocytes circulating in the blood can migrate out of blood vessels into various tissues and differentiate into **macrophages** and **dendritic cells** (**Fig. 23.15**). Macrophages are phagocytic and form a major part of the amorphous **reticuloendothelial system**, which is widely distributed throughout the body. The reticuloendothelial system is a group of cells with the ability to take up and sequester inert particles. In addition to macrophages, the system is composed of specialized endothelial cells lining the sinusoids of the liver, spleen, and bone marrow as well as reticular cells of lymphatic tissue (macrophages) and of bone marrow (fibroblasts). The function of macrophages and the reticuloendothelial system is to phagocytize microorganisms. Macrophages are the cells most likely to make first contact with invading pathogens. They have two functions. As part of innate immunity, they kill invaders directly. Protrusions from the macrophage surface extend and clasp nearby bacteria, pulling them into the cell (see **Fig. 23.15C**). Once ingested by the macrophage, the bacteria are destroyed.

Subsequently, the remnants are, as the first step in adaptive immunity, processed (degraded) into smaller peptides (called antigens), which are then presented on the macrophage cell surface. Specific white blood cells called T cells (a type of lymphocyte) can bind the displayed antigens and become activated as part of the adaptive immune response. Thus, macrophages are considered **antigen-presenting cells** (see Chapter 24).

Dendritic cells are also antigen-presenting cells. Present in the spleen and lymph nodes, they, like macrophages, can take up, process, and present small antigens on their cell surface. The Chapter 24 opening photo shows the interaction between a dendritic cell and a T cell. Dendritic cells are different from macrophages in structure and in the fact that dendritic cells primarily take up small soluble antigens from their surroundings rather than through phagocytosis and degradation of whole bacteria.

Lymphoid Organs

Lymphocytes, which are the main participants in adaptive immunity, are present in blood at about 2,500 cells per microliter, accounting for over one-third of all peripheral white blood cells (**Fig. 23.13A**). However, an individual lymphocyte spends most of its life within specialized solid tissues (lymphoid organs) and enters the bloodstream only periodically, where it migrates from one place to another, surveying tissues for possible infection or foreign antigens. No more than 1% of the total lymphocyte population can be found in the blood at any one time.

The tissues of the immune system where the great majority of lymphocytes are found are classified as primary or secondary lymphoid organs or tissues, depending on their function (**Fig. 23.16**). The primary lymphoid organs and tissues are where immature lymphocytes mature into antigen-sensitive B cells and T cells. **B cells**, which ultimately produce antibodies (see Chapter 25), develop in bone marrow tissue, whereas T cells, which modulate various facets of adaptive immunity, develop in the thymus, an organ located above the heart.

The secondary lymphoid organs serve as stations where lymphocytes can encounter antigens. These encounters lead to the differentiation of B cells into antibody-secreting plasma cells, and T cells into antigen-specific helper cells, as will be discussed in Chapter 24. The spleen is an example of a secondary lymphoid organ. It is designed to filter blood directly and detect microorganisms. Macro-

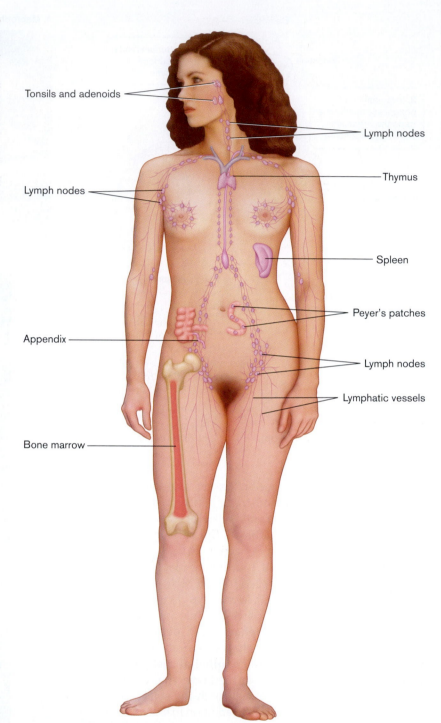

Figure 23.16 Lymphoid organs.

phages in the spleen engulf these organisms and destroy them, and then migrate to other secondary lymph organs and present pieces of the microbe (called antigens) to the B and T cells, which then become activated.

The **lymph nodes** are another kind of secondary lymphoid organ. Lymph nodes are arranged to trap organisms from local tissues, not from blood. Lymph nodes are situated at various sites in the body where lymphatic vessels converge (for example, under the armpits). There are also lymphoid tissues in the mucosal regions of the gut and respiratory tracts (for example, Peyer's patches and gut-associated lymphoid tissue or GALT, discussed later). Other secondary lymphoid organs are the tonsils, adenoids, and appendix.

TO SUMMARIZE:

- **The immune system** consists of both innate and adaptive mechanisms that recognize and eliminate pathogens.
- **Myeloid bone marrow stem cells** differentiate to form cells of the innate immune system—phagocytic PMNs, monocytes, macrophages, antigen-presenting dendritic cells, and mast cells.
- **Lymphoid stem cells** differentiate into natural killer cells (part of the innate immune system), and lymphocytes (cells of the adaptive immune system). Lymphocytes are classified as B cells, which ultimately produce antibodies, and T cells, which regulate adaptive immunity.

23.4 Barbarians at the Gate: Innate Host Defenses

When defending a castle in medieval times, the first lines of defense included physical barriers (the castle wall), chemical barriers (boiling oil tossed onto invaders trying to scale the wall), and finally hand-to-hand combat once the wall was breached. Similarly, the body's initial defenses against infectious disease are composed of physical, chemical, and cellular barriers designed to prevent a pathogen's access to host tissues. Although generally described as nonspecific, some innate defense systems are more specific than others.

Physical Barriers to Infection

The first line of defense against any potential microbial invader (either commensal or pathogenic) occurs where parts of the body interface with the environment. These interfaces (skin, lung, gastrointestinal tract, genitourinary tract, and oral cavities) have similar defense strategies, although each has unique characteristics.

Skin. Few microorganisms can penetrate skin due to the thick **keratin** armor produced by closely packed cells called **keratinocytes**. Keratin protein is a hard substance (hair and fingernails are made of this) that is not degraded by known microbial enzymes. An oily substance (sebum) produced by the sebaceous glands also covers and protects the skin. Its slightly acidic pH inhibits bacterial growth. Competition between species also limits colonization by pathogens, and microorganisms that manage to adhere are continually removed by the constant shedding of outer epithelial skin layers.

There are other, more specialized cells just under the skin that can recognize microbes managing to slip through the physical barrier. They are part of a consortium of cells called **skin-associated lymphoid tissue (SALT)**. **Langerhans cells** make up a significant portion of SALT. They are specialized dendritic cells that can phagocytize microbes. Once a lymphoid Langerhans cell has ingested a microbe, the cell migrates by ameboid movement to nearby lymph nodes and "presents" parts of the microbe to the immune system to activate antimicrobial immunity.

NOTE: Do not confuse phagocytic Langerhans cells with the pancreatic "islets of Langerhans" that secrete insulin.

Mucous membranes. Mucosal surfaces form the largest interface (200–300 m^2) between the human host and the environment (the intestine alone is 23–26 ft long). Mucous membranes in general present a containment problem. They must be selectively permeable in order to exchange nutrients as well as to export products and waste components. At the same time, they must constitute a barrier against invading pathogens. Mucosal membranes are covered with special tight-knit epithelial layers that support this barrier function. The mucus secreted from stratified squamous epithelial cells coats mucosal surfaces and traps microbes. Compounds within the mucus can serve as a food source for some commensal organisms, but other secreted compounds like the enzymes lysozyme, which cleaves cell wall peptidoglycan, and lactoperoxidase, which produces superoxide radicals, can kill an organism trapped in the mucus.

There are also semispecific innate immune mechanisms associated with mucosal surfaces. Host cells, even epithelial cells, in mucosa have evolved mechanisms to distinguish harmless compounds and commensal microorganisms from dangerous pathogens. Patterns of conserved structures on pathogenic microbes are recognized by host cell surface receptors such as various Toll-like receptors and CD14.

First discovered in insects and named Toll receptors, **Toll-like receptors (TLRs)** are evolutionarily conserved cell surface glycoproteins present on the cells of many eukaryotic genera. They are transmembrane proteins with an extracellular leucine-rich repeat domain and an intracellular Toll/interleukin 1 receptor domain (TIR domain). Humans have numerous Toll-like receptors, each of which recognizes different pathogen-associated molecular patterns (PAMPs) present on pathogenic microorganisms. This makes them an innate defense mechanism with some degree of specificity. For example, TLR2 binds to lipoarabinomannan from mycobacteria, zymosan from yeast, lipopolysaccharide (LPS) from spirochetes, and peptidoglycan. TLR4, on the other hand, binds to LPS as well as host proteins released at sites of infection (for example, heat-shock protein 60). CD14, mentioned earlier, serves as a coreceptor for LPS. Note that these receptors bind fragments of structures after they are released from the microbe. They don't interact with the whole organism.

Once bound to their ligands, the TLRs trigger an intracellular regulatory cascade via their TIR domain that causes the host cell to release proteins called cytokines that diffuse away from the site, bind to receptors on various cells of the immune system, and direct them to engage the invader. Cytokines are discussed further in Chapter 24.

Like skin, the gastrointestinal system possesses an innate mucosal immune system, in this case called **gut-associated lymphoid tissue (GALT)**. GALT includes tonsils, adenoids, and Peyer's patches (**Fig. 23.17A**). These tissues include specialized **M cells** that dot the intestinal surface and are wedged between epithelial cells. M stands for "microfold," which describes their appearance (**Fig. 23.17B**). These are fixed cells that take up microbes from the intestine and release them, or pieces of them, into a pocket formed on the opposite, or basolateral, side of the cell. Other cells of the innate immune system, such as macrophages (a type of white blood cell that migrates through tissues), gather here and collect the organisms that emerge. Macrophages engulf and try to kill the organism and then place small, degraded components of the organism on the macrophage cell surface where other immune system cells can recognize them. As a result, M cells are extremely important for the development of mucosal immunity to pathogens. However, M cells can also serve as a portal for some pathogens to gain entry to the body.

The lungs. The lung also has a formidable defense. In addition to the respiratory ciliary elevator discussed in Section 23.1, microorganisms larger than 100 μm become trapped by hairs and cilia lining the nasal cavity and trigger the forceful expulsion of air from the lungs (a sneeze). The sneeze is designed to clear the organism from the respiratory tract. Organisms that make it to the alveoli are met by phagocytic cells called **alveolar macrophages**. These cells can ingest and kill most bacteria.

Lung epithelia can also internalize and destroy organisms. This mechanism is thought to explain why *Pseudomonas aeruginosa* does not typically infect healthy lungs but readily infects the lungs of cystic fibrosis patients, whose cells cannot bind and kill the bacteria.

Cystic fibrosis is a genetic disease in which a membrane chloride channel called the **cystic fibrosis transmembrane conductance regulator (CFTR)** is defective. CFTR is important for regulating chloride movement across the membrane, an essential part of hydrating mucus in healthy individuals. Localization of the protein at the lung cell surface is depicted in **Figure 23.18**. In addition to its role in regulating chloride channels, CFTR appears

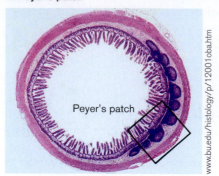

A. Peyer's patch

Peyer's patch

www.bu.edu/histology/p/12001oba.htm

B. M cell

Antigens or bacteria are taken up by the M cell.

M cell

Lymphocyte

Macrophage

Antigens or bacteria that exit the M cell are taken up by macrophages.

Figure 23.17 Gut-associated lymphoid tissue (GALT). A. A Peyer's patch located on the intestine. **B.** Diagram of an M cell (microfold cell).

to serve as a receptor for internalizing *P. aeruginosa* by lung epithelial cells, which helps clear the organism from the lung. While normal lung epithelial cells prevent *Pseudomonas* infection, cells defective in CFTR cannot bind and clear the organism, leading to life-threatening disease. The strains of *P. aeruginosa* that are the most serious threat produce a thick slimy material (alginate) that retards other lung clearance mechanisms (**Fig. 23.19**).

Chemical Barriers to Infection

Some examples of chemical barriers were mentioned previously, such as the acid pH of the stomach, lysozyme in tears, and generators of superoxide. In addition, a vari-

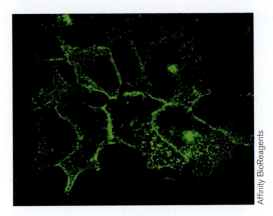

Figure 23.18 **Plasma membrane location of the cystic fibrosis gene product (CFTR).** Laser scanning confocal fluorescence microscopy was used to image living epithelial cells of the human airway. The cells were transfected with DNA expressing a cystic fibrosis transmembrane conductance regulator (CFTR) protein tagged with green fluorescent protein (GFP-CFTR). Note that the fluorescence is heaviest at the edge of the cell, indicating that the GFP-CFTR protein is localized at the cell surface.

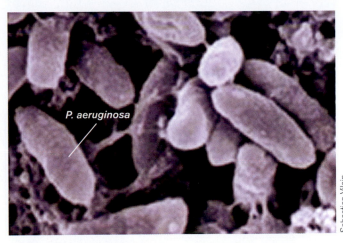

Figure 23.19 **Mucoid strains of *Pseudomonas aeruginosa* that infect cystic fibrosis patients.** The organism is shown growing in a biofilm with strings of dehydrated exopolysaccharide connecting cells. This EPS material gives colonies of these strains a very mucoid appearance. (Cells 2–3 μm long, SEM.)

ety of human cells generate small antimicrobial, cationic (positively charged) peptides called **defensins** (**Table 23.2**). These antimicrobial peptides are important components of innate immunity against microbial infections.

Defensins. Defensins range in length from 29 to 47 amino acids and are found in mammals, birds, amphibians, and plants. They kill by destroying the invader's cytoplasmic membrane and are effective against gram-positive and gram-negative bacteria, fungi, and even some viruses (those with membranes, such as HIV). To kill gram-negative bacteria, the peptides must first bind outer membrane lipopolysaccharides, which are negatively charged, and move into the periplasm. The defensins then insert into the cytoplasmic membrane of either gram-positive or

gram-negative bacteria. Insertion is driven by the transmembrane electrical potential, which in bacteria is about –150 mV (cell interior more negative than exterior). The negative charge pulls the small cationic peptides into the membrane, where they assemble into channels that destroy the cytoplasmic membrane barrier, killing the bacterial cell. Defensins generally do not affect eukaryotic cells because these cells have a very low membrane potential (–15 mV) across their plasma membrane. Defensins are produced by many human cells, including cells of the skin, lungs, genitourinary tract, and gastrointestinal tract (**Fig. 23.20**).

There are two classes of vertebrate defensins, alpha (α) and beta (β). The alpha defensins are stored in membrane-bounded granules within neutrophils and in Paneth cells

Table 23.2	**Categories of natural antimicrobial peptides.**		
Class	**Examples**	**Major presence in humans**	**Source**
Alpha defensins	-1, -2, -3, -4[a]	Neutrophils	Stored in granules
Alpha defensins	-5, -6	Paneth cells, small intestine	Stored in granules
Cathelicidins	LL-37, hCAP-18	Neutrophils	Secreted
Histatins		Saliva	Secreted
Beta defensins	HBD-1, -2	Epithelia	Secreted
		Other species	
Maganins		Frogs	
Protegrins		Pigs	
Indolicidin		Cattle	

[a]Alpha defensins are named alpha defensin-1, alpha defensin-2, etc.

A.

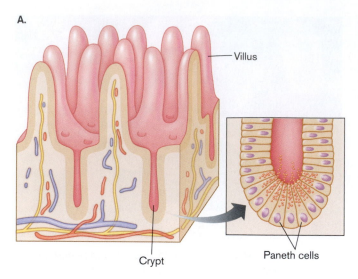

B.

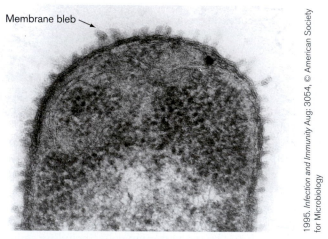

1995. *Infection and Immunity* Aug: 3054, © American Society for Microbiology

Figure 23.20   Defensins.   A. Certain defensins are produced in the crypts of the intestine. The crypts contain granule-rich Paneth cells (inset) that discharge their granules into the crypt lumen in response to the entry of bacteria or as a result of food-related stimulation by acetylcholine. **B.** Effect of cationic peptides on *E. coli* 0111. Polymixin B is a small cationic peptide antibiotic that mimics the action of defensins. In this micrograph, polymixin B causes blebs of membrane to ooze from the surface of the cell. *Source*: A. C. L. Bevins, Research Institute, and D. Schumick, Department of Medical Illustration, Cleveland Clinic Foundation. ©2000, Cleveland Clinic Foundation.

in the small intestine (see **Fig. 23.20A**). When stimulated, these cells **degranulate** (release their granule contents) by fusing their granule membranes to cytoplasmic or vacuolar membranes, dumping their contents into the surroundings or into phagocytic vacuoles, where the alpha defensins can destroy engulfed microbes (**Fig. 23.20B**). In contrast, the beta defensins are not stored in cytoplasmic granules. Their synthesis is activated only after contact with bacteria or their products.

Cathelicidins. Cathelicidins are also cationic antimicrobial peptides, but their structure differentiates them from defensins. Initially synthesized as inactive precursor proteins, cathelicidins contain a 100-amino-acid N-terminal domain called cathelin that is attached to a smaller antimicrobicidal domain at the N-terminus. Extracellular proteolytic cleavage of the large proprotein releases the now-active antimicrobial peptide. As with the defensins, most cathelicidins are thought to function by assembling within bacterial membranes, where they disrupt membrane integrity. Some cathelicidins, however, penetrate to the cell interior and disrupt protein synthesis.

Other antimicrobial peptides are listed in Table 23.2.

> **THOUGHT QUESTION 23.6**   Why do defensins have to be so small?

TO SUMMARIZE:

- **Skin defenses** against invading microbes include closely packed keratinocytes and a SALT lymphoid system made up largely of phagocytic Langerhans cells.
- **Mucous membrane defenses** involve secreted enzymes, cytokines, and GALT tissues, such as Peyer's patches, that contain phagocytic M cells.
- **Phagocytic alveolar macrophages** inhabit lung tissues, contributing to nonspecific defense.
- **Chemical barriers against disease** include cationic defensins, acid pH in the stomach, and superoxide produced by certain cells.

23.5 Innate Immunity: The Acute Inflammatory Response

The boil shown in **Figure 23.21** is an inflammatory response triggered by infection with the organism *Staphylococcus aureus*. Inflammation is a critical innate defense in the war between microbial invaders and their hosts. It provides a way for phagocytic cells (such as neutrophils) normally confined to the bloodstream to gain access to infected sites within tissues. Movement of these cells out of blood vessels is called **extravasation**, which we discuss later. Once at the infection site, the neutrophils begin engulfing microbes. The white pus associated with an infection is teeming with these white blood cells. The five cardinal signs of inflammation, first described over 2,000 years ago, are redness, warmth, pain, swelling, and altered function at the affected site.

Although many things can trigger inflammation, our focus is on how microbes cause the response. The process begins with the infection itself. Microorganisms introduced into the body, for example on a wood splinter, will begin to grow and produce compounds that damage host

Special Topic 23.1   Do Defensins Have a Role in Determining Species Specificity for Infection?

Differences in cationic peptides between host species may explain why some bacteria cause devastating disease in certain animals and only mild disease in others. *Salmonella enterica* serovar Typhimurium, for example, usually kills mice, but most humans survive. Insects, completely lacking an immune system, resist invading pathogens altogether. Could these differences be due to the type of defensin made?

Studies suggest that differences in immunity across species may indeed reflect species-specific defensins. Scientists placed the gene for human defensin-5 (made in human intestinal Paneth cells) into mice, which do not normally produce this peptide. When exposed to high doses of *S.* Typhimurium, the engineered mice did not get sick, while normal mice became profoundly ill (**Fig. 1**).

Studies of defensins and their selectivity offer the promise of a new class of antibiotics. Frog skin, for example, is rich with antimicrobial peptides called maganins. The effectiveness of these peptides is impressive. Frog skin maganins are so powerful that frogs do not become infected when put into natural pond water teeming with bacteria. Harnessing these peptides as antibiotics would be a tremendous advance in medicine. The potential to discover new antibiotics is another reason for protecting our rain forests, which harbor many undiscovered species, including frogs, whose unique chemical defenses have yet to be studied.

A.

Transgenic mice   Wild-type mice

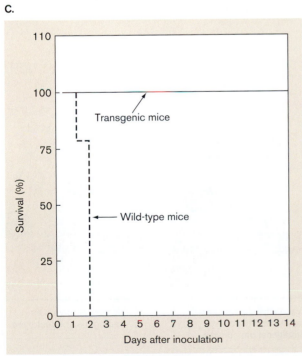

B.

Transgenic mice   Wild-type mice

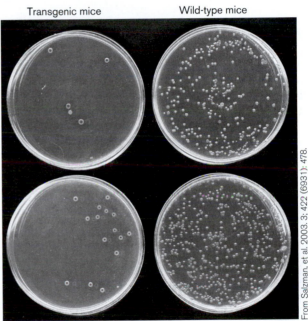

C.

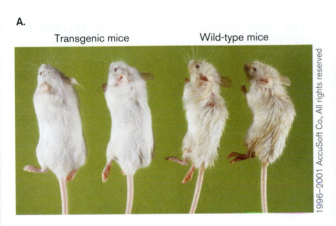

Figure 1 **Role of defensins in the host range of *Salmonella enterica* serovar Typhimurium infection.**
A. Comparison of human defensin-5 transgenic (left) and wild-type (right) mice 12 hours after oral challenge with 1.5×10^8 cfu of *S.* Typhimurium. The wild-type mice are clearly sick. **B.** Comparison of bacterial burden in terminal ileum in human defensin-5 transgenic and wild-type control mice 6 hours after oral challenge with 1×10^8 cfu of *S.* Typhimurium. The *Salmonella* grew better in the wild-type mice and produced more colonies on an agar plate. **C.** Survival curves comparing age-matched HD-5 transgenic mice with wild-type control mice after oral challenge with 1.5×10^9 cfu of *S.* Typhimurium ($n = 6$ for each group). All wild-type mice died by 2 days, while the human defensin-5 mice lived at least 14 days.

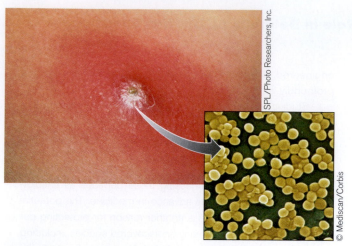

Figure 23.21   Inflammation caused by infection.   Boil resulting from infection of a hair follicle by *Staphylococcus aureus* (size 0.5–1.0 μm, colorized SEM).

cells (**Fig. 23.22**●▌). Resident macrophages that wander into the infected area engulf these organisms and then release inflammatory mediators (chemoattractants) that "call" for more help. These mediators include **vasoactive factors** such as leukotrienes, platelet-activating factor, and prostaglandins, which act on blood vessels of the microcirculation, increasing blood volume and capillary permeability to help deliver white blood cells to the area. In addition, the small protein molecules called cytokines are secreted, diffusing to the vasculature and stimulating expression of specific receptors (selectins) on the endothelial cells of capillaries and venules.

Vasoactive Factors and Cytokines Begin the Extravasation Process

The eventual passage of neutrophils through vascular walls (extravasation; **Fig. 23.23**) requires a relaxation of endothelial cell adhesion. The process is initiated by vasoactive factors released by macrophages; these vasoactive factors increase vascular permeability and stimulate vasodilation. Vasodilation slows blood flow and, as a result, increases blood volume in the affected area. The more permeable vessel also allows the escape of plasma into tissues. Both events cause localized swelling, redness, and heat. Vasoactive factors also stimulate local nerve endings, causing pain, which draws awareness to the infected area.

There are many kinds of cytokines, all of which are involved in regulating the immune response. Cytokines such as **interleukin 1 (IL-1)** and **tumor necrosis factor alpha (TNF-α)** are released by macrophages and stimulate the production of adhesion molecules (**selectins**) on the inner lining of the capillaries. P-selectin is produced first, followed by E-selectin. The selectins snag neutrophils zooming by in the bloodstream, slow them down, and cause them to roll along the endothelium (see **Fig. 23.23**). Rolling neutrophils that encounter inflammatory mediators are activated to produce and display integrin adhesion molecules on their surface. The integrins on the neutrophils lock onto endothelial adhesion molecules ICAM-1 (intracellular adhesion molecule 1) and VCAM-1 (vascular cell adhesion molecule 1). This stops the neutrophils from rolling and initiates extravasation, where the

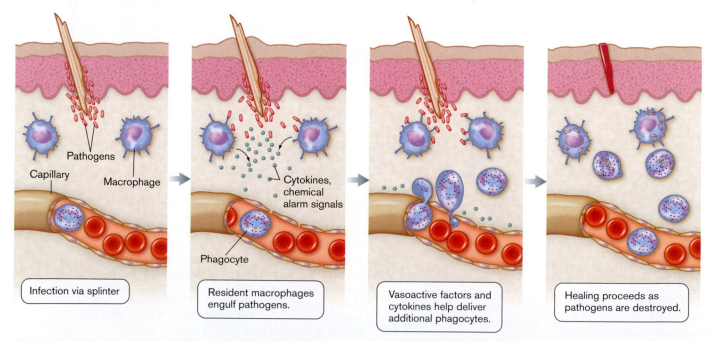

Figure 23.22   Basic inflammatory response.   Neutrophils (a type of phagocyte) circulate freely through blood vessels and can squeeze between cells in the walls of a capillary (extravasation) to the site of infection. They then engulf and destroy any pathogens they encounter. ●▌

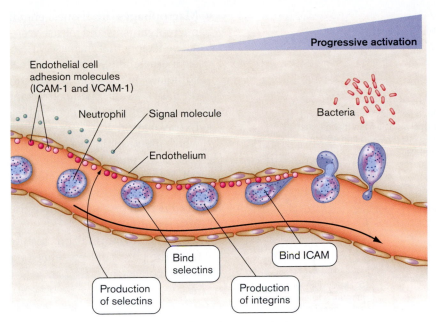

Figure 23.23 Mechanism of extravasation. Extravasation is the process by which leukocytes move from the bloodstream into surrounding tissues. Signal molecules produced by damaged tissue cells induce the production of selectins (produced early in process) on the surfaces of endothelial cells and integrins on the surface of the white blood cell (selectins and integrins not shown). Selectins capture leukocytes traveling through blood vessels. Leukocytes begin to roll along the vessel wall, and the integrins on their surface lock onto the endothelial cell's adhesion molecules (ICAM-1 or VCAM-1). The leukocytes are progressively activated while rolling. The neutrophil ultimately squeezes through the wall between endothelial cells (extravasation).

have f-met (*N*-formylmethionine) as their N-terminal amino acid, but mammalian proteins, in general, do not. Bacteria will often cleave off the f-met peptide, which can then diffuse away from the bacterium. The f-met peptide binds to neutrophil receptors and stimulates pseudopod projections aimed toward the microbe. This causes the white blood cell to migrate in the direction of the infection. In addition, **chemokine** peptides (IL-8 and MCP-1) produced by damaged tissues can serve as chemoattractants for these white blood cells. Thus, much of what takes place at the site of an infection is not due directly to substances produced by the microbe, but is the result of the body's reaction to the presence of the intruding organism. Once the cascade of inflammatory events delivers phagocytic cells to the site of infection, they begin devouring the microbes in an attempt to cure the infection.

Chronic Inflammation Causes Permanent Damage

Chronic inflammation is a response of long duration provoked by the persistent presence of a causative stimulus. This type of inflammation inevitably causes permanent tissue damage, even though the body attempts repair. The causes of chronic inflammation are many. For example, infectious organisms such as *Mycobacterium tuberculosis*, *Actinomycetes bovis*, and various protozoan parasites can avoid or resist host defenses (**Fig. 23.25A**). As a result, they persist at the site and continually stimulate the basic inflammatory response. Nonliving, irritant material like wood splinters, inhaled asbestos particles, or surgical implants can also lead to chronic inflammatory responses. Autoimmune diseases are another important cause. Autoimmunity (reaction against self) occurs when there is a failure to regulate some aspect of adaptive immunity (see Chapter 24). As a result, the immune system does not recognize one specific component of the body as self and begins to react against it. Rheumatoid arthritis is one example of the body attacking itself.

Whatever causes autoimmune disease, there is continual recruitment of macrophages and lymphocytes from the circulation. The body attempts to "wall off" the site of inflammation by forming a **granuloma**. A granuloma begins as an aggregation of the mononuclear inflammatory cells surrounded by a rim of lymphocytes. The body then deposits fibroconnective tissue around the lesion, causing tissue hardening known as fibrosis.

white blood cells squeeze through the endothelial wall and into the tissues, where they can help macrophages attack the invading microbes.

Damaged tissue cells in the area of inflammation will release **bradykinin**, a nine-amino-acid polypeptide that helps loosen the tight junctions between endothelial cells to promote extravasation (**Fig. 23.24**). Bradykinin molecules also bind to mast cells in the area, causing a calcium influx that triggers degranulation. The histamine released from mast cells further loosens the endothelial cell junctions, which allows more fluid and cells to move out (increased vascular permeability), adding to fluid accumulation (edema). Bradykinin will attach to capillary cells and induce synthesis of prostaglandins that in turn stimulate nerve endings, causing pain in the area. A key enzyme involved in prostaglandin synthesis is cyclooxygenase (COX). Aspirin as well as the popularly prescribed anti-inflammatory agents Vioxx and Celebrex are COX inhibitors that prevent the synthesis of prostaglandins and thus reduce inflammatory pain.

Once neutrophils have passed through the vascular wall, chemotactic factors are needed to lure them to the proper location. One of these, f-met-leu-phe peptide, is made by the bacterium itself. Many bacterial proteins

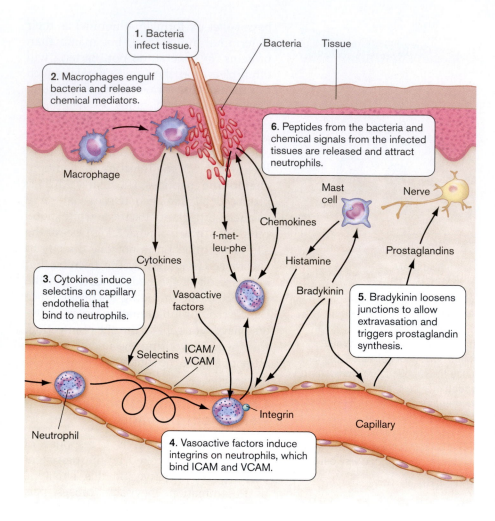

1. Bacteria infect tissue.

2. Macrophages engulf bacteria and release chemical mediators.

Bacteria   Tissue

6. Peptides from the bacteria and chemical signals from the infected tissues are released and attract neutrophils.

Macrophage

Mast cell

Nerve

Chemokines

f-met-leu-phe

Cytokines

Histamine

Prostaglandins

3. Cytokines induce selectins on capillary endothelia that bind to neutrophils.

Vasoactive factors

Bradykinin

5. Bradykinin loosens junctions to allow extravasation and triggers prostaglandin synthesis.

Selectins ICAM/ VCAM

Integrin

Capillary

Neutrophil

4. Vasoactive factors induce integrins on neutrophils, which bind ICAM and VCAM.

Figure 23.24   Summary of the inflammatory response.

Several forms of granulomas are shown in **Figure 23.25**. Mycobacteria are resistant to killing by macrophages and can be found living within these cells (**Fig. 23.25A**). This resistance to host defense can lead to long-term chronic infections and the production of granulomas. For example, skin infections caused by *Mycobacterium marinum* will produce skin granulomas, while *M. tuberculosis* can cause liver granulomas (**Figs. 23.25B and C**). Crohn's disease, which commonly manifests as abdominal pain, frequent bowel movements, and rectal bleeding, has been attributed to an autoimmune reaction possibly activated in some way by intestinal microflora. In this case, the intestinal bacteria are thought to cause a chronic inflammation resulting in characteristic granulomas (**Figs. 23.25D and E**).

TO SUMMARIZE:

■ **Inflammation** begins with a mechanism (extravasation) that moves neutrophils from the bloodstream into infected tissues.

■ **Macrophages** in tissues engulf microorganisms and release vasoactive factors that increase vascular permeability, cytokines that stimulate production of selectin receptors, and chemoattractant molecules that call in neutrophils.

■ **Extravasation** is the process by which neutrophils pass between cells of the endothelial wall. Once out of the circulation, the neutrophils travel to the site of infection.

■ **Bradykinin** produces pain in the affected area.

■ **Chronic inflammation** results from the persistent presence of a foreign object.

23.6 Phagocytosis

Phagocytosis is an effective nonspecific immune response. However, it would be disastrous were white blood cells to indiscriminately phagocytize host as well as nonhost structures. Fail-safe controls built into our immune defenses prevent this from happening. Without these restraints, the mammalian immune system would constantly attack itself, causing more problems than it solves. When one of these fail-safe mechanisms fails, autoimmune diseases can arise.

Phagocytes Recognize Alien Cells and Particles

For phagocytosis to proceed, macrophages and neutrophils must first "see" the surface of a particle as being foreign. Only then is phagocytosis possible (**Fig. 23.26**). When a phagocyte surface interacts with the surface of another body cell, the phagocyte becomes temporarily paralyzed (inept at pseudopod formation). Paralysis allows the phagocyte to evaluate whether the other cell is friend or foe, self or nonself. Self-recognition involves glycoproteins located on the white blood cell membrane binding inhibitory glycoproteins present on all host cell membranes. The inhibitory glycoprotein on human cells is called CD47. Because invading bacteria lack these inhibitory surface molecules, they can readily be engulfed.

In addition to the method of self-recognition, macrophages have a molecular recognition apparatus specific for bacteria—the TLRs. Toll-like receptors specifically recognize bacterial structures such as lipopolysaccharide (LPS),

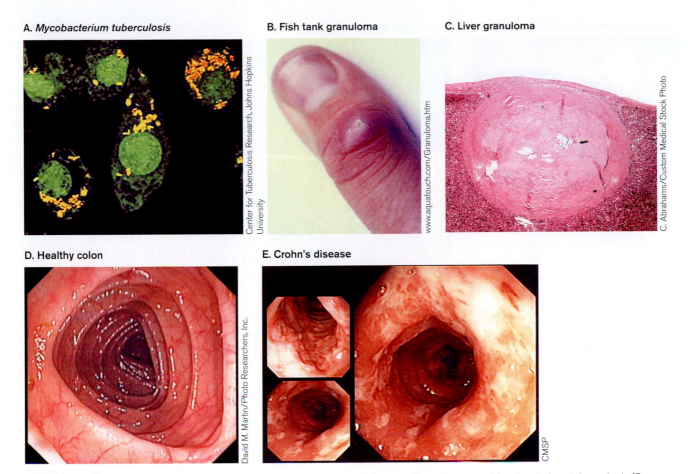

A. *Mycobacterium tuberculosis*

B. Fish tank granuloma

C. Liver granuloma

D. Healthy colon

E. Crohn's disease

Figure 23.25 **Chronic inflammation.** **A.** The fluorescent orange organisms shown here are *Mycobacterium tuberculosis* (2 μm long, fluorescence microscopy) within macrophages in a tuberculosis abscess. Mycobacteria have a thick, waxy cell wall that protects them against the mechanisms used by macrophages to destroy microorganisms. As a result, they survive for prolonged periods within macrophages. This sequestration also protects the bacteria from other cellular and humoral mechanisms that deal with invading organisms. The continual stimulation of an inflammatory response leads to chronic inflammation. **B.** Fish tank granuloma. *Mycobacterium marinum* can cause tuberculosis-like infection in fish. The organism can accidentally enter the human body through an open wound or abrasion during cleaning of an aquarium containing infected fish. The infection is first noted as a lesion that heals very slowly on the hand or forearm. Once it does heal, it often forms a granuloma at the site that contains live organisms. **C.** Cross section of liver showing a necrotizing granuloma caused by *M. tuberculosis* that spread through the bloodstream to this site after escaping the lung. **D.** Healthy colon. **E.** Intestinal granulomas of Crohn's disease.

flagella, or peptidoglycan. There are different TLRs for each structure. Interaction between one of these receptors and its ligand initiates a signal transduction cascade in the phagocyte that stimulates phagocytosis and triggers release of chemokines and cytokines to call for more help.

Although many bacteria, such as *Mycobacterium* or *Listeria* species, are easily recognized and engulfed by phagocytosis (**Figs. 23.26A–C**), others, such as *S. pneumoniae*, possess polysaccharide capsules that are too slippery for pseudopods to grab (**Fig. 23.26D**). This is where innate immunity and adaptive immunity join forces. Adaptive immunity produces anticapsular antibodies that aid the innate immune mechanism of phagocytosis through a process known as **opsonization** (**Fig. 23.27**). The anticapsular antibodies coat the surface of the bacterium, leaving the tail ends of the antibodies, called the

Fc region (see Section 24.3), pointing outward. The Fc regions of these antibodies are recognized and bound by specific receptors on phagocyte cell surfaces. As a result, the antibodies bind the bacteria to phagocytic Fc receptors, allowing the phagocyte to engulf the invader.

Phagocytes Kill through Oxygen-Independent and Oxygen-Dependent Mechanisms

During phagocytosis, the cytoplasmic membrane of the phagocyte flows around, and then engulfs, the bacterium, producing an intracellular phagosome, as described earlier. Subsequent phagosome-lysosome fusion (producing a phagolysosome) results in both oxygen-independent and oxygen-dependent killing pathways. Oxygen-independent

A. White blood cell attacking bacteria

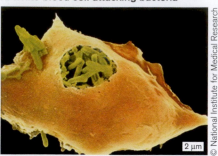

C. Contacts between phagocyte and target microbe

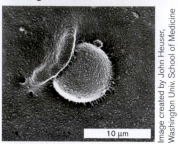

D. *Streptococcus pneumoniae* and capsule

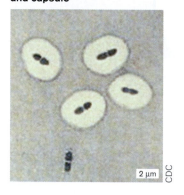

B. Macrophage engulfing bacterium

Figure 23.26   Images of phagocytosis.   A. A white blood cell attacking bacteria. *Mycobacterium* (green, 2 μm long) being phagocytosed by a white blood cell (colorized SEM). **B.** A macrophage engulfing bacteria on the outer surface of a blood vessel (SEM, magnification ×1,315). **C.** Contacts between phagocyte and target microbe, illustrating how phagocytes "grab" the target bacterium. **D.** *Streptococcus pneumoniae* and capsule. The slippery nature of the polysaccharide capsule makes phagocytosis more difficult.

Myeloperoxidase, present only in neutrophils, converts hydrogen peroxide and chloride ions to hypochlorous acid (HOCl).

Macrophages, mast cells, and neutrophils also generate reactive nitrogen intermediates that serve as potent cytotoxic agents. Nitric oxide (NO) is synthesized by NO synthetase. Further oxidation of NO by oxygen yields nitrite (NO_2^-) and nitrate (NO_3^-) ions. All of these reactive oxygen species attack bacterial membranes and proteins. These mechanisms are responsible for the large increase in oxygen consumption noted during phagocytosis, called the **oxidative burst**. The reactive chemical species formed during the oxidative burst do little to harm the phagocyte because the burst is limited to the phagosome and because the various reactive oxygen species, such as superoxide, are very short-lived.

TO SUMMARIZE:

- **Phagocytosis** is selective for particles recognized as foreign to the body.
- **Different Toll-like receptors (TLRs)** on macrophage surfaces recognize different microbial structures.
- **Oxygen-independent** and **oxygen-dependent mechanisms of killing** are initiated by fusion between a lysosome and a bacteria-containing phagosome.
- **The oxidative burst**, a large increase in oxygen consumption during phagocytosis, results in the production of superoxide ions, nitric oxide, and other reactive oxygen species.

mechanisms include enzymes like lysozyme to destroy the cell wall, compounds such as lactoferrin to sequester iron away from the microbe, and defensins, small cationic antimicrobial peptides (described in Section 23.4).

Oxygen-dependent mechanisms are probably activated through the TLRs. They kill through the production of various oxygen radicals. NADPH oxidase, myeloperoxidase, and nitric oxide synthetase in the phagosome membrane are extremely important. NADPH oxidase yields superoxide ion (O_2^-), hydrogen peroxide, and, ultimately, hydroxyl (•OH) radicals.

$$\text{NADPH} + 2O_2 \xrightarrow{\text{NADPH oxidase}} \text{NADP}^+ + H^+ + 2O_2^- \xrightarrow{\text{Superoxide dismutase}}$$

$$H_2O_2 + Cl^- \xrightarrow{\text{Myeloperoxidase}} OH^- + HOCl$$

23.7 Innate Defenses by Interferon and Natural Killer Cells

When a community is threatened by a thief, a neighbor who has been robbed alerts others to take precautions. And if the thief is discovered, police are called to arrest him. Similarly, interferon peptides and natural killer cells are two innate mechanisms of defense that, respectively, warn healthy host cells of a nearby infection and seek out and then destroy cells already infected.

Interferons Are Host Specific, Not Virus Specific

In 1957, it was discovered that cells exposed to inactivated viruses produce at least one soluble factor that can "inter-

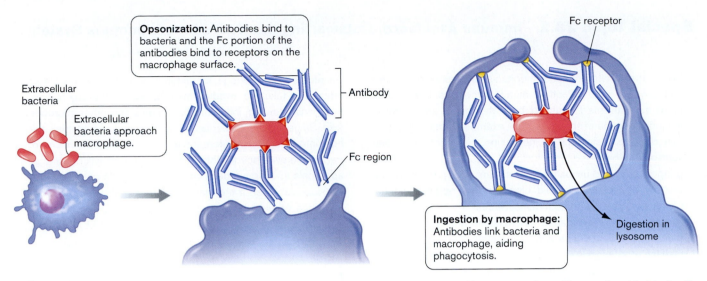

Figure 23.27   Opsonization.   Opsonization is a process that facilitates phagocytosis. Here, macrophage Fc receptors bind to the Fc region of antibodies binding to bacteria.

fere" with viral replication when applied to newly infected cells. The term **interferon** was coined to represent these molecules. Actually, several different macromolecules have the property of interfering with viral replication. Interferons are low-molecular-weight cytokines (14–20 kDa) produced by many eukaryotic cells in response to intracellular infection—that is, infection by viruses or by bacterial pathogens, such as *Listeria*, that can grow intracellularly. The action of interferons is usually species specific (that is, interferon from mice will not work on human cells) but virus nonspecific (human interferon will help protect against both poliovirus and influenza virus, for example).

There are two general types of interferons, which differ in the receptors they bind and the responses they generate (**Table 23.3**). Type I interferons have high antiviral potency; they consist of IFN-alpha (IFN-α), IFN-beta (IFN-β), and IFN-omega (IFN-ω). Type II interferon, IFN-gamma (IFN-γ), has more of an immunomodulatory function that will be discussed in the next chapter.

Type I interferons can bind to specific receptors on uninfected host cells and render those cells resistant to viral infection. The host cell becomes resistant because interferon induces the intracellular production of two classes of proteins. One class encompasses double-stranded RNA-activated endoribonucleases that can cleave viral RNA. Proteins in the second class are protein kinases (PKRs) that phosphorylate and inactivate eukaryotic initiation factor eIF2, which is required to translate viral RNA. These mechanisms affect both RNA and DNA viruses because protein synthesis is required for the propagation of all viruses. However, RNA viruses are better than DNA viruses at inducing interferon.

Type II interferon functions by activating various white blood cells—for example, macrophages, natural killer cells, and T cells—to increase the number of **major histocompatibility complex (MHC)** antigens on their surfaces. MHC proteins are important for recognizing self and for presenting foreign antigens to the adaptive immune system. They will be discussed more fully in Chapter 24.

Natural Killer Cells Recognize Infected Cells and Cancer Cells

Body cells that are infected or that are cancerous can be a major problem for the host. Infected cells can harbor the pathogenic microbe, hiding it from the immune system. Cancer cells, on the other hand, can take over and kill the

Table 23.3   Classes of interferon.

Type	Examples	Function	Mechanism of action
Type I	IFN-alpha IFN-beta IFN-omega	Antiviral	Induce dsRNA endonuclease, eIF2 kinase
Type II	IFN-gamma	Immunomodulatory	Triggers signal cascades in macrophages, NK cells, and T cells; increases MHC II on cell surfaces

Special Topic 23.2 Immune Avoidance: Outsmarting the Host's Innate Immune System

While the innate immune mechanisms described in this chapter are quite effective, many pathogens have developed mechanisms to subvert both the innate and adaptive immune systems of their host. For example, many organisms can develop resistance to defensins, *Legionella* can prevent phagosome-lysosome fusion, *Listeria* can escape from phagosomes, and uropathogenic strains of *E. coli* can invade bladder epithelium to avoid clearance by defense mechanisms associated with host cell surfaces.

Some other avoidance mechanisms are quite clever in their design. *Salmonella enterica*, a major cause of diarrheal disease, can engage in close-contact duels with phagocytic macrophages to see which can kill the other first. The macrophage, of course, tries to engulf the microbe and destroy it via oxygen-dependent and oxygen-independent mechanisms within a phagolysosome. For its part, *Salmonella* sidles up to the macrophage and uses a syringe-like type III secretion system (see Section 25.5) to penetrate the eukaryotic cell membrane and then inject the bacterial protein SipB directly into the cytoplasm. SipB binds to and activates a protein in the macrophage called caspase 1. Caspase 1 is part of a cascade of caspase enzymes that, when activated, cause the macrophage to commit suicide in a programmed response called apoptosis. Apoptosis is a process normally used by the host to kill unwanted host cells. For example, the differentiation of human fingers requires the cells between the fingers to initiate apoptosis so that the fingers can separate. However, apoptosis is also triggered as a last resort by a host cell that is heavily compromised by a bacterial or viral infection.

An infected macrophage, for instance, will sacrifice itself in an attempt to take the invading microbe with it, thereby halting, or at least slowing, infection. *Salmonella* and other microbes (like *Shigella*) inject caspase-activating proteins to trick the host into killing itself before it can engulf and kill the microbe. Part of the evidence for this is that *Salmonella* will not trigger apoptosis in caspase-deficient cells in Peyer's patches (**Fig. 1**). However, even if the macrophage wins this duel, *Salmonella* has other subversive mechanisms at its disposal that can protect it after phagocytosis (see Section 25.5).

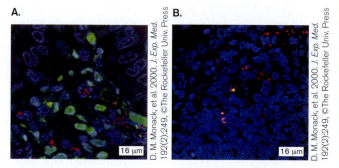

Figure 1 *Salmonella* causes apoptosis in caspase-positive but not caspase-negative cells. Cells undergoing apoptosis are colored green by a fluorescent dye that binds to damaged host DNA. *Salmonella* is labeled red. Normal host cell nuclei are blue. **A.** Caspase-positive cells from Peyer's patches in the mouse intestine undergo apoptosis. **B.** Caspase-negative mouse cells remain viable when infected by *Salmonella*.

host. A class of lymphoid cells called **natural killer (NK) cells** identifies and handles these situations. Instead of killing microbes, the mission of these cells is to destroy host cells that harbor microorganisms or that have been transformed into cancer cells (**Fig. 23.28**). Natural killer cells recognize changes in cell surface proteins of compromised cells and then degranulate to release chemicals that kill those cells.

Natural killer cells recognize their targets in two basic ways. One involves MHC class I molecules, and the other utilizes Fc receptors. A normal host cell displays two classes of MHC molecules on the outside of the cell membrane. MHC I is an indicator of "self." (MHC II molecules will be discussed in Chapter 24.) NK cells have specific receptors that bind to self MHC I molecules on the surfaces of other cells in the body. An NK cell that "touches" a self MHC I–containing cell from the same person will not attack that cell. However, if a host cell lacks MHC class I molecules (or has nonself MHC), NK cells perceive

the target as foreign and a potential threat. Host cells can lose their MHC molecules during infection or as a result of malignant transformation. When an NK cell encounters a host cell lacking these markers, the NK cell inserts a pore-forming protein (**perforin**) into the membrane of the target cell, through which cytotoxic enzymes are delivered.

Natural killer cells also contain Fc receptors on their cell surface. The second killing mechanism, called **antibody-dependent cell-mediated cytotoxicity (ADCC)**, occurs when the Fc receptor on the NK cell links to an antibody-coated host cell. The part of an antibody that does not bind to a target molecule (antigen) is called the Fc region. Why would host cells be coated with antibodies? During their replication, many viruses place viral proteins in the membrane of the infected cell. Antibodies to those viral proteins will coat the compromised cell, tagging it for ADCC. Once the compromised cell is targeted, it is killed by the NK cell in the same way as

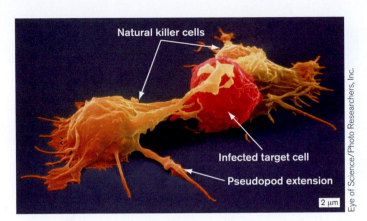

Natural killer cells

Infected target cell

Pseudopod extension

2 μm

Eye of Science/Photo Researchers, Inc.

Figure 23.28   Natural killer cells.   Natural killer (NK) cells attack eukaryotic cells infected by microbes, not the microbes themselves. Perforin produced by the NK cell punctures the membrane of target cells, causing them to burst.

described for cells lacking MHC, by insertion of a perforin molecule. This killing mechanism is another example of cooperativity between innate (NK cells) and adaptive immunity (antibody producing lymphocytes).

> **THOUGHT QUESTION 23.7**   If NK cells can attack infected host cells coated with antibody, why won't neutrophils?

TO SUMMARIZE:

- **Interferons are species-specific** molecules that can nonspecifically interfere with viral replication (type I) and modulate the immune system (type II).
- **Natural killer (NK) cells** are a class of white blood cell that destroy cancer cells or cells harboring microorganisms.
- **Natural killer cells target host cells** that have lost MHC class I receptors as a result of infection or cancer, or host cells that are coated with antibody (antibody-dependent cell-mediated cytotoxicity, ADCC).
- **Natural killer cells kill** by inserting perforin pores into the membranes of target cells.

23.8 Complement's Role in Innate Immunity

Bacterial invaders can also be attacked by a mechanism known as the complement cascade. Complement was first discovered as a heat-labile component of blood that enhances (or complements) the killing effect of antibodies on bacteria. The complement system consists of up to 20 proteins, several of which are proteases that sequentially cleave each other. Once a complement cascade is triggered, a number of things can happen. Pores are inserted into membranes and cause cytoplasmic leaks; in addition, pieces of the complement system can attract white blood cells and also facilitate phagocytosis (opsonization).

Pathways of Complement Activation Produce a Membrane Attack Complex (MAC)

The three routes to complement activation are officially known as the classical pathway, the alternative pathway, and the lectin pathway. The classical complement pathway depends on antibody, so it is part of adaptive rather than innate immunity and is discussed in Chapter 24 (Section 24.7). The lectin pathway requires synthesis of mannose-binding lectin by the liver in response to macrophage cytokines. This lectin coats the surface of invading microbes and activates complement without antibody. We focus in this chapter on the alternative pathway because it is a well-characterized part of innate resistance. Like the lectin pathway, the alternative pathway is a nonspecific defense mechanism (does not require antibody for activation). The alternative complement pathway can attack invading microbes long before a specific immune response can be launched.

The goal of the complement cascade is to insert pores into target microbial membranes. The pores destroy membrane integrity, which kills the cell. The alternative complement pathway begins with the complement factor C3 (**Fig. 23.29**). In blood, C3 slowly cleaves into C3a and C3b (step 1). C3b, under normal circumstances, is rapidly degraded, a process that thwarts inadvertent complement activation. However, if C3b meets LPS on an invading gram-negative microbe, the bound C3b becomes stable and binds another factor, designated Factor B (step 2), and makes Factor B susceptible to cleavage by yet another protein, Factor D (step 3). The resulting complex, called C3bBb, is changed into what is called C5 convertase by properidin, another serum protein (step 4). C5 convertase cleaves C5 in serum to C5a and C5b (step 5). (C3bBb is also called C3 convertase because it can quickly cleave more C3 and amplify the cascade.) C5b now starts to form a prepore complex by binding to C6 and C7 (step 6). The resulting C5bC6C7 complex binds to membranes. Finally, C8 and C9 factors join in to form the **membrane attack complex (MAC)**, creating a destructive pore in the membrane of the target cell (step 7).

If the target of complement is a eukaryotic cell, Na^+ and H_2O enter and the cell osmotically lyses. Eukaryotic cells infected with viruses and coated with antibody can activate complement via the classical complement pathway (see Section 24.7). When complement affects gram-negative cells, lysozyme (present in serum) enters through outer membrane pores and cleaves peptidoglycan, making the cell membrane more susceptible to the

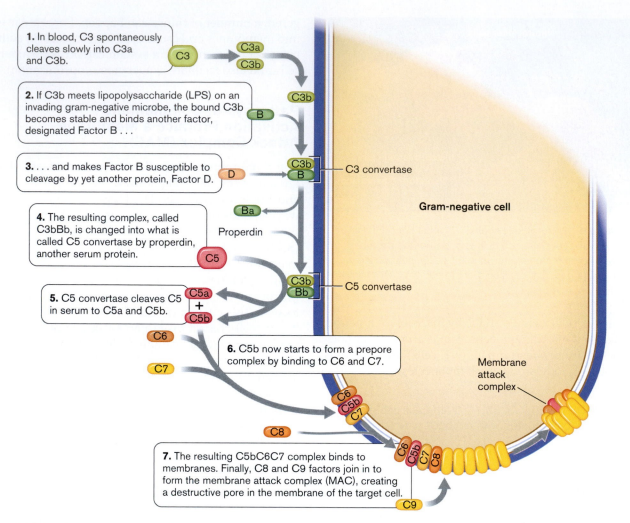

1. In blood, C3 spontaneously cleaves slowly into C3a and C3b.

2. If C3b meets lipopolysaccharide (LPS) on an invading gram-negative microbe, the bound C3b becomes stable and binds another factor, designated Factor B . . .

3. . . . and makes Factor B susceptible to cleavage by yet another protein, Factor D.

4. The resulting complex, called C3bBb, is changed into what is called C5 convertase by properdin, another serum protein.

5. C5 convertase cleaves C5 in serum to C5a and C5b.

6. C5b now starts to form a prepore complex by binding to C6 and C7.

7. The resulting C5bC6C7 complex binds to membranes. Finally, C8 and C9 factors join in to form the membrane attack complex (MAC), creating a destructive pore in the membrane of the target cell.

C3 convertase

C5 convertase

Gram-negative cell

Properdin

Membrane attack complex

Figure 23.29 **The alternative complement pathway.** Although called "alternative," this complement cascade is part of the first-line innate defense.

membrane attack complex. Teichoic acid on gram-positive cell walls can induce the complement cascade but is less effective than LPS. Nevertheless, gram-positive bacteria are resistant to complement because they lack an outer membrane (no LPS to efficiently start the cascade) and have a thick peptidoglycan layer that hinders access of complement components. However, even in the absence of LPS, there are ways to activate complement involving antigen-antibody complexes (see Section 24.7).

The Association between Acute Phase Reactants, Complement, and Heart Disease

As noted earlier, inflammation is associated with the production of various cytokines by macrophages. Some of these cytokines (such as IL-1, TNF-α, and IL-6) travel to the liver, where they stimulate synthesis of several so-called acute phase reactant proteins, including **C-reactive**

protein. This protein, named for its ability to activate complement, is an acute phase reactant that will bind to components of bacterial cell surfaces but not to host cell membranes (**Fig. 23.30**). Once tied to the bacterial cell surface, C-reactive protein will bind complement factor C3 and convert it to C3b, initiating the complement cascade. Although C3 can spontaneously produce C3b, C-reactive protein accelerates C3b production at the bacterial surface, where it can do the most damage.

Elevated levels of C-reactive protein have been linked to an increased risk of cardiovascular disease. The hypothesis is that fatty deposits in the arteries trigger inflammation in the area of deposition. Inflammation then triggers an increase in C-reactive protein, which can be measured as an indicator of cardiovascular disease. Taking an aspirin a day may reduce the risk of heart attack by preventing inflammation and, in turn, reducing levels of C-reactive protein. Whether C-reactive protein has a direct role in cardiovascular disease is not known.

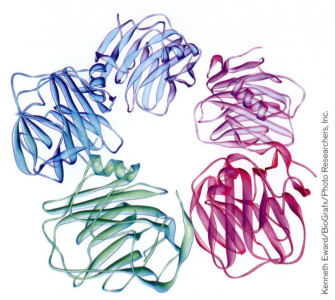

Kenneth Eward/BioGrafx/Photo Researchers, Inc.

Figure 23.30   C-reactive protein.   The three-dimensional structure of human C-reactive protein (CRP) is unique. CRP is synthesized as a 206-amino-acid polypeptide that folds to form a flattened jellyroll structure, which then assembles into a radially symmetrical pentamer that circulates in serum. This structure can bind bacterial cell surfaces and C3 complement factor.

TO SUMMARIZE:

- **Complement** is a series of 20 proteins naturally present in serum.
- **Activation of the complement cascade** results in a pore being introduced into target membranes.
- **The three pathways for activation** are the classical pathway, the alternative pathway, and the lectin pathway.
- **The alternative activation pathway** begins when complement factor C3b is stabilized by interaction with the LPS of an invading microbe.
- **The cascade of protein factors** C3b $\longrightarrow$ B $\longrightarrow$ Factor D $\longrightarrow$ properidin $\longrightarrow$ C5 $\longrightarrow$ C6 $\longrightarrow$ C7 $\longrightarrow$ C8 and C9 results in formation of a membrane attack complex in target membranes.
- **C-reactive protein** in serum is activated when bound to microbial structures and will convert C3 to C3b, which can start the complement cascade.

23.9 Fever

In a healthy individual, body temperature is kept constant within a very small range (36–38°C), despite large differences in surrounding temperature and physical activity. Body temperature is normally regulated by blood flow through the skin and subcutaneous areas. Vasoconstriction (tightening of blood vessel diameter) allows the increased accumulation of heat, while vasodilation

secures its quick release. If body temperature increases above 42–43°C, irreversible damage to the brain occurs.

Information about the body's skin temperature is relayed through the nervous system to the hypothalamus, which acts as a thermostat. If body temperature is too high, the hypothalamus directs increased blood flow through the skin to accelerate heat release. If body temperature is too low, blood flow will decrease to conserve heat, and shivering increases to generate heat.

Fever (elevated body temperature) is a natural reaction to infection. Fever is usually accompanied by general symptoms, such as sweating, chills, sensation of cold, and other subjective sensations. Substances that cause fever are known as **pyrogens**. Pyrogens fall into two classes: exogenous and endogenous. Exogenous pyrogens are those that originate outside the body, such as bacterial toxins. Endogenous pyrogens, on the other hand, are formed by the body's own cells in response to an outside stimulus (such as a bacterial toxin). Exogenous pyrogens generally cause fever by inducing the release of endogenous pyrogens. Endogenous pyrogens include cytokines (interferon, tumor necrosis factor, IL-6, others). These cytokines indirectly cause the hypothalamus to reset the normal temperature level. Cytokine receptors are located on neurons in proximity to the anterior hypothalamus, the location of the thermoregulatory center. The cytokine-receptor interaction stimulates production of phopholipase A2, an enzyme required to make prostaglandins. As a result, prostaglandin E2 is made and crosses the blood-brain barrier, where it changes the responsiveness of the thermosensitive neurons that make up the thermoregulatory center. In other words, it turns up the thermostat.

Because the ideal growth temperature for many microbes is 37°C, elevated temperature can place the organism outside its "comfort zone" of growth. There is also evidence that fever reduces iron availability to bacteria. Slower growth of the pathogen allows the body's immune system time to subdue the infection before it is too late. Consequently, interventions that reduce a moderate fever caused by infection may be counterproductive to a speedy recovery.

> **THOUGHT QUESTION 23.8**   If increased fever limits bacterial growth, why would bacteria make pyrogenic toxins?

TO SUMMARIZE:

- **The hypothalamus** acts as the body's thermostat.
- **Exogenous and endogenous pyrogens** elevate body temperature by stimulating production of prostaglandins.
- **Prostaglandins** change the responsiveness of thermosensitive neurons in the hypothalamus.

Concluding Thoughts

The various "hardwired" innate immune mechanisms described in this chapter keep our normal microbial flora at bay and provide an effective first line of defense against potential pathogens. The next chapter addresses what happens when microbes breach these innate defenses. Unlike the general protective mechanisms discussed in this chapter, the system of acquired immunity generates a molecular defense specifically tailored to a given pathogen.

CHAPTER REVIEW

Review Questions

1. Name some sterile body sites.
2. What body sites are colonized by normal microbial flora?
3. Under what circumstances can commensal organisms cause disease?
4. Why are commensal organisms beneficial to the host?
5. Name and describe various types of innate immunity.
6. What are probiotics? How do they help maintain health?
7. How does the lung avoid being colonized?
8. What is a gnotobiotic animal?
9. Describe GALT and SALT.
10. Describe some chemical barriers to infection.
11. Discuss the different types of white blood cells.
12. What is a lymphoid organ?
13. Outline the process of inflammation.
14. Explain why phagocytes do not indiscriminately phagocytize body cells.
15. What is interferon?
16. Describe antibody-dependent cell cytotoxicity.
17. How does complement kill bacteria?
18. Why might fever be helpful in fighting infection?

Key Terms

adaptive immunity (873)
alveolar macrophage (878)
antibody-dependent cell-mediated cytotoxicity (ADCC) (888)
antigen (873)
antigen-presenting cell (876)
B cell (876)
bacteremia (867)
basophil (874)
bioburden (864)
bradykinin (883)
cathelicidin (880)
chemokine (883)
commensal organism (864)
complement (873)
compromised host (871)
conjunctiva (866)
C-reactive protein (890)
cystic fibrosis transmembrane conductance regulator (CFTR) (878)
cytokine (872)
defensins (879)
degranulate (880)
dendritic cell (875)

enterotoxins (872)
eosinophil (874)
epidermis (864)
extravasation (880)
gnotobiotic animal (872)
granuloma (883)
gut-associated lymphoid tissue (GALT) (878)
immune system (873)
immunomodulin (872)
innate immunity (873)
interferon (887)
interleukin 1 (IL-1) (882)
keratin (877)
keratinocyte (877)
Langerhans cell (877)
lymph node (876)
M cell (878)
macrophage (875)
major histocompatibility complex (MHC) (887)
mast cell (874)
membrane attack complex (MAC) (889)
monocyte (875)

mucociliary elevator (868)
nasopharynx (867)
natural killer (NK) cell (888)
neutrophil (874)
nonadaptive immunity (innate immunity) (873)
normal flora (869)
opportunistic pathogen (871)
opsonization (885)
oropharynx (867)
oxidative burst (886)
perforin (888)
phagosome (874)
probiotics (870)
pyrogen (891)
reticuloendothelial system (875)
skin-associated lymphoid tissue (SALT) (877)
Toll-like receptor (TLR) (877)
tumor necrosis factor (TNF) (882)
urinary tract infection (UTI) (871)
vasoactive factor (882)

Recommended Reading

Chen, Ching-Chen, Michelle A. Grimbaldeston, Mindy Tsai, Irving L. Weissman, and Stephen J. Galli. 2005. Identification of mast cell progenitors in adult mice. *Proceedings of the National Academy of Science USA* **102**:11408–11413.

Corcos, Maurice, Olivier Guilbaud, Sabrina Paterniti, Marlène Moussa, Jean Chambry, et al. 2003. Involvement of cytokines in eating disorders: A critical review of the human literature. *Psychoneuroendocrinology* **28**:229–249.

Ganz, Tomas, and Robert I. Lehrer. 1998. Antimicrobial peptides of vertebrates. *Current Opinion in Immunology* **10**:41–44.

Gunn, Jon S. 2000. Mechanisms of bacterial resistance and response to bile. *Microbes and Infection* **2**:907–913.

Hathaway, Lucy J., and Jean-Pierre Kraehenbuhl. 2000. The role of M cells in mucosal immunity. *Cellular and Molecular Life Science* **57**:323–332.

Hewetson, J. T. 1904. The bacteriology of certain parts of the human alimentary canal and of the inflammatory processes arising therefrom. *British Medical Journal* **2**:1457–1460.

Hornef, Mathias W., Mary J. Wick, Mikael Rhen, and Staffan Normark. 2002. Bacterial strategies for overcoming host innate and adaptive immune responses. *Nature Immunology* **3**:1033–1040.

Menezes, Juscilene da Silva, Daniel De Sousa Mucida, Denise C. Cara, Jaqueline I. Alvarez-Leite, Momtchilo Russo, et al. 2003. Stimulation by food proteins plays a critical role in the maturation of the immune system. *International Immunology* **15**:447–455.

Niedergang, Florence, and Jean-Pierre Kraehenbuhl. 2000. Much ado about M cells. *Trends in Cell Biology* **10**:137–141.

Ogawa, Michinaga, and Chihiro Sasokawa. 2006. Bacterial evasion of the autophagic defense system. *Current Opinion in Microbiology* **9**:62–68.

Peekhaus, N., and Tyrrell C. Conway. 1998. What's for dinner? Entner-Doudoroff metabolism in *E. coli. Journal of Bacteriology* **180**:3495–3502.

Salzman, Nita H., Dipankar Ghosh, Kenneth M. Huttner, Yvonne Paterson, and Charles L. Bevins. 2003. Protection against enteric salmonellosis in transgenic mice expressing a human intestinal defensin. *Nature* **422**:522–526.

Stetson, Daniel B., and Ruslan Medzhitov. 2006. Type I interferons in host defense. *Immunity* **25**:373–381.

Stuart, Lynda M., and R. Alan B. Ezekowitz. 2005. Phagocytosis; elegant complexity. *Immunity* **22**:539–550.

Sullivan, Ann, and Carl E. Nord. 2002. The place of probiotics in human intestinal infections. *International Journal of Antimicrobial Agents* **20**:313–319.

Tannock, Gerald W. 2002. Analysis of the intestinal microflora using molecular methods. *European Journal of Clinical Nutrition* **56** (Suppl 4):S44–S49.

Tlaskalova-Hogenova, H., L. Tuckova, R. Lodinova-Zadnikova, R. Stepankova, B. Cukrowska, et al. 2002. Mucosal immunity: Its role in defense and allergy. *International Archives of Allergy and Immunology* **128**:77–89.

Chapter 24

The Adaptive Immune Response

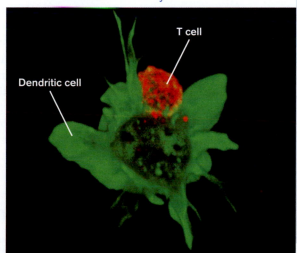

Memory does not reside solely in the brain. Adaptive immunity also depends on a kind of memory that does not rely on nerves and neurons. Our immune system's memory relies on special lymphocytes in the circulation called memory B cells and T cells that form during an infection and "remember" it for years. Although a single member of these cells can recall exposure to only one part of an infecting agent, collectively memory cells remember every microbe that has managed to breach our innate defenses. Armed with this knowledge, they circulate throughout the body like tiny sentries, ready to detect and quickly respond to a second attack.

From birth, the naive immune system is able to recognize billions of possible foreign antigens. How is this possible? Our chromosomes seem too small. The system relies on genetic shuffling of a limited number of gene cassettes to produce each antibody.

But will a system this flexible not attack itself? Or does it? Type 1 diabetes, hypothyroidism, and arthritis are three examples of the immune system reacting against self. So why don't we all suffer from these autoimmune diseases? This chapter will answer these questions and demystify the phenomenon of adaptive immunity.

A naive, antigen-specific T cell is seen (red) contacting an antigen-presenting dendritic cell (DC, green). Ultimately, the T cell will be activated by this interaction. Red staining shows the distribution of the T-cell receptor (TCR) by staining with an antibody. The TCR binds to foreign antigens placed on the surface of antigen-presenting cells. Green shows staining for actin by phalloidin-FITC. Both figures show the same cell pair. The lower image is rotated 180° on the vertical axis to show the DC membrane veils more precisely. Note the homogeneous distribution of the TCR on the T-cell surface despite its interaction with the DC. The TCR signal is not enriched at the T-cell-DC interface called the immunological synapse interface. *Source*: Cover. *Blood*. 2007. Vol. 110, No. 5.

David Philip Vetter was born in 1971 without an immune system. Better known as the "bubble boy," he was the only human known to live in a plastic, germ-free bubble for his entire 12 years of life (1971–1984; **Fig. 24.1A**). His predicament stemmed from a genetic disease known as severe combined immunodeficiency (SCID). In the most serious form of this disease, the patient harbors no T cells and has dysfunctional (or sometimes no) B cells, so the body cannot launch a meaningful immune defense against any invading microbe—whether it be pathogen or normal flora. Because David's elder brother had died of SCID before David was born, physicians were alerted that he might have the disorder, too. Consequently, David was transferred to a sterile environment within seconds after birth to await a bone marrow transplant. Water, air, food, diapers, clothes all were disinfected with special cleaning agents before entering his sterile plastic bubble. He was handled only through special plastic gloves attached to the wall. He lived for 12 years physically isolated in this plastic, sterile environment, venturing out only in a NASA-designed sterile space suit. A bone marrow transplant was attempted in 1984, but his defective immune system failed to protect him from Epstein-Barr virus undetected in the transplanted cells. He died just before his 13th birthday.

One in every million people develops SCID. The two most common forms of the disease result from genetic abnormalities linked to the X chromosome. Patients with X-linked SCID either lack an enzyme called adenosine deaminase (resulting in the toxic accumulation of deoxyguanosine triphosphate) or cannot produce an important cytokine receptor that T cells need for their development and to communicate with B cells. Cytokines are discussed in Section 23.2.

A dramatic therapeutic attempt to cure the disease took place in 2004 using stem cells. Stem cells are undifferentiated, but are capable of changing into many different cell types, including B cells and T cells (see Section 23.3). A team of scientists headed by Dr. Adrian Thrasher of University College, London, removed stem cells from the bone marrow of four SCID children who lacked the gene for the cytokine receptor. They inserted the gene encoding the normal cytokine receptor into a severely defective leukemia virus that can infect cells without causing disease. The debilitated leukemia virus served as a gene vector to deliver the normal cytokine receptor gene into the patient's stem cells. Once the genetic material entered the nucleus, the healthy copy of the gene began to function and the corrected stem cells were reintroduced into the patients. After this gene replacement therapy, all four children started making T cells with the correct receptor and produced functional B cells. The patients developed an immune system, were discharged, and are now living at home.

SCID dramatically illustrates the importance of our immune system and how fragile is its development. All it takes is a small defect in a single gene to subvert the entire process.

This chapter begins by describing the two types of adaptive immunity and the factors that influence the immunogenicity of foreign proteins, lipids, and so on. We will explore how B cells ultimately differentiate into plasma cells and make antibodies (defining what is called humoral, or circulating, immunity) and how certain T-cell lymphocytes develop to directly kill infected host cells in what is called cell-mediated or cellular immunity. You will also learn that another type of T cell controls the balance between humoral and cell-mediated responses to a given infection. Ultimately, you will appreciate that adaptive immunity is a major reason the human race still exists.

A.

B.

Courtesy of Texas Children's Hospital

Courtesy of UCL News

Figure 24.1 **Living without an immune system.** **A.** David Vetter, the bubble boy, inside his environmental bubble. The tube behind him is a port that was used to introduce sterile food, clothes, etc. **B.** Adrian J. Thrasher, University Hospital London, conducted a successful gene therapy trial on children with severe combined immunodeficiency.

24.1 Adaptive Immunity

As Chapter 23 points out, the immune system has both nonadaptive and adaptive mechanisms. Nonadaptive (innate, or nonspecific) immune mechanisms are present from birth. Adaptive immunity, in contrast, develops as the need arises. For instance, adaptive immunity against malaria is not developed until the individual has encountered the plasmodial parasite responsible for the disease. The **adaptive immune response** is a complex, interconnected, and cross-regulated defense network.

NOTE: The terms *adaptive immune response* and *immune response* are often used interchangeably.

Two types of adaptive immunity are recognized: humoral immunity and cell-mediated immunity. In **humoral immunity**, **antibodies** are produced that directly target microbial invaders. The term *humoral* means "related to bodily fluids." Thus, antibodies are proteins that circulate in the bloodstream and recognize foreign structures called antigens. An **antigen** (also called an **immunogen**) is any molecule that will, when introduced into a person, elicit the synthesis of antibodies that specifically bind the antigen. Antigens stimulate B cells (B lymphocytes) to differentiate into antibody-producing cells. **Cell-mediated immunity**, the second type of adaptive immunity, employs teams of T cells (T lymphocytes) that can also recognize antigens and then destroy host cells infected by the microbe possessing the antigen. In truth, the humoral and cellular immune responses are intertwined, each relying on some facet of the other to work efficiently. T cells serve a central role in adaptive immunity by determining whether humoral or cell-mediated mechanisms predominate in response to a specific antigen.

Adaptive immunity develops over a three- to four-day period after exposure to an invading microbe. The immune system does not recognize the *whole* microbe, but innumerable tiny *pieces* of it. Each small segment of an antigen that is capable of eliciting an immune response is called an **antigenic determinant** or **epitope**. Many single-protein antigens are recognized when the larger antigen is broken into smaller segments upon being phagocytized. Even distinct tertiary (three-dimensional) shapes within a protein may be counted as antigenic determinants if they produce a specific response. This happens when two or more stretches of amino acids are far removed from each other in a protein's primary sequence, yet protein folding aligns them side by side in three-dimensional space (**Fig. 24.2**). Such a three-dimensional structure may be recognized by the immune system as a single entity or antigen. Besides proteins, other structures in the cell, such as complex polysaccharides, can have linear and three-dimensional epitopes. So the immune response to a microbe is really a composite of responses to different epitopes by thousands of individual B cells, which are the cells responsible for producing antibodies. The response to each individual epitope is **clonal**; that is, it gives rise to a population of cells that originate from a single cell. This means that each clone of immune host cells will target a unique epitope.

As mentioned previously, the humoral response starts when an antigen triggers the differentiation of B cells into antibody-producing cells. In contrast, the cell-mediated immune response occurs when certain types of

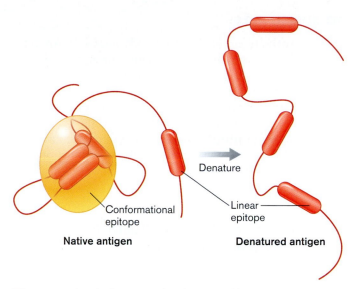

Figure 24.2 Antigens and epitopes. Native proteins fold into a three-dimensional shape, where several regions separated in the linear sequence can reside next to each other to form a conformational epitope. Denaturing the protein with the detergent sodium dodecyl sulfate (SDS) and reducing agents like dithiothreitol (to remove disulfide bonds) will unfold the protein and separate the various amino acid stretches that formed the conformational epitope.

T cells become activated by microbial antigens placed on infected host cell surfaces or on the surface of phagocytic cells that have engulfed the microbe. In cellular immunity, the activated T cell can directly kill the infected host cell in an attempt to kill the invading microbe. In addition, T cells synthesize soluble growth factors called cytokines (introduced in Section 23.5) that incite nearby macrophages to indiscriminately attack cells in the local area. Thus, cellular immunity, in general, is critical for dealing with intracellular pathogens such as viruses, whereas humoral immunity is most effective against extracellular bacterial pathogens like *Streptococcus pneumoniae*, one cause of pneumonia.

As we proceed through the chapter, we will reveal how the immune system functions in layers, with each layer building upon the previous one. As a result, we will periodically return to a particular aspect of the immune response—for example, B-cell differentiation into plasma cells—to integrate seemingly distinct parts of the immune system into a unified concept of immunity.

THOUGHT QUESTION 24.1   Two different stretches of amino acids in a single protein form a three-dimensional antigenic determinant. Will the specific immune response to that three-dimensional antigen also respond to one of the two amino acid stretches alone?

TO SUMMARIZE:

- **An antigen** can elicit an antibody response. An antigen is usually made of many different epitopes (antigenic determinants), each of which binds to a different, specific antibody.
- **Humoral immunity** against infection is the result of antibody production originated by B cells.
- **Cellular immunity** involves a type of lymphocyte called T cells, which control antibody production and can directly kill host cells.

24.2 Factors That Influence Immunogenicity

Immunogenicity measures the effectiveness by which an antigen elicits an immune response. One antigen can be more immunogenic than another. For example, proteins are the strongest antigens, but carbohydrates can also elicit immune reactions. Nucleic acids and lipids are usually weaker antigens. This is, in part, because these molecules are very flexible and present a variable three-dimensional structure that does not easily interact with antibodies. Nucleic acids and lipids are also weak antigens because they are both made of relatively uniform repeating units. Proteins are more effective antigens for three reasons: They form a variety of shapes, they maintain their tertiary structure, and they are made of many different amino acids that can be assembled in many different combinations. These features provide stronger interactions with antibodies in the bloodstream and enable better recognition by lymphocytes, the cellular workhorses of the immune system.

Several other factors contribute to the immunogenicity of proteins (**Table 24.1**). For example, the larger the antigen, the more likely it is that phagocytic cells will "see" and engulf it. This is important because for an immune response to occur, phagocytic cells such as macrophages and dendritic cells must first engulf large antigens and degrade them, presenting the epitopes on their cell surface. Phagocytic cells that degrade larger antigens and thus expose antigenic determinants are called **antigen-presenting cells (APCs)**. The immune response begins once a B cell binds to a foreign peptide and a T lymphocyte binds to the same foreign peptide displayed on the surface of an antigen-presenting cell.

Presentation of antigens on APCs requires that the antigen be placed on a membrane surface protein structure called the **major histocompatibility complex**, or **MHC** (discussed later). The better an antigen can bind to these MHC surface proteins, the more immunogenic it is. The stronger the binding, the easier it is for T cells to recognize the complex.

Each specific antigen shows a different **threshold dose** needed to generate an optimal response. A dose higher or lower than that threshold will not generate as strong an immune response. Lower doses activate only a few B cells. Exceedingly high doses of antigen can cause **B-cell tolerance**, a state in which B cells have been overstimulated to the point at which they do not respond to subsequent antigen exposures and make antibody. Tolerance is part of the reason your immune system does not react against your own protein antigens.

As you might expect, the body must regulate the immune system carefully so that a response is not leveled against itself. In effect, the immune system must become "blind" to its own antigens; as a result, the host will often be blind to foreign antigens that resemble epitopes of its own cells. Therefore, the more complex the foreign protein is, the more likely it will possess antigenic determinants that a lymphocyte can recognize as nonself. The further an antigen is from "self," the greater its immunogenicity will be.

Specific Interactions between Antibodies and Antigens Create Immunological Specificity

Smallpox is a devastating disease that caused enormous suffering and killed millions of people in the seventeenth and eighteenth centuries (**Fig. 24.3**). There was no cure,

---

Table 24.1   Factors that influence immunogenicity of antigens.

Parameter	High immunogenicity	Low immunogenicity
Size	Large	Small
Dose	Intermediate	High or low
Inoculation route	Subcutaneous > intraperitoneal > intravenous >	intragastric[a]
Composition	Complex	Simple
Form	Particulate	Soluble
	Denatured	Native
Differences from self proteins	Many	Few
Interactions with host MHC proteins	Effective	Ineffective

[a]Route of inoculation affects immunogenicity of a given antigen. In general, subcutaneous inoculation gives the strongest immune response while intragastric yields the weakest response.

A. Smallpox patient

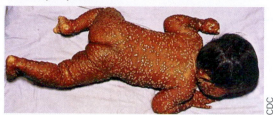

CDC

Figure 24.3 Immunological specificity is the basis of vaccination. **A.** Photo of a smallpox patient, showing the white pox pustules. **B.** The smallpox virus, variola major (300 nm long, TEM). **C.** The vaccinia virus that causes cowpox (360 nm long, electron micrograph). Edward Jenner recognized the similarity between the deadly smallpox and less severe cowpox diseases and used cowpox scrapings to vaccinate humans against smallpox.

B. Variola major

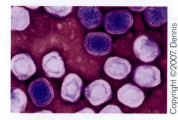

Copyright ©2007. Dennis Kunkel Microscopy, Inc.

C. Varicella

CDC/PHIL/Corbis

proteins critical to the pathogenesis of the two different microorganisms share key antigenic determinants. No cross-protection occurs if these determinants differ significantly. A good example is the common cold, which is caused by hundreds of closely related rhinovirus strains (rhinitis, a runny nose, is one of the symptoms of this viral disease). Infection with one strain will not immunize the victim against a second strain. The reason is that the structures of the viral proteins used to attach to the ICAM-1 protein on host cells differ dramatically between different strains of rhinovirus (**Fig. 24.4A**). Antibodies called neutralizing antibodies, which bind to the attachment protein on one strain of rhinovirus, will prevent infection by the same virus strain (**Fig. 24.4B**) but will not bind a similar but antigenically distinct ICAM-1 receptor protein from a different strain. A key to one lock will not work on a different lock.

> **THOUGHT QUESTION 24.2** How does a neutralizing antibody that recognizes a viral coat protein prevent infection by the associated virus?

and the only available preventive treatment was to take dried material from lesions of a previous smallpox sufferer, place it on a healthy person, and hope the person survived. Those who survived were protected from subsequent bouts of smallpox but were still susceptible to other diseases. This early observation gave rise to the idea of **immunological specificity**, which means that an immune response to one antigen is not effective against a different antigen. In other words, the immune response to smallpox will not protect someone against the plague bacillus (*Yersinia pestis*), which is antigenically different from smallpox.

While immunological specificity is important, it is not absolute. As described in Chapter 1, an English country physician named Edward Jenner (see Fig. 1.19) in the late eighteenth century (long before viruses were discovered) learned to protect townsfolk from deadly smallpox disease by inoculating them with scrapings from lesions produced by a tamer disease, cowpox. By this process, Jenner unwittingly transferred the vaccinia virus (which we now know causes cowpox; see **Fig. 24.3C**) to villagers susceptible to smallpox (caused by the related but more dangerous, variola virus). The resulting immune reaction to vaccinia produced an effective cross-protection against variola and thus prevented smallpox. This illustrates that an immune reaction against one organism or virus may be sufficient to protect against an antigenically related, if not identical, organism. The technique of exposing individuals to "tame" microbes to protect them against pathogens, now generally called **vaccination**, has been used to protect humans against many microbial and viral pathogens (**Table 24.2**). Most vaccinations today involve administering crippled (attenuated) strains of the pathogenic microbe or inactivated microbial toxins (for example, diphtheria toxin).

Cross-protection, where immunization against one microbe protects against a second, will work only if two

Antigens, Immunogens, and Haptens

Karl Landsteiner (1868–1943) discovered the ABO blood group system in 1901, for which he received the 1930 Nobel Prize in Physiology or Medicine (**Fig. 24.5A**). He observed that red blood cells (RBCs) from a type A individual, whose RBCs contained the A antigen, were destroyed (lysed) when transfused into a type B person, whose RBCs contained the B antigen. However, type A RBCs remained intact when introduced into another type A individual, and type B RBCs were undamaged when introduced into a type B individual (**Fig. 24.5B**). This is the phenomenon of blood group incompatibility, which occurs when specific antibodies in the serum of an individual of one blood type bind to antigens on red blood cells of a different type (that is, foreign RBCs). We now know that type B individuals carry specific anti-A antibodies in their bloodstream, while type A individuals carry specific anti-B antibodies. (Type AB people carry neither antibody, and type O individuals, whose RBCs contain neither A nor B antigen, carry both anti-A and anti-B antibodies.) The discovery by Landsteiner defined the basic concept of immunological specificity and explains why a type A person cannot donate blood to a type B individual, and vice versa. It also explains why a type O person (called a universal donor) can donate to type A, B, AB, or O individuals.

Table 24.2 Vaccines against viral and bacterial pathogens.

Disease	Vaccine	Vaccination recommended for:
Viral		
Chickenpox	Attenuated strain (will still replicate)	Children 12–18 months
Hepatitis A	Inactivated virus (will not replicate)	Travelers to endemic areas
Hepatitis B	Viral antigen	Medical personnel, children 1–18 months
Influenza	Inactivated virus or antigen	Adults over 65 years
Measles	Attenuated viruses; MMR combined vaccine	Children 15–19 months
Mumps		
Rubella		
Polio	Attenuated (oral, Sabin)	Children 2–3 years
	Inactivated (injection, Salk)	
Rabies	Inactivated virus	Persons in contact with wild animals
Yellow fever	Attenuated virus	Military personnel
Bacterial		
Anthrax	*B. anthracis*, components of toxin	Agricultural and veterinary personnel, key
	Unencapsulated strain	health care workers
Cholera	*Vibrio cholerae,* components	Travelers to endemic areas
Diphtheria	Toxoid (inactivated toxin)	Children 2–3 months
Pertussis	Acellular *Bordetella pertusis*	
Tetanus	Toxoid	
Haemophilus influenzae type b (meningitis)	Bacterial capsular polysaccharide	Children under 5 years
Lyme disease	Lipoprotein OspA surface antigen, *Borrelia burgdorferi*	Individuals in endemic areas
Meningococcal disease	Bacterial capsular polysaccharides, *Neisseria meningitidis*	Military and high-risk individuals
Pneumococcal pneumonia	Bacterial capsular polysaccharides, *Streptococcus pneumoniae*	Adults over 50 years
Tuberculosis (*Mycobacterium tuberculosis*)	Attenuated *Mycobacterium bovis* (BCG vaccine)	Exposed individuals
Typhoid fever	Killed *Salmonella typhi*	Individuals in endemic areas
Typhus fever	Killed *Rickettsia prowazekii*	Medical personnel in endemic areas and scientists

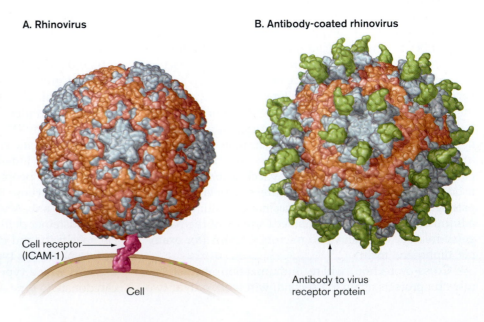

Figure 24.4   Antibodies prevent rhinovirus attachment to cell receptors.   A. This figure illustrates the complexity of the rhinovirus capsid and shows the attachment of the virus to the cell surface molecule ICAM-1 (intercellular adhesion molecule, shown in reddish brown). (PDB code: 1rhi) **B.** This figure shows rhinovirus coated with protective (neutralizing) antibodies (green) that block the ICAM-1 receptors on the virus. As a result, the virus fails to attach to and infect the host cell. (PDB code: 1rvf)

A. Rhinovirus

B. Antibody-coated rhinovirus

Cell receptor (ICAM-1)

Cell

Antibody to virus receptor protein

A.

Karl Landsteiner. *The Specificity of Serological Reactions.* Dover Publications.

B.

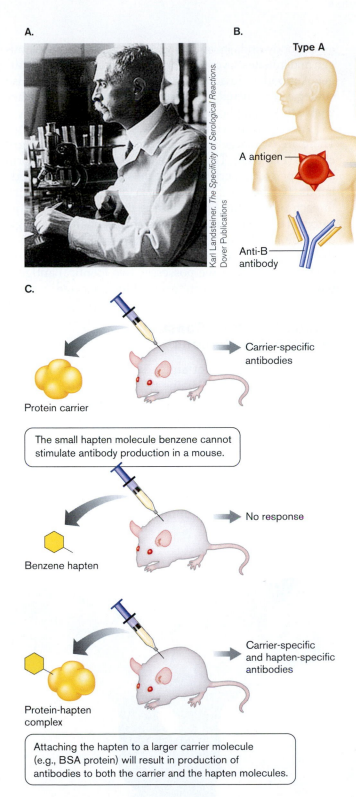

Type A

RBCs from a type A person are coated with A antigen. RBCs from a type B person are coated with B antigen.

A antigen

Transfusion

Type B

B antigen

Type A blood contains anti-B antibodies. Type B blood contains anti-A antibodies.

Anti-B antibody

Anti-A antibody

The anti-A antibodies in a type A person will attack and destroy transfused type B red blood cells.

Type A RBC is lysed

C.

Carrier-specific antibodies

Protein carrier

The small hapten molecule benzene cannot stimulate antibody production in a mouse.

No response

Benzene hapten

Carrier-specific and hapten-specific antibodies

Protein-hapten complex

Attaching the hapten to a larger carrier molecule (e.g., BSA protein) will result in production of antibodies to both the carrier and the hapten molecules.

Figure 24.5 Karl Landsteiner, ABO blood groups, and haptens. A. Karl Landsteiner discovered ABO blood groups and haptens. **B.** Individuals with type A antigens on their red blood cells possesses antibodies to B antigen, while type B individuals have B antigens on RBCs and carry anti-A antibodies. Transfusing RBCs from a type B individual into a type A individual will lead to the anti-B antibodies attacking and destroying the type B red blood cells. **C.** Basic concept of a hapten.

Landsteiner wanted to better understand this idea of immunological specificity. Because at that time nothing was known about antigen or antibody structure, he chose to study very simple molecules that, by themselves, were too small to elicit antibodies. Molecules of molecular weight less than 1,000 are generally not immunogenic (we now know the reason is that they do not bind MHC molecules);

however, Landsteiner discovered that these small molecules would elicit production of specific antibodies if they were covalently attached to a larger carrier protein or other molecule. He called these types of small antigens **haptens** (derived from the Greek word meaning "to fasten" and the German word for "stuff"). Haptens can be thought of as small incomplete antigens. An example of a hapten is the antibiotic penicillin, a serious cause of immune hypersensitivity reactions in some individuals (see Section 24.8).

Landsteiner chose to work with benzene haptens with sulfate bound at different positions of the ring. As illustrated in **Figure 24.5C**, a protein carrier like bovine serum albumin (BSA) injected into a mouse would elicit antibodies that react against BSA. Because BSA antigen elicits an immune response to itself, it is called an immunogen. In contrast, a mouse injected with the benzene-sulfate hapten alone fails to produce antibodies to the hapten. However, when the hapten is attached to BSA and the hapten-BSA complex is injected, the mouse produces antibodies that react to the carrier (BSA) as well as other antibodies that react against the benzene hapten. The reason for this carrier effect will be evident later as we describe how antigens are processed by cells of the immune system. Thus, antigens include immunogens that elicit an immune response by themselves, and haptens that must be attached to an immunogen in order to generate an immune response.

TO SUMMARIZE:

- **Proteins are better immunogens** than nucleic acids and lipids, because proteins have more diverse chemical forms.
- **Antigen-presenting cells** are cells, such as phagocytes, that degrade microbial pathogens and present distinct pieces on their cell surface MHC proteins.
- **Immunological specificity** means that antibody made to one epitope will not bind to different epitopes (although some weak cross-binding can happen).
- **A hapten** is a small compound that must be conjugated to a larger carrier antigen to elicit production of an antibody.

24.3 Antibody Structure and Diversity

Antibodies, also called **immunoglobulins**, are members of the larger immunoglobulin superfamily of proteins. Antibodies are the keys to immunological specificity. They are glycoproteins made by the body in response to an antigen. The immunoglobin superfamily of proteins has in common a 110-amino-acid domain with an internal disulfide bond. The immunoglobulin superfamily includes antibodies and other important binding proteins, such as the major histocompatibility proteins and B-cell receptors described later.

Like miniature "smart bombs," antibody immunoglobulins individually circulate through blood, ignoring all antigens except those for which they were designed. When an antibody finds its antigenic match, it binds to the antigen and initiates several events designed to destroy the target. Antibodies, in addition to being free-floating, are also strategically situated on the surfaces of B cells, where they enable these lymphocytes to recognize specific antigens.

An antibody consists of four polypeptide chains. There are two large **heavy chains** and two smaller **light chains** (**Fig. 24.6**). The four polypeptides combine to form a Y-shaped tetrameric structure held together by disulfide bonds. Two bonds connect the two identical heavy chains to each other. One light chain is then attached near its carboxyl end to the middle of each heavy chain by a single disulfide bond. The antigen-binding sites are formed at the amino-terminal ends of the light and heavy chains. One antibody molecule possesses two identical antigen-binding sites, one on each "arm" of the molecule.

Antibodies Have Constant and Variable Regions

There are five classes of antibodies, defined by five different types of heavy chains called alpha (α), mu (μ), gamma (γ), delta (δ), and epsilon (ϵ). The heavy-chain classes are distinguished one from another by regions of highly conserved amino acid sequences, known as **constant regions** (denoted C, **Fig. 24.7**). Antibodies containing gamma heavy chains are called IgG; those with alpha, delta, mu, and epsilon heavy chains are called IgA, IgD, IgM, and IgE, respectively. Each antibody class serves a specific purpose in the immune system.

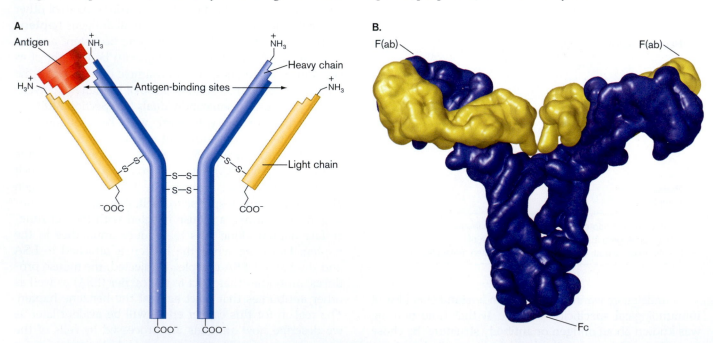

Figure 24.6 Basic antibody structure. **A.** Each antibody contains two heavy chains and two smaller, light chains held together by disulfide bonds. The Y-shaped structure contains two antigen-binding sites, one at each arm of the molecule. The two antigen-binding sites are formed by the amino-terminal regions of the heavy and light chain pairs. **B.** Three-dimensional structure of an antibody. The heavy chains are shown in blue, the light chains in yellow. The F(ab) regions represent the antigen-binding sites. The Fc portion points downward and is used to attach the antibody to different cell surface molecules. (PDB code: 1r70)

In contrast to heavy chains, there are only two classes of light chains: kappa (κ) and lambda (λ). An antibody of any heavy-chain class may contain two kappa chains or two lambda chains, but never one of each. Two-thirds of all antibody molecules carry kappa chains; the rest have lambda chains.

The antigen-binding part of an antibody is formed by highly variable amino acid sequences situated at the amino-terminal ends of the light and heavy chains. These **variable regions** are referred to as the V_L region and the V_H regions (see **Fig. 24.7**). The rest of the immunoglobulin chains are composed of the highly conserved constant regions. $C_H 1$, $C_H 2$, and $C_H 3$ denote the three different constant regions in each heavy chain. Each light chain also has a constant region, designated C_L. Section 24.5 discusses the genes that code for these regions and the combinatorial possibilities underlying the formation of different antibody molecules.

In addition to the two "arms" that bind antigen, every antibody contains a "tail" that serves a different function. Intact antibodies can be dissected into their two functional parts through digestion with the protease papain. This endoprotease cleaves the molecule at the hinge region, which releases the tail of the antibody (called the **Fc region**) from the antigen-binding portion of the molecule (called the **F(ab)$_2$ region**) (see **Fig. 24.7**). The c in Fc refers to the ease with which this fragment can be crystallized. The Fc region is not involved in antigen recognition but is important for anchoring antibodies to the surface of certain host cells and for binding components of the complement system.

Isotypes, Allotypes, and Idiotypes Represent Different Levels of Antibody Diversity

As noted earlier, antibodies are classified based on variations in amino acid sequences in regions of the light and heavy chains. Certain differences in the constant region give rise to **isotypes**, which are common to all members of a particular species. IgG, IgM, IgA, IgD, and IgE are antibody isotypes. The IgG isotype is identical among humans but is different from the IgG isotype of monkeys.

Within an isotype, there are amino acid differences in the constant region that are shared by some, but not all, members of a species. These are called *allo-*

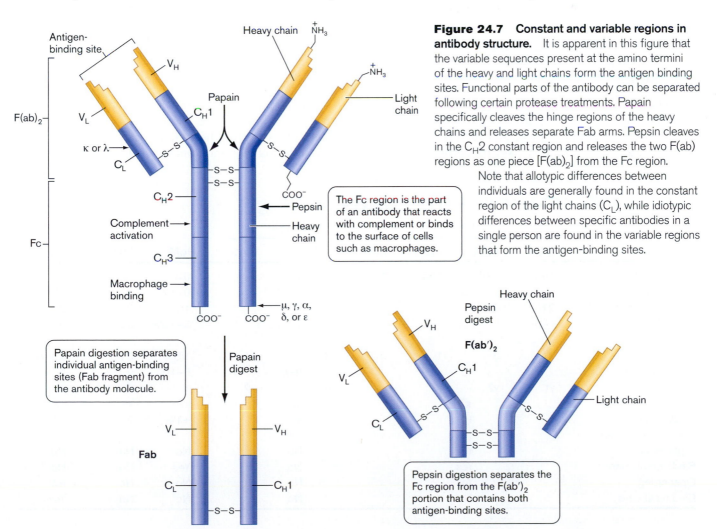

Figure 24.7 Constant and variable regions in antibody structure. It is apparent in this figure that the variable sequences present at the amino termini of the heavy and light chains form the antigen binding sites. Functional parts of the antibody can be separated following certain protease treatments. Papain specifically cleaves the hinge regions of the heavy chains and releases separate Fab arms. Pepsin cleaves in the $C_H 2$ constant region and releases the two F(ab) regions as one piece [F(ab)$_2$] from the Fc region.

Note that allotypic differences between individuals are generally found in the constant region of the light chains (C_L), while idiotypic differences between specific antibodies in a single person are found in the variable regions that form the antigen-binding sites.

The Fc region is the part of an antibody that reacts with complement or binds to the surface of cells such as macrophages.

Papain digestion separates individual antigen-binding sites (Fab fragment) from the antibody molecule.

Pepsin digestion separates the Fc region from the F(ab')$_2$ portion that contains both antigen-binding sites.

typic differences. Allotypic differences in an amino acid sequence usually occur in the light chains. For example, John possesses circulating IgG molecules that have allotypic differences from the IgG antibodies circulating in Sherrie. (John and Sherrie have different IgG **allotypes**.)

On another level, alterations in *hyper*variable regions within a single antibody class in a single person are referred to as *idiotypic* differences. These differences occur in the antigen-binding sites of the various antibodies of a single individual. Thus, the IgG molecules in Sherrie that bind to epitope A possess idiotypic differences from Sherrie's IgG molecules that bind epitope B. In sum, the human IgG isotype is composed of different allotypes (constant region differences between individuals), and each allotype is comprised of different **idiotypes** (variable region differences within an individual).

Note that antibodies are proteins and as such are themselves antigens. Thus, the amino acid differences found within a single class of antibody—IgG for example—also represent different *epitopes* within that class. Isotypic, allotypic, and idiotypic differences define epitopes in antibody molecules that can be detected using immunological techniques such as immunoprecipitation and Western blotting assays described in **Special Topic 24.1**.

Different Antibody Isotypes Have Different Functions and "Super" Structures

All antibody isotypes have the same basic structure. However, each isotype has a unique "super" structure (for example, monomer, dimer), and each is designed to carry out a different task. Some key properties of the five different immunoglobulin classes are listed in **Table 24.3**. IgG is the simplest and most abundant antibody in blood and tissue fluids. It is made as a monomer but has four subclasses. Each subclass varies in its amino acid composition and by the number of interchain cross-links. IgG molecules carry out several missions for the immune system. First, they bind and **opsonize** microbes; that is, they make the microbe more susceptible to phagocytes. Opsonizing IgG antibodies coat the microbe with their Fc portions protruding outward. Phagocytes possess surface Fc receptors that can attach to the Fc region of the antibody to gain a firmer "grip" on the microbe, facilitating phagocytosis (discussed in Section 23.6). IgG can also directly neutralize viruses by binding to virus attachment sites and is one of only two antibody types that can activate complement by the classical pathway (to be described in Section 24.7).

IgA is secreted across mucosal surfaces and is most commonly found as a dimer (**Fig. 24.8A**). This explains why IgA can bind four molecules of antigen (each monomer can bind two molecules of antigen). The components of the IgA dimer are linked by disulfide bonds to a protein called the J chain, which joins two IgAs by their Fc regions. A sixth protein, the secretory piece, is wrapped around the IgA dimer during the secretion process. The secreted molecule, now called **sIgA** (secretory IgA), is found in tears, breast milk, and saliva and on other mucosal surfaces. The molecule sIgA is important for mucosal immunity against pathogens.

Circulating **IgM** is a huge, Ferris wheel–shaped molecule formed from five monomeric immunoglobulins tethered together by the J-chain protein (**Fig. 24.8B**). It can also be found in monomeric form on the surfaces of B

Table 24.3 Properties of human immunoglobulins.

Property	IgG				IgA		IgM	IgD	IgE
	IgG1	IgG2	IgG3	IgG4	IgA1	IgA2			
Mol. wt. (kDa)	146	146	165	146	160		970	184	188
% Carbohydrate		3			7–11		9–12	9–11	12
Serum half-life (days)	21	20	7	21	6		10	3	2
% Total serum Ig		70%			15–20%		5–10%	0.2%	0.002%
Avr. concentration (mg/ml)	9	3	1	0.5	3	0.5	1.5	0.03	5×10^{-5}
Ag-binding sites		2			2–4		5–10	2	2
Heavy chain	γ1	γ2	γ3	γ4	α1	α2	μ	δ	ε
Light chain		κλ			κλ		κλ	κλ	κλ
Produced by fetus		+/−			+/−		+++	?	+/−
Transmitted across placenta		Yes			No		No	No	No
Bind complement		Yes			No		Yes	No	No
Opsonizing		Yes			No		No	No	No
Bind mast cells		No			No		No	No	Yes

cells, where it forms part of the B-cell receptor. IgM is the first antibody isotype detected during the early stages of an immune response. Unlike the smaller IgG immunoglobulins, IgM is so large that it cannot cross the placenta (see **Table 24.3**).

Two other antibody isotypes are present at very low levels in the blood. **IgD** is a monomer that can neither bind complement nor cross the placenta. These molecules are, however, abundantly found on the surface of B cells.

Attached to the cell surface by their Fc regions, IgD, along with monomeric IgM, can bind antigen and signal B cells to differentiate and make antibody.

IgE is also present in trace amounts in the blood, but is found more prominently bound to the surface of mast cells and basophils, where it has potent biological activity. Mast cells and basophils contain granules loaded with inflammatory mediators. The primary role of IgE is to amplify the body's response to invaders. Once secreted into serum, IgE

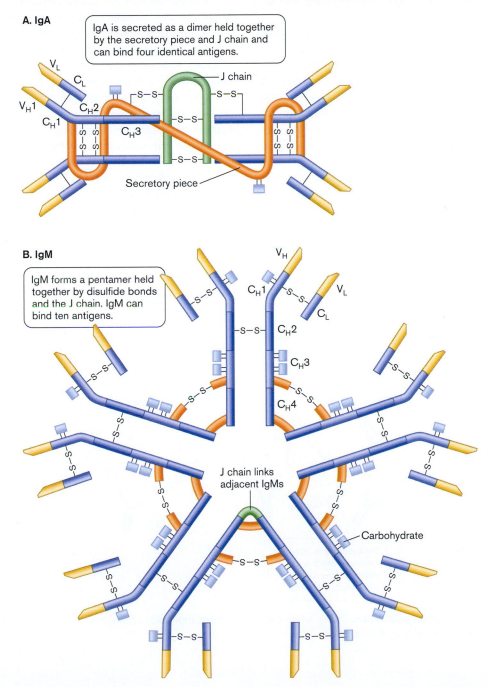

Figure 24.8 Structures of IgA and IgM. The antibodies are made as multimers of two (IgA) or five (IgM) immunoglobulin molecules.

Special Topic 24.1   Applications Based on Antigen-Antibody Interactions

A typical antibody molecule will bind only one type of antigen but has two antigen-binding sites. Because of the ability to bind to more than one antigen molecule, antibodies can cross-link antigens in solution, ultimately forming complexes too large to remain soluble (**Fig. 1**). The phenomenon, called **immunoprecipitation**, is normally only observed in vitro, where the concentration of antigen and antibody can be manipulated experimentally.

Immunoprecipitation occurs only with appropriate ratios of antigen and antibody molecules. Too many antigen molecules (antigen excess; **Fig. 1A**) or too few antigen molecules (antibody excess; **Fig. 1B**) result in complexes too small to immunoprecipitate. Large complexes are formed only at an appropriate antigen:antibody ratio called **equivalence (Fig. 1C)**. Equivalence is the point where the number of antigenic sites is roughly equal to the number of antigen-binding sites.

Immunoprecipitation is the basis for many experimental immunological techniques. For example, the concentration of an antigen can be determined in vitro, or antibodies can be used to identify and remove specific antigens from a complex mixture because the specific antigen-antibody complex falls out of solution (**Figs. 2A** and **B**). **Figure 2B** shows how specific proteins can be isolated from a complex cell extract using immunoprecipitation. Antibody is added to a complex mix of different antigens. The antibody, however, can only bind to its specific antigen. Beads coated with a molecule known as protein A (derived from *Staphylococcus aureus*)

are added to specifically purify that antigen-antibody complex. Protein A binds to the Fc portion of an IgG antibody molecule. Consequently, all the antigen-antibody complexes will become bound to the bead. Because the bead is heavy, centrifugation will remove the desired antigen from the complex mixture.

Another application, called **radial immunodiffusion**, allows the concentration of an antigen in a solution to be determined (**Fig. 2C**). This technique involves visualizing a ring of precipitation in an agarose gel impregnated with antibody. Antigen placed within a well will diffuse outward until reaching a zone of equivalence with the embedded antibody. At this point, antigen-antibody complexes precipitate and form a ring a certain distance from the well. Antigen concentration progressively decreases the farther the antigen diffuses away from the well. Consequently, the higher the concentration of antigen that is originally present in the well, the farther it will have to diffuse before the zone of equivalence is reached and a ring of precipitation forms. The concentration of antigen in an unknown solution is determined by comparing the radius of immunoprecipitation formed against a standard curve in which known concentrations of antigen are plotted against the radius of the ring of precipitation.

Another important research technique involving antibodies is the **Western blot**, which is used to detect the presence of a specific protein in cell extracts. As described in Section 12.1, proteins are separated using SDS PAGE

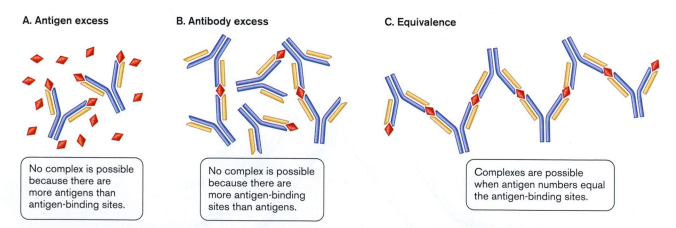

A. Antigen excess

No complex is possible because there are more antigens than antigen-binding sites.

B. Antibody excess

No complex is possible because there are more antigen-binding sites than antigens.

C. Equivalence

Complexes are possible when antigen numbers equal the antigen-binding sites.

Figure 1   Basis of immunoprecipitation.   Only when the number of epitopes and antigen-binding sites are roughly equivalent will a large complex form and fall out of solution.

electrophoresis. The proteins are transferred (that is, blotted) from the gel onto a nitrocellulose or other membrane. The membrane is then probed with an antibody directed against a specific protein. Antibody sticks to the protein in ques-

tion, and because that antibody is labeled in some way (for example, a radioactive or fluorescent probe has been added), the protein band can be visualized by autoradiography or phosphoimaging.

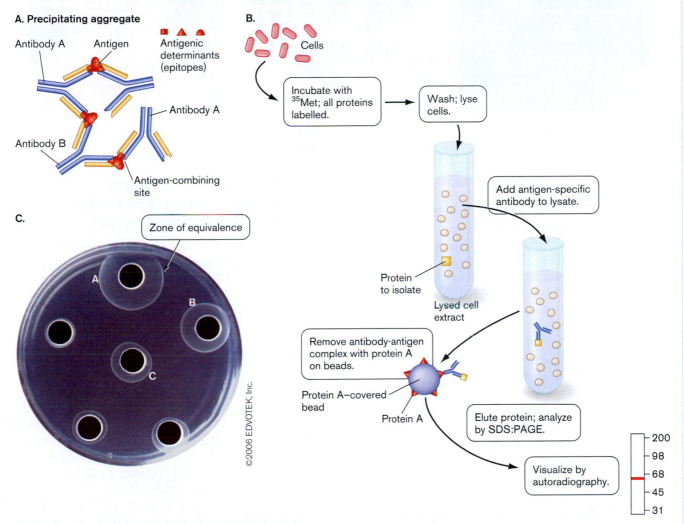

Figure 2   Applications based on antigen-antibody behavior.   A. Precipitating aggregate. Antibodies to different antigenic determinants residing on antigen molecules can cross-link the antigens to make a huge, insoluble complex that precipitates. **B.** Purifying proteins. Removing specific proteins from a complex cell extract using immunoprecipitation. In the experiment, antibody to a specific cellular protein (shown as yellow box) is added to a lysed cell extract. Protein A beads are then added. Protein A will bind to the Fc region of the antigen-antibody complex. Centrifugation then is used to pull down the beads and the protein with it. The protein is eluted from the antibody and analyzed by SDS PAGE (Protein A remains covalently attached to the bead). **C.** Radial immunodiffusion. The agarose plate shown is embedded with antibodies specific for a certain antigen. Different concentrations of the antigen are placed in each well. The antigen diffuses into the agarose. A ring of precipitation occurs when the concentration of the diffusing antigen reaches a zone of equivalence with the antibody in the plate. The more antigen placed in the well, the larger the diameter of the precipitation ring. Well A was loaded with more antigen than wells B or C.

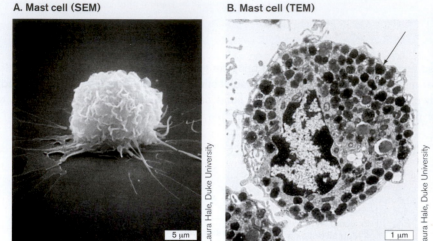

A. Mast cell (SEM)

B. Mast cell (TEM)

C. Hayfever

5 µm

1 µm

Laura Hale, Duke University

Laura Hale, Duke University

Mark Clarke

Figure 24.9   Mast cells.   A. Scanning EM of a mast cell. **B.** Electron transmission EM showing granules (arrow). **C.** Hayfever is the result of degranulation of IgE-coated mast cells, which release histamine and other pharmacological mediators.

attaches to mast cells (**Figs. 24.9A** and **B**), again by way of its Fc region, and, like a Venus flytrap, waits until its matched antigen binds to its antigen-binding site. When surface IgE molecules of two mast cells are cross-linked by antigen, a signal is sent internally that triggers degranulation (see hypersensitivity, Section 24.8). The subsequent release of histamine and other pharmacological mediators from these granules helps orchestrate the acute inflammation that takes place during early host responses to microbial infection (that is, while the antibody response is gearing up). The system is also responsible for severe allergic hypersensitivities, such as anaphylaxis, and milder forms like hay fever (see **Fig. 24.9C**).

TO SUMMARIZE:

- **Antibodies, or immunoglobulins**, are members of the immunoglobulin superfamily.
- **Antibodies are Y-shaped** molecules that contain two heavy chains and two light chains.
- **There are five classes (isotypes) of heavy chains.** Each antibody isotype is defined by the structure of the heavy chain.
- **Each antibody molecule contains two antigen-binding sites.** Each binding site is formed by the hypervariable ends of a heavy and light chain pair.
- **The Fc portion** of an antibody can bind to specific receptors on host cells. This binding is antigen independent.

THOUGHT QUESTION 24.3   There are immune disorders in which an individual overproduces a specific class of antibody, for example hypergammaglobulinemia. How could the radial immunodiffusion technique be used to identify what class of antibody is in excess?

24.4 Humoral Immunity: Primary and Secondary Antibody Responses

After a lag period of several days following primary immunization or natural infection, antibodies begin to appear in the **serum** (the fluid that remains after blood clots). During the lag period, a series of molecular and cellular events occur that cause a distinct subset of B cells to proliferate and differentiate into antibody-secreting **plasma cells** and **memory B cells**. This process is known as the **primary antibody response**, the events of which will be discussed later. A secondary exposure to the antigen, which can take place months or years after the initial encounter, will trigger a rapid, almost instantaneous increase in the production of antibodies and is called the **secondary antibody response** (**Fig. 24.10**). This quick response occurs thanks to the memory B cells formed during the primary response. Once stimulated, memory B cells rapidly differentiate into plasma cells and secrete antibody.

The net result of the primary antibody response is the early synthesis and secretion of IgM molecules specifically directed against the antigen, also called the immunogen. Later during the primary response, a process known as **isotype switching (class switching)** occurs, by which the predominant antibody type produced becomes IgG rather than IgM (discussed shortly). Antibodies made during this primary phase, while specific for the immunogen, are actually *not* of the highest affinity. Mechanisms to increase antibody affinity occur later.

As the immunogen is cleared from the body, the levels of both IgG and IgM decline because the plasma cells that produced them die. Plasma cells have a life span of only 100 days. However, the immune system has been primed to respond more aggressively to the immunogen should it encounter it again. This is because the memory B cells produced as a result of the primary response react to anti-

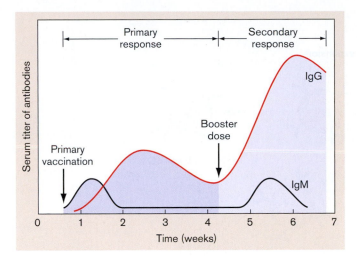

Figure 24.10   Primary versus secondary antibody response.
Primary vaccination or infection leads to the early synthesis of IgM
followed by IgG. Reinfection or a second, booster dose of a vaccine
results in a more rapid antibody response comprised mainly of IgG
due to memory B cells formed during the primary response.

gen more quickly and take less time to make antibody than
naive B cells that had no prior exposure to the antigen.

B cells are maintained in the body because, unlike
plasma cells, they continue to divide. If a person encounters the same antigen at a later time, memory B cells
quickly proliferate and differentiate into plasma cells with
no lag phase. Thus, memory B cells quickly initiate the secondary antibody response (or anamnestic response, from
the Greek *anamnesis*, meaning "remembrance"). During
the secondary response, copious amounts of IgG antibody
are secreted from plasma cells made from memory B cells
that have undergone isotype switching. These antibodies
have a higher specificity for the antigen than the antibodies produced during the primary response. Small amounts
of IgM are also produced from the few memory cells that
did not undergo isotype switching during the primary
response. (Most memory B cells undergo isotype switching and produce IgG rather than IgM, but a few do not.)

The speedy production of antibody during the secondary response is the basis of immunization. Pathogens
can do considerable harm during the lag phase of the primary response. To avoid this, an innocuous version of a
pathogen, or a harmless piece of it, can be injected into a
person to trigger the primary response without producing
disease (or, at worst, producing only a mild form of the illness). Immunization thus primes the immune system to
respond efficiently and without delay upon encountering
the real pathogen. **Table 24.2** lists a variety of viral and
bacterial diseases for which immunizations are available.

> **THOUGHT QUESTION 24.4**   The mother of a
> newborn was found to be infected with rubella, a viral
> disease. Infection of the fetus could lead to serious
> consequences for the newborn. How could you deter-
> mine if the newborn was infected while in utero?

B-Cell Differentiation into Plasma Cells Occurs by Clonal Selection

As mentioned earlier, each B cell circulating throughout the body or ensconced in a lymphoid organ is programmed to synthesize antibody that reacts with a *single*
epitope. In a process called **clonal selection**, an invading antigen will inadvertently select which B-cell clone
will proliferate to large numbers and differentiate into
antibody-producing plasma cells or memory B cells. In
this way, large amounts of antibody specific for the antigen are made. The mechanism of clonal selection begins
with the antigen binding to a matching B cell (a B cell
preprogrammed to bind to that antigen) (**Fig. 24.11**).

Mature naive B cells (those that have not previously encountered antigen) can produce only IgM and
IgD, both of which have identical antigen specificities.
These two antibody classes are displayed like tiny satellite dishes on the B-cell surface, anchored by their Fc
regions through hydrophobic transmembrane segments.
These surface antibodies are the keys to stimulating the
proliferation and differentiation of B cells into antibody-
producing plasma cells or memory B cells. Upon binding
to its corresponding antigen via these surface antibodies,
the B cell is said to become **activated**, whereby it multiplies and differentiates into a plasma cell that ultimately
synthesizes only one antibody isotype (for example, IgG1
or IgA2). Clonal selection has begun. (Remember that
some of the activated B cells become memory B cells and
do not become antibody-secreting plasma cells.)

Each membrane-bound antibody on the B cell is associated with two other membrane proteins called Igα and
Igβ (these are not immunoglobulins but are designated
Ig because they *associate* with the surface antibody). The
complex is called the **B-cell receptor** (**Fig. 24.12**). Each B
cell may have upwards of 50,000 B-cell receptors. When B
cells bind antigen, they often, but not always, differentiate into plasma cells (some become memory cells). During this transformation process, antibody heavy-chain
isotype switching (class switching) occurs. Class, or isotype, switching changes the class of antibodies produced.
After binding to the antigen, the surface B-cell receptors
begin to cluster in a process called **capping**. Capping
activates Igα and Igβ to initiate a phosphorylation signal
cascade directed into the nucleus (**Fig. 24.13**). The transcription factors at the end of the cascade stimulate transcription of genes that contribute to cellular activation
and DNA recombination events involved in heavy-chain
class switching. In addition, B-cell receptors trigger endocytosis of a bound antigen. Once inside the B cell, the
antigen is degraded, processed, and repositioned on the
B-cell membrane, a step that will ultimately enable T cells
to directly bind and communicate with the B cell, giving
the B cell "permission" to become an antibody-secreting
plasma cell. How these antigens are placed on the B-cell
membrane will be discussed in Section 24.6.

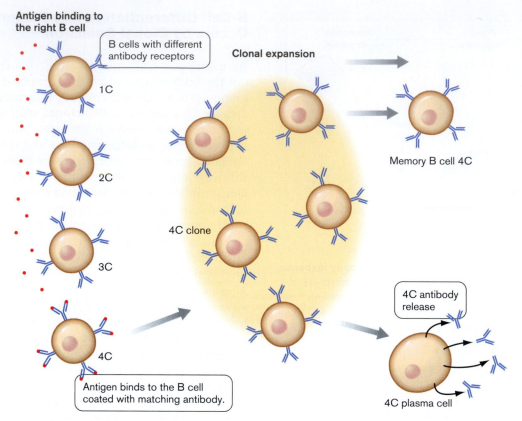

Figure 24.11   Clonal selection theory. The B-cell population is composed of individuals that have specificity for different antigens. When a B cell contacts its cognate antigen, an intracellular signal is generated, leading to proliferation and differentiation of that clone (clonal expansion). Plasma cells and memory B cells result.

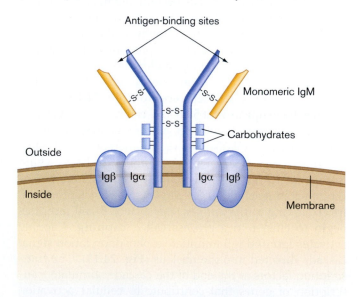

Figure 24.12   B-cell receptor. The B-cell receptor is formed as a complex between a monomeric IgM and Igα and Igβ in the membrane. Igα and Igβ are not immunoglobulins.

There are two routes by which antigens can stimulate B cells to differentiate into plasma cells. In one, called the T-cell-independent route, antigens that possess multiple repeating epitopes can directly cross-link B-cell receptors (the capping process), a step essential for triggering differentiation (see **Fig. 24.13**). Proteins, however, which are the largest group of antigens, do not contain multiple repeating units. Proteins possess many small, discrete, single epitopes, making cross-linking of B-cell receptors difficult. B-cell responses to these types of antigens require help from specific T cells, and this constitutes the second route to B-cell activation. Thus, B cells usually require multiple signals to initiate a primary response. How T cells help foster B-cell activation will be discussed in Section 24.6.

TO SUMMARIZE:

- **The primary antibody response** to an antigen begins when B cells differentiate into antibody-producing plasma cells and memory B cells. IgM antibodies are generally the first class of antibodies made during the primary response.
- **Isotype switching** occurs during the primary response when a subclass of B cells switches during differentiation from making IgM to making other antibody isotypes.
- **The secondary antibody response** occurs during subsequent exposures to an antigen and arises because memory B cells are activated. IgG is the predominant antibody made.

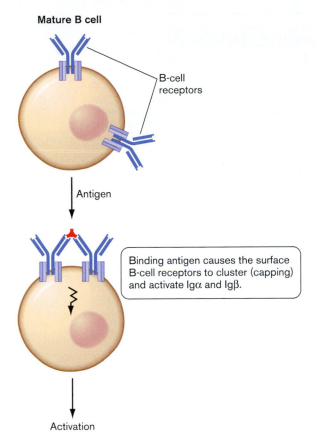

Mature B cell

B-cell receptors

Antigen

Binding antigen causes the surface B-cell receptors to cluster (capping) and activate Igα and Igβ.

Activation

Figure 24.13 **Capping and activation of the B cell.** The capping process initiates a signal cascade that activates differentiation and proliferation of the B cell.

- **Clonal selection** is the rapid proliferation of a subset of B cells during the primary or secondary antibody response.
- **A B-cell receptor** consists of a membrane-bound antibody in association with the Igα and Igβ proteins. Binding of antigen to the B-cell receptor triggers B-cell proliferation and differentiation.

24.5 Genetics of Antibody Production

It is estimated that each human can synthesize 10^{11} different antibodies. Given that each B cell displays antibodies to only one antigenic determinant, it follows that there are 10^{11} different B cells in the body. We have learned, however, that each person possesses only about 1,000 genes or gene segments involved in antibody formation. How are 10^{11} different antibodies made from only 10^3 genes? Susumu Tonegawa was awarded the 1987 Nobel Prize in Physiology or Medicine for discovering that antibody genes can move and rearrange themselves within the genome of a differentiating cell. Three steps are involved: rearrangement of antibody gene segments (or

cassettes), the random introduction of somatic mutations, and the generation of different codons during antibody gene splicing. In humans, the process of generating antibody diversity takes place constantly over a lifetime.

DNA Rearrangements in Gene Splicing and Hypermutation Generate Antibody Diversity

The first step in making a specific antibody occurs during the formation of a B cell from a progenitor stem cell in bone marrow. Immunoglobulin genes in a bone marrow stem cell consist of many gene segments that can rearrange in many possible combinations. During differentiation of a stem cell into a mature B cell, DNA segments are deleted in a process called **gene splicing**, decreasing the number of gene segments in the mature B-cell DNA. The process starts at the 5′ end of an immunoglobulin gene cluster, which corresponds to the variable (V) end of the ultimate peptide (**Fig. 24.14**). The 5′ end encodes the N-terminal end of the protein that ultimately binds antigen (that is, the antigen-binding site).

In both the heavy- and light-chain genes, there are a number of tandem gene cassettes encoding potential variable regions separated by **recombination signal sequences (RSS)**. These sequences allow recombination to bring two widely separated gene segments together. There are approximately 135 V gene segments for the heavy and light chains. The light-chain variable-region gene cluster lies upstream of a cluster of J (joint or joining)-region genes, which are ultimately used to join the variable region to the light-chain constant regions (C), whose genes reside farther downstream in the DNA. The arrangement of the heavy-chain genes is slightly more complex. In this case, the heavy-chain V cluster is followed by a D (diversity) cluster, then the J region.

> **NOTE:** Do not confuse the J-region gene segments used to make heavy- and light-chain proteins with the J-chain protein that holds together IgM and IgA multimers. They are completely different and unrelated. The J-region gene segments do not encode the J chain.

During the formation of each mature B cell, recombination between the various recombination signal sequences deletes all but one heavy-chain variable region (see **Fig. 24.14**). As a result of the deletion, a single V segment is placed next to one D region. The structure of the particular peptide encoded by the selected V region begins to limit antigen recognition properties. At the same time, all but one D segment and one J segment are deleted. The result is the VDJ portion of the heavy chain. A similar sequence of events occurs for the light chains, only the product is VJ. **Table 24.4** illustrates the amount of

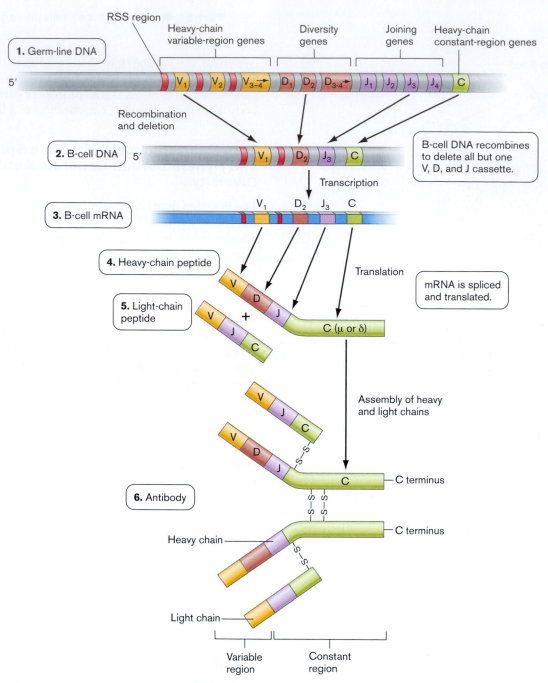

Figure 24.14 **Formation of the VDJ regions of heavy chains.** Note that only a small subset of the V, D, and J genes listed in Table 24.4 are actually shown in this model.

Table 24.4 **Antibody diversity attributed to combinatorial joining in the human germ line.**

Chain type	Number of			Number of combinations
	V regions	D regions	J regions	
λ light chains	30	0	4	$30 \times 4 = 120$
κ light chains	40	0	5	$40 \times 5 = 200$
Heavy chains	65	27	6	$1,000 \times 30 \times 4 = 120,000$
Number of possible antibodies	$7,800 \times 200 = 1.56 \times 10^6$			$65 \times 27 \times 6 = 7,800$
	$7,800 \times 120 = 0.94 \times 10^6$			
	$1.56 \times 10^6 + 0.94 \times 10^6 = 2.5 \times 10^6$ combinations			

antibody diversity in humans that can be achieved simply by this combinatorial rejoining (a total of about 2.5×10^6 antigens can be recognized).

Where does the rest of the diversity arise? In addition to recombination, the V regions of the germ lines are susceptible to high levels of somatic mutation, resulting in the hypervariable regions. Hypermutation happens every time a memory B cell is exposed to the antigen: The memory B cells divide and the hypervariable regions mutate. Additional diversity comes from the junctions of VJ and VDJ, where recombinational splicing can occur between different nucleotides. Each gene splice event can generate additional codons, so the resulting peptides will differ by one or more amino acids. The interactions between the light- and heavy-chain hypervariable regions in an antibody form the antigen-binding sites.

In sum, a combination of gene splicing and random mutations creates the remarkable level of antibody diversity we each possess. As noted earlier, the human body is capable of responding to 10^{11} antigens; yet there are only 10^8 estimated antigens in nature. This apparent overkill suggests that the immune system is well prepared to cope with any possible antigen it could encounter. Unfortunately for humans, enterprising microbes, such as the trypanosomes that cause sleeping sickness, can stay one step ahead of the immune system by changing the structure of key surface antigens. Changing the antigenic structure of a protein renders useless those antibodies made to the previous structure.

It isn't hard to understand why multicellular organisms, like humans, need the capacity to make any one of billions of different antibodies quickly. Pathogens can undergo many generations of growth within the single life span of a human host. So humans have to generate recombinant clones of cells quickly to overcome rapidly dividing pathogens.

Isotype Class Switching Is the Result of Gene Splicing

The first stage in isotype switching involves the transition of the immature B cell in bone marrow to a mature B cell, which ends up in circulation or in lymph nodes. The immature B cell (a B cell that has not yet "seen" the antigen but has already assembled its immunoglobulin VDJ-binding site) produces only monomeric IgM as part of the B-cell receptor on its cell surface. The immature B cell ultimately differentiates into a mature B cell (a naive B cell), which then makes both IgM and IgD B-cell receptors (the naive B cell is referred to as IgM$^+$ IgD$^+$). This type of immunoglobulin class switching does not involve recombination at the DNA level, but occurs through the splicing of RNA. For instance, the immunoglobulin mRNA transcript in these cells includes both the mu- and delta-region sequences (the sequences that

encode the IgM and IgD isotypes, respectively). Unprocessed, the transcript will yield IgM. However, the part of the transcript containing the mu constant region can be spliced out, in which case the processed transcript will make IgD.

The primary immune response is launched when a mature (naive) B cell receptor finds its matched antigen. As mentioned, in the early stages of the response, plasma cells produced from these B cells secrete only IgM, and an antigen-activated mature B cell will make IgM and IgD (IgM$^+$ IgD$^+$). If this activated B cell also receives appropriate signals from a certain type of T cell known as a **helper T cell (T$_H$ cell)**, further immunoglobulin isotype switching will occur; the switched B cell may then make either IgG, IgA, or IgE. The type of switch is influenced by small peptides called **cytokines** that are secreted by helper T cells.

How does the switch occur? Notice in **Figure 24.15** that the constant-region gene segments defining different antibody heavy-chain classes are arranged in tandem *after* a VDJ region. The mechanism by which a B cell switches to make IgG (C regions gamma 1–4), IgE (C-epsilon), or IgA (C-alpha 1 and 2) is very similar to VDJ formation. Each constant segment, except delta, contains a repeating DNA base sequence called a **switch region**. Recombination between these switch regions will delete the intervening DNA between the VDJ region and one of the constant regions. The primary RNA transcript produced after this recombination event has all of its introns spliced out before translation. Because the VDJ region is the same regardless of which C$_H$ gene is selected, whatever antibody is produced will have the same antigenic specificity as the original IgM. However, the heavy-chain switch selection process is *not* random. The type of cytokine present at the time of the switch will influence which C$_H$ gene is selected.

Note that any heavy-chain peptide can combine with any light chain. However, a single, mature B cell can only make one type of heavy chain and one type of light chain.

> **THOUGHT QUESTION 24.5**   Why does the delta region have no switch region?
>
> **THOUGHT QUESTION 24.6**   Why do individuals with type A blood have anti-B and not anti-A antibodies?

Making Memory B Cells

As illustrated in **Figure 24.11**, an antigen-activated mature B cell may become a memory B cell. These are B cells that have already undergone class switching (that is, they are committed to making IgG) but are

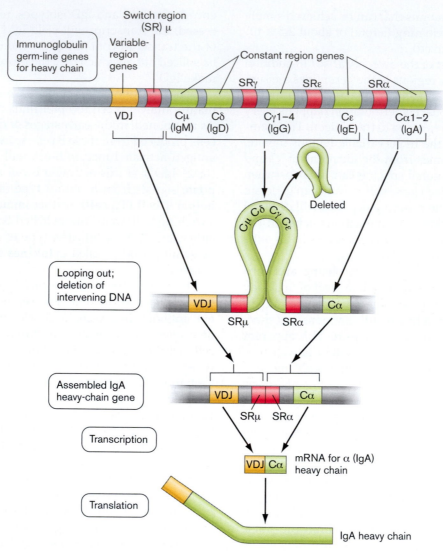

Figure 24.15 Heavy-chain class switching. As a B cell becomes activated, a switch in antibody isotype will occur. The switch involves recombination between isotype cassettes that brings one heavy-chain constant region (Cα in the example, for IgA) in tandem with a VDJ sequence. RSS = recombination signal sequences.

very long-lived. Normally, chromosome ends (called telomeres) in every cell progressively shorten with each round of DNA replication. When they become too short, the cell dies. Memory B-cell production is associated with an increase in telomerase activity. Recall that telomerase is a reverse transcriptase that shares common ancestry with the reverse transcriptase of the virus HIV. Telomerase activity maintains the length of the telomeres and prolongs the life of the cell.

Memory B cells hypermutate the VJ regions of antibody genes (as already described). These mutations increase the affinity of the antibody produced during the secondary response. Hypermutation helps fine-tune the immune response, making it even more specific toward a given microbe. Note, too, that as an immune response progresses, antigen becomes scarce. This means that only B cells with the highest affinity antibodies as part of

their B cell receptors will remain activated. The process is known as affinity maturation.

THOUGHT QUESTION 24.7   Why do immunizations lose their effectiveness over time?

TO SUMMARIZE:

- **Each B cell is programmed** to make primarily a single antibody isotype that specifically binds one epitope.
- **Antigens bound** to the surface IgM or IgD antibodies on mature B cells are internalized, degraded, and redisplayed on MHC molecules at the B-cell surface.
- **B cells differentiate** into antibody-secreting plasma cells after a helper T cell interacts with the MHC-antigen complex.

- **During the primary antibody response**, IgM is the predominant class of antibody secreted.
- **Memory B cells** rapidly proliferate and differentiate into antibody-secreting plasma cells upon a second encounter with an antigen/epitope. Most antibody made during this secondary response is IgG.
- **Diversity of idiotype** occurs via a complex series of splicing events between adjacent DNA cassettes as well as mutational events in DNA sequences encoding the hypervariable regions of heavy and light chains.

24.6 T Cells, Major Histocompatibility Complex, and Antigen Processing

The following section describes how antigens are recognized by T cells and how T cells then influence both humoral and cellular immunities. In the process, we will expand on how B cells become activated once they bind antigen, and will begin to explain why a host usually does not react against itself.

The T Cell Links Humoral and Cell-Mediated Immunity

Recall that there are two general types of immune response: humoral (antibody-based) immunity and cell-mediated immunity. Different types of infection tilt the immune response toward one or the other. B cells are clearly linked to the humoral immune system because they ultimately become plasma cells that make antibodies. T cells, on the other hand, serve as a nexus, or link, between humoral and cell-mediated responses. They play integral roles in both antibody production and cell-mediated immunity. Although derived from the same progenitor stem cell as B cells, T cells develop in the thymus (rather than in the bone marrow, where B cells develop) and contain surface antigens different from those of B cells. In general, T cells can be divided into two broad groups differentiated by the type of cellular differentiation (CD) proteins present on their cell surfaces (**Table 24.5**).

One of the two major types of T cells has already been mentioned: the so-called helper T cells (T_H cells). The second major type of T cells are known as **cytotoxic T cells (T_C cells)**. Helper T cells display the surface antigen CD4, while cytotoxic T cells display CD8.

Helper T cells come in three types: T_H1 cells, which assist in the activation of cytotoxic T cells; T_H2 cells, which stimulate B-cell differentiation into plasma cells; and T_H0 cells, which are really undifferentiated precursors of T_H1 and T_H2 cells. Cytotoxic T (T_C) cells are the "enforcers" of the cell-mediated immune response. They destroy the membranes of host cells infected with viruses or bacteria. The ratio between the number of T_H1 cells and T_H2 cells produced during an infection affects whether the immune response will lean more toward humoral (antibody, T_H2) or cell-mediated (T_H1) immune mechanisms. As will be described, different cytokines produced during infection influence whether an activated T_H0 cell develops into a T_H1 or T_H2 cell.

Why are T_H and T_C cells different? The reasons are many but it all starts with their relative abilities to recognize two different protein complexes that are present on antigen-presenting cells. These complexes are called the major histocompatibility proteins.

The Major Histocompatibility Complex (MHC) Is Critical to the Immune System

The factors affecting immune responsiveness are many, but the major histocompatibility complex (MHC) proteins encoded by a set of genes on a single chromosome, play a particularly important role. MHC proteins differ between species and among individuals within a species. They help determine whether a given antigen is recognized as coming from the host (a self antigen) or from another source (a foreign antigen), in a phenomenon called histocompatibility. Thus, the name major histocompatibility complex. There are two classes of MHC molecules found on cell surfaces. Both classes belong to the immunoglobulin family of proteins, but they are not immunoglobulins. Class I MHC receptors (**Fig. 24.16A**) are found on all nucleated cells. Class II MHC molecules (**Fig. 24.16B**), on the other hand, have a more limited distribution, principally on antigen-presenting cells.

Table 24.5 Major classes of T cells.

T-cell type	Coreceptor	MHC restriction	Cytokines produced	Major function
Helper T cell				
T_H1	CD4	Class II	IFN-γ, IL-2, TNF-β	Activating cytotoxic T cells
T_H2	CD4	Class II	IL-4, IL-5, IL-6	B-cell helper
Cytotoxic T cell	CD8	Class I	IFN-γ, TNF	Killing virus-infected and cancer cells

A. Major histocompatibility proteins

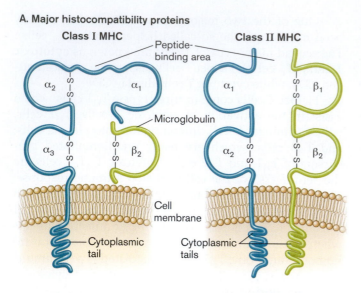

B. MHC binding to antigen

C. MHC-antigen binding

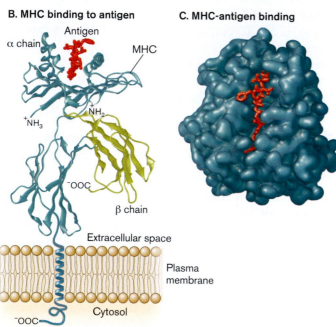

Figure 24.16   Major histocompatibility proteins.
A. MHC Class I molecules are composed of a 45-kDa chain and a small peptide called β_2-microglobulin (12 kDa). MHC Class II molecules contain an alpha chain (30-34 kDa) and a beta chain (26–29 kDa). The peptide-binding regions of both classes show variability in amino acid sequence that yield different shapes and grooves. Peptide antigens nestle in the grooves and are held there awaiting interaction with T-cell receptors. CD8 T cells recognize antigen peptides associated with class I molecules, while CD4 T cells recognize peptides bound to class II molecules. **B.** Antigen binding to an MHC. **C.** Top view of antigen (red) nestled in the MHC peptide-binding site. (PDB code: 1bii)

They are always expressed by the antigen-presenting dendritic cells and B cells, for example, but can also be made by macrophages under the influence of cytokines such as interferon-gamma (IFNγ). The salient feature of MHC molecules is that they bind antigen, presenting it on the cell surface. The antigen-binding clefts of different MHC molecules differ markedly and have distinct binding affinities for different antigenic peptides.

MHC molecules are critical to the immune system because T cells only recognize antigens associated with MHC proteins; they do not recognize free-floating antigens. (B cells, on the other hand, *do* recognize free-floating antigens.) Once a T cell recognizes an antigen-presenting cell with a foreign antigen attached to an MHC molecule, the T cell becomes activated and in turn activates the immune system. It is important to note that not all potentially antigenic peptides can be recognized in one animal. An animal will not respond to an antigen if its antigen-presenting cells lack the MHC molecule needed to bind that antigen. This is the basis of tolerance to the self antigens present on an animal's own tissues. Tolerance means your immune system does not launch an immune response to your own antigens. How tolerance to self-antigens develops is described in **Special Topic 24.2**.

Antigen Processing and Presentation Occur by Two Paths

How are foreign antigens placed, or presented, on host cell surfaces? In the initial stages of an immune response, microbes either infect (as with viruses) or are engulfed by (as with bacteria) antigen-presenting cells. Inside the APCs, foreign proteins are degraded into smaller pieces (for example, converted to peptides). These smaller pieces are placed within MHC-binding clefts and transported back to the cell surface. Whether an antigen peptide binds to class I or class II molecules generally depends on how the antigen initially entered the cell (**Fig. 24.17**). Antigens synthesized within the cytoplasm of an APC (called endogenous antigens), as would occur during infections with viruses and intracellular bacteria, will attach to class I MHC molecules on the endoplasmic reticulum and are moved to the cell surface. Antigens produced outside of the APC (called exogenous antigens), as are most bacterial antigens, will enter the cell via phagocytosis and attach to class II MHC molecules transported to the acidic phagosome or lysosome (**Fig. 24.17**, right side). The MHC class II–peptide complex is then carried to the cell surface. Once the antigen is presented on the APC cell surface, T cells can interact with it via the T-cell receptors. Interactions between antigen-presenting cells and naive T cells occur within the lymph nodes, the spleen, or Peyer's patches in the gut. So the APCs must make their way to those locations. How T cells then distinguish between MHC I and MHC II presentation is explained later.

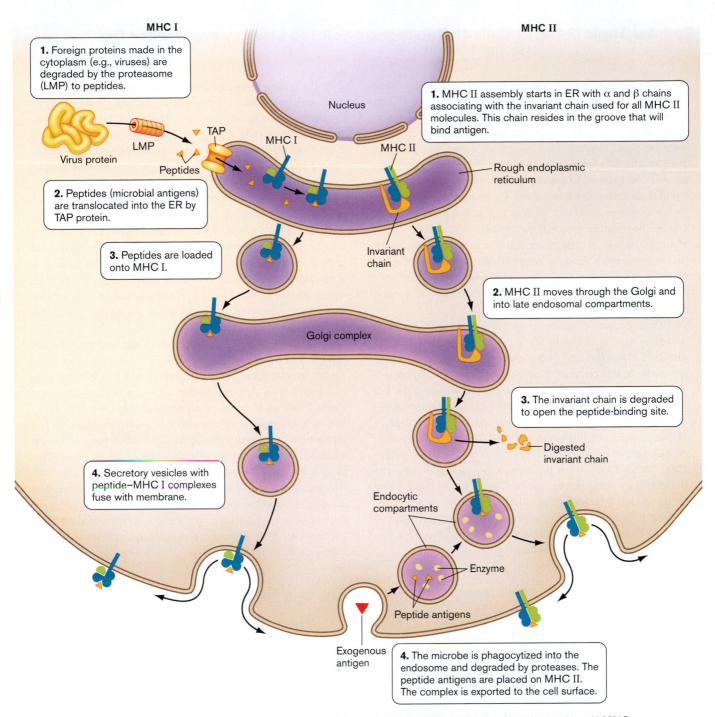

MHC I

1. Foreign proteins made in the cytoplasm (e.g., viruses) are degraded by the proteasome (LMP) to peptides.

Nucleus

Virus protein

LMP

TAP

Peptides

MHC I

MHC II

MHC II

1. MHC II assembly starts in ER with α and β chains associating with the invariant chain used for all MHC II molecules. This chain resides in the groove that will bind antigen.

Rough endoplasmic reticulum

2. Peptides (microbial antigens) are translocated into the ER by TAP protein.

3. Peptides are loaded onto MHC I.

Invariant chain

2. MHC II moves through the Golgi and into late endosomal compartments.

Golgi complex

3. The invariant chain is degraded to open the peptide-binding site.

Digested invariant chain

4. Secretory vesicles with peptide–MHC I complexes fuse with membrane.

Endocytic compartments

Enzyme

Peptide antigens

Exogenous antigen

4. The microbe is phagocytized into the endosome and degraded by proteases. The peptide antigens are placed on MHC II. The complex is exported to the cell surface.

Figure 24.17   Processing and presentation of antigens by antigen-presenting cells on class I and class II MHC proteins.   Microbial proteins made in the host cytoplasm (upper left) are degraded and peptides are placed on MHC class I molecules in the ER. Microbial proteins made outside the cell (lower center) are endocytosed, degraded in the endosome, and placed on MHC class II molecules. TAP = transporter of antigen peptides; LMP = low-molecular-mass polypeptide component of the proteasome.

Greek mythology tells of Narcissus, a young man punished by the gods for scorning the women who fell in love with him. One day Narcissus saw his own beautiful reflection in a pool of water. He unwittingly fell in love with his own reflection, fell into the pool, and drowned. Narcissus's inability to identify his own image is tantamount to the inherent danger posed by an immune system. To avoid disaster, immune systems must be able to distinguish what is self (meaning antigens present in our own tissues) from what is not self. Chapter 23 discusses how innate immunity accomplishes this task, in part through the use of pattern recognition proteins such as the Toll-like receptors. Equally important are the ways adaptive immune mechanisms avoid recognizing self. At the outset of development, the immune system is fully capable of reacting against self.

One protective mechanism the body uses to avoid attacking itself is to delete any T cells that react against self antigens. In humans, this occurs throughout life as new T cells are made. In this process, T cells undergo a two-stage selection or "education" process in the thymus to recognize self versus nonself. This education process involves the use of self antigens, not foreign antigens. T cells bearing T-cell receptors (TCRs) that *weakly* recognize self MHC proteins displayed on thymus epithelial cells are allowed to survive (**positive selection**). These T cells leave the thymus to seed secondary lymphoid organs such as the spleen. T cells in the thymus that recognize self MHC peptides *too strongly*, however, are killed, or deleted from the population (**negative selection**). Almost 95% of T cells entering the thymus die during these positive and negative selection processes.

The positive selection process is needed because T cells must be able to recognize self MHC to "see" the associated antigen. Antigen bound to MHC actually increases the affinity to the matched TCR. If our T-cell repertoire included cells that bound self MHC too tightly, then our T cells would constantly react to our own MHC molecules, regardless of what antigen peptides were attached.

In contrast to negative selection for B cells, which occurs in bone marrow, T-cell education is limited to the thymus. But, if the thymus only expresses thymus antigens, how can T cells that respond to antigens expressed on other host cells (for example, heart cells) be removed? The answer is a special gene activator in thymus cells that allows them to synthesize *all* human proteins in small amounts. This expression is necessary to complete T-cell education within the thymus.

You might ask how someone who has had their thymus removed (a treatment for myasthenia gravis) can live if the organ is critical for T-cell maturation and for deleting self-reactive T cells. Actually, within a few years after birth, the thymus begins to lose its utility such that very little function remains in adults. Research has shown that some T-cell maturation can occur extrathymically in secondary lymphoid tissues, although not as efficiently as in the thymus. This is at least part of the reason why children and adults with thymectomies are able to live relatively normal lives. The situation is more serious with newborns who have had their thymuses removed. Newborns have not had sufficient time to populate their secondary extrathymic lymphoid organs with T cells.

T-Cell Receptors Bind Antigens on Antigen-Presenting Cells (APCs)

T-cell receptors (TCRs) are antigen-binding molecules (not immunoglobulins) present on the surfaces of T cells (**Fig. 24.18**). These receptors do not bind *soluble* antigen. A TCR, in general, will only bind to antigens attached to MHC surface proteins that are present on antigen-presenting cells. However, the TCRs of cytotoxic T cells can also bind viral antigens present on any virus-infected cells, whether or not that cell is normally considered an APC.

The T-cell receptor is composed of several transmembrane proteins. The part used to recognize antigen is composed of two molecules, alpha and beta. Much like the immunoglobulins, the alpha and beta proteins of the TCR are formed from gene clusters that undergo gene rearrangements. The diversity of TCR antigen receptors comes from random recombinations of various V, D, and J segments (analogous to, but different from, the immunoglobulin genes) and to variability in the precise joining of segments. There is no hypermutation, however.

The TCR alpha and beta proteins are found in a complex with four other peptides; together these form the CD3 complex. When stimulated, these ancillary CD3 complex proteins recruit and activate intracellular protein kinases and launch a phosphorylation cascade that triggers proliferation of the T cell.

Activation of T$_H$0 Helper Cells

Figure 24.19 summarizes the steps that lead to T-cell activation in the lymph node and highlights how acti-

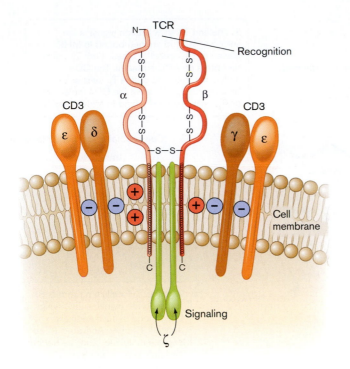

Figure 24.18   The T-cell receptor (TCR) and CD3 complex. The T-cell receptor proteins are associated with CD3 proteins at the cell surface. Antigen binds to the alpha and beta subunits. The positive and negative charges holding the complex together come from amino acids in the peptide sequences. Once bound to antigen, the complex transduces a signal into the cell. This signal triggers T-cell proliferation.

vated T cells influence the two arms of adaptive immunity: humoral and cellular. We will refer to **Figure 24.19** often within the next sections.

Two molecular signals are required to activate T cells. The first step, regardless of the T-cell type, involves the cross-linking of T-cell receptors by antigen-MHC proteins on antigen-presenting cells (**Fig. 24.19**●, step 1). In the case of T_H0 cells, this involves class II MHC complexes. But this first step alone is not enough to activate a T cell. A second signal is required by T_H cells, which involves the binding of a CD28 molecule present on the T-cell surface to a molecule called B7 protein on the APC cell surface (**Fig. 24.19**, step 2). Different cytokines produced during infection will then convert activated T_H0 cells to T_H1 or T_H2 cells (**Fig. 24.19**, steps 3 and 4). Activation of cytotoxic T cells (T_C cells) will be examined later.

B-Cell Activation Revisited

As with T cells, B cells need two signals to become activated. The first signal, described earlier, occurs when antigen cross-links B-cell receptors on B-cell membranes (in the process called capping). But to become activated

during a primary immune response, B cells usually require helper T cells. But not just any helper T cell can stimulate a B cell; specificity is involved. How does a B cell gain specific T-cell help? T cells (the T_H2 type) know which B cells to help via their T-cell receptors. As part of the B-cell capping mechanism, some antigen bound by B-cell receptors becomes internalized, to be processed and presented back on the B cell's surface MHC receptor (see **Fig. 24.17**, step 4). T_H2 cells that are activated by dendritic cells presenting antigen A, for instance, can use their T-cell receptor to bind to antigen A presented on the matched B cells. This contact allows CD40 on the B cell to bind CD154 on the T cell, an interaction that triggers the second intracellular signal needed for B-cell activation (**Fig. 24.19**, step 5).

During the secondary response, B cells still need help to become plasma cells but do not need direct contact with helper T cells. Memory B cells that have antigen bound to their B-cell receptors can respond to the soluble IL-4 and IL-6 cytokines secreted by activated helper T cells without having direct contact with the T_H cell. IL-4 stimulates B-cell proliferation, while IL-6 directs differentiation into antibody-secreting plasma cells (**Fig. 24.19**, step 6).

Production of Cytotoxic T cells

As with helper T cells, the first signal needed to activate cytotoxic T cells is the binding of the TCR to an antigen-MHC complex (in this case, MHC class I) on an antigen-presenting cell (see **Fig. 24.19**, step 1). **Figure 24.20** shows a closer look at the interaction between an antigen-presenting cell and a CD8 cytotoxic T cell. This interaction occurs in a lymph node. As with the production of helper T cells, this interaction alone will not activate the cytotoxic T cell. The second signal needed here is actually a cytokine called IL-2 produced by the T_H1 class of helper T cells (**Fig. 24.19**, step 7). The first signal (the TCR–antigen–MHC I complex) initiates the synthesis of IL-2 receptors on the T_C cell surface. Upon subsequent binding of IL-2, the T_C cell gains cytotoxic activity, after which it can leave the lymph node, travel to the site of infection, and kill any cell bearing the same peptide–MHC class I complex (for example, cells infected with the same virus that triggered T_C production) (**Fig. 24.21**). The B7-CD28 interaction that was needed to activate helper T cells is not required for T_C cell activation but can help.

Superantigens Do Not Require Processing to Activate T Cells

Normal antigens require processing by antigen-presenting cells. Each peptide produced by antigen processing is ultimately recognized in the context of

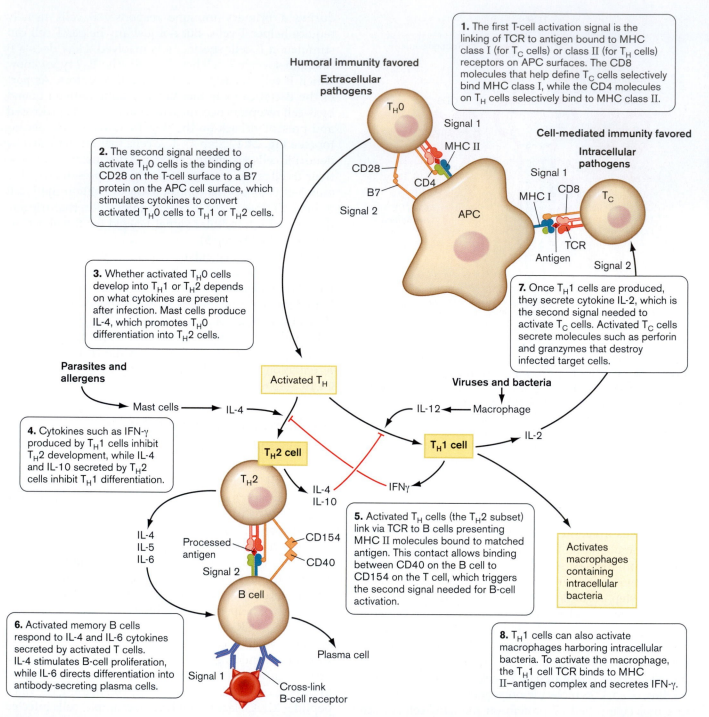

1. The first T-cell activation signal is the linking of TCR to antigen bound to MHC class I (for T_C cells) or class II (for T_H cells) receptors on APC surfaces. The CD8 molecules that help define T_C cells selectively bind MHC class I, while the CD4 molecules on T_H cells selectively bind to MHC class II.

2. The second signal needed to activate T_H0 cells is the binding of CD28 on the T-cell surface to a B7 protein on the APC cell surface, which stimulates cytokines to convert activated T_H0 cells to T_H1 or T_H2 cells.

3. Whether activated T_H0 cells develop into T_H1 or T_H2 depends on what cytokines are present after infection. Mast cells produce IL-4, which promotes T_H0 differentiation into T_H2 cells.

4. Cytokines such as IFN-γ produced by T_H1 cells inhibit T_H2 development, while IL-4 and IL-10 secreted by T_H2 cells inhibit T_H1 differentiation.

5. Activated T_H cells (the T_H2 subset) link via TCR to B cells presenting MHC II molecules bound to matched antigen. This contact allows binding between CD40 on the B cell to CD154 on the T cell, which triggers the second signal needed for B-cell activation.

6. Activated memory B cells respond to IL-4 and IL-6 cytokines secreted by activated T cells. IL-4 stimulates B-cell proliferation, while IL-6 directs differentiation into antibody-secreting plasma cells.

7. Once T_H1 cells are produced, they secrete cytokine IL-2, which is the second signal needed to activate T_C cells. Activated T_C cells secrete molecules such as perforin and granzymes that destroy infected target cells.

8. T_H1 cells can also activate macrophages harboring intracellular bacteria. To activate the macrophage, the T_H1 cell TCR binds to MHC II–antigen complex and secretes IFN-γ.

Humoral immunity favored
Extracellular pathogens

Cell-mediated immunity favored
Intracellular pathogens

Parasites and allergens

Viruses and bacteria

Activates macrophages containing intracellular bacteria

Figure 24.19   Summary of the activation of humoral and cell-mediated pathways.   Intracellular pathogens generally activate cell-mediated immunity by stimulating cytotoxic T cells. Extracellular pathogens tend to activate humoral immunity. The balance between cell-mediated and humoral immune responses to a given infection is regulated by the balance between the production of T_H1 (cell-mediated) versus T_H2 (antibody) helper cells. This balance is influenced by whether the foreign antigen was made by intracellular pathogens (T_H1-favored) or extracellular pathogens (T_H2-favored), and whether antigen presentation (APC) occurred through a macrophage (T_H1-favored) or a mast cell (T_H2-favored). T_H1 cells will encourage activation of cytotoxic T cells (cell-mediated immunity; best for killing intracellular pathogens), while T_H2 promotes antibody production (humoral immunity, best for attacking extracellular pathogens). See text for more details. ▶❚

self MHC molecules by a specific cognate T-cell receptor. However, when an antigen is introduced into a host, there are only a few T cells with the proper TCR needed to recognize that antigen—in the range of 1–100 cells in a million. Proliferation of the T cells through antigen-dependent activation increases that number and increases the immune response to that antigen.

Some proteins, called **superantigens**, stimulate T cells much more rapidly by bypassing the normal route of antigen processing. In fact, recognition as an antigen is not even involved. Certain microbial toxins, such as staphylococcal toxic shock syndrome toxin (TSST), are actually superantigens. These proteins, as illustrated in **Figure 24.22**, simultaneously bind to the outside of T-cell receptors on T cells and to the MHC molecules on antigen-presenting cells (for example, macrophages). This joining of the T cell and macrophage activates more T cells than a typical immune reaction, and stimulates release of massive amounts of inflammatory cytokines from both cell types.

The effect of superantigens can be devastating because cytokines like tumor necrosis factor can overwhelm the host immune system regulatory network and cause severe damage to tissues and organs. The result is disease and sometimes death. Jim Henson, the creator of Kermit the frog and other Muppet characters, died in 1990 from complications of pneumonia caused by a potent superantigen produced by *Streptococcus pyogenes*. This example is but one of many ways that overreaction of the immune system causes morbidity (disease) and mortality (death), more so than the direct effects of the pathogen involved.

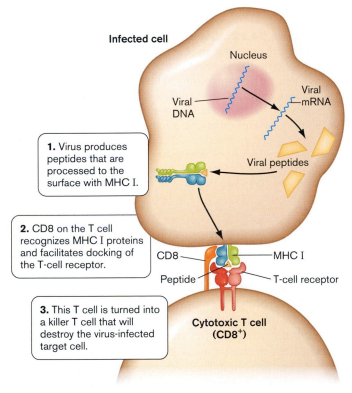

Figure 24.20 Presentation of a viral antigen to a T cell. CD8 protein on a T cell directs the interaction between a T-cell receptor and a viral antigen bound to a class I MHC protein on an antigen-presenting cell.

Infected cell
Nucleus
Viral DNA
Viral mRNA
1. Virus produces peptides that are processed to the surface with MHC I.
Viral peptides
2. CD8 on the T cell recognizes MHC I proteins and facilitates docking of the T-cell receptor.
CD8
MHC I
Peptide
T-cell receptor
3. This T cell is turned into a killer T cell that will destroy the virus-infected target cell.
Cytotoxic T cell (CD8⁺)

The CD4 and CD8 Surface Proteins Help T Cells Differentiate between MHC I and MHC II Surface Proteins

One question not yet addressed is how do the TCRs on T_C and T_H0 cells "know" to respectively bind MHC class I and class II antigen complexes? The answer is based on the different clusters of cellular differentiation (CD) molecules displayed on T_C and T_H0 cell surfaces. The surface CD proteins CD8 and CD4 help T cells distinguish between MHC class I and class II molecules on antigen-presenting cells. The CD8 molecules on T_C cells selectively bind MHC class I while CD4 molecules on T_H cells selectively bind to MHC class II (see **Fig. 24.19**, step 1). Antigens presented on class I MHC molecules generally arise from intracellular pathogens such as viruses and some bacteria, situations that will require cell-mediated immunity for resolution. In contrast, antigen peptides presented on class II MHC molecules originate from extracellular infections and are only recognized by CD4 T_H cells.

Once activated, cytotoxic CD8 T_C cells secrete molecules such as **perforin** and **granzymes** that destroy infected target cells (similar to natural killer cells; discussed in Section 23.7). Perforin forms a pore in target cells that is used to deliver toxic enzymes (granzymes).

Figure 24.21 Cytotoxic T-cell action.
A. An antigen-presenting cell presenting viral peptide on MHC I surface molecules interacts with and activates a cytotoxic T cell in a lymph node. **B.** The activated T_C cell then leaves the lymph node and migrates to the site of infection where it can recognize the viral peptide presented on the MHC class I receptor on an infected cell. This interaction authorizes the cytotoxic T cell to kill the infected cell.

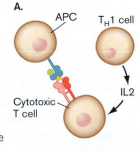

A.
APC
T_H1 cell
Cytotoxic T cell
IL2

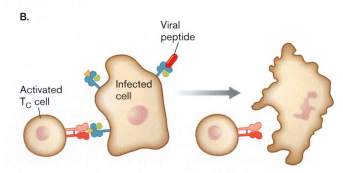

B.
Viral peptide
Activated T_C cell
Infected cell

A.

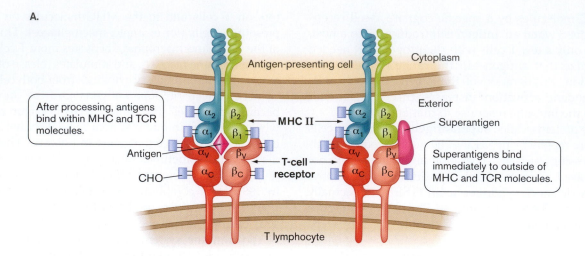

After processing, antigens bind within MHC and TCR molecules.

Antigen-presenting cell

Cytoplasm

MHC II

Exterior

Superantigen

Superantigens bind immediately to outside of MHC and TCR molecules.

Antigen

CHO

T-cell receptor

T lymphocyte

B.

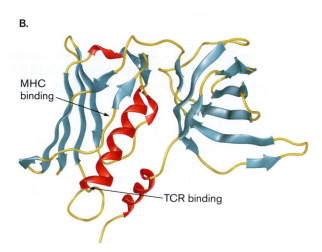

MHC binding

TCR binding

Figure 24.22 Superantigens. A. The difference between presentation of antigen and superantigen. Antigen presentation requires the antigen to bind within the binding pockets of MHC and TCR molecules. Superantigens do not require processing. They can bind directly to outer aspects of the TCR and MHC proteins, linking and activating the two cell types. **B.** Staphylococcal toxic shock syndrome toxin is one example of a potent superantigen. Alpha helices are shown as red ribbons, while sections of the beta sheet are shown as blue ribbons. (PDB code: 2qil)

Because they directly attack host cells, CD8 T_C cells are the major "enforcers" of the cellular immune system, along with macrophages and NK cells.

Why does the presentation of antigens on MHC I molecules activate cytotoxic T cells rather than B cells and antibody production? A major reason is that intracellular pathogens (for example, viruses) hide inside host cells, where they are protected from antibody. Consequently, these pathogens are best killed when the harboring host cell is also sacrificed (via cellular immunity). Because class I MHC molecules are found on all nucleated cells, any infected cell can potentially activate T_C cells, converting them to cytotoxic T cells that ultimately kill the infected cell (T_C cells are not actually cytotoxic until they are activated). The infected host cell is sacrificed for the good of the whole animal, human or otherwise.

Whereas antigens presented on class I MHC molecules elicit only cell-mediated immunity, antigens presented on class II molecules can actually stimulate either arm of the immune system. In contrast to CD8 T_C cells, CD4 helper T cells do not directly kill host cells. Instead, activated CD4 T_H cells release various cytokines called lymphokines that, among other things, attract additional white cells to the area. **Table 24.6** lists a fraction of the many different cytokines produced by various cell types and their influence over the immune system. Activated T_H1 cells secrete cytokines that stimulate the cytotoxic T_C cells of the cellular immune system (see **Fig. 24.19**, step 7), whereas activated T_H2 cells directly interact with the class II MHC–antigen peptide complexes on B cells, as described earlier, and stimulate those cells to differentiate and produce antibody (see **Fig. 24.19**, step 5).

AIDS (acquired immunodeficiency syndrome) provides a vivid and tragic illustration of the importance of CD4 T cells in immunity. The human immunodeficiency virus (HIV) binds to CD4 molecules and thus is able to invade and infect CD4$^+$ T cells. As the disease progresses, the number of CD4 T cells declines below its normal level of about 1,000 per microliter. When the number of CD4 T cells drops below 400 per microliter, the ability of the patient to mount an immune response declines dangerously. Not only does the patient become hypersusceptible to infections by pathogens, but the individual becomes susceptible to infection by commensal organisms as well. Most patients actually die from these opportunistic infections.

Table 24.6 Select cytokines that modulate the immune response.

Cytokine	Sample sources	General functions
IL-1	Many cell types, including endothelial cells, fibroblasts, neuronal cells, epithelial cells, macrophages	Affects differentiation and activity of cells in inflammatory response; central nervous system; acts as endogenous pyrogen
IL-2	T_H1 cells	Stimulates T-cell and B-cell proliferation
IL-3	T cells, mast cells, keratinocytes	Stimulates production of macrophages, neutrophils, mast cells, others.
IL-4	T_H2 cells, mast cells	Promotes differentiation of CD4 and T cells into T_H2 helper T cells; promotes proliferation of B cells
IL-5	T_H2 cells	Acts as a chemoattractant for eosinophils; activates B cells and eosinophils
IL-6	T_H2 cells, macrophages, fibroblasts, endothelial cells, hepatocytes, neuronal cells	Stimulates T-cell and B-cell growth; stimulates production of acute-phase proteins
IL-8	Monocytes, endothelial cells, T cells, keratinocytes, neutrophils	Chemoattracts PMNs; promotes migration of PMNs through endothelium
IL-10	T_H2 cells, B-cells, macrophages, keratinocytes	Inhibits production of IFN-γ, IL-1, TNF-α, IL-6 by macrophages
IFN-α/β	T cells, B cells, macrophages, fibroblasts	Promotes antiviral activity
IFN-γ	T_H2 cells, cytotoxic T cells, NK cells	Activates T cells, NK cells, macrophages
TNF-α	T cells, macrophages, NK cells	Exerts wide variety of immunomodulatory effects
TNF-β	T cells, B cells	Exerts wide variety of immunomodulatory effects

The MHC Restriction Rule Limits T-Cell Reactivity

As noted previously, T cells can only recognize antigens complexed to an MHC molecule. Actually, it must be a self, MHC molecule. The MHC molecules of each person are unique. Thus, T cells from one individual normally do not recognize peptides placed on antigen-presenting cells from another individual, a property known as **MHC restriction**. CD4 T_H cells are class II MHC restricted, while CD8 T_C cells are class I restricted. Rolf Zinkernagel (University of Zurich) and Peter Doherty (St. Jude Children's Hospital, Memphis) discovered this phenomenon in 1976 with a simple yet profoundly insightful experiment. They demonstrated that mouse cytotoxic T cells would kill only virus-infected target cells that possessed the same MHC surface protein. If the infected target cell came from a different strain of mouse containing a different MHC protein, the infected target cell was spared. Thus, there are two components to T-cell recognition: self and nonself. The self component comes from the MHC receptor, while the nonself component comes from the pathogen. This concept won Zinkernagel and Doherty the 1996 Nobel Prize in Physiology or Medicine.

Cytokines Tip the Balance between Cell-Mediated Immunity and Antibody Production

We have already discussed how cytokines made from T_H1 cells help activate cells involved with cell-mediated immunity (the CD8 T_C cells, macrophages, and natural killer cells) and that other cytokines from T_H2 cells trigger B cells to class-switch their immunoglobulin. But what determines whether T_H1 or T_H2 cells predominate during an infection? Part of the answer is found in the contingent of cytokines generated by macrophages or mast cells during different types of infection. Infections by viruses and bacteria favor the production of T_H1 cells, whereas allergens and parasites favor differentiation to T_H2. Early in an infection, bacteria and viruses stimulate cells of the innate immune system (macrophages and natural killer cells) to produce cytokines that tilt development of T_H0 into T_H1 (one such cytokine is IL-12). During interactions with an allergen, on the other hand, mast cells produce IL-4, which promotes T_H0 differentiation into T_H2 cells (see **Fig. 24.19**, step 3).

There is also cross-inhibition of T_H1 and T_H2 development. Cytokines such as interferon-gamma (IFN-γ) produced by T_H1 cells inhibit T_H2 development, while interleukins IL-4 and IL-10 secreted by T_H2 cells inhibit T_H1 differentiation (**Fig. 24.19**, step 4). As a result, once the immune system starts to go down one pathway, the other pathway is held back. But this is not an all-or-none response. For instance, a viral infection will also result in T_H2 cells that stimulate antibody production. The tilt toward T_H1 or T_H2 predominance occurs because the body realizes which pathway, cell mediated (T_H1) or humoral (T_H2), will more effectively clear a particular infection.

Microbial Evasion of Adaptive Immunity

Efficient as our immune system is, numerous viral and bacterial pathogens have developed effective means for avoiding the adaptive immune response. Many viruses produce proteins that downregulate production of class I MHC on infected cell surfaces. This will limit antigen presentation, since MHC I is needed for that process. On the other hand, losing MHC I should expose the infected host cell to natural killer cells because NK cells attack peers that lack MHC I (discussed in Section 23.7). To surmount this obstacle, it appears that human cytomegalovirus places a decoy MHC I-like molecule on the surface of infected cells. These decoys are thought to bind inhibitory receptors on NK surfaces that block NK cell cytotoxicity.

Bacteria, like viruses, are masters of illusion when it comes to the immune system. A major cause of gastric ulcers, *Helicobacter pylori* expresses proteins from a cluster of pathogenicity genes that trigger apoptosis of T cells. (Apoptosis is a programmed cell death.) Other bacteria have evolved mechanisms that interfere with signal transduction pathways controlling the expression of cytokines. For example, YopP from *Yersinia enterocolitica*, one cause of gastroenteritis, inhibits a specific signal transduction pathway needed to produce TNF, IL-1, and IL-8. Thus, *Yersinia* avoids the detrimental effects of those pro-inflammatory cytokines. Another method of immune avoidance is employed by various mycobacteria, some of which cause tuberculosis and leprosy (**Fig. 24.23**). These bacteria induce the production of anti-inflammatory cytokines, which dampen the immune response. Mycobacterium-infected macrophages produce IL-6, which inhibits T-cell activation, and IL-10, which downregulates the production of MHC II molecules needed for specific activation of T cells. Fortunately for humans, in most cases the immune system catches on to these tricks and through redundant humoral and cellular mechanisms manages to resolve these infections.

Activated T$_H$1 Cells Can Also Activate Macrophages

Macrophages, too, must be activated to become highly effective killers of microbes. Activation of macrophages also requires two signals. One activation pathway involves interferon-gamma (IFN-γ) produced by nearby infected or damaged cells, followed by binding of the macrophage to lipopolysaccharide (LPS) or other microbial components. Once activated, the macrophage becomes highly aggressive in terms of phagocytosis and increases production of numerous antimicrobial reactive oxygen intermediates.

As noted earlier, some microorganisms such as mycobacteria, the causative agents of tuberculosis and leprosy, are intracellular pathogens that grow primarily in the phagolysosomes of macrophages. There they are shielded from the effects of both antibodies and cytotoxic T cells. These pathogens live in the usually hostile environment of the phagocyte either by inhibiting the fusion of lysosomes to the phagosomes in which they grow or by preventing the acidification of the vesicles needed to activate lysosomal proteases. However, macrophages can rid themselves of these pathogens if the macrophage is activated by a T$_H$1 cell (see **Fig. 24.19**, step 8). Macrophages, even unactivated, are able to process some of the intracellular bacteria and place antigen from them on their class II MHC molecules. T$_H$1 cells then activate these macrophages though binding of their TCR molecules to the macrophage MHC II–antigen complex and by secreting interferon-gamma. The activated macrophages can then kill any bacteria that may be growing within them.

Because activated macrophages are such effective assassins of microbial pathogens, why are macrophages not always kept in an active state? A major reason is that once activated, macrophages also damage nearby host tissue through the release of reactive oxygen radicals and proteases. Thus, effective killing of microbial pathogens comes at the expense of host tissue damage.

TO SUMMARIZE:

■ **The major histocompatibility complex** consists of membrane proteins with variable regions that can bind antigens. Class I MHC molecules are on all nucleated cells, while antigen-presenting cells contain both class I and class II MHC molecules.

■ **Antigen-presenting cells (APCs)** such as dendritic cells present antigens synthesized during an intra-

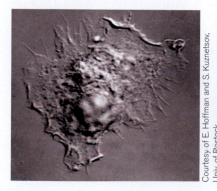

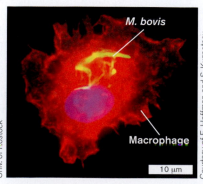

Courtesy of E. Hoffman and S. Kuznetsov, Univ. of Rostock

M. bovis

Macrophage

10 µm

Figure 24.23 *Mycobacterium bovis* **growing within cultured macrophages (cell line J774).** Left panel is a differential interference contrast image of an infected macrophage. Right image is the same cell viewed by fluorescence microscopy. *M. bovis* is carrying green fluorescent protein. The macrophage was stained for F-actin (red) and nucleus (blue).

cellular infection on their surface class I MHC molecules, but place antigens from engulfed microbes or allergens on their class II MHC molecules.

- **Activation of a T_H0 cell** requires two signals: TCR/CD4 binding to an MHC II–antigen complex on an antigen-presenting cell, and CD20-B7 interaction.
- **T_H0 cell differentiation** to T_H1 or T_H2 cells is influenced by different cytokine "cocktails" secreted by macrophages and NK cells during an infection.
- **Activation of a B cell** into an antibody-producing plasma cell usually requires two signals: antigen binding to BCR and binding to a T_H2 cell activated by the same antigen.
- **Activation of cytotoxic T cells** requires two signals: TCR/CD8 molecules that recognize MHC I–antigen complexes and cytokines such as IL-2 made from activated T_H1 cells. The activated cytotoxic T cells in turn destroy infected host cells.
- **Superantigens** stimulate T cells by directly linking TCR on a T cell with MHC on an APC without undergoing APC processing and surface presentation.

> **THOUGHT QUESTION 24.8** Transplant rejection is a major consideration when transplanting most tissues because host T_C cells can recognize allotypic MHC on donor cells. So, why are corneas easily transplanted from a donor to just about any other person?
>
> **THOUGHT QUESTION 24.9** Why does attaching a hapten to a carrier protein allow production of anti-hapten antibodies?

24.7 Complement as Part of Adaptive Immunity

Chapter 23 describes how complement can attack invading microbes before a specific immune response is launched. In what is called the alternative pathway (see Section 23.8), C3b binding to LPS sets off a cascade of reactions resulting in the formation of a membrane attack complex (MAC) pore composed of C5b, C6, C7, C8, and C9 proteins (see **Fig. 23.29**). In addition to the alternative pathway, antibodies made during the humoral response to a pathogen offer another route to activate complement called the **classical complement pathway**. Called "classical" because it was the first complement pathway to be discovered, it requires a few additional proteins before reaching C3, the linchpin factor connecting the two pathways.

The classical cascade begins with the binding of the complement protein C1 complex to the Fc region of an antibody that is part of an antigen-antibody complex, as you would find when antibody binds to a bacterial or viral pathogen (**Fig. 24.24**). The bound C1 complex then cleaves

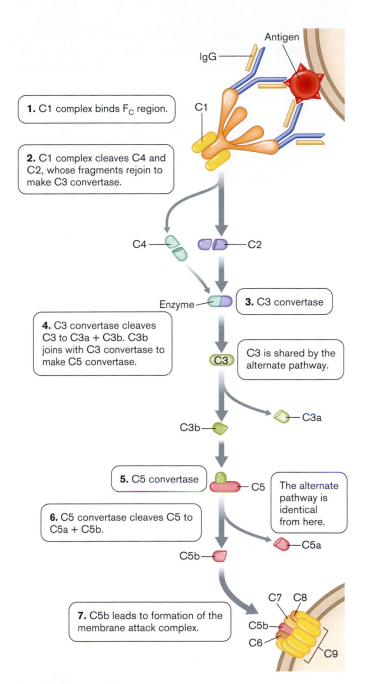

1. C1 complex binds F_C region.

2. C1 complex cleaves C4 and C2, whose fragments rejoin to make C3 convertase.

3. C3 convertase

4. C3 convertase cleaves C3 to C3a + C3b. C3b joins with C3 convertase to make C5 convertase.

C3 is shared by the alternate pathway.

5. C5 convertase

The alternate pathway is identical from here.

6. C5 convertase cleaves C5 to C5a + C5b.

7. C5b leads to formation of the membrane attack complex.

Figure 24.24 Classical complement cascade.

two other complement factors, C2 and C4, not used in the alternative pathway. Two fragments, one each for C2 and C4, unite to form another protease called C3 convertase. This enzyme cleaves C3 into C3a and C3b. C3b, which in the alternative pathway is stabilized by interacting with LPS, combines in the classical pathway with the C3 convertase to make a C5 convertase (note that this C5 convertase is different from the C5 convertase formed in the alternative pathway; see **Fig. 23.31**). The subsequent steps leading to formation of a membrane attack complex are the same as in the alternative pathway. As before, C5b binds to a target membrane and is joined by C6, C7, and C8. Multiple C9 proteins then assemble around the membrane attack complex and form a pore that compromises the integrity of the target cell.

One might wonder why we need complement with all the other immune responses at our disposal, but the need becomes evident in individuals with complement deficiencies. These patients are extremely susceptible to recurrent septicemic (blood) infections by organisms such as *Neisseria gonorrhoeae* (the cause of gonorrhea) that normally do not survive forays into the bloodstream because they are killed by complement. Why does reinfecting *Neisseria* not quickly succumb to all the other defenses, such as antibodies and cytotoxic T cells? Because key antigens on the bacterial surface undergo phase variations (see Sections 10.6 and 27.7), many antibodies made to cells of *N. gonorrhoeae* participating in the initial infection are useless during a second infection.

Regulation of Complement Activation

How do normal body cells prevent self-destruction following complement activation? The sequential assembly of the membrane attack complex provides several places where regulatory factors can intervene. One such factor is the host cell surface protein CD59. This protein will bind any C5b-C8 complex trying to form in the membrane and prevent C9 from polymerizing. Thus, no pore is formed and the host cell is spared. (Complement will not normally attack normal or infected host cells, since both contain CD59.)

Another regulatory mechanism hinges on a normal serum protein called **factor H**. This protein prevents the inadvertent activation of complement in the absence of infection. In the absence of factor H, the uncontrolled activation of complement could damage host cells despite the presence CD59. To short-circuit the cascade, factor H binds to C3bBb, displaces Bb from the complex, and acts as a cofactor for factor I protease that then cleaves C3b. Without C3b, the complement cascade stops. Some bacteria, such as some strains of *Neisseria gonorrhoeae,* have learned to protect themselves from complement by binding factor H.

Normal flora can also assist host cells in resisting complement. The intestinal microbe *Bacteroides thetaiotamicron* protects host cells from complement-mediated cytotoxicity by upregulating a **decay-accelerating factor** present in host cell membranes. Decay-accelerating factor stimulates decay of complement factors and prevents their deposition at the cell surface.

Other Roles for Complement Fragments

The C3a, C4a, and C5a cleavage fragments generated by the complement cascade do not participate in MAC for-

mation but are important for amplifying the immune reaction. These peptides act as chemoattractants to lure more inflammatory cells into the area. C3b, which does participate in MAC formation, also can act as an opsonin when it is bound to a bacterial cell. An opsonin is a protein factor that can facilitate phagocytosis. Because phagocytes contain C3b receptors on their surfaces, it is easier for them to grab and engulf cells coated with C3b.

> **THOUGHT QUESTION 24.10**   Do bone marrow transplants in a patient with severe combined immunodeficiency require immunosuppressive chemothereapy?
>
> **THOUGHT QUESTION 24.11**   How can a stem cell be differentiated from a B cell at the level of DNA?

TO SUMMARIZE:

- **The classical pathway for complement activation** begins with an interaction between the Fc portion of an antibody bound to an antigen and C1 factor in blood. (The alternative pathway does not need antibody but begins when C3b binds to LPS on a bacterium.)
- **The Fc-C1 complex** reacts with C2 and C4, leading to production of C3b and a novel C5 convertase specific to the classical pathway.
- **After C5 convertase cleaves C5**, the classical pathway is the same as the alternative pathway, resulting in the formation of a membrane attack complex (MAC).
- **CD59 and/or factor H** prevent inappropriate MAC formation in host cells.

24.8 Failures of Immune System Regulation: Hypersensitivity and Autoimmunity

Sometimes the immune system overreacts to certain foreign antigens and causes more damage than the antigen (microbe) alone might cause. Furthermore, some foreign antigens possess structures similar in shape to host structures and can trick the immune system into reacting against self. These immune miscues are called allergic hypersensitivity reactions, and the antigen causing the reaction is called an **allergen**. There are four types of hypersensitivity reactions (**Table 24.7**). The first three types of hypersensitivity are antibody mediated; the fourth is cell mediated.

Case History: Type I Hypersensitivity

A bee stings a 9-year-old boy walking with his mother at the zoo. Within minutes the boy begins sweating and itching. His

Table 24.7 Summary of hypersensitivity reactions.

Type	Description	Time of onset	Mechanism	Manifestations
I	IgE-mediated hypersensitivity	2–30 min	Ag induces cross-linking of IgE bound to mast cells with release of vasoactive mediators	Systemic anaphylaxis, local anaphylaxis, hay fever, asthma, eczema
II	Antibody-mediated cytotoxic hypersensitivity	5–8h	Ab directed against cell surface antigens mediates cell destruction via ADCC or complement	Blood transfusion reactions, hemolytic disease of the newborn, autoimmune hemolytic anaemia
III	Immune complex–mediated hypersensitivity	2–8h	Ag-Ab complexes deposited at various sites induce mast cell degranulation via Fc receptor; PMN degranulation damages tissue (Arthus reaction, localized)	Systemic reactions, disseminated rash, arthritis, glomerulonephritis
IV	Cell-mediated hypersensitivity	24–72h	Memory T_H1 cells release cytokines that recruit and activate macrophages	Contact dermatitis, tubercular lesions

Ag = antigen; Ab = antibody; ADCC = antibody-dependent cell cytotoxicity; PMN = polymorphonuclear leukocyte

chest then starts to tighten and he has tremendous difficulty breathing. Terrified, he looks to his equally frightened mother for help.

This is a classic and severe example of **type I (immediate) hypersensitivity**, sometimes called **anaphylaxis**. (Anaphylaxis is defined as the contraction of smooth muscle and dilation of capillaries due to release of pharmacologically active substances.) Type I hypersensitivity occurs within minutes of a *second* exposure to an allergen when the allergen reacts with IgE-coated mast cells. Recall that the primary role of IgE is to amplify the body's response to invaders by causing mast cells to release inflammatory mediators (see Section 24.3). On initial exposure, an allergen elicits the production of IgE antibodies specific to the allergen (bee venom in this example). The Fc portions of these antibodies bind to Fc receptors on the surfaces of mast cells. IgE-coated mast cells are then said to be *sensitized*. Most allergen molecules contain several identical antigenic sites. Thus, a single allergen particle during a second exposure can bind to adjacent surface IgE molecules. The result is a bridge between adjoining binding sites (**Fig. 24.25**). This cross-linked complex inhibits the mast cell enzyme **adenylate cyclase**, so that intracellular levels of cyclic adenosine monophosphate (cAMP) fall. The lower levels of intracellular cAMP cause the mast cell granules to quickly migrate to the cell surface and release their contents in a process called degranulation. Understanding this process is clinically important because anaphylaxis inhibitors such as epinephrine (also known as adrenaline) stimulate adenylate cyclase activity, stopping the anaphylaxis cascade in its tracks.

Mast cell degranulation releases chemicals with potent pharmacological activities. The most important of these is histamine, which binds to histamine receptors (H1 receptors) present on most body cells. Antihistamines have a structure similar to histamine and work by binding H1 receptors to prevent histamine from binding. When histamine binds H1, it activates an enzyme (phospholipase A) that releases from the cell membrane a 20-carbon fatty acid called arachidonic acid. Arachidonic acid is then converted to leukotrienes by 5-lipooxygenase A and to prostaglandins by cyclooxygenase. Another important chemical released during degranulation is **tumor necrosis factor**. All of these mediators (leukotrienes, prostaglandins, and tumor necrosis factor) cause the smooth muscles nearby to contract. This constricts small blood vessels and expands capillary pores. Fluid is then forced from the circulation into the tissues. The immediate consequence is swelling (**edema**) in the joints and around the eyes and a rash (similar to hives), with burning and itching of the skin due to nerve involvement. In the clinical example involving the bee sting, the contraction of lung smooth muscles led to breathing difficulties.

What we described was a severe type I allergic reaction caused by a bee sting. Type I hypersensitivity reactions do not usually involve the whole body. Most type I reactions are more localized and cause what is called **atopic** ("out of place") disease. Hay fever, or allergic rhinitis, is a common manifestation of atopic disease that can be caused by the inhalation of dust mite feces (**Fig. 24.26**), animal skin or hair (dander), and certain types of grass or weed pollens. This disease affects the eyes, nose, and upper respiratory tract. Atopic asthma, another form of type I hypersensitivity resulting from inhaled

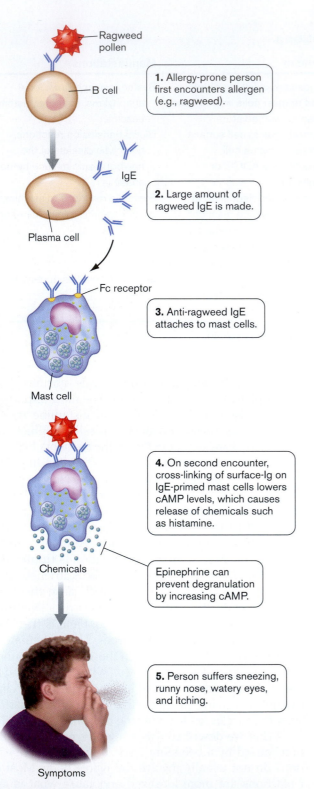

1. Allergy-prone person first encounters allergen (e.g., ragweed).

2. Large amount of ragweed IgE is made.

3. Anti-ragweed IgE attaches to mast cells.

4. On second encounter, cross-linking of surface-Ig on IgE-primed mast cells lowers cAMP levels, which causes release of chemicals such as histamine.

Epinephrine can prevent degranulation by increasing cAMP.

5. Person suffers sneezing, runny nose, watery eyes, and itching.

Ragweed pollen

B cell

IgE

Plasma cell

Fc receptor

Mast cell

Chemicals

Symptoms

Figure 24.25 Events leading to type I hypersensitivity reactions.

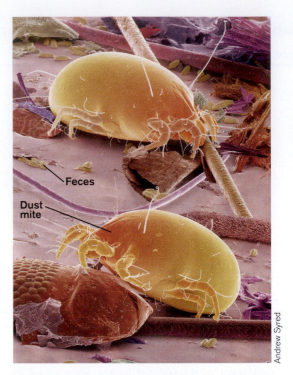

Feces

Dust mite

Andrew Syred

Figure 24.26 Inhaled allergen. Dust mite (200 to 600 micrometers) and dust mite feces (colorized SEM).

allergens, affects the lower respiratory track and is characterized by wheezing and difficulty breathing.

The administration of antihistamines is an effective treatment of allergic rhinitis, but the more important chemical mediators of asthma are the leukotrienes produced dur-

ing what is called **late-phase anaphylaxis**. In late-phase anaphylaxis, mast cells release chemotactic factors that call in eosinophils. Eosinophils entering the affected area produce large amounts of leukotrienes that, like histamine, cause vasoconstriction and inflammation. At this point, antihistamines have little effect. Effective treatment includes inhaled steroids to minimize inflammation and bronchodilators (for example, albuterol) to widen bronchioles and facilitate breathing. A more recently developed medication, SINGULAIR™, works by blocking leukotriene receptors in the lung. During severe asthma attacks, injected epinephrine is critical. Epinephrine will open airways to ease breathing through direct hormonal action.

People sensitized to allergens are not necessarily doomed to suffer with the allergy their entire life. A clinical treatment called **desensitization** can sometimes be used to prevent anaphylaxis. Desensitization involves injecting small doses of allergen over a period of months. This process is thought to produce IgG molecules that circulate, bind, and neutralize allergens before they contact sensitized mast cells. Desensitization has been useful in cases of asthma, food allergies, and bee stings, although results are variable and by no means guaranteed.

There is a relatively new treatment for allergic asthma, omalizumab (Xolair), that can actually prevent sensitization. Omalizumab is a monoclonal antibody (an antibody preparation of one antibody type that only binds one epitope), administered by injection that selectively binds IgE. It prevents IgE from binding to Fc receptors on mast cells and, thus, prevents sensitization.

Why don't all antigens generate type I hypersensitivity? The answer is that most antigens do not elicit high levels of IgE antibodies. It is unclear what makes certain antigens capable of this and what makes different individuals prone to different allergies.

Case History: Type II Hypersensitivity

A hospitalized male patient with blood group B accidentally receives a transfusion with type A blood. Within hours, the individual experiences chills and lowered blood pressure and has blood in his urine, all symptoms of ABO blood group incompatibility.

This is a form of **type II hypersensitivity**. Type II hypersensitivity starts with antibody binding to cell surface antigens (donor blood cells, in this case). This differentiates it from type III hypersensitivity (considered next), which involves antibody binding to *soluble* antigens. The classic type II hypersensitivity reaction is illustrated by blood group incompatibilities resulting from transfusions. Destruction of the foreign red blood cells can occur by two routes. In the first route, antibody-dependent cell-mediated cytotoxicity is triggered when NK cells or macrophage surface Fc receptors bind to the Fc regions of antibodies bound to foreign red blood cell antigens. The second route involves complement activation, in which complement (C1) binds to the Fc regions of circulating antibodies affixed to foreign red blood cells. The complement cascade then creates a membrane attack complex (MAC) that destroys the RBC.

Another form of type II hypersensitivity is Rh incompatibility disease of the fetus and newborn (**Fig. 24.27**). Similar to ABO antigens, Rh antigens are chemical groups on red blood cells. Incompatibility between mother (Rh$^-$) and fetus (Rh$^+$) can lead to a hypersensitivity reaction in the fetus, especially in a second pregnancy. During birth of an Rh$^+$ child to an Rh$^-$ mother, fetal red blood cells enter the mother's bloodstream. The Rh$^-$ mother's immune system will make anti-Rh$^+$ IgG antibodies that can pass through the placenta. This is a problem during a subsequent pregnancy because these antibodies will attack an Rh$^+$ fetus. Rh disease can be prevented if the mother is passively given anti-Rh$^+$ immunoglobulin by intramuscular injection within a day after birth. This antibody binds the fetal Rh$^+$ antigen and prevents the mother's immune system from making Rh$^+$ antibody and the attendant memory B cells. Passive immunity only lasts four to six weeks, long enough to clear the fetal antigens from the mother's bloodstream.

Case History: Type III Hypersensitivity

*A 16-year-old girl is treated for acne with minocycline (a tetracycline derivative). Two weeks later she develops a fever and a urticarial rash (**Fig. 24.28B**) and experiences muscle and joint pain. (Urticaria are hives, an itchy rash caused by tiny amounts of fluid that leak from blood vessels just under the skin surface.) Her symptoms resolve after five days of anti-inflammatory steroid treatment.*

This is an example of immune complex disease, also called **type III hypersensitivity**. In contrast to type II hypersensitivity, type III hypersensitivity is initiated by IgG antibody binding to *soluble* antigen, which in this case is minocycline. Penicillin sensitivity, usually associated with type I anaphylaxis, can also involve a type III hypersensitivity reaction. Normally, antigen-antibody complexes are readily cleared by macrophage engulfment, but an excessive amount of complexes may overwhelm the system and circulate in the blood, free to bind complement (C1 factor, classical pathway). The binding of C1 triggers a cascade in which complement fragments recruit and activate polymorphonuclear leukocytes (PMNs). The complement-activated PMNs release lysosomal proteases and reactive oxygen species that damage host cells in the area.

A more recent model for type III hypersensitivity is shown in **Figure 24.28A**. In this model, type III hypersensitivity is very similar to type I hypersensitivity in that mast cells are involved. The difference is that the immune complexes involve IgG molecules that are not prebound to mast cells. Once made, the soluble IgG-antigen complexes bind to Fc receptors on mast cells, causing degranulation and PMN recruitment. The PMNs use their Fc receptors to bind the complexes formed or deposited near tissues. The PMNs then release enzymes that damage cells in the area. Tissue deposition is important for disease because PMNs can more easily engage the deposited complex than if it is floating in blood, and the factors released will maintain a higher concentration in the local area. Because antigen-antibody complex deposition occurs throughout the body, the result is still a systemic disease. Systemic type III hypersensitivity will resolve once the immune complexes are cleared. In the case history described, steroid treatment was administered to minimize inflammation as the complexes were cleared.

In addition to systemic effects, there are also localized type III reactions. In 1903, a Frenchman named Maurice Arthus (1862–1945) discovered a phenomenon now

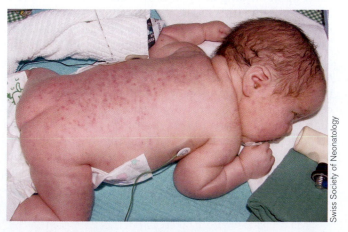

Figure 24.27   Type II hypersensitivity. Targeting foreign red blood cells in Rh disease of the newborn.

Swiss Society of Neonatology

A.

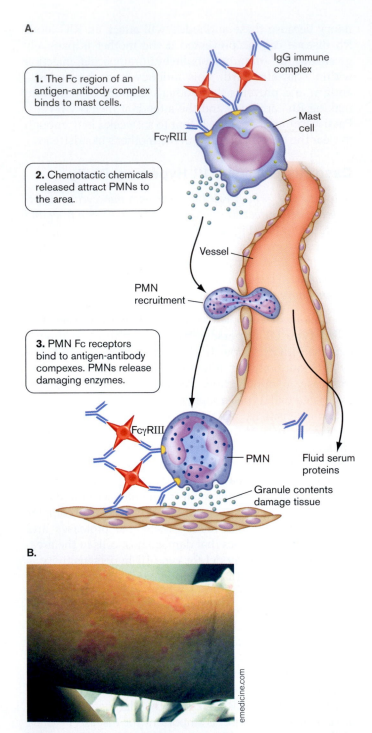

1. The Fc region of an antigen-antibody complex binds to mast cells.

IgG immune complex

FcγRIII

Mast cell

2. Chemotactic chemicals released attract PMNs to the area.

Vessel

PMN recruitment

3. PMN Fc receptors bind to antigen-antibody compexes. PMNs release damaging enzymes.

FcγRIII

PMN

Fluid serum proteins

Granule contents damage tissue

B.

emedicine.com

Figure 24.28   Type III hypersensitivity.   A. Development of hypersensitivity. **B.** Urticarial rash caused by serum sickness, a form of type III hypersensitivity that can follow the administration of serum. The rash is similar to that formed after the administration of nonprotein drugs (for example, minocycline).

named for him—the Arthus reaction. Every few weeks, he would inject the same set of rabbits with horse serum. With each succeeding injection, he noted an increasingly severe reaction at the injection site. On early rounds only redness was noted, but after the fifth round, the lesions

became necrotic and slow to heal. This is a *localized* type III reaction, where antigen-antibody complexes form at the vessel walls.

Case History: Type IV Hypersensitivity

The patient is a 19-year-old woman from Argentina. As a child she received the BCG vaccine for tuberculosis. Upon entering university in the United States, she is required to take a skin test for tuberculosis. She tries to refuse but is told it is a requirement and allows the nurse to apply the test to her arm. Three days later, the test site has a large red lesion and the skin is starting to slough (peel off).

Type IV hypersensitivity, also known as **delayed-type hypersensitivity (DTH)**, is the only class of hypersensitivity to be triggered by antigen-specific T cells. Because T cells have to react and proliferate to cause a response, it generally takes 24–48 hours before a reaction is noticed. This delay distinguishes type IV hypersensitivity from the more rapid antibody-mediated allergic reactions (types I–III). The woman in the case study was initially sensitized with BCG (an attenuated strain of *M. bovis,* a close relative of *M. tuberculosis*) as a result of childhood vaccination. Because the organism is intracellular, the vaccination produced a cell-mediated immunity, complete with preactivated memory T cells (similar to memory B cells). When she was reinoculated by the skin test, the memory T cells activated and elicited a localized reaction at the site of injection.

There are two stages to the development of type IV sensitivity. The first stage, sensitization, involves processing and presentation of antigen on cutaneous dendritic cells (called Langerhans cells). These APCs travel to the lymph nodes, where T_H0 cells can react to them as described earlier (see **Figure 24.19**), generating activated T cells and a subset of memory T cells. On second exposure, two routes leading to delayed-type hypersensitivity (DTH) are possible. In the first pathway, memory T_H1 cells, upon binding antigen that is complexed to class II MHC receptors (this happens at the site of infection), will release IFN-gamma, TNF-beta, and IL-2 (**Fig. 24.29A**). These cytokines recruit macrophages and PMNs to the site and will activate macrophages and natural killer cells to release a number of other inflammatory mediators that will damage innocent, uninfected bystander host cells. In the second pathway, memory T_C cells can recognize antigen on class I MHC receptors, become activated and directly kill the host cell presenting the antigen (**Fig. 24.29B**). A hallmark of type IV hypersensitivity is that transferring white blood cells, not serum, can transfer the sensitivity to a naive animal. This is because T cells, not serum antibody, are responsible for the reaction.

Forms of delayed-type hypersensitivity reactions include contact dermatitis (such as poison ivy) and

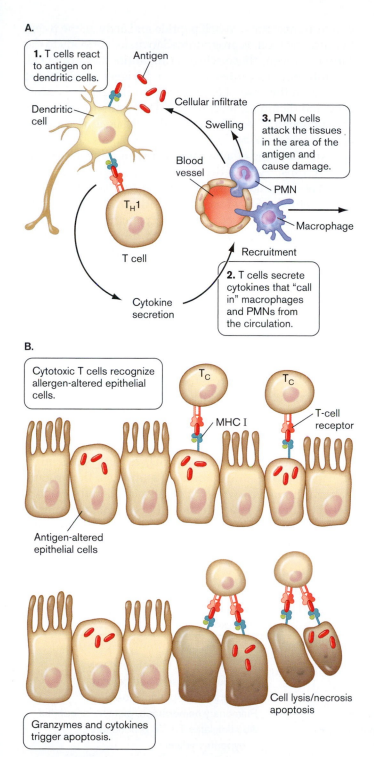

A.

1. T cells react to antigen on dendritic cells.

Antigen

Dendritic cell

Cellular infiltrate

Swelling

3. PMN cells attack the tissues in the area of the antigen and cause damage.

Blood vessel

T_H1

PMN

T cell

Macrophage

Recruitment

Cytokine secretion

2. T cells secrete cytokines that "call in" macrophages and PMNs from the circulation.

B.

Cytotoxic T cells recognize allergen-altered epithelial cells.

T_C T_C

MHC I

T-cell receptor

Antigen-altered epithelial cells

Cell lysis/necrosis apoptosis

Granzymes and cytokines trigger apoptosis.

Figure 24.29 Mechanisms of damage in delayed-type hypersensitivity (type IV). A. T_H1 cells react to antigen on dendritic cells. **B.** Antigen-sensitized cytotoxic T cells recognize allergen haptens attached to proteins of other cells. This recognition triggers granzyme release and cytokine production that triggers apoptosis of the target cell.

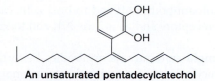

An unsaturated pentadecylcatechol

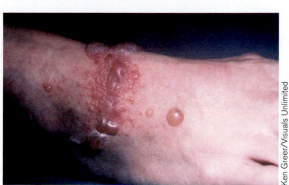

Ken Greer/Visuals Unlimited

Figure 24.30 Contact dermatitis. Pentadecylcatechols are chemicals present on the surface of poison ivy leaves. They are haptens that bind to proteins in dermal cells, where they can activate T cells. The result of the ensuing delayed-type hypersensitivity is contact dermatitis.

allograft rejection. The antigens involved with contact dermatitis are usually small haptens that have to bind and modify normal host proteins to become antigenic (**Fig. 24.30**). In this case, the hapten-modified protein is processed, and fragments containing the bound hapten are presented on the surfaces of antigen-presenting cells.

Autoimmunity Occurs When the Immune System Is Tricked into Recognizing Self as Nonself

The ability to distinguish between self antigens and foreign antigens is crucial to our survival; without this ability, our immune system would constantly attack us from within. Normally, the body develops tolerance to self; occasionally, however, an individual loses immune tolerance against some self antigens and mounts an abnormal immune attack against his or her own tissues. The attack can involve antibodies or T cells and is called an **autoimmune response**. Autoimmune responses may or may not be associated with pathological changes (autoimmune disease). Almost 30% of the population will have an autoimmune antibody by age 65, but many will not exhibit disease. The mechanisms that lead to autoimmune disease are essentially hypersensitivity reactions.

An autoimmune response results in autoimmune disease when an autoantibody or autoimmune lymphocyte causes damage to tissue components. Tissue damage can ensue when antibody to self activates antibody-dependent cell cytotoxicity (see Section 24.5). In this scenario, autoantibodies affixed to host cells can bind to NK cell Fc receptors and cause the NK cell to kill the tissue cell.

How might autoimmune antibodies be formed? One proposed mechanism starts with the occasional escape of self-reacting B cells from the negative selection process. The negative selection process normally kills autoreactive B cells. Self-reacting B cells that escape this process are usually not a problem because the specific helper T cells needed to activate them were deleted from the T-cell population. However, the self-reacting B cell can still be activated. The B cell can, of course, take up and process a self antigen (or a foreign antigen that mimics a self antigen). The problem comes if the antigen *also* contains a nonself epitope. This nonself epitope will also be presented on a B-cell receptor. When this happens, T cells specific to the nonself epitope will recognize the nonself peptide and activate that B cell. However, the B cell is programmed to make antibody to the self antigen, not the nonself antigen. Those antibodies will begin to attack the host antigen.

A good example of this is the M protein (pili) of *Streptococcus pyogenes*, a microbe that causes "strep throat" and the autoimmune disease rheumatic fever. M pili contain an epitope that resembles cardiac antigen, but the cardiac-like epitope is flanked by epitopes that are not related to the human host. The cardiac-like epitope can bind surface antibody on an escaped self-reacting B cell and be taken up. Because the nonself epitope and the heart-like epitope are contained within the same protein, the nonself epitope will "piggyback" its way into the B cell. The self epitope is processed and placed on the B-cell surface MHC. When this happens, T cells specific to the nonself epitope

will recognize the non-self peptide and activate the B cell. Because the B cell is programmed to make antibody to the cardiac antigen, those self-reacting antibodies are made, begin to bind to cardiac tissue, and trigger autoimmune disease. In the case of *S. pyogenes*, cardiac tissue is damaged and rheumatic fever results. This similarity between the epitope in M protein and cardiac tissue is an example of antigenic mimicry.

Another type of autoimmune reaction involves cytotoxic T cells. Type 1 diabetes, also known as juvenile diabetes, is an autoimmune disease that usually occurs during childhood, when T cells attack and destroy the insulin-producing islet cells in the pancreas. (In contrast to type 1 diabetics, type 2 diabetics continue to produce insulin, but their cells no longer respond to it.) A theoretical series of events has been proposed that involves molecular mimicry and starts with a viral infection of some sort. In this unproven model, an infecting virus (one study suggests Coxsackie virus), possesses an antigen with a shape similar to the glutamic acid decarboxylase in pancreatic beta islet cells. The cytotoxic T cells produced in response to the viral infection will also attack the beta cells, destroying them and the ability to make insulin.

Other autoimmune diseases involve the production of autoantibodies that bind and block certain cell receptors. Graves' disease (hyperthyroidism), for example, occurs when an autoantibody is made that binds to the thyroid-stimulating hormone (TSH) receptor. This mimics TSH binding and continually stimulates the production of T3 and T4, which generally increases metabolic rate. The thyroid is not destroyed but in fact enlarges to form a goiter. Examples of other important autoimmune diseases are listed in **Table 24.8**.

Table 24.8 Examples of autoimmune diseases.

Autoimmune disease	Autoantigen	Pathology
Type II hypersensitivity—mediated by antibody to cell surface antigens		
Acute rheumatic fever	Streptococcal M protein, cardiomyocytes	Myocarditis, scarring of heart valves
Autoimmune hemolytic anemia	Rh blood group	Destruction of red blood cells by complement, phagocytosis
Goodpasture's syndrome	Basement membrane collagen	Pulmonary hemorrhage, glomerulonephritis
Graves' disease	Thyroid-stimulating hormone receptor	Ab stimulates T3, T4 production, hyperthyroidism
Myasthenia gravis	Acetylcholine receptor	Interrupts electrical transmission, progressive muscular weakness
Type III hypersensitivity—mediated by antibody complexes with small soluble antigens (immune complex)		
Systemic lupus erythematosus	DNA, histones, ribosomes	Arthritis, vasculitis, glomerulonephritis
Type IV hypersensitivity—mediated by antigen specific T cells		
Type I diabetes	Pancreatic β cell antigen	β cell destruction
Multiple sclerosis	Myelin protein	Demyelination of axons

Special Topic 24.3   Organ Donation and Transplantation Rejection

It is now possible to donate organs from one person to another because we can minimize the risk of the recipient's immune system rejecting the donated tissue. As a result of their "education" process, T cells that survive the selection process by and large do not have TCRs that "see" isogenic MHC proteins on other cells of the body. However, about 5% of the surviving T cells can recognize and bind to allotypic MHC proteins on cells from other individuals (donors). This is because the allotypic MHC resembles a self MHC-foreign antigen complex. For this reason, transplanting organs from a donor with one type of MHC protein into a recipient with a different type of MHC (called an **allograft**) typically ends badly, with the recipient rejecting the graft. The allotypic MHC proteins of the donor play a major role in this transplantation rejection process.

To understand transplant rejection, it is important to remember that during normal antigen presentation, the T-cell receptor actually recognizes part of the presenting MHC molecule (self) as well as the peptide (antigen) presented. Consequently, allograft rejection is initiated in one of two ways. TCR on a recipient's T cells can *directly* bind the allotypic MHC (that is, the nonself MHC) on donor target cells that are presenting peptides (**Fig. 1A**). Alternatively, T cells from the recipient can *indirectly* recognize *pieces* of the donor allotypic MHC presented on the recipient's host cell through normal antigen-processing pathways (**Fig. 1B**). In the direct route, it does not matter whether the donor's antigen-presenting cells are presenting self peptides (from the recipient) or foreign peptides (from the donor). The allotypic differences in MHC are sufficient to trigger a response. The indirect route, however, requires that foreign donor peptides be presented on recipient APCs. Once activated by one of these routes, cytotoxic T cells will destroy all the donor cells containing the allotypic MHC proteins. The tissue is rejected.

Transplant rejection due to allotypic differences in MHC proteins might appear to contradict the MHC restriction rule for T-cell responses that was noted earlier. By this rule pre-activated T cells are restricted to act only on other host cells, that is, cells with compatible MHC proteins. Transplant rejection, however, is the result of T-cell action on foreign cells with incompatible MHC proteins and so appears to violate the MHC restriction rule. This paradox can be understood if we look at the time scales involved. In a typical laboratory experiment, when virus-infected cells of one MHC allotype are mixed with already activated T cells of a different MHC allotype, the in vitro cytotoxic reaction is monitored only for an hour. In this time period, activated cytotoxic T cells will not kill virus-infected cells of a different allotype but could kill virus-infected cells of the same isotype MHC. In transplant rejection, however, an unactivated T cell initially recognizes the donor cell allotypic MHC protein as nonself. The subsequent in vivo response to nonself will take days before rejection occurs. There are probably a few CD8 T_C cells in the in vitro experiment that could recognize the allotypic MHC antigen, but because they are so few in number, activation would take a long time.

MHC type matching is an important consideration for a successful transplant. Then how, you may ask, can we successfully carry out blood transfusions, which do not require MHC matching? Because human red blood cells, which are not nucleated, do not contain MHC proteins, tissue rejection via MHC incompatibility does not occur. However, the A or B antigens present on donor red blood cells can be recognized as foreign.

A. Direct route

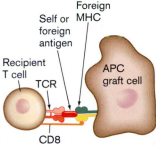

B. Indirect route

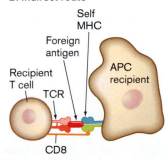

Figure 1   Transplantation rejection.   A. In the direct route to tissue rejection, T cells from the recipient recognize the foreign MHC on a donor (graft) cell and become activated. **B.** The indirect route occurs when pieces of foreign donor MHC (allo-MHC) are presented on the surface of recipient APCs. Recipient T_C cells then respond and become activated.

TO SUMMARIZE:

- **Allergens** cause the host immune system to over-respond or react against self.
- **Type I hypersensitivity** involves IgE antibodies bound to mast cells by the antibody Fc region. Binding of antigen to the bound IgE causes mast cell degranulation. Can occur within minutes of an exposure.
- **Type II hypersensitivity** (for example, Rh incompatibility) involves antibodies binding to cell surface antigens. This can trigger cell-mediated cytotoxicity or activate the complement cascade. Can occur within hours of an exposure.
- **Type III hypersensitivity** is an immune complex disease involving antibody complexes with small, soluble antigens. The complexes can activate complement or bind to mast cells. The result is the recruitment of PMNs that damage host cells in the area. May take weeks to form.
- **Type IV hypersensitivity** (delayed-type hypersensitivity) involves antigen-specific T cells. T_H1 cells release cytokines that activate macrophages and NK cells. T_C cells can directly kill cells that present the antigen. Reaction seen within a few days of exposure.
- **Autoimmune disease** is caused by the presence of lymphocytes that can react to self.
- **Autoreactive B cells** (B cells that make antibody directed against self epitopes) are activated when they present a nonself epitope on their surface MHC. T cells that recognize the nonself epitope activate the B cell, which then secretes antibody against the self epitope.
- **Cytotoxic T cells can produce autoimmune disease** by killing host cells that make a self protein that closely resembles a foreign antigen.

Concluding Thoughts

Immunity is a remarkable feat of nature. The realization that any one person's immune system can recognize and respond to virtually any molecular structure and yet remain selectively "blind" to his or her own antigens is hard to comprehend. But even more remarkable is that pathogenic microbes have managed ways to outmaneuver the interconnected redundancies and safeguards of the immune system. In some cases, the pathogen evolves to become less harmful and coexists with the host; indeed, pathogens in nature are far outnumbered by closely related strains that are harmless or even beneficial to their hosts. In other cases, however, evolution generates a never-ending arms race between pathogen and host. The next chapter describes some of the microbial strategies that contribute to the success of a pathogen.

CHAPTER REVIEW

End of Chapter Questions

1. Define antigen, epitope, hapten, and antigenic determinant.
2. What is the basic difference between humoral immunity and cellular immunity?
3. Why are proteins better immunogens than nucleic acids?
4. What makes IgA antibody different from IgG?
5. What is immunoprecipitation and radial immunodiffusion?
6. Explain isotypic, allotypic, and idiotypic differences in antibodies.
7. How is IgE involved in allergic hypersensitivity?
8. Discuss differences in the primary and secondary antibody responses.
9. Outline the basic steps that turn a B cell into a plasma cell.
10. What is isotype switching, and how is antibody diversity achieved?
11. What signals are needed to activate helper T cells?
12. What signals activate cytotoxic T cells?
13. Discuss the differences between transplant rejections of nucleated cells and rejection of blood cells.
14. How do superantigens activate T cells?
15. Discuss the differences between the alternative and classical pathways of complement activation.
16. How does the host prevent membrane attack complexes from being formed in host cells?
17. Describe the differences between hypersensitivity types I, II, III, and IV.
18. How does a B cell programmed to make an antibody against self become activated in the absence of specific T-cell help?

Key Terms

activated (909)
adaptive immune response (896)
adenylate cyclase (927)
allergen (926)
allograft (933)
allotype (904)
anaphylaxis (927)
antibody (897)
antigen (897)
antigenic determinant (897)
antigen-presenting cell (APC) (898)
atopic (927)
autoimmune response (931)
B-cell receptor (909)
B-cell tolerance (898)
capping (909)
cell-mediated immunity (897)
classical complement pathway (925)
clonal (897)
clonal selection (909)
constant region (902)
cytotoxic T cell (T$_C$ cell) (915)
cytokine (913)
decay-accelerating factor (926)
desensitization (928)
edema (927)
epitope (897)
equivalence (906)

F(ab)$_2$ region (903)
factor H (926)
Fc region (903)
gene splicing (911)
granzyme (921)
hapten (901)
heavy chain (902)
helper T cell (T$_H$ cell) (913)
humoral immunity (897)
idiotype (904)
immunogen (897)
immunogenicity (898)
immunoglobulin (902)
  IgA (904)
  IgD (905)
  IgE (905)
  IgM (904)
immunological specificity (899)
immunoprecipitation (906)
isotype (903)
isotype switching (class switching) (908)
late-phase anaphylaxis (928)
light chain (902)
major histocompatibility complex (MHC) (898)
memory B cell (908)
MHC restriction (923)

negative selection (918)
opsonize (904)
perforin (921)
plasma cell (908)
positive selection (918)
primary antibody response (908)
radial immunodiffusion (906)
recombination signal sequence (RSS) (911)
secondary antibody response (908)
serum (908)
sIgA (904)
superantigen (921)
switch region (913)
threshold dose (898)
tumor necrosis factor (927)
type I (immediate) hypersensitivity (927)
type II hypersensitivity (929)
type III hypersensitivity (929)
type IV (delayed-type) hypersensitivity (DTH) (930)
vaccination (899)
variable regions (903)
Western blot (906)

Recommended Reading

Hornef, Mathias W., Mary Jo Wick, Mikael Rhen, and Staffan Normark. 2002. Bacterial strategies for overcoming host innate and adaptive immune responses. *Nature Immunology* **3**:1033–1040.

Kaufmann, Stefan H., and Ulrich E. Schaible. 2005. Antigen presentation and recognition in bacterial infections. *Current Opinion in Immunology* **17**:79–87

Manz, Rudolf A., Anja E. Hauser, Falk Hiepe, and Andreas Radbruch. 2005. Maintenance of serum antibody levels. *Annual Review of Immunology* **23**:367–386

Medzhitov, Ruslan, and Charles A. Janeway, Jr. 2002. Decoding the patterns of self and nonself by the innate immune system. *Science* **296**:298–300.

Müller-Alouf, Heidi, Christophe Carnoy, Michel Simonet, and Joseph E. Alouf. 2001. Superantigen bacterial toxins: State of the art. *Toxicon* **39**:1691–1701

Nestle, Frank O., Arpad Farkas, and Curdin Conrad. 2005. Dendritic-cell-based therapeutic vaccination against cancer. *Current Opinion in Immunology* **17**:163–169.

Orange, Jordan S., Marlys S. Fassett, Louise A. Koopman, Jonathan E. Boyson, and Jack L. Strominger. 2002. Viral evasion of natural killer cells. *Nature Immunology* **3**:1006–1011.

Robinson, Harriet L., and Rama Rao Amara. 2005. T-cell vaccines for microbial infections. *Nature Medicine* **11**:s25–s32.

Savino, Wilson. 2006. The thymus is a common target organ in infectious diseases. *PLoS Pathogens* **2**:472–483

Chapter 25

Microbial Pathogenesis

Mammals have developed elaborate physical, chemical, and immunological defenses that protect against disease-causing microbes, or pathogens. Yet every fortress has its weakness. Pathogenic microbes have identified those weaknesses and exploit them to gain entry. The result is disease.

What makes a pathogen different from a commensal organism? The answer varies, depending on the pathogen. Some bacterial pathogens can avoid being phagocytosed by host cells, whereas others actively encourage it. Why do some microbes prefer to live intracellularly? Even more mysterious are those bacterial pathogens that develop a latent stage in the host, where they are undetectable, but still emerge and cause disease. Pathogens also differ widely with respect to their effects on a host. Some pathogens efficiently slay their hosts. Others persist for many years without killing. Which is the more successful pathogen?

The fundamental question of microbial pathogenesis is how an organism too small to be seen with the naked eye can kill a human that is 1 million times larger. This chapter will explore the varied methods that different bacterial and viral pathogens use to accomplish this feat.

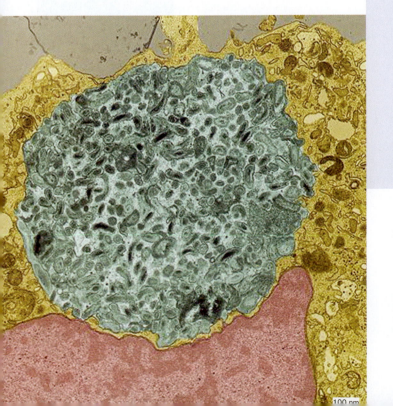

Pseudocolored transmission electron micrograph (magnification ×4,000) showing a Vero cell infected for 8 days with *Coxiella burnetii*, an obligate intracellular bacterium and the agent of human Q fever. *C. burnetii* multiplies within a large vacuole, called a parasitophorous vacuole (green). The organism directs development of this vacuole, a process presumably aided by the potent antiapoptotic properties of the organism, which keeps the host cell from killing itself. *Source*: Elizabeth Fischer of the Research Technologies Branch, Rocky Mountain Laboratories Microscopy Unit. From Daniel E. Voth, et al. 2007. *Infection and Immunity* 75:4263.

A 23-year-old Hispanic mother brought her 3-year-old daughter into the emergency room. The child was lethargic, had a fever of 40°C (104°F), and was having difficulty breathing. The mother explained that the family arrived in the United States from El Salvador the previous week. The attending physician noted an extreme swelling of the child's cervical lymph nodes, giving the girl a thick, "bull-neck" appearance. She also noticed the beginnings of a membranous growth at the back of the child's throat that was beginning to obstruct the trachea. It was grayish in color and bled when scraped. When asked, the distraught mother admitted that the child had not received any vaccinations before arriving in New Mexico. Suspecting the nature of the child's illness, the physician granted immediate admission to the hospital and ordered administration of penicillin and a specific antitoxin. The results of a throat swab sent to the microbiology lab confirmed the physician's suspicion. The root of the child's disease was *Corynebacterium diphtheriae*, which causes diphtheria.

Corynebacterium diphtheriae is a deadly pathogen able to kill humans by attacking the respiratory and cardiac systems. When untreated, death often comes by suffocation when the gray membrane completely covers the trachea. This pathogen is a gram-positive, nonspore-forming rod identical in appearance to many common commensal throat microbes, such as the more docile *C. striatum*. So, two gram-positive rods, identical in appearance, are both found in the human throat. Why is one a killer and the other not? The child in the case history most likely came in contact with the pathogen before leaving El Salvador, where vaccinations in some areas are difficult to obtain. The result was diphtheria. The question we want to ask in this chapter is *not* how this could have been prevented, but what genetic distinctions separate closely related disease-causing pathogens from innocuous nonpathogens. In this instance, the difference between friend and foe is a bacteriophage genome embedded in the genome of the organism. This phage carries the gene for diphtheria toxin, whose properties will be described later.

Pathogens, such as *C. diphtheriae*, that kill their hosts or only transiently reside there must also be prepared for life outside the host. These microbes possess an alternate physiology that allows survival in nonhost environments such as a lake or soil. However, some pathogens are not as versatile and die when separated from their host. They must be passed directly from person to person. If the host dies before transmission, the pathogen dies with it. Thus, to kill or not to kill is an important question each pathogen must address. This chapter will discuss the various relationships that occur between pathogens and their hosts and the factors that contribute to **pathogenesis**, the processes by which microbes cause disease in a host. The degree of harm that is caused depends on the mechanisms the pathogen has at its disposal.

25.1 Host-Pathogen Interactions

Parasites, in the broadest sense, include bacteria, viruses, fungi, and protozoa that colonize and harm their hosts. However, the term **pathogen** is typically used to refer to bacterial, viral, and fungal agents of disease. Disease-causing protozoans and worms, on the other hand, are normally called parasites. Pathogens and parasites infect their animal and plant hosts in a variety of ways and enter into different host pathogen relationships, depending on the site of colonization. For example, organisms that live on the surface of a host are called **ectoparasites**. The fungus *Trichophyton rubrum*, one cause of athlete's foot, is an ectoparasite (**Fig. 25.1**). *Wuchereria bancrofti*, the worm parasite that causes elephantiasis, is an **endoparasite** because it lives inside the body (**Fig. 25.2**).

Before examining the mechanisms microbes use to cause disease, it is helpful to know the terminology of pathogenesis. An **infection** occurs when a pathogen or parasite enters or begins to grow on a host. Be aware that

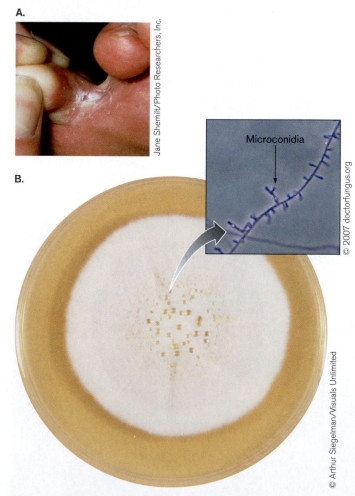

Figure 25.1   An ectoparasite.   A. Athlete's foot is caused by the fungus *Tricophyton rubrum*. **B.** Colony morphology and microscopic, branching conidia (blowup) of *T. rubrum*. Conidia are described in Chapter 20.

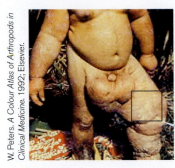

W. Peters, A Colour Atlas of Arthropods in Clinical Medicine. 1992; Elsevier.

CDC

Figure 25.2   An endoparasite.   The disease filariasis, commonly known as "elephantiasis" for obvious reasons, is caused by the worm *Wuchereria bancrofti* (blowup), which enters the lymphatics and blocks lymphatic circulation. Adult worms are threadlike and measure 4–10 cm in length. The young microfilaria (shown) are approximately 0.5 mm in length. Though not a problem in the United States, it is found throughout middle Africa, Asia, and New Zealand.

the term *infection* does not necessarily imply overt disease. Any potential pathogen growing in or on a host is said to cause an infection, but that infection may be only transient because immune defenses kill the pathogen before noticeable disease results. Indeed, most infections go unnoticed. For example, every time you have your teeth cleaned by a dentist your gums bleed and your oral flora transiently enter the bloodstream, but you rarely suffer any consequences.

Primary pathogens, also called **frank pathogens**, are disease-causing microbes possessing the means to breach the defenses of a healthy host. For example, *Shigella flexneri*, the cause of bacillary dysentery, is a frank pathogen. When ingested, it can survive the natural barrier of an acidic (pH 2) stomach, enter the intestine, and begin to replicate. **Opportunistic pathogens**, on the other hand, only cause disease in a compromised host. *Pneumocystis jiroveci* (previously *P. carinii*) is an opportunistic pathogen that causes life-threatening infections in AIDS patients, whose immune systems have been eroded (**Fig. 25.3A**). Some microbes even enter into a **latent state** during infection, where the organism cannot be found by culture. Herpes virus, for instance, can enter the peripheral nerves and remain dormant for years, then suddenly emerge to cause cold sores (**Fig. 25.3B**). The bacterium *Rickettsia prowazekii* causes epidemic typhus, but it can also enter a latent phase and months or years later cause a disease relapse called recrudescent typhus.

The term **pathogenicity** refers to an organism's ability to cause disease. It is defined in terms of how easily an organism causes disease (infectivity) and how

severe that disease is (virulence). Pathogenicity, overall, is shaped by the genetic makeup of the pathogen. In other words, an organism is more—or less—pathogenic depending upon the tools at its disposal (such as toxins), and their effectiveness.

Virulence is a measure of the degree or severity of disease. For instance, Ebola virus and the closely related Marbury virus have case fatality rates near 70%. This means that they are highly virulent (**Fig. 25.4**). On the other hand, rhinovirus, the cause of the common cold, is very effective at causing disease but never kills its victims. So it is highly infective but has a low virulence. Both organisms are pathogenic, but with one you live and the other you die.

One way to measure virulence is to determine how many bacteria or virions are required to kill 50% of an experimental group of animal hosts. This is called the LD_{50} (lethal dose 50%). An organism with a low LD_{50}, in which very few organisms are required to kill 50% of the hosts, is more virulent than one with a high LD_{50} (**Fig. 25.5**). For organisms that colonize but do not kill the host, the **infectious dose** needed to colonize 50% of the experimental hosts (ID_{50}) can be measured. Infectious dose is measured by determining how many microbes are required to cause disease symptoms in half of an experimental group of hosts.

Although one might be able to measure the infectious dose, rather than lethal dose, for a lethal pathogen, it is not typically done. Because it gives a clear end point, LD_{50} is much easier to use when trying to determine the effectiveness of a given treatment (an antibiotic, for example) or quantitate the role of a given gene in pathogenesis.

> **THOUGHT QUESTION 25.1**   Is a microbe with an LD_{50} of 5×10^4 more or less virulent than a microbe with an LD_{50} of 5×10^7?

A.

B.

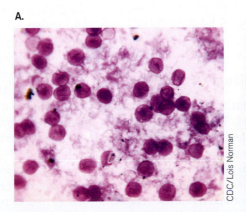
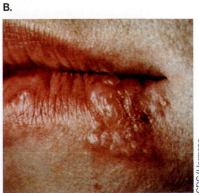

CDC/Lois Norman

CDC/Hermann

Figure 25.3   Opportunistic and latent infections.   **A.** *Pneumocystis jiroveci* cysts (5 to 10 μm diameter) in broncheoalveolar material. Notice how the fungi look like crushed Ping-Pong balls. **B.** Cold sore produced by a reactivated herpes virus hiding latent in nerve cells.

A.

B.

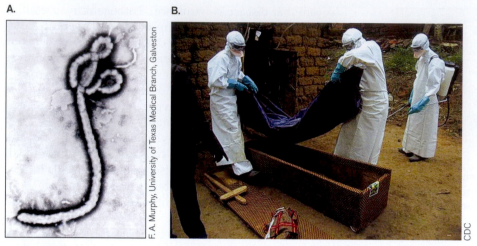

F. A. Murphy, University of Texas Medical Branch, Galveston

CDC

Figure 25.4 Highly virulent Ebola virus.   A. Ebola virus (approx. 1 μm long, TEM).
B. The body of a victim of Marburg virus is placed in a coffin for safe burial in Angola.
Marburg and Ebola cause hemorrhagic infections in which patients bleed from the mouth,
nose, eyes, and other orifices. They have a 70–80% mortality rate.

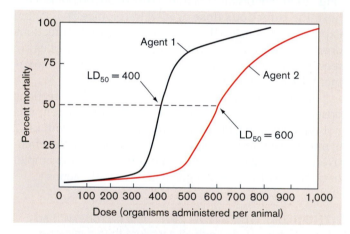

Figure 25.5 Measurement of virulence.   Each LD_{50}
measurement requires infecting small groups of animals with
increasing numbers of infectious agent and viewing how many
animals die. The number of microbes that kill half the animals is
called the LD_{50} dose. In this example, agent 1 is more virulent
than agent 2.

Infection Cycles Can Be Direct or Indirect

Somehow, pathogens must pass from one person or animal
to another if a disease is to spread. The route an organism takes to accomplish this is called the **infection cycle**.
A cycle of infection can be simple or complex (**Fig. 25.6**).
Organisms that spread directly from person to person, such
as the rhinovirus or *Shigella*, have simple infection cycles.
Inanimate objects through which pathogens can be relayed
to hosts are called **fomites**. More complex cycles often
involve **vectors**, usually insects, as intermediaries. Vectors
serve to carry infectious agents from one animal to another.
A mosquito vector, for example, transfers the virus caus-

ing yellow fever from infected to
uninfected individuals (**Fig. 25.7**)
in what is called **horizontal transmission**. The mosquito can also
bequeath this particular virus to
its offspring via infected eggs in
a form of **vertical transmission**
called **transovarial transmission**. Although yellow fever is not a
problem in the United States today,
West Nile virus, a flavivirus closely
related to yellow fever, currently
claims several victims each year in
this country.

Because insects are instrumental in transmitting pathogens, killing insect vectors is an
important way to halt the spread
of disease. Interventions include
spraying insecticide in a community during egg-hatching season
or using other microbes as assassins "trained" to kill the
vector. For example, *Bacillus thuringiensis* will kill many
types of insects that carry infectious agents. More recently,
an insect virus called baculovirus has been developed that
kills the *Culex* mosquito vectors carrying the West Nile
virus. The advantage of these vector-targeting microbes is
that they do not kill other insects or animals, as do many
chemical insecticides.

Another critical factor in an infection cycle is the
reservoir of infection. A reservoir is an animal, bird, or
insect that normally harbors the pathogen. In the case of
yellow fever, the mosquito is not only the vector, but the
reservoir as well, because the insect can pass the virus to
future generations of mosquitoes through vertical transmission. The virus causing eastern equine encephalitis
(EEE), however, uses birds as a reservoir. The microbe is
normally a bird pathogen and is transmitted from bird to
bird via a mosquito vector. The virus does not persist in
the insect, but transmission by the insect vector keeps the
virus alive by passing it to new avian hosts. Humans or
horses entering geographic areas harboring the disease
(called **endemic areas**) can also be bitten by the mosquito. When this happens, they become accidental hosts
and contract disease. The virus does not replicate to high
titers in mammals, which means that horses and humans
are poor reservoirs for the virus. However, the virus does
replicate to high numbers in the avian host. Reservoirs
are critically important for the survival of a pathogen and
as a source of infection. If the eastern equine encephalitis
virus had to rely on humans to survive, the virus would
cease to exist because of limited replication potential
and limited access to mosquitoes. It is important to note
that the reservoir of a given pathogen might not exhibit
disease.

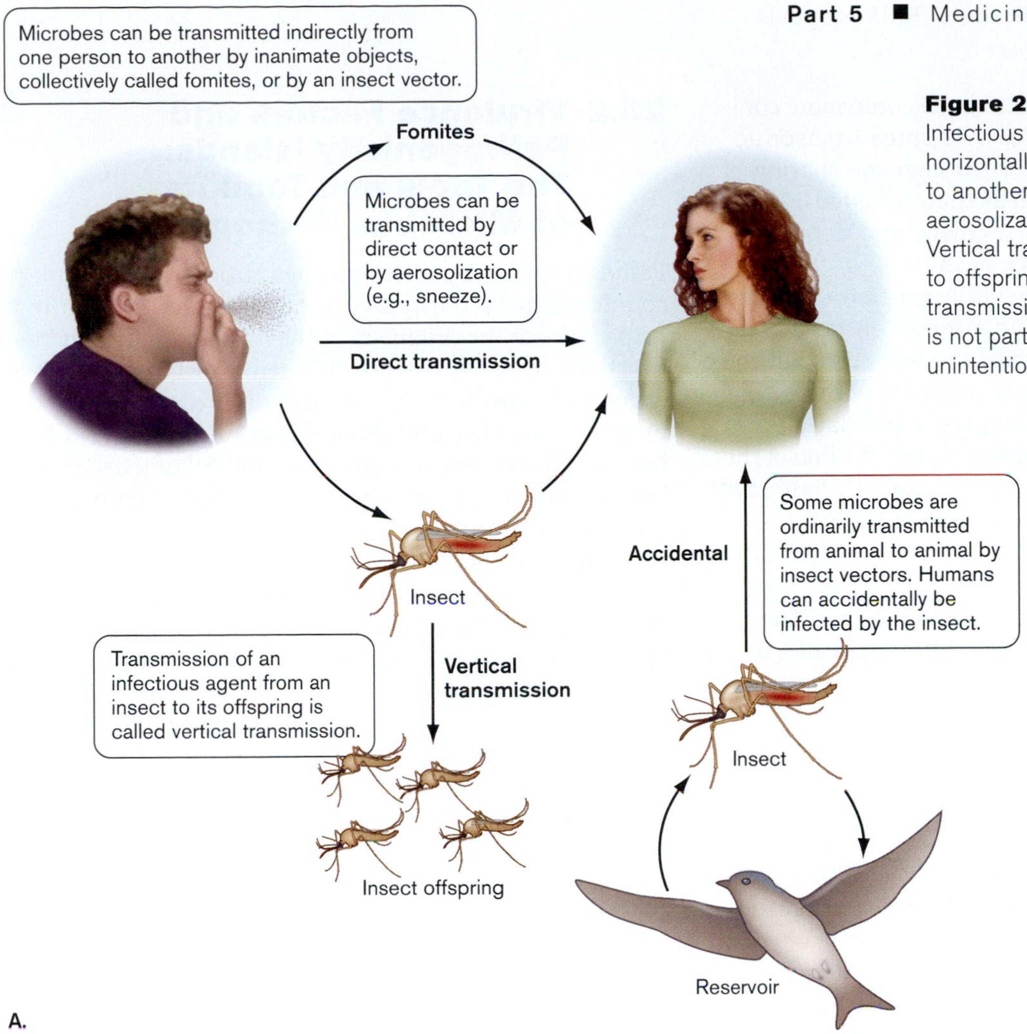

Microbes can be transmitted indirectly from one person to another by inanimate objects, collectively called fomites, or by an insect vector.

Fomites

Microbes can be transmitted by direct contact or by aerosolization (e.g., sneeze).

Direct transmission

Insect

Transmission of an infectious agent from an insect to its offspring is called vertical transmission.

Vertical transmission

Insect offspring

Accidental

Some microbes are ordinarily transmitted from animal to animal by insect vectors. Humans can accidentally be infected by the insect.

Insect

Reservoir

Figure 25.6   Infection cycles.
Infectious agents can be transmitted horizontally from one member of a species to another by a variety of means: fomites, aerosolization, direct contact, or insect vector. Vertical transmission is passage from parent to offspring during birth, while accidental transmission happens when a host that is not part of the normal infectious cycle unintentionally encounters that cycle.

A.

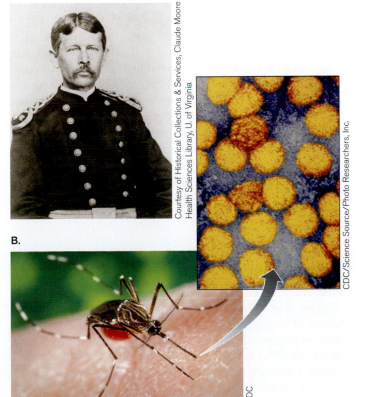

Courtesy of Historical Collections & Services, Claude Moore Health Sciences Library, U. of Virginia

CDC/Science Source/Photo Researchers, Inc.

B.

CDC

Figure 25.7   Insect vector and yellow fever.   A. Walter Reed, a member of the University of Virginia Medical School class of 1869, proved in 1901 that the mosquito *Aedes aegypti* was the vector of transmission for yellow fever, a disease named for the jaundice produced by liver damage. **B.** *Aedes aegypti.* Yellow fever is caused by a flavivirus (inset, TEM) carried by this mosquito. The virus varies in size from 50 to 90 nm. Yellow fever remains endemic in the northern part of South America and in central Africa.

Even a simple infectious cycle can become more complex. For example, rhinovirus can be spread person-to-person by a sneeze (airborne) or through the sharing of inanimate objects (fomites) such as contaminated utensils (fork, pen), towels, cloth handkerchiefs, and doorknobs. Handshaking is also an efficient means of transferring some pathogens. Imagine that one person in a city of 100,000 people has a cold and sneezes on her hands. Then, without washing her hands, she goes through the day shaking hands with 10 people, and each of those people shake hands with another 10 people per day, and so on. If there were no repeat handshakes, and if none of the contacts washed their hands, it would take only four days to spread the virus throughout the population.

In this example, eventually the entire populace of the city would come in contact with the virus, but not everyone would actually contract disease. Additional factors influence whether the virus successfully replicates in a given individual.

Portals of Entry

How do infectious agents gain access to the body? Each organism is adapted to enter the body in different ways. Food-borne pathogens (for example, *Salmonella*, *E. coli*, *Shigella*, and rotavirus) are ingested by mouth and ultimately colonize the intestine. They have an oral portal of entry. Airborne organisms, in contrast, infect through the respiratory tract. Some microbes enter through the conjunctiva of the eye, others through the mucosal surfaces of the genital and urinary tracts. Agents that are only transmitted by mosquitoes or other insects enter their human hosts via the parenteral route, meaning injection into the bloodstream. Wounds and needle punctures can also serve as portals of entry for many microbes. For instance, shared needle use between drug addicts has been an important factor in the spread of HIV.

TO SUMMARIZE:

- **Infection** with a microbe does not always lead to disease.
- **Primary pathogens** have mechanisms that help the organism circumvent host defenses.
- **Opportunistic pathogens** cause disease only in a compromised host.
- **Pathogenicity** refers to the mechanisms a pathogen uses to produce disease and how efficient the organism is at causing disease, whereas virulence is a measure of disease severity.
- **Diseases can be spread** by direct or indirect contact between infected and uninfected persons/animals or by insect vectors.
- **Pathogens use portals of entry** best suited to their mechanisms of pathogenesis.

25.2 Virulence Factors and Pathogenicity Islands: The Tools and Toolkits of Microbial Pathogens

Pathogens can be distinguished from their avirulent counterparts by the presence of **virulence factors** that help establish the organism in the host and that alter host functions to cause disease. Virulence factors, which are encoded by virulence genes, include toxins, attachment proteins, capsules, and other devices used to avoid host innate and adaptive immune systems. All of these factors enhance the disease-producing capability, or pathogenicity, of the pathogen.

Extensive sequencing efforts have allowed us to compare genomes of many pathogens and expose some "footprints" of their evolution. For example, in bacterial pathogens, most chromosomes are dotted with clusters of pathogenicity genes that encode virulence functions. These gene clusters, called **pathogenicity islands**, can be considered the toolboxes of pathogens (originally discussed in Section 9.7). Many virulence genes reside in pathogenicity islands, although many others do not. Some virulence genes reside on plasmids (for example, the genes for the diarrhea-producing labile toxin of certain *E. coli* strains) or in phage genomes (such as the genes encoding the diphtheria toxin of *Corynebacterium diphtheriae*).

Many of the genes in pathogenicity islands were originally inherited through horizontal transmission from other organisms by, for example, conjugation or transduction (discussed in Section 9.2). But what do the pathogenicity genes do? Some genes encode molecular "grappling hooks," such as pili that attach to host cells. Once attached, microbes can secrete toxins that injure the host cell. Other bacteria wall themselves off to prevent damage by host inflammatory responses. Some bacterial pathogens are even capable of what could be called "host cell reprogramming." These organisms inject proteins directly into the host cell to disrupt normal signaling pathways. This reprogramming causes the target cell to either engulf the bacterium, commit suicide (undergo apoptosis), or provide an even more intimate attachment platform at the cell surface.

Pathogenicity Islands Are Different from the Rest of the Genome

How do new pathogens evolve? DNA sequencing efforts suggest that gene swapping followed by divergent evolution is a major force in the development of emerging pathogens. Horizontal gene transfers move whole blocks of DNA (more than 10 kbp) from one organism to another, placing the blocks directly in the chromosome in what is called a **genomic island**. If the island increases the "fitness" of a microorganism (pathogen) that interacts with a host, it is called a pathogenicity island. Genomic islands generally reveal themselves by several anomalies that they possess with respect to the rest of the host genome:

■ They are often linked to a tRNA gene and generally have a GC/AT ratio very different from the rest of the chromosome (**Fig. 25.8**). For example, a plot of GC content along the length of a chromosome may reveal that most of the genome has a 50% GC content. But somewhere in the middle, a 50-kbp region sticks out on the graph, showing a content of 40%. This probably reflects the GC content of the microbe that donated the island. The reason tRNA genes are often

the targets for insertion of pathogenicity islands is not known. One hypothesis is that the conserved secondary structure of tRNA genes provides a structural motif that facilitates integration by an integrase.

■ They are typically flanked by genes with homology to phage or plasmid genes. This is thought to reflect the transfer vector used to move the island from one organism to another.

■ They carry gene clusters with specific functions, such as protein export systems that secrete toxins (for example, type III secretion systems that inject toxic proteins directly into target host cells; see Section 25.5). These are the operons that contribute to the fitness of the organism in pathogenic circumstances.

Table 25.1 lists several pathogenicity islands and their functions. It provides examples of pathogenicity islands present in different bacteria, names a key function of the products of the island, and gives the disease caused. Various pathogenicity island functions will be described as the chapter proceeds.

Shigella and *Escherichia* Are Examples of Evolution through Horizontal Transmission

Figure 25.9 shows a schematic comparison of the circular *Shigella flexneri* genome with those of *E. coli* K-12 (a commonly used avirulent lab strain) and *E. coli* O157:H7 (a virulent strain of enterohemorrhagic *E. coli*). All three gram-negative rods are closely related but differ greatly in terms of pathogenic potential and mechanisms. *S. flexneri* and *E. coli* O157:H7 cause bloody diarrhea, while *E. coli* K-12 has a commensal origin. DNA sequence analysis has revealed chromosome regions that are common among these organisms, as well as genomic islands specific to individual species and strains (Fig. 25.9). The core genes needed for sustaining growth (that is, for transcription, translation, replication, and so on) are common to all

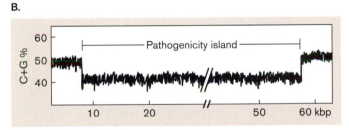

A.

B.

Figure 25.8   Model pathogenicity island.   A. Schematic model of a pathogenicity island. The DNA block is linked to a tRNA gene and flanked by direct repeats (DR) that may be "footprints" of a transposon or viral-mediated transfer. The integrase gene (int) and insertion sequences (IS) may also be remnants of transposition. **B.** The guanine + cytosine (GC) content of the island is different from that of the core genome.

Table 25.1	Examples of pathogenicity islands.		
Pathogenicity island	**Function**	**Organism**	**Disease**
HPI (high pathogenicity island)	Iron uptake	*Yersinia* spp.	Plague, enterocolitis
VPI (*Vibrio* pathogenicity island)	Toxin production	*Vibrio cholerae*	Cholera
PAI III (pathogenicity island III)	Encodes adhesins	Uropathogenic *E. coli*	Urinary tract infection
SPI-1 and SPI-2			
(*Salmonella* pathogenicity islands)	Type III secretion	*Salmonella enterica*	Gastroenteritis
SHI-1 and SHI-2 (*Shigella* islands)	Type III secretion	*Shigella flexneri*	Bloody diarrhea
YSA (*Yersinia* secretion apparatus)	Type III secretion	*Yersinia* spp.	Plague, enterocolitis
Cag PI (cytotoxin-associated gene)	Type IV secretion	*Helicobacter pylori*	Gastric ulcers, gastric cancer
icm/dot (intracellular multiplication)	Type IV secretion	*Legionella pneumophila*	Legionnaires' disease

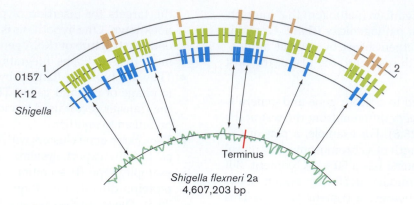

Figure 25.9   Comparison of the *Shigella flexneri* 2a chromosome with those of *E. coli* K-12 and O157:H7 (EDL933).
Segments of the three genomes between 1 MB and 2 MB are shown. Gray line indicates DNA sequences shared among the organisms (O157, top arc; K-12, second arc; *Shigella*, third arc). Colored boxes depict genomic islands (including pathogenicity islands) present in each organism. The bottom arc illustrates the GC content of the *Shigella* genome. Each data point along the graph indicates GC content relative to AT content as averaged over a sliding 10 kbp window. Note how major differences in *Shigella* GC content shown above or below the center line (indicating 50% GC content) often correlate with the genomic islands depicted in the third arc. Arrows indicate some islands with obvious correlation to GC content.

three organisms, whereas other genes may be present in only one. These unique genes and islands are thought to be the result of horizontal gene transfers originating from widely different genera. Within the unique DNA segments of *S. flexneri* and *E. coli* O157:H7 are genes encoding host cell attachment, toxin secretion, and the toxins themselves—all of which are absent from K-12. The degree of gene shuffling that was required to separate these otherwise similar bacteria is remarkable.

The following sections describe some of the specific tools pathogens use to undermine the integrity of the body. From attachment, to toxins, to intracellular invasion, the infection process is like a chess match, with each side, human and microbe, trying to outmaneuver the other.

TO SUMMARIZE:

■ **Pathogenicity islands** are DNA sequences within a species that are acquired by horizontal gene transfer from a different species.
■ **Virulence genes** encode genes whose products enhance the disease-causing ability of the organism. Many virulence genes can be found within pathogenicity islands, but some are located outside of an obvious genomic island or reside in plasmids.
■ **Pathogenicity islands** contain distinct features, such as GC content and the remnants of phage or plasmids, that mark them as being different from the rest of the genome.

25.3 Virulence Factors: Microbial Attachment

Regardless of the disease, pathogens must reach a colonization site either through their own motility or by

hitchhiking with a vector. Once at the site, attachment mechanisms are needed to stay there.

The human body has many ways to exclude pathogens. The lungs use a mucociliary elevator (see **Fig. 23.5**) to rid themselves of foreign bodies, the intestine uses peristaltic action to ensure that its contents are constantly flowing, and the bladder uses contraction to propel urine through the urethra with tremendous force. How do bacteria ever manage to stay around long enough to cause problems? Like a person grasping a telephone pole during a hurricane, successful pathogens moving through the body manage to grab onto host cells and tenaciously hold on. Thus, the first step toward infection is attachment, also called adhesion. An **adhesin** is the general term for any microbial factor that promotes attachment.

Viruses attach to the host through their capsid or envelope proteins, which bind to specific host cell receptors discussed in Chapters 6 and 11. Bacteria have a variety of similar strategies. They can use hairlike appendages called **pili** (also called **fimbriae**), whose tips contain receptors for mammalian cell surface structures. Or they can use a variety of adherence proteins or other molecules (adhesins) that are not part of a pilus. Sometimes they use both. **Table 25.2** summarizes bacterial attachment strategies.

Types of Pili

Different pili from different bacterial species have been classified historically based on phenotype. In some cases the phenotype classes have turned out to be inconsistent with sequence-based homologies. We will consider three groups in this chapter.

Table 25.2 **Specific attachments of bacteria to cell or tissue surfaces.**

Bacterium	Adhesin	Host receptor	Attachment site	Disease
Streptococcus pyogenes	Protein F	Amino terminus of fibronectin	Pharyngeal epithelium	Sore throat
Streptococcus mutans	Glucan	Salivary glycoprotein	Pellicle of tooth	Dental caries
Streptococcus salivarius	Lipoteichoic acid	Unknown	Buccal epithelium of tongue	None
Streptococcus pneumoniae	Cell-bound protein	*N*-acetylhexosamine galactose disaccharide	Mucosal epithelium	Pneumonia
Staphylococcus aureus	Cell-bound protein	Amino terminus of fibronectin	Mucosal epithelium	Various
Neisseria gonorrhoeae	*N*-methylphenylalanine pili	Glucosamine galactose carbohydrate	Urethral/cervical epithelium	Gonorrhea
Enterotoxigenic *E. coli*	Type I fimbriae (pili)	Species-specific carbohydrate(s)	Intestinal epithelium	Diarrhea
Uropathogenic *E. coli*	Type I fimbriae (pili)	Complex carbohydrate	Urethral epithelium	Urethritis
Uropathogenic *E. coli*	P pili (pyelonephritis-associated pili)	P-blood group	Upper urinary tract	Pyelonephritis
Bordetella pertussis	Pili ("filamentous hemagglutinin")	Galactose on sulfated glycolipids	Respiratory epithelium	Whooping cough
Vibrio cholerae	*N*-methylphenylalanine pili	Fucose and mannose carbohydrate	Intestinal epithelium	Cholera
Treponema pallidum	Peptide in outer membrane	Surface protein (fibronectin)	Mucosal epithelium	Syphilis
Mycoplasma	Membrane protein	Sialic acid	Respiratory epithelium	Pneumonia
Chlamydia	Unknown	Sialic acid	Conjunctival or urethral epithelium	Conjunctivitis or urethritis

Type I pili are a group that, in general, adhere to mannose residues on host cell surfaces. Because adding free mannose will inhibit attachment of most type I pili, this binding is called mannose sensitive. Another group of pili is called mannose resistant because the pili do not bind to mannose residues. There are at least two types of mannose-resistant pili. Members of one type, sometimes called **type III pili**, bind to red blood cells treated with tannic acid. The other more commonly studied type is called **type IV pili**. Unlike type I pili, which simply stick out from the cell surface, type IV pili are more dynamic and are assembled on the cell surface through a very different pathway.

There are other types of attachment pili that do not fall neatly into one of these three groups. For clarity, it is important to note that the primary classification of pili is now based largely on protein sequence information (deduced from DNA sequence) and may contradict earlier phenotype-based schemes. For instance, the pyelonephritis adhesion pili (Pap) of the uropathogenic *E. coli* are mannose resistant, since they bind to a digalactoside on host surfaces (called the P blood group antigen). However, Pap is very similar to type I pili based on amino acid sequence homology.

Pilus Assembly

How bacteria assemble pili on their cell surfaces is an engineering marvel. The shafts of pili are cylindrical structures composed of identical pilin protein subunits. Several different proteins adorn the tip, including one at the very apex that binds to host receptors (**Fig. 25.10A**). In addition to these structural components, numerous other proteins work together as a machine to assemble the structure. Genes encoding a given pilin protein and the associated assembly apparatus are typically arranged on the chromosome as an operon (**Fig. 25.10B**).

In **Figure 25.11**, the assembly of a type I pilus is shown, using uropathogenic *E. coli* Pap pili as a model. The mechanism is representative of other type I pili; only the names of the proteins will differ for each system. Protein components are secreted into the periplasm by the SecA-

A. FimH adhesion tip (TEM)

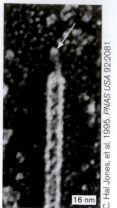

C. Hal Jones, et al. 1995. *PNAS USA* 92:2081

16 nm

Figure 25.10 Attachment pili and encoding operon. **A.** High-resolution micrograph showing a type I pilus (TEM). The FimH adhesin at the tip is the protein that binds to the cell receptor. **B.** Genetic organization of the type I gene cluster, which includes genes involved in pilus assembly. The genes are designated *fim A–H.*

B. Type I pilus gene cluster

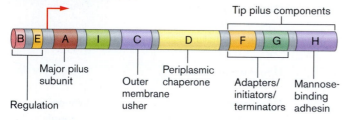

Tip pilus components

B E A I C D F G H

Regulation
Major pilus subunit
Outer membrane usher
Periplasmic chaperone
Adapters/ initiators/ terminators
Mannose-binding adhesin

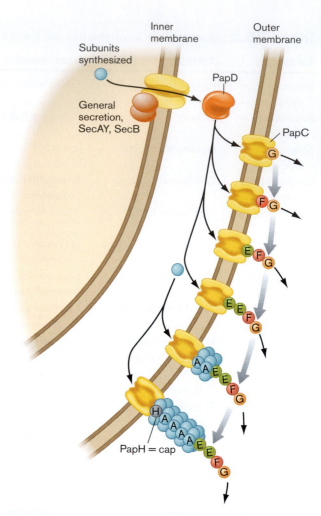

Inner membrane — Outer membrane

Subunits synthesized

General secretion, SecAY, SecB

PapD

PapC

PapH = cap

Figure 25.11 Assembly of type I pili. The figure illustrates pyelonephritis-associated pilus (Pap) assembly but is representative of other type I pili. Only the names of the proteins will differ. Proteins are secreted by the Sec system to the periplasm, where they are chaperoned by PapD to the site of assembly. PapC, also called the usher, assembles the individual proteins in the proper order. Assembly starts with the tip protein, PapG, marked at the far right, which ultimately binds to carbohydrates on host membranes. The subunits fit together like pieces of a jigsaw puzzle. The arrow at the head of the elongating pilus indicates direction of pilus growth.

dependent general secretory system (discussed in Section 8.4). Once in the periplasm, the subunits are chaperoned by PapD to the membrane site of assembly, which is marked by the presence of the usher protein PapC. PapC proteins form channels in the outer membrane large enough to accommodate individual pilus subunits and, like an usher in a theater, direct the subunits to their proper places. Chaperoning of the pilin building blocks by PapD is necessary to prevent pilin subunits from inadvertently assembling in the periplasm. As illustrated in Figure 25.11, appropriate assembly of pili at the usher site starts with the tip protein, PapG, which will ultimately bind to carbohydrates on host membranes after the pilus is complete. After PapG, the ushers add PapF and PapE, forcing PapG farther away from the surface. Then a series of identical PapA pilin subunits are strung together to form the shaft. The PapA subunits assemble by swapping domains, essentially linking themselves together like pieces of a jigsaw puzzle.

Interestingly, shear forces generated by fast-moving urine in the urethra may actually tighten bacterial adhesion to target cells. The tip protein, FimH, which is analogous to PapG but for a different type I pilus, contains two domains, the pilin domain (pale yellow) and the lectin domain (blue), as shown in **Figure 25.12.** The pilin domain integrates FimH into the tip of the pilus. The lectin domain binds the host receptor. Scientists have described a mechanism whereby a shear force, such as might be experienced by uropathogenic *E. coli* bound to bladder cells during urination, extends the interdomain

linker between the FimH lectin and pilin domains and *strengthens* the hold of the lectin domain on the mannose receptor. It is not clear whether shear ends up changing the existing site from a low-affinity to a high-affinity binding site or if the conformational shift exposes a second, hidden, mannose-binding site. Regardless, it becomes harder to pry *E. coli* from its target host cell. Thus, fluid forces in the human body, as occur with salivating, swallowing, sneezing, urinating, and weeping, may in some cases strengthen bacterial adhesion instead of detaching and flushing away the infectious agents.

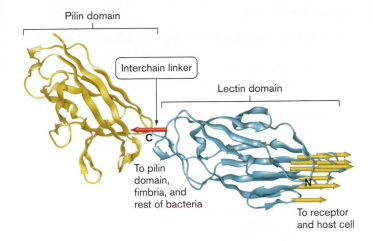

Pilin domain

Interchain linker

Lectin domain

C

To pilin domain, fimbria, and rest of bacteria

N

To receptor and host cell

Figure 25.12   Structure of type I pili tip protein.   FimH contains two domains, the pilin domain (yellow) and the lectin domain (blue). The pilin domain integrates FimH into the tip of the pilus. The lectin domain binds the host receptor. When shear is created, the receptor-binding site on the lectin domain is pulled in one direction (yellow arrows), and residue T158 (orange arrow) is pulled in the opposite direction. (PDB codes: 2co4, 1tr7)

Other pili with an important role in pathogenesis are the type IV pili. Type IV pili are found in a broad spectrum of gram-negative bacteria and share amino acid homology in their major pilin structure (**Fig. 25.13**). Species with type IV pili include *V. cholerae, Pseudomonas aeruginosa* (**Fig. 25.13B**), certain pathogenic strains of *E. coli, Neisseria meningitidis,* and *N. gonorrhoeae.* All type IV pili use similar secretion and assembly machinery involving at least a dozen proteins. One major difference between the assembly of type IV and type I pili is that type IV pilus proteins are never free in the periplasm; they are transported directly from the cytoplasm through a channel in the outer membrane (**Fig. 25.13A**). Thus, type IV pilus assembly is SecA independent. As described here, the blueprints for this mechanism were adapted and modified through evolution to secrete other proteins via what is generally called type II protein secretion (discussed in Section 25.5).

A.

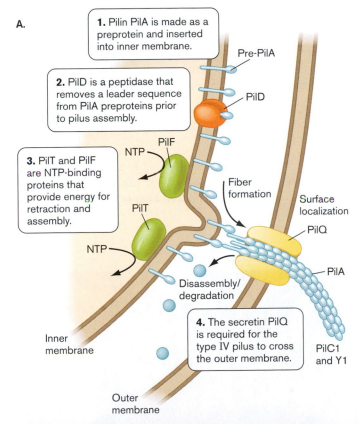

1. Pilin PilA is made as a preprotein and inserted into inner membrane.

Pre-PilA

2. PilD is a peptidase that removes a leader sequence from PilA preproteins prior to pilus assembly.

PilD

PilF

NTP

3. PilT and PilF are NTP-binding proteins that provide energy for retraction and assembly.

PilT

NTP

Fiber formation

Surface localization

PilQ

PilA

Disassembly/ degradation

Inner membrane

4. The secretin PilQ is required for the type IV pilus to cross the outer membrane.

PilC1 and Y1

Outer membrane

Figure 25.13   Type IV pili.   A. Model of pilus assembly and disassembly. In this example, PilA is the pilin protein, and PilC1 and Y1 form the attachment tip. Diameter of the filament is approximately 6 nm. Note that assembly/disassembly requires hydrolysis of nucleoside triphosphate and takes place at the inner membrane, not in the periplasm. **B.** Photographic evidence of type IV pilus extension and retraction in cells of *P. aeruginosa.* Filament b retracts, then filament d extends at 6 seconds and retracts. Filament c attaches briefly at its distal tip (note straightening at 24 seconds), then begins to retract. Time in seconds. Fluorescent microscopy. **C.** Type IV pili are essential for the interaction of *Neisseria meningitidis* with brain endothelial cells. Type IV pili are green in this SEM. Diplococcal cells are approx. 1.6 μm in diameter. *Sources:* A. Bardy, Ng, and Jarrel. 2003 *Microbiol.* 149:295–304. B. J. M. Skerker and H. C. Berg. 2001. *PNAS* 98:6901–6904.

B. Extension and contraction of Type IV pili

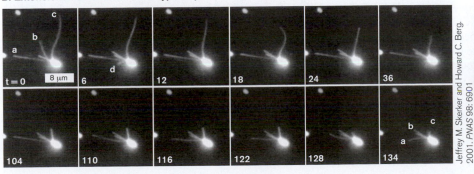

c

b

a

t = 0   8 μm   6

d

12

18

24

36

104

110

116

122

128

134

b

c

a

Jeffrey M. Skerker and Howard C. Berg. 2001. *PNAS* 98: 6901

C. Type IV pili of *Neisseria meningitidis*

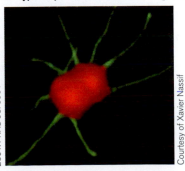

Courtesy of Xavier Nassif

Type IV pili can actually make cells move because the assembly process involves the reiterative elongation and retraction of the pili. The process, called "twitching motility," occurs when the pilus elongates, attaches to a surface, then depolymerizes from the base, which shortens the pilus and pulls the cell forward. This mechanism is akin to using a grappling hook to scale a building. The gliding motility of the slime mold *Myxococcus xanthas* is also due to type IV pili. The type IV pili of *Neisseria meningitidis*, shown in **Figure 25.13C**, are essential for crossing the blood-brain barrier and causing bacterial meningitis.

Bacteria also carry afimbriate adhesins (proteins that aid in attachment but do not form pili) that mediate binding to host tissues (**Fig. 25.14**). Some examples include *Bordetella* pertactin (which binds to integrin), *Streptococcus* protein F (binds to fibronectin), *Streptococcus* M protein (binds to fibronectin and complement regulatory factor H), and intimin of enteropathogenic *E. coli* (binds to Tir; discussed in Section 25.5). Fimbriae (pili) often mediate the initial binding between bacterium and host, after which a more intimate attachment is created by an afimbriate attachment protein. In the case of *Neisseria gonorrhoeae*, once the type IV pilus has attached to the surface of the mucosal epithelial cell, the filamentous pilus contracts, pulling the bacterium down onto the host cell membrane. Tight secondary interactions are then mediated by the neisserial Opa membrane proteins, another example of an afimbriate adhesin (Opa is named because of the *opa*city it adds to colony appearance).

TO SUMMARIZE:

- **Bacteria use pili and nonpilus adhesins** to attach to host cells.
- **Type I pili** produce a static attachment to the host cell, whereas type IV pili continually assemble and disassemble.
- **Nonpilus adhesins** are bacterial surface proteins, or other molecules, that can tighten interactions between bacteria and target cells.

25.4 Toxins: A Way to Subvert Host Cell Function

Following attachment, many microbes secrete protein toxins (called **exotoxins**) that kill host cells and unlock their nutrients (because dead host cells ultimately lyse). Bacterial pathogens have developed an impressive array of toxins that take advantage of different key host proteins, or structures. All gram-negative bacteria also possess a nonprotein yet toxic compound called **endotoxin** that can hyper-activate host immune systems to harmful levels.

Microbial Exotoxins Have Many Modes of Action

Microbial exotoxins fall into five broad categories based on their mechanisms of action (**Table 25.3**). Three of these classes are illustrated in **Figure 25.15**.

A. M protein

Maria Fazio and Vincent A. Fischetti, Rockefeller University

M protein fibrils

Figure 25.14 Nonpilus adhesins. A. M protein surface fibrils on *Streptococcus pyogenes* (TEM). Cell size 0.5 to 1 μm diameter. **B.** Colonization of tracheal epithelial cells by *Bordetella pertussis* (SEM). This organism uses a surface protein called pertactin as well as a pilus called filamentous hemagglutinin (FHA) to bind bronchial cells.

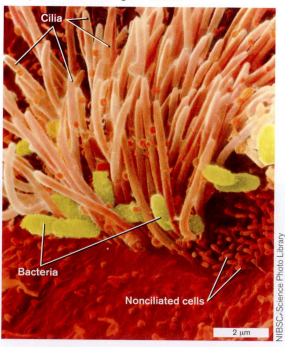

B. *B. pertussis* colonizing the trachea

Cilia

Bacteria

Nonciliated cells

2 μm

NIBSC/Science Photo Library

- **Cell membrane disruption.** Members of the first class, exemplified by the alpha (α) toxin of *Staphylococcus aureus*, disrupt the cell membrane and cause leakage of cell constituents (**Fig. 25.15A**).
- **Protein synthesis disruption.** A second class, exemplified by diphtheria and Shiga toxins, targets eukaryotic ribosomes and destroys protein synthesis (**Fig. 25.15B**).
- **Second messenger pathway disruption.** The third broad mechanism of action involves the toxin subverting host cell secondary messenger pathways. Cholera toxin and *E. coli* ST (stable toxin), for instance, cause runaway synthesis of cAMP (described later) and cGMP (**Fig. 25.15C**), respectively, in target cells. Elevated cAMP or cGMP levels, in turn, trigger critical changes in ion transport and fluid movement.
- **Superantigens.** The fourth class includes toxins that act as superantigens and activate the immune system without being processed by antigen-presenting cells (discussed in Section 24.6). The pyrogenic toxins of *Staphylococcus aureus* (toxic shock syndrome toxin) and *Streptococcus pyogenes* are examples of superantigenic toxins.
- **Proteases.** The fifth class of toxins consists of proteases. One example is tetanus toxin, a protease that cleaves host components involved in nerve signal transmission.

This section focuses on the key concepts of microbial toxin activity. Section 25.5 discusses how bacteria secrete these toxins.

A common structural theme among many, but not all, bacterial toxins is that they have two subunits, usually called A and B. These two-subunit complexes are termed **AB toxins**. The actual toxic activity in AB toxins resides within the A subunit. The role of the B subunit is limited to binding host cell receptors. Thus, the B subunit for each toxin delivers the A subunit to the host cell. Many AB toxins have five B subunits arranged as a ring, in the center of which is nestled a single A subunit (**Fig. 25.16A**).

A major AB toxin subclass comprises toxins that have an **ADP-ribosyltransferase** enzymatic activity. These enzyme toxins transfer the ADP-ribose group from an NAD molecule to a target protein (**Fig. 25.16B**). The ADP-ribosylated protein has an altered function. Sometimes the function is destroyed (for example, protein synthesis is destroyed by diphtheria toxin); other times, the protein is locked into an active form insensitive to regulatory feedback control (for example, cAMP synthesis continues unchecked in the presence of cholera toxin).

Mechanisms of selected toxins representing the various classes are described in this section.

Alpha toxin. The hemolytic alpha toxin is produced by *Staphylococcus aureus*, an organism that causes boils and blood infections. Alpha toxin forms a transmembrane, oligomeric (seven member) beta barrel pore in target cell membranes. It is easy to see how the resulting leakage of cell constituents and influx of fluid cause the target cell to burst. To form the pore, hydrophobic areas of each monomer face the lipids of the membrane and hydrophilic residues face the channel interior.

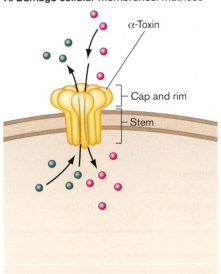

A. Damage cellular membranes/matrices

α-Toxin

Cap and rim

Stem

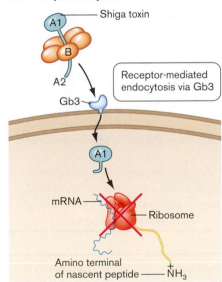

B. Inhibit protein synthesis

Shiga toxin

A1

B

A2

Receptor-mediated endocytosis via Gb3

Gb3

A1

mRNA

Ribosome

Amino terminal of nascent peptide — $\overset{+}{N}H_3$

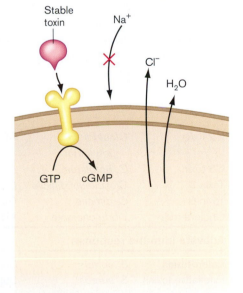

C. Activate secondary messenger pathways

Stable toxin

Na⁺

Cl⁻

H₂O

GTP   cGMP

Figure 25.15 Three classes of microbial toxins. These classes are defined by mode of action. **A.** Pore-forming toxins assemble in target membranes and cause leakage of compounds into and out of cells. **B.** Shiga toxin attaches to ganglioside Gb3, enters the cell, and cleaves 28S rRNA in eukaryotic ribosomes to stop translation. **C.** Enterotoxigenic *E. coli* heat-stable toxin affects cGMP production. The result is altered electrolyte transport: inhibition of Na⁺ uptake and stimulation of Cl⁻ transport. Water follows the resulting electrolyte imbalance and leaves the cell.

Table 25.3 Characteristics of bacterial toxins.[a]

Toxin	Organism	Mode of action	Host target	Disease	Toxin implicated in disease[b]
Damage membranes					
Aerolysin	*Aeromonas hydrophila*	Pore former	Glycophorin	Diarrhea	(Yes)
Perfringolysin O	*Clostridium perfringens*	Pore former	Cholesterol	Gas gangrene[c]	Unknown
Hemolysin[d]	*Escherichia coli*	Pore former	Plasma membrane	UTIs	(Yes)
Listeriolysin O	*Listeria monocytogenes*	Pore former	Cholesterol	Food-borne systemic illness, meningitis	Yes
Alpha toxin	*Staphyloccocus aureus*	Pore former	Plasma membrane	Abcesses[c]	(Yes)
Pneumolysin	*Streptococcus pneumoniae*	Pore former	Cholesterol	Pneumonia[c]	(Yes)
Streptolysin O	*Streptococcus pyogenes*	Pore former	Cholesterol	Strep throat Scarlet fever	Unknown
Inhibit protein synthesis					
Diphtheria toxin	*Corynebacterium diphtheriae*	ADP-ribosyltransferase	Elongation factor 2	Diphtheria	Yes
Shiga toxins	*E. coli/Shigella dysenteriae*	*N*-glycosidase	28S rRNA	HC and HUS	Yes
Exotoxin A	*Pseudomonas aeruginosa*	ADP-ribosyltransferase	Elongation factor 2	Pneumonia[c]	(Yes)
Activate second messenger pathways					
CNF	*E. coli*	Deamidase	Rho G proteins	UTIs	Unknown
LT	*E. coli*	ADP-ribosyltransferase	G proteins	Diarrhea	Yes
ST[d]	*E. coli*	Stimulates guanylate cyclase	Guanylate cyclase receptor	Diarrhea	Yes
EAST	*E. coli*	ST-like?	Unknown	Diarrhea	Unknown
Edema factor	*Bacillus anthracis*	Adenylate cyclase	ATP	Anthrax	Yes
Dermonecrotic toxin	*Bordetella pertussis*	Deamidase	Rho G proteins	Rhinitis	(Yes)
Pertussis toxin	*B. pertussis*	ADP-ribosyltransferase	G protein(s)	Pertussis (whooping cough)	Yes
C2 toxin	*Clostridium botulinum*	ADP-ribosyltransferase	Monomeric G-actin	Botulism	Unknown
C3 toxin	*C. botulinum*	ADP-ribosyltransferase	Rho G protein	Botulism	Unknown
Toxin A	*Clostridium difficile*	Glucosyltransferase	Rho G protein(s)	Diarrhea/PC	(Yes)
Toxin B	*C. difficile*	Glucosyltransferase	Rho G protein(s)	Diarrhea/PC	Unknown
Cholera toxin	*Vibrio cholerae*	ADP-ribosyltransferase	G protein(s)	Cholera	Yes
Activate immune response					
Enterotoxins	*S. aureus*	Superantigen	TCR and MHC II	Food poisoning[c]	Yes
Exfoliative toxins	*S. aureus*	Superantigen (and serine protease?)	TCR and MHC II	Scalded skin syndrome[c]	Yes
Toxic shock toxin	*S. aureus*	Superantigen	TCR and MHC II	Toxic shock syndrome[c]	Yes
Pyrogenic exotoxins	*S. pyogenes*	Superantigens	TCR and MHC II	Toxic shock syndrome Scarlet fever	Yes Yes

Table 25.3 Characteristics of bacterial toxins[a] (*continued*)

Toxin	Organism	Mode of action	Host target	Disease	Toxin implicated in disease[b]
Protease					
Lethal factor	*B. anthracis*	Metalloprotease	MAPKK1/MAPKK2	Anthrax	Yes
Neurotoxins A–G	*C. botulinum*	Zinc metalloprotease	VAMP/synaptobrevin SNAP-25 syntaxin	Botulism	Yes
Tetanus toxin	*Clostridium tetani*	Zinc metalloprotease	VAMP/synaptobrevin	Tetanus	Yes

[a]**Abbreviations:** CNF, cytotoxic necrotizing factor; LT, heat-labile toxin; ST, heat-stable toxin; CLDT, cytolethal distending toxin; EAST, enteroaggregative *E. coli* heat-stable toxin; TCR, T-cell receptor; MHC II, major histocompatibility complex class II; MAPKK, mitogen-activated protein kinase kinase; VAMP, vesicle-associated membrane protein; SNAP-25, synaptosomal-associated protein; UTI, urinary tract infection; HC, hemorrhagic colitis; HUS, hemolytic uremic syndrome; PC, antibiotic-associated pseudomembranous colitis.

[b]Yes, strong causal relationship between toxin and disease; (yes), role in pathogenesis has been shown in animal model or appropriate cell culture.

[c]Other diseases are also associated with the organism.

[d]Toxin is also produced by other genera of bacteria.

Source: C. K. Schmidt, K. C. Meysick, and A. O'Brien. 1999. Bacterial toxins: Friends or foes? *Emerging Infectious Diseases* 5:224–234.

A completed pore and a cutaway view exposing the channel are illustrated in **Figure 25.17A** and **B**. Diagnostic microbiology laboratories visualize hemolysins (proteins that lyse red blood cells) such as alpha toxin by inoculating bacteria onto agar plates containing sheep red blood cells (**Fig. 25.17C**). The clear, yellow zones around the *S. aureus* colonies growing on blood agar indicate that the microbe secretes a hemolysin.

Cholera and *E. coli* labile toxins. *Vibrio cholerae* (**Fig. 25.18A**) produces a severe diarrheal disease called cholera that generally afflicts malnourished people populating poor countries like Bangladesh, or countries in which access to clean water has been disrupted by war or natural disasters. This microbe produces a gastrointestinal enterotoxin nearly identical to one produced by some strains of *E. coli* associated with what is known as "traveler's diarrhea." Enterotoxins specifically affect the intestine. The *E. coli* enterotoxin is called **labile toxin (LT)** because it is easily destroyed by heat. Cholera toxin (CT) and labile toxin are both AB toxins with identical modes of action, which is to increase the level of cAMP made inside the host cell.

After the bacteria attach to the cells lining the intestinal villa (**Figs. 25.18C** and **D**), they secrete these AB toxins, the structure of which is shown in **Figure 25.18B**. Both toxins have five B subunits arranged as a ring around a single A subunit. The B subunits bind to ganglioside GM1 on eukaryotic cell membranes and deliver the A subunit to the target cell (**Fig. 25.19A**, steps 1 to 4). The A subunit possesses the toxic part of the molecule, an ADP-ribosyltransferase, which must be activated by the host.

Figure 25.16 AB toxins.
A. A typical AB toxin consists of an A subunit and a pentameric B subunit joined noncovalently.
B. Many AB toxins are ADP-ribosyltransferase enzymes that modify protein structure and function.

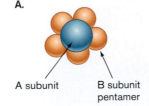

A subunit   B subunit pentamer

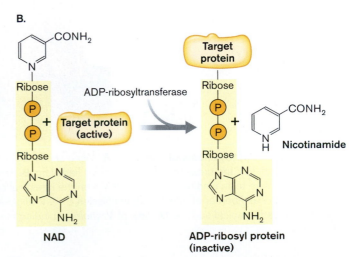

A. Alpha hemolysin

B. Cross section of alpha hemolysin

C. Hemolysis by *S. aureus*

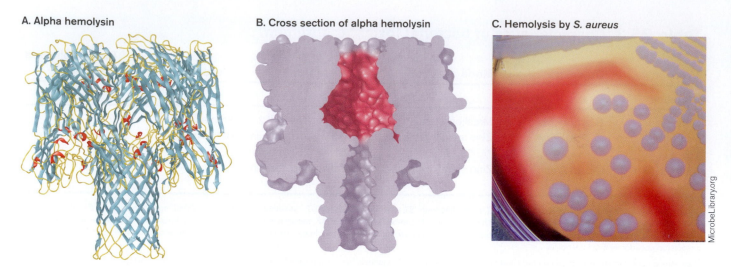

Figure 25.17   Alpha hemolysin of *Staphylococcus aureus*.   A. Three-dimensional figure of the pore complex comprising seven monomeric proteins. (PDB code: 7ahl) **B.** Cross section showing the channel. **C.** A blood agar plate inoculated with *S. aureus*. The alpha toxin is secreted by the organism and diffuses away from the producing colony. It forms pores in the red blood cells embedded in the agar, causing them to lyse. This is visible as a clear area surrounding each colony.

The binding of CT or LT to GM1 triggers endocytosis and the formation of a toxin-containing vacuole. The phagosome is then transported to the endoplasmic reticulum. During this time, the A subunit is cleaved by a host protease into two fragments called A1 and A2, which are still held together by a disulfide bond. The reducing environment at the cell surface reduces that bond and frees the A1 peptide containing active ADP-ribosylase into the endoplasmic reticulum, which exports the toxin into the cytoplasm.

The mission of A1 peptide is to modify (ADP-ribosylate) a membrane-associated GTPase (called a G protein) that binds to adenylate cyclase and controls its activity (**Fig. 25.19A**, step 5). To understand how the toxin produces diarrhea, it is helpful to observe what happens in the absence of toxin (**Fig. 25.19B**).

A. *Vibrio cholerae* (SEM)

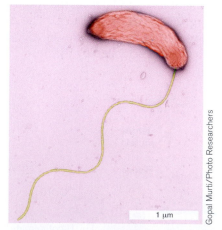

B. Cholera toxin

A subunit

B₅ subunit

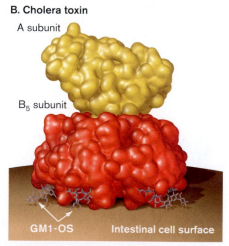

GM1-OS   Intestinal cell surface

C. Brush border of intestine

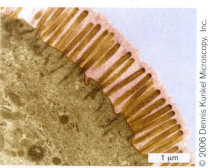

D. *V. cholerae* attachment

Figure 25.18   Pathogenesis of cholera.   A. *Vibrio cholerae* (SEM). Note the slight curve of each cell and the presence of a single polar flagellum. **B.** Three-dimensional structure of cholera toxin binding ganglioside GM1 on the intestinal cell surface. (PDB code: 1s5f) **C.** Brush border of intestine (transmission EM). *V. cholerae* binds to the fingerlike villi on the apical surface. **D.** View of *V. cholerae* binding the surface of a host cell (SEM).

A.

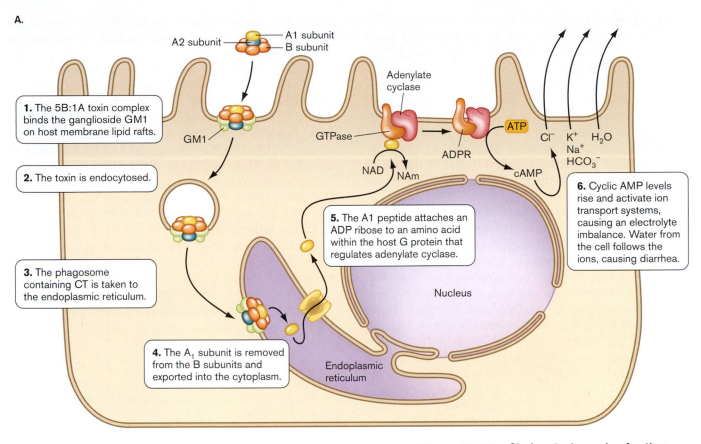

1. The 5B:1A toxin complex binds the ganglioside GM1 on host membrane lipid rafts.

2. The toxin is endocytosed.

3. The phagosome containing CT is taken to the endoplasmic reticulum.

4. The A₁ subunit is removed from the B subunits and exported into the cytoplasm.

5. The A1 peptide attaches an ADP ribose to an amino acid within the host G protein that regulates adenylate cyclase.

6. Cyclic AMP levels rise and activate ion transport systems, causing an electrolyte imbalance. Water from the cell follows the ions, causing diarrhea.

A2 subunit — A1 subunit — B subunit

Adenylate cyclase

GM1

GTPase

ADPR

NAD NAm

ATP

cAMP

Cl⁻ K⁺ H₂O Na⁺ HCO₃⁻

Nucleus

Endoplasmic reticulum

B.

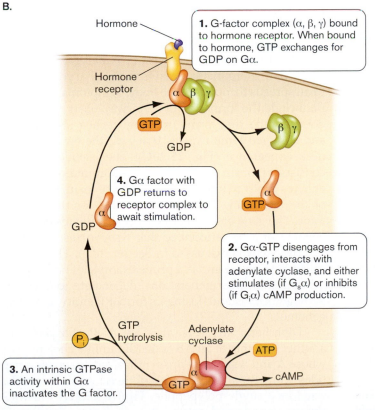

Hormone

Hormone receptor

1. G-factor complex (α, β, γ) bound to hormone receptor. When bound to hormone, GTP exchanges for GDP on Gα.

α β γ

GTP

GDP

β γ

GTP α

4. Gα factor with GDP returns to receptor complex to await stimulation.

α GDP

2. Gα-GTP disengages from receptor, interacts with adenylate cyclase, and either stimulates (if Gₛα) or inhibits (if Gᵢα) cAMP production.

GTP hydrolysis

Pᵢ

Adenylate cyclase

ATP

3. An intrinsic GTPase activity within Gα inactivates the G factor.

α GTP

cAMP

Figure 25.19   Cholera toxin mode of action. A. Delivery of cholera toxin into target cells and deregulation of adenylate cylase activity. **B.** Normal regulation of human cell adenylate cyclase. Human cells have two G-factor complexes, Gₛ and Gᵢ, that stimulate and inhibit adenylate cyclase, respectively. ▶❚❚

The intestinal epithelium has both absorptive and secretory functions. Transport is characterized by a net absorption of NaCl, short-chain fatty acids (SCFA), and water, allowing extrusion of a feces with very little water and salt content. In addition, the epithelium secretes mucus, bicarbonate, and KCl. A major player in the secretion of Cl⁻ is the cystic fibrosis transmembrane conductance regulator (CFTR). Chloride export via CFTR is activated indirectly by cAMP. (CFTR is so named because a genetically based defect in the CFTR transporter/regulator manifests as the lung disease cystic fibrosis.)

Human cells have two types of G-factor complexes called G_s and G_i that stimulate and inhibit adenylate cyclase, respectively. The G-factor complexes share beta and gamma subunits but have distinct alpha subunits (G_s-alpha and G_i-alpha). The complexes are normally bound to specific membrane hormone receptors (**Fig. 29.19B**). Stimulation of the hormone receptor allows GTP to bind the associated G-alpha subunit (replacing a bound GDP). Subsequently, the G-alpha–GTP protein leaves the receptor complex and binds adenylate cyclase, either stimulating cAMP production (if G_s-alpha) or inhibiting it (if G_i-alpha).

Normally, an intrinsic GTPase activity within the G-alpha protein degrades GTP to GDP. G-alpha protein bound to GDP is inactive and cannot control adenylate cyclase. G-factor-associated GTP turnover controls cAMP levels and thus CFTR activity in normal cells.

In the disease state (**Fig. 25.19A**, step 5), G-alpha protein is ADP-ribosylated by cholera or labile toxin. Modified G-alpha protein can still activate adenylate cyclase, but it has a defective GTPase. Because the altered G factor cannot degrade GTP, the protein continually stimulates cAMP production. This increases the levels of cAMP tremendously. Elevated levels of cAMP stimulate a host enzyme called protein kinase A that activates various ion transport channels (**Fig. 25.19**, step 6). One of these channels is the cystic fibrosis transmembrane conductance regulator that controls chloride transport (discussed in Section 24.4). As a result, chloride, sodium, and other ions leave the cell and, in an attempt to equilibrate osmolarity, water leaves as well. Because the affected cells line the intestine, the escaping water enters the intestinal lumen, leading to watery stools, or diarrhea.

How is diarrhea a benefit to the microorganism? A major benefit to the organism is to propagate the species. The more diarrhea that is produced and expelled, the greater the number of organisms are made and disseminated throughout the environment. This increases the chance that another host will ingest the organism, ensuring survival of the species. Diarrhea can also benefit a pathogen by decreasing competition with other organisms as they are swept away. In fact, the vast majority of organisms found in the diarrheal fluids of cholera patients are *V. cholerae* bacteria.

Different pathogens have discovered alternative ways to alter host cAMP levels. The gram-negative pathogen *Bordetella pertussis* causes a childhood respiratory infection called whooping cough, so named for the whooping sound made as a child tries to take a breath after a fit of coughing. This microbe also secretes an ADP-ribosylating toxin, but one that modifies G_i-alpha (the inhibitory factor that normally downregulates adenylate cyclase activity). ADP-ribosylation in this instance prevents that inhibition, with the result (again) of runaway cAMP synthesis. Pertussis toxin, however, is structurally different from cholera toxin. In addition, the organism makes a "stealth" adenylate cyclase that is secreted from the bacterium but remains inactive until it enters host cells, where it becomes active by binding the calcium-binding protein calmodulin. The resulting increase in cAMP levels causes inappropriate triggering of certain host cell signaling pathways.

> **THOUGHT QUESTION 25.2** Antibodies to which subunit of cholera toxin will best protect a person from the toxin's effects?

Diphtheria toxin. Other exotoxins can affect host protein synthesis. The microbe *Corynebacterium diphtheriae* produces a potent toxin that causes the disease diphtheria by inhibiting protein synthesis in eukaryotic cells (**Fig. 25.20A**). The organism remains in the pharynx while the toxin spreads throughout the body. Symptoms of diphtheria include the formation of an airway-obstructing pseudomembrane that forms over the trachea (**Fig. 25.20B**) and enlarged lymph nodes that give the patient a bull-neck appearance (**Fig. 25.20C**). The pseudomembrane consists of fibrin, bacteria, and inflammatory cells; it adheres firmly to underlying tissue. The toxin, not growth of the organism, is solely responsible for these symptoms. The gene encoding the toxin actually resides in the genome of a bacteriophage that lysogenizes *C. diphtheriae* (lysogeny is discussed in Section 10.7). The phage is called Beta phage and is a prime example of how a phage genome may, over generations, degenerate to become a pathogenicity island, losing its former identity as a phage genome.

Diphtheria toxin is also an ADP-ribosylase and another example of an AB toxin. But unlike cholera toxin, whose A and B subunits are synthesized as separate peptides from separate genes, diphtheria toxin is synthesized as a single polypeptide. **Figure 25.21A** illustrates the monomeric and natural dimeric forms of diphtheria toxin. There are three domains within the peptide. A

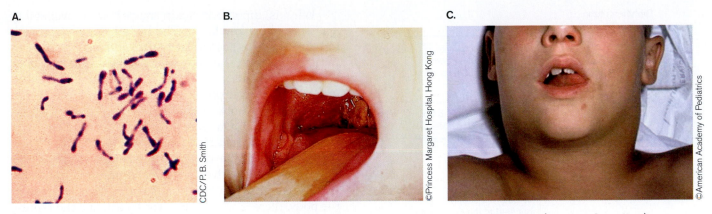

Figure 25.20 Pathogenesis of diphtheria. **A.** Methylene blue stain of *Corynebacterium diphtheriae* (size 1 to 8 µm long). **B.** Pseudomembrane formed across trachea. **C.** Bull-neck appearance due to enlarged cervical lymph nodes.

central **receptor-binding domain** (the B subunit of the toxin) binds to the target cell receptor on the membrane. There is also a C-terminal **transmembrane domain** that allows the protein to embed in the host membrane. The third domain, located at the amino terminus, carries the ADP-ribosylase activity and is the A subunit of the peptide toxin.

As with cholera toxin, diphtheria toxin must be cleaved to become active. As illustrated in **Figure 25.22**, the receptor-binding domain (B) binds the host membrane, and the toxin (A + B) then enters the cell by endocytosis. Once the toxin enters the cell, its peptide backbone

is cleaved, but the halves are held together by disulfide bonds. When the inside of the endosome vesicle acidifies (by H⁺-pumping ATPases), the disulfide bonds are reduced (cleaved), allowing the transmembrane domain of subunit A to embed in the host endosome membrane and then to pass through the vesicle membrane. The transmembrane domain then facilitates passage of the A peptide (that is, the catalytic domain) through the vesicle membrane. Once inside the cytoplasm, the catalytic domain (A subunit) ADP-ribosylates an important protein synthesis factor called elongation factor 2 (EF2). This effectively halts protein synthesis and kills the cell.

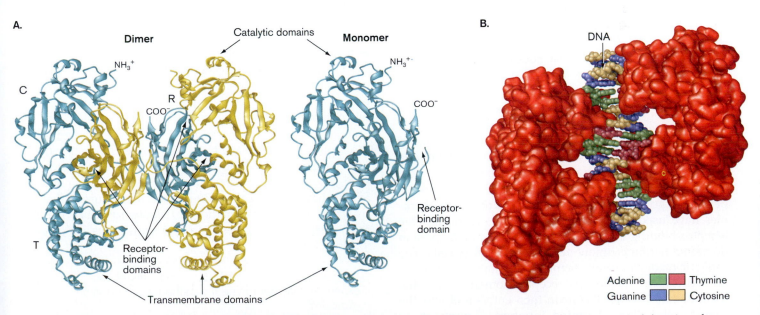

Figure 25.21 Diphtheria toxin and the diphtheria toxin repressor. **A.** View of diphtheria toxin showing swapped domains of a dimer and a monomer of the toxin. To form a dimer, the receptor-binding domains of two monomers swap to produce a stable structure. (PDB code: 1SGK) **B.** Diphtheria toxin repressor (DtxR) bound to the *dtx* promoter DNA. When bound to iron, the repressor binds to the *dtx* gene in the beta phage to prevent toxin synthesis. (PDB code: 1C0W)

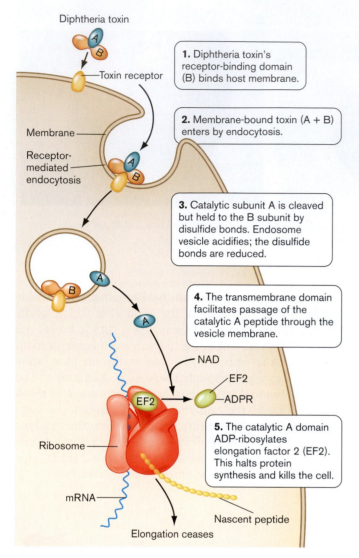

Diphtheria toxin

Toxin receptor

1. Diphtheria toxin's receptor-binding domain (B) binds host membrane.

Membrane

Receptor-mediated endocytosis

2. Membrane-bound toxin (A + B) enters by endocytosis.

3. Catalytic subunit A is cleaved but held to the B subunit by disulfide bonds. Endosome vesicle acidifies; the disulfide bonds are reduced.

4. The transmembrane domain facilitates passage of the catalytic A peptide through the vesicle membrane.

NAD

EF2

ADPR

EF2

Ribosome

5. The catalytic A domain ADP-ribosylates elongation factor 2 (EF2). This halts protein synthesis and kills the cell.

mRNA

Nascent peptide

Elongation ceases

Figure 25.22   Diphtheria toxin mode of action.
ADP-ribose, ADPR.

Today, thanks to intense vaccination efforts using inactivated toxin (called a toxoid), the disease diphtheria is rarely seen in this country. However, scientists are now learning to modify diphtheria toxin to serve new therapeutic purposes. Some experimental anticancer therapies have used gene-splicing techniques to replace the receptor-binding domain of diphtheria toxin with protein domains that bind unique receptors on cancer cells. The catalytic (toxic) and transmembrane domains of the toxin are left intact, but the new receptor domain targets the toxin to cancer cells. The toxin then enters and kills the cancer cell without harming innocent (noncancerous) bystander cells. Though showing promise, recombinant diphtheria toxin has not yet been proven effective in clinical trials.

Why is diphtheria toxin made? Iron availability appears to be a key factor. The body holds its iron very tightly in proteins such as lactoferrin and ferritin. So to an invading microbe, the body seems to be a very iron-poor environment. This makes it difficult for the pathogen to grow. Diphtheria toxin offers the organism a way to rob the host's iron stores.

Diphtheria toxin genes (*dtx*) are controlled by a repressor (DtxR) that binds iron (see **Fig. 25.21B**). When iron is plentiful, the repressor binds as a dimer to the promoter region of the *dtx* gene and prevents transcription. However, when free iron is scarce, as is the case in the body, there is not enough iron to activate the repressor and the *dtx* gene is expressed. As a result, diphtheria toxin is made, target cells are killed, and iron is released for bacterial use. This quest for iron is another common theme used by many pathogens to control pathogenic mechanisms.

THOUGHT QUESTION 25.3   Would patients with iron overload (excess free iron in the blood) be more susceptible to infection?

Shiga toxin: *Shigella* and *E. coli* O157:H7. *S. flexneri* and *E. coli* O157:H7 (also known as enterohemorrhagic *E. coli*) cause food-borne diseases whose symptoms include bloody diarrhea. These organisms produce an important toxin known as Shiga toxin (or Shiga-like toxin). The toxin has five B subunits for binding and one A subunit imbued with toxic activity. The A subunit, upon entry, destroys protein synthesis by cleaving 28S rRNA in eukaryotic ribosomes. Strains that produce high levels of this toxin are associated with acute kidney failure known as hemolytic uremic syndrome.

E. coli O157:H7 is a recently emerged pathogen. The organism can colonize cattle intestines without causing bovine disease; as a result, undetected bacteria can easily contaminate meat products following slaughter. The organism is sometimes referred to as the "Jack-in-the-Box" microbe, a reference to the first large U.S. outbreak associated with fast-food hamburgers at a Washington state Jack-in-the-Box restaurant in 1993. Both *E. coli* and *Shigella* have remarkable acid resistance mechanisms that rival that of the gastric pathogen *Helicobacter pylori*. This acid resistance makes these pathogens infectious at a very low infectious dose.

THOUGHT QUESTION 25.4   How might you experimentally determine if a pathogen secretes an exotoxin?

Anthrax. A century ago, anthrax (caused by *Bacillus anthracis*; **Fig. 25.23A**) was mainly a disease of cattle

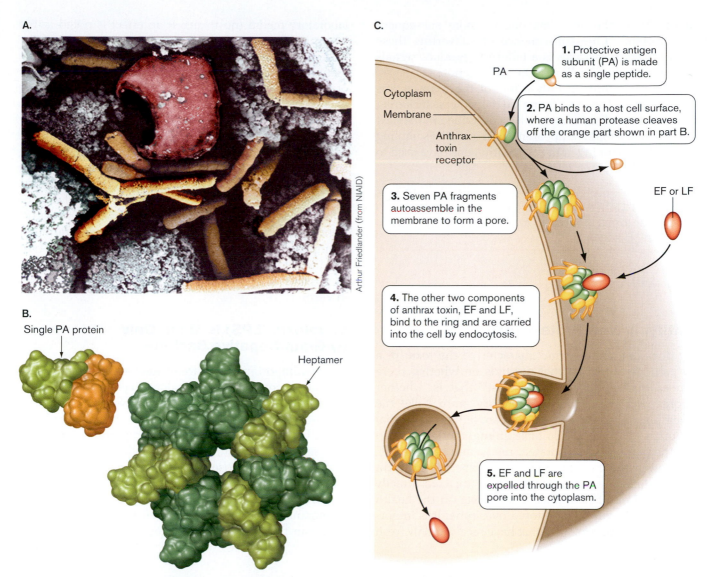

A.

Arthur Friedlander (from NIAID)

B.

Single PA protein

Heptamer

C.

Cytoplasm

Membrane

PA

Anthrax toxin receptor

1. Protective antigen subunit (PA) is made as a single peptide.

2. PA binds to a host cell surface, where a human protease cleaves off the orange part shown in part B.

3. Seven PA fragments autoassemble in the membrane to form a pore.

EF or LF

4. The other two components of anthrax toxin, EF and LF, bind to the ring and are carried into the cell by endocytosis.

5. EF and LF are expelled through the PA pore into the cytoplasm.

Figure 25.23 *Bacillus anthracis* and anthrax toxin.   A. *Bacillus anthracis* (SEM). Approx. 2 μm in length. Splenic tissue from a monkey. Spores are not visible. **B.** Single subunit and heptamer of protective antigen. (PDB code: 1TZO) **C.** Mechanism of toxin entry.

and sheep. Humans only acquired the disease accidentally. Today we fear the deliberate shipment of *B. anthracis* through the mail or its dispersion from the air ducts of heavily populated buildings. What makes this grampositive, spore-forming microbe so dangerous? In large part, its lethality is due to the secretion of a plasmid-encoded tripartite toxin. The core subunit of the toxin is called **protective antigen (PA)** because immunity to this protein *protects* hosts from disease. Protective antigen is made as a single peptide but then binds to the host cell surface, where a human protease cleaves off a fragment (**Fig. 25.23B**). The remaining part of PA can autoassemble in the membrane to form a heptameric (seven-membered) pore. The other two components of anthrax toxin, **edema factor (EF)** and **lethal factor (LF)**, bind to separate rings and are carried into the cell (see **Fig.**

25.23C). The complex is endocytosed, and the two proteins carried in are passed through the pore into the host cytoplasm.

Edema factor and lethal factor are the toxic parts of anthrax toxin. Both are enzymes that attack the signaling functions of the cell. Edema factor is an adenylate cyclase that remains inactive until entering the cytoplasm, where it binds calmodulin. This binding activates adenylate cyclase, resulting in a huge production of cAMP, and inactivates calmodulin from its normal function in the cell.

Lethal factor is actually a protease that cleaves several host protein kinase kinases, each of which is part of a critical regulatory cascade affecting cell growth and proliferation. A protein kinase kinase is an enzyme that phosphorylates, and thereby activates, another protein kinase that can then phosphorylate one or more subsequent target proteins. One consequence of subverting these phosphorylation cascades is a failure to produce signals that recruit immune cells to fight the infection.

We have examined only a few of the many toxins employed by pathogens. Some of the others, including tetanus and botulism toxins, will be described in the next chapter. What should be apparent from our brief sampling is the evolutionary ingenuity that pathogens have used to try to tame the human host.

> **THOUGHT QUESTION 25.5**   Internet problem: What other toxins are related to the cholera enterotoxin A subunit? B subunit?

www | Anthrax tutorial

Identifying New Protein Toxins

How does one identify and characterize the toxin of a new pathogen? In part that depends on whether there is an animal model for the disease (that is, whether the organism will infect a laboratory animal like the guinea pig or mouse and cause disease similar to that of humans). If there is a suitable animal model, one can grow the pathogen in laboratory media and harvest cell-free growth media into which the suspected exotoxin was secreted. The supernatant can be injected directly into a small number of mice, and effects on the health of the animals can be monitored over time. It is important to also use control mice that have received only fresh laboratory media (no toxin). If an effect is noted in the test mice, the exotoxin-containing supernatant can be treated with proteinase. If the toxin is protein, the treatment will destroy toxic activity. Subsequently, a variety of protein purification techniques, such as ion-exchange and molecular sieve chromatography, can be used to purify the protein.

One can also utilize tissue culture cells to test for the presence of a cell-free toxin in growth media. In using this method, one thing to keep in mind is that the effect may be tissue specific: The toxin may work on one tissue but not another. Another concern is whether the laboratory medium used to grow the bacterial cells is sufficient to elicit toxin production. A high-iron medium, for example, will prevent synthesis of diphtheria toxin. Thus, it is important to mimic the host environment as closely as possible when coaxing a pathogen to make exotoxin in vitro.

Endotoxin (LPS) Is Made Only by Gram-Negative Bacteria

Another important virulence factor common to all gram-negative microorganisms is endotoxin present in the outer membrane (discussed in Chapter 3). Not to be confused with secreted exotoxins, endotoxin is an embedded part of the bacterial cell surface and an important contributor to disease. Endotoxin, otherwise called lipopolysaccharide (LPS), is composed of lipid A, core glycolipid, and a repeating polysaccharide chain (**Fig. 25.24**). LPS molecules form the outer leaflet of the gram-negative outer membrane. As bacteria die, they release endotoxin. Endotoxin can then bind macrophages or B cells and trigger

A. LPS membrane

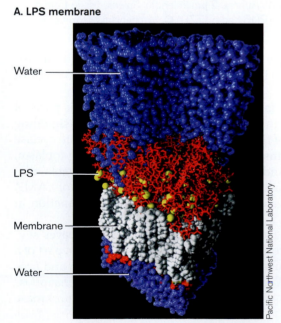

Water

LPS

Membrane

Water

Pacific Northwest National Laboratory

B. Gram-negative bacterial endotoxin (lipopolysaccharide, LPS)

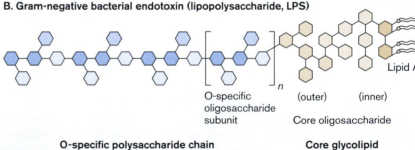

O-specific oligosaccharide subunit

(outer)   (inner)

Core oligosaccharide

Lipid A

O-specific polysaccharide chain   **Core glycolipid**

Figure 25.24   Endotoxin.   A. Model of a lipopolysaccharide membrane of *Pseudomonas aeruginosa*, consisting of 16 lipopolysaccharide molecules (red) and 48 ethylamine phospholipid molecules (white). **B.** Basic structure of endotoxin, showing the repeating O antigen side chain that faces out from the microbe and the membrane proximal core glycolipid and lipid A (contains endotoxic activity).

the release of TNF-alpha, interferon, IL-1, and other cytokines. The release of these active agents causes a variety of symptoms, such as:

- Fever
- Activation of clotting factors, leading to disseminated intravascular coagulation
- Activation of the alternate complement pathway
- Vasodilation, leading to hypotension (low blood pressure)
- Shock due to hypotension
- Death when other symptoms are severe

The lipid A moiety of LPS possesses endotoxic activity (**Fig. 25.24**).

> **NOTE:** Lipopolysaccharides are also the outer membrane structures called O antigens that are used to classify different strains of *E. coli* as well as other gram-negative organisms (for example, *E. coli* O157 versus *E. coli* O111).

The role of endotoxins can be seen in infections with the gram-negative diplococcus *Neisseria meningitidis* (**Fig. 25.25A**), a major cause of bacterial meningitis. *N. meningitidis* has, as part of its pathogenesis, a septicemic phase where the organism can replicate to high numbers in the bloodstream. The large amount of endotoxin present causes a massive depletion of clotting factors that leads to internal bleeding, most prominently displayed to a physician as small pinpoint hemorrhages called **petechiae** on the patient's hands and feet. (**Fig. 25.25B**). Capillary bleeding near the surface of the skin causes petechiae. One danger of treating massive gram-negative sepsis with antibiotics is that the enormous release of endotoxin from dead bacteria could well kill the patient. Untreated gram-negative sepsis is, however, almost always fatal, so its treatment, albeit risky, is imperative.

TO SUMMARIZE:

- **There are five categories of protein exotoxins** based on mode of action. These include toxins that cause membrane disruption, inhibition of protein synthesis, or alteration of host cell signal molecule synthesis, as well as superantigens and target-specific proteases.
- **AB subunit toxins** are common. The B subunit promotes penetration through host cell membranes, and the A subunit has toxic activity.
- *S. aureus* **alpha-toxin** forms pores in host cell membranes.
- **Cholera toxin, *E. coli* labile toxin, and pertussis toxin** are AB toxins that alter host cAMP production by adding ADP-ribose groups to different G-factor proteins.

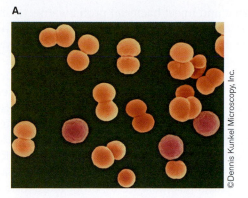

A.

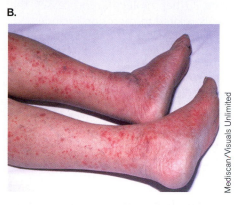

B.

©Dennis Kunkel Microscopy, Inc.

Mediscan/Visuals Unlimited

Figure 25.25   Effect of *Neisseria meningitidis* endotoxin.
A. *Neisseria meningitidis* (cell 0.8 to 1 µm diameter, SEM).
B. Petechial rash caused by *N. meningitidis.*

- **Diphtheria toxin** is an AB toxin that adds an ADP-ribosyl group to eukaryotic elongation factor 2—a modification that stops protein synthesis.
- **Shiga toxin** is an AB toxin that cleaves host cell 28S rRNA in host cell ribosomes.
- **Anthrax toxin** is a three-part AB toxin with one B subunit (protective antigen) and two different A subunits that affect cAMP levels (edema factor) and cleave host protein kinases (lethal factor).
- **Lipopolysaccharide**, known as endotoxin, is an integral component of gram-negative outer membranes and an important virulence factor that triggers massive release of cytokines from host cells. The indiscriminate release of cytokines can trigger fever, shock, and death.

25.5 Protein Secretion and Pathogenesis

A recurring theme among bacterial pathogens is the secretion of proteins that destroy, cripple, or subvert host target cells. The bacterial toxins described in Section 25.4 are secreted into the surrounding environment, where they float randomly until chance intervenes and they hit a membrane-binding site. However, many pathogens attach to a tissue and inject bacterial proteins directly into the host cell cytoplasm. The proteins may not kill the cell but redirect host signaling pathways in ways that benefit the microbe.

Protein secretion pathways were introduced in Section 8.5, focusing on ATP-binding cassette (ABC) proteins as a model. Additional secretion models are described here in their critical role of delivering pathogenicity proteins such as toxins. A particularly interesting aspect of these secretory systems is that many of them evolved from, and bear structural resemblance to, other more innocuous systems. Other molecular processes that are evolutionarily related to secretion include:

- Type IV pilus biogenesis (homologous to type II protein secretion)
- Flagellar synthesis (homologous to type III protein secretion)
- Conjugation (homologous to type IV protein secretion)

Table 25.4 gives examples of virulence proteins associated with these export systems.

Type II Secretion Systems Resemble Pilus Assembly

Type II secretion offers a clear example of how nature has modified the blueprints of one system to do a very different task. DNA sequence analysis has revealed that the genes used for type IV pilus biogenesis (see Section 25.3) were duplicated at some point during evolution and redesigned to serve as a protein secretion mechanism. Type IV pili have the unusual ability to extend and retract from the outer membrane, a property that produces the gliding motility of *Myxococcus* (see Section 4.7) and the twitching motility of *Neisseria* and *Pseudomonas*. As you might guess, assembly/disassembly of these appendages is quite complex. Type II protein secretion mechanisms mirror this complexity. Proteins to be secreted first make their way, via the Sec-dependent general secretion pathway, to the

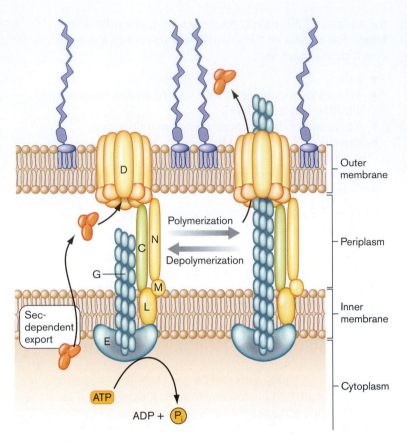

Figure 25.26   Type II secretion.   C, D, E, G, L, M, and N are protein components of the secretion system. *Source:* Modified from Moat, Foster, and Spector. 2002. *Microbial Physiology,* 4th ed.

periplasm, where they then encounter the appropriate type II secretion system. Because of this periplasmic "layover," the proteins are folded *before* secretion. Type II secretion systems use the pilus-like structure as a piston to ram the folded proteins through an outer membrane pore structure and into the surrounding void (**Fig. 25.26**). Piston action occurs via cyclic assembly and disassembly of pilus-like proteins, driving the pilus-like structure through an outer membrane pore and then retracting. Cholera toxin is a well-known example of a protein expelled by a type II secretion mechanism.

Table 25.4	Secretion systems for bacterial toxins.	
Secretion type	**Features**	**Examples**
I	SecA dependent	*E. coli* alpha-hemolysin, *Bordetella pertussis* adenyl cyclase
II	SecA dependent, similar to type IV pili	*Pseudomonas aeruginosa* exotoxin A, elastase
Autotransporters	SecA dependent to periplasm, self-transport through outer membrane	Gonococcal and *Haemophilus influenzae* IgA proteases
III	SecA independent, syringe	*Yersinia* Yop proteins, *Salmonella* Sip proteins, EPEC EspA proteins, TirA
IV	Related to conjugational DNA transfers	*B. pertussis* toxin, *Helicobacter* CagA

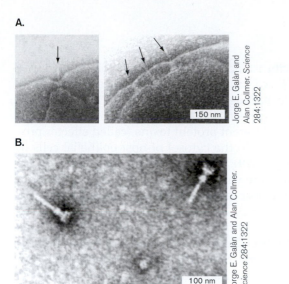

A.

150 nm

Jorge E. Galán and Alan Collmer. *Science* 284:1322

B.

100 nm

Jorge E. Galán and Alan Collmer. *Science* 284:1322

C.

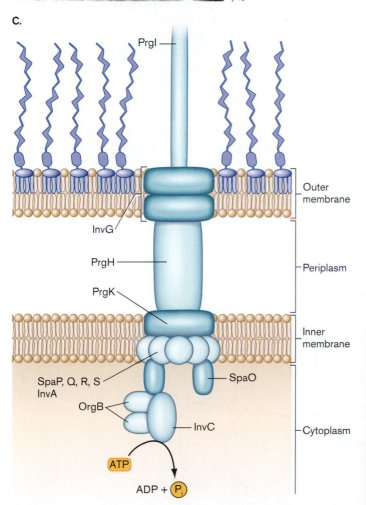

- PrgI
- Outer membrane
- InvG
- PrgH
- Periplasm
- PrgK
- Inner membrane
- SpaP, Q, R, S
  InvA
- SpaO
- OrgB
- InvC
- Cytoplasm
- ATP
- ADP + P$_i$

Figure 25.27   The needle complex of the *S. typhimurium* type III secretion system.   A. Unlike other secretion systems, the type III mechanism injects proteins directly from the bacterial cytoplasm into the host cytoplasm. The proteins in these systems are related to flagellar assembly proteins. Shown here are TEMs of osmotically shocked *S. typhimurium* with needle complexes visible in the bacterial envelope (arrows). **B.** Purified needle complexes (electron micrograph). **C.** A schematic representation of the *S. typhimurium* needle complex and its putative components. *Source:* Modified from Moat, Foster, and Spector. 2002. *Microbial Physiology,* 4th ed.

Type III Secretion Injects Protein into Host Cells

In the last decade of the previous century, it was discovered that the etiological agents of Black Death and various forms of diarrhea (caused by species of *Yersinia, Salmonella,* and *Shigella*) could somehow take their bacterial virulence proteins made in the cytoplasm and drive them directly into the eukaryotic cell cytoplasm without the protein ever getting into the extracellular environment. Direct delivery is a good idea because it eliminates the dilution that happens when a toxin is secreted into media. Another advantage of this strategy is that it avoids the need to tailor the toxin to fit a preexisting host receptor.

What kind of molecular machine can directly deliver cytoplasmic bacterial proteins into target cells? Research has shown that some microbes use tiny molecular syringes embedded in their membranes to inject proteins directly into the host cytoplasm; this mechanism of delivery is called **type III secretion**. **Figure 25.27** shows electron micrographs of actual type III secretion needles and a model of the complex. Genes encoding type III systems are actually related to flagellar genes, whose products export the flagellin proteins through the center of a growing flagellum (discussed in Chapter 3). It appears that flagella genes were evolutionarily reengineered to act more like molecular syringes (see **Fig. 25.27C**). The bacterial virulence proteins secreted by type III systems subvert normal host cell signaling pathways and cause dramatic rearrangements of host membranes that often lead to engulfment of the microbe (**Fig. 25.28**). The genes encoding type III systems are usually located within pathogenicity islands inherited from other microbial sources.

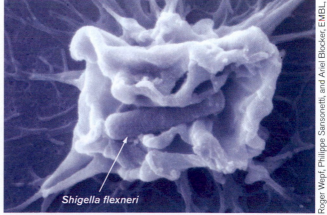

Roger Wepf, Philippe Sansonetti, and Ariel Blocker, EMBL, Heidelberg

Shigella flexneri

Figure 25.28   *Shigella* invades a host cell ruffle produced as a result of type III secretion. *Shigella flexneri* entering a HeLa cell (scanning transmission microscopy). The bacterium (small diameter approx. 1 μm) interacts with the host cell surface and injects (via its type III secretion apparatus) its invasin proteins, which act to choreograph a local actin-rich membrane ruffle at the host cell. The ruffle engulfs the bacterium and eventually disassembles, internalizing the bacterium.

Many bacterial pathogens use this type of secretion system, including plant pathogens such as *Pseudomonas syringae* (the cause of blight, a disease of many plants in which leaves or stems develop brown spots). Secretion is normally triggered by cell-cell contact between host and bacterium.

The use of type III secretion will be examined during our discussion of *Salmonella* and *E. coli* pathogenesis (**Special Topic 25.1**). Because of their importance to virulence, type III secretion systems are the subject of intensive research designed to exploit them as potential drug targets.

***Salmonella* pathogenesis.** The enteric pathogen *S. enterica* uses type III secretion systems to invade the eukaryotic host cell and become an intracellular parasite. Following ingestion of contaminated food or water, this bacterium attaches to and invades M cells that are interspersed along the intestinal wall. M cells are specialized intestinal epithelial cells (see Figure 23.17) that sample the intestinal flora and transfer pathogens across the epithelial barrier for recognition by the immune system. *Salmonella* subverts the normal function of M cells and causes an inflammatory response that leads to diarrhea. During its evolutionary journey toward becoming a pathogen, *Salmonella* has acquired at least five and as many as ten

pathogenicity islands that are absent from related, but harmless, *E. coli* strains. Only two will be discussed here. ***Salmonella* pathogenicity island 1 (SPI-1)** encodes a type III protein secretion system that delivers a cocktail of at least 13 different protein toxins (called effector proteins) directly into the cytosol of host epithelial cells in the gut (**Fig. 25.29**). Inside epithelial cells, these effector proteins interfere with signal transduction cascades and modulate the host response. One mission of these effectors is to induce cytoskeletal rearrangements that cause ruffling of the eukaryotic membrane around the microbe (see **Fig. 25.28**). The membrane ruffling starts the process of engulfment. *Salmonella* induces this response as a way to avoid the normal endocytic process.

Once *Salmonella* enters an epithelial cell or a macrophage, it finds itself in a vacuole. In the normal course of events, an enzyme-packed lysosome would then fuse with the phagosome and release its contents in an effort to kill the invader. *Salmonella*, however, possesses a second pathogenicity island called **SPI-2** that subverts this host response. SPI-2 uses another type III secretion system to inject proteins that alter vesicle trafficking, thereby reducing phagosome-lysosome fusion so that the intracellular bacteria are spared. The pathogenesis of *Salmonella* is even more complex than just outlined. There are 12 recognized pathogenicity islands present in the genome.

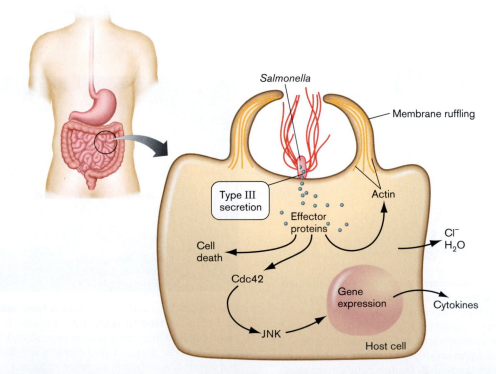

Figure 25.29 Schematic overview of *Salmonella* pathogenesis. Effector proteins injected by *Salmonella* into a host M cell affects the activity of host proteins that trigger cell death (apoptosis), influence gene expression (Cdc42 and JNK), influence electrolyte movements, and induce actin rearrangement. Actin rearrangement produces the ruffling of host membrane to engulf the organism.

Many pathogens rely on key host surface structures such as gangliosides to recognize and attach to the correct host cell. But the host species can evolve to become resistant to infection when the gene encoding the receptor (or receptor synthesis) mutates. The mutation could lead to complete loss of the protein or a change in its shape to prevent recognition or alter its function. An example is the T-cell surface protein CCR5, which acts as a receptor for HIV virus (see Section 11.5). Individuals with a genetic defect that eliminates CCR5 are immune to HIV infection, so absent methods for preventing and curing HIV infection, humans would evolve a level of resistance to HIV.

Some microbes circumvent possible loss of a host receptor by not relying on the natural array of host receptors in the first place. Instead, these bacterial pathogens use a type III secretion system to insert their own receptors into target cells (**Fig. 1A**). One such group of enterprising pathogens is enteropathogenic *E. coli* (EPEC). The adhesion molecule on the bacterial cell is called **intimin**. Intimin is a 94-kDa integral outer membrane bacterial protein needed for intimate adherence to host cells. However, there is no natural receptor for intimin on host membranes. As discovered by Brett Finley and his colleagues in Vancouver, British Columbia, EPEC must deliver its own intimin receptor, a 57-kDa bacterial protein called Tir (for translocated intimin receptor) that the bacteria inject into the host cell. The genes that encode intimin, Tir, and the secretion apparatus are all part of an EPEC pathogenicity island.

Once injected and placed in the host membrane, Tir binds intimin on the bacterial surface (**Fig. 1B**). The result is the more intimate adherence between the bacterium and host cell surface required for infection to proceed. In addition, host protein kinases phosphorylate Tir at tyrosine residue 474. Phosphorylated Tir directly triggers a remarkable reorganization of host cellular cytoskeletal components (actin, alpha-actinin, ezrin, talin, and myosin light chain) such that a membrane "pedestal" is formed, raising the microbe up (**Fig. 1C**). The result of this attachment is the characteristic attaching and effacing (A/E) lesion, characterized by destruction of the microvilli and pedestal formation.

Figure 1 *E. coli* **type III secretion and cell-cell interaction.** **A.** EspA filaments form a bridge between enteropathogenic *E. coli* (EPEC) and an epithelial cell during the early stages of attaching and effacing lesion formation (SEM). These filaments are part of the type III secretion apparatus, functioning as a molecular syringe to inject proteins from pathogenic *E. coli* into the host cell. **B.** Model of cytoskeletal components within the EPEC pedestal. EPEC injects Tir protein into the host cell, where it moves to the membrane and acts as a receptor for intimin. Tir also communicates through phosphorylation with other host proteins to cause a change in actin cytoskeleton, which leads to pedestal formation. **C.** Pedestal formation (SEM).

A.

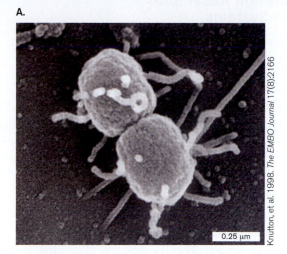

Knutton, et al. 1998. *The EMBO Journal* 17(8):2166

B.

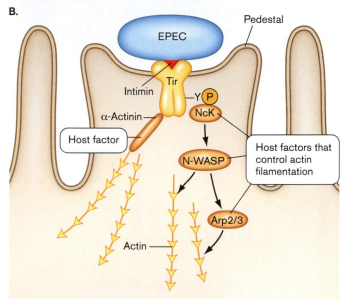

C.

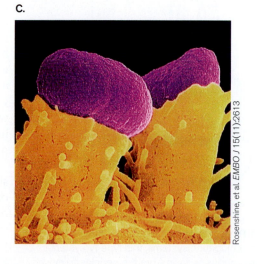

Rosenshine, et al. *EMBO J* 15(11):2613

Type IV Secretion Resembles Conjugation Systems

As Chapter 9 describes, many bacteria can transfer DNA from donor to recipient cells via a cell-cell contact system known as conjugation (discussed in Section 9.2). The conjugation systems of some pathogens have, through evolution, been modified into new systems that transport proteins, or proteins plus DNA, directly into target cells. *Agrobacterium tumefaciens*, for example, uses its Vir system to transfer the tumor-producing Ti plasmid and some effector proteins into plant cells. The result is a plant cancer called crown gall disease. The bacterium responsible for whooping cough in humans, *Bordetella pertussis*, also uses a type IV secretion system to export pertussis toxin, but simply exports the toxin without injecting it into the host (**Fig. 25.30**). Type IV systems also differ with respect to whether the protein is taken directly from the cytoplasm, like CagA from *Helicobacter,* or from the periplasm, as with pertussis toxin. In the latter case, the SecA-dependent general secretory system first delivers the toxin to the periplasmic space.

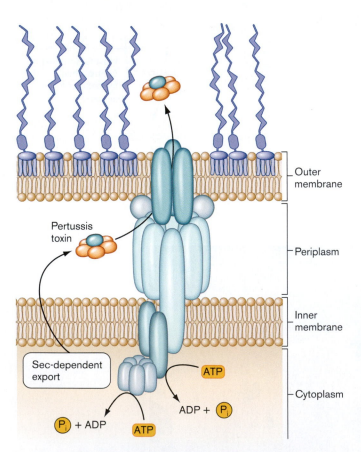

Figure 25.30   Type IV secretion of pertussis toxin.
Evolutionarily related to conjugation systems, this type IV system in *Bordetella pertussis* takes pertussis toxin from the periplasm (moved there by SecA-dependent transport) across the outer membrane.

Labels in figure: Outer membrane; Periplasm; Inner membrane; Cytoplasm; Pertussis toxin; Sec-dependent export; ATP; ADP + P_i; P_i + ADP; ATP

THOUGHT QUESTION 25.6   Protein and DNA have very different structures. Why would a protein secretion system be derived from a DNA-pumping system?

TO SUMMARIZE:

- **Many pathogens** use specific protein secretion pathways to deliver toxins.
- **Type II secretion** systems use a pilus-like extraction/retraction mechanism to push proteins out of the cell.
- **Type III secretion** uses a molecular syringe to inject proteins from the bacterial cytoplasm into the host cytoplasm.
- **Type IV secretion** utilizes an entourage of proteins that resemble conjugation machinery to secrete proteins from either the cytoplasm or periplasm.

25.6 Finding Virulence Genes

All pathogens must enter a host; find their unique niche; avoid, circumvent, or subvert normal host defenses; multiply; and eventually be transmitted to a new susceptible host. Although certain common pathogenic tactics have come to be appreciated, each microbe has a unique "pathogenic signature" that permits survival and leads to its freedom to multiply. When an emerging pathogen causes disease, how do we figure out its mechanisms of virulence? Virulence constitutes a measurable phenotype of an organism. Genes encoding virulence factors are expressed just like those for other traits (for example, expression of the *lac* operon confers the ability to metabolize lactose; discussed in Section 10.2). For metabolic or biosynthetic pathways, identifying the requisite genes is relatively straightforward. For instance, mutations in genes for amino acid biosynthesis produce a clear phenotype that can be observed on an agar plate. If the pathway is defective, the amino acid is not made and must be added to the medium or the mutant will not grow. Because virulence is a consequence of growth in a host, there is no way to identify a phenotype for virulence gene mutants in vitro. Virulence mutants fail to survive or fail to cause disease in animals. So how are these genes identified?

If the genome has been sequenced and a gene bears similarity to a known virulence gene present in another organism, that gene can be selectively mutated and the resulting mutant strain tested for virulence, usually by measuring LD_{50}. But what of genes not assigned an annotated function? They, too, could encode virulence factors. And what of genes that are expressed only during an infection? How do we find those?

In Vivo Expression Technologies Can Identify Virulence Genes

Bacterial genes that are exclusively expressed by the pathogen only during growth within an animal host can now be discovered using what is called **in vivo expression technology (IVET)**. The first step in IVET is to cripple a pathogen so that it cannot grow in the host animal unless a "rescue" gene that produces the missing factor is expressed. Next, the rescue gene is placed without its promoter into a plasmid to make what is called a promoter trap plasmid. The promoter of the plasmid-borne rescue gene is then replaced with random promoters cloned from the microbial genome. If one of the cloned promoters drives expression of the bacterial rescue gene after the organism is injected into the animal, the organism will flourish and kill the host. The trick is to find the promoters that express in vivo but that do not express, or that very poorly express, in laboratory media. Promoters that fit these criteria respond only to unknown in vivo conditions within the host and can be used to determine what those in vivo conditions are.

The IVET procedure was first described in 1995 by J. Mekalanos and colleagues at Harvard University using a promoter trap plasmid in which random fragments of *Salmonella* DNA were inserted (that is, cloned) upstream of a promoterless *cat-lacZ* fusion in the plasmid (**Fig. 25.31**). The *cat* gene encodes chloramphenicol acetyltransferase, an enzyme that inactivates chloramphenicol, making the strain resistant to the drug. The collection of random clones was then injected into mice that were immediately injected with chloramphenicol. When a promoter allowed in vivo expression of the *cat* gene, the microbe became resistant to the drug, grew in the animal, and traveled to the spleen.

The researchers knew that some of the in vivo expressed promoters would also be expressed in vitro (for example, promoters that drive genes encoding ribosome proteins). However, the goal was to identify promoters *only* expressed in the host, a characteristic that would indicate the presence of a downstream gene dedicated to virulence. So, the next task was to identify which of the in vivo expressed promoters was *not* expressed in vitro. To do this, the bacteria extracted from the spleen were plated on artificial media containing X-gal (an indicator of beta-galactosidase enzyme made from the *lacZ* gene). A colony that was blue indicated that the promoter was expressed in vitro. A white colony indicated that the promoter was *not* expressed in vitro, but since the strain grew in the mouse, that same promoter must be exclusively expressed in vivo. DNA sequence analysis of the insert identified the promoter and, based on the promoter's location on the bacterial genome, led to the conclusion that downstream genes were expressed specifically in vivo. This approach has now been used to identify potential virulence genes in several pathogens.

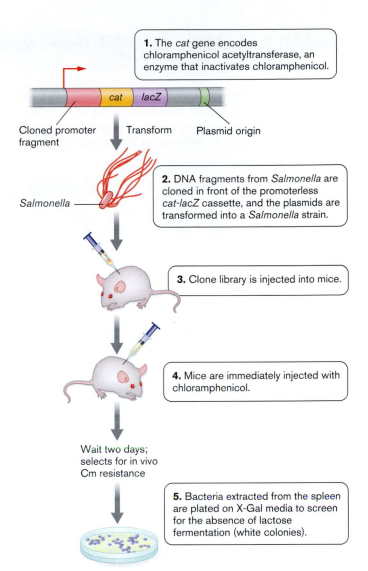

1. The *cat* gene encodes chloramphenicol acetyltransferase, an enzyme that inactivates chloramphenicol.

Cloned promoter fragment   Transform   Plasmid origin

Salmonella

2. DNA fragments from *Salmonella* are cloned in front of the promoterless *cat-lacZ* cassette, and the plasmids are transformed into a *Salmonella* strain.

3. Clone library is injected into mice.

4. Mice are immediately injected with chloramphenicol.

Wait two days; selects for in vivo Cm resistance

5. Bacteria extracted from the spleen are plated on X-Gal media to screen for the absence of lactose fermentation (white colonies).

Figure 25.31   In vivo expression technology. A library of promoter trap plasmids containing randomly cloned fragments of *Salmonella* DNA is transferred into *Salmonella* by transformation. The cloned pool is fed or injected into mice that have themselves been injected with chloramphenicol. Plasmids containing promoters expressed in vivo will yield Cm-resistant bacteria that can survive for two days and travel to the spleen. Bacteria recovered from the spleen are tested for an inability of the promoter fragment to drive *lacZ* expression in vitro. Lactose-fermenting colonies will turn blue when grown on X-Gal medium.

You could ask why these virulence or pathogenicity genes are needed for in vivo growth of pathogens if commensal microbes, which also grow in vivo, do not need them. The reason is that commensal microbes exist in distinct niches within or on the body. Pathogens have mechanisms to subvert or bypass host defenses to gain entrance and grow in privileged sites normally unassailable by commensal bacteria. Finding these genes is the goal of IVET and other gene selection technologies (**Special Topic 25.2**).

Special Topic 25.2 | Signature-Tagged Mutagenesis

An especially elegant approach to identifying virulence genes is signature-tagged mutagenesis. Signature-tagged mutagenesis is a negative selection procedure that will allow the researcher to find mutants that do *not* replicate in hosts. The process, as applied to *Salmonella*, starts with a pool of mutants (**Fig. 1**, step 1). Each member of the pool is mutated (randomly) in a different gene and each mutant gene is tagged with a different oligonucleotide (like a bar code)

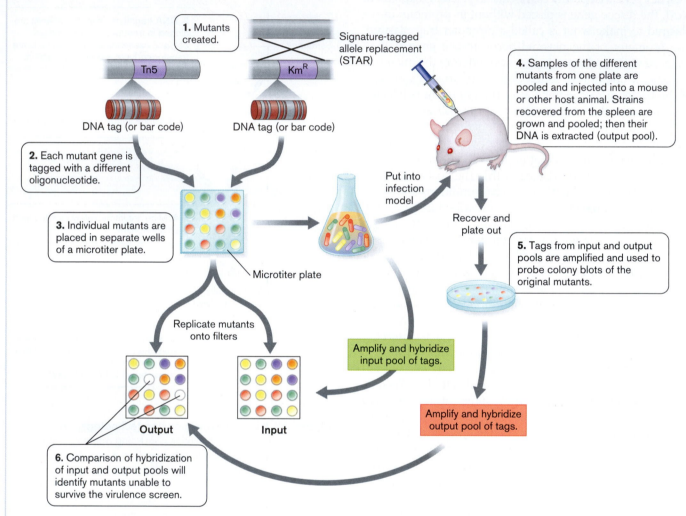

1. Mutants created.

Signature-tagged allele replacement (STAR)

4. Samples of the different mutants from one plate are pooled and injected into a mouse or other host animal. Strains recovered from the spleen are grown and pooled; then their DNA is extracted (output pool).

Tn5

Km^R

DNA tag (or bar code)   DNA tag (or bar code)

2. Each mutant gene is tagged with a different oligonucleotide.

3. Individual mutants are placed in separate wells of a microtiter plate.

Microtiter plate

Put into infection model

Recover and plate out

5. Tags from input and output pools are amplified and used to probe colony blots of the original mutants.

Replicate mutants onto filters

Amplify and hybridize input pool of tags.

Output   Input

Amplify and hybridize output pool of tags.

6. Comparison of hybridization of input and output pools will identify mutants unable to survive the virulence screen.

Figure 1   Signature-tagged mutagenesis.   DNA on the filter from mutants unable to survive in mouse will not hybridize to anything amplified from the output pool.

Using Modern Tools to Define Pathogenesis

***Helicobacter pylori* and gastric ulcers.** *Helicobacter pylori* is a gram-negative microbe, transmitted orally, that causes a variety of stomach diseases, including ulcers and cancer (**Fig. 25.32**). *H. pylori* has a genome size of 1.6 megabases (Mb) encoding 1,590 genes. The organism, however, is extremely difficult to grow in vitro and does not easily exchange genetic material. Because of this, classical approaches for studying its pathogenicity that rely on mutagenesis, clonal selection, and gene transfer are nearly impossible.

However, cutting-edge gene array technology, which can identify the repertoire of genes expressed under any condition (see Section 10.9), and signature-tagged mutagenesis have successfully identified numerous genes involved in the virulence of *Helicobacter*. A basic scheme of

(step 2). Each individual mutant with its unique sequence tag is placed in a separate well in a 96-well microtiter plate (step 3). Samples of the 96 different mutants from one plate are pooled and injected into a mouse or other host animal. Strains with mutations in pathogenicity genes will not grow in the mouse and will not be recovered from the spleen. These defective virulence mutants are identified using what is called a colony blot.

For the colony blot, strains recovered from the spleen are grown and pooled, and then their DNA is extracted (step 4). The various tagged genes in this pool of DNA are then amplified and radiolabeled using PCR (all of the unique tag sequences are flanked by common sequences to which PCR primers are made). The individual clones originally arrayed in the microtiter dish are blotted onto a nitrocellulose membrane (step 5). This colony blot membrane is treated to separate DNA strands and submerged into a solution containing the pooled radioactive PCR tag fragments made from the bacteria that survived in the mice and made it to the spleen (see step 4). Hybridization is spotted when a colony spot becomes radioactive. The conclusion is that representative bacteria from that colony survived in vivo. In contrast, a colony on the nitrocellulose membrane that does *not* hybridize to any of the probes in the mix (and thus is not radioactive) represents an *attenuated* strain that did not make it through the mouse selection (step 6). This strain contains a mutation (with a linked unique tag) within a virulence gene needed for in vivo survival of the bacterium. The gene can then be identified through various DNA-sequencing strategies. The SPI-2 pathogenicity island of *Salmonella enterica* was discovered using this technique (**Fig. 2**).

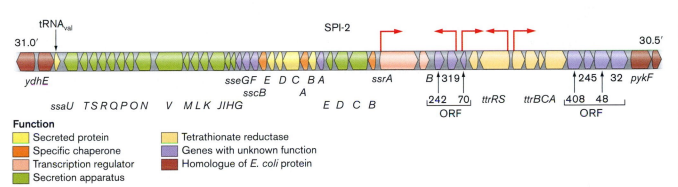

Figure 2 **The *Salmonella* pathogenicity island SPI-2 was identified using signature-tagged mutagenesis.** The island is located between 30.5 and 31 centisomes. ORFs are indicated by colored block arrows pointing in the direction of transcription. Gene designations are positioned below each ORF. The functional classes of SPI-2 genes are represented by different colors.

pathogenesis has been suggested as a result (**Fig. 25.33**). As with most microbial diseases the first step is colonization. Following ingestion of the pathogen, *H. pylori* flagella propel the organism toward the mucosa (step 1). As it approaches the mucosa, the pathogen produces intracellular and perhaps extracellular urease, which converts urea to CO_2 and ammonia, locally reducing the acidic pH of the stomach (step 2). Two other enzymes, collagenase and mucinase, soften the mucus lining, which allows the bacteria to reach the stomach's epithelial lining (step 3). The epithelial lining is much less acidic than the lumen, permitting the organism to grow and divide. Once at the epithelium, *Helicobacter* produces various adhesins, such as BapA or HpaA, to bind host cells (step 4). Once the organism has adhered, tissue damage occurs following the release of vacuolating cytotoxin (VacA) and neutrophil-activating protein (NAP), which activates neutrophils and mast cells that further damage local tissue (step 5).

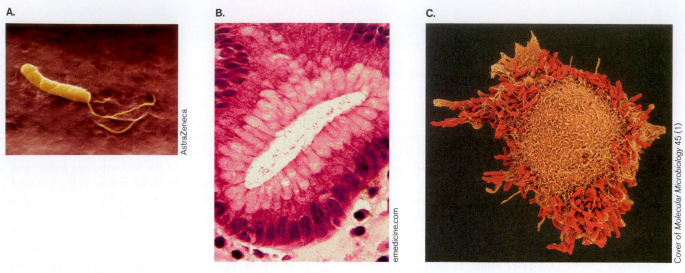

Figure 25.32 *Helicobacter* **pathogenesis.** **A.** *H. pylori* (SEM). Note the tuft of flagella at one pole. Cell length approx. 2 μm. **B.** *Helicobacter* attached to gastric mucosa (SEM). **C.** *Helicobacter* (red) binding to human gastric epithelial cells (colorized SEM). Intimate binding facilitates translocation of the bacterial effector protein CagA into the gastric cell via type IV secretion.

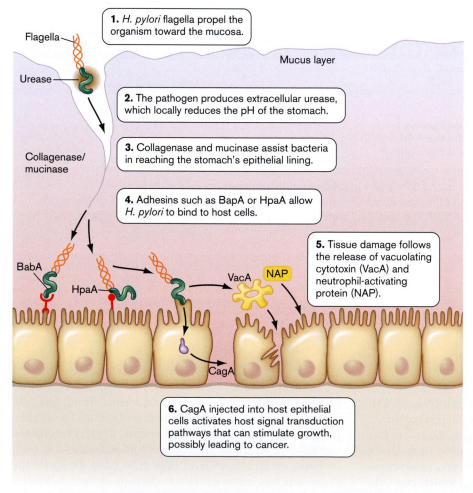

1. *H. pylori* flagella propel the organism toward the mucosa.

Mucus layer

2. The pathogen produces extracellular urease, which locally reduces the pH of the stomach.

3. Collagenase and mucinase assist bacteria in reaching the stomach's epithelial lining.

4. Adhesins such as BapA or HpaA allow *H. pylori* to bind to host cells.

5. Tissue damage follows the release of vacuolating cytotoxin (VacA) and neutrophil-activating protein (NAP).

6. CagA injected into host epithelial cells activates host signal transduction pathways that can stimulate growth, possibly leading to cancer.

Flagella
Urease
Collagenase/mucinase
BabA
HpaA
VacA
NAP
CagA

Figure 25.33 **Summary of steps involved in *Helicobacter* colonization.**

VacA forms a hexameric pore in the host membrane and induces apoptosis (programmed cell death). Another protein, CagA, is injected into host epithelial cells, where it is phosphorylated; CagA then interacts with host signaling proteins and activates host signal transduction pathways that can stimulate growth, possibly leading to cancer (step 6).

As you can see, genomics has been a powerful tool in the study of *Helicobacter*. From DNA sequence alone, a fairly detailed picture of *Helicobacter* physiology and pathogenesis was proposed based on homologies to known enzymes from other organisms (see **Fig. 8.42**). Several aspects of these models have been confirmed experimentally.

Whipple's disease. It is difficult to characterize a pathogen if you cannot even grow it in culture. The microorganism *Tropheryma whipplei* causes a gastrointestinal illness called Whipple's disease, whose symptoms include diarrhea, intestinal bleeding, abdominal pain, loss of appetite, weight loss, fatigue, and weakness. Though identified, the causative agent had never been grown outside of fibroblast cells. This made it nearly impossible to study its physiology. Once the complete genome sequence of two strains of *T. whipplei* became available, however, Didier Raoult and his colleagues at the Marseille School of Medicine discovered how to feed this fastidious bacterium. The genome (less than 1 Mb) contained a great deal of functional information. Computer modeling studies used this information to discover that the organism lacks the machinery to make several amino acids. With this knowledge, the investigators were able to design a cell-free culture medium that supported the growth of three different strains of *T. whipplei* as well as a new strain isolated from a heart valve. Similar applications of modern genomic techniques should lead to other successes in growing previously unculturable intracellular pathogens.

TO SUMMARIZE:

- **In vivo expression technology (IVET)** is a positive selection tool to identify genes required for a pathogen to grow in vivo.
- **Signature-tagged mutagenesis** is a technique that uses negative selection to identify genes required for in vivo growth.
- **Genomic tools** were used to predict mechanisms of *Helicobacter pylori* pathogenesis and the growth requirements for the Whipple's disease microbe.

25.7 Surviving within the Host

Once inside a host, a successful pathogen must avoid detection and destruction for as long as possible. Many products of the virulence factor arsenal help the microbe escape or resist innate immune mechanisms. Others are dedicated to stealth, that is, hiding from the immune system.

Intracellular Pathogens

In an effort to escape both nonspecific and humoral immune mechanisms (see Chapter 24), many pathogens, called **intracellular pathogens**, seek refuge by invading host cells. Antibodies and phagocytic cells will not penetrate live host cells, so hiding there provides temporary safe harbor for the pathogen. Some bacteria dedicate their entire lifestyle to intracellular parasitism. *Rickettsia*, for example, for reasons unknown, will not grow outside a living eukaryotic cell. Other microbes, such as *Salmonella* and *Shigella*, are considered facultative intracellular pathogens. **Facultative intracellular pathogens** can live either inside host cells or free. We have discussed how intracellular parasites get into cells, but how do they withstand intracellular attempts to kill them?

Once inside the phagosome, intracellular pathogens have three options to avoid being killed by the phagolysosome. They can escape the phagosome, prevent phagosome-lysosome fusion, or survive the result.

Escape from the phagosome. The gram-negative bacillus *Shigella dysenteriae* and the gram-positive bacillus *Listeria monocytogenes*, both of which cause food-borne gastrointestinal disease, use hemolysins to break out of the phagosome vacuole before fusion. In this way, they completely avoid lysozymal enzymes. Once free in the cytoplasm, they are thought to enjoy unrestricted growth. Yet even in the cytoplasm, these microbes have found a way to redirect host cell function to their own ends.

A fascinating aspect of escaping the phagosome involves motility. *Shigella* and *Listeria* are both nonmotile at 37°C in vitro; however, they both move around inside the host cell, even though they have no flagella. How do they move? These species are equipped with a special device at one end of the cell that mediates host cell actin polymerization. The polymerizing actin, called a "rocket tail," propels the organism forward through the cell (**Fig. 25.34A**) until it reaches a membrane. The membrane is then pushed into an adjacent cell, where the organism once again ends up in a vacuole. This strategy allows the microbe to spread from cell to cell without ever encountering the extracellular environment, where it would be vulnerable to attack.

Inhibit phagosome-lysosome fusion. Other intracellular pathogens avoid the hazard of lysosomal enzymes by preventing lysosomal fusion with the phagosome. *Salmonella*, *Mycobacterium*, *Legionella*, and *Chlamydia* are good examples. For example, *Legionella pneumophila* grows inside alveolar macrophage phagosomes and produces the potentially fatal Legionnaires' disease, so named for the veterans group that suffered the first recognized outbreak in 1976. The organism, once inside a phagosome, secretes proteins through the vesicle membrane and into the cytoplasm. These bacterial proteins interfere with the cell-signaling pathways that cause phagosome-lysosome fusion. The result is that *L. pneumophila* can grow in a friendlier vesicle.

Thrive under stress. In what could be called the "grin and bear it" strategy, some intracellular pathogens prefer the harsh environment of the phagolysosome. *Coxiella burnetii*, for example, grows well in the very acidic phagolysosome environment (**Fig. 25.34B**). This obligate intracellular organism causes a flu-like illness called Q fever (Query fever). The symptoms of Q fever include sore throat, muscle aches, headache, and high fever. It has a mortality rate of about 1%, so most people recover to good health. The organism allows phagosome-lysosome fusion because the acidic environment that results is needed for it to survive and grow.

> **THOUGHT QUESTION 25.7** How can one determine if a bacterium is an intracellular parasite?

Why some bacteria are obligate intracellular pathogens is unclear. One intracellular bacterium, *Rickettsia prowazekii*, a cause of epidemic typhus, appears to be an "energy parasite" that can transport ATP from the host

A. *Shigella flexneri* (red) and actin tails (green)

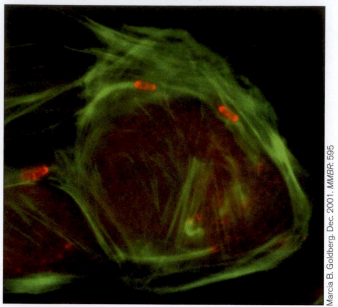

B. *Coxiella burnetii*

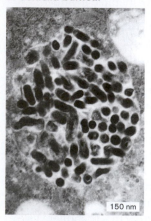
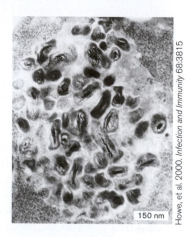

150 nm   150 nm

Figure 25.34   Intracellular pathogens: *Shigella* motility and *Coxiella* development.   A. Intracellular *Shigella flexneri* (fluorescence microscopy), fluorescently stained red (1 μm in length), are propelled through the host cytoplasm by actin tails stained green. **B.** TEM image showing a typical vacuole in J774A.1 mouse macrophage cells infected with *C. burnetii* at 2 hours and 6 hours postinfection. The organism, which prefers to live in the acidified vacuole, undergoes a form of differentiation that changes its shape and alters its interactions with the host cell.

cytoplasm and exchange it for spent ADP in the bacterium's cytoplasm. But this does not explain its obligate intracellular status, since giving rickettsia ATP outside a host does not allow the bacterium to grow. Other factors remain to be discovered.

THOUGHT QUESTION 25.8   Why would killing a host be a bad strategy for a pathogen?

Extracellular Immune Avoidance

This topic was first discussed in the chapters on host defense (Chapter 23) and immunology (Chapter 24). Many bacteria, such as *Streptococcus pneumoniae* and *Neisseria meningitidis*, produce a thick polysaccharide capsule that envelopes the cell. Capsules help organisms resist phagocytosis in several ways. Recall that phagocytes must recognize bacterial cell surface structures or surface-bound C3b complement factor to begin phagocytosis (see Section 23.6). Capsules cover bacterial cell wall components and mannose-containing carbohydrates that phagocytes normally use for attachment. The uniformity and slippery nature of capsule composition makes it difficult for phagocytes to lock onto the bacterial cell. But what about complement factor C3b, which can bind to the bacterial cell? Phagocytes have surface C3b receptors that latch onto C3b molecules. Capsules will envelop any C3b complement factor that binds to the bacterial surface, thereby hiding it from the phagocyte. Fortunately, immune defense mechanisms can circumvent this avoidance strategy by producing opsonizing antibodies (IgG) against the capsule itself (see Section 24.3). The Fc regions of antibodies that bind to the capsule stick out away from the bacterium, where they can be recognized by the Fc receptors on phagocyte membranes. Binding of the Fc region to the phagocyte's Fc receptor can then trigger phagocytosis.

Pathogens can also use proteins on the cell surface to avoid phagocytosis. *Staphylococcus aureus* has a cell wall protein called **protein A** that binds to the Fc region of antibodies, hiding the bacteria from phagocytes. This works in two ways. Protein A can bind to the Fc region of an antibody before the antibody binding site ever finds its bacterial target. Thus, the business end of the antibody—that is, the part with antigen-binding sites—is pointing *away* from the microbe. However, even if the antibody finds its antigen target on one bacterium, protein A from a second bacterium can bind to that Fc region, once again blocking phagocyte recognition.

Other microbes can trigger apoptosis in target host cells. Proteins made by the pathogen enter the host cell and trigger this programmed cell death. If a macrophage is targeted for self-destruction, it cannot destroy the microbe.

Another immune avoidance strategy used by microorganisms is to shift their antigenic structure. Genes encoding flagella, pili, and other surface proteins often use, as part of their regulatory features, site-specific inversions to express alternative proteins (for example, *Salmonella* phase variation) or slipped-strand mispairing to add or remove amino acids from a sequence (see Section 10.6).

Many bacteria and viruses also use mimicry to confuse the immune system (discussed in Section 24.6). In some cases, microbial proteins are made that look like cytokines.

These factors manipulate the balance of T helper cells and send immunity down the wrong path for combating the microbe. All of these strategies are designed to buy the microbe more time to overwhelm the host.

Chapter 26 presents additional mechanisms bacteria use to survive in vivo.

Regulating Virulence: How Do Pathogenic Bacteria Know Where They Are?

Many pathogens can grow either outside or inside a host. To accommodate the needs imposed by these different environments, microbes will employ alternate physiologies appropriate to each. Why make a type III secretion system if there are no host cells around? So how do microbial pathogens know whether they are in a host or in a pond? And what bacterial genes are exclusively expressed while in a host?

The same types of regulatory mechanisms that sense environmental conditions in a pond are used by the microbe to intuit its whereabouts in a host. That is, various sensing systems act in concert to recognize any specific environmental niche. Two-component signal transduction systems, discussed in Section 10.1, are used to monitor magnesium concentrations, which are characteristically low in a host cell vacuole. Other regulators measure pH, which will be acidic in the same vacuole. There are many other examples. The point is that there is no one in vivo sensor system. The various regulators collaborate to trigger the expression of virulence genes. As one example, the concentration of free iron, which is typically very low in the host, is an important signal used to induce the synthesis of virulence proteins whose purpose is to liberate iron bound up in host proteins. However, regulators sensitive to other in vivo signals must also be activated to achieve a successful infection.

Cell-cell communication is also important during infections. *Pseudomonas aeruginosa*, for example, has at least two quorum-sensing systems that detect secreted autoinducers. As the number of bacteria in a given space increases, so, too, does the concentration of the chemical autoinducer. When the autoinducer reaches a critical concentration, it diffuses back into the bacterium or binds to a surface receptor and triggers the expression of bacterial target genes. Genes included in the *P. aeruginosa* quorum-sensing regulons encode *Pseudomonas* exotoxin A and other secreted proteins, such as elastase, phospholipase, and alkaline protease. Why would a pathogen employ quorum sensing to regulate virulence factors? One reason may be to prevent alerting the host that it is under attack before enough microbes can accumulate through replication. Tripping the host's alarms too early would make eliminating infection easy. However, releasing toxins and proteases once a large number of bacteria have amassed could overwhelm the host.

Knowledge of quorum sensing has also provided an opportunity for treating disease. We know that bacteria can communicate with each other by secreting, and sensing, autoinducer chemicals. We also know that these molecules will accumulate in the growth environment and trigger collaborative responses, such as biofilm formation, among members of the population. A new chemical has been developed that interferes with quorum sensing and could serve as a new type of antibiotic. The compound, N-(2-oxocyclohexyl)-3-oxododecanamide, is an analog of the homoserine lactone autoinducers used by *P. aeruginosa*. The compound can bind, but not activate, the regulatory proteins LasR and RhlR. When tested in vitro, this compound reduced biofilm formation and decreased production of virulence factors.

TO SUMMARIZE:

- **Intracellular bacterial pathogens** attempt to avoid the immune system by growing inside host cells. Different mechanisms are used to avoid intracellular death.
- **Hemolysins** are used by certain pathogens to escape from the phagosome and grow in the host cytoplasm.
- **Actin tails** are used by some microbes to move within and between host cells.
- **Inhibiting phagosome-lysosome fusion** is one way pathogens can survive in phagosomes.
- **Specialized physiologies** enable some organisms to survive in the normally hostile environment of fused phagolysosomes.
- **Extracellularly, pathogens** evade the immune system by hiding in capsules, by changing their surface proteins, by triggering apoptosis, or by using molecular mimicry.
- **Two-component signal transduction systems** can regulate virulence gene expression in response to the environment.
- **Quorum sensing** may prevent pathogens from releasing toxic compounds too early during infection.

25.8 Viral Pathogenesis

The pathogenic strategies of viruses are at once similar to and yet distinctly different from those of bacteria. On their own, viruses are inert objects, lacking motility and unable to replicate. Like tiny molecular mines, they float in the environment until bumping into a suitable target cell. There is no need to transport food, expel waste, or generate energy, because the host does all that. But each virus must still attach to a host cell and subvert its biochemistry, directing it to make more virus. Because of the differences between bacteria and viruses, important insights into pathogenicity can be derived by comparing infection strategies used by different viruses.

The Common Cold Versus Influenza

You have probably wondered why you repeatedly get colds if the immune system is so effective. Why are you not protected from recurring infections after having one cold? The common cold is caused by over 100 known serotypes of rhinovirus, a small (30 nm), nonenveloped single-strand RNA virus. Serotypes of a given organism differ from one another in the antigenic makeup of a specific protein.

The host receptor for these viruses is ICAM-1 (see **Fig. 24.3**), which is also important for the extravasation of leukocytes through blood vessels (see Section 23.5). Many of the differences among the rhinovirus strains can be traced to the structure of one of the capsid proteins, VP1, which is thought to bind ICAM-1. Because of these differences, neutralizing antibodies to one strain of rhinovirus will not neutralize a different strain. Thus, we repeatedly contract a cold because we are infected with a different strain.

Unlike many viruses, rhinoviruses prefer to replicate at 33°C. As a result, they grow in the cooler regions of the respiratory tract, such as nasal passages. The reason these viruses cause a runny nose is that virus-infected cells release bradykinin and histamine, both of which affect fluid loss from local blood vessels (discussed in Section 24.5).

Certain strains of coronavirus also cause common cold symptoms. It is significant to note, however, that a recent variant of coronavirus is also responsible for the newer, more deadly disease known as severe acute respiratory syndrome (SARS).

> **THOUGHT QUESTION 25.9** Why do rhinovirus infections fail to progress beyond the nasopharynx?

Influenza infection proceeds very differently from rhinovirus infections, even though the two are often confused by the public. Influenza virus first establishes a local upper respiratory tract infection, where it targets and kills mucus-secreting, ciliated epithelial cells, destroying this primary defense. If the virus spreads down to the lower respiratory tract, it can cause shedding of bronchial and alveolar epithelium down to the basement membrane. As a result, oxygen and CO_2 exchange is compromised and breathing becomes difficult. Systemic symptoms, such as muscle aches, are due to the release of interferon and lymphokines in response to the infection. Because natural defenses are compromised, persons infected with flu are very susceptible to superinfection with bacterial pathogens such as *Haemophilus influenzae*. In fact, these bacterial infections are so prevalent in persons infected with influenza virus that *Haemophilus influenzae* was originally suspected as the causative agent of flu.

What makes flu so dangerous is that every 10 to 15 years the virus undergoes a major genetic alteration called an **antigenic shift** that results in a new pandemic strain. The change takes place in the gene encoding hemagglutinin and produces a change in the protein's antigenic form. (Hemagglutinin binds to and agglutinates red blood cells. It is used by the virus for attachment to host cells.) As a result of the change in form, antibodies that recognized the old hemagglutinin will not recognize and neutralize the new one. Antigenic shift occurs when two different strains of influenza infect the same cell, usually in animals like pigs or chickens. Recall that influenza has a segmented RNA genome. Thus, as the two viruses disgorge their eight nucleic acid segments, recombinations can occur between them, generating a chimeric, and thus new, form of the hemagglutinin gene (see Section 11.4). This new influenza virus can then jump to humans, and a new pandemic begins. During the time between major shifts, minor changes in antigen structure also occur. Because they are minor, the process is called antigenic drift. Antigenic shifts and drifts are the reasons why a new vaccine (the flu shot) is required each year. These shots are highly recommended for the elderly, who are very susceptible to the disease, and are a good idea for everyone else.

An estimated 21 million people worldwide died of the flu in the pandemic of 1918 (**Fig. 25.35**). No one then knew what caused the disease. Influenza was originally thought to be of bacterial etiology, and not until 1933 was it established that a virus was at fault. Many flu epidemics have taken place since then, but none have been so deadly. Though different from influenza, what was so frightening about the SARS outbreak that began in 2003 is that it had a fatality rate of greater that 10%, making it more deadly than the 1918 influenza strain, which had a fatality rate of 2–4%.

The 1918 influenza virus was thought lost, and with it the opportunity to learn why it was so deadly. Such knowledge could prove important as we face the threat of another deadly strain, which, as of this writing, is the avian flu. However, we now have a new opportunity to learn about this deadly strain. In February 1997, Ann Reid, Jeffery Taubenberger, and their colleagues at the Armed Forces Institute of Pathology reported the partial sequences of five influenza genes recovered from the preserved lung tissue of a U.S. soldier who died from influenza in 1918. Additional sequences of this virus were obtained from an Alaskan influenza victim who was buried in the permafrost in November 1918. The freezing helped preserve the integrity, if not the viability, of the flu genome. In 2005, a group of scientists at the CDC, using these sequences, actually regenerated a live version of the 1918 pandemic virus, which is kept under tight security and containment. Its resurrection is expected to reveal secrets of its pathogenesis that may help us combat the inevitable emergence of a new, equally deadly flu virus.

A.

National Museum of Health and Medicine

B.

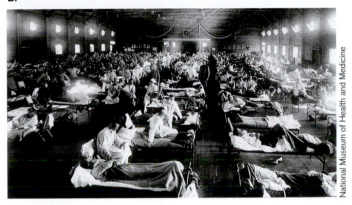

National Museum of Health and Medicine

C.

Figure 25.35   Flu epidemic of 1918.   A. Influenza unit, U.S. Army hospital in Luxembourg. Notice the mouth covering designed to prevent spread of the disease. **B.** A Kansas emergency hospital trying to cope with the 1918 flu. **C.** Life expectancy in the United States in the twentieth century. Notice the dramatic dip in 1918, the result of deaths from influenza.

Human Immunodeficiency Virus

Acquired immunodeficiency syndrome (AIDS) became an enormously important disease in the late twentieth century and continues to grip us today. There are between 30 to 40 million cases worldwide. As a result of infection by the retrovirus HIV (human immunodeficiency virus), the patient becomes severely immunocompromised and susceptible to a wide variety of infectious diseases. Death in patients with AIDS is usually the result of these secondary infections. Chapter 11 discusses the detailed molecular biology of HIV; here we will deal with the pathogenesis of the disease.

In contrast to rhinovirus, which exclusively binds to one cell receptor, ICAM-1, HIV binds multiple cell receptors via its glycoprotein 120 (gp120, also called the spike protein; see Section 11.5 and **Fig. 25.36A**). For HIV, the basic receptor unit on host cells is CD4 (meaning T cells are the usual targets), in combination with a chemokine receptor CCR-5 (see Fig 11.29A). By binding to the chemokine receptor, HIV can block chemokine binding and disconnect communication between the target cell and the immune system. You might still expect that the infection would be stopped because another aspect of the viral infection should, on its own, trigger apoptosis of the infected cell. Viral protein gp120 cross-links CD4 surface molecules on T cells, which, coupled with *antigen* binding to the T-cell receptor, should trigger apoptosis. The ensuing programmed cell death would simultaneously kill the infecting virus. However, the virus makes a protein (negative factor, NEF) that hinders apoptosis. NEF is one of several HIV proteins that affect pathogenesis (**Table 25.5**). As a result, the virus can replicate and bud from these lymphocytes (**Fig. 25.36B**).

Despite the presence of NEF, however, the lymphocytes eventually do die. Because CD4$^+$ T cells are the primary target of HIV, the levels of these T cells decline dramatically in an infected patient and cause an immunodeficiency, particularly in immune regulation. Monitoring T-cell numbers, therefore, is a critical diagnostic tool.

Comparing HIV infection and AIDS, the disease it causes, underscores the difference between infection and disease. Detection of HIV in an individual does not equate with that person having overt disease. The individual can go for years without developing symptoms. But even if infected individuals are not showing signs of disease, they are fully able to transmit the virus to another person sexually, through blood transfusion, or through the use of shared hypodermic needles.

As with many viral diseases, AIDS begins with flu-like symptoms—fever, headache, fatigue, sore throat, and sometimes a rash. Consequently, symptomology is never a reliable way to diagnose HIV infection. The detection of antibodies is a good initial screen to gauge whether a person has been exposed to the virus. Once exposure has been confirmed by a serum test (a change in test result

A.

env
Surface Glycoprotein SU
gp 120

env
Transmembrane
Glycoprotein TM
gp41

RNA
(2 molecules)

gog
Capsid CA
(Core Shelf)
p24

pol
Protease PR p9
Polymerase RT
RNAse H RNH p66
Integrase IN p32

gog
Membrane Associated
(Matrix) Protein MA
p17

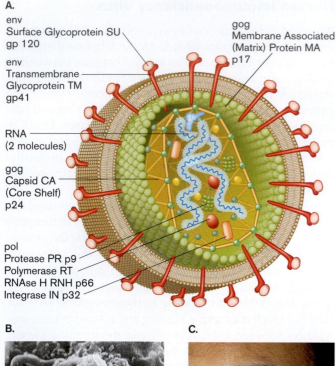

D.

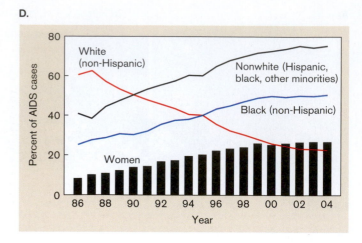

Figure 25.36   HIV and AIDS.   A. HIV. **B.** HIV-1 budding from cultured lymphocyte (SEM). **C.** Kaposi's sarcoma (oval spots) and periorbital cellulitis infection. **D.** AIDS incidence in the United States by race and age.

B. **C.**

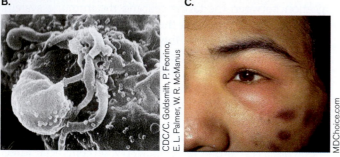

CDC/C. Goldsmith, P. Feorino,
E. L. Palmer, W. R. McManus

MDChoice.com

from negative to positive is called seroconversion), it is important to use PCR techniques to measure actual HIV viral load in a patient's blood. The higher the viral load the faster the disease will progress. Thus, for proper disease management, it is important to know viral load in addition to monitoring T-cell numbers.

The symptoms of AIDS itself begin to manifest once T-cell numbers fall to around 300 per microliter of blood (the normal range is 500–1,500). It can often take four or

five years (or more) before a patient starts to experience this decline. With this immunocompromised state come secondary infections such as shingles (herpes zoster virus), Pneumocystis pneumonia, tuberculosis, and oral thrush (a fungal infection of the mouth caused by the yeast *Candida albicans*). Late-stage AIDS is defined by CD4$^+$ counts below 200, at which point a malignancy called **Kaposi's sarcoma** can develop (**Fig. 25.36C**). Kaposi's sarcoma is a malignancy originating in endothelial or lymphatic cells, and tumors can arise anywhere—gastrointestinal tract, mouth, lungs, skin, or brain.

An important cause of AIDS symptoms is the HIV Tat protein (trans activator of the HIV promoter). A transactivator is a protein that can diffuse through the cytoplasm to bind and regulate a target gene sequence. The Tat protein is needed for viral replication. In addition, it is released from infected monocytes and enters other cells, where it alters host regulatory cascades and gene expression. The result is an increase in the production of nuclear factor kappa beta (NF κβ), which helps further disassemble the

Table 25.5 HIV proteins that affect pathogenesis.

Proteins	Role in pathogenesis
TAT (transcriptional trans-activator)	Secreted from infected cells; accelerates HIV gene transcription by host polymerase, has chemokine-like properties, acts as growth factor for endothelial cells, alters gene expression in target cells
VPR (viral protein R)	Induces G_2 cell cycle arrest and nuclear import of the preintegration complex
NEF (negative factor)	Downregulates cell surface CD4 and MHC class I (important for virus release); enhances virion infectivity by inhibiting apoptosis.
VIF (virion infectivity factor)	Counteracts a host cellular protein that naturally inhibits HIV replication
VPU (viral protein U)	Enhances HIV release from infected cells and CD4 degradation by targeting CD4 to the proteasome

A.

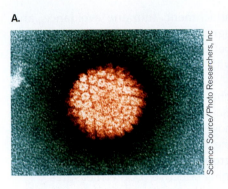

Science Source/Photo Researchers, Inc

B.

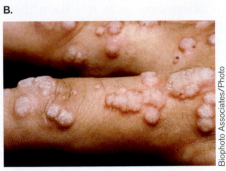

Biophoto Associates/Photo Researchers, Inc

Figure 25.37 Human papillomavirus.
A. Human papillomavirus (40–50 nm diameter, TEM). **B.** View of hand warts.
C. Part of the strategy HPV uses to deregulate cell division in infected cells. Protein p53 normally helps limit cell division by inactivating EF2. Virus E6 protein, however, removes p53, so the cell will begin to divide uncontrollably and become cancerous.

C.

Nondisease state: Normally, p53 neutralizes EF2, a protein that is required to activate genes encoding essential cell division proteins. EF2 stimulates cell division.

Disease state: HPV viral protein E6 chaperones human protein p53 toward degradation.

p53

HPV virus

Inactivate p53

E6

EF2

E6
p53

EF2

Proteosome degradation of p53

c-myc   *c-fos*

Leads to uncontrolled cell division

Cell division

numbers for women, African Americans, and Hispanic minorities continue to rise.

Human Papillomavirus

Human papillomavirus (HPV) is a very common virus that causes warts, an abnormal growth of skin tissue (**Figs. 25.37A** and **B**). HPV can produce warts on the feet, hands, vocal cords, mouth, and genital organs. Over 60 types of HPV have been identified so far. Each type infects different parts of the body. Some strains of HPV have also been associated with cervical and penile cancer. Genital HPV is one of the most common sexually transmitted diseases among college students.

How does this virus cause excessive growth of tissues? Once inside host cells, HPV synthesizes a protein called E6 (**Fig. 25.37C**). E6 binds to a host protein, p53, that normally prevents cells from completing a cell cycle if their DNA is not properly replicated. Protein p53 inhibits cell division by binding to a transcription factor called EF2, preventing it from activating two oncogenes *c-myc* and *c-fos*. Transcription of *c-myc* and *c-fos* is needed for mitosis, so blocking their transcription prevents cell division and controls growth. The HPV protein E6 targets p53 for destruction by the cell proteosome (discussed in Section 8.6). With p53 thus removed, *c-myc* and *c-fos* proteins are made continually. As a result, cell division proceeds unabated to produce warts or, in the worst-case scenario, cervical cancer. After the initial infection has resolved, the virus may remain latent in tissues and can reactivate if immune system function is impaired.

An effective HPV vaccine that will protect against cervical cancer has been developed, although guidelines for its use remain controversial. The problem is that it must be administered before a girl becomes sexually active (and potentially exposed) in order to work. The FDA and CDC have recommended its use in women and girls between ages 9 and 26.

TO SUMMARIZE:

- **Multiple reinfection** of the same person by a virus can occur because different strains of the virus have minor sequence differences in attachment proteins

immune system, and activation of endothelial cell growth, which can lead to Kaposi's sarcoma. Some believe that a vaccine using inactivated Tat will be useful in preventing AIDS. Other HIV proteins important to the pathogenesis of HIV and AIDS are presented in **Table 25.5**.

The relative incidence of AIDS by race since 1986 is examined in **Figure 25.36D**. Since 1990, the overall incidence, approximately 40,000 annual cases, has declined in the United States, mainly as a result of improvements in antiviral therapies. However, the decrease is also due to a drop in the numbers of white males afflicted. The

(for example, capsid). Because of these differences, neutralizing antibodies made during a previous infection are ineffective against the new strain.

- **Viral infection** can increase a patient's susceptibility to other, less virulent microbes.
- **Antigenic shifts** in a viral antigen (as in the hemagglutinin of influenza) can lead to a new pandemic of the disease.
- **Animals coinfected with two viruses** can serve as incubators for antigenic shifts or the evolution of new viruses.
- **Binding to host cells** can involve a single host receptor (as in rhinovirus) or multiple receptors (as in HIV).
- **Virus-infected cells** can secrete proteins that enter uninfected cells and disrupt signaling pathways.

Concluding Thoughts

This chapter has only scratched the surface of all that is known about microbial pathogenesis. It should be clear at this point that attachment, immune avoidance, and subversion of host signaling pathways are common goals of most successful pathogens, be they bacterial, viral, or parasitic. But even with all we know, there remains much to learn.

The remaining chapters deal with the basic principles used to diagnose disease and eradicate offending pathogens, but many pathogens remain hard to detect and difficult, if not impossible, to kill. So the more we know about the mechanisms microbes use to cause disease, the better we will be at developing effective countermeasures.

As you will learn in the coming chapters, new pathogens are constantly emerging. Over the last few decades, we have seen the development of HIV, SARS, avian flu, hantavirus, West Nile virus, *E. coli* O157:H7, and the re-emergence of flesh-eating streptococci, to name but a few. Can we ever stop pathogens from emerging? Probably not. For every countermeasure we develop, nature designs a counter-countermeasure. Our hope is that continuing research into the molecular basis of pathogenesis and antimicrobial pharmacology will keep us one step ahead.

CHAPTER REVIEW

Review Questions

1. Describe the differences between infection versus disease; pathogenicity versus virulence; LD_{50} versus ID_{50}.
2. What is meant by direct versus indirect routes of infection?
3. What are the characteristics of a good reservoir for an infectious agent?
4. Name the various portals of entry for infectious agents and name a disease associated with each.
5. Describe the basic features of a pathogenicity island.
6. Explain various ways bacteria can attach to host cell surfaces.
7. Describe the basic steps by which pili are assembled on the bacterial cell surface. How do type I and type IV pili differ?
8. Explain the five broad categories of toxin mode of action.
9. What is ADP-ribosylation and how does it contribute to pathogenesis?
10. Explain the differences between exotoxins and endotoxins.
11. Explain the mechanisms of secretion carried out by type II and type III protein secretion systems. What are the paralogous origins of these systems?
12. Describe the key features of *Salmonella* pathogenesis.
13. How can genomic approaches help identify pathogens in an infection?
14. What different mechanisms do intracellular pathogens use to survive within the infected host cell?
15. Describe different molecular strategies that microbes use to avoid the immune system.
16. How do bacteria determine whether or not they are in a host environment?
17. What is antigenic shift?
18. How is attachment of influenza virus different from HIV? Why does HIV produce an immune deficiency while influenza does not?
19. Explain the basics of how human papilloma virus can trigger cancer.

Key Terms

AB toxin (949)
adhesin (944)
ADP-ribosyltransferase (949)
antigenic shift (972)
ectoparasite (938)
edema factor (EF) (957)
endemic area (940)
endoparasite (938)
endotoxin (948)
exotoxin (948)
facultative intracellular pathogen (969)
fimbria (944)
fomite (940)
frank pathogen (939)
genomic island (942)
horizontal transmission (940)
in vivo expression technology (IVET) (965)
infection (938)

infection cycle (940)
infectious dose (939)
intimin (963)
intracellular pathogen (969)
Kaposi's sarcoma (974)
labile toxin (LT) (951)
latent state (939)
LD$_{50}$ (939)
lethal factor (LF) (957)
opportunistic pathogen (939)
parasite (938)
pathogen (938)
pathogenesis (938)
pathogenicity (939)
pathogenicity island (942)
petechia (959)
pilus (944)
primary pathogen (939)
protective antigen (PA) (957)

protein A (970)
receptor-binding domain (955)
reservoir (940)
Salmonella pathogenicity island 1 (SPI-1) (962)
SPI-2 (962)
transmembrane domain (955)
type I pilus (945)
type II secretion (960)
type III pilus (945)
type III secretion (961)
type IV pilus (945)
vector (940)
vertical transmission (transovarial transmission) (940)
virulence (939)
virulence factor (942)

Recommended Reading

Aktories, Klaus, and Joseph T. Barbieri. 2005. Bacterial cytotoxins: Targeting eukaryotic switches. *Nature Reviews in Microbiology* 3:397–410.

Autret, Nicolas, and Alain Charbit. 2005. Lessons from signature-tagged mutagenesis on the infectious mechanisms of pathogenic bacteria. *FEMS Microbiology Reviews* 27:703–717.

Burrows, Lori L. 2005. Weapons of mass retraction. *Molecular Microbiology* 57:878–888.

Campellone, Kenneth G., Andrew Giese, Donald J. Tipper, and John M. Leong. 2002. A tyrosine-phosphorylated 12-amino-acid sequence of enteropathogenic *Escherichia coli* Tir binds the host adaptor protein Nck and is required for Nck localization to actin pedestals. *Molecular Microbiology* 43:1227–1241.

Dean, Paul, Marc Maresca, and Brendan Kenny. 2005. EPEC's weapons of mass subversion. *Current Opinion in Microbiology* 8:28–34.

Dietrich, Guido, Sebastian Kurz, Claudia Hübner, Christian Aepinus, Stephanie Theiss, et al. 2003. Transcriptome analysis of *Neisseria meningitidis* during infection. *Journal of Bacteriology* 185:155–164.

Dobrindt, Ulrich, Franziska Agerer, Kai Michaelis, Andreas Janka, Carmen Buchrieser, et al. 2003. Analysis of genome plasticity in pathogenic and commensal *Escherichia coli* isolates by use of DNA arrays. *Journal of Bacteriology* 185:1831–1840.

Dobrindt, Ulrich, Bianca Hochhut, Ute Hentschel, and Jörg Hacker. 2004. Genomic islands in pathogenic and environmental microorganisms. *Nature Reviews Microbiology* 2:414–424.

Groisman, Eduardo, and Josep Casadesús. 2005. The origin and evolution of human pathogens. *Molecular Microbiology* 56:1–7.

Hamon, Mélanie, Hélène Bierne, and Pascale Cossart. 2006. *Listeria monocytogenes* a multifaceted model. *Nature Reviews in Microbiology* 4:423–434.

Hensel, Michael, Jill E. Shea, Colin Gleeson, Michael D. Jones, Emma Dalton, and David W. Holden. 1995. Simultaneous identification of bacterial virulence genes by negative selection. *Science* 269:400–403.

Hensel, Michael. 2004. Evolution of pathogenicity islands of *Salmonella enterica*. *International Journal of Medical Microbiology* 294:95–102.

Jin, Qi, Zhenghong Yuan, Jianguo Xu, Yu Wang, Yan Shen, et al. 2002. Genome sequence of *Shigella flexneri* 2a: Insights into pathogenicity through comparison with genomes of *Escherichia coli* K12 and O157. *Nucleic Acid Research* 30:4432–4441.

Liu, Tie F., Kimberley A. Cohen, Jason G. Ramage, Mark C. Willingham, Andrew M. Thorburn, and Arthur E. Frankel. 2003. A diphtheria toxin-epidermal growth factor fusion protein is cytotoxic to human glioblastoma multiforme cells. *Cancer Research* 15:1834–1837.

Mahan, Michael J., John W. Tobias, James M. Slauch, Philip C. Hanna, R. John Collier, and John J. Mekalanos. 1995. Antibiotic based selection for bacterial genes that are specifically induced during infection of a host. *Proceedings of the National Academy of Science* 92:669–673.

Mota, Luís Jaime, Isabel Sorg, and Guy R. Cornelis. 2005. Type III secretion: The bacteria-eukaryotic cell express. *FEMS Microbiology Letters* **252:**1–10.

Pallen, Mark J., and Brendan W. Wren. 2007. Bacterial pathogenomics. *Nature* **449:**835–842.

Renesto, Patricia, Nicolas Crapoulet, Hiroyuki Ogata, Bernard La Scola, Guy Vestris, et al. 2003. Genome-based design of a cell-free culture medium for *Tropheryma whipplei*. *Lancet* **362:**447–449.

Saenz, Henri L., and Christoph Dehi. 2005. Signature-tagged mutagenesis: Technical advances in a negative selection method for virulence gene identification. *Current Opinion in Microbiology* **8:**612–619.

Schilling, Joel D., Matthew A. Mulvey, and Scott J. Hultgren. 2001. Structure and function of *Escherichia coli* type 1 pili: New insight into the pathogenesis of urinary tract infections. *Journal of Infectious Diseases* **183:**S36–S40.

Smith, Kristina M., Yigong Bu, and Hiroaki Suga. 2003. Induction and inhibition of *Pseudomonas aeruginosa* quorum sensing by synthetic autoinducer analogs. *Chemistry & Biology* **10:**81–89.

Taubenberger, Jeffrey K., Ann H. Reid, and Thomas G. Fanning. 2005. Capturing a killer flu. *Scientific American* **292:**48–57.

Taubenberger, Jeffrey K., Ann H. Reid, A. E. Krafft, K. E. Bijwaard, and Thomas G. Fanning. 1997. Initial genetic characterization of the 1918 "Spanish" influenza virus. *Science* **275(5307):**1793–1796.

Thomas, Wendy E., Elena Trintchina, Manu Forero, Viola Vogel, and Evgeni V. Sokurenko. 2002. Bacterial adhesion to target cells enhanced by sheer force. *Cell* **109:**913–923.

Thompson, Lucinda J., and Hilde de Reuse. 2002. Genomics of *Helicobacter pylori*. *Helicobacter* **7:**1–17.

Wiles, Siouxsie, William P. Hanage, Gad Frankel, and Brian Robertson. 2006. Modelling infectious disease—time to think outside the box? *Nature Reviews in Microbiology* **4:**307–312.

Chapter 26

Microbial Diseases

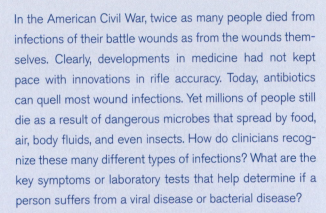

In the American Civil War, twice as many people died from infections of their battle wounds as from the wounds themselves. Clearly, developments in medicine had not kept pace with innovations in rifle accuracy. Today, antibiotics can quell most wound infections. Yet millions of people still die as a result of dangerous microbes that spread by food, air, body fluids, and even insects. How do clinicians recognize these many different types of infections? What are the key symptoms or laboratory tests that help determine if a person suffers from a viral disease or bacterial disease?

Microbial diseases are with us daily and continue to be one of the main contributors to global mortality and morbidity. The emergence of new pathogens, the increasing development of drug resistance, and the threat of bioterrorism have combined to create an increased demand for investigations into disease mechanisms and the body's ability to combat the organisms responsible. Along with those basic studies, effective diagnostic algorithms are needed to quickly identify infectious diseases and prevent their spread. This chapter explores the general classes and causes of microbial diseases and introduces the art of diagnosis.

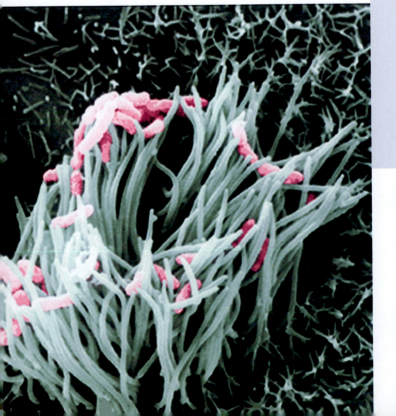

Scanning electron micrograph of primary cultured rabbit tracheal epithelial cells following 5 minutes of coincubation with virulent wild-type *Bordetella bronchiseptica*, a close relative of *B. pertussis*, the cause of human whooping cough. In this model for early host-pathogen interactions, wild-type *B. bronchiseptica* preferentially adheres to cilia and is rarely seen bound to aciliated cells or aciliated portions of ciliated cells. This adherence pattern impedes the ciliary elevator that normally removes bacteria from the lung. *B. bronchiseptica* is a common respiratory pathogen of cats and dogs but can cause whooping cough in immuno-compromised humans. Bacteria (approx. 1 μm in length) have been pseudocolored pink for easier visualization.

979

A visibly ill 16-year-old girl was taken to her pediatrician, where she complained of diarrhea, high fever (39.9°C; 102°F), and vomiting. She told the physician that she was healthy until two days prior. Upon physical exam, the doctor noticed that the girl had a much lower than normal blood pressure (76/48 mmHg), a rapid heart rate of 120 beats per minute, and an ery-thematous (red) rash on her trunk. Because of her deteriorating condition, the patient was admitted to the pediatric intensive care unit at the local hospital, and cultures were obtained. She was immediately given intravenous fluids and IV antibiotics, yet she died within a week. Patient history taken upon admission revealed that the girl had started her menstrual period four days before becoming ill, providing an important clue as to the cause of the disease.

This tragedy was part of a larger story that emerged in the late 1970s and early 1980s, when women started dying from this dangerous, new (emerging) disease. Now known as toxic shock syndrome, scientists and physicians learned that it was caused by certain strains of *Staphylococcus aureus* that produced a superantigen type of toxin (toxic shock syndrome toxin, or TSST; see Section 24.6). Why was the disease not seen in previous decades? The answer, surprisingly, turned out to be the use of one brand of superabsorbent tampons (since removed from the market). The tampons produced a rich growth environment for *S. aureus*. So if the patient were colonized by a TSST-producing strain, huge amounts of toxin would be released to circulate in the bloodstream. Today we recognize that toxic shock syndrome is a possible consequence of any *S. aureus* infection and can occur in both men and women. We now have tools to diagnose it quickly.

In Chapter 25, we discussed the arsenal of weapons that microbes use to cause disease (including TSST). But what of the diseases themselves? Microbial diseases are often presented from the microbe's point of view—for example, where in the body does *Staphylococcus aureus* cause disease? Another approach, which we take in this chapter, is to look at infections from the vantage point of the infected organ. We might ask, for example, what organisms cause vaginal infections versus lung infections, and how are they differentially diagnosed? Are the patient's complaints consistent with gastrointestinal tract disease or a lung infection? This approach is more attuned to the practices of health care workers and the foundation by which a physician interacts with a patient. An individual who goes to a clinician complaining of a fever, cough, and chest pain does not typically report having encountered a particular bacterium. The physician must determine from the patient that the disease is localized to the chest and then intuit that it may be a respiratory tract infection. Appropriate samples are then collected and sent to a clinical microbiology laboratory, where data are collected to confirm or dispute the presence of a specific etiologic agent such as *Streptococcus pneumoniae* or other lung pathogen.

In this chapter, microbial diseases are classified as respiratory, gastrointestinal, genitourinary, nervous system, skin and soft-tissue, bone and joint, cardiovascular, and generalized (disseminated) infections. We do not present an exhaustive compendium of microbial illnesses but a representative sampling of infections that illustrate key aspects of microbial disease. The material is arranged and presented so as to integrate your knowledge of microbiology and immunology within the framework of the practice of medicine and the study of infectious disease.

26.1 Characterizing and Diagnosing Microbial Diseases

Although this chapter will classify representative microbial diseases by organ system, there are other very instructive ways to group and study microbial infections. These include the organism and portal-of-entry approaches. Each has clear benefits and pitfalls. The organism approach is useful when examining the different ways a given species can cause disease. For instance, *E. coli* can cause urogenital tract infections, gastrointestinal infections, and meningitis. How do the various strains causing these diseases differ from one another? Are virulence factors present in the uropathogenic strain that are absent or different in the diarrhea-producing strain? (Virulence factors are discussed in Chapter 25; see Sections 25.2 and 25.3.)

Pathogens are also commonly classified by their route of infection (or portal of entry; see Chapter 23). In this approach, pathogens are classified as food-borne, airborne, blood-borne, or sexually transmitted. The port-of-entry approach has its merits, but it sometimes seems at odds with the disease itself. *Salmonella typhi*, for example, is categorized as a food-borne pathogen. Food-borne pathogens typically have a fecal-oral route of infection and cause pronounced gastrointestinal disease. Yet *S. typhi* does not actually cause gastrointestinal disease until very late in the disease process. After ingestion, it quickly enters the bloodstream from the intestine to produce a serious septicemia (blood infection) called enteric (typhoid) fever. At this stage, the organism cannot be found in feces. (Much later the organism can re-enter the intestine via the gall bladder and cause mild gastrointestinal symptoms.) The obvious value of knowing the route of infection is that society can design effective measures to avoid disease. Hand washing and proper food preparation, for instance, will effectively prevent the septicemia of typhoid fever. But clearly, all organisms entering via the gastrointestinal tract do not cause gastrointestinal disease.

The organ systems approach has its drawbacks as well, given that organisms that initially affect one organ system can disseminate to infect other organ systems. *Neisseria gonorrhoeae*, for example, initially infects the genitourinary tract, but can disseminate via the bloodstream, causing a blood infection. *Yersinia enterocolitica*, a cause of gastroenteritis, can disseminate from the intestine and cause abscesses elsewhere in the body. Yet we favor the organ systems approach because it is illuminating to know how microbial diseases present themselves to physicians treating patients. The physician's examination usually begins not by taxonomic evaluation of the agent or knowledge of the portal of entry, but with symptoms emanating from the organ system that has been affected. You will benefit, however, from studying infectious disease from various perspectives, and you should keep in mind factors such as the portal of entry and the taxonomy of the infectious organism.

Patient Histories Provide Important Diagnostic Clues

A problem often faced by clinicians is that many infectious diseases display similar symptoms, making diagnosis difficult. *Vibrio cholerae* and enterotoxigenic *E. coli*, for instance, both produce diarrheal diseases characterized by cramps, lethargy, and liters of watery stool each day. Cholera, however, is not commonly seen in the United States. Nevertheless, a clinician might suspect cholera if a patient recently traveled to or arrived from an endemic area (a specific geographical locale such as India or Bangladesh) where the disease is regularly observed. This travel information is not gained by examining the patient but comes from taking a "patient history."

Knowledge of the patient's hobbies can be equally important for diagnosing a disease. This is another critical part of taking a patient's history. For example, a person who is suffering from enlarged glands, fever, and headaches and who was recently rabbit hunting may have been exposed to the gram-negative rod *Francisella tularensis*, an intracellular pathogen that infects various wild animals and is the cause of tularemia, an illness also known as "rabbit fever" (**Fig. 26.1A**). A hunter with these symptoms could have accidentally infected a cut while cleaning the kill. Tularemia is an example of a **zoonotic disease**, an infection that normally affects animals but that can be transmitted to humans.

How can knowing a person's occupation be important? Consider the case of a man with acute pneumonia. The clinician discovers that the man is a sheep farmer. Knowing this, *Coxiella burnetii*, a pathogenic bacterial species that infects sheep and is shed in large quantities in placental materials, would be considered one of the possible sources of infection. The organism grows intracellularly only (it is an obligate intracellular pathogen). It infects sheep, cattle, and goats but does not usually cause clinical disease in these animals. *C. burnetii* is secreted in body fluids and shed in high numbers in the animal's amniotic fluids and placenta, heavily contaminating the soil. When the contaminated soil dries, the microbes become aerosolized whenever the dirt is disturbed. Humans (such as farmers) that inhale the dried particles can develop the lung infection called Q fever, another type of zoonotic disease (**Fig. 26.1B**).

One more example of an "occupational hazard" is illustrated by a woman with severe respiratory disease who works in a leather factory, handling animal hides. Knowing her occupation, a clinician would investigate whether she might have contracted anthrax (woolsorter's disease) from inhaling spores present on the animal hide.

Because patient histories are so helpful in diagnosing infectious diseases, we will present several case histories in the following sections and segue into discussions of various microbes that can infect each organ system. Key aspects of infectious diseases will be revealed as we proceed.

It should be stressed that further insight into pathogenesis and new innovations in diagnosis will come from the remarkable number of microbial genomes whose sequences have been completed as well as from those that are still under way. A sampling of the pathogens whose genomes have already been sequenced is presented in **Table 26.1**. Many of the ORFs identified have no known function and

Figure 26.1 **Examples of bacteria that cause zoonotic diseases.** **A.** *Francisella tularensis* (colorized SEM), the cause of tularemia. Tularemia is a highly infectious disease, usually spread via a tick vector but also through cuts. **B.** *Coxiella burnetii* (colorized TEM), the cause of Q fever. This irregularly-shaped organism undergoes developmental stages that range in size from 0.2 μm to 1 μm.

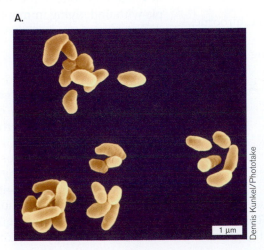

A.

1 μm

Dennis Kunkel/Phototake

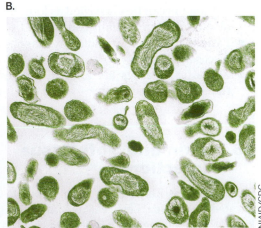

B.

NIAID/CDC

Table 26.1 Examples of pathogens whose genomes have been sequenced.[a]

Pathogen	Disease
Escherichia coli O157:H7	Diarrhea, hemolytic uremic syndrome (HUS)
Salmonella typhi	Typhoid fever
Yersinia pestis	Plague
Haemophilus influenzae	Meningitis, otitis media
Haemophilus ducreyi	Chancroid (sexually transmitted)
Vibrio cholerae	Cholera (severe diarrhea)
Vibrio vulnificus	Septicemia, wound infections
Pseudomonas aeruginosa	Wound infections, septicemia, respiratory infections
Coxiella burnetii	Q fever (respiratory)
Neisseria meningitidis	Meningitis, septicemia
Bordetella pertussis	Whooping cough (respiratory)
Helicobacter pylori	Stomach ulcers, gastric cancer
Campylobacter jejuni	Diarrhea
Rickettsia prowazekii	Epidemic typhus
Brucella suis	Flu-like disease, brucellosis
Bacillus anthracis	Anthrax (respiratory or wound)
Staphylococcus aureus	Boils, toxic shock syndrome, wound infections, osteomyelitis (bone)
Streptococcus pyogenes	Strep throat, wound infections
Listeria monocytogenes	Flu-like, diarrhea, stillbirths
Clostridium tetani	Tetanus
Clostridium perfringens	Wound infections
Mycoplasma pneumoniae	Respiratory
Mycobacterium tuberculosis	Tuberculosis (respiratory)
Mycobacterium leprae	Leprosy
Chlamydia trachomatis	Eye infection, urethritis (sexually transmitted)
Borrelia burgdorferi	Lyme disease
Treponema pallidum	Syphilis (sexually transmitted)
Leptospira interrogans	Septicemia

[a]To view genomes of pathogens, access the "Kyoto Encyclopedia of Genes and Genomes" site on the Internet or the "Genome" site at Entrez PubMed.

may eventually explain differences in host and tissue specificity and reveal new toxins or other pharmacologically active molecules produced by virulent microbes.

TO SUMMARIZE:

- **Pathogens can be classified** as food-borne, airborne, blood-borne, or sexually transmitted.
- **Patient histories** are vital in diagnosing microbial diseases.
- **Zoonotic diseases** are animal diseases accidentally transmitted to humans.

26.2 Skin and Soft-Tissue Infections

Skin infections range from simple boils to severe, complicated so-called "flesh-eating" diseases (**Table 26.2**). Recall that the integrity of the skin as well as the presence of normal skin flora prevent infection. However, even minor insults to the skin (such as a paper cut) can result in infections, most of which are caused by the gram-positive pathogen *Staphylococcus aureus*. Healthy individuals develop infections of the skin only rarely; however, people with underlying immunosuppressive diseases such as diabetes are at much higher risk.

Boils

Staphylococcus aureus is a common cause of the painful skin infections called boils, or carbuncles (**Fig. 26.2**). This gram-positive organism, often a normal inhabitant of the nares (nostrils), can infect a cut or gain access to the dermis via a hair follicle. It possesses a number of enzymes that contribute to disease, including **coagulase**, which helps coat the organism with fibrin, thereby walling off the infection from the immune system and antibiotics. As a result, boils generally require surgical drainage as well as antibiotic therapy. As noted earlier, some strains of *S.*

Figure 26.2
Staphylococcus aureus.
A. *S. aureus* (colorized SEM).
B. Exfoliative toxin from some strains of *S. aureus* cause scalded skin syndrome.

A.

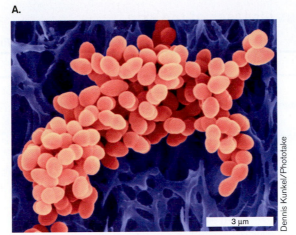

3 µm

Dennis Kunkel/Phototake

B.

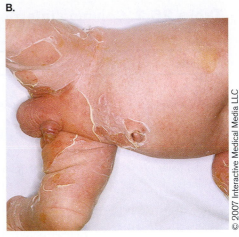

© 2007 Interactive Medical Media LLC

aureus also produce toxic shock syndrome toxin, a superantigen that can lead to serious systemic symptoms (see Chapters 24 and 25).

A particularily dangerous strain of *S. aureus* called methicillin-resistant *S. aureus* (MRSA) has recently emerged. *S. aureus* infections are commonly treated with penicillin-like drugs, such as methicillin. MRSA has developed resistance to methicillin and many other penicillin-like drugs through a mutation that alters one of the proteins (called a penicillin-binding protein; PBP) involved in cell wall synthesis. Because methicillin is normally used as a first line of defense against staphylococcal infections, treatment failure of MRSA can be life-threatening, and alternative drugs, such as vancomycin, need to be used. MRSA first appeared as agents of nosocomial infections—infections that occur after a patient enters a hospital. However, these antibiotic-resistant organisms are no longer contained only in the hospital. Individuals who have not been in a hospital are being infected with MRSA in what are called community-acquired infections. This is occuring at an epidemic rate in the United States (an incidence of approximately 20 per 100,000 population). Seventy percent of Staphylococcal skin infections are now caused by MRSA. Thus, a physician can no longer assume that a patient walking into the office with a staphylococcal infection will respond to methicillin. The doctor must assume that it may be MRSA. As a result, treatment regimens around the country are being forced to change. Today, vancomycin and linezolid are the antibiotics typically used to treat staph infections. Antibiotics are discussed in Chapter 27.

Other staphylococcal diseases are caused by toxin-producing strains in which the organism remains localized but the toxin disseminates. We have already mentioned TSST, but there are other toxins. For example, some strains of *Staphylococcus aureus* produce a toxin called **exfoliative toxin** that causes a blistering disease in children called staphylococcal scalded skin syndrome (**Fig. 26.2C**). Exfoliative toxin, like TSST, is a superanti-

gen, but it also cleaves a skin cell adhesion molecule that when inactivated results in blisters.

Table 26.2 presents other common infections of the skin and soft tissues.

Case History: Necrotizing Fasciitis by "Flesh-Eating" Bacteria

*One weekend in June, Cassi was camping with her three children. She suffered a minor cut on her finger, which she bandaged properly. She also injured the left side of her body while playing sports with her kids. Not thinking much of either of her minor injuries, she went to bed. Two days later, Cassi was extremely ill. Her symptoms included vomiting, diarrhea, and a fever. She was also in severe pain where she had injured her side, and the area had begun to bruise (the skin was not broken). By the next day, she could barely get out of bed, and by the end of the night, she was breathing with difficulty and could not see. Her side began to leak fluid and blood. Cassi was admitted to the hospital in shock, with no detectable blood pressure. An infectious disease specialist diagnosed the problem as necrotizing fasciitis, and she was rushed into surgery. In an effort to save her life, about 7% of her body surface was removed. Because the large wound infection in her side would need to resolve before a skin graft could be performed to repair it, the hole in Cassi's body was left wide open (**Fig. 26.3A**). After nearly three months and several operations, Cassi recovered.*

What kind of organism can cause this type of devastating disease? The disease **necrotizing fasciitis**, also known as flesh-eating disease, is rare and is often caused by the gram-positive coccus *Streptococcus pyogenes*, a microbe normally associated with sore throat infections (pharyngitis). Although sometimes described as a recently emerging infectious disease, necrotizing fasciitis was first discovered in 1783, in France. Its incidence may have risen recently owing to the increased use of anti-inflammatory drugs, which increase a person's susceptibility to infection. In this case history, Cassi probably had this organism on her skin

Table 26.2 Common infectious diseases of the skin.

Disease	Symptoms	Etiological agent	Virulence factors
Bacterial			
Folliculitis	Boils	*Staphylococcus aureus* (G+ cocci); fibrin wall around abscess renders it poorly accessible to antibiotics	Coagulase, protein A, TSST, leucocidin, exfoliative toxin
Scalded skin syndrome	Peeling skin on infants, systemic toxin		
Impetigo	Skin lesions on the face, mostly children		
Scarlet fever	Sore throat, fever, rash	*Streptococcus pyogenes* (G+ cocci)	M-protein pili, C5a peptidase, hemolysin, pyrogenic toxins, others
Erysipelas	Skin lesions, usually facial, that spread to cause systemic infection		
Necrotizing fasciitis	Rapidly progressive cellulitis		
Viral			
Rubella[a]	Discolored, pimply rash; mild disease unless congenital	Rubella virus [ssRNA(+)]	
Measles[a]	Severe disease, fever, conjunctivitis, cough, rash	Rubeola virus [ssRNA(−)]	
Chickenpox[a]	Generalized discolored lesions	Varicella-zoster (dsDNA)	
Shingles	Pain and skin lesions, usually on trunk in adults		
Smallpox	Raised, crusted skin rash, highly contagious	Variola major (dsDNA)	
Warts	Rapid growth of skin cells	Papillomaviruses (dsDNA)	
Fungal			
Dermatomycosis	Dry, scaly lesions like athlete's foot	Dermatophytes	
Sporotrichosis	Granulomatous, pus-filled lesions; can disseminate to lungs or other organs	*Sporothrix schenckii*	
Blastomycosis	Granulomatous, pus-filled lesions; can disseminate to lungs or other organs	*Blastomyces dermatitidis*	BAD1 adherence
Candidiasis	Patchy inflammation of mouth (thrush) or vagina; can disseminate in immunocompromised patients	*Candida albicans*	Proteinase, phospholipase, Ssn6/Tup1 regulators, others
Aspergillosis	Infects wounds, burns, cornea, external ear	*Aspergillus* spp.	PacC/fos1 regulators, gliotoxin
Zygomycosis	Mainly affects diabetic patients; can rapidly disseminate	*Mucor* and *Rhizopus* spp.	

[a] Vaccine is available against this agent.

when the injury to her side occurred. The injured area probably suffered an invisible microabrasion, providing a good growth environment for the organism, leading to the secretion of potent toxins and death of surrounding tissues.

Rapid, aggressive antibiotic treatment is required in these extreme cases, even before the clinical microbiology lab has had time to identify the organism. Therapy can include several antibiotics, such as clindamycin and metronidazole, which act against anaerobes and gram-positive cocci, and gentamicin, a drug particularly effective against gram-negative microbes. (Chapter 27 further discusses these and other antibiotics.) Often, antibiotic treatment of patients with necrotizing fasciitis is ineffective because of insufficient blood supply to the affected tissues.

Figure 26.3 Flesh-eating *Streptococcus pyogenes*. A. Flesh removed from a patient in an effort to stop the spread of necrotizing fasciitis. **B.** Gangrene of the fingers caused by the same infection in the same patient.

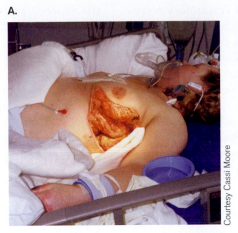

A.

Courtesy Cassi Moore

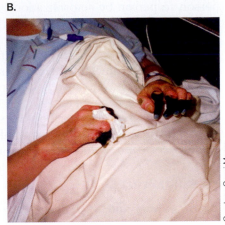

B.

Courtesy Cassi Moore

> **THOUGHT QUESTION 26.1** Why would treatment of an infection sometimes require multiple antibiotics?

WWW | Antibiotic treatment

Is There an Immunological Predilection to Necrotizing Faciitis?

S. pyogenes is best known for causing sore throats and for the immunological sequelae that can develop, such as rheumatic fever and glomerulonephritis. **Sequelae** are secondary diseases that develop after resolution of a primary infection. They are often the result of a cross-reactivity between bacterial and host antigens. The immune system, activated by the bacterial antigen, begins to attack the cross-reacting self antigens and damages tissues (see Section 24.8). Why some patients infected with *S. pyogenes* develop necrotizing faciitis while others do not may be due more to the immunological status of the patient than to the virulence arsenal of the microbe, but its arsenal is vast nonetheless.

The source of many established putative virulence factor genes in *S. pyogenes* are the many prophages present in its genome. Prophages constitute approximately 10% of the organism's genome. One study found that a soluble factor produced by human pharyngeal cells can facilitate activation of at least some of these phages and cause horizontal transfer of the associated virulence factors between strains of this pathogen.

Viral Diseases Causing Skin Rashes

Several viruses can produce skin rashes, although their route of infection is usually through the respiratory tract. Measles, for example (see Section 6.1), is a highly contagious viral infection caused by a paramyxovirus, whose hallmark symptom is skin rash (Fig. 6.2). Also known as rubeola, the first signs are fever, cough, runny nose, and red eyes occurring 9–12 days after exposure. A few days later, spots in the mouth appear (Koplik's spots) along with a sore throat. Then a skin rash develops that typically starts on the face and spreads down the body. The virus replicates in the lymph nodes and spreads to the bloodstream (viremia), where it can infect endothelial cells of the blood vessels. The rash occurs when T cells begin to interact with these infected cells.

Although skin rash is the main symptom of measles, infection can also cause respiratory symptoms and complications, including pneumonia, bronchitis, croup, and even a fatal encephalitis in immunocompromised patients. In the United States, measles has been almost completely eliminated by the measles, mumps, and rubella (MMR) multivalent vaccine (one exception is the AIDS patient, where measles can be fatal). Worldwide, however, measles is still a serious problem in places where vaccinations are not routine.

Rubella virus, a togavirus, causes a maculopapular rash known as German measles, or three-day measles. The rash is similar but less red than that of measles (**Fig. 26.4**). German measles is an infection that primarily affects the skin and lymph nodes and is usually transmitted from

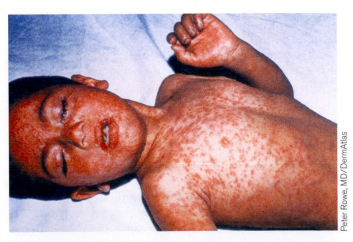

Peter Rowe, MD/DermAtlas

Figure 26.4 German measles. Skin rash caused by rubella virus.

person to person by aerosolization of respiratory secretions. It is not dangerous in adults or children; the virus can, however, cross the placenta in a pregnant woman and infect her fetus. If the virus crosses the placenta within the first trimester, the result is congenital rubella syndrome, which can cause death or serious congenital defects in the developing fetus.

Other viruses affecting the skin, such as chickenpox, the related disease shingles, and smallpox, are discussed in Chapters 6 and 11 and included in **Table 26.2**.

TO SUMMARIZE:

- **S. aureus** and **S. pyogenes** are common bacterial causes of skin infections. The organisms usually infect through broken skin.
- **Methicillin-resistant S. aureus** (MRSA) has become an important cause of community-acquired staphylococcal infections.
- **Necrotizing fasciitis** is usually caused by S. pyogenes but can be the result of other infections.
- **Infections of the skin** can disseminate via the bloodstream to other sites in the body.
- **Rubeola** and **Rubella viruses** infect through the respiratory tract, but their main manifestation is the production of similar maculopapular skin rashes.

26.3 Respiratory Tract Infections

Lung and upper respiratory tract infections are among the most common diseases of humans. Many different bacteria, viruses, and fungi are well adapted to grow in the lung. Successful lung pathogens come equipped with appropriate attachment mechanisms and countermeasures to avoid various lung defenses (such as alveolar macrophages). Although many microbes can cause lung infection, most respiratory diseases are of viral origin, and most viral infections (such as the common cold) do not spread beyond the lung. Fortunately, viral diseases by and large are self-limiting and typically resolve within two weeks; however, the damage caused by a primary viral infection can lead to secondary infections by bacteria.

Bacterial infections of the lung, whether of primary or secondary etiology, require intervention. Today this means antibiotic therapy. Before the advent of antibiotics, the only recourse was to insert a tube into the patient's back to drain fluid accumulating in the pleural cavity around the lung (a pathological process known as pleural effusion). Unless released, the pressure on the lung will collapse the alveoli and make breathing difficult.

Bacterial infections of the lung that arise secondarily to viral disease occur, in part, because the patient dehydrates. The resulting increase in mucus viscosity hampers the mucociliary elevator (described in Section 23.1), as will growth of the bacterium itself. Because a compromised mucociliary elevator makes it difficult to expel the microbe, the patient's susceptibility to secondary bacterial infection increases. Hence, cold sufferers are advised to drink plenty of fluids, which help decrease the viscosity of mucus and consequently improve mucociliary elevator function.

Case History: Bacterial Pneumonia

In March, an 80-year-old resident of a New Jersey nursing home had a fever accompanied by a productive cough with brown sputum (mucous secretions of the lung that can be coughed up). He reported to the attending physician that he had pain on the right side of his chest and suffered from night sweats. Blood tests revealed that his white blood cell (WBC) count was 14,000/μl with a makeup of 77% segmented forms (polymorphonuclear leukocytes, PMNs) and 20% bands (immature PMNs). The chest radiograph revealed a right upper lobe infiltrate with cavity formation (Fig. 26.5A). From this information, the clinician made a diagnosis of pneumonia. Microscopic examination of the patient's sputum revealed gram-positive cocci in pairs and short chains surrounded by a capsule (Fig. 26.5B). Bacteriological culture of his sputum and blood yielded Streptococcus pneumoniae.

Note that pneumonia is a disease, not a specific infection. Many different microbes can cause pneumonia (**Table 26.3**). The pneumococcus S. pneumoniae accounts for about 25% of community-acquired cases of pneumonia, but pneumococcal pneumonia occurs mostly among the elderly and immunocompromised, including smokers, diabetics, and alcoholics. A breakdown of pneumonia cases by causative organism is shown in **Figure 26.5C**.

The noses and throats of 30–70% of a given population can contain S. pneumoniae. The microbe can be spread from person to person by sneezing, coughing, or other close personal contact. Pneumococcal pneumonia may begin suddenly, with a severe shaking chill usually followed by high fever, cough, shortness of breath, rapid breathing, and chest pains.

After being aspirated into the lung, the microbe will grow in the nutrient-rich edema fluid of the alveolar spaces. Neutrophils and alveolar macrophages then arrive to try to stop the infection. They are called into the area from the circulation by chemoattractant chemokines released by damaged alveolar cells. The thick polysaccharide capsule of the pneumococcus, however, makes phagocytosis very difficult. In an otherwise healthy adult, pneumococcal pneumonia usually involves one lobe of the lungs; thus, it is sometimes called lobar pneumonia. The infiltration of PMNs and fluid lead to the typical radiological findings of diffuse cloudy areas. In contrast, infants, young children,

A. Lobar pneumonia

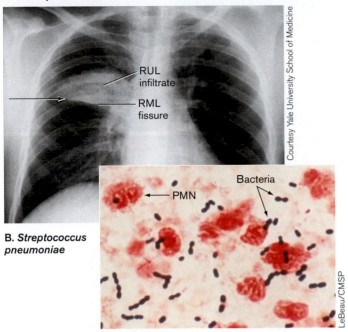

Courtesy Yale University School of Medicine

RUL
infiltrate

RML
fissure

B. *Streptococcus pneumoniae*

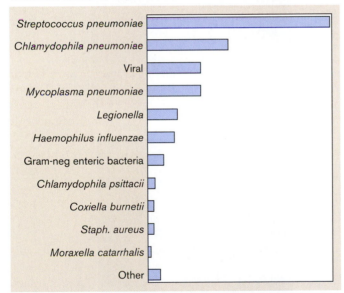

Bacteria

PMN

LeBeau/CMSP

C. Relative incidence of pneumonia

Streptococcus pneumoniae

Chlamydophila pneumoniae

Viral

Mycoplasma pneumoniae

Legionella

Haemophilus influenzae

Gram-neg enteric bacteria

Chlamydophila psittacii

Coxiella burnetii

Staph. aureus

Moraxella catarrhalis

Other

Figure 26.5   Pneumonia caused by *Streptococcus pneumoniae*.   A. X-ray view of a patient with lobar pneumonia. Infiltration in the right upper lobe is caused by *S. pneumoniae*. The sharp lower border represents the upper boundary of the middle lobe fissure (arrow). **B.** Micrograph of *S. pneumoniae*. Sputum sample showing numerous PMNs and extracellular diplococci in pairs and short chains. Bacteria range from 0.5 μm to 1.2 μm in diameter. **C.** Relative incidence of pneumonia caused by various microorganisms.

and elderly people more commonly develop an infection in other parts of the lungs, such as around the air vessels (bronchi), causing bronchopneumonia.

The white cell count in the case history is telling. The patient had an elevated WBC count (normal is 5,000 to 10,000/μl) and elevated band cells (normal is 0–8%). These increases are indicative of a bacterial, not viral, infection. Neutrophils (PMNs), the front line combatants against infection, rise in response to bacterial infections and are first released from bone marrow as immature band cells, whose presence is a sure sign of bacterial infection.

Several outbreaks of pneumococcal pneumonia have occurred over recent years in nursing homes, where numerous residents have been affected. This underscores the importance of elderly people receiving the pneumococcal polysaccharide vaccine (PPV) as a hedge against infection. While there are over 80 antigenic types of pneumococcal capsular polysaccharide, the injected vaccine contains only the 23 types that are most often associated with disease. A vaccine that is formulated to respond to multiple antigens is termed a **multivalent vaccine**. The immune response that results will protect vaccinated individuals against infection by those antigenic types. The vaccine is recommended for individuals over 65 as well as for those who are immunocompromised. The patient in this case history failed to receive the vaccine.

In addition to causing serious infections of the lungs (pneumonia), *S. pneumoniae* can invade the bloodstream (bacteremia) and the covering of the brain (meningitis). The death rates for these infections are about one out of every 20 who get pneumococcal pneumonia, about two out of 10 who get bacteremia, and three out of 10 who get meningitis. Individuals with special health problems, such as liver disease, AIDS (caused by HIV), or organ transplants, are even more likely to die from the disease because of their compromised immune systems.

An emerging infectious disease problem throughout the United States and the world is the increasing resistance of *S. pneumoniae* to antibiotics. At least 30% of the strains isolated are already resistant to penicillin, the former drug of choice for treating the disease. Chapter 27 discusses why antibiotic resistance is on the rise for this and other microbes.

Case History: Disseminated Disease from a Fungal Lung Infection

A 35-year-old male boxer named Tyrrell, recently admitted to a Maryland hospital, was in good health until six months ago, when he developed a chronic cough that produced blood-tinged white sputum. He also experienced flu-like symptoms, a decrease in appetite, and weight loss. One month prior to admission, he became so short of breath that he could no longer continue boxing. At that time, an X-ray taken in the emergency room showed right upper lobe infiltrate, indicating pneumonia (**Fig. 26.6A**). *A tuberculosis skin test was negative. He was given a prescription for the antibiotic azithromycin (a commonly used macrolide antibiotic to treat bacterial infections of the respiratory tract) and discharged. Despite antibiotic treatment, the cough never diminished and he developed several painless subcutaneous nodules. The largest nodule was on his left leg, contained pus, caused pain, and eventually hindered walking* (**Fig. 26.6B**). *This prompted his current admission to the hospital. The patient's history revealed that he installed home insulation for a living and had not traveled outside the area for the past year. He complained of fevers, chills, night sweats, and a 10-kg (22-lb) weight loss. His right tibia was tender to the touch, indicating bone involvement. He denied having prior pneumonia, sinus infection, arthritis, hematuria (blood in the urine), numbness, or muscle weakness. He had no history of intravenous drug use and had been in a monogamous relationship for four years. His white cell count was 10,500/mm³ with a normal differential of PMNs and band cells.*

In this case, the expression of frankly purulent material (pus) from the left leg nodule suggested that this was an infectious process. The infection in this patient probably started in the lung (clued by the cough), after which the organism disseminated throughout the body via the bloodstream. Fungus is a probable cause, given the chronic nature of the patient's symptoms. Tuberculosis, caused by the bacterium *Mycobacterium tuberculosis*, also causes a chronic lung infection and might have been suspected except for the negative TB skin test. The tuberculin skin test involves injecting a small amount of mycobacterial antigen called PPD (purified protein derivative) under the skin of the lower arm. A person who has been infected with *M. tuberculosis* will exhibit a localized delayed-type hypersensitivity reaction at the site of injection, although this does not equate to currently active disease.

The most likely fungal causes of infection in this case history are the endemic mycoses, such as histoplasmosis, blastomycosis, and coccidiomycosis. This patient had never traveled to the western United States, where coccidioidomycosis is endemic. This ruled out exposure to *Coccidioides*. Histoplasmosis most commonly presents as a flu-like pulmonary illness, with erythema nodosum (tender bumps on skin) and arthritis (swollen joints) or arthralgias (joint pain), none of which the patient had. Blastomycosis can disseminate to the lung, skin, bone, and genitourinary tract, which is consistent with the pattern of organ involvement seen in this patient. *Cryptococcus*, an encapsulated yeast, is not the likely cause, since it typically requires an immunocompromised host to cause disease, not the case in this instance. (*Cryptococcus*, which causes cryptococcosis, is an opportunistic pathogen that commonly infects AIDS patients.) The most prevalent clinical form of cryptococcosis is meningoencephalitis, although

A. Pneumonia infiltrate

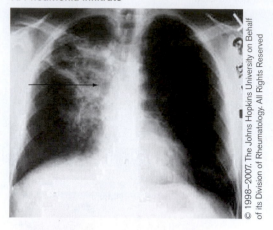

B. Leg lesion

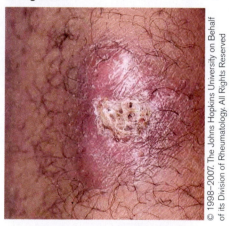

C. Colony of *Blastomyces dermatitidis*

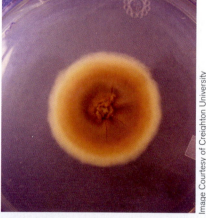

Figure 26.6 **Pneumonia and metastatic disease caused by *Blastomyces dermatitidis*.** **A.** Diffuse infiltrate in the right lung (arrow). **B.** Metastatic leg lesion at tibia. **C.** Fungal colony of *Blastomyces dermatitidis*.

disease can also involve the skin, lungs, prostate gland, urinary tract, eyes, myocardium, bones, and joints.

Amphotericin B, an antifungal agent, was finally given to this patient. His fever lowered almost immediately, and the skin nodules diminished. After two weeks, a fungus was found in the cultures of the nodule biopsy, bronchoalveolar lavage (washes), and urine (**Fig. 26.6C**). This fungus was identified by a DNA probe as *Blastomyces dermatitidis*, confirming the diagnosis of blastomycosis.

Blastomyces dermatitidis is a dimorphic fungus that resides in the soil of the Ohio and Mississippi River valleys and the southeastern United States. The portal of entry is the respiratory tract, and infection is usually associated with occupational and recreational activities in wooded areas along waterways, where there is moist soil with a high content of organic matter and spores. The incubation period ranges from 21 to 106 days. This patient most likely inhaled conidia (fungal spores) from the soil while crawling underneath houses installing insulation. The physician learned that he used only a T-shirt to cover his mouth and nose, not an effective method of keeping spores from entering the respiratory tract. He should have worn a respirator.

Several critical features of this case help differentiate it from the preceding case of pneumococcal pneumonia. First, the initial macrolide antibiotic, azithromycin, should have killed most bacterial sources of infection. Second, the X-ray finding of diffuse infiltrate is more indicative of fungal lung infection than bacterial infection, which in a patient of this age would likely be confined to one lobe. The patient was young and in good health prior to the infection, making it unlikely to be pneumococcal pneumonia. The blood count was also a clue. Fungal infections do not usually cause an increase in WBCs or an increase in band cells. Finally, the **metastatic lesions** (infectious lesions that develop at a secondary site away from the initial site of infection) on the leg were in no way consistent with *S. pneumoniae*. They arise when the organism moves through the bloodstream from the primary site of infection to another body site, where it can begin to grow. Many infectious diseases start out as a localized infection but end up disseminating throughout the body to cause metastatic lesions.

NOTE: The term "metastasis" means to spread disease from one organ to another noncontiguous organ. Only microbial infections and malignant cancer cells can metastasize.

Tuberculosis as a Reemerging Disease

Until recently, tuberculosis, caused by the acid-fast bacillus *Mycobacterium tuberculosis*, was considered of passing historical significance to physicians practicing in the developed world. In 1985, however, owing primarily to the newly recognized HIV epidemic and a growing indigent population, TB resurfaced, especially in inner-city hospitals. In 1991, highly virulent multidrug-resistant (MDR) strains of *M. tuberculosis* were reported. These strains not only produced fulminant (rapid onset) and fatal disease among patients infected with HIV (the time between TB exposure to death is 2–7 months), but also proved highly infectious. Tuberculin skin test conversion rates of up to 50% are reported in exposed health care workers. A positive tuberculin skin test is seen as a delayed-type hypersensitivity reaction to *M. tuberculosis* proteins [called partially purified derivative (PPD)] injected under the skin. More information on mycobacteria can be found in Section 18.4.

M. tuberculosis primarily causes a respiratory infection, but can disseminate through the bloodstream to produce abscesses in many different organ systems. Disseminated disease is called miliary TB because the size of the infected nodules, called tubercles, approximates the size of millet seeds. The organisms are spread from person to person (no animal reservoir) through aerosolization of respiratory secretions. Once in the lung, the bacteria are phagocytized by macrophages and survive ensconced within modified phagolysosomes. A delayed-type hypersensitivity response results, and small hard tubercles form. Over time, the tubercles develop into caseous lesions that have a cheese-like consistency and can calcify into the hardened Ghon complexes seen on typical X-ray findings (**Fig. 26.7**).

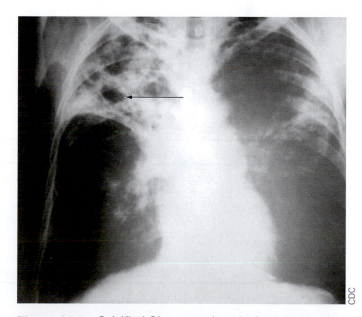

Figure 26.7   Calcified Ghon complex of tuberculosis. The arrow points to a complex in a patient's right upper lobe. Note the difference in appearance compared to Figure 26.6A.

Table 26.3 Microbes causing respiratory tract diseases.

Disease	Symptoms	Etiologic agent
Bacterial		
Inhalation anthrax	Fever, muscle aches, hypotension, respiratory failure	*Bacillus anthracis* (G+ rod)
Diphtheria	Tracheal pseudomembrane	*Corynebacterium diphtheriae* (G+ rod)
Whooping cough	Fever, runny nose, sneezing, violent cough followed by inhalation "whoop"	*Bordetella pertussis* (G– coccobacilli)
Tuberculosis	Fever, chills, cough, bloody sputum, fatigue, weight loss	*Mycobacterium tuberculosis* (acid-fast bacillus)
Pneumonia	Infects cystic fibrosis patients; fever, chills, cough, chest pain	*Pseudomonas aeruginosa* (G– rod)
Pneumonia	Fever, chills, cough, chest pain	*Streptococcus pneumoniae* (G+ diplococcus)
Psittacosis	Fever, sore throat, chest pain, runny nose	*Chlamydophila psittaci*
Pneumonia	Fever, sore throat, chest pain, runny nose (also associated with cardiovascular disease)	*Chlamydophila pneumoniae*
"Walking pneumonia"	Fever, sore throat, nonproductive cough, chills	*Mycoplasma pneumoniae* (wall-less microbe)
Legionnaires' disease (legionellosis)	Fever, chest pain, chills, cough, muscle pain, vomiting, diarrhea	*Legionella pneumophila* (G– rod)
Viral		
CMV disease	Fever, chills, cough, chest pain	Cytomegalovirus (CMV); dsDNA
RSV disease	Fever, chills, cough, chest pain	Respiratory syncytial virus; ssRNA (–)
Influenza	Fever, chills, cough, chest pain, sore throat, muscle pain	Influenza and parainfluenza viruses; ssRNA (–), segmented
	Sore throat, fever, runny nose	Adenovirus; dsDNA
Severe acute respiratory syndrome	Fever, chills, cough, chest pain	SARS virus; ssRNA (+)
Fungal		
Aspergillosis	Lung, sinuses, fever, chills, breathing difficulty	*Aspergillus* spp.
Histoplasmosis	Flu-like; pulmonary infiltrate	*Histoplasma capsulatum*
Coccidioidomycosis	Flu-like; pulmonary infiltrate	*Coccidioides immitis*
Blastomycosis	Chest pain, cough, skin lesion, pulmonary infiltrate	*Blastomyces dermatitidis*
Pneumocystosis	Infects AIDS patients; cough, fever, weight loss	*Pneumocystis jiroveci* (formerly *carinii*)

[a]Bacillus Calmette-Guérin (a weakened strain of the bovine tuberculosis strain).

Virulence properties	Source	Treatment	Vaccine
Peptide capsule, PA, LF, and EF toxins	Airborne	Ciprofloxacin	+, Military
Diphtheria toxin	Airborne; localizes to nasopharynx	Penicillin	+
Tracheal cytotoxin, adenylate cyclase toxin, filamentous hemagglutinin (adhesin)	Respiratory droplets	Erythromycin	+
Cord factor, wax D Intracellular growth	Human	Combination therapy (rifampin, isoniazid, ethambutol, pyrazinamide)	+ BCG[a]
Exotoxin A, phospholipase C, exopolysaccharide, others	Water, soil	Quinolones, aminoglycosides	
Capsule, pneumolysin	Inhalation	Macrolides, quinolones, ceftriaxone	Multivalent from capsule antigens
Obligate intracellular growth; prevents phagolysosome fusion	Bird droppings, dust; inhalation	Tetracycline, erythromycin	
Intracellular, obligate; prevents phagolysosome fusion	Person-to-person	Quinolone	
Adhesin tip	Inhalation, person-to-person	Erythromycin and other macrolides	
Intracellular growth, hemolysin, cytotoxin, protease	Inhalation	Erythromycin	
	Saliva, tears, breast milk	Gangcyclovir	Experimental
	Respiratory droplets	Treat symptoms; ribovarin	
	Respiratory droplets	Relenza	+
	Respiratory droplets	Treat symptoms	+, Military
	Respiratory droplets		
	Inhalation	Amphotericin B, voriconazole	
	Inhalation; bird, chicken, bat droppings	Amphotericin B, itraconazole	
	Inhalation	Amphotericin B, fluconazole	
	Inhalation	Amphotericin B, itraconazole	
	Inhalation	Trimethoprim sulfamethoxazole	

After a period of 4–12 weeks, patients with disease exhibit a productive cough generating sputum and experience fever, night sweats, and weight loss. They also become tuberculin test positive owing to delayed-type hypersensitivity. However, it is important to note that a positive tuberculin skin test does not signify *active* disease, only that the person was infected at one time. The bacterium may have been killed by the immune system without having caused disease. Treatment of active disease is aggressive and involves a four-drug regimen including isoniazid, rifampin, pyrazinamide, and ethambutol given over a course of several months. MDR strains are treated with a nine-drug regimen.

Viral Diseases of the Lung

Numerous viruses can cause lung infections (see **Table 26.3**). Influenza, rhinovirus, and SARS are discussed in Chapters 6, 11, and 25. An important viral lung infection not discussed elsewhere is respiratory syncytial disease, caused by **respiratory syncytial virus (RSV)**. A negative-sense, single-strand RNA, enveloped virus, RSV is the most common cause of bronchiolitis and pneumonia among infants and children under 1 year of age. Illness begins most frequently with fever, runny nose, cough, and sometimes wheezing. RSV is spread from respiratory secretions through close contact with infected persons or by contact with contaminated surfaces or objects. Infection can occur when the virus contacts mucous membranes of the eyes, mouth, or nose and possibly through the inhalation of droplets generated by a sneeze or cough. Unlike rubella or rubeola, which infect the respiratory tract and disseminate though the body, RSV remains localized in the lung.

The majority of children hospitalized for RSV infection are under 6 months of age. RSV can cause repeated infections throughout life, usually associated with moderate-to-severe cold-like symptoms. Severe lower respiratory tract disease may occur at any age, especially among the elderly or people with compromised cardiac, pulmonary, or immune systems. As yet, a vaccine to control this disease is not available.

Table 26.3 presents many other bacterial, fungal, and viral microbes that can cause respiratory tract infection. Be aware that very different diseases can produce similar symptoms. For instance, people constantly confuse influenza (the flu) with the common cold. Symptomatically, they may start out similarly, but there are telling differences. **Influenza** is characterized by fever, myalgia (muscle aches), pharyngitis (sore throat), and headache. A runny nose is *not* one of the symptoms. The common cold, however, manifests as a runny nose, nasal congestion, sneezing, and throat irritation. No myalgia. The observant clinician will note the difference.

TO SUMMARIZE:

- **The mucociliary elevator** is a primary defense mechanism used by the lung to avoid infection.
- **Pneumonia** is a disease that can be caused by many microorganisms.
- **An elevated white cell count** in blood is an indicator of bacterial infection.
- **Pneumococcal vaccine** should be administered to the elderly because they are often immunocompromised.
- **Fungal agents** commonly cause long-term, chronic infections.
- **Tuberculosis** is an ancient yet reemerging bacterial disease with an increasing mortality rate caused by the development of multidrug-resistant strains, the susceptibility of HIV patients, and an increasing indigent population.
- **Localized infections in the lung can disseminate** via the bloodstream to form metastatic lesions at other body sites.
- **Respiratory syncytial virus** is one of several viruses that can cause lung disease but rarely disseminates.

26.4 Gastrointestinal Tract Infections

Nearly everyone has experienced diarrhea—a condition characterized by frequent loose bowel movements accompanied by abdominal cramps. Hundreds of millions of cases occur each year in the United States. The loose stools usually result from inflammation (called gastroenteritis) due to viral growth, bacterial growth, or toxin production, causing large amounts of water and electrolytes to leave the intestinal cells and enter the intestinal lumen. As a result, the patient not only suffers diarrhea, but can become dangerously dehydrated. As with respiratory tract infections, most diarrheal disease is viral in origin, with rotavirus being the primary culprit. Among the bacteria, the gram-negative curved bacillus *Campylobacter* is the most frequent cause of self-limiting diarrheal disease.

The main symptoms of gastroenteritis, whether bacterial or viral, are watery diarrhea and vomiting. A more severe form of gastroenteritis is called **dysentery**, whose symptoms include abdominal pain, a persistent desire to empty the bowels, and diarrhea with passage of blood or mucus. Bacterial causes of dysentery include several *Shigella* species and some strains of *E. coli*. An ameba, *Entamoeba histolytica*, can also cause a form of the disease called amebic dysentery. Features of bacterial dysentery are discussed in more detail later in this section.

Remarkably, *Staphylococcus aureus* causes gastrointestinal disease without ever producing infection. Everyone has heard of the local church picnic where scores of people became violently ill within hours of eating unrefrigerated potato salad. *S. aureus* is the usual cause of these disasters. Not all strains of *S. aureus* cause food poisoning, but certain strains can secrete enterotoxins into tainted foods such as pies, turkey dressing, or potato salad. After ingestion, the toxin travels to the intestine, where it enters the bloodstream and stimulates nerves leading to the vomit center in the brain. Because the toxin is preformed, symptoms occur quickly after ingestion. Within 2–6 hours, the poisoned patient will begin vomiting and may also experience diarrhea. The disease, though violent, is not life-threatening and usually resolves spontaneously within 24–48 hours. In contrast, diarrhea caused by infectious agents, such as *Salmonella enterica*, that must first grow in the victim do not occur until 12–24 hours after ingestion, sometimes longer. A clinician noting quick onset of symptoms in a patient will immediately suspect staphylococcal food poisoning. Obviously, antibiotic treatment is not needed for staph food poisoning but may be indicated for other gastrointestinal infections.

Antibiotics Are Often Inappropriate When Treating Gastroenteritis

Although antibiotic treatment of infectious gastroenteritis seems intuitive, it is rarely used. Most gastrointestinal infections are viral, and so antibiotics are ineffective. Likewise, gastroenteritis caused by bacteria usually resolves spontaneously without antibiotic treatment. Of course, severe systemic disease stemming from gastroenteritis can develop, often in the young or elderly. Diseases such as typhoid fever (*Salmonella typhi*) or bacillary dysentery (*Shigella dysenteriae*) will respond well to antibiotics.

In some cases, antibiotic treatment can actually *trigger* gastrointestinal disease. For example, the antibiotic clindamycin can kill most normal intestinal flora except the naturally resistant gram-positive anaerobe *Clostridium difficile*, the causative agent of pseudomembranous enterocolitis. Unrestrained by microbial competition, *C. difficile* will grow in the intestine and produce specific toxins that can damage intestinal cells. The organism's growth leads to inflammation and the formation of pseudomembrane structures along the intestinal wall (refer to **Fig. 23.8**). Because they block the intestinal mucosa, pseudomembranes cause malabsorption of nutrients and water, which results in diarrhea. As the pseudomembrane enlarges, it begins to slough off and pass into the stool. Diagnosis of this disease involves PCR identification of the organism or immunological identification of the toxin in fecal samples.

Case History: Diarrhea and Dysentery

In April, a 6-year-old girl from Montgomery County, Pennsylvania, arrived at the ER with bloody diarrhea, a temperature of 39°C, abdominal cramping, and vomiting. Hospital admission was five days after a kindergarten field trip to the local dairy farm. The child's health history was otherwise unremarkable. At the time of hospital admission, the parents were questioned as to the child's activities during the trip. They confirmed that the child purchased a snack while at the farm. Upon laboratory analysis, a fecal smear was positive for leukocytes, and isolation of organisms confirmed the presence of gram-negative rods that produced Shiga toxins 1 and 2. Subsequent testing of the isolate by pulsed-field gel electrophoresis was indistinguishable from E. coli O157:H7. By this time, the child developed further problems. Symptoms included puffy face and hands as well as neurological abnormalities. Initial examination suggested renal failure, and laboratory analyses supported this diagnosis with thrombocytopenia (reduced blood platelet count) and hemolytic uremic syndrome (HUS; renal failure). The child was treated by fluid and electrolyte replacement (intravenous). Antibiotics were not administered.

In this case history, the presence of leukocytes in a fecal smear is a sign that the intestinal pathogen has invaded the epithelial mucosa of the intestine. Breaching this barrier sends out a chemical call to neutrophils, which then enter the area. *Shigella dysenteriae, Salmonella enterica*, and enteroinvasive *E. coli* (EIEC) are invasive; because they actually enter enterocytes, they are considered intracellular pathogens. Enterohemorrhagic *E. coli* (EHEC), which also produces leukocytes and blood in stools, is not an intracellular parasite (it is not invasive), but causes damaging attachment and effacing lesions, described in Section 25.2, that causes destruction of the mucosal epithelium. The resulting inflammation, in conjunction with damage caused by Shiga toxin, leads to blood and white cells in the stool. *E. coli* O157:H7, the etiological agent in the case history, is a common serotype of EHEC.

There are at least six different classes of pathogenic *E. coli* that differ based on their repertoire of pathogenicity islands, plasmids, and virulence factors. They include EIEC and EHEC, already mentioned, enterotoxigenic *E. coli* (ETEC) and uropathogenic *E. coli*, (UPEC), both described in Chapter 25, as well as enteropathogenic *E. coli* (EPEC) and enteroaggregative *E. coli* (EAEC). All but UPEC cause gastrointestinal disease. To tell them apart, each group has telltale O and H antigens that can be identified using serology.

NOTE: "O antigen" is part of the bacterium's LPS, while "H antigen" is flagellar protein. Thus, O157:H7 denotes the specific version of LPS (O157) and flagellar protein (H7) found on *E. coli* O157:H7. Other pathogenic strains of *E. coli* have different O and H antigens.

Shiga toxin. *Shigella* and EHEC, the agent in the preceding case history, both produce toxins called Shiga toxin 1 and 2 that are encoded by genes of bacteriophage genomes embedded in the bacterial chromosome. These toxins inhibit host protein synthesis and, in the process, damage endothelial cells in the kidney and brain. Endothelial damage triggers the formation of platelet-fibrin microthrombi (clots) that occlude blood vessels in the various organs, leading to two major syndromes—hemolytic uremic syndrome (HUS) and thrombotic thrombocytopenic purpura (TPP). HUS occurs when the microthrombi are limited to the kidney. The microclots clog the tiny blood vessels in this organ and cause decreased urine output, ultimately leading to kidney failure and death. In TPP, the clots occur throughout the circulation. This will cause reddish skin hemorrhages called petichiae and purpura. Neurological symptoms (for example, confusion, severe headaches, possibly coma) then arise from microhemorrhages in the brain. The hemorrhaging occurs because platelets needed for normal clotting have been removed from the circulation as they form the microthrombi. The decreased number of platelets is called thrombocytopenia.

The toxins, which are absorbed through the intestine and disseminated via the bloodstream, have five B subunits used to bind to target cell membranes and one A subunit imbued with toxic activity (see Section 25.4). The A subunit, upon entry, destroys protein synthesis by cleaving an adenine from 28S rRNA in eukaryotic ribosomes.

Enterohemorrhagic *E. coli* (EHEC). *E. coli* O157:H7 is a recently emerged pathogen that can colonize cattle intestines without causing disease and as a result can contaminate meat products following slaughter. The organism is sometimes referred to as the "Jack-in-the Box" microbe, a reference to the first documented large-scale U.S. outbreak in 1993, linked to fast-food hamburgers purchased at a Washington state Jack-in-the-Box restaurant. As many as 600 people were sickened in that outbreak, and 3 children died. Though 1993 marked the first large-scale outbreak in the United States, the first report of the disease in this country occurred 11 years earlier, in 1982.

E. coli O157:H7 rarely affects the health of the reservoir animal. But when an infected steer is slaughtered, fecal contamination of the carcass can happen, despite manufacturers' considerable efforts to prevent it. Grinding the tainted meat into hamburger will distribute the microbe throughout. As a result, cooking burgers to 160°C is essential to kill any existing EHEC. Also be aware that cross-contamination between foods is possible. Using the same cutting board to prepare meat and salad is a great way to contaminate the salad, which will not be cooked.

Despite its common association with hamburger, vegetarians are not safe from this organism. During heavy rains, waste from a cattle farm can easily wash into nearby vegetable fields unless precautions are taken. If the cattle waste contains *E. coli* O157:H7, the crops are contaminated. One such outbreak occurred in 2006, when spinach from certain areas of California became contami-

nated with this pathogen, prompting a nationwide recall of bagged spinach and a month without spinach salad.

Early on, the remarkably low infectious dose of *E. coli* O157:H7 mystified researchers. However, we have since learned that *E. coli* has an impressive level of acid resistance that rivals that of the gastric pathogen *Helicobacter pylori*. Acid resistance mechanisms permit survival of *E. coli* in the acidic stomach and enable a mere 10–100 individual organisms to cause disease.

As already noted, EHEC strains produce two toxins that are identical to the Shiga toxins produced by *Shigella* species. The toxins cleave host ribosomal RNA, thereby halting translation. As discussed earlier, one consequence of Shiga toxin is HUS. The development of HUS, as in the case history described, is a common consequence of *E. coli* O157:H7 infection. Unfortunately, HUS can only be treated with supportive care, such as blood transfusions and dialysis throughout the critical period until kidney function resumes. Antibiotic treatment can increase the release of Shiga toxins from the organisms and actually trigger HUS. Thus, antimicrobial therapy is not recommended.

In contrast to the case just described, many gastrointestinal infections do not produce fecal leukocytes or blood in the stool. Diarrheal diseases caused by *V. cholerae* (cholera) or enterotoxigenic *E. coli* (ETEC), which produces a cholera-like disease, do not involve invasion of the intestinal lining by the microbe and yield copious amounts of watery diarrhea. In these two toxin-driven diseases, the bacteria attach to cells lining the intestine and secrete toxins that are imported into the target cells (see Section 25.4).

Epidemiology of EHEC. Epidemiology is the study of factors and mechanisms involved in the spread of disease. The father of epidemiology was the British physician John Snow, who, even before there was a clear connection between bacteria and disease, managed to trace the source of a cholera epidemic in London in 1854 to a single well in the city. He found that the addresses of all the cholera patients clustered around just one of the city's numerous water pumps that had, as it turns out, been contaminated with human feces. Sealing the well ended the epidemic.

The U.S. Centers for Disease Control (CDC) used the same basic strategy to identify the risk factors associated with the case study presented earlier. Fifty-one infected patients and 92 controls (children who visited the farm but did not become ill) were interviewed. Infected patients were more likely than controls to have had contact with cattle, an important reservoir for *E. coli* O157:H7. All 216 cattle on the farm were sampled by rectal swab, and 13% yielded *E. coli* O157:H7 with a DNA restriction pattern indistinguishable from that isolated from the patients. This finding indicated that the cattle were the source of infection. Activities that promoted hand-mouth contact, such as nail-biting and purchasing food from an outdoor concession were more common among the children that contracted disease (fecal-oral route of infection). Furthermore, separate areas were not established for eating and interactions with farm animals. Visitors could touch cattle, calves, sheep, goats, llamas, chickens, and a pig while eating and drinking. Hand-washing facilities were unsupervised and lacked soap, and disposable hand towels were out of the children's reach. All of these situations provided opportunity for infection.

Type III secretion and diarrhea. Type III secretion was first described in Section 25.5, where we discussed the pathogenesis of *Salmonella enterica*, but these secretion systems are present in numerous gram-negative pathogens, such as EHEC in our case history. Recall that type III protein secretion systems directly inject proteins from the cytoplasm of a bacterial pathogen into the cytoplasm of a target eukaryotic host cell. The system delivers proteins across three membranes—two for the gram-negative bacterial pathogen and one for the target cell.

In addition to stimulating bacterial entry into host cells, bacterial proteins injected by type III transport systems cause host cells to secrete pro-inflammatory cytokines. The cytokines then "call in" inflammatory cells and alter ion transport through the epithelial membrane. Excessive export of ions such as chloride causes water to leave the cell in an attempt to equilibrate the internal and external ionic concentrations. The water entering the intestine results in diarrhea. We are just beginning to understand how the type III–translocated effector proteins induce these host cellular responses and how this leads to intestinal inflammation during an infection.

Although most gastrointestinal disease is intestinal in locale, specialized microbes can also target the stomach, with its harsh acidic environment.

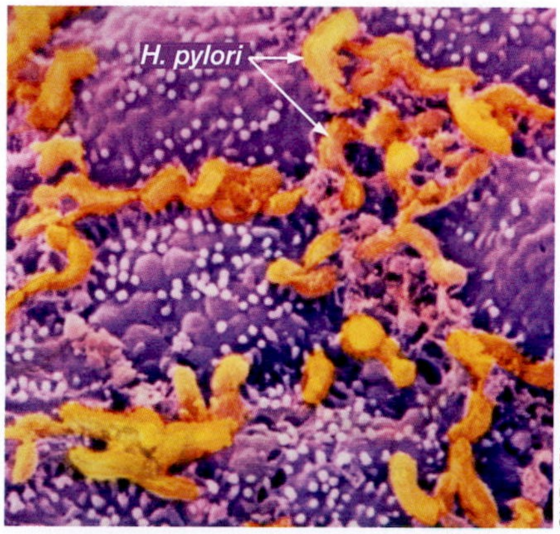

Case History: Ulcers—It's Not What You Eat

Gary is a 34-year-old accountant who immigrated to Nebraska from Poland seven years ago. Since his teenage years, he has been bothered periodically by episodes of epigastric pain (pain around the stomach), nausea, and heartburn. Antacids usually alleviated the symptoms. Over the years, he received several courses of treatment with Tagamet or Pepcid to reduce acid secretion and provide relief. Recently, an upper-GI endoscopy was performed, in which a long, thin tube tipped with a camera and light source was inserted into Gary's mouth and into his stomach. The view through the endoscope showed some reddened areas in the antrum (bottom part) of the stomach. The endoscope was also equipped with a small clawlike structure that obtained a small tissue sample from the lining of Gary's stomach. A urease test performed on the antral biopsy turned positive in 20 minutes. Histological examination of the biopsy confirmed moderate chronic active gastritis (inflammation of the stomach lining) and revealed the presence of numerous spiral-shaped organisms. Cultures of the antral biopsy were positive for H. pylori.

Painful and sometimes life-threatening gastric ulcers were for many years blamed on spicy foods and stress. These factors were believed to cause increased acid production that ate away at the stomach lining, even though the gastric mucosa is normally well protected from stomach acid, which can fall as low as pH 1.5. This argued against the model but was ignored. In the 1980s, based on their discovery of odd, helical-shaped bacteria present in the biopsies of gastric ulcers, Australians Robin Warren and Barry Marshall (a medical intern at the Royal Perth Hospital at the time; **Fig. 26.8A**) proposed that bacteria, not pepperoni, caused ulcers (**Fig. 26.8B**). Their hypothesis was viewed with skepticism and declared as heresy by the established medical community. Faced with disbelief bordering on ridicule, the young intern drank a vial of the helical organisms and waited. A week later he began vomiting and suffered other painful symptoms of gastritis. Barry Marshall could not have been happier. He had proved his point. We now know that this curly-shaped microbe causes the vast majority of stomach ulcers.

The discovery of *Helicobacter pylori* and its association with gastric ulcer disease led to an upheaval

A.

B.

Figure 26.8 A bacterial cause of gastric ulcers. A. Australian physician Barry Marshall was so sure he was right about the cause of stomach ulcers, he swallowed bacteria to prove his point. **B.** View of *H. pylori* in stomach crypts (colorized SEM). Bacteria are approximately 2 μm in length.

in gastroenterology. Prior to this discovery, treatment focused on suppressing acid production, which did not provide long-term relief. Within one year after acid-suppressive therapy, up to 80% of patients suffer a relapse of their ulcer. Therapy now includes antimicrobial treatment to kill the bacteria and acid suppression therapy to prevent further inflammation while the ulcer heals. Warren and Marshall, who recovered from his gastritis, received the 2005 Nobel Prize in Physiology or Medicine for their groundbreaking work.

The exact mechanism by which the organism causes gastric ulcers is not known, although a variety of virulence factors have been described (see Section 25.6). The urease enzyme of *H. pylori* is a characteristic virulence feature. Urease converts urea, produced by the gastric lining, to ammonia and carbon dioxide. The ammonia neutralizes acid in and around the organism and allows the microbe to survive in the stomach. In an animal model, the urease enzyme of *H. pylori* was found to be important for bacterial colonization in the stomach. Treatment with urease inhibitors such as acetohydroxamic acid, however, does not lead to eradication of *H. pylori*. This failure is likely due to protection afforded the organism once it reaches the mucous layer that blankets the gastric epithelium (see **Fig. 26.8B**). The pH of this environment is closer to neutral, so urease may no longer be needed. Tools useful for diagnosing *H. pylori* include histology, rapid urease testing, and serology (for example, enzyme-linked immunosorbent assay (ELISA) to detect antibody to the CagA antigen).

Enzyme-linked immunosorbent assay (ELISA). ELISA is a common immunological tool to detect the presence, in serum, of antibodies to a specific organism—an indication of infection. The assay is performed by coating wells in a plastic dish with an antigen (for example, *Helicobacter* CagA). Serum from the patient is then added to the well. If antibodies to CagA are present, they will bind to the CagA antigen. Unbound antibody is removed by washing, and a secondary antibody that binds human IgG is added to the well. These anti-antibodies have an enzyme linked to them. A sandwich is formed as follows: [plastic dish]–[CagA protein]–[anti-CagA antibody]–[anti-IgG antibody]–[enzyme]. When the appropriate enzyme substrate is added to the well, the enzyme acts on it to produce light or a chromogenic product that can be detected. The more anti-CagA antibody present in the serum, the more light or product is produced by the linked enzyme. Applications of ELISA and further discussion of the methodology involved are described in Section 28.2.

***H. pylori* infection, cancer, and bad breath.** Problems caused by *H. pylori* are not limited to gastric ulcers. The microbe has also been associated with gastric cancer. The evidence is not yet conclusive, but many individuals with gastric cancer are also colonized by *H. pylori*. The most compelling experimental proof is that gerbils infected with this organism develop gastric cancer. In addition, when the CagA protein of *H. pylori* is injected into gastric epithelial cells, it is phosphorylated on a tyrosine residue and activates a regulatory cascade that causes the gastric cell to proliferate (**Fig. 25.33**). The connection between *H. pylori* and gastric cancer is sobering when you consider that *H. pylori* can be detected in about two-thirds of the world's population, especially in impoverished countries.

There is also the unusual case of a patient with a 60-year history of halitosis (extreme bad breath) that was resistant to all standard therapies but was cured by a triple-drug therapy regimen used for *H. pylori*. This case is an illustration of how microorganisms can cooperate to cause disease. *H. pylori* did not cause the smell. It appears that the patient did not produce much stomach acid. As a result, the urease produced by *H. pylori* easily neutralized the remaining stomach acid, allowing anaerobes ingested with food to start putrefying the stomach contents, leading to the foul-smelling breath. Other examples of disease caused by microbial cooperation will be described later.

Rotavirus Is the Single Greatest Cause of Gastroenteritis

Many people wrongly think that most cases of diarrhea are caused by a bacterial agent. Actually, a virus, **rotavirus**, causes more intestinal disease than any bacterial species. It is highly infectious; it is endemic around the globe; and it affects all age-groups, although children between 6 and 24 months are most severely affected. Spread by the fecal-oral route, it is estimated that by age 3, all children have had a rotavirus infection.

The incubation period is approximately two days, after which the victim suffers frequent watery, dark green, explosive diarrhea. All of this may be accompanied by nausea, vomiting, and abdominal cramping. Severe dehydration and electrolyte loss due to the diarrhea will cause death unless supportive measures, such as fluid replacement, are undertaken. There is no cure, but most patients recover if rehydrated properly. Few deaths from rotavirus occur in the United States, but each year more than 600,000 children worldwide die from this viral diarrhea, which is why efforts to develop a safe, effective vaccine are so important. Success in this endeavor seems close at hand.

Protists (Protozoa) Are Another Major Cause of Diarrheal Diseases

Most students of biology are familiar with protozoa (more properly called protists) such as paramecia and amebas, but many are surprised to learn that some species of protists cause serious human diseases. For instance,

Figure 26.9 *Giardia lamblia.* This protist is a major cause of diarrhea in the world. **A.** Cysts (7 to 14 μm) present in fecal matter (colorized SEM). **B.** Trophozoite form (colorized SEM, 5 to 15 μm in length).

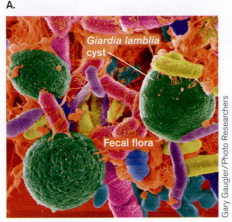

A.

Giardia lamblia cyst

Fecal flora

Gary Gaugler/Photo Researchers

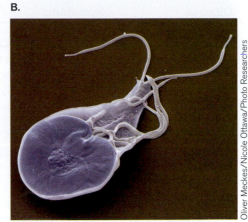

B.

Oliver Meckes/Nicole Ottawa/Photo Researchers

Entamoeba histolytica and *Cryptosporidium parvum* cause the diarrheal diseases amebic dysentery and cryptosporidiosis, respectively. In 2005, the CDC tallied 2,640 cases of cryptosporidiosis, a reportable disease in the United States. Two other amebas, *Nagleria* and *Acanthamoeba*, cause amebic menigoencephalitis. Because some species of *Acanthamoeba* can infect the eye, soft contact wearers should take precautions to prevent contamination of their lenses.

The flagellated protozoan *Giardia lamblia* is a major cause of diarrhea throughout the world. In the United States alone, *G. lamblia* was responsible for over 11,000 reported cases of giardiasis diarrhea in 2005 and likely caused thousands more that were not reported. *Giardia lamblia* enters a human or other host as a cyst present in drinking water contaminated by feces (**Fig. 26.9A**). Aside from humans, *G. lamblia* can be found in various rodents, deer, cattle, and even household pets. It is very infectious. Ingestion of as few as 25 cysts can lead to disease. Following ingestion, the hard, outer coating of the cyst is dissolved by the action of digestive juices to produce a trophozoite, which attaches itself to the wall of the small intestines and reproduces (**Fig. 26.9B**). Offspring quickly encyst and are excreted out of the host's body.

Asymptomatic carriers of *G. lamblia* are common—it has been estimated that anywhere from 1 to 30% of children in U.S. day-care centers are carriers. Disease usually manifests as greasy stools alternating between a watery diarrhea, loose stools, and constipation. However, some patients will experience explosive diarrhea. Diagnosis is usually done by observing the cysts or trophozoite forms of the protozoan in feces. Metronidazole is a drug often used to cure the disease. As for prevention, proper treatment of community water supplies is essential.

We have examined only a handful of the microbes that cause gastrointestinal infection. Others are presented in **Table 26.4**.

TO SUMMARIZE:

- **Diarrhea** leads to dehydration, for which fluid replacement is a critical treatment. Antibiotic treatment is usually not recommended.
- **Staphylococcal food poisoning** is not an infection. It is a toxigenic disease.
- **Antibiotic treatments** can sometimes cause gastrointestinal disease (for example, pseudomembranous enterocolitis by *Clostridium difficile*).
- **Bacteria that invade intestinal epithelial mucosal cells** lead to the presence of blood and white blood cells in fecal contents. This occurs with intracellular pathogens such as *Shigella*, *Salmonella*, and EIEC.
- **Bacteria that do not invade intestinal cells** usually produce watery diarrhea. EHEC is an exception because the attachment and effacing lesions it produces result in bloody stools.
- **Bacterial toxins** produced by bacterial enteric pathogens can cause systemic symptoms.
- **John Snow** founded the science of epidemiology while studying a cholera outbreak in London.
- **The bacterium *H. pylori***, a common cause of gastric ulcers, lives in the stomach and is highly acid resistant.
- **Rotavirus** is the single greatest cause of diarrhea worldwide.
- *Giardia lamblia* is a major protozoan cause of diarrhea worldwide.

26.5 Genitourinary Tract Infections

Although the genital and urinary tracts are different organ systems, their close association in the body has led them to be grouped together when discussing infections.

Table 26.4 Microbes causing diseases of the gastrointestinal tract.

Disease	Symptoms	Etiologic agent	Virulence factors	Source	Treatment
Bacterial					
Staphylococcal food poisoning	Symptoms within 4 h of ingestion; nausea, vomiting, diarrhea	*Staphylococcus. aureus* (G+)	Enterotoxin	Preformed toxin in foods	Supportive
Botulism	Symptoms begin quickly; flaccid paralysis	*Clostridium botulinum* (G+, anaerobe)	Neurotoxin	Preformed toxin in food	Antiserum
Salmonellosis	Symptoms after 18 h; abdominal pain, diarrhea; invade intestinal M cells	*Salmonella enterica* [gram negative (G–)]	Type III secretion, intracellular growth	Chickens, other animals; fecal-oral	Oral rehydration; antibiotics if severe
Typhoid fever	Headache, fever, chills, abdominal pain, rash (rose spots), hypotension, diarrhea in late stages	*Salmonella typhi* (G–)	Type III secretion, intracellular growth, PhoPQ regulators, Vi antigen capsule	Human carriers (gall bladder reservoir), food, water	Quinolones
Traveler's diarrhea	Watery diarrhea	Enterotoxigenic *Escherichia coli*	Labile and stable toxins	Humans; food, water	Oral rehydration
Gastroenteritis	Bloody diarrhea, HUS	Enterohemorrhagic *E. coli*	Intimin, Tir, type III secretion, Shiga toxin	Contaminated foods (hamburger) and crops	Oral rehydration; antibiotics if severe
Shigellosis	Bloody diarrhea, HUS	*Shigella* spp. (G–)	Shiga toxin, type III secretion, intracellular growth, actin-based motility, escape phagosome	Human fecal-oral route	Oral rehydration; antibiotics if severe
Cholera	Watery diarrhea	*Vibrio cholerae* (G–)	Cholera toxin, TCP pili, ToxR regulator	Human waste-contaminated water	Oral rehydration, antibiotics
Gastroenteritis	Diarrhea, blood in stool	*V. parahaemolyticus* (G–)	Enterotoxin	Raw seafood	Self-limiting
Gastroenteritis	Watery diarrhea, nausea	*Clostridium perfringens* (G+)	Alpha toxin	Soil, food	Self-limiting
Pseudomembranous enterocolitis	Fever, abdominal pain, diarrhea, pseudomembrane in colon	*C. difficile* (G+)	Cytotoxin, antibiotic resistance	Animals, normal flora	Vancomycin
Gastroenteritis	Fever, muscle pain, watery diarrhea, blood in stool, headache	*Campylobacter jejuni* (G–)	Cytotoxin, enterotoxin, adhesin	Poultry, unpasteurized milk	Erythromycin
Gastric ulcers	Abdominal pain, bleeding, heartburn	*Helicobacter pylori* (G–)	Adhesin, urease CagA, vacuolating toxin	??	Triple drug (omeprazole, clarithromycin, metronidazole)
Viral					
Stomach "flu"	Nausea, vomiting, diarrhea (Most common cause)	Noroviruses (Norwalk virus)	??	Fecal-oral route	Oral rehydration
		Rotavirus	??	Fecal-oral	Oral rehydration

Urinary Tract Infections

The urinary tract includes the kidneys, ureters, urinary bladder, and urethra. Infections anywhere along this route are called urinary tract infections (UTIs). Urinary tract infections are the second most common type of bacterial infection in humans, ranking in frequency just behind respiratory infections such as bronchitis or pneumonia. Along with numerous office visits, bladder infections and other urinary tract infections result in 100,000 hospital admissions and $1.6 billion in medical expenses each year. Most sufferers are women. One in five women will get a urinary tract infection at some point in her life, and 20–40% of those infected will develop recurrent infections.

Urine, as produced in the kidneys and stored in the bladder, is normally sterile. Microorganisms must be introduced into the bladder to cause infection. Active infection of the urinary tract occurs in one of three basic ways:

- **Descending infection from the kidney.** Descending infection occurs when an infected kidney sheds bacteria that descend via the ureters into the bladder. Kidney infections arise when microorganisms are deposited in the kidneys from the bloodstream.
- **Ascending infection to the kidney.** This is when an established infection in the bladder ascends along the ureter to infect the kidney.
- **Infection from the urethra to the bladder.** Bacteria residing along the superficial urogenital membranes of the urethra can ascend to the bladder. This is more common in women than men. Microorganisms can also be introduced into the bladder by means of mechanical devices such as catheters or cystoscopes that are passed through the urethra into the bladder.

> **THOUGHT QUESTION 26.2** Why do you think most urinary tract infections occur in women?

Urine is bacteriostatic to most of the commensal organisms inhabiting the perineum and vagina, such as *Lactobacillus*, *Corynebacterium*, diphtheroids, and *Staphylococcus epidermidis*. In contrast, many gram-negative organisms thrive in urine. As a result, most urinary tract infections are caused by facultative gram-negative rods from the GI tract. The most common etiological agents of UTI are:

- Certain serotypes of *E. coli* that comprise the uropathogenic *E. coli* (75% of all UTI)
- *Klebsiella*, *Proteus*, *Pseudomonas*, *Enterobacter* (20%)
- *S. aureus*, *Enterococcus*, *Chlamydia*, fungi, *Staphylococcus saprophiticus*, other (5%)

Case History: Classic Urinary Tract Infection

Lashandra is 24 years old and has been experiencing back pain, increased frequency of urination, and dysuria (painful urination) *over the past three days. She consulted her general practitioner, who requested that a mid-stream specimen of urine be examined. This was the first time Lashandra had ever suffered from persisting dysuria. Upon microscopic examination, the urine was found to contain more than 50 leukocytes per µl and 35 red blood cells per µl. No epithelial squamous cells were seen. The urine culture plated on agar medium yielded more than 10^5 colonies per milliliter (meaning more than 10^5 organisms per milliliter in the urine) of a facultative anaerobic gram-negative bacillus capable of fermenting lactose.*

The first question to ask in this case is whether the patient had a significant UTI. The purpose of the midstream urine collection is to provide laboratory data to make this determination. Even though urine in the bladder is normally sterile, urine becomes contaminated with normal flora that have adhered to the urethral wall. In a mid-stream collection, the patient urinates briefly, stops to position a collection jar, and resumes urinating to collect a midstream sample. This minimizes the number of organisms in the sample by washing away organisms clinging to the urethra before actually collecting the sample. Nevertheless, the collected sample will still contain low numbers of organisms representing normal flora of the urethra. A diagnosis of UTI is made when the number of bacteria in the sample becomes greater than 10^5/ml. (However, that number can be as low as 1,000/ml in symptomatic women.) The patient in the case history is symptomatic and has sufficient numbers of bacteria in her urine to indicate a UTI.

The laboratory found the organism to be a gram-negative bacillus that ferments lactose, suggesting *E. coli* as the likely culprit. Given that the normal habitat of *E. coli* is the gastrointestinal tract and this is the first UTI suffered by the patient, the infection is likely the result of an inadvertent introduction of the microbe into the urethra. The organism makes its way up the urethra and into the bladder. If the patient had a recurring UTI, it could result from organisms entering the bladder from a descending route along the ureter from an infected kidney (also see **Special Topic 26.1**).

Gram-negative rods not only thrive in urine, but may be adapted to cause urinary tract infections through specialized pili. These pili have terminal receptors for glycolipids and glycoproteins present on urinary tract epithelial cells (**Fig. 26.10A**). Uropathogenic strains of *E. coli*, for example, typically have P-type pili, with a terminal receptor for the P antigen (a rather appropriate name for a bladder-specific virulence factor). The P antigen is a blood group marker found on the surface of cells lining the perineum and urinary tract; it is expressed by approximately 75% of the population. These individuals are particularly susceptible to UTIs. The P antigen can also be shed and found in vaginal and prostate secretions. When this happens, the secreted P antigens can be protective by acting as a decoy.

Doctors used to think that women suffering from recurrent bladder infections were repeatedly infected as a result of sexual activity or poor hygiene, a presumption many women with chronic infections found frustrating and offensive. New evidence indicates that bacteria may actually hide inside bladder cells, forming a reservoir for reinfection.

Specialized colonies of *E. coli* have been found living in the surface layer of cells lining the bladders of infected mice (**Fig. 1A**). Though the mouse served as a model system, the bacteria used were originally isolated from humans. These strains of *E. coli* cause about 80% of all urinary tract infections. The bacteria use pili to latch onto proteins coating the bladder cells. Those proteins form a protective substance, known as uroplakin, that strengthens the epithelial cells and shields them from toxins that may build up in the urine (**Fig. 1B**). The bacteria can slip under the protective barrier and penetrate superficial cells lining the bladder. Once inside the cell, the microbes begin to multiply, and the epithelial cell fills with bacteria also coated with uroplakin, taking on the appearance of a pod (**Fig. 1C**).

The pod-like *E. coli* biofilms we describe here are notoriously resistant to antibiotic treatment and attacks by the immune system. Living inside host cells and surrounded by uroplakin, the *E. coli* in the bladder pods have devised a clever mechanism for surviving. Bacteria living on the edges of the biofilm may break free, leading to subsequent rounds of infection. Understanding the cycle of infection and pod formation may help researchers design drugs to prevent new urinary tract infections or stop established ones.

Although pods have so far been found only in mice, researchers are tracking women with persistent infections to see whether pods are also present in humans. Biofilms hiding inside cells elsewhere in the body also may act as reservoirs for other hard-to-treat chronic or recurring infections, such as ear infections.

Source: G. G. Anderson, J. J. Patermo, J. D. Schilling, R. Roth, J. Heuser, and S. J. Hultgren. 2003. Intracellular bacterial biofilm-like pods in urinary tract infections. *Science* 301:105–107.

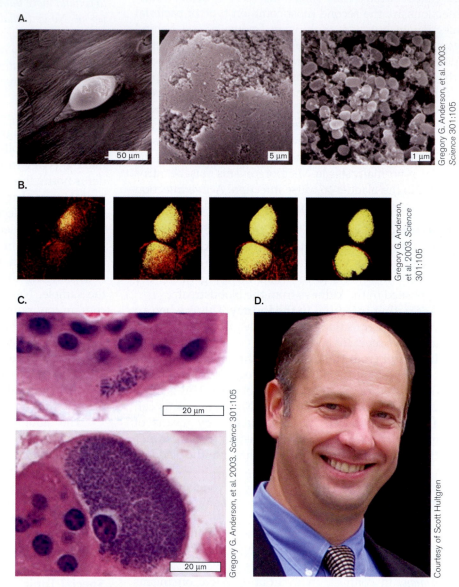

Figure 1   Bladder pods of uropathogenic *E. coli*.   Scott Hultgren's laboratory (Washington University, St. Louis) has discovered that intracellular bacterial communities extend like pods into the bladder lumen. **A.** Increasing magnifications of large intracellular communities of uropathogenic *E. coli* (UTI89) inside pods on the surface of a mouse bladder infected for 24 hours (SEM). **B.** Confocal photos of a whole-mounted bladder infected with UTI89 expressing green fluorescent protein from the plasmid pcomGFP. Uroplakin on the surface of the two pods is revealed by treatment with antibody to uroplakin (primary antibody) and tetramethyl rhodamine isothiocyanate–labeled secondary antibody (red). The series starts on the left with the lumenal surface and progresses toward the right in sections that move downward through the epithelium. Optical section thickness, 1 μm. **C.** Hematoxylin- and eosion-stained sections of UTI89-infected mouse bladders show a bacterial factory 6 hours after inoculation (top panel) and a pod 24 hours after inoculation (bottom panel). Bacteria in the pod were densely packed and shorter and completely filled the host cell. **D.** Scott Hultgren.

A.

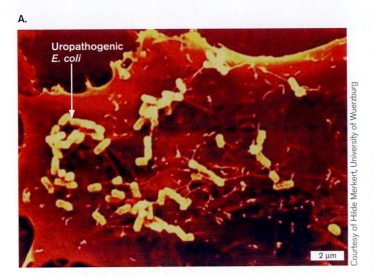

Uropathogenic
E. coli

2 μm

Courtesy of Hilde Merkert, University of Wuerzburg

B.

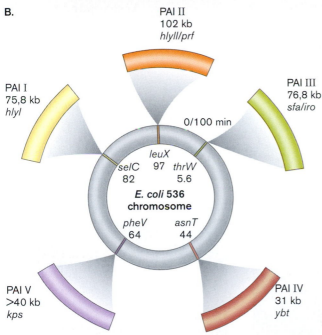

PAI II
102 kb
hlyII/prf

PAI I
75,8 kb
hlyI

PAI III
76,8 kb
sfa/iro

0/100 min

leuX 97

selC *thrW*
82 5.6

**E. coli 536
chromosome**

pheV *asnT*
64 44

PAI V
>40 kb
kps

PAI IV
31 kb
ybt

Figure 26.10   Uropathogenic *E. coli*.   A. Bladder cell
with adherent uropathogenic *E. coli* (SEM). Bacteria are approx.
1 μm in length. **B.** Distribution of pathogenicity islands (PAI) in
uropathogenic *E. coli*. The location of each insert is given in map
units within the circle representing the genome. The 0–100 map
units are called centisomes. Each centisome is approximately
44 kb of DNA. 0 is arbitrarily placed at the *thr* (threonine) gene.
The origin of replication on this map is near 82 centisomes. A
chromosomal gene flanking the insert is also provided. The size
of each island is provided above the insert. A key virulence gene
for each island is listed.

They bind to the bacterial receptor, preventing binding of
the organism to the surface epithelium. Individuals most
susceptible to UTIs are those who express P antigen on
their cells but lack P antigen in their secretions.

Urinary tract infections are among those most frequently acquired during a hospital stay (so-called noso-

comial infections). In these cases, the causal organism is
less likely to be *E. coli* and more likely to be another gram-
negative bacterium or *Staphylococcus*. Many UTIs resolve
spontaneously, but others can progress to destroy the kidney or, via gram-negative septicemia, the host. As a result,
antibiotic therapy is recommended. In older patients, UTIs
frequently show atypical symptoms, including delirium,
which disappears when the UTI is treated.

> **THOUGHT QUESTION 26.3**   Urine samples col-
> lected from six hospital patients were placed on a table
> at the nurses' station awaiting pickup from the micro-
> biology lab. Several hours later, a courier retrieved the
> samples and transported them to the lab. The next day,
> the lab reported that four of the six patients had UTI.
> Would you consider these results reliable? Would you
> start treatment based on these results?

What makes uropathogenic *E. coli* different from other
E. coli? This is a question still under investigation, but
genomic analysis has exposed five pathogenicity islands
unique to these strains (**Fig. 26.10B**). The functions of
these pathogenicity islands are still under investigation.

Sexually Transmitted Diseases

Sexually transmitted diseases (STDs) are defined as
infections transmitted primarily through sexual contact,
which may include genital, oral-genital, or anal-genital
contact. The organisms or viruses involved are generally
very susceptible to drying and require direct physical contact with mucous membranes for transmission. Because
sex can take many forms in addition to intercourse, these
microbes can initiate disease in the urogenital tract, rectum, or oral cavities. Examples of common sexually transmitted diseases are given in **Table 26.5**.

Case History: Secondary Syphilis

*A pregnant 18-year-old woman came to the county urgent-
care clinic with a low-grade fever, malaise, and headache. She
was sent home with a diagnosis of influenza. She again sought
treatment seven days later after she discovered a macular rash
(flat, red) developing on her trunk, arms, palms of her hands,
and soles of her feet. Further questioning of the patient when
serology results were known revealed that one year ago, she
had a painless ulcer on her vagina that healed spontaneously.*

The vaginal ulcer, the long latent period, and secondary development of rash on the hands and feet described
in the case history are classic symptoms of syphilis.
Syphilis was recognized as a disease as early as the sixteenth century, but the organism responsible, the spirochete *Treponema pallidum*, was not discovered until 1905
(**Fig. 26.11A**). The illness has several stages. The disease

Table 26.5 Microbes causing sexually transmitted diseases.

Disease	Symptoms	Etiologic agent	Virulence factors	Treatment
Gonorrhea	Purulent discharge, burning urination; can lead to sterility	*Neisseria gonorrhoeae* (G−)	Type IV pili, phase variation	Ceftriaxone
Syphilis	1°Chancre, 2°Joint pain, rash 3°Gummata, aneurism, CNS damage	*Treponema pallidum* (spirochete)	Motility	Penicillin
Nongonococcal urethritis	Watery or mucoid urethral discharge, burning urination	*Chlamydia trachomatis*	Intracellular growth; prevents phagolysosome fusion	Azithromycin
Trichomoniasis	Vaginal itching, painful urination, strawberry cervix	*Trichomonas vaginalis* (protozoan)	Cytotoxin	Metronidazole
Chancroid	Painful genital lesion	*Haemophilus ducreyi* (G−)	?	Erythromycin
Acquired immunodeficiency syndrome (AIDS)	Fever, diarrhea, cough, night sweats, fatigue, opportunistic infections	HIV	gp120, rev and nef proteins, tat protein	Azidothymidine, AZT, protease inhibitors, Zidovudine
Genital herpes	Painful ulcer on external genitals, painful urination	Herpes simplex 2	Cell fusion protein, complement-binding protein, latency	Acyclovir, iododeoxyuridine
Genital warts	Warts on external genitals	Human papillomavirus	E6, E7 proteins	Vaccine now available

has often been called the great imitator because its symptoms in the second stage, as exhibited in the case history, can mimic many other diseases. The incubation stage can last from two to six weeks after transmission, during which time the organism multiplies and spreads throughout the body. **Primary syphilis** is an inflammatory reaction at the site of infection called a **chancre**. About a centimeter in diameter, the chancre is painless and hard, and it contains spirochetes. Patients are usually too embarrassed to seek medical attention, and because it is painless, hope that it will "go away." It does go away after several weeks, and without scarring. The disease has now entered the primary latent stage. Over the next five years, symptoms may be absent, but at any time, as described in the case history, the infected person can develop the rash typical of **secondary syphilis** (**Fig. 26.11B**). The rash can be similar to rashes produced by

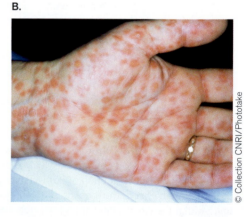

Figure 26.11   **Syphilis.**   **A.** *Treponema pallidum* (darkfield microscopy). Organisms are 10–25 μm long. **B.** Rash of secondary syphilis.

many different diseases, which contributes to the "great imitator" label. The patient remains contagious in this stage. Some patients eventually progress over years to **tertiary syphilis**, in which a great number of symptoms can develop, mostly cardiovascular and nervous systems. The patient can develop dementia and eventually dies from the disease.

The presence of the organism in tissues can be detected with fluorescent antibody, but the initial screen is usually serological (that is, patient serum is tested for antibodies). Antibiotics are useful for eradicating the organism, but there is no vaccine, and cure does not confer immunity.

The disease is particularly dangerous in pregnant women. The treponeme can cross the placental barrier and infect the fetus to cause **congenital syphilis**. At birth, infected newborns will have notched teeth (visible on X-rays), perforated palates, and other congenital defects. Women should be screened for syphilis as part of their prenatal testing to prevent these congenital infections.

Columbus and the New World theory of syphilis. An outbreak of syphilis that spread throughout Europe soon after Columbus and his crew returned from the Americas (1493) led to the theory that Columbus brought the treponeme to Europe from the New World. However, the theory that Columbus brought syphilis to Europe is most likely wrong. Native Americans buried *before* Columbus arrived in America showed no signs of syphilis. Furthermore, in recent years, pre-Columbian skeletons—such as those unearthed at the Hull friary in England—have been found with distinctive signs of syphilis. While some historians disagree, the progenitor organism was probably brought to Europe by African slaves long before the discovery of America, and it was in Europe that it developed into venereal syphilis. Thus, the crew of the Columbus voyages may actually have brought the disease to America along with smallpox and measles.

The Tuskegee experiment. Unfortunately, much of what we know of syphilis is the result of the infamous Tuskegee experiment conducted in the 1930s in Alabama. The study was entitled "Untreated Syphilis in the Negro Male." Through dubious means and deception, a group of African-American males were enlisted in a study that promised treatment but whose real purpose was to observe how the disease progressed without treatment. Today, such experiments are barred thanks to strict institutional review board (IRB) oversights in which human subjects must sign informed consent forms. An interesting treatise on the Tuskegee experiment can be found on the Web at the National Center for Case Study Teaching in America (Search for Bad Blood).

Chlamydial Infections Are Often Silent

Chlamydia is the most frequently reported sexually transmitted infectious disease in the United States, according to the Centers for Disease Control and Prevention, but many people are completely unaware that they are infected. Three-fourths of infected women, for instance, have no symptoms.

The chlamydiae are unusual gram-negative organisms with a unique developmental cycle. They are obligate intracellular pathogens that start as a small, nonreplicating infectious elementary body that enters target eukaryotic cells. Once inside vacuoles, they begin to enlarge into replicating reticulate bodies (**Fig. 26.12**). As the vacuole fills, the reticulate bodies divide to become new nonreplicating elementary bodies. *Chlamydia trachomatis* and *Chlamydophila pneumoniae* can both cause STDs, as well as other diseases, such as trachoma of the eye or pneumonia.

People most at risk of developing genitourinary tract infections with chlamydia are young, sexually active men and women; anybody who has recently changed sexual partners; and anybody who has recently had another sexually transmitted disease. The astute clinician knows that when one STD is discovered, others may also be present. A recent report indicates that *Chlamydia* can bind to sperm in a process called hitchhiking. It is thought that this interaction helps the organism spread to females.

Left untreated, chlamydia can cause serious health problems. In women, the organism can produce pelvic

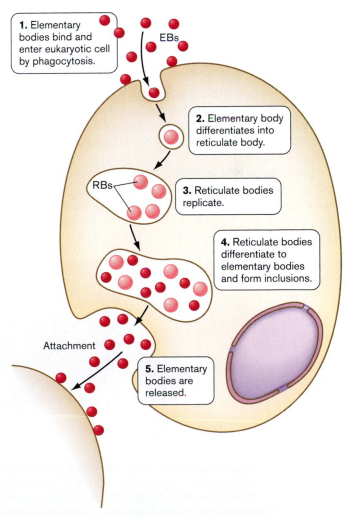

1. Elementary bodies bind and enter eukaryotic cell by phagocytosis.

EBs

2. Elementary body differentiates into reticulate body.

RBs

3. Reticulate bodies replicate.

4. Reticulate bodies differentiate to elementary bodies and form inclusions.

Attachment

5. Elementary bodies are released.

Figure 26.12 Replication cycle of *Chlamydia*.

inflammatory disease. Pelvic inflammatory disease is a damaging infection of the uterus and fallopian tubes that can be caused by several different microbial species. The damage produced can lead to infertility, tubal pregnancies, and chronic pelvic pain. Men left untreated can suffer urethral and testicular infections and a serious form of arthritis.

Case History: Gonorrhea

A 22-year-old mechanic saw his family doctor for treatment of painful urination and urethral discharge. His medical history was unremarkable. The patient was sexually active, with three regular and several "one time–good time" partners. Physical examination was unremarkable except for prevalent urethral discharge. The discharge was Gram stained and sent for culture. The Gram stain revealed many pus cells, some of which contained numerous phagocytosed gram-negative diplococci (Fig. 26.13A). Blood was drawn for syphilis serology, which proved negative. The patient was given an intramuscular injection of ceftriaxone (250 mg), and oral tetracycline (500 mg, four times a day) was prescribed for seven days. The bacteriology lab was able to recover the bacteria seen in the Gram-stained smear of the urethral discharge. The organism produced characteristic colonies on chocolate agar (agar plates containing heat-lysed red blood cells that turn the medium chocolate brown; Fig. 26.13B). The case was subsequently reported to the state public health department. Upon his return visit, the patient's symptoms had resolved, and a repeat culture was negative. The patient confirmed that all three of his regular sexual contacts were seen at the STD clinic.

The disease here is classic gonorrhea caused by *Neisseria gonorrhoeae*. A characteristic distinguishing *Neisseria* infections from *Chlamydia* infections is that bacterial cells are seen in gonorrheal discharges but not in chlamydial discharges. Gonorrhea has been a problem for centuries and remains epidemic in this country today. Symptoms generally occur 2–7 days after infection but can take as long as 30 days to develop. Most infected men exhibit symptoms; only about 10–15% of men do not. Symptoms include painful urination, yellowish white discharge from the penis, and in some cases swelling of the testicles and penis. The Greek physician Galen (AD 129–ca199) originally mistook the discharge for semen. This led to the name gonorrhea, which means "flow of seed."

In contrast to men, most infected women (80%) do not exhibit symptoms and constitute the major reservoir of the organism. If they are asymptomatic, they have no reason to seek treatment and thus can spread the disease. When symptoms are present, they are usually mild. A symptomatic woman will experience a painful burning sensation when urinating and will notice vaginal discharge that is yellow or occasionally bloody. She may also complain of cramps or pain in her lower abdomen, sometimes with fever or nausea. As the infection spreads throughout the reproductive organs (uterus and fallopian tubes), pelvic inflammatory disease occurs (see earlier discussion of chlamydia). It is important to note that there is no serological test or vaccine for gonorrhea because the organism frequently changes the structure of its surface antigens.

Although *N. gonorrhoeae* is generally serum sensitive owing to its sensitivity to complement, certain serum-resistant strains can make their way to the bloodstream and carry infection throughout the body. As a result, both sexes can develop purulent arthritis (joint fluid containing pus), endocarditis, or meningitis. An infected mother can also infect her newborn during parturition (birth), leading to a serious eye infection called ophthalmia neonatorum. Because of this risk and because most infected

A.

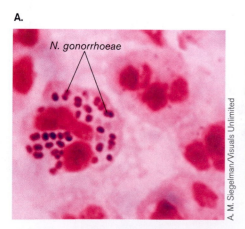

A. M. Siegelman/Visuals Unlimited

B.

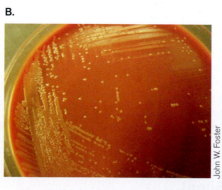

John W. Foster

C.

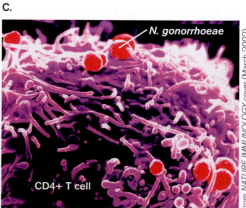

from *NATURE IMMUNOLOGY* cover (March 2002). Photo Courtesy of Ian C. Boulton and Gray-Owen

Figure 26.13 *Neisseria gonorrhoeae.* **A.** Within pus-filled exudates, the gram-negative diplococci are found intracellularly inside PMNs. The intracellular bacteria (approx. 0.6 to 1 μm in diameter) in this case are no longer viable, having been killed by the antimicrobial mechanisms of the white cell. **B.** Colonies of *N. gonorrhoeae* growing on chocolate agar (agar plates containing heat-lysed red blood cells that turn the medium chocolate brown). **C.** *N. gonorrhoeae* binding to CD4+ T cells (colorized SEM), inhibiting T-cell activation and proliferation, which may explain the ease of reinfection.

women are asymptomatic, all newborns receive antimicrobial eyedrops at birth.

Because adults engage in a variety of sexual practices, *N. gonorrhoeae* can also infect the anus or the pharynx, where it can develop into a mild sore throat. These infections generally remain unrecognized until a sex partner presents with a more typical form of genitourinary gonorrhea. Because no lasting immunity is built up, reinfection with *N. gonorrhoeae* is possible. This is due in part to the phase variation in various surface antigens that takes place and because the organism can apparently bind to CD4+ T cells, inhibiting their activation and proliferation to become memory T cells (**Fig. 26.13C**).

THOUGHT QUESTION 26.4 Considering that *N. gonorrhoeae* is exquisitely sensitive to ceftriaxone, why do you suppose the patient was also treated with tetracycline? And why can one person be infected repeatedly with *N. gonorrhoeae*?

Human Immunodeficiency Virus (HIV) Causes Sexually Transmitted and Blood-Borne Disease

Discovered in 1981, HIV caused the greatest pandemic of the late twentieth century and remains a serious problem today, especially in Africa, where it is estimated that 10–30% of the population are infected with HIV (prevalence in the United States is less than 1%). HIV has claimed the lives of more than 22 million people worldwide, over half a million in the United States alone. The molecular biology and virulence of HIV are discussed in Chapters 11 and 25. This section focuses on the disease that HIV causes—acquired immunodeficiency syndrome (AIDS).

HIV, a lentivirus in the retroviral family, is a prominent example of viruses that can be transmitted either sexually (vaginally, orally, anally, homosexually, or heterosexually) or through direct contact with body fluids, such as occurs with blood transfusion or the sharing of hypodermic needles by intravenous drug users. HIV is *not* transmitted by kissing, tears, or mosquito bite. It can, however, be transferred from mother to fetus through the placenta (transplacental transfer). **Figure 26.14A** illustrates the incidence of AIDS cases by year. The drop-off beginning in 1990 corresponds to the development of more effective treatment regimens. Starting in 1999, however, the trend began climbing upward once again. The proportion of AIDS cases by sex and ethnicity is described in Chapter 25 (see **Fig. 25.37**).

Having entered the bloodstream, HIV infects CD4+ T cells and replicates very rapidly, producing a billion particles per day. The subsequent decrease in CD4+ T cells leads to the disease symptoms collectively called AIDS. Though the symptoms of AIDS may take years to

develop (the average time between infection and AIDS is 11 years), nearly all HIV-infected individuals eventually become ill and die from the disease.

The first stage of the disease, previously known as AIDS-related complex (ARC), can include fever, headache, a macular rash, weight loss, and the appearance of antibodies to HIV in serum. These early, relatively mild symptoms can manifest within a few months of infection, resolve within a few weeks, and then may recur. The debilitated immune system leads to secondary infections by the yeast *Candida albicans* (candidiasis) and related species (**Fig. 26.14B**).

After several years, the disease can progress to the second stage, AIDS, marked by a significant depletion of the CD4+ T cell population. In 1993, the CDC revised its

A.

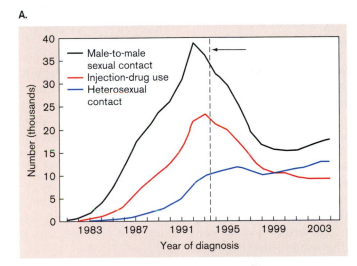

B.

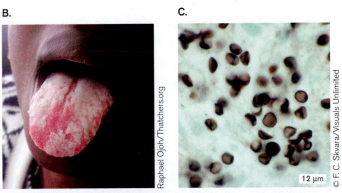

Raphael Ojoh/Thatchers.org

C.

12 µm

© F. C. Skvara/Visuals Unlimited

Figure 26.14   Acquired immunodeficiency syndrome.
A. Number of AIDS cases in the United States by major transmission category and year. In 1993, government agencies developed a specific definition of HIV infection that is used for public health surveillance only (vertical dotted line). It is estimated that 250,000–300,000 HIV-infected individuals are unaware of their HIV infection. **B.** Oral candidiasis. The white patches are caused by secondary infection by the yeast *Candida albicans*. **C.** *Pneumocystis jiroveci* infection of the lung. Note the cuplike appearance of the fungus, almost like crushed Ping-Pong balls. Organisms range from 2 to 6 µm in diameter.

definition of AIDS to include all persons with a CD4+ cell count of less than 200/μl. Once the CD4+ T cell population falls below 400 cells/μl, opportunistic infections begin to arise and disease processes begin. Opportunistic infections include pneumonia by *Mycobacterium avium–intracellulare* or *Pneumocystis jiroveci* (**Fig. 26.14C**), cryptococcal meningitis, *Histoplasma capsulatum* infection, and tuberculosis.

The third stage of AIDS includes changes in mental cognition, muscular action, and reflexes. Cardiovascular disease and brain tumors are also common. The neurological changes appear coincident to inflammation and demyelination of neurons.

The fourth stage of disease is marked by the appearance of various cancers triggered when the depressed immune system cannot detect and destroy cancers initiated by secondary agents. For instance, Kaposi's sarcoma (**Fig. 25.37C**) is a common cancer seen in AIDS patients caused by human herpes virus type 8 (HHV8).

Diagnosis of AIDS involves detecting anti-HIV antibodies and determining CD4+ cell count in a patient. Assay for HIV to determine viral load is done via quantitative PCR to detect HIV-specific genes such as *gag*, *nef*, or *pol*. Remember, a person who is HIV-positive does not necessarily have AIDS; it may take years to develop. A vaccine is not yet available to prevent AIDS, in part because the envelope proteins of the virus (see **Fig. 11.29**) typically change their antigenic shape. However, progression of the disease can be controlled by antimicrobials that inhibit two HIV enzymes critical to the replication of the virus—reverse transcriptase and protease; this is discussed further in Chapter 27.

The Protozoan *Trichomonas vaginalis* Causes a Common Vaginal Infection

Trichomonas vaginalis is a flagellated protozoan (**Fig. 26.15**) that causes an unpleasant sexually transmitted vaginal disease called trichomoniasis. Approximately 2–3 million infections occur each year in the United States. It can occur in men or women; however, men are usually asymptomatic. Even among infected women, 25–50% are considered asymptomatic carriers.

There is no cyst in the life cycle of *T. vaginalis*, so transmission is via the trophozoite stage only (discussed in Chapter 20). The female patient with trichomoniasis may complain of vaginal itching and/or burning and a musty vaginal odor. An abnormal vaginal discharge also may be present. Males will complain of painful urination (dysuria), urethral or testicular pain, and lower abdominal pain.

Owing to colonization by lactobacilli (which produce large amouts of acidic lactic acid), the normal, healthy vagina has a pH of less than 4.5. However, since *Trichomonas vaginalis* feeds on bacteria, the pH of the vagina rises as the numbers of lactobacilli decrease. Definitive diagnosis

Figure 26.15 *Trichomonas vaginalis* (SEM), a protist that causes a common sexually transmitted disease (size 7 μm × 10 μm).

David M. Phillips/Visuals Unlimited

requires demonstrating the flagellated protozoan in secretions by microscopy. PMNs, which are the primary host defense against the organism, are also usually present. As with giardiasis, this disease is treated with metronidazole.

THOUGHT QUESTION 26.5   Human immunodeficiency virus was discussed in this section and in Chapter 25. Like the plague, it is a blood-borne disease. Why, then, do fleas and mosquitoes fail to transmit HIV?

TO SUMMARIZE:

- **Urinary tract diseases** can result from ascending (to the kidney) or descending (from the kidney) routes of infection. The most common route leading to bladder infection, however, is through the urethra.
- *E. coli* is the most common cause of UTI.
- **Syphilis, gonorrhea, and chlamydia** are the most common sexually transmitted diseases.
- **A patient with one sexually transmitted disease** often has another sexually transmitted disease.
- **Complement sensitivity** prevents dissemination by *N. gonorrhoeae*. In contrast, *N. meningitidis*, a cause of meningitis, frequently disseminates in the bloodstream because it is complement resistant.
- **HIV depletion of CD4+ T cells** results in lethal secondary infections and cancers.
- *Trichomonas vaginalis* is a flagellated protozoan that causes a sexually transmitted vaginal disease. The reservoir for this organism is the male urethra and female vagina.

26.6 Infections of the Central Nervous System

Microbes cannot gain easy access to the brain in large measure because of the **blood-brain barrier**, a filter mechanism that allows only selected substances into the brain. The blood-brain barrier works to our advantage when harmful substances, such as bacteria, are prohibited from entering. However, it works to our disadvantage when substances we want to enter the brain, such as antibiotics, are kept out. The barrier is not a single structure but is a function of the way blood vessels, especially capillaries, are organized in the brain. Furthermore, the endothelial cells in those vessels have tight junctions that do not allow most compounds or microbes to cross. And yet, brain infections do occur.

Case History: Meningitis

*In April 2001, a 4-month-old infant from Saudi Arabia was hospitalized with fever, tender neck, and purplish spots (purpuric spots) on her trunk (**Fig. 26.16A**). Suspecting meningitis, the clinician took a cerebral spinal fluid (CSF) sample and examined it by Gram stain. The smear revealed gram-negative diplococci inside PMNs. The CSF was turbid with 900 leukocytes per μl, and* Neisseria meningitidis *was confirmed by culture. The child was treated with cefotaxime and made a full recovery. Her father, the person who brought her in, was clinically well. However, meningococcus was isolated from his oropharynx, as well as from the throat of the patient's 2-year-old brother. Isolates from the patient, her father, and her brother were positive by agglutination with meningococcus A, C, Y, W135 polyvalent reagent. The father's vaccination certificate confirmed that he had received a quadrivalent meningococcal vaccine. All three isolates were sent to the WHO (World Health Organization) Collaborating Center, which confirmed meningococcus serogroup W135. The three isolates were examined by pulsed-field gel electrophoresis and were found to be indistinguishable (that is, they had identical DNA restriction patterns).*

Meningitis is an inflammation of the meninges—the membrane that surrounds the brain and spinal cord. Meningitis can be either bacterial or viral in origin. Sinus and ear infections can extend directly to the meninges, while septicemic spread requires passage through the blood-brain barrier. Viral meningitis is serious but rarely fatal in people with a normal immune system. The symptoms generally persist for seven to ten days and then completely resolve. Bacterial meningitis is usually caused by *Streptococcus pneumoniae, Neisseria meningitidis,* or *Haemophilus influenzae.* Symptoms of bacterial meningitis can include sudden onset of fever, headache, neck pain or stiffness, painful sensitivity to strong light (photophobia), vomiting (often without abdominal complaints), and irritability. Prompt medical attention is extremely important because the disease can quickly progress to convulsions and death.

The meningococcus *Neisseria meningitidis* (**Fig. 26.16B**) can colonize the human oropharynx, where it causes mild, if any, disease. At any given time, 10–20% of the healthy population can be colonized and asymptomatic. The organism spreads directly by person-to-person contact or indirectly via droplet nuclei from sneezing or fomites. The problem arises when this organism enters the bloodstream. Unlike the gonococcus, *N. meningitidis* is very resistant to complement owing to its production of a polysaccharide capsule. The capsule allows the microbe to produce a transient blood infection (bacteremia) and reach the blood-brain barrier.

N. meningitidis uses type IV pili to adhere to these endothelial cells and becomes intracellular. Then, utilizing the process of **transcytosis**, in which a cell moves from one side of a host cell to another, the microbe moves to the brain side of the capillary, where it exits. Once in the cerebral spinal fluid, microbes can multiply almost at will. **Figures 26.16C** and **D** shows the remarkable damage

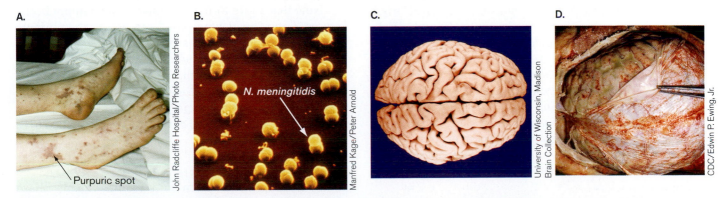

A. Purpuric spot — John Radcliffe Hospital/Photo Researchers

B. *N. meningitidis* — Manfred Kage/Peter Arnold

C. University of Wisconsin, Madison Brain Collection

D. CDC/Edwin P. Ewing, Jr.

Figure 26.16 Bacterial meningitis. Meningococcal disease is dreaded by parents and medical practitioners alike for its rapid onset and the difficulty of obtaining a timely and accurate diagnosis. **A.** Purpuric spots produced by local intravascular coagulation due to *Neisseria meningitidis* endotoxin. The rash in meningitis typically has petechial (small) and purpuric (large) components. **B.** *N. meningitidis* (SEM, approx. 1 μm diameter). **C.** Normal brain. **D.** Autopsy specimen of meningitis due to *Streptococcus pneumoniae.* Note the greening of the brain, compared with the pink normal brain.

that *N. meningitidis* and other microbes, such as *S. pneumoniae*, cause in the brain.

Several antigenic types of capsules, called type-specific capsules, are produced by different strains of pathogenic *N. meningitidis*: types A, B, C, W135, and Y. Types A, C, Y, and W135 are usually associated with epidemic infections seen among people kept in close proximity, such as college students or military personnel. Type B meningococcus is typically involved in sporadic infections. Antibodies to these capsular antigens are used to classify the capsular types of the organisms causing an outbreak. Knowing the capsular type of organism involved in each case helps determine whether the disease cases are related and where the infection may have started.

Meningococcal meningitis is highly communicable. As a result, close contacts, such as the parents or siblings of any patient with meningococcal disease, should receive antimicrobial prophylaxis within 24 hours of diagnosis; a single dose of ciprofloxacin (a quinolone antibiotic, discussed in Section 27.4) can be given to adults, and two days of rifampicin given to children.

Highly susceptible populations can be immunized with a vaccine containing the four capsular structures. Generally, large numbers of people housed together are susceptible to this disease. For instance, it is a significant problem in college dormitories, prompting several colleges to request that incoming students be vaccinated.

> **THOUGHT QUESTION 26.6**   Normal cerebral spinal fluid is usually low in protein and high in glucose. The protein and glucose content does not change much during a viral meningitis, but bacterial infection leads to greatly elevated protein and lowered glucose levels. What could account for this?

Case History: Botulism—It Is What You Eat

In June, a 47-year-old resident of Oklahoma was admitted to the hospital with rapid onset of progressive dizziness, blurred vision, slurred speech, difficulty swallowing, and nausea. Findings on examination included drooping eyelids, facial paralysis, and impaired gag reflex. He developed breathing difficulties and required mechanical ventilation. The patient reported that during the 24 hours before onset of symptoms, he had eaten home-canned green beans and a stew containing roast beef and potatoes. Analysis of the patient's stool detected botulinum type A toxin, but no Clostridium botulinum organisms were found. The patient was hospitalized for 49 days, including 42 days on mechanical ventilation, before being discharged.

Bacterial Neurotoxins That Cause Paralysis

Imagine a disease that causes complete loss of muscle function. Using secreted exotoxins, two microbes cause lethal paralytic diseases. In one instance, the victim suffers a flaccid paralysis in which the muscles go limp, as in the case history just given, causing paralysis and respiratory difficulty. Voluntary muscles fail to respond to the mind's will because botulinum toxin interferes with neural transmission. The disease (called botulism) is typically food-borne and is caused by an anaerobic, gram-positive, spore-forming bacillus named *Clostridium botulinum*.

In striking contrast to botulism is tetanus, a very painful disease in which muscles continuously and involuntarily contract (called tetany or spastic paralysis). Tetanus is caused by **tetanospasmin**, a potent exotoxin made by another anaerobic, gram-positive spore-forming bacillus called *Clostridium tetani* (**Fig. 26.17A**). Tetanospasmin interferes with neural transmission, but in contrast to botulism toxin, it causes *excessive* nerve signaling to muscles, forcing the victim's back to arch grotesquely while the arms flex and legs extend. The patient remains locked this way until death. Spasms can be strong enough to fracture the patient's vertebrae. **Figure 26.17B** shows the result of injecting a mouse's hind leg with just a tiny amount of tetanus toxin. In both botulism and tetanus, death can result from asphyxiation.

Botulism is typically caused by ingesting preformed toxin, although infected wounds or germination of ingested spores can also occasionally produce disease. The microbe germinates in the food and produces toxin. Because the organism is an anaerobe, an anaerobic environment must be present. Home-canning processes are designed to remove oxygen in the canned food as well as sterilize the food, but if the sterilization process is not complete, surviving spores germinate and produce toxin. The toxin is susceptible to heat, but will remain active in improperly cooked food. After ingestion, the toxin is absorbed from the intestine. As in the case history given here, it is not unusual for the organism to be absent from stool samples. This is why serological identification of the toxin is important.

Note that a rare form of botulism, called infant botulism or "floppy head syndrome," can occur when infants are fed honey. Honey can harbor *C. botulinum* spores that can germinate in the gastrointestinal tract after which the growing vegetative cells will secrete toxin.

> **THOUGHT QUESTION 26.7**   Knowing the symptoms of tetanus, what kind of therapy would you use to treat the disease?

Toxin structure. Botulism and tetanus toxins share 30–40% identity and have similar structures and nearly identical modes of action. Botulinum and tetanus toxins are each composed of two peptides—a large, or heavy, fragment analogous to a B subunit of AB toxins and a small, or light, fragment analogous to an A subunit. Both

A. *Clostridium tetani*

Spores

8 μm

A. M. Siegelman/Visuals Unlimited

B. Spastic paralysis due to tetanus toxin

Tetany

Courtesy American Society for Microbiology

D. Tetanus toxin structure

Binding domain

Catalytic domain

Translocation domain

C. Basic structure of tetanus and botulinum toxins

COO⁻

Ganglioside binding domain

Heavy chain

Translocation domain

Proteolytic cleavage

Light chain

Zn Catalytic domain

NH₂

1. The binding domain binds to receptor molecules (gangliosides) of the nerve cell.

2. The translocation domain makes a pore for passage of the toxin.

3. The protease toxin disrupts release of neurotransmitter.

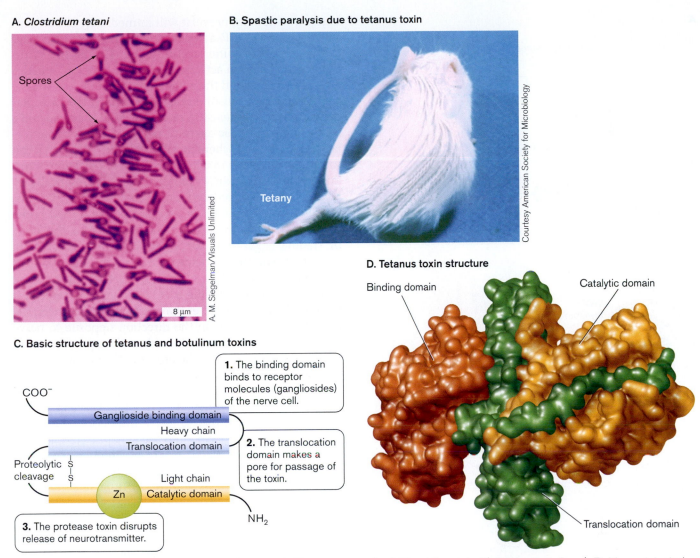

Figure 26.17   Tetanus and botulinum toxins.   A. Photomicrograph of *Clostridium tetani* (cell length 4–8 μm). **B.** Mouse injected with tetanus toxin in left hind leg. **C.** Schematic diagram of tetanus and botulinum toxins. **D.** Three-dimensional representation of the tetanus neurotoxin with the domains marked. (PDB code: 3BTA)

toxins are initially made as single peptides (about 150 kDa) that are cleaved after secretion to form two fragments (heavy and light chains) that remain tethered by a disulfide bond (**Fig. 26.17C**). Each heavy chain includes the binding domain, which binds to receptor molecules (gangliosides) on the nerve cell membrane, and a translocation domain that makes a pore in the nerve cell through which the toxin passes. The light chains (catalytic domains) are proteases that disrupt the movement of exocytic vesicles containing neurotransmitters needed for contraction or relaxation. **Figure 26.17D** shows a three-dimensional rendition of tetanus toxin with the three domains marked.

Mechanism of action. Both toxins bind to target membranes and are brought into the cell by endocytosis. The low pH that forms in the endosome reduces the disulfide bonds holding the two halves of the toxin together and causes the heavy chain to assemble as a channel through the membrane. That channel allows release of the proteolytic light chain into the cytoplasm. The toxic subunit then cleaves key host proteins such as synaptobrevin (VAMP), involved in the exocytosis of vesicles containing neurotransmitters (**Fig. 26.18**).

With such a high degree of mechanistic similarity, why do tetanus and botulism toxins have such drastically different effects? The answer is based on where each toxin acts in the nervous system. Both toxins enter at peripheral nerve ends where nerve meets muscle. The disease botulism results from the ingestion of preformed toxin in poorly prepared, contaminated foods. Whether the organism is also ingested is immaterial. Once botulism toxin

A.

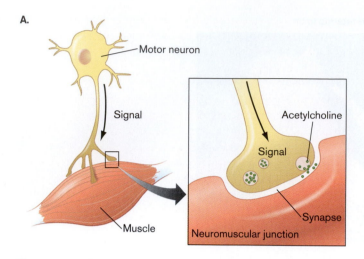

B.

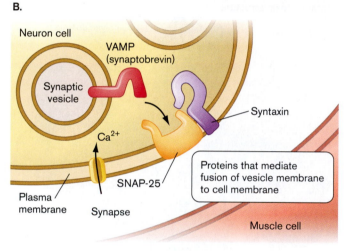

C.   **D.**

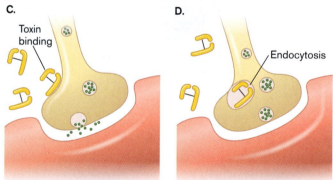

Figure 26.18 Mechanism of action of botulism toxin.
A. The neuromuscular junction. Inset shows vesicles filled with neurotransmitters. **B.** A series of proteins within the nerve are needed to allow synaptic vesicles to bind to the nerve endings. Fusion of the membranes releases acetylcholine into the neuromuscular junction. Botulinum toxin types A and E cut SNAP-25. Botulinum toxins B, D, F, and G cut VAMP. Botulinum toxin C1 cuts syntaxin and SNAP-25. **C, D.** Toxin binds via the heavy chain and is endocytosed into the nerve terminal. Once the toxin cleaves its target, the nerve terminal is no longer able to release acetylcholine.

enters a peripheral nerve, it will immediately cleave peptides, such as VAMP, associated with the exocytic release of the neurotransmitter acetylcholine (**Fig. 26.18**). Without acetylcholine to activate nerve transmission, muscles will not contract and the patient is paralyzed.

Although botulism disease is now rare, the toxin has gained renewed interest in recent years. It is considered a select biological toxic agent of potential use to bioterrorists, but that is not where most interest lies. Because it can safely relax muscles in a localized area if injected in small doses, botulism toxin, or Botox, is used cosmetically by plastic surgeons to reduce facial wrinkles in their patients.

Tetanus, in contrast to botulism, is not a food-borne disease. *C. tetani* spores are introduced into the body by trauma (such as by stepping on a dirty, rusty nail). Necrotic tissue then provides the anaerobic environment required for germination. Growing, vegetative cells release tetanospasmin, which enters the peripheral nerve cells at the site of injury. But rather than cleaving targets here, the toxin travels up axons in the direction opposite to nerve signal transmission until it reaches the spinal column, where it becomes fixed at the presynaptic inhibitory motor neuron. There the toxin also cleaves proteins like VAMP, but the function of vesicles in these nerves is to release *inhibitory* neurotransmitters that dampen nerve impulses. Tetanus toxin blocks release of the inhibitory neurotansmitters GABA and glycine into the synaptic cleft, leaving nerve impulses unchecked. As a result, impulses come too frequently and produce the generalized muscle spasms characteristic of tetanus. **Figure 26.19** illustrates the mechanism of action of tetanus toxin.

> **THOUGHT QUESTION 26.8** How do the actions of tetanus toxin and botulinum toxin actually help the bacteria colonize or obtain nutrients?
>
> **THOUGHT QUESTION 26.9** If *C. botulinum* is an anaerobe, how might botulism toxin get into foods?

Case History: Eastern Equine Encephalitis

In August, Mr. C brought his 21-year-old son, Rich, to a New Jersey emergency room. Rich appeared dazed and had trouble responding to simple commands. When questioned about his son's activities over the past few months, Mr. C told the physician that Rich planned to enter veterinary school in the fall and had spent the month of July relaxing and sunning himself on New Jersey beaches. Aside from the shore, his favorite locale was a pond in a wooded area near a horse farm. On the afternoon prior to admission, Rich became lethargic and tired. He returned home and went to bed. That evening, his father woke him for supper but Rich was confused and had no appetite. By 11 p.m., Rich had a fever of 40.7°C and could not respond to questions. A few hours later, when his father had

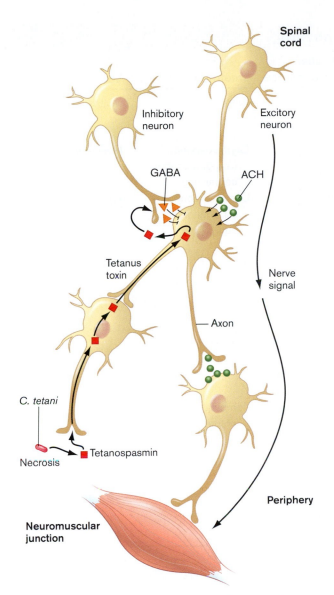

Figure 26.19 Retrograde movement of tetanus toxin to an inhibitory neuron. Tetanus toxin enters the nervous system at the neuromuscular junction and travels retrogradely up the axons until reaching an inhibitory neuron located in the central nervous system. There it cleaves VAMP protein (Fig. 26.18B) associated with exocytosis of vesicles containing inhibitory neurotransmitters. GABA = gamma-aminobutyric acid; ACH = acetylcholine. ▶❚❚

trouble rousing him, he brought Rich to the ER. Over the next week, Rich's condition worsened to the point where his limbs were paralyzed. Two weeks later, he died.

Serum samples taken upon entering the hospital and a few days before he died showed a sixfold rise in antibody titer to eastern equine encephalitis (EEE) virus. Brain autopsy showed many small foci of necrosis in both the gray and white matter.

The EEE virus, a member of the family Togaviridae, is transmitted from bird to bird by mosquitoes. Horses contract the disease in the same way. Rich contracted it inadvertently while visiting the pond. The disease, an encephalitis, is often fatal (35% mortality) but fortunately rare. One reason human disease is rare is that the species of mosquito that usually transmits the virus between marsh birds does not prey on humans. But sometimes a human-specific mosquito bites an infected bird and then transmits EEE virus to humans. Another reason human disease is rare is that the virus generally does not fare well in the body. Many persons infected with EEE virus have no apparent illness because the immune system thwarts viral replication. However, as already noted, mortality is high in those individuals that do develop disease.

An interesting point in the case history is that diagnosis relied on detecting an increase in antibody titer to the virus. Typically, at the point that disease symptoms first appear, the body has not had time to generate large amounts of specific antibodies. After a week or so, often when the patient is nearing recovery (convalescence), antibody titers have risen manyfold. The rule of thumb is that a greater than fourfold rise in antibody titer between acute disease and convalescence (or in this case death) indicates that the patient has had the disease. While this knowledge could not help the patient in this case, it was valuable in terms of public health and prevention strategies.

Several bacterial, fungal, and viral causes of encephalitis and meningitis are given in **Table 26.6**.

Prion Diseases Are Caused by Proteins

An unusual infectious agent called the prion has been implicated as the cause of a series of relatively rare but invariably fatal brain diseases (**Table 26.7**). Prions are infectious agents that do not have a nucleic acid genome. It seems that a protein alone can mediate an infection. The **prion** has been defined as a small proteinaceous infectious particle that resists inactivation by procedures that modify nucleic acids. The discovery that proteins alone can transmit an infectious disease has come as a surprise to the scientific community. Diseases caused by prions are especially worrying, since prions resist destruction by many chemical agents and remain active after heating at extremely high temperatures. There have even been documented cases in which sterilized surgical instruments, originally used on a prion-infected person, still held infectious agent and transmitted the disease to a subsequent surgery patient.

How can a nonliving entity without nucleic acid be an infectious agent? Prions associated with human brain disease are thought to be aberrantly folded forms of a normal brain protein. The theory is that when a prion is introduced into the body and manages to enter to the brain, it will cause normally folded forms of the protein to refold incorrectly. The improperly folded proteins fit together like Lego blocks to produce damaging aggregated structures within brain cells.

Table 26.6 Microbes causing meningitis/encephalitis.

Type of meningitis	Etiologic agent	Virulence factors	Typical initial infection or source	Treatment	Vaccine
Bacterial (septic)	*Streptococcus pneumoniae* (G+)	Capsule, pneumolysin	Lung	Ampicillin	Multivalent, capsule
	Neisseria meningitidis (G–)	Capsule IgA protease, endotoxin	Throat	Cephalosporin (3rd generation), ceftriaxone	Multivalent, capsule
	Haemophilus influenzae type B (G–)	Polyribitol capsule IgA protease Endotoxin	Ear infection	Cephalosporin (3rd generation), ceftriaxone	Type b polysaccharide
	Other G– bacilli		Septicemia		
	Group B streptococci (G+)	Sialic acid capsule, streptolysin, inhibition of alternate complement path	Neonate infected during parturition	Ampicillin	Capsular
	Listeria monocytogenes (G+)	Intracellular growth, PrfA regulator, actin-based motility	Mild GI disease of mother	Ampicillin plus gentamicin	
	Mycobacterium tuberculosis	Cord factor, wax D, intracellular growth	Lung	Combination therapy (rifampin, isoniazid, ethambutol, pyrazinamide)	BCG
	Staphylococcus aureus (G+)	Coagulase, protein A, TSST, leucocidin	Septicemia	Methicillin, vancomycin	
	Staphylococcus epidermidis (G+)	Biofilm, slime	Complication of surgical procedure	Vancomycin, methicillin	
Aseptic[a]	Fungi (e.g., *Coccidioides*, *Cryptococcus*)		Sinusitis, direct spread to meninges	Ketoconezole, fluconazole	
	Amebas (e.g., *Naegleria*)		Swimming in contaminated waters	Amphotericin B	
	T. pallidum		Syphilis		
	Mycoplasmas	Adhesin tip	Respiratory	Erythromycin	
	Leptospira	Burrowing motility	Septicemia, water contaminated with animal urine	Erythromycin	
Viral	Viruses (90% caused by enteroviruses)			Self-limiting	
Eastern equine encephalitis	EEE virus	?	Mosquito bite	None; often fatal	
West Nile disease	West Nile virus	?	Mosquito bite	Supportive therapy[b]	

[a]Aseptic meningitis refers to an inability to isolate bacterial sources by ordinary means.

[b]Supportive therapy means hospitalization, intravenous fluids, airway management, respiratory support, prevention of secondary infections.

Table 26.7 Prion diseases.

Disease	Susceptible animal	Incubation period	Disease characteristics
Creutzfeldt-Jakob [sporadic, familial, new variant (vCJD)]	Human	Months (vCJD) to years	Spongiform encephalopathy (degenerative brain disease)
Kuru	Human	Months to years	Spongiform encephalopathy
Gerstmann-Straussler-Scheinker syndrome	Human	Months to years	Genetic neurodegenerative disease
Fatal familial insomnia	Human	Months to years	Genetic neurodegenerative disease with untreatable insomnia
Mad cow disease	Cattle	5 years	Spongiform encephalopathy
Wasting disease	Deer	Months to years	Spongiform encephalopathy

Some investigators suggest that living, infectious agents, such as *Spiroplasma* (a spiral-shaped genus whose members lack cell walls), can somehow precipitate a conformational change in the normal brain protein, converting it to an infectious prion protein. This proposal is not, however, universally accepted and has yet to be proved.

Prion diseases are often called **spongiform encephalopathies** because the postmortem appearance of the brain includes large, spongy vacuoles in the cortex and cerebellum. These are visible in a brain sample from a victim of one of these diseases, called Creutzfeldt-Jakob disease (CJD) (**Figs. 26.20A** and **B**). Most mammalian species appear to develop these diseases.

Since 1996, mounting evidence points to a causal relationship between outbreaks in Europe of a disease in cattle called bovine spongiform encephalopathy (BSE, or "mad cow disease") and a disease in humans called variant Creutzfeldt-Jakob disease (vCJD). Both disorders are invariably fatal brain disorders. They have unusually long incubation periods (measured in years) and are caused by an unconventional transmissible agent.

As of this writing, only four cases of BSE have been detected in the United States. Due to aggressive surveillance efforts in the United States and Canada (where only nine cases have been found), it is unlikely, but not impossible, that BSE will be a food-borne hazard to humans in this country. The CDC monitors the trends and current incidence of typical and variant CJD in the United States (**Fig. 26.20C**).

TO SUMMARIZE:

- *N. meningitidis* is resistant to serum complement because it produces a type-specific capsule. This allows the organism to reach and then cross the blood-brain barrier.
- **A vaccine for *N. meningitidis*** is available, but none exists for *N. gonorrhoeae*.
- **Tetanospasmin** causes spastic paralysis.
- **Botulinum toxin** causes flaccid paralysis.

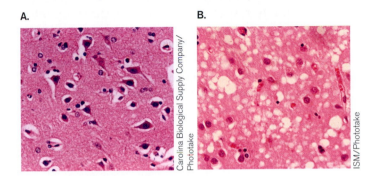

A. B.

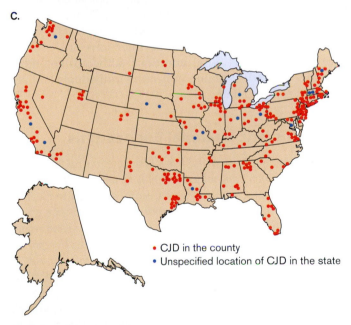

C.

- CJD in the county
- Unspecified location of CJD in the state

Figure 26.20   Spongiform encephalopathies.   A. Normal brain section. **B.** Section of brain taken from a CJD victim. Note the "Swiss-cheese" appearance, indicating brain damage. **C.** Locations of CJD (not vCJD) cases in the USA.

- **Serological diagnosis of an infectious disease** can be done if a fourfold rise in specific pathogen antibody titer occurs between the acute and convalescent stages.
- **Spongiform encephalopathies** are believed to be caused by nonliving proteins called prions.

26.7 Infections of the Cardiovascular System

Infections of the cardiovascular system include septicemia, endocarditis (inflammation of the heart's inner lining), pericarditis (inflammation of the heart's outer lining) and, possibly, atherosclerosis (the deposition of fatty substances along the inner lining of arteries). These are all life-threatening diseases.

Septicemia is, by strict definition, the presence of bacteria or viruses in the blood. The presence of viruses is a condition called **viremia**, while bacteria in the circulation is more specifically called **bacteremia**. In practice, however, the terms *septicemia* and *bacteremia* are often used interchangeably. Septicemia can develop from a local infection situated anywhere in the body, although blood factors such as complement can nonspecifically kill many types of bacteria that enter the blood. Nevertheless, gram-positives, gram-negatives, aerobes, and anaerobes can all produce septicemia under the right conditions.

Endocarditis can be either viral or bacterial in origin. It can be a consequence of many bacterial diseases, such as brucellosis, gonorrhea, psittacosis, staphylococcal and streptococcal infections, candidiasis, and Q fever. Among the many viral causes are Coxsackie virus, ECHO virus, Epstein-Barr, and HIV. Bacterial infections of the heart are always serious. Viral infections, although common, are rarely life-threatening in healthy individuals and are usually asymptomatic.

Case History: Bacterial Endocarditis

Elizabeth is 38 years old and has a history of mitral valve prolapse (a common congenital condition in which a heart valve does not close properly). She was recently admitted to the hospital complaining of fatigue, intermittent fevers for five weeks and headaches for three weeks, symptoms the physician recognized as possible indications of endocarditis. Elizabeth reported having a dental procedure a few weeks prior to the onset of symptoms. A sample of her blood placed in a liquid bacteriological media grew gram-positive cocci, which turned out to be Streptococcus mutans, *a member of the viridans streptococci. As a result, the diagnosis of bacterial endocarditis was confirmed. The patient began a one-month course of intravenous penicillin G and gentamicin therapy and eventually recovered to normal health.*

Endocarditis (inflammation of the heart) is traditionally classified as acute or subacute, based on the pathogenic organism involved and the speed of clinical presentation. Subacute bacterial endocarditis (SBE) has a slow onset with vague symptoms. It is usually caused by a bacterial infection of a heart valve (**Fig. 26.21**). Subacute bacterial endocarditis infections are usually (but not

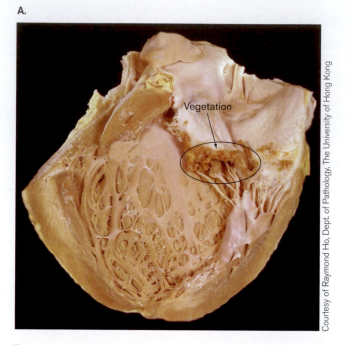

A.

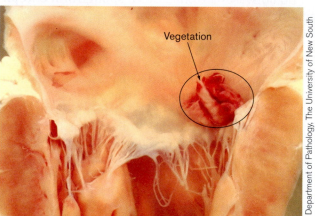

B.

Courtesy of Raymond Ho, Dept. of Pathology, The University of Hong Kong

© Department of Pathology, The University of New South Wales, Sydney Australia

Figure 26.21 Views of bacterial endocarditis.
A. Whole-heart view, showing vegetations on the mitral valve.
B. Close-up of mitral valve endocarditis, showing vegetation (arrow).

always) caused by a viridans streptococcus from the oral flora (for example, *Streptococcus mutans*, a common cause of dental caries). Viridans streptococci is a general term used for normal flora streptococci whose colonies produce green alpha hemolysis on blood agar (viridans from the Greek viridis, to be green). Most patients who develop infective endocarditis have mitral valve prolapse (90%), although this is frequently not the case when patients are intravenous drug abusers or have hospital-acquired (nosocomial) infections.

As in the case history presented here, bacterial endocarditis often begins at the dentist's office. Following a dental procedure (such as tooth restoration), oral flora can transiently enter the bloodstream and circulate.

S. mutans, which is not normally a serious health problem, can become lodged onto damaged heart valves, grow as a biofilm, and secrete a thick glycocalyx coating that encases the microbes and forms a vegetation on the valve, damaging it further. If untreated, the condition can be fatal within six weeks to a year.

When more virulent organisms, such as *Staphylococcus aureus*, gain access to cardiac tissue, a rapidly progressive (acute) and highly destructive infection ensues. Symptoms of acute endocarditis include fever, pronounced valvular regurgitation (back flow of blood through the valve), and abscess formation.

Most patients with subacute bacterial endocarditis present with a fever that lasts several weeks. They also complain of nonspecific symptoms, such as cough, shortness of breath, joint pain, diarrhea, and abdominal or flank pain. Endocarditis is suspected in any patient who has a heart murmur and an unexplained fever for at least one week. It should also be considered in an intravenous drug abuser with a fever, even in the absence of a murmur. In either case, a definitive diagnosis requires blood cultures that grow bacteria. Blood cultures involve taking samples of a patient's blood from two different locations (such as two different arms). Growth of the same organism in cultures taken from two body sites rules out inadvertent contamination with skin flora, which would likely yield growth only in one culture. The blood is then added to liquid culture medium and incubated at 37°C. Incubation should be done aerobically and anaerobically.

Curing endocarditis is difficult because the microbes are usually ensconced in a nearly impenetrable glycocalyx. Consequently, eradicating microorganisms from the vegetations almost always requires hospitalization, where high doses of intravenous antibiotic therapy can be administered and monitored. Antibiotic therapy usually continues for at least a month, and in extreme cases, surgery may be necessary to repair or replace the damaged heart valve.

Although at increased risk, patients with heart valve prolapse should not shy away from the dentist. Prophylactic treatment with penicillin or erythromycin taken an hour before a procedure will kill any oral flora that enter the bloodstream and prevent the development of endocarditis.

Viruses such as adenovirus and some enteroviruses can also cause endocarditis as well as a condition known as myocarditis, an inflammation of the heart muscle.

> **THOUGHT QUESTION 26.10**   A patient presenting with high fever and in an extremely weakened state is suspected of having a septicemia. Two sets of blood cultures are taken from different arms. One bottle from each set grows *Staphylococcus aureus*, yet the laboratory report states that the results are inconclusive. New blood cultures are ordered. Why might this be?

Atherosclerosis and Coronary Artery Disease

Atherosclerosis can occur in most blood vessels. Coronary artery disease is the result of atherosclerotic deposits in the arteries of the heart, which can produce a heart attack if blood flow is occluded. The causes of atherosclerosis have been studied and debated for many years. It now appears that microbes may have an important role in the development of this disease. *Chlamydophila pneumoniae*, for example, is suspected of involvement. Serological studies indicate that a large number of patients with coronary heart disease have antibodies to this organism. Although it is still not clear if *Chlamydophila* truly causes atherosclerosis, one possible model is presented in **Figure 26.22.** Infection of alveolar macrophages leads to circulating "activated" macrophages that generate inflammatory cytokines and factors that result, among other things, in lowered high-density cholesterol (the "good" cholesterol) and increased clotting—all of which can damage endothelial cells and form plaques. In support of an infection model, some studies indicate that antibiotic treatment will reduce the number of recurrent heart attacks in patients with acute coronary artery disease.

Malarial Parasites Feed on Blood

Malaria is the most devastating infectious disease known. Each year 300–500 million people develop malaria worldwide, and 1–3 million of these people, mostly children, die. Fortunately, the disease is relatively rare in the United States; only around 1,000 cases occur annually and almost all are acquired as a result of international travel to endemic areas (**Fig. 26.23A**). Once again, this illustrates the diagnostic importance of knowing a patient's history.

The disease is caused by four species of *Plasmodium*; *P. falciparum* (the most deadly), *P. malariae*, *P. vivax*, and *P. ovale*. The life cycle of *Plasmodium*, discussed in detail in Chapter 20, is complex and involves two cycles—an asexual erythrocytic cycle in the human and a sexual cycle in the mosquito (discussed in Chapter 20; **Fig. 20.37**).

In the erythrocytic cycle, the organisms enter the bloodstream through the bite of an infected female *Anopheles* mosquito (the mosquito injects a small amount of saliva containing *Plasmodium*). The haploid sporozoites travel immediately to the liver, where they undergo asexual fission to produce merozoites. Released from the liver, the merozoites attach to and penetrate red blood cells, where *Plasmodium* consumes hemoglobin and enlarges into a trophozoite. The protist nucleus divides, so that the cell, now called a schizont, contains up to 20 or so nuclei. The schizont then divides to make the smaller, haploid merozoites (**Fig. 26.23B**). The glutted red blood cell eventually lyses, releasing merozoites that can infect new red blood cells (**Fig. 20.36**).

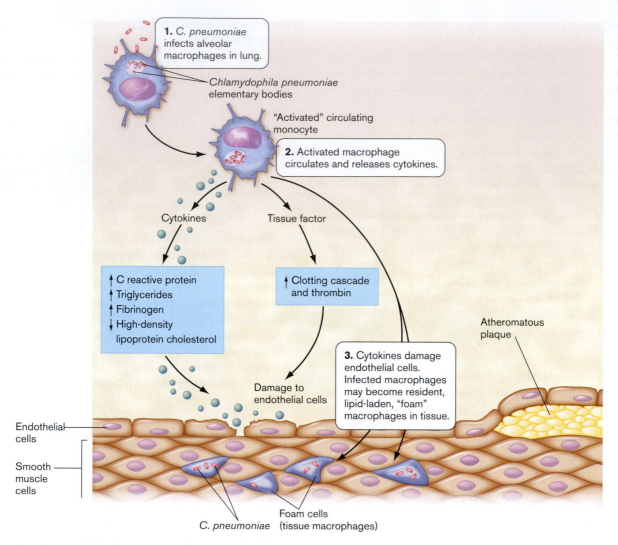

Figure 26.22 **Model outlining a possible role for *Chlamydophila pneumoniae* in cardiovascular disease.** Upward and downward arrows indicate increases or decreases in factors. Atheromatous plaques consist of fatty deposits of cholesterol that narrow blood vessels.

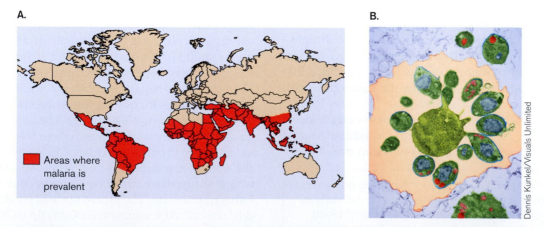

Figure 26.23 **Malaria is a major disease worldwide.** **A.** Endemic areas of the world where malaria is prevalent. **B.** *Plasmodium falciparum* (TEM). Schizont after completion of division. A residual body of the organism (yellow-green) is left over after division. The erythrocyte has lysed and only a ghost cell remains; no cytoplasm is seen surrounding the merozoites just being released. Free merozoites are seen outside the membrane.

Sudden, synchronized release of the merozoites and red cell debris triggers the telltale symptoms of malaria—violent, shaking chills followed by high fever and sweating. The erythrocytic cycle, and thus the symptoms, repeats every 48–72 hours. After several cycles, the patient goes into remission lasting several weeks to months, after which there is a relapse.

Much of today's research focuses on why malarial relapse happens. Why does the immune system fail to eliminate the parasite after the first episode? When *Plasmodium* invades the red blood cells, it lines the blood cells with a protein, PfEMP1, that causes them to stick to the sides of blood vessels. This removes the parasite from circulation, but the protein cannot protect the parasite from patrolling macrophages, which eventually detect the invader and recruit other immune cells to fight it. So, during a malarial infection, a small percentage of each generation of parasites switches to a different version of PfEMP1 that the body has never seen before. In its new disguise, *P. falciparum* can invade more red blood cells and cause another wave of fever, headaches, nausea, and chills.

These sticky surface proteins are the antigens that the body's immune system recognizes and attacks. Once the immune system fights off one version of falciparum malaria, the parasite alters which gene is expressed. The resulting antigenic variation blinds the immune system, allowing a new wave of illness. The body now has to repeat the recognition and attack responses all over again. The parasite has 60 cloaking genes, called *var*, that can be turned on and off individually, changing the organisms' antigenic structure, like a criminal repeatedly changing his disguise to elude police.

In April 2005, Australian scientists Alan Cowman, Brendan Crabb, and colleagues (Walter and Eliza Hall of Institute of Medical Research) showed that *var* genes are regulated by chromosome packaging, which unwraps one gene to be expressed at a time and literally packs away the inactive genes. DNA can be encased so securely by some proteins that other proteins cannot access the nucleic acid for transcription, a process known as epigenetic silencing. Becoming immune to all the types of malaria can take upwards of five years and requires constant exposure; otherwise the immunity is lost. Many children do not live long enough to gain immunity to malaria in all its forms.

Diagnosis involves microscopic demonstration of the protist within erythrocytes (Wright stain) or through serology to identify antimalarial antibodies. Treatment regimens include chloroquine or mefloquine, which kill the organisms in their erythrocytic asexual stages, and primaquine, effective in the exoerythrocytic stages. The chloroquine family of drugs acts by interfering with the detoxification of heme generated from hemoglobin digestion. Malaria parasites accumulate the hemoglobin released from red blood cells in plasmodial lysosomes, where digestion occurs. The parasites use the amino acids from hemoglobin to grow but find free heme toxic. To prevent eating themselves to death from accumulating too much heme, the organism detoxifies heme via polymerization, which produces a black pigment. Many antimalarial drugs, such as chloroquine, prevent polymerization by binding to the heme. As a result, the increased iron (from heme) level kills the parasite.

Chloroquine is given prophylactically to persons traveling to endemic areas. Unfortunately, *Plasmodium* has been developing resistance to these drugs, forcing development of new ones. The antigenic shape-shifting carried out by this parasite has so far stymied development of an effective vaccine.

TO SUMMARIZE:

- **Blood cultures** are useful in diagnosing septicemia and endocarditis.
- **Endocarditis** can have acute or subacute onsets.
- **Subacute bacterial endocarditis** is usually an endogenous infection caused by *S. mutans*.
- ***Chlamydophila pneumoniae* infection** may contribute to atherosclerosis.
- **Malaria, caused by *Plasmodium* species**, manifests as repeated episodes of chills, fever, and sweating owing to the organism's ability to alter the antigenic appearance of its surface proteins and evade the immune response.

26.8 Systemic Infections

Many pathogenic bacteria can produce a septicemia as a way to disseminate throughout the body and infect other organs. These organisms cause what are considered systemic infections.

Case History: The Plague

*A 25-year-old New Mexico rancher was admitted to an El Paso hospital because of a two-day history of headache, chills, and fever (40°C). The day before admission, he began vomiting. The day of admission, an orange-sized, painful swelling in the right groin area was noted (**Fig. 26.24A**). A lymph node aspirate and a smear of peripheral blood were reported to contain gram-negative rods that exhibited bipolar staining (**Fig. 26.24B**). The patient's white blood cell count was 24,700/μl (normal is 4,300–10,800/μl), and platelet count was 72,000/μl (normal is 130,000–400,000/μl). In the two weeks prior to becoming ill, the patient had trapped, killed, and skinned two prairie dogs, four coyotes, and one bobcat. The patient had cut his left hand shortly before skinning a prairie dog. PCR and typical biochemical testing of a gram-negative rod isolated from blood cultures identified the organism as* Yersinia pestis, *the organism that causes plague. The patient received an antibiotic cocktail of gentamicin and tetracycline. He eventually recovered after six weeks in intensive care.*

Plague is caused by the bacterium *Yersinia pestis*, which can infect both humans and animals. During the Middle Ages, the disease, known as the Black Death, decimated over a third of the population of Europe. Such was the horror it evoked that invading armies would actually catapult dead plague victims into embattled fortresses. This was probably the first use of biowarfare.

Contrary to popular belief, *Y. pestis* is present in the United States. The organism is endemic in 17 western states. It is normally transmitted from animal to animal, typically rodents like rats and even prairie dogs (**Fig. 26.24C**) by the bite of infected fleas. **Figure 26.25** illustrates the various infective cycles of the plague bacillus. Humans are not typically part of the natural infectious cycle. However, in the absence of an animal host, the flea can take a blood meal from humans and thereby transmit the disease to them. During the Middle Ages, urban rats venturing back and forth to the countryside became infected by the fleas of wild rodents that served as a res-

ervoir. Upon returning to the city, the rat flea passed the organism on to other rats, which then died in droves. The rat fleas, deprived of their normal meal, were forced to feed on city dwellers, passing the disease on to them.

Individuals bitten by an infected flea or accidentally infected through a cut while skinning an infected animal first exhibit the symptoms of **bubonic plague**. Bubonic plague emerges as the organism moves from the site of infection to the lymph nodes, producing characteristically enlarged nodes called buboes (see **Fig. 26.24A**). From the lymph nodes, the pathogen can enter the bloodstream, causing **septicemic plague**. In this phase, the patient can go into shock from the massive amount of endotoxin in the bloodstream. Neither bubonic nor septicemic plague is passed from person to person. As the organism courses through the bloodstream, however, it will invade the lungs and produce **pneumonic plague**, which can be easily transmitted from person to person through aerosol droplets generated by coughing (**Fig. 26.24D**). Pneu-

monic plague is the most dangerous form of the disease because it can kill quickly and spread rapidly through a population. Pneumonic plague is so virulent that an untreated patient can die within 24–48 hours. Identification of the organism is usually achieved postmortem.

The organism has numerous virulence factors, including so-called V and W cell surface lipoprotein antigens that inhibit phagocytosis. Another factor, the F1 protein surface antigen, is partly responsible for *Y. pestis* survival in the gut of the flea. It blocks flea digestion, making the flea feel "starved" even after a blood meal. Therefore, the flea jumps from host to host in a futile effort to feel full. This curious effect on the insect vector is another unique aspect of how plague spreads so quickly.

Y. pestis also uses type III secretion systems to inject virulence proteins (YopB and YopD) into host cell membranes. Unlike *Salmonella*, which uses type III–secreted proteins to gain entrance into host cells, *Y. pestis* is not primarily an intracellular pathogen. Injection of the Yop proteins disrupts the actin cytoskeleton and so helps the organism evade phagocytosis. By evading phagocytosis, the organism avoids triggering an inflammatory response and produces massive tissue colonization.

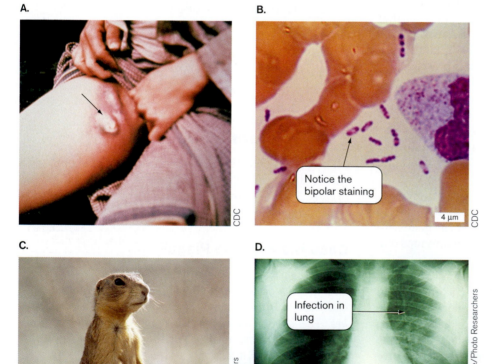

Figure 26.24 The plague. **A.** Classic bubo (swollen lymph node) of bubonic plague. **B.** *Yersinia pestis*, bipolar staining (length 1 to 3 μm). **C.** Prairie dogs are often hosts to fleas that carry plague bacilli. **D.** X-ray of pneumonic plague, showing bilateral pulmonary infection.

Plague has disappeared from Europe; the last major outbreak occurred in 1772. The reason for its disappearance is not known but is thought to be the result of human interventions. Although it wasn't until the nineteenth century that doctors understood how germs could cause disease, Europeans recognized by the sixteenth century that plague was contagious and could be carried from one area to another. Beginning in the late seventeenth century, governments created a medical boundary, or *cordon sanitaire*, between Europe and the areas to the east from which epidemics came. Ships traveling west from the Ottoman Empire were forced to wait in quarantine before passengers and cargo could be unloaded. Those who attempted to evade medical quarantine were shot.

Case History: Lyme Disease

Brad, a 9-year-old from Connecticut, developed a fever and a large (8-cm) reddish rash with a clear center (Erythema migrans) on his trunk (**Fig. 26.26A**). *He also had some left facial nerve palsy. Brad had returned a week previously from a Boy Scout camping trip to the local woods, where he did a lot of hiking. When asked by his physician, Brad admitted finding a tick on his stomach while in the woods, but thought little of it. The doctor ordered serological tests for* Borrelia burgdorferi *(the organism responsible for Lyme disease),* Rickettsia rickettsii *(which produces Rocky Mountain spotted fever), and* Ehrlichia equi *(which causes ehrlichiosis). The ELISA test for* B. burgdorferi *came back positive, confirming a diagnosis of Lyme disease. The boy was given a three-week regimen of doxycycline (a tetracycline derivative), which led to resolution of the rash and palsy.*

Lyme arthritis was first reported in Lyme, Connecticut, in the 1970s, but the causative organism, *B. burgdorferi*, was not identified until 1982. Since then, Lyme disease (aka borreliosis) has become the most common vector-borne illness in the United States and is considered an emerging infectious disease. The main endemic areas are the northeastern coastal area from Massachusetts to Maryland, Wisconsin and Minnesota, and northern California and Oregon. Lyme disease is also common to parts of Europe, such as Sweden, Germany, Austria, Switzerland, and Russia.

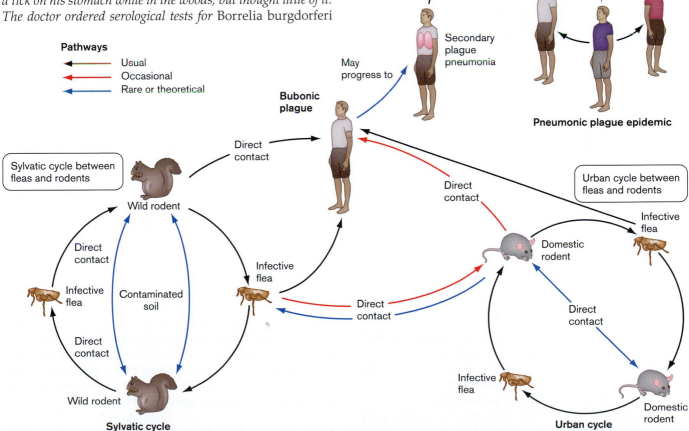

Figure 26.25   The cycles of plague.   The sylvatic cycle occurs in the wild, where fleas transmit the organism between rodents. An accidental interaction with urban rats can trigger a similar urban cycle. Humans can be infected through contact with infected fleas coming from either cycle. This starts bubonic plague symptoms that can progress to pneumonic plague. Pneumonic plague is highly infectious, which can cause epidemic spread of the disease.

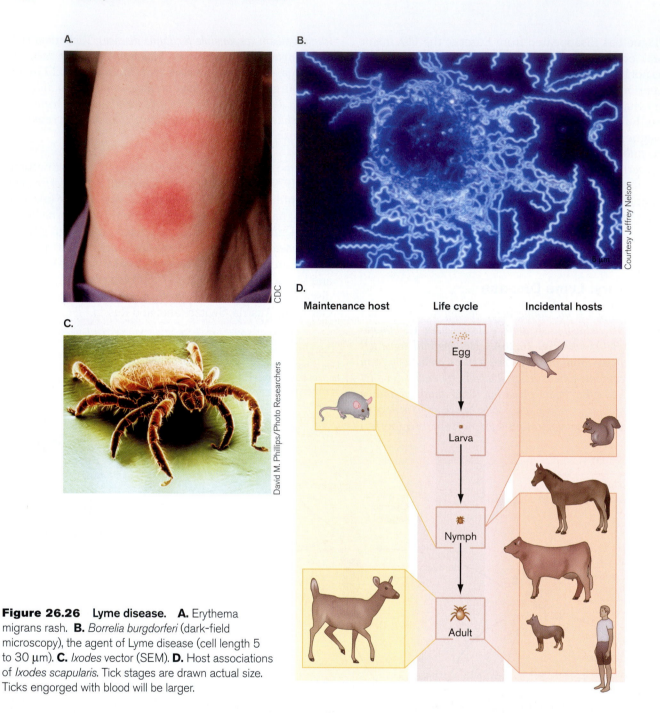

Figure 26.26   Lyme disease.   A. Erythema migrans rash. **B.** *Borrelia burgdorferi* (dark-field microscopy), the agent of Lyme disease (cell length 5 to 30 μm). **C.** *Ixodes* vector (SEM). **D.** Host associations of *Ixodes scapularis*. Tick stages are drawn actual size. Ticks engorged with blood will be larger.

B. burgdorferi is a spirochete (**Fig. 26.26B**) transmitted to humans by ixodid ticks (hard ticks; **Figs. 26.26 C** and **D**). In the northeast and central United States, where most cases occur, the deer tick *Ixodes scapularis* transmits the spirochete, usually during the summer months. In the western United States, *I. pacificus* is the tick vector.

During its nymphal stage (the stage after taking its first blood meal), *I. scapularis* is the size of a poppy seed. Its bite is painless, so it is easily overlooked. Infection takes place when the tick feeds because the spirochete is regurgitated into the host. However, the organism grows in the tick's digestive tract and takes about two days to

make its way to the tick's salivary gland, so if the tick is removed before that time, the patient will not be infected. Once it is transferred to the human, the microbe can travel rapidly via the bloodstream to any area in the body, but prefers to grow in skin, nerve tissue, synovium (joint lining), and the conduction system of the heart.

Lyme disease has three stages. Three general stages of Lyme disease are recognized. Stage 1 infections occur 3–30 days after the initial exposure. Approximately 75% of patients experience an erythema migrans rash, usually at the site of the tick bite, that varies in appearance but is classically erythematous with central clearing ("bulls-eye" rash). This stage is often associated with constitutional symptoms such as fever, myalgias, arthralgias, and headache.

Stage 2 occurs weeks to months after the initial infection. In this stage, the patient can be quite ill with malaise, myalgias (muscle pain), arthralgias (joint pain), or neurological or cardiac involvement. Common neurological manifestations include Bell's palsy (facial paralysis), inflammation of spinal nerve roots, and chronic meningitis. The most common cardiac manifestation is an irregular heart rhythm.

Stage 3 borreliosis occurs months to years later and can involve the synovium, nervous system, and skin, though skin involvement is more common in Europe than in North America. Arthritis occurs in the majority of previously untreated patients; it is usually intermittent, involves the large joints, particularly the knee, and lasts from weeks to months in any given joint. Joint fluid analysis typically shows a WBC count of 10,000 to 30,000/μl. Late neurological involvement may include peripheral neuropathy and encephalopathy, manifested by memory, mood, and sleep disturbances.

Treatment with antibiotics is recommended for all stages of Lyme disease, but is most effective in the early stages. Treatment for early Lyme disease (stage 1) is a course of doxycycline for 14–28 days. Lyme arthritis is typically slow to respond to antibiotic therapy. Despite antimicrobial drug treatment, patients with persistent active arthritis and persistently positive PCR tests may have incomplete microbial eradication and may be the most likely to benefit from repeated treatment with injected antibiotics.

Curiously, 25% of people infected with *B. burgdorferi* never experience erythema migrans, and many infected individuals are also unsure of tick bites. Hence, patients with Lyme disease may present with arthritis as their first complaint. Although this makes diagnosis extremely difficult, knowing that the patient lives or recently traveled to an endemic area can provide a critical clue.

Many other bacteria can cause septicemia and systemic illness (**Table 26.8**). Gram-negative organisms like *E. coli*, *S. typhi*, and *Francisella* and gram-positive microbes like *S. aureus*, *Enterococcus*, and *Bacillus anthracis* can grow in the bloodstream if they can gain entrance. Even anaerobes that are normal inhabitants of the intestine (for example, *Bacteroides fragilis*) can be lethal if they escape the intestine and enter the blood, as might happen following surgery. This is one reason surgical patients are given massive doses of antibiotics immediately before and after surgery.

Hepatitis Viruses Target the Liver

Hepatitis is a general term meaning inflammation of the liver. Hepatitis is caused by several viruses, including hepatitis A, B, and C viruses. Although these viruses are members of very different families, they all target the liver. We include them under the section on systemic infections because their infectious routes take them to the bloodstream before arriving at the liver.

Hepatitis A virus (HAV) is a single-stranded RNA picornavirus (**Fig. 26.27A**) that causes an acute infection spread person-to-person by the fecal-oral route, but can also result from eating undercooked shellfish collected from contaminated waters. The organism replicates in the intestinal endothelium and is disseminated via the bloodstream to the liver. After replicating in hepatocytes, the progeny enter the bile and are released into the small intestine, explaining why stools are so infectious. Though the virus has an early viremic stage after leaving the intestine, it is rarely transmitted by transfusion because the viremic stage is transient, ending after development of liver symptoms. In contrast, hepatitis B and C viruses produce persistent viremia and are readily transmitted by transfusion.

Many people who are infected with HAV are asymptomatic or exhibit very mild symptoms that include nausea, vomiting, diarrhea, low-grade fever, and fatigue. As the virus attacks the liver, the patient may become jaundiced (from accumulation of bilirubin in the skin), and their urine will turn dark brown. There is no specific treatment, but the disease usually lasts for only a few months and then resolves without establishing a carrier state. Disease can, however, be prevented if immunoglobulin is given to someone who has had contact with an infected individual. There is an inactivated hepatitis A vaccine available called HepA vaccine, which is administered after 1 year of age. For those not vaccinated, frequent hand washing is important for preventing the spread of the disease because it interrupts the fecal-oral cycle.

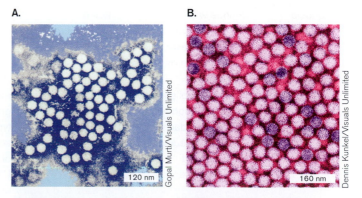

Figure 26.27   Structures of hepatitis A and hepatitis B viruses.   Both viruses are icosahedral in shape. **A.** Hepatitis A (TEM), spread by fecal-oral route, is a single-stranded RNA virus in the Picornaviridae family (30 nm in diameter). **B.** Hepatitis B (TEM) is an enveloped, double-stranded DNA virus in the Hepadnaviridae family (40 nm in diameter).

Table 26.8 Microbes causing systemic disease.

Disease	Symptoms	Etiologic agent	Virulence properties	Source	Treatment	Vaccine
Lyme disease	Stage 1: rash Stage 2: chills, headache, malaise, systemic involvement Stage 3: neurological changes	*Borrelia burgdorferi* (spirochete)	Antigenic variation, OspE (binds complement)	Deer tick	Penicillin, tetracycline	+
Brucellosis	Fever, weakness, sweats, splenomegaly, osteomyelitis, endocarditis, others	*Brucella abortis* (G– rod)	Intracellular, growth in monocytes	Animal products, unpasteurized milk	Doxycycline	
Leptospirosis	Fever, photophobia, headache, abdominal pain, skin rash, liver involvement, jaundice,	*Leptospira interrogans* (spirochete)	Burrowing motility	Urine of infected animals	Erythromycin, penicillin	
Epidemic typhus	Chills, fever, headache, muscle pain, splenomegaly, coma	*Rickettsia prowazekii* (G– rod)	Obligate intracellular growth, escapes phagosome	Human louse, flying squirrel flea	Tetracycline, chloramphenicol	
Tularemia	Fever, chills, headache, muscle pain, rash, bacteremia	*Francisella tularensis* (G– rod)	Intracellular	Rabbits, rodents, insect vectors	Gentamicin, streptomycin	
Typhoid fever	Septicemia, chills, fever, hypotension, rash (rose spots)	*Salmonella typhi* (G– rod)	Type III secretion, intracellular growth, PhoPQ regulators, Vi antigen capsule	Gallbladder of human carrier	C. profloxacin, ceftriaxone	Vi antigen
Typhoid fever	Septicemia, chills, fever, hypotension	*Salmonella cholerasuis* (G–) rod	Intracellular growth, invasin	Animals, poultry	Ceftriaxone	
Vibriosis	Serious with immunocompromised patients; fever, chills, multi-organ damage, death	*Vibrio vulnificus* (G– curved rod)	Cytolysin, capsule	Seawater, raw oysters	Tetracycline plus aminoglycoside	
Bubonic plague	Buboes (swollen lymph glands), high fever, chills, headache, cough, pneumonia, septicemia	*Yersinia pestis* (G– rod)	Intracellular growth, type III secretion of YOPs (*Yersinia* outer proteins), phospholipase D, toxin	Rodents, rodent fleas, human respiratory aerosol, potential bioterrorism agent	Streptomycin or tetracycline	

In contrast to HAV, hepatitis B virus is a partially double-stranded circular DNA virus (family Hepadna-viridae) that causes diseases of varying severity. These include acute and chronic hepatitis; cirrhosis; and hepa-tocarcinoma. The virus also wears a membrane envelope donned when progeny viruses are released from infected cells. The virion coat protein, a surface antigen, is called HBsAg. The virus makes an excess amount of HBsAg, so it is sometimes extended as a tubular tail on one side of the virus particle and is often found in the blood of infected individuals in the form of noninfectious filamen-tous and spherical particles (**Fig. 26.27B**). The presence of HBsAg in blood in an indicator of HBV infection.

HBV is primarily transferred via blood transfusions, con-taminated needles shared by IV drug users, and any human body fluid (saliva, semen, sweat, breast milk, tears, urine, feces). It can be transferred transplacentally to a fetus and can be sexually transmitted. Infection by HBV has two stages: a short-term acute phase and a long-term chronic phase that, if it extends beyond six months, may never resolve. Symp-toms resemble the flu but with jaundice and brown urine. Liver damage caused by HBV infection is in large part due to an efficient cell-mediated immune response. Cytotoxic T cells and natural killer cells cause immune lysis of infected liver cells. Over the long term, chronic hepatitis will lead to a scarred and hardened liver (cirrhosis), the only recourse being a liver transplant. Fortunately, about 90% of those infected are able to fight off infection and never proceed to the chronic stage. A HepB vaccine (made from recombinant HBsAg) is available. Its administration is recommended after birth followed by booster shots administered by 2 months and 18 months of age.

Hepatitis C virus (HCV) causes another form of hep-atitis. HCV is a single-stranded linear DNA virus with a lipid coat; it is a member of the Flaviviridae family. It is transmitted by blood transfusions and is, in fact, responsi-ble for 90% of transfusion-related cases of hepatitis. It can also be transmitted by needle sticks and, less frequently, by sex. Over 100 million people worldwide are infected with HCV. Most (80%) HCV-infected individuals do not exhibit symptoms, and in those that do, symptoms may not appear for 10–20 years. At least 75% of patients that exhibit symptoms ultimately progress to chronic hepati-tis requiring a liver transplant. Fortunately, infection can be detected using an ELISA assay. Liver biopsies of HCV patients are used to determine the extent of liver damage, which in turn helps establish the stage of disease.

Prevention of HBC or HCV infection for health care personnel includes avoiding inadvertent needle sticks and, if one should occur, the administration of immu-noglobulin within seven days. Chronic hepatitis can be treated with interferon but only with modest success. Though vaccines have been developed for HAV and HBV, no vaccine is yet available for HCV. It is important to note that since hepatitis viruses can be spread via contaminated blood products, all blood donations collected by the Red Cross and other agencies are tested for the presence of these viruses, as well as for HIV. Thus, the blood supply is safe.

TO SUMMARIZE:

- **Septicemia** is caused by many gram-positive and gram-negative bacterial pathogens. It can start with the bite of an infected insect, introduction via a wound, escape from an abscess, or by the pathogen penetrat-ing mucosal epithelium (as through the intestine or vagina); it can lead to disseminated, systemic disease.
- **Plague** has sylvatic and urban infection cycles involv-ing transmission between fleas and rats.
- **Y. pestis-infected flea** bites lead to bubonic plague. Bubonic plague can progress to septicemic and pneu-monic stages.
- **Aerosolized respiratory secretions** will directly spread *Y. pestis* pneumonic plague from person to person (no insect vector).
- **Lyme disease** is caused by the spirochete *Borrelia burgdorferi*, which is transmitted from animal reser-voirs to humans by the bite of *Ixodes* ticks.
- **Three stages of Lyme disease**: stage 1, a bull's-eye rash (erythema migrans); stage 2, joint, muscle, and nerve pain; stage 3, arthritis with WBCS in the joint fluid.
- **Hepatitis** is caused by several unrelated viruses; among them, HAV, HBV, and HCV account for most disease.
- **HAV** is transmitted by the fecal-oral route, does not establish chronic infection, and can be prevented by a vaccine.
- **HBV and HCV** can be transmitted by blood prod-ucts (such as transfusions) and shared hypodermic needles and can lead to chronic hepatitis.
- **Vaccine for HBV**, but not HCV, is available.

26.9 Immunization

In our discussion of various infections caused by bacteria and viruses, we often noted the availability of vaccines that prevent disease. In Chapter 24, we also listed many of the available vaccines and their basic makeup (see Table 24.2). As noted in that table, some vaccines utilized killed organisms (examples are the hepatitis A vaccine or the Salk inactivated polio vaccine), while others contained attenuated microbes (BCG for tuberculosis or the Sabin live polio vaccine) or consist of purified components of an infectious agent (*Streptococcus pneumoniae* and *Haemophi-lus influenzae b* capsular antigens). Some vaccines are injected in combination as polyvalent vaccines to control multiple diseases (measles, mumps, and rubella), while others are given individually but may change from year to year (influenza viral capsid proteins).

Table 26.9   Recommended childhood and adolescent immunization schedule, by vaccine and age—United States, 2007.

Vaccine	Age ⟶ Birth	1 month	2 months	4 months	6 months	12 months
Hepatitis B[a]	HepB (dose1)	HepB (dose 2)			HepB	
Rotavirus			Rota (dose1)	Rota (dose 2)	Rota (dose 3)	
Diphtheria, tetanus, pertussis[b]			DTaP (dose1)	DTaP (dose 2)	DTaP (dose 3)	
Haemophilus influenzae type b[c]			Hib (dose1)	Hib (dose 2)	Hib[c] (dose 3)	Hib (dose 4)
Inactivated poliovirus			IPV (dose1)	IPV (dose 2)	IPV (dose 3)	
Measles, mumps, rubella[d]						MMR (dose 1)
Varicella[e]						Varicella (dose 1)
Meningococcal[f]						
Pneumococcal[g]			PCV (dose 1)	PCV (dose 2)	PCV (dose 3)	PCV
Influenza[h]					Influenza (yearly)	
Hepatitis A[i]						HepA series

This schedule indicates the recommended ages for routine administration of currently licensed childhood vaccines, as of December 1, 2005.

▉ Indicates age groups that warrant special effort to administer those vaccines not previously administered.

▉ Range of recommended ages   ▉ Catch-up immunization   ▉ Assessment at age 11–12 years

[a]Hepatitis B vaccine (HepB). *At birth:* All newborns should receive monovalent HepB soon after birth and before hospital discharge.

[b]Diphtheria and tetanus toxoids and acellular pertussis vaccine (DTaP).

[c]*Haemophilus influenzae* type b conjugate vaccine (Hib).

[d]Measles, mumps, and rubella vaccine (MMR).

[e]Varicella vaccine.

[f]Meningococcal vaccine (MCV4). Meningococcal conjugate vaccine (MCV4) should be administered to all children at age 11–12 years as well as to unvaccinated adolescents at high school entry (age 15 years). Vaccine contains four types of capsule.

[g]Pneumococcal vaccine. The heptavalent pneumococcal conjugate vaccine (PCV) is recommended for all children 2–23 months and for certain children aged 24–59 months. The final dose in the series should be administered at age ≥12 months. Pneumococcal polysaccharide vaccine (PPV) is recommended in addition to PCV for certain high-risk groups. PPV contains 23-capsule antigens.

[h]Influenza vaccine. Influenza vaccine is recommended annually for children aged ≥6 months with certain risk factors (including, but not limited to, asthma, cardiac disease, sickle-cell disease, human immunodeficiency virus infection, diabetes, and conditions that can compromise respiratory function or handling of respiratory secretions or that can increase the risk of aspiration), health-care workers, and other persons (including household members) in close contact with persons in groups at high risk (see *MMWR* 2005; 54[No. RR-8]).

[i]Hepatitis A vaccine (HepA). HepA is recommended for all children at age 1 year (i.e., 12–23 months).

Source: http://www.immunize.org/cdc/child-schedule.pdf

Contrary to what you may think, multiple vaccines can be given simultaneously and safely. Most vaccines are administered during childhood, when the diseases can be most devastating. **Table 26.9** provides the current immunization schedule recommended for children and adolescents by the Centers for Disease Control and Prevention. As you will notice, all of these vaccines are given in multiple doses, called booster doses. The exception to this rule is the influenza vaccine, which is given in a single dose but changes every year. The reason for multiple doses, as noted in Chapter 24, is that secondary exposure to an antigen provides a more robust and long-lasting immunity. However, most vaccines are not administered until 2 months of age because maternal antibody crossing the placenta to the fetus (or to a newborn through breast milk) will persist for a short time in the newborn, temporarily protecting the baby from disease and possibly dampening the response to an antigen administered during that time.

You might wonder whether all members of a community must be vaccinated against a given microbe in order to lower the risk of disease for every individual in that community. It depends on the disease. For infections spread by person-to-person contact, the risk of disease to an unvaccinated person can actually be lowered dramatically if only around two-thirds of the community are vaccinated. This concept, called community (or herd) immunity, works because it interrupts the transmission

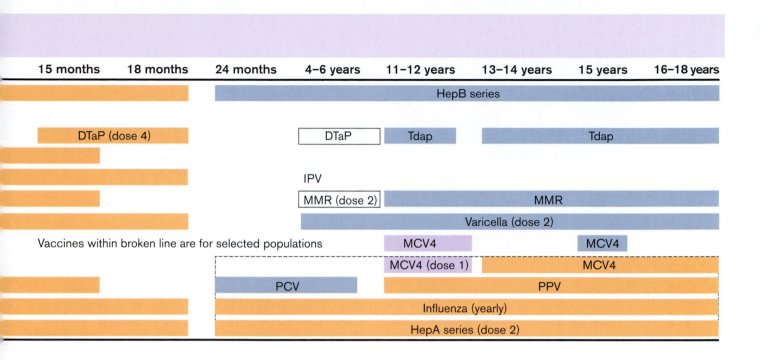

15 months	18 months	24 months	4–6 years	11–12 years	13–14 years	15 years	16–18 years

HepB series

DTaP (dose 4)   DTaP   Tdap   Tdap

IPV

MMR (dose 2)   MMR

Varicella (dose 2)

Vaccines within broken line are for selected populations

MCV4   MCV4

MCV4 (dose 1)   MCV4

PCV   PPV

Influenza (yearly)

HepA series (dose 2)

of the disease. If one individual contracts the disease, the chance that he or she will come into contact with another unvaccinated person and transmit the disease is much reduced. Thus, the risk of disease to any one unvaccinated person is lessened as a result of community immunity.

Herd immunity works well for diseases such as diphtheria, whooping cough (pertussis), measles, and mumps. However, herd immunity will not lower the risk of an unvaccinated person contracting tetanus, which is not spread by person-to-person contact. *Clostridium tetani*, the agent whose toxin causes tetanus, is a ubiquitous soil organism transmitted through punctured skin. The risk of tetanus for an unvaccinated person doesn't change even if every other person in the community is vaccinated against tetanus.

It is important to note that the vast majority of people who receive vaccines suffer no, or only mild, reactions, such as fever or soreness at the injection site. Very rarely do more serious side effects, such as allergic reactions, occur. Vaccines are *extremely* safe. You may find some unsettling, erroneous information on the Internet purporting a link between vaccinations and other diseases, such as diabetes or autism, but no well-controlled scientific study supports these claims. Clearly, the risks, including death, associated with these preventable infectious diseases are far greater than the minimal risk associated with being vaccinated against them. As proof of this point, the undervaccination of children in the United States in the 1980s and 1990s has led to an increase in cases of whooping cough (25,616 in 2005), caused by *Bordetella pertussis*. Nevertheless, the successful vaccination programs car-

ried out in the United States have come close to eradicating many diseases once feared, such as polio, rubella, and diphtheria, and have dramatically reduced the morbidity and mortality of many others.

The one caveat to this is that the current risk of infection for some diseases (because they are so rare) is lower than the risk of an adverse reaction to immunization. However, failing to vaccinate, as just mentioned, will result in a population susceptible to the microbe and a resurgence of serious disease.

TO SUMMARIZE:

- **Vaccines** can be made from live attenuated organisms, killed organisms, or purified microbe components.
- **Herd immunity** can help protect unimmunized persons from diseases transmitted from person to person.
- **Serious side effects** very rarely result from immunizations.

Concluding Thoughts

This chapter described key concepts of infectious disease using just a small sample of disease-causing pathogens. Many more pathogens can be explored in further studies. The principal goal with this and the previous chapter was to illustrate how microbial metabolism can undermine the physiology of a human host. The next chapter will describe how humans fight back using pharmacology to sabotage the physiology of the infecting microbes.

CHAPTER REVIEW

Review Questions

1. How are pathogens classified as to their portal of entry?
2. Discuss some common skin infections.
3. What causes boils?
4. What are the symptoms of necrotizing fasciitis?
5. What is the difference between primary and secondary infections?
6. How do pneumococci avoid engulfment by phagocytes?
7. What causes the clouding seen in X-rays of infected lungs?
8. What are the key features of the pneumococcal vaccine?
9. Name the common fungal causes of lung disease.
10. What is a metastatic lesion?
11. Why is diarrhea watery?
12. What is the most common microbial cause of diarrhea?
13. List common bacterial agents that cause diarrhea.
14. Would you suspect *Salmonella* infection in a cluster of nauseated patients rushed to the hospital directly from a church picnic? Why or why not?
15. What is the significance of finding leukocytes in stool?
16. What is one reservoir of *E. coli* O157:H7?
17. How is a diagnosis of a UTI made?
18. What is an important virulence determinant of uropathogenic *E. coli*?
19. What is the most common sexually transmitted disease?
20. How is gonorrhea different in men and women?
21. Why will *N. gonorrhoeae* not usually disseminate in the bloodstream, while *N. meningitidis* will?
22. Name the major causes of bacterial meningitis. What are the two routes of infection?
23. If tetanus and botulism toxins have the same mode of action, why do they cause opposite effects on muscles?
24. What are some virulence factors of *Y. pestis*?

Key Terms

bacteremia (1014)
blood-brain barrier (1007)
bubonic plague (1018)
chancre (1002)
chlamydia (1003)
coagulase (982)
congenital syphilis (1003)
dysentery (992)
epidemiology (994)
exfoliative toxin (983)
influenza (992)

meningitis (1007)
metastatic lesion (989)
multivalent vaccine (987)
necrotizing fasciitis (983)
pneumonic plague (1018)
primary syphilis (1002)
prion (1011)
respiratory syncytial virus (RSV) (992)
rotavirus (996)
rubella virus (985)
secondary syphilis (1002)

septicemia (1014)
septicemic plague (1018)
sequela (985)
sexually transmitted disease (STD) (1001)
spongiform encephalopathy (1013)
tertiary syphilis (1002)
tetanospasmin (1008)
transcytosis (1007)
viremia (1014)
zoonotic disease (981)

Recommended Reading

Broudy, Thomas, and Vincent Fischetti. 2003. In vivo lysogenic conversion of Tox⁻ *Streptococcus pyogenes* to Tox⁺ with lysogenic streptococci or free phage. *Infection and Immunity* **71:**3782–3786.

Clark, Ian A., Alison C. Budd, Lisa M. Alleva, and William B. Cowden. 2006. Human malarial disease: A consequence of inflammatory cytokine release. *Malaria Journal* **5:**85.

Duerr, Ann, Judith N. Wasserheit, and Lawrence Corey. 2006. HIV vaccines: New frontiers in vaccine development. *Clinical Infectious Diseases* **43:**500–511.

Eiden, Martin, Anne Buschmann, Leila Kupfer, and Martin H. Groschup. 2006. Synthetic prions. *Journal of Veterinary Medicine* **53:**251–256.

Eugene, Emmanuel, Isabelle Hoffmann, Celine Pujol, Pierre-Oliver Couraud, Sandrine Bourdoulous, and Xavier Nassif. 2002. Microvilli-like structures are associated with the internalization of virulent capsulated *Neisseria meningitidis* into vascular endothelial cells. *Journal of Cell Science* **115:**1234–1241.

Kim, Kwang Sik. 2006. Microbial translocation of the blood-brain barrier. *International Journal for Parasitology* **36:**607–614.

Perrin, Agnès, Stéphane Bonacorsi, Ettiene Carbonnelle, Driss Talibi, Philippe Dessen, et al. 2002. Comparative genomics identifies the genetic island that distinguish *Neisseria meningitidis,* the agent of cerebrospinal meningitis, from other *Neisseria* species. *Infection and Immunity* **70:**7063–7072.

Petersen, Andreas M., and Karen A. Krogfelt. 2003. *Helicobacter pylori*: an invading microorganism? A review. *FEMS Immunology and Medical Microbiology* **36:**117–126.

Russell, David G. 2007. Who puts the tubercle in tuberculosis? *Nature Reviews in Microbiology* **5:**39–47.

Trevejo, Rosalie T., Margaret C. Barr, and Robert Ashley Robinson. 2005. Important emerging bacterial zoonotic infections affecting the immunocompromised. *Veterinary Research* **36:**493–506.

Chapter 27

Antimicrobial Chemotherapy

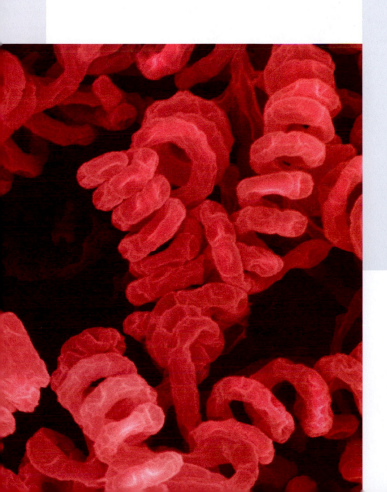

The discovery of antibiotics less than 80 years ago has played a major role in increasing life expectancy throughout the world. Prior to 1918, the average life expectancy in the United States was 45–50 years. That number is now between 70 and 75 years, thanks in part to antibiotics. Yet we face the possibility that antibiotics will become useless. Over several decades, antibiotics have been used indiscriminately to treat patients and are often included in animal feed—not to treat animals but to grow larger ones. These abuses, combined with the ability of bacteria to become antibiotic resistant, have led experts to predict an impending crisis where the human race is vulnerable to infectious diseases we once thought had been conquered.

Important questions about chemotherapeutic agents will be addressed in this chapter, including: Why do antimicrobials inhibit the growth of bacteria, but not humans or animals? What is antibiotic resistance? How do clinicians know which antibiotic to use to treat an infection? We will also discuss what makes a good antibiotic target, and how new antibiotics are discovered.

The genome of *Streptomyces avermitilis*, shown here as an SEM, may yield new antibiotics through unrecognized pathways of secondary metabolism. *S. avermitilis* produces avermectins, a group of antiparasitic agents used in human and veterinary medicine. The genome was found to contain 9,025,608 bases that encode at least 7,574 potential open reading frames (ORFs). Thirty gene clusters related to secondary metabolite biosynthesis were identified, corresponding to 6.6% of the genome. It is thought that some of these pathways may yield other useful antibiotics. *Source: Cover of Nature Biotechnology. May 2003. 21(5).*

Antibiotics undeniably have been a benefit to modern society. We live longer and contribute more to society than in the past, in large measure because of antibiotics. Infections we consider minor today often killed their victims just 50 or 60 years ago. But there has been a downside. Consider the case of a 56-year-old man with diabetes who entered the hospital for a heart transplant. The operation went well, but one week after the surgery, he developed a severe chest wound infection notable for exuded pus. Treatment with methicillin, a penicillin derivative commonly used to treat infections, failed and the patient became comatose. The diagnostic microbiology laboratory ultimately identified the agent as a methicillin-resistant *Staphylococcus aureus* (MRSA), prompting the surgeon to immediately place the patient on intravenous vancomycin for six weeks. Vancomycin is an antibiotic, structurally different from methicillin, that can usually kill methicillin-resistant bacteria. Fortunately, this patient recovered; too often, they do not.

Where did the infection come from? Surprising as it may seem, this drug-resistant pathogen was a resident of the hospital itself. To find the source, nasal swabs were taken of all hospital personnel and screened for the presence of *S. aureus*. The results revealed that several members of the surgery team actually harbored this organism as part of their normal flora. But which individual was the actual source of the patient's infection? In what could be described as an example of forensic microbiology, each strain was subjected to pulsed-field gel electrophoresis, and the resulting genomic restriction patterns were compared with that of the isolate from the infected patient. A match pointed to the source. The unwitting culprit turned out to be the perfusionist who manipulated the tubing used for cardiopulmonary bypass.

Hospital-acquired, or **nosocomial**, infections are not unusual. As many as 5–10% of all patients admitted to acute-care hospitals acquire nosocomial infections. This leads to over 80,000 deaths each year. The deaths are due, in part, to the poor health of the patient and in part to the antibiotic-resistant nature of bacteria lurking in hospitals. In fact, as much as 60–70% of staph infections that develop in a hospital setting are the result of methicillin-resistant *S. aureus*.

We begin this chapter with a discussion of the golden age of antibiotic discovery (1940–1960) and move on to describe the basic concepts of antibiotics and their use and misuse. We will also delve into the ways genomic and proteomic approaches broaden our ability to search for new antibiotics and identify new antibiotic targets—all part of our attempt to stay one step ahead of evolving, antibiotic-resistant pathogens.

27.1 The Golden Age of Antibiotic Discovery

Antibiotics (from the Greek, meaning "against life") are compounds produced by one species of microbe that can kill or inhibit the growth of other microbes. While

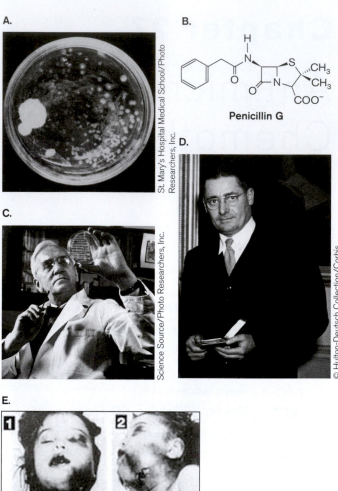

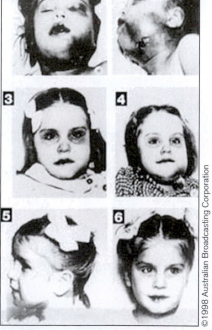

Figure 27.1   The dawn of antibiotics.   A. Alexander Fleming's photo of the dish with bacteria and penicillin mold. **B.** The chemical structure of penicillin G. **C.** Alexander Fleming at work in his laboratory. **D.** Howard Florey. **E.** Pictures taken in 1942, shortly after the introduction of penicillin, show the improvement in a child suffering from an infection four days (panel 3) and nine days (panel 5) after treatment. Panels 5 and 6 show her fully recovered.

we think of antibiotics as being a recent biotechnological development, their use has historical precedent. Ancient remedies called for cloths soaked with organic material to be placed on wounds to allow them to heal faster. This organic material likely contained "natural antibiotics" that killed bacteria and prevented further infection. The medicinal properties of molds were also recognized for centuries. Historical accounts refer to the ancient Chinese successfully treating boils with warm soil and molds scraped from cheeses, and in England, a paste of moldy bread was a home remedy for wound infections up until the beginning of the twentieth century.

The modern antibiotic revolution began with the discovery of penicillin in 1928 by Sir Alexander Fleming (1881–1955). His discovery was actually a rediscovery, and was arguably one of the greatest examples of serendipity in science. Although Fleming generally receives the credit for discovering penicillin, a French medical student, Ernest Duchesne (1875–1912), originally discovered the antibiotic properties of *Penicillium* in 1896.

Duchesne observed that Arab stable boys at the nearby army hospital kept their saddles in a dark and damp room to encourage mold to grow on them. When asked why, they told him the mold helped heal saddle sores on the horses. Intrigued, Duchesne prepared a solution from the mold and injected it into diseased guinea pigs. All recovered. Although he submitted his work as a dissertation to the Pasteur Institute, it was ignored because of his young age and because he was unknown.

Penicillium was forgotten in the scientific community until Fleming rediscovered it one day in the late 1920s. Petri dishes were glass in those days and could be rewashed and sterilized. Fleming was preparing to wash a pile of old Petri dishes he'd used to grow the pathogen *S. aureus*. He opened and examined each dish before tossing it into a cleaning solution. He noticed that one dish had grown contaminating mold, which in and of itself was not unusual in old plates, but all around the mold the staph bacteria had failed to grow (**Fig. 27.1A**). He took a sample of the mold and found it was from the penicillium family, later identified as *Penicillium notatum*. The mold appeared to have synthesized a chemical, now known as penicillin (**Fig. 27.1B**), that diffused through the agar, killing cells of *S. aureus* before they could form colonies. Fleming (**Fig. 27.1C**) presented his findings in 1929, but they raised little interest, since penicillin appeared to be unstable and would not remain active in the body long enough to kill pathogens.

As World War II began, an Oxford professor named Howard Florey (**Fig. 27.1D**), together with his colleague Ernst Chain, rediscovered Fleming's work, thought it held promise, and set about purifying penicillin. To their amazement, when the purified penicillin was injected into mice infected with staphylococci or streptococci, the majority of mice survived. Subsequent human trials also proved successful (**Fig. 27.1E**), and penicillin gained wide

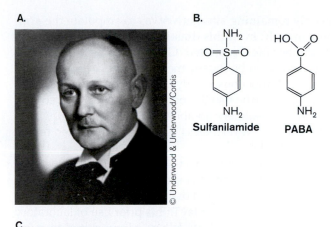

Sulfanilamide   PABA

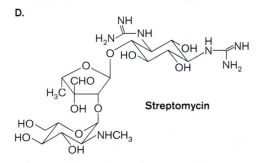

Streptomycin

Figure 27.2   The discoverers of sulfanilamide and streptomycin.   A. Gerhard Domagk discovered sulfanilamide. **B.** Chemical structure of sulfanilamide, an analog of para-aminobenzoic acid (PABA), a precursor of the vitamin folic acid, necessary for growth. Sulfonilamide inhibits one of the enzymes that converts PABA into folic acid. **C.** Selman Waksman discovered streptomycin in 1944. **D.** Chemical structure of streptomycin.

use, saving countless lives during the war. As a fitting tribute to serendipity, Fleming, Florey, and Chain received the 1945 Nobel Prize in Physiology or Medicine for their work. Duchesne's work, however, went unrecognized by the scientific community for decades.

The next landmark discovery in antibiotics was made by Gerhard Domagk (1895–1964), a German physician at the Institute of Experimental Pathology and Bacteriology who investigated antimicrobial compounds in the 1930s (**Fig. 27.2A**). In 1935, Domagk's 6-year old child was afflicted with a serious streptococcal infection induced by an innocent pinprick to the finger. The infection spread to the lymph nodes under her arm and became so severe that lancing and draining the pus 14 times did little to help.

The only remaining alternative was to amputate the arm. Unfortunately, even this drastic measure would probably not save her life. Frustrated, Gerhard Domagk took what would appear to be drastic measures. He administered a dose of an innocuous red dye he was investigating that, on agar plates (the usual medium for testing antibiotics), had shown absolutely no ability to inhibit the growth of streptococcus. Nevertheless, Domagk's daughter recovered completely.

How did Domagk conceive of such a therapy when the conventional method of screening for antibiotic activity on agar plates indicated that this compound (Prontosil) was useless? The answer lay in his prior use of laboratory animals to study the drug. When administered to mice, Prontosil was very effective at preventing infection. Had he not used live animals, Domagk would never have discovered that Prontosil was metabolized by the body into another compound, sulfanilamide, clearly lethal to the streptococcus. This finding led to an entire class of drugs, the sulfa drugs, which saved hundreds of thousands of lives, including that of his own child. Sulfanilamide is an analog of para-aminobenzoic acid (PABA), a precursor of folic acid, a vitamin necessary for nucleic acid synthesis (**Fig. 27.2B**). The drug stops bacterial growth by inhibiting the conversion of PABA to folic acid.

Folic acid is not synthesized by humans, but is a dietary supplement. Furthermore, bacteria do not transport folic acid, which explains why the sulfa drugs can inhibit bacterial growth in humans without affecting human cells.

In a dark turn, the pressing need for effective antibiotics during World War II dictated a change in the in vivo screening methods used by Domagk's German employer. Animal testing was abandoned in favor of human testing, but the humans were not volunteers. They were concentration camp prisoners intentionally infected with bacterial diseases, such as gangrene, and then treated with new chemical compounds. Domagk's possible participation in these experiments, even if unwilling, was a source of controversy that haunted him long after the war ended. Yet his contributions to medicine continued as he also developed the first effective chemotherapy for tuberculosis via the thiosemicarbazones and isoniazid, still used today.

During the same period of history, Selman Waksman (1888–1973) at Rutgers University began screening 10,000 strains of soil bacteria and fungi for an ability to inhibit growth or kill bacteria (**Fig. 27.2C**). In 1944, this Herculean effort paid off with the discovery of streptomycin, an antibiotic produced by the actinomycete *Streptomyces griseus* (**Fig. 27.2D**). Waksman's discovery of streptomycin triggered the antibiotic gold rush that is still under way.

TO SUMMARIZE:

- **The importance of antibiotics** in treating disease was recognized in the early 1940s.

- **Some antimicrobial agents are initially inactive** until converted by the body to an active agent.
- **Florey and Chain purified penicillin** and capitalized on Fleming's discovery of penicillin.

27.2 Basic Concepts of Antimicrobial Therapy

Antibiotics comprise the vast majority of chemotherapeutic agents used to treat microbial diseases. As already noted, the term *antibiotic* originally referred to any compound produced by one species of microbe that could kill or inhibit the growth of other microbes. Today, the term is also used for synthetic chemotherapeutic agents, such as sulfonamides, that are clinically useful but chemically synthesized. There are a large number of natural and synthetic compounds that affect microbial growth. However, their utility in a clinical setting is dictated by certain key characteristics.

Antibiotics Exhibit Selective Toxicity

As early as 1904, the German physician Paul Ehrlich (1854–1915) realized that a successful antimicrobial compound would be a "magic bullet" that selectively kills or inhibits the pathogen but not the host. This seemingly obvious premise was innovative at the time. Ehrlich made several discoveries based on this concept, the most celebrated of which was the arsenical compound known as salvarsan. Salvarsan proved to be quite active in killing the syphilis agent, *Treponema pallidum*. Syphilis, a sexually transmitted disease, had been untreatable and the source of considerable long-term suffering. Ehrlich's "magic bullet" concept is now known as **selective toxicity**.

Selective toxicity is possible because key aspects of a microbe's physiology are different from those of eukaryotes. For example, suitable bacterial antibiotic targets include peptidoglycan, which eukaryotic cells lack, and ribosomes, which are structurally distinct between the two domains of life. Thus, chemicals like penicillin, which prevents peptidoglycan synthesis, and tetracycline, which binds to bacterial 30S ribosomal subunits, inhibit bacterial growth but are essentially invisible to host cells, since they do not interact with them at low doses.

While their intended targets are bacterial cells, some antibiotics, particularly at high doses, can interact with elements of eukaryotic cells and cause side effects that harm the patient. For example, chloramphenicol, a drug that targets bacterial 50S ribosomal subunits, can interfere with the development of blood cells in bone marrow, a phenomenon that may result in aplastic anemia (failure to produce red blood cells). The toxicity of an antibiotic can also depend on the age of the patient. Tetracycline, for instance, can cause defects in human bone growth plates and should not be administered to children. Problems can even arise if the drug does not directly impact mammalian physiology. For example, many people develop an extreme allergic sensitivity toward penicillin, a situation in which the treatment of an infection may end up being worse than the infection itself. Physicians must be aware of these allergies and use alternative antibiotics to avoid harming their patients.

As a student of microbiology, it is important that you distinguish between drug *susceptibility* and drug *sensitivity*. A microbe is susceptible to the drug's action, but a human can develop an allergic sensitivity to the drug.

Antimicrobials Have a Limited Spectrum of Activity

No one antimicrobial drug affects all microbes. As a result, antimicrobial drugs are classified based on the type of organisms they affect. Thus, we have antifungal, antibacterial, antiprotozoan, and antiviral agents. The term antibiotic is usually reserved for antimicrobial compounds that affect bacteria. Even within a group, one agent might have a very narrow **spectrum of activity**, meaning it only affects a few species, while another antibiotic will inhibit many species. For instance, penicillin has a relatively narrow spectrum of activity, primarily killing gram-positive bacteria. However, ampicillin is penicillin with an added amino group that allows the drug to more easily penetrate the gram-negative outer membrane. As a result of this chemically engineered modification, ampicillin kills gram-positive and gram-negative organisms and is described as having a spectrum of activity broader than penicillin. There are also

antimicrobials that exhibit extremely narrow activities. One example is isoniazid, which is clinically useful only against *Mycobacterium tuberculosis*, the agent of tuberculosis. **Table 27.1** presents the spectrum of activity of select antibiotics.

Antibiotics Are Classified as Bacteriostatic or Bactericidal

Nonscientists typically believe that all antibiotics kill their intended targets. This is a misconception. Many drugs simply prevent the growth of the organism and let the body's immune system dispatch the intruding microbe. Thus, antimicrobials are also classified on the basis of whether or not they kill the microbe. An antibiotic is **bactericidal** if it kills the target microbe; it is **bacteriostatic** if it merely prevents bacterial growth.

TO SUMMARIZE:

- **Antimicrobial agents** may be produced naturally or artificially.
- **Selective toxicity** refers to the ability of an antibiotic to attack a unique component of microbial physiology that is missing or distinctly different from eukaryotic physiology.
- **Antibiotic side effects** on mammalian physiology can limit the clinical usefulness of an agent.
- **Antibiotic spectrum of activity** refers to the range of microbes that a given drug affects.
- **Bactericidal antibiotics** kill microbes; **bacteriostatic antibiotics** inhibit their growth.

Table 27.1　Antibacterial activity (spectrum) of antimicrobial agents.

Spectrum	Aerobic bacteria		Anaerobic bacteria		Examples
	Gram (+)	Gram (−)	Gram (+)	Gram (−)	
Broad	+	+	+	+	Cefoxitin, chloramphenicol, imipenem, tetracyclines
Intermediate	+	+	+	±	Carbenicillin, ticarcillin, ceftiofur, amoxicillin/ clavulanic acid, cephalosporins
	+	±	+	±	Ampicillin, amoxicillin
Narrow		+			Aztreonam, polymyxin
	+	±	+	±	Benzyl penicillin G
	+	+			Aminoglycosides, spectinomycin, sulfonamides, trimethoprim
	+	+			Fluoroquinolones
	+		+	+	Lincosamides, macrolides, pleuromutilins
	+		+		Bacitracin, vancomycin
			+	+	Nitroimidazoles

± variable activity, depending on species/strain

27.3 Measuring Drug Susceptibility

One critical decision a clinician must make when treating an infection is what antibiotic to prescribe for the patient. There are several factors to consider, including:

■ The relative effectiveness of different antibiotics on the organism causing the infection. This includes learning whether the organism isolated from a specific patient has developed resistance to the drug.

■ The average attainable tissue levels of each drug. An antibiotic may appear to work on an agar plate, but the concentration at which it affects bacterial growth may be too high to be safe in the patient. In fact, an important aspect of designing new antibiotics is to enhance the pharmacological activity of an existing drug—for example, modifying it so that the body does not break it down or quickly secrete it in urine.

Minimal Inhibitory Concentration Reflects Antibiotic Efficacy

The in vitro effectiveness of an agent is determined by measuring how little of it is needed to stop growth. This is classically measured in terms of an antibiotic's **minimal inhibitory concentration (MIC)**, defined as the lowest concentration of the drug that will prevent the growth of an organism. But the MIC for any one drug will differ among different bacterial species. For example, the MIC of ampicillin needed to stop the growth of *S. aureus* will be different from that needed to inhibit *Shigella dysenteriae*. The reasons that a drug may be more effective against one organism than another include the ease with which the drug penetrates the cell and the affinity of the drug for its molecular target.

So how do we measure MIC? As shown in **Figure 27.3**, an antibiotic is serially diluted along a row of test tubes containing nutrient broth. After dilution, the organism to be tested is inoculated at low, constant density into each tube, and the tubes are usually incubated overnight. Growth of the organism is seen as turbidity. Note that in **Figure 27.3** the tubes with the highest concentration of drug were clear, indicating no growth. The tube containing the MIC is the tube with the *lowest* concentration of drug that shows no growth. However, the MIC does not tell whether a drug is bacteriostatic or bactericidal.

> **THOUGHT QUESTION 27.1**   The drug tobramycin is added to a concentration of 1,000 µg/ml in a tube of broth from which serial twofold dilutions were made. Including the initial tube (tube 1), there are a total of ten tubes. Twenty-four hours after all the tubes are inoculated with *Listeria monocytogenes*, turbidity is observed in tubes 6–10. What is the MIC?
>
> **THOUGHT QUESTION 27.2**   What additional test performed on an MIC series of tubes will tell you whether a drug is bacteriostatic or bactericidal?

MIC determinations are very useful for estimating a single drug's effectiveness against a single bacterial pathogen isolated from a patient but not very practical when trying to screen 20 or more different drugs. Dilutions take time—time that the technician, not to mention the patient, may not have. The time required to evaluate antibiotic effectiveness can be reduced by using a strip test (like the Etest® shown in **Fig. 27.4**) that avoids the need for dilutions. The strip, containing a gradient of antibiotic, is placed over a fresh lawn of bacteria spread on an agar plate. While the bacteria are trying to grow, the drug diffuses out of the disk and into the media. Drug emanating from the more concentrated areas of the strip will travel faster and farther through the agar than will drug from the less concentrated areas of the strip. Thus, the higher concentrations will kill or inhibit the growth of cells farther from the strip than the lower concentration areas. The result is a **zone of inhibition** where the antibiotic has stopped bacterial growth. The MIC is the point at which the elliptical zone of inhibition intersects with the strip.

µg/ml

| 0.06 | 0.125 | 0.25 | 0.5 | 1.0 | 2.0 | 4.0 | 8.0 |

John W. Foster

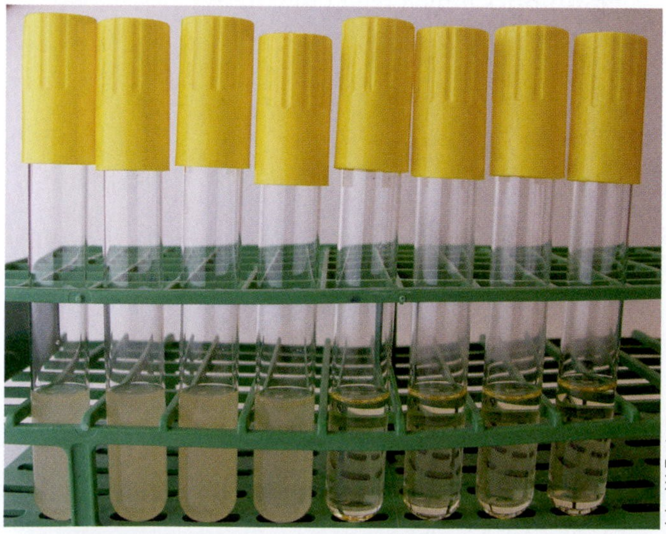

Figure 27.3   Determining minimal inhibitory concentration.   In this series of tubes, tetracycline was diluted serially starting at 8 µg/ml (far right tube). Each tube was then inoculated with an equal number of bacteria. Turbidity indicates that the antibiotic concentration was not sufficient to inhibit growth. The MIC in this example is 1.0 µg/ml.

Kirby-Bauer Disk Susceptibility Test

Although the strip test eliminates the time and effort needed to make dilutions, it would take 20 or more plates to test an equal number of antibiotics for just one bacterial

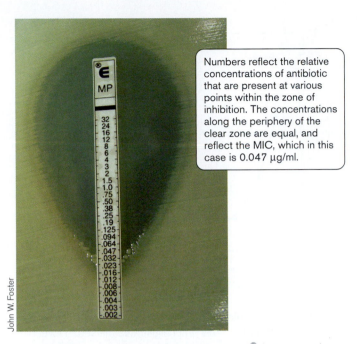

Numbers reflect the relative concentrations of antibiotic that are present at various points within the zone of inhibition. The concentrations along the periphery of the clear zone are equal, and reflect the MIC, which in this case is 0.047 µg/ml.

John W. Foster

Figure 27.4 An MIC strip test. The Etest® (AB Biodisk) is a commercially prepared strip that creates a gradient of antibiotic concentration (µg/ml) when placed on an agar plate. The MIC corresponds to the point where bacterial growth crosses the numbered strip.

isolate. Clinical labs can receive up to 100 or more isolates in one day, so individual MIC determinations are not practical. A simplified agar diffusion test, however, which can test 12 antibiotics on one plate, makes evaluating antibiotic susceptibility a manageable task.

Named for its inventors, the **Kirby-Bauer assay** uses a series of round filter paper disks impregnated with different antibiotics. A dispenser (**Fig. 27.5**) delivers up to 12 disks simultaneously to the surface of an agar plate covered by a bacterial lawn. Each disk is marked to indicate the drug used. During incubation, the drugs diffuse away from the disks into the surrounding agar and inhibit growth of the lawn to different distances. The zones of inhibition vary in width, depending on which antibiotic is used, the concentration of drug in the disk, and the susceptibility of the organism to the drug. The diameter of the zone correlates to the MIC of the antibiotic against the organism tested.

Correlations between MIC values and Kirby-Bauer zone sizes are made empirically. Every disk containing a given antibiotic is impregnated with a standard concentration of drug, and every antibiotic has a quantifiable MIC that will differ when tested against different bacterial species and strains. The outermost ring of the no-growth zone in a Kirby-Bauer disk test must, by

A.

John W. Foster

Figure 27.5 The Kirby-Bauer disk susceptibility test.
A. Device used to deliver up to 12 disks to the surface of a Mueller-Hinton plate. The device is placed over the plate, and the plunger is depressed to deliver the disks. **B.** Disks impregnated with different antibiotics are placed on a freshly laid lawn of bacteria and incubated overnight. The clear zones around certain disks indicate growth inhibition. Shown are the results with normal *Staphylococcus aureus*. **C.** Methicillin-resistant *Staphylococcus aureus* (MRSA). Note the lack of inhibition by the oxacillin disk. This strain is resistant to both methicillin and oxacillin. **D.** *Streptococcus pneumoniae*. The brownish tint of the blood agar plates outside the zones of bacterial inhibition is caused by a hemolysin secreted by the lawn of pneumococci. Penicillin (P); oxacillin (OX); erythromycin (E); clindamycin (CC); rifampin (RA); vancomycin (VA); tetracycline (TE); chloramphenicol (C); cephazolin (CZ); sulbactam-ampicillin (SAM); norfloxacin (NDR); sulfa-trimethoprim (SXT).

B.

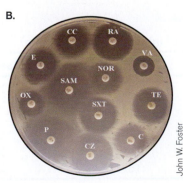

John W. Foster

C.

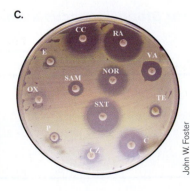

John W. Foster

D.

John W. Foster

definition, contain the minimal concentration of drug needed to prevent growth on agar. Thus, if species A and B have MIC values for penicillin of 4 µg/ml and 40 µg/ml, respectively, then species A will exhibit a proportionally larger zone of inhibition than species B in the disk test. A graph plotting MIC on one axis and zone diameter on the other provides the correlation.

After incubating agar plates in the Kirby-Bauer test, the diameters of the zones of inhibition around each disk are measured, and the results are compared with a table listing whether a zone size is wide enough (meaning the MIC is low enough) to be clinically useful. **Table 27.2** shows susceptibility data for *S. aureus*. The concentration of antibiotic used and the zone size that is considered clinically significant are correlated with the average attainable tissue level for each antibiotic. For the antibiotic to remain effective in vivo, it is important that the tissue concentration of the drug remain above the MIC; otherwise, invading bacteria will not be affected.

Correlating antibiotic MIC with tissue level. The average attainable tissue level for a drug depends on how quickly the antibiotic is cleared from the body via secretion by the kidney or destruction in the liver. It also depends on when side effects of the drug start to appear. As shown in **Figure 27.6**, as long as the concentration of the drug in tissue or blood remains higher than the MIC, the drug will be effective. The concentration can be kept at sufficient levels either by administering a higher dose, which runs the risk of side effects, or by giving a second dose at a time when the levels from the first dose have declined. This is why patients are told to take doses of some antibiotics four times a day and other antibiotics only once a day.

To ensure reproducibility, the Kirby-Bauer test was standardized a half century ago. Reproducibility means that results from a laboratory in California will match those in Alabama, Ohio, or any other state. The following are standardizations used to make the test reproducible and easier.

■ **Size of the agar plate.** The plates used (150 mm) are larger than standard agar plates (100 mm) to accommodate more disks and to maintain sufficient dis-

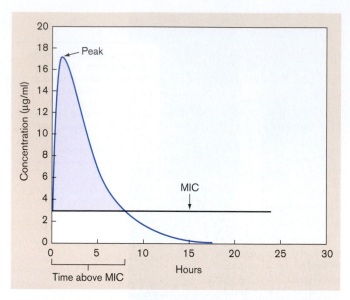

Figure 27.6   Correlation between MIC and serum or tissue level of an antibiotic. This graph illustrates the serum level of ampicillin over time. The important consideration here is how long the serum level of the antibiotic remains higher than the MIC. Once the concentration falls below the MIC, owing to destruction in the liver or clearance through the kidneys and secretion, the infectious agent fails to be controlled by the drug—in this case about 7–8 hours after the initial dose. To maintain a serum level higher than the MIC, a second dose would be taken. The shaded area of the curve represents time above MIC.

tance between disks so that zones of inhibition do not overlap.

■ **Depth of the media.** Antibiotics diffuse out of impregnated disks in not two, but three dimensions. Because diffusion cannot occur very far downward in a thinly poured agar, the drug is forced to move more laterally. Thus, the zone of inhibition measured from a thinly poured agar plate will be larger than the zone from a thick agar plate.

■ **Media composition.** Media composition can also affect results. For example, most common laboratory media contain para-aminobenzoic acid (PABA), which is used by the cell to make folic acid, a vitamin needed for purine and pyrimidine synthesis.

Table 27.2   Susceptibility results for *Staphylococcus aureus*.

Antibiotic	Quantity in disk (µg)	Zone of inhibition diameter (mm)		
		Resistant	Intermediate	Susceptible
Ampicillin	10	<12	12–13	>13
Chloramphenicol	30	<13	13–17	>17
Erythromycin	15	<14	14–17	>17
Gentamicin	10	≤12.5		>12.5
Streptomycin	10	<12	12–14	>14
Tetracycline	30	<15	15–18	>18

Sulfonamides are analogs of PABA that competitively inhibit one of the enzymes needed to convert PABA to folic acid. Media containing PABA, such as standard nutrient agar, will flood the bacterial cell with PABA and limit the ability of the sulfa drugs to compete for the enzyme. Thus, even though the drug might be effective in vivo, on the agar plate there would be no zone of inhibition. The standardized media used for the Kirby-Bauer test is called **Mueller-Hinton agar**, which contains no PABA.

- **The number of organisms spread on the agar plate.** The number of organisms placed on an agar surface is inversely proportional to the size of the zone of inhibition. The more organisms there are, the smaller the zone. The reason for this is that there is a time lag between dropping a disk on a plate and the diffusion of the antibiotic. The more organisms there are on a plate, the less time it takes for them to form visible growth; so by the time the antibiotic gets to them, visible growth has already formed. As a result, a standard optical density solution of each organism is prepared, and a cotton swab is used to spread the entire agar surface.
- **Size of the disks.** A standard diameter of 6 mm means that all antibiotics start diffusing into agar at the same point.
- **Concentrations of antibiotics in the disks.** The higher the concentration of antibiotic in a disk, the faster the drug can diffuse through the plate and kill bacteria (or at least inhibit growth) before replicating cells are able to form visible growth. To avoid differences between labs, the concentration of each given drug impregnating a disk has been standardized.
- **Incubation temperature.** Incubation temperature will not affect growth and diffusion equally. To avoid differences, a temperature of 37°C is standard.

THOUGHT QUESTION 27.3 You are testing whether a new antibiotic will be a good treatment choice for a patient with a staph infection. The Kirby-Bauer test using the organism from the patient shows a zone of inhibition of 15 mm around the disk containing this drug. Clearly, the organism is susceptible. But you conclude from other studies that the drug would not be effective in the patient. What would make you draw this conclusion?

TO SUMMARIZE:

- **The spectrum of an antibiotic and the susceptibility** of the infectious agent are critical points of information required before prescribing antibiotic therapy.
- **Minimal inhibitory concentration (MIC)** of a drug, when correlated with average attainable tissue levels of the antibiotic, can predict the effectiveness of an antibiotic in treating disease.
- **MIC is measured** using tube dilution techniques but can be approximated using the Kirby-Bauer disk diffusion technique.

27.4 Mechanisms of Action

As noted earlier, selective toxicity of an antibiotic depends on enzymes or structures unique to the bacterial target cell. The following aspects of a microbe's physiology are classic targets:

- Cell wall
- Cell membrane
- DNA synthesis
- RNA synthesis
- Protein synthesis
- Metabolism

Table 27.3 summarizes the general targets of common antibiotics. Chapters 3, 7, and 8 describe these cellular components and provide the tools needed to understand how antibiotics work.

Cell Wall Antibiotics

Bacterial cell walls are an obvious structure that could be the basis for selective toxicity because peptidoglycan does not exist in mammalian cells; thus, antibiotics that target the synthesis of these structures should selectively kill bacteria. The following case history illustrates the use of two-cell-wall targeting antibiotics and also reveals how bacteria can evolve to escape destruction.

Case History: Meningitis

A 3-year-old child was brought to the emergency room crying, with a stiff neck and high fever. Gram stain of cerebral spinal fluid revealed gram-positive cocci, generally in pairs. The diagnosis is

---

Table 27.3 **Mechanisms of action of antimicrobial agents.**

Target	Antibiotic examples
Cell wall synthesis	Penicillins, cephalosporins, bacitracin, vancomycin
Protein synthesis	Chloramphenicol, tetracyclines, aminoglycosides, macrolides, lincosamides
Cell membrane	Polymyxin, amphotericin, imidazoles vs. fungi
Nucleic acid function	Nitroimidazoles, nitrofurans, quinolones, rifampin; some antiviral compounds, especially antimetabolites
Intermediary metabolism	Sulfonamides, trimethoprim

meningitis. The physician immediately prescribed intravenous ampicillin. Unfortunately, the child's condition worsened, so antibiotic treatment was changed to a third-generation cephalosporin (which will cross the blood-brain barrier). The patient began to improve within hours and was released after two days. A report from the clinical microbiology laboratory identified the organism as Streptococcus pneumoniae.

Both of the antibiotics used in this case kill bacteria by targeting cell wall synthesis. Synthesis of peptidoglycan is a complex process but is represented simply in **Figure 27.7**. Basically, sugar molecules called *N*-acetylglucosamine (NAG) and *N*-acetylmuramic acid (NAM) are made by the cell and linked together by a **transglycosylase** enzyme into long chains assembled at the cell wall. *N*-Acetylmuramic acid contains a short side chain of amino acids that is assembled enzymatically, not by a ribosome. Rigidity of the macromolecular structure, essential for maintaining cell shape, is achieved by cross-linking the side chains from

adjacent strands. The enzyme **transpeptidase** (D-alanyl-D-alanine carboxypeptidase/transpeptidase) catalyzes the cross-link. **Figure 27.7** also indicates where several antibiotics target various stages of this assembly process.

Peptidoglycan is assembled outside the cell membrane. Before we can explain how cell wall antibiotics work, it is important to know how the cell wall is made. Synthesis of peptidoglycan starts in the cytoplasm with a uridine diphosphate (UDP)-NAM molecule. The amino acids L-alanine, D-glutamic acid, and L-lysine (or diaminopimelic acid in gram-negative organisms) are individually and sequentially added to NAM (**Fig. 27.7**, step 1); and then a dipeptide of D-alanine is attached (step 2). Next, the NAM-pentapeptide is transferred to a membrane-situated, 55-carbon, lipid molecule called bactoprenol (step 3) and UMP is released. Another sugar molecule, NAG, is then linked to NAM, once again through a UDP intermediate (step 4). All of this takes place on the cytoplasmic

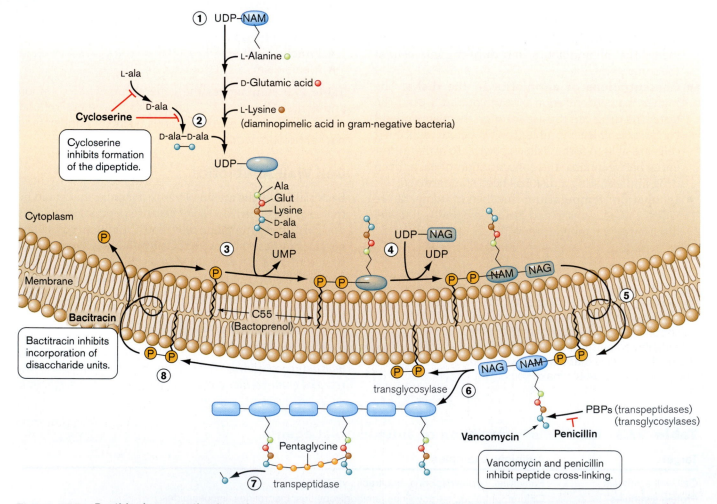

Figure 27.7   Peptidoglycan synthesis and targets of antibiotics.   A series of small-molecular-weight compounds are joined to form a disaccharide unit that will be added to preexisting chains of this unit. Steps in the synthesis are described in the text. Red lines indicate inhibition. Cycloserine inhibits ligation of the two D-alanines (step 2). Bacitracin inhibits the linking of the disaccharide units, while vancomycin and the beta-lactams, such as penicillin, inhibit the peptide cross-linking of peptidoglycan side chains.

side of the membrane. Bactoprenol carrying NAM-NAG then moves to the outer side of the cytoplasmic membrane (step 5) where transpeptidases and transglycosylases (two so-called **penicillin-binding proteins**) bind to the D-ala-D-ala part of the pentapeptide. Transglycosylase attaches the new disaccharide unit to an existing peptidoglycan chain (step 6). Transpeptidase then links two peptide side chains with a pentaglycine cross-link (in *Staphylococcus aureus*). The pentaglycine connects L-lysine on one side chain and the penultimate D-ala on the other side chain (step 7). The terminal D-ala is removed in the process. Other bacteria do not use a pentaglycine cross-link but directly form a peptide bond between L-lysine (or DAP) and the penultimate D-ala. Cross-linking, which strengthens the cell wall, can take place between either the same strand or adjacent strands. In the last part of the cycle, one of the phosphates on the now liberated bactoprenol is removed and the lipid moves back to the cytoplasmic side of the membrane, ready to pick up and taxi another unit of peptidoglycan to the growing chain (step 8).

Penicillin and other beta-lactam antibiotics target penicillin-binding proteins. Penicillin is an antibiotic derived from cysteine and valine, which combine to form

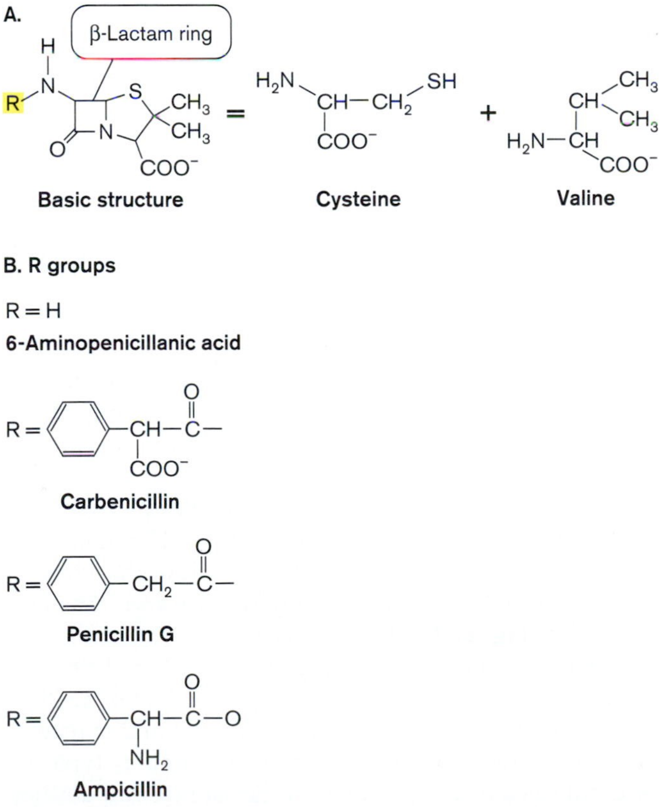

the beta-lactam ring structure shown in **Figure 27.8A**. Different R groups can be added to the basic ring structure to change the antimicrobial spectrum and stability of the derivative penicillin (**Fig. 27.8B**). Penicillin itself chemically resembles the D-ala-D-ala piece of peptidoglycan. This molecular mimicry allows the drug to bind transpeptidase and transglycosylase (which is why they are called penicillin-binding proteins), preventing their activities and halting synthesis of the chain. The consequence is a disaster for bacteria that are trying to grow larger and larger. Eventually, the growing cell bursts for lack of cell wall restraint. Penicillin, then, is a bactericidal drug (unless the treated organism is suspended in an isotonic solution). Penicillin is most effective against gram-positive organisms because the drug has difficulty passing through the gram-negative outer membrane. Ampicillin, which was used in the case history, is a modified version of penicillin that more easily penetrates this membrane and is more effective than penicillin against gram-negative microbes. Thus, ampicillin has a broader spectrum of activity than penicillin.

As noted earlier, antibiotic resistance is a growing problem throughout the world. Bacteria develop resistance to penicillin in two basic ways. The first is through inheritance of a gene encoding one of the beta-lactamase enzymes, which cleave the critical ring structure of this class of antibiotics. Beta-lactamase is transported out of the cell and into the surrounding media (for gram-positives) or the periplasm (for gram-negatives), where it can destroy penicillin before the drug even gets to the cell. Bacteria that produce beta-lactamase are still susceptible to certain modified penicillins and cephalosporins engineered to be poor substrates for the enzyme. Methicillin, for example, works well against beta-lactamase-producing microbes.

The second way a microbe can become resistant is through mutations in genes encoding key penicillin-binding proteins. Resistance occurs when the mutated gene produces an altered protein that no longer binds to the antibiotic. Methicillin-resistant bacteria use this strategy. Hospitals take special interest in methicillin-resistant *S. aureus* (MRSA) because very few drugs can kill it. One of the few remaining antibiotics effective against MRSA is vancomycin. Unfortunately, resistance to this drug is also developing. The penicillin-resistant *S. pneumoniae* in the preceding case history actually had an altered penicillin-binding protein. No beta-lactamase producing *S. pneumoniae* has ever been found.

Cephalosporins are another type of beta-lactam antibiotic originally discovered in nature but modified in the laboratory to fight microbes that are naturally resistant to penicillins (especially *Pseudomonas aeruginosa*). Over the years, the basic structure of cephalosporin has undergone a series of modifications to improve its effectiveness against penicillin-resistant pathogens. Each modification is increasingly more complex and produces what is referred to as a new "generation" of cephalosporins.

Figure 27.8 The structure of penicillins. **A.** Penicillanic acid (R = H) is derived from cysteine and valine. **B.** The R group marked in (A) can be any one of a number of different groups, some of which are shown here. Modifying this group changes the pharmacological properties and antimicrobial spectrum.

There are currently four generations of this semisynthetic antibiotic (**Fig. 27.9**). Unfortunately, the microbial world constantly adapts and eventually becomes resistant to new antibiotics. In the case of the cephalosporins, new beta-lactamases evolve that can attack the sterically buried beta-lactam rings in these molecules. It is also important to note that since the core feature of these drugs is the beta-lactam ring, persons who are sensitive to penicillins may also suffer a hypersensitivity reaction to cephalosporins.

Treatment note: The infecting strain of *S. pneumoniae* in the preceeding case history turned out to be resistant to ampicillin. If the patient had been an adult, a fluoroquinolone (discussed in Chapter 7) might have been the best secondary drug of choice because its target, a type II topoisomerase, is unrelated to cell wall synthesis. However, quinolones are not recommended for children, as in this case, due to potential side-effects. Other beta-lactam

antibiotics, such as the third-generation cephalosporins, may still work on penicillin-resistant *S. pneumoniae* since the modified antibiotic often can still bind the altered PBP. Nevertheless, vancomycin or an oxazolidinone are the best last choices, since cephalosporin-resistant strains of *S. pneumoniae* are now appearing.

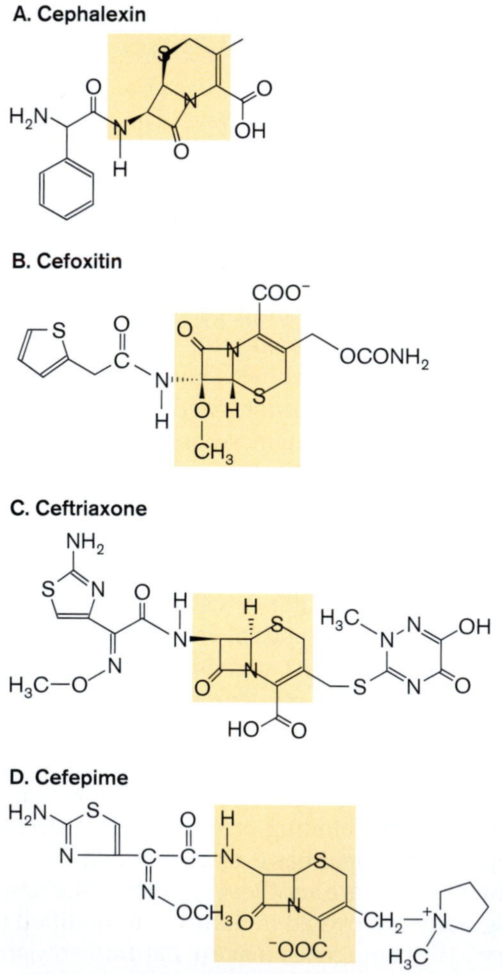

NOTE: Archaea peptidoglycan contains talosaminuronic acid instead of muramic acid and lacks the D-amino acids found in bacterial peptidoglycan. Because of this, archaea are insensitive to penicillins, which interfere with bacterial transpeptidases.

Cell wall antibiotics target other steps in peptidoglycan synthesis. Another antibiotic that affects cell wall synthesis is **bacitracin**, a large polypeptide molecule produced by *Bacillus subtilis* and *Bacillus licheniformis* (**Fig. 27.10A**). The antibiotic inhibits cell wall synthesis by binding to the bactoprenol lipid carrier molecule that normally transports monomeric units of peptidoglycan across the cell membrane and to the growing chain (see **Fig. 27.7**). Bacitracin binds to and inhibits dephosphorylation of the carrier, thereby preventing the carrier from accepting a new unit of UDP-NAM. Resistance to bacitracin can develop if the organism can rapidly recycle the phosphorylated lipid carrier molecule through dephosphorylation or if the organism possesses an efficient drug export system (discussed in Section 27.6). Normally, bacitracin is used only topically because of serious side effects, such as kidney damage, that can occur if it is ingested.

Cycloserine (made by *Streptomyces garyphalus*) is one of several antimicrobials used to treat tuberculosis (**Fig. 27.10B**). Relative to bacitracin, it acts at an even earlier step in peptidoglycan synthesis. Cycloserine inhibits the two enzymes that make the D-ala-D-ala dipeptide. As a result, the complete pentapeptide sidechain on *N*-acetylmuramic acid cannot be made (see **Fig. 27.7**). Without these alanines, cross-linking cannot occur and peptidoglycan integrity is compromised.

Vancomycin, a very large and complex glycopeptide produced by *Amycolatopsis orientalis* (**Fig. 27.10C**), binds to the D-ala-D-ala terminal end of the disaccharide unit and prevents the action of transglycosylase and transpeptidases (see **Fig. 27.7**). The mechanism of resistance is very different for vancomycin and penicillin. This difference makes vancomycin particularly useful against penicillin-resistant bacteria. To prevent development and spread of vancomycin-resistant bacteria, this antibiotic is typically used only as a drug of last resort. Resistance can develop when products from a cluster of *van* genes collaborate to make D-lactate and incorporate it into the ester D-ala-D-lactate to which vancomycin cannot bind. Another enzyme in the *van* gene cluster prevents the accumulation of D-ala-D-ala; as a result, the lactate version is incorporated into peptidoglycan. Peptidoglycan containing D-ala-D-lactate

Figure 27.9 Cephalosporin generations. Representative examples. **A.** First generation: cephalexin (Keflex). **B.** Second generation: cefoxitin. **C.** Third generation: ceftriaxone. **D.** Fourth generation: cefepime. Note that with each successive generation, the side groups become more complex. Highlighted areas indicate the core structure of each of the cephalosporins, with beta-lactam rings.

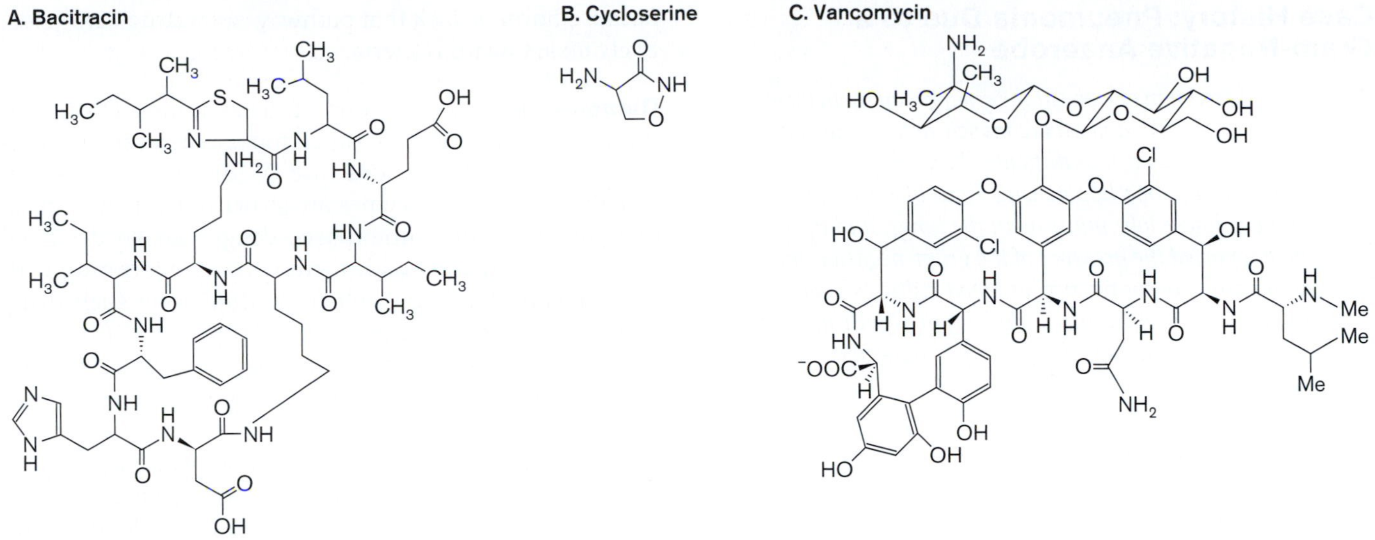

A. Bacitracin

B. Cycloserine

C. Vancomycin

Figure 27.10   Other antibiotics that affect peptidoglycan synthesis.   A. Bacitracin is produced by *Bacillus subtilis*. It is generally used only topically to prevent infection. **B.** Cycloserine, an analog of D-alanine, is one of several drugs used to treat tuberculosis. **C.** Vancomycin is a cyclic polypeptide made by *Amycolatopsis orientalis*, previously classified as a streptomycete. These antibiotics, especially bacitracin and vancomycin, are synthesized by exceedingly complex biochemical pathways in the producing organisms.

functions just fine, but since vancomycin cannot bind the D-lactate form, the organism is resistant to the antibiotic.

It is important to note that antibiotics targeting cell wall biosynthesis generally only kill *growing* cells. These drugs do not affect static or stationary phase cells because in this state the cell has no need for new peptidoglycan.

> **THOUGHT QUESTION 27.4**   When treating a patient for an infection, why would combining a drug such as erythromycin with a penicillin be counterproductive?

Drugs That Affect Bacterial Membrane Integrity

Poking holes in a bacterial cytoplasmic membrane is an effective way to kill bacteria. There are a few compounds useful in this regard, among them a group called the peptide antibiotics, of which **gramicidin** is an example. Gramicidin, produced by *Bacillus brevis*, is a cyclic peptide composed of 15 alternating D- and L-amino acids. It inserts into the membrane as a dimer, forming a cation channel that disrupts membrane polarity (**Fig. 27.11**). Polymyxin (from *Bacillus polymyxa*), another polypeptide antibiotic, has a positively charged polypeptide ring that binds to the outer (lipid A) and inner membranes of bacteria, both of which are negatively charged. Its major lethal effect seems to be to destroy the inner membrane, much like a detergent. These antibiotics are used only topically to treat or prevent infection. Because they can also form channels across human cell membranes, they are never ingested. Polymyxin has been fused to some bandage materials used to treat burn patients who are particularly susceptible to gram-negative infections (e.g., *P. aeruginosa*).

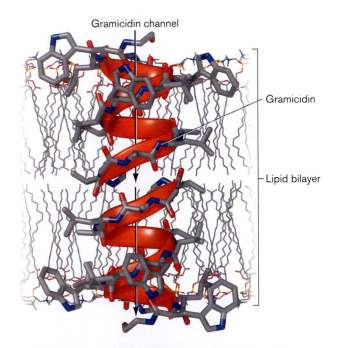

Figure 27.11   Gramicidin is a peptide antibiotic that affects membrane integrity.   Gramicidin forms a cation channel through cell membranes through which H$^+$, Na$^+$, or K$^+$ can freely pass. (PDB code: 1grm)

Drugs That Affect DNA Synthesis and Integrity

Bacteria generally make and maintain their DNA using enzymes that closely resemble those of mammals. Thus, you might consider it impossible to selectively target bacterial DNA synthesis. As you will see, this is not the case.

Case History: Pneumonia Due to a Gram-Negative Anaerobe

A 23-year-old woman arrived at the emergency room by ambulance with fever, chills, and severe muscle aches. She developed a nonproductive cough, had difficulty breathing, had pleuritic chest pain, and became hypotensive (low blood pressure). An X-ray showed lower lobe infiltrate in the lungs, and the clinical laboratory reported the presence of the gram-negative anaerobe Fusobacterium necrophorum *in blood cultures. The patient was diagnosed with pneumonia and treated with metronidazole, a DNA-damaging agent specific for anaerobes. She fully recovered.*

There are several classes of drugs, including sulfa drugs, quinolones, and metronidazole, that selectively affect the synthesis or integrity of DNA in microorganisms.

Sulfa drugs. The sulfa drugs, originally discovered by Domagk, belong to a group of drugs known as antimetabolites because they interfere with the synthesis of metabolic intermediates. Ultimately, the sulfa drugs act to inhibit the synthesis of nucleic acids. Drugs such as sulfamethoxazole work at the metabolic level to prevent the synthesis of tetrahydrofolic acid (THF), an important cofactor in the synthesis of nucleic acid precursors (**Fig. 27.12**). All organisms use THF to synthesize nucleic acids, so why are the sulfa drugs selectively toxic to bacteria? The selectivity occurs because mammalians do not synthesize a precursor of THF called folic acid. Higher mammals generally rely on bacteria and green leafy vegetables as sources of folic acid. Bacteria make folic acid from the combination of PABA, glutamic acid, and pteridine. Sulfamethoxazole, a structural analog of PABA, competes for one of the enzymes in the bacterial folic acid pathway and inhibits both folic acid and THF production (**Fig. 27.12C**).

Because humans lack that pathway, sulfa drugs are selectively toxic toward bacteria.

Quinolones. Another group of drugs inhibit DNA synthesis by targeting microbial topoisomerases such as DNA gyrase and topoisomerase IV (discussed in Section 7.2). Because these enzymes are structurally distinct from their mammalian counterparts, drugs can be designed to selectively interact with them without interfering with mammalian DNA metabolism. In 1963, one such drug, nalidixic acid, was discovered as a by-product of the synthesis of chloroquine, an antimalarial drug. Nalidixic acid, which targets DNA gyrase, has a very narrow antimicrobial spectrum, covering only a few gram-negative organisms. However, various chemical modifications, such as adding fluorine and amine groups, have increased its antimicrobial spectrum and its half-life in the bloodstream. The result is the class of drugs known as the **quinolones**. (The mode of action of quinolones and fluoroquinolones was discussed in Section 7.2.)

Metronidazole. Also known as Flagyl, metronidazole is an example of a drug that is harmless until activated, known as a prodrug. Metronidazole is activated after receiving an electron (is reduced) from the microbial protein cofactors flavodoxin and ferredoxin, found in microaerophilic and anaerobic bacteria such as *Bacteroides* (**Fig. 27.13**). Once activated, the compound begins nicking DNA at random, thus killing the cell. Because the etiological agent in our case history was an anaerobe, metronidazole was an effective therapy. Metronidazole is also effective against protozoans such as *Giardia* and *Entamoeba*. Aerobic microbes, although in possession of ferredoxin, are incapable of reducing metronidazole, presumably because oxygen is reduced in preference to metronidazole.

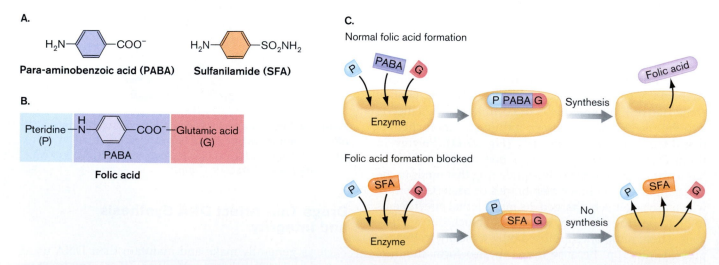

Figure 27.12   Mode of action of sulfanilamides.   A. The structures of PABA and sulfanilamide are very similar. **B.** PABA, pteridine, and glutamic acid combine to make the vitamin folic acid. **C.** Normal synthesis of folic acid requires that all three components engage the active site of the biosynthetic enzyme. The sulfa drugs replace PABA at the active site. The sulfur group, however, will not form a peptide bond with glutamic acid, and the size of sulfanilamide sterically hinders binding of pteridine so folic acid cannot be made.

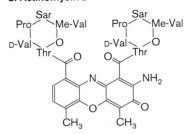

Figure 27.13   Activation of metronidazole. Single-electron transfers are made by ferredoxin and flavodoxin from anaerobes. Ferredoxin and flavodoxin are reducing agents capable of reducing very low redox potential compounds.

THOUGHT QUESTION 27.5 The enzyme DNA gyrase, a target of the quinolone antibiotics, is an essential protein in DNA replication. The quinolones bind to and inactivate this protein. Research has proved that quinolone-resistant mutants contain mutations in the gene encoding DNA gyrase. If the resistant mutants contain a mutant DNA gyrase and DNA gyrase is essential for growth, why are these mutations not lethal?

RNA Synthesis Inhibitors

Antibiotics that inhibit transcription, such as rifampicin and actinomycin D (**Fig. 27.14**), were described in Special Topic 8.1. These drugs are considered bactericidal in action and are most active against growing bacteria. The tricyclic ring of actinomycin D binds DNA from any source. As a result, it is not selectively toxic and not used to treat infections. Rifampin (also called rifampicin), on the other hand, does exhibit selective toxicity and is often prescribed for treatment of tuberculosis or meningococcal meningitis. Curiously, because rifampin is reddish orange, it turns bodily secretions, including breast milk, orange. The astute physician will warn the patient of this highly visible but harmless side effect to avoid unnecessary anxiety when the patient's urine changes color.

Protein Synthesis Inhibitors

The differences between prokaryotic and eukaryotic ribosomes account for the selective toxicity of antibiotics that specifically inhibit bacterial protein synthesis. How various antibiotics inhibit protein synthesis was discussed in Section 8.3. We recommend reviewing Chapter 8 for details. As a brief review, protein synthesis inhibitors can be classified into several groups based on structure and function (**Fig. 27.15**). Note that most of these antibiotics work by binding and interfering with the function of bacterial rRNA, which differs from eukaryotic rRNA. Also recall that protein synthesis inhibitors are, by and large, bacteriostatic.

Case History: Erysipelas in a Penicillin-Sensitive Patient

Sixteen-year-old Jamal arrived at the emergency room after two days of fever, malaise, chills, and neck stiffness. His most notable symptom was a painful, red, rapidly spreading rash covering the right side of his face. The rash covered his entire cheek, which was swollen, and extended into his scalp. About seven days earlier, Jamal had a severe sore throat. However, since it subsided in two days, he was not clinically evaluated. Throat cultures taken on admission revealed group A Streptococcus pyogenes, suggesting that the rash was a case of erysipelas caused by this organism. Although penicillin would be the drug of choice, Jamal is known to be allergic to this antibiotic.

A. Rifampin

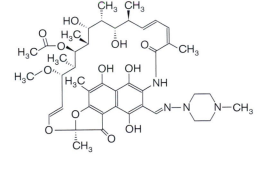

B. Rifampin and RNA polymerase

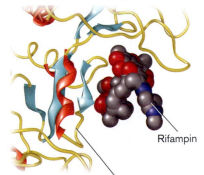

C. Actinomycin D

D. Actinomycin D and DNA

Figure 27.14   Antibiotics that inhibit transcription. A. Rifampin. **B.** Rifampin-binding site on the RNA polymerase beta subunit. (PDB code: 1i6v) **C.** Actinomycin D. **D.** Actinomycin D (yellow and red) interacting with DNA. Covalent intercalation of actinomycin interferes with DNA synthesis and transcription. (PDB code: 1DSC)

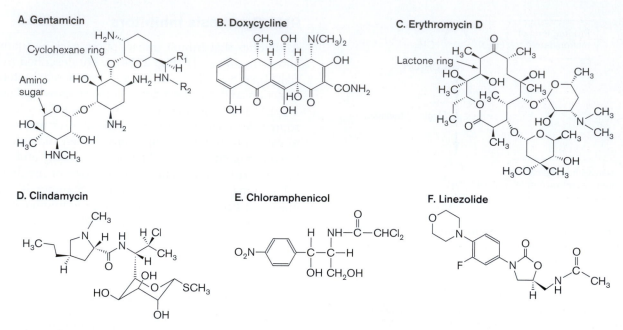

Figure 27.15   Protein synthesis inhibitors.   A. The aminoglycoside gentamicin. **B.** The tetracycline doxycycline. **C.** The macrolide antibiotic erythromycin. **D.** The lincosamide antibiotic clindamycin. **E.** Chloramphenicol. **F.** Linezolide.

When a patient is known to be immunologically sensitive to the usual drug of choice, a structurally distinct drug is best. Often, that drug will be one that inhibits protein synthesis; in this case, the drug chosen was the macrolide azithromycin. Drugs that inhibit protein synthesis can be subdivided into different groups based on their structures and on what part of the translation machine is targeted.

Drugs That Affect the 30S Ribosomal Subunit

The classification of antibiotics affecting protein synthesis is initially based on the bacterial ribosomal subunit targeted. Thus, one class of antibiotics interferes with 30S subunit function, and the other scrambles 50S subunit activities.

Aminoglycosides. There is considerable variation in structure among different aminoglycosides, but all contain a cyclohexane ring and amino sugars (**Fig. 27.15A**). The aminoglycosides are unusual among protein synthesis inhibitors in that they are bactericidal rather than bacteriostatic. Most of them bind 16S rRNA and cause translational misreading of mRNA, which is why these drugs are bactericidal. The resulting synthesis of jumbled peptides wreaks havoc with physiology and kills the cell. Streptomycin and gentamicin (see **Fig. 27.15A**) are two widely used drugs in this class. Ototoxicity (hearing damage) is a major, but uncommon, side effect of these antibiotics (approximately 0.5–3% of patients treated with gentamicin suffer from this toxicity). Hearing is generally affected at frequencies above 4,000 Hz.

Tetracyclines. Tetracycline antibiotics are characterized by a structure with four fused cyclic rings; hence the name. **Figure 27.15B** shows one frequently used example, called doxycycline. Tetracyclines are bacteriostatic and work by binding to and distorting the ribosomal A site that accepts incoming charged tRNA molecules. Doxycycline is used to treat early stages of Lyme disease (*Borrelia burgdorferi*), acne (*Corynebacterium acne*), and other infections. An important adverse side effect of tetracyclines is that they can interfere with bone development in a fetus or young child. Tetracycline use by pregnant mothers will also cause yellow discoloration of the infant's teeth. As a result, this drug is not recommended for pregnant women or nursing mothers.

Drugs That Affect the 50S Subunit

Five classes of drugs subvert translation by binding to the 50S ribosomal subunit. Most of these drugs were discussed in Chapter 8 and will be recapped here only briefly.

- **Macrolides**, all of which contain a 16-member lactone ring (**Fig. 27.15C**), inhibit translocation of the growing peptide (bacteriostatic action). Commonly prescribed examples are erythromycin and azithromycin. Azithromycin was the antibiotic used to treat the *S. pyogenes* infection in our case history, although other drugs could have been used. Because it is structurally dissimilar to any of the beta-lactam antibiotics, such as penicillin, it can be used safely in patients who are penicillin sensitive.

- **Lincosamides** (**Fig. 27.15D**), such as clindamycin, are similar to macrolides in function but have a different structure.
- **Chloramphenicol** (**Fig. 27.15E**) inhibits peptidyl transferase activity (bacteriostatic). Bone marrow depression leading to aplastic anemia is the most common serious side effect and limits its clinical use.
- **Oxazolidinones** (**Fig. 27.15F**) are a recently discovered class of antibiotics effective against many antibiotic-resistant microbes. In fact, this is the first new class of antibiotics discovered in over 35 years. Oxazolidinones such as linezolid bind to the 23S rRNA in the 50S subunit of the prokaryotic ribosome and prevent formation of the protein synthesis 70S initiation complex. This is a novel mode of action; other protein synthesis inhibitors either block polypeptide extension or cause misreading of mRNA. Linezolid binds to the 50S subunit near where chloramphenicol binds, but does not inhibit peptidyl transferase. Resistance is limited because most bacterial genomes have multiple operons encoding 23S rRNA. Usually more than one of these genes must mutate to confer high-level resistance. The more unmutated 23S rRNA genes there are, the more ribosomes susceptible to the antibiotic will be present. Oxazolidinones are useful primarily against gram-positive bacteria. Gram-negative bacteria are intrinsically resistant because of multidrug efflux pumps and decreased permeability due to the outer membrane.
- **Streptogramins** (**Fig. 27.16**), produced by some *Streptomyces* species, fall into two groups, A and B. Streptogramins belonging to group A have a large nonpeptide ring. Streptogramin B members are cyclic peptides. The two groups differ in their modes of action, although both inhibit bacterial protein synthesis by binding to the peptidyl transferase site. Group A streptogramins bound to the peptidyl transferase site distort the ribosome to prevent binding of tRNA to the ribosome A site. In contrast, group B streptogramins are thought to narrow the peptide exit channel, preventing exit of the peptide and thereby blocking translocation.

Natural streptogramins are produced as a mixture of A and B, the combination of which is more potent than either individual compound alone (an example of synergy). In tribute to this synergistic action, the drug combination is marketed under the name Synercid. Synergy between the two drugs occurs because the A type streptogramin alters the binding site for the B type drug, increasing its affinity. Bacteria can develop resistance either through ribosomal modification (the modification in 23S rRNA is the same one that provides resistance to macrolides), via the production of inactivating enzymes, or by active efflux of the antibiotic.

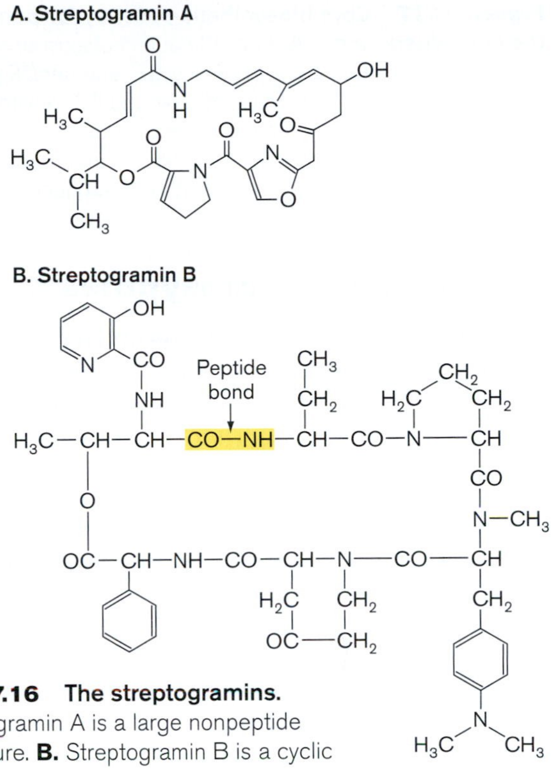

Figure 27.16 The streptogramins.
A. Streptogramin A is a large nonpeptide ring structure. **B.** Streptogramin B is a cyclic peptide.

TO SUMMARIZE:

- **Antibiotic specificity** for bacteria can be achieved by targeting a process that occurs only in bacteria, not host cells; by targeting small structural differences between components of a process shared by bacteria and hosts; or by exploiting a physiological condition such as anaerobiosis present only in certain bacteria.
- **Antibiotic targets** include cell wall synthesis, cell membrane integrity, DNA synthesis, RNA synthesis, protein synthesis, and metabolism.
- **Antibiotics targeting the cell wall** bind to the transglycosidases, transpeptidases, and lipid carrier proteins involved with peptidoglycan synthesis and cross-linking.
- **Antibiotics interfering with DNA** include the antimetabolite sulfa drugs that inhibit nucleotide synthesis, quinolones that inhibit DNA topoisomerases, and a drug, metronidazole, that, when activated, randomly nicks the phosphodiester backbone.
- **Inhibitors of RNA synthesis** target RNA polymerase (rifampin) or bind DNA and inhibit polymerase movement (actinomycin D).
- **Aminoglycosides and tetracyclines** bind the 30S subunit of the prokaryotic ribosome.
- **A variety of antibiotics** bind the 50S ribosomal subunit and inhibit translocation (macrolides, lincosamides), peptidyl transferase (chloramphenicol), formation of 70S complex (oxazolidinones), or peptide exit through the ribosome exit channel (streptogramins).

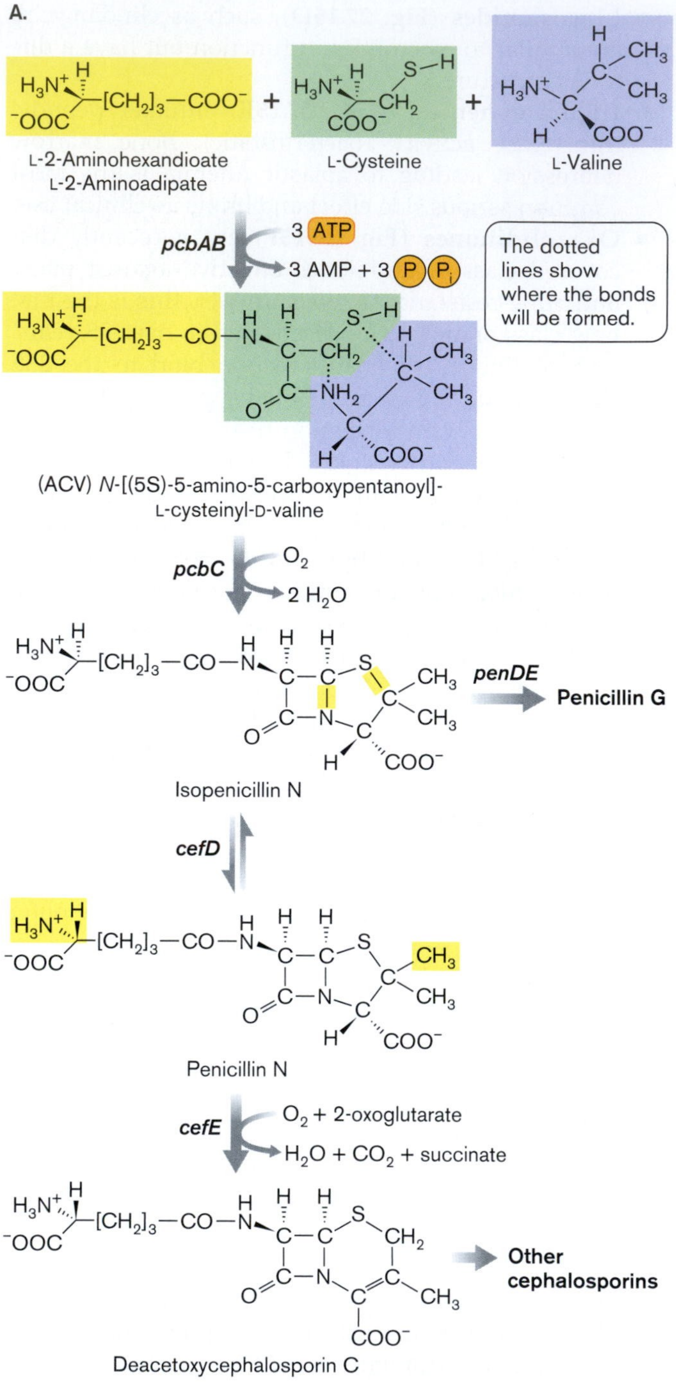

Figure 27.17 **Core biosynthetic pathway for penicillin and the cephalosporins.** **A.** Penicillin and cephalosporins share a biochemical pathway. The *pcbAB*, *pcbC*, and *penDE* genes encode ACV synthetase, IPN synthase, and IPN acyltransferase, respectively. CefD is required to make penicillin N, a precursor of the cephalosporins, and CefE is a synthase that makes the first cephalosporin structure in the biosynthetic sequence.

27.5 Antibiotic Biosynthesis

Antibiotics are considered **secondary metabolites** because they often have no apparent use in the producing organism. This most likely means a purpose has yet to be identified. Although the production of antibiotics is thought to help one microbe compete with another in nature, growth inhibition may not have been the original purpose of secondary metabolite production. The relative complexity of the biosynthetic pathways suggests a more immediate purpose—for example, cell-cell signaling—that evolved into cross-species inhibition. Antibiotic biosynthetic pathways are sometimes relatively simple (as in penicillin) and sometimes quite complex. **Figure 27.17A** outlines the relatively straightforward biosynthetic pathway used to make penicillin and the related cephalosporins, all of which contain a beta-lactam ring. Note that the ring results from the joining of two amino acids, cysteine and valine. Several different types of fungi can produce these drugs. All use similar pathways and evolutionarily related genes (**Fig. 27.17B**).

Polyketide antibiotics such as erythromycin (**Fig. 27.15D**) are synthesized in a modular fashion by modular polyketide synthase enzymes (discussed previously in Special Topic 15.2). The synthesis strategy is reminiscent of the way fatty acids are made. Malonyl-ACP units containing different R groups are successively condensed by modular subunits of the polyketide synthase complex until elongation is terminated by the thioesterase. The repeating forms of polyketides enable these long-chain molecules to be synthesized by a small number of enzymes.

Curiously, the peptide antibiotics such as bacitracin or the gramacidins are not synthesized using ribosomes. Complex biosynthetic pathways are involved that do not rely on mRNA. Nonribosomal synthesis of peptide antibiotics is discussed in Section 15.6 and Special Topic 15.3.

One theory of why certain microbes make antibiotics is that they prevent the growth of competing organisms, but how do microbes that make antibiotics avoid committing suicide? Fungi that make penicillin do not face any consequence for having done so because the organism does not contain peptidoglycan. Actinomycetes that produce compounds such as streptomycin or chloramphenicol, however, could be susceptible to their own secondary metabolite. Ribosomes isolated from *Streptomyces griseus*, for example, are fully sensitive to the streptomycin pro-

A.

(ACV) *N*-[(5S)-5-amino-5-carboxypentanoyl]-L-cysteinyl-D-valine

Isopenicillin N

Penicillin N

Deacetoxycephalosporin C

duced by this organism. *S. griseus* avoids killing itself in two ways. First, the organism synthesizes an inactive precursor of streptomycin, 6-phosphorylstreptomycin, which is secreted from the cell and, once outside the mycelium, becomes activated by a specific phosphatase. In addition, this streptomycete has an enzyme that inactivates any streptomycin that may leak back into the mycelium. Other organisms protect themselves by methylating key residues on their rRNA to prevent drug binding or set up permeability barriers that thwart reentry of the antibiotic.

B.

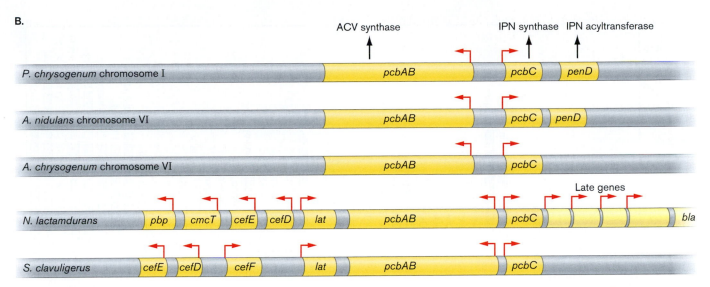

Figure 27.17 (*continued*) **B.** Penicillin gene clusters in various fungi that synthesize penicillins and cephalosporins. The arrows indicate the orientation of the genes. Note that the *pcbAB-pcbC* promoter region is transcribed bidirectionally. Genes involved in cephalosporin synthesis are labeled *cef*. The *bla* gene encodes a beta-lactamase that protects the producing organism *Nocardia lactamdurans* and appears to serve as an ancestral homolog to *bla* genes found in clinical pathogens.

THOUGHT QUESTION 27.6 Why might a combination therapy of an aminoglycoside antibiotic and cephalosporin be synergistic?

THOUGHT QUESTION 27.7 Could genomics ever predict the drug resistance phenotype of a microbe? If so, how?

TO SUMMARIZE:

- **Antibiotics are synthesized as secondary metabolites.**
- **Microbes may make antibiotics** to eliminate competitors in the environment.
- **Antibiotic producers prevent self-destruction** by various antibiotic resistance mechanisms.

27.6 The Challenges of Antibiotic Resistance

In this section, we discuss the various ways microbes can become resistant to antibiotics and explore the evolutionary road to resistance. In addition, we address how resistance moves between species and how we can deal with the spread of resistance throughout the world.

Case History: Multiple-Drug-Resistant Pneumonia

A 14-year-old boy with fever (39°C), chills, and left-sided pleuritic chest pain was referred to a hospital emergency department by his general practitioner. A chest X-ray showed left *lower lobe pneumonia. The boy reported that he was allergic to amoxicillin and cephalosporins (as a child he had developed a rash to these agents) and had been taking daily doxycycline (tetracycline) for the previous three months to treat mild acne. He was admitted to the hospital and treated with intravenous erythromycin because of his reported beta-lactam allergies, but he continued to feel sick. The day after admission, both sputum and blood cultures grew* Streptococcus pneumoniae. *After 48 hours, antibiotic susceptibility results indicated that the microbe was resistant to penicillin, erythromycin, and tetracycline. Armed with this information, the clinician immediately changed antibiotic treatment to vancomycin. The boy's fever resolved over the next 12 hours, and he made a slow but full recovery over the next week.*

Unfortunately the scenario presented in this case is far too common and has become an extremely serious concern. **Figure 27.18** illustrates the rapid rise of penicillin resistance among *S. pneumoniae* in the world. Another instance of emerging antibiotic resistance is unfolding in Europe and the Far East. The non-Enterobacteriaceae gram-negative rod *Acinetobacter baumanii* is increasingly seen as a dangerous cause of nosocomial infections. It commonly colonizes hospitalized patients, particularly those in intensive care units. Before 1998, there were almost no cases of multidrug-resistant *A. baumanii*. The rate is now as high as 8%. The organism is resistant to drugs as diverse as ciprofloxacin (a quinolone), amikacin (an aminoglycoside), penicillins, third-generation cephalosporins, tetracycline, and chloramphenicol. Imipenem, a relatively new class of beta-lactam drugs, is currently useful, but resistance to it is also likely to develop.

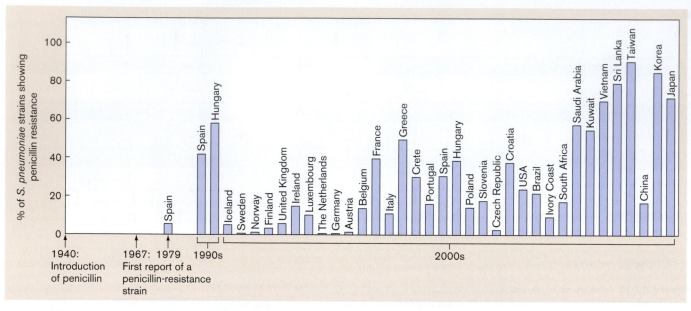

Figure 27.18   The rise of penicillin-resistant *Streptococcus pneumoniae* throughout the world.   Numbers reflect the number of penicillin-resistant strains among clinical isolates (strains of disease-causing bacteria isolated from patients).

There are four basic forms of antibiotic resistance. The resistant organism can:

- **Modify the target so that it no longer binds the antibiotic.** Mutations in key penicillin-binding proteins and ribosomal proteins, for instance, can confer resistance to methicillin and streptomycin, respectively. These mutations occur spontaneously and are not typically transferred between organisms.
- **Destroy the antibiotic before it gets into the cell.** One example is the enzyme beta-lactamase (or penicillinase), which is made exclusively to destroy penicillins. The sites of ring cleavage and structure of the enzyme are illustrated in **Figure 27.19**.
- **Add modifying groups that inactivate the antibiotic.** For instance, there are three classes of enzymes that modify and inactivate aminoglycoside antibiotics. The results of these types of enzyme modifications are illustrated for kanamycin in **Figure 27.20**.
- **Pump the antibiotic out of the cell using specific (for example, tetracycline export) and nonspecific transport proteins.** This strategy works because the pumps bail drugs out of the cell faster than the drugs can get in. Some are single-component pumps present in the cytoplasmic membrane of gram-negative and gram-positive bacteria (for example, NorA in *S. aureus*, PmrA in *S. pneumoniae*, and the TetA and B proteins in gram-negatives). Other drug efflux pumps are multicomponent systems present in gram-negative bacteria only (discussed shortly). Efflux in either case is usually energized by proton motive force.

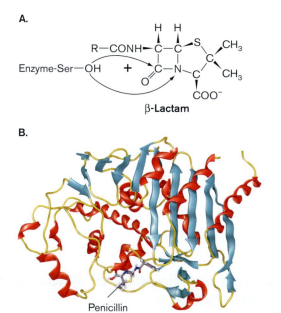

Figure 27.19   Destroying penicillin.   A. Beta-lactamase (or penicillinase) cleaves the beta-lactam ring of penicillins and cephalosporins. There are two types of penicillinases, based on where the enzyme attacks the ring. In either type, a serine hydroxyl group launches a nucleophilic attack on the ring. **B.** Structure of a beta-lactamase and location of the penicillin-binding site. (PDB code: 1XX2)

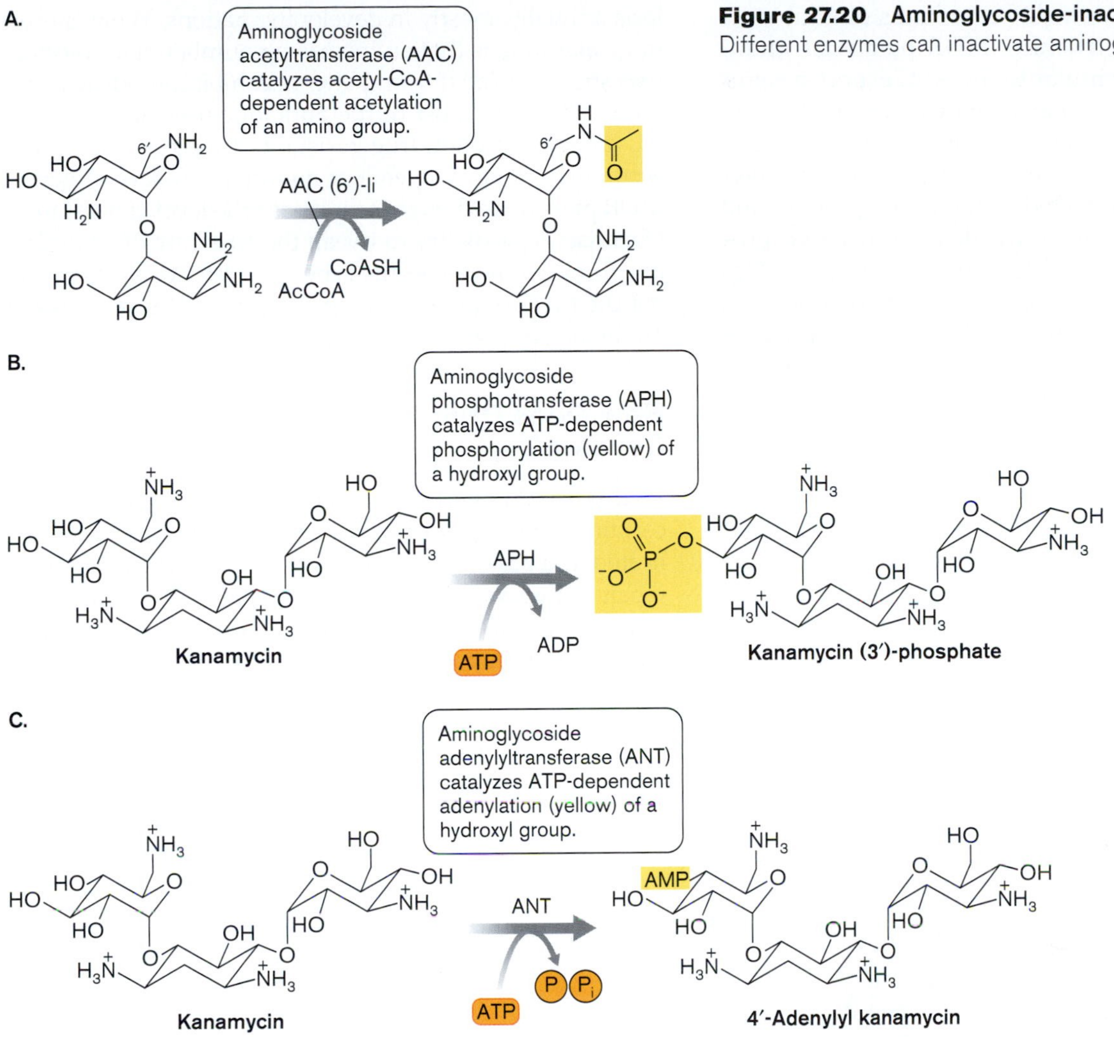

A.

Aminoglycoside acetyltransferase (AAC) catalyzes acetyl-CoA-dependent acetylation of an amino group.

AAC (6')-li

CoASH
AcCoA

Figure 27.20 Aminoglycoside-inactivating enzymes. Different enzymes can inactivate aminoglycoside antibiotics.

B.

Aminoglycoside phosphotransferase (APH) catalyzes ATP-dependent phosphorylation (yellow) of a hydroxyl group.

Kanamycin

APH

ATP   ADP

Kanamycin (3')-phosphate

C.

Aminoglycoside adenylyltransferase (ANT) catalyzes ATP-dependent adenylation (yellow) of a hydroxyl group.

Kanamycin

ANT

ATP   P P$_i$

4'-Adenylyl kanamycin

THOUGHT QUESTION 27.8 Fusaric acid is a cation chelator that normally does not penetrate the *E. coli* membrane, which means *E. coli* is typically resistant to this compound. Curiously, cells that develop resistance to tetracycline become *sensitive* to fusaric acid. Resistance to tetracycline is usually the result of an integral membrane efflux pump that pumps tetracycline out of the cell. What might explain the development of fusaric acid sensitivity?

THOUGHT QUESTION 27.9 Mutations in the ribosomal protein S12 (encoded by *rpsL*) confer resistance to streptomycin. Given a cell containing both *rpsL*$^+$ and *rpsL*^{R} genes, would the cell be streptomycin resistant or sensitive? (Recall that genes encoding <u>r</u>ibosome proteins for the <u>s</u>mall subunit are designated *rps*. A + indicates the wild-type allele, while R indicates a gene whose product is resistant to a certain drug.)

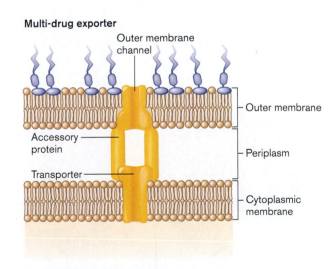

Multi-drug exporter

Outer membrane channel

Outer membrane

Accessory protein

Periplasm

Transporter

Cytoplasmic membrane

Figure 27.21 Basic structure of a multidrug efflux pump in gram-negative bacteria. These efflux systems have promiscuous binding sites that can bind and pump a wide range of drugs out of the bacterial cell.

A particularily dangerous type of drug resistance is mediated by what are called **multidrug resistance (MDR) efflux pumps** (**Fig. 27.21**). A single pump in this class can export many different kinds of antibiotics with little regard for structure. MDR pumps of gram-negative microbes are similar to the ABC export systems described in Section 4.2. They include three proteins: an inner membrane pump protein (fueled by proton motive force, a distinction from true ABC exporters), an outer membrane channel connected to the pump protein, and an accessory protein that may link the other two proteins. For instance, the ArcB transporter, shown in **Figure 27.22**, almost indiscriminately binds antibiotics in a large central cavity (a promiscuous binding site) and uses proton motive force to move those compounds through a pore and out of a funnel that connects to an outer membrane channel, TolC.

Antibiotic efflux pumps are now believed to contribute significantly to bacterial antibiotic resistance because of the very broad variety of substrates they recognize and because of their expression in important pathogens. Strains of the pathogen *M. tuberculosis*, for instance, have developed multiple drug resistance phenotypes due in part to MDR pumps. Approximately 2 million people die from tubercu-losis annually, mostly in developing nations. What is even more alarming is that an increasing number of *M. tuberculosis* strains isolated from patients exhibit multidrug resistance. Although most of the antibiotic resistance in most *M. tuberculosis* multidrug resistant strains is due to the accumulation of independent mutations in several genes, MDR pumps are thought to increase the level of resistance. Chemists typically try to tweak the structure of an antibiotic to overcome a specific type of resistance mechanism; but the MDR pumps act on an exceptionally wide range of antibiotics, almost without regard to structure.

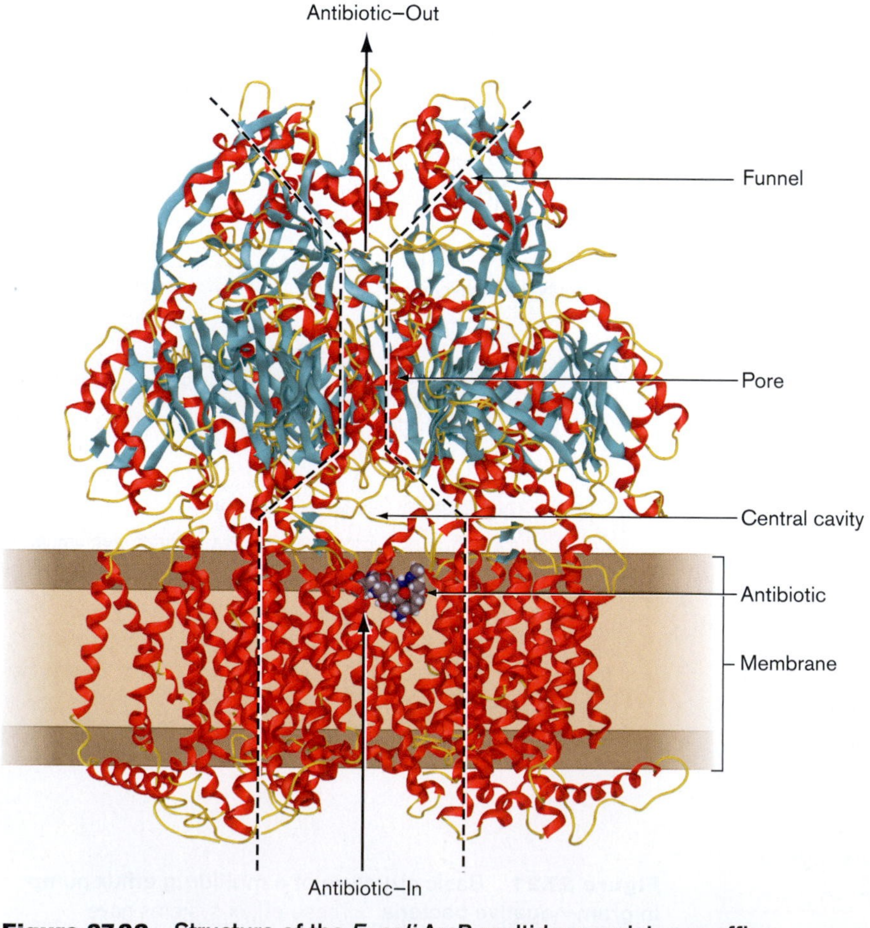

How Does Drug Resistance Develop?

As discussed in earlier chapters, nature has engineered a certain degree of flexibility in the way genomes are replicated and passed from one generation to the next. DNA repair pathways involving lesion bypass polymerases (for example, UmuDC; see Section 9.5) are thought to play a large role in randomized adaptive and evolutionary processes. For instance, at some point during evolution, gene duplication and mutational reshaping generated a gene product able to cleave the beta-lactam ring, creating an organism resistant to penicillin.

However, antibiotic resistance in all species does not occur de novo through gene duplication and/or mutation. Why reinvent the wheel—or, in this case, drug resistance? Gene transfer mechanisms such as conjugation, described in Chapter 9, can move antibiotic resistance genes from one organism to another and from one species to another. In fact, several drug resistance genes found in pathogenic bacteria actually had their start in the chromosomes of the drug-producing organisms and were passed on through gene swapping. For instance, *Streptomyces clavuligerus* produces a beta-lactamase (encoded by the *bla* gene) that protects this organism from the penicillin it produces. Transfer of a drug resistance gene is particularly evident if the gene has been incorporated into a plasmid and that plasmid is found in a new species.

An interesting case study of antibiotic resistance is provided by the gram-positive bacterium *Enterococcus faecalis*, a natural inhabitant of the mammalian gastrointestinal tract that can cause life-threatening disease if granted access to other body sites (as in subacute bacterial endocarditis; see Section 26.7).

Figure 27.22   Structure of the *E. coli* ArcB multidrug resistance efflux pump. The structure is trimeric and driven by proton motive force. Dotted lines outline the central cavity, pore, and efflux funnel through which the drug is pumped. Connection to TolC in the outer membrane is not shown. (PDB code: 2GIF)

E. faecalis is naturally resistant to numerous antibiotics, which makes disease treatment particularly challenging.

Vancomycin is one of the last lines of defense for treating serious *E. faecalis* infections. Unfortunately, increasing numbers of vancomycin-resistant strains have arisen in recent years. The completed genome sequence of one vancomycin-resistant strain illustrates the reason. The organism has an incredible propensity for incorporating mobile genetic elements that encode drug resistance. About a quarter of the genome consists of mobile or exogenously acquired DNA. These include 7 probable phages, 38 insertion elements, numerous transposons, and integrated plasmid genes. One such mobile element encodes vancomycin resistance in the sequenced strain.

Recently, multidrug resistance in various microbes (for example, *Salmonella enterica*) has also been attributed to the presence of integrons. Integrons are gene expression elements that account for rapid transmission of drug resistance because of their mobility and ability to collect resistance gene cassettes (see Special Topic 9.2).

We should note that the development of antibiotic resistance is not without consequence to the bacterium. Altering one aspect of an organism's physiology for the better may weaken another area. Bacteria may become resistant to a certain antibiotic, but that resistance comes at a price. For example, the altered DNA gyrase that affords resistance to quinolones may not function as well as the "normal" gyrase. Thus, in a situation where both resistant and susceptible organisms cohabit the same environment, the wild-type (sensitive) strain may grow faster and eventually overwhelm the mutant strain—unless fluoroquinolone is present, of course.

How Did We Get into This Mess?

Consider the following case: A young mother brings her 4-year-old child to the physician. The child is screaming because he has an extremely painful sore throat. Simply looking at the throat is not diagnostic. The raw tissue could mean the child is suffering from a bacterial infection, in which case antibiotics are needed. Alternatively, a virus could be the cause, a situation where antibiotics do nothing but pacify the parent. More often than not, the clinician will prescribe an antibiotic without ever knowing the cause of disease. The problem is this: the more an antibiotic is used, the more opportunities there are to select for an antibiotic-resistant organism. The presence of drug does not *cause* resistance but will kill off or inhibit the growth of competing bacteria that are sensitive, thereby allowing the resistant organism to grow to high numbers.

When *should* antimicrobials be administered? Certainly, in life-threatening situations where time is of the essence, antibiotics should be administered even before knowing the cause of the infection. On the other hand, the most prudent course to take when a patient has a simple infection is to confirm a bacterial etiology, and then prescribe. An exception to this may be in an elderly or otherwise immunocompromised individual, who may be more susceptible to secondary bacterial infections that can occur subsequent to viral disease.

Another proposed source of antibiotic resistance is the widespread practice of adding antibiotics to animal feed. No one quite knows why, but feeding animals antibiotics in their food makes for larger, and therefore more profitable, animals. Some estimates suggest that 70% of all antibiotics used in the United States is fed to healthy livestock. The consequence of this is that the animals may serve as incubators for the development of antibiotic resistance. Even if the resistance develops in nonpathogens, the antibiotic genes produced can be transferred to pathogens. Efforts have been under way since the 1960s to curb this practice.

Fighting Drug Resistance

Several strategies are being used to stay one step ahead of drug-resistant pathogens. In some instances, dummy target compounds that inactivate resistance enzymes have been developed. Clavulanic acid, for example, is a compound sometimes used in combination with penicillins such as amoxicillin. Clavulanic acid, a beta-lactam compound with no antimicrobial effect, competitively binds to beta-lactamases secreted from penicillin-resistant bacteria. Because the enzyme releases bound clavulanic acid very slowly, the amoxicillin remains free to enter and kill the bacterium. Another strategy is to alter the structure of the antibiotic in a way that sterically hinders the access of modifying enzymes. **Figure 27.23** illustrates how adding a side chain to gentamicin, converting it to amikacin, blocks the activity of various aminoglycoside-modifying enzymes. Of course, we are now seeing resistance to amikacin develop (this resistance is ribosome based, involving mutational alteration of the S12 protein or 16S rRNA).

Linking antibiotics is another strategy currently used to limit resistance. Recent advances have been made in linking a quinolone to an oxazolidinone to form a hybrid antibiotic with dual modes of action (**Fig. 27.24**). Because it has two modes of action, this hybrid antibiotic may limit the development of antibiotic resistance. Here's why: The rate of spontaneous resistance to a given antibiotic is roughly 1 out of 10^7 cells. For spontaneous resistance to develop to two antibiotics, that probability rises to 1 out of 10^{14}, making it very unlikely that an organism can become doubly drug resistant. However, multidrug resistance efflux pumps, integron cassettes, and plasmids carrying multiple antibiotic resistance genes can overwhelm that approach.

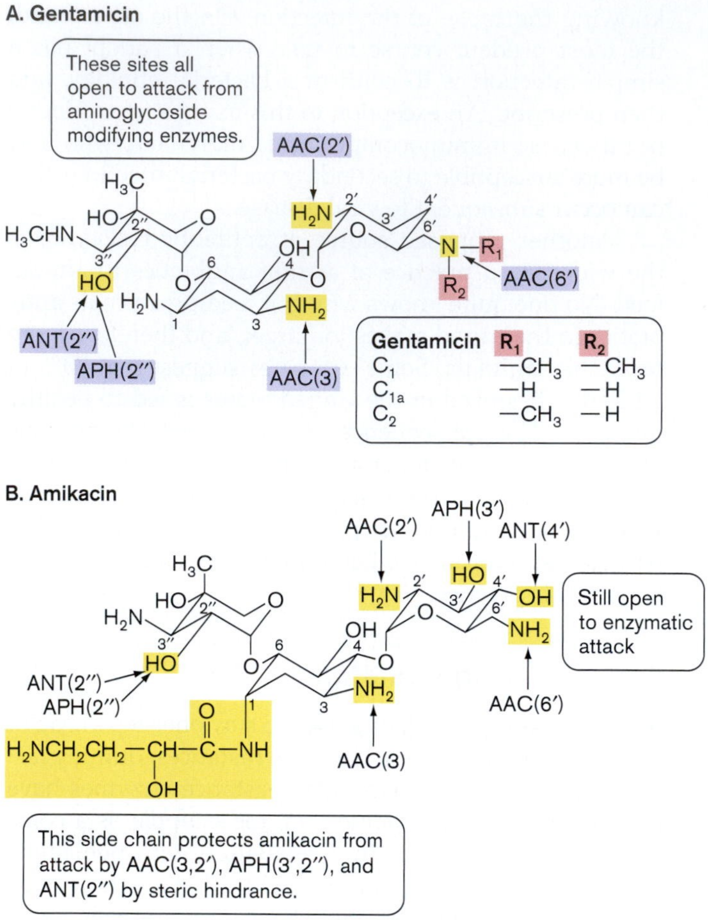

A. Gentamicin

These sites all open to attack from aminoglycoside modifying enzymes.

Gentamicin	R_1	R_2
C_1	—CH_3	—CH_3
C_{1a}	—H	—H
C_2	—CH_3	—H

B. Amikacin

Still open to enzymatic attack

This side chain protects amikacin from attack by AAC(3,2′), APH(3′,2′), and ANT(2″) by steric hindrance.

Figure 27.23 Fighting drug resistance. A. Sites where gentamicin is vulnerable to enzymatic inactivation. AAC = aminoglycoside acetyltransferase; ANT = aminoglycoside adenylyltransferase; APH = aminoglycoside phosphotransferase. Inset shows the R groups for different gentamicin compounds C_1, C_{1a}, and C_2. **B.** Gentamicin can be chemically modified at the highlighted sites to prevent loss of activity due to enzyme action. The side groups block access to enzyme active sites by steric hindrance (that is, the added group prevents the active site from interacting with its target structure), but does not inactivate the antibiotic.

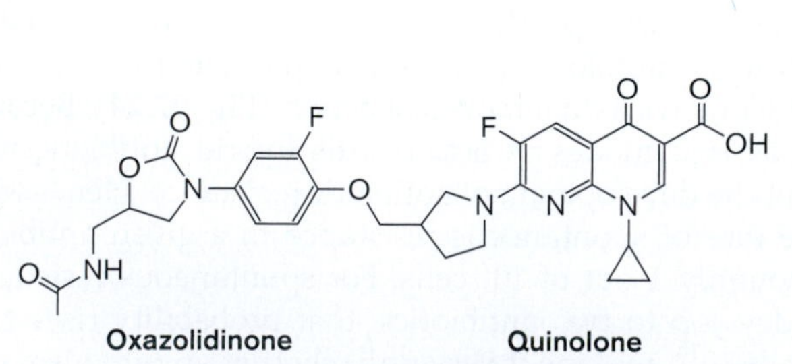

Oxazolidinone   **Quinolone**

Figure 27.24 Quinolone-oxazolidinone hybrid. Physically combining two different antibiotics may help reduce the emergence of drug-resistant bacteria.

TO SUMMARIZE:

- **Antibiotic resistance** is a growing problem worldwide.
- **Mechanisms of antibiotic resistance** include modifying the antibiotic, destroying the antibiotic, altering the target to reduce affinity, and pumping the antibiotic out of the cell.
- **Multidrug resistance pumps** use promiscuous binding sites to bind antibiotics of diverse structure.
- **Antibiotic resistance can arise spontaneously** through mutation, can be inherited by gene exchange mechanisms, or can arise de novo through gene duplication and mutational reengineering.
- **Indiscriminate use of antibiotics** has significantly contributed to the rise in antibiotic resistance.
- **Measures to counter antibiotic resistance** include chemically altering the antibiotic, using combination antibiotic therapy, and adding a chemical decoy.

27.7 The Future of Drug Discovery

Is humankind doomed to a future where antibiotics will no longer work? The general consensus is no. Through prudent use of current antibiotics and innovative approaches for finding new ones, humans should continue to effectively control evolving bacterial pathogens.

How does one screen for new drugs that target specific proteins? Certainly the classic approach in which microbes, plants, and even animals collected from around the world are screened for their abilities to make new antibiotics is still valid and remains the most fruitful source of new drugs. However, the ability to search for novel bacterial drug targets or validate theoretical targets has been revolutionized by genome sequence analysis and associated genetic techniques. Powerful computational and bioinformatic methods are required in the initial identification and selection of molecular targets, followed by a series of postgenomic approaches to validate and characterize the targets, devise screens for effective inhibitor molecules, and pursue structure-based drug design to handle the inescapable emergence of future resistance.

The modern drug discovery process can be outlined as follows:

1. Identify new targets (such as unique essential enzymes) based on genomics.
2. Find or design compounds that inhibit the target in vitro and show that the inhibitory compound has actual antibacterial activity.
3. Show that the target within the bacterial cell is the same as the in vitro target.

4. Optimize the MICs against susceptible species by altering the compound's structure.

5. Examine the compound's spectrum of activity (gram-positive, gram-negative, aerobe, anaerobe, etc.) and the rate at which antibiotic resistance develops in pathogens.

6. Determine the new drug's toxicity to animals and humans, and its pharmacological properties (for instance, how long does it stay at therapeutic levels in a patient).

The functions of genes that constitute potential new drug targets fall into three broad categories: those required only for growth of bacteria in laboratory media (in vitro expressed genes), those required only for bacterial infection in vivo (the in vivo induced genes, discussed in Section 25.6), and those required for bacterial growth both in vitro and in vivo (housekeeping genes). Products of novel and essential genes falling into the second and third categories constitute new potential drug targets.

Take, for example, the case of *S. pneumoniae*. Twenty years ago, *S. pneumoniae* was uniformly sensitive to penicillin. Today, a growing number of strains have become resistant (as illustrated in the multiple-drug-resistant pneumonia case history presented earlier; also see **Fig. 27.18**). The discovery of proteins within this microbe that could potentially be targeted by new antibiotics would offer hope of future treatments. Recently, a computer-assisted strategy has helped identify several potential drug targets in this pathogen. The search consisted of scanning the pneumococcal genome for sequence motifs commonly found in cell-surface-exposed or virulence-related proteins of other bacteria.

One such motif, called the choline-binding domain (CBD), is used by a variety of bacteria to interact with host cells. These domains consist of repeats of 2–10 amino acids. The rationale used to find potential drug targets in the pneumococcus was to look for pneumococcal proteins that contain these domains. It was already known that various streptococcal species had one such protein, CbpA, which functions as a surface adhesin and plays an important role in nasopharyngeal colonization. By performing genome-wide searches with the C-terminal choline-binding region of *cbpA*, Khoosheh Gosink and his collaborators at St. Jude Children's Hospital (Memphis, TN) identified six genes predicted to encode CBD-containing proteins ranging from 20 to 80 kDa (CbpD, E, F, G, I, and J). They constructed mutations in each gene and tested the mutants for pathogenicity. Mutants defective in CbpG adhered poorly to nasopharyngeal cells and were less virulent in a mouse infection model. Therefore, CbpG appears to play an important role in invasion and infection of the mucosa and the bloodstream. Its structural similarity to proteases suggests that it may be an excellent candidate for target-based drug development because numerous chemical inhibitors of proteases already exist.

Another advance in drug design utilizes a combination of genomic and proteomic approaches. The basis of this technique is to develop fluorescent molecules that bind only to *active* proteins of a given family of proteins. The probes can be created by rational design in which knowledge of the active site structure allows chemists to engineer chemicals that interact with that site. A true inhibitor of that protein and thus, a potential antibiotic, will prevent binding of the tagged probe, an event that can be viewed after directly blotting cell extracts on a membrane filter, much like a Western blot. In the absence of inhibitor, the probe will bind the spot, which will fluoresce. However, the presence of an inhibitor that binds well to the active site will prevent binding of the probe and the spot will not fluoresce.

Alternatively, if the probe binds to more than one protein, the same test can be performed using gel electrophoresis separation techniques. Proteins are separated on a nondenaturing gel so that the protein in question will retain its shape and its ability to bind probe. Tagged probe, with or without a potential inhibitor, is added to duplicate gels or a blot of those gels. The protein band will fluoresce in the absence of inhibitor or in the presence of a poor inhibitor. The band will not light up if an effective inhibitor is present.

The inhibitors are often made using more or less random combinatorial chemistry techniques, in which random variations of a molecule are created and tested. A library of these related compounds, which could number in the thousands, is robotically screened for inhibitory activity.

Genomic and combinatorial chemistries are certainly the cutting edge of antibiotic discovery. However, the old brute force screening of natural products was recently used in a novel way to discover a promising new class of antibiotic. Merck scientists screened 250,000 natural product extracts for an ability to specifically inhibit bacterial fatty acid biosynthesis. Fatty acid biosynthesis is an attractive target for antibiotic development because the process and proteins involved are different from those of eukaryotes. The strategy specifically targeted the protein FabF because it is an essential component of fatty acid synthesis and is conserved among key pathogens such as *S. aureus*.

Scientists engineered a strain of *S. aureus* to contain a gene expressing an antisense RNA to *fabF* mRNA. When the antisense RNA was induced, it bound to *fabF* mRNA and prevented efficient translation. As a result, the level of FabF protein in the cell decreased. The strain could still grow but would be exquisitely sensitive to any compound that targeted the remaining FabF protein. Fewer molecules of FabF present per cell means that fewer molecules of an active ingredient in a natural product are needed to stop fatty acid synthesis. Thus, zones of inhibition on an agar plate will be wider than they would be if cells produced normal amounts of FabF.

This novel targeted screening method led to the discovery of the antibiotic platensimycin, made by *Streptomyces platensis,* an organism isolated from South African soil. Platensimycin was subsequently shown to bind FabF and exhibit bacteriostatic, broad-spectrum activity, acting on gram-positive and gram-negative bacteria. It is only the third entirely new antibiotic developed in the last four decades. The novel chemical structure of platensimycin and its unique mode of action provide a great opportunity to develop a new class of critically needed antibiotics—a class that selectively targets fatty acid synthesis. More information on fatty acid biosynthesis and inhibitors can be found in Chapter 15.

Another type of rationale for developing new antimicrobial compounds is presented in **Special Topic 27.1**.

TO SUMMARIZE:

- **Potential targets for rational antimicrobial drug design** include proteins expressed only in vivo or proteins expressed both in vivo and in vitro.
- **Candidate antimicrobial compounds** can be designed to interact at the active site of a known enzyme and inhibit its activity.
- **Combinatorial chemistry** is used to make random combinations of compounds that can be tested for enzyme inhibitory activity and antimicrobial activity.

27.8 Antiviral Agents

A father pleading with a physician to give his child antibiotics when the infant is suffering with a cold is an all too common dilemma faced by the general practitioner,

but there is nothing of substance the physician can do. The common cold is caused by the rhinovirus, and no antibiotic designed for bacteria can touch it. So why are there so few antiviral agents in the clinician's arsenal? The reason is that applying the principle of selective toxicity is much harder to achieve for viruses than it is for bacteria. Viruses routinely usurp host cell functions to make copies of themselves. Thus, a drug that hurts the virus is likely to also harm the patient. Nevertheless, there are several useful antiviral agents in which selective viral targets have been found and exploited. Some of these agents are listed in **Table 27.4**. Select examples are discussed in this section. Note that all of the molecular mechanisms of viruses presented in Chapter 11 are studied as potential drug targets.

Antiviral Agents That Prevent Virus Uncoating or Release

Membrane-coated viruses are vulnerable at two stages: when the virus is invading the host cell and after viral propagation, when the progeny viruses release from the host cell. The flu virus presents a good example of both.

Case History: Antiviral Treatment of Infant Influenza

A 9-month-old infant arrived at the Johns Hopkins Hospital with an acute onset of fever, cough, regurgitation from his gastrostomy feeding tube, and dehydration. This illness occurred following a series of chronic problems, including respiratory syncytial virus bronchiolitis (infection and inflammation of the bronchioles) and neonatal group B streptococcal sepsis. [Neonatal sepsis is often caused by Lancefield group B Streptococcus agalactiae *described in Chapter 28.] Physical*

Table 27.4   Examples of antiviral agents.

Virus	Agent	Mechanism of action	Result
Influenza	Amantadine	Inhibits viral M2 protein	Prevents viral uncoating
	Zanamivir	Neuraminidase inhibitor	Prevents viral release
Herpes simplex virus and varicella-zoster virus (shingles)	Acyclovir	Guanosine analog	Halts DNA synthesis
	Famciclovir	Prodrug of penciclovir a guanosine analog	Halts DNA synthesis
Cytomegalovirus	Gancyclovir	Similar to acyclovir	
	Foscarnet	Analog of inorganic phosphate	Binds and inhibits virus-specific DNA polymerase
Respiratory synticial virus and chronic hepatitis C	Ribavirin	RNA virus mutagen	Causes catastrophic replication errors
HIV	Zidovudine (AZT)	Nucleoside analog, resembles thymine	Inhibits reverse transcriptase
	Nevirapine	Binds to allosteric site	Inhibits reverse transcriptase
	Nelfinavir	Protease inhibitor	Prevents viral maturation

Reza Ghadiri and his colleagues at the Scripps Research Institute have designed tiny peptide molecules that will self-assemble within bacterial membranes to form lethal nano-tubes that puncture the cell surface (**Fig. 1A**). The component parts of the tube are six- and eight-residue cyclic D, L alpha peptides. Built with alternating right- and left-handed amino acids joined end to end, the peptides expose their amino acid side chains outside of the ring. (In nature, the great majority of peptides contain only left-handed L-amino acids, but synthetic peptides can use both right- and left-handed forms.) This alternating D and L configuration allows adoption of a planar ring structure and stacking of rings to form a tube. Ghadiri and his colleagues were able to design the amphipathic pep-tides to insert themselves selectively into bacterial cell mem-branes (differences in the phospholipid content of mammalian and bacterial membranes enables selective toxicity). Once the peptides enter the charged environment of the bacterial membrane, they become sticky and form hollow, antiparallel, hydrogen-bonded stacks (**Fig. 1B**).

These nanotubes effectively poke holes in the membrane and disrupt the normal electrical potential and ion gradients that bacteria use to maintain homeostasis, generate energy, and carry out chemical reactions necessary for survival. By forming nano-tubes and poking holes in the cells, cyclic peptides disrupt these gradients and kill the cells. This is not just a theoretical whimsy. The authors have found these peptides to be highly effective in curing methicillin-resistant *S. aureus* infections in mice. Ques-tions of selectivity and toxicity have yet to be addressed, so it will be years before the use of nanotubes in humans can be realized.

Sources: Fernandez-Lopez, S., H.-S. Kim, E. C. Choi, M. Delgado, J. R. Granja, et al. 2001. Antibacterial agents based on the cyclic D, L-α-peptide architecture. *Nature* 412:452–455. Dartois, V., J. Sanchez-Quesada, E. Cabezas, E. Chi, C. Dubbelde, et al. 2005. Systemic antibacterial activity of novel synthetic cyclic peptides. *Antimicrobial Agents and Chemotherapy* 49:3302–3310.

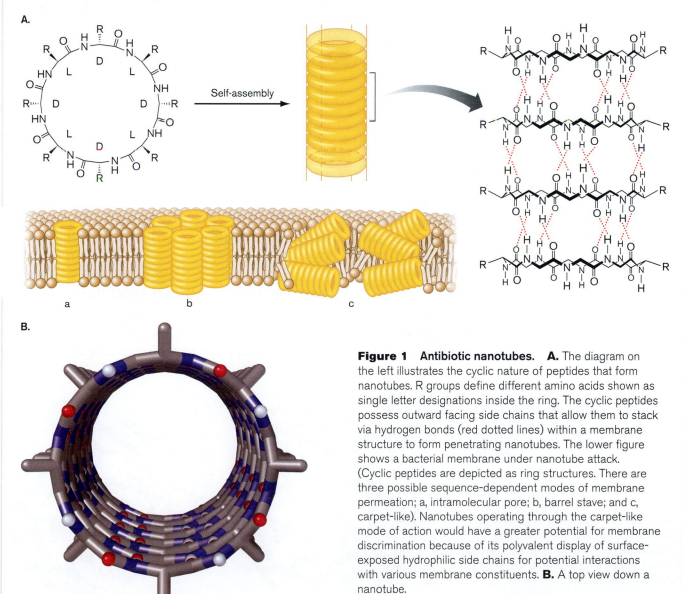

A.

Self-assembly

a   b   c

B.

Figure 1 Antibiotic nanotubes. A. The diagram on the left illustrates the cyclic nature of peptides that form nanotubes. R groups define different amino acids shown as single letter designations inside the ring. The cyclic peptides possess outward facing side chains that allow them to stack via hydrogen bonds (red dotted lines) within a membrane structure to form penetrating nanotubes. The lower figure shows a bacterial membrane under nanotube attack. (Cyclic peptides are depicted as ring structures. There are three possible sequence-dependent modes of membrane permeation; a, intramolecular pore; b, barrel stave; and c, carpet-like). Nanotubes operating through the carpet-like mode of action would have a greater potential for membrane discrimination because of its polyvalent display of surface-exposed hydrophilic side chains for potential interactions with various membrane constituents. **B.** A top view down a nanotube.

exam revealed fever, a severe cough resulting in respiratory distress, a rapid heart rate, and moderate dehydration. Naso-pharyngeal aspirate was positive for influenza A antigen. The patient was treated with amantadine when influenza was diagnosed. He gradually improved and was discharged home four days after admission.

An unusually severe form of influenza spread across the United States in 2003. Most states reported a higher than normal number of influenza-related deaths of children and young adults. Several factors, however, helped keep the outbreak from becoming an epidemic of larger proportions. First, the administration of flu vaccine afforded the population what is called herd immunity (discussed in Section 26.9). Herd immunity occurs even when only a fraction of a population is immunized. This fraction of individuals will not become infected and so cannot spread the disease to others. At the very least, this slows progression of infection throughout the population. (It is impossible to immunize all humans against any given disease. However, it is estimated that immunizing two-thirds of a population can eliminate the disease by cutting off transmission. Even this, however, is rarely achieved.)

The second factor that prevented an influenza pandemic was the availability of antiviral agents that can limit the disease course. As a result of studying the molecular biology of influenza, scientists discovered two selective targets. Influenza virus (200 nm) is encased in a membrane envelope donned when the virion buds from an infected cell. As described in Section 11.4, the envelope contains the viral proteins neuraminidase and hemagglutinin. Spikes of hemagglutinin bind to glycoprotein receptors on the host cell and trigger receptor-mediated endocytosis. The virus ends up inside the resulting endosome. After the endosome is formed, proton pumps in the membrane acidify endosomal contents. The drop in pH changes the structure of hemagglutinin on the viral membrane so it can now bind to receptors on the endocytic membrane. The result is fusion of the two membranes and release of the virion into the cytoplasm. For this change in

hemagglutinin structure to occur, it is essential that the *interior* of the enveloped virion also be acidified, and this is facilitated by the formation of a channel by the virus-encoded M2 envelope protein. Amantadine (**Fig. 27.25A**), a drug that has been used to treat some strains of the flu virus, is a specific inhibitor of the influenza M2 protein; however, it is useful against the influenza type A viral strains only. It is also effective against the avian H5N1 strain that has occasionally infected humans and is feared as the source of a future influenza pandemic. Unfortunately, amantadine-resistant strains are circulating in Asia due in part to the widespread use of amantadine by Chinese poultry farmers.

The second target of the influenza virus is the envelope protein neuraminidase. The newer antiflu drugs, such as zanamivir (Relenza), are **neuraminidase inhibitors** that act against types A and B influenza strains (**Figs. 27.25B** and **C**). Neuraminidase on the viral envelope allows virus particles to leave the cell in which they were made. Neuraminidase inhibitors prevent this release and cause the virus particles to aggregate at the cell surface, reducing the number of virus particles released. The contributions of different NA and HA genes on the severity of influenza are discussed in **Special Topic 27.2**.

Amantadine and the neuraminidase inhibitors, when used within 48 hours of disease onset, decrease shedding and reduce the duration of influenza symptoms by approximately one day. However, flu symptoms generally last only from three to ten days. While this does not sound like a substantial benefit, in the elderly, shortening the course of the flu can minimize damage to the lungs, which in turn reduces the chance of developing life-threatening secondary bacterial infections such as pneumonia and bronchitis.

Antiviral DNA Synthesis Inhibitors

Most antiviral agents work by inhibiting viral DNA synthesis. These drugs chemically resemble normal DNA nucleosides, molecules containing deoxyribose and analogs of

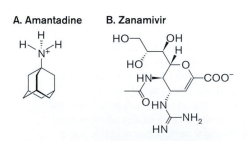

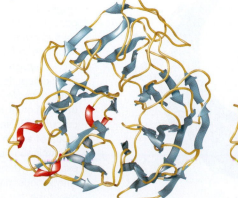

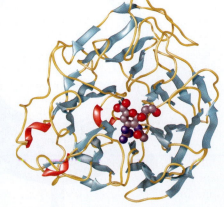

Figure 27.25 Inhibitors of influenza proteins. A. Amantadine inhibits the M2 protein. **B.** Neuraminidase inhibitor zanamivir. **C.** Neuraminidase without (left) and with (right) bound inhibitor. (PDB: 2HTQ)

Special Topic 27.2   Critical Virulence Factors Found in the 1918 Strain of Influenza Virus

As Chapter 26 describes, the influenza virus that swept the globe from 1918 to 1919 took the lives of 20–40 million people, making it a killer greater than the first World War. One might wonder whether today's drugs could prevent a similar disaster should that virus strain reemerge after all these years. This question may now have been answered.

A team of scientists working in a fortresslike biohazard research facility has used forensic molecular biology techniques to partially re-create that deadly virus. Their initial goal was to determine why the 1918 virus was so lethal compared with typical influenza strains, but they also wanted to learn if the antivirals on hand today could avert a 1918-like disaster.

Christopher F. Basler, of the Mount Sinai School of Medicine in New York, and his colleagues incorporated several genes from the 1918 flu into an influenza strain that was adapted to kill mice. The viral genes introduced included those that encode hemagglutinin (HA), neuraminidase (NA), and protein M2. The team actually expected that those genes would *reduce* the virulence of the mouse-adapted virus. This

was because flu viruses isolated from people rarely prove lethal to rodents, and genes from human-adapted strains typically weaken the rodent-adapted viruses. Not so for the HA and NA genes of the 1918 flu.

The engineered virus containing both of the 1918 HA and NA genes efficiently killed mice, but the presence of only one or the other of the two genes lowered virulence. This suggests that the *specific combination* of HA and NA may underlie the 1918 flu disaster. Even more remarkable was that treating the recombinant virus-infected mice with an NA inhibitor prevented 90% of the mice from dying. Furthermore, all mice infected with a recombinant flu strain carrying the 1918 gene for M2 survived when treated with M2 inhibitors, even if treatment began 6 hours after infection. The conclusion is that HA and M2 are critical components of influenza virulence. Thus, the currently available flu drugs targeting HA and M2 may help prevent a new pandemic of the 1918 influenza strain or a similar flu, such as the H5N1 avian flu.

adenine, guanine, cytosine, or thymine. Viral enzymes then add phosphate groups to these nucleoside analogs to form DNA nucleotide analogs. The DNA nucleotide analogs are then inserted into the growing viral DNA strand in place of a normal nucleotide. Once inserted, however, new nucleotides cannot attach to the nucleoside analogs, and DNA synthesis stops (**Fig. 27.26**). These DNA chain-terminating analogs are selectively toxic because viral polymerases are more prone to incorporate nucleotide analogs into their nucleic acid than are the more selective host cell polymerases. Antiviral DNA synthesis inhibitors work on DNA viruses or retroviruses, but not on viruses such as influenza with its RNA genome.

Case History: Treatment of HIV

A married couple come to the community clinic for prenatal care. He is 20 years old. She is 19 and reportedly two months pregnant with her first child. She denies intravenous (IV) drug use or a history of other sexual partners and has no history of sexually transmitted disease; however, a routine prenatal HIV antibody screen is reported as positive for HIV-1. Careful questioning of the patient and her husband elicits from him a history of IV drug use five years earlier. An HIV antibody screen for him is also positive. The laboratory results indicate that the wife may not yet require therapy (she has a low viral load—that is, less than 1,000 copies per milliliter of blood—and a high CD4 T-cell count), but since she is pregnant, a short

course of antiretroviral therapy (AZT) will be helpful in preventing the transmission of the virus to her child. The husband had an HIV viral load of 10,000 copies per milliliter and was started on protease inhibitor treatment.

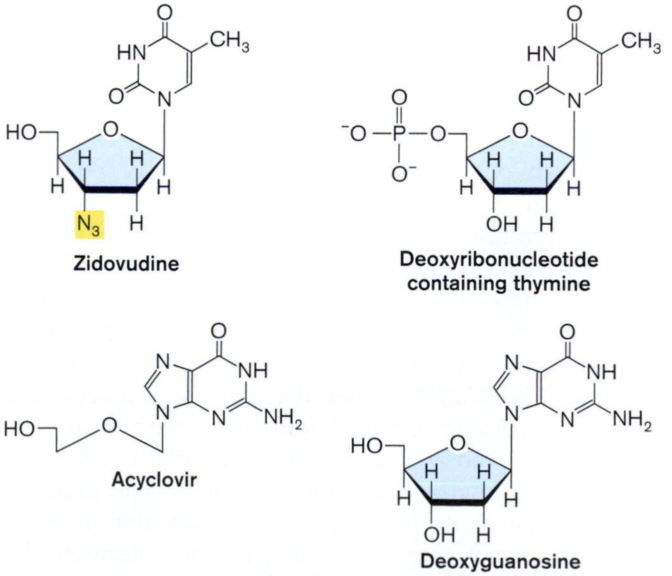

Figure 27.26   Antiviral inhibitors that prevent DNA synthesis. Zidovudine (AZT) and acyclovir are analogs of thymine and guanine nucleotides, respectively. Because the analogs have no 3′ OH to which another nucleotide can add, chain elongation ceases.

Nucleoside and Nonnucleoside Reverse Transcriptase Inhibitors

Human immunodeficiency virus (HIV) is an RNA retrovirus that uses a reverse transcriptase to make DNA that then integrates into host nuclear DNA to form a provirus (discussed in Section 11.5). The antiretroviral drug zidovudine (abbreviated ZDV or AZT), which was used in the case history, is a nucleoside analog recognized by reverse transcriptase. Once incorporated into a replicating HIV DNA molecule, the DNA chain-terminating property of AZT prevents further DNA synthesis. Because HIV transmission from the mother to the neonate can occur at delivery or by breast-feeding, treatment of the mother and the child are important steps to prevent transmission.

There are also nonnucleoside reverse transcriptase inhibitors. For example, the drug delavirdine binds directly to reverse transcriptase and allosterically inactivates the enzyme.

Because HIV can mutate rapidly and become resistant to single-drug therapies, treatment usually involves administering combinations of three or more antiretroviral drugs.

Protease Inhibitors

To make optimum use of its limited provirus DNA sequence, HIV makes long nonfunctional polypeptide chains that are proteolytically cleaved to make the actual proteins and enzymes used to replicate and produce new virions. For example, the *gag* and *pol* genes reside next to each other in the HIV genome and are transcribed as a single mRNA molecule (discussed in Chapter 11). The Gag and Pol open reading frames overlap but are offset from one another by one base. This mRNA produces two polyproteins called Gag and Gag-Pol, the latter being the result of a shift in reading frames that takes place during translation. Once made, both polyproteins are cleaved by HIV protease. The Gag protein is proteolytically cleaved to make different capsid components (p17, p24, and p15, which is further cleaved to make nucleocapsid protein p7), while Gag-Pol is cleaved to make reverse transcriptase and integrase (**Fig. 27.27A**). Protease inhibitors such as Viracept® and Lopinavir® are a powerful new class of drugs that block the HIV protease (**Fig. 27.27B**). When the protease is inactivated, even though new virus particles are made, the polyproteins remain uncleaved and the virus cannot mature. Because immature HIV particles cannot infect other cells, progress of the disease stalls. It is important to note that protease inhibitors do not cure AIDS; they can only decrease the number of infectious copies of HIV.

Antiviral therapy with protease inhibitors is recommended for patients with symptoms of AIDS and for

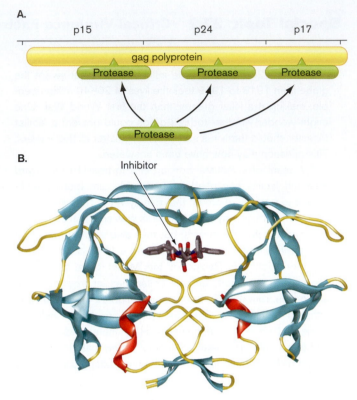

Figure 27.27   HIV protease inhibitor.   A. Representation of HIV protease cleavage of a single gag polyprotein into multiple, smaller proteins. **B.** The protease enzyme is shown as a ribbon structure, while the protease inhibitor BEA 369 is shown as a stick model. (PDB code: 1EBY)

asymptomatic patients with viral loads above 30,000 copies per milliliter. Treatment should be considered even for patients with viral loads above 5,000 copies per milliliter, as for the husband in the case history presented earlier.

TO SUMMARIZE:

- **Fewer antiviral agents** are available when compared to antibacterial agents because it is harder to identify viral targets that provide selective toxicity.
- **Preventing viral attachment to, or release from, host cells** are mechanisms of action for antiviral agents such as amantadine and Zanamivir used to treat influenza virus.
- **Inhibiting DNA synthesis** is the mode of action for most antiviral agents, although they only work for DNA viruses and retroviruses.
- **HIV treatments** include reverse transcriptase inhibitors that prevent synthesis of DNA and protease inhibitors that prevent the maturation of viral polyproteins into active forms.

27.9 Antifungal Agents

Fungal infections are much more difficult to treat than bacterial infections in part because fungal physiology is more similar to that of humans than is bacterial physiology. The other reason is that fungi have an efficient drug detoxification system that modifies and inactivates many antibiotics. Thus, to have a fungistatic effect, repeated applications of antifungal agents are necessary to keep the level of unmodified drug above MIC levels.

Case History: Blastomycosis

A 37-year-old male presented to the emergency department of a Florida hospital with persistent fever, malaise, and a painful right arm mass. He denied trauma to the arm. White blood cell count was elevated at 27,000/µl, and chest X-ray revealed a left lung infiltrate. Bronchoscopy revealed granulomatous inflammation containing a single yeast-like mass. Incision and drainage was performed on the arm mass and cultures obtained. Serum cryptococcal antigen tests were negative, as were tests for Bartonella henselae *and* Toxoplasma. *Cultures from the right arm grew out a fungal form similar to that identified from the bronchoscopy specimens. A tentative diagnosis of* Blastomyces dermatitides *was confirmed using PCR. The patient was placed on amphotericin B, and his fevers and leukocytosis subsequently subsided. His medication was changed to fluconazole for a recommended duration of six months.*

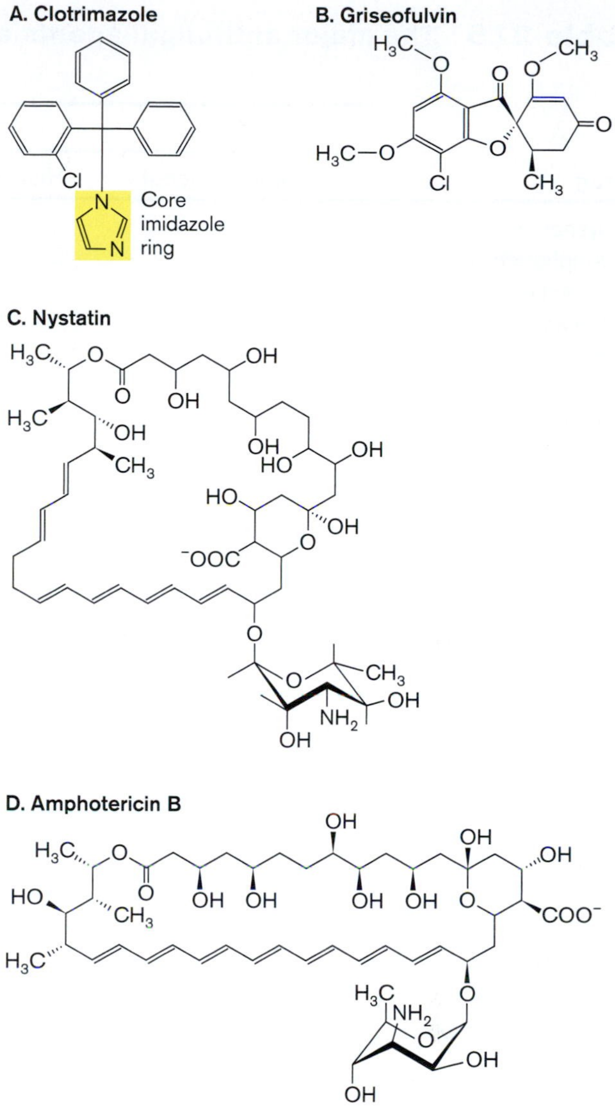

Superficial **mycoses** (fungal infections), such as athlete's foot, and systemic mycoses, such as blastomycosis, require very different treatments. Imidazole-containing drugs (clotrimazole, miconazole) are often used topically in creams for superficial mycoses (**Fig. 27.28A**). Others, such as itraconazole, are administered orally. Superficial mycoses include infections of the skin, hair, and nails, as well as *Candida* infections of moist skin and mucous membranes (for example, vaginal yeast infections). The imidazole-containing drugs appear to disrupt the fungal membrane by inhibiting sterol synthesis. More chronic dermatophytic infections typically require another antifungal agent called **griseofulvin**, produced by a *Penicillium* species (**Fig. 27.28B**). Griseofulvin disrupts the mitotic spindle and derails cell division (called metaphase arrest). This does not kill the fungus, but as the hair, skin, or nails grow and are replaced, the fungus is shed. Vaginal yeast infections caused by *Candida* are often treated with **nystatin**, a polyene antifungal agent synthesized by *Streptomyces* that forms membrane pores (**Fig. 27.28C**). The name *nystatin*, by the way, came about because two of the people who discovered it worked for the New York State Public Health Department.

The serious, sometimes fatal consequences of systemic mycoses require more aggressive therapy. The

Figure 27.28 Examples of antifungal agents.
A. Clotrimazole belongs to the group of imidazole antifungals, which are so named because they all contain an imidazole ring.
B. Griseofulvin is produced by *Penicillium griseofulvum*.
C. Nystatin is a polyene macrolide produced by *Streptomyces noursei*. **D.** Amphotericin B is a polyene produced by *Streptomyces nodosus*.

drugs used in these instances include **amphotericin B** (produced by *Streptomyces*; see **Fig. 27.28D**) and fluconazole. Amphotericin B binds to the sterols in fungal membranes and destroys membrane integrity. It has a high affinity for ergosterol, which is prevalent in fungal, but not mammalian, membranes. Fluconazole, on the other hand, inhibits the synthesis of ergosterol. Thus, fungal cells grown in the presence of fluconazole make defective membranes. Typically, curing systemic fungal infections

Table 27.5   The major antifungal agents and their common uses.

	Clinical application			
	Systemic mycoses			
Drug	Coccidioidomycosis	Histoplasmosis	Blastomycosis	Paracoccidioidomycosis
Polyenes				
Amphotericin B	+	+	+	+
Nystatin	–	–	–	–
Pimaricin	–	–	–	–
Imidazoles				
Clotrimazole	–	–	–	–
Miconazole	–	–	–	–
Ketoconazole	+	+	+	+
Triazoles				
Itraconazole	+	+	+	+
Fluconazole	+	?[b]	?	?
Antimetabolite				
5-Fluorocytosine[c]		–	–	–

[a.] mc, Mucocutaneous but not systemic candidiasis.

[b.] Insufficient data.

[c.] Used only in combination with amphotericin B.

requires long-term treatment to prevent disease relapse. **Table 27.5** lists a number of other commonly used antifungal agents.

TO SUMMARIZE:

- **Fungal infections** are difficult to treat because of similarities in human and fungal physiologies.
- **Imidazole-containing** antifungal agents inhibit sterol synthesis.
- **Griseofulvin** inhibits mitotic spindle formation.
- **Nystatin** produces membrane pores.
- **Amphotericin B** binds to membranes and destroys membrane integrity.

Concluding Thoughts

Antibiotics have done much to improve the health and well-being of people and animals around the globe. Although they have successfully kept at bay infectious diseases dreaded for centuries (such as the plague and tuberculosis), a crisis of antibiotic resistance looms because of the irresponsible use of antimicrobial agents. The fact that very few new, clinically useful antimicrobials have been discovered over the last quarter century should give us pause. Redoubling efforts to find new drugs combined with the responsible use of existing antibiotics are required to maintain our advantage over constantly evolving pathogens.

CHAPTER REVIEW

Review Questions

1. What is selective toxicity? Provide examples.
2. Explain the difference between antibiotic susceptibility and antibiotic sensitivity.
3. What does the term *spectrum of antibiotic activity* mean?
4. Provide examples of bacteriostatic and bactericidal antibiotics.
5. What is the Kirby-Bauer test? Does it tell whether a drug is bacteriostatic or bactericidal?
6. Give examples of drugs that target cell wall synthesis; RNA synthesis; protein synthesis; DNA replication. What are their modes of action?
7. What is the mechanism by which peptide antibiotics are synthesized by the producing organism?

Clinical application				
	Opportunistic mycoses			
Aspergillosis	Candidiasis	Cryptococcosis	Dermatophytosis	Other
+	+	+	−	
−	mc[a]	−	−	
−	−	−	−	Mycotic keratitis
−	mc	−	+	
−	mc	−	+	
−	+	−	+	
+	mc	+	+	Sporotrichosis
−	+	+	+	Sporotrichosis
+	+	+	−	Phaeohyphomycosis

8. How do antibiotic-producing microorganisms prevent suicide?
9. Why is antibiotic resistance a growing problem?
10. What are the four basic mechanisms of antibiotic resistance?
11. Explain the basic concept of an MDR efflux pump.
12. Discuss the current concepts of the origin of antibiotic resistance.
13. What are some mechanisms used to combat the development of drug resistance?
14. Why are there few antiviral agents available to treat disease?
15. What is herd immunity?
16. How does amantadine inhibit influenza?
17. Discuss the general modes of action of antifungal agents.

Key Terms

amphotericin B (1059)
antibiotic (1030)
bacitracin (1040)
bactericidal (1033)
bacteriostatic (1033)
chloramphenicol (1045)
cycloserine (1040)
gramicidin (1041)
griseofulvin (1059)
Kirby-Bauer assay (1035)
lincosamide (1045)

macrolide (1044)
minimal inhibitory concentration (MIC) (1034)
Mueller-Hinton agar (1037)
multidrug resistance (MDR) efflux pump (1049)
mycosis (1059)
neuraminidase inhibitor (1056)
nosocomial (1030)
nystatin (1059)
oxazolidinone (1045)

penicillin-binding protein (1039)
quinolones (1042)
secondary metabolite (1046)
selective toxicity (1032)
spectrum of activity (1033)
streptogramin (1045)
transglycosylase (1038)
transpeptidase (1038)
vancomycin (1040)
zone of inhibition (1034)

Recommended Reading

Cohen, Mitchell L. 2000. Changing patterns of infectious disease. *Nature* **406**:762–767.

Dibner, Julia J., and James D. Richards. 2005. Antibiotic growth promoters in agriculture: History and mode of action. *Poultry Science* **84**:634–643.

Fernandez-Lopez, Sara, Hui-Sun Kim, Ellen C. Choi, Mercedes Delgado, Juan R. Granja, et al. 2001. Antibacterial agents based on the cyclic D,L-α-peptide architecture. *Nature* **412**:452–455.

Fernandez-Tornero, Carlos, Rubens Lopez, Ernesto Garcia, Guillermo Gimenez-Gallego, and Antonio Romero. 2001. A novel solenoid fold in the cell wall anchoring domain of the pneumococcal virulence factor LytA. *Nature Structural Biology* **8**:1020–1024.

Freiberg, Christoph, and Heike Brötz-Oesterhelt. 2005. Functional genomics in antibacterial drug discovery. *Drug Discovery Today* **10**:927–935.

Gosink, Khoosheh K., Elizabeth R. Mann, Chris Guglielmo, Elaine I. Tuomanen, and H. Robert Masure. 2000. Role of novel choline binding proteins in virulence of *Streptococcus pneumoniae*. *Infection and Immunity* **68**:5690–5695.

Hubschwerlen, Christian, Jean-Luc Specklin, Daniel K. Baeschlin, Yves Borer, Sascha Haefeli, et al. 2003. Structure–activity relationship in the oxazolidinone–quinolone hybrid series: Influence of the central spacer on the antibacterial activity and the mode of action. *Bioorganic & Medicinal Chemistry Letters* **13**:4229–4233.

Johnson, Kirk W., Denene Lofland, and Heinz E. Moser. 2005. PDF inhibitors: An emerging class of antibacterial drugs. *Current Drug Targets—Infectious Disorders* **5**:39–52.

Kloss, Patricia, Liqun Xiong, Dean L. Shinabarger, and Alexander S. Mankin. 1999. Resistance mutations in 23S rRNA identify the site of action of the protein synthesis inhibitor linezolid in the ribosomal peptidyl transferase center. *Journal of Molecular Biology* **294**:93–101.

Marquez, Beatrice. 2005. Bacterial efflux systems and efflux pumps inhibitors. *Biochimie* **87**:1137–1147.

Paulsen, I. T., L. Banerjee, G. S. Myers, K. E. Nelson, R. Seghadri, et al. 2003. Role of mobile DNA in the evolution of vancomycin-resistant *Enterococcus faecalis*. *Science* **299**:2071–2074.

Shah, P. M. 2005. The need for new therapeutic agents: What is the pipeline? *Clinical Microbiology and Infection* **11 (Suppl 3)**:36–42.

Swaney, Steve M., Hiroyuki Aoki, M. Clelia Ganoza, and Dean L Shinabarger. 1998. The oxazolidinone linezolid inhibits initiation of protein synthesis in bacteria. *Antimicrobial Agents and Chemotherapy* **42**:3251–3255.

Turnidge, John, and David J. Paterson. 2007. Setting and revising antibacterial susceptibility breakpoints. *Clinical Microbiology Reviews* **20**:391–408.

Walsh, Christopher. 2003. *Antibiotics: Actions, Origins, Resistance.* ASM Press, Washington, DC.

Wang, Jun, Stephen M. Soisson, Katherine Young, Wesley Shoop, Srinivas Kodalil, et al. 2006. Platensimycin is a selective FabF inhibitor with potent antibiotic properties. *Nature* **441**:358–361.

Chapter 28

Clinical Microbiology and Epidemiology

28.1 Principles of Clinical Microbiology

28.2 Approaches to Pathogen Identification

28.3 Specimen Collection

28.4 Biosafety Containment Procedures

28.5 Principles of Epidemiology

28.6 Detecting Emerging Microbial Diseases

Throughout history, infectious diseases have killed more people than all of our wars combined. Our present success in controlling the spread of disease is due in large part to worldwide surveillance agencies that are equipped to detect outbreaks quickly, before major epidemics develop. These agencies rely on smaller clinical microbiology laboratories scattered throughout the world. These laboratories help clinicians diagnose infectious diseases.

This chapter discusses the principles of clinical microbiology and epidemiology that are used to identify, treat, and contain outbreaks. How do clinical laboratories know what organisms to suspect in a given case? And what tests will unequivocally identify the right pathogen? How are the *real* pathogens found among the normal flora? Finding the *source* of an outbreak is also critical. How do epidemiologists identify "patient zero" and recognize and contain emerging diseases? Last but not least, we'll consider bioterrorism. Clinical scientists and epidemiologists are trained to detect bioterrorist attacks using the same principles they employ to detect naturally occuring infectious diseases.

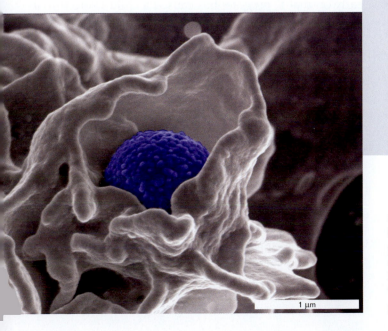

Community-acquired methicillin-resistant *S. aureus* (in blue) destroys a neutrophil attempting to phagocytose it, thus breaching the body's first line of defense. Recent outbreaks of community-acquired MRSA have become a serious health problem.
Source: Jovanka Voyich, et al. 2006. *The Journal of Infectious Diseases* 194:1761. Photo from David W. Dorward/Rocky Mountain Laboratories/NIAID.

Wilfrid Saintilus lay immobile on a straw mat when first seen by Dr. Paul Farmer in 1996. The 33-year-old peasant farmer had been sent home without hope by health care workers in a small clinic near his village in Haiti. He was pronounced "paralyzed from the waist down"; it was said that "nothing more could be done." Farmer, however, quickly realized that Wilfrid's problem was not paralysis but a long-term *Salmonella* infection, very likely contracted from an unclean drinking water supply—a problem for 80% of Haiti's population. Farmer realized that, while normally a gastrointestinal pathogen, *Salmonella* can occasionally disseminate via the bloodstream and infect bones, causing a painful disease called osteomyelitis. The microbe had infected Wilfrid's hip, causing pain so severe he could not move. His leg muscles had atrophied over time, and he now weighed under 100 pounds. Contrary to prediction, Wilfrid did not die. Dr. Farmer's diagnostic skills and antibiotics saved his life.

Thousands of stories like this, most with more tragic outcomes, underscore the integral roles that clinical microbiology and epidemiology have in health care. In Wilfrid's case, a simple laboratory test would have revealed a fully treatable *Salmonella* infection and prevented his enormous physical and emotional pain.

As in Chapters 24–27, we use case histories in this chapter to introduce various strategies and methodologies used by modern medicine to diagnose infectious diseases. The approach is, by necessity, selective. Not all techniques, nor all diseases, can be described. Our goal is not to catalog the many kinds of infectious diseases but to demonstrate general principles and problem-solving approaches used in identifying disease-causing microbes. It will become apparent in the process that modern tools of clinical microbiology are indispensable in the diagnosis and treatment of infectious disease.

28.1 Principles of Clinical Microbiology

As in any good detective mystery, the first step in investigating an infectious disease is to identify the most likely suspects. This can be accomplished, in part, from observing the disease symptoms in a patient and knowing what organisms typically produce those symptoms. It also helps if the physician is aware of a similar disease outbreak under way in the community. Beyond these clues, the etiological agent must be identified through biochemical, molecular, serological, or antigen detection strategies.

Why Take the Time to Identify an Infectious Agent?

This is the first question many students ask when contemplating the effort and expense required to identify the genus and species of an organism causing an infection. Why not simply treat the patient with an antibiotic and be done with it? While this approach sounds appealing, there are several compelling reasons for identifying an infectious agent.

Many bacteria are resistant to certain antibiotics. As discussed in Chapter 27, antibiotic resistance is an increasingly serious global problem. Characterizing a microbe's antibiotic resistance profile is thus part of any microbial identification process. Understanding which antibiotics are effective in treating an infectious agent will help the physician avoid prescribing an inappropriate drug. Furthermore, knowing which drugs are ineffective enables health organizations to track the spread of antibiotic-resistant strains. For example, before 1970, most strains of *Neisseria gonorrheae* were susceptible to penicillin. Today, most are penicillin resistant, due in part to the widespread use of penicillin to treat gonorrhea during and after the Vietnam War, when U.S. soldiers first contracted penicillinase-producing *N. gonorrhoeae* (PPNG).

There are strain-specific disease complications. Many diseases have serious complications that are common to a given organism or strain of organism. For example, children whose sore throats are caused by certain strains of *Streptococcus pyogenes* can develop serious complications affecting the heart and kidney long after the infection has resolved. These complications are the immunological consequence of bacterial and host antigen cross-reactivity; they are called **sequelae** because they occur *after* the infection itself is over. Life-threatening sequelae such as rheumatic fever and acute glomerular nephritis caused by certain strains of *S. pyogenes* produce severe damage to the heart and kidney, respectively. Knowing early on that *S. pyogenes* has caused a child's sore throat allows the physician to prescribe antibiotics that will quickly eradicate the infection and prevent development of the sequelae.

NOTE: Penicillin or a penicillin-like antibiotic is the usual treatment for *S. pyogenes* infection. Contrary to what you might expect, *S. pyogenes* has not developed any penicillin-resistant strains.

Tracking the spread of a disease can lead to its source. Consider a situation in which ten infants scattered throughout a city develop bloody diarrhea. The clinical laboratory identifies the same strain of *Shigella sonnei*, a gram-negative bacillus, as the cause in each case. Finding the same strain in all cases suggests they probably originated from the same source. Shigellosis is transferred from person to person by what is called the fecal-oral route. Carriers of *Shigella* will shed this organism in their feces. Inadequate hand washing after defecation will leave bacteria on their hands, which can then transfer the pathogen to foods or utensils or to another person by touching. Uninfected persons can become infected after eating the contaminated food or placing their fingers in their mouth.

Public health officials, armed with the knowledge that all the patients had the same strain of *Shigella*, then question the parents and learn that all of the children attend the same day-care center. By testing the other children and workers in that center, officials can stop the infection, called shigellosis, from spreading. This investigative process, called epidemiology, is covered more completely later in this chapter.

TO SUMMARIZE:

Identifying a pathogen enables us to:

- Use appropriate **antibiotics**, if necessary.
- Anticipate possible **sequelae**.
- **Track the spread** of the disease.

28.2 Approaches to Pathogen Identification

Pathogens can be identified on the basis of a variety of observations, including the patient's symptoms, finding organisms in stained clinical specimens, biochemical clues, and serology (the presence in a patient's serum of antibodies reactive against a specific microbe).

Bacterial Pathogen Identification Requires Knowledge of Microbial Physiology and Genetics

Thousands of bacterial species are capable of causing disease. How can a laboratory quickly, sometimes within hours, identify which one causes a given infection? The solution, in part, comes from knowing the biochemical and enzymatic features of these bacteria. Because no two species have the same biochemical "signature," the clinical lab can look for reactions, or combinations of reactions, that are unique to a given species.

The same is now true at the genetic level. In the current genomics-dominated era, DNA sequences of many pathogens have been completely elucidated. This has enabled design of polymerase chain reaction (PCR) primers that rapidly detect species-specific genes or even genes that are unique to highly pathogenic strains within a species. One example of such a gene is the attaching and effacing locus (*eae*) that is present in enterohemorrhagic, but not commensal, strains of *E. coli*. Finding *E. coli* in a fecal stool sample is not unusual. Finding *E. coli* with the *eae* gene, however, indicates that it is a pathogen.

Case History: Medical Detective Work

A 38-year-old woman with no significant previous medical history came to the emergency room complaining of a mild sore throat persisting for three days. Her symptoms included arthralgia (joint pain), myalgia (muscle pain), and low-grade fever. The day before, she had a severe headache with neck stiffness, nausea, and vomiting. She was not taking any medications, had no known drug allergies, and did not smoke. She lived with her husband and two children, all of whom were well. Cerebral spinal fluid (CSF) was collected from a spinal tap. The CSF appeared cloudy (it should be clear) and contained 871 white blood cells per microliter (normal is 0–10/µl); the glucose level was 1 mg/dl (normal is 50–80 mg/dl); and the total protein level was 417 mg/dl (normal is under 45 mg/dl). Gram stain of a CSF smear revealed gram-negative rods. The CSF sample was sent to the diagnostic laboratory for microbial identification.

As discussed in Thought Question 26.6, low glucose and elevated protein levels in CSF are indicators of *bacterial* infection, not viral. The increase in white blood cells revealed that the woman's immune system was trying to fight the disease. The presence of gram-negative rods in the CSF smear confirmed a diagnosis of bacterial meningitis, since CSF should be sterile. Now it was up to the clinical laboratory to determine the etiological agent.

Clinical Laboratories Use Problem-Solving Algorithms to Identify Bacteria

How does a clinical microbiology laboratory handle incoming specimens such as the one considered here? What are the first steps toward identifying the etiological agent of a disease? Over the years, clinical microbiologists have developed algorithms (step-by-step problem-solving procedures) that expose the most likely cause of a given infectious disease. For instance, there are only a limited number of microbes known to cause meningitis. The microbiologist poses a series of binary "yes" or "no" questions about the clinical specimen in the form of biochemical or serological tests. Typical questions in this case might include: "Is an organism seen in the CSF of a patient with symptoms of meningitis?" "Is the organism gram-positive or gram-negative?" "Does it stain acid-fast?" Answers to a first round of questions will then dictate the next series of tests to be used.

Because speed is of the essence in deciding how to treat the patient, a series of tests is not always carried out sequentially. To save time, a slew of tests are carried out simultaneously, but the results are *interpreted* sequentially based on the algorithm. In our case history, for instance, consider the most common causes of bacterial meningitis: *Neisseria meningitidis*, *Streptococcus pneumoniae*, *Haemophilus influenzae*, and *Escherichia coli*. The CSF sample was

Gram-stained by a microbiologist and simultaneously plated onto three media—chocolate agar, blood agar, and Hektoen agar. Chocolate agar is an extremely rich medium that looks brown owing to the presence of heat-lysed red blood cells (**Figs. 28.1A** and **B**). Because it is so nutrient-rich, all four organisms will grow on chocolate agar. However, nutritionally fastidious organisms such as *N. meningitidis* and *H. influenzae* will not grow well, if at all, on ordinary blood agar because these bacteria cannot lyse red blood cells and release required nutrients. Less fastidious organisms, such as *S. pneumoniae* and *E. coli*, will grow on blood agar, but only *E. coli* can grow on Hektoen agar (**Figs. 28.1C–E**), which is a selective and differential medium for enteric gram-negative rods. Differential and selective media are described in Section 4.3.

In our case history, the Gram stain of the CSF revealed gram-negative rods, which ruled out *Neisseria* (a gram-negative diplococcus) and *Streptococcus pneumonia* (a gram-positive diplococcus). The organism in CSF did grow on blood agar, which eliminated *Haemophilus influenzae* (a gram-negative, nonenteric rod) as a candidate. It also grew on Hektoen, where it produced orange lactose-fermenting colonies. Thus, the organism was a gram-negative, enteric rod and likely *E. coli*. Additional biochemical tests confirming the identity of the organism had to be carried out, but this simple example shows how simultaneous tests can be interpreted.

Identifying gram-negative bacteria. The gram-negative bacterium in this case was subjected to a battery of biochemical tests packaged as 20 separate chambers in a patented analytical profile index (API 20E) strip that can be used for pathogen identification (**Fig. 28.2**). The strip

A. Chocolate agar

B. Colonies of *N. gonorrhoeae*

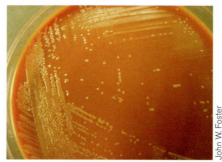

C. Hektoen agar

D. *E. coli* colonies on Hektoen

E. *S. enterica* colonies on Hektoen

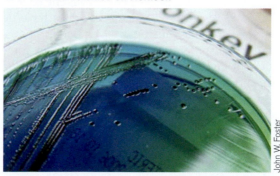

Figure 28.1   Chocolate agar and Hektoen agar; two widely used clinical media.   A. Uninoculated chocolate agar. Its color is due to gently lysed red blood cells that provide a rich source of nutrients for fastidious bacteria. **B.** Chocolate agar inoculated with a species of *Neisseria*. This organism will not grow well on typical blood agar because important nutrients remain locked within intact red blood cells. **C.** Uninoculated Hektoen agar, which contains lactose, peptone, bile salts, thiosulfate, an iron salt, and the pH indicators bromthymol blue and acid fuchsin. The bile salts prevent growth of gram-positive microbes. **D.** Hektoen agar inoculated with *Escherichia coli*. This organism ferments lactose to produce acidic fermentation products that give the medium an orange color owing to the pH indicators acid fuchsin and bromphenol blue. **E.** Hektoen agar inoculated with *Salmonella enterica*. This organism does not ferment lactose but instead grows on the peptone amino acids. The amines produced are alkaline and produce a more intense blue color with bromthymol blue. *Salmonella* species also produce hydrogen sulfide gas from the thiosulfate. Hydrogen sulfide reacts with the medium's iron salt to produce an insoluble, black iron sulfide precipitate visible in the center of the colonies.

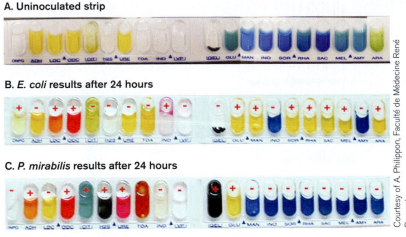

A. Uninoculated strip

B. *E. coli* results after 24 hours

C. *P. mirabilis* results after 24 hours

Courtesy of A. Philippon, Faculté de Médecine René Descartes

Figure 28.2   API 20E strip technology for the biochemical identification of Enterobacteriaceae.   A. Uninoculated API strip. Each well contains a different medium that tests for a specific biochemical capability. The color of the media after 24-hour incubation indicates a positive or negative reaction (see Table 28.1). **B.** API results for *E. coli*. Plus (+) and minus (−) indicate positive and negative reactions, respectively. **C.** API results for *Proteus mirabilis*.

requires overnight incubation and tests whether the organism can ferment a series of carbon sources. It also provides evidence of specific end products produced as a result of fermentation. The results appear as different-colored reactions in each chamber and are scored as positive or negative, depending on the color (**Table 28.1**). For

example, in the indole chamber (ninth well from the left in **Fig. 28.2B**), a red reaction is positive and indicates that the organism can produce indole from tryptophan. A colorless chamber would be a negative result. Similar API strips are available for the identification of gram-positive bacteria and yeasts.

The results of the API chambers can be interpreted in two ways. First, as a dichotomous key, a stepwise interpretation can be done by lab personnel starting with a key reaction, such as lactose fermentation. The reaction is read as positive or negative based on the color. Then, following a printed flowchart, the technician would go to the next key reaction, indole production, and read it as positive or negative. If the organism was a lactose fermentor and the indole test was positive, the choices have been narrowed to *E. coli* or *Klebsiella* species. Another reaction is read to distinguish between the remaining choices. The process continues until a single species is identified. A simplified example is shown in **Figure 28.3** using a limited number of enteric gram-negative species. In reality, many more reactions than the 11 shown have to be used to make a definitive identification because of species differences with respect to a single reaction. For example, Figure 28.3 shows that *K. pneumoniae* and *K. oxytoca* exhibit opposite indole reactions,

Table 28.1 Reading the API 20.

Tests	Substrate	Reaction tested	Negative results	Positive results
ONPG	ONPG	Beta-galactosidase	Colorless	Yellow
ADH	Arginine	Arginine dihydrolase	Yellow	Red/orange
LDC	Lysine	Lysine decarboxylase	Yellow	Red/orange
ODC	Ornithine	Ornithine decarboxylase	Yellow	Red/orange
CIT	Citrate	Citrate utilization	Pale green/yellow	Blue-green/blue
H2S	Na thiosulfate	H_2S production	Colorless/gray	Black deposit
URE	Urea	Urea hydrolysis	Yellow	Red/orange
TDA	Tryptophan	Deaminase	Yellow	Brown-red
IND	Tryptophan	Indole production	Yellow	Red (2 min)
VP	Na pyruvate	Acetoin production	Colorless	Pink/red (10 min)
GEL	Charcoal gelatin	Gelatinase	No diffusion of black	Black diffuse
GLU	Glucose	Fermentation/oxidation	Blue/blue-green	Yellow
MAN	Mannitol	Fermentation/oxidation	Blue/blue-green	Yellow
INO	Inositol	Fermentation/oxidation	Blue/blue-green	Yellow
SOR	Sorbitol	Fermentation/oxidation	Blue/blue-green	Yellow
RHA	Rhamnose	Fermentation/oxidation	Blue/blue-green	Yellow
SAC	Sucrose	Fermentation/oxidation	Blue/blue-green	Yellow
MEL	Melibiose	Fermentation/oxidation	Blue/blue-green	Yellow
AMY	Amygdalin	Fermentation/oxidation	Blue/blue-green	Yellow
ARA	Arabinose	Fermentation/oxidation	Blue/blue-green	Yellow

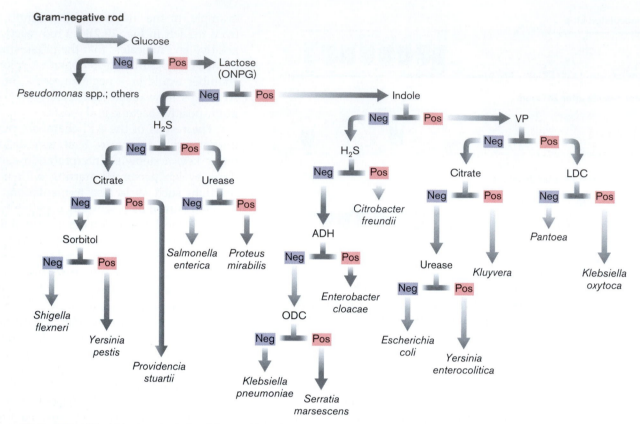

Figure 28.3 Simplified biochemical algorithm to identify gram-negative rods. The diagram presents a dichotomous key using a limited number of biochemical reactions and selected organisms to illustrate how species identifications can be made using biochemistry. Abbreviations and reactions: Lactose fermentation to produce acid, ONPG (orthonitrophenyl beta-D-galactoside, cleaved by beta-galactosidase); sorbitol fermentation to produce acid, indole production from tryptophan; VP (Voges-Proskauer test) indicates production of acetoin or 2,3-butanediol; LDC (lysine decarboxylase) cleaves lysine to produce CO_2 and cadaverine; ODC (ornithine decarboxylase) cleaves ornithine to make CO_2 and putrescine; ADH (arginine dihydrolase) is the stepwise degradation of arginine to citrulline and ornithine; citrate utilization as a carbon source; H_2S is the production of hydrogen sulfide gas; urease produces CO_2 and ammonia from urea. (*Note:* "Neg" for *Pseudomonas* in terms of glucose indicates an inability to ferment glucose. *Pseudomonas* can still use glucose as a carbon source.)

even though they are of the same genus. Even within a given species, only a certain percentage of strains might be positive for a given reaction. The inherent danger in using a dichotomous key is that one anomolous result can lead to an incorrect identification.

In a second approach, the API strips can be used to generate a seven-digit identification number that points to the identity of the bacterium. To generate this number, individual reactions are given numerical values based on whether they give a positive or negative result, and then the number values from three reactions are added to produce a single digit. Because there are 21 reactions (the 21st reaction is an oxidase test performed on colonies), the end result will be a seven-digit number. The example in **Figure 28.4A** shows how this is done. Once the number is generated, it is compared to a database of numbers generated by computer. The database was compiled by taking all possible test results and applying them to all species of Enterobacteriaceae. Each species has a known probability of being positive for a given reaction. As discussed in Section 17.3, taking each spe-

cies and multiplying the probabilities for the results generated by the test organism will generate a probability score. The test organism is identified as the species with the highest score. This probability approach is considerably more accurate than the dichotomous key.

Most clinical laboratories in the United States and Europe now use automated identification systems that are even more sophisticated than API. These include BioMerieux's *VITEC*, Dade Behring's *WalkAway*, and Becton, Dickinson and Company Phoenix systems (**Fig. 28.4B**). All of them use cards or plates with 30 or more biochemical or enzymatic reactions. As with the API chambers, the results appear as colored reactions that can be read by computer. The BD Phoenix system, for example, will convert the results into a 10-digit code that is used to identify genus and species.

Identifying nonenteric gram-negative bacteria. The procedures we have outlined will accurately identify members of Enterobacteriaceae, but pathogenic gram-negative bacilli can also be found in other phylogenetic

A.

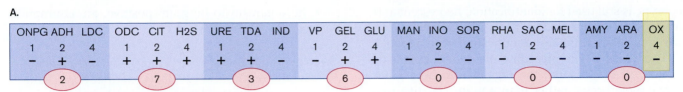

ONPG	ADH	LDC	ODC	CIT	H2S	URE	TDA	IND	VP	GEL	GLU	MAN	INO	SOR	RHA	SAC	MEL	AMY	ARA	OX
1	2	4	1	2	4	1	2	4	1	2	4	1	2	4	1	2	4	1	2	4
−	+	−	+	+	+	+	+	−	−	+	+	−	−	−	−	−	−	−	−	−

2 7 3 6 0 0 0

B.

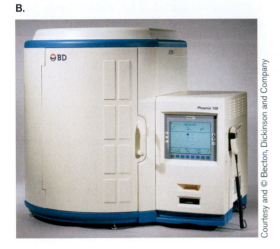

Courtesy and © Becton, Dickinson and Company

Figure 28.4 Generating an identification number from the API 20E strip. A. Results from Figure 28.2C were used to generate an identification number. The 20 reactions in the API plus an oxidase reaction performed separately are divided into seven reaction triplets. Each reaction within a triplet is given a value (1, 2, or 4). If the reaction is positive, that value is added to the value of any other positive reaction in that group, and the resulting number is placed in the oval. The result is a seven-digit number that will be unique to a given genus and species of enteric microorganism. Note that because of biochemical diversity within different strains of a species, several different numbers may be generated; but because of the way the test is designed, the numbers will still be unique to the species. **B.** Automated microbiology system. The BD Phoenix™ system uses plates with numerous reaction wells and a computerized plate reader to automatically identify pathogenic bacteria. The numbers and their evaluation can be done automatically using this type of instrument.

families. One possibility in the preceding case history is that the gram-negative bacillus seen in CSF smears would *not* grow on blood agar or on the other selective media, but would grow as small, glistening colonies on chocolate agar. This result would implicate the gram-negative rod *Haemophilus influenzae*. Meningitis caused by *H. influenzae* was a major problem prior to 1988, before the introduction of vaccinations with *H. influenzae* type b capsule material. Growth of *H. influenzae* requires hemin (X factor) and NAD (V factor), so confirming the iden-

tity *H. influenzae* involves growing the organism on agar medium containing hemin and NAD. This can be done by placing small filter paper disks containing these compounds on a nutrient agar surface (not blood agar) that has been covered with the organism. *H. influenzae* will only grow around a strip containing both X and V factors. Alternatively, X and V factors can be incorporated into Mueller-Hinton agar, as shown in **Fig. 28.5A**. The organism grows on chocolate medium because the lysed red blood cells release these factors. Although XV growth

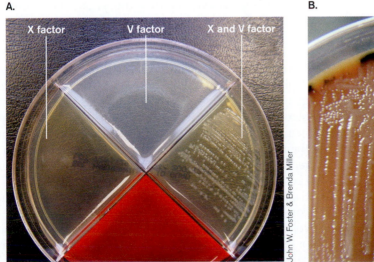

Figure 28.5 *Haemophilus influenzae* **growth factors and** *Neisseria meningitidis* **oxidase reaction. A.** *H. influenzae* will grow on an agar plate (here, Mueller-Hinton agar) only when the medium has been fortified with both X factor (hemin) and V factor (NAD), but not either one alone. **B.** Oxidase positive reaction for *Neisseria meningitidis*. Oxidase reagent was dropped onto colonies of *N. meningitidis* grown on chocolate agar. The test is called the cytochrome oxidase test, but it really tests for cytochrome *c*.

phenotype is still used for identification, fluorescent antibody staining is more specific (discussed shortly).

A completely different identification scheme would have been used in the preceding meningitis case if the laboratory had discovered the organism to be a gram-negative diplococcus (rather than a gram-negative rod). A gram-negative diplococcus would suggest *Neisseria meningitidis*. *N. gonorrhoeae* is also possible, but less likely because the gonococcus lacks the protective capsule meningococci used to survive in the bloodstream. Without this capsule, *N. gonorrhoeae* is not as resistant to serum complement and cannot disseminate to the meninges.

The first test to determine if the organism is a species of *Neisseria* is the cytochrome oxidase test (**Fig. 28.5B**). In this test, a few drops of the colorless reagent N,N,N′,N′-tetramethyl-p-phenylenediamine dihydrochloride are applied to the suspect colonies. The reaction, which only takes place if the organism possesses both cytochrome oxidase and cytochrome *c*, turns the p-phenylenediamine reagent (and the colony) a deep purple/black. While many bacteria possess cytochrome oxidase, *Neisseria* is one of only a few genera that also contain cytochrome *c* in their membranes. Oxidase-positive organisms use cytochrome oxidase to oxidize cytochrome *c*, which then oxidizes p-phenylenediamine. Other oxidase-positive bacteria include *Pseudomonas*, *Haemophilus*, *Bordetella*, *Brucella*, and *Campylobacter* species—all of them gram-negative rods. None of the Enterobacteriaceae, however, are oxidase-positive because they lack cytochrome *c*.

An oxidase-positive, gram-negative diplococcus is very likely a member of *Neisseria*. Differentiation between species of *Neisseria* is based on their ability to grow on certain carbohydrates; or it can be determined using immunofluorescent antibody staining tests that test for the presence of different capsule antigens in *N. meningitidis*.

Identifying gram-positive pyogenic cocci. Recall the case of the woman with necrotizing fasciitis (see Section 26.2). How did the laboratory determine that the etiological agent was *Streptococcus pyogenes*? A sample algorithm, or flowchart (**Fig. 28.6**), shows how it was done. The physician sends a cotton swab containing a sample from a lesion to the clinical laboratory. The laboratory technician streaks the material onto several media: (1) blood agar (which will grow both gram-positive and gram-negative organisms), (2) blood agar containing the inhibitors colistin and naladixic acid (called a CNA plate, this agar will grow *only* gram-positives), and (3) MacConkey agar (which will grow only gram-negative organisms (see Section 4.3). The suspect organism in this case grows on the CNA and blood plates. Because they grow in the presence of the gram-negative inhibitory compounds in CNA, one would immediately sus-

pect the organism to be gram-positive, an assumption borne out by the Gram stain.

> **NOTE:** Though the skin is normally populated by many different microorganisms (normal flora), samples from an infected lesion are overwhelmingly populated by the etiological agent. This occurs because the pathogenic microbe outgrows normal flora. Using selective media to isolate the infectious agent will further simplify diagnosis by reducing growth of any normal flora that may still be present.

The algorithm tells the laboratory technician that since the organism is a gram-positive coccus, the next step is to test for catalase production. Catalase, which converts hydrogen peroxide (H_2O_2) to O_2 and H_2O, clearly distinguishes staphylococci from streptococci (remember, Gram stain morphology alone is insufficiently reliable to make that distinction). The catalase test is performed by mixing a colony with a drop of H_2O_2 on a glass slide. Effusive bubbling due to the release of oxygen indicates catalase activity (see **Fig. 28.6**). Staphylococci are catalase-positive, while the streptococci are catalase-negative. Note that many other organisms possess catalase activity, including the gram-negative rod *E. coli*. However, based on the algorithm, *E. coli* would not be considered, since it does not grow on CNA agar and is not gram-positive.

> **NOTE:** When performing a catalase test from colonies grown on blood agar, be sure not to transfer any of the agar, since red blood cells also contain catalase.

Having established that the organism is catalase-negative, the technician examines the blood plate for evidence of hemolysis. Three types of colonies are possible: nonhemolytic, alpha-hemolytic, and beta-hemolytic. Nonhemolytic streptococci do not produce any lytic zone. Alpha-hemolytic strains produce large amounts of hydrogen peroxide that oxidize the heme iron within intact red blood cells to produce a green product. As a result, alpha-hemolytic streptococci produce a green zone around their colonies called alpha hemolysis—even though the red blood cells remain intact. (For example, *S. mutans*, a cause of dental caries and subacute bacterial endocarditis, is alpha-hemolytic.) Still other streptococci produce a completely clear zone of true hemolysis surrounding their colony. This is called beta hemolysis. Complete hemolysis of red blood cells occurs owing to the export of enzymes, called hemolysins, that lyse red cell membranes. The flowchart indicates that the organism from the case history was beta-hemolytic.

The final relevant test in this flowchart involves susceptibility to the antibiotic bacitracin, which identifies the

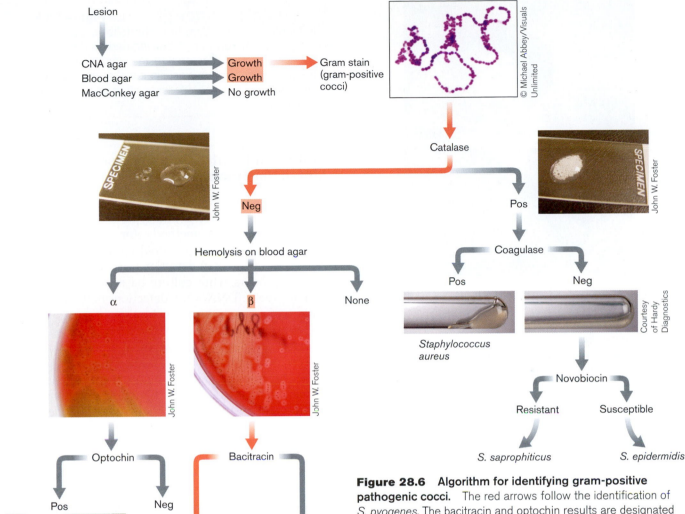

Figure 28.6 Algorithm for identifying gram-positive pathogenic cocci. The red arrows follow the identification of *S. pyogenes*. The bacitracin and optochin results are designated "positive" if the organism is susceptible and "negative" if the organism is resistant to the agent.

cocci (also called GAS), defined as the species *Streptococcus pyogenes*.

Unfortunately, the Lancefield classification procedure is somewhat time-consuming and not readily amenable as a rapid identification method. However, the group A beta-hemolytic streptococci are uniformly susceptible to the antibiotic bacitracin. Thus, a simple antibiotic disk susceptibility test can be used to indicate the group A beta-hemolytic streptococci (that is, *S. pyogenes*). But beware; many bacteria are bacitracin sensitive, so like the catalase test, the bacitracin test must be used in conjunction with an algorithm to be useful for identification. The technician must follow the appropriate algorithm before assigning importance to this or any other test result. It is irrelevant, for instance, if an alpha-hemolytic organism is bacitracin susceptible. Some of these may exist, but they are not associated with disease.

most pathogenic group of beta-hemolytic streptococci. The beta-hemolytic streptococci are subdivided into many different groups, known as the Lancefield groups, based on differences in the composition of their carbohydrate peptidoglycans. These cell wall differences are distinguished from each other immunologically and divide the streptococci into Lancefield groups A through U. Rebecca Lancefield, for whom the classification scheme is named, was the first to use immunoprecipitation to group the streptococci (**Fig. 28.7**). The vast majority of streptococcal diseases are caused by group A beta-hemolytic strepto-

Marine Biological Laboratory

Figure 28.7   Rebecca Lancefield. In 1918, Dr. Lancefield joined the Rockefeller Institute for Medical Research in New York City, where she studied the hemolytic streptococci, known then as *Streptococcus haemolyticus*. She was the first to use serum precipitation methods to classify *S. haemolyticus* into groups according to differences in cell wall carbohydrate antigens. The basic technique is still used today and is known as the Lancefield classification scheme in her honor.

The organism in our case of necrotizing fasciitis, however, was beta-hemolytic, so bacitracin susceptibility indicated that the organism was *S. pyogenes*.

The other tests named in Figure 28.6 are equally important for the identification of gram-positive infectious agents. For example, *Streptococcus pneumoniae* is an important cause of pneumonia. Like *S. pyogenes*, *S. pneumoniae* is a catalase-negative, gram-positive coccus; but unlike *S. pyogenes*, it is alpha-hemolytic. Optochin susceptibility is a property closely associated with *S. pneumoniae* while other alpha-hemolytic strains of streptococci are resistant to this compound. Thus, an optochin susceptibility disk test is a useful tool for identifying *S. pneumoniae*.

Coagulase is a key reaction used to distinguish the pathogen *Staphylococcus aureus*, which causes boils and bone infections, from other staphylococci, such as the normal skin species *S. epidermidis*. To conduct a coagulase test, a tube of plasma is inoculated with the suspect organism. If the organism is *S. aureus*, it will secrete the enzyme coagulase. Coagulase will convert fibrinogen to fibrin and produce a clotted, or coagulated, tube of plasma. Coagulase-negative staphylococci can still be medically important, however. *S. saprophiticus*, for instance, is an important cause of urinary tract infections. It can be distinguished from *S. epidermidis* based on resistance to novobiocin.

THOUGHT QUESTION 28.1   Use Figure 28.6 to identify the organism from the following case. A sample was taken from a boil located on the arm of a 62-year-old man. Bacteriological examination revealed the presence of gram-positive cocci that were also catalase-positive, coagulase-positive, and novobiocin resistant.

Pathogen Identifications Based on Molecular Genetics

Many diagnostic laboratories have implemented DNA-based methods to detect and type bacteria and viruses. Molecular detection methods for bacteria are often more rapid than the traditional culture-based methods described earlier. DNA detection (for example, PCR) takes only a few hours, while culture-based identification takes days to weeks. DNA/RNA detection methods are especially useful for viruses, which otherwise require elaborate electron microscopy to view morphology or serology to detect an increased presence of antiviral antibodies. The problem with serology is that by the time these antibodies become detectable in blood, the patient is already recovering from the disease. It is also important to note that molecular detection techniques may provide the *only* means of identifying newly recognized, or emerging, pathogens.

The polymerase chain reaction (PCR) is the most widely used molecular method in the clinical laboratory's diagnostic arsenal. DNA primers that bind to unique genes in a pathogen's genome can be used to specifically amplify DNA or RNA present in a clinical specimen. Successful amplification is visualized as an appropriately sized fragment in agarose gels following electrophoresis.

Why is PCR needed to detect the presence of these nucleic acids? Without the PCR amplification steps, clinical samples usually provide too little nucleic acid from infecting microorganisms to be detected. For example, *Mycobacterium tuberculosis*, the cause of tuberculosis, can take weeks to grow on standard bacteriological media. Detecting the presence of its DNA in the specimen, however, would allow a very rapid diagnosis. Unfortunately, the amount of bacterial DNA present in the sample is infinitesimal, too little to detect by standard techniques such as Southern blot. By amplifying *M. tuberculosis* nucleic acid by PCR, however, one copy of DNA will become billions of copies.

PCR is very quick. Preparing a clinical specimen for PCR by extracting the DNA or RNA usually takes less than an hour. PCR itself is completed in 2–3 hours. Detection of the PCR-amplified DNA by DNA gel electrophoresis takes an additional 1–2 hours. So, what might take two to three days (or sometimes weeks) using biochemical algorithms may take less than a day by molecular strategies. Most clinical laboratories are now equipped with thermocyclers (the instrument used to precisely and rapidly cycle temperature for PCR reactions) and can carry out this type of identification protocol for organisms for which specific primers are available.

Table 28.2 lists several instances in which DNA detection tests are useful. A specific example presented in **Figure 28.8** illustrates the use of PCR to type different strains of the anaerobic pathogen *Clostridium botulinum*, the cause of food-borne botulism. These organisms are not typed serologically, as is the case for *S. pneumoniae*. *C. botulinum* is divided into different types based on what neurotoxin genes they possess. In this example, **multiplex PCR** was used to simultaneously search for these toxin genes. Multiplex PCR uses multiple sets of primers, one pair for each gene, combined in a single tube with a specimen. Care must be taken to be sure that the primers chosen make different-sized products, do not interfere

with each other, and do not produce artifactual products that can confuse interpretation. Multiplex PCR can help identify sets of specific genes present in a single species or can screen for the presence of multiple pathogens in a clinical sample. In the latter, primer sets are designed to amplify genes unique to each pathogen.

Case History: Outbreak of Hospital-Acquired Wound Infections in New Delhi

From August through December, 45 patients in the pediatric surgery unit of a New Delhi hospital developed postoperative wound infections. Of these, 42 were outpatients, while 3 were inpatients who had undergone major surgery. The diseases ranged from chronic ear infections to bacteremia associated with the use of hemodialysis equipment. Thirty-two clinical samples of pus and wound exudates were tested for acid-fast bacilli by the Ziehl-Neelsen method (**Fig. 28.9A**).

Table 28.2   **Some DNA-based detection tests.**

Test	Intended use	Specimen for PCR	Transport conditions
Detection of HIV proviral DNA	Diagnosis of HIV infection in newborns; to resolve indeterminate serologic results	>2 ml whole blood in EDTA tube. EDTA prevents coagulation by chelating (binding) ions	Room temperature, within 24 hours after collection
Detection of *Bordetella pertussis*	Diagnosis of whooping cough (pertussis)	Nasopharyngeal swab	Room temperature, within 24 hours after collection
Detection of *Borrelia burgdorferi*	Diagnosis/surveillance of Lyme disease	Tick	Wrap in moist tissue in small plastic bag
		Skin biopsy; >1 ml spinal fluid; or urine	4°C (cold packs); within 24 hours after collection
Detection of *Rickettsia rickettsii*	Diagnosis/surveillance of Rocky Mountain spotted fever	Tick	Wrap in moist tissue in small plastic bag
		>2 ml whole blood in EDTA tube Skin biopsy	Blood: room temperature Skin lesion biopsy within 24 hours after collection; skin: 4°C
Detection of *Ehrlichia chaffeensis*	Diagnosis/surveillance of human monocytic ehrlichiosis	Tick	Wrap in moist tissue in small plastic bag
		>2 ml whole blood in EDTA tube	Room temperature, within 24 hours after collection
Detection of Norwalk and other small round-structured viruses	Diagnosis of viral gastroenteritis; investigation of food-borne and waterborne outbreaks	>1 ml diarrheal stool in sterile container	4°C (cold packs) within 72 hours after collection
Typing of *Mycobacterium tuberculosis*	Determine relatedness of *M. tuberculosis* isolates/ investigation of suspected outbreaks	Sputum, urine	Room temperature
Typing of various gram-negative and gram-positive bacteria	Determine relatedness of isolates/investigation of suspected nosocomial or food-borne outbreaks	Food, feces	4°C (cold packs), within 24 hours after collection (if *Campylobacter* suspected, transport at room temperature)

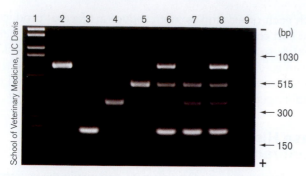

Figure 28.8 Multiplex PCR identification of *Clostridium botulinum*. *C. botulinum* cells are not typed serologically based on surface antigens, but on the basis of what toxin genes a strain possesses. Typing isolates can be done by a single PCR reaction that uses primer pairs specific for each of the four major toxin genes: Type A, B, E, and F. Because multiple products are sought in a single reaction, this is called multiplex PCR. Each lane in the agarose gel was loaded with multiplex products from different isolates of *C. botulinum* and subjected to electrophoresis. The slower-moving fragments (toward top of gel) are larger than those moving farther down the gel toward the positive pole. Lane 1, DNA size markers; lane 2, type A (*cntA*); lane 3, type B (*cntB*); lane 4, type E (*cntE*); lane 5, type F (*cntF*); lane 6, type A, B, and F; lane 7, type B, E, and F; lane 8, type A, B, E, and F.

*The same smear samples were cultured on Lowenstein-Jensen slants (**Fig. 28.9B**) and examined for growth over a course of several weeks. Biochemical tests indicated that the organism was a mycobacterium in each case. The slow-growing organism,* Mycobacterium abscessus, *was identified by PCR. Based on the DNA fingerprint, the source of the outbreak was traced to the tap water in the operating room and to a defective autoclaving process (the result of a leaking vacuum pump and faulty pressure gauge in the autoclave).*

This case history points out an underlying problem with some biochemical identifications. It took weeks to grow enough organisms to perform these biochemical tests. An alternative PCR test that probed for mycobacterium-specific genes encoding 16S rDNA could have been performed directly on the clinical specimen and would have confirmed the diagnosis immediately, much sooner than by biochemical methods. The PCR test that was done was not performed until *after* the cells were grown in the laboratory.

M. tuberculosis can be presumptively diagnosed in the clinical laboratory by the acid-fast stain, a technique first described in 1882 (called the Ziehl-Neelsen stain) that is used to find the tubercle bacillus in a patient's sputum (**Fig. 28.9A**). The acid-fast stain enables a technician to visualize bacteria such as *Mycobacterium* species that are *not* stained by the Gram stain. Mycobacteria have a very

waxy outer coat composed of mycolic acid that resists penetration by most dyes, an obstacle the acid-fast stain was designed to overcome. The original Ziehl-Neelsen acid-fast stain used phenol and heat to drive carbol fuchsin (a red dye) into mycobacterial cells on glass slides. Destaining with an acid alcohol solution removes the stain from all cell types *except* mycobacteria. The slide is subsequently counterstained with methylene blue, after which the mycobacteria will be seen as curved, red rods (called acid-fast bacilli), while everything else will appear blue. A more modern version of the acid-fast stain uses the fluorochrome auramine O to stain the mycolic acid. This dye also resists removal by an acid alcohol wash, so when observed under a fluorescence microscope, the mycobacteria will fluoresce bright yellow.

Although the acid-fast stain is very useful, sometimes organisms cannot be found in a sputum sample; and even if they are found, confirmatory tests are needed for a definitive diagnosis. Relying on growth of the organism means that diagnosis may not be clear for several weeks. For our case history, a rapid DNA-based method would have made the job of determining species much easier. This test probes a 383-bp DNA sequence located at the end of a heat-shock gene that is highly conserved among all mycobacterial species. As a result, the sequence can be amplified by PCR regardless of the strain. However, fragments amplified from different species will have somewhat different DNA sequences. The alterations in DNA sequence can be exposed by restriction digestion. Sequences from different species will produce unique digestion fragment patterns that can then be used for identification. The clinical laboratory compared the restriction patterns produced from mycobacteria isolated from different areas throughout the hospital in the case history. The results allowed the infection control staff to trace the source of all the infections to the tap water in the operating room.

In addition to PCR, restriction fragment length polymorphisms (RFLPs), a form of DNA fingerprinting, can be used to track a given strain during the course of an epidemic. This technique relies on small differences in sequence that occur between strains of a single species. Because of these small differences in sequence, the number and location of restriction enzymes sites in the chromosome will differ. In the example in Figure 28.9, DNA from a strain of *Mycobacterium tuberculosis* is cut with restriction enzymes, and the digested fragments are separated by gel electrophoresis (**Figs. 28.9C** and **D**). The gel is then probed with a radioactively labeled fragment of insertion sequence IS6110 (a fluorescent probe can also be used). IS6110 is a repetitive DNA sequence present at multiple sites in the chromosome. The radioactive probe can hybridize only to DNA fragments that contain IS6110 sequences. Strains of

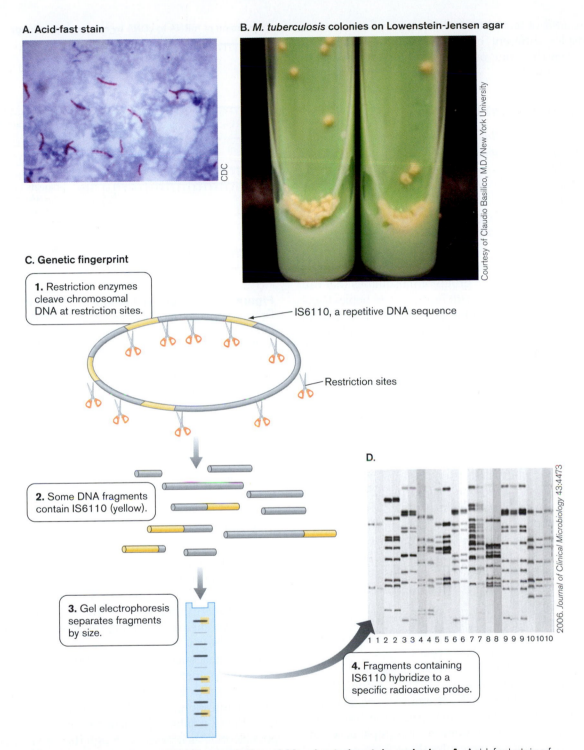

A. Acid-fast stain

CDC

B. *M. tuberculosis* colonies on Lowenstein-Jensen agar

Courtesy of Claudio Basilico, M.D./New York University

C. Genetic fingerprint

1. Restriction enzymes cleave chromosomal DNA at restriction sites.

IS6110, a repetitive DNA sequence

Restriction sites

2. Some DNA fragments contain IS6110 (yellow).

3. Gel electrophoresis separates fragments by size.

D.

2006. *Journal of Clinical Microbiology* 43:4473

1 1 2 2 3 3 4 4 5 5 6 6 7 7 8 8 9 9 9 101010

4. Fragments containing IS6110 hybridize to a specific radioactive probe.

Figure 28.9 Morphology, growth and DNA fingerprinting of *Mycobacterium tuberculosis*. A. Acid-fast stain of *M. tuberculosis*. **B.** Lowenstein-Jensen medium enables growth of mycobacterial species, some of which grow extremely slowly. The colonies have a "bread-crumb-like" appearance. **C.** Genetic fingerprinting of *Mycobacterium tuberculosis* isolates. Fragments containing IS6110 are marked yellow. **D.** Fragments containing IS6110 hybridize to the specific radioactive probe. A characteristic banding pattern (fingerprint) appears for each isolate. Isolates with similar banding patterns are assigned the same number in this example.

M. tuberculosis obtained from different outbreaks would likely display different hybridization patterns. Strains isolated from the same outbreak would look very similar, if not identical.

Case History: West Nile Virus

A 55-year-old man was admitted to a local hospital complaining of headache, high fever, and neck stiffness. The man appeared confused and disoriented. He also complained of muscle weakness. History indicated he had received several mosquito bites approximately two weeks previously. A blood specimen was sent to the laboratory. The report the following day indicated that the patient was suffering from West Nile virus.

West Nile virus is primarily an infection of birds and culicine mosquitoes (a group of mosquitoes that can transmit human diseases), with humans and horses serving as incidental, dead-end hosts. Replication of virus in this bird-mosquito-bird cycle begins when adult mosquitoes emerge in early spring and continues until fall. Among humans, the incidence of disease peaks in late summer and early fall. Birds provide an efficient means of geographical spread of the virus. As a result, over the past several years, the virus has spread throughout much of the United States.

Isolating a disease-causing virus is extremely challenging. Most laboratories are not equipped for the special tissue culture techniques required to grow viruses. Consequently, most viral infections, including human West Nile virus infections, are usually diagnosed by measuring the antibody response of the patient. For instance, the presence of West Nile virus–specific IgM in cerebrospinal fluid is a good indicator of current West Nile virus infection, but it is indirect and not conclusive. Real-time PCR is a molecular test that can quickly reveal the presence of the virus itself.

Real-time quantitative PCR (RTQ-PCR, or qPCR) is used routinely for the high-throughput diagnosis of viral pathogens such as West Nile virus. Because West Nile virus (Flaviviridae family) contains single-stranded RNA, its RNA must first be converted to DNA using reverse transcriptase before PCR can be attempted. The quantitative advantage of RTQ-PCR is that you can estimate the number of virus particles present in the sample based on the number of viral RNA molecules there.

The basic technique is as follows: RNA is first extracted from the sample. A subsequence of viral RNA is then converted to cDNA using a single primer and reverse transcriptase (**Fig. 28.10**). The more virus particles there are in the sample, the more viral RNA will be present and the more cDNA product is made. The cDNA is then amplified by PCR using two specific primers and Taq polymerase (see Section 12.3; Fig. 12.15). The trick

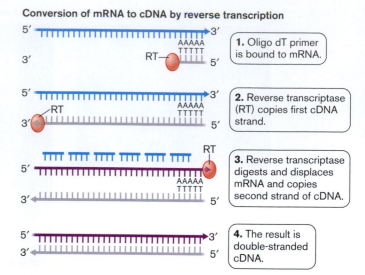

Conversion of mRNA to cDNA by reverse transcription

1. Oligo dT primer is bound to mRNA.

2. Reverse transcriptase (RT) copies first cDNA strand.

3. Reverse transcriptase digests and displaces mRNA and copies second strand of cDNA.

4. The result is double-stranded cDNA.

Figure 28.10 Reverse transcriptase synthesis of DNA. Reverse transcriptase (RT) uses mRNA as a template to make DNA. The DNA can then be amplified using standard PCR methods. Molecules of eukaryotic and viral mRNA usually contain poly-A tails at their 3′ ends. An oligonucleotide poly-T primer added to the reaction tube will anneal to the poly-A tail and allow RT to synthesize the first strand of a complementary DNA (cDNA). A second primer specific to the viral gene is then used to prime RT synthesis of the opposite strand (second-strand synthesis), using the first DNA strand as template. Subsequent amplification by PCR requires the addition of a thermostable DNA polymerase (Taq) that can withstand the denaturing and DNA synthesis temperatures required for amplifying the cDNA.

for quantitation is in the method used to detect amplification. In one method, a third, fluorescent oligonucleotide (called the probe) is added to the PCR reaction (see Fig. 12.15A). The probe contains a fluorescent dye at the 3′ end and a chemical dye at the 5′ end that quenches (that is, absorbs) energy emitted from the fluorescent dye. As long as the two chemicals are kept in close proximity by the intact probe, no light is emitted. The probe is designed to anneal to a sequence in between the binding sites of the two other primers (modifications on the ends of the probe prevent it from being used as a primer). So in a successful amplification, Taq polymerase will synthesize DNA from the two outside primers and degrade the probe oligonucleotide as it passes through that area. This cleavage separates the dye from the quencher, and the dye begins to fluoresce. The greater the amount of cDNA there was to begin with, the fewer cycles it takes to register a fluorescence increase over background (**Fig. 28.11**).

THOUGHT QUESTION 28.2 Why does finding IgM to West Nile virus indicate current infection? Why wouldn't finding IgG do the same?

A.

Cycle number	Amount of DNA
0	1
1	2
2	4
3	8
4	16
5	32
6	64
7	128
8	256
9	512
10	1,024
11	2,048
12	4,096
13	8,192
14	16,384
15	32,768
16	65,536
17	131,072
18	262,144
19	524,288
20	1,048,576
21	2,097,152
22	4,194,304
23	8,388,608
24	16,777,216
25	33,554,432
26	67,108,864
27	134,217,728
28	268,435,456
29	536,870,912
30	1,073,741,824
31	1,400,000,000
32	1,500,000,000
33	1,550,000,000
34	1,580,000,000

B.

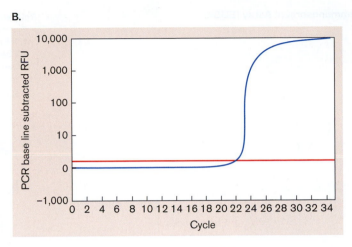

Figure 28.11   Results of real-time PCR.   A. The exponential increase in PCR products after each cycle of hybridization and polymerization. The switch to yellow indicates the point where the increase in product plateaus because the primers have been exhausted. **B.** Increase in relative fluorescence units (RFU) as the fluorescent dye is released from the dual-labeled probe during real-time PCR. The blue curve representing PCR product is flat for the first 22 cycles because the amount of DNA made, and therefore level of fluorescent dye released, remains below background level (red line). The more starting DNA there is, the sooner RFU values will increase over background (that is, fewer cycles are needed to see the increase over background). The slope eventually decreases because the fluorescent probe has become limiting.

Identifications Based on Serology

Case History: Ebola

Between October and November 2000, 62 residents of a small village north of Gulu, a town in Uganda, became ill with high fever, diarrhea, headache, vomiting, and gastrointestinal bleeding from the rectum. As many as 36 patients died. By January, 425 cases were reported, of whom 224 died. The presumptive diagnosis was Ebola. Laboratory confirmation tests included viral antigen detection and antibody ELISA tests. Laboratory-confirmed Ebola patients were defined as patients who were either positive for Ebola virus antigen or Ebola IgG antibody. Once identified, rigorous quarantine mechanisms were implemented to limit spread of the disease to other villages.

What are these tests?

The ELISA test detects antibodies indirectly or antigens directly. The **enzyme-linked immunosorbent assay (ELISA)** can detect antigens or antibodies present in nanogram and picogram quantities. One form of ELISA detects serum antibodies. It is carried out in a 96-well microtiter plate, which allows multiple patient serum samples to be tested simultaneously. An antigen from the virus (Ebola in this case) is attached (or adsorbed) to the plastic of the wells (**Fig. 28.12**). Albumin or powdered milk is used to block the remaining sites on the plastic that could result in false positives. Patient serum is then added. Ebola-specific antibodies present in the serum will react with the antigen attached to the microtiter plate. The antigen-antibody complex is then reacted with goat-antihuman IgG to which an enzyme has been attached, or conjugated (for example, horseradish peroxidase). This forms an antibody "sandwich" that attaches the enzyme to the well. The chromogenic substrate for the enzyme is added next (for example, tetramethylbenzidine). If enzyme-conjugated antibody has bound to any human IgG, the enzyme will convert the substrate to a colored product (blue for tetramethybenzidine). Enzyme activity can be measured with an ELISA plate reader. The amount of colored product formed, detected as absorbance with a spectrophotometer, will be an indication of the amount of anti-Ebola antibody present in the patient sample.

> **THOUGHT QUESTION 28.3**   Why does adding albumin or powdered milk prevent false positives in ELISA?

Enzyme-Linked Immunosorbent Assay (ELISA)

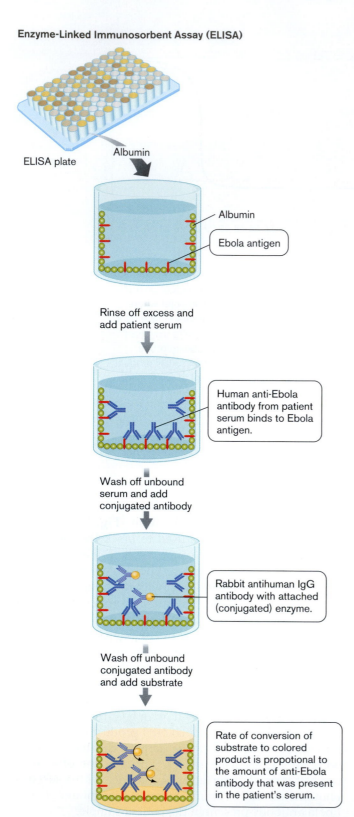

Figure 28.12 on the left; Figure 28.13 on the right.

Antigen capture

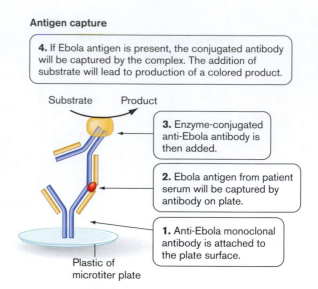

4. If Ebola antigen is present, the conjugated antibody will be captured by the complex. The addition of substrate will lead to production of a colored product.

Substrate   Product

3. Enzyme-conjugated anti-Ebola antibody is then added.

2. Ebola antigen from patient serum will be captured by antibody on plate.

1. Anti-Ebola monoclonal antibody is attached to the plate surface.

Plastic of microtiter plate

Figure 28.13   Antigen capture ELISA.   ELISA technique to capture Ebola antigens circulating in patient serum.

Antigen capture is another ELISA technique, but in this instance, anti-Ebola antibody, not viral antigen, is absorbed to the wells of a microtiter plate (**Fig. 28.13**). Patient serum is then added to the wells. If the serum contains Ebola antigen, the antigen will be captured by the antibody in the well. Then a second, enzyme-conjugated antibody against the Ebola antigen is added. The more antigen present in the serum, the more enzyme-linked antibody will affix to the well. Addition of the appropriate chromogenic substrate will produce a colored product that can be measured.

Antibody against Ebola may be easier to detect than viral antigen because antibodies will be present at higher levels than the virus itself. But because there is a delay between the time when the virus is first present in serum and when the body manages to make antibody, a speedier diagnosis can be made by directly detecting viral antigen.

While bacterial infections are commonly diagnosed by growing the infecting organism on artificial medium in the clinical laboratory, viral diseases are usually diagnosed by immunological means. Viruses are more difficult to grow than bacteria and do not exhibit the biochemical diversity so useful for identifying different bacterial species. In addition to immunological tests, viral diseases can be diagnosed using molecular approaches, such as PCR, to identify viral DNA or RNA sequences.

ELISA plate

Albumin

Albumin

Ebola antigen

Rinse off excess and add patient serum

Human anti-Ebola antibody from patient serum binds to Ebola antigen.

Wash off unbound serum and add conjugated antibody

Rabbit antihuman IgG antibody with attached (conjugated) enzyme.

Wash off unbound conjugated antibody and add substrate

Rate of conversion of substrate to colored product is propotional to the amount of anti-Ebola antibody that was present in the patient's serum.

Figure 28.12   Enzyme-linked immunosorbent assay (ELISA).   ELISA to detect anti-Ebola antibodies circulating in patient serum. The 96-well plate can be used to make dilutions of a single patient's serum to more precisely determine the amount of anti-Ebola antibody, or it can be used to test samples from multiple patients.

THOUGHT QUESTION 28.4   Specific antibodies against an infectious agent can persist for years in the bloodstream. So how is it possible that antibody titers can be used to diagnose diseases such as infectious mononucleosis?

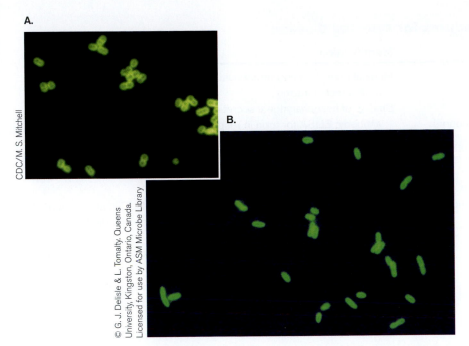

Figure 28.14   Fluorescent antibody stain.   A. *S. pneumoniae* capsule (cells approx. 0.8 μm, fluorescence microscopy). The capsule is the green halo. The center of the halo is the cell. **B.** *Legionella pneumophila* (approx. 1 μm in length) from a respiratory tract specimen.

Fluorescent Antibody Staining

Chapter 27 presents a case history involving an 80-year-old nursing home resident who contracted pneumonia caused by *Streptococcus pneumoniae*. The laboratory diagnosis was probably made using the biochemical algorithm previously described. However, there are over 80 serological types of *S. pneumoniae*, each one containing a different capsular antigen. How can the lab identify which antigenic type has caused the infection? One way is to stain the organism with antibodies.

Figure 28.14A illustrates the result of staining a smear of the isolated streptococcus with fluorescently tagged antibodies directed against a specific antigenic type of capsule. Viewed under a fluorescence microscope, the organism is "painted" green when the right antibody binds to the capsule. In the case of pneumonia, this knowledge probably will not help in the treatment of the individual patient, but its broader value is in determining if a *single type* of organism is responsible for an outbreak of pneumonia, which in turn is of epidemiological value for identifying the source of the bacterium.

On the other hand, fluorescent antibody staining techniques are critically important for rapidly identifying organisms that are difficult to grow. Infected tissues can be subjected to direct fluorescent antibody staining. **Figure 28.14B**, for example, shows a direct fluores-cent antibody stain of pleural fluid from a patient with Legionnaires' disease.

Other Microbes

Space does not permit a complete listing of the various diseases and the methods used to identify the etiological agents, but **Table 28.3** presents some additional examples. In general, bacteria that are easily cultured are grown in the laboratory, after which biochemical tests are performed. Bacterial, viral, and fungal species that are difficult to grow are typically identified using immunological techniques. These tests either identify antigen from the microbe in infected tissues or measure a rise in antibody titer. As noted earlier, DNA-based methodologies are gaining increasing acceptance. Eukaryotic microbial parasites such as *Plasmodium* species (the cause of malaria), *Giardia lamblia* (which causes giardiasis, a diarrheal disease), and *Entamoeba histolytica* (the cause of amebic dysentery) can be identified via their telltale morphologies under the microscope, making biochemical tests unnecessary.

TO SUMMARIZE:

- **Selective media** are used to inhibit growth of one group of organisms while permitting the growth of others (such as gram-positive bacteria versus gram-negative bacteria). This technique is often used to prevent growth of normal flora while permitting the growth of pathogens.
- **Differential media** exploit the unique biochemical properties of a pathogen to distinguish it from similar-looking nonpathogens.
- **Identification of bacterial species** can be accomplished with biochemical analyses, molecular techniques (for example, PCR), and/or immunological methods (for example, ELISA).
- **Viral diseases are often diagnosed using immunological tests,** such as ELISA, that measure the presence of antibody or antigen, or by real-time quantitative PCR.
- **Fluorescent antibody staining** can rapidly identify organisms or antigens present in tissues.

Table 28.3 Identification procedures for selected diseases.

Agent	Disease	Identification
Corynebacterium diphtheriae	Diphtheria	Material from nose and throat cultured on a special medium; in vivo or in vitro tests for toxin
Bordetella pertussis	Pertussis (whooping cough)	Smears of nasopharyngeal secretions stained with fluorescent antibody; ELISA for toxin in respiratory secretions; culture on special media
Legionella pneumophila	Legionellosis, Legionnaires' disease, Pontiac fever	Culture on special medium is preferred method; antigen detection by ELISA; *Legionella* nucleic acid can be identified in clinical material using PCR and nucleic acid probes
Campylobacter spp.	Campylobacteriosis	Isolate bacteria on selective media incubated at 42°C in atmosphere of nitrogen containing 5% oxygen and 10% carbon dioxide; then biochemical testing performed
Leptospira interrogans	Leptospirosis	Serological tests early in the illness and after 2–3 weeks to detect rise in antibody titer
Listeria monocytogenes	Listeriosis	Culture of blood and spinal fluid; selective and enrichment cultures performed on food samples to grow potential pathogens; DNA probe for rapid identification of colonies
Chlamydia trachomatis	Chlamydial genital infections	Identification of *C. trachomatis* antigen in urine or pus using monoclonal antibody; nucleic acid probes
Treponema pallidum	Syphilis	Direct fluorescent antibody staining; serological tests
Francisella tularensis	Tularemia	Cultures using cysteine-containing media; fluorescent antibody stain of pus; detection of rise in antibody titer
Yersinia pestis	Plague	Identification of capsular antigen using fluorescent antibody or ELISA
Viruses		
Rhinovirus	Common cold	Strain identification requires use of specific antibodies
Influenza	Flu	Tests comparing influenzal antibody levels in blood samples taken during acute and convalescent stages of illness
Hantavirus	Hantavirus pulmonary syndrome	Antigen detection in tissues using electron microscopy or monoclonal antibody; ELISA and Western blot tests for IgG and IgM antibodies in victim's blood
Herpes simplex	Different strains cause cold sores, ocular lesions, and genital lesions	Identifying the viral antigen in clinical material using fluorescent antibody or DNA probes
Mumps	Mumps	Rise in antibody titer, or presence of IgM antibody to mumps virus in victim's blood
Rotavirus	Diarrhea	Electron microscopy or ELISA of diarrheal stool for virus
HIV	AIDS	Detection of antibody to HIV-1 in patient's blood
Fungi		
Coccidioides immitis	Coccidioidomycosis, infection of lung; can disseminate to almost any tissue	Observation of large, thick-walled, round spherules from clinical specimens; PCR identification
Histoplasma capsulatum	Histoplasmosis, intracellular infection of lung; sometimes disseminates	Stained material from pus, sputum, tissue, etc., examined for intracellular *H. capsulatum* yeast phase; blood tests for antibody to the organism

28.3 Specimen Collection

Physicians must collect, and the laboratory must process, a wide variety of clinical specimens. Types of specimens range from simple cotton swabs of sore throats, in which the swab is placed into a liquid transport medium before being sent to the laboratory, to urine and fecal samples that are transported directly. Tables 28.2 and 28.3 outline some of these samples and the techniques used to collect them.

Case History: Abdominal Abscess

A 4-year-old boy was admitted to the hospital for evaluation and treatment of persistent pain in the rectal area. His prob- *lem began about one week earlier with ill-defined pain in the same area. He had a white blood cell count of 24,900 with 87% granulocytes. An abdominal computed tomography scan revealed an abscess adjacent to his rectum. A needle aspiration drained 20 mL of yellowish, foul-smelling fluid from the abscess. Aerobic cultures of this specimen plated on blood and MacConkey agars were negative. Why didn't the infectious agent grow?*

The problem in this instance is related to specimen collection and processing. Internal abscesses located near the gastrointestinal tract are often anaerobic infections, in this case caused by the gram-negative rod *Bacteroides fragilis*, a strict anaerobe (**Fig. 28.15A**).

A.

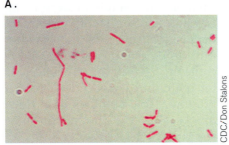

CDC/Don Stalons

B.

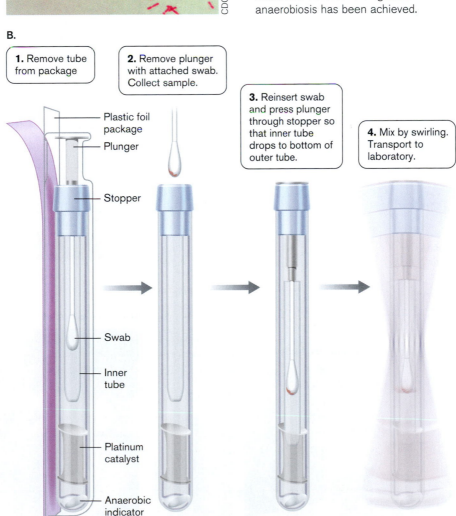

Figure 28.15 **Anaerobic infection.**
A. Gram stain of *Bacteroides fragilis* (1.5 to 4 μm in length). **B.** Vacutainer anaerobic specimen collector. Plunging the inner tube to the bottom will activate a built-in oxygen elimination system. The anaerobic indicator changes color when anaerobiosis has been achieved.

1. Remove tube from package

2. Remove plunger with attached swab. Collect sample.

3. Reinsert swab and press plunger through stopper so that inner tube drops to bottom of outer tube.

4. Mix by swirling. Transport to laboratory.

Plastic foil package

Plunger

Stopper

Swab

Inner tube

Platinum catalyst

Anaerobic indicator

Intestinal microbes, the majority of which are anaerobic, can sometimes escape the intestine if the organ is damaged in some way. The specimen in this instance should have been collected under anaerobic conditions by aspiration into a nitrogen-filled tube prior to transport to the clinical laboratory. Alternatively, a swab of the abcess material can be inserted into a special transport tube that has a built-in oxygen elimination system (**Fig. 28.15B**). Because it was collected, transported, and handled in air, many anaerobic microbes were probably killed by the oxygen.

Because *B. fragilis* has a stress response system that permits survival of this anaerobe for one or two days in oxygen, some of the bacteria may have survived transport. The laboratory still had a chance to find the organism, which raises the second problem in the case. Once the lab received the specimen, they cultured it only under aerobic conditions. The laboratory should have also incubated a series of plates anaerobically (see Section 5.6; Figure 5.22). This case, therefore, illustrates the importance of both proper specimen collection and proper processing.

Some body sites should not contain any microorganisms when collected from a healthy individual. These include blood, cerebral spinal fluid, and urine from the bladder. Because these sites are sterile, specimens can be plated onto nonselective agar media as well as selective media. Nonselective media, such as blood or chocolate agars, can be used because any organism found in these specimens is considered significant.

Urine collection can be problematic, however. When collected from a catheterized patient, urine should be sterile (**Fig. 28.16A**). Catheterization involves passing a thin, sterile tubing through the urethra and directly into the bladder. (Catheterization is primarily used to assist urination by immobilized patients, but it also provides a convenient way to collect urine for bacteriological examination.) Collections made from the tube should be sterile unless an infection is present. Unfortunately, the simple process of inserting the catheter through the non-sterile urethra can sometimes introduce organisms into the bladder and precipitate an infection. Also, the urine should be collected from the catheter, never from the collection bag. Urine may sit for hours in the collection bag, so organisms initially present at low numbers have time to replicate to high numbers even if the patient does not have a urinary tract infection (UTI).

When a catheter is not in place, urine is most commonly collected by what is called a midstream clean-

Figure 28.16 Specimen collection. **A.** Urinary catheter showing placement in the urethra. **B.** Throat swab. **C.** Sputum collection. A TB patient has coughed up sputum and is spitting it into a sterile container. (The patient is sitting in a special sputum collection booth that prevents the spread of tubercle bacilli. The booth is decontaminated between uses.) **D.** Lumbar puncture to obtain cerebral spinal fluid.

A. Urinary catheter

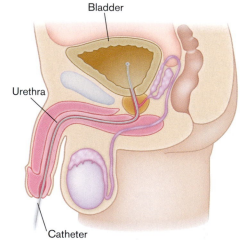

B. Throat swab

Will & Deni McIntyre/Photo Researchers

C. Sputum collection

Courtesy of the CDC

D. Lumbar puncture

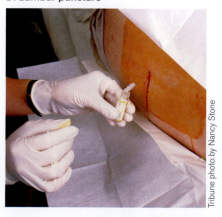

Tribune photo by Nancy Stone

catch technique, which is performed by the patient. In this procedure, the external genitals are first cleaned with a sterile wipe containing an antiseptic. The patient then partially urinates to wash as many organisms as possible out of the urethra and then collects 5–15 ml of the midstream urine in a sterile cup. This urine sample will not usually be sterile because of urethral contamination, but the number of bacteria will be low. The clinical laboratory determines how many organisms per milliliter are present in the midstream catch and informs the physician as to whether an infection is present. Finding more than 100,000 organisms per milliliter of urine from a midstream clean catch is considered indicative of an infection even in an asymptomatic patient; 10,000 or fewer is considered normal. If the patient actually has symptoms of a UTI, however, then any count greater than 1,000 CFUs/ml of a single species is considered significant and should be treated.

Identifying pathogens present at sites that contain normal flora is more challenging. A stool or fecal sample, for instance, is normally teaming with microbial flora. These specimens are typically plated onto selective media (for example, MacConkey, Hektoen, CNA agar) to eliminate or decrease the number of normal flora that might contaminate the specimen. The following represent techniques used to collect specimens from sterile and nonsterile body sites:

Collections from sites with normal flora:

Swabs	Throat swabs (**Fig. 28.16B**), for example, should be placed in specialized liquid nutrient transport medium
Sputum	Deep lung secretions expectorated for oral collection (**Fig. 28.16C**)
Stool samples	Cup or rectal swab; for identifying diarrhea-causing microbes
Abscesses	Needle aspirations

Collections from normally sterile sites:

Cerebral spinal fluid	Lumbar punctures (spinal taps; **Fig. 28.16D**); testing for meningitis
Urine samples	Midstream clean catch or from catheters placed in the bladder; testing for urinary tract infections
Blood samples	Generally taken by syringe from two body sites and placed in liquid media for aerobic and anaerobic culture. The same organism isolated from blood samples taken from the two sites is considered the likely etiological agent.

THOUGHT QUESTION 28.5 Two blood cultures, one from each arm, were taken from a patient with high fever. One culture grew *Staphylococcus epidermidis*, but the other blood culture was negative (no organisms grew out). Is the patient suffering from septicemia caused by *S. epidermidis*?

THOUGHT QUESTION 28.6 A 30-year-old woman with abdominal pain went to her physician. After the examination, the physician asked the patient to collect a midstream urine sample that they would send to the lab across town for analysis. The woman complied and handed the collection cup to the nurse. The nurse placed the cup on a table at the nurse's station. Three hours later, the courier service picked up the specimen and transported it to the laboratory. The next day, the report came back "greater than 200,000 CFUs/ml; multiple colony types; sample unsuitable for analysis." Why was this determination made?

TO SUMMARIZE:

■ **Special collection precautions** must be taken when collecting specimens from abscesses of suspected anaerobic etiology.
■ **Common specimens** include blood, pus, urine, sputum, throat, stool, and CSF.
■ **Specimens** from sites containing normal flora must be handled differently than specimens taken from normally sterile body sites.

28.4 Biosafety Containment Procedures

Case History: Fatal Meningitis

On July 15, an Alabama microbiologist was taken to the emergency room with acute onset of generalized malaise, fever, and diffuse myalgias. She was given a prescription for oral antibiotics and released. On July 16, she became tachycardic and hypotensive and returned to the hospital. She died 3 hours later. Blood cultures were positive for N. meningitidis *serogroup C. Three days before the onset of symptoms, the patient had prepared a Gram stain from the blood culture of a patient subsequently shown to have meningococcal disease; the microbiologist also handled agar plates containing cerebrospinal fluid (CSF) cultures from the same patient. Coworkers reported that in the laboratory, aspirating fluids from blood culture bottles was typically performed at the open laboratory bench. No biosafety cabinets, eye protection, or masks were used for this procedure. Testing at CDC indicated that the isolates from both patients were indistinguishable. The laboratory at the hospital infrequently processed isolates of* N. meningitidis *and had not processed another meningococcal isolate during the previous four years.*

Medical and laboratory personnel are exposed to extremely dangerous pathogens on a daily basis. The microbiologist in this case did not take appropriate measures to protect herself and ended up with a laboratory-acquired infection leading to meningitis. The CDC has published a series of regulations designed to protect workers at risk of infection by human pathogens. Infectious agents are ranked by the severity of disease and ease of transmission. Based on this ranking, four levels of containment are employed (**Table 28.4**).

Table 28.4   Biological safety levels and select agents.

	Containment level			
	Level 1	**Level 2**	**Level 3**	**Level 4**
Class of disease agent	**Category I** Agents not known to cause disease	**Category II** Agents of moderate potential hazard; also required if personnel may have potential contact with human blood or tissues	**Category III** Agents may cause disease by inhalation route	**Category IV** Dangerous and exotic pathogens with high risk of aerosol transmission; only six such labs in the United States
Recommended safety measures	Basic sterile technique; no mouth pipetting	Level 1 procedures plus limited access to lab; biohazard safety cabinets used; hepatitis vaccination recommended	Level 2 procedures plus ventilation providing directional airflow into room, exhaust air directed outdoors; restricted access to lab (no unauthorized persons)	Level 3 procedures plus one-piece positive-pressure suits; lab completely isolated from other areas present in the same building or is in a separate building
Representative organisms in class	*Bacillus subtilis* *E. coli* K12 *Saccharomyces*	*Bordetella pertussis* *Burkholderia mallei* (glanders) *Campylobacter jejuni* *Chlamydia* spp. *Clostridium* spp. *Corynebacterium diphtheriae* *Cryptococcus neoformans* *Cryptosporidium parvum* Dengue Diarrheagenic *E. coli* *Entamoeba histolytica* *Francisella tularensis* (tularemia) *Giardia lamblia* *Haemophilus influenzae* *Helicobacter pylori* Hepatitis *Legionella pneumophila* *Listeria monocytogenes* *Mycoplasma pneumoniae* *Neisseria* spp. Pathogenic *Vibrios* *Salmonella* *Shigella* spp. *Staphylococcus aureus* *Toxoplasma* *Yersinia pestis* *Yersinia enterocolitica*	*Bacillus anthracis* (anthrax) *Brucella* spp. (brucellosis) California encephalitis *Coxiella burnetii* (Q fever) EEE (Eastern equine encephalitis) Japanese encephalitis virus La Crosse encephalitis LCM (lymphocytic choriomeningitis) *Mycobacterium tuberculosis* Rabies *Rickettsia prowazekii* (typhus fever) Rift Valley fever SARS (severe acute respiratory syndrome) Variola major (smallpox) and other poxviruses VEE (Venezuelan equine encephalitis) West Nile virus Yellow fever	Ebola Guanarito virus Hantavirus Junin virus Kyasanur Forest virus Lassa fever Machupo virus Marburg virus Tick-borne encephalitis viruses

Organisms in blue are on the list of CDC select agents that are considered possible agents of bioterrorism.

Category I organisms have little to no pathogenic potential and require the lowest level of containment (biosafety level 1). Standard sterile techniques and laboratory practices are sufficient. Category II agents have greater pathogenic potential but are not generally transmitted by the respiratory route. They require more rigorous containment procedures, such as limiting laboratory access when experiments are in progress and the use of biological laminar flow cabinets if aerosolization is possible (biosafety level 2). Category III pathogens can be transmitted via the respiratory route and present a high risk of infection. To safely handle these organisms, level 2 procedures are supplemented with a lab design that ensures that ventilation air flows only *into* the room (negative pressure) and that exhaust air vents directly to the outside. Negative pressure will keep any organism that may aerosolize from escaping into hallways. In addition, access to the lab is strictly regulated and includes double-door air locks at the entrance (biosafety level 3).

By law, extremely dangerous pathogens such as the Ebola virus may only be studied at a biosafety level 4 containment facility. Practices here dictate that lab personnel wear positive-pressure lab suits connected to a separate air supply (**Fig. 28.17**). The positive pressure ensures that if the suit is penetrated, organisms will be blown away from the breach and not sucked into the suit. As reasonable as these regulations may seem, they were not always in effect. Prior to 1970, scientists had an almost cavalier approach toward handling pathogens. For instance, culture material was routinely transferred from one vessel to another by mouth pipetting (essentially using a glass or plastic pipette as a straw). This is now forbidden for obvious reasons.

TO SUMMARIZE:

- **Various levels** of protective measures are used when handling potentially infectious biological materials.
- **Category I** agents are generally not pathogenic and require the lowest level of containment.
- **Category II** agents are pathogenic but not typically transmitted via the respiratory tract. Laminar flow hoods are required.
- **Category III** agents are virulent and transmitted by respiratory route. They require laboratories with special ventilation and air-lock doors.
- **Category IV** agents are highly virulent and require the use of positive-pressure suits.

28.5 Principles of Epidemiology

Case History: Inhalation Anthrax

On October 16, a 56-year-old African-American U.S. Postal Service worker became ill with a low-grade fever, chills, sore throat, headache, and malaise. This was followed by minimal dry cough, chest heaviness, shortness of breath, night sweats, nausea, and vomiting. On October 19, he arrived at a local hospital, where he presented with a normal body temperature and normal blood pressure. He was not in acute distress but had decreased breath sounds and rhonchi (dry sounds in lungs due to congestion). No skin lesions were observed, and he did not smoke. Total WBC count was normal, but there was a left shift in the differential—that is, more polymorphonuclear lymphocytes (PMNs; see Section 26.3). A chest X-ray showed bilateral pleural effusions (accumulation of fluids in the lung) and a small right lower lobe air space opacity. Within 11 hours, blood cultures taken upon admission grew Bacillus anthracis. *Ciprofloxacin, rifampin, and clindamycin antibiotic treatments were initiated, and the patient recovered. His job at the post office was to sort mail.*

From October 4 to November 2, 2001, the Centers for Disease Control and Prevention (CDC) and various state and local public health authorities reported 10 confirmed cases of inhalational anthrax and 12 confirmed or suspected cases of cutaneous anthrax in persons who worked in the District of Columbia, Florida, New Jersey, and New York. Many of them were postal workers. It was clear that a biological attack was in progress.

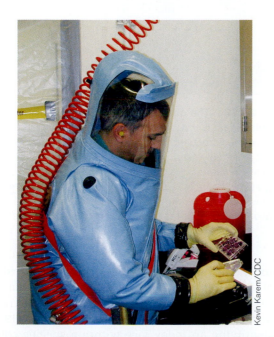

Figure 28.17 Biosafety level 4 containment. Dr. Kevin Karem at the CDC performs viral plaque assays to determine the neutralization potential of serum from smallpox vaccination trials. He is protected by a positive pressure suit working in a biosafety level 4 laboratory. The airflow into his suit is so loud he must wear earplugs to protect his hearing.

The word *epidemiology* is derived from the Greek, meaning "that which befalls man." In scientific parlance, epidemiology examines the distribution and determinants of disease frequency in human populations. Put more simply, epidemiologists determine the source of a disease outbreak and the factors that influence how many individuals will succumb to the disease. Epidemiological principles are also used to determine the effectiveness of therapeutic measures and to identify new syndromes, such as SARS (severe acute respiratory syndrome) and Lyme disease. Some of the basic concepts of epidemiology were already covered in Chapter 25 when we discussed infection cycles. Now we will explore how those principles are used to track disease.

Epidemiological early-warning systems require an extensive organization that coordinates information from many sources. In the United States, that duty falls to the Centers for Disease Control and Prevention (CDC). On the world stage, it is the World Health Organization (WHO). Any disease considered highly dangerous or infectious is first reported to local public heath centers, usually within 48 hours of diagnosis. The local centers forward that information to their state agencies, which then report to the CDC in Atlanta. This is how authorities in 2001 quickly recognized that an outbreak of anthrax was under way.

The terms **endemic** and **epidemic** are often used when referring to disease outbreaks. A disease is endemic if it is always present in a population at a low frequency. For example, Lyme disease, caused by the spirochete *Borrelia burgdorferi*, is endemic to the Northeastern United States because the organism has found a **reservoir** in deer and ticks. Recall that a reservoir is an animal, bird, or insect that harbors the infectious agent and is indigenous to a geographical area. Humans become infected only when they come in contact with the reservoir. Thus, the disease incidence is low but relatively constant. An epidemic disease, on the other hand, is when large numbers of individuals in a population become infected over a short time. This is due, in part, to rapid and direct human-to-human transmission.

Figure 28.18A illustrates the difference in the frequency of cases observed between endemic and epidemic disease. An endemic disease can become epidemic if the population of the reservoir increases, which allows for more frequent human contact, or if the infectious agent evolves to spread directly from person to person, bypassing the need for a reservoir. This is the concern with the H5N1 avian flu virus, which is endemic in animals and birds in Asia (see Sections 11.4 and 25.8). A **pandemic** is an epidemic that occurs over a wide geographical area, usually the world. The bubonic plague pandemic in the fourteenth century and the AIDS pandemic in the late twentieth and early twenty-first centuries are good examples.

Finding Patient Zero

When trying to contain the spread of an epidemic, it is vital to track down the first case of the disease (known as the **index case** or **patient zero**) and then identify everyone who has had contact with the individual so that they can be treated or separated from the general population (that is,

A. Endemic versus epidemic

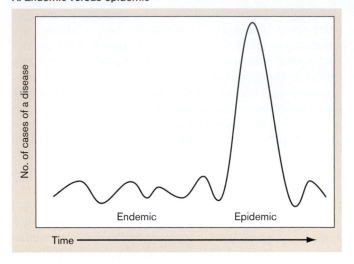

B.

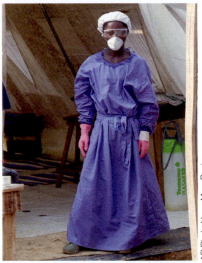

AP Photo/Jean-Marc Bouju

Figure 28.18 The difference between endemic and epidemic disease.   A. An endemic disease is continuously present at a low frequency in a population. If the disease frequency suddenly rises, this constitutes an epidemic. **B.** A health care worker stands outside a quarantined area housing Ebola patients in the Ivory Coast. Epidemics can be minimized if infected persons are kept segregated from the general population (quarantined) to avoid spread of the infectious agent.

quarantined). When a new disease arises, the epidemiological search for the index case starts only after a number of patients have been diagnosed and a new disease syndrome declared. This is what happened with AIDS in the 1970s.

Identifying an index case within a specific community is easier if the disease syndrome is already recognized, as was the case with the 2003 SARS outbreak in Singapore. According to the World Health Organization, a suspected case of SARS is defined as an individual who has a fever greater that 38°C, who exhibits lower respiratory tract symptoms, and who has traveled to an area of documented disease or has had contact with a person afflicted with SARS. The index case in Singapore was a 23-year-old woman who had stayed on the ninth floor of a hotel in Hong Kong while on vacation. A physician from southern China who stayed on the same floor of the hotel during this period is believed to have been the source of her infection, as well as that of the index patients who precipitated subsequent outbreaks in Vietnam and Canada.

During the last week of February, the woman, who had returned to Singapore, developed fever, headache, and a dry cough. She was admitted to Tan Tock Seng Hospital, Singapore, on March 1 with a low white blood cell count and patchy consolidation in the lobes of the right lung. Tests for the usual microbial suspects (*Legionella*, *Chlamydia*, *Mycoplasma*) were negative. Electron microscopy of nasopharyngeal aspirations showed virus particles with widely spaced club-like projections (**Fig. 28.19A**). At the time of her admission to the hospital, the clinical features and highly infectious nature of SARS were not known. Thus, for the first six days of hospitalization, the patient was in a general ward, without barrier infection control measures. During this period, the index patient

infected at least 20 other individuals, including hospital staff, nearby patients, and visitors.

Within weeks, the WHO named the disease in China SARS and issued travel alerts (discussed further in Section 28.6). These alerts allowed Singapore health officials to rapidly identify the index patient and her contacts. As a result, they were able to limit spread of the illness. When all was said and done, SARS killed fewer than 1,000 victims worldwide, although thousands more became ill and recovered (**Fig. 28.19B**). Today, SARS remains a threat but one of lesser concern. Methicillin-resistant *Staphylococcus aureus* (MRSA) and H5N1 avian flu are considered more pressing dangers.

Identifying Disease Trends Depends on Physician Notification of Health Organizations

How do epidemiologists first recognize that an epidemic is under way, and then identify the agent and its source? Certain diseases, because of their severity and transmissibility, are called **reportable** or **notifiable diseases** (**Table 28.5**). Physicians are required to report instances of these diseases to a central health organization, such as the CDC in the United States and the WHO. As a result, the incidences of certain diseases within a population can be tracked and upsurges noted. An emerging disease not on the list of notifiable diseases can be detected as a cluster of patients with unusual symptoms or combinations of symptoms. This is possible because diseases of unknown etiology are also reported to health authorities. A new disease could manifest with common symptoms (for example, the cough and fever of SARS) that cannot be linked to a known disease agent by clinical tests. An upsurge in cases, either of a reportable disease or an emerging disease, will set off institutional "alarms" that initiate epidemiological efforts to determine the source and cause of the outbreak.

John Snow (1813–1858), the Father of Epidemiology

The first case in which the source of a disease outbreak was methodically investigated took place in the mid-nineteenth century. During a serious outbreak of cholera in London in 1854, John Snow (**Fig. 28.20A**) used a map to plot the locations of all the diarrheal cases he learned about. The source of the infection was unknown, but Snow thought that if the cases clustered geographically, he might gain a clue as to the source. Water in that part of London was pumped from separate wells located in the various neighborhoods. Snow's map revealed a close association between the density of cholera cases and a single well located on Broad Street (**Fig. 28.20B**). Simply removing the pump handle of the Broad Street well put an

Courtesy of CDC/Dr. Fred Murphy

A.

B.

Anat Givon/AP Photo

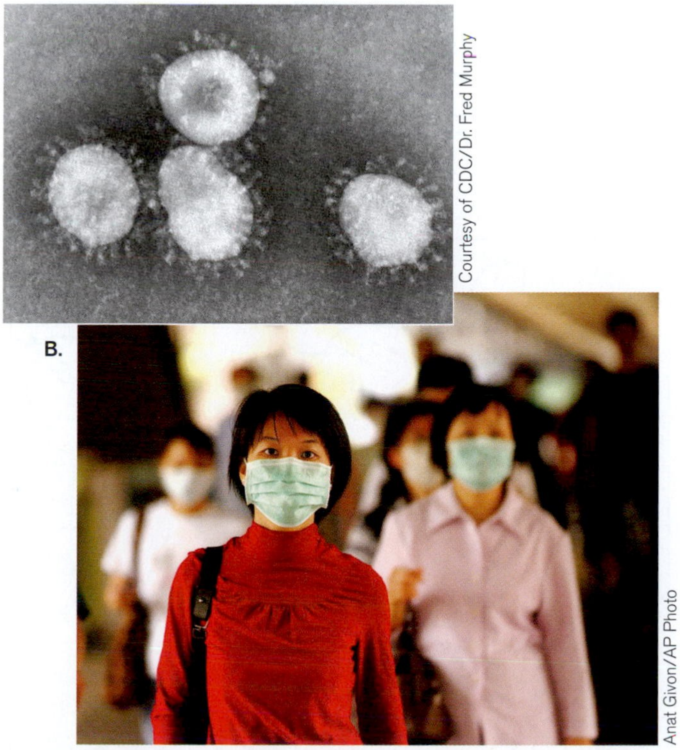

Figure 28.19　Severe acute respiratory syndrome (SARS). **A.** The coronavirus SARS (TEM). **B.** Citizens of China, including the military, donned surgical masks in 2003 to slow the spread of SARS.

Table 28.5 Notifiable infectious diseases.

Bacterial

Anthrax	Hansen's disease (leprosy)	Rocky Mountain spotted fever
Botulism	Legionellosis	Salmonellosis (nontyphoid fever types)
Brucellosis	Leptospirosis	Shigellosis
Campylobacter infection	Listeriosis	Staphylococcal enterotoxin
Chlamydia infection	Lyme disease	Streptococcal invasive disease
Cholera	Meningitis, infectious	*Streptococcus pneumoniae*, invasive disease
Diphtheria	Pertussis	Syphilis
Ehrlichiosis	Plague	Tuberculosis
E. coli O157:H7 infection	Psittacosis	Tularemia
Gonorrhea	Q fever	Typhoid fever
Haemophilus influenzae, invasive disease		Vancomycin-resistant *Staphylococcus aureus* (VRSA)

Viral

Dengue fever	Mumps	Varicella (chickenpox), fatal cases only (all types)
Hantavirus infection	Poliomyelitis	Yellow fever
Hepatitis, viral	Rabies	
Human immunodeficiency virus infection	Rubella and congenital rubella syndrome	
Measles	Smallpox	

Fungal

Coccidioidomycosis

Parasitic

Amebiasis	Giardiasis	Malaria
Cryptosporidiosis	Trichinosis	
Cyclosporiasis	Microsporidiosis	

Figure 28.20 Early epidemiology. A. John Snow. **B.** The map of London that Snow used to pinpoint the source of the cholera outbreak in 1854. The Broad Street well is found within the red circle. Each black bar represents a death from cholera.

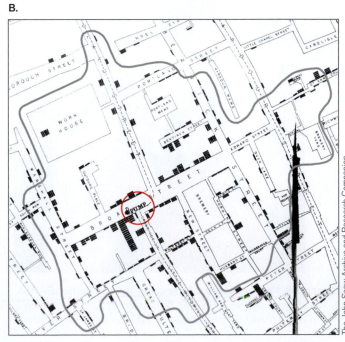

end to the epidemic, proving that the well water was the source of infection. This approach succeeded brilliantly despite the fact that the infectious agent that causes cholera, *Vibrio cholerae*, was not recognized until 1905, over 50 years later. Identifying clusters of patients afflicted with a given disease is still used to identify potential sources of infectious disease outbreaks.

> **THOUGHT QUESTION 28.7** Methicillin, a beta-lactam antibiotic, is very useful in treating staphylococcal infections. The development of methicillin-resistant strains of *Staphylococcus aureus* (MRSA) is a very serious development because there are few antibiotics that can kill these strains. Imagine a large metropolitan hospital in which there have been eight serious nosocomial infections with MRSA and you are responsible for determining the source of infection so it can be removed. How would you accomplish this using common bacteriological and molecular techniques?

Genomic Strategies Help Identify Nonculturable Pathogens

Microbiologists have successfully developed many strategies to identify the causes of infectious diseases. Robert Koch in the late 19th century devised a set of postulates that, when followed, can identify the agent of a new disease (discussed in Section 1.3). One important feature of Koch's postulates is that the suspected organism be grown in pure culture. However, there are bacterial diseases for which the agent could not be cultured. How might they be identified?

Case History: Whipple's Disease

In April, a 50-year-old man had an abrupt onset of watery diarrhea with a stool frequency of up to ten times every 24 hours. This was the latest episode in a six-year history of illness beginning with recurring fevers, flu-like symptoms, profuse night sweats, and painful joint swelling. The current bout of diarrhea was associated with gripping lower abdominal pain, especially after meals. No blood or mucus was found in the stools. Blood and stool cultures tested negative for known infectious agents. Serology was also negative for syphilis, brucellosis, toxoplasmosis, and leptospirosis. His weight fell rapidly from 83 kg to 73 kg within four weeks of onset. A flexible sigmoidoscopy showed only diffuse mild erythema in the bowel. However, the appearance of the small bowel was consistent with malabsorption, a disease in which nutrients are poorly absorbed by the intestine. A duodenal biopsy to look for the organisms of Whipple's disease showed large macrophages in the lamina propria (a layer of loose connective tissue beneath the epithelium of an organ). Bacterial rods characteristic of Whipple's disease were seen with electron microscopy. PCR analysis of tissue samples confirmed the diagnosis.

First diagnosed in 1907 by George Whipple, the symptoms of this disease are malabsorption, weight loss, arthralgia (joint pain), fevers, and abdominal pain. Any organ system can be affected, including the heart, lungs, skin, joints, and central nervous system. The cause of Whipple's disease went undiscovered for 85 years but was suspected to be of bacterial etiology even though an organism was never successfully cultured. The agent was finally identified in 1992 not by culturing, but by blindly amplifying 16S rDNA sequences from biopsy tissues.

All bacterial 16S rRNA genes have some sequence regions that are highly homologous across species and other sequences that are unique to a species. Tissues from numerous patients diagnosed with Whipple's disease were subjected to PCR analysis using the common 16S rDNA primers. If a bacterial agent were present, it was predicted that PCR should successfully amplify a DNA fragment corresponding to the agent's 16S rRNA gene. All tissues produced such a fragment, indicating that there were bacteria in the tissues. DNA sequence analysis of these fragments indicated that the organism was similar to actinomycetes but was unlike any of the known species.

The organism is actually a gram-positive soil-dwelling actinomycete that has been named *Tropheryma whipplei* in honor of the physician who first recognized the disease. Because of its bacterial etiology, Whipple's disease can be treated with antibiotics, usually trimethoprim sulfamethoxazole (Bactrim, Septra, Cotrim). **Figure 28.21** shows an in situ hybridization for *T. whipplei* RNA in a tissue biopsy.

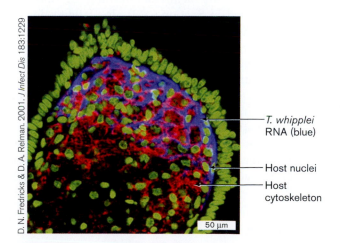

D. N. Fredricks & D. A. Relman. 2001. *J Infect Dis* 183:1229

T. whipplei RNA (blue)

Host nuclei

Host cytoskeleton

50 μm

Figure 28.21 Whipple's disease. Fluorescent in situ hybridization of a small intestinal biopsy in a case of Whipple's disease (confocal laser scanning microscopy). In this test, a fluorescently tagged DNA probe that specifically hybridizes to *T. whipplei* RNA is added to the tissue. Other fluorescent probes are used to visualize host nuclei and cytoskeleton. *T. whipplei* rRNA is blue, nuclei of human cells are green and the intracellular cytoskeletal protein vimentin is red. Magnification approximately 200×.

This story reveals that Koch's postulates (see Figure 1.18) must sometimes be modified when identifying the cause of a new disease. In this instance, the organism could not be cultured in pure form in the laboratory but was found, via molecular techniques, in all instances of the disease. Note that more recent studies have successfully cultured *T. whipplei* in vitro. The medium used was formulated based on nutritional requirements deduced from knowing the DNA sequence of the *T. whipplei* genome (discussed in Section 25.6).

> **THOUGHT QUESTION 28.8**   What are some reasons why some diseases spread quickly through a population while others take a long time?

Molecular Approaches Can Be Used for Disease Surveillance

A worldwide pandemic of pulmonary tuberculosis currently affects over 2 billion people. Many of the *Mycobacterium tuberculosis* infections are caused by multidrug-resistant strains that are difficult, if not impossible, to kill with existing antibiotics (see Section 26.3). This problem is especially serious among refugee populations attempting to flee war-torn countries. As a result, it is important to screen these refugees as they enter neighboring countries with low incidences of tuberculosis. Although chest X-rays are mandatory in many cases, a positive image will be obtained only if the disease is at a relatively advanced stage. Actively infected individuals who have not developed the characteristic lung tubercles seen on X-ray will not be identified.

Unfortunately, the acid-fast staining of sputum samples (discussed in Section 28.2) also fails to detect individuals at an early stage of infection. Studies have shown, however, that PCR techniques are much more sensitive for detecting these individuals. As time progresses, PCR surveillance strategies will be used more often to track the worldwide ebb and flow of microbial diseases. PCR and restriction fragment length polymorphism (RFLP) strategies are already used for epidemiological purposes to type (that is, determine relatedness of) different microbial isolates by generating a complex DNA profile that is specific for a particular strain (Section 28.2).

Bioterrorism

"A wide-scale bioterrorism attack would create mass panic and overwhelm most existing state and local systems within a few days," said Michael T. Osterholm, director of the Center for Infectious Disease Research and Policy at the University of Minnesota. "We know this from simulation exercises." These words were spoken in October 2001.

Less than a month after the September 11 attack on the World Trade Center, some unknown person(s) sent weapons-grade anthrax spores through the U.S. mail. Thankfully, only 5 persons died and a mere 25 became ill. But even though the efficiency of the attack was poor, the impact was enormous. Over 10,000 people took a two-month course of antibiotics after possible exposure, and mail deliveries throughout the country were affected. The simple act of opening an envelope suddenly became a risky endeavor.

As a result of this attack and other events, the CDC and the National Institutes of Health (NIH) have assembled a list of select agents (marked in blue in **Table 28.4**) that could potentially be used as **bioweapons**. A bioweapon is considered to be any infectious agent or toxin that has high virulence and/or mortality rate. Microorganisms considered bioweapons can be used to conduct biowarfare, in which the intent is to inflict massive casualties, or bioterrorism, in which there may be a few casualties but widespread psychological trauma.

Although the list of select agents is recent, biowarfare is not new. In the Middle Ages, victims of the Black Death (plague caused by *Yersinia pestis*) were flung over castle walls using catapults; during the French and Indian War, in the eighteenth century, Jeffery Amherst distributed smallpox-infected blankets to Native Americans; and during World War II, the Japanese Imperial Army experimented with infectious disease weapons using Chinese prisoners as guinea pigs. Even the United States has participated through the development of weapons-grade anthrax spores. It was originally thought that the strain of *B. anthracis* used in the 2001 anthrax attack was the same as the strain developed by the military. We now know that the strain used in the attack had a more ordinary pedigree.

The first documented act of bioterrorism in the United States occurred in 1984, when followers of the cult leader Bhagwan Shri Rajneesh tried to control a local election in Dalles, Oregon, by infecting salad bars with *Salmonella*. Over 900 people became ill. Rajneesh was sentenced to 20 years in prison.

How effective are bioweapons? The method by which a biological agent is dispersed plays a large role in its effectiveness as a weapon. Only a few people became ill during the 2001 anthrax attack not only because of the epidemiological surveillance, but because anthrax is inherently difficult to disperse. Thousands of spores must be inhaled to contract disease, which means that effective dispersal of the spores is critical for the use of anthrax as a weapon. Once spores hit the ground, the threat of infection is limited. Weapons-grade spores are very finely ground so that they stay airborne longer. But as we saw, the letter-borne dispersal system was not very effective in terms of generating large numbers of victims. Nevertheless, the potential threat of weapons-grade anthrax on the battlefield led the U.S. military in 2006 to resume vaccinating all soldiers serving in Iraq, Afghanistan, and South Korea.

An effective bioweapon would be one that capitalizes on person-to-person transmission. In an easily transmitted disease, one infected person could disseminate disease to scores of others within one or two days. So in terms of generating massive numbers of deaths, anthrax was a poor choice. But the goal of most terrorists is not to kill large numbers of people, but to terrorize them. In that regard, the anthrax attack succeeded.

The most effective bioweapon in terms of inflicting death (biowarfare) would have a low infectious dose, be easily transmitted between people, and be one to which a large percentage of the population is susceptible. Smallpox fits these criteria and would be the bioweapon of choice (**Fig. 28.22**). Fortunately, however, smallpox has been eradicated (almost) from the face of the Earth and is not easily obtained. Two laboratories still harbor the virus—one in the United States and one in Russia. It is believed that the virus has been destroyed in all other laboratories. Since smallpox is the perfect biowarfare agent, it is imperative that the last two smallpox repositories remain secure.

While the good news is that smallpox disease has been eradicated, the bad news is that no individual born after 1970 has been vaccinated. As a result, anyone under 35 years of age is susceptible to smallpox. Even those of us who received the smallpox vaccination over 35 years ago are at risk, since our protective antibody titers have diminished. A terrorist attack with smallpox would cause terrible numbers of deaths. The vaccine does, however, exist. Were a smallpox attack to be launched, the vaccine would be rapidly administered to limit the spread of disease. Nevertheless, the economic and psychological impact of a smallpox epidemic would be devastating.

The CDC has assembled a list of so-called "select agents" that are considered possible bioterrorist weapons (see Table 28.4). Research with organisms considered to be select agents is tightly regulated. Because *Yersinia pestis*, for instance, is a select agent, laboratory personnel working with it must now possess security clearance with the Department of Justice (even though the organism itself can be handled under biosafety level 2 conditions). The laboratory must also register with the CDC to legally possess this pathogen and access to the lab and the organism must be tightly controlled.

Much has improved since Michael Osterholm offered his dire assessment of a wide-scale bioterrorism attack. Education and surveillance procedures have been bolstered, and new detection technologies are being developed (**Special Topic 28.1**). We will probably never be fully protected from attack, biological or otherwise, but recent efforts have improved the situation.

WWW | Bioterrorism Preparedness Act

TO SUMMARIZE:

- **John Snow** founded the discipline of epidemiology.
- **Epidemiology** is a field that examines factors that determine the distribution and source of disease.
- **Endemic, epidemic, and pandemic** are terms for different frequencies of disease in different geographical areas.
- **Finding patient zero** is important for containing the spread of disease.
- **Molecular approaches using PCR and nucleic acid hybridization** are used to identify nonculturable pathogens and to track disease movements.
- **Bioweapons**, when they have been used, typically kill few people but create great fear.
- **The CDC** has assembled a list of select agents with bioweapon potential.

Figure 28.22 Dealing with bioterrorism. A. Members of a hazardous materials team near Capitol Hill during the anthrax attacks in 2001. **B.** An Illinois man suffering from smallpox, 1912.

A.

Kenneth Lambert/AP

B.

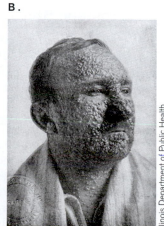

Illinois Department of Public Health

The emergence of pathogenic infectious diseases resulting from both natural and intentional acts is a growing global concern. Acts of bioterrorism, particularly, represent a formidable challenge that dictates the need for rapid and efficient means of pathogen detection. The ideal detection system would integrate sample processing and pathogen characterization into a single automated device that would eliminate laborious, time-consuming sample processing and costly detection. One promising detection system involves a hybridization-based bioelectronic DNA detection chip that is in the testing phases.

The basic design, illustrated in **Figure 1**, requires two hybridization events to generate a detection signal. First, a DNA capture probe is affixed to a gold electrode chip. The capture probe is complementary to one part of a gene that is specific to a given pathogen. Next, a solution containing unknown DNA is injected into the chip. If the target gene is present, the capture probe will hybridize to a single-stranded part of the gene's DNA and "capture" it. The actual detection signal is then generated following a second hybridization. The second probe used is complementary to the downstream region of the same captured gene. This probe is not affixed to the gold support but contains a ferrocene label at one end. If the target DNA is captured, the second probe will anneal

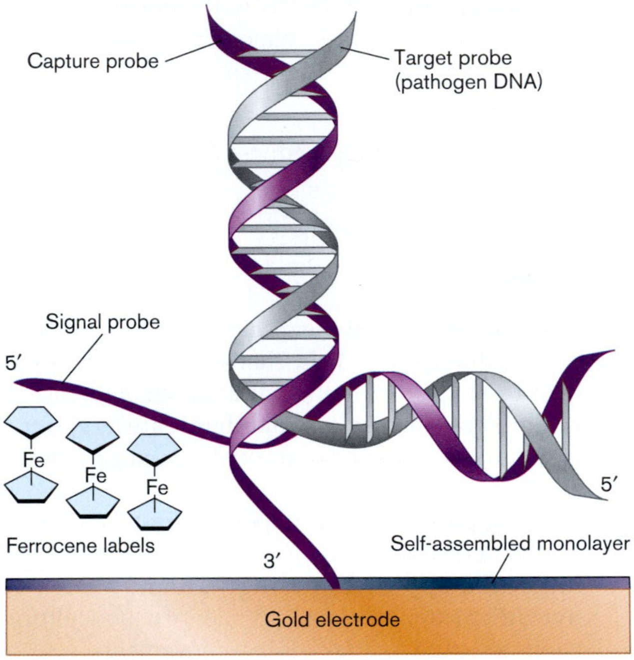

Figure 1 Electronic detection of pathogens. Schematic representation of the eSensor DNA Detection System. A self-assembled monolayer is generated on a gold electrode. For electrochemical detection to occur across the monolayer, two hybridization events must occur: the first between the capture probe (shown in purple perpendicular to the electrode) and the target (for example, HPV DNA, shown in gray), and the second between an adjacent region of the target HPV DNA and a second, ferrocene-labeled signal probe (shown in purple parallel to the monolayer).

28.6 Detecting Emerging Microbial Diseases

Case History: SARS

A 48-year-old man was hospitalized in a Dutchess County, New York, hospital with a 101°F fever, headache, and body aches. He also had difficulty breathing. He had just returned from a business trip to China.

SARS: An Epidemiological Success Story

Within seven weeks in early 2003, epidemiologists armed with global technologies and rapid DNA-sequencing techniques tracked, named, identified, completely sequenced and contained a newly emerging disease with a scary death rate: severe acute respiratory syndrome (SARS). This was a remarkable feat, especially when we consider that after

the first cases of AIDS appeared in 1981, it took over three years just to identify the virus, and it took seven years to track down the Lyme disease spirochete, *Borrelia burgdorferi*, after Lyme disease was first recognized in 1975.

We now know that SARS first developed in Asia, then began to spread by jet to other countries, such as Canada. While the death rate seemed modest at about 5%, it was higher than the 4% reached during the 1918 flu epidemic that killed 25–40 million people worldwide. SARS clearly posed a formidable threat. Unprecedented cooperation between the WHO, the CDC, and numerous other health organizations around the world played a major role in tracking and containing the disease. Patient information, case histories, possible treatment regimens flooded into a secure WHO website. The most exciting posting was the 30,000-bp sequence of the SARS genome, taken from one-millionth of a gram of genetic

and bring the ferrocene molecules in contact with the gold electrode, generating an electrochemical signal. These signals are then detected by a scanning AC voltammetry technique (voltammetry is an electroanalytical method whereby information about an analyte is obtained by measuring current as potential is varied). As shown in **Figure 2**, Suzanne Vernon and colleagues at the Centers for Disease Control (CDC) have used this approach to detect the presence of human papillomavirus (HPV, a cause of genital warts) in a dramatic test of the system.

Of course, an ideal system to use in the field would also include an automated DNA extraction technology, a development that is yet to come. Nevertheless, the elegance of this design promises to be invaluable for pathogen detection under a variety of circumstances.

Source: S. D. Vernon, D. H. Farkas, E. R. Unger, V. Chan, D. L. Miller, et al. 2003. Bioelectronic DNA detection of human papillomaviruses using eSensor™: A model system for detection of multiple pathogens. *BMC Infectious Diseases* 3:12.

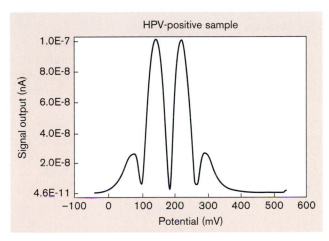
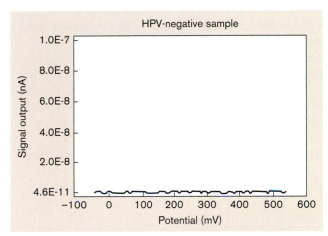

Figure 2   Bioelectronic signal output.   The eSensor bioelectronic representative signal output. The peaks in the first panel reveal the electrochemical signal generated when the ferrocene from the DNA probe comes in contact with the gold electrode. The second panel represents a negative control sample.

material isolated from a Toronto patient. The following 2003 timeline illustrates the remarkable speed with which all this took place:

February 14—China first reports 305 cases of atypical pneumonia to WHO.

March 15—WHO issues emergency travel alert, names illness SARS.

March 17—Leading epidemiologists join WHO in conference call, agree to unprecedented cooperation.

April 4—U.S. President George Bush authorizes quarantine of SARS patients.

April 12—Canadian scientists in Vancouver post the completed genome on the Internet.

April 16—WHO announces that SARS is caused by a pathogen never before seen in humans, a new coronavirus.

Technology Helps New Infectious Agents Emerge and Spread

Despite all we know of microbes, despite the many ways we have to combat microbial diseases, our species, for all its cleverness, still lives at the mercy of the microbe. Lyme disease, MRSA, SARS, Ebola, *E. coli* O157:H7, HIV, "flesh-eating" streptococci, hantavirus—all of these and many other new diseases have emerged over the last 30 years. Worse yet, forgotten scourges, such as tuberculosis, have reappeared. Yet in the 1970s, medical science was claiming victory over infectious disease. What happened?

Part of the equation has been progress itself. Travel by jet, the use of blood banks, and suburban sprawl have all opened new avenues of infection. People unwittingly infected by a new disease in Asia or Africa can, traveling by jet, bring the pathogen to any other country in the world within hours. The person may not even show symptoms until days or weeks after the trip. This means that diseases can spread faster and farther than ever before. In addition, newly emerged blood-borne pathogens can spread by transfusion. This was a major problem with HIV before an accurate blood test was developed to screen all donated blood.

Although human encroachments into the tropical rain forests have often been blamed for the emergence of new pathogens, one need go no farther than the Connecticut woodlands to find such developments. *Borrelia burgdorferi*, the spirochete that causes Lyme disease, lives on deer and white-footed mice and is passed between these hosts by the deer tick (Fig. 26.26). This infectious cycle has been going on for years. Humans have crossed paths with these animals long before the disease erupted in our communities. Why have we suddenly become susceptible? The answer appears to be suburban development. In the wild, foxes and bobcats hunt the mice that carry the Lyme agent. These predators disappear as developers clear land and build roads and houses, leaving the infected mice and ticks to proliferate. Humans in these developed areas are more likely to be bitten by an infected tick and contract the disease than in prior decades. Luckily, many diseases that successfully leap from animals to humans find the new host to be a dead end, unable to spread the disease to others.

There are numerous examples in which technology and progress have had the unintended consequence of breeding disease. Here are just a few:

- **Mad cow disease:** Modern farming practices (North America and Europe) of feeding livestock the remains of other animals helped spread transmissible spongiform encephalopathies similar to Creutzfeldt-Jakob disease associated with prions. Because the prion is infectious, the brain matter from one case of mad cow disease could end up infecting hundreds of other cattle, which in turn increases the chance that the disease could spread to humans.

- **Lyme disease:** Suburban development in the northeastern United States destroys predators of the mice that carry *Borrelia burgdorferi*.

- **Hepatitis C:** Transfusions and transplants spread this blood-borne disease.

- **Influenza:** Live poultry markets in Asia serve as breeding grounds for avian flu viruses that can jump to humans.

- **Enterohemorrhagic *E. coli* (for example, *E. coli* O157:H7):** Modern meat-processing plants can accidentally grind trace amounts of these acid-resistant, fecal organisms into beef while making hamburgers.

Natural environmental events can also trigger upsurges in the incidence of unusual diseases. A good example of this involves the hantavirus (also known as Sin Nombre virus). An unprecedented outbreak of hantavirus pulmonary disease occurred in the Four Corners

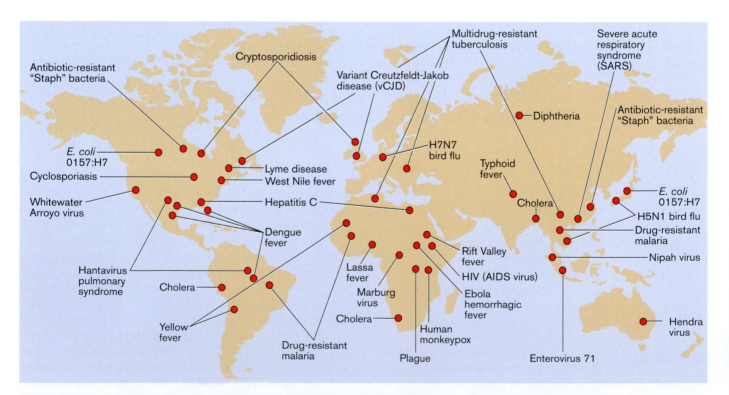

Figure 28.23 **Locations of some emerging and reemerging infectious diseases.** The examples given represent extreme increases in the reported cases. Many of these diseases, such as HIV and cholera, are widespread but show alarming increases in the areas indicated.

area of Arizona, New Mexico, Colorado, and Utah in 1993 when rain lead to greater than normal increases in plant and animal numbers. The resulting tenfold increase in deer mice, which carries the virus, made it more likely that infected mice and humans would come in contact.

Detecting Emerging and Reemerging Pathogens

A world map showing the general locations of **emerging** and **reemerging diseases** is shown in **Figure 28.23.** A reemerging disease is one thought to be under control but whose incidence has risen. For example, the incidence rate of tuberculosis dropped sharply in the 1950s, but the number of recent cases has increased just as dramatically. Tuberculosis is a reemerging disease. The trigger for its reemergence was the AIDS pandemic beginning in 1980. Immunocompromised AIDS patients are highly susceptible to infection by many organisms, including *Mycobacterium tuberculosis.*

Mycobacterium tuberculosis is also reemerging among non-AIDS patients owing to the development of drug-resistant strains. Drug resistance in *M. tuberculosis* developed largely because of noncompliance on the part of patients to complete their full courses of antibiotic treatment. Treatment usually involves three or more antibiotics to reduce the risk that resistance to any one drug will develop. Many patients failed to take all three drugs simultaneously, allowing the organism to develop resistance to one drug at a time until it became resistant to all of them. These highly drug-resistant strains are almost impossible to kill. The link between noncompliance and the development of drug resistance is the primary reason why tuberculosis patients are now required to take the multiple antibiotics prescribed in the presence of the physician or a member of the medical staff.

As described in Section 28.5, highly infectious diseases are identified using aggressive epidemiological surveillance. Communications between local, national, and world health organizations are critical and helped expose the rise in tuberculosis. To see for yourself how emerging diseases are monitored, search the Internet for a website titled ProMed-mail. There you will find daily reports posted from around the world that describe new outbreaks of infectious diseases.

For example, a posting on June 21, 2006, reports that "11 students at a Franklin elementary school (Boston) have confirmed cases of salmonella, and state health officials are investigating whether owl pellets used in a class project were the source of the bacterial infection." The source was eventually confirmed to be the owl pellets. Another posting from the Congo in Africa notes reports of 100 cases of suspected pneumonic plague, including 19 deaths in the Ituri district, Oriental province. The site also posts reports of undiagnosed illnesses that *may* be the initial signs of emerging diseases. For instance, on June 23, 2006, there is a report of an undiagnosed die-off of wild birds in Zambia, which may be due to avian flu. By tracking these reports, trends of disease spread can be determined.

> **THOUGHT QUESTION 28.9.** Using the "Maps of Outbreaks" provided on the ProMed-mail website, identify the countries considered to have anthrax epidemics.

TO SUMMARIZE:

- **Emerging diseases** can spread quickly around the world as a result of air travel.
- **Modern technology** and urban growth have provided opportunities for new diseases to emerge.

Concluding Thoughts

This chapter has examined the basic principles used to collect, detect, and track pathogenic microorganisms. As you can see, the task of controlling the spread of disease is daunting and ever changing because microorganisms continue to evolve. Known pathogens change, eluding eradication efforts by the immune system and antibiotics, while new pathogens keep emerging. The world is depending on the next generation of microbiologists to face the growing threat of these evolving microbes.

CHAPTER REVIEW

Review Questions

1. Why is it important to identify the genus and species of a pathogen?
2. What is an API strip, and what is its use in clinical microbiology?
3. Describe three examples of selective media.
4. If a colony on a nutrient agar plate is catalase-positive, does this mean it is made up of gram-positive microorganisms? Why or why not?
5. Describe the types of hemolysis visualized on blood agar.

6. What is the clinical significance of a group A, beta-hemolytic streptococcus?
7. How does one distinguish *S. aureus* from *S. epidermidis*?
8. Why are PCR identification tests preferable to biochemical approaches?
9. How is RTQ-PCR performed?
10. Describe an ELISA.
11. Name some sterile and nonsterile body sites.
12. List seven common types of clinical specimens collected for bacteriological examination.
13. Describe key features of the four levels of biological containment.
14. How is a pandemic different from an epidemic?
15. How can genomics help identify nonculturable pathogens?
16. List and provide short descriptions of four emerging diseases.
17. Name four select agents and the diseases they cause.

Key Terms

bioweapon (1090)
emerging disease (1095)
endemic (1086)
enzyme-linked immunosorbent assay (ELISA) (1077)

epidemic (1086)
index case (1086)
multiple PCR (1073)
pandemic (1086)
patient zero (1086)

quarantined (1086)
reemerging disease (1095)
reportable (notifiable) disease (1087)
reservoir (1086)
sequela (1064)

Recommended Reading

Baker, Edward L., Margaret A. Potter, Deborah L. Jones, Shawna L. Mercer, Joan P. Cioffi, et al. 2005. The public health infrastructure and our nation's health. *Annual Review of Public Health* **26**:303–318.

Bengis, R. G., F. A. Leighton, J. R. Fischer, M. Artois, T. Morner, and C. M. Tate. 2004. The role of wildlife in emerging and re-emerging zoonoses. *Reviews in Science and Technology* **23**:497–511.

Espy, Mark, James Uhl, Lynne Sloan, Seanne Buckwalter, Mary Jones, et al. 2006. Real time PCR in clinical microbiology: Application for routine laboratory testing. *Clinical Microbiology Reviews* **19**:165–256.

Fournier, Pierre-Edouard, Michel Drancourt, and Didier Raoult. 2007. Bacterial genome sequencing and its use in infectious diseases. *Lancet Infectious Diseases* **7**:711–723.

Gaydos, Charlotte A., Mellisa Theodore, Nicholas Dalesio, Billie J. Wood, and Thomas C. Quinn. 2004. Comparison of three nucleic acid amplification tests for detection of *Chlamydia trachomatis* in urine specimens. *Journal of Clinical Microbiology* **42**:2041–3045.

Kuiken, Thijs, Ron Fouchier, Guus Rimmelzwaan, and Albert Osterhaus. 2003. Emerging viral infections in a rapidly changing world. *Current Opinion in Biotechnology* **14**:241–246.

Madoff, Lawrence C., and John P. Woodall. 2005. The Internet and global monitoring of emerging diseases: Lessons from the first 10 years of ProMed-mail. *Archives of Medical Research* **36**:724–730.

Molinari, J. A. 2003. Diagnostic modalities for infectious diseases. *Dental Clinics of North America* **47**:605–621.

Pejcic, Bobby, Roland De Marco, and Gordon Parkinson. 2006. The role of biosensors in the detection of emerging infectious diseases. *Analyst* **131**:1079–1090.

Raman, Lakshmy Anantha, Noman Siddiqi, Mohammed Shamim, Monorama Deb, Geeta Mehta, and Seyed Ehtesham Hasnain. 2000. Molecular characterization of *Mycobacterium abscessus* strains from a hospital outbreak. *Emerging Infectious Diseases* **6**:561–562.

Relman, David A., Thomas M. Schmidt, Richard P. MacDermott, and Stanley Falkow. 1992. Identification of the uncultured bacillus of Whipple's disease. *New England Journal of Medicine* **327**:283–301.

Shi, Pei-Yong, Elizabeth Kauffman, Ping Ren, Andy Felton, Jennifer Tai, et al. 2001. High throughput detection of West Nile Virus RNA. *Journal of Clinical Microbiology* **39**:1264–1271.

Shvartzman, Pesach, and Yussuf Nasri. 2004. Urine culture collected from gel-based diapers: Developing a novel experimental laboratory method. *The Journal of the American Board of Family Practice* **17**:91–95.

Appendix 1

Biological Molecules

This appendix reviews information typically covered in an introductory biology course. We first cover the chemical bonding principles needed to understand biological molecules, with an emphasis on the special properties of water. We then discuss organic molecules, paying particular attention to four important classes of organic biomolecules—proteins, polysaccharides, nucleic acids, and lipids. Finally, we explore common chemical principles, such as concentrations, thermodynamics, equilibrium, pH, and oxidation-reduction reactions.

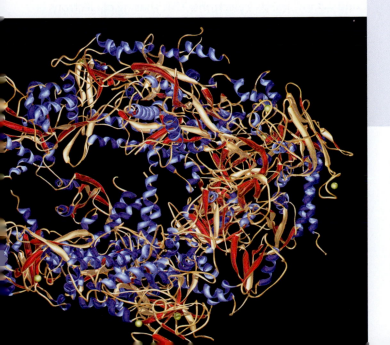

A computer model showing the structure of the enzyme RNA polymerase II. The molecule comprises 12 subunits. This enzyme synthesizes a complementary mRNA strand from a strand of DNA during a process called transcription. It recognizes a start sign on the DNA strand and then moves along the strand building the mRNA until it reaches a stop sign. mRNA is the intermediary between DNA and its protein product. *Source*: Mark J. Winter/Photo Researchers, Inc.

Living cells are remarkably complex machines, able to integrate and respond to multiple stimuli, to catalyze reactions, and to replicate themselves. Yet despite all the various tasks that cells perform, 98% of the mass of living organisms consists of only six elements: hydrogen (H), oxygen (O), nitrogen (N), phosphorus (P), sulfur (S), and carbon (C); and 90% of the mass is accounted for by just C, H, and O. Of the compounds formed from these elements, the most abundant in cells is water. The remainder of the cell consists, for the most part, of just four different kinds of organic (carbon-based) macromolecules: proteins, nucleic acids, carbohydrates, and lipids. In one respect, cells can be thought of as compartments that orchestrate chemical reactions between these organic molecules. It is clear that to understand life, we must understand the properties of water and organic molecules and of their building blocks, the chemical elements.

A1.1 Elements, Bonding, and Water

Cells consist mostly of water and organic molecules. These cellular components are formed from atoms of various elements. An atom consists of a positively charged nucleus that contains protons and neutrons, surrounded by negatively charged electrons. The hydrogen nucleus consists of a single proton. Protons, neutrons, and electrons differ in their mass and charge as summarized in **Table A1.1**.

The elements can be organized into a periodic table as in **Figure A1.1**, indicating each element's **atomic number** (the number of protons) and **atomic mass** (the mass in grams of 1 mole of the element). The defining characteristic of an element is the atomic number. For example, all carbon atoms have six protons in their nucleus. To maintain neutrality, atoms have negatively charged electrons equal in number to the positively charged protons.

The atomic mass is found by adding together the number of protons and neutrons. The atomic mass is an average that takes into account the relative abundance of each isotope. **Isotopes** are atoms of an element that differ in the number of neutrons. For example, the most abundant isotope of carbon is carbon-12 (with six neutrons), but there are naturally occurring isotopes of carbon-13 (seven neutrons) and carbon-14 (eight neutrons). Because carbon-13 and carbon-14 are rare, the average atomic mass is close to but not exactly 12. While some isotopes are stable, others decay at a known rate and give off radioactivity. Carbon-14 has a **half-life** (the amount of time it takes for half of the sample to decay) of 5,700 years and is used in radiocarbon dating to determine the age of organic material. Carbon-14 and shorter-lived isotopes such as tritium (hydrogen with a mass of 3—one proton and two neutrons) are used by scientists as tracers to follow specific atoms in metabolic pathways.

Bonds Hold Elements Together to Form Molecules

Atoms combine by sharing electrons to form molecules. For example, two atoms of oxygen combine to form molecular oxygen, O_2. Molecules may also contain more than one kind of element; an example is water, H_2O. The symbols O_2 and H_2O are examples of **molecular formulas**, a shorthand notation indicating the number and type of atoms present in a molecule.

Each column (group) in the periodic table contains elements of similar reactivity as a result of the similarity in their electronic configurations, particularly of electrons in the outermost shell. The shell closest to the nucleus can hold a maximum of two electrons and the next shell a maximum of eight electrons. **Figure A1.2A** shows all the electrons for hydrogen, carbon, nitrogen, and oxygen. Each unpaired electron is capable of participating in a bond. If we examine **Figure A1.2A**, it is clear that H can form one bond, C four bonds, N three bonds, and O two bonds. For example, carbon has four unpaired electrons in its outermost shell; each can form a bond with an unpaired electron from another atom. The two atoms involved in this bond "share" the two electrons. This sharing of electrons is termed a *covalent bond*. Covalent bonds are very strong and difficult to break. In methane (CH_4), carbon forms four covalent bonds with four hydrogen atoms (**Fig. A1.2B**). Methane is stable because both carbon and hydrogen have filled their outer shells. Atoms can also share more than one pair of electrons with another atom, forming double or triple bonds. In carbon dioxide, CO_2, each oxygen shares two pairs of electrons with carbon; and in diatomic nitrogen, N_2, the nitrogen atoms each share three pairs of electrons (see **Fig. A1.2B**). The bonding in molecules can be represented in a shorthand representation by a **structural formula**, in which covalent bonds are shown as a line between two atoms (**Fig. A1.2C**).

Another way atoms can obtain full outer shells is by gaining or losing electrons. A complete transfer of electrons can occur between two atoms that have a large

Table A1.1	The mass and charge of atomic particles.	
Particle	**Mass** (atomic mass unit)	**Charge** (electronic charge unit)
Proton	1	+1
Neutron	1	0
Electron	0.0005	−1

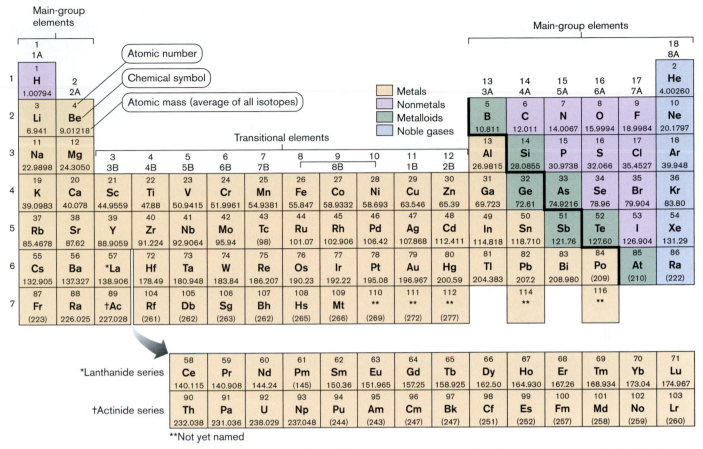

Figure A1.1 Periodic table of the elements. The atomic number (number of protons) and atomic mass are shown for each element.

Symbol	Name
Ac	Actinium
Al	Aluminum
Am	Americium
Sb	Antimony
Ar	Argon
As	Arsenic
At	Astatine
Ba	Barium
Bk	Berkelium
Be	Beryllium
Bi	Bismuth
Bh	Bohrium
B	Boron
Br	Bromine
Cd	Cadmium
Ca	Calcium
Cf	Californium
C	Carbon
Ce	Cerium
Cs	Cesium
Cl	Chlorine
Cr	Chromium

Symbol	Name
Co	Cobalt
Cu	Copper
Cm	Curium
Db	Dubnium
Dy	Dysprosium
Es	Einsteinium
Er	Erbium
Eu	Europium
Fm	Fermium
F	Fluorine
Fr	Francium
Gd	Gadolinium
Ga	Gallium
Ge	Germanium
Au	Gold
Hf	Hafnium
Hs	Hassium
He	Helium
Ho	Holmium
H	Hydrogen
In	Indium
I	Iodine

Symbol	Name
Ir	Iridium
Fe	Iron
Kr	Krypton
La	Lanthanum
Lr	Lawrencium
Pb	Lead
Li	Lithium
Lu	Lutetium
Mg	Magnesium
Mn	Maganese
Mt	Meitnerium
Md	Mendelevium
Hg	Mercury
Mo	Molybdenum
Nd	Neodymium
Ne	Neon
Np	Neptunium
Ni	Nickel
Nb	Niobium
N	Nitrogen
No	Nobelium
Os	Osmium

Symbol	Name
O	Oxygen
Pd	Palladium
P	Phosphorus
Pt	Platinum
Pu	Plutonium
Po	Polonium
K	Potassium
Pr	Praseodymium
Pm	Promethium
Pa	Protactinium
Ra	Radium
Rn	Radon
Re	Rhenium
Rh	Rhodium
Rb	Rubidium
Ru	Ruthenium
Rf	Rutherfordium
Sm	Samarium
Sc	Scandium
Sg	Seaborgium
Se	Selenium
Si	Silicon

Symbol	Name
Ag	Silver
Na	Sodium
Sr	Strontium
S	Sulfur
Ta	Tantalum
Tc	Technetium
Te	Tellurium
Tb	Terbium
Tl	Thallium
Th	Thorium
Tm	Thulium
Sn	Tin
Ti	Titanium
W	Tungsten
U	Uranium
V	Vanadium
Xe	Xenon
Yb	Ytterbium
Y	Yttrium
Zn	Zinc
Zr	Zirconium

difference in **electronegativity**, a measure of the affinity of an atom for electrons. A large electronegativity indicates a strong attraction for electrons. Of the elements listed in **Table A1.2**, oxygen has the greatest attraction for electrons and sodium the weakest. If two elements with greatly different electronegativities come into close contact, one element can "steal" an electron from the other. For example, sodium (Na) and chlorine (Cl) interact to form table salt, NaCl. The electronegative Cl strips an electron away from Na, and both Cl^- and Na^+ now

A.

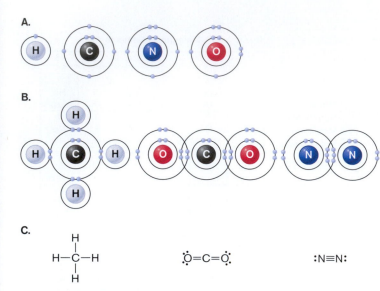

B.

C.

Figure A1.2   Covalent bonding of hydrogen, carbon, nitrogen, and oxygen.   A. The electrons present in hydrogen (H), carbon (C), nitrogen (N), and oxygen (O). **B.** Electron sharing in the four single bonds present in methane, CH_4; the two double bonds present in carbon dioxide, CO_2; and the triple bond present in diatomic nitrogen, N_2. **C.** Structural formulas for methane, carbon dioxide, and diatomic nitrogen. Each line represents a covalent bond. Lone pairs of electrons in the outermost shell are represented by dots.

Table A1.2	Electronegativities of some common elements.

Element	Electronegativity
Oxygen	3.44
Chlorine	3.16
Nitrogen	3.04
Sulfur	2.58
Carbon	2.55
Hydrogen	2.10
Sodium	0.93

have full outer shells (**Fig. A1.3A**). Both Cl^- and Na^+ are charged atoms called ions, in which the number of electrons and protons are unequal. **Anions** are negatively charged ions, and **cations** are positively charged ions. Anions and cations can form ionic crystals held together by **ionic bonds**, electrostatic attractions between anions and cations (**Fig. A1.3B**). In the absence of water, ionic crystals maintain their integrity, but they are destabilized in the presence of water. To understand why, we need to understand the structure of water.

A.

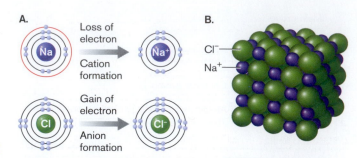

B.

Figure A1.3   Formation of ions and ionic crystals.   A. The loss of an electron from an atom of sodium to form the cation Na^+ and the gain of an electron by a chlorine atom to form the chloride anion Cl^-. **B.** Oppositely charged anions and cations—in this case, Cl^- and Na^+—are attracted to one another and form crystals of table salt (sodium chloride).

Water Is the Solvent of Life

Living organisms consist mostly of water, and water has many unique properties that render it particularly suitable for sustaining life. To appreciate these properties, we must understand the forces that hold water together: polar covalent bonds and hydrogen bonds. For molecules such as H_2 and O_2, there is no charge distribution within the molecule because both atoms in the molecule have the same electronegativity. Therefore, the electrons in the covalent bond are shared equally and form **nonpolar covalent bonds**. In contrast, the shared electrons in H_2O spend more of their time around the highly electronegative oxygen than in the vicinity of the less electronegative hydrogen (**Fig. A1.4A**). A bond with unequal electron sharing is a **polar covalent bond**, so called because the molecule has partial positive and negative poles. Polar covalent bonds occur in individual water

A.   **B.**

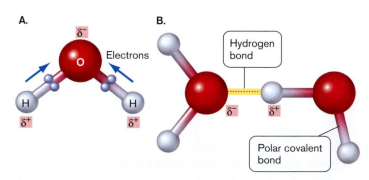

Figure A1.4   Polar covalent and hydrogen bonds in water. A. The polar covalent bonds in an individual water molecule. Displacement of electrons toward oxygen causes oxygen to have a partial negative charge and the hydrogen atoms to have a partial positive charge. Hence, water is polar. **B.** A hydrogen bond between two water molecules.

Table A1.3 **Bond types and strengths.**

Type of bond	Description	Bond strength in water (kJ/mol)[a]
Covalent	Sharing of electrons	210–418
Ionic	Electrostatic attraction between anion and cation	12.5
Hydrogen	Electrostatic attraction between a hydrogen bound to a nitrogen or oxygen and a second nitrogen or oxygen	4
van der Waals	Electrostatic attraction between temporary, shifting electron clouds	0.42 (per atom)

[a]Bond strengths are given in water, which is similar to the environment in the aqueous cytoplasm. Anhydrous bond strengths differ from those listed.

molecules, creating a partial charge separation within the molecule.

Water molecules experience a strong electrostatic attraction for one another as a result of the charge separation present in individual molecules. This electrostatic attraction, known as a **hydrogen bond**, occurs between a hydrogen bound to an oxygen or nitrogen and a second oxygen or nitrogen, either in the same or a different molecule. Hydrogen bonds are short-lived and constantly break and re-form in liquid water. **Figure A1.4B** shows the polar covalent and hydrogen bonds present in water.

The hydrogen bonds in water contribute to its unique properties that enable it to support life. Water is a liquid over a large temperature range because hydrogen bonds cause water molecules to associate, favoring the liquid state over the gas. The hydrogen bonds that keep water a liquid over a wide range of temperatures also endow water with a high specific heat, the amount of energy needed to raise the temperature of 1 gram (g) of a substance by 1°C. The high specific heat of water moderates the temperature of all aqueous environments, oceans and cells alike.

The polar nature of water is also responsible for its properties as a solvent. Compounds that are ionic or polar themselves tend to dissolve in water and are termed **hydrophilic**. For example, the ionic bonds in NaCl are very strong in the absence of water, but are easily dissolved in the presence of water. This is because the polar water can surround and interact with the sodium and chloride ions and shield them from each other (**Fig. A1.5A**). Water also dissolves polar compounds. In this case, water does not break the polar covalent bonds; rather, individual, intact polar molecules are surrounded by water owing to electrostatic interactions. In contrast, compounds that are mostly nonpolar do not dissolve in water and are termed **hydrophobic**. Nonpolar molecules have no partial charges to attract water. Because water hydrogen-bonds with other water molecules, it tends to exclude nonpolar compounds, forcing them together (**Fig. A1.5B**). The aggregated nonpolar compounds can be further stabilized by **van der Waals forces**, weak temporary

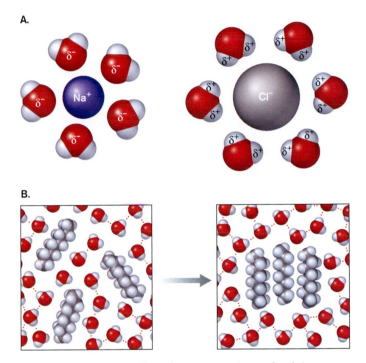

Figure A1.5 **Interactions between water and solutes.** **A.** Water surrounds and interacts with individual ions or polar molecules, causing them to dissolve in water. **B.** Nonpolar molecules do not dissolve in water. Instead, to minimize the disruption of hydrogen bonding among water molecules, nonpolar molecules aggregate in water.

electrostatic attractions between molecules caused by random movements of their electron clouds. **Table A1.3** lists the bonds we have discussed and indicates the strength of the bonds in water.

A1.2 Common Features of Organic Molecules

Now that we have discussed water, let's take a look at some of the organic molecules found in cells. **Organic molecules** are those that contain a carbon-carbon bond.

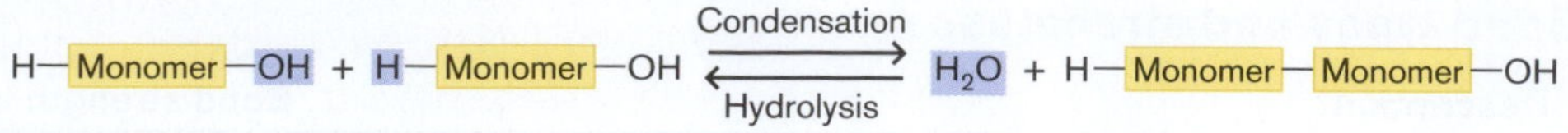

Condensation

$$\text{H}\!-\!\boxed{\text{Monomer}}\!-\!\text{OH} + \text{H}\!-\!\boxed{\text{Monomer}}\!-\!\text{OH} \rightleftharpoons \text{H}_2\text{O} + \text{H}\!-\!\boxed{\text{Monomer}}\!-\!\boxed{\text{Monomer}}\!-\!\text{OH}$$

Hydrolysis

Figure A1.6   Condensation and hydrolysis reactions.   Monomers can be covalently bonded to form polymers through a condensation reaction that liberates a water molecule. Polymers can be broken apart into monomers through hydrolysis (water-splitting) reactions.

The major macromolecular components of cells are proteins, polysaccharides, nucleic acids, and lipids. Although the macromolecules differ from each other in their structures and cellular functions, they all share some common features:

■ As polymers, the macromolecules found in cells are composed of smaller units called monomers. The monomers are joined to one another by a common reaction, called a **condensation,** that involves splitting out a molecule of water for each monomer unit added. Conversely, the polymers can be broken apart into monomers by **hydrolysis**, the addition of a molecule of water, as depicted in **Figure A1.6.**

■ The three-dimensional shape, or structure, of a molecule is critical for proper function. The structure is determined by which atoms are present and how they are bonded together. Structural formulas like those in **Figure A1.2C** represent the sequence of bonds in a molecule; however, this does not adequately convey information about the three-dimensional shape of the molecule. It is particularly important to understand the shape of carbon-containing molecules because carbon is the backbone for most cellular macromolecules. The bonding orbitals in the second shell of carbon are arranged so that they point to the vertices of a tetrahedron; methane, the simplest hydrocarbon, actually has the shape of tetrahedron, with the hydrogen atoms at the vertices (**Fig. A1.7A**). **Figure A1.7B** and **C** show two different models used to depict the three-dimensional shape of molecules. The **space-filling model** shows the volume filled by the outer shell of the atoms. The **stick model** shows the length and orientation of interatomic bonds between the nuclei of each pair of atoms.

There are inherent difficulties in presenting three-dimensional representations, such as the models just described, on a two-dimensional page. Images on a computer screen in conjunction with special glasses can give a sense of the actual three-dimensional structures of molecules. Viewers can use the mouse to rotate the molecule to see it from any angle. Often the viewer can choose whether to view the molecule as a space-filling or stick representation.

■ An important feature of any organic molecule is the number and type of functional groups present. **Functional groups** are small groups of atoms with characteristic bonding, shape, and reactivity. A few of the more common ones are shown in **Table A1.4.** Know-

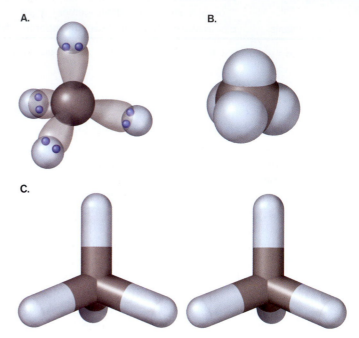

Figure A1.7   Molecular models of methane.   A. A molecule of methane showing the tetrahedral arrangement of the electrons in the second (valence) shell of carbon. **B.** A space-filling model of methane. **C.** A stereo-image view of a stick model of methane. To view the stereo-image, place a piece of cardboard vertically between the two images. Position your eyes on opposite sides of the cardboard and force them to focus behind the images. The images will merge and produce a 3-D image.

ing which functional groups are present in a molecule allows us to infer something about the structure and reactivity of the molecule. Information about the functional groups present in a molecule is often indicated by a molecule's name; for example, amino acids contain an amino group and a carboxyl (carboxylic acid) group.

We will discuss different classes of molecules individually, but in living cells, molecules of different types are frequently found in combination. For example, sugars decorate some proteins (glycoproteins) and lipids (glycolipids), and the ribosome is a complex of proteins and ribonucleic acids. Let us now examine the structure of the fundamental biological macromolecules in detail.

A1.3 Proteins

Cells express thousands of different proteins and may contain more than 2 million protein molecules. Proteins perform many functions, such as catalyzing reactions,

Table A1.4 Common functional groups.

Functional group	General structure	Example	Comments
Aldehyde	R—C(=O)—H	Acetylaldehyde	Can react with alcohols $R-C(=O)H + HO-R' \rightarrow R-C(OH)(H)-OR'$
Alkane	R—CH₂—H	Ethane	Nonpolar, tends to make molecules containing it hydrophobic; nonreactive
Amino	R—N(H)(H)	Methylamine	Acts as a base by accepting a proton; found in amino acids $R-NH_2 + H^+ \rightarrow R-NH_3^+$
Carboxyl	R—C(=O)—O—H	Acetic acid	Acts as an acid by donating a proton; the ionized form name ends in *ate* (i.e., acetate) $CH_3-C(=O)-O^- + H^+$ Acetate Proton
Ester	R—C(=O)—O—R	Phosphodiester $R_1-O-P(=O)(O^-)-O-R_2$	Common linkage found in lipids
Hydroxyl	R—O—H	Ethanol	Polar, makes compounds more soluble through hydrogen bonding; found in alcohols and sugars
Ketone	R—C(=O)—R	Dihydroxyacetone	Found in many intermediates of metabolism
Phosphate	R—O—P(=O)(O⁻)—O⁻	Glyceraldehyde-3-phosphate	When more than one phosphate is linked together, a high-energy bond forms due to repulsion of the negative oxygens

serving as receptors and transporters, providing structure, and aiding movement. Proteins can carry out such diverse functions because they can fold into a variety of three-dimensional structures.

The Building Blocks of Proteins Are Amino Acids

Although different proteins can have vastly different structures, all proteins are composed of the same building blocks: amino acids. All 20 of the amino acids commonly occurring in nature have the same general structure. Each **amino acid** contains a central carbon atom (the alpha carbon) covalently bound to four different moieties—a hydrogen atom, an amino group, a carboxyl group, and a side chain (R, residue) that is unique for each amino acid (**Fig. A1.8A**). The alpha carbon is an example of a **chiral carbon**, a carbon with four different groups attached to it. Chiral carbons exist in two different forms called optical isomers, or enantiomers.

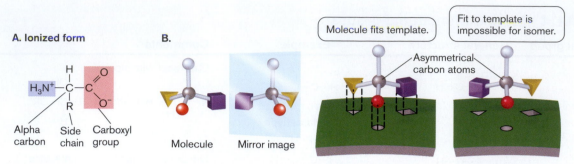

A. Ionized form

Alpha carbon | Side chain | Carboxyl group

B.

Molecule   Mirror image

Molecule fits template.

Fit to template is impossible for isomer.

Asymmetrical carbon atoms

Figure A1.8 Amino acid structure and chiral carbons. **A.** All amino acids contain a central carbon (the alpha-carbon) bonded to an amino group, a carboxyl group, a hydrogen, and a variable group or side chain, designated R. At cellular pH, the amino and carboxyl groups are ionized. Because the alpha-carbon is bound to four different molecules, it is a chiral carbon. **B.** Chiral molecules exist in two different forms that are mirror images of each other. The two forms cannot be superimposed, and only the correct isomer can interact with a template (for example, an enzyme).

Optical isomers have the same molecular formula and the same order of bonds within the molecules, but a different arrangement of their atoms in space. Optical isomers are actually mirror images of each other, like your left and right hands, and the two forms cannot be super-

imposed (**Fig. A1.8B**). Each of the amino acid isomers is designated L or D, based on the configuration at the alpha carbon. The amino acids used to make proteins all have the L configuration at the alpha carbon, and all of these amino acids can be derived by substituting the different

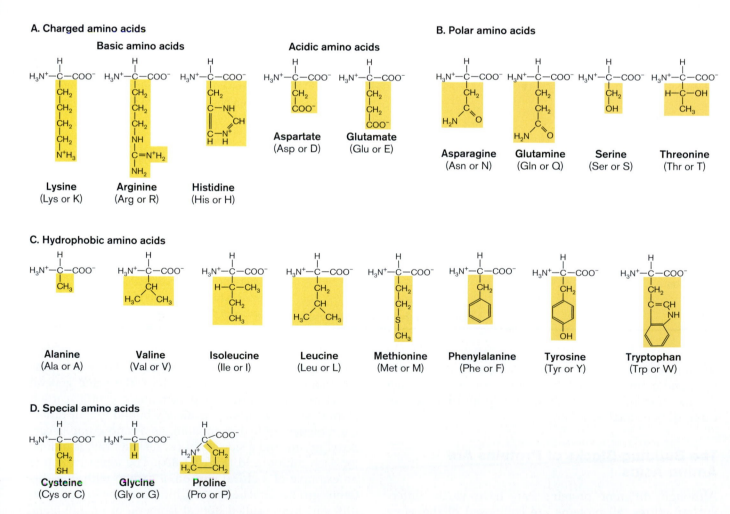

A. Charged amino acids

Basic amino acids

Lysine (Lys or K)   Arginine (Arg or R)   Histidine (His or H)

Acidic amino acids

Aspartate (Asp or D)   Glutamate (Glu or E)

B. Polar amino acids

Asparagine (Asn or N)   Glutamine (Gln or Q)   Serine (Ser or S)   Threonine (Thr or T)

C. Hydrophobic amino acids

Alanine (Ala or A)   Valine (Val or V)   Isoleucine (Ile or I)   Leucine (Leu or L)   Methionine (Met or M)   Phenylalanine (Phe or F)   Tyrosine (Tyr or Y)   Tryptophan (Trp or W)

D. Special amino acids

Cysteine (Cys or C)   Glycine (Gly or G)   Proline (Pro or P)

Figure A1.9 Twenty common amino acids.   The grouping of amino acids is based on their side chains, highlighted in yellow.

R groups for one another. Some unusual D-amino acids, with the opposite configuration at the alpha carbon, are found in bacterial cell walls.

The side chains of amino acids can be grouped according to their hydrophobicity, charge, or presence of specific functional groups. **Figure A1.9** depicts the 20 common amino acids, along with their three-letter abbreviations and single-letter codes. Several amino acids have unusual features: Glycine does not contain a chiral carbon and therefore does not have optical isomers; proline has a ring structure that interrupts the regular geometry of the poly-peptide chain; and the side chain of cysteine is capable of forming inter- and intramolecular disulfide (–S–S–) bonds.

There Are Four Levels of Protein Organization

The first level of organization, the **primary structure**, is simply the linear sequence of amino acids. This is genetically determined; the DNA code specifies the amino acid sequence. During protein synthesis, amino acids are connected together by a condensation reaction to form a covalent **peptide bond**. The peptide bond forms between the carboxyl group of one amino acid and the amino group of a second amino acid, as shown in **Figure A1.10A**. The portion of the amino acid incorporated into the peptide after condensation is called an amino acid residue. Note that the peptide chain has an amino terminus (N-terminus) and a carboxyl terminus (C-terminus); numbering of the amino acid residues starts at the amino terminus (**Fig. A1.10B**). The bonds on either side of the alpha carbon can rotate, but there is no rotation around the peptide bond, so the backbone of a protein does not rotate freely.

The **secondary structures** of proteins are regular patterns that repeat over short regions of the polypeptide chain. **Figure A1.11A** depicts two common secondary structures, the alpha helix and the beta sheet. The alpha helices are flexible, coiled structures formed by a series of hydrogen bonds between an oxygen in a carboxyl group and a hydrogen in an amino group five amino acids down the chain. The beta-sheets are more rigid than the alpha helices. Beta sheets form as a result of extensive hydrogen bonding between regions of the protein lying next to each other. Both alpha helix and beta sheet secondary structures are stabilized by hydrogen bonding between the oxygen and hydrogen atoms of the peptide bonds in the main chain; the side chains do not directly participate. Proline, however, acts as a helix breaker because its ring structure does not allow the proper rotation around the alpha carbon that is necessary to form an alpha helix.

The **tertiary structure** of a protein is its unique three-dimensional shape. At the tertiary level of organization, regions distant in the primary structure may be brought close together. The nature and order of the side chains (the primary structure) determines how a protein folds into its final tertiary structure. The large number

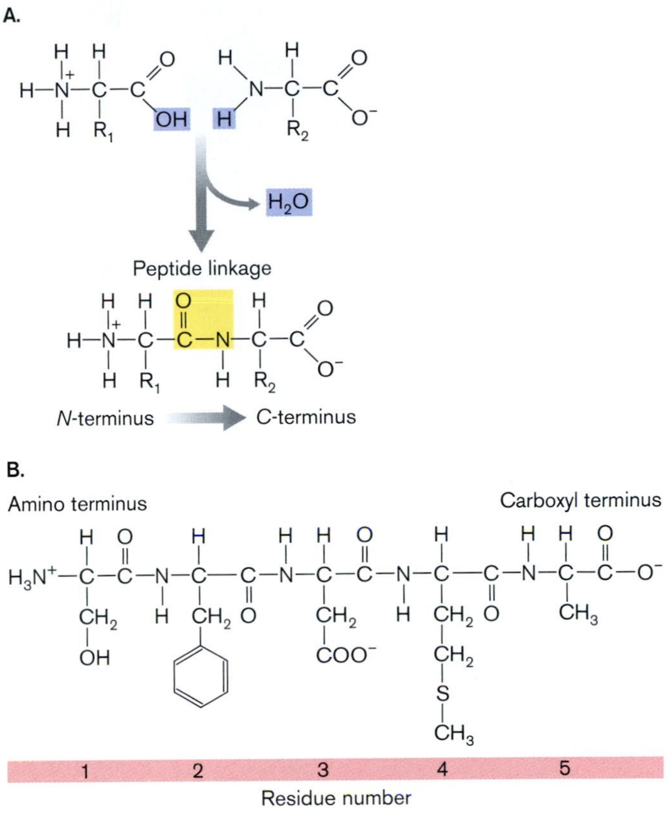

Figure A1.10   The peptide bond and primary protein structure.   A. Formation of a peptide bond. Amino acids are joined by a condensation reaction between the carboxyl group of one amino acid and the amino group of another to form a covalent linkage called a peptide bond. **B.** The primary structure of a pentapeptide, a chain of five amino acids. This pentapeptide has the sequence SFDMA. A protein can be formed of hundreds or thousands of amino acids.

of amino acid side chains is responsible for the diversity of protein structures and hence functions. As proteins emerge from the ribosome, they are bound by chaperone proteins that help them fold into their functional shape, known as the **native conformation**. In soluble proteins, amino acids with hydrophobic side chains tend to cluster inside the protein to minimize reactions with water, while polar amino acids are exposed on the surface (**Fig. A1.11B**). Tertiary structure is stabilized by hydrogen bonding between side chains and between side chains and the main chain. Ionic interactions between acidic and basic side chains also contribute to tertiary structure. The one covalent interaction found stabilizing tertiary structure is disulfide bonding between two cysteine residues. The interactions that contribute to protein tertiary structure are shown in **Figure A1.11C,** and a ribbon representation of the three-dimensional structure of a protein is illustrated in **Figure A1.11D**.

Some proteins form stable, functional complexes with other proteins. These multisubunit proteins exhibit the fourth level of protein organization, **quaternary structure** (**Fig. A1.11E**). The forces that hold interacting polypeptide

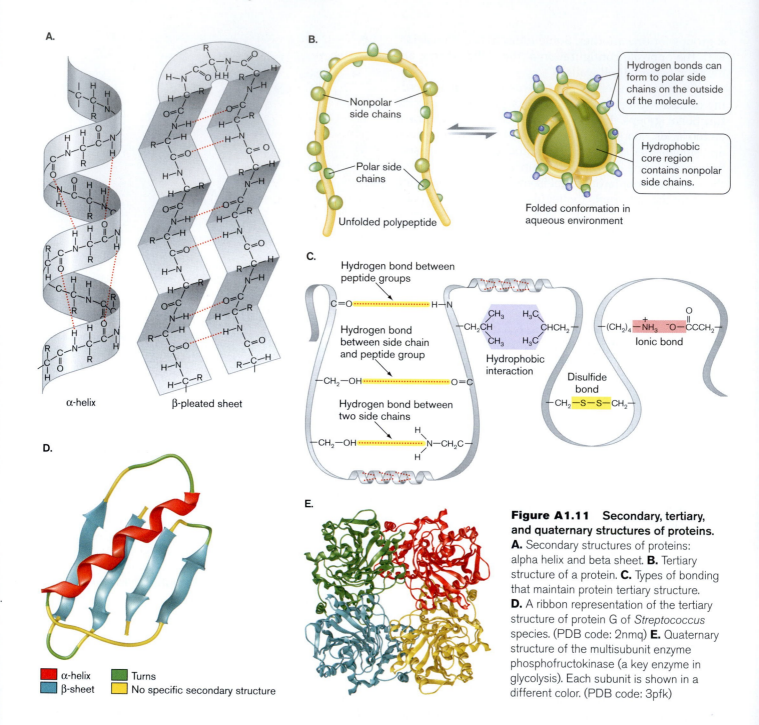

Figure A1.11 **Secondary, tertiary, and quaternary structures of proteins.** **A.** Secondary structures of proteins: alpha helix and beta sheet. **B.** Tertiary structure of a protein. **C.** Types of bonding that maintain protein tertiary structure. **D.** A ribbon representation of the tertiary structure of protein G of *Streptococcus* species. (PDB code: 2nmq) **E.** Quaternary structure of the multisubunit enzyme phosphofructokinase (a key enzyme in glycolysis). Each subunit is shown in a different color. (PDB code: 3pfk)

chains together are the same noncovalent interactions and disulfide bonds responsible for protein tertiary structure.

It is important to note that proteins are not static structures. The individually weak hydrogen bonds, ionic interactions, and van der Waals forces that contribute to protein structure are constantly being broken and re-formed. Because these noncovalent bonds do not break all at once, the large number of such bonds present act collectively to maintain protein integrity. Some proteins are dynamic entities that may permute through a number of semistable states. Such conformational changes may

be stochastic, random fluctuations due to the intrinsic thermal (kinetic) energy of the protein. Alternatively, the conformational changes may be influenced by external factors. For many proteins, alterations in conformation are critical for protein function.

A1.4 Polysaccharides

Polysaccharides are complex carbohydrates, composed of monosaccharide (simple-sugar) monomers. Carbohy-

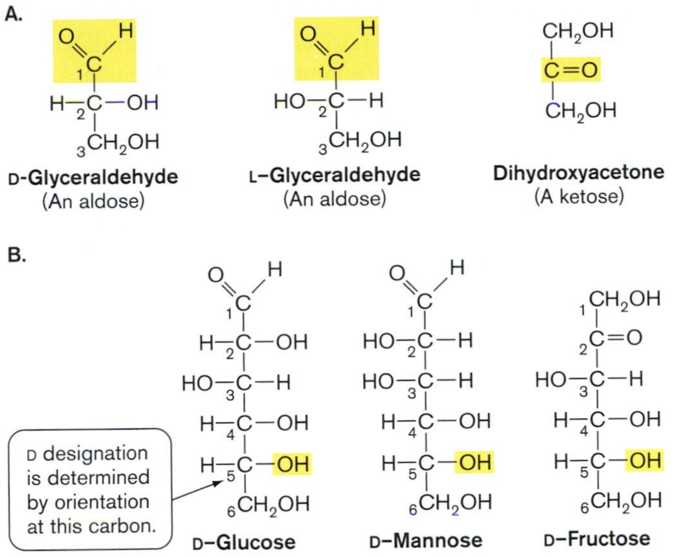

A.

D-Glyceraldehyde
(An aldose)

L-Glyceraldehyde
(An aldose)

Dihydroxyacetone
(A ketose)

B.

D designation is determined by orientation at this carbon.

D-Glucose

D-Mannose

D-Fructose

Figure A1.12   Structural formulas for some monosaccharides.   A. The three-carbon sugars dihydroxyacetone and glyceraldehyde. The aldehyde and ketone functional groups are shown in yellow. Glyceraldehyde contains a chiral carbon and has two optical isomers. **B.** The hexose sugars glucose, mannose, and fructose. The numbering of the carbons starts at the carbon of the aldehyde group or at the free carbon closest to the ketone group. The D designation is determined from the orientation of the hydroxyl group on a chiral carbon farthest from the ketone or aldehyde functional group (highlighted).

drates are so named because their molecular formulas are multiples of CH_2O (hydrated carbon). Carbohydrates are a preferred energy source for many cells and also serve to store energy. They also play a structural role, as in cell walls. Moreover, when attached to proteins or lipids, they can serve an informational role.

Monosaccharides Are Aldehydes or Ketones with Multiple Hydroxyl Groups

Monosaccharides differ from each other in the number of carbons present, in whether they have an aldehyde or a ketone group, and in the orientation of the hydroxyl

groups. The smallest sugars, glyceraldehyde and dihydroxyacetone, have three carbons (**Fig. A1.12A**). These two sugars have the same molecular formula, $C_3H_6O_3$, but since their atoms are arranged differently, they are **structural isomers** of each other. In addition, because glyceraldehyde contains a chiral carbon at C-2, it also has an optical isomer. In contrast to proteins, where the L-amino acids predominate, it is the D-sugars that are of biological importance.

The six-carbon sugars (hexoses) are particularly important in the cell. Hexose sugars include glucose, mannose, and fructose (**Fig. A1.12B**). Hexoses all have the molecular formula $C_6H_{12}O_6$ and are structural isomers of each other. While fructose is a keto sugar (ketose), mannose and glucose are both aldehydes (aldoses). The difference between mannose and glucose lies in the orientation of the hydroxyl group on one of the chiral carbons, C-2. The carbons in sugars are numbered starting with the carbon of the aldehyde group or the free carbon closest to the ketone group so the aldehyde group is always C-1.

In sugars of five or more carbons, one of the hydroxyl groups can react with the aldehyde or ketone group, forming a ring structure. The ring form predominates because it is more energetically favorable than the straight-chain sugar. In the formation of the cyclic form of glucose, two products can be formed (**Fig. A1.13**). Depending on how the bond between C-1 and C-2 was oriented during the formation of the ring, the cyclic form obtained will be alpha-glucose or beta-glucose. Both forms interconvert rapidly in aqueous solution, and both forms are found as components of biological structures.

The hydroxyl groups of sugars can react with other functional groups to form modified sugars. Examples include fructose 1,6-bisphosphate, an intermediate in glycolysis, and N-acetylglucosamine, a component of the peptidoglycan cell wall (**Fig. A1.14**).

Two Monosaccharides Condense to Form a Disaccharide

Disaccharides form when two monosaccharides undergo a condensation reaction to form a covalent glycosidic bond. The glycosidic bond often occurs between the C-1 carbon of one monosaccharide and the C-4 carbon of another

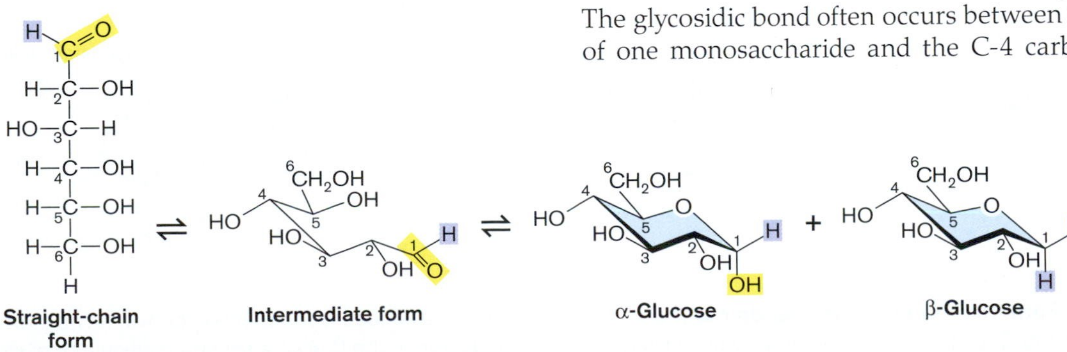

Straight-chain form

Intermediate form

α-Glucose

β-Glucose

Figure A1.13   Straight-chain and cyclic forms of glucose.   The straight-chain form of glucose can cyclize to form two different ring structures, alpha-glucose and beta-glucose. In this process, a new chiral carbon is formed at C-1, the anomeric carbon.

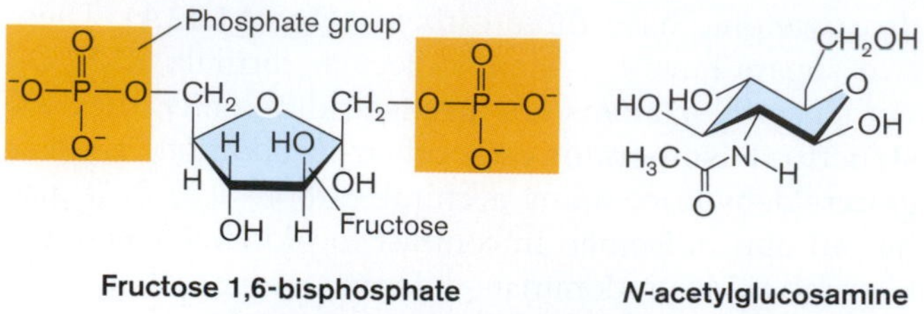

A. Sugar phosphate

B.

Fructose 1,6-bisphosphate

N-acetylglucosamine

Figure A1.14 Modified sugars. A. Phosphorylated sugars, such as fructose 1,6-bisphosphate, are intermediates in glycolysis. **B.** *N*-acetylglucosamine is a component of bacterial cell walls.

monosaccharide (**Fig. A1.15**). The structure of the disaccharide varies, depending on which monosaccharides are involved and whether the C-1 carbon participating in the bond is in the alpha or beta form. Oligosaccharides consist of a few monosaccharides linked together, while polysaccharides are much longer chains of monosaccharides. Polysaccharides are unbranched if all the linkages are 1,4. In branched polysaccharides, additional sugars are attached to other hydroxyl groups (for example, at C-6).

A1.5 Nucleic Acids

Cells contain two kinds of nucleic acids: deoxyribonucleic acid (DNA) and ribonucleic acid (RNA). DNA is the hereditary material of the cell; it encodes the information to make proteins. Three different types of RNA play key roles in protein synthesis: Messenger RNA (mRNA) is transcribed from DNA and is the template for protein synthesis; transfer RNAs (tRNAs) bind amino acids and deliver them to the ribosome; and ribosomal RNA (rRNA) is a catalytic component of the ribosome.

A Chain of Nucleotides Forms the Primary Structure of Nucleic Acids

Nucleic acids are polymers of nucleotides. **Nucleotides** have three components—a pentose sugar (ribose in RNA, 2-deoxyribose in DNA), a nitrogenous base, and a phosphate group (**Fig. A1.16A**). The nitrogenous base is attached to the C-1 of the sugar. Nitrogenous bases come in two structural classes: purines (adenine, A; and guanine, G), and pyrimidines (cytosine, C; thymine, T, found only in DNA; and uracil, U, found only in RNA). The phosphate group is esterified to the 5'-hydroxyl group of the sugar. (The prime indicates that the numbering refers to the sugar portion of the nucleotide; the base is numbered without primes.)

A nucleoside is similar to a nucleotide, except that it has only two components—a sugar and a base (that is, it lacks a phosphate group). The names of the nucleosides are listed in **Table A1.5**. *Nucleotides* are formed when a phosphate group forms a covalent ester bond to the C-5 carbon of the sugar portion of a *nucleoside*. The nomenclature for nucleotides uses the abbreviation for the nitrogenous base and adds the abbreviation MP, DP, or TP to designate the phosphate group as a monophosphate, diphosphate, or triphosphate, respectively (**Fig. A1.16B**). A small "d" in front of the nucleotide indicates that the sugar is 2-deoxyribose. For example, adenosine diphosphate is ADP, and deoxycytosine triphosphate is dCTP. In addition to being the precursors for nucleic acid synthesis, the ribose nucleotides have cellular functions of their own. For example, ATP is an energy carrier in the cell, and GTP and cyclic adenosine monophosphate (cAMP) are signaling molecules.

Table A1.5 Bases and nucleosides.

Base (abbreviation)[a]	Nucleoside (base plus ribose or deoxyribose)
Adenine (A)	Adenosine
Guanine (G)	Guanosine
Cytosine (C)	Cytidine
Thymine (T)	Thymidine
Uracil (U)	Uridine

[a]A,T,G, C, and U are abbreviations for the nucleotide bases; however, they are also used when referring to the nucleotides in a DNA or RNA strand to indicate which base is present in the primary structure.

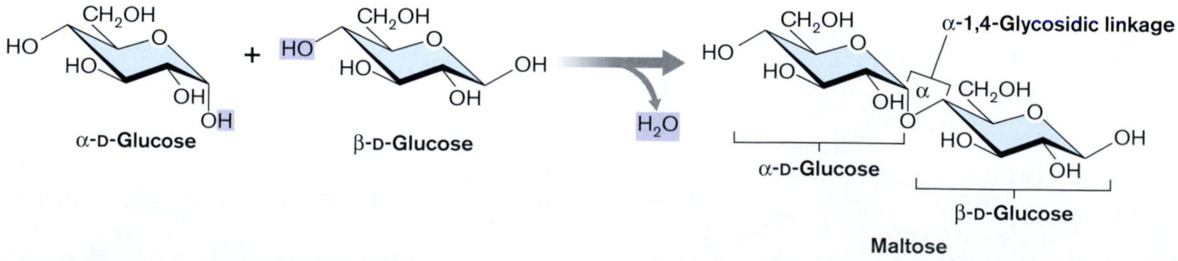

Figure A1.15 Formation of the disaccharide maltose. Maltose forms by a condensation reaction between two molecules of D-glucose, so that the C-1 of one molecule of glucose is linked by an oxygen atom to the C-4 of a second molecule of glucose. The hydroxyl groups of the second glucose can be either alpha or beta (the beta form is shown). The covalent glycosidic bond is called an α-1,4-glycosidic linkage because the oxygen on the C-1 carbon of the glycosidic bond is in the α position.

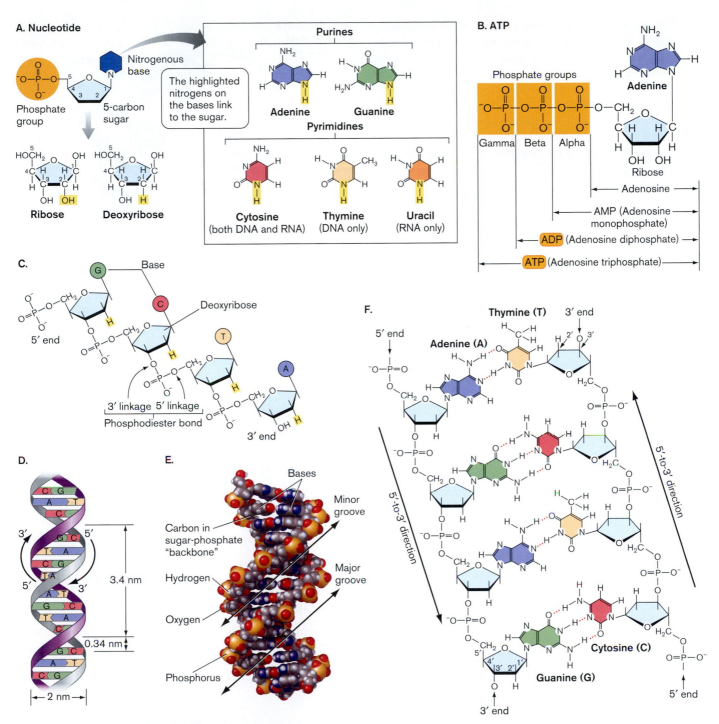

Figure A1.16 Nucleotide and DNA structure.   A. A nucleotide consists of a five-carbon sugar (ribose or deoxyribose), a phosphate group, and a nitrogenous base. **B.** A nucleotide may be a monophosphate, diphosphate, or triphosphate. The phosphate attached to the sugar is designated alpha, the beta phosphate is in the middle, and the gamma phosphate is distal to sugar. **C.** The primary structure of DNA. The orientation of the strand is determined by the presence of a free phosphate on the 5′ end and a free hydroxyl on the 3′ end. The sequence of a nucleic acid is read from the 5′ end to the 3′ end, so for this portion of the DNA strand, the primary structure is GCTA. **D, E, F.** The secondary structure of DNA. **D.** A ribbon structure of the DNA double helix showing the purine-pyrimidine base pairs. **E.** A space-filling model of DNA. Note that the base pairs stack on top of each other like the rungs of a ladder. **F.** Hydrogen bonding between the base pairs. Two hydrogen bonds form between A and T, three hydrogen bonds between G and C. Note the antiparallel orientation of the two strands.

The nucleotide triphosphate monomers (collectively designated NTPs or dNTPs) form nucleic acids through reactions catalyzed by nucleic acid polymerase enzymes. The phosphate at the C-5 position forms a diester linkage with the hydroxyl group at C-3, releasing pyrophosphate. At one end of the nucleic acid is a free phosphate group; at the other end is a free C-3 hydroxyl group. The terminus with a free phosphate is called the 5′ end, and the terminus with the free C-3 hydroxyl is called the 3′ end (**Fig. A1.16C**). Repeating sugar-phosphate linkages form the "backbone" of the molecule, while the information content of nucleic acids resides in the sequence of bases attached to the backbone. This linear sequence of nucleotides, from the 5′ end to the 3′ end, is the primary structure of the nucleic acid.

The Secondary Structure of DNA Is the Famous Double Helix

Two strands of DNA, held together by hydrogen bonding between a purine and a pyrimidine on opposite strands, twist around each other to form a helix (**Figs. A1.16 D** and **E**). Two hydrogen bonds form between adenine and thymine, and three hydrogen bonds form between cytosine and guanine (**Fig. A1.16F**). The sugar-phosphate backbone of DNA is on the outside of the helix. The base pairs are inside the helix, but portions of them are accessible to proteins such as transcription factors at two grooves in the helix—a wide major groove and a narrower minor groove. The double-stranded structure of DNA is critical to its function as the hereditary material. Each strand of DNA encodes the information to make a new double helix based on strict base-pairing rules.

The Secondary Structure of RNA Is Diverse

In contrast to the invariant double-helix structure of DNA, RNA molecules have a variety of secondary structures that allow RNA molecules to perform a number of different functions in the cell. Secondary structures are possible because RNA is single-stranded and capable of complementary base pairing. Base pairing in RNA is similar to the base pairing exhibited by DNA, except that in RNA, uracil replaces thymine. Base pairing is used in making mRNA from a DNA template and also in protein synthesis, where a tRNA charged with an amino acid matches its anticodon with the codon on the mRNA (**Fig. A1.17**). Because RNAs are single-stranded, they can undergo *intramolecular* as well as *intermolecular* base pairing. Base pairing within a single RNA molecule can lead to interesting three-dimensional shapes that may allow the RNA to function as a catalyst. The activity of catalytic RNAs (ribozymes) may be enhanced by the reactive hydroxyl group at the C-2 of the ribose sugar, which is lacking in DNA.

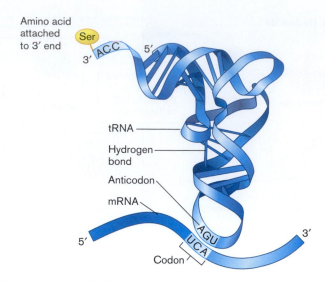

Figure A1.17   RNA secondary structure.   A tRNA folds up into a secondary structure stabilized by the hydrogen bonds shown in the figure. The tRNA forms complementary base pairs with an mRNA.

A1.6 Lipids

Lipids are a structurally diverse class of molecules that have diverse functions. They are major structural components of cell membranes, they store energy, and they act as cellular signals. The common structural feature of lipid molecules is that they contain a substantial number of nonpolar C–H and C–C covalent bonds. Because lipids are nonpolar they are hydrophobic and do not dissolve in water.

Lipids Can Be Hydrophobic or Amphipathic

An example of a hydrophobic lipid is isoprene (**Fig. A1.18**), which functions as a building block for more complex cellular lipids such as squalene. Other lipid

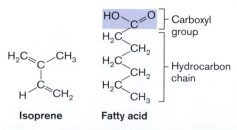

Figure A1.18   Lipid building blocks.   Isoprene is used to make squalene and cholesterol. Fatty acids are components of triglycerides and phospholipids.

building blocks include the fatty acids (**Fig. A1.18**). In contrast to isoprene, the fatty acids have both a hydrophobic portion (the hydrocarbon tail) and a hydrophilic component (the carboxylic acid). Molecules with both hydrophobic and hydrophilic portions are described as **amphipathic**.

Fatty Acids May Be Saturated or Unsaturated

The cell contains many different fatty acids, which differ in the number of carbons (usually even) and their saturation (**Fig. A1.19**). In saturated fatty acids, the carbon-carbon bonds are all single bonds, and the carbons are bonded to the maximum number of hydrogen atoms. Unsaturated fatty acids contain one or more double bonds between adjacent carbons in the hydrocarbon tail. Monounsaturated fatty acids have one double bond; polyunsaturated fatty acids have multiple double bonds. Saturated fatty acids can pack together in an arrangement where they can be stabilized by van der Waals forces between adjacent hydrocarbon chains. Thus, saturated fatty acids (including animal fats such as lard) tend to be solids at room temperature. In unsaturated fatty acids, the double bond causes a kink in the chain that prevents tight packing, so unsaturated fatty acids, such as vegetable oils, tend to remain fluid at room temperatures.

Glycerol Is an Important Building Block of Many Lipids

The three hydroxyl groups on glycerol can undergo a condensation reaction with the carboxylic acid portion of fatty acids to form triesters, known as triglycerides.

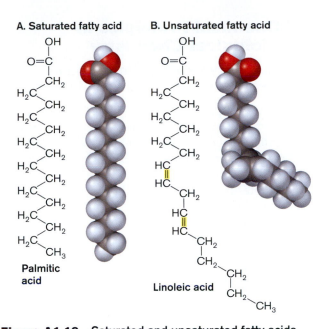

Figure A1.19   Saturated and unsaturated fatty acids.
A. Palmitic acid is a saturated fatty acid; there are no double bonds between the carbon atoms. **B.** Linoleic acid is an example of a polyunsaturated fatty acid (it has more than one double bond). Note that the double bonds create kinks in the hydrocarbon tail.

Triglycerides are a compact energy source for cells (**Fig. A1.20**). Phospholipids, key components of the cell membrane, contain fatty acids attached to two of the hydroxyl groups of glycerol and a phosphate covalently attached to the third hydroxyl. The structure of phospholipids and how they function in cell membranes is fully explored in Appendix A2.1.

Figure A1.20   Formation of a triglyceride.   Condensation reactions form covalent ester linkages between the carboxyl groups of three fatty acids and the three hydroxyl groups of glycerol.

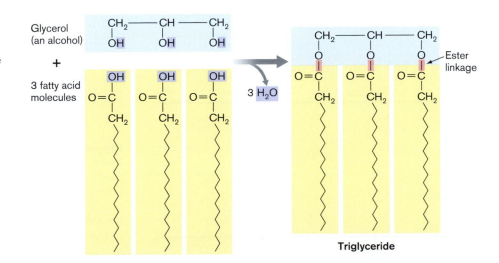

A1.7 Chemical Principles in Biological Chemistry

To Express Very Large or Very Small Quantities, We Use Scientific Notation

Scientists often express very large numbers (such as the number of microorganisms in a liter of seawater) or very small numbers (such as the diameter of a bacterium) with the help of exponents. For example, a million, or 1,000,000, is often written as 1×10^6. The positive exponent 6 indicates how many decimal places to the right of the number 1 our number is. One-millionth, or 0.000001, is written as 1×10^{-6} to indicate six decimal places to the left of the number 1. Special prefixes can be used to modify the magnitude of units. These are listed in **Table A1.6**. For example, 1×10^{-6} is designated by the symbol μ (which stands for "micro"), so a length of 0.2 micrometer (μm) is equivalent to 2×10^{-7} meter (m).

When scientists work with a set of numbers that range over many orders of magnitude, they often use a logarithmic scale. For example, the pH scale is based on base 10 logarithms. Base 10, or common, logarithms are abbreviated log, as opposed to natural logarithms, abbreviated ln. Base 10 logarithms are defined as $\log 10^x = x$. For example the log of $1 \times 10^{-4} = -4$. The log of 10 = 1, and the log of 1 = 0.

Molarity Is a Unit Commonly Used to Measure Concentration

Scientists often measure concentrations. Concentration refers to how much of something is present in a given volume. A frequent way to report concentration is in units of molarity. **Molarity** is defined as the number of moles of substance per liter of solution (usually water). One mole is Avogadro's number (6.02×10^{23}) of molecules. If you wanted to make a 1-molar solution of NaCl, it would be impossible to know when you had added 6.02×10^{23} mol-

ecules to a liter of water. It's more convenient to weigh out a mole of NaCl on a balance. The weight of a mole of a substance is Avogadro's number times the weight of a molecule. For example, NaCl has a mass of 58.5 atomic mass units (amu), 23 from sodium and 35.5 from chlorine (see **Fig. A1.1**). Each amu is equal to 1.66×10^{-24} grams (g), so one molecule of NaCl weighs 9.71×10^{-23} g. A mole of NaCl then weighs 9.7×10^{-23} g times 6.021×10^{23} (Avogadro's number) = 58.5 g. (Note that this weight is the same as just adding together the atomic masses of Na and Cl and expressing it in grams.) So to make a 1-molar solution of NaCl, you would take 58.5 g of NaCl and add water up to a liter.

The Change in Free Energy Determines Whether a Reaction Proceeds Spontaneously

Reactions in organisms, such as the synthesis of proteins from amino acids by condensation, can occur only if they are energetically favorable. The study of energy and matter changes is called thermodynamics. All systems (including living systems, such as cells) must obey the laws of thermodynamics. The first law of thermodynamics states that energy is neither created nor destroyed. In a closed system (a system that is not exchanging energy with its environment), the total amount of energy remains constant. Although the amount of energy remains constant, energy can be converted from one form to another. The second law of thermodynamics states that in energy transformations, some energy becomes unavailable to do work and is lost as disorder, or entropy. In other words, entropy tends to increase. The laws of thermodynamics are summarized in **Figure A1.21**.

The total energy in a system, called enthalpy, can be expressed in the following equation, where H stands for

Table A1.6 Common numerical prefixes.		
Prefix	**Symbol**	**Factor**
kilo	k	10^3
deci	d	10^{-1}
centi	c	10^{-2}
milli	m	10^{-3}
micro	μ	10^{-6}
nano	n	10^{-9}
pico	p	10^{-12}
femto	f	10^{-15}

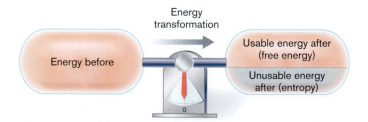

Figure A1.21   The first and second laws of thermodynamics. The first law states that in an energy transformation, the total amount of energy remains constant. Both sides have the same amount of energy (the balance reads zero). The second law states that in an energy transformation, the amount of usable energy, or free energy, decreases and the amount of unusable energy, or entropy, increases.

enthalpy, G stands for free energy, S stands for entropy, and T stands for temperature:

$$H = G + T \cdot S$$

Because scientists are interested in the amount of free energy—that is, the amount of energy available, or free, to do work—the energy equation is usually written in terms of G:

$$\Delta G = \Delta H - T \cdot \Delta S$$

The triangle is the Greek letter delta, which signifies "change in." Here, a change in energy is defined as the energy of the products minus the energy of the reactants. Thus, ΔG means $G_{products} - G_{reactants}$, and ΔS indicates $S_{products} - S_{reactants}$.

Systems Tend Toward the Most Stable, Lowest Free Energy State

Reactions are favored when the free energy of the products is less than the free energy of the reactants, that is, when ΔG is negative. Reactions with a negative ΔG are termed spontaneous, or **exergonic**, because they are energetically favorable and can occur without an input of energy. Reactions with a positive ΔG are **endergonic**. They are not energetically favorable, and energy needs to be added for them to proceed. Reactions with a ΔG of zero are at equilibrium, and neither products nor reactants are favored.

The sign of ΔG is determined by changes in both enthalpy and entropy. For example, the following reaction, the oxidation, or "burning," of methane to release carbon dioxide and water, is spontaneous and has a negative change in free energy:

$$CH_4 + O_2 \rightleftharpoons CO_2 + H_2O + energy$$

The bonds in methane and oxygen have more energy than the bonds in CO_2 and H_2O. Because the products CO_2 and H_2O have less energy than the reactants, the change in enthalpy for this reaction, ΔH ($H_{products} - H_{reactants}$), is negative and the reaction is said to be exothermic.

A reaction does not need to be exothermic to be spontaneous. Consider an ice cube placed at room temperature. The ice cube will spontaneously melt even though this reaction is endothermic, absorbing heat from the environment. Ice melting is spontaneous because of the increase in entropy in going from a solid to a liquid. Remember that entropy, S, is a measure of the disorder of a system. Gases are more disordered than liquids, and liquids are more disordered than solids. Thus, the change in entropy, ΔS ($S_{products} - S_{reactants}$), from solid water (ice) to liquid water is positive because the product, liquid water, has more entropy than ice. Positive changes in entropy may result in a negative change in free energy, especially at high temperatures.

There is a Relationship between the Standard Free Energy Change and Chemical Equilibrium

To compare changes in free energy among different reactions, a standard change in free energy, $\Delta G°$ (the circle is pronounced naught), is defined as the change in free energy when the concentrations of all reactants and products are at 1 molar. $\Delta G°$ is a constant and depends on the nature of the products and reactants. Another constant, $\Delta G°'$ refers to the standard change in free energy at pH 7. A standard free energy change can be related to an equilibrium constant, K_{eq}, that indicates whether products or reactants will be favored at equilibrium. Reactions between biomolecules can be written as chemical equations. For example, in the following reaction, reactants A and B form products C and D:

$$A + B \underset{k_r}{\overset{k_f}{\rightleftharpoons}} C + D$$

The double arrow indicates that the reaction is reversible; A and B can be forming C and D at the same time that C and D are reverting to A and B. What determines whether the forward or reverse reaction will predominate? The forward and reverse rates depend on rate constants k_f and k_r, respectively. The rate constants depend on the nature of the interacting molecules and are an indication of how easily they will react. The forward and reverse rates also depend on concentrations of reactants and products:

Rate forward $= [A][B]k_f$

Rate reverse $= [C][D]k_r$

If the forward and reverse reactions are allowed to proceed, eventually equilibrium will be reached. All chemical reactions have a preferred state called **equilibrium**, where there is no *net* change in the reaction. Equilibrium is not static; rather, at equilibrium, the rate of the forward reaction equals the rate of the reverse reaction:

$$[A][B]k_f = [C][D]k_r$$

For every reaction, there is an equilibrium constant, K_{eq}, that indicates the relative ratios of products and reactants at equilibrium:

$$K_{eq} = k_f/k_r = [C][D]/[A][B]$$

A K_{eq} of 1 means that neither products nor reactants are favored; a K_{eq} greater than 1 indicates that products are favored at equilibrium; and a K_{eq} less than 1 means that reactants predominate. K_{eq} and $\Delta G°$ are related as follows where R is the gas constant (8.3×10^{-3} kJoule/mole·degree) and T is the temperature in kelvins:

$$\Delta G° = -RT \ln K_{eq}$$

A value of 298 K is often used (this is roughly 25°C). The relationship between K_{eq} and $\Delta G°$ is usually shown with the base 10 logarithm instead of the natural log:

$$\Delta G° = -2.303 \, RT \log K_{eq}$$

This equation defines a relationship between $\Delta G°$ and K_{eq} that is discussed in Chapter 13. When K_{eq} is 1 (neither products nor reactants favored), $\Delta G°$ (defined as the change in free energy when concentrations of all substances are at 1 molar) is zero because the reaction is at equilibrium at that point. A K_{eq} of less than 1 (reactants favored) corresponds to a positive $\Delta G°$ because at 1-molar concentrations of reactants and products, the reaction will move to the left. A K_{eq} of more than 1 (products favored) corresponds to a negative $\Delta G°$ because at 1-molar concentrations, the reaction will move to the right.

Free Energy in Cells and the Law of Mass Action

Although the standard changes in free energy are useful for comparing the energetics and equilibria of different reactions, they do not reflect what is happening in the cell where concentrations of substances are probably not 1 molar. While K_{eq} and $\Delta G°$ are constants and unique for a particular reaction, ΔG depends on the actual concentrations of products and reactants as shown in the equation relating ΔG to $\Delta G°$:

$$\Delta G = \Delta G° + 2.303 \, RT \log ([C][D]/[A][B])$$

In this equation, the concentrations of products and reactants are not their equilibrium concentrations, but the actual concentrations present at a given point in time. Because the concentrations of products and reactants will change as a reaction proceeds, the value of ΔG will change over time.

At equilibrium, where the total energy on each side of the reaction is equivalent, ΔG is zero. For an exergonic reaction at equilibrium, the total energy of reactants and products is equivalent because the amount of products is greater than the amount of reactants. If, at equilibrium, reactants (A, B) are added or products (C, D) are removed, ΔG will become negative. A negative ΔG causes the reaction to move to the right to reestablish equilibrium. As A and B convert to C and D, ΔG becomes less and less negative until it reaches zero again at equilibrium. If, at equilibrium, products are added or reactants are removed, ΔG will become positive and the reaction will proceed to the

left to reestablish equal energy on both sides. This **law of mass action** is the tendency of a reaction to reestablish equilibrium after perturbations in the concentrations of products or reactants.

The law of mass action means that a reaction with a positive $\Delta G°$ can be made spontaneous and driven to the right by keeping the concentration of reactants high or the concentration of products low. The cell employs this strategy in metabolic pathways, where the product of one reaction is constantly removed by a subsequent reaction.

The Rate at Which a Reaction Proceeds Depends on the Activation Energy

It is important to realize that although a reaction with a large positive K_{eq} and negative $\Delta G°$ is spontaneous, it may be slow. The value of $\Delta G°$ says nothing about the rate of a reaction. Everyday examples of spontaneous but slow reactions are the aging process in humans and the rusting of metal. A reaction will be slow, even though it is spontaneous, if it must pass through an unstable, high-energy transition state on the way to forming products. The **activation energy** (E_a) is the energy needed to reach this transition state (**Fig. A1.22**). For reactions with low activation energies, random collisions between reactants may provide enough energy to boost them up and over the E_a. For reactions with high activation energies, collisions between molecules may not provide enough energy for them to reach the transition state.

Enzymes are biological catalysts that can speed up reaction rates by stabilizing transition states and lowering the activation energy. Most enzymes are proteins that specifically bind reactants and provide an environment

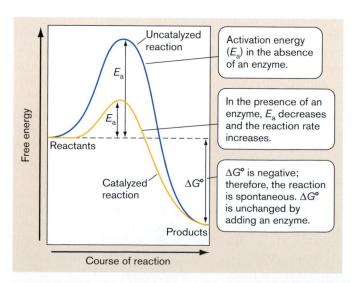

Figure A1.22   Enzymes, activation energies, and reaction rates.

that facilitates product formation. Although enzymes increase reaction rates, they do not change the $\Delta G°$ or K_{eq} of a reaction. **Figure A1.22** illustrates how the difference in free energy between products and reactants is unchanged even in the presence of an enzyme that lowers the activation energy.

Many Biological Processes Are Sensitive to Changes in pH

Many biological processes occur only within a narrow range of hydrogen ion concentrations. Hydrogen ions (H^+) are also referred to as protons because they have lost their single electron and consist of only a proton. Hydrogen ion concentration is reported using a pH (power of hydrogen) scale, where pH is the negative logarithm of the hydrogen ion concentration: $pH = -\log [H^+]$. In pure water, which is considered neutral, $[H^+]$ is 1×10^{-7} molar (M), a pH of 7. Because the pH scale is logarithmic, every pH unit corresponds to a tenfold change in hydrogen ion concentration. A solution with a pH of 6 has a hydrogen ion concentration of 1×10^{-6} M, ten times the hydrogen ion concentration at pH 7. Hydrogen ion concentration $[H^+]$ multiplied by hydroxide ion concentration $[OH^-]$ always equals 1×10^{-14}. Hence, the concentrations of H^+ and OH^- are reciprocally related (if one goes up, the other goes down). If the pH is less than 7, a solution is acidic (more H^+, less OH^-). If the pH is greater than 7, the solution is basic, or alkaline (less H^+, more OH^-).

Acids (for example, carboxyl groups) donate protons, and bases (for example, amino groups) accept protons (**Fig. A1.23**). At intracellular pH (near 7), the carboxylic acid and amino group of amino acids are both ionized (charged), the carboxyl group carrying a negative charge and the amino group a positive charge. The ionization of these groups can change if the pH changes. At lower pH (more acidic solution) protons move back onto the ionized carboxylic acid. For example, the side chain of an acidic amino acid such as glutamate (see **Fig. A1.9**) is ionized at normal cell pH but may regain a proton and become

glutamic acid at a lower pH. At higher pH values (lower proton concentrations), amino groups lose protons to become uncharged. Disturbances in the ionization state of carboxyl or amino groups on the side chains of amino acid residues can disrupt ionic bonds between these groups and lead to protein denaturation. This effect of pH on protein tertiary structure is one reason why cells can tolerate only a narrow range of intracellular pH.

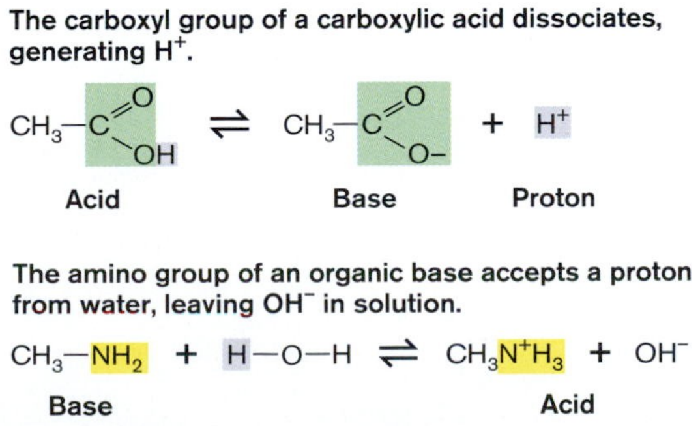

Oxidation-reduction (Redox) Reactions Are an Important Way of Transferring Energy in Biological Systems

Biomolecules may undergo an important class of chemical reactions termed redox reactions. Redox reactions involve the transfer of electrons from one molecule to another or from one atom to another. The molecule that gains electrons becomes reduced, and the molecule that loses electrons becomes oxidized (**Fig. A1.24**). (A useful mnemonic device to remember this is "LEO the lion says GER." LEO: lose electrons oxidation; GER: gain electrons reduction.)

Redox reactions are always coupled. If one atom loses electrons, another atom must gain electrons.

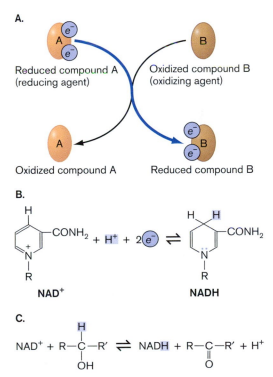

Figure A1.24 Oxidation-reduction reactions. A. A reducing agent A donates electrons to reduce compound B. Because A loses electrons, it becomes oxidized itself in the process. B is the oxidizing agent. **B.** NADH is a common biological reducing agent. **C.** NAD$^+$ is reduced to NADH as an alcohol is oxidized to an aldehyde.

The carboxyl group of a carboxylic acid dissociates, generating H^+.

Acid ⇌ Base + Proton

The amino group of an organic base accepts a proton from water, leaving OH^- in solution.

$CH_3-NH_2 + H-O-H \rightleftharpoons CH_3N^+H_3 + OH^-$

Base   Acid

Figure A1.23 Organic acids and bases.

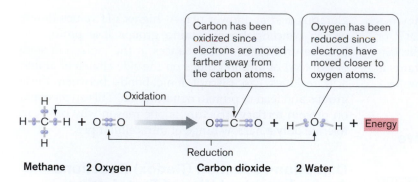

Figure A1.25   **The oxidation of methane.** Bonding electrons were shared equally by carbon and hydrogen in methane, but are closer to oxygen in carbon dioxide and water. Carbon and hydrogen have been oxidized; oxygen has been reduced.

Oxygen, with its large electronegativity, usually gains electrons and becomes reduced in redox reactions. Molecules that become reduced are called oxidizing agents because they cause something else to become oxidized. In contrast, reducing agents can donate electrons, reduce other molecules, and become oxidized themselves in the process. Molecules with reduced carbons contain more energy than their oxidized counterparts. Common reducing agents in the cell are NADH, NADPH, and FADH$_2$. These are all high-energy molecules that can donate electrons.

In addition to achieving a complete transfer of electrons, redox reactions can also occur if electrons are shifted toward or away from an atom. For example, in the burning (oxidation) of methane, electrons have moved away from the carbon and toward the oxygen (**Fig. A1.25**). Oxygen usually acts as an oxidizing agent; thus, we expect it to get reduced (gain electrons). By default, then, methane is oxidized; indeed, while the bonding electrons were shared fairly equally between C and H in methane, they are now close to the O in both carbon dioxide and water and farther away from the C and H. Thus, methane has been oxidized and oxygen has been reduced. This reaction releases energy because CO$_2$ and H$_2$O are the most stable lowest-energy forms available when carbon, hydrogen, and oxygen are combined.

Appendix 2

Introductory Cell Biology: Eukaryotic Cells

This appendix presents a review of cell biology principles generally covered in an introductory-level biology class. Cell biology encompasses fundamental principles of cell structure and function. Because all cells are enclosed by a cell membrane and need to regulate the transport of materials across the membrane, we cover this topic first. We then go on to discuss structures unique to eukaryotic cells.

All eukaryotic cells contain a nucleus, an organelle that houses the DNA. We describe the structure of the nucleus and the process of mitosis—a process that ensures the accurate separation of replicated chromosomes during cell division. We then explore the challenges encountered by eukaryotic cells owing to their large size compared to most prokaryotic cells. The endomembrane system, a network of internal membranes is discussed next, followed by a discussion of the cytoskeleton, a collection of proteins that form the basis of cell architecture and are responsible for cell movement. Finally, we discuss mitochondria and chloroplasts, the energy-producing organelles of eukaryotic cells.

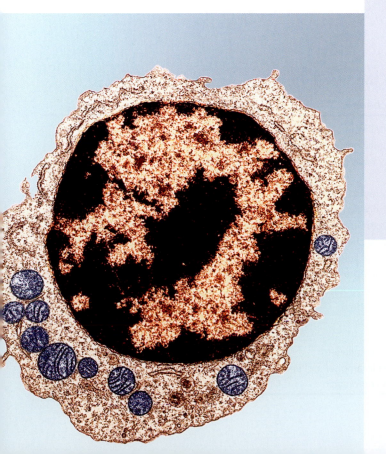

This eukaryotic cell (lymphocyte white blood cell) shows numerous major organelles such as a large nucleus (orange) and multiple mitochondria (blue). (TEM ×20,550) *Source*: ©Gopal Murti/Visuals Unlimited

The cell is the basic unit of life. All organisms consist of either a single cell or a collection of cells. All cells can potentially perform a common set of tasks, including replication, catalysis, and regulation. The structures that enable these functions are found in all cells and include the cell membrane, DNA, and ribosomes (**Fig. A2.1**). Although the detailed organization of these structures differs in cells from the three domains (**Bacteria**, **Archaea**, and **Eukarya**; see **Table A2.1**), the fundamental structure and function of the cell membrane, DNA, and ribosomes are the same in all cells. This appendix describes the cell structures that are found across all three domains of life, then focuses on those that are unique to the eukaryotic domain, specifically organelles and the cytoskeleton.

A2.1 The Cell Membrane

All cells are enclosed by a **cell membrane** (sometimes called the **plasma membrane** or **cytoplasmic membrane**) that creates an internal environment distinct from the external environment. The aqueous fluid inside the membrane is called **cytoplasm** (or **cytosol**). Major functions of the cell membrane include the regulated transport of substances into and out of the cell and the reception of signals from the external environment. Membranes are also critical for energy production.

As shown in **Figure A2.2**, membranes consist mainly of lipids and proteins, but also contain some carbohydrates found in hybrid structures such as glycolipids and glycoproteins (sugars joined to lipids or to proteins, respectively).

Although the membrane contains more lipid molecules than protein molecules, proteins may contribute most of the mass. The number and nature of proteins within a membrane depend on the membrane under consideration. For example, the inner membrane of the mitochondrion contains a rich array of proteins involved in energy production

Membrane proteins can be classified based on how they interact with membranes (**Fig. A2.2A**). **Transmembrane proteins** (integral proteins) span the bilayer. Transmembrane proteins are **amphipathic** meaning they have both hydrophobic and hydrophilic portions; the hydrophilic portions face the cytoplasm and the extracellular environment, and the hydrophobic components span the membrane. The transmembrane domains are often alpha-helices containing 15–20 amino acid residues (**Fig. A2.2B**). The portions of the protein that are intracellular, in the membrane, and extracellular depend on the protein and can be quite different for different proteins.

Peripheral membrane proteins are associated with the cell membrane but are not directly inserted into the bilayer. Some peripheral membrane proteins are retained at membranes through noncovalent interactions with transmembrane proteins, others by noncovalent interactions with the lipids of the membrane.

Lipids Are Responsible for Forming Membranes

The predominant lipids in membranes are phospholipids. **Phospholipids** consist of a core of glycerol, to which two fatty acids and a modified phosphate group are attached

Table A2.1	**Comparison of cell structures in the three domains.**		
Feature	*Bacteria*	*Archaea*	*Eukarya*
Genome	Usually circular DNA	Usually circular DNA	Linear DNA
	Usually one chromosome	Usually one chromosome	Multiple chromosomes, in pairs
	May have plasmids	May have plasmids	Plasmids rare
	Usually lacks introns		May have introns
Location of DNA	Nucleoid region in cytoplasm	Nucleoid region in cytoplasm	Contained within membrane-bound nucleus
Cell membrane	Glycerol bonded to straight-chain fatty acids via ester linkages	Glycerol bonded to branched fatty acids via ether linkages	Glycerol bonded to straight-chain fatty acids via ester linkages
Cell wall	Usually present, composed of peptidoglycan	Usually present, composed of pseudopeptidoglycan	If present composed of cellulose (algae) or chitin (fungi)
Internal membranes	Uncommon	Uncommon	Extensive membranous organelles
	May have mesosomes or energy-transducing lamellae		
Ribosomes	Sensitive to chloramphenicol and streptomycin	Not sensitive to chloramphenicol and streptomycin	Not sensitive to chloramphenicol and streptomycin

A. Prokaryotic cell

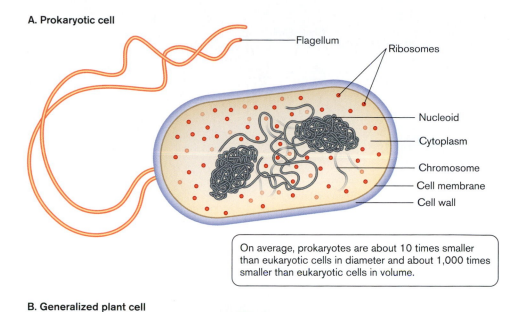

Flagellum
Ribosomes
Nucleoid
Cytoplasm
Chromosome
Cell membrane
Cell wall

On average, prokaryotes are about 10 times smaller than eukaryotic cells in diameter and about 1,000 times smaller than eukaryotic cells in volume.

B. Generalized plant cell

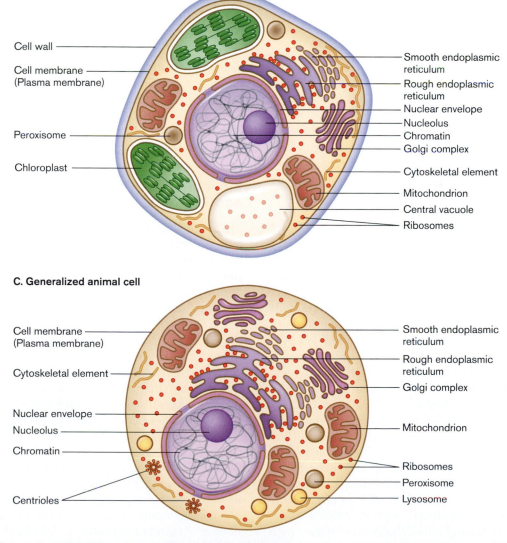

Cell wall
Cell membrane (Plasma membrane)
Peroxisome
Chloroplast

Smooth endoplasmic reticulum
Rough endoplasmic reticulum
Nuclear envelope
Nucleolus
Chromatin
Golgi complex
Cytoskeletal element
Mitochondrion
Central vacuole
Ribosomes

C. Generalized animal cell

Cell membrane (Plasma membrane)
Cytoskeletal element
Nuclear envelope
Nucleolus
Chromatin
Centrioles

Smooth endoplasmic reticulum
Rough endoplasmic reticulum
Golgi complex
Mitochondrion
Ribosomes
Peroxisome
Lysosome

Figure A2.1   The prokaryotic cell and the eukaryotic cell.   A. The prokaryotic cell typically contains a single compartment, and its DNA is organized in the nucleoid region. **B, C.** Eukaryotic cells are typically much larger than prokaryotic cells and contain organelles.

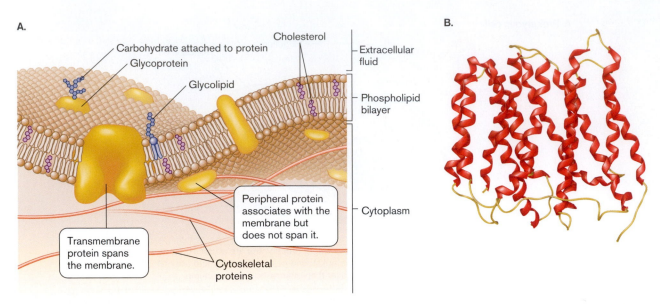

A.

Carbohydrate attached to protein

Glycoprotein

Cholesterol

Extracellular fluid

Glycolipid

Phospholipid bilayer

Transmembrane protein spans the membrane.

Peripheral protein associates with the membrane but does not span it.

Cytoplasm

Cytoskeletal proteins

B.

A2.2   The cell membrane.   A. A cutaway view of the cell membrane. **B.** Ribbon diagram from X-ray crystallographic data of the *E. coli* EmrD protein, a multi-drug transporter. Red portions are transmembrane alpha helices and yellow portions are intracellular and extracellular loops. (PDB code: 2gfp) Source: Yin, et al. 2006. *Science* 312:741.

via ester linkages (**Fig. A2.3A**). Unlike eukaryotes and bacteria, archaea have phospholipids with ether linkages. Phospholipids are amphipathic; the fatty acid hydrocarbon tails are hydrophobic, and the phosphate head group is hydrophilic. Amphipathic lipids are most stable in water when the hydrophilic portions interact with water and the hydrophobic portions cluster together away from water. One way phospholipids can achieve stability is by forming a bilayer. Indeed, the cell membrane is a **phospholipid bilayer**, two layers of phospholipids whose hydrocarbon fatty acid tails face the interior of the bilayer and whose charged phospholipid head groups face the aqueous cytoplasm and extracellular environment (**Fig. A2.3B**). The phospholipid layer in contact with the cytoplasm is called the **inner leaflet**, and the layer in contact with the environment is called the **outer leaflet**. Cells contain a number of different phospholipids that differ in how the phosphate head group is modified (**Fig. A2.3C**). Phospholipids can also differ in the length and saturation of the fatty acids chains (see Section A1.6 for fatty acid structures), and, as we shall see, these structural variations have functional consequences.

In addition to phospholipids, eukaryotic membranes contain a variable amount of the steroid cholesterol (**Fig. A2.3D**). Like phospholipids, cholesterol is amphipathic, with a hydrophilic hydroxyl group and hydrophobic hydrocarbon rings and tail. In membranes, cholesterol is oriented so that the hydrophilic hydroxyl group interacts with the phosphate head groups of phospholipids, while the hydrophobic rings and tail of cholesterol interact with the phospholipid hydrocarbon tails. The amount of cho-

lesterol present in membranes varies among cells and also among organelles within a cell. Cholesterol contributes to membrane integrity by providing mechanical stability to the phospholipid bilayer. Bacterial membranes do not contain cholesterol but do contain hopanoids that serve similar roles.

Movement of Membrane Lipids and Proteins

Membranes are not static structures; rather, many membrane components (a mosaic of various lipids and proteins) can undergo rapid movement. The **fluid mosaic model** of membranes states that membrane components are free to diffuse in the plane of the membrane. Recent evidence indicates that certain membrane proteins may be restricted to specific regions of the membrane by interactions with cytoskeletal proteins and that these membrane regions may be enriched with specific lipids (for discussion of cytoskeletal proteins, see Section A2.5).

Although many phospholipids and membrane proteins can move laterally within a leaflet, they do not "flip-flop" from one leaflet of the bilayer to the other (**Fig. A2.3E**). Flip-flop of phospholipids is rare owing to the highly unfavorable interactions of charged head groups moving through the hydrophobic interior of the membrane. Thus, the inner and outer leaflets of the membrane may be made up of different phospholipids. It is thought that such phospholipid asymmetry may be important for the correct functioning of membrane proteins—that is, that proteins may work best when surrounded by particular phospholipids.

A. Phospholipid

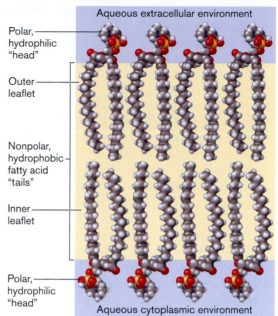

Phosphate

Glycerol

Fatty acids

B. Phospholipid bilayer

Aqueous extracellular environment

Polar, hydrophilic "head"

Outer leaflet

Nonpolar, hydrophobic fatty acid "tails"

Inner leaflet

Polar, hydrophilic "head"

Aqueous cytoplasmic environment

C. Selected phospholipids

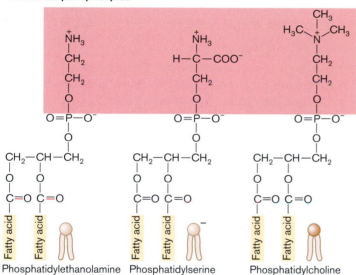

Phosphatidylethanolamine   Phosphatidylserine   Phosphatidylcholine

D. Cholesterol structures

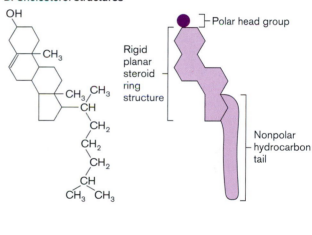

Rigid planar steroid ring structure

Polar head group

Nonpolar hydrocarbon tail

E. Phospholipid motions

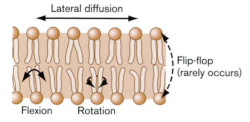

Lateral diffusion

Flip-flop (rarely occurs)

Flexion   Rotation

F. Membrane fluidity

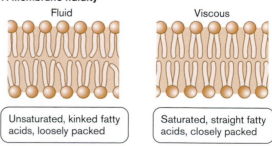

Fluid   Viscous

Unsaturated, kinked fatty acids, loosely packed

Saturated, straight fatty acids, closely packed

Figure A2.3 **Phospholipids, cholesterol, the lipid bilayer, and membrane fluidity.** **A.** Generic (saturated) phospholipid. **B.** Orientation of phospholipids in the bilayer. **C.** Some phospholipids present in cell membranes. **D.** Structural formula and schematic drawing of cholesterol. **E.** Motions of phospholipids in membrane bilayers. **F.** The ratio of saturated and unsaturated fatty acids in the phospholipids affects membrane fluidity.

Glycolipids and glycoproteins also contribute to membrane asymmetry because the carbohydrate moieties always face the extracellular environment.

Membrane fluidity refers to the movement of membrane phospholipids within the plane of the membrane, and this fluidity is important for proper membrane function. For example, transport across the membrane is affected by membrane fluidity. Decreased fluidity is associated with decreased transport rates. Because a drop in temperature decreases fluidity, cold temperatures may slow transport processes across the membrane.

The composition of the membrane, especially the types of phospholipids present, can have a dramatic effect on membrane fluidity. For example, saturated fatty acids decrease membrane fluidity because the linear hydrocarbon tails pack together well. In contrast, unsaturated fatty acids have kinks in the hydrocarbon chains that limit packing and increase fluidity (**Fig. A2.3F**). The length of the fatty acid chains also affects fluidity. Phospholipids with longer hydrocarbon chains have increased hydrophobic interactions with neighboring lipids and thus decreased membrane fluidity. Organisms can alter membrane fluidity in response to temperature stress by changing the length and degree of saturation of fatty acids present in membrane phospholipids. For example, as environmental temperatures drop, both eukaryotes and prokaryotes maintain membrane fluidity by replacing long-chain fatty acids with shorter chains and increasing the percentage of unsaturated fatty acids in their membranes.

Cholesterol also influences membrane fluidity. The effects of cholesterol on membrane fluidity are complicated and depend on factors such as the ratio of saturated to unsaturated fatty acids in the membrane. Cholesterol may prevent packing of saturated fatty acids, thus increasing fluidity. In membranes with unsaturated fatty acids, cholesterol may fill in the spaces between adjacent phospholipids, stabilizing them and decreasing fluidity. In this case, cholesterol can decrease the permeability of the membrane to hydrophobic substances by packing in between the hydrocarbon chains and preventing substances from slipping through.

Membranes Are Semipermeable: Some Substances Pass through Them Easily; Others Do Not

The major functions of membranes (such as containing cytoplasmic components, regulating what substances enter and leave cells and organelles, and producing energy) depend on the semipermeable nature of membranes. **Semipermeable** (also called **selectively permeable**) **membranes** are permeable to some substances but not to others. In general, the cell membrane is permeable to hydrophobic molecules and impermeable to charged molecules (**Fig. A2.4**). Diffusion across the membrane also depends on the size of the molecule. The membrane is freely permeable to small nonpolar molecules such as O_2. Larger nonpolar molecules can also diffuse across the membrane, albeit more slowly. Molecules that are polar but small (such as ethanol and water) can also diffuse across the membrane. The membrane is impermeable to large polar molecule such as glucose and to charged molecules, regardless of their size. The impermeability of the membrane to charged substances such as ions is important for energy production at membranes because the proton motive force depends on the ability of the membrane to separate compartments of different ion concentrations.

Figure A2.4 Selective permeability of cell membranes.

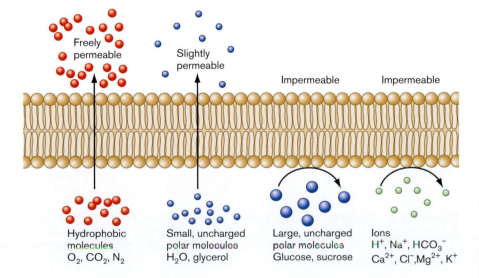

Freely permeable

Slightly permeable

Impermeable

Impermeable

Hydrophobic molecules
O_2, CO_2, N_2

Small, uncharged polar molecules
H_2O, glycerol

Large, uncharged polar molecules
Glucose, sucrose

Ions
H^+, Na^+, HCO_3^-
Ca^{2+}, Cl^-, Mg^{2+}, K^+

Transport Proteins Move Many Substances across Cell Membranes

Although ions and large polar molecules such as glucose cannot diffuse directly across the cell membrane, they do need to enter cells. In both eukaryotes and prokaryotes, movement of molecules across the cell membrane is accomplished through specific transmembrane proteins such as channels and transporters. Channels can also increase the diffusion of molecules that move across the membrane too slowly on their own to supply the cell's needs. For example, aquaporins can increase the rate of water movement across the membrane. There are many different types of transporters, each differing in its energy requirements and in the types of molecules that it transfers across the membrane.

Diffusion Moves Substances across Cell Membranes and within Cells

Diffusion is the net movement of molecules from an area of high concentration to one of low concentration. It is a spontaneous process because it is accompanied by an increase in entropy (positive ΔS) that results in a negative free energy change (ΔG). The process requires no energy input and is brought about by the random, thermal movement of molecules.

Many factors can influence the *rate* of diffusion of a molecule across a membrane. Diffusion rates can be determined with an artificial membrane system as depicted in **Figure A2.5**. Molecules are added to the left side of the beaker, and samples from the right side are analyzed at various time points for the presence of the test molecule.

Factors that influence diffusion of molecules across a membrane include:

- **Temperature.** Increased temperatures mean faster motion. The faster the molecules are moving, the faster they will arrive at the membrane and cross it.
- **Solubility of the molecules in the membrane.** To cross the membrane, the molecules must penetrate it. Hydrophobic molecules will dissolve in the membrane and cross it; charged molecules will not.
- **Surface area of the membrane.** To cross the membrane, molecules must first encounter it. The chances of this happening are increased with an increase in membrane surface area.
- **Concentration gradient of the dissolved molecules.** A larger concentration gradient speeds up diffusion because the more molecules there are, the more will encounter the membrane and cross.
- **Thickness of the membrane.** Diffusion rates are inversely proportional to the square of the distance the solute must travel across the membrane. The thinner the membrane, the faster the molecules can get across.
- **Mass of the molecule.** Friction between a molecule and its medium is a source of resistance that slows down motion. Larger molecules with more mass experience more resistance and cross the membrane more slowly.

These factors can be expressed as follows:

$$\text{Diffusion rate} \propto \frac{\text{temperature} \times \text{surface area} \times \text{concentration gradient}}{\text{mass} \times \text{distance}^2}$$

Conditions for the diffusion of gases and other substances across the cell membrane will be most favorable when the surface area of the membrane is large, the concentration gradient across the membrane is high, and the membrane is thin.

Transport of Water Across the Cell Membrane Must Be Tightly Controlled

Osmosis is the diffusion of water across a selectively permeable membrane from regions of high water concentration (low solute) to regions of low water concentration

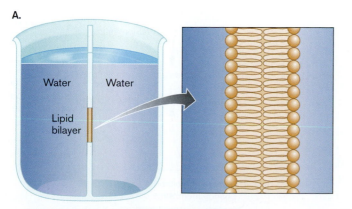

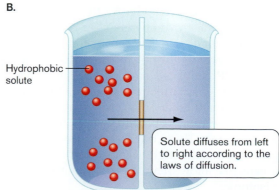

Figure A2.5   Diffusion across a phospholipid bilayer.   A. An artificial bilayer system. **B.** Diffusion across an artificial membrane.

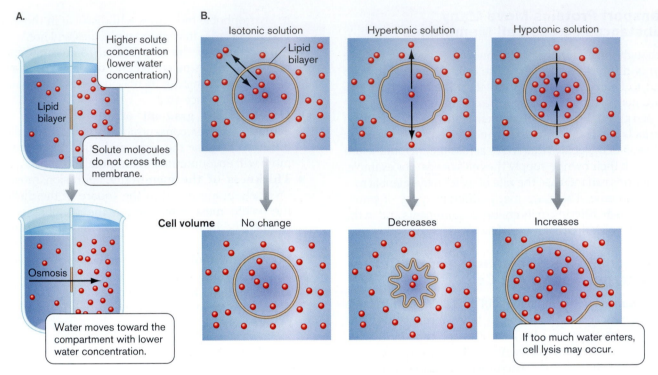

Figure A2.6 **Osmosis and water balance.** **A.** Osmosis. **B.** Movement of water across the cell membrane, and shrinkage or expansion of the membrane in isotonic, hypertonic, and hypotonic environments. Arrows indicate net water movement.

(high solute) (**Fig. A2.6A**). Cells must maintain osmotic balance with the surrounding environment. The direction of water movement depends on the concentration of dissolved solutes in the cell relative to the cell's environment. When a cell is in an **isotonic** environment (equal concentrations of dissolved solutes inside the cell and out), there is osmotic balance, and water will enter and exit the cell at equal rates (that is, there is no net movement of water). In a **hypertonic** environment (higher concentration of solutes outside the cell), there is a net loss of water from a cell. The cell shrinks, and the concentration of cell contents increases. In a **hypotonic** environment (lower concentration of solutes outside the cell), there is a net uptake of water by the cell; the cell swells, and the cell components are diluted (**Fig. A2.6B**). If enough water enters, the cell is destroyed by **lysis**—a rupturing of the cell membrane and dispersal of cell contents. Both hypertonic and hypotonic environments can cause other problems for cells. Proteins have specific salt requirements, and intracellular environments that have salt concentrations that are either higher or lower than normal for a cell can cause denaturation of proteins, with potentially fatal results for the cell. Thus, transport of water across the cell membrane must be tightly controlled, and cells need mechanisms that allow them to live in environments that are not isotonic.

Most cells live in environments that are hypotonic. To deal with this challenge, most prokaryotes and many eukaryotes have a cell wall external to the cell membrane. As water enters by osmosis and pushes against the cell wall (turgor pressure), the wall resists the tension and pushes back with an equal but opposite force known as wall pressure. The wall pressure is an inward pressure exerted by the cell wall against the cell membrane (**Fig. A2.7**). When turgor pressure and wall pressure are equal in magnitude, the cell is at equilibrium with respect to

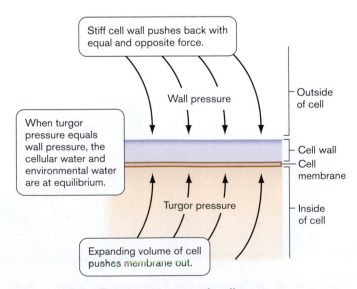

Figure A2.7 **Turgor pressure and wall pressure.**

water movement. Organisms that lack a cell wall employ other strategies to deal with hypotonic environments. For example, some freshwater protists that lack a cell wall expel excess water through a contractile vacuole.

In contrast to freshwater microbes, ocean-dwelling organisms face the problem of water loss. Many of these organisms accumulate solutes known as osmolytes to ensure that they are isotonic to the external environment, thus preventing water loss.

Eukaryotes Can Move Substances across the Cell Membrane by Endocytosis and Exocytosis

All cells use diffusion and transport proteins to move molecules into or out of the cell, but some eukaryotes can, in addition, use endocytosis and exocytosis to achieve this end. In **endocytosis**, parts of the cell membrane bud into the cytoplasm and eventually separate from it to form **endosomes** (**Fig. A2.8**). Endosomes are a type of **vesicle**, a small membranous sphere found within a cell. The interior of these endosomes contains extracellular material. **Phagocytosis** (cell eating) is a form of endocytosis in which large extracellular particles are brought into the cell. **Pinocytosis** (cell drinking) is endocytosis of the extracellular fluid. Endocytosis is a controlled, energy-requiring process that relies on many proteins, including cytoskeletal proteins. **Exocytosis** is the reverse of endocytosis. In exocytosis, intracellular vesicles fuse with the cell membrane, and the contents of the vesicles are released to the extracellular environment. Cells can use exocytosis to release wastes.

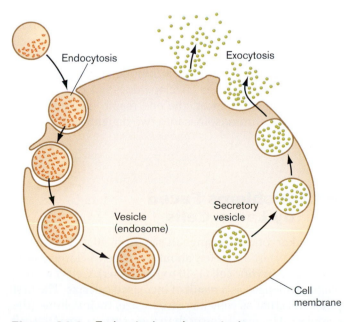

Figure A2.8 Endocytosis and exocytosis.

A2.2 The Nucleus and Mitosis

All cells need to synthesize proteins, and all cells contain DNA, the genetic material that encodes the information needed to specify protein primary structure. The central dogma of molecular biology holds that DNA is transcribed into messenger RNA (mRNA), and mRNA is translated into protein on ribosomes. Although this central dogma of molecular biology holds for all cells, details in the structure of DNA and ribosomes vary among the three domains of life (see **Table A2.1**). For example, ribosomes always function in protein synthesis, but ribosomes from organisms in different domains differ in their sedimentation rates and their sensitivity to various antibiotics. DNA always encodes the information needed for protein synthesis, but whereas bacterial chromosomes are usually circular, eukaryotic chromosomes are typically linear. Another difference between DNA in prokaryotes and DNA in eukaryotes is the location of DNA within the cell. Prokaryotic DNA is found in an area of the cytoplasm known as the nucleoid, while eukaryotic DNA is enclosed by a membrane-bound nucleus.

Eukaryotic DNA Is Housed in the Membrane-bound Nucleus

Eukaryotes derive their name (*eukaryote* means "true kernel") from the fact that they possess a **nucleus**, and, indeed, the nucleus is often the most prominent feature of eukaryotic cells viewed under a microscope (**Fig. A2.9A**). The nucleus is an **organelle**, an intracellular membrane-bound compartment with a specific function. The nucleus contains **chromatin**, a complex of DNA and proteins. The nuclear membrane (envelope) consists of two concentric phospholipid membranes. The outer nuclear membrane is continuous with the membrane of the endoplasmic reticulum (ER), and the space between the two nuclear membranes is continuous with the lumen (inside) of the ER (**Fig. A2.9B**). Nuclei contain a region called the **nucleolus**, where ribosome assembly begins. At the nucleolus, multiple rRNA (ribosomal RNA) genes are transcribed, and the resulting rRNA combines with ribosomal proteins imported into the nucleus from the cytoplasm to form the ribosomal subunits. The ribosomal subunits then need to exit the nucleus.

The nuclear membrane contains nuclear pore complexes (NPCs) that allow for transport of material into and out of the nucleus. Metabolites and small proteins can diffuse through the NPCs, but larger proteins and organelles cannot enter by diffusion. Large proteins that need to enter the nucleus are actively transported in through the NPCs. These selectively imported proteins contain a nuclear localization signal, a sequence of amino acids that acts like a zip code to direct them into the nucleus. In addition to their role in protein import, NPCs also function in exporting mRNAs out of the nucleus.

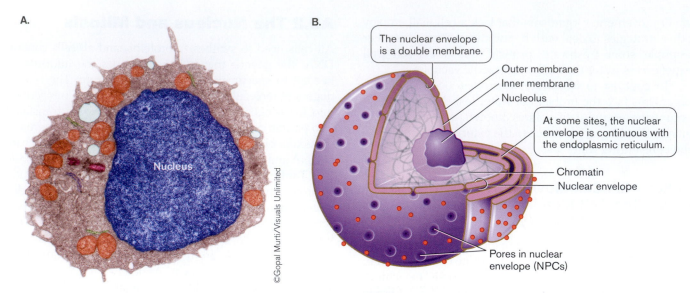

A.

Nucleus

©Gopal Murti/Visuals Unlimited

B.

The nuclear envelope is a double membrane.

Outer membrane
Inner membrane
Nucleolus

At some sites, the nuclear envelope is continuous with the endoplasmic reticulum.

Chromatin
Nuclear envelope

Pores in nuclear envelope (NPCs)

Figure A2.9   The nucleus.   A. An electron micrograph of a eukaryotic yeast cell showing the prominent nucleus. **B.** Diagram of a nucleus.

Eukaryotic Cells Replicate by Mitosis

Cells need to ensure an accurate replication and division of their DNA. Prokaryotes replicate by fission, as described in Chapter 3. Eukaryotic cells replicate their nuclear DNA and divide by mitosis. **Mitosis** is a series of steps that segregates duplicated chromosomes and ensures that each daughter cell receives a copy of the genetic material. When the cell is not undergoing mitosis, it is in interphase (**Fig. A2.10A**). During interphase, the individual chromosomes are long and thin and not visible by a light microscope. Interphase can be divided into three phases: G_1, S, and G_2. Cells that are not committed to dividing are in G_1, the first gap phase. If cell division is to occur, then the chromosomes are replicated during S phase (S for "synthesis"). The duplicated chromosomes, called sister chromatids, remain attached to each other at the centromere (**Fig. A2.10B**). After chromosome replication, a second gap phase, G_2, occurs, after which mitosis can proceed.

Mitosis is divided into a number of steps: prophase, metaphase, anaphase, and telophase (**Fig. A2.10C**). During *prophase*, the chromosomes condense and become visible by light microscopy. The nuclear membrane may break down. The mitotic spindle, responsible for separating the sister chromatids to opposite poles of the cell, begins to form. The mitotic spindle is a network of microtubules (see Section A2.5) that originate from centrosomes. There are two centrosomes, and these migrate to opposite sides of the cell. Each centrosome contains two centrioles, and each centriole contains nine sets of microtubules in a radial formation (see Section A2.5). The free ends of the microtubules establish connections with the sister chromatids at a structure called the kinetochore.

At *metaphase*, the spindle apparatus is complete and each sister chromatid is connected to a microtubule. The chromosomes are arranged along an imaginary plane in the middle of the cell. In the next phase of mitosis, *anaphase,* the microtubules shorten and pull the sister chromatids apart, separating the replicated chromosomes. At the end of anaphase, each set of chromosomes is located on opposite sides of the cell. During, *telophase*, the final step in mitosis, the nuclear membrane re-forms around the chromosomes and the chromosomes become long and thin again. Cytokinesis also occurs during telophase and partitions the original cell into two daughter cells by the formation of a cell membrane between them.

The preceding description of mitosis forms a synopsis of this process as it occurs in eukaryotic cells from multicellular organisms, where it has been particularly well studied. However, some unicellular eukaryotes exhibit differences from the canonical mitotic phases. In yeast, for example, the nuclear envelope does not break down. Despite these differences, the end result, the separation of duplicated chromosomes into two daughter cells, is the same.

A2.3 Problems Faced by Large Cells

Most eukaryotic cells range in size from 10 to 100 micrometers (μm) in diameter, about ten times as large as the typical prokaryotic cell (1–10 μm diameter). Cells face two major challenges as a result of increased cell size. The first problem is that as cells increase in size, their volume (the cytoplasm) increases faster than their surface area (the cell

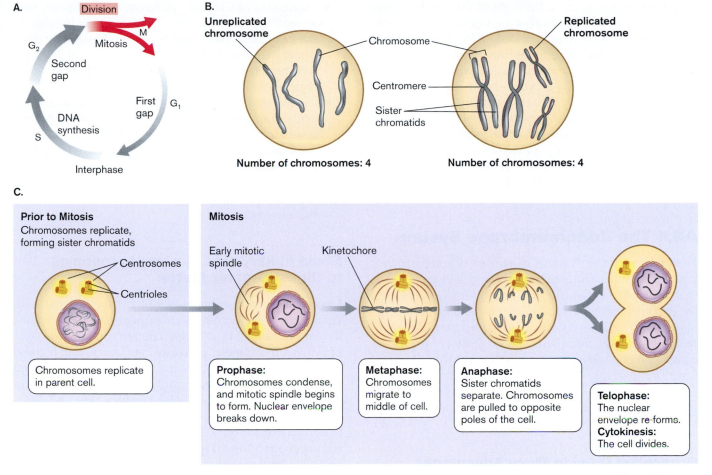

Figure A2.10   The cell cycle and mitosis.   A. Stages of the eukaryotic cell cycle. G_1, S, and G_2 make up interphase (in gray). **B.** Duplication of the chromosomes during S phase. **C.** The phases of mitosis. See text for details.

membrane) (**Fig. A2.11**). Cells are filled with metabolically active cytoplasm that requires nutrients and energy and produces wastes. Energy production, nutrient import, and waste disposal are events that take place at the cell membrane. As cells increase in size, the cell membrane area may not be able to keep up with the demands placed on it by a proportionally larger cytoplasm. The eukaryotic cell's answer to this problem is the endomembrane system, an extensive network of internal membranes that effectively increases the membrane surface area without an increase in cell volume.

The second problem associated with an increase in size is related to diffusion. The amount of time it takes a molecule to diffuse a given distance is proportional to the distance squared. For example, if it takes a particular molecule 1 second to diffuse 1 μm, then it takes that same molecule 100 seconds to diffuse 10 μm. Many biochemical reactions depend on the partners in the reaction finding each other by diffusion (that is, their rate is diffusion controlled), so longer diffusion times result in slower reactions. Furthermore, signals received at the cell membrane

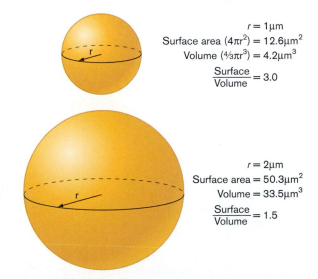

$r = 1\mu m$
Surface area ($4\pi r^2$) $= 12.6\mu m^2$
Volume ($\frac{4}{3}\pi r^3$) $= 4.2\mu m^3$
$\dfrac{\text{Surface}}{\text{Volume}} = 3.0$

$r = 2\mu m$
Surface area $= 50.3\mu m^2$
Volume $= 33.5\mu m^3$
$\dfrac{\text{Surface}}{\text{Volume}} = 1.5$

Figure A2.11   Cell volume increases faster than surface area. Because the surface area increases by the radius squared and the volume increases by the radius cubed, the volume increases faster than the surface area.

need to be communicated throughout the cell. In a large cell, the amount of time it takes for a molecule to traverse the cell by diffusion may be too slow for the cell to rapidly adjust to signals it receives from the environment or from other sites within the cell. To deal with this problem, eukaryotic cells possess a cytoskeleton, a group of proteins that, among other things, maintains cell shape and moves molecules around the cell, relieving the cell from the need to rely on diffusion for transport.

As we shall see in the next two sections, the endomembrane system and the cytoskeleton have other advantages for the eukaryotic cell.

A2.4 The Endomembrane System

The endomembrane system is a series of compartments found inside eukaryotic cells and bounded by membranes that are separate from the cell membrane. Organelles of the endomembrane system include the endoplasmic reticulum (ER), the Golgi complex (also called the Golgi apparatus), lysosomes, and peroxisomes. Different organelles contain unique subsets of proteins that contribute to their function. To a large extent, the function of the ER and the Golgi complex is to direct to their proper cellular location proteins destined for lysosomes, for the cell membrane, or for secretion from the cell.

Organelles Provide Many Advantages to the Eukaryotic Cell

- Endomembranes increase the membrane surface area without increasing the cell volume.
- Separating cellular contents in small, enclosed compartments increases the concentrations of enzymes and their substrates, allowing reactions to proceed faster (**Fig. A2.12**).

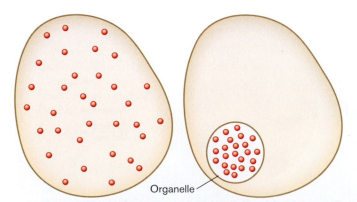

Figure A2.12   Advantages of organelles. Solutes (red dots) are concentrated in organelles, reactive molecules are separated from the cytoplasm, and membrane surface area is increased without an increase in cell volume.

- Organelles can provide different environments that allow disparate reactions to occur simultaneously. For example, proteins are being synthesized by ribosomes in the neutral pH cytoplasm, while at the same time proteins are being hydrolyzed within the acidic organelles known as lysosomes.
- Compartmentalization protects cytoplasmic components from harmful substances. For example, hydrogen peroxide (H_2O_2), a product of cellular oxidation reactions, is produced and converted to water within **peroxisomes**. Localizing the reaction within the peroxisome keeps the toxic peroxide away from other cell components, such as proteins and DNA, that are sensitive to oxidative stress.

Some Eukaryotic Cells Use Lysosomes to Digest Organic Matter

Lysosomes are membrane-bound organelles that help eukaryotic cells obtain nourishment from macromolecular nutrients. Lysosomes contain many hydrolytic enzymes (for example, proteases, nucleases, and lipases) and have an acidic pH of around 5. Lysosomes are formed when vesicles containing hydrolytic enzymes and proton pumps bud off from the Golgi complex (**Fig. A2.13A**). Lysosomes then fuse with vesicles containing the material to be digested. Often this material comes from outside the cell via phagocytosis (**Figs. A2.13A** and **B**). Phagocytosis and lysosomal digestion help the eukaryotic cell because they effectively increase the membrane surface area over which nutrients can be absorbed. Bacteria lack these processes; to obtain nutrition from large molecules in their environment, bacteria must secrete digestive enzymes. The extracellularly digested materials are subsequently transported across the bacterial cell membrane through specific transporters (see Section 4.2). In contrast, lysosomes allow for intracellular digestion, and digested material crosses the lysosomal membrane into the cytoplasm. Any waste products left in the lysosome can leave the cell via exocytosis.

The Area Inside the Endoplasmic Reticulum Is Separate from the Cytoplasm

The endoplasmic reticulum is continuous with the outer nuclear membrane, and the lumen (interior) of the ER is continuous with the space between the two nuclear membranes (**Fig. A2.14A**). As well as being continuous with the space between the nuclear membranes, the lumen of the ER is spatially equivalent to the interior spaces of other endomembrane components and to the outside of the cell. This means that material in the ER does not need to cross a membrane to enter these other spaces, and

A.

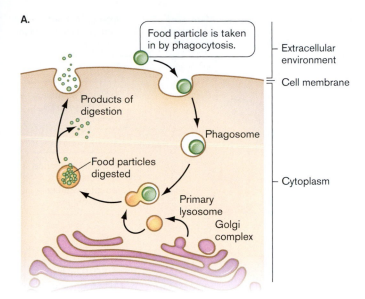

Food particle is taken in by phagocytosis.

Extracellular environment

Cell membrane

Products of digestion

Phagosome

Food particles digested

Cytoplasm

Primary lysosome

Golgi complex

B.

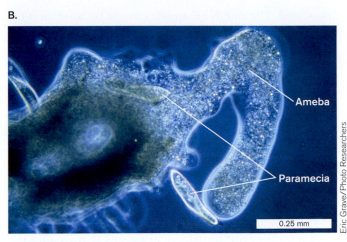

Ameba

Paramecia

0.25 mm

Eric Grave/Photo Researchers

Figure A2.13 **Lysosomes.** **A.** Lysosomes contain hydrolytic enzymes to digest material brought into the cell by phagocytosis. **B.** Amobas engulfing paramecia (light microscopy). The paramecia will be digested with the aid of lysosomes.

mixing can occur via vesicle fusion. For example, material contained within the lumen of the ER can mix with the contents of the Golgi complex or with the extracellular milieu by fusion of vesicles from the ER with Golgi membrane or the plasma membrane. These topologically equivalent areas, indicated by a common color in **Figure A2.14A**, are completely separated from the cytoplasm by endomembranes, so that the ER can be used to sequester substances that must be held at low concentrations in the cytoplasm, for example, calcium ions.

Smooth ER and Rough ER Have Different Structures and Functions

There are two morphologically and functionally distinct types of endoplasmic reticulum—smooth ER and rough ER, as shown in the micrograph of **Figure A2.14C**. The smooth ER is the site of lipid synthesis and some detoxification of noxious compounds. The rough ER is the site where transmembrane proteins, secreted proteins, and resident proteins of the ER, Golgi, or lysosomes are

A.

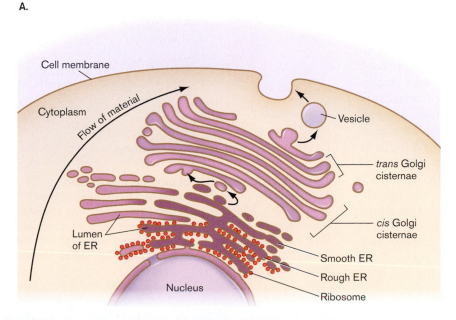

Cell membrane

Cytoplasm

Flow of material

Vesicle

trans Golgi cisternae

cis Golgi cisternae

Lumen of ER

Smooth ER

Rough ER

Nucleus

Ribosome

B. Golgi

SPL/Photo Researchers

C. Endoplasmic reticulum

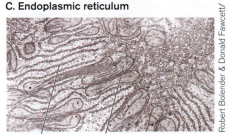

Rough ER   Smooth ER

Robert Bolender & Donald Fawcett/ Visuals Unlimited

Figure A2.14 **The endoplasmic reticulum and Golgi complex.** **A.** The relationship of the ER to other cellular membranes and the flow of material through vesicles from the rough ER to the cell membrane. **B.** Electron micrograph of Golgi cisternae. **C.** Electron micrograph showing rough and smooth ER.

translated. The rough ER appears rough because its cytoplasmic surface is studded with ribosomes. These ribosomes are located on the rough ER because the protein being synthesized by the ribosome has a signal sequence on its amino terminus (the first part of the protein translated from the mRNA). A **signal sequence** is a specific sequence of amino acids that directs proteins to a specific cellular location. The signal sequence that directs proteins to the ER recognizes a receptor (the **signal recognition particle**) on the rough ER membranes (**Fig. A2.15A**), so the protein-ribosome complex is directed to the surface of the ER membrane. Note that the ribosomes attached to the rough ER are identical to cytoplasmic ribosomes and only attach to the ER transiently because of the type of protein they are translating, proteins that contain the correct signal sequence.

The Translation of Proteins across Rough Endoplasmic Reticulum

After the ribosome docks with the signal recognition particle, the protein is threaded through the ER membrane as it is translated. Secreted proteins and proteins destined for the lumen of an organelle are threaded completely through the ER membrane and end up in the lumen of the ER (see **Fig. A2.15A**). In contrast, transmembrane proteins are not threaded completely through, and part of the protein spans the membrane (**Fig. A2.15B**). The membrane-spanning regions of transmembrane proteins usually contain a continuous stretch of about 20 hydrophobic amino acids that form an alpha helix with the hydrophobic side chains facing out toward the hydrophobic hydrocarbons of the membrane.

As the nascent polypeptide chains are threaded across the ER membrane, the unfolded proteins are bound by chaperonins. Chaperonins (also known as heat-shock proteins) prevent partially folded proteins from clumping together and help proteins attain their correct tertiary structure. In the ER, proteins may be modified. The signal sequence that directed the proteins to the ER is usually cleaved off. The environment inside the ER allows disulfide bonds to form between cysteine residues of some proteins. Other ER proteins have oligosaccharide groups covalently attached, a posttranslational modification known as glycosylation. The enzymes that attach the sugars are found only in the ER lumen. In transmembrane proteins of the cell membrane, sugars always face the extracellular environment because the extracellular environment is topologically equivalent to the lumen of the ER (see **Fig. A2.14A**).

Resident ER proteins, those that will stay and function in the ER, are retained inside the ER because they contain a sequence of amino acids that acts as an ER retention signal.

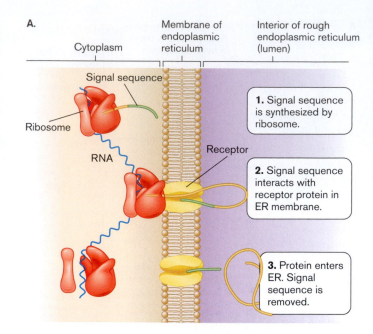

A.

Cytoplasm | Membrane of endoplasmic reticulum | Interior of rough endoplasmic reticulum (lumen)

Signal sequence

Ribosome

RNA

Receptor

1. Signal sequence is synthesized by ribosome.

2. Signal sequence interacts with receptor protein in ER membrane.

3. Protein enters ER. Signal sequence is removed.

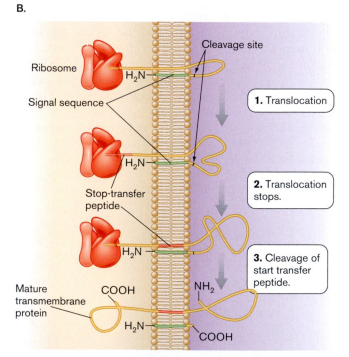

B.

Ribosome

Signal sequence

Stop-transfer peptide

Mature transmembrane protein

Cleavage site

H_2N

H_2N

H_2N

COOH

NH_2

H_2N

COOH

1. Translocation

2. Translocation stops.

3. Cleavage of start transfer peptide.

Figure A2.15   Insertion of proteins into the ER.
A. Proteins are targeted to the ER by a signal sequence on the growing polypeptide chain. Secreted proteins and proteins destined for the lumen of an organelle end up in the lumen of the ER. **B.** Transmembrane proteins are not completely inserted through the ER membrane, and a portion of the protein remains within the membrane.

The Golgi Complex Directs the Transport of Proteins

Proteins not retained in the ER pass on to the Golgi complex by way of vesicles. The **Golgi complex** consists of separate membrane stacks (cisternae) that each contain unique enzymes. The *cis* face of the Golgi complex is nearest to the ER, and the *trans* face is farthest from the ER, closest to the cell membrane (see **Figs. A2.14A** and **B**). As proteins pass through the cisternae, the carbohydrates on them may be trimmed and modified. These modified carbohydrates can serve as address tags to target proteins to particular organelles. For example, proteins tagged with mannose-6-phosphate (mannose phosphorylated on its number 6 carbon) are selectively sent to lysosomes; that is, vesicles enriched in proteins containing mannose-6-phosphate bud off from the Golgi and are directed to lysosomes. Proteins not targeted to lysosomes or marked for retention in the Golgi complex, may be sent to the cell membrane. Vesicles leaving the Golgi complex may fuse with the cell membrane, releasing their contents to the extracellular environment (see **Fig. A2.14A**). Transmembrane proteins in these vesicles can then become part of the cell membrane. Regions of transmembrane proteins that were inside vesicles will face the extracellular environment, and cytoplasmic portions will remain cytoplasmic.

A2.5 The Cytoskeleton

Eukaryotic cells contain proteins called intermediate filaments, microfilaments, and microtubules that are collectively termed the **cytoskeleton**. As their name implies, these proteins serve as a cell skeleton and impart specific shapes to eukaryotic cells. However, cytoskeletal proteins are multifunctional and are also involved in whole-cell movements and movements of substances within the cell.

Intermediate Filaments Are Formed of Various Proteins

Intermediate filaments (**Fig. A2.16A**) consist of various fibrous proteins that have a diameter of about 10 nm. Intermediate filaments often form a meshwork under the cell membrane and, in cells that lack a cell wall, help impart and maintain cell shape. Intermediate filaments also strengthen the cell by resisting tension placed on the cell membrane. The proteins that make up intermediate filaments vary with cell type. Intermediate filaments are fairly stable and are not thought to undergo acute changes in length the way microfilaments and microtubules do.

Microfilaments Are Polymers of the Protein Actin

Microfilaments, also known as actin filaments, have a diameter of 7 nm (**Fig. A2.16B**). They are formed when individual actin monomers (globular actin, or G-actin) polymerize, in a process fueled by ATP hydrolysis, to form chains of filamentous actin (F-actin). Two F-actin chains twist around each other to form microfilaments that have a plus end and a minus end. Microfilaments are dynamic structures, growing and shrinking in a controlled manner. New monomer units add to the plus end and dissociate from the minus end. Whether actin will polymerize

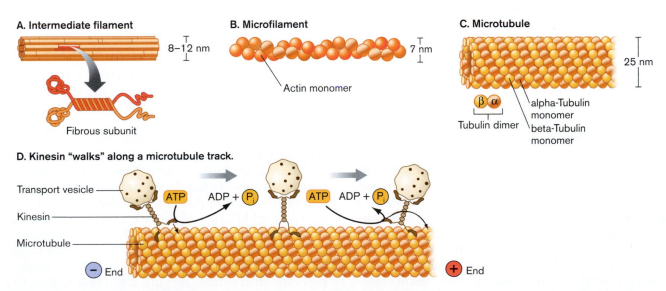

A. Intermediate filament   8–12 nm

B. Microfilament   7 nm   Actin monomer

C. Microtubule   25 nm

Fibrous subunit

β α   Tubulin dimer   alpha-Tubulin monomer   beta-Tubulin monomer

D. Kinesin "walks" along a microtubule track.

Transport vesicle   ATP   ADP + P$_i$ ATP   ADP + P$_i$

Kinesin

Microtubule

– End   + End

Figure A2.16   Cytoskeletal proteins.   A. Intermediate filaments are ropelike assemblages of various proteins. **B.** Microfilaments consist of two strands of actin polymers twisted together. **C.** Microtubules are polymers of tubulin dimers. **D.** Fueled by the hydrolysis of ATP, the motor protein kinesin can move vesicles or organelles toward the plus end of microtubules.

or depolymerize depends on a number of factors, including the concentration of G-actin. The critical concentration is a measure of the ability of actin to polymerize. At G-actin concentrations below the critical concentration, F-actin will depolymerize, and at concentrations greater than the critical concentration, G-actin will polymerize. The plus end of microfilaments has a critical concentration less than that of the minus end, so actin is preferentially added on to the plus end and removed from the minus end.

Some microfilaments play a structural role in the cell and work in conjunction with intermediate filaments to maintain cell shape. These structural microfilaments have protein caps at both ends to prevent changes in microfilament length. Other microfilaments have functions that require dynamic changes in length. For example, the pseudopod movement of an ameba depends on the polymerization of actin at the leading edge of growth. The plus end of the microfilament is located underneath the cell membrane of the extending pseudopod, and polymerization is enhanced by the actin-binding protein profilin. Microfilaments are also responsible for cytoplasmic streaming, a mixing of the cytoplasm that aids diffusion. The protein myosin works with microfilaments to generate the forces needed for cell streaming, pseudopod formation, and cytokinesis, the separation of daughter cells after nuclear division.

Microtubules Are Polymers of Tubulin

Microtubules have a larger diameter (25 nm) than microfilaments and intermediate filaments (**Fig. A2.16C**). The hollow microtubule structure consists of 13 tubulin dimers; one alpha-tubulin protein plus one beta-tubulin protein forms one tubulin dimer. Like microfilaments, microtubules have plus (faster-growing) and minus (slower-growing) ends and are dynamic structures that can polymerize and depolymerize. Polymerization is an energy-requiring process, and the necessary energy is obtained by coupling polymerization to GTP hydrolysis. Microtubules aid movement of substances within the cell and are also involved in powering whole-cell movement by cilia and flagella.

Traffic of proteins through the endomembrane system (see Section A2.4) relies on the controlled movement of vesicles from one cellular compartment to the next. Microtubules provide tracks that can move vesicles from one organelle to the next in an efficient, directed fashion. Working with microtubules to accomplish this are motor proteins. Motor proteins such as kinesin and dynein can capture cargo (for example, vesicles or organelles) and walk them along microtubule tracks in an ATP-dependent process (**Fig. A2.16D**). Kinesin moves cargo toward the plus end of microtubules, while dynein moves

cargo toward the minus end. In addition to moving vesicles, microtubules are responsible for the segregation of duplicated chromosomes during mitosis.

Eukaryotic Cilia and Flagella

Cilia and flagella are thin extensions of the cell membrane that can move in a whiplike fashion, driven by interactions between microtubules and the motor protein dynein. Flagella (singular, flagellum) are relatively long, and cells usually have only one or two of them; cilia (singular, cilium) are shorter and more numerous. Both flagella and cilia can move cells through space. Cilia may also aid in food capture—for example, by sweeping extracellular fluid into the gullet of a paramecium. Eukaryotic cilia and flagella differ from bacterial cilia and flagella; bacterial flagella depend on the proton motive force to rotate a motor that causes flagellar movement, whereas eukaryotic flagella rely on ATP hydrolysis by dynein to move the flagella in a whiplike fashion.

The study of protists has played a key role in determining the structure and function of flagella and cilia. Much information about flagella structure has come from studies of a *Chlamydomonas* species, a unicellular alga that has two long flagella (**Fig. A2.17A**). Dynein was first discovered in the cilia of the unicellular eukaryote *Tetrahymena* (**Fig. A2.17B**). Flagella and cilia have the same structure and mechanism of action, so for convenience we will restrict the following discussion to flagella.

A cross section through a flagellum reveals a central bundle of microtubules called the axoneme (**Fig. A2.17C**). The axoneme originates at a microtubule-organizing center called the basal body, which is similar to the centriole (see Section A2.2). The minus ends of the microtubules are at the basal body; the plus ends at the tip of the flagellum are capped to prevent changes in length.

As shown in the electron micrograph in **Figure A2.17D,** the axoneme has a characteristic arrangement of two central microtubules and nine microtubule doublets around the periphery. The microtubule doublets are connected to each other and to the central microtubules through protein cross-links called bridges and spokes. (**Fig. A2.17E**). The motor protein dynein connects adjacent microtubule doublets. If the cell membrane is removed from the surface of a flagellum and the protein cross-links (but not dynein) are dissolved, the microtubule doublets are observed to lengthen in the presence of ATP. This is due to the action of dynein, which slides the microtubule doublets past each other (**Fig. A2.17F**). In intact flagella, the tension imparted as dynein tries to slide microtubules past each other is translated into a bending motion. Controlled cycles of dynein activation and inactivation on opposite sides of the axoneme lead to the whiplike motion of flagella.

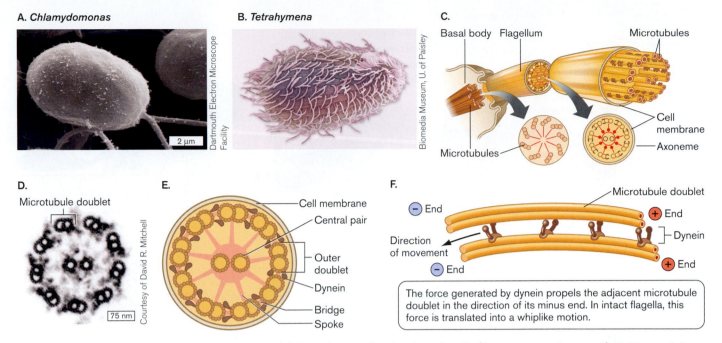

Figure A2.17   Flagella and cilia.   A. The protist *Chlamydomonas* has two long flagella (fluorescence micrograph). **B.** The protist *Tetrahymena* has numerous cilia (scanning electron micrograph). **C.** Structure of flagella and cilia. **D.** Cross section of axoneme (TEM). **E.** Cross section of axoneme indicating bridges and spokes formed by cross-linking proteins. **F.** Force generated by dynein on a microtubule doublet.

A2.6 Mitochondria and Chloroplasts

Mitochondria and **chloroplasts** are organelles involved in cellular energy production. Mitochondria perform oxidative respiration and are found in nearly all eukaryotes. Chloroplasts perform photosynthesis and are found only in photosynthetic eukaryotes, such as green algae. Both organelles are thought to have become part of eukaryotic cells through a process of endosymbiosis. The **endosymbiosis theory** states that mitochondria and chloroplasts were once free-living prokaryotes that became ingested, but not digested, by a larger (possibly eukaryotic) cell (**Fig. A2.18A**). A symbiotic relationship developed, with the larger eukaryotic cell providing protection to the prokaryote and the prokaryote providing energy to the eukaryote. As we shall see, the structure of mitochondria and chloroplasts strongly supports the endosymbiosis theory.

Mitochondria Produce ATP by Oxidative Respiration

Mitochondria are the powerhouses of the eukaryotic cell. A cell may contain tens or hundreds of mitochondria, depending on its energy needs. Mitochondria have two membranes, an outer membrane and an inner membrane (**Fig. A2.18B**). The inner membrane has numerous infoldings called cristae that increase its surface area. It is thought that as a result of the endocytosis of the prokaryote, the inner mitochondrial membrane is derived from the prokaryote and the outer membrane is derived from the larger eukaryote. Supporting this idea is the fact that the inner membrane has structural characteristics of a prokaryotic cell membrane, while the outer membrane is similar to the host eukaryotic membrane. For example, 20% of the phospholipids in the inner membrane are cardiolipin, a phospholipid largely absent from eukaryotic membranes.

Mitochondria contain two distinct compartments: the intermembrane space between the two membranes and the matrix inside the inner membrane. Different stages of oxidative respiration occur in specific compartments. As predicted by the endosymbiosis theory, the topology of these processes is similar in prokaryotes and mitochondria (**Table A2.2**). For example, in prokaryotes, the citric acid cycle occurs in the cytoplasm; in mitochondria, the citric acid cycle takes place inside the matrix, the metabolic equivalent of the prokaryotic cytoplasm.

The fact that mitochondria contain their own DNA and ribosomes lends further support to the endosymbiosis theory, since these cell features would be a necessary part of any free-living organism. Both the DNA and ribosomes of mitochondria show similarities with the DNA and ribosomes of prokaryotes. For example, like prokaryotic DNA, mitochondrial DNA is circular, and mitochondrial ribosomes are sensitive to antibiotics that disrupt

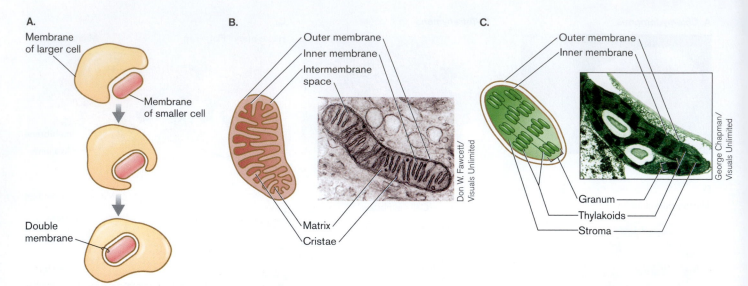

Figure A2.18 **Endosymbiosis theory, mitochondria, and chloroplasts.** **A.** Origin of organelles according to endosymbiosis theory. **B.** Structures in a mitochondrion. **C.** Chloroplast structure.

Table A2.2	**Location of oxidative respiration components in prokaryotes and mitochondria.**	
Feature	Prokaryotes	Mitochondria
Electron transport chain (ETC)	Cell membrane	Inner membrane
ATP synthase	Cell membrane	Inner membrane
Citric acid cycle	Cytoplasm	Matrix
Proton motive force	Protons diffuse from lower-pH extracellular environment into the cytoplasm across the cell membrane.	Protons diffuse from the lower-pH intermembrane space into the matrix across the inner membrane.

prokaryotic ribosomes. Although nuclear DNA encodes some mitochrondrial proteins, phylogenetic analysis has shown that these nuclear genes are related to prokaryotic genes. Furthermore, the generation of new mitochondria is not tied to the replication of the cell, and the division of mitochondria within the cell is similar to the fission seen in prokaryotic cells.

Chloroplasts Allow Eukaryotes to Perform Photosynthesis

Chloroplasts are organelles found only in photosynthetic eukaryotes. In the light reactions of photosynthesis, chloroplasts convert light energy from the sun to ATP and reduced NADPH. In the subsequent light-independent reactions, the ATP and NADPH are used to reduce CO_2 to sugar.

Like mitochondria, chloroplasts are probably the result of endosymbiosis. Bacteria related to cyanobacteria are thought to be the prokaryotic partners that gave rise to chloroplasts. Like eukaryotic chloroplasts, cyanobacteria contain chlorophylls a and b and perform aerobic photosynthesis. In addition, cyanobacteria contain extensive internal membranes called thylakoids that contain chlorophyll and participate in the light reactions of photosynthesis.

Chloroplasts have three membranes whose topology can be understood in light of the endosymbiosis theory (**Fig. A2.18C**). The outer membrane appears to be derived from the host eukaryotic cell; the inner membrane is equivalent to the bacterial cell membrane; and the thylakoid membrane is derived from the prokaryotic thylakoid membranes. The region inside the inner membrane is called the stroma and is equivalent to the bacterial cytoplasm. The thylakoid membrane is packed with the chlorophyll pigments that give chloroplasts their green color. ATP and NADPH are produced in the stroma and used there in the light-independent reactions. As would be expected as a result of endosymbiosis, chloroplasts, like mitochondria, contain their own circular DNA and their own ribosomes.

Answers to Thought Questions

Chapter 1

1.1 The minimum size of known microbial cells is about 0.2 μm. Could even smaller cells be discovered? What factors may determine the minimum size of a cell?

ANSWER: The smallest cells known, about 0.2 μm in length, are cell wall-less bacteria called mycoplasmas; for example, *Mycoplasma pneumoniae*, a causative agent of pneumonia. Bacteria might be discovered that are smaller than 0.2 μm, but it is hard to see how their cell components such as ribosomes (about a tenth this size) could fit inside such a small cell. The volume required for DNA and the apparatus of transcription and translation probably sets the lower limit on cell size.

1.2 If viruses are not functional cells, are they truly "alive"?

ANSWER: A traditional definition of a life form includes an entity with the capability for metabolism and homeostasis (maintaining internal conditions of its cytoplasm) as well as reproduction and response to its environment. Viruses reproduce themselves indefinitely, and respond to the environment of the host cell, but lack metabolism or homeostasis outside their host cell. Nevertheless, viruses such as herpesviruses contain numerous metabolic enzymes that participate in the metabolism of their host. Certain large viruses such as the mimivirus appear to have evolved from cells. Some microbiologists argue that viruses should be considered "alive" if reproduction is the main criterion, and if the viral "environment" is considered the inside of the host cell.

1.3 Why do you think it took so long for humans to connect microbes with infectious disease?

ANSWER: For most of human history we were unaware of the existence of microbes. Even after microscopy had revealed their existence, the incredible diversity of the microbial world and the difficulties in isolating and characterizing microbial organisms made it difficult to discern the specific effects of microbes. All healthy people contain microbes; and most disease-causing microbes are indistinguishable from normal flora by light microscopy. Not all microbial diseases are directly transmittable from human to human; they may require complex cycles with intermediate hosts, such as the fleas and rats that carry bubonic plague.

1.4 How could you use Koch's postulates to demonstrate the causative agent of influenza? What problems would you need to overcome that were not encountered with anthrax?

ANSWER: In order to use Koch's postulates to demonstrate the causative agent of influenza, an animal model host would be needed. Secretions from diseased patients could be applied to different animal species, such as monkeys and mice, in order to find an animal that shows signs of the disease. To determine the causative agent of disease, the patient's secretions could be filtered in order to separate bacteria and viruses. Only the filtrate would cause disease, as it contains viruses (relevant to Koch's postulates 1 and 3). Viruses, however, are more difficult to isolate in pure culture than bacteria (postulate 2), a problem Koch did not address. Furthermore, some viruses, such as HIV (human immunodeficiency virus) have no animal model; they grow only in human cells. Today, viruses are usually isolated in a tissue culture. Once isolated, the virus could be used to inoculate a new host animal (if an animal model exists) or a tissue culture, and determine whether infection results (postulates 3 and 4). Another problem Koch did not address was the detection of infectious agents too small to be observed under a microscope. Today, antibody reactions are used to determine whether an individual has been exposed to a putative pathogen. An antibody test could be used to determine whether healthy and diseased individuals had been exposed to the isolated virus.

1.5 Why do you think some pathogens generate immunity readily, whereas others evade the immune system?

ANSWER: Some pathogens (microbes that cause disease) have external coat proteins that strongly stimulate the immune system and induce production of strong antibodies. Other pathogens have evolved to avoid the immune system by changing the identity of their external proteins. Immunity also varies greatly with the host's status. The very young and very old generally have weaker immune systems than people in the prime of their lives. Some pathogens, such as HIV, will directly attack the immune system, limiting the immune response to the pathogen.

1.6   How do you think microbes protect themselves from the antibiotics they produce?

ANSWER: Microbes protect themselves from the antibiotics they produce by producing their own resistance factors. As discussed in later chapters, microbes may synthesize pumps to pump the antibiotics out; or they make altered versions of the target macromolecule, such as the ribosome subunit; or they make enzymes to cleave the antimicrobial substance.

1.7   Why don't all living organisms fix their own nitrogen?

ANSWER: Nitrogen fixation requires a tremendous amount of energy, about thirty molecules of ATP per dinitrogen molecule converted to ammonia (discussed in Chapter 15). In a community containing adequate nitrogen sources, organisms that lose the nitrogen fixation pathway would make more efficient use of their energy reserves than those that spend energy to fix nitrogen from the atmosphere. Another consideration is that nitrogenase is an oxygen-sensitive enzyme, whereas plants, animals, and fungi are aerobes. In order to fix nitrogen, aerobic organisms need to develop complex mechanisms to exclude oxygen from nitrogenase.

1.8   What arguments support the classification of Archaea as a third domain of life? What arguments support the classification of archaea and bacteria together, as prokaryotes, distinct from eukaryotes?

ANSWER: The sequence of 16S rRNA (small-subunit rRNA) and other fundamental genes differs as much between archaea and bacteria as it does between archaea and eukaryotes. The composition of archaeal cell walls and phospholipids is completely distinct from those of bacteria and eukaryotes. Some aspects of gene expression, such as the RNA polymerase complex, are more similar between archaea and eukaryotes than between archaea and bacteria. On the other hand, archaeal and bacterial cells are prokaryotic; they both lack nuclei and complex membranous organelles. Archaeal metabolism and lifestyles are more similar to those of bacteria than eukaryotes. Some archaea and bacteria sharing the same environment, such as high-temperature springs, have undergone horizontal transfer of genes encoding traits such as heat-stable membrane lipids.

1.9   Do you think engineered strains of bacteria should be patentable? What about sequenced genes or genomes?

ANSWER: Microbes—as well as multicellular organisms, such as transgenic mice—have been patented, and the patents have stood up in court. DNA sequence per se is not patentable, but specific plans for use of DNA sequence can include the sequence as part of the patent. The reason for granting these patents is to encourage medical research by companies that need to earn a profit. The disadvantage of patents is that they restrict information flow and under-cut competition. Furthermore, religious and philosophical arguments have been made that patenting live organisms cheapens life. Current laws attempt to reach a balance among these concerns.

Chapter 2

2.1   You have discovered a new kind of microbe, never observed before. What kind of questions about this microbe might be answered by light microscopy? What questions would be better addressed by electron microscopy?

ANSWER: Light microscopy could answer questions such as: What is the overall shape of this cell? Does it form individual cells or chains? Is the organism motile? Only light microscopy can visualize an organism alive. Electron microscopy can answer questions about internal and external subcellular structures. For example, does a bacterial cell possess external filamentous structures such as flagella or pili? If the dimensions of the unknown microbe are smaller than the lower limits of a light microscope's resolution, EM may be the only way to observe the organism. Viruses are often characterized by shape, and this shape is observed by electron microscopy.

2.2   Explain what happens to the refracted light wave as it emerges from a piece of glass of even thickness. How does its new speed and direction compare with its original (incident) speed and direction?

ANSWER: The part of the wave front that emerges first travels faster than the portion still in the glass, resulting in the wave front bending toward the surface of the glass. Ultimately, the wave travels in the same direction and with the same speed as before it entered the glass. The path of the emerging light ray is parallel to the path of the light ray entering the glass and is shifted over by an amount dependent on the thickness of the glass. This refraction will alter the path of the beam of light and decrease the amount reaching the lens of the microscope. Immersion oil has the same refractive index as glass and will limit the amount of light lost in this way.

2.3   Parabolic lenses are generally "biconvex," that is, curving outward on both sides. What would happen to parallel light rays that pass through a lens that is concave on both sides? Or a lens that is convex on one side and concave with equal curvature on the other?

ANSWER: When light passes through a lens that is concave (curving inward) on both sides, the light rays diverge within the lens material, then diverge again more steeply on their way out. If the lens is concave on one side and convex with equal curvature on the opposite side, then the light rays will diverge within the lens but will emerge parallel, although slightly farther apart than when they entered.

2.4   In theory, what angle θ would produce the highest resolution? What practical problem would you have in designing a lens to generate this light cone?

ANSWER: In theory, an angle theta (θ) of 90 degrees would produce the highest resolution. However, a 90-degree angle of theta generates a cone of 180 degrees, which would require the object to sit in the same position as the objective lens; in other words, to have a focal distance of zero. In practice, the cone of light needs to be somewhat less than 180 degrees, in order to allow room for the object and to avoid substantial aberrations (light-distorting properties) in the lens material.

2.5   Under starvation conditions, bacteria such as *Bacillus thuringiensis*, the biological insecticide, repackage their cytoplasm into spores, leaving behind an empty cell wall. Suppose, under a microscope, you observe what appears to be a hollow cell. How can you tell if the cell is indeed hollow or if it is simply out of focus?

ANSWER: You could tell whether the cell is out of focus or actually hollow by rotating the fine-focus knob to move the objective up and down while observing the specimen carefully. If the hollow shape appears to be the sharpest image possible, it is probably a hollow cell. If the hollow shape turns momentarily into a sharp, dark cell, it was probably out of focus before. You could also use a confocal microscope to visualize the center of the hollow cell.

2.6   Some early observers claimed that the rotary motions observed in bacterial flagella could not be distinguished from whiplike patterns, comparable to the motion of eukaryotic flagella. Can you imagine an experiment to distinguish the two and prove that the flagella rotate? *Hint:* Bacterial flagella can get "stuck" to the microscope slide or coverslip.

ANSWER: To prove that flagella rotate, you can "tether" the bacteria to the microscope slide by getting one of its flagella stuck to the slide. A simple way to tether bacteria is by using a slide coated with anti-flagellin antibody. When the flagellum is stuck to the slide, its motor continues to rotate; thus, the entire cell now rotates. The rotation of the cell body can easily be seen by video microscopy. If the flagella moved in a whiplike fashion, the tethered cell would move back and forth, not rotate.

2.7   What experiment could you devise to determine the actual order of events in DNA movement toward the pole during formation of an endospore?

ANSWER: One way to track the movement of DNA during sporulation would be to stain the DNA with a dye such as propidium iodide at various stages of sporulation. Another way to determine the order of events in sporulation could be to observe mutant strains of bacteria that contain defects at different points in the sporulation process. (Sporulation is discussed further in Chapter 4.)

2.8   Compare and contrast fluorescence microscopy with dark-field microscopy. What similar advantage do they provide, and how do they differ?

ANSWER: Both dark-field and fluorescence enable detection (but not resolution) of objects whose dimensions are below the wavelength of light. Dark-field technique is based on light scattering, which detects all small objects without discrimination. Fluorescence, however, provides a means to label specific parts of cells, such as cell membrane or DNA, or particular species of microbes, using fluorescent antibody tags.

2.9   An electron microscope can be focused at successive powers of magnification, as in a light microscope. At each level, the image rotates at an angle of several degrees. Based on the geometry of the electron beam, as shown in Figure 2.34, why do you think this rotation occurs?

ANSWER: The image rotates because the electron beam is not straight, as for photons, but travels in a spiral through the magnetic field lines. As magnification increases, the spiral expands, and it reaches the image plane at a slightly different angle than before.

2.10   What kind of research questions could you investigate using SEM? What questions would be answered using TEM?

ANSWER: SEM could be used to examine the surface of cells: Do the cells possess a smooth surface, or does their surface contain protein complexes or bulges that serve special functions? How do pathogens attach to the surface of cells? TEM can be used to determine the intracellular structure of attachment sites as well as of internal organelles. TEM can also visualize the shape of macromolecular complexes such as flagellar motors or ribosomes.

2.11   What kind of experiments could prove or disprove the interpretations of the images of "nanobacteria" in blood plasma?

ANSWER: Attempt to observe the proposed "nanobacteria" in the presence of a general antibacterial agent such as sodium azide. If the particles still appear despite the sodium azide, they cannot be alive and are probably inorganic in nature. Another approach is to use PCR amplification of 16S ribosomal RNA sequences to establish the existence of a novel isolate. So far, the PCR sequences obtained from "nanobacteria" in blood plasma have been identified as those of an environmental microbe that commonly contaminates PCR samples.

Chapter 3

3.1   Which molecules occur in the greatest number in a prokaryotic cell? The smallest number? Why does a cell contain 100 times as many lipid molecules as strands of RNA?

ANSWER: Inorganic ions occur in the largest number in a prokaryotic cell (250 million/cell). They are also the smallest in size. DNA molecules are found in the smallest number (one large molecule, branched during replication). A prokaryotic cell contains a hundred times as many lipid molecules as strands of RNA because lipids are small structural molecules, highly packed. RNA molecules are long macromolecules that are either packed into complexes (such as ribosomal RNA), or are temporary information carriers (messenger RNA), present only as needed to make proteins.

3.2   Why does the Svedberg coefficient of the intact ribosome (70S) differ from the sum of the individual subunits of the ribosome (30S and 50S)?

ANSWER: The Svedberg coefficient is a nonlinear function. A particle's mass, density, and shape will determine its S value. It depends on the frictional forces retarding the particle's movement, which in turn are based on the average cross-sectional area of the particle. The sum of the cross-sectional areas of the two separate subunits is greater than the cross-sectional area of the complex; thus, the sum of their S units (30S + 50S) is greater than that of the intact complex (70S).

3.3   What are the advantages and limitations of biochemical and genetic approaches to cell structure and function?

ANSWER: Biochemistry is useful to separate particles and identify their functions in isolation. In some cases, particles can also be combined together to study the function of a complex such as the cell-free translation system. On the other hand, biochemistry may miss aspects of function that are observed only in the intact cell. Genetics reveals the functions of specific genes in the phenotype of a living cell. On its own, however, genetics reveals little about the spatial relationships of gene products working together. Together, genetics and biochemistry reveal a more complete view of how cellular structures function.

3.4   Amino acids have acidic and basic groups that can dissociate. Why are they *not* membrane-permeant weak acids or weak bases? Why do they fail to cross the phospholipid bilayer?

ANSWER: At neutral pH, amino acids each have both a positively charged amine and a negatively charged carboxylate, that is, they can act as either a weak acid or a weak base. Charged ions, no matter what their size, will not freely pass a plasma membrane. If either charged group becomes neutralized by acid or base, the other group remains charged, so the molecule as a whole will never cross the membrane.

3.5   The actual thickness of the cell wall is difficult to determine based solely on electron microscopic observation of the envelope layers. Devise a biochemical experiment that can show the number of layers of peptidoglycan in cells of a given species.

ANSWER: To demonstrate the number of layers of peptidoglycan in a cell wall, determine the percentage by weight of peptidoglycan in the cell sample. Estimate the surface area of cells based on scanning EM. Given the known thickness of a layer of peptidoglycan, calculate how many layers would be needed to yield the amount of peptidoglycan present in the cell.

3.6   Why would laboratory culture conditions select for evolution of cells lacking an S-layer?

ANSWER: Degeneration of protective traits is a common problem when conducting research on microbes that can produce thirty generations overnight. Their rapid reproductive rate gives ample opportunity for spontaneous mutations to accumulate over an experimental time scale. In the case of the S-layer, in a laboratory test tube free of predators or viruses, mutant bacteria that fail to produce the thick protein layer would save energy compared to S-layer synthesizers, and would therefore grow faster. Such mutants would quickly take over a rapidly growing population.

3.7   Suppose that one cell out of a million has a mutant gene blocking S-layer synthesis, and suppose that the mutant strain can grow twice as fast as the S-layered parent. How many generations would it take for the mutant strain to constitute 90% of the population?

ANSWER: After each generation, the parent strain increases two-fold, whereas the mutant numbers increase by a factor of four. The mutant fraction is equal to $(4^N)/[4^N + (10^6 \times 2^N)]$, where N is the number of mutant cells after a given generation. Plotting this out on an Excel spreadsheet shows that by the twenty-third generation, the mutant fraction approximates 90%.

3.8   Why would the proteins listed in **Table 3.2** be confined to specific cell fractions? Why could a protein not function throughout the cell?

ANSWER: Each protein has evolved a specific function optimized for a specific part of the cell. For example, water-conducting porins are found solely in the inner membrane (cell membrane), which is otherwise impermeable to water. The outer membrane, which is water-permeable, is the sole location for specific porins transporting small peptides and sugars. The sugars then need to be taken across the inner membrane by transport proteins that evolved to function best in this location. Similarly, different chaperones (proteins that aid peptide folding) have evolved to function best in the cytoplasmic environment or in the periplasmic environment, membrane-enclosed regions that differ substantially in pH and ion concentrations. In a different chemical environment of the cell, the protein denatures and loses its functional structure.

3.9 What do you think are the advantages and disadvantages of a contractile vacuole, compared with a cell wall?

ANSWER: The disadvantage of a contractile vacuole is that it requires continual input of energy to bail out the water. On the other hand, the contractile vacuole permits the existence of a cell that is flexible enough to engulf other cells as prey. A cell wall does not require energy (other than the initial energy to synthesize it). A cell wall is inflexible and does not allow another cell to be engulfed as prey.

3.10 How would a magnetotactic species have to behave if it were in the Southern Hemisphere instead of in the Northern Hemisphere?

ANSWER: In the Northern Hemisphere, the field lines for magnetic north point downward; in the Southern Hemisphere, the opposite is true. Thus, if downward direction is the aim of magnetotaxis, bacteria existing in the two hemispheres would have to respond oppositely to the magnetic field; in the Southern Hemisphere, anaerobic magnetobacteria swim toward magnetic south. Near the equator, the proportions of north-seeking and south-seeking bacteria are roughly equal.

3.11 Most strains of *E. coli* and *Salmonella* commonly used for genetic research actually lack flagella entirely. Why do you think this is the case? How can a researcher maintain a motile strain?

ANSWER: The motility apparatus requires fifty different genes generating different protein components. Cells that acquire mutations eliminating expression of the motility apparatus gain an energy advantage over cells that continue to invest energy in motors. In a natural environment, the nonmotile cells lose out in competition for nutrients despite their energetic advantage; but in the laboratory, cells are cultured in isotropic environments such as a shaking test tube, where motility confers no advantage. These culture conditions lead to evolutionary degeneration of motility, as they do for the S-layer (see Thought Question 3.7). In order to maintain a motile strain, bacteria are cultured on a soft agar medium containing an attractant nutrient. As cells consume the attractant, they generate a gradient and chemotaxis leads them to swim outward. By subculturing only bacteria from the leading edge of swimming cells, one can maintain a motile strain.

Chapter 4

4.1 In a mixed ecosystem of autotrophs and heterotrophs, what happens when a heterotroph allows the autotroph to grow and begin to make excess organic carbon?

ANSWER: At first the growth of the heterotroph might outpace the growth of the autotroph, using the carbon sources faster than the autotroph can make them. As the organic carbon sources diminish through consumption, growth of the heterotroph decreases, but the CO_2 formed by the heterotroph will allow the autotroph to grow and make more organic carbon. Ultimately, the ecosystem comes into balance.

4.2 In what situation would antiport and symport be passive rather than active transport?

ANSWER: Antiport and symport would be passive when both molecules are moving down their concentration gradients. A symport or an antiport system can do work to move a molecule from a low concentration to a high (against a concentration gradient) as long as the co-transported molecule is moving from high to low concentration. When this happens, it is a form of active transport. If a symport or antiport moves both molecules with the concentration gradient (from high concentration to low), it is passive transport, assuming the molecules are moving no faster than the rate of diffusion.

4.3 What kind of transporter, other than an antiporter, could produce electroneutral coupled transport?

ANSWER: Electroneutral coupled transport can occur by symport if molecules of opposite charge are symported; for example Na^+ flux together with Cl^-.

4.4 What would be the phenotype (growth characteristic) of a cell that lacks the *trp* genes (genes required for the synthesis of tryptophan)? What would be the phenotype of a cell missing the *lac* genes (genes whose products catabolize the carbohydrate lactose)?

ANSWER: The difference lies in the function of the two pathways. The *trp* operon is a biosynthetic operon. Errors in the biosynthetic pathway will lead to a failure to produce tryptophan. Therefore, a *trp* auxotrophic mutant will *only* grow on defined medium if tryptophan is added. The lactose operon involves the catabolism of a carbon source, lactose. If any of these genes are damaged, the cells are no longer able to use lactose as a carbon source. A *lac* mutant will *not* grow on defined medium with lactose as the sole carbon source.

4.5 The addition of sheep blood to agar produces a very rich medium called blood agar. Do you think blood agar can be considered a selective medium? A differential medium? *Hint:* Some bacteria can lyse red blood cells.

ANSWER: Blood agar can be considered differential, because different species growing on blood have different abilities to lyse the red blood cells in the agar. Some do not lyse, others completely lyse red blood cells (secreted hemolysin produces complete clearing around a colony), while still others only partially lyse the blood (the secreted hemolysin produces a greening around the colony). It will therefore differentiate between hemolytic and non-hemolytic bacteria. The medium is very rich and does not prevent the growth of any organism so is not considered selective.

4.6   A virus such as influenza virus might produce 800 progeny virus particles from one infected host cell. How would you mathematically represent the exponential growth of the virus? What practical factors might limit such growth?

ANSWER: In theory, the growth rate of the virus would be proportional to 800^n. In practice, however, all the virus particles would never find 800 different host cells to infect. Furthermore, it turns out that only a small proportion of the influenza virus progeny are viable (see Chapter 11).

4.7   Suppose 1,000 bacteria are inoculated in a tube of minimal salts medium, where they double once an hour; and 10 bacteria are inoculated into rich medium, where they double in 20 minutes. Which tube will have more bacteria after 2 hours? After 4 hours?

ANSWER: After 2 hours, the minimal medium will contain $1,000 \times 2^2 = 4,000$ bacteria, whereas the rich medium will contain only $10 \times 2^{(2\,h)(3/h)} = 640$ bacteria. After 4 hours, the minimal medium will contain 1.6×10^5 bacteria, whereas the rich medium will have surpassed this count, reaching nearly 4.1×10^5.

4.8   An exponentially growing culture has an optical density at 600 nm (OD_{600}) of 0.2 after 30 minutes and an OD_{600} of 0.8 after 80 minutes. What is the doubling time?

ANSWER: Because the OD_{600} has increased from 0.2 to 0.8, there are four times as many cells after 80 minutes as after 30 minutes, and the culture has doubled twice. Two doublings occurred in 50 minutes; therefore the doubling time is 25 minutes.

4.9   **Figure 4.20C** shows growth curves for different population densities of *Thiobacillus thiooxidans* when the concentration of sulfur in the medium is constant. Draw the growth curves you would expect to see if the initial population density was constant but the concentration of sulfur varied.

ANSWER:

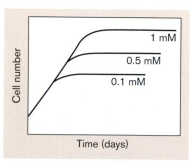

4.10   It takes 40 minutes for a typical *E. coli* cell to completely replicate its chromosome and about 20 minutes to prepare for another round of replication. Yet the organism enjoys a 20-minute generation time growing at 37°C in complex medium. How is this possible?

ANSWER: After the DNA is replicated about halfway around the chromosome, each daughter half-chromosome initiates a second round of replication, so the time needed to divide from one cell to two is effectively halved. Most cells in a log-phase culture in rich medium actually have four copies of the DNA origin of replication, each with a separate attachment site on the cell envelope, the future midpoint of a cell two generations ahead (see Chapter 3).

4.11   What can happen to the growth curve when a culture medium contains two carbon sources, one a preferred carbon source of growth-limiting concentration and a second, nonpreferred source? (see Section 10.3)

ANSWER: There are two possibilities. If the enzyme systems needed to utilize both carbon sources are always made, the growth curve will look normal because both will be used simultaneously. Usually the enzyme system for the nonpreferred carbon source is not produced until the preferred source is used up. In this case, a second lag phase will interrupt the exponential phase. This is called a diauxic growth curve and is commonly seen when cells are grown on both glucose and lactose. Lactose is the nonpreferred carbon source and is used second. The second lag phase marks the exhaustion of one nutrient and the gearing up by the cell to use the other (see Chapter 10).

4.12   How would you modify the equations describing microbial growth rate to describe the rate of death?

ANSWER: The death rate applies to a period of declining cell numbers. Therefore, the logarithm of the cell number ratio of N_1 to N_0 will be a negative number, and this factor will need to be preceded by a negative sign to convert it to a positive "halving time," or half-life of the culture.

4.13   Why are cells in log phase larger than cells in stationary phase?

ANSWER: Cells strive to maintain a certain DNA/mass ratio. In so doing, they balance the amount of biochemical processes needed to sustain viability. If the cell mass becomes large relative to the number of copies of a given, critical gene, the amount of enzyme produced may not be sufficient to keep the cell alive and growing. In addition, the DNA/mass ratio serves as a signal to trigger cell division. Thus, when a cell divides faster than it replicates its chromosome, it must start a second round of replication before it finishes the first. This type of replication ensures that at least one chromosome duplication will be complete at the time of division. Because fast-growing cells contain more than one chromosome, they will increase in size to maintain the desired DNA/mass ratio. If the ratio were not maintained, cell division would not occur when needed.

4.14   How might *Streptomyces* and *Actinomyces* species avoid committing suicide when they make their antibiotics?

ANSWER: Bacteria that produce antibiotics need to make defenses against the antibiotic within their own cytoplasm.

For example, their genes can express an altered form of the target molecule, such as a ribosomal subunit; or they can make pumps to pump the antibiotic out of the cell.

Chapter 5

5.1 Why haven't cells evolved so that all their enzymes have the same temperature optimum? If they did, wouldn't they grow even faster?

ANSWER: An enzyme's function is not determined by temperature alone. There are other physicochemical constraints based on the variety and complexity of functions different enzymes must carry out. The thousands of different enzyme molecules must work in a coordinated fashion to support the basic functions of life. By having some enzymes work below or above their optimum temperatures, the rate of the reactions they catalyze will be altered. A population will evolve based on the entire organism's ability to reproduce, not the speed at which each individual chemical reaction is carried out. The primary goal of a microbe is not just to grow fast but also to survive. Growing too fast could deplete food sources and produce toxic by-products too quickly.

5.2 If microbes lack a nervous system, how can they sense a temperature change?

ANSWER: Most bacteria respond to outside stimuli, such as heat, by altering their gene expression. They sense heat by monitoring the level of misfolded proteins, a consequence of too high a temperature. The mechanism does not perceive heat per se, but recognizes the deleterious effects of moving outside the optimum growth temperature range so that the cell can launch an emergency response. The same mechanisms can sense other environmental stresses that misfold proteins, such as acid stress. See Chapter 10.

5.3 What could be a relatively simple way to grow barophiles in the laboratory?

ANSWER: High pressures can be maintained using a syringe. Scientists would place medium in a stainless steel syringe and maintain pressure with a calibrated vise.

5.4 How might the concept of water availability be used by the food industry to control spoilage?

ANSWER: Food preservation traditionally includes water exclusion by salt, as seen in hams, back bacon, salted fish, etc., or by high concentration of sugar, as in canned fruit or jellies. The lower a_w prevents microbial growth. Dehydrating foods will also prevent microbial growth.

5.5 In Chapter 4, we stated that an antiporter couples movement of one ion down its concentration gradient with movement of another molecule uphill, against its gradient. If this is true, how could a Na^+/H^+ antiporter work to bring protons into a haloalkaliphile growing in high salt at pH 10? Since the Na^+ concentration is lower inside the cell than outside and H^+ concentration is higher inside than outside (see **Fig. 5.16**), both ions are moving against their gradients.

ANSWER: In this situation the cell has to expend some energy to make the antiporter work. The energy involved is rooted in the charge difference between the inside of the cell (negative charge) and the outside of the cell (positive charge, called delta psi, $\Delta\psi$. See Chapter 13). The antiporter in this case must also exchange a different number of NA^+ and H^+ ions, maintaining an electrical charge difference across the membrane. For example, export $2NA^+$ and import $1H^+$.

5.6 If anaerobes cannot live in oxygen, how do they incorporate oxygen into their cellular components?

ANSWER: They incorporate oxygen from their carbon sources (e.g., CO_2 and carbohydrates such as glucose), all of which contain oxygen. This form of oxygen will not damage the cells.

5.7 How can anaerobes grow in the human mouth when there is so much oxygen there?

ANSWER: A synergistic relationship occurs between facultatives and anaerobes within a tooth biofilm. The facultatives consume oxygen within the biofilm microenvironment, which allows the anaerobes to grow underneath them.

5.8 What evidence led people to think about looking for anaerobes? *Hint:* Look up Spallanzani, Pasteur, and spontaneous generation on the web.

ANSWER: The priest Lazzaro Spallanzani (1729–1799), during his quest to disprove spontaneous generation, said, "Every beast on Earth needs air to live, and I am going to show just how animal these little animals are by putting them in a vacuum and watching them die." He dipped a glass tube into a culture, sealed one end and attached the other to a vacuum. He was astonished to find that the microbes lived for weeks. He then wrote: "How wonderful this is. For we have always believed there is no living being that can live without the advantages air offers it. " Fifty years later, Pasteur observed that air could kill some organisms. After looking at a drop of liquid from a fermentation culture, he wrote, "There is something new here—in the middle of the drop they are lively, going every which way, but on the edge they were stiff as pokers."

5.9 How would you test the killing efficacy of an autoclave?

ANSWER: Construct a death curve by measuring survival of a known quantity of spores (e.g., *Bacillus stearothermophilus*) after autoclaving for various lengths of time. Spores should be used because they are more resistant to heat than any vegetative cell. Typically, autoclaves are regularly checked with spore strips that change color once the endospores are no longer viable.

Chapter 6

6.1   Which viruses do you know that have a narrow host range, and which have a broad host range?

ANSWER: Examples of viruses with a narrow host range include poliovirus (poliomyelitis), which infects only humans and chimpanzees; smallpox, which infects only humans; and feline T-cell leukemia virus, which infects only cats. Examples of viruses with a broad host range include rabies, which infects numerous species of mammals, and influenza strains, which show preference for particular species but can jump between various mammals and birds.

6.2   What would happen if a virus particle remained intact within a host cell instead of releasing its genome?

ANSWER: The virus would be unable to reproduce, because DNA polymerases could not reach its genome for reproduction and RNA polymerase could not transcribe its genes to make gene products.

6.3   Why do viral capsids take the form of an icosahedron instead of some other polyhedron?

ANSWER: The icosahedron is a polyhedron of 20 triangular faces, the largest number possible. Thus, the icosahedron turns out to be the largest and most economical form to enclose space based on a small repeating unit. Natural selection probably favored viruses that could build the largest capsid based on the smallest amount of genetic information.

6.4   How can viruses with different kinds of genomes (RNA versus DNA) combine and exchange genetic information?

ANSWER: DNA viruses require messenger RNA intermediates to express their proteins. It is possible that a DNA virus could acquire the ability to package its RNA transcript in a capsid, rather than its DNA. Alternatively, RNA retroviruses form DNA intermediates within their host cell; these DNA intermediates might recombine with the DNA genome of another virus.

6.5   What are the relative advantages and disadvantages (to a virus) of the slow-release strategy, compared with the strategy of a temperate phage, which alternates between lysis and lysogeny?

ANSWER: A disadvantage of slow release is that the phages can never reproduce progeny phage as rapidly as in a lytic burst. The drain on resources of the host cell infected by a slow-release virus causes it to grow more slowly compared with uninfected cells; in contrast, a lysogenized cell suffers little or no reproductive deficit compared with uninfected cells. An advantage of reproduction by slow release is the continual release of phage, while avoiding the possibility of releasing all particles into an environment where no other host cell exists.

6.6   How could humans evolve to resist rhinovirus infection? Is such evolution likely? Why or why not?

ANSWER: Resistance to all rhinovirus infections might evolve through a mutation in the host gene encoding ICAM-1. The mutation would have to prevent rhinovirus binding without impairing the protein's ability to bind integrin. Such evolution is unlikely because of the importance of integrin binding and because rhinovirus infection is rarely fatal; thus, there is little selection pressure to evolve inherited resistance. Note, however, that the immune system rapidly generates immunity to particular strains of rhinovirus. Over a lifetime, most individuals acquire immunity to many rhinovirus strains but remain susceptible to others.

6.7   What are the advantages and disadvantages to the virus of replication by the host polymerase, compared with using a polymerase encoded by its own genome?

ANSWER: An advantage of using the host polymerase is energetic; the virus avoids the energetic cost of manufacturing a polymerase to package with each virion. This is an advantage to the virus, since its reproductive potential is limited by the energy resources of its host cell. Furthermore, because DNA and RNA polymerases are so central to cell function, the host species is unlikely to evolve a mutant form of the polymerase that resists the virus. On the other hand, the advantage of the virus making its own polymerase is that the viral polymerase can evolve to better meet the needs of its own replication, such as high speed and low accuracy to generate frequent variants.

6.8   Why does bacteriophage reproduction give a step curve, whereas cellular reproduction generates an exponential growth curve? Could you design an experiment in which viruses generate an exponential curve? Under what conditions does the growth of cellular microbes give rise to a step curve?

ANSWER: Lytic viruses appear to make a step curve because the number of progeny per infected cell is 100 or more, released simultaneously. After two or three generations the cell cycles would fall out of synchrony, and the curve would smooth out, but the later cell cycles are rarely observed in practice because by then the supply of host cells is exhausted. If, however, an extremely low ratio of viruses to host cells is provided, the growth of virus particles will eventually generate an exponential curve. By contrast, the growth of cellular microbes is rarely observed during the first few doublings. By the time we measure the population, the cells are all undergoing different stages of division, and the population growth overall generates a smooth exponential curve. If, however, we observe the growth of a synchronized population of cells, we see a step curve of cell division two.

Chapter 7

7.1   What do you think happens to two single-stranded DNA molecules isolated from *different* genes when they are mixed together at very high concentrations of salt?

ANSWER: In high salt conditions, the stacking of hydrophobic bases is so strongly favored that two single strands of DNA will form a duplex no matter what the sequence of base pairs.

7.2 Do you think the kinetics of denaturation and renaturation are dependent on DNA concentration?

ANSWER: The speed of denaturation does not depend on DNA concentration, but the speed of renaturation does. The higher the concentration of ssDNA, the more likely it is that complementary sequences will find each other and the faster the duplex can re-form.

7.3 DNA gyrase is essential to cell viability. Why, then, are nalidixic acid-resistant cells that contain mutations in *gyrA* still viable?

ANSWER: The *gyrA* mutations alter only the nalidixic acid binding site on GyrA, not its gyrase activity. In other words, active DNA gyrase is still made, but the drug cannot bind to it.

7.4 Bacterial cells contain many enzymes that can degrade DNA. How, then, do linear chromosomes in organisms like *Borrelia burgdorferi* (the causative agent of Lyme disease) avoid degradation?

ANSWER: DNA-digesting exonucleases act on free 5′ or 3′ ends. The *Borrelia* linear chromosomes possess covalently closed hairpin ends called telomeres and do not possess free 5′ or 3′ groups.

7.5 Suppose you have the following capabilities: You can label DNA in a bacterium by growing cells in medium containing either nitrogen ^{14}N or the heavier isotope ^{15}N; you can isolate pure DNA from the organism; you can subject DNA to centrifugation in a cesium chloride solution, a solution that forms a density gradient when subjected to centrifugal force, thereby separating the light (^{14}N) and heavy (^{15}N) forms of DNA to different locations in the test tube. Given these capabilities, how might you prove that DNA replication is semiconservative?

ANSWER: Grow bacteria in medium containing heavy ^{15}N, so that all the DNA made by the cells will be in the heavy form. Transfer the cells to medium containing only ^{14}N and allow the cells to divide for one generation. Extract the DNA and centrifuge the preparation in cesium chloride. If replication is semiconservative, a hybrid DNA band will be seen at a position located between and equidistant from where heavy DNA and light DNA would be located. The hybrid band will be composed of one heavy strand and one light strand, which makes it of intermediate weight (density). This, in fact, was how Meselson and Stahl proved semiconservative replication in 1958, five years after the discovery that DNA was double-stranded.

7.6 How fast does *E. coli* DNA polymerase synthesize DNA (in nucleotides per second), given that the genome is 4.6 million base pairs, replication is bidirectional, and the chromosome completes a round of replication in 40 minutes?

ANSWER: It takes *E. coli* 40 minutes (2,400 seconds) to complete the replication of its 4,639,221-bp chromosome. Since Pol III works as a dimer to synthesize both strands simultaneously, we will only consider one strand in the calculation. The rate is 4,639,221 nt per 2,400 seconds = 1,933 nt per second. But there are two replication forks, so each polymerase dimer only synthesizes one-half of the chromosome, which means that each polymerase still synthesizes a remarkable 800–1000 nt per second.

7.7 Individual cells in a population of *E. coli* typically initiate replication at different times (asynchronous replication). However, depriving the population of a required amino acid can synchronize reproduction of the population. What happens is that ongoing rounds of DNA synthesis finish, but new rounds do not begin. Replication stops until the amino acid is once again added to the medium, an action that triggers simultaneous initiation in all cells; that is, reproduction of the population becomes synchronized. Why?

ANSWER: Initiation requires synthesis of the initiator protein DnaA. Consequently, depriving the population of an amino acid prevents protein synthesis, which precludes synthesis of DnaA. Because DnaA is not required to complete already initiated rounds of replication, all rounds already started are completed, but reinitiation cannot occur. Adding the amino acid once again will allow all cells to simultaneously make DnaA so that initiation is triggered in all cells at the same time.

7.8 The antibiotic rifampin inhibits transcription by RNA polymerase, but not by primase (DnaG). What happens to DNA synthesis if rifampin is added to a synchronous culture?

ANSWER: Initiation of DNA synthesis requires primer transcription at the origin by RNA polymerase, an enzyme sensitive to rifampin. Primase (DnaG), which synthesizes RNA primers in the lagging strand throughout DNA synthesis, is resistant to rifampin. So adding rifampin to a synchronized culture will prevent new rounds of DNA replication but will not affect already initiated rounds.

7.9 Knowing that *E. coli* possesses restriction enzymes that cleave DNA from other species, how is it possible to clone a gene from one organism to another without the cloned gene sequence being degraded?

ANSWER: The best way to overcome the restriction barrier is to use a mutant strain of *E. coli* in which its restriction system has been inactivated. DNA entering the cell will not be degraded and, if the modification system is still in place, will be modified. Once modified, that DNA can be safely transferred to other *E. coli* strains that still have their restriction enzymes.

7.10   PCR is a powerful technique, but one can easily contaminate a sample and amplify the wrong DNA—perhaps sending an innocent person to jail. What might you do to minimize this possibility?

ANSWER: Use sterile, filtered micropipet tips. The filters prevent contamination of the micropipet and subsequent reactions. Also, perform a control PCR reaction without adding a DNA sample. If the water, buffers, or tubes are contaminated with a given DNA, a PCR product will appear even when DNA template has not been added deliberately to the reaction.

Chapter 8

8.1   If each sigma factor recognizes a different promoter, how does the cell manage to transcribe genes that respond to multiple stresses, each involving a different sigma factor?

ANSWER: In these situations, a given gene will have multiple promoters. Each promoter will be recognized by a different sigma factor and will begin transcription at different distances from the ORF start codon.

8.2   With respect to two different sigma factors with different promoter recognition sequences, predict what would happen to the overall gene expression profile in the cell if one sigma factor were artificially overexpressed. Could there be a detrimental effect on growth?

ANSWER: Since sigma factors compete for the same site on core polymerase, overexpressing one sigma factor could displace the other sigma factor from the RNA polymerase population and compromise expression of those target genes. If those genes were important to survival, the cell could die.

8.3   Why might some genes contain multiple promoters, each one specific for a different sigma factor?

ANSWER: The gene might need to be expressed under multiple conditions at different levels. If a given condition increases expression of an alternate sigma factor, the target gene will need a promoter that the new sigma can recognize. As the need disappears and the sigma factor diminishes in concentration, a promoter that uses the housekeeping sigma will be needed. For example, the gene for DnaK heat shock protein has promoters for RpoH (sigma-32) and sigma-70, the housekeeping sigma factor. The level of protein needed during normal growth is supplied by sigma-70. Upon encountering heat stress, the RpoH sigma factor level increases and mediates an increase in DnaK production.

8.4   How might the redundancy of the genetic code be used to establish evolutionary relationships?

ANSWER: The codon preferences of different microorganisms are based in part on their GC content. Thus, an organism with an AT-rich genome will preferentially use codons for a given amino acid that have As and Ts over those with Gs and Cs. Evolutionarity, finding a long AT-rich region encoding mRNA with AT bias within a chromosome that is otherwise GC-rich, suggests that the AT-rich region was horizontally inherited from some illicit DNA exchange with another species. (See Chapter 9.)

8.5   Synthesis of ribosomal RNAs and ribosomal proteins cause a major energy drain on the cell. How might the cell regulate synthesis of these molecules when growth rate slows as a result of amino acid limitation?

ANSWER: As energy levels fall, fewer amino acids are made, which means that fewer charged tRNA molecules are assembled. This will cause ribosomes to pause during translation until finding the right charged tRNA. This pause causes synthesis of a signal molecule called guanosine tetraphosphate (ppGpp), which interacts with RNA polymerase and selectively stops transcription of the rRNA genes. This also stops new ribosome synthesis. As cells divide, the ribosomes will dilute to a lower steady-state level. Since less rRNA is made, fewer partners for ribosomal proteins are available and unassembled ribosomal proteins accumulate. These proteins bind sequences in their own mRNA molecules that are similar to the target sequences on rRNA. By binding to their mRNA, these ribosomal proteins can prevent their own translation. (See Chapter 10.)

8.6   While working as a member of a pharmaceutical company's drug discovery team, you have found that a soil microbe snatched from the jungles of South America produces an antibiotic that will kill even the most deadly, drug-resistant form of *Enterococcus faecalis,* which is an important cause of heart valve vegetations in bacterial endocarditis. Your experiments indicate that the compound stops protein synthesis. How could you more precisely determine the antibiotic's mode of action?

ANSWER: One way is to take a culture of sensitive bacteria and isolate resistant mutants (bacteria that are not killed by the antibiotic), purify their ribosomes, and separate the 30S and 50S ribosomal subunits. Cross-mix subunits from sensitive and resistant cells (for example, mix 30S subunits from sensitive cells with 50S subunits from resistant cells). Then measure protein synthesis by the hybrid ribosomes with and without the drug. If resistance is due to an altered ribosomal protein or RNA, then the subunit mix containing the altered component will make protein regardless of whether the drug is present. Once identified, the responsible ribosomal subunits from resistant and sensitive cells can be broken down further into their component parts, reconstituted in hybrid form, and again tested for an ability to make protein in the presence of drug. This reductive approach will likely, but not always, uncover the target ribosomal protein or rRNA.

8.7   How might one gene code for two proteins with different amino acid sequences?

ANSWER: By having two different translation start sites in different reading frames. While this is not a common occurrence, it happens.

8.8 Why involve RNA in protein synthesis? Why not translate directly from DNA?

ANSWER: Because transcription enables the cell to amplify the gene sequence information into multiple copies of RNA. Amplification means that more ribosomes can be engaged in translating the same protein, causing the concentration of the protein to rise more quickly than if only a single gene were used. The transcriptional process also presents an opportunity to regulate production of a protein.

8.9 Codon 45 of a 90-codon gene was changed into a translation stop codon. This produced a shortened, truncated protein. What kind of mutant cell could produce a full-length protein from the gene *without* removing the stop codon?

ANSWER: A mutant in which a tRNA gene has been altered by mutation so that the anticondon of the tRNA "sees" the stop codon as an amino acid codon. This mutated tRNA molecule will transfer its amino acid to the peptide chain. The attached amino acid can be used to bridge the gap caused by the stop condon, and a full-length protein is made. These modified tRNAs are called suppressor tRNAs because they suppress the mutant phenotype.

8.10 An ORF 1,200 bp in length could encode a protein of what size and molecular weight?

ANSWER: There are three bases per codon, so the ORF can encode 400 amino acids. The average molecular weight of an amino acid is 110 Da. Therefore, the hypothetical protein will be approximately 44,000 Da (44 kDa).

Chapter 9

9.1 Transfer of an F factor from an F^+ cell to an F^- cell converts the recipient to F^+. Why does transfer of an Hfr not do the same?

ANSWER: The last piece of an Hfr to transfer is the F factor and *oriT*. It is a rare event that an entire chromosome will transfer from one cell to another, so most Hfr transfers do not result in transfer of *oriT* and so cannot initiate conjugation.

9.2 Would it be easier to demonstrate generalized transduction using a temperate phage (which generates a lysogen) or a virulent phage?

ANSWER: Both can be used, but more care must be taken with a virulent phage. The reason is that a lytic phage can kill a potential recipient through superinfection. Superinfection occurs when a normal phage particle and a transducing particle infect the same cell. The normal lytic phage will lyse the cell before a recombinant can be formed. Generalized trans-

duction usually occurs more easily with a temperate phage (like P22 or P1) because a superinfecting temperate phage will usually just lysogenize the cell and not kill it. Therefore, the host DNA delivered by the transducing particle has time to recombine into the recipient genome. Lytic phage, like the variant of the lysogenic P1 phage called P1 vir, commonly used for *E. coli* transduction, can be used in the lab if care is taken to avoid superinfection. (The vir designation stands for *virulent* because the mutant P1 phage lost the ability to lysogenize.) If citrate is added moments after mixing P1 vir transducing lysate with a recipient, the ions needed for phage absorption are chelated, thereby limiting superinfection. In nature, where dilution of phage may be very great, there is less chance of superinfection and death. In this situation, transduction can easily be performed by lytic phage.

9.3 How do you think phage DNA containing restriction sites evade the restriction-modification screening systems of its host?

ANSWER: Phage DNA will survive because sometimes the modification enzyme gets to the foreign DNA before the restriction enzyme. Once methylated, the phage DNA will be shielded from restriction and the methylated molecule will replicate unchallenged. If, on the other hand, a foreign DNA fragment (not necessarily a phage) has been modified and the conditions are right, it might recombine into the host chromosome and convey a new character to the strain.

9.4 Type I restriction-modification system genes are often designated *hsdR, hsdM,* and *hsdS,* which code, respectively, for the restriction (HsdR), modification (HsdM), and sequence recognition (HsdS) proteins. Can a plasmid grown in a wild-type strain be used to transform a strain defective in *hsdR*? An *hsdS* mutant? Can a plasmid grown in an *hsdM* mutant strain be transformed into an *hsdR* defective strain?

ANSWER: A plasmid grown in a wild-type strain will be methylated, but methylated or not, it can survive in a restrictionless *hsdR* mutant or an *hsdS* mutant that lacks the recognition protein needed by both the restriction and modification subunits. Since an *hsdM* mutant lacks the modification protein, it cannot protect its genes from restriction enzymes and would commit suicide, so this last experiment cannot be done.

9.5 In a transductional cross between an $A^+B^+C^+$ genotype donor and an $A^-\,B^-\,C^-$ genotype recipient, 100 A^+ recombinants were selected. Of those 100, 15% were also B^+, while 75% were C^+. Is gene B or C closer to gene A?

ANSWER: Gene C, because it was cotransduced with A at the highest frequency.

9.6 Calculate the mutation rate for a gene in which the number of mutations increases from 0 to 50 as cell number increases from 10^1 to 10^8.

ANSWER: One calculation that quickly estimates mutation rate is $m[\ln2/N_t]$ where m = number of mutation in the culture and N_t = the number of cells in the culture. Because bacterial cultures grow logarithmically, $\ln2/N_t$ estimates the number of generations that went into making the culture. Thus:

$50[\ln2/(10^8 - 10^1)] =$

$50[0.69/(10^8 - 10^1)] =$

$50[0.69/\text{approx. }10^8] =$

$50/1.44(10^8) = \text{mutation rate}$

mutation rate approx. equals 3.47×10^{-7}

9.7   It has been reported that hypermutable bacterial strains are overrepresented in clinical isolates. Out of 500 isolates of *Haemophilus influenzae,* for example, 2–3% were mutator strains having mutation rates 100–1,000 times higher than lab reference strains. Why might mutator strains be beneficial to pathogens?

ANSWER: The mutator strains may speed microbial evolution, which could help the microbe outwit the immune system or escape the effects of administered antibiotics.

9.8   Diagram how a composite transposon like Tn*10* can generate inversions or deletions in target DNA during transposition. *Hint*: This occurs when the transposon tries to jump from one site to another in the same chromosome using the ends of the inverted repeats closest to the tetracycline resistance gene (see **Fig. 9.35**). If you draw it right, you will see that in the end, the tetracycline gene is lost.

ANSWER: From the diagram below you can see that inversions or deletions occur when the inner rather than outer ends of the inverted repeats are the targets of the transposase. The tetracycline resistance gene is lost, and the chromosome will contain a deletion or inversion depending on the orientation of the chromosome (looped or unlooped) at the time of transposase action.

9.9   The heat-shock gene *dnaK* in *E. coli* encodes a chaperone (Hsp70), homologs of which are found in all three domains of life. However, there is no other phylogenetic evidence that a *dnaK*-like gene was present in the most recent common ancestor of all organisms. For instance, some species of archaea do not encode a *dnaK* homolog, while other archaeal species do. In fact, molecular phylogenies indicate that the archaeal homologs are quite closely related to those of bacteria. It seems that the most recent common ancestor of all *dnaK* (or *hsp70*) genes existed more recently than the common ancestor of all organisms. That being the case how do you suppose *hsp70* genes arose in archaea?

ANSWER: The gene is thought to have been moved by some type of horizontal gene transfer mechanism from domain Bacteria to some members of domain Archaea.

DIAGRAM FOR ANSWER TO THOUGHT QUESTION 9.8

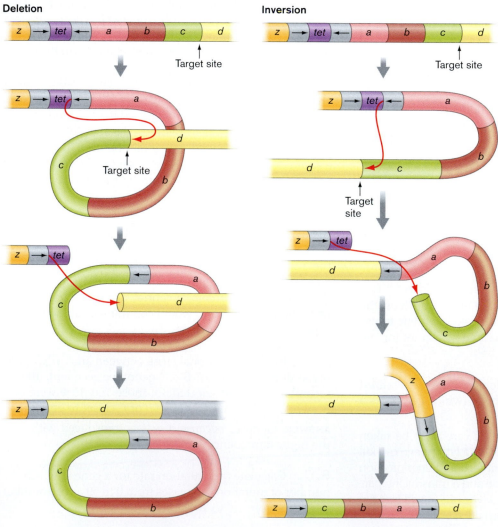

Chapter 10

10.1   If the gene *lacZ* has a nonsense mutation in its open reading frame, will *lacY* be translated?

ANSWER: *LacY* mRNA will still be translated because each gene in this polycistronic mRNA (*lacZYA*) has its own ribosome binding site and the *lacY* ribosome binding site is functional. A transitional stop codon in an upstream gene does not usually affect transcription of the downstream RNA.

10.2   Null mutations completely eliminate the function of a mutated gene. Predict the effect of the following null mutations on the induction of β-galactosidase by lactose and predict whether the *lacZ* gene is expressed at high or low levels: *lacI, lacO, lacP, crp, cya* (the gene encoding adenylate cyclase). Also, what effect will those mutations have on catabolite repression?

ANSWER: Loss of LacI repressor will lead to constitutive expression of *lacZ* but will not affect catabolite repression. A *lacO* mutant will not bind LacI repressor, so the phenotype will mimic that of a *lacI* mutation. A *lacP* mutation will prevent expression of *lacZYA* because RNA polymerase will not bind. Mutations in *crp* or *cya* will prevent catabolite repression, but *lacZYA* induction by lactose will be normal. However, without the cAMP-CRP complex, expression can never achieve maximal levels.

10.3   Predict what will happen to the expression of *lacZ* under the following partial diploid conditions (see Section 9.1). The genotypes of these strains are presented as follows: chromosomal genes/plasmid gene. (a) *lacI lacO⁺P⁺Z⁺Y⁺A⁺*/plasmid *lacI⁺*; (b) *lacO lacI⁺P⁺Z⁺Y⁺A⁺*/plasmid *lacO⁺*; (c) *crp lac I⁺O⁺P⁺ Z⁺Y⁺A⁺*/plasmid *crp⁺*.

ANSWER: (a) The *lacI⁺* gene on the plasmid will produce LacI repressor protein that can diffuse through the cytoplasm, bind chromosomal *lacO*, and repress the *lacZYA* operon. Because the complementing gene is on a different DNA molecule than the mutation, the gene is said to work in *trans*. (b) Because the *lacO* gene does not produce a diffusible product (e.g., protein or RNA), the plasmid *lacO⁺* cannot complement a *lacO* mutation in trans, the strain will not make β-galactosidase. Thus, the *lacO* gene only functions in *cis*, that is, when it resides next to the gene it regulates. (c) The *crp* gene produces a diffusible protein product, so it can function in *trans* and complement a *crp* mutation. The strain will make β-galactosidase to the highest level, in the presence of inducer lactose.

10.4   Researchers often use isopropyl-β-D-thiogalactopyranoside (IPTG) rather than lactose to induce the *lacZYA* operon. IPTG structure resembles lactose, which is why it can interact with the LacI repressor, but it is not degraded by β-galactosidase. Why do you think the use of IPTG is preferred in these studies?

ANSWER: There are at least two reasons. First, the level of IPTG inducer will not change, but the level of lactose inducer will continually decrease as it is consumed. This can affect the kinetics of induction. Second, the act of degrading lactose produces glucose and galactose. Glucose, as the preferred carbon source, will catabolite repress the *lacZYA* operon, once again affecting the kinetics of induction.

10.5   Predict the phenotype of a *spoIIAA* mutant that completely lacks this anti-anti-sigma factor (see **Fig. 10.16**).

ANSWER: The *spoIIAA* mutant will have no way to efficiently remove anti-sigma F factor from sigma F. As a result, sigma F will almost never be active, even in the forespore. The cell can get as far as asymmetrical cell division to make what would be the forespore but since sigma F and anti-sigma F equally distribute in both compartments, sigma F will not be activated.

10.6   What is the phenotype of a *fljA* mutant? A *fliC* mutant?

ANSWER: A *fljA* mutant lacks the repressor needed to turn off FliC. Thus, a cell will switch back and forth from making H2 flagellin to making both H1 and H2 flagellin. A *fliC* mutant, however, will switch from being motile to nonmotile. In one orientation, the invertible element will allow H2 flagellin to be made, but in the opposite orientation, no flagellin will be made, at which point the cell will not be motile.

10.7   The plaques of normal phage lambda appear as clear rings with cloudy centers of lysogenized cells. What do you think the plaque will look like in a *cI* mutant? A *cII* mutant? A *cro* mutant?

ANSWER: Both the *cI* and *cII* mutant plaques will be completely clear (no cloudy center) because without CI, the virus can only go down the lysis pathway. A *cro* mutant may not even make a plaque, since Cro is needed to stop CI transcription and allow lysis.

10.8   Predict why UV irradiation can activate lambda prophage. *Hint:* What other host system is activated by UV?

ANSWER: The RecA coprotease activated by the ssDNA molecules resulting from excessive UV irradiation not only stimulates autocleavage of the LexA repressor that controls the SOS response but also facilitates autocleavage of the lambda CI repressor. This allows transcription from lambda promoters P_L and P_R and sends the integrated prophage down the lytic path.

10.9   Predict the phenotype of a *glnB* mutant. Will it be a glutamine auxotroph or not? What about a *glnD* or a *glnE* mutant?

ANSWER: A *glnB* mutant will not dephosphorylate NtrB or remove AMP from GlnA. The cell will continue to make GlnA and fail to inactivate it. Thus, the cell will overproduce glutamine. A *glnD* mutant, however, will not add UMP to GlnB, so GlnB will not direct GlnE to remove AMP from GlnA (remember, GlnA without AMP is active). As a result, GlnA will remain inactive and the cell will likely require glutamine. The same will occur with a *glnE* mutant.

10.10   Genes encoding luciferase can be used as "reporters." What do you think would happen if the promoter for an SOS response gene were fused to the luciferase open reading frame?

ANSWER: It could serve as a real-time biosensor for an environmental stress response. You could fuse the luciferase gene to the *recA* promoter and examine a real-time increase in fluorescence after ultraviolet irradiation by inserting the whole culture into a spectrofluorometer. The opening figure in Chapter 9 illustrates this strategy, using green fluorescent protein rather than luciferase.

10.11   What do you think would happen if a culture were coinoculated with a *V. fischeri luxI* mutant and a *luxA* mutant, neither of which produces light?

ANSWER: The *luxI* mutant would glow. The *luxA* mutant would still make autoinducer as it grew. This autoinducer would accumulate in the culture medium, then diffuse and enter the *luxI* mutant cells, where it would trigger induction of the *lux* operon and production of luciferase.

10.12   Why might the mass of a protein excised from a gel fail to match any of the ORF molecular weights predicted by the genome sequence?

ANSWER: The protein might be altered by phosphorylation, acetylation, removal of a leader sequence, or otherwise proteolytically processed. Identification might still be made if certain unmodified fragments of the protein match up with predicted fragments.

Chapter 11

11.1   Why would phage T4 production require a tenfold higher rate of DNA replication than does the host cell? Why would the phage substitute all of its cytosine with an unusual base that requires greater energy to synthesize?

ANSWER: The phage T4 genome needs to replicate itself and generate progeny as fast as possible, in order to use as much of the host cell's resources as possible before the cell deteriorates. By replacing cytosine with a modified base, the phage can avoid degradation of its chromosome by its own endonucleases (which cleave host DNA) or by those of the host (which are made to cleave phage DNA).

11.2   The phage T4 chromosome is a linear piece of DNA; yet the genomic map of T4 is circular. Why?

ANSWER: Each headpiece contains just enough space to package slightly more than the length of a genome. The head-filling mechanism of DNA packaging means that successive genomes on a concatemer each get cut at different positions within the genome sequence. Thus, any map of gene order based on recombination linkages between genes, or based on the sequencing of a population of genome molecules, will generate a circular map.

11.3   Which numbered genes from Figure 11.8 might be responsible for each of the defective phage populations presented in Figure 11.9D?

ANSWER: Mutant 1 may have a defect in a gene whose product is needed for tail fiber attachment, such as gene 63. Mutant 2 may be defective for tail fiber assembly (genes 34 or 57). Mutant 3 could be defective for sheath assembly around the internal tube (gene 18).

11.4   Compare and contrast the reproductive strategies of phages T4 and M13.

ANSWER: Both species of phage aim to maximize production of progeny and sustain an unbroken chain of infection. Both recognize and attach to proteins of the cell envelope that are essential for host cell structure, and both are specific for *E. coli;* T4 attaches to OmpC, whereas M13 attaches to the F pilus. T4 injects its DNA only, discarding its capsid, whereas M13 injects its DNA and recycles its coat proteins into progeny phage. T4 immediately destroys its host genome and reshapes its host's biosynthetic apparatus to accelerate production of phage, whereas M13 moderates its rate of production so that the host will continue to grow and acquire nutrients, perhaps generating more phage in the long run. Most of T4 assembly occurs in the cytoplasm, whereas most of M13 assembly occurs in the cell membrane. T4 produces an enzyme that lyses the host to liberate all phage progeny, whereas M13 exits through a capped pore that maintains the integrity of the host, enabling gradual sustained production of phage.

11.5   Why does influenza virus have to provide its polymerase ready-made, whereas poliovirus does not?

ANSWER: The influenza viral chromosome is (–) strand RNA. When it enters the host cytoplasm, it cannot be translated by ribosomes, and no host polymerase makes RNA from RNA. Therefore, the virion must provide an RNA-dependent RNA polymerase to generate a (+) strand RNA for translation to proteins. Poliovirus, however, injects (+) strand RNA, which can immediately be translated by host ribosomes to generate RNA-dependent RNA polymerase.

11.6   Explain why the expression $(8!/8^8)$ approximates the proportion of infective particles of influenza virus. What assumption might be changed to make the proportion greater or less?

ANSWER: An infective particle can start by packaging any one of the eight chromosomes. Once the first is chosen, seven possibilities remain for the second, six for the third, and so forth. Thus there are 8! ways to obtain a perfect set out of 8^8 total ways to pick eight segments at random. This expression assumes, however, that exactly eight segments are packaged in every virion. If the average number of segments packaged is less than eight, the proportion of infectives falls off. If the number of segments is usually more than eight, the proportion of virions containing at least one of each segment increases.

11.7 How do attachment and entry of HIV resemble attachment and entry of influenza virus? How do attachment and entry differ between these two viruses?

ANSWER: Attachment and entry of HIV requires the envelope spike proteins to bind receptors in the host plasma membrane, just as the influenza envelope protein hemagglutinin binds the sialic acid protein in the host membrane. In both cases, initial receptor binding signals major rearrangement of a viral envelope protein so as to insert a fusion peptide into the host membrane. The processes differ between these two virus species, in that influenza virus induces formation of an endocytic vesicle, whose acidification triggers membrane fusion and release of the core contents into the host cytoplasm. HIV virions, however, do not induce endocytosis and do not require acidification to induce membrane fusion and release of the core into the cytoplasm.

11.8 Compare and contrast the priming of chromosome replication in HIV with the priming mechanisms of poliovirus and influenza virus.

ANSWER: Poliovirus uses a viral protein (Vpg) to provide the 3′ OH needed to prime synthesis of RNA. In influenza virus, RNA synthesis is primed by "capped" fragments of host mRNA. Instead of host mRNA, HIV utilizes tRNA for primers.

11.9 Compare and contrast the fate of the HSV chromosome with that of the chromosome of HIV.

ANSWER: HSV-1 contains a DNA chromosome, which is transported to the nuclear membrane within an intact capsid. HIV contains twin RNA chromosomes, which are released in the cytoplasm upon dissolution of the capsid. The RNA chromosomes are copied to double-stranded DNA for transport into the nucleus. In HSV-1, the DNA chromosome circularizes and generates concatemeric duplicates by rolling-circle replication. In HIV, the replicated DNA circularizes but immediately integrates into the host chromosome. In both cases, the viral DNA can persist for decades in latent infection.

Chapter 12

12.1 In this postgenomic era, how might you generally design a strategy to construct a strain of *E. coli* that cannot synthesize histidine?

ANSWER: The genomic sequence of *E. coli* is known. We also know which genes encode the steps of histidine biosynthesis. The strategy today would be to engineer a plasmid containing DNA sequences (about 200 bp) that flank the histidine biosynthesis gene you want to delete, but in place of the target gene, you would instead insert an antibiotic resistance gene (e.g., kanamycin resistance). The plasmid used would

be a "suicide" plasmid that cannot replicate in the strain you wish to mutate but which can replicate in other *permissive* strains (strains that produce a specific protein the plasmid needs to replicate). The permissive strain allows you to make a lot of the plasmid. The recombinant plasmid is then moved into the nonpermissive target strain by transformation, and the transformation mixture is plated on nutrient agar medium containing the antibiotic. Since the plasmid cannot replicate in this strain, the only way the target strain can become resistant to the antibiotic is if the chromosomal sequences flanking the drug marker in the plasmid undergo homologous recombination with the same DNA sequences on the actual chromosome. This double recombination event "surgically" removes the target gene from the chromosome and replaces it with the antibiotic resistance marker. In our example, the newly constructed mutant can grow in the presence of antibiotic, but only if histidine is present.

12.2 You have monitored expression of a gene fusion where *lacZ* is fused to the glutamate decarboxylase gene, *gadA*. You find that there is a 50-fold increase in expression of *gadA* (as measured for β-galactosidase) when the cells carrying this fusion are grown in media at pH 5.5 as compared to cells grown at pH 8. What additional experiment must you do to tell if the control is transcriptional or translational?

ANSWER: You must compare these results with those obtained with an operon fusion to the same gene. If both fusions show an equal 50-fold increase, then the gene is controlled at the transcriptional level. If the gene fusion shows a 50-fold increase in induction but the transcriptional fusion shows no induction, then regulation occurs postranscriptionally, most likely at the translational level.

12.3 What would you conclude if the Northern blot of the *gad* genes showed no difference in mRNA levels in cells grown at pH 7.7 and pH 5.5 but the Western blot showed more protein at pH 5.5 than at pH 7.7?

ANSWER: Regulatory control would likely be at the posttranscriptional level, either through increasing mRNA stability, translational efficiency, or protein stability.

12.4 Another regulator of glutamate decarboxylase (Gad) production does not affect the production of *gad* mRNA but is required to accumulate Gad protein. What two regulatory mechanisms might account for this phenotype?

ANSWER: Control of translation or control of protein degradation.

12.5 How could you use two-hybrid analysis to determine if *three* proteins can interact? You suspect that proteins A and C can simultaneously bind to protein B and that this interaction then allows A to interact with C in the complex.

ANSWER: Place three plasmids in the cell. One makes protein A fused to the *Gal*4 DNA-binding domain, the second plasmid makes protein C fused to the *Gal*4 activator domain, and the third plasmid simply makes protein B. If the yeast cell only contains the first two plasmids, no interaction will occur and activation of *lacZ* will not take place. If all three plasmids are in the same yeast cell, protein B will be the center of a sandwich, linking the A and C fusion proteins. The fusion proteins can then interact and activate *lacZ*.

12.6   How would you have to modify real-time PCR to quantitate the level of a specific mRNA?

ANSWER: You would have to first convert the mRNA into cDNA using the enzyme reverse transcriptase. An oligonucleotide primer hybridizing to the to the 3′ end of the mRNA is acted on by reverse transcriptase to synthesize DNA (called complementary DNA, or cDNA). Normal real-time PCR techniques can then quantitate the cDNA.

12.7   You have just employed phage display technology to select for a protein that tightly binds and blocks the eukaryotic cell receptor targeted by anthrax toxin. When this receptor is blocked, the toxin cannot get into the target cell. You suspect that the protein may be a useful treatment for anthrax. How would you recover the gene following phage display and then express and purify the protein product?

ANSWER: The gene can be cut out of the phage genome using restriction enzymes or amplified by PCR using known phage DNA sequences that flank the gene. The fragment can then be cloned into a His$_6$ tag expression vector. This requires that the sequence of the gene be known so that the DNA encoding the His$_6$ tag can be placed in-frame with the open reading frame of the receptor-blocking protein. The plasmid containing the His$_6$-tagged protein gene is then induced to overexpress the protein in *E. coli*. The vector will contain an inducible promoter (e.g., *lacP*) to drive expression of the gene. The His$_6$-tagged protein can then be purified by pouring cell extracts over a nickel column, as described in the text.

12.8   Would insect resistance to an insecticidal protein be a concern when developing a transgenic plant? Why? How would you design a transgenic plant to limit the possibility of insects developing resistance?

ANSWER: Insects have, in fact, developed resistance to single insecticidal proteins. The most common resistance mechanism involves a change in the membrane receptors in the midgut to which activated *Bt* toxins bind. Resistance can be due to a reduced number of *Bt* toxin receptors or to a reduced affinity of the receptor for the toxin. Some insects, such as the spruce budworm, can inactivate specific toxins by precipitating them with a protein complex present in the midgut.

While developing a transgenic plant, several steps could be taken to limit the development of resistance in an insect population. The insecticidal gene could be fused to a promoter that is only expressed at a time when the plant is most susceptible to attack. Alternatively, the gene could be fused to a promoter that is only expressed in a tissue of the plant that is most vulnerable to attack. This would limit the time during which the insects can develop resistance. For instance, cotton plants attacked by bollworms could produce toxin only in young boll tissues, the most important part of the plant. In addition to specifically protecting the critical plant tissue, this strategy would only affect one generation of bollworms, avoiding the constant selection pressure that hastens evolution of resistance. Another technique would be to engineer two different insecticidal proteins into the plant genome that will not exhibit cross-resistance. In other words, even if an insect develops resistance to one toxin, it will still remain susceptible to the second.

Chapter 13

13.1   Why can't photosynthesis be driven by solar microwaves or by X-rays?

ANSWER: The energy of microwaves is a hundred times lower than that of visible light and near infrared. Microwaves have too little energy to form or break chemical bonds. On the other hand, photons of higher energy such as X-rays break molecular bonds indiscriminately, in a manner that enzymes cannot control.

13.2   Methanogens are a small proportion of the microbial community in soil, and the ΔG of methanogenesis is small. Yet methanogens produce large volumes of methane, large enough to contribute significantly as a greenhouse gas to global warming. Why would methanogens produce a relatively large quantity of waste product?

ANSWER: For methanogenesis, the ΔG per reaction cycle is small. Thus, the organism must run numerous cycles in order to store sufficient energy to build biomass.

13.3   The soil bacterium *Geobacter* can metabolize acetic acid by the following reaction:

$$CH_3COOH + 2H_2O \rightarrow 2CO_2 + 4H_2$$

Yet the $\Delta G°′$ is +95 kJ/mol. What could happen in the soil enabling *Geobacter* to grow?

ANSWER: *Geobacter* must grow in the presence of different species of bacteria that oxidize the H_2 produced by *Geobacter*. Removal of H_2 drives forward the acetate catabolism. The sum of the acetate reaction plus the hydrogen oxidation yields a negative $\Delta G°′$. This joint metabolism of two species is called syntrophy. The rate of reaction in this syntrophy is further enhanced at high temperature, which magnifies the term $-\Delta RT \ln K$.

13.4   After ATP has been dephosphorylated to ADP, how else might ADP provide energy for the cell?

ANSWER: Within the cell, some enzymes can hydrolyze ADP to AMP, breaking the second high-energy phosphodiester bond.

13.5 Linking an amino acid to its cognate tRNA is driven by ATP hydrolysis to AMP plus pyrophosphate. Why release PP_i instead of P_i?

ANSWER: The formation of aminoacyl-tRNA must be irreversible until the ribosome is ready to release the tRNA. The pyrophosphate from ATP is immediately cleaved to $2P_i$, preventing the reversal of aminoacyl-tRNA formation.

13.6 In the microbial community of the bovine rumen, the actual ΔG value has been calculated for glucose fermentation to acetate:

$$C_6H_{12}O_6 + 2H_2O \rightarrow 2C_2H_3O_2^- + 2H^+ + 4H_2 + 2CO_2$$
$$\Delta G = -318 \text{ kJ/mol}$$

If the actual ΔG for ATP formation is 44 kJ/mol and each glucose fermentation yields four molecules of ATP, what is the thermodynamic efficiency of energy gain? Where does the lost energy go?

ANSWER: The energy efficiency is (4 × 44 kJ/mol)/(318 kJ/mol) × 100 = 55%. The remaining energy is dissipated as heat. (Kohn and Boston, 2000.)

13.7 What would happen to the cell if pyruvate kinase catalyzed PEP conversion to pyruvate but failed to couple this reaction to ATP production?

ANSWER: If PEP conversion to pyruvate occurred without ATP production, then the energy liberated by this reaction would be lost as heat. No energy would be channeled into cell function and growth.

13.8 Some bacteria make an enzyme, dihydroxyacetone kinase, that phosphorylates dihydroxyacetone to dihydroxyacetone phosphate. Why would this enzyme be useful?

ANSWER: Bacteria can obtain dihydroxyacetone from their environment, using a transporter protein, then phosphorylate the substrate and direct it into glycolysis. Some bacteria can grow on dihydroxyacetone as a sole carbon source.

13.9 In the EMP pathway, how are the two water molecules generated?

ANSWER: Each water molecule is removed from 2-phosphoglycerate, converting $R-CH_2OH$ to $R=CH_2$. The water molecule accounts for the water equivalent from incorporation of P_i during oxidation by NAD^+. The P_i ultimately is used to convert one ADP to ATP, a reaction that includes water formation.

13.10 Explain why the ED pathway generates only one ATP, whereas the EMP pathway generates two.

ANSWER: The EMP pathway primes the six-carbon sugar with two phosphoryl groups. The sugar then splits into two three-carbon units (glyceraldehyde 3-phosphate), each of which generates two ATPs for one of the original ATPs. By contrast, the ED pathway only phosphorylates the sugar once before it splits in two. The phosphorylated end yields glyceraldehyde 3-phosphate, which enters the EMP pathway to generate ATP, ending up as pyruvate. The unphosphorylated three-carbon unit yields pyruvate directly, with no ATP.

13.11 How does the structure of coenzyme A resemble that of NADH? How does it differ?

ANSWER: Coenzyme A and NADH each consist of ADP esterified to a nitrogenous compound. The linked compound for coenzyme A, however, is not nicotinamide (as in NADH) but a linear molecule terminating in a thiol (SH).

13.12 Compare the reactions catalyzed by pyruvate dehydrogenase and pyruvate formate lyase (see Section 13.4). What conditions favor each reaction, and why?

ANSWER: The pyruvate dehydrogenase complex (PDC) is favored in the presence of oxygen because the electrons transferred to NADH can enter the electron transport chain, eventually combining with oxygen to release energy. In the absence of oxygen, pyruvate formate lyase is favored to yield fermentation products that can be excreted from the cell without reducing more energy carriers. At high pH, formate and acetate production is especially favorable because the extra acid counteracts alkalinity.

13.13 Suppose a cell is pulse-labeled with ^{14}C-acetate (fed the ^{14}C label briefly, then "chased" with unlabeled acetate). Can you predict what will happen to the level of radioactivity observed in isolated TCA intermediates? Plot a curve showing your predicted level of radioactivity, as a function of number of rounds of the cycle.

ANSWER: The amount of radioactivity measured in TCA intermediates will rise steeply as labeled acetate is incorporated, then will decrease by half with each succeeding cycle, as the order of the carbons is randomized by succinate.

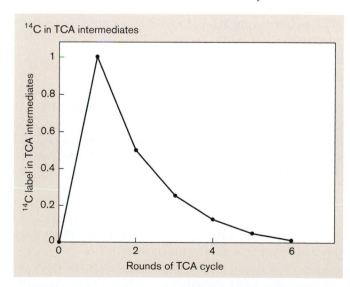

13.14   How might the enzymes that catalyze acenaphthene catabolism be discovered?

ANSWER: A genomic library of *Pseudomonas* strain A2279 might be used to transform another species that lacks the ability to degrade acenaphthene. If a transformant capable of degradation is found, the genes can be mapped and the deduced amino acid sequences compared with known enzymes to provide clues as to their mode of action. Other genes in the same operons might be found to catalyze other steps in the catabolic pathways.

Chapter 14

14.1   *Pseudomonas fluorescens,* a soil bacterium at neutral pH, oxidizes NADH with nitrate (NO_3^-) at the neutral pH. What is the value of $E°'$?

ANSWER: To calculate the reduction potential $E°'$:

$$NADH + H^+ + NO_3^- \rightarrow NAD^+ + NO_2^- + H_2O$$

$$E°' = 320 \text{ mV} + 420 \text{ mV} = 740 \text{ mV}$$

14.2   Could a bacterium obtain energy from succinate as an electron donor with nitrate (NO_3^-) as an electron acceptor?

ANSWER: For nitrate reduction:

$$\text{Succinate} + NO_3^- \rightarrow \text{fumarate} + NO_2^- + H_2O$$

$$E°' = -33 \text{ mV} + 420 \text{ mV} = 387 \text{ mV}$$

Although succinate is a relatively poor electron donor, nitrate is a strong electron acceptor. This reaction should provide energy for bacterial metabolism.

14.3   What do you think happens to $\Delta\psi$ as the cell's external pH increases or decreases? What happens to Δp in bacteria growing at a pH two or three units above their internal pH?

ANSWER: As external pH changes, ΔpH increases or decreases, affecting the magnitude of Δp. Some bacteria, such a *Bacillus* species, regulate their electrical potential with counter ions in order to maintain overall Δp at about 200 mV. Other species, such as *E. coli,* seem to tolerate a range of Δp from near zero to 250 mV. Alkaliphiles growing at high pH appear to spend much of their Δp to maintain the inverted ΔpH. How they stay energized is not understood.

14.4   Suppose that de-energized cells ($\Delta p = 0$) with an internal pH 7.6 are placed in a solution at pH 6. What do you predict will happen to the cell's flagella? What does this demonstrate about the function of Δp?

ANSWER: The flagella will rotate, driven solely by the ΔpH component of Δp. This result is consistent with the hypothesis that the trans-membrane potential Δp drives flagellar rotation.

14.5   What is the advantage of the oxidoreductase transferring electrons to a pool of mobile quinones, which then reduce the terminal reductase (cytochrome complex)? Why does each oxidoreductase not interact directly with a cytochrome complex?

ANSWER: The mobile quinone pool serves to connect diverse electron donors with diverse electron acceptors. If each oxidoreductase had to interact specifically with a different terminal oxidase, the pathways of electron transport would be limited; for example, NADH might only donate electrons to O_2, whereas succinate might only donate electrons to nitrate. Instead, all potential electron donors can be coupled with all potential acceptors.

14.6   Why are most electron transport proteins fixed within the cell membrane? What would happen if they "got loose" in aqueous solution?

ANSWER: If the electron transport proteins came away from the membrane into water solution, they could carry their energized electrons back into the cytoplasm or lose them outside the cell. In either case, they could no longer convert the electron energy to a proton gradient.

14.7   Why would bacteria evolve dehydrogenases that fail to pump protons even when sufficient donor potential exists?

ANSWER: Dehydrogenases that fail to transport protons despite sufficient reduction potential are needed for cases in which reoxidation of the donor is required but the proton gradient is already high enough to drive generation of ATP. Too high a transmembrane voltage can disrupt the membrane.

14.8   Experiments suggest that NADH dehydrogenase 2 is more often coupled with cytochrome *bo,* and dehydrogenase 1 more often with cytochrome *bd.* Why might this be the case?

ANSWER: The non-proton-pumping oxidoreductase NDH-2 may be used to moderate the amount of proton gradient generated with abundant oxygen. The proton-pumping NDH-1 may be useful under low-oxygen conditions, where the cell is forced to use the high-affinity oxidase (cyt *b/d*), which fails to pump the extra $2H^+$.

14.9   How might the bacterial cell compensate for environmental conditions in which Δp falls below 100 mV, a suboptimal level for ATP synthesis?

ANSWER: To generate sufficient ATP when Δp is low, the bacteria could synthesize a greater number of F_1F_o complexes for the membrane. The overall rate of ATP synthesis would thus be increased.

14.10   Would *E. coli* be able to grow in the presence of an uncoupler that eliminates the proton potential supporting ATP synthesis?

ANSWER: Yes, *E. coli* can grow with the proton gradient eliminated, but only with a rich supply of nutrients for substrate phosphorylation to generate ATP (for example, from glycolysis). Also, the external pH and salt levels must be maintained close to those of the cytoplasm, to minimize the need for ion transport.

14.11 The proposed scheme for uranium removal requires injection of acetate under highly anoxic conditions, with less than 1 part per million (ppm) dissolved oxygen. Why must the acetate be anoxic?

ANSWER: Oxygen is the strongest terminal electron acceptor. If O_2 is present, bacteria will use it preferentially (instead of U^{6+}) to oxidize the acetate to CO_2.

14.12 Hydrogen gas is so light that it rapidly escapes from Earth. Where does all the hydrogen come from to be used for hydrogenotrophy and methanogenesis?

ANSWER: Hydrogen is produced in substantial quantities as a by-product of fermentation. It may seem surprising that organisms would readily excrete quantities of energy-rich H_2, but in the absence of a good electron acceptor (or the enzymes to utilize electron acceptors), hydrogen may be just another waste product. Hydrogen gas trapped underground supports large communities of methanogens and hydrogenotrophs. The human colonic bacteria generate so much hydrogen that all parts of the body show traces of hydrogen gas.

14.13 Suppose you discover bacteria that require a high concentration of Fe^{2+} for photosynthesis. Can you hypothesize what the role of Fe^{2+} may be? How would you test your hypothesis?

ANSWER: The organism uses reduced iron as an electron donor for its photosystem ($Fe^{2+} \rightarrow Fe^{3+}$). To test this, grow the organism on a defined concentration of Fe^{2+}. Measure the amount of iron oxidized and the amount of carbon fixed into biomass; if the Fe^{2+} is an electron donor for the photosystem, the two numbers should show a linear correlation.

Chapter 15

15.1 Propose a simple experiment to reveal the key intermediate to receive CO_2.

ANSWER: Add a very short pulse of $[^{14}C]\ CO_2$ to the cells. The pulse must contain sufficient CO_2 to begin fixation, but insufficient to continue beyond one "turn" of the cycle. In this case, the radiolabel will accumulate in the intermediate that needs to assimilate CO_2 for the next round. This kind of experiment ultimately identified the key intermediate, ribulose 1,5-bisphosphate.

15.2 Speculate why rubisco catalyzes a competing reaction with oxygen. Why might researchers be unsuccessful in attempting to engineer rubisco without this reaction?

ANSWER: The oxygenation reaction might have an essential function in regulation of metabolism. For example, it might help prevent excessive reduction of cell components or fixation of too much carbon to be used in biosynthesis. Given the universal existence of the oxygenation reaction in bacterial and chloroplast rubiscos, it seems unlikely that oxygenation serves no purpose. For this reason, the attempt to engineer rubisco without oxygenation may not succeed.

15.3 Why does ribulose 1,5-bisphosphate have to contain two phosphoryl groups, whereas the other intermediates of the Calvin cycle contain only one?

ANSWER: Only ribulose 1,5-bisphosphate needs to split into two molecules (3-phosphoglycerate). Each of the two products needs to have its own phosphate as a tag for the enzymes to recognize it within the cycle.

15.4 Which catabolic pathway (see Chapter 13) includes some of the same sugar-phosphate intermediates as those of the Calvin cycle? What might this suggest about the evolution of the two pathways?

ANSWER: The pentose phosphate pathway includes ribulose 5-phosphate, erythrose 4-phosphate, and sedoheptulose 7-phosphate in a similar series of carbon exchanges. Perhaps these pathways evolved from a common amphibolic pathway of sugar consumption and biosynthesis. Alternatively, the one pathway evolved earlier, then the sugar intermediates were available for evolution of the second pathway.

15.5 For a given species, uniform thickness of a cell membrane requires uniform chain length of its fatty acids. How do you think chain length may be regulated?

ANSWER: In *E. coli,* the chain length of a growing fatty acid appears to be limited by beta-ketoacyl-ACP synthase, which binds only precursor acyl-ACPs shorter than 18 carbons.

15.6 Suggest two reasons why transamination is advantageous to cells.

ANSWER: Transamination enables cells to store amine groups in nontoxic form, readily available for biosynthesis. The availability of multiple enzymes of transamination from different amino acids enables cells to quickly recycle existing resources into the amino acids most needed by the cell in a given environment. For example, if a sudden supply of glutamine appears, cells can immediately distribute its amines into all twenty amino acids.

15.7 Which energy carriers (and how many) are needed to make arginine from 2-oxoglutarate? (See **Fig. 15.24**.)

ANSWER: Arginine biosynthesis requires three ATP molecules and three NADPH molecules (including two for converting two molecules of 2-oxoglutarate to glutamate). An additional ATP is spent converting acetate to acetyl-CoA.

15.8 Why are purines and pyrimidines not synthesized separately from the sugar phosphate?

ANSWER: Purines and pyrimidines are highly hydrophobic, insoluble in the cytoplasm. The ribose phosphate component solubilizes the molecule, enabling synthesis to occur in the cytoplasm, where needed to make RNA and DNA.

15.9 Why are the ribosyl nucleotides synthesized first, then converted to deoxyribonucleotides as necessary? What does this suggest about the evolution of nucleic acids?

ANSWER: Ribonucleic acid is believed to be the original chromosomal material of cells. Cells evolved to synthesize RNA first; then later, as DNA was used, pathways evolved to synthesize it by modification of RNA, which the cell already had the ability to make.

Chapter 16

16.1 Why do the lipid components of food experience relatively little breakdown during fermentation?

ANSWER: Lipids are highly reduced molecules, largely hydrocarbon with relatively low oxidizing potential. Thus, lipids cannot undergo as many intramolecular redox reactions as do sugars, which readily generate energy through anaerobic fermentation.

16.2 Why does oxygen allow excessive breakdown of food, compared with anaerobic processes?

ANSWER: Oxygen functions as the terminal electron acceptor for the complete breakdown of all kinds of organic molecules to water and CO_2.

16.3 In an outbreak of listeriosis from unpasteurized cheese, only the refrigerated cheeses were found to cause disease. Why would this be the case?

ANSWER: In the cheeses kept at room temperature, other naturally occurring bacteria outgrew the pathogenic *Listeria*, whereas in the refrigerator, only the *Listeria* could grow. (Note, however, that many other potential pathogens, such as *Salmonella*, are inhibited by refrigeration.)

16.4 Cow's milk contains 4% lipid (butterfat). What happens to the lipid during cheese production?

ANSWER: Lipids undergo little catabolism, because the fermentation conditions are anaerobic. During coagulation, lipid droplets become trapped in the network of denatured protein and are largely retained in the bulk of the cheese. "Low-fat" cheeses are made from skim milk, which eliminates the lipids before fermentation.

16.5 In traditional fermented foods, without pure starter cultures, what determines the kind of fermentation that occurs?

ANSWER: The fermentation can be controlled by introducing a crude starter culture obtained from a previous batch of the food product or from a natural source of a particular microbe; for example, rice straw is a source of *Bacillus natto* for natto production. The fermentation type can be manipulated by the addition of factors, such as brine, that retard growth of all but a few strains. In pidan, for example, the high concentration of sodium hydroxide limits bacterial growth to alkali-tolerant strains of *Bacillus*.

16.6 Compare and contrast the role of fermenting organisms in the production of cheese and bread.

ANSWER: In cheese production, fermentation causes major biochemical changes in the food, such as the buildup of acids and the breakdown of proteins to smaller peptides and amino acids. Minor by-products, such as methanethiols and esters accumulate to levels that confer flavors. In yeast bread, by contrast, the only significant product of fermentation is the carbon dioxide that leavens the dough. The small amount of ethanol produced evaporates during cooking. A form of bread in which extended fermentation does generate flavor is injera, for which the dough ferments for three days.

16.7 Compare and contrast the role of low-concentration by-products in the production of cheese and beer.

ANSWER: In both cheese and beer, minor by-products such as esters contribute flavor. In both cases, oxidation leads to off-flavors. In cheese, however, the exclusion of oxygen usually prevents off-flavors. In beer, the yeast requires a low level of oxygen; thus, significant amounts of acetaldehyde and diacetyl are produced and must be eliminated by a secondary fermentation.

16.8 Why would bacteria convert trimethylamine oxide (TMAO) to trimethylamine? Would this kind of spoilage be prevented by exclusion of oxygen?

ANSWER: TMAO acts as a terminal electron acceptor—that is, an alternative to oxygen for anaerobic respiration, as discussed in Chapter 14. Exclusion of oxygen inhibits only aerobic bacteria; TMAO respirers continue to grow and can spoil the fish.

16.9 Is it possible for physical or chemical preservation methods to completely eliminate microbes from food? Explain.

ANSWER: Preservation methods either slow microbial growth or induce microbial death. Microbial death follows a negative exponential curve, as discussed in Chapter 5. In theory, the exponential curve never reaches zero, so total exclusion of microbes is impossible. In practice, there is a high probability of totally eliminating microbes if the treatment time extends several "half-lives" beyond the time at which microbial concentration declines to less than one per total volume.

16.10   Why would different industrial strains or species be used to express different kinds of cloned products?

ANSWER: Different industrial strains have biochemical systems that favor different products. Some fungi naturally possess the highly complex pathways to generate antibiotics, as well as regulatory timing to turn on these pathways after the culture has grown to high population density. On the other hand, bacteria such as *B. subtilis* are the most genetically tractable and predictable in their growth cycles and the easiest to manipulate to express recombinant products such as human genes.

16.11   Why would an herbicide resistance gene be desirable in an agricultural plant? What long-term problems might be caused by microbial transfer of herbicide resistance genes into plant genomes?

ANSWER: Introduction of an herbicide resistance gene allows application of higher amounts of herbicide to crops in order to control growth of weeds. But the higher concentrations of herbicide may also have greater side effects on animals and on human consumers of the crop. In the long run, the herbicide resistance gene is likely to escape into weed plants through natural genetic transfer mechanisms. Thus, eventually the weeds may require still higher concentrations of herbicide. While the costs versus benefits of new gene modifications remain poorly understood, it must be recognized that all modern crops today are the product of many generations of genetic manipulation.

Chapter 17

17.1   What would have happened to life on Earth if the sun were a different stellar class, substantially hotter or colder than it is now?

ANSWER: If the sun were hotter, too much ultraviolet and gamma radiation would reach the Earth, breaking chemical bonds of living organisms so rapidly that life could not be sustained. If the sun were colder, too little radiation with sufficient energy would be available to drive photosynthesis. In either case, life as we know it could not have evolved on Earth.

17.2   Outline the strengths and limitations of each model of the origin of living cells. Which aspects of living cells does each model explain?

ANSWER: The three models are complementary in that each explains aspects of modern cells not addressed by the others. The prebiotic soup model accounts for the major classes of compounds used by cells, such as nucleosides, TCA cycle intermediates, amino acids, and fatty acids. It also suggests the origin of membranes as soap bubble-like micelles. It does not, however, account for the evolution of metabolic pathways and replication of genetic informa-

tion. The metabolist model accounts for the prevalence of major cellular reactions such as carbon fixation and the TCA cycle. It does not explain the evolution of membranes and genetic material. The RNA world accounts for the central role of RNA in living cells; of all molecular classes, RNA and ribonucleotides probably serve the widest range of functions as information carriers, agents of catalysis, and genetic regulators. The origin of membranes is not addressed.

17.3   Suppose living organisms were to be found on Mars. How might such a find shed light on the origin and evolution of life on Earth?

ANSWER: If life on Mars were to show a completely different basis than that of Earth—for example, based on silicon polymers instead of carbon—such a find would support the view that life originated independently on each planet, rather than traveling from one planet to the other; or that both planets were seeded from somewhere else. If life on Mars is based on similar macromolecules, perhaps even showing the same genetic code, this would support the view that life arose on Mars first; or that both planets were seeded from the same source.

17.4   Based on **Figure 17.18,** what would be the identification of a straight, nonsheathed gram-negative bacterium that has sulfur granules, is motile, and is 1.0 μm wide? What would happen if you happened to assign the bacterium a width of 0.9 μm?

ANSWER: The bacterium would key out as *Beggioatoa* sp. If the cell width were measured as 0.9 μm, however, the identification would proceed down a completely different, wrong track, at the first step of the key. This is one disadvantage of the dichotomous key.

17.5   In **Figure 17.19B**, were any test results atypical for *Salmonella enterica*? What would happen if another atypical result had been obtained? What would happen if the actual isolate had not been one of those in the database?

ANSWER: The negative result for dulcitol fermentation is atypical. If you multiply the probabilities, for most of the tests a second atypical result will decrease the score for *S. enterica* by about an order of magnitude but will still leave it at the top of the list. Thus the probabilistic indicator is relatively tolerant of atypical results for species within the database. On the other hand, if an organism outside the database is isolated it cannot be correctly identified.

17.6   What kind of DNA sequence changes have no effect on gene function?

ANSWER: Base substitutions that do not change the amino acid specified by the codon have no effect on gene function. For example, CUA → CUG still encodes leucine. Additionally, it turns out that a majority of the amino acids in any given protein can be replaced by an amino acid of similar form (for example leucine → valine) without significantly affecting function of the gene.

17.7   Solve the three alignment problems shown in **Figure 17.22**. How could errors or ambiguities in alignment lead to mistakes in interpreting the molecular clock?

ANSWERS:
A. Add the new sequence from organism F to this alignment:

Seq A	C C C C A G C U U C G G C U G G G G G A G G
Seq B	C C U U A G C G A A A G C U – A A G G A G G
Seq C	C C U C A G C G U G A G C U – G A G G A G G
Seq D	C C C A A G C U U U – G C U A U G G G A G G
Seq E	C C – – A G C U U U G G C U – – – G G A G G
Seq F	**C C – A A G C G A G A G C U – U G G – A G G**

B. Make an alignment of these RNAs:

RNA 1	– G G G G G U U C G C C C C C A A –
RNA 2	– G G G A G U U U G C U C C C A A –
RNA 3	– G A G G G U U C G C C C U C A U –
RNA 4	– G G G A G G A A A C U C C C A C A
RNA 5	– G A G G G G U G A C U C U C A C A
RNA 6	G G A G G G U U U – C C C U C A U –

C. Which group of Archaea does 'Isolate X' most likely belong to based on the RNA shown?

ANSWER:

Position Number	Isolate X	Crenarchaea	Euryarchaea	Korarchaea
4	C	(C or U)	C	U
5	A	U	U	A
8	G	A	G	C
11	G	C	G	G
15	C	R (A or G)	C	U
18	U	U	U	C
19	G	A	G	Y (C or U)
20	A	A	A	U
21	A	A	A	Y (C or U)
22	G	G	G	A

Isolate X shows 9/10 key positions in common with the Euryarchaea, as compared with 5/10 Crenarchaea and 2/10 Korarchaea. Thus Isolate X most likely belongs to the Euryarchaean clade.

Errors in alignment could lead to overestimate of the time of divergence between organisms. For example, if position 15 were aligned by counting bases in Isolate X without skipping the loop extension, base 15 could be given as G, which would lead to only 8/10 positions in common with Euryarchaea.

17.8   What are the major sources of error in constructing phylogenetic trees?

ANSWER: Phylogenetic trees are affected by variability in the number of substitutions, or rate of mutation in different strains. The tree is distorted by errors in sequence alignment, and by systematic errors due to failure of the fundamental assumptions of the molecular clock. These assumptions include the constant rate of mutation for all branches; constant generation time; and true orthology of the gene chosen (that is, the encoded product has the same function and hence the same degree of selection pressure in all taxa under consideration).

17.9   What are the limits of evidence for horizontal gene transfer in ancestral genomes? What alternative interpretation might be offered?

ANSWER: Horizontal gene transfer is inferred based on the appearance of genes in clade A that are absent from other members of the clade but present in clade B. The degree of similarity between genes in the two clades, however, must be high enough to exclude the possibility that the genes in question were retained from a common ancestor of the two clades, but lost from other members of clade A. This possibility is difficult to exclude in the case of deep-branching clades, where all genes have had a long time to diverge. For example, the large number of archaean genes present in deep-branching thermophilic bacteria such as *Thermatoga* may include some inherited from the last common ancestor.

17.10   Besides mitochondria and chloroplasts, what other kinds of entities within cells might have evolved from endosymbionts?

ANSWER: Some of the large "megaplasmids" found in bacteria and protists are as large as genomic chromosomes and contain numerous housekeeping genes. These megaplasmids may have originated as endosymbiotic cells that lost all their membranes through reductive evolution. Similarly, some of the giant viruses such as Mimivirus and smallpox, as well as phages such as T4, possess a wide spectrum of housekeeping genes. These viruses may have originated as cellular parasites that underwent reductive evolution.

Chapter 18

18.1   Which taxonomic groups in **Table 18.1** stain gram-positive and which gram-negative? Which group contains both gram-positive and gram-negative species? For which groups is the Gram stain undefined, and why?

ANSWER: See **Table 18.1**. Most Firmicutes and Actinobacteria stain gram-positive, as do the Cyanobacteria. These bacteria have relatively thick cell walls that retain the stain. The Proteobacteria, Nitrospira, and Bacteroides/Chlorobi groups stain gram-negative. The Thermus/Deinococcus group includes both gram-positive and gram-negative staining members. For Chlamydia and Planctomyces, the

Gram stain is irrelevant because they lack the cell wall that retains the stain. For Spirochetes, many species are too narrow to observe the stain under light microscopy.

18.2 Which groups of species share common structure and physiology within the group? Which groups show extreme structural and physiological diversity?

ANSWER: Cyanobacteria all share oxygenic photosynthesis through thylakoid membranes. Their overall cell structure and organization, however, takes diverse forms. Spirochetes all share the common structure of sheathed flexible spiral with internal flagella; most share anaerobic or facultative heterotrophy. Chlamydia and Planctomyces groups of species each share general structural features. Other groups, particularly Firmicutes and Actinomycetes, show considerable diversity of form and physiology. The Proteobacteria display more extreme diversity of metabolism than any other division.

18.3 What taxonomic questions are raised by the apparent high rate of gene transfer between archaea and thermophilic bacteria?

ANSWER: If gene transfer results in a species containing a quarter of its genes from organisms outside its domain, such a mosaic genome raises questions of how to define the species and the domain. How can a species be defined if its genome contains large portions from distantly related sources? Other interesting questions relate to the means of gene transfer. How do such distantly related organisms as bacteria and archaea maintain a compatible mechanism of gene transfer?

18.4 What are the relative advantages and disadvantages of propagation by hormogonia, as compared with akinetes?

ANSWER: Hormogonia are motile, and thus capable of active chemotaxis toward a more favorable environment. On the other hand, hormogonia have active metabolism that requires nutrition; if the environment lacks nutrients, the hormogonia will die. Akinete cells can persist until environmental conditions improve, but they cannot actively seek out a new location.

18.5 What are the relative advantages and disadvantages of the different strategies for maintaining separation of nitrogen fixation and photosynthesis?

ANSWER: Temporal separation has the advantage that all cells possess both capacities for nitrogen fixation and photosynthesis. On the other hand, it eliminates the ability of a chain of cells to conduct both processes simultaneously—the benefit of heterocysts. Heterocysts face the problem of operating in close proximity to photosynthetic cells generating toxic oxygen; this problem may be solved by associated respiring bacteria. Globular clusters of cells can bury their nitrogen fixers within the cluster; this arrangement effectively excludes oxygen, but it may lack flexibility during environmental change. Endosymbiotic nitrogen fixation within a respiring eukaryote is probably the most effective strategy of all, as the host provides oxygen-removing proteins such as leghemoglobin. Endosymbiosis, however, requires the presence of an appropriate host organism.

18.6 Why would *Streptomyces* produce antibiotics targeting other bacteria?

ANSWER: *Streptomyces* species may produce antibiotics to curb the growth of bacterial competitors with smaller genomes and faster rates of reproduction. The lysed cells release nutrients that feed growing mycelia of *Streptomyces*.

18.7 Why might genes for the proteorhodopsin retinal photopump be more likely to transfer horizontally than the bacteriochlorophyll-based photosystems PS I and PS II?

ANSWER: Proteorhodopsin requires only the one gene encoding the pump, plus one or two genes to produce retinal. This involves a relatively small amount of sequence to transfer, and the encoded products generate proton potential on their own, without requiring interaction with recipient enzymes. By contrast, PSI and PSII each involve multiple electron carriers that must function together and interact with the recipient electron transport chain.

18.8 Why do you think it took many years of study to realize that *Escherichia coli* and other Proteobacteria can grow as a biofilm?

ANSWER: *E. coli* and its relatives grow exceptionally well in liquid culture. Liquid culture is attractive because it enables quantitative measurement of defined aliquots of microbial population. However, repeated subculturing in liquid medium selects for planktonic (non-biofilm) cells. Eventually, the biofilm-forming property may be lost if mutants outgrow the original genotype in liquid medium.

18.9 Compare and contrast the formation of cyanobacterial akinetes; firmicute endospores; actinomycete arthrospores; and myxococcal myxospores.

ANSWER: An endospore forms as the daughter product (forespore) of a single cell. Within the same cell, endospore development is supported by the motherspore, which disintegrates after release of the endospore. Endospores have tough coatings of calcium picolinate; they are heat-resistant. By contrast, arthrospores and myxospores are less durable and are not heat-resistant, although they can persist in the environment for an extended period. Arthrospores form through binary fission of actinomycete filaments. Myxospores are formed by a multicellular fruiting body. In all three cases, spore formation can be induced by depletion of nutrients; and the spore-producing entity is left behind to die.

Chapter 19

19.1   If two deeply diverging clades each show a wide range of growth temperature, what does this suggest about the evolution of thermophily or psychrophily?

ANSWER: The two clades diverged before temperature adaptation occurred, and adaptations to high or low temperature must have evolved independently in the two clades.

19.2   What might be the advantages of flagellar motility for a hyperthermophile living in a thermal spring or in a black smoker vent? What would be the advantages of growth in a biofilm?

ANSWER: Flagellar motility enables isolated cells to detect a new nutrient source, or an approximate temperature range, and approach it through chemotaxis. Growth in a biofilm attached to a substrate prevents the microbes from floating away from the nutrient source, or from being carried away in the flow from the vent.

19.3   What problem with cell biochemistry is faced by acidophiles that conduct heterotrophic metabolism?

ANSWER: Heterotrophic metabolism generates fermentation products such as acetate and lactate, which act as permeant acids. Permeant acids become protonated outside the cell, at low pH; the protonated forms then permeate the membrane, returning into the cell. Given the high transmembrane pH difference maintained by *Sulfolobus,* one would expect even small traces of fermentation acids to cross the membrane in the protonated form, then dissociate and accumulate to toxic levels of organic acids. It is unknown how *Sulfolobus* solves this problem.

19.4   What conclusions might be drawn if viruses of mesophilic archaea are found to have RNA genomes? What if they all have DNA genomes only?

ANSWER: If mesophilic viruses show RNA genomes, then it is likely that only double-stranded DNA is sufficiently stable for viruses to persist in the environment of hyperthermophiles. If only double-stranded DNA viruses are found throughout the archaea, this would suggest that all archaeal viruses evolved from viruses infecting a common ancestral cell that was a thermophile. The latter hypothesis would require supporting evidence from archaeal cell physiology and phylogeny.

19.5   What do the multiple metal requirements suggest about the evolution of methanogens?

ANSWER: The requirement for so many different metals may suggest that methanogens evolved in habitats such as geothermal vents where superheated water carries up high concentrations of dissolved metal ions.

19.6   Compare and contrast the metabolic options available for *Pyrococcus* and for the crenarchaeote *Sulfolobus.*

ANSWER: *Sulfolobus* catabolizes sugars and amino acids aerobically, using O_2 as terminal electron acceptor. *P. abyssi* catabolizes sugars and amino acids anaerobically, using S^0 as terminal electron acceptor. *Sulfolobus* oxidizes sulfur lithotrophically, from S^{2-} to S^0, SO_3^{2-}, and ultimately SO_4^{2-}. *P. abyssi,* however, reduces sulfur lithotrophically with H_2 to form H_2S.

19.7   Compare and contrast sulfur metabolism in *Pyrococcus* and in *Ferroplasma.*

ANSWER: *Pyrococcus* species reduce S^0 with hydrogens from organic substrates, forming HS^- and H_2S. *Ferroplasma* species oxidize sulfur in the form of FeS_2 (using oxidant Fe^{3+}) forming sulfuric acid. The result is extreme acidification of their environment.

Chapter 20

20.1   Why would yeasts remain unicellular? What are the relative advantages and limitations of hyphae?

ANSWER: Yeast grow in environments with sufficient dissolved nutrients to absorb from the medium. The advantage of forming hyphae is that they enable penetration of other organisms, and hence provide access to nutrients. On the other hand, hyphae formation limits the rate of dispersal of progeny cells. Since yeasts grow in environments where dissolved nutrients can be absorbed from the medium, they do not need to produce hyphae and can proliferate more rapidly than mycelial fungi.

20.2   Why would some fungi have lost their sexual life cycle? What are the advantages and limitations of sexual reproduction?

ANSWER: The sexual life cycle involves significant genetic and metabolic costs to the organism. Reductive evolution leading to loss of sexual reproduction might eliminate an energy drain and perhaps enable greater proliferation with fewer resources. On the other hand, the sexual life cycle provides a valuable means of generating diversity through genetic recombination, so that the population may respond to environmental change. Asexual fungi, like prokaryotes, must rely on mutation and gene transfer by viruses and mobile sequence elements to generate genetic diversity.

20.3   What are the advantages and limitations of motile gametes, as compared to non-motile spores?

ANSWER: Motile gametes have the advantage of rapid dispersal on their own and the potential for chemotaxis toward a food source or toward a gamete of the opposite mating type. On the other hand, motility uses up energy that could alternatively be invested in production of a greater number of non-motile gametes. Motile gametes are especially useful in a watery habitat, but are little use in a terrestrial habitat, where air currents or animal hosts must be used for dispersal.

20.4 Compare the life cycle of an oomycete (**Figure 20.17C**) with that of a chytridiomycete (**Figure 20.11C**). How are they similar, and how do they differ?

ANSWER: Both chytridiomycetes and oomycetes undergo alternation of generations. Each possesses the alternative route of an asexual diploid cycle between mycelium, motile zoospores and cysts. At the cellular and molecular level, however, the two kinds of organism differ radically. The chytridiomycete flagellum differs from the oomycete paired flagella, one of which has brush-like side hairs. The oomycete gametes have highly specialized forms that fuse by a process completely different from that of chytridiomycete zoospore fusion.

20.5 How are coralline algae able to grow at greater depths than coral?

ANSWER: Corals generally grow symbiotically with green algae, which contain chlorophyll but lack accessory photopigments. Corallines are red algae, possessing both chlorophyll and phycoerytherins. The latter absorb blue and green light that are not absorbed by green algae, and thus coralline algae penetrate deeper in the water column.

20.6 What might happen when an ameba phagocytoses algae?

ANSWER: If light is available, the algae may be retained as endosymbionts providing energy through photosynthesis. For example, *Chlorachnion* possesses obligate chloroplast-bearing endosymbionts descended from green algae. Alternatively, the ameba may digest all but the algal chloroplast, which persists for some time providing photosynthetic products.

20.7 What kind of habitats would favor a flagellated ameba?

ANSWER: A dilute watery habitat would favor flagella, which allow more rapid propulsion than pseudopods. Pseudopod motility requires a solid substrate.

20.8 Compare and contrast the processes of conjugation in ciliates and in bacteria.

ANSWER: Conjugation in ciliates is a completely different process from conjugation in bacteria, although the purpose (genetic exchange) is similar. In ciliates, two cells form a bridge allowing cytoplasm to flow directly between them, whereas in bacteria, a donor cell attaches to another (by pili in some cases), then a protein complex transfers DNA across both cell envelopes, without direct cytoplasmic contact. In bacterial conjugation, DNA is transferred unidirectionally from the donor cell to the recipient, whereas in ciliates there is reciprocal exchange of DNA. A donor bacterium generally transfers only part of its genome, whereas ciliates exchange entire copies of theirs.

20.9 For ciliates, what are the advantages and limitations of conjugation, as compared with gamete production?

ANSWER: The process of conjugation avoids the necessity of dissolving the intricate cell structure of the ciliate in order to form gametes that fuse or fertilize each other. On the other hand, conjugation requires two diploid organisms to find each other and make contact for several hours, during which time feeding is suspended and the pair is vulnerable to predation.

Chapter 21

21.1 From Chapters 13–15, give examples of microbial metabolism that fit patterns of assimilation and dissimilation.

ANSWER: Microbes can assimilate carbon either by reducing carbon dioxide or by oxidizing methane. Dissimilation of organic carbon occurs by fermentation and respiration. Nitrogen is assimilated by N_2 fixation, and by incorporation of NH_4^+ into glutamine and glutamate. Nitrogen is dissimilated by deamination of amino acids, and by lithotrophic oxidation.

21.2 Does "viable but nonculturable" mean that growth is permanently shut down?

ANSWER: Nonculturable bacteria might be induced to grow again, perhaps when the correct growth conditions are discovered. Some "unculturable" isolates have been made to grow by new methods after decades of attempts. Alternatively, some nonculturable cells may come from multicellular assemblages such as biofilms in which reproductive growth is limited to a portion of the cells. In this case, some of the population is permanently prevented from growth, but other cells continue to grow.

21.3 How do viruses select for increased diversity of microbial plankton? Why would this be so?

ANSWER: Since viruses tend to infect only a narrow host range, their existence favors the evolution of a large number of different species with highly dispersed populations. Highly dispersed populations minimize the chance of viral transmission from one host to another.

21.4 Design an experiment to test the hypothesis that the presence of mycorrhizae enhances plant growth in nature.

ANSWER: Such an experiment requires a control based on the natural environment, where various unknown factors may be very different than in the laboratory. One possibility is to compare the growth of seedlings in natural soil versus sterilized natural soil. However, this experiment would not prove that fungi are the cause of enhanced growth in unsterile soil. The sterilization procedure (usually involving heat and pressure) could break down key nutrients in the soil. A follow-up experiment might be to grow the plants in the presence of a fungus inhibitor in sterilized and unsterilized soil.

21.5   High levels of nitrate or ammonium ion corepress the *nod* genes through NodR. Why?

ANSWER: Nitrate and ammonia are the main forms of nitrogen assimilated by plants. If they are abundant in the soil, the plant does not need rhizobial symbionts; therefore, development of rhizobia-legume symbiosis is inhibited.

21.6   One unanswered question is, how do symbiotic rhizobia reproduce? Why do bacteroids develop if they cannot proliferate?

ANSWER: Various answers have been proposed. Not all of the invading bacteria become bacteroids; some continue to undergo cell division, particularly within senescing tissues of the plant. These bacteria benefit from plant growth, which is sustained by the bacteroids whose genes they share. Alternatively, the entire plant-bacteroid system may benefit rhizobia that grow just outside the plant, in the rhizosphere.

21.7   Compare and contrast the processes of plant infection by rhizobia and by fungal haustoria.

ANSWER: Both rhizobial bacteria and fungal haustoria penetrate the volume of a plant cell, but they keep the plant cell membrane intact, its invagination always surrounding the invading cell. Rhizobia establish a complex, highly regulated exchange of nutrients with the host, receiving catabolites and oxygen in exchange for ammonium and cycling the components of amino acids. By contrast, haustoria establish one-way removal of nutrients such as sucrose, while providing no nutrients in return. Fungal pathogens weaken the structure of the host plant and decrease or halt its growth.

21.8   Why does ruminant fermentation leave food value for the animal host? How is the animal able to obtain nourishment from waste products that the microbes could not use?

ANSWER: The rumen interior is anaerobic. In the absence of oxygen as a terminal electron acceptor, microbes are forced to generate waste products in which the electrons are put back onto the electron donors (fermentation; see Chapter 13). When the short-chain fatty acid wastes enter the animal's bloodstream, the blood is full of oxygen, which enables complete digestion to CO_2 and water.

21.9   How do you think cattle feed might be altered or supplemented to decrease methane production?

ANSWER: Several methods have been proposed to limit methanogenesis. One is to feed cattle the antibiotic monensin, an inhibitor of sodium transport, which is required for methanogenesis. Another approach is to feed cattle an electron acceptor for H_2, such as fumarate, which bacteria use to generate short-chain acids instead of methane. These approaches have been used with only partial success—not surprising given the complexity of the system.

Chapter 22

22.1   Why is oxidation state critical for the acquisition, usability, and potential toxicity of cycled compounds? Cite examples based on your study of microbial metabolism.

ANSWER: Many examples can be cited. In the case of carbon, CO_2 can be fixed by many species with substantial input from photosynthesis or hydrogen donors. The reduced form methane, however, can be assimilated only by methanotrophic bacteria, usually with oxygen as electron acceptor. Nitrogen gas can be assimilated only by nitrogen-fixing bacteria and archaea, whereas NH_4^+ can be assimilated by many plants and microbes. But NH_4^+ can also be oxidized by lithotrophs to NO_3^-, a substance potentially toxic to humans.

22.2   What would happen if wastewater treatment lacked microbial predators? Why would the result be harmful?

ANSWER: Without predators, too many planktonic bacteria would remain in the wastewater after sedimentation of the sludge. The bacteria could be killed by chlorination, but the treated water would have significant BOD because the bacterial remains provide an organic carbon source for respirers.

22.3   How many kinds of biomolecules can you recall that contain nitrogen? What are the usual oxidation states for nitrogen?

ANSWER: Amino acids, nucleotide bases, polyamines for DNA stabilization, peptidoglycan (both amino sugar and peptide chains), and the heme derivatives of cytochromes, chlorophyll, and vitamin B_{12} all include nitrogen (as do many other biochemicals). The oxidation states of nitrogen in living organisms are nearly always reduced, either $R-NH_2$, $R=NH$, or $R-N=R$. An exception is the neurotransmitter NO (nitric oxide).

22.4   In the laboratory, which kind of bacteria would likely grow on artificial medium including NH_2OH as the energy source: *Nitrosomonas* or *Nitrobacter*?

ANSWER: *Nitrosomonas* is more likely to utilize NH_2OH, since it performs the intermediate oxidation of NH_2OH during nitrification of ammonia.

22.5 The nitrogen cycle has to be linked with the carbon cycle, since both contribute to biomass. How might the carbon cycle of an ecosystem be affected by increased input of nitrogen?

ANSWER: One hypothesis is that the injection of nitrogen into an ecosystem accelerates growth of producers (phytoplankton in the ocean or trees in a forest) and therefore facilitates net removal of CO_2 from the atmosphere. Overall, however, the additional fixed carbon may end up dissipated by consumers and decomposers.

22.6 Compare and contrast the cycling of nitrogen and sulfur. How are the cycles similar? How are they different?

ANSWER: Both nitrogen and sulfur cycling involve interconversion between different oxidation states. Most of these interconversion reactions are performed solely by microbes, many of them solely by bacteria. Examples include nitrification of ammonia and denitrification to N_2, as well as sulfide oxidation and photolysis. In both cases, oxidation produces strong acids (HNO_3, H_2SO_4). The major sources and sinks differ; nitrogen is obtained primarily from the atmosphere as N_2, whereas sulfur is at high levels in the ocean and soil. Sulfur is rarely limiting, whereas nitrogen frequently is. Sulfur participates extensively in phototrophy; nitrogen shows little involvement in phototrophy; phototrophy based on nitrate reduction has been observed.

22.7 Compare and contrast the cycling of nitrogen and phosphorus. How are the cycles similar? How are they different?

ANSWER: Nitrogen and phosphorus are both limiting nutrients in many ecosystems—marine, aquatic, and terrestrial. Addition of either element into an aquatic system may cause algal bloom and eutrophication. On the other hand, the two elements differ in their major sources: the atmosphere for nitrogen, and crustal rock for phosphate. Within biomass, nitrogen exists almost entirely in reduced form, whereas phophorus is entirely oxidized. Phosphorus cycles through the biosphere mainly as inorganic or organic phosphates, whereas nitrogen cycles through a broad range of oxidation states, from NH_3 to NO_3^-.

22.8 Do you think that iron fertilization would succeed in maintaining lowered atmospheric CO_2? What consequences might you predict for widespread iron fertilization?

ANSWER: The algal bloom could occur in a population of toxin-producing algae, which would cause massive die-off of fish. Alternatively, if iron fertilization were widespread, the increased population density of algae would lead to increased viral predation, which would mineralize the algae, returning their carbon to the atmosphere. Other scenarios are possible, but it appears unlikely that iron fertilization would lead to a sustained decrease in atmospheric CO_2.

Chapter 23

23.1 How can an anaerobic microorganism grow on skin or in the mouth, both of which are exposed to air?

ANSWER: Facultative organisms living in proximity to the anaerobes will deplete oxygen in the environment, especially around nooks and crannies (e.g., between teeth and gums, gingival pockets) that would ordinarily prevent anaerobes from growing. These small spaces have limited access to oxygen.

23.2 What can a person with mitral valve prolapse do to prevent formation of subacute bacterial endocarditis when visiting the dentist?

ANSWER: Take antibiotics prophylactically. A high dose of antibiotic (usually amoxicillin or acephalosporin) taken 1 hour before the procedure will produce a high enough blood level to kill bacteria. A patient that is hypersensitive to β-lactam antibiotics such as amoxicillin will be prescribed an alternate antibiotic such as azithromycin.

23.3 Why do many gram-positive microbes that grow on the skin, such as *S. epidermidis*, grow poorly or not at all in the gut?

ANSWER: Bile salts present in the intestine (not on the skin) easily gain access to and destroy cytoplasmic membranes of gram-positive organisms (unless the organism possesses bile hydrolases). Gram-negative microbes have an extra protection in the form of an outer membrane and so can survive better in the intestine.

23.4 How might normal flora escape the intestine and cause disease at other body sites?

ANSWER: Normal flora can escape through intestinal perforations resulting from gunshot or knife wounds, surgery, or cancer.

23.5 Why can the colon be considered a fermenter?

ANSWER: The contents are continually flowing through the intestinal tube, with food containing substrates for fermentation ingested at one end and waste containing fermentation products removed from the other.

23.6 Why do defensins have to be so small?

ANSWER: They need to be small so they can get through the outer membrane of gram-negative organisms and the thick peptidoglycan maze of gram-positive organisms.

23.7 If NK cells can attack infected host cells coated with antibody, why won't neutrophils?

ANSWER: Actually, neutrophils (PMNs) *can* attack infected cells coated with antibody—but the killing mechanism is different than ADCC. Human neutrophils do not make perforin nor the other ADCC-related compounds, called granzymes, used by NK cells to kill target cells. In addition to that difference,

NK cells possess a type of Fc receptor not found on neutrophils, which means the intracellular signaling pathways are different between NK cells and neutrophils. Neutrophils can, however, be activated when their Fc receptors bind antibody. Activated neutrophils make reactive oxygen products and can release a variety of peptides including defensins, cathelicidin, and myeloperoxidase, which can all damage target cells.

23.8 If increased fever limits bacterial growth, why would bacteria make pyrogenic toxins?

ANSWER: The pyrogenic toxins have other effects that compromise and damage the host. The toxins can induce cytokines that damage local host cells or confuse the immune system. This can provide the pathogen with nutrients and help it hide from the immune system longer. Pyrogenic toxins include lipopolysaccharide and protein toxins such as toxic shock syndrome toxin (see Chapter 25).

Chapter 24

24.1 Two different stretches of amino acids in a single protein form a three-dimensional antigenic determinant. Will the specific immune response to that three-dimensional antigen also respond to one of the amino acid stretches alone?

ANSWER: Most likely no. It is the three-dimensional shape formed by the two stretches that is recognized as an antigen. A denatured protein that contains both amino acid stretches will not possess the three-dimensional shape of the antigen. On the other hand, there can be other specific immune responses that involve different subsets of lymphocytes that individually recognize each of the amino acid stretches that form the three-dimentional antigenic determinant. As an analogy, take a computer image of a friend's face and shuffle the facial features. Turn the nose upside down, exchange the eyes with the mouth, and lower the ears. Since you were programmed to respond to the original facial configuration, you likely would not recognize the rearranged face as a whole. But you might find that the nose looks familiar.

24.2 How does a neutralizing antibody that recognizes a viral coat protein prevent infection by the associated virus?

ANSWER: Neutralizing antibodies usually bind attachment proteins on the virus and sterically prevent them from binding to host cell receptors (see **Fig. 24.4**). Some antibodies to enveloped viruses might trigger the complement cascade (see later in the chapter), destroying the membrane.

24.3 There are immune disorders in which an individual overproduces a specific class of antibody, for example hypergammaglobulinemia. How could the radial immunodiffusion technique be used to identify what class of antibody is in excess?

ANSWER: Because they are proteins, immunoglobulin antibodies are also antigens. Therefore, antibodies can be made that react to epitopes in the heavy chains. Antibodies to human IgG can be raised in rabbits, for example. Radial immunodiffusion using agarose impregnated with rabbit anti-human IgG antibodies can be used to measure the concentration of human IgG in serum loaded into a well. The larger the diameter of the resulting ring, the more IgG there is in the serum sample. A patient that has an unusually high amount of IgG antibody relative to other antibody isotypes, as compared to a normal healthy person, would be diagnosed with IgG hypergammaglobulinemia.

24.4 The mother of a newborn was found to be infected with rubella, a viral disease. Infection of the fetus could lead to serious consequences for the newborn. How could you determine if the newborn was infected while in utero?

ANSWER: Since maternal IgM antibodies cannot cross the placenta, finding IgM antibodies to rubella antigens in the newborn's circulation indicates that the fetus was infected and initiated its own immune response. If the newborn has only IgG antibodies to rubella (no IgM antibodies), this indicates that the child was not infected and that maternal IgG crossed the placenta.

24.5 Why does the delta region have no switch region?

ANSWER: Because B cells at the early stages have both IgM and IgD surface antibodies. Recombination at the DNA level is not involved because alternative RNA-splicing events after transcription determine whether an IgM or IgD molecule is made.

24.6 Why do individuals with type A blood have anti-B and not anti-A antibodies?

ANSWER: Because the B-cell population that would react to type A antigen was deleted during B-cell maturation.

24.7 Why do immunizations lose their effectiveness over time?

ANSWER: Because memory B cells eventually die. Without some exposure to antigen, those memory cells will not be replaced.

24.8 Transplant rejection is a major consideration when transplanting most tissues because host T_C cells can recognize allotypic MHC on donor cells. So why are corneas easily transplanted from a donor to just about any other person?

ANSWER: The cornea is not normally vascularized. So even though corneal cells express MHC proteins, circulating host T cells do not have an opportunity to interact with them. The cornea will not be rejected. This is called an immune privileged site.

24.9   Why does attaching a hapten to a carrier protein allow production of anti-hapten antibodies?

ANSWER: B cells with anti-hapten surface antibody (as part of the B-cell receptor) can take up hapten but cannot present the hapten to a helper T cell. The same B cell can also take up the hapten bound to a carrier molecule, and because the carrier molecule is larger than the hapten, the B cell will present the carrier epitope to the helper T cell. The helper T cell stimulates the B cell, which was already programmed to make anti-hapten antibody, to differentiate into plasma and memory B cells.

24.10   Do bone marrow transplants in a patient with severe combined immunodeficiency require immunosuppressive chemothereapy?

ANSWER: No. Since the patient has no T cells to recognize foreign antigens, the transplant is not rejected.

24.11   How can a stem cell be differentiated from a B cell at the level of DNA?

ANSWER: Gene splicing will have taken place in the B cell but not the stem cell. Thus, the B cell will have fewer cassettes for each of the V, D, and J regions, while the stem cell will have all of them. PCR techniques can be used to view those differences.

Chapter 25

25.1   Is a microbe with an LD_{50} of 5×10^4 more or less virulent than a microbe with an LD_{50} of 5×10^7?

ANSWER: Because it takes fewer cells to cause disease, the microbe with the smaller LD_{50} (5×10^4) is the more virulent.

25.2   Antibodies to which subunit of cholera toxin will best protect a person from the toxin's effects?

ANSWER: Antibodies to the B subunit will be more protective. Inactivating the B subunit will prevent binding of the toxin to cell membranes. The A subunit active site is typically sequestered in these toxins and inaccessible to antibody. Furthermore, once the A subunit has entered a host cell, antibodies cannot enter and neutralize it.

25.3   Would patients with iron overload (excess free iron in the blood) be more susceptible to infection?

ANSWER: Withholding iron from potential pathogens is a host defense strategy because when iron is plentiful, the microbe does not have to expend energy to get it and so can readily grow. On the other hand, low iron can also be a signal to express various virulence genes, so for some organisms, high iron might hinder infection.

25.4   How might you experimentally determine if a pathogen secretes an exotoxin?

ANSWER: The microbe can be grown in liquid culture and the cells removed either by centrifugation or filtration. If the organism makes an exotoxin, it may well be present in the cell-free supernatant. The presence of a toxin can be determined by injecting the supernatant into an animal model (e.g., mice) and examining the result (death or altered function; see tetanus toxin below). Alternatively, the supernatant can be administered to a layer of tissue culture cells and the health of the monolayer noted.

25.5   Internet problem: What other toxins are related to the cholera enterotoxin A subunit? B subunit?

ANSWER: Go to Pub-med (http://www.ncbi.nlm.nih.gov/PubMed/) In the search pop-down window mark "Protein." Then type "cholera enterotoxin" or "cholera B subunit." Click on the appropriate result. Highlight and copy the amino acid sequence at the bottom of the document. Then, back at the NCBI home page, click on "Blast", select "Protein-Protein Blast," paste the B subunit sequence into the Search box and click "Blast." After clicking "Format" on the next page, all the homology results will be displayed. Items with E-values having large negative numbers represent the most significant homologies. Note that the toxin of *Citrobacter freundii*, another gram-negative rod, possesses a B subunit similar to that of cholera, but not a similar A subunit (you must perform a second "Blast" for this). Scroll down the page to find the actual alignments between the query protein and homolog.

25.6   Protein and DNA have very different structures. Why would a protein secretion system be derived from a DNA-pumping system?

ANSWER: Conjugation systems actually move DNA that is attached to a pilot protein at the 5′ end. A pilot protein is made by the conjugation system, then binds to the 5′ end of the DNA to be transferred, and "pilots" the DNA through the conjugation pore. So a modified conjugation system that moves protein only is not as much of a leap as might initially be thought.

25.7   How can one determine if a bacterium is an intracellular parasite?

ANSWER: Microscopic examination to see if bacteria are found within cultured mammalian cells is usually not satisfactory. The difficulty lies in determining if the organism is inside the host cell or just bound to its surface—or, if it is inside, whether it is a live or dead bacterium. One commonly used approach is to add to infected cell monolayers an antibiotic that can kill the microbe but will not penetrate the mammalian cells. The protein synthesis inhibitor gentamicin is typically used. A bacterium that invades a host cell will gain sanctuary from gentamicin and grow intracellularly. Extracellular bacteria and bacteria attached to the outside of the host cell are killed. Counting viable colony-forming units of bacteria released from the mammalian cells by gentle detergent treatments at various times will reveal if the organism grew intracellularly. Of course, this will only work if the microorganism is not an obligate intracellular parasite.

25.8   Why would killing a host be a bad strategy for a pathogen?

ANSWER: The goal of any microbe is to maintain its species. If a microbe does not have an opportunity to easily spread to a new host, killing its host would be tantamount to suicide.

25.9   Why do rhinovirus infections fail to progress beyond the nasopharynx?

ANSWER: For one thing, rhinoviruses are susceptible to acid pH (pH 3.0), so they are unable to replicate in the gastrointestinal tract. They also grow best at 33°C, which may help explain their predilection for the cooler environs of the nasal mucosa.

Chapter 26

26.1   Why would treatment of an infection sometimes require multiple antibiotics?

ANSWER: Sometimes treatment is initiated before knowing the infecting microbe or its susceptibility to different drugs. Without knowing the antibiotic susceptibility of the infecting agent, a physician will use multiple antibiotics to ensure that one will kill or inhibit the growth of the microbe. Combinations of antibiotics are also used because different antibiotics are effective against different sets of microbes (e.g., gram-positive vs. gram-negative, or aerobe vs. anaerobe).

26.2   Why do you think most urinary tract infections occur in women?

ANSWER: The major reason is anatomy. Most bladder UTIs come from access through the urethra. Since the urethra in men is longer than in women, it usually takes a catheter to introduce bacteria into a male bladder. The trip for the infecting organism in women is much shorter. However, in people older than 50, UTIs become more common in both men and women, with less difference between the sexes. The reason is not clear.

26.3   Urine samples collected from six hospital patients were placed on a table at the nurse's station awaiting pickup from the microbiology lab. Several hours later, a courier retrieved the samples and transported them to the lab. The next day, the lab reported that four of the six patients had UTI. Would you consider these results reliable? Should you start treatment based on these results?

ANSWER: Because urine is good growth media for many bacteria, the delay of several hours in picking up the sam-

ples gave the organisms time to replicate and increase their numbers. Consequently, the lab results should be viewed with suspicion.

26.4   Considering that *N. gonorrhoeae* is exquisitely sensitive to ceftriaxone, why do you suppose the patient was also treated with tetracycline? And why can one person be infected repeatedly with *N. gonorrhoeae*?

ANSWER: Because STDs often travel in pairs and because the initial symptomolgy is similar, the clinician will want to "cover" the patient for possible chlamydia infection. Chlamydiae are not susceptible to ceftriaxone. The large, single dose of ceftriaxone (as opposed to smaller multiple injections given over days) was given because patients with gonorrhea are typically poorly compliant and fail to return for subsequent injections. A person who experiences gonorrhea is not protected from reinfection because the organism's surface antigens undergo phase variations (see Section 10.6) and may downregulate the immune system (see **Fig 26.13C**).

26.5   Human immunodeficiency virus was discussed in this section and in Chapter 25. Like the plague, it is a blood-borne disease. Why, then, do fleas and mosquitoes fail to transmit HIV?

ANSWER: When insect vectors take a blood meal, they typically defecate or regurgitate simultaneously. So theoretically, they could serve as a vector for HIV. Pathogens such as the West Nile virus that are transmitted by insect vectors actually grow in their insect hosts; however, HIV does not. Because it is unusually fragile, HIV will die quickly. There have been no cases of HIV transmitted by insect vectors.

26.6   Normal cerebral spinal fluid is usually low in protein and high in glucose. The protein and glucose content does not change much during a viral meningitis, but bacterial infection leads to greatly elevated protein and lowered glucose levels. What could account for this?

ANSWER: There are several explanations. Bacteria and infiltrating PMNs will consume glucose and alterations in the blood-brain barrier can lead to decreased transport of glucose into the spinal fluid. As a result of these factors, glucose levels plummet. Growth of the bacteria and infiltration of PMNs accounts for the increase in CSF protein levels. Viruses are very small and consist mostly of nucleic acids, so even at high numbers, they will not significantly add to CSF protein content. Since viruses do not grow in CSF directly, they will not consume glucose. In addition, viral meningitis does not cause a great infiltration of PMNs into the CSF, another reason glucose levels remain high and protein levels remain low.

26.7   Knowing the symptoms of tetanus, what kind of therapy would you use to treat the disease?

ANSWER: At the first sign of muscle spasm, antitoxin should be given. If tetany is severe, muscle relaxants can relieve spasms.

26.8   How do the actions of tetanus toxin and botulinum toxin actually help the bacteria colonize or obtain nutrients?

ANSWER: This is a difficult question to answer. Few scientists have speculated. Recall that the toxins are encoded by genes in resident bacteriophages that became part of the clostridial genome through horizontal transfer from some other source. Since the organisms normally reside in soil, the actual function of these toxins may have something to do with survival in that habitat. The toxin's effects on humans may simply be an unfortunate accident. However, with tetanus toxin, there may be benefit in that muscle spasms could limit oxygen delivery to infected tissues, enabling a more anaerobic environment for growth. Cell death may also release iron or other nutrients useful to *C. tetani*.

26.9   If *C. botulinum* is an anaerobe, how might botulism toxin get into foods?

ANSWER: The most common way botulism toxin gets into food today is via home canning. The canning process involves heating food in jars to very high temperatures. The heat destroys microorganisms and drives out oxygen; both processes help to preserve foods. If the jars are not heated to sterilization conditions, spores of *C. botulinum* will survive. When the jars are cooled for storage at room temperature, the spores germinate; the organism then grows in the anaerobic medium and releases the toxin. When the food is eaten, the toxin is eaten, too.

26.10   A patient presenting with high fever and in an extremely weakened state is suspected of having a septicemia. Two sets of blood cultures are taken from different arms. One bottle from each set grows *Staphylococcus aureus* yet the laboratory report states that the results are inconclusive. New blood cultures are ordered. Why might this be?

ANSWER: There must have been something different about the two strains of staphylococci grown in the separate bottles. For example, they may have different antibiotic susceptibility patterns when tested against a battery of antibiotics. One strain may be sensitive to penicillin while the other strain is resistant. Since the expectation is that a single strain initiated the infection, both isolates should exhibit the same drug susceptibility pattern. The laboratory suspects contamination from separate sources.

Chapter 27

27.1   The drug tobramycin is added to a concentration of 1,000 µg/ml in a tube of broth from which serial two-fold dilutions were made. Including the initial tube (tube 1), there are a total of ten tubes. Twenty-four hours after all the tubes are inoculated with *Listeria monocytogenes,* turbidity is observed in tubes 6–10. What is the MIC?

ANSWER: 62.5 µg/ml, the concentration in tube 5, the last tube with no growth. (Relative to tube 1, tube 5 has been diluted 2^4 (16-fold dilution: 1,000/500/250/125/62.5.)

27.2   What additional test performed on an MIC series of tubes will tell you whether a drug is bacteriostatic or bactericidal?

ANSWER: Streaking a portion of the broth from the dilution tubes showing no growth. If the drug is bacteriostatic, colonies will form on the agar plate because during streaking, the bacteria are removed from the presence of the drug. If the antibiotic is bactericidal, no colonies will form because the organisms are dead before plating. This method determines the minimum bactericidal concentration (MBC) of an antibiotic. MBC would be defined as the lowest dilution that did not yield viable cells.

27.3   You are testing whether a new antibiotic will be a good treatment choice for a patient with a staph infection. The Kirby-Bauer test using the organism from the patient shows a zone of inhibition of 15 mm around the disk containing this drug. Clearly, the organism is susceptible. But you conclude that the drug would not be effective in the patient. What would make you draw this conclusion?

ANSWER: If the average attainable tissue level of the drug is below the MIC, the drug will not be effective.

27.4   When treating a patient for an infection, why would combining a drug such as erythromycin with a penicillin be counterproductive?

ANSWER: Erythromycin, a bacteriostatic drug, will stop growth, which indirectly stops cell wall synthesis and renders the microbe insensitive to penicillin.

27.5   The enzyme DNA gyrase, a target of the quinolone antibiotics, is an essential protein in DNA replication. The quinolones bind to and inactivate this protein. Research has proved that quinolone-resistant mutants contain mutations in the gene encoding DNA gyrase. If the resistant mutants contain a mutant DNA gyrase and DNA gyrase is essential for growth, why are these mutations not lethal?

ANSWER: The mutations cause changes in DNA gyrase that have little to no effect on function but that do prevent the drug from binding. Thus, the mutant organism will continue to twist its DNA and grow well with or without the drug.

27.6   Why might a combination therapy of an aminoglycoside antibiotic and cephalosporin be synergistic?

ANSWER: The two drugs given together could act synergistically because the cephalosporin can weaken the cell wall and allow the aminoglycoside easier access to the cell interior, where it can attack ribosomes. This is especially useful in organisms that have some resistance to both drugs.

27.7   Could genomics ever predict the drug resistance phenotype of a microbe? If so, how?

ANSWER: Yes. If the organism's genome possesses genes whose deduced protein sequences harbors significant similarity to antibiotic resistance proteins from other organisms, one can predict a similar drug resistance. Definitive proof of drug resistance requires actual in vitro testing.

27.8   Fusaric acid is a cation chelator that normally does not penetrate the *E. coli* membrane, which means *E. coli* is typically resistant to this compound. Curiously, cells that develop resistance to tetracycline become *sensitive* to fusaric acid. Resistance to tetracycline is usually the result of an integral membrane efflux pump that pumps tetracycline out of the cell. What might explain the development of fusaric acid sensitivity?

ANSWER: Fusaric acid is imported by the tetracycline efflux pump. This phenomenon can be used to isolate mutants with deletions of transposons encoding tetracycline resistance. Transposons, such as Tn*10*, which carries tetracycline resistance, can spontaneously delete at around a frequency of 10^{-6}, but finding one out of a million tetracycline-susceptible cells is impossible without a positive selection. Fusaric acid provides that positive selection, since the cell with the Tn*10* deletion will be resistant to fusaric acid.

27.9   Mutations in the ribosomal protein S12 (encoded by *rpsL*) confer resistance to streptomycin. Given a cell containing both *rpsL*⁺ and *rpsL*ᴿ genes, would the cell be streptomycin resistant or sensitive? (Recall that genes encoding ribosome proteins for the small subunit are designated *rps*. A + indicates the wild-type allele, while R indicates a gene whose product is resistant to a certain drug.

ANSWER: This merodiploid cell would contain two sets of ribosomes, one set, containing normal S12, would be sensitive to streptomycin; another set, containing the resistant S12, would be resistant. Because streptomycin causes mistranslation of mRNA on sensitive ribosomes, inappropriate proteins that can kill the cell would still be synthesized. Thus, the cell would remain sensitive to streptomycin. Note, however, that the recessive nature of antibiotic resistance seen in this case is not the norm. Resistance is a dominant trait in a majority of cases.

Chapter 28

28.1   Use **Figure 28.6** to identify the organism from the following case. A sample was taken from a boil located on the arm of a 62-year-old man. Bacteriological examination revealed the presence of gram-positive cocci that were also catalase-positive, coagulase-positive, and novobiocin resistant.

ANSWER: *Staphylococcus aureus.* The novobiocin test is irrelevant in this situation.

28.2   Why does finding IgM to West Nile virus indicate current infection? Why wouldn't finding IgG do the same?

ANSWER: Upon infection with any organism, IgM antibodies are the first to rise. After a short time the levels of IgM decline as IgG levels rise. IgG, however, can remain in serum for years, making it a poor prognosticator of current infection.

28.3   Why does adding albumin or powdered milk prevent false positives in ELISA?

ANSWER: Antibodies are proteins. They can stick to plastic just as easily as the antigen being tested. Without blocking all the possible binding sites on plastic with albumin, any antibody from the patient's serum could stick to the plastic instead of to the antigen and react with the secondary enzyme-conjugated antihuman antibody.

28.4   Specific antibodies against an infectious agent can persist for years in the bloodstream. So how is it possible that antibody titers can be used to diagnose diseases such as infectious mononucleosis?

ANSWER: During the course of a disease, the body's immune system will increase the amount of antibody made specifically against the infectious agent. Thus, one compares the antibody titer in a blood sample taken from a patient in the active, or acute, phase of disease with the antibody titer several weeks later, when the patient is in the recovery, or convalescent, stage. Seeing a greater than fourfold rise in a specific antibody titer (for example, in mononucleosis) indicates that the patient's immune system was responding to the specific agent. Remember, simply finding IgG against an organism or virus in serum only indicates that the patient was exposed to that microbe at some time in the past.

28.5   Two blood cultures, one from each arm, were taken from a patient with high fever. One culture grew *Staphylococcus epidermidis,* but the other blood culture was negative (no organisms grew out). Is the patient suffering from septicemia caused by *S. epidermidis*?

ANSWER: Probably not. *S. epidermidis* is a common inhabitant of the skin and could easily have contaminated the needle when blood was taken from the patient. The fact that only one of the two cultures grew this organism supports this conclusion. If the patient had really been infected with *S. epidermidis,* both blood cultures would have grown this organism.

28.6   A 30-year-old woman with abdominal pain went to her physician. After the examination, the physician asked the patient to collect a midstream urine sample that they would send to the lab across town for analysis. The woman complied and handed the collection cup to the nurse. The nurse placed the cup on a table at the nurse's station. Three hours later, the courier service picked up the specimen and transported it to the laboratory. The next day, the report came back "greater than 200,000 CFUs/ml; multiple colony types; sample unsuitable for analysis." Why was this determination made?

ANSWER: Although the CFU number is high enough to consider relevant, UTIs are typically caused by a single organism. The fact that the lab found many different colony types suggests a problem with specimen collection. In this case, not refrigerating the sample allowed the small number of urethral contaminants to overgrow the specimen.

28.7 Methicillin, a beta-lactam antibiotic, is very useful in treating staphylococcal infections. The development of methicillin-resistant strains of *Staphylococcus aureus* (MRSA) is a very serious development because there are few antibiotics that can kill these strains. Imagine a large metropolitan hospital in which there have been eight serious nosocomial infections with MRSA and you are responsible for determining the source of infection so it can be removed. How would you accomplish this using common bacteriological and molecular techniques?

ANSWER: Samples from all the affected patients and from the hospital staff would be screened for the presence of *S. aureus* resistant to methicillin. Each strain would then be analyzed for restriction fragment length polymorphisms (RFLP). Strains from all the patients will likely have identical restriction patterns if they came from the same source. The source is then identified by determining which staff member possesses MRSA with the same pattern. The source may also be inanimate, such as surgical equipment or ventilation apparatus. A connection between patients and specific staff members or instruments would also have to be demonstrated.

28.8 What are some reasons why some diseases spread quickly through a population while others take a long time?

ANSWER: There are several factors. One is mode of transmission; airborne diseases can spread more quickly than food-borne disease, for instance. Sexually transmitted diseases spread slower still. Herd immunity is another factor. Herd immunity is based on the number of individuals within a population that are resistant to a disease. Someone immune to the disease cannot pass it on. The more immune people there are in a population (or herd), the slower the spread of the epidemic to susceptible people.

28.9 Using the "Maps of Outbreaks" provided on the ProMed-mail website, identify the countries considered to have anthrax epidemics.

ANSWER: The latest map available at the time this was written was drawn from reported occurrences in 2003. At that time, 14 countries had epidemic occurrences of anthrax (Turkey, Niger, Chad, Guinea, Sierra Leone, Liberia, Ivory Coast, Ghana, Togo, Ethiopia, Zambia, Zimbabwe, Myanmar, and Commonwealth of Independent States). Many others, such as Mexico, are endemic. The United States has sporadic cases of anthrax. Note that the ProMed-mail website as of 2008 has an interactive map that shows the locations of specific disease outbreaks that have taken place during the past 30 days across the globe.

Glossary

AB toxin A bacterial toxin containing two protein subunits, A and B. Subunit A is the toxic protein and subunit B binds host cell receptors.

ABC transporter An ATP powered transport system that contains an ATP binding cassette.

aberration An imperfection in a lens.

abiotic Produced without organisms; occurring in the absence of life.

absorption In optics, the capacity of a material to absorb light.

acceptor end The amino acid attachment site at the 3' end of tRNAs.

acceptor site (A site) The region of the ribosome that binds an incoming charged tRNA.

accessory protein A protein found in the viral capsid or tegument that is needed early in the viral life cycle.

acid-fast stain A diagnostic stain for mycobacteria, which retain the dye fuchsin due to the presence of mycolic acids in the cell wall.

acidophile An organism that grows fastest in acid (generally defined as below pH 5).

acquired immunodeficiency syndrome (AIDS) A disease caused by HIV that leads to the destruction of T cells and the inability to fight off opportunistic infections.

Actinobacteria A phylum of high GC content gram-positive bacteria.

activated sludge Organic material concentrated from wastewater, containing microbes that digest the material to inorganic compounds.

activation energy The energy needed for reactants to reach the transition state between reactants and products.

activator A protein that increases gene transcription.

active transport An energy-requiring process that moves molecules across a membrane against their electro-chemical gradient.

acyl carrier protein (ACP) A protein that can carry an acetyl group for anabolic pathways such as fatty acid synthesis.

adaptive immune response See **adaptive immunity**.

adaptive immunity Immune responses activated by a specific antigen and mediated by B cells and T cells.

adenosine triphosphate (ATP) A ribonucleotide with three phosphates and the base adenine. It has many functions in the cell including precursor for RNA synthesis and energy carrier.

adenylate cyclase An enzyme that converts ATP into cyclic adenosine monophosphate, cAMP.

adherence The ability of an organism to attach to a substrate.

adhesin Any cell surface factor that promotes attachment of an organism to a substrate.

ADP-ribosyltransferase A bacterial toxin that enzymatically transfers the ADP-ribose group from NAD^+ to target proteins, altering the target protein's structure and function.

adsorb The attachment of virions to a host cell.

aerated (A) horizon The layer of soil below the organic horizon, containing decomposed organic particles and minerals.

aerial mycelium A hypha that extends above the surface and produces spores at its tip.

aerobic respiration The use of oxygen as the terminal electron acceptor in an electron transport chain. A proton gradient is generated and used to drive ATP synthesis.

agar A polymer of galactose that is used as a gelling agent.

Airy disk A bright central point surrounded by rings of light and dark caused by the pattern of interference of spherical wavefronts converging at the focal point.

akinete A specialized spore cell formed by some filamentous cyanobacteria.

alga A microbial eukaryote that contains chloroplasts.

algal bloom An overgrowth of algae on the water surface, caused by an increase in a limiting nutrient.

alkaline fermentation Bacterial fermentation in conjunction with proteolysis and amino acid catabolism that generates ammonia in amounts that raise pH.

alkaliphile An organism with optimal growth in alkali (generally defined as above pH 9).

allergen An antigen that causes an allergic hypersensitivity reaction.

allograft The transplantation of tissue from a donor with one type of major histocompatibility complex (MHC) protein into a recipient with a different type of MHC.

allosteric site A regulatory site on a biological molecule distinct from the ligand/substrate binding site.

allotype An amino acid sequence in the antibody constant region that is shared by some, but not all, members of a species.

alternation of generations A life cycle that alternates between a haploid cell population and a diploid cell population.

alveolar macrophage Macrophages, located in the lung alveoli, that phagocytose foreign material.

Alveolata (alveolate)   A group of ciliated or flagellated protists with complex cortical structure.

ameba (or amoeba)   A protist that moves via pseudopods.

amensalism   An interaction between species that harms one partner but not the other.

amino acid   The monomer unit of proteins. Each amino acid contains a central carbon covalently bound by a hydrogen, an amino group, a carboxylic acid group and a side chain. An exception is proline, in which the side chain is cyclized with the central carbon.

aminoacyl-tRNA transferase   An enzyme that attaches a specific amino acid to the correct tRNA thereby charging the tRNA.

aminoglycosides   Bacteriocidal protein synthesis inhibitors used as antibiotics.

ammonification   The generation of ammonia from organic nitrogen.

Amoebozoa   A taxonomic class of unshelled amebas that move via lobed pseudopods. Also known as Lobosea.

amphibolic   Metabolic pathways that are reversible and can be used for both catabolism and anabolism.

amphipathic   A molecule with both hydrophilic and hydrophobic portions.

amphotericin B   An anti-fungal drug that binds the fungal specific sterol ergosterol and destroys membrane integrity.

anabolism   The building up of complex biomolecules from smaller precursors.

anaerobic photosystem I   A protein complex that harvests light from a chlorophyll, splits an electron from a small molecule such as H_2S or H_2O, and stores energy in the form of NADPH.

anaerobic photosystem II   A protein complex that splits an electron from bacteriochlorophyll and stores energy in the form of a proton potential.

anaerobic respiration   The use of a molecule other than oxygen as the final electron acceptor of an electron transport chain.

anammox reaction   The anaerobic oxidation of ammonium to nitrogen gas (using nitrate as electron acceptor); yields energy.

anaphylaxis   A severe hypersensitivity reaction caused by chemically induced contraction of smooth muscles and dilation of capillaries.

anaplerotic reaction   Metabolic reactions, occurring in all organisms, that fix small amounts of CO_2 to regenerate TCA cycle intermediates.

angle of aperture   The width of a light cone (theta, θ) that projects from the midline of a lens. Greater angles of aperture increase resolution.

anion   A negatively charged ion.

annotation   Deciphering genome sequences, including identification of genes and prediction of gene function.

antenna complex   A complex of chlorophylls and accessory pigments in the photosynthetic membrane that collects photons and funnels them to a reaction center.

anthropogenic   Caused by humans.

anti-anti-sigma factor   A protein that inhibits an anti-sigma factor, allowing the target sigma factor to participate in initiating transcription.

anti-attenuator stem loop   An mRNA secondary structure whose formation prevents assembly of a downstream transcriptional termination stem loop. The anti-attenuator stem permits transcription of the downstream structural genes.

antibiotic   A molecule that can kill or inhibit the growth of selected microorganisms.

antibody   A host defense protein produced by B cells in response to a specific antigenic determinant. Antibodies bind to their corresponding antigenic determinant.

antibody stain   The attachment of a stain to an antibody to visualize cell components recognized by the antibody with high specificity.

antibody-dependent cell-mediated cytotoxicity (ADCC)   The process by which natural killer cells destroy viral protein expressing antibody-coated host cells.

anticodon   Three nucleotides in the middle loop of a tRNA that base pair with a codon in mRNA.

antigen   A compound, recognized as foreign by the cell, that elicits an adaptive immune response.

antigenic determinant   A small segment of an antigen that is capable of eliciting an immune response. An antigen can have many different antigenic determinants.

antigenic shift   A genetic change in a pathogen's surface protein that prevents recognition by antibodies and host immune cells produced in response to the previous version of the protein.

antigen-presenting cell (APC)   An immune cell that can process antigens into antigenic determinants and display those determinants on the cell surface for recognition by other immune cells.

antimicrobial agent   A chemical substance that can kill microbes or slow their growth.

antiparallel   The orientation of the two strands in opposite directions. Commonly refers to a nucleic acid double helix with one strand in the 5' to 3' orientation and the other strand in the 3' to 5' orientation.

antiport   A transport protein in which the molecules being transported move in opposite directions across the membrane.

antisense RNA   A non-coding RNA that binds to a complementary sequence of protein-coding RNA and (usually) prevents its translation.

antisepsis   The removal of pathogens from living tissues.

antiseptic   A chemical that kills microbes.

anti-sigma factor   A protein that inhibits a specific sigma factor, preventing transcription initiation.

AP endonuclease   An enzyme that cleaves the DNA backbone at regions missing a nitrogenous base.

AP site   A position in DNA where there is no base attached to the sugar of the backbone.

apical complex   A specialized structure that facilitates entry of Apicomplexan parasites into host cells.

Apicomplexa   A taxonomic group of parasitic alveolates that possess an apical complex used for entry into a host cell.

apicoplast   An organelle unique to Apicomplexans, derived from genetic reduction of a chloroplast, that no longer performs photosynthesis but provides an essential function in fatty acid metabolism.

apurinic site   A DNA site missing a purine base due to hydrolysis of the bond linking the base to the sugar.

aquaporin   A membrane-embedded channel that increases the rate of water diffusion across the membrane.

arbuscules   The structure formed by penetrating hyphae of endomycorrhizae.

archaea   Prokaryotic organisms that are members of the domain *Archaea*.

Archaea   One of the three domains of life, consisting of organisms with a last common ancestor not shared with members of *Bacteria* or *Eukarya*. Organisms are prokaryotic (lacking nuclei, unlike eukaryotes) and possess ether-linked phospholipid membranes (unlike bacteria).

Archaean eon   The second eon (major time period) of Earth's existence, from 3.8 to 2.5 gigayears (Gyr, 10^9 years) before the present. The earliest geological evidence for life dates to this eon.

archaeon   See *pl.* **archaea**.

aromatic   A ring-shaped organic molecule with π electrons delocalized equally around the ring.

arthrospore   A small vegetative cell, produced by mature mycelia, that gets dispersed.

artifact   A structure viewed through a microscope that is incorrectly interpreted.

Ascomycota (ascomycete)   A taxonomic group of fungi whose mycelia form paired nuclei. Haploid ascospores are produced in pods called asci.

ascospore   The spore produced by an Ascomycete fungus.

ascus *pl.* **asci**   A spore-containing pod produced by Ascomycete fungi.

aseptic   An environment that is free of microbes.

assembly   The packaging of a viral genome into the capsid to form a complete virion.

assimilation   An organism's acquisition of an element, such as carbon from CO_2, to build into body parts.

assimilatory nitrate reduction   The uptake of nitrate by plants and bacteria for use in biosynthetic pathways.

atomic force microscopy (AFM)   A technique that maps the three dimensional topography of a object using van der Waals forces between the object and a probe.

atomic mass   The mass (in grams) of one mole of an element.

atomic number   The number of protons in an atom, it is unique for each element.

ATP synthase   A protein complex that synthesizes ATP from ADP and inorganic phosphate using energy derived from the transmembrane proton potential. It is located in the prokaryotic cell membrane and in the mitochondrial inner membrane.

attach   See **adsorb**.

attenuator stem loop   An intramolecular mRNA structure consisting of a base-paired stem connected by a single-strand loop. The stem loop structure causes transcription to terminate. Its formation requires efficient translation of a leader peptide sequence.

autoclave   An appliance that uses pressurized steam to sterilize materials by raising the temperature above the boiling point of water at standard pressure.

autoimmune response   A pathology caused by lymphocytes that can react to self antigens.

autoinducer   A secreted molecule that induces quorum sensing behavior in bacteria.

autoradiography   The visualization of a radioactive probe by exposing the probed material to X-ray film, followed by photographic development of the film.

autotroph   An organism that can reduce carbon dioxide to produce organic carbon for biosynthesis.

axenic growth   The ability of an organism to grow in the absence of any other species, as, for example, in a pure culture.

azidothymidine (AZT)   A nucleotide analog that inhibits reverse transcriptase and was the first drug clinically used to fight HIV infections.

B cell   An adaptive immune cell that can give rise to antibody-producing cells.

bacillus *pl.* **bacilli**   A bacterium with a linear, rod-like shape.

bacitracin   A topical antibiotic that affects cell wall synthesis.

bacteremia   A bacterial infection of the blood.

bacteria *sing.* **bacterium**   Prokaryotic organisms that are members of the domain

Bacteria   One of the three domains of life, consisting of organisms with a last common ancestor not shared with members of *Archaea* or *Eukarya*. Organisms are prokaryotic (lacking nuclei, unlike eukaryotes) and possess ester-linked phospholipid membranes (unlike archaea).

bactericidal   An agent that kills bacterial cells.

bacteriochlorophyll   The chlorophylls of anaerobic phototrophs; they absorb photons most strongly in the far red end of the light spectrum.

bacteriophage (phage)   A virus that infects bacteria.

bacteriorhodopsin (BR)   An archaeal membrane-embedded protein that contains retinal and acts as a light-driven proton pump; homologous to the bacterial proteorhodopsin.

bacteriostatic   An agent that inhibits the growth of bacterial cells.

bacterium   See *pl.* **bacteria**.

bacteroid   Cell-wall-less, undividing, differentiated rhizobial cell within a plant cell. The bacteroid provides fixed nitrogen for the plant.

Bacteroidetes   A phylum of gram-negative bacteria; nearly all their members are obligate anaerobes.

banded iron formation   A geological formation consisting of layers of oxidized iron (Fe^{3+}) which indicates formation under oxygen-rich conditions.

barophile   See **piezophile**. An organism that requires high pressure to grow.

base excision repair   A DNA repair mechanism that cleaves damaged bases off the sugar-phosphate backbone. After endonuclease activity at the AP site, a new correct DNA strand is synthesized complementary to the undamaged strand.

Basidiomycota (basidiomycete)   A clade of fungi that form mushrooms.

basidiospore   A haploid spore formed by a basidiomycete through meiosis of a basidium.

basidium A diploid cell, formed by the fusion of paired nuclei, that lines the gills of mushrooms (Basidiomycota).

basophile A white blood cell, stained by basic dyes, that secretes compounds that aid innate immunity.

batch culture The growth of bacteria in a closed system without inputs of nutrients.

B-cell receptor A B-cell membrane protein complex containing an antibody in association with the Igα and Igβ immunoglobulins.

B-cell tolerance The exposure of B cells to a high antigen dose, preventing future antibody production against that antigen.

bedrock The region of the Earth's crust where the soil layer ends.

benthic organism Organisms that live on the ocean floor or within the sediment.

beta barrel A cylinder of beta-sheets, found for example in a porin protein.

binary fission The process of replication in which one cell divides to form two daughter cells of equal size.

bioburden The microorganisms that normally inhabit a particular body ecosystem.

biofilm A community of microbes growing on a solid surface.

biogeochemical cycles The recycling of elements needed for life (such as carbon or nitrogen) through the biotic and abiotic components of the biosphere.

biogeochemistry The metabolic interactions of microbial communities with the abiotic (mineral) components of their ecosystems.

bioinformatics A discipline at the intersection of biology and computing that analyzes gene and protein sequence data.

biological oxygen demand (BOD) The amount of oxygen removed from an environment by aerobic respiration.

biological signature See **biosignature**.

biomass The mass found in bodies of living organisms.

bioprospecting The search for organisms with potential commercial applications.

bioremediation The use of microbes to detoxify environmental contaminants.

biosignature (biological signature) A chemical indicator of life.

biosphere The area containing the sum total of all life on Earth.

biosynthesis See **anabolism**. The building up of complex biomolecules from smaller precursors.

biotic Referring to processes caused by living organisms.

black smoker An oceanic thermal vent containing high concentrations of minerals such as iron sulfide.

borreliosis A tick-borne disease caused by *Borrelia burgdorferi*, which may involve skin lesions and arthritis; also known as Lyme disease.

botulism A food borne disease caused by a *Clostridium botulinum* toxin, involving muscle paralysis.

bradykinin A cell signaling molecule that promotes extravasation, activates mast cells, and stimulates pain perception.

bright-field microscopy A type of light microscopy in which the specimen absorbs light and appears dark against a light background.

bubonic plague A disease caused by the bacterium *Yersinia pestis*; characterized by swollen lymph nodes that often turn black.

budding A form of reproduction in which mitosis of the mother cell generates daughter cells of unequal size.

burst size The number of virus particles released from a lysed host cell.

calorimetry A technique to measure the amount of heat released or absorbed during a reaction.

Calvin cycle (Calvin-Benson cycle, Calvin-Benson-Bassham cycle, CBB cycle) The metabolic pathway of carbon fixation in which the CO_2 condensing step is catalyzed by rubisco. Found in chloroplasts and in many bacteria.

candidate species A newly described microbial isolate that may become accepted as an official species.

capsid A protein shell that surrounds a virion's nucleic acid.

capsule A slippery outer layer composed of polysaccharides that surrounds the cell envelope of some bacteria.

carbon dioxide fixation The enzymatic covalent incorporation of inorganic carbon dioxide (CO_2) into an organic compound.

carbon monoxide reductase pathway The carbon fixation process in methanogenic archaea, so called because the key enzyme can fix CO as well as CO_2.

carbon sink An ecosystem that removes carbon dioxide from the atmosphere, for example by fixation into biomass or by dissolving into marine water.

carboxysome A protein-bounded compartment containing rubisco to fix CO_2.

cardiolipin (diphosphatidylglycerol) A double phospholipid linked by glycerol.

carotenoids Accessory photosynthetic pigments that absorb photons in the green end of the spectrum.

catabolism The cellular breakdown of large molecules into smaller molecules, releasing energy.

catabolite repression The inhibition of transcription of an operon encoding catabolic proteins in the presence of a more favorable catabolite, such as glucose.

catalytic RNA An RNA capable of enzymatic reactions. Also known as a ribozyme.

catenane Linked rings of DNA found immediately after replication of circular chromosomes.

cathelicidin An antimicrobial peptide that is synthesized as an inactive precursor and activated extracellularly.

cation A positively charged ion.

cell membrane (plasma membrane, cytoplasmic membrane) The phospholipid bilayer that encloses the cytoplasm.

cell surface receptor A transmembrane protein that senses a specific extracellular signal and may be the docking site for a specific virus.

cell wall A rigid structure external to the cell membrane. The molecular composition depends on organism; composed of peptidoglycan in bacteria.

cell-mediated immunity   A type of adaptive immunity, employing mainly T cell lymphocytes.

cellular slime mold   A slime mold in which the individual cells retain their own cell membranes.

cellulolytic bacteria   Bacteria that catabolyze cellulose; found in the rumen of cattle fed a high cellulose diet.

Cercozoa   A taxonomic group of shelled amebas that have thin, filamentous pseudopods.

chain of infection   The serial passage of a pathogenic organism from an infected individual to an uninfected individual, thus transmitting disease.

chancre   A painless, hard lesion due to an inflammatory reaction at the site of infection with *Treponema pallidum*, the causative agent of syphilis.

chaperone   A protein that helps other proteins fold into their correct tertiary structure.

cheddared curd   Curd that has been cut and piled in order to remove the liquid whey.

cheese   A solid or semi-solid food product prepared by coagulating milk proteins. Its production commonly involves microbial fermentation.

chemoautotroph   An organism that oxidizes inorganic compounds to yield energy and reduce carbon dioxide.

chemoautotrophy   The oxidation of inorganic compounds to yield energy used to reduce carbon dioxide.

chemoheterotroph   An organism that oxidizes organic compounds to yield energy.

chemoheterotrophy   The oxidation of organic compounds to yield energy without the use of light.

chemokine   An attractant for white blood cells that is produced by damaged tissues.

chemokine receptor (CCR)   A human T-cell membrane protein that binds chemokine hormones, but is also used by HIV for attachment and infection.

chemolithotroph (lithotroph, chemoautotroph)   See **chemoautotroph**.

chemostat   A continuous culture system in which the introduced media contains a limiting nutrient.

chemotaxis   The ability of organisms to move towards or away from specific chemicals.

chemotrophy   Metabolism that yields energy from oxidation-reduction reactions without using light energy.

Chlamydia   A phylum of intracellular parasitic bacteria that lose most of their cell envelope during intracellular growth and generate multiple spore-like structures that escape to infect the next host.

Chlorobi   A phylum of gram-negative bacteria. They are obligate anaerobes, "green sulfur" phototrophs that photolyze sulfides or H_2.

chlorophyll   A magnesium-containing pigment that captures light energy at the start of photosynthesis.

chloroplast   An organelle of endosymbiotic origin that conducts oxygenic photosynthesis; found in algae and plant cells.

chlorosome   A membranous photosynthetic organelle found in bacterial groups such as *Chloroflexus*.

choanoflagellate   A microbial opisthokont closely related to animals and fungi.

cholesterol   A sterol lipid found in eukaryotic cell membranes.

chromatin   Chromosomal DNA complexed with proteins. Usually refers to a eukaryotic chromosome.

chromophore   A light-absorbing redox cofactor.

chrysophyte   A clade of flagellated heterokont protists possessing chloroplasts as secondary endosymbionts; also known as "golden algae."

Chydridomycota (chytridiomycete)   A taxonomic group of fungi that produce flagellated zoospores. Includes ruminal endosymbionts and parasites of amphibians.

ciliate   An Alveolate that has paired cilia.

cistron   A functional unit of RNA, containing the information from a single gene.

clade   See **monophyletic**.

class switching   See **isotype switching**.

classical complement pathway   An antibody-mediated pathway for complement activation.

classification   The recognition of different forms of life and their placement into different categories.

clonal   A population of genetically identical cells, all descendants of a single cell.

clonal selection   The rapid proliferation of a subset of B cells during the primary or secondary antibody responses.

cloning vector   A small genome that can carry specific genes for cloning.

cloning   The insertion of DNA into a plasmid where it can be replicated.

CO_2-concentrating mechanism (CCM)   The inducible expression and insertion of CO_2 and bicarbonate (HCO_3^-) transporters into carboxysomes to enhance CO_2 levels near rubisco.

coastal shelf   Shallow regions of the ocean less than 200 meters deep.

coccus *pl.* **cocci**   A spherically shaped microbial cell.

codon   A set of three nucleotides that encodes a particular amino acid.

coenzyme A (CoA)   A non-protein cellular organic molecule that can carry acetyl groups and participates in metabolism.

coevolution   The evolution of two species in response to one another.

cofactor   A metallic ion or a coenzyme required by an enzyme to perform normal catalysis.

cointegrate   A DNA molecule formed by a single site recombination event joining two participating circular DNA molecules.

colony   A visible cluster of microbes on a plate, all derived from a single founding microbe.

commensal organism   An organism that benefits from, but neither helps nor harms the host. In medical usage, some commensals benefit the host. Commensal organisms are normally found at various nonsterile host body sites.

commensalism   An interaction between two different species that only benefits one partner.

compatible solute A small non-toxic molecule that can accumulate in a cell to prevent cell water loss in hypertonic environments.

competence factor A species-specific secreted bacterial protein that induces competence for transformation.

competent A cell that is able to take up DNA from the environment.

complement Innate immunity proteins in the blood that form holes in bacterial membranes, killing the bacteria.

complementary DNA (cDNA) A DNA synthesized complementary to an RNA template via reverse transcriptase.

complex medium A nutrient-rich growth solution including undefined chemical components such as beef broth.

complex transposon A transposon containing a gene for the transposase enzyme which is needed for replicative transposition.

composite transposon A transposon containing genes in addition to those of transposition, such as antibiotic resistance or catabolic functions.

compound microscope A microscope with multiple lenses to compensate for lens aberration and increase magnification.

compromised host An animal with a weakened immune system.

condensation A chemical reaction that joins two molecules and produces a water molecule.

condenser A lens that focuses parallel light rays from the light source onto a small area of the specimen to improve the resolution of the objective lens.

conditional lethal mutation A mutation that leads to death under one growth condition but permits growth under a second condition.

confluent A lawn of organisms that have completely covered a surface.

congenital syphilis Syphilis contracted in utero.

conjugation Horizontal gene transmission involving cell-cell contact. In bacteria, pili draw together the donor and recipient cell envelopes, and a protein complex transmits DNA across. In ciliated eukaryotes, a conjugation bridge forms between two cells connecting their cytoplasm, through which micronuclei are exchanged.

conjugative transposon A transposon that can be transferred from one cell to another via conjugation.

consensus sequence A sequence of nucleotides or amino acids with a common function at many nucleic acid or protein positions. Consists of the base pair or amino acid most frequently found at each position in the sequence.

constant region The region of an antibody that defines the class of a heavy chain or a light chain.

consumers Organisms that acquire nutrients from producers, either directly or indirectly.

contig Overlapping fragments of cloned DNA that are contiguous along a chromosome.

continuous culture A culture system in which new medium is continually added to replace old medium.

contractile vacuole An organelle in eukaryotic microbes that pumps water out of the cell.

contrast Differential absorption or reflection of electromagnetic radiation between an object and background that allows the object to be distinguished from the background.

coral bleaching The death or expulsion of coral algal symbionts. One cause is an increase in temperature.

coralline alga An algal species that calcifies its fronds into hardened shapes similar to corals.

core particle A viral capsid enclosing its nucleic acid genome.

core polysaccharide A sugar chain that attaches to the glucosamine of lipopolysaccharides and extends outside the cell.

coreceptor A cell surface receptor needed for viral entry along with a primary receptor.

corepressor A small molecule that must bind to a repressor to allow the repressor to bind operator DNA.

cortical alveolus One of the vesicles that forms a network in the outer covering of an alveolate.

cotransduced Genes that are transferred together to a recipient cell during transduction.

Coulter counter A device to count cells based on increasing electrical resistance as cells pass through a small hole.

counterstain A secondary stain used to visualize cells that do not retain the first stain.

coupled transport The movement of a substance against its electrochemical gradient (from lower to higher concentration, or from opposite charge to like charge) using the energy provided by the simultaneous movement of a different chemical down its electrochemical gradient.

C-reactive protein A peptide that stimulates the complement cascade, induced by cytokines in the liver. Elevated levels in the blood may be associated with heart disease.

Crenarchaeota One of the two major divisions of Archaea, containing sulfur thermophiles and marine mesophiles.

cross-bridge An attachment that links parallel molecules such as the peptide link between glycan chains in peptidoglycan.

cryocrystallography X-ray crystallography on crystals that are flash-cooled in liquid nitrogen.

cryoelectron microscopy (cryo-EM) Electron microscopy in which the sample is cooled rapidly in a cryoprotectant medium that prevents freezing. The sample does not need to be stained.

cryptogamic crust A low-growing ground cover consisting of an algal-fungal symbiont, similar to lichens.

curd Coagulated milk proteins produced by the combined action of lactic acid-producing bacteria and stomach enzymes of certain mammals such as cattle.

Cyanobacteria A phylum of photoautotrophic bacteria containing chlorophylls a and b. They are closely related to chloroplasts.

cyclic photophosphorylation A photosynthetic process in which chlorophyll serves as both the initial electron donor and the final electron acceptor. ATP is produced via the proton potential from an electron transport system, but no NADPH is generated.

cycloserine A polypeptide antibiotic that inhibits peptidoglycan synthesis.

cystic fibrosis transmembrane conductance regulator (CFTR) A chloride channel found in respiratory epithelia. Mutations in the CFTR gene lead to cystic fibrosis.

cytochrome   A membrane protein that donates and receives electrons.

cytokine   A small, secreted host protein that binds to receptors on various endothelial and immune system cells, regulating the cells responses.

cytoplasm (cytosol)   The aqueous solution contained by the cell membrane in all cells and outside the nucleus (in eukaryotes).

cytoskeleton   A collection of filamentous proteins that impart structure to and aid movement of cells; in a eukaryote, these include intermediate filaments, microtubules and microfilaments.

cytotoxic T cell (T_C cell)   T cells that express CD 8 on their cell surface and can secrete toxic proteins such as perforin and granzymes.

dark-field microscopy   The detection of microbes too small to be resolved by light rays by observing the light they scatter.

dead zone   An anoxic region of an ocean, devoid of most fish and invertebrates.

death phase   The period of cell culture following stationary phase, in which cells die faster than they replicate.

death rate   The rate at which cells die; exponential during the death phase.

decay-accelerating factor   A host cell membrane protein that stimulates decay of complement factors and prevents their deposition at the cell surface.

decimal reduction time (D-value)   The length of time it takes for a treatment to kill 90% of a microbial population, and hence a measure of the efficacy of the treatment.

decolorizer   A chemical that removes loosely bound stain from a specimen.

decomposer   Organism that consumes dead biomass.

defensins   Small, positively charged peptides, produced by animal tissues, that destroy the cell membranes of invading microbes.

defined minimal medium   A solution of known compounds for organismal growth that contains only the minimal components required for growth.

degradosome complex   A multi-enzyme complex that degrades RNA.

degranulate   The release of antimicrobial granule contents by the fusion of granule membranes to cytoplasmic or vacuolar membranes.

degron   A degradation signal contained within a protein.

dehalorespiration   The reduction of halogenated organic molecules by H_2.

deletion   The loss of nucleotides from a DNA sequence.

denature   The loss of secondary and tertiary structure in a protein or nucleic acid due to high temperature or chemical treatment.

dendritic cell   An antigen-presenting white blood cell that primarily takes up small soluble antigens from its surroundings.

denitrification   Energy-yielding metabolism that involves the reduction of nitrate (NO_3^-) to nitrite (NO_2^-), diatomic nitrogen, N_2, and in some cases ammonia (NH_3).

denitrifying bacterium   A heterotrophic bacterium that uses nitrate (NO_3^-) as the final electron acceptor in an electron transport chain and produces nitrogen gas, N_2, or ammonia.

depth of field   A region of the optical column over which a specimen appears in reasonable focus.

derepression   An increase in gene expression caused by the decrease in concentration of a corepressor.

desensitization   A clinical treatment to reduce allergic reactions by exposing patients to small doses of the allergen.

detection   The ability to determine the presence of an object.

detritus   Discarded biomass that can be consumed by decomposers.

diaphragm   A device in a microscope to vary the diameter of the light column, changing the amount of light admitted.

diatom   A taxonomic group of protists (Bacillariophyta) known for intricate, silica-containing bipartite shells.

diauxic growth   A biphasic cell growth curve caused by the depletion of the favored carbon source and a metabolic switch to the second carbon source.

dichotomous key   A tool for identifying organisms, in which a series of yes/no decisions successively narrows down the possible categories of species.

differential medium   A growth medium that can distinguish between various bacteria based on metabolic differences.

differential stain   A stain that differentiates among objects by staining only particular types of cells or specific sub-cellular structures.

diffusion   The energy independent net movement of a substance from a region of high concentration to a region of lower concentration.

dilution streaking   A method of spreading of bacteria on a plate in order to obtain colonies arising from an individual bacterium.

Dinoflagellata (dinoflagellate)   A taxonomic group of tertiary endosymbiont alveolates with two flagella, one of which is wrapped distinctively around the cell equator.

dioxygenase   An enzyme that coordinately oxygenates two adjacent ring carbons.

diphosphatidylglycerol   See **cardiolipin**.

diplococcus   The paired cocci of *Neisseria* species.

direct repeat   Two identical sequences in a DNA molecule, aligned in the same direction.

Discicristata (discicristate)   A protist taxonomic group whose members have mitochondria with distinctive disc-shaped cristae.

disinfection   The removal of pathogenic organisms from inanimate surfaces.

dissemination   The movement of virions from the initial site of infection to other regions of the body.

dissimilation   An organism's catabolism or oxidation of nutrients to inorganic minerals that are released into the environment.

dissimilatory denitrification   Metabolic reduction of nitrate or nitrite to yield energy; anaerobic respiration of nitrate or nitrite.

dissimilatory metal reduction   A type of anaerobic respiration that uses metal cations as terminal electron acceptors.

dissimilatory nitrate reduction   The reduction of nitrate and nitrite through a series of decreased oxidation states back to atmospheric nitrogen and/or ammonia.

dissimilatory nitrate reduction to ammonia (DNRA) A metabolic pathway in which nitrate serves as an electron acceptor and hydrogen gas (H_2) as the electron donor forming water and ammonia.

D-loop formation A triplex DNA molecule that forms as an intermediary structure during generalized recombination.

DNA control sequence A region of DNA, such as the promoter region, that controls the expression of structural genes but is not itself transcribed to RNA.

DNA ligase An enzyme cells use to create a covalent bond at a nick in the phosphodiester backbone. Also used in molecular biology laboratories to join pieces of DNA.

DNA microarray (or microchip) A microchip containing short DNA sequences corresponding to all the open reading frames in an organism affixed to specific locations. It can be used to measure the amount of specific mRNA molecules transcribed in cells.

DNA probe A labeled DNA sequence that hybridizes only to a particular complementary DNA sequence.

DNA reverse-transcribing virus A virus with a double-stranded DNA genome that generates an RNA intermediate and thus requires reverse transcriptase to generate progeny DNA genomes.

DNA sequencing A technique to determine the order of bases in a DNA sample.

DNA shuffling A technique in which fragments of similar genes are combined to generate new genes with potentially new or improved functions.

domain 1. One of the three major taxonomic groups of living organisms. 2. A region of a protein with a particular structure and function.

dormant cell A vegetative cell that has entered a physiological state where it remains metabolically active but fails to replicate or form colonies. Also called the viable-but-nonculturable state, it is distinct from spore forms such as endospores.

doubling time The generation time of bacteria in culture. The amount of time it takes for the population to double.

downstream processing The recovery and purification of the commercial product produced by industrial microbes.

dry weight The weight of a population of cells after the water has been removed.

early gene A viral gene expressed early in the infectious cycle.

eclipse period The time after viral genome injection into host cell but before complete virions are formed.

ecosystems Communities of species plus their environment (habitat).

ectomycorrhizae Mycorrhizae that colonize the surface of plant roots. Their mycelia do not penetrate the root cells.

ectoparasite A harmful organism that colonizes the surface of a host.

edema Tissue swelling due to fluid accumulation.

edema factor (EF) A component of anthrax toxin with adenylate cyclase activity.

electrochemical potential A type of potential energy formed by the combined concentration gradient of a molecule and the electrical potential across a membrane.

electrogenic A transport system that results in a net movement of charged molecules across a membrane.

electron acceptor An oxidized molecule (e.g., NAD^+) that can accept electrons.

electron donor A reduced molecule (e.g., NADH) that can donate electrons.

electron microscope A microscope that obtains high resolution and magnification by focusing electron beams on samples using magnetic lenses.

electron transport chain See **electron transport system**.

electron transport system (ETS) (electron transport chain) A collection of membrane proteins that converts the energy of redox reactions into a proton potential.

electronegativity The affinity of an atom for electrons. The greater the electronegativity, the stronger the attraction for electrons.

electroneutral A transport system that does not result in any net change in charge across the membrane.

electrophoresis A technique to separate charged proteins and nucleic acids based on how rapidly they migrate in an electric field through a gel.

electrophoretic mobility shift assay (EMSA) A technique to observe DNA-protein interactions based on the ability of a bound protein to slow the voltage-driven migration of DNA through a gel.

electroporation A laboratory technique that temporarily makes the cell membrane more leaky to allow the uptake of DNA.

elementary body The endospore-like form of Chlamydia transmitted outside host cells.

eluviated horizon The soil layer below the aerated horizon that periodically experiences water saturation from rain.

Embden-Myerhof-Parnas (EMP) pathway A glycolytic pathway in which glucose-6-phosphate isomerizes to fructose-6-phosphate, ultimately yielding 2 pyruvate, 2 ATP, and 2 NADH.

emission wavelength The wavelength of light emitted by a fluorescent molecule. It is of a lower energy and longer wavelength than the excitation wavelength.

empty magnification Magnification without an increase in resolution.

endemic A disease that is always present in a population, although the frequency of infection may be low.

endemic area The geographical region where an pathogenic organism or virus is found, usually colonizing or infecting animals indigenous to the area.

endergonic A reaction that requires an input of energy to proceed.

endocarditis An inflammation of the heart's inner lining.

endocytosis The budding in of the cell membrane to form a vesicle that contains extracellular material.

endogenous retrovirus A retroelement that contains *gag, env,* and *pol* genes.

endoliths Bacteria that grow within the crystals of solid rock.

endomycorrhizae Mycorrhizae in which fungal hyphae penetrate plant root cells.

endoparasite A parasite that lives inside the host.

endophyte   An endosymbiont of vascular plants.

endosome   A vesicle formed from the pinching in of the cell membrane.

endospore   A durable, inert, heat-resistant spore that can remain viable for thousands of years.

endosymbiont   An organism that lives as a symbiont inside another organism.

endosymbiosis   An intimate association between different species in which one partner population grows within the body of another organism.

endosymbiosis theory   The theory that mitochondria and chloroplasts were originally free-living prokaryotes that formed an internal symbiosis with early eukaryotes.

endotoxin   Lipopolysaccharides in the outer membrane of gram negative bacteria that become toxic to the host after the bacterial cell has lysed.

energy   The ability to do work.

energy carrier   Molecules in the cell, such as ATP and NADH, that serve as energy currency. They are produced during catabolic reactions, and can be used to drive energy-requiring reactions.

enhancer   A non-coding DNA region in eukaryotes that can lead to activation of transcription when bound by the appropriate transcription factor. Its location on the chromosome can be far removed from the regulated gene.

enriched medium   A growth solution for fastidious bacteria, consisting of complex media plus additional components.

enrichment culture   The use of selective growth media to allow only certain microbes to grow.

enterotoxins   Proteins that damage the intestine of the host and cause diarrhea, produced by some gram-negative pathogens.

enthalpy   A measure of the heat energy in a system.

Entner-Doudoroff (ED) pathway   A glycolytic pathway in which glucose-6-phosphate is initially oxidized to 6-phospho-gluconate, and ultimately yields 1 pyruvate, 1 ATP, 1 NADH and 1 NADPH.

entropy   A measure of the disorder in a system.

envelope   Structures external to the cell membrane such as a cell wall or outer membrane.

enzyme   A biological catalyst; a protein or RNA that can speed up the progress of a reaction without itself being changed.

eosinophil   A white blood cell that stains with the acidic dye eosin, and secretes compounds that facilitate innate immunity.

epidemic   A disease outbreak in which large numbers of individuals in a population become infected over a short time.

epidermis   The outer protective cell layer in most multicellular animals.

epilimnion   Warm upper layer of a freshwater lake, above the thermocline, a region where temperature drops steeply.

episome   A DNA element that can exist as part of the chromosome or independently, as a plasmid.

epitope   See **antigenic determinant**.

equilibrium   A dynamic state in which there is no net change in a reaction.

equilibrium density gradient centrifugation   A technique to separate cell components based on their differential densities.

During centrifugation, cell fractions migrate to locations in the centrifuge tube that match their density.

equivalence   The antigen:antibody ratio that leads to immuno-precipitation of large, insoluble complexes.

error-prone repair   Low accuracy DNA repair mechanisms that allow mutations.

error-proof repair   DNA repair mechanisms that minimize the formation of mutations.

erythema migrans   A bulls'eye rash characteristic of borreliosis (Lyme disease).

essential nutrient   A compound that an organism cannot synthesize and must acquire from the environment in order to survive.

ethanolic fermentation   A fermentation reaction yielding 2 ethanol and 2 CO_2 as products.

ether link   A covalent attachment of two organic molecule through an oxygen, C—O—C. Found in archaeal phospholipids between the glycerol and the fatty acids.

Eukarya   One of the three domains of life, consisting of organisms with a last common ancestor not shared with members of *Archaea* or *Bacteria*. Cells possess nuclei, unlike cells of bacteria and archaea.

eukaryote *pl.* **eukaryotes**   A cell that contains a nucleus and is a member of the domain *Eukarya*.

Eumycota   True fungi, a taxonomic group of opisthokont eukaryotes with chitinous cell walls; the group most closely related to animals.

euphotic (photic) zone   The region of the ocean that receives light capable of supporting photosynthesis. Also known as the photic region.

Euryarchaeota   One of the two major divisions of Archaea, containing methanogens, halophiles, and extreme acidophiles.

eutrophic lake   A lake in which overgrowth of heterotrophic microbes has eliminated oxygen, leading to a decrease in animal life.

eutrophication   A sudden increase of a formerly limiting nutrient in an aquatic environment, leading to overgrowth of algae and grazing bacteria and subsequent oxygen depletion.

Excavata   A protist group that shows extensive evolutionary reduction and lacks mitochondria.

excitation wavelength   The wavelength of light that must be absorbed by a molecule in order for the molecule to fluoresce.

exergonic   A spontaneous reaction that releases free energy.

exit site (E site)   The region of the ribosome that holds the uncharged, exiting tRNA.

exocytosis   Fusion of vesicles with the cell membrane to release vesicle contents extracellularly.

exon   An expressed or protein-coding portion of a eukaryotic gene.

exonuclease   An enzyme that cleaves DNA from the end.

exopolysaccharide (EPS)   A thick extracellular matrix of poly-saccharides and entrapped materials that forms around the microbes in a biofilm.

exotoxin   Protein toxin, secreted by bacteria, that kills or damages host cells.

exponential   A mathematical function that is raised to a power. It gives a curve whose slope continually increases.

exponential phase   A phase in bacterial cell culture when bacteria are growing at their maximal possible rate given the conditions, usually exponentially. Same as log phase.

extravasation   The movement of immune cells out of blood vessels and into surrounding infected tissue.

extremophile   An organism that only grows in an extreme environment; that is, an environment including one or more conditions that are "extreme" relative to the conditions for human life.

F⁻ cell   The DNA recipient cell in conjugation

F(ab)₂ region   The amino-terminus variable "arms" of an antibody that bind to a specific antigen.

F⁺ cell   The DNA donor cell that transmits the fertility factor F⁺ to an F⁻ cell during conjugation.

factor H   A normal serum protein that prevents the inadvertent activation of complement in the absence of infection.

facultative   An organism that can grow in the presence or absence of a given environmental factor, such as oxygen.

facultative aerobe (facultative anaerobe)   An organism that can grow either in the presence or absence of oxygen.

facultative anaerobe   See **facultative aerobe**.

facultative intracellular pathogen   A pathogen that can live either inside host cells or outside host cells.

fatty acid synthase complex   A collection of all the enzymes and binding proteins necessary for fatty acid synthesis.

Fc region   The region of an antibody that binds to specific receptors on host cells in an antigen-independent manner. It is found in the carboxy-terminal "tail" region of the antibody.

feline leukemia virus (FeLV)   A retrovirus that is a major cat pathogen.

femtoplankton   Marine or aquatic viruses.

fermentation   The production of ATP via substrate level phosphorylation, using organic compounds as both electron donors and electron acceptors.

fermentative metabolism   See **fermentation**.

fermented food   Food products that are biochemically modified by microbial growth.

ferredoxin   An iron- and sulfur-containing protein that transfers electrons in electron transport systems.

fertility factor (F factor)   A specific plasmid (transferred by an F⁺ donor cell) that contains the genes needed for pilus formation and DNA export.

field   The background observed in microscopy as opposed to the specimen of interest.

filamentous virus   A viral structure type consisting of a helical capsid surrounding a single stranded nucleic acid.

filopodium   A needle-shaped pseudopod.

fimbria *pl.* **fimbriae**   See **pilus**.

Firmicutes   A phylum of low GC content gram-positive bacteria.

fixation   The adherence of cells to a slide by a chemical or heat treatment.

flagellate   An alveolate that has paired flagella.

flagellum *pl.* **flagella**   A filamentous structure for motility. In prokaryotes, a helical protein filament attached to a rotary motor; in eukaryotes, an undulating cell membrane-enclosed complex of microtubules and ATP driven motor proteins.

flavin adenine dinucleotide (FADH₂, FAD⁺)   An energy carrier in the cell that can donate (FADH₂) or accept (FAD⁺) electrons.

flavonoids   Signaling molecules released from legumes to attract nitrogen-fixing rhizobia.

flocs   Biofilms formed on microbial filaments and the soluble organic content of wastewater.

fluid mosaic model   A model of the cell membrane in which proteins are free to diffuse laterally within the membrane.

fluorescence   The emission of light from a molecule that absorbed light of a shorter, higher energy wavelength.

fluorescence resonance energy transfer (FRET)   The detectable transfer of fluorescent energy from one molecule to another. Since the participating molecules must be near each other, FRET can be used to monitor protein-protein interactions in cells and is also used in real time PCR.

fluorescence-activated cell sorter (FACS)   A device that can count cells and sort them based on differences in fluorescence.

fluorescent-focus assay   An assay to detect viruses that do not kill host cells, based on intracellular detection of viruses using anti-viral antibodies or green fluorescent protein modified viruses.

fluorophore   A fluorescent molecule used to stain specimens for fluorescent microscopy.

focal plane   A plane that contains the focal point for a given lens.

focal point   The position at which light rays that pass through a lens intersect.

focus   A group of cells infected by a virus.

folate (folic acid)   A heteroaromatic cofactor, that is required by some enzymes.

fomite   An inanimate object on which pathogens can be transmitted from one host to another.

food contamination   See **food poisoning**.

food irradiation   The exposure of food to ionizing radiation, sterilizing the food for long-term storage.

food poisoning   The presence of human disease-causing microbial pathogens or toxins in food.

food spoilage   Microbial changes that render a food unfit or unpalatable for consumption.

food web   A network of interactions in which organisms obtain or provide nutrients for each other, for example by predation or by mutualism.

foraminiferan   An ameba with a calcium carbonate shell and a helical arrangement of chambers.

forespore   In sporulation of gram-positive bacteria, the smaller cell compartment formed through asymmetrical cell division; it develops into the endospore.

fossil fuels   Ancient organismal remains that have been converted to hydrocarbons through microbial digestion followed by reduction under high pressure underground; extracted and burned by humans for energy.

F-prime (F′) factor or F′ plasmid   A fertility factor plasmid that contains some chromosomal DNA.

frameshift mutation   A gene mutation involving the insertion or deletion of nucleotides that cause a shift in the codon reading frame.

frank pathogen   See **primary pathogen**.

freeze-drying   The removal of water from food, by freezing under vacuum, to limit microbial growth.

fruiting body   A multicellular fungal or bacterial reproductive structure.

frustule   The silica bipartite shell produced by diatoms.

functional group   A cluster of covalently bound atoms that behaves with specific properties and functions as a unit.

fungus *pl.* **fungi**   A heterotrophic opisthokont eukaryote with chitinous cell walls. Mainly includes Eumycota, but traditionally may refer to fungus-like protists such as the oomycetes.

fusion peptide   A portion of a viral envelope protein that changes shape to facilitate envelope fusion with the host cell membrane.

gain-of-function mutation   A mutation that enhances the activity or allows new activity of a gene product. Contrast with **loss-of-function mutation**.

gamogony   The differentiation of parasitic haploid cells into male and female gametes.

gas chromatography   A technique to separate and quantify gasses in a sample.

gas vesicle   An organelle that traps gasses to increase buoyancy of aquatic microbes.

GC content   The proportion of an organism's genome consisting of guanine-cytosine base pairs.

gene fusion   A technique to measure control of gene transcription or translation by inserting a reporter gene into a target gene. The reporter relies on both the promoter and the ribosome binding site of the target gene. Also known as reporter fusion.

gene silencing   The inhibition of transcription from an mRNA by a complementary small interfering RNA. Also known as RNA interference or RNAi.

gene splicing   The cutting and pasting of DNA to produce new gene variants.

generalized recombination   Recombination between two DNA molecules that share long regions of DNA homology.

generalized transduction   A phage-mediated gene transfer process in which any donor gene can be transferred to a recipient cell.

generation time   The species-specific time period for doubling of a population (for example, by bacterial cell division) in an given environment, assuming no depletion of resources.

genetic analysis   Determination of the function of cell RNAs and proteins based on the phenotype of cells in which the gene encoding the RNA or protein is mutated.

genome   The complete genetic content of an organism. The sequence of all the nucleotides in a haploid set of chromosomes.

genomic island   A region of DNA sequence whose properties indicate that it has been transferred from another genome. Usually comprises a set of genes with shared function, such as pathogenicity or symbiosis support.

genus name   A taxonomic rank consisting of closely related species.

geochemical cycling   The global interconversion of various inorganic and organic forms of elements.

geomicrobiology   See **biogeochemistry**.

geosmin   A molecule released from decaying *Streptomyces* cells; it causes the characteristic odor of soil and can affect the taste of drinking water.

germ theory of disease   The theory that many diseases are caused by microbes.

germicidal   A substance that kills cells but not spores.

germination   The activation of a dormant spore to generate a vegetative cell.

Gibbs free energy change (ΔG)   In a chemical reaction, a measure of how much energy available to do work is released or required as the reaction proceeds.

gliding motility   The movement of cells individually or as a collective over surfaces using pili.

global warming   The overall rise in temperature of our biosphere over the past hundred years.

glucosamine   A glucose modified with an amine group.

glutamate dehydrogenase (GDH)   An enzyme that condenses NH_4^+ with 2-oxoglutarate to form glutamate. The condensation requires reduction by NADPH.

glutamate synthase (GOGAT)   An enzyme that converts 2-oxoglutarate plus glutamine into two molecules of glutamate.

glutamine synthetase (GS)   An enzyme that condenses NH_4^+ with glutamate to form glutamine.

glycan   See **polysaccharide**.

glycolysis   The catabolic pathway of glucose oxidation to pyruvate, generating ATP and NADH.

glyoxalate bypass   An alternative to the tricarboxylic acid cycle, induced under low glucose conditions.

gnotobiotic animal   An animal that is germ-free or colonized by a known set of microbes.

Golgi complex   A series of membrane stacks that modifies proteins and helps sort them to the correct eukaryotic cell compartment.

Gram stain   A differential stain that distinguishes between cells that possess a thick cell wall and retain a positively-charged stain (gram-positive) from cells with a thin cell wall and outer membrane that fail to retain the stain (gram-negative).

gramicidin   A peptide antibiotic that disrupts membrane integrity.

gram-negative   Cells that do not retain the Gram stain.

gram-negative bacteria   Bacteria that fail to retain the Gram stain because they have a thin cell wall.

gram-positive   Cells that do retain the Gram stain and appear dark purple after staining.

gram-positive bacteria   Bacteria that retain the Gram stain by virtue of their thick cell wall reinforced by teichoic acids.

granuloma   A thick lesion formed around a site of infection.

granzyme   A cytotoxic T cell-secreted enzyme that damages target cells.

grazers   The first level of consumers that feed directly on producers.

green alga (chlorophyte)   Microbial eukaryotes containing chloroplasts; primary endosymbionts, closely related to plants (Viridiplantae).

greenhouse effect   The trapping of solar radiation heat in the atmosphere by CO_2; a cause of global warming.

griseofulvin   An anti-fungal antibiotic that inhibits cell division.

group translocation   A form of active transport in which the transported molecule is modified after it enters the cell, thus keeping a favorable inward concentration gradient for the unmodified extracellular molecule.

growth factor   A compound needed for the growth of only certain cells.

growth rate   The rate of increase in population number or biomass.

gut-associated lymphoid tissue (GALT)   Lymphatic tissues such as tonsils and adenoids that are found in conjunction with the gastrointestinal tract and contain immune cells.

Haber process   Industrial nitrogen fixation, in which dinitrogen is hydrogenated by methane (natural gas) under extreme heat and pressure to form ammonia.

habitat   The environment where a particular species grows.

Hadean eon   The first period of Earth's existence, from 4.5 to 3.8 gigayears (10^9) ago.

half-life   The amount of time it takes for one half of a radioactive sample to decay.

haloarchaeon   An archaeal species that inhabits high salt environments.

Halobacteriales   An order of Euryarchaeota that contains haloarchaea.

halophile   An organism that requires a high extracellular sodium chloride concentration for optimal growth.

halorhodopsin (HR)   An archaeal light-driven chloride pump.

hapten   A small compound that must be conjugated to a larger carrier antigen in order to elicit production of an antibody that binds to it.

haustorium   A bulbous hyphal extension of a fungal plant pathogen into the host cell.

heat-shock protein   A chaperone protein that is induced by high temperature stress.

heat-shock response   A coordinated response of cells to higher than normal temperatures. It includes changes in the membrane and expression of heat shock genes.

heavy chain   The larger protein that comprises an antibody. An antibody contains two identical heavy chain and two light chain proteins.

helicase   A protein that unwinds the DNA helix.

helper T cell (T$_H$ cell)   A T lymphocyte that secretes cytokines that modulate B-cell class switching.

heme   An organic molecule containing a ring of conjugated double bonds surrounding an iron atom. It is involved in redox reactions and oxygen binding.

hepatitis   An inflammation of the liver, caused by infection or by exposure to a toxic substance.

herd immunity   The slowing of disease spread due to some population members being immune and thus unable to infect other members.

heteroaromatic   Aromatic rings containing non-carbon atoms such as the nitrogenous bases found in nucleotides.

heterocyst   A specialized nitrogen-fixing cell in filamentous cyanobacteria that maintains a reduced environment and excludes O_2.

Heterokonta (heterokont)   A taxonomic group of protists whose flagellated members possess two flagella of unequal length.

heterolactic fermentation   A fermentation reaction in which the products are lactic acid, ethanol and CO_2.

heterotroph   An organism that relies on external sources of organic carbon compounds for biosynthesis.

Hfr   A high frequency recombination bacterial strain, caused by the presence of a chromosomally integrated F factor.

high-performance liquid chromatography (HPLC)   A technique for the high-resolution separation of chemical products through columns packed with beads of various physical properties.

histone   A protein that helps compact eukaryotic chromosomes in nucleosomes.

holdfast   Adhesion factors secreted by the tip of a stalk to firmly attach an organism to a substrate.

Holliday structure (junction)   A cross-like configuration of recombining DNA molecules that forms during generalized recombination.

homologous   Genes that derived from a common ancestral gene.

hopane (hopanoid)   Five-ringed hydrocarbon lipids found in bacterial cell membranes.

horizontal gene transfer   The passage of genes from one cell into another mature cell.

horizontal transfer   See **horizontal gene transfer**.

horizontal transmission   The transfer of a pathogen from one organism into another, non-progeny organism.

hormogonium   A short motile chain of three to five cells produced by filamentous bacteria to disseminate their cells.

host range   The species that can be infected by a given pathogen.

human immunodeficiency virus (HIV)   A human-specific retrovirus that is the causative agent of AIDS.

humic material   Phenolic molecules, derived from lignin, that are resistant to degradation and hence very stable in soil.

humoral immunity   A type of adaptive immunity mediated by antibodies.

humus   Organic breakdown products of lignin that are found in the soil.

hybridization   The annealing of a nucleic acid strand with another nucleic acid strand containing a complementary sequence of bases. The binding of one nucleic acid strand with a complementary strand.

hydric soil Soil that undergoes periods of anoxic water saturation.

hydrogen bond An electrostatic attraction between a hydrogen bound to an oxygen or nitrogen and a second, nearby oxygen or nitrogen.

hydrogenosome Found in some eukaryotes, an organelle that ferments carbohydrates in a pathway generating H_2; may have evolved from mitochondria.

hydrogenotrophy The use of molecular hydrogen (H_2) as an electron donor for a variety of electron acceptors.

hydrological cycle The cyclic exchange of water between atmospheric water vapor and the Earth's bodies of liquid water.

hydrolysis The cleaving of a bond by the addition of a water molecule.

hydrophilic Soluble in water. Includes ionic and polar molecules.

hydrophobic Non-polar; insoluble in water.

3-hydroxypropionate cycle A carbon fixation pathway that condenses hydrated CO_2 and acetyl-CoA to form 3-hydroxypropionate.

hyperthermophile An organism with optimum growth at extremely high temperatures (generally considered as above 80°).

hypertonic An environment that has more solutes than another environment separated by a membrane. Water will tend to flow towards the hypertonic solution.

hypha Thread like filaments forming the mycelium of a fungus.

hypolimnion The lower, colder region of a body of freshwater, below the epilimnion.

hypotonic An environment that has fewer solutes than another environment separated by a membrane.

hypoxia A state of lower than normal oxygen concentration.

identification The recognition of the class of a microbe isolated in pure culture.

idiotype Amino acid differences in hypervariable regions (N-terminus of heavy and light chains) within a single antibody class in an individual that allow recognition of different antigens.

IgA An antibody class containing the alpha heavy chain. It can be secreted and found in tears, saliva, breast milk, etc.

IgD An antibody isotype that contains the delta heavy chain and is found on B cell membranes.

IgE An antibody isotype that contains the epsilon heavy chain and is involved in degranulation of mast cells.

IgM The first antibody isotype detected during the early stages of an immune response. It contains the Mu heavy chain and is found as a pentamer in serum.

immersion oil An oil with a refractive index similar to glass that minimizes light ray loss at wide angles, thereby minimizing wavefront interference and maximizing resolution.

immune avoidance The changing of cell surface proteins by pathogens to prevent antibody detection and prolong infection.

immune system An organism's cellular defense system against pathogens.

immunity The resistance to a specific disease.

immunization The stimulation of an immune response by deliberate inoculation with a weakened pathogen, in hopes of providing immunity to disease caused by the pathogen.

immunogen See **antigen**.

immunogenicity A measure of the effectiveness of an antigen in eliciting an immune response.

immunoglobulin A member of a family of proteins that contain a 110-amino-acid domain with an internal disulfide bond. Members include antibodies and major histocompatibility proteins.

immunological specificity The ability of antibodies produced in response to a particular epitope to bind that epitope almost exclusively. Antibodies made to one epitope bind only weakly, if at all, to other epitopes.

immunomodulin A protein made by normal bacterial flora that influences the host immune response by modifying the secretion of host proteins, such as a cytokine.

immunoprecipitation The antibody mediated cross-linking of antigens to form large, insoluble complexes. Immunoprecipitation is used in research labs and is normally only seen in vitro.

in vivo expression technology (IVET) A laboratory technique to uncover bacterial genes that are only expressed during growth within a host.

incubation stage The time after infection when an organism multiplies but the patient remains asymptomatic.

index case Also known as patient zero, the first case of an infectious disease and an important piece of data for helping contain the spread of disease.

indigenous flora Microbes found naturally in a particular location, often in association with a food substrate.

inducer A molecule that binds to a repressor and prevents repressor binding to the operator sequence.

inducer exclusion The ability of glucose to cause metabolic changes that prevent the cellular uptake of less favorable carbon sources that could cause unnecessary induction.

induction Increased transcription of target genes due to an inducer binding to a repressor and preventing repressor-operator binding.

industrial fermenter The equipment used to grow microbes on an industrial scale.

industrial microbiology The commercial exploitation of microbes.

industrial strain A microbial strain whose characteristics are optimized for industrial use.

infection The growth of a pathogen or parasite in or on a host.

infection cycle The route a pathogen takes as it passes from one host into another.

infection thread A column of rhizobial cells that projects down a tube into a plant root cell during the early stages of the rhizobia-legume symbiosis.

infectious dose (ID$_{50}$) The number of microbes required to cause disease symptoms in half of an experimental group of hosts.

informational gene A DNA sequence that encodes a product essential for transcription or translation.

infrared chemical analysis A technique based on infrared spectrometry that can identify small molecules in a sample; it can be used to measure levels of gasses in the upper atmosphere.

initiation factors Proteins required for initiation of protein synthesis in bacteria.

injera A highly fermented Ethiopian flat bread using the grain teff.

innate immunity See **non-adaptive immunity**.

inner leaflet The layer of the cell membrane phospholipid bilayer that faces the cytoplasm.

inner membrane In gram-negative bacteria, the membrane in contact with the cytoplasm, equivalent to the cell membrane.

insertion The addition of nucleotides into the middle of a DNA sequence.

insertion sequence (IS) A simple transposable element that consists of a transposase gene flanked by short, inverted repeat sequences that are the target of transposase.

integrase An enzyme that catalyzes the integration, via a double cross-over, of one DNA molecule into another at a specific sequence.

integron A large transposon that can contain many different antibiotic resistance genes.

interference The interaction of two wave fronts. Interference can be additive (amplitudes in phase, constructive) or subtractive (amplitudes out of phase, destructive).

interference microscopy Observation of an object using contrast enhanced by superimposing interference bands upon an image to accentuate small differences in refractive indexes.

interferon A host-secreted immuno-modulatory protein that inhibits viral replication.

interleukin 1 (IL-1) A cytokine release by macrophages.

intermediate filament A eukaryotic cytoskeletal protein that is composed of various proteins depending on the cell type.

intimin A pathogenic *E. coli* adhesion protein that binds tightly to an *E. coli*-produced receptor injected into host cells.

intracellular parasite A parasite that lives within a host cell.

intron In eukaryotic genes, an intervening sequence that does not code for protein and is spliced out of the mRNA prior to translation.

inversion A flipping of a DNA fragment within a chromosome. Inversion may allow or repress the transcription of a particular gene.

inverted repeat A DNA sequence that is found in an identical but inverted form at two sites on the same double helix (e.g., 5′ ATCGATCGnnnnnnnCGATCGAT 3′).

ionophore A small organic molecule that binds to an ion and solubilizes it in the membrane, allowing the ion to move down its electrochemical gradient.

irradiation The exposure of a substance to high-energy electromagnetic radiation, usually for the purpose of sterilization.

isoelectric focusing A technique that separates proteins based on their charge.

isoelectric point The pH at which there is no net charge on an amino acid or a protein.

isolate A microbe that has been obtained from a specific location and grown in pure culture.

isoprenoid Condensed isoprene chains, found in archaeal membrane lipids.

isotonic A state in which two solutions separated by a semi-permeable membrane are in osmotic balance. A cell in an isotonic environment will neither gain nor lose water.

isotope An atom of an element with a specific number of neutrons. For example carbon-12, carbon-13, and carbon-14 are all isotopes of carbon.

isotope ratio The ratio of amounts of two different isotopes of an element. The isotope ratio may serve as a biosignature if the ratio between certain isotopes of a given element is altered by biological activity.

isotype A species-specific antibody class, defined by the structure of the heavy chain. IgG, IgA, and IgE are examples of isotypes.

isotype switching (class switching) A change in the predominant antibody isotype produced by a cell.

joules (J) The standard SI unit for energy.

Kaposi's sarcoma A malignancy originating in endothelial or lymphatic cells that is often found in late stage AIDS patients.

kelp A heterokont protist, also known as "brown algae," that grows as multicellular sheets at the water's surface.

kimchi A popular Korean food based on brine-fermented cabbage.

Kirby-Bauer test Method for determining antibiotic susceptibility. Antibiotic-impregnated disks are placed on an agar plate whose surface has been confluently inoculated with a test organism. The antibiotic diffuses away from the disk and inhibits growth of susceptible bacteria. The width of the inhibitory zone is proportional to the susceptibility of the organism.

knockout mutation A mutation that completely eliminates the activity of the gene product.

Koch's postulates Four criteria that should be met for a microbe to be designated the causative agent of an infectious disease.

Korarchaeota A deeply branching division of Archaea.

labile toxin (LT) An *E. coli* enterotoxin, destroyed by heat, that increases cellular cAMP concentrations.

lactate fermentation A fermentation reaction that generates lactic acid from reduction of pyruvic acid.

lactic acid fermentation See **lactate fermentation**.

lag phase A phase of slow growth or no growth right after bacteria are inoculated into new media.

lamellar pseudopod A pseudopod with a sheet-like morphology.

laminar flow biological safety cabinet An air filtration appliance that removes pathogenic microbes from within the cabinet.

Langerhans cell Specialized, phagocytic dendritic cells that are the predominant cell type in skin-associated lymphatic tissue.

laser scanning confocal microscopy A type of fluorescence microscopy in which the excitation and emitted light are focused together, producing high resolution images.

late gene A gene that is expressed late in the viral infectious cycle, such as genes for capsid proteins or lysozyme.

latent period The time in the viral life cycle when progeny virions have formed but are still within the host cell.

latent state When a pathogenic agent is dormant in the host and can not be cultured.

late-phase anaphylaxis Anaphylaxis caused by leukotrienes released by eosinophils recruited by mast cells.

lateral transfer See **horizontal gene transfer**.

law of mass action The tendency of a reaction at equilibrium to return to equilibrium after the concentrations of products or reactants is altered.

leaching The process of metal dissolution from ores.

leader sequence A short DNA sequence preceding a structural gene. In amino acid operons it contains multiple codons for the amino acid synthesized by the downstream structural genes. The leader sequence and translating ribosome help determine whether the structural genes are transcribed.

leaflet One of the two lipid layers in a phospholipid bilayer. The inner leaflet of the cell membrane faces the cytoplasm.

leaven The generation of air spaces in bread due to carbon dioxide production by yeast fermentation.

leghemoglobin An iron-bearing plant protein that sequesters oxygen to maintain an anoxic environment for nitrogenase within cells containing bacteroids.

lentivirus A member of a family of retroviruses that propagate slowly. An example is HIV.

lethal dose (LD$_{50}$) A measure of virulence, it is the number of bacteria or virions required to kill 50% of an experimental group of hosts.

lethal factor (LF) A component of anthrax toxin that cleaves host protein kinases.

lichen An organism formed by a mutualistic relationship between algae and fungi.

light chain The smaller of the two different proteins comprising an antibody. Each antibody contains two heavy chains and two light chains.

light microscopy Observation of a microscopic object based on light absorption and transmission.

lignin A complex aromatic organic compound that forms the key structural support for trees and woody stems.

limiting nutrient A nutrient that is in short supply and limits growth.

lipopolysaccharide (LPS) Structurally unique phospholipids found in the outer leaflet of the outer membrane in gram-negative bacteria. Many are endotoxins.

lithotroph (chemolithotroph) An autotrophic organism that gains energy by oxidizing inorganic compounds. See **chemoautotroph**.

lithotrophy (chemolithotrophy) Energy-yielding metabolism that uses an inorganic electron donor; usually includes fixation of CO_2 into biomass.

littoral zone The upper water layer of aquatic habitats that light can penetrate.

logarithmic (log) phase See **exponential phase**.

long terminal repeat (LTR) A repeated nucleic acid sequence at the 5′ and 3′ ends of a provirus.

loss-of-function mutation A mutation that eliminates or decreases the function of the gene product.

lumen The interior of an intracellular membrane-bound compartment.

lymph node A secondary lymphatic organ, formed by the convergence of lymphatic vessels, that traps foreign particles from local tissue and presents them to resident immune cells.

lyophilization A method to freeze-dry microbes or food for long term storage.

lysate The contents of broken cells; may include virus particles.

lyse To break open cells by disruption of their cell membrane.

lysis The rupture of the cell by a break in the cell membrane.

lysogen A bacterial cell that harbors a complete, yet quiescent, phage genome. Depending on the phage, the phage genome may be integrated into the bacterial chromosome or exist as an autonomous plasmid in the cell.

lysogeny A viral life cycle in which the viral genome integrates into and replicates with the host genome, but retains the ability to initiate host cell lysis.

lysosome An acidic eukaryotic organelle that aids digestion of molecules. Not found in plant cells.

lytic A viral life cycle in which the virus produces new virions and lyses the cell, releasing virions.

M cell A phagocytic innate immune cell (microfold cell) found between intestinal epithelial cells.

MacConkey medium A differential, selective medium that selects for gram-negative bacteria and can differentiate between lactose fermenters and non-fermenters.

macronucleus A form of nucleus found in ciliates, derived from gene amplification and rearrangement of micronuclear DNA, that contains actively transcribed genes.

macronutrient A nutrient an organism needs in large quantities.

macrophage A mono-nuclear, phagocytotic, antigen presenting cell of the immune system.

magnetosome An organelle containing the mineral magnetite that allows microbes to sense a magnetic field.

magnetotaxis The ability to direct motility along magnetic field lines.

magnification An increase in the apparent size of a viewed object as an optical image.

major histocompatibility complex (MHC) Transmembrane cell proteins important for recognizing self and for presenting foreign antigens to the adaptive immune system.

malaria A disease caused by the apicomplexan *Plasmodium falciparum*, transmitted by mosquitoes.

malolactic fermentation Fermentation of L-malate (a side product of glucose fermentation) by *Oenococcus oeni* bacteria, an important process in wine making.

marine snow Microbial biofilms on inorganic particles suspended in marine water.

mast cell White cells that secrete proteins that aid innate immunity. Mast cells reside in connective tissues and mucosa and do not circulate in the bloodstream.

matrix protein A protein found in some viruses that is located between the capsid and the membrane envelope.

mean generation time The reciprocal of the mean growth rate constant; the mean time period for doubling of a population.

mean growth rate constant The number of organismal generations per unit time, k.

mechanical transmission A nonspecific mode of viral entry into damaged tissue.

membrane attack complex (MAC) A cell-destroying pore produced in the membrane of invading bacteria by the host cell complement cascade.

membrane potential Energy stored as an electrical voltage difference across a membrane.

membrane-permeant organic acid See **membrane-permeant weak acid**.

membrane-permeant weak acid An acid that exists in charged and uncharged forms such as acetic acid. The uncharged form can cross the membrane.

membrane-permeant weak base A base that exists in charged and uncharged forms such as methylamine. The uncharged form can dissolve in the membrane.

memory B cell A long-lived type of lymphocyte preprogrammed to produce a specific antibody. After encountering its activating antigen, memory B cells differentiate into antibody-producing plasma cells.

merodiploid A partial diploid strain with chromosomal and F-prime factor copies of particular genes.

merozoite The form of *Plasmodium falciparum*, the causative agent for malaria, which invades red blood cells.

mesocosm A small controlled model ecosystem.

mesophile An organism with optimal growth between 20°C and 40°C.

messenger RNA (mRNA) An RNA molecule that encodes a protein.

metabolist model A model of early life in which the central components of intermediary metabolism arose from self-sustaining chemical reactions based on inorganic chemicals.

metastatic lesions Lesions of infection, or of cancerous cells, that develop at secondary sites away from the initial site of infection.

metazoan A multicellular animal with cells organized into tissues.

methane gas hydrate A crystalline material in which methane molecules are surrounded by a cage of water molecules. This molecular configuration is found in the deep ocean.

methanogen An organism that uses hydrogen to reduce CO_2 and other single-carbon compounds to methane, yielding energy.

methanogenesis An energy-yielding metabolic process that produces methane. It is unique to archaea.

methanotroph An organism that oxidizes methane to yield energy.

methanotrophy The metabolic oxidation of methane to yield energy.

methyl mismatch repair A DNA repair system that fixes misincorporation of a nucleotide after DNA synthesis. The unmethylated daughter strand is corrected to complement the methylated parent strand.

methylene blue A cationic dye commonly used as a simple stain for bacteria.

methylotrophy The ability of an organism to oxidize single-carbon compounds such as methanol, methylamine, or methane.

metranidazole An antibiotic activated by anaerobic bacteria that kills cells by nicking the DNA.

MHC restriction The ability of T lymphocytes to only recognize antigens complexed to self-MHC molecules.

microaerophilic An organism that requires oxygen at a concentration lower than that of the atmosphere, and can not grow in high oxygen environments.

microbe An organism or virus too small to be seen with the unaided human eye.

microcolony A small colony of bacteria only visible with the aid of a microscope.

microfilament A eukaryotic cytoskeletal protein composed of polymerized actin.

microfossil A microscopic fossil in which calcium carbonate deposits have filled in the form of ancient microbial cells.

micronucleus A form of nucleus found in ciliates that contains a diploid set of chromosomes and undergoes meiosis for sexual exchange by conjugation.

micronutrient A nutrient that an organism needs in small quantity, typically a vitamin or a mineral.

microplankton Plankton consisting of cells 20–200 μm in diameter.

microscope A tool that increases the magnification of specimens to enable viewing at higher resolution.

microtubule A eukaryotic cytoskeletal protein composed of polymerized tubulin.

minicell A small spherical cell fragment lacking DNA, generated by improper septation near a pole.

minimal inhibitory concentration (MIC) The lowest concentration of a drug that will prevent the growth of an organism.

miso A Japanese condiment, made from ground soy and rice, salted and fermented by the mold *Aspergillus oryzae*.

missense mutation A point mutation that alters the sequence of a single codon, leading to a single amino acid substitution in a protein.

mitochondrion A eukaryotic organelle that produces ATP through the use of an electron transport chain to produce a proton motive force. O_2 is the final electron acceptor to produce H_2O.

mitosis The orderly replication and segregation of eukaryotic chromosomes, usually prior to cell division.

mitosporic fungus A species of fungus that generates spores by mitosis and lacks a known sexual cycle.

mixotrophic An organism that can switch among metabolic strategies, such as heterotrophy and phototrophy, depending on the environmental conditions.

mixotrophs Organisms capable of both photosynthetic and heterotrophic metabolism.

modulon A group of genes, operons and regulons that is coordinately activated in response to a particular stimulus.

molarity A unit of concentration measured as the number of moles of solute per liter of solution.

molecular clock The use of DNA or RNA sequence information to measure the time of divergence among different species.

molecular formula   A notation that indicates the number and type of atoms in a molecule. For example H_2O is the molecular formula for water.

molecular mimicry   A structural similarity between two different molecules.

monocistronic   The RNA produced from a single gene.

monocyte   A white cell with a single nucleus that can differentiate into macrophages or dendritic cells.

monophyletic   A group of organisms that includes an ancestral species and all of its descendents.

monophyletic group   See **monophyletic**.

monosaccharide   The monomer unit of sugars. Monosaccharides have a molecular formula of CH_2O.

mordant   A chemical binding agent that causes specimens to retain stains better.

mother cell   The larger cell that forms during the asymmetric cell division leading to spore formation. The mother cell will engulf the forespore.

motherspore   See **mother cell**.

motility   The ability of a microbe to generate self-directed movement.

mucociliary elevator   The ciliated mucus lining of the trachea, bronchi, and bronchioles that sweeps foreign particles up and away from the lungs.

Mueller-Hinton agar   A specialized, standardized, para-amino benzoic acid-free media used for the Kirby-Bauer test.

multidrug resistance (MDR) efflux pump   A transmembrane protein pump that can export many different kinds of antibiotics with little regard for structure.

multiplex PCR   A polymerase chain reaction that uses multiple pairs of oligonucleotide primers to amplify several different DNA sequences simultaneously.

murein   See **peptidoglycan**.

murein lipoprotein   The major lipoprotein that connects the outer membrane of gram-negative bacteria to the peptidoglycan cell wall.

mutagen   A chemical that damages DNA and leads to mutations.

mutation   A heritable change in the DNA sequence.

mutation frequency   The fraction of mutant cells (defective in a given gene) within the total cell population.

mutation rate   The number of mutations introduced into DNA per generation (cell doubling).

mutator strain   A strain of cells with a high mutation rate, usually due to a mutation in a DNA repair enzyme.

mutualism   A symbiotic relationship in which both partners benefit.

mycelia   See **mycelium**.

mycelium *pl.* **mycelia**   A fungal hypha that projects into the air (aerial mycelium) or into the growth substrate (substrate mycelium).

mycolic acid   One of a diverse class of sugar-linked fatty acids found in the cell envelopes of mycobacteria.

mycology   The study of fungi.

mycorrhizae   A symbiotic relationship between plant roots and fungi.

myxospore   Durable spherical cells produced by the fruiting body of Myxobacteria.

Nanoarchaeota   A deeply branching division of Archaea; includes thermophilic cells of extremely small size.

nanoplankton   Plankton consisting of cells 2–20 μm in diameter.

nasopharynx   The area leading from the nose to the oral cavity.

native conformation   The fully folded, functional form of a protein.

natto   A soybean product, similar to tempeh, produced by alkaline fermentation.

necrotizing fasciitis   A severe skin infection, also know as flesh-eating disease, usually caused by the gram-positive coccus *Streptococcus pyogenes*.

negative selection   The destruction of T cells bearing T-cell receptors (TCRs) that bind strongly to self-MHC proteins displayed on thymus epithelial cells.

negative-sense (–) strand (see also template strand) Single-stranded RNA whose sequence is complementary to that of mRNA. Also the DNA strand that is complementary to an mRNA.

negative stain   A stain that colors the background and leaves the specimen unstained.

neuraminidase inhibitors   Anti-influenza drugs that target neuraminidase on the viral envelope and decrease the number of virus particles released.

neutralophile   An organism with an optimal growth range in environments between pH 5 and 8.

neutrophil   An innate immune system white blood cell that can phagocytose and kill microbes.

niche   A description of an organism's environmental requirements for existence and its relations with other members of the ecosystem.

nicotinamide adenine dinucleotide (NADH, NAD$^+$)   An energy carrier in the cell that can donate (NADH) or accept (NAD$^+$) electrons.

nitrification   The oxidation of reduced nitrogen compounds to nitrite or nitrate.

nitrifier   An organism that converts reduced nitrogen compounds to nitrite or nitrate.

nitrifying bacterium   A bacterium that performs nitrification, gaining energy by oxidizing ammonium (NH_4^+) to nitrate (NO_3^-).

nitrogen fixation   The ability of some prokaryotes to reduce inorganic diatomic nitrogen gas (N_2) to two molecules of ammonium ion ($2\ NH_4^+$).

nitrogenase   The enzyme that catalyzes the conversion of dinitrogen (N_2) to two molecules of ammonium ion (NH_4^+).

nitrogen-fixing bacterium   A bacterium that can reduce diatomic nitrogen gas, N_2, into ammonium ion (NH_4^+).

Nitrospirae   A phylum of gram-negative bacteria, many of which are lithotrophs, oxidizing nitrite to nitrate.

Nod factors   Signaling molecules released by rhizobial cells and sensed by legumes that help initiate the rhizobia-legume symbiosis.

nomenclature   The naming of different taxonomic groups of organisms.

non-adaptive immunity (innate immunity)   Non-specific mechanisms for protecting against pathogens.

nonpolar covalent bond   A covalent bond in which the electrons in the bond are shared equally by the two atoms.

non-ribosomal peptide antibiotic   Secondary metabolite with antimicrobial activity synthesized by modular enzymes and not by ribosomes.

nonsense mutation   A mutation that changes an amino acid codon into a premature stop codon.

nonsense suppressor   An altered (mutant) tRNA molecule whose anticodon will bind to a nonsense codon (premature stop codon) in a mutant mRNA and allow translation to proceed.

nori   A Japanese food obtained from the red algae *Porphyra* sp.

normal flora   The resident population of harmless organisms normally present at nonsterile body sites (e.g. the intestine).

Northern blot   A technique to detect specific RNA sequences. Sample RNA is subjected to gel electrophoresis, transferred to a blot, and probed with a labeled cDNA that will hybridize to target RNA sequences.

nosocomial   Hospital-acquired.

N-terminal rule   The tendency of the N-terminal amino acid of a protein to influence protein stability.

nuclear magnetic resonance (NMR)   A technique that provides structural information based on the absorption and emission of electromagnetic radiation resulting from changes in the spin state of atomic nuclei.

nucleocapsid protein (NP)   A protein that coats a viral genome.

nucleoid   The looped coils of a bacterial chromosome.

nucleolus   A region inside the nucleus where ribosome assembly begins.

nucleotide   The monomer unit of nucleic acids, consisting of a five-carbon sugar, a phosphate and a nitrogenous base.

nucleotide excision repair (NER)   A DNA repair mechanism that cuts out damaged DNA. New correct DNA is synthesized by DNA polymerase I.

nucleus   A eukaryotic organelle that contains the DNA.

numerical aperture (NA)   The product of the refractive index of the medium and sin θ. As NA increases the magnification increases.

O polysaccharide   A sugar chain that connects to the core polysaccharide of lipopolysaccharides.

objective lens   In a compound microscope, the lens closest to the specimen that generates the initial magnification.

ocular lens   In a compound microscope, the lens situated closest to the observer's eye. It is also called the eyepiece.

Okazaki fragments   Short fragments of DNA that are synthesized on the lagging strand during DNA synthesis.

oligotroph   An organism that can only grow in environments containing extremely low concentrations of organic nutrients.

oligotrophic lake   A lake having low concentration of organic nutrients, the opposite of eutrophic.

oncogene   A gene that through mutation or inappropriate expression can lead to cancer.

oncogenic virus   A virus that causes cancer.

Oomycetes   A taxonomic group of heterkont protists whose life cycle resembles that of fungi; they were formerly classified as fungi.

open reading frame (ORF)   A DNA sequence predicted to encode a protein.

operational gene   A DNA sequence that encodes a product not essential for transcription or translation but involved in cell functions such as metabolism, stress response, or pathogenicity.

operator   A DNA sequence that binds a repressor protein, preventing transcription of a gene downstream. (Less common, refers to a sequence binding an activator protein.)

operon   A collection of genes that are in tandem on a chromosome and are transcribed into a single RNA.

operon fusion   A technique to monitor transcriptional regulation by fusing a reporter gene containing its own ribosome binding site to the 3′ end of an operon. Unlike the "gene fusion" technique, only the promoter of the target gene directs expression of the reporter.

opine   A specialized amino acid used as a carbon and nitrogen source by the plant pathogen *Agrobacterium tumefasciens*.

Opisthokonta   The eukaryotic clade that contains fungi, multicellular animals, and some protist-like organisms.

opportunistic pathogen   A microbe that normally is not pathogenic but can cause infection or disease in an immune compromised host organism.

opsonization   The coating of pathogens with antibodies that aid pathogen phagocytosis by innate immune cells.

opsonize   The binding of IgG antibodies to microbes to enhance microbial phagocytosis by host immune cells.

optical density   A measure of how many particles are suspended in a solution based on light scattering by the suspended particles.

optical isomer   Also known as a stereoisomer, a molecule that has a mirror image. Molecules that contain a chiral carbon can have optical isomers.

oral groove   A ciliate structure for food uptake.

organelle   A membrane-bound compartment within eukaryotic cells that serves a specific function.

organic horizon   The top, surface layer of the soil.

organic molecule   A molecule that contains a carbon-carbon bond.

organotrophy   Metabolism of organic compounds to yield usable energy.

origin (*oriC*)   The region of a bacterial chromosome where DNA replication initiates.

origin of replication (*ori*)   A DNA sequence at which DNA replication initiates. In a bacterial chromosome this site is also attached to the cell envelope.

oropharynx   The area between the soft palate and the upper edge of the epiglottis.

ortholog   See **orthologous gene**.

orthologous Genes present in more than one species that derived from a common ancestral gene and encode the same function.

orthologous gene A gene present in more than one species that derived from a common ancestral gene and encodes the same function.

osmolarity A measure of the number of solute molecules in solution.

osmosis The diffusion of water from regions of high water concentration (low solute) to regions of low water concentration (high solute) across a semi-permeable membrane.

osmotic pressure Pressure exerted by the osmotic flow of water through a semipermeable membrane.

outer leaflet The layer of the cell membrane phospholipid bilayer that faces away from the cytoplasm.

outer membrane A membrane external to the cell wall in gram-negative bacteria.

oxidative burst A large increase in the oxygen consumption of immune cells during phagocytosis of pathogens as the immune cells produce oxygen radicals to kill the pathogen.

oxidative phosphorylation An electron transport chain that uses diatomic oxygen as a final electron acceptor and generates a proton gradient across a membrane for the production of ATP via ATP synthase.

oxidized, oxidation An oxidized compound is one that has lost electrons. Oxidation is the loss of electrons.

oxidoreductase An electron transport system protein that accepts electrons from one molecule (oxidizing that molecule), and donates electrons to a second molecule, thereby reducing the second molecule.

oxygenic Z pathway An ATP-producing photosynthetic pathway consisting of photosystems I and II. Water serves as the initial electron donor (generating O_2) and $NADP^+$ is the final electron acceptor, generating NADPH.

palindrome A DNA sequence in which the top and bottom strands have the same sequence in the 5′ to 3′ direction.

pandemic An epidemic that occurs over a wide geographic area.

panspermia The hypothesis that life forms originated elsewhere and "seeded" life on Earth.

paper chromatography A technique to separate compounds based on their differential migration in a solvent that wicks up through paper. Compounds are separated based relative solubility in the solvent.

paralogous Genes that arise by gene duplication within a species and evolve to carry out different functions.

parasite Bacteria, viruses, fungi, or protozoan (protist) that colonizes and harm its host; the term usually refers to protozoa.

parasitism A symbiotic relationship in which one member benefits and the other is harmed.

parfocal In a microscope with multiple objective lenses, having the objective lenses set at different heights that maintain focus when switching among lenses.

passive transport Net movement of molecules across a membrane without energy expenditure by the cell.

pasteurization The heating of food at a temperature and time combination that will kill spores of *Mycobacterium tuberculosis* and *Coxiella burnetii*.

pathogen A bacterial, viral, or fungal agent of disease.

pathogenesis The processes through which microbes cause disease in a host.

pathogenicity The ability of a microorganism to cause disease.

pathogenicity island A type of **genomic island**, a stretch of DNA that contains virulence factors and may have been transferred from another genome.

pelagic Referring to the open ocean.

pellicle A thick, flexible cell covering found in protists.

penicillin An antibiotic, produced by the *Penicillium* mold, that blocks cross-bridge formation during peptidoglycan synthesis.

penicillin-binding proteins Bacterial proteins, involved in cell wall synthesis, that are targets of the penicillin antibiotic.

pentose phosphate shunt (PPS) An alternate glycolytic pathway in which glucose-6-phosphate is first oxidized, then decarboxylated to ribulose-5-phosphate, ultimately generating 1 ATP and 2 NADPH.

peptide bond The covalent bond that links two amino acid monomers.

peptidoglycan (murein) A polymer of peptide-linked chains of amino sugars; a major component of the bacterial cell wall.

peptidyltransferase The rRNA enzymatic ability to form peptide bonds.

peptidyl-tRNA site (P site) The ribosomal site that contains the growing protein attached to a tRNA.

perforin A cytotoxic protein, secreted by T cells, that forms pores in target cell membranes.

peripheral membrane protein A protein that is associated with a membrane but does not transverse the phospholipid bilayer.

permease A substrate-specific carrier protein in the membrane.

permissive temperature A temperature at which a temperature-sensitive mutation in a gene is masked, permitting growth of the organism.

peroxisome A eukaryotic organelle that converts hydrogen peroxide to water.

petechiae Pinpoint capillary hemorrhages due to the absence of clotting factors; may indicate the presence of endotoxin.

petri dish A round dish with vertical walls covered by an inverted dish of slightly larger diameter. The smaller dish can be filled with a substrate for growing microbes.

phage display A technique in which a phage particle contains recombinant coat proteins expressed by genes encoding a coat protein fused to the protein of interest, such as a vaccine antigen.

phagocytosis A form of endocytosis in which large extracellular particles are brought into the cell.

phagosome A large intracellular vesicle that forms as a result of phagocytosis.

phase variation A gene regulatory mechanism that changes the amino acid sequence of a protein from one antigenic type to another. One mechanism involves site specific recombination that flips a DNA sequence in a chromosome.

phase-contrast microscopy Observation of a microscopic object based on the differences in the refractive index between cell components and the surrounding medium. Contrast is generated as the difference between refracted light and transmitted light shifted out of phase.

phenol coefficient test A test of the ability of a disinfectant to kill bacteria; the higher the coefficient the more effective the disinfectant.

phenol red broth test A clinical test for particular bacterial fermentation pathways, indicating the presence or absence of particular species, based on fermentative acids changing a pH indicator.

phosphatidate A negatively charged phosphate head group of a phospholipid.

phosphatidylethanolamine A type of phospholipid with a positively charged ethanolamine attached to the phosphate group.

phosphatidylglycerol A type of phospholipid with a glycerol attached to the phosphate group.

phosphodiester bond The bond that covalently attaches to adjacent nucleotides in a nucleic acid.

phospholipid The major component of membranes. A typical phospholipid is composed of a core of glycerol to which two fatty acids and a modified phosphate group are attached.

phospholipid bilayer Two layers of phospholipids; the hydrocarbon fatty acids tails face the interior of the bilayer, and the charged phosphate groups face the cytoplasm and extracellular environment.

phosphorylation The enzyme-catalyzed addition of a phosphoryl group onto a molecule.

phosphotransferase system (PTS) A group translocation system that uses phosphoenol pyruvate to transfer phosphoryl groups onto the incoming molecule.

photoautotroph Organisms that perform photosynthesis, using light energy to reduce carbon dioxide.

photoautotrophy The reduction of carbon dioxide using light as an energy source.

photoheterolithotroph An organism that can use heterotrophy, lithotrophy, or photosynthesis for energy production, depending on environmental conditions.

photoheterotrophy The production of energy by the photolysis of organic compounds.

photolysis The first energy-yielding phase of photosynthesis, the light-driven separation of an electron from a molecule coupled to an electron transport system.

photoreactivation A light-induced, photolyase-catalyzed repair of pyrimidine dimers.

photosynthesis The metabolic ability to absorb and convert solar energy into chemical energy for biosynthesis; a precise definition includes CO_2 fixation.

photosystem I and II See **anaerobic photosystem I and II**.

phototroph A general term for organisms that can obtain metabolic energy through light absorption, with or without CO_2 fixation.

phototrophy Obtaining energy from chemical reactions triggered by the absorption of light.

phycobilisome A protein complex that captures light in photosynthetic bacteria.

phycoerythrin A photopigment that give red algae their characteristic color.

phylogenetic tree A diagram depicting estimates of the relative amounts of evolutionary divergence among different species.

phylogeny A measurement of genetic relatedness. The classification of animals based on their genetic relatedness.

phylum The taxonomic rank one level below domain; a group of organisms sharing a common ancestor that diverged early from other groups.

phytoestrogens See **flavonoids**.

phytoplankton Phototrophic marine bacteria, algae, and protists, the primary producers in pelagic food webs.

picoplankton Plankton consisting of cells 0.3–2 μm in diameter.

picornavirus A class of icosahedral RNA viruses such as poliovirus.

piezophile See **barophile**. An organism that requires high pressure to grow.

pilin The protein monomer that polymerizes to form a pilus.

pilus *pl.* **pili (fimbria,** *pl.* **fimbriae)** A straight protein filament composed of a tube of protein monomers that extend from the bacterial cell envelope.

pinocytosis A form of endocytosis in which only extracellular fluid and small molecules are brought into the cell.

Planctomycetes A phylum of free-living bacteria that have stalked cells and reproduce by budding. Their nucleoid is surrounded by a membrane.

plankton Organisms that float in water.

planktonic cells Isolated cells, growing individually in a liquid without connections to other cells.

plaque A cell-free zone on a lawn of bacterial cells caused by viral lysis.

plaque assay An assay to determine the presence of bacteriophages based on their ability to form plaques.

plaque-forming unit (PFU) A measure of the concentration of phage particles in liquid culture.

plasma cell A short-lived antibody-producing cell.

plasmid An extrachromosmal genetic element that may be present in some cells.

plasmodesma *pl.* **plasmodesmata** A membrane channel in plants that connects adjacent plant cells.

plasmodial slime mold A slime mold in which the mass of aggregating cells becomes a multinucleate single cell.

plasmodium The giant multinucleate cell formed by plasmodial slime molds.

pneumonic plague A highly virulent and contagious *Yersinia pestis* lung infection.

point mutation A change in a single nucleotide within a nucleic acid sequence.

poliomyelitis (polio) The paralytic disease caused by the poliovirus.

poliovirus A picornavirus that is the causative agent of poliomyelitis.

poliovirus receptor (Pvr) The cell surface receptor unique to humans and simians through which poliovirus gains entry to cells.

polyamine A positively charged molecule containing multiple amine groups.

polycistronic An RNA produced from an operon containing several genes and hence containing several functional sequences (usually encoding different proteins).

polymerase chain reaction (PCR) A method to amplify DNA in vitro using many cycles of DNA denaturation, primer annealing, and DNA polymerization with a heat-stable polymerase.

polyphyletic An organism or group of organisms that have multiple evolutionary origins.

polysaccharide A polymer of sugars.

polysome A cell structure consisting of multiple ribosomes performing translation on the same mRNA molecule.

porin A transmembrane protein complex that allows movement of specific molecules across the cell membrane or the outer membrane.

positive selection Growth of a population under a condition that favors a particular genotype and prevents growth of other genotypes. In immunology: the survival of T cells bearing T-cell receptors (TCRs) that don't recognize self-MHC proteins displayed on thymus epithelial cells.

positive-sense (+) strand A molecule of of DNA that has the same sequence as mRNA (except for T replacing U). In virology: a molecule of RNA (mRNA) that can be directly translated into viral proteins.

pour plate Also known as a Petri dish, a round, shallow-sided container into which molten agar is poured and subsequently cooled.

prebiotic soup A model for the origin of life based on the abiotic formation of fundamental biomolecules and cell structures such as membranes out of a "soup" of nutrients present on early Earth.

predator A consumer that feeds on grazers.

preliminary mRNA transcript (pre-mRNA) A eukaryotic messenger RNA prior to intron removal.

primary antibody response The production of antibodies upon first exposure to a particular antigen. B cells become activated and differentiate into plasma cells and memory B cells.

primary endosymbiont A lineage of organisms derived from a single endosymbiotic event.

primary latent stage The period of syphilis infection between the chancre of primary syphilis and the rash of secondary syphilis.

primary pathogen A disease-causing microbe that can breach the defenses of a healthy host. Same as a frank pathogen.

primary producer An organism that produces biomass (reduced carbon) from inorganic carbon sources such as CO_2.

primary recovery The initial isolation of commercial product from industrial microbes.

primary structure The first level of organization of polymers, consisting of the linear sequence of monomers, for example the sequence of amino acids in a protein or nucleotides in a nucleic acid.

primary syphilis The initial inflammatory reaction (chancre) at the site of infection with *Treponema pallidum*.

primase An RNA polymerase that synthesizes short RNA primers complementary to a DNA template to launch DNA replication.

primer extension A technique to determine the 5′ end of an RNA transcript.

prion An infectious agent that causes propagation of misfolded host proteins; usually consists of a defective version of the host protein.

probabilistic indicator A means of quickly identifying microbes in the clinical setting, based on a battery of biochemical tests performed simultaneously on an isolated strain.

probiotic A food or nutritional supplement that contains live microorganisms and aims to improve health by promoting beneficial bacteria.

prokaryote *pl.* **prokaryotes** Organism whose cell or cells lack a nucleus; includes both bacteria and archaea.

promoter A non-coding DNA regulatory region immediately upstream of a structural gene that is needed for transcription initiation.

promoter trap plasmid A laboratory technique to discover promoters that only drive gene expression under particular circumstances.

proofreading An enzymatic activity of some nucleic acid polymerases that attempts to correct mispaired bases.

prophage A phage genome integrated into a host genome.

propionic acid fermentation The fermentation of lactic acid to propionic acid by *Propionibacterium* species; used in the production of Swiss cheese.

prosthetic group A non-protein component of an enzyme required for catalytic ability.

protease inhibitor A molecule that inhibits a protease enzyme; some are used as anti-HIV drugs to block the virally encoded protease needed to complete HIV assembly.

protective antigen (PA) The core subunit of anthrax toxin, so called because immunity to this protein protects against disease.

protein A A *Staphylococcus aureus* cell-wall protein that binds to the Fc region of antibodies, hiding the *S. aureus* cells from phagocytes.

Proteobacteria A large, metabolically and morphologically diverse, phylum of gram-negative bacteria.

proteome All the proteins expressed in a cell at a given time. The "complete proteome" includes all the proteins the cell can express under any condition. The "expressed proteome" represents the set of proteins made under a given condition.

proteomics The biological field of proteome analysis.

proteorhodopsin A bacterial membrane protein that contains retinal and acts as a light-driven proton pump; homologous to the archaeal protein bacteriorhodopsin.

protist A motile single-celled eukaryote.

proton motive force See **proton potential**.

proton potential The potential energy of the concentration gradient of protons (hydrogen ions, H^+) plus the charge difference across a membrane.

provirus   A viral genome that is integrated into the host cell genome.

pseudogene   A gene that is no longer functional.

pseudomurein   See **pseudopeptidoglycan**.

pseudopeptidoglycan   A peptidoglycan-like molecule composed of sugars and peptides that is found in some archaeal cell walls.

pseudopod   A locomotory extension of cytoplasm bounded by the cell membrane.

psychrophile   An organism with optimal growth at temperatures below 20°C.

pure culture   A culture containing only a single strain or species of microorganism. A large number of microorganisms all descended from a single individual cell.

purine   A nitrogenous base with fused rings found in nucleotides, such as adenine and guanine.

putrefaction   Food spoilage due to the decomposition of proteins and amino acids.

pyrimidine   A single-ring nitrogenous base found in nucleotides such as cytosine, thymine, or uracil.

pyrogen   Any substance that induces fever.

pyruvate dehydrogenase complex (PDC)   The multi-subunit enzyme that couples the oxidative decarboxylation of pyruvate to acetyl-CoA and NADH production.

quarantine   The separation of infectious individuals from the general population to limit the spread of infection.

quaternary structure   The highest level of protein structure, in which more than one polypeptide chain interact and function together.

quinol   A reduced electron carrier that can diffuse laterally within membranes.

quinolones   A group of antibiotic drugs that inhibit DNA synthesis by targeting bacterial topoisomerases such as DNA gyrase.

quinone   An oxidized electron carrier that can diffuse laterally within membranes.

quinone pool   The oxidized quinones and reduced quinols that diffuse freely within the phospholipid membrane and are able to transfer electrons between many different redox enzymes.

quorum sensing   The ability of bacteria to sense the presence of other bacteria via secreted chemical signals called autoinducers.

radial immunodiffusion   A laboratory technique to determine the concentration of an antigen in solution.

Radiolaria   A taxonomic group of amebas with a silicate shell penetrated by filamentous pseudopods.

rancidity   Food spoilage due to the oxidation of fats; may or may not involve microbial activity.

rank   A level of taxonomic hierarchy, such as phylum, class, order or family.

reaction center   The chlorophyll molecule that donates its excited electron to an electron transport system.

reading frame   The position in a nucleic acid sequence from which triplet codons encode amino acids.

real-time PCR   A technique using fluorescence to detect the products of PCR amplification as the reaction progresses, in order to quantify the amount of DNA in a sample.

receptor-binding domain   A region of a protein responsible for binding to a receptor.

recombinant DNA   DNA that has been combined with other DNA to create novel DNA sequences.

recombination   The replacement of host DNA with donor DNA.

recombination signal sequence (RSS)   A DNA region downstream of antibody heavy and light chain genes that allows recombination between widely separated gene segments.

recombinational repair   A DNA repair mechanism that relies on recombination between an undamaged chromosome and a gap that occurred during replication of damaged DNA.

red alga   See **Rhodophyta**.

redox couple   The oxidized and reduced states of a compound. For example, NAD^+ and NADH form a redox couple.

reduced, reduction   A reduced compound is one that has gained electrons. Reduction is the gain of electrons.

reduction potential   See **voltage potential**.

reductive acetyl-CoA pathway   A carbon assimilation pathway in which two CO_2 molecules are condensed and reduced by $2 H_2$ to form an acetyl group.

reductive evolution   The loss or mutation of DNA encoding unselected traits.

reductive pentose phosphate cycle   See **Calvin cycle**.

reductive (reverse) TCA cycle   A CO_2 fixation pathway that generates acetyl-CoA through reversal of TCA cycle reactions. It requires ATP and NADPH.

reflection   Deflection of an incident light ray by an object, at an angle equal to the incident angle.

refraction   The bending and slowing of light as it passes through a substance.

refractive index   The degree to which a substance causes the refraction of light, a ratio of the speed of light in a vacuum to its speed in another medium.

regolith   Loose rock fragments covering bedrock; commonly refers to the surface layer of Mars or of the moon.

regulatory protein (regulator)   A protein that can bind DNA and modulate transcription in response to a metabolite.

regulon   A group of genes and operons that is co-ordinately regulated and shares a common biochemical function.

release factor   A molecule that enters a ribosome A site containing an mRNA stop codon and initiates protein cleavage from the tRNA.

replication fork   During DNA synthesis the region of the chromosome that is being unwound.

replisome   A complex of DNA polymerase and other accessory molecules that performs DNA replication.

reporter gene   A gene, such as *lacZ* (β-galactosidase) or *gfp* (green fluorescent protein), whose protein product can be easily quantified; commonly used in a gene fusion.

repression   The down-regulation of gene transcription.

repressor   A regulatory protein that can bind to a specific DNA sequence and inhibit transcription of genes.

reservoir   1. An organism that maintains a virus in an area by serving as a high titer host. 2. A part of the biosphere containing significant amounts of an element needed for life.

resolution The smallest distance that two objects can be separated and still be distinguished as separate objects.

respiration The oxidation of reduced electron donors through a series of membrane-embedded electron carriers to a final electron acceptor. The energy derived from the redox reactions is stored as an electrochemical gradient across the membrane, which may be harnessed to produce ATP.

response regulator A cytoplasmic protein that is phosphorylated by a sensor kinase and modulates gene transcription depending on its phosphorylation state.

restriction endonuclease A bacterial enzyme that cleaves double-stranded DNA within a specific short sequence, usually a palindrome. Often called a restriction enzyme.

restriction enzyme See **restriction endonuclease**.

restriction site A DNA sequence recognized and cleaved by a restriction enzyme.

restrictive temperature A temperature at which a temperature-sensitive mutation in a gene leads to the mutant phenotype, which generally includes failure to grow.

reticulate body The metabolically and reproductively active form of Chlamydia.

reticuloendothelial system A collection of cells that can phagocytize and sequester extracellular material.

retinal A vitamin A related cofactor in opsin proteins; it undergoes a conformational change in response to photon absorption.

retroelements Non-virion-producing, ancient retroviral genomes found in host genomes.

retrotransposon Retroelements that contain only partial retroviral sequences but may contain reverse transcriptase to allow further movement into the host genome.

retrovirus A virus containing a positive single-stranded RNA genome; uses reverse transcriptase to generate a double-stranded DNA.

reverse transcriptase An enzyme that produces a double-stranded DNA molecule from a single-stranded RNA template.

reversion test A way to screen the mutagenic potential of compounds by their ability to fix a gene that is necessary for the biosynthesis of a needed metabolite.

rhinovirus A picornavirus that causes the common cold.

rhizobia Bacterial species of the order Rhizobiales that form highly specific mutualistic associations with plants in which the bacteria fix nitrogen for the plant.

rhizoplane The surface of plant roots.

rhizosphere The soil environment surrounding plant roots.

Rho-dependent A bacterial transcription termination mechanism that requires Rho protein.

Rhodophyta (rhodophyte) A taxonomic group of unicellular primary algae containing red pigment; also known as "red algae."

Rho-independent A bacterial transcription termination mechanism that requires a GC-rich region near the transcript terminus.

ribosomal RNA (rRNA) RNA molecules that include the scaffolding and catalytic components of ribosomes.

ribosome-binding site See **Shine-Dalgarno sequence**. A stretch of nucleotides upstream of the start codon in an mRNA that binds to a specific location of the ribosome.

riboswitch A regulatory RNA sequence (upstream of an operon) that forms a stem-loop structure stabilized by the operon's product and modulates transcription and translation.

ribozyme A catalytic RNA molecule.

ripening The aging of cheese.

rise period During viral culture, the time when cells lyse and viral progeny enter the media.

RNA reverse-transcribing virus See **retrovirus**.

RNA world A model of early life in which RNA performed all the informational and catalytic roles of today's DNA and proteins.

RNA-dependent RNA polymerase An enzyme that produces an RNA complementary to a template RNA strand.

rod 1. A bacteria with a linear shape (also referred to as a bacillus). 2. A photoreceptor cell.

rooted tree A phylogenetic tree showing the relative distances between different species, including the earliest time of divergence from a common ancestor.

Rous sarcoma virus (RSV) A retrovirus that was the first virus shown to cause cancer.

rubisco, ribulose 1,5-bisphosphate carbon dioxide reductase/oxidase The enzyme that catalyzes the carbon fixation step in the Calvin cycle.

rumen The first chamber of the digestive tract of ruminant animals such as cattle; the main site for microbial digestion of feed.

sacculus The single covalent molecule that comprises the bacterial cell wall.

saprophytes Fungal decomposers.

sarcina *pl.* **sarcinae** A cubical octad cluster of cells formed by septation at right angles to the previous cell division.

sargassum weed Unrooted kelp forests that float in marine water.

scanning electron microscopy (SEM) Electron microscopy in which the electron beams scan across the specimen's surface to reveal the three-dimensional topology of the specimen.

scattering Interaction of light with an object resulting in propagation of spherical light waves at relatively low intensity.

schizogony Mitotic reproduction of parasitic cells to achieve a large population within a host tissue.

secondary antibody response A memory B cell-mediated rapid increase in the production of antibodies in response to a repeat exposure to a particular antigen.

secondary endosymbiont An organism evolved through engulfment of a primary endosymbiont.

secondary metabolite See **secondary product**.

secondary product Biosynthetic products that are not essential nutrients but enhance nutrient uptake or inhibit competing species (e.g. antibiotics).

secondary structure The second level of organization of polymers, consisting of regular patterns that repeat, such as the double helix in DNA or the beta-sheet in proteins.

secondary syphilis A rash that may appear at some point after the primary latent stage of syphilis.

sedimentation rate The rate at which particles of a given size and shape travel to the bottom of a tube under centrifugal force. The rate depends on the particle's mass and cross-sectional area.

segmented genome A viral genome that consists of more than one nucleic acid molecule.

selective medium A medium that allows the growth of certain species or strains of organisms but not others.

selective toxicity The ability of a drug, at a given dose, to harm the pathogen and not the host.

semiconservative The mode of DNA replication whereby each new double helix contains one old, parental strand and one newly synthesized daughter strand.

semipermeable membrane (selectively permeable membrane) A membrane that is permeable to some substances but impermeable to other substances.

sensor kinase A transmembrane protein that phosphorylates itself in response to an extracellular signal, and transfers the phosphoryl group to a receiver protein.

septation The formation of a septum, a new section of cell wall and envelope to separate two prokaryotic daughter cells.

septicemia An infection of the bloodstream.

septicemic plague Infection of the bloodstream by *Yersinia pestis*.

septum A plate of cell wall and envelope that forms to separate two daughter cells.

sequelae A serious, harmful immunological consequence of bacterial and host antigen cross-reactivity that occur after the infection itself is over. An example is rheumatic fever.

serovar A strain of a microbial species that expresses envelope proteins with an antigenicity distinct from other strains.

serum The non-cell, liquid component of the blood.

sex pilus A pilus specialized for DNA transfer between bacteria.

Shine-Dalgarno sequence See **ribosome-binding site**. In bacteria, a stretch of nucleotides upstream of the start codon in an mRNA that hybridizes to the 16S rRNA of the ribosome, correctly positioning the mRNA for translation.

shuttle vector A plasmid with origins of replication for recognized by both bacteria and eukaryotes.

side effects Drug effects that harm the patient.

siderophore A high-affinity iron binding protein used to scavenge iron from the environment and deliver it to the siderophore-producing organism.

sigma factor A protein needed to bind RNA polymerase for the initiation of transcription in bacteria.

signal recognition particle (SRP) A receptor that recognizes the signal sequence of peptides undergoing translation. The complex attaches to the cell membrane of prokaryotes (or the rough endoplasmic reticulum of eukaryotes), where it docks the protein-ribosome complex to the membrane for protein membrane insertion or secretion.

signal sequence A specific amino acid sequence on the amino-terminus of proteins that directs them to the endoplasmic reticulum (of a eukaryote) or the cell membrane (of a prokaryote).

signature-tagged mutagenesis A technique to discover pathogenicity genes by creating randomly mutated genes, each tagged with a different oligonucleotide. Mutant strains unable to infect an organism contain mutations in pathogenicity genes, which can be isolated.

silencer A eukaryotic DNA element that can lead to decreased transcription when bound by an appropriate transcription factor.

silent mutation A mutation that does not change the amino acid sequence encoded by an open reading frame. The changed codon encodes the same amino acid as the original codon.

simple stain A stain that makes an object more opaque, increasing its contrast with the external medium or surrounding tissue.

single-celled protein Edible microbes of high food value such as Spirulina or some yeasts.

siphonous alga A form of alga consiting of a single cell that forms extended siphon-like tubes and fronds without cell partitions.

site-specific recombination Recombination between DNA molecules that do not share long regions of homology but do contain short regions of homology specifically recognized by the recombination enzyme.

skin-associated lymphoid tissue (SALT) Immune cells, such as dendritic cells, located under the skin that help eliminate bacteria that have breached the skin surface.

S-layer A crystalline protein surface layer replacing or external to the cell wall in many species of archaea and bacteria.

sliding clamp A protein that keeps DNA polymerase affixed to DNA during replication.

slime mold An organism in which unicellular amebas can aggregate into a fruiting body.

sludge The solid products of wastewater treatment.

"small RNA" (smRNA) Non-protein coding regulatory RNAs that modulate translation.

small subunit rRNA (in bacteria, **16S rRNA**) A ribosomal RNA found in the small subunit of the ribosome. Its gene is often sequenced for phylogenetic comparisons.

soil A complex mixture of decaying organic and mineral matter that covers the terrestrial portions of the planet.

solute Any dissolved molecule.

sorbitol MacConkey agar A media that can detect whether a bacterial strain can ferment sorbitol. Used clinically to test for *E. coli* O157:H7.

SOS response A coordinated cellular response to extensive DNA damage. It includes error-prone repair.

sourdough An undefined yeast population, derived from a previous batch of dough, that is used in bread production.

Southern blot A technique to detect specific DNA sequences. Sample DNA segments are separated by gel electrophoresis, transferred to a blot, and probed with a labeled DNA that will hybridize to complementary DNA sequences.

specialized transduction (restricted transduction) Transduction in which the phage can transfer only a specific, limited number of donor genes to the recipient cell.

species A single, specific type of organism, designated by a genus and species name.

spike protein A viral glycoprotein that connects the membrane to the capsid or the matrix and may be involved in viral binding to host cell receptors.

spirochete A bacterium with a tight, flexible spiral shape; a species of the phylum Spirochetes (Spirochaetes).

Spirochetes (Spirochaetes) A phylum of bacteria with a unique morphology, a flexible, extended spiral that twists via intracellular flagella.

spongiform encephalopathies Brain-wasting diseases caused by prions.

spontaneous generation The theory, much debated in the nineteenth century, that under current Earth conditions life can arise spontaneously from non-living matter.

sporangiospore A haploid spore that can germinate to form a haploid mycelium.

sporangium A fungal organ that disburses non-motile spores via ballistic expulsion.

spore stain A type of differential stain that is specific for the endospore coat of various bacteria, typically a firmicute species.

sporogony The development of a parasite's zygote into a spore-like form transmissible to the next host.

spread plate A method to grow separate bacterial colonies by plating serial dilutions of a liquid culture.

staining The process of treating microscopic specimens with a stain to enhance their detection or to visualize specific cell components.

stalk An extension of the cytoplasm and envelope that attaches a microbe to a substrate.

stalked ciliate A ciliate that adheres to a substrate and uses its cilia to obtain prey.

standard Gibbs free energy change ($\Delta G°$) The free energy change under standard conditions of 1 atmosphere pressure, 298°K and 1 M concentrations of all products and reactants.

standard reduction potential ($E°$) A standard value of E, at standard temperature and pressure and assuming initial 1 M concentrations of all reactants and products.

staphylococcus *pl.* **staphylococci** A hexagonal arrangement of cells formed by septation in random orientations.

start codon A codon (usually AUG) that signals the first amino acid of a protein.

starter culture A mixture of fermenting microbes added to a food substrate to generate a fermented product.

stationary phase A period of no net increase in replication that follows the exponential growth phase.

sterilization The destruction of all cells, spores and viruses on an object.

stop codon One of three codons (UAA, UAG, UGA) that do not encode an amino acid and trigger the end of translation.

strand invasion During generalized recombination, the enzyme-catalyzed pairing of single stranded DNA with its complementary strand in the DNA duplex.

strict aerobe An organism that performs aerobic respiration and can only grow in the presence of oxygen.

strict anaerobe An organism that cannot grow in the presence of oxygen.

stringent response A cellular response to idle ribosomes (often indicating low carbon and energy stores) that includes a decrease in rRNA and tRNA production.

stroma The compartment contained by the inner chloroplast membrane where the light-independent reactions of photosynthesis occur (CO_2 fixation).

stromatolite A mass of sedimentary layers of limestone produced by a marine microbial community over many years.

structural formula A representation of molecular structure in which covalent bonds are shown as a line between atoms.

structural gene A string of nucleotides that encodes a functional RNA molecule.

structural isomer A molecule with the same molecular formula as a different molecule but a different arrangement of atoms.

subcellular fractionation A procedure to separate cell components; often includes ultracentrifugation.

substrate level phosphorylation Formation of ATP by the enzymatic transfer of phosphate from a substrate molecule onto ADP.

substrate mycelium A long multinucleate filament (hypha) emanating from the germinating spore of some species that grows along a food source.

substrate-binding protein An extra-cytoplasmic protein that binds specific substrates and delivers them to their cognate uptake ABC transporters.

sugar acid Sugars containing a carboxylic acid group; carboxylate sugars

sulfa drugs Antibiotics that that inhibit nucleotide synthesis.

sulfur granule Elemental sulfur generated by metabolism and stored inside the cell. May refer to extracellular sulfur deposits, also called sulfur globules.

superantigen Molecules that directly stimulate T cells without undergoing antigen presenting cell processing and surface presentation.

supercoil An extra twist or turn found in DNA, either positive (increases DNA winding) or negative (decreases DNA winding).

superoxide radical A highly reactive oxygen species, O_2^-, that is toxic to cell components.

Svedberg coefficient A measure of particle size based on the particle's sedimentation rate in a tube subjected to a high g force.

swarming A behavior in which some microbial cells differentiate into large swarmer cells and swim together as a unit.

switch region A repeating DNA sequence in antibody constant segment genes that serves as a recombination site during isotype switching.

symbiogenesis An evolutionary process by which two or more species become intimately associated.

symbiont An organism that lives in a close association with another organism.

symbiosis The intimate association of two unrelated species.

symbiosome Bacteroids sequestered within a sac of plant-derived membrane.

symport A transport protein where the molecules being transported move in the same direction across the membrane.

synergism Cooperation between species in which both species benefit but can grow independently. The cooperation is less intimate than symbiosis.

synthetic medium A bacterial growth solution that contains defined, known components.

syntrophy Metabolic cooperation between two different species.

tailed phage   A phage such as T4 that contains a genome delivery device called the tail.

tandem repeat   A stretch of directly repeating DNA sequence (direct repeats) without any intervening DNA.

taxon *pl.* **taxa**   A category of organisms grouped together based on genetic relatedness.

taxonomy   The description of distinct life forms and their organization into different categories.

tegument   The contents of a virion between the capsid and the envelope.

teichoic acid   Chains of phosphodiester-linked glycerol or ribitol that thread through and reinforce the cell wall in gram-positive bacteria.

telomerase   A eukaryotic enzyme with reverse transcriptase activity that is required to replicate the ends of linear chromosomes to avoid chromosome shortening.

telomere   The end of a linear chromosome, composed of a repeated DNA sequence.

tempeh   A mold-fermented soy product, popular as a food in parts of Asia.

temperate phage   Phage capable of lysogeny.

template (–) strand   A DNA strand (or an RNA strand in some viruses) that is used as a template for the synthesis of mRNA.

terminal electron acceptor   The final electron acceptor at the end of an electron transport system.

termination (*Ter*) site   A sequence of DNA that halts replication of DNA by DNA polymerase.

terpenoid   A branched lipid derived from isoprene that is found in hydrocarbon chains of archaeal membranes.

terraforming   The idea of transforming the environment of a planet to make it suitable for life from Earth.

tertiary structure   The third level of polymer structure, the unique three-dimensional shape of a polymer.

tertiary syphilis   A final stage of syphilis, manifested by cardiovascular and nervous system symptoms.

tetanospasmin   The tetanus-causing potent exotoxin produced by *Clostridium tetani*.

tetracyclines   A class of antibiotics with four fused cyclic rings, that inhibits translation.

tetraether   A molecule containing four ether links. An example is found in archaeal membranes, when two lipid side chains form ether linkages with a pair of side chains from the other side of the bilayer.

tetrapyrrole   An essential primary product that contains four pyrroles (five-membered rings containing nitrogen), each with one or two double bonds. A precursor for many important cell cofactors such as chlorophylls and vitamin B_{12}.

thermocline   A region of the ocean where temperature decreases steeply with depth, and water density increases.

thermophile   An organism adapted for optimal growth at high temperatures, usually 55°C or higher.

Thermoproteales   An order of crenarchaeote containing hyperthermophilic organisms.

threshold dose   The concentration of antigen needed to elicit adequate antibody production.

thylakoid   A chlorophyll-containing membrane folded within a phototrophic bacterium or a chloroplast.

Ti plasmid   A plasmid found in tumorigenic strains of *A. tumefaciens* that can be used as a vector to introduce DNA into plant cells.

tmRNA   A molecule resembling both tRNA and mRNA that rescues ribosomes stalled on damaged mRNAs lacking a stop codon.

tobacco mosaic virus   An RNA virus that is the causative agent of tobacco mosaic disease.

Toll-like receptor (TLR)   A member of a eukaryotic transmembrane glycoprotein family that recognizes a particular pathogen-associated molecular pattern (PAMP) present on pathogenic microorganisms.

tomography   The acquisition of projected images of a transparent specimen from different angles that are digitally combined to visualize the entire specimen.

topoisomerase   A enzyme that can change the supercoiling of DNA.

total magnification   The magnification of the ocular lens multiplied by the magnification of the objective lens.

transamination   The transfer of an ammonium ion between two metabolites.

transcript   An RNA copy of a DNA template.

transcription   The synthesis of RNA complementary to a DNA template.

transcriptional attenuation   A transcriptional regulatory mechanism in which translation of a leader peptide affects transcription of downstream structural genes.

transcriptome   The set of transcribed genes in a cell at given time. The "complete trancriptome" includes all the possible RNA transcription products from a given genome. The "expressed transcriptome" represent the set of RNAs present during a given condition.

transcytosis   The movement of a cell or substance from one side of a polarized cell to the other side, using an intracellular route.

transduction   The transfer of host genes between bacterial cells via a phage head coat.

transfer RNA (tRNA)   An RNA that carries an amino acid to the ribosome. The anticodon on the tRNA matches the codon on the mRNA.

transform   The conversion of cultured cells into cancer cells.

transformation   The internalization of free DNA from the environment into bacterial cells.

transformed-focus assay   The detection of oncogenic viruses based on their ability transform cells, generating foci of unrestricted cell growth.

transglycosylase   An enzyme that links N-acetyl glucosamine and N-acetyl muramic acid into chains during bacterial cell wall synthesis.

transition mutation   A type of point mutation in which a purine is replaced by a different purine or a pyrimidine is replaced by a different pyrimidine.

translation   The ribosomal synthesis of proteins based on triplet codons present in mRNA.

translational control   A regulatory mechanism that modulates protein production by influencing the translation of mRNA.

translocasome   A bacterial cell membrane protein complex that imports external DNA during transformation.

translocation   The energy-dependent movement of the ribosome to the next triplet codon along an mRNA.

transmembrane domain   A region of a protein that spans a membrane.

transmembrane protein   A protein with a membrane-spanning region.

transmission electron microscopy (TEM)   A type of electron microscopy in which electron beams are transmitted through a thin specimen to reveal internal structure.

transovarial transmission   The passage of genes or pathogens from parent to offspring via the egg cell.

transpeptidase   An enzyme that cross-links the side-chains from adjacent peptidoglycan strands during bacterial cell wall synthesis.

transport protein (transporter)   A membrane protein that moves specific molecules across a membrane.

transposable element   A segment of DNA that can move from one DNA region to another.

transposase   A transposable element-encoded enzyme that catalyzes the transfer of the transposable element from one DNA region to another.

transposition   The process of moving a transposable element from one DNA region to another.

transposon   A transposable DNA element that contains genes in addition to those required for transposition.

transversion   A point mutation in which a purine is replaced by a pyrimidine or vice versa.

tricarboxylic acid (TCA) cycle   A metabolic cycle that catabolizes the acetyl group from acetyl-CoA to 2 CO_2 with the concomitant production of NADH, $FADH_2$, and ATP.

trophic level   A level of the food web representing the consumption of biomass of organisms from a lower trophic level.

tropism   The tissue types infected by a specific virus in a given host.

trypanosome   A parasitic discicristate protist that has a cortical skeleton of microtubules culminating in a long flagellum.

tumor necrosis factor (TNF)   A cytokine released by several cell types (e.g., macrophages) in response to cell damage.

turbidostat   A continuous culture device that can measure optical density and through changing culture flow rates, maintain a specific cell density.

twitching motility   A type of bacterial movement on solid surfaces where a specific pilus extends and retracts.

two-component signal transduction system   A message relay system composed of a sensor kinase protein and a response regulator protein that regulates gene expression in response to a signal (usually an extracellular signal).

two-dimensional polyacrylamide gel electrophoresis (2-D PAGE, 2-D gels)   A technique to separate proteins based on differences in charge and molecular weight.

type I (immediate) hypersensitivity   An IgE mediated allergic reaction that causes degranulation of mast cells within minutes of exposure to the antigen.

type I pilus   A pilus that adheres to mannose residues on host cell surfaces.

type I protein secretion   An ATP-binding cassette transport system that moves molecules directly into or out of bacterial cells.

type II hypersensitivity   An immune condition in which antibodies bind to cell surface antigens, triggering cell-mediated cytotoxicity or activation of the complement cascade.

type II secretion   A bacterial protein secretion system that uses a type-IV pilus-like extraction/retraction mechanism to push proteins out of the cell.

type III hypersensitivity   An immune over-reaction triggered by IgG antibody binding to soluble antigens.

type III pilus   A bacterial pilus that does not bind to mannose but binds to tannic acid.

type III secretion   A bacterial protein secretion system that uses a molecular syringe to inject bacterial proteins into the host cytoplasm.

type IV (delayed-type) hypersensitivity (DTH)   An immune response that develops 24 to 72 hours after exposure to an antigen that the immune system recognizes as foreign. DTH is triggered by antigen-specific T cells. The response is delayed because the T cells need time to proliferate after being activated by the allergen.

type IV pilus   A dynamic pilus that can repeatedly assemble and disassemble; it mediates twitching motility.

ubiquitin (Ub)   A small eukaryotic protein that can be attached to other proteins to target them for degradation via the proteasome.

ultracentrifuge   A machine that exposes samples to high centrifugal forces and can be used to separate subcellular components.

uncoupler   A molecule that makes a membrane permeable to protons, dissipating the proton motive force and uncoupling electron transport from ATP synthesis.

unrooted tree   A phylogenetic tree showing only the relative distances between different species, without indicating which of these diverged earliest from the common ancestor.

upstream processing   The culturing of industrial microbes to produce large quantities of desired product.

vaccination   Exposure of an individual to a weakened version of a microbe to provoke immunity and prevent development of disease upon re-exposure.

van der Waals force   Weak, temporary electrostatic attraction between molecules caused by shifting electron clouds.

vancomycin   A glycopeptide antibiotic that inhibits bacterial cell wall synthesis in a mechanism distinct from penicillin inhibition.

variable region   The amino-terminal regions of antibody light and heavy chains that confer specificity to antigen-binding and define the antibody idiotype.

vasoactive factor   A cell signaling molecule that increases capillary permeability.

vector   An organism (e.g., insect) that can carry infectious agents from one animal to another. In molecular biology, a molecule of DNA into which exogenous DNA can be inserted to be cloned.

vegetative cell   A metabolically active, replicating bacterial cell.

vegetative mycelium   A branched filament produced by vegetative cells that expands into the substrate.

Verrucomicrobia   A phylum of free-living aquatic bacteria with wart-like protruding structures containing actin.

vertical transfer   See **vertical transmission**.

vertical transmission   The passage of genes from parent to offspring.

vesicle   A small membrane-bound sphere found within a cell.

vesicular-arbuscular mycorrhiza (VAM)   A mutualistic association between plant roots and certain fungi, involving hyphal penetration of plant root cells.

viable   An organism that can replicate, for instance by forming a colony on an agar plate.

viable but nonculturable (VBNC)   An organism that is metabolically active but can not replicate to form a colony on a plate by current means of culture. Also called a dormant cell.

viral envelope   A host-derived membrane that surrounds a virus capsid.

viremia   The presence of large numbers of virions in the bloodstream.

Viridiplantae   A taxonomic group of eukaryotes that includes the land plants and primary endosymbiont algae.

virion   A virus particle

viroid   An infectious naked nucleic acid.

virulence   A measure of the severity of a disease caused by a pathogenic agent.

virulence factor   A trait of a a pathogen that enhances the pathogen's disease-producing capability.

virus   An acellular particle containing a genome that can replicate only inside a cell.

virus shedding   The release of virions from the host organism into the environment.

voltage potential difference ($E°$)   The difference in electrical potential between the oxidized and reduced forms of a molecule; the tendency of a molecule to accept electrons.

wastewater treatment   A series of wastewater transformations designed to lower biological oxygen demand and eliminate human pathogens before water is returned to local rivers.

water activity   A measure of the water that is not bound to solutes and is available for use by organisms.

water table   The layer of soil that is permanently saturated with water.

Western blot   A technique to detect specific proteins. Proteins are subjected to gel electrophoresis, transferred to a blot, and probed with enzyme-linked or fluorescently-tagged antibodies that specifically bind the protein of interest.

wet mount   A technique to view living microbes with a microscope by placement of the microbes in water on a slide under a coverslip.

wetlands   Regions of land that undergo seasonal fluctuations in water level and aeration.

whey   The liquid portion of milk after proteins have precipitated out of solution, usually during cheese production.

Winogradsky column   A tube containing a stratified environment that causes specific microbes to grow at particular levels; a type of enrichment culture for the growth of microbes from wetland environments.

X-ray crystallography   A technique to determine the positions of atoms (atomic coordinates) within a molecule or molecular complex, based upon the diffraction of X-rays by the molecule.

X-ray diffraction analysis   See **X-ray crystallography**.

yeast   A unicellular fungus.

yeast two-hybrid system   An *in vivo* technique to determine protein-protein interactions in which DNA sequences encoding proteins of interest of are fused separately to the DNA-binding and transactivation domains of a yeast transcription factor. The recombinant yeast is then tested for expression of a reporter gene.

yogurt   A semi-solid food produced through acidification of milk by lactic acid-producing bacteria.

zone of hypoxia   See **dead zone**.

zone of inhibition   A region of no bacterial growth on an agar plate due to the diffusion of a test antibiotic. Correlates to the **minimal inhibitory concentration**.

zoonotic disease   An infection that normally affects animals but can be transmitted to humans.

zooplankton   Heterotrophic marine microbes.

zoospore   A flagellated reproductive cell produced by chytridiomycete fungi.

zooxanthella   A phototrophic dinoflagellate that is a coral endosymbiont.

Zygomycota (zygomycete)   A clade of fungi forming nonmotile haploid gametes that grow toward each other, fusing to form the zygospore.

zygospore   In zygomycetes, the diploid structure formed by the fusion of two gamete-bearing hyphae.

Index

Genomes of representative bacteria and archaea.

Species (strain)	Genomic chromosome(s)* (kilobase pairs, kbp)	Plasmid(s)* (kbp)	Total (kbp)
Bacteria			
Mycobacterium tuberculosis Tuberculosis	4,400		4,400
Mycoplasma genitalium Normal flora, human skin	580		580
Burkholderia cepacia	3,870 + 3,217 + 876	93	8,056
Escherichia coli K-12 (W3110) Model strain for *E. coli* proteomics	4,600		4,600
Anabaena species (PCC 7120) Cyanobacteria: Major photosynthetic producer of carbon source for aquatic ecosystems	6,370	110 + 190 + 410	7,080
Borrelia burgdorferi Lyme disease	911	21 plasmids with sizes between 9–58	>1,250
Agrobacterium tumefaciens Tumors in plants; genetic engineering vector	2,840 + 2,070	214 + 542	5,666
Archaea			
Methanocaldococcus jannaschii Methanogen from thermal vent	1,660	16 + 58	1,734
Haloarcula marismortui Halophile from volcanic vent	3,130 + 288	33 + 33 + 39 + 50 + 155 + 132 + 410	4,270

1,000 kbp

500 kbp

*Colored schematics indicate relative sizes of genomic elements and whether these are circular or linear. Size bars are provided under each column.